CHEMICAL SYNONYMS

AND

TRADE NAMES

CHEMICAL SYNONYMS

AND

TRADE NAMES

A DICTIONARY AND COMMERCIAL HANDBOOK
CONTAINING OVER 35,500 DEFINITIONS

BY

WILLIAM GARDNER

EIGHTH EDITION
REVISED & ENLARGED

BY

EDWARD I. COOKE
M.A.(Cantab.), B.Sc.(Lond.), C.Chem., M.R.I.C.

AND

RICHARD W. I. COOKE
M.B.B.S.(Lond.), M.R.C.P.(U.K.), D.C.H.

OXFORD
TECHNICAL PRESS

ISBN 0 291 39678 X

Library of Congress Card Number 77–85232

MADE AND PRINTED IN GREAT BRITAIN BY
WILLIAM CLOWES & SONS, LIMITED
LONDON, BECCLES AND COLCHESTER

PREFACE

To THIS EIGHTH EDITION of the late Mr. William Gardner's *Chemical Synonyms and Trade Names* there have been added some 3,300 new entries, principally in the fields of plastics, alloys and pharmaceuticals. A number of entries describing products known to the Editors to be no longer commercially available have been deleted, with the principal object of keeping the bulk of the book within reasonable bounds; but it has been possible to add nearly 400 names to the *Index of Manufacturers* to be found at the end of the book. The sum of these additions and deletions represents a net increase of about 10 per cent. in the scope of this Eighth Edition as compared with its predecessor published in 1971.

As in earlier editions, all entries known to the Editors to be proprietary trade names have been distinguished by a reference number appearing in brackets in, or at the end of, the definition itself, thus " keying " the entry to the *Index of Manufacturers* mentioned above. Only in a few cases, however, has it been possible to add this reference number to the many older entries in the book, principally because the source of these older entries can no longer be traced and the existence of possible proprietary rights in them be thereafter investigated. To the regret of the Editors, many of the new entries in this Edition also lack their appropriate reference number, for the reason that mention of proprietary rights is frequently not made in the trade and professional literature from which the Editors have perforce drawn the bulk of their primary information.

Arising from this very point, the Editors and their Publishers wish to repeat with emphasis a warning given in the Prefaces to most of the earlier editions of this book. It is a matter of regret to them that the presence of a name in " Gardner " has frequently caused difficulty for manufacturers and patent agents trying to register that name as a trade mark in certain countries of the world. Registrars sometimes take the view that, because a word appears in " Gardner ", it must necessarily be a word of such general currency in the chemical or pharmaceutical industry that it would be wrong to restrict the right to use it in advertisements or other promotional literature to a single person or body corporate. Alternatively, they sometimes resist the plea that a word appearing in " Gardner " is already a registered trade name elsewhere in the world by pointing out that the word in question, though defined in " Gardner ", is not there specifically stated to be a proprietary trade name.

It is hoped that these difficulties may be obviated if the following facts are borne in mind.

Words and names appear and are defined in " Gardner " principally because they are words and names in common usage. But many words and names *are* in common usage precisely because they are trade names which have been vigorously promoted as such by the person or corporation owning them. Nearly every entry in " Gardner ", old or new, has been abstracted from published technical literature. Whenever it has been possible to deduce from the source of the information that the name in question is owned and used by a person or body corporate to describe one or more of their products, the fact that the name is claimed as a proprietary one has been recorded. The same rule has been applied in cases where the fact that the word is a trade name is either a matter falling within the personal knowledge of one of the Editors, or is obviously deducible on other grounds.

Many entries in " Gardner " may be, and some doubtless are, proprietary trade names in one or more parts of the world without that fact being known to the Editors, and it by no means follows that an entry defined in " Gardner " without any indication of proprietary status is not in fact a registered trade name.

To those manufacturers and patent agents who have provided details of, and draft definitions for, a sizeable proportion of the new entries in this Eighth Edition of " Gardner ", the Editors and their Publishers express grateful appreciation. Communications from interested parties seeing defined herein, unqualified by a proprietary reference, a word or name in which proprietary rights are claimed are always welcomed by the Publishers at their offices in Freeland, Oxford OX7 2AP. The claim will be submitted to the Editors, and an appropriate entry will then normally be made in future editions of the book.

CHEMICAL SYNONYMS AND TRADE NAMES

A

A.A.A. Spray. A proprietary preparation of benzocaine and cetalkonium chloride in an aerosol. A throat spray. (327).

A.—1. Thiocarbanilide. A registered trade mark for an accelerator. (57).

A.7. A proprietary rubber accelerator used in the manufacture of ebonite and rubber thread. It is an acetaldehyde-aniline condensate, specific gravity 1·04 at 25° C., soluble in acetone, benzene, chloroform and toluene. (559).

A.16. A proprietary rubber accelerator for use with reclaim. It is an acetaldehyde-butyraldehyde-aniline condensate, specific gravity 1·06 at 25° C., soluble in acetone, benzene and chloroform. (559).

A.32. A proprietary rubber accelerator giving a high modulus, and tensile strength producing resilient compounds. It is a butyraldehyde aniline condensate specific gravity 1·005 at 25° C., soluble in acetone, benzene, chloroform and naphtha. (57).

A. 80 Resin. An alpha methyl styrene resin of m.p. 80° C., sp. gr. 1·07; a 70 per cent w/w resin solution in toluene has a viscosity of 3 poise. Has a wide range of compatibility and solubility and is of low colour (0·25 wax Barretts). Used as a film former in coatings, rubber tile and paper formulations. (566).

A.A.A. Ointment. A proprietary ointment containing 2 per cent. ammoniated mercury, 1·25 per cent. salicylic acid, 1 per cent. boric acid and 15 per cent. zinc oxide, used to treat skin irritations. (437).

A.—100. A proprietary vulcanisation accelerator. The reaction product between butyraldehyde, acetaldehyde and aniline. (57).

A-acid. 2 : 5 : 1-Amino-naphthol sulphonic acid.

Aanerödite. A mineral, synonymous with Annerödite.

Aarane. Cromolyn sodium. A proprietary inhibitor of bronchial asthma. (809).

Aarite (Arite). A mineral, Ni(As.Sb).

A.A.T.P. PARATHION.

A.B. Arsenobillon. See SALVARSAN.

Abaca. Manila hemp, the inner fibre of *Musa textilis*.

Abacid. A proprietary preparation containing aluminium hydroxide gel and magnesium trisilicate. An antacid. (434).

Abacid Plus. A proprietary preparation containing aluminium hydroxide, magnesium trisilicate, dimethicone and dicyclomine hydrochloride. An antacid. (434).

Abalak (Kopan, Jewelith). Trade names for synthetic resin varnishes, made in Austria.

Abalyn. A registered trade name for a synthetic resin composed of the methyl ester of abietic acid. A plasticiser. A liquid used for the manufacture of varnishes. (93).

Abanone. Magnesium phospho-tartrate.

Aba-odo. An African term for a mixture of rubber latices, probably those from *Funtumia elastica* and *Ficus Vogelii*.

Abasin. A proprietary preparation of acetyl carbromal. A sedative. (112).

Abassi Cotton. An Egyptian cotton, white in colour, and known in trade as White Egyptian.

Abatia. The leaves of *Abatia rugosa*, used as a black dye.

Abavit B. Organo-mercury seed dressing. (501).

Abavit S. Organo-mercurial dip. (501).

Abbcite No. 2. An explosive for coal mines. It consists of 56·5–59·5 parts ammonium nitrate, 22–24 parts sodium chloride, 7·5–9·5 parts nitro-glycerin, 0·5–2·5 parts dinitro-toluol, and 7–9 parts dried meal.

A.B.C. Liniment. Compound liniment of aconite. (*Linimentum aconiti compositum B.P.C.*)

Abecedin. A proprietary preparation containing Vitamin A, thiamine, riboflavine, ascorbic acid and cholecalciferol. Vitamin supplement. (334).

Abeku Nuts. See NJAVE BUTTER.

Abel Alloy. See FRENCH SILVER.

Abelite. A proprietary name for blasting explosives containing ammonium nitrate, trinitro-toluene, sodium chloride, etc.

Abel's Reagent. A 10 per cent. solution of normal chromic acid. It is used in the micro-analysis of carbon steels, for etching.

Aberel. A proprietary preparation used in the treatment of acne. Retinoic acid. Vitamin A acid. (853).

Abernethy's Mixture. Similar to Black draught (*q.v.*).

Aberoid. A proprietary casein plastic material.

Abfoil. A proprietary trade name for cellulose acetate butyrate. (71).

Abichite. See CLINOCLASTITE.

Abicol. A proprietary preparation of reserpine and bendrofluazide used for control of hypertension. (253).

Abidec. A proprietary preparation containing Vitamin A, Vitamin D, thiamine, riboflavine, nicotinamide and ascorbic acid. Vitamin supplement. (264).

Abies Bark. Hemlock spruce bark.

Abietene. A hydrocarbon obtained by distilling the exudation of *Pinus sabiniana*. It consists almost entirely of normal heptane, C_7H_{16}, and is sold under the names of Abietene, Aurantine, Erasine, and Thioline for removing grease-spots and paint-stains from clothing.

Abietic Acid. Synonymous with Sylvic acid.

Abietic Anhydride. Synonymous with Colophony.

Abir. An Indian aromatic powder containing curcuma, cloves, sandalwood, etc.

Abisol. A 40 per cent. solution of sodium hydrogen sulphite, $NaHSO_3$. A disinfectant and preservative.

Abkhazite. A mineral, a variety of Tremolite having the formula $2[Ca_3Mg_4Si_8O_{22}(OH)_2]$.

Ablick Clay. A mineral, synonymous with Ablykite.

Ablikite. A mineral, synonymous with Ablykite.

Ablykite. A mineral. It is an aluminosilicate of Mg, Ca and K.

Abopon. A proprietary liquid inorganic resinous product which forms films in a few minutes on drying in air. It is recommended as an adhesive, a suspended medium for pigments and abrasives, and for sealing of surfaces to be lacquered or painted. It is stated to be a boro-phosphate.

Abrac. A trade mark for plasticisers. Rosin glycerides. Fine chemicals and other products. (503).

Abrac A. Epoxidised oil stabiliser. Non-migratory plasticiser for PVC and chlorinated rubber. (503).

Abrac C. A proprietary plasticiser consisting of *n*-butyl epoxy-stearate. (503).

Abrac M. Epoxidised ester plasticiser. (503).

Abracol. A registered trade mark for toluene sulphonic acid esters, used as plasticisers. Emulsifying agents. (504).

Abracol 203. A registered trade mark for *p*-toluene sulphanilide.

Abracol 234. A registered trade mark for glyceryl dicresyl ether.

Abracol 777. A registered trade mark for glyceryl mono-cresyl ether diacetate.

Abracol 789. A registered trade mark for toluene sulphonamide.

Abracol 888. A registered trade mark for glyceryl dicresyl ether acetate.

Abracol 1001. A registered trade mark for tertiary butyl cresol.

Abrasite. See GISMONDITE.

Abrastol (Asaprol, Asaprol-etrasol). Calcium - β - naphthol - α - sulphonate, $[C_{10}H_6(OH)SO_3]_2Ca + 3H_2O$. It is used as a clarifier for wines, and is recommended as an antithermic in typhoid fever and articular rheumatism.

Abraum. A red ochre used to stain new mahogany.

Abraum Salts (Stripping Salt, Stassfurt Salts, Potash Salts). The names applied to the upper layers of mixed chlorides of magnesium, potassium, and sodium, overlying the beds of rock-salt at Stassfurt.

Abrazite. Gismondite (*q.v.*).

Abriachanite. A mineral. It is an amorphous earthy variety of crocidolite.

Abril. Synthetic waxes. (505).

Abrodil. A proprietary preparation of sodium monoiodo-methane sulphonate.

Abros. A heat-resisting alloy containing 88 per cent. nickel, 10 per cent. chromium, and 2 per cent. manganese.

A.B.S. Abbreviation for acrylonitrile butadiene styrene, an impact resistant moulding material.

Absaglas. A registered trade name for flame retardant acrylonitrile-butadiene-styrene.

Absinthe. A cordial prepared by distilling rectified alcohol or brandy in which wormwood, star-anise, green anise seed, fennel, coriander, angelica root, or other aromatics have been macerated for about a week.

Absinthol. See THUJOL.

Absolute Acetic Acid. See ACETIC ACID, GLACIAL.

Absolute Alcohol. See ALCOHOL, ABSOLUTE.

Abson A.B.S. 213. A proprietary general-purpose grade of A.B.S. used in injection moulding and extrusion. (414).

Abson A.B.S. 300. A grade of A.B.S. having medium impact resistance. Its good gloss makes it suitable for refrigeration applications. (414).

Abson A.B.S. 230. A proprietary general-purpose grade of A.B.S. possessing more toughness. Used in applications requiring a higher gloss. (414).

Abson A.B.S. 500. A grade of A.B.S. having high-to-medium impact resistance and good strength over a wide temperature range. (414).

Absorbent Cotton. Cotton obtained from linters, card fly, and comber waste which has been purified and bleached.

Absorption Oils. See HEAVY OILS.

Abstem. A proprietary preparation of citrated calcium carbamide, used in the treatment of alcoholism. (306).

Abstrene. A trade mark for acrylonitrile butadiene styrene (A.B.S.) terpolymer moulding compounds. M100 has the best impact strength but the lowest softening point. M300 has the highest softening point and best chemical resistance. (12).

Abukumalite. A mineral. Its formula is $2[(Ca,Y)_5(P,Si)_3O_{12}(OH,F)]$. Also quoted as $(Ca,Y,Th)_{10}[PO_4,SiO_4AlO_4][F,O]$.

Aburana. A yellow oil from *Brassica campretris*. Used by the Japanese for culinary and lighting purposes.

Abus Ha Heree. An inferior kind of asafœtida in the Bombay market. It is derived from *Ferula alliacea*.

Abyssinian Gold (Talmi Gold, Cuivre Poli). A yellow alloy of copper and zinc. It usually consists of about 91 per cent. copper and 8 per cent. zinc, but sometimes contains 86 per cent. copper, 12 per cent. zinc, and 1 per cent. tin. Employed in the manufacture of trinkets.

Abyssinian Gutta. A sticky elastic gum, similar in appearance to gutta-percha.

Abyssinian Tea. See ARABIAN TEA.

AC-1S. A proprietary electrically-conductive, silver-filled epoxy resin with low heat cure. Its resistivity is 0·0001–0·0003 ohm-cm. (418).

AC-2H. A proprietary electrically-conductive, silver-filled epoxy resin having high temperature resistance. Its resistivity is 0·0001–0·0003 ohm-cm. (418).

AC-2LV. A proprietary electrically-conductive, silver-filled epoxy resin. A high-viscosity liquid with a resistivity of 0·002 ohm-cm. (418).

AC-2S. A proprietary electrically-conductive, silver-filled epoxy resin possessing good overall properties. Its resistivity is 0·0001–0·0003 ohm-cm. (418).

AC-4V. A proprietary electrically-conductive, silver-filled epoxy resin whose main characteristics are non-critical mix, room-temperature cure and resistivity of 0·0007 ohm-cm. (418).

A.C.A. A proprietary name for aspirin, caffeine and acetophenetidin tablets. (754).

Acabel. See BEVONIUM METHYLSULPHATE.

Acacetin. 5 : 7-dihydroxy-4′-methoxy-flavone.

Acacia Gum. See AMRITSAR GUM. AUSTRALIAN GUM, HABBAK, and GUM ARABIC.

Acadialite. A mineral. It is a variety of chazabite.

Acadialith. A mineral, synonymous with Acadialite.

Acadiolite. A mineral, synonymous with Acadialite.

Acagine. A mixture of lead chromate and bleaching powder. Used to purify acetylene.

A-Caine. A proprietary ointment for the relief of haemorrhoids, composed of 2 per cent. benzocaine, balsam Peru, bismuth subgallate, cod-liver oil, ephedrine sulphate, lanolin and zinc oxide. (855).

Acajou Balsam (Cardol). A material obtained from the fruits of *Anacardium occidentale* (mahogany nuts, elephant nuts) by the extraction of the powdered nuts. The chief constituent of the balsam is cardol, a non-volatile oil, the balsam being known as raw cardol. It is used medicinally as a blister, and is employed in the preparation of indelible inks and colours for die-sinking work. From the fruits of *A. orientale* a similar product is obtained.

Acanthicone. A mineral, synonymous with Epidote.

Acanthite. A mineral. It is a variety of argentite (silver sulphide, Ag_2S), and occurs at Erzgebirge, in Hungary.

Acanticone. A mineral, synonymous with Epidote.

Acanticonite. A mineral, synonymous with Epidote.

Acarbodavyne. It is an aluminosilicate, sulphate and chloride of calcium and alkali metals; a variety free from CO_2.

Acaroid Balsam (Acaroid Resin, Resin of Botany Bay, Resin Lutea). A yellow resin which exudes from the tree *Xanthorrhœa Hastile* and other species. It is used in the preparation of sealing-wax, lacquers, and japanner's gold size. The balsam contains a resin, also benzoic and cinnamic acids. It melts at about 97° C., yields an ash of from 1–3 per cent., and about 87 per cent. is soluble in alcohol. It has an acid value of 125–140, and a saponification value of 200–

220. There is also a red balsam of Xanthorrhœa or black boy gum, from *X. Australis* of Australia. It is a red resin resembling dragon's blood, and does not contain cinnamic acid. It melts at about 110° C., gives an ash of about 0·02 per cent.; and 96 per cent. is soluble in alcohol. The acid value is 60–100, and the saponification value 160–200. The red gum is also called grass-tree gum or red yacca gum, and both the yellow and red varieties have been called gum accroides, Xanthorrhœa balsams, and earth shellac. The commercial products vary considerably and contain insoluble matter (vegetable products and sand) from 5–10 per cent., and the amount of sand may be as high as 5 per cent.

Acaroid Resin. See ACAROID BALSAM.

Accelemal. A proprietary rubber vulcanisation accelerator. It is possibly thiocarbamide.

Accelerated Cement. A Portland cement containing a high proportion of lime.

Accelerator A1. A proprietary rubber vulcanisation accelerator. It is thiocarbanilide.

Accelerators A5–10. Proprietary rubber vulcanisation accelerators. They consist of formaldehyde-anilines.

Accelerator A7. A proprietary rubber vulcanisation accelerator. It is made from 2 molecules of ethylidene aniline condensed with 1 molecule of acetaldehyde.

Accelerators A11, 16E. Proprietary rubber vulcanisation accelerators. They are aldehyde derivatives of a Schiff's base.

Accelerator A17. A proprietary rubber vulcanisation accelerator. It consists of methylene-*p*-toluidine.

Accelerator A19. A proprietary rubber vulcanisation accelerator. It is a modified ethylidene-aniline.

Accelerator A20. A proprietary rubber vulcanisation accelerator. It consists of aldehyde-amine diluted with mineral oil.

Accelerator A22. A proprietary rubber vulcanisation accelerator. It consists of di-*o*-tolyl-thiourea.

Accelerator A32. A proprietary rubber vulcanisation accelerator. It consists of condensation products of aldehydes and Schiff's base, *e.g.* butyraldehyde and butylidine aniline.

Accelerator A50. A proprietary rubber vulcanisation accelerator. It consists of aldehyde-amine.

Accelerator A1010. A proprietary rubber vulcanisation accelerator consisting of formaldehyde-aniline.

Accelerator A77. Condensation product of acetaldehyde and aniline.

Accelerator 100. Aldehyde derivative of a Schiff's base, made from both butyraldehyde and acetaldehyde.

Accelerator 108. A proprietary rubber vulcanising accelerator. Tetramethyl thiuram disulphide 2/3, 2-mercapto-benzthiazole 1/3. (435).

Accelerator 182. A non-metallic accelerator giving clear polyester castings. (564).

Accelerator 808. A proprietary rubber vulcanising accelerator. Butyraldehyde aniline.

Accelerator BB. A proprietary rubber vulcanisation accelerator. It is butyraldehyde *p* - amino - dimethyl - aniline.

Accelerator D. A proprietary rubber vulcanisation accelerator. It is diphenyl-guanidine.

Accelerator DBA. A proprietary rubber vulcanisation accelerator. It is dibenzyl-amine.

Accelerator DT. A French proprietary rubber vulcanisation accelerator. It is di-*o*-tolyl-guanidine.

Accelerator E-A. A proprietary rubber vulcanisation accelerator. It consists of ethylidene-aniline.

Accelerators F-A. Proprietary rubber vulcanisation accelerators. The high melting one is formaldehyde-aniline, and the low melting one consists of methylene-dianilide.

Accelerator G.M.F. A proprietary rubber vulcanising accelerator. Quinone dioxime.

Accelerator L. A French proprietary rubber vulcanisation accelerator. It is thiocarbanilide.

Accelerator Mercapto. A German rubber vulcanisation accelerator. It is mercapto-benzthiazole.

Accelerator PTX. A proprietary rubber vulcanising accelerator. It is phenyl-tolyl-xylyl-guanidine.

Accelerator R2. A proprietary rubber vulcanisation accelerator. It is a condensation product from methylene-dipiperidine and carbon disulphide.

Accelerator R3. A proprietary rubber vulcanisation accelerator. It is the zinc salt of a dithio-carbaminic acid.

Accelerator R5. A proprietary rubber vulcanisation accelerator. It is dithio-carbamate.

Accelerator W29. A proprietary rubber vulcanisation accelerator. It is a compound of diphenyl-guanidine with dibenzyl-dithio-carbaminic acid.

Accelerator W80. A proprietary rubber vulcanisation accelerator. It is the diphenyl-guanidine salt of mercapto-benzthiazole.

Accelerator X28. A proprietary rubber vulcanisation accelerator. It is stated to be impure tarry diphenylguanidine.

Accelerator XLM. Rubber vulcanising accelerator. Condensation product butyraldehyde - p - amino - dimethyl - aniline.

Accelerator XLO. Rubber vulcanising accelerator. Two parts diphenyl-guanidine and 1 part magnesium oxide.

Accelerator Z88. A proprietary rubber vulcanisation accelerator. It is the ammonia salt of mercapto-benzthiazole mixed with a softener.

Accelerator ZBX. A proprietary rubber vulcanisation accelerator. It is zinc butyl xanthate.

Accelerator ZPD. Proprietary rubber vulcanising accelerator. Zinc penta-methylene-dithio-carbamate.

Accelerator 2P. A proprietary rubber vulcanising accelerator. Piperidinium pentamethylene dithiocarbamate. (290).

Accelerator 4P. A proprietary rubber vulcanising accelerator. Dipenta-methylene thiuram disulphide. (290).

Accelerators. The term used for those substances which effect a reduction in the time necessary for the vulcanisa-tion of rubber. Certain oxides, such as litharge and magnesia, have been known to have this effect for a long time, but more recently, organic accelerators have come into prominence, such as amino-compounds, xanthates and other bodies. The action is stated to be due to an ac-tive form of sulphur being produced by means of the accelerator. See ACCEL-ERENE, VULCAFOR, VULCONE, SUPARAC anb SULZIN.

Accelerene (Accinelson, Vulcafor I, Vulcaniline A). Trade names for a rubber vulcanisation accelerator. It is p-nitroso-dimethyl-aniline $C_6H_4(NO)$ $N.(CH_3)_2$.

Accelerene V 1. A proprietary rubber vulcanisation accelerator. It is stated to consist of equi-molecular proportions of p-nitroso-dimethyl-aniline and β-naphthol.

Accinelson. See ACCELERENE.

Accoine. An aqueous solution of 14 per cent. sodium benzoate, and 2 per cent. sodium carbonate. It is sold as a preservative for milk and cream.

Accomet. A proprietary chromium-based product used in the treatment of metal. (503).

Accotile. A proprietary flooring material made from asbestos, pigments, and asphalt.

Accra Copal. See COPAL RESINS.

Accra Gum. See COPAL RESINS.

Accroides. See ACAROID BALSAM.

Accromix. Chrome plating solutions. (503).

Accrosol. Chrome plating solutions. (503).

Accrotan. A proprietary self-basifying chrome tanning material. (503).

Accrovert. A proprietary grade of chromite. (503).

Accumulator Metal. An alloy called by this name contains 90 per cent. lead, 9·25 per cent. tin, and 0·75 per cent. antimony.

Acderm. A proprietary preparation of propylene glycol, phenazone, triethanol-amine, sulphestolis and sulphur. Acne lotion. (814).

Acdrile. A proprietary preparation of methylcysteine hydrochloride. An expectorant. (137).

Ace. A proprietary trade name for hard rubber. High performance liquid de-tergents. (506).

Acebutolol. A beta adrenergic receptor blocking agent. ($\pm$)-1-(2-Acetyl-4-butyramidophenoxy) - 3 - isopropylam-inopropan-2-ol. "M&B 17803A" is the hydrochloride currently under clini-cal trial.

Acecoline. A proprietary preparation of acetylcholine chloride, used to treat arterial spasm. (815).

Acedapsone. An anti-malarial and anti-leprotic drug currently undergoing clinical trial as "CI556." It is di(4-acetamidophenyl) sulphone.

Acedicone. Acetyl-dihydro-codeinone.

Acedronoles. See AZONINES.

Ace-ite. A proprietary bituminous or asphalt composition.

Aceito de Abeto. A turpentine of sharp acrid flavour, from the Mexican *Pinus religiosa*.

Acele. A proprietary cellulose acetate (yarn).

Aceloid. A proprietary cellulose acetate material used as a moulding composi-tion.

Acelon. Cellulose acetate coated fabric. (507).

Acelose. A proprietary cellulose acetate material.

A.C.E. Mixture. Compound chloroform inhalation. It is a mixture of alcohol 1, chloroform 2, and ether 3 parts by volume.

Acenocoumarol. Nicoumalone.

Acepifylline. Piperazine theophyllin-7-ylacetate. Acefylline Piperazine.

Aceplus. A proprietary cellulose acetate material.

Acepromazine. 2 - Acetyl - 10 - (3 - di-methylaminopropyl) phenothiazine. The maleate of the latter is called Notensil.

Acerado (Fierroso). Names used for mercurial earths.

Acerdese. A mineral, synonymous with Manganite.

Acerdol. Calcium permanganate, $CaMn_2O_8$. Used in gastro-enteritis and diarrhœa.

Acerilla. A mineral, synonymous with Galena.

Acertannin. The tannin of *Acer ginnula*, the Korean maple tree. It is probably digalloyl-aceritol. The crystallised tannin melts at 164–165° C.

Ace-Sil. A proprietary trade name for microporous rubber. It is used for battery separators and filters.

Aceta. A proprietary brand of cellulose rayon. The name is also applied to a French nitro-cellulose lacquer.

Acetaldehyde-naphthylamine. See V.G.B.

Acetaloid. A proprietary cellulose acetate material in the form of rods, sheet, and moulding composition.

Acetamide, Phenyl. See ANTIFEBRIN.

Acetamino-benzoyl-eugenol. See ACETAMINOL.

Acetamino-ethyl-salicylic Acid. See BENZACETIN.

Acetaminol. p-Acetamino-benzoyl-eugenol.

Acetaminophen. Acetaminophenol. See JALUPAP; KORUM.

Acetaminosalol. Salophene (*q.v.*).

Acetanil. Dyestuffs. (508).

Acetanilide. See AMMONAL, ANTIFEBRIN, and ANTINERVINE.

Acetanisidine. See METHACETIN.

Acetanizidin. Methacetin (*q.v.*).

Acetannin (Diacetyl tannin). Acetyl-tannic acid $(CH_3CO)_2C_{14}H_{18}O_9$.

Acetargol. A mixture of formic and acetic acids.

Acetarsol. See ACETARSONE.

Acetarsone. 3-Acetylamino-4-hydroxy-phenyl-arsonic acid. (311). Also ACETARSOL, ACETPHENARSINE, AMARSAN, DYNARSAN, EHRLICH 594, LIMARSOL, ORASAN, OSARSAL, OSVARSAN, PAROSCYL, STOVARSOL.

Acetasem. A proprietary analgesic containing phenacetin, aspirin, caffeine and gelsemium extract. (857).

Acetate, Green. A chromium pigment made from a chromium salt and lead acetate.

Acetate of Lime. The commercial name for the calcium acetate, $(C_2H_3O_2)_2Ca$, prepared from crude pyroligneous acid.

Acetate Silk. A material made from the esters of cellulose and organic acids, cellulose triacetate, $C_6H_7(C_2H_3O_2)_3O_2$ + nH_2O, being the ester mainly used. It dissolves in glacial acetic acid, chloroform, ethyl benzoate, and nitro-benzene to form a syrupy solution, which may be squirted through holes into alcohol or ammonium chloride, to form silk-like threads. Such acetate is non-inflammable, and is used for coating very fine copper wires for electrical purposes, also as a dope for aeroplane wings.

Acetazide. A proprietary preparation of acetazolamide. A diuretic. (279).

Acetazolamide. 5 - acetamido - 1,3,4 - thiadiazole-2-sulphonamide. Diamox. (436).

Acet-Dia-Mer-Sulphonamide. A preparation of sulphacetamide, sulphadiazine and sulphamerazine. Buffonamide, Baf-Sal, Cetazine, Cocodiazine, Incorposul, Trizyl.

Acetest. A mixture of sodium nitroprusside, aminoacetic acid, disodium phosphate, and lactose, in tablet form, used to test for the presence of ketones in blood and urine. (807).

Acetex. A proprietary safety-glass.

Acetic Acid. See ALCOHOL OF VINEGAR; AROMATIC VINEGAR ; DISTILLED VINEGAR ; PICKLING ACID ; PYROLIGNEOUS ACID ; SPIRIT OF COPPER ; SPIRIT OF VERDIGRIS ; SPIRIT OF VINEGAR ; TOILET VINEGAR ; VINEGAR, ARTIFICIAL ; VINEGAR ESSENCE.

Acetic Acid, Absolute. See ACETIC ACID, GLACIAL.

Acetic Acid, Anhydrous. Acetic anhydride, $(CH_3CO)_2O$.

Acetic Acid Bacteria. See MYCODERMA ACETI.

Acetic Acid, Glacial (Absolute Acetic Acid). So-called because it crystallises at ordinary temperatures. The technical acid contains 97 per cent. acetic acid, CH_3COOH.

Acetic Alcohol. A fixing agent used in microscopy. It consists of 6 parts absolute alcohol, 3 parts chloroform, and 1 part glacial acetic acid.

Acetic Ester (Acetic Ether, Vinegar Naphtha). Ethyl acetate, $CH_3COOC_2H_5$.

Acetic Ether. See ACETIC ESTER.

Acetic Ethers. Alkyl acetates.

Acetic-gelatin Cement. A cement obtained by dissolving gelatin in glacial acetic acid. It remains semi-fluid, and is made quite fluid by warming. It is used for uniting glass and china. Diamond cement, Hercules cement, and Giant cement are trade names for cements of this type.

Acetidin. Ethyl acetate, $CH_3COOC_2H_5$.

Acetin Blue. Solutions of indulines in acetins (acetic esters of glycerol). See INDULINE, SPIRIT SOLUBLE.

Acetinduline R. See INDULINE.

Acetine. A mixture of various acetyl derivatives of glycerol. Used in dyeing, either in a neutral form, or in one containing about 20 per cent. acid.

Aceto-acetic Ester. Ethyl-aceto-acetate, $CH_3.CO.CH_2.COO.C_2H_5$ or $CH_3.C(OH):CH.COO.C_2H_5$.

Acetobromal. See NEURONAL.

Aceto-carmine. To a 45 per cent. acetic acid is added carmine until no more dissolves, and filtered. A microscopic stain.

Aceto-caustin. Trichlor-acetic acid, $CCl_3.COOH$. A caustic.

Acetodin. Aspirin (q.v.).

Acetoform. A registered trade mark currently awaiting re-allocation by its proprietors. (983)

Acetohexamide. N - 4 - acetylbenzene-sulphonyl-N'-cyclohexylurea. Dimelor. An oral hypoglycaemic agent.

Acetoin. 3 - Hydroxy - 2 - butanone, $C_4H_8O_2$.

Acetol. A proprietary cellulose acetate material in the form of flake.

Acetol. Acetal or diethyl-aldehyde, $C_6H_{14}O_2$. Used as a hypnotic.

Acetol. Hydroxyacetone $CH_3.CO.CH_2.OH$.

Acetol. A liquid fuel. It contains 80 per cent. alcohol, and 20 per cent. other materials, including acetylene. Used for internal combustion engines.

Acetol Ethyl Ether. See ETHOXY-ACETONE.

Acetol Salicylic Ether. See SALACETOL.

Acetomesal. See SALACETOL.

Aceto-morphine. See HEROIN and MORPHACETIN.

Acetonal. A 10 per cent. solution of aluminium and sodium acetates, used in medicine.

Acetone. See OIL OF ACETONE; PYROACETIC SPIRIT.

Acetone Alcohol. Greenwood spirits; Manhatten spirits; Standard wood spirits (q.v.).

Acetone Bisulphite. See ACETONE SULPHITE and KALLE'S SALT.

Acetone Bromoform. See BROMETONE.

Acetone Chloroform. A name for chloretone (q.v.). The term is also used for chloroform prepared from acetone.

Acetone Oils. The by-products obtained in the commercial preparation of acetone. The oils consist of ketones, aldehydes, and condensation products of acetone, and are sold in two grades. Light, having a boiling-point of 75–130° C., and heavy, with a boiling-point of 130–270° C. They are used in the purification of anthracene.

Acetone Sulphite (Acetone Bisulphite). Acetone - sodium - bisulphite, $CH_3.CO.CH_3.NaHSO_3$. Used in photography.

Acetonitrile. Methyl cyanide, $CH_3.CN$.

Acetonyl. A proprietary analgesic. It is the sodium salt of aspirin with potassium and sodium tartrates. (325).

Acetophen. A proprietary preparation of aspirin, phenacetin and caffeine in capsule form. (437).

Acetophenazine Maleate. A proprietary tranquilliser. Tindal maleate. 2 - acetyl - 10 - {3 - [4 - (- β - hydroxyethyl)-piperazinyl] propyl} phenothiazine dimaleate. OR 10-(3-(4-(2-hydroxymethyl) - 1 - piperazinyl) propyl)-phenothiazin-2-yl. methyl ketone maleate (1 : 2). (438).

Acetophenetidin. An antipyretic and analgesic. Phenedidin. Ethoxyacetanilide.

Acetophenone. See MITTEL HNA; PHENYL - METHYL - KETONE; HYPNONE.

Acetophenone - acetyl -p-aminophen-olester. See HYPNOACETIN.

Acetophenone-phenetidine. See MALARINE.

Acetopiperon. p-Cumarhydrin, $C_6H_8O_2$ $CH_2.CO.CH_3$.

Acetopurpurin. See DIANOL BRILLIANT RED.

Acetopyrin. See ACOPYRIN.

Acetorphine. A narcotic analgesic. O^3 - Acetyl - 7, 8 - dihydro - 7α - [1(R) - hydroxy - 1 - methylbutyl] - O^6 - methyl - 6, 14 - endoethenomorphine.

Acetosal. See ASPIRIN.

Acetosalic Acid. See ASPIRIN.

Acetosalin. See ASPIRIN.

Acetose. Acetyl-cellulose. Used in the manufacture of artificial silk.

Acetosol (Westrol). A trade name for tetrachlor-ethane.

Acetosulphone. 4,4' - diaminodiphenyl-sulphone - 2 - N - acetylsulphonamide sodium. (N' - acetyl - 6 - sulphanilyl - metanilamido) sodium. Promacetin. (264).

Acetoxane. See NEURODINE.

Acetoxyphenyl Mercury. Phenyl-mercuric oxide.

Acetozone (Benzoxate, Benzozone). A mixture of equal parts of acetyl-benzoyl-peroxide, $C_6H_5CO.O.O.C_2H_3O$, and kieselguhr. A germicide.

Acetphenarsine. See ACETARSONE.

Acetphenolisatin. Oxyphenisatin acetate. ISALAX.

Acet-theocin-sodium (Soluble Theocin). Theophyllin sodium acetate, a diuretic.

Acetryptin Hydrochloride. A proprietary anti-hypersensitive agent. 5-Acetyltryptamine hydrochloride. (858).

Acetycol. A proprietary analgesic containing colchicine salicylate, aspirin, p-aminobenzoic acid, vitamin C and niacin. (859).

Acetyl Adalin. ACETYL CARBROMAL.

Acetyl-amino-azo-toluene. See AZOVERNIN.

Acetyl-amino-ethoxy-benzene. See PETONAL.

Acetyl-amino-phenyl-salicylate. See SALOPHENE.

Acetyl-amino-salol. See ACET-AMINOSOLOL ; PHENESAL.

Acetylarsan. A proprietary preparation. It is diethylamine-oxyacetyl-amino-phenyl-arsinate. Diethylamine acetarsol. (507).

Acetylatoxyl. See ARSACETIN.

Acetyl-benzene. See PHENYL-METHYL-KETONE.

Acetyl-benzoyl-peroxide. See ACETOZONE.

Acetyl-bromo-salol. See PHENYL-SEDASPIRIN.

p-**Acetylaminobenzaldehyde Thiosemicarbazone.** A pharmaceutical used in the treatment of tuberculosis. AMITHIOZONE, ANTIB, BERCULON A, BENZOTHIOZON, CONTEBEN, MYCIZONE, NEUSTAB, TEBETHION, THIOMICID, THIOPARAMIZONE, THIACETAZONE.

Acetyl Carbromal. Acetyl adalin. Acetylbromodiethylacetyl - carbamide. (860).

Acetyl-cellulose. See ACETOSE.

Acetyl-choline Hydrochloride. See ACECHOLINE.

Acetyl-coumaric Acid. See TYLMARIN.

Acetyl-cresotic Acid. See AMATIN.

Acetyl-cresotinic Acid. See CRESO-PIRINE ; ERVASIN.

Acetylcysteine. N - acetyl - 1 - cysteine. Airbron.

Acetyl-Digitoxin. α-digitoxin monoacetate. Acylanid.

Acetyldihydrocodeinone. Thebacon.

Acetyl-dinitro-butyl-xylene. See MUSK KETONE.

Acetylmethadol. Methadyl Acetate.

Acetylene Black (Shawinigan Black). A carbon black made by exploding acetylene, C_2H_2, with insufficient oxygen for complete combustion. It has a specific gravity of 1·9 and is used in rubber mixings.

Acetylene Blue 6B, BX, 3R. Dyestuffs. These colours dye cotton from a neutral bath containing sodium sulphate.

Acetylene Stones. Acetylith (*q.v.*).

Acetyl-iodo-salicyclic Acid. See ASPROIDINE.

Acetylith. The name given to a preparation of calcium carbide, CaC_2, coated with sugar.

Acetyloid. A proprietary cellulose acetate plastic.

Acetylon. A proprietary cellulose acetate material.

Acetysal. See ASPIRIN.

Acetyl-salol. See VESIPYRINE.

Acetyltrypoxyl. Sodium-*p*-acetyl-amino-phenyl-arsinate.

Achates. A mineral, synonymous with Agate.

Achavalite. A mineral. It is iron selenide, FeSe.

Acheson's Deflocculated Graphite (Dag). The trade mark for a lubricant obtained by macerating graphite with a solution of tannin for several weeks, whereby it is so finely divided that it forms a permanent emulsion, containing a very small quantity of carbon, with water. The graphite with water is called Aquadag. Oildag is prepared by pouring oil over the filtered dag, then freeing the material from moisture. See AQUADAG and OILDAG.

Achete Juice (Amole Juice). An alkaline decoction made from the juice of a South American Tree called Achete or Coasso. It is used to coagulate rubber latex.

Achiardite. A mineral, synonymous with Dachiardite.

Achirite. A name which has been used for dioptase.

Achlusite. A mineral. It is a cloudy alteration product of clear topaz.

Achmatite. A mineral, synonymous with Epidote.

Achmit. A mineral, synonymous with Acmit.

Achondrite. A class of meteorite mineral.

Achrematite. A mineral. It is a molybdo-arsenate of lead found in Mexico, and consists of the oxides of arsenic, molybdenum, and lead, together with some lead chloride.

Achroite. A mineral. It is a colourless tourmaline (an alkali-calcium-aluminium-silicate).

Achromaite. A mineral. It is a variety of horneblend.

Achromycin. A proprietary preparation containing Tetracycline Hydrochloride. An antibiotic. (306).

Achroodextrins. Intermediate products formed by the hydrolysis of starch or dextrin.

Achtaragdite. A mineral. It is an aluminosilicate of Ca, Fe and Mg.

Achtarandite. A mineral, synonymous with Achtaragdite.

Achtaryndit. A mineral, synonymous with Achtaragdite.

Acicular Bismuth. A mineral, synonymous with Aikinite.

Aciculite (Needle Ore, Patrinite). A mineral. It is a sulphide of bismuth, copper, and lead.

Acid 810. A mixture of branched-chain carboxylic acids containing 8–10 carbon atoms. An intermediate for high quality general purpose paint driers. (2).

Acid, Acetosalic. See ASPIRIN.

Acid Alizarin. See CHROME.

Acid Alizarin Black. A dyestuff prepared from diazotised 6-nitro-2-amino-phenol-4-sulphonic acid, and β-naphthol.

Acid Alizarin Blue BB, GR (Alizarin blue BB, GR, Alizarin Acid Blue BB). A dyestuff. It is the sodium salt of hexaoxy - anthraquinone - disulphonic acid, $C_{14}H_6O_{14}S_2Na_2$. Dyes wool red from an acid bath, and blue from subsequent chroming. The mark GR gives duller shades.

Acid Alizarin Brown B. An acid mordant dyestuff, which dyes wool from an acid bath.

Acid Alizarin Green B, G. A dyestuff. It is the sodium salt of di-sulphydro-tetraoxy - anthraquinone - disulphonic acid, $C_{14}H_6O_{12}S_4Na_2$. Dyes wool greenish-blue from an acid bath.

Acidax. A stearic acid for use in the rubber industry.

Acid Black 6B. See NAPHTHOL BLACK 6B.

Acid Black HA. A dyestuff. It is a British equivalent of Naphthol blue black.

Acid Black N. A dyestuff. It is a British equivalent of Naphthylamine black D.

Acid Blue B. A dyestuff for wool, of a greener shade than Acid Blue R.

Acid Blue 6G. See CYANOL EXTRA.

Acid Blue R. A dyestuff for wool, which it dyes from a bath of sodium sulphate containing a little sulphuric acid. Employed as a groundwork for logwood black.

Acid Blue Black. A dyestuff. It is a British equivalent of Naphthol blue black.

Acid Blue Black 428. A dyestuff. It is a British equivalent of Naphthol blue black.

Acid Bricks. Bricks which contain an excess of silica. They are used for lining furnaces.

Acid Bronze. Alloys containing from 82–88 per cent. copper, 8–10 per cent. tin, 2–8 per cent. lead, and 0–2 per cent. zinc. One alloy consists of 73·5 per cent. copper, 17 per cent. lead, 8 per cent. tin, and 1·5 per cent. nickel. A metal containing 90 per cent. copper and 10 per cent. aluminium is also known as Acid Bronze.

Acid Brown. See FAST BROWN N.

Acid Brown (Dahl & Co.). See FAST BROWN G.

Acid Brown G. An azo dyestuff, $C_{18}H_{16}N_6O_3S$, formed by the action of the hydrochloride of diazo-benzene upon chrysoidine-sulphonic acid. Dyes wool brown from an acid bath.

Acid Brown G (Brotheron & Co.). A dyestuff. It is a British equivalent of Fast brown N.

Acid Brown R. (Vega brown R). A dyestuff. It is the sodium salt of p-sulpho-naphthalene - azo - phenylene-diamine-azo-benzene $C_{22}H_{17}N_6O_3SNa$. Dyes wool brown from an acid bath.

Acid Brown R (British Dyestuffs Corporation). A dyestuff. It is an equivalent of Fast brown N.

Acid Brown R (Brotherton & Co.). A dyestuff. It is equivalent to Fast brown 3B.

Acid Carbonate (Bicarbonate). Hydro-carbonate, of which sodium-hydrogen-carbonate, $NaHCO_3$, is a type.

Acid Carmoisine B. See FAST RED E.

Acid Cerise. See MAROON S.

Acid, Chryseinic. See CAMPOBELLO YELLOW.

Acid Clay, Japanese. See JAPANESE ACID CLAY.

Acid Colours. Dyestuffs which have an acid character and dye wool readily from an acid bath. They are usually divided into three groups : (1) Sulphonated basic colours, of which acid magenta, $C_{20}H_{11}(SO_3H)_3(OH)$ (NH_2), is a type; (2) Azo-sulphonic acids, of which azo yellow (amino-azo-benzene-disulphonate of sodium), $C_6H_4(SO_3Na)$ $N : NC_6H_3(SO_3Na)NH_2$, is a type; and (3) Nitro compounds, such as picric acid and naphthol yellow S.

Acid, Consolidated Sulphuric. See SOLIDIFIED SULPHURIC ACID.

Acid, Cresylic. See CRESYL HYDRATE.

Acid, Croceinic. See BAYER'S ACID.

Acid, Crocic. See BAYER'S ACID.

Acid, Disulphuric. See OLEUM.

Acid, Elixir, Haller's. Consists of alcohol, 1 part, and sulphuric acid, 1 part, by weight.

Acid Elixir of Vitriol. Aromatic sulphuric acid B.P. It contains tincture of ginger, spirit of cinnamon, sulphuric acid, and alcohol.

Acid Eosin. Tetrabrom-fluoresceine, $C_{20}H_8Br_4O_5$. The potassium salt forms soluble eosin, the sodium salt, Eosin C, and the ammonium salt, Eosin B. See EOSIN.

Acid, F. See DELTA ACID.

Acid Fuchsine. See ACID MAGENTA.

Acid Green (Vert Sulpho J). (The Bayer Co.; Meister, Lucius, and Brüning; Chemische Fabriken; K. Oehler.) See LIGHT GREEN SF, YELLOWISH.

Acid Green (Vert Sulpho B). (The Bayer Co. ; Meister, Lucius and Brüning ; Durand, Huguenin & Co. ; Farbwerk Griesheim ; Notzel, Istel & Co.) See LIGHT GREEN SF, BLUISH.

Acid Green. See HELVETIA GREEN, GUINEA GREEN B, FAST GREENS, and above.

Acid Green, Bluish. See LIGHT GREEN SF, BLUISH.

Acid Green D. See LIGHT GREEN SF, YELLOWISH.

Acid Green, Extra Conc. See LIGHT GREEN SF, YELLOWISH.

Acid Green G. A dyestuff It is a British equivalent of Guinea green B.

Acid Green GG. See LIGHT GREEN SF, YELLOWISH.

Acid Green JJ. See LIGHT GREEN SF, YELLOWISH.

Acid Green M. See LIGHT GREEN SF, BLUISH.

Acid Green SOF. See LIGHT GREEN SF, YELLOWISH.

Acid Green, Yellowish. See LIGHT GREEN SF, YELLOWISH.

Acid Hæmatoxylin. See EHRLICH'S ACID HÆMATOXYLIN.

Acid, Iso-anthraflavic. See ANTHRAFLAVIC ACID.

Acid, Laurent's Naphthalidinic. See LAURENT'S ACID.

Acid, Lepargylic. See ANCHOIC ACID.

Acid Levelling Red 2B. A dyestuff. It is a British equivalent of Azofuchsine B.

Acid Magenta (Acid Fuchsine, Fuchsine S, Magenta S, Cardinal S, Acid Roseine, Acid Rubine, Rubine S). A dyestuff. It is a mixture of the sodium or ammonium salts of the trisulphonic acids of rosaniline and pararosaniline. The sodium salts have the formulæ $C_{19}H_{16}N_3O_{10}S_3Na_3$, and $C_{20}H_{18}N_3O_{10}S_3Na_3$. Dyes wool and silk red from an acid bath. Impure Acid magenta is sold as Acid maroon, Acid cerise, Maroon S, Grenadine S, and Cerise S.

Acid Magenta II, Conc, L, N, N Extra. Dyestuffs. They are British equivalents of Acid magenta.

Acid Maroon. See MAROON S.

Acid Mauve B. A sulphonated mauvaniline. It is an acid dyestuff, and dyes wool and silk mauve from an acid bath.

Acid Metaborate. See BORAX.

Acid Milling Scarlet. A dyestuff. It is the sodium salt of azoxy-toluenedisazo - α - naphthol - monosulphonic - β-naphthol-disulphonic acid, $C_{34}H_{23}N_6O_{12}S_3Na_3$. Dyes wool scarlet from an iron bath.

Acid, Nordhausen. See FUMING SULPHURIC ACID.

Acid, NW. See NEVILE AND WINTHER'S ACID.

Acid of Amber. Succinic acid, COOH. CH_2CH_2COOH.

Acid of Lemons. Citric acid, $C_3H_4(OH)(COOH)_3.H_2O$.

Acid of Sugar. Oxalic acid, COOH. COOH.

Acidol (Bakusin, Myloin). Sodium salts of the mixed acids obtained from the alkaline refining lyes of Russian petroleum. They are used in soap-making. The name "Acidol" is also used for betaine hydrochloride, $CH_2.N(CH_3)_3$. Cl.COOH. It is also called Lycin, and is sold in the form of pastilles for stomach complaints.

Acidol-Pepsin. A proprietary preparation of Betaine hydrochloride and pepsin, used in the treatment of achlorhydria. (439).

Acidolamin. A German preparation. It is acidol-hexamethylene-tetramine, and has similar properties to acidol.

Acidol Chromate Violet B. An acid chrome-developing dyestuff.

Acidol Chrome Brilliant Blue BL. An acid chrome-developing dyestuff, giving navy blue shades on wool.

Acidol Fast Violet BL. An acid colour suitable for wool printing.

Acidol Grounding Yellow. An acid grounding dyestuff for leather.

Acidol-Pepsin. A proprietary preparation of betaine hydrochloride, and pepsin, used in the treatment of achlorhydria gastritis. (112).

Acid Orange. See ORANGE II.

Acid Orange G. A dyestuff. It is a British equivalent of Orange II.

Acid, Pararosilic. Aurine (q.v.).

Acid, Polygallic. See STRUTHIIN.

Acid Ponceau. See FAST ACID SCARLET.

Acid-proof Cements. See CEMENT, ACID-PROOF.

Acid-proof Ultramarine. See ULTRAMARINE, ACID-PROOF.

Acid, Pyrosulphuric. See FUMING SULPHURIC ACID.

Acid Reclaim. See RECLAIMED RUBBER.

Acid Red. A dyestuff. It is a British equivalent of Fast red.

Acid-resisting Alloys. These alloys vary considerably in composition, but many are alloys of nickel, chromium, and iron, often with small additions of molybdenum and manganese. One alloy contains nickel, zinc, and molybdenum, and another cobalt, zinc, and molybdenum. A few are made from nickel and silicon with small amounts of copper and aluminium.

Acid-resisting Bricks. See BLUE BRICKS.

Acid-resisting Wood. Planks of deal or pine which have been dipped into melted naphthalene, or have been painted with sodium silicate.

Acid, Rheic. See RHAPONTICIN.

Acid, Rhubarbaric. See RHAPONTICIN.

Acid Rosamine A (Violamine G). A dyestuff. It is the sodium salt of dimesidyl-m-amino-phenol-phthalein-sulphonic acid, $C_{38}H_{32}N_2O_6SNa_2$. Dyes wool and silk pink.

Acid Roseine. See ACID MAGENTA.

Acid, Rosilic. See AURINE.

Acid Rubine. See ACID MAGENTA.

Acid, Rumpff's. See BAYER'S ACID.

Acid Scarlet. A dyestuff. It is a British equivalent of Ponceau 2G.

Acid Scarlet R. A dyestuff. It is a British equivalent of New Coccine.

Acid, Selenhydric. See SELENIETTED HYDROGEN.

Acid Soaps. Soaps obtained by treating a dissolved soap with acid just insufficient to produce separation of the fatty acids. They are used to replace Turkey red oil in dyeing.

Acid, Tannic. See GALLOTANNIC ACID.

Acid Tar. The waste acid from the washing of crude light oils of coal tar. It is also the name applied to the acid which has been used to purify petroleum oil. It consists of sulphuric acid of specific gravity 1·74 which has been agitated with the once-distilled crude oil, allowed to settle, and drawn-off. It is used as a source of oil fuel by dilution, when the oil forms a top layer, and the diluted acid is used for the preparation of ammonium sulphate.

Acid, Thymic. See THYME CAMPHOR.

Acid Violet. A general name applied to a series of dyestuffs. They are the sodium salts of the sulphonated methyl and ethyl colours.

Acid Violet 6B (Berlin Aniline Co.). A dyestuff. It is the sodium salt of dimethyl-dibenzyl-triamino-triphenyl-carbinol-disulphonic acid, $C_{39}H_{41}N_3O_7$ S_2Na_2. Dyes wool bluish-violet from an acid bath.

Acid Violet 6B (J. R. Geigy). See FORMYL VIOLET S4B.

Acid Violet 6B (Bayer & Co.). See ACID VIOLET 4BN.

Acid Violet 7B (Société pour l'Industrie Chemique ; The Badische Co.). A dyestuff. It is the sodium salt of diethyl-dimethyl-diphenyl-triamino-triphenyl-carbinol-disulphonic acid, $C_{37}H_{37}N_3O_7S_2Na_2$. Dyes wool bluish-violet from an acid bath.

Acid Violet 7B (Farbwerk Muhlheim). See ACID VIOLET 4BN.

Acid Violet 4B Extra. See FORMYL VIOLET S4B.

Acid Violet 3BN. An acid colour, which dyes wool a less blue shade than 4BN.

Acid Violet 4BN (Acid Violet 6B, Acid Violet 7B, Acid Violet N). A dyestuff. It is the sodium salt of benzyl-pentamethyl-triamino-triphenyl-carbinol-sulphonic acid, $C_{31}H_{34}N_3O_4SNa$. Dyes wool bluish-violet.

Acid Violet 6BN. A dyestuff. It is the sodium salt of tetramethyl-p-tolyl-triamino-ethoxy-triphenyl-carbinol-sulphonic acid, $C_{32}H_{36}N_3O_5SNa$. Dyes wool and silk bluish-violet.

Acid Violet 7BN. A dyestuff. It is the sodium salt of tetramethyl-diphenyl-rosaniline-disulphonic acid, $C_{35}H_{33}N_3$ O_7S_2Na. Dyes wool and silk bluish-violet from an acid bath.

Acid Violet 7BS, 5BNS, BNS. Dyestuffs. They are the sulphonic acids of β-naphthyl-penta-alkyl-rosaniline, and dye wool bluish-violet.

Acid Violet N. See ACID VIOLET 4BN.

Acid Violet 4R. See FAST ACID VIOLET A2R.

Acid Violet 4RS (The Bayer Co.). A dyestuff. It is a mixture of Acid violet with Fuchsine S. See RED VIOLET 4RS.

Acid Violet 5RS. See RED VIOLET 5RS.

Acid Vitriolated Tartar. Potassium bisulphate, $KHSO_4$.

Acid, Vitriolic. See OIL OF VITRIOL.

Acid Whey. Cow's milk deprived of cream, boiled with a little cream of tartar, and coagulated with vinegar or lemon-juice.

Acid Wool Dyes. Colours which dye wool directly from an acid bath.

Acid Yellow (Fast Yellow G, Fast Yellow GL, Acid Yellow G, Fast Yellow, Fast Yellow Extra, New Yellow L, Fast Yellow, Greenish, Fast Yellow S, Solid Yellow S). A dyestuff. It is a mixture of sodium amino-azobenzene-disulphonate with some sodium amino-azobenzene - monosulphonate. Dyes wool and silk yellow from an acid bath.

Acid Yellow D. See ORANGE IV.

Acid Yellow G. See ACID YELLOW.

Acid Yellow 2G. See METANIL YELLOW S.

Acid Yellow OO. See BRILLIANT YELLOW.

Acid Yellow R. See FAST YELLOW R.

Acid Yellow RS. See TROPÆOLINE O.

Acid Yellow S. See NAPHTHOL YELLOW S.

Acid Yellow 79210. A dyestuff. It is a British equivalent of Tartrazine.

Acieral. A silver-white alloy of aluminium. It consists mainly of aluminium, with from 2·3–6·4 per cent. copper, 1·0–1·5 per cent. manganese, 0·7–1·4 per cent. iron, 0–0·98 per cent. nickel, 0·1–0·5 per cent. magnesium, 0–0·4 per cent. silicon, and often contains traces of silver, cobalt, tungsten, cadmium, and tin. It has been used in the manufacture of helmets.

Aci-jel. A proprietary preparation of acetic, boric, ricinoleic acids and oxyquinoline sulphate in a gel. A vaginal antiseptic. (369).

Acinitrazole. 2 - Acetamido - 5 - nitrothiazole. Aminitrazole. Trichorad. Tritheon.

Acitophosan. Calcium aspirin with calcium cinchophen.

Acitrin. The ethyl ester of 2-phenyl-quinoline-4-carboxylic acid, (Atophan), $C_9H_5N(C_6H_5)CO_2C_2H_5$. An analgesic used in the treatment of gout and rheumatism.

Aclobrom. A proprietary preparation containing propantheline bromide. An antispasmodic. (279).

Acme Rubber. A fine grade of American reclaimed rubber.

Acme Yellow. See TROPÆOLINE O.

Acmite. A pyroxene mineral, NaFe $(SiO_3)_2$.

Acmite-Augite. A mineral, synonymous with Aegirine-Augite.

Acnil. A proprietary preparation of resorcinol, precipitated sulphur, and cetrimide. An acne treatment. (188).

Acoin (Guanicaine). Di-p-anisil-mono-phenetyl-guanidine-hydrochloride, HCl. C (NHC$_6$H$_4$O (CH$_3$)) (NC$_6$H$_4$O (C$_2$H$_5$)) (NHC$_6$H$_4$O(CH$_3$)). Employed as a local anæsthetic.

Aconite Liniment. See A.B.C. LINIMENT.

Acoomo Seeds. See OTE SEEDS.

Acopyrin (Acetopyrin, Pyrosal, Phenazopirin). Antipyrine-acetyl-salicylate, $C_{11}H_{12}N_2O.C_2H_3O.OC_6H_4.COOH$. Prescribed in cases of acute rheumatism, also for fever and headache.

Acorite. A mineral. It is a variety of azorite.

Acorn Galls. See GALLS AND KNOPPERN.

Acorn Sugar. Quercitol, $C_6H_7(OH)_5$, found in acorns.

Acote. A trade name for a one-coat airdrying wood primer for oil paints. (254).

Acquerite. A native alloy of silver and mercury, found in Chile, approximating to the formula $Ag_{12}Hg$.

Acrafil. A registered trade name for flame retardant styrene-acrylonitrile.

Acraglas. Moulded acrylic synthetic resins.

Acratamine Dyestuffs. Acid dyestuffs suitable for wool.

Acrawax. A proprietary synthetic wax for lubrication and blending with other waxes to increase their melting-point.

Acrawax C. N,N'-ethylene. bis stearamide. A proprietary lubricant used in the manufacture of A.B.S., PVC and polystyrene. (440).

Acree-Rosenheim Reagent. A 1 : 6000 solution of commercial formalin. A proteid reagent.

Acrest. A proprietary polyester laminating resin. (441).

Acrex. Dinobuton. (501).

Acridan. An abbreviation of dihydroacridine.

Acridine Orange. See PHOSPHINE.

Acridine Orange L, LP. Dyestuffs. They are British equivalents of Acridine orange.

Acridine Orange NO. A dyestuff. It is the zinc double chloride of tetramethyl-diamino-acridine, $C_{17}H_{19}N_3HCl$. ZnCl$_2$. Dyes silk orange with greenish fluorescence, cotton mordanted with tannin, orange ; also leather.

Acridine Orange R Extra. A dyestuff. It is a salt of tetramethyl-diamino-phenyl-acridine, $C_{23}H_{23}N_3HCl$. Dyes tannined cotton, orange red.

Acridine Orange RS. A basic dyestuff for use in calico-printing.

Acridine Red. A dyestuff. It is a mixture of Acridine orange and Pyronine.

Acridine Red B, 2B, 3B. Dyestuffs. The chloride of tetraethyl-diamino-oxy-diphenyl-carbinol. Dyes silk or mordanted cotton, yellow shades of red.

Acridine Scarlet. A dyestuff. It is a mixture of Acridine orange and Pyronine.

Acridine Scarlet R, 2R, 3R. Dyestuffs. They are mixtures of Acridine orange with Pyronine.

Acridine Yellow. A dyestuff. It is the hydrochloride of diamino-dimethyl-acridine, $C_{15}H_{16}N_3Cl$. Dyes silk greenish-yellow with green fluorescence and cotton mordanted with tannin, yellow. See PHOSPHINE.

Acridine Yellow GR, R, 2R. Dyestuffs. They are British equivalents of Acridine yellow.

Acriflavine (Trypaflavine). A yellow dyestuff. It is 3 : 6-diamino-methyl-acridine-chloride-hydrochloride, NH$_2$. $C_6H_3.CH.N(CH_3)Cl.NH_2HCl$. A powerful antiseptic.

Acriflavine, Neutral. See EUFLAVINE.

Acriflex. 5-Aminoacridine, oil in water emulsion. (510).

Acrochordite. A mineral, As_2O_5. 2MnO.MgO.6H$_2$O.

Acrodel. An insecticide preparation. (511).

Acrolite. A proprietary synthetic resin. It is a condensation product of glycerol and phenol or homologues of phenol. It is soluble in alcohol and acetone, and melts at 300° C. It is used in varnishes and in the preparation of electrical insulating materials. Acrolite is also the trade mark for abrasive and refractory materials, the essential constituent of which is crystalline alumina.

Acronal. A registered trade name for polyacrylates. (49).

Acronal D. A registered trade mark for a polyacrylic ester dispersion. (49).

Acronal I. High viscosity methyl acrylate polymer sold in solid form or as a 27 per cent. solution in ethyl acetate. (3).

Acronal I-D. A polymer from methylacrylic ester sold as 25 per cent. and 40 per cent. emulsions. (3).

Acronal II. High viscosity ethyl acrylate polymer sold in solid form and as a 30 per cent. solution in ethyl acetate. Also a low viscosity ethyl acrylate polymer sold as 60 per cent. solution in toluene or 50 per cent. solution in polysolvan E (*q.v.*). (3).

Acronal II-D. A polymer from ethyl acrylic ester sold as 25 per cent. and 40 per cent. emulsions. (3).

Acronal IV. Low viscosity butyl acrylate polymer sold as 50 per cent. solution in ethyl acetate or 60 per cent. solution in Ligroin (*q.v.*). (3).

Acronal 14D. A plasticiser-free acrylic copolymer in the form of a 55 per cent. dispersion. (49).

Acronal 21D, 27D and 30D. Plasticiser-free dispersions of thermosetting copolymers of various acrylic asters. (49).

Acronal 160D. A plasticiser-free acrylic copolymer in the form of a 40 per cent. dispersion. (49).

Acronal 250-D. A copolymer from 80 parts of methyl acrylate, 20 parts polyvinyl benzoate and 2 parts of acrylic acid sold as 40 per cent. emulsion. (3).

Acronal 300-D. A copolymer from $\frac{1}{3}$ butyl acrylate, $\frac{1}{3}$ vinyl acetate and $\frac{1}{3}$ vinyl chloride sold as 50 per cent. emulsion. (3).

Acronal 350-D. A plasticiser-free acrylic type copolymer in the form of a 45 per cent. dispersion. (49).

Acronal 430-D. A copolymer from 50 parts vinyl isobutyl ether, 30 parts methyl acrylate and 20 parts acrylonitrile butyl ether sold as 30 per cent. and 50 per cent. emulsions. (3).

Acronal 450-D. A copolymer from 66 parts ethyl acrylate, 20 parts acrylonitrile butyl ether, 12 parts styrene and 2 parts acrylic acid sold as 40 per cent. emulsion. (3).

Acronal 500-D. A copolymer from 49 parts butyl acrylate, 49 parts vinyl acetate and 2 parts acrylic acid sold as 50 per cent. emulsion. (3).

Acronal 700. A low viscosity polymer composed of 75 per cent. polybutyl acrylate and 25 per cent. polyvinylisobutyl ether sold as 50 per cent. solution in ethyl acetate. (3).

Acronol. Basic dyestuff. (512).

Acronol Brilliant Blue. A dyestuff. It is a British equivalent of Setocyanine.

Acronol Yellow T. A dyestuff. It is a British equivalent of Thioflavine T.

Acrosyl. A saponified cresol disinfectant. It is a British lysol.

Acry-Ace. Polymethyl methacrylate. A proprietary moulding powder. (442).

Acrylafil G-40/20/FR and G-40/30/FR. Proprietary flame-retardant grades of styrene-acrylonitryle, reinforced with long glass. (415).

Acrylite. A trade mark for methyl methacrylate polymer. A moulding compound. (56).

Acrylite H, Type 11. Proprietary acrylic pellets used for injection moulding or in the extrusion of parts requiring maximum service temperature. (416).

Acrylite H, Type 12. Proprietary acrylic pellets used in the making of critical mouldings where better flow and maximum service temperatures are required. (416).

Acrylite H, Type 15. Proprietary acrylic pellets of extrusion grade, for use at maximum service temperature and possessing extra toughness. (416).

Acrylite M, Type 30. Proprietary acrylic pellets providing medium heat resistance, used in injection moulding and extrusion where greater flow is required. (416).

Acryloid. A proprietary trade name for acrylic ester resins used for coatings. Made by polymerising acrylic compounds. Pale in colour and suitable as a constituent of lacquers, etc.

Acrymul AM 123R. A proprietary trade name for a self cross-linking acrylic emulsion.

Acrysol. A proprietary trade name for acrylic varnish and lacquer resins for rubber and rubberised surfaces, etc.

Actal. A proprietary preparation containing sodium polyhydroxy aluminium monocarbonate hexitol complex. An antacid. Activated aluminas. (112).

Acterol. An irradiated ergosterol, a vitamin D concentrate.

Acthar. A proprietary preparation of corticotrophin for injection. (327).

Acthar Gel. A proprietary preparation of corticotrophin gelatin for injection. (327).

Acticarbone. A proprietary trade name for a peat based activated elemental carbon having a special micropore structure. Used for the decolorization of glucose, glycerine phthalate esters etc. (206).

Actidil. A proprietary preparation of triprolidine hydrochloride. (277).

Actifed. Triprolidine and pseudo-ephedrine. (514).

Actifed Compound Linctus. A proprietary preparation containing triprolidine hydrochloride, pseudo-ephedrine hydrochloride and codeine phosphate. A cough linctus. (277).

Actinac. A proprietary preparation of CHLORAMPHENICOL, HYDROCORTISONE acetate, butoxyethyl nicotinate, allantoin and sulphur, used in the treatment of acne. (443).

Actinolin. A mineral, synonymous with Actinolite.

Actinolite (Spanish Chalk). A mineral. It is a silicate of magnesium, calcium, and iron, and is an amianth variety of asbestos.

Actinomycin. Antimicrobial substances with antitumour activity produced by streptomyces antibioticus and streptomyces chrysomallus.

Actinote. A mineral, synonymous with Actinolite.

Activated Alumina. Partially dehydrated aluminium trihydrate in the form of hard porous lumps. Has a strong affinity for water and will remove it from air.

Activated Carbon. A carbon specially treated and prepared by distillation and the action of steam to remove hydrocarbons during manufacture, so that it is very highly adsorptive.

Activated Sludge. A material obtained by allowing the growth of microorganisms in the sludge deposited by sewage. It is used in the treatment of sewage (after it has been oxidised) by circulating it with the material, allowing it to stand, when the coagulated solids settle out.

Activax. Fowl pox vaccine. (514).

Active Diamyl. See DIAMYL, ACTIVE.

Active Oxygen. See OXYGEN, ACTIVE.

Activit. A proprietary rubber vulcanisation accelerator. It is thiocarbanilide.

Activol. Colour developer. (515).

Activox. Zinc oxide—rubber fast curing activator. (516).

Actol. Silver lactate, $AgC_3H_5O_3 + H_2O$. An antiseptic injection.

Actomol. A proprietary preparation of mebanazine. A monoamine oxidase inhibitor antidepressant. (2).

Acton. Ethyl-o-formate, $CH(OC_2H_5)_3$.

Actonorm. A proprietary preparation containing papaverine and aneurine hydrochlorides, atropine sulphate, malt diastase, phenobarbitone, pancreatin, aluminium hydroxide, calcium carbonate, kaolin, magnesium trisilicate and oil of peppermint. An antacid. (255).

Actonorm-Sed. A proprietary antacid preparation containing aluminium hydroxide, magnesium hydroxide, simethicone, phenobarbitone and belladonna. (255).

Actos 50. A proprietary preparation of prolonged action adrenocorticotrophic hormone. (256).

Actrapid. A proprietary preparation of neutral pork insulin. (182).

Actril. Selective weed killer. (507).

Actriol. A proprietary trade name for Epioestriol. (316).

Actynolite. A mineral, synonymous with Actinolite.

Acykal. A German preparation. It is a silver cyanide containing 54 per cent. silver, and used as a germicide.

Acylanid. ACETYL DIGITOXIN. α-digitoxin monoacetate. (267).

Ac-zol. A mixture of ammonia, copper, zinc salts, and phenol. A wood preservative.

Adabee. A proprietary preparation containing Vitamins A and C, thiamine, riboflavine, pyridoxine and nicotinamide. Vitamin supplement. (258).

Adalin (Uradal, Planadalin). Diethylbromo - acetyl - urea $(C_2H_5)_2.CBr.CO.NH_2$. A proprietary preparation of carbromal. A sedative. (112).

Adalox. A proprietary trade name for coated abrasives for sanding metal or plastics.

Ad-aluminium. A proprietary trade name for an alloy of 82 per cent. copper, 15 per cent. zinc, 2 per cent. aluminium, and 1 per cent. tin.

Ad. A.M. A proprietary preparation of ephedrine hydrochloride and butethamate citrate. A bronchial antispasmodic. (273).

Adamac. Specially prepared tar for roads. (517).

Adamant. Synonymous term for either diamond or corundum.

Adamanta. An American name for a German substitute for rubber, made from linseed oil, sulphur, lime, and resin.

Adamantine. See DIAMOND-BORON. Also a proprietary brand of fireproof and alkali-proof material, used in sulphur burners.

Adamantine Cement. See CEMENT, ADAMANTINE.

Adamantine Spar. A mineral, synonymous with corundum.

Adamine. A mineral from Chile. It is a hydrated zinc arsenate, $Zn_3(AsO_3)_2H_2O$.

Adamite. A trade mark applied to certain abrasive and refractory materials the essential constituent of which is crystalline alumina.

Adamite. A proprietary high-carbon nickel-chromium iron alloy used for dies.

Adamon. (1) Dibromo-dihydro-cinnamic acid bornyl ester. A sedative.

Adamon. (2) The registered trade name for an antispasmodic. It is β-diethylaminoethyl - (α - methyl - 2,5 - endomethylene - Δ^3 - tetrahydrobenzhydryl) ether bromomethylate. (844).

Adamsite (D.M.). 10-chloro-5 : 10-dihydro-phenarsazine. Diphenylamine-arsenious chloride, a poison gas. A greenish-black variety of mica is also known by this name. A mineral. It is a variety of Muscovite.

Adanon. A proprietary trade name for Methadone.

Adansonia Fibre. A fibre obtained from the bark of *Adansonia digitata*. Used for making rope and sacking, and also for special paper.

Adaphax 758. A proprietary factice containing neither sulphur nor chlorine used as a processing aid in the manufacture of polyurethanes and PVC. It enables good electrical properties to be maintained in PVC and tackiness to be reduced. It can also be used in white nitrile mixes. Acetone extract 16 per cent. specific gravity 1·01. (417).

Adaprin. A proprietary preparation of nicotinamide and acetomenaphthone, used to treat chilblains. (319).

Adarola. A proprietary casein plastic.

Adcora. A proprietary trade name for coating materials as follows:—

　P6. A neoprene coating with good chemical and abrasion resistance and good flexibility.

　H2. A hypalon (*q.v.*) coating for tanks etc. Possesses good resistance to halogens.

　V. A Viton (fluoroelastomer) coating with very good heat stability resistant to temperatures up to 260° C.

　A3. A hard coal tar epoxy coating for tanks and pipes for effluent and for resistance to fumes and weathering.

Adcora SP. A solventless epoxide-based coating giving a hard finish resembling enamel.

Adcortyl. A proprietary preparation of triamcinolone. (326).

Adcortyl-A. A proprietary name for the acetonide of Triamcinolone.

Add-add. The leaves of *Celastraceæ serratus*, of Abyssinia. Used as an antiperiodic.

Adder Oil. See OIL OF VIPERS.

Adelan. Wool fat.

Adelfolite. A mineral, synonymous with Adelpholite.

Adelite. A mineral. It is a calcium magnesium hydroxy - arsenate, Ca (MgOH)AsO$_4$.

Adelphane. A proprietary preparation of reserpine, and dihydrallazine sulphate. An antihypertensive. (18).

Adelpholite. A mineral. It is a niobate of iron and manganese.

Adenos Cotton. The finest type of cotton from the Orient.

Adenosine Phosphate. Adenosine 5-(dihydrogenphosphate). Adenosine monophosphate. A M P. My-B-Den.

Adenotriphos. A proprietary preparation of adenosine triphosphoric acid sodium. (323).

Adenyl. A proprietary preparation of ADENOSINE mono-phosphoric acid used in the treatment of rheumatic diseases. (323).

Adepsine. See OZOKERINE.

Adepsine Oil. See PARAFFIN, LIQUID.

Adexolin. A proprietary preparation containing Vitamin A palmitate and calciferol. (335).

Adhæsol (Copal Lotion). A lotion containing 350 parts copal, 30 parts benzoin, 30 parts tolu, 20 parts oil of thyme, 3 parts α-naphthol, and 1000 parts ether. Used as an antiseptic varnish in diphtheritic affections of the throat.

Adheso. A proprietary trade name for a synthetic wax consisting of a modified polymerised terpene.

Adhesor. A sticky rubber substitute, used to a certain extent in friction.

Adicillin. 6 - [D(+) - 5 - Amino - 5 - carboxyvaleramide] penicillanic acid. Penicillin N.

Adigan. A trade mark for a digitalis preparation containing all the active principles except digitonin, this having been removed.

Adigeite. A mineral. Its formula is Mg$_5$Si$_3$O$_{11}$4H$_2$O.

Adinol. Textile auxiliaries based on triethyl citrate. (521).

Adinole. A mineral consisting of metallic silver and albite.

Adipite. A mineral. It is an aluminosilicate of Ca, Mg and K, similar to chabizite.

Adiplantin. A German preparation used for application to wounds.

Adipocere. A wax-like mass left when animal bodies decompose in the earth. It is a mixture of the palmitates of calcium and potassium.

Adipocerite. A mineral, synonymous with Hatchettite.

Adipocire. A mineral, synonymous with Hatchettite.

Adipol XX. A proprietary trade name for di-isodecyl adipate. A vinyl plasticiser. (73).

Adipol 2EH. A proprietary trade name for di-2-ethyl hexyl adipate. A vinyl plasticiser. (73).

Adipol 10A. A proprietary trade name for di-iso-octyl phthalate. A vinyl plasticiser. (73).

Adipol BCA. A proprietary trade name for dibutoxy ethyl adipate. A vinyl plasticiser. (73).

Adipol ODY. A proprietary trade name for N-octyl n-decyl adipate. A vinyl plasticiser. (73).

Adipol 810. A proprietary trade name for isodecyl octyl adipate. A vinyl plasticiser. (73).

Adiprene. A trade mark for urethane rubbers, L-245, L-265, L-700, L-767, resistant to oil and radiation. (10).

Adirondackite. A rubber substitute made from sulphurised oils. Used in the proofing of cloth, and as an insulator.

Adjab Fat. Njave butter (q.v.).

Adjective Dyestuffs (Mordant colours). Dyestuffs which require a mordant for fixing them are called adjective dyestuffs or mordant colours.

Admex 301. A proprietary trade name for a plasticiser manufactured from Soy bean. (74).

Admex 710. A proprietary trade name for an epoxy polyester plasticiser. (74).

Admiralty Antifriction Metal. Usually an alloy of 85 per cent. tin and 15 per cent. antimony.

Admiralty Bell Metal. An alloy of 78 per cent. copper and 22 per cent. tin.

Admiralty Brass. See BRASS.

Admiralty Brazing Metal. An alloy of 90 per cent. copper and 10 per cent. zinc.

Admiralty Gun Metal. Alloys. Some contain from 87–90 per cent. copper and from 10–13 per cent. tin, whilst others consist of from 86–88 per cent. copper, 6–10 per cent. tin, and 2–6 per cent. zinc.

Admiralty Nickel. See ADNIC.

Admiralty White Metal. A bearing alloy containing 86 per cent. tin, 8·5 per cent. antimony, and 5·5 per cent. copper.

Admos Alloys. Brass alloys of varying composition containing small amounts of tin, nickel, lead, and iron.

Admul. Glycerol monostearate. (522).

Admune. A proprietary preparation of inactivated influenza virus used to confer immunity to the disease. (444).

Adnephrin. See ADRENALIN.

Adnic (Admiralty Nickel). An alloy containing 70 per cent. copper, 29 per cent. nickel, and 1 per cent. tin. It is resistant to corrosion and heat.

Adobe. Clay-like masses used for making bricks.

Adocain. Adonidin and cocaine hydrochloride.

Adolyl. A German commercial preparation containing 0·1 gram of the sodium salt of diethyl-barbiturate and 0·1 gram acetyl-salicylic acid in each cubic centimetre. An analgesic.

Adorin. A powder containing a polymeric form of formaldehyde. Used in cases of perspiration of the feet.

Adovern. A German cardio-tonic, consisting of a mixture of the two active glucosides present in *Adonis vernalis*.

AD-Pilo. A proprietary preparation of pilocarpine hydrochloride, epinephrine bitartrate, used to treat glaucoma. (329).

Adraganthin. See BASSORIN.

Adrenal. See ADRENALIN.

Adrenalin (Adnephrin, Adrenal, Adrenamine, Adrenin, Adrin, Epinephran, Epinephrine, Epirenan, Hæmostasin, Hemisine, Nephridine, Paranephrine, Renalina, Renoform, Renostypticin, Renostyptin, Styptirenal, Supradin, Suprarenaline, Suprarenine, Surrenine, Vasoconstrictin). o-Dihydroxy-phenyl methyl-amino-methyl-carbinol hydrochloride, $(OH)_2.C_6H_3.CH(OH).CH_2.$ $NH(CH_3)HCl$. It is the active principle of the suprarenal gland, and is used as a styptic.

Adrenamine. See ADRENALIN.

Adrenapax. A proprietary preparation of adrenaline in a cream base. An embrocation. (385).

Adrenin. See ADRENALIN.

Adrenocorticotrophic Hormone. Corticotrophin.

Adrenocortrophin ACTH. Corticotrophin.

Adrenolone. Methyl-amino-acetocatechol, $C_6H_3(OH)_2.CO.CH_2NH.CH_3$.

Adrenoxyl. A proprietary preparation of adrenochrome monosemicarbazone dihydrate. (269).

Adreson. A proprietary preparation of cortisone acetate. (316).

Adreucaine. See EUDRENINE.

Adriamycin. A proprietary pharmaceutical used in the chemotherapy of malignant disease. (353).

Adrin. A proprietary brand of Adrenalin (q.v.).

Adroit. Cutting oils. (523).

Adronal. See CYCLOHEXANOL.

Adronal Acetate (Hexalin Acetate). Cyclohexanol acetate, a resin solvent.

Adroyd. A proprietary preparation of oxymetholone. An anabolic agent (264).

Adsol. See JAPANESE ACID CLAY.

Adularia. A mineral. It is a clear crystalline variety of orthoclase (q.v.).

Adurol. Monochlor and monobromhydroquinones, $C_6H_3Cl(OH)_2$, and $C_6H_3Br(OH)_2$. Photographic developers.

Advagum. A proprietary plasticiser. A terpene resin.

Advance Alloy. An alloy of copper with from 44–46 per cent. nickel, and small quantities of iron and manganese.

Advastab A70. A proprietary trade name for a stabiliser for PVC suitable for use in contact with food. It is an aminocrotonic ester.

Advastab A80. A trade name for powdered aminocrotonic ester used as a stabiliser for unplasticised PVC film and moulded bottles. (14).

Advawax 280. A proprietary range of synthetic amide compositions used in paints, adhesives, asphalts and resins. (419).

Advita. A proprietary concentrate of vitamins A and D.

Advitacon. Oil soluble vitamins. (522).

Advitagel. Flour and confectionery emulsifier. (522).

Advitagent. Detergent powder. (522).

Advitamix. Animal feed supplements. (522).

Advitamul. Bakery release agent. (522).

Advitaroma. Butter and meat flavouring. (522).

Advitastat. Bakery release agent. (522).

Adwin Balm. A proprietary preparation of adrenalin, methyl salicylate and capsicum in a cream base. An embrocation. (342).

Aedelforsite. A mineral It is a variety of wollastonite.

Aedelite. A mineral, variously regarded as synonymous with Edelforsite, Laumontite, Natrolite or Prehnite.

Aegerite. A trade name for Elaterite.

Aegirine. A mineral. Its composition is $4[NaFe^{\cdots}Si_2O_6]$.

Aegirine-Augite. A mineral. Its composition is $8[(Na,Ca,Mg,Fe^{\cdot\cdot},Fe^{\cdots},Al)(Si,Al,Fe^{\cdots})O_3]$.

Aegirinediopside. A mineral, synonymous with Aegirine-Augite.

Aegirine-Hedenbergite. A mineral. Its composition is $4[(Ca,Fe^{\cdot\cdot},NaFe^{\cdot\cdot\cdot})S_2O_6]$.

Aegirite. Acmite (q.v.).

Aegrosan. A German preparation. It consists of an alcoholic solution of 0·8 per cent. ferrous saccharate and 0·4 per cent. calcium phosphate, and is used in the treatment of anæmia and rachitis.

Aenigmatite. A mineral. It is a titanosilicate of ferrous iron and sodium. $4[Na_4(Fe^{\cdot\cdot},Fe^{\cdots},Ti)_{13}S_{12}O_{42}]$.

Aeonite. A nickel silver containing 20 per cent. nickel. A trade name for Elaterite.

Aerated Butter. See BUTTER, RENOVATED.

Aerial Cement. A term applied to cements which set in air, the setting being due to desiccation and carbonation.

Aerialite. A proprietary synthetic resin.

Aerinite. A mineral which is a mixture of calcium-bearing leptochlorite with pyroxene, quartz, and spinel. It is an aluminosilicate of $Fe^{\cdot\cdot}Fe^{\cdots}Mg$ and Ca.

Aero. A proprietary trade name for rosin-glycerol varnish and lacquer resins.

Aero DOP. A proprietary trade name for di-2-ethyl hexyl phthalate. A plasticiser for vinyl compounds. (56).

Aero DOPI. A proprietary trade name for di-iso-octyl phthalate. A plasticiser for vinyl compounds. (56).

Aerodag 504. A proprietary trade name for a colloidal dispersion of graphite in white spirit, supplied as an aerosol. (207).

Aerogene Gas. A variety of air gas obtained by mixing air and gasoline, or benzine.

Aerolite. An alloy of aluminium with small amounts of copper, iron, magnesium, silicon, and zinc. It is also a proprietary name for a safety-glass.

Aeromatt. Precipitated calcium carbonate for cosmetics. (524).

Aero Metal. An aluminium alloy, consisting mainly of aluminium with from 2·1–2·9 per cent. magnesium, 0·3–1·3 per cent. iron, and 0·2–0·6 per cent. copper.

Aeromin. An alloy of 91·6 per cent. aluminium and 8·4 per cent. magnesium.

Aeron. See SELERON.

Aeroplex. A proprietary safety-glass.

Aero Quality Lumarith. A proprietary trade name for cellulose acetate transparent sheeting for use in aviation.

Aerosil. A trade name for finely divided silicon dioxide. There are several grades the particle sizes of which range from 7 to 40 millimicrons. Used as a thixotropic agent for paints, polishes, epoxy resin, cosmetics and adhesives. (6).

Aerosil COK 84. A trade name for a mixture of Aerosil and alumina in 5 : 1 ratio. It is suited particularly for thickening aqueous and other polar systems. (6).

Aerosil Composition. A trade name for a mixture of aerosil with 15 per cent. starch, specially designed for tableting. (6).

Aerosite. A name which has been applied to the mineral pyrargyrite.

Aerosols. A proprietary trade name for a series of wetting agents. Usually sulphonated esters.

Aerosporin. A proprietary trade name for the sulphate of Polymyxin B. An antibiotic. (277).

Aerotrol. A proprietary trade name for the hydrochloride of Isoprenaline. An aerosol bronchodilator. (525).

Aero X. A proprietary rubber vulcanisation accelerator. It is aniline-diisopropyl-dithio-phosphate.

Aerox. Aero X.

Aerozonin-gala. A fluid containing formaldehyde, alcohol, and aromatic oils. Used as a perfume to overcome tobacco smoke.

Aerugite. A nickel ore, $2NiO.Ni_3(AsO_4)_3$.

Aeschynite. A mineral, $Ce_2(Ca.Fe)_2(Ti.Th)_8O_2.2Ce(CbO_3)_4$.

Aescorcin. See ESCORCIN.

Æ Super Nickel. A proprietary copper-nickel alloy containing 80 per cent. copper and 20 per cent. nickel. It is used for the manufacture of condenser tubes.

Aeternol. A proprietary synthetic resin.

Aethiops Mineral. A mineral, synonymous with Metacinnabarite.

Æthone. A proprietary preparation. It is ethyl formate, $HCOOC_2H_5$, used as a sedative.

Aethrol. A plastic of the pyroxylin-cellulose acetate type.

Afcolene. A proprietary polystyrene. (445).

Afdigel. A proprietary preparation of digitoxin and aminophylline, used in cardiac failure. (815).

Afenil. Calcium chloride urea.

Afghan Yellow. A dyestuff produced by warming p-nitro-toluene-sulphonic acid with sodium hydroxide. It dyes unmordanted cotton in the presence of sodium chloride, and silk and wool from an acid bath.

African Phosphates. Mineral phosphates found in Tunis and Algeria. They contain from 55–65 per cent. calcium phosphate. Others found at Safaga and Kosseir contain 60–70 per cent. calcium phosphate. Fertilisers.

African Saffron. Carthamus.

Afridi Wax. See ROGHAN.

Afridol (Mercuritol). The sodium salt of hydroxy-mercury-o-toluic acid, $CH_3.C_6H_3(CO_2Na)HgOH$. Said to be valuable as an antiseptic in the treatment of skin diseases.

Afridol Violet. A dyestuff. $C_{33}H_{22}O_{15}N_8S_4Na_4$, produced by coupling di-amino-diphenyl-urea with 2 molecules of H acid. It has been used in the treatment of trypanosomes.

Afrin. A proprietary trade name for the hydrochloride of oxymetazoline.

Afrodit. A mineral, synonymous with Aphrodite.

Afrol. Timber insecticide. (511).

Afror Tyne Powder. A low-freezing explosive containing 8–10 per cent. of a nitrated mixture of glycerin and ethylene glycol, 56·5–59·5 per cent. ammonium nitrate, 7·5–9·5 per cent. wood meal, 21–23 per cent. sodium chloride, 0–2 per cent. magnesium carbonate, and 0–2 per cent. nitro-naphthalene.

A.F.S. A Canada balsam substitute made from aniline, formaldehyde and sulphur.

Afsal. See URASOL.

Aftalosa. A mineral, synonymous with Aphthitalite.

Aftalosia. A mineral, synonymous with Aphthitalite.

Afterschörl. A mineral, synonymous with Axinite.

Afwillite. A mineral of Kimberley. It is a hydrated calcium silicate and approximates to the formula, $3CaO.2SiO_2.3H_2O$.

Agalite (Mineral pulp, Asbestine Pulp). A New Jersey variety of talc (hydrated magnesium silicate). Used by paper-makers.

Agalma Black 10B. A dyestuff prepared from H-acid (8-amino-1-naphthol-3 : 6-disulphonic acid) by adding it to an equal amount of diazotised p-nitraniline, then coupling with diazotised aniline. It dyes wool and silk a greenish-black from an acid bath. It is the basis of many black dye mixtures on the market. Vegetable fibres are unstained by it.

Agalmatolite (Pagodite). A soft stone used in China for carving. Generally, the mineral pyrophyllite is known by this name, but sometimes it is applied to steatite and pinite.

Agalyn. A proprietary pyroxylin material used for dentures.

Agaphite. A Persian variety of turquoise.

Agapite. A synonym for Turquoise.

Agar-agar (Bengal Gelatin or Isinglass, Ceylon Gelatin or Isinglass, Chinese Gelatin or Isinglass, Japanese Gelatin or Isinglass, Japan Agar, Layor Carang, Vegetable Glue). The material obtained from certain varieties of algæ by boiling water. Used as a sizing for cloth, and as a culture medium for bacteria.

Agarase. An agar-agar preparation.

Agaricin. Agaric acid, a resin acid, obtained by extraction with alcohol of the fruit bodies of *Polyporus officinalis* and *Agaricus albus*. A febrifuge.

Agaric Mineral. A mineral. It is a white earthy variety of calcite.

Agar, Japan. See AGAR-AGAR.

Agarol. A proprietary preparation containing liquid paraffin, phenolphthalein and agar. A laxative. (262).

Agate. Silica, SiO_2, in crystalline, crypto-crystalline, and colloidal form. Also see POTATO STONES and SCOTCH PEBBLES.

Agate-Jasper. An impure variety of agate.

Agate Ware (Enamelled Iron-granite) Enamelled iron.

Agate, White. Chalcedony (q.v.).

Agathine (Salizone, Cosmin). Salicyl-α-methyl-phenyl-hydrazide,$C_6H_5(CH_3)$. N.NHCO(C_6H_4OH). Used in the treatment of neuralgia and rheumatism.

Agatine. A proprietary phenol-formaldehyde resin in the form of sheet, rods, tubes, etc.

Agatit. A German reclaimed rubber. It is sold as a solid and as an emulsion, and has been used as a substitute for leather.

Agavin. Thiosolucin-dihydrostreptomycin preparation for the veterinary field. (507).

Agchem V. A proprietary trade name for a ketone-aldehyde condensate used as an accelerator for rubber and as an antioxidant. (161).

Age (Axin). The fat of *Coccus Axin*, growing in Mexico. It consists of the glycerides of lauric and axinic acids.

Agene. A patented bleaching agent for flour. It is nitrogen trichloride, NCl_3.

Age-Rite AK. A trade name for an antioxidant. Polymethyldihydroquinoline. (290).

Age-Rite Alba. A proprietary trade name for a rubber antioxidant. It is para-benzyl-oxyphenol. (162).

Age-Rite Gel. A proprietary product. It is a combination or composition, consisting of ditolylamines and a selected petroleum wax. An antioxidant for rubber.

Age-Rite Hipar. A proprietary trade name for a mixture of isopropoxy diphenylamine, diphenyl-para-phenylamine and phenyl-beta-naphthylamine. An antioxidant for rubber. (162).

Age-Rite HP. A proprietary trade name for a mixture of phenyl-beta-naphthylamine and diphenyl-para-phenylene diamine. (162).

Age-Rite Powder. A proprietary antioxidant for rubber. It is phenyl-β-naphthylamine, and has a melting-point of 100° C. and a specific gravity of 1·19.

Age-Rite Resin. A proprietary product obtained by the interaction of acetaldol with α-naphthylamine. It is a cherry-red resin, brittle when cold but soft when warmed, flowing at 100° C. It is sparingly soluble in petroleum ether, but freely soluble in benzene, has a specific gravity of 1·16, and from 0·5–1 per cent. is used as an antioxidant in rubber to increase the length of life of this material.

Age Rite Resin D. A proprietary trade name for trimethyl-dihydroquinoline in a polymerised state. (162).

Age-Rite Spar. Styrenated phenol. A proprietary anti-oxidant. (290).

Age-Rite Stalite. A proprietary trade name for polymerised trimethyl-dihydro-quinoline. An antioxidant. (162).

Age-Rite Stalite S. Alkylated diphenylamines. A proprietary anti-oxidant. (290).

Age-Rite White. A proprietary anti-oxidant for rubber. It is di-β-naphthyl-p-phenylene-diamine.

Agesulf. Sulphosalicylic acid and colloidal silver proteinate.

Agfa. A proprietary intensifier for photographic negatives. It is mercuric thiocyanate dissolved in a solution of ammonium thiocyanate.

Agglomerated Coal. See BRIQUETTES.

Aggrepox. A proprietary range of epoxy resin mastics. (420).

Agidex. Glucoamylase for conversion of starch into dextrose. (520).

Agit. A German product. It is a lime preparation, each dose containing calcium salicylate, calcium lactate, and acetylene. It has the same use as acetyl-salicylic acid.

Aglaite. A mineral. It is a silicate of potassium, sodium, and aluminium.

Aglaurite. A mineral. It is a variety of orthoclase felspar.

Aglutella Gentili. A proprietary group of pasta food preparations—gluten-free and low in electrolyte and protein—used in diets in cases of phenylketonuric and renal failure. (446).

Agmatine. Guanidine-butyl-amine, NH_2C:(NH).NH.$(CH_2)_4NH_2$. Found in ergot.

Agnesite. A mineral of Cornwall. It is a carbonate of bismuth.

Agnin (Agnolin). Purified wool fat.

Agnolin. Lanolin substitute. (526). See AGNIN.

Agnolite. A mineral. It is a silicate of manganese.

Agnowax. Wool wax alcohols. (526).

Ago. A proprietary leather cement consisting of (a) a solution of celluloid in acetone and (b) pure acetone.

Agobilin. Strontium cholate, mixed with small quantities of strontium salicylate and diacetyl-phenol-phthalein.

Agocholine. A proprietary preparation of peptine and magnesium sulphate. A laxative. (802).

Agoleum. A German preparation. It is a 1 per cent. suspension of colloidal silver in vegetable oil. Used in Cystitis.

Agomensin. The hydrosoluble ovarian substance which provokes hyperæmia of the female sexual organs, stimulates ovarian function, and activates menstruation.

Agonol. Santalol benzoate.

Agotan. See ATOPHAN.

Agprosol. A proprietary preparation containing acetylsalicylic acid, codeine phosphate, phenacetin, phenolphthalein, citric acid and calcium carbonate. An analgesic. (259).

Agprotane. A proprietary preparation of bismuth subgallate, resorcinol, zinc oxide, balsam of Peru and benzocaine. Suppositories. (259).

Agral. Wetting agent. (511).

Agramite. A mineral. It is a variety of meteoric iron.

Agricol. A proprietary range of alginates used for root dipping. (536).

Agricolite (Ahelestite). A mineral. It is found in Saxony, and contains 57 per cent. Bi_2O_3, 12 per cent. $Fe_2P_2O_6$, and 30 per cent. As_2O_3.

Agricultural Copper Sulphate. The commercial copper sulphate used in agriculture. It contains iron sulphate as an impurity.

Agricultural Paints. These are generally enamels made from copal varnish, but sometimes quick-drying colours which require varnishing are used.

Agrilite Alloy. A copper-lead alloy containing small amounts of tin.

Agripol. A proprietary trade name for a synthetic vulcanisable rubber made from soya bean and ethyl cellulose compound.

Agrisol A and B. Cresol preparations. Used as preservatives for rubber latex.

Agritol. A proprietary ammonium dynamite explosive.

Agritox. Selective weed killers. (507).

Agrocide. Insecticides. (511).

Agro Cotto. Concentrated lemon juice from Sicily.

Agrosan. Organo mercury seed dressing. (511).

Agrosol. Liquid mercury seed dressing. (511).

Agrothion. Liquid insecticide. (511).

Agroxone. Selective weed killer. (511).

Agucarina. Saccharine (q.v.).

Ague Salt. Quinine sulphate.

Aguilarite. A mineral approximating to Ag (SSe).

Agulin. Sheep vaccine. (520).

Agurin. Theobromine-sodium-acetate, $C_7H_7N_4O_2Na.C_2H_3O_2Na$. A diuretic.

Agustite. A mineral. It is a variety of apatite from Saxony.

Aguttan. Oxy-quinoline-sulphonic acid.

Agyrol. See LARGIN.

Ahelestite. See AGRICOLITE.

Aiathesin. o-Oxy-benzoyl-alcohol (salicyl - alcohol), $C_6H_4(OH)CH_2OH$. Antirheumatic and antiseptic.

Aich Metal (Gedge's Metal, Sterro Metal). An alloy similar to Delta metal, except that it contains iron. It usually consists of 60 per cent. copper, 38 per cent. zinc, and 1·5–2 per cent. iron, and is used for sheathing ships.

Ai Hao. A Chinese drug prepared from the leaves of *Artemisia vulgaris*. A remedy for hæmorrhage and diarrhœa.

Aikinite. A mineral $Bi_2S_3.2PbS.Cu_2S$.

Ailanthus Silk. See FAGARA SILK.

Aimatolite. Synonym for Hematolite.

Aimax. Methallibure. (512).

Ainalite (Tantalum Cassiterite). A cassiterite containing about 9 per cent. tantalic oxide found in Finland.

Air, Alkaline. See VOLATILE ALKALI.

Air Blue. A pigment. It is basic copper carbonate, $2CuCO_3.Cu(OH)_2$. Also see VERDITER BLUE and CHESSYLITE.

Airbron. A proprietary preparation of acetylcysteine. A bronchial inhalant. (182).

Air, Dephlogisticated. See VITAL AIR.

Airedale. Dyestuffs for leather. (508).

Air Gas. A gaseous fuel obtained by treating heated carbonaceous matter (coal or coke) with an air blast. It contains from 30–33 per cent. carbon monoxide, 1–5 per cent. hydrogen, and 61–65 per cent. nitrogen. The name is also applied to a mixture of air and some volatile inflammable hydrocarbon, usually petrol or carburine, but sometimes benzene. See PRODUCER GAS.

Airglow. Bright nickel plating process. (528).

Air-hardening Steel. A trade term applied to a manganese tool steel containing some tungsten, which hardens when cooled in air.

Air, Nitrous. See NITROUS GAS OR AIR.

Air, Pure. See VITAL AIR.

Air Saltpetre (Norwegian Saltpetre). A mixture of calcium nitrite and nitrate. It is produced by passing air over a series of high intensity alternating arcs, absorbing the gases produced by lime water, and evaporating.

Airstrip. Waterproof non-occlusive plaster. (529).

Aithalite. Asbolane (q.v.).

Aix Oil (Nice Oil, Var Oil, Riviera Oil, Bari Oil). Commercial varieties of edible olive oil.

Aizen. A semi-synthetic rubber made from chlorinated rubber. Japanese origin.

Ajacol. See GUAIACOL ETHYL.

Ajax Alloy. A bearing metal. It contains 30–70 per cent. iron, 25–50 per cent. nickel, and 5–20 per cent. copper.

Ajax Phosphor Bronze. An alloy containing 81 per cent. copper, 11 per cent. tin, 7 per cent. lead, and 0·4 per cent. phosphorus.

Ajax Plastic Bronze. See PLASTIC BRONZE.

Ajax Powder. An explosive consisting of potassium perchlorate, nitroglycerol, ammonium oxalate, wood meal, and small quantities of collodion cotton, and nitro-toluenes.

Ajkaite (Ajkite). A fossil resin found in Hungary. It has a saponification value of approximately 160 and an acid value of 0.

Ajkite. See AJKAITE.

Ajowan. An Indian name for the fruit of a plant similar to Caraway. It is a source of thymol, and is used as a spice.

Ajuin. Synonym for Haüyne.

A.K. A proprietary plastic of the phenol-formaldehyde type, with asbestos.

Akakia. An extract of the pods of acacia. It is sold in Indian bazaars.

Akalbir. The root of *Datisca cannabina*. It has been extensively used in Kashmir, and throughout the Himalayas as a yellow dyestuff. It dyes alum mordanted silk.

Akailt. A proprietary casein material.

Akanda. See AK MUDAR.

Akarittom Fat. A solid fat from *Parinarium laurinum*. It melts at 49–50° C. and has an iodine value of 214.

Akbar. A rubber vulcanisation accelerator. It is a condensation product of formaldehyde and *p*-toluidine.

Akco Resins. A proprietary synthetic phenolic resin for varnish manufacture.

Akee Oil. See OIL OF AKEE.

Akerite. A mineral. It is a pale-blue variety of spinel.

Akermanite. A mineral, $4CaO.SiO_2$.

Akesol. Valeri-amido-quinine.

Akineton. The hydrochloride or lactate of Biperiden. Used to treat Parkinsonism. (85).

Akinetone. A trade mark for a benzyl compound, $C_6H_5CH_2NH_2.CO.C_6H_4.CO_2H$, an anti-spasmodic. (227).

Aklar. A modified copolymer of monochlorofluoroethylene. (424).

Ak Mudar (Akanda, Akra Rui, Erukku Erukkam). The bark of *Calotropis gigantea* and *C. procera*. An important Indian drug.

Akra Rui. See AK MUDAR.

Akrite. A tool alloy. It contains 38 per cent. cobalt, 30 per cent. chromium, 10 per cent. nickel, 4 per cent. molybdenum, and 2–5 per cent. carbon.

Akroflex A. A proprietary trade name for an antioxidant. It is composed of secondary aromatic amines. (2).

Akroflex B. A proprietary trade name for an antioxidant composed of secondary amines. (2).

Akroflex C. A proprietary trade name for a rubber antioxidant consisting of a mixture of 65 parts of phenyl-alpha-naphthylamine and 35 parts of diphenyl-para-phenylene diamine. (2).

Akroflex CD. A proprietary trade name for an antioxidant composed of diphenyl-para-phenylene-diamine and NEOZONE D. (10).

Akroflex D. A proprietary trade name for an antioxidant consisting of aromatic secondary amines. (2).

Akroflex DAZ. A proprietary amine blend used as a combined anti-oxidant and anti-ozonant. (10).

Akrol. A proprietary synthetic resin.

Akron. An alloy of 63 per cent. copper, 36 per cent. zinc, and 1 per cent. tin.

Akrotherm. A proprietary preparation of histamine acid phosphate, acetylcholine chloride, and oxycholesterol in a cream base. Skin stimulant. (383).

Aksel. A rubber vulcanisation accelerator. It is stated to be the zinc salt of an allyl dithiocarbamic acid.

Aktiplast F. A proprietary preparation of zinc salts of unsaturated high carboxylic acids, used as a peptiser for rubber. For use in natural rubber, isoprene rubber, SBR and BR, and in blends of these polymers. (421).

Aktivin. See CHLORAMINE T.

Akto Protein. A preparation of casein for use in the treatment of gonorrhœa.

Akulon. Nylon 6 and 66. (294).

Akulon B2-65. A high-viscosity grade of Nylon 6. (426).

Akulon B.10. A grade of Nylon 6 possessing increased flexibility and toughness. (426).

Akulon B 10-X355. A grade of Nylon 6 possessing higher viscosity than AKULON B10. (426).

Akulon K and M. Proprietary grades of Nylon 6. (5).

Akulon K2-S. A grade of Nylon used in injection moulding, giving a transparent product. (426).

Akulon K2 Special. A grade of nylon 6 used in the extrusion of products in the food processing industry. (426).

Akulon K10 2G340. A grade of injection-moulded nylon used in the production of snap-on caps and catches. (426).

Akulon M2. An extrusion grade of Nylon 6 possessing good impact strength and resistance to abrasion and fatigue. (426).

Akulon M2-2G340. A grade of Nylon 66 which crystallises rapidly and thus reduces processing time. (426).

Akulon M 10. A grade of Nylon 6 possessing greater flexibility and toughness than AKULON M2. (426).

Akulon R2. A grade of Nylon 66. (5).

Akulon R-600. A grade of Nylon 66 of medium viscosity, having greater rigidity and surface hardness than Nylon 6 grades, and a higher softening temperature. (426).

Akund. A vegetable down of the kapok class, from *Asclepias* species of South Africa.

Alabandin (Alabandite). A mineral. It is manganese sulphide MnS.

Alabandite. See ALABANDIN.

Alabaster. A form of gypsum, $CaSO_4.2H_2O$ used for ornamental carvings. Also see ONYX OF TECALI.

Alabaster, Oriental. A compact form of marble, $CaCO_3$.

Alabrastine. A water-mixing cement made by calcining plaster of paris and immersing in a solution of alum. Also see CARBON.

Alacet (Alaret). A preservative for fruit preparations, made from carbon monoxide and caustic soda. It contains from 50–60 per cent. formic acid.

Alacetan. Aluminium aceto-lactate.

Alacreatine. Lactyl guanidine, $C_4H_9O_2N_3$.

Aladar. See ALUDUR.

Aladdinite. A material which is stated to be identical with Galalith (*q.v.*).

Alaite. A mineral, $V_2O_5.H_2O$.

Alalite. Diopside (*q.v.*).

Alamask. Industrial reodorants. (507).

Alamosite. A mineral. It is lead silicate, $PbSiO_3$.

Alan-gilan. Cananga oil, a neutral oil from *Cananga odorata*.

Alanin. Propionic-glycocoll, $CH_3.CH(NH_2)COOH$.

Alanin-mercury. Mercury-amino-propionate. Proposed as a substitute for corrosive sublimate in hypodermic injections.

Al Anodised. See ELOXAL.

Alant Camphor. Helenin, C_6H_8O.

Alantin. See HELENIN.

Alapurine. A special name given to the purest kinds of lanoline (wool fat).

Alaret. See ALACET.

Alargan. An alloy of aluminium and silver, the surface having been dusted with platinum black and hammered or subjected to pressure. A platinum substitute.

Alasil. A proprietary preparation of calcium acetyl-salicylate and colloidal aluminium hydroxide.

Alaskaite. A mineral. It is an argentiferous galeno-bismuthite, $Bi_2S_3.(Cu_2.Ag_2.Pb)S$. The name is also applied to a rock containing quartz and felspar.

Alathon. A registered trade name for polyethylene and other copolymer resins. (10). See also ALKATHENE.

Alavac. A group of proprietary preparations of alum-precipitated extracts of allergenic materials used in the desensitisation treatment of allergic conditions. (272).

Albacar. Highly refined calcite. (85).

Albacer. A proprietary synthetic wax for increasing melting-point of waxes.

Albalith. A white, light-resisting lithopone, used in the paint and rubber industries.

Albamycin. A proprietary trade name for the calcium salt of Novobiocin.

Albamycin G.U. A proprietary preparation of novobiocin calcium and sulphamethizole. Urinary antiseptic. (325).

Albamycin T. A proprietary preparation containing Novobiocin and Tetracycline. An antibiotic. (325).

Albane. A resin, $C_{10}H_{16}O$, insoluble in alcohol, which is a constituent of guttapercha.

Albanite. A bituminous material found in Albania.

Albanol. See OIL WHITE. Also a plasticiser made by condensing a viscous ester from balsam. (3).

Albaphos Dental Na 211. A proprietary trade name for sodium monofluorophosphate. A fluorine component for toothpastes, the toxic effects of which are only $\frac{1}{3}$ of those of sodium fluoride. (9).

Albanose. A leucite rock.

Albargin. Silver-gelatose, a compound of gelatose and silver nitrate. Used in the treatment of wounds, and as an injection for gonorrhœa.

Albarium. A lime obtained by burning marble. Used for stucco.

Albasan. A proprietary antioxidant for rubber. It is composed of a naphthol mixed with a polyalkylene polamine. (163).

Albata. A nickel-brass or low nickel-silver containing about 8 per cent. nickel.

Albatex B8. A proprietary trade name for sodium *m*-nitrobenzene sulphonate.

Albatex OR. A proprietary trade name for a non-foaming polyvalent amide. Used as a levelling agent for vat dyes. (18).

Albatra Metal. A nickel silver. It contains 57·5 per cent. copper, 22·5 per cent. zinc, 18·75 per cent. nickel, and 1·25 per cent. lead.

Albegal CL. A proprietary trade name for an ester of sulphonated fat. It is used as a levelling agent in wool dyeing. (18).

Albene. See VEGETABLE BUTTER.

Albene. An acetate silk deadened with pigments in the spinning process.

Alberene. A blue-grey soapstone mined in Virginia.

Alberit MF. A proprietary melamine formaldehyde thermosetting moulding compound used in the manufacture of tracking-resistant mouldings for the electrical industry. (427).

Alberit MP. A proprietary melamine/phenol - formaldehyde thermosetting moulding compound. (427).

Alberit PF. A proprietary phenol-formaldehyde resin thermosetting moulding compound. (427).

Alberit VP. A proprietary unsaturated polyester thermosetting moulding compound used in the production of impact-resistant mouldings. (427).

Albert. Basic slag for fertilising purposes. (509).

Albertan. A compound of aluminium with phenolic bodies. An antiseptic.

Albertat. A proprietary range of chemical fillers, extenders and additives for products containing synthetic resins, such as additives for thickening and preventing setting in paints and varnish. (4).

Albert Coal. Albertite (*q.v.*).

Albertite. A jet-black mineral resembling asphalt. It is used in America for producing oil and coke, and contains 62 per cent. volatile matter, and 37 per cent. carbon. It is fusible with difficulty only slightly soluble in carbon disulphide, and has a specific gravity of 1·07–1·10.

Albertol. A trade mark for a series of alkyd-phenolic resin combinations. (4).

Albertol IIIL. A phenol-resin condensation product melting at 106–133° C. It has a saponification value of 15·8, is insoluble in alcohol, but soluble in linseed oil. It is stated to be a good substitute for kauri gum in the manufacture of oil varnishes. Rosin glycerine - diane - formaldehyde condensate in the presence of alkali. (4).

Albertol 142-R. Butyl phenol formaldehyde. (4).

Albertol 175-A. The aluminium salt of unesterified Albertol III L (*q.v.*). (4).

Albertol 237-R. Di-*iso* butylphenolformaldehyde. (4).

Albertol 326-R (387L). A rosin modified phenolic resin used in aircraft primers made from 1 part Diane (*q.v.*), 1 part rosin and 0.1 part paraformaldehyde. (4).

Albertol 347Q. An ester gum-phenolic combination made from xylenol-formaldehyde rosin, pentaerythritol and glycerogen (*q.v.*). (4).

Albertol 369-Q (209-L). An ester gumphenolic combination made from phenol-formaldehyde, rosin, pentaerythritol and glycerogen (*q.v.*). (4).

Albichthol. See ICHTHYOL-ALBUMEN.

Albiogen. Tetramethyl-ammonium oxalate, $[CO_2N(CH_3)_4]_2$.

Albion Metal. A sheet of metal containing tin and lead. It is formed by pressing together sheets of these metals.

Albion Oil. A compressed lubricating oil used for looms. It has a specific gravity of 0·914 at 60° C. and a flash-point of 380° F.

Albit. A German explosive.

Albite (Pericline, Soda Felspar). A mineral. It is a silicate of sodium and aluminium, $Al_2O_3.3SiO_2.Na_2O.3SiO_2$.

Albocarbon. See CARBON.

Alboferine. See ALBOFERRIN.

Alboferrin (Alboferine). A phospho-albumen preparation of iron, a tonic.

Albolene. See PARAFFIN, LIQUID.

Alboleum. Oil insecticide. (511).

Alboline. See PARAFFIN, LIQUID.

Albolit. A proprietary phenol-formaldehyde synthetic resin.

Albondur. A Bondur alloy (see BONDUR) coated on each side with pure aluminium to improve corrosion resistance.

Albone C. A proprietary trade name for a brand of hydrogen peroxide, 100 vols., containing 13 per cent. available oxygen.

Albor Die Steel. A proprietary steel containing small amounts of chromium, molybdenum, and carbon.

Alboresin. A proprietary urea-formaldehyde synthetic resin. Moulding composition.

Alborubol. See PELLIDOL.

Albral. Aluminium bronze flux. (531).

Albroman. Isopropyl-brom-acetyl-carbamide. A German soporific.

Albron. A proprietary name for atomised powder, flake powder and pastes of pure aluminium. (131).

Albucid. A proprietary preparation of sulphacetamide sodium and ephedrine hydrochloride. Nasal spray. (265).

Albumaid Preparations. A group of proprietary preparations of beef protein hydrolysate, amino acids, carbohydrates, vitamins and minerals used in the treatment of aminoacidaemias and mal-absorption syndromes. (447).

Albumen Paper. A photographic printing-out paper, the sensitive substance being silver chloride, and the ground, albumen.

Albumen Powder (Aleuronate). Dried and powdered gluten.

Albustix. A prepared test strip of tetra-bromphenol blue with a citrate buffer, used to detect protein in urine. (807).

Albutannin (Protan). Albumen tannate.

Alcacement. See ALCEMENT.

Alcacit. A Czechoslovakian acid and alkali-proof covering compound.

Alcad. A duralumin coated with pure aluminium.

Alcan. A trade mark for alloys of aluminium coded as follows:

    GB-99·8 per cent. Wrought, non heat-treatable high purity aluminium.

GB-1S. Wrought, non heat-treatable 99·5 per cent. aluminium.

GB-2S. Wrought, non heat-treatable commercially pure aluminium.

GB-3S. Wrought, non heat-treatable aluminium. Stronger and harder than 2S.

GB-B53S, 54S, D54S, A56S, M57S. Wrought non heat-treatable aluminium alloys in which magnesium is the main additive.

GB-50S. Wrought heat-treatable aluminium alloy. Forms well in the " W " (solution heat-treated) condition.

GB-B51S, 65S. Wrought, heat-treatable medium strength alloys of aluminium.

GB-100. Cast, non heat-treatable commercially pure aluminium.

GB-160. Cast, non heat-treatable aluminium alloy.

GB-B320. Cast, non heat-treatable medium strength aluminium alloy.

GB-B116. Cast, heat-treatable alloy available in four conditions " M " (as cast), " P " (precipitation treated), " W " (solution treated) and " WP " (fully heat treated).

GB-350. Cast, heat-treatable, aluminium alloy specially impact resistant. Good resistance to marine conditions. (132).

Alcement (Alcacement). Fused cement prepared in the electric furnace from bauxite and lime. It contains approximately 40 per cent. CaO, 40 per cent. Al_2O_3, 10 per cent. SiO_2, and 10 per cent. Fe_2O_3.

Alchemy. Paint driers and esters and plasticisers. (532).

Alcho. A term used to denote a substance containing aluminium carbonate. It is used medicinally. It is also the name for an ester gum. See ESTER GUMS.

Alcian. Dyestuffs, blues, greens and yellow dyes formed by introducing chloromethyl groups into phthalocyanin and its derivatives by means of dichlorodimethyl ether in pyridine containing aluminium chloride. (512).

Alcian Dyes. Blue, green and yellow dies formed by introducing chloromethyl groups into phthalocyanine and its derivatives by means of dichlorodimethyl ether in pyridine containing aluminium chloride.

Alcin. A proprietary preparation containing sodium magnesium aluminium silicate and basic magnesium aluminate. An antacid. (243).

Alcoa 2-S. A proprietary trade name for a commercially pure aluminium.

Alcoa 3-S. A proprietary alloy of aluminium, containing small amounts of copper, iron, silicon, and zinc.

Alcoa 32-S. A proprietary alloy of aluminium with 12 per cent. silicon, 0·8 per cent. nickel, 1 per cent. magnesium, and 0·8 per cent. copper.

Alcoa 43. A proprietary alloy of aluminium and silicon. Contains 5 per cent. silicon.

Alcoa 47. A proprietary alloy of aluminium with 12·5 per cent. silicon.

Alcoa 108. A proprietary aluminium alloy with 3 per cent. silicon and 4 per cent. copper.

Alcoa 112. A proprietary aluminium alloy containing 7–8·5 per cent. copper, 1–2 per cent. zinc, and up to 1·7 per cent. of other metals, mostly iron.

Alcoa 122. A proprietary alloy of aluminium with 10 per cent. copper, 0·2 per cent magnesium, and 1·2 per cent. iron.

Alcoa 145. A proprietary aluminium alloy containing 10 per cent. zinc, 2·5 per cent. copper, and 1·2 per cent. iron.

Alcoa 195. A proprietary alloy containing aluminium with 4 per cent. copper.

Alcoa 220-TA. A proprietary aluminium alloy containing aluminium with 10 per cent. magnesium.

Alcoa 356. A proprietary alloy of aluminium containing 4 per cent. silicon, 0·3 per cent. magnesium.

Alcoa 515. A proprietary alloy containing aluminium with magnesium, silicon, and iron.

Alcoa 535. A proprietary alloy of aluminium with 0·25 per cent. chromium, 1·25 per cent. magnesium, and 0·7 per cent. silicon.

Alcoa 24-S. A proprietary wrought aluminium alloy containing 93·7 per cent. aluminium, 1·5 per cent. magnesium, 4·2 per cent. copper, 0·6 per cent. manganese.

Alcobon. A proprietary preparation of FLUCYTOSINE. A systemic anti-fungal agent. (314).

Alcobronze. An alloy of copper and aluminium of golden colour. Very hard.

Alcoform. Low water content formaldehyde solutions in a variety of alcohols. (89).

Alcogas. A trade mark for a mixture of alcohol (anhydrous) and hydrocarbons in varying proportions.

Alcohol. Ethyl alcohol, C_2H_5OH. Ordinary alcohol contains 90 per cent. ethyl alcohol.

Alcohol, Absolute. Ethyl alcohol, containing not more than 1 per cent. water. It has a flash-point of 55° F.

Alcohol, Denatured. See METHYLATED SPIRIT.

Alcoholene. A mixture of alcohol and ether.

Alcohol-Hydrocarbon Gas. A mixture of alcohol and hydrocarbon vapours. Used for lighting and motive power.

Alcohol, Industrial. An ethyl alcohol of 66–67 over proof, containing 95–96 per cent. ethyl alcohol. Used for the manufacture of esters, chloroform, iodoform, dye intermediates, etc.

Alcohol of Vinegar. Glacial acetic acid.

Alcohol, Phenic. See PHENIC ACID.

Alcohol, Wood. See METHANOL.

Alcolec 532. A proprietary preparation of vinyl-based resin in bead form. A carboxylated vinyl copolymer used in the formulation of flexographic printing inks and paper lacquers. (296).

Alcolite. A proprietary pyroxylin product used for denture purposes.

Alcolube CL. A cationic polyethylene/ lanolin emulsion used as a substantive softener for polyamide and polyester fibres. (565).

Alcon. A proprietary name for an acetal copolymer used as a moulding material. (2).

Alcopal FA. A foaming agent for aqueous systems used for carpet backing. (296).

Alcopan. Bephenium hydrochloride. (514).

Alcopar. A proprietary preparation containing bephenium hydroxynaphthoate. An anthelmintic. (277).

Alcos-Anal. A proprietary preparation of sodium morrhuate, benzocaine and chlorothymol used in the treatment of haemorrhoids. (370).

Alcosperse. A range of speciality dispersing and levelling agents. They are used in the dyeing of synthetic fibres. (296).

Alcovar. A proprietary trade name for fast dyes for spirit and cellulose varnishes. (533).

Alcresta Ipecacuanha. A trade mark for a medicinal preparation containing the alkaloids of ipecacuanha for human use.

Alcumite. A proprietary corrosion-resisting alloy containing 87·5 per cent. copper, 7·5 per cent. aluminium, 3·5 per cent. iron, and 1·5 per cent. nickel.

Alcuronium Chloride. Diallyldinor-toxiferin dichloride. Alloferin.

Aldacol. Textile stripping aid. (534).

Aldactide. A proprietary preparation of spironolactone and hydroflumethiazide. A diuretic. (285).

Aldactone. A proprietary preparation of spironolactone. A diuretic. (285).

Aldal. A new French aluminium-copper alloy of the duralumin type, but containing less copper.

Aldamine. A proprietary trade name for acetaldehyde-ammonia CH_3COCH.

OH.NH_2. The material is used in the manufacture of plastics and as a pickling inhibitor for steel, and is a rubber vulcanising accelerator.

Aldehol A. An oxidised kerosene to be used in U.S.A. for denaturing methylated spirit.

Aldehyde. Acetaldehyde, CH_3CHO.

Aldehyde C14. A proprietary flaming material. Undecalactone.

Aldehyde Green (Aniline Green, Usébe Green). A diphenyl-methane dyestuff. It is a quinoline derivative of rosaniline, and dyes silk green from an acid bath.

Alder Bark. The bark of *Alnus glutinosa*. Used for fixing yellow dyes and as a tanning material.

Alderlin. A proprietary trade name for the hydrochloride of Pronethalol.

Alderton's Solution. A solution of ammonium ichthosulphonate (1 oz.), in glycerol (10 oz.).

Aldiphen. See DINITRA.

Aldocorten. A proprietary preparation of aldosterone. (18).

Aldoform. A formaldehyde preparation.

Aldogen. A mixture of trioxymethylene and bleaching powder. An antiseptic.

Aldol. β-Oxy-butyric-aldehyde, CH_2 CH(OH).CH_2.COH. A hypnotic.

Aldomet. A proprietary preparation of methyldopa. An antihypertensive. (310).

Aldomet Injection. The hydrochloride of Methyldopate.

Aldones. Flavour bases. (504).

Aldosterone. 11β,21 - dihydroxy - 3, 20 - dioxopregn - 4 - en - 18 - al. Aldocorten; Electrocortin.

Aldoxim "AX". A proprietary trade name for butyraldoxime used as anti-skinning agent in paints and printing inks. (190).

Aldrey. A new aluminium alloy of Swiss origin. It is used in the form of wire and cables for electrical conductors. It contains 98·7 per cent. aluminium, 0·6 per cent. silicon, 0·4 per cent. magnesium, and 0·3 per cent. iron. It has the same specific gravity as aluminium (2·7) with only 8 per cent. less electrical conductivity. It has a tensile strength of 35 kilos per sq. mm. and a resistance to flexure 80 per cent. higher than aluminium wire.

Aldrin. Hexachlorooctahydronaphthalene. A powerful insecticide.

Aldur. A condensation product of formaldehyde and urea, a new window-glass substitute.

Al-dur-bra. A patented alloy of 76 per cent. copper, 22 per cent. zinc, and 2 per cent. aluminium.

Aldydale. A proprietary phenol-formaldehyde synthetic resin.

Ale. A fermented drink, prepared by the action of yeast upon malted barley. It contains from 6-8 per cent. alcohol.

Alegar. Vinegar.

Alembicol-D. A proprietary preparation of mainly C_8 & C_{10} triglycerides derived from cocoanut oil, used in the treatment of mal-absorption. (448).

Alentina. A German preparation. It consists of a material containing albumen and sugar ; a nutrient.

Alepol. A proprietary preparation of selected sodium salts of hydnocarpus oil acids.

Alepsal. A preparation of pheno-barbitone with belladonna and caffeine.

Aletodin. See ASPIRIN.

Aleudrin (Dichloramal). The carbamic acid ester of α-dichloro-isopropyl alcohol $(CH_2Cl)_2.CH.O.NH_2$. An analgesic and sedative. A proprietary trade name for the sulphate of Isoprenaline. A bronchodilator. (252).

Aleuronate. See ALBUMEN POWDER.

Aleurone Grains. Granules deposited in the cells of seeds of plants. They consist mainly of globulins, and are used by the plant as food material during germination.

Alevaire. A proprietary preparation containing Tyloxapol. A mucolytic. (112).

Alexander Green. See MOUNTAIN GREEN.

Alexandrian Laurel Oil. See LAUREL NUT OIL.

Alexandrite. Beryllium aluminate, a precious stone.

Alexandrolite. A mineral containing hydrated oxides of chromium, aluminium, and silicon.

Alexipon. Ethyl-acetyl-salicylate.

Alexite. A proprietary trade name for an aluminium oxide abrasive.

Alfa. A variety of esparto grass used in the manufacture of paper. It is also the term for a synthetic tannin, a red-brown liquid containing 23 per cent. tanning substance, 11 per cent. nontannins, 66 per cent. water, and a trace of sulphuric acid.

Alfa-841. A proprietary aliphatic polyurethane elastomeric polymer used as an adhesive and as a coating. It is serviceable at temperatures up to 300° F. and flexible down to −60° F. It is especially used as an adhesive for flexible substances. (76).

Alfa-antar. See YXIN.

Alfenide. See NICKEL SILVERS.

Alferium. A proprietary alloy of aluminium with 2·5 per cent. copper, 0·62 per cent. magnesium, 0·5 per cent. manganese, and 0·3 per cent. silicon.

Alferric. Aluminium sulphate. (513).

Alficetyn. Preparation of chloramphenicol. (510).

Alfloc. Water treatment chemicals. (512).

Alflorone. A proprietary trade name for the 21-acetate of Fludrocortisone.

Alfodex. Fungal-α-amylase. (520).

Alfol. A proprietary trade name for an aluminium foil in a crumpled condition used for heat insulation.

Alforder. A proprietary synthetic resin.

Alformin. A 16 per cent. solution of basic aluminium formate, $Al_2(OH)_2(HCO_2)_4$. An astringent and antiseptic. It is also the name for a German preparation of finely divided aluminium oxide used for tooth powder.

Alfrax. A proprietary trade name for refractory material produced from electrically fused crystalline alumina. Used as a catalyst carrier.

Alfrax B301. A commercial grade of bubble aluminium oxide. (292).

Alftalat. A proprietary name for glyceryl phthalate resins. (4).

Algalex 104. A solution of the sodium salt of a chlorinated phenyl derivative of methane containing surface active and antifoam agents. A non-toxic non-corrosive bactericide for water systems. (1).

Algalith. A proprietary algine plastic.

Algamine. Horticultural algæcide. (535).

Algarobilla. A vegetable tanning material. It consists of the pods of *Cæsulpinias brevifolia* of Chile, and contains about 60 per cent. tannin.

Algarobillin. A dye product obtained from the carob tree, *Ceratonia siliqua*, found in the Argentine. It is employed for dyeing cloth, khaki.

Algaroth, Powder of. See POWDER OF ALGAROTH.

Algarotti Powder. See POWDER OF ALGAROTH.

Algarovilla. A Columbian name for a copal resin obtained there.

Algerite. A mineral. It is an altered scapolite.

Alger Metal. An alloy of 90 per cent. tin and 10 per cent. antimony. A silvery-white alloy used in making jewellery.

Algesal. A proprietary preparation of diethylamine salicylate. A rubefacient. (822).

Algestone Acetonide. A progestational steroid. 16α,17α - Isopropylidenedioxypregn - 4 - ene - 3,20 - dione.

Algier's Metal. A jeweller's alloy. (a) Consists of 90 per cent. tin and 10 per cent. antimony, and is used for the manufacture of forks and spoons ;

(b) contains 94·5 per cent. tin, 5 per cent. copper, and 0·5 per cent. antimony. It is used for making handbells.

Algin. A gelatinous substance which is the residue from the water maceration of seaweed, during the process of obtaining iodine. It is used as a substitute for isinglass.

Alginade. Alginate ice cream stabiliser. (536).

Algine. A proprietary seaweed celluloid. A moulding composition.

Alginoid Iron. See ALGIRON.

Algipan Balm. A proprietary preparation of methylnicotinate, glycol salicylate, histamine and capsicin. An embrocation. (245).

Algiron (Alginoid iron). An iron compound of alginic acid (from seaweed). It contains 11 per cent. iron.

Algistat. Preparations for water treatment. (527).

Algocratine. Phenyl-amido-xanthine-citrate.

Algodon. Cotton wool.

Algodon de Seda. The fibre of *Calotropis gigantea* is known in Venezuela by this name.

Algodonite (Mohawkite). A mineral. It is an arsenide of copper Cu_6As.

Algoflon. A registered trade mark for PTFE resins. (61).

Algolan (Algolave). The salicylic ester of propyl-dihydroxy-butyric acid, $C_6H_4(OH)CO.O.CH_2.C(CH_3)(OH).CO_2 C_3H_7$. An antipyretic.

Algolave. See ALGOLAN.

Algol. A trade mark for a range of dyestuffs. (983).

Algol Blue C. A dyestuff. It is equivalent to Indanthrene blue GC.

Algol Blue CF. A dyestuff. It is the chloro derivative of indanthrene.

Algol Blue 3G. A dyestuff. It is a dihydroxy-indanthrene.

Algol Blue K. A dyestuff. It is *n*-dimethyl-indanthrene.

Algol Blue 3R (Algol Brilliant Violet 2B). A dyestuff. It consists of dibenzoyl-diamino-anthrarufin.

Algol Bordeaux 3B. A dyestuff. It is an anthraquinone-imide.

Algol Brilliant Orange FR. A dyestuff. It is benzoyl-1 : 2 : 4-triamino-anthraquinone.

Algol Brilliant Red 2B. A dyestuff. It is 1 : 5-dibenzoyl-diamino-4-hydroxy-anthraquinone.

Algol Brilliant Violet 2B. See ALGOL BLUE 2B.

Algol Green B. A dyestuff. It consists of dibromo-diamino-indanthrene.

Algol Grey B. A dyestuff. It is 1 : 5-diamino-anthraquinone condensed with α-chlor-anthraquinone. The product is reduced with alkali sulphide, and then nitrated.

Algol Olive R. A dyestuff. It is prepared by the action of chloro-sulphonic acid upon dibenzoyl-diamino-anthraquinone.

Algol Orange R. A dyestuff. It is α-β-dianthraquinonyl-amine.

Algol Pink. A dyestuff. It consists of benzoyl-4-amino-1-hydroxy-anthraquinone.

Algol Red 5G. A dyestuff. It is dibenzoyl-1 : 4-diamino-anthraquinone.

Algol Red R Extra. A dyestuff. It is dibenzoyl-1 : 5-diamino-8-hydroxy-anthraquinone.

Algol Scarlet G. A dyestuff. It consists of benzoyl-1-amino-4-methoxy-anthraquinone.

Algol Violet B. A dyestuff. It is benzoyl-amino-4 : 5 : 8-trihydroxy-anthraquinone.

Algol Yellow 3G. A dyestuff. It is succinyl-1-amino-anthraquinone.

Algol Yellow R (Hydranthrene Yellow ARNAG New, Indanthrene Yellow GK, Caledon Yellow 3G). A dyestuff. It is dibenzoyl-1 : 5-diamino-anthraquinone.

Algol Yellow WG (Hydranthrene Yellow AGR). A dyestuff. It is benzoyl-1-amino-anthraquinone.

Algrol. Textile detergent. (537).

Algulose. A very pure cellulose used in paper-making. It is obtained from kelp.

Alibated Iron. Iron coated with aluminium to form a protective covering.

Alidine. The dihydrochloride of Anileridine.

Alimemazine. Trimeprazine.

Alipamide. A diuretic currently undergoing clinical trial as "CI-546". 4-Chloro - 2′,2′ - dimethyl - 3 - sulpha-moylbenzohydrazide.

Alipite. A mineral. It is a silicate of nickel and magnesium.

Aliso. A proprietary trade name for an aluminium isopropoxide. (190).

Aliso "B". A proprietary trade name for an aluminium isopropoxide/sec. butoxide. (190).

Alisonite. A mineral of Chile. It is a sulphide of copper and lead, PbS. $3Cu_2S$.

Alite. The chief constituent of Portland cement. It consists mainly of calcium ortho-silicate, $3CaO.SiO_2$, with calcium oxide, CaO, and a certain amount of calcium aluminate, and ferrite.

Alival. 3-Iodo-1 : 2-dihydroxy-propane, $CH_2I.CHOH.CH_2OH$. Used as a substitute for iodides in medicine.

Alizanthrene Blue GC. A dyestuff. It is a British equivalent of Indanthrene blue GC.

Alizanthrene Blue GCP. A dyestuff. It is a British equivalent of Indanthrene blue GCD.

Alizanthrene Dark Blue O. A dyestuff. It is a British equivalent of Indanthrene dark blue BO.

Alizanthrene Olive G. A dyestuff. It is a British equivalent of Indanthrene olive G.

Alizarin (Alizarin No. 1, Alizarin No. 1B new, Alizarin I^e, Alizarin P, Alizarin V, Alizarin VI, Alizarin Blue Shade, Alizarin for Violet, Alizarin Red NB). A dyestuff. It is 1 : 2-dioxy-anthraquinone, $C_6H_4(CO)_2.C_6H_2(OH)_2$. Commercial alizarin is sold in the form of a yellow paste, usually containing 20 per cent. alizarin, but it is sometimes met with as dry powdery lumps containing starch and glycerol. It dyes cotton mordanted with alumina, scarlet red ; with tin, bluish red ; with iron, violet ; and with chrome, puce brown. It is used for dyeing " Turkey red " upon cotton mordanted with " sulphated oil " and alumina.

Alizarin Acid Blue BB. See ACID ALIZARIN BLUE, BB, GR.

Alizarin Acid Blues. Dyestuffs prepared from anthrachrysone. They consist of polyoxy-anthraquinone-disulphonic acids, and dye wool from an acid bath.

Alizarin Acid Green. A dyestuff prepared by reducing dinitro-anthrachrysone-disulphonic acid.

Alizarin Astrol. A dyestuff. It is 1 - methyl - amino - 4 - sulpho-toluidine-anthraquinone, $C_6H_4(CO)_2.C_6H_2(NH.C_7H_6SO_3H)(NH.CH_3)$.

Alizarin Black Blue. See ALIZARIN CYANINE BLACK G.

Alizarin Black P. A dyestuff. It is flavopurpurin-quinoline, $C_{17}H_9NO_5$. It dyes chromed wool violet-grey to black. Used also in calico printing.

Alizarin Black S (Alizarin black SW, SRW, WR, Alizarin Blue Black RW, Alizarin Blue Black SW, Naphthazarin S). A dye-stuff. It is the sodium bisulphite compound of naphthazarin (di-oxynaphthoquinone), $C_{10}H_7SO_7Na$. It dyes chromed wool black, and is used in printing.

Alizarin Black S. A dyestuff. It is the sodium bisulphite compound of Alizarin Black P, $C_{17}H_9NO_5.2NaHSO_3$. It dyes chromed wool grey to black, and is also used in calico printing.

Alizarin Black SRA. See ALIZARIN BLACK S.

Alizarin Black SW. See ALIZARIN BLACK S.

Alizarin Black WR. See ALIZARIN BLACK S.

Alizarin Black WX. A dyestuff. It is tetrahydroxy-naphthalene.

Alizarin Blue (Alizarin Blue A, AB, ABI, BSS, DNW, F, GW, R, RR). A dyestuff. It is dioxy-anthraquinone-β-quinoline, $C_{17}H_9NO_4$. Dyes mordanted fabrics blue. The mark BSS is the sodium salt.

Alizarin Blue BB, GR. See ACID ALIZARIN BLUE BB, GR.

Alizarin Blue Black RW. See ALIZARIN BLACK S.

Alizarin Blue Black SW. See ALIZARIN BLACK S.

Alizarin Blue Green. A dyestuff. It is the sulphonic acid of a trihydroxy-anthraquinoline-quinone.

Alizarin Blue, Insoluble Paste. A dyestuff. It is a British equivalent of Alizarin blue.

Alizarin Blue S (Alizarin Blue ABS, Anthracene Blue S). A dyestuff. It is the bisulphite compound of Alizarin blue (dioxy-anthraquinone-β-quinoline), $C_{17}H_{11}NO_{10}S_2Na_2$. Dyes chromed fabrics. Used in printing.

Alizarin Blue SNG, SNW, SWN. Dyestuffs identical or isomeric with anthracene blues (q.v.).

Alizarin Blue, Soluble Powder. A dyestuff. It is a British equivalent of Alizarin blue S.

Alizarin Blue WA. See ALIZARIN BLUE XA.

Alizarin Blue XA (Alizarin Blue WA). The soluble sodium salt of Alizarin blue (q.v.).

Alizarin Bordeaux B, BD (Alizarin Cyanine 3R, Quinalizarin). A dyestuff. It is tetraoxy-anthraquinone, $C_{14}H_8O_6$. It dyes wool mordanted with alumina, bordeaux, and with chrome, dark violet blue.

Alizarin Bordeaux BA, BAY. Dyestuffs. They are British equivalents of Alizarin bordeaux B.

Alizarin Bordeaux G, GG. Dyestuffs isomeric with Alizarin bordeaux B.

Alizarin Brilliant Blue B. A dyestuff. It is a British equivalent of Alizarin saphirol B.

Alizarin Brilliant Blue GS. An equivalent of Alizarin saphirol B.

Alizarin Brilliant Violet R. A dyestuff. It is a British equivalent of Alizarin irisol R.

Alizarin Brown. A dyestuff. It is α-nitro-alizarin, $C_{14}H_7NO_6$, and dyes wool yellow-brown with a chrome mordant.

Alizarin Brown. See ANTHRACENE BROWN.

Alizarin Brown M. A dyestuff. It is a British equivalent of Metachrome brown B.

Alizarin Cardinal. See ALIZARIN GARNET R.

Alizarin Carmine. See ALIZARIN RED S.

Alizarin Celestol. See FORMYL VIOLET S4B.

Alizarin Chestnut. A dyestuff. It is amino-alizarin, obtained by the reduction of Alizarin orange (β-nitroalizarin).

Alizarin Claret. See ALIZARIN GARNET R.

Alizarin Cyanine AC. A dyestuff. It is a British equivalent of Alizarin cyanine R.

Alizarin Cyanine Black G. An anthracene dyestuff. It is probably the sulphonic acid of purpurin-toluidide, and dyes wool with chrome mordants.

Alizarin Cyanine G. A dyestuff. It is the imide of tri- or tetroxy-anthraquinone, and dyes wool mordanted with alumina, blue, and with chrome, bluishgreen.

Alizarin Cyanine Greens (Quinizarin Greens, Alizarin Viridine, Kymrie green). Dyestuffs obtained by treating quinizarin with excess of aniline and sulphonating. The sodium salt is Quinizarin green. They dye wool mordanted with chrome, green.

Alizarin Cyanine 3G. A dyestuff. It is a sulphonic acid of a polyamino-oxyanthraquinone. Dyes wool.

Alizarine Cyanine R. A dyestuff consisting chiefly of $1:2:4:5:8$-pentaoxyanthraquinone, $C_{14}H_8O_7$. It dyes wool mordanted with alumina, violet, and with chrome, blue.

Alizarin Cyanine 3R. See ALIZARIN BORDEAUX B, BD.

Alizarin Dark Blue. A mordant dyestuff. It gives indigo blue shades on chrome mordanted wool.

Alizarin Dark Green. A dyestuff obtained by the treatment of naphthazarin melt with phenols. It dyes chromed wool grey-green to greenish-black.

Alizarin Dark Red B.BB.G.5G. Acid dyestuffs suitable for dyeing wool.

Alizarin DCA, JCA, CAF, YAR. Dyestuffs. They are British equivalents of Flavopurpurin.

Alizarin Delphinol. A blue anthracene dyestuff for wool or silk.

Alizarin Delphinol B. A dyestuff. It is a British equivalent of Alizarin saphirol B.

Alizarin Direct Blue A. An acid dyestuff suitable for dress goods and curtains, also for silk and acetate silk.

Alizarin Direct Blue A3G. An acid dyestuff suitable for silk.

Alizarin Direct Brilliant Light Blue R. An acid dyestuff suitable for silk.

Alizarin Emerald. An acid dyestuff for wool or silk.

Alizarin FA. See FLAVOPURPURIN.

Alizarin Fast Blue, BHG. A mordant dyestuff for wool.

Alizarin for Violet. See ALIZARIN.

Alizarin Garnet R (Alizarin Cardinal, Alizarin Claret, Alizarin Granat R). A dyestuff. It is α-amino-alizarin, $C_{14}H_9NO_4$. It dyes wool mordanted with alumina, bluish-red. It is also used in calico-printing.

Alizarin GB. See FLAVOPURPURIN.

Alizarin GD. See ANTHRAPURPURIN.

Alizarin GI. See FLAVOPURPURIN.

Alizarin Green. See CŒRULEINE and CŒRULEINE S.

Alizarin Green B. A dyestuff. It is dioxy - naphthoxazonium - sulphonate, $C_{20}H_{11}NO_6S$. It dyes chromed fabrics green.

Alizarin Green S. A dyestuff. It is a mixture of the bisulphite compounds of tri- and tetraoxy-anthraquinonequinolines and their sulphonic acids, chiefly $C_{17}H_9NO_6 + 2NaHSO_3$. It dyes chromed wool bluish-green, and is also used for cotton printing.

Alizarin Green S. A dyestuff. It is the bi-sulphite compound of α-alizarinquinoline, and consists of a dark violet paste or powder. It gives green lakes with nickel or chromium mordants, and dyes chrome mordanted wool or cotton bluish-green. Employed in printing.

Alizarin Heliotrope. A mordant dyestuff, which dyes unmordanted wool bluish-red from an acid bath, and chromed wool, bluish-violet. Used in calico-printing.

Alizarin Indigo Blue. A dyestuff which consists of $1:2:5:7:8$-penta-oxyanthraquinoline.

Alizarin Indigo Blue S. A dyestuff. It is a mixture of the bisulphite compounds of tetra- and penta-oxy-anthraquinone-quinolines, and their sulphonic acids, chiefly $C_{17}H_9NO_7 + 2NaHSO_3$. Dyes wool (chromed) indigo blue.

Alizarin Irisol (Solway purple). An anthracene dyestuff. It is 1-hydroxy-4 - sulpho - toluidine - anthraquinone, $C_6H_4(CO)_2.C_6H_2(OH).(NH.C_7H_6SO_3H)$. Dyes wool bluish-violet from an acid bath.

Alizarin Maroon. A mordant dyestuff. It consists of amino-alizarin mixed with amino-purpurins, and dyes wool mordanted with alumina, garnet-red, and with chrome, maroon. It is used in wool, silk, and cotton dyeing.

Alizarin No. 6. See PURPURIN.

Alizarin No. 10CA. See FLAVOPURPURIN.

Alizarin OG. See ALIZARIN ORANGE.

Alizarin Oil. See TURKEY RED OILS.

Alizarin OK. See ALIZARIN ORANGE.

Alizarin OR. See ALIZARIN ORANGE.

Alizarin Orange (Alizarin Orange A, AO, D, N, OG, OK, OR). A mordant dyestuff. It is β-nitro-alizarin (nitro-dioxy-anthraquinone), $C_{14}H_7NO_6$. Dyes cotton mordanted with alumina, orange; with iron, reddish-violet ; and with chrome, brown. Also employed for wool.

Alizarin Orange Powder (Alizarin Orange AOP). A dyestuff. It is the sodium salt of Alizarin orange.

Alizarin Orange AOP. See ALIZARIN ORANGE POWDER.

Alizarin Orange G. A dyestuff. It is 1 : 2 : 6 : 3-nitro-flavopurrin, $C_{14}H_7NO_7$. Dyes alumina mordanted wool or cotton orange.

Alizarin 1P, 1PS, 1PX, 2PX, 3PX, BAR. Dyestuffs. They are British equivalents of Alizarin.

Alizarin Paste. See ALIZARIN.

Alizarin Powder. See ALIZARIN.

Alizarin Powder SA. See ALIZARIN RED S.

Alizarin Powder W. See ALIZARIN RED S.

Alizarin Puce. A colour obtained in calico-printing by using a mixture of aluminium and iron mordants with alizarin.

Alizarin Pure Blue. See ALIZARIN SKY BLUE.

Alizarin Purple. See GALLOCYANINE DH, BS.

Alizarin Red S (Alizarin S, 2S, 3S, SA, W, WS, Alizarin Carmine, Alizarin Powder SA, W). A mordant dyestuff. It is the sodium salt of alizarin-monosulphonic acid, $C_{14}H_7O_7SNa$. Dyes wool mordanted with alumina, scarlet, and with chrome, bordeaux red.

Alizarin Red WGG. Flavopurpurin for wool.

Alizarin Red 3WS. A dyestuff. It is the sodium salt of flavopurpurin-sulphonic acid, $C_{14}H_7O_8SNa$.

Alizarin RF. See ANTHRAPURPURIN.

Alizarin RG. See FLAVOPURPURIN.

Alizarin RT. See ANTHRAPURPURIN.

Alizarin RX. See ANTHRAPURPURIN.

Alizarin S, 2S, 3S. See ALIZARIN RED S.

Alizarin SA. See ALIZARIN RED S.

Alizarin Saphirol B (Solway blue, Durasol acid blue B, Alizarin brilliant blue GS, Alizarin brilliant blue B, Alizarin delphinol B, Alizurol sapphire). A dyestuff. It is the sodium salt of diamino-anthrarufin-disulphonic acid, $C_{14}H_8N_2O_{10}S_2Na_2$. Dyes wool blue from an acid bath.

Alizarin SAR. A dyestuff. It is a British equivalent of Anthrapurpurin.

Alizarin SC. See ANTHRAPURPURIN.

Alizarin SDG. See FLAVOPURPURIN.

Alizarin Sky Blue (Alizarin Pure Blue). A dyestuff which is the mono-sul-

phonic acid of 1-amino-4-p-toluidine-2-bromo-anthraquinone.

Alizarin SSA. See ANTHRAPURPURIN.

Alizarin SSS. Sodium 1 : 2 : 6-tri-hydroxyanthraquinonesulphonate. An acid dyestuff.

Alizarin SX. See ANTHRAPURPURIN.

Alizarin SX Extra. See ANTHRAPURPURIN.

Alizarin Violet. See GALLEINE. In calico printing and cotton dyeing, this term is applied to the colour produced by alizarin red with an iron mordant.

Alizarin Viridine. See ALIZARIN CYANINE GREENS.

Alizarin W. See ALIZARIN RED S.

Alizarin WG. See ANTHRAPURPURIN

Alizarin WS. See ALIZARIN RED S.

Alizarin X. See FLAVOPURPURIN.

Alizarin YCA. See FLAVOPURPURIN.

Alizarin Yellow. See ANTHRACENE YELLOW.

Alizarin Yellow A. A mordant dyestuff. It is trioxy-benzophenone, $C_{10}H_{13}O_4$. Dyes cotton mordanted with alumina and lime, a golden yellow. Used in printing.

Alizarin Yellow C. A mordant dyestuff. It is gallacetophenone (trioxy-acetophenone), $C_8H_8O_4$. Dyes cotton mordanted with alumina, yellow ; with chrome, brown ; and with iron, black.

Alizarin Yellow FS. A mordant and acid dyestuff. It is diphenyl-tolyl-carbinol-trisazo-trisalicylic acid, $C_{41}H_{30}N_6O_{10}$. Dyes chromed wool yellow.

Alizarin Yellow GG, Paste (Alizarin Yellow GGW Powder). A mordant or acid dyestuff. It is m-nitro-benzene-azo-salicylic acid, $C_{13}H_9N_3O_3$. Dyes chromed wool yellow.

Alizarin Yellow 5G (Azoalizarin Yellow 6G, Tartrachromin GG). A dyestuff prepared from p-phenetidine and salicylic acid.

Alizarin Yellow L WS. Dyestuffs. They are British equivalents of Milling yellow.

Alizarin Yellow, Paste. A mordant dyestuff. It is ellagic acid, $C_{14}H_8O_9$. Dyes chromed wool sulphur yellow.

Alizarin Yellow R (Orange R, Terracotta R). A dyestuff. It is p-nitro-benzene-azo-salicylic acid, $C_{13}H_9N_3O_5$. Dyes chromed wool yellowish-brown.

Alizarin Yellow RW. A dyestuff. It is the sodium salt of Alizarin yellow R.

Alizarin Yellow W. An oxyketone dyestuff obtained by the condensation of β-naphthol with gallic acid in the presence of zinc chloride. It dyes chrome mordanted wool.

Alizurol Sapphire. A dyestuff. It is a British equivalent of Alizarin saphirol B.

Alk. See CHIAN TURPENTINE.

Alka-Donna. A proprietary preparation containing magnesium trisilicate, aluminium hydroxide and belladonna extract. An antacid. (246).

Alka-Donna P. A proprietary preparation of Alka-Donna with phenobarbitone. An antacid and sedative. (246).

Alkagel. A trade name for alginates made for waterproofing.

Alkali Blue (Nicholson's Blue, Fast Blue, Soluble Blue, Guernsey Blue). An acid dyestuff. It consists of a mixture of the sodium salts of triphenyl-rosaniline-mono-sulphonic acid, and triphenyl-pararosaniline-mono - sulphonic acid, $C_{38}H_{32}N_3O_4SNa$, and $C_{37}H_{30}N_3O_4$ SNa. Dyes wool blue from a bath made alkaline with borax.

Alkali Blue 6B. See METHYL ALKALI BLUE.

Alkali Blue D. See METHYL ALKALI BLUE.

Alkali Blue XG (Soluble Blue XG, Non-mordant Cotton Blue). A dyestuff consisting of the sulphonic acids of β-naphthylated-rosaniline. Dyes cotton and silk from an acid bath.

Alkali Brown D. See INGRAIN BROWN.

Alkali Cellulose (Hydrated Cellulose). The product of the combination between cotton and caustic soda. When this is decomposed by water it gives hydrated cellulose, the first form of mercerised cotton.

Alkali Dark Brown G, V (Alkali Red Brown 3R). A dyestuff which consists of mixed disazo compounds from benzidine, tolidine, or dianisidine, with one molecule of the bisulphite compound of nitroso-β-naphthol, and one molecule of α-amino-naphthol-sulphonic acid. Dyes cotton and half-wool dark brown.

Alkali Fast Red B, R. An acid colour which dyes wool bluish-red.

Alkali Green (Viridine). An acid dyestuff. It is sodium-diphenyl-diamino - triphenyl - carbinol - sulphonate, $C_{31}H_{23}N_2O_3SNa$. Dyes wool and silk green from an acid bath.

Alkalin. See METHYL ALKALI BLUE.

Alkaline Air. See VOLATILE ALKALI.

Alkali Reclaim. Rubber, reclaimed.

Alkali Red. A tetrazo dyestuff prepared from benzidine, naphthionic acid, and α-naphthylamine-disulphonic acid. Dyes cotton from a salt bath.

Alkali Red Brown 3R. See ALKALI DARK BROWN G, V.

Alkalised Mercury. See GREY POWDER.

Alkalit. A proprietary synthetic resin obtained by heating the sodium salt of phenol-phthalein with toluoyl chloride. The resin is very resistant to acids.

Alkali Violet. An acid dyestuff. It is the sodium salt of tetraethyl-monomethyl - phenyl - pararosaniline - monosulphonic acid, $C_{34}H_{40}N_3O_4SNa$. Dyes

wool from an alkaline, neutral, or acid bath, bluish-violet.

Alkali, Waste (Tank Waste, Soda Waste). A mixture of calcium sulphide, unchanged coal, and calcium carbonate, left after the lixiviation of the black ash in the Leblanc soda process for sodium carbonate.

Alkali Yellow. See ORIOL YELLOW.

Alkali Yellow R. A dyestuff. It is the sodium salt of diphenyl-disazo-salicylic-dehydro-thio-toluidine-sulphonic acid, $C_{33}H_{22}N_6O_6S_2Na_2$. Dyes cotton yellow.

Alkalsite. An explosive containing 25–32 per cent. potassium perchlorate, ammonium nitrate, trinitro-toluene, and other constituents.

Alkanet (Alkanna, Anchusin). Terms applied to two different plants, *Lawsonia inermis* and *Anchusa tintoria*, whose roots are the source of a red dye, anchusine (alkannin), $C_{15}H_{14}O_4$. The name is applied to the dye as well as to the plant. The dye is used for colouring tinctures, oils, pomades, varnishes, and cheese, for tinting marble, and for dyeing corks and wines. It dyes cotton mordanted with iron, grey, and with alumina, lilac-violet.

Alkanna. See ALKANET.

Alkannin. Anchusine, $C_{15}H_{14}O_4$. See ALKENET.

Alkanol. A proprietary trade name for a wetting agent. A sodium naphthalene sulphonate.

Alkargen. Bunsen's name for cacodylic acid.

Alkarsin. Bunsen's name for cacodyl (*q.v.*). Also see CADET'S FLUID.

Alkasal (Alkasol). A combination of aluminium salicylate and potassium acetate, used in medicine.

Alkasit. A proprietary cellulose adhesive.

Alkasol. See ALKASAL.

Alkasperse 25. A proprietary series of pigment dispersions based on a short-oil xylol-thinned alkyd used in the colouring of medium to fast air-drying surface coatings. (422).

Alkathene (Polyethylene, Polythene). The registered trade mark for solid polymers of ethylene prepared by subjecting ethylene to extremely high pressures under carefully controlled conditions of temperature. They are translucent white thermoplastic products, having a softening point of 108° C.

Polyethylene is a saturated straight chain hydrocarbon in which the individual molecules are of the order of 1,000 carbon atoms long. This chemical structure is associated with great resistance to water and other chemicals. The non-polar nature of the molecule gives a very low power factor and dielectric constant at high frequencies. The following Alkathenes are coded by the manufacturer (512) according to their nature and properties: *XJF80/52* has a melt flow index of 1·5 and is used in the extrusion of tubular film in general-purpose applications; *Powder 11/04/00*. A polyethylene used in rotational moulding; *Powder 11/04/01H Black 909*. A polyethylene containing a nominal 2·5 per cent. furnace black used in rotational moulding. It possesses good weathering properties; *Powder 11/04/ 15H Black 904*. A polyethylene powder for rotational moulding used in the production of chemical tanks and general-purpose containers; *Powder 19/ 04/00A*. A polyethylene for rotational moulding and carpet backing. It possesses good melt flow properties; *Powder 19/04/01H Black 904*. A polyethylene similar to Alkathene 19/04/00A but containing 2·5 per cent. furnace black to impart weathering properties; *WSM78*. A fast-cycling, high-gloss polyethylene, easy to process and used for injection moulding.

Alkazene. A proprietary trade name for several compounds. They are chlorethyl benzenes or polyisopropyl benzenes.

Al-kenna. The powdered roots and leaves of *Lawsonia inermis*. It is used in the East for dyeing the nails, teeth, and hair. See ALKANET.

Alkeran. A proprietary preparation of melphalan. An antimitotic. (277).

Alkermes. See KERMES.

Alkolite. A proprietary phenophthalein resin.

Alkydal. Proprietary glyceryl phthalate resins. (3).

Alkydal L45U. 57 per cent. glyceryl phthalate, 43 per cent. linseed and synthetic fatty acids.

Alkydal STK. Phthalic anhydride-trimethylolethane and castor oil-synthetic fatty acids.

Alkydal RD. Adipic acid non drying alkyd. No oil.

Alkydal RD25. Phthalic anhydride, trimethylol propane and synthetic fatty acids from paraffin oxidation and a small amount of maleic acid. (3).

Alkydal ST. 42 per cent. glyceryl phthalate, 58 per cent. castor oil acids. (3).

Adkydal ST. 75 per cent. Alkydal ST with 25 per cent. xylol. (3).

Alkydal T. 51 per cent. glyceryl phthalate, 49 per cent. linseed oil acids. (3).

Alkydal TO. 31 per cent. anhydride.

18 per cent. glycerine and 60 per cent, linseed oil acids. (3).

Alkydal TN. Low viscosity Alkydal T. (3).

Alkydharz 2521. A long oil alkyd containing 16·4 per cent. phthalic anhydride. (31).

Alkydon. A sardine oil type varnish.

Alkyphen. An oil soluble, heat reactive, 100 per cent. alkyl phenolic resin. (3).

Allaco 1215. A proprietary epoxy dipcoating compound for use with heat-sensitive electronic components. Its cure temperature is 125° C. (423).

Allaco 1295. A proprietary epoxy-based compound for potting and encapsulating. (423).

Allactite. A mineral. It is a basic arsenate of manganese.

Allactol. Aluminium lacto-tartrate.

Allagite. A mineral. It is allied to rhodonite.

Allanite. See BAGRATIONITE.

Allan Red Bronze. An alloy of 62·5 per cent. copper, 30 per cent. lead, and 7·5 per cent. tin.

Allan Red Metal. An alloy of 50 per cent. copper, and 50 per cent. lead.

Allbee with C. A proprietary preparation of Vitamins B and C. (258).

Allercur. A proprietary preparation of clemizole hydrochloride. (265).

Allegheny 33, 44, 55, 66. Corrosion-resisting alloys (formerly Ascoloy 33, 44, 55, 66). They contain iron and chromium : 33 contains 12–16 per cent. chromium, and 55 contains 26–30 per cent. chromium.

Allegheny Metal. A proprietary corrosion-resisting alloy. It contains iron with 17–20 per cent. chromium and 7–10 per cent. nickel.

Allegron. A proprietary preparation of nortriptyline hydrochloride. An antidepressant. (309).

Allemontite (Arsenical Antimony). A mineral. It is a native alloy of arsenic and antimony, $SbAs_3$.

Allenite. Synonym for Pentahydrite.

Allenoy. A proprietary molybdenum steel.

Allen's Metal. An alloy of 55·3 per cent. copper, 44·6 per cent. lead, and 0·1 per cent. tin.

Allercur. A proprietary name for the hydrochloride of clemizole.

Allergen Extracts. A proprietary preparation of pollen allergen extracts.

Allethrin. It is 2-allyl-4-hydroxy-3-methyl-2-cyclo-penten-l-one ester of chrysanthimummono carboxylic acid. A synthetic insecticide structurally similar to pyrethrin.

Alletorphine. An analgesic currently

undergoing clinical trial as "R & S 218-M". N-Allyl - 7,8 - dihydro - 7α - (1(R) - hydroxy - 1 - methylbutyl) - O^6 - methyl - 6,14 - *endo*ethenonormorphine. N - Allylnoretorphine.

Alley Stone. Aluminite ($q.v.$).

Alliance Chrome Blue-black B. A dyestuff. It is a British equivalent of Chrome fast cyanine G.

Alliance Chrome Blue-black R. A dyestuff. It is a British equivalent of Palatine chrome blue.

Alliance Fast Red A Extra. A dyestuff. It is a British equivalent of Fast red.

Allico Varnish. Urethane varnish. (539).

Alligator Wood. The wood of *Guarea grandifolia*, of West India.

Allingite. A commercial variety of amber. It melts at 300° C.

Alliols. Crude sulphur-bearing spirits obtained from the first distillation fraction of light tar oils. They contain impure sulphides of carbon.

Allisan. Horticultural fungicide. (502).

Allisatin. A German preparation. It contains all the active principles of garlic. An antithelmintic.

Allmatein (Almatein). A condensation product of hæmatoxylin and formaldehyde, $CH_2O_2 . (C_6H_{12}O_5)_2 : CH_2$. Used as an internal disinfectant.

All-Mine Pig. See STAFFORDSHIRE ALL-MINE PIG.

All-O. A proprietary special quality soap in standard strength liquid form for use as a rubber lubricant in moulding rubber.

Allocaine Lumière. Ethocaine hydrochloride.

Allocaine S. Ethyl-mydriatine hydrochloride. A local anæsthetic.

Allochroite. A mineral. It is a variety of andradite.

Allochrysine Lumière. Sodium auropropanol-sulphonate.

Alloclase. A mineral regarded as cobalt glance ($q.v.$).

Alloclasite. A mineral. It is a cobalt ferrous sulpho-arsenito-bismuthite.

Alloferin. A proprietary preparation of alcuronium chloride, a muscle relaxant. (314).

Allogonite. See HERDERITE.

All Oil Water Gas. An illuminating gas produced by passing oil and steam into heated chambers. It consists mainly of hydrogen, methane, carbon monoxide, hydrocarbons, and nitrogen.

Allomorphite. A mineral consisting mainly of barium sulphate.

Allopalladium. A mineral. It is a rare hexagonal modification of palladium.

Allophane (Riemannite). A mineral, $Al_2O_3.SiO_2.nH_2O$.

Allophite. A mineral. It is a silicate of aluminium and magnesium.

Alloprene. A trade mark. It is a white fibrous chlorinated rubber derivative resistant to acids and oxidising agents, and soluble in a wide range of solvents at ordinary temperatures. Its composition approximates to $C_{10}H_{13}Cl_{17}$. For use in the paint and varnish industry. Other trade names for chlorinated rubber are Pergut, Tegofan, and Dartex.

Allopurinol. 1 - H - Pyrazolo[3,4- d] - pyrimidin-4-ol. Zyloric. An uricosuric drug.

Allosan. Santalyl-allophanate, NH_2. $CO.NH.COO.C_{15}H_{23}$. Used in the treatment of gonorrhœa.

Alloxan. Mesoxalyl-urea, $C_4H_4O_5N_2$.

Alloy ADR. An alloy possessing between 0–500° a coefficient of expansion lower than that of the added metals.

Alloy AM4·4. An aluminium-magnesium alloy used in aircraft construction. It contains magnesium with 4 per cent. aluminium, 0·4 per cent. manganese, and 0·15 per cent. silicon.

Alloy AM7·4. An alloy used in aircraft construction. It contains magnesium with 7 per cent. aluminium, 0·4 per cent. manganese, and 0·15 per cent. silicon.

Alloy AMF. An alloy containing 50–60 per cent. nickel, which does not become brittle at very low temperatures.

Alloy AP33. An aluminium alloy containing 4·5 per cent. copper and 0·4 per cent. titanium.

Alloy JL. An aluminium alloy containing 4·5 per cent. copper, 0·41 per cent. iron, and 0·35 per cent. silicon.

Alloy L5. An aluminium alloy containing, in addition to aluminium, 13 per cent. zinc and 2·8 per cent. copper. It has a specific gravity of 3.

Alloy 2L5. An aluminium alloy with 12·5–14·5 per cent. zinc and 2·5–3 per cent. copper.

Alloy 2L8. An aluminium alloy containing 11–13 per cent. copper. It is chiefly used for motor pistons.

Alloy 3L11. An aluminium alloy containing 6–8 per cent. copper and tin may be added up to 1 per cent.

Alloy L7. An alloy containing aluminium with 14 per cent. copper, and 1 per cent. manganese. It has a specific gravity of 2·9.

Alloy L8. An alloy containing aluminium with 12 per cent. copper. It has a specific gravity of 2·9.

Alloy L10. An alloy containing aluminium with 10 per cent. copper and

1 per cent. tin. It has a specific gravity of 2·95.

Alloy L11. An alloy containing aluminium with 7 per cent. copper and 1 per cent. tin. It has a specific gravity of 2·9.

Alloy L24. See ALLOY Y.

Alloy MG7. An alloy consisting mainly of aluminium with magnesium and manganese. It has mechanical properties similar to Duralmin, and is stated to be highly resistant to corrosion. It has a specific gravity of 2·63.

Alloy N. An alloy of 91 per cent. aluminium, 6 per cent. copper, and 3 per cent. manganese.

Alloy NCT3. An alloy containing 44·5 per cent. iron, 37·5 per cent. chromium, 17·5 per cent. nickel, and 0·5 per cent. manganese.

Alloys RR. A series of aluminium alloys containing aluminium with 0·5–5 per cent. copper, 0·2–2·5 per cent. nickel, 0·05–5 per cent. magnesium, 0·6–1·5 per cent. iron, 0·05–0·5 per cent. titanium, and 0·2–5 per cent. silicon.

Alloy Steel. The term applied to a steel containing one or more elements in addition to carbon. Also see STEELS V. and W.

Alloy T. An alloy containing aluminium with 3·8 per cent. magnesium, 0·5 per cent. iron, 0·5 per cent. silicon, and 0·1 per cent. copper.

Alloy-treated Steel. A term used for a steel to which one or more elements have been added in small quantities, for the treatment of the steel, and not to form an alloy.

Alloy W.7. 1 per cent. silicon, 4·5 per cent. manganese, balance nickel with controlled zirconium addition. (143).

Alloy W.9. 1 per cent. silicon, 4·5 per cent. manganese, balance nickel with controlled zirconium addition. (143).

Alloys Wm. German white-bearing metals. Wm5 contains 78·5 per cent. lead, 15 per cent. antimony, 5 per cent. tin, and 1·5 per cent. copper, with a specific gravity of 10·1. Wm10 consists of 73·5 per cent. lead, 15 per cent. antimony, 10 per cent. tin, and 1·5 per cent. copper, and has a gravity of 9·7. Wm42 contains 42 per cent. tin, 41 per cent. lead, 14 per cent. antimony, and 3 per cent. copper. Its gravity is 8·5. Wm80 consists of 80 per cent. tin, 10 per cent. antimony, 10 per cent. copper. It has a specific gravity of 7·5.

Alloy Y (Alloy 24). An alloy of aluminium with 4 per cent. copper, 2 per cent. nickel, and 1·5 per cent. magnesium. It has a specific gravity of 2·79.

Alloy 39. An alloy containing aluminium with 3·75–4·25 per cent. copper, 1·2–1·7 per cent. magnesium, and 1·8–2·3 per cent. nickel.

Alloy 109. An aircraft alloy containing 88 per cent. aluminium and 12 per cent. copper.

Alloy 122. An aircraft alloy containing 88·6 per cent. aluminium, 10 per cent. copper, 1·2 per cent. iron, and 0·25 per cent. magnesium.

Alloy 142. An aircraft alloy containing 92·5 per cent. aluminium, 4 per cent. copper, 2 per cent. nickel, and 1·5 per cent. magnesium.

Alloy 145. An aluminium alloy containing 10–11 per cent. zinc, 2–3 per cent. copper, and 1–1·5 per cent. iron.

Alloy 195. An aluminium alloy containing 4–5 per cent. copper, and not more than 1·2 per cent. silicon, 1·2 per cent. iron, 0·35 per cent. magnesium, and 0·35 per cent. zinc.

Alloy 2129. A special nickel-iron alloy having fairly high permeability and excellent mechanical properties enabling it to be drawn into boxes for magnetic shielding.

Allpyral Allergen Extracts. A proprietary group of preparations containing standardized allergen extracts, used in the hyposensitisation therapy of asthma and in allergic rhinitis. (272).

Allspice (Jamaica Pepper). Pimento.

Allspice Oil. Oil of Pimento.

Alluman. An alloy of aluminium with 10–20 per cent. tin and 4–6 per cent. copper.

Allyl Mustard Oil. See OIL OF MUSTARD.

Allyloestrenol. 17α - allyloestr - 4 - en-17β-ol. Gestanin.

Allylprodine. 3 - Allyl - 1 - methyl - 4 - phenyl - 4 - propionyloxypiperidine.

Almacarb. Tablets of aluminium hydroxide- magnesium carbonate co-dried gel, antacid. (182).

Almadina. See POTATO GUM.

Almadur. A proprietary moulding compound based on urea formaldehyde. (449).

Almag. A proprietary aluminium alloy similar in composition to Alferium (q.v.).

Almagrerite. A mineral consisting of anhydrous zinc sulphate, found in Spain.

Almandine. A mineral. It consists of an iron-aluminium silicate, $Fe_3Al_2Si_3O_{12}$. It is a precious stone, and when cut and polished is called carbuncle.

Almandine Ruby. Spinel (q.v.).

Almandite. A variety of garnet, $Fe_3Al_2(SiO_4)_3$, used as a coating for abrasive paper and cloth.

Almasilium. An aluminium alloy containing 1 per cent. magnesium and 2 per cent. silicon.

Almatein. See ALLMATEIN.

Almeidina Gum. See POTATO GUM.

Almelec. A proprietary alloy of aluminium containing 0·7 per cent. magnesium, 0·5 per cent. silicon, and 0·3 per cent. iron. Its specific gravity is 2·5 grams per c.c.; breaking strain, 35 kg. per sq. mm.; elongation 6–8 per cent.; coefficient of expansion, 23 × 10⁻⁶. It is used for electrical conduction.

Almen's Reagent. A solution containing 5 grams tannic acid in 240 c.c. of 50 per cent. alcohol, to which has been added 10 c.c. of a 25 per cent. solution of acetic acid. A precipitate is given with nucleo-proteins.

Almeraite. A mineral. It is a chloride of potassium, sodium, and magnesium.

Almevax. A proprietary preparation of live rubella vaccine. (277).

Almex. A proprietary trade name for a compound of powdered lead and chemical-quality lead sheet capable of standing up to varied conditions of corrosion, temperature and vibration. (133).

Almond Emulsion. *Mistura amygdalæ* B.P. Compound of almonds in distilled water.

Almond Meal. The cake obtained after the expression of the oil from sweet almonds.

Almond Oil, French (Almond Oil, Persic). The commercial oil consists either of pure apricot-kernel oil or a mixture of this oil with peach-kernel oil, or the latter alone.

Almond Oil, Persic. See ALMOND OIL, FRENCH.

Almond Paste. Ground almond cake, from which the oil has been expressed, is mixed with an equal quantity of whiting and made into a stiff paste with water. It is used for luting stills and other purposes.

Almonds, Milk of. See MILK OF ALMONDS.

Almo Steel. Proprietary chrome-molybdenum steels.

Alnagon. Aluminium sulphonaphthalate in glycerin and water.

Alneon. An aluminium alloy for castings. It contains 10–20 per cent. zinc, a small quantity of copper, and a little nickel. It has a tensile strength of 27 kg. per sq. mm. and elongation 4 per cent.

Alnovol. A series of spirit soluble pure phenolic resins. (4).

Alnovol 429-K. Acid condensed cresylic acid-formaldehyde+5 per cent. maleic acid.

Alnovol 276-K. Acid condensed Xylenol - formaldehyde+5 per cent. maleic acid.

Alnuoid. An extract of *Alnus serulata* or American alder bark. Used as a tonic.

Alobrent. A trade mark used for articles of abrasive and refractory type. Its main constituent is crystalline alumina.

Alocol. A proprietary preparation of colloidal aluminium hydroxide.

Alocol-P. A proprietary preparation containing dried aluminium hydroxide gel with oil of peppermint. An antacid. (244).

Alocrom. Chemical pretreatment for aluminium. (512).

Aloe. The inspissated juice of the leaves of the aloe. Socotrine aloe is derived from *Aloe Perryi* : Zanzibar aloe is a hepatic variety from the same source. The Curaçao aloe is obtained from *Aloe vera* and Cape aloe from *Aloe ferox*. Natal aloe is probably derived from *Aloe succotina*, and Uganda or Crown aloe is a brand of Cape aloe formerly produced by allowing it to ferment slightly and then drying it. Hepatic, callabine, fetid, and horse aloes are inferior qualities of aloes from various sources. Other aloes are Jafferabad or Mocha aloe from *Aloe abyssinica*, and Musambra aloe from *Aloe vera* of India.

Aloe Fibre (Mauritius Hemp). A fibre obtained mainly from the leaves of *Fourcroya fœtida*.

Aloesol. The mono-acetyl derivative of Cape or Uganda aloes, $C_{11}H_4O_3Cl_4$.

Aloisiite. A mineral. It is a hydrated silicate of calcium, iron, magnesium, and sodium.

Alomite. A trade name applied to a variety of sodalite used as an ornamental stone.

Alon. A registered trade mark. Fumed alumina.

Alophen. A proprietary preparation of aloin, phenolphthalein, ipecacuanha, strychnine and belladonna. A laxative. (264).

Alopon. A preparation containing the hydrochlorides of opium alkaloids.

Alotrichite. A mineral. It is a hydrated silicate of aluminium and ferrous iron.

Alowalt. A trade mark for articles of the abrasive and refractory type made essentially of alumina.

Alox. A proprietary trade name for a series of methyl esters of higher alcohols.

Aloxidone. 3 - Allyl - 5 - methyl - oxazolidine - 2,4 - dione. Allomethadione. Malidone.

Aloxiprin. A polymeric condensation product of aluminium oxide and O-acetyl salicylic acid. Palaprin. Present in Placodin. (190).

Aloxite. A trade mark for abrasive and refractory materials consisting essentially of alumina.

Alpaca (Vigogne, Llama). Fibres made from the hair of different kinds of goat.

Alpacca (Alpakka). An alloy of 64 per cent. copper, 19 per cent. zinc, 14·5 per cent. nickel, 2 per cent. silver, 0·4 per cent. iron, and 0·12 per cent. tin. It is a nickel silver.

Alpakka. See ALPACCA.

Alpax. See SILUMIN.

Alperox C. A registered trade mark for lauroyl peroxide. A catalyst for polymerisation.

Alpex 4505. Cyclised rubber used for surface coatings and printing inks.

Alpha C. See ALPHA NK.

Alphacaine. See α-EUCAINE.

Alphadrol. A proprietary name for Fluprednisolone.

Alphacetylmethadol. α - 3 - acetoxy - 6 - dimethylamino - 4,4 - diphenyl - heptane. O-Acetate of α-6-dimethylamino - 4,4 - diphenylheptan - 3 - ol.

Alphachroic. Chrome dyestuffs. (541).

Alphachym. A proprietary preparation of alphachymotrypsin. Anti-inflammatory agent. (279).

Alpha Daphnone. Specially pure α-isomethylionone. (504).

Alphaderm. A proprietary preparation of HYDROCORTISONE and urea used as a skin ointment in the treatment of eczema and related disorders. (818).

Alphadolone. An anæsthetic component present in ALTHESIN. 3α,21-Dihydroxy - 5α - pregnane - 11,20 - dione.

Alphameprodine. α - 3 - Ethyl - 1 - methyl - 4 - phenyl - 4 - propionyl - oxypiperidine.

Alphamethadol. α - 6 - Dimethyl - amino-4,4-diphenylheptan-3-ol.

Alphamint. Exquisite peppermint blend. (504).

Alpha-Naphthol Orange. See ORANGE I.

Alpha NK. A new synthetic tannin made by bringing certain products obtained in the manufacture of cellulose into reaction with a mixture of phenols and homologues, instead of β-naphthol, which latter yields Alpha C. The tanning liquids contain 20 per cent. tanning substance.

Alphanol. Medium chain length alcohols. (512).

Alphaprodine. α - 1,3 - Dimethyl - 4 - phenyl - 4 - propionyloxypiperidine. Nisentil is the hydrochloride.

Alphasol OT. A proprietary trade name for the sodium salt of an alkyl ester of sulpho-succinic acid.

Alphaxalone. An anæsthetic component present in ALTHESIN. 3α-Hydroxy-5α - pregnane - 11,20 - dione.

Alphenate. A series of phenolic-alkyd plasticised resins. (4).

Alphide. A proprietary trade name for a cold moulded refractory ceramic.

Alphidine. A proprietary preparation. It is an allotropic form of iodine. It is also stated to be an iodine-tannic acid complex. (366).

Alphogen (Alphozone, Succinoxate). Succinyl peroxide, $(COOH.CH_2.CH_2CO)_2O_2$. An antiseptic.

Alphol. α-Naphthol-salicylate, $C_6H_4(OH)(COOC_{10}H_7)$. An antiseptic and antirheumatic.

Alphosyl. A proprietary preparation of allantoin and coal tar extract for dermatological use. (376).

Alphozone. See ALPHOGEN.

Alpine Blue (Fast Wool Blue). A dyestuff. It is the sodium salt of tetra - methyl - ethyl - benzyl - triamino-triphenyl-carbinol-trisulphonic acid. Dyes wool and silk blue from an acid bath.

Alplate. A proprietary aluminium-coated steel.

Alpon. Fibre reinforced polyester moulding material. (542).

Alprenolol. A beta adrenergic receptor blocking agent. 1 - (2 - Allylphenoxy) - 3-isopropylaminopropan-2-ol.

Alprokyds. Proprietary drying oil and non-drying oil modified alkyd resins.

Alquifon (Black Lead Ore, Potter's Ore). A mineral. It consists of zinc sulphide, and is used in pottery to give a green glaze.

Alquifou. See ALQUIFON.

Alresat. A series of proprietary ester gum-maleic treated resins. (4).

Alresat 313-C. Maleic anhydride-rosin-glycerine (m.p. 95° C). (4).

Alresen. Oil soluble phenolic resins. (4).

Alresen 260-R. Ortho cresol-turpentine condensate cooked with butylated diane (q.v.) condensate. (4).

Alrheumat. A proprietary preparation of ketoprafen. An anti-rheumatic. (112).

Alromin Ru 1000. Antistatic agent. (543).

Al Root. See CHAY ROOT.

Alsace Green. See DINITROSO-RESORCIN.

Alsace Green J. See GAMBINE Y.

Alsace Grey. See NEW GREY.

Alsace Gum. Dextrin (q.v.).

Alshedite. A mineral. It is a variety of sphene.

Alsi. A pigment consisting of finely ground aluminium-silicon alloy used to give durable and rust-preventative paints.

Alsibronz. A trade mark for micaceous powders for use as parting materials in the manufacture of pneumatic tyres and other uses.

Alsica Alloys. Aluminium-silicon-copper alloys.

Alsifer. A proprietary alloy of 40 per cent. silicon, 40 per cent. iron, and 20 per cent. aluminium. A hardener alloy for adding silicon to aluminium alloys.

Alsifilm. A material made from bentonite, used in place of mica.

Alsimag. A registered trade mark for ceramic materials. Used for insulation and for the dielectric of condensers. (35).

Alsimag 754. A beryllia ceramic.

Alsimag 779. A trade mark for leachable ceramic cores for precision metal castings. (35).

Alsimin. A similar alloy to Alsifer.

Alsol. Aluminium-aceto-tartrate, a non-poisonous germicide. It is used as a 50 per cent. solution, for the treatment of wounds, and as a mouth wash.

Alsol Liquid. A 50 per cent. solution of Alsol (q.v.).

Alstonia (Australian Bark, Dita Bark, Bitter Bark). The dried bark of *Alstonia scholaris*. Used in the treatment of malarial and other fevers.

Alstonite (Baryto-calcite, Bromlite). A mineral. It is a barium-calcium-carbonate, $(Ba.Ca)(CO_3)_2$.

Alsynate. Paint driers. (532).

Altacite. A proprietary preparation of HYDROTALCITE. An antacid. (443).

Altaite. A mineral, PbTe.

Altal. A proprietary trade name for triphenylphosphate.

Altan. Veterinary laxative. (507).

Altax (Captax Disulphide). A proprietary rubber vulcanisation accelerator. It is essentially benzo-thiazyl-disulphide.

Althein. See ASPARAGIN.

Althesin. A proprietary preparation of ALPHAXALONE and ALPHADOLONE acetate used in the induction of anaesthesia. (335).

Altrex. Histological dehydrating agent. (544).

Alubarb. A proprietary preparation of aluminium hydroxide, phenobarbitone and belladonna. A gastrointestinal sedative. (350).

Alu-Cap. A proprietary preparation of aluminium hydroxide gel used as an antacid and in the treatment of hypophosphataemia in cases of renal failure. (275).

Alucetol. Aluminium acetate in the form of tablets. A German preparation.

Aludrox. A proprietary preparation containing aluminium hydroxide gel. An antacid. (245).

Aludrox CO. A proprietary preparation containing alumina-sucrose powder and magnesium hydroxide. An antacid. (245).

Aludrox SA. A proprietary antacid preparation containing aluminium hydroxide gel, magnesium hydroxide, *sec.* butobarbitone and ambutonium bromide. (245).

Aludur (Aladar). An alloy of aluminium and silicon, containing from 5–20 per cent. silicon. Also see SILUMIN.

Alugan. A proprietary preparation of BROMOCYCLEN. A veterinary pesticide.

Aluhyde. A proprietary preparation of aluminium hydroxide gel, magnesium trisilicate, sodium quinabarbitone and belladonna. An antacid. (385).

Alujet. Aluminium cleaner. (545).

Alum. Potassium aluminium sulphate, $K_2SO_4.Al_2(SO_4)_3 24H_2O$, ammonium aluminium sulphate $(NH_4)_2SO_4.Al_2(SO_4)_3.24H_2O$, and aluminium sulphate, $Al_2(SO_4)_3.9H_2O$, are all known by this name, but it is usually applied to the potassium salt.

Aluman. An alloy of 88 per cent. aluminium, 10 per cent. zinc, and 2 per cent. copper.

Alumantine. A proprietary refractory containing 60–65 per cent. Al_2O_3, tested to fuse above 2,000° C. It is stated to resist the action of lime at high temperatures.

Alumbro. A trade mark for a 2 per cent. aluminium-brass alloy having good resistance to corrosion by sea water and marine atmospheres. (134), (2).

Alum Cake (Patent Alum). Produced from kaolin by calcining, then treating it with sulphuric acid, whereby the aluminium oxide is converted into aluminium sulphate, $Al_2(SO_4)_3.9H_2O$. When this is powdered, it is known as alum cake.

Alum Curd. When white of egg is shaken with a lump of alum, it is coagulated as a curd, at the same time a part of the alum is dissolved. The curd is used between folds of gauze, over the eye, in some cases of ophthalmia.

Alum, Dried. Potassium alum, from which the water of crystallisation has been removed.

Alumel. An electrical resistance alloy containing 94 per cent. nickel, 1 per cent. silicon, 2 per cent. aluminium, 2·5 per cent. manganese, and 0·5 per cent. iron.

Alumetised Steel. A steel which has been sprayed with aluminium and then heat-treated to give a surface of alloy.

Alum, Exsiccated. See BURNT ALUM.

Alum-hæmatoxylin Solution. A microscopic stain. It consists of 1 gram hæmatoxylin dissolved in a saturated solution of ammonium alum (100 c.c.) and 300 c.c. distilled water. The solution is filtered after it has assumed a dark red colour, and 0·5 gram thymol added as preservative.

Alumian. A mineral, Al (AlO) (SO₄)₂.

Alumilite. A proprietary trade name for chemical coatings applied to aluminium electrically.

Alumin. Sodium aluminate, $Na_2Al_2O_4$.

Alumina. Aluminium oxide, Al_2O_3.

Aluminac. A similar alloy to Alpax (*q.v.*), for making die castings.

Alumina Cream. A cold saturated solution of alum is divided into two parts, a slight excess of ammonia added to the larger portion, and the other alum solution added by degrees until the whole is faintly acid. It is used to clarify sugar solutions by the removal of proteins.

Aluminage. A term used for coating and protecting metals (usually iron), by spraying them with metallic aluminium.

Aluminate Cement. A cement made from bauxite. It usually contains about 40 per cent. Al_2O_3, 40 per cent. lime, 15 per cent. iron oxide, silica, and magnesia.

Aluminite (Websterite). A mineral. It is a hydrated sulphate of aluminium, $Al_2O_3.SO_3.9H_2O$. It is also the name for an aluminium alloy containing 73 per cent. aluminium, 23 per cent. zinc, 3 per cent. copper, and traces of iron and silicon. Also name of proprietary alumina refractory.

Aluminium Alloys for Aircraft Construction. The main constituent of these alloys is aluminium, usually from 87–99 per cent. The other metals vary from 0–10·75 per cent. copper, 0–1·2 per cent. silicon, 0–1·75 per cent. manganese, 0–0·75 per cent. magnesium, 0–0·25 per cent. zinc, and 0–1·5 per cent. iron.

Aluminium Amalgam. An alloy of aluminium and mercury, formed when aluminium foil is immersed in a solution of mercuric chloride. Used as a reducing agent.

Aluminium Brass. Alloys of from 59–70 per cent. copper, 26–40 per cent. zinc, 0·3–5·2 per cent. aluminium, and sometimes a little iron. The alloy containing 59 per cent. copper, 39 per

cent. zinc, 1 per cent. aluminium, and 1 per cent. iron has a melting-point of 900° C. and a specific gravity of 8·4.

Aluminium Bronze. There are various alloys under this name. Those containing a high percentage of aluminium are termed light, and have from 83–89 per cent. aluminium and 11–17 per cent. copper. The other type is called heavy, and contains from 85–95 per cent. copper and 5–15 per cent. aluminium. The alloy containing 95 per cent. copper and 5 per cent. aluminium has a tensile strength of from 53,000–69,000 lb. per sq. in., that consisting of 92 per cent. copper and 8 per cent. aluminium from 60,000–83,000 lb. per sq. in., and the one containing 90 per cent. copper and 10 per cent. aluminium from 80,000–96,000 lb. per sq. in. The 10 per cent. aluminium alloy has a specific gravity of 7·6, melts at 1050° C., and has a specific resistance at 0° C. of 12 micro-ohms per cm.[3] An alloy containing 88·7 per cent. copper and 11·3 per cent. aluminium has a specific heat of 0·104 calories per gram at from 20–100° C. The heavy alloys are sometimes used in the manufacture of jewellery, as they resemble gold. Also see COPROR, CUPRO-ALUMINIUM, and MANHARDT'S ALUMINIUM BRONZE.

Aluminium Cadmium. Alloys of 90–93 per cent. aluminium, 2·5–3·5 per cent. cadmium, and 4–6 per cent. copper.

Aluminium Clofibrate. A pharmaceutical used in the treatment of arteriosclerosis. Di - [2 - (4 - chlorophenoxy) - 2 - methylpropionato]-hydroxyaluminium.

Aluminium Iron (Ferro-aluminium). An alloy of iron and aluminium. It is used for refining iron, also as a permanent ingredient for increasing the strength. A 15 per cent. alloy has been used for crucibles exposed to high temperatures.

Aluminium Iron Brass. An alloy containing 61·1 per cent. copper, 35·3 per cent. zinc, 1·1 per cent. iron, and 2·3 per cent. aluminium.

Aluminium Iron Bronze. An alloy containing 85–89 per cent. copper, 6–9 per cent. aluminium, and 3–7 per cent. iron.

Aluminium Magnesium Bronze. An alloy of 89–94 per cent. copper, 5–10 per cent. aluminium, and 0·5 per cent. magnesium.

Aluminium Manganese. An alloy of aluminium with 2–3 per cent. manganese.

Aluminium Manganese Brass. An alloy consisting of 56·3 per cent. copper,

40 per cent. zinc, 2·7 per cent. manganese, and 1 per cent. aluminium.

Aluminium Manganese Bronze. An alloy of 89 per cent. copper, 9·6 per cent. aluminium, and 1·2 per cent. manganese.

Aluminium Nickel. An alloy containing varying amounts of nickel with aluminium. One alloy consists of 76·4 per cent. nickel, and 23·6 per cent. aluminium.

Aluminium Nickel Bronze. An alloy containing 85 per cent. copper, with from 5–10 per cent. aluminium and 5–10 per cent. nickel.

Aluminium-Nickel-Titanium. An alloy of 97·6 per cent. aluminium, 2 per cent. nickel, and 0·4 per cent. titanium.

Aluminium Nickel Zinc. An alloy consisting of 85 per cent. aluminium, 10 per cent. nickel, and 5 per cent. zinc.

Aluminium Ore. The varieties of bauxite used for the preparation of aluminium.

Aluminium Paper. A paper which has been coated with powdered aluminium. It is used for packing purposes.

Aluminium Silicon Alloy C. A British Chemical Standard alloy. It contains 12·74 per cent. silicon, 0·34 per cent. iron, 0·005 per cent. manganese, 0·020 per cent. zinc, 0·006 per cent. titanium, and 0·010 per cent. copper.

Aluminium Silver (Silver Metal). An alloy consisting of 57 per cent. copper, 20 per cent. nickel, 20 per cent. zinc, and 3 per cent. aluminium. Used for the spacing levers of typewriters. The name is also applied to an alloy of 95 per cent. aluminium, and 5 per cent. silver.

Aluminium Solders. Various. They are usually alloys of from 50–85 per cent. tin, 15–50 per cent. zinc, and aluminium. Higher melting alloys contain from 73–87 per cent. tin, 8–15 per cent. zinc, and 5–12 per cent. aluminium. Some alloys contain copper instead of tin, others have copper and lead as well as zinc, tin and aluminium, whilst a few contain cadmium. The following are some types of aluminium solders : (a) contains 55 per cent. tin, 33 per cent. zinc, 1 per cent. copper and 11 per cent. aluminium ; (b) consists of 89·5 per cent. zinc, 6 per cent. aluminium, and 4·5 per cent. copper ; (c) contains 23 per cent. zinc, 46 per cent. tin, 8 per cent. copper, 8 per cent. silver, and 15 per cent. aluminium.

Aluminium Steel. Aluminium is usually added to steel on account of its oxidising properties. Steels containing more than 2 per cent. aluminium are somewhat brittle, but 5 per cent. can be added without affecting the tensile strength.

Aluminium Tin Bronze. An alloy of 85 per cent. copper, 10 per cent. tin, 2·5 per cent. aluminium, and 2 per cent. zinc.

Aluminium Titanium Bronze. An alloy of from 89–90 per cent. copper, 9–10 per cent. aluminium, 1 per cent. iron, and a trace of titanium.

Aluminium, Victoria. See PARTINIUM.

Aluminoferric. Consists of crude aluminium sulphate, and contains some iron sulphate. It is used as a precipitating agent in sewage and refuse liquids treatment, and also for removing suspended matter from boiler feed water. (513)

Aluminoid. A trade mark for goods of the abrasive and refractory type, the essential constituent being crystalline alumina.

Aluminol. A solder for aluminium. It is stated to be a mixture of rare metals.

Aluminon. Aurine tricarboxylic acid, a dyestuff which is used as a test for aluminium in solution.

Alumino-Vanadium. An alloy of aluminium and vanadium, obtained by adding a mixture of vanadium pentoxide and powdered aluminium to liquid aluminium. Used as a deoxidising agent.

Aluminox. A trade mark for articles as abrasives and refractories. The essential constituent is crystalline alumina.

Aluminum. Alloys of from 50–86 per cent. tin, 0–47 per cent. lead, 0–66 per cent. zinc, and 0–12 per cent. aluminium. One alloy contains 14 per cent. bismuth, and another 5 per cent. copper.

Alumite. See ALUMSTONE.

Alum Meal. Crystals of potassium alum, containing small quantities of iron.

Alumnasol. See ALUMNOL.

Alumnol (Alumnasol). The aluminium salt of β-naphthol-disulphonic acid, $[C_{10}H_6OH(SO_3)_2]_3Al_2$. An antiseptic and astringent. It is used for wounds, but more especially for gonorrhœa.

Alumohydrocalcite. A mineral of Siberia. It corresponds to the formula $CaO.Al_2O_3.2CO_2.5H_2O$, is brittle, and has a specific gravity of 2·231.

Alum Ore. See ALUM SHALE.

Alum, Paperhanger's. See PAPERHANGER'S ALUM.

Alum, Rock. See ALUMSTONE AND ROCHE ALUM.

Alum Schist. See ALUM SHALE.

Alum Shale (Alum Ore, Alum Schist). Minerals. They are mixtures of aluminium silicates, iron pyrites, and bituminous matter.

Alumstone (Alunite, Alum Rock). A natural double sulphate of aluminium and potassium. It is a basic alum $(SO_4)_2Al.K.Al(OH)_3$.

Alumyte. An alum clay from Ireland.

Alundum. A registered trade mark for various types of goods, such as grinding wheels, abrasive and refractory grain, refractory articles and cement, porous plates, crucibles, and other articles made from crystalline alumina, or alumina which has been electrically fused and crystallised. The alumina may vary in purity from substantially 100 per cent. aluminium oxide to grades of material containing a considerable per centage of other substances, such as titania, silica, iron oxide, etc.

Aluni. An aluminium-nickel alloy used as an anode for deposition of the alloy coating.

Alunite. See ALUMSTONE.

Alunogen (Halotrichite, Hair Salt, Keramohalite). A mineral. It is a hydrated sulphate of aluminium, $Al_2O_3.SO_3.18H_2O$.

Alunozal. A proprietary preparation. It is stated to be a basic aluminium salicylate, and to be used as an internal antiseptic.

Alupent. A proprietary preparation of orciprenaline sulphate. A bronchial antispasmodic. (278).

Alupent Expectorant. A proprietary preparation of orciprenaline sulphate and Bisolvon. A bronchial antispasmodic. (278).

Alupent Obstetric. A proprietary preparation of ORCIPRENALINE used in the management of premature labour. (278).

Alupent-Sed. A proprietary preparation of amylobarbitone and orciprenaline sulphate. A bronchial antispasmodic. (278).

Aluphos. A proprietary preparation containing aluminium phosphate. An antacid. (219).

Alusac. A proprietary preparation containing sucralox and polymethyl siloxane. An antacid. (247).

Alushtite. A mineral. It is a hydrated aluminium silicate.

Alusil. Aluminium silicate. (546).

Alusil ET. A proprietary trade name for a synthetic aluminium silicate of controlled particle size used as an extender in emulsion paint. (205).

Aluzone. A proprietary preparation of phenylbutazone. An antirheumatic. (371).

Aluzyme. A proprietary preparation of aneurine hydrochloride, riboflavin, pyri-

doxine, niacin, pantothenic acid and folic acid. Vitamin supplement. (823).

Alva Marina. A seaweed specially treated for use as a stuffing material.

Alvar. A trade mark for compound derived from Gelvas (q.v.) by hydrolysing partially or completely and treating the products with acetaldehyde. Used for injection moulding.

Alvesco. An insecticide. It contains chlorinated hydrocarbons, and is used in vapour form.

Alvex. Blended alkaline detergent. (546).

Alvite (Anderbergite). A mineral. It is a silicate of zirconium, cerium, yttrium, calcium, magnesium, beryllium aluminium, and zinc.

Alvyl. A French polyvinyl alcohol.

Alypin (Amydricaine). Benzoyl-1 : 3-tetramethyl - diamino - 2 - isopropyl - alcohol-hydrochloride,$C_6H_5CO.O(C_2H_5)$ $C[CH_2.N(CH_3)_2]_2HCl$. A substitute for cocaine as a local anæsthetic.

Alzen. An alloy of 66 per cent. aluminium, and 33 per cent. zinc.

Alzogur. A proprietary trade name for an agricultural chemical based on cyanamide having special effects as a fertilizer and plant protective, as a defoliation agent, and as a disinfectant and deodorant. (899).

Alzomin. A proprietary trade name for an artificial fertilizer based on calcium cyanamide, containing s-triazine herbicides. (899).

Alzotrin. See ALZOMIN. (899).

AM 80 RESIN. An alphamethylstyrene resin of m.p. 80° C., sp. gr. 1·07. A 70 per cent. w/w resin in toluene solution has a viscosity of 3 poise. Used as a film former in coatings, rubber, tile and paper formulations. (196).

AM 100 Resin. An alphamethylstyrene resin of m.p. 100° C. and sp. gr. 1·07. A 70 per cent. w/w resin in toluene solution has a viscosity of 30 poise. Used as a film former in coatings, rubber, tile and paper formulations. (196).

Amadou (German tinder). A spongy combustible substance prepared from a species of fungus, *Boletus igniarius*.

Amalgam. A name applied to alloys of metals with mercury. It is also used as a term for a native alloy of mercury and silver with a formula varying between AgHg and Ag_2Hg_3.

Amalgams. Alloys of mercury with other metals. A typical one is sodium amalgam.

Amalgol. A proprietary trade name for a sulphonated fatty alcohol. A detergent used in the textile industry.

Amalith. See BAKELITE. The name is also applied to an American grade of vulcanised fibre used for electrical insulation.

Amaloy. A proprietary product. It is a single continuous coating of practically pure lead applied by a hot dipping process. It is corrosion-proof and adheres to base metal better, without alloying or impregnating, than galvanising. The name appears to be applied also to an alloy containing tungsten, chromium, and nickel, and which is used for cutting tools and surgical instruments.

Amandol. A proprietary grade of benzaldehyde. (507).

Amang. The heavy sands obtained in the Chinese process of tin-mining in Malaya.

Amantadine. A drug used in the treatment of the Parkinsonian syndrome. 1 - Adamantanamine. SYMMETREL is the hydrochloride.

Amapa Milk. The milky sap of the bark of *Plumeria Fallax*, a Brazilian tree. Used as a vermifuge, also externally.

Amaranth. See FAST RED D, and SCARLET 6R.

Amaranth A. A dyestuff. It is a British equivalent of Fast red D.

Amaranth Spirit. See TIN SPIRITS.

Amarantite. A mineral from Chile. It is ferric sulphate, $Fe_2(SO_4)_3$.

Amargosite. A trade name for a clay of the Bentonite type.

Amargyl. A proprietary preparation of chlorpromazine hydrochloride and amylobarbitone. A sedative. (336).

Amarin. Triphenyl-dihydro-glyoxaline, $(C_6H_5.C.NH)_2 : CH.C_6H_5$. It is also the name for a German salve containing formaldehyde.

Amaron. Tetraphenyl - p - pyrimidin, $C_{28}H_{20}N_2$.

Amarsan. See ACETARSONE.

Amatin. A German proprietary antipyretic and analgesic. It is stated to be acetyl-m-cresotic acid.

Amatols. Mixtures of trinitro-toluene and ammonium nitrate. An 80/20 amatol contains 80 parts ammonium nitrate and 20 parts trinitro-toluene. They are British high explosives.

Amax. A proprietary rubber accelerator. 2-(4 - morpholinyl mercapto) - benzthiazole. (450).

Amax No. 1. A proprietary accelerator, consisting of 90 per cent. 2-(4-morpholinyl mercapto) benzthiazole + 10 per cent. dibenzthiazyl disulphide. (450).

Amax Metal. An alloy of 81·3 per cent. copper, 11 per cent. tin, 7·4 per cent. lead, and 0·3 per cent. phosphorus.

Amazonite. Amazon stone (*q.v.*).

Amazon Stone. A mineral. It is a green variety of microcline (*q.v.*).

Ambara. A Columbian copal resin is known by this name.

Ambari Hemp. An East Indian fibre obtained from the bast of *Hibiscus cannabinus*.

Ambatoarinite. A mineral allied to ancylite (*q.v.*).

Ambazone. 1,4-benzoquinone amidinohydrazone thiosemicarbazone hydrate. Iversal.

Ambazyme. A proprietary preparation of amyloglucosidase. (550).

Ambenonium Chloride. NN' - Di - (2-diethylaminoethyl) oxamide bis-2-chlorobenzylochloride. Mytelase.

Amber. A fossil resin formed in certain beds of clay and sand, stated to be derived from *Pinites succinifer*, and to have the formula $C_{40}H_{64}O_4$. Amber has a specific gravity of 1·05–1·1, a melting-point of 250–325° C., a saponification value of 85–150, an iodine value of 62, an acid value of 15–35, and an ash of 0·3 per cent. The following are varieties of amber : Succinite (m.pt. 250–300° C.), Gedanite (m.pt. 150–180° C.), Glessite (m.pt. 250–300° C.), Beckerite, an opaque variety.

Amber, Artificial. Mellite (*q.v.*).

Amber, Baltic. The best kinds of amber.

Amberdeen. See BAKELITE.

Amberex. A factis of amber shade for light rubber stocks.

Amber, Friable. See GEDANITE.

Amber Glass. The colour is produced by the addition of carbonaceous matter, such as grain cake or sawdust, to a lime glass.

Amberglow. A proprietary phenolformaldehyde synthetic resin.

Ambergris. A grey, wax-like product found in the sea. It occurs in certain conditions of the intestines of the sperm whale. The chief constituent is ambrein, $C_{23}H_{40}O$, and it is used in perfumery. There is also a very dark, viscid product of inferior quality.

Amber-Guaiacum Resin. A variety of guaiacum resin. It is not an amber.

Amberine. A mineral. It is a variety of chalcedony.

Amberite. A smokeless powder consisting of 71 per cent. nitro-cotton, 18·6 per cent. barium nitrate, 1·3 per cent. potassium nitrate, 1·4 per cent. wood meal, and 5·8 per cent. petroleum jelly. It is a 42 grain powder. Also see BAKELITE.

Amberlac. A proprietary trade name for a synthetic varnish and lacquer resin of the alkyd class. It is used for air drying and baking finishes. (411.)

Amberlite. A proprietary trade name

for phenol-formaldehyde synthetic resin adhesives.

Amber Mica. Phlogopite (*q.v.*), a mica containing less than 2 per cent. iron.

Amberoid. See NICKEL SILVERS.

Amber Oil. See OIL OF AMBER.

Amberol. A phenol-formaldehyde resin combined with rosin or other resin. Used in the varnish industry.

Amber, Pressed. See AMBROID.

Amber, Prussian. Is Baltic amber, one of the best kinds.

Amber Salt. See SALT OF AMBER.

Amber Seed. The seed of musk mallow.

Amber, Soft. See GEDANITE.

Ambilhar. A proprietary preparation of niridazole. An antibilharzial agent. (18).

Ambiteric. Ampholytic surface active agents. (547).

Amblerlite. A proprietary packing made from asbestos with binders.

Ambloid. A proprietary casein plastic.

Amblygonite (Montebrasite). A mineral, $Li.Al(F.OH)PO_4$.

Amblystegite (Enstatite). A mineral, $MgO.SiO_2$, or $(MgO.FeO)SiO_2$.

Ambodryl. A proprietary preparation of bromodiphenhydramine hydrochloride. (264).

Ambor. Amine borates. (548).

Amboyna Wood. An Indian wood from *Pterospermum indicum*.

Ambra. A phenol-formaldehyde synthetic resin.

Ambrac Metal. A nickel silver. One alloy contains 74·5 per cent. copper, 20 per cent. nickel, 5 per cent. zinc, and 0·5 per cent. manganese, and another 64 per cent. copper, 25·4 per cent. nickel, 5·67 per cent. zinc, and 0·42 per cent. manganese. They are corrosion-resistance alloys suitable when exposed to salt solutions, dilute sulphuric acid, and chloride bleach solutions.

Ambraloy 927. A proprietary trade name for an alloy of 76 per cent. copper, 22 per cent. zinc, and 2 per cent. aluminium.

Ambraloy 928. A proprietary trade name for an aluminium bronze containing 92 per cent. copper and 8 per cent. aluminium.

Ambraloy 930. A proprietary trade name for an alloy of copper with 8 per cent. aluminium and 2·5 per cent. iron.

Ambramycin. A proprietary trade name for Tetracycline.

Ambrasite. See BAKELITE.

Ambrene (Musk Ambrette). An artificial musk perfume. It is dinitro-ter-butyl-*m*-cresol-methyl ether, $C_6H(OCH_3)(CH_3)(NO_2)_2C(CH_3)_3$.

Ambretta Seeds. Musk seeds.

Ambrine (Mulene, Parresine, Thermozine, Hyperthermine). Names for a proprietary preparation for burns. It consists of solid paraffin with an antiseptic, often resorcinol. A substance resembling cholesterol, which occurs in ambergris to the extent of 85 per cent., is also known as Ambrine.

Ambrite. A resin found in the lignite of Auckland, New Zealand.

Ambroid (Pressed Amber). A product consisting of small fragments of amber heated under pressure.

Ambroin. A German substitute for ebonite. It consists of asbestos which has been soaked in a solution of copal. When dry it is as hard as stone and can be ground up and moulded to any shape. Also see BAKELITE.

Ambrol. A phenol-formaldehyde synthetic resin.

Ambrosia Oil. An oil distilled from seeds of *chimopodium ambrosiodes*. Used in perfumery.

Ambucetamide. α - Dibutylamino - 4-methoxyphenyl acetamide. Present in Femerital.

Ambutonium Bromide. (3 - Carbamoyl - 3,3 - diphenylpropyl) ethyldi - methylammonium bromide.

AMC Antiseptic Solution. A proprietary trade name for a germicide. It is amyl-*m*-cresol.

Amcide. Weed killer. (503).

Amcosite. A proprietary ebonite or hard rubber.

Amechol. A proprietary trade name for Methacholine Chloride.

Ameisine. The trade name for aluminium formate.

Ameletite. A mineral. It is $Na_{37}Al_{24}Si_{24}O_{102}Cl$.

Amenorone. A proprietary preparation of ethisterone and ethinyloestradiol, used to treat amenorrhœa. (307).

Amenyl. Methyl-hydrastimide-hydrochloride, $CH_8O_2.C_6H_2(CO_2)N(CH_3).HCl$. Used in the treatment of amenorrhœa.

Amer-Glo. A proprietary trade name for pyroxylin (cellulose nitrate) and cellulose acetate.

American Antifriction Metal. See ANTIFRICTION METAL.

American Ashes. Crude potassium carbonate.

American Blue. Prussian blue (*q.v.*).

American Chrome Yellow. See CHROME YELLOW.

American Cloth (Leather Cloth). This material usually consists of a mixture of oxidised linseed oil, rosin, fillers, and colouring matter, spread on a cotton cloth. Other qualities are made from formaldehyde and gelatin.

American Copal. A copal from *Hymenæa courbaril.*

American Mace Butter. See OTOBA BUTTER.

American Olibanum. The oleo-resin from *Juniperus phœnicea.*

American Sienna. See INDIAN RED.

American Silver. A nickel silver. It contains from 49–59 per cent. copper, 20–24 per cent. zinc, 11–24 per cent. nickel, 0–1·3 per cent. iron, 0–2·9 per cent. lead, 0–0·69 per cent. tin, and sometimes small quantities of aluminium and phosphorus.

American Storax (Sweet Gum). A resin obtained from *Liquidambar styraciflua.* Used for chewing.

American Turpentine (English Turpentine). The oleo-resin from *Pinus australius*, *P. abies*, *P. pinaster*, and *P. iæda.* It contains the terpene australene, $C_{10}H_{16}$.

American Vermilion. See CHROME RED.

American Wormseed Oil. Chenopodium oil, from the fruits of *Chenopodium ambrosioides.*

Ameripol. A trade mark for synthetic rubber. A butadiene co-polymer. It is oil resistant.

Ameripol 1903. A trade mark for a high styrene synthetic rubber prepared from Ameripol 1511 cold SBR type latex and 85 per cent. bound styrene reinforcing resin. (38).

Amerite. A proprietary trade name for rubber derivatives and rubber-like resins in aqueous dispersion.

Amerith. A proprietary pyroxylin plastic in the form of sheet, rod, etc. Also a proprietary cellulose nitrate plastic.

Ameroid. A proprietary casein plastic.

Amerol. The methyl ester of saccharin.

Amerone. Perfumery base. (549).

Amesec. A proprietary preparation of aminophylline, ephedrine hydrochloride and amylobarbitone. A bronchial antispasmodic. (333).

Amesite. A chloritic mineral.

" A " Metal. A nickel-iron-copper alloy containing 6–8 per cent. copper. Used for making audio-frequency transformers.

Ametazole. 3 - (2 - Aminoethyl) pyrazole. Betazole. Histalog is the hydrochloride.

Amethopterin. A proprietary trade name for Methotrexate.

Amethyst. Purple or violet varieties of quartz. An analysis of an amethyst gave 97·5 per cent. SiO_2, 0·5 per cent. F_2O_3, 0·25 per cent. Al_2O_3, and 0·25 per cent. manganese oxide. It is also the name for the safranine dyestuffs, tetramethylsafranine, and tetraethyl-safranine, which dye violet. See MAGENTA.

Amethystine. A mineral. It is ferric sulphate, $Fe_2(SO_4)_3$.

Amethyst, Oriental. A green variety of corundum.

Amethyst Violet (Iris Violet). A dyestuff. It is tetraethyl-diamino-phenyl-phenazonium-chloride, $C_{26}H_{31}N_4Cl$. It is a basic dye, and dyes silk violet, with red fluorescence.

Ametox. A proprietary trade name for a specially purified and sterilised sodium thiosulphate for use in metallic poisoning. (507).

A.M.F. (A Nickel-Iron Alloy). It contains from 55–60 per cent. nickel, 1–3 per cent. manganese, and 0·2–0·4 per cent. carbon.

Amfac. A proprietary preparation of antimenorrhagic factor, used in the treatment of functional uterine hæmorrhage. (327).

Amfaid. Surface active agents and detergents. (550).

Amfecloral. An appetite suppressant. α - Methyl - N - (2,2,2 - trichloroethylidene) - phenethylamine.

Amfipen. A proprietary preparation of AMPICILLIN. An antibiotic. (317).

Amfix. High speed fixer for photographic processing. (507).

Amfonelic Acid. A stimulant for the central nervous system. 7 - Benzyl - 1 - ethyl - 4 - oxo - 1,8 - naphthyridine - 3 - carboxylic acid.

Amianite. A trade name for a specially hardened asbestos product.

Amianth. See AMIANTHUS.

Amianthoid. Asbestos (*q.v.*).

Amianthus (Amianth, Mountain Flax). A white and satiny variety of asbestos.

Amicar. 6-Aminohexanoic acid.

Amicarbalide. An anti-protozoan for veterinary use. 3,3′-Diamidinocarbanilide. DIAMPRON is the isethionate.

Amidated Cotton. See COTTON, AMIDATED.

Amidephrine. A vasoconstrictor and nasal decongestant. 3 - (1 - Hydroxy - 2 - methylaminoethyl)methanesulphonanilide. DRICOL is the mesylate.

Amide Powder. A blasting powder similar to gunpowder. It contains 40 per cent. potassium nitrate, 38 per cent. ammonium nitrate, and 22 per cent. charcoal.

Amidin. A solution of starch in water.

Amido-G-Acid (Amido-G-Salt). β-Naphthylamine-disulphonic acid G or γ. (2 : 6 : 8) $C_{10}H_5(SO_3H)_2(NH_2)$. Amido-G-salt is the sodium salt.

Amidogen. The group NH_2 is known by this name.

Amidogene. An explosive consisting of 73 per cent. potassium nitrate, and 1 per cent. magnesium sulphate in $\frac{1}{3}$ the weight of water ; 8 per cent. wood charcoal, 8 per cent. bran, and 10 per cent. sulphur is added.

Amido-G-Salt. See AMIDO-G-ACID.

Amidol. 1 : 2 : 4-Diamino-phenol, C_6H_4. $OH(NH_2)_2$. Used as a photographic developer, as the sulphate or chloride, in the proportion of 20 grams amidol, 200 grams crystallised sodium sulphite, and 1000 grams water.

Amidonaphthol Red GL. An acid dyestuff for dress materials.

Amidone. A proprietary trade name for Methadone.

Amidopyrine. See PYRAMIDONE.

Amido-R-Acid (Amido-R-salt). 2-Naphthylamine-3 : 6-disulphonic acid, R or α, $C_{10}H_5(SO_3)_2(NH_2)$. Amido-R-salt is the sodium salt.

Amido-R-Salt. See AMIDO-R-ACID.

Amiesite. A proprietary asphalt-rubber product used in road surfacing.

Amigren. A German preparation containing 0·2 gram acetyl-salicylic acid, 0·2 gram sodium citrate, and 0·05 gram caffeine. An analgesic.

Amikacin. An antibiotic currently undergoing clinical trial as BBK-8 and " Amikin ". $(2S)$ - 4 - Amino- N - $[(1R,\ 3S,3S,4R,5S)$ - 5 - amino - 2 - (3 - amino - 3 - deoxy - α - D - glucopyranosyloxy) - 4 - (6 - amino - 6 - deoxy - α - D - glucopyranosyloxy - 3 - hydroxycyclohexyl] - 2 - hydroxybutyramide.

Amikapron. A proprietary trade name for Tranexamic acid.

Amilee. See PARAFFIN, LIQUID.

Amiloride. A diuretic. N-Amidino-3,5-diamino - 6 - chloropyrazine - 2 - carboxamide. " MK-870 " is currently under clinical trial as the hydrochloride.

Aminacyl. A proprietary preparation of calcium or sodium para-aminosalicylic acid. An antituberculous agent. (244).

Amine O. Dewatering agent and corrosion inhibitor. (543).

Aminic Acid. Formic acid, H.COOH.

Aminocaproic Acid. 6-Aminohexanoic acid. Amicar. Epsicapron.

Aminoform. Hexamine (q.v.).

Aminoformic Acid. Carbamic acid, NH_2COOH.

Aminofusin. A proprietary preparation of l-aminoacids, sorbitol, vitamins and electrolytes used for parenteral nutrition. (451).

Amino-G Acid. 2-Naphthylamine-6, 8-disulphonic acid.

Aminogen I. A rubber vulcanisation

accelerator. It is a-naphthylamine, $C_{10}H_9N$.

Aminogen II. A rubber vulcanisation accelerator. It is p-phenylene-diamine, $C_6H_4(NH_2)_2$.

Amino-Glaucosan. β-Iminazolylethyl-amino-chlor-hydrate.

Aminogran. A proprietary PHENYLALANINE - free food used in the treatment of phenylketonuria. (284).

Aminomed. A proprietary preparation of theophylline sodium glycinate used in the treatment of bronchospasm. (250).

Aminomed Compound. A proprietary preparation of theophylline sodium glycinate, ephedrine hydrochloride and amylobarbitone used in the treatment of asthma. (250).

Aminometradine. 1 - Allyl - 6 - amino - 3 - ethylpyrimidine - 2,4 - dione. Mictrine.

Amino Plastics. A term applied to moulded materials made from urea-formaldehyde condensation products.

Aminoplex. A proprietary preparation of amino-acids, sorbitol, ethanol, vitamins and electrolytes used for parenteral nutrition. (386).

Aminopterin Sodium. Sodium N-[4 - (2,4 - diaminopterid - 6 - ylmethyl) - amino-benzoyl]-1-glutamate.

Amino-R Acid. 2-Naphthylamine-3, 6-disulphonic acid.

Aminorex. An appetite suppressant. 2 - Amino - 5 - phenyl - 2 - oxazoline. Apiquel is the fumarate.

Aminosol-Vitrum. A proprietary preparation of enzymatic casein hydrolysate containing 0·25 mEq. of sodium and potassium. Intravenous electrolyte. (329).

Aminox. A proprietary antioxidant for use in rubber at 0·25–0·35 per cent. It is composed of the reaction product of a ketone and an arylamine. (163).

Aminosol. A proprietary preparation of aminoacids and peptides of low molecular weight, used for parenteral nutrition. (330).

Aminutrin. A proprietary preparation of aminoacids for oral nutrition. (386).

Amiphenazole. 2,4 - Diamino - 5 - phenylthiazole. Daptazole.

Amisometradine. 6 - Amino - 1 - methallyl - 3 - methylpyrimidine -2,4 - dione. Rolicton.

Amisyn. A proprietary preparation of acetomenaphthone and nicotinamide, used in the treatment of chilblains. (327).

Amithiozone. A preparation used in the treatment of tuberculosis. P-acetyl-aminobenzaldelhyde thiosemicarbazide.

Amitraz. A veterinary ascaricide. N-

Methylbis - (2, 4 - xylyliminomethyl) - amine. TAKTIC; TRIATRIX; TRIATOX.

Amitryptyline. 3 - (3 - Dimethyl - aminopropylidone) - 1,2 : 4,5 - di - benzocyclohepta-1,4-diene. Laroxyl, Saroten and Tryptizol are the hydrochloride.

Ammiolite. A copper-antimony mineral, $CuSb_2O_5$, found in Chile.

Ammoket. A proprietary product. It is described as an elixir of ammonium mandelate for the treatment of urinary infections.

Ammonal (Antikamnia, Antisepsin, Antitoxine, Asepsin, Phenalgin). Preparations containing acetanilide or some allied body as the chief constituent. Antisepsin, or asepsin, is mono-bromacetanilide, $C_6H_4BrNH(C_2H_3O)$, and was formerly used as an antipyretic and analgesic. Another substance called antisepsin, or antiseptin, is iodo-borothymolate of zinc. Ammonal is also the name for an Austrian explosive. It consists of 30 per cent. trinitro-toluene, 47 per cent. ammonium nitrate, 22 per cent. aluminium powder, and 1 per cent. charcoal.

Ammonaldehyde. Hexamine (q.v.).

Ammon-Aspirin. Ammonium acetylsalicylate.

Ammon-Carbonite. An explosive. It contains ammonium nitrate, flour, nitroglycerin, and collodion wool.

Ammon-Dynamite. A French explosive. It contains 40 per cent. nitroglycerin, 10 per cent. wood meal, 10 per cent. sodium nitrate, and 40 per cent. ammonium nitrate.

Ammondyne. A coal-mine explosive containing 9–11 per cent. of nitro-glycerin, 45–51 per cent. of ammonium nitrate, 8–10 per cent. of sodium nitrate, 17–19 per cent. of ammonium oxalate, and 11–13 per cent. of wood meal.

Ammon-Fœrdite I. An explosive. It contains ammonium nitrate, flour, nitro-glycerin, collodion wool, glycerin, diphenylamine, and potassium chloride.

Ammon-Gelatin-Dynamite. A blasting explosive. It consists of 50 per cent. nitro-glycerin, 2·5 per cent. collodion cotton, 45 per cent. ammonium nitrate, and 2·5 per cent. rye meal.

Ammon-Halalit. A German explosive containing nitro-glycerin, ammonium nitrate, vegetable meal, nitro-compounds, and potassium perchlorate.

Ammoniacal Cochineal. See COCHINILLE AMMONIACALE.

Ammoniacal Turpethum. Hydrated dimercuri-ammonium sulphate, $(NHg_2)_2$ $SO_4.2H_2O$.

Ammonia Carmine. A solution of carmine in ammonia.

Ammonia, Copperised. See WILLESDEN FABRICS.

Ammoniacum. A gum resin obtained from *Dorema ammoniacum*, and used in medicine internally as an expectorant and externally as plasters. It contains from 18–28 per cent. gum, 50–75 per cent. resin, and 1·5–6 per cent. oil. It has a specific gravity of about 1·2, a saponification value of 160–177, an acid value of 100–106, and an ash of 2–7 per cent. African ammoniacum (gum galbanum), obtained from *Ferula tingitana*, containing 56–68 per cent. resin, having a specific gravity 1·11–1·21, and an acid value 69–80 is used to adulterate the Persian variety.

Ammonia Dynamite No. 1. A French explosive. It contains 40 per cent. nitro-glycerin, 45 per cent. ammonium nitrate, 5 per cent. sodium nitrate, and 10 per cent. wood meal or wheat flour.

Ammonia Dynamite No. 2. A French explosive. It consists of 20 per cent. nitro-glycerin, 75 per cent. ammonium nitrate, and 5 per cent. wood meal.

Ammonia Dynamite, Pulverent. An explosive containing 20 per cent. nitroglycerin, 25 per cent. ammonium nitrate, 36 per cent. sodium nitrate, and 19 per cent. rye flour.

Ammonia Dynamites. Explosives. They usually contain nitro-glycerin, wood pulp, ammonium nitrate, sodium nitrate, and calcium or magnesium carbonate. One consists of 75 per cent. ammonium nitrate, 4 per cent. paraffin, 3 per cent. charcoal, and 18 per cent. nitro-glycerin.

Ammonia Gelatin A. An explosive. It consists of 30 per cent. nitro-glycerin, 3 per cent. nitro-cotton, and 67 per cent. ammonium nitrate.

Ammonia Gelignite. An explosive containing 29·3 per cent. nitro-glycerin, 0·7 per cent. nitro-cotton, and 70 per cent. ammonium nitrate.

Ammonia, Hepatised. Ammonium sulphide.

Ammonia Nitrate Powder. An explosive. It consists of 80 per cent. ammonium nitrate, 5 per cent. potassium chlorate, 10 per cent. nitroglycerin, and 5 per cent. coal-tar.

Ammonia-Olein. The trade name for a form of sulphonated castor oil.

Ammonia Soda Ash. See SODA ASH, AMMONIA.

Ammonia-superphosphate. See NITROPHOSPHATE.

Ammoniated Alcohol. Made by passing ammonia into alcohol, until the latter increases in weight by about $\frac{1}{9}$.

Ammoniated Mercury. See WHITE PRECIPITATE.

Ammoniated Peat. An American fertiliser obtained by heating ammonia and peat under pressure. About two-thirds of the ammonia is changed into combinations that are not soluble in water. The material can contain as much as 20 per cent. nitrogen.

Ammonia Turpeth. Mercuric hydroxysulphato-amide, $(HO.Hg.NH.Hg)_2SO_4$. It is the sulphate of Millon's base.

Ammonia Water. A solution of ammonia, NH_4OH.

Ammonioformaldehyde. Hexamine (q.v.).

Ammoniojarosite. A mineral of Utah containing 49·3 per cent. Fe_2O_3, 1·56 per cent. K_2O, 4·23 per cent. $(NH_4)_2O$, 34·5 per cent. SO_3, 9·86 per cent. H_2O, and small amounts of Al_2O_3, CuO, MgO, and Na_2O.

Ammonio-Mercuric Chloride. See MERCURAMMONIUM CHLORIDE.

Ammonite. An explosive. Originally it consisted of ammonium nitrate, and dinitro-naphthalene. It now contains ammonium nitrate, trinitro-toluene, and sodium chloride.

Ammonium Iron Alum. Iron alum (ammonium-ferric sulphate).

Ammonium Muriate. See SAL-AMMONIAC.

Ammonium Superphosphate. See NITROPHOSPHATE.

Ammonium-Syngenite. Synonym for Koktaite.

Ammonocarbonous Acid. Hydrocyanic acid, HCN.

Ammo-Phos. The trade mark for a highly concentrated fertiliser containing nitrogen and phosphorus in an available form, and usually containing ammonium phosphate.

Amodiaquine. 7 - Chloro - 4 - (3 - diethylaminomethyl - 4 - hydroxy - anilino)-quinoline. Camoquin is the hydrochloride.

Amois. An explosive. It contains 90 per cent. ammonium nitrate, 5 per cent. chlor-dinitro-benzene, and 5 per cent. wood pulp.

Amole Juice. See ACHETE JUICE.

Amorphous Carborundum. See CARBORUNDUM FIRESAND.

Amorphous Digitalin. See DIGITALIN, AMORPHOUS.

Amorphous Hyoscyamine. Hyoscine, $C_{17}H_{23}NO_3$.

Amorphous Phosphorus. See RED PHOSPHORUS.

Amosite. An asbestos mineral. It is a magnesium silicate rich in iron. An average analysis gives 49·6 per cent. SiO_2, 39·6 per cent. FeO and Fe_2O_3, 4·8 per cent. MgO, 3·2 per cent. H_2O, and 2·2 per cent. Al_2O_3. The fibre is harsher than chrysotile and rather dirty.

Amoxal. A proprietary preparation of pentyloxybenzamide, pentyloxyacetophenone, and salicylic acid. A skin fungicide. (824.)

Amoxil. A proprietary preparation of Amoxycillin. An antibiotic. (272).

Amoxycillin. An antibiotic. 6 - [(—) - α - Amino - 4 - hydroxyphenylacetamido]penicillanic acid.

Amp. Adenosine phosphate.

Ampacet. A proprietary trade name for cellulose acetate moulding compound.

Ampangabeite. A rare earth mineral. It is a tantalo-columbate of uranium.

Ampco. An aluminium bronze. It contains from 86—92 per cent. copper, 7—11 per cent. aluminium, and 1—3 per cent. iron.

Amphactil. Chlorpromazine-dexamphetamine tablets. (507).

Amphenol. A proprietary trade name for polystyrene products.

Amphibole. A mineral, $Ca(Mg.Fe)_3(SiO_3)_4$, with $Na_2Al_2(SiO_3)_4$, and $(Mg.Fe)_2.(Al.Fe)_4.Si_2O_6$. The term is more particularly applied to the group of metallic metasilicates.

Amphibole, White. See TREMOLITE.

Amphigene. See LEUCITE.

Amphilogite. See DIDYMITE.

Amphionic. Ampholytic surface active agents. (547).

Amphithalite. A mineral. It is a phosphate of aluminium.

Amphodelite. A mineral. It is a variety of anorthite.

Amphomycin. An antibiotic produced from Streptomyces canus.

Amphotericin. Polyene antibiotics isolated from a strain of a *Streptomyces* species known as *Streptomyces nodosus*. Specific substances are designated by a terminal letter; thus, Amphotericin B (of which FUNGILIN and FUNGIZONE are examples).

Amphotropin. Hexamethylene-tetramine - camphorate, $(C_6H_{12}N_4)_2.C_8H_{14}(COOH)_2$. A urinary antiseptic.

Ampicillin. 6 - [D(—) - α - Amino - phenylacetamido]penicillanic acid. [D(—)α-Aminobenzyl]penicillin. Penbritin.

Ampiclox. A proprietary preparation of ampicillin and cloxacillin. An antibiotic. (364).

Amplex. A proprietary preparation of chlorophyll. A deodorant. (452).

Amplex-Sol. A proprietary preparation of sodium copper chlorophyllin, p-chloro-meta-xylenol, terpinol, iso propanol and sodium sulphoricinate. A deodorant. (452).

Amprol. A proprietary preparation of AMPROLIUM. An anti-protozoan for veterinary use.

Amprolium. An anti-protozoan present as the hydrochloride in AMPROL, AMPROLMIX and PANCOXIN. It is 1 - (4 - amino - 2 - propylpyrimidin - 5 - ylmethyl) - 2 - picolinium chloride.

Amprolmix. A proprietary preparation of AMPROLIUM. An anti-protozoan for veterinary use.

Amprolmix-UK. A proprietary preparation of ETHOPABATE. A veterinary anti-protozoal.

Amprol-Plus. A proprietary preparation of ETHOPABATE. A veterinary antiprotozoal.

Amrad Gum (Amra Gum, Oomra Gum). East African gums of a reddish colour. Used as substitutes for gum arabic.

Amritsar Gum. See GUM, AMRITSAR.

Amron. A proprietary vinylite base plastic for coatings.

Amsco Steel. A proprietary high-manganese steel containing 12–13 per cent. manganese and 1·2 per cent. carbon.

Amsil. A trade mark for silver copper alloy containing oxygen-free copper and 0·05 per cent. silver (3 oz. per ton). Conductivity 101 per cent. IACS. (33).

Amsulf. A trade mark for sulphur-copper alloy containing oxygen-free copper and 0·3 per cent. sulphur. Conductivity 96 per cent. IACS. (33).

Amtel. A trade mark for tellurium-copper alloy containing oxygen-free copper and 0·5 per cent. tellurium. Conductivity 90 per cent. IACS. (33).

Amvis. An explosive containing ammonium nitrate.

Amycaine. A French local anæsthetic, described as the hydrochloride of amylene and novocaine.

Amydricaine. See ALYPIN.

Amygdophenin. p-Phenetidine-amygdalate. Used in medicine.

Amyl Acetate. Isoamyl acetate, $(CH_3)_2.CH.CH_2.CH_2O.C_2H_3O$.

Amyl Alcohol (Fermentation amyl alcohol, Fusel Oil). Isoamyl alcohol, $C_5H_{12}O$.

Amylarine. Isoamyl-trimethyl-ammonium-hydroxide, $C_5H_{11}(CH_3)_3N.OH$.

Amylase. Diastase, an enzyme which renders starch soluble, by converting it into maltose.

Amylenol. Amyl salicylate, $C_6H_4(OH)$ $(CO_2.C_5H_{11})$. Sometimes used in the place of sodium salicylate in cases of rheumatism.

Amylit. A diamalt compound, which is an enzymic product. It is used for desizing in the textile industry.

Amylmetacresol. An anti-infective present in STREPSILS. It is 6-pentyl-m-cresol.

Amyl Mustard Oils. Amyl thiocarbimides, $C_5H_{11}N.CS$.

Amylocaine. See STOVAINE.

Amylocarbol. The trade name for a disinfectant, said to consist of 9 parts carbolic acid, 160 parts amyl alcohol, 150 parts soft soap, and 690 parts water.

Amylodextrin. See SOLUBLE STARCH.

Amyloform (Formamylum). A compound of formaldehyde and starch. Used as a specific for colds, also for the treatment of wounds, in the place of iodoform.

Amylogen. A soluble starch.

Amyloid. Concentrated sulphuric acid dissolves cellulose, gradually converting it into dextrin, and ultimately into dextrose. If the solution as soon as it is made is diluted with water, a gelatinous hydrate is produced. This substance is known as amyloid, since it resembles starch. The term is also applied to a substance found in the seeds of pæonies and certain cresses.

Amylomet. A proprietary preparation of amylobarbitone and ipecacuanha. (360).

Amylon. See MALTOBIOSE.

Amylopsin. Pancreatic diastase.

Amylozine Spansule. A proprietary preparation of trifluoperazine and amylobarbitone. A sedative. (281).

Amylozyme. Amylase starch converting enzyme. (550).

Amylum. Starch $(C_{12}H_{20}O_{10})n$.

Amysal. Amyl salicylate $C_6H_4(OH)$ $COOC_5H_{11}$

Amytal. Isoamyl-ethyl-barbituric acid. An anæsthetic applied intravenously. A proprietary preparation of amylobarbitone. A hypnotic. (333).

Amytin. The ammonium salt of a mixture of sulphonated hydrocarbons, prepared by the action of sulphuric acid upon various hydrocarbons. Under this name has been introduced a 33 per cent. aqueous solution of ichthyol-sulphonic acid. Used in skin diseases.

Amzirc. A trade mark for zirconium copper alloys containing oxygen-free copper and 0·15 per cent. zirconium. Conductivity 37 per cent. IACS. (33).

Anabalm. A proprietary preparation of oleoresin capsicum, menthol, histamine hydrochloride, beta-chloroethyl salicylate and squalene. An embrocation. (280).

Anabolex. A proprietary preparation of stanolone. Anabolic agent. (347).

Anacal. A proprietary preparation containing mucopolysaccharide, polysulphuric acid ester, prednisolone lauromacrogol - 400 and hexachlorophene, used in the treatment of haemorrhoids. (453).

Anacardic Acid. Obtained from *Anacardium occidentale* (cashew nut). The ammonium salt is used as a vermifuge.

Anacobin. A proprietary trade name for Cyanocobalamin.

Anacot Pastilles. A German product. Each pastille is stated to contain 0·01 gram formaldehyde, 0·001 gram menthol, and 0·005 gram citric acid. A mouth and throat antiseptic.

Anadonis Green. A pigment. It is hydrated chromium sesquioxide.

Anæstheform. A compound of diiodophenol - sulphonate (sozoiodol), and ethyl-amino-benzoate (anæsthesin). It is in the form of crystals melting at 225° C., and is used in the treatment of wounds.

Anæsthesin (Benzocaine, Anæsthone). Ethyl-*p*-amino-benzoate, $NH_2.C_6H_4.CO_2.C_2H_5$. Used as a substitute for cocaine as a local anæsthetic, also to check vomiting.

Anæsthesin, Soluble. Subcutin (*q.v.*).

Anæsthol. An anæsthetic containing ether and chloroform with 17 per cent. ethyl chloride.

Anæsthone. See ANÆSTHESIN.

Anæsthyl. See ANESTILE.

Anaflex. A proprietary preparation of polynoxylin. (386).

Anafranil. A proprietary preparation of chlomipramine hydrochloride. An antidepressant drug. (17).

Anagenite. Chrome ochre (*q.v.*).

Anahæmin. A proprietary preparation of a solution of an active erythropoietic fraction of liver. A hæmatinic. (182).

Analar. Laboratory reagents and chemicals. (527).

Analbite. A mineral. It is $4[(Na,K)AlSi_3O_8]$.

Analcine. See ANALCITE.

Analcite (Analcine). A zeolite mineral. It is a hydrated silicate of aluminium and sodium, $Na_2O.Al_2O_3.4SiO_2.2H_2O$.

Analexin. A proprietary trade name for the hydrochloride of Phenyramidol.

Analgen (Benzanalgen, Benzalgen, Chinalgen, Quinanalgen, Labordin). *o*-Ethoxy- ana - benzoyl - amino - quinoline, $C_9H_5(O.C_2H_5)(NH.CO.C_6H_5)N$. It is taken internally for rheumatism and gout.

Analgesco. A patented hypnotic composition. It is stated to be a molecular compound of amidopyrin and ethyl-isopropyl-barbituric acid.

Analgesic Balsam. Compound methyl salicylate ointment. (*Unguentum Methylis salicylatis compositum B.P.C.*)

Analgesine. See ANTIPYRINE.

Analgin. A proprietary preparation of aspirin, phenacetin, codeine, caffeine and phenolphthalein. An analgesic. (350).

Analoam. Soil testing reagent. (501).

Analoids. A proprietary article, consisting of tablets containing the exact amount of reagent for use in analysis.

Analutos. See KALMOPYRIN.

Ananas Oil (Essence of Pineapple Oil). Ethyl butyrate dissolved in alcohol. Used for flavouring rum and confectionery.

Ananase. A proprietary preparation of bromelains, used as an anti-inflammatory agent. (266).

Anapaite. A mineral. It is a hydrated ferrous calcium phosphate.

Anaphe Silk. See SILK, ANAPHE.

"A" Naphtha. See BENZINE.

Anaphylline. A proprietary preparation of theophylline, and niacinamide. Bronchial antispasmodic. (376).

Anapolon. A proprietary preparation of oxymetholone. An anabolic agent. (805).

Anaprel 500. A proprietary preparation of rescinnamine. An antihypertensive. (313).

Anaprotin. A proprietary preparation of stanolone. An anabolic agent. (825).

Anapyralgin. Diacetyl-*p*-amino-phenol, an antipyretic and analgesic.

Anarcotine. Narcotine.

Anaroids. A proprietary preparation containing resorcinol, powdered gall, bismuth subgallate, titanium dioxide, zinc oxide, boric acid, balsam of peru and kaolin. An antipruritic. (273).

Anasite. An explosive. It consists of ammonium perchlorate and myrabolans, usually with some sodium or potassium nitrate, and a small quantity of agar-agar.

Anaspalin. A form of wool fat.

Anasthol. A mixture of ethyl and methyl chlorides. A local anæsthetic.

Anatase (Octahedrite). A mineral. It is titanium dioxide, TiO_2.

Anatomical Alloy. An alloy of 53·5 per cent. bismuth, 19 per cent. tin, 17 per cent. lead, and 10·5 per cent. mercury, is known by this name. It melts at 60 °C.

Anatto. See ANNATTO.

Anauxite. A mineral, $3Al_2O_3.10SiO_2 8H_2O$.

Anaxeryl. A proprietary preparation of dioxanthranol, resorcin, salicylic acid, icthiol, birch tar oil, and balsam of Peru. Keratolytic agent. (802).

Anazol. Ethyl phthalate.

Ancaflex. A proprietary trade name for a series of boron trifluoride based polymers. Used as curing agents for epoxy resins to give flexible products. (162).

Ancaflex 70 and 150. A proprietary range of polymeric hardeners based on boron trifluoride. (454).

Ancamide 280, 400. A proprietary trade name for a fluid complex polyamide used as a curing agent for epoxy resins. (162).

Ancamine LO. An epoxy hardener for use at low temperatures. Free from phenolic odour and possessing a low irritation index. (290).

Ancamine LT. A proprietary trade name for an activated aromatic amine curing agent for epoxy resins. (162).

Ancamine MCA. A modified cycloaliphatic polyamine used on floorings and coatings. (290).

Ancamine XT. A faster version of LT above.

Ancaris. A preparatory compound of thenium for veterinary use. (277).

Ancaster Stone. A limestone employed in building.

Ancatax. A proprietary rubber accelerator. Dibenzthiazyl disulphide. (290).

Ancazate BU. A proprietary trade name for a self dispersible zinc butyl dithiocarbamate. Used as an accelerator for vulcanisation and an antioxidant for rubbers. (162).

Ancazate EPH. Zinc ethyl phenyl dithiocarbamate. (290).

Ancazate ET. Zinc diethyl dithiocarbamate. (290).

Ancazate ME. A proprietary accelerator. Zinc dimethyl dithiocarbamate. (290).

Ancazate Q. A proprietary complex of zinc dithiocarbamate used as an accelerator. (290).

Ancazate XX. A proprietary rubber accelerator. Butyl dithiocarbamate. (290).

Ancazide ET. A proprietary rubber accelerator. Tetraethyl thiuram disulphide. (290).

Ancazide IS. A proprietary rubber accelerator. Tetramethyl thiuram monosulphide. (290).

Ancazide ME. A proprietary rubber accelerator. Tetramethyl thiuram disulphide. (290).

Anchoic Acid (Lepargylic Acid). Azelaic acid, $C_9H_{16}O_4$.

Anchor. A proprietary trade name for a vanadium tool steel.

Anchor 1040, 1115, 1170, 1171 and 1222. A proprietary group of boron-trifluoride epoxy hardener curing agents. They consist of modified amine complexes of boron trifluoride. (454).

Anchoracel. A proprietary rubber vulcanisation accelerator. It is thiocarbanilide.

Anchorite. A light rubber product used in rubber mixings.

Anchred. A red oxide used in rubber mixings.

Anchusin. See ALKANET.

Ancofen. Tablets of meclozine hydrochloride, ergotamine tartrate and caffeine for relief of migraine. (182).

Ancolan. A proprietary trade name for the dihydrochloride of Meclozine-antihistamine and anti-emetic. (182).

Ancoloxin. Tablets of meclozine hydrochloride and pyridoxine hydrochloride taken to combat nausea and vomiting of pregnancy. (182).

Ancovert. Tablets of meclozine hydrochloride with nicotinic acid—vertigo and Meniere's disease. (182).

Ancrod. An anti-coagulant. The active principle is obtained from the venom of the Malayan pit-viper *Agkistrodon rhodostoma*, acting specifically on fibrinogen. See ARVIN.

Ancudite. Synonym for Kaolinite.

Ancylite. A mineral. It is a hydrated basic carbonate of cerium, lanthanum, didymium, and strontium, $4Ce(OH)CO_3.3SrCO_3.3H_2O$.

Andalusite. A mineral. It is a silicate of aluminium, Al_2SiO_5.

Andeer's Solution. A 9 per cent. aqueous solution of resorcinol.

Anderbergite. See ALVITE.

Andersonite. A mineral. It is $6[Na_2CaVO_2(CO_3)_2.6H_2O]$.

Anderson's Solution. A solution containing 100 c.c. of potassium hydroxide of 1·55 specific gravity, and 15 grams of pyrogallol. It is used in the Hempel burette for absorption of oxygen.

Andesine. A mineral. It is a soda-lime felspar.

Andesite. A volcanic rock containing about 60 per cent. silica.

Andiroba Oil. See CRAB WOOD OIL.

Andorite. A mineral. It is a sulphantimonite of silver and lead, $Ag.Pb.Sb_3S_6$.

Andradite (Demantoid, Uralian Emerald). A mineral. It is a calcium-iron-garnet, $Ca_3Fe_2Si_3O_{12}$. When cut, it is called Demantoid or Uralian emerald. It is sometimes erroneously called olivene.

Andreolite. Harmotone (*q.v.*).

Andreson's Acid. 1-Naphthol-3 : 8-disulphonic acid.

Andrewsite. A mineral. It is a phosphate of iron and copper. $(Cu,Fe'')_3Fe_6'''(PO_4)_4(OH)_{12}$.

Andriole. A German preparation consisting of bismuth iodide and an iodine-uranium compound.

Androcur. A proprietary preparation of CYPROTERONE acetate used in the control of hypersexual and deviant behaviour in males. (438).

Androgeston. A proprietary preparation of methyltestosterone, and ethisterone, used to control menorrhagia. (265).

Androstalone. A proprietary preparation of methylandrostanolone, used to treat premenstrual tension. (307).

Androstin. A total testicular extract physiologically standardised, containing all the active principles of the male genital glands.

Androx 3961. Shield. A water displacing fluid meeting DTD 900/4942.

Andursil. A proprietary preparation of aluminium hydroxide gel, magnesium hydroxide and carbonate, and simethicone. An antacid. (17).

Anectine. A proprietary preparation of suxamethonium chloride. A short acting muscle relaxant. (277).

Anemolite. A mineral. It is a variety of calcium carbonate.

Anesin. A solution of chloretone (tertiary - trichloro - butyl - alcohol), CCl_3.$(CH_3)_2$.COH. Used in medicine.

Anesone. Acetone chloroform (q.v.).

Anestile (Anæsthyl, Anesthyl, Anesthol). A mixture of ethyl and methyl chlorides. An anæsthetic.

Anesthol. See ANESTILE.

Anesthyl. See ANESTILE.

Anethaine. A proprietary preparation of amethocaine hydrochloride cream. Antipruritic. (335).

Anethol. p-Allyl-phenyl-methyl ether, $C_{10}H_{12}O$, a constituent of fennel and anise oil.

Anethol-Borneol. Anethol hexahydride, $C_{10}H_{18}O$.

Anethol-Camphor. Anethol tetrahydride, $C_{10}H_{16}O$.

Aneurin. Vitamin B_1. See THIAMINE.

Aneurone. A proprietary preparation of Vitamin B_1, strychnine hydrochloride, caffeine, compound infusion of gentian and sodium acid phosphate. A tonic. (387).

Angaralite. A mineral. It is a silicate of iron, aluminium, and magnesium.

Angarite. A Russian cast basalt (q.v.) used as an electrical insulator.

Angelardite. A mineral. It is a phosphate of iron.

Angel Red. See INDIAN RED.

Angico (Bahia Bark). The bark of *Peptedenia columbrina* of Brazil. A tanning material containing 18–20 per cent. of tannins.

Angico Gum. See BRAZILIAN GUM.

Anginos. A German preparation. It consists of tablets containing formaldehyde and menthol, which are used for mouth disinfection.

Angio-Conray. A proprietary trade name for the sodium salt of Iotholamic acid.

Angioneurosin. Nitro-glycerin, $C_3H_5(O.NO_2)_3$.

Angiotensin Amide. A hypertensive. Val^5 - hypertensin II - Asp - β - amide. Asn - Arg - Val - Tyr - Val - His - Pro - Phe.

Angioxyl. A proprietary preparation of pancreatic extract insulin free.

Angised. A proprietary preparation of glyceryl trinitrate in stabilised base. A vasodilator used for angina pectoris. (277).

Anglarite. A mineral, $FeS.Sb_2S_3$.

Anglesite (Lead Vitriol). A mineral. It is a lead sulphate, $PbSO_4$.

Anglopyrin. See ASPIRIN.

Angola Copal. See COPAL RESINS.

Angolon. A proprietary trade name for the hydrochloride of Imolamine.

Angomensin. A preparation containing lipamin, a chemical substance from the corpus luteum.

Angustura Bark. Cusparia bark.

Anhalamine. The methyl ester of 3 : 4 : 5-trimethyoxy-phenyl-ethylamine. Used in medicine.

Anhaline. See HORDENINE.

Anhydrite (Muriacite). A mineral. It occurs in the Stassfurt deposits, and is calcium sulphate, $CaSO_4$.

Anhydrite, Soluble. See SOLUBLE ANHYDRITE.

Anhydrokainite (Basaltkainite). A mineral, $KMgClSO_4$, found in Prussia.

Anhydrone. A proprietary name for a perchlorate of magnesium, a drying agent for gases. It is stated to be capable of absorbing as high as 35 per cent. of its weight of water.

Anhydrous Acetic Acid. See ACETIC ACID, ANHYDROUS.

Anidoxime. An analgesic undergoing clinical trial as " BRL 11870 " and " E-142 ". It is 3 - diethylaminopropiophenone O - (p - methoxyphenyl-carbamoyl) - oxime.

Anilazone. See ANILIPYRIN.

Anileine. See MAUVEINE.

Anileridine. Ethyl 1 - (4 - amino - phenethyl) - 4 - phenylpiperidine - 4 - carboxylate.

Anilin. See KYANOL.

Aniline Black (Aniline Black in Paste, Fine Black, Oxidation Black). A black dyestuff produced on the fibre by the oxidation of an aniline salt. Salts of nigraniline, $(C_6H_5N)n$, are produced on the fibre. Used for dyeing and printing black.

Aniline Black in Paste. See ANILINE BLACK.

Aniline, Blue. See ANILINE OIL.

Aniline Blue, Spirit Soluble (Rosaniline Blue, Opal Blue, Spirit Blue O, Gentian Blue 6B, Hessian Blue, Fine Blue, Lyons Blue, Light Blue, Parma, Bleu de Nuit, Bleu Lumière). The hydrochloride, sulphate, or acetate of triphenyl-rosaniline and triphenyl-pararosaniline. Hydrochlorides. $C_{37}H_{30}N_3Cl$ and $C_{38}H_{32}N_3Cl$. Sulphates, $C_{74}H_{60}N_6$ SO_4, and $C_{76}H_{64}N_6SO_4$. Acetates, $C_{39}H_{33}N_3O_2$ and $C_{40}H_{35}N_3O_2$. Dyes wool and silk greenish-blue.

Aniline Brown. See BISMARCK BROWN.

Aniline for Blue. See ANILINE OIL.

Aniline for Red. See ANILINE OIL FOR RED.

Aniline for Safranine. See ANILINE OIL FOR SAFRANINE.

Aniline Green. See ALDEHYDE GREEN.

Aniline, Light. See ANILINE OIL.

Aniline Mauve. See MAUVEINE.

Aniline Oil (Aniline for Blue, Aniline Oil for Blue, Blue Aniline, Blue Oil, Light Aniline). Almost pure aniline, free from toluidines. Used for phenylating rosaniline. See KYANOL.

Aniline Oil for Blue. See ANILINE OIL.

Aniline Oil for Red (Aniline for Red, Red Aniline). A mixture of about 32 per cent. aniline, 27 per cent. *p*-toluidine, and 41 per cent. *o*-toluidine. Used principally for the preparation of fuchsine.

Aniline Oil for Safranine (Aniline for Safranine). A mixture of 35–50 per cent. aniline, and 65–50 per cent. *o*-toluidine. Used in the preparation of safranines.

Aniline Orange. See VICTORIA YELLOW.

Aniline Pink. Safranine (*q.v.*).

Aniline Purple. See MAUVEINE.

Aniline, Red. See ANILINE OIL FOR RED.

Aniline Red. See MAGENTA.

Aniline Rose. Safranine (*q.v.*).

Aniline Salt. Aniline hydrochloride, $C_6H_5.NH_2.HCl$. Used for dyeing black on cotton.

Aniline Spirits. See TIN SPIRITS.

Aniline Tailings. A name applied to the least volatile portion of aniline oils. They contain little or no aniline, some toluidine, xylidine, and cumidine ; nitrobenzene and its homologues.

Aniline Tartar Emetic. Aniline-antimonyl-tartrate, $C_4H_5O_6(SbO)C_6H_7$ $N.H_2O$. It is used in the treatment of trypanosomiasis.

Aniline Violet. See MAUVEINE.

Aniline Yellow (Spirit Yellow). A dyestuff. It is the hydrochloride of amino-azo-benzene, $C_{12}H_{12}N_3Cl$. It is not used as a dye, but for the preparation of acid yellow and indulines.

Anilipyrin (Anilazone). A mixture of antipyrine and acetanilide.

Anilotic Acid. Nitro-salicylic acid.

Animal Black. A pigment prepared by carbonising all kinds of animal refuse (leather, horn, hoofs, and skin).

Animal Cellulose. Tunicin, $C_6H_{10}O_5$, prepared from the outer covering, or mantle, of the molluscs belonging to the class *Tunicata*.

Animal Charcoal (Bone Charcoal). This term is used for all charcoal produced by the ignition of animal substances with exclusion of air, but more particularly to that obtained from bones. This material contains approximately 10 per cent. carbon, and 90 per cent. mineral matter, mainly calcium phosphate. Used for absorbing dyes and other purposes.

Animal Cholesterin. Cholesterin, $C_{26}H_{44}O$.

Animal Glycerin. Neatsfoot oil.

Animalised Cotton. Cotton which has been treated with casein or other nitrogenous matter, such as milk and sodium stannate, then with alum or calcium chloride, and ammonia. The material is more easily dyed after such treatment.

Animal Oil. See BONE OIL.

Animal Oil, Dippel's. See BONE OIL.

Animal Rouge. Carmine (*q.v.*).

Animal Soap. Curd soap (*q.v.*).

Animal Starch. Glycogen, found in the blood and liver of mammals.

Anime Resin (Gum Anime). A fossil copal resin from South America.

Anime Resin, West Indian. See TACAMAHAC RESIN.

Animikite. A mineral. It is a silver antimonide, Ag_2Sb, and is considered a variety of dyscrasite.

Aniodol. A solution of paraform in glycerol, mixed with oil of mustard. An internal anæsthetic.

Anise Camphor. Anethol, $CH_3O.$ $C_6H_4.CH : CH.CH_3$.

Aniseed Cordial. Consists of aniseed water, 4 parts, and sugar, 1 part.

Anisette. A liqueur similar to Chartreuse.

Anisidine. Ortho-aminophenol methyl ether.

Anisidine Ponceau. See ANISOL RED.

Anisidine Scarlet. See ANISOL RED.

Anisil. The dimethyl derivative of dioxy-benzil, $C_6H_4(OCH_3).CO.COC_6H_4$ (OCH_3).

Anisindione. 2-(4-Methoxyphenyl) indane-1,3-dione. Miradon.

Anisoin. Anethoin, $(C_{10}H_{12}O)n$.

Anisoline. See RHODAMINE 3B.

Anisol Red (Anisidine Ponceau, Anisidine Scarlet). An acid dyestuff. It is anisol-azo-β-naphthol-disulphonic acid, $CH_3 . O . C_6H_4 . N_2 . C_{10}H_4(HSO_3)_2OH$. Dyes wool and silk red from an acid bath.

Anisotheobromine. An addition product of theobromine-sodium and sodium anisate, $C_7H_7N_4O_2.Na.C_6H_4(OCH_3) CO_2Na$.

Anisotropine Methylbromide. Octatropine Methylbromide.

Ankara. An artificial butter consisting of a mixture of cocoa butter with 10 per cent. milk and coloured with yolk of egg.

Anka. A proprietary trade name for a range of austenitic stainless steels. See WELDANKA. (136).

Anka E. A proprietary trade name for a low carbon stainless steel containing 18 parts chromium and 8 parts nickel. (135).

Anka M. A deep drawing steel. (135).

Anka Steel (V2A Steel). Nickel-chromium steels containing from 15–16 per cent. chromium and from 7–10 per cent. nickel.

Ankerite. A mineral. It is a carbonate of calcium, magnesium, and iron.

Ankylotaphine. A disinfectant containing about 15 per cent. creosol.

Annabergite (Nickel Bloom). A mineral. It is a hydrated arsenate of nickel, $Ni_3As_2O_8.8H_2O$.

Annalidin. A thymol compound, $C_{20}H_{22}O_3I_3$. A substitute for iodoform.

Annaline. See GYPSUM.

Annatto (Anatto, Annotto, Arnatto, Arnotto, Orleans, Rocou). A natural dyestuff, obtained from the fleshy covering of the seeds of the ruccu tree, *Bixa orellana*. It contains a yellow colouring matter, bixen, $C_{28}H_{34}O_5$. The extract is used for colouring butter and cheese. It dyes cotton, orange.

Annealing Carbon (Carbon of Cementation, Carbon of Normal Carbide). The names given to the carbon found combined, in definite composition, in steel.

Annerodite. A mineral. It is a niobate of thallium and cerium.

Annex. A rubber vulcanisation accelerator. It is hexamethylene-tetramine mixed with other organic material.

Annidalin. Thymol iodide.

Annite. A mineral. It is a variety of mica, from Cape Ann.

Annivite. A mineral. It is tennantite containing bismuth.

Annotto. See ANNATTO.

Annovax. A vaccine for Newcastle disease, infectious bronchitis and epidemic tremor. (514).

Anobolex. A proprietary trade name for Stanolone.

Anodesyn. A proprietary preparation containing ephedrine hydrochloride, lignocaine hydrochloride and allantoin. An antipruritic. (253).

Annulex BHT B.P. grade. A proprietary trade name for 2,6-di-tertbutyl *p*-cresol. An antioxidant for packaging materials for food and drugs. (180).

Anonaid. Anionic surfactants and detergents. (550).

Anode Mud (Anode Slime). The material which falls to the bottom of the electrolysing vessel during the electrolytic refining of copper. It contains copper (10–25 per cent.), gold (0·7–2 per cent.), silver (5–40 per cent.), tellurium, antimony, and arsenic.

Anode Slime. See ANODE MUD.

Anodised Aluminium. See ELOXAL, ALUMINIUM ANODISED.

Anodyne Balsam (Anodyne Liniment). Opium liniment.

Anodyne Colloid. Anodyne collodion. (*Collodium anodynum B.P.C.*)

Anodyne Drops. See HOFFMANN'S ANODYNE.

Anodyne, Hoffmann's. See HOFFMANN'S ANODYNE.

Anodyne Liniment. See ANODYNE BALSAM.

Anodynin. See ANTIPYRINE.

Anodynon. Ethyl chloride, C_2H_5Cl.

Anogon. The mercury salt of 2 : 6-diiodo-phenol-4-sulphonic acid, $C_6H_2I_2 (OHg)(SO_3Hg)$.

Anol. *p*-Propenyl alcohol.

Anolineum. A coal-tar product used as a wood preservative.

Anomite. A mineral containing lithium. It is similar to biotite (*q.v.*).

Anon. See CYCLOHEXANON.

Anone. Cyclohexanol, a resin solvent.

Anophorite. A mineral. It is $2[Na_3Mg_3Fe^{..},Fe^{..}(Ti,Si)_8O_{22}(OH)_2]$.

Anorthite (Lime Felspar). A mineral, $Al_2O_3.SiO_2+CaO.SiO_2$.

Anorthoclase. A mineral, $Na_2O.2SiO_2 +Al_2O_3.4SiO_2$.

Anorvit. Tablets of ferrous sulphate, ascorbic acid and acetomenaphthone—hypochromic anaemia. (182).

Anotal. A proprietary preparation of phenyl-cinchonyl-urethane.

Anotex. Dyes for anodised metal. (551).

Anovlar. A proprietary preparation of norethisterone acetate and ethinyl œstrodiol. Oral contraceptive. (898).

Anozol. The trade name for a preparation of iodoform. It contains from 10 to 20 per cent. thymol.

Anquil. A proprietary preparation of BENPERIDOL used in the control of deviant sexual behaviour. (356).

Ansa. Alkyl naphthaline sodium sulphonate. (503).

Ansal. Phenazone salicylate.

Anschutz Chloroform. A very pure form of chloroform, prepared from salicylide chloroform.

Anscor-P. A proprietary preparation of magnesium oxide in powder form, derived from sea-water. It is used in compounding polychloroprene and related elastomers. (428).

Anserine. A natural peptide from muscle. It is β-alanyl methyl histidine.

Ansol A. A proprietary blend of anhydrous denatured ethyl alcohol with small percentages of esters, other alcohols, and hydrocarbons. A nitrocellulose and resin solvent. It has a specific gravity of 0·8058 at 20° C.; none boils below 70° C., and not more than 50 per cent. above 80° C. It contains a small amount of amyl alcohol. A trademark.

Ansol B. A preparation similar to A, except that B contains a small amount of normal butyl alcohol in the place of amyl alcohol. It has a specific gravity and boiling-point similar to A. A trademark.

Ansol E-121. A trade mark for ethylene glycol dimethyl ether. (897).

Ansol E-161. A trade mark for triethylene glycol dimethyl ether. Dimethoxy triethylene glycol. TRIGLYME. (897).

Ansol E-181. A trade mark for tetraethylene glycol dimethyl ether. TETRA-GLYME. (897).

Ansol M. A preparation similar to A but containing no higher alcohols. It has a specific gravity of 0·7982 at 20° C., none boils below 70° C., and not more than 10 per cent. above 80° C. A trademark.

Ansol PR. A preparation similar to M but with a higher ester content. It has a specific gravity of 0·8414 at 20° C., not more than 10 per cent. boils below 70° C., and at least 80 per cent. below 80° C. A trademark.

Ansolysen. A proprietary preparation of pentolinium tartrate. An antihypertensive. (336).

Antabuse. A proprietary preparation of tetraethylthiuram disulphide, used in the treatment of alcoholism. (808).

Antacedin. Calcium saccharate, $(C_{12}H_{22}O_{11})_3CaO$. Recommended in the treatment of dyspepsia and flatulence.

Antacidin. Antacedin (q.v.).

Antaciron. A proprietary silicon containing 14·5 per cent. silicon.

Antalby. A proprietary preparation of theophylline-para-amino benzoate. Bronchial antispasmodic. (802).

Antalcrol. Amido-pyrin-quino-salicylate.

Antalgine. The trade name for salicylaldehyde-α-methyl-phenyl-hydrazone. An anti-neuralgic and anti-rheumatic.

Antamokite. A mineral. It is a gold-silver telluride.

Antas. A proprietary preparation containing pepsin, aluminium hydroxide, magnesium trisilicate, belladonna, magnesium hydroxide. An antacid. (256).

Antazoline. 2-N-Phenylbenzylamino-methyl-2-imidazoline.

Antelope. Acid sodium pyrophosphate. (503).

Antemin. A proprietary preparation of sodium lauryl sulphate, and trisopropylphenoxypolyethoxyethanol. Spermicidal cream. (361).

Antepan. Piperazine anthelmintic. (514).

Antepar. A proprietary preparation of piperazine citrate. Antihelmintic. (277).

Anthical. A proprietary preparation of mepyramine maleate and zinc oxide. An antihistamine skin cream. (336).

Anthiomaline. Hexa-lithium antimony tri(mercaptosuccinate). (507).

Anthion. Potassium persulphate, used as a hypo eliminator in photography.

Anthiphen. A proprietary preparation of dichlorophen. An antihelmintic. (336).

Anthisan. A proprietary name for the hydrogen maleate of Mepyramine. (336).

Anthium Dioxcide. A proprietary additive based on stabilised chlorine dioxide. A deodorant designed to remove odours caused by residual monomers in resins. It works by oxidation. (429).

Anthochroite. A mineral. It is a silicate of calcium, magnesium, and manganese. Synonym for Violan.

Anthocyans. The red, blue, and violet colouring matters of plants.

Anthoinite. A mineral. It is $AlWO_4OH.H_2O$.

Antholite. Anthophyllite (q.v.).

Anthophyllite. A mineral, $Mg.Fe.O$ (SiO_2).

Anthosiderite. A mineral, $2Fe_2O_3.$ $9SiO_2.3H_2O$.

Anthracene Acid Black. A dyestuff obtained by coupling amino-salicylic acid with α-naphthylamine-sulphonic acids (1 : 6 and 1 : 7), the product being diazotised, and coupled with β-naphthol-disulphonic acid R.

Anthracene Blue. 1 : 2 : 4 : 5 : 6 : 8-Hexaoxy-anthraquinone.

Anthracene Blue New WG. A dyestuff obtained by the long heating of anthracene blue with caustic soda. It dyes chromed wool greenish-blue.

Anthracene Blue S. See ALIZARIN BLUE S.

Anthracene Blue SWB, SWG, SWR.
Dyestuffs. They are brands of anthracene blue, soluble in water.

Anthracene Blue WG, WB. Dyestuffs obtained by the action of fuming sulphuric acid, and ordinary sulphuric acid upon 1 : 5-dinitro-anthraquinone. WB is hexaoxy-anthraquinone, and is probably identical with Brilliant alizarin cyanine 3G. It dyes wool mordanted with alumina, blue, and chromed wool, bluish-green. WG is isomeric with WB, and dyes wool mordanted with alumina, greenish-blue.

Anthracene Blue WGG, WGG Extra. A dyestuff which is the sodium salt of diamino-dioxy-anthraquinone-sulphonic acid. It dyes wool bluish-violet from an acid bath, and chromed wool bluish-green.

Anthracene Blue WR. A dyestuff. It is hexaoxy-anthraquinone, $C_{14}H_8O_8$, and is isomeric with WB. It dyes wool mordanted with alumina, violet, and with chromium, blue.

Anthracene Brown Powder. A dyestuff. It is the sodium salt of anthracene brown, and dyes wool with chromium mordants, and silk with aluminium, iron, or chromium mordants.

Anthracene Brown R, G (Anthragallol, Alizarin Brown). A dyestuff. It is trioxy-anthraquinone, $C_{14}H_8O_5$. Dyes cotton mordanted with chromium, brown. Also see CHROME FAST BROWN FC.

Anthracene Chrome Black F, FE, 5B. An acid azo dyestuff from 2 : 3-amino-naphthol-6-sulphonic acid, and alkylised 1 : 8-amino-naphthol-3 : 6-disulphonic acid. It dyes with chromium mordants.

Anthracene Chrome Yellow BN. A dyestuff. It is a British equivalent of Milling yellow.

Anthracene Dark Blue W. A mordant dyestuff. It is similar to Anthracene blue WR.

Anthracene Green. See CŒRULEINE AND CŒRULEINE S.

Anthracene Oil. That fraction from coal-tar distilling over between 270° and 400° C., of specific gravity 1·085 to 1·095. Approximately 38 gallons are obtained from a ton of tar. It is a yellowish-green mass containing about 3 per cent. anthracene.

Anthracene Red. An acid dyestuff, which is the sodium salt of nitro-diphenyl - disazo - salicylic - α - naphtholsulphonic acid, $C_{28}H_{17}N_5O_9SNa_2$. It dyes wool fast red from an acid bath.

Anthracene Violet. See GALLEINE.

Anthracene Yellow (Alizarin Yellow). A dyestuff. It is dibromo-dioxy-β-methyl-coumarine, $C_{10}H_6Br_2O_4$. It dyes chromed wool greenish-yellow.

Anthracene Yellow BN. See MILLING YELLOW.

Anthracene Yellow C (Fast Chrome Yellow C, Fast Mordant Yellow). An acid dyestuff. It is the sodium salt of thio-dibenzene-disazo-disalicylic - acid, $C_{26}H_{16}N_4O_6SNa_2$. It dyes chromed wool, yellow.

Anthracene Yellow GG. An acid dyestuff giving greener shades than the C brand.

Anthracene Yellow Paste. A mordant dyestuff. It is dibromo-dioxy-methyl-coumarine, and dyes chrome mordanted wool greenish-yellow.

Anthrachrysone. 1 : 3 : 5 : 7-Tetroxy-anthraquinone.

Anthracite. A hard coal, containing 85–95 per cent. carbon. It burns with little smoke.

Anthracite Black B (Phenylene Black). An acid dyestuff. It is the sodium salt of disulpho-naphthalene-azo-α-naphthalene - azo - diphenyl - m - phenylene - diamine, $C_{38}H_{26}N_6O_6S_2Na_2$. It dyes wool black from an acid bath.

Anthracoal. A metallurgical fuel composed of a mixture of small particles of anthracite and practically pure carbon, formed from the distillation of coal-tar pitch or similar material.

Anthraconite. A mineral. It is calcium carbonate, with bituminous matter.

Anthracoxene. A mixture of resins found in Bohemia.

Anthracyl Chrome Green. A dyestuff obtained from diazotised picramic acid and naphthionic acid.

Anthraflavic Acid. 2 : 6-Dioxyanthraquinone, $C_{14}H_6O_2(OH)_2$.

Anthraflavone. A dyestuff which is prepared by treating 2-methyl-anthraquinone with condensing agents. It dyes cotton yellow shades.

Anthragalanthranol. Trioxyanthranol, $C_{14}H_{10}O_4$.

Anthragallol. See ANTHRACENE BROWN.

Anthrahydroquinone. Oxanthranol $C_{14}H_{10}O_2$.

Anthramethane Colours. Dyestuffs for yarns and dress materials.

Anthranilic Acid. o-Amino-benzoic acid, $C_6H_4(NH_2).COOH$.

Anthrapuranthranol. Trioxyanthranol, $C_{14}H_6(OH)_4$.

Anthrapurpurin (Isopururin, Alizarin GD, Alizarin RF, Alizarin RT, Alizarin RX, Alizarin SC, Alizarin SSA, Alizarin SX, Alizarin SX Extra, Alizarin WG). A mordant dyestuff. It is 1 : 2 : 7-trioxy-anthraquinone, $C_{14}H_5O_2(OH)_3$, and dyes red shades on alumina mordanted wool. Also see ALIZARIN SAR.

Anthraquinone Black. A dyestuff produced by the fusion of 1 : 5-dinitro-anthraquinone with sodium polysulphide. It dyes cotton black from an alkaline or sulphide bath.

Anthraquinone Blue. A dyestuff. It is 2 : 4 : 6 : 8-tetrabromo-1 : 5-diamino-anthraquinone, heated with p-toluidine, and sulphonated.

Anthraquinone Green GX. An anthracene dyestuff closely related to Alizarin viridine.

Anthraquinone Violet. An anthracene dyestuff, $C_{14}H_6O_2(HSO_3.C_7H_6NH)_2$.

Anthraquinone Violet 3R. A dyestuff. It is a British equivalent of Anthraquinone violet.

Anthrarobin. Dioxy - anthranol, $C_{14}H_{10}O_2$. Used in the treatment of skin diseases.

Anthrarufin. 1 : 5-Dioxy-anthraquinone, $C_{14}H_6O_2(OH)_2$.

Anthrasol (Colourless Coal-tar). A coal-tar freed from bases such as pyridine and chinoline by acid treatment, and from pitch by distillation. Juniper tar and oil of peppermint are added. It is an antiseptic and antiparasiticide having the colour and consistency of olive oil, and is employed for itching skin diseases. A glycerol containing 20 per cent. anthrasol is sold under the name of anthrasolin.

Anthrasol Brown IVD. A dyestuff. The sulphuric ester of the reduced form of the condensation product of 6 : 7-dichloro-5-methoxythioindoxyl and the 4 : 5-benz-derivative of thioindoxyl.

Anthrasol Yellow 129. The di-o-sulphuric ester of 1 : 5-di (m-trifluoro-benzamido) anthraquinone in the reduced form.

Anthrasolin. See Anthrasol.

Anthratetrol. Tetrahydroxy-anthracene, $C_{14}H_6(OH)_4$.

Anthratriol. Trihydroxy-anthracene, $C_{14}H_7(OH)_3$.

Anthra Wool Red CR. A new vat dyestuff.

Anthraxolite. A variety of anthracite.

Anthropic Acid. A mixture of palmitic and stearic acids.

Anti-Age 33. A proprietary trade name for an antioxidant. It is an amine reaction product. (164).

Anti-Age 34. A proprietary trade name for an antioxidant. It is a naphthol-amine reaction product. (164).

Anti-Age 44. A proprietary trade name for an antioxidant. It is a naphthol-amine reaction product. (164).

Anti-Age 55. A proprietary trade name for an antioxidant. It is an acetaldehyde-aniline reaction product. (164).

Anti-Age 66. A proprietary name for an antioxidant. It is an amine reaction product. (164).

Anti-Age 77. A mixture of Anti-age 33 with Anti-age 66 ($q.v.$). (164).

Anti-Age 99. A proprietary trade name for a wax preparation of an antioxidant. (164).

Antiar. The milky juice of the upas tree. Used as an arrow poison.

Antiarthrin. A condensation product of saligenin and tannin, probably the hydroxy-benzyl ester of tannic acid. Used in medicine.

Antib. A preparation used in the treatment of tuberculosis. p-acetylamino-benzaldehyde thiosemicarbazone.

Antibacterin. A suggested preservative for bread. It is sodium-p-toluene sulpho-chloramide (chloramine T).

Antibenzine-Pyrine. The commercial name for the magnesium salt of oleic acid. Used as a preventative of spontaneous combustion of benzine.

Antibortin. A proprietary preparation of hesperidin ascorbate. (332).

Antichlor. The name given to any substance which removes free chlorine from materials. Sodium bisulphite is a typical example. Used to remove the last traces of chlorine from goods bleached with it.

Anticorodal. An aluminium alloy containing 1 per cent. silicon, 0·06 per cent. magnesium, and 0·06 per cent. manganese.

Anti-corrosion Paints. Various types of bituminous preparations are used for this purpose. Some are made from asphalt, coal-tar pitch, and boiled oil. Red lead and graphite paints are employed for covering ironwork.

Antidiabetin. The trade name for a preparation of saccharin and mannite.

Antidine. The phenyl ether of glycerol.

Antidol. A proprietary preparation of ethosalamide, paracetamol and caffeine. An analgesic. (252).

Antidust 2. A proprietary preparation similar to Antidust F, but containing a particularly fine zinc stearate in aqueous dispersion. (421).

Antidust F. A proprietary group of surface-active agents in combination with polyvalent alcohols, used in a concentrated aqueous solution to prevent undesirable surface tackiness in sheets and extrudates of plastics and rubber materials. (421).

Antiethylin. The active principle of a serum from horses treated regularly with alcohol. Recommended for the treatment of alcoholism.

Antifebrin (Phenylacetamide). Acetanilide, $C_6H_5.NH.COCH_3$. Used as a febrifuge in cases of typhoid fever, smallpox, phthisis, rheumatism, and erysipelas.

Antifekt. A German preparation containing thymol, tribrom-β-naphthol, and neutral ground soap. Used as an antiseptic washing agent.

Antifoam ET. An emulsified form of Antifoam T. (112).

Antifoam T. Tri-n-butyl-phosphate. An antifoaming agent used in the manufacture of paper coatings. (112).

Antiformin. A disinfectant. It consists of sodium hypochlorite and caustic soda. Used in breweries. Also see RADOFORM.

Anti-fouling Compositions. These usually consist of iron oxide paints containing some poisonous substance such as white arsenic, mercury oxide, or copper salt, in a medium of oil varnish and shale naphtha. Used for painting ships' bottoms.

Anti-freezing Solutions. These vary considerably, but the following are common ones : (a) calcium chloride in water, (b) glycerin in water, and (c) methylated spirit in water.

Antifriction Metal. Also called White Metal (q.v.). There are several varieties which are used for bearings. (a) Used for rapid working, consists of 77 per cent. antimony, 17 per cent. tin, and 6 per cent. copper ; (b) Extra hard, containing 82 per cent. antimony, 12 per cent. tin, 2 per cent. zinc, and 4 per cent. copper ; (c) Medium hard, consisting of 26 per cent. antimony, 72 per cent. tin, and 2 per cent. copper ; (d) American, containing 20 per cent. antimony, 78 per cent. lead, 1 per cent. zinc, and 1 per cent. iron ; (e) Babbitt's, consisting of 7 per cent. antimony, 89 per cent. tin, and 4 per cent. copper. Other Babbitt's bearing metals contain from 55–90 per cent. zinc, 1–29 per cent. tin, 3–8 per cent. copper, and 1–3 per cent. antimony, and still another class consists of from 10–80 per cent. lead, 1–75 per cent. tin, 8–20 per cent. antimony, and 0–5 per cent. copper.

Anti-frost Gelammonite No. 1. A low-freezing explosive containing a nitrated mixture of glycerin and ethylene glycol, nitro-cotton, ammonium nitrate, wood meal, sodium chloride, and dinitro-toluol.

Anti-frost Penrhyn Powder. A low-freezing explosive containing a nitrated mixture of glycerin and ethylene glycol, ammonium nitrate, wood meal, and sodium chloride.

Antifungin. The trade name for magnesium borate, a fungicide.

Antigalta. A liniment consisting of camphor, ethereal oils, and animal fats, which has proved an active medicament against acute and chronic mastitis.

Antigelde Sûreté. A Belgian explosive containing 25 per cent. nitroglycerin, 15 per cent. nitro-toluene, sodium nitrate, wood meal, and a little ammonium sulphate.

Antigen A. A proprietary trade name for phenyl-alpha-naphthylamine. An antioxidant. (165).

Antigen B.C. Thought to be a proprietary trade name for a condensation product between aldol and alpha-naphthylamine. (165).

Antigen C. A proprietary trade name for a mixture of phenyl-alpha-naphthylamine and meta-toluylene diamine. An antioxidant. (165).

Antigermin. The commercial name for a preparation of a copper salt of a phenol-carboxylic acid, mixed with lime. A fungicide.

Antigorite. A mineral. It is a dark-green variety of serpentine (q.v.).

Antigrison. An explosive. It consists of 27 per cent. nitro-glycerin, 1 per cent. nitro-cotton, and 72 per cent. ammonium nitrate.

Antihypo. Potassium percarbonate, $K_2C_2O_6$. Used as a bleaching agent and hypo eliminator in photography.

Antikamnia. See AMMONAL.

Anti-knock Fluids. A fluid for admixture with petrol to prevent or reduce knocking in internal combustion engines. A widely-known anti-knock fluid is that sold under the trade mark Ethyl (q.v.). See also MOTYL.

Anti-leprol. A trade name for certain esters of the fatty acids of chaulmoogra. Used in the treatment of leprosy.

Antiliton. A boiler compound for the prevention of scale. It contains 20 per cent. solids and 12 per cent. tannins.

Antillite. A mineral. It is $(Mg,Fe)_8Si_2O_5(OH)_4$.

Antiluetin. Potassium-ammonium-antimonyl-tartrate, $[SbO(C_4H_4O_6)_2KNH_4]$ H_2O. A trypanocide.

Antilusin. A concentrated preparation of normal serum. A proprietary trade name for Pentamethonium iodide.

Antilux AOL. A proprietary blend of paraffinic hydrocarbons used as an anti-weathering agent in natural and synthetic rubbers. (421).

Antimalum. A mixture of essential oils used for rheumatism by outward application.

Antimicrobium. Sodium amino-o-benzoyl - sulpho - iso - amido - hydro - cupron-nucleo-formate.

Antimildioidium. A fungicide containing 39 per cent. copper sulphate and 61 per cent. sodium carbonate.

Antimonial Copper. See ANTIMONIAL COPPER GLANCE.

Antimonial Copper Glance (Antimonial Copper, Wolfsbergite). A mineral. It is cuprous sulphantimonite, $CuSbS_2$.

Antimonial Lead. An alloy of 87 per cent. lead and 13 per cent. antimony.

Antimonial Nickel. A mineral, NiSb.

Antimonial Nickel Ore (Ullmanite). A mineral, NiSbS.

Antimonial Powder. (*Pulvis Antimonialis*, B.P.) It contains 25 parts antimonious oxide, and 50 parts calcium phosphate.

Antimonial Silver. See DYSCRASITE.

Antimonic Acid. Antimonic oxide, Sb_2O_5.

Antimonin. See LACTOLIN.

Antimonite (Stibnite, Grey Antimony Ore, Antimony Glance). A mineral. It consists of antimony sulphide, Sb_2S_3.

Antimoniuretted Hydrogen (Stibine). Antimony hydride, SbH_3.

Antimonous Acid. Antimonous oxide, Sb_4O_6.

Antimony, Arsenical. See ALLEMONTITE.

Antimony Ash. Obtained by roasting the grey sulphide in air. It consists chiefly of Sb_2O_4, and is used for the preparation of antimony compounds.

Antimony Blende. See RED ANTIMONY.

Antimony Bloom (Senarmontite, Valentinite, White Antimony). A mineral. It is mainly Sb_2O_3.

Antimony Caustic. Liquor antimony chloride, B.P.

Antimony Cinnabar. See RED ANTIMONY.

Antimony Crocus. See SULPHURATED ANTIMONY.

Antimony Flux. Crude potash melted with liquated antimony sulphide. It is used in refining antimony.

Antimony Glance. See ANTIMONITE.

Antimony Glass. Produced in the preparation of antimony from its ores. It is antimony oxysulphide, and is used for colouring glass yellow.

Antimony, Japanese. See BLACK ANTIMONY.

Antimony Lead. See HARD LEAD.

Antimony Mordant. See TARTAR EMETIC POWDER.

Antimony Needle. See BLACK ANTIMONY.

Antimony Ochre. A mineral. It contains Cervantite, $Sb_2O_3.Sb_2O_5$, and Stiblite, $Sb_2O_3.Sb_2O_5.2H_2O$.

Antimony Ore. See CRUDE ANTIMONY.

Antimony Ore, Grey. See ANTIMONITE.

Antimony Regulus. See REGULUS OF ANTIMONY.

Antimony Saffron. See SULPHURATED ANTIMONY.

Antimony Salt. Different mixtures which are double salts of antimony fluoride, SbF_3, with alkali sulphates, or with alkali fluorides. One is a double salt of antimony fluoride and ammonium sulphate, $SbF_3(NH_4)_2SO_4$. Used as mordants in dyeing.

Antimony Silver Blende. A silver ore, $Sb_2S_3.5Ag_2S$, and $Sb_2S_3.3Ag_2S$.

Antimony, Tartrated. See TARTAR EMETIC.

Antimony Vermilion. A red pigment. It is antimony oxysulphide, $Sb_6S_6O_3$.

Antimony, White. See ANTIMONY BLOOM.

Antimony Yellow (Mérimée's Yellow). A pigment. It is bismuth antimonate mixed with ammonium chloride, and litharge, then fused.

Antinevene. An antivenomous serum.

Antinervine. A mixture of 50 per cent. acetanilide, 25 per cent. salicylic acid, and 25 per cent. ammonium bromide. Prescribed for neuralgia and headache.

Antinonnin. See VICTORIA YELLOW.

Antinosine. Sodium tetra-iodo-phenolphthalein, (the sodium salt of nosophene (*q.v.*)), $C_6H_2I_2ONa_2.C.O.C_6H_4O$. It is used for treating wounds, and is given internally as a stomach disinfectant.

Antinutisin. Pyrogallol-mono-ethylester.

Anti-oxidant 425. A proprietary preparation of 2,2-methylene-bis(4-ethyl-6-tertiary-butyl) phenol. (416).

Anti-oxidant 2246. A proprietary preparation of 2,3 - methylene - bis (4 - methyl - 6 tertiary - butyl) phenol. (416).

Anti-oxidant 4010. A proprietary preparation of N-phenyl-N'cyclohexyl-*p*-phenylene diamine.

Anti-Oxidant AH. A proprietary aldol-alpha-naphthylamine resin. (112).

Antioxidant AN. A proprietary antioxidant. It is aldol-α-naphthylamine paste.

Anti-Oxidant AP. A proprietary aldol-alpha-naphthylamine powder. (112).

Anti-Oxidant DDA. A proprietary derivative of DIPHENYLAMINE. (112).

Anti-Oxidant DNP. Di - β - naphthyl - p - phenylenediamine. (112).

Anti-Oxidant DOD. 4 - 4' - dioxydiphenyl. (112).

Anti-Oxidant EM. A 30 per cent. aqueous emulsion of a diphenylamine derivative. (112).

Anti-oxidant K1. A mixture of 60 per cent. dihydroxydiphenyl and 40 per cent. magnesium. (3).

Anti-oxidant MB. Phenylenethiourea. A proprietary antioxidant. (3). 2-mercaptolbenzimidazole. (112).

Anti-Oxidant PAN. Phenyl - alpha-naphthylamine. (112).

Antioxidant PBN. See also NEOZONE D. A proprietary antioxidant. It is phenyl-β-naphthylamine.

Antioxidant RA. A proprietary antioxidant. It is aldol-α-naphthylamine powder.

Antioxidant RES. A proprietary antioxidant. It is aldol-α-naphthylamine resin.

Antioxidant RRIO. A condensation product of styrene with resorcinol. (3).

Anti-Oxidant RR 10 N. A proprietary range of alkylated phenols. (112).

Anti-Oxidant SP. A proprietary styrenated phenol. (112).

Antioxygene A. A proprietary trade name for phenyl-alpha-naphthylamine. An antioxidant. (166).

Antioxygene AFL. A proprietary trade name for a Ketone-amine reaction product. (166).

Antioxygene AN. A proprietary trade name for aldol-alpha-naphthylamine paste. An antioxidant. (166).

Antioxygene BN. A proprietary trade name for β-naphthol. An antioxidant. (166).

Antioxygene CAS. A proprietary trade name for a mixture of phenyl-alpha-naphthylamine and meta-toluylene-diamine. (166).

Antioxygene INC. A proprietary trade name for aldol-alpha-naphthylamine in powder form. An antioxidant. (166).

Antioxygene MC. A proprietary trade name for phenyl-beta-naphthylamine. An antioxidant. (166).

Antioxygene RA. A proprietary trade name for aldol naphthylamines. Antioxidants. (166).

Antioxygene RES. A proprietary trade name for an aldol-alpha-naphthylamine resin. An antioxidant. (166).

Antioxygene RM. A proprietary trade name for aldol naphthylamine. An antioxidant. (166).

Antioxygene RO. A proprietary trade name for aldol naphthylamine. An antioxidant. (166).

Antioxygene STN. A proprietary trade name for phenyl-alpha-naphthylamine mixed with meta-toluylene-diamine and stearic acid. An antioxidant. (166).

Antioxygene WBC. A proprietary trade name for an antioxidant. (166).

Antiozone. Appears to be atomic or ionic oxygen, liberated from ozone.

Antiperiostin. Mercury-iodo-cantharidate, $C_8H_{11}IO(COO)_2Hg$.

Antiphlogin. A trade name for a mixture of boric acid, ammonium phosphate, and acetic acid. It is stated to render certain artificial silks non-inflammable.

Antiphlogistine. A trade mark for a medicated poultice and cold remedy products.

Antiputrol. A commercial name given to a thick yellowish-brown disinfectant containing 65–70 per cent. phenols, made soluble with soap.

Antipynin. Sodium metaborate $NaBO_2$. H_2O. Antiseptic.

Antipyoninum. Neutral sodium tetra-borate, prepared by fusing together borax and boric acid.

Antipyreticum. See ANTIPYRINE.

Antipyrine (Analgesine, Anodynin, Antipyreticum, Methozin, Metozin, Paradyn, Phenazone, Phenylon, Pyrazine, Pyrazoline, Sedatine). 1-Phenyl-2 : 3-dimethyl-pyrazolone, $C_6H_5(CH_3)_2 \cdot C_3HN_2O$. A febrifuge and analgesic.

Antipyrinum. Antipyrine of the German Pharmacopœia.

Antique Purple. A colouring matter. It is 6 : 6'-dibrom-indigo.

Antiquinol. Allyl-phenyl-cinchonic ester. An antiseptic and analgesic.

Antirheumatin. A compound of sodium salicylate and methylene blue. An antirheumatic.

Antirheumol. A 20 per cent. solution of the glycerol esters of salicylic acid in glycerol and alcohol. Used in the treatment of rheumatism.

Antirhinol. A mixture of tannin, salol, and sandal oil.

Antisapros. An oil containing 70–80 per cent. limonene, extracted from *Pinus pinea*. It is used in the treatment of chronic bronchitis.

Antisclerosine. Consists of the mixed inorganic constituents of normal blood.

Antisepsin (Bromanilide). See AMMONAL.

Antiseptin. A mixture of boric acid, zinc iodide, and thymol. An antiseptic dusting powder.

Antiseptol. A registered trade mark. Cinchonine-iodo-sulphate. Used as a substitute for iodoform.

Antisin. The hydrochloride of Antazoline.

Antisoil Agent A. An aqueous anionic dispersion which is not readily removed from treated fabric. Gives an anti-static antisoil finish for carpets and upholstery of either natural or synthetic fibres. (2).

Antisol. A proprietary preparation of lobeline hydrochloride used as a smoking remedy. (389).

Antispasmin. The trade name for a combination of narceine and sodium salicylate. A narcotic and sedative.

Antistaphin. Methyl-hexamethylene-tetramine-pentaborate.

Antistat A. An antistatic agent for plasticised PVC compounds. (503).

Antistatic 812 and 813. A proprietary range of antistatic compounds for plastics. They are 100 per cent. active phenolic ethoxylates, used in proportions of 5–7 per cent. (431).

Antistatic 816. A proprietary antistatic compound for plastics, 95 per cent. active lauric imide, used in polyethylene, PVC and polystyrene. (431).

Antistin. A proprietary preparation of antazoline hydrochloride. (18).

Antistitin-Privine. A proprietary preparation of ANTAZOLINE sulphate and NAPHAZOLINE nitrate used in the treatment of allergic rhinitis. (18).

Antitan. A tannin remover. (523).

Antitartaric Acid. Lævo-tartaric acid, COOH.CH(OH).CH(OH).COOH.

Antithermin. Phenyl-hydrazine-levulinic acid, $CH_3.C(N_2H.C_6H_5).CH_2.CH_2.COOH$. An antipyretic, now almost out of use, being toxic.

Antitode. A mixture of magnesium oxide, charcoal, and ferric hydroxide. An antidote for poisoning.

Antitoxine. See AMMONAL.

Antitryx. An antiskinning agent. (552).

Antituman. Sodium-chondritin-sulphate. Used in the treatment of cancers.

Antitussin. A 5 per cent. ointment of difluor-diphenyl, $(C_6H_4F)_2$. Used in the treatment of whooping cough.

Anti-white Lead. See OIL WHITE.

Antlerite. A mineral. It is a variety of Brochantite, $4CuO.SO_3.3H_2O$, or $7CuO.2SO_3.5H_2O$.

Antoban. A veterinary anthelmintic. (514).

Antodin (Antodyne). The phenyl ether of glycerol, $C_3H_5(OH)_2O.C_6H_5$. An analgesic.

Antodyne. See ANTODIN.

Ant Oil. See OIL OF ANTS.

Antoin. A proprietary preparation of aspirin, calcium carbonate, citric acid, phenacetin, codeine phosphate and caffeine citrate. An analgesic. (248).

Anton N. A trade mark for a general-purpose antioxidant. It is an octylated diphenylamine used in many elastomers. (10).

Antox. This name is used for a proprietary antioxidant for rubber, consisting of a powder containing p-aminophenol dispersed on barium sulphate. It is also applied to a pigment consisting of antimony oxide, and to an aluminium-zinc alloy which resists corrosion. A proprietary trade name for a derivative of butyraldehyde-aniline. An antioxidant. (2).

Antox N. Octylated Diphenylamine. A registered trade name for a general

purpose antioxidant with only mild discolouring and staining characteristics. (10).

Antoxyl. A preparation consisting mainly of sodium hyposulphite. A preservative for perfumed soaps.

Antozonite. A mineral. It is a variety of fluorite.

Antrenyl Duplex. A proprietary preparation of oxyphenonium bromide. Gastrointestinal sedative. (18).

Antrimolite. A mineral. It is a variety of mesolite.

Antrycide. Slats of quinapyramine. (512).

Antrypol. A proprietary preparation of suramin. A tryanoside. (2).

Anturan. A proprietary preparation of sulphinpyrazone. An antirheumatic. (17).

Antwerp Blue. A pigment, which is a low grade Prussian blue. Alum and zinc are used in its manufacture, and it is practically a mixture of iron and zinc ferrocyanides. See PRUSSIAN BLUE.

Antwerp Brown. Asphalt or bitumen used as a pigment is known by this name.

Anugesic. A proprietary preparation of boric acid, balsam of Peru, bismuth oxide, bismuth subgallate, resorcinol, zinc oxide and pramoxine. Analgesic suppositories. (262).

Anugesic-HC. A proprietary preparation of PRAMOXINE hydrochloride, HYDROCORTISONE acetate, zinc oxide, Peru balsam, benzyl benzoate, bismuth oxide and resorcinol, used in the treatment of haemorrhoids. (262).

Annulex. An antioxidant. (552).

Anusol. A proprietary preparation containing boric acid, zinc oxide, bismuth subgallate, bismuth oxide, Peru balsam and resorcinol. An antipruritic. (262).

Anusol-HC. A proprietary preparation of benzyl benzoate, bismuth subgallate, bismuth oxide, resorcinol, Peru balsam, zinc oxide and hydrocortisone acetate, used in the treatment of haemorrhoids. (262).

Anvico. A rubber vulcanisation accelerator. It is stated to be an impure thiocarbanilide.

Anvil Brass. See BRASS.

Anxine. Tablets of dexamphetamine sulphate, cyclobarbitone and mephenesin. (510).

Anysin (Anytin). Aqueous solutions of ichthyol-sulphonic acid (33 per cent.). Germicidal.

Anytin. See ANYSIN.

Anytols. Solutions of phenol in ichthyol.

Aolan. A proprietary albumin milk preparation.

AOR/GR. A proprietary name for technically-comminuted rubber. (456).

Apagallin. Tetraiodo-phenol-phthalein. An antiseptic and indicator.

Apallagin. See Nosophene.

Aparatine. See Vegetable Glue.

Apatelite. A mineral. It is a hydrated sulphate of iron.

Apatite (Eupyrochroite). A mineral. It is a phosphate of lime combined with either calcium fluoride or calcium chloride. See Fluorapatite and Chlorapatite.

Aperione. Phenol-phthalein, $C_{20}H_{14}O_4$.

Aperitol (Valothalein). A mixture of equal parts of isovaleryl- and acetyl-phenol-phthalein. A purgative.

Apex. A proprietary brand of cellulose rayon.

Apex 400. A proprietary alloy of aluminium with silicon.

Aphalerite. A mineral. It is zinc sulphide.

Aphanesite (Clinoclase). A mineral, $2CuO.As_2S_3.4H_2O$.

Aphit. An alloy of 20–21 per cent nickel, 70–75 per cent. copper, 2–5 per cent. zinc, and 2–4·5 per cent. cadmium.

Aphrite (Earth Foam, Foam Spar). A mineral which is a variety of calcite.

Aphrizite. A mineral. It is a black tourmaline.

Aphrodine. A proprietary preparation of corynine.

Aphrodite. A mineral, $H_6Mg_4Si_4O_{15}$, of Sweden.

Aphrogen. A saponin. It is a German material obtained from plant products.

Aphrosiderite. A chloritic mineral found at Ouvaly. It contains 24·5 per cent. SiO_2, 22 per cent. Al_2O_3, 7·4 per cent. Fe_2O_3, 31·3 per cent. FeO, 3·6 per cent. MgO, and traces of CaO, MnO, and TiO_2. It agrees closely with Thuringite.

Aphrosol. Foaming agents. (512).

Aphthisin. A mixture of potassium-guaiacol-sulphate and petro-sulphol-ammonium. Recommended in lung and bronchial diseases.

Aphthitalite (Arcanite, Glaserite). A mineral. It is a sulphate of potassium and sodium, $(KNa)_2SO_4$.

Aphthite. An alloy of 800 parts copper, 25 parts platinum, 10 parts tungsten, and 170 parts gold.

Aphthonite. A silver ore consisting mainly of tetrahedrite.

Aphtite. Zinc and cadmium-containing nickel bronzes. Used for high-grade imitation silver products.

Apicosan. A German preparation. It is an aqueous solution of the natural poison in bees. Used for subcutaneous or intramuscular injection for rheumatic pains.

Apiezon. High vacuum oils, greases, waxes. (553).

Apigenine. A yellow dyestuff obtained by decomposing the glucoside apiine found in parsley.

Apinol. A product obtained by the distillation of the wood of *Pinus palustris*. It consists mainly of lævo-menthone, $C_{10}H_{18}O$, and is used locally as an antiseptic, and as an anæsthetic.

Apiol (Apiol, White). The crystalline constituent of parsley oil. Liquid apiol is essential oil derived from an apiol-bearing variety of parsley. Sometimes used as a diuretic.

Apisate. A proprietary preparation containing diethylpropion, aneurine hydrochloride, riboflavine, pyridoxine hydrochloride and nicotinamide. An anti-obesity agent. (245).

Apitong. The wood of the tree *Dipterocarpus Grandiflorus*, etc.

Apjohnite. A mineral. It is manganous aluminium sulphate, $MnSO_4.Al_2(SO_4)_3.24H_2O$, found in Algoa Bay, South Africa.

Apla. A proprietary trade name for Saran tubing and fittings.

Aplexil. A proprietary preparation. It is an anti-influenza vaccine.

Apochin. A German compound of acetyl-salicylic acid and quinine. An antispasmodic.

Apoconquinine. Apoquinidine, $C_{19}H_{22}N_2O_2$.

Apollinaris. A mineral water containing 217 parts sodium bicarbonate, 4·5 parts sodium chloride, 2·8 parts sodium silicate, 29 parts calcium chloride, 42·8 parts magnesium carbonate, 16·7 parts magnesium sulphate, 3 parts ferrous sulphate, and 1·7 parts hydrochloric acid in 100,000 parts water.

Apollo Red (Archil Substitute Extra, Orchil Extract N Extra, Orchil Substitute N Extra). An acid dyestuff. It is the sodium salt of *p*-nitro-benzene-azo-α-naphthylamine-disulphonic acid, $C_{16}H_{10}N_4O_8S_2Na_2$. Dyes wool red from an acid bath.

Apolloy. A proprietary copper-iron alloy containing 0·25 per cent. copper and 0·08 per cent. carbon.

Apolomine. A proprietary preparation of hyoscine hydrobromide, benzocaine, riboflavine, pyridoxine hydrochloride, and nicotinamide. An antinauseant. (112).

Apolysin. Mono-phenetidine-citrate, $C_6H_4(O.C_2H_5)NH(CO.C_3H_4(OH)(COOH)_2$. Used medicinally as an internal remedy for neuralgia.

Aponal. Amyl carbamate, $NH_2.COOC_5H_{11}$. A somnifacient.

Apophyllite. A zeolite mineral, $4(CaO.2SiO_2.H_2O)KF$, a source of potassium.

Aposafranine. A dyestuff. It is diazotised safranine boiled with alcohol.

Aposet 707. A proprietary trade name for a Ketone peroxide catalyst for polyesters.

Apothesine. The ester of γ-diethyl-amino-propyl-alcohol and cinnamic acid. $C_6H_5.CH : CH.CO.OCH_2.CH_2.CH_2N.(C_2H_5)_2$. An anæsthetic.

Appalachian Tea (Carolina Tea, Black Drink). Leaves from two plants, *Prinos glaber* and *Viburnum cassinoides*, of America. Used as a substitute for tea.

Appallagin. Mercury of nosophen (tetraiodo-phenol-phthalein).

Apparatine. A strong solution of starch, prepared by means of caustic soda in which the alkali has been subsequently neutralised by sulphuric acid. It has been sold as a sizing preparation under this name.

Apperitive Saffron of Iron. Ferric subcarbonate.

Appertol. The trade name for a preparation of sodium bisulphite. Used as a preservative and disinfectant.

Appetrol. A proprietary preparation of meprobamate and amphetamine sulphate. An antiobesity agent. (255).

Apple Acid. Malic acid, $COOH.CH_2.CH(OH).COOH$.

Apple Essence. See ESSENCE OF APPLES.

Apple Flour. Consists of washed and peeled apples dried at 30–35° C. The ground product is bleached and sterilised by ozonisation. It keeps well and contains all the constituents of fresh apples.

Apple Oil. The amyl ester of valeric acid, $C_4H_9COOC_5H_{11}$. Used in confectionery, and to some extent for catching night-moths. See ESSENCE OF APPLES.

Apple Sugar. Originally a preparation of apple juice and sugar, but is more generally a preparation without apples, and flavoured with lemon. It contains 63–80 per cent. sucrose, 14–31 per cent. invert sugar, and 0·5–1·2 per cent. levulose.

Apple Wine (Cider). A drink produced by the fermentation of apple juice. It contains from 5–6 per cent. alcohol.

Appretan Ant. A proprietary dispersant surfactant for finishing woven fabrics. An acrylate-based copolymer. (432).

Appretan CPF. A proprietary polyvinyl acetate dispersion surfactant used for finishing woven, non-woven and knitted fabrics. (432).

Appretan GM. A proprietary polyvinyl-acetate-based dispersion surfactant used in the finishing of especially light-weight fabrics, knitted fabrics and non-wovens. (432).

Appretan H. A shellac substitute from 95 per cent. polyvinyl acetate and 5 per cent. crotonaldehyde.

Appretan TN. A proprietary dispersion surfactant for finishing woven, non-woven and knitted fabrics. A vinyl acetate copolymer. (432).

Appreteen. Water soluble size. (523).

A.P.P. Stomach Powder. A proprietary preparation containing calcium carbonate, magnesium carbonate, magnesium trisilicate, bismuth carbonate, aluminium hydroxide gel, homatropine methylbromide, papaverine and phenobarbitone. An antacid. (256).

APR 2730, 5958, 8178, 9276 and 10344. A proprietary range of polyester resins. (433).

Apramycin. An anti-bacterial currently undergoing clinical test as " EL-857/820 ". Produced by *Streptomyces tenebrarius*, it is o^4 - [3α - Amino - 6α - (4 - amino - 4 - deoxy - α - D - glucopyranosyloxy) - 2, 3, 4, 4aβ, 6, 7, 8, 8aα - octahydro - 8β - hydroxy - 7β - methylaminopyrano[3, 2 - b]pyran - 2α - yl] - 2-deoxy-D-streptamine.

Apresoline. A proprietary preparation of hydralazine hydrochloride. Antihypertensive. (18).

Apricot Essence. See ESSENCE OF APRICOTS.

Apricotine. A mineral. It is a variety of quartz.

Aprindine. An anti-arrhythmic currently undergoing clinical test as " 99 170 ". It is 3 - [N - indan - 2 - yl) - N - phenylamino]propyldiethylamine.

Aprinox. A proprietary preparation of bendrofluazide. A diuretic. (253).

Aproten. A range of proprietary low-protein, gluten-free foods used in the treatment of renal failure and phenylketonuria. (365).

Aprotinin. A polypeptide proteinase inhibitor. TRASYLOL.

Apsin VK. Phenoxymethylpenicillin.

Aptocaine. A local anaesthetic. 2 - Pyrrolidin - 1 - ylpropiono - o - toluidide. Pirothesin is the hydrochloride.

Apyonin. See PYOCTANIN.

Apyre. Synonym for Andalusite.

Apyrite. Rubellite (*q.v.*).

Apyrogen. A proprietary preparation of pyrogen-free water for injection. (285).

Apyron (Grifa, Hydropyrine, Litmopyrine, Tyllithin). Lithium-acetyl-salicylate, $C_2H_3O.OC_6H_4COOLi$. Used in medicine.

Aquacide. A diquat-based herbicide. (2).

Aquaclene. Tablets containing chloramine. (527).

Aquacrepite. A mineral. It is a silicate of aluminium, iron, and magnesium.

Aquadag. A registered trade mark for colloidal graphite in water used as a lubricant for drawing tungsten and molybdenum filament wires, for metal forming operations such as extrusion, as an aid to cutting and for forming electrically conducting coatings. (207).

Aquadrate. A proprietary preparation of urea in a cream base, used in the treatment of ichthyosis and related skin disorders. (818).

Aqua Dulcis. Chloroform water.

Aqua Fortis. Nitric acid. HNO_3.

Aquagel. A proprietary trade name for a colloidal bentonite. Also a proprietary hydrated silicate of alumina for waterproofing cement.

Aqualac. A proprietary shellac.

Aqualite. A proprietary phenol-formaldehyde synthetic resin laminated product and bearing material requiring only water as lubricant.

Aqualose. A proprietary trade name for Poloxyl Lanolin.

Aqualube. A proprietary plasticiser for water soluble materials.

Aquamarine. See BERYL.

Aqua-Mephyton. A proprietary preparation of phytomenadione. (310).

Aquamol. See CARBOSIL.

Aquamox. A proprietary preparation of quinethazone. A diuretic. (306).

Aquapearl. A proprietary phenol-formaldehyde resin with a pearl effect used for decorative work, buttons, etc.

Aquaplex. A proprietary trade name for alkyd resin dispersed in an aqueous medium. An emulsion varnish and lacquer resin vehicle for stucco, etc.

Aqua Regia (Nitro-muriatic Acid). Nitro-hydrochloric acid, a mixture of one volume of nitric acid and three volumes of hydrochloric acid.

Aquaresin GB. A proprietary plasticiser. It is glyceryl boriborate.

Aquaresin M. A proprietary product. It is glycol boriborate, a water soluble non-drying resin more viscous than glycerine and strongly hygroscopic. It is soluble in water, ethyl alcohol, methyl alcohol, and cellosolve, but insoluble in hydrocarbons.

Aquarium Cement. A cement of this type contains 1 part dry white lead, 1 part litharge, 1 part silver sand, and ½ part finely powdered rosin.

Aquaseal. A proprietary trade name for a cold applied plastic bitumen for waterproofing and sealing.

Aquasol. A proprietary trade name for a sulphonated castor oil.

Aquatec. A proprietary trade name for a waterproofing material. It is a paraffin wax emulsion sometimes with aluminium acetate.

Aqua Veija. The mother liquor from the crystallisation of sodium nitrate from raw caliche.

Aquavit (1). A liqueur used as an appetiser. Alcohol from grain mash is rectified, and an addition of carraway seeds and orange peel is made to flavour it.

Aquavit. (2) A proprietary preparation of vitamin A, calciferol, aneurine hydrochloride, riboflavin, nicotinamide, pyridoxine hydrochloride, calcium pantothenate, ascorbic acid and tocapheryl acetate. Vitamin supplement. (338).

Aquila Alba. Calomel (q.v.).

Aquinite (Klop). Chloropicrin (nitrotrichloromethane), CCl_3NO_2, a poison gas.

Aquinol. (1) A preparation of formalin, a potash soap, and thymol. Antiseptic. (2) A proprietary brand name for invert emulsions.

Aquirin (Theonacet). A preparation of theobromine and sodium acetate, $C_7H_7O_2N_4Na + NaC_2H_3O_2$. A diuretic.

Arabian Gum. Khordofan gum (q.v.).

Arabian Tea (Abyssinian Tea, Kat, Kath). Catha, the leaves of *Catha edulis*.

Arabin. See GUM ARABIC.

Arabiose. See DIARABINOSE.

Arachis Butter. A preparation made by natives from the slightly roasted kernels of arachis nuts. Commercial samples contain from 45–56 per cent. fat, having an iodine value of 88–94.

Arachis Nuts. See GROUND NUTS.

Arachis Oil. Earthnut oil (q.v.).

Aradolene. A proprietary preparation of diethylamine salicylate, capsicum, oil of camphor and menthol. An embrocation. (380).

Araeoxene. Dechenite (q.v.).

Aragonite. A mineral. It is calcium carbonate, $CaCO_3$.

Arakawaite. A mineral, $(Cu.Zn)_6$ $P_2O_{11}.7H_2O$, of Japan.

Araldite. A proprietary trade name for a series of epoxy resins used for casting, encapsulating, laminating, surface coating and as an adhesive. (18).

Aralen. A proprietary trade name for the diphosphate of chloroquine.

Aramide. An aromatic polyamide. See ARENKA.

Aramayoite. A mineral. It is $6[Ag(Bi,Sb)S_2]$. A Bolivian silver sulphoantimonio bismuthite.

Aramid Fibre. See KEVLAR.

Aramine. A proprietary preparation of metaraminol tartrate. A vasoconstrictor used in resuscitation. (310).

Arandisite. A mineral. It is $Sn_5Si_3O_{16}$. $4H_2O$.

Aranox. A proprietary trade name for p-(p-tolyl sulphonylamido)-diphenylamine. A rubber antioxidant. (163).

Arappo. The cake produced after the expression of the oil (Illipe butter) from the seeds of *Illipe malabrorum*.

Araroba. See GOA POWDER.

Arasina Gurgi. An impure variety of gamboge from Camara.

Arbeflex. Plasticisers. (554).

Arbeflex 489. A heat and light and secondary plasticiser for PVC. Oxirane oxygen 3.3 per cent. min. Iodine number 3 max; viscosity 98 cS/25° C. (554).

Arbeflex 550. A low priced moderate performance plasticiser for PVC. Sap. value 380 mg KOH/gm. Acid value 1 mg KOH/gm. (554).

Arbutin Benzoic Ester. Cellotropin (*q.v.*).

A.R.C. A trade mark for goods of the abrasive and refractory type, the essential constituent of which is crystalline alumina.

Arcanite. See APHTHITALITE. A mineral. It is $4[K_2SO_4]$.

Arcanol. A German preparation containing atophan and aspirin, used in the treatment of colds and influenza. The name is also applied to a lead pigment which contains the lead partly in a very finely divided condition, and partly in the form of oxide. It differs from white or red lead in being electrically conductive, and is able to combine chemically with oxygen and moisture of the air and thus protect surfaces from these effects.

Archangel Pitch. A pine pitch from the distillation of the resinous woods of Russian forests.

Archiardite. Synonym for Dachiardite.

Archibald's Stain. A microscopic stain. Solution A contains 0.5 gram thionin, 2.5 grams phenol, 1 c.c. formalin, and 100 c.c. distilled water, and solution B contains 0.5 gram methylene blue, 2.5 grams phenol, 1 c.c. formalin, and 100 c.c. distilled water. For use mix equal parts of A and B, and filter.

Archibromin. Mono-bromo-isovaleryl-glycolyl-urea, $CH_2(OH) . CO . NH . CO.NH.CO.C_4H_8Br$.

Archil (Orchil, Orseille, Persio, Cudbear, Orchellin). A natural colouring matter obtained from *Roccela tinctoria* and other lichens. The colouring principle is orcin, $C_6H_3(CH_3)(OH)_2$, which, in the presence of air and ammonia,

oxidises to a violet dye, orcein, $C_{28}H_{24}N_2O_7$. The alkali salts dye wool and silk. It is sold in three forms : (*a*) A pasty mass called archil ; (*b*) a mass of drier character named persio ; and (*c*) as a reddish powder known as cudbear or French indigo. Commercial brands of cudbear are Cudbear O, I, II, extra, fine, violet, red-violet, blue-violet and red. Orchellin is a product very rich in the dyestuff.

Archil Brown. An azo dyestuff prepared by the action of diazotised amine-azo-benzene-disulphonic acid upon α-naphthylamine, $HSO_3.C_6H_4NH.C_6H_3.HSO_3.N : N.C_{10}H_6NH_2$.

Archil Carmine. A dyestuff containing archil colouring matter in a state of great purity.

Archil, Earth. See EARTH ARCHIL.

Archil Purple. A dyestuff similar to archil carmine (*q.v.*).

Archil Red A. An azo dyestuff, prepared by the action of diazotised amino-azo-xylene upon β-naphthol-disulphonic acid, $C_8H_9.N_2.C_8H_8N_2.C_{10}H_4(HSO_3)_2OH$.

Archil Substitute. See APOLLO RED and NAPHTHIONIC RED.

Archil Substitute G (Orchil substitute G, Orchil Extract G). An azo dyestuff from p-nitraniline and β-naphthylamine, $C_6H_4(NO_2).N : N.C_{10}H_5.NH_2.SO_3Na$. Dyes wool orchil red from an acid bath.

Archil Substitute N Extra. See APOLLO RED.

Archil Substitute V. Société Anonyme des Matières Colorantes, Claus & Rée, Read Holiday & Sons (Orchil Extract V, Orchil Substitute V, Naphthindone Red). A dyestuff which is the sodium salt of p-nitro-benzene-azo-α-naphthylamine-p-sulphonic acid, $C_{16}H_{11}N_4O_5SNa$. Dyes wool red from an acid bath.

Archil Substitute V. (The Berlin Aniline Co. Orchil Substitute V, Orchil Extract V). See ARCHIL SUBSTITUTE 3VN.

Archil Substitute 3VN (Archil Substitute V, Orchil Substitute V, Orchil Extract 3VN, Orchil Substitute 3VN). A dyestuff which is the sodium salt of p-nitro-benzene-azo-α-naphthylamine-sulphonic acid, $C_{16}H_{11}N_4O_5SNa$. Dyes wool red from an acid bath.

Archil Violet. See FRENCH PURPLE.

Archiodin. Mono-iodo-isovaleryl-glycolyl-urea, $CH_2(OH).CONH.CO.NH.CO.C_4H_8I$.

Arcolite. A proprietary phenolformaldehyde synthetic resin. A moulding composition.

Arcolloy. A non-magnetic alloy of iron with 12–16 per cent. chromium, and less than 0.12 per cent. carbon, 0.5 per

cent. manganese, 0·025 per cent. phosphorus, 0·025 per cent. sulphur, and 0·5 per cent. silicon.

Arcoloy. A proprietary copper-silicon casting alloy containing 97·25 per cent. copper, 2·63 per cent. silicon, 0·12 per cent. iron, and 0·01 per cent. phosphorus.

Arcovite. An asbestos-silica product, used as an insulator.

Arctolite. A mineral. It is a silicate of aluminium, calcium, and magnesium.

Arcton. A proprietary trade name for fluorinated hydrocarbon refrigerants. (fluorocarbons) (q.v.). (2).

Ardealite. A mineral. It is $2[Ca_2HPO_4SO_4.4H_2O]$.

Ardeer Powder. A Nobel explosive of the dynamite class. It contains 31–34 per cent. nitro-glycerin, 11–14 per cent. kieselguhr, 47–51 per cent. magnesium sulphate, 4–6 per cent. potassium nitrate, and 0·5 per cent. ammonium or calcium carbonate.

Ardena. A proprietary preparation of zinc oxide, titanium dioxide, kaolin, calcium carbonate, menthyl salicylate and iron oxide pigments. A covering cream for the skin. (457).

Ardenite. A proprietary synthetic resin. A moulding composition.

Ardennite. A mineral. It is a vanadiosilicate of aluminium and manganese, $H_{10}Mn_8Al_8V_2Si_8O_{46}$.

A.R.D. Gum. A proprietary anti-dry-rot compound containing glue, resin, boiled oil, and water.

Ardil. A name for a textile fibre resembling wool made from peanut protein. It is not attacked by moths but lacks the abrasion resistance of wool.

Ardinex. A proprietary preparation of guaiphenesin, methaqualone hydrochloride and ephedrine hydrochloride. A bronchial antispasmodic. (253).

Ardmorite. A variety of Bentonite (q.v.), found in the Pierre shales at Ardmore, South Dakota.

Arecaidine. N-methyl-△-3-tetrahydropyridine-3-carboxylic acid.

Areca Red. A red colouring matter extracted from the areca nut.

Arecoline. The methyl ester of N-methyl-△ -3-tetrahydro-pyridine -3 - carboxylic acid.

Arecolineserine. A mixture of arecoline hydrobromide and eserine. Used for hypodermic injection.

Arendalite. A mineral which is a fine crystalline variety of epidote (q.v.).

Arenka. A proprietary high-strength yarn manufactured from aramides (aromatic polyamides). (458).

Arenolite. An artificial siliceous-argillaceous-calcareous stone.

Are-o-Sol. A rubber compounding filler made from oil impregnated shale.

Arequipite. A mineral. It is a lead-antimony silicate.

Areskap. A proprietary trade name for a butyl-phenyl-phenol sodium sulphonate.

Aresket. A proprietary trade name for a wetting agent. It is stated to be a butyl-diphenyl sodium sulphonate.

Aresklene. A proprietary trade name for a dibutyl-phenyl-phenol sulphonate. A mould lubricant and emulsifier.

Aretan. A mercurial fungicide. (511).

Aretone 270. A proprietary trade name for a fine mica used as a pigment in paints.

Arfoedsonite. A mineral. It is amphibole (q.v.), with soda.

Arfonad. A proprietary trade name for the (+)-camsylate of Trimetaphan. A vasodilator. (314).

Argacid. A silver-glucoside-boron aluminate containing 21 per cent. silver. A remedy for gonorrhœa.

Argal. See ARGOL.

Argalbose. See HEGONON.

Argaldin. A combination of albumen-silver and hexamethylene-tetramine. It contains 10 per cent. silver.

Argaldon. A combination of protein and hexamethylene-tetramine. It is recommended for gargling, as formaldehyde is liberated by the action of the alkaline saliva.

Argasoid. A low-grade nickel silver.

Argatoxyl. Silver atoxylate, NH_2. $C_6H_4.As.O.AgO.OH$. Used in medicine, chiefly to increase bodily resistance to infection.

Argein. A colloidal silver protein compound used in the treatment of gonorrhœa.

Argenol. A silver albuminate used as an antiseptic.

Argental. An alloy of 85 per cent. copper, 10 per cent. tin, and 5 per cent. cobalt.

Argentalium. An aluminium alloy containing antimony. It has a specific gravity of 2·9.

Argentamine (Dimargyl). Ethylene-diamine-silver-nitrate, $CH_2.CH_2.NH_2$. $NH_2.AgNO_3$. A substitute for lunar caustic, used for gonorrhœa.

Argentan. A nickel silver. It consists of 56 per cent. copper, 26 per cent. nickel, 18 per cent. zinc, and 1 per cent. iron, and is used as an electrical resistance alloy.

Argentan solder. An alloy of 57 per cent. zinc, 35 per cent. copper, and 8 per cent. nickel. Also see NICKEL SILVERS.

Argentarsyl. A mixture of colloidal silver and cacodylic acid.

Argentichthol. See ICHTHARGAN.

Argentide. A silver iodide solution containing 100 grains in 1 oz.

Argentiferous Copper Glance. Strohmeyerite (*q.v.*).

Argentin. See NICKEL SILVERS. It is also the name for a jeweller's alloy containing 85·5 per cent. tin, and 14·5 per cent. antimony, used for the manufacture of spoons and forks.

Argentine. A finely divided spongy tin, used for tin-plating. The term is also applied to a special powder prepared from zinc, which, when combined with casein, gives a metallic finish to yarn or other fabrics. It is also the name for a mineral, which is a white, lamellar, shining variety of calcite.

Argentis Quinaseptol. See ARGENTOL.

Argentite (Glance Ore, Glass Ore, Silver Glance, Soft Ore). A mineral. It is a silver ore, and consists of silver sulphide, Ag_2S.

Argento-Domeykite. An argentiferous domeykite, $(Cu.Ag)_3As$.

Argentojarosite. A silver mineral found in Utah. It contains 18 per cent. Ag_2O, 43 per cent. Fe_2O_3, 28 per cent. SO_3, and 10 per cent. H_2O.

Argentol (Argentis quinaseptol). Silver-oxychinolin-sulphonate, $C_9H_5N.OH.SO_3Ag$. An antiseptic used in the treatment of gonorrhœa.

Argento-pyrites. A mineral, $3FeS.3FeS_2.Ag_2S$.

Argentorat. The trade name for a flash powder, consisting of a mixture of potassium and aluminium. Used in photography.

Argentous Tungstate. A mixture of normal silver tungstate and metallic silver.

Argentum. Silver, Ag.

Argentum Casein. A silver caseinate.

Argentum Metal. See METAL ARGENTUM.

Arghan Fibre (Honduras Silk). A fibre obtained from *Pita floji*.

Arghan Tissue. A material prepared from a species of wild pineapple. It consists of pure cellulose, and is used for fine thread.

Argilla. See CHINA CLAY.

Arginine. α-Amino-s-guanido-valeric acid, $(H_2N).C(NH).NH.CH_2.CH_2.CH_2.CH(NH_2)COOH$.

Arginine Glutamate. A nutrient. It is the L-arginine salt of L-glutamic acid.

Arginine-Sorbitol. A proprietary preparation of L-arginine monohydrochloride, L-arginine base, sodium hydrosulphite, and sorbitol, used in the treatment of hepatic coma. (809).

Argipressin. An anti-diuretic hormone.

Argirin. A bactericide. It is a silver preparation.

Argiroide. See NICKEL SILVERS.

Argochrom. A proprietary preparation of silver methylene blue (*q.v.*).

Argoferment. Colloidal silver, a germicide.

Argofil. A proprietary trade name for copper alloy rods for the argon arc welding of copper. (134).

Argoflavin. The silver salt of acriflavine. A bactericide.

Argol (Argal, Crude Tartar). A crystalline crust deposited on the sides of the vat in which grape juice has been fermented. It contains 40–70 per cent. tartaric acid, principally as potassium hydrogen tartrate. It is known as red or white argol, according to whether it is deposited from the red or white grapes.

Argolaval. A German preparation. It consists of a solution of silver nitrate and hexamethylene-tetramine (Ag = 0·635–0·762). Used in the treatment of catarrh of the bladder.

Argolit. A proprietary casein product.

Argol, Red. See ARGOL.

Argol, White. See ARGOL.

Argonin. A preparation of casein silver, obtained by precipitating a solution of casein with silver nitrate and alcohol. It contains 4·2 per cent. silver, and is used in cases of gonorrhœa. See LARGIN.

Argonin L. A silver caseinate containing 10 per cent. silver.

Argophol. See SOPHOL.

Argotone. A proprietary preparation of ephedrine hydrochloride, and silver vitellin. Nasal drops. (323).

Argozie (Arguzoid, Argozoil). An alloy of from 54–56 per cent. copper, 23–38 per cent. zinc, 2–4 per cent. tin, 2–3·5 per cent. lead, and 13·5–14 per cent. nickel.

Argulan. A preparation which is stated to be antipyrine-sulphamino-mercury, $C_{11}H_{11}N_2O.NH.SO_3.HgOH$. It is used in the treatment of syphilis.

Argus Powder. An explosive powder containing 81 per cent. potassium nitrate, and 18 per cent. charcoal.

Arguzoid. An alloy of 56 per cent. copper, 23 per cent. zinc, 4 per cent. tin, 3·5 per cent. lead, and 13·5 per cent. nickel.

Arguzoil. Argozie (*q.v.*).

Argyn. A silver proteinate.

Argyrite. Argentite (*q.v.*).

Argyroceratite. Cerargyrite (*q.v.*).

Argyrodite. A mineral. It is a sulphide of silver and germanium, $GeS_2.3Ag_2S$.

Argyroide. See NICKEL SILVERS.

Argyrol Brand Silver Vitellin. Used in the treatment of all accessible inflammations of the eye, ear, nose, throat, and genito-urinary tract; non-toxic, non-irritant, contains approximately 20 per cent. silver.

Argyrolith (China Silver, Electroplate). Names given to alloys containing 50–70 per cent. copper, 10–20 per cent. nickel, and 5–30 per cent. zinc. Alfenide and Argentan are similar alloys. They are nickel silvers (German silvers).

Argyro-pyrites. A silver ore containing sulphides of silver and iron. It is $Ag_2Fe_7S_{11}$.

Arhemapectyl. A proprietary preparation. It is a pectin isotonic colloidal solution.

Arhéol. Santalol, $C_{15}H_{24}O$.

Arhoin (Arhovin). An addition product of diphenylamine and ethyl-thymyl-benzoate, $(C_6H_5)_2NH . C_{10}H_{13} . C_6H_4 . COOC_2H_5$. An antiseptic prescribed in cases of gonorrhœa, also in the form of small sticks as a local disinfectant.

Arhovin. See ARHOIN.

Aricyl. A proprietary preparation of sodium aceto-arsinate.

Aridex. A retexturing and reproofing aid. (513).

Arigal PMP. A proprietary trade name for a solution of an organic mercuric compound which, when used together with Arigal C, imparts a mildew resistant and rot-proof finish on cellulosic fibres; fast to water, washing and dry cleaning. (18).

Ariloks. A preparation of polyphenylene oxide. (424).

Arilvax. A proprietary vaccine against yellow fever. (277).

Aristar. Ultrapure reagents and solvents. (527).

Aristochin. See ARISTOQUININE.

Aristocort. A proprietary trade name for Triamcinolone.

Aristol. Dithymyl-diiodide, $(C_6H_2OI . CH_3.C_3H_7)_2$. Used as a substitute for iodoform, for the treatment of wounds.

Aristol, Naphthol. See NAPHTHOL ARISTOL.

Aristoquin. See ARISTOQUININE.

Aristoquinine (Aristochin, Aristoquin). The carbonic ester of quinine, $CO(OC_{20}H_{23}N_2O)_2$. A practically tasteless quinine body, used for whooping cough.

Aristosan. A German product. It is a solution of glucose, methylene blue, and formic acid, and is injected intramuscularly for rheumatism and gout.

Arite. See AARITE. A mineral. It is $2[Ni(Sb,As)]$.

Arizona Ruby. A mineral. It is a pyrope, $Mg_3Al_2Si_3O_{12}$ (q.v.), from Arizona.

Arizona Shellac. See SONORA GUM.

Arizonite. A mineral. It is a ferric metatitanate, $Fe_2(TiO_3)_3$.

Arkady Yeast Food. A mixture of salts, chiefly ammonium chloride, calcium sulphate, and potassium bromide.

Arkansas Stones. Whetstones used for sharpening tools.

Arkansite. A mineral, which is a variety of Brookite (q.v.).

Arkite. Explosives used in coal mines. (a) Contains nitro-glycerin, collodion cotton, potassium nitrate, wood meal, and ammonium oxalate ; (b) Consists of nitro-glycerin, collodion cotton, potassium nitrate, sodium nitrate, wood meal, and ammonium oxalate.

Arklone P. A proprietary trade name for trichlorotrifluoroethane, a solvent for cleaning. (2).

Arklone W. A proprietary trade name for a mixture of Arklone P, a surface active agent and water. Used for cleaning off lyophobic and lyophilic contamination simultaneously. (2).

Arko Metal. An alloy of 80 per cent. copper, and 20 per cent. zinc.

Arksutite. Chiolite (q.v.).

Arlacel C, 83. A proprietary trade name for Sorbitan Sesquioleate.

Arlef-100. A proprietary preparation of flufenamic acid. An antirheumatic. (264).

Arlin. A proprietary name for polyethylene film. (430).

Armac. Long chain aliphatic amine acetates. (555).

Armangite. A mineral, $Mn_3(AsO_3)_2$, found in Sweden.

Armco. A trade mark for ingot iron and stainless steel.

Armco Ingot Iron. A trade mark for a very pure iron. It is guaranteed to be 99·84 per cent. pure, and the annealed wire has an ultimate tensile strength of 25 tons per sq. in.

Armeen. Long chain aliphatic amines (555).

Armenian Bole. See INDIAN RED.

Armenian Cement. A jeweller's cement containing gum mastic, isinglass, gum ammoniac, alcohol, and water. It is made by soaking the isinglass in water and mixing it with the spirit containing the gums.

Armenian Salt. A very early name for ammonium chloride.

Armide. See CAMITE.

Armids. Long chain aliphatic amides. (555).

Armite. A proprietary trade name for thin fish paper insulation.

Armogard. Fuel oil additive. (555).

Armogel. Thickening agent. (555).

Armohib. Acid inhibitors. (555).

Armor-ply. A proprietary trade name for metal bonded plywood.

Armoured Concrete. See REINFORCED CONCRETE.

Armowax. Synthetic waxes. (555).

Armowax EBS. A proprietary trade

name for N-N'ethylene bis-stearamide. (86).

Armstrong Acid. See SCHAEFFER'S ACID.

Armstrong and Wynne's Acid. α-Naphthol-sulphonic acid (1 : 3), $C_{10}H_6$(OH)(SO$_3$H).

Armstrong's δ-Acid. Naphthalene-disulphonic acid (1 : 5), $C_{10}H_6$(SO$_3$H)$_2$.

Armyl. A proprietary trade name for Lymecycline.

Arnatto. See ANNATTO.

Arnaudon Green. A green pigment prepared by stirring up 128 parts ammonium phosphate and 149 parts potassium bichromate with water to a paste, and heating the mixture from 170–180° C. See CHROMIUM GREEN.

Arneel DN. A proprietary trade name for the dimerised product of octa-decene and octadecadiene nitriles. A vinyl plasticiser. (75).

Arneel HF. A proprietary trade name for the 18, 20 and 24 carbon atom fatty acid nitriles. Vinyl plasticisers. (75).

Arneel S. A proprietary trade name for a derivative of octadecene and octa-decadiene nitriles. A vinyl plasticiser. (75).

Arneel TOD. A proprietary trade name for a derivative of octadecene and octa-decadiene nitriles. A vinyl plasticiser. (75).

Arnica. The dried flower-heads of *Arnica montana*. The tincture is used for bruises and sprains.

Arnica Yellow. An azo dyestuff prepared by condensing *p*-nitro-toluene-sulphonic acid with *p*-amino-phenol, in the presence of aqueous caustic soda. Dyes cotton golden-yellow from a salt bath.

Arnimite. A mineral, 5CuO.2SO$_3$.6H$_2$O.

Arnite. A proprietary trade name for a polyethylene terephthalate polymer characterised by extreme hardness. (Rockwell M106). (187).

Arnite A.K.U. A trade name for a polyethylene glycol terephthalate injection moulding material. (5).

Arnite G. Thermoplastic polyester grade for injection moulding and extrusion. (5).

Arnotto. See ANNATTO.

Arobon. A proprietary preparation of carob flour, starch and cocoa. A gastro-intestinal sedative. (282).

Arocan. Proprietary products. They consist of procaine hydrochloride and adrenaline preparations.

Arochem (Aroplax). A proprietary trade name for soft oil modified alkyds.

Aroclor. A trade mark. Polychlor-diphenyls varying from water-white mobile oils through oils and soft waxes

to brittle resins. Used as ingredients in nitro-cellulose lacquers. They are proprietary products, and the following are some qualities: No. 1219, a liquid with b.pt. of 275° C. and a f.pt. of 127° C., for use as a solvent; No. 1255, a syrupy material suggested as a substitute for resins; No. 2615, a black solid softening at 79° C. and fluid at 170° C.; No. 2265, a light yellow wax suggested for use as a varnish gum. Nos. 1242, 1248 and 1254 are used as vinyl plasticisers. (57).

Arofene. Proprietary phenolic synthetic resins for varnishes and enamels.

Aromaplas. Range of perfumes for plastics. (549).

Aromasol. Solvents. (512).

Aromasol N. A aromatic solvent consisting of a mixture of C9 aromatic hydrocarbons. (2).

Aromatic Tincture. An alcoholic solution of cinnamon, ginger, cloves, and cardamon. An aromatic.

Aromatic Vinegar. Acetic acid with odorants.

Aromatin. A hop substitute made from gentian root.

Aromex. Powdered perfumery compounds. (504).

Aromite. A mineral. It is a sulphate of aluminium and magnesium.

Aromix. Solvent emulsifier concentrates for pesticide formulations. (511).

Aromox. Long chain aliphatic amine oxides. (555).

Aron Alpha. A proprietary cyano-acrylate adhesive. (459).

Aroplax. See Arochem.

Aroplaz 6065. A proprietary trade name for a short oil alkyd resin used in fast drying industrial paints. (195).

Aropol. Water emulsifyable polyester resins. (293).

Arothane 185-ML-60. A 60 per cent. solution of a polyisocyanate safflower/DHC oil. Used for floor varnishes. (297).

Aroxine. A proprietary trade name for Forminitrazole.

Arphoalin. An arsenic and phosphorus albumin compound.

Arpylene 1325. A proprietary self-extinguishing polystyrene injection-moulding compound containing 25 per cent. asbestos reinforcement. (152).

Arpylene 2045. A proprietary asbestos-reinforced polypropylene proof against distortion at temperatures below 150° C. (152).

Arpylene APPN 2025. A proprietary brand of polypropylene containing 25 per cent. by weight of asbestos reinforcement. A moulding material. (152).

Arpylene APPN 2040 and 2240. A proprietary range of asbestos-reinforced polypropylenes. The reinforcement is 40 per cent. by weight. (152).

Arpylene APSN 1025. A polystyrene reinforced with 25 per cent. asbestos, used in injection moulding. (152).

Arquad. Quaternary ammonium salts. (555).

Arquad DMHTB-75 per cent. A proprietary trade name for hydrogenated tallow and benzyl dimethyl and di-stearyl benzyl quaternary ammonium chloride. An emollient and conditioning agent in toiletries and shampoos. A demulsifier and dispersing agent. (86).

Arquerite. See ACQUERITE.

Arrack (Raki). A liqueur of the Arabs and Indians. It is a type of brandy prepared from rice, cane sugar, and coconuts, and contains from 48–54 per cent. alcohol.

Arragonite. A mineral. It is a crystalline calcium carbonate, $CaCO_3$.

Arrconox AHT, DNL and DNP. A proprietary range of anti-oxidants used in the manufacture or processing of rubber. (556).

Arrconox GP. Proprietary name for a general-purpose staining anti-oxidant for rubber.

Arrconox S.P. A non-staining anti-oxidant. (556).

Arrcopep. A rubber-reclaiming agent.

Arrcorez 16. A butyl rubber curing resin. (556).

Arrcorez 17. A tackifying resin. (556).

Arrehbor Tragacanth. A tragacanth gum which exudes from the cut branches of certain trees in Persia.

Arrhenal (Arsinyl, New Cacodyl). Sodium - methyl - arsinate, $AsO(CH_3)(ONa)_2.6H_2O$.

Arrhenalic Acid. Methyl-arsenious acid.

Arrhenite. A complex rare earth mineral containing from 3–4 per cent. ZrO_2.

Arrojadite. A mineral. It is $12[Na_2(Fe,Mn)_5(PO_4)_4]$.

Arrope. Sherry boiled to a syrup. Used for colouring wines.

Arrow Black. A proprietary gas black used in rubber mixings.

Arrow Poison, Kombe's. See KOMBE'S ARROW POISON.

Arrowroot. Starch from *Maranto arundinacea* and *Indica*, of West Indies and Brazil.

Arrowroot, Queensland. See TOUS-LES-MOIS STARCH.

Arrow Tool Steels. Proprietary steels containing 0·9–1·02 per cent. chromium, 0·16–0·20 per cent. vanadium, 0·5–0·6

per cent. manganese, and 0·20–0·30 per cent. carbon.

Arsacetin (Acetylatoxyl). Sodium-*p*-acetyl-amino-phenyl-arsinate, $C_2H_3O.NH.C_6H_4.AsO(ONa)(OH)_3-4H_2O$. Employed in sleeping-sickness, skin diseases, and syphilis.

Arsacetin-quinine. A medicinal compound containing 43 per cent. arsacetin and 54 per cent. quinine.

Arsalyt. Bismethyl-amino-tetra-amino-arseno-benzene.

Arsamin. See ATOXYL.

Arsaminol. The trade name for a brand of salvarsan (*q.v.*).

Arsamon. A solution of sodium mono-methyl-arsonate.

Arsan. A glidin (*q.v.*), preparation containing silver.

Arsanilic Acid. A preparation used in the treatment of enteritis and in the promotion of growth. It is *p*-amino-phenylarsonic acid.

Arsenargentite. A mineral containing 81 per cent. silver, and 19 per cent. arsenic.

Arsenated Petroleum Oil. Petroleum oil containing such arsenic compounds as phenarsazine chloride and oxide, diphenyl-chloro-arsine, and nitroso-dimethyl - aniline - chloro - arsine. It is stated to be very effective for the preservation of timber.

Arsen-glidin. A combination of 4 per cent. arsenious acid with glidin (*q.v.*).

Arsenhemol. See HEMOL.

Arsenic. See ARSENIOUS ACID.

Arsenic Acid. Arsenic oxide, As_2O_5.

Arsenical Antimony. See ALLEMONTITE.

Arsenical Cobalt. See SMALTINE.

Arsenical Copper. Domeykite (*q.v.*).

Arsenic Alginate. A combination of sodium alginate and arsenic chloride.

Arsenicafahlerz. Synonym for Tetrahedrite.

Arsenical Green. Copper aceto-arsenite.

Arsenical Iron (Leucopyrite). A mineral, $FeAs_2$.

Arsenical Iron Ore. See MISPICKEL.

Arsenical Nickel. See NICCOLITE.

Arsenical Nickel Ore. A mineral, NiAsS.

Arsenical Pyrites. See MISPICKEL.

Arsenical Soot. Arsenious acid (*q.v.*).

Arsenic, Black. See BLACK ARSENIC.

Arsenic Bronze. An alloy of 80 per cent. copper, 10 per cent. tin, 9·2 per cent. lead, and 0·8 per cent. arsenic.

Arsenic, Butter of. See BUTTER OF ARSENIC.

Arsenic, Flaky. See FLAKY ARSENIC.

Arsenic, Flowers of. See ARSENIOUS ACID.

Arsenic Glass (Vitreous White Arsenic). Prepared by the volatilisation of arsenious oxide under slight pressure.

Arsenic Glass, Red. See REALGAR.

Arsenic Greens. See SODA GREENS.

Arsenic Meal. Arsenious acid (q.v.).

Arsenic Orange. See REALGAR.

Arsenic, Red. Realgar (q.v.).

Arsenic, Ruby. See REALGAR.

Arsenic Silver Blende. A silver ore, $Ag_2S_3.Ag_2S$.

Arsenic, Vitreous White. See ARSENIC GLASS.

Arsenic, White. See ARSENIOUS ACID.

Arsenic, Yellow. See YELLOW ARSENIC.

Arseniopleite. A mineral, $(Mn.Fe)_2$ $(Mn.Ca.Pb.Mg)_3(MnOH)_6(AsO_4)_6$.

Arseniosiderite (Arsenocrocite). A mineral, $3Ca_3(AsO_4)_2 . 6FeAsO_4 . 8Fe(OH)_3$.

Arsenious Acid (White Arsenic, Flowers of Arsenic, Arsenic). Arsenious oxide, As_4O_6.

Arsenite. See ARSENOLITE.

Arseno-Argenticum. Sodium-silver-arsenobillon.

Arsenobenzol. See SALVARSAN.

Arsenobillon. See SALVARSAN.

Arsenobismite. A mineral. It is a bismuth arsenate.

Arsenocrocite. See ARSENOSIDERITE.

Arsenoferratin. A compound of arsenic and albumen, with ferratin. Used in the treatment of anæmia.

Arsenoferratose. A similar preparation to Arsenoferratin.

Arsenogen. A combination of iron, phosphorus, and arsenic, with paranucleinic acid. Used in the treatment of anæmia.

Arsenolite (Arsenite). A mineral. It is arsenious oxide, As_4O_6.

Arsenolsolvine. The sodium salt of amino-phenyl-arsenic acid.

Arsenomelane. A mineral, $2PbS.As_2O_3$.

Arseno-pyrite. See MISPICKEL.

Arsenosiderite. A mineral, $Fe_4Ca_3(OH)_9(AsO_4)_3$.

Arsenotellurite. A mineral. It is a sulphide and telluride of arsenic, $As_2S_3.2TeS_2$.

Arsenotriferrin. Iron-arseno-p-nucleinate.

Arsen-phenolamine. A synonym for salvarsan (q.v.).

Arsenuretted Hydrogen. See ARSINE.

Arsine (Arsenuretted Hydrogen). Arsenic trihydride, AsH_3.

Arsinette. Arsenate insecticides. (511).

Arsinyl. See ARRHENAL.

Arsitriol. Calcium glycero-arseniate.

Arsobin. Strontium chlor-arseno-benolate.

Arsoferrin. A compound of arsenic,

iron, and glycerophosphoric acid. A remedy used for anæmia and debility.

Arsol. A compound of phosphorus and calcium arsenate. A nerve tonic.

Arsphen. Salvarsan (q.v.).

Arsphenamine. Salvarsan (q.v.).

Arsulin. A German product. It is a combination of arsenic and insulin (10 c.c. contain 10 clinical units of insulin and 1 mg. of arsenic in the form of mono-methyl-arsenic acid).

Arsybismol. Bismuth-3-acetyl-amino-4-oxyphenyl-arsinate.

Arsycodile. A proprietary preparation of sodium cacodylate.

Arsykodyls. The salts of cacodylic acid.

Arsylene. Sodium-propenyl-arsenate.

Arsynal. Sodium methyl-arsenate.

Artane. A proprietary preparation of BENZHEXOL hydrochloride, used in the treatment of Parkinson's disease. (306).

Artane Sustets. A proprietary preparation of benzhexol hydrochloride, in a sustained release form, used in Parkinsonism. (306).

Art Bronze. An alloy of 80–90 per cent. copper, and 5–8 per cent. tin.

Arterenol (Homadrenine). o-Dihydroxy - phenyl - amino - methyl - carbinol-hydrochloride, $(OH)_2C_6H_3.CH(OH).CH_2.NH_2.HCl$. An adrenaline substitute.

Artesin. A German preparation. It contains phenyl-cinchonyl-anthranilic acid.

Arthene. A proprietary preparation of phenyl ethyl iodide, terpineol, phenyl ethyl isocyanate, in a cream base. A rubefacient. (826).

Arthripax Cream. A proprietary preparation of benzyl salicylate, glycol salicylate, terebene, menthol, ephedrine hydrochloride and capsicin. An embrocation. (271).

Arthrisin. See ERTHRISIN.

Arthriticin. A synthetic product. It is a compound of diethylene-amine and the nitrile of the ethyl cresol of amino-acetic acid, $C_6H_4(O.C_2H_5)N(N(CH_2)_4NH)(CH_2NH_2CO)$. Used in the treatment of gout.

Arthrytin. A proprietary preparation. It is stated to be ammonium iodoxybenzoic acid. Used in America for the treatment of certain forms of chronic arthritis.

Artic. A proprietary trade name for methyl chloride used in refrigeration.

Artificial Amber. Mellite (q.v.).

Artificial Asphalt. See ASPHALT, ARTIFICIAL AND GERMAN ASPHALT.

Artificial Barytes. Artificial heavy spar. Precipitated barium sulphate.

Artificial Bitumen. See GOUDRON.

Artificial Caoutchouc. See CAOUT-CHOUC, ARTIFICIAL.

Artificial Chappe. See STAPLE FIBRE.

Artificial Gold. See GOLD, ARTIFICIAL.

Artificial Gum. See BRITISH GUM.

Artificial Gutta-Percha. See GUTTA-GENTZSCH, NIGRITE, VELVRIL, and SOREL'S GUTTA-PERCHA SUBSTITUTE.

Artificial Hair. See GLANZSTOFF.

Artificial Heavy Spar. Barium white (q.v.).

Artificial Hemp Bast. See HEMP BAST, ARTIFICIAL.

Artificial Honey (Herzfeld's Honey). An invert sugar syrup similar in composition to honey. It is obtained by heating a mixture of cane sugar and water with a little tartaric or citric acid.

Artificial Horse-hair. Prepared from certain Mexican grasses by treatment with sulphuric acid.

Artificial Ivory. See IVORY, ARTIFICIAL.

Artificial Musk. See MUSK, ARTIFICIAL.

Artificial Ochre. See MARS YELLOW and SIDERINE YELLOW.

Artificial Pineapple Oil. See PINE-APPLE OIL, ARTIFICIAL.

Artificial Pine-needle Oil. See PINE-NEEDLE OIL, ARTIFICIAL.

Artificial Rose Oil. See ROSE OIL, ARTIFICIAL.

Artificial Rubber. See CAOUTCHOUC, ARTIFICIAL, LAKE'S INDIA-RUBBER COMPOUND, DAVY'S SUBSTITUTE, and THINOLINE.

Artificial Saltpetre. See GERMAN SALTPETRE.

Artificial Turpentine. See TURPENTINE, ARTIFICIAL.

Artificial Ultramarine. See ULTRAMARINE.

Artificial Vinegar. See VINEGAR, ARTIFICIAL.

Artificial Wool. See STAPLE FIBRE and KOSMOS FIBRE.

Artinite. A mineral, $2MgCO_3.Mg(OH)_2.3H_2O$.

Artist's Linseed Oil. See LINSEED OIL, ARTIST'S.

Artosin. o-Carboxy-anilide of 2-phenyl-4-quinoline-carboxylic acid, $C_{10}H_5N.C_6H_5.CONHC_6H_4CO_2H$. It has a pharmacological reaction similar to atrophan. A proprietary trade name for Tolbutamide. An oral hypoglycæmic agent. (275).

Artrite. A proprietary polyester laminating resin. (460).

Arvetane. A proprietary adhesive containing polyurethane. (461).

Arvin. A proprietary preparation of ANCROD. An anti-coagulant. (137).

Arvitin. A proprietary preparation of colloidal silver with vitellinic acid.

Arvynol. A proprietary preparation of ethchlorvynol. A hypnotic. (315).

Arylan. Alkyl aryl sulphonates. (558).

Arylene. A proprietary trade name for a wetting agent. It is a sulphonated compound.

Arylon T. A proprietary polyaryl ether moulding compound used in the manufacture of gears, valves, etc. (287).

Arylon T-3198. A polyaryl ether. A high melting point thermoplastic resin. (287).

Arzrunite. A mineral. It is $Cu_4Pb_2SO_4Cl_6(OH)_4.2H_2O$.

Asa Dulcis. Benzoin.

Asafœtida. A gum-resin. It is the dried juice of the roots of Ferula narthex and F. Fœtida. A nerve stimulant and antispasmodic.

Asagran. Acetyl salicylic acid P.B.C. granules. (559).

Asahi Promoloid. A by-product from the manufacture of glass in Japan. It contains 88 per cent. water, and the dried material 17·72 per cent. MgO, 0·85 per cent. K_2O, and 0·38 per cent. P_2O_5. A fertiliser.

Asaprol. See ABRASTOL.

Asaprol-Etrasol. See ABRASTOL.

Asarabacca Oil. See OIL OF ASARA-BACCA.

Asarite. Impure asarone, obtained from the root of Asarum europæum.

Asarol (Asaronic Camphor, Asarum Camphor). Propenyl-trimethoxy-benzene, $C_6H_2(CH : CHCH_3)(OCH_3)_3$. An emetic and cathartic.

Asaron. The trade name for a camphor derivative from Asarum europæum, $(CH_3)_3.C_6H_2(CH)_2.CH_2.CH_3$. Recommended as a tonic and antiseptic.

Asaronic Camphor. See ASAROL.

Asarum Camphor. See ASAROL.

Asbarg. The dried flowers and flowering stems of Delphinium zalil, found in Afghanistan. Used as a dyeing material for silk all over India.

Asbedex. A proprietary asbestos compound.

Asbeferrite. A mineral. It is a variety of dannemorite.

Asbeskin. A proprietary product consisting of asbestos and vulcanised oils.

Asbestic. Waste asbestos mixed with powdered serpentine. Largely employed in fireproof buildings.

Asbestine. An asbestos in powder form. It has a specific gravity of 2·7–2·8 and consists mainly of magnesium silicate. It is used as a rubber filler.

Asbestine Pulp. See AGALITE.

Asbestite. A Swedish synthetic resin varnish-asbestos-paper product. Used for electrical insulation.

Asbestolite. A trade name for a synthetic varnish-asbestos-paper product. The term is also applied to an asbestos cement material.

Asbestone. A registered trade mark for a fibrous cement product. (981).

Asbestonite. An asbestos product used as a heat insulator.

Asbestos. A natural magnesium silicate mixed with varying amounts of lime. The term is applied to two distinct minerals, Chrysotile and Actinolite.

Asbestos, Bastard. See BASTARD ASBESTOS.

Asbestos, Blue. See CROCIDOLITE.

Asbestos, Canadian. See CHRYSOTILE.

Asbestos, Elastic. See MOUNTAIN CORK.

Asbestos Magnesia Mixture. A mixture of asbestos and magnesium oxide. Used for steam packings.

Asbestos, Micro. See MICRO ASBESTOS.

Asbestos, Platinised. See PLATINISED ASBESTOS.

Asbestos Shingles. Boards usually made from asbestos and Portland cement.

Asbestos Sponge. An asbestos impregnated with a platinum salt and ignited. A catalyser.

Asbolane (Earthy Cobalt, Asbolite, Black Cobalt, Wad). A mineral. It is an earthy variety of manganese dioxide, containing oxide of cobalt, generally $(Co.Mn)O.2MnO_2.4H_2O$.

Asbolin. Obtained by the aqueous infusion of lampblack. It consists mainly of pyrocatechin and homopyrocatechin, and has been used in the treatment of tuberculosis.

Asbolite. See ASBOLANE.

Ascabiol. A proprietary preparation of benzyl benzoate, used in the treatment of scabies and pediculosis. (336).

Ascarite. Caustic soda deposited upon an inert substance. An absorbent.

Ascharite. A mineral, $MgHBO_3$. Synonym for Szájbelyite.

Aschtrekker. Tourmaline (q.v.).

Asciatine. See TRIGEMIN.

Ascoloy. See ALLEGHENY METAL.

Ascon. A proprietary preparation containing dried aluminium hydroxide gel, magnesium trisilicate and hyoscyamine hydrobromide. An antacid. (248).

Ascorbic Acid. Vitamin C.

Ascoxal. A proprietary preparation of ascorbic acid, sodium percarbonate and copper sulphate. (338).

Asepsin. See AMMONAL.

Aseptic Acid. Boric acid.

Aseptisil. An alkaline bottle washing detergent. (560).

Aseptin. The trade name for a mixture of hydrogen peroxide, boric acid, and salicylic acid. Antiseptic.

Aseptoforn. A proprietary trade name for an alkyl ester of hydroxy-benzoic acid. An antiseptic.

Aseptol. A 33 per cent. solution of phenol-sulphonic acid, $C_6H_4(OH)SO_3H$, (Sozolic acid). It contains about 27 per cent. of the p-acid, and 6 per cent. of the o-acid. Antiseptic.

Aseptolin. Pilo-carp-phenol, $C_{11}H_{16}N_2O_2OH.C_6H_5$. A 0·02 per cent. solution with a 2·75 per cent. solution of phenol is used in the treatment of malaria and tuberculosis.

Aseptosol. A new phenol obtained from the essential oil derived from the leaves of *Chavica betle*. A powerful antiseptic.

Aserbine. A proprietary preparation of malic acid ester of propylene glycol, malic acid, benzoic acid and salicylic acid used as a desloughing agent. (269).

Asferryl. Said to be a combination of an iron salt and arseno-tartaric acid. Used in the treatment of anæmia.

Ash, Antimony. See ANTIMONY ASH.

Ashberry Metal (Ashbury Metal). An alloy of 80 per cent. tin, 14 per cent. antimony, 2 per cent. copper, and 1 per cent. zinc.

Ash, Black. See BLACK ASH.

Ash, Blue. See SMALT.

Ash, Bone. See BONE ASH.

Ashbury Metal. See ASHBERRY METAL.

Ashby Lime. Usually contains 46 per cent. CaO, 23 per cent. MgO, and 10 per cent. SiO_2.

Ash, Canadian. See CANADIAN ASH.

Ashes, Blue. See CHESSYLITE and MOUNTAIN BLUE.

Ashes, Green. See BREMEN.

Ashi Fibre. A fibre resembling straw in its physical and chemical character. It is obtained from a Japanese reed.

Ash, Kelp. See KELP ASH.

Ash, Pearl. See PEARL ASH.

Ash, Soda. See SODA ASH.

Ash, Tin. See TIN ASH.

Ash, Ultramarine. See ULTRAMARINE ASH.

Asiatic Pill. Contains $\frac{1}{12}$ gr. arsenic, and $\frac{3}{4}$ gr. black pepper.

Asilone. A proprietary preparation containing polymethylsiloxane and aluminium hydroxide. An antacid. (137).

Asilone Pædiatric. A proprietary preparation containing polymethylsiloxane and carob flower. An antacid. (137).

Asiphenin. Acetyl-salicylic acid and phenacetin.

Asiphyl. See ATOXYL MERCURY.

Askol. A proprietary phenol-formaldehyde synthetic resin.

Asmac. A proprietary preparation of allobarbitone, ipeccacuana, ephedrine, caffeine and aminophylline. A bronchial antispasmodic. (244).

Asmacort. A proprietary preparation of prednisolone, ephedrine, theophylline and phenobarbitone. Bronchial antispasmodic. (307).

Asmal. A proprietary preparation of theophylline, ephedrine and phenobarbitone. A bronchial antispasmodic. (350).

Asmapax. A proprietary preparation of ephedrine resinate, theophylline and bromvaletone. A bronchial antispasmodic. (271).

Asma-Vydrin. A proprietary preparation of adrenaline, atropine, papaverine, pituitary extract and chlorbutol. A bronchial inhalation. (252).

Asparagin (Althein). Amino-succinamic acid, $C_4H_8N_2O_3$, found in many vegetable juices.

Asparagin Mercury. Mercuric-amino-succinimate, $Hg(C_4H_7N_2O_3)_2$. Used in the treatment of syphilis.

Asparaginase. An anti-neoplastic. L-asparagine aminohydrolene.

Asparagus Stone. A mineral. It is a pale yellowish-green variety of apatite ($q.v.$).

Asparol. A German product. It is caffeine-calcium-acetyl-salicylate. An analgesic used in cases of rheumatism, influenza, and dysmenorrhœa.

Aspartame. A sweetening agent. 3-Amino - N - (α - methoxycarbonylphenethyl)succinamic acid. L - Aspartyl - L - phenylalanine methyl ester.

Aspartic Acid. Amino-succinic acid, $HOOC.CH_2.CH(NH_2)COOH$.

Aspellin. A proprietary preparation containing menthol, camphor, aspirin, methyl salicylate, glycerin ammonia, citronella oil and methylated spirit. A liniment. (390).

Aspergum. A proprietary preparation of aspirin, phenacetin, codeine phosphate and caffeine. An analgesic. (462).

Asperolite. A copper silicate which occurs at Tagilsk.

Asphaline. An explosive containing 54 per cent. potassium chlorate, 42 per cent. bran, and 4 per cent. potassium nitrate, and potassium sulphate.

Asphalt. Species of natural bitumen, sometimes mixed with mineral matter.

Asphalt, Artificial. Many mixtures are made up to imitate asphalt. The following are some in common use : (*a*) Made by the fusion of colophony with sulphur, and (*b*) by mixing chalk, lime, or sandstone with the pitchy residues of gas tar. See GERMAN ASPHALT.

Asphaltenes. Constituents of bitumen insoluble in hexane, but soluble in carbon tetrachloride.

Asphalt, French. See AUVERGNE BITUMEN.

Asphalt, German. See GERMAN ASPHALT.

Asphaltine. See PETROLINE.

Asphaltites. See PETROLITE.

Asphalt Mastic. Calcareous bituminous stones, containing from 5–20 per cent. bitumen. These stones are powdered and fused with a certain quantity of bitumen, and used for paving.

Asphalt, Natural. See BITUMEN.

Asphalt, Retin. See RETINITE.

Asphalt Rock. See ROCK ASPHALT.

Asphalt, Rubbered. See RUBBERED ASPHALT.

Asphalt Sludge. A sludge consisting of oxygenated bodies found in electrical transformers.

Asphalt, Syrian. See SYRIAN ASPHALT.

Asphalt Tar. See GOUDRON.

Asphalt, Trinidad. See TRINIDAD ASPHALT.

Asphaltum, Liquid. A solution of asphalt in turpentine. Used sometimes as a pigment.

Asphaltum Oil. See OIL OF ASPHALTUM.

Aspidelite. A mineral. It is a variety of sphene.

Aspidolite. A mineral. It is an olive-green mica.

Aspiphenyl. A German preparation of aspirin (acetyl-salicylic acid) and phenacetin. An antirheumatic.

Aspirgan. A pure acetyl-salicylic acid (Aspirin).

Aspirgran. A proprietary preparation of aspirin (acetyl-salicylic acid). (559).

Aspirin (Acetodin, Acetosal, Acetosalin, Acetysal, Acetosalic Acid, Helicon, Salacetin, Coxpyrin, Anglopyrin, Salaspin, Empirin, Regipyrin, Genasprin, Aspirgran, Asposal, Aspro, Atonin, Atylin, Nupyrin, Salantin). Trade names for acetyl-salicylic acid, $C_2H_3O.O.C_6H_4COOH$. Used for rheumatism, neuralgia, and headache.

Aspirin, Methyl. See METHYL RHODIN.

Aspirin, Quinine. See QUININE ACETOSALATE.

Aspirin, Soluble. See KALMOPYRIN.

Aspiriodine. A combination of acetyl-salicylic acid (aspirin), and iodine. A medicinal preparation.

Aspirochyl. See ATOXYL MERCURY.

Aspirolithite. A mineral, $CuSiO_3$. $3H_2O$.

Aspirophen (Phenaseptin). Amino-acet - phenetidide - acetyl - salicylate, $C_2H_5O.C_6H_4NH.C_2H_2O.N H_2.C_2H_3O.OC_6H_4.COOH$. An antirheumatic and antineuralgic.

Asplit. A synthetic resin adhesive. (561).

Asplosal. See Aspirin.

Aspon. Acid sodium orthophosphate for laundry use. (503).

Aspriodine. A proprietary preparation of acetyl-iodo-salicylic acid.

Aspro. See ASPIRIN.

Aspulum. A mercury derivative of chlorophenol. A seed preservative.

Asquirrol. Mercury dimethoxide, $(CH_3O)_2.Hg$.

Assam White. See GUTTA-SUSU.

Astaki. See MAZUT.

Asterite. Acrylic sheet for vacuum forming. (512).

Asteroite. A mineral. It is a variety of pyroxene.

Asterol. A registered trade mark applied to antifungal preparations containing 2-dimethylamino-6-(beta-diethylaminoethoxy)-benzothiazole and its salts. A skin fungicide. (314).

Asthmatussin. A proprietary preparation of guaiphenesin, ephedrine sulphate and phenobarbitone. A bronchial antispasmodic. (258).

Asthmatussin-T. A proprietary preparation of guaiphenesin, ephedrine sulphate, phenobarbitone and theophylline. A bronchial antispasmodic. (258).

Asthmeudin. A German product containing camphor, menthol, thymol, turpentine, eucalyptus, and ol. pumilion. Used for the prevention and cure of asthma.

Astiban. A proprietary trade name for Stibocaptate.

Astochite. A mineral. It is a variety of amphibole.

Astradur A and T. A registered trade mark for high impact PVC.

Astrafer. A proprietary preparation of dextriferron. A hæmatinic. (338).

Astrakanite (Blodite, Simonyite). A mineral, $Na_2SO_4.MgSO_4.4H_2O$. Occurs in the Stassfurt deposits.

Astralex. A proprietary range of chemicals used in the bright plating of nickel. (24).

Astraline. A fraction of Russian petroleum of specific gravity 0·832–0·835. It flashes at 40–45° C., and is used for burning.

Astralit. (1) A German explosive containing ammonium nitrate, vegetable meal, nitro-compounds, potassium perchlorate, and nitro-glycerin.

Astralit. (2) A registered trade mark for unclorinated PVC.

Astralite. A glass resembling aventurine, made by fusing silica, lead oxide, sodium carbonate, and borax, with lime, copper oxide, and ferric oxide. A cupreous compound is formed.

Astral Oil. See KEROSENE.

Astralon. A registered trade mark for PVC polymers in sheet form.

Astraphloxine FF. A sensitiser and dyestuff obtained from trimethyl indolenine methiodide and alkyl orthoformate in pyridine solution.

Astrictum. A compound containing cotton, pitch, asphalt, ground granite, resin, coal-tar, and mastic.

Astrinite. A proprietary bitumen plastic.

Astrolatum. A specially purified and bleached petroleum jelly which yields no brown colour with sulphuric acid and no fluorescence.

Astrolin (Glycopyrin). Antipyrine-methyl-ethyl-glycollate, $C_5H_{10}O_3.C_{11}H_{12}ON_2$. Proposed as a substitute for antipyrine in the treatment of headache.

Astrolite. A mineral. It is a silicate of aluminium, iron, potassium, and sodium.

Astrolith. A proprietary trade name for a special lithopone. A pigment.

Astrophyllite. A mineral, $(Fe.Mn)_4(Na.KH)_4(Si.Ti.Zr)_5O_{16}$.

Astroplax. A hydraulic gypsum cement.

Astryl. Sodium p-glycollylarsanilate. (507).

Asuntol. A proprietary preparation of COUMAPHOS. A veterinary insecticide.

Asurol (Hydurol). An addition product of mercury salicylate and sodium-amino-hydroxy-isobutyrate. It contains 40 per cent. mercury, and is recommended in the treatment of syphilis.

Asyphil. See ATOXYL MERCURY.

Asyphyl. Atoxyl-mercury ($q.v.$).

Atacamite (Remolinite). A native copper oxychloride from Peru and Saxony, $CuCl_2.3Cu(OH)_2$. It formerly constituted the pigment, Brunswick green.

A.T. 10. A proprietary preparation of dihydrotacbysterol. Used in vitamin D deficiency. (112).

Atarax. A proprietary preparation of hydroxyzine hydrochloride. A sedative. (85).

Atasorb. A proprietary preparation of activated attapulgite. An anti-diarrhœal. (463).

Atasorb-N. A proprietary preparation of activated attapulgite, neomycin sulphate, and pectin. An antidiarrhœal. (333).

AT Cellulose B. Ethylcellulose. (3).

Atebrin (Erion, Mepacrine). A proprietary anti-malarial stated to be the hydrochloride of an allyl-amino-alkyl-amino-acridine derivative.

Atelestite. A mineral. It is a bismuth arsenate containing iron phosphate.

Atelite. A mineral. It is a basic copper chloride.

Atempol. A proprietary trade name for Methylpentynol.

Atenolol. A beta adrenergic blocking agent currently undergoing clinical test as " ICI 66,082 ". It is 4-(2-hydroxy-3-isopropylaminopropoxy) phenylacetamide.

Atensine. A proprietary preparation of DIAZEPAM. A tranquilliser. (137).

Atephen. A phenolic coating possessing extreme chemical resistance. (3).

Aterite. A nickel silver. It usually contains from 47–68 per cent. copper, 17–38 per cent. zinc, 10–14 per cent. nickel, 1·5–1·9 per cent. iron, and 0·16–2·2 per cent. manganese.

Atgard. A proprietary preparation of DICHLORVOS. An insecticide.

Atinosol. A proprietary thallium acetate solution.

Atiran. A potato fungicide. (511).

Ativan. A proprietary preparation of LORAZEPAM. A tranquilliser. (245).

Atlac 382. A proprietary bisphenol-A-fumarate polyester resin. (113).

Atlas. A proprietary phenol-formaldehyde synthetic resin.

Atlas 10 Bronze. A proprietary trade name for an aluminium bronze containing 9·0 per cent. aluminium, 7·0 per cent. lead with copper.

Atlas 89. A proprietary trade name for an alloy of copper with 9·0 per cent. aluminium and 1·0 per cent. iron.

Atlas 90. A proprietary trade name for an aluminium bronze containing 90·0 per cent. copper with 10·0 per cent. aluminium.

Atlasite. A mineral. It is a carbonate of copper containing chlorine.

Atlas Orange. See ORANGE II.

Atlas Powders. Explosives. (a) Contains 75 per cent. nitro-glycerin, 2 per cent. sodium nitrate, 21 per cent. wood pulp or sawdust, and 2 per cent. magnesium carbonate ; (b) contains 50 per cent. nitro-glycerin.

Atlas Red. A dyestuff. It is the sodium salt of primuline-azo-tolylene-diamine. Dyes cotton terracotta-red from an alkaline bath.

Atlas Red, Patent. See GERANINE.

Atlas Scarlet No. 3. An azo dyestuff. It is sulpho-xylene-azo-β-naphthol-di-sulphonic acid, $HSO_3.C_8H_8N_2.C_{10}H_4.(HSO_3)_2.OH$. The sodium salt is the commercial product.

Atlas Silk. The product of the Atlas caterpillar, *Attacus atlas*, East Indies. It is similar to Tussah silk.

Atlas Steel. A proprietary hot die steel containing 9–11 per cent. tungsten, 3·25–3·5 per cent. chromium, and a little vanadium.

Atmido. A siliceous earth, used as a filtering medium, also as a rubber filler.

Atmoid. A mineral filler for rubber goods. It consists of almost pure silica.

Atochinol (Atoquinol). The allyl ester of atophan (*q.v.*). Used in the treatment of gout.

Atole (Chica). A South American liqueur.

Atoleine. See PARAFFIN, LIQUID.

Atoline. See PARAFFIN, LIQUID.

Atomol. Nasal decongestant with prolonged action. (510).

Atonin. See ASPIRIN.

Atomised Glue. A fine glue powder produced by spraying liquid glue into a heated evacuated chamber. It easily dissolves in water.

Atomite. A proprietary trade name for water-ground whiting.

Atophan (Quinophan, Agotan). 2-Phenyl-quinoline-4-carboxylic acid, $C_9H_5N(C_6H_5)COOH$. Used in the treatment of gout and sciatica.

Atophanyl. A German preparation. It is a solution of the sodium salt of Atophan and salicylic acid. Used in the treatment of rheumatism.

Atopite. A mineral, $(Ca.Na_2.Fe.Mn)_2 Sb_2O_7$.

Atoquinol. Allylphenylcinchoninic ester. A powerful uric acid solvent and eliminator.

Atoxycocaine. Ethocaine hydrochloride.

Atoxyl (Arsamin, Soamin). The mono-sodium salt of *p*-amino-phenyl-arsenic acid, $NH_2.C_6H_4.AsO(OH)(ONa)$. It contains 37 per cent. As_2O_5, and is used as a subcutaneous injection for sleeping-sickness and syphilis. Also used in the treatment of tuberculosis.

Atoxyl Mercury (Aspirochyl, Asyphil, Asyphyl, Asiphyl, Atyrosyl). Mercury-*p*-amino-phenyl-arsenate, $C_{12}H_{14}O_6N_2 As_2Hg$. Used in the treatment of syphilis.

Atrabilin. A pale fluid derived from a glandula suprarenalis.

Atramentum Stone. A mixture of ferric and ferrous sulphates with ferric oxide. Used in the manufacture of inks.

Atreol. Ammonium atreolate, an aqueous solution containing the ammonium salts of a mixture of organic acids obtained from petroleum distillation. A mild antiseptic.

Atrix. A German product. It is a depilatory cream containing a sulphur compound and chalk.

Atrixo. A silicone hand cream. (529).

Atro-Glyceric Acid. α-Phenyl-glyceric acid.

Atrolactic Acid. α-Oxy-α-phenyl-propionic acid, $C_6H_5.C(CH_3)(OH)$ COOH.

Atromid-S. A proprietary preparation of clofibrate, used to reduce blood cholesterol levels. (2).

Atroscine. Optically inactive scopolamine (d-l-hyoscine, $C_{17}H_{21}O_4N$).

Atroxindol. The anhydride of o-amino-α-phenyl-propionic acid, $C_6H_4(NH)$ $(CH.CH_3)CO$.

Atta Beans. See OWALA BEANS.

Attacolite. A mineral. It is $(Ca,Mn)AlPO_4(OH)_2$.

Attar. Otto or Essential Oil.

Attawa Nuts. See OWALA BEANS.

Aturbane. A proprietary preparation of phenglutarimide hydrochloride, used in Parkinsonism. (18).

A.T.V. Steel. A complex nickel-chromium austenitic steel.

Atyek. A name given in the Cameroons to the bark of *Pausinystalia brachythyrsus*.

Atyeng. A name applied in the Cameroons to the bark of a species of *Canthium*.

Atylin. A proprietary brand of acetyl-salicylic acid (aspirin).

Atyrosyl. See ATOXYL MERCURY.

Aubepine. Anisaldehyde, $C_6H_4.(COH)$. $O.CH_3$. Used in perfumery.

Audax. A proprietary preparation of choline salicylate, ethylene oxide polyoxy propylene condensate. Ear drops. (334).

Audicort. A proprietary preparation of TRIAMCINOLONE acetonide, NEOMYCIN undecylenate and benzocaine used as ear-drops in the treatment of otitis externa. (306).

Auel Solder. An alloy of 63 per cent. tin, 35 per cent. zinc, 1·7 per cent. copper, and 0·3 per cent. aluminium.

Auerbachite. A mineral, $ZrSiO_4$.

Auerlite. A mineral. It is a silico-phosphate of thorium, $3ThO_2.(3SiO_2. P_2O_5).6H_2O$.

Auer Metal (Ferro-cerium). A pyrophoric alloy, containing iron with cerium earth metals. It forms the sparking material for automatic gas and cigar lighters. See MISCH METAL.

Augelite. A mineral, $Al_2(OH)_3PO_4$.

Augite (Coccolite). A mineral. It is essentially $CaO.MgO.2SiO_2$, a variety of pyroxene ($q.v.$), and occurs in basalt, a felspathic rock.

Augite-Syenite. A mineral. It is a coloured variety of tourmaline ($q.v.$).

Augsburg Metal. A brass containing 72 per cent. copper, and 28 per cent. zinc.

Auina. Synonym for Haüyne.

Aula Salt. A proprietary mixture containing nitrite, for curing meat.

Auligan. A German product. It is stated to be bis-ethyl-xanthogen, and

consists of yellowish needles insoluble in water and easily soluble in ether, petroleum ether, and oils. It contains 52 per cent. of sulphur, and an ointment containing from 2–6 per cent. is used in cases of parasitic skin diseases.

Aulin. A parasiticide. It is stated to be a yellow oily liquid " bis-ethyl-xanthogen " in an organic solvent.

Aulinox. Polyoxethylene derivative of ricinoleic acid for veterinary use. (512).

Auracet. A proprietary preparation of aluminium acetotartrate and lead subacetate. Ear drops. (319).

Auralgicin. A proprietary preparation of ephedrine, benzocaine, chlorbutol, potassium hydroxyquinoline sulphate, phenazone and glycerine. Ear drops. (188).

Auraltone. A proprietary preparation of benzocaine and phenazone in glycerine. Ear drops. (340).

Auramine (Auramine O, Auramine I, II, and conc., Aureum, Pyoctanin). A diphenylamine basic dyestuff, which is the hydrochloride of imino-tetramethyl-diamino-diphenyl-methane, $N(CH_3)_2$. $C_6H_4.C(NH), C_6H_4.N(CH_3)_2HCl$. Dyes silk and cotton direct, and tannin mordanted cotton greenish-yellow. It is used for staining paper. It has also come into prominence as a powerful antiseptic for use in nose and ear surgery, and in gonorrhœa. Also see GLAURAMINE.

Auramine I, II. Auramine ($q.v.$), mixed with dextrine.

Auramine, Conc. See AURAMINE.

Auramine G. A dyestuff. It is the hydrochloride of imino-dimethyl-dia-mino-ditolyl-methane, $C_{17}H_{22}N_3Cl$. Dyes tannined cotton greenish-yellow.

Auramine O. See AURAMINE.

Auramine, Medicinal. See PYOCTANIN and AURAMINE.

Aurantia (Imperial Yellow). An acid dyestuff. It is the ammonium salt of hexanitro-diphenyl-amine, $C_{12}H_8N_8O_{12}$. Dyes wool and silk orange from an acid bath.

Aurantiène. A residue containing terpenes left behind in the refining of orange oil. Used as a perfume for soaps.

Aurantiin (Hesperidine). Naringin, $C_{21}H_{26}O_{11}$.

Aurantine. The trade name for Osage orange extract (from the bark of a shrub), used in the textile industry for tanning. Also see ABIETENE.

Auremetine. A proprietary preparation of auramine emetine periodide.

Aureocort. A proprietary preparation of TRIAMCINOLONE acetomide and chlortetracycline hydrochloride used in the treatment of skin disorders. (306).

Aureolin (Cobalt Yellow, Indian Yellow). A yellow pigment. It is a double nitrite of potassium and cobalt, $K_6Co_2(NO_2)_{12}.3H_2O$.

Aureoline. See PRIMULINE.

Aureomycin. A proprietary preparation of chlortetracycline. An antibiotic. (306).

Aureosin. A chlorinated fluoresceine.

Aureum. See AURAMINE.

Aurichalcite (Green Calamine). A mineral, $2(Zn.Cu)CO_3.3(Zn.Cu)(OH)_2$.

Auricome (Golden Hair Water). A dilute solution of hydrogen peroxide, H_2O_2, mixed occasionally with nitric acid, for imparting a light colour to the hair.

Aurine (Rosilic Acid). A dyestuff. It consists of mixtures of aurine (trioxytriphenyl-carbinol), oxidised aurine, methyl-aurine, and pseudo-rosilic acid. Pseudo-rosilic acid is the chief constituent of commercial aurine. It dyes wool and silk, and is also used for colouring spirit varnishes and lacquers. Aurine, $C_{19}H_{14}I_3$. Oxyaurine, $C_{19}H_{16}O_6$. Methylaurine, $C_{20}H_{16}O_3$. Pseudorosilic acid, $C_{20}H_{16}O_4$.

Aurine R. See PÆONINE.

Aurine Red. See PÆONINE.

Aurocantan. Cantharidyl-ethylene-diamine-aurocyanide.

Aurochin. Quinine-p-amino-benzoate.

Aurocyanase. A colloidal gold and potassium double cyanide.

Auromet 55. A proprietary trade name for an alloy containing 76–80 per cent. copper, 10–12 per cent. aluminium, 4–6 per cent. iron, and 4–6 per cent. nickel.

Auronal Black. A dyestuff obtained by the fusion of dinitro-p-amino-diphenyl-amine with sodium polysulphide. It dyes cotton a blue-black from a sodium sulphide bath.

Aurophos. A sodium-gold double salt of an amino-aryl-phosphorous acid and hyposulphurous acid. It is a German product and consists of a white, odourless, easily soluble substance containing 25 per cent. of gold. It is stated to be not very poisonous, and is used for treating tubercular pains and lupus.

Auro-Protasin. Colloidal gold.

Aurora Yellow. See CADMIUM YELLOW.

Aurotine. An acid or mordant dyestuff for wool. It is the sodium salt of tetranitro-phenol-phthalein, $C_{16}H_8O_{12}N_4Na_2$.

Aurum. Gold, Au.

Austenite. A characteristic constituent of very highly carbonised steel, containing more than 1.1 per cent. carbon. It can be scratched with a needle.

Austinite. A mineral. It is $4[CaZnAsO_4.OH]$.

Australene. Dextro-pinene.

Australian Bark. See ALSTONIA.

Australian Caoutchouc. See COO-RONGITE.

Australian Copal (Kowree Gum, Cowdee Gum, Cawree Gum, Kauri Gum, Kawrie Gum, Cowdie Gum). Australian Dammara resin, obtained from *Dammara australis*. It is employed in the manufacture of oil varnishes.

Australian Gold. A gold-silver alloy containing 8.33 per cent silver, used for coinage.

Australian Gum (Wattle Gum). A variety of gum acacia, from *Acacia pycnantha*. It is usually sold as large reddish tears or lumps.

Australian Sandarac. See SANDARAC RESIN.

Australin. A fraction of Russian petroleum of specific gravity 0.850, of a pale-yellow colour, with a flash-point of not less than $122°$ F.

Austrapol. A trade name for a group of styrene-butadiene polymers. (464).

Austrian Cinnabar. See CHROME RED.

Autan. A disinfectant consisting of 1 part formaldehyde, and 2 parts barium peroxide, together with a little alkali carbonate. When mixed with water it evolves formaldehyde, and kills organisms in air.

Autex. A proprietary trade name for cotton fibre and phenolic resin.

Autodyne. Phenoxy-glycerin, $C_9H_{12}O_3$, an anodyne.

Autogal. A trade mark for a flux used for soldering and welding aluminium. It is a mixture of the halogen salts of the alkali metals.

Autogene Black. A dyestuff produced by the condensation of amino-oxy-diphenyl-amine, with the products of the action of sulphur chloride upon phenol, cresol, or amines, and finally fusing with sodium sulphide. Dyes cotton black.

Autogene Grey. A sulphide dyestuff.

Autol Red GL. A dyestuff prepared from diazotised p-chlor-o-anisidine. Used for lakes.

Automolite. A mineral, $ZnO.Al_2O_3$.

Autunite (Uranium Mica, Calcouranite, Uranite, Sideranatrite). A mineral. It is an anhydrous phosphate of calcium and uranium, $Ca(UO_2)_2(PO_4)_28H_2O$.

Auvergne Bitumen (French Asphalt). A species of natural bitumen found in the province of Auvergne, France. It is similar to Trinidad bitumen, but contains clay, silica, iron, etc.

Auxite. See LUCIANITE.

Avalite. A mineral containing hydrated

oxides of chromium, aluminium, and silicon.

Avantine. A brand of isopropyl alcohol. An anæsthetic. (289).

Avanturine Felspar. Oligoclase (*q.v.*).

Avaram Bark. A tanning material. It is the product of *Cassia auriculata* of Southern India.

Avazyme. A proprietary preparation of chymotrypsin, used for treatment of phlebitis and bruises. (382).

Avecolite. A proprietary phenol-formaldehyde synthetic resin.

Aveeno. A proprietary preparation of colloidal oatmeal. Used in dermatology. (320).

Aveeno Oilated. A proprietary preparation of colloidal oatmeal and emollient oils. Used in dermatology. (320).

Avenose. A mixture of oatmeal with soluble acorn malt extract.

Aventurine. See ROUGE FLAMBÉ.

Aventurine Glass. A glass containing either finely divided copper, or cuprous oxide. It is deep yellow, and full of metallic-looking scales.

Aventyl. A proprietary preparation of nortriptyline hydrochloride. An antidepressant. (333).

Avenyl. A proprietary product. It is stated to be a solution of 2-myristoxymercuri - 3 - hydroxy - benzaldehyde in hydnocarpus oil. Employed in the treatment of leprosy complicated by syphilis.

Aveonal. An alloy of aluminium with 4 per cent. copper, 0·05 per cent. magnesium, 0·05 per cent. manganese, and 0·05 per cent. silicon.

Avertin. A proprietary preparation of bromethol, used to control eclampsia in toxæmia of pregnancy. (112).

Avialite. A proprietary trade name for an alloy of copper with about 9·0 per cent. aluminium and 1·0 per cent. iron.

Avicalm. See METOSERPATE.

Avicell-RC. A proprietary trade name for a chemically pure colloidal cellulose. Forms thixotropic dispersions both mechanically and thermally stable. Edible and metabolically alert. Used as a thickening agent. (209).

Avignon Grains. Persian berries, the yellow colouring matter of which is rhamnin. See PERSIAN BERRIES.

Avilon. Metal complex dyes. (530).

Avional D. An alloy of aluminium with 3·9 per cent. copper, 0·5 per cent. nickel, 0·5 per cent. magnesium, 0·55 per cent. silicon, and 0·3 per cent. iron.

Avirol. See MONOPOLE SOAP.

Avisol. Neutral soluble sulphathiazole for poultry. (507).

Avitex. A proprietary trade name for textile finishing materials. Consists of sulphated higher alcohols.

Avivan (Tetrapol, Koloran, Flerhenol). Wetting out agents consisting of sulphonated oils with organic solvents.

Avlinox. A proprietary polyoxyethylene derivative of ricinoleic acid for veterinary use. (2).

Avlochlor. A proprietary preparation of chloroquine phosphate. Anti-inflammatory agent. (2).

Avlodex. Pharmaceutical products. (512).

Avloprocil. Procaine penicillin preparation for injections. (512).

Avloprocil A.S. A proprietary preparation containing procaine penicillin. An antibiotic. (2).

Avlosulphon. A proprietary preparation of dapsone, used in the treatment of leprosy. (2).

Avlothane. A proprietary grade of hexachloroethane. (2).

Avogadrite. A mineral. It is $4[(K,Cs)BF_4]$.

Avoleum. A proprietary preparation. It is a concentrated preparation of vitamin A.

Avomine. A proprietary preparation of promethazine theoclate. (Promethazine 8-chlorotheophyllinate). An antiemetic. (336).

Avoparcin. A growth promoter. A glycopeptide antibiotic obtained from cultures of *Streptomyces candidus* or the same substance obtained by any other means.

Avothane. A proprietary polyurethane elastomer. (465).

Awaruite. A mineral, $FeNi_2$, of New Zealand.

Axerophthol. Vitamin A.

Axestone. A mineral which is a variety of jade.

Axf Plastic (Shinkle Plastic). A proprietary synthetic compounding ingredient for rubber. It is a condensation product of ethylene dichloride and other bodies.

Axin. See AGE.

Axinite. A mineral. A silicate of aluminium, calcium, magnesium, and iron, with some boron, $H_2(Ca.Fe.Mn)_4(BO)Al_2(SiO_4)_5$.

Axite. An explosive. It is a smokeless powder which contains guncotton, nitro-glycerin, pet. jelly, and a little potassium nitrate.

Az. See AZOTE.

Azacyclonal. Diphenyl - 4 - piperidyl - methanol.

Azadirach, Indian. Neem or Margosa bark.

Azafran. Saffron (*q.v.*).

Azaleine. See MAGENTA.

Azalin. A mixture of quinoline red and quinoline blue. Used for making orthochromatic plates.

Azalomycin. A mixture of related antibiotics produced by *Streptomyces hygroscopicus*, var. *azalomyceticus*.

Azamethonium Bromide. 3 - methyl - 3 - azapentamethylenedi (ethyldimethyl ammonium bromide).

Azamine 4B. See BENZOPURPURIN 4B.

Azaniles. See AZONINES.

Azanol Brown N (Azanol Dark Brown RR, Orange Brown RY, Red Brown R). New acid brown dyestuffs.

Azaperone. A tranquilliser. 4' - Fluoro-4 - [4 - (2 - pyridyl)piperazin - 1 - yl)-butyrophenone. STRESNIL, SUICALM.

Azapetine. 1 - Allyl - 2,7 - dihydro - 3, 4 : 5,6-dibenzazepine.

Azapropazone. An analgesic and anti-inflammatory. 5 - Dimethylamino - 9 - methyl - 2 - propyl - 1 *H* - pyrazolo - [1, 2 - *a*] [1, 2, 4] - benzotriazine - 1, 3 (2*H*) - dione.

Azaribine. A preparation used in the treatment of psoriasis. 2-β-D-Ribofuranosyl - 1, 2, 4 - triazine - 3, 5 - (2*H*, 4*H*)dione 2', 3', 5' - triacetate. 6 - Azauridine 2', 3', 5', - triacetate.

Azarine R. A dyestuff. It is the ammonium bisulphite compound of the product of the reaction between β-naphthol and diazotised diamino-oxy-sulpho-benzide. It resembles Azarine S.

Azarine S. A mordant dyestuff. It is the ammonium bisulphite compound of dichloro - phenol - azo - β - naphthol, $C_{16}H_{15}Cl_2N_3SO_5$. Used for pinks and reds in calico printing, and for dyeing silk.

Azatadine. An anti-histamine. It is 5, 6 - dihydro - 11 - (1 - methyl - 4 - piperidylidene) benzo(h)cyclohepta(b) - pyridine.

Azathioprine. 6 - (1 - methyl - 4 - nitro - imidazol-5-ylthio)purine.

Azeosin. See AZO-EOSIN.

Azeta. A proprietary brand of low-protein biscuits used in diets in cases of renal failure and phenylketonuria. (466).

Azetepa. PP - Di(1 - aziridinyl) - N - ethyl - 1,3,4 - thiadiazol - 2 - ylphosphinamide.

Azidine Brilliant Yellow 6G Extra. A direct cotton colour suitable for artificial silk.

Azidine Chrome Black Blue BL. A direct cotton blue suitable for artificial silk.

Azidine Chrome Brown BB, BG, BR. Direct cotton browns after treatment with potassium bichromate and copper sulphate. Suitable for artificial silk.

Azidine Fast Red F. Diamine fast red (*q.v.*).

Azidine Fast Scarlets. Dyestuffs obtained by passing carbonyl chloride into a mixture of *m*-toluylene-diamine-4-sulphonic acid and J-acid.

Azidine Fast Yellow G. A dyestuff obtained by warming *p*-nitro-toluenesulphonic acid with caustic soda. It gives golden shades on cotton in the presence of sodium chloride, and on wool and silk from an acid bath.

Azidine Wool Blue R. A tetrazo dyestuff obtained by coupling diazotised tolidine with β-naphthol-8-sulphonic acid and 8-amino-α-naphthol-5-sulphonic acid.

Azidocillin. An antibiotic currently undergoing clinical test as " BRL 2534 " and " SPC 297D ". It is 6-[D (—) - α - azidophenylacetamido] - penicillanic acid.

Azilex. A proprietary preparation of azulene. An antipruritic. (827).

Azindone Blue G (Anzindone Blue R). Indigo blue dyestuffs closely allied to the naphthyl dyes.

Azine Blue, Spirit Soluble. See INDULINE, SPIRIT SOLUBLE.

Azine Green. A safranine dyestuff, $C_{30}H_{25}N_4Cl$, the product of the reaction between nitroso-dimethyl-aniline hydrochloride and 2 : 6-diphenyl-naphthylene-diamine.

Azine Green GB (Azine Green TO). A basic dyestuff. It is dimethyl-aminophenyl-amino - phenyl - pheno - naphthazonium chloride, $C_{30}H_{25}N_4Cl$. Dyes tannined cotton dark green.

Azine Green S. An acid dyestuff. It is the sodium salt of azine green sulphonic acid. Dyes wool bluish-green from an acid bath.

Azine Green TO. See AZINE GREEN GB.

Azine Scarlet G. A dyestuff. It is dimethyl - diamino - methyl - toluphenazonium-chloride, $C_{16}H_{19}N_4Cl$. Dyes yellow-scarlet.

Azlocillin. An antibiotic currently undergoing clinical trial as " BAYe 6905 ". It is 6 - [D - 2 - (2 - oxoimidazolidine - 1 - carboxamido) - 2 - phenylacetamido] - penicillanic acid.

Azo Acid Black B, G, R, TL, 3BL. Acid dyestuffs employed in wool dyeing.

Azo Acid Black TL Extra. A mixture of a blackish-blue, a green, a violet-blue, and an orange dye. It dyes wool black from an acid bath containing sodium sulphate.

Azo Acid Blue B. A dyestuff belonging to the same class as Victoria violet BS. It dyes wool pure blue.

Azo Acid Blue, 4B, 6B. Acid dyestuffs for wool.

Azo Acid Brown. An acid colour

for wool, which it dyes from an acid bath.

Azo Acid Carmine. An acid dyestuff for wool.

Azo Acid Magenta B, G. Acid dyestuffs for wool.

Azo Acid Rubine. See AZO RUBINE S.

Azo Acid Rubine 2B. See FAST RED D.

Azo Acid Violet 4R. An acid colour, dyeing wool reddish-violet from an acid bath.

Azo Acid Yellow. See CITRONINE B.

Azoalizarin Black. A dyestuff. It is the sodium salt of benzene-disazo-salicylic acid-dioxy-naphthalene-mono- or disulphonic acid, $C_{23}H_{13}N_4S_2O_{11}Na_3$. Dyes chromed wool black.

Azoalizarin Bordeaux W. A dyestuff. It is the sodium salt of benzene-disazo-salicylic acid - α - naphthol - sulphonic acid, $C_{23}H_{14}N_4SO_7Na_2$. Dyes chromed wool bordeaux, and is used for wool printing.

Azoalizarin Yellow 6G. See ALIZARIN YELLOW 5G.

Azo-aniline. Amino-benzene-azo-aniline, $C_6H_4(NH_2)N_2C_6H_4(NH_2)$.

Azobenzene. Benzene-azo-benzene, $C_6H_5.N_2.C_6H_5$.

Azobenzene Red. See BIEBRICH SCARLET.

Azo Black. See BLUE BLACK B.

Azo Black Blue (Azo Navy Blue). A dyestuff. It is the sodium salt of ditolyl - disazo -m- oxy - diphenylamine-amino - naphthol - disulphonic acid, $C_{36}H_{25}N_6S_2O_8Na_2$. Dyes cotton grey to dark violet from a boiling salt bath.

Azo Black O. See BLUE BLACK B.

Azo Blue. A dyestuff. It is the sodium salt of ditolyl-disazo-bi-α-naphthol-p-sulphonic acid, $C_{34}H_{24}N_4O_8S_2Na_2$. Dyes cotton greyish-violet.

Azo Bordeaux. See BUFFALO RUBINE.

Azo Brown O. See FAST BROWN N.

Azo Camphor. Monoketazo-camphor quinone, $C_{10}H_{16}N_2O$.

Azo Cardinal G. An acid dyestuff for wool.

Azocarmine B (Rosinduline 2B). A dyestuff. It is the acid sodium salt of phenyl-rosinduline-trisulphonic acid, $C_{28}H_{17}N_3S_3O_9Na_3$. Dyes wool bluish-red, and is used as an archil substitute.

Azocarmine BX Powder and G Paste. Acid dyestuffs similar to azo-carmine B and G.

Azocarmine G (Rosazine, Rosarin). A dyestuff. It is the sodium salt of phenyl - rosinduline - disulphonic acid, $C_{28}H_{17}N_3S_2O_6Na_2$. Dyes wool bluish-red. Employed as a substitute for archil.

Azochromine. A mordant dyestuff. It is tetroxy-azo-benzene, $C_{12}H_{10}O_4N_2$. Dyes chromed wool and cotton dark brown. Used in cotton printing.

Azococcin G. See TROPÆOLINE OOOO.

Azococcin 2R (Double Scarlet R, Xylidine Scarlet, Jute Scarlet R). An acid dyestuff. It is the sodium salt of xylene - azo - α - naphthol - p - sulphonic acid, $C_{18}H_{15}N_2O_4SNa$. Dyes wool red from an acid bath.

Azococcin 7B. See CLOTH RED G.

Azocochineal. A dyestuff. It is the sodium salt of anisol-azo-α-naphthol-disulphonic acid, $C_{17}H_{12}N_2S_2O_8$. Dyes wool red from an acid bath.

Azocoralline. A dyestuff. It is the sodium salt of p-acetamino-benzene-azo-β-naphthol-disulphonic acid, $C_{18}H_{13}N_3S_2O_8Na$. Dyes wool from an acid bath.

Azo Corinth. A dyestuff. It is the sodium salt of sulpho-naphthalene-azo-resorcin-azo-ditolyl-azo-amino - phenol-sulphonic acid, $C_{38}H_{27}N_7O_9S_2Na_2$. Dyes cotton brown from a soap bath.

Azodermin. Acetylised amino-azo-toluene.

Azodiphenyl Blue. See INDULINE SPIRIT SOLUBLE.

Azodolen. A mixture of equal parts of pellidol (q.v.), and an indol compound called indolen (q.v.). Used in surgery.

Azoene. Fast red salt (ponceau fast L salt). A pinkish cream powder for use in automatic SGO-T assays. (527).

Azoeosin (Azeosin). An acid dyestuff. It is the sodium salt of anisol-azo-α-naphthol-p-sulphonic acid, $C_{17}H_{13}N_2O_5$ SNa. Dyes wool eosin red from an acid bath.

Azofix. Azoic colours. (521).

Azoflavin. See CITRONINE B.

Azoflavin 3R Extra Conc. Citronine B or 2B (q.v.).

Azoflavin S or 2. See CITRONINE B.

Azoform. Monoacetyl-azoxy-toluene.

Azofuchsine B (Acid Levelling Red 2B). An acid and acid mordant dyestuff. It is the sodium salt of toluene-azo-1 : 8-dioxynaphthalene-sulphonic acid, $C_{17}H_{13}N_2SO_5Na$. Dyes wool from an acid bath magenta red, becoming violet-black on chroming.

Azofuchsine G. An acid and acid mordant colour. It is the sodium salt of p-sulpho-benzene-azo-dioxy-naphthalene-sulphonic acid, $C_{16}H_{10}N_2S_2O_8Na_2$. Dyes wool magenta red from an acid bath.

Azofuchsine GN Extra. A dyestuff, similar to Azofuchsine G.

Azo Galleine. A mordant dyestuff. It is p-dimethyl-amino-benzene-azopyrogallol, $C_{14}H_{15}N_3O_3$. Dyes chromed wool blackish-violet. It is used for calico printing, giving dark violet shades.

Azogen R. A dyestuff similar to Azophor red PN.

Azo Green. A dyestuff. It is tetra-methyl - diamino - triphenyl - carbinol - azo-salicylic acid, $C_{30}H_{30}N_4O_4$. Dyes chromed wool green.

Azo Green T. A dyestuff of the same class as Basle blue R.

Azogrenadine L. A dyestuff from p-amino-acetanilide and 3 : 4-naphthol-sulphonic acid.

Azoground. Azoic colours. (521).

Azogrenadine S. See SORBINE RED.

Azoguard. A stabiliser for diazo compounds. (512).

Azol. 4 - aminophenol-hydrochloride in solution. (515).

Azolith. A proprietary pigment containing 71·0 per cent. $BaSO_4$ and 29·0 per cent. ZnS.

Azolitmin. A specially pure litmus colouring matter, soluble in water.

Azolone. Belgian phenolic synthetic resins.

Azomagenta G. A dyestuff obtained by diazotising sulphanilic acid, and treating the product with S acid.

Azoman. A proprietary trade name for Hexazole.

Azo Mauve B. A dyestuff. It is the sodium salt of ditolyl-disazo-α-naphthyl amine - amino - naphthol - disulphonic acid, $C_{34}H_{26}N_6O_7S_2Na_2$. Dyes cotton blackish-blue-violet from an alkaline bath.

Azo Mauve R. The benzidine derivative corresponding to Azo mauve B.

Azonaphthol Red S. See FAST RED D.

Azo Navy Blue. See AZO BLACK BLUE.

Azonigrin. An acid dyestuff from di-sulpho-phenol-azo-naphthylamine and β-naphthol. Dyes wool from an acid bath brownish black.

Azonines (Acedronoles, Silkons, Aza-niles). Trade names for various amino and diamino compounds such as nitrani-line, amino-azo-benzene, and dianisi-dine. They are stated to be suitable for dyeing acetyl silk.

Azo Orange R. A dyestuff. It is the sodium salt of bi-tolyl-tetra-disazo-dioxy - diphenyl - methane - bi - naphth-ionic acid, $C_{61}H_{48}N_{10}O_8S_2Na_2$. Dyes cotton orange.

Azo Orchil R. An acid dyestuff, giving brownish-red colour on wool.

Azo Orseillin. An azo dyestuff pre-pared by the action of diazotised ben-zidine upon α-naphthol-sulphonic acid, $(C_6H_4)_2(N_2)_2.(C_{10}H_5)_2(OH)_2(HSO_3)_2$. The commercial product is the sodium salt. It is a direct cotton dye, giving brownish-red shades from a soap bath.

Azophor Blue. See DIANISIDINE BLUE.

Azophor Orange. A dyestuff. It is m - nitro - diazo - benzene - sulphate mixed with zinc sulphate.

Azophor Red. See PARANITRANILINE RED.

Azophor Red PN. A dyestuff. It consists of diazotised paranitraniline and desiccated Glauber's salt or alu-minium sulphate.

Azophosphine GO. A basic dyestuff. It is the chloride of m-trimethyl-amino-benzene - azo - resorcin, $C_{14}H_{18}N_3O_2Cl$. Dyes cotton direct from an acid bath orange-red.

Azoprint. Azoic printing colours. (521).

Azopurpurin 4B. A direct cotton dye-stuff, which it dyes orange from a salt bath.

Azorapid. Azoic printing colours. (521).

Azo Red A. A dyestuff allied to Palatine red ($q.v.$).

Azo Resins. Artificial resins made by the condensation of phenols and form-aldehyde, in the presence of a diazo compound, such as diazotised p-nitrani-line. They are stated to be suitable as dyes, fat or oil-soluble varnishes, and for insulation.

Azoresorcin. See DIAZORESORCIN.

Azoresorufin (Diazoresorufin). Re-sorufin, $C_{12}H_7NO_3$.

Azorite. A mineral, $ZrSiO_4$.

Azorubine. See Azorubine S.

Azorubine A. See AZORUBINE S.

Azorubine S (Azo acid rubine, Azoru-bine, Azorubine A, Brilliant carmoisine O, Brilliant crimson, Carmoisine, Chro-motrope FB, Crimson B, Fast red C, Mars red, Nacarat, Rainbow Red RB, NB, Cardinal Red 3B, Carmoisine L.LAS.S.WS). A dyestuff. It is the sodium salt of p-sulpho-naphthalene-azo-α-naphthol-p-sulphonic acid, $C_{20}H_{12}N_2O_7S_2Na_2$. Dyes wool red from an acid bath.

Azorubine 2S. An azo dyestuff pre-pared by the action of diazotised amino-azo-benzene-sulphonic acid upon α-naphthol-mono-sulphonic acid, HSO_3. $C_6H_4.N_2.C_6H_4.N_2.C_{10}H_5.HSO_3.OH$.

Azosan. A fungicide. (507).

Azostix. A prepared test strip of urease, bromothymol blue and buffers, for the semi-quantitative determination of blood urea levels. (807).

Azote (Az). Lavoisier's name for nitrogen.

Azothioprine. 6-(1-methyl-4-nitro imi-dazol-5-ylthio)purine. Available as Imuran.

Azotic Acid. Nitric acid, HNO_3.

Azotine. An Austrian explosive.

Azo Turkey Red. An azo dyestuff pro-duced in a similar way to Paranitrani-line red ($q.v.$), by using diazotised β-naphthylamine, instead of diazotised p-nitraniline.

Azovan Blue. The tetrasodium salt of 4,4'-di-[7-(1-amino-8-hydroxy-2, 4-disulpho) naphthylazo]-3,3'-bitolyl. Evans blue.

Azovernin. Acetyl-amino-azo-toluene, $CH_3.C_6H_4.N_2.C_6H_3(CH_3).NH.C_2H_3O$.

Azo Violet. A dyestuff. It is the sodium salt of dimethoxy-diphenyl-disazo-naphthionic-α-naphthol-p-sulphonic acid, $C_{34}H_{25}N_5O_9S_2Na_2$. Dyes cotton bluish-violet from a soap bath.

Azovskite. A mineral. It is $Fe_3(PO_4)(OH)_6$.

Azo Yellow. See CITRONINE B.

Azo Yellow 3G Extra Conc. See CITRONINE B.

Azo Yellow M. See CITRONINE B.

Azudimidine. A proprietary trade name for Salazosulphadimidine.

Azufrol. Colloidal sulphur.

Azulin. See Azuline.

Azuline (Azulin). A blue compound produced when the essential oil of the Patchouli plant is distilled.

Azuline (Azurine). A basic blue dye-stuff obtained by the action of aniline upon aurine. It is now obsolete.

Azuline Yellow. A colouring matter obtained by nitrating azuline.

Azure Blue. See MOUNTAIN BLUE and COBALT BLUE.

Azuresin. Prepared from carbacrylic cation exchange resin and azure A dye (3-amino-7-dimethylaminophenaza-thionium chloride). Diagrex Blue.

Azurine. See AZULINE.

Azurite. See CHESSYLITE.

Azurmalachite. A mineral. It is a mixture of blue and green copper carbonates.

Azymil. Vitamin preparations. (563).

B

BA 21. A non-heat-treatable alloy of aluminium containing 2·25 per cent. magnesium and 0·3 per cent. manganese. Suitable for pressure vessels up to 204° C. (136).

BA 25. A heat-treatable alloy of aluminium containing 0·7 per cent. magnesium, 1 per cent. silicon and 0·5 per cent. manganese. (136).

BA 27. A non-heat-treatable alloy of aluminium containing 3·5 per cent. magnesium and 0·4 per cent. manganese. Possesses good corrosion resistance but cannot be used for pressure vessels above 66° C.

BA 60. A non-heat-treatable alloy of aluminium containing 1·2 per cent. manganese. (136).

BA 271. A non-heat-treatable alloy of aluminium stronger than BA 21 containing 2·7 per cent. magnesium, 0·8

per cent. manganese and 0·1 per cent. chromium. Otherwise similar to BA 21. (136).

BA 281. A non-heat-treatable alloy stronger than but otherwise similar to BA 27 containing 4·5 per cent. magnesium, 0·7 per cent. manganese and 0·15 per cent. chromium. (136).

B.A.A. A special proprietary sulphuric acid, usually prepared from natural brimstone and distilled water.

Bababurdanite. A mineral. It is $2[Na_{2.5}(Fe'',Mg)_3Fe_{2.5}'''Si_8O_{22}(O,OH)_2]$.

Babal Gum (Gond-babel, Bengal gum, Gond babul, Babul gum). An inferior kind of gum arabic, from *Acacia verek*.

Babassu Cake. A cattle cake prepared from the seeds of the South American palm, *Attalea funifera*.

Babbington's Solution. A solution used for preserving animal tissue. It consists of 1 part wood naphtha and 7 parts water, by volume.

Babbitt's Metals. White bearing metals, which usually contain 83–90 per cent. tin, 7·3–11 per cent. antimony, and 3·7–8·3 per cent. copper. Hard alloys contain 83·3 per cent. tin, 8·3 per cent. antimony, and 5·5–8·3 per cent. copper, whilst an ordinary Babbitt consists of 89 per cent. tin, 7·3 per cent. antimony, and 3·7 per cent. copper. This alloy has a specific gravity of 7·5 and melts at 230° C. Lead often replaces the dearer tin in these alloys, and a marine Babbitt contains 72 per cent. lead, 21 per cent. tin, and 7 per cent. antimony. A soft Babbitt metal contains 84 per cent. tin, 7·4 per cent. antimony, 5·6 per cent. lead, and 3 per cent. copper. Another alloy, also known as Babbitt, consists of 90 per cent. tin and 10 per cent. copper, and still another contains 69 per cent. zinc, 19 per cent. tin, 5 per cent. lead, 4 per cent. copper, and 3 per cent. antimony.

Babel Quartz. A mineral. It is a variety of quartz.

Babesan. Quinuronium sulphate used in veterinary preparations. (512).

Babe's Solution. A mixture of equal parts of a concentrated alcoholic solution and an aqueous solution of safranine. Microscopic stain.

Babingtonite. A mineral, (Ca.Fe.Mn) $O.SiO_2.Fe_2O_3.3SiO_2$. Also described as $2[Ca_2Fe''Fe'''Si_5O_{14}OH]$.

Bablah (Neblah, Web-neb). Commercial names for the fruit of certain acacias, used in the East for calico printing. They contain from 15 to 20 per cent. tannin.

Babool. *Acacia arabica*, the bark of which contains 12 to 20 per cent. tannin. Used considerably as a tanning material.

Babul Bark. Acacia bark.

Babul Gum. See BABAL GUM.

Babulum Oil. Neatsfoot oil (*q.v.*).

Bacchusstein. Synonym for amethyst.

Bacdip. A proprietary preparation of QUINTIOFOS. A veterinary insecticide.

Bacher's Pills. Consist mainly of black helebore. Recommended for dropsy.

Bachite. A specially prepared carbon made from anthracite.

4B Acid. An intermediate used in the production of azo pigments. It is $CH_3.C_6H_3.SO_3HNH_2$. (568).

β-Acid. Anthraquinone-2-sulphonic acid. A dyestuff intermediate.

Bacillarite. Synonym for Leverrierite.

Bacillol. A cresol soap solution, used as a disinfectant.

Bacitracin. An antibiotic produced by a strain of Bacillus subtilis.

Bäckströmite (Baeckstroemite, Pseudopyrochroite). A mineral. It is manganese hydroxide, $Mn(OH)_2$, found in Sweden.

Baclofen. A muscle relaxant. 4 - Amino - 3 - (4 - chlorophenyl)butyric acid β - aminomethyl - *p* - chlorohydrocinnamic acid. LIORESAL.

Bacnelo. A proprietary mixture of 3 per cent. sulphur, 2 per cent. coal tar solution, zinc oxide and calcium hydroxide, used in the treatment of skin disorders. (861).

Bacon Oil. Lard oil, the olein of lard.

Bacterase. Bacterial enzyme or diastase. (550).

Bacterised Peat. A culture of bacteria used to inoculate soils.

Bacterium Aceti. Mycoderma aceti (*q.v.*).

Bacterol (Creolin-Pearson, Cresoleum, Germol, Kresolig, Kreso, Lysol, Saponified cresol, Trikresol). Cresol preparations. Disinfectants.

Bactoform. A neutral soap solution of formaldehyde and coal-tar derivatives. Used for rendering antiseptic the instruments used in surgery.

Bactox. A proprietary antiseptic and disinfectant, made from certain phenolic bodies in a form miscible with water.

Bactrim. A proprietary preparation of trimethoprim, and sulphamethoxazole. An antibiotic. (314).

Bactylan. A proprietary preparation of sodium para-amino salicylic acid. Antituberculous agent. (307).

Badamier Bark (Hatafa Bark). The bark of *Terminalia catappu*. It contains 19 per cent. pyrogallol-tannin.

Badam Kohee. See CHOOLI-KI-TEL.

Baddeckite. A mineral. It is a variety of muscovite.

Baddeleyite (Brazilite, Zircite). A mineral. It consists of zirconium oxide, ZrO_2.

Baden. A Siberian tanning material obtained from *Saxifrage crassifolia*. It is stated to contain 20 per cent. tannin.

Baden Acid (α-Acid). 2 : 8-Naphthylamine-sulphonic acid. See BADISCHE ACID.

Baden Hemp. A fine variety of hemp fibre, prepared by stripping the bast from the retted stalks by hand.

Badenite. A mineral. Possible composition $(Co,Ni)_3(As,Bi)_4$.

Badin Metal. A name used for an alloy of iron with 8–10 per cent. aluminium, 18–20 per cent. silicon, and 4–6 per cent. titanium. Used for adding silicon to steel.

Badional. A proprietary trade name for Sulphathiourea.

Badische Acid. β-Naphthylamine-α-sulphonic acid, $C_{10}H_6(NH_2)(SO_3H)$.

Badouin's Reagent. A 1 per cent. solution of furfural in alcohol. It is used to test for sesamé oil, 5 c.c. of which is shaken with 5 c.c. hydrochloric acid (conc.) and 2 drops reagent. The acid layer becomes rose-coloured if sesamé oil is present.

Baeckstroemite. See BÄCKSTRÖMITE.

Bael (Bela, Bel). The fruit of *Ægle marmelos*.

Baeumlerite. A mineral. It is a potassium-calcium chloride, $KCl.CaCl_2$.

Bagasse. See BEGASSE.

Bagasses Oil. See PYRENE OIL.

Bagotite. A mineral. It is a variety of thomsonite.

Bagrationite (Allanite). A mineral, cerium epidote.

Bahia Arrowroot. Tapioca.

Bahia Bark. See ANGICO.

Bahia Powder. See GOA POWDER.

Bahia Wood. A wood obtained from *Mitragyna macrophylla*, of Gaboon. The dry wood contains 0·6 per cent. fats and waxes, 70 per cent. cellulose, and 28·62 per cent. lignin.

Bahn Metal. A lead base bearing alloy containing copper, 0·7 per cent. calcium, 0·6 per cent. sodium, and 0·04 per cent. lithium.

Baicalite. Synonym for Baikalite.

Baiculein. 5 : 6 : 7-trihydroxy-flavone.

Baierine. Synonym for Columbite.

Baierite. Columbite (*q.v.*).

Baikalite. A mineral. Diopside (*q.v.*).

Baikarite. Synonym for Baikerite.

Baikerinite. A tarry substance separated from Baikerite.

Baikerite. A variety of Ozocerite, a paraffin wax.

Bailey Solder. An alloy containing 70 per cent. tin, 16 per cent. zinc, 10 per cent. aluminium, and 4 per cent. phosphor tin.

Baird's Pills. Compound aloes pills. (*Pilulæ aloes compositæ B.P.C.*)

Baize. A rough woollen material. It has a nap on one side, and is used as a covering material.

Bajuvarin. A yeast preparation used internally and locally in veterinary practice.

Baka Gum. A latex from *Ficus obliqua*. Used as a bird-lime.

Bakalaque. A trade name for a proprietary synthetic resin varnish.

Bakdura. An insulating composition made from synthetic resin, asbestos, pulverised paper, and other materials. See BAKELITE.

Bakelite. The original trade name for synthetic resins of the phenol-formaldehyde type, usually prepared by heating together phenol and formaldehyde with a small amount of an alkaline condensing agent. When heated under pressure, these products become hard and solid. They are used for a variety of purposes, but especially as heat and electric insulators. The following are trade names, many of which are proprietary ones for the resins either in the hardened condition or in the form of press mouldings with the resin as binder. Abalack, Amalith, Amberdeen, Amberite, Ambrasite, Ambroin, Bakdura, Bakelaque, Bucheronium, Bosch-Bakelite, Cérite, Condensite, Dekorite, Durax, Elastolith, Elo, Estralite, Faturan, Formite, Gummon, Herolith, Invelith, Isolierstahl, Issolith, Juvelith, Kopan, Laccain, Marbolith, Melusite, Metakalin, Nuloid, Ornalith, Phenoform, Redmanol, Resan, Ricolite, Sibolite, Sipilite, Tenacite, Tenalan, Trolite, Trolon, Vigorite. The transparent, moulded, and laminated materials are unaffected by organic solvents and dilute acids, but are all affected by alkalis. They stand a continual heat up to 300° F. and soften and char at 500° F. The Bakelite Corporation of America now appear to market polystyrene, urea-formaldehyde, and polybasic plastics under this name as a trade mark. (95).

Bakelite A and B (Resinite). Bakelite (a phenol-formaldehyde resin), soluble in certain solvents. (95).

Bakelite C-9. A proprietary trade name for soft oil modified alkyds. (95).

Bakelite, Continental. See CONTINENTAL BAKELITE.

Bakelite Dilecto. A laminated product consisting of paper or fibre cemented together with a phenol-formaldehyde resin. (95).

Bakelite DQD-3269. A proprietary trade name for an ethylene-vinyl acetate copolymer. A plastics material

which retains its flexibility and toughness at low temperatures. (95).

Bakelite Micarta. A similar product to BAKELITE DILECTO. (95).

Bakerite. A mineral. It is a hydrated boro-silicate of calcium, $8CaO.5B_2O_3.6SiO_2.6H_2O$.

Baker Plasticizers. Coded as follows. (76).

Baker F-1, F-2. A proprietary trade name for vinyl plasticisers. Acetylated castor oil derivations. (76) (104).

Baker P-1C. A proprietary trade name for a vinyl plasticiser. Methyl cellosolve ricinoleate. (76).

Baker P-2C. A proprietary trade name for cellosolve ricinoleate. A vinyl plasticiser.

Baker P-3E. A proprietary trade name for ethyl butyl ricinoleate. A vinyl plasticiser. (76).

Baker P-4. A proprietary trade name for methyl acetyl ricinoleate. A vinyl plasticiser. (76).

Baker P-6. A proprietary trade name for butyl acetyl ricinoleate. A vinyl plasticiser. (76).

Baker P-6C. A proprietary trade name for butyl cellosolve acetyl recinoleate. A vinyl plasticiser. (76).

Baker P-6E. A proprietary trade name for ethyl butyl acetyl ricinoleate. A vinyl plasticiser. (76).

Baker P-8. A proprietary trade name for glyceryl triacetyl ricinoleate. A plasticiser for vinyl polymers and GR/S, neoprene GN, ethyl cellulose, and perbunan.

Baker PG-16. A proprietary trade name for butyl diethyl polyricinoleate. A vinyl plasticiser. (76).

Baker PY-3E. A proprietary trade name for the ethyl butyl ester of soya bean fatty acid. A vinyl plasticiser. (76).

Baker PY-16. A proprietary trade name for butyl diethyl polyricinoleate. A vinyl plasticiser. (76).

Baker's Anæsthetic Ether. Diethyl ether or anæsthetic ether. (507).

Baker's P.8. A proprietary plasticiser for polyvinyl chloride and copolymers of PVC, GR/S, Neoprene GN, ethyl cellulose, and perbunan. It is glyceryl triaceto-ricinoleate.

Baker's P and S Liquid. A proprietary mixture of 1 per cent. phenol, sodium chloride, liquid paraffin and water. (862).

Baker's Salt. Ammonium carbonate, $(NH_4)_2CO_3$.

Bakhar. An enzymic substance prepared from rice, or powdered roots, in

India. Used in the production of Hindu rice beer.

Bakhilite. A trade name for refined Asphaltite.

Baking Powder (Yeast Powder). Usually a mixture of bicarbonate of soda and tartaric acid or cream of tartar. A common one, however, contains 40 per cent. starch. Used as a substitute for yeast, as a chemical generator of carbon dioxide gas, causing the dough to rise.

Baking Soda. Sodium hydrogen carbonate, $NaHCO_3$.

Bakoby Oil. See BAKOLY OIL.

Bakoly Oil (Kekuna Oil). Candlenut oil, from *Aleurites triloba*, of West Indies. " Bakoly oil " is the Madagascar term, and " Kekuna oil " the name it is known by in India. It is used in soap manufacture, and sometimes as a substitute for linseed oil, but it has poor drying properties. The term " Bakoly oil " or " Bakoby oil " is also used in Madagascar for Chinese wood oil (Tung oil).

Bako Nuts. See NJAVE BUTTER.

Bakuin. Russian mineral machine oils, used for lubrication.

Bakuol. A name given by Mendeleef to an illuminating oil, prepared from the crude oils of Baku, by mixing the intermediate oil with kerosene.

Bakurin. A lubricating oil, consisting of 100 parts of crude Baku oil, 25 parts of castor oil, and 60–70 parts of sulphuric acid of 66° B.

Bakurol. See PARAFFIN, LIQUID.

Bakusin. See ACIDOL.

B.A.L. A proprietary preparation of dimercaprol, used to treat heavy metal intoxication. (253).

Balam Tallow. A fat obtained from the seeds of *Palaquium pisang*, of Sumatra.

Balanced Salt Solution. A proprietary solution of 0·49 per cent. sodium chloride, 0·075 per cent. potassium chloride and 0·036 per cent. calcium chloride. (854).

Balas. Oil of Apitong, obtained from the Apitong tree.

Balas Ruby (Spinel Ruby). A mineral. It is spinel (magnesium aluminate, $MgAl_2O_3$), a precious stone.

Balata. The coagulated milky juice of *Mimusops globosa* (the balata or bullet tree) of Venezuela, Dutch, British, and French Guiana, the Amazon, Brazil, Bolivia, and Peru. Panama balata is obtained from *Mimusops darienensis*. The material from British, French, and Dutch Guiana is usually in the form of thin sheets obtained by drying thin layers of the milky juice in the sun. The latex contains 65 per cent. of balata,

and this upon analysis gives approximately 50 per cent. of balata hydrocarbon, 44 per cent. of resin, 4 per cent. of impurities, and 2 per cent. of moisture. Block balata from Panama, Venezuela, and the Amazon districts is made by heating the latex and forming it into blocks. The average analysis of this material gives 39 per cent. of balata hydrocarbon, 43 per cent. of resin, 12 per cent. of impurities, and 6 per cent. of moisture. Venezuelan block balata is also known as Angostura block balata. Peruvian balata is derived from the quinilla tree, and is in two grades : (*a*) Rosada (red), a superior product, and (*b*) Blanca (white), an inferior product, usually a mixture. Balata, when cleaned and purified, is used for a variety of purposes, but mainly as a substitute for gutta-percha in electrical insulation, for transmission belts, and as soles for shoes.

Balbiani's Ointment. An agricultural insecticide. It is prepared by slaking 120 lb. quicklime and melting by means of the heat produced 20 lb. coaltar and 60 lb. naphthalene, and stirring into a paste.

Balbiano's Acid. A dibasic acid, $C_8H_{12}O_5$, obtained by oxidising camphoric acid with permanganate very slowly at ordinary temperatures. It is mixed with small quantities *of camphanic, camphoronic, and trimethylsuccinic acids, and an equivalent amount of oxalic acid. The acid upon reduction with hydriodic acid gave an acid, $C_8H_{14}O_4$, which was identified as *a*-β-β-trimethyl-glutaric acid.

Balchaschit. Synonym for Balkhashite.

Baldaufite. A mineral. It is a phosphate of ferrous iron.

Baldwin's Phosphorus. Fused calcium nitrate, $Ca(NO_3)_2$. When heated it becomes phosphorescent.

Bale Blue. A dyestuff prepared by the condensation of nitroso-dimethyl-aniline with diphenyl - 2 : 7 - naphthylene - diamine. Used on tannined goods.

Balenite. A product intermediate between soft rubber and vulcanite, obtained by heating a mixture of 10 parts caoutchouc, 2 parts shellac, calcined magnesite, sulphur, and antimony sulphide.

Balipramine. A drug used in the treatment of Parkinson's disease. It is 3 - dibenz[*b*, *f*]azepin - 5 - ylpropyldimethylamine. DEPRAMINE.

Balkeneisen. Synonym for Kamacite.

Balkhashite. A mineral. It is a variety of Elaterite.

Ballas. A mineral. It is a variety of Bort.

Ball-bearing Steel. See CHROME STEELS.

Ball Clay. A white plastic clay.

Ballesterosite. A mineral. It is a variety of Pyrite, FeS_2, containing zinc and tin.

Ballistite. A trade mark for a smokeless powder containing nitro-glycerin and collodion cotton.

Balm Drops. Friar's balsam (q.v.).

Balm of Gilead. Mecca balsam, an oleo-resin, the product of *Balsamodendron gileadense.*

Balm Oil. See OIL OF BALM.

Balmex Medicated Lotion. A proprietary protective lotion of lanolin, hexachlorophene, allantoin, balsam Peru and silicone oil in a non-mineral oil base. (863).

Balmosa. A proprietary preparation of sodium iodide, menthol, camphor and methyl salicylate. An embrocation. (366).

Bal-nela. An alloy of 28 per cent. nickel, 66 per cent. copper, with 6 per cent. manganese, silicon, iron, and zinc.

Balnetar Liquid. A proprietary preparation used to treat skin disorders. It is a mixture of 2·5 per cent. crude coal tar, a lanolin derivative, mineral oil and a non-ionic emulsifier. (864).

Balsam, Canada. See CANADA TURPENTINE.

Balsam, Carpathian. Riga Balsam (q.v.).

Balsam, Copalm. See LIQUIDAMBAR.

Balsam, Earth. See EARTH OIL.

Balsam, Friar's. See TURLINGTON BALSAM.

Balsam, Litaner. See DAGGET.

Balsam of Capivi. See OLEO-RESIN COPAIBA.

Balsam of Captivo. A balsam resin, used as a substitute for copiaba.

Balsam of Cascara (Balsam of Tacuasonte). An inferior grade of Balsam of tolu.

Balsam of Copaiba (Balsam of Capivi). An oleo-resin from the tree *Copaifera landsdorfi* and others. It contains 41–62 per cent. oil, has a specific gravity of 0·92–0·99, and an acid value of 33–36. Used in medicine. It is often adulterated with a balsam called African balsam of copaiba or gurgun balsam.

Balsam of Copaiva. See OLEO-RESIN COPAIBA.

Balsam of Fern. The liquid extract of male fern.

Balsam of Fir. See CANADA TURPENTINE.

Balsam of Life. Compound decoction of aloes.

Balsam of Peru. An oleo-resin obtained from the bark of *Myroxylon pereinæ.* It has a specific gravity of 1·14–1·16, an acid value of 27–77, a saponification value of 215–243, and a refractive index of 1·48–1·4855. It contains from 55–65 per cent. cinnamein (a mixture of the esters of cinnamic and benzoic acids). It is antiseptic, and is used in medicine and perfumery.

Balsam of Soap. Soap liniment.

Balsam of Storax. Prepared storax.

Balsam of Sulphur. Consists of 1 part sulphur and 4–9 parts olive oil heated together.

Balsam of Tacuasonte. See BALSAM OF CASCARA.

Balsam of Tolu. The product of the tree *Myroxylon toluifera.* A good variety has 90 per cent. soluble in alcohol and 3–8 per cent. in petroleum ether. It has an acid value of 105–140 and a saponification value of 170–202. Used in medicine and in perfumery.

Balsams. Resins containing benzoic or cinnamic acids.

Balsam Turpentine. The term applied to the natural or artificial exudation of living conifers. Also see WOOD TURPENTINE.

Balsam Wool. A heat insulator made from pulp-wood bark and sulphite screenings.

Balsam, Xanthorrhœa. See ACAROID BALSAM.

Balsan. Proprietary name for a specially purified preparation of balsam Peru. (863).

Balsan-Katel. See MECCA BALSAM.

Baltic or Prussian Amber. The best kind of amber.

Baltimore Chrome Yellow. See CHROME YELLOW.

Baltimorite. A mineral. It is a silicate of aluminium, calcium, and magnesium. See CHRYSOTILE.

Balvraidite. A mineral. It is an Aluminosilicate of magnesium, calcium and potassium.

Bamate. A proprietary name for MEPROBAMATE. A tranquilliser. (865).

Bambuk Butter. See SHEA BUTTER.

Bamethan. 2 - *n* - Butylamino - 1 - (4-hydroxyphenyl) ethanol. Vasculit is the sulphate.

Bamethan Sulphate. A vasodilator. It is o - {(butylamino)methyl} - *p* - hydroxybenzyl alcohol sulphate.

Bamifylline. 8 - Benzyl - 7 - [2 - (N - ethyl - 2 - hydroxyethylamino)ethyl] theorphylline. Trentadil is the hydrochloride.

Bamipine. 4 - (N - Benzylanilino) - 1 - methylpiperidine. Solventol.

Bamlite. A mineral. It is a variety of sillimanite.

Bamo. A proprietary name for MEPRO-BAMATE. A tranquilliser. (866).

BA-Mordenite. A variety of Mordenite of composition $4(BaAl_2Si_{10}O_{24}.7H_2O)$.

Banacid. A proprietary antacid. A mixture of aluminium hydroxide, magnesium hydroxide and tribasic calcium phosphate. (868).

Banalg. A proprietary preparation of camphor, eucalyptus oil, menthol and methyl salicylate in a greaseless base, used as a counter-irritant for external use. (869).

Banalsite. A mineral. It is $4(Na_2BaAl_4Si_4O_{16})$.

Banana Oil (Bronzing Liquid). A solution of pyroxylin in amyl acetate. Used with a bronze powder, usually aluminium bronze. Banana oil is also the name given to amyl acetate, $CH_3COOC_5H_{11}$, a solvent for lacquers and varnishes.

Banana Rubber (Musa Rubber). A rubbery material contained in green bananas to the extent of about 3 per cent. An investigation by H. von Loesecke (*Rubber age*, N.Y., 22, 129), shows the material to be more closely related to gum chicle than to rubber. It has no technical importance.

Bandalasta. A proprietary trade name for an urea-formaldehyde plastic.

Bandeisen. Synonym for Tænite.

Bandjaspis. Synonym for Ribbon Jasper.

Bandolin (Fixature, Clysphitique). A thick mucilage of algæ (Carrageen, Irish, or pearl moss), scented with a spirit, and used for stiffening silk.

Bandrowski's Base. Tetra-amino-diphenyl-*p*-azo-phenylene.

Bandylite. A mineral. It is $2(CuBO_2Cl.2H_2O)$.

Banilloes. Vanilla pods.

Banistyl. A proprietary preparation of dimethothiazine mesylate. (336).

Banket. A matrix of gold. It is a hard, compact conglomerate, consisting of pebbles of quartz, cemented together by a siliceous material.

Bankoil. A variety of menhaden oil. It is rich in fatty acids, and is used in the oil-tanning of skins to make a chamois leather.

Bank Paper. Usually a high-grade rag and chemical wood pulp writing-paper.

Banks Oil. Cod-liver oil.

Banocide. A proprietary preparation of diethylcarbamazine citrate. Filaricide. (277).

Banrock. A proprietary trade name for a rock wool (*q.v.*) made from high silica limestone.

Bantex. Zinc mercaptobenzothiazole. A rubber accelerator. (559).

Bantex DN. A proprietary rubber accelerator. It is zinc mercaptobenzthiazole + diphenyl guanidine. (559).

Banthine Bromide. A proprietary trade name for Methanthelinium Bromide.

BA-Priderite. A mineral. It is a variety of priderite.

B.A.R. A proprietary trade name for butyl acetyl ricinoleate. A vinyl plasticiser. (77).

BAR-15, BAR-25, BAR-30, BAR-100. Proprietary tablets (containing respectively 15, 25, 30 and 100 mg. phenobarbital each) used as anti-convulsants, hypnotics and sedatives. (754).

Bararite. A mineral. It is the hexagonal, low temperature phase of $[(NH_4)_2SiF_6]$.

Baratol. A proprietary trade name for Trimethidinium Methosulphate.

Barbadoes Nuts. Purging nuts from *Jatropha curcas*.

Barbadoes Tar. See GREEN TAR.

Barbarit. A German explosive containing potassium chlorate and mineral oil.

Barbasprin. A proprietary preparation of allobarbitone and aspirin. (250).

Barbatimao Bark. A tanning material. It is the bark of *stryphnodedron barbatimao* of Brazil, and contains 24–26 per cent. tannin.

Barbenyl. Phenobarbital.

Barberite. A corrosion-resisting alloy containing 88·5 per cent. copper, 5 per cent. nickel, 5 per cent. tin, and 1·5 per cent. silicon. It is stated to have a high tensile strength in addition to good corrosion-resisting properties.

Barberry Root. From *Berberis vulgaris*. It contains a natural dyestuff berberine, $C_{20}H_{17}NO_4.4H_2O$. Used for dyeing leather and silk yellow.

Barbertonite. A mineral. It is $[Mg_6Cr_2CO_3(OH)_{16}.4H_2O]$.

Barbierite. A mineral. It is a soda-felspar.

Barbital. See VERONAL, VESPERAL, URONAL, SEDEVAL MALONAL, HYPNOGENE, DORMONAL, DEBA, BARBITONE. All are sedatives and hypnotics.

Barbidex. A proprietary preparation of dexamphetamine resinate, and phenobarbitone. Antiobesity agent. (271).

Barbitone. See VERONAL.

Barbitone, Soluble. Veronal sodium (*q.v.*).

Barbœstryl. A proprietary preparation of ethynylœstradiol, phenobarbitone and papaverine. Used to treat menopausal disturbances. (307).

Barbosalite. A mineral of the Lazulite group. It is $2[Fe''Fe_2'''(PO_4)_2(OH)_2]$.

Barbouze's Alloy. An alloy of aluminium with 10 per cent. tin.

Barbul Bark (Barbura Bark). Acacia.

Barbura Bark. See BARBUL BARK.

Barc Cream. A proprietary preparation of isobornyl thiocyanoacetate and other terpenes with dioctylsodium sulphosuccinate, used for combating head lice, nits and crabs. (871).

Barcenite. A mineral, $3HgO.Sb_2O_3$.

Bardase. A proprietary preparation containing phenobarbitone, hyoscyamine sulphate, hyoscine hydrobromide, atropine sulphate and aspergillus oryzæ enzymes. Gastric sedative. (264).

Bardiglio. A variety of calcite.

Bardiglione. Synonym for Anhydrite.

Bardol. A new dispersing oil for rubber compounding. It is a coal-tar product having a specific gravity of 1·08 and a flash-point of 212° F.

Bardolite. A mineral found at Barda, in Poland. Its formula approximates to $(H.K)_3(Fe . Al)_3Si_3O_{12}.3H_2MgSiO_4.6\frac{1}{2}H_2O$.

Barege. Sulphurated potash (q.v.).

Barettite. A mineral. Its approximate composition is $(Ca,Mg,Fe)_6Si_5O_{16}\frac{1}{2}H_2O$.

Barff Boroglycerin. A saturated solution of boric acid in glycerin. A preservative for animal and vegetable specimens.

Barfoed's Reagent. Copper acetate, 1 part, is dissolved in water, 15 parts. To 200 c.c. of this solution, 50 c.c. of 68 per cent. acetic acid are added. Used to distinguish glucose and other monosaccharides from disaccharides, such as lactose and maltose. With glucose red cuprous oxide is produced.

Baribux. Waterproofed ground limestone. (512).

Baricalcite. A mineral. It is a variety of calcite containing some barium carbonate.

Baridur. Barium carbonate. (571).

Bariform. Barium sulphate for X-ray diagnosis. (513).

Barilith. Barium green, yellow and blue pigments. (571).

Barilla (Barillor). The commercial name for an impure sodium carbonate prepared from the ash of certain plants (soda). Kelp is sometimes called British barilla.

Barilla, Copper. See COPPER BARILLA.

Barillor. See BARILLA.

Bariohitchcockite. Synonym for Gorceixite.

Bari Oil. See AIX OIL.

Bario Metal. An alloy containing 90 per cent. nickel, 1·22 per cent. tungsten, 0·29 per cent. silicon, and 4·25 per cent. chromium. It is stated to be acid and heat-resisting. A soft Bario metal is stated to contain 60 per cent. cobalt, 20 per cent. chromium, and 20 per cent. tungsten, and a hard one 30 per cent. cobalt, 30 per cent. chromium, 25 per cent. tungsten, 10 per cent. manganese, and 5 per cent. titanium.

Bario-Muscovite. Synonym for Œllacherite.

Bario-Phlogopite. Synonym for Barium Phlogopite.

Bariostrontianite. Synonym for Stromnite.

Barisol. A water repellent for concrete. (571).

Baritite. Synonym for Baryte.

Barite (Bolognian spar, Michel-Levyite). A mineral. It is barium sulphate, $BaSO_4$.

Barium-Anorthite. A mineral. Its composition is $BaCa_2Al_8Si_8O_{28}2H_2O$.

Barium Chrome. See BARIUM YELLOW.

Barium Aragonit. Synonym for Alstonite.

Barium Autunite. Synonym for Uranocircite.

Barium-Hamlinite. A mineral. It is $BaAl_4(PO_4)_2(OH)_8.H_2O$.

Barium-Heulandite. A mineral. It is $[(Na,Ca,Ba)_4Al_6(Al,Si)_4Si_{23}O_{72}.24H_2O]$.

Barium-Hydroxylapitite. A mineral. It is $2[Ba_5(PO_4)_3OH]$.

Barium Muscovite. Synonym for Œllacherite.

Barium-Natrolite. An artificial base exchange product. Its composition is $2[BaAl_2Si_8O_{10}.nH_2O]$.

Barium-Nepheline. An artificial barium aluminosilicate of composition $BaAl_2Si_2O_8$.

Barium Orthoclase. A mineral. It is $4[(KSi,BaAl)AlSi_2O_8]$.

Barium-Phlogopite. A mineral. It is $4[KMg_3AlSi_3O_{10}(OH)_2]$ with about 1 per cent BaO.

Barium-Phosphoruranite. Synonym for Uranocircite.

Barium Sanidine. A mineral. It is a variety of sanidine with about 5 per cent. BaO.

Barium Nickelite. A term used for a compound of nickel oxide and barium oxide, $BaO.2NiO$, formed when nickel oxide, Ni_2O_3, is heated with barium carbonate, $BaCO_3$, in the electric furnace.

Barium Parisite. See CORDYLITE.

Barium Rat Poison. This usually contains 32·25 per cent. barium carbonate, 32·25 per cent. sugar, 32·25 per cent. oatmeal, and 3·25 per cent. aniseed. It is harmful to other animals.

Bariumuranite. See URANOCIRCITE.

Barium White (Artificial heavy spar, Baroxlenite, Terra ponderosa, Permanent White, Mineral White, New White, Snow White, Blanc Fixe, Constant White, Barytes White, Baryta White). A pigment. It is barium sulphate, $BaSO_4$.

Barium Yellow (Barium Chrome, Chrome Yellow, Lemon Chrome, Lemon Yellow, Ultramarine Yellow, Yellow Ultramarine). A pigment. It is barium chromate, $BaCrO_4$.

Bark. See QUERCITRON.

Bark, Australian. See ALSTONIA.

Bark, Bahia. See ANGICO.

Bark, Bitter. See ALSTONIA.

Bark Eleuthera. See SWEET BARK.

Barkevikite. A mineral. It is a variety of amphibole.

Barkite B. Di(dimethylcyclohexyl)oxalate. A plasticiser for cellulose lacquers. (513).

Bark Liquor. The aqueous extract of quercitron bark, used in calico printing.

Barklyite. A mineral. It is a variety of corundum.

Bark Rubber. The rubber recovered from the strip of bark removed in tapping. It is obtained from the inner part of the bark, also from the residue of rubber remaining on the bark after removal of scrap rubber. It is usually finely ground under a stream of water, the bark particles floated away and the rubber pressed together. It has a lower value than scrap rubber.

Bark, Whitewood. See CANELLA.

Barley Gum. A gum obtained in the nitrogen-free extractive material of cereals.

Barley Sugar. Made by heating cane sugar with water, then boiling off the added water. There is a change from the crystalline to the vitreous condition.

Barm. The yeasty top formed on fermenting beer, used as leaven for bread.

Barnhardtite. A mineral. It is a sulphide of copper and iron.

Barol. An oil prepared from coal-tar. It is antiseptic, and is used for preserving timber.

Barolite. Witherite (q.v.).

Baros. A heat-resisting alloy containing 90 per cent. nickel and 10 per cent. chromium.

Baros Camphor. See CAMPHOL.

Baroselenite. Barium white (q.v.).

Bar-o-Sil. A barium based stabilizer for PVC. (572).

Barosil. A proprietary preparation of zinc oxide and dimethicone in a cream base. A barrier cream. (328).

Barosperse. A proprietary preparation of barium sulphate. (819).

Barotrast. A proprietary preparation of barium sulphate. (873).

Barquinol H.C. A proprietary preparation of hydrocortisone and clioquinol. An antibacterial dermatological agent. (188).

Barracanite. A mineral, $CuFe_2S_4$.

Barrandite. A Nevada mineral, $(Fe.Al)_2O_3.P_2O_5.4H_2O$.

Barras. A pine resin.

Barreswill's Solution. Fehlings' solution (q.v.).

Barroisite. A mineral. It is a variety of amphibole. Synonym for Carinthine.

Barronia Metal. An alloy of from 80–83 per cent. copper, 11–17 per cent. zinc, 4–4·5 per cent. tin, 0·4–1·15 per cent. iron, and 0·5–0·65 per cent. lead.

Barsilowsky's Base. Amino-ditolyl-p-toluquinone-di-imine.

Barsovite. A mineral. It is an aluminium-calcium silicate.

Barsowite. Synonym for Anorthite.

Barthite. A mineral, $3Zn(AsO_3)_2.Cu(OH)_2.H_2O$.

Bartholomite. A mineral. It is a sulphate of iron and sodium.

Barth's Blue. Indigo sulphate.

Baru. A cotton-like fibre obtained from a sago palm in India. It is used as a substitute for tow for caulking boats.

Barutin. The double salt of barium-theobromine and sodium salicylate. Used medicinally. It is the corresponding barium salt to Diuretine.

Barwood. See REDWOODS.

Barwood Spirits. See TIN SPIRITS.

Barylite. A mineral, $3BaO.SiO_2$.

Barysaltpeter. Synonym for Nitrobarite.

Barysilit. A mineral, $3PbO.2SiO_2$.

Baryta. Barium monoxide, BaO.

Baryta, Caustic. See CAUSTIC BARYTA.

Baryta Felspar. A mineral. It is Hyalophane, $BaO.K_2O.Al_2O_3.8SiO_2$.

Baryta Green. See ROSENTHIEL'S GREEN.

Baryta Mixture. A mixture of 1 volume of a saturated solution of barium nitrate and 2 volumes of a saturated solution of barium hydrate.

Barytapatite. A mineral, $3Ba_3(PO_4)_2BaCl_2$.

Baryta Saltpetre. A mineral. It is barium nitrate, $Ba(NO_3)_2$.

Baryta Water. Made by heating 100 grams barium hydrate, $Ba(OH)_2.8H_2O$ in 1,500 c.c. water, filtering hot, allowing solution to cool, and pouring off the baryta water. It is put into a bottle with a trap containing moist soda lime to prevent the entrance of carbon dioxide.

Baryta White. See BARIUM WHITE.

Barytbiotite. A mineral. It is $2[(K,Ba)_2(MgAl)_{4-6}(AlSi)_8O_{20}(OH)_4]$.

Barytes (Heavy Spar, Heavy White). A mineral. Barium sulphate, $BaSO_4$. Specific gravity, 3·9–4·53.

Barytes, Cockscomb. See CRESTED BARYTES.

Barytes White. See BARIUM WHITE.

Baryt-Flusspath. Synonym for Fluobaryt.

Barytglimmer. Synonym for Œllacherite.

Baryt-Hedyphane. A mineral. It is $2[(Pb,CaBa)_5(AsO_4)_3Cl]$.

Barytocalcite. See ALSTONITE.

Barytocelestine (Barytocelestite). A mineral. It consists of the mixed sulphates of barium, strontium, and calcium.

Barytocelestite. See BARYTOCELESTINE.

Barytophyllite. See CHLORITOID.

Barytsaltpeter. Synonym for Nitrobarite.

Basafrit. A name applied to a dolomite.

Basalt. A felspathic rock containing crystals of augite (q.v.). Used as a building material.

Basaltic Hornblende. A mineral. It is $2[(Ca,Na,K)_{2-3}(MgFe'''Al)_5(SiAl)_8O_{22}(O,OH)_2]$.

Basaltine. Synonym for Augite.

Basaltkainit. See ANHYDROKAINITE.

Basaluminite. A mineral. It is $Al_4SO_4(OH)_{10}.5H_2O$.

Basanite. A mineral. It is a variety of jasper, used for the streak plates for testing gold alloys. Synonym for Lydian stone.

Basanomelan. A mineral. It is a variety of menaccanite.

Base Bullion. Argentiferous lead containing up to 0·1 per cent. silver.

Base Metal. A name applied to the commoner metals as contrasted with the rare ones.

Base, Millon's. See MILLON'S BASE.

Base Oil. See BLOWN OILS.

Basex. A proprietary base-exchange material for water softening. It has an approximate composition $Na_2Al_2O_3$ $14SiO_2$. It is stated to have high softening capacity.

Basic Bricks. Bricks composed of magnesia, or magnesia and lime. Employed for lining furnaces.

Basic Chloride. See POWDER OF ALGAROTH.

Basic Cinder. See BASIC SLAG.

Basic Dyes. Dyestuffs containing salts of colourless bases which contain chromophoric group. They require mordants for cotton, but dye animal fibres direct.

Basic Heliotrope B. A dyestuff. It is a British equivalent of Tannin heliotrope.

Basicin (Corticin). A mixture of quinine hydrochloride and caffeine. Employed externally for malaria, and internally for neuralgia and influenza.

Basic Lead Sulphate. See BLUE BASIC LEAD SULPHATE.

Basic Phosphate. See BASIC SLAG.

Basic Pig. A pig iron made for use in the basic Bessemer process. It contains very little silicon (1–1·5 per cent.), and from 1–3 per cent. phosphorus.

Basic Slag (Basic Cinder, Basic Phosphate, Belgian Slag, Thomas Meal, Thomas Phosphate, Thomas Slag). A by-product of the Thomas-Gilchrist process for the conversion of phosphatic pig-iron into steel. It consists mainly of tetracalcium phosphate, $Ca_4P_2O_9$, calcium silicate, lime, and ferric oxide. It usually contains from 25–40 per cent. phosphate, and is used as a fertiliser. Also see SLAG A.

Basilicon (Resin Cerate, Basilicon Ointment). A mixture of oil, wax, and resin.

Basilicon Ointment. See BASILICON.

Basilite. A wood preservative. It consists of 89 per cent. sodium fluoride and 11 per cent. aniline and dinitrophenol. It is also the name for a mineral, $11(Mn_2O_3.Fe_2O_3).Sb_2O_5.21H_2O$.

Basilüte. A mineral. It is an antimonate of manganese.

Basinetto Silk. See GALETTAME SILK.

Basle Blue. A safranine dyestuff. It is tolyl-dimethyl-amino-pheno-tolyl-imino-naphthazonium chloride, $C_{32}H_{29}N_4Cl$.

Basle Blue BB. A dyestuff, which is the compound corresponding to Basle blue R, only obtained from nitrosodiethyl-aniline. Dyes cotton mordanted with tannin and tartar emetic blue from a neutral bath.

Basle Blue R. A dyestuff. It is dimethyl-amino-tolyl-amino-tolyl-pheno-naphthazonium chloride, $C_{32}H_{29}N_4Cl$. Dyes cotton mordanted with tannin and tartar emetic, blue.

Basle Blue RS. An acid dyestuff. It is a sulphonated derivative of Basle blue R.

Basle Blue S. A dyestuff. It is the sodium salt of Basle blue sulphonic acid. Dyes wool and silk blue from an acid bath.

Basobismutite. A mineral. It is a basic bismuth carbonate.

Basofor. A proprietary trade name for a specially precipitated barium sulphate.

Basolit. A cement for porcelain.

Bassanite. A mineral. It is calcium sulphate.

Bassetite. A uranium mineral found in Cornwall.

Bassio Tallow (Mohrah Butter, Illipé Butter). The fat from the seeds of *Bassia latifolia* and *B. longifolia*.

Bassoba Nuts (Caco Babassu Nuts, Curcia Nuts, Coquilho Nuts, Coquilla Nuts, Cognito Nuts, Vaua-assu Nuts). The fruits of a species of *Attalea*, probably *A. funifera*. The kernels contain 65–68 per cent. of Babassu-kernel oil.

Bassora Gum (Caramania Gum, Hog Gum, India Gum, False Tragancanth, Kutera Gum). A gum said to be the product of plum and almond trees, from Bassora.

Bassorin (Adraganthin). Tragacanthin, $C_{12}H_{20}O_{10}$, the mucilage from gum tragacanth.

Bast. A fibre obtained from certain trees, the lime-tree being a particular type for producing this material. Jute, flax, and hemp are bast fibres of other plants. Used for making mats.

Bastanet. A decolorising carbon.

Bastard Asbestos. Picrolite, a fibrous variety of serpentine, is known as bastard asbestos.

Bastard Saffron. See SAFFLOWER.

Bast Hemp. Hemp fibre produced by breaking is known by this name.

Bastinite. A mineral. It is a phosphate of lithium with some manganese and iron.

Bastite. See SCHILLERSPAR.

Bastnasite (Hamartite). A mineral, $[(Cd.La.Di)F]CO_3$.

Bastonite. A mineral. It is a variety of mica.

Bastose. The cellulose of jute.

Basudin. An insecticide. A proprietary preparation containing DIMPYLATE.

Batana Oil. See PATAVA OIL.

Batavian Dammar. See DAMMAR RESIN.

Batavite. A mineral. It is a silicate of aluminium and magnesium.

Batchelorite. A mineral, $Al_2O_3.2SiO_2. H_2O$, of Tasmania.

Batchite. An activated carbon produced from coco-nut shells.

Bate's Alum Water. Made by dissolving 1 oz. alum and 1 oz. zinc sulphate in 48 oz. water and filtering.

Bath Brick. A silico-calcareous deposit found in the estuary at Bridgwater, and other places. Used as an abrasive.

Bathite. A trade mark for articles made as abrasives and refractories. The essential constituent is crystalline alumina.

Bath Metal. A brass containing 83 per cent. copper and 17 per cent. zinc. Another alloy consists of 55 per cent. copper and 45 per cent. zinc.

Bath Stone. A calcium carbonate stone, used for building purposes.

Bathvillite. Name of a naturally occuring wood resin.

Batist. A cotton material impregnated on one or both sides with rubber. Used for antiseptic dressings.

Batrachite. A mineral. It is a variety of monticellite.

Batterium Metal. A high copper alloy containing 89 per cent. copper, 9 per cent. aluminium, and 2 per cent. nickel, etc.

Battery Acid. Sulphuric acid and potassium bichromate.

Battery Copper. A brass containing 94 per cent. copper and 6 per cent. zinc.

Battery Manganese. See MANGANESE BLACK.

Battery Plate Metal. Various. They usually contain from 73–89 per cent. copper, 2–18 per cent. zinc, and 0–22 per cent. tin.

Battery Plates. Various. One alloy contains 63·5 per cent. zinc, 21·5 per cent. tin, 12 per cent. lead, and 3 per cent. copper ; and another consists of 94 per cent. lead and 6 per cent. antimony.

Batu Gum. An East Indian Dammar.

Baudische's Reagent. Cupferron (q.v.).

Baudoin's Metal. Complex nickel silvers. One contains 72 per cent. copper, 16·6 per cent. nickel, 1·8 per cent. cobalt, 2·25 per cent. tin, and 7·1 per cent. zinc, and another 75 per cent. copper, 16 per cent. nickel, 2·25 per cent. zinc, 2·75 per cent. tin, 2 per cent. cobalt, 1·5 per cent. iron, and 0·5 per cent. aluminium.

Bauer Oil. The high-boiling residue from molasses fusel oil. It is found on the lower plate of the column of the continuous beer still, and contains the ethyl esters of capric acid, $C_{10}H_{20}O_2$, and lauric acid, $C_{12}H_{24}O_2$.

Baume's Flux. A mixture of 6 parts potassium nitrate, 2 parts sulphur, and 2 parts sawdust. It will deflagrate with sufficient intensity to fuse a small silver coin into a globule.

Baumhauerite. A mineral, $4PbS. 3As_2S_3$.

Baum's Acid. See SCHAEFFER'S ACID.

Baurach. Synonym for Borax.

Bauxite. A mineral. It consists mainly of hydrated alumina, $Al_2O_3.2H_2O$, mixed with silica and iron hydroxide, and is the most important source of aluminium.

Bavalite. A mineral. It is an aluminium-iron silicate.

Bavarian Blue DBF. See METHYL BLUE.

Bavarian Blue DSF (Methyl Blue, Water Soluble, Navy Blue B, Methyl Blue for Silk MLB, Marine Blue B). A dyestuff. It is the sodium salt of triphenyl - pararosaniline - disulphonic acid, $C_{37}H_{27}N_3O_6S_2Na_2$, with some trisulphonic acid. Dyes silk blue from a soap bath.

Bavarian Blue, Spirit Soluble. See DIPHENYLAMINE BLUE, SPIRIT SOLUBLE.

Bavarite. An explosive containing nitrated solvent naphtha.

Bave. Double silk fibres as existing in the cocoon. See BRINS.

Bavenite. A mineral. It is a calcium-aluminium silicate.

Bavon D. A leather-waterproofing agent consisting of $C18$ to $C36$ dibasic acids solubilised at 50 per cent. solids, which is fully water soluble, enabling water-proofing treatments to be made during normal tannery procedure. Applicable to a wide range of grain and suede leathers. (569).

Bayate. A local name for a ferruginous jasper found in Cuba.

Bay-berries. The fruit of *Laurus nobilis.*

Bay - berry Camphor. Laurin, $C_{22}H_{30}O_3$.

Bayberry Tallow. See LAUREL WAX.

Baycaine. A proprietary trade name for the hydrochloride of Tolycaine.

Baycaron. A proprietary preparation of metruside. A diuretic. (112).

Bayee Balsam. A gum-resin obtained from *Balsamodedron pubescens.* A myrrh substitute.

Bayer 205. See FOURNEAU 309.

Bayerite. Aluminium oxide, $Al_2O_3.H_2O$.

Bayer's Acid (Cassella's Acid F). β-naphthylamine-δ-sulphonic acid (2 : 7).

Bayer's Acid (Rumpff's Acid, Croceinic Acid, Crocic Acid, Crocein-sulphonic Acid). β-Naphthol-sulphonic acid, $C_{10}H_{16}(OH)(SO_3H)$ (2 : 8).

Bayer's Hypo Destroyer. Potassium percarbonate+potassium persulphate.

Bayko Metal Yarn. Consists of a core of cotton, silk, or other thread, which is coated with cellulose acetate solution containing finely divided metals. Used as a trimming material for fabrics.

Bayldonite. A mineral, $(Cu.Pb)AsO_4.$ $CuOH.\frac{1}{2}H_2O$.

Bayleyite. A mineral. It is $4[Mg_2UO_2(CO_3)_3.18H_2O]$.

Bayliss's Solution. Contains sodium chloride and gum acacia in water. Used for intravenous injection.

Bay Oil. See OIL OF BAY.

Bayolin. A proprietary preparation of heparinoid, glycolester of monosalicylic acid and benzylester of nicotinic acid. Local analgesic cream. (341).

Bay Rum (Spirit of Myrcia). Made by distilling rum with the leaves of *Myrcia acris,* and other plants, or by dissolving their oil in alcohol.

Bay Salt. A coarser grained variety of salt than fishing salt, NaCl, produced by the evaporation of sea water.

Bay Wood. A mahogany from Honduras.

Bazzite. A mineral. It is a silicate of scandium with cerium, yttrium, iron, and sodium.

BB Accelerator. A rubber vulcanisation accelerator. It is a derivative of dimethyl-p-phenylene-diamine. (102).

BB Alloys. A range of corrosion and acid resisting stainless steels comprising, BBOK containing 18 per cent. chromium, 8 per cent. nickel and $1\frac{1}{2}$ per cent. molybdenum; ; BB2K containing 18 per cent. chromium, 8 per cent. nickel and 2 per cent. molybdenum ; BB4K containing 18 per cent. chromium, 8 per cent. nickel and 3 per cent. molybdenum ; and BBMK containing 18 per cent. chromium, 82 per cent. molybdenum and 2 per cent. copper. (135).

B.B.D.C. Standard Alloy. An alloy of 88·5 per cent. copper, 10 per cent. tin, 1 per cent. nickel, 0·25 per cent. lead, and 0·25 per cent. phosphorus.

B.C. 500. A proprietary vitamin supplement containing thiamine, riboflavine, pyridoxine, calcium partothenate, cyanocobalamin and ascorbic acid. (467).

B.C.G. Bacillus Calmette-Guerin. An attenuated strain of *Mycobacterium bovis* used in the prophylaxis of tuberculosis in man.

β-Corlan. A proprietary preparation of betamethazone for use in local oral ulceration. (335).

B.D.C. A mixture of phenyl arsenious oxide, C_6H_5AsO and phenyl arsenious chloride, $C_6H_5As.Cl_2$.

Bdellium. A gum resin resembling myrrh. There are African and East Indian varieties obtained from various species of *Commiphora* and *Balsamodedron.* An aromatic. The African variety has an acid value of 10–20, and the East Indian 35–37.

Beaconite. A mineral. It is a variety of talc.

Beacon Red. A proprietary trade name for a red lake using an aluminium base.

Beading Oil. A mixture of almond oil and ammonium sulphate. Used to produce beading in distilled liquids.

Beale's Carmine (Beale's Stain). A microscopic stain. It contains 1 gram carmine, 1·5 c.c. solution of ammonia, 80 c.c. glycerin, 25 c.c. water, and 120 c.c. alcohol.

Bean Ore. A variety of Limonite.

Bean's Alloy. An alloy used in dental work. It contains 95 per cent. tin and 5 per cent. silver.

Bean Shot Copper (Feather Shot Copper). Varieties of copper for brass-making. They are prepared by pouring the molten metal into hot or cold water.

Bearing Bronze. See BRONZE BEARING METAL.

Bearing Metal, Krupps. An alloy of 87 per cent. aluminium, 8 per cent. copper, and 5 per cent. tin.

Bearing Metals (Antifriction Metals, White Metals). There are many varieties of these alloys. Some of these have lead in the highest proportion with tin and antimony as the other main constituents. They contain from 42–86 per cent. lead, 7–20 per cent. antimony, 2–42 per cent. tin, and 0–2 per cent. copper. Some have tin as the predominating metal with copper and antimony as the other ingredients. These have tin from 38–91 per cent., antimony from 4–26 per cent., and copper from 2–21 per cent. Another type have copper in high proportion with zinc, tin, and sometimes lead in addition. Such alloys contain 72–90 per cent. copper, 0–22 per cent. zinc, 2–20 per cent. tin, and 0–15 per cent. lead. Finally, a smaller class contain 47–90 per cent. zinc, 1–38 per cent. tin, 0·5–8 per cent. copper, 0–4 per cent. lead, and 0–6 per cent. antimony. The constituents of some varieties of bearing metal are listed below.

Bearium. Proprietary alloys of copper with 17·5–28 per cent. lead and 10 per cent. tin. Bearing metal.

Bears. Infusible masses of a titanium compound formed in the blast furnace during the smelting of iron ores containing titanium oxide.

Beatin. A trade mark used in Germany for the sale of Famel Syrup (*q.v.*). (228).

Beaudouin's Reagent. A 1 per cent. furfural in alcohol.

Beaumontique. A wood stopping composed of shellac, rosin, and beeswax. It is coloured to suit the wood.

Beaumontite. A mineral. It is a variety of heulandite.

Beaverite. A mineral, $CuSO_4.PbSO_4.(Fe.Al)_2O_3.4H_2O$, from Utah.

Beaver Steel. A proprietary nickel-chromium steel containing 1·5 per cent. nickel, 0·75 per cent. chromium, 0·6 per cent. manganese, and 0·55 per cent. carbon.

Bebate. A proprietary preparation containing betamethasone 17-benzoate. A steroid skin preparation. (262).

Bebeeru Bark. The bark of *Nectandra rodæi*.

Bébé Oil. Inukaya oil, obtained from the seeds of *Cephalotaxes drupacea*.

Bebirine. Bebeerine, $C_{19}H_{21}NO_3$.

Bebrit. A proprietary moulding compound.

B.E.C. A proprietary trade name for dibutoxy ethyl diglycol carbonate. A vinyl plasticiser. (78).

Becantyl. A proprietary preparation containing sodium dibunate. A cough syrup. (269).

Beccarite. A mineral, ZrO_2.

Bearing Metal	Cu.	Zn.	Sn.	Pb.	Sb.	P.Sn.
	Per cent.	Per cent.	Per cent.	Per cent.	Per cent.	Per cent.
American automobile	2–6	...	87–89	...	7–9	...
,, railway	...	...	8–21	68–73	11–18	...
Automobile	77–88	0–14	7–10	0–12	...	0–2
Belgian railway	80	...	16	2	2	...
Babbitt's (original)	4	69	19	5	3	...
,, (hard)	8	...	88	...	4	...
,, (marine)	...	...	21	72	7	...
,, (normal)	3	...	90	...	7	...
,, (soft)	3	...	84	6	7	...
English	7·8	...	76·7	...	15·5	...
Dutch locomotive	85	2	13	...	...	...
Extra hard	4	2	12	...	82	...
French car	5·5	...	83·3	...	11·1	...
,, railway	0–16	...	20–82	0–70	10–22	...
,, automobile	...	...	10	75	15	...
German	5–7	...	71–76	0–12	9–17	...
,, railway	6	...	83	...	11	...
	21·4	...	71·4	...	7·2	...
Kamarsch	9·5	...	70·7	...	19·7	...
Katzenstein	...	...	7·3	75·6	16·8	0·4 Cr
Lafond's	80–84	2–14	2–18	...	...	...
Locomotive	89	7·8	2·4	...	...	...
,, piston	74	22	4	...	...	...
Medium hard	2	...	72	...	26	...
Piston	6·5	...	81	...	12·5	...
Valve rod	8	...	82	...	10	...

Bechilite. See PRICITE.

Beckacite. A proprietary trade name for modified phenolic varnish and lacquer resins. Also alkyd resin varnish and lacquers. Also applied to a synthetic resin. (32)

Beckamine. A proprietary trade name for urea-formaldehyde synthetic lacquer and varnish resins.

Beckasol. A proprietary trade name for a synthetic resin.

Beckalide 50 H.V. A proprietary trade name for a polyamide solution for surface coating applications. (181).

Beckalide 51. A proprietary trade name for a polyamide solution for surface coating applications. (181).

Beckelite. A mineral. It is a zircon-silicate of rare earths and lime.

Beckerite. An opaque variety of amber.

Beckol. A proprietary trade name for synthetic alkyd resin.

Beck-o-lac. A proprietary trade name for a synthetic resin.

Beckolin. A proprietary trade name for synthetic oil lacquer and varnish resins.

Beckolloid. A proprietary synthetic resin plastic.

Beckophen. A proprietary trade name for a series of heat reactive resins. Used for moulding. (32).

Beckopol. A proprietary trade name for phenolated copal varnish and lacquer resins.

Beckosol. A proprietary trade name for synthetic oil lacquer and varnish resins. Also a trade name for synthetic alkyd resin used in paints.

Beckosol 1424. A proprietary trade name for a long oil soya alkyd for surface coatings. (181).

Beckosol 1425. A proprietary trade name for a long oil safflower alkyd for surface coatings. (181).

Becksol. A proprietary trade name for alkyd synthetic resins.

Beckton White. See LITHOPONE.

Beckurane M-190. A proprietary trade name for a moisture-curing isocyanate prepolymer for surface coating applications. (181).

Beckurane 2-293. A proprietary trade name for an isocyanate prepolymer for use with a catalyst in surface coating applications. (181).

Beclamide. N - β - chloropropionyl - benzylamine. Nydrane.

Beclomethasone. A corticosteroid. 9α - Chloro - 11β, 17α, 21 - trihydroxy - 16β - methylpregna - 1, 4 - diene - 3, 20 - dione. 9α - Chloro - 16β - methyl-prednisolone. BECOTIDE and PROPA-DERM are the $17,21$-dipropionate.

Becmulse. A proprietary emulsified tar for cold application. (573).

Becomel. A proprietary preparation containing nicotinamide, thiamine, pyridoxine and riboflavine. Vitamin supplement. (280).

Beconase. A proprietary preparation of BECLOMETHASONE dipropionate as an aerosol, used in the treatment of allergic rhinitis. (284).

Becosed. A proprietary preparation of phenobarbitone, aneurine, riboflavin and nicotinamide. (350).

Becosym. A proprietary preparation containing thiamine, riboflavine, nicotinamide and pyridoxine. A Vitamin B complex. (314).

Becosym Forte. A proprietary preparation containing thiamine, riboflavine, nicotinamide and pyridoxine. (314).

Becotide. A proprietary preparation of BECLOMETHASONE dipropionate as an aerosol, used in asthma therapy. (284).

Becovite. A proprietary preparation of the Vitamin B complex. Vitamin supplement. (272).

Becquerelite. A mineral. It is uranium hydroxide, $UO_3.2H_2O$, found in the pitchblende at Katanga. It is also stated to be $4[7UO_3 . 11H_2O]$.

Becspray. Specially prepared tar for spraying. (573).

Bectaphalt. Tar-bitumen compound for spraying and grouting. (573).

Bedacryl. An acrylic resin preparation. (512).

Bedafin. Textile resin finishes. (512).

Beda Nuts. Myrobalans (*q.v.*).

Bedenite. A mineral of approximate composition $Ca_2Mg_4(Fe\cdots,Al)(Si,Al)_8 . O_{22}(OH)_2$.

Bedesol. Synthetic resins. (570).

Beechwood Sugar. Wood sugar (*q.v.*).

Beegerite. A mineral. It is a sulphide of bismuth and antimony, found in Colorado. It is $Pb_6Bi_2S_9$.

Been Oil. See OIL OF BEEN.

Beer. An alcoholic beverage obtained by fermenting the aqueous decoction of barley malt and hops. It is then charged with carbon dioxide. It contains from 3–8 per cent. alcohol.

Beer Vinegar. A variety of vinegar containing less acetic acid than wine vinegar. It contains 4·5–6 per cent. of extract rich in maltose, dextrin, albuminoids, and phosphates.

Beeswax. A wax produced by the bee, *Apis mellifica*, which secretes the wax internally and uses it for making the honeycomb. The wax has a specific gravity of 0·96–0·963, and melts at 62–64° C. The crude wax has a saponification value of 93·5 and an acid value of 20·3.

Beet Molasses. See MOLASSES.

Beet Sugar. See SUCROSE.

Beetle. A proprietary trade name for resin systems, e.g. Beetle foundry resins, laminating resins, polyester resins, synthetic resin, moulding powders, synthetic lacquers, adhesives, textile, paper and leather resins. (570).

Beetle Resin BT 333. A cyclic reactant. Recommended for soft mechanical finishes on cellulosic fabrics, also soft crease resistant and shrink resistant finishes on cellulose/synthetic fibre blends. (570).

Beetle Resin BT 334. A cyclic reactant-modified melamine cross linking agent. Recommended for chlorine resistant finishes on cotton fabrics and easy care finishes on cellulose/synthetic fibre blends. (570).

Beetle Resin W69. A modified urea-formaldehyde resin. Recommended for glueing " Vac-Vac " treated timbers where difficulties may be experienced with standard adhesives. (570).

Beetle Ware. A proprietary trade name for urea-formaldehyde synthetic resin mouldings.

Befanamite. A mineral. A variety of Thortveitite of composition $2(Sc_2Si_2O_7)$.

Beflavit. A proprietary preparation of riboflavin (Vitamin B_2). (314).

Befortiss. A proprietary preparation containing thiamine, riboflavine, pyridoxine and nicotinamide. (272).

Begasse (Megass, Bagasse). The woody residue left after the removal of the juice from the sugar cane. Used as a fuel.

Behn Oil. See OIL OF BEEN.

Beidellite. A mineral, previously described as Leverrierite. It is a meta-silicate, $Al_2O_3.3SiO_2.nH_2O$, of Beidell, Colorado.

Beilstein. Synonym for Nephrite.

Beinstein. Synonym for Osteocolla.

Bel. See BAEL.

Bela. See BAEL.

Belco. Trade name for a cellulose product.

Beldongrite. A mineral. It is a manganese iron oxide.

Beleck Porcelain. A porcelain made in Ireland. It is glazed Parian ware (q.v.).

Belgian Phosphates. They consist of phosphatic chalk and carbonate, and contain from 48–60 per cent. calcium phosphate and 40–45 per cent. lime. A fertiliser.

Belgian Slag. See BASIC SLAG.

Belgite. Willemite (q.v.).

Beligno Seeds. See OLIVES OF JAVA.

Belite. A constituent of Portland cement and cement clinker.

Bellabarbitol. A proprietary preparation of phenobarbitone and hyoscyamine. A sedative. (340).

Belladenal. A proprietary preparation of phenobarbitone and belladonna leaf. A sedative. (267).

Belladonna, Extract of Belladonna Alkaloids. A preparation of constitutionally related alkaloids, notably atropine and hyoscyamine, obtained from the Solanaceæ, especially Atropa belladonna.

Bell Brass. See BRASS.

Bellergal. A proprietary preparation of belladonna, ergotamine tartrate and phenobarbitone. A sedative. (267).

Belleroid. A proprietary trade name for a plastic with a rubber base.

Bellingerite. A mineral. It is $[3Cu(IO_3)_2.2H_2O]$.

Bellite. A mineral. It is a chromo-arsenate of lead, $PbCrO_4.XAs_2O_3$. The name is also used for certain coal mine explosives.

Bellite. Explosives for coal mine (a) Consists of 83·5 per cent. ammonium nitrate and 16·5 per cent. dinitro-benzene ; (b) contains ammonium nitrate, dinitro-benzene, and sodium chloride ; (c) consists of 93·5 per cent. ammonium nitrate, and 6·5 per cent. dinitro-benzene ; (d) contains ammonium nitrate, dinitro-benzene, and sodium chloride.

Bell Metal. Alloys usually consisting of from 78–80 per cent. copper and 20–22 per cent. tin, but sometimes lead and zinc are added. The alloy containing 78 per cent. copper and 22 per cent. tin has a specific gravity of 8·7, and melts at 890° C. Another bell metal consists of 83 per cent. aluminium, 10 per cent. manganese, and 7 per cent. cadmium. Used for casting bells and other purposes. An alloy containing 80 per cent. copper and 20 per cent. tin has a melting point of 890° C., specific gravity of 8·7, and specific heat 0·0862 cal./gm./°C. between 14–98° C.

Bell Metal Ore. See TIN PYRITES.

Bellobarb. A proprietary preparation of phenobarbitone, belladonna and magnesium trisilicate. An antacid. (259).

Belloform. Identical with Bactoform (q.v.).

Belloid. Dispersing agents. (543).

Bell Pepper. The fruit of *Capsicum grossum*.

Belmontine Oil. See MINERAL BURNING OILS.

Belonesite. Found in Vesuvian lava. It is essentially a magnesium molybdate —$MgMoO_4$.

Belonesite. See BELONITE.

Belonite (Belonesite). A mineral. It is identical with Sellaite, MgF_2.

Belovite. A mineral. It is $Ca_2(Ca,Mg)(AsO_4)_2.2H_2O$.

Belsol. Wetting and penetrating agents. (541).

Bemal. An alloy of 70 per cent. copper, 29 per cent. zinc, and traces of lead and iron.

Bemasulph. A proprietary preparation of chloroquin sulphate. Anti-inflammatory agent. (828).

Bemberg Silk. See BEMBERG YARN.

Bemberg Yarn (Bemberg Silk, Cupra Yarn). A proprietary cuprammonium silk made by forcing a cuprammonium solution of cellulose prepared from cotton linters, through orifices, and drawing the resulting product through water, treating with sulphuric acid, washing, and winding.

Bemegride. β - ethyl - β - methyl - glutarimide. Megimide.

Bementite. A mineral. It is a manganese silicate. Variously formulated: $MnSiO_3 . \frac{1}{2}H_2O$ and $Mn_5Si_4O_{10}(OH)_{16}$.

Benacine. A proprietary preparation of diphenhydramine and hyoscine hydrobromide. Antinauseant. (264).

Benactyzine. 2 - Diethylaminoethyl benzilate. Cevanol, Lucidil, Nutinal and Suavitil are hydrochlorides of the above.

Benadon. A proprietary preparation of pyridoxine. An antinauseant. Pyridoxine—Vitamin B$_6$. (314).

Benadryl. A proprietary preparation of diphenhydramine hydrochloride. (264).

Benafed. A cough suppressant. A proprietary preparation of diphenhydramine hydrochloride, dextramethorphan hydrobromide, pseudoephedrine hydrochloride, ammonium chloride, sodium citrate, chloroform and menthol in a syrup. (264).

Ben-a-Gel. A gelling agent for water systems. (572).

Benalite. A proprietary trade name for a lignin plastic. Cured lignin sheets.

Benaloid. A proprietary trade name for uncured lignin sheets.

Benapen. Benethamine Penicillin. (520).

Benapryzine. A drug used in the treatment of the Parkinsonian syndrome. 2 - (N - Ethylpropylamino)ethyl benzilate.

Benbows. Dog health products. (535).

Benatol. Specially pure benzyl alcohol. (504).

Benazma. A proprietary preparation of pseudoephedrine hydrochloride, stramonium extract, syrup of prunes and syrup of codeine. Bronchial antispasmodic. (272).

Bendalite. A proprietary trade name for a cast styrene synthetic resin.

Bendazac. International recommended non-proprietary name (I.N.N.) for BINDAZAC.

Bendrofluazide. 3 - Benzyl - 3,4 - dihydro - 6 - trifluoromethylbenzo -1, 2,4 - thiadiazine - 7 - sulphonamide 1,1 - dioxide. Bendroflumethiazide. Aprinox. Berkozide. Centyl. Neo-Naclex.

Bendroflumethiazide. Bendrofluazide.

Benedict-Hitchcock Standard Uric Acid Solution. Disodium orthophosphate, Na_2HPO_4, 9 grams, and monosodium orthophosphate, NaH_2PO_4, 1 gram, are dissolved in 250 c.c. distilled water at 60° C. Filter and make up to 500 c.c. Pour the warm solution over 200 mg. uric acid suspended in a few cubic centimetres of water. Agitate, cool, and add exactly 1·4 c.c. glacial acetic acid and 5 c.c. chloroform, and dilute to 1,000 c.c. 5 c.c. = 1 mg. uric acid.

Benedict Metal. An alloy containing 84–86 per cent. copper and 14–16 per cent. nickel. Also see NICKEL-COPPER ALLOYS.

Benedict-Nash Reagent. For the ammonia content of the blood. Potassium carbonate, 8 grams, is placed in a 100 c.c. flask and 50 c.c. of distilled water added, then 12 grams of potassium oxalate crystals. The solution is boiled to about 30 c.c. and then diluted to 50 c.c. with ammonia-free distilled water. This is repeated and finally diluted to 80 c.c.

Benedict Plate. A nickel silver. It contains 57 per cent. copper, 28 per cent. zinc, and 15 per cent. nickel.

Benedict's Creatinin Solution. Creatinin, 100 mg., is placed in a 100 c.c. flask with some distilled water, 1 c.c. of concentrated hydrochloric acid added, and the whole made up to 100 c.c. with distilled water. 1 c.c. = 1 mg. creatinin.

Benedict's Decoloriser for Creatin Determination. Powdered lead, chemically pure.

Benedict's Molybdic Acid Reagent. For blood phosphorus. Pure molybdic acid, 20 grams, is placed in 25 c.c. of 20 per cent. caustic soda solution and warmed to 50° C. Dilute to 200 c.c. with distilled water, mix, and filter. Add 200 c.c. concentrated sulphuric acid, keeping the solution cool during its addition.

Benedict's Picric-picrate Solution. Normal solution of caustic soda, 125 c.c., is added to 700 c.c. distilled water at 80° C., and then 36 grams of dry or 40 grams of moistened picric acid added and dissolved by shaking. It is finally made up to 1,000 c.c.

Benedict's Reagent for Blood Phosphorus. Sodium hydrogen sulphite, 30 grams, is dissolved in 100 c.c. distilled water, 1 gram hydro-quinone added, and the whole diluted with water to 200 c.c.

Benedict's Solution for Glucose (Qualitative). A solution of 173 grams sodium citrate and 100 grams anhydrous sodium carbonate is made using 600 c.c. water. The solution is filtered into a litre measuring cylinder and diluted to 850 c.c. Copper sulphate, $CuSO_4.5H_2O$, 17·3 grams, is dissolved in 100 c.c. water, diluted to 150 c.c., and added to the citrate-carbonate solution. The mixture is used to test for glucose by adding 5–10 drops of the glucose solution to 5 c.c. of the reagent and boiling. A red, yellow, or green precipitate is produced, the final colour depending upon the amount of glucose present.

Benedict's Solution for Glucose (Quantitative). A solution is made by dissolving 200 grams of sodium carbonate, $Na_2CO_3.10H_2O$, or 100 grams Na_2CO_3, 200 grams sodium citrate, and 125 grams potassium thiocyanate in enough water to make 800 c.c., and filtered. Copper sulphate, $CuCO_4.5H_2O$, 18 grams, is dissolved in 100 c.c. water and poured into the above solution, and 5 c.c. of a 5 per cent. solution of potassium ferrocyanide are added, and the whole diluted to 1 litre. Used for the determination of glucose, 25 c.c. of the reagent=0·05 gram of glucose.

Benedict's Standard Glucose Solution. (a) For urine : pure glucose, 167 mg. is dissolved in 400 c.c. water, 5 c.c. toluene added, and the whole mixed and diluted to 500 c.c. with water. (b) For blood : pure glucose, 0·1 gram, is dissolved in 400 c.c. of a 2 per cent. picric acid in a 500 c.c. flask. It is mixed and diluted to the mark with the saturated picric acid solution. 3 c.c. =0·6 mg. glucose.

Benedict's Standard Phenol Solution. Resorcinol, 11·62 mg. in 100 c.c. 0·1 normal hydrochloric acid solution. 5 c.c. =0·5 mg. phenol.

Benedict's Standard Phosphorus Solution. Potassium di-hydrogen phosphate, 0·4394 gram, is dissolved in 800 c.c. distilled water, and made up to 1,000 c.c. with chloroform water (10 c.c. with 200 c.c. water). 5 c.c. contains 0·025 mg. phosphorus.

Benedict's Sulphur Reagent. Used to test for sulphur in urine. It consists of 200 grams of sulphur-free copper nitrate and 50 grams of sodium or potassium chlorate dissolved in distilled water to 1 litre.

Benefing Oil. A drying oil derived from a species of *Hyptis spicigera*. Suggested as a substitute for linseed oil.

Benemid. A proprietary preparation of probenecid. An antirheumatic. (310).

Benerva. A proprietary preparation of thiamine hydrochloride (Vitamin B_1). (314).

Benerva Compound. Vitamin B_1, B_2 and nicotinamide. (540).

Benethamine Penicillin. N-Benzyl-phenethylamine salt of benzylpenicillin. Benapen.

Bengal Blue. See INDULINE, SOLUBLE.

Bengal Blue G. See GENZOAZURINE G.

Bengal Catechu. See CUTCH.

Bengal Fire. A mixture of realgar, nitre, and sulphur.

Bengal Gram (Chick Pea). The seed of *Cicer arietinum*, of India. It is used as a food for cattle. It is also employed by the natives as a cooling agent in fever, as the plant exudes oxalic acid.

Bengal Gum. See BABAL GUM.

Bengaline. See INDOINE BLUE R.

Bengaline Blue. A dyestuff. It is a mixture of indoine and methylene blue.

Bengal Isinglass. See AGAR-AGAR.

Bengal Kino. See KINO, BENGAL.

Bengal Lights. See BENGAL FIRE.

Bengal Red. See ROSE BENGAL.

Bengue's Balsam. A proprietary preparation of menthol and methyl salicylate. A bronchial inhalation. (802).

Benicke's Cement. A luting cement made from rubber, linseed oil, pipeclay, and litharge.

Benin Copal. See COPAL RESIN.

Beni Oil. See GINGELLY OIL.

Beniseed Oil. See GINGELLY OIL.

Benit. A German alloy containing 96·91 per cent. aluminium 2 per cent. copper, 0·45 per cent. manganese, and 0·27 per cent. tungsten.

Benitoite. A mineral. It is an acid titano-silicate of barium, $BaTiSi_3O_5$.

Benjamin. Benzoin.

Benjamin Gum. See GUM BENJAMIN.

Benjaminite. A mineral, $Pb_2(Ag.Cu)_2B_4S_9$.

Benjamin Oil. See OIL OF BENJAMIN.

Bennatate. A proprietary trade name for a cellulose acetate plastic.

Benné Oil. See GINGELLY OIL.

Ben Oil. See OIL OF BEEN.

Benoral. A proprietary preparation of BENORYLATE. An anti-rheumatic drug. (439).

Benorylate. An analgesic. 4 - Acetamidophenyl *O* - acetylsalicylate. BENORAL.

Benoxaprofen. An anti-inflammatory and analgesic, currently under clinical trial as " LRCL 3794 ". It is 2 - (2 - *p* - chlorophenylbenzoxazol - 5 - yl) - propionic acid.

Benoxyl. A proprietary preparation of benzoyl peroxide used in dermatology as an antibacterial agent. (379).

Benperidol. A tranquilliser. 1 - {1 - [3 - (4 - Fluorobenzoyl)propyl] - 4 - piperidyl} benzimidazolin - 2 - one ANQUIL.

Bensapol. A proprietary trade name for a wetting agent consisting of sulphonated oils and a solvent.

Benserazide. A preparation used in the treatment of the Parkinsonian syndrome. It is DL - 2 - amino - 3 - hydroxy - 2' - (2, 3, 4 - trihydroxy-benzyl) propionohydrazide. " Ro 04-4062 " is currently under clinical trial as the hydrochlorate.

Bensuccin. A proprietary preparation of benzyl succinate.

Bensuldazic Acid. A fungicide. (5 - Benzyl - 6 - thioxo - 1, 3, 5 - thiadiazin - 3 - yl) acetic acid. DEFUNGIT is the sodium salt.

Benteine. Specially pure benzyl acetate. (504).

Bentene. Butyraldehyde - aniline. A proprietary rubber accelerator. (435).

Bentone. An organic gelling agent. (572).

Bentonite (Denver Mud, Denver Clay, Paper Clay, Soap Clay, Gumbo, Ardmonite). A mineral. It is a variety of bedded clay which has a great affinity for water. It consists mainly of silica, with some alumina, and smaller quantities of ferric oxide, calcium oxide, magnesium oxide, and alkalis. The material has been called Taylorite, and has been used as a soap. Employed as a filler for soap and paper, and for decolorising oils.

Benuride. A proprietary preparation of pheneturide. An anticonvulsant. (802).

Benvil. A proprietary preparation of tybamate. A sedative. (272).

Benylin Expectorant. A proprietary preparation containing diphenhydramine hydrochloride, ammonium chloride, sodium citrate, chloroform and menthol. A cough linctus. (264).

Benzacetin. Acet-amino-ethyl-salicylic acid, $C_6H_3(OC_2H_5)(NH . CO . CH_3)$ COOH. Recommended as an analgesic.

Benzadone. Anthraquinone. Vat dyestuffs. (508).

Benzaldehyde Copper. Diphenyl-cupriphen, $C_6H_5CHOCuOCHC_6H_5$.

Benzaldehyde Green. See MALACHITE GREEN.

Benzaldehyde Violet. A dyestuff. It is a diamino-triphenyl-methane derivative, $C_{19}H_{17}N_2Cl$. It has no technical importance.

Benzalgen. See ANALGEN.

Benzal Green. See MALACHITE GREEN.

Benzalkonium Chloride. Alkylbenzyl-dimethylammonium chlorides. P.R.Q. Antiseptic; Roccal; Zephiran. Present in Tonsillin, Capitol, Drapolene.

Benzamine. See EUCAINE.

Benzanalgen. See ANALGEN.

Benzanthronequinolines. Violet blue dyestuffs from α-amino-anthraquinone.

Benzanil. Direct cotton colour dyestuffs. (508).

Benzarene Blue 2B. A dyestuff. It is a British equivalent of Diamine blue 2B.

Benzarene Brown G. A dyestuff. It is a British equivalent of Benzo brown G.

Benzarene Green B. A dyestuff. It is a British equivalent of Diamine green B.

Benzathine Penicillin. NN'-Dibenzyl-ethylenediamine di(benzypenicillin). Dibencil; Neolin; Penidural; Permapen.

Benzbromarone. A drug used in the treatment of hyperuricæmia. 3 - (3, 5 - Dibromo - 4 - hydroxybenzoyl) - 2 - ethylbenzofuran. MINURIC.

Benzedrex. A proprietary preparation containing Propylhexedrine.

Benzene for Red. A mixture of benzene and toluene.

Benzethidine. Ethyl 1-(2-benzyl-oxyethyl) - 4 - phenylpiperidine - 4 - carboxylate.

Benzethonium Chloride. Benzyl-dimethyl - 2 - {2 - [4 - (1,1,3,3 - tetra - methylbutyl)-phenoxy] ethoxy} ethyl-ammonium chloride. Phemeride; Phemerol Chloride.

Benzets. A proprietary preparation of benzalkonium chloride and benzocaine. Throat lozenges. (350).

Benzex. A proprietary product. Benzyl cellulose.

Benzhexol. 1 - Cyclohexy - 1 - phenyl - 3-piperidinopropan-1-ol Trihexypheni-dyl. Artane and Pipanol are the hydrochloride. Present in Pathibamate.

Benzhydrol. Diphenyl - carbinol, $(C_6H_5)_2.CH.OH$.

Benzidam. Zinin's name for aniline (1842). See KYANOL.

Benzidine. Di-p-amino-diphenyl, $C_6H_4(NH_2)_2C_6H_4$.

Benzidine Blue. A dyestuff prepared from diazotised benzidine and B-naphthol-disulphonic acid, $(C_6H_4)_2(N_2)_2(C_{10}H_4)_2(HSO_3)_2.(OH)$. Obsolete.

Benzidine Red. An azo dyestuff from diazotised benzidine-disulphonic acid and naphthionic acid, $(C_6H_3)_2(HSO_3)_2(N_2)_2(C_{10}H_5)_2(HSO_3)_2(NH_2)_2$. Obsolete.

Benzilonium Bromide. 3-Benziloyl-oxy-1,1-diethylpyrrolidinium bromide. Portyn.

Benzine (Petroleum Naphtha, Benzo-line, " A " Naphtha). That fraction obtained in the refining of petroleum which boils from 70–120° C., of specific gravity 0·725–0·737. Heptane, C_7H_{16}, is the principal constituent of the Pennsylvanian oil.

Benzine-Collas. Benzine introduced by Collas, for cleaning purposes.

Benzine, Heavy. See LIGROIN.

Benzine, Petroleum. See C-PETROLEUM NAPHTHA.

Benzine, Standard. See STANDARD BENZINE.

Benzinoform. Carbon tetrachloride, CCl_4.

Benziodarone. 2-Ethyl-3-(4-hydroxy-3, 5-di-iodobenzoyl)coumarone. Caridvix.

Benznaphthanthrene. A reddish-blue dyestuff from naphthanthraquinone.

Benzo-aniline. p-Amino-benzophenone.

Benzoated Lard. Lard heated with benzoin to give it an agreeable odour, and to retard rancidity (*Adeps Benzoatus B.P.*).

Benzoated Suet. Similar to benzoated lard, except that suet is in the place of lard.

Benzoazurin. A dyestuff formed from 1 molecule of phenyl-chloroform and 2 molecules of phenol.

Benzoazurin B. A dyestuff. It is the homologue of benzoazurin, from diazotised diamino-diphenetol.

Benzoazurin G (Benzoazurin 3G, Bengal Blue G). A dyestuff. It is the sodium salt of dimethyoxy-diphenyl-disazo-bi-α-naphthol-p-sulphonic acid, $C_{34}H_{24}N_4O_{10}S_2Na_2$. Dyes cotton blue from an alkaline bath.

Benzoazurin 3G. See BENZOAZURIN G.

Benzoazurin R. A dyestuff. It is a mixture of Benzoazurin G and Azo blue.

Benzobis. An emulsion containing the bismuth basic salt of dioxy-benzoic acid monomethyl ester.

Benzo-bismuth. Tribismuthyl-benzoic acid. An antisyphilitic.

Benzo Black. A direct cotton dyestuff.

Benzo Black-blue G. A dyestuff. It is the sodium salt of disulpho-diphenyl-disazo - α - naphthalene - azo - bi - α - naphthol-p-sulphonic acid, $C_{42}H_{24}N_6O_{14}S_4Na_4$. Dyes cotton black-blue from an alkaline bath.

Benzo Black-blue 5G. A dyestuff. It is the sodium salt of disulpho-diphenyl-disazo - α - naphthalene - azo - bi - di - oxynaphthalene-sulphonic acid, $C_{42}H_{24}N_6O_{16}S_2Na_2$. Dyes cotton greenish-black.

Benzo Black-blue R. A dyestuff. It is the sodium salt of ditolyl-disazo-α - naphthalene - azo - bi - α - naphthol-sulphonic acid, $C_{44}H_{30}N_6O_8S_2Na_2$. Dyes cotton dark bluish-violet from a soap bath.

Benzo Black-brown. A direct cotton dyestuff, or for union material.

Benzo Blue (Diamine Blue, Benzo Cyanine, Congo Blue, Congo Cyanine, Chicago Blue, Columbia Blue). Tetrazo dyestuffs from toluidine or dianisidine and 1 : 8-amino-naphthol-4-sulphonic acid, or mixed peri-amino-naphthol-sulphonic acid dyestuffs, mostly prepared with 1 : 8-amino-naphthol-4- and 2 : 4-disulphonic acids.

Benzo Blue BB. See DIAMINE BLUE BB.

Benzo Blue BX. Diamine blue BX (*q.v.*).

Benzo Blue 3B. See DIAMINE BLUE 3B.

Benzo Blue 3G. See DIAMINE BLUE 3B.

Benzo Blue 2R. See COLUMBIA BLUE G.

Benzo Blue 4R. See CHICAGO BLUE 4R.

Benzo Blue RW. A dyestuff identical with Chicago blue RW and Diamine blue RW. It is a similar dye to Chicago blue 2R.

Benzo Brown B (Orion Brown B, Enbico Direct Brown R). A dyestuff. It is the sodium salt of sulpho-naphthalene-azo-phenylene-brown, $C_{38}H_{28}N_{12}O_6S_2Na_2$. Dyes cotton brown from a neutral salt bath.

Benzo Brown BX and BR. Dyestuffs, which are derivatives of chrysoidine. They are identical with Cotton brown A and N.

Benzo Brown G (Benzarene Brown G, Orion Brown G, Enbico Direct Brown BB, Chlorazol Brown GM). A dyestuff. It is the sodium salt of sulpho-benzene-azo-phenylene-brown, $C_{30}H_{24}N_{12}O_6S_2Na_2$. Dyes cotton yellowish-brown from a neutral salt bath.

Benzo Brown 5R. See INGRAIN BROWN.

Benzocaine. See ANÆSTHESIN.

Benzo Chrome-black B. A dyestuff similar to Benzo brown.

Benzo Chrome-brown 3R. A polyazo dyestuff of the same class as Benzo indigo blue. Dyes unmordanted cotton.

Benzoctamine. A tranquilliser. N - Methyl - 9, 10 - ethanoanthracene - 9 (10H) - methylamine. Tacitin is the hydrochloride.

Benzocyanine 3B, B, R. Substantive cotton dyestuffs, identical with the corresponding marks of Congo cyanine and Diamine cyanine. Employed in calico printing.

Benzo Dark Brown. A direct cotton dyestuff.

Benzo Dark Green B. A polyazo dyestuff, similar to Diamine green G and Columbia green. Dyes cotton bluish- and yellowish-olive.

Benzo Fast Grey. A direct cotton dyestuff, giving bluish-grey shades from an alkaline bath.

Benzo Fast Orange S. See BENZO FAST SCARLET GS.

Benzo Fast Pink 2BL (Chlorazol Fast Pink BK). A dyestuff. It is the sodium salt of diphenyl-urea-disulphonic acid-disazo - bi - amino - naphthol-sulphonic acid, $C_{33}H_{20}N_6O_{14}S_4Na_4$. Dyes cotton pink.

Benzo Fast Red. See BENZO FAST SCARLET GS.

Benzo Fast Scarlet 4BS. See BENZO FAST SCARLET GS.

Benzo Fast Scarlet 8BS. See BENZO FAST SCARLET GS.

Benzo Fast Scarlet GS (Benzo Fast Scarlet 4BS, 8BS, Benzo Fast Orange S, Benzo Fast Red, Chlorazol Fast Scarlet 4BS, Direct Fast Scarlet SE, Paramine Fast Scarlet 4BS, Direct Fast Orange SE, Paramine Fast Orange S, Chlorazol Fast Orange R). Dyestuffs. They are the sodium salts of bi-benzene-(or homologue)-disazo-di-oxynaphthyl-urea-disulphonic acid. Cotton dyes.

Benzoflavine. A dyestuff. It is the hydrochloride of diamino-phenyl-dimethyl-acridine, $C_{21}H_{19}N_3HCl$. Dyes silk, wool, and mordanted cotton yellow.

Benzoflex 2-45. A proprietary trade name for diethylene glycol dibenzoate. (79).

Benzoflex 9-88. A proprietary trade name for the dibenzoic ester of a dipropylene glycol. A vinyl plasticiser. (78).

Benzo-furane Resin. See PARA-COUMARONE RESIN.

Benzo Grey. A dyestuff. It is the sodium salt of diphenyl-disazo-α-naphthalene - azo - α - naphthol - sulphonic - salicylic acid, $C_{39}H_{24}N_6O_7SNa_2$. Dyes cotton grey.

Benzoic Acid, English. The acid made from the neutral resin benzoin.

Benzoic Acid, German. The acid made by synthetic methods. The term was used for that prepared from hippuric acid.

Benzo Indigo Blue. A dyestuff. It is the sodium salt of ditolyl-disazo-α-naphthalene - azo - bi - dioxy - naphthalene-sulphonic acid, $C_{44}H_{30}N_6O_{10}S_2Na_2$. Dyes cotton indigo blue.

Benzoin (Gum Benjamin). A resin obtained from species of *Stryax*. Siam benzoin is one of the best varieties, others being Sumatra and Palembang benzoin, the latter being of the poorest quality. The main constituent is benzoic acid, and vanillin is present in Siam and Sumatra benzoin. Good qualities contain from 12–20 per cent. benzoic acid. Benzoin has a specific gravity of 1·2 and melts at 75–100° C. It has a saponification value of 155–270, the mineral matter varies from 0·24–2·85 per cent., the solubility in 90 per cent. alcohol from 86–96 per cent., and the acid value from 98–158.

It is partly soluble in amyl alcohol, petroleum, ether, benzene, and chloroform. Benzoin is used as a constituent of Friar's balsam, as a perfuming agent, and in the manufacture of incense.

Benzoin, Flowers of. See FLOWERS OF BENJAMIN.

Benzoin Yellow. A dyestuff obtained by the condensation of benzoin with gallic acid in the presence of cold sulphuric acid. Dyes chromed wool yellow.

Benzol. Commercial benzene, C_6H_6.

Benzol 30 per cent. A benzol with a boiling range of 80–130° C., 30 per cent. of which distils over below 100° C. It has a specific gravity at 15° C. of 0·870–0·872, and a refractive index at 15° C. of about 1·49.

Benzol 50 per cent. A benzol with a boiling range of 80–125° C., 50 per cent. of which distils over below 100° C. It has a specific gravity at 15° C. of 0·875–0·877, and a refractive at 15° C. of 1·49–1·50.

Benzol 90 per cent. Benzol containing 80–85 per cent. benzene, 13–15 per cent. toluene, and 2–3 per cent. xylene, sometimes with traces of olefines, paraffins, and sulphuretted hydrogen. It is called 90 per cent. benzol because in its distillation 90 per cent. distils over at below 100° C. It has a boiling range of 80–120° C., a specific gravity of 0·88–0·886, a refractive index at 15° C. of 1·5–1·502, and an ignition point of about —8° C. It is used as a solvent for extraction, in the linoleum and varnish industries, and in rubber manufacture. It is also employed for extracting the fat from bones, and as a fuel.

Benzol, Crude. A mixture of benzene, toluene, and xylene.

Benzol, Crystallisable. Nearly pure benzene. It contains traces of thiophene, carbon disulphide, toluene, and paraffins. It flashes below —13° F. Used for the production of mono- and dinitro-benzene.

Benzoline. The more volatile fraction of benzine, of boiling point 70–95° C. The term is often applied to benzine.

Benzol Oils. A mixture of equal parts of Diesel engine oil and benzol. A fuel for motor vehicles.

Benzol, Standard. Benzene containing 7 per cent. toluene. Employed in the determination of water in tar, naphthalene, butter, and sausages.

Benzonaphthol. Benzoyl-β-naphthol, $C_{10}H_7.O.COC_6H_5$. An antiseptic used internally for abdominal diseases.

Benzonatate. 2 - (ω - Methoxypoly-ethyleneoxy)ethyl 4-butylamino-benzoate. Tessalon.

Benzone. See DIPHENYL-KETONE.

Benzo Nitrol. The name applied to diazotised para-nitraniline, when used as a developer for direct cotton dyes.

Benzo Olive. A dyestuff. It is the sodium salt of diphenyl-disazo-α-naphthalene azo - amino - naphthol - disul - phonic-salicylic acid, $C_{39}H_{24}N_7O_{10}S_2$ Na_3. Dyes cotton greenish-olive from a neutral salt bath.

Benzo Orange R (Paramine Orange G, R, Chlorazol Orange RN). A dyestuff. It is the sodium salt of diphenyl-disazo-salicylic-naphthionic acid, C_{29} $H_{19}N_5O_6SNa_2$. Dyes cotton orange from an alkaline bath, also chromed wool.

Benzoparal. An oily preparation of benzoin containing 16 parts of liquid paraffin and 1 part of benzoin.

Benzophenol. Phenol, C_6H_5OH.

Benzo-piperaz. Piperazine benzoate.

Benzo Pure Blue. Identical with Diamine pure blue and Congo pure blue. See DIAMINE SKY BLUE.

Benzo Pure Blue A. Identical with Diamine pure blue A and Congo pure blue A.

Benzo Pure Blue 4B. Identical with Diamine blue 4B and Chicago blue 4B.

Benzopurpurin B. An azo dyestuff. It is the sodium salt of ditolyl-disazo-bi-β-naphthylamine-β-sulphonic acid, $C_{34}H_{26}N_6O_6S_2Na_2$. Dyes cotton red from an alkaline bath.

Benzopurpurin 4B (Cotton Red 4B, Sultan Red 4B, Erie Red B Conc., Imperial Red, Eclipse Red, Victoria Red, Fast Scarlet, Azamine 4B). A dyestuff. It is the sodium salt of ditolyl-disazo-bi-naphthionic acid, C_{34} $H_{26}N_6O_6S_2Na_2$. Dyes cotton red from an alkaline, and wool from a neutral bath.

Benzopurpurin 6B. A dyestuff. It is the sodium salt of ditolyl-disazo-bi-α-naphthylamine-sulphonic acid, $C_{34}H_{26}$ $N_6O_6S_2Na_2$. Dyes cotton red from an alkaline bath.

Benzopurpurin 10B (Cotton Red 10B). A dyestuff. It is the sodium salt of dimethoxy - diphenyl - disazo - bi - naphthionic acid, $C_{34}H_{26}N_6O_8S_2Na_3$. Dyes cotton carmine red from an alkaline bath.

Benzopyrocatechol. Dihydroxybenzophenone.

Benzoquinol. Hydroquinone, C_6H_4 $(OH)_2$.

Benzoquinone. Quinone, $C_6H_4O_2$.

Benzo Red Blue G, R. Dyestuffs identical with Columbia blue G, R, and Diamine blue LG, LR.

Benzoresorcin. Dioxy-benzophenone, $C_6H_5.CO.C_6H_3(OH)_2$.

Benzorhoduline. A dyestuff derived from the carbamides of 2 : 5 : 7-amino-naphthol-sulphonic acid.

Benzosalin. Methyl-benzoyl-salicylate, $C_6H_4(O.CO.C_6H_5)COOCH_3$. Prescribed for gout and rheumatism.

Benzo Sky Blue. See DIAMINE SKY BLUE.

Benzosol. Benzoyl-guaiacol, $C_6H_4(O.$ $CH_3)(O.CO.C_6H_5)$. Recommended as a substitute for creosote in the treatment of intestinal tuberculosis.

Benzo Violet. A tetrazo dyestuff prepared from diazotised benzidine coupled with 1 molecule of α-naphthol-sulphonic acid NW and 1 molecule of α-naphthol-3 : 6-disulphonic acid.

Benzoxate. See ACETOZONE.

Benzoyl Green. See MALACHITE GREEN.

Benzoyl Pink (Rose de Benzoyl). A dyestuff. It is the sodium salt of benzoyl - amino - ditolyl - azo - α - naphthol-sulphonic acid, $C_{31}H_{27}N_3O_5SNa$. Dyes unmordanted cotton pink.

Benzoyl Superoxide. Lucidol (q.v.).

Benzozone. See ACETOZONE.

Benzphetamine. (+) - N - Benzyl - Nα-dimethylphenethylamine. Didrex is the hydrochloride. An appetite suppressant.

Benzquinamide. 2 - Acetoxy - 3 - diethylcarbamoyl - 1,3,4,6,7,11b - hexahydro - 9,10 - dimethoxy - 2H - benzo[a]quinolizine. A tranquilliser.

Benzthiazide. 3 - Benzylthiomethyl - 6 - chlorobenzo - 1,2,4 - thiadiazine - 7 - sulphonamide 1,1-dioxide. Fovane. A diuretic. (307).

Benztropine. 3 - Diphenylmethoxy - tropane. Cogentin is the methanesulphonate.

Benzyanil. Benzyl valerianate.

Benzycin. Sodium-benzyl-succinate. Used in pharmacology.

Benzydamine. 1 - Benzyl - 3 - (3 - dimethylaminopropoxy)indazole.

Benzyl. Acid wool dyestuffs. (530).

Benzyl Blue. A dyestuff made from rosaniline, which dyes wool, silk, and cotton. The term is also used for a mixture of methyl violet and malachite green.

Benzyl Green B. A dyestuff obtained by combining o-chloro-benzaldehyde with ethyl - benzyl - aniline - sulphonic acid, and oxidising the product.

Benzylidene Rubber. A product obtained by treating crepe rubber in carbon tetrachloride with benzyl chloride in the presence of aluminium chloride. It is almost insoluble in organic solvents.

Benzyl Violet (Paris Violet 6B, Methyl Violet 6B, Methyl Violet 5B, Methyl Violet 6B Extra, Violet 5B, Violet 6B). A dyestuff, which is chiefly a mixture of the hydrochlorides of benzyl-penta-methyl - pararosaniline, $C_{31}H_{34}N_3Cl$, and hexamethyl - pararosaniline. It dyes silk and wool violet, and cotton after mordanting with tannin and tartar emetic.

Benzyl Violet 5BN. A dyestuff. It is a British equivalent of Formyl violet S4B.

Benzyphos. Sodium-dibenzyl-phosphate. Used in pharmacology.

Beogex. A proprietary preparation containing sodium acid phosphate and sodium bicarbonate. A laxative. (330).

Bepanthen. A proprietary preparation of pantothenic alcohol or dexpanthenol, used to treat paralytic ileus. (314).

Bephenium Hydroxynaphthoate. Benzyldimethyl-2-phenoxyethylammonium 3-hydroxy-2-naphthoate. Alcopar.

Beplete. A proprietary preparation of phenobarbitone, thiamine hydrochloride, riboflavine, pyridoxine hydrochloride, nicotinamide and alcohol. (245).

Beplex. A proprietary preparation of thiamine, riboflavine and nicotinamide. (245).

Beprochin. See PLASMOQUIN.

Beraunite. A mineral. It is a ferric phosphate, $Fe_2(PO_4)_2.4FePO_4(OH)_3. 3H_2O$.

Berberine. $C_{20}H_{17}NO_4$, a natural dyestuff, used to a small extent for dyeing silk and leather yellow.

Berculon A. A proprietary trade name for Thiacetazone, used in the treatment of tuberculosis. See p-ACETYLAMINO-BENZALDEHYDE THIOSEMICARBAZONE.

Berdo. A proprietary trade name for nickel lead bronzes with hard wearing properties and possessing the ability to function with little or no lubrication. (225).

Berelex. Gibberellic acid. (511).

Berengelite. A pitch-like material found in Peru, used for caulking. A resin asphalt high in oxgyen.

Berenil. A proprietary preparation of DIMINAZENE. A veterinary anti-proto-zoan and bactericide.

Beresowite (Beresovite, Berezovite). A mineral. It is a lead carbonate-chromate, $6PbO.3CrO_3.CO_2$.

Berex. A proprietary preparation of calcium succinate and aspirin. An analgesic. (388).

Berezovskite. A mineral. It is $8[(Fe,Mg)Cr_2O_4]$.

Bergamaskite. A mineral. It is a variety of amphibole.

Bergaminol (Bergamiol). Artificial oil of bergamot. Linalyl acetate, $(CH_3)_2 C : CH(CH_2)_2CCH_3 . (OCOCH_3)CH : CH_2$. Used in perfumery.

Bergamiol. See BERGAMINOL.

Bergamot Essence. Oil of Bergamot.

Berger Mixture. A mixture of zinc, potassium chlorate, a chlorinating agent, such as carbon tetrachloride, and a filler, such as kieselguhr. A smoke screen material.

Bergmannite. A mineral. It is a variety of natrolite.

Bergmann's Powder. An explosive, consisting of 50 per cent. potassium chlorate, 5 per cent. pyrolusite, and 45 per cent. bran, sawdust, or tar.

Bergmehl. Diatomaceous earth or diatomite ($q.v.$).

Berkalloy. A proprietary trade name for a zinc/aluminium alloy in powder form. Used for metal spraying to resist marine corrosion. (137).

Berkalon. A proprietary trade name for a polyamide plastics material produced in powder form suitable for flame-spraying. (137).

Berkatekt. Protective coatings for heat treatment. (572).

Berkathene. A proprietary trade name for polyethylene produced in powder form for flame-spraying. (137).

Berkazide. Bendrofluazide.

Berkbent. Binder for feed pelletizing. (572).

Berkbent CE. A proprietary name for Bentonite ($q.v.$) used in civil engineering. (137).

Berkbent F. A proprietary trade name for Bentonite ($q.v.$) used in animal foodstuffs as a hardening additive. (137).

Berkbond. Bonding agent for foundry sands. (572).

Berkdopa. A proprietary preparation of LEVODOPA used in the treatment of Parkinson's disease. (137).

Berkefeld Insulating Material. A heat insulator. It is a mixture of kieselguhr and cowhair.

Berkeyite. A mineral. It is a variety of lazulite.

Berkfurin. A proprietary preparation containing nitrofurantoin. A urinary antiseptic. (137).

Berkmycen. A proprietary preparation of oxytetracycline. An antibiotic. (137).

Berkomine. A proprietary preparation of imipramine hydrochloride. An antidepressant. (137).

Berkonite. White sodium bentonite. (572).

Berkozide. A proprietary preparation of bendrofluazide. A diuretic. (137).

Berkshire Sand. A purified sea sand used for filtering.

Berkstat D. Internal antistatic agent. (572).

Berkstat L. Internal antistatic agent. (572).

Berkwhite. Paper filler grade of bentonite. (572).

Berlese Mixture. A mixture of molasses and sodium arsenite. It is recommended for poisoning the olive fly.

Berlin Alloy. An alloy containing from 52–63 per cent. copper, 26–31 per cent. zinc, and 6–22 per cent. nickel.

Berlin Black. A mixture of asphalt, boiled oil, and turpentine. A paint.

Berlin Blue. See PRUSSIAN BLUE and CHESSYLITE.

Berlin Brown. A pigment produced by charring Prussian blue. It is a mixture of ferroso-ferric oxide and charcoal, and is used as an artist's colour.

Berlinite. A mineral. It is an aluminium phosphate.

Berlin Red. See INDIAN RED.

Berlin White. White Lead (q.v.).

Bermanite. A mineral. It is $(Mn, Mg)_5(Mn^{...}, Fe^{...})_8(PO_4)_8(OH)_{10}. 15H_2O]$.

Bernit. A proprietary trade name for a cellulose acetate plastic.

Bernthsen's Violet (Isothionine). A dyestuff made from β-dinitro-diphenyl-amine-sulphoxide. It is β-amino-di-phenyl-imide, and is isomeric with Lauth's violet. Dyes reddish-violet.

Beroliet. A proprietary casein plastic.

Berries, Yellow. See PERSIAN BERRIES.

Bersch Bearing Metal. An alloy of 93 per cent. aluminium and 7 per cent. nickel.

Berthierine. A mineral. It is $(Fe^{...},Fe Mg,Al)_{3-x}(Si,Al)_2O_5(OH)_4$.

Berthierite. A mineral containing 58·6 per cent. antimony, 10 per cent. iron, and 29 per cent. sulphur, having a probable formula, $FeS.Sb_2S_3$.

Berthier's Alloy. A copper-nickel alloy containing approximately 32 per cent. nickel.

Bertholite. A name applied to chlorine used as a poison gas.

Berthollet's Neutral Carbonate of Ammonia. Ammonium bicarbonate, NH_4HCO_3.

Berthonite. A mineral of Tunis. It approximates to $2PbS.9Cu_2S.7Sb_2S_3$.

Bertoni's Ether. Tertiary amyl nitrite, $C_5H_{11}NO_2$.

Bertrandite. A mineral. It is beryllium hydro-silicate, $Be_2(BeOH)_2Si_2O_7$.

Beryl (Aquamarine). A mineral. It is a double silicate of aluminium and beryllium, $Al_2Be_3(SiO_3)_6$. A precious stone.

Berylla. Beryllium oxide.

Beryllium Bronze. An alloy of 97·5 per cent. copper with 2·5 per cent. beryllium.

Beryllium Humite. A mineral. It is $4[(Mg,Ba),Si_3O_{12}(OH)_2]$.

Beryllium-Leucite. A mineral. It is $K_2Be_3Si_4O_{12}$.

Beryllonite. A mineral. It is sodium beryllium phosphate, $NaBePO_4$.

Berzelianite. A mineral. It is copper selenide, Cu_2Se.

Berzeliite. A mineral. It is an arsenate of calcium, magnesium, and manganese.

Berzeline. Synonym for Haüyne.

Berzelite. See PETALITE.

Besankura. A preparation consisting mainly of sodium bicarbonate, sodium chloride, sodium sulphate, and yeast, with additions.

Bessemer Pig (Hæmatite Pig). A pig-iron made from hæmatite or other ores free from phosphorus. It contains not more than 0·05 per cent. phosphorus, and 0·1 per cent. sulphur, and is usually high in silicon (3–3·5 per cent.). It also contains manganese.

Bessemer Steel. Steel made in the Bessemer converter. The carbon content can be made from 0·2–0·6 per cent.

Bestanino. A trade name for a sulphited quebracho extract. Used for tanning.

Bestorite. A trade name for a water-proof and fireproof asbestos fabric resembling leather, made from asbestos and vulcanised rubber.

Bestuscheff's Tincture. An ethereal tincture of iron chloride.

Betabeet. A weed killer preparation. (574).

Beta-borocaine. A proprietary preparation of benzamine borate.

Betacaine. See EUCAINE.

Betacaine Lactate. Benzamine lactate. See EUCAINE.

Beta-cardone. A proprietary preparation of sotalol hydrochloride used in the treatment of cardiac dysrhythmias. (444).

Betacetylmethadol. β - 3 - Acetoxy-6 - dimethylamine - 4,4 - diphenyl-heptane O-Acetate of β-6-dimethyl-amino-4,4-diphenylheptan-3-ol.

Betadine. A proprietary preparation of povidone iodine. Skin disinfectant. (137).

Beta-eucaine. See EUCAINE.

Betafite. A mineral. It is a hydrated uranylcolumbate containing 35·5 per cent. Cb_2O_5.

Betaform. See NOVIFORM.

Betaine. Trimethyl-glycine, $O.CO(CH_2).N(CH_3)_3$, found in molasses.

Betaloc. A proprietary preparation of METAPROBOL tartrate used in the treatment of angina pectoris. (477).

Betameprodine. β - 3 - Ethyl - 1 - methyl - 4 - phenyl - 4 - propionyl - oxypiperidine.

Betamethadol. β - 6 - Dimethylamino - 4,4 - diphenylheptan - 3 - ol.

Betamethasone. 9α - Fluoro-11β,17α, 21 - trihydroxy - 16β - methylpregna - 1. 4 - diene - 3,20 - dione. 9α - Fluoro - 16β - methylprednisolone. Betnelan ; Betnesol is the disodium phosphate ; Betnovate is the 17-valerate.

Betamethasone Acibutate. A corticosteroid currently undergoing clinical trial as " GR2/541 ". 21 - Acetoxy - 9α - fluoro - 11β - hydroxy - 16β- methyl - 17 - (2 - methylpropionyloxy) - pregna - 1, 4 - diene - 3, 20 - dione.

Beta-naphthol Orange. See ORANGE II.

Beta-naphthol Sodium. See MICRO-CIDIN.

Betanol. An oil used in the production of para-reds, giving bluer shades.

Betanox. A proprietary trade name for a rubber antioxidant. A ketone-amine reaction product. (163).

Betaprodine. A narcotic analgesic. β - 1, 3 - Dimethyl - 4 - phenyl - 4 - propionyloxypiperidine.

Betaprone. A proprietary trade name for Propiolactone.

Betaquinine. Quinidine, $C_{20}H_{24}O_2N_2$.

Betasol Ot-A. A proprietary trade name for a wetting agent. It is a sulphonated ester.

Beta-sulphopyrin (Sulphopyrin). A compound of sulphanilic acid with antipyrin. Prescribed for influenza and colds.

Betazole. Ametazole.

Betel. Leaves of *Piper betel*.

Betel Nuts. Areca nuts.

Betex. See HYDREX.

Bethanid. A proprietary preparation of bethanidine sulphate. An antihypertensive. (247).

Bethanidine. N - Benzyl - N′N″ - dimethylguanidine. Esbatal is the sulphate.

Betilon. A German product. It is a new derivative of mandelic acid, $CH.C_6H_5(O.SO_3Na).CO_2.CH_2C_6H_5$, containing the benzyl and sulphonate groups. It is stated to be efficacious in the treatment of colic, asthma, and angina pectoris.

Betite. A refined and vulcanised bitumen. An insulating material.

Betnelan. A proprietary preparation of betamethasone. (335).

Betnesol. A proprietary preparation of betamethasone disodium phosphate. (335).

Betnesol-N. A proprietary preparation of betamethasone sodium phosphate and neomycin sulphate. (335).

Betnovate. A proprietary preparation of betamethazone 17-valerate. A dermatological corticosteroid. (335).

Betnovate-C. A proprietary preparation of betamethasone valerate and clioquinol used in dermatology. (335).

Betnovate-N. A proprietary preparation of betamethasone valerate and neomycin sulphate used in dermatology. (335).

Betnovate Scalp. Betamethasone 17-valerate in alcoholic solution for scalp application. (620).

Beto. A material containing 97 per cent. magnesium sulphate, $MgSO_4.7H_2O$, 2 per cent. talc, and traces of cinnamon oil.

Betol (Naphthosalol, Naphtholsalol, Salinaphthol). The salicylic ester of β-naphthol, $C_{10}H_7.O.CO.C_6H_4OH$. An internal antiseptic for catarrh of the bladder.

Beton. See CONCRETE.

Betonet. A trade name for a rigid PVC quick air drying solution coating for concrete and masonry. (254).

Betrox. Sodium chloride fertilizer. (512).

Betsolan. Betamethasone 21 phosphate veterinary preparations. (520).

Bettendorf's Reagent. A concentrated solution of stannous chloride in fuming hydrochloric acid. Used for the detection of arsenic.

Betula Camphor. Betulinol, $C_{24}H_{40}O_2$, an alcohol from *Betula* species.

Beudantite. A mineral, $FePO_4.PbSO_4.Fe_2(OH)_6$.

Beustite. A mineral. It is a variety of epidote.

Bevantolol. A beta adrenergic blocking agent. It is 1 - (3, 4 - dimethoxyphenethylamino) - 3 - *m* - tolyloxypropan - 2 - ol.

Bevonium Methysulphate. An antispasmodic. 2 - Benziloyloxymethyl - 1, 1 - dimethylpiperidinium methylsulphate. ACABEL.

Bewax. Rosin wax emulsion. (575).

Bex. A proprietary phenol-formaldehyde plastic.

Bexoid. A proprietary trade name for a cellulose acetate plastic.

Bextasol. A proprietary preparation of betamethasone valerate as an aerosol, used in asthma therapy. (335).

Bextrene. A proprietary brand of polystyrene. (468).

Beyerite. A mineral. It is $2[Ca(BiO)_2(CO_3)_2]$.

Beyrichite. See MILLERITE.

Bezitramide. A narcotic analgesic. 4 - [4 - (2 - Oxo - 3 - propionylbenzimidazolin - 1 - yl) piperidino] - 2, 2 - diphenylbutyronitrile.

B.F.I. A proprietary preparation of bismuth formic iodine. A skin antiseptic. (310).

Bhang (Ganja, Charas, Hashish). A drug. It is the resinous exudation of the hemp plant, *Cannabis sativa*, " ganja " being the resin of the flowers, " charas " of the young shoots, and " bhang " of the mature leaves. Hashish, an intoxicating drink used by the Arabs and Indians, is prepared from bhang. An extract of the resin has been used in medicine as an anodyne, hypnotic, and antispasmodic.

BHC. γ-hexachlorocyclohexane. A powerful insecticide.

BHC 238 and BHC 318. A plant detergent and steriliser. (576).

BHC 350. A quarternary ammonium compound detergent. (576).

BHC 2008. A fully rinsable quarternary steriliser. (576).

BHT. A proprietary anti-oxidant. 2, 6 - di - tertiary - butyl - p - cresol. (57).

Bhimsiam Camphor. See CAMPHOL.

Biakmetals. A group of alloys some of which are zinc-copper alloys with small amounts of nickel or manganese or both, and others are aluminium-zinc-copper alloys, or zinc-aluminium alloys.

Bialamicol. 3,3' - Diallyl - 5,5' - bisdi - ethylaminomethyl - 4,4' - dihydroxy - biphenyl. Camoform is the dihydrochloride.

Bial's Reagent. A reagent used for the detection of pentoses, giving a green colour. It consists of 1 gram orcinol in 500 c.c. of 30 per cent. hydrochloric acid to which 30 drops of 10 per cent. ferric chloride have been added.

Bianchite. A mineral. It is $8[(Zn,Fe)SO_4 . 6H_2O]$.

Biasbeston. A synthetic varnish-asbestos product.

Biaxial Mica. See MUSCOVITE.

Bibenzonium Bromide. 2 - (1,2 - Diphenylethoxy)ethyltrimethylammonium bromide. Thoragol.

Biberite. A mineral. It consists of normal cobaltous sulphate, $CoSO_4 . 7H_2O$.

Bibiru Bark. Bebeeru bark.

Biborate of Soda. See BORAX.

Bibra Alloy. An alloy of lead with smaller amounts of bismuth and tin.

Bicarton. A product made with synthetic resin and pressboard.

Bice. Pigments. There are two varieties, blue and green. The former is obtained from native copper carbonate, and the latter by mixture of the blue material with yellow orpiment.

Bice, Blue. See BICE.

Bice, Green. See BICE.

Bichloride of Tin. A solution of tin tetrachloride of 50° Be. It is a misleading term.

Bichromate. See BICHROME.

Bichrome (Chrome, Bichromate). Terms used generally for potassium bichromate, but occasionally for sodium bichromate. Used as a mordant for wool dyeing.

Bicillin. A proprietary preparation of penicillin G. An antibiotic.

Bicol. A proprietary laxative comprising 5 mg. tablets of BISACODYL. (875).

Bicreol. A proprietary cream containing precipitated metallic bismuth in creocamph (*q.v.*). An antisyphilitic.

Biddery Ware. An alloy of from 84–90 per cent. zinc, 6–11 per cent. copper, 2·5–3 per cent. lead, and 0·1–1·5 per cent. tin, with a small quantity of iron.

Bidesyl. Diphenyl-diphenylethylene-diketone, $CH.C_6H_5.C_7H_6O.CH.C_6H_5.C_7H_6O$.

Bidetal. A preparation whose active constituent is sodium bicarbonate.

Bidimazium Iodide. An anthelmintic currently undergoing clinical trial as " 65 - 318 ". 4 - (Biphenyl - 4 - yl) - 2 - (4 - dimethylaminostyryl) - 3 - methylthiazolium iodide.

Bidizole. A proprietary trade name for Sulphasomizole.

Bidormal. A proprietary preparation of pentobarbitone sodium and butobarbitone. A hypnotic. (284).

Bidry (Vidry). An Indian alloy, containing zinc, copper, lead, and occasionally tin.

Bieberite. A mineral. It is a sulphate of cobalt, $CoSO_4.7H_2O$.

Bieber's Reagent. Consists of fuming nitric acid, concentrated sulphuric acid, and an equal volume of water. One part of this solution mixed with 5 parts of almond oil gives a yellow-white colour, and with apricot or cherry-kernel oil a cherry-red colour.

Biebrich Acid Blue. A triphenyl-methane dyestuff. It is an acid wool dye.

Biebrich Acid Red 4B. See CHROMOTROPE 2R.

Biebrich Black AO, 4AN, 6AN, 3BO, 4BN, RO. Acid dyestuffs, producing shades of black on wool from an acid bath, sometimes with the addition of copper sulphate.

Biebrich Patent Black 4AN. A dyestuff. It is produced by coupling diazotised naphthionic acid with α-naphthylamine-sulphonic acids (1 : 6 and 1 : 7), diazotising the product, and coupling it with α-naphthylamine.

Biebrich Patent Black BO. A dyestuff produced by coupling diazotised α-naphthylamine-disulphonic acid with α-naphthylamine-sulphonic acids (1 : 6 and 1 : 7), diazotising the product, and coupling it with β-naphthol-disulphonic acid R.

Biebrich Scarlet (Ponceau 3R or 3RB, Ponceau B, Fast Ponceau B, New Red L, Imperial Scarlet, Azobenzene Red, Old Scarlet, Ponceau B Extra, Scarlet Red, Scarlet B, New Red, Scarlet EC). A dyestuff. It is the sodium salt of sulpho - benzene - azo - sulpho - benzene - azo - β - naphthol $C_{22}H_{14}N_4O_7S_2Na_2$. Dyes wool scarlet from an acid bath.

Biebrich Scarlet R, Medicinal. See SUDAN IV.

Bielzite. A resin-asphalt containing sulphur and nitrogen.

Bifacton. A proprietary preparation of Vitamin B_{12} with the intrinsic factor. (316).

Bife. A proprietary mixture of thiamine hydrochloride and ferrous sulphate, used in the treatment of iron deficiency. (877).

Bifluranol. A preparation currently undergoing clinical test as " BX 341 ", for use in the treatment of benign hypertrophy of the prostate. It is $(2R, 3S)$ - 4, 4′ - pentane - 2, 3 - diyldi - (2 - fluorophenol).

Biformyl. Glyoxal, $(CHO)_2$.

Bigall. Bismuth subgallate, $Bi(OH)_2$. $C_7H_5O_5$.

Bigarade Essence. Oil of Bitter Orange Peel.

Bigarré. A mixture of white and black seeds of sesamê. See SUFFETTIL and TILLIE.

Bigatren. Basic bismuth oxyquinoline sulphonate.

Bigeston. A proprietary preparation containing magnesium oxide, aluminium hydroxide and magnesium carbonate as a co-dried gel with simethicone. An antacid. (316).

Bigwood-Ladd Nitroprusside Reagent. For acetone. To 10 c.c. glacial acetic acid add 10 c.c. of a 10 per cent. solution of sodium nitroprusside.

Bikarton. A proprietary trade name for a phenol-formaldehyde synthetic resin. A laminated product.

Bikh. See BISH.

Bikorund. A trade mark for materials of the abrasive and refractory type. They contain crystalline alumina.

Bilagit. A German product containing the sodium salts of all the gallic acid acids, oil of menthol, phenol phthalein, hexamethylene tetramine, novatropine, and papaverine. Tablets are used in cholelithiasis.

Bilbao Iron Ores. Hæmatites from Bilbao, in Spain.

Bilein. Sodium tauroglycocholate.

Bilevon. A proprietary preparation of NICLOFOLAN. A veterinary anthelmintic.

Bilgen Bronze. An alloy of 97 per cent. copper, 1·9 per cent. tin, 0·52 per cent. iron, and 0·24 per cent. lead.

Biligrafin. A proprietary trade name for the bis-meglumine salt of Iodiparnide. (898).

Biligram. See IOGLYCAMIC ACID.

Biliodyl. A proprietary preparation of phenobutiodil, used in oral cholangiography. (336).

Biliposol. A proprietary preparation. It is a solution of bismuth-camphocarbonate in oil.

Bilival. A German preparation. It is a lecithin-sodium-cholate.

Billietite. A mineral. It is $4[BaU_6O_{13}.11H_2O]$.

Billroth's Mixture. An anæsthetic containing 3 parts chloroform, 1 part alcohol, and 1 part ether.

Bilopaque. A proprietary preparation of sodium tyropanoate, used as a contrast medium in radiography.

Bilopoid. A preparation containing the bismuth and lecithin salts of tricamphocarbonic acid.

Biloptin. A proprietary trade name for Sodium Ipodate. (898).

Bilostat. A proprietary trade name for dehydrocholic acid.

Bilston. A basic slag for fertilizing purposes. (509).

Biluen. A bismuth preparation of German origin. It is a suspension of bismuth lactate in olive oil. An anti-syphilitic.

Bima's Redwood. See REDWOODS.

Bimexkitt. A mixture of tar preparations and bituminous asphalts, used in Germany for motor roads. It is stated to have a long life and to produce no dust.

Bimez. A proprietary trade name for Sulphamoprine.

Bimli Hemp. See GAMBO HEMP.

Binarite. Marcasite $(q.v.)$.

Binazine. See TODRALAZINE.

Bindazac. An anti-inflammatory drug. 1 - Benzylindazol - 3 - yloxyacetic acid. See BENDAZAC.

Bindheimite. A mineral. It is a hydrous antimoniate of lead, an oxidation product of Jamesonite.

Bindschedler's Green. A dyestuff. It is a tetramethyl-indamine hydro-chloride, $C_{16}H_{19}N_3HCl$.

Binnite (Cuprobinnite). A mineral, $2As_2S_3.3Cu_2S_2$.

Biobor JF. A proprietary trade name for a biocide comprising a mixture of 2,2′-oxybis-(4,4,6-trimethyl 1,3,3-dioxaborinane) and 2,2′-(1-methyltrimethylenedioxy) bis-(4-methyl 1,3,2-dioxaborinane). It is used to control micro-organisms in jet aircraft fuels.

Biodal. Mono-iodo-dibismuth-methyl-ene-dicresotinate. A powder suggested as an odourless substitute for iodo-form.

Bioferrin. A liquid hæmoglobin pre-paration, obtained by the treatment of ox blood. It is prescribed for chlorosis.

Biogaicol. Guaiacol phosphate, $(C_6H_4O.CH_3O)_3.PO$. Used for administering guaiacol.

Biogastrone. A proprietary prepara-tion of carbenoxolone sodium. A gas-tro-intestinal sedative. (137).

Biogen (Magnesium Perhydrol). A mix-ture of magnesium oxide and mag-nesium peroxide. It contains from 15–25 per cent. magnesium peroxide, MgO_2, and is used as a bleaching agent and antiseptic. Pure magnesium per-oxide is also called magnesium per-hydrol, and is known by the names Novozone and Magnodat.

Biojodin. Tablets containing 40 per cent. sucrose, 25 per cent. lactose, 15 per cent. talc, 11·5 per cent. sodium chloride, 8·5 per cent. sodium bicarbon-ate, and 0·04 per cent. potassium iodide.

Biolactyl. A preparation of lactic acid bacilli.

Biomer. A proprietary segmented poly-ether - urethane (originally Du Pont T - 126), now licensed for the manu-facture of heart valves. (469).

Biomydrin. A proprietary preparation of neomycin, gramicidin, thonzylamine hydrochloride and phenylephrine hydro-chloride. Nasal spray. (262).

Biopar Forte. A proprietary prepara-tion of Vitamin B_{12} and intrinsic factor. Hæmatinic. (327).

Bioral. A proprietary preparation of carbenoxolone sodium. Used to treat peptic ulceration. (137).

Bios. A vitamin. It is Wildier's name for Funk's vitamin D.

Bioson. A nutritious preparation con-taining iron, albuminoids, and lecithin. It consists of 80 per cent. albuminoids, 0·24 per cent. iron, 1·27 per cent. lecithin, 15 per cent. cocoa, and 3 per cent. salts.

Biosone. A proprietary preparation of enoxolone. Antipruritic skin ointment. (137).

Biostat A.1. Oxytetracycline for fish preparation. (577).

Biotexin. A proprietary trade name for the sodium salt of Novobiocin for resis-tant staphylococci for veterinary use. (520).

Biotexin P. Novobiocin and penicillin veterinary preparation. (520).

Biotin. Part of the Vitamin B complex. Not known to be essential to man.

Biotite (Black Mica). A mineral. It is a magnesium-iron mica, $(KH)_2O.(Mg.Fe)_2O_3.Al_2O_3$. A source of aluminium.

Biotrase. A proprietary preparation of trypsin and bithional. Treatment of ulcers and wounds. (347).

Biotren. A proprietary preparation of glycine, zinc bacitracin, neomycin sul-phate, l - cysteine and di-threonine. An anti-bacterial powder. (246).

Biperiden. 1 - (Bicyclo[2,2,1]hept - 5 - en - 2 - yl) - phenyl - 3 - piperidino - propan-1-ol. Akineton is the hydro-chloride or the lactate.

Biphasic Insulin Injection. A hypo-glycæmic agent. It is a suspension of insulin crystals in a solution of insulin buffered at pH 7.

Biphenamine. A proprietary trade name for Xenysalate.

Biquinyl. Quinine-bismuth-iodide. An antisyphilitic.

Birch Bark Rubber. A black gum from the outer layers of the birch tree.

Birch Camphor. Betulinal.

Birch Oil. See OIL OF BIRCH.

Birch Tar Oil. The lighter oils from birch tar.

Bird Glue. See BIRD LIME.

Bird Lime (Bird Glue). A viscid sub-stance existing in various plants, par-ticularly in the bark of *Liscum album*, the mistletoe.

Bird Manure. See GUANO.

Bird Pepper. Capsicum devoid of pungency.

Bireez. The Persian name for gum galbanum.

Birlane. Registered trade-mark (553) for an insecticide. It is 2 chloro-1-(2, 4 dichlorophenyl vinyl) diethyl phosphate. CHLORFENVINFOS.

Birmabright. A corrosion-resisting alloy containing 96 per cent. aluminium, 3·5 per cent. magnesium, and 0·5 per cent. manganese.

Birmasil Alloy. A special alloy which is a nickel-aluminium-silicon alloy con-taining up to 3·5 per cent. nickel and from 8–13 per cent. silicon. It has high tensile strength.

Birmidium. A proprietary trade name for an alloy of aluminium with smaller amounts of copper, nickel, and magnesium. It is similar to Y-alloy.

Birmingham Nickel Silver. See NICKEL SILVER.

Birmingham Platina. A brass con-taining 47 per cent. copper, 53 per cent. zinc, and 0·25 per cent. iron.

Birmite. See BURMITE.

Birmite. A proprietary trade name for urea-formaldehyde synthetic resin.

Bisabol Myrrh. A variety of myrrh, a gum-resin, obtained from *Balsamea erythrea*. Used in medicine as an aromatic.

Bisacodyl. Di - (4 - acetoxyphenyl) - 2 - pyridylmethane. Dulcolax, Bicol.

Bisantol. Bismuth salicylate in oil suspension.

Bisbeeite. Cupric metasilicate, $CuSiO_3.H_2O$, found in the Bisbee Mine, Arizona.

Biscam. Bismuth camphorate in olive oil.

Bischofite. A mineral. It is a chloride of magnesium, $MgCl_2.6H_2O$, and occurs in the Stassfurt deposits.

Bischlorol. A preparation of bismuth oxychloride. An antisyphilitic.

Bisciniod. A combination of cinchonidine hydriodide and bismuth iodide, $C_{19}H_{22}N_2O.HI.BiI_3$.

Bisedia. A proprietary preparation of liquor bismuth, pepsine, morphine hydrochloride, hydrocyanic acid, and tincture of nux vomica. Gastro-intestinal sedative. (829).

Bisermol. A bismuth-mercury amalgam in an oil suspension.

Bisformasal. See FORMASAL.

Bisglucol. A proprietary preparation. It is a suspension of metallic bismuth in isotonic solution of glucose. Used for intramuscular injection in the treatment of syphilis.

Bish (Bikh). The tuberous roots of *Aconitum ferox*, and other species.

Bislumina. A proprietary preparation containing bismuth aluminate and magnesium oxide. An antacid. (249).

Bislumina Suspension. A proprietary preparation containing bismuth aluminate. An antacid. (249).

Bismal. Bismuth-methylene-gallate, $4(C_{15}H_{12}O_{10})+3Bi(OH)_3$. Used as an astringent in diarrhœa.

Bismarck Brown (Aniline Brown, Gold Brown, Manchester Brown, Phenylene Brown, Vesuvine, Leather Brown, Cinnamon Brown, English Brown). A dyestuff. It is the hydrochloride of benzene - disazo - phenylene - diamine, $NH_2.C_6H_4.N_2C_6H_3(NH_2)_2$. Dyes wool, leather, and tannined cotton reddish-brown.

Bismarck Brown R. See MANCHESTER BROWN EE.

Bismarck Brown T. See MANCHESTER BROWN EE.

Bismarsen. Bismuth sulpharsphenamine.

Bismite. (See BISMUTH OCHRE.) A mineral. It is $4[Bi_2O_3]$.

Bismoclite. A mineral. It is $2[BiOCl]$.

Bismocoral. A German preparation.

It is oxy-bismuth-tartramide. An antiluetic.

Bismocymol. Basic bismuth camphorcarboxylate.

Bismogenol. A German preparation. It is a bismuth double compound used in the treatment of syphilis.

Bismolinide. See CRURIN.

Bismon. Colloidal bismuth oxide, Bi_2O_3, produced by the reaction between bismuth salts and an alkaline solution of sodium lysalbinate or protalbinate. Recommended as a mild intestinal astringent in cases of gastro-intestinal catarrh.

Bismosan. An antisyphilitic consisting of a suspension of bismuth hydroxide.

Bismosol. Sodium - potassium - bismuthyl-tartrate in solution.

Bismostab. A preparation of finely divided bismuth metal suspended in glucose solution. An antisyphilitic.

Bismoxyl. A preparation of protein-bismuth obtained by treatment of a bismuth salt with liver extract. It is used in the treatment of syphilis.

Bismucyn. A proprietary preparation of bismuth sodium tartrate. Throat lozenges. (273).

Bismutan (Isutan). A yellow powder consisting of bismuth, resorcinol, and tannin.

Bismuthal. See BISMUTHOL.

Bismuth Amalgam. This usually contains 80 per cent. mercury and 20 per cent. bismuth.

Bismuthan. A combination of bismuth with albumin.

Bismuthaurite. A natural alloy of gold and bismuth, AuBi.

Bismuth Barley (Graupen). The residue from bismuth smelting.

Bismuth Brass. A nickel silver. It contains from 47–52 per cent. copper, 30–31 per cent. nickel, 12–21 per cent. zinc, 0–5 per cent. lead, 0–1 per cent. tin, and 0·1–1 per cent. bismuth.

Bismuth Bronze. An alloy. (*a*) Consists of 1 part bismuth with 16 parts tin ; (*b*) contains 1 part bismuth, 63 parts copper, 21 parts spelter, and 9 parts nickel. Resists sea-water.

Bismuth, Cosmetic. See COSMETIC BISMUTH.

Bismuth Cream. Bismuth metal in a creo-camph base.

Bismuth Glance. A mineral. It is bismuth sulphide, Bi_2S_3.

Bismuth Glycollylarsanilate. Bismuthyl N-glycolloylarsanilate. Glycobiarsol. Milibis.

Bismuth Gold. A mineral, Au_2Bi.

Bismuth Hypoloid. A proprietary preparation of bismuth metal in liquid suspension.

Bismuthine. Bismuth trihydride, BiH_3.

Bismuthinite. A mineral, bismuth sulphide, Bi_2S_3.

Bismuthite. A mineral. It is a basic bismuth carbonate, $3(BiO)_2CO_3.2Bi(OH)_3.3H_2O$, found in South Carolina. The term is also used for the sulphide Bi_2S_3.

Bismuth Magister. See MAGISTER OF BISMUTH.

Bismuth Nickel. A mineral. It is bismuth sulphide, Bi_2S_3, with some nickel sulphide, NiS.

Bismuth Ochre (Bismite). A mineral. It is bismuth trioxide, Bi_2O_3, together with some ferric oxide, water, and carbon dioxide. Synonym for Sillénite.

Bismuth Oil. Bismuth hydroxide in a 1 per cent. solution of camphor in sesamé oil.

Bismuthol (Bismuthal). A mixture of bismuth phosphate and sodium salicylate. Recommended as an antiseptic in the treatment of wounds, and for perspiring feet.

Bismuthose. A bismuth-iron compound containing from 21·5–22 per cent. bismuth. An intestinal astringent.

Bismuthosmaltite. A mineral, Co $(As.Bi)_2$.

Bismuth, Phenol. See PHENOL BISMUTH.

Bismuth Silver. See CHILENITE.

Bismuth Solder. Alloys of from 25–50 per cent. tin, 25–40 per cent. bismuth, and 25–40 per cent. lead. One containing 33 per cent. tin, 33 per cent. bismuth, 33 per cent. lead, melts at 284° F. A solder containing 40 per cent. bismuth, 40 per cent. lead, and 20 per cent. tin melts at 111° C. Also see ROSE'S, KRAFT'S, and HOMBERG'S METALS.

Bismuth Spar. See BISMUTOSPHAERITE.

Bismuth Subcarbonate (Oxycarbonate of bismuth). Bismuth carbonate.

Bismuth, Telluric. See TETRADYMITE.

Bismuth Ternitrate. Bismuth trinitrate. Bismuth nitrate.

Bismuth White (Paint White, Pearl White, Spanish White, Fard's Spanish White). Basic bismuth nitrate, $BiO(NO_3)$, a pigment.

Bismuth Yellow. Bismuth chromate, $3Bi_2O_3.2CrO_3$, a pigment.

Bismuthyl. A preparation of precipitated bismuth in a glucose solution. An antisyphilitic.

Bismutite. A mineral, $Bi_2H_2CO_6$.

Bismutoferrite. A mineral, $Bi_2Fe_4Si_4O_{27}$.

Bismutol. Sodium-potassium-bismuthyl-tartrate.

Bismutoplagionite. A mineral, $5PbS.4Bi_2S_3$.

Bismutose. Bismuth albuminate, a combination of bismuth and albumin. It contains about 22 per cent. bismuth and 66 per cent. albumin.

Bismutosmaltine. See BISMUTHOSMALTITE.

Bismutosphaerite (Bismuth Spar). A mineral. It is bismuthyl carbonate, $(BiO)_2CO_3$.

Bismutotantalite. A mineral. It is $BiTaO_4$.

Bisnene. Sodium-urea-p-amino-phenyl-bismuthate.

Bisoflex. A trade mark for a group of benzoate plasticisers. (470).

Bisoflex DNA. A trade mark. Dinonyl adipate. A vinyl plasticiser. (288).

Bisoflex ODN. A trade mark for an adipic ester plasticiser. (211).

Bisoflex 8N. A trade mark for the condensation product of 2-ethyl hexyl urethane and formaldehyde. A vinyl plasticiser. (288).

Bisoflex L79. A higher straight chain phthalate plasticiser. (288).

Bisoflex 79A. A trade mark. The adipate of mixed C_7–C_9 alcohols. A vinyl plasticiser. (288).

Bisoflex 81. A trade mark. Dioctyl phthalate. A vinyl plasticiser. (288).

Bisoflex 82. A trade mark. Dioctyl phthalate. A vinyl plasticiser. (288).

Bisoflex 88. A trade mark. The phthalate of a C_8 alcohol. A vinyl plasticiser. (288).

Bisoflex 91. A trade mark. Dinonyl phthalate. A vinyl plasticiser. (80).

Bisoflex 100. A trade mark for a plasticiser. Diisodecyl phthalate. (470).

Bisoflex 102a and DOA. Dioctyl adipate. A vinyl plasticiser.

Bisoflex 104. A trade mark for an adipic ester plasticiser. (470).

Bisoflex 106. A primary low temperature plasticiser for PVC and synthetic rubber. (288).

Bisoflex 130. A trade mark for ditridecyl phthalate. A plasticiser. (470).

Bisoflex 610. A trade mark for a plasticiser. C_6 - C_{10} aliphatic alcohol phthalate. (470).

Bisoflex 619. A trade mark for a plasticiser. C_6 - 6_{10} trimellitates. (470).

Bisoflex 791. A trade mark for a plasticiser. Dialkyl (C_7 - C_9) phthalate. (288).

Bisoflex 799. A trimellitate plasticiser for higher temperature resistant PVC for cable use. (288).

Bisoflex 810. A trade mark for a plasticiser. C_8 - C_{10} aliphatic phthalate. (288).

Bisoflex 819. A trade mark for a plasticiser. C_8 - C_{10} trimellitates. (470).

Bisoflex 1002. A trade mark for a polymeric plasticiser. (470).

Bisoflex 1007. A trade mark for a polymeric plasticiser. (470).

Bisoflex L911. A higher straight chain phthalate plasticiser. (288).

Bisol. (1) Soluble bismuth phosphate.

Bisol. (2) Solvents, plasticisers and intermediates. (578).

Bisolene. Proprietary liquid fuels. (578).

Bisolite. Solid fuels, e.g. metaldehyde. (578).

Bisolube. Oil additives. (578).

Bisolvomycin. A proprietary preparation of Bisolvon and oxytetracycline hydrochloride. An antibiotic. (278).

Bisolvon. A proprietary preparation of bromhexine hydrochloride. A mucolytic agent. (278).

Bisomer DALP. A trade name for a diallylphthalate plasticiser. (470).

Bisomer DAM. A trade name for a plasticiser. Dialkyl maleate. (C_7 - C_9 alcohols). (470).

Bisomer DBF. A trade name for a dibutyl fumarate plasticiser. (470).

Bisomer DBM. A trade name for a dibutyl maleate plasticiser. (470).

Bisomer D10M. A trade name for a Diisooctyl maleate plasticiser. (470).

Bisomer DNM. A trade name for a dinonyl maleate plasticiser. (470).

Bisomer DOM. A trade name for a diethyl hexyl maleate plasticiser. (470).

Bisomer 2HEA. A proprietary trade name for 2-hydroxy-ethyl-acrylate. A monomer which permits the production of polymers with side chain hydroxyl groups suitable for further reaction (cross-linking) and the production of thermosetting acrylic adhesives. (288).

Bisomer 2HEMA. A proprietary trade name for 2-hydroxy-ethyl methacrylate. A monomer which permits the production of polymers with side chain hydroxyl groups suitable for cross-linking and the production of thermosetting acrylic surface coating adhesives. (288).

Bisomer 2HPMA. As above except that the monomer is 2-hydroxy-propyl-methacrylate. (288).

Bisoprufe. Preparations for waterproofing cements. (578).

Bisoxatin. A laxative. 2, 3 - Dihydro - 2, 2 - di - (4 - hydroxyphenyl) - 1, 4 - benzoxazin - 3 one. Laxonalin is the diacetate.

Bisoxol. A registered trade mark for antioxidants.

Bisoxyl. Bismuth oxychloride in liquid suspension. An antisyphilitic. (182).

Bissa Bol. An Indian trade name for Habbak haddi, a gum resin from *Commiphora Erythorœa*, now used in Europe as opoponax.

Bissy Nuts. Kola seeds.

Bistabillin. Penicillin preparation. (502).

Bistose. The wool obtained by shearing twice a year.

Bistoval. An arsenic-bismuth compound used in the treatment of syphilis.

Bistovol. A proprietary preparation consisting of an oily suspension of bismuth acetyl-oxy-amino-phenyl-arsenate. Used for intramuscular injection in the treatment of syphilis.

Bistre (Brom Lake). A bituminous brown pigment prepared from the soot of wood. Used in water-colours.

Bisulphol. A French preparation of colloidal sulphur, recommended for rheumatism and skin diseases.

Bisupen-Heyden. A German preparation. It is an oil suspension of bismuth salicylate, containing 6 per cent. bismuth. Used in the treatment of syphilis.

Bisuprol. A proprietary preparation consisting of 7 per cent. colloidal bismuth in oil. Used in the treatment of syphilis.

Bitaco. Black bituminous paint. (580).

Bi-Tarco. Tar compounds for roads. (517, 573, 580).

Bithionol. 2,2′ - Thiobis - (4,6 - dichlorophenol). Present in Biotrase.

Bitit. See ESBENIT.

Bitose. A proprietary trade name for a refined bitumen.

Bitran. Adhesion agents. (547).

Bitran H. A proprietary long-chain cyclic polyamine. (194).

Bitrex. A bitter principle for denaturing alcohol. It has been described as denatonium benzoate. (579).

Bitter Almond Essence. See ESSENCE OF BITTER ALMONDS.

Bitter Almond Oil. Benzaldehyde, $C_6H_5.CHO$. See OIL OF BITTER ALMONDS.

Bitter Almond Oil Camphor. Benzoin (*q.v.*).

Bitter Almond Oil Green. See MALACHITE GREEN.

Bitter Apple. Colocynth pulp.

Bitter Ash. Quassia.

Bitter Bark. See ALSTONIA.

Bittern. The mother liquor remaining after the crystallisation of sodium chloride from sea-water. It is a source of magnesium, and also contains bromides and iodides.

The same term is used for a mixture of equal parts of quassia extract and sulphate of iron, 2 parts of extract of *Cocculus indicus*, 5 parts of Spanish liquorice, and 8 parts of treacle. Used to sophisticate beers.

Bitter Salt. See EPSOM SALTS.

Bitter Salt Spar. See DOLOMITE.

Bitter Spar. A mineral. It is a ferruginous variety of dolomite (*q.v.*).

Bitter Waters. Waters containing either sodium or magnesium sulphates, or both.

Bitter Wood. Quassia.

Bituba. A synthetic resin varnish canvas product.

Bitumarnish. Bituminous varnish. (581).

Bitumastic. A proprietary trade name for a spirit paint made from refined coal-tar pitch, etc.

Bitumen (Mineral Pitch, Natural Asphalt, Compact Bitumen, Jew's Pitch, Bitumen of Judæa, Naphthine). A hard pitchy material found at the surface of the Dead Sea, and in the pitch lake at Trinidad. The terms are applied to a number of mineral substances containing mainly hydrocarbons. Asphalt mineral, mountain or rock tallow, or hatchettine, and mineral caoutchouc or elaterite, are all bitumens. A mixture of asphalt with drying oil, used as a pigment, is also known as bitumen.

Bitumen, Artificial. See GOUDRON.

Bitumen, Elastic. See ELATERITE.

Bitumen of Judæa. See BITUMEN.

Bituminol. Ammonium - sulpho - bituminolate. A German manufacture, used as a substitute for ichthyol.

Bitumoid. A soft mineral rubber with a melting-point of 80° C.

Bituvar. Anticorrosive paint. (541).

Bituvia. A proprietary trade name for a tar for road purposes.

Bityite. A mineral. It is $Li_6(Ca,Be)_{11}Al_{32}Si_{20}O_{105}.14H_2O$.

Biuret Reagent (Gies). A 10 per cent. potassium hydroxide solution to which 25 c.c. of a 3 per cent. copper sulphate solution per litre has been added. A purple-violet or pinkish-violet colour is obtained with $CONH_2$ groups, such as are contained in biuret and oxamide.

Bivatol. Basic bismuth-*a*-carboxethyl-β-methyl-nonoate.

Bixbite. A mineral. It is a variety of beryl.

Bixbybite. A mineral. It is $16[(Mn,Fe)_2O_3]$.

Bjelkite (Cosalite). A mineral containing 43 per cent. bismuth, 40 per cent. lead, iron, and sulphur.

Black Albumen. A lower quality of blood albumen. It is almost black in colour, and is obtained from blood serum. Used in sugar refining.

Black, Animal. See ANIMAL BLACK.

Black Antimony (Antimony Needle, Iron Black, Japanese, Antimony). Mainly consists of the native sulphide of antimony, Sb_2S_3 (stibnite), freed from impurities by fusion. Used in pharmacy.

Black Arsenic. The arsenic obtained by the sublimation of arsenic in a tube of hydrogen. It condenses on the cooler part of the tube.

Black Ash. The product obtained by heating salt cake, Na_2SO_4, limestone, $CaCO_3$, and powdered coal together, in the Leblanc soda process.

Black Balsam. Peru Balsam.

Black-band Ironstones. Iron ores in Lanarkshire. They consist essentially of ferrous carbonate, mixed with considerable quantities of coaly matter, and contain about 30 per cent. iron.

Black Bellite. An explosive, containing 61 per cent. ammonium nitrate, 12 per cent. trinitro-toluene, 24 per cent. sodium chloride, and 3 per cent. plumbago.

Black-boy Gum. See ACAROID BALSAM.

Black Bronze. An alloy of 50 per cent. lead, 40 per cent. tin, and 10 per cent. antimony. Used in the cutlery trade.

Black Catechu. See CUTCH.

Black Chalk. A clay containing carbon, found in Carnarvonshire.

Black Cobalt. See ASBOLANE.

Black Copper. See BLISTER COPPER and MELACONITE.

Black Cummin. Nigella seeds.

Black Dammar. A resin which exudes from *Canarium strictum*, a South Indian tree. An average grade has an acid value of 81·5, a saponification value of 94, and melts at 123° C. There is also a black dammar of Borneo and Sumatra with an acid value of 13 and a saponification value of 34·3. Black dammar is partially soluble in acetone and alcohol, and completely soluble in benzene and turpentine.

Black-damp. See CHOKE-DAMP.

Black Diamond. See CARBONADO.

Black Draught. Compound mixture of senna (*Mistura sennæ composita B.P.*).

Black Drink. See APPALACHIAN TEA.

Black Drop (*Acetum opii*). A vinegar.

Black Earthy Cobalt. See ASBOLANE.

Black Fish Oil. Malon oil.

Black Flux. A reducing compound, produced by heating potassium bitartrate. It consists of a mixture of carbon and potassium carbonate, and is used for the reduction of arsenic compounds in testing for arsenic.

Black Gip. Indian ink (*q.v.*).

Black Glass. Produced by excess of colouring matter, such as manganese, cobalt, or iron.

Black Gold. See MALDONITE.

Black Grain. The cochineal insect killed at three months old, in hot water, is called black grain.

Black Hypo. See BURNT HYPO.

Blacking. Mixtures of varying composition, mostly bone black, wood oil, hog fat, syrup, and sulphuric acid, or without sulphuric acid, but with turpentine or shellac. See BOOT POLISHES.

Black Iron. See PYROLIGNITE OF IRON.

Black Iron Liquor. See PYROLIGNITE OF IRON.

Black Iron Oxide. Magnetic ferric oxide, $Fe_2O_3(OH)_2$.

Black Jack (False Lead). The miner's name for zinc blende or zinc sulphide, ZnS. The term Black Jack is also applied to black draught (q.v.).

Black Jam. Confection of senna.

Black Japan. A mixture of asphalt, boiled oil, terebene, and turpentine. A paint.

Black Lead. See GRAPHITE.

Black Lead Ore. See ALQUIFON.

Blackley Blue. A dyestuff. It is the sulphonation product of a peculiar kind of phenylated rosaniline, prepared by acting upon crude magenta with aniline. Used for dyeing paper pulp. See SOLUBLE BLUE.

Black Liquor. See PYROLIGNITE OF IRON.

Black Manganese. See MANGANESE BLACK.

Black Mercurial Lotion. See BLACK WASH.

Black Metal. An electrolytic deposit in a black form of certain metals, such as platinum and palladium.

Black Mica. See BIOTITE.

Black Mordant. See PYROLIGNITE OF IRON.

Blackmorite. A mineral. It is a yellow variety of opal.

Black Mustard Seed. The ripe seed of *Brassica nigra*.

Black Nickel. An electro-deposited mixture of nickel, sulphur, and zinc.

Black Ore. See KUROKO.

Black Oxide of Bismuth. Hypobismuthous oxide, Bi_2O_2.

Black Oxide of Cobalt. Cobalto-cobaltic oxide, Co_3O_4.

Black Oxide of Copper. Copper monoxide, CuO.

Black Oxide of Iron. See MAGNETIC IRON ORE.

Black Oxide of Manganese. Manganese dioxide, MnO_2.

Black Paste. See FAST BLACK.

Black Phosphorus (Hittorf's Phosphorus). Prepared by heating yellow phosphorus with lead in a sealed tube.

Black Powder. See GUNPOWDER.

Black Precipitate. A black compound having the formula $Hg_2O.NH_2.Hg_2NO_3$. Used in medicine.

Black Prussiate of Potash. Potassium perferricyanide, $K_2[Fe(CN)_5H_2O]$.

Black Rouge. Black ferric oxide (magnetic iron oxide), $FeO.Fe_2O_3$.

Black Ruthenium Hydroxide. Hydrated ruthenium sesquioxide, $Ru_2O_3.3H_2O$.

Black Salt. Obtained in the Leblanc soda process. It consists of crystals of sodium carbonate, $Na_2CO_3.H_2O$.

Black Sassafras. Oliver Bark.

Black Silver. See PYRARGYRITE.

Black Solder. Usually an alloy of 58 per cent. zinc, 40 per cent. copper, and 2 per cent. tin.

Black Spodium. Animal charcoal (q.v.).

Blackstrap. A low grade of cane molasses.

Black Sugar. Liquorice extract in sticks.

Black Sulphur. Impure native sulphur.

Black Tea. A class of tea prepared by exposing the leaf to sun and air, so that changes due to oxidation and fermentation take place, and drying.

Black Tellurium. See MAGYAGITE.

Black Tin. Tinstone after the removal of sand, iron, and copper.

Black Tung Oil. Tung oil obtained by the hot pressing method.

Black Varnish, Burma. See THITSI.

Black Varnishes. These are usually asphalt varnishes or stearin pitch varnishes, using asphalt or pitch with benzine, with or without the addition of boiled oil. Cheaper kinds are made from coal-tar pitch and benzene.

Black Vitriol. Bluish-black crystals, $(Cu.Mg.Fe.Mn.Co.Ni)SO_4.7H_2O$, obtained from the mother liquor of the copper sulphate, at Mansfield. It is isomorphous with green vitriol, $FeSO_4.7H_2O$.

Black Wash (Black Mercurial Lotion) (*Lotio Hydrargyri Nigra B.P.*). It contains mercuric chloride, glycerin, mucilage of tragacanth, and lime-water.

Bladder Green. See SAP GREEN.

Blakeite. Coquimbite (q.v.).

Blanc D'Argent. See FLAKE WHITE.

Blanc de Baleine. Spermaceti cetaceum.

Blanc de Ceruse (Blanc de Plomb). Lead carbonate, $PbCO_3$.

Blanc de Fard (Blanc D'espagne). Bismuth subnitrate, $Bi(NO_3)_3.Bi_2O_3.3H_2O$. Employed in medicine, as a flux for enamels, and as a cosmetic.

Blanc de Meudon. See BLANC DE TROYES.

Blanc de Neige. Zinc oxide, ZnO.

Blanc de Perle. Precipitated bismuth oxychloride, or a mixture of 1 part zinc oxide and 1 part bismuth oxychloride. See PEARL WHITE.

Blanc de Plomb. See BLANC DE CERUSE.

Blanc D'Espagne. See BLANC DE FARD.

Blanc de Troyes (Blanc de Meudon). White chalk.

Blanc de Zinc. Zinc oxide, ZnO.

Blanc Fixe. See BARIUM WHITE.

Blanchite. See HYDROSULPHITE.

Blancol. An optical brightening agent. (582).

Blandine. See PARAFFIN, LIQUID.

Blandite. A patented rubber substitute made from oxidised linseed oil, chloride of sulphur, and asphalt.

Blandlax. A laxative. (502).

Blandola. Pure vegetable gelatine.

Blandthax. A blackquarter/anthrax vaccine. (502).

Blanfordite. A mineral. It is a variety of pyroxene.

Blankit. See HYDROSULPHITE.

Blanko-Blech. An alloy of 80 per cent. copper and 20 per cent. nickel.

Blanquette (Salicor, Soude douce). A crude soda. It is the ash from soda plants growing in France.

Blapsin. A German product. It consists of ceramin and chrysarobin. Used as a soap and ointment for psoriasis.

Blast Furnace Gas. The waste gas derived from furnaces used in the smelting of iron and other ores. It consists of from 25–30 per cent. carbon monoxide, 2–5 per cent. hydrogen, 2–4·4 per cent. methane, 8–12 per cent. carbon dioxide, and 53–59 per cent. nitrogen. When coal is used, tarry matter and hydrocarbons are present.

Blast Furnace Slag. Consists of the earthy material of the ore, the ash of the coke, and the lime of the limestone. Essentially, it is a silicate of lime and alumina.

Blastine. An explosive. It is a mixture of ammonium perchlorate, and sodium nitrate, with dinitro-toluene and paraffin wax.

Blasting Gelatin. A blasting explosive. It is the jelly-like mass obtained when nitro-cotton is added to nitroglycerin, and contains from 90–93 per cent. nitro-glycerin, and 7–10 per cent. dry nitro-cotton.

Blasting Oil (Nitroleum). "Nitroglycerine"—Glyceryl trinitrate.

Blatt Gold. A brass containing 77 per cent. copper and 23 per cent. zinc.

Blatt Silver. An alloy of 91·1 per cent. tin, 8·25 per cent. zinc, 0·35 per cent. lead, and 0·23 per cent. iron.

Blaud's Pill. Iron pill (q.v.).

Blau Gas (Blue Gas). An illuminating gas made by decomposing oil, purifying it, and subjecting it to a pressure of 100 atmospheres. It contains 52 per cent. hydrogen, 44 per cent. methane,

and some ethylene. It has a calorific value of 1,800 B.T.U. per cubic foot, and a specific gravity of 1·08.

Ble. A proprietary antioxidant for rubber. It is a ketone-diarylamine-type reaction product. It is a liquid with a specific gravity of 1·087, and is insoluble in water. (163).

B.L.E. 25. A ketone-amine reaction product. An antioxidant. (163).

Bleached Shellac. Shellac which has been bleached in solution by means of sodium hypochlorite. The lac is precipitated by addition of sulphuric acid.

Bleaching Earth (Florida Earth). An aluminium - magnesium - hydro - silicate, $4MgO.3Al_2O_3.25SiO_2$. Used for bleaching fats.

Bleaching Liquid. See BLEACH LIQUOR.

Bleaching Powder (Calcium oxymuriate, Chloride of lime, Chlorinated calcium oxide, Chlorinated lime, Oxymuriate of lime, Calcium hypochlorite). Prepared by saturating slaked lime with chlorine, $Ca(OCl)Cl$. An oxidising and bleaching agent.

Bleach Liquor (Bleaching Liquid, Liquid Chloride of Lime). Prepared by passing chlorine through milk of lime. It is a solution of chlorinated lime.

Blei (Plomb, Plumbum). Lead, Pb.

Bleiantimonglanz. Synonym for Zinckenite.

Bleiarsenit. Synonym for Dufrenoysite.

Bleicyanamid. Lead cyanamide. A rust inhibitor. (3).

Bleinierite. A mineral. It is basic lead antimonate, $Pb_3(SbO_3)_2(OH)_4.2H_2O$, found in Siberia, and in Cornwall.

Bleiwismuthglanz. Synonym for Galenobismutite.

Blenal. The carbonic acid ester of santalol, $CO(OC_{15}H_{23})_2$. Employed as an internal remedy in the treatment of gonorrhœa.

Blenda. See OIL WHITE.

Blende. Native zinc sulphide, ZnS. It usually contains iron sulphide, which gives it a black colour, hence the name, Black Jack, applied to it. See ZINC BLENDE.

Blende, Cadmium. See GREENOCKITE.

Blende, Mercury. See CINNABAR.

Blende, Nickel. See MILLERITE.

Blenderm. A proprietary preparation of polythene adhesive tape for surgical use. (471).

Blennostasine. Cinchonine-dihydrobromide. A sedative for the brain and spinal cord.

Bleomycin. An anti-neoplastic antibiotic produced by *Streptomyces verticillus*.

Bleu D'Azure. See SMALT.

Bleu de Garance. Artificial ultramarine (*q.v.*).

Bleu de Lyon. See ANILINE BLUE, SPIRIT SOLUBLE.

Bleu de Nuit. See ANILINE BLUE, SPIRIT SOLUBLE.

Bleu de Paris. See METHYL BLUE.

Bleu de Saxe. See SMALT.

Bleu Direct. See METHYL BLUE.

Bleu Fluorescent. A colouring matter made from dicyoresorufin. It dyes silk and wool blue, with a brownish fluorescence. See FLUORESCENT BLUE.

Bleu Lumière. See ANILINE BLUE, SPIRIT SOLUBLE.

Bliabergite. A mineral. It is a variety of mica.

Blind Coal. A Scotch term for anthracite.

Blister Copper (Black Copper). A copper produced from copper matte by oxidation, then heating with coal in the furnace, when the metallic oxides are reduced. It contains 90–95 per cent. copper, 1–2·5 per cent. iron, and 0·5–2·5 per cent. sulphur.

Blister Gas. Mustard gas (*q.v.*).

Blistering Beetles or Flies. Cantharides.

Blistering Collodion. (*Collodium vesicans B.P.*) A solution of pyroxylin in acetic ether containing the active principle of cantharides.

Blister Steel. Steel after its withdrawal from the chests of charcoal in the cementation process.

Blocadren. A proprietary preparation of trinolol maleate used in the treatment of angina pectoris. (472).

Block Brass. See BRASS.

Blockite. A selenium ore from Bolivia, approximating to $(Ni.Cu)Se_2$.

Block Tin. Tin oxide, called mine tin, is reduced, and the fused metal is run off. A less fusible tin alloy containing small quantities of arsenic, copper, iron, or lead, remains, known as block tin.

Blodite. See ASTRAKANITE.

Blomstrandine. A rare earth mineral consisting of a titano-columbate of the yttrium metals.

Blood Albumen. Serum albumin, used as a mordant in dyeing.

Blood Charcoal. Obtained by the evaporation of blood with potassium hydroxide and carbonising. Used for decolorising substances.

Bloodit. A synthetic gum for process engraving. (515).

Blood Meal. A nitrogenous fertiliser prepared by coagulating blood, drying and grinding the product. It contains on an average, from 11–14 per cent. nitrogen, and 0·75 per cent. phosphorus.

Blood Molasses. A manufactured product from beet molasses.

Blood Red. See INDIAN RED.

Bloodstone (Heliotrope). A mineral. It is a variety of chalcedony of a dark green colour with bright red spots. The term bloodstone is also applied to a hard compact variety of hæmatite, Fe_2O_3, used for polishing.

Bloom. The term given to a mass of iron after it leaves the puddling furnace.

Bloom, Cobalt. See ERYTHRITE.

Bloom, Nickel. See ANNABERGITE.

Bloom, Zinc. See HYDROZINCITE.

Blown Alumina. Molten alumina as it issues from the furnace is met by a high pressure air or steam blast from a nozzle situated just below the spout where the alumina emerges. The material is blown into fragments which are in the form of hollow globules ranging in size from very fine ones to those of $\frac{3}{16}$ in. or more. The wall thickness as a rule does not exceed 0·01 in.

Blown Oils (Oxidised Oils, Polymerised Oils, Soluble Castor Oils, Base Oil). When semi-drying vegetable oils, marine animal oils, and liquid waxes are warmed at from 70–120° C., and a current of air blown through them, the oils oxidise to viscid fluids, miscible with mineral oils. Such oils are known as soluble castor oils. They are rich in triglycerides of the hydroxy-acids, and are used as lubricants. If the blowing is continued for a long time, thickened oils result.

Blown Pitch. Usually a petroleum pitch produced by blowing air through residual oils. Hydrolene and Byerite are trade names for pitches of this type.

Blue 1900 (Deep Blue Extra R, Violet Moderne). Leucogallocyanines, giving blue to violet shades upon a chrome mordant.

Blue Acid. Sulpho-nitronic acid, H_2SNO_5, sometimes found in the Gay-Lussac towers in the lead chamber process for sulphuric acid.

Blue Aniline. See ANILINE OIL.

Blue Asbestos. See CROCIDOLITE.

Blue Ashes. See CHESSYLITE and MOUNTAIN BLUE.

Blue Basic Lead Sulphate. A basic lead sulphate containing lead oxide, sulphite, sulphide, zinc oxide, and carbon, produced from lead ore by volatilisation. It is used in rubber mixing and in priming paint.

Blue Bice. See BICE.

Blue Billy (Burnt Ore). The residual ferric oxide after treatment of iron pyrites, FeS_2, for sulphur. The name " Blue Billy " is also applied to a dark blue rosin oil.

Blue-black (Vine Black). A black pigment obtained by carbonising vine twigs and the like.

Blue-black B (Azo Black O, Azo Black, Naphthol Black). A dyestuff. It is the sodium salt of β-sulpho-naphthalene - azo - α - naphthalene - azo - β - naphthol-disulphonic acid, $C_{30}H_{17}N_4O_{16}S_3Na_3$. Dyes wool bluish-violet from an acid bath.

Blue-black R. A dyestuff similar to Blue-black B.

Blue Bricks (Acid-resisting Bricks). Bricks made from clay containing a considerable quantity of oxide of iron. They are fired at a high temperature, under such conditions as to reduce ferric oxide to the ferrous condition.

Blue, Brunswick. See BRUNSWICK BLUE.

Blue Butter (Blue Unction). Ung. Hydrarg. Mit. B.P.

Blue Carmine. Indigo carmine (q.v.).

Blue CB. See INDULINE, SOLUBLE.

Blue CB, Spirit Soluble. See INDULINE, SPIRIT SOLUBLE.

Blue Cobalt Colours. These essentially consist of glass fluxes or alumina (cobalt aluminates with alumina) coloured blue with cobalt oxide.

Blue Copper. See COVELLITE.

Blue Copperas (Bluestone, Blue Vitriol, Roman Vitriol). Copper sulphate, $CuSO_4.5H_2O$.

Blue Cross Gas (Sneezing Gas). Diphenyl-chloro-arsine, $(C_6H_5)_2.AsCl$. A military poison gas.

Blue Felspar. See CHESSYLITE.

Blue Flax. Flemish flax.

Blue Gas. See BLAU GAS.

Blue Glass. The colour is produced by cobalt or copper.

Blue Gold. A jeweller's alloy, consisting of 75 per cent. gold and 25 per cent. iron.

Blue Ground. A mineral formed in volcanic pipes. Essentially it is a hydrated silicate of magnesium. Diamonds are found in this earth.

Blue Gum. Eucalyptus gum.

Blue Iron Earth (Blue Iron Ore, Mullicite, Vivianite). A mineral. It is hydrated triferrous phosphate, $Fe_3(PO_4)_2.8H_2O$, and occurs in North America.

Blue Iron Ore. See BLUE IRON EARTH.

Blueite. A mineral. It is pyrites containing nickel.

Blue Jack. A Scotch mineral oil.

Blue John. A dark purplish, fibrous variety of fluorspar (calcium fluoride, CaF_2), found in Derbyshire.

Blue Lead. A blue sulphide of lead produced in the sublimation of galena, PbS. It is used as a pigment for the production of priming coats. The term is also applied to galena, to distinguish it from white lead ore, or carbonate.

Blue Malachite. See CHESSYLITE.

Blue Metal. In the treatment of coarse metal to produce matte copper, if the iron sulphide is insufficiently oxidised, " blue metal " results. It contains 50–55 per cent. copper.

Blue Mould. *Penicillium Glaucum*, a fungus. See PENICILLIN.

Blue of Borrel. A dyestuff prepared by the action of silver oxide upon methylene blue. Used in microscopy.

Blue of Manson. A dyestuff prepared by the action of sodium borate upon methylene blue. Used in microscopy.

Blue Oil. Cœrulignol, $C_9H_{10}(OCH_3)(OH)$. The term is also used for linseed oil, thickened by boiling, with the addition of 6–10 per cent. Paris blue. Used for making asphalt varnishes. Also see ANILINE OIL.

Blue Ointment. See MERCURY OINTMENT.

Blue Oxide of Tungsten. Ditungsten pentoxide, W_2O_5.

Blue Pill. See MERCURY PILL.

Blue Powder (Zinc Dust, Zinc Fume). A by-product in the smelting of zinc. It consists of a mixture of finely divided zinc and zinc oxide. It has the power of absorbing hydrogen, and is, therefore, used in the chemical industries. Employed to discharge locally the colour of dyed cotton goods. It is also used for the recovery of gold from the cyanide solution of the metal.

Blue PRC. See CHROMOCYANINE V AND B.

Blue Pyoctanin. See PYOCTANIN.

Blue Ridge. See Catalpo.

Blue Rosin Oil. A type of rosin oil having a dark blue colour.

Blue Salt. See INDIGOSOL O.

Blue Salts. Nickel sulphate, $NiSO_4.7H_2O$.

Blue Silver. See NIELLO SILVER.

Blue Size. Originally it was size dyed heavily with logwood with the addition of alum or ferrous sulphate to produce a blue-black, but aniline dyes are now used for the colour. It has been used by shoemakers to fill up pores in leather.

Blues, Laundry. Ultramarine, indigo, Prussian blue, methylene blue, and aniline blue, together with starch, clay, gypsum, chalk, and sometimes glycerin, are all sold under this name. They are used to colour fabrics with a blue tint, to cover the yellowish tint produced by alkali in washing.

Blue Spar. See CHESSYLITE.

Blue Spinel. A mineral. It is magnesium aluminate, Al_2MgO_4.

Bluestone. See BLUE COPPERAS. Also the name for a greyish-blue felspathic rock.

Blue-stripe. A proprietary trade name for silicon carbide paper and cloth.

Blue Sulphur (Green Sulphur). Formed by mixing a solution of ferric chloride with sulphuretted hydrogen water, when the liquid becomes blue or green in colour.

Blue Unction. See BLUE BUTTER.

Blue Verdigris. See VERDIGRIS. It consists principally of the basic copper acetate, $(C_2H_3O_2)_2Cu_2O$.

Blue Vitriol. See BLUE COPPERAS.

Blue, Washing. See BLUES, LAUNDRY.

Blue Water. A solution of copper sulphate obtained from copper-bearing earths.

Blue Water Gas. Water gas, not carburetted. It is an impure mixture of hydrogen and carbon monoxide. A typical analysis gave 52 per cent. hydrogen, 38·5 per cent. carbon monoxide, 4·5 per cent. carbon dioxide, 4 per cent. nitrogen, and 1 per cent. methane.

Bluish Eosin. See ERYTHROSIN.

Blumenbachite. Alabandin (q.v.).

Blümner's Benzine. A motor spirit prepared by Blümner's method for the continuous distillation of tar and oils. It is a " cracked " benzine.

Blythite. A mineral. It is $8[Mn_3'''Mn_2'''Si_3O_{12}]$.

B-mixture. A term used for the product obtained by pouring rubber latex into an equal volume of boiling water. The rubber is in a flocculated form in the mixture.

B.M. Mixture. A mixture for producing smoke screens. It is an American preparation containing 35 per cent. zinc, 42 per cent. carbon tetrachloride, 9 per cent. sodium chlorate, 5 per cent. ammonium chloride, and 8 per cent. magnesium carbonate.

" B " Naphtha. A fraction of American petroleum oil, of specific gravity 0·714–0·718.

B.N. Powder. A French military powder similar to Poudre B.

Boas Reagent. A solution of 5 grams resorcinol and 3 grams sugar in 100 c.c. dilute alcohol. It is used to test for hydrochloric acid in the gastric juices, giving a rose-red colour.

Bobbinite. A blasting explosive used in coal mines. There are two varieties, one of which contains 62–65 per cent. potassium nitrate, 17–19·5 per cent. wood charcoal, 1·5–2·5 per cent. sulphur, together with 13–17 per cent. of a mixture of ammonium sulphate and copper sulphate. The other variety contains potassium nitrate, charcoal, sulphur, starch, and paraffin wax.

Bobierre Metal. A brass containing 58–66 per cent. copper and 34–42 per cent. zinc. Ships' sheathing metal.

Bobierrite (Hautefeullite). A mineral, $Mg_3(PO_4)_2.8H_2O$.

Bobrovkite. A natural nickel-iron alloy found in the Urals.

Bocasan. A proprietary preparation of sodium perborate and sodium hydrogen tartrate used in the treatment of mouth and gum infections. (473).

Bodenbenderite. A new mineral of Argentina, $(Mn.Ca)_4.Al[(Al.Yt)O][(Si.Ti)O_4]_3$.

Bodenite. A mineral. It is a variety of Allanite, containing yttrium and cerium metals.

Bodenstein. Synonym for amber.

Bodryl. A proprietary preparation of bromodiphenhydramine hydrochloride, aspirin, phenacetin, phenylephrine, caffeine and aluminium hydroxide dried gel. Cold remedy. (264).

Boea Gum. A fossil resin of the Manila copal class. A varnish gum.

Boehmite. A mineral. It is a form of bauxite, $Al_2O_3.H_2O$.

Boehringer's Coagulating Powder. A powder consisting chiefly of aluminium lactate. It is used for coagulating rubber latex.

Bœumlerite. A mineral, $KCl.CaCl_2$.

Boffinite. An explosive. It is the same as Bobbinite.

Bog Butter (Butyrellite). A substance resembling adipocere, occasionally found in peat.

Boghead Mineral. A bituminous shale, found in Scotland.

Bog Iron Ore (Pea Iron Ore, Clay Iron Stones, Lake Ore, Brown Iron Ore). Earthy varieties of Limonite (q.v.).

Bog Manganese. See WAD.

Bogoslovskite. Chrysocolla (q.v.).

Bohemian Earth (Veronese Earth, Tyrolean Earth, Seladon Green, Cyprian Earth, Terre Verte, Stone Green). Green earths, which are products of the disintegration of minerals, chiefly of the hornblende type. Stone green is a mixture of ground green earth and white clay, and has been used for the manufacture of waterproofing paints.

Bohemian Garnet. A mineral. It is pyrope, or magnesium-aluminium-garnet, $Mg_3Al_2Si_3O_{12}$.

Bohemian Glass. A potash glass (a silicate of potassium and calcium), used for hollowware, being less fusible than soda glass, and less easily attacked by acids. See POTASH-LIME GLASS.

Bohemian Ruby. Synonym for Rose Quartz.

Bohemian Topaz. Synonym for Citrine.

Bohnalite B. A proprietary trade name for an aluminium alloy containing 4·5 per cent. copper and 0·3 per cent. magnesium.

Bohnalite J. A proprietary trade name for an alloy of aluminium with 10 per cent. copper.

Bohnalite S43. A proprietary trade name for an alloy of aluminium with about 5 per cent. silicon.

Bohnalite S51. A proprietary trade name for an alloy of aluminium with small amounts of magnesium, silicon, and iron.

Bohnalite U. A proprietary trade name for an alloy of aluminium with 13 per cent. silicon.

Bohnalite Y. A proprietary aluminium alloy containing small quantities of copper, nickel, and magnesium.

Boiled Butter. See BUTTER, RENOVATED.

Boiled-off Silk (Degummed Silk). Raw silk which has been deprived of 20–30 per cent. of its weight by removing sericin or silk gum by means of a dilute hot solution of soap.

Boiled Oil. Linseed oil, which has been boiled with litharge to render the oil more " drying." The term is also applied to linseed oil which has been heated for some time. The tendency to oxidise is thereby increased.

Boiler Plug Alloy. An alloy of 8 parts bismuth, 5–30 parts lead, and 3–24 parts tin.

Boiled Turpentine. The residue left after distilling turpentine in steam.

Boiled Varnish. Produced by heating linseed oil from 220–300° C. for from 2–3 hours, in the presence of minium, litharge, or manganese dioxide. See BOILED OIL.

Boiler Scale. The incrustation formed in boilers. It consists mainly of calcium sulphate, $CaSO_4.2H_2O$, with some magnesium hydrate, $Mg(OH)_2$, and alumina, Al_2O_3.

Boksputite. A mineral. It is $Pb_6Bi_2O_6(CO_3)_3$.

Boldenone. An anabolic steroid. 17β-Hydroxyandrosta - 1, 4 - dien - 3 - one. " CIBA-29038 " is the undec-10-enoate currently under clinical trial.

Boldo. Leaves and young twigs of *Peumus fragrans*.

Bole. See INDIAN RED.

Bole, Armenian. See INDIAN RED.

Boleite. A mineral, $3PbCuCl_2(OH)_2$. AgCl.

Bole, Red. See INDIAN RED.

Bole, Venetian. Indian red (*q.v.*).

Bole, White. See CHINA CLAY.

Boliformin. A condensation product of formalin with an aluminium salt. An antiseptic dusting-powder used in veterinary practice.

Bolidenite. Synonym for Falkmanite.

Bolivarite. A mineral. It is a hydrated aluminium phosphate, $AlPO_4.Al(OH)_3$. H_2O, found in Spain.

Bolivianite. A mineral. It is an antimonial silver sulphide, $Ag_2S.6Sb_2S_3$. The name appears to be also applied to a mineral containing 35 per cent. tin, 25 per cent. copper, 33 per cent. sulphur, and 0–10 per cent. iron found in the Bolivian tin deposits.

Bolivite. An oxysulphide of bismuth, $Bi_2S_3.Bi_2O_3$, found naturally.

Bolley's Green. A pigment. It is a copper borate, dehydrated by heating. Used as an oil colour.

Bolmantalate. 17β-Hydroxyoestr-4-en-3-one adamantane-1-carboxyl-ate.

Bolognian Phosphorus. A sulphide of barium, prepared by heating barium sulphate with powdered carbon. It contains some sulphate as well as sulphide, and is phosphorescent in the dark.

Bolognian Spar. Barite (*q.v.*).

Boloretin. A resin found in Danish peat bogs.

Bolster Silver. A nickel silver. It is an alloy of 65·5 per cent. copper, 16 per cent. zinc, 18 per cent. nickel, and 0·5 per cent. lead.

Bolting Cloth. An unsized silk used for making sieves.

Boltonite. A mineral. It is stated to be a variety of Forsterite.

Boluphen. A phenol-formaldehyde condensation product.

Bolusal. A German intestinal astringent containing bismuth, calcium, and aluminium salts.

Bolus Alba. Kaolin (*q.v.*).

Bolvidon. See MIANSERIN.

Bombay Catechu. See CUTCH.

Bombay Copal (Salem Copal). Names applied to Zanzibar copal resin, obtained from *Trachylobium verrucosum*.

Bombay Gum. See EAST INDIA GUM.

Bombycin. A glucoside occurring in the green silkworm cocoon. It resembles in character the flavone glucoside.

Bonamite. A trade name for a calamine of apple-green colour.

Boncarine. Carbon tetrachloride, CCl_4.

Bondelite. A proprietary trade name for a laminated phenol formaldehyde synthetic resin product.

Bonderised Iron and Steel. Iron and steel which has been treated to impart a rust-resisting surface to the metal. A layer of iron phosphate is often used.

Bonderite 67. A trade mark for a corrosion inhibiting treatment for iron, steel, aluminium and zinc which converts the metal surface into a zinc phosphate coating. (217).

Bond-etched. A proprietary trade name for an etched laminated thermoplastic sheet.

Bondex. A proprietary trade name for a laminated phenol-formaldehyde synthetic resin.

Bond Paper. A superior writing-paper. Rag bond is made from linen and cotton rags.

Bondur. An aluminium alloy containing 4·2 per cent. copper, 0·3–0·6 per cent. manganese, and 0·5–0·9 per cent. magnesium. It is a corrosion-resisting alloy.

Bone Ash (Bone Earth). The ash obtained by heating bones in air. It contains from 67–85 per cent. basic calcium phosphate, 2–3 per cent. magnesium phosphate, 3–10 per cent. calcium carbonate, a little caustic lime, and a little calcium fluoride. Used as a cleaning and polishing material, as a manure, and for the manufacture of phosphorus.

Bone Black (Char). A black pigment prepared by carbonising bones. The specific gravity is 2·8. Also see IVORY BLACK.

Bone Brown. A pigment obtained by partially charring ivory dust.

Bone Charcoal. See ANIMAL CHARCOAL.

Bone, Dry. See CALAMINE.

Bone, Earth. See BONE ASH.

Bone Fat. The fat obtained from the marrow in hollow bones.

Bone Glue. Made from bones by first treating them with hydrochloric acid to remove mineral matter (calcium phosphate), then heating in water to dissolve the glue. It can be also obtained by treating the bones with solvent for bone fat, then with steam and boiling water for glue.

Bone Meal. Bones from which the fat and gelatin have been removed. When powdered, it is used as a phosphatic manure.

Bone Meal, Steamed. See STEAMED BONE MEAL.

Bone Oil (Animal Oil, Dippel's Oil, Oil of Hartshorn, Bone Tar). A dark-brown oil, rich in pyridine bases, obtained by the distillation of bones. Used for denaturing spirits. The liquid part of bone fat is also called bone oil.

Bone-phosphate. Synonym for Osteolite.

Bone Pitch. A pitch obtained by the distillation of bone oil. It is not very soluble in solvents.

Bone Superphosphate. The superphosphate prepared from bones. It contains all the phosphoric acid in a water-soluble condition. Used as a manure.

Bone Tar. See BONE OIL.

Bone Turquoise (Odontolite). A mineral. It consists of fossil bones or teeth, coloured with ferrous phosphate, $Fe_3P_2O_8$.

Bone, Vitriolised. See VITRIOLISED BONES.

Bonjela. A proprietary preparation containing choline salicylate and cetyldimethyl benzyl-ammonium chloride. (358).

Bonner L-894. A proprietary trade name for a polyester resin.

Bonnyware. Proprietary synthetic resins of the phenol-formaldehyde and urea-formaldehyde types.

Bonoform. A name for tetrachlorethane.

Bonsor's Black. Logwood extracts with ferrous sulphate or copper sulphate. Used in dyeing.

Bontex. Range of synthetic blended detergents. (506).

Bontid. A proprietary preparation of carboxyphen. An antiobesity agent. (830).

Boothal. A material containing aluminium sulphate. It is used as a coagulant for fine paper pulp and clay.

Boothite. A mineral, $8[CuSO_4.7H_2O]$, of California.

Boot Polishes. These vary greatly. The essential constituents are shellac, wax, lampblack or aniline black, with turpentine, paraffin, or alcohol, as vehicle. Sometimes a little spermaceti is used, and occasionally glycerin, or chestnut tannin extract is included. Also see BLACKING.

Boracic Acid. Boric acid, H_3BO_3. A preservative.

Boracite. Crystalline boracite, $2Mg_3B_8O_{15}.MgCl_2$, hydroboracite, $CaMgB_4O_8.6H_2O$, and pinnoite, $MgB_2O_4.3H_2O$, are classed under this name, but the term is usually applied to a mixture of crystalline boracite and hydroboracite. Crystalline boracite is soluble in water, but hydroboracite is nearly insoluble. The former is also known as Stassfurtite. Used for the preparation of boric acid. The term is also applied to a mixture of magnesium borocitrate and sugar.

Boracite, Crystalline. See BORACITE and STASSFURTITE.

Boral. Aluminium-boro-tartrate. An antiseptic and astringent prescribed in skin diseases, and in inflammation of the ear.

Borascu. General non-selective weedkillers. (548).

Borated Benzol. A mixture of 10 per cent. boric acid with 90 per cent. benzol.

It is stated to improve the cellulose acetate varnishes used for painting aeroplane wings.

Borateem. Borax. (548).

Borate of Lime. See BORONATRO-CALCITE.

Borate Spar. See COLEMANITE.

Boratto. A silk and wool fabric.

Borax (Acid Metaborate, Biborate of Soda). Sodium dimetaborate, $Na_2B_4O_7.10H_2O$.

Borax, Burnt. See BORAXOSTA.

Borax Carmine. A microscopic stain. It consists of 1 drachm carmine dissolved in 2 drachms strong ammonia, and adding 12 ounces of a saturated solution of borax.

Borax Glass (Nitrified Borax). Fused borax.

Borax Honey. (*Mel boracis B.P.*) Consists of 2 parts borax and 16 parts honey. Used in the treatment of thrush in infants' mouths.

Borax Lime. A similar material to Tiza. It is Boronatrocalcite mixed with aluminium, magnesium, and calcium and sodium sulphates.

Borax, Methylene Blue. See SAHLI'S STAIN.

Borax, Nitrified. See BORAX GLASS.

Boraxo. Industrial hand cleaners. (548).

Boraxosta (Burnt Borax). Crystalline borax with most of its water of crystallisation removed.

Borax, Tartarised. Potassium boro-tartrate.

Boraxusta. Calcined borax.

Borcher's Metal. Non-corrosive alloys. One alloy contains 64·6 per cent. nickel, 32·3 per cent. chromium, 1·8 per cent. molybdenum, and 0·5 per cent. silver. Others are stated to contain 30 per cent. chromium, 34–35 per cent. cobalt, and 34–35 per cent. nickel ; whilst alloys consisting of 60 per cent. iron, 36 per cent. chromium, 4 per cent. molybdenum, and 65 per cent. chromium and 35 per cent. iron respectively, are also known as Borcher's metal.

Bordeaux B. See FAST RED B.

Bordeaux BL. See FAST RED B.

Bordeaux BX. Dyestuffs. (*a*) Is the sodium salt of sulphoxylene-azo-xylene-azo - β - naphthol - β - sulphonic acid, $C_{26}H_{22}N_4O_7S_2Na_2$. Dyes wool red from an acid bath. (*b*) Consists of the sodium salt of xylene-azo-xylene-azo-β - naphthol - β - mono - sulphonic acid, $C_{26}H_{23}N_4O_4SNa$.

Bordeaux Cov. See CONGO VIOLET.

Bordeaux D. See FAST RED B.

Bordeaux Developer. Benzyl-naphthylamine. Used for developing

bordeaux on fibre treated with primuline yellow.

Bordeaux DH. See FAST RED D.

Bordeaux Extra. See CONGO VIOLET.

Bordeaux G (Claret G). A dyestuff. It is the sodium salt of sulpho-toluene-azo - toluene - azo - β - naphthol - β - sulphonic acid, $C_{24}H_{18}N_4O_7S_2Na_2$. Dyes wool red from an acid bath.

Bordeaux G, BD, GG, GD, GDD. Dyestuffs allied to Fast red B.

Bordeaux Mixture. A fungicide for plant diseases, especially for those which attack potatoes. (*a*) Consists of 5 lb. copper sulphate, 5 lb. lime, and 50 gallons water. (*b*) Contains 6 lb. copper sulphate, 4 lb. lime, and 50 gallons water. It is made by the addition of the lime to the copper sulphate solution. Generally the substance, $10CuO.SO_3.4CaO.SO_3$, is formed. Also see PEACH BORDEAUX MIXTURE, SODA BORDEAUX MIXTURE, and POTASH BORDEAUX MIXTURE.

Bordeaux Mixture, Improved. This mixture contains $12\frac{3}{4}$ oz. copper sulphate, $12\frac{3}{4}$ oz. lime, $\frac{3}{4}$ oz. nicotine, and 8 oz. lead arsenate in 10 gallons water.

Bordeaux Mixture, Stabilised. Granulated sugar has been found to prevent Bordeaux mixture from deteriorating. Sugar, $\frac{1}{4}$ oz. dissolved in water is added for each 1 lb. copper sulphate used.

Bordeaux Red. A dyestuff. It is azo-naphthyl - naphthol - sulphonic acid. Used for colouring Italian wines.

Bordeaux R Extra. See FAST RED B.

Bordeaux R or B (Claret Red). A dyestuff isomeric with Bordeaux G.

Bordeaux S. See FAST RED D.

Bordeaux Turpentine. See FRENCH TURPENTINE.

Bordite. A mineral. It is a hydrated calcium silicate.

Bordosite. A mineral, $AgCl.2HgCl$.

Borester. Organic boron compounds. (548).

Borester N. Trinonyl borate (tri-3,5,5-trimethylhexyl borate) $C_{27}H_{57}O_3B$.

Borester 2. Tri-*n*-butyl $C_{12}H_{27}O_3B$.

Borester 7. Trihexylene glycol biborate (tri-(2-methyl-2,4-pentadiol)-diborate) $C_{18}H_{36}O_6B_2$.

Borester 8. Tri-*m,p*-cresyl borate (tri-*m,p*-tolyl borate) $C_{21}H_{21}O_3B$.

Borester 25. Trimethoxyboroxine (methyl metaborate) $(CH_3OBO)_2$. This is used for preventing polymerisation of liquid sulphur dioxide.

Borester A. Triaryl borate $C_{21}H_{21}O_3B$.

Boresters. A series of esters of boric acid. They are used as a convenient means of introducing boron into organic media such as paints and plastics.

Borethyl. Ethyl boride, $(C_2H_5)_3B$.

Borgströmite. A mineral. It is a basic ferric sulphate.

Borickite. A mineral, $Fe_4(OH)_6 . Ca(PO_4)_2 . 3H_2O$.

Borite. A trade mark for goods used as abrasives and refractories. They consist essentially of crystalline alumina.

Borium. A fused tungsten carbide diamond substitute formed by exposing tungsten at high temperatures to carbon monoxide or hydrocarbon gases.

Bornaprine. A spasmolytic. 3-Diethylaminopropyl 2 - phenylbicyclo[2.2.1] - heptane - 2 - carboxylate. SORMODREN.

Borneo Camphor. See CAMPHOL.

Borneo Tallow. A fatty material obtained from the fruit of the family of *Dipterocarpus*, also from *Shorea stenoptera*. Used as a substitute for cacao butter, under the name of green butter. Tangkawang fat is a native name for Borneo tallow. The seed is imported into England under the name of Pontianak illipé nuts.

Bornesite. Monomethyl-inositol, $C_6H_6 (OH)_5(OCH_3)$.

Bornite. See PURPLE COPPER ORE.

Bornjosal. Bornyl-di-iodo-salicylate, $C_6H_2I_2(OH)_2COOC_{10}H_{17}$. It is soluble in ether, chloroform, and benzene.

Bornolin. A proprietary preparation of cod liver oil, halibut liver oil and Vitamin D. Used in treatment of burns and bedsores. (802).

Bornyval. The trade name for bornyl-iso-valerianate, $C_{10}H_{17}O.C_5H_9O$. Used medicinally as a sedative in the treatment of nervous diseases.

Bornyval, New (Glycoborval, Neobornyval). Bornyl-isovaleryl-glycollate, $C_4H_9COOCH_2.COOC_{10}H_{17}$. Used for the same purposes as Bornyval.

Borocaine. A proprietary local anæsthetic. It is the borate of diethyl-aminoethyl-*p*-amino-benzoic acid (ethocaine), and is used as a substitute for cocaine.

Borocalcite. Calcium dimetaborate, CaB_4O_7, found naturally in South America. Synonym for Ulexite.

Borocarbone. An artificial corundum. See ALUNDUM.

Borocil. Non-selective weed killer. (548).

Borocil 3. A granular herbicide based on sodium borate combined with bromacil. It is a root-killer needing rain to activate it and carry it into the topsoil. It is non-selective. Usual application rate 200 kg per hectare. (183).

Borofil. A proprietary trade name for copper rods erosion and corrosion resistant for welding H.C. copper. (134).

Borofluorin. A mixture of boric acid, benzoic acid, sodium fluoride, and formaldehyde. Antiseptic and germicide.

Boroflux. A mixture of boron suboxide with boric anhydride and magnesia. Used to the extent of about 1 per cent. to deoxidise copper during purification.

Boroform. A solution of formaldehyde in sodium glyceroborate. Antiseptic.

Borogen. The ethyl ester of boric acid, $B(O.C_2H_5)_3$. Has been suggested as a local remedy in diseases of the nasal and respiratory mucous membrane.

Boroglyceride (Boroglycerine). Glyceryl borate, $C_3H_5BO_3$. Used as a preservative for wines and fruits. The aqueous solution was formerly used as a preservative for meat and milk, and has been sold under the name of Glacialin.

Boroglycerine. See BOROGLYCERIDE.

Borol. A fused mixture of boric acid and sodium bisulphate, used as an antiseptic.

Borolon. A trade mark for articles made for abrasive and refractory purposes. They consist essentially of crystalline alumina.

Boromagnesite (Szaibeyite). A mineral, $4Mg(BO_2)OH.Mg(OH)_2$.

Boromethyl (Trimethylborine). Methyl boride, $B(CH_3)_3$.

Boromica. A Belgian laboratory glass. It is an acid glass in which silica is partly replaced by boric acid. It resists chemical action, and has a low coefficient of expansion.

Boronatrocalcite (Ulexite, Tiza, Cotton Balls, Hayesin, Raphite). A soft, fibrous, silky mineral, from Chile and Peru, $Ca_2B_6O_{11}Na_2B_4O_7.16H_2O$. Also known as borate of lime.

Boronised Copper. Copper which has been purified when in the molten condition by the addition of 0·03–0·1 per cent. boron. The product is practically free from boron.

Boron, Manganese. See MANGANESE BORON.

Borophenol. Borax, in which phenol has been absorbed. A disinfectant.

Borophenylic Acid. A mixture of phenylborate, $C_6H_5BO_2$, and phenyltriborate, $C_6H_5B_3O_5$. Antiseptic.

Borosalicylic Acid. A solution containing 4 per cent. each boric and salicylic acids. Has been used as an antiseptic.

Borosalyl. See BORSALYL.

Borosodine. Sodium borotartrate solution.

Borovertin. Hexamethylene-tetramine-triborate, $(CH_2)_6N_4.3HBO_2$. Used as an antiseptic in urinary troubles.

Boroxo. A heavy-duty soap powder hand-cleanser, based on borax. (183).

Borra. A mass of material consisting of stones, clay, and sand, mixed with some sodium nitrate solution, obtained during the lixiviation of raw caliche.

Borrel Blue. See BLUE OF BORREL.

Borromite. An artificial zeolite similar to permutit.

Borsalyl (Borosalyl). The sodium salt of borosalicylic acid, $B.OH(O.C_6H_4.COOH)_2$. An antiseptic.

Borsyl. A mixture of cetyl alcohol, $C_{16}H_{33}OH$, with boric acid or borates. A cosmetic.

Bort. Cloudy, rounded crystals of diamond, used for rock drills.

Bortin " 45 ". Brucella abortus vaccine for veterinary purposes. (520).

Borussic Acid. Dilute hydrocyanic acid, HCN.

Borvicote. A proprietary range of emulsions of vinyl homopolymers and copolymers. (474).

Boryckite. A mineral, $Fe_4(OH)_6.Ca(PO_4)_2.3H_2O$.

Boryl. The trade name for ethyl-borosalicylate. Used as a substitute for ethyl salicylate in medicine.

Bosa. A Turkish cooling drink preparation, obtained by the limited fermentation of millet.

Bosanga. A juice obtained from the seeds of an African plant. Used to coagulate rubber latex.

Bosch. An inferior butter prepared in Holland. The term is sometimes used for margarine.

Bosch-bakelite. See BAKELITE.

Bosjemanite. A manganese alum.

Bosphorite. A mineral. It is a ferric phosphate. $Fe_3(PO_4)_2(OH)_3.7H_2O$.

Bostonite. The purest asbestos, found in Canada.

Botallackite. An oxychloride of copper, found in Cornwall. A mineral. It is $Cu_2Cl(OH)_3.nH_2O$.

Botany Bay Kino. See KINO, AUSTRALIAN.

Botany Bay Resin. See ACAROID BALSAM.

Botargo. A delicacy prepared in Greece. It is the salted, sun-dried, and compressed roe of certain fish preserved in wax.

Botrilex. A horticultural fungicide. (511).

Botryogen. (1) See RED VITRIOL.

Botryogen. (2) A mineral. It is $MgFe'''(SO_4)_2OH.7H_2O$.

Botryolite. A mineral. See DATOLITE.

Bottled Gas (Philgas, Pyrofax). Liquefied petroleum gas distributed in cylinders. It is sold under the names of Philgas and Pyrofax. It consists principally of propane and butane.

Bottle Glass. Usually composed of soda, lime, alumina, and sand. It owes its colour to a silicate of iron.

Bottle Green. See BRONZE GREEN.

Bottle-nose Oil. An inferior sperm oil from the blubber of the bottle-nose whale used in soap making. The term is also used for a cheap grade of olive oil also used in soap making.

Bottom Salts. See LEAD BOTTOMS.

Bouazzerite. Synonym for Stichtite.

Bouchardt's Reagent. A solution of 1 part iodine and 2 parts potassium iodide in 20 parts water. An alkaloidal reagent giving a brown precipitate.

Boudin's Solution. An aqueous solution of arsenic trioxide with wine. One fluid ounce contains $\frac{1}{4}$ grain arsenic trioxide.

Bougault's Reagent. Phosphorous acid in hydrochloric acid solution.

Bougival White. A clay similar to China clay found at Bougival, in France. It has been used as a rubber filler.

Bouglisite. A mineral. It is a mixture of gypsum and anglesite.

Bouillon Noir. See PYROLIGNITE OF IRON.

Bouin's Picro-formol. See PICRO-FORMOL.

Bouisol. A proprietary product. It is a colloidal copper compound for horticultural purposes.

Boulangerite. See ZINKENITE.

Boulion's Solution (French Mixture). A solution containing iodine, phenol, glycerin, and water.

Bourbonal (Vanillal). A proprietary preparation of ethyl protocatechuic aldehyde, $C_2H_5O.C_6H_3.OH.CHO$. Used in perfumery.

Bourbonne's Metal. An alloy of 50·48 per cent. tin, 48·8 per cent. aluminium, 0·25 per cent. copper, and 0·33 per cent. iron.

Bourboulite. A native ferroso-ferric sulphate, found at Puy-de-Dôme.

Bourbouze Aluminium Solder. An alloy of 47 per cent. zinc, 37 per cent. tin, 10 per cent. aluminium, and 5 per cent. copper.

Bourbouze Solder. An alloy of 83 per cent. tin and 17 per cent. aluminium.

Bourette Silk. The shorter fibres of waste silk after spinning and cording.

Bourgeoisite. A calcium silicate mineral.

Bourghoul. A Lebanon food product. It is grain which is boiled for a short time, dried, and ground to a paste. It contains 61 per cent. starch, 14·3 per cent. water, 0·79 per cent. phosphoric oxide, and 0·6 per cent. fat.

Bournonite (Endellionite). A mineral. It is a sulphide of antimony, lead, and copper, $3(Pb.Cu_2)S.Sb_2S_3$.

Boussingaultite (Cerbolite). A mineral. It is a hydrated ammonium magnesium sulphate, $(NH_4)_2SO_4.MgSO_4.6H_2O$.

B.O.V. See BROWN ACID.

Bovey Coal. A brown coal, consisting chiefly of bitumen and wood.

Bowenite. An exceptionally hard variety of serpentine (*q.v.*).

Bowhill's Stain. A microscopic stain. It contains 15 c.c. of a saturated alcoholic solution of orcin, 10 c.c. of 20 per cent. tannic acid solution, and 30 c.c. water.

Bowlingite. Steatite (*q.v.*).

Bowl Spirit. See SCARLET SPIRIT.

Bowmanite. See HAMLINITE.

Bowranite. A proprietary bituminous preparation resistant to acid fumes. It is used as a coating for tanks and structural steel-work.

Bow-wire Brass. See BRASS.

Boxite. A trade mark for articles of the abrasive and refractory class. They consist essentially of crystalline alumina.

Box Oil. See OIL OF BOX.

Boydite. Synonym for Proberite and Austenite.

Boyle's Fuming Liquor. A yellow liquid obtained by distilling ammonium chloride with sulphur and lime. Ammonium polysulphides are the chief constituents.

Bozetol. A proprietary trade name for a wetting agent. It is a sulphonated castor oil product.

B-Pasinah. A proprietary preparation of calcium benz amidosalicylate and isoniazid. An antituberculous agent. (244).

B.P.D. Selectron. A proprietary polyester laminating resin. (475).

Brackebuschite. A mineral, (Cu.Pb. Fe.Zn.Mn)$_3$V$_2$O$_8$.H$_2$O.

Bradford Blue. An obsolete triphenylmethane dyestuff.

Bradilan. A proprietary preparation of tetranicotinoylfructose in enteric coated tablets. (339).

Bradleyite. A mineral. It is Na$_3$MgPO$_4$CO$_3$.

Bradosol. A proprietary preparation of domiphen bromide. Throat lozenges (18).

Braemer's Reagent. A tannin reagent. It consists of 1 gram sodium tungstate and 2 grams sodium acetate in 10 c.c. of water.

Braga. An alcoholic beverage used in Romania, prepared by the fermentation of millet. It contains 1·3 per cent. alcohol.

Bragite. A mineral. It is a variety of fergusonite.

Brain Sugar. Cerebrose.

Brakol. A water treatment alkali. (513).

Bramma Orange and Bramma Red B, 2B, 6B. Azo dyestuffs prepared from benzidine. Direct cotton colours.

Bram's Powder. An explosive. It contains 60 per cent. of a mixture of potassium chlorate, potassium nitrate, wood charcoal, and oak sawdust, saturated with 40 per cent. of trinitroglycerin.

Bran (Pollards, Sharps, Middlings). Materials obtained from the outer coats of wheat.

Branalcane. A boric acid preservative.

Brandãosite. A mineral. It is (Mn,Fe)$_4$(Al,Fe)$_2$Si$_4$O$_{15}$.

Branderz. Synonym for Idrialite.

Brandish's Solution. An impure solution of potassium hydroxide.

Brandisite. A mineral. It is a variety of seybertite.

Brandite. A mineral, Mn.Ca$_2$(AsO$_4$)$_2$. 2H$_2$O.

Brandol. A 1 per cent. solution of picric acid, containing 0·4 per cent. of undissolved acid. Used for burns.

Brandy (Spirit of French Wine). A spirit. French cognac generally consists of blends of pot-still brandy with patent-still spirit produced from grapes in the Midi districts of France. British brandy is a compounded spirit prepared by redistilling compounded spirits made from grain, with flavouring materials. Hamburg brandy, manufactured in Germany, is made from potato or beet spirit, and flavoured to imitate grape brandy.

Brandy Bitters. Tr. Gent. Co.

Brandy, French. See COGNAC and BRANDY.

Brandy, Mint. Peppermint.

Brandywine. A proprietary trade name for vulcanised fibre tubes used for electrical insulation.

Brannerite. A mineral. It is a titanite of uranium with rare earths. Approximate composition is (U,Yt,Ca,Fe,Th)$_3$. Ti$_5$O$_{16}$.

Bran Oil. Furfural, C$_3$H$_3$CO.COH.

Brasivol. A proprietary preparation of hexachlorophane in a skin cleanser base. (382).

Bravit. A proprietary preparation of thiamine hydrochloride, riboflavine, pyridoxine hydrochloride, nicotinamide and ascorbic acid. A vitamin supplement. (476).

Brass. A copper-zinc alloy of varying proportions. Usually it contains more than 18 per cent. zinc, and lead is sometimes added to the extent of 1–2 per cent. Ordinary brass contains 67 per cent. copper and 33 per cent. zinc. This alloy has a specific gravity of 8·4 and melts at 940° C. The alloy containing 60 per cent. copper and 40 per cent. zinc has a specific heat of 0·0917 at from 20–100 C., and melts at

900° C., and that containing 72 per cent. copper and 28 per cent. zinc 0·094 at 14–98° C. The alloy consisting of 90 per cent. copper and 10 per cent. zinc melts at 1040° C., and that containing 70 per cent. copper and 30 per cent. zinc at 930° C. The specific resistance of brass at 0°C. varies from 7–8 micro-ohms per cm.[3] The following are varieties of brass made for specific purposes:

Brass	Copper.	Zinc.	Tin.	Lead.	Iron.
	per cent.	per cent.	per cent.	per cent.	per cent.
Admiralty brass .	70	29	1	...	...
Anvil brass . .	62·5	37·5	...	...	...
Bell brass . .	64·25	35	0·75	...	...
Block brass . .	66·5	32	...	1·5	...
Bow-wire brass .	93	2	5	...	...
Brazing brass . .	75–85	15–25	...	...	...
Bristol brass . .	60–76	24–39	...	...	...
Bronze powder brass	84	16	...	...	...
Bullet brass . .	90	9	...	1	...
Burr brass . .	62	38	...	...	...
	90	10	...	...	...
Button brass . .	89·5	10	0·5	...	...
	43	57	...	...	...
Button brass (Bristol)	61	36	3	...	...
,, ,, (Gold) .	58·7	33	5·5	2·75	...
,, ,, (Jackson)	63·9	30·5	5·5	...	...
,, ,, (Luden-scheidt)	20	80	...	...	...
Cap gilding brass .	89·7	9·9	...	0·4	...
Cartridge brass .	67–72	28–33	...	0–0·05	0–0·1
Check brass . .	62	38	...	...	...
Clock brass . .	61–62·5	36–36·5	...	1·75–2·5	...
Collet brass . .	61	36·5	...	2·5	...
Cymbal brass . .	78	22	...	...	...
Door-plate brass .	63	35	...	2	...
Drawing brass . .	66·7	33·3	...	...	...
Drill-rod brass .	62	35·5	...	2·5	...
Dutch brass . .	79·5	20·5	...	...	...
Electrical castings .	84	13	3	...	...
English brass . .	70·3	29·3	0·17	0·26	...
Engraver's brass .	66	33	...	1	...
Escutcheon pin brass	64·5	35·1	...	0·43	...
Eyelet brass . .	68	32	...	...	...
Fan blade brass .	61·5	37	...	1·5	...
Flush plate brass .	65·75	32·75	...	1·5	...
Fob metal . .	87·5	12	0·5	...	...
Forging brass . .	57–60	40–43	...	...	...
,, ,, (Russian)	53·5	42	...	...	4·5Mn
German brass . .	50	50	...	...	...
Gold leaf (imitation).	77–78	22–23	...	...	...
Grommet brass .	70	30	...	...	...
Hard brass . .	60	40	...	...	...
Hardware brass .	88	9·5–12	0–1·5	0–1	...
Helmet brass . .	72·3	27·6	...	0·1	...
High brass . .	61·5	38·5	...	...	...
Hooker brass . .	61	37	...	2	...
Japanese brass . .	66·5	33·4	...	...	0·1
Jeweller's brass .	83–91·5	6·5–13	0·75–2	...	...
Journal box brass .	92·5	7·5	...	...	...
Kick plate brass .	84	15	...	1	...
Lancashire brass .	73	25	...	2	...
Low brass . .	80	20	...	...	...
Machinery brass .	83	16	1	...	...
Matrix brass . .	62	36·5	...	1·5	...
Naval brass . .	62	37	1	...	...
Pen metal . .	85	13	2	...	...
Percussion cap brass.	95	5	...	...	...
Primer gilding brass .	97	3	...	...	...
Red brass . .	90	10	...	...	...
Reed brass . .	69	30	1	...	...

Brass—*continued.*

Brass	Copper.	Zinc.	Tin.	Lead.	Iron.
	per cent.	per cent.	per cent.	per cent.	per cent.
Rule brass .	62·5	35	2·5	...	...
Russian cast brass .	78	21	1	...	...
Screen plate brass .	58	41	0·75	0·25	...
Screw brass . .	62–78	16–38	0–4·5	0–1·5	...
Sheet brass . .	55–72	27–44	0–3·3	0·25–2·0	...
Shell-head brass .	75	25	...	...	...
Shoe nail brass .	63	37	...	...	...
Spinning brass. .	67	...	33	...	...
Spring brass . .	72	28	...	...	...
Stamping brass .	65	35	...	...	...
Sterling brass . .	66	33	...	...	0·7
Thurston's brass .	55	45·5	0·5	...	...
Tube brass . .	60–70	29–38	0–1	...	...
Turbine brass . .	67–76	24–32	...	0–0·23	...
Valve brass . .	90	6	...	4	...
Washer brass . .	60–62	38–40	...	...	...
Wheel brass . .	68	30	...	2	...
Wire brass (common)	65·4	35·6	...	...	...
,, ,, (English)	70·3	29·4	0·28	0·17	...
,, ,, (German)	72	27·3	...	0·85	...
Yellow brass . .	70	30	...	...	...

Brassidic Acid. Erucic acid, $C_{22}H_{42}O_2$.

Brassil. Iron Pyrites (*q.v.*).

Brass, Iron. A brass containing from 1–9 per cent. iron.

Brass, Iserlohn. An alloy of 64 per cent. copper, 33·5 per cent. zinc, and 2·5 per cent. tin.

Brass, Japanese. See SIN-CHU.

Brass, Leaded. Alloys of from 71–79 per cent. copper, 4·5–9·5 per cent. lead, 8·5–23 per cent. zinc, and traces to 3 per cent. tin.

Brassoline. See ZAPON VARNISH.

Brass Solders. For soft soldering a tin-lead solder is used, and for hard soldering, a copper-zinc alloy, poorer in copper than the pieces to be soldered, is employed. A solder suitable for copper, brass, or iron contains 67 per cent. copper and 33 per cent. zinc, and one consisting of 66 per cent. zinc and 33 per cent. copper is usually known as brass solder.

Brattice Cloth. A very coarse, tightly woven jute cloth.

Braunite. A native sesquioxide of manganese, Mn_2O_3.

Braunmanganeez. Synonym for Manganite.

Bravaisite. A mineral similar to Glauconite.

Bravilite. A trade name for an asbestos cement board.

Bravoite. A mineral. It is an iron-nickel sulphide containing vanadium.

Brazilian Arrowroot. Manihot, the farina of *Manihot itilissima*.

Brazilian Balsam. An oleo-resin from *Myroxylon peruiferum*. It resembles balsam of Peru, but is harder and more red in colour.

Brazilian Elemi. Elemi from *Icica icariba*.

Brazilian Emerald. Tourmaline (*q.v.*).

Brazilian Gum (Para Gum, Gum Angico). A gum which is stated to be the product of *Acacia angico*. Brazilian gum is the term applied to a gum resin from *Hymenæa courbaril*, which is used in varnishes, and a gum obtained from *Piptadenia rigida* is also known as Brazilian gum.

Brazilianite. A mineral. It is $NaAl_3(PO_4)_2(OH)_4$.

Brazilian Pebble. A variety of quartz (*q.v.*), used formerly for spectacle lenses.

Brazil Indian Lake. A lake which is analogous to carmine lake.

Brazilite. See BADDELEYITE.

Brazil Powder. See GOA POWDER.

Brazil Wax. Carnauba Wax.

Brazil Wood. See FUSTIC and RED-WOODS.

Brazing Solder. Alloys of from 35–45 per cent. copper, 45–57 per cent. zinc, and 8–10 per cent. nickel. One alloy, containing 50 per cent. copper and 50 per cent. zinc, is used for brazing, and for acetylene soldering an alloy of 90 per cent. copper and 10 per cent. zinc is used.

Brazing Spelter. An alloy of 50 per cent. copper and 50 per cent. zinc is known by this name.

Brea. A road dressing usually consisting of sand impregnated with petroleum oil.

Breadalbanite. A mineral. It is a variety of hornblende.

Bredbergite. A mineral. It is $8[(Ca,Mg)_3Fe_2Si_3O_{12}]$.

Breecht's Double Salt. Potassium bimagnesium sulphate, produced in the treatment of kainit, and schonite, at Stassfurt. Used as a manure.

Breezes. The dust of coke or charcoal.

Breislakite. A mineral. It is a variety of amphibole.

Breithaupite. A mineral, NiSb.

Bremen (Green Ashes, Paul Veronese Green, Mineral Green). Copper pigments.

Bremen Blue. See VERDITER BLUE.

Bremen Green. See VERDITER BLUE.

Brentacet. Azoic dyestuffs for acetate rayon and nylon. (512).

Brentamine. Azoic dyestuffs. (512).

Brenthol. Azoic dyestuffs. (512).

Brentogen. Azoic printing colours. (512).

Brentosyn. Azoic components for polyester fibres. (512).

Brenzcaine. Guaiacol-benzyl ester, $C_6H_4OCH_3.OCH_2.C_6H_5$. A local anæsthetic.

Breon. A trade mark. Vinyl materials, nitrile and acrylic rubbers. (584).

Breon GA 301A. A proprietary high-quality insulation grade of polyvinyl chloride suitable for use at temperatures up to 105° C. (288).

Breon GA 302A. A hard insulation grade of polyvinyl chloride suitable for use at 85° C. (288).

Breon GA 304A. A soft insulation and sheathing grade of heat-resisting polyvinyl chloride, suitable for use at 85° C.

Breon GA 314A. A high-quality soft insulation and sheathing grade of polyvinyl chloride able to withstand high temperatures. It is typically used to sheathe the insulated wiring in electric blankets. (288).

Bresille Wood. See REDWOODS.

Bresin. A proprietary trade name for a phenol-formaldehyde synthetic resin.

Bresk. Jelutong (q.v.).

Bretonite. Iodoacetone, $CH_3CO.CH_2I$.

Bretylate. A proprietary preparation of bretylium tosylate, used in the treatment of cardiac arrhythmias. (277).

Bretylium Tosylate. N - 2 - Bromo - benzyl - N - ethyl - NN - dimethyl - ammonium p-toluenesulphonate. Darenthin.

Breunerite. A mineral, $(Mg.Fe)CO_3$.

Brevicite. A mineral. It is a variety of natrolite.

Brevidil. A proprietary preparation of suxemethonium bromide, a muscle relaxant. (336).

Brevidil E. A short acting muscle relaxant. (507).

Brevidil M. Suxamethonium salt (507).

Brevidil M. A proprietary trade name for Suxamethonium Bromide o Chloride.

Brevital. A proprietary trade name for Methohexitone.

Brewing Glucose. See GLUCOSE CHIPS.

Brewsterite. A mineral, $Al_2O_3.H$ $(Ba.Sr)O_3.(SiO_2)_6.3H_2O$.

Brewsterlin. A name applied to a liquid found enclosed in a variety of topaz.

Bricanyl. A proprietary preparation of terbutaline sulphate used in asthma therapy. (477).

Bricanyl Expectorant. A proprietary preparation of terbutaline sulphate and guaiphenesin. An expectorant cough medicine. (477).

Brickerite. Synonym for Austinite.

Brick Oil. See OIL OF BRICK.

Bricks, Acid-resisting. See BLUE BRICKS.

Brietal. A proprietary preparation of methohexitone sodium. An intra-venous barbiturate anæsthetic. (333).

Brightray Alloy B. A trade mark for an electrical resistance alloy of 59 per cent. nickel, 16 per cent. chromium, 0·3 per cent. silicon and the balance iron. (143).

Brightray Alloy C. A trade mark for an electrical resistance alloy of 19·5 per cent. chromium, 1·5 per cent. silicon, 0·04 per cent. rare earth elements and the balance nickel. (143).

Brightray Alloy F. A trade mark for an electrical resistance alloy of 37 per cent. nickel, 18 per cent. chromium, 2·2 per cent silicon and the balance iron. (143).

Brightray Alloy H. A trade mark for an electrical resistance alloy of 19·5 per cent. chromium, 3·6 per cent. aluminium and the balance nickel. (143).

Brightray Alloy S. A trade mark for an electrical resistance alloy of 20 per cent. chromium and the balance nickel. (143).

Bright Red Oxide. See INDIAN RED.

Briklens. Laundry detergents. (513).

Briline. Diazo compounds and coupling agents. (585).

Brilliant Acid Bordeaux. A dyestuff. It is a British equivalent of Fast red D.

Brilliant Alizarin Blue G and R (Indochromine T). A dyestuff produced by the condensation of β-naphthol-quinone-disulphonic acid with dimethyl - p - phenylene - diamine - thio - sulphonic acid. Dyes chromed wool, cotton, and silk blue.

Brilliant Alizarin Cyanine G. An analogous product to Anthracene blue WR.

Brilliant Alizarin Cyanine 3G. An analogous product to Anthracene blue WR.

Brilliant Archil C. A dyestuff. It is the sodium salt of the azimide of p - nitro - benzene - azo - 1 : 8 - naphthylene-diamine-sulphonic acid, $C_{16}H_8 N_6O_8S_2Na_2$. Dyes wool red.

Brilliant Azofuchsine 2G. A dyestuff. It is a British equivalent of Brilliant sulphone red B.

Brilliant Azurin B. A direct cotton dyestuff.

Brilliant Azurin 5G (Hexamine Azurine 5G). An azo dyestuff from dianisidine and dioxy - naphthalene - mono - sul - phonic acid, $C_{34}H_{24}N_4O_{12}S_2Na_2$. Dyes cotton blue from a salt bath, and wool from a neutral bath.

Brilliant Benzo Blue 6B. See DIAMINE SKY BLUE.

Brilliant Benzo Green B. A direct cotton dyestuff.

Brilliant Black B. See NAPHTHOL BLACK B.

Brilliant Black 5G. A dyestuff. It is the sodium salt of dimethyoxy-diphenyl - disazo - bi - dioxy - naphthalene, $C_{34}H_{24}N_4O_{12}S_2Na_2$. Dyes cotton bluish-black.

Brilliant Blue. Naphthazine blue mixed with Naphthol black.

Brilliant Bordeaux B, 2B. Dyestuffs. They are British equivalents of Fast red D.

Brilliant Bordeaux S. An acid dyestuff.

Brilliant Carmoisine O. See AZO RUBINE S.

Brilliant Chrome Red Paste. A mordant dyestuff used in calico printing.

Brilliant Cochineal 2R, 4R. Acid colours belonging to the same group as Palatine scarlet. They dye wool and silk red from an acid bath.

Brilliant Congo G. A dyestuff. It is the sodium salt of diphenyl-disazo-β-naphthylamine - sulphonic - β - naph - thylamine-disulphonic acid, $C_{32}H_{21}N_6 O_9S_3Na_3$. Dyes cotton red from a soap bath.

Brilliant Congo R (Vital Red). A dyestuff. It is the sodium salt of ditolyl-disazo - β - naphthylamine - mono - sul - phonic - β - naphthylamine - disulphonic acid, $C_{34}H_{25}N_6O_9S_3Na_3$. Dyes cotton red from a soap bath.

Brilliant Congo 2R. A dyestuff ob-

tained from tetrazotised tolidine, amino-R-salt, and F-acid.

Brilliant Cotton Blue, Greenish. See METHYL BLUE.

Brilliant Cresyl Blue 9B. A dyestuff prepared from nitroso-dimethyl-m-amino-cresol and benzyl-m-amino-dimethyl-p-toluidine. Dyes tannined cotton and silk.

Brilliant Crimson. See AZO RUBINE S.

Brilliant Crimson O. A dyestuff. It is a British equivalent of Fast red D.

Brilliant Croceine 3B. See BRILLIANT CROCEINE M.

Brilliant Croceine 9B. A dyestuff. It is the sodium salt of disulpho-β-naphthalene - azo - benzene - azo - β - naph - thol-disulphonic acid, $C_{26}H_{14}N_4O_{13}S_4 Na_4$. Dyes wool red from an acid bath.

Brilliant Croceine, Bluish. See BRILLIANT CROCEINE M.

Brilliant Croceine LBH. A dyestuff. It is a British equivalent of Brilliant croceine M.

Brilliant Croceine M (Brilliant Croceine 3B, Brilliant Croceine, Bluish, Cotton Scarlet, Cotton Scarlet 3B Conc., Ponceau BO Extra, Paper Scarlet, Blue Shade, Croceine Scarlet 3B, 9187K, Brilliant Croceine LBH). A dyestuff. It is the sodium salt of benzene-azo-benzene-β-naphthol disulphonic acid, $C_{22}H_{14}N_4O_7S_2Na_2$. Dyes wool and silk red from an acid bath. Also used in paper staining.

Brilliant Double Scarlet G (Orange Red I). A dyestuff prepared from β-naphthylamine-sulphonic acid and β-naphthol.

Brilliant Fast Red G. A dyestuff obtained from diazotised α-naphthyl-amine-5-sulphonic acid and β-naphthol.

Brilliant Gallocyanine. See CHROMOCYANINE V AND B.

Brilliant Gelatin. A mixture containing potassium carbonate and carnauba wax incorporated into molten gelatin, and allowed to solidify.

Brilliant Geranine. See GERANINE.

Brilliant Glacier Blue. See STETOCYANINE.

Brilliant Green (Malachite Green G, New Victoria Green, Ethyl Green, Emerald Green, Fast Green J, Diamond Green C, Smaragdgreen, Solid Green J, Solid Green TTO, Fast Green S). A dyestuff, which is the sulphate or double zinc chloride of tetra-ethyl-diamino-triphenyl-carbinol. Sulphate, $C_{27}H_{34}N_2 O_4S$. Dyes wool, silk, jute, leather, and cotton mordanted with tannin and tartar emetic, green. A specially pure form of this dye, and free from zinc, is used as a powerful antiseptic in medicine and dentistry.

Brilliant Heliotrope 2R Conc. A dyestuff. It is a British equivalent of Methylene violet 2RA.

Brilliant Hessian Purple. A dyestuff. It is the sodium salt of disulpho-stilbene-disazo - bi - β - naphthylamine - β - sul - phonic acid, $C_{34}H_{22}N_6O_{12}S_4Na_4$. Dyes cotton bluish-red.

Brilliant Indigo B. 5 : 6 : 7-trihydroxy-flavone.

Brilliant Induline. A dyestuff, which is a sulphonic acid of an induline.

Brilliant Milling Green B. A diphenyl-naphthyl-methane dyestuff.

Brilliant Oil Crimson. A dyestuff. It is a British equivalent of Rosaniline.

Brilliant Orange. See PONCEAU 4GB.

Brilliant Orange G. See PONCEAU 4GB.

Brilliant Orange GL. A dyestuff. It is a British equivalent of Ponceau 4GB.

Brilliant Orange O. See ORANGE GT.

Brilliant Orange OL. A dyestuff. It is a British equivalent of Orange GT.

Brilliant Orange R. See SCARLET GR.

Brilliant Orchil C. An azo dyestuff, which dyes wool and silk.

Brilliant Paraffin. A mixture of 75 per cent. paraffin and 25 per cent. carnauba wax.

Brilliant Phosphine 5G. An acridine dyestuff.

Brilliant Ponceau. See NEW COCCINE.

Brilliant Ponceau G. See PONCEAU R.

Brilliant Ponceau GG. See PONCEAU 2G.

Brilliant Ponceau 4R. See DOUBLE SCARLET EXTRA S and NEW COCCINE.

Brilliant Ponceau 5R. A dyestuff from diazotised naphthionic acid and G-acid.

Brilliant Ponceau 5R. A dyestuff. It is a British brand of New coccine.

Brilliant Purpurin R. A dyestuff. It is the sodium salt of ditolyl-disazo-naphthionic - β - naphthylamine - disul - phonic acid, $C_{34}H_{25}N_6O_9S_3Na_3$. Dyes cotton red from an alkaline bath.

Brilliant Red. See FAST RED.

Brilliant Rhoduline Red. See RHODULINE REDS B and G.

Brilliant Scarlet. See NEW COCCINE and IODINE RED.

Brilliant Scarlet Extra. See DOUBLE SCARLET EXTRA S.

Brilliant Scarlet 2R. A dyestuff similar to Palatine scarlet. It gives a light scarlet colour.

Brilliant Scarlet 4R. A dyestuff similar to Palatine scarlet.

Brilliant Scarlet 4R. A dyestuff. It is a British equivalent of New coccine.

Brilliant Sulphonazurine R. A direct cotton dyestuff.

Brilliant Sulphone Red (Brilliant Azo-fuchsine 2G, Vega Red S). A dyestuff of the same class as Fast sulphone violet 5BS.

Brilliant Ultramarine. See ULTRA-MARINE.

Brilliant Yellow. A pigment. It is a mixture of cadmium sulphide and white lead.

Brilliant Yellow. (Acid Yellow OO). A dyestuff prepared from the sulphonic acids of mixed toluidines.

Brilliant Yellow. A dyestuff. It is the sodium salt of disulphostilbene - disazo - bi - phenol, $C_{26}H_{18}N_4O_8S_2Na_2$. Dyes cotton yellow from an acid bath. Used for paper-staining.

Brilliant Yellow (Naphthol Yellow RS). A dyestuff. It is the sodium salt of dinitro -α- naphthol -α- monosulphonic acid, $C_{10}H_5N_2O_8SNa$. Dyes wool and silk yellow from an acid bath.

Brilliant Yellow S (Yellow WR, Brilliant Yellow, Curcumine). A dyestuff. It is the sodium salt of p-sulpho-benzene - azo - diphenylamine - disul - phonic acid, $C_{18}H_{13}N_3O_6S_2Na_2$. Dyes wool and silk yellow.

Brimstone. Lumps or blocks of sulphur, obtained in the refining of sulphur. It collects on the floor of the condensing chambers, when it is cast into sticks (roll sulphur).

Brimstone Acid. Sulphuric acid prepared from sulphur. It is a fairly pure acid, being free from arsenic.

Brimstone, Flowers of. See FLOWERS OF SULPHUR.

Brinaldix. A proprietary preparation of CLOPAMIDE with potassium chloride and bicarbonate. A diuretic. (478).

Brins. Single silk fibres as they exist in the cocoons. See BAVE.

Briquettes (Agglomerated Coal). Made by compressing a mixture of small coal slack, or coal dust, with pitch or other similar material, into blocks.

Bristocycline. A proprietary preparation of tetracycline. An antibiotic. (324).

Bristol Brass. See BRASS.

Bristrex. See TETRACYCLINE PHOSPHATE COMPLEX.

Britannia Metal. An alloy of from 74–91 per cent. tin, 5–24 per cent. antimony, and 0·15–3·68 per cent. copper, sometimes with small quantities of zinc, lead, and bismuth. A Britannia metal containing 90 per cent. tin and 10 per cent. antimony has a specific gravity of 7·9, and melts at 260° C. Used in the manufacture of cheap tableware, such as teapots and spoons, also as an antifriction metal.

Britholite. A mineral. It is a basic phospho-silicate of iron, calcium, magnesium, sodium, and cerium metals.

British Barilla. See BARILLA.

British Gum (Starch Gum, Vegetable

Gum, Gommeline, Artificial Gum). Dextrin, $n(C_6H_{10}O_5)$, obtained by the action of diastase on starch paste, or by heating starch with a trace of acid. Used as an adhesive.

British Gum, Crystallised (Crystallised Gum). Dextrin which has been decolorised with animal charcoal, and evaporated down to give a brittle mass.

Britonite. An explosive containing nitro-glycerin, potassium nitrate, wood meal, and ammonium oxalate.

Britsulite. A proprietary trade name for a phenol-formaldehyde synthetic resin. A moulding composition. Also for laminating.

Brittalite. See HYDROSULPHITE NF.

Brittle Silver Ore. See PYRARGYRITE.

Britulite. See HYDRALDITE C EXT.

Brix Metal. Alloys of from 60–75 per cent. nickel, 15–20 per cent. chromium, 5 per cent. copper, 1–4 per cent. tungsten, 4 per cent. silicon, 3 per cent. titanium, 2 per cent. aluminium, and 1 per cent. bismuth. Stated to be non-corrosive.

Briz. A scouring powder. (506).

Brizin. A proprietary preparation of BENAPRYZINE hydrochloride used in the treatment of Parkinson's disease. (364).

Broad Salt. Ground rock salt.

Brocade Colours. See BRONZE COLOURS.

Brocatello. A variety of Calcite.

Brocadopa. A proprietary preparation of LEVODOPA used in the treatment of Parkinson's disease. (317).

Brocchite. Chondrodite (q.v.).

Brochantite. A native basic copper sulphate, $CuSO_4.3Cu(OH)_2$, found in Cumberland. A synonym for Warringtonite. Formula also given as $4[CuSO_4(OH)_6]$.

Brocillin. A proprietary preparation containing propicillin potassium. An antibiotic. (364).

Brockite. An explosive. It is a mixture of barium chlorate and aluminium powder.

Brocresine. A histidine decarboxylase inhibitor. O - (4 - Bromo - 3 - hydroxybenzyl) hydroxylamine. CONTRAMINE is the phosphate.

Brodie's Graphite. An inferior graphite treated with 2 parts of sulphuric acid and $\frac{1}{15}$ part potassium chlorate.

Broenner's Acid. β-Naphthylaminesulphonic acid, $(NH_2.SO_3H. 2 : 6)$.

Brofezil. An anti-inflammatory preparation. 2 - (4 - p - Bromophenyl - thiazol - 2 - yl) propionic acid. "ICI 54, 594" is currently under clinical trial as the sodium salt.

Brofo. A proprietary trade name for a cumarone resin.

Broggerite. See CLEVITE. A mineral. It is $4[(U,Th)O_2]$.

Brojosan. A German product. It contains bromine, iodine, and albumen, and is used against arterial sclerosis.

Brolene. A proprietary preparation of propamidine isethionate. An ocular antiseptic. (336).

Bromacetol. Methyl bromacetol, $CH_3.CBr_2CH_3$.

Bromal. Tribrom-acetic aldehyde, $CBr_3.CHO$. Used in medicine.

Bromalbacid. A brominated egg albumin. It has similar therapeutic properties to bromides.

Bromalbin. A bromine derivative of protein. Used as a substitute for alkaline bromides in medicine.

Bromaline (Bromethylformin). Hexamethylene - tetramine - ethyl - bromide, $[(CH_2)_6N_4]C_2H_5Br$. Used as a substitute for bromine salts as a sedative in neurasthenia and epilepsy.

Bromamide. The hydrobromide of tribrom-aniline, $C_6H_4Br_3.NHBr$. An antipyretic and antineuralgic.

Bromamil. Brom-ethyl-formin.

Bromanex. A German product. It is a yeast-bromine preparation, tasting like meat extract. Used for nervous troubles.

Bromanil. Tetrabrom-quinone, $C_6Br_4O_2$.

Bromanilide. See ANTISEPSIN.

Bromargyrite (Bromite). A mineral. It is silver bromide, $4[AgBr]$.

Bromatacamite. A mineral. It is $4[Cu_2Br(OH)_3]$.

Bromazepam. A tranquilliser currently undergoing clinical trial as "Ro 05-3350". It is 7 - bromo - 1, 3 - di - hydro - 5 - (2 - pyridyl) - $2H$ - 1, 4 - benzodiazepin - 2 - one.

Bromazine. Bromodiphenhydramine.

Brom-Bischofite. A mineral. It is $2[MgBr_2.6H_2O]$.

Brombutol. See BROMETONE.

Bromcamphor. Camphor monobromide, $C_{10}H_{15}BrO$. Used in medicine as a sedative.

Bromcellite. A mineral. It is $2[BeO]$.

Bromchlorargyrite. Synonym for Embolite.

Bromchlorenone. A proprietary preparation of 6-bromo-5-chloro-2-benzoxazolinone. VINYZENE. (878).

Bromchlorphenol Blue. Dibromodichloro - phenol - phenol - sulphon - phthalein.

Bromchlorphenol Red. Dibromodichloro-phenol-sulphon-phthalein. An indicator.

Brom-Cornallite. A mineral. It is $12[KMgCl_3.6H_2O]$.

Bromcresol Green. Tetrabromo-m-cresol-sulphon-phthalein. An indicator.

Bromcresol Purple. Dibromo-*o*-cresol-sulphon-phthalein. A substitute for litmus for use in milk cultures.

Bromebric Acid. A cytotoxic agent. *cis* - 3 - Bromo - 3 - *p* - anisolyacrylic acid. Cytembena is the sodium salt.

Bromeigon. See EIGONES.

Bromeikon. Sodium - tetrabromo - phenol-phthalein.

Bromelains. A concentrate of proteolytic enzymes derived from Ananas comosus Merr. Ananase. Recommended as a substitute for pancreatin. Prepared from pineapple juice. A pale brown powder. (182).

Bromelia. See NEROLIN II.

Bromellite. A mineral. It is beryllium oxide.

Bromethyl. Ethyl bromide, C_2H_5Br.

Bromethylformin. See BROMALINE.

Brometone (Acetone-bromoform, Brombutol). Tribromo-tertiary-butyl-alcohol, $CBr_3.C(CH_3)_2OH$. Used in medicine as a sedative.

Bromhemol. See HEMOL.

Bromhexine. Approved name for N-cyclohexyl-N-methyl-(2-amino-3, 5-dibrombenzyl)amine. Available as Bisolvon

Bromhydric Ether. See HYDROBROMIC ETHER.

Bromide Solution, Rice's. See RICE'S BROMIDE SOLUTION.

Bromidia. A proprietary preparation óf potassium bromide, chloral hydrate, and extract of hyacyamus. A sedative. (391).

Bromidine. A dry mixture of sodium bisulphate with sodium or potassium bromide and bromate. A disinfectant.

Bromidsodalith. A mineral. It is $Na_4Al_3Si_3O_{12}Br$.

Bromile. A combination of bromine with an unknown organic base. Used as a substitute for bromides.

Brominated Lime. Lime which has been treated with bromine. It has bleaching properties.

Bromindigo FB (Indigo MLB/4B, Ciba Blue 2B, Durindone Blue 4B). 5 : 7 : 5′ : 7′-Tetrabrom-indigo.

Bromindione. 2 - (4 - Bromophenyl) indane-1,3-dione. Circladin.

Bromine Fat. See BROMIPIN.

Bromine Salt. A mixture made by saturating caustic soda with bromine, draining off the mother liquor, and adding sodium bromate, $NaBrO_3 + 2NaBr$. Used in the extraction of gold ores.

Bromine, Solidified. Diatomaceous earth with suitable binders saturated with bromine and formed into sticks.

It contains from 50–72 per cent· bromine by weight.

Bromine Water. Bromine (3·2 grams) in water (100 grams). This solution is known as bromine water.

Brominol. See BROMIPIN.

Bromipin (Bromine Fat, Brominol). A bromine addition product of sesamé oil, containing 10–33 per cent. bromine. Used internally as a sedative, and externally as an ointment.

Bromite. See BROMARGYRITE.

Bromlite. See ALSTONITE.

Bromlost. A blister gas. It is dibromo-ethyl sulphide, $S(CH_2CH_2Br)_2$.

Bromobehenate. See SABROMINE.

Bromochinol. Acid dibromo-salicylate of quinine, $C_{20}H_{24}N_2O_2.2C_6H_2Br_2(OH)$ COOH. A feeble antipyretic, also a soporific.

Bromochloral. Dichloro-bromo-acetic-aldehyde, $CCl_2.Br.CHO$.

Bromochloroform. Dichloro-bromo-methane, $CHCl_2.Br$.

Bromocoll. A registered trade mark currently awaiting re-allocation by its proprietors to cover a range of pharmaceutical products. (983).

Bromocriptine. A prolactin inhibitor currently undergoing clinical trial as " CB 154 ". 2 - Bromo - α - ergocryptine.

Bromocyclen. An insecticide and acaricide. 5 - Bromomethyl - 1, 2, 3, 4, 7, 7 - hexachloronorborn - 2 - ene. ALUGAN.

Bromodan. A brominating agent. (509).

Bromodeine. A proprietary preparation containing bromoform, codeine hydrochloride, liquid extract of krameria, wild cherry and senega. A cough linctus. (280).

Bromodiphenhydramine. NN - Dimethyl - 2 - (4 - bromodiphenyl methoxy)ethylamine. Bromazine. Ambodryl.

Bromo-ethane. Ethyl bromide, C_2H_5Br.

Bromoform. Tribrom-methane, $CHBr_3$. Used in the treatment of whooping cough.

Bromoformin. Bromaline (*q.v.*).

Bromoglidin. A wheat gluten preparation, containing 10 per cent. bromine. Used as a substitute for alkaline bromides in medicine.

Bromogluten (Bromoprotein). A vegetable albumin preparation, containing 8 per cent. bromine.

Bromol. Tribrom-phenol, $C_6H_2(OH)$ Br_3. Used as a caustic and disinfectant for wounds, and internally as an intestinal disinfectant.

Bromolaurionite. A mono-hydrated lead oxy-dibromide, $PbO.PbBr_2.H_2O$, obtained by heating a solution of lead acetate and sodium bromide for 12 hours.

Bromolecithin. An addition product of bromine and lecithin, containing 25–30 per cent. bromine. Used in medicine.

Bromolein. The brominated unsaturated fatty acids of almond oil, containing 20 per cent. bromine. Used hypodermically, or upon the skin. A sedative.

Bromo-mangan. A ferromangan, with the addition of 3 per cent. brom-peptone, which contains 11 per cent. bromine. Said to be a tonic.

Bromomethane. Methyl bromide, CH_3Br.

Bromomimetite. Lead bromo-tri-o-arsenate, $(PbBr)Pb_4(AsO_4)_3$.

Bromonitroform. Bromo-trinitro-methane, $CBr(NO_2)_3$.

Bromophenol Blue. Tetrabromo-phenol-sulphon-phthalein. An indicator.

Bromophenol Red. Dibromo-phenol-sulphon-phthalein. An indicator.

Bromophin. Apomorphine-bromo-methylate. Used in medicine.

Bromophor. Stated to be dibromo-ricinoleate. Recommended as an external application in pruritus and erysipelas.

Bromopicrin. Tribromo-nitro-methane, $CBr_3(NO_2)$.

Bromopin. A compound of bromine and oil. Used as a substitute for potassium bromide in medicine.

Bromoprotein. See BROMOGLUTEN.

Bromo-purpurin. Bromo-trioxy-anthraquinone, $C_{14}H_4O_2(OH)_3Br$.

Bromopyrin. Monobrom-antipyrine, $C_{11}H_{11}BrN_2O$. Medicinal. A mixture of caffeine, antipyrine, and sodium bromide is also sold under this name.

Bromopyromorphite. Lead bromo-triortho-phosphate, $PbBr_2.3Pb_3(PO_4)_2$, obtained by fusing a mixture of 12 parts lead ortho-phosphate and 5 parts lead bromide with excess of sodium bromide.

Bromoquinal. Quininic acid dibromo-salicylate.

Bromosalizol. Bromosaligenin, an antiseptic and local anæsthetic.

Bromosin. A similar preparation to Bromalbacid (*q.v.*), containing 10 per cent. bromine.

Bromotan. Brom-tannin-methylene-urea. Used in combination with talc and zinc oxide, in the treatment of eczema.

Bromovaletin. See PHENOVAL.

Bromovalidol. Validol with sodium bromide. A sedative.

Bromowagnerite. A compound of

calcium bromide and phosphate, $Ca_3(PO_4)_3.CaBr_2$.

Bromoxylenol Blue. An indicator, $C_{23}H_{20}O_5Br_2S$, prepared by adding bromine to xylenol blue dissolved in glacial acetic acid.

Bromoxynil. 1,4 - Dibromo - 3 - cyno phenol. A specific herbicide for use in cereal crops.

Brompheniramine. 1 - (4 - Bromo - phenyl) - 3 - dimethylamine - 1 - (2 - pyridyl) - propane. Dimotane is the hydrogen maleate.

Bromstrontiuran. A German product. It is the bromine compound of stron-tiuran (10 per cent. strontium chloride-urea solution). Used in cases of itching and skin diseases.

Brom-tetragnost. Sodium tetra-bromo-phenol-phthalein.

Bromthymol Blue. Dibromothymol-sulphon-phthalein. An indicator.

Bromum Solidificatum. Kieselguhr made plastic with molasses, pressed into sticks, which sticks are heated, then saturated with bromine. It contains 75 per cent. bromine.

Bromural (Dormigene, Uvaleral). α-Bromo-isovaleryl-urea, $C_4H_8Br.CO.NH.CO.NH_2$. A nervous sedative, and a mild somnifacient.

Bromurea. Bromo-isovaleryl-urea.

Bromvaletin. α-Bromo-isovaleryl-p-phenetidine.

Bromyrite. A mineral, AgBr.

Bronce Amarillo. Synonym for Chalcopyrite.

Bronce Blanco. Synonym for Arseno-pyrite.

Bronchilator. A proprietary preparation of isoetharine methanesulphonate, phenylephrine hydrochloride and thenyldiamine hydrochloride. A bronchial antispasmodic. (112).

Bronchotone. A proprietary preparation of ephedrine hydrochloride, caffeine and sodium salicylate, sodium iodide and tincture of belladonna. A bronchial antispasmodic. (340).

Brongniardite. A mineral. It is a lead-silver-antimonide, $PbAg_2.Sb_2S_5$.

Bronner's Acid. 2-Naphthylamine-6-sulphonic acid.

Bronopol. 2 - Bromo - 2 - nitropro-pane-1,3-diol.

Brontina. A proprietary preparation of deptropine dihydrogen citrate. A bronchial antispasmodic. (317).

Brontisol. A proprietary preparation of deptropine citrate and isoprenaline hydrochloride. A bronchial antispasmodic. (317).

Brontrium. A proprietary preparation of chlordiazepoxide, theophylline and ephedrine hydrochloride used in the treatment of bronchospasm. (314).

Brontyl. A proprietary preparation of proxyphylline. Bronchial antispasmodic. (347).

Bronyrite. A mineral. Silver bromide, AgBr.

Bronze. Alloys usually consisting of copper and tin in varying proportions, often with zinc, and occasionally with lead. The copper varies from about 74–95 per cent., the tin from 1–20 per cent., zinc from 0–17 per cent., and the lead from 0–18 per cent. The alloys containing from 70–91 per cent. copper and 9–30 per cent. tin vary in melting-point from 750–1030° C. and in specific gravity from 8·79–8·93.

Bronze " A." A British chemical standard. It contains 85·5 per cent. copper, 9·96 per cent. tin, 1·86 per cent. zinc, 1·83 per cent. lead, 0·25 per cent. phosphorus, 0·24 per cent. antimony, 0·07 per cent. iron, 0·04 per cent. nickel, and 0·06 per cent. arsenic.

Bronze, Acid. See ACID BRONZE.

Bronze, Ajax. See AJAX BRONZE.

Bronze, Allan Red. See ALLAN RED BRONZE.

Bronze, Aluminium. See ALUMINIUM BRONZE.

Bronze, Aluminium Iron. See ALUMINIUM IRON BRONZE.

Bronze, Aluminium Magnesium. See ALUMINIUM MAGNESIUM BRONZE.

Bronze, Aluminium Manganese. See ALUMINIUM MANGANESE BRONZE.

Bronze, Aluminium Tin. See ALUMINIUM TIN BRONZE.

Bronze, Aluminium Titanium. See ALUMINIUM TITANIUM BRONZE.

Bronze, Arsenic. See ARSENIC BRONZE.

Bronze, Artificial. See TOMBAC RED, and MANNHEIM GOLD.

Bronze, Atlas. See ATLAS BRONZE.

Bronze Bearing Metals. Very variable alloys. One type contains from 70–91 per cent. tin, 7–26 per cent. antimony, and 2–22 per cent. copper ; whilst another class contains from 70–86 per cent. copper, 4–20 per cent. tin, 0–30 per cent. zinc, and 0–15 per cent. lead. Also see KOCHLIN'S BEARING BRONZE.

Bronze, Beryllium. See BERYLLIUM BRONZE.

Bronze, Bilgen. See BILGEN BRONZE.

Bronze, Black. See BLACK BRONZE.

Bronze Blues. Types of Prussian blue.

Bronze Browns. Mixtures of unburnt umber and chrome yellow, toned with Cassel brown or chrome orange. Pigments.

Bronze, Cadmium. See CADMIUM BRONZE.

Bronze, Calsun. See CALSUN BRONZE.

Bronze, Carbon. See CARBON BRONZE.

Bronze, Carloon. See CARLOON BRONZE.

Bronze, Caro. See CARO BRONZE.

Bronze, Chamet. See CHAMET BRONZE.

Bronze, Chinese. See CHINESE BRONZE.

Bronze, Chromax. See CHROMAX BRONZE.

Bronze, Chromium. See CHROMIUM BRONZE.

Bronze, Cobalt. See COBALT BRONZE.

Bronze, Coin. See COINAGE BRONZE.

Bronze Colours (Brocade Colours). Powdered metals or metallic alloys mixed with linseed oil varnish. Brocade colours are not so finely powdered.

Bronze, Conductivity. See CONDUCTIVITY BRONZE.

Bronze, Cornish. See CORNISH BRONZE.

Bronze, Cowles' Aluminium. See COWLES' ALUMINIUM BRONZE.

Bronze, Cyprus. See CYPRUS BRONZE.

Bronze, Damar. See DAMASCUS BRONZE.

Bronze, Dawson. See DAWSON BRONZE.

Bronze, Durbar. See DURBAR BRONZE.

Bronze, Eclipse. See ECLIPSE BRONZE.

Bronze, Eisen. See EISEN BRONZE.

Bronze, Eisler's. See EISLER'S BRONZE.

Bronze, Elephant. See ELEPHANT BRONZE.

Bronze, Emerald. See EMERALD BRONZE.

Bronze, False Gold. See FALSE GOLD BRONZE.

Bronze, File. See FILE BRONZE.

Bronze, Gold. See SAFFRON BRONZE.

Bronze, Graney. See GRANEY BRONZE.

Bronze Greens (Bottle Green, Olive Green). Pigments. They usually consist of Brunswick greens mixed with ochre and umber, or of Chrome green with black and yellow colours.

Bronze, Gurney. See GURNEY BRONZE.

Bronze, Harmonia. See HARMONIA BRONZE.

Bronze, Harrington. See HARRINGTON BRONZE.

Bronze, Helmet. See HELMET BRONZE.

Bronze, Herbohn. See HERBOHN BRONZE.

Bronze, Hercules. See HERCULES BRONZE.

Bronze, Holfos. See HOLFOS BRONZE.

Bronze, Hydraulic. See HYDRAULIC BRONZE.

Bronze, Instrument. See INSTRUMENT BRONZE.

Bronze, Japanese. See JAPANESE BRONZE.

Bronze, Kern's Hydraulic. See KERN'S HYDRAULIC BRONZE.

Bronze, Kochlin's Bearing. See KOCHLIN'S BEARING BRONZE.

Bronze, Kuhne's Phosphor. See KUHNE'S PHOSPHOR BRONZE.

Bronze, Lafond. See LAFOND BRONZE.

Bronze, Lead. See LEAD BRONZE.

Bronze, Leaded. See LEADED BRONZE.

Bronze, Liquid. See LIQUID BRONZES.

Bronze, Litnum. See LITNUM BRONZE.

Bronze, Lowroff Phosphor. See LOWROFF PHOSPHOR BRONZE.

Bronze, Lumen. See LUMEN BRONZE.

Bronze, Machine. See MACHINE BRONZE.

Bronze, Magenta. See MAGENTA BRONZE.

Bronze, Manganaluminium. See MANGANALUMINIUM BRONZE.

Bronze, Manganese. See MANGANESE BRONZE.

Bronze, McKechnie's. See McKECHNIE'S BRONZE.

Bronze, Medal. See MEDAL BRONZE.

Bronze, Mirror. See MIRROR BRONZE.

Bronze, Naval. See NAVAL BRONZE.

Bronze, Needle. See NEEDLE BRONZE.

Bronze, Nickel. See NICKEL BRONZE.

Bronze, Nickel Aluminium. See NICKEL ALUMINIUM BRONZE.

Bronze, Nickel Manganese. See NICKEL MANGANESE BRONZE.

Bronze, Olympic. See OLYMPIC BRONZE.

Bronze, Optical. See OPTICAL BRONZE.

Bronze, Oranium. See ORANIUM BRONZE.

Bronze, Phono. See PHONO BRONZE.

Bronze, Phosphor. See PHOSPHOR BRONZE.

Bronze, Plastic. See PLASTIC BRONZE.

Bronze, Platinum. See PLATINUM BRONZE.

Bronze Powders. These are usually finely divided alloys of copper and zinc (brass). The following are common :

	Copper per cent.	Zinc per cent.
Rich gold	80	20
Rich pale gold	85	15
Pale gold	90	10
Red gold	95	5

Bronze, Reich's. See REICH'S BRONZE.

Bronze, Resistance. See RESISTANCE BRONZE.

Bronze, Roman. See ROMAN BRONZE.

Bronze, Saffron. See SAFFRON BRONZE.

Bronze, Screw. See SCREW BRONZE.

Bronze, Sea-water. See SEA-WATER BRONZE.

Bronze, Sheathing. See SEA-WATER BRONZE.

Bronze, Silicon. See SILICON BRONZE.

Bronze, Silver. See SILVER BRONZE.

Bronze, Silzin. See SILZIN BRONZE.

Bronze, Statuary. See STATUARY BRONZE.

Bronze, Steel. See UCHATIUS BRONZE.

Bronze, Stone's. See STONE'S BRONZE.

Bronze, Sun. See SUN BRONZE.

Bronze, Telegraph. See TELEGRAPH BRONZE.

Bronze, Tin. See TIN BRONZE.

Bronze, Tobin. See TOBIN BRONZE.

Bronze, Tungsten. See TUNGSTEN BRONZE.

Bronze, Turbadium. See TURBADIUM BRONZE.

Bronze, Turbiston. See TURBISTON BRONZE.

Bronze, Uchatius. See UCHATIUS BRONZE.

Bronze, Valve. See VALVE BRONZE.

Bronze, Vanadium. See VANADIUM BRONZE.

Bronze Varnishes. These are usually prepared by mixing suitable dyes with spirit varnishes. Used for colouring articles to give iridescence.

Bronze, Vulcan. See VULCAN BRONZE.

Bronze Wire. An alloy of 98·75 per cent. copper, 1·2 per cent. tin, and 0·05 per cent. phosphorus.

Bronze, Wolfram. See TUNGSTEN BRONZE.

Bronze, Zinc. See ADMIRALTY GUN METAL.

Bronzing Liquids. Consist of volatile liquids which will hold up the metal and some material which will keep the metallic powder from rubbing off after it has been applied. The best one contains pyroxylin dissolved in amyl acetate to which the metallic powder is added. A cheaper form consists of rosin dissolved in benzene with metallic powder added, and a very cheap one is a solution of sodium silicate with metal powder. Also see BANANA OIL.

Bronzing Powder. See MOSAIC GOLD.

Bronzing Solder. An alloy of 50 per cent. zinc and 50 per cent. copper.

Bronzite. A pyroxene mineral. It is $16[(Mg,Fe)SiO_3]$.

Bronzoke. A brass and copper cleaning compound. (545).

Brookite. A mineral. It is titanium dioxide, TiO_2.

Brooksite. An insulating material made from rosin and rosin oils.

Brophenin. Bromo-isovaleryl-phenocoll, a brominated phenetidine derivative. An antipyretic.

Broprins. A proprietary preparation of aspirin, citric acid, calcium carbonate and saccharin sodium. An analgesic. (354).

Brossite. A mineral. It is a variety of Dolomite.

Brostenite. A mineral. It is a manganite of iron and manganese.

Brotianide. An anthelmintic currently undergoing clinical trial as " FB 4059 ". It is 2 - bromo - 4 - chloro - 6 - (4 - bromophenylthiocarbamoyl) phenyl acetate.

Brotonat. A German product. This was formerly Brotropon, a bromine-albumen preparation containing 40 per cent. bromine. A sedative.

Brotox. A registered trade mark for a special tar for application to roads. It is stated to be non-toxic and for use on roads near streams. (580).

Brotropon. See BROTONAT.

Brovalol (Eubornyl, Valisan). Bornyl-bromo-valerate, $CH_3.CH(CH_3).CHBr.CO.O.(C_{10}H_{17})$. A nerve sedative, containing 25·2 per cent. bromine.

Brovolin Cough Syrup. Codeine phosphate, ephedrine hydrochloride ammonium chloride and potassium guaiacolsulphonate. A cough linctus. (339).

Brovon. A proprietary preparation of adrenaline, atropine methonitrate,papaverine hydrochloride and chlorbutol. Bronchial inhalant. (339).

Brovonex. A proprietary preparation of caffeine sodium iodide, sodium iodide and theophylline monoethanolamine. Bronchial antispasmodic. (339).

Brown Acetate. A calcium acetate prepared from crude pyroligneous acid and lime. It contains from 60–70 per cent. acetate. The name is also applied to an impure variety of lead acetate prepared from the same acid.

Brown Acid (Brown Oil of Vitriol, B.O.V.). Sulphuric acid of specific gravity 1·72. It is usually a tower acid, and contains about 80 per cent. sulphuric acid, and is largely employed for making superphosphate, and in other rough chemical manufactures. It is technically called brown acid, brown oil of vitriol, or simply, B.O.V.

Brown Barberry Gum. See MOROCCO GUM.

Brown, Bone. See BONE BROWN.

Brown, Catechu. See CATECHU BROWN.

Brown Hæmatite. A hydrated oxide of iron, an iron ore, $2Fe_2O_3.3H_2O$.

Browning. See BURNT SUGAR.

Brown Iron Ore. Limonite (*q.v.*).

Brown Lake. Bistre (*q.v.*).

Brown Lead Ore (Linnets). A mixture of phosphate of lead and lead chloride, $3[Pb_3(PO_4)_2]+PbCl_2$.

Brown Lead Oxide. Lead dioxide, PbO_2.

Brown Madder. A pigment. It is a lake prepared from madder root.

Brown Manganese Ore. See MANGANITE.

Brown NP and NPJ. A dyestuff from diazotised *p*-nitraniline and pyrogallol, $C_{12}H_9N_3O_5$. Dyes chrome mordanted wool.

Brown Ochre. See YELLOW OCHRE.

Brown Oil of Vitriol. See BROWN ACID.

Brown Ointment. A mixture of camphorated brown plaster, suet, and olive oil.

Brown Ore. A variable mixture of hydrated oxides of iron, usually $2Fe_2O_3.3H_2O$.

Brown Oxide of Chromium. See CHROMATE OF CHROMIUM.

Brown Oxide of Iron. Peroxide of iron.

Brown Oxide of Tungsten. Tungsten dioxide, WO_2.

Brown Pink (Stil de Grain). A pigment prepared from Turkish or Persian berries. It is a lake precipitated by alum from a decoction of the colouring matter.

Brown PM. A dyestuff. It is the hydrochloride of *p*-amino-benzene-azo-*m*-phenylene-diamine, $C_{12}H_{14}N_5Cl$. Dyes tanned cotton dark brown.

Brown Powder (Cocoa Powder). Brown gunpowder, consisting of approximately 79 per cent. nitre, 18 per cent. charcoal, and 3 per cent. sulphur.

Brown Precipitate. Iodine dissolved in potassium iodide.

Brown Red. See INDIAN RED.

Brown Red Antimony Sulphide. See KERMES MINERAL.

Brown Rice. See RED RICE.

Brownspar. A mineral. It is triferrous carbonate of calcium and magnesium.

Brown Sugar of Lead. A lead acetate prepared from crude pyroligneous acid and litharge.

Brown Terre Verte. See BURNT SIENNA.

Brown Ultramarine. Ultramarine in which the sulphur constituent is replaced by selenium.

Brown Vinegar (Malt Vinegar). A vinegar which is seldom produced by the acetous fermentation of malt wort. It is now nearly all derived from cereals, the starch of which has been saccharified.

Broxil. A proprietary preparation containing phenethicillin potassium. An antibiotic. (364).

Brozgerite. See CLEVITE.

Brucite (Nematolite, Nemalite). A mineral. It is magnesium hydroxide, $Mg(OH)_2$.

Brücke's Reagent. This reagent consists of 50 grams potassium iodide in 500 c.c. water saturated with mercuric iodide (120 grams) and made up to 1 litre. A precipitating agent for proteins.

Brufen. A proprietary preparation containing IBUPROFEN. An anti-rheumatic drug. (502).

Brugère Powder. An explosive. It is a priming composition, containing 54 per cent. ammonium picrate, and 46 per cent. potassium nitrate.

Brugnatellite. A mineral. It is [$Mg_6FeCO_3(OH)_{13}.4H_2O$].

Bruiachite. A mineral. It is a variety of fluorite.

Brulidine. A proprietary preparation of dibromopropamidine isethionate. Antiseptic skin cream. (336).

Brulon. Nylon brush mono filaments. (512).

Brunnerite. A mineral. It is a variety of calcite.

Brunner's Salt. Obtained by dissolving vermilion in potassium monosulphide. It has the formula, $HgS.K_2S.5H_2O$. The method is used to test vermilion for iron and metallic mercury.

Brünnichite. Apophyllite (q.v.).

Brunol. A proprietary preparation of n-butyl-salicylate.

Brunolinum. See CARBOLINEUM.

Brun's Paste. A mixture of airol (q.v.), mucilage, and glycerin. Used for skin diseases.

Brunsvigite. A mineral. It is $2[(Fe^{...},Mg,Al)_6(Si,Al)_4O_{10}(OH)_8]$.

Brunswick Black. Asphalt or pitch mixed with turpentine and linseed oil, and heated. Some preparations contain benzoline, and the pigment is bone pitch.

Brunswick Blue (Celestial Blue). A pigment produced by mixing 50–90 per cent. barytes with Prussian blue.

Brunswick Green. A pigment. It was originally obtained by treating copper sulphate with sodium chloride, and precipitating with milk of lime. In this case a basic copper chloride is produced. When precipitated with sodium carbonate, it possesses the composition, $CuCO_3.Cu(OH)_2$.

At the present time the name has been transferred to mixtures of chrome yellow and blue, to which the names Prussian green and Victoria green are also applied. The term Brunswick green is, however, also used for Bremen blue, for an arsenical copper green, and for a variety of Schweinfurth green. Also see CHROME GREEN and MOUNTAIN GREEN.

Brunswick Green, New. See CHROME GREENS.

Brunswick Green, Old. A basic chloride of copper, now rarely manufactured. Emerald green and Bremen blue are both known by this name.

Brushite. A mineral. It is a hydrated phosphate of lime, $HCaPO_4.2H_2O$.

Brush Wire. A brass wire containing 64·25 per cent. copper, 35 per cent. zinc, and 0·75 per cent. tin.

Brussels System. A systemic insecticide. (501).

B.R.V. A coal-tar distillate consisting chiefly of high boiling constituents. It is used as a rubber softener.

Bryta. Automatic dishwashing machine detergent. (586).

B.S. Copper. The purest form of copper on the market, containing less than 0·05 per cent. arsenic.

B.S. Sea Water Alloy. An aluminium alloy containing 7·5–9·5 per cent. magnesium, 0·2 per cent. silicon, and 0·2–0·6 per cent. manganese. It has a tensile strength of 45–55 kg. per sq. mm.

B.T.G. Alloy. A heat-resisting alloy containing 60 per cent. nickel, 12 per cent. chromium, 1–4 per cent. tungsten, and balance iron.

Buban. A proprietary preparation of BUNAMIDINE. A veterinary anthelmintic.

Bubblefil. A proprietary trade name for regenerated cellulose.

Bucar. A proprietary isoprene-isobutylene rubber. (479).

Bucarpolate. A pyrethrum synergist. (504).

Buccosperin. A German preparation. The chief constituents are buchu extract, oil of menthæ, acetyl-salicylic acid, salol, and hexamethylene-tetramine. An antiseptic.

Bucetin. N - β - Hydroxybutyryl - p - phenetidine.

Bucheronium. A proprietary trade name for a phenol-formaldehyde synthetic resin.

Bucholzite. See FIBROLITE.

Buchu (Bucku). The dried leaves of Barosma species. A diuretic.

Buchu Camphor. Diosphenol, $C_{10}H_{15}O_2$, the chief constituent of the essential oil obtained from Buchu leaves from Barosma betulina. It is antiseptic.

Bückingite. A mineral. It is a sulphate of iron.

Bucklandite. A mineral. It is a variety of allanite.

Buckland's Cement. A label cement consisting of 50 per cent. gum arabic, 37·5 per cent. starch, and 12·5 per cent. white sugar. It is mixed with a little water for use.

Buckram. A coarse open fabric resembling cheesecloth but heavily sized with gum or other stiffener.

Buckroid. A very tough form of pure vulcanised rubber. Used for making mats, and for other purposes.

Buckthorn Berries. The dried unripe fruit of certain species of *Rhamnus*. The berries contain a glucoside, xanthorrhamnin, which splits up into rhamnetin and isodulcite. The dyestuff, rhamnetin, gives a fine yellow tin lake.

Buckthorn Green. See SAP GREEN.

Buclamase. A proprietary preparation of alpha amylase. (320).

Buclizine. 1 - (4 - *t* - Butylbenzyl) - 4 - (4 - chloro - α - phenylbenzyl) - piper - azine. Longifene and Vibazine are the hydrochloride.

Buclosamide. N - Butyl - 4 - chloro - salicylamide. Present in Jadit.

Bucrea. A proprietary trade name for urea-formaldehyde mouldings.

Bucrol. Carbutamide.

Buctril. A selective weed killer. (507).

Bucurit. A Romanian phenol-formaldehyde synthetic resin.

Budale. A proprietary preparation of paracetamol, codeine phosphate and butobarbitone. A sedative. (320).

Bufa. A proprietary trade name for dibutyl phthalate. (81).

Bufexamac. An anti-inflammatory. 4-Butoxyphenylacetohydroxamic acid.

Buffalo Rubine (Azo Bordeaux). A dyestuff. It is the sodium salt of α-naphthalene - azo - α - naphthol - disulphonic acid, $C_{20}H_{12}N_2O_7S_2Na_2$. Dyes wool red from an acid bath.

Bufferin. A proprietary preparation of aspirin, magnesium carbonate and aluminium glycinate. An analgesic. (324).

Buffo. A dilute form of toril, a variety of beef extract.

Bufrolin. A preparation for the treatment of allergic airway obstruction. 6 - Butyl - 1, 4, 7, 10 - tetrahydro - 4 10 - dioxo - 1, 7 - phenanthroline - 2, 8 - dicarboxylic acid. " ICI 74,917 " is currently under clinical trial as the disodium salt.

Bufuralol. A beta adrenergic blocking agent. 2 - *tert* - Butylamino - 1 - (7 - ethylbenzofuran - 2 - yl) ethanol. " Ro 03-4787 " is currently under clinical trial as the hydrochloride.

Bufylline. Theophylline 2 - amino - 2 - methylpropan-1-ol.

Buhrstone. See BURRSTONE.

Bulbocapnin. An alkaloid obtained from *Corydalis cava*. It acts against tremors or trembling in paralysis.

Bull-dog. A mixture of ferric oxide and silica obtained by roasting tap cinder in the puddling furnace.

Buller's Glue Compound. This is a compound of glue, sulphur, and Paris white, and is used to secure adhesion to metallic surfaces.

Bullet Brass. See BRASS.

Bullet-proof Glass. A glass made from several layers of glass with a transparent plastic between the layers.

Bull Metal. An alloy similar to Delta metal in composition.

Bultfonteinite. A mineral. It is $Ca_2SiO_2(OH,F)_4$

Bumetanide. A diuretic. 3 - Butylamino - 4 - phenoxy - 5 - sulphamoylbenzoic acid. See BURINEX.

Buna Rubbers. German synthetic rubbers obtained from butadiene by means of polymerisation with sodium as catalyst. They are vulcanisable.

Buna AP. A trade name for a group of ethylene - propylene rubbers. AP147 is used for injection moulding; AP447 for improving the properties of plastics. (480).

Buna S. A proprietary trade name for a vulcanisable synthetic rubber. It is a butadiene-styrene co-polymer.

Buna SS. Contains a larger percentage of styrene. An important series of synthetic products is made by copolymerising butadiene with 10–30 per cent. of another polymerisable substance such as styrene or acetonitrile, *e.g.* Buna N, etc.

Bunamidine. An anthelmintic. *NN*-Dibutyl - 4 - hexyloxy - 1 - naphthamidine. SCOLABAN is the hydrochloride; DUBAN the hydroxynaphthoate.

Buniodyl. 3 - (3 - Butyramido - 2,4,6 - tri-iodiphenyl)-2-ethylacrylic acid. Orabilix is the sodium salt.

Bunsenine. See WHITE TELLURIUM.

Bunsenite. A mineral. It is nickel oxide, NiO, found in Saxony.

Bunte's Salt. Sodium-ethyl thiosulphate.

Bunting. A plain woven worsted fabric employed for making flags, etc. A cotton bunting is also made for the manufacture of cheap flags, etc.

Buntkupfererz. Bornite (*q.v.*).

Buphenine. 1 - (4 - Hydroxyphenyl) - 2 - (1 - methyl - 3 - phenylpropyl amino)propanol. Perdilatal is the hydrochloride.

Bupivacaine. A local anæsthetic. 1 - Butyl - 2 - (2, 6 - xylylcarbamoyl) - piperidine. MARCAIN is the hydrochloride.

Buprenorphine. An analgesic. *N*-Cyclopropylmethyl - 7, 8 - dihydro - 7α - (1 - (*S*) - hydroxy - 1, 2, 2 - trimethylpropyl) - *O*⁶ - methyl - 6, 14 - *endo*ethanonormorphine. " RX 6029-M " is currently under clinical trial as the hydrochloride.

Buratite. A mineral. It is a variety of aurichalcite.

Burdeo. A proprietary deodorant containing 25 per cent. aluminium subacetate and 75 per cent. boric acid. (879).

Burgess-Hambuechen Solution. A solution containing 275 grams ferrous ammonium sulphate and 1000 c.c. water. Used for the electro-deposition of iron, using a current density of 6–10 amps. per sq. ft. at 30° C.

Burgess Solder. An alloy of 76 per cent. tin, 21 per cent. zinc, and 3 per cent. aluminium.

Burgundy Lake. A proprietary trade name for a red lake containing organic colours, aluminium hydroxide, and Blanc fixé.

Burgundy Mixture. For the prevention of blight or potato disease, and its cure. (a) Consists of 4 lb. copper sulphate dissolved in 5 gallons water, and made up to 35 gallons. (b) Contains 5 lb. washing soda dissolved in 5 gallons water. (b) is added to (a) and stirred.

Burgundy Pitch. The resinous exudation of the European silver fir. Artificial burgundy pitch is usually a compound made from yellow pine-resin, and is practically a solution of paraffin oil in common rosin. An ordinary mixing consists of 95 per cent. rosin and 5 per cent. mineral oil. The melting-point varies from 176–212° F. It is used as a rubber softener. Another mixture consists of common pitch, rosin, and turpentine.

Burinex. A proprietary preparation of bumetanide. A diuretic. (308).

Burinex K. A proprietary preparation of bumetanide with potassium chloride in slow-release form. A diuretic. (308).

Burkeite (Gauslinite). A double salt of sodium carbonate and sulphate.

Burlap. A coarse jute fabric used for wrapping and as a backing for linoleum.

Burma Black Varnish. See THITSI.

Burmite (Birmite). Burmese amber of a reddish-brown colour.

Burmol (Laundros). Names for a preparation of sodium hydrosulphite. A bleaching agent. Also see HYDRO-SULPHITE.

Burn-a-lay. A proprietary mixture of 0·75 per cent. chlorobutanol, 0·025 per cent. oxyquinoline benzoate, 2 per cent., zinc oxide and 0·5 per cent. thymol used in the treatment of burns. (881).

Burnett's Disinfecting Fluid. A solution of zinc chloride.

Burnojel. A proprietary preparation of copper guaicol sulphonate, 5-amina-crine hydrochloride and benzamine borate. Burn ointment. (826).

Burnol Acriflavine Cream. A proprietary product. It is an emulsified cream containing neutral acriflavine for use in cases of burns, wounds, etc. (502).

Burnt Alum (Exsiccated Alum). Potash alum, which has been heated at low redness.

Burnt Borax. See BORAXOSTA.

Burnt Carmine. A pigment obtained by calcining carmine.

Burnt Hypo (Black Hypo, Eureka Compound). A mixture of lead thiosulphate and sulphide, and sulphur. Used in the vulcanisation of rubber.

Burnt Iron. Iron which has been heated to a high temperature for a long time. It is brittle, also cold and hot short.

Burnt Lime. See LIME.

Burnt Magnesia. Magnesium oxide MgO.

Burnt Nickel. A term used for a grey pulverulent nickel precipitated by too strong a current during its electro-deposition.

Burnt Ochre. See INDIAN RED.

Burnt Ore. See BLUE BILLY.

Burnt Pyrites. Pyrites which have been burnt until 70 per cent. of ash is left. It consists of iron oxide, Fe_2O_3.

Burnt Roman Ochre. An orange pigment obtained by calcining Roman ochre.

Burnt Sienna (Brown Terre Verte, Raw and Burnt Umber, Cappagh Brown, Mars Brown, American Sienna). Earths or ochres, raw or calcined, or artificial ochre. They are brown pigments containing iron or iron and manganese. Burnt sienna usually contains 65–75 per cent. Fe_2O_3, and 13–20 per cent. SiO_2. Also see INDIAN RED.

Burnt Steel. Steel which has been heated to near its point of fusion.

Burnt Sugar (Browning, Gravy Salt, Gravy Colour, Caramel, Sugar Colouring). Obtained by making a dextrose solution alkaline with soda, evaporating, and heating to 220° C.

Burnt Topaz. When Brazilian topaz is heated, it changes from a cherry-yellow to a rose-pink, being then known as burnt topaz.

Burnt Umber. Umber which has been heated, whereby its colour is somewhat reddened. Raw umber is a brown earthy variety of ochre, coloured by oxides of iron. Also see BURNT SIENNA and UMBER.

Bur Oil. Burdock oil, obtained from the seeds of *Lappa minor*.

Burow's Solution. A 7·5–8 per cent. solution of aluminium acetate. (*Liquor alumini acetatis, B.P.C.*)

Burr Brass. See BRASS.

Burrstone (Buhrstone). A siliceous rock, used for millstones.

Bursoline. Sulphonated oil for tanners. (530).

Burveen. A proprietary preparation of colloidal oatmeal with aluminium sulphate and calcium acetate. Used in dermatology. (320).

Burwood. A proprietary trade name for a cold moulded wood fibre plastic.

Buscopan. A proprietary preparation containing hyoscine-N-butylbromide. An antispasmodic. (278).

Busheled Iron. A low grade iron or steel made from scrap in a hearth furnace.

Bushmanite. A mineral. It is $(Mn,Mg)Al_2(SO_4)_4.22H_2O$.

Bush Metal. An alloy of 72 per cent. copper, 14 per cent. tin. and 14 per cent. yellow brass.

Bu-shol. See RAR.

Bush Salt. A mixture of 78 per cent. potassium chloride and 19 per cent. potassium sulphate. It is an ash prepared by the lixiviation of the ashes of the sedge.

Bustamentite. A mineral. It is a lead iodide.

Bustamite. A pyroxene mineral, $(Mn.Ca)SiO_2$.

Busulphan. Tetramethylene di(methanesulphonate). An anti-neoplastic agent. See MYLERAN.

Butabarbital. A sedative and hypnotic. 5 - ethylbarbituric acid. BUTISOL, DA-SED, EXPANSATOL, MEDASED.

Butabarbitone. A proprietary trade name for Secbutobarbitone.

Butachlor. A proprietary trade name for a polychloroprene moulding compound. (12).

Butacite (Butvar). A proprietary trade name for a polyvinyl-butyral synthetic resin used for interleaving safety glass and in adhesive preparations.

Butacote. A proprietary preparation of phenyl butazone. An anti-rheumatic drug. (17).

Butakon. Butadiene copolymers. (512).

Butakon A. A trade mark for butadiene acrylonitrile copolymers (nitrile rubbers) having excellent resistance to oils. (2).

Butakon A2554. A proprietary butadiene copolymer rubber. (481).

Butakon ML 577/1. A proprietary butadiene/methyl methacrylate latex. (481).

Butalamine. A vasodilator. 5 - (2 - Dibutylaminoethyl) amino - 3 phenyl - 1, 2, 4 - oxadiazole. "LA 1221" is currently under clinical trial as the hydrochloride.

Butamin. p-Amino-benzoyl-α-dimethyl-amino-β-methyl-γ-butanol hydrochloride.

Butamyrate. A cough suppressant. 2 - (2 - Diethylaminoethoxy) ethyl 2 - phenylbutyrate. Sinecod is the citrate.

Butane Tetrol. See TETRA-NITROL.

Butanilicaine. N - Butylaminoacetyl - 6-chloro-o-toluidine. Present in Hostacain as the phosphate.

Butanol. Normal butyl alcohol, C_4H_9OH. It has a specific gravity of 0·81, boils at 114–117° C., and flashes at 35° C. It is now obtained from corn mash by fermentation and distillation (100 parts corn yield 10·7 butanol). It is used as a solvent in the lacquer industry.

Butanox. A proprietary trade name for methyl ethyl ketone peroxides. (90).

Butaphen. A proprietary preparation of phenylbutazone. (279).

Butasan. A proprietary trade name for a rubber vulcanisation ultra accelerator. It is the zinc salt of dibutyl dithiocarbamic acid.

Butazate. A proprietary preparation of zinc dibutyl dithiocarbamate. (435).

Butazolidin. A proprietary preparation of phenylbutazone. (17).

Butazolidin Alka. A proprietary preparation of aluminium hydroxide gel, magnesium trisilicate and Butazolidine. (17).

Butazolidin with Xylocaine. A proprietary preparation of phenylbutazone and lignocaine. (17).

Butazone. A proprietary preparation of phenylbutazone. (363).

Butazolidin. A proprietary trade name for Phenylbutazone.

Butea Gum. See KINO, BENGAL.

Butene. A reaction product of butyl aldehyde and aniline. It is an amber liquid, and is used as an accelerator for rubber vulcanisation.

Butesin (Scuroforme). Trade mark for butyl-p-amino-benzoate.

Butethamate Citrate. 2 - Diethyl - aminoethyl 2-phenylbutyrate citrate. Tephamine. An anti-spasmodic agent.

Butex. (1) A proprietary trade name for a synthetic rubber.

Butex. (2) Esters of 4 hydroxybenzoic acid. (504).

Buthalitone Sodium. A mixture of 100 parts by weight of the mono-sodium derivative of 5-allyl-5-isobutyl-2-thiobarbituric acid and 6 parts by weight of dried sodium carbonate. Transithal; Ulbreval.

Butipyrine. The name given to an imitation of trigemin (q.v.).

Butirosin Sulphate. An anti-bacterial. It is a mixture of the sulphates of the A and B forms of an antibiotic produced by *Bacillus circularis*.

Butisol. See BUTABARBITAL.

Butlerite. A mineral. It is an iron sulphate.

Butoben. A proprietary trade name for p-hydroxy-n-butyl benzoate.

Butocresiol. See EUROPHENE.

Butofan D. A registered trade mark for a polybutadiene dispersion. (49).

Butolan. A proprietary preparation. It is p-oxy-diphenyl-methane carbamic ester.

Butomet. A proprietary preparation of butobarbitone and ipecacuanha. A hypnotic. (360).

Butorphanol. An analgesic and anti-tussive, currently undergoing clinical trial as " levo-BC-2627 ". (—) - 17 - (Cyclobutylmethyl) morphinan - 3, 14 - diol.

Butox. A proprietary preparation of polyisobutylene-isoprene. (482).

Butoxamine. ($\pm$) - erythro - 1 - (2,5 - Dimethoxyphenyl)2 - t - butylamino - propan-1-ol.

Butoxone. A selective weed killer. (512).

Butoxyl. 1-methoxybutyl acetate.

Butriptyline. An anti-depressant. DL - 3 - (10, 11 - Dihydro - 5H - dibenzo[a, d]-cyclohepten - 5 - yl) - 2 - methylpro-pyldimethylamine. "AY 62014" is currently under clinical trial as the hydrochloride.

Bütschliite. A mineral. It is a variety of calcium carbonate. $CaCO_3$. Also: $K_6Ca_2(CO_3)_5.6H_2O$.

Butt Brass. See BURR BRASS.

Butte 15. A proprietary preparation of sodium butabarbital in quarter-grain tablets. (754).

Butter, American Mace. See OTOBA BUTTER.

Butter, Abjab. See NJAVE BUTTER.

Butter, Arachis. See ARACHIS BUTTER.

Butter, Artificial. See MARGARINE.

Butter, Bambuk. See SHEA BUTTER.

Butter, Blue. See BLUE BUTTER.

Butter, Bog. See BOG BUTTER.

Butter, Boiled. See BUTTER, RENO-VATED.

Butter, Cacao. See COCOA BUTTER.

Butter, Cocoa. See COCOA BUTTER.

Butter, Coco-nut. See COCO-NUT BUTTER.

Butter Colour. Annatto (q.v.).

Buttercup Yellow. See ZINC YELLOW.

Butter, Dika. See DIKA BUTTER.

Butter, Djave. See NJAVE BUTTER.

Butter, Dutch. See MARGARINE.

Butter, Galam. Shea butter (q.v.).

Butter, Gamboge. See GAMBOGE BUTTER.

Butter, Ghé. See INDIAN BUTTER.

Butter, Ghee. See INDIAN BUTTER.

Butter, Goa. Kokum butter (q.v.).

Butter, Green. See BORNEO TALLOW.

Butter, Illipé. See BASSIO TALLOW.

Butter, Indian. See INDIAN BUTTER.

Butterine. See MARGARINE.

Butterine Oil. An inferior oil obtained from the peanut.

Butter, Irvingia. See COCHIN-CHINA WAX.

Butter, Kanga. See LAMY BUTTER.

Butter, Kanja. See LAMY BUTTER.

Butter, Kokum. See KOKUM BUTTER.

Butter, Lamy. See LAMY BUTTER.

Butter, Macaja. See MOCAYA OIL.

Butter, Macassar Nutmeg. See MACASSAR NUTMEG BUTTER.

Butter, Mace. See MACE BUTTER.

Butter, Mahwa. See BASSIO TALLOW.

Butter Milk. Milk from which the fat has been removed.

Buttermilk Ore. A mixture of clay and silver chloride, found in the Hartz.

Butter, Mineral. See MINERAL BUTTERS.

Butter, Mohrah. See BASSIO TALLOW.

Butter, Mohwrah. See BASSIO TALLOW.

Butter, Njave. See NJAVE BUTTER.

Butter, Nutmeg. See NUTMEG BUTTER.

Butter Nuts. The nuts from *Caryocar tomentosum*, imported from South America, where they are known as Suari or Surahwa nuts.

Butter of Antimony. An old name for antimony trichloride, $SbCl_3$. Used in surgery as a caustic.

Butter of Arsenic (Caustic Oil of Arsenic). Arsenic chloride, $AsCl_3$.

Butter of Paraffin. Soft paraffin.

Butter of Sulphur. Precipitated sulphur.

Butter of Tin (Oxymuriate of Tin). A crystalline hydrated chloride of tin, obtained by exposing stannic chloride to air, $SnCl_4$, $5H_2O$. Used in dyeing.

Butter of Zinc. Hellot's name for zinc chloride, $ZnCl_2$.

Butter Oil (Sweet Nut Oil). Cotton-seed oil has been sold under these names, and sometimes from 2–5 per cent. palm oil is used with it. Used in the manufacture of margarine and lard compound.

Butter, Palm. See PALM BUTTER.

Butter, Papua Nutmeg. See MACASSAR NUTMEG BUTTER.

Butter, Para. See PARA PALM OIL.

Butterpits. The seeds of *Acanthosicyos horrida*. Eaten as dessert nuts, or used in confectionery.

Butter Powder. Sodium bicarbonate, $NaHCO_3$.

Butter, Process. See BUTTER, RENO-VATED.

Butter, Renovated (Process Butter, Boiled Butter, Aerated Butter, Sterilised Butter). Prepared from rancid butter, which is kneaded with a solution of sodium, bicarbonate, then washed with water, and mixed with milk. Often air is blown through it, while hot.

Butter Rock. A soft compound of iron and aluminium exuded from certain rocks.

Butter, Shea. See SHEA BUTTER.

Butter, Sierra Leone. See LAMY BUTTER.

Butter, Sterilised. See BUTTER, RENOVATED.

Butter Surrogate. See MARGARINE.

Butter, Vegetable. See VEGETABLE BUTTER.

Butter, Wax. See WAX BUTTER.

Butter Yellow (Oil Yellow, Butyro-flavine). An azo dyestuff. It is di-methyl-amino-azo-benzene, $C_6H_5.N_2$. $C_6H_4.N(CH_3)_2$. Used for colouring butter and oils.

Buttgenbachite. A mineral. It is a chloride and nitrate of copper.

Buttirol. An artificial butter prepared by emulsifying oleomargarine or other fat with milk, and separating the arti-ficial butter by centrifugalising after slight fermentation.

Button Brass. See BRASS.

Button Lac. See LAC.

Button Metal. An alloy of 57 per cent. zinc and 43 per cent. copper.

Button Solder (White Solder). Usually contains 50 per cent. tin, 30 per cent. copper, and 20 per cent. brass ; or 33 per cent. copper, 27 per cent. brass, and 40 per cent. zinc.

Butvar (Butacite). The trade mark for an acetal type of synthetic resin made from Gelva (*q.v.*)., and butyraldehyde.

Butvar B-79. A trade mark for poly-vinyl butyral resin specially designed for surface coatings and adhesive applications. It has a low molecular weight and is readily soluble in a wide range of solvents. (57).

Butylated Hydroxyanisole. A mixture of 2-*t*-butyl-4-methoxyphenol and 3-*t*-butyl-4-methoxyphenol.

Butylated Hydroxytoluene. 2,6 - Di - *t*-butyl-*p*-cresol.

Butyl Carbitol. Diethylene-glycol-monobutyl ether. A lacquer solvent boiling at 240° C.

Butyl Cellosolve. An American brand of ethylene-glycol-monobutyl ether. It is a nitrocellulose solvent, and is used in the manufacture of brushing lac-quers. It boils at 170° C., has a specific gravity of 0·9–0·905, and no acidity.

Butyl-Hypnal. Butyl-chloral-anti-pyrine, $C_{11}H_{12}N_2O + C_4H_5Cl_3H_2O$. An-algesic and hypnotic.

Butylite. A proprietary butyl rubber sealant. (483).

Butyl Rubber. A proprietary trade name for a co-polymer of isobutylene with a small percentage of diolefine such as butadiene. An unsaturated synthetic rubber possessing the mini-mum unsaturation required for vul-canisation.

Butyn. A trade mark for the sulphate of

p - amino - benzoate of γ - di - *n* - butyl-amino-propyl alcohol [$NH_2.C_6H_4$. $COO(CH_2)_3.N(C_4H_9)_2]_2.H_2SO_4$. A local anæsthetic.

Butyrellite. See BOG BUTTER.

Butyric Ether. Ethyl butyrate, $C_4H_7O_2.C_2H_5$. Also see PINEAPPLE OIL.

Butyrin. The glyceryl ester of butyric acid. It is found in butter.

Butyroflavine. See BUTTER YELLOW.

Butyroin. Propyl-oxy-butyl ketone, $C_3H_7.CO.CH(OH)C_3H_7$.

Butyrone. Di-propyl-ketone, $CO(CH_2. CH_2.CH_3)_2$.

Buxine. Berberine.

Buxton Lime. A lime containing 98 per cent. lime and 0·5 per cent. magnesia.

B.V.B. An oxygenated brightener for white textiles. (588).

B.V. Monopolin. A German motor fuel stated to be a mixture of alcohol with a grade of light gasoline and motor benzol.

B-X-A. A proprietary anti-oxidant. A product of the reaction diarylamine-ketone-aldehyde. (556).

Bydolax. A proprietary preparation of oxyphenisatin diacetate. A laxative. (339).

Byerite. See BLOWN PITCH.

Byerlite. An asphaltic product ob-tained by the oxidation with the oxygen of the air of petroleum residue at 230° C. then at 340° C. It is soluble in carbon disulphide, and about 60 per cent. in ligroin, 75–110° C.

Byssolite. A mineral. A variety of amphibole.

Byssus Silk (Sea Silk). A silk found in tufts protruding from the shells of a mollusc, *Pinna nobilis*. Used in Italy for ornamental braid.

Byne or Bynes. Malt.

Bynin. A trade mark. A liquid malt extract.

Bynin Amara. A proprietary prepara-tion of iron phosphorus and nux vomica in Bynin liquid malt. Tonic. (284).

Byroline. A mixture of boric acid, wool fat, glycerin, and water. Used for skin troubles.

Bytownite. A mineral similar to Anorthite (*q.v.*).

BZ 55. Carbutamide.

Bz. Cellulose. The benzyl ether of cellulose. Used in the manufacture of plastic materials, films, and varnishes. It is stable towards acids, alkalis, water and weathering, alcohol, and petrol. It is soluble in benzene, toluene, xylene, naphtha+10 per cent. alcohol, and carbon tetrachloride+10 per cent. alcohol. (3)

C

Caapi. A proprietary preparation of atropine sulphate, phenacetin, caffeine, quinine alkaloid, cinnamon and ascorbic acid. Cold remedy. (334).

Cabardine Musk. An inferior musk from Thibet.

Cabbage Oil. See OIL OF CABBAGE.

Cabbage Red. The red colouring matter of red cabbage. It is called cauline, and is used for colouring wines.

Cabflex DIOA. A trade mark. Diisooctyl adipate. A vinyl plasticiser. (82).

Cabflex DIOZ. A trade mark. Diisooctyl azelate.

Cabflex DOA. A trade mark. Didecyl adipate. A vinyl plasticiser. (82).

Cabflex DOP, DDP. Trade marks for dioctyl phthalate and didecyl phthalate. Vinyl plasticisers. (82).

Cabflex HS-10. A trade mark. An alkyl aryl phthalate. A vinyl plasticiser. (82).

Cablinol. A speciality perfumery chemical. (549).

Cabol 100. A trade mark. A hydrocarbon oil type vinyl plasticiser.

Cab-O-Sil. A registered trade name for fire dried fumed silica having a surface area between 200 and 400 m^2/gm. For rendering resins thixotropic. Refractive index 1·46. (63).

Cabrerite. A mineral, $(NiMg.Co)_3 (AsO_4)_2.8H_2O$.

Cabsol. A trade name for a cellulose acetate butyrate air drying solution coating for woodwork. (254).

Cabtyrit. A proprietary preparation of compounded rubber in the unvulcanised state. It softens and becomes plastic, and can be easily moulded when slightly warmed. It is stated to resist the action of hydrochloric acid up to a temperature of 110° C., and to be suitable for lining tanks. It can also be employed as a paint.

Cabufix. An adhesive for cellulose acetate butyrate. (507).

Cabulite. Cellulose acetate butyrate film and sheet. (507).

Cacao Butter. See COCOA BUTTER.

Cacao Oil. See OIL OF COCOA.

Cacao Red. The red pigment, $(C_{17}H_{16} O_7)_2$, of cacao.

Cacao Seeds. The seeds of *Theobroma cacao*.

Cacao Shell. A by-product of the cocoa and chocolate industry, 100 lb. yielding 10½ lb. cacao shell in a factory. It contains 4–11 per cent. moisture 2·5–4·5 per cent. fat, and 7–12 per cent. ash. It is mainly used as an ingredient in cattle food and to a limited extent in the preparation of a mildly stimulating drink "Cascarilla" in South America. It is a source of theobromine and furfural.

Cacheutaite. A mineral. A variety of zorgite.

Cacholong. A mineral. It is a variety of Chalcedony or opal.

Cachou. See CUTCH.

Cachou de Lavel (Cachou de Lavel S, Katigene black brown N). A sulphide dyestuff, $(C_4H_2S_3)n$, prepared by heating sawdust or bran, with sodium polysulphide. The S mark is the bisulphite compound. Dyes cotton brown.

Cachou de Lavel S. See CACHOU DE LAVEL.

C Acid. 2-Naphthylamine-4 : 8-disulphonic acid.

Cacoclastite. A mineral. It is a silicate of aluminium and calcium.

Cacodyl (Alkarsin). Tetramethyl-diarsenide, $As_2(CH_3)_4$.

Cacodyliacol. See CACODYLLIAGOL.

Cacodylic Acid. Dimethyl-arsinic acid, $As(CH_3)_2O(OH)$.

Cacodylliagol (Cacodyliacol, Guaiacodyl). Guaiacol-cacodylate, $(CH_3)_2 AsO.O.C_6H_4(OCH_3).H_2O$. An antiseptic used in medicine.

Cacodyl, New. See ARRHENAL.

Cacodyl Oxide. Tetramethyl-diarsineoxide, $[As(CH_3)_2]_2O$.

Cacoxenite. A mineral. It is $[Fe_4'''(PO_4)_3(OH)_3.12H_2O]$.

Cadaverine. Pentamethylene-diamine, $NH_2(CH_2)_5NH_2$. A base found in ergot.

Cade Oil. See JUNIPER TAR OIL.

Caderite. A tear gas. It is benzyl bromide and stannic chloride.

Cadet's Fluid. A fluid containing small quantities of cacodyl, and much cacodyl oxide.

Cadie Gum. See CAMBOGIA.

Cadmia. See CADMIUM YELLOW.

Cadmiol. A 10 per cent. suspension of cadmium salicylate in liquid paraffin containing a small quantity of cresol. A German preparation.

Cadmium Blende. See GREENOCKITE.

Cadmium Bronze. A copper alloy containing 0·5–1·2 per cent. cadmium used for telephone and trolley wire.

Cadmiumised Zinc. Zinc metal placed in a 2 per cent. cadmium sulphate solution for five minutes, then well washed. It is used as a reducing agent for the determination of iron.

Cadmium Lithopone (Cadmopone). A pigment analogous to lithopone, in which cadmium replaces zinc. It is made by the precipitation of cadmium sulphate solution with barium sulphide, and contains 38 per cent. cadmium sulphide.

Cadmium Orange (Cadmium Red). Cadmium sulphide, CdS, with smaller amounts of cadmium selenide and heavy spar.

Cadmium, Pale. See CADMIUM YELLOW.

Cadmium Red. See CADMIUM ORANGE.

Cadmiumspat. Synonym for Otavite.

Cadmium Yellow (Cadmia, Jaune Brilliant, Pale Cadmium, Orient Yellow, Radiant Yellow, Aurora Yellow, Daffodil). A pigment. It is cadmium sulphide, CdS. Specific gravity, 4·9.

Cadmium-zinc Spar. A mineral. It is a zinc carbonate containing cadmium.

Cadmiumzinkspath. A mineral. It is $(Zn,Cd)CO_3$.

Cadmopone. A registered trade mark. See CADMIUM LITHOPONE.

Cadogel. That portion of oil of cade which distils between 220° and 300° C. It is employed in the treatment of eczema.

Cadox BFF-50. A granular product containing 50 per cent. benzoyl peroxide and 50 per cent. of an inert phlegmatizer. (405).

Cadwaladerite. A mineral. It is $Al(OH)Cl.4H_2O$.

Caen Stones. Limestones used for building purposes.

Cæsium-Beryl. Synonym for Vorobbyevite.

Cafadol. A proprietary preparation of paracetamol and caffeine citrate. An analgesic. (355).

Cafaspin. A proprietary preparation. It contains acetyl-salicylic acid and caffeine.

Cafedrine. An analeptic. L-7-[2-(β-Hydroxy - α - methylphenethylamino)-ethyl] theophylline.

Caffelite. A proprietary trade name for moulding compound derived from coffee beans.

Cafergot. A proprietary preparation of ergotamine tartrate and caffeine, used for migraine. (267).

Cafformasol. See FORMASAL.

Cafinal. A proprietary preparation of phenobarbitone and caffeine.

Cagamite. A mineral. It is a hydrated zinc carbonate.

Cagunite. A mineral. It is ferric-mono-metaborate, $Fe_2O_3 . 3B_2O_3 . 3H_2O$. Found in Tuscany.

Cahnite. A new mineral from Franklin, N.J. It is a boro-arsenate of calcium, $4CaO.B_2O_3.As_2O_5.4H_2O$.

Caincic Acid. Obtained from the root of *Chiococca anguifuga* or *C. racemosa.* Used as a diuretic cathartic.

Cairngorm. A brown variety of quartz, SiO_2.

Cajuelite. Rutile (*q.v.*).

Cajuputol. Cineol (Eucalyptol), $C_{10}H_{18}O$.

Cake Lac. See LAC.

Calaba Oil. See LAUREL NUT OIL.

Calaband. A proprietary name for bandages impregnated with zinc paste, urethane and calamine used in the treatment of eczema. (489).

Calabar Bean (Physostigma, Esere Nut). Ordeal bean, the seed of *Physostigma venenosum.*

Calac. Calcium acetate. (589).

Caladryl. A proprietary preparation of diphenhydramine, camphor and calamine in an aerosol spray, for dermatological use. (264).

Calafatite. A mineral. It is a variety of alumite (*q.v.*).

Calaghanite. A mineral. It is $2[Cu_4Mg_4Ca(CO_3)_2(OH_{14}2H_2O].$

Calaite. Turquoise, $AlPO_4.Al(OH)_2$ H_2O.

Calamine (Zinc Spar, Spathic Zinc Ore, Smithsonite). A term applied to both the silicate and carbonate of zinc, usually the carbonate, $ZnCO_3$, found native. In mineralogy, the name calamine is employed for the silicate, while the term smithsonite is given to the carbonate. Zinc carbonate is known among miners as " dry bone." The term calamine is also used for a brass, made by melting together copper, zinc carbonate, and charcoal.

Calamine, Cupriferous. See TYROLITE.

Calamine, Green. See AURICHALCITE.

Calamine, Prepared. See PREPARED CALAMINE.

Calamine, Red. See RED CALAMINE.

Calamine, Siliceous. See SILICEOUS CALAMINE.

Calamine, White. See ZINC SULPHIDE GREY and WHITE CALAMINE.

Calamita. Synonym for Magnetite.

Calamite. A mineral. It is tremolite.

Calamus Root. The rhizome of *Acorus calamus.*

Calaroc. A resin for paper and textile finishes. (512).

Calaroc PG. An aqueous solution of dimethylol di-hydroxymethylene urea used in the production of durable crease resistant, dimensionally stable and mechanical (embossed, glazed, pleated or schreinered) finishes on cellulosic fibres or blends of these with synthetics (particularly polyester fibres). (512).

Calaroc PK. An aqueous solution of dimethylol dihydroxymethylene urea used in the production of " deferred cure " finishes on garments of cellulosic fibres and blends of these with polyamide or polyester fibres. (512).

Calasec. A thickening agent for resin dispersions. (512).

Calatac. Leather and textile finishing agents. (512).

Calaton. A textile finishing agent. (512).

Calaverite. See SYLVANITE.

Calavite. A proprietary preparation containing vitamin A, ascorbic acid, CYANOCOBALAMIN, nicotinamide, thiamine and CALCIFEROL. A vitamin supplement. (246).

Calbonite. A proprietary trade name for an imitation marble product.

Calbor. Calcium borate minerals. (548).

Calbux. A proprietary ground limestone. (2)

Calcaona. A German preparation. It is a mixture of cocoa and chalk.

Calcareobarite. A mineral. It is $(Ba,Ca)SO_4$.

Calcareous Tufa. See TRAVERTINE.

Calcaroni. Circular heaps of Sicilian sulphur ore which have been covered with the moistened ash of burnt ore, and having internal chambers in which the fused sulphur collects. The heaps are heated by means of burning wood.

Calcars. Calcium arsenate. (589).

Calcas. A casein preparation to replace Larosan. It is a calcium caseinate.

Calc-Clinobronzite. A mineral. It is $8[(Mg,Fe,Ca)SiO_3]$.

Calcedonyx. A mineral. It is a variety of Onyx.

Calcene. A trade name applied to a precipitated calcium carbonate with 2 per cent. organic material. It is prepared for use as a rubber filler.

Calcetal. Calcium-acetyl-salicylate.

Calcibronat. A proprietary preparation of calcium bromido-lactobionate. A sedative. (267).

Calcichrome. Cyclo - tris - 7 - (1 - azo - 8 - hydroxy - naphthalene - 3 : 6 - disulphonic acid). A mauve crystalline powder. A sensitive and specific reagent for calcium giving a red colour with Ca^{2+} ions in alkaline solution. (182).

Calcic Liver of Sulphur. Sulphurated lime (q.v.).

Calcidine. Calcium iodide, CaI_2.

Calcidrine. A proprietary preparation containing calcium iodide, ephedrine hydrochloride, codeine phosphate and pentobarbitone sodium. An expectorant bronchodilator sedative cough syrup. (311).

Calciferol. Synthetic Vitamin D. Vitamin D_2. Available as Sterogyl-15, and is present in many multivitamin preparations.

Calcified Milk. Prepared by adding 10 c.c. of normal calcium chloride (5·55 per cent.) to 50 c.c. of fresh milk. Used for the comparison of clotting of milk by enzymes.

Calciglycin. Calcium chloride glycocoll.

Calcimangite. A mineral. It is composed of the minerals Franklinite and Zincite. Synonym for Spartaite.

Calcimalt. A German product. It is a malt extract and silicic acid in colloidal form. Used in the treatment of tuberculosis.

Calcimax. A proprietary preparation of calcium glycine hydrochloride, Vitamin D_2, Vitamin B_1, Vitamin B_2, Vitamin B_6, Vitamin B_{12}, Vitamin C, nicotinamide and calcium pantothenate. A vitamin supplement. (255).

Calcimine. Whitewash.

Calcinaphthol. Calcium-naphthol-sulphonate, $[C_{10}H_6(OH)SO_3]_2Ca$.

Calcined Gypsum. Plaster of Paris (q.v.).

Calcined Magnesia. Magnesium oxide, MgO.

Calcined Mercury. Red mercuric oxide, HgO.

Calcinite. See CARBORA.

Calcinol. Calcium iodate, $Ca(IO_3)_2$.

Calcio-Ancylite. A mineral. It is $(Sr,Ca)_3Ce_4(CO_3)_7(OH)_4.3H_2O$.

Calciobiotite. A mineral. It is a variety of biotite rich in calcium.

Calciocelestine. A mineral. It is $4[(Sr,Ca)SO_4]$.

Calciocelsian. A mineral. It is $4[(Ba,Ca)Al_2Si_2O_8]$.

Calcio-coramine. Double crystalline hydrosoluble salt of Coramine (pyridine-β-carboxylic acid diethylamide) and calcium sulphocyanate. A potent cardiac and respiratory stimulant with an expectorant action.

Calcioferrite. A mineral, $(Fe.Al)_3(OH)_3(Ca.Mg)_3(PO_4)_4.8H_2O$.

Calciogadolinite. A mineral. It is $2[Be_2(Fe''',Fe'')(Yt,Ca)_2Si_2(O,OH)_{10}]$.

Calciomalachite. See LIME MALACHITE.

Calcio-Olivine. A mineral. It is $4[(Mg,Ca,Mn)_2SiO_4]$.

Calcioscheelite. Tungstenite (q.v.).

Calciostab. A proprietary preparation consisting of calcium thiosulphate solution in ampoules.

Calciostrontianite. A mineral, $(Sr.Ca)CO_3$.

Calciotantalite. A mineral. It is $4[(Fe,Mn,Ca)(Ta,Cb)_2O_6]$.

Calciothorite. A mineral. It is a variety of thorite containing lime, $5ThSiO_4.2Ca_2SiO_4.10H_2O$, of Norway.

Calciotite. A mineral. It is a biotite rich in calcium, found in Italy.

Calciovolborthite. A mineral. It is a vanadate of calcium and copper.

Calcipen-V. Phenoxymethylpenicillin.

Calcitare. A proprietary preparation of porcine CALCITONIN used in the treatment of Paget's disease. (75).

Calcite (Calc-spar, Marble, Limestone). A mineral. It is calcium carbonate, $CaCO_3$.

Calcitonin. A polypeptide hormone of ultimobranchial origin, extractable from the thyroid gland of mammalian species or the ultimobranchial gland of non-mammals. It lowers the calcium concentration in the plasma of mammals. THYROCALCITONIN

Calcium-Akermanite. A mineral. It is $Ca_3Si_2O_7$.

Calcium Aluminium. An alloy of calcium and aluminium.

Calcium Aspirin. Kalmopyrin (q.v.).

Calcium Benzamidosalicylate. Calcium - 4 - benzamido - 2 - hydroxy - benzoate. Aminacyl. B-PAS. Therapas.

Calcium-B-Pas. A proprietary preparation of calcium benzamidosalicylate. An antituberculous agent. (244).

Calcium Caseinate. Casein containing about 2 per cent. calcium oxide.

Calcium Disodium Versenate. Sodium Calciumedetate.

Calcium Diuretin. Theobromine and calcium salicylate.

Calcium-Eisenspessartine. A mineral. It is $8[(Mn,Ca,Fe)_3Al_2Si_3O_{12}]$.

Calcium Eosolate. The calcium salt of trisulpho-acetyl-guaiacol.

Calcium Gummate. See GUM ARABIC.

Calcium Hydrochlorphosphate. A mixture of calcium chloride and phosphate.

Calciumjarosite. A mineral. It is $CaFe_6^{\cdots}(SO_4)_4(OH)_{12}$.

Calcium Lactophosphate. A mixture of calcium lactate, calcium acid lactate, and calcium acid phosphate. It contains about 2 per cent. P_2O_5.

Calcium-Larsenite. A mineral. It is $(Pb,Ca)ZnSiO_4$.

Calcium Lazulite. A variety of lazulite, containing about 3 per cent. lime, found in Canada.

Calcium-Melitite. A mineral. It is $Ca_3Al_2Si_4O_{14}$.

Calcium Nitrogen. The name given by Frank to raw calcium cyanamide.

Calcium Oxymuriate. See BLEACHING POWDER.

Calcium Phytinate. The calcium salt of anhydro-methylene-phosphoric acid.

Calcium-Psilomelane. Synonym for Ranciéte.

Calcium Resonium. A proprietary preparation of calcium polystyrene sulphonate used in the treatment of hyperkalæmia. (439).

Calcium-Sandoz. A German product. It is calcium gluconate, and is a tasteless calcium preparation for oral, intramuscular, and intravenous application.

Calcium Trisodium Penetate. Calcium chelate of the trisodium salt of diethylenetri - amine - NNN'N''N'' - penta - acetic acid.

Calcium White. Calcium carbonate, $CaCO_3$.

Calclacite. A material found as a efflorescence on museum specimens. It is $CH_3.COO.CaCl.5H_2O$.

Calcotephroite. Synonym for Glaucochroite.

Calcouranite. See AUTUNITE.

Calcowulfenite. A mineral. It is $4[(Pb,Ca)MoO_4]$.

Calc-Pyralmandite. A mineral. It is $8[(Mg,Fe^{\cdots},Ca)_3Al_2Si_3O_{12}]$.

Calcreose. Calcium creosotate.

Calc Sinter. See TRAVERTINE.

Calcspar. See CALCITE.

Calc-Spessartite. A mineral. It is $8[(Mn,Ca)_3Al_2Si_3O_{12}]$.

Calcusol. An effervescent preparation of potassium bicarbonate and piperidine - p - sulphamine - benzoate, $SO_2.(NH_2)C_6H_4COOH.C_5H_{11}N$.

Calcutta Hemp. A name applied to jute.

Calcydic. Calcium monohydrogen phosphate and Vitamin D. (510).

Caldasite. A mineral. It is a mixture of baddeleyite with silicates of Brazil.

Calderite. A mineral. It is a variety of garnet.

Caldiox. Calcium plumbate. (590).

Caldo. The solution of sodium nitrate obtained from raw caliche. The nitrate is crystallised from this solution.

Caldura. A high temperature (240° C) resistant resin, containing aromatic hydrocarbon groups linked by oxygen and methylene bridges. (409).

Caledon Blue. A dyestuff synonymous with Indanthrene blue.

Caledon Blue GC. A dyestuff. It is a British equivalent of Indanthrene blue GC.

Caledon Blue GCD. A dyestuff. It is a British brand of Indanthrene blue GCD.

Caledon Blue R. A British dyestuff equivalent to Indanthrene X.

Caledon Brown. Indanthrene brown BB.

Caledon Dark Blue O. A dyestuff. It is a British equivalent of Indanthrene dark blue BO.

Caledon Green. Indanthrene green B.

Caledonian Brown. A permanent native pigment used in oil. It is an earth composed chiefly of manganese and iron oxides and hydroxides.

Caledonite. A mineral, $(Pb.Cu)O.(Pb.Cu)SO_4.H_2O$.

Caledon Jade Green. The dimethyl ether of $12:12'$ dihydroxy-dibenzanthrone. A dyestuff giving blue-green shades.

Caledon Pink. Indanthrene pink B.

Caledon Purple. Indanthrene dark blue BO (q.v.).

Caledon Red. Indanthrene red BN.

Caledon Violet. Indanthrene violet R extra.

Caledon Yellow. Indanthrene yellow G.

Caledon Yellow 3G. A dyestuff. It is a British brand of Algol yellow R.

Calfax. A catalyst for flash-age printing. (512).

Calgon. A proprietary trade name for sodium hexametaphosphate. Used as a means of dispersing lime soaps in textiles. It also has some detergent properties.

Calgonite. Detergents for dish washing machines. (503).

Caliatour-wood. A dyewood. See REDWOODS.

Calibrite. A French trade name for ground calcite (26). See also OMYA.

Calibrite. A trade mark for series of aluminium pigments for plastics.

Caliche. Crude nitrate of soda (Chile saltpetre). The soluble salts which cement together the sand, clay, and stones in the mixture known as caliche are mainly sodium nitrate and sodium chloride and the following is an analysis of the salts of a typical caliche :

Sodium nitrate	.	17·6 per cent.
,, chloride	.	16·1 ,,
,, sulphate	.	6·5 ,,
Calcium sulphate	.	5·5 ,,
Magnesium sulphate		3·9 ,,
Potassium nitrate	.	1·3 ,,
Sodium borate	.	0·94 ,,
Potassium per-chlorate	.	0·23 ,,
Sodium iodide	.	0·11 ,,

It is used as a fertiliser.

Calido. An alloy of 64 per cent. nickel with from 15–25 per cent. iron, 8–12 per cent. chromium, and 3–8 per cent. manganese.

Calido Brass. An alloy of 70 per cent. copper, 30 per cent. zinc, and traces of iron.

Calido-elalco. A heat-resisting alloy containing 60 per cent. nickel, 24 per cent. iron, and 16 per cent. chromium.

Californite. A mineral. It is idocrase (q.v.), used for making small ornaments.

Calingastite. A mineral. It is $8[(Fe,Zn,Cu)SO_4.7H_2O]$.

Cali Nuts. See KALI NUTS.

Calioben. See SAIODINE.

Calisaya Bark. Yellow cinchona bark.

Calite. An alloy of 50 per cent. iron, 35 per cent. nickel, 10 per cent. aluminium, and 5 per cent. chromium. It is very resistant to heat, and melts at 2777° F.

Calkinsite. A mineral. It is $4[(La,Ce)_2(CO_3)_3.4H_2O]$.

Callabine Aloe. See ALOE.

Callaica. Synonym for Turquoise.

Callainite. A mineral, $AlPO_4.2H_2O$.

Callaquol. An aromatic milky fluid, which is said to be a suspension of the acid ester of oxycarb-triallyl in oil of thyme. Used in the treatment of pleuritis and cystitis.

Callilite. A mineral, $(Ni.Co.Fe)(Sb.As.Bi)S$.

Callochrome. Crocoisite (q.v.).

Callusolve. A proprietary preparation of alkyl dimethyl benzyl ammonium halide dibromide. Keratolytic agent. (377).

Calmet. An additive for degreasing in hard water. (586).

Calmine. A mixture of dimethyl-phenyl-pyrazolone (antipyrine), and diacetyl-morphine (heroin). Used medicinally for coughs.

Calmitol. A proprietary preparation of chloral hydrate, camphor, menthol, iodine, hyoscyamine and zinc oxide in an ointment base. Antipruritic. (269).

Calmoden. A proprietary preparation of CHLORDIAZEPOXIDE. A tranquilliser. (137).

Calmonal. Calcium-methyl-bromide.

Calmurid. A proprietary preparation of urea in a cream base used in the treatment of ichthyosis. (337).

Calmurid H.C. A proprietary preparation of urea and hydrocortisone in a cream base used in the treatment of eczema. (337).

Calodal. A pure albumin preparation from meat. Used for feeding through the skin.

Calofil A4 and B1. A trade name for precipitated calcium carbonate fillers. (23). See also CALOPAKE PC, CALO-FORT-S HAKUENKA CCR, WINNOFIL S, VEDAR.

Calofort S. Precipitated calcium carbonate. (23).

Calofort U50. A proprietary calcium carbonate filler for rubbers. Its fine particles are formed by precipitation. (23).

Calogerasite. Synonym for Simpsonite.

Calomel (Calomel, Sublimed, Precipitated, Calomel). Mercurous chloride, Hg_2Cl_2.

Calomel, Precipitated. See CALOMEL.

Calomel, Sublimed. See CALOMEL.

Calomic. A nickel-iron-chromium resistance alloy. It is used for electric heater elements operating up to 900° C.

Calonutrin. A proprietary preparation of mono-, poly- and di-saccharides used in the tube feeding of invalids. (386).

Calopake EP. A proprietary trade name for a fine uncoated precipitated calcium carbonate of extremely high whiteness of good opacity and ease of mixing. Used for p.v.a. homo-polymer and copolymer based emulsion paints. (184).

Calopake PC. Precipitated calcium carbonate with a high extinction coefficient and good opacity. (23).

Caloreen. A proprietary preparation of a polyglucose polymer used in high-calorie diets. (447).

Calorene. A proprietary trade name for a gas mainly composed of ethylene obtained by breaking down alcohol.

Caloric PG. An aqueous solution of dimethylol dihydroxymethylene urea, used in the production of durable crease-resistance and dimensional stability on cellulosic fibres or blends of these with polyester fibres. (2).

Caloride. A proprietary trade name for cakes of calcium chloride for drying purposes.

Calorite. Alloys. One contains 65 per cent. nickel, 15 per cent. iron, 12 per cent. chromium, and 8 per cent. manganese ; and another 65 per cent. nickel, 23 per cent. iron, and 12 per cent. chromium.

Caloroc PK. Similar to Calaroc PG but used in the production of deferred cure finishes on blends of cellulosic and synthetic fibres. (2).

Caloxol. A dispersion of calcium oxide in oil. (524).

Caloxal CLP 45. A dispersion of high-grade calcium oxide in chlorinated paraffin used as a desiccant for rubber and plastics. (23).

Caloxol CP2. A proprietary calcium-oxide desiccant for rubber. (23).

Calphos. A handwashing detergent. (586).

Calpol. A proprietary preparation of paracetamol. An analgesic. (247).

Calsept. A proprietary preparation of calamine, menthol, cetrimide, camphor and zinc oxide. Antiseptic skin cream. (350).

Calsize. An emulsified rosin size. (575).

Calsolene. A wetting agent. (512).

Calsun Bronze. A proprietary trade name for an aluminium bronze containing copper with 2·5 per cent. aluminium and 2 per cent. tin.

Calsynar. A proprietary preparation of synthetic salmon CALCITONIN used in the treatment of Paget's disease. (327).

Calurea. A nitrogenous fertiliser. It is a mixture of urea and calcium nitrate containing 34 per cent. nitrogen, and is stated to be suitable for tobacco and as a high-grade vegetable fertiliser.

Calvert's Carbolic Acid. A disinfecting fluid consisting chiefly of cresylic acid.

Calvert's Carbolic Acid Powder. Made by adding carbolic acid to the siliceous residue resulting from the manufacture of aluminium sulphate from shale or kaolin.

Calvonigrite. Psilomelane (*q.v.*).

Calx. Lime, CaO.

Calyptol. A proprietary preparation of eucalyptol, terpineol and oil of pine needles, thyme and rosemary. A bronchial inhalant. (268).

Calyptolite. A mineral. It is a variety of zircon, $ZrSiO_4$.

C.A.M. A proprietary preparation of ephedrine hydrochloride and butethamate citrate. A bronchial antispasmodic. (273).

Cambe Wood. See REDWOODS.

Cambilene. A selective weedkiller. (509).

Cambison. A proprietary preparation of prednisolone, neomycin and quinolyl derivatives used in dermatology as an antibacterial agent. (312).

Cambogia (Cadie Gum). Gamboge, a gum resin.

Cambrian Essence. See ESSENCE OF SMOKE.

Cambrite. A smokeless powder, containing 22–24 per cent. nitro-glycerin, 3–4·5 per cent. barium nitrate, 26–29 per cent. potassium nitrate, 32–35 per cent. dried wood meal, 1 per cent. calcium carbonate, and 7–9 per cent. calcium chloride.

Camcolit. A proprietary preparation of lithium carbonate, used in the treatment of mania and psychoses. (370).

Camcopot. A proprietary preparation of potassium chloride in soluble base. Potassium supplement. (370).

Camelia Metal. An alloy of 70·2 per cent. copper, 14·7 per cent. lead, 10·2 per cent. zinc, 4·2 per cent. tin, and 0·5 per cent. iron.

Camelina Oil. German sesamé oil.

Cameline Oil (Dodder Oil). German sesamé oil obtained from the seeds of *Camelina sativa*. It has a specific gravity of 0·922–0·926, an acid value of 13, a saponification value of 185–188, and an iodine value of 135–142.

Camenthol. A mixture of camphor and menthol for inhalation.

Cameos. Onyx, a variety of chalcedony.

Camerantite. A mineral. It is [K_2SiF_6].

Camite (Dimondite, Haystellite, Armide, Straus metal, Purdurum, Phoran, Hartmetall). Proprietary trade names for tungsten carbide materials.

Camoform. Bialamicol dihydrochloride.

Camoquin. A proprietary preparation containing amodiaquine and hydrochloride. An antimalarial drug. (264).

Campanil. Red hæmatite from Bilbao, in Spain.

Campanuline. See MUSCARINE.

Campeachy Wood. See LOGWOOD.

Camphacol. See CAMPHACOLLASIS.

Camphacollasis (Camphacol). The camphoric acid ester of methylene-diguaiacol. Used as an expectorant in phthisis, and as an antiseptic in cystitis.

Camphaquin. A German preparation. It is a solution of Japanese camphor for intramuscular injection. An analeptic and a cardiac.

Camphellite. A proprietary trade name for a phenol-formaldehyde synthetic resin. A laminated product.

Camphemol. A 10 per cent. emulsion of camphor oil in emulsamin (a urethane derivative). An analeptic.

Camphine. Decinene, $C_{10}H_{18}$. The name is sometimes applied to rectified oil of turpentine.

Camphire. Camphor.

Camphochol. An additive compound of Japan camphor and apocholic acid. A medicinal preparation.

Camphoid. A collodion substitute. It consists of 1 part pyroxylin in 20 parts each camphor and absolute alcohol, by weight.

Camphol (Baros Camphor, Bhimsiam Camphor, Borneo Camphor, Dryobalanops Camphor, Malay Camphor, Sumatra Camphor). Borneol, $C_{10}H_{17}OH$, a terpene from *Dryobalanops camphora*. It has a specific gravity of 1·009, melts at 203° C., and boils at 212° C., and is used in perfumery, in celluloid manufacture, and in medicine as an antiseptic and stimulant. The term Camphol has also been applied to oxanilide (diphenyl-oxamide), $CO(NHC_6H_5)(NHC_6H_5)CO$, a softening agent for cellulose esters.

Campholene. Euninene, C_9H_{16}.

Campho-phenique. Consists of 1 part camphor, and 1 part phenol.

Camphoral. The camphoric acid ester of santalol, $C_8H_{14}(COO)_2.(C_{15}H_{23})_2$.

Camphor Aldehyde. Formyl camphor.

Camphor, Anise. Anethole, $CH_3CH : CH.C_6H_4OCH_3$.

Camphor, Asaronic. See ASAROL.

Camphor, Asarum. See ASAROL.

Camphorated Chalk. Chalk containing one-tenth of its weight of camphor.

Camphorated Gelatin. Blasting gelatin containing 4 per cent. camphor. Used in explosives.

Camphorated Oil. (*Linimentum camphoræ B.P.*) It consists of 200 parts camphor (in flowers), and 800 parts olive oil.

Camphor, Baros. See CAMPHOL.

Camphor, Bhimsiam. See CAMPHOL.

Camphor, Borneo. See CAMPHOL.

Camphor, Bromated. Mono-brom-camphor.

Camphor, Carbolated. Camphor phenique (*q.v.*).

Camphor, Dryobalanops. See CAMPHOL.

Camphor Gum. Camphor.

Camphor Ice. A mixture of synthetic camphor and paraffin wax. One brand contains white wax, almond oil, spermaceti, and flowers of camphor.

Camphor Julep or Mixture. Camphor water (*q.v.*).

Camphor, Laurel. See JAPAN CAMPHOR.

Camphorline. Naphthalene, $C_{10}H_{18}$.

Camphorloid. A proprietary pyroxylin plastic.

Camphor, Malay. See CAMPHOL.

Camphor Naphthol. A product obtained by the condensation of camphor with β-naphthol. Poisonous. See NAPHTHOL-CAMPHOR.

Camphorodyne. A mixture of morphine, chloroform, ether, camphor, hydrocyanic acid, and peppermint oil.

Camphor Oil. See LIQUID CAMPHOR.

Camphor Ointment. A mixture of 22 per cent. camphor with lard and white wax. An anodyne and antipyretic.

Camphoroxol. A compound of camphor with hydrogen peroxide. Antiseptic.

Camphor, Sumatra. See CAMPHOL.

Camphor Water. (*Aqua camphoræ B.P.*) It consists of 1 gram camphor, 2,000 c.c. 90 per cent. alcohol, and 1,000 c.c. water.

Camphosal (Camphossil). The camphoric ester of santalol, $C_8H_{14}(COO)_2$ $(C_{15}H_{23})_2$. Used in catarrh of the bladder. The term camphossil is also used for camphor salicylate. Also see CAMPHORAL.

Camphosan. A solution of 15 parts camphoric acid methyl ester in 85 parts santal oil. A urinary antiseptic.

Camphossil. See CAMPHOSAL.

Camphre de Persil. See PARSLEY CAMPHOR.

Camphresic Acid (Camphretic Acid). A mixture of camphoric and camphoronic acids.

Camphretic Acid. See CAMPHRESIC ACID.

Camphrosal. *p*-Toluene-sulphamide. A softening agent for cellulose esters.

Camphrosalyl. Camphor and benzyl salicylate in oil.

Camphylene. See CARBON.

Camphylite. A mineral. It is mimetite (*q.v.*), found in small balls or barrels.

Campillit. A poison gas. It is cyanogen bromide, CN.Br.

Campiodol. Iodised rape-seed oil.

Campobello Yellow (French Yellow, Chryseinic Acid). A dyestuff. It is the sodium salt of 1 : 4-nitro-α-naphthol.

Campolon. A proprietary preparation of liver extract with Vitamin B_{12}. A hæmatinic. (112).

Camsellite. A mineral. It is a hydrated borate of magnesium, $2MgO.B_2O_3.H_2O$, found in British Columbia. The name is also applied to a mineral, $2(MgO.$ $FeO)(Bi_2O_3.SiO_2)H_2O$, of California.

Camwood (Cambe Wood). See RED-WOODS.

Camyna. A proprietary preparation of thioxolone. (278).

Canaanite. A mineral. It is a variety of pyroxene.

Canacert. Food-stuffs colouring matter meeting Canadian regulations. (551).

Canada Balsam. See CANADA TURPEN-TINE.

Canada Pitch (Hemlock Pitch). The exudation of hemlock spruce, *Tsuga canadensis*, sometimes incorrectly called hemlock gum. It is analogous to Burgundy pitch, and is used for similar purposes.

Canada Turpentine (Canada Balsam, Balsam of Fir). An oleo-resin, the product of gilead fir, *Pinus balsamea* The genuine variety is soluble in alcohol, amyl acetate, benzene, chloroform, and ether, and has a specific gravity of 0·985–0·995, and an acid value of 84–87. The saponification value varies from 90–96.

Canadian Asbestos. See CHRYSOTILE.

Canadian Ash. Commercial potassium carbonate, K_2CO_3.

Canadium. An alloy of 1 part palladium, 2 parts platinum, and 6 parts nickel. Used as a substitute for platinum.

Canadol. See GASOLINE.

Canadrast. A proprietary preparation of hydrastinine hydrochloride.

Canaigre. The ground roots of *Rumex hymenosepalum* of North America. The root contains 17–23 per cent. tannin and is used for tanning leather. The root is also known as Raiz del Indio.

Canarin (Persulphocyanogen Yellow). A yellow colouring matter obtained by the action of bromine upon potassium or ammonium thiocyanate. It is probably perthiocyanogen, $C_3HN_3S_3$. Used in calico printing.

Canary Glass. The colour is produced by uranium.

Canary Seed. The seed of *Phalaris canariensis*, an annual grass originally grown in the Canary Islands, but now in Europe. A bird seed.

Canary Yellow. See CHROME YELLOW.

Canbyite. A mineral, $H_4Fe_2Si_2O_9.2H_2O$.

Cancer Serum. See CANCROIN.

Cancrinite. A mineral, $(Na_2K_2)_4O_4.$ $4Al_2O_3.9SiO_2$.

Cancroin (Cancer Serum). These were called phenol-neurin-citrates, and had at one time a wide use in the treatment of cancer.

Candelilla Wax. A substance deposited on the surface of the candelilla plant of

Mexico and Texas. The parts of the plant containing the material yield 2·5–6 per cent. of crude wax when extracted with chloroform. The crude wax contains about 6 per cent. water and from 3–15 per cent. dirt, and gives an ash of about 1 per cent. The melting-point of the wax varies from 66–68° C., has a specific gravity of 0·9825–0·9850, an acid value of 12–16, a saponification value of 46–65, and an iodine value of 16–37. It is used in the manufacture of boot polishes and in inferior varnishes.

Candelite (Talgol, Coryphol, Duratol, Synthetic Tallow). Hydrogenised oils, prepared by subjecting fatty oils to hydrogen in the presence of a catalyst. The unsaturated liquid fatty acids of the glycerides of the oils are transformed into solid stearic acid.

Candeptin. A proprietary preparation of candicidin in a petrolatum base. Antimonial vaginal ointment. (804).

Candeptin-N. A proprietary preparation of candicidin and neomycin, in a petrolatum base. Antibiotic skin ointment. (804).

Candicidin. One or more of a mixture of heptaenes with antifungal activity produced by Streptomyces griseus and other Streptomyces species. Candeptin.

Candlenut Oil. See BAKOLY OIL.

Candle Pitch. Stearine pitch (*q.v.*).

Candles. The best are made from stearine (solid stearic and palmitic acids). They are also made from tallow, and contain wax or paraffin.

Candle Tar. See STEARINE PITCH.

Cane Cream. A sugar food product made from the juice of the sugar cane. It has the colour of cane syrup.

Canella. The dried inner bark of *Canella Winterana*. In commerce it is known as cinnamon bark, or whitewood bark. A mild aromatic tonic.

Canelle. An old name for Bismarck brown.

Cane Sugar. See SUCROSE.

Canesten. A proprietary preparation of CLOTRIMAZOLE used in the treatment of vaginal infections. (112).

Canfieldite. A mineral, Ag_8SnS_6, similar to Argyrodite. It is also given as $32[Ag_8(Sn,Ge)S_6]$.

Canilep. Vaccines for dogs. (520).

Canilep D.D. A combined double-dose vaccine for dogs. (520).

Canilin D. A canine distemper vaccine. (520).

Cannabine Tannate. The tannic acid salt of the crude alkaloid of Indian hemp.

Cannabinol. 6, 6, 9 - Trimethyl - 3 - pentylbenzo[*c*]chromen - 1 - ol.

Cannabis. The dried flowering and

fruiting tops of the pistillate plant of *Cannabis Sativa*. It is poisonous and contains a resin which is also poisonous.

Canna Starch. Starch from the root of *Canna edulis*.

Cannel Coal. A variety of bituminous coal of fine grained texture. Used for production of gas and oil.

Cannel Powder. An explosive containing 80 per cent. ammonium perchlorate, and 20 per cent. cannel coal.

Cannizzarite. A mineral, $PbS.2Bi_2S_3$, found in a volcano. It is also given as $Pb_3Bi_5S_{11}$.

Cannonite. A smokeless powder. It consists of 86 per cent. nitro-cotton, 6 per cent. barium nitrate, 3 per cent. vaseline, 1 per cent. lampblack, and 2 per cent. potassium ferrocyanide.

Cannon Spar. A mineral. It is a variety of calcite.

Canogard. A proprietary preparation of DICHLORVOS. An insecticide for veterinary use.

Canquoin's Paste. Zinc chloride mixed with flour or dried gypsum and water. A caustic.

Cantharene. Dihydro-*o*-xylene, C_8H_{12}.

Cantharidin Camphor. Cantharidine, $C_{10}H_{12}O_4$, the lactone of cantharidic acid.

Cantharidol. A liquid mixture the essential constituents of which are mercuric chloride, mercurous chloride, potassium chloride, sodium chloride, methyl acetate, and cantharidin.

Cantil. A proprietary preparation containing mepenzolate bromide. An antispasmodic. (249).

Cantonite. A mineral. It is a variety of covellite.

Canton's Phosphorus. Calcium sulphide, CaS. It is phosphorescent.

Canutillos. Synonym for Emerald.

Canvas. A name for a dense cloth used for sails, etc. It was formerly made from hemp or flax, but is now made from cotton.

Canvasite. A trade name for a synthetic resin varnish-paper product.

Canzler Wire. An alloy of 98·8 per cent. copper, 1 per cent. silver, and 0·2 per cent. phosphorus.

Caoucho. A grade of rubber from *Castilloa Ulei*, of Bolivia.

Caoutchene. A French rubber substitute made from sunflower-seed oil, sulphur, and gums.

Caoutchin. See DIPENTENE.

Caoutchine. See CAOUTCHOUCINE.

Caoutchite. Vulcanised rubber exposed to heat for some days, and devulcanised and recovered by this means.

Caoutchol. A plasticising agent for rubber. It is a brown-black oil with an odour resembling pine tar. It has antioxidant properties and is a mild accelerator.

Caoutchouc. Rubber.

Caoutchouc, Artificial. Vulcanised linseed oil, cottonseed oil, and fish oils have been used as indiarubber substitutes, but more success has been obtained with vulcanised maize oil, which products are known as Factis, in Germany. White rubber substitute is prepared by treating fatty oils, notably rape and cottonseed oils, with sulphur chloride. The finest are made from castor oil (French substitutes). Brown rubber substitutes are made by heating fatty oils as such, or blown with sulphur. Oxidised or blown oils, and sulphonated oils are also used. Solutions of nitro-celluloses in heavy oil solvents are used as rubber substitutes in varnishes. Tar, glycerin, resins, glues, asphalt, cellulose, and seaweed, with or without mineral fillers, have all been used in different substitutes.

Caoutchouc, Australian. See COORONGITE.

Caoutchoucine (Caoutchine). A crude oil obtained by the dry distillation of rubber.

Caoutchouc, Mineral. See ELATERITE.

Capa Homen. See CIPÔ DO CABODÔ.

Caparsolate. Thiacetarsamide sodium. (525).

Capastat. A proprietary preparation containing Capreomycin (as the sulphate). An antibiotic. (309).

Cap Cements. Used for fitting brass caps or stopcocks on to glass apparatus. See FARADAY'S, VARLEY'S, and SINGER'S CEMENTS.

Cap Composition. A mixture of mercury cyanate, antimony sulphide, and potassium chlorate.

Cap Copper. An alloy of 95–97 per cent. copper and 3–5 per cent. zinc. It is used for deep drawing and stamping.

Cape Aloe. See ALOE.

Cape Blue. See CROCIDOLITE.

Cape Gum. A variety of gum arabic, from *Acacia horrida*. It is not strongly adhesive.

Cape Ruby. A mineral. It is a pyrope (*q.v.*), from the diamond mines of South Africa.

Capexco. A coal mine explosive consisting of 32–34 per cent. nitro-glycerin, 0·5–1·5 per cent. nitro-cotton, 24–25 per cent. sodium nitrate, 30–32 per cent. ammonium oxalate, and 8–10 per cent. wood meal.

Cap Gilding Brass. See BRASS.

Capillose. Millerite (*q.v.*).

Capillary Pyrites. See MILLERITE.

Capillitite. A mineral. It is $2[(Mn,Zn,Fe)CO_3]$.

Capitol. A proprietary preparation of benzalkonium chloride. A keratolytic scalp cleanser. (377).

Capivi. See OLEO-RESIN COPAIBA.

Capivi Balsam. See OLEO-RESIN COPAIBA.

Capla. A proprietary preparation of mebutamate. An antihypertensive. (269).

Capnite (Kapnite, Zinc-iron Spar). A mineral. It is a ferruginous variety of zinc carbonate.

Caponets. Implants of silbœstrol dipropionate. (591).

Caporcianite. Laumonite.

Caporiet. A calcium hypochlorite. It yields 50 per cent. available chlorine.

Caposil. A trade mark for calcium silicate and asbestos based insulating materials. Caposil 1400 withstands 1400° F. (760° C.) and Caposil HT withstands 1850° F. (1000° C.). (138).

Cappagh Brown (Euchrome, Mineral Brown). A natural pigment used in oil-painting. It is a species of bog earth, containing hydrated oxide of iron and 27 per cent. manganese peroxide. See BURNT SIENNA.

Cappelenite. A mineral. It is a borosilicate of rare earth metals and barium, of Norway.

Capramin. Decongestant and antipyretic tablets. (510).

Capreomycin. An antibiotic produced by Streptomyces capreolus, present as the sulphate in CAPASTAT.

Capreomycin GMC. An antitubercular agent. (538).

Capri Blue Gon. A dyestuff. It is the zinc double chloride of dimethyl-diethyl - diamino - toluphenazonium chloride, $C_{17}H_{20}N_3OCl$. Dyes tannined cotton greenish-blue.

Caprin. A proprietary preparation of aspirin. An analgesic. (137).

Caprine. a-Amino-caproic acid, $(CH_3)(CH_2)_3CH(NH_2)COOH$.

Capriton. A proprietary preparation of chlorpheniramine maleate, phenylephine hydrochloride, aspirin, caffeine. Decongestant and antipyretic tablets. (284).

Caprokol. A proprietary name for hexyl-resorcinol (n-hexyl-2 : 4-dihydrobenzene). An efficient urinary antiseptic.

Caprolan. A registered trade name for a range of thermoplastic polyurethenes. (485).

Capron 8270. A proprietary modified nylon 6 compound used in blow moulding. (30).

Caprosem. A proprietary preparation

containing Testosterone, 17-chloral hemi-acetal acetate. Male sex hormone. (308).

Caproxamine. An anti-depressant. (E) - 3' - Amino - 4' - methylhexanophenone O - (2 - aminoethyl)oxime. " DU 22550 " is currently under clinical trial as the sulphate.

Capsicum (Spanish Pepper, Cayenne Pepper). German capsicum is the fruit of *Capsicum annum*. British and American capsicum is the fruit of *Capsicum minimum*.

Capsifor. A gelatinous product containing methyl salicylate, menthol, ammonia, camphor, and capsicum. It is applied externally for rheumatism.

Capsolin. A proprietary preparation of oleoresin capsicum, camphor, oil of turpentine and eucalyptus. A rubefacient. (264).

Capsule Metal. An alloy of 92 per cent. lead and 8 per cent. tin.

Captan-50. A foliage fungicide. (511).

Captan-Col. A captan fungicide. (511).

Captax. A trade name for mercapto-benzo-thiazole, a product obtained from carbon disulphide, aniline, and sulphur. A rubber vulcanisation accelerator.

Captax-disulphide. See ALTAX.

Captodiame. 4 - Butylthio - α - phenyl-benzyl 2-dimethylaminoethyl sulphide. Covatin is the hydrochloride.

Captol. A trade mark for a certain condensation product of hydrated chloraltannin. A hair tonic.

Caput Mortuum. See INDIAN RED.

Caracolite. A mineral, $Na_2Pb(OH)SO_4Cl$.

Caradol C1, C2, E100, E101. A proprietary trade name for polyether polyol blends. Used as a rigid polyurethane foam component used with Caradate 30 for cavity filling. (99).

Carajura. See CHICA RED.

Caramania Gum. See BASSORA GUM.

Caramel. See BURNT SUGAR.

Caramelan. An amorphous constituent of caramel.

Caramiphen. 2-Diethylaminoethyl 1-phenylcyclopentane - 1 - carboxylate. Parpanit is the hydrochloride; Taoryl is the edisylate.

Carana. A brittle resinous body from *Urotium carana*.

Caraway. The seeds of *Carum carvi*. Used as an aromatic and stimulant.

Carbacaine. A local anæsthetic. It is the diethyl-amino-ethyl ester of carbazole-n-carboxylic acid.

Carbadox. An anti-bacterial. Methyl 3 - quinoxalin - 2 - ylmethylenecarbazate N^1N^4 - dioxide. MECADOX.

Carbagel. A proprietary trade name for

a drying agent. It is an activated carbon containing calcium chloride.

Carbamazepine. 2,3 : 6,7 - Dibenzazepine-1-carboxamide. Tegretol.

Carbamide. Urea, $CO(NH_2)_2$.

Carbanilic Ether. Phenyl-urethane.

Carbanilide. Diphenyl-urea, $C_6N_5NH.CO.NHC_6H_5$.

Carbarsone. 4-Carbamino-phenylarsonic acid.

Carbaryl. An insecticide. 1 - Naphthyl methylcarbamate.

Carbathene. A proprietary trade name for an ethylene N-vinyl carbazole copolymer. A moulding compound form stable over 240° C. (186).

Carbazene Blue. Sulphurised blue vat dyestuffs. (582).

Carbazol. Diphenyl-imide, $C_{12}H_9N$.

Carbazole Blue. A dyestuff, $(C_{12}H_7NH)_2C : C_{12}H_7NH.CO_2H$, obtained by fusing carbazole with oxalic acid.

Carbazole Violet. A dyestuff obtained by fusing 9-ethyl-carbazole with oxalic acid.

Carbazole Yellow. A dyestuff. It is the sodium salt of carbazole-disazo-bisalicylic acid, $C_{26}H_{15}N_5O_6Na_2$. Dyes cotton from a boiling alkaline bath.

Carbazotic Acid. Picric acid (tri-nitrophenol), $C_6H_2(NO_2)_3OH$. Dyes wool and silk from an acid bath. Also used in explosives.

Carbellon. A proprietary preparation containing charcoal, magnesium hydroxide, belladonna dry extract and peppermint oil. An antacid. (250).

Carben. A proprietary trade name for an acetylene condensation product.

Carbenes (Carboids). Names applied to the oxidation products of mineral oils.

Carbenicillin. An antibiotic. 6 - (α - Carboxyphenylacetamido) penicillanic acid. PYOPEN is the disodium salt.

Carbenoxolone. 3β - (3 - Carboxy - propionyloxy) - 11 - oxo - olean - 12 - en-30-oic acid. Biogastrone and Bioral are the disodium salt.

Carbenzol. A distillate from shale tar containing phenols. Used in skin diseases.

Carbergan. Carbon tetrachloride and promethazine in oil. (507).

Carbex. See CARBORA.

Carbic. A calcium carbide which has been mixed with other substances, and coated with a fat.

Carbicon. See CARBORA.

Carbide. Calcium carbide, CaC_2.

Carbide Black. A dyestuff. It is a British brand of Columbia black FF extra.

Carbide Black BO. A polyazo dyestuff. It dyes cotton bluish-black in the presence of sodium chloride.

Carbide Black D. A dyestuff. It is a British equivalent of Zambesi black D.

Carbide Black RT. A dyestuff similar to Carbide black BO.

Carbide Carbon. A definite carbide of iron, Fe_3C. It is the combined carbon present in steels.

Carbide Italia. A lime fertiliser giving, upon analysis, 30 per cent. CO_2, 2 per cent. SiO_2, and 9 per cent. H_2O.

Carbidopa. A dopa-decarboxylase inhibitor, currently under clinical trial as " MK-486 ". It is L - 2 - (3, 4 - dihydroxybenzyl) - 2 - hydrazinopropionic acid.

Carbimazole. Ethyl 3-methyl-2-thiomidazoline - 1 - carboxylate. Neo - Mercazole.

Carbinol. See METHANOL.

Carbinoxamine. 4 - Chloro - α - 2 - pyridylbenzyl 2 - dimethylaminoethyl ether. Clistin is the maleate.

Carbion. A proprietary preparation of charcoal in potassium chloride solution.

Carbiphene. An analgesic. α - Ethoxy - N - methyl - N - [2 - (N - methylphenethylamino) - ethyl]diphenylacetamide.

Carbite. An explosive containing nitroglycerin, potassium nitrate, flour, barium nitrate, bark, and sodium carbonate. The name is also used for both diamond and graphite.

Carbitol. A proprietary solvent. It is diethylene - glycol - monoethyl ether, $C_2H_5O.CH_2.CH_2.O.CH_2.CH_2.OH$. It boils at 175–200° C. and is a solvent for cellulose nitrate, shellac, copal, rosin, etc. It is used in dopes for artificial leather and is added to brushing lacquers.

Carbo Alumina. A trade mark for goods of the abrasive and refractory class. They consist mainly of crystalline alumina.

Carboazotine. A mixture of nitre, lampblack, sawdust, and sulphur. An explosive.

Carbobrant. See CARBORA.

Carbocaine. A proprietary preparation of MEPIVACAINE hydrochloride used in dentistry as a local anaesthetic. (486).

Carbocloral. Ethyl (2,2,2-trichloro-1-hydroxyethyl)carbamate. Chloralurethane. CI 336.

Carbochrome. Non toxic Indian ink for injections. (544).

Carbocoal. Briquettes of practically pure carbon, prepared by distilling bituminous coal at a low temperature, the residue being carbocoal, containing from 1–4 per cent. of volatile matter only.

Carbocort. A proprietary preparation of hydrocortisone and coal tar solution for dermatological use. (372).

Carbo-corundum. A trade mark for articles made from crystalline alumina. They are refractories and abrasives.

Carbodavyne. A mineral. It is $Na_3CaAl_3Si_3O_{12}CO_3$.

Carbo-Dome. A proprietary preparation of coal tar solution for dermatological use. (372).

Carbodynamite. A dynamite containing 90 per cent. nitro-glycerin, absorbed by 10 per cent. cork charcoal.

Carbofilm. A proprietary trade name for starch covered with cellulose acetate.

Carbofrax. A carborundum refractory containing more than 90 per cent. silicon carbide.

Carbofuce. Heat-treatment salt. (512).

Carbo-gel. Silica gel impregnated with carbon, for use to absorb such vapours as benzene and toluene.

Carbogelatin. A mixture of nitrogelatin, nitre, and wood meal. An explosive.

Carboids. See CARBENES.

Carbokaylene. A proprietary preparation of vegetable charcoal with colloidal aluminium silicate.

Carbolan. Acid milling dyestuffs. (512).

Carbolated Camphor. See CAMPHO-PHENIQUE.

Carbolated Creosote. A similar preparation to Jeyes' fluid.

Carbolfuchsine (Ziehl's stain). A microscopic stain for bacteria. It contains 5 parts fuchsine, 25 parts phenol, 50 parts alcohol, and 500 parts water.

Carbol-gentian-violet Solution. A microscopic stain. It contains 10 c.c. of a saturated alcoholic solution of gentian violet and 90 c.c. of a 5 per cent. phenol solution.

Carbolic Acid. Phenol, C_6H_5OH. In trade, the term is used for pure phenol, the cresols and their mixtures with phenol, and also for crude tar oils.

Carbolic Acid, Commercial Powder 15 per cent. This usually consists of chalk impregnated with 15 per cent. by weight of phenol.

Carbolic Acid Crystals. Pure phenol.

Carbolic Camphor. See PHENOL CAMPHOR.

Carbolic Oils. See MIDDLE OILS.

Carbolineum (Brunolinum). Proprietary coal-tar preparations. They usually consist of the higher fractions of coal-tar distillation (the anthracene oils). Some are treated with chlorine, and others also contain zinc chloride or colophony. Used for preserving wood. They are also used for the treatment of

bark diseases of rubber trees. Also see HEAVY OILS.

Carbolised Iodine Solution. A mixture of tincture of iodine, phenol, and glycerin, in hot water. Used as a gargle for diphtheria.

Carbolite. See KARBOLITE and CARBORA.

Carbol Methyl Violet. A microscopic stain. It contains 10 parts of an alcoholic solution of methyl violet 6B with 90 parts of a 5 per cent. aqueous solution of phenol.

Carboloid. A proprietary phenol-formaldehyde synthetic resin moulding composition.

Carbolon. A trade mark for abrasive articles consisting essentially of silicon carbide.

Carbolonium Bromide. Hexamethyl-enedi(carbamoylcholine bromide). Imbretil. Hexacarbacholine Bromide.

Carbolox. See CARBORA.

Carboloy. A trade mark for hard metal compounds consisting of tungsten carbide and cobalt for cutting glass and for high-speed tools. It scratches sapphire, which comes just below the diamond on the scale of hardness.

Carboluphen. A German product. It is a condensation product of formaldehyde and phenol with carbo-medicinalis and bolus alba. An antiseptic, astringent, and absorbent.

Carbol Xylene. A microscopic clearing solution containing 3 parts xylene and 1 part phenol.

Carbolysin. Pastilles of phenol. They contain 51·8 per cent. phenol, 2·74 per cent. sodium bicarbonate, and 46·06 per cent. cream of tartar.

Carbomang. A proprietary trade name for a steel containing 1–1·25 per cent. manganese, 0·45 per cent. chromium, 0·5 per cent. tungsten, and 1 per cent. carbon.

Carbomer. A suspension agent. A polymer of acrylic acid cross-linked with allyl sucrose. CARBOPOL.

Carbomucil. A proprietary preparation of charcoal, magnesium carbonate, and sterculia. A antidiarrhœal. (286).

Carbon (Albocarbon, Disinfecting Carbon, Alabrastine, Camphylene). Names applied to naphthalene, $C_{10}H_8$, an insecticide.

Carbona. A mixture of benzine and carbon tetrachloride, a proprietary cleaning liquid.

Carbonado (Black Diamond). A black, compact variety of diamond, used in the steel crowns of rock-drills.

Carbonan. A liquid motor fuel prepared from coal-tar and tar oils.

Carbonas. Pipe-like deposits of tin ore (cassiterite), found at St. Ives.

Carbonate-Marialite. A mineral. It is $Na_5Al_3Si_9O_{24}CO_3$.

Carbonate-Meionite. A mineral. It is $2[Ca_4Al_6Si_6O_{24}CO_3]$.

Carbonate-Sodalite. A mineral. It is $Na_5Al_3Si_3O_{12}CO_3$.

Carbon Black. Modern carbon black is obtained by burning natural gas in a regulated air supply, and the black deposited on a metal surface. It is a dry powder. B.S.S. No. 284 (1927) states that loss in weight 95–98° C. not be more than 3 per cent. Ash <0·2 per cent., not more than 0·5 per cent. soluble in methylated ether and 0·25 per cent. soluble in water. Also see CHARCOAL BLACK, ACETYLENE BLACK, GASTEX, GASTONEX, MICRONEX, KOSMOS F4, MODULEX, NEOSPECTRA.

Carbon Bronze. An alloy containing 75·5 per cent. copper, 9·75 per cent. tin, and 14·5 per cent. lead. A white metal used for bearings.

Carbon Carbonite. An explosive. It consists of 25 per cent. nitro-glycerin, 34 per cent. potassium nitrate, 1 per cent. barium nitrate, 38·5 per cent. wheaten flour, 1 per cent. ground tan, and 0·5 per cent. sodium carbonate.

Carbondale Silver. An alloy of 66 per cent. copper, 18 per cent. nickel, and 16 per cent. zinc. See NICKEL SILVERS.

Carbon, Disinfecting. See CARBON.

Carbonet. Non-adherent gauze dressing. (529).

Carbonex. A flaky product made from coal-tar residue after the removal of the light fractions. The average specific gravity is 1·25 and the melting-point 215° F. It contains 40–50 per cent. free colloidal carbon, the remainder being hydrocarbons. It is used as a rubber softener and is suitable material for introducing carbon into rubber.

Carbon, Gas. See RETORT CARBON.

Carbonic Acid. The name applied to carbon dioxide, CO_2.

Carbonic Snow. Solid carbon dioxide, CO_2.

Carboniet. Kamforite (q.v.).

Carbonised Clay. Clay carbonised by carbon-charged gases. It is produced by heating together ground raw clay and coal.

Carbonite. A trade mark for an explosive used in coal mines. It contains usually nitro-glycerin, sodium nitrate, flour, and potassium bichromate. Sometimes, however, potassium nitrate, barium nitrate, and sodium carbonate are used, also sulphur, kieselguhr, cane sugar, and nitro-benzene. Carbonite is also the name for a natural coke found in Ayrshire. The name is also applied to an activated charcoal made from anthracite, pitch, and sulphur. It is also a proprietary trade name for a silicon carbide abrasive.

Carbo-nite. A cotton-seed preparation, a hard rubber substitute.

Carbonite, Coal. See COAL CARBONITES I. and II.

Carbon, Metallic. See RETORT CARBON.

Carbon Nitride. Cyanogen $(CN)_2$.

Carbon of Cementation. See ANNEALING OF CARBON.

Carbon of Normal Carbide. See ANNEALING CARBON.

Carbon, Silicised. See SILFRAX.

Carbon Steels. See STEEL A2, E, etc.

Carbonyle. A charcoal specially prepared for use as a fuel.

Carbophenothion. An insecticide. S-(4 - Chlorophenylthiomethyl) oo - diethyl phosphorodithioate. GARRATHION, TRITHION.

Carboplastic. A proprietary fireproof plastic cement, the main constituent of which is carborundum. It is suitable for repairing furnaces.

Carbopol P. A proprietary name for CARBOMER. (102).

Carbora (Carbonite, Carbolite, Carbolox Mimico, Carbobrant, Carbicon, Corex, Silexon, Crystolite, Sterbon, Storabon, Natalon, Natrundum, Lotens, Calcinite, Idilon, Carbex). Proprietary names for silicon carbide materials. Abrasives.

Carboraffin. A decolorising black, made by mixing peat with zinc chloride, and heating. The zinc is afterwards removed by means of water and hydrochloric acid.

Carboresin A, B, C. A polymerisation product from the distillation of butadiene. (3).

Carboresin R and RR. A by-product from the polymerisation of ethylene in the presence of aluminium chloride. (3).

Carborite. See CARBORA.

Carborundum. The name applied to particular goods the main constituent of which is silicon carbide, SiC. The term carborundum is used as a generic term for silicon carbide in Great Britain only. It is a registered trade mark in all other English-speaking countries. Used as an abrasive in the form of sharpening tools and grinding-wheels.

Carborundum, Amorphous. See CARBORUNDUM FIRESAND.

Carborundum Firesand (Amorphous Carborundum, Whitestuff). A mixture of amorphous silicon carbide, SiC, with siloxicon (q.v.), formed in the preparation of crystalline carborundum.

Carbosant. Santalol carbonate, $(C_{15}H_{23}).O.CO.O.(C_{15}H_{23})$, produced by the action of santalol with carbonyl chloride, COCl₂, and an alkali. It contains 94 per cent. santalol, and is recommended to relieve pain in the urinary organs.

Carboserin. A German product. It is an absorption charcoal in the form of tablets. It is used against stomach and intestinal disturbances and for moderating the so-called narcosis sickness. The carbon in the stomach combines with the ether and makes it harmless.

Carbosil (Aquamol). Proprietary preparations of soda and sodium silicate. Used as detergents for aluminium ware.

Carbosolite. A German silicon carbide, used as an abrasive for granite finishing.

Carbostyril. 2-Hydroxy-quinoline, C₉H₆(OH)N.

Carbotex. A brand of plantation rubber and gas carbon black. Used in rubber mixings.

Carbowax. Polyethylene glycols. (592).

Carboxide. A proprietary fumigant containing 1 part ethylene oxide and 8 parts carbon dioxide. It is a liquefied mixture.

Carbox Metal. An alloy of 84 per cent. lead, 14 per cent. antimony, 1 per cent. iron, and 1 per cent. zinc.

Carboxymethylcysteine. A mucolytic agent. S - Carboxymethylcysteine. MUCODYNE; THIODRIL.

Carbrital. A proprietary preparation of carbromal and pentobarbitone sodium. A hypnotic. (264).

Carbromal. Adalin (diethyl-bromo-acetyl-urea).

Carbrox. A decolorising carbon used for treating sugar juices.

Carbuncle. See ALMANDINE.

Carburante Nacional. A motor fuel of Uruguay. It contains 50 per cent. alcohol (98 per cent.), and 50 per cent. naphtha.

Carburet of Iron. Graphite (q.v.).

Carburet of Sulphur. Carbon disulphide, CS₂.

Carburetted Hydrogen, Heavy. See OLEFIANT GAS.

Carburetted Water Gas. Water gas which has been carburetted with the decomposition products of petroleum.

Carburine. A light petroleum oil used for carburetting coal gas, also as a motor spirit.

Carburite. A mixture of pure iron and carbon. It is used for the production of high carbon steels. The term Carburite or Karburite is also used for a smokeless fuel made from lignite. It contains 4 per cent. water, 12 per cent. ash, 11 per cent. volatile matter, and gives no tar.

Carbutamide. N - Butyl - N' - sulpha - nilylurea. Bucrol; BZ.55; Invenol; Nadisan.

Carbuterol. A drug used as a bronchodilator. It is [5 - (2 - *tert* - butylamino - 1 - hydroxyethyl) - 2 - hydroxyphenyl] - urea.

Carchedony. A mineral. It is a variety of carbuncle, a precious stone.

Carclazite. A rock from which china clay is prepared.

Carcola. A trade name for a moulded asbestos product.

Cardaissin. An extract from the suprarenal gland of the cow. A heart stimulant.

Cardamist. A proprietary preparation of glyceryl trinitrate in aerosol form, used in angina pectoris. (271).

Cardamon. The seeds of *Ellettaria cardamomum*, an aromatic and carminative.

Carcanol G23. A proprietary trade name for a fraction of cashew nut shell liquid.

Cardiacap. A proprietary preparation of pentaerythritol tetranitrate. A vasodilator used for angina pectoris. (256).

Cardiacap A. A proprietary preparation of pentærythritol tetranitrate and amylobarbitone in a sustained release form, used in angina pectoris. (256).

Cardiazole. Pentamethylene-tetrazole, C₆H₁₀N₄. A water-soluble analeptic suitable for intravenous application.

Cardibaine. Crystallised strophanthin.

Cardinal. A dyestuff. It is a mixture of chrysoidine and fuchsine. Used for dyeing cotton red.

Cardinal Red 3B. A dyestuff. It is a British equivalent of Azorubine S.

Cardinal Red J. A dyestuff. It is a British brand of Fast red.

Cardinal Red S. See MAROON S.

Cardinal S. See ACID MAGENTA.

Cardine. A proprietary trade name for Visnadine.

Cardioquin. A proprietary preparation of quinidine polygalacturonate used in the treatment of cardiac arrhythmias. (487).

Cardiuix. Benziodarone.

Cardobarb. A proprietary preparation of digitalis, aminophylline and phenobarbitone. (259).

Cardol. See ACAJOU BALSAM.

Cardolite. A proprietary cashew nut derivative of the phenol aldehyde polymer class. Also proprietary trade name for plasticisers, resins, rubber-like polymers, and solvents.

Cardophylin. Aminophylin tablets, ampoules and suppositories. (509).

Cardox. An explosive utilising liquid carbonic acid.

Cardura. A proprietary trade name for non-drying alkyd resins with excellent weathering properties used for surface coatings. (99).

Cardyl. Bismuth camphor-carboxylate in oil.

Carelinite. A mineral, Bi_4O_3S.

Careystone. A proprietary material made from Portland cement and asbestos. Manufactured into blocks.

Carfecillin. An antibiotic. 6 - (α - Phenoxycarbonylphenylacetamido) - penicillanic acid. "BRL 3475" is currently under clinical trial as the sodium salt.

Cargentos. Colloidal silver. Germicidal.

Cariflex. A proprietary trade name for a range of synthetic rubbers (styrene butadiene, cis-polyisoprene and poly-butadiene). (99).

Cariflex S. A proprietary styrene - butadiene rubber. (303).

Carin. Hexamine (*q.v.*).

Carina. A proprietary trade name for polyvinyl chloride (*q.v.*). (99).

Carindacillin. An antibiotic. 6 - (α - Indan - 5 - yloxycarbonylphenylaceta - mido) penicillanic acid. "CP - 15,464" is currently under clinical trial as the sodium salt.

Carinex. A proprietary trade name for a series of polystyrenes and toughened polystyrenes. (99).

Carinex SB41. A proprietary trade name for an easy processing poly-styrene. (99).

Carinex SI 73. A proprietary high-impact grade of polystyrene possessing good flow properties in moulding. (488).

Carisoma. A proprietary preparation of carisoprodol, a muscle relaxant. (255).

Carisoprodol. 2-Carbamoyloxymethyl-2 - N - isopropylcarbamoyloxy-methyl-pentane. Carisoma.

Caritrol. Diethylcarbamazine citrate. (507).

Carletti's Indicator. Phenol-phthalein which has been reduced by caustic soda and zinc dust.

Carlona. A proprietary trade name for high and low density polyethylenes. (99). See also Telcothene, Alkathene, Rigidex.

Carlona LB 157. A proprietary low-density polyethylene used in the pro-duction of film. (99).

Carlona LF 456. A proprietary poly-ethylene having high impact strength, used in the production of film. (99).

Carlona LF 459. A proprietary low-den-sity polyethylene possessing good opti-cal and mechanical properties for the production of film. (99).

Carlona P. A proprietary trade name for polypropylene. (99). See also Pro-pathene, Noblen.

Carlona P PLZ 532. A proprietary talc-filled, heat-stabilised polyethylene used in the production of rigid mould-ings. (99).

Carlona P PY 61. A proprietary poly-propylene homopolymer used in the production of film, especially of fibrilla-ted stretched tapes for use as plastic string. (99).

Carlona 55-004. A proprietary high-density polyethylene possessing high resistance to stress cracking, used in the blow-moulding of containers. (99).

Carlona 60-010. A proprietary high-density polyethylene used in the extru-sion blow-moulding of thin-walled bot-tles. (99).

Carlona 60-060. A proprietary high-density polyethylene used in the injec-tion-moulding of heavy-duty containers. (99).

Carlona 60-120. A proprietary high-density polyethylene used in the injec-tion-moulding of thin sections. (99).

Carlona 460. A proprietary polyethylene used in the production of film. It has high impact strength but contains a slip additive. (99).

Carlona 462. A proprietary polyethylene containing slip additives, for optical and mechanical use. (99).

Carlona 463. A proprietary polyethy-lene containing slip additives for optical and mechanical film. (99).

Carloon Bronze. An alloy of 75.5 per cent. copper, 14.5 per cent. lead, and 10 per cent. tin.

Carlsbad Salts, Artificial. Made by mixing 45 per cent. sodium sulphate, 18 per cent. sodium chloride, and 36 per cent. sodium bicarbonate.

Carlsonite. A high explosive consisting of ammonium perchlorate, trinitro-toluene, and dinitro-naphthalene.

Carlton Suspension N.K. A propriet-ary preparation containing neomycin sulphate and light kaolin. An anti-diarrhœal. (246).

Carmalum. A microscopic stain. It consists of 1 gram carminic acid, 10 grams alum, and 200 c.c. water. Filtered and antiseptic added to keep it.

Carmenite. A name which has been applied to a mineral, Cu_2S.

Carminaph. See Sudan I.

Carminaph Garnet (Scarlet 2R, Naph-thylamine Bordeaux on Fibre). A dye-stuff. It is α-naphthalene-azo-β-naphthol, $C_{20}H_{14}N_2O$. Employed in dyeing and printing.

Carminaph J. See Sudan G.

Carminaphtha. See Sudan I.

Carminaphtha Garnet. See CAR-MINAPH GARNET.

Carmine. A pigment. It is that preparation of cochineal which contains most colouring matter, and least aluminium base. See COCHINEAL.

Carmine Base. See COCHINEAL.

Carmine Lake (Lac Lake, Indian Lake). Cochineal carmine prepared by precipitating a decoction of cochineal with alum or stannic chloride, with addition of acid oxalate or tartrate of potassium. See COCHINEAL.

Carmine Red. $C_{14}H_{12}O_7$, obtained by boiling a dilute aqueous solution of carminic acid with a few drops of a mineral acid. It gives coloured lakes.

Carmine, Safflower. See SAFFLOWER.

Carmine Spar. See CARMINITE.

Carminette Blue. A pigment. It is a red lead coloured with a bluish eosin.

Carminette Blue-red (Carminette Reddish-yellow, Carminette Warm Red, Carminette Warm Dark Red). Pigments, which are similar to Carminette blue and yellow.

Carminette Reddish-yellow. See CARMINETTE BLUE-RED.

Carminette Warm Dark Red. See CARMINETTE BLUE-RED.

Carminette Warm Red. See CAR-MINETTE BLUE-RED.

Carminette Yellow. A pigment. It consists of red lead coloured with eosin, extra yellowish.

Carminite (Carmine Spar). A mineral, $Pb_3(AsO_4)_2.10FeAsO_4$.

Carmoisine. See AZORUBINE S.

Carmoisine L, LAS, S, WS. Dyestuffs. They are British brands of Azorubine S.

Carnahuba Wax. Carnauba wax.

Carnalithe. A proprietary trade name for a casein product.

Carnallite. A double chloride of potassium and magnesium, $MgCl_2.KCl.6H_2O$, found in the Stassfurt deposits. It is a source of potassium chloride and potash manures. Crude carnallite contains on an average, 15·7 per cent. potassium chloride, and consists of a mixture of carnallite, rock salt, and kieserite.

Carnat. A mineral. It is a variety of kaolinite containing iron.

Carnauba Wax. A wax derived from the carnauba palm, *Corypha cerifera*. The wax is found on the young leaves. It is greyish, yellowish, or greenish in colour, and when freshly purified melts at 85–86° C. Old wax melts at 90–91° C. The specific gravity of the wax varies from 0·995–0·999, the saponification value is 80, the acid value 2·8, the iodine value 12·7, and the ash 0·27 per cent.

Carnegieite. A soda felspar, $Na_2Al_2Si_2O_8$.

Carnelian. An orange-red variety of chalcedony, which is a crypto-crystalline variety of quartz.

Carniferrin. Iron-phospho-sarcolactate. Used in medicine for chloritic condition.

Carnitine. See NOVAIN.

Carnoid. A proprietary trade name for a casein plastic.

Carnotine. See PRIMULINE.

Carnotite. A mineral. It consists of a vanadate of uranium and potassium, sometimes containing small quantities of calcium, barium, and iron. Usually, 15–18 per cent. vanadium pentoxide is present. It is $2[K_2(UO_2)_2(VO_4)_2.3H_2O]$.

Carnotite, Silico. See SILICO-CARNO-TITE.

Carnot's Reagent. Basic bismuth nitrate (100 grams) is dissolved in hot, concentrated hydrochloric acid, and diluted to a litre with 92 per cent. alcohol.

Carnoy's Fluid. A mixture of absolute alcohol and glacial acetic acid. Used for fixing animal tissue.

Caroa Fibre. A fibre from *Neoglazovia variegata*, of Brazil. It is suggested as suitable for the manufacture of paper.

Carobel. A proprietary preparation of carob flour. A gastro-intestinal sedative. (283).

Caro Bronze. An alloy of 92 per cent. copper, 8 per cent. tin, and 0·25 per cent. phosphorus. A bearing metal.

Carob-seed Gum (Locust-Kernel Gum). A gum obtained from the hard brown seeds contained in the locust bean, or the fruit of the carob tree, *Ceratonia siliqua*. It is used as a sizing material for rayon, cotton, etc.

Carofax. A proprietary brand of solidified carron oil.

Carolina Tea. See APPALACHIAN TEA.

Carolite. A mineral, Co_2CuS_4.

Carom. A proprietary butadiene-styrene copolymer available in three grades: 1500, 1700 and 1712. (489).

Caro's Acid. Oxypersulphuric acid, $S_2O_9H_2$.

Caro's Reagent. Obtained by dissolving ammonium or potassium persulphate in concentrated sulphuric acid. It is a pasty mass having great oxidising power. Used for testing alkaloids.

Carotine. The yellow-red colouring matter of carrots. Used for colouring butter and cheese.

Caroubier. An acid dyestuff, giving crimson shades on wool.

Carovax. Pasteurella vaccine for veterinary use. (514).

Carpathian Balsam. Riga balsam (*q.v.*).

Carpene. Nonone, C_9H_{14}.

Carperidine. Ethyl 1-(2-carbamoylethyl) - 4 - phenylpiperidine - 4 - carboxylate.

Carphenazine. 10 - {3 - [4 - (2 - Hydroxyethyl)piperazine - 1 - yl]pro - pyl} - 2 - propionylphenothiazine. Proketazine is the dimaleate.

Carpholite. A mineral, (Mn.Fe)O.(Fe. Al)$_2$O$_3$.2SiO$_2$.2H$_2$O.

Carphosiderite. A mineral. It is a basic ferric sulphate.

Carphostilbite. A mineral. It is a variety of thomsonite.

Carpolite. A clay of Bohemia.

Carrol-Dakin Solution. A solution of chlorinated soda.

Carrol Gum. A rubber substitute. It is a sulphurised oil.

Carrollite. A mineral. It is a sulphide of cobalt and copper, Co$_2$CuS$_4$.

Carron Oil (Lime Liniment). Linseed or olive oil, mixed with an equal volume of lime water. Used for burns.

Carsalam. Benz - 1,3 - oxazine - 2,4 - dione. O-Carbamoylsalicylic acid lac-tam.

Casamids. A registered trade mark for polyamides for use as epoxy resin curing agents, adhesive and thixotropic agents. (218).

Carstanjen's Compound. Potassium sulphite-thymoquinone.

Carstone. An iron stone of Norfolk.

Carterite. A proprietary trade name for resin-pulp moulding compound.

Carter's Solution. An eye lotion. It contains zinc sulphate, boric acid, camphor water, and water.

Carthamin. The red colouring matter of safflower, C$_{14}$H$_{16}$O$_7$. See SAFFLOWER.

Carticaine. A local anæsthetic cur-rently undergoing clinical test as " 40 045 " and " Hoe 045 ". Methyl 4 - methyl - 3 - (2 - propylaminopro - pionamido) thiophene - 2 carboxylate.

Cartogene. A trade name for a synthetic resin varnish-paper product.

Carton Pierre. A material for roofing. It is made either by mixing paper pulp with asphalt, coal tar, brown coal tar, lignite, or by impregnating common cardboard with boiling asphalt.

Cartridge Brass. See BRASS.

Cartridge Paper. Drawing-paper made usually from cotton and linen rags and wood pulp.

Carvacrol. 2-Methyl-5-isopropyl-phenol, C$_{10}$H$_{14}$O.

Carvene. See HESPERIDENE.

Carvol. Carvone, C$_{10}$H$_{14}$O.

Caryinite (Karyinite). A mineral, (Mn.Ca.Pb.Mg)$_3$(AsO$_4$)$_2$.

Caryocerite (Karyocerite). A mineral. It is a fluo-silicate of common metals with cerium and yttrium metals.

Caryopilite. A mineral. It is a hydrated silicate of manganese.

Casan Pink. See PYRONINE G.

Casbis. A proprietary preparation. It is an activated oil suspension of bismuth hydroxide, and is used in the treatment of syphilis.

Casca. Sassy bark.

Cascade. A photographic wetting agent. (507).

Cascalote. A Mexican tanning material. It contains 55 per cent. tannin, and is obtained from *Cæsalpinia cacolaco*.

Cascamid. A proprietary trade name for a group of polyamide resins. (490).

Cascamite. A proprietary trade name for a urea-formaldehyde resin glue in the form of a dry powder.

Cascara Balsam. See BALSAM OF CASCARA.

Cascaras. Cascara tablets.

Cascara Sagrada. The dried bark of *Rhamnus purshiana*. A laxative.

Cascarilla. The dried bark of *Croton elateria*.

Cascarin. Stated to be the active prin-ciple of the laxative cascara sagrada, but it is in all probability an impure product.

Casco-Dur. A proprietary trade name for a synthetic rubber based material suitable as a liquid applied roof coating. (139).

Cascophen. A proprietary trade name for a liquid and powdered phenol-formaldehyde synthetic resin glue.

Casco Resin. A proprietary liquid urea-formaldehyde synthetic resin glue.

Cascote. A proprietary trade name for furane resins and caralysts used for flooring and sealing in chemically resistant tiles. (139).

Cascovin. A proprietary trade name for a group of vinyl adhesives. (490).

Casein. A nucleo-albumin, which is contained in the colloidal condition in milk. Also see CASUMEN, LACTARIN, and NUTROSE.

Casein Ammonia. See EUCASEIN.

Casein-cement. Made by mixing casein with milk of lime. Used by joiners and bookbinders, also for fixing together glass and porcelain.

Casein Colours. Colours for painting, in which the cementing material is casein, quicklime, and water.

Casein Glue. Made by stirring up casein with 25 per cent. distilled water, and 1–4 per cent. sodium bicarbonate, add-ing another 25 per cent. distilled water, standing, and adding an antiseptic to prevent mould. It can be applied cold. Borax is sometimes used instead of the bicarbonate. A substitute for glue. Another preparation consists of 100 parts casein, 10 parts alum, and 3–5 parts soda ash, mixed with 500 c.c. water.

Casein Magnesia. A preparation which consists of powdered casein, water, and magnesia. It will fix mineral pigments to stand washing.

Casein Paints. Paints formed by the addition of a powder containing casein and alkali, to water. The colouring matter and lime are added.

Casein Plastics. See under AMBLOID, BEROLIET, DEFIANCE, FANTASIT, GALA, LACRINITE, LACTOLOID, LUPINIT PAPYRUS.

Casein, Serum. See GLOBULIN.

Casein Silk. Casein dissolved in alkali or zinc chloride solution, and forced through tiny holes into an acid bath.

Casein, Technical. See LACTARIN.

Caseogum. A solution of casein in lime water. Used as an adhesive, more particularly for impregnating linen and cotton fabrics. The casein on the fibre assists the latter to absorb dyestuffs.

Caseon. See NUTROSE.

Cashew Lake. A brown sienna, on to which organic dyes are precipitated.

Cashew Nut. The seed of *Anacardium occidentale*.

Cashmere. The fibre from the hair of the Cashmere goat.

Casilan. Calcium caseinate. (520).

Casmalon. A proprietary name for cyclarbamate.

Casoid. A casein plastic.

Casolithe. Gallatite (*q.v.*).

Casper Wax. A mineral wax made from crude petroleum found in the Casper Wyoming field. It is yellow, harder than paraffin, and has a specific gravity of 0·933 and melts at 160–166° F.

Cassareep. A syrup used in conjunction with capsicums, in the preparation of " pepper-pot " of the West Indies. It is produced by boiling down the juice from Cassava meal. The juice contains hydrocyanic acid, and is obtained by pressing the pulp in bags. Cassareep is an antiseptic, and is used as a meat preservative.

Cassava. A food product. It consists of the starch obtained from the roots of the Manioc, *Manihot utilissima*.

Cassava Meal. Ground cassava root. Also see FARINE.

Cassel Black. See IVORY BLACK.

Cassel Brown. See VANDYCK BROWN and UMBER.

Cassel Earth. Vandyck brown (*q.v.*).

Cassel Green. Rosenthiel's green (a manganate of barium). Used as a pigment.

Cassella's Acid. β-Naphthol-sulphonic acid δ or F.

Cassella's Acid F. See BAYER'S ACID.

Casselmann's Green. A pigment made by mixing boiling solutions of copper sulphate and an alkaline acetate. It consists of $CuSO_4.3Cu(OH)_{24}H_2O$.

Cassel Yellow. See TURNER'S YELLOW.

Cassia Bark (Chinese Cinnamon). The bark obtained from a species of *Cinnamonum*, grown in China.

Cassia Oil. See OIL OF CASSIA.

Cassinite. A mineral. It is an aventurine-felspar.

Cassiterite. See TINSTONE.

Castaloy. A proprietary trade name for high carbon, high chromium steel.

Castanite. A mineral. It is a basic ferric sulphate, $Fe_2O_3.2SO_3.8H_2O$, of California.

Cast Brass. Brass which is not required to be spun, rolled, drawn or hammered. It is made by melting together the copper and zinc. This usually contains 66 per cent. copper with zinc, and the lead varies from 1–3 per cent.

Castellanos Powder. A dynamite containing nitro-glycerin, nitro-benzene, fibrous material, and kieselguhr.

Castellite. A mineral. It is a silicate of titanium and calcium.

Castelnaudite. A mineral. It is a variety of Xenotine, containing zirconium, from Brazil.

Castile Soap. A pure olive oil-soda soap. Sometimes the olive oil is replaced by pea-nut or cotton-seed oil. A recent Spanish order defines this product. It is produced from the saponification of good quality olive oil by means of caustic soda. It must not contain more than 20 per cent. water, or more than 0·3 per cent. free alkali by weight. The fatty substances liberated by means of mineral acids must show an iodine number (Hubl) of from 69–82.

Castillite (Frenzelite). A mineral. It is guanajuatite.

Casting Copper. An American copper. It contains 98·5–99·75 per cent. copper. Used for casting.

Cast Iron D2. A British chemical standard. It is a grey phosphoric cast iron in the form of fine turnings. It contains 1·31 per cent. silicon, 1·07 per cent. phosphorus, and 1·64 per cent. manganese, also about 2·5 per cent. graphitic carbon, 0·8 per cent. combined carbon, and 0·03 per cent. sulphur.

Castomer A-G. A trade name for a polylactone based isocyanate-terminated polyurethane elastomer prepolymer. Used for casting and moulding. Shore D.80 and tensile strength

9000 p.s.i. Elongation 170 per cent. (57).

Castomer ES 725. A proprietary three-component all-liquid cast-urethane elastomer system free of toluylene diisocyanate and diamine-type cross-linking agents. (539).

Castomer MU 14C33. A proprietary polyurethane elastomer for curing at room temperature. Used in the production of foam. (539).

Castor Blend. A lubricating oil obtained by mixing distilled castor oil with pale filtered mineral oil, under pressure.

Castor Cake (Castor Meal). The cake obtained from castor beans after separation of oil. The ground cake is the meal and is used as a fertiliser.

Castordag. A registered trade-mark for colloidal graphite in castor oil used principally as a lubricant. (207).

Castorite. A mineral. It is a variety of petalite.

Castor Oil Lozenges. Calomel lozenges.

Castor Oil Meal. The residue obtained after the extraction of the oil from the castor oil bean. A fertiliser.

Castor Oil, Soluble. See BLOWN OILS.

Cast Steel (Crucible Steel). A steel of high quality, used for making heavy guns, turrets, and similar objects. It is obtained by melting Bessemer steel, or mixtures of other steels, in refractory crucibles in the furnace, and pouring into moulds.

Cast Yellow Brass. Usually an alloy of 67 per cent. copper, 31 per cent. zinc, and 2 per cent. lead. It melts at 895° C.

Casumen. A soluble form of casein (flocculent casein). It contains 90 per cent. protein.

Catabond. A proprietary trade name for a phenol-formaldehyde liquid resin used for plywood manufacture where a waterproof bond is required.

Catafor. Antistatic agents. (547).

Cataid. Surface active agents and detergents. (550).

Catalase. An oxidising enzyme.

Catalazuli. A proprietary trade name for a phenol-formaldehyde synthetic resin product.

Catalex. A proprietary trade name for an expanded phenol formaldehyde plastic.

Cataldi's Chlor-cellulose. A chlor-cellulose prepared from esparto grass by the action of electrolytically prepared chlorine.

Catalin. A registered trade mark. (302).

Catalin (95-1). A proprietary grade of Nylon 6. (302).

Catalin CAO-1 DBPC. 2,6-di-tert. butyl-para-cresol (butylated hydroxy toluene). An antioxidant for polymers especially polyethylene, polystyrene and polyurethane. (302).

Catalin CAO-3. 2,6-di-tertiary butyl-para-cresol (butylated hydroxy toluene). Food grade of BHT an antioxidant for plastics that come into contact with food. (302).

Catalin CAO-4. 2,2'-thiobis-(4-methyl-6 tertiary-butylphenol). A non-staining and non-discolouring synergistic antioxidant (when used with carbon black) for rubber, latex compounds, high polymers (polyolefins, polyamides, polyurethanes) adhesives, paraffin waxes, fats, oils and petroleum hydrocarbons. (302).

Catalin CAO-5. 2,2'-methylenebis-(4-methyl-6-tertiary-butyl phenol). A high molecular weight antioxidant for use at high temperatures. (302).

Catalin CAO-6. A high purity version of CAO-4. (302).

Catalin CAO-7. Butylated hydroxyanisole (BHA). A mixture of 2-tertiary-butyl-4-methoxyphenol with a small amount of 3-tertiary-butyl-4-methoxy-phenol. A stabiliser for animal, mineral and vegetable oils, fats and waxes. (302).

Catalin CAO-14. A high purity grade of CAO-5. (302).

Catalin CAO-30. 1,1'-thiobis-(2-naphthol). The highest molecular weight CAO antioxidant for polymeric materials. It contains a strong thio bond giving a synergistic effect with carbon black fillers. Used for high temperature applications. (302).

Catalin CAO-42. A low volatility, heat-stable, non-staining, hindered-phenolic antioxidant. Suitable for gasolines (petrol), lubricants and fuels. Also suitable for plastics and rubber. (302).

Catalinite. A trade name for pebbles from Santa Catalina Island, California.

Catalin Melamine. Proprietary melamine (q.v.), synthetic resin.

Catalpo. A British mineral earth treated by a new chemical process. It is a colloidal china clay, prepared from Cornish china clay, and is said to be a fine pigment to take up colours. It is also used in rubber mixings. Other proprietary trade names for china clays are : Blue Ridge, Dixie, Devolite, Langford, Par, Stockalite, and Suprex. Dixie is a hard clay while Catalpo is soft.

Catalysol. An oxidising agent used to purify acetylene. It oxidises the phosphine and arsine. It contains mercurial salts, manganese oxide, and oxidising agents such as cupric chloride or chromium trioxide.

Catalyst AC. A curing agent for amino textile resins. (559).

Catalytic Oleine. Commercial oleic acid, obtained by enzymic or catalytic decomposition.

Cataplas 930R. A group of proprietary phenolic resins used in the reinforcement of natural and synthetic rubbers. (62).

Cataplas 940R. A proprietary range of light-coloured phenolic resins. (62).

Cataplas 942R. CATAPLAS 940R with the addition of 10 per cent. hexamine—the whole ground to a powder. (62).

Catapleiite. A mineral. It is a hydrated silicate and zirconate of sodium and calcium, containing from 30–40 per cent. zirconium oxide.

Catapol SR. A proprietary range of polyurethane elastomers ranging in hardness from 20–60 Durometer A. (491).

Catapres. A proprietary preparation of clonidine hydrochloride. An antihypertensive. (278).

Catavar. A proprietary phenol-formaldehyde surface coating or laminating varnish.

Catechine. A colouring matter used as a substitute for catechu. Used for dyeing cotton cloth, sail-cloth, and fishingnets.

Catechin Red. The red substance remaining in alcohol, after the removal of catechin from catechu.

Catechol. See PYROCATECHIN.

Catechol, Methyl. See METHYL CATECHOL.

Catechu. See CUTCH.

Catechu, Bengal. See CUTCH.

Catechu, Black. See CUTCH.

Catechu, Bombay. See CUTCH.

Catechu Brown. Prepared from Bismarck brown. A direct cotton dyestuff, $C_{30}H_{28}N_{14}$.

Catechu, Cubical. See CUTCH.

Catechu, Gambier. See CUTCH.

Catechu, Pale. See CUTCH.

Catechu, Pegu. See CUTCH.

Catechu, Yellow. See CUTCH.

Catgut. A material obtained from freshly soaked sheep intestine. It is washed, treated with alkali, bleached, dried, polished, and oiled.

Cathartine. An extract of senna leaves containing a mixture of cathartic acid and its salts.

Catheter Oil. See LUND'S OIL.

Catheter Paste. Lubricant paste. (*Pasta lubricans B.P.C.*)

Cathkinite. A mineral. It is a variety of saponite.

Cathomycin. A proprietary preparation containing Novobiocin (as Novobiocin Sodium). An antibiotic. (310).

Cathopen. A proprietary preparation containing Benzylpenicillin (as the potassium salt) and Novobiocin. An antibiotic. (310).

Catigene Black Brown N. A dyestuff obtained by sulphurising sawdust or bran.

Catigene Brown N (Cold Blacks B and R). Dyestuffs of the same class as thion blue B. They dye cotton in a cold bath, containing sodium sulphate and soap.

Catigene Indigo R Extra. A sulphide dyestuff, recommended for cop-dyeing.

Catigene Red-brown. A dyestuff obtained by the action of alkali sulphides and sulphur upon amino-hydroxyphenazines.

Cativo Balsam. Balsam of Captivo (*q.v.*).

Cativo Gum. A gum obtained from the sap of the mangrove. It mixes with rubber.

Catlinite (Indian Pipestone). A red clay used by the Indians for making pipes, found in Minnesota.

Cat's Eye. Quartz, containing, enclosed in the crystal, fine fibres of asbestos parallel to each other. Synonym for Katoptrite.

Cat's Eye Dammar. See DAMMAR RESIN.

Cat's-eye Resin. An East Indian dammara resin obtained from *Pinus dammara* or *Dammara alba*.

Cat's Gold (Cat's Silver). Very finely powdered mica is sometimes called by these names. The term " cat's gold " is also used for mosaic gold (a tin sulphide).

Cat's Silver. See CAT'S GOLD. Synonym for Mica.

Cattierite. A mineral. It is $4[CoS_2]$.

Cattimandu Gum. A latex obtained from a species of *Euphorbia*. It resembles gutta-percha in that it softens when heated.

Cattu Italiano. A dyestuff obtained by sulphurising sawdust or bran.

Catylac. A cold curing wood finish. (512).

Cauchillo Gum. A gum resembling chicle obtained from a tree in South America. It is used as a chewing gum.

Caucho Blanco. A rubber obtained from different species of the genus *Sapium* which belongs to the *Enphorbiaceæ* family and found in the northern part of South America.

Caucho Rubber. A name usually applied to rubber derived from species of Castilloa.

Caustic Alcohol. (*Liquor sodii ethylates B.P.*) An alcoholic solution of 18 per cent. sodium ethylate (C_2H_5ONa).

Caustic Baryta. Barium hydroxide, $Ba(OH)_2$.

Causticised Ash. A mixture of soda ash and caustic soda containing 15–45 per cent. sodium hydrate. A water softener and cleaner.

Caustic Lime. See LIME.

Caustic Oil of Arsenic. See BUTTER OF ARSENIC.

Caustic Potash. Potassium hydroxide, KOH.

Caustic Soda. Sodium hydroxide, NaOH.

Caustic Soda Ash. A sodium carbonate containing caustic soda.

Caustic Soda Liquor. A sodium hydrate (caustic soda) solution of 90° or 100° Tw.

Caustic, Toughened. See TOUGHENED CAUSTIC.

Caustic, Vienna. See VIENNA CAUSTIC.

Caustic, White. See WHITE CAUSTIC.

Causul. A nickel-copper-chromium cast iron of marked acid and alkaline resistance. It is used in the manufacture of valves intended particularly for handling sulphuric acid and caustic soda.

Causyth. A German product. It is cyclo-hexo-cuperidine, and is used in the treatment of rheumatism.

Cavalite. A proprietary trade name for a rubber-coated silk.

Caved-S. A proprietary preparation containing glycyrrhizinic acid as powdered block liquorice, bismuth subnitrate, aluminium hydroxide gel, light magnesium carbonate, sodium bicarbonate and powdered frangula bark. (294).

Cavodil. A proprietary trade name for the hydrochloride of Pheniprazine.

Cawk. Aggregates of crystals of barytes, $BaSO_4$, found in the lead mines in Derbyshire.

Cawree Gum. See AUSTRALIAN COPAL.

Cayenne Pepper. See CAPSICUM.

Cayota. A reddish-brown bark from Mexico, where it is used for tanning leather. It contains from 22–30 per cent. tannin.

Caytine. A proprietary trade name for the hydrochloride of Protokylol.

Caytur 4. A proprietary partial complex of zinc chloride with benzothiazyl disulphide used as a cross-linking agent in sulphur-curable urethane elastomers. Formerly known as LD-55. (10).

Caytur 21 & 22. A trade mark for a range of curing agents used in urethane elastomers. (10).

Cazeline Oil. See MINERAL BURNING OILS.

Cazin. An alloy of cadmium and zinc, containing 82·6 per cent. cadmium.

C.E. A mixture of 2 parts chloroform and 3 parts ether.

Ceanal. A proprietary preparation of phenylethyl alcohol, cetrimide, undecenoic acid and lanolin. Skin antiseptic. (831).

Ceara Cotton. A Brazilian cotton.

Ceara Wax. Carnauba wax (q.v.).

Cearin. An ointment vehicle capable of taking up a large proportion of water. Usually it consists of 1 part carnauba wax with 3 parts paraffin or ceresin, mixed with four times its weight of liquid paraffin.

Cebollite. A mineral. It is $Ca_5Al_2Si_3O_{12}(OH)_4$.

CE-Cobalin. A proprietary preparation of Vitamin B_{12} and Vitamin C. Vitamin supplement. (329).

Cecolene 1. A proprietary trade name for trichlorethylene.

Cecolene 2. A proprietary trade name for perchlorethylene.

Cedarite. See CHEMAWINITE.

Cedilanid. A proprietary preparation of lanatoside C. (267).

Cedilanid Ampoules. A proprietary preparation of deslanoside. (267).

Cedocard. A proprietary preparation of sorbide nitrate used in the treatment of angina pectoris. (251).

Cedrarine. See OREXINE.

Cedrat Oil. See OIL OF CEDRAT.

Cedrene Camphor. Cedrol, $C_{15}H_{26}O$, obtained from the oil of red cedar.

Cedriret. Cœrulignone, $C_{16}H_{16}O_6$.

Cedroc. The oil of *Cedrus Atlantica*.

Cedronol. A microscopic immersion oil. (544).

Ceduran. A proprietary preparation of NITROFURANTOIN with deglyceryrrhizinated liquorice. An antibiotic. (251).

Cefapirin. An antibiotic. 7 - [α - (4 - Pyridylthioacetamido)] cephalo - sporanic acid.

Cefazedone. An antibiotic. 7 - [2 - (3, 5 - Dichloro - 4 - oxo - 1 - pyridyl) - acetamido] - 3 - 5 - methyl - 1, 3, 4 - thiadiazol - 2 - ylthiomethyl) - 3 - cephem - 4 - carboxylic acid.

Cefuroxime. An antibiotic. $(6R, 7R)$ - 3 - Carbamoyloxymethyl - 7 -](2Z) - 2 - (2 - furyl) - 2 - methoxyiminoace - tamido]ceph - 3 - em - 4 - carboxylic acid. " 640/359 " is currently under clinical trial as the sodium salt.

Cegamite. Hydrozincite (q.v.).

Cehasol. A German preparation. It is ammonium sulpho-ichthyfossilicum. A product similar to ichthyol.

Ceilinite. A proprietary asbestos preparation.

Cekas. A heat-resisting alloy containing 59·7 per cent. nickel, 11·2 per cent. chromium, 28 per cent. iron, and 2 per cent. manganese.

Ce-K-Sal. A proprietary preparation of aspirin, ascorbic acid and acetomenaphthone. (329).

Celacol. A proprietary trade name for a cellulose ether. Used as a boxing material in the firing of porcelain. (213).

Celadonite. A mineral. It is a silicate of iron, magnesium, and potassium.

Celafuse. A proprietary range of polyamide resins. Celafuse 100 is a terpolyamide with good melt flow properties (melting point 103–108° C.). Celafuse T is a terpolyamide with a higher viscosity (melting point 115–125° C.). Celafuse CP has a melting point of 145–150° C. with good hydrophobic properties. Celafuse SG is a plasticised co-polyamide with a melting point of 110–120° C. (213).

Celanese (Lustron). Celanese is an English name for cellulose acetate silk. The same type of silk is known in America as Lustron.

Celanex 917 and J101. A proprietary range of glass-reinforced polyester thermoplastics. (493).

Celastic. A proprietary trade name for a pyroxylin product.

Celastoid. A non-inflammable celluloid substitute, made from cellulose acetate.

Celasyl. Fast dyestuffs for artificial fibres. (541).

Celatene Colours. Proprietary colours which are amino-anthraquinone derivatives.

Celbenin. A proprietary preparation containing methicillin sodium. An antibiotic. (364).

Celbenin. A proprietary trade name for the sodium salt of Methicillin.

Celcon (11). A trade name for acetal plastics. See also DELRIN (10) and HOSTAFORM C (9). POLYFYDE (12).

Celcon GC-25. A glass-coupled variant of CELCON M 90. It is an acetal copolymer used in injection moulding, and is capable of bearing high loads. (493).

Celcon M25. A proprietary range of easily-processed acetal copolymers. (493).

Celcon M90. A proprietary range of general-purpose acetal copolymers used in injection moulding. (493).

Celcon M90-07. A proprietary range of acetal copolymers available in different colours for tumble blending. (493).

Celcon M90-08. An acetal copolymer stabilised for exposure to ultraviolet light. (493).

Celcon M270. A proprietary acetal copolymer possessing good moulding properties for small wall-sections. (493).

Celcon U10-01. A proprietary acetal copolymer with high melt strength used in extrusion moulding. (493).

Celcon U10-11. An easily-processed proprietary acetal copolymer giving off little odour, used in the making of aerosols. (493).

Celeron. A proprietary trade name for a synthetic resin.

Celescot. A proprietary trade name for a pyroxylin product.

Celeste. See WILLOW BLUE.

Celestial Blue. See BRUNSWICK BLUE.

Celestialite. A variety of Ozocerite found in certain iron meteorites.

Celestine (Celestite). Native strontium sulphate, $SrSO_4$.

Celestine Blue B. See COREINE 2R.

Celestite. See CELESTINE.

Celestol. A proprietary trade name for an alkyd synthetic resin.

Celestols. A proprietary trade name for polybasic acid—polyhydric alcohol fatty acid type synthetic resins.

Celestron. A condensation product of phenol and formaldehyde with fillers. Used as an insulator.

Celevac. A proprietary preparation containing oxyphenisatin diacetate. A laxative. (264).

Celite. A registered trade mark for a material for separating impurities and other matters from fluids, and used as an aid in filtering, dehydrating, and demulsifying of liquids. It is also used as a filler in the plastics industry. The term is also applied to a constituent of Portland cement and clinker. It consists of a solution of dicalcium silicate in dicalcium aluminate.

Cellacephate. A partial mixed acetate and hydrogen phthalate ester of cellulose, used as an enteric coating.

Cellactite. A material prepared by dispersing hard pitch in a ball mill with water, to which 0·1–0·25 per cent. of a resin soap has been added as the dispersing agent. A colloidal fluid is obtained, to which asbestos or cellulose pulp is added. The pitch coagulates upon the fibre, and the material is converted into soft, pliable sheets on a paper machine. These sheets are compressed and called Cellactite. It is said to be acid-proof, and capable of being nailed, punched, or screwed. It has high insulating properties. A registered trade mark. (981).

Cellanite. A proprietary trade name for a synthetic resin paper product.

Cellase. Cytase, an enzyme which attacks cellulose.

Cellastine. A proprietary trade name for a cellulose-acetate product.

Cellemoil. A French nitro-cellulose lacquer.

Cellesta. A proprietary trade name for a pyroxylin product.

Cellestron Silk. Silk made from acetyl cellulose. Used for insulating wires.

Cellit B. Cellulose acetobutyrate. (3).

Cellit L and K. Cellulose acetate. (3).

Cellit TP. Cellulose propionate. (3).

Cellite. A material made from a solution of the lower acetylated products of cellulose in alcohol. It consists of cellulose acetates. Cellite films for the cinematograph are made by mixing these acetates with camphor. They are much less inflammable than celluloid.

Cellobiose. Cellose (*q.v.*).

Cellobond. A registered trade mark for thermohardening and thermoplastic adhesives for general and specific purposes. (288).

Cellocaps. A proprietary viscose product.

Cellofas A. A trade mark. Methyl Ethyl Cellulose. (2).

Cellofas B. A trade mark. Sodium Carboxymethyl Cellulose. (2).

Celloidin (Photoxylin). A substance which is obtained from collodion by precipitating it from its solution in alcohol and ether. It consists of pure nitro-cellulose. Used in microscopy and in surgery.

Cellolyn. Synthetic resins for lacquers. (593).

Cellomold. A registered trade mark for a cellulose acetate moulding material and other plastics.

Cellon. A proprietary safety glass made from Trolitul, a polystyrene thermoplastic. Also a trade name for tetrachlorethane.

A trade mark for a German product resembling but much less inflammable than celluloid. It is a mixture of cellulose acetate and camphor.

Cellon-Caoutchouc. An elastic product made by mixing solutions of rubber and cellulose ester in hexalin (cyclohexanol). It is stated to be used in the manufacture of photographic films, and for other purposes.

Celloncrete. A Danish preparation of sand and cement mixed with a froth liquid. It has a specific gravity of 0·25–0·94, and is used as a heat insulator.

Cellon-drahtglas. A proprietary cellulose acetate coated netting.

Cellophane (Registered Trade Mark). A brand of regenerated cellulose film produced from viscose by treatment with sulphuric acid and/or ammonium salts. The film is non-inflammable and is insoluble in water, alcohol, and oils ; is used as a transparent wrapping material and also for decorations and in millinery, and has been suggested as a gutta-percha and celluloid substitute. See also VISCOSE and VISCOSE SILK.

Celloresin. A proprietary trade name for a synthetic resin.

Cellose (Cellobiose). A glucose-β-glucoside obtained in the hydrolysis of cellulose.

Cell-o-silver. A proprietary product. A silvered vinyl plastic mirror.

Cellosolve. Ethanediol ethers and ether esters. (592).

Cellosolve Acetate. Ethylene-glycol-monoethyl ether acetate. It is used as a lacquer solvent. It boils at 154° C., has an acidity of less than 0·01 per cent., calculated as acetic acid, and a specific gravity of 0·973–0·982.

Cellotropin. The benzoic ester of arbutin, $C_6H_4O.C_6H_{11}O_5.O.COC_6H_5$. Prescribed in cases of tuberculosis and scrofula.

Cell Substance. Cellulose.

Cellucon. A proprietary preparation of methylcellulose used as a laxative and as an antiobesity agent. (250).

Cellucraft. A proprietary trade name for nitro-cellulose spray coating.

Celluflex CEF. A trade mark. Trichloroethyl phosphate. A vinyl plasticiser. (83).

Celluflex 112. A trade mark. Cresyl diphenyl phosphate. A vinyl plasticiser. (83).

Celluflex M142. A trade mark. An aromatic phosphate. A vinyl plasticiser. (83).

Celluflex 179C. A trade mark. Tricresyl phosphate. A vinyl plasticiser. (83).

Celluflex 179EG. A trade mark. A phosphate plasticiser for vinyl plastics. (83).

Celluflex 180. A trade mark. Isopropyl *o*-tolyl phosphate. A vinyl plasticiser. (83).

Celluflex M179. A proprietary trade name for alkyl-aryl plasticiser.

Cellulak. A proprietary trade name for a laminated product made from kraft paper coated with resin, e.g. shellac laminated under heat and pressure.

Cellulase. See CYTASE.

Cellulate. A proprietary trade name for a cellulose acetate plastic.

Cellulith. A material made by grinding cellulose in water until a jelly is produced, boiling, when it hardens to the consistency of bone. Used as a binding agent for carborundum wheels, also as a packing material.

Celluloid (Xylonite). Composed of a soluble nitro-cellulose mixed with camphor, obtained by gelatinising nitrocellulose by means of a solution of camphor in ethyl alcohol. It can be moulded, and is used for making toys, combs, and other articles.

Celluloid-caoutchouc. A material prepared by dissolving rubber and celluloid in hexalin (cyclohexanol) and mixing the solutions. An elastic substance is produced which is said to be suitable for photographic films, and for other purposes.

Cellulon. A fibre made in Germany, from wood pulp.

Cellulose Acetate Plastics. See under ACEYLOID, CRAYONNE, ERINOFORT, LANOPLAST, LANZOID, RHODIALITE, SERACELLE, SUPEREZ, TENITE, VITREOCOLLOID.

Cellulose Enamels. Usually consist of cellulose nitrate, resin, and plasticide.

Cellulose, Hydrated. See ALKALI CELLULOSE.

Cellulose Pitch. The residue obtained from the evaporation of the waste sulphite lye, from the treatment of wood in the sulphite process. Used for making briquettes.

Cellulose, Starch. See FARINOSE.

Cellulose Turpentine Oil. See SULPHITE TURPENTINE OIL.

Cellulo Silk. Silk made by Thiele's process (cuprammonium cellulose solution).

Cellulosine. A proprietary trade name for a bleached celluloid.

Cellulube. A proprietary trade name for a lubricant for mixing.

Celluplastic. A proprietary trade name for a plasticised cellulose used for containers.

Cellushi. A proprietary viscose product used for packing.

Celluvarno. A proprietary trade name for cellulose nitrate surfacing material.

Celmar. A registered trade mark for polypropylene/glass fibre reinforced structures. (213).

Celmontite. A coal mine explosive containing 65·5–68·5 per cent. ammonium oxalate, 10·5–12·5 per cent. trinitrotoluene, and 19·5–21·5 per cent. sodium chloride.

Cel-o-glass. A proprietary trade name for a fine wire screen whose interstices are covered with cellulose acetate. Used as a glass to transmit the ultraviolet solar rays.

Celontin Kapseals. A proprietary preparation of methsuximide. An anticonvulsant. (264).

Celoron, Condensite. See CONDENSITE CELORON.

Celotex. A registered trade mark for fibre board, insulation board, ceiling tile and panels, gypsum products, roofing products and other building products. (845).

Celsian. A mineral. It is a silicate of aluminium and barium. $4[BaAl_2Si_2O_8]$.

Celsit. An alloy similar in composition to Stellite.

Celta. See LUFTSEIDE.

Celtid. A proprietary trade name for a pyroxylin product.

Celtite. An explosive. It consists of 56–59 per cent. nitro-glycerin, 2–3·5 per cent. nitro-cotton, 17–21 per cent. potassium nitrate, 8–9 per cent. wood meal, and 11–13 per cent. ammonium oxalate.

Celulon. A proprietary trade name for a series of cellular polyurethane coating materials suitable for application by brush or spray. (140).

Celvaloid. A proprietary trade name for a cellulose acetate-wood product.

Cement, Acid-proof. Usually mixtures of glass and sodium silicate or of asbestos, barium sulphate, and sodium silicate.

Cement, Adamantine. A mixture of powdered pumice and silver amalgam. Used in dentistry.

Cement, American. A rubber cement made from 10 parts rubber, 6 parts chloroform, and 2 parts mastic.

Cementation Copper. See CEMENT COPPER.

Cementation Steel. Obtained by heating bars of good malleable iron, packed with nitrogenous matter, or wood charcoal.

Cement, Clinker. The granules of cement produced in the manufacture of Portland cement by heating limestone and clay. It is ground up to supply the Portland cement of commerce.

Cement Copper (Cementation Copper, Copper Precipitate). Copper produced from copper liquors and mine liquors, by means of iron (pig iron, scrap iron, or spongy iron). It is usually contaminated with arsenic, antimony, and iron.

Cement, Electrical. See FARADAY'S CEMENT, and SINGER'S ELECTRIC CEMENT.

Cement, Giant. See ACETIC-GELATIN CEMENT.

Cement, Hercules. See ACETIC-GELATIN CEMENT.

Cement, Iron. See IRON CEMENT and EISEN-PORTLAND CEMENT.

Cementite. Triferrous carbide, Fe_3C, the hardest component of steel.

Cementite, Independent. Cementite in rectilinear lamellæ.

Cementkote. See CERESIT.

Cement, Le Farge. Grappier cement (q.v.).

Cement, Lehner's. See ELASTIC CEMENT.

Cement, Magnesia. See SOREL CEMENT.

Cement Mortar. A mixture of natural

slag or Portland cement, sand, and water. Lime is also added.

Cement, Pozzolana. See EISEN-PORTLAND CEMENT.

Cement Prodor. An acid resisting cement. (594).

Cement, Puzzolana. See EISEN-PORTLAND CEMENT.

Cement Rock. A clay-limestone found in America.

Cement, Roman. See ROMAN CEMENT.

Cement, Slag. See EISEN-PORTLAND CEMENT.

C.E. Mixture. See C.E.

C-enamel. A patented lacquer consisting of gum and oils which have been specially treated. Used as a lacquer on tinplate for packing foods.

Cendevax. A proprietary rubella vaccine. (281).

Ceneg. A proprietary trade name for dinonyl phthalate. A vinyl plasticiser (81).

Cenomassa. A mixture of yeast and yeast extract. Recommended as a pill or tablet constituent.

Cenosite. See KAINOSITE.

Censedal. A proprietary preparation of nealbarbitone. A sedative. (336).

Centralite. Dimethyl-diphenyl-urea. Used in explosive powders, by dissolving in the surface layer and making the powder burn more slowly at first.

Centrallassite. A calcium silicate mineral of Nova Scotia.

Centrals. The rubber obtained from *Castilloa elastica*, of South America, is known by this name.

Centricast Mark 17. A proprietary trade name for an iron containing 33 per cent. chromium made to the requirements of DTD.462. It is exceptionally resistant to corrosion, heat and abrasion. (141).

Centricast Mark 18. A proprietary trade name for an austenitic cast iron which contains 16 per cent. nickel, 2 per cent. chromium and 7 per cent. copper. It has low magnetic permeability. (141).

Centricast Mark 21. A proprietary trade name for cast iron alloy containing 1 per cent. aluminium and 1·5 per cent. chromium developed for a cylinder lining material which could be nitrided. (141).

Centrifugal Syrup. See GREEN SYRUP. A selective weedkiller. (596).

Control. A selective weedkiller. (596).

Centyl. A proprietary preparation of bendrofluazide. A diuretic. (308).

Centyl K. A proprietary preparation of bendrofluazide and potassium chloride in a slow release core. A diuretic. (308).

C.E. Powders. Explosive powders, containing tetryl- or tetranitro-methyl-aniline.

Cephaldol. A substance prepared by the action of citric acid and sulphuric acid upon phenetidine, with an addition of quinine and sodium carbonate. Used as a mild antipyretic and antineuralgic.

Cephalin. A mono-amino-mono-phosphatide found in egg-yolk.

Cephaloram. 7 - Phenylacetamido - cephalosporanic acid.

Cephaloridine. 7 - [(2 - Thienyl)acet - amido] - 3 - (1 - pyridylmethyl) - 3 - cephem - 4 - carboxylic acid betaine. Ceporin.

Cephalexin. An antibiotic. 7 - (D - α - Aminophenylacetamido) - 3 - methyl - 3 - cephem - 4 - carboxylic acid. CEPOREX, KEFLEX.

Cephaloglycin. An antibiotic. 7 - [D - (—) - α - Aminophenylacetamido] - cephalosporanic acid. KEFGLYCIN.

Cephalonium. An antibiotic currently under clinical trial as " 87/90 ". It is 3 - (4 - carbamoyl - 1 - pyridiniomethyl) - 7 - (2 - thienylacetamido) - 3 - cephem - 4 - carboxylate.

Cephalosporin C. 7 - (5 - Amino - 5 - carboxyvaleramido)cephalosporanic acid.

Cephalothin. An antibiotic. 7 - (2 - Thienylacetamido)cephalosporanic acid. KEFLIN is the sodium salt.

Cephamandole. An antibiotic currently undergoing clinical trial as " 83405 ". 7 - D - Mandelamido - 3 - [(1 - methyl - tetrazol - 5 - yl)thiomethyl] - 3 - cephem - 4 - carboxylic acid.

Cephazolin. An antibiotic currently under clinical trial. 3 - [(5 - Methyl - 1, 3, 4 - thiadiazol - 2 - yl)thiomethyl] - 7 - (tetrazol - 1 - ylacetamido) - 3 - cephem - 4 - carboxylic acid.

Cephoxazole. An antibiotic currently undergoing clinical trial as " 291/1 ". It is 7 - (3 - o - chlorophenyl - 5 - methylisoxazole - 4 - carboxamido) - cephalosporanic acid.

Cephradine. An antibiotic. 7 - [D - 2 - Amino - 2 - (cyclohexa - 1, 4 - dienyl) - acetamido] - 3 - methyl - 3 - cephem - 4 - carboxylic acid. ESKASEF, VELOSEF.

Cephreine. Pure citronellyl acetate. (504).

Cephrol. Specially pure dextro citronellol. (504).

Cephthalothin. 7 - (2 - Thienylacet - amido)cephalosporanic acid.

Ceporex. A proprietary preparation containing cephalexin. An antibiotic. (335).

Ceporin. A proprietary preparation containing Cephaloridine. An antibiotic. (335).

Cerabrit. Synthetic waxes. (505).

Ceracine. See FAST RED.

Ceralumin B. An aluminium alloy containing 1·5 per cent. copper, 1·5 per cent. nickel, 0·2 per cent. magnesium, 0·7 per cent. iron, 1·5 per cent. silicon, and 0·1 per cent. cerium. Used for pistons, cylinder heads, etc.

Ceralumin " C." A proprietary alloy of aluminium with 2·5 per cent. copper, 1·5 per cent. nickel, 1·2 per cent. iron, 1·2 per cent. silicon, 0·8 per cent. magnesium, and 0·15 per cent. cerium.

Ceramic Glazes. These are usually compounds of silicates, consisting of silica, bases, and metallic oxides.

Ceramite. A solution of fluo-silicates, used as a disinfectant, as a preservative for wood, and for hardening cements.

Ceramocoll. The pasty mass formed when Japanese acid clay is treated with an oxalic acid solution. It is used to purify naphthalene.

Ceramyl. A solution of hydrofluo-silicic acid, and its salts (chiefly of iron and aluminium). It contains 19·8 per cent. free and combined hydrofluo-silicic acid. A disinfectant.

Cerapatite. A mineral. It is $2[(\mathrm{Ca,Ce})_5(\mathrm{PO_4})_3(\mathrm{F,O})]$.

Cerargyrite (Argyroceratite, Horn Silver, Kerargyrite). Minerals. They are compounds of silver with chlorine, bromine, or iodine.

Cerasin (Ceresine). A wax obtained by the treatment of ozokerite with concentrated sulphuric acid. Also, a term used to describe the insoluble part of cherry-tree gum.

Cerasin(e). A registered trade mark currently awaiting re-allocation by its proprietors. (983).

Cerasine. See KERASITE.

Cerasine Orange G. See SUDAN G.

Cerasine Red. See SUDAN III.

Cerasite. An iolite of Japan.

Cerate. A hard ointment containing wax.

Cerate, Resin. See BASILICON.

Ceratex. A trade mark for wax and rubber containing impregnating compounds for textiles and papers.

Cerbolite. See BOUSSINGAULTITE.

Cereal Rubber. A patented rubber substitute prepared from wheat by treatment with ptyalin.

Cereal Soaps. These are obtained by the treatment of starchy and albuminous materials such as are found in bran of maize and oats, with a strong solution of caustic soda. Sodium salts of amino-acids are formed, the excess of acid being neutralised by free fatty acid.

Cerebrose. Galactose, $C_6H_{12}O_6$.

Cereclor. A registered trade mark for a series of secondary plasticisers manufactured from chlorinated waxes. The percentage of chlorine is indicated by the number after the name, e.g. Cereclor 70. (2).

Ceredin. See CEROLIN.

Cerelose. A commercial glucose.

Cerere. Tricresyl-mercuro-acetate. It is a mixture of mono- and diacetate derivatives of the three cresols, with about 75 per cent. of the mono-derivative, and containing about 57 per cent. mercury. It accelerates the germination of grain and affords protection against animal and vegetable parasites.

Ceres. A fungicide. It consists of potassium sulphide.

Ceresine. See CERASIN.

Ceresit (Driwal, Impervite, Cementkote). Waterproofing compounds for cement, consisting mainly of calcium carbonate, alum, and calcium soap, sometimes with more or less free oil or fat. Sold in the form of powder to be mixed with dry cement, or as a paste to be mixed with water.

Cerevisin. A protein obtained from yeast. It resembles leguminin of peas. A dried brewer's yeast.

Cerevon. A proprietary preparation of ferrous gluconate. A hæmatinic. (247).

Cerex. A proprietary thermoplastic stated to be resistant to deformation at 100° C. It is a copolymer containing carbon, hydrogen, and nitrogen—probably of the acrylonitrile type.

Cerfluorite. A compound, $(\mathrm{Ca_3Ce_2})\mathrm{F_6}$, prepared artificially.

Cergadolinite. A mineral. It is $2[\mathrm{Ba_2Fe(Yt,Ce)_2Si_2O_{12}}]$.

Ceridin. See CEROLIN.

Ceriform. An antiseptic compound of certain ceric double sulphates.

Cerin. See OZOKERITE.

Cerinite. A mineral. It is a silicate of aluminium and calcium.

Cerirouge. Optical glass polisher. (595).

Cerise. See MAGENTA.

Cerise B, BB, G, 2YS. Dyestuffs. They are British brands of Magenta.

Cerit. An alloy similar in composition to Stellite.

Cerite. A cerium mineral, $2(\mathrm{Ca.Fe})\mathrm{O}$ $3\mathrm{Ce_2O_3.6SiO_2.3H_2O}$.

Cérite. A French synthetic resin of the phenol-formaldehyde type.

Ceritone. An aromatic flavouring chemical. (522).

Cerium Copper. An alloy containing 10 per cent. cerium metals and 90 per cent. copper. Used as a deoxidiser in the preparation of non-ferrous metals.

Cerolin (Ceredin). The solid extract of yeast, containing fat. Used in medicine.

Cerolite. A mineral, $H_2Mg_3Si_2O_8H_2O$.

Cerol S. A proprietary trade name for a textile waterproofing material. It is a paraffin wax emulsion.

Ceromel. A mixture of beeswax and honey.

Cerosiline (Ceroxylin). Palm-tree wax, from *Ceroxylon andicola*.

Cerosin. See OZOKERITE.

Cerotin. Ceryl cerotate, $C_{26}H_{53}COOC_{27}$ H_{55}. Occurs in Chinese wax.

Cerotine Orange C Extra. See CHRYSOIDINE R.

Cerotine Yellow R. A dyestuff obtained from diazotised aniline and resorcinol.

Ceroxin GL. A proprietary trade name for 12-hydroxy stearic acid. A lubricant for plastics processing. (212).

Ceroxin GMO. A proprietary trade name for a partially esterified fatty acid made from naturally occurring saturated or unsaturated fatty acids and a polyhydric alcohol. A lubricant for plastics processing. (212).

Ceroxin GMR. A proprietary trade name for naturally occurring saturated or unsaturated fatty acids partially esterified with a polyhydric alcohol. A lubricant for p.v.c. processing. (212).

Ceroxin GMSI. A proprietary trade name for a solid lubricant for p.v.c. processing made from naturally occurring saturated or unsaturated fatty acids partially esterified with a polyhydric alcohol. (212).

Ceroxin TKI. A proprietary trade name for hydroxystearic acid glyceride. A lubricant for p.v.c. which does not cause discoloration and which is particularly suitable for compounds for electrical purposes. (212).

Ceroxylin. See CEROSILINE.

Cerrobase. An alloy of lead and bismuth used as a pattern metal in foundry work. Melting-point 124° C.

Cerrobend. An alloy of lead, bismuth, tin, and cadmium used for tube and section bending in foundry work. Melting-point 70° C.

Cerromatrix. A proprietary alloy of bismuth, lead, tin, and antimony. It expands on cooling.

Certicol. Foodstuff colours of guaranteed specification. (533).

Certolake. Foodstuff colour lakes. (533).

Cerubidin. A proprietary preparation of DAUNORUBICIN hydrochloride. A cytotoxic used in the treatment of leukaemia. (494).

Cerulean Blue. See CŒRULEUM.

Ceruleite. A mineral. It is $CuAl_4(AsO_4)_2(OH)_8.4H_2O$.

Ceruleofibrite. A mineral. It is a basic chloro-arsenate of copper, $CuCl_2$. $\frac{1}{3}Cu_3As_2O_8.6Cu(OH)_2$, found in Arizona.

Cerumol. A proprietary preparation of *p*-dichlorbenzene, benzocaine, chlorbutol and oil of terebinth. Ear drops. (352).

Ceruse. See WHITE LEAD.

Cerussite (White Lead Ore, Wheatstone, Lead Earth). A mineral. It is lead carbonate, $PbCO_3$.

Cervantite. See ANTIMONY OCHRE.

Cervitan. A proprietary preparation. It is a yeast-tannin compound.

Cesarolite. A mineral. It is an acid lead manganate, $H_2PbMn_3O_8$, found in Tunisia.

Cesol. Chlor-methyl-pyridine-β-carbonic acid methyl ester. A registered trade name. (896).

Cestarsol. Arecoline acetarsol. (507).

Cetaceum. Spermaceti.

Cetal. An alloy of 87 per cent. aluminium and 13 per cent. silicon.

Cetalkonium Chloride. Benzylhexadecyldimethylammonium chloride. Present in Bonjela.

Cetaped. Veterinary preparation. (512).

Cetavlex. A proprietary preparation of cetrimide in a cream base. Antiseptic skin cream. (2).

Cetavlon. A proprietary preparation of cetrimide. Skin disinfectant and wound cleanser. (2).

Cetec. A proprietary trade name for a cold moulding bituminous compound. (Non-refractory.)

Cetec-refractory. A proprietary trade name for a cold moulding inorganic compound.

Ceteprin. A proprietary preparation of emepronium bromide, used in the control of frequency of micturition. (318).

Cetiacol. Sec-cetylguaiacyl.

Cetine. Cetyl palmitate, $C_{15}H_{31}COOC_{16}$ H_{33}. Occurs in spermaceti.

Cetol. A fatty alcohol.

Cetomacrogol 1000. Polyethylene glycol 1000 monocetyl ether. Polyoxyethylene glycol 1000 monocetyl ether.

Cetosalol. See SALOPHENE.

Cetosan. A mixture of the higher alcohols of spermaceti, mainly cetyl and octodecyl alcohols, with petroleum jelly. It gives a creamy emulsion with water. It is an ointment.

Cetoxime. N - Benzylanilinoacetamidoxime. Febramine is the hydrochloride.

Cetrarin. Cetrarinic acid, $C_{30}H_{30}O_{12}$.

Cetrimonium Chloride. Hexadecyl - trimethylammonium chloride.

Cetylguaiacyl (Cetiacol, Palmiacol). Pyrocatechin methyl-cetyl-ether, C_6H_4 $(OCH_3)(C_{16}H_{23})$. A remedy recommended for consumption.

Cetylpyridinium Chloride. 1 - Hexa - decylpyridinium chloride. Ceepryn.

Cevadilla. The seed of *Schœnocaulon officinale*.

Cevadilline (Sabadilline, Cevadine). Veratrine, $C_{32}H_{49}NO_9$, an alkaloid.

Cevadine. See CEVADILLINE.

Cevedine. Cevine, $C_{27}H_{43}NO_8$, an alkaloid.

Ceylon Cinnamon. Cinnamon (q.v.).

Ceylon Hyacinth. Garnet (q.v.).

Ceylon Isinglass or Gelatin. See AGAR-AGAR.

Ceylonite. A mineral. It is an iron-magnesium spinel.

Ceyssatite. A white earth consisting of almost pure silica. An absorbent powder.

C-Film. A proprietary preparation of nonoxinol in a film of dried polyvinyl alcohol. A contraceptive. (495).

Chabazite (Phacolite). A zeolite, CaO. $Al_2O_3.4SiO_2.6H_2O$.

Chagual Gum (Magyey Gum). A Chilean gum of uncertain origin.

Chalcanthite (Copper Vitriol, Cyanosite). A native copper sulphate, found at Cape Calamita, as a crust on iron ores. A mineral. It is $2[CuSO_4.5H_2O]$.

Chalcanthum. Ferrous sulphate, $FeSO_4$.

Chalcedony. A mineral. It is a variety of crypto-crystalline quartz.

Chalcedonyx. An agate with white and grey layers.

Chalchuite. A mineral. It is a variety of turquoise.

Chalcoalumite. A mineral, $CuSO_4$. $4Al(OH)_3.3H_2O$, of Arizona.

Chalcochlore. Goethite (q.v.).

Chalcocite. A mineral. It is cuprous sulphide.

Chalcodite. A mineral. It is a variety of stilpnomelane.

Chalcolamprite. A mineral. It is a silico-niobate of cerium and yttrium metals with Zr, Ca, Fe, Na, K, of Greenland.

Chalcolite. A mineral, $CuO.UO_5.P_2O_5$. $8H_2O$. It contains radium.

Chalcomenite. A mineral, $CuSeO_3$-$2H_2O$.

Chalcophanite. A mineral, Mn_2O_3ZnO. H_2O.

Chalcophyllite (Euchlorose). See TAMARITE.

Chalcopyrite. See COPPER PYRITES.

Chalcopyrrhotite. A mineral. It is a sulphide of copper and iron.

Chalcosiderite. A mineral, $(Al.Fe)_3$ $(OH)_3(Ca.Mg)_3.(PO_4)_4.8H_2O$.

Chalcosine. See REDRUTHITE.

Chalcostibnite. A mineral, $CuSbS_2$.

Chalcotrichite (Hair Copper). Native copper suboxide, Cu_2O, a variety of cuprite.

Chalcuite. A new Mexican bluish-green turquoise.

Chaldinite. Synonym for Enstatite.

Chalilite. A mineral. It is a variety of thomsonite.

Chalk (Orleans White, Whiting, English White, Spanish White, Paris White). Pure amorphous calcium carbonate, $CaCO_3$.

Chalkomorphite. A mineral. It is a calcium silicate.

Chalkone. Benzal-aceto-phenone, C_6H_5. CH : $CH.CO.C_6H_5$.

Chalk, Prepared. See PREPARED CHALK.

Chalk, Red. See INDIAN RED.

Chalk, Spanish. See ACTINOLITE.

Chalk, Tailor's. See TAILOR'S CHALK.

Challenger. A proprietary trade name for an acrylic denture material.

Chalmersite. A mineral. It is a sulphide of copper and iron, $CuFe_2S_3$.

Chalybeated Tartar. Tartrated iron (q.v.).

Chalybeate Waters. Iron waters. They usually contain from 0·03–0·15 per cent. ferrous carbonate.

Chalybite (Spathic Iron Ore, Siderite, Gyrite). A mineral. It is a carbonate of iron, $2[FeCO_3]$.

Chamber Acid. Sulphuric acid, which condenses on the floors in the lead chamber process for sulphuric acid. It contains from 62–70 per cent. acid.

Chamet Bronze. A brass containing 62 per cent. copper and 38 per cent. zinc.

Chamois (Rouille). A yellowish-brown hydrated sesquioxide of iron, precipitated on calico by the action of caustic soda on ferrous sulphate.

Chamois Mordant. Pure ferrous acetate.

Chamomile Flowers. Flowers of *Anthemis nobilis.* Also see GERMAN CHAMOMILE.

Chamomillysatum Bürger. A German proprietary preparation of camomile blossoms used as a disinfectant and carminative. It is supplied as ointment, liquid, and powder.

Chamosite. A natural silicate of iron, $Al_2O_3.3(Fe.Mg)O.2SiO_2.3H_2O$.

Chamotan. Leather dressing oils. (563).

Chamotte. A mixture of fireclay and burnt pottery. Used for making fire-bricks.

Chanarcillite. A mineral, $Ag_2(As.Sb)_3$.

Chance Mud (Lime Mud). The mud left after the treatment of alkali waste with carbon dioxide to recover the sulphur. It contains approximately 33 per cent. calcium carbonate, and is used for agricultural purposes.

Chance's Stone. An artificial stone made by melting a basaltic rock found in Staffordshire (Rowley rag), and casting it into the shapes required.

Channel Black. A gas black. See CARBON BLACK.

Chapmanite. A mineral of Ontario. It is a hydrous ferrous-silico-antimonate, and approximates to the formula, $5FeO.5SiO_2.Sb_2O_52H_2O$.

Chappe. A yarn made from waste silk.

Chappe, Artificial. See STAPLE FIBRE.

Chapped Silk. Silk which has been partially degummed by a fermentation process.

Char. See BONE BLACK.

Charas. See BHANG.

Charcoal Black (Carbon Black). A pigment. It consists of carbon, obtained by heating wood in closed retorts to a high temperature, then washing and grinding. Also see CARBON BLACK.

Charcoal, Bone. See ANIMAL CHARCOAL.

Charcoal Iron. A high grade iron free from sulphur.

Charcoal Mixture. Meal powder (500 parts), mixed with from 6–8 parts charcoal. Used in fireworks.

Charcolite. A charcoal made from coal by a special process.

Chardonnet Silk. A collodion silk prepared from nitrated cellulose dissolved in alcohol and ether. It is afterwards denitrated.

Charklets. Small briquettes of powdered charcoal with a binder. Used for coal.

Charleston Phosphate. A mineral, consisting mainly of calcium phosphate.

Char, Lignite. See GRUDEKOK.

Charlton White. See LITHOPONE.

Charmour. A proprietary trade name for cellulose acetate materials. Used for lampshades, etc.

Chartreuse. A liqueur prepared from balm mint, cinnamon, saffron, hyssop, angelica, sugar, alcohol, and other ingredients.

Chateauagt Iron. A low phosphorous copper-free pig-iron made from magnetite ore. Used for casting gears, etc.

Chathamite. A mineral. It is $(Ni,Fe)As_2$.

Chatelier Solder. An alloy of 70 per cent. tin, 25 per cent. zinc, 2 per cent. aluminium, and 1·5 per cent. phosphorus.

Chats. A name for hard sand tailings from certain lead ores, used for sawing stone.

Chatterton's Compound. Mixtures of tar, rosin, and gutta-percha. Used for cementing gutta-percha to wood and metals.

Chaubert's Oil. Consists of 75 per cent. oil of turpentine, and 25 per cent. oil of hartshorn.

Chaulgrol. A proprietary preparation. It is ethyl chaulmoograte.

Chaulmartin. Chaulmoogra fatty acid esters.

Chaulmestrol. Consists of the ethyl esters of the fatty acids of chaulmoogra oil.

Chaulphosphate. Sodium dichaulmoogryl-β-glycerophosphate. Used in leprosy.

Chaux de Theil. A hydraulic lime, containing more than 20 per cent. clayey matter. Used by French engineers for marine work.

Chaval Nickel Silver. See NICKEL SILVERS.

Chavicol. p-Allyl-phenol, $C_6H_4(OH)C_3H_5$.

Chavosol (Chavosote). A dental antiseptic containing p-allyl-phenol, $C_6H_4(OH)C_3H_5$.

Chavosote. See CHAVOSOL.

Chay. See CHAY ROOT.

Chay-aver. See CHAY ROOT.

Chay Root (Chay-aver, Indian Madder). The root of *Oldenlandia umbellata*, known in India as turbuli, cheri-vello, ché or chay, sayawer, and imburel. The root contains a dyestuff, which dyes similar to Turkey red.

Chazellite. A mineral, $3FeS.2Sb_2S_3$.

Ché. See CHAY ROOT.

Cheaply-black. A mineral carbon for rubber. It has a specific gravity of 2·2–2·3.

Check Brass. See BRASS.

Cheddites. Blasting explosives, containing perchlorates and chlorates of ammonium, potassium, or sodium, the explosive nature of which is reduced by coating them with castor oil having a nitro-compound (nitro-naphthalene or nitro-toluene) dissolved in it. They are manufactured at Chedde, in France.

Cheesecloth. A thin cotton cloth originally used for wrapping cheese, now employed for a variety of purposes.

Chel. Chelating agents. (543).

Chelafrin. Adrenalin (*q.v.*).

Chelen. Ethyl chloride, C_2H_5Cl.

Chelevtite. A mineral. It is a variety of smaltite.

Cheltenham Salts. A mixture of 34 parts sodium sulphate, 23 parts magnesium sulphate, and 50 parts sodium chloride.

Chemawinite (Cedarite). A pale yellow Canadian amber.

Chemglaze. A trade mark for a polyurethane coating material. (36).

Chemical Lead. Lead usually containing only a small amount of copper The British Engineering Standards Association (334, 1928) states that it should contain 99·99 per cent. lead, but if copper is present the amount of lead may be reduced by an amount not exceeding the copper present. The total impurities excluding copper should not exceed 0·01 per cent. The maximum amount of any single impurity should not exceed the following : silver,

0·002 per cent. ; bismuth, 0·005 per cent. ; iron, 0·003 per cent. ; antimony, 0·002 per cent. ; zinc, 0·002 per cent. ; copper, 0·050 per cent. ; nickel and cobalt, 0·001 per cent. together, and tin, cadmium, and arsenic as traces. Other tests are the aqua regia test and the flash test.

Chemical Red. See INDIAN RED.

Chemical Sand. The iron salt used to neutralise the waste soap lyes for glycerin recovery. It consists of a mixture of ferric sulphate, ferric oxide, and a little free acid, and is made by treating powdered iron ore with concentrated sulphuric acid, breaking it up, and heating.

Chemical Stoneware. Essentially an aluminium silicate partly combined and partly mixed with other silicates. It usually contains about 73 per cent. silica and 22 per cent. alumina.

Chemical Wood Pulp. Pure cellulose.

Chemical Yarn. Artificial silk.

Chemick. A dilute solution of bleaching powder, used in the textile industry.

Chemigum. A patented synthetic rubber derived from petroleum. It is a butadiene copolymer.

Chemloc. A proprietary chelating agent for bonding polyaminocarboxylic acids to complex metal ions with an organic compound, to form a closed ring structure and a non-ionic complex. Used in the paper industry. (496).

Chemlok. Adhesives; rubber to metal bonding agents. (516).

Chemlube 353 and 1109. A proprietary range of barium-lead stabiliser lubricants. (492).

Chemol. A registered trade mark currently awaiting re-allocation by its proprietors. (983).

Cheneviscite (Chenevixite). A mineral. It is an arsenate of copper and iron.

Chenevixite. See CHENEVISCITE.

Chenocoprolite. Ganomatite (q.v.).

Chenzinsky-Plehn's Solution. A microscopic stain. It contains 0·25 gram eosin, 50 grams of 70 per cent. alcohol, 100 grams of a saturated solution of methylene blue, and 100 c.c. distilled water.

Cheri-vello. See CHAY ROOT.

Cherokine. A mineral. It is a variety of pyromorphite.

Cherry Essence. Usually consists of 15 parts amyl acetate, 10 parts benzaldehyde, 1 part sweet oil of oranges, 2 parts cassia oil, 2 parts lemon oil, 2 parts oil of cloves, and 8 parts amyl benzoate.

Cherrypit Oil. Cherry oil. Used as an edible oil.

Chert. A mineral, consisting of a massive form of silica SiO_2.

Chessy Copper. See CHESSYLITE.

Chessylite (Chessy Copper, Berlin Blue, Native Smalt, Blue Ashes, Klaprothite, Mollite, Blue Malachite, Azurite, Lasurite, Lazurite, Lazulite, Smalt Blue, Blue Felspar, Blue Spar). A blue carbonate of copper, found near Lyons, $2CuCO_3.Cu(OH)_2$.

Chesterlite. A mineral. It is a variety of microcline.

Chestnut. See MAGENTA and UMBER.

Chestnut Brown. Umber (q.v.).

Chestnut Extract. An extract obtained from the wood of Spanish chestnut, *Castanea vesca*. The extract contains 26–32 per cent. tannin.

Chevkinite. See TSCHEFFKINITE.

Chewing Gum. Gum chicle, washed and dried, and mixed with flavouring materials, sugar, and filling substances.

Chhilua. Nitrate soils or earths found in the East. They are used for obtaining calcium and potassium nitrates by extraction with water.

Chian Turpentine (Alk, Chio Turpentine, Chios Turpentine, Cyprian Turpentine, Scian Turpentine). Names applied to the oleo-resin obtained from the bark of *Pistachi terebinthus*, a tree in the Mediterranean and Asia Minor.

Chia Oil. An oil obtained from the Mexican plant *Salvia hispanica*. The raw oil dries slowly, but the boiled oil is a good drying oil.

Chiastolite (Cross-stones). A mineral. It is a variety of andalusite (a silicate of aluminium, Al_2SiO_5). Worn as charms.

Chica. See ATOLE.

Chicago Acid. 2S Acid.

Chicago Blue. Benzo blue (q.v.).

Chicago Blue B. A dyestuff. It is the sodium salt of dimethoxy-diphenyl-disazo-bi-amino-naphthol-sulphonic acid, $C_{34}H_{26}N_6S_2O_{10}Na_2$. Dyes cotton blue.

Chicago Blue 4B and RW (Chlorazol Blue RW). Mixed azo dyestuffs from dianisidine, 1 : 8 : 4 : 6-amino-naphthol-disulphonic acid, and another component.

Chicago Blue 6B. See DIAMINE SKY BLUE.

Chicago Blue R. A dyestuff. It is the sodium salt of ditolyl-disazo-bi-amino-naphthol-sulphonic acid, $C_{34}H_{26}N_6O_8S_2Na_2$. Dyes cotton blue.

Chicago, Blue 2R. See COLUMBIA BLUE G.

Chicago Blue 4R (Columbia Blue R, Benzo Blue 4R, Diamine Blue C4R). Mixed diazo compounds from benzidine, amino-naphthol-sulphonic acid (1 : 8 : 4) or disulphonic acid (1 : 8 : 2 : 4), and α-naphthol-sulphonic acid. Dyes cotton blue.

Chicago Grey. An azo dyestuff from Chicago orange and amino-naphthol-sulphonic acid G. A direct cotton dye.

Chicago Orange RR. A dyestuff produced by the condensation of *p*-nitro-toluene-sulphonic acid with benzidine, in the presence of caustic soda. Dyes cotton orange from a salt bath.

Chica Red (Carajura). A rare pigment prepared by the Indians of Central America, from *Bigonia chica*. It is obtained from the leaves, and is used for painting their bodies. It is also called chika.

Chick Pea. See BENGAL GRAM.

Chicle. A gum obtained from *Achras* species and others of Mexico, British Honduras, and Venezuela. It has been used in the manufacture of chewing-gum, but has now been considerably superseded by other gums, e.g. Jelu-tong. Other names are Tunogum, Zapoto gum.

Chicory. The root of *Cichorium intybus*. It is mixed with coffee.

Chiendent. See RAIZ DE ZACATON.

Chierite. A proprietary moulding material of urea formaldehyde. (497).

Chierol. A proprietary phenolic moulding material. (497).

Chika. See CHICA RED.

Childrenite. A mineral, $FeAl(OH)_2PO_4$. H_2O.

Chileite. A mineral. It is a copper-lead vanadate.

Chilenite (Bismuth Silver). A mineral, Ag_5Bi or Ag_6Bi.

Chili Bar. A crude copper resembling blister copper, and containing 98-99 per cent. of the metal.

Chili Nitre. See CHILI SALTPETRE.

Chili Saltpetre (Chili Nitre, Soda Salt-petre, Peru Saltpetre, Cubic Saltpetre, Soda Nitre, Cubic Nitre, Nitratine). Sodium nitrate, $NaNO_3$, found as deposits in Chile and Peru.

Chillagite. A mineral, $PbWO_4.PbMoO_4$.

Chillies or Chillie Pods. The fruit of *Capsicum minimum*.

Chilworth Powders. Explosives. They are mixtures of potassium nitrate and chlorate, sulphur, and charcoal.

China Blue. See SOLUBLE BLUE.

China Clay (Porcelain Clay, Cornish Clay, Kaolin, Argilla, White Bole, Pipe-clay, Porcelain Earth, Catalpo). The purest form of clay (a silicate of aluminium, $Al_2O_3.2SiO_2.2H_2O$), found naturally. It is used in the manufacture of china, in distemper work, and for other purposes.

China Clay Rock. The disintegrated

granite from which China clay is washed is known by this name.

Chinadone. A proprietary preparation of potassium-oxychinoline-sulphonate.

China Grass (White Ramie). A fibre from *Bœhmeria nivea*.

China Green. See MALACHITE GREEN.

Chinaldine. 2-Methyl-quinoline.

Chinalgen. See ANALGEN.

Chinamin. A solution of the neutral salt of quinine, in combination with adrenaline and hamamelis. Used as a nasal spray, and as an instillation in the eyes.

China Oil. Peru balsam.

Chinaphenin. See QUINAPHENIN.

Chinaphthol (Quinaphthol). Quinine-β-naphthol-α-mono-sulphonate, $C_{20}H_{24}$ $N_2O_2(C_{10}H_6.OH.SO_3H)_2$. Used in rheumatism and typhoid fever.

China Silver. See ARGYROLITH and NICKEL SILVERS.

China Stone. See CORNISH STONE.

Chineonal (Quineonal). Quinine-di-ethyl-barbiturate, containing 64 per cent. quinine. Has been used in the treatment of whooping cough.

Chinese Bean Oil. Soja bean oil, extracted from the bean of *Soja hispida, S. japonica*, or *Phaseolus hispidus*. Used in soap-making.

Chinese Blue. See PRUSSIAN BLUE. Several mixtures are also sold under this name, such as ultramarine and flake white, or cobalt blue and white lead.

Chinese Bronze. Alloys of from 72·5-74 per cent. copper, 15-18·5 per cent. lead, 10-14 per cent. zinc, and 1-5 per cent. tin. One alloy, also known as Chinese bronze, contains 78 per cent copper and 22 per cent. tin.

Chinese Carmine. Said to be prepared by extracting cochineal with a boiling solution of alum, heating the filtered decoction with the addition of tin in nitric and hydrochloric acids, and finally allowing it to stand for the carmine to separate.

Chinese Cinnamon. See CASSIA BARK.

Chinese Cinnamon Oil. See OIL OF CHINESE CINNAMON.

Chinese Galls. See GALLS.

Chinese Glue. Shellac dissolved in alcohol. Used for joining wood, earthenware, or glass.

Chinese Green. See SAP GREEN.

Chinese Ink. See INDIAN INK.

Chinese Isinglass. See AGAR-AGAR.

Chinese Jute. A commercial fibre derived from *Abutilon avicennœ* (Indian mallow). It is not a jute.

Chinese Nickel Silver. See NICKEL.

Chinese Orange. A brownish-orange lake pigment made from an alizarin derivative precipitated on an aluminium base.

Chinese Red. See CINNABAR and CHROME RED.

Chinese Rice Paper. This is prepared by cutting out in spiral form the pith of *Aralia papyrifera*, and pressing.

Chinese Scarlet. See CHROME RED.

Chinese Silver (Peru Silver). A German silver, containing a little aluminium.

Chinese Tallow (Vegetable Tallow). A waxy substance obtained from the outer coating of the fruit of *Stillingia sebifera*, of China. It has a specific gravity of 0·918–0·922, a solidifying point of 24–34° C., a saponification value of 179–206, and an acid value of 2·4. See also STILLINGIA OIL.

Chinese Vermilion. See SCARLET VERMILION.

Chinese Wax (Vegetable Spermaceti, Tree Wax). Also wrongly called Japanese wax. Insect wax obtained from the insect *Coccus ceriferus*, or *C. pela*, which deposits wax on certain trees. The wax has a melting-point of 65° C., a saponification value of 92·9, an acid value of 13, and an iodine value of 15·2.

Chinese White. See ZINC WHITE. The name is sometimes applied to barium sulphate.

Chinese White Copper. An alloy of 40 per cent. copper, 31 per cent. nickel, 25 per cent. zinc, and 2 per cent. iron.

Chinese Wood Oil (Wood Oil). Tung oil obtained from the seeds of *Aleurites* species in China and Japan. It has a specific gravity of 0·936–0·944 at 15° C., a refractive index 1·510–1·525 at 15° C., a saponification value of 188–197, and an iodine value 154–176. A drying oil which has been used in the manufacture of linoleum.

Chinese Yellow (Persian Yellow, Spanish Yellow, Royal Yellow). Trisulphide of arsenic, As_2S_3. A pigment. Also see OCHRE.

Chinic Acid. See KINIC ACID.

Chinidine (Chinotin). Quinidine, $C_{20}H_{24}O_2N_2$.

Chinine. Quinine.

Chinkolobwite (Sklodowskite). A radio-active material found at Katanga in the Belgian Congo. The main constituents are uranium oxide and silica.

Chinn Mineral. A snow-white calcium carbonate mined in Kentucky. It has a specific gravity of 2·64, and is used as a rubber filler.

Chinocol. A compound of hydroxy-quinoline and potassium sulphate. An antiseptic.

Chinoform. A proprietary name for cliquinol. See QUINOFORM.

Chinoidin. See QUINOIDINE.

Chinol. Quinoline-mono-hypochlorite,

$C_9H_6N.ClO$. Said to be antipyretic and analgesic.

Chinolin. See LEUCOLINE.

Chinone. Quinone, $C_6H_4O_2$.

Chinoral. A compound of quinine and chloral. A hypnotic and antiseptic.

Chinosol. See QUINOSOL.

Chinotheine. A mixture of quinine, caffeine, and antipyrine.

Chinotin. See CHINIDINE.

Chinotoxine. Dichinoline-dimethyl-sulphate-quinotoxin. Toxic.

Chinotropine. Formamine quinate (hexamine quinate), used in medicine.

Chiolite (Arksatite). A mineral. It is a double fluoride of aluminium and sodium, $3AlF_3.5NaF$.

Chios Turpentine. See CHIAN TURPENTINE.

Chio Turpentine. See CHIAN TURPENTINE.

Chirol. A solution of resins and fatty acids in ether-alcohol.

Chirosoter. A solution of waxy or balsam-like bodies in carbon tetra-chloride. Recommended as an antiseptic.

Chirt. A material consisting of a silicate of aluminium and magnesium found with the mineral barite.

Chisso-rite. A proprietary synthetic resin made by condensing formaldehyde and acid oils from low temperature coal carbonisation.

Chiviatite. A mineral, $2PbS.3Bi_2S_3$.

Chizeuilite. Andalusite (*q.v.*).

Chladnite. A mineral. It is Enstatite.

Chloanthite (White Nickel). A mineral. It is nickel diarsenide, $NiAs_2$.

Chlophedianol. 1 - (2 - Chlorophenyl) - 3 - dimethylamino - 1 - phenylpropan - 1-ol.

Chloractil. A proprietary preparation of CHLORPROMAZINE hydrochloride. A tranquilliser. (498).

Chloral. Chloral hydrate, $CCl_3.CHO$.

Chloralamide. Chloral formamide, $CCl_3.CH(OH).NH.CHO$. A mild hypnotic and sedative.

Chloral Betaine. Chloral hydrate - betaine adduct. Somilan.

Chloral Iodine. A solution of chloral hydrate (50 grams in 20 c.c. water) saturated with iodine. Used for the detection of starch grains.

Chloralose. Anhydro-gluco-chloral, $C_8H_{11}CCl_3O_6$. A hypnotic.

Chloral-tannin. See CAPTOL.

Chloralum. An impure solution of aluminium chloride, $AlCl_3$, of specific gravity 1·15. It contains calcium and sodium salts as impurities. A disinfectant.

Chloraluminite. A mineral. It is $2(AlCl_3.6H_2O)$.

Chloralurethane. Carbochloral.

Chlorambucil. γ - 4 - Di(2 - chloro - ethyl)aminophenylbutyric acid. Leukeran.

Chloramide. See CHLORALAMIDE.

Chloramide of Mercury. See WHITE PRECIPITATE.

Chloramine B. Sodium-benzene-sulpho-chloro-amide, $C_6H_5.SO_2.NNaCl+2H_2O$. Used like Chloramine T.

Chloramine Black N. A dyestuff which dyes unmordanted cotton, half wool, and half silk.

Chloramine Blue 3G and HW. Polyazo dyestuffs which dye mixed fabrics. The brand 3G gives a greenish-blue on unmordanted cotton, and HW, a blackish-blue.

Chloramine Green B. A dyestuff. It is the sodium salt of diphenyl-disazo-phenol (or salicylic)-disulpho-amino-naphthol-azo-dichloro-benzene, $C_{34}H_{21}N_7O_8S_2Na_2Cl_2$. Dyes cotton green.

Chloramine-Heyden. See CHLORAMINE T.

Chloramine Orange. See MIKADO ORANGE G and 4R.

Chloramine T (Tochlorine, Tolamine, Chloramine-Heyden, Pyrgos, Mianin, Aktivin). Sodium-p-toluene-sulpho-chloramide, $C_6H_4(CH_3)(SO_2 . NCl . Na) 3H_2O$. An antiseptic used in medicine. Solutions of Chloramine T are also used as detergents and bleaching agents. Zauberin, Mannolit, Gansil, Glekosa, Purus, and Washington Bleach are trade names for washing and bleaching materials containing Chloramine T as the active principle.

Chloramine Yellow (Chlorophenine Y, Diamine Yellow A, Diamine Fast Yellows B, C, and FF, Oxyphenine, Oxyphenine Gold, Thiophosphine J, Columbia Yellow). Oxidation products of dehydro-thio-toluidine-sulphonic acid or the latter and primuline together. Dye cotton from a neutral bath, and silk from an acid bath, yellow.

Chloramphenicol. D - ($-$) - threo - 2 - Dichloroacetamido - 1 - (4 - nitro - phenyl)-propane-1,3-diol. Alficetyn ; Chloromycetin ; Paraxin.

Chloranil. Tetrachlor-quinone, $C_6Cl_4O_2$.

Chloranil Violet. A dyestuff which is probably identical with methyl violet, obtained by the action of dimethyl-aniline upon chloranil.

Chloranol. A photographic developer. It is a combination of methyl-p-amino-phenol and chloro-hydroquinone.

Chloranthrene Blue. Indanthrene blue.

Chloranthrene Dyes. Chloranthrene is a proprietary trade name applied to certain dyestuffs.

Chlorantine. Dyestuffs fast to light. (530).

Chlorantine Red. See DIANOL BRILLIANT RED.

Chlorapatite. A mineral. It is a variety of apatite, $3Ca_3.2PO_4.CaCl_2$.

Chlorargyrite. See HORN SILVER. A mineral. It is 4[AgCl].

Chlorarsine. Cacodyl, $As_2(CH_3)_4$.

Chlorasol. A proprietary trade name for a mixture of carbon tetrachloride and ethylene dichloride.

Chlorastrolite. A variety of jade.

Chlorat-Baldurit. A German explosive.

Chlorat-Rivalit. A German explosive.

Chlorazene. A preparation of CHLORAMINE-T and sodium p-toluene-sulpho-chloramide. (884).

Chlorazol. Direct cotton dyestuffs. (512).

Chlorazol Black BH. A dyestuff. It is a British brand of Diamine black BH.

Chlorazol Black FFH. A dyestuff. It is a British equivalent of Columbia black FF extra.

Chlorazol Black SD. A dyestuff. It is a British brand of Zambesi black D.

Chlorazol Blue B. A dyestuff. It is a British brand of Diamine blue 2B.

Chlorazol Blue 3B. A British dyestuff. It is equivalent to Diamine blue 3B.

Chlorazol Blue 6G. See DIAMINE SKY BLUE.

Chlorazol Blue R and 3G. A dyestuff. It is dimethoxy-diphenyl-disazo-bi-chloro-α-naphthol-sulphonic acid, $C_{34}H_{22}N_4Cl_2O_{10}S_2Na_2$. Dyes cotton blue from a salt bath.

Chlorazol Blue 2R. A dyestuff. It is a British brand of Diamine blue BX.

Chlorazol Blue RW. A dyestuff. It is a British equivalent of Chicago blue RW.

Chlorazol Brown B. A British dyestuff. It is equivalent to Diamine brown B.

Chlorazol Brown 2G. A dyestuff. It is a British brand of Trisulphone brown 2G.

Chlorazol Brown GM. A dyestuff. It is a British equivalent of Benzo brown G.

Chlorazol Brown LF. A dyestuff. It is a British equivalent of Trisulphone brown B.

Chlorazol Brown M. A dyestuff. It is a British brand of Diamine brown M.

Chlorazol Deep Brown. A dyestuff identical with Trisulphone brown B, except that the end component is m-tolylene-diamine.

Chlorazol Diazo Blue 2B. A dyestuff. It is a British equivalent of Diaminogen blue BB.

Chlorazol Dyes. A registered trade mark applied to certain dyestuffs.

Chlorazol Fast Orange D. A dyestuff. It is a British brand of Mikado orange.

Chlorazol Fast Orange R. A British dyestuff. It is equivalent to Benzo fast orange S.

Chlorazol Fast Pink BK. A dyestuff. It is a British equivalent of Benzo fast pink 2BL.

Chlorazol Fast Red FG. A dyestuff. It is a British equivalent of Diamine fast red.

Chlorazol Fast Scarlet 4BS. A dyestuff. It is a British brand of Benzo fast scarlet 4BS.

Chlorazol Fast Yellow 5GK. A British equivalent of Cotton yellow (The Badische Co.)

Chlorazol Green BN. A dyestuff. It is a British equivalent of Diamine green B.

Chlorazol Green G. A dyestuff which is the same as Diamine green G (Cassella). Dyes unmordanted cotton, green.

Chlorazol Orange RN. A British dyestuff. It is equivalent to Benzo orange R.

Chlorazol Pink Y. A dyestuff. It is a British brand of Rosophenine 10B.

Chlorazol Sky Blue FFS. A dyestuff identical with Benzo sky blue, Chicago blue 6B, and Diamine sky blue FF. See DIAMINE SKY BLUE.

Chlorazol Violet N. A dyestuff. It is a British equivalent of Diamine violet N.

Chlorazol Violet WBX. A dyestuff. It is a British equivalent of Trisulphone violet B.

Chlorazol Yellow GX. A dyestuff. It is a British brand of Direct yellow R.

Chlorazone. See CHLORAMINE T.

Chlorbetamide. Dichloro - N - (2,4 - dichlorobenzyl) - N - (2 - hydroxyethyl)-acetamide. Mantomide ; Pontalin.

Chlor-Buna. Chlorinated synthetic rubber.

Chlorbutol. See CHLORETONE.

Chlorcahücit. A German explosive.

Chlor-cellulose, Cataldi's. See CATALDI'S CHLOR-CELLULOSE.

Chlorcenthite. A mineral, $NiAs_2$. The nickel is frequently replaced by cobalt and iron.

Chlorcosan. A liquid chlorinated paraffin wax. It is used as a solvent for Dichloramine T in 8 per cent. solution. The chlorine content varies from 27–35 per cent.

Chlor Cresol Green. Tetrachlor-*m*-cresol-sulphon-phthalein.

Chlorcyclizine. 1 - (4 - Chloro - α - phenylbenzyl) - 4 - methylpiperazine. Di-paralene and Histantin are the hydrochloride.

Chlordantoin. 5 - (1 - Ethylpentyl) - 3 - (trichloromethylthio)hydantoin.

Chlordiazepoxide. 7-Chloro-2-methyl-amino - 5 - phenyl - 3H - benzo - 1,4 - diazepine 4-oxide. Librium is the base or the hydrochloride. Present in Pentrium.

Chlorethyl. Ethyl chloride, C_2H_5Cl.

Chloretone (Chlorbutanol, Methaform, Chlorbutol, Acetone Chloroform). Tri-chloro-butyl-alcohol, $(CH_3)_2 . C(OH) . CCl_3.\frac{1}{2}H_2O$. Used as a hypnotic, soporific, as an inhalation anæsthetic, and as a remedy for sea-sickness.

Chlorfenvinphos. An insecticide. 2 - Chloro - 1 - (2, 4 - dichlorophenyl) - vinyl diethyl phosphate. BIRLANE.

Chlor-Fluorapatite. A mineral. It is $2[Ca_5(PO_4)_3(F,Cl)]$.

Chlorhexadol. 2 - Methyl - 4 - (2,2,2 - trichloro - 1 - hydroxyethoxy) - pen - tan-2-ol.

Chlorhexidine. 1,6 - Di(4 - chloro - phenyldiguanido)hexane. Hibitane.

Chloric Ether. Spirit of chloroform (*Spiritus chloroformi, B.P.*). It contains 50 c.c. chloroform and 1000 c.c. alcohol.

Chloride-Meionite. A mineral. It is $Ca_4Al_6Si_6O_{24}Cl_2$.

Chloride of Lime. See BLEACHING POWDER.

Chloride of Lime, Liquid. See BLEACH LIQUOR.

Chloride of Soda. See EAU DE JAVELLE.

Chlorin. See DINITROSO-RESORCIN.

Chlorina. Chloramine.

Chlorinated Calcium Oxide. See BLEACHING POWDER

Chlorinated Lime. Bleaching powder (*q.v.*).

Chlorinated Rubber. See DETEL, TORNESIT.

Chlorinated Soda. See EAU DE JAVELLE.

Chlorinated Train Oil. Chlorinated whale oil. A binding and adhesive agent.

Chlorinated Wool. Wool which has first been treated with hydrochloric acid, then with bleaching powder or sodium hypochlorite solution. It is stated to give a silk-like gloss to the material, and to aid dyeing.

Chlorisondamine Chloride. 4,5,6,7 - Tetrachloro - 2 - (2 - dimethylamino - ethyl) - isoindoline dimethochloride. Ecolid.

Chlorite (Greenite). A mineral, $5MgO.Al_2O_3.3SiO_2.4H_2O$.

Chlorite Spar. See CHLORITOID.

Chloritoid (Masonite, Phyllite, Chlorite Spar, Barytophyllite). A mineral, $Al_2O_3.FeO.SiO_2.H_2O$.

Chlorkautschuk. Chlorinated natural rubber.

Chlormadinone. 6 - Chloro - 17α - hydroxypregna - 4,6 - diene - 3,20 - dione. Clordion is the acetate.

Chlormagnesite. A mineral. It consists of magnesium chloride, $MgCl_2$.

Chlormanganokalite. A mineral. It is $2[K_4MnCl_6]$.

Chlormerodrin. 3 - Chloromercuri - 2 - methoxypropylurea. Mercloran ; Neohydrin.

Chlormethane. Methyl chloride, CH_3Cl.

Chlormethiazole. 5 - (2 - Chloroethyl) - 4-methylthiazole. Heminevrin is the edisylate.

Chlormethine. Mustine.

Chlormethylencycline. A proprietary name for clomocycline.

Chlormezanone. 2 - (4 - Chlorphenyl) tetrahydro - 3 - methyl - 1,3 - thiazin - 4-one. 1,1-dioxide. Trancopal. Present in Gerison and Lobak.

Chlormidazole. 1 - (4 - Chlorobenzyl) - 2-methylbenzimidazole.

Chlormytol. A proprietary preparation of chloramphenicol and prednisolone used in dermatology as an antibacterial agent. (264).

Chloro-acetene. A mixture of aldehyde, paraldehyde, and carbonyl chloride.

Chloroben. o-Dichlorobenzene. Used in sewage treatment.

Chlorobrom. A solution of 6 parts potassium bromide, 6 parts chlorobromamide, and 58 parts water.

Chlorobromal. Chloro-dibromo-acetic-aldehyde, $CCl.Br_2.CHO$.

Chlorobromhydrin. α-Chloro-α-bromo-isopropyl-alcohol, $CH_2Br.CH(OH).CH_2Cl$.

Chlorobromoform. Chloro-dibromo-methane, $CHCl.Br_2$.

Chlorochroite. A mineral, $K_2SO_4.CuCl_2$.

Chlorodyne. A medicinal compound containing prussic acid, morphia, chloroform and hemp. An opiate and antispasmodic.

Chlorofin 42. A proprietary trade name for a chlorinated paraffin extender for vinyl plastics containing 40 per cent. chlorine. (93).

Chlorofin 70. A proprietary trade name for a chlorinated paraffin extender for vinyl plastics containing 70 per cent. chlorine.

Chloroform, Nitro. See CHLORO-PICRIN.

Chloroform, Pictet. Chloroform prepared from chloral-chloroform, by freezing this compound to −75° C.

Chloroform Water. (*Aqua chloroformi, B.P.*) It consists of 2·5 c.c. chloroform, made up with water to 1000 c.c.

Chlorogen. A bleaching liquid similar to Chlorozone.

Chlorogenine. Alstonine, $C_{21}H_{20}N_2O_4$, an alkaloid.

Chloroiodolipol. Stated to be a chlorine substitution product of phenol, creosote, and guaiacol. Used in chronic affections of the respiratory tract.

Chlorol. A disinfectant containing mercuric chloride, sodium chloride, hydro-

chloric acid, and copper sulphate, dissolved in water.

Chlorolin. A disinfectant said to contain 20 per cent. mono-chlor-phenol, and trichlor-phenol. Applied to wounds as a 0·5–3 per cent. solution.

Chloromelane. Cronstedite (*q.v.*).

Chloromelanite. Jadeite (*q.v.*).

Chloromethane (Chloro - methyl). Methyl chloride, $CHCl_3$.

Chloro-methyl. See CHLOROMETHANE.

Chloromycetin Palmitate Suspension. A proprietary preparation containing chloramphenicol. An antibiotic. (264).

Chloromycetin Intramuscular. A proprietary preparation containing Chloramphenicol. An antibiotic. (264).

Chloromycetin Succinate. A proprietary preparation containing chloramphenicol, (as the sodium salt of the monosuccinic ester). An antibiotic. (264).

Chloromycetin Kapseals. A proprietary preparation containing Chloramphenicol. An antibiotic. (264).

Chloromycetin Pure. A proprietary preparation containing chloramphenicol. An antibiotic. (264).

Chloromycetin Suppositories. A proprietary preparation containing chloramphenicol. An antibiotic. (264).

Chloronitrous Gas. Nitrosyl chloride, $NOCl$.

Chlorophaeite. A mineral. It is a hydrated silicate of iron and magnesium.

Chlorophane (Pyro-emerald). A variety of fluorspar with a greenish fluorescence.

Chlorophanerite. A mineral. It is a variety of glauconite.

Chlorophenine G. A dyestuff similar to Oxyphenine, Oxyphenine gold, and Thiophosphine J. Its properties and application are the same as Chloramine yellow.

Chlorophenine Orange R and GO. Intermediate reduction products of curcuphenine.

Chlorophenine Orange RR and RO. A dyestuff. It is the sodium sulphonate of the dehydro-thio-toluidide of azo-stilbene-aldehyde, obtained by the reduction of curcuphenine with glucose and caustic soda. Dyes cotton bright orange.

Chlorophenine Y. See CHLORAMINE YELLOW.

Chlorophœnicite. A mineral formed in the Franklin furnace. It contains 34·46 per cent. MnO, 29·72 per cent. ZnO, 19·24 per cent. As_2O_5, 3·36 per cent. CaO, 1·34 per cent. MgO, 0·48 per cent. FeO, and 11·6 per cent. H_2O.

Chlorophyll (Leaf Green, Chromule).

The green colouring matter of plants, leaves, and stalks. It is a magnesium compound, and is used for colouring confectionery and liqueurs. Chlorophyll as imported into America is a copper compound and not the magnesium one, the copper compound being more stable. The water-soluble chlorophyll is usually a soapy mass containing copper chlorophyll salts, and the alcohol and oil-soluble products consist of wax and fat mixed with copper chlorophyll.

Chloro-picrin (Nitro-chloroform). Trichloro-nitro-methane, $CCl_3(NO_2)$.

Chloroprene. 1 : 3 chlor-2-butadiene. Used as a protective coating and in synthetic rubber manufacture by polymerisation. The polymerised product bears the name of Neoprene.

Chloropryl. See Isopral.

Chloropyramine. Halopyramine.

Chloropyrilene. N-5-Chloro-2-thenyl-N'N' - dimethyl - N - 2 - pyridylethylenediamine.

Chloroquine. 7 - Chloror - 4 - (4 - diethylamine - 1 - methylbutylamino) - quinoline. Aralen, Avloclor and Resochin are the diphosphatel Nivaquine is the sulphate. Present in Nivembin as the sulphate.

Chloros. A registered trade mark for a disinfectant containing sodium hypochlorite. It is stated to contain 10 per cent. available chlorine. (512).

Chlorosalol. Chlor-phenol-salicylate, $C_6H_4(OH)COOC_6H_4Cl$. An external antiseptic.

Chlorosene. Bleaching powder for disinfectants. (512).

Chlorosoda. A proprietary form of solidified sodium hypochlorite for use as a bleaching agent. A small proportion of a saturated fatty acid, such as lauric acid, is incorporated.

Chlorosonin. A compound of chloral hydrate with hydroxylamine.

Chlorospinel. A mineral. It is a magnesium-iron spinel.

Chlorostab. A proprietary preparation of bismuth oxychloride in liquid suspension.

Chlorotex. A reagent for estimating residual chlorine in water. (527).

Chlorothene VG. An inhibited grade of 1-1-1-trichlorethane. A registered trade name for a chlorinated vapour degreasing solvent. (64).

Chlorothiazide. 6 - Chlorobenzo - 1,2,4 - thiadiazine-7-sulphonamide 1,1-dioxide. Chlotride. Diuril. Saluric.

Chlorothionite. A mineral, $(CuK_2)(SO_4.Cl_2)$.

Chlorothorite. Thorogummite (q.v.).

Chlorotile. A mineral, $(CuOH)_3.Cu(OH)_2.AsO_4$.

Chlorotrianisene. Chlorotri-(4-methoxy-phenyl)ethylene. TACE.

Chlorovin. A registered trade mark for polyvinyl chloride.

Chlorowax. A trade mark for chlorinated paraffins. (84).

Chlorowax 40. Liquid chlorinated paraffin. A vinyl plasticiser. (84).

Chlorowax 50. Chlorinated paraffin (50 per cent. chlorine). A vinyl plasticiser. (84).

Chlorowax 70. Resinous chlorinated paraffin. A vinyl plasticiser. (84).

Chlorowax LV. A chlorinated paraffin. A vinyl plasticiser. (84).

Chloroxethose. Hexachloro-divinyloxide, $(CCl_2 : CCl)_2O$.

Chloroxiphite. A mineral. It is an oxychloride of copper and lead.

Chloroxyl. Quinophan hydrochloride.

Chlorozone. A bleaching liquor prepared by passing chlorine into caustic soda.

Chlorphenamine. A proprietary trade name for chlorphenivamine.

Chlorphenarsine. 3-Amino-4-hydroxyphenyl - dichlorarsine - hydrochloride solution.

Chlorphenesin. 3 - (4 - Chlorophenoxy) propane-1,2-diol. Mycil.

Chlorpheniramine. 1 - (4 - Chloro - phenyl) - 3 - dimethylamino - 1 - (2-pyridyl)-propane. Chlorphenamine. Chlorprophenpyridamine. Chlor-Trimeton and Piriton are the hydrogen maleate.

Chlorphenoctium Amsonate. 2,4 - Dichlorophenoxymethyldimethyl - n - octylammonium, 4,4'-diaminostilbene - 2,2'-disulphonate.

Chlorphenol Red. An indicator. It is dichlor-phenol-sulphon-phthalein.

Chlorphenoxamine. 1 - (4 - Chloro - phenyl) - 1 - phenylethyl 2 - dimethylaminoethyl ether. Clorevan is the hydrochloride.

Chlorphentermine. 4 - Chloro - $\alpha\alpha$ - dimethylphenethylamine. Lucofen is the hydrochloride.

Chlorproguanil. N' - 3,4 - Dichloro - phenyl-N^5-isopropyldiguanide. Lapudrine is the hydrochloride.

Chlorpromazine. 2 - Chloro - 10 - (3 - dimethylaminopropyl)phenothiazine. Largactil is the hydrochloride. Present in Amargyl as the hydrochloride.

Chlorpropamide. N - 4 - Chloro - benzenesulphonyl-N'-propylurea. Diabinese.

Chlorprophenpyridamine. A proprietary trade name for Chlorphenivamine.

Chlorprothixene. α - 2 - Chloro - 9 -

(3 - dimethylaminopropylidene) - thi - axanthen. Taractan.

Chlorpyrifos. An insecticide currently undergoing clinical trial as " Dowco 179 ", " Dursban " and " Lorsban ". It is OO - diethyl O - 3, 5, 6 - trichloro - 2 - pyridyl phosphorothioate.

Chlorpyrin. See HYPNAL.

Chlorquinaldol. 5,7 - Dichloro - 8 - hydroxy-2-methylquinoline. Steroxin.

Chlorquinol. See HALQUINOL.

Chlorsapphir. A deep-green variety of sapphire.

Chlor-Tabs. Effervescent chlorine tablets. (522).

Chlor-tetragnost. Sodium-tetra-chlor-phenol-phthalein.

Chlorthalidone. 3 - (4 - Chloro - 3 - sulphamoylphenyl) - 3 - hydroxy - isoindolin-1-one. Hygroton.

Chlorthenoxazin. 2 - (2 - Chloroethyl) - 2,3-dihydro-4-oxobenz-1,3-oxazine.

Chlorthymol. A German product. It is a chlor-thymol preparation containing iodine. A yellowish-brown liquid soluble in water and having a slight smell of thymol. A disinfectant. It comes to the market as Lavasan.

Chlor-Trimeton. A proprietary trade name for the hydrogen maleate of chlorpheniramine.

Chloryl (Chloryl Anæsthetic). Ethyl chloride, C_2H_5Cl, used as an anæsthetic.

Chlorylen. Trichlorethylene, C_2HCl_3.

Chlorzoxazone. 5-Chlorobenzoxazolin-2-one. Paraflex. Present in Parafon.

Chlotride. A proprietary trade name for chlorothiazide.

Chlumin. A new aluminium alloy resistant to sea water. It contains chromium, a few per cent. of magnesium, and iron.

Chobile. A proprietary extract of ox bile with oxidised ox-bile acids, used as a stimulant to bile secretion. (819).

Chocolate. A mixture of cocoa and sugar with flavouring materials. A good sample contains from 40–60 per cent. cocoa, and the rest sugar. Ordinary qualities contain from 10–15 per cent. starch.

Chocolate Fat. See VEGETABLE BUTTER.

Chocolate Nuts. Theobroma seeds.

Chocolate Varnish. A liquid preparation, used in Spain, for hardening chocolate, to prevent it becoming white on keeping. It contains gum benzoin, a resin, and alcohol.

Chocolate Whetstone. A mica schist from New Hampshire, used for sharpening tools.

Chocolite. A mineral. It is a silicate of iron, nickel, and magnesium.

Chocovite. A proprietary preparation

of calcium gluconate and Vitamin D. Calcium supplement. (250).

Chodneffite. A mineral. It is a fluoride of sodium and aluminium. Synonym for Khodnevite.

Chodnewite. Synonym for Khodnevite.

Choke-damp (Black-damp). A gas which issues from soils into mines and wells. It usually contains from 85–95 per cent. nitrogen and 5–15 per cent. carbon dioxide.

Cholartosan. A blood cholesterol reduction preparation. (563).

Cholazyl. A chloro-acetyl-choline-chloride-urea compound. It causes lowering of blood pressure.

Cholecalciferol. 9,10 - Secocholesta - 5,7,10(19)-trien-3β-ol. Vitamin D_3.

Choledyl. A proprietary preparation of choline theophyllinate. A bronchial antispasmodic. (284).

Choleflavin. A German preparation. It consists of trypaflavin, papaverine, podophyllin, and oil of menthol. A gall antiseptic.

Cholelysin. A pure potassium oleate.

Choleol. An oily solution, said to contain the palmitic and oleic esters of cholesterin.

Cholestol. Oxy-quino-terpene, $C_{30}H_{48}O_2$.

Cholestrophane. Dimethyl-parabanic acid, $C_5N_6N_2O_3$.

Cholestyramine. A styryl-divinylbenzene copolymer (about 2 per cent. divinyl-benzene) containing quaternary ammonium groups. Cuemid.

Choleval. A preparation of colloidal silver and sodium tannate. A bactericide used in the treatment of gonorrhœa.

Choline (Sincaline). Neurine (dimethyl-oxy - ethyl - amine - methylo - hydrox - ide, $CH_2(OH).CH_2.N(CH_3)_3OH)$.

Choline Salicylate. An analgesic and antipyretic. Choline salt of salicylic acid.

Choline Theophyllinate. A bronchodilator. Choline salt of theophylline. CHOLEDYL.

Choloxin. A proprietary preparation of sodium D-thyroxine, used to treat hypercholesterolæmia. (832).

Cholumbrin. A proprietary preparation of sodium tetraiodo-phenol-phthalein.

Chondrarsenite. A mineral, (Mn.Ca. Mg).$(MnOH)_4.(AsO_4)_2.\frac{1}{2}H_2O$.

Chondrille Rubber. A rubber-like substance obtained from a plant probably a species of *Chondrilla* occurring in Central Asia. The material is found at the root and mixed with sand. The product freed from sand and amounting to about 20 per cent. is stated to contain rubber and a pitchy material in the proportion of about 50 per cent. of each.

Chondrodite (Brocchite, Humite). A mineral, $8MgO.3SiO_2$. Iron and fluorine are present.

Chonicrite. A mineral. It is a silicate of aluminium, magnesium, and calcium.

Chooli-ki-tel (Badam Kohee). Apricot oil, expressed from the kernels of *Prunus Armeniaca*.

Chop Nut. Se ORDEAL BEAN.

Chorzon-Salpeter-nitrofas. A Polish fertiliser containing 15·5 per cent. nitrogen and 9 per cent. phosphoric acid.

Chrisma. See OZOKERINE.

Chrismaline. See OZOKERINE.

Chrismol. Liquid paraffin.

Christianite. Anorthite (*q.v.*).

Christobalite. A mineral. It is a variety of silica.

Christofle. See NICKEL SILVERS.

Christolit. A proprietary plastic of the phenol-formaldehyde type.

Christophite. Synonym for Marmatite.

Chroatol. Terpin-iodo-hydrate, $C_{10}H_{16}.2HI$. Has been used externally in the treatment of psoriasis and other skin diseases, as a dusting-powder or ointment.

Chrogo U42. An alloy of 40 per cent. gold, 45 per cent. copper, 14 per cent. nickel, 1 per cent. chromium, and traces of platinum. A dental alloy.

Chromacetin Blue. A dyestuff obtained by the condensation of gallo-cyanines with aromatic alkylated amines.

Chromagan. Nickel-chromium steel for the manufacture of table ware. It is of German manufacture.

Chromalay. Materials for thinlayer chromatography. (507).

Chromaline. A chrome mordant made by reducing chromic acid with glycerin. Used for printing chrome colours on wool.

Chromaloid. A proprietary trade name for a dual metal in which zinc is bonded with chromium (plating).

Chromaloy. Nickel-chromium-iron alloys. The specific gravity varies from 8·15–8·35, and the melting-point 1360–1390° C.

Chromaluminium. An alloy of aluminium, chromium, and other metals. It has a specific gravity of 2·9.

Chromammonite. An explosive containing ammonium nitrate.

Chroman B. A Rohn alloy containing 64 per cent. nickel, 20 per cent. iron, 15 per cent. chromium, and 1 per cent. manganese.

Chroman Co. A Rohn alloy containing 79 per cent. nickel, 20 per cent. chromium, and 1 per cent. manganese.

Chromanil Black BF, 2BF, 3BF, RF, 2RF. Direct cotton blacks obtained after treatment with potassium bichromate.

Chromar. A proprietary trade name for a chromium based maraging steel. An ultra-high tensile strength steel with high corrosion and stress corrosion resistance. Suitable for the production of pressure vessels subjected to corrosive conditions. (142).

Chromargan VA. A non-oxidising steel. It contains from 18–25 per cent. chromium. It is non-magnetic, and is used for the manufacture of turbine blades, valves, cutlery, etc.

Chromargan VM. A non-oxidising steel. It contains from 13–15 per cent. chromium and a small amount of nickel. It has similar uses to VA.

Chromate, Copper. See COPPER CHROMATE.

Chromate of Chromium (Brown Oxide of Chromium). Chromium sesqui-oxide, Cr_2O_3.

Chromatised Gelatin. Made by adding 1 part potassium bichromate to 5 parts of a solution (5–10 per cent.) of gelatin. A cement for glass.

Chromax. An electrical resistance alloy containing 75 per cent. nickel and 25 per cent. chromium.

Chromax Bronze. An alloy of 15 per cent. nickel, 67 per cent. copper, 12 per cent. zinc, 3 per cent. aluminium, and 3 per cent. chromium.

Chromazol Violet. A trade mark. A dyestuff. It is a British equivalent of Modern violet. (2).

Chromazone Blue R. A dyestuff. It is the sodium salt of the ethyl-phenyl-hydrazone of benzaldehyde-azo-1 : 8-di-hydroxy-naphthalene-3 : 6-disulphonic acid. Gives blue shades on chromed wool.

Chromazone Red A. A dyestuff. It is the sodium salt of benzaldehyde-azo-1 : 8-dihydroxy-naphthalene-3 : 6-di-sulphonic acid. Dyes wool from an acid bath.

Chromazurines. Dyestuffs similar to Delphine blue.

Chromazurol S. A dyestuff. It is o-chloro-benzaldehyde-sulphonic acid, condensed with salicylic acid, and oxidised.

Chrombral. Flux for copper/chromium alloys. (531).

Chrome. See BICHROME.

Chrome (Acid Alizarin, Acid Anthracene, Diamond Salicine). Acid chrome colours for wool. The wool is dyed and mordanted with bichromate in the same vessel.

Chrome-Acmite. A mineral. It is $NaCrSi_2O_6$.

Chrome Alum. Potassium chrome alum, $Cr_2(SO_4)_3.K_2SO_4.24H_2O$.

Chrome Amalgam. An alloy of chromium and mercury obtained by the electrolysis of chromic chloride, using a mercury cathode. It has the formula Hg_3Cr. Another alloy, $HgCr$, is obtained from the above by pressure.

Chrome Black. Anhydrous copper chromate.

Chrome Black A. A dyestuff. It is a British equivalent of Eriochrome black A.

Chrome Black I. A dyestuff. It is the sodium salt of sulpho-carboxy-phenol-azo-α-naphthalene - azo - α - naphthol - p-sulphonic acid, $C_{27}H_{15}N_4O_{10}S_2Na_3$. Dyes chromed wool.

Chrome Blue (Neochrome Blue-black R). A dyestuff. It is tetra-methyl-diamino-oxy - diphenyl - naphthyl carbinol - carboxylic acid, $C_{28}H_{27}N_2O_3Cl$. Dyes chromed wool blue, but is chiefly used for cotton printing.

Also the name for a chromium-silicon-phosphate, obtained by fusing a mixture of potassium bichromate, fluorspar, and silica.

Chrome Bordeaux. A mordant dye-stuff, used for printing with acetate of chromium mordants, giving claret-red shades.

Chrome Bordeaux 6B Double. A dye-stuff similar to Chrome bordeaux.

Chrome Bronze. Crystalline chromium oxide obtained by heating potassium bichromate with sodium chloride or in a stream of hydrogen. It has a metallic sheen, a specific gravity of 5·61, and cuts glass.

Chrome Brown. Manganese chromate.

Chrome Brown P. A dyestuff prepared by the action of picramic acid upon m-amino-phenol.

Chrome Brown PA. A dyestuff. It is p-nitro-benzene-azo-pyrogallol.

Chrome Brown RO. See FAST BROWN N.

Chrome Brown RR. A dyestuff. It is the sodium salt of disulpho-oxy-ben-zene-azo-pyrogallol, $C_{12}H_8N_2O_{10}S_2Na_2$. Dyes chrome wool brown, and cotton reddish-brown with a chrome mordant.

Chrome Carmine. See CHROME RED.

Chrome Cement. A strong solution of gelatin, to which has been added 1 part of a solution of acid chromate of lime to every 5 parts gelatin. Used as a glass cement.

Chrome Cinnabar. See CHROME RED.

Chrome Die Steel. See CHROME STEELS.

Chrome-Diopside. A mineral. It is $4[MgCaSi_2O_6]$.

Chromeduol. A concentrated chrome tanning powder. (598).

Chrome Emerald Green. See CHROMIUM GREEN.

Chrome Fast Black A. A dyestuff. It is a British brand of Eriochrome black A.

Chrome Fast Cyanine G (Neochrome Blue-black B, Solochrome Black 6B, Fast Chrome Cyanine, Alliance Chrome Blue-black B, Eriochrome Blue-black B). A dyestuff prepared from diazo-tised 1 - amino - 2 - naphthol - sulphonic acid, and α-naphthol.

Chrome Fast Black FW. A British dyestuff. It is equivalent to Diamond black F.

Chrome Fast Brown FC. A dyestuff. It is a British equivalent of Anthracene brown R.

Chrome Fast Cyanine B, BN. Dye-stuffs. They are British brands of Palatine chrome blue.

Chrome Fast Yellow. See MILLING YELLOW.

Chrome Fast Yellow G. A dyestuff. It is the sodium salt of diphenyl-pheno-triazine - azo - salicylic - di (and tri-)-sulphonic acid. Dyes chromed wool yellow.

Chrome Fast Yellow O. A dyestuff. It is a British equivalent of Milling yellow.

Chrome Garnet. See CHROME RED and UVAROVITE.

Chrome Glue. Either a mixture of glue with potassium or ammonium bi-chromate, or of glue with chrome alum. On drying these mixtures, they become very insoluble, and are used as cements for glass, for waterproofing materials, and for photographic purposes.

Chrome Greens. Tetramethyl-diamino-triphenyl-carbinol-m-carboxylic acid, $C_{24}H_{25}N_2O_3$, is known as Chrome green. It dyes chromed wool green, also used in cotton printing.

Also, under the names of Chrome green, Milory green, German green, Green vermilion, Printing green, Satin green, Silk green, Hooker's green, Milori green, Brunswick green, Cinna-bar green, and New Brunswick greens, are mixtures of Prussian blue with Chrome yellow, and sometimes with barytes sold as pigments. Used for wall-paper colour, and for lithographic colour.

Formerly, chromium sesquioxide, Cr_2O_3, was known as Chrome green. It was used as a pigment, and for imparting a green colour to glass. Other names for this oxide are Ultramarine green, Green cinnabar, Oil green, and Leaf green. It is now sold as chrome oxide green. Also see CHROMIUM GREEN.

Chrome IA. An alloy of 80 per cent. nickel and 20 per cent. chromium.

Chrome Iron (Ferrochrome). An alloy of iron and chromium, usually containing from 62–68 per cent. chromium. Used in the manufacture of chrome steel.

Chrome Iron Ore (Chromite, Ferro-chromite, Chrome Iron Stone, Chromo-ferrite, Siderochrome). An ore analo-gous to magnetic iron ore. Its formula is Cr_2FeO_3, but the chromium oxide is always partially replaced by ferric oxide and alumina, and the iron oxide by magnesium oxide.

Chrome Iron Stone. See CHROME IRON ORE.

Chromel. Alloys made in three qualities, (a) containing 80 per cent. nickel and 20 per cent. chromium ; (b) consisting of 85 per cent. nickel and 15 per cent. chromium ; and (c) containing 64 per cent. nickel, 11 per cent. chromium, and 25 per cent. iron. They are used for heating elements, and as moulds for glass.

Chromel D. An alloy of 66 per cent. iron, 26 per cent. nickel, and 8 per cent. chromium.

Chromel P. A heat-resisting alloy con-taining 90 per cent. nickel and 10 per cent. chromium.

Chrome-nickel Steel, High. These alloys usually contain 70–75 per cent. iron, 17–20 per cent. chromium, and 8–10 per cent. nickel.

Chrome Ochre (Araginite). A yellowish-green earthy deposit found in the Shet-land Islands.

Chrome Orange. See CHROME RED.

Chrome Oxide Green. See CHROME GREENS.

Chrome Patent Blacks TG, TB, T, TR. Wool dyestuffs which are em-ployed in an acid bath, the fabric being then treated with a 1·5 per cent. solu-tion of potassium bichromate.

Chrome Patent Green A and N. A dyestuff. It is the sodium salt of ben-zene - azo - amino - naphthol - disul - phonic acid - azo - α - naphthalene - azo - salicylic acid, $C_{33}H_{21}N_7O_9S_2Na_2$. Dyes wool by the chrome method, a dark bluish-green.

Chrome-Pistazite. A mineral. It is $Ca_2(Fe,Al,Cr)_3Si_3O_{12}OH$.

Chrome Prune. A mordant dyestuff. It gives claret shades with chrome mor-dants.

Chrome Red (Austrian Cinnabar, Chinese Red, Persian Red, Victoria Red, Vienna Red, Derby Red, Chrome Cinnabar, Chrome Orange, American Vermilion, Chrome Garnet, Chrome Ruby, Chrome Carmine, Chinese Scarlet). Pigments consisting of basic lead chromate, $PbCrO_4.Pb(OH)_2$, or a mixture of this substance with chrome yellow.

Chrome Red R. A mordant dyestuff, used in calico printing, with chromium acetate. It gives red shades.

Chrome Ruby. See CHROME RED.

Chrome Steels. These steels represent a range of alloys containing from 0·2–2 per cent. carbon and from 0·5–15 per cent. chromium. Ball-bearing steel contains 1 per cent. carbon and 1 per cent. chromium, chrome die steel con-tains 2 per cent. carbon and 12 per cent. chromium, and stainless steel 0·2–0·4 per cent. carbon and 11–13 per cent. chromium. Stainless steel is un-affected by acetic, boric, benzoic, picric, oleic, pyrogallic, tannic, and stearic acids, by fruit juices, milk, paraffin, petrol, camphor, inks, benzol, and ammonia. It is affected by carbolic, chloracetic, lactic, oxalic, sulphuric, tartaric, citric, hydrochloric, hydro-fluoric, hydrocyanic, and malic acids, also by ammonium sulphate, bromine water, iodine, mercuric chloride, potas-sium bromide, sodium oxide, sodium sulphate, ferric chloride, ammonium chloride, and aluminium sulphate.

Chrome Steels, High. These usually contain from 64–81 per cent. iron and 18–35 per cent. chromium.

Chrometan. A chrome used in tanning. It contains oxide of chromium.

Chrome-tin Pink. Obtained by calcin-ing a mixture of stannic oxide and a small amount of chromic oxide.

Chrome Violet. See MAUVEINE.

Chrome Violet. The Bayer Co. A dye-stuff. It consists of tetramethyl-dia-mino - oxy - triphenyl - carbinol - m - carboxylic acid, $C_{24}H_{25}N_2O_4$. Dyes chromed wool violet, and is used in calico printing.

Chrome Violet. Geigy. A dyestuff. It is the sodium salt of aurine-tricar-boxylic acid, $C_{22}H_{13}O_{10}Na_3$. Gives a reddish-violet colour in calico printing with chrome mordants.

Chrome Yellow. A pigment. It is normal lead chromate, $PbCrO_4$, specific gravity of 5·8. The compounds, $PbCrO_4.PbSO_4$ and $PbCrO_4.2PbSO_4$, are also known as chrome yellow.

The lower qualities of Chrome yellow, also called Lemon yellow, Paris yellow, Leipsic yellow, Lemon chrome, Royal yellow, New yellow, Cologne yel-low, American chrome yellow, French chrome yellow, Baltimore chrome yel-low, Imperial yellow, King's yellow, Canary yellow, and Zwickau yellow, are mixtures of Chrome yellow with gypsum, heavy spar, china clay, whit-ing, and kieselguhr. Also see BARIUM YELLOW.

Chrome Yellow D. See MILLING YELLOW.

Chrome Yellow GG. A dyestuff pre-pared from diazotised o-anisidine and salicylic acid.

Chromglaserite. A double salt, $3K_2CrO_4.Na_2CrO_4$.

Chromic Anhydride. Chromic acid is sometimes called by this name.

Chromidium. A nickel-chromium cast iron.

Chromiform. A reagent used for the preservation of milk samples. It consists of pastilles containing 0·25 gram potassium bichromate and 0·25 gram trioxymethylene.

Chromine G. See THIOFLAVINE S.

Chromite. A mineral. It is $8[FeCr_3O_4]$. See CHROME IRON ORE.

Chromitite. A mineral, $(Fe.Al)_2O_3.2Cr_2O_3$, of Serbia. It has magnetic properties.

Chromium Bronze. The term applied to copper-zinc or copper-tin alloys to which chromium has been added up to 5 per cent.

Chromium - cobalt - molybdenum Steel. See COBALT-CHROMIUM-MOLYBDENUM STEEL.

Chromium Copper. An alloy of copper and chromium, containing 10 per cent. chromium. Used in the manufacture of hard steels. Also added to increase elasticity.

Chromium Green (Guignet's Green, Pannetier Green, Arnaudon Green, Matthieu-Plessy Green, Chrome Emerald Green, Mittler's Green, Permanent Green, Emerald Green, Veridian, Chrome Green, Grüne's chromoxyd, French Veronese green). Green pigments consisting of hydrated sesquioxide of chromium, with phosphate or borate of chromium. Also see CHROME GREENS.

Chromium Manganese. An alloy containing 30 per cent. chromium and 70 per cent. manganese. Used in the manufacture of hard steels. Also added to copper to increase elasticity.

Chromium Mica. Fuchsite (q.v.).

Chromium Molybdenum. An alloy of 50 per cent. chromium and 50 per cent. molybdenum. Used in the manufacture of hard steels.

Chromium-molybdenum Steel. Alloys containing from 0·06–1·2 per cent. carbon, traces to 15 per cent. molybdenum, and traces to 6 per cent. chromium.

Chromium Nickel. An alloy of 10 per cent. chromium and 90 per cent. nickel, also 50 per cent. chromium, and 50 per cent. nickel. Used in the manufacture of hard steels.

Chromium - nickel - copper. See NICKEL-CHROMIUM-COPPER.

Chromium - nickel - molybdenum Steels. See NICKEL-CHROMIUM-MOLYBDENUM STEELS.

Chromium-nickel Steels. See NICKEL-CHROMIUM STEELS.

Chromium Oxide, Green. See GREEN OXIDE OF CHROMIUM.

Chromium Oxide, Red. See RED OXIDE OF CHROMIUM.

Chromium Sandoz. A proprietary preparation of chromium sesquioxide. (267).

Chromium Steels. See CHROME STEELS.

Chromium - vanadium -molybdenum Steels. Alloys of iron containing 0·1–0·55 per cent. carbon, 0·22–1·45 per cent. molybdenum 0·8–1·5 per cent. chromium, and 0·15–0·45 per cent. vanadium.

Chromium-vanadium Steel. An alloy of this type contains 0·3–0·4 per cent. carbon, 1–1·5 per cent. chromium, and 0·15–0·25 per cent. vanadium. Used in the manufacture of torpedo tubes and driving axles. Also for making springs, tools, and locomotive forgings.

Chromo-acetic Acid Solution. A microscopic fixing solution for plant material. It contains 1 gram chromic acid, 0·5 c.c. glacial acetic acid, and 98·5 c.c. water.

Chromo-aceto-osmic Acid. Fleming's solution (q.v.).

Chromobrugnatellite. Stichtite (q.v.).

Chromochlorite (Rhodochrome). A mineral. It is a variety of Pennine.

Chromocyanines. See INDALIZARIN.

Chromocyanine V and B (Blue PRC, Brilliant Gallocyanine). Sulphonic acids of leuco-gallocyanines. They dye violet to blue shades with chrome mordants. Employed in calico printing.

Chromocylite. A mineral. It is a variety of apophyllite.

Chromoferrite. Chrome iron ore (q.v.).

Chromoform. A combination of methylhexamethylene-tetramine and dichromic acid, $(C_6H_{12}N_4CH_3)_2Cr_2O_3$. A urinary antiseptic.

Chromogen C, LL. See CHROMOTROPE ACID.

Chromogen I. See CHROMOTROPE ACID.

Chromohercynite. A spinel mineral, $FeCr_2O_4(Fe.Mg.Mn)Al_2O_3$, found in Madagascar.

Chromol. Metachrome dyestuffs. (599).

Chromol Paints. Paints prepared from green oil (made by adding chromyl chloride, dissolved in carbon tetrachloride, to linseed oil), siccatives, and pigments.

Chromo-nitric Acid. See PERENYI'S FLUID.

Chromopicotite. A mineral. It is a variety of chromite (chrome iron ore).

Chromo Santonin (Golden Santonin). A modification of ordinary santonin, formed by exposing it to sunlight. Given in sprue.

Chromosone Red. Polychrome dye for cytology. (544).

Chromospinel. See PICOTITE.

Chromotrope Acid (Chromogen I, Chromogen C, LL). Dioxy-naphthalene-disulphonic acid. An acid dye for wool.

Chromotrope 2B. A dyestuff. It is the sodium salt of p-nitro-benzene-azo-1 : 8-dioxy-naphthalene-3 : 6-disulphonic acid, $C_{16}H_9N_3O_{10}S_2Na_2$. Dyes wool from an acid bath, bluish-red, becoming blue to black on chroming.

Chromotrope 6B. A dyestuff. It is the sodium salt of p-acetamino-benzene-azo-1 : 8-dioxy-naphthalene-disulphonic acid, $C_{17}H_{13}N_3O_9S_2Na_2$. Dyes wool violet-red from an acid bath.

Chromotrope 8B. A dyestuff. It is the sodium salt of p-sulpho-naphthalene-azo - dioxy - naphthalene - disulphonic acid, $C_{20}H_{11}N_2O_{11}S_3Na_3$. Dyes wool reddish-violet from an acid bath.

Chromotrope 10B. A dyestuff. It is the sodium salt of α-naphthalene-azo-1 : 8-dioxy - naphthalene - disulphonic acid, $C_{20}H_{12}N_2O_8S_2Na_2$. Dyes wool reddish-violet from an acid bath.

Chromotrope FB. See AZORUBINE S.

Chromotrope 2R (Biebrich Acid Red 4B, Lighthouse Chrome Blue B, XL Carmoisine 6R). A dyestuff. It is the sodium salt of benzene-azo-1 : 8-dioxy-naphthalene - 3 : 6 - disulphonic acid, $C_{16}H_{10}N_2O_8S_2Na_2$. Dyes wool bluish-violet.

Chromovan Steel. A proprietary trade name for a non-sparking tool steel containing 12·5 per cent. chromium, 0·8 per cent. molybdenum, 1 per cent. vanadium, and 1·6 per cent. carbon.

Chromoxan Colours. Dyestuffs. They are aldehydes of the naphthalene series, condensed with hydroxy-acids of the benzene series, and oxidised.

Chromule. See CHLOROPHYLL.

Chronin. An alloy of 84 per cent. nickel and 15 per cent. chromium.

Chronite. Heat-resisting alloys containing 63–67 per cent. nickel, 13–16 per cent. chromium, 12–20 per cent. iron, 0–0·4 per cent. silicon, 0–1 per cent. manganese, and 0–0·8 per cent. aluminium.

Chrysamin (Flavophenin). An azo dyestuff from diazotised benzidine and salicylic acid, $(C_6H_4)_2(N_2)_2(C_6H_3)_2$ $(OH)_2(COOH)_2$. The commercial product is the sodium salt, and is used for dyeing cotton goods yellow.

Chrysamin G (Erie Yellow XC, Enbico Direct Yellow G). A dyestuff. It is the sodium salt of diphenyl-disazo-bi-salicylic acid, $C_{26}H_{16}N_4O_6Na_2$. Dyes cotton yellow from a soap bath. See CHRYSAMIN.

Chrysamin R. A dyestuff. It is the sodium salt of ditolyl-disazo-bi-salicylic acid, $C_{28}H_{20}N_4O_6Na_2$. Dyes cotton yellow from a soap bath.

Chrysaniline. See PHOSPHINE.

Chrysarobin, Crude. See GOA POWDER.

Chrysaureine. See ORANGE II.

Chrysazin. Dioxy - anthraquinone, $C_6H_3(OH) : C_2H_2 : C_6H_3OH$.

Chrysazol. Dioxy-anthracene, $C_{14}H_8$ $(OH)_2$.

Chryseine. An iodine derivative of urotropine. An antiseptic.

Chryseinic Acid. See CAMPOBELLO YELLOW.

Chryseoline. See TROPÆOLINE O.

Chryseoline Yellow. See TROPÆOLINE O.

Chryseone (Silicone). A compound, $Si_4H_4O_3$, obtained by digesting calcium silicide with concentrated hydrochloric acid in the absence of light.

Chrysil. A rubber obtained from *Chrysothamus nauseosus*.

Chrysin. 1 : 3-Dihydroxy-flavone. A pigment found in poplar buds.

Chrysine. See OZOKERINE.

Chrysinic Acid. Chrysin, $C_{15}H_{10}O_4$.

Chrysitin. Massicot (*q.v.*).

Chrysoberyl. A mineral. It is beryllium aluminate, $Al_2O_3.BeO$.

Chrysocale. A jeweller's alloy, containing 9 parts copper, 8 parts zinc, and 2 parts lead.

Chrysochalk (Gold-copper). An alloy containing 59–93 per cent. copper, 8–39 per cent. zinc, and 1·6–1·9 per cent. lead. A jeweller's alloy.

Chrysocolla (Cyanochalcite, Bogoslovskite). A copper ore. It is a hydrated silicate, $CuO.SiO_2.2H_2O$.

Chrysofluorine. Naphthylene-phenylene-methane, $C_{17}H_{12}$.

Chrysoform. Dibromo-diiodo-hexamethylene-tetramine, $C_6H_8Br_2I_2N_4$. Antiseptic.

Chrysoharmine. Nitro-harmaline, $C_{13}H_{13}(NO_2)H_2O$.

Chrysoidine (Chrysoidine G, Pure Chrysoidine RD, YD, Chrysoidine R, R Powder, R Crystals, Y Powder, Y Crystals, YRP, YL, RL, Y, 1606, Pure Crystals, Supra Crystals). Diamino-azo-benzene, $C_6H_5.N_2.C_6H_3.NH_2.$ NH_2. Used in combination with safranines to produce a scarlet on mordanted cotton.

Chrysoidine Crystal. A dyestuff. It consists of the hydrochloride of benzene-azo-m-phenylene-diamine, with some of the homologues from o- and p-toluidine. Dyes wool and silk orange.

Chrysoidine G. See CHRYSOIDINE.

Chrysoidine R. Durant, Hugenin & Co. A dyestuff. It is the hydrochloride of toluene-azo-m-tolylene-diamine, $C_{14}H_{17}N_4Cl$. Dyes tannined cotton brown.

Chrysoidine R (Cerotine Orange C Extra, Gold Orange for Cotton). Read, Holliday & Sons, Williams Bros., Levenstein, Ltd., Geigy, Société pour l'Industrie Chimique. A dyestuff. It is the hydrochloride of benzene-azo-*m*-tolylene-diamine, $C_{13}H_{15}N_4Cl$.

Chrysoidine R, R Powder, R Crystals, Y Powder, Y Crystals, YRP, YL, Y, RL, Y, 1606, Pure Crystals, Supra Crystals. Dyestuffs. They are British equivalents of Chrysoidine.

Chrysoidine Y. A dyestuff. It is the hydrochloride of diamino-azo-benzene, $C_{12}H_{13}N_4Cl$. Dyes wool, silk, and tannined cotton orange.

Chrysoine. See TROPÆOLINE O.

Chrysoine Extra. A dyestuff. It is a British equivalent of Tropæoline O.

Chrysoline. The sodium salt of benzylfluoresceine. Dyes silk yellow. See TROPÆOLINE O.

Chrysolite. See OLIVINE.

Chrysophanic Acid. Dioxy-methylanthraquinone, $C_{15}H_{10}O_4$.

Chrysophenine. A dyestuff. It is the sodium salt of disulpho-stilbene-disazo-phenetol - phenol, $C_{28}H_{22}N_4O_8S_2Na_2$. Dyes cotton or wool yellow from a neutral or acid bath, and silk from an acetic acid bath.

Chrysoprase. An apple-green or grass-green variety of quartz.

Chrysorin. An alloy of 66 per cent. copper and 34 per cent. zinc.

Chrysosulphite. A mixture of magnesium picrate with sodium sulphate. A red filter for photographic purposes.

Chrysotile (Chrysotile Asbestos, Canadian Asbestos). An asbestos mineral, the average composition of which is 40·5 per cent. silica, 41·5 per cent. magnesium oxide, 14 per cent. water, 2·6 per cent. iron oxides, and 1·3 per cent. aluminium oxide. It yields the best fibre for spinning, and is considered the best type of asbestos. It is a fibrous form of serpentine (hydrous magnesium silicate).

Chrysotile Asbestos. See CHRYSOTILE.

Chrysyl. Zinc-boro-picrate. Antiseptic.

Chubutite. A mineral, $14PbO.2PbCl_2$.

Chufa Oil. An oil obtained from the tubers of *Cyperus esculentus*. It contains 73 per cent. glycerol oleate and 9 per cent. glycerol palmitate.

Chunam. A rendering made in India by burning Kunkur (*q.v.*), mixing with sand, water, and a little molasses.

Chu-Pax. A proprietary preparation of aspirin in a chewable base. An analgesic. (385).

Churchill's Iodine Caustic. This solution contains iodine, potassium iodide, and water.

Churchite. A mineral. It is a hydrated phosphate of cerium, containing 52 per cent. cerium oxide, Ce_2O_3. It is also given as $4[(Yt,Er)PO_4.2H_2O]$.

Chute's Solvent. A U.S. patented rubber resin solvent consisting of a mixture of methyl acetone and acetone.

Chymacort. A proprietary preparation of hydrocortisone acetate, neomycin palmitate and pancreatic enzymes, used in dermatology. (327).

Chymar. A proprietary preparation of alpha-chymotrypsin. Treatment of wounds. (327).

Chymar Ointment. A proprietary preparation of hydrocortamate hydrochloride, neomycin palmitate and pancreatic enzymes, used in dermatology. (327).

Chymocyclar. A proprietary preparation of tetracycline, trypsin and chymotrypsin. Antibiotic. (327).

Chymoral. A proprietary preparation of trypsin and chymotrypsin. Digestive enzymes. (327).

Chymosin. Pepsin.

Chymotrypsin. An enzyme obtained from chymotrypsinogen by activation with trypsin. α-Chymotrypsin. Avazyme; Chymar; Chymo-Trypure. Present in Chymacort and Chymoral.

Chymo-Trypure. A proprietary name for chymotrypsin.

CI 336. Carbochloral.

Cialit. The sodium salt of 2-ethyl mercury mercapto benzoxazol-5-carboxylic acid. An antifouling pigment.

Ciba. Indigoid vat dyestuff. (530).

Ciba 1906. A proprietary preparation containing thiambutosine. (18).

Ciba Blue B. A dyestuff. It is 5 : 7 : 5'-tribrom-indigo.

Ciba Blue 2B. A dyestuff. It is tetra-brom-indigo.

Ciba Bordeaux. A dyestuff. It is 5 : 5'-dibrom-thio-indigo.

Cibacet. Dispersive dyes. (530).

Cibacrolan. Reactive dyestuffs. (530).

Ciba Green G. A dyestuff. It is dibrom-β-naphthindigo.

Ciba Heliotrope B. A dyestuff. It is tetrabrom-indirubin.

Cibalan. Wool dyestuffs. (530).

Cibalgin. Dial (diallylbarbituric acid) and amidopyrine. Analgesic and sedative without narcotic alkaloids.

Cibanite. A proprietary trade name for an aniline-formaldehyde resin resistant to water, oil, alkalis, and organic solvents.

Cibanoid. A proprietary trade name for urea-formaldehyde synthetic resins.

Cibanone Black B, 2B. A dyestuff prepared by fusing 2-methyl-benzanthrone with sulphur.

Cibanone Blue 2G. A dyestuff prepared in a similar way to Cibanone black.

Cibanone Brown. A dyestuff. It is amino-2-methyl-anthraquinone fused with sulphur.

Cibanone Green. A dyestuff prepared as Cibanone black.

Cibanone Orange R. A dyestuff prepared from dichloro-methyl-anthraquinone.

Cibanone Yellow R. A dyestuff prepared from mono-chloro-methyl-anthraquinone.

Cibaphasol 6042. A proprietary trade name for a fatty acid amide derivative giving improved quality and appearance to continuous dyeings. Used as a dyeing auxiliary for polyacrylonitrile fibres. (18).

Ciba Red B. A dyestuff. It is 6 - 6'-dichloro-thio-indigo.

Ciba Red G. A dyestuff. It is dibromo-thio-indigo-scarlet.

Ciba Scarlet G. A dyestuff obtained by the condensation of thio-indoxyl with acenaphthene-quinone.

Cibatex PA. A proprietary trade name for a synthetic tanning agent of the phenol sulphonic acid type used as an improver of wet fastness in the dyeing of nylon. (18).

Cibatex 248. A proprietary trade name for a phenol sulphonic acid derivative which gives improved wet fastness properties of dyeings. Used also as a synthetic tanning agent for dyes on nylon 66. (18).

Ciba Violet A, B, and R. A dyestuff. It is halogenated thio-indigo.

Ciba Yellow G. A dyestuff. It is the dibromo derivative of Indigo yellow 3G.

Ciba Yellow 5R. A dyestuff prepared from Ciba yellow G by reduction.

Cibrola. A preparation of milk and glycerophosphate, similar to Sanatogen.

Cicatricine. A solution containing 20 parts thiosinamine, 33 parts antipyrine, and 0·65 part eucaine lactate per 100 parts.

Cicatrin. A proprietary preparation containing neomycin sulphate, bacitracin, di-threonine, l-cysteine and glycine. Local antibiotic. (247).

Ciclazindol. An anti-depressant. 10 - m - Chlorophenyl - 2, 3, 4, 10 - tetra - hydropyrimido[1, 2 - a] - indol - 10 - ol. "Wy 23409" is currently under clinical trial as the hydrochloride.

Ciclopirox. A fungicide. 6 - Cyclo - hexyl - 1 - hydroxy - 4 methyl - 2 - pyridone.

Cicloprofen. An anti-inflammatory currently undergoing clinical trial as "SQ

20824". It is 2 - (2 - fluorenyl)propionic acid.

Cicloxolone. A preparation for the treatment of gastric ulcers. 3β - (cis - 2 - Carboxycyclohexylcarbonyloxy) - 11 oxo - 18β - olean - 12 en - 30 - oic acid. "BX 363A" is currently under clinical trial as the disodium salt.

Cicuta. Conium.

Cicutine. See CONIA.

Cider. See APPLE WINE.

Cider and Perry. Jam of apples and pears formerly allowed to ferment with yeasts occurring naturally on the surface of the fruits. Modern practice uses pure cultures of appropriate yeasts to give standard products.

Cider Vinegar. A vinegar containing a proportion of malic acid, with usually 3–6 per cent. acetic acid.

Cidomycin Injectable. A proprietary preparation containing gentamicin sulphate. An antibiotic. (307).

Cidrase. Cider yeast.

Cieddu. A Sardinian fermented milk, similar to Yogurt ($q.v.$).

Ciempozuelite. A mineral. It is a sulphate of calcium and sodium.

Cignolin. 1 : 8-Dihydroxy-anthranol. Used in medicine.

Ciment Fondu. A trade mark. A typical cement contains 44 per cent. alumina, 40 per cent. lime, 10 per cent silica and 3·5 per cent. metallic iron with small amounts of magnesia and ferrous oxide. It is stated to be resistant to sea water. (234).

Cimet. Stainless alloys containing about 48 per cent. chromium with iron.

Cimetidine. An H_2-receptor histamine antagonist currently undergoing clinical trial as "SK & F 92334". 1 - Methyl - 3 - [2 - (5 - methylimidazol - 4 - yl-methylthio)ethyl]guanidine - 2 - car - bonitrile.

Cimmol. Cinnamyl hydride.

Cimolite. Purified Fuller's earth ($q.v.$).

Cinchaine. Isopropyl-hydro-cupreine. A local anæsthetic.

Cinchona Bark. See COUNTESS POWDER, JESUITS' BARK, LOXA BARK, LUGO'S POWDER, PERUVIAN BARK, and TALBOR'S POWDER.

Cinchona, Pale. The bark of *Cinchona officinalis*.

Cinchona, Red. The bark of *Cinchona succirubra*.

Cinchona, Yellow. The bark of *Cinchona calisaya*.

Cinchonidene. Cinchene, $C_{19}H_{20}N_2$.

Cincophen (Phenyl-Cinchoninic Acid). 2-Phenyl-quinoline-4-carboxylic acid $C_{16}H_{11}O_2N$, melting-point 210° C.

Cindal. An aluminium alloy containing zinc and small amounts of magnesium and chromium.

nder Pig. A pig iron made from tap cinder, and other forms of slag. It is usually high in phosphorus.

nelin. A proprietary trade name for cellulose acetate and nitrate moulding powder bases.

nene. See DIPENTENE.

neol. Eucalyptol, $C_{10}H_{18}O$.

nepazet. A drug used in the treatment of angina pectoris. Ethyl 4 - (3, 4, 5 - trimethoxycinnamoyl)piperazin - 1 - ylacetate.

inepazide. A peripheral vasodilator. 1 - Pyrrolidin - 1 - ylcarbonylmethyl - 4 - (3, 4, 5 - trimethoxycinnamoyl) - piperazine.

inereine. An induline dyestuff obtained from azoxy-aniline, aniline hydrochloride, and p-phenylene-diamine.

innabar (Chinese Red, Mercury Blend, Vermilion, Liver Ore, Patent Red, Cinnabarite). A mineral. It is mercuric sulphide, HgS, used as a pigment.

innabar, Austrian. See CHROME RED.

innabar, Chrome. See CHROME RED.

innabar Green. See CHROME GREENS.

innabar, Imitation (Bluish and Yellowish Cinnabar Substitutes). Red lead coloured with Rose bengal, or a mixture of this dye with some Cochineal scarlet 2R.

innabarite. See CINNABAR.

innabar of Antimony. See RED ANTIMONY.

innabar Red. See COTTON SCARLET.

innamal. Cinnamic aldehyde, C_6H_5 CH : CHCHO.

innamein. A term applied to benzyl cinnamate. It is, however, also used for the mixture of ester and alcohol in the Balsams of Tolu and Peru, which are not extractable by alkalies from an ethereal solution.

innamene. See STYROL.

innamol. See STYROL.

innamon (Ceylon Cinnamon). The dried inner bark of the shoots of *Cinnamomum Zeylanicum*.

innamon Bark, Wild. See CANELLA.

innamon Brown. See BISMARCK BROWN.

innamon, Chinese. See CASSIA BARK.

innamon Leaf Oil. See OIL OF CINNAMON LEAF.

innamon Oil. See OIL OF CINNAMON.

innamon Stone. See HESSONITE.

innamon Water. (*Aqua cinnamomi B.P.*) Cinnamon bark (100 grams), is added to 2,000 c.c. water, and the solution distilled to 1,000 c.c.

innarizine. 1 - trans - cinnamyl - 4 - diphenylmethylpiperazine. Mitronal.

inoxacin. An anti-bacterial currently undergoing clinical trial as " 64 716 ".

It is 1 - ethyl - 4 - oxo [1, 3] dioxolo - [4, 5 - g] cinnoline - 3 - carboxylic acid.

Cinoxolone. A preparation for the treatment of gastric ulcers, currently undergoing clinical trial as " BX 311 ". It is cinnamyl 3β - acetoxy - 11 - oxo - 18β - olean - 12 - en - 30 - oate.

Cinquinolite. See WHITE COPPERAS.

Ciodrin. A proprietary preparation of CROTOXYFOS. An insecticide.

Ciose. A dry soluble protein product of beef, containing 83–85 per cent. protein.

Ciplyte. A mineral, $4CaO.2P_2O_5.SiO_2$.

Cipô do Cabodô (Capa Homen). A preparation of *Davilla rugosa*, a Brazilian plant. It is produced by extracting the leaves with water and evaporating. It has analgesic properties.

Circladin. Bromindione.

Cirin. A proprietary preparation of aspirin and ascorbic acid 6–1. An analgesic. (885).

Cirrasol. A textile softening agent, fibre lubricant or antistatic agent. (512).

Cirrasol PAC. An aqueous cationic dispersion used for the production of full soft finishes on natural and man-made fibres, when application is by padding and exhaustion methods. (512).

Cirrosol PB. As above. Particularly recommended for the finishing of dyed nylon/terylene half hose which are later post boarded to minimize reductions in the rubbing, wet and perspiration fastness of shades. (503).

Cirrolite. A mineral, $Al_2Ca_3(OH)_3$ $(PO_4)_3$.

Cisdene. A proprietary synthetic rubber. (499).

Cisium. An alloy in which aluminium, zinc, tin, copper, as well as traces of antimony and arsenic are found.

Citalo. A proprietary safety glass.

Citanest. A proprietary preparation of prilocaine hydrochloride. A local anæsthetic. (338).

Citarin (Goutine, Citromin). Sodium-anhydro - methylene - citrate, CH_2O. $O.CO.C(CH_2COONa)_2$. It dissolves uric acid, and is used as a remedy for gout.

Cithrol. Surface active agents. (526).

Citobaryum. A proprietary preparation of special barium sulphate.

Citragan. A proprietary preparation of silver-sodium citrate.

Citra-Gran. A proprietary preparation containing tartaric acid sodium bicarbonate, citric acid, potassium bicarbonate, magnesium sulphate, lithium benzoate, anhydrous sodium phosphate and calcium lactophosphate. An antacid. (294).

Citralka. A proprietary preparation containing sodium acid citrate. (264).

Citramine. See HELMITOL.

Citraminoxyphen. See HELMITOL.

Citranova. Synthetic citrus oils. (600).

Citrate of Magnesia, Effervescent. See EFFERVESCENT CITRATE OF MAGNESIA.

Citrene. See HESPERIDENE.

Citrex. Trade mark for a range of surfactants derived from sulphopolycarboxylic acids. (500).

Citridic Acid. See EQUISETIC ACID.

Citrin. Vitamin P. See QUERCITIN.

Citrine (Yellow Quartz, Oriental Topaz, False Topaz). A mineral. It is a clear yellow variety of quartz and contains ferric oxide.

Citrine Ointment (Citron Ointment). Mercuric nitrate ointment.

Citrocoll. Claimed to be a neutral citrate of amino-acet-p-phenetidin, $(C_6H_4 . OC_2H_5 . NH . CO . CH_2 . NH_2)_3 . C_6H_8O_7$. Antipyretic and analgesic.

Citroflex 2. A trade mark for triethyl citrate. A vinyl plasticiser. (85).

Citroflex 4. A proprietary trade name for tributyl citrate. A vinyl plasticiser. (85).

Citroflex A-2. A proprietary trade name for acetyl triethyl citrate. A vinyl plasticiser. (85).

Citroflex A-4. A proprietary trade name for acetyl tributyl citrate. A vinyl plasticiser. (85).

Citroflex A-8. A proprietary trade name for acetyl tri-2-ethyl hexyl citrate. A vinyl plasticiser. (85).

Citromin. See CITARIN.

Citronella Oil. See OIL OF CITRONELLA.

Citronine. See NAPHTHOL YELLOW S.

Citronine A. See NAPHTHOL YELLOW S.

Citronine B or 2B (Azo Yellow, Azo Yellow M, Azo Yellow 3G Extra Conc., Azo Acid Yellow, Azoflavine S or 2, Azoflavine 3R Extra Conc., Jaune Yellow, Indian Yellow, Indian Yellow G, New Yellow, Citronine 2AEJ, Citronine NE, Helianthin, Jasmin). An azo dyestuff. It consists of a mixture of the nitro derivatives of p-sulphobenzene-azo-diphenylamine, and mono-di-tri-nitro-phenylamine. Dyes wool and silk yellow from an acid bath.

Citronine 2AEJ. See CITRONINE B.

Citronine NE. See CITRONINE B.

Citron Ointment. See CITRINE OINTMENT.

Citron Yellow. See ZINC CHROME.

Citrophan Tablets. The chief constituents of these tablets are tetra-iodophenol-phthalein and lactose. Stated to be fat-reducing.

Citrophen (Phenocitrin). Phenetidine citrate, $C_3H_4(OH)(CO . NH . C_6H_4O$

$C_2H_5).H_2O$. Used medicinally in cases of fever and neuralgia.

Citrosalic Acid. Methylene-citryl salicylic acid.

Citrosodine. Sodium citrate, $Na_3C H_5O_7$.

Citrozone. Vanadium-sodium-citro chloride.

Citrullin. A resinoid from *Citrullu Colocynthis*. An external purgative.

Clair de Lune. A proprietary trade name for cellulose acetate used for decorative purposes.

Clairine. A motor fuel containing 79 per cent. alcohol (90 per cent.), 0·4 per cent. pyridine, 0·4 per cent. Simonser oil, and 20 per cent. ether. It is coloured with methyl violet.

Clamidoxic Acid. 2 - (3,4 - Dichloro benzamido)phenoxyacetic acid. SNI 1804.

Clamoxyquin. 5-Chloro-7-(3-diethyl aminopropylaminomethyl)-8-hydroxy quinoline. CI 433 is the di-hydro chloride.

Claradin. A proprietary preparation o aspirin in an effervescent base. An analgesic. (271).

Clarain (Durain, Fusain, Vitrain). Term applied to coal ingredients having diff ferent structure or appearance. Clarai consists of a band of material with a subdued lustre and containing bande dull and bright streaks. With vitrai it is considered equivalent to " bright " or " glance " coal. The ash of clarai is generally low (1–1·5 per cent.), but i sometimes as high as 3 per cent. It ha an alkaline reaction. The volatile matter from clarain is generally highe than that of other ingredients, and the percentage of clarain in the seam of coa is higher than any other. Durain is a firm, hard, dull coal with fine hair-like streaks, and is equivalent to dull, har coal. It has an ash usually from 4– per cent. which has a high value of alu minium silicate. The percentage o durain in the coal is usually somewha lower than that of clarain. Fusain i a brittle powdery material obtained by scraping at the cleavage surfaces con taining it. It is equivalent to " mothe of coal " or " mineral charcoal." I gives the highest ash of any ingredien (10–14 per cent.) which contains con siderable amounts of lime. The vola tile matter is the lowest of any ingre dient. Fusain amounts to not less tha 2–3 per cent. and may be 5–8 per cent in certain layers. Vitrain is brilliantly glossy. The ash is usually low. The volatile matter is high, but the amoun in the seam is low.

lar-Apel. A proprietary viscose plastic used as a packing material.

larax. A washing powder. It is a mixture of sodium phosphate, borax, and sodium perborate.

larenite. Black bituminous varnish. (601).

laret G. See BORDEAUX G.

laret Red. See BORDEAUX R or B.

larite. A mineral. It is cuprous sulpharsenate, Cu_3AsS_4, found in the Black Forest.

lark 1. An irritant poison gas. Diphenyl-chloro-arsine, $(C_6H_5)_2AsCl$.

lark 2. An irritant poison gas. Diphenyl-cyano-arsine, $(C_6H_5)_2AsCN$. Strongly irritating to the mucous membrane of the nose.

larkeite. A new uranium mineral containing $1 \cdot 1$ per cent. CaO, $3 \cdot 7$ per cent. PbO, $1 \cdot 4$ per cent. K_2O, $2 \cdot 6$ per cent. Na_2O, $0 \cdot 04$ per cent. BaO, $82 \cdot 7$ per cent. UO_3, $0 \cdot 5$ per cent. Fe_2O_3 and Al_2O_3, $1 \cdot 12$ per cent. rare earths, $5 \cdot 2$ per cent. H_2O, and $0 \cdot 3$ per cent. SiO_2.

lark-Lubs Indicators. Phenol red. The dry powder (10 dg.) is ground with $5 \cdot 7$ c.c. $1/20N$ caustic soda, and diluted to 25 c.c. with distilled water. Bromophenol blue : as for phenol red, but use 3 c.c. of alkali. Cresol red : as for phenol red, but use $5 \cdot 3$ c.c. alkali. Bromocresol purple : as for phenol red, but use $3 \cdot 7$ c.c. alkali. Thymol blue : as for phenol red, but use $4 \cdot 3$ c.c. alkali. Bromthymol blue : as for phenol red, but use $3 \cdot 2$ c.c. alkali. Methyl red : as for phenol red, but use $7 \cdot 4$ c.c. alkali.

lark's Patent Alloy. An alloy of 75 per cent. copper, $7 \cdot 2$ per cent. zinc, $14 \cdot 5$ per cent. nickel, $1 \cdot 9$ per cent. tin and $1 \cdot 9$ per cent cobalt.

larodene. Thermoplastic coumaroneindene resins. (601).

larostene. Thermoplastic coumaronestyrene resin. (601).

larus Metal. A light aluminium alloy used for sheets and tubes.

laudetite. A mineral. It is a monoclinic form of arsenic oxide, As_2O_3.

laudilithe. Gallalith (q.v.).

laussenite. Hydrargilite (q.v.).

lausthalite. A mineral. It is lead selenide, PbSe.

lay. The name applied to a large class of minerals, differing considerably in composition. The basis of all clay is aluminium silicate, $Al_2O_3.2SiO_2.2H_2O$.

lay Band Ores. Iron ores consisting of carbonate of iron, $FeCO_3$, mixed with considerable quantities of clayey matter. They may be regarded as clays or shales, which have become impregnated with iron carbonate.

Clay, Cornish. See CHINA CLAY.

Clay, Fat. Ball clay (q.v.).

Clay Iron Stone. See BOG IRON ORE.

Clayite. The non-crystalline part of China clay. Synonym for Kaolinite.

Clay, Pipe. See PIPECLAY.

Clay, Porcelain. See CHINA CLAY.

Clay, Potter's. Pipeclay (q.v.).

Claysil. A silicate adhesive. (546).

Clayton Black D. A dyestuff prepared from nitroso-phenol by solution in caustic soda, dilution, then mixing successively with sodium thiosulphate and 38 per cent. sulphuric acid. Filtered, and solution mixed with an alkaline solution of nitroso-phenol, then heated. The dyestuff is precipitated. It dyes cotton black in a sodium sulphide bath.

Clayton Cloth Red (Stanley Red). A dyestuff. It is the ammonium or sodium salt of sulpho-benzenyl-amino-thio-cresol-azo-β-naphthol, $C_{24}H_{21}N_4O_4S_2$. Dyes wool and silk red from an acid bath.

Clayton Cotton Brown. An azo dyestuff obtained from dehydro-thio-toluidine, naphthylamine-sulphonic acid, and phenylene-diamine. Dyes cotton brown from a salt bath.

Clayton Fast Blacks (Clayton Fast Greys). Dyestuffs which are probably sulphides or thiosulphonic acids of aniline black and its analogues. They dye cotton from a sodium sulphide bath, and when applied with caustic soda upon glucose-prepared calico, give black prints.

Clayton Fast Greys. See CLAYTON FAST BLACKS.

Clayton Wool Brown. An azo dyestuff obtained from aniline, sulphanilic acid, naphthionic acid, and m-phenylene-diamine. Dyes wool and silk brown from an acid bath.

Clayton Yellow (Turmerine, Thiazol Yellow, Mimosa, Thiazol Yellow G, Titan Yellow). A dyestuff. It consists of the sodium salt of the diazo-amino compound of dehydro-thio-toluidine-sulphonic acid, or of the mixed diazo-amino compounds of dehydro-thio-toluidine-sulphonic acid, and primuline. Dyes cotton greenish-yellow from a salt bath.

Clay, White. See CHINA CLAY.

Cleaning Oil. A fraction of Russian petroleum distillation. See B-PETROLEUM NAPHTHA and A-PETROLEUM NAPHTHA.

Clearpol. Alginate for brewing. (536).

Clearsite. A proprietary trade name for cellulose acetate used for transparent containers.

Cleavelandite. A mineral. It is a white lamellar albite.

Clebrium Alloys. Heat-resisting alloys. One contains 76·5 per cent. iron, 13 per cent. chromium, 3·6 per cent. molybdenum, 2·6 per cent. carbon, 2 per cent. nickel, 1·5 per cent. silicon, and 0·75 per cent. manganese. Another consists of 70 per cent. iron, 18·5 per cent. chromium, 4·6 per cent. nickel, 2 per cent. copper, and 2 per cent. carbon.

Cleero. A proprietary cleaning preparation for the hair. It contains coconut oil soap, glycerin, ammonium carbonate, and water.

Clefamide. $\alpha\alpha$ - Dichloro - N - (2 - hydroxyethyl) - N - [4 - (4 - nitrophen - oxy)benzyl] acetamide. Mebinol.

Clelands Reagent. Dithiothreitol. A white powder, m.p. 37° C. A reagent for protecting —SH groups in certain biochemical systems. (182).

Clemantine. See METHYLENE VIOLET 2RA.

Clemantine Girofle. Safranine MN (*q.v.*).

Clemastine. An anti-histamine. (+) - 2 - [2 - (4 - Chloro - α - methylbenz - hydryloxy) ethyl] - 1 - methylpyrrolidine. TAVEGIL is the hydrogen fumarate.

Clemen's Solution. A solution of potassium arsenate and bromide. (*Liquor Potassi arsenatis et bromidi. B.P.C.*)

Clemizole. 1 - (4 - Chlorobenzyl) - 2 - pyrrolidinomethylbenzimidazole. Allercur is the hydrochloride.

Clemizole Penicillin. Benzylpenicillin combined with 1-(4-chlorobenzyl)-2-pyrrolidinomethylbenzimidazole. Neopenyl.

Clenesco. A proprietary trade name for anhydrous sodium metasilicate.

Clenicose. A proprietary preparation of TRYPSIN and SHYMOTRYPSIN, NEOMYCIN and hydrocortisone acetate. A topical steroid skin cream. (633).

Clenpyrin. An insecticide currently undergoing clinical trial as " FBb 6896 ". It is 1 - butyl - 2 - (3, 4 - dichlorophenylimino)pyrrolidine.

Clerici Solution. A molecular mixture of thallium malonate and formate. Used for floating mineral specimens to determine the specific gravity.

Clerit. A horticultural fungicide. (511).

Clerite. A horticultural fungicide. (511).

Cleritex. A textile cleaning agent. (508).

Cleroxide. A stabilizer for polyvinyl chloride. (583).

Cletoquine. An anti-inflammatory. 7 - Chloro - 4 - [4 - (2 - hydroxyethyl - amino) - 1 - methylbutyl]amino - quinoline.

Cleveland Pig. A pig iron made in the Middlesbrough district, from Cleveland ores. It contains about 1 per cent phosphorus and 2·5 per cent. silicon.

Cleveite (Broggerite). Names applied to certain varieties of pitch-blende rich in rare earths.

Cleve's Acid. Sulphonic acids of α - naphthylamine, α-naphthol, and nitro naphthalene. β-Acid is α-naphthyl amine-sulphonic acid, $C_{10}H_6(SO_3H)(NH_2)$ 1 : 6, also nitro-naphthalene sulphonic acid, $C_{10}H_6(SO_3H)(NO_2)$ γ-Acid is α-naphthylamine-sulphonic acid, $C_{10}H_6(SO_3H)(NH_2)$ 1 : 3 ; θ-Acid or J-acid is α-naphthylamine-sulphonic acid 1 : 7 ; and θ- or δ-acid is nitro naphthalene-sulphonic acid 1 : 7.

Clevite. A registered trade mark. (843)

Cliché Metal. An alloy of 33 per cent tin, 46 per cent. lead, and 21 per cent cadmium. Another alloy called by this name contains 48 per cent. tin 32·5 per cent. lead, 9 per cent. bismuth and 10·5 per cent. antimony.

Clichy White. A white lead made at Clichy.

Clidinium Bromide. 3-Benziloyloxy 1-methylquinuclidinium bromide.

Cliftonite. A cubic form of graphitic carbon, found in meteoric iron.

Climatone. A proprietary preparation of ethinoloestrodiol and methyl testosterone used in the treatment of pre-menstrual menopausal symptoms (330).

Climax. A magnetic alloy containing 30 per cent. nickel and 70 per cent.iron. An alloy stated to contain 73 per cent. iron, 24·4 per cent. nickel, and 2·6 per cent manganese has also been called Climax.

Climax 193. An alloy of 68 per cent iron, 28 per cent. nickel, 2 per cent chromium, and 1 per cent. manganese.

Climax Sugars. A dark-coloured syrup mainly consisting of dextrose. Used by brewers for colouring porter.

Climeline. A sodium phosphate used as a water softener.

Clindamycin. An antibiotic. Methyl 7 - chloro - 6, 7, 8 - trideoxy - 6 - (*trans* 1 - methyl - 4 - propyl - L - 2 - pyrro lidinecarboxamido) - 1 - thio - L *threo* - α - D - galacto - octopyranoside DALACIN C.

Clinimycin. A proprietary preparation of OXYTETRACYCLINE. An antibiotic (335).

Clinistix. A prepared test strip of glucose oxidase, peroxidase and *o*-toluidine used for the detection of glucose in urine. (807).

Clinitest. A prepared test tablet containing copper sulphate, sodium hydroxide, citric acid, and sodium carbonate used for the detection of reducing substances in urine. (807).

Clinitetrin. A proprietary preparation

of tetracycline hydrochloride. An anti-biotic. (335).

linium. See LIDOFLAZINE.

linker. Burnt cement produced in cement kilns is called by this name. See CEMENT CLINKER.

linoclase. See APHANESITE.

linoclastite (Albichite, Aphanesite, Siderochalcite). A mineral. It is a copper hexahydroxy-*o*-arsenate, $Cu_3(AsO_4)_2.3Cu(OH)_2$.

linoclore. A mineral. It is a hy-drated silicate of aluminium and mag-nesium, with varying amounts of iron. See CHLORITE.

linocrocite. A mineral. It is a sul-phate of aluminium, iron, sodium, and potassium.

linohedrite. A mineral, $(CaOH)(ZnOH)SiO_3$.

linostrengite. Synonym for Phospho-siderite.

linovariscite. Synonym for Metavaris-cite.

lintonite (Holmesite). A mineral, $10(Mg.Ca.Fe)O.5Al_2O_3.4SiO_2.3H_2O$.

lioquinol. 5 - Chloro - 8 - hydroxy - 7 - iodoquinoline. Iodochlorhydroxyquin; Iodochlorhydroxyquinoline. Barquinol; Chinoform; Vioform.

lioxanide. An anthelmintic. 2 - (4 - Chlorophenylcarbamoyl) - 4, 6 - di - iodophenyl acetate. TREMERAD.

liradon. See KETOBEMIDONE.

listin. The maleate of Carbinoxamine.

lobazam. A tranquilliser. 7 - Chloro - 1 - methyl - 5 phenyl - 1, 5 - benzodia - zepine - 2, 4 - dione.

Clobetasol. A corticosteroid. 21 - Chloro - 9α - fluoro - 11β, 17α - di - hydroxy - 16β - methylpregna - 1, 4 - diene - 3, 20 - dione. " GR2/925 " is currently under clinical trial as the 17 - propionate.

Clobetasone. A corticosteroid. 21 - Chloro - 9α - fluoro - 17α - hydroxy - 16β - methylpregna - 1, 4 - diene - 3, 11, 20 - trione. " GR/1214 " is currently under clinical trial as the 17-butyrate.

Clocel. Prepolymer systems for rigid polyurethane foams. (539).

Clociguanil. An anti-malarial drug. 4, 6 - Diamino - 1 - (3, 4 - dichloro - benzyloxy) - 1, 2 - dihydro - 2, 2 - di - methyl - 1, 3, 5 - triazine. " BRL 50216 " is currently under clinical trial as the hydrochloride.

Clock Brass. See BRASS.

Clofazimine. A preparation used in the treatment of leprosy. 3 - (4 - Chloro-anilino) - 10 - (4 - chlorophenyl) - 2, 10 - dihydro - 2 - isopropyliminophenazine. LAMPRENE.

Clofibrate. Ethyl-α-(4-chlorophenoxy)-α-methylpropionate). Atromid-S.

Clofluperol. A neuroleptic currently

undergoing clinical trial as " R.9298 ". It is 4 - (4 - chloro - 3 - trifluoromethyl - phenyl) - 1 - [3 - (4 - fluorobenzoyl) - propyl] piperidin - 4 - ol.

Clogestone. A progestational steroid. 6 - Chloro - 3β, 17 - dihydroxypregna - 4, 6 - diene - 20 - one. " AY 11440 " is currently under clinical trial as the 3, 17-diacetate.

Cloguanamile. An anti-malarial drug currently under clinical trial as " 57C65 ". It is 1 - amidino - 3 - (3 - chloro - 4 - cyanophenyl)urea.

Cloisonette. A proprietary trade name for cellulose acetate used for toilet ware.

Clomacran. A tranquilliser. *NN*-Di-methyl - 3 - (2 - chloro - 9, 10 - dihydro - acridin - 9 - yl) propylamine. " SK&F 14,336 " is currently under clinical trial as the phosphate.

Clomid. A proprietary preparation of clomiphene citrate. (263).

Clomiphene. 1 - Chloro - 2 - [4 - (2 - diethylaminoethoxy)phenyl] - 1,2 - diphenylethylene.

Clomipramine. An anti-depressant. 3 - Chloro - 5 - (3 - dimethylamino - propyl) - 10, 11 - dihydrodibenz[*b*,*f*] - azepine. ANAFRANIL is the hydro-chloride.

Clomocycline. N^2 - (Hydroxymethyl) chlorotetracycline. Chlormethylen - cycline. Megaclor.

Clonazepam. An anti-convulsant. 5 - (2 - Chlorophenyl) - 1, 3 - dihydro - 7 - nitro - $2H$ - 1, 4 - benzodiazepin - 2 - one. RIVOTRIL.

Clonidine. A drug used in the treatment of low blood pressure. 2-(2,6-Dichloro-anilino) - 2 - imidazoline. CATAPRES and DIXARIT are the hydrochloride.

Clonaline. A proprietary French con-denser wax. It is chlorinated naph-thalene.

Clonitazene. 2 - (4 - Chlorobenzyl) - 1 - (2 - diethylaminoethyl) - 5 - nitro - benzimidazole.

Clopaine. A proprietary trade name for the hydrochloride of cyclopentamine.

Clopamide. 4 - Chloro - N - (2,6 - dimethylpiperidino) - 3 - sulphamoyl-benzamide. Brinaldix.

Clopenthixol. A psychotropic. 2 - Chloro - 9 - {3 - [4 - (2 - hydroxyethyl) - piperazin - l - yl] - propylidene}thiaxan-thene. SORDINOL is the dihydro-chloride.

Clophen A60. Chlorinated diphenyl (60 per cent. chlorine). A plasticiser.

Clophenharz W60. Chlorinated di-phenyl or polyphenyl (60 per cent. chlorine).

Clopidol. An anti-protozoan. 3, 5 - Dichloro - 4 - hydroxy - 2, 6 - dimethyl-pyridine. COYDEN 25.

Clopirac. An anti-inflammatory currently under clinical trial as " BRL 13856 " and " CP 172 AP ". 1 - p - Chlorophenyl - 2, 5 - dimethylpyrrol - 3 - ylacetic acid.

Cloponone. β,4 - Dichloro - α - dichloro - acetamidopropiophenone. An antiseptic present in GINETRIS.

Cloprostenol. A preparation currently undergoing clinical trial for use in cases of infertility, the aim being to achieve synchronisation of œstrus. ($\pm$) - 7 - {(1R, 2R, 3R, 5S) - 2 - [(3R) - 4 - (3 - Chlorophenoxy) - 3 - hydroxybut - 1 - (E) - enyl] - 3, 5 - dihydroxycyclopent - 1 - yl}hept - 5 - (Z) - enoic acid. ICI 80,996 is the sodium salt.

Cloquinate. Chloroquine di-(8-hydroxy-7 - iodoquinoline - 5 - sulphonate). Resotren.

Cloran. A trade mark for a chlorinated anhydride thermal and chemical stabiliser for polymers. (51).

Clorazepic Acid. A sedative. 7 - Chloro - 2, 3 - dihydro - 2, 2 - dihydroxy-5 - phenyl - 1H - 1, 4 - benzodiazepine - 3 - carboxylic acid. TRANXENE is the dipotassium salt.

Clorcetin. A proprietary preparation of chloramphenicol. An antibiotic. (279).

Clordion. The acetate of chlormadinone.

Clorevan. A proprietary preparation of chlorphenoxamine hydrochloride, used for control of Parkinsonism. (833).

Clorexolone. 5 - Chloro - 2 - cyclohexyl - 6 - sulphamoylisoindolin - 1 - one. Nefrolan.

Clorgyline. A mono-amide oxidase inhibitor and anti-depressant, currently undergoing clinical trial as " M&B 9302 ". N-3-(2,4-Dichlorophenoxy) - propyl - N - methylprop - 2 - ynylamine.

Clorindione. 2 - (4 - Chlorophenyl)in - dane-1,3-dione. Indalitan.

Clorprenaline. 1 - (2 - Chlorophenyl) - 2-isopropylaminoethanol.

Clostebol Acetate. An anabolic steroid. 17β - Acetoxy - 4 - chloroandrost - 4 - en - 3 - one. It is present in STERANABOL.

Clostrin. Combined vaccine for sheep. (520).

Cloth Brown G. A dyestuff. It is the sodium salt of diphenyl-disazo-dioxy-naphthalene-salicylic acid, $C_{29}H_{19}N_4O_5Na$. Dyes chromed wool brownish-yellow.

Cloth Brown R. A dyestuff. It is the sodium salt of diphenyl-disazo-sali-cylic-naphthol-salicylic acid, $C_{29}H_{18}N_4O_7SNa_2$. Dyes chromed wool brownish-red.

Cloth, Leather. See AMERICAN CLOTH.

Cloth Oil. That fraction obtained from

the residue of Russian petroleum b distillation, of specific gravity 0·875.

Cloth Oils. See WOOL OILS.

Cloth Orange. A dyestuff. It is th sodium salt of diphenyl-disazo-reso cinol - salicylic acid, $C_{25}H_{17}N_4O_5Na$ Dyes chromed wool brownish orange.

Cloth Red B. The Bayer Co., Dahl Co. A dyestuff. It is the sodium sa of toluene-azo-toluene-azo-α-naphtho mono-sulphonic acid, $C_{24}H_{19}N_4O_4SNa$ Dyes chromed wool red.

Cloth Red B. K. Oehler. A dyestuf It is the sodium salt of toluene-azc toluene - azo - β - naphthol - disulphoni acid, $C_{24}H_{18}N_4O_7S_2Na_2$. Dyes chrome wool brownish-red from an acid bath Other names for this colour are Clot red O, Fast bordeaux O, Cloth red BA and Fast milling red B.

Cloth Red BA. See CLOTH RED B.

Cloth Red 3B Extra. A dyestuff. It i the sodium salt of toluene-azo-toluene azo - ethyl - β - naphthylamine - sul phonic acid, $C_{26}H_{24}N_5SO_3Na$. Dye wool and silk bluish-red from an acic bath.

Cloth Red G. The Bayer Co. A dye stuff. It is the sodium salt of benzene azo - benzene - azo - α - naphthol - p sulphonic acid, $C_{22}H_{15}N_4O_4SNa$. Dye wool red from an acid bath. Othe names for this dyestuff are Cloth red R and Fast red 7B.

Cloth Red G. K. Oehler. A dyestuff It is the sodium salt of toluene-azo toluene-azo-β-naphthol-mono-sulphonic acid, $C_{24}H_{19}N_4O_4SNa$. Dyes chromed wool dark red from an acid bath. Another name for this dyestuff is Cloth red G extra.

Cloth Red G Extra. See CLOTH RED G.

Cloth Red 3GA. See CLOTH RED 3G EXTRA.

Cloth Red 3G Extra (Cloth Red 3GA). A dyestuff. It is the sodium salt of toluene-azo-toluene-azo-β-naphthyla-mine-mono-sulphonic acid, $C_{24}H_{20}N_5O_5$ SNa. Dyes chromed wool red.

Cloth Red O. See CLOTH RED B.

Cloth Red R. See CLOTH RED G.

Cloth Scarlet G. A dyestuff. It is the sodium salt of sulpho-benzene-azo-ben-zene-azo-β-naphthol, $C_{22}H_{15}N_4O_4SNa$. Dyes wool red from an acid bath.

Cloth Scarlet R. A dyestuff. It is the sodium salt of sulpho-toluene-azo-toluene-β-naphthol, $C_{24}H_{19}N_4O_4SNa$.

Clothiapine. A tranquilliser. 2 - Chloro - 11 - (4 - methylpiperazin - 1 - yl)dibenzo[b,f][1, 4]thiazepine.

Clotrimazole. An anti-mycotic agent. 1 - (α - 2 - Chlorophenylbenzhydryl) - imidazole. CANESTEN.

Cloudy Ammonia. This contains soap,

strong ammonia, and sometimes is perfumed with lavender water.

Cloustonite. A variety of asphalt.

Clovean. A detergent for bakehouse usage. (513).

Clowes' Solution. A solution containing 160 grams potassium hydroxide in 200 c.c. water and 10 grams pyrogallol. Used in the Hempel pipette for the absorption of oxygen.

Cloxacillin. 3 - (2 - Chlorophenyl) - 5 - methylisoxazol - 4 - ylpenicillin. Orbenin is the sodium salt.

Clozapine. A sedative currently undergoing clinical trial as " HF 1854 ". 8-Chloro - 11 - (4 - methylpiperazin - 1 - yl) - 5H - dibenzo[b, e][1, 4] - diazepine.

Club Moss. Lycopodium ($q.v.$).

Clumina. Humus treated with chlorine. A fertiliser.

Clysphitique. See BANDOLIN.

' C " Naphtha. A fraction of American petroleum oil, of specific gravity 0·700.

Coagol. A colloidal preparation of aluminium hydroxide, $Al(OH)_3$. It precipitates colloidal solutions.

Coagulatex. A coagulating agent consisting of sulphuric acid. It was formerly used for coagulating rubber latex.

Coagulating Powder, Boehringer's. See BOEHRINGER'S COAGULATING POWDER.

Coagulen-Ciba (Coagulin, Kocherfonio). A physiological hæmostatic prepared according to the method of Kocher-Fonio. It contains the natural coagulating elements of animal blood suitable for oral, subcutaneous, intramuscular and intravenous administration in 3 per cent. solution.

Coagulin. See COAGULEN-CIBA.

Coagulose (Hemostyl). The latter is a trade mark for normal serum preparations (hæmoplastin).

Coal, Agglomerated. See BRIQUETTES.

Coalatex. A proprietary coagulating material for use in coagulating rubber latex. It primarily consists of oxalic acid and oxalate, and is a good coagulating agent in 10 per cent. solution.

Coal Bitumen. A bituminous material which is soluble in pyridine, and insoluble in carbon tetrachloride.

Coal Black. See MINERAL BLACK.

Coal Brasses. Iron pyrites, FeS_2, as met with in coal.

Coal Carbonites I and II. Explosives containing respectively 25 and 30 per cent. nitro-glycerin, 30·5 and 24·5 per cent. sodium nitrate, 39·5 and 40·5 per cent. rye flour, and 5 per cent. potassium bichromate.

Coalite. A smokeless coal, prepared by distilling coal at a low temperature, so that it contains sufficient matter to enable it to burn without much smoke.

Coalite N.T.P. Non-toxic trixylenyl phosphate plasticiser. (602).

Coal Motor Spirit. Motor spirit derived from the volatile products from the distillation of coal and cannels.

Coal Oil. See KEROSENE.

Coal Tar, Colourless. See ANTHRASOL.

Coarse Metal (Matte). An impure mixture of ferrous and cuprous sulphides produced in copper smelting. It usually contains from 20–75 per cent. copper, 12–45 per cent. iron, and 19–25 per cent. sulphur. The impurities consist of arsenic, antimony, bismuth, manganese, nickel, cobalt, zinc, lead, selenium, tellurium, gold, and silver.

Coarse Para. See SERNAMBY.

Coarse Solder. See PLUMBER'S SOLDER and TINSMITH'S SOLDER.

Cobadex Ointment. A proprietary preparation of hydrocortisone in a silicone base for dermatological use. (248).

Cobalamin. Vitamin B_{12}. See CYANOBALAMIN.

Cobalin. A proprietary preparation of cyanocobalamin. (329).

Cobalin-H. A proprietary preparation of HYDROXOCOBALAMIN. (330).

Cobalt, Arsenical. See SMALTINE.

Cobalt Autunite. A mineral. It is $Co(UO_2)(PO_4)_2.8H_2O$.

Cobalt, Black. See ASBOLANE.

Cobalt Bloom. See ERYTHRITE.

Cobalt Blue (Cobalt Ultramarine, Thenard's Blue, Leithner's Blue, Leyden Blue, Gahn's Ultramarine, Vienna Blue, Azure Blue, Hoxner's Blue, Hungary Blue, King's Blue). A blue pigment consisting of an aluminate of cobalt, prepared by calcining cobalt compounds with clay or alumina. Phosphoric acid and zinc oxide are often added. Used as an oil colour.

Cobalt Brass. Acid-resisting alloys of 52 per cent. copper, 17–25 per cent. zinc, and 22–30 per cent. cobalt.

Cobalt Bronze. A phosphate of cobalt and ammonium.

Cobalt Brown. A pigment prepared by calcining a mixture of ammonium sulphate, cobalt sulphate, and ferrous sulphate.

Cobalt-Cabrerite. A mineral. It is $2[(Co,Mg)_3(AsO_4)_2.8H_2O]$.

Cobalt-Chalcanthite. A mineral. It is $2[CoSO_4.5H_2O]$.

Cobalt - chromium - molybdenum Steels. Usually an alloy of iron with 3·05 per cent. molybdenum, 2·16 per cent. chromium, 1·33 per cent. cobalt, and 0·65 per cent. carbon.

Cobaltcrome (Kobaltchrom). Alloys of from 60–70 per cent. iron, 25–30 per cent. chromium, and 5–10 per cent. cobalt. One alloy contains 79·5 per cent. iron, 13·6 per cent. chromium, 3·7 per cent. cobalt, 0·8 per cent. molybdenum, 0·8 per cent. silicon, 1·5 per cent. carbon, and 0·2 per cent. manganese. They are permanent magnet steels.

Cobalt, Earthy. See ASBOLANE.

Cobalt Enamels. Alkali silicates coloured blue by cobalt oxide, or in other colours, due to mixtures of cobalt oxide with other oxides, such as zinc oxide.

Cobalt Glance. See COBALTINE.

Cobalt Green (Rinmann's Green, Zinc Green). A pigment prepared by heating the precipitated oxides of cobalt and zinc. Also see GELLERT GREEN.

Cobalt, Grey. See COBALTINE.

Cobaltine (Cobalt Glance, Grey Cobalt, Cobaltite). A mineral. It is a sulpharsenide of cobalt, frequently containing iron, $CoAs_2.CoS_2$.

Cobalting Solutions. Usually consist of solutions of cobalt chloride and ammonia in water. Used for the electro-deposition of cobalt.

Cobaltite. See COBALTINE.

Cobalt Manganese Pink. A pigment obtained by mixing a thin paste of magnesium carbonate with cobalt nitrate solution, drying, and heating.

Cobalt Nickel Blende (Cobalt Nickel Pyrites). A mineral, $2RS.3R_2S_3$. (R is nickel or cobalt). It usually contains 11–40 per cent. cobalt, and 14–42 per cent. nickel, and iron.

Cobalt Nickel Pyrites. See COBALT NICKEL BLENDE.

Cobalt, Nitro. See NITRO-COBALT.

Cobaltoadamite. A mineral. It is $4[(Zn,Co)_2AsO_4OH]$.

Cobalt Ochre. Cobalt bloom (q.v.).

Cobaltomenite. A mineral. It is a selenide of cobalt.

Cobalto-Sphærosiderite. A mineral. It is $2[(Fe,Mg,Mn,Co)CO_3]$.

Cobalt Oxide, Prepared. See PREPARED COBALT OXIDE.

Cobalt Pink. Cobalt phosphate, $Co_3(PO_4)_2$.

Cobalt Pyrites. A mineral. It is cobaltic sulphide, Co_2S_3.

Cobalt Red (Cobalt Rose). Cobalt oxide, Co_2O_3.

Cobaltron Steel Alloy. A proprietary alloy containing iron with about 11 per cent. chromium, 2·25 per cent. cobalt, 1·5 per cent. carbon, 1·25 per cent. molybdenum, and 0·25 per cent. tungsten.

Cobalt Rose. See COBALT RED.

Cobalt-speiss. A cobalt ore (Co.Ni. Fe)As.

Cobalt Steel. Alloys of steel with cobalt, used for certain parts of electrical machinery requiring a high permeability. An alloy containing 34·5 per cent. cobalt, at high inductions, has a higher permeability than pure iron. Also see PERMANITE and K.S. MAGNET STEEL.

Cobalt, Tin-white. See TIN-WHITE COBALT.

Cobalt Ultramarine. See COBALT BLUE.

Cobalt Violet. A pigment prepared by mixing solutions of cobalt sulphate and sodium phosphate. A precipitate of hydrated cobaltous phosphate is formed, which upon fusing produces a violet pigment. When broken up, washed, and dried, it forms the cobalt violet of commerce.

Cobalt Violet, Pale. See PALE COBALT VIOLET.

Cobalt Vitriol. Cobaltous sulphate, $CoSO_4.7H_2O$.

Cobalt Yellow. See AUREOLIN.

Coban. A proprietary preparation of MONENSIN sodium. A veterinary antiprotozoan.

Cobastab. A proprietary preparation of cyanocobalamin for injection. (253).

Cobbler's Wax. This usually contains about 63 per cent. Swedish pitch, 31 per cent. rosin, and 6 per cent. tallow. These are melted together and the mass poured into water and pulled when it becomes partly cooled.

Cobione. A proprietary trade name for cyanocobalamin.

Cobolt. Smaltite, a mineral.

Cobrol. A photographic developer. (507).

Coca. The dried leaves of *Erythroxylon Coca.*

Cocæthylin. The ethyl ester of benzoyl-ecgonine, $C_6H_5CO.C_8H_{13}NO.COOC_2H_5$.

Co-Carboxylase. Pyrophosphoric ester of aneurine.

Cocceryl Coccerate. Coccerin, $C_{30}H_{60}(O.C_{31}H_{61}O_2)_2$.

Coccine 2B. See CROCEINE 3BX.

Coccinin. See PHENETOLE RED.

Coccinine B. A dyestuff. It is the sodium salt of *p*-methoxy-toluene-azo-β-naphthol-disulphonic acid, $C_{18}H_{14}N_2O_8S_2Na_2$. Dyes wool red from an acid bath.

Coccinite. A mineral. It is mercuric iodide, HgI_2.

Coccolite. See AUGITE.

Coccolith. Very tiny calcareous skeletons of floating marine algæ.

Cocculin. Picrotoxin.

Coccus. Cochineal.

Cochin-China Wax (Irvingia Butter). Cáy-cáy fat (candle-tree fat), obtained from *Irvingia Oliveri*. It is used in the manufacture of candles.

:ochineal (Carmine, Carmine Lake). The dried bodies of the female shield louse, *Coccus Cacti*. It is the source of carmine, a red colouring matter, the dyeing principle of which is carminic acid. Used for dyeing scarlet on tin mordanted wool, but chiefly for the preparation of pigments. Carmine lake consists chiefly of the aluminium salt of carminic acid. Two qualities of cochineal are white or silver-grey, and black cochineal or Zaccatila.

:ochineal Cake. Cochineal pressed into the form of cakes.

:ochineal Lakes. Pigments. They consist of the colouring matter of the cochineal insect, precipitated by an aluminium salt. See COCHINEAL.

:ochineal, Liquid. Made by digesting the insects or carmine with alkali and adding glycerin and water or alcohol.

:ochineal Red A. See NEW COCCINE.

:ochineal Scarlet G. A dyestuff. It is the sodium salt of benzene-azo-α-naphthol-mono-sulphonic acid, $C_{16}H_{11}N_2O_4SNa$. Dyes wool brick-red from an acid bath.

:ochineal Scarlet PS. See PALATINE SCARLET.

:ochineal Scarlet 2R. A dyestuff. It is the sodium salt of toluene-azo-α-naphthol-sulphonic acid, $C_{17}H_{13}N_2O_4SNa$. Dyes wool red from an acid bath.

:ochineal Scarlet 4R. A dyestuff. It is the sodium salt of xylene-azo-α-naphthol-sulphonic acid, $C_{18}H_{15}N_2O_4SNa$. Dyes wool red from an acid bath.

:ochinille Ammoniacale. Ammoniacal cochineal, a colouring matter consisting of dark-brown tablets, which are obtained by digesting cochineal with ammonia (excluding air), and precipitating the solution with aluminium nitrate.

:ochranite. A prepared titanium dicyanide.

:ochrome. An alloy of 60 per cent. cobalt, 12 per cent. chromium, 24 per cent. iron, and 2 per cent. manganese. It has been used in the place of Nichrome for the elements of electrical heating apparatus.

:ocinerite. A mineral, Cu_4AgS, found in Mexico.

:ock Metal. An alloy containing 20 lb. copper, 8 lb. lead, 1 oz. lead oxide, and 3 oz. antimony.

:ockscomb Barytes. See CRESTED BARYTES.

:ockscomb Pyrites (Spear Pyrites). Minerals. They are groups of crystals of Marcasite (disulphide of iron, FeS_2).

Coclopet. Copper chloride (petroleum refining grade). (589).

Coclor. Industrial cupric chloride (35/36 per cent. Cu). (589).

Cocloran. Anhydrous copper chloride (46–47 per cent. Cu). (589).

Coco. An aqueous extract of liquorice root, used as a polishing agent in the micro-analysis of carbon steels. Also the name for a liqueur from Bologna.

Cocoa Butter (Cacao Butter). The fat extracted from the seeds of *Theobroma cacao*. The seeds contain from 35–45 per cent. of the butter. It has an acid value of 0·6–1·3 (oleic acid), a saponification value of 192–193, and melts at 33–34° C. It contains the glycerides of arachic, palmitic, oleic, stearic, and lauric acids.

Cocoaline. See VEGETABLE BUTTER.

Cocoa Oil. See OIL OF COCOA.

Cocoa Powder. See BROWN POWDER.

Coconucite. A mineral. It is a variety of travertine.

Coco-nut Butter (Coco-nut Oil, Coco-nut Milk, Copra). The sweet watery liquid contained in the coco-nut is called coco-nut milk. This disappears and gives place to a soft edible pulp, which hardens in the air, and is sold as copra. Copra contains from 60–70 per cent. oil, which is extracted as coco-nut oil or butter.

Coco-nut Cake. The compressed material after the expression of coco-nut oil.

Coco-nut Fibre. Coir (*q.v.*).

Coco-nut Milk. See COCO-NUT BUTTER.

Coco-nut Oil. See COCO-NUT BUTTER.

Cocose. See VEGETABLE BUTTER.

Cocuiza Fibre. Cuban hemp is known in Venezuela by this name.

Codactide. A corticotrophic peptide currently undergoing clinical trial as "CIBA41 795-Ba". D - Ser[1] - Lys -[17,18] - β[1–18] - corticotrophin amide.

Codeine. Methyl-morphine, $C_{17}H_{18}(CH_3)NO_3$.

Codel Cortone. A proprietary preparation of prednisolone. (310).

Codel Cortone T.B.A. A proprietary preparation of prednisolone butyl-acetate. (310).

Codelsol. A proprietary preparation of prednisolone sodium phosphate. (310)

Codeltra. A proprietary preparation of prednisolone, magnesium trisilicate and aluminium hydroxide. Anti-inflammatory agent. (310).

Codemprin. Acetylsalicylic acid, phenacetin, caffeine and codeine compound. (514).

Codeonal. A mixture of 1 part Medinal (*q.v.*), and 2 parts Codeine-veronal. Used as a somnifacient, as a sedative, and as an analgesic.

Codethyline. Ethyl-morphine, $C_{17}H_{18}$ $(C_2H_5)NO_3$.

Codinyl. A proprietary preparation of ephedrine hydrochloride, menthol, codeine phosphate, syrup of prunes and kola. Cough linctus. (834).

Codis. A proprietary preparation of aspirin and codeine phosphate. An analgesic. (243).

Codite. A proprietary trade name for a vulcanised fibre and pure cotton cellulose plastic tubing.

Codoil. See RETINOL.

Codol. Retinol (q.v.).

Codur. A proprietary trade name for a synthetic clear or coloured baking enamel.

Codyl Syrup (Ingelheim). A German product. It contains the hydrochlorides of narcotine, papaverine, narceine, thebaine, and codeine. It is taken when the respiratory organs are inflamed.

Cœline. See CŒRULEUM.

Co-Elorine Pulvules. A proprietary preparation of amylobarbitone and tricyclamol chloride. (333).

Cœrulean Blue. See CŒRULEUM.

Cœruleine (Alizarin Green, Anthracene Green, Cœruleine A). A dyestuff obtained by heating galleine with concentrated sulphuric acid, $C_{20}H_{10}O_6$. Dyes chromed wool, silk, or cotton, green. Used in cotton printing.

Cœruleine A. See CŒRULEINE.

Cœruleine S (Alizarin Green, Anthracene Green S, Cœruleine SW). The bisulphite compound of cœruleine, $C_{20}H_{10}$ $O_6.2NaHSO_3$. Dyes chromed wool, silk, or cotton, green. Used in calico printing.

Cœruleine SW. See CŒRULEINE S.

Cœruleite. A mineral, $CuO.2Al_2O_3$. $As_2O_5.8H_2O$.

Cœrulene. Calcium carbonate coloured green and blue by malachite and chessylite.

Cœruleofibrite. A mineral. It is a basic chloro-arsenate of copper.

Cœruleolactite. A mineral. It is a hydrated aluminium phosphate.

Cœruleum (Cœline, Sky Blue, Cœrulean Blue). A blue pigment prepared by calcining a mixture of cobaltous sulphate, tin salt, and chalk, or by treating tin with nitric acid, adding a solution of cobalt nitrate, evaporating, and heating. Calcium sulphate is also usually found in the pigment.

Cofectant. A proprietary disinfectant, containing phenols in a form soluble in water.

Co-Ferol. A proprietary preparation of ferrous fumarate and folic acid. A haematinic. (248).

Coffee Essence. Various kinds of sugar, which are burnt, together with aromatics and fillings. Used to cheapen coffee.

Coffeine. Caffeine.

Coffeminal. A 50 per cent. mixture of luminal (q.v.) and caffeine. An antispasmodic. It is also used in the treatment of migraine.

Coffetylin. A preparation of caffeine and acetyl-salicylic acid.

Cofil. A rubber fabric adhesive. (534).

Cofill 11. An additive for direct bonding of rubber and of polyvinyl chloride to textiles and metals. (190).

Cogebi. A proprietary phenol-formaldehyde resin.

Cogentin. A proprietary preparation of benztropine methanesulphonate, used in Parkinsonism. (310).

Cognac (French Brandy). A brandy obtained by distilling weak wines of special vintages, refining and maturing the product in casks of Angoulême or Limousin oak. Also see BRANDY.

Cognac Oil. See OIL OF COGNAC.

Cognito Nuts. See BASSOBA NUTS.

Cohardite. A proprietary trade name for a hard rubber.

Cohenite. An iron carbide, Fe_3C, found in meteorites.

Co-Hydeltra. A proprietary preparation of prednisolone, magnesium trisilicate, and aluminium hydroxide. Antiinflammatory agent. (310).

Cohydrol. A colloidal graphite solution.

Coinage Bronze. An alloy of 95 per cent. copper, 4 per cent. tin, and 1 per cent. zinc. It has a specific gravity of 8·9 and melts at 900° C.

Coinage, Nickel. See NICKEL COINAGE.

Coinage, Silver. See SILVER COINAGE.

Coir (Coco-nut Fibre). The fibre from the shell of the coco-nut. Used as a cordage, and as a filling material.

Cojene. Analgesic and antipyretic tablets. (509).

Coke. The residue from the distillation of coal. It amounts to from 70–80 per cent.

Coke Oven Gas. The average composition of this gas is 53 per cent. hydrogen, 26 per cent. methane, 11 per cent. nitrogen, 7 per cent. carbon monoxide, and 3 per cent. heavier hydrocarbons. It is used for the production of hydrogen.

Colac. A preparation for use in the treatment of alimentary disorders. It comprises equal parts of lactic acid and calcium lactate.

Colace. Dioctyl sulphosuccinate. A proprietary laxative. (886).

Colalin. Cholalic acid, $C_{24}H_{40}O_5.H_2O$.

Colamine. Amino-ethyl-alcohol.

Colaspase. L - Asparagine amidohydrolase obtained from cultures of *Escherichia coli* A.T.C.C. 9637. CRASNITIN.

Colasta. A proprietary trade name for a phenol-formaldehyde synthetic resin.

Colastex. A patented mixture of Colas (cold asphalt) and rubber latex for improving and hardening roads.

Colbenemid. A proprietary preparation of probenecid and colchicine, used in the treatment of gout. (310).

Colcar D. A proprietary modified diethyl-phthalate range of carriers for the dying of triacetate fibres. (565).

Colcemid. A proprietary trade name for demecolcine.

Colchisal. A trade name for a preparation of colchicine salicylate, $C_{22}H_{25}O_6N.C_7H_6O_3$. Used in medicine for treatment of arthritis, rheumatism, and gout.

Colcothar. See INDIAN RED.

Cold Blacks B, R. See CATIGENE BROWN N.

Cold Blast Pig. A pig iron made in a furnace worked with a cold blast. It is usually low in silicon.

Cold Drawn Oil. See VIRGIN OIL.

Cold-short Iron. Iron which is brittle in the cold owing to its phosphorus content.

Cold Varnishes. Varnishes obtained by heating linseed oil to 105° C. for four and a half hours, adding manganese borate, linoleate, or resinate, and stirring the mass with compressed air.

Colemanite (Borate spar). A mineral. It is calcium borate, $Ca_2B_6O_{11}.5H_2O$, found in California. A source of boron and boric acid.

Colerainite. A mineral. It is a hydrated silicate of magnesium and aluminium.

Coles Solder. An alloy of 82 per cent. tin, 11 per cent. aluminium, 5 per cent. nickel, and 2 per cent. manganese.

Coletyl. Veterinary carbachol solution. (591).

Colfite. A proprietary trade name for a graphited laminate bearing material.

Coliacron. A proprietary preparation of glutamine synthetase, acetyl-Co A-kinase and oxidative phosphorylising enzymes used in the treatment of psychosomatic disorders. (887).

Colifoam. A proprietary preparation of hydrocortisone acetate in an aerosol foam base. For rectal application. (374).

Colistin. A mixture of polypeptides produced by strains of Bacillus polymyxa var. colistinus. Coloimycin is the sulphate.

Colistin Sulphomethate. An antibiotic obtained from Colistin Sulphate by sulphomethylation with formaldehyde and sodium bisulphite.

Collactivit. A proprietary trade name for the product obtained by the action of sulphuric acid on sawdust. It is a colloidal carbon product and is used as a decoloriser and purifier for sugar juices.

Collagenase. Clostridiopeptidase A; 3.4.4.19. Digests native collagen at about physiological pH. (182).

Collagen Sugar. Glycocoll, $C_2H_5O_2N$.

Collargol (Crede's Silver). Colloidal silver, used in medicine for many septic diseases.

Collasan. Colloidal kaolin.

Collatex. Ammonium alginate. (536).

Collaurin. A form of colloidal gold.

Collene. Colloidal silver.

Collet Brass. See BRASS.

Collet Steel. A manganese steel with small amounts of chromium and 0·75 per cent. carbon.

Collex. Aqueous oil and grease remover. (545).

Collidine. 2-Methyl-4-ethyl-pyridine, b.pt. 178° C., is α-modification. 4-Methyl-3-ethyl-pyridine, b.pt. 195–196° C., is β-modification. 2:4:6-Trimethyl-pyridine, b.pt. 171° C., is γ-modification.

Colliery Steelite. A chlorate explosive, consisting of 74 per cent. potassium chlorate, 25 per cent. oxidised resin, and 1 per cent. castor oil.

Collin. A preparation made by heating hide powder or gelatin with caustic soda, and neutralising with acetic acid. Used in the analysis of tanning materials.

Collinsonite. A mineral. It is a hydrated phosphate of calcium, magnesium, and iron.

Colliron. A proprietary preparation of ferric hydroxide colloid. A hæmatinic. (182).

Collocal-D. A proprietary preparation of calcium oleate and stearate and calciferol. (Vitamin D.) (280).

Collo-calamine. Colloidal calamine.

Collodion. Nitro-cotton, a weakly nitrated cellulose, usually regarded as dinitro-cellulose, $C_6H_8(NO_2)_2O_5$. Its solution in alcohol-ether is also known as collodon. Employed in medicine and photography.

Collodion Cotton. Guncotton (*q.v.*).

Collodion, Elastic. Flexible collodion (*q.v.*).

Collodion Pyroxyline. Cellulose tetra- and tri-nitrates.

Collodion Silk. A weakly nitrated cellulose (collodion cotton), is dissolved in a mixture of alcohol and ether, to form a thick syrup, subjected to pressure, and forced out in fine threads to form silk. It is passed through a reducing bath to remove NO groups, to render it non-inflammable and non-explosive. Sometimes the nitro-cellulose is mixed with resin, oil, or fishglue.

Collodion Wool. A weakly nitrated cellulose, usually a dinitro-cellulose. Used for the preparation of celluloid and blasting gelatin.

Collodiumwolle A. Alcohol-soluble nitrocelluloses of varying viscosities.

Collodiumwolle E. A series of estersoluble nitrocelluloses of varying viscosities.

Colloidal Aluminium. See VAGINTUS.

Colloidal Fuel (Coalinoil, Liquid Fuel). A mixture of finely pulverised coal or other solid fuel and a liquid fuel such as petroleum, oil, coal-tar naphtha, lignite tar or shale oil fraction, in such a condition that the solid particles will not separate out on standing. One such mixture contains 30 per cent. coal, 10 per cent. coal tar distillate, and 60 per cent. mineral oil.

Colloidox. Colloidal copper containing 20 per cent. Cu. (589).

Collone. Emulsifying agents. (547).

Collones. A proprietary trade name (547) for a group of emulsifying waxes comprising the following:

AC—A saturated fatty alcohol/ethylene oxide condensation product. Non-ionic.

HV—A fatty alcohol modified by added saponifiable fats. Anionic.

NI—A saturated fatty alcohol containing a proportion of a cetyl polyoxyethylene glycol ether. Non-ionic. (Cetomacrogol emulsifying wax BPC.)

QA—Higher saturated fatty alcohols incorporating a proportion of a quaternary ammonium compound. Cationic. (Cetrimide emulsifying wax BPC.)

SE—A saturated fatty alcohol containing a proportion of a sulphated alcohol. Anionic.

SEC—Anionic emulsifying wax BP.

Collongite. Phosgene or carbonyl chloride, $COCl_2$. A poison gas.

Collophane (Morite). A mineral, $Ca_3(PO_4)_2H_2O$.

Collophanite. See COLLOPHANE.

Colloresin D. A proprietary product. It is a methyl cellulose soluble in cold water.

Colloresin DK. A proprietary product. An alkyl ether of cellulose soluble in cold water insoluble in hot water. For use as a thickening agent in textile printing.

Collosin. See FILMOGEN.

Collosol Argentum. A proprietary preparation of colloidal silver, used as an ocular antiseptic. (280).

Collosol Calamine. A proprietary preparation of colloidal calamine. A skin cream. (280).

Collosol Manganese. A proprietary preparation of colloidal copper, chromium and manganese. Skin antiseptic. (280).

Collotone. A proprietary preparation of iron and ammonium citrate, iron and manganese citrate, potassium glycerophosphate, sodium glycerophosphate, thiamine hydrochloride, tincture of nux vomica and caffeine citrate. A tonic. (280).

Colloxylinum. Pyroxylin (*q.v.*).

Collozets. A proprietary preparation of tyrothricin and cetylethyldimethylammonium ethyl sulphate, used for sore throat. (280).

Collozine. Colloidal zinc hydroxide.

Collubarb. A proprietary preparation of aluminium hydroxide gel, phenobarbitone and atropine sulphate. An antacid. (182).

Collumina. A proprietary preparation of colloidal aluminium hydroxide. (591).

Collyrite. A white clay found in Hungary.

Collyrium. Synonym for Kaolinite.

Colmonoy. A proprietary trade name for a chromium boride, an abrasive.

Colmonoy No. 6. A corrosion-resisting alloy with about 75 per cent. nickel base. An essential constituent is chromium boride.

Colmenthol. Menthol-ethyl-glycollate.

Colodol Lubricant. Conveyor and band lubricant. (576).

Colofac. A proprietary preparation containing mebeverine hydrochloride. An antispasmodic. (280).

Cologel. A proprietary preparation containing methylcellulose. A laxative. (333).

Cologne Brown. See VANDYCK BROWN and UMBER.

Cologne Earth. A brown pigment. Originally it was a native bituminous earth, but it is now prepared by calcining Vandyck brown. See UMBER.

Cologne Glue. The best and purest skin glue.

Cologne-Rottweiler Powder. An explosive. It consists of 93 per cent. ammonium nitrate, 0·9 per cent. barium nitrate, 1·2 per cent. sulphur, and 4·9 per cent. vegetable oil.

Cologne Spirit. Spirit of wine, used in pharmacy.

Cologne Yellow. See CHROME YELLOW.

Colombo Root. Calumba root.

Colomycin. A proprietary preparation containing colistin sulphate. An antibiotic. (330).

Colona Steel. A proprietary trade name for a nickel-chromium steel containing some manganese.

Colonial Spirit. A purified methyl alcohol, CH_3OH.

Colonia Powder. An explosive, consisting of a mixture of from 30–35 per cent. nitro-glycerin, with from 65–70 per cent. ordinary gunpowder.

Colophonite. A mineral. It is a calcium-iron-silicate, a garnet.

Colophony (Rosin, Resin). The residue which remains after the volatile oils have been removed by the distillation of crude turpentine. It is essentially abietic anhydride, $C_{44}H_{62}O_4$. French rosin is obtained from *Pinus maritima*, by tapping the trees to yield the gemme containing 19–21 per cent. essential oil and 66–69 per cent. dry products. This is purified by sedimentation and then distilled for turpentine and rosin. The latter is purified by filtration through phosphor-bronze sieves. The best grades of French rosin are highly valued. The lightest is called " crystal," and the remaining ones range from AA to AAAAA, the more A's the lighter the colour. American rosin is obtained from several species of *Pinus*, mainly from *Pinus palustris* and *P. Australis*. The first year's or pale varieties are marked WW and K. The lightest in colour (almost colourless) is known as " Excelsior." WW is slightly yellow and WG yellow. Others range from A–N (A being the darkest), H is light brown, G dark brown, and F–A very dark. Spanish rosin is obtained mainly from *Pinus maritima*, but *Pinus larix* and the Aleppo pine are also used. Indian rosin is obtained from several varieties including *Pinus longifolia*. The essential oil amounts to 20 per cent. and the rosin to 70–72 per cent. The rosin has an acid value of 174, a saponification value of 184, and softens at 74° C. Colophony is soluble in alcohol, amyl alcohol, amyl acetate, acetone, benzene, chloroform, carbon tetrachloride, ether, carbon disulphide, and acetic acid. The specific gravity varies from 1·045–1·09, the melting-point 120–150° C. (it softens at about 75° C.), the saponification value 150–200, the acid value 145–185, the iodine value 112–120, and the ash from 0·02–0·05 per cent. It is used in the manufacture of soap, in varnishes, and for a variety of other purposes.

Coloradoite. A mineral, HgTe.

Colorado Silver. An alloy of 57 per cent. copper, 25 per cent. nickel, and 18 per cent. zinc. It is a nickel silver (German silver). Also see NICKEL SILVERS.

Colorol 20. A proprietary trade name for a standardised modified lecithin used as a wetting and suspending agent for use in Nitrocellulose pigmental lacquers containing 10 per cent. polar solvent. (189).

Col-o-tex. A proprietary trade name for lacquer coated fabrics.

Coloured Inks. Inks which are usually prepared from water-soluble aniline dyes.

Colouring, Sugar. See BURNT SUGAR.

Colourless Coal Tar. See ANTHRASOL.

Colours, Mordant. See ADJECTIVE DYESTUFFS.

Colour Wash. A mixture of lime and water with pigments.

Col-o-vin. A proprietary trade name for polyvinyl synthetic resin coated fabrics.

Colsil. A German product. It consists of cholesterin with a small amount of lecithin.

Colsonite. A proprietary smokeless fuel. It is a semi-coke and has an ash of not more than 7·5 per cent.

Coltapaste. Zinc paste and coal tar bandage. (529).

Coltrock. A proprietary trade name for a phenolic plastic.

Coltwood. A proprietary trade name for a phenolic plastic.

Colugel. A proprietary aluminium hydroxide gel used as an antacid. (889).

Coluitrin. See PITUITRIN.

Columbia Black B. A dyestuff. It is the sodium salt of dimethoxy-diphenyl-disazo-naphthol-disulphonic acid-azo-bi-*m*-tolylene-diamine, $C_{38}H_{34}N_{10}O_9S_2Na_2$. Dyes cotton black.

Columbia Blacks 2BX, 2BW. Dyestuffs belonging to the same group as Columbia black B.

Columbia Black FB, FF Extra (Paramine Brilliant Black FB, FR, Carbide Black, Chlorazol Black FH). A dyestuff. It is the sodium salt of benzene-disazo-α-naphthylamine-sulphonic acid-naphthol-sulphonic acid-azo-*m*-phenylene-diamine, $C_{32}H_{23}N_9O_7S_2Na_2$. Dyes cotton black.

Columbia Black R. A dyestuff. It is the sodium salt of ditolyl-disazo-naphthol-disulphonic acid-azo-bi-*m*-tolylene-diamine, $C_{38}H_{34}N_{10}O_7S_2Na_2$. Dyes cotton black.

Columbia Blue G (Chicago Blue 2R, Benzo Blue 2R, Diamine Blue C4R). Mixed disazo dyestuffs from tolidine, amino-naphthol-sulphonic acids, and a third component. Dyes cotton blue.

Columbia Blue R. See CHICAGO BLUE 4R.

Columbia Brown R. A direct cotton dyestuff.

Columbia Chrome Black 2B. A direct cotton dyestuff.

Columbia Fast Blue 2G. A dyestuff similar to Chicago blue 2R.

Columbia Fibre. A fibre originating from Africa and South America. It is used for the production of yarns similar to cotton and wool, and also silky yarns.

Columbia Green (Direct Green CO). A dyestuff. It is the sodium salt of diphenyl-disazo-salicylic acid-amino-naphthol-sulphonic acid-azo-benzene-sulphonic acid, $C_{35}H_{22}N_7O_{13}S_2Na_4$. Dyes cotton green.

Columbian Spirit. A brand of wood alcohol of a higher quality than the ordinary commercial kind. It contains only a trace of acetone.

Columbia Orange R. A direct cotton dyestuff.

Columbia Red 8B. See GERANINE.

Columbia Resin. A proprietary trade name for a transparent thermosetting synthetic resin.

Columbia Yellow. See CHLORAMINE YELLOW.

Columbite (Baierite). A mineral. It is essentially a niobate of iron and manganese, $(Fe.Mn)O.Nb_2O_5$. Part of the niobium in the mineral is often replaced by tantalum.

Columbium. Niobium, Nb.

Coluval. A proprietary preparation of diperodon hydrochloride, methylthionin chloride, naphazoline nitrate and sodium chloride. Eye drops. (391).

Colza Oil (Sweet Oil). Rapeseed oil from *Brassica campestris*.

Colzarine Oil. See MINERAL BURNING OILS.

Combestrol. Hexœstrol and stilbœstrol. (514).

Combizym. A proprietary preparation of lipase, amylase, proteases, hemicellulases and cellulase. (391).

Combustol. A liquid antiseptic plaster containing salicylic acid in 4 per cent. solution. Used in skin diseases.

Comet. (1) A proprietary trade name for hard rubber used for pipe bits.

Comet. (2) Liquid soaps. (506).

Comet Metal. An alloy of 67 per cent. iron, 30 per cent. nickel, 2·2 per cent. chromium, with small quantities of manganese and copper.

Commander's Balsam. Tinct. Benzoin Co.

Common Mica. See MUSCOVITE.

Common Salt. Sodium chloride, NaCl.

Comosal. Antimalarial compound. (512).

Comox. Cobalt and molybdenum oxides on alumina. (513).

Compact Bitumen. See BITUMEN.

Compazine. A proprietary trade name for the dimaleate or edisylate of Prochlorperazine.

Compitox. A selective weedkiller. (507).

Complac. A proprietary trade name for a shellac compound.

Complamin. (1) A proprietary trade name for Xanthinol Nicotinate.

Complamex. A proprietary preparation of XANTHINOL NICOTINATE. A peripheral vasodilator. (462).

Complamin. (2) A German preparation. It contains phenazolone, the propyl ester of phenyl-cincholin-carboxylic acid, and calcium citrate. Analgesic, antipyretic, and antirheumatic.

Complevite. A proprietary preparation of Vitamins A, D, C, thiamine hydrochloride, ferrous sulphate and calcium phosphate. (272).

Compocillin-UK. Phenoxymethyl-penicillin. Also stated to be Potassium penicillin V. (525).

Compocillin V-K. A proprietary preparation containing Phenoxymethylpenicillin potassium. An antibiotic. (311).

Compocillin V-K with Sulphas. A proprietary preparation of phenoxymethylpenicillin, sulphadiazine and sulphadimadine. An antibiotic. (311).

Compoglas. A proprietary trade name for a thermoplastic composition containing glass fibres.

Compo-site. A proprietary trade name for a shellac compound.

Composition Lava. A steatite scrap which is powdered and re-pressed.

Compounded Oil. A mixture of essential oils made to resemble certain flower odours and used in perfumery.

Compound Liquorice Powder. Contains powdered liquorice, powdered senna, sulphur, sugar, and oil of fennel.

Compound, Lake's. See LAKE'S INDIARUBBER COMPOUND.

Compral. A German preparation. It is a combination of voluntal and pyramidone. An analgesic.

Compregnite. A proprietary trade name for an impregnation solution containing phenol-formaldehyde synthetic resin.

Comprena. Food colour and flavour compounds. (522).

Compron. Veterinary compressed products. (507).

Comptonite. A mineral. It is a variety of thomsonite.

Comuccite. A mineral. It is a sulphantimonite of lead and iron.

Concavit. A proprietary preparation of

Vitamins A, B_1, B_2, B_6, B_{12}, C, D_2, E, nicotinamide, panthenol and folic acid. Vitamin supplement. (255).

Concentrated Alum. Aluminium sulphate, $Al_2(SO_4)_3$.

Concentrated Crystal Soda. A soda used in wool washing, Na_2CO_3. $NaHCO_3.2H_2O$.

Concentrated Size. Powdered glue.

Conchinine. Quinidine, $C_{20}H_{24}O_2H_2$.

Conchite. A mineral. It is calcium carbonate.

Conchlor. A water chlorinating agent. (512).

Concordin. A proprietary preparation of protriptyline hydrochloride. An antidepressant. (310).

Concrete (Béton). A conglomerate of pebbles, stones, gravel, blast furnace slag, or cinders, embedded in a matrix of mortar or cement.

Concrete, Armoured. See REINFORCED CONCRETE.

Concrete Oil of Mangosteen. See KOKUM BUTTER.

Condensed Milk. Milk from which some water has been removed and sugar added.

Condenser Foil. An alloy of 90 per cent. lead, 9·25 per cent. tin, and 0·75 per cent. antimony.

Condensite. See BAKELITE.

Condensite Celoron. A laminated product consisting of fibre and canvas cemented together by means of a phenol-formaldehyde resin.

Condor. See OIL WHITE.

Condres de Varech. The ash from seaweed.

Conductivity Bronze. A bronze containing copper with 0·8 per cent. cadmium and 0·6 per cent. tin.

Condurango. The dried bark of *Marsdenia condurango*.

Condy's Fluid. A solution made by dissolving potassium permanganate and aluminium sulphate in water. The potash alum which crystallises out is separated. The liquid left is a solution of permanganate of aluminium, with some aluminium sulphate. It is an oxidising agent, and is used as a disinfectant and deodoriser.

Condy's Green Fluid. A disinfecting liquid. It contains chiefly sodium manganate, $NaMnO_4$.

Conephrine. A solution of cocaine and paranephrine. Used as a local anæsthetic.

Confectionery Red. A dyestuff. It is a British equivalent of New coccine.

Con-Fer. A proprietary preparation containing ethinyloestrodiol and norethisterone acetate with ferrous fumarate tablets. An oral contraceptive with an iron supplement. (264).

Confit Wood. A similar preparation to Oropon, used for bating.

Confolensite. A variety of the mineral Montmorillonite.

Congluten. A vegetable substance obtained from almonds, containing 18 per cent. nitrogen, and 0·6 per cent. sulphur. Used in medicine as a 6 per cent. solution.

Congo Blue 2B (Benzo Blue 2B, Diamine Blue 2B). A dyestuff. It is the sodium salt of dimethoxy-diphenyl-disazo-α-naphthol-*p*-sulphonic-β-naphthol-disulphonic acid, $C_{34}H_{23}N_4O_{12}S_3Na_3$. Dyes cotton blue.

Congo Blue 3B. See DIAMINE BLUE 3B.

Congo Blue BX. See DIAMINE BLUE BX.

Congo Blue 2BX. See DIAMINE BLUE 2B.

Congo Brown G. A dyestuff. It is the sodium salt of sulpho-benzene-azo-resorcinol-azo-diphenyl-azo-salicylic acid, $C_{31}H_{20}N_6O_8SNa_2$. Dyes cotton brown.

Congo Brown R. A dyestuff. It is the sodium salt of sulpho-naphthalene-azo-resorcinol-azo-diphenyl-azo-salicylic acid, $C_{35}H_{22}N_6O_8SNa_2$. Dyes cotton brown.

Congo Copal. See COPAL RESIN.

Congo Corinth B. A dyestuff. It is the sodium salt of ditolyl-disazo-naphthionic-α-naphthol-*p*-sulphonic acid, $C_{34}H_{25}N_5O_7S_2Na_2$. Dyes cotton brownish-violet from a soap bath.

Congo Corinth G (Cotton Corinth G, Congo Corinth GW). A dyestuff. It is the sodium salt of diphenyl-disazo-naphthionic-α-naphthol sulphonic acid, $C_{32}H_{21}N_5O_7S_2Na_2$. Dyes cotton brownish-violet from a soap bath.

Congo Corinth GW. A dyestuff. It is a British brand of Congo Corinth G.

Congo Cyanine. Benzo blue (*q.v.*).

Congo Fast Blue B. A dyestuff. It is the sodium salt of dimethoxy-diphenyl-disazo-α-naphthalene-azo-bi-α-naphthol-disulphonic acid, $C_{44}H_{28}N_6O_{16}S_4Na_4$. Dyes cotton blue.

Congo Fast Blue R. A dyestuff. It is the sodium salt of ditolyl-disazo-α-naphthalene-azo-bi-α-naphthol-disulphonic acid, $C_{44}H_{28}N_6O_{14}S_4Na_4$. Dyes cotton blue.

Congo GR. A dyestuff. It is the sodium salt of diphenyl-disazo-*m*-amino-benzene-sulphonic acid-naphthionic acid, $C_{22}H_{20}N_6O_6S_2Na_2$. Dyes cotton red from a soap bath.

Congo Orange G. A dyestuff. It is the sodium salt of diphenyl-disazo-phenetol-β-naphthylamine-disulphonic acid, $C_{30}H_{23}N_5S_2O_7Na_2$. Dyes cotton orange.

Congo Orange R. A dyestuff. It is the sodium salt of ditolyl-disazo-phenetol-β-naphthylamine - disulphonic acid, $C_{32}H_{27}N_5O_7S_2Na_2$. Dyes cotton orange.

Congo Pure Blue. The same as Diamine pure blue.

Congo 4R. An azo dyestuff from diazotised tolidine, resorcinol, and naphthionic acid, $C_{30}H_{25}N_5O_5S$. See CONGO RED 4R.

Congo Red (Congo Red Conc., L, R, Extra, W Conc.). A dyestuff. It is the sodium salt of diphenyl-disazo-binaphthionic acid, $C_{32}H_{22}N_6O_6S_2Na_2$. Dyes wool or cotton red from a neutral or alkaline bath.

Congo Red Conc., L, R Extra, W Conc. Dyestuffs. They are British equivalents of Congo Red.

Congo Red 4R (Congo 4R). A dyestuff. It is the sodium salt of ditolyl-disazo-resorcinol-naphthionic acid, $C_{30}H_{24}N_5O_5SNa$. Dyes cotton red from a soap bath.

Congo Rubine. A dyestuff. It is the sodium salt of diphenyl-disazo-naphthionic acid-β-naphthol-sulphonic acid, $C_{32}H_{21}N_5O_7S_2Na_2$. Dyes cotton bluish-red.

Congo Sky Blue. See DIAMINE SKY BLUE.

Congo Violet (Bordeaux COV, Bordeaux Extra, Direct Violet). A dyestuff. It is the sodium salt of diphenyl-disazo-bi-β-naphthol-β-sulphonic acid, $C_{32}H_{20}N_4O_8S_2Na_2$. Dyes wool bordeaux red from an acid bath, and cotton violet from a salt bath.

Congo Yellow (Yellow Paste). An azo dyestuff, $C_{24}H_{19}N_5O_4S$, prepared by acting with diazotised benzidine upon phenol, and then combining the product with sulphanilic acid.

Conia (Conylia, Conicine, Cicutine). Conine (α-propyl-piperidine), $C_5H_{10}N$ (C_3H_7), an alkaloid which occurs in hemlock.

Conichalcite. A mineral, (Cu.Ca) (CuOH)(As.P.V)O_4.$\frac{1}{2}H_2O$.

Conicine. See CONIA.

Conima Resin. See HIAWA GUM.

Conistonite. An oxalate of calcium from Coniston, Cumberland. Synonym for Whewellite.

Conite. A variety of Dolomite, MgCa $(CO_3)_2$.

Connarite. A mineral. It is a hydrated nickel silicate.

Connellite. A mineral, $CuSO_4.CuCl_2$.

Connemara Marble. A crystalline limestone, containing admixed serpentine.

Conn's Stain. A microscopic stain. It

contains 1 gram rose bengal, 5 grams phenol, and 100 c.c. distilled water.

Conotrane. A proprietary preparation of penotrane and silicone cream. Antiseptic skin barrier cream. (319).

Conovid. A proprietary preparation of norethynodrel and mestranol. Oral contraceptive. (285).

Conovid-E. A proprietary preparation of norethynodrel and mestranol. Oral contraceptive. (285).

Conox. Ethylene oxide condensates and sulphates, nonionic and anionic surfactants. (597).

Conpernick. A proprietary trade name for an alloy of nickel and iron.

Conquinine. Quinidine, $C_{20}H_{24}N_2O_2$.

Conradite. The residue left in a tin ore found in New South Wales after the extraction of most of the metal. The remaining tin is very resistant to reagents. The residue approximates to the formula $7SnO_2.Fe_2O_3$.

Conray. A proprietary trade name for the meglumine salt of Iothalamic acid.

Consolidated Sulphuric Acid. See SOLIDIFIED SULPHURIC ACID.

Constantan. An alloy of 60 per cent. copper and 40 per cent. nickel. It has a specific gravity of 8·9, and melts at 1290° C. Used as an electrical resistance. It has a specific resistance of 48 micro-ohms per cm.[3], and a specific heat of 0·098 calories per gram at 0° C.

Constantin. Electrical resistance alloys. One contains 54 per cent. copper and 46 per cent. nickel, and another consists of 54 per cent. copper, 44 per cent. nickel, 1·3 per cent. manganese, and 0·4 per cent. iron.

Constant White. A pigment. It is barium tungstate. Also see BARIUM WHITE.

Constructal. An aluminium alloy containing 3 per cent. of alloying elements, chiefly zinc.

Contact Acid. Dipping acid, Tower acid, Sulphuric acid.

Contact T. A product similar to the Twitchell reagent. It is made by treating a fraction of petroleum distillate with a little sulphuric acid and then sulphonating. The sulphonated product is separated, and is called Contact T. Used in the decomposition of fats.

Contax. A proprietary preparation containing the diacetate of oxyphenisatin. A laxative. (248).

Conteben. P - acetylaminobenzalde - hyde thiosemicarbazone.

Continental Bakelite. A laminated product consisting of canvas cemented together by means of a phenol-formaldehyde resin.

Contracid. A corrosion-resisting alloy stable to nitric acid, hydrochloric acid, sulphuric acid, and other reagents. It contains from 50–60 per cent. nickel, 15–20 per cent. chromium, 0–20 per cent. iron, and up to 10 per cent. molybdenum or tungsten.

Contradet. A laundry contra-flow detergent. (513).

Contraluesin. A colloidal emulsion containing 0·0001 gram gold amalgam, 0·00001 gram iodine, and 0·001 gram arsenic per c.c. Used in the treatment of syphilis.

Contramine. Diethyl - ammonium - di - ethyl - dithio - carbamate. An anti-syphilitic.

Contrapar. A proprietary preparation of gloxazone. A veterinary anti-protozoan.

Controvlar. A proprietary preparation of norethisterone acetate, and ethinyl œstradiol, used for control of menstrual cycle. (265).

Conversion Saltpetre. See GERMAN SALTPETRE.

Conylia. See CONIA.

Conyrine. α-Propyl-pyridine, $C_8H_{11}N$.

Cookeite. A mineral, $Al_3LiH(SiO_4)_2$ $(OH_3)H_2O$.

Cook's Alloys. One contains 68·5 per cent. antimony and 31·5 per cent. zinc, and another consists of 57 per cent. antimony and 43 per cent. zinc.

Cooksons. White oxide of antimony. (590).

Coolgardite (Kalgoorlite). A mineral containing gold, silver, mercury, and tellurium.

Coolspray. A proprietary preparation of fluoro-chloro-hydrocarbons. Soothing aerosol spray. (802).

Coomassie. Acid milling dyestuffs. (512).

Coomassie Black B. A dyestuff. It is the sodium salt of sulpho-naphthalene-disazo - β naphthylamine - β - naphthol - disulphonic acid, $C_{30}H_{18}N_6O_{10}S_3Na_3$. Dyes wool deep black.

Coomassie Blue. A proprietary trade name for Sodium Anoxynaphthonate.

Coomassie Blue-Black. A dyestuff. It is a British brand of Naphthol blue-black.

Coomassie Navy Blue. A dyestuff. It is the sodium salt of sulpho-naphthalene - disazo-β-naphthol-p-naphthol - di-sulphonic acid, $C_{30}H_{17}M_4O_{11}S_3Na_3$. Dyes wool navy blue.

Coomassie Scarlet 9012K. A dyestuff. It is a British equivalent of Double scarlet extra S.

Coomassie Union Blacks. The sodium salt of sulpho - naphthalene - disazo - naphthol - sulphonic acid - azo - bi - m - phenylene diamine (or m-tolylene-diamine or resorcinol). Dye mixed and union goods black.

Coomassie Violet. See FORMYL VIOLET S4B.

Coomassie Violet AV. A British dyestuff which is equivalent to Victoria violet 4BS.

Coomassie Violet R. A dyestuff. It is a British brand of Formyl violet S4B.

Coomassie Wool Black R. A dyestuff. It is the sodium salt of amino-benzene-azo-naphthalene-azo - β - naph - thol-mono-sulphonic acid S, $C_{26}H_{18}N_5O_4SNa$. Dyes wool violet black.

Coomassie Wool Black S. A dyestuff. It is the sodium salt of amino-benzene-azo-naphthalene-azo-β-naphthol - disul - phonic acid R, $C_{26}H_{17}N_5O_7S_2Na_2$. Dyes wool black.

Coopal Powder No. 1 and KS Explosives. They are 42-grain powders.

Cooperite. (1) An alloy of 80 per cent. nickel, 14 per cent. tungsten, and 6 per cent. zirconium. Used for cutting tools. A modified alloy contains tantalum. It is also a suggested name for a new platinum mineral which is a sulpharsenide of platinum containing 64·2 per cent. platinum, 17·7 per cent. sulphur, and 7·7 per cent. arsenic.

Cooperite. (2) A mineral. It is 2[PtS].

Cooper's Gold. An alloy of 19 per cent. platinum and 81 per cent. copper. It resembles gold.

Cooper's Pen Metal. An alloy of 50 per cent. platinum, 37·5 per cent. silver, and 12·5 per cent. copper.

Cooper's Speculum Metal. An alloy of 58 per cent. copper, 27 per cent. tin, 10 per cent. platinum, 4 per cent. zinc, and 1 per cent. arsenic.

Coorongite (Australian Caoutchouc). A material resembling rubber, first discovered in South Australia. It is regarded as a petroleum product.

Copaiva (Copivi). Copaiba. See OLEO-RESIN COPAIBA.

Copaiva Balsam. See OLEO-RESIN COPAIBA.

Copalm Balsam. See LIQUIDAMBAR.

Copal Oils. Oils obtained by the dry distillation of copal. Used for the preparation of oil varnishes.

Copaloy. A proprietary platinum-tin-antimony alloy with a small percentage of copper. A bearing metal.

Copal, Pontianac. See MACASSAR COPAL.

Copal Resin (Gum Copal). A resin comprising a number of different types of recent and fossil origin. There are five principal types: East African, West African, Manila, Kauri, and South American copal. The East African copals are fossil resins, and are marketed at Zanzibar after sorting, being sold as Zanzibar copal or anime. It is a hard copal with acid value 115–123. West African copals are of lower value than the East African grades, and they are recent and fossil products, with acid value 122–143. Manila copal is sold as Macassar, Pontianac, or Singapore copal, and has an acid value 120–135. Kauri copal is a fossil resin derived from *Dammara Australis*, with acid value 65–85. South American copal is the product of *Hymenœa* species, and is known as Demerara animi. It is obtained from living trees. Accra copal melts at 120–180° C., has an acid value of 124–134, and gives an ash of 0·2–2 per cent. Angola copal, a fossil resin, is found in two qualities, red and white. Benin copal melts at 120–140° C., has an acid value of 110–116, and gives an ash of 0·5 per cent. Congo copal has an acid value of 132. Demerara copal, obtained from *Hymenœa verrucosa*, has an acid value of 97. Manila copal, from *Agathis alba*, is obtained in two qualities, hard and soft. The hard has an acid value of 72–130, and the soft 51–83. A bright amber variety melts at 103° C., and has an acid value of 142. Madagascar copal, from *Hymenœa verrucosa*, has an acid value of 76–93. Sierra Leone copal, from *Copaifera guibourtiana*, has an acid value of 110, and Zanzibar copal, from *Trachylobium* species, has acid value 60–123.

Copal, Salem. See BOMBAY COPAL.

Copal, Singapore. See MACASSAR COPAL.

Copel Alloy. An alloy of 55 per cent. copper and 45 per cent. nickel.

Copene. A proprietary trade name for a polyterpene copolymer resin used in lacquers, paints, and varnishes.

Coperflex. A proprietary synthetic rubber. (635).

Coperite. A name applied to a mineral consisting of cuprous sulphide, Cu_2S.

Copernick. An alloy similar to Hypernick. The permeability is constant over a wide range of flux densities.

Copholco. A proprietary preparation containing pholcodine, menthol, cineole, chloroform, oil of aniseed and terpin hydrate. A cough linctus. (340).

Copholcoids. A proprietary preparation containing pholcodine, menthol, terpin hydrate, chloroform, cineole and oil of aniseed. Cough pastilles. (340).

Copiapite (Misylite). A mineral, $Fe_4S_5O_{18}.12H_2O$.

Copivi. See COPAIVA.

Copper. A Bohme dyestuff. It contains catechu, safranine, and a copper salt.

Copper Abietinate. $Cu(C_{19}H_{27}O_2)$. Green scales, soluble in oils, imparting an emerald green colour. It has been used as a vermifuge in veterinary practice, also for impregnating and preserving wood.

Copper Alanine. Copper amino-propionate.

Copper Aluminate Cakes. A fused mixture of copper sulphate, potash alum, potassium nitrate, and camphor.

Copper, Antimonial. See ANTIMONIAL COPPER GLANCE.

Copper Arsenical. Domeykite (*q.v.*).

Copperas. See IRON VITRIOL.

Copper Aseptol. Copper sulpho-carbolate.

Copperas, Green. See IRON VITRIOL.

Copper Barilla. A copper sand containing from 60–80 per cent. copper, the remainder being quartz, found in Chile.

Copper Black. Copper blue B (*q.v.*), shaded with a yellow dye.

Copper Blue. See MOUNTAIN BLUE and COVELLITE.

Copper Blue B. A disazo dyestuff, which dyes wool reddish-blue shades in the presence of copper sulphate.

Copper Blues. Carbonates, silicates, and other salts of copper.

Copper, Cementation. See CEMENT COPPER.

Copper, Chessy. See CHESSYLITE.

Copper, Chili Bars. A variety of copper imported into England. It contains 95–99 per cent. copper.

Copper Chromate. A solution of copper ammonium chromate is known commercially under this name.

Copper Clad. A proprietary trade name for a roofing material consisting of copper sheet bonded to asbestos felt.

Copper Clude. A proprietary trade name for a copper steel used in the form of tubing.

Copper, Electrolytic. Contains 99·93 per cent. copper or over this amount.

Copper, Euflavine. See EUFLAVINE CADMIUM.

Copper, Feather Shot. See BEAN SHOT COPPER.

Copper Froth. See TYROLITE.

Copper Glance. See REDRUTHITE.

Copper, Gold. See CHRYSOCHALK.

Copper Green. A term applied to the mineral Malachite.

Copper, Grey. See TETRAHEDRITE.

Copper, Hair. See CHALCOTRICHITE.

Copper Hemi-oxide. See HEMI-OXIDE OF COPPER.

Copper Hemol. See HEMOL.

Copper, Indigo. See COVELLITE.

Copper Inhibitor X-872. Disalicylalethylene diamine. (2).

Copper Inhibitor X-872A. Disalicylal-ethylene diamine 50 parts, cumar 25 parts and stearic acid 25 parts. (2).

Coppering Solutions. These generally consist of solutions of copper acetate and potassium cyanide, sometimes ammonia, in water. Used for the electro-deposition of copper.

Copperised Ammonia. See WILLESDEN FABRICS.

Copper Mica. See TAMARITE.

Copper-nickel. An alloy of 50 per cent. copper and 50 per cent. nickel, used in the manufacture of nickel-copper alloys, is known by this name. See NICKEL COPPER ALLOYS. The term is also applied to the mineral Niccolite, NiAs.

Copperone. Cupferron (q.v.).

Copper Ore, Grey. See TETRAHEDRITE.

Copper Ore, Red. See CUPRITE.

Copper Ore, Yellow. See COPPER PYRITES.

Copper Oxide, Red. See HEMI-OXIDE OF COPPER.

Copper Paints. Mixtures of cupric and cuprous oxides. Used for painting ships' bottoms.

Copper Precipitate. The trade name for copper used in electrolytic processes.

Copper Protoxide. See HEMI-OXIDE OF COPPER.

Copper Pyrites (Chalcopyrite, Towanite, Yellow Copper Ore). A mineral. It is a sulphide of iron and copper, $CuFeS_2$.

Copper, Red. See VIOLET COPPER.

Copper, Red Ore. See CUPRITE.

Copper, Ruby. See CUPRITE.

Copper Rust. See MALACHITE.

Copper Salvarsan. Made by mixing salvarsan with a copper salt, adding caustic soda, and precipitating the mixture. Used in sleeping-sickness.

Copper Scale. The substance formed on copper when it is heated. It consists of a mixture of cupric and cuprous oxides.

Copper-silicon Alloy. An alloy usually containing 1·0–5·0 per cent. silicon, 1·5 per cent. manganese, 5·0 per cent. zinc (maximum), 2·5 per cent. iron (maximum), 2·0 per cent. tin (maximum), 0·5 per cent. impurities (maximum, remainder copper. One alloy contains 95 per cent. copper, 4 per cent. silicon, 1 per cent. manganese. Melting-point 1000° C. Specific gravity 8·15. Tensile strength 50,000 lb. per sq. in. Elongation on 2 in. is 20 per cent.

Copper Silumin. An alloy of 85·6 per cent. aluminium, 13 per cent. silicon, 0·8 per cent. copper, and 0·6 per cent. iron.

Copper Silver Glance. See STROHMEYERITE.

Copperskin. A proprietary construc-

tural material consisting of bitumised paper faced with copper metal.

Copper Soap. Copper resinate.

Copper Solder. An alloy of 2 parts lead and 5 parts tin.

Copper Sponge. A porous copper made by mixing with a copper powder a non-metallic volatile substance, pressing and heating in hydrogen, whereby the volatile matter is expelled. The sponge is then saturated with lead by immersion in molten lead. The material is used for bearings.

Copper, Standard. British copper containing not less than 97 per cent. of the metal.

Copper Steel. An alloy of steel with up to 1 per cent. copper, usually 0·5 per cent. It resists corrosion.

Copper Trisaylt. A copper sulphate preparation used in electro-plating.

Copper, Vitreous. See REDRUTHITE.

Copper Vitriol. See CHALCANTHITE.

Copper Water. Iron sulphate, $FeSO_4$.

Copper, White. See NICKEL SILVER and TOMBAC, WHITE.

Copper Wool. A form of metallic copper in a very finely divided condition.

Copper-Zinc-Epsomite. A mineral. It is $4[(Mg,Zn,Cu)SO_4 . 7H_2O]$.

Copper-Zinc-Melanterite. A mineral. It is $(Cu,Zn)SO_4 . 7H_2O$.

Coppesan. A copper fungicide. (502).

Coppite. A mineral, $Ca_3(CaF)(Na.K)(CeO)(Cb_2O_7)_3$.

Copra. See COCO-NUT BUTTER.

Coprah Oil. Coco-nut Oil. See COCO-NUT BUTTER.

Coprantine. Cotton dyestuffs. (530).

Copraol. A fat obtained from coco-nut oil.

Coprol. A proprietary trade name for dioctyl sodium sulphosuccinate.

Coprola. A proprietary preparation containing dioctyl sodium sulphosuccinate. A laxative. (332).

Coprolax. A proprietary preparation containing dioctyl sodium sulphosuccinate and dihydroxyanthraquinone. A laxative. (332).

Coprolites. Deposits formerly used as phosphatic manures. They consist mainly of about 25 per cent. phosphoric oxide and about 42 per cent. calcium oxide, and are apatites of organic origin.

Copro Yeast. A yeast containing a high proportion of coproporphysin.

Copying Ink Pencils. Made by mixing a paste of powdered graphite and kaolin with a concentrated methyl violet solution. Pressed into sticks and dried.

Co-Pyronil. A proprietary preparation containing pyrrobutamine, methapyriline and cyclopentamine hydrochloride. (333)

Coquilho Nuts. See BASSOBA NUTS.

Coquilla Nuts. See BASSOBA NUTS.

Coquimbite (Blakeite). A mineral. It is a hydrated ferric sulphate, $Fe_2(SO_4)_3$. $9H_2O$.

Coquina. A coarse limestone, composed of marine shells.

Coracite. A mineral. It is an uraninite partly altered to gummite, and occurs north of Lake Superior.

Corajo. Vegetable ivory (q.v.).

Coral. Solid mesodermal calcareous skeletons of the coral polyps, *Anthozoa*. They consist chiefly of calcium carbonate, coloured with ferric oxide.

Coralite. A vulcanite having the colour of coral.

Coralline. See PÆONINE.

Coralline Earth. A mineral. It is a mixture of cinnabar, bituminous matter, and apatite.

Coralline Limestone. Limestone $CaCO_3$, formed from the deposition of grains of coral.

Coralline Phthalein (Phenolcoralline). Pseudo-rosilic acid, $C_{20}H_{16}O_4$.

Coralline Red. See PÆONINE.

Coralline Yellow (Yellow coralline). The sodium salt of aurine (q.v.).

Coramine. A 25 per cent. aqueous solution of pyridine-β-carboxylic acid diethylamide. A powerful cardiac and respiratory stimulant.

Corangil. A proprietary preparation of glyceryl trinitrate, pentaerythritol tetranitrate, diprophylline and papaverine hydrochloride, used in angina pectoris. (182).

Corapel. Pipeline enamels and primers. (603).

Corba Oil (Sphagnol). A black tarry substance distilled from peat. An external antiseptic.

Corbasil. A proprietary preparation. It is o-dihydroxy-phenyl-propanolamine.

Corbestos. A proprietary material consisting of sheet metal covered with graphited asbestos. Used for gaskets to resist heat and pressure.

Cordal. See CORDOL.

Cordec. A series of plastisol products based on a dual dispersion of vinyl and a crosslinkable resin. (403).

Cordeine. Methyl-tribrom-salol. An intestinal astringent.

Cordex. A proprietary preparation of prednisolone and aspirin. Anti-inflammatory agent. (325).

Cordierite. See IOLITE.

Cordilox. A proprietary preparation of verapamil. (85).

Cordite (Cordite MD, Maximite). An explosive used as powder and filaments. It is a mixture of 65 per cent. guncotton, gelatinised by means of acetone,

30 per cent. nitro-glycerin, and 5 per cent. petroleum jelly.

Cordite MD. See CORDITE.

Cordite SC. Cordite. SC stands for sliced.

Cordocel. A proprietary preparation of alum and zinc powder used in preventive treatment for sepsis of the umbilical cord. (486).

Cordol (Cordal). Tribrom-salol, C_6H_4OH. $COO.C_6H_2Br_3$. An intestinal astringent.

Cordran. A proprietary trade name for Flurandrenolone.

Cordyl. Acetyl-tribrom-salol, an intestinal astringent. Also see CORDOL.

Cordylite (Barium-parasite, Pseudo-parasite). A mineral. It is a carbonate and fluoride of the cerium metals and barium, $(CeF)_2.Ba(CO_3)_3$.

Core Gum. Usually linseed oil products. They are used as binders for the sand used for cores in casting.

Coreine AR, AB. A dyestuff obtained by heating Coreine 2R with aniline and sulphonating. Dyes chromed wool blue. Chiefly used for printing.

Coreine 2R (Celestine Blue B). A dyestuff. It is the amide of diethyl-gallo-cyanine, $C_{17}H_{18}N_3O_4Cl$. Dyes chromed wool bluish-violet.

Corekal A. Sodium alkyl naphthalene sulphonate.

Core Oils. Liquid binders used for sand cores in foundrywork. They are usually linseed oil mixed with vegetable oils, etc.

Corex. Carbora (q.v.).

Corferrol. Suprarenal extract with the addition of iron and pyrrol blue. It is suggested for the treatment of cancer.

Coriandrol. d-Linalool.

Coriban. A proprietary preparation of diamphenethide. A veterinary anthelmintic.

Corichrome. Titanium lactates. Employed as mordants and " strikers " in the leather industry.

Coridine. n-Propyl-lutidine.

Cori Ester. Glucopyraminose-1-monophosphate.

Corilene. An aqueous degreasing agent for leather. (512).

Corinac A. A chromate based corrosion inhibitor having sp. gr. 1·04. Used as an additive to priming paints. (503).

Corinac C. A chromate based corrosion inhibitor having an sp. gr. 1·04. Used as an additive to priming paints. (503).

Corinal. A trade mark for a synthetic tannin prepared by condensing heavy tar oils with formaldehyde and then making the aluminium salt. Also see ESCO EXTRACT and NERADOL. (236).

Corindite. The trade name for an abrasive and refractory of the carborundum type. It is obtained from bauxite by heating it with anthracite, and contains 69 per cent. Al_2O_3, with SiO_2 and Fe_2O_3.

Corioflavines. Red or reddish-brown dyestuffs, used for leather.

Coriphosphines. Dyestuffs. They are alkylated amino-acridines.

Corisol. A proprietary preparation of suprarenal gland hormone.

Cork. The outer bark of a tree, *Quercus suber*.

Corkaline. A material made from ground cork, glue, glycerin, tannin, and chromic acid. Used in the manufacture of mats.

Cork Black. A pigment. It is the soot obtained by charring cork.

Cork Carpets. Made from cork grains and linoleum cement.

Corkite. A mineral $2PbO.3Fe_2O_3.3SO_3.P_2O_5.6H_2O$. It is a variety of beudantite.

Corkstone. A material made from finely divided cork, and a mineral cementing substance. Used in building construction as a non-conducting and fireproof material.

Corlan. A proprietary preparation of hydrocortisone hemi-succinate sodium for mouth ulcers. (335).

Cornalith. See GALLATITE.

Cornelian. Carnelian (*q.v.*).

Cornetite. A mineral, $Cu_3(PO_4)_2.3Cu(OH)_2$.

Cornish Bronze. An alloy of 78 per cent. copper, 9·5 per cent. tin, and 12·5 per cent. lead. A white bearing metal.

Cornish Clay. See CHINA CLAY.

Cornish Stone (China Stone). A granite rock occurring in Cornwall. It is rich in felspathic material, and is used in earthenware manufacture.

Cornite. A hard vulcanite.

Corn Oil (Maize Oil). The oil pressed from maize grains, used in soap making.

Cornox. A selective weedkiller. (502).

Corn Paint. Compound salicylic collodion. (*Collidium salicylicum Compositum, B.P.C.*)

Corn Silk. Maize stigmas.

Corn Sugar. See GLUCOSE.

Cornuite. A protein-like mineral found in the diatomaceous earth from Neu-Ohe. It is a gelatinous albuminous material with a 3 per cent. dry residue.

Cornutine. Impure ergotoxine.

Cornutol. A liquid extract of ergot, containing the water soluble and the alcohol soluble constituents of ergot, with about 10 per cent. alcohol.

Cornwallite. A mineral. It is a basic copper arsenate, $Cu(CuOH)_4(AsO_4)_2.3H_2O$.

Corodenin. A German product. It is a quinoline derivative with suprarenin, and 100 c.c. contains 0·3 gram 8-ethoxy-quinoline-5-sulphonic acid neutralised with 0·15 gram sodium tetraborate and 0·02 gram suprarenin hydrochloride. It is a yellow liquid, and is recommended for protecting the eyes from ultra-violet rays, and in eye trouble such as conjunctivitis.

Co-ro-felt. A proprietary trade name for a phenolic sisal plastic (unimpregnated).

Co-ro-lite. A proprietary trade name for a phenolic sisal plastic (impregnated).

Corolox. A trade mark for abrasive and refractory materials. The essential constituent is crystalline alumina.

Corona. Anhydrous lanolin. (526).

Coronadite. A mineral. It is a manganate of lead and manganese, $(Mn.Pb)Mn_3O_7$.

Coronation. A proprietary trade name for a casein plastic.

Corondite. An abrasive obtained from bauxite.

Coronguite. A mineral. It is a silver-lead antimonate.

Coronite. An explosive, containing from 38–40 per cent. nitro-glycerin, 1–1·5 per cent. soluble guncotton, 26–28 per cent. ammonium nitrate, 3–5 per cent. potassium nitrate, 11–14 per cent. aluminium stearate, 8–11 per cent. rye flour, 2–4 per cent. sawdust, and 2–4 per cent. liquid paraffin.

Coronium. An alloy containing 80 per cent. copper, 15 per cent. zinc, and 5 per cent. tin.

Coronium Bromide. Strontium bromide, $SrBr_2$.

Corotox. A trade mark for goods made for abrasive and refractory purposes. The essential constituent is crystalline alumina.

Corowalt. A trade mark for materials of the abrasive and refractory type, the essential constituent of which is crystalline alumina.

Corox. A new proprietary insulating material having a great thermal conductivity and high electrical insulating power. It consists essentially of magnesium oxide.

Corozo. Vegetable ivory, the seeds of *Phytelephas macrocarpa*.

Corprene. A proprietary trade name for a synthetic rubber made from ground cork, rubber derivatives and rubber-like resins.

Corronel 220. A trade mark for a nickel - molybdenum - vanadium alloy with good resistance to hydrochloric, sulphuric and phosphoric acids under reducing conditions. (143).

Corronel Alloy 230. A trade mark for an alloy of 35 per cent. chromium and 65 per cent. nickel. It is resistant to nitric and nitric/hydrochloric acid mixtures. (143).

Corronil. An alloy of 70 per cent. nickel, 26 per cent. copper, and 4 per cent. manganese. It is a corrosion-resisting alloy, melting at 1400° C., having a specific gravity of 8·8 and a specific resistance of 50 micro-ohms per cm.³ at 0° C.

Corrosalloy. A proprietary trade name for stainless steels coded as follows :
   S. 18 per cent. chromium, 8 per cent. nickel.
   N.D.P. 10 per cent. titanium or niobium. 18 per cent. chromium, 8 per cent. nickel.
   18. 18 per cent. chromium.
   These alloys have wide applications in the food, brewing, petrochemical and chemical industries. (141).
   M. 8 per cent nickel, 18 per cent. chromium, 3 per cent. molybdenum.
   DU. 29 per cent. nickel, 20 per cent. chromium, 2·5 per cent. molybdenum, 4 per cent. copper.
   13. 13 per cent. chromium.

Corrosiron. A Swedish iron-silicon alloy containing 12 per cent. silicon. It is stated to be very resistant to the corrosive action of acids.

Corrosist. A proprietary trade name for nickel base corrosion resistant alloys with compositions as follows :
   1. 60 per cent. nickel, 30 per cent. molybdenum, 6 per cent. iron.
   2. 50 per cent. nickel, 15 per cent. chromium, 20 per cent. molybdenum, 3 per cent. tungsten, 8 per cent. iron.
   3. 80 per cent. nickel, 10 per cent. silicon, 3 per cent. copper, 5 per cent. iron.
   4. 60 per cent. nickel, 24 per cent. chromium, 4 per cent. molybdenum, 2 per cent. copper, 7 per cent. iron.
   5. 5 per cent. nickel, 23 per cent. chromium, 8 per cent. copper, 4 per cent. molybdenum, 2 per cent. tungsten.
   Used for valves in chemical plant subject to contact with chlorine.

Corrosive-acetic Fluid. A fixing agent used in microscopy. It consists of 100 c.c. of a saturated corrosive sublimate solution with 2 c c. glacial acetic acid.

Corrosive Sublimate. Mercuric chloride, $HgCl_2$.

Corsite (Napoleonite). A mineral. It is a variety of diorite.

Corsodyl. A proprietary preparation of chlorhexidine gluconate used in the treatment of gingivitis. (2).

Cortacream. A proprietary preparation of hydrocortisone acetate and silicon fluid on a dressing, used for ecxematous skin disorders. (268).

Cor-Tar-Quin. A proprietary preparation of hydrocortisone, coal tar solution and di-iodohydroxy quinolone used in dermatology as an antibacterial agent. (372).

Cortef. A proprietary preparation of hydrocortisone for local use on skin. (325).

Cortelan. A proprietary preparation of cortisone acetate. (335).

EF-Cortelan. A proprietary trade name for Hydrocortisone.

Corten. A proprietary trade name for a chromium steel containing 0·5 per cent. chromium, 0·10 per cent. carbon, 0·1 per cent. manganese, 0·5 per cent. silicon, and 0·3 per cent. copper.

Cortenema. A proprietary preparation of hydrocortisone in saline. A steroid enema. (802).

Corticin. See BASICIN.

Cortico-Gel. A proprietary preparation of corticotrophin gelatin. (280).

Corticotrophin. Adrenocorticotrophic hormone. Adrenocorticotrophin; ACTH. Cortrophin.

Cortifoam. A proprietary preparation of hydrocortisone in aerosol form for dermatological use. (85).

Cortiphenicol. A proprietary preparation of PREDNISOLONE metasulphonobenzoate and chloramphenicol, used as ear drops. (320).

Cortipix. A proprietary preparation of coal tar and hydrocortisone acetate, used in the treatment of ecxema. (188).

Cortisone. 17α,21 - Dihydroxypregn - 4-ene-3,11,20-trione. 17-Hydroxy-11-dehydrocorticosterone. Adreson, Cortelan, Cortistab and Cortisyl are the 21-acetate.

Cortisporin. A proprietary preparation of polymyxin B sulphate, neomycin sulphate, and hydrocortisone alcohol. Antibiotic steroid cream. (277).

Cortistab. A proprietary preparation of cortisone acetate. (253).

Cortistan. Cortisone in a solution of 25 mg/10 c.c. (890).

Cortisyl. A proprietary preparation of cortisone acetate. (307).

Cortitrane. A proprietary preparation of phenylmercuric dinaphthylmethane disulphonate and prednisolone. Antibiotic steroid skin cream. (319).

Cortocaps. A proprietary preparation of hydrocortisone acetate and neomycin sulphate, used in eye infections. (280).

Cortoderm. A proprietary preparation of hydrocortisone acetate for dermatological use. (280).

Cortoderm-N. A proprietary preparation of hydrocortisone and neomycin used in dermatology as an antibacterial agent. (280).

Cortodoxone. A corticosteroid. $17,21$-Dihydroxypregn - 4 - ene - 3, 20 - dione.

Cortomycin. A proprietary preparation of hydrocortisone acetate and neomycin sulphate used in dermatology as an antibacterial agent. (380).

Cortril. A proprietary preparation of hydrocortisone for dermatological use. (85).

Cortrophin. Corticotrophin.

Cortrophin-Zn. A proprietary preparation of corticotrophin zinc hydroxide. (316).

Cortrosyn Depot. A proprietary preparation of TETRACOSACTRIN adsorbed on to zinc hydroxide. (316).

Cortucid. A proprietary preparation of sodium sulphacetamide, and hydrocortisone acetate. Eye cream. (265).

Corubin. An artificial corundum Al_2O_3. It is the alumina which constitutes the slag formed in the reaction between aluminium and metallic oxides (Thermit). Used for polishing purposes, and in the manufacture of fireproof stones.

Corundellite. A mineral. It is a variety of margarite.

Corundite. A trade mark used for abrasive and refractory materials the essential constituent of which is crystalline alumina.

Corundophilite. A mineral. It is a variety of Clinochlore.

Corundum. A mineral. It is alumina, Al_2O_3. Used as an abrasive in the form of wheels and emery paper.

Corundum, Artificial. An abrasive made by fusing bauxite.

Corvic. A registered trade mark for polymers of polyvinyl chloride for moulding and extrusion purposes. (2).

Coryfin. Methyl-ethyl-glycollate, CH_2O $(C_{10}H_{19}).COOC_2H_5$. Used as a local anæsthetic, and a sedative.

Coryl. A mixture of methyl and ethyl chlorides. Used in dentistry and minor surgery, as an anæsthetic.

Corynite. A mineral, (Ni.Fe)(As.Sb)S.

Coryphol. See CANDELITE.

Coryzomed. A proprietary preparation of codeine phosphate, papaverine hydrochloride and phenolphthalein. (137).

Cosaldon. A proprietary preparation of hexyl theobromine and nicotinic acid. (137).

Cosalite. See BJELKITE.

Cosaprin. Sodium-acetyl-sulphanilate, $C_6H_4(NH.CO.CH_3)(SO_3Na)$. Resembles phenacetin in its medicinal action.

Coscopin. A proprietary preparation containing noscapine. A cough linctus. (182).

Coscopin Pædiatric. A proprietary preparation containing noscapine. A cough linctus. (182).

Coslettised Steel. A steel whose surface has been rust-proofed by dipping in a solution of iron phosphate and phosphoric acid.

Cosmegin Lyovac. A proprietary preparation of ACTINOMYCIN D. A cytotoxic drug. (636).

Cosmetic Bismuth. Bismuth oxychloride, BiOCl.

Cosmetic Mercury. See WHITE PRECIPITATE.

Cosmic. Emulsifying agents. (505).

Cosmin. Agathin $(q.v.)$.

Cosmoline. A motor fuel consisting of 70 per cent. benzene and 30 per cent. naphthalene with the addition of $1 \cdot 5$ per cent. cresol, $0 \cdot 25$ per cent. naphthylamine, and $0 \cdot 25$ per cent. nitro-naphthalene. It is also the name for a variety of soft paraffin. See OZOKERINE.

Cosmos Alloy. A proprietary lead base alloy with small amounts of tin and antimony. A bearing metal.

Cosmos Fibre. A material obtained by breaking up fabrics of flax, jute, and hemp into fibre.

Cossaite. A mineral. It is a variety of Paragonite.

Cossyrite. A mineral. It is an iron and columbium titanosilicate. The name is also applied to a petroleum jelly article.

Costan. A proprietary preparation of tablets of tin metal and oxide.

Coster's Paste. Iodine dissolved in light oil of wood tar. Used for ringworm.

Costra. A lower grade of Chilean nitrate, the higher grade being called Caliche $(q.v.)$.

Cosulid. A proprietary preparation of SULPHACHLORPYRADAZINE. A veterinary anti-microbial.

Cosylan. A proprietary preparation containing tincture of cocillana, liquid extract of squill and senega, antimony potassium tartrate, cascarin, ethylmorphine hydrochloride, menthol and syrup. A cough linctus. (264).

Cosyntropin. See TETRACOSACTRIN.

Cotacord. A proprietary trade name for a synthetic resin-coated cord.

Cotargite. A double salt of ferric chloride and cotarine hydrochloride.

Cotatape. A proprietary trade name for a synthetic resin-coated tape.

Cotazym. A proprietary preparation

of pancreatic extract, used in pancreatic insufficiency. (316).

Cotazym B. A proprietary preparation of PANCRELIPASE, ox-bile extract and cellulase. (316).

Coterpin. A proprietary preparation of codeine, terpin, menthol, pine oil and eucalyptus. A cough linctus. (344).

Cothias Metal. An alloy of 67 per cent. copper and 33 per cent. tin. Used as a hardener for zinc alloys.

Cotinazin. A proprietary trade name for Isomazid.

Cotinin. See YOUNG FUSTIC.

Coto Bark. Used in medicine for diarrhœa. The active principle is cotoin, $C_{14}H_{12}O_4$.

Cotopa. A form of textile insulating material composed of acetylated cotton yarn.

Co-trimoxazole. Compounded preparations of trimethoprim and sulphamethoxazole in the proportions of 1 part to 5 parts, used as anti-bacterials. BACTRIM, SEPTRIN.

Cottaite. A mineral. It is a variety of Orthoclase.

Cotterite. A mineral. It is a variety of quartz.

Cotton, Amidated. Cotton which has had amino groups introduced into its molecule. This is accomplished by treating immunised cotton (q.v.) with aqueous ammonia. Amidated cotton has a great affinity for acid dyestuffs.

Cotton Balls. See BORONATROCALCITE.

Cotton Black. A dyestuff obtained by the fusion of o-p-dinitro-diphenylamine-sulphonic acid with sodium polysulphides. Dyes cotton brownish-black.

Cotton Black Grease. A grease obtained when cotton-seed mucilage is acidified. It contains oil and fatty acids.

Cotton Blue. Many of the direct cotton blues, such as Diamine, Benzo, and Congo blues, are sold under this name. See SOLUBLE BLUE and METHYL BLUE.

Cotton Blue R. See MELDOLA'S BLUE.

Cotton Bordeaux. A dyestuff from diamino - diphenylene - ketoxime and naphthionic acid, $C_{33}H_{21}N_7O_7S_2Na_2$. Dyes cotton from a soap bath.

Cotton Brown A, N. A tetrazo dyestuff. See BENZO BROWN BX, BR.

Cotton Brown R. See INGRAIN BROWN.

Cotton, Ceara. See CEARA COTTON.

Cotton Corinth G. See CONGO CORINTH G.

Cotton Fast Blue 2B. See MELDOLA'S BLUE.

Cotton Gum (Wood Gum). The name given to the product extracted from cotton by boiling alkali. It contains cotton wax and pectic and fatty matters.

Cottonin. Cottonised flax (q.v.).

Cottonised Flax. A product made by disintegrating flax by means of caustic soda, followed by acid, and carbon dioxide. A cotton-like material is produced.

Cotton, Mineral. See SLAG WOOL.

Cotton, Nitro. Guncotton (q.v.).

Cotton Orange 6305. A dyestuff. It is equivalent to Ingrain brown.

Cotton Orange R. A dyestuff. It is the sodium salt of primuline-azo-disulpho-m-phenylene-diamine-azo-benzene-m-sulphonic acid. Dyes cotton orange from a boiling bath.

Cotton Pitch. See COTTON-SEED PITCH.

Cotton Ponceau. An azo dyestuff from diamino-dixylyl-methane and naphthol-disulphonic acid R.

Cotton Red. See BENZOPURPURIN 4B.

Cotton Red 4B. See BENZOPURPURIN 4B.

Cotton Red 10B. A dyestuff. It is a British brand of Benzopurpurin 10B.

Cotton Rhodine BS. A dyestuff. It is the methylene derivative of dimethyl-homo-rhodamine ester, $C_{51}H_{50}N_4O_6Cl_2$. Dyes tannined cotton violet-red.

Cotton Root Bark. The root bark of *Gossypium* species.

Cotton Scarlet. The Badische Co. See BRILLIANT CROCEINE M.

Cotton Scarlet. Leipziger Anilinfabrik Beyer & Kegel. A dyestuff. It is the sodium salt of phenyl-dixylyl-me-thane-disazo-bi-β-naphthol-disulphonic acid, $C_{43}H_{32}N_4O_{14}S_4Na_4$. Dyes cotton red from a boiling alkaline bath, and is used for preparing lakes. Another name is Cinnabar red.

Cotton Scarlet 3B Conc. See BRILLIANT CROCEINE M.

Cotton-seed Meal. Obtained from cotton seed, by grinding the cake which is left when the oil is pressed out. Used as a nitrogenous fertiliser, and as a cattle-food.

Cotton-seed Mucilage (Cotton Soap Stock). The soap which separates out when cotton-seed oil is refined by treatment with caustic soda. It contains about 35 per cent. cotton-seed oil and 55 per cent. soap.

Cotton-seed Pitch (Cotton Pitch). A pitch material obtained in the refining of cotton-seed oil. It is obtained by the steam distillation of cotton black grease (q.v.). It is variable in hardness, has a specific gravity of from 0·9–1·2, and a saponification value of from 90–120.

Cotton-seed Stearin. Usually consists of impure stearic acid, obtained from cotton-seed oil. Used in the manufacture of soap and candles.

Cotton, Silicate. See SLAG WOOL.

Cotton Soap Stock. See COTTON-SEED MUCILAGE.

Cotton Spirits. See TIN SPIRITS.

Cotton Stearin. The solid stearin deposited when cotton-seed oil is chilled.

Cotton, Sthenosised. See STHENO-SISED COTTON.

Cottonstone. A mineral. It is a variety of mesolite.

Cotton Wax. The wax found in cotton fibre. It melts at 82–86° C., and appears similar to carnauba wax.

Cotton Wool. Gossypium.

Cotton Yellow 6307. A British dyestuff which is an equivalent of Diphenyl citronine G.

Cotton Yellow G. A dyestuff. It is the sodium salt of primuline-azo-m-phenylene-diamine-disulphonic acid, and dyes cotton orange-yellow.

Cotton Yellow G. The Badische Co. (Chlorazol Fast Yellow 5GK). A dyestuff. It is the sodium salt of diphenyl-urea-disazo-bi-salicylic acid, $C_{27}H_{18}N_6O_7Na_2$. Dyes cotton yellow from a boiling alkaline bath.

Cotton Yellow R. See ORIOL YELLOW.

Cotunnite. A mineral. It is lead chloride, $PbCl_2$.

Couac. A coarse kind of farina ($q.v.$), made in Brazil. It is a food product, obtained from the dried pulp of the roots of the plant Manioc, *Manihot utilissima*. Couac jaune is obtained by the addition of turmeric.

Coulsonite. A mineral. It is $(Fe,V)_3O_4$.

Coumadin. A proprietary trade name for Warfarin.

Coumalic Acid. α-Pyrone-3-carboxylic acid.

Coumaphos. An insecticide. o-3-Chloro-4'-methylcoumarin-7 yl oo-diethyl phosphorothioate. ASUNTOL.

Coumarin. o-Benzo-pyrone.

Coumarone Resin. See PARACOUMARONE RESIN.

Coumassie Union Blacks. See COOMASSIE UNION BLACKS.

Countess Powder. Cinchona bark in powder.

Count Palmer's Powder. Magnesium carbonate, $MgCO_3$.

Coupier's Blue. See INDULINE, SPIRIT SOLUBLE.

Courline. A proprietary trade name for polypropylene monofil yarns. (213).

Cournova. A proprietary trade name for polypropylene oriented slit film for the manufacture of strings, twines and ropes. (213).

Courto. A proprietary brand of artificial silk.

Court Plaster. Isinglass dissolved in water, alcohol, and glycerin, and painted on taffeta.

Courtzilite. A variety of asphalt.

Couvar. Bitumen varnish. (603).

C.O.V. Concentrated oil of vitriol (sulphuric acid), containing from 93–95 per cent. acid.

Covatin. The hydrochloride of captodiame.

Covellite (Blue Copper, Indigo Copper). A mineral. It is copper sulphide, CuS.

Coveral. Aluminium alloy cleansing flux. (531).

Coverite. A wetting agent. (501).

Covermark. A proprietary preparation of titanium dioxide with different pigments in a cream base. A skin masking cream. (379).

Covexin. Combined sheep vaccine. (514).

Cowdee Gum. See AUSTRALIAN COPAL.

Cowdie Gum. See AUSTRALIAN COPAL.

Cowhage. The hairs on the fruit of *Mucuma pruriens*. A mechanical irritant used to promote the expulsion of worms.

Cowles' Aluminium Bronze. An alloy containing 89–98·75 per cent. copper and 1·25–11 per cent. aluminium. One alloy contains small amounts of iron and silicon.

Cowrie Gum. Gum dammar.

Coxcomb Pyrites. See COCKSCOMB PYRITES.

Coxine. The name given to a red solution of coal-tar dyes, which absorb the actinic rays.

Coxpyrin. See ASPIRIN.

Coxycle. Copper oxychloride (56/58 per cent. Cu). (589).

Coxydust. Copper oxychloride dust. (589).

Coxyoil. Copper oxychloride in oil. (589).

Coxysan. Copper oxychloride (50 per cent. Cu) with wetting agent and spreader. (589).

Coxytrol. Cyproquinate. (682).

Coyden. See CLOPIDOL. (613).

Coyden 25. A proprietary preparation of CLOPIDOL. A veterinary antiprotozoan.

Coyuntla Juice. A juice obtained from a Mexican plant. It is used to coagulate rubber latex.

Cozirc 69. A proprietary trade name for a metal organic compound containing cobalt and zirconium. An improved high metal, lead free dryer used in paints. (190).

C.P.D. A rubber vulcanisation accelerator. It is cadmium-pentamethylene-dithio-carbamate.

C-Quens. A proprietary preparation of mestranol (14 yellow tablets) and mestranol and chlormadinone (7 pink tablets). Oral contraceptive. (333).

Crab Ointment. See MERCURY OINTMENT.

Crab-orchard Salt. A mild saline purgative obtained by evaporating the water of the springs at Crab Orchard, Lincoln County, Kentucky. It consists mainly of magnesium sulphate, together with some sodium chloride, sodium sulphate, potassium sulphate, and calcium sulphate, also a little iron, calcium, and magnesium carbonates, and silica.

Crab's Eyes. Found in crayfish. They consist of calcium carbonate and phosphate, with some animal matter. They are also called Crabstones. A prepared chalk is called Crab's eyes.

Crabstones. See CRAB'S EYES.

Crab Wood Oil (Andiroba Oil, Toulou-couna Oil). Carapa oil, obtained from *Carapa* species

Cracked Motor Spirit. Motor oils obtained by cracking oils of less volatile mineral oils, or tar oils.

Craig Gold. An alloy of 80 per cent. copper, 10 per cent. nickel, and 10 per cent. zinc. It is a nickel silver (German silver). Also see NICKEL SILVERS.

Cramp Bark. The bark of *Viburnum opulus*.

Cranco. Lacquers and enamels. (512).

Crandallite. A mineral. It is a hydrated phosphate of calcium and aluminium.

Crasnitin. A proprietary preparation of COLASPASE used as a cytotoxic drug. See ASPARAGINASE. (112).

Crategine. Anisaldehyde, $CH_3O.C_6H_4.CHO$.

Crayonne. A proprietary cellulose acetate moulding compound.

Crealbin. Egg albumin in combination with creolin. Used for the internal administration of creolin. It contains 50 per cent. creolin.

Cream. Cream of tartar (*q.v.*).

Cream Caustic. An inferior caustic soda from tank liquor.

Cream, Formalin. See FORMALIN CREAM.

Cream, Lambkin's. See LAMBKIN'S CREAM.

Cream of Tartar (Refined Tartar). Acid potassium tartrate, $CO_2K.CH(OH).CH(OH).CO_2H$.

Cream Powder. A name applied to a commercial calcium phosphate.

Creatine. Methyl-guanino-acetic acid.

Creatinine. Glycocoll-methyl-guanidine.

Crede's Silver. See COLLARGOL.

Crednerite. A Californian ore, $Cu_3Mn_4O_9$.

Creedite. A mineral. It is a hydrated fluoride and sulphate of calcium and aluminium.

Cremalgex. A proprietary preparation of methyl nicotinate, capsicum oleoresin, glycol salicylate and histamine hydrochloride. A rubefacient. (462).

Cremalgin. A proprietary preparation of methyl nicotinate, histamine dihydrochloride, capsicin and glycol salicylate. An embrocation. (137).

Cremathurm R. A proprietary preparation of methyl nicotinate, ethyl salicylate, histamine and capsicum oleoresin. A rubefacient. (637).

Crème de Cacao. Made from an infusion of cocoa, alcohol, and sugar.

Crème de Menthe. A sweetened liqueur. The flavour is obtained from fresh mint leaves, usually peppermint.

Crème de Roses. A sweetened liqueur. It has the odour of oil of roses.

Crème de Vanilla. A liqueur.

Crème de Violet. See CRÈME DE YVETTE.

Crème de Yvette (Crème de Violet). A liqueur which has an odour of violets, and has a violet colour.

Cremnite. A building material made from clay, sand, and fluorspar.

Cremnitz. See FLAKE WHITE.

Cremnitz White. See FLAKE WHITE.

Cremodiazine. A proprietary preparation of sulphadiazine. An antibiotic. (310).

Cremoglycic Acid. A preparation used in the treatment of allergic airway obstruction. 1, 3 - Di - (2 - carboxy - 4 - oxochromen - 5 - yloxy)propan - 2 - ol. It is present in INTAL and RYNACROM as the disodium salt.

Cremomycin. A proprietary preparation of succinylsulphathiazole, kaolin and neomycin sulphate. An anti-diarrhœal. (310).

Cremonite. An explosive consisting of 48·85 per cent. ammonium perchlorate, and 51·15 per cent. ammonium picrate.

Cremor Antispasmodic. A proprietary preparation of benzyl benzoate, camphor, bismuth carbonate, anise oil, chloroform and glycerin. (340).

Cremostrep. A proprietary preparation of streptomycin sulphate, succinyl-sulphathiazole and kaolin. An anti-diarrhœal. (310).

Cremosuxidine. A proprietary preparation of succinylsulphathiazole, and kaolin. An antidiarrhœal. (310).

Cremotresamide. A proprietary preparation of sulphadiazine, sulphameraxine and sulphadimidine. (310).

Crems White. See FLAKE WHITE.

Creocamph. A mixture of equal parts of creosote and camphoric acid. Used to minimise pain at the seat of an injection.

Creocide. A proprietary disinfectant containing creosote.

Creolin. (Proprietary Name). A coal tar disinfectant, germicide, antiseptic, deodorant. A dark brown liquid with a characteristic odour. Specific gravity 1·02–1·04. Approximate composition. Tar acids and oils 75–77 per cent.; emulsifying soaps 15–17 per cent.; water 8–10 per cent.; Forms stable emulsions when diluted with water.

Creoline-iodoform. A compound of creoline and iodoform. An antiseptic with only a faint aromatic odour.

Creolin-Pearson. See BACTEROL.

Creosal. See TANNOSAL.

Creosoform. A condensation product of formaldehyde and creosote. An internal antiseptic.

Creosol. Homocatechol methyl ester, $C_6H_3CH_3O.CH_3O$.

Creosol Soap. A potash soap soluble in water, containing 50 per cent. crude cresols. Used in midwifery as an antiseptic.

Creosotal. A carbonate of creosote, prepared by passing carbonyl chloride into a solution of creosote in caustic soda. Used as an internal antiseptic.

Creosote. A term used in reference to the mixed phenols and phenoloid bodies obtained from wood tar, coal tar, and other sources. The name is usually applied to the purified portion of beech-wood tar distilling between 200° and 220° C. Impure phenol is sometimes sold under this name.

Creosote Albuminate (Nutrincreosote). A compound of creosote and albumin. It is said to contain 40 per cent. creosote.

Creosote Carbonate. A mixture of neutral carbonic esters produced from a mixture of phenols.

Creosote-guaiacol (Homoguaiacol). Creosol.

Creosote Oils. See HEAVY OILS.

Creosote Phosphite (Phosphatol). Names given to preparations of guaiacol phosphite, $P(O.O_6H_4.OCH_3)$. Antiseptics.

Creosote Tannate. See TANNOSAL.

Creosotide. Creosote iodide, containing 25 per cent. iodine. Prescribed in incipient tuberculosis, also in asthma, and as a gastric and intestinal antiseptic.

Cresalols (Cresol salols). Cresol salicylate, $C_6H_4(OH)(COOC_6H_4CH_3)$. Generally the m-cresol compound is referred to by the above names.

Cresamol. See KRESAMINE.

Cresantol. Germicides. (559).

Cresaprol. See CRESIN.

Cresatin. m-Cresol-acetate, $CH_3.C_6H_4 OC_2H_3O$. External antiseptic and analgesic.

Cresaurin. The anhydride of trioxy-tritolyl-carbinol, $C_{22}H_{20}O_3$.

Cresavon. Hospital antiseptic soap. (513).

Crescormon. See GROWTH HORMONE.

Cresegol. See EGOLS.

Cresilite. An explosive. It is trinitro-cresol.

Cresin (Cresaprol). Made by dissolving 25 per cent. cresol in 75 per cent. of a solution of sodium-cresoxyl-acetate. Used in ½–1 per cent. solution as a disinfectant for wounds.

Cresineol. A proprietary preparation of cineol and o-cresol.

Cresival. Calcium cresol-sulphonate.

Cresochin. Said to be a neutral sulphonate of quinoline and tricresol. A disinfectant for surgical instruments.

Cresoform. A condensation product of formaldehyde and creosote. An antiseptic.

Cresol. Crude cresol contains approximately 35 per cent. o-, 40 per cent. m-, and 25 per cent. p-cresol, $C_6H_4(CH_3)OH$. Antiseptic.

Cresoleum. See BACTEROL.

Cresoline. See CREOLINE.

Cresol Purple. m-Cresol-sulphon-phthalein. An indicator.

Cresol Red. o-Cresol-sulphon-phthalein. An indicator.

Cresol-resorcin. See CRESORCIN.

Cresol Salols. See CRESALOLS.

Cresol, Saponified. See BACTEROL.

Cresol, Solid. See PARALYSOL.

Cresolution. Tar oil disinfectant. (601).

Cresopirine. Acetyl-o-cresotinic acid.

Cresorcin (Cresol-resorcin, α-Isorcin, Lutorcin, Cresorcinol). Dioxy-toluene, $C_6H_3(CH_3)(OH)_2.(1:2:4)$.

Cresorcinol. See CRESORCIN.

Cresosteril (Kresosteril). The acid o-oxalic ester of m-cresol, $CH_3.C_6H_4O. C(OH)_2.C(OH)_2.OC_6H_4.(CH_3)$. A germicide.

Cresotic Acids. Cresol carboxylic acids.

Cresotine Yellow G. A dyestuff. It is the sodium salt of diphenyl-disazo-bi-o-cresol-carboxylic acid, $C_{28}H_{20}N_4O_6 Na_2$. Dyes cotton yellow.

Cresotine Yellow R. A dyestuff. It is the sodium salt of ditolyl-disazo-bi-o-cresol-carboxylic acid, $C_{34}H_{24}N_4O_6 Na_2$. Dyes cotton yellow.

Cressylite. A mixture of picric acid and trinitrocresol. Used in explosives.

Crestalkyd. Alkyd resins. (542).

Crestanol. Modified phenolic resins. (604).

Crestapol RE 290, 292 and 294. A proprietary trade name for polyester plasticisers for vinyl plastics. (94).

Crestavin. A proprietary trade name for vinyl acetate resins.

Crested Barytes (Cockscomb Barytes). Aggregates of crystals of barytes, $BaSO_4$.

Crestmoreite. A mineral. It is a hydrated calcium silicate.

Creston. Solventless coating resins. (542).

Crestophen. Phenolic synthetic resins. (604).

Cresyl Blue BB, BBS. A dyestuff. It is possibly dimethyl-diamino-toluphenazonium chloride and homologues. Dyes tannined cotton blue.

Cresyl Hydrate (Cresylic acid). Cresol, $C_6H_4(CH_3)OH$.

Cresylic Acid. See CRESYL HYDRATE.

Cresylite. A French explosive. It is trinitro-cresol. Also see CRESILITE.

Cresylol. Cresol, $C_6H_4OH.CH_3$.

Cresyl Violet B, BB. A dyestuff closely related to Capri blue GN.

Cresyntan. See NERADOL.

Cretaform. Oxy-methyl-cresol-tannin.

Creto. A cationic surface-active agent. (526).

Crex. A registered trade name for a sodium sesquicarbonate, Na_2CO_3. $NaHCO_3.2H_2O$.

Crichtonite. A mineral. It is a variety of ilmenite, and contains from 50–53 per cent. of titanium dioxide.

Crill. Surface-active agents. (526).

Crillets. Surface-active agents. (526).

Crimson Lake. A cochineal lake containing more aluminium base than carmine.

Crimson Spirits. See TIN SPIRITS.

Crinagen. A proprietary preparation of 3,4,4-trichlorocarbanilide, allantoin, salicylic acid, and thymol in gel base. Antifungal, keratolytic skin application. (366).

Crinol. A proprietary artificial horse-hair made by the cuprammonium process.

Crinoline. An open weave cotton fabric which has been heavily sized.

Crin Végétal. A palm fibre used as a substitute for horsehair in upholstery.

Crisalbine. A proprietary preparation of gold-sodium thiosulphate for the treatment of tuberculosis.

Cristal. See FLINT GLASS.

Cristalbine. See CRISALBINE.

Cristalite. A microscopic mountant. (544).

Cristalline. See ZAPON VARNISH.

Cristaloid. A proprietary phenol-formaldehyde resin.

Cristite. A proprietary trade name for a steel containing 10 per cent. chromium, 17 per cent. tungsten, 3·5 per cent. carbon and 2·5 per cent. molybdenum.

Cristobalite. A mineral. It is a form of silica, SiO_2.

Cristolax. A combination of malt extract and liquid paraffin. Recommended as a laxative.

Crocalite. A red zeolite mineral.

Crocein B. A dyestuff. It is the sodium salt of benzene-azo-benzene-azo-α-naphthol-disulphonic-acid,$C_{22}H_{14}$ $N_4O_7S_2Na_2$. Dyes wool red from an acid bath.

Crocein 3B. A dyestuff. It is the sodium salt of toluene-azo-toluene-azo-α-naphthol-disulphonic acid, $C_{24}H_{18}N_4$ $O_7S_2Na_2$. Dyes wool red from an acid bath.

Crocein 3BX (Coccine 2B, Scarlet OOO). A dyestuff. It is the sodium salt of p-sulpho - naphthalene - azo - β - naphthol-mono-sulphonic acid,$C_{20}H_{12}N_2O_7S_2Na_2$. Dyes wool red from an acid bath.

Croceine. Benzene-azo-β-naphthol-6-sulphonic acid.

Croceine. 2-naphthol-8-sulphonic acid.

Croceinic Acid. See BAYER'S ACID.

Crocein Orange. See PONCEAU 4GB.

Crocein Orange (Brotherton & Co., Ltd.). A dyestuff. It is an equivalent of Orange GT.

Crocein Scarlet B. A mixture of Crocein scarlet 3B with Orange 3B.

Crocein Scarlet 2B. A mixture of Crocein scarlet 3B with Orange 7B.

Crocein Scarlet 3B (Ponceau 4RB). A dyestuff. It is the sodium salt of sulpho - benzene - azo - benzene - azo - β - naphthol-sulphonic acid, $C_{24}H_{14}N_4O_7S_2$ Na_2. Dyes wool scarlet from an acid bath, and cotton from an alum bath.

Crocein Scarlet 3BX. An azo dyestuff isomeric with Fast red E, prepared from diazotised naphthionic acid and β-naphthol-α-sulphonic acid. The commercial product is the sodium salt.

Crocein Scarlet 3B, 9187K. Dyestuffs. They are British brands of Brilliant crocein M.

Crocein Scarlet 4BX. See NEW COCCINE.

Crocein Scarlet 7B (Ponceau 6RB, Crocein Scarlet 8B). A dyestuff. It is the sodium salt of sulpho-toluene-azo-toluene - azo - β - naphthol- α - sulphonic acid, $C_{24}H_{18}N_4O_7S_2Na_2$. Dyes wool red from an acid bath.

Crocein Scarlet 8B. See CROCEIN SCARLET 7B

Crocein Scarlet O Extra. A dyestuff. It is the sodium salt of sulpho-benzene-azo-sulpho-benzene-azo-β-naphthol-sulphonic acid, $C_{22}H_{13}N_4O_{10}S_3Na_3$. Dyes wool and silk scarlet.

Crocein Scarlet R. A mixture of Crocein scarlet 3B with Orange 5B.

Crocein Sulphonic Acid. See BAYER'S ACID.

Crocein Yellow. A dyestuff. It is the potassium salt of dinitro-β-naphthol-sulphonic acid, $OK.C_{10}H_4(NO_2)_2.SO_3K$.

Crocell. A strippable coating composition. (526).

Crocic Acid. See BAYER'S ACID.

Crocidolite (Blue Asbestos). A fibrous mineral. It is a member of the group of minerals known as soda-amphiboles. It is a hydrated silicate of sodium and iron approximating to the formula $(Fe.Na_2)_4Si_4O_{12}.FeSiO_3$. A pale blue asbestos found in Bolivia and South Australia differing chemically and in quality, is also termed crocidolite. Good commercial crocidolite is found in South Africa and Western Australia, and the African varieties are marketed as Cape blue or Transvaal blue. The name crocidolite is also incorrectly applied to a variety of quartz, known as Tiger's eye.

Crocoisite (Crocoite). A mineral. It is a chromate of lead, $PbCrO_4$.

Crocoite. See CROCOISITE.

Crocq's Solution. A nerve stimulant injected hypodermically. It is a solution of sodium phosphate.

Crocus. Saffron, a colouring matter from the dried and powdered flowers of the saffron plant, *Crocus sativus*, used for colouring confectionery, and ferric oxide having a bluish tint used for polishing metals, are both known as crocus. See SAFFRON and INDIAN RED.

Crocus Martius. A hydrated ferric oxide. It is used as a pigment for pottery and other purposes.

Crocus of Antimony. See SULPHURATED ANTIMONY.

Crocus Saturni. Red lead (*q.v.*).

Croda Fluid. Lanolin rust preventatives. (526).

Crodafos. A range of phosphate anionic surfactants. (526).

Crodalan. Lanolin. (526).

Crodapur. A proprietary trade name for lanoline (pure). Croda is crude lanoline.

Crodax. Rust preventatives. (526).

Crodimyl. A proprietary trade name for Methylchromone.

Crodol. A barrier cream. (526).

Crodon. The trade name for a type of chromium plate that is claimed to be about ten times as hard as nickel plate. It does not oxidise below 700° F., and protects sheet from scaling up to 1500° F. It has been utilised by manufacturers of pyrometer parts, oil burner parts, oil cracking equipment, etc. It is not affected by salt water.

Crolax. A proprietary preparation containing dioctyl sodium sulphosuccinate and dihydroxyanthaquinine. A laxative. (280).

Croloy. A proprietary trade name for high chromium steel containing molybdenum and vanadium.

Cromal. An alloy of aluminium with 2–4 per cent. chromium and smaller amounts of nickel and manganese. It melts at 700° C., and is specially suitable for castings.

Cromaloy II. An alloy of 80 per cent. nickel, 15 per cent. chromium, and 5 per cent. iron. Its specific gravity is 8·15, and its specific resistance 110 micro-ohms per cm.[3]

Cromaloy III. An alloy of 85 per cent. nickel and 15 per cent. chromium. It melts at 1380° C., has a specific gravity of 8·25, and a specific resistance of 93 micro-ohms per cm.[3]

Cromaloy IV. An alloy of 80 per cent. nickel and 20 per cent. chromium. It melts at 1380° C., has a specific gravity of 8·35, and a specific resistance of 100 micro-ohms per cm.[3]

Cromfordite. See KERASITE.

Cromodined Iron and Steel. A process for forming a rust-preventing coat to iron and steel. It employs an aqueous solution of chromic acid and and activating agent.

Cromophtals. High performance organic pigments for plastics. A registered trade name. (18).

Cromo Steel. A proprietary trade name for a chrome-molybdenum steel.

Cronetal. A proprietary trade name for Disalfiram.

Cronite. An alloy of nickel and chromium. No. 1 contains 85 per cent. nickel, and 15 per cent. chromium.

Cronstedite. A mineral. It is Chlorite (*q.v.*), containing manganese.

Crookesite. A mineral, $(Cu.Tl.Ag)Se$.

Cropropamide. N - [α - (N - Crotonoyl - N - propylamino)butyryl]dimethyl - amine.

Cross Dye Black FNG. See KHAKI YELLOW C.

Cross Dye Blacks (Sulpho Blacks, Cross Dye Navy). Dyestuffs produced by the fusion of a variety of amino compounds and phenols with sodium polysulphide. They dye cotton dark blue to black.

Cross Dye Navy. See CROSS DYE BLACKS.

Crossite. A mineral. It is a variety of amphibole.

Cross-stones. See CHIASTOLITE.

Crotal. See CROTTLE.

Crotaline. The venom of the rattlesnake. It has been recommended in the treatment of phthisis and asthma.

Crotamiton. N - Crotonoyl - N - ethyl - o-toluidine, Eurax.

Crotethamide. N - [α - (N - Crotonoyl - N - ethylamino)butyryl]dimethyl - amine.

Croton Chloral Hydrate. Butylchloral-hydrate, $C_3H_4Cl_3CH(OH)_2$.

Crotonitrile. Allyl cyanide, CH_2:$CH.CH_2CN$.

Crotorite. A cupro-manganese alloy. It contains 68 per cent. copper, 30 per cent. manganese, and 2 per cent. iron.

Crotothane. A fungicide. (507).

Crotoxyfos. An insecticide. α - Methyl-benzyl 3 - (dimethoxyphosphinyloxy) - isocrotonate. CIODRIN.

Crottle (Crotal). A colouring matter obtained from certain lichens.

Cro-tung. A proprietary trade name for a chromium-tungsten steel.

Croweacin. An oil obtained from a shrub, *Eriostemon crowei* or *Crowea salique,* of New South Wales.

Crown Aloe. See ALOE.

Crown Glass. A glass composed of soda, lime, and sand. Used for sheet glass, plate glass, and optical work.

Crown Solder. An alloy of 63 per cent. tin, 18 per cent. zinc, 13 per cent. aluminium, 1 per cent. lead, 3 per cent. copper, and 2 per cent. antimony.

CR Resins. A proprietary trade name for allyl resins.

Crucible Cast Steel. See CAST STEEL.

Crude Antimony (Antimony Ore). Antimony trisulphide, Sb_2S_3, the liquated sulphide obtained from the ore.

Crude Potashes. When the wood or other parts of plants are burnt to ashes, the ashes lixiviated with water, and the solution evaporated to dryness, the residue when fused constitutes crude potashes.

Crude Tartar. See ARGOL.

Crude Wood Spirit. Contains 80 per cent. methyl alcohol, CH_3OH.

Crude Wood Vinegar. See PYROLIG-NEOUS LIQUOR.

Crufomate. An insecticide and anthelmintic. 4 - *tert* - Butyl - 2 - chloro - phenyl methyl methylphosphoramidate. RUELENE.

Crumpsall Direct (Fast Brown O). A dyestuff. It is the sodium salt of diphenyl - disazo - benzene - azo - phenyl-amino - naphthol - sulphonic - salicylic - acid, $C_{41}H_{27}N_7O_7SNa_2$. Dyes cotton olive brown.

Crumpsall Direct Fast Brown B. A dyestuff. It is the sodium salt of diphenyl - disazo - benzene - azo - amino-naphthol - sulphonic - salicylic acid, $C_{35}H_{23}N_7O_7SNa_2$. Dyes cotton dark brown.

Crumpsall Yellow. A dyestuff. It is the sodium salt of disulpho-naphtha-lene-azo-salicylic acid, $C_{17}H_{10}N_2O_9S_2$ Na_2. Dyes wool yellow.

Crurin (Bismolinide). Quinoline-bis-muth-thiocyanate, $(C_9H_7.N.HSCN)_3Bi$ $(SCN)_3$. Prescribed for internal use, also as an injection for gonorrhœa.

Cruverlite. A proprietary trade nam for a luminous plastic moulding powde

Crylene. A proprietary rubber vulcar isation accelerator. It is an acetald hyde-aniline condensation product (form of ethylidene-aniline).

Crylene Paste. A proprietary rubbe vulcanisation accelerator. It consist of aldehyde-amine and stearic acid.

Cryofine. See KRYOFINE.

Cryogen Blue GR. A sulphide dye stuff prepared from 1 : 8-dinitro-naph thalene.

Cryogen Brown. A sulphide dyestu prepared from 1 : 8-dinitro-naphthalene Dyes cotton brown from a salt bath.

Cryogenin (Kryogenin, Semi-benzide) Phenyl - semicarbazide, $C_6H_5.NH.NH$ $CONH_2$. It has antipyretic propertie and is employed in the treatment o phthisis. It has been described a Benzamino-semicarbazide.

Cryolite (Ice-spar, Ice-stone, Cryolith) A mineral. It is a fluoride of alumi nium and sodium, $AlF_3.3NaF$, anc occurs as deposits in Greenland.

Cryolith. See CRYOLITE.

Cryolithionite. A mineral. It is a fluoride of lithium, sodium, and alu minium, $Li_3Na_3Al_2F_{12}$.

Cryophylite. A mineral. It is a lithium mica of an emerald green colour.

Cryphiolite. A mineral. It is a phos-phate of magnesium and calcium.

Cryptargol. Sodium argento - thio - glycerin-sulphonate.

Cryptoclase. A mineral. It is a variety of albite.

Cryptohalite. A mineral, $(NH_4)_2SiF_6$.

Cryptol. A mixture of carbon, graphite, and carborundum. Used as a resisting material for heating in electric furnaces.

Cryptolite. A mineral. It is a variety of monazite.

Cryptomorphite. A mineral. It is a borate of calcium and sodium.

Cryptone (Titanox, Duolith, Titanolith, Tidolith). Proprietary trade names for white pigments containing lithopone and TiO_2.

Cryptopyrrole. 2-4-Dimethyl-3-ethyl-pyrrole.

Crysmalin. See PARAFFIN, LIQUID.

Crysolgan. A proprietary preparation. It is sodium-4-amino-2-aurothiophenol-carbonate.

Crysta. A bright zinc process. (605).

Crystal Carbonate. The name given to a pure monohydrate carbonate of sodium, $Na_2CO_3.H_2O$.

Crystalex. A proprietary trade name for an acrylic denture base.

Crystal Glass. A glass composed of lead and potassium silicates.

Crystalite. A proprietary trade name for an acrylic moulding powder.

Crystallin. A solution of 1 part pyroxylin, in 4 parts methyl alcohol, and 15 parts amyl acetate. A vehicle for remedial agents, such as chrysarobin and iodoform. Also a proprietary trade name for a phenol-formaldehyde synthetic resin.

Crystalline. Di-iodo-*p*-phenol-sulphonic acid.

Crystalline Boracite. See STASSFURTITE and BORACITE.

Crystalline Phosphorus. Yellow phosphorus.

Crystallised Gum. See BRITISH GUM, CRYSTALLISED.

Crystallised Verdigris. See CRYSTALS OF VENUS.

Crystallose. See SACCHARIN, EASILY SOLUBLE.

Crystal Mineral. Sal prunella.

Crystal Minium (True Red Lead). Commercial red lead, prepared by calcining lead oxide, PbO, in the form of massicot, not litharge.

Crystal Naphthylene Blue R. See MELDOLA'S BLUE.

Crystal Ponceau. See CRYSTAL SCARLET 6R.

Crystal Ponceau 6R. See CRYSTAL SCARLET 6R.

Crystal Scarlet 6R (Crystal Ponceau 6R, New Coccine R). A dyestuff. It is the sodium salt of α-naphthalene-azo-β-naphthol-disulphonic acid, $C_{20}H_{12}N_2O_7S_2Na_2$. Dyes wool scarlet from an acid bath.

Crystals of Venus (Crystallised verdigris). Copper acetate, $Cu(C_2H_3O_2)_2H_2O$.

Crystal, Tin. See TIN SALTS.

Crystal Vinegar. See DISTILLED VINEGAR.

Crystal Violet (Crystal Violet 5BO, Crystal Violet O, Violet C, Violet 7B Extra, Methyl Violet 10B). A dyestuff. It is the hydrochloride of hexamethyl-para-rosaniline, $C_{25}H_{30}N_3Cl$. Dyes silk and wool violet, also cotton mordanted with tannin and tartar emetic. It has also been used as an antiseptic for malignant growths.

Crystal Violet 5BO. See CRYSTAL VIOLET.

Crystal Violet O. See CRYSTAL VIOLET.

Crystamycin. A proprietary preparation containing Benzylpenicillin sodium and Streptomycin. An antibiotic. (335).

Crystamycin Forte. A proprietary preparation containing Benzylpenicillin Sodium and Streptomycin. An antibiotic. (335).

Crystapen. A proprietary preparation containing Benzylpenicillin Sodium. Sodium penicillin G. An antibiotic. (335).

Crystapen G. A proprietary preparation containing Benzylpenicillin (as the potassium salt). An antibiotic. (335).

Crystapen V. A proprietary preparation containing Phenoxymethylpenicillin (as the calcium salt). An antibiotic. (335).

Crystex. A proprietary trade name for a high-polymer form of sulphur. It is insoluble and not subject to blooming in rubber.

Crystic 193. A proprietary trade name for a methyl methacrylate modified unsaturated polyester resin, light stabilized, water clear and suitable for machine mouldings. (94).

Crystic 213. A proprietary trade name for a special high-impact unsaturated polyester resin used with glass fibre. (94).

Crystic 217. A proprietary trade name for an unsaturated bisphenol polyester resin. Used with glass fibre for industrial mouldings. (94).

Crystic 326. A proprietary trade name for a filled unsaturated polyester resin. (94).

Crystolite. See CARBORA.

Crystolite Paint. A proprietary trade name for a corrosion-resisting paint made from refined coke-oven pitch and certain oils.

Crystex. A proprietary viscose product used for packing.

Crystolon. A registered trade name for various types of goods such as grinding-wheels, abrasives, and refractory grain, etc., made from silicon carbide.

Crystopurpurin. Hexamethylene-tetramine-sodium acetate, $(CH_2)_6N_4.2CH_3COONa.6H_2O$. A urinary antiseptic.

Csiklovaite. A mineral. It is Bi_2TeS_2.

CT-680. A polyurethane elastomeric resin possessing good heat resistance. (434).

CT-690, CT-700. Polyurethane elastomeric resins possessing greater heat resistance than CT-680. (434).

Cuba Black (Dianil Black). A dyestuff. It is the sodium salt of sulpho-diphenyl-disazo-bi-sulpho-naphthol disazo-bi-*m*-phenylene-diamine, $C_{44}H_{31}N_{12}O_{11}S_3Na_3$. Dyes cotton black from an alkaline bath.

Cubanite. A mineral, $CuFe_2S_4$.

Cuba Orange. A dyestuff prepared by the action of sodium sulphite upon diazo-naphthalene-sulphonic acid. Dyes wool orange.

Cuba Wood. See FUSTIC.

Cubeb Camphor. The solid part of cubeb oil.

Cubeb Paste. Powdered cubebs mixed with copaiba.

Cubeite. A mineral. It is a hydrated sulphate of iron and magnesium. Synonym for Rubrite.

Cube Ore. Tyrolite (q.v.).

Cube Powder. A similar explosive powder to Imperial Schultze (q.v.).

Cubical Catechu. See CUTCH.

Cubic Nitre. See CHILE SALTPETRE.

Cubic Saltpetre. See CHILE SALTPETRE.

Cucar. Copper carbonate. (589).

Cuclat. Kerosene sweeting catalyst. (589).

Cudbear. See ARCHIL.

Cudinoc. A proprietary trade name for the copper derivative of dinitro-o-cresol. A herbicide.

Cuemid. A proprietary trade name for cholestyramine.

Cufenium. An alloy of 22 per cent. nickel, 72 per cent. copper, and 6 per cent. iron. Used for table ware.

Cufor. Copper formate. (589).

Cuite. Natural silk freed from the silk gum or sericin (q.v.). Raw silk consists of about 66 per cent. fibroin, $C_{15}H_{23}N_5O_6$, forming the real silk substance, and silk gum or sericin. The silk gum is removed by treating with hot neutral soap solution. The de-gummed silk is called cuite.

Cuivre Poli. See ABYSSINIAN GOLD.

Cullen Earth. Cologne Earth (q.v.).

Cullet. Broken glass, used in the manufacture of glass.

Culm. A term used for anthracite.

Culsageeite. A mineral. It is a variety of jefferisite.

Culver's Root. Leptandra U.S.P.

Cumaline. A rubber solution said to be suitable for use with paints and varnishes.

Cumar. See PARACOUMARONE RESIN.

Cumar Gum. See PARACOUMARONE RESIN.

Cumar Resin. See PARACOUMARONE RESIN.

Cumate. A proprietary trade name for a vulcanising accelerator containing copper.

Cumengeite. A mineral, $4PbCl_2.4CuO.5H_2O$.

Cumetharol. 4,4' - Dihydroxy - 3,3' - (2 - methoxyethylidene)di - coumarin. Dicumoxane.

Cumidine Ponceau. See PONCEAU 3R.

Cumidine Red. See PONCEAU 3R.

Cuminol. Paracuminic aldehyde.

Cummingtonite. A mineral, $(Fe.Mg)SiO_3$.

Cumol. Cumene, $C_6H_5CH(CH_3)_2$.

Cumopyran. A proprietary trade name for cyclocumarol.

Cunerol (Kunerol). A margarine made from cocoa-butter, treated with a saline solution containing yolk of egg.

Cuniloy. An alloy of nickel, manganese, copper, and small quantities of lead.

Cunitex. Bird and animal repellant. (507).

Cuoxam. A solution containing 2·5 grams copper carbonate 50 c.c. strong ammonia, and 50 c.c water. It dissolves cellulose.

Cupac. Copper acetate. (589).

Cupal. A German copper-plated aluminium.

Cupar. Copper arsenate. (589).

Cuperatin. Copper albuminate.

Cupferron (Copperone). Nitroso-phenyl-hydroxylamine, $C_6H_5.N(NO).OH$. The ammonium salt is used as a precipitating agent for copper, in the determination of copper.

Cupoline. A refractory for cupolas. It contains silica, iron oxide, alumina, and magnesia.

Cupolloy. Ferro-alloy briquettes for cupola. (531).

Cupragol. A compound of copper with proteic acid. An external antiseptic.

Cupralgin. Copper alginate.

Cuprammonium Silk (Cuprate Silk). Artificial silks made by dissolving cotton in a cuprammonium solution (copper hydrate dissolved in ammonia), and precipitating it in a fine thread.

Cuprammonium Solution. See CUPRAMMONIUM SILKS and WILLESDEN FABRICS.

Cupranil Brown B. A dyestuff. It is a British equivalent of Diamine brown B.

Cupranium. A name for certain brass and bronze alloys. The bronze contains tantalum and vanadium.

Cuprargol. A silver-copper albuminate. Used as an astringent in ophthalmology in a 20 per cent. solution.

Cuprase. A form of colloidal copper. A proprietary trade name for a vulcanising accelerator containing copper.

Cuprate Silk. See CUPRAMMONIUM SILK.

Cupratin. Copper albuminate, used as a germicide.

Cuprea Bark. The bark of *Remijia species*.

Cuprein. A name applied to mineral copper sulphide, Cu_2S.

Cupren. Polymerisation products of acetylene, obtained by the action of copper oxide, CuO, upon acetylene at 230–260° C. By repeated treatments, products containing only very small amounts of copper are obtained. Used as an explosive constituent. A certain product of this type is employed as a cork substitute.

Cuprene. See CUPREN.

Cuprex. (1) A solution of copper abietate in a mixture of benzene and paraffin oil. An insecticide.

Cuprex. (2) Copper alloy general purpose flux. (531).

Cupri-adeptol. See CUPRI-ASEPTOL.

Cupri-aseptol (Cupri-adeptol). Copper m-phenol-sulphonate, $(HO.C_6H_4.SO_3)_2$ $Cu.6H_2O$. A $\frac{1}{2}$–1 per cent. solution is used as a styptic and antiseptic.

Cupricine (Cuprisin). A preparation of copper and potassium cyanides. It has been recommended medicinally for trachoma.

Cupriferous Calamine. See TYROLITE.

Cuprimine. A proprietary preparation of PENICILLAMINE. (472).

Cuprinol. A copper naphthenate or sodium pentachlorphenate preparation used as a preservative for timber and fabrics.

Cuprisin. See CUPRICINE.

Cuprit. Copper alloy cleansing flux. (531).

Cuprite (Ruby Copper, Red Copper Ore). A mineral. It is cuprous oxide, Cu_2O.

Cuproadamite. A mineral. It is $4[(Zn,Cu)_2AsO_4.OH]$.

Cupro-aluminiums. Alloys of aluminium and copper containing 1–20 per cent. aluminium. They are also wrongly called aluminium bronzes.

Cuprobinnite. Binnite ($q.v.$).

Cuprobismuthite. A mineral, $Cu_6Bi_8S_{15}$.

Cuprocalcit. A fungicide consisting of 20–25 per cent. copper sulphate and 75–80 per cent. calcium carbonate.

Cuprocalcite. A mineral containing calcium carbonate and cuprous oxide.

Cuprocassiterite. A mineral, $4SnO_2$. $Cu_2Sn(OH)_6$.

Cuprocide. A proprietary trade name for red copper oxide.

Cuprocitrol. Copper citrate, $2Cu_2C_6$ $H_4O_7.5H_2O$. Similar to Cusylol ($q.v.$).

Cuprocollargol. A combination of electro-colloidal silver solution and electro-colloidal copper solution. A useful medicine in veterinary work for septic illnesses.

Cuprocyan. A double cyanide of potassium and copper.

Cuprocyanid. Copper cyanide.

Cuprodesclozite. A mineral, $(PbOH)$ $(Pb.Cu.Zn)VO_4$.

Cuprodine. A medium for coating steel with copper. (512).

Cuprodust. Cuprous oxide dust. (589).

Cuproiodargyrite. A mineral. It is a copper-silver iodide.

Cuprol. The copper salt of nucleic acid, containing 6 per cent. copper. An

external antiseptic, and has been recommended in the treatment of trachoma.

Cupromagnesite. A mineral, $(Cu.Mg)$ $So_4.7H_2O$.

Cupromagnesium. An alloy of 90 per cent. copper and 10 per cent. magnesium. It has a specific gravity of 8·4, melts at 1290° C., and is used as a deoxidiser.

Cupro-manganese. See MANGANESE COPPER.

Cupron. A proprietary trade name for a copper nickel alloy with a low temperature coefficient of resistance.

Cupron. A name suggested for benzoin-oxime, $C_6H_5.CNOH.CNOH.C_6H_5$, a spot reagent for copper. Dilute ammonia is added to the copper solution, then an alcoholic solution of benzoin-oxime, when a greenish precipitate is formed. This is filtered off and dried at 105–115° C., and weighed. Copper = 22·02 per cent. of the weight of the precipitate.

Cupro-nickel. See NICKEL-COPPER ALLOYS.

Cupronova. Yellow cuprous oxide 50 per cent. wettable powder. (589).

Cuprophenyl. An after coppering direct dye. (543).

Cuproplumbite. A mineral. It is a basic copper-lead arsenate, found in South-West Africa.

Cuproplumbite (Plumbocuprite). A mineral, $Cu_2S.2PbS$.

Curopyrite. An impure form of the mineral Chalcopyrite.

Cupror. An aluminium bronze. It contains 94·2 per cent. copper and 5·8 per cent. aluminium.

Cuproscheelite (Cuprotungstite). A mineral, $(Cu.Ca_2(WO_4)_3)$.

Cupro-silicon. Copper silicide. Cu_4Si.

Cuprosklodowskite. A mineral. It is $Cu(UO_2)_2Si_2O_7.6H_2O$.

Cupro-steel. An alloy of steel with copper up to 4 per cent. Occasionally used for printing rollers and projectiles.

Cupro-titanium. Usually an alloy of copper with 10 per cent. titanium. Used as a deoxidiser in making brass and bronze castings.

Cuprotungstate. A tungsten mineral, $CuWO_4$.

Cuprotungstite. See CUPROSCHEELITE.

Cuprouranite (Torberite). A mineral. It is a hydrated phosphate of uranium and copper, $CuO.2UO_2.P_2O_5.8H_2O$.

Cuprous 80 and 50. Cuprous oxide 80 per cent. and 50 per cent. wettable powder. (589).

Cuprovanadinite. A mineral. It is $2[(Pb,Cu)_5(VO_4)_3Cl]$.

Cupro-vanadium. An alloy usually containing from 10–15 per cent. vanadium, 60–70 per cent. copper, 10–15 per cent. aluminium, and 2–3 per cent. nickel.

Cuprozincite. A mineral. It is a basic carbonate of copper and zinc, $(Cu.Zn)CO_3(Cu.Zn)(OH)_2$.

Cuprylox. Colloïdal cuprous oxide. (589).

Curaçao. A liqueur prepared from oranges in the Island of Curaçao.

Curaçao Aloe. See ALOE.

Curacit. The sodium salt of cholic or cholloic acid. A solvent used in the treatment of wool for dyeing.

Curacit-natron. The sodium salt of the crude mixture of the gall acids of ox-gall. Used as a soap for washing delicate fabrics.

Curare (Woorari, Woorara, Woorali, Urari). A South American poison.

Curaril. A strong and stable preparation of curare, the arrow poison of the South American Indians.

Curathane. A proprietary trade name for a moisture cured polyurethane resin used for clear floor coatings. (189).

Curcia Nuts. See BASSOBA NUTS.

Curcuma. See TURMERIC.

Curcumeine. See CITRONINE B and FAST YELLOW N.

Curcumine. Geigy. See BRILLIANT YELLOW S.

Curcumine S. See SUN YELLOW.

Curcuphenine. A dyestuff. It is the sodium sulphonate of the dehydro-thio-toluidide of azoxy-stilbene-aldehyde. Dyes cotton yellow.

Curd Soap. A soap made with caustic soda and a purified animal fat, consisting mainly of stearin. It is therefore a sodium stearate. See also STEARIN SOAP, TALLOW SOAP, ANIMAL SOAP.

Curetard. A retarding agent for rubber. (559).

Curgon. A proprietary trade name for naphthenate driers.

Curite. A radio-active mineral, found in the Belgian Congo. It is a hydrated uranate of lead and uranium, $2PbO.5UO_3.4H_2O$.

Curodex. Rubber odorants and deodorants. (504).

Curry. The powdered leaves of *Murraya kœnigii* of India.

Curtisite. A new organic mineral from California. Its formula approximates to $C_{24}H_{18}$, and it is associated with sandstone and silica.

Cusamon. Copper sulphate monohydrate. (589).

Cusatrib. Tribasic copper sulphate. (589).

Çus-cus Oil. Vetiver oil, derived from the roots of *Vetiveria zizanioides*. Used in perfumery.

Cusiloy A. A proprietary trade name for an alloy containing 95·5 per cent. copper, 3 per cent. silicon, 1 per cent. iron, and 0·5 per cent. tin.

Cusisa. A copper-arsenic preparation used as a fungicide.

Cusol. Cusylol (*q.v.*).

Cusso. The flowers of *Brayera anthelmintica*.

Custerite. A mineral. It is a hydrated fluosilicate of calcium.

Cusyd. Copper thiocyanate. (589).

Cusylol. Soluble copper citrate, $2Cu_2C_6H_4O_7.5H_2O$.

Cutal. Cutol (*q.v.*).

Cutch (Kutch, Catechu, Katechu, Gambier, Japan earth, Cachou, Cutt). Natural dyestuff. It consists of a brown or reddish-brown amorphous extract, obtained by boiling with water the wood of certain kinds of *Acacia*. Cutch, Catechu, Black catechu, or Pegu catechu, is obtained by boiling in water the wood of *Acacia catechu*, or the wood and fruit of *betel* or *arecanus*, and evaporating.

Gambier, Pale catechu, Terra japonica, Yellow catechu, or Cubical cutch, is obtained by boiling the leaves of *Uncaria gambir*.

Bombay catechu is obtained from the fruit and heartwood of *Areca catechu*.

Bengal catechu is made from the unripe fruit pods and twigs of *Mimosa catechu*.

Katechu preparations are made by heating brown or yellow catechu with potassium bichromate, aluminium sulphate, or copper sulphate.

The dyeing principles of cutch are catechin, $C_{15}H_{14}O_6$, catechu-tannic acid, together with some quercetin. It is used for the production of browns on cotton, fixed with bicarbonate, or a copper salt. Used in preparation of some leathers, also for stains.

Borneo cutch is extracted from a species of Borneo mangrove, the so-called tengah bark or bastard mangrove. It contains 58·2 per cent. tannin. Burma cutch is from wood of *Acacia* species.

Cutisan. A proprietary preparation of trichloro-3,4,4'-carbanalide. (320).

Cutlanego. An alloy of 50 per cent. bismuth and 50 per cent. tin. Used for tempering steel tools.

Cutol (Lutol, Cutal). Aluminium-borotannate. An antiseptic and astringent.

Cutrene. A German preparation. It is a 10 per cent. suspension of the bismuth salt of *o*-oxy-quinoline-sulphonic acid. Antileutic.

Cutt. See CUTCH.

Cutting Oils (Soluble Oils, Cutting

Compounds). Materials used for the lubrication of the cutting tools in the machining of metals. They are either oils such as lard or rape, or a mixture of these with mineral oil, or they are emulsions containing oil, soap, and water. These liquids lubricate, but also carry away heat from the work and prevent oxidation. The oils are known as cutting oils, and mixtures for preparing emulsions as soluble oils or cutting compounds.

C.V.K. Compocillin VK (q.v.). (512).

Cyacetazide. An anthelmintic. Cyano-acetohydrazide.

Cyanacryl. A proprietary trade name for a synthetic rubber with excellent resistance to oils and high temperatures. (162).

Cyananthrene B Double. An anthracene dyestuff prepared by melting benzanthrone-quinoline with alkali hydroxides.

Cyanaprene A-8, A-9, D-5, D-6, D-7 and 4590. Trade names for urethane elastomers used with curatives CYAMA-SET M.S.H. and B to obtain various properties. Trade marks. (56).

Cyanegg. Sodium cyanide, NaCN, in egg-shaped lumps.

Cyangran. Sodium cyanide, NaCN, in granular form.

Cyanide Wool. Mercury and zinc cyanide wool.

Cyanine (Leitch's Blue, Quinoline Blue). A blue pigment consisting of a mixture of Cobalt blue and Prussian blue. A quinoline dyestuff, also known as Cyanine or Quinoline blue, is lepidine-quinoline-amyl-cyanine-iodide, $C_{29}H_{35}N_2I$. It is used as a panchromatic sensitiser, and also dyes silk.

Cyanine B. A dyestuff obtained by the oxidation of Patent blue with ferric salts or chromic acid. Dyes wool indigo blue.

Cyanine Blue. See CYANINE.

Cyanine Moderns. Dyestuffs, which are condensation products of gallo-cyanines and allyl-diamines.

Cyanite. See KYANITE.

Cyanite, Fireproof. See FIREPROOF CYANITE.

Cyanochalcite. Chrysocolla (q.v.).

Cyanochroite (Cyanochrome). A mineral $K_2SO_4.CuSO_4.6H_2O$.

Cyanochrome. See CYANOCHROITE.

Cyanocobalamin. α - (5,6 - Dimethyl - benzimidazolyl)cobamide cyanide. Vitamin B_{12} Anacobin ; Cobastab ; Cobione ; Cytacon ; Cytamen.

Cyanogas. A proprietary calcium cyanide for use as an ant killer.

Cyanogen Mud. Formed in gas purifiers, where coal gas free from tar is brought into contact with a saturated solution of ferrous sulphate, when the mud is produced. Prussian blue, to the extent of 30 per cent. is contained in this mud.

Cyanogran. A proprietary trade name for a granulated form of sodium cyanide (cf. CYANEGG).

Cyanoids. Sodium cyanide. (512).

Cyanol Extra (Acid Blue 6G). A dyestuff. It is the sodium salt of m-oxy-diethyl-diamino-phenyl-ditolyl-carbinol disulphonic acid, $C_{25}H_{28}N_2O_8S_2Na_2$. Dyes wool and silk blue from an acid bath.

Cyanol FF. A dyestuff of the same class as Cyanol extra. It dyes wool and silk from an acid bath.

Cyanol Green B. A triphenyl-methane dyestuff, which dyes wool from a bath containing sodium sulphate and sulphuric acid.

Cyanolime. Calcium cyanide. (589).

Cyanolit. An alpha cyanoacrylate. An adhesive containing no solvent and forming strong bonds on pressure alone. (261).

Cyanolit-Hitemp. A proprietary range of high-temperature cyanoacrylate adhesives. (434).

Cyanolite. A calcium silicate mineral, of Nova Scotia.

Cyanosin. See METHYL PHLOXIN and PHLOXINE.

Cyanosin A (Cyanosin, Spirit Soluble). The alkali salt of tetrabromo-dichloro-fluoresceine-methyl-ether, $C_{21}H_7Cl_2Br_4O_5K$. Used in silk dyeing.

Cyanosin B. A dyestuff. It is the sodium salt of tetrabromo-tetrachloro-fluoresceine-ethyl-ether, $C_{22}H_7Cl_4Br_4O_5Na$. Dyes wool bluish-red.

Cyanosin Spirit Soluble. See CYANO-SIN A.

Cyanosite. See CHALCANTHITE.

Cyanotrichite. A mineral. It is a hydrated sulphate of aluminium and copper.

Cyan Salt. Produced by the fusion of ferrocyanide with sodium carbonate.

Cyanthrene. See INDANTHRENE DARK BLUE BO.

Cyanthrol R. A dyestuff. C_6H_4COCO $C_6H.CH_3(NH_2).NH(NaSO_3.CH_3.C_6H_3)$ from 2-methyl-anthraquinone nitrating to the 1-nitro compound, reducing to the 1-amino compound, brominating and replacing the bromine by p-tolyl-amino group and finally sulphonating.

Cyarsal. Potassium-p-cyan-mercuri-salicylate.

Cyasorb UV-9. A registered trade mark for 2 - hydroxy - 4 - methoxybenzophe - none. An ultraviolet absorber for plastics. (56).

Cyasorb UV-24. A registered trade mark for 2,2'-dihydroxy-4-methoxybenzophenone. An ultraviolet absorber for plastics. (56).

Cyasorb UV-531. A trade mark for an ultraviolet light absorbing additive for plastics. (56). See TINUPAL also.

Cybond WD-4517 and WD-4521. Proprietary two-component polyurethane adhesives of the solvent type. (434).

Cyclamic Acid. N-Cyclohexylsulphamic acid. Hexamic acid.

Cyclamine. A dyestuff, obtained by the bromination of the product of the action of sodium sulphide upon dichloro-fluoresceine. It dyes wool and silk bluish-red from a neutral bath.

Cyclamycin. A proprietary preparation of triaceto-oleandromycin. An antibiotic. (245).

Cyclandelate. 3,3,5 - Trimethylcyclo - hexyl mandelate. Cyclospasmol.

Cyclarbamate. 1,1 - Di(phenylcarbamoyloxymethyl)cyclopentane. Casmalon.

Cyclatex. A proprietary cyclised rubber. (417).

Cycleweld. A proprietary trade name for synthetic adhesives of the thermosetting type.

Cyclimorph. Cyclizine tartrate and morphine tartrate. (514).

Cycline Oil. A trade mark for a blend of vegetable and mineral oils. Used as a softener in rubber mixings (0·5–2 per cent.).

Cyclite. A proprietary cyclised rubber. (223).

Cyclizine. 1 - Methyl - 4 - α - phenyl - benzylpiperazine. Marzine is the hydrochloride; Valoid is the hydrochloride or the lactate.

Cyclobarbital. Ethyl-cyclohexyl-barbituric acid.

Cyclobarbitone. 5 - (Cyclohex - 1 - enyl) - 5 - ethylbarbituric acid. Phanodorm.

Cyclocaine. See CYCLOFORM.

Cyclochem. Self emulsifiable wax. (606).

Cyclocoumarol. 5′6′ - Dihydro - 6′ - methoxy - 6′ - methyl - 4′ - phenyl - pyrano - (3′,2′; 3,4)coumarin. Cumopyran.

Cyclodex. Dispersible driers for emulsion paints. (607).

Cyclofenil. A drug used in the treatment of sterility. 4, 4′ - Diacetoxy - benzhydrylidenecyclohexane. ONDONID, SEXOVID.

Cyclofor. A paint additive. (606).

Cycloform (Cyclocaine). The isobutyl ester of *p*-amino-benzoic acid, C_6H_4 $(NH_2)COOC_4H_9$. Recommended as a local anæsthetic.

Cyclogenin. *m*-Benzamino-semi-carbazide, $NH_2.CO.C_6H_4.NH.NH.CO.NH_2$. An antipyretic.

Cyclogol. A non-ionic emulsifier. (606).

Cycloguanil Embonate. 4,6-Diamino-1 - (4 - chlorophenyl) - 1,2 - dihydro -2, 2 - dimethyl - 1,3,5 - triazine compound (2 : 1) with 4,4′-methylenedi-(3-hydroxy-2-naphthoic acid).

Cyclogyl. A proprietary trade name for Cyclopentolate.

Cyclohexanol (Hexalin, Sextol, Adronal). Hexahydro-phenol, $C_6H_{11}OH$. A solvent for gums, oils, waxes, rubber, and nitro-cellulose. It is added to lower-boiling solvents for lacquers, as it produces glossy films. It is also used as an emulsifier in detergents. The pure alcohol boils at 160° C., has a density of 0·94 and a flash-point of 155° F. The commercial material boils from 160–180° C., has a density of 0·93–0·94 and flash-point of 155° F.

Cyclohexanone (Hexanon, Anon, Sextone). Keto-hexamethylene, CH_2CH_2. $CH_2.CH_2.CH_2.CO$. A solvent for cellulose acetate, rosin, ester gum, rubber, shellac, oils, waxes, etc. It has a specific gravity of 0·93–0·96, a boiling-point of 150–165° C., a flash-point of 117–148° C., and a refractive index of 1·443–1·451.

Cyclolac. Tough hard rigid polymers. (608).

Cyclomet. A proprietary preparation of cyclobarbitone and ipecacuanha. A hypnotic. (360).

Cyclomethycaine. 3 - (2 - Methyl - piperidino)propyl 4-cyclohexyloxybenzoate. Surfathesin is the sulphate.

Cyclomorph. A proprietary preparation of morphine tartrate and cyclizine tartrate. An analgesic. (277).

Cyclon. A 30 per cent. aqueous solution of hydrocyanic acid to which has been added 10 per cent. ethyl-chloro-carbonate. A fumigator and insecticide.

Cyclonal. A proprietary preparation of hexobarbitone sodium. Intravenous anæsthetic. (336).

Cyclonal Sodium. A soluble hexabarbitone. (507).

Cyclonite. Cyclo-trimethylene-tetramine.

Cyclonox. Cyclohexanone peroxide. (587).

Cyclopentamine. 1 - Cyclopentyl - 2 - methylaminopropane. Clopane is the hydrochloride.

Cyclopenthiazide. 6 - Chloro - 3 - cyclopentylmethyl - 3,4 - dihydrobenzo - 1,2,4-thiadiazine-7-sulphonamide 1,1-dioxide. Navidrex.

Cyclopentolate. 2 - Dimethylamino - ethyl α - 1 - hydroxycyclopentyl - α - phenylacetate. Cyclogyl; Mydrilate.

yclophosphamide. 2 - [Di(2 - chloro - ethyl)amino] - 1 - oxa - 3 - aza 2 - phosphacyclohexane 2-oxide. Cytoxan; Endoxanal.

yclopol. A detergent composition. (606).

yclops Metal. A nickel-chromium-iron alloy, containing 18 per cent. nickel, 18 per cent. chromium, and the rest iron. It resists corrosion.

yclorans. Wetting-out agents for textiles. They contain an alcohol of high boiling-point emulsified with a potassium olein soap. Also see TERPURILE.

yclorubbers. Thermoplastic products made by heating a mixture of rubber sheet with about 10 per cent. of its weight of an organic sulphonyl chloride or an organic sulphonic acid. These products resemble gutta-percha or can be made to resemble shellac according to treatment. They are termed Thermoprene GP if like gutta-percha or Thermoprene SL if like shellac.

yclosal. Methyl-isopropyl-cyclo-hexanone and sodium salicylate.

yclosan. 4 per cent. Calomel dust. (507).

ycloserine. D - 4 - Aminoisoxazolid - 3-one. Seromycin.

yclospasmol. A proprietary preparation of cyclandelate. A vasodilator. (317).

yclothiazide. A diuretic currently under clinical trial as " MDi 193 ". 3 - (Bicyclo[2.2.1]hept - 5 - en - 2 yl) - 6 - chloro - 3, 4 - dihydro - 1, 2, 4 - benzothiadiazine - 7 - sulphonamide 1, 1 - dioxide. 6 - Chloro - 3, 4 - di - hydro - 3 - (norborn - 5 - en - 2 - yl) - 1, 2, 4 - benzothiadiazine - 7 - sulphon - amide 1, 1 - dioxide.

yclotropin. A urotropine-salicyl-caffeine preparation of German manufacture.

yco. A material made from vegetable gums. It is a puncture fluid.

ycolac. A proprietary range (434) of A.B.S. moulding materials classified as follows:

CIT: clear; DF: medium impact; GT-4502: giving high gloss on thin cross-sections; JH: cellular material. (434).
JP: Cellular A.B.S. used in injection moulding. (434).
JS: Cellular A.B.S. used in expansion casting. (203).
KL: High-impact, self-extinguishing A.B.S. injection-moulding material.
KJ: Self-extinguishing A.B.S. injection-moulding material, easy to process.
KT: High-impact A.B.S. of self-extinguishing extrusion grade.

LTH and LTHP: Moulding grades of A.B.S. which retain their toughness at low temperatures.
T-4501-G: A moulding grade of A.B.S.
X-17: A high-flow moulding grade of A.B.S. subject to lower residual strains.

Cycoloy KHE. A proprietary A.B.S./polycarbonate mixture used as a sheet-moulding compound. (433).

Cycostat. A proprietary preparation of ROBENIDINE hydrochloride. A veterinary anti-protozoan.

Cycrimine. 1 - Cyclopentyl - 1 - phenyl - 3-piperidinopropan-1-ol.

Cydril. A proprietary trade name for the succinate of Levamphetamine.

Cyfol. A proprietary preparation of ferrous gluconate, folic acid and Vitamin B_{12}. A hæmatinic. (273).

Cylert. A proprietary preparation of PEMOLINE used to treat hyperkinesis in children. (311).

Cylinder Oils. Petroleum lubricating oils. Dark cylinder oils are the undistilled residues which remain in the stills after the steam distillation of non-asphaltic base crude oils. Filtered cylinder oils are made from dark cylinder oils by filtration through ignited fuller's earth. It is sometimes mixed with from 3–10 per cent. acid-free tallow oil to increase oiliness.

Cylindrite. A mineral. It is a sulphide ore of tin.

Cylline. A disinfectant. It is a coal-tar oil, made soluble by the help of fatty acids and resin acids.

Cymag. A powder for fumigation evolving hydrocyanic acid. (511).

Cymanol. A proprietary trade name for methyl-isopropyl-benzene.

Cymarose. A glucoside occurring in Canadian Hemp. It is 3-methyl ether of digitoxose.

Cymbal Brass. See BRASS.

Cymbal Metal. A brass containing 78 per cent. copper and 22 per cent. zinc.

Cymbilide. Cyclamen aldehyde. (507).

Cymenol. Carvacrol, $C_6H_3CH_3.CH(CH_3)_2OH$.

Cymogen (Rhigolene). The liquid obtained by submitting the gases, originally dissolved in crude American petroleum, to cold and pressure. It consists of nearly pure butane, having a boiling-point of 10° C. Used in refrigerators.

Cymogran. Phenylalanine-free casein hydrolysate. (510).

Cymol. Cymene, $CH_3.C_6H_4.CH_2.CH_2.CH_3$.

Cymophane. A precious stone. It is beryllium aluminate.

Cymrite. A mineral. It is $64[BaAlSi_3O_8OH]$.

Cymyl Orange. A new indicator. It is an azo dye obtained by combining diazotised sulphonated amino-cymene with dimethyl-aniline.

Cynomel. A proprietary trade name for the sodium derivative of Liothyronine.

Cynorex. A bright cyanide copper plating process. (605).

Cyprenorphine. 12 - Cyclopropyl - methyl - 1,2,3,3a,8,9 - hexahydro - 5 - hydroxy - 2a - (1 - hydroxy - 1 - methyl - ethyl) - 3 - methoxy - 3,9a - etheno - 9,9b - iminoethanophenanthro[4,5 - bcd]-furan. N-Cyclopropylmethyl-7,8-dihydro - 7a(1 - hydroxy - 1 - methyl - ethyl)O methyl - 6ml4 - endoethenonor-morphine. N-Cyclopropylmethyl-7a(1-hydroxy - 1 - methylethyl) - 6,14 - endoethenotetrahydro-oripavine. N-Cyclopropylmethyl - 19 - methylnoro - vinol. M.285 is the hydrochloride.

Cypress Pine Resin. See PINE GU.

Cyprian Earth. Bohemian earth (q.v.).

Cyprian Turpentine. See CHIAN TURPENTINE.

Cyprian Vitriol. Prepared at Chessy. It contains copper sulphate, zinc sulphate, and water, $CuSO_4.3ZnSO_4.28H_2O$.

Cypridol (Hydriodol). Mercuric iodide (1 per cent.) in sterilised oil.

Cyprine. A mineral. It is a sky-blue variety of Vesuvianite or Idocrase.

Cyproheptadine. 4 - (1,2 : 5,6 - Diben-zocycloheptatrienylidene) - 1 - methyl - piperidine. Periactin is the hydro-chloride.

Cyprol. A proprietary preparation containing dextromethorphan hydrobromide, ephedrine hydrochloride, ammonium chloride, ipecacuanha liquid extract, tolu syrup and glycerin. A cough mixture. (334).

Cyproterone. An anti-androgen. 6 - Chloro - 17α - hydroxy - 1α, 2α - methylenepregna - 4, 6 - diene - 3, 20 - dione.

Cyprus Bronze. This contains 65 per cent. copper, 30 per cent. lead and 5 per cent. tin.

Cyprusite. A mineral. It is a hydrated ferric sulphate.

Cyrene. A resin used in adhesives, coatings, and moulding compounds.

Cyrogene Brown. A dyestuff prepared by sulphurising sawdust.

Cyrtolite. A beryllium-zirconium mineral. It usually contains about 15 per cent. of the oxide of beryllium.

Cystamin. See HEXAMINE.

Cystazol. An addition product of hexa-mine (q.v.), and sodium benzoate. Antiseptic.

Cystine. The disulphide of α-amino-β-sulphhydro-propionic acid, $(S.CH_2.CH(NH_2)COOH)_2$.

Cystogen. See HEXAMINE.

Cystopurin. A compound formed by the interaction of formaldehyde, ammonia, and sodium acetate, $(CH_2)_6N_2.2CH_3COONa.6H_2O$. Recommended in the treatment of cystitis and gonor-rhœa.

Cytacon. A proprietary preparation of cyanocobalamin. Oral Vitamin B_{12} preparations. (335).

Cytamen. A proprietary preparation of cyanocobalamin Vitamin B_{12}. (335).

Cytarabine. An anti-viral preparation. 1 - β - D - Arabinofuranosylcytosine. CYTOSAR.

Cytarsan. A sodium-bismuth caco-dylate. An antisyphilitic.

Cytase (Cellulase). An enzyme which decomposes cellulose.

Cythion. A proprietary preparation of MALATHION. An insecticide.

Cytosar. A proprietary preparation of CYTARABINE. A cytotoxic drug. (325).

Cytoxan. A proprietary trade name for Cyclophosphamide.

Cytrel. A proprietary tobacco substitute. (11).

D

D40. Sodium-3 : 5-di-iodo-4-pyridoxyl-N-methyl-2 : 6-carboxylic acid.

DA. Diphenylchlorarsine $(C_6H_5)_2As.Cl$.

DAB. A proprietary intermediate for various high-temperature plastics, used to make polypyrones and poly-quinoxalines. It is 3, 3' - diamino-benzidine. (325).

Dacarbazine. An anti-neoplastic. 5-(3, 3 - Dimethyltriazeno)imidazole - 4 - carboxamide.

Dachiardite. A mineral. It is $[(Ca,K_2,Na_2)_3Al_4Si_{18}O_{45}.14H_2O]$.

Dacortilene. A proprietary trade name for Prednylidene.

Dacrene. A name suggested for a crystalline diterpene found in the oil from *Dacrydium biforme*.

Dacron. A polyethylene terephthalate fibre having high strength and low water absorption. See also Terylene. (10).

Dactil. A proprietary preparation containing piperidolate hydrochloride. An antispasmodic. (249).

Dacuronium Bromide. A neuromuscular blocking agent. 3α-Acetoxy-17β-hydroxy - 5α - androstan - 2β, 16β - di - (1 - methyl - 1 - piperidinium) dibrom-ide.

Dadhi. An Indian fermented milk, similar to Yogurt (q.v.).

Daffodil. A pigment. It is cadmium sulphide, CdS. (Cadmium Yellow).

Dag. A registered trade mark covering colloidal dispersions of graphite and other products. (207).

Dagenite. A proprietary trade name for a bituminous asbestos-filled thermoplastic for accumulator cases.

Dagget (Doggert, Litaner Balsam). Birch tar, a component of leather finishes.

Dahlia. See HOFMANN's VIOLET and METHYL VIOLET B. Also a mixture of methyl violet and fuchsine.

Dahlin. Helenin. An alanta lactone.

Dahllite (Podolite). A mineral, $Ca_7(PO_4)_4.CO_3$.

Dahl's Acids. Acid II, α-naphthylamine-disulphonic acid, (1 : 4 : 6). Acid III, α-naphthylamine-disulphonic acid (1 : 4 : 7).

Dahmenite A. An explosive containing 91·3 per cent. ammonium nitrate, naphthalene, and potassium bichromate.

Daintex. A proprietary trade name for a wetting agent. It contains miscible terpene alcohols.

Dairos. A dairy detergent. (513).

Dairozon. A dairy sterilization agent. (513).

Daiton-sulphur. A prismatic sulphur found in the Daiton volcano.

Dakamballi starch. A starch prepared from the fruit of *Aldina insignis*, a tree of British Guiana.

Daka-ware. A proprietary trade name for urea moulded products.

Dakin's Oil. Another name for Chlorcosane, the solvent for Dichloramine T.

Dakin's Solution. A mixture of hypochlorite and perborate of sodium, with small amounts of hypochlorous and boric acids. Antiseptic.

Daktarin. A proprietary preparation of MICONAZOLE nitrate used as an antifungal agent. (356).

Dalacide. An herbicide. (548).

Dalacin C. A proprietary preparation of CLINDAMYCIN. An antibiotic. (325).

Daleminzite. A mineral. It is a silver sulphide.

Dalmane. A proprietary preparation of FLURAZEPAM. A hypnotic. (314).

Dalmatian Insect Powder. See INSECT POWDER.

Dalspray. An herbicide. (548).

Daltocel. Polyesters or polyethers for flexible foams. (512).

Daltocel HF. A highly branched polyester-based resin containing a small amount of water. Used for high modulus foams for use in safety padding and packaging. With tolylene-diisocyanate it gives low density foams of

greater hardness than is obtainable with currently available resins. (512).

Dal-Tocol. A proprietary preparation of d-α-tocopheryl succinate. A vasodilator. (272).

Daltoflex. A polyurethane raw rubber. (512).

Daltoflex 535, 540H, 540L and 745, 845. A proprietary range of solution-based urethane polymers. (512).

Daltogard. A polyurethane foam additive. (512).

Daltogard F. A proprietary halogenated phosphate ester used as a flame retardant in flexible urethane foams. (512).

Daltogen. A catalyst for polyurethane lacquers. (512).

Daltolac. Polyester and polyether rigid foam surface coatings. (512).

Daltolac 83. A proprietary polyether blend developed for the production of rigid urethane foams for use in insulation. (512).

Daltolac 1560. A highly branched saturated polyester resin in *m*-cresol used as a component of stable single pack high temperature—curing wire enamels, chemically resistant primers and finishes on metal, glass cloth and rubber. (2).

Daltolite. Pigments for flexible polyurethane foams. (512).

Daltomold. Trade mark (512) for a range of plastic moulding compounds of which some examples follow.

135:—A thermoplastic polyurethane elastomer of the polyester type. It has a Shore D hardness of 35.

140:—Similar to DALTOMOLD 135 but with a Shore D hardness of 40.

150:—Similar to DALTOMOLD 135 but with a Shore D hardness of 50.

160:—Similar to DALTOMOLD 135 but with a Shore D hardness of 60.

230:—A thermoplastic urethane elastomer of the polyether type. It has a Shore D hardness of 30.

238 and 338:—Thermoplastic polyurethane elastomers of the polyether type, with a Shore D hardness of 38.

245:—A thermoplastic polyurethane elastomer of the polyether type, with a Shore D hardness of 45.

Daltorol. Polyester for printers rollers. (512).

Dalyite. A mineral. It is $[K_2ZrSi_6O_{15}]$.

Dalysep. A proprietary preparation of sulphametapyrazine. An antibiotic. (809).

Damar Bronze. See DAMASCUS BRONZE.

Damascenised Steel. A steel made by repeatedly welding, drawing out, and doubling up a bar composed of a mixture of steel and iron, the surface of which has been treated with an acid.

The steel is left with a black coating of carbon, and the iron retains its metallic lustre.

Damascus Bronze (Damar bronze). An alloy of 76 per cent. copper, 10·5 per cent. tin, and 12·5 per cent. lead. A white bearing metal.

Dambonite. Dimethyl-i-inositol, $C_6H_6(OH)_4(OCH_3)_2$. It occurs in Gaboon rubber.

Dame's Violet Oil. Garden rocket oil, obtained from the seeds of *Hesperis matronalis*.

Damiana. The dried leaves of a Mexican plant, *Turnera diffusa*. A tonic.

Dammar Resin. A resin obtained from *Hopea*, *Shorea*, and *Balanocarpus* species, mainly of Federated Malay States. The melting-point usually varies from 90–200° C., acid value from 33–72, and the ash from 0·04–0·52 per cent. Rock dammar from Burma is derived from *Hopea odorata*, and has a specific gravity of from 0·98–1·013, a melting-point of 90–115° C., a saponification value of 31–37, acid value 31, and ash 0·55–0·68 per cent. A Borneo dammar, from *Retinodedron rassak*, melts at 130–150° C., has an acid value of 140–150, and saponification value 159–165. Perak dammar, obtained from *Balanocarpus heimii* of Malay States, is pale yellow and amber, has an acid value of 34–37, and a melting-point of 80–100° C. Batavian dammar is a good variety, melting at 100° C., and having an acid value of 35·5. Dammar Mata Kuching or Cat's-Eye dammar is a high-grade dammar from species of *Hopea*, and has a melting-point of 80–100° C. and an acid value of 21–24. Dammar Sengai, from trees of *Busseraceæ* species of Malay. It is dark brown in colour, has a melting-point of 120–135° C., and a low acid value. Dammar Siput, is obtained from *Shorea Ridleyana* of Malay. It is hard and dark in colour, has a melting-point of 190–220° C., and an acid value of 26. Dammar Temak from *Shorea Crassifolia* of Malay. A pale resin with a melting-point of 82–85° C. Has an acid value of 17–25. Dammar Hiroe, a type of Borneo dammar having an acid value of 13·5, a saponification value of 57–61, melting-point 190–200 °C., and is soluble in benzene, turpentine, and chloroform. It is suitable for use in lacquers and varnishes. Dammar Hitaru, from *Balanocarpus Penangianus* of Malay. It melts at 140° C., has an acid value of 16, and a saponification value of 34. Dammar Kapur, from Kapur, the Borneo camphor wood tree. It has an acid value of 52 and a saponification value of 78. Dammar Kedon-dong from Kedondong, a name for a species of *Canarium* of Malay. It has a low acid value. Dammar, Meranti Tembaga, from a tree of same name of Malay. It has a melting point of 180–210° C., and an acid value of 30–41. Dammar Meranti Jerit, from a tree of same name, a species of *Shorea* of Malay. It melts at 60–70° C. and has an acid value of 11. Dammar Minyak from *Ayathio Alba* of Malay. A milky-white resinous liquid with acid value of 130. Also see BLACK DAMMAR. Other varieties of dammar resin mainly of low grade are : Dammar Hitaru, acid value 14·2, melting-point 140–170° C.; Dammar Kepong, acid value 10·1, melting-point 160–180° C. ; Dammar Saraya, acid value 24·2, melting-point 135–175° C.; Dammar Batu, acid value 18·1, melting-point 140–180° C. ; Dammar Daging, acid value 23·3, melting-point 120–160° C. Dammar is soluble in amyl acetate, benzene, carbon tetrachloride, and chloroform, and partly soluble in acetone and alcohol. It is used in the manufacture of spirit varnishes.

Damourite. A mineral. It is a yellowish variety of muscovite.

Danaite. A mineral. It is a variety of mispickel (*q.v.*), in which part of the iron is replaced by an equivalent amount of cobalt.

Danalite. A mineral, $2(Mg.Fe)O.SiO_2$. It also contains zinc, beryllium, iron, and manganese.

Danazol. An anterior pituitary suppressant currently under clinical trial as " WIN 17/757 ". 17α - Pregna - 2, 4 - dien - 20 - yno[2, 3 - *d*]isoxazol - 17 - ol.

Danbar. A proprietary preparation of veratrum viride alkaloids, for scalp application. (392).

Danburite. A mineral, $CaO.B_2O_3.2SiO_2$.

Dandelion Metal. An alloy of 72 per cent. lead, 18 per cent. antimony, and 10 per cent. tin.

Dandricide. A proprietary anti-seborrheic. Benzalkonium chloride, alkyl isoquinilinium bromide, sorbutol and polysorbates. (867).

Daneral. A proprietary preparation of pheniramine p-aminosalicylate. (312).

Daneral-SA. A proprietary preparation of pheniramine maleate. (312).

Daneral. A proprietary trade name for the 4-aminosalicylate of Pheniramine.

Danforth's Oil. See NAPHTHA.

Dankwerth's Substitute. A rubber substitute made from hemp oil, wood-tar oil, coal-tar oil, ozokerite, and spermaceti.

Dannemorite. A mineral, $(Fe.Mg.Mn)SiO_3$.

Danol. A proprietary preparation of DANAZOL, used as a suppressant of gonadotrophins. (439).

Danthron. 1,8 - Dihydroxyanthra - quinone.

Dantrium. A proprietary preparation of DANTROLENE sodium. A skeletal muscle relaxant. (818).

Dantrol. A proprietary preparation of cholesterol, lecithin, chlorocresol, salicylic acid. A scalp lotion. (814).

Dantrolene. A skeletal muscle relaxant. 1 - (5 - p - Nitrophenylfurfurylidene-amino)imidazoline - 2, 4 - dione. "F440" and "Dantrium" are used as the sodium salt, with the former still under clinical trial.

Dantyl. A proprietary preparation of p-aminosalicylic acid phenylester, p-aminosalicylic acid and sucrose. Antituberculous agent. (308).

Dantyl-Inah. A proprietary preparation of phenyl-p-aminosalicylic acid, p-aminosalicylic acid, isoniazid and sucrose. Antituberculous drug. (308).

Daonil. A proprietary preparation of GLIBENCLAMIDE used in the treatment of late-onset diabetes. (312).

Daphnetin. 7 : 8 - Dioxy - coumarin, $C_8H_6O_4$.

Daphnin. See EOSIN BN.

Daphnite. A mineral. It is an iron-aluminium silicate, $H_{56}Fe_{27}Al_{29}Si_{18}O_{121}$.

Daphyllite. A mineral. It is tetradymite.

Dapicho (Zaspis). The South American name for the caoutchouc from the roots of *Hevea guianensis*.

Dapon M. A trade mark for diallyl isophthalate. A moulding material. (130).

Dapon 35. A trade mark for diallyl phthalate. A moulding material. (130).

Daprisal. A proprietary preparation of dexamphetamine sulphate, amylobarbitone, aspirin and phenacetin. (281).

Dapsetyn. A chloramphenicol/dapsone veterinary preparation. (510).

Dapsone. Di - (4 - aminophenyl)sulphone. Diaphenylsulphone. Avlosulfon.

Dapsyvet. A chloramphenicol/dapsone veterinary preparation. (510).

Daptazole. A proprietary preparation of amiphenazole. A central nervous stimulant. (271).

Darachlor. An antimalarial compound. (514).

Daranide. A proprietary preparation of dichlorphenamide. Respiratory stimulant. (310).

Daraprim. A proprietary preparation containing pyrimethamine. An antimalarial drug. (277).

Darapskite. A mineral, Na_2SO_4. $NaNO_3.H_2O$.

Daratac SP 1025. A PVC emulsion adhesive for sticking PVC film to a substrate. (404).

D'Arcet's Alloy. (*a*) Consists of 50 per cent. bismuth, 25 per cent. lead, and 25 per cent. tin, melting-point 93° C.; (*b*) contains 50 parts bismuth, 25 parts lead, 25 parts tin, and 250 parts mercury.

Darcil. A proprietary preparation of potassium phenethicillin. An antibiotic. (245).

Darco. A decolorising and refining carbon. A substitute for bone char.

Darbid. A proprietary trade name for Isopropamide iodide.

Darenthin. A proprietary preparation of bretylium tosylate. (277).

Daricon. A proprietary preparation containing oxyphencyclimine hydrochloride. An antispasmodic. (85).

Dark Cylinder Oils. See CYLINDER OILS.

Dark Green. See DINITROSO-RESORCIN.

Dark Oxide of Iron. See INDIAN RED.

Dark Red Gold. An alloy of 50 per cent. gold and 50 per cent. copper.

Dark-red Silver Ore. See PYRARGYRITE.

Darlingite. A mineral. It is a variety of lydian stone.

Dartalan. A proprietary preparation of thiopropazate dihydrochloride. A sedative. (285).

Dartex. See ALLOPRENE, a German chlorinated rubber.

Darurnite. See DARWINITE.

Darvic. A proprietary trade name for unplasticised p.v.c. sheet. (2).

Darvisul. A proprietary preparation containing DIAVERIDINE. A veterinary anti-protozoan.

Darvon. A proprietary trade name for the hydrochloride of Dextropropoxyphene.

Darwin Glass. Fused rock produced by the impact of a meteorite from Mt. Darwin, Tasmania.

Darwinite (Darurnite, Whitneyite). A mineral. It is an arsenide of copper, Cu_9As.

Darwinol Acetate. Acetic ester obtained from *Darwinia grandiflora*.

Darwins AR3. A nickel-chromium-copper-iron alloy designed to withstand the action of sulphuric, nitric, phosphoric and acetic acids. (144).

Darwins AR 654A. A nickel-molybdenum-iron alloy designed to withstand corrosion from hydrochloric and sulphuric acids up to 70° C. (144).

Darwins AR 655B. Similar to 654A but richer in molybdenum. It will resist attack up to boiling point.

Darwins AR 656C. A proprietary trade name for nickel-molybdenum chromium iron alloy for handling acids of an oxidising nature such as nitric acid. (144).

Da-Sed. A proprietary preparation of BUTABARBITAL. (870).

Dasag. Surophosphate (q.v.). A fertiliser.

Dashkesanite. A mineral. It is $2[(Na, K)Ca_2(Fe^{..}, Mg, Fe^{...})_5(Si,Al)_8.O_{22}Cl_2]$.

Daster. A semi-refined cholesterol.

Datem. A proprietary trade name for di-acetyl tartaric esters of mono-glycerides. Edible emulsifiers for use in lipsticks and similar products. (191).

Date Sugar (Date-Tree Sugar). Palm sugar, obtained from the liquid extract of certain palms.

Date-tree Sugar. See DATE SUGAR.

Datholite. See DATOLITE.

Datiscetine. A yellow dyestuff occurring as a glucoside (datiscine).

Datolite (Datholite, Botryolite, Esmarkite, Palacheite). A mineral, $2CaO.B_2O_3.2SiO_2.H_2O$.

Daturine. See DUBOISINE.

Daua-daua Cakes. The kernels of *Parkia africana*, sold as a food and condiment in the Sudan.

Dauberite. A mineral. It is a basic uranium sulphate.

Daubréelite. A mineral. It is a ferrous sulphochromite, $FeS.Cr_2S_3$.

Daubreite. A mineral. It is a bismuth oxychloride, $Bi_2O_3.BiOCl$, found in Bolivia.

Daudelin Solder. An alloy of 65·6 per cent. tin, 12·2 per cent. zinc, 1 per cent. aluminium, 17·4 per cent. lead, 3·1 per cent. copper, and 0·4 per cent. phosphorus.

Daufresne's Solution. Chlorinated soda and sodium bicarbonate, B.P.C.

Daunorubicin. An antibiotic produced by *Streptomyces cæruleorubidus*. 3-Acetyl - 1, 2, 3, 4, 6, 11 - hexahydro - 3, 5, 12 - trihydroxy - 10 - methoxy - 6, 11 - dioxonaphthacen - 1 - yl - 3 - amino - 2, 3, 6 - trideoxy - β - D - galactopyranoside.

Davenol. A proprietary preparation containing carbinoxamine maleate, ephedrine hydrochloride and pholcodine. A cough linctus. (245).

Davey's Gray. A pigment prepared from siliceous earths. Used principally in mixtures with other colours to reduce tones.

Davidite. A mineral. It is a titanate of iron, uranium, vanadium, cerium, and yttrium metals, of Australia.

Davidsonite. A mineral. It is a variety of beryl.

Daviesite. A mineral. It is an oxychloride of lead.

Davis Metal. An alloy of 67 per cent.

copper, 29 per cent. nickel, 2 per cent. iron, and 1·5 per cent. manganese.

Davisonite. A mineral. It is $Ca_3Al(PO_4)_2(OH)_3$.

Davreuxite. A mineral. It is a silicate of aluminium and manganese.

Davyne. A mineral. It is $[(Na,K,Ca)_6Al_6Si_6O_{24}(SO_4,CO_3Cl_2)_2]$.

Davy's Cement. A cement consisting of 4 parts pitch and 4 parts gutta-percha.

Davy's Substitute. An indiarubber substitute consisting of partially saponified oils heated with sulphur.

Dawson Bronze. An alloy of 83·9 per cent. copper, 15·9 per cent. tin, and traces of antimony, lead, iron, arsenic, and zinc.

Dawsonite. A mineral, $Al_2(CO_3)_3.Na_2CO_3H_2O$.

Dayamin. A proprietary preparation of Vitamins A, D, C, thiamine, riboflavine, nicotinamide and pyridoxine. Vitamin supplement. (311).

Daycollan. A proprietary polyurethane elastomer. (638).

D.B.A. See ACCELERATOR D.B.A.

DB Gran. A fertilizer. (548).

DBP. Dibutylphthalate. A plasticiser for vinyl and other plastics.

DBPC. A proprietary anti-oxidant. Di-*tert.*-butyl-para-cresol. (639).

Dchit (Jaft). A Persian tanning material, which is supposed to be an oak product, containing 40 per cent. tannin.

DCP. Dicapryl phthalate. A plasticiser for vinyl plastics.

D.D.D. An accelerator for rubber vulcanisation. It is dimethylamine dimethyl-dithio-carbamate.

D.D.T. Dichlorodiphenyltrichlorethane or 2,2-bis (parachlorphenyl) 1,1,1-trichlorethane. A powerful insecticide.

De-Acidite. An anion exchange material. (609).

Dead Borneo. Jelutong (q.v.).

Dead Cotton. Cotton fibre from unripe plants. They do not spin well and are difficult to dye.

Dead Oil. The high boiling-point fraction of shale oil, from which the greater part of the paraffin has been crystallised out. Also see HEAVY OILS.

Dead Silver (Frosted Silver). Silver, whitened by heating in air, and immersed in dilute sulphuric acid.

Deanase. A proprietary preparation of desoxyribonuclease for treatment of ulcers, bruises and abscess. (256).

Deanase D.C. A proprietary preparation of delta chymotrypsin. Digestive enzyme. (256).

Deaner. A proprietary trade name for the 4-acetamido-benzoate of Deanol.

Deanol. 2-Dimethylaminoethanol. ANP 235 is the 4-chlorophenoxyacetate hydrochloride; Deaner is the 4-acetamidobenzoate.

Deba. Veronal (q.v.).

Debendox. A proprietary preparation of dicyclomine hydrochloride, doxylamine succinate and pyridoxine hydrochloride. Antiemetic. (263).

Debrisoquine. 2 - Amidino - 1,2,3,4 - tetrahydroisoquinoline. Declinax.

Debron 711. A proprietary coating compound manufactured from polyphenylene sulphide. (640).

Decadron. A proprietary preparation of dexamethasone sodium phosphate. (310).

Deca-Durabolin. A proprietary preparation of nandrolone decanoate. An anabolic agent. (316).

Decalex. A photographic developer. (507).

Decalin. See DEKALIN.

Decalite. A proprietary phenolic moulding material. (641).

Decamethonium Iodide. Decamethylenedi(trimethylammonium iodide) Eulissin; Syncurine.

Decamianto. A proprietary phenolic moulding material. (641).

Decamphorised Oil of Turpentine (Oxidised Oil of Turpentine). The residue from the manufacture of camphor. It consists mainly of dipentene.

Decapryn. A proprietary trade name for Doxylamine.

Decaserpyl. A proprietary trade name for Methoserpidine.

Decaserpyl Plus. A proprietary preparation of methoserpidene and benzthiazide. (307).

Decaspray. A proprietary preparation of dexamethasone and neomycin sulphate used in dermatology as an antibacterial agent. (310).

Deccox. A proprietary preparation of DECOQUINATE. A veterinary antiprotozoan.

Decelox. A zinc oxide slow curing rubber. (610).

Dechenite (Aræoxene). A mineral. It is a lead meta-vanadate, $Pb(VO_3)_2$.

Dechlorane A-O. A proprietary synergistic agent for fire-retardant plastics. It is antimony oxide and contains halogens. (201).

Decholin. A German preparation. It is the sodium salt of dehydro-cholic acid. Used in the treatment of cholelithiasis and other diseases of the liver and gall-ducts.

Deciquam 222. A sterilizer for breweries, pipe lines and stoppers. (576).

Deciquam 223. An ice cream servers sterilizer. (576).

Deciquat. An aqueous solution of didecyldimethylammonium bromide. (576).

Deckor. PVC pastes. (542).

Declinax. A proprietary preparation of dibrisoquine sulphate. An antihypertensive. (314).

Declomycin. A proprietary preparation of diethylchlortetracycline. An antibiotic. (306).

Declonal. A proprietary preparation of diethyl-chlor-acetamide.

Decolite. A trade name for an asbestos cement used for flooring purposes.

Decolorising Powder. A by-product in the manufacture of prussiate of potash. It contains from 30–40 per cent. animal charcoal, silica, and silicates, with a little iron oxide. Used for decolorising paraffin.

Deconyl. A proprietary trade name for a weatherproof nylon coating. (146).

Decoquinate. An anti-protozoan. Ethyl-6 - decyloxy - 7 - ethoxy - 4 - hydroxy - quinoline - 3 - carboxylate. DECCOX.

Decorpa. A proprietary preparation of guar gum granules used as an anti-obesity agent. (286).

Decortilen. A proprietary trade name for Prednylidene.

Decortisyl. A proprietary preparation of prednisone. (307).

Decroline. See HYDROSULPHITE.

Decrose. A proprietary preparation of glucose with calcium glycero-phosphate and Vitamin D.

Decrysil. 4 : 6-Dinitro-o-cresol.

De De Tane. D.D.T. products. (q.v.). (501).

Deegrol. A textile and hard surface peroxide. (537).

Deeline. See PARAFFIN, LIQUID.

Deenax. A proprietary anti-oxidant. 2, 6 di - tert. - butyl 4 - methyl - phenol. (642).

Deep Blue Extra R. See BLUE 1900.

Deep Chrome. A pigment. It is a chromate of lead.

Defencin. A proprietary preparation of ISOXUPRINE resinate. A peripheral vasodilator. (324).

Defiance. A proprietary casein plastic.

Defirust. A proprietary trade name for a rustless iron containing 12–15 per cent. chromium and 0·1 per cent. carbon.

Defloc Compound (Tablets). A water scale reducer for bottling machines. (576).

Defolia. Defoliant for hops. (501).

Defungit. A proprietary preparation of BENSULDAZIC ACID. A veterinary fungicide.

Defolup. A proprietary trade name for an agricultural chemical based on cyanamide used for cleaning hop-plants and for weed-control in hop-gardens. (899).

Degalan V. An acrylic modifier for p.v.c.

Degalan 6. A proprietary polymethyl injection moulding compound based on methacrylate. (643).

Degalan 6E. A proprietary polymethyl extrusion compound based on methacrylate. (643).

Degalan 7. A proprietary polymethyl injection moulding compound based on methacrylate. (643).

Degalan 7E. A proprietary polymethyl compound based on methacrylate. (643).

Degalan 8. A proprietary injection compound based on polymethyl methacrylate. (643).

Degalan 8E. A proprietary extrusion compound based on polymethyl methacrylate. (643).

Degalol. Menthadioxy-cholic acid.

Degasser. A degassing agent for aluminium alloys. (531).

Degeröite. A mineral. It is a variety of hisingerite.

Dégragène. See DÉGRAS.

Degranol. A proprietary preparation of mannomustine hydrochloride. A carcino-chemotherapeutic agent.

Dégras (Tanning Grease, Leather Grease, Sod Oil). A material consisting of rancid fish oil, resinous substances called dégragène or dégras-former (about 20 per cent.), from the oxidation of the oil, mineral matter (consisting of lime, soda, and sulphates), a considerable quantity of water, and the residues of skin, membranes, and hair. It is obtained in the chamoising process, and is used for tanning other skins. At the present time the term dégras is used for a mixture of moellon (see DÉGRAS, ARTIFICIAL), with wool fat, tallow, and other fats.

Dégras is a *wool* grease. Not to be confused with moellon which is the grease extracted from the *skin* of sheep. Extracted dégras has melting-point of 97–100° F. Moisture 2–2·5 per cent., ash 0·1–0·5 per cent. Free fatty acid (as oleic) 7·8 per cent. Saponifiable matter 60 per cent. Specific gravity at 60° F. 13·15–13·88. Wool grease is used in synthetic rubber extrusion.

Dégras, Artificial. Prepared by kneading the refuse and clippings of skins with fish oil, exposing the mass to air to oxidise it, and pressing out the dégras or moellon. Modern moellon is an aqueous emulsion of fish oil.

Dégras Former. See DÉGRAS.

Degrasine. A concentrated glandular extract, sold in the form of tabloids.

Degreroite. A mineral, $Fe_2O_3.SiO_2.3H_2O$.

Degummed Silk. See BOILED-OFF SILK.

De Haën Salt. A double salt of antimony trifluoride and ammonium sulphate, $SbF_3(NH_4)_2SO_4$. A mordant.

Dehybor. Trade name for borax from which the water of crystallization has been removed by heat. (Anhydrous borax.) It is widely used in glass, vitreous enamel, ceramic glaze and metallurgical industries where borax has to be melted. (183).

Dehydrite. A registered trade name for magnesium perchlorate trihydrate. A drying agent.

Dehydrocholic Acid. 3,7,12-Trioxo-5β-cholanic acid. Bilostat; Decholin : Dehydrocholin.

Dehydrocholin. A proprietary preparation of dehydrocholic acid. A laxative. (182).

Dehydroemetine. 3 - Ethyl - 1,6,7,11b-tetrahydro - 9,10 - dimethoxy - 2 - (1,2,3,4 - tetrahydro - 6,7 - dimethoxy - isoquinol - 1 - ylmethyl)4H - benzol(a) quinolizine. 2,3-Dehydroemetine. Mebadin.

Dehydromorphine (Oxydimorphine, Oxymorphine, Dimorphine). Pseudomorphine, $C_{17}H_{17}NO_3$.

Deicke's Mixture. Obtained by boiling spent residues containing sulphur, with iron filings, and then regenerating the Fe_2S_3 formed by oxidation. Used as a purifying agent for coal gas. It contains 66 per cent. ferric oxide, a little lime, and sawdust.

Dekalin (Decalin). Deca-hydro-naphthalene, $C_{10}H_{18}$, a paint and resin solvent. The commercial variety contains 80 volumes decahydro-naphthalene and 20 volumes tetra-hydro-naphthalene. It has a specific gravity of 0·895 and a flash-point of 52° C. Used as turpentine substitute.

Dekhotinsky. A proprietary cement for porcelain, plastics, and glass.

Dekorit. A German synthetic resin of the phenol-formaldehyde type.

Dekrysil. A proprietary preparation of 4 : 6-dinitro-cresol.

DEL. A proprietary preparation of methenamine mandelate. (754).

Delac-S. A proprietary rubber accelerator. It is 2-cyclohexyl-2-benzthiazyl sulphenamide. (435).

Delafield's Hæmatoxylin. A microscopic stain prepared by adding 4 grams hæmatoxylin dissolved in 25 c.c. absolute alcohol to 400 c.c. of a saturated aqueous solution of ammonium alum. This solution is exposed for several days, filtered, and 100 c.c. glycerin and

100 c.c. methyl alcohol added. Allowed to stand and filtered.

)elafila. A proprietary powdered slate used as a filler, base, loader, etc., in the manufacture of paints, paper, asphalt road construction, etc. The best grade contains 99 per cent. passing through 300-mesh sieve (0·0018-in. opening), and contains 59·7 per cent. silica and 26·9 per cent. alumina.

)elafossite. A mineral. It is an oxide of copper, iron, and aluminium.

)elalot's Alloy. An alloy containing 80 per cent. copper, 18 per cent. zinc, 2 per cent. manganese, and 1 per cent. calcium phosphate.

)elan-Col. A dithionon fungicide. (511).

)elanium. A proprietary trade name for carbon and graphite materials highly resistant to all chemicals except some oxidising agents. (145).

)elanolite. A variety of the mineral Montmorillonite.

)elatynite. An amber from Delatyn in the Galician Carpathians. Low in succinic acid and free from sulphur.

)elavilie. Zinc dust. (611).

Delawarite. A mineral. It is a variety of orthoclase.

)elchowyte. A decolorising carbon, prepared from peat.

)elessite (Iron-chlorite). A mineral, $4(Mg.Fe)O.2(Al_2Fe_2)O_3.4SiO_2.5H_2O$.

Delexin Expectorant. A proprietary preparation containing guaiphenesin and phenylpropanolamine hydrochloride. A cough linctus. (244).

)elfen. A proprietary preparation of nonoxinol. A contraceptive cream and foam. (369).

)elft Blue. A pigment. It is a mixture of indigo and ultramarine.

)elhi Rustless Iron. A corrosion-resisting alloy containing 18 per cent. chromium, 1·5 per cent. silicon, and not more than 0·08 per cent. carbon.

)elimon. A proprietary preparation of 1-phenyl-2,3-dimethyl-4-(2-phenyl-3 - methyl - hydroxazino - methyl) - pyrazolone-(5)-hydrochloride, paracetamol and salicylamide. An analgesic. (256).

)elint. A proprietary fibrous material from the hulls of cotton seed. Used for the production of fine paper and artificial silk.

)ellerite. A trade name for a waterproof and fireproof asbestos product made from asbestos and vulcanised rubber. It resembles leather.

)ellite. A Swiss synthetic resin varnish-paper product used for electrical insulation.

Delmadinone. A progestational steroid.

6 - Chloro - 17α - hydroxypregna - 1, 4, 6 - triene - 3, 20 - dione. " RS 1301 " is currently under clinical trial as the 17-acetate.

Delnav. A proprietary preparation of DIOXATHION. A veterinary insecticide.

Delorenzite. A mineral, $2FeO.UO_2$. $2Y_2O_3.24TiO_2$, of Italy.

Delphen. A proprietary preparation of nonylphenoxypolyethoxyethanol cream. A spermicide. (369).

Delphine Blue. A dyestuff. It is the ammonium salt of the sulphonic acid of dimethyl-phenyl-diamino-oxy-phenoxazone, $C_{20}H_{20}N_4SO_6$. Dyes chromed wool indigo blue. Employed in calico printing.

Delphinite. A mineral. It is a variety of epidote.

Delrin. A proprietary trade name for a stiff strong engineering plastic of the acetal resin type. It has excellent fatigue resistance and is used as a replacement for die cast parts in gears, bearings and housings. (10). Special grades are used, or possess characteristics, as follows: *Delrin 100*, for mouldings in which toughness is a prerequisite; *Delrin 150*, for general extrusions, blown bottles, tubing and rod; *Delrin 507* contains a light stabiliser; *Delrin 550*, in general-purpose moulding and wire-coating applications; *Delrin 900*, a special purpose high-flow resin. See CELCON (11) and HOSTAFORM C (9). POLYFYDE (12).

Delta Acid (F-acid). β-Naphthylamine-sulphonic acid ($NH_2 : SO_3H = 2 : 7$).

Delta-Butazolidine. A proprietary preparation of phenylbutazone and prednisolone. (17).

Delta Cortef. A proprietary preparation of prednisone. (325).

Delta Cortelan. A proprietary preparation for the acetate of prednisone. (335).

Deltacortil DA. A proprietary trade name for the hydrochloride of Prednisolamate.

Delta Cortone. A proprietary preparation of prednisone acetate. (310).

Delta Cortril. A proprietary preparation of prednisolone. (85).

Delta-Fenox. A proprietary preparation of prednisolone, phenylephrine hydrochloride, and naphazoline nitrate. Nasal spray. (253).

Deltaform. See THIOFORM.

Delta-Genacort. A proprietary preparation of prednisolone. An anti-inflammatory agent. (188).

Deltaite. A mineral. It is $[Ca_2Al_2(PO_4)_2(OH)_4.H_2O]$.

Delta Metals. The registered trade mark for a variety of metals, metallic alloys and metal articles.

Deltamin. A German preparation containing pyrazolone - phenyl - diamino - methylum, atophan, and codeine phosphate. An antineuralgic.

Deltanephrin. Glaucosan (*q.v.*).

Deltapurpurin 5B. See DIAMINE RED B.

Deltapurpurin 7B. See DIAMINE RED 3B.

Deltapurpurin G. An azo dyestuff, $(C_6H_4)_2(N_2)_2 \cdot (C_{10}H_5)_2(NH_2)_2(HSO_3)_2$. It is isomeric with Congo red, and is obtained from diazotised benzidine acting upon β-naphthylamine-δ-sulphonic acid.

Deltastab. Proprietary preparations of prednisolone, e.g. the acetate. (253).

Deltyl. A proprietary trade name for a plasticiser. It is a fatty acid ester.

Delvauxite. A mineral. It is a hydrated iron phosphate containing vanadium, $Fe_4(OH)_6(PO_4)_2 \cdot 17H_2O$.

Delvex. A proprietary trade name for the iodide of dithiazanine.

Delvinal. A proprietary trade name for Vinbarbitone.

Delysid. A proprietary trade name for Lysergide.

Dema. A proprietary brand of TETRACYCLINE. (872).

Demantoid. See ANDRADITE.

Demecarium Bromide. Decamethylenedi - (3 - dimethylaminophenyl N-methylcarbamate methobromide). HUMORSAL; TOSMILEN.

Demeclocycline. An antibiotic. 7-Chloro - 4 - dimethylamino - 1, 4, 4a, 5, 5a, 6, 11, 12a - octahydro - 3, 6, 10, 12, 12a - pentahydroxy - 1, 11 - di - oxonaphthacene - 2 - carboxamide. It is present in DETECLO and LEDERSTATIN.

Demecolcine. N - Methyl - N-deacetylcolchicine. Colcemid.

Demerara Anime. South American copal resin. It is the product of *Hymenœa* species.

Demerara Animi. See COPAL RESIN.

Demerara Copal. See COPAL RESIN.

Demethylchlortetracycline. 7-Chloro-4-dimethylamino - 1,4,4a,5,5a,6,11,12a-octahydro-3,6,10,12,12a-pentahydroxy-1,11 - dioxonaphthacene - 2 - carboxyamide. Declomycin; Ledermycin.

N-Demethylcodeine. Norcodeine.

N-Demethylmorphine. Normorphine.

Demidovite. A mineral found in the Urals. It is a copper silico-phosphate.

Demineralised Gelatin. A pure gelatin obtained by dialysing a solution of gelatin or allowing it to set to a jelly and steeping this repeatedly in cold water to remove salts and impurities.

Demulen. A proprietary preparation of MESTRANOL and ETHYNODIOL diacetate. An oral contraceptive. (644).

Demulen 50. A proprietary preparation of ethinyloestrodiol and ETHYNODIOL diacetate. An oral contraceptive. (644).

Denatonium Benzoate. Benzyldiethyl-(2,6-xylylcarbamoylmethyl) ammonium benzoate. Bitrex.

Denatured Alcohol. See METHYLATED SPIRITS.

Dendrid. A proprietary preparation of IDOXURIDINE used as eye-drops. (374).

Denigés' Acid Mercuric Sulphate Solution. Mercuric oxide (5 grams) is added to the hot solution produced when 20 c.c. of concentrated sulphuric acid is added to 100 c.c. water. Used to test for acetone.

Dennisonite. Synonym for Davisonite.

De-Nol. A proprietary preparation of tri-potassium di-citratobismuthate used in the treatment of peptic ulcers. (317).

Denol. A mixture of aliphatic higher alcohols, used as a denaturant.

Densite. Alloying additions of nickel and nickel–tin. (531).

Densites. Mining explosives. They contain ammonium nitrate, sodium or potassium nitrate, and trinitro-toluene.

Densithene. A proprietary trade name for lead powder-loaded polythene. Used for radi-opaque screening. (133).

Dental Alloys. Usually consist of 65 per cent. gold, 11–18 per cent. platinum, 4–16 per cent. palladium, 1·5–6 per cent. silver, and 7 per cent. copper. Other alloys used in dentistry contain 66–75 per cent. silver and 25–33 per cent. platinum. Also see VON ECKART'S, MELLOTA'S, KINGSLEY'S, REESE'S, and BEAN'S ALLOYS.

Dental Amalgam. Usually an alloy of silver and tin with a little copper. It is used with mercury.

Dental Gold. Alloys usually containing 65–90 per cent. of gold, 5–12 per cent. silver, 4–12 per cent. copper, and sometimes small amounts of platinum.

Dentalone. A solution of chloretone in essential oils.

Dentelles (Spectacles). The ossein made from button-makers' refuse.

Dentist's Amalgam. This is often an alloy of 70 per cent. mercury and 30 per cent. copper.

Dentplus Special. A proprietary trade name for dicalium phosphate dihydrate. Used as a thickening agent, cleaning agent, and carrier in toothpaste. (9).

Denver Clay. See BENTONITE.

Denver Mud. See KAOLIN CATAPLASM and BENTONITE.

Deodorised Alcohol. Rectified spirit, ethyl alcohol, C_2H_6OH.

Deodorised Iodoform. See IODOFORM, DEODORISED.

Deodorised Oils. Oils which have been

hydrogenised, or subjected to other treatment, to remove objectionable odours.

Dephlogisticated Air. See VITAL AIR.

Dephlogisticated Muriatic Acid Gas. Chlorine, Cl.

Depixol. A proprietary preparation of FLUPENTHIXOL used in the treatment of psychoses. (645).

Deplet. A proprietary preparation of potassium teclothiazide. A diuretic. (271).

Depomedrone. A proprietary preparation of methyl prednisolone acetate. (325).

Depo-Provera. A proprietary preparation containing medroxyprogesterone acetate. (325).

Depostat. A proprietary preparation of gestronal hexanoate used to treat benign prostatic hypertrophy and endometrial carcinoma. (438).

Depot Glumorin. A proprietary preparation of kallikrein bound to a high molecular weight steroid. (341).

Depramine. See BALIPRAMINE.

Deprodone. A corticosteroid. 11β, 17α-Dihydroxypregna - 1, 4 - diene - 3, 20 - dione. " R.D. 20,000 " is currently under clinical trial as the 17α-propionate.

Depronal S.A. A proprietary preparation of dextropropoxyphene hydrochloride. An analgesic. (262).

Depropanex. A proprietary preparation of deprotienated pancreatic extract. Urinary antispasmodic. (310).

Deptropine. 3 - (1,2:4,5 - Dibenzocyclo-heptadien-3-yloxy)tropane. Brontina is the dihydrogen citrate.

Dequadin. A proprietary preparation of dequalinium chloride. (284).

Dequalinium Chloride. Decamethyl-enedi-(4-aminoquinaldinium chloride). Dequadin.

Dequalone. Dequadin and predniso-lone. (510).

Dequaspon. A proprietary preparation of gelatin sponge impregnated with dequadin. Dental packing for hæmorrhage. (284).

Derakane. A trade mark for vinyl ester resins for use with styrene in the moulding of reinforced plastics. (64).

Derakane 470-45. A proprietary polyvinyl ester resin possessing good chemical resistance to chlorinated solvents. (613).

Derakane 510-40. A proprietary polyvinyl ester resin used in fire-retardant laminates. It contains 20 per cent. bromine. (613).

Derbac. A proprietary preparation of MALATHION used to treat infestation by lice. (802).

Derbac Shampoo. A proprietary shampoo containing carbaryl used to treat infestation by lice on the head and in the hair. (802).

Derbylite. A mineral of Brazil, Fe $(SbO_3)_2.5FeTiO_3$.

Derby Red. See CHROME RED.

Derbyshire Spar. See FLUORSPAR.

Dercolyte. A range of proprietary alpha and beta pinene resins. (874).

Dericin. See FLORICIN.

Deriphyllia. A compound of theophyllin with an oxamine used for the treatment of angina pectoris. It is superior to theophyllin itself.

Derizine. See FLORICIN.

Dermacaine. A proprietary preparation of cinchocaine. Local anæsthetic cream. (250).

Dermal Salve. This contains as the essential ingredients 16 per cent. hydrophile fatty material, 11–12 per cent. alum, 0·05 per cent. active chlorine (as hypochlorite), together with a little sodium chloride.

Dermalex. A proprietary preparation of squalene, hexachlorophene and allantoin. A protective skin lotion. (646).

Dermamed. A proprietary preparation of bacitracin and neomycin. An antibacterial skin ointment. (250).

Dermasulph. A proprietary preparation of polythionates. Sulphur skin ointment. (280).

Dermatin. A mineral. It is a hydrated silicate of iron and magnesium.

Dermative. A mineral. It is $(Mg,Fe)SiO_3.2H_2O$.

Dermatol. A basic bismuth salt of gallic acid, $C_6H_2(OH)_3.CO.Bi(OH)_2$. Used medicinally for the treatment of wounds and skin diseases, also as a remedy for perspiring feet. Sometimes employed internally as a preventative of gonorrhœa.

Dermevan. Solubilised surface activated iodine. (591).

Dermiforma. A patent bating material composed of whey, lactic acid, and other organic acids.

Dermocaine. A proprietary preparation of cinchocaine used as a skin cream. (250).

Dermogen. See EKTOGAN.

Dermogesic. A proprietary preparation of calamine, benzocaine, and hexylated m-cresol. An antipruritic. (310).

Dermol. Bismuth chrysophanate, Bi $(C_{15}H_9O_4)_2Bi_2O_3$. Used in the treatment of skin diseases as a 5–20 per cent. ointment.

Dermoline. A glycerin liniment containing camphor, alcohol, rose water, glycerin, cajuput oil, soap, and liq. hamamelidis.

Dermonistat. A proprietary preparation of MICONAZOLE nitrate. An anti-fungal agent. (356).

Dermoplast. A proprietary preparation of benzocaine, benzethonium chloride, menthol, hydroxyquinalone benzoate and methyl paraben in the form of an aerosol, used as a soothing skin spray. (467).

Dermovate. A proprietary preparation of CLOBETASOL propionate used in the treatment of eczema and psoriasis. (335).

Dernbachite. A mineral of Dernbach. It is a variety of beudantite.

Deronil. A proprietary trade name for Dexamethasone.

Derosne's Salt. Narcotine.

Deroton. Polytetramethylene tereph-thalate. A proprietary thermoplastic polyester. (512). See ARNITE.

De Rossi's Stain. A microscopic stain. It consists of two solutions : (a) Tannic acid 25 grams, distilled water 100 c.c. ; (b) fuchsin 0·25 gram, phenol 5 grams, alcohol 10 grams, and distilled water 100 grams.

Derrid. An acid resin obtained from *Derris elliptica*. It is poisonous.

Desalgin. See DISALOIN.

Desaloin. See DISALOIN.

Desaulesite. A mineral. It is a hydrated silicate of nickel and zinc.

Desavin. Di-(phenoxy-ethyl)formal. A plasticiser.

Desbutal. A proprietary preparation of methylamphetamine hydrochloride and pentobarbitone sodium. (311).

Descloizite. A mineral. It is a basic vanadate of lead and zinc, $3PbO.V_2O_5.Zn(OH)_2$, found in Mexico and Chile.

Deschenite. A mineral. It is a variety of lead-vanadium mica.

Desencin. A German preparation containing 19 parts of the benzyl ester of iodo-ethoxy-benzoic acid and 1 part of pseudo-sulphimido-benzazide. It contains 31·6 per cent. iodine, and has the property of lowering blood pressure.

Deseril. A proprietary preparation of methysergide. Treatment of migraine. (267).

Deserpidine. 11 - Demethoxyreserpine. Harmonyl.

Desferal. A proprietary preparation of desferrioxamine mesylate used to treat hæmochromatosis and acute iron poisoning. (18).

Desferrioxamine. 30 - Amino - 3,14, 25 - trihydroxy - 3,9,14,20,25 - penta-azatriacontane - 2,10,13,21,24 - pen-taone. Deferoxamine. Desferal is the mesylate.

Desibyl Kapseals. A proprietary preparation of dried whole bile. (264).

Desicchlora. A proprietary name for a perchlorate of barium, a drying agent to replace calcium chloride, sulphuric acid, and potassium hydroxide. It absorbs 20 per cent. of its weight of water.

Desichthyol (Desichtol). The oil ichthyol, deodorised by distilling with steam, or by treating it with hydrogen peroxide.

Desichtol. See DESICHTHYOL.

Designolle's Powders. Explosives for torpedoes. They contain from 50–55 per cent. potassium picrate, 45–50 per cent. potassium nitrate, and charcoal.

Desipramine. 4,5 - Dihydro - 1(3-methylaminopropyl) - 2,3 : 6,7 - di-benzazepine. Petrofan is the hydrochloride.

Desklon. An inhibited acid descalant. (576).

Deslanoside. Deacetyl - lanatoside C. Cedilanid (for injection).

Desmine. A mineral, $(Ca.Na_2)O.Al_2O_3.6SiO_2.6H_2O$.

Desmodur (Formerly **Desmosit**). Di-isocyanates designed for use with hydroxyl containing materials. (3).

Desmodur H. Hexamethylene diiso-cyanate. (3).

Desmodur T. Toluene diisocyanate. (3).

Desmodur 1L. A proprietary trade name for an isocyanate prepolymer. Used with Desmophen (*q.v.*) to provide improved polyurethane finishes. (112).

Desmodur TH. Desmodin T partially reacted with trimethylolpropane and 1,4 hexanediol.

Desmodur 15. 1-5 Naphthalene diiso-cyanate. (3).

Desmopan. A proprietary polyester-based polyurethane thermoplastic used in injection moulding. (112).

Desmophen. A trade name for a series of alkyd resins designed for use with Desmodur (*q.v.*). (3).

Desmophen 300. Phthalic anhydride-trimethylolpropane. (3).

Desmophen 800. ½ mol. phthalic acid, 2½ mols adipic acid and 4·1 mols tri-methylolpropane. (3).

Desmophen 900. Adipic acid-trimethyl-olpropane. (3).

Desmophen 1100. Adipic acid-tri-methylolpropane-1,3 butanediol. (3).

Desmophen 1200. Adipic acid-tri-methylolpropane-1,4 butanediol. (3).

Desmophen 2000. A proprietary linear polyester based on ethylene glycol and adipic acid. It possesses terminal OH groups and is used in the production of polyurethane elastomers. (3).

Desmophen 2001. A proprietary linear

polyester based on butane diol 1-4 and ethane diol, used in the production of polyurethane elastomers. (3).

Desmopressin. A preparation used in the treatment of diabetes insipidus.

Desogen. A proprietary preparation of dodecanoyl N methylaminoethyl(phenyl carbamyl methyl)dimethyl ammonium chloride. Throat lozenges. (17).

Desomorphine. 7,8 - Dihydro - 6 - deoxymorphine.

Desonide. An anti-inflammatory. 11β, 21 - Dihydroxy - 16α, 17α - isopropylidene - dioxypregna - 1, 4 - diene - 3, 20 - dione. TRIDESILON.

Desoxymethasone. A topical corticosteroid. 9α - Fluro - 11β, 21 - dihydroxy - 16α - methylpregna - 1, 4 - diene - 3, 20 - dione. ESPERSON.

Despyrin. Salicyl-tartaric ester.

Destinezite. A mineral. It is a variety of diadochite.

Destral PCPL. Trade name for pentachlorphenyl laurate produced by the esterification of pentachlorphenol with mixed fatty acids of molecular weight approximately 200. It is used to render textiles and other organic materials resistant to attack of bacteria, fungi and insects. (183).

Desulfex. A ladle desulphuriser for iron. (531).

Detarex. Organic chelating agents. (572).

Detarol. Organic chelating agents. (572).

Deteclo. A proprietary preparation of chlortetracycline, tetracycline and democlocycline hydrochlorides. An antibiotic. (306).

Detel A and A(d/H). A proprietary trade name for chlorinated rubber based coating materials with good corrosion resistance. (148).

Detel EP. A proprietary trade name for an epoxy-pitch coating material for tank lining. (148).

Detel H. A proprietary trade name for a cyclised rubber surface coating material for lining galvanised hot water tanks. (148).

Detel HB Epoxy. A proprietary trade name for an epoxy based surface coating material with a high film build allowing single coats to achieve 0·008 in. thickness. (148).

Detel Thixochlor. A proprietary trade name for a thixotropic chlorinated rubber paint with good chemical resistance giving up to 0·005 in. per coat. Used in chemical works, plating shops, swimming pools and bottling plants. (148).

Deterlex. A surface active detergent. (576).

Dethlac. An insecticidal lacquer. (535).

Dethmor. A rodenticide.

Detigon. The hydrochloride of Chlophedianol.

Detigon Linctus. A proprietary preparation containing chlophedianol citrate and potassium guaiacol sulphonate. A cough linctus. (341).

Detonal. Diethyl - acetyl - urethane, $(C_2H_5)_2CH.CO.NH.COOC_2H_5$. A hypnotic.

Detoxin. A German preparation. It is sodium keratinate containing 1·9 per cent. sulphur. It is used for subcutaneous injection in cases of chronic rheumatism of the joints.

Detrustex. A rust remover for ferrous metals. (545).

Dettol. A proprietary trade name for a germicide containing chloroxylenols and terpineol. It is very powerful in action and yet non-poisonous.

De Valangin's Solution. Arsenious oxide (1$\frac{1}{4}$ grains), dissolved in 100 fluid parts dilute hydrochloric acid, then made up with water to 1 fluid ounce.

Devarda's Alloy. An alloy of 45 per cent. aluminium, 50 per cent. copper, and 5 per cent. zinc.

Devegan. A proprietary preparation. It is 3-acetyl-amino-4-hydroxy-phenyl-arsonic acid.

Devilline. A mineral. It is a variety of langite.

Devil's Dung. Ascefetida.

Devitrite. A mineral. It is $Na_2Ca_3Si_6O_{16}$.

Devitsky's Solder. An aluminium solder. It consists of tin.

Devolite (Catalpo). A refined china clay used as a filler for rubber, floorings, tiles, road blocks, etc.

Devonshire Batts. Whetstones used for sharpening scythes.

De Vry's Reagent. An alkaloidal reagent prepared by precipitating a solution of ammonium molybdate with excess of sodium phosphate at 40–50° C. It is left for 24 hours, filtered, and the precipitate washed with water and dissolved in the minimum quantity of sodium carbonate solution. The solution is evaporated and the residue ignited to free it from ammonia. Dissolve in hot water and add sufficient nitric acid to dissolve the precipitate at first formed.

Dewalquite. A mineral. It is ardennite.

Deward Steel. A proprietary non-shrinking steel containing 1·55 per cent. manganese, 0·3 per cent. molybdenum, and 0·9 per cent. carbon.

Deweylite (Gymnite). A mineral, 3MgO. $H_2O.2SiO_2H_2O$.

Dewindtite. A mineral, $4PbO.8UO_3$. $3P_2O_5.12H_2O$. A hydrated phosphate of uranium and lead, of the Belgian Congo.

De Wint's Green (Olive Green). A green pigment of variable composition.

Dew of Death. See LEWISITE.

Dewrance Metal. See DURANCE'S METAL.

Dexa Cortisyl. A proprietary preparation of dexamethasone acetate. (307).

Dexacaine. A proprietary preparation of dexivacaine. (876).

Dexamed. A proprietary preparation of dexamphetamine sulphate. An anti-obesity agent. (250).

Dexamethasone. 9α - Fluoro - $11\beta,17\alpha$, 21 - trihydroxy - 16α - methylpregna 1,4 - diene - 3,20 - dione. 9α - Fluoro-16α-methylprednisolone. Decadron ; Deronil ; DexaCortisyl ; Dextelan ; Millicorten ; Oradexon.

Dexamphetamine. $(+)$-α-Methylphenethylamine $(+)$-2-Aminopropylbenzene $(+)$-Amphetamine. Dexedrine is the sulphate.

Dexa-Rhinaspray. A proprietary preparation of neomycin, dexamethasone and tramazoline. (278).

Dexedrine. A proprietary preparation of dexamphetamine sulphate. Appetite control and anti-depressant agent. (281).

Dexetimide. A preparation used in the treatment of the Parkinsonian syndrome. It is $(+)$-3-(1-benzyl-4-piperidyl) - 3 - phenylpiperidine - 2, 6 - dione.

Dexdale. A proprietary preparation of dexamphetamine, codeine phosphate and paracetamol. (320).

Dexine 521. A polyisobutylene material with a good resistance to chemical attack from oxidising liquors up to 110° C. Used for lining and covering metal tanks. (149).

Dexine 656. A natural rubber based compound with good abrasion resistance used for lining metal tanks. (149).

Dexine 687. A natural rubber based material with good resistance to chemicals especially sodium hypochlorite. (149).

Dexine 759. A Hypalon (*q.v.*) based lining and covering material with very good resistance to chemical attack. It can be used with sulphuric acid at concentrations up to 95 per cent. (149).

Dexine 779. A polyurethane lining and covering material with a very good resistance to abrasion. (149).

Dexobarb. A proprietary preparation of amylobarbitone and dexamphetamine sulphate. Anti-depressive agent. (250).

Dexocodene. A proprietary preparation of aspirin, aluminium hydroxide, phenacetin, codeine phosphate, phenolphthalein, and dexamphetamine sulphate. (250).

Dexon. See POLYGLYCOLIC ACID.

Dexonite. A trade name for a proprietary hard rubber moulded material for electrical insulation. A proprietary trade name for ebonite. (149).

Dexoplas. A proprietary trade name for a butadiene-styrene plastics material used for constructing corrosion resistant fittings. (149).

Dexpanthenol. $(+)$-2,4-Dihydroxy-N-(3 - hydroxypropyl) - 3,3 - dimethylbutyramide. Bepanthen.

Dexpropranolol. An anti-arrhythmic. $(+)$ - 1 - Isopropylamino - 3 - (1 - naphthyloxy)propan - 2 - ol.

Dextelan. A proprietary trade name for Dexamethasone.

Dexten. A proprietary preparation of dexamphetamine resinate. (271).

Dextonite. A super-ebonite for chemical plant, having good corrosion resistance to alkalis and most gases and acids.

Dextran. A proprietary trade name for α 1-6 polyglucose or polyanhydroglucose. (219).

Dextran (Viscose). Fermentation gum, formed in the lactic fermentation of cane sugar.

Dextranomer. DEXTRAN cross-linked with epichlorohydrin. A promoter of wound-healing.

Dextraven. A proprietary preparation of dextrans in normal saline, used to restore blood volume. (188).

Dextriferron. A colloidal solution of ferric hydroxide in complex with partially hydrolysed dextrin.

Dextrin. See BRITISH GUM, LIQUID GLUE, and MAZAM.

Dextrin-Maltose. See GLUCOSE SYRUP.

Dextrinozole. See OZOLE.

Dextroform. A condensation product of dextrin and formaldehyde. An antiseptic used in medicine as a substitute for iodoform.

Dextro-glucose. See GLUCOSE.

Dextro-mannose. Mannose, CHO. $CH(OH).CH(OH).CH(OH).CH(OH).CH_2OH.$

Dextromethorphan. $(+)$-3-Methoxy-N-methylmorphinan. Romilar is the hydrochloride.

Dextromoramide. $(+)$ - 1 - (β - Methyl - γ - morpholine - $\alpha\alpha$ - diphenylbutyryl) - pyrrolidine. Jetrium ; Palfium ; R.875.

Dextro-pinene. Australene, a terpene.

Dextropropoxyphene. $(+)$ - 4 - Dimethylamino - 3 - methyl - 1,2 - diphenyl - 2 - propionyloxybutane (α-form). Darvon and Doloxene are the hydrochloride.

Dextrorphan. (+) - 3 - Hydroxy - N - methylmorphinan.

Dextrose. See GLUCOSE and CLIMAX SUGAR.

Dextrostix. A proprietary test strip impregnated with a buffered mixture of glucose oxidase, peroxidase, and a chromogen system, used to estimate blood glucose. (807).

Dextro-tartaric Acid. Common tartaric acid, $CO_2H.CH(OH).CH(OH).CO_2H$.

Dextrothyroxine. D - α - Amino - β-[4 - (4 - hydroxy - 3,5 - di-iodophenoxy) - 3,5-di-iodophenyl] acid. Choloxin is the sodium derivative.

Dexytal. A proprietary preparation of sodium amylobarbitone and dexamphetamine sulphate. An antidepressive agent. (333).

D.F. 118. A proprietary preparation of dihydrocodeine bitartrate. An analgesic. (182).

D.F.P. A proprietary preparation of dyflos, used in the treatment of glaucoma. (253).

Dhak gum. See KING, BENGAL.

Dhil Mastic. A mixture of 1 part massicot with 10 parts brickdust, and enough linseed oil to form a paste. Used by builders for repairing stone.

Dhobies' Earth. A saline earth found in India. It is used for washing, and in soap-making.

Dhurrin. A glucoside occurring in millet. It is p-hydroxy-mandelonitrile-glucoside.

Diabase. A dark green crystalline rock consisting chiefly of plagioclase, augite, and magnetite.

Diabantite. A mineral. It is $2[(Mg,Fe^{..},Al)_6(Si,Al)_4O_{10}(OH)_8]$.

Diabetic Cough Mixture. A proprietary preparation containing codeine phosphate, pholcodine, butethamate citrate, ipecacuanha liquid extract and squill liquid extract. A cough mixture. (273).

Diabetic Sugar. See GLUCOSE.

Diabetin. See LEVULOSE.

Diabinese. A proprietary preparation of chlorpropamide. An oral hypoglycæmic agent. (315).

Diaboleite. A mineral. It is $[Pb_2CuCl_2(OH)_4]$.

Diacetamate. An analgesic. 4-Acetamidophenyl acetate.

Diacetin. Glycerol diacetate, $(CH_3 COOCH_2)_2CHOH$.

Diacetone Alcohol. A compound, $(CH_3)_2C(OH)CH_2COCH_3$, obtained from diacetonamine and nitrous acid. It boils at 164° C. It is a solvent for cellulose acetate and nitrate.

Diacetylnalorphine. oo' - Diacetyl - N-allynormorphine.

Diacetyl Tannin. See ACETANNIN.

Diachylon. Lead plaster (q.v.).

Diadelphite. Synonym for Hematolite.

Diadem Chrome. Acid mordant wool dyestuffs. (583).

Diadem Chrome Black A. A dyestuff. It is a British equivalent of Eriochrome black A.

Diadem Chrome Black F New. A British dyestuff. It is equivalent to Diamond black F.

Diadem Chrome Black PV. A dyestuff. It is a British brand of Diamond black PV.

Diadem Chrome Blue-black P6B. A dyestuff. It is equivalent to Palatine chrome blue.

Diadem Chrome Red L3B. A dyestuff. It is a British equivalent of Eriochrome red B.

Di-Ademil. A proprietary trade name for Hydroflumethiazide.

Diadochite. A mineral. It is a hydrated phosphate and sulphate of iron, $Fe_4O(OH)_2.(HSO_4)_2.(PO_4)_2$.

Di-Adreson. A proprietary preparation of prednisone. (316).

Di-Adreson-F. A proprietary preparation of prednisolone. (316).

Diadzein. A glucoside. It is 7 : 4'-dihydroxy-isoflavine.

Diafor. Urea acetyl-salicylate.

Diaginol. Sodium acetrizoate. (507).

Diagnex Blue. An azuresin diagnostic test containing two tablets of caffeine sodium benzoate for stimulation of gastric secretion, and 2 gm. of Diagnex blue granules equivalent to 100 mg. azure A carbacrylic resin. Used in the diagnosis of achlorhydria. (326).

Diagnothorine. A proprietary preparation. It is a colloidal preparation of thorium dioxide.

Diakon. A proprietary trade name for an acrylic moulding powder. See also Perspex. (512).

Diakon Apai. A proprietary acrylic copolymer used to improve the processing of P.V.C. (512).

Dial. Diallylbarbituric acid. A powerful sedative and hypnotic.

Dialacetin. Tablets containing Dial (diallyl-barbituric acid) and allyl-p-aceto-amino-phenol. A hypnotic.

Dialin. Dihydro-naphthalene, $C_{10}H_{10}$.

Diall. A proprietary polyester laminating resin. (647).

Diallage. A mineral, $(Mg.Fe)O.CaO.2SiO_2$, with some alumina.

Diallogite. See RHODOCROSITE.

Dialozite. A mineral. It is a manganese carbonate, $MnCO_3$.

Dialuramide (Murexan). Uramil, $C_4H_5N_3O_3$.

Dialyl. Methylamine and lithium citrate.

Dialysed Iron. A solution containing ferric oxide and acetic acid, made by dialysing ferric acetate. Used medicinally.

Diamalt. A registered trade mark for a malt preparation for use in bread-making.

Diamantin. Alundum (q.v.), manufactured at Rheinfelden, under this name.

Diamazon. Diacetyl-amino-azo-toluene $CH_3.C_6H_4.N : N.C_6H_3(CH_3)N(CH_3CO)_2$. Used in the treatment of old wounds and ulcers.

Diamethine. A proprietary trade name for the bromide of dimethyltubocurarine.

Diamido - anthrachrysone - disulphonic Acid. A dyestuff. It is the sodium salt of diamino-tetroxy-anthraquinone-disulphonic acid, $C_{14}H_3N_2O_{12}S_2Na_2$. Dyes wool violet from an acid bath ; with alumina, violet blue ; and with chrome, blue.

Diamidogen. Hydrazine, $NH_2.NH_2$.

Diamine Azo Blue RR. A direct cotton dyestuff.

Diamin(e). A trade mark for a range of dyestuffs. (983).

Diamine Black BH (Melanthrene BH, Diazo Black DHL, Paramine Black BH, Chlorazol Black BH). A dyestuff. It is the sodium salt of diphenyl-disazo-amino-naphthol-sulphonic acid-amino-naphthol-disulphonic acid, $C_{32}H_{21}N_6O_{11}S_3Na_3$. Dyes cotton blue-black.

Diamine Black BO. A dyestuff. It is the sodium salt of ethoxy-diphenyl-disazo - bi - amino - naphthol - sulphonic acid, $C_{34}H_{26}N_6O_9S_2Na_2$. Dyes cotton blue-black. Used for ingrain blacks.

Diamine Black HW (Paramine Black HW). A dyestuff. It is the sodium salt of diphenyl-disazo-sulpho-amino-naphthol - disulpho - amino - naphthol - azo-nitro-benzene, $C_{38}H_{24}N_9O_{13}S_3Na_3$. Dyes cotton greenish-black.

Diamine Black RO. A dyestuff. It is the sodium salt of diphenyl-disazo-bi-amino-naphthol-sulphonic acid, $C_{32}H_{22}N_6O_8S_2Na_2$. Dyes cotton greyish-violet.

Diamine Black-blue B (Diamine Dark Blue B, Oxydiamine Deep Black N, NR, SOOO, Diazo Blue-Black RS Direct Deep Black R, T, Direct Blue-Black B, Pluto Black, Tabora Black, Zambesi Black BR). Tetrazo dyestuffs of a similar character to Diamine black HW.

Diamine Blue B. A dyestuff. It is the sodium salt of ethoxy-diphenyl-disazo - β - naphthol - δ - disulphonic -α-naphthol-mono-sulphonic acid, $C_{34}H_{23}N_4O_{12}S_3Na_3$. Dyes cotton blue.

Diamine Blue 2B (Benzo Blue BB, Congo Blue 2BX, Diazine Blue 2B, Orion Blue 2B, Benzarene Blue 2B, Paramine Blue 2B new, Direct Blue 2B, 2BL, 3B supra, Chlorazol Blue B). A dyestuff. It is the sodium salt of diphenyl - disazo - bi - amino - naphthol - disulphonic acid, $C_{32}H_{20}N_6O_{14}S_4Na_4$. Dyes cotton blue.

Diamine Blue 3B (Benzo Blue 3B, Congo Blue 3B, Diamine Blue 3G, Chlorazol Blue 3B, Direct Blue 3B). A dyestuff. It is the sodium salt of ditolyl-disazo-bi-amino-naphthol-disulphonic acid, $C_{34}H_{24}N_6O_{14}S_4Na_4$. Dyes cotton blue.

Diamine Blue-black E. A dyestuff. It is the sodium salt of ethoxy-diphenyl-disazo - amino - naphthol - mono - sulphonic-β-naphthol-δ-disulphonic acid, $C_{34}H_{24}N_5O_{12}S_3Na_3$. Dyes cotton a black-blue.

Diamine Blue BX (Benzo Blue BX, Congo Blue BX, Chlorazol Blue 2R). A dyestuff. It is the sodium salt of ditolyl - disazo - α - naphthol - mono - sulphonic - amino - naphthol - disulphonic acid, $C_{34}H_{24}N_5O_{11}S_3Na_3$. Dyes cotton deep blue.

Diamine Blue C4R. See COLUMBIA BLUE G.

Diamine Blue 3G. See DIAMINE BLUE 3B.

Diamine Blue 6G. A dyestuff. It is the sodium salt of disulpho-β-naphthalene - azo - ethoxy - α - naphthalene - azo-β-naphthol, $C_{32}H_{22}N_4O_8S_2Na_2$. Dyes cotton blue.

Diamine Blue 3R. A dyestuff. It is the sodium salt of ethoxy-diphenyl-diazo-bi-α-naphthol - p - sulphonic acid, $C_{34}H_{24}N_4O_9S_2Na_2$. Dyes cotton reddish-blue.

Diamine Blue RW, RG (Diamine New Blue R and G, Chicago Blues, Chicago Grey, Diazo Blue, Columbia Blue R and G). Analogous dyestuffs to Diamine sky blue. They dye cotton from a neutral salt bath.

Diamine Bordeaux B. A dyestuff allied to Diamine scarlet B. It dyes cotton red from an alkaline salt bath.

Diamine Bordeaux S. A dyestuff allied to Diamine scarlet B. It dyes cotton as above.

Diamine Brilliant Blue. A dyestuff. It is the sodium salt of dimethoxy-diphenyl - disazo - bi-1 : 8-chloro-naphthol-disulphonic acid, $C_{34}H_{20}N_4O_{16}S_4Cl_2Na_2$. Dyes cotton blue.

Diamine Brilliant Blue G. See DIAMINE SKY BLUE.

Diamine Bronze G. A dyestuff. It is the sodium salt of diphenyl-disazo-naphthol - disulphonic - azo - m - phenylene-diamine-salicylic acid, $C_{35}H_{23}N_8$ $O_{10}S_2Na_3$. Dyes cotton yellowish-brown of metallic appearance.

Diamine Brown B (Cupranil Brown B, Paramine Fast Brown B, Chlorazol Brown B). A dyestuff. It is the sodium salt of diphenyl-disazo-salicylic acid-phenyl-amino-naphthol-sulphonic acid γ, $C_{29}H_{19}N_5O_7SNa_2$. Dyes cotton dark brown.

Diamine Brown 3G. An azo dyestuff similar in composition and application to Diamine brown B.

Diamine Brown M (Chlorazol Brown M, Paramine Brown M, Direct Brown M). It is the sodium salt of diphenyl-disazo-salicylic acid-amino-naphthol-sulphonic acid γ, $C_{29}H_{19}N_5$ O_7SNa_2. Dyes cotton brown.

Diamine Brown V. A dyestuff. It is the sodium salt of diphenyl-disazo-phenylene-diamine-amino-naphthol-sulphonic acid, $C_{28}H_{22}N_7O_4SNa$. Dyes cotton dark violet-brown.

Diamine Catechin B and G. Direct cotton dyestuffs, used in conjunction with potassium bichromate.

Diamine Cutch. See NAPHTHYLENE VIOLET.

Diamine Dark Blue B. See DIAMINE BLACK-BLUE B.

Diamine Deep Black. A disazo dyestuff derived from diazotised di-p-amino-diphenylamine, coupled with 1 molecule of amino-naphthol-sulphonic acid G, and 1 molecule of m-tolylene-diamine.

Diamine Deep Blue B and R. A direct cotton dyestuff.

Diamine Fast Red (Azidine Fast Red F, Direct Fast Red F, Paramine Fast Red F, Chlorazol Fast Red FG). A dyestuff. It is the sodium salt of diphenyl-disazo-salicylic-amino-naphthol-sulphonic acid, $C_{20}H_{19}N_5O_7SNa$. Dyes cotton and chromed wool red.

Diamine Fast Yellows, B, C, FF. See CHLORAMINE YELLOW.

Diamine Gold (Diamine Golden Yellow). A dyestuff. It is the sodium salt of disulpho-naphthalene-disazo-phenetol-phenol, $C_{24}H_{18}N_4O_8S_2Na_2$. Dyes cotton yellow from a salt bath.

Diamine Golden Yellow. See DIAMINE GOLD.

Diamine Green B (Diazine Green B, Enbico Direct Green B, B Extra Conc., Direct Green B, BL, Orion Green B, Chlorazol Green BN, Benzarene Green B). A dyestuff. It is the sodium salt of diphenyl-disazo-phenol-disulpho-amino-naphthol-azo-nitro-benzene, $C_{34}H_{22}N_8O_{10}S_2Na_2$. Dyes cotton green.

Diamine Green G (Paramine Green B, G, Orion Green G, Direct Green BG, G). A dyestuff. It is the sodium salt of

diphenyl-disazo-salicylic acid-disulpho-amino-naphthol-azo-nitro-benzene, $C_{35}H_{20}N_8O_{12}S_2Na_2$. Dyes cotton green from a neutral salt bath, also wool and silk.

Diamine Grey. An azo dyestuff dyeing mordanted cotton from an alkaline bath.

Diamine Jet Black OO. An azo dyestuff, dyeing cotton from an alkaline bath.

Diamine Jet Black OR. A direct cotton dyestuff, after-treated with potassium bichromate.

Diamine Jet Black RB, SS. Azo dyestuffs, having the same application as Diamine jet black OO.

Diamine New Blue G. An azo dyestuff, which dyes cotton from an alkaline salt bath, and wool from an acid bath.

Diamine New Blue R. A dyestuff similar to the G mark.

Diamine Orange G and B. A tetrazo dyestuff, which dyes cotton reddish-orange shades from an alkaline salt bath, and wool from a sodium sulphate bath.

Diamine Pink. See DIAMINE ROSE.

Diamine Pure Blue. See DIAMINE SKY BLUE.

Diamine Pure Blue FF. See DIAMINE SKY BLUE.

Diamine Red B (Delta purpurin 5B). A dyestuff. It is the sodium salt of ditolyl-disazo-bi-β-naphthylamine-sulphonic acid, $C_{34}H_{26}N_6O_6S_2Na_2$. Dyes cotton red from an alkaline bath.

Diamine Red 3B (Delta purpurin 7B). A dyestuff. It is the sodium salt of ditolyl-disazo-bi-β-naphthylamine-δ-sulphonic acid, $C_{34}H_{26}N_6O_6S_2Na_2$. Dyes cotton red from an alkaline bath.

Diamine Red NO. A dyestuff. It is the sodium salt of ethoxy-diphenyl-disazo-β-naphthylamine-β-sulphonic β-naphthylamine-δ-sulphonic acid, $C_{34}H_{26}N_6O_7S_2Na_2$. Dyes cotton red from an alkaline bath.

Diamine Rose (Diamine Pink). A dyestuff. It is the sodium salt of benzenyl-amino-thiophenol-azo-chloro-naphthol-disulphonic acid, $C_{24}H_{15}N_3ClS_3O_7Na_2$. Dyes unmordanted cotton pink shades.

Diamine Scarlet B (Direct Red B). A dyestuff. It is the sodium salt of diphenyl-disazo-phenetol-β-naphthol-γ-disulphonic acid, $C_{30}H_{22}N_4O_8S_2Na_2$. Dyes wool and silk from an acid or neutral bath, and cotton from an alkaline bath.

Diamine Scarlet 3B. A dyestuff belonging to the same group as Diamine scarlet B.

Diamine Sky Blue (Benzo Sky Blue, Congo Sky Blue, Chlorazol Blue 6G,

Chicago Blue 6B, Benzo Pure Blue, Brilliant Benzo Blue 6B, Diamine Pure Blue FF, Diamine Brilliant Blue G, Chlorazol Sky Blue FF, Orion Sky-blue, Direct Sky-blue GS). A dyestuff. It is the sodium salt of dimethoxy-diphenyl-disazo-bi-amino-naphthol-disul-phonic acid, $C_{34}H_{24}N_6O_{16}S_4Na_4$. Dyes cotton blue from an alkaline bath.

Diamine Violet N (Chlorazol Violet N, Paramine Fast Violet N). A dyestuff. It is the sodium salt of diphenyl-disazo-bi-amino-naphthol-sulphonic acid, $C_{32}H_{22}N_6O_8S_2Na_2$. Dyes cotton violet, also wool and silk from a neutral bath.

Diamine Yellow A. See CHLORAMINE YELLOW.

Diamine Yellow N. A dyestuff. It is the sodium salt of ethoxy-diphenyl-disazo-phenetol-salicylic acid, $C_{29}H_{25}N_4O_5Na$. Dyes cotton yellow.

Diaminogen. A registered trade mark currently awaiting re-allocation by its proprietors to cover a range of dyestuffs. (983).

Diaminogen Black. A dyestuff prepared from diazotised acetyl-1 : 4-naphthylene-diamine-7-sulphonic acid, and α-naphthylamine, the product being diazotised, coupled with 8-amino-β-naphthol-6-sulphonic acid G, and then hydrolysed.

Diaminogen Blue BB (Substitute for Indigo, Chlorazol Diazo Blue 2B). A dyestuff prepared in a similar way to Diaminogen black, but containing β-naphthol-6-sulphonic acid, instead of amino-naphthol-sulphonic acid G.

Diaminogen Blue G. A dyestuff containing β-naphthol-3 : 6-disulphonic acid as the final component.

Diaminogen Extra. A dyestuff similar to Diaminogen.

Diammonphos. A pure diammonium phosphate, $(NH_4)_2HPO_4$. It contains 21 per cent. nitrogen and 35 per cent. soluble phosphoric acid. It is a fertiliser with its phosphate in an available form.

Diamocaine. A local anæsthetic. 1 - (2 - Anilinoethyl) - 4 - (2 - diethylaminoethoxy) - 4 - phenylpiperidine. " R 10,948 " is currently under clinical trial as the dicyclamate.

Diamol. 2 : 4 - Diamino - phenol - hydrochloride.

Diamond. One of the two allotropic forms of carbon, owing its popularity as a gem to its hardness and high refractive index which render it imperishable and lustrous.

Also the second smallest type used in printing.

Diamond, Black. See CARBONADO.

Diamond Black 2B. A mixture of a violet-black and a bluish-green dye-

stuff. Dyes wool bluish-black in the presence of sodium sulphate, acetic acid, and a little potassium bichromate.

Diamond Black F (Chrome Fast Black FW, Durochrome Black, Enbico Chrome Black F, Diadem Chrome Black F new, Solochrome Black F). A dyestuff. It is the sodium salt of carboxy-phenol-azo-α-naphthalene-azo - α - naphthol-*p*-sulphonic acid, $C_{27}H_{16}N_4O_7SNa_2$. Dyes chrome mordanted wool bluish-black.

Diamond Black PV (Diadem Chrome Black PV). A dyestuff prepared from *o*-amino-phenol-sulphonic acid, and 1 : 5-dihydroxy-naphthalene.

Diamond-boron (Adamantine). A crystalline variety of boron.

Diamond Brown Paste. An acid mordant dyestuff, giving brown shades on wool mordanted with bichromate. Used in calico printing.

Diamond Cement. A cement containing 8 parts isinglass, 1 part gum ammoniacum, 1 part galbanum, and 4 parts alcohol. Used for mending china and glass. Also see ACETIC-GELATIN CEMENT.

Diamond Fibre. A proprietary trade name for a vulcanised fibre ; a laminated acid-treated cotton cellulose.

Diamond Flavine G. A dyestuff. It is *p*-oxy-diphenyl-azo-salicylic acid, $C_{19}H_{14}N_2O_4$. Dyes chromed wool yellow.

Diamond Fuchsine. The largest crystals of fuchsine.

Diamond Green. A dyestuff. It is the sodium salt of carboxy-phenol-azo-α-naphthalene-azo-dioxy - naphthalene-sulphonic acid, $C_{27}H_{16}N_4O_8SNa$. Dyes chrome mordanted wool dark bluish-green.

Diamond Green B. See MALACHITE GREEN.

Diamond Green C. See BRILLIANT GREEN.

Diamond Grey. See ZINC GREY.

Diamondite. An alloy of 95·65 per cent. tungsten with 3·91 per cent. carbon.

Diamond Magenta. See MAGENTA.

Diamond Orange. A dyestuff similar to Diamond yellow.

Diamond Salicine. See CHROME.

Diamond Vinegar. A vinegar containing 3·5 per cent. acetic acid.

Diamond White. See OIL WHITE.

Diamond Yellow G. A dyestuff. It is the sodium salt of *m*-carboxy-benzene-azo-salicylic acid, $C_{14}H_{10}N_2O_5$. Dyes chrome mordanted wool reddish-yellow.

Diamond Yellow R. A dyestuff. It is the sodium salt of *o*-carboxy-benzene-azo-salicylic acid, $C_{14}H_{10}N_2O_5$. Dyes chrome mordanted wool reddish-yellow.

Diamorphine. See HEROIN.

Diamox. A proprietary preparation of acetazolamide. A diuretic. (306).

Diamoxalic Acid. Oxy-dodecoic acid, $(C_5H_{11})_2.C(OH).CO_2H$.

Diamphenethide. An anthelmintic. $\beta\beta'$ - Oxydi(aceto - p - phenetidide). Coriban.

Diampromide. An analgesic. N - [2 - (N - Methylphenethylamino) - propyl]propionanilide.

Diampron. A proprietary preparation of Amicarbalide. An anti-protozoan for veterinary use.

Diamthazole. 6 - (2 - Diethylamino-ethoxy) - 2 - dimethylaminobenzothia-zole. Dimazole. Asterol is the dihydro-chloride.

Diamyl, Active. Decane, $CH(CH_3)$ $(C_2H_5)CH_2.CH_2.CH(CH_3)(C_2H_5)$.

Diana. Silver, Ag.

Dianabol. A proprietary preparation of methandienone. An anabolic agent. (18).

Diandrone. A proprietary preparation of dehydro-iso-androsterone, used in the treatment of psychoneuroses. (316).

Diane. Diphenylolpropane used as the phenolic reactant in resin manufacture. (4).

Dianil Black. See Cuba Black.

Dianil Black HW. A dyestuff which dyes mixed fabrics.

Dianil Black R. A dyestuff. It is the sodium salt of diphenyl-disazo-m-pheny-lene - diamine - disulpho-dioxy-naphtha-lene - azo - naphthalene - sulphonic acid, $C_{38}H_{27}N_8O_{11}S_3Na_3$. Dyes cotton black.

Dianil Blue B. A dyestuff obtained by combining tetrazotised paradiamine with 2 molecules of 1 : 8-dioxy-naphtha-lene-3 : 6-disulphonic acid. Dyes cotton blue from a salt bath.

Dianil Yellow. A dyestuff obtained by coupling diazotised primuline with aceto-acetic ester.

Dianisidine Blue (Azophor Blue). A dyestuff. It is a copper derivative of dimethoxy-diphenyl-disazo-bi-β-naph-thol. A reddish-blue on cotton.

Dianite. A name applied to the mineral columbite.

Dianol Black Brown. A direct cotton dyestuff, fixed by treatment with potas-sium bichromate.

Dianol Brilliant Red (Toluylene Red, Acetopurpurin, Chlorantine Red). A dyestuff. It is the sodium salt of dichloro-benzidine - disazo - bi - β - naph-thylamine-disulphonic acid, $C_{32}H_{18}N_6$ $O_{12}S_4Na_4Cl_2$. Dyes cotton bluish-red.

Dianol Brown T, Y, YY, R. Direct cotton browns, fixed with potassium bichromate.

Dianol Olive. A direct cotton dyestuff.

Dianol Red B. A dyestuff. It is the sodium salt of dichloro-diphenyl-dis-azo-bi-β-naphthylamine-sulphonic acid, $C_{32}H_{20}N_6O_6S_2Na_2Cl_2$. Dyes cotton yellowish-red.

Dianol Red 2B. A dyestuff. It is the sodium salt of dichloro-diphenyl-disazo-bi-naphthionic acid, $C_{32}H_{20}N_6O_6S_2Na_2$ Cl_2. Dyes cotton bluish-red.

Dianthine. A trade mark for a range of dyes. (236).

Dianthine. See St. Denis Red.

Dianthine B. See Erythrosin.

Dianthine G. See Erythrosin G.

Diapente. Consists of 8 parts powdered gentian root, and 1 part powdered bay berries.

Diaphan Oil. A mixture of methyl-hexalin and sodium oleate. Used in the preparation of transparent soaps.

Diaphoretic Antimony. This term ap-pears to have been applied to both antimony oxide and potassium anti-monate.

Diaphorite. A name applied to the mineral freieslebenite.

Diaphorm. Diamorphine hydrochloride.

Diaphtherine (Oxychinaseptol). A com-pound of 1 molecule of o-phenol-sul-phonic acid, and 2 molecules of oxy-quinoline. Used externally as an anti-septic.

Diapthol (Quinaseptol). o-Oxy-quino-line-m-sulphonic acid,$C_9H_5N.OH.SO_3H$. An antiferment and urinary disinfec-tant.

Diarabinose (Arabiose). A sugar. It is arabinose, $C_{10}H_{18}O_9$.

Diarex. A proprietary polystyrene. (663).

Diarsenal. A proprietary brand of salvarsan ($q.v.$).

Diasal. Sodium - di - iodo - salicylate, a germicide.

Diaspirin (Salicylsuccinate). Succinyl-salicylic acid, $(CH_2.COO)_2(C_6H_4.$ $COOH)_2$. Used in medicine for rheu-matism and neuralgic troubles.

Diaspore. A mineral. It is hydrated aluminium oxide, $Al_2O_3.H_2O$.

Diastase. See Amylase, Diastin, Dias-tofor, Maltine, French, and Tex-tase.

Diastatite. A mineral. It is a variety of hornblende.

Diastin. A form of diastase for use in certain kinds of dyspepsia.

Diastix. A proprietary test strip im-pregnated with glucose oxidase and peroxidase, plus potassium iodide, used to detect glycosuria. (807).

Diastofor. A registered trade mark for a size for cotton and other threads and yarns.

Diathesin. A name for Saligenin (q.v.).

Diatite. A vulcanite substitute made from shellac and infusorial earth.

Diatol. A registered trade name for diethyl-carbonate, $(C_2H_5)_2CO_3$. It is a solvent for nitro-cellulose, with not more than 15 per cent. boiling at or below 100° C., and 85 per cent. at or below 125° C.

Diatomaceous Earth. See INFUSORIAL EARTH.

Diatomite. A German name for a preparation of infusorial earth (q.v.), obtained by heating it. Bricks of this material are not affected by water, steam, acids, or alkalis.

Diatrin. A proprietary trade name for the hydrochloride of Methaphenilene.

Diatrizoic Acid. A radio-opaque substance present in Angiografin and Urografin as the meglumine salt, and in Urovison as the sodium salt. It is 3, 5 - diacetamido - 2, 4, 6 - tri - iodo - benzoic acid. HYPAQUE is the sodium salt.

Diaveridine. An anti-protozoan. 2, 4 - Diamino - 5 - (3, 4 - dimethoxybenzyl) - pyrimidine. It is present in DARVISUL.

Diax. A proprietary product used as a diastatic ferment.

Diazamine Fast Yellow H. A dyestuff. It is a British equivalent of Mikado yellow.

Diazepam. 7 - Chloro - 2,3 - dihydro - 1 - methyl - 5 - phenyl - 1,4 - benzodiazepin-2-one. Valium.

Diazine. Direct duestuffs for cotton and artificial silk. (541).

Diazine Black. A dyestuff. It is safranine-azo-phenol. Dyes tannined cotton black.

Diazine Blue. See INDOINE BLUE R.

Diazine Blue 2B. A British dyestuff. It is equivalent to Diamine blue 2B.

Diazine Blue BR. See INDOINE BLUE R.

Diazine Green (Janus Green B and G). A dyestuff. It is the chloride of safranine-azo-dimethyl-aniline. Dyes cotton dull bluish-green.

Diazine Green B. A dyestuff. It is a British brand of Diamine green B.

Diazine Yellow R. A dyestuff. It is a British equivalent of Direct yellow R.

Diazinon. A proprietary preparation of DIMPYLATE. An insecticide.

Diazitol. Diazonon insecticide. (501).

Diazo Black B. A tetrazo dyestuff, obtained by coupling diazotised benzidine with 2 molecules of α-naphthylamine-5-sulphonic acid L. Dyes cotton grey, becoming darker upon developing with β-naphthol.

Diazo Black 3B, G, H, BHN, R. Dyestuffs similar to Diazo black B.

Diazo Black DHL. A dyestuff. It is a British equivalent of Diamine black BH.

Diazo Blue-black B. See DIAMINE BLACK-BLUE B.

Diazo Blue-black RS. Diamine Black-blue B.

Diazo Blue 3R. A dyestuff having a similar application to Diazo black B.

Diazo Bordeaux. A dyestuff allied to primuline, having the same application as primuline.

Diazo Brilliant Black B. A tetrazo dyestuff, obtained by coupling diazotised tolidine with 2 molecules of α-naphthylamine - 5 - sulphonic acid L. Dyes cotton the same as Diazo black B.

Diazo Brilliant Black R. A similar dyestuff to the B mark.

Diazo Brown. A tetrazo dyestuff, dyeing cotton direct.

Diazo Brown R Extra. A dyestuff similar to Diazo brown.

Diazo Fast Black (Diazo Fast Black H). Direct cotton dyestuffs which are developed on the fibre.

Diazo Indigo Blue. An analogous product to Diamidogen (q.v.).

Diazone. Fast dyestuffs for cotton. (541).

Diazoresorcin (Azoresorcin, Resazoin). Resazurin, $C_{12}H_9NO_4$.

Diazoresorufin. See AZORESORUFIN.

Diazoxide. 7 - Chloro - 3 - methyl-benzo-1,2,4-thiadiazine 1,1-dioxide.

Diazurin B. A dyestuff. It is the sodium salt of dimethoxy-diphenyl-disazo-bi-α-naphthylamine-5-sulphonic acid, $C_{34}H_{26}N_6O_8S_2Na_2$. Dyes cotton direct, being further developed on the fibre by β-naphthol.

Diazurin G. A similar dyestuff to the B brand.

Dibencil. A proprietary preparation containing Benzathine Penicillin. An antibiotic. (2).

Dibenyline. A proprietary preparation of phenoxybenzamine hydrochloride. Vascular antispasmodic. (281).

Dibexin. A proprietary preparation of Vitamin B. (264).

Dibenzepin. 4-(2-Dimethylaminoethyl)-1,4 - dihydro - 1 - methyl - 2,3 : 6,7 - dibenzo-1,4-diazepin-5-one. Noveril is the hydrochloride.

Dibenzyline. A proprietary preparation containing phenoxybenzamine hydrochloride. (281).

Dibistin. A proprietary preparation of antazoline hydrochloride and tripelennamine hydrochloride. Antihistamine. (18).

Dibnal. Dibutyl-aceturethane.

Dibotin. A proprietary preparation of

phenformin hydrochloride. An oral hypoglycæmic agent. (112).

Dibrogan. Dibromopropamidine isethionate/promethazine cream. (507).

Dibromin. Dibromo - barbituric acid. An antiseptic.

Dibupyrone. Sodium 2,3-dimethyl-1-phenyl - 5 - pyrazolon - 4 *yl* - N - isobutylaminomethanesulphonate.

Dicalite 14, 14B, and 14W. Proprietary trade names for diatomaceous silica fillers. Used for heat insulating and as a filler for plastics, etc.

Dicarburetted Hydrogen. See OLEFIANT GAS.

Dicene. A dyeing assistant for polyester fibre. (508).

Dice Ore. See GALENA.

Dicestal. Dichlorophen. (507).

Dichlofenthion. An insecticide currently undergoing clinical trial as " V-C13 ". O - 2, 4 - Dichlorophenyl OO - diethyl phosphorothioate.

Dichlofuanide. A proprietary paint fungicide. N - Dimethylamino - N' - phenyl - N' - (fluorodichloromethylthio) sulphamide. (112).

Dichlone. Dichloronaphthaquinone. An organic fungicide used as a seed dressing.

Dichloralphenazone. A complex of chloral hydrate and phenazone. Welldorm.

Dichloramal. See ALEUDRIN.

Dichloramine - M. Methyl - diphenyl-methyl-dichloramine.

Dichloramine T. *p*-Toluene-sulphon-dichlor-amide, $C_6H_4(CH_3)(SO_2.NCl_2)$. A disinfectant.

Dichloraminet. See DICHLORAMINE T.

Dichlorene. Dichlor-ethylene. A narcotic used in conjunction with ether.

Dichloroditane. A proprietary trade name for *p*-dichlorodiphenylmethane. (95).

Dichlorophen. Di - (5 - chloro - 2 - hydroxyphenyl)methane. Anthiphen.

Dichlorphenamide. 4,5 - Dichloro-benzene-1,3-disulphonamide. Daranide; Oratrol.

Dichlorophenarsine. 3 - Amino - 4 - hydroxyphenyldichloroarsine.

Dichlorvos. An anthelmintic and insecticide. 2, 2 - Dichlorovinyl dimethyl phosphate. ATGARD, CANOGARD, EQUIGARD, VAPONA.

Dichlotride. A proprietary trade name for Hydrochlorothiazide.

Dichroite. See IOLITE.

Dichromium Trioxide. A diagnostic aid. Chromium sesquioxide.

Dick. A military poison gas. It is a nose gas, ethyl-dichloroarsine C_2H_5 $AsCl_2$. It produces asthma.

Dickerlite. A trade name for a synthetic resin-asbestos composition used as an electrical insulation.

Dickinsonite. A mineral. It is $4[H_2Na_6(Mn\cdot\cdot,Fe\cdot\cdot)_{14}(PO_4)_{12}.H_2O]$.

Dickite. A mineral. It is $4[Al_2Si_2O_5(OH)_4]$.

Dicköl Varnish. See STAND OIL.

Dicksbergite. A mineral. It is a variety of rutile.

Diclofenac. An anti-inflammatory. [2 - (2, 6 - Dichloroanilino)phenyl]acetic acid. Voltarol is the sodium salt.

Dicodethine (Dicodethylene). Ethylene-dimorphine, $C_2H_4(C_{17}H_{18}NO_3)_2$.

Dicodethylene. See DICODETHINE.

Dicodid. A German codeine preparation. It is hydro-codeinone, and is used as a morphine substitute.

Diconal. A proprietary preparation of dipipanone hydrochloride and cyclizine hydrochloride. An analgesic. (277).

Dicopin. A proprietary preparation of dihydrocodeine tartrate and aspirin. An analgesic. (633).

Dicotox. A selective weedkiller. (507).

Dicrasite. A mineral. It is an antimonide of silver, Ag_4Sb.

Dicrylan 270. A proprietary trade name for an aqueous emulsion of an acrylic resin for increasing the abrasion resistance of crease-resistant fabrics. (18).

Dicumoxane. A proprietary preparation of cumetharol. An anticoagulant. (249).

Di-Cup. Trade name for dicumyl peroxide, used as a crosslinking agent for vinyl, acrylics and polyester resins. (93).

Dicyanin. A red sensitiser, prepared by the action of potash and oxygen upon α-γ-dimethyl-quinolinium salts.

Dicyclomine. 2 - Diethylaminoethyl bicyclohexyl-1-carboxylate. Dicyclo-verine. Merbentyl and Wyovin are the hydrochloride.

Dicycloverine. Dicyclomine.

Dicynene. A proprietary preparation of ethamsylate. A clotting agent. (349).

Didandin. A proprietary trade name for Diphenadione.

Didial. Dial (diallylbarbituric acid) and diallylbarbiturate of ethylmorphine. A highly active hypnotic and sedative for the treatment of grave insomnia.

Didi-Col. A D.D.T. insecticidal spray. (511).

Didigram. An oil insecticide. (511).

Didimac. A D.D.T. insecticide. (511).

Didrex. A proprietary preparation of benzphetamine hydrochloride. An antiobesity agent. (325).

Didymite. A mineral. It is a variety of mica.

Didymolite. A mineral. It is a silicate of calcium and aluminium.

Die-casting Alloys. These are usually zinc-base alloys containing 86 per cent. zinc, 7–10 per cent. tin, 4–7 per cent. copper, 0·5–1 per cent. aluminium. Some alloys have a tin base, and a typical one contains 90 per cent. tin, 4 per cent. copper, and 6 per cent. antimony.

Di-el. A material having a specific gravity of 1·35, used as an insulating material for high voltage engineering as a substitute for hard paper, also for porcelain.

Dieldrin. Product containing 85 per cent. of 1,2,3,4,10,10-hexachloro-6,7-epoxy - 1,4,4a,5,6,7,8,8a - octahydro-exo - 1,4 - endo - 5,8 - dimethano-naphthalene.

Dieline. A proprietary solvent. It is sym-dichlorethylene, $CHCl = CHCl$. Its specific gravity is 1·25–1·278, boiling-point 48–60° C., flash-point 36° F. It is a solvent for cellulose acetate, rubber, and oils.

Dielmoth. A moth proofing agent for wool. (553).

Diemenal. Colloidal manganese and iron.

Diene. A proprietary synthetic rubber. (648).

Dienerite. A mineral, Ni_3As_2.

Dienol. Colloidal manganese.

Di-esterex N. A proprietary trade name for a rubber vulcanising accelerator stated to be 60 per cent. of the dinitrophenyl ester of mercapto-benzthiazole and 40 per cent. of the acetate of diphenyl-guanidine.

Diethadione. 5,5 - Diethyloxazine-2,4-dione. Toce.

Diethazine. 10 - (2 - Diethylamino-ethyl)phenothiazine. Diparcol is the hydrochloride.

Diethylcarbamazine. 1 - Diethyl-carbamoyl - 4 - methylpiperazine. Danocide, Ethodryl and Hetrazan are the dihydrogen citrate.

Diethylin. The diethyl ether of glycerin. Boiling-point 191° C.

Diethylpropion. α - Diethylamino-propiophenone. Tenuate is the hydrochloride.

Diethyl-sulphonal. Tetronal (q.v.).

Diethylthiambutene. 3 - Diethylamino-1,1 - di - (2 - thienyl)but - 1 - ene. Themalon.

Diethyltoluamide. NN - Diethyl - m - toluamide.

Dietzeite. A mineral. It is a double iodate and chromate of calcium, $Ca(IO_3)_2.CaCrO_4$.

Di-Farmon. A herbicide. (503).

Difenoxin. An anti-diarrhœal. 1 - (3 - Cyano - 3, 3 - diphenylpropyl) - 4 - phenylpiperidine - 4 - carboxylic acid.

Difenzoquat. An ion pesticide. 1, 2 - dimethyl - 3, 5 - diphenylpyrazolium ion.

Difetarsone. An arsenical. NN' - Ethylene - 1, 2 - diarsanilic acid.

Diflon. A Russian trade name for a polycarbonate resin for moulding purposes.

Diflubenzuron. A pesticide. 1 - (4 - chlorophenyl) - 3 - (2, 6 - difluoro-benzoyl) urea.

Diflucortolone. A glucocorticoid. 6α, 9 - Difluoro - 11β, 21 - dihydroxy - 16α - methylpregna - 1, 4 - diene - 3, 20 - dione.

Diflumidone. An anti-inflammatory. 3' - Benzoyldifluoromethanesulphon - anilide.

Diformyl. Glyoxal, $(CHO)_2$.

Digalin. A solution of digitoxin (0·3 milligramme in 1 millilitre).

Digallic Acid. Tannic acid.

Diganox. A proprietary preparation of digoxin. (781).

Digenite. A name applied to a cuprous sulphide mineral.

Digibaine. Crystallised strophanthin combined with digitalin.

Digifoline. A physiologically standardised extract of the cardio-active glucosides of digitalis leaf.

Digifortis Kapseals. A proprietary preparation of digitalis. (264).

Digilanid. A proprietary preparation of a glycosidal complex of lanatosides, used for treatment of heart failure. (267).

Diginutin. A proprietary preparation. It is a stable standardised solution of the total glucosides of digitalis, free from inert matter.

Digipotene. A preparation containing the whole of the glucosides of digitalis leaves.

Digipuratum. See DIGITAN.

Digistrophena. A German product. It is the dialysate from *Digitalis purp.* and *Straphanthus kombé.*

Digitalin, Amorphous (Homolle's digitalin). This consists largely of digitoxin with some digitalin.

Digitaline. A proprietary preparation of a cardiotonic glycoside of digitalis purpurea; used for the treatment of heart failure. (835).

Digitalin, Homolle's. See DIGITALIN, AMORPHOUS.

Digitalin, Nativelle's. See NATIVELLE'S DIGITALIN.

Digitalis. The dried leaves of the flowering plants of *Digitalis purpurea.*

Digitan (Digipuratum). An American preparation containing digitoxin and digitalin, in the form of tannates.

Digitin. German crystallised digitalin.

Digitsaponin. Amorphous digitonin.

Diglyme. The dimethyl ether of diethylene glycol. A solvent for polystyrene, PVC/PVA copolymer (*q.v.*) and polymethyl methacrylate. (2).

Diguanil. A proprietary preparation of metformin hydrochloride. An oral hypoglycæmic agent. (279).

Dihydrallazine. 1,4 - Dihydrazinophthalazine. Nepresol is the mesylate or the sulphate.

Dihydrite. See PSEUDO-MALACHITE.

Dihydrohydroxycodeinone. Oxÿcodone.

Dihydrohydroxymorphinone. Oxymorphone.

Dihydroisophorone. 3,5,5 - triethylcyclohexanone-1. A high boiling point ketone solvent for surface coatings.

Dihydrotachysterol. 9,10 - Secoergosta - 5,7,22 - trien - 3β - ol. A.T.10.

Diiodoform. Ethylene periodide, C_2I_4.

Di-iodoform. Ethylene periodide.

Di-iodohydroxyquinoline. 8-Hydroxy-5,7-di-iodoquinoline. Diodoquin; Embequin; Floraquin.

Di - iodo - salol. Phenyl-di-iodo-salicylate, $C_6H_2I_2(OH)CO_2 \cdot C_6H_5$. An antiseptic used in skin diseases.

Di-isoprene. See DIPENTENE.

Dijex. A proprietary preparation containing aluminium hydroxide and magnesium carbonate as a co-dried gel. An antacid. (253).

Dijodil. Ricinstearolic di-iodide.

Dijozol. A German product. It is an alcoholic solution of a di-iodated salt of phenol-sulphonic acid with homogeneously combined iodine. It is used in diluted solution as an antiseptic for disinfecting the operative field and the hands of the operator.

Dika Bread (Dika Chocolate, Gaboon Chocolate). Cakes made from the kernels of various kinds of *Irvingia*, of the West Coast of Africa. They are food products, and contain Dika fat.

Dika Butter (Dika Oil, Oba Oil, Wild Mango Oil). Dika fat, obtained from the kernels of *Irvingia* species, of the West Coast of Africa.

Dika Chocolate. See DIKA BREAD.

Dika Oil. See DIKA BUTTER.

Dikegulac. A pesticide. 2 : 3 : 4, 6-di-O-isopropylidene - α - L - xylo - 2 - hexulofurano - sonic acid.

Diketo - camphor. 3-Keto-camphor, $C_{10}H_{14}O_2$.

Dilatin NA Liquid. A proprietary preparation used in the dyeing of 100 per cent. polyester goods. It is based on unchlorinated aromatic hydro-carbons. (267).

Dilaudid. A proprietary preparation of dihydromorphone hydrochloride. (393).

Dilavase. A proprietary preparation of isoxsaprine hydrochloride. A vasodilator. (316).

Dilectene. A proprietary trade name for an aniline-formaldehyde synthetic resin.

Dilecto. A proprietary trade name for a phenol-formaldehyde synthetic resin laminated product.

Dillnite. A mineral. It is $Al_6Si_2O_{13} \cdot 6H_2O$.

Dilosyn. A preparation of methdilazine hydrochloride. (527).

Diloxanide. 4 - (N - Methyldichloroacetamide)phenol. Entamide; Furamide is the 2-furoate.

Dillenburgite. A mineral. It is a variety of chrysocolla.

Dill Water. (*Aqua anethi B.P.*) It is obtained by adding 100 parts of dill fruit to 1,000 parts of water, and distilling.

Dilo Oil. See Laurel nut oil.

Dilver. An alloy containing 42 per cent. nickel. It is used as the leading-in wire for filament lamps, as it has the same coefficient of expansion as glass.

Dimagel. A proprietary preparation containing dimagnesium aluminium trisilicate. An antacid. (252).

Dimagel-Belladonna. A proprietary preparation containing dimagnesium trisilicate and belladonna alkaloids. An antacid. (252).

Dimargyl. See ARGENTAMINE.

Dimatos. Infusorial earth (*q.v.*).

Dimazole. Diamthazole.

Dimazon. Diacetyl-amino-azo-toluene, $C_{18}H_{19}O_4N_3$, a red dye used in ointment or as a dusting powder.

Dimedon. Dimethyl-dihydro-resorcinol.

Dimefline. 8 - Dimethylaminomethyl - 7 - methoxy - 3 - methylflavone. Remeflin is the hydrochloride.

Dimelor. A proprietary preparation of acetohexamide. An oral hypoglycæmic agent. (333).

Dimenformon. A proprietary preparation of œstradiol monobenzoate, used to arrest lactation and to prevent abortion. (316).

Dimenhydrinate. Diphenhydramine salt of 8-chlorotheophylline. Dramamine.

Dimenoxadole. 2-Dimethylaminoethyl-α - ethoxy - αα - diphenylacetate.

Dimephtanol. 6 - Dimethylamine - 4,4-diphenylheptan-3-ol. Methadol.

Dimepregnen. An anti-œstrogen currently undergoing clinical trial as " St 1411 ". 3β - Hydroxy - 6α, 16α - dimethylpregn - 4 - en - 20 - one.

Dimepropion. An anorexigenic. α-

Dimethylaminopropiophenone. Metamfepramone is the I.N.N.

Dimer X. See Iocarmic Acid.

Dimesone. An anti-inflammatory ster-
oid. 9α - Fluoro - 11β, 21 - dihydroxy -
16α, 17 - dimethylpregna - 1, 4 - diene -
3,20-dione.

Dimethachlor. A pesticide. α-chloro-
N - (2 - methoxyethyl)acet - $2'$, $6'$ -
xylidide.

Dimethicone. A polydimethylsiloxane.

Dimethindene. 2 - {1 - [2 - (2 - Dimethylaminoethyl)inden - 3 - yl]ethyl}-
pyridine. Fentostil is the hydrogen
maleate.

Dimethisoquin. 3 - Butyl - 1 - (2 -
dimethylaminoethoxy) isoquinoline.
Quinisocaine. Quotane is the hydrochloride.

Dimethisterone. 17β - Hydroxy - 6α,
21 - dimethyl - 17α - pregn - 4 - en - 20-
yn - 3 - one. 6a,21 - Dimethylethis-
terone. Secrosteron.

Dimethothiazine. 10 - (2 - Dimethylaminopropyl) - 2 - dimethylsulphamoyl-
phenothiazine. 8599 R.P. is the mesylate.

Dimethoxanate. 2 - (2 - Dimethyl -
aminoethoxy)ethyl phenothiazine - 10-
carboxylate.

Dimethylaniline Orange. See Orange
III.

Dimethyl Carbinol. Isopropyl alcohol,
$CH_3.CH_3.CH(OH)$.

Dimethyl-phenylene-green. The tetramethyl derivative of Phenylene blue,
$C_{16}H_{20}N_3Cl$. Dyes silk and other
fabrics.

Dimethylthiambutene. 3 - Dimethylamino - 1,1 - di(2 - thienyl)but - 1 - ene.

Dimethyltubocurarine. Dimethyl
ether of (+)-tubocurarine. Diamethine
is the bromide.

Dimethyl-vinyl-carbinol. Pentenyl
alcohol, $CH_2 : CH.C(CH_3)_2OH$.

Dimetridazole. An anti-protozoan. 1,
2 - Dimethyl - 5 - nitroimidazole.
Emtryl.

Diminazene. An anti-protozoan and
bactericide. pp' - Diamidinodiazoam-
inobenzene. It is present in Berenil
as the aceturate.

Dimipressin. A proprietary preparation of Imipramine. An anti-depressant. (368).

Dimol. A trade mark for a preparation
of dimethylo-methoxy-phenol. An antiseptic.

Dimopyrane. The trade name for a
preparation intended to replace pyramidone.

Dimorphine. See Dehydromorphine.

Dimorphite. A mineral. It is an
arsenic sulphide.

Dimotane Expectorant. A proprietary
preparation containing bromphenira-

mine maleate, guaiphenesin, phenylephrine hydrochloride and phenylpropanolamine hydrochloride. A cough
mixture. (258).

Dimotane Expectorant DC. A proprietary preparation containing di-
hydrocodeinone bitartrate, brompheniramine maleate, guaiphenesin, phenylephrine hydrochloride and phenylpropanolamine hydrochloride. A cough
mixture. (258).

Dimotapp Elixir. A proprietary preparation containing brompheniramine
maleate, phenylephrine hydrochloride
and phenylpropanolamine hydrochloride. A cough linctus. (258).

Dimpylate. An insecticide. oo-Diethyl
o - (2 - isopropyl - 6 - methylpyrimidin -
4 - yl) phosphorothioate. Diazinon.
It is present in Basudin.

Dimundite. See Camite.

Dimycin. Streptomycin and dihydrostreptomycin for veterinary purposes.
(520).

Dimyril. A proprietary preparation containing isoaminile citrate. A cough
suppressant. (188).

Dinamene. Selective weedkiller. (501).

Dinamit No. 3. An explosive containing
25 per cent. nitro-glycerin, 54 per cent.
sodium nitrate, 19 per cent. wood meal,
2 per cent. soda, and yellow ochre.

Dinas Bricks (Flintshire Stones). Fireproof stones made from sand, milk of
lime, and a cementing material.

Dindevan. A proprietary preparation
of phenindione. An anticoagulant.
(182).

Dingler's Green. A pigment. It is a
chromium phosphate.

Dinitolmide. An anti-protozoan. 3,5-
Dinitro-o-toluamide. Zoalene. It is
also present in Zoamix.

Dinitra (Aldiphen, Dinitrenal, Nitraphen, Diphen). 2 : 4-Dinitro-phenol.

Dinitrenal. See Dinitra.

**Dinitro - anthrachrysone - di - sul-
phonic Acid.** A dyestuff. It is the
sodium salt of dinitro-tetroxy-anthra-
quinone-disulphonic acid, $C_{14}H_4N_2O_{16}$
S_2Na_2. Dyes wool brown.

Dinitro-naphthol Yellow. See Martius Yellow.

Dinitrophenol Black (Immedial Black
N, Sulphur Black T Extra, Thiophenol
Black T Extra, Thiol Black, Cross Dye
Black BX). Dyestuffs obtained by
heating 2 : 4-dinitrophenol with sodium
sulphide and sulphur under reflux.
They dye unmordanted cotton black.

Dinitrosoresorcin (Fast Green O, Dark
Green, Chlorin, Russian Green, Fast
Myrtle Green, Alsace Green, Resorcin
Green, Resorcinol Green, Solid Green
O). Dinitroso-resorcinol, $C_6H_4N_2O_4$.

Dyes iron mordanted cotton, green, and iron mordanted wool, dark green.

)inner Pills. Aloes and mastic pills. (*Pilulæ aloes et mastiche B.P.C.*)

)inocillon-RT. A proprietary preparation of sodium methicillin. An antibiotic. (326).

)inopol. A proprietary trade name for di-*n*-octyl phthalate. A vinyl plasticiser. (73).

)inopol 235. A proprietary trade name for *n*-octyl *n*-decyl phthalate. A vinyl plasticiser. (73).

)inopol I.D.O. A proprietary trade name for isodecyl octyl phthalate. A vinyl plasticiser. (73).

)inopol MOP. A proprietary trade name for a mixed octyl phthalate vinyl plasticiser. (73).

Dinoprost. A smooth muscle activator. 7 - [3α, 5α - Dihydroxy - 2β - [(3*S*-) - hydroxy - *trans* - oct - 1 - enyl]cyclo - pent - 1 - yl] - *cis* - hept - 5 - enoic acid.

Dinoprostone. A smooth muscle activator. 7 - [3α - Hydroxy - 2β - [(3*S*) - hydroxy - *trans* - oct - 1 - enyl] - 5 - oxocyclopent - 1 - yl] - *cis* - hept - 5 - enoic acid.

Diocaine. A German product. It is diallyl- oxy - ethenyl - diphenyl - diamidene-hydrochloride. A local anæsthetic used in eye treatment.

Dioctyl. A proprietary trade name for dioctyl sodium sulphosuccinate.

Dioctyl Ear Capsules. A proprietary preparation of dioctyl sodium sulphosuccinate. (250).

Dioctyl Forte. A proprietary preparation containing dioctyl sodium sulphosuccinate. A laxative. (250).

Dioctyl Medo. A proprietary preparation containing dioctyl sodium sulphosuccinate. A laxative. (250).

Dioctyl Sodium Sulphosuccinate. Di - (2-ethylhexyl)sodium sulphosuccinate. Coprol. Dioctyl. Exomic OT, Manoxol OT. Norval.

Diocroma. A mineral. It is a variety of zircon.

Dioderm. A proprietary preparation of hydrocortisone in a cream base. A steroid skin cream. (377).

Dioderm C. A proprietary preparation of HYDROCORTISONE and CLIOQUINOL in a cream base. A steroid skin cream. (377).

Diodoquin. A proprietary preparation of di-iodohydroxyquinoline. An antiamœbic drug. (285).

Diofan. See also KUROFAN. A registered trade name for polyvinylidene chloride. (49).

Diofan D. A registered trade mark for a polyvinylidene chloride dispersion. (49).

Dioform. Acetylene dichloride, $CHCl : CHCl$. Recommended as an anæsthetic.

Diogen. Sodium-amino-naphthol-disulphonate, $C_{10}H_4(NH_2)(OH)(SO_3Na)$. It forms the basis of the photographic developer of this name.

Diogenal. Dibromo - propyl - diethyl - barbituric acid, $(C_2H_5)_2C.CO.NH.CO.N C_3H_5Br_2).CO$. Used as a somnifacient and as a general nerve sedative.

Diol. Glycol, $CH_2OH.CH_2OH$.

Diolpate. A proprietary trade name for a polymeric adipate plasticiser with non-migratory characteristics in P.V.C. (96).

Diolpate 150. A proprietary polymeric plasticiser used to render P.V.C. resistant to oil. (96).

Diolpate 190. A proprietary general-purpose plasticiser used in P.V.C. plastisols. (96).

Diolpate 195. A proprietary trade name for a linear polyester vinyl plasticiser. (96).

Diolpate 582. A proprietary polymeric plasticiser combining good plasticising properties with flexibility at low temperatures and a high level of resistance to extraction. (96).

Diolpate PPA. Polypropylene adipate. A proprietary polymeric plasticiser. (96).

Diolpate PPA 400. A proprietary plasticiser. Polypropylene adipate. (96).

Dional 55. A solvent mixture of 70 parts of Losingsmittle E33 (*q.v.*) and 30 parts polysolvan E (17·5 per cent. methanol, 52·5 per cent. methyl acetate, 30·0 per cent. acetates of propyl isobutyl and amyl alcohols).

Dionine. Ethyl-morphine-hydrochloride, $C_{19}H_{23}NO_3$. Used in medicine as a substitute for morphine.

Dionosil. Propyliodone suspensions or powders. (520).

Diopside. A mineral. It is a calcium-magnesium - silicate, $Ca . Mg(SiO_3)_2$. Often a small amount of ferrous iron replaces an equivalent amount of magnesium. Clear crystals are sometimes cut as gem stones.

Diopside-jadeite. A mineral intermediate between jadeite and diopside in composition.

Dioptase. A mineral, $CuO.SiO_2.H_2O$.

Diorez 514. A proprietary polyester containing a very low number of end OH groups, used in the manufacture of polyurethane thermoplastics. (96).

Diorez 550. A proprietary liquid linear polyester used in conjunction with a diisocyanate and a cross-linking agent in the manufacture of polyurethane elastomers, especially for automobile tyres and shoe soles. (96).

Diorez 620. A proprietary slightly-branched liquid polyester with a high proportion of end groups. DIOREZ GY. (96).

Diorez 750 and 770. A proprietary range of soft wax-like polyesters similar to DIOREZ 550. (96).

Diorez GY. See DIOREZ 620. (96).

Diorite. A rock composed of felspar and hornblende.

Diosal. Sodium di-iodo-salicylate.

Diothane. The hydrochloride of piperidino - propanediol - diphenyl - urethane, $C_5H_{10}N.CH_2.CH(O.CO.NHC_6H_5).CH_2.$ $O.CO.NHC_6H_5$. Local anæsthetic.

Diothyl. Veterinary insecticide preparations. See PYRIMITHATE. (512).

Diotroxin. L-Thyroxine and L-triiodo-thyronine mixture. (520).

Dioval. A proprietary preparation containing aluminium hydroxide, magnesium hydroxide and dimethyl polysiloxane. An antacid. (255).

Diox DR 22. A trade name for a rutile (titanium dioxide) type white pigment with a blue tone and good dispersability. (58).

Dioxamate. 4 - Carbamoyloxymethyl-2 - methyl - 2 - nonyl - 1,3 - dioxolan.

Dioxan. Tetramethyl-1 : 4-oxide, CH_2 $OCH_2.CH_2OCH_2$. It is a solvent for cellulose acetate, etc., boiling at 101° C., melting at 11° C., and having a specific gravity of 1·0338 at 20° C. It is miscible with water in all proportions.

Dioxaphetyl Butyrate. Ethyl 4-morpholino-2,2-diphenylbutyrate.

Dioxathion. An insecticide and acaricide. A mixture consisting essentially of *cis*- and *trans*-SS'-1,4-dioxan-2,3-diyl bis(*oo*-diethyl phosphorodithioate). DELNAV.

Dioxine (Gambine R). A dyestuff. It is nitroso-dioxy-naphthalene, $C_{10}H_7$ NO_3. Dyes iron mordanted fabrics, green, and chrome mordanted materials, brown.

Dioxitol. Ethyldigol. (553).

Dioxogen. A 3 per cent. solution of hydrogen peroxide, H_2O_2. A disinfectant.

Dioxyanthranol. Anthrarobine, C_{14} $H_{10}O_3$. Used externally for skin diseases.

Dioxylite. See LANARKITE.

Dipar. A proprietary preparation of PHENFORMIN hydrochloride. A hypoglycæmic agent. (312).

Diparalene. A proprietary preparation of chlorcyclizine hydrochloride. (311).

Diparcol. A proprietary trade name for the hydrochloride of diethazine.

Dipasic. A proprietary preparation of aminosalicylate of isonicotinic acid hydrazide. An antituberculous drug. (386).

Dipaxin. A proprietary trade name for Diphenadione.

Dipenine Bromide. 2 - Dicyclopentyl acetoxyethyltriethylammonium bromide HL267. Diponium Bromide.

Dipentek. A technical grade of dipenta erythritol.

Dipentene (Di-isoprene, Cinene, Caoutchin). *i*-Limonene, $C_{10}H_{16}$.

Dipentrol. Dipentaerythritol. (612).

Diperodon. 1,2 - Di(phenylcarbamoyl oxy)-3-piperidinopropane.

Dipex. A proprietary molud lubricant for rubber. It is a water-soluble sodium sulphonate obtained from petroleum-acid sludges.

Diphanite. See MARGARITE.

Diphemanil Methylsulphate. 4-Diphenylmethylene - 1,1 - dimethyl-piperidinium methylsulphate. Diphenatil.

Diphen. A trade name for phenolic resins. See DINITRA. (43).

Diphen 60-B. A phenol-urea formaldehyde resin.

Diphenadione. 2 - Diphenylacetylindane-1,3-dione. Didandin ; Dipaxin ; Oragulant.

Diphenal. The sodium salt of diamino-oxy-diphenyl. A photographic developer.

Diphenatil. A proprietary trade name for diphemanil methylsulphate.

Diphenhydramine. NN-Dimethyl-2-diphenylmethoxyethylamine. Benadryl is the hydrochloride.

Diphenidol. 1,1 - Diphenyl - 4 - piperidinobutan-1-ol.

Diphenoxylate. Ethyl 1-(3-cyano-3,3-diphenylpropyl) - 4 - phenyl - piperidine-4-carboxylate.

Diphenylamine Blue. See METHYL BLUE.

Diphenylamine Blue, Spirit Soluble (Spirit Sky Blue, Diphenylamine Opal Blue, Bavarian Blue, Spirit Soluble, XL Opal Blue). The hydro-chloride of triphenyl-pararosaniline, $C_{37}H_{30}N_3$ Cl.

Diphenylamine Opal Blue. See DIPHENYLAMINE BLUE, SPIRIT SOLUBLE.

Diphenylamine Orange. See ORANGE IV.

Diphenylamine Yellow. See ORANGE IV.

Diphenyl Black. A dyestuff. It is amino-diphenylamine.

Diphenyl Black Oil DO. Consists of 1 part amino-diphenylamine (diphenyl black), in 3 parts aniline oil.

Diphenyl Blue-black. A dyestuff. It is the sodium salt of diphenyl-disazo-ethyl-amino-naphthol-sulphonic-amino-naphthol-disulphonic acid, $C_{34}H_{25}N_6$ $O_{11}S_3Na_3$. Dyes cotton blue-black.

)iphenyl Brown BN. A dyestuff. It is the sodium salt of diphenyl-disazo-salicylic acid-dimethyl-amino-naphthol-sulphonic acid, $C_{31}H_{23}N_5O_7SNa_2$. Dyes cotton dark brown.

)iphenyl Brown 3GN. A dyestuff. It is the sodium salt of ditolyl-disazo-salicylic acid-dimethyl-amino-naphthol-sulphonic acid, $C_{33}H_{27}N_5O_7SNa_2$. Dyes cotton dark yellowish-brown.

)iphenyl Brown RN. A dyestuff. It is the sodium salt of diphenyl-disazo-salicylic acid-methyl-amino-naphthol-sulphonic acid γ, $C_{30}H_{21}N_5O_7SNa_2$. Dyes cotton dark reddish-brown.

)iphenyl Catechine G. A dyestuff obtained by the diazotisation of the alkaline condensation product of dinitro-dibenzyl-disulphonic acid with aniline, and the combination of the diazo compound with dimethyl-amino-naphthol-sulphonic acid γ, $C_{32}H_{23}N_6O_{11}S_3Na_3$. Dyes cotton cutch brown.

)iphenyl Chrysoine G. A dyestuff obtained by the ethylation of the product of the condenstion of p-nitro-toluene-sulphonic acid with p-amino-phenol, in the presence of caustic soda. Dyes cotton golden yellow.

)iphenyl Chrysoine RR. A dyestuff obtained by the diazotisation of the alkaline condensation product of di-nitro-dibenzyl-disulphonic acid and aniline, and the combination of the diazo compound with phenol, and ethylation, $C_{28}H_{21}N_5O_8S_2Na_2$. Dyes cotton reddish-orange.

Diphenyl Citronine G (Cotton Yellow 6370). A dyestuff obtained by the condensation of dinitro-dibenzyl-disulphonic acid with aniline, in the presence of caustic soda. Dyes cotton greenish-yellow.

Diphenyl Fast Black. A dyestuff. It is the sodium salt of ditolyl-amine-disazo-m-tolylene-diamine-amino-naphthol-sulphonic acid, $C_{31}H_{29}N_8O_4SNa$. Dyes cotton black.

Diphenyl Fast Brown G. A dyestuff obtained by the diazotisation of the alkaline condensation product of di-nitro-dibenzyl-disulphonic acid and aniline, and the combination of the diazo compound with phenyl-amino-naphthol-sulphonic acid γ, $C_{36}H_{23}N_6O_{11}S_3Na_3$. Dyes cotton dark yellowish-brown.

Diphenyl Fast Yellow. A dyestuff obtained by the condensation of di-nitro-dibenzyl-disulphonic acid with primuline, in the presence of caustic soda. Dyes cotton yellow.

Diphenyl Green G. A dyestuff. It is the sodium salt of diphenyl-disazo-phenol-disulpho-amino-naphthol-azo-chloro-nitro-benzene, $C_{34}H_{21}N_8O_{10}S_2Na_2Cl$. Dyes cotton green.

Diphenyl Green 3G. A dyestuff. It is the sodium salt of diphenyl-disazo-salicylic-disulpho-amino-naphthol-azo-chloro-nitro-benzene, $C_{35}H_{20}N_9O_{12}S_2Na_3Cl$. Dyes cotton green.

Diphenyl Ketone (Benzone). Benzophenone, $C_6H_5CO.C_6H_5$.

Diphenyl Orange. See ORANGE IV.

Diphenyl Orange RR. A dyestuff. It is the sodium salt of nitroso-stilbene-disulphonic acid-azo-aniline.

Diphenylpyraline. 4-Diphenylmethoxy-1-methylpiperidine. Histryl is the hydrochloride.

Diphosgene. Superpalite ($q.v.$).

Diphthosan. 2 : 7-Dimethyl-3-dimethyl-amino-6-amino-methyl-acridinium chloride.

Dipidolor. A proprietary preparation of PIRITRAMIDE. An analgesic. (356).

Dipipanone. 4,4-Diphenyl-6-piperidinoheptan-3-one. Pipadone is the hydrochloride.

Diplomycin. A proprietary preparation of zinc bacitracin, polymixin B sulphate in a starch base. Antibiotic powder. (279).

Diplosal. Salicyl-salicylic acid, HO.$C_6H_4.CO.O.C_6H_4.COOH$. Used in medicine in chronic and acute rheumatism.

Dip Oil. A 25 per cent. crude phenol, used in the manufacture of cattle dips and disinfectants.

Dipolymer. A coumarone-indene resin.

Diponium Bromide. Diperine bromide.

Dippel's Acid Elixir. (*Acidum sulphuricum aromaticum B.P.*) It is sulphuric acid diluted with alcohol, and mixed with tincture of ginger and spirit of cinnamon.

Dippel's Animal Oil. See BONE OIL.

Dippel's Oil. See BONE OIL.

Dipping Acid. A mixture of nitric and sulphuric acids. Also see CONTACT ACID.

Dipping Metal. A jeweller's alloy containing 48 parts of copper and 15 parts of zinc.

Dipping Oil. Sulphuric acid, used for pickling iron and steel.

Dippol. Sheets of cardboard impregnated with quinosol.

Diprenorphine. A narcotic antagonist currently under clinical trial as " M. 5050 ". N-Cyclopropylmethyl-7,8-dihydro-7α-(1-hydroxy-1-methylethyl)-O^6-methyl-6, 14 *endo*ethano-normorphine.

Dipronal. Dipropyl-aceto-urethane.

Dipropæsin. A urea derivative containing the residues of 2 molecules of propæsin, CO(NH.C_6H_4.CO_2.C_3H_7)$_2$. A local anæsthetic.

Diprophylline. 7 - (2,3 - Dihydroxy-propyl)theophylline. Isophylline; Neutraphylline; Silbephylline.

Dipropyl Carbinol. Secondary heptyl alcohol, $C_7H_{15}.OH$.

Dipsanil. A waterproofing agent. (512).

Dipsar. A proprietary solvent-based organotin wood preservative. (425).

Dipterex. A proprietary preparation of METRIPHONATE. An insecticide.

Dipyridamole. 2,6 - Di - [di - (2 - hydroxyethyl)amino] - 4,8 - dipiperidinopyrimido[5,4 - d]pyrimi dine 2,6 - Di - [di - (2 - hydroxyethyl)amino] - 4,8 - dipiperidino - 1,3,5,7 - tetra-azanaphthalene. Persantin.

Dipyridyl Oil. An oil prepared from pyridine. It is stated to be more effective than nicotine as an insecticide.

Dipyrone. Sodium 2,3 - dimethyl - 1 - phenyl - 5 - pyrazolon - 4 - yl - N - methylaminomethane sulphonate. Sodium noramidopyrine methanesulphonate. Methampyrone.

Direct Black (Imperial Black, Nigrosalin). Dyestuffs. They are mixtures of logwood with iron and copper sulphates.

Direct Black for Cotton. A dyestuff composed of 50 per cent. water, 45 per cent. of either hæmatoxylin or hæmatein, and 3·5–7 per cent. copper sulphate.

Direct Black V. A dyestuff. It is the , sodium salt of diphenyl-disazo-naphtho - disulphonic - azo - α - naphthylamine - amino - naphthol - sulphonic acid, $C_{42}H_{27}N_8O_{11}S_3Na_3$. Dyes cotton a violet black.

Direct Blue B. A dyestuff. It is the sodium salt of dimethoxy-diphenyl-disazo - dioxy - naphthoic - sulphonic - α-naphthol-p-sulphonic acid, $C_{35}H_{23}N_4O_{18}S_2Na_3$. Dyes cotton steel-blue to black-blue.

Direct Blue 2B, 2BL, 2B Supra. Dyestuffs. They are British equivalents of Diamine blue 2B.

Direct Blue 3B. A dyestuff. It is a British brand of Diamine blue 3B.

Direct Blue-black B. See DIAMINE BLACK-BLUE B.

Direct Blue-black 2B. A polyazo dyestuff which dyes unmordanted cotton, in the presence of sodium sulphate, and some sodium carbonate.

Direct Blue R. A dyestuff. It is the sodium salt of ditolyl-disazo-dioxy-naphthoic - sulphonic - α - naphthol - p - sulphonic acid, $C_{35}H_{23}N_4O_{11}S_2Na_2$. Dyes cotton black-violet.

Direct Brown J. A dyestuff. It is the sodium salt of carboxy-benzene-azo-phenylene-brown, $C_{32}H_{24}N_{12}O_4Na_2$. Dyes cotton brown.

Direct Brown M. A dyestuff. It is a British equivalent of Diamine brown M.

Direct Brown R. See POLYCHROMINE B.

Direct Deep Black E. A polyazo dyestuff which dyes cotton with the addition of salt to the bath.

Direct Deep Black G. A polyazo dyestuff, suitable for dyeing cotton, half-wool, and jute. Cotton is dyed in a boiling bath containing 10 per cent. salt, and half-wool in a 15 per cent. salt bath.

Direct Deep Black R, T. See DIAMINE BLACK-BLUE B.

Direct Deep Black RW. A polyazo dyestuff, which dyes cotton violet-black in a boiling bath containing 5–15 per cent. salt. Also recommended for dyeing linen and jute.

Direct Dyestuffs. Dyestuffs which are used for cotton, and dye the fibre direct by combination with it. The Congo colours are types.

Direct Fast Orange D2G, D2R. British colours equivalent to Ingrain brown.

Direct Fast Orange 4R, RL, 2RL. Dyestuffs. They are British brands of Mikado orange.

Direct Fast Orange SE. A dyestuff. It is a British equivalent of Benzo fast orange S.

Direct Fast Red F. A British equivalent of Diamine fast red.

Direct Fast Scarlet SE. A dyestuff. It is a British equivalent of Benzo fast scarlet 4BS.

Direct Fast Yellow GL, 2GL, 3GL, R, RL. Dyestuffs. They are British brands of Direct yellow R.

Direct Fast Yellow 2GLO. A British equivalent of Mikado yellow.

Direct Green B, BL. Dyestuffs. They are British equivalents of Diamine green B.

Direct Green BG, G. Dyestuffs. They are British brands of Diamine green G.

Direct Green CO. See COLUMBIA GREEN.

Direct Grey. See NEW GREY.

Direct Grey B. A dyestuff. It is the sodium salt of ditolyl-disazo-bi-dioxy-naphthoic-sulphonic acid, $C_{36}H_{22}N_4O_{14}S_2Na_2$. Dyes cotton steel-grey to bluish-black.

Direct Grey R. A dyestuff. It is the sodium salt of diphenyl-disazo-bi-dioxy-naphthoic-sulphonic acid, $C_{34}H_{18}N_4O_{14}S_2Na_2$. Dyes cotton reddish-grey to bluish-black.

Direct Heliotrope B. A dyestuff. It is the sodium salt of sulpho-α-naphthol-azo-diphenyl-azo-toluene-disazo-bi-α-naphthol-*p*-sulphonic acid, $C_{49}H_{31}N_8O_{12}S_3Na_3$. Dyes cotton violet.

Direct Indigo Blue A. A dyestuff. It is the sodium salt of diphenyl-disazo-cresol-ether-azo-amino-naphthol-disulphonic-amino-phenol-disulphonic acid, $C_{36}H_{26}N_8O_{15}S_4Na_4$. Dyes cotton indigo from an alkaline bath.

Direct Indigo Blue BK. A dyestuff. It is the sodium salt of diphenyl-disazo-cresol-ether-azo-amino-naphthol isulphonic-amino-naphthol-sulphonic acid, $C_{40}H_{29}N_8O_{12}S_3Na_3$. Dyes cotton indigo blue from an alkaline bath.

Direct Indigo Blue BN. A dyestuff. It is the sodium salt of diphenyl-disazo-dioxy - naphthoic - sulphonic - amino-naphthol-disulphonic acid, $C_{33}H_{19}N_5O_{14}S_3Na_4$. Dyes cotton from a slightly alkaline bath.

Direct Indone Blue R. A dyestuff. It is the sodium salt of diphenyl-disazo-naphthol - disulphonic - α - naphthyl-amine - amino - naphthol - disulphonic acid, $C_{42}H_{26}N_{10}O_{14}S_4Na_4$. Dyes cotton grey-blue to indigo-blue.

Direct Jet Black R and T. Direct cotton dyestuffs.

Direct Orange. A direct cotton dye-stuff.

Direct Orange G. See MIKADO ORANGE G and 4R.

Direct Orange 2R. See MIKADO ORANGE G AND 4R.

Direct Red. An azo dyestuff obtained from diamino-phenyl-tolyl, and naphthionic acid, $C_{33}H_{24}N_6O_6S_2Na_2$. Dyes cotton red from an alkaline salt bath.

Direct Red B. Diamine scarlet B (*q.v.*).

Direct Scarlet. See GERANINE.

Direct Sky Blue. Diamine sky blue (*q.v.*).

Direct Sky-Blue. A British dyestuff. It is equivalent to Diamine sky blue.

Direct Sky-Blue GS. A British brand of Diamine sky blue.

Direct Violet. A dyestuff. It is a British equivalent of Congo violet.

Direct Violet. See METHYL VIOLET B.

Direct Violet BB. A dyestuff. It is the sodium salt of dimethoxy-diphenyl-disazo-*m*-tolylene-diamine-dioxy-naphthalene-sulphonic acid, $C_{31}H_{27}N_6O_7$ SNa. Dyes cotton bluish-violet.

Direct Violet R. A dyestuff. It is the sodium salt of diphenyl-disazo-*m*-tolylene - diamine - dioxy - naphthoic - sulphonic acid, $C_{30}H_{22}N_6O_7$SNa. Dyes cotton reddish-violet.

Direct Yellow G (Direct Yellow R, (Chlorazol Yellow GX, Diazine yellow R, Enbico Direct Yellow R, Direct Fast Yellow GL, 2GL, 3GL, R, RL, Paramine Yellow R, 2R, Y). A dyestuff. It is the sodium salt of the so-called dinitroso-stilbene-disulphonic acid. It dyes cotton yellow from a salt bath, and wool from an acid bath.

Direct Yellow 2G, 4G. See MIKADO YELLOW.

Direct Yellow R. See DIRECT YELLOW G.

Direma. A proprietary preparation of hydrochlorothiazide. A diuretic. (309).

Dirigold. See ORANIUM BRONZE.

Disaloin (Desaloin, Desalgin). A combination of egg albumin and chloroform (chloroform = 25 per cent.). Used externally as a skin disinfectant, and internally in stomachic catarrh.

Disalol. Phenyl-salicyl-salicylate, HO. $C_6H_4.CO.O.C_6H_4COOC_6H_5$. Used in medicine.

Disamide. A proprietary trade name for disulphamide. An oral diuretic. (182).

Discelite. A proprietary trade name for a diatomaceous silica filler.

Discenite. A mineral $(MnOH)_2Mn_3$ $SiO_3.(AsO_3)_2$.

Discus. A machine dishwashing detergent. (537).

Disecron. A proprietary preparation of progesterone and œstradiol monobenzoate. (265).

Discharge Lake R and RR. See PARANITRANILINE RED.

Disipal. A proprietary preparation of orphenadrine hydrochloride. An antidepressive agent. (317).

Discol. A proprietary preparation containing 50 parts 95 per cent. denatured alcohol, 25 parts benzol, 25 parts hydrocarbon spirit. A motor fuel.

Discolite (Formopon, Hydrosulphite AW, Rongalite C, Hydros). NaH $SO_2.CH_2O.2H_2O$, a reducing agent for stripping in dyeing.

Discrasite. See DYSCRASITE.

Diseptal. *p*-Aminobenzene-sulphonyl-*p*′-amino-benzene sulphone methylamine, an anæsthetic.

Disfico. A trade name for a vulcanised fibre used for electrical insulation.

Disilane. Silico-ethane (disilicon-hexahydride), Si_2H_6.

Disinfectant, Macdougall's. See MACDOUGALL'S DISINFECTING POWDERS.

Disinfecting Carbon. See CARBON.

Disinfecting Powders. Usually consist of mixtures of inorganic materials, with varying amounts of cresol, crude carbolic acid, or tar oils. Slaked lime, gypsum, and chalk are the ordinary mineral constituents.

Disinfection Oil. See SAPROL.

Disipal. A proprietary preparation of ORPHENADRINE hydrochloride used in the treatment of Parkinson's disease. (317).

Di-Sipidin. A proprietary preparation of posterior pituitary extract. Antidiuretic hormone. (329).

Dismenol. A proprietary preparation of *p*-aminobenzoic acid and phenazone. An analgesic. (391).

Disodium Edetate. A chelating agent. Disodium dihydrogen ethylenediamine-*NNN'N'*-tetra-acetate.

Disogrin. A proprietary polyester-based polyurethane elastomer cross-linked with with diols. (649).

Disopyramide. An anti-arrhythmic. 4 - Di - isopropylamino - 2 - phenyl - 2 - (2 - pyridy)butyramide. RYTHMODAN.

Dispargen. A form of colloidal mercury.

Disparin. A preparation of lead arsenate containing 49 per cent. lead oxide, 16 per cent. arsenic acid, 4 per cent. tar, and 31 per cent. water. An agricultural insecticide.

Disparit A. A disinfecting cleaning compound. It is a solution of tar oils in crude petroleum distillate.

Disparit B. Trichlorethylene, C_2HCl_3. A disinfecting cleaning compound.

Dispensing Syrup. Consists of equal volumes of glycerin, syrup, alcohol, and mucilage of acacia.

Dispermin. Piperazine, $NH(CH_2.CH_2)$ $(CH_2CH_2)NH$. Used in medicine as a uric acid solvent.

Dispersite. A proprietary trade name for a dispersion of rubber, rubber-like and film-forming resins in water.

Dispersol. Dispersed dyestuffs for synthetic fibres. (512).

Dispersol OS. An 8 per cent. solution of a polyethenoxy compound in isopropyl alcohol. A tailor made dispersant for oil slicks. (2).

Dispex G40. A proprietary dispersant for high-gloss paints. It contains 40 per cent. by weight of an acrylic-based material. (565).

Dispolac. Intramammary creams for veterinary use. (514).

Dissolved Guano. Natural guano, which has been treated with an acid in order to obtain a more soluble product.

Dissolvine. Sequestering agents. (587).

Dista G. Procaine penicillin. (538).

Distalgesic. A proprietary preparation of dextropropoxyphene hydrochloride and paracetamol. An oral analgesic. (309).

Distamine. A proprietary preparation of penicillamine hydrochloride, used to treat heavy metal poisoning and Wilson's disease. (309).

Distaquaine Fortified. A proprietary preparation containing Procaine potassium Penicillin and Benzylpenicillin. An antibiotic. (309).

Distaquaine G. A proprietary preparation containing Procaine Penicillin. An antibiotic. (309).

Distaquaine Suspension. A proprietary preparation containing Procaine Penicillin. An antibiotic for injection. (309).

Distaquaine V. A proprietary preparation containing Phenoxymethylpenicillin. An antibiotic. (309).

Distaquaine V-K. A proprietary preparation containing Phenoxymethylpenicillin Potassium. An antibiotic. (309).

Distaquaine V Sulpha. A proprietary preparation of phenoxymethylpenicillin, sulphadimidine, and sulphamerazine. An antibiotic. (309).

Distavone. A proprietary preparation containing Benzylpenicillin, Procaine penicillin and Streptomycin. An antibiotic for injection. (309).

Distaval. A proprietary trade name for Thalidomide.

Distec. Fatty acids and glycerides. (555).

Distempers. Dry distempers or kalsomines are mixtures of whiting, china clay, or lime, with binding agents such as casein, glue, or dextrose, with an alkali or alkaline earth, and the tinting colour. The paste distempers contain a little linseed oil or wood oil, glue, and pigments.

Disthene. See KYANITE.

Distigmine Bromide. NN' - Hexamethylenedi - [1 - methyl - 3 - (methylcarbamoyl - oxy)pyridinium bromide] Ubretid.

Distillation Glycerin. That glycerin obtained by the direct saponification of fats with sulphuric acid.

Distillation Oleine. Commercial oleic acid, obtained by steam distillation, after treatment of sulphuric acid.

Distilled Grease. Yorkshire grease (*q.v.*), which has been distilled.

Distilled Verdigris. A crystalline variety of verdigris, obtained by dissolving copper oxide in pyroligneous acid, or by precipitating copper sulphate with calcium acetate, and evaporating the solution obtained in either case to the crystallising point.

Distilled Vinegar (Crystal Vinegar). Dilute acetic acid, obtained by the distillation of malt vinegar.

Distiller's Wash (Vinasse). The residue of fermented mash, after the alcohol is distilled off. Used as a feeding stuff for animals.

Distivit. Oral Vitamin B preptide. (538).

Distivit Oral. A proprietary preparation of cyanocobalamin for oral use. (309).

Distoline. A proprietary trade name for commercial oleic acid obtained from vegetable oils.

Distrene. A proprietary trade name for a polystyrene synthetic resin. (12).

Disulfiram. Tetraethylthiuram disulphide. Antabuse; Cronetal.

Disulphine. An acid wool dyestuff. (512).

Disulphine Blue. A preparation of sulphan blue used as a visual diagnostic agent for circulatory disorders. (2).

Disulphine Blue A. A dyestuff. It is a British equivalent of Patent blue A.

Disulphine Green B. See NEPTUNE GREEN.

Disulphamide. 4 - Chloro - 6 - methylbenzene - 1,3 - disulphonamide. Disamide.

Disulpho Acid S. 1-Naphthylamine-4 : 8-disulphonic acid.

Disulphuric Acid. See FUMING SULPHURIC ACID.

Dita Bark. The bark of *Alstonia scholaris*

Dithan. Trional (*q.v.*).

Dithiazanine. 3,3' - Diethylthiadicarbocyanine. Delvex and Telmid are the iodide.

Dithion. Sodium-dithio-salicylate, $S(C_6H_3(OH)CO.ONa)_2$. There are two isomeric salts, one of which is used in veterinary practice as an antiseptic dusting-powder, whilst the other finds employment as an antirheumatic and antipyretic.

Dithizone. Diphenyl - thiocarbazone. Used for the detection of heavy metals.

Ditonal. A German preparation. It contains alsol, acetonal, and pyramidone. Analgesic.

Ditophal. Diethyl dithiolisophthalate. Etisul.

Dittmarite. A mineral, $Mg(NH_4)PO_4$. $2MgH_2(PO_4)_2.8H_2O$, found in bat guano.

Diufortan. A German product. It contains 17 per cent. iodine, 11 per cent. calcium, and 50 per cent. theobromine, and is free from potassium and salicylic acid. A diuretic.

Diuretic Salt. Potassium acetate, $(C_2H_3O_2)K$.

Diuretine. Theobromine - sodium - salicylate. Prescribed in gout, dropsy, and diseases of the heart and kidneys. It acts as a diuretic.

Diuril. A proprietary trade name for chlorothiazide.

Divanillin. The dimethyl derivative of tetra-oxy-diphenyl-dicarboxylic-aldehyde, $C_6H_2(OCH_3)$ (OH) (CHO).C_6H_2 (OCH$_3$)(OH)CHO.

Diver's Liquid. A liquid formed by absorbing ammonia in solid ammonium nitrate. It is capable of dissolving ammonium nitrate.

Divi-divi. The dried pods of *Cæsalpinia coriaria*, containing 40–45 per cent. tannin. Used as a partial substitute for gambier.

Divinyl (Erythrene). 1 : 3-Butadiene.

Dixarit. A proprietary preparation of clonidine hydrochloride used in the prophylaxis of migraine and for menopausal flushing. (650).

Dixenite. A mineral. It is a silicate and arsenite of manganese, $(MnOH)_2$ $Mn_3SiO_3(AsO_3)_2$, found in Sweden.

Dixie. See CATALPO.

Dixie 5 and Dixie Special 102. Proprietary carbon-black pigments.

D.M. See ADAMSITE.

D.M. 10-chloro-5 : 10 - dihydro - phenarsazine.

DMP. Dimethylphthalate. Used in insect repellants.

DNOC. Dinitro-*o*-cresol. A herbicide. Its derivatives are soluble. See CUDINOC and SODINOC.

Djave Butter. See NJAVE BUTTER.

D.N.P. Dinitro-phenol (1 : 2 : 4). Used as a fungicide in the rubber industry.

D.N.T. Dinitro-toluene, $CH_3.C_6H_3$ $(NO_2)_2$.

DOA. Di iso octyl adipate. A vinyl plasticiser.

Dobane. Detergent alkylates. (553).

Dobanol 25. A proprietary trade name for a synthetic linear C_{12}–C_{15} primary alcohol. Used as an intermediate in the manufacture of ethoxylates, sulphates and ethoxy sulphates. (99).

Dobatex. An anionic detergent. (553).

Dobbin's Reagent. Prepared by adding mercuric chloride solution to a solution of potassium iodide until a permanent precipitate is obtained. The solution is filtered and 1 gram of ammonium chloride added, then dilute caustic soda until a precipitate is formed. Filtered and made up to 1 litre. Used for detecting traces of caustic alkalis in soap.

Dobell Solution. An aqueous solution containing 1·5 per cent. sodium borate, 1·5 per cent. sodium bicarbonate, and 0·3 per cent. phenol and glycerin. An alkaline antiseptic.

Dobschauite. A mineral. It is a variety of gersdorffite

Dobutamine. A cardiac stimulant currently under clinical trial as " 81929 ". ($\pm$) - 4 - [2 - (3 - *p* - Hydroxyphenyl - 1 - methylpropylamino)ethyl]pyrocatechol.

Doca. A proprietary preparation of deoxycortone acetate. (316).

Doctojonan. A German preparation. It is potassium-arsenite and manganese iodate preparation, and is used for intramuscular injection for tuberculosis.

Dr. Haller's Tutorol. A fluosilicate preparation stated to increase the hardness of concrete.

Doctor Metal. An alloy of 88 per cent. copper, 9·5 per cent. zinc, and 2·5 per cent. tin.

Dodder Oil. See CAMELINE OIL.

Dodicin. 2,5,8 - Triazaeicosane - 1 - carboxylic acid. Dodecyl - di(aminoethyl) glycine.

Doebner's Violet. Amino - fuchsoneimonium-chloride. A dyestuff of no technical importance.

Dofamium Chloride. An antiseptic. 2 - (N - Dodecanoyl - N -methylamino) - ethyldimethyl(phenylcarbamoylmethyl) ammonium chloride.

Doggert. See DAGGET.

Dogmatyl. See SULPIRIDE.

Dognacskaite. A mineral. It is $Cu_2Bi_4S_7$. Also given as $Pb_4Bi_{10}S_{19}$.

Dog-tooth Spar. A mineral. It is a variety of calcite, $CaCO_3$.

Dogwood Oil (Sanguinella Oil). Cornel oil obtained from the seeds of *Cornus sanguinea*. The oil has a specific gravity of 0·922, a saponification value of 192, and an iodine value of 100.

Dohyfral. A proprietary preparation of vitamin D tablets.

Dolalgin. A proprietary preparation of butobarbitone, codeine phosphate, and paracetamol. An analgesic. (336).

Dolantin. A German preparation. It consists of 4-β-methoxy-methyl-aminobenzoic acid-β-piperidine-methyl ester mono-hydrochloride and 1 per cent. suprarenin. A non-irritating anæsthetic.

Dolasan. A proprietary preparation of DEXTROPROPOXYPHENE napsylate and aspirin. An analgesic. (463).

Doler Brass. A proprietary alloy. It is a silicon brass.

Dolerite. A mineral. It is a basalt.

Dolerofano. Synonym for Dolerophane.

Dolerophane. A mineral. It is $4[Cu_2SO_5]$.

Dolo-Adamon. The registered trade name for a strong pain relieving agent. It is a mixture of sodium phenyldimethyl pyrazolone methylaminomethanesulphonate(Noramidazophenum), Codeine phosphate and 5-ethyl-5-crotyl barbituric acid (Crotarbital) and Adamon (*q.v.*). (844).

Dolerophanite. A mineral, $CuSO_4$.CuO.

Dolomite (Magnesium Limestone, Bitter Salt Spar). A mineral. It is a double carbonate of calcium and magnesium, $MgCa(CO_3)_2$.

Dolomite Spar. A name applied to Dolomite, $MgCa(CO_3)_2$.

Dolomitic Limestone. Limestone containing more than 5 per cent. magnesium carbonate.

Dolomol. A white insoluble powder, consisting mainly of magnesium stearate, with small amounts of magnesium oleate and palmitate. Used as a dusting powder in affections of the skin.

Dolomol-acid Boric. A combination of magnesium stearate and palmitate, with boric acid.

Dolomol-calomel. A combination of magnesium stearo-palmitate and calomel.

Dolomol-iodine. A stearo-palmitate of magnesium containing small quantities of iodine.

Dolordon. A German product. It consists of tablets containing 0·3 gram pyramidone, 0·02 gram dipropyl-barbituric acid, and 0·03 gram caffeine. Used in the treatment of headache and neuralgia.

Doloresum tophiment. A German preparation. It contains phenyl-cinchonic acid, methyl salicylate, chloroform, menthol, mustard oil, and oil of turpentine. Used for rheumatism and gout.

Doloxene. A proprietary preparation of dextropropoxyphene hydrochloride. An analgesic. (333).

Doloxene Compound-65. A proprietary preparation of dextropropoxyphene hydrochloride, phenacetin, aspirin, and caffeine. An analgesic. (333).

Doloxytal. A proprietary preparation of dexopropoxyphene hydrochloride and amylobarbitone. An analgesic and sedative. (333).

Dolphin Blue. A dyestuff, obtained from gallocyanine and aniline, and sulphonation. A mordant dye used in calico printing.

Dolviran. A proprietary preparation of aspirin, phenacetin, codeine phosphate, caffeine, and phenobarbitone. (341).

Domba Oil. See LAUREL NUT OIL.

Dome-Acne. A proprietary preparation of sulphur and resorcinol acetate used in the treatment of acne vulgaris. (651).

Dome-Cort Cream. A proprietary preparation of hydrocortisone for dermatological use. (372).

Domestos. Stabilised sodium hypochlorite. (506).

Domeykite. A mineral. It is an arsenide of copper, Cu_3As.

Domibrom. A proprietary preparation of domiphen bromide. A throat lozenge. (279).

Domical. A proprietary preparation of AMITRYPTILINE hydrochloride. An anti-depressant. (137).

Domingite. See WARRENITE.

Dominit 18. A similar explosive to Donarit V.

Domiphen Bromide. Dodecyldimethyl - 2 - phenoxyethylammonium bromide. Phenododecinium Bromide. Bradosol.

Donacargyrite. Freieslebenite (q.v.).

Donar. A Sprengel explosive. It is a mixture of potassium chlorate, and potassium permanganate, acted upon by a mixture of turpentine with nitrobenzene or phenol.

Donarite. An explosive containing 80 per cent. ammonium nitrate, 12 per cent. trinitro-toluene, 4 per cent. flour, 3·8 per cent. nitro-glycerin, and 0·2 per cent. collodion wool.

Donarit V. A German explosive containing nitro-glycerin, ammonium nitrate, vegetable meal, nitro-compounds, and potassium perchlorate.

Donbassite. A mineral. It is $Al_8Si_5O_{22} . 7H_2O$.

Dongola Leather. A leather which has been made by handling the skins in a bath containing gambier, alum, and salt.

Donnagel. A proprietary preparation containing kaolin, pectin, hyoscyamine sulphate, atropine sulphate and hyoscine hydrobromide. An anti-diarrhœal. (258).

Donnagel-PG. A proprietary preparation containing kaolin, pectin, hyoscyamine sulphate, atropine sulphate, hyoscyamine hydrobromide and opium. An anti-diarrhœal. (258).

Donnar. See DONAR.

Donnatal. A proprietary preparation containing hyoscyamine sulphate, atropine sulphate, hyoscine hydrobromide and phenobarbitone. An antacid. (258).

Donnazyme. A proprietary preparation containing hyoscyamine sulphate, atropine sulphate, hyoscine hydrobromide, phenobarbitone, pepsin, pancreatin and bile salts. An antacid. (258).

Donovan's Solution. (Liquor arsenii et Hydrargyri iodidi B.P.) A solution of 1 part by weight of arsenious iodide, and 1 part by weight of mercuric iodide, in 100 fluid parts of water.

Doom Bark. Sassy Bark.

Door Plate Brass. See BRASS.

DOP. Di-2-ethyl hexyl phthalate. A vinyl plasticiser.

Dopa. Dioxy-phenyl-alanine.

Dopamet. A proprietary preparation of METHYLDOPA. A drug used in the treatment of hypertension. (137).

Dopamine. A sympathomimetic. 4-(2-Aminoethyl)pyrocatechol.

Dope. The name given to various solutions of cellulose or cellulose compounds in acetone, amyl alcohol, amyl acetate, and other solvents. Used for painting aeroplane wings, and other purposes.

Dope. Emaillite (q.v.).

Doplex. A proprietary cellophane-faced fabric.

Dopplerite. The humus in peat.

Dopram. A proprietary preparation of DOXAPRAM hydrochloride. A respiratory stimulant. (258).

Dor. A proprietary preparation of aromatic oils and essences. A deodorant. (652).

Dorbanex. A proprietary preparation containing danthron and poloxalkol. A laxative. (275).

Doré Silver. A silver containing small amounts of gold.

Dorevane. A proprietary trade name for Propiomazine.

Doriden. A proprietary preparation of glutethimide. A hypnotic. (18).

Dorfit. A German explosive containing ammonium nitrate, with 5 per cent. potassium nitrate, trinitro-toluene, flour, and sodium chloride.

Dormalgen. A German product. It is a combination of pyramidone and noctal.

Dormen. A German proprietary narcotic. It is diallyl-acetyl-bromo-isovaleryl-urea.

Dormigene. See BROMURAL.

Dormiole. Amylene-chloral, $CCl_3.CH(OH)O.C_5H_{11}$. A soporific prescribed in heart disease and nervous troubles.

Dormiral. A German preparation. It is chemically pure ethyl-phenyl-barbituric acid. A hypnotic, sedative, and antepileptic.

Dormonal. A proprietary preparation of veronal.

Dormupax. A proprietary preparation of calcium-n-butyl-allyl-barbiturate and carbromal. A hypnotic. (396).

Dormwell. Dichloralphenazone tablets. (519).

Dorsite. An activated carbon prepared from coco-nut shells. It is used for clarifying and decolorising liquids.

Doryl. Trimethyl - amino - formyl - β - hydroxy-ethyl-amino-chloride.

Doss. The Japanese name for a dyewood from Ilex mertensii.

Dosulphin. A proprietary preparation of sulphaproxyline and sulphamerazine. An antibiotic. (17).

D.O.T.G. Di-o-tolyl-guanidine. A rubber vulcanisation accelerator.

Dothiepin. An anti-depressant. 11-(3-Dimethylaminopropylidene) - 6H - di - benzo[b,e]thiepin. PROTHIADEN is the hydrochloride.

Dotolite. A mineral. It is calcium

boro-silicate, found in California and Nova Scotia.

D.O.T.T. Di-o-tolyl-thiourea.

Double-boiled Oils. Linseed oil heated to 220–250° C., and after addition of driers (metallic oxides) it is boiled for a considerabletime. They are boiled oxidised oils.

Double Brilliant Scarlet G. See FAST RED B.

Double Brilliant Scarlet 3R. See DOUBLE SCARLET EXTRA S.

Double Green S.F. Methyl green ($q.v.$).

Double Nickel Salt. Nickel ammonium sulphate, $Ni(NH_4)_2.(SO_4)_2.6H_2O$. Used in the plating trade.

Double Ponceau. A dyestuff obtained from α-naphthylamine and $1 : 5$-naphthol-sulphonic acid.

Double Scarlet. See FAST SCARLET.

Double Scarlet Extra S (Brilliant Scarlet Extra, Double Brilliant Scarlet 3R, Brilliant Ponceau 4R, Coomassie Scarlet 9012K). A dyestuff. It is the sodium salt of sulpho-naphthalene-azoα-naphthol-mono-sulphonic acid, $C_{20}H_{12}N_2O_7S_2Na_2$. Dyes wool scarlet from an acid bath.

Double Scarlet R. See AZOCOCCINE 2R.

Double Seidlitz Powder. See SEIDLITZ POWDER, DOUBLE.

Double Soluble Glass. Consists of about equal quantities of potassium and sodium silicates.

Double Stearine. See STEARINE.

Double Superphosphate. A monocalcium phosphate usually prepared by treating phosphate rock with phosphoric acid. It contains from 45–50 per cent. phosphoric acid in a soluble form.

Double Twitchell Reagent. The barium salt of the sulphonated mixture of naphthalene and fatty acid. Used in the decomposition of fats. See TWITCHELL REAGENT.

Double White. A proprietary trade name for a general purpose potassium silicate cement for acid conditions, e.g., as a bedding and jointing material for tiles. (150).

Doucil. A water softener. It is stated to be a compound having the formula $Al_2O_3.Na_2O.5SiO_2$. Sodium aluminosilicate base exchange material. (546).

Doughtyite. A mineral. It is a hydrated basic aluminium sulphate.

Douglasite. A mineral, $2KCl.FeCl_2.2H_2O$. Occurs in the Stassfurt deposits.

D.O.V. Distilled oil of vitriol or double oil of vitriol (sulphuric acid containing 95–96 per cent. acid).

Dover's Powder. (*Pulvis Ipecacuanhæ compositus B.P.*) It consists of 1 part of powdered ipecacuanha, 1 part of powdered opium, and 8 parts of potassium sulphate. An emetic.

Dowco 179. A proprietary preparation of CHLORPYRIFOS. An insecticide. (64).

Dow DBR. A registered trade mark for dibenzoyl resorcinol. An ultraviolet absorber for plastics. (64).

Dow H Alloy. See S.A.E. No. 50 ALLOY.

Dowicide 1. A proprietary trade name for o-phenyl-phenol. An antiseptic and fungicide.

Dowicide 2. A proprietary trade name for $2 : 4 : 5$-trichlorphenol. An antiseptic and fungicide.

Dowicide 3. A proprietary trade name for chloro-o-phenyl-phenol. An antiseptic and fungicide.

Dowicide 5. A proprietary trade name for brom-p-phenyl-phenol. A germicide.

Dowicide 6. A proprietary trade name for tetrachlor-phenol. An antiseptic.

Dowicide 7. A proprietary trade name for pentachlor-phenol. An antiseptic and fungicide.

Dowicide A. A proprietary trade name for sodium-o-phenyl-phenate. An antiseptic and germicide.

Dowicide B. A proprietary trade name for sodium $2 : 4 : 5$-trichlorphenate. An antiseptic and germicide.

Dowicide C. A proprietary trade name for sodium-chloro-o-phenyl-phenate. An antiseptic and germicide.

Dowicide F. A proprietary trade name for sodium-tetrachlor-phenate. An antiseptic and germicide.

Dowmetal Alloys. Aircraft alloys containing magnesium with small amounts of aluminium and manganese, sometimes with the addition of small quantities of copper, cadmium, and zinc. They have low specific gravity.

Dow Plasticiser No. 5. A proprietary trade name for diphenyl monoortho xenyl phosphate. A vinyl plasticiser. (64).

Dow Plasticiser No. 55. A proprietary trade name for a technical grade of diphenyl mono ortho xenyl phosphate. A vinyl plasticiser. (64).

Dowpon. Systemic grass killer. (613).

Dowson Gas (Suction Gas, Generator Gas, Mixed Gas, Semi-Water Gas). Gases produced by passing air and steam over red-hot coal. It is cheaper than water gas, and contains about 30 per cent. carbon monoxide, 15 per cent. hydrogen, and 50 per cent. nitrogen.

Dowtherm A. A proprietary trade name for a product consisting of a mixture of diphenyl and diphenyl oxide. It is used for heating industrial machinery to high temperatures (e.g. 200° C.) in place of steam.

Dow 276-V2. A proprietary trade name for an alpha methyl styrene derivative. A vinyl plasticiser. (64).

Dow V9. A proprietary trade name for an alpha methyl styrene derivative. A vinyl plasticiser. (64).

Doxapram. 1 - Ethyl - 4 - (2 - morpholinoethyl) - 3,3 - diphenyl - 2 - pyrrolidone.

Doxepin. An anti-depressant. 11-(3-Dimethylaminopropylidene) - 6H - di - benz[b, e]oxepin. SINEQUAN is the hydrochloride.

Doxorubicin. 14-Hydroxydaunorubicin. An antibiotic produced by *Streptomyces peuceticus*, var. *cæsius*. ADRIAMYCIN.

Doxybetasol. A corticosteroid. 9α-Fluoro - 11β, 17α - dihydroxy - 16β - methylpregna - 1, 4 - diene - 3, 20 - dione. " GR2/443 " is currently under clinical trial as the 17-propionate.

Doxycycline. 6 - Deoxy - 5 - hydroxy-tetracycline.

Doxylamine. 2 - (α - 2 - Dimethyl-aminoethoxy - α - methylbenzyl) - pyridine. Decapryn.

DP. 250. A polyester vinyl plasticiser.

DP 324. A proprietary silicone rubber used in the food industry and in prostheses. (512).

DP/4137-16. A proprietary fast-curing polyester resin with low viscosity. A fire-retardant. (653).

D.P.G. Diphenyl-guanidine. A rubber vulcanisation accelerator.

D.P.G. Salt. A form of diphenyl-guanidine made by a special process. A rubber vulcanisation accelerator. Also see VULCAFOR II.

DR 19549. A proprietary phenolic resol dispersed in water. (468).

Draco Mitigatus. Calomel.

Draconyl. A hydrocarbon, $C_{14}H_7$, obtained by the distillation of dragon's blood.

Dracyl. Toluene, $C_6H_5CH_3$.

Dragées. Sugar-coated pills.

Dragendorf's Reagent (Kraut's Reagent). Potassium - iodo - bismuthate. Used for testing alkaloids.

Dragon Green. An old name for Malachite green.

Dragon Gum. Gum tragacanth.

Dragon's Blood. A red resin. The two varieties are Palm Dragon's Blood, obtained from the rattan palm, *Dæmon-orops draco*, of Sumatra and Borneo, and Socotra Dragon's Blood, from *Dracæna cinnabari*, of Socotra, and the West Indies. Employed as a pigment, for the preparation of red lakes, and varnishes. Specific gravity is 1·2. Melting-point is 120° C. Saponification value is 150–160. Ash is 3–4 per cent.

Dramamine. A proprietary preparation of dimenhydrinate. An antiemetic. (285).

Drapex. A registered trade mark for an epoxy plasticiser. (97).

Drapex 3.2. A registered trade mark for a vinyl plasticiser derived from an epoxide. (97).

Drapolene. A proprietary preparation of benzalkonium chloride in a cream base, used to treat nappy rash. (247).

Dravite. A mineral. It is a brown variety of tourmaline (*q.v.*).

Drawing Brass. See BRASS.

Drazine. A proprietary trade name for the hydrogen maleate of Phenoxy-propazine.

Dreadnought Powder. An explosive containing 73–77 per cent. ammonium nitrate, 14–17 per cent. sodium nitrate, 4–6 per cent. ammonium chloride, and 3–5 per cent. trinitrotoluene.

Dreelite. A mineral containing barium and calcium sulphates.

Dreft. A proprietary trade name for a washing material consisting of sodium lauryl sulphate.

Drenamist. A proprietary preparation of adrenaline hydrochloride in an aerosol form. Bronchial antispasmodic. (271).

Drene. A proprietary trade name for a shampoo containing sodium lauryl sulphate.

Drenison. A proprietary preparation of flurandrenalone for dermatological use. (333).

Dresden Thick Oil. The trade name for a thick turpentine or oleo-resin. It is similar to Venice turpentine, and is used as a vehicle for colours for painting.

Dresinate. Resin soap emulsifiers. (593).

Dricol. A proprietary preparation of amidephrine mesylate. A nasal decongestant. (324).

Drierite. A proprietary trade name for anhydrous calcium sulphate used for drying gases.

Driers. A trade term for those substances which are added during the process of boiling linseed oil, to accelerate its drying properties. Driers appears to absorb the oxygen from the air and transfer it to the oil, thereby aiding its oxidation. The term is used for the oxides of lead, manganese, and cobalt, which were formerly used as driers. More recently, the oxalate, acetate, and borate of manganese have been employed, and at present, the metallic salts of fatty acids, such as lead and manganese linoleates are much used. The metallic resinates are also employed as driers. Also see BOILED OIL, TEREBINE, and LEAD TUNGATE.

Drikold. A trade mark for solid carbon dioxide. A refrigerating agent.

Drill Rod Brass. See BRASS.

Drimol. A wax obtained from the leaves of *Drimys granatensis*.

Drinamyl. A proprietary preparation of dexamphetamine sulphate and amylobarbitone. (281).

Dri-Sil. Silicone water repellents. (614).

Drittel Silver. An alloy of 67 per cent. aluminium and 33 per cent. silver.

Driwal. See CERESIT.

Drocarbil. The acetarsone salt of ARECOLINE.

Droleptan. A proprietary preparation of droperidol. Used for neuroleptanalgesia and premedication. (369).

Dromoran. A proprietary preparation of levorphanol tartrate. An analgesic. (314).

Dromostanolone. A proprietary trade name for Drostanolone.

Droogmansite. A mineral which occurs with sklodowskite and curite as small yellow globules consisting of needle-like crystals.

Drop Black. See FRANKFORT BLACK.

Droperidol. A neuroleptic. 1-{1-[3-(4- Fluorobenzoyl)propyl] - 1, 2, 3, 6 - tetra - hydro - 4 - pyridyl}benzimidazol-2 - one. DROLEPTAN.

Dropped Tin. See GRAIN TIN.

Dropropizine. An anti-tussive. 1-(2,3-Dihydroxypropyl)-4-phenylpiperazine.

Drops, Haarlem. See HAARLEM OIL.

Drostanolone. 17 - Hydroxy - 2α - β methyl - 5(α) - androstan - 3 - one. Dromostanolone.

Drotebanol. An analgesic and antitussive. 3, 4 - Dimethoxy - 17 - methyl-morphinan-6β,14-diol.

Drott. A proprietary pyroxylin plastic.

Droxalin. A proprietary preparation containing polyhydroxy aluminium sodium carbonate and magnesium trisilicate. An antacid. (257).

Droxychrome. A photographic colour developer. (507).

Droxypropine. 1 - [2 - (2 - Hydroxyethoxy)ethyl] - 4 - phenyl - 4 - propionylpiperidine.

Dry Bone. See CALAMINE.

Dry Copper. Metal obtained in the refining of copper. It contains 6–7 per cent. Cu_2O.

Dry Ice. Solid carbon dioxide. Its specific gravity is 1·56.

Dryobalanops Camphor. See CAMPHOL.

Dryptal. A proprietary preparation containing polymethyl siloxane and alu-minium hydroxide and magnesium carbonate as a co-dried gel. An antacid. (137).

Dry Soaps. See SOAP POWDERS.

Dry Wash. Dried spent wash. See SPENT WASH.

Dry Wines. See WINE.

DS-207. A trade mark for dibasic lead stearate. A stabiliser and lubricant. (47).

D.S.P. A chlorinated rubber-based spreading paste used for pointing acid and alkali resisting tiles. (148).

D-steel. A steel containing 1·1–1·4 per cent. manganese, 0·33 per cent. carbon and 0·12 per cent. silicon.

D.T.S. Dehydro-thio-*p*-toluidine-sulphonic acid.

Dualin. An explosive of the dynamite class, containing 80 per cent. nitroglycerin, 10 per cent. sawdust, and 10 per cent. potassium nitrate.

Dubbin. Mixtures of waxes and tallow with colouring matter, sometimes with the addition of rosin. Used to render leather waterproof, and to preserve it.

Dubois' Hydrocarbon. Carotine, a natural colouring matter. See CAROTINE.

Duboisine (Daturine). Hyoscyanine, $C_{17}H_{23}NO_3$.

Duck. A heavy fabric usually made from cotton and employed for a variety of purposes such as tents, bags, aprons, sails, etc. There is also a linen duck.

Duco. A proprietary trade name for pyroxylin lacquers, containing cellulose nitrate.

Dudgeonite. A mineral. It is a variety of annabergite.

Dudley Metal. An alloy of 98 per cent. tin, 1·6 per cent. copper, and 0·25 per cent. lead.

Dufox. A potato fungicide. (501).

Dufrenite. A mineral. It is a ferric phosphate, $Fe_2(OH)_3PO_4$.

Dufrenoisite. A mineral, $2PbS.As_2S_3$.

Duftite. A mineral. It is a basic copper-lead arsenate, $2Pb_3(AsO_4)_2$ $Cu_3(AsO_4)_2.4Cu(OH)_2$, found in South-West Africa.

Duhnul-balasan. See MECCA BALSAM.

Duhudu Oil. An oil obtained from *Celatrus paniculatus*. It is used in Ceylon as a nerve stimulant, also externally for sores.

Duka (Hurka, Karu). Names for an alkaline deposit found in Arabia. It contains sodium carbonate and sodium chloride.

Duke's Metal. A heat-resisting alloy, containing 81 per cent. iron, 12 per cent. chromium, 4 per cent. cobalt, 1·5 per cent. carbon, and small quantities of manganese, tungsten, and silicon.

Dulcine (Sucrol, Sucrene). Mono - *p* -

phenetol-carbamide, $NH_2.CO.NH.C_6H_4$ OC_2H_5. A sweetening substance. It is 200 times sweeter than cane sugar.

Dulcodos. A proprietary preparation containing bisacodyl and dioctyl sodium sulphosuccinate. A laxative. (278).

Dulcolax. A proprietary preparation containing bisacodyl. A laxative. (278).

Dulenza. A proprietary viscose silk.

Dullray. A heat-resisting alloy containing 60 per cent. iron, 34 per cent. nickel, and 5 per cent. chromium.

Du-Lustre. Polychromatic stoving car finishes. (512).

Dulux. A proprietary trade name for alkyd varnish and lacquer resins.

Dumasin. A mixture of methyl - α - butenyl-ketone, $CH_3.COCH : CH.C_2H_5$, and cyclo-pentanone, obtained by the distillation of wood spirit oil. The material occurs in the fraction boiling at 120°–130° C.

Dumet. A copper-clad nickel-iron alloy.

Dumonite. A radio-active mineral found in the Belgian Congo. It has the formula, $2PbO.3UO_3.P_2O_5.5H_2O$.

Dumortierite. A mineral, $2Al_2O_3.$ $3SiO_2$.

Dumoulin's Glue. A liquid glue, obtained by dissolving 8 oz. of best glue in ½ pint of water, adding 2½ oz. of nitric acid of specific gravity 1·330, and when all action ceases, corking it up.

Dumreicherite. A mineral. It is a hydrated silicate of magnesium and aluminium.

Duncaine. A proprietary trade name for Lignocaine.

Dunclad CE. A proprietary trade name for a laminate of Penton (q.v.) and a synthetic rubber. Offers a highly corrosion resistant lining at temperatures up to 125–130° C. (151).

Dunclad VN. A laminate of unplasticised P.V.C. and synthetic rubber designed to enable metal tanks to be lined using special adhesives, giving a highly corrosive resistant surface to prevent attack from oxidising and other acids up to 85° C.

Dundasite. A mineral. It is an aluminium-lead carbonate.

Dunder. The wash obtained after the alcohol has been distilled off from fermented sugar-cane molasses. It contains gluten and gums, and is used for the fermentation of molasses.

Dunging Salt. Sodium silicate (water-glass), used as a substitute for cow-dung in dyeing and calico printing.

Dung Salt. Sodium arsenate, Na_2 $HAsO_4$.

Dunhamite. A mineral. It is $PbTeO_3$.

Dunhill's Solution. An anæsthetic.

It contains ethocaine, adrenalin hydrochloride, and saline solution.

Dunite. A rock containing the minerals olivine and chromite.

Dunlop PL. A laminate of polypropylene and synthetic rubber for tank lining. (151).

Dunlop Grade 6167. A first quality butyl rubber compound which can be used up to 110° C. in corrosive conditions. (151).

Dunlop 6593. A high grade neoprene compound giving high chemical and abrasion resistance at elevated temperatures. (151).

Dunnite. An American explosive. The main constituent is picric acid.

Duo-Autohaler. A proprietary preparation of ISOPRENALINE hydrochloride and PHENYLEPHRINE bitartrate. A bronchospasm-relaxant inhaler. (275).

Duodenin (Pansecretin). Secretin.

Duogastrone. A proprietary preparation of carbenoxolone sodium in a delayed release capsule, for the treatment of duodenal ulcers. (137).

Duogen. A proprietary preparation of hydrogen peroxide. It is stated to be specially pure.

Duolith. See CRYPTONE.

Duomac. Aliphatic diamine acetates. (555).

Duomeen. Long chain aliphatic diamines. (555).

Duoquad T 50 per cent. A proprietary trade name for a liquid di-quaternary ammonium compound. A bactericide, sterilising agent, algaecide, aqueous corrosion inhibitor and foaming agent. (86).

Duotal. Guaiacol-carbonate, $(C_6H_4.O.$ $CH_3)_2CO_3$. Used in medicine as a remedy for tuberculosis and typhus.

Duoterics. Special blends of surface active agents. (547).

Duphalac. A proprietary preparation of LACTULOSE. A laxative. (654).

Duphaspasmin. A proprietary trade name for the hydrochloride of Mebeverine.

Duphaston. A proprietary preparation of dydrogesterone. (280).

Duplate. A proprietary trade name for a safety glass (q.v.).

Duplexite. An explosive. It contains 70 per cent. potassium chlorate, 10 per cent. charcoal, 10 per cent. dinitrobenzene, and 10 per cent. coal tar.

Duponol. A proprietary trade name for a wetting and emulsifying agent containing the sodium salt of sulphated higher fatty alcohols.

Duponol LS. A proprietary trade name for sodium-oleyl-sulphate, a wetting agent.

Duponol ME. A proprietary trade name for sodium-lauryl-sulphate, a wetting agent.

Duponol WA. A proprietary trade name for a mixture of sodium salt of sulphated lauryl alcohol and lauryl alcohol.

Du Pont Accelerator No. 19. See VULCONE.

Du Pont Accelerators. Du Pont is a registered trade mark for proprietary products. Accelerator No. 1 is composed of *p*-nitroso-dimethyl-aniline ; No. 4 consists of aniline ; No. 5 is a formaldehyde-aniline product ; No. 6 is methylene dianilide ; No. 8 consists of anhydro-formaldehyde-*p*-toluidine ; No. 11 is triphenyl-guanidine ; No. 12 is diphenyl-guanidine ; No. 15 is composed of thiocarbanilide ; No. 17 is di-*o*-tolyl-thiourea ; No. 18 is di-*o*-tolyl-guanidine ; No. 19 is a synthetic resin produced by the condensation of aliphatic aldehydes with aniline. The above are all used in the vulcanisation of rubber.

Du Pont Permissible No. 1. A trade mark for an explosive containing nitroglycerin, ammonium nitrate, wood pulp, and sodium chloride.

Du Pont Powder. A trade mark for an explosive containing nitro-cellulose, nitro-glycerin, and ammonium picrate.

Duporthite. A mineral. It is a silicate of aluminium, iron, and magnesium.

Duprene. A proprietary trade name for a synthetic rubber made by the polymerisation of chloroprene (chlorobutadiene) which 'is obtained by the action of hydrogen chloride upon monovinylacetylene. The latter is obtained by polymerising acetylene in the presence of a catalyst. This synthetic rubber is resistant to heat, oil, ozone, and most other chemicals. See also NEOPERNE.

Dupuytren's Caustic Paste. A caustic containing small amounts of arsenious acid and calomel with gum arabic, using water to make the paste.

Durabolin. A proprietary preparation of nandrolone phenylpropionate. An anabolic agent. (316).

Duracillin. A proprietary preparation of procaine penicillin. An antibiotic. (333).

Duracon. A proprietary range of polyacetals. (880).

Duracreme. A proprietary preparation of propylene glycol, glycerin, sodium alginate, boric acid and hexyl resorcinol. A spermicidal jelly. (804).

Durafil. A proprietary synthetic fibre resembling Nylon.

Duraform. A proprietary trade name for asbestos reinforced thermoplastics. They have greater stiffness, lower coefficient of expansion, higher heat distortion point, lower creep, higher tensile and flexural strengths than the basic resins.

Durafur. Bases for fur dyeing. (512).

Durabit. An Austrian trade name for hard rubber and similar products. Used for electrical insulation.

Durabolin. A proprietary trade name for the phenylpropionate of Nandrolone.

Duradent. A proprietary plastic used as a denture.

Duragel. A proprietary preparation of nonoxinol. A contraceptive gel. (655).

Duragen. A proprietary synthetic rubber. (656).

Duraglas. A proprietary trade name for glass fibre reinforced materials such as polyesters, phenolic resins, epoxide, silicone and melamine resins. Used for producing tanks, pipes, ducting and pressure vessels. (152).

Durain. See CLARAIN.

Dural. Duralumin (*q.v.*).

Dural. A trade mark for a moulding and extrusion compound of acrylic resin modified with polyvinyl chloride. (657).

Duralium. An alloy of aluminium with from 3.5–5.5 per cent. copper and small amounts of magnesium and manganese.

Duralon. A proprietary trade name for a vinyl plasticiser. A furan resin. (98).

Duraloy. A proprietary corrosion-resisting alloy of iron with 27–35 per cent. chromium.

Also a proprietary trade name for a phenol-formaldehyde synthetic resin laminated product.

Duralum. An alloy of 79 per cent. aluminium, 10 per cent. copper, zinc, and tin.

Duralumin. Alloys of aluminium with from 3–5.5 per cent. copper, 0.5–1 per cent. manganese, 0.5 per cent. magnesium, and small quantities of silicon and iron. The alloy, containing 95.5 per cent. aluminium, 3 per cent. copper, 1 per cent. manganese, and 0.5 per cent. magnesium, has a specific gravity of 2.8 and melts at 650° C. These alloys are resistant to sea water and dilute acids.

Duramold. A proprietary trade name for a synthetic resin-impregnated wood product used in aeroplane construction.

Duranalium. An aluminium alloy similar in composition and properties to hydronalium and B.S. sea-water alloy.

Durana Metal. An alloy of 65 per cent. copper, 30 per cent. zinc, 2 per cent. tin, 1.5 per cent. iron, and 1.5 per cent. aluminium, of a golden yellow colour.

Durance's Metal (Dewrance Metal).

A bearing metal, consisting of about 33 per cent. tin, 23 per cent. copper, and 45 per cent. antimony.

Durand's Metal. An alloy of 66·6 per cent. aluminium, and 33·3 per cent. zinc.

Durangite. A mineral, $2NaF.Al_2O_3.As_2O_5$.

Durango. See GUAYULE.

Duranic. An alloy containing aluminium with from 2–5 per cent. nickel and 1·5–2·5 per cent. manganese.

Duranite. A proprietary trade name for a fast-baking synthetic enamel.

Duranol. Dispersed dyestuffs for synthetic fibres. (512).

Duranthrene Blue GCD. A dyestuff. It is a British equivalent of Indanthrene blue GCD.

Duranthrene Blue RD Extra. A British dyestuff. It is equivalent to Indanthrene X.

Duranthrene Dark Blue BO. A dyestuff. It is a British brand of Indanthrene dark blue BO.

Duranthrene Dyes. A registered trade name for certain British dyestuffs.

Duranthrene Gold Yellow Y. A British brand of Pyranthrone.

Duranthrene Olive GL. A dyestuff. It is a British equivalent of Indanthrene olive G.

Duraplaz 2114 and 2115. A proprietary range of phthalate ester plasticisers. (516).

Duraplay DNP. A proprietary plasticiser. Dinonyl phthalate. (516).

Duraplex. A proprietary trade name for alkyd varnish and lacquer resins.

Durasol Acid Blue B. A dyestuff. It is a British brand of Alizarin saphirol B.

Durastic. A bitumen compound stated to be composed of high grade bitumen freed from organic acids and used as a protective coating.

Durate. Paint driers. (508).

Duratol. See CANDELITE.

Durax. A moulded insulation made from paper pulp mixed with a synthetic resin. See BAKELITE.

Durbar Bronze. A proprietary trade name for an alloy of copper with 24 per cent. lead and 4 per cent. tin.

Durbar Hard Bronze. A proprietary trade name for an alloy of copper with 10 per cent. tin and 20 per cent. lead.

Durcoton. A proprietary phenol-formaldehyde resin impregnated textile.

Durdenite. A mineral. It is a hydrated iron telluride. Synonym for Emmonsite.

Durecol. A proprietary French glycerophthalic synthetic resin.

Durehete 900. A proprietary trade name for steel containing 1 per cent. chromium and ½ per cent. molybdenum.

It is suitable for studs and bolts for service at temperatures up to 900° F. (482° C.). (153).

Durehete 950. A proprietary trade name for a steel containing 1 per cent. chromium, ½ per cent. molybdenum and ¼ per cent. vanadium for studs and bolts for service at temperatures up to 950° F. (510° C.). (153).

Durehete 1050. A proprietary trade name for a steel containing 1 per cent. chromium, 1 per cent. molybdenum and ¾ per cent. vanadium for bolting materials capable of operating at metal temperatures up to 1050° F. (153).

Durelast 100. A proprietary thermoplastic polyurethane elastomer used to improve the toughness of P.V.C. and its resistance to flex cracking and abrasion. (96).

Durena Metal. See DURANA METAL.

Durenate. A proprietary preparation of sulphamethoxydiazine. An antibiotic. (265).

Durene. Tetramethyl - benzene, $C_6H_2(CH_3)_4$.

Duresco. See LITHOPONE.

Durestos. A proprietary trade name for an asbestos/phenolic moulding material. (152).

Durethan B. A proprietary brand of Nylon 6. (112).

Durethan BKV. A registered trade name for a glass-filled polyamide 6. (112).

Durethane. A proprietary range of polyurethane thermoplastics used in injection moulding. (112).

Durex. A proprietary trade name for an alloy of 83 per cent. copper, 10 per cent. tin, and 4–5 per cent. carbon. Also a proprietary trade name for phenol-formaldehyde synthetic resin.

Durex White. Barium carbonate, $BaCO_3$, used as a pigment in paints.

Durez. A synthetic resin of the phenol-formaldehyde type. It is oil soluble. There are other Durez phenol-formaldehyde thermosetting synthetic resins and moulding powders. Also the trade name for a range of furfuryl alcohol resins used for acid proof tile cements. (37).

Durez 18783. A trade mark for a glass fibre filled, high impact diallyl phthalate moulding compound. (37).

Durfeldite. (1) A mineral, $3(Pb.Ag_2)S.Sb_2S_3$.

Durfeldtite. (2) A mineral. It is $MnPbAgSb_2S_6$.

Durham Lime. A good lime, containing 90 per cent. CaO, 6 per cent. MgO, and 1 per cent. SiO_2.

Durham's Stain. A microscopic stain. It contains a saturated solution of stannous chloride and a 15 per cent. solution of tannic acid in equal parts, with a few drops of an alcoholic solution of methylene blue.

Durichlor. A hydrochloric acid resisting alloy containing 81 per cent. iron, 14·5 per cent. silicon, 3·5 per cent. molybdenum, and 1 per cent. nickel.

Duridine. Tetramethyl - phenylamine, $C_6H(CH_3)_4.NH_2$.

Durimet Alloys. Proprietary alloys for acid resistance. Alloy A is stated to contain iron with 25 per cent. nickel, no chromium, and 5 per cent. silicon ; B contains iron with more nickel than A, and with chromium content about one-third of the nickel, and 5 per cent. silicon ; alloy D contains iron with 15 per cent. nickel, and smaller amounts of chromium and silicon.

Durindone. Indigoid vat dyestuffs. (512).

Durindone Dyes. Registered trade names for certain British dyestuffs.

Durindone Blue 4B. A dyestuff. It is a British equivalent of Bromindigo FB.

Durindone Blue 6B. A dyestuff. It is a British brand of Indigo MLB/6B.

Durindone Red B. A British equivalent of Thioindigo red B.

Durindone Red 3B. A dyestuff. It is equivalent to Helindone red 3B, and is of British manufacture.

Durine. A formalin preparation.

Duriron. An acid resisting alloy, which is a silicon-iron alloy. It contains 15·5 per cent. silicon, 82 per cent. iron, 0·66 per cent. manganese, 0·83 per cent. carbon, and 0·57 per cent. phosphorus. It has a specific gravity of 7 and melts at 1200° C.

Durisol. A trade mark. See PERMALI.

Durite. A trade mark for a coal-tar product in which the lighter oils are absent. It is stated to be suitable as a road-surfacing material.

Durite. A registered trade mark (U.S.) for phenol-formaldehyde and phenol-furfural synthetic resins and moulding compositions of the thermosetting type for mechanical and electrical purposes. See also BAKELITE. (168).

Duro Cement. A cement used in the manufacture of acid towers. It contains 96 per cent. silica and 4 per cent. sodium silicate.

Durocer. A proprietary trade name for a wax.

Durochrome. Fast chrome mordant dyes. (508).

Durochrome Black. A dyestuff. It is a British equivalent of Diamond black F.

Durodi Steel. A proprietary trade name for a nickel-chromium-molybdenum steel.

Duroftal. A proprietary synthetic resin.

Duroftal 293-E. See DUROPHEN 330V.

Duroglass. A proprietary borosilicate resistance glass for chemical use.

Duroid. Special road-dressing material. (603).

Durol. Monochlor-hydroquinone.

Duromine. (1) A proprietary trade name for Phentermine. An ion-exchange resin complex. (2) A proprietary preparation of phenyl tertiary butyl-amine resin complex. An antiobesity agent. (275)

Duromorph. A proprietary preparation of morphine in a long acting form. A hypnotic. (352).

Duronze. A proprietary trade name for a high-silicon copper alloy.

Durophen. A trade name for a series of plasticised phenolic resins widely used in baking finishes. (4).

Durophen 127-B. See KUNSTHARZ HM or DUROPHEN 373U. An ammonia condensed phenol formaldehyde resin melting at 55° C. (4).

Durophen 170W. A butylated diane formaldehyde condensate cooked with trimethylene glycol maleate (65 per cent. solids). (6).

Durophen 218V. A butylated diane formaldehyde castor oil resin. (4).

Durophen 287W. A butylated phenol urea formaldehyde resin sold at 58 per cent. solids. (4).

Durophen 309W. A butylated xylenol formaldehyde resin containing butyl glyceryl adipate. (4).

Durophen 330V. A butylated diane formaldehyde resin cooked with glyceryl phthalate and synthetic fatty acids. (4)

Durophen 373U. See KUNSTHARZ HM or DUROPHEN 127-B.

Durophet. A proprietary preparation of lævo- and dextro-amphetamine in a resin base. (275).

Durophet-M. A proprietary preparation of amphetamine, dexamphetamine, and methaqualone. (275).

Duroplaz. Plasticizers for PVC. (516).

Duroplaz 610, 810, 911. Proprietary trade names for phthalate esters of straight chain alcohols.

Durophenine Brown. A dyestuff prepared by boiling nitroso-phenol with dilute sulphuric acid. It dyes cotton dark violet-brown from a sodium sulphide bath.

Duroprene. A registered trade name for a halogen product of polyprene, obtained by the exhaustive chlorination of natural rubber. It can be moulded, and is soluble in benzene, coal-tar naphtha, and carbon tetrachloride. It is resistant to chemical action and is used in paints and varnishes.

Duroseal. Aluminium soaps for waterproofing. (516).

Durostabe 2201. A proprietary barium-cadmium-zinc complex stabiliser for the high-speed calendering of flexible and semi-rigid P.V.C. (516).

Durostabes. Stabilizers for PVC. (516).

Durosil. A glass free from lead and antimony, suitable for chemical work.

Durox. A mullite (q.v.) made by fusing kyanite and alumina. A proprietary trade name for an ammonium dynamite. An explosive.

Duroxide. Zinc oxides for outdoor paints. (610).

Duroxyn W. Epoxy resins based upon bisphenol A and epichlorhydrin.

Dursban. A proprietary preparation of CHLORPYRIFOS. An insecticide.

Durundum. See METALITE.

Dussertite. A mineral of Algeria. It approximates to the formula $(Ca.Mg)_3.(Fe.Al)_3.(AsO_4)_2.(OH)_9$. It is also given as $[BaFe_3(AsO_4)_2(OH)_5.H_2O]$.

Dusting Powder. Consists of 3 parts of zinc oxide, 1 part of salicylic acid, and 12 parts of starch.

Dust, Rape. See RAPE CAKE.

Dutch Boy Red Lead. A paint pigment. It is an improved form of red lead, Pb_3O_4, and is stated to be chemically pure.

Dutch Brass. See BRASS.

Dutch Butter. See MARGARINE.

Dutch Camphor. Obtained from the wood of the Japanese camphor laurel, *Cinnamomun camphora.*

Dutch Drops. See HAARLEM OIL.

Dutch Liquid. Ethylidene dichloride, $CH_2Cl.CH_2Cl$.

Dutch Liquid Monochlorinate. β-trichlorethane, a solvent for fats, oils, and waxes. It is also used in conjunction with alcohol as a solvent for synthetic resins such as cellulose acetate, etc.

Dutch Metal. An alloy of 80 per cent. copper and 20 per cent. zinc.

Dutch Oil. See HAARLEM OIL.

Dutch Pink. A pigment. It is a yellow colour made by absorbing quercitron on barytes or alumina Specific gravity 2·5.

Dutch Varnish. A solution of rosin in turpentine.

Dutch White. A pigment consisting of 1 part white lead with 3 parts heavy spar.

Dutch Yellow. A dyestuff obtained by the action of sodium bisulphite upon the intermediate product from tetrazotised benzidine and 1 molecule of sali-cylic acid. Dyes chromed wool brownish-yellow. Also see PERSIAN BERRY CARMINE.

Duthane. A proprietary polyester-based polyurethane elastomer cross-linked with diols. (151). VULKOLLAN.

Dutral. A proprietary trade name for an ethylene-propylene synthetic rubber copolymer suitable for tank linings, seals, hose, cables. (99).

Dutral-Co. A proprietary range of ethylenepropylene copolymers. (636).

Dutral-Ter. A proprietary range of ethylene-propylene-diene terpolymers (EPDM). (636).

Dutrex 20, 25. A proprietary trade name for an extender-plasticiser for vinyl resins. A petroleum derivative. (99).

Duty Oil. See OIL OF RHODIUM.

Duvadilan. A proprietary preparation of isoxsuprine hydrochloride. A vasodilator. (280).

Duxalid. A proprietary trade name for a synthetic resin.

Duxite. An explosive containing nitroglycerin, collodion cotton, sodium nitrate, wood meal, and ammonium oxalate. Duxite is also the name applied to a resin from lignite.

Duxol. A proprietary trade name for a synthetic resin.

D-Vac Pollens. A proprietary preparation of grass pollen vaccine used in the hyposensitization treatment of hay-fever and asthma. (272).

DX-830. A proprietary grade of TPX methyl pentene polymer used in injection moulding. It is opaque white in colour. (512).

DX-836. A proprietary opaque white TPX methyl pentene polymer used in the extrusion of sheets. (512).

D.X.L. High boiling tar acids. (602).

Dyamul. Textile scouring and levelling agents. (508).

Dyazide. A proprietary preparation of TRIAMPTERINE and hydrochlorthiazide. A diuretic. (658).

Dyclasite. A mineral. It is a hydrated calcium silicate.

Dycote. Foundry die coating. (531).

Dydrogesterone. 9β,10α - Pregna - 4, 6-diene-3,20-dione. 6-Dehydro-9β,10α-progesterone. Duphaston.

Dyer's Broom. *Genista tinctoria,* used for dyeing yellow on leather.

Dyer's Lac. See LAC DYE.

Dyer's Saffron. Safflower (q.v.).

Dyer's Woodruff. *Asperula tinctoria,* employed as a substitute for madder. Its roots contain a small amount of alizarin.

Dyflor L90. A proprietary polyvinyl fluoride. It is processed in dispersion form. (882).

Dyflos. Di-isopropyl phosphorofluoridate. Di-isopropyl fluorophosphonate. DFP.

Dylene. A proprietary polystyrene. (639).

Dymal. Didymium salicylate, used as a non-irritant antiseptic dressing for wounds.

Dymsol 38C. A proprietary anionic, bio-degradable, polymerisation emulsifier for improving the processing characteristics of S.B.R., nitrile rubber and neoprene. (659).

Dynalkol. A name given in Czecho-slovakia to a motor fuel consisting of a mixture of 60 per cent. benzene and 40 per cent. alcohol (96·6 per cent.). Sometimes 1 per cent. naphthalene, 1 per cent. tetralin, and 5 per cent. ether are added.

Dynamagnite. See NITROMAGNITE.

Dynamin. A mixture of benzene and stellin (q.v.). It contains 40 per cent. by volume of the former and 60 per cent. by volume of the latter.

Dynamine. A rubber vulcanisation accelerator. It is diphenyl-guanidine.

Dynamite. See KIESELGUHR DYNA-MITE.

Dynamite Acid. Concentrated nitric acid, used for making 96 per cent. mixed acids (34 per cent. nitric acid + 62 per cent. sulphuric acid).

Dynamite, Ammonia, Pulverent. See AMMONIA DYNAMITE, PULVERENT.

Dynamite Antigrisouteuse V. An explosive similar to Grisoutite, except that sodium sulphate replaces magnesium sulphate.

Dynamite de Trauzl. An explosive containing 75 per cent. nitro-glycerin, 23 per cent. guncotton, and 2 per cent. charcoal.

Dynamite Glycerin. Glycerin of specific gravity 1·263, containing 98–99 per cent. It contains no lime, sulphuric acid, chlorine, or arsenic.

Dynamite No. 1. English contains 75 per cent. nitro-glycerin. Austrian contains 65½ per cent. nitro-glycerin, 2 per cent. collodion wool, 7½ per cent. wood meal, 25 per cent. nitre, and ¼ per cent. soda.

Dynamite No. 2. Austrian. It contains 46 per cent. nitro-glycerin.

Dynamite No. 2A. Austrian. It contains 38 per cent. nitro-glycerin.

Dynammon. An Austrian explosive containing 87–88 per cent. ammonium nitrate, and 12–13 per cent. charcoal.

Dynapol L. A proprietary range of linear saturated polyesters of high molecular weight, based on terephthalic acid. They are used for weather-resistant coatings. (883).

Dynapol P. A proprietary terephthalic polyester coating material, applied electrostatically. (882).

Dynarsan. See ACETARSONE.

Dynasil. A trade mark for synthetic fused silica. (660).

Dynastite. An explosive containing 94 per cent. potassium chlorate, and 6 per cent. barium nitrate, dipped in nitro-toluene.

Dynat. A proprietary brand of mechanically comminuted rubber. (661).

Dynat W. A proprietary brand of mechanically comminuted rubber from Malaysia.

Dynobel. An explosive containing potassium perchlorate, nitro-glycerin, ammonium oxalate, wood meal, and a little collodion cotton.

Dynoform. A proprietary phenolic moulding material. (662).

Dyphos XL. A trade mark for modified dibasic lead phosphite, a stabiliser for vinyl plastics. (47).

Dypolene. A textile dyeing and printing assistant. (508).

Dysanalyte. A mineral, $(Ca.Ce.Fe)O.Nb_2O_5$.

Dyscrasite (Dyskrasite, Discrasite, Antimonial silver). A silver ore. It is silver antimonide, Ag_2Sb to Ag_6Sb.

Dyskrasite. See DYSCRASITE.

Dysluite. A mineral, $(Zn.Mn)(Al.Fe)_2O_4$.

Dyslysin. Cholalic anhydride.

Dyslytite. See SCHREIBERSITE.

Dysodil (Paper Coal). A carbonaceous material found in Sicily. It burns with an odour resembling burning rubber.

Dysoid. A bearing bronze containing 62 per cent. copper, 18 per cent. lead, 10 per cent. tin, and 10 per cent. zinc.

Dyspastol. A proprietary preparation of ephedrine hydrochloride, potassium bromide and extract of hyoscyamus. A urinary antispasmodic. (334).

Dyssnite. A mineral. It is a black silicate of manganese.

Dystome Spar. A mineral. It is a variety of datolite.

Dysyntribite. A mineral. It is a hydrated silicate of aluminium and potassium.

Dytac. A proprietary preparation of triamterene. A diuretic. (281).

Dythal. A dibasic lead phthalate stabilizer for PVC. (572).

Dytide. A proprietary preparation of triamterene and benzthiazide. A diuretic. (281).

Dytransin. A proprietary trade name for Ibufenac.

Dyvon. A proprietary preparation of METRIPHONATE. An insecticide.

E

E45 Cream. A proprietary preparation of white soft paraffin, light liquid paraffin and wool fat used as a skin cream. (502)

E107. See AVERTIN.

Eagle Silk. A German proprietary artificial silk made by the cuprammonium process.

Eagle-stone. A mineral. It is a variety of clay iron ore.

Eakinsite. A mineral It is $Pb_5Sb_4S_{11}$.

E Alloy. An alloy of 76 per cent. aluminium, 20 per cent. zinc, 2·5 per cent. copper, 0·2 per cent. iron, 0·5 per cent. manganese, 0·5 per cent. magnesium, and 0·2 per cent. silicon.

Earth Archil. Archil (q.v.), contaminated with mineral matter. Used for the preparation of litmus.

Earth Balsam. See EARTH OIL.

Earth, Bone. See BONE ASH.

Earth, Cassel. Vandyck Brown (q.v.).

Earth, Cullen. Cologne Earth (q.v.).

Earth, Diatomaceous. See INFUSORIAL EARTH.

Earth, Florida. See BLEACHING EARTH.

Earth Foam. See APHRITE.

Earth, Gold. An ochre. See YELLOW OCHRE.

Earth Green. See LIME GREEN.

Earth, Infusorial. See INFUSORIAL EARTH.

Earth, Japan. See CUTCH.

Earth, Lemnos. See LEMNIAN EARTH.

Earthnut Oil (Peanut Oil). Arachis oil, obtained from the seeds of *Arachis hypogœa*.

Earth Nuts. See GROUND NUTS.

Earth Oil (Rock Oil, Naphtha, Mineral Oil, Ohio Oil, Lima Oil, Earth Balsam, Rangoon Oil). Crude petroleum.

Earth, Porcelain. See CHINA CLAY.

Earthquake Powder. A safety explosive containing 79 per cent. nitre, and 21 per cent. charcoal.

Earth, Red. See INDIAN RED.

Earth Rubber. Rubber latex which drops from the tree or is spilled from the cups on to the ground. It is contaminated with sand and has low value.

Earth, Santorin. See SANTORIN EARTH.

Earth Shellac. See ACAROID BALSAM.

Earth Wax. See OZOKERITE.

Earth Worm Oil. See OIL OF EARTH WORMS.

Earthy Cobalt. See ASBOLANE.

Earth, Yellow. See OCHRE.

East India Gum (Bombay Gum). A variety of gum arabic, pale amber or pinkish in colour.

East Indian Balsam of Copaiba. A name given to Gurjun balsam (the oleo-resin from the stems of *Dipterocarpus* species).

Eastman 910. A proprietary trade name for a cyano-acrylate adhesive which sets with the application of pressure. Variants are: *910 EM* for vinyls; *910 FS* for quicker setting; *910 MHT* for applications involving high temperatures. (214).

Eastman Inhibitor DOBP. A proprietary trade name for 4-dodecyloxy-2-hydroxybenzophenone. A U.V. light inhibitor suitable for use in P.V.C., polyesters, polystyrene and butyrate-acrylic coatings.

Eastman Inhibitor OPS. A proprietary trade name for *p*-octylphenyl salicylate. A U.V. light inhibitor suitable for polyolefins.

Eastman Inhibitor RMB. A proprietary trade name for a U.V. light inhibitor for polar resins (cellulosics). It is resorcinol monobenzoate.

Eastman Yellow. A yellow colouring matter used as a corrective filter in photography. It is the sodium salt of glucose-phenyl-osazone-*p*-carboxylic acid.

Eastonite. A mineral. It is $K_2Mg_5Al_4Si_5O_{20}(OH)_4$.

Easton's Syrup. (*Syrupus Ferri Phosphatis cum Quinina et Strychnina B.P.*) Quinine sulphate and strychnine are added to the solution obtained by dissolving iron in concentrated phosphoric acid, then the syrup is added.

Eastozone 32. A proprietary anti-oxidant. N, N'-dimethyl-N, N'-di-(1 methylpropyl)-p-phenylenediamine. (242).

Easy-Flo. Fluxes for silver alloy brazing. (615).

Eau de Brouts. Petitgrain water.

Eau de Goudron. Tar water.

Eau de Javelle (Eau de Labarraque, Chlorinated Soda, Chloride of Soda). The name Eau de Javelle, was first applied to potassium hypochlorite, but is now used for the liquor obtained by passing chlorine into a solution of sodium carbonate, when sodium hypochlorite is formed. Used for bleaching.

Eau de Labarraque. See EAU DE JAVELLE.

Eau de Luce. Oil of amber mixed with rectified spirit and ammonia. Formerly used in medicine.

Eau de Naphe. Orange flower water.

Eau de Surean. Elderflower water.

Eau de Vie. Brandy.

Eau Mountant. Water miscible microscopic mounting medium. (544).

Ebelmenite. A mineral. It is a variety of psilomelane.

Ebert and Merz's α-Acid. Naphthalene-disulphonic acid, $C_{10}H_6(SO_3H)_2$, 2 : 7.

Ebert and Merz's β-Acid. Naphthalene-disulphonic acid, $C_{10}H_6(SO_3H)_2$, 2 : 6.

Ebivit. A Vitamin D_3 sterile injection. (591).

Ebner's Fluid. A mixture of 2·5 c.c. hydrochloric acid, 2·5 grams sodium chloride, 100 c.c water, and 500 c.c. alcohol. Used for decolorising in bacteriological work.

Eboli Green. A polyazo dyestuff which is a similar product to Diamine green G.

Ebonestos. A trade name for a series of proprietary moulded products for electrical and heat insulation.

Ebonised Monel. A monel metal with a fine finish produced by an oxidising process.

Ebonite (Vulcanite, Hardened Rubber). A material prepared by vulcanising rubber with up to 75 per cent. sulphur or metallic sulphides, with the addition of chalk, gypsum, or other filling and colouring substances. Used as an insulating material. The dielectric strength varies from 10,000–38,000 volts per mm., the tensile strength from 3,500–6,500 lb. per sq. in., the density from 1·12–1·4. The water absorbed in 24 hours is 0·02 per cent. It is attacked by acetone and alcohol, partly attacked by carbon disulphide and ether, also softened by aniline and benzene. Alkalis do not attack it, and ammonium chloride has no effect upon it.

Ebontex. A proprietary trade name for an emulsified asphalt and used for waterproofing tanks.

Ebony Black. A blackish-brown dyestuff mixed with a blue dyestuff. It is used for dyeing cotton from a bath containing sodium sulphate and sodium carbonate, and half-wool. Also see GAS BLACK.

Ebony, Green. See GREEN EBONY.

Eborex. A proprietary preparation containing about 65–70 per cent. sodium fluosilicate. It is a light fluosilicate for use as an insecticide.

Ebrok. A proprietary trade name for a bituminous plastic.

Ebucin. Calcium gluconate solutions. (591).

E.C. See ELECTROLYTIC CHLOROGEN.

E.C.A. Cresylic acids. (501).

Ecarite, Ecaron. A German cellulose acetate.

Ecboline. Ergotinine, $C_{35}H_{40}N_4O_5$.

Eccaine. n - Benzoyl - hydroxy - propyl-novec-gonidene ester. It is allied to cocaine.

Eccobond 114. A proprietary one-part, filled epoxy adhesive. (410).

Eccobond Paste 99. A proprietary one-part thixotropic epoxy adhesive of high thermal conductivity. It is used in heat-sink applications. (410).

Eccobond SF40. A proprietary low-density, two-component epoxy-based adhesive and rigid filler. (410).

Eccocoat SJB. A proprietary epoxy resin used as a coating for semiconductor junctions. (410).

Eccofloat EG35. A proprietary epoxy-resin-bound syntactic foam material used in deep-sea applications. (410).

Eccofloat Encapsulant 1421. A proprietary epoxy-resin-bound encapsulant used to protect under-sea components. (410).

Eccofloat HG452. A proprietary polyester-resin-bound low-density float material for use in deep-sea applications. (410).

Eccofloat PC61. A proprietary polyester-resin-bound castable material for use in deep-sea applications. (410).

Eccofloat PG23. A proprietary polyester-resin-bound syntactic foam used to fill voids in submarine hulls. (410).

Eccofloat PP22 and 24. Proprietary grades of polyester-bound syntactic foam which can be packed *in situ* to fill voids and to make buoys. (410).

Eccofloat SP 12, 20. A proprietary polyester-bound low-density syntactic foam for use where buoyancy is required in harbour and off-shore applications. (410).

Eccofloat SS40. A proprietary polyurethane rubber-bound material used in the making of deep-sea diving suits. (410).

Eccofloat UG 36. A proprietary polyurethane-bound semi-flexible non-compressible material. (410).

Eccofloat US 35. A proprietary polyurethane - bound material — flexible, compressible and usable down to about 1000 ft depth of water. (410).

Eccofoam PP. A proprietary group of hydrocarbon resin closed-cell foams used in high-frequency electrical applications. (410).

Eccofoam PS-A. A proprietary polystyrene-based foam having a controlled dielectric constant. (410).

Eccomold 1099. A proprietary epoxy moulding compound filled with glass microballoons giving it a very low density. (410).

Eccosorb Coating 268E. A proprietary epoxy coating which is brushed on to surfaces to increase their electrical loss in the L-band of the high-frequency range. (410).

Eccosorb 269E. A proprietary epoxy coating having properties similar to those of Eccosorb Coating 268E but for use in the S to the K bands inclusive. (410).

Eccosorb MF. A proprietary range of magnetically-loaded epoxy resins. (410).

Eccospheres. Trade name for small hollow glass or silica spheres of diameter ranging from 10–250 microns. Used as a loading material for plastics to impart lightness and reduced permittivity which it does by virtue of the large air space. (410).

Ecedemite (Ekedemite). A mineral, $2PbCl_2.Pb_2As_2O_7$.

Ecepox PB1 and PB2. Epoxidised esters used as plasticisers and stabilisers for P.V.C. compounds. Registered trade names.

Echappes. o-Toluidine, recovered as a distillate on fusing fuchsine.

Échappe Silk. A name for floss or waste silk.

Echellite. A zeolite mineral, (Ca.Mn) $O.2Al_2O_3.3SiO_2.4H_2O$, found in Ontario.

Echicaoutchin. A low-grade gutta-like material from *Alstonia scholaris*.

Echurin. A mixture of picric acid and nitro-flavin.

Eckermannite. A mineral. It is $2[Na_3(Mg,Li)_4(Al,Fe)Si_8O_{22}(OH,F)_2]$.

Eclipse Black. A sulphide dyestuff.

Eclipse Blue. A dyestuff. The indophenol from dimethyl-p-phenylene-diamine and phenol when treated with alkali sulphites gives a sulphonic acid. The sodium salt of this acid is heated with sulphur and sodium sulphide to give Eclipse Blue.

Eclipse Bronze. A proprietary trade name for a nickel-bronze containing 60–65 per cent. nickel, 24–27 per cent. copper, 9–11 per cent. tin, and small amounts of iron, silicon, and manganese.

Eclipse Brown B. A dyestuff obtained by heating m-tolylene-diamine and oxalic acid with polysulphides.

Eclipse Green G. A sulphide dyestuff.

Eclipse Red. See BENZOPURPURIN 4B. Also Geigy's Eclipse red, by heating amino-hydroxy-phenazines with alkali sulphides and sulphur.

Eclipse Yellow. A dyestuff obtained by heating diformyl-m-tolylene-diamine and sulphur at 240° C.

Ecolac. A proprietary trade name for an air drying lacquer and adhesive for plastics.

Ecolid. Chlorisondamine chloride.

Economycin. A proprietary preparation of tetracycline hydrochloride. An antibiotic. (363).

Ecomytrin. A proprietary preparation of amphomycin and neomycin. Antibiotic skin cream. (262).

Econopen V. A proprietary preparation

containing Phenoxymethylpenicillin Potassium. An antibiotic. (137).

Econosil V-K. A proprietary preparation of phenoxymethyl penicillin potassium. An antibiotic. (363).

Ecoro. A proprietary packaging material of polypropylene loaded with calcium carbonate to ease disposal by incineration. (663).

Ecothiopate Iodide. S-2-Dimethylaminoethyl diethyl phosphorothioate methiodide. Phospholine Iodide.

E.C. Powder. A smokeless sporting powder. It consists of grains of nitrocotton (79 per cent.), hardened by treatment with a solvent (ether-alcohol), and mixed with 4·5 per cent. potassium nitrate, 7·5 per cent. barium nitrate, 4·1 per cent. camphor, and 3·8 per cent. wood meal.

Ecrasite. An Austrian explosive. It is the ammonium salt of trinitro-cresol. See LYDDITE.

Ecru Silk. Silk which has lost about 3–4 per cent. of its weight of sericin or silk gum.

Ectimar. A proprietary preparation of ETISAZOLE. A veterinary fungicide.

Ectylurea. α - Ethylcrotonoylurea. Nostyn.

Edecrin. A proprietary preparation of ethacrynic acid. A diuretic. (310).

Edelfeka. A nickel-containing silver-copper-cadmium alloy.

Edelmist. A manure prepared from farmyard manure by speeding up fermentation by piling loosely and allowing fermentation to take place until the temperature reaches 55–65° C., then compacting the mass by tramping, covering with a layer of fresh dung and leaving it for 3–4 months. The manure is in a more decomposed form, and the temperature caused by fermentation is fatal to most weed seeds. It can be stored longer than ordinary manure.

Edelresanol. A proprietary synthetic resin.

Edelweiss. See OIL WHITE.

Edenite. A mineral, $Na_2Al_2(SiO_3)_4$.

Edenol. A proprietary trade name for a neutral ester of adipic acid and a special compound of synthetic alcohols.

Edenol 74. A proprietary alkyl-epoxy stearate-type plasticiser. (664).

Edenol B35. A proprietary alkyl-epoxy stearate-type plasticiser for plastisols. (664).

Edenol B316. A proprietary epoxy plasticiser made of an epoxidised linseed oil and used in rigid P.V.C. (664).

Edenol D72. A proprietary alkyl-epoxy stearate-type plasticiser used for plastisols. (664).

Edenol D82. A proprietary epoxy plasticiser. An epoxidised soya bean oil, it is used in both rigid and soft P.V.C. (664).

Edenol HS 235. A proprietary alkyl-epoxy stearate-type plasticiser for use in plastisols. (664).

Eder's Solution. A solution of mercuric chloride and ammonium oxalate used in photometric determinations.

Edetic Acid. Ethylenediamine - NNN′N′-tetra-acetic acid. Versene Acid.

Edicol. Foodstuff colours of guaranteed purity. A registered trade name. (512).

Edifas A. A trade mark. Methyl Ethyl Cellulose for use in foods. (2).

Edifas B. A trade mark. Sodium Carboxymethyl Cellulose for use in foods. (2).

Edifor. Vinyl-based thermoplastic alloy. (61).

Edimet. Polymethylmethacrylate.

Edingtonite. A mineral, $BaO.Al_2O_3. 3SiO_2.3H_2O$.

Edinol. A photographic developer. It contains p-amino-saligenin, acetone sulphite, potassium hydroxide, and potassium bromide.

Edisonite. A mineral. It is dioxide of titanium, TiO_2.

Edistir. Polystyrene. General purpose and high impact grades.

Edogestrone. A progestational steroid currently undergoing clinical trial as " P.H. 218 ". It is 17-acetoxy-3,3-ethylenedioxy - 6 - methylpregn -5 - en - 20-one.

Edosol. A proprietary food product containing fat, protein, lactose and minerals used in low-sodium diets. (665).

E.D.P. Bismuth subiodide dusting powder. (591).

Èdrédon Végétale (Pattes de Lièvre). Vegetable down, the hair fibres of *Ochroma lagopus*.

Edrisal. A proprietary preparation of amphetamine sulphate, aspirin, and phenacetin. (281).

Edrophonium Chloride. Ethyl - (3 - hydroxyphenyl)dimethylammonium chloride. Tensilon.

Edward's Speculum. A zinc and arsenic bearing bronze containing 63·3 per cent. copper, 32·2 per cent. tin, 2·9 per cent. zinc, and 1·6 per cent. arsenic.

Eel Antifriction Metal. An alloy of 75 per cent. lead, 15 per cent. antimony, 6 per cent. tin, 1·5 per cent. cadmium, 0·5 per cent. arsenic, and 0·1 per cent. phosphorus.

Efamil. A proprietary preparation of ephedrine, aminophylline, and pheno-

barbitone. A bronchial antispasmodic. (279).

Efcortelan. A proprietary preparation of hydrocortisone for dermatological use. (335).

Efcortelan-N. A proprietary preparation of hydrocortisone and neomycin used in dermatology as an antibacterial agent. (335).

Efcortelan Soluble. A proprietary preparation of HYDROCORTISONE, for injection in cases of shock and adrenal crisis. (335).

Efcortesol. A proprietary preparation of hydrocortisone as the 21-disodium phosphate ester, used to treat adrenal insufficiency and shock. (335).

Effervescent Citrate of Magnesia. Generally, it is a mixture of sodium bicarbonate, citric acid, tartaric acid, sugar, and a small quantity of either magnesium carbonate or magnesium sulphate, or of both, and flavouring materials.

Effervescent Epsom Salts. (*Magnesii sulphas effervescens B.P.*) Magnesium sulphate dried, until nearly half of its water of crystallisation has been removed, then mixed with citric and tartaric acids, sodium bicarbonate, and sugar.

Effervescent Magnesia. Prepared by mixing magnesium citrate with sodium bicarbonate, a small quantity of citric acid, and sugar. Used as a refreshing drink, and as a purgative.

Effervescent Magnesium Sulphate. See EFFERVESCENT EPSOM SALTS.

Effervescent Tartrated Soda Powder. See SEIDLITZ POWDER.

Effesay. Sulphonated alcohols and detergents. (550).

Effico. A proprietary preparation of thiamine hydrochloride, nicotinamide, tincture of nux vomica and caffeine hydrate. A tonic. (330).

Efflorescent Pyrites. See WHITE PYRITES.

Efpa Gel. A proprietary preparation of di-isobutylphenoxypolyethoxyethanol, and ricinoleic acid. A spermicide. (369).

EFTE. A commonly-used " shorthand " expression for ethylene tetrafluoroethylene. See (e.g.) TEFZEL 200.

Egeran. A mineral. It is a variety of idocrase.

Egg Cement. White of egg beaten up and mixed with sufficient slaked lime to form a thin paste. Used for cementing earthenware, china, and glass.

Egg Oil. Yolk of eggs, employed in the preparation of glove leather.

Egg Oils. See WHITE OILS.

Eggonite. A mineral, $CdO.SiO_2$. Synonym for Sterrettite.

Egg Yellow. Both the dyestuffs Naphthol yellow S and Tartrazine are sold under this name. Used for colouring foodstuffs.

Eglantine. A name which has been applied to both isobutyl benzoate and to isobutyl-phenol acetate. Used in perfumery.

Eglestonite. A mineral. It is mercury oxychloride, Hg_4Cl_2O.

Egols. The name applied by E. Gautrelet to a series of substances produced by him. They are o-nitro-p-sulphonates of phenol, cresol, and thymol, in chemical combination with mercury and potassium. Phenegol from phenol, Cresegol from cresol, and Thymegol from thymol. They are said to be powerfully bactericidal.

Egueiite. A mineral. It is $CaFe_{14}(PO_4)_{10}(OH)_{12} \cdot 21H_2O$.

Egyptian Blue (Vestorian Blue). CaO. $CuO.4SiO_2$, is the formula which corresponds to this ancient blue, used by the Romans.

Egyptian Fibre. See VULCANISED FIBRE.

Egyptianised Clay. Clay rendered more plastic by the addition of tannin.

Ehlite. See PSEUDO-MALACHITE.

Ehrenbergite. A mineral. It is a variety of cimolite.

Ehrhard's Metal. An alloy of 89 per cent. zinc, 4 per cent. copper, 4 per cent. tin, and 3 per cent. lead.

Ehrlich 418. See SPIRARSYL.
Ehrlich 594. See ACETARSONE.
Ehrlich 606. See SALVARSAN.
Ehrlich 914. See NEOSALVARSAN.

Ehrlich-biondi Stain. A microscopic stain containing 100 c.c. of a saturated solution of Orange G, 30 c.c. of a saturated solution of acid fuchsin, and 50 c.c. of a saturated solution of methyl green.

Ehrlich-hata 606. See SALVARSAN.

Ehrlich's Acid Hæmatoxylin. A stain used in microscopy. It contains 100 c.c. water, 100 c.c. absolute alcohol, 100 c.c. glycerin, 10 c.c. glacial acetic acid, 2 grams hæmatoxylin, and alum in excess.

Ehrlich's Diazo Reagent. (a) For indol : 4 grams p-dimethyl-amino-benzaldehyde in 380 c.c. alcohol and 80 c.c. concentrated hydrochloric acid. One volume of the solution to be tested is used with 1 volume of the reagent, a positive colour being red. (b) Used in the diagnosis of typhoid ; the reagents are (1) 5 grams sulphanilic acid and 50 c.c. hydrochloric acid in 1 litre of water and (2) 5 grams sodium nitrate in 1 litre of water. A pink froth gives a positive test on shaking the reagents with the unknown solution.

Ehrlich's Hæmatoxylin. A microscopic stain. It consists of 30 grains hæmatoxylin, 100 c.c. absolute alcohol, 100 c.c. glycerin, 30 grains ammonium alum, and 100 c.c. distilled water.

Ehrlich's Triple Stain. A microscopic stain for blood corpuscles. It contains 135 parts of a saturated aqueous solution of Orange G, 100 parts of a saturated aqueous solution of Methyl green, 100 parts of a saturated solution of Acid fuchsine, 100 parts of glycerin, 200 parts of absolute alcohol, and 300 parts water.

Eichbergite. A mineral, $(Cu.Fe)_2S$. $3(Bi.Sb)_2S_3$.

Eichrome Red B. A dyestuff obtained from 1 - amino - 2-naphthol-4-sulphonic acid and 1 - phenyl - 3 - methyl - 5 - pyrazolone.

Eichwaldite. See EREMEYEVITE.

Eicosylene. Icosinene, $C_{20}H_{38}$, prepared from ozokerite.

Eigones. The name given to certain albuminous compounds of iodine and bromine. Bromeigone contains 11 per cent. bromine, and is used as a substitute for bromides. Iodine eigones are used in syphilis, rheumatism, and asthma. Peptoneigone is a peptonised bromine-albumin compound, containing 11 per cent. bromine.

Eikonogen. A photographic developer. Sodium-α-amino-β-naphthol-β-sulphonate, $C_{10}H_5(OH)(NH_2)(SO_3Na)$, forms the basis of this material. It is also used in the treatment of ivy poisoning.

Eisenamianth. A fibrous form of silica, found in the bottom of iron smelting vessels.

Eisenanthophyllite. Synonym for Ferro-anthophyllite.

Eisen Bronze. An alloy of 82·5 per cent. copper, 8·55 per cent. tin, 4·45 per cent. zinc. and 4 per cent. iron.

Eisenbrucite. A mineral. It is $Mg_6Fe_2\cdot\cdot CO_3(OH)_{14} \cdot 4H_2O$.

Eisencordierit. Synonym for Iron-cordierite.

Eisenepidot. Synonym for Ferriepidote.

Eisenglimmer. Synonym for Vivianite.

Eisengymnite. A mineral. It is $(Mg,Fe)_4Si_3O_8(OH)_4$.

Eisenoxydrot. A synthetic red iron oxide pigment. See also OXYDROT F-140.

Eisenphyllit. Synonym for Vivianite.

Eisen-Portland Cement (Iron cement, Slag Cement, Iron Portland Cement, Pozzolana Cement, Puzzolana Cement). This material consists of 70 per cent. Portland Cement (made from slag and limestone), and 30 per cent. granulated slag.

Eisenschefferit. Synonym for Iron schefferite.

Eisen, Silicon. See SILICON-EISEN.

Eisenstassfurtit. Synonym for Huyssenite.

Eisler's Bronze. A bronze containing 5·9 per cent. tin. Used for art castings.

Eistan Firnis. Boiled linseed oil in mineral spirits (55 per cent. solids). Also 30 per cent. linseed oil and 20 per cent. oil-linseed alkyd in mineral spirits.

Eka-aluminium. Gallium, Ga.

Eka-boron. Scandium, Sc.

Ekammon. A proprietary preparation of aspirin, and vitamins C and K. An analgesic. (319).

Ekanda Rubber. A rubber obtained from the shrub, *Raphionacme utilis* in Angola.

Eka-silicon. Germanium, Ge.

Ekdemite. A mineral. It is $32[Pb_3As'''O_{4-n}Cl_{2n-1}]$.

Ekebergite. A mineral. It is a compound of silica, alumina, lime, and soda.

Ekedemite. See ECEDEMITE.

E-Kote 3042. A trade name for a silver filled air drying epoxy coating material soluble in isobutyl ketone. (16).

Ektobrom. A German product. It is a 10 per cent. solution of sodium bromate in 4 per cent. calorose solution. Used in the treatment of eczema and other skin diseases.

Ektogan (Ektogen, Zinc Perhydrol, Zinconal). A preparation of zinc oxide, containing from 40–60 per cent. zinc peroxide. An antiseptic used for dressing wounds and burns, also as an astyptic.

Ektogen. See EKTOGAN.

Ektropite. A mineral. It is $Mn_3Si_2O_7$. $1\frac{3}{4}H_2O$.

Ekzebrol. A German preparation. It is a 20 per cent. solution of grape sugar (10 c.c. contain 1 gram strontium bromate). Used in the treatment of dermatitis.

Elæolite. Nepheline (*q.v.*).

Elaine. See OLEIN.

Elaite. A mineral. It is a variety of copiapite.

Elamol. A proprietary preparation of TOFENACIN hydrochloride. An antidepressant. (317).

Elaol. A proprietary plasticiser. It is stated to be dibutyl-phthalate, C_6H_4 $(COO.C_4H_9)_2$. Its specific gravity is 1·05, boiling-point 325° C., and flashpoint 328° F. It dissolves cellulose nitrate, ester gum, coumarone, etc.

Elaol 1. A trade name for a plasticiser made from C_4 to C_6 paraffin fatty acids and hexanetriol.

Elaol 2. A trade name for a plasticiser made from C_6 to C_9 paraffin fatty acids and hexanetriol.

Elaol 3. A trade name for a plasticiser made from C_4 to C_6 paraffin fatty acids and pentaerythritol.

Elaol 4. A trade name for a plasticiser made from C_6 to C_9 paraffin fatty acids and pentaerythritol.

Elargol. A silver finish for mica and plastics. (518).

Elarsan. Strontium-chloro-arseno-behenolate $(C_{22}H_{39}O_3AsCl)_2S_2$. It has the same use in medicine as arsenious oxide.

Elase. A proprietary preparation of fibrinolysin and desoxyribonuclease. Dermatological stimulant. (264).

Elastacast. A proprietary polyurethane elastomer. (666).

Elasteine. A patented French rubber substitute made by the treatment of copals with oleic acid.

Elastes. A rubber substitute belonging to the class of mixtures of glue or gelatin, glycerin, and oils, treated with tannic acid, chromates, or formaldehyde.

Elastex. A mineral rubber used in rubber mixings. Also a proprietary material consisting of emulsified asphalt and used for floor surfacing.

Elastex. A proprietary trade name for vinyl plasticisers as follows:
DCHP. dicyclohexyl phthalate.
IOP. diisooctyl phthalate.
20A. diisodecyl adipate.
28P. di-2-ethyl hexyl phthalate.
40P. butyl isodecyl phthalate.
50B. butyl cyclohexyl phthalate. (30).

Elastic Asbestos. See MOUNTAIN CORK.

Elastic Bitumen. See ELATERITE.

Elastic Cement (Lehner's Cement). A mixture of 5 parts rubber, 3 parts chloroform, and 1 part gum mastic. It is elastic and transparent.

Elastic Collodion. Flexible Collodion (*q.v.*).

Elastic Glue. A mixture of glue and glycerin. See GLYCERIN GLUE.

Elasticite. A rubber substitute made from corn oil.

Elasti-glass. A proprietary trade name for a vinyl copolymer used for belts, braces, raincoats, tobacco pouches, etc.

Elastite. A sulphurised oil rubber substitute. Also a proprietary flooring block made from asphalt, fibre, and fillers.

Elastolac. A proprietary trade name for a shellac derivative. It is water and alcohol soluble.

Elastolith. A synthetic resin. See BAKELITE.

Elastollan. A registered trade name for a range of thermoplastic polyurethanes. (485).

Elastomag 20. A trade mark for low-activity magnesium oxide. (888).

Elastomag 170. A trade mark for high-activity magnesium oxide. (888).

Elaterite (Elastic Bitumen, Mineral Caoutchouc, Helenite). A fossil resin, resembling asphaltum, found in some of the lead mines in Derbyshire. It contains 6–7 per cent. mineral matter, and is slightly soluble in ether.

Elaterite, Artificial. A proprietary product made from liquid bitumen and vegetable oils, then treatment with heat and pressure with sulphur chloride, saltpetre, and sulphur. Used for water-proofing and insulation.

Elaterium. The dried sediment from the juice of *Ecballium elaterium*. The active principle is elaterin.

Elayl. See OLEFIANT GAS.

Elbelan. Metal complex wool dyestuffs. (582).

Elbenyl. Dyestuffs for nylon. (582).

Elbon " Ciba." Cinnamolyl - *p* - oxy-phenylurea. Antipyretic and pulmonary disinfectant for the treatment of tuberculosis, hayfever, and affections of the respiratory tract.

Elcomet. A proprietary trade name for a steel containing chromium, silicon, copper, and nickel.

Elcor. A proprietary polyvinyl chloride. (667).

Elder Berries. The fruit of *Sambucus nigra*.

Elder Oil. See OIL OF CABBAGE.

Eldoform. Albumen tannate.

Eldoradoite. A trade name for a blue variety of quartz.

Electrargol. A form of electrolytic colloidal silver. A germicide.

Electraurol. A form of colloidal gold.

Electrical Castings Brass. See BRASS.

Electric Amalgam. Consists of 1 part each of tin and zinc, with 3 parts of mercury. It is the exciting material which is rubbed against the glass plate of an electrical machine.

Electric Bronze. An alloy of 87 per cent. copper, 7 per cent. tin, 3 per cent. zinc, and 3 per cent. lead.

Electric Calamine. Crystalline zinc silicate, $Zn_2SiO_4 + H_2O$. Synonym for Hemimorphite.

Electric Cement. An aluminous cement manufactured in the electric furnace. It consists essentially of $Al_2O_3.CaO$. The term is also used for Faraday's cement (*q.v.*).

Electricidal. Electro-colloidal iridium.

Electric Metal. See TELEGRAPH BRONZE.

Electriridol. Colloidal iridium.

Electrisil. A proprietary silicone rubber composition used for insulating conductors. (59).

Electrisil 758. A proprietary flame-retardant silicone rubber compound used to insulate high-voltage cables. (59).

Electrisil 9025. A proprietary silicone rubber compound used in applications where radiation and high temperatures may be encountered. (59).

Electrit. A trade mark for goods of the abrasive and refractory class, the essential constituent of which is crystalline alumina.

Electro-ammon. Ammonium sulphate produced in Germany by the interaction of ammonia and sulphur trioxide. The sulphur trioxide is produced by oxidising the sulphur obtained from coke-oven gas.

Electro-collargol. Colloidal silver.

Electrocortin. Aldosterone.

Electrocuprol. A form of colloidal copper.

Electro-filtros. A diaphragm material. It consists of grains of pure crystalline silica cemented together with a fused siliceous binding substance. Used in electrolytic processes.

Electrofine. A registered trade name for chlorinated paraffins for use as extenders in plasticised materials.

Electro-fused Cement. See FUSED CEMENT.

Electro-granodised Iron and Steel. A process for forming a rust-preventing coat on iron and steel. An alternating current plates a continuous coating of zinc phosphate.

Electrokali. A potash fertiliser introduced in Sweden. It is obtained as a residue from the manufacture of ferro-silicon. When felspathic rocks are mixed with iron turnings and graphite, and heated to 1800° C., ferro-silicon and a slag consisting mainly of potassium silicate is obtained. This slag, finely ground, is the article sold as " Electrokali."

Electrolon. See CARBORA.

Electrolytic Chlorogen (E.C.). A chlorinated soda prepared by the electrolysis of brine.

Electrolytic Copper. See COPPER, ELECTROLYTIC.

Electromartiol. A form of colloidal iron.

Electromercurol. A form of colloidal mercury.

Electronite. A safety explosive containing 75 per cent. ammonium nitrate, 5 per cent. barium nitrate, with wood meal and starch.

Electronite No. 2. An explosive consisting of 95 per cent. ammonium nitrate, and 5 per cent. wood meal and starch.

Electropalladiol. A form of colloidal palladium.

Electroplate. See ARGYROLITH and NICKEL SILVERS.

Electroplatinol. A form of colloidal platinum.

Electrorhodiol. A form of colloidal rhodium.

Electrorubin. A trade mark for abrasive and refractory materials. The essential constituent is crystalline alumina.

Electrose. A proprietary trade name for a shellac plastic.

Electroselenium. A form of colloidal selenium.

Electrotype Metal. An alloy of 93 per cent. lead, 4 per cent. antimony, and 3 per cent. tin.

Electrox. Zinc oxide for photocopying paper. (610).

Electrozone. A similar preparation to Chloros (sodium hypochlorite solution). A disinfectant.

Electrum. See NICKEL SILVERS.

Electrundum. A trade mark for materials of the abrasive type and consisting essentially of alumina.

Eledon. A German product. It consists of buttermilk in powder form obtained by means of an atomising apparatus. Used in dyspepsia.

Elektron. A registered trade mark used in connection with certain magnesium alloys containing up to about 10 per cent. of various alloying constituents, such as aluminium, zinc, and manganese. Its specific gravity is about 1·8. It is used in cast and wrought forms for aero engines and other purposes.

Elemi Resins. These are somewhat soft oleo-resins from species of *Canarium*, principally *Canarium luzonicum*. The chief elemi of commerce is Manila elemi. The fresh resin contains from 20–30 per cent. of essential oil which is composed mainly of hydrocarbons, the main one being the terpene phellandrene. The soft resin contains 15–20 per cent., volatile oil, an ash of 0·02–0·2 per cent., and an acid value of 17–25. The hard resin has 8–9 per cent. volatile matter, 0·2–1 per cent. ash, and an acid value of 15–28. Other elemi resins are Yucatan elemi from *Amyris plumieri*, Mexican elemi from *Amyris elemifera*, Brazilian elemi from *Protium heptaphyllum*, African elemi from *Boswellia freriana*, and East Indian elemi from *Canarium zephyrenum*.

Elemite. A proprietary trade name for a wetting agent and detergent. It is a combination of sulphonated oils and solvents.

Eleonorite. A mineral, $3Fe_2O_3.2P_2O_5.8H_2O$.

Elephant Bronze. An alloy of 85 per cent. copper, 10·5 per cent. tin, 2·75 per cent. zinc, 1·5 per cent. lead, and 0·1–0·2 per cent. phosphorus.

Elephant Nuts. See ACAJOU BALSAM.

Elephant-S Bronze. An alloy of 80·5 per cent. copper, 10·2 per cent. tin, 9 per cent. antimony, and 0·1–0·3 per cent. phosphorus.

Elestol. A proprietary preparation of chloroquine phosphate, prednisolone and aspirin. An anti-inflammatory agent. (341).

Eleuthera Bark. See SWEET BARK.

Elexar. A proprietary range of thermoplastic rubbers designed for use in the cable industry. (99).

Elhuyarite. A red allophane mineral.

Elianite I. An acid-resisting alloy, containing 82 per cent. iron, 15 per cent. silicon, and 0·6 per cent. manganese.

Elianite II. An acid-resisting alloy, consisting of 81 per cent. iron, 15 per cent. silicon, 0·5 per cent. manganese, 2·2 per cent. nickel, 0·8 per cent. carbon, and 0·06 per cent. phosphorus.

Eliasite. A mineral. It is a variety of gummite found near Joachimstahl.

Elie Ruby. A mineral. It is a Pyrope (*q.v.*).

Elimit. A proprietary preparation of orphenadrine hydrochloride and reserpine. A tranquilliser. (317).

Elinvar. A nickel steel containing 36 per cent. nickel, 46 per cent. iron, 12 per cent. chromium, 4 per cent. tungsten, and 1–2 per cent. manganese. It has a very low temperature coefficient of the elasticity modulus, and is used for the more delicate parts of watches.

Elite Fast. Wool dye stuffs, fast to washing. (582).

Elityran. A proprietary preparation of throid extract. (112).

Elixir Gabail. A proprietary preparation of valerian, strontium bromide, chloral hydrate. A sedative. (815).

Elixir Virvina. A proprietary preparation of thiamine hydrochloride, riboflavine, pyridoxine hydrochloride, nicotinamide, calcium glycerophosphate, potassium glycerophosphate, and magnesium glycerophosphate. A tonic. (310).

Elixir of Vitriol. See ACID ELIXIR OF VITRIOL.

Elkonite. A copper-tungsten alloy used for making welding dies. It has a Brinnell hardness of 225, a compression strength of 208,000 lb. per sq. in., and is not annealed at red heat.

Elkosin. A proprietary trade name for Sulphasomidine.

Ellagitannin. A variety of tannin found in divi-divi, knoppern, and myrobalans.

It resembles gallotannin, and is probably galloyl-ellagic acid.

Ellagite. A mineral. It is a variety of natrolite.

Ellestadite. A mineral. It is $2[Ca_5(S,Si)_3O_{12}(O,OH,Cl,F)]$.

Ellsworthite. A mineral. It is a hydrated calcium-uranium niobate, found in Ontario.

Elmarid. An alloy of 89 per cent. tungsten, 4·5 per cent. cobalt, 5·9 per cent. carbon, and 0·4 per cent. iron.

Elner's German Silver. A nickel-silver containing 57·4 per cent. copper, 26·6 per cent. zinc, 13 per cent. nickel, and 3 per cent. iron.

Elo. A phenolic moulding material. (668).

Elonite. A mineral. It is a hydrated magnesium silicate.

Elorine Chloride. A proprietary trade name for Tricyclamol chloride.

Eloxal. A proprietary trade name for an aluminium alloy which has been treated in an electrolytic chemical bath to give it a surface of oxide to render it resistant to corrosion. It is also called anodised aluminium.

Elpasolite. A mineral. It is cryolite, with potassium instead of sodium, $3KF.AlF_3$.

Elpidite. A mineral, $Na_2O.ZrO_2.6SiO_2.3H_2O$.

El Sixty. A proprietary trade name for bis-N, N′(2-benzothiazyl thiomethyl) urea. A thiozole accelerator.

Elsner Green. A pigment. It is a mixture of Genteles green (copper stannate), with fustic decoction.

Elsner's Reagent. A basic zinc chloride solution obtained by dissolving 500 grams zinc chloride and 20 grams zinc oxide in 425 c.c. water and warming. A solvent for silk.

Elterco. A proprietary preparation of terpin hydrate, eucalyptol, codeine phosphate and menthol. A cough linctus. (399).

Eltesol. Aromatic sulphonates for laundry detergents. (503).

Eltroxin. A proprietary preparation of thyroxine sodium. Thyroid hormone. (335).

Eludril. A proprietary preparation of CHLORHEXIDINE digluconate, CHLOR-BUTOL and chloroform. An antiseptic mouthwash. (633).

Elvacite 6026. A proprietary trade name for a low molecular weight copolymer of methyl methacrylate and *n*-butyl methacrylate as a 50 per cent. solution in toluene.

Elvanol. A proprietary polyvinyl alcohol. (10).

Elvax D. Proprietary dispersions of ionomers and vinyl resins. (10).

Elverite. A proprietary trade name for charcoal iron used for crushing mills.

El Varnish. A varnish containing 12 per cent. rosin ester, 16 per cent. alkyd resin, 21 per cent. linseed stand oil, 1 per cent. drying agent, and 50 per cent. benzoline.

El Zair. A mixture of 42 parts magnesium sulphate, 28 parts acetic acid, and oil of bergamot and water to make 100 parts.

Emanosal. A preparation containing radium. It is added to the bath to give a certain proportion of radium emanation. Used for rheumatism.

Emarex. A mineral rubber. See BITUMEN. The name is also applied to a German preparation containing pulsatilla, ignatia amara, cyclamen, cinifugar acimosa. Used in dysmenorrhœa.

Embacel. Kieselguhr for gas chromatography. (507).

Embacoid. A cement for cine films. (507).

Embadol. A proprietary trade name for Thiomesterone.

Embafume. Methyl bromide fumigant. (507).

Embamix. Potassium iodide mixtures. (507).

Embanox. A food grade antioxidant. (507).

Embarin. A preparation said to contain 6·6 per cent. sodium-mercury-salicyl-sulphonate and 0·5 per cent. dipara - anisyl - mono - phenetyl - guanidine-hydrochloride. Used as an injection in cases of syphilis.

Embatex. A cellulose acetate coated fabric. (507).

Embathion. An insecticide. (507).

Embazin. A proprietary preparation of sulphaquinoxaline sodium. A veterinary coccidiostat. (507).

Embedyne. Chlorodyne. (507).

Embelic Acid. An acid obtained from the fruit of *Embelia ribes*. Used in medicine for worms.

Embequin. A proprietary preparation of di-iodohydroxyquinoline. An anti-amœbic. (336).

Embolite. A mineral. It is a double chloride and bromide of silver, $Ag(Cl.Br)$.

Embramine. 2 - [1 - (4 - Bromodiphenyl)ethoxy] - NN - dimethylethylamine. Mebryl is the hydrochloride.

Embrithite. A mineral, $Pb_3Sb_2S_6$.

Embutox. A selective weedkiller. (507).

Embutramide. N - [2 - Ethyl - 2 - (3 - methoxyphenyl)butyl] - 4 - hydroxybutyramide.

Emdite. A proprietary trade name for a 50 per cent. w/w aqueous solution of ethyl-ammonium ethyl-dithio-carbonate, an alternative to hydrogen sulphide in qualitative inorganic analysis. (527).

Emepronium Bromide. Ethyl(3,3-diphenyl - I - methyl - propyl dimethyl-ammonium bromide). Ceteprin.

Emerald. A mineral. It is a variety of beryl with a rich green colour, a beryllium aluminium silicate. A precious stone.

Emerald, Brazilian. See BRAZILIAN EMERALD.

Emerald Bronze. An alloy of 50 per cent. copper, 49·7 per cent. zinc, and 0·3 per cent. aluminium.

Emerald Copper. Dioptase, $CuO.SiO_2$. H_2O, a mineral silicate of copper.

Emerald Green (Schweinfurth Green, Mitis Green, Vienna Green, Paris Green, Verdigris Green, Emperor Green, New Green, Mineral Green, Original Green, Vert Paul, Veronese Green, Parrot Green, Imperial Green, Kaiser Green, Meadow Green, English Green, Patent Green). Formerly the name Emerald green was applied to the hydrous oxide, $Cr_2(OH)_6$, but it is now given to cupric aceto-arsenite, $Cu(C_2H_3O_2)_3.3CuAs_2O_4$, made by mixing a hot solution of white arsenic in sodium carbonate, with the calculated quantity of copper sulphate solution. Also see CHROMIUM GREEN, CHROME GREEN, and OIL GREEN.

Emerald Green. See BRILLIANT GREEN.

Emerald Nickel (Zaratite). A mineral, $NiCO_3.6H_2O$.

Emerald Powder. An explosive. It is a 33-grain powder.

Emerald, Prismatic. See EUCLASE.

Emerald, Pyro. See CHLOROPHANE.

Emerald, Uralian. See ANDRADITE.

Emery. A trade mark for abrasive materials the essential constituent of which is alumina.

Emeryllite. See MARGARITE.

Emery's L–110. A proprietary form of azelaic acid, $C_7H_{14}(COOH)_2$. Melting-point 96° C. Decomposition greater than 380° C. It is used as a softener for alkyd resins.

Emery's L–114. A proprietary mixture of low molecular weight aliphatic acids in which pelargonic acid, C_8H_7COOH, predominates. It is used in the oil modification of alkyd resins.

Emeside. A proprietary preparation of ethosuximide. An anticonvulsant. (352).

Emetic Tartar. See TARTAR EMETIC.

Emetol. An emetine solution in olive oil.

Emetrol. A proprietary preparation of laevulose, dextrose and orthophosphoric acid. An anti-emetic. (330).

Emfor. An emulsifying and clouding agent. (591).

Emge. Magnesium hyposulphite in ampoules and tablets.

EMI-24. A trade mark. 2-ethyl-4-methylimid-azole. A curing agent for epoxy resins used in low proportions thus improving chemical resistance.

Emildine. A mineral. It is $8[Mn_3Al_2Si_3O_{12}]$.

Emilite. Synonym for Emildine.

Emin Red. A dyestuff. It is the sodium salt of methyl-benzenyl-amino-thioxylenol-azo - β - naphthol-sulphonic acid, $C_{20}H_{20}N_3O_4S_2Na$. Dyes wool red from an acid bath.

Emko. A proprietary preparation of BENZETHONIUM CHLORIDE and non-oxinol. A contraceptive foam. (809).

Emmensite. An explosive consisting of 5 parts Emmens acid, 5 parts ammonium nitrate, and 6 parts picric acid.

Emmonite. A mineral. It consists of a double carbonate of calcium and strontium.

Emmonsite. A mineral. It is $Fe_2(TeO_3)_3.2H_2O$.

Emodin. Trioxy-methyl-anthraquinone $C_{15}H_{10}O_5$. Used in medicine.

Emol Keleet. A purified fuller's earth.

Empee PP. A proprietary flame-retardant polypropylene. (669).

Emperor Alloy. A nickel-chromium alloy. It will resist a temperature of 1750°–1800° F.

Emperor Brass. An aluminium bronze. It consists of 60 per cent. copper, 20 per cent. aluminium, and 20 per cent. zinc.

Emperor Green. See EMERALD GREEN.

Empesin. A dentist's local anæsthetic of German origin.

Empholite. A mineral. It is a variety of diaspore.

Empicol. A wetting agent. It is an aliphatic alcohol sulphate and aliphatic alcohol alkyl ether sulphate. (503).

Empicol LX. A series of products manufactured from primary cut lauryl alcohol containing sodium lauryl sulphate of high solubility. Used for carpet shampoos.

Empicol LZ. A proprietary trade name for a series of sodium salts of lauryl alcohol. (210).

Empilan. Fatty acid esters, emulsifiers and foam stabilisers. A trade mark for an aliphatic ethoxylate. (503).

Empiphos. A synthetic detergent. It is an aliphatic phosphate. (503).

Empiquat. An alkyl dimethylbenzyl-ammonium chloride. (503).

Empire Cloth. A woven fabric, usually

cotton, impregnated with varnish. Used as an insulating material.

Empire Powder. An explosive. It is a 33-grain powder containing insoluble and soluble nitro-cotton, 10 per cent. metallic nitrate, and 7 per cent. pet. jelly,

Empire Stone. A stone which is similar to Victoria Stone (*q.v.*).

Empirin. Acetylsalicylic acid. See Aspirin. (514).

Empirin Compound. A proprietary preparation of aspirin, phenacetin and caffeine. An analgesic. (277).

Empiwax. A self-emulsifying wax. (503).

Emplecite. A mineral, $CuBiS_2$.

Emprazil. A proprietary preparation of pseudoephedrine hydrochloride, aspirin, and caffeine. A cold remedy. (277).

Empressite. A mineral. It is a silver telluride, AgTe.

Empyreal (Fire-air). The name given to oxygen by Scheele.

Empyroform. A condensation product of birch tar and formaldehyde. Used externally in the treatment of skin diseases, such as acute and chronic eczema.

Emser Salt, Artificial. Consists of 90 parts sodium chloride, 220 parts sodium bicarbonate, 2 parts sodium sulphate, and 4 parts potassium sulphate.

Emtryl. A proprietary preparation of DIMETRIDAZOLE. A veterinary antiprotozoan. (507).

Emugenol. A German preparation. It is a 2 per cent. emulsion of bromoform in cod-liver oil. Used in the treatment of coughs.

Emulax. A proprietary preparation of dioctyl sodium sulphosuccinate and casanthranol. A laxative. (338).

Emulgeen P. A proprietary trade name for potassium ricinoleate. (202).

Emulgeen S. A proprietary trade name for sodium ricinoleate. (202).

Emulgen. A jelly-like mass used for the rapid emulsification of oils and resins. It contains tragacanth, gum arabic, pittoporad, glycerin, alcohol, and water.

Emulphor. A proprietary trade name for a condensation product of ethylene oxide and an organic acid.

Emulsamin. A proprietary trade name for menthol diurethane, a wetting agent and detergent.

Emulsene. Emulsifying agents. (504).

Emulsifier L.W. Cyclohexyl ammonum oleate. (513).

Emulsin. A ferment. It decomposes the glucoside, amygdalin, into grape sugar, benzaldehyde, and hydrocyanic acid.

Emultex 307 and 328. A proprietary range of unplasticised vinyl acetate homopolymer emulsions stabilised with polyvinyl alcohol, used in the manufacture of adhesives. (481).

Emultex AC 431. A proprietary vinyl acetate/acrylic ester copolymer emulsion used in general-purpose emulsion paints. (481).

Emylcamate. 1 - Ethyl - 1 - methylpropyl carbamate. Striatran.

Enal. A proprietary epoxy-novolac resin. (424).

Enalite. A mineral. It is $(Th,U)O_2 . nSiO_2 . 2H_2O$.

Enamelled Iron-granite. See AGATE WARE.

Enameloid Cloisonné. A proprietary trade name for cellulose acetate and methyl methacrylate (engraved plastic).

Enamels. Usually consist of oil varnishes, into which pigments have been ground.

Enamel White. See LITHOPONE.

Enanth. Nylon 7.

Enargite (Guayacanite, Luzonite). A copper ore, $4CuS.Cu_2S.As_2S_3$.

Enavid 5mg. A proprietary preparation containing norethynodrel and mestranol. (285).

Enavid-E. A proprietary preparation containing norethynodrel and mestranol. (285).

Enbico Chrome Black F. A dyestuff. It is a British equivalent of Diamond black F.

Enbico Direct Brown B. A British equivalent of Benzo brown G.

Enbico Direct Brown G. A dyestuff. It is a British brand of Benzo brown B.

Enbico Direct Fast Pink Y Conc. A dyestuff. It is equivalent to Salmon red (Berlin Aniline Co.), and is of British manufacture.

Enbico Direct Green B, B Extra Conc. Dyestuffs. They are British equivalents of Diamine green B.

Enbico Direct Yellow G. A British brand of Chrysamine G.

Enbico Direct Yellow R. A dyestuff. It is a British equivalent of Direct yellow R.

Enbucrilate. A surgical tissue adhesive. Butyl 2-cyanoacrylate. HISTOACRYL.

Enceladite. Synonym for Warwickite.

Encem Steel. A proprietary trade name for a nickel-chromium-molybdenum steel.

Encephabol. A proprietary trade name for the dihydrochloride of Pyritinol.

Encynex. A proprietary preparation of *p*-aminobenzoic acid, sodium salicylate, sodium bicarbonate and phenacetin. An analgesic. (815).

Endellionite. See BOURNONITE.

Endermol. A compound ointment

vehicle, containing hydrocarbons of the paraffin series, and stearic acid amide.

Endlichite. A mineral, $Pb_5Cl[(As.V)O_4]_3$.

Endocaine. A proprietary trade name for Pyrrocaine.

Endocrocine. The orange-yellow colouring matter isolated from *Nephioniopsis endocrocea*, a lichen growing in Japan. It is a hydroxyanthraquinone, $C_{16}H_{10}O_7$.

Endotryptase. A proteolytic enzyme.

Endoxana. A proprietary preparation of cyclophosphamide. An antimitotic. (319).

Endrine. A proprietary preparation of ephedrine, menthol, camphor, and eucalyptol. A nasal decongestant. (245).

Enduro Alloys. Proprietary corrosion-resisting alloys of iron with chromium, or with nickel and chromium. Enduro A contains iron with from 16·5–18·5 per cent. chromium ; Enduro KA2 contains iron with 17–20 per cent. chromium and 7–10 per cent. nickel ; Enduro S has iron with 12·5–14·5 per cent. chromium.

Endurol. Vat dye colours. (599).

Enduron. A proprietary preparation of methylclothiazide. A diuretic. (311).

Enduronyl. A proprietary preparation of methylclothiazide and deserpidine. (311).

Energin. An organic preparation of proteids and acids obtained from dried cod livers. Recommended as a tissue reconstructor.

Energine. A specially pure refined gasoline, said to be free from paraffin, mineral grease, sulphur, and dirt, also to be less inflammable than ordinary gasoline. It has a specific gravity of 0·709 at 60° F.

Eneril. A proprietary preparation of paracetamol. An analgesic. (271).

Enesol. The mercury salt of arsenic-salicylic acid, $(C_6H_4(OH)COO)_2AsHg$, containing 38 per cent. mercury and 14 per cent. arsenic. It is employed in the arseno-mercurial treatment of syphilis.

Engelhardtite. Zircon (*q.v.*).

Engenamel. An oil-resisting finish. (512).

England, Salts of. See EPSOM SALTS.

Englate. A proprietary preparation of theophylline sodium glycinate. A bronchial antispasmodic. (279).

English Bearing Metal. Anti-friction and fitting metal. It contains usually 53 per cent. tin, 33 per cent. lead, 10·5 per cent. antimony, and 2·5 per cent. copper.

English Blue. See MOUNTAIN BLUE.

English Brown. See BISMARCK BROWN.

English Green. See EMERALD GREEN and GREEN VERMILION.

Englishite. A mineral. It is $KCa_2Al_4PO_4)_4(OH)_5.4\frac{1}{2}H_2O$.

English Metal. A jeweller's alloy containing 88 parts tin, 2 parts copper, 2 parts brass, 2 parts nickel, 1 part bismuth, 8 parts antimony, and 2 parts tungsten.

English Powder. See POWDER OF ALGAROTH.

English Red. See INDIAN RED.

English Salt (Smelling Salt). Ammonium carbonate, $(NH_4)_2CO_2$. Smelling salts are usually mixtures of ammonium carbonate or sometimes ammonium chloride with essential oils. White smelling salts consist of a coarsely powdered ammonium carbonate perfumed with oils of bergamot, lavender, clove, rose, or cinnamon. Violet smelling salts consist of ammonium carbonate moistened with a tincture of orris root, ammonia, and violet extract.

English Turpentine. See AMERICAN TURPENTINE.

English Vinegar. A preparation containing acetic acid, camphor, cinnamon oil, clove oil, and lavender oil. Used for smelling as a restorative.

English Vitriol. Ferrous sulphate, $FeSO_4$.

English White. See CHALK.

English White Bearing Metal. An antifriction metal, containing 77 per cent. tin, 15 per cent. antimony, and 8 per cent. copper.

English Yellow. See TURNER'S YELLOW and VICTORIA YELLOW.

Engobe. A fusible mixture of clay, felspar, and silica. Used for the manufacture of glazes on pottery.

Engraver's Brass. See BRASS.

Enhydros. See WATER STONE.

Enlactol. A white powder prepared by evaporating milk. It contains 33 per cent. albumin, 46 per cent. fat, 14 per cent. carbohydrates, and 5 per cent. saline matter.

Enlax. A proprietary preparation. It is a phenol-phthalein preparation.

Enophite. A serpentine mineral. It is $(Mg,Ca,Fe'',Fe''')_4(Si,Al)_3O_{10}.4H_2O$.

Enoxolone. 3 - Hydroxy - 11 - oxo-olean-12-en-30-oic acid. Biosone.

Enpac. A proprietary preparation containing resistant strains of Lactobaccillus Acidophilus. (394).

Enpiprazole. A psychotrophic drug. 1 - (2 - Chlorophenyl) - 4 - [2 - (1 -methyl-pyrazol - 4 - yl)ethyl]piperazine. "H 3608" is currently under clinical trial as the dihydrochloride.

Ensecote S. A proprietary trade name for a modified epoxy resin coating material for high temperature stoving or spraying.

Enstatite. See also AMBLYSTEGITE. A mineral. It is $16[MgSiO_3]$.

Entacyl. A proprietary preparation of piperazine adipate. An antihelmintic. (182).

Entair. Capsules and syrup containing theophylline and guaiphenesin. Reduces secretion of mucus in chronic bronchitis. (182).

Entair-A. A proprietary preparation of theophylline, guaiphenisin and ephedrine hydrochloride. A bronchial antispasmodic. (182).

Entair Expectorant. A preparation of diphenhydramine hydrochloride and guaiphenesin. (527).

Entamide. A proprietary trade name for diloxamide. An amoebicide. (502).

Enterfram. A proprietary preparation containing framycetin sulphate and light kaolin. An antidiarrhœal. (188).

Enteromide. A proprietary trade name for the calcium salt of Sulphaloxic Acid. An antibiotic. (256).

Enterorose. A preparation of vegetable albumin, impregnated with beef extract and diastase. It contains 18 per cent. albumin, 11 per cent. fat, 3·8 per cent. salts, and 59 per cent. carbohydrates, and is used in the treatment of chronic constipation.

Enterosan. A proprietary preparation of di-iodohydroxyquinoline chlorodyne, tincture of belladonna and kaolin. An antidiarrhœal. (836).

Entero-vioform. A specific therapeutic agent for the treatment of acute and chronic amœbic dysentery and other gastro-enteric disturbances occurring in tropical and subtropical countries. The active principle is iodochlorohydroquinolene.

Endotin. A proprietary preparation. It is a 20 per cent. solution of hexamethylene-isopropanol-biniodide.

Entobex. A proprietary trade name for Phanquone.

Entomycin. A proprietary preparation of neomycin sulphate and kaolin. An antidiarrhœal. (344).

Entramin. 2-Amino-5 nitrothiazole premix. (507).

Entramin A. Acetamidonitrothiazole premix. (507).

Entrosalyl. A proprietary preparation of sodium salicylate in an enteric coated tablet. Antirheumatic drug. (248).

Entrosalyl Standard. A proprietary preparation of enteric coated sodium salicylate. (248).

Envacar. A proprietary preparation of guanoxan sulphate. An antihypertensive. (85).

Enysite. A mineral. It is a basic sulphate of aluminium and copper. It is $Cu_2Al_6SO_4(OH)_{20}$.

Enzopride. See NADIDE.

Enzypan. A proprietary preparation of pepsin, pancreatin, and bile. Digestive enzyme supplement. (286).

Enzytol. A 10 per cent. aqueous solution of choline borate.

Eolite. Synonym for Realgar.

Eosamine B. A dyestuff. It is the sodium salt of p-cresol-methyl-ether-azo-α-naphthol-disulphonic acid, $C_{18}H_{14}N_2O_8S_2Na_2$. Dyes wool and silk bluish-red.

Eosin (Eosin A, Eosin B, Eosin C, Eosin A Extra, Eosin DH, Eosin GGF, Eosin G Extra, Eosin 3J, Eosin JJS, Eosin G, Eosin KS, Eosin Yellowish, Water Soluble Eosin, Acid Eosin, Eosin 4J Extra). The alkali salts of tetra-bromo-fluoresceine, $C_{20}H_6O_5Br_4Na_2$. Dyes wool and silk yellowish-red. Used for making pigments.

Eosin A. See EOSIN.

Eosin A Extra. See EOSIN.

Eosin B. The ammonium salt of tetra-bromo-fluoresceine. Also see EOSIN and EOSIN BN.

Eosin 10B. See Phloxine.

Eosin BB. See SPIRIT EOSIN.

Eosin BS. A dyestuff. It is a British equivalent of Eosin BN.

Eosin, Bluish. See ERYTHROSIN.

Eosin BN (Eosin B, Eosin BW, Eosin DVH, Eosin Scarlet, Eosin Scarlet BB, Eosin Scarlet B, Daphnin, Scarlet J, JJ, and V, Nopalin, Imperial Red, Methyl Eosin, Saffrosine, Lutecienne, Kaiser Red, Kaiserroth, Eosin BS). The potassium or sodium salt of di-bromo-dinitro-fluoresceine, $C_{20}H_6N_2O_9Br_2K_2$. Dyes silk and wool bluish-red.

Eosin BW. See EOSIN BN.

Eosin C. See EOSIN.

Eosin DH. See EOSIN.

Eosin DVH. See EOSIN BN.

Eosin G. See EOSIN.

Eosin 3G. See EOSIN ORANGE.

Eosin G Extra. See EOSIN.

Eosin GGF. See EOSIN.

Eosin J. See ERYTHROSIN.

Eosin 3J and 4J Extra. See EOSIN.

Eosin JJS. See EOSIN.

Eosin KS. See EOSIN.

Eosin Orange (Eosin 3G, Salmon Pink). Varying mixtures of di- and tetra-bromo-fluoresceine, used in dyeing.

Eosin S. See SPIRIT EOSIN.

Eosin Scarlet. See EOSIN BN.

Eosin Scarlet B and BB. See EOSIN BN.

Eosin Soluble in Spirit. See ERY-THRINE and SPIRIT EOSIN.

Eosin, Water Soluble. See EOSIN.

Eosin, Yellowish. See EOSIN.

Eosin YS. A British brand of Eosin.

Eosite. A mineral. It is a vanado-molybdate of lead.

Eosolate. Silver acetyl-guaiacol-tri-sulphonate. An antiseptic.

Eosolsaures Salz. A preparation similar to thiocoll. It is probably a salt of acetyl-creosote-trisulphonic acid.

Eosote. Creosote valerinate, a mixture of valeric acid esters of the phenols contained in creosote. An internal antiseptic.

Eosphorite. A mineral, $(Fe.Al.Mn)_2$ $O_3P_2O_5.5H_2O$.

Epanutin. A proprietary preparation of phenytoin sodium. An anticonvulsant. (264).

Epargol. A form of colloidal silver in ampoules.

Eparseno (Pomaret's 132). A stabilised solution of amino-arseno-phenol. An antisyphilitic.

EPDM. A commonly-used "shorthand" expression for the ethylene propylene diene methylene group.

Ephedrine. A new alkaloid obtained from the Chinese plant Ma Huang. It is α-phenyl-β-methyl-amino-propanol, $C_6H_5.CHOH.CH.CH_3NH.CH_3$. It is used as a mydriatic and as an atropine substitute in ophthalmology. It melts at 38–40° C., and is soluble in alcohol, ether, and chloroform.

Ephetonin. A German product. It is the hydrochloride of phenyl-methyl-amino-propanol. Its action is similar to that of ephedrine, but it is made synthetically. Ephedrine is læva-rotatory, but ephetonin is optically inactive.

Ephos. A basic phosphate, containing 60–65 per cent. tricalcium phosphate.

Ephpect Elixir. A proprietary preparation containing ephedrine hydrochloride, ammonium chloride, sodium citrate and extract of ipecacuanha. A cough linctus. (342).

Ephrelix. Ephedrine elixir. (591).

Ephretuss. Ephedrine syrup. (591).

Ephynal. A proprietary preparation of Vitamin E. A vasodilator. (314).

Epibond 1522 A and B. A proprietary range of epoxy adhesives used in repair work. (670).

Epiboulangerite. A mineral, $Pb_3Sb_2S_3$.

Epicaine. A proprietary preparation. It is a solution of epinine and cocaine hydrochloride.

Epi-camphor. 3-Keto-camphane, C_{10} $H_{16}O$.

Epicarin (Naphoxytol). β-Oxy-naphthyl o-oxy-m-toluic acid, $HO.C_{10}H_6$.

$CH_2.C_6H_4(OH)COOH$. Used as an antiseptic in skin diseases, such as scabies and mange.

Epichlorhydrin. Chloro-propylene oxide.

Epichlorite. A chloritic mineral found in Harzberg.

Epicillin. An antibiotic currently under clinical trial. 6-(D-α-Aminocylcohexa-1, 4-dien-1-ylacetamido)penicillanic acid.

Epicure. A trade name for polyamide curing agents for epoxy resins. (99).

Epi-Cure 8515. A proprietary amido-amine curing agent for epoxy resins. (432).

Epidermin. An ointment made from white wax, water, gum arabic, and glycerin. The term is also applied to a mixture of fluor-pseudo-cumene and difluor-diphenyl, in the form of an ointment for burns.

Epididymite. A mineral, $HNaBeSi_3O_8$.

Epidote. A mineral. It is a complex silicate of aluminium, iron, and calcium, $4CaO.3(Al.Fe)_2O_3.6SiO_2.H_2O$.

Epiethylin. Ethyl-glycide ether.

Epiflex. A proprietary trade name for an epoxy resin expansion jointing material. (150).

Epigenite. A mineral, $Cu_7As_2S_{12}$.

Epiglaubite. An impure calcium phosphate.

Epiglo. A proprietary trade name for a cold cure liquid epoxy resin coating material. (154).

Epihydrin. Propylene oxide, C_3H_6O.

Epiianthinite. A mineral. It is $8[UO_3.2H_2O]$.

Epikote. A proprietary trade name for a series of epoxy resins whose characteristics may be modified by hardeners and other additives. (99).

Epikote DX-209-B-80. A proprietary 80% solution of epoxy resin in methyl ethyl ketone. It is used with EPIKURE 3400 as a curing agent. (99).

Epikote DX-210-B-80. A proprietary 80% solution of epoxy resin in methyl ethyl ketone for use in work involving carbon fibres. It is cured with EPIKURE 3400. (99).

Epikote DX-231-B-91. A proprietary solution containing 91% epoxy resin in methyl ethyl ketone for use in work involving carbon film. It is cured with EPIKURE 3400. (99).

Epikure 3400. A proprietary curing agent for epoxy resins. (99).

Epilim. A proprietary preparation of sodium valproate. An anti-convulsant. (243).

Epilink. Epoxy curing agents. (583).

Epilok. Epoxy curing agents. (583).

Epilon. A proprietary trade name for an epoxy resin cement. (150).

Epimillerite. A mineral. It is moreno-site.

Epinalin. A proprietary preparation of adrenaline and ephedrine in solution.

Epinephram. Adrenalin (q.v.).

Epinephrine. Adrenalin (q.v.).

Epinine. A proprietary preparation of 3 : 4 - dihydroxy - phenyl - ethyl-methyl-amine, $(OH)_2C_6H_3.CH_2.CH_2.NH.CH_3$. It is allied to adrenaline in formula and physiological action.

Epioestriol. OEstra - 1,3,5(10) - triene - 3,16,17-triol 16-epiOEstriol. Actriol.

Epiosine. 1-Methyl-4 : 5-diphenylene-imidazole, $CH(C_6H_4.C.N(CH_3)_3)(C_6H_4.C.N)$.

Epiphassol. A viscous oily liquid containing naphthene-sulphonic acids. It is a similar preparation to Kontakt, and is used for cleaning cotton fabrics.

Epiphen. A proprietary range of epoxy resins. (168).

Epirenan. See ADRENALIN.

Epirez 501. Butyl glycidyl ether. A proprietary reactive diluent for epoxy resin systems. (432).

Epirez 502. A proprietary aliphatic diepoxide. (432).

Epirez 520C. A proprietary epoxy resin of the bisphenol A type. (432).

Episol. A proprietary preparation of chlorodiethylaminoethoxyphenylbenz-thiazole. (Halethazole). An antifungal agent. (280).

Epistilbite. A mineral, $CaO.Al_2O_3.6SiO_2.5H_2O$.

Epistolite. A mineral. It is a sodium titano-silico-columbate, $Na_7Ti(CbO)_3.(SiO_4)_5.3\frac{1}{2}H_2O$.

Epitar. An epoxy resin additive. (616).

Epitate. An epoxy resin additive. (616).

Epithelan. A German ointment substance containing pure carbon and vaseline.

Epithermol. Amino-azo-toluene. Used for healing the skin.

Epithiazide. 6 - Chloro - 3,4 - dihydro - 3 - (2,2,2 - trifluoroethylthiomethyl) - benzo - 1,2,4 - thiadiazine - 7 - sulphona-mide 1,1-dioxide.

Epithol Gold (Epithol Silver). Finely powdered copper and tin, which have been used in veterinary practice as a local application for wounds.

Epithol Silver. See EPITHOL GOLD.

Epivax. A canine distemper vaccine. (514).

Epocast 8408. An epoxy resin system giving tough rubber-like castings.

Epocuprol. A form of colloidal copper in ampoules.

Epodyl. A proprietary preparation of ethoglucid. An antimitotic. (2).

Epok. A proprietary trade name for synthetic resins and solutions of the phenolic and cresylic type.

Epok V8007. The proprietary trade name for a vinyl acetate/acrylate copolymer emulsion used to manufacture emulsion paints. (12).

Epolac. A proprietary range of epoxy resins. (671).

Epolite. A proprietary epoxy resin. (714).

Epon. A proprietary epoxy resin. (303).

Epon 8280. A proprietary liquid epoxy resin for use in filled compounds. (99).

Eponite. A decolorising material for sugar juices. It consists of carbon made from woods, such as pine or cedar, and poplar or willow.

Epontol. A proprietary preparation of propanidid. Short duration general anæsthetic. (112).

Epophen. A proprietary trade name for epoxy resins and hardeners, hot and cold setting. (139).

Eposir. A proprietary range of epoxy resins. (715).

Eposis. Epoxy resins.

Epotuf. A registered trade name for a hardener for epoxy resins. (48).

Epoxol G-5. Linseed oil epoxy resin.

Epoxol 7-4. Soybean oil epoxy resin.

Epoxol 80, 130. A proprietary trade name for vinyl plasticisers manufactured from soya bean. (107).

Epox-S. A proprietary trade name for an epoxidised triester vinyl plasticiser. (106).

Eppy. A proprietary preparation of adrenaline used to treat glaucoma. (824).

Epronal. Ethyl-propyl-acetyl-urethane. A hypnotic.

Epsikapron. A proprietary preparation of 6-aminocaproix acid, used in the treatment of fibrinolysis. (318).

Epsilon Acid (E. Acid). 1-naphthol-3, 8-disulphonic acid. 1-naphthylamine-3, 8-disulphonic acid. Dyestuff intermediates.

Epsoline. Phenol-phthalein, $C_6H_4(OC)C.(C_6H_4.OH)_2O$.

Epsomite (Richardite). Magnesium sulphate, $MgSO_4.7H_2O$. It occurs naturally.

Epsom Salts (Salts, Salts of England, Hair Salt, Bitter Salt). Magnesium sulphate, $MgSO_4.7H_2O$.

Epsom Salts, Effervescent. See EFFERVESCENT EPSOM SALTS.

Epsom Salts, Mock. Needle crystals of sodium carbonate.

Eptoin. Soluble phenytoin. (502).

Epurite. A mixture of bleaching powder, iron sulphate, and copper sulphate. It is used for the production of oxygen, which gas is obtained by action of water.

Equadiol. A proprietary preparation of meprobamate and ethinylœstradiol, used for control of menopause. (245).

Equagesic. A proprietary preparation of ethoheptazine citrate, meprobamate, asprin, and calcium carbonate. An analgesic. (245).

Equalised Guano. Natural guanos, blended or mixed with ammonium salts, to obtain definite proportions of nitrogen and phosphorus. A fertiliser.

Equanil. A proprietary preparation of meprobamate. A sedative. (245).

Equaprin. A proprietary preparation containing meprobamate and aspirin. An analgesic. (245).

Equatrate. A proprietary preparation of meprobamate and pentaerythrityl tetranitrate, used in treatment of angina pectoris. (245).

Equigard. A proprietary preparation of DICHLORVOS. A veterinary insecticide.

Equimate. A proprietary preparation of FLUPROSTENOL sodium used to treat infertility in animals.

Equionic. Sanitizer/detergent compounds. (547).

Equipose. A proprietary preparation of hydroxyzine pamoate. A sedative. (85).

Equisetic Acid (Citridic Acid). Aconitic acid, $C_3H_3(COOH)_3$.

Equivert. A proprietary preparation of buclizine hydrochloride. An antinauseant. (85).

Era 147. A proprietary trade name for a steel containing 0·22 per cent. carbon, 0·20 per cent. silicon, 0·04 per cent. sulphur, 0·04 per cent. phosphorus, 0·50 per cent. manganese, 5·00 per cent. chromium and 0·50 per cent. molybdenum. It is used for forging steel pressure vessels for service with hydrogen. (155).

Era 164. A proprietary trade name for a steel containing 0·20 per cent. carbon, 0·25 per cent. silicon, 0·04 per cent. sulphur, 0·04 per cent. phosphorus and 1·5 per cent. manganese. It is used in the manufacture of forged steel pressure vessels for use with intermediate pressures. (155).

Era CR1. A proprietary trade name for a steel containing 0·06 per cent. carbon, 0·30 per cent. silicon, 0·04 per cent. sulphur, 0·04 per cent. phosphorus, 1·0 per cent. manganese, 18·50 per cent. chromium and 9·00 per cent. nickel. It is used in the manufacture of forged steel pressure vessels with good corrosion resistance. (155).

Era CR15 (CB). A proprietary trade name for a steel containing 0·06 per cent.

carbon, 0·50 per cent. silicon, 0·04 per cent. sulphur, 0·04 per cent. phosphorus, 1·00 per cent. manganese, 19·00 per cent. chromium, 10·00 per cent. nickel and 0·6 per cent. niobium. (155).

Era Chrome Dark Blue B. A dyestuff. It is a British equivalent of Palatine chrome blue.

Eraclene. A registered trade mark for low density polyethylene. (19).

Eradite. Sodium hyposulphite, $Na_2S_2O_4$, incorrectly termed "hydrosulphite."

Era Dyes. These are registered trade names for certain British dyestuffs.

Era Metal. A steel containing 21 per cent. chromium and 7 per cent. nickel.

Eranol. A form of colloidal iodine.

Erasine. See ABIETENE.

Erasol. Stripping and printing agents. (582).

Eraydo. A proprietary trade name for a zinc alloy containing copper and silver for use as radio screens.

Erbolin. A proprietary preparation. It is a standardised ergot powder.

Ercal. A proprietary name for ergotamine tartrate. (709).

Ercedylate. Bismuth cacodylate, a compound of bismuth and arsenic. An antisyphilitic.

Ercerhinol. A colloidal silver.

Erdmannite (Michaelsonite). A mineral. It is a silicate of cerium and yttrium metals and calcium, with zirconium, beryllium, thorium, aluminium, and iron, of Norway.

Erdmann's Reagent. Made by adding 40 c.c. of concentrated sulphuric acid to 20 drops of a solution containing 10 drops of nitric acid (specific gravity 1·153) and 20 c.c. of water. Used in testing for alkaloids.

Erdol. Quinoline-sulpho-salicylate.

Eremeyevite (Eichwaldite). A mineral. It is an aluminium borate. It is 12[$AlBO_3$].

Erepton. A meat preparation, in which the proteins are completely resolved into amino acids. It contains 12 per cent. nitrogen, and has been recommended for rectal feeding.

Ereugol. A German product. It contains atropine, papaverine, caffeine, camphor, and ether in the form of an emulsion. It is used in asthma by intramuscular injection.

Ergamine (Histamine). A constituent of ergot. It is β-iminazolyl-ethyl-amine, CH : N.NH.CH.C.CH$_2$.CH$_2$.NH$_2$. Used in medicine to decrease blood pressure.

Erganol. Dibenzyl ether. A softening agent for cellulose esters.

Ergine. A trade name for the liquid hydrocarbons (homologues of benzene), obtained by the dry distillation of black and brown tar oils.

Ergodex. A proprietary ergot preparation containing 0·03 per cent. ergotoxine.

Ergodryl. A proprietary preparation of ergotamine tartrate, caffeine citrate and diphenhydramine hydrochloride, used for migraine. (264).

Ergol. A proprietary plasticiser. It is stated to be benzyl benzoate, C_6H_5.COO.CH$_2$.C$_6$H$_5$. It has a specific gravity of 1·2–1·26 and a boiling-point of 323° C. It dissolves ester gum, coumarone, and hard resins, and is used as a softening agent for cellulose esters.

Ergometrine. A water-soluble alkaloidal substance isolated from ergot.

Ergon Carbon. A special kind of carbon, used for arc-lights which are used in various light cures.

Ergorone. An irradiated ergosterol. A vitamin D preparation.

Ergosterol. A German product. It is a body related to cholesterol, which after radiation with ultra-violet light has as good an effect in cases of rickets as cod-liver oil. It comes into trade as Vigantol.

Ergot. A dark coloured fungus, which attacks damp rye and other grasses, and when contained in flour, causes ergotism. It is a mixture of the alkaloids, ergotoxine, p-hydroxy-phenylethylamine, isoamylene, and β-aminoethyl-glyoxal, and is used in midwifery for causing contractile action of the pregnant uterus.

Ergotamine. p-Hydroxy-phenyl-ethylamine.

Ergotine. *Extractum ergotæ B.P.*

Ergot Oil. See OIL OF ERGOT.

Ergotoxine. Purified cornutine.

Ericin. See MESOTAN.

Ericol. Guaiacol acetate.

Ericon. A proprietary trade name for a phenol-formaldehyde synthetic resin.

Erie Blue GG. A dyestuff which dyes cotton or wool, afterwards treated with copper sulphate.

Erie Red B Conc. See BENZOPURPURIN 4B.

Erie Yellow XC. See CHRYSAMIN G.

Erika B. A dyestuff. It is the sodium salt of methyl-benzenyl-amino-thioxylenol - azo - α - naphthol - disulphonic acid, $C_{26}H_{19}N_3O_7S_3Na_2$. Dyes unmordanted cotton rose pink.

Erika G. An azo dyestuff from dehydrothio-m-xylidine and β-naphthol-γ-sulphonic acid. It dyes in a similar way to Erika B.

Erikite. A mineral. It is a phosphosilicate of cerium, calcium, aluminium, potassium, sodium, with thorium dioxide, of Greenland.

Erinite. A mineral, $2CuO.As_2O_3.4H_2O$.

Erinofort. A proprietary cellulose acetate plastic.

Erinoid. A proprietary trade name for casein-formaldehyde synthetic resin insulating material. (716).

Erio. Acid dyes. (543).

Eriochalcite. A mineral. It contains cupric chloride, $CuCl_2$.

Eriochlorine A, B, CB, BB. Green dyestuffs similar to Erika B.

Eriochrome Azurole B. A dyestuff obtained by the condensation of o-chloro-benzaldehyde and o-cresotinic acid, then oxidation.

Eriochrome Black A (Chrome Black A, Chrome Fast Black A, Diadem Chrome Black A). A dyestuff prepared from diazotised 8-nitro-1-amino-β-naphthol-4-sulphonic acid and β-naphthol.

Eriochrome Black T. A dyestuff obtained by the diazotisation of 8-nitro-1-amino-β naphthol-4-sulphonic acid and β-naphthol.

Eriochrome Blue-black B. A dyestuff. It is equivalent to Chrome fast cyanine G.

Eriochrome Blue-black R. See PALATINE CHROME BLUE.

Eriochrome Cyanine. A dyestuff obtained by the condensation of benzaldehyde, o-sulphanilic acid, and o-cresotinic acid, and oxidation.

Eriochrome Phosphine. A dyestuff prepared from diazotised p-nitraniline-o-sulphonic acid and salicylic acid.

Eriochrome Red B (Diadem Chrome Red L3B). A dyestuff prepared from diazotised 1 - amino - β - naphthol - sulphonic acid and phenyl-methyl-pyrazolone.

Eriochrome Verdone A. A dyestuff prepared by diazotising sulphanilic acid, combining it with m-amino-p-cresol, diazotising the product, and combining it with β-naphthol. It dyes wool claret red shades from an acid bath, which upon chroming becomes blue-green.

Eriocyanine A. A dyestuff. It is the sodium salt of tetramethyl-dibenzyl-rosaniline-disulphonic acid, $C_{37}H_{37}N_3O_7S_2Na_2$. Dyes wool reddish-blue from an acid bath.

Eriocyanine B. A similar dyestuff to the above.

Erioglaucine A. A dyestuff. It is the acid ammonium salt of the trisulphonic acid of diethyl-dibenzyl-diamino-triphenyl-carbinol, $C_{35}H_{38}N_4O_9S_3$. Dyes wool and silk greenish-blue from an acid bath.

Erioglaucine RB, BB, B, J, GB. Dyestuffs similar to above.

Erionyl. Acid dyes for polyamide fabrics. (543).

Eriopon. Surface active agents. (543).

Eri Silk. The product of the caterpillar of *Attacus rincini*, of Assam.

Erlangen Blue. See PRUSSIAN BLUE.

Erlau Green. A pigment prepared in a similar manner to the old Brunswick green, by adding a mixture of copper sulphate and sodium chloride to milk of lime, washing the precipitate, and treating with a solution of neutral potassium chromate. Vienna white or heavy spar is added to it.

Erlicki's Solution. A hardening agent used in microscopy. It consists of potassium bichromate 2·5 parts, calcium sulphate 1 part and water 100 parts.

Ermite. A proprietary trade name for a synthetic resin.

Ernite. Synonym for Grossular.

Ernolith. A product obtained by treating waste yeast with formaldehyde, drying, and grinding with or without the addition of tar, tar oils, sulphur, and pigments, and compressing in into moulds. It is used as a substitute for Bakelite and Celluloid.

Ernutin. A proprietary medicinal product. It is synthetic ergot.

Erodin. A culture medium of peptonised gelatinous tissue, with a special mixed culture of selected bacteria. Used for " bating " skins.

Errite. A mineral. It is a variety of parsettensite.

Ersbyite. A mineral. It is a silicate of calcium and aluminium.

Erthrisin (Arthrisin). Acetyl-salicyl-amide. Used in the treatment of rheumatism.

Erubescite. See PURPLE COPPER ORE.

Erukku Erukkam. See AK MUDAR.

Ervadiol 220S. A proprietary unsaturated polyester resin. (717).

Ervamine. A proprietary trade name for melamine-formaldehyde.

Ervamix. A proprietary trade name for a fibrous glass reinforced polyester moulding compound.

Ervasin. Acetyl-cresotinic acid. An anti-rheumatic, analgesic, and antipyretic.

Erycen. A proprietary preparation of ERYTHROMYCIN. An antibiotic. (137).

Eryophylite. A mineral, $3(Li.Na)F.Al_2F$.

Erysilin. An erysipelas vaccine for swine. (520).

Erythrene. See DIVINYL.

Erythrin (Spirit Eosin, Methyl Eosin, Primrose Soluble in Alcohol, Erythrin Methyl Eosin, Eosin Soluble in Spirit). A dyestuff. It is the potassium salt of tetrabromo-fluoresceine - methyl - ether,

$C_{21}H_9Br_4O_5K$. Dyes silk bluish-red, with a reddish fluorescence.

Erythrin Methyl Eosin. See ERYTHRIN.

Erythrin X. See PONCEAU 5R.

Erythrite (Cobalt Bloom). A mineral. It is a hydrated arsenate of cobalt, $Co_3As_2O_8.8H_2O$. Also see ERYTHROMANNITE.

Erythrobenzine. See MAGENTA.

Erythrocin. A proprietary preparation containing erythromycin as the lactobionate. An antibiotic. (311).

Erythrocin I.M. A proprietary preparation containing Erythromycin Ethyl Succinate. An antibiotic. (311).

Erythroglucin. See ERYTHROMANNITE.

Erythromannite (Erythrite, Phycite, Pseudorcin, Erythroglucin, Eryglucin). Erythrol, $C_4H_{10}O_4$.

Erythromid. A proprietary preparation containing Erythromycin. An antibiotic. (311).

Erythromycin. An antibiotic produced by a strain of Streptomyces erythreus. Erythrocin; Ilotycin. Erythroped is the ethyl succinate.

Erythromycin Estolate. Erythromycin propionyl ester lauryl sulphate. Ilosone.

Erythroped. A proprietary preparation containing erythromycin (as the ethyl succinate). An antibiotic. (311).

Erythroretin. A mixture of chrysophanic acid, emodin, and rhein. Obtained from rhubarb.

Erythrosiderite. A mineral, $2KCl.FeCl_3.H_2O$.

Erythrosin (Erythrosin B, Erythrosin D, Pyrosin B, Iodeosin B, Eosin Bluish, Eosin J, Rose B, Dianthine B, Primrose Soluble, Soluble Primrose). The sodium or potassium salt of tetraiodofluoresceine, $C_{20}H_6O_5I_4Na_2$. Dyes silk and wool bluish-red. Used for paperstaining, as a sensitiser of silver bromide in the photographic industry, and in the production of orthochromatic dry plates.

Erythrosin A. The sodium or potassium salt of tri-iodo-fluorescein.

Erythrosin B. See ERYTHROSIN.

Erythrosin BB. See PHLOXINE P.

Erythrosin D. See ERYTHROSIN.

Erythrosin G (Dianthine G, Pyrosin G, Iodeosin G). The sodium or potassium salt of diiodo-fluoresceine, $C_{20}H_8O_5I_2Na_2$. Dyes wool yellowish-red, with yellowish-red fluorescence.

Erythroxyanthraquinone. *o*-Hydroxyanthraquinone, $C_6H_4(CO)_2.C_6H_3OH$.

Esbach's Reagent. Picric acid, 10 grams, is dissolved in 600 c.c. distilled water. Citric acid, 10 grams, is added, and water to 1,000 c.c. It distinguishes gelatin, which is precipitated from peptone and similar bodies.

Esbatal. A proprietary preparation of benthanidine sulphate. An antihypertensive. (277).

Esbenit (Bitit). Ebonite substitutes made from asbestos with a pitch, tar, or asphalt binder. Esbenit is made from a mixture of cellulose, asbestos dust, magnesia, and lime.

Esbenite. A material made from cellulose, powdered mica, and magnesium silicate.

Escalin. Consists of 2 parts aluminium powder mixed with 1 part glycerin into a paste. It has been used as a substitute for bismuth salts in the treatment of gastric ulcers.

Esch. See ESCHBODEN.

Eschboden (Esch). The name applied to the heavy sandy soil of Western Hanover.

Eschel. A fine-grained light coloured smalt (q.v.).

Eschenite. A mineral containing titanium, columbium, thorium, cerium, and other rare earths.

Escherite. A mineral. It is a variety of epidote.

Eschka Mixture. A mixture of 2 parts by weight pure calcined magnesia and 1 part pure anhydrous sodium carbonate. Used for the determination of sulphur in coal by heating the coal with the mixture, then adding hydrochloric acid and barium chloride, when barium sulphate is precipitated.

Eschwegeite. A mineral. It is a hydrated yttrium - tantalo - columbo - titanite, $2Ta_2O_5.4Cb_2O_5.10TiO_2.5Y_2O_3.7H_2O$.

Eschweg Soaps. Blue mottled soaps, usually made from 20–25 per cent. tallow, 25–30 per cent. bone-fat, 10–15 per cent. cotton-seed oil, 20–40 per cent. palm-kernel oil, and 20–30 per cent. coco-nut oil.

Esco Extract. A synthetic tannin prepared from sulphonated heavy tar oils, by condensation with formaldehyde, and then forming the chromium salt. Also see NERADOL.

Escorcin (Escorcinal, Æscorcin). It has the formula $C_9H_8O_4$, and is prepared from Æsculetin by the action of sodium amalgam. Used for discovering defects in the cornea.

Escorcinal. See ESCORCIN.

Escorto. A proprietary brand of artificial silk.

Esculetin. 6 : 6-dihydroxy-coumarin.

Escutcheon Pin Brass. See BRASS.

Esdeform. Stain removers. (523).

Esdesol. A paint stripper. (523).

Esdogen. Wetting agent and solvent soaps. (523).

Eseré Nut. See CALABAR BEAN.

Eserine. Physostigmine, $C_{15}H_{21}N_3O_2$. An alkaloid obtained from the calabar bean.

Eserin Oil. A solution of 0·2 gram physostigmine salicylate in 40 grams olive oil. Used in the treatment of eye diseases.

Eshalit. A bakelite (phenol-formaldehyde resin) having a very high dielectric resistance even in moist air.

Esidrex. A proprietary preparation of hydrochlorothiazide (6-chloro-3,4-dihydro-7-sulphamyl-1,2,4-benzothiadiazine 1,1-dioxide). A diuretic. (18).

Esidrex-K. A proprietary preparation of hydrochlorothiazide and potassium chloride in slow release core. A diuretic. (18).

Esien Andradit. Synonym for Skiagite.

Eskacef. A proprietary preparation of cephradine. An antibiotic. (658).

Eskacillin V. A proprietary preparation containing phenoxymethyl penicillin.

Eskacillin 100. A proprietary preparation containing Benzylpenicillin (as the potassium salt). An antibiotic. (281).

Eskacillin 200. A proprietary preparation containing procaine penicillin. An antibiotic. (281).

Eskacillin 100 Sulpha. A proprietary preparation of benzylpenicillin potassium and sulphadimidine. An antibiotic. (281).

Eskacillin 200 Sulpha. A proprietary preparation of procaine penicillin and sulphadimidine. An antibiotic. (281).

Eskamel. A proprietary preparation of resorcinol, sulphur and hexachlorophane. Acne treatment. (281).

Eskornade. A proprietary preparation of ISOPROPAMIDE, DIPHENYLPYRALINE and PHENYLPROPANOLAMINE hydrochloride. A nasal decongestant. (658).

Esmaillite. A dope consisting of a mixture of cellulose acetate and volatile solvent, usually ethyl formate.

Esmarch's Caustic. A caustic containing arsenious acid, morphine sulphate, calomel, and gum arabic.

Esmarkite. See DATOLITE.

Esmeraldaite. A mineral. It is a hydrated ferric oxide.

Esmodil. A proprietary trade name for Meprochol.

Esoderm. A proprietary preparation of gamma benzene hexachloride and D.D.T. in a shampoo, used for removal of head lice. (383).

Esperson. See DESOXYMETHASONE.

Essar (W). A proprietary trade name for a general purpose acid and alkali resistant furane resin cement for bedding in acid resisting tiles. (150).

Esshete CML. A proprietary trade

name for a steel containing 1 per cent. chromium and $\frac{1}{2}$ per cent. molybdenum possessing high creep strength and corrosion resistance. It is suitable for operating up to 1000 °F. (153).

Esshete CRM2. A steel containing $2\frac{1}{4}$ per cent. chromium and 1 per cent. molybdenum and superior properties to CML. It can be used up to 1100 °F. (153).

Esshete CRM5. A steel containing 5 per cent. chromium and $\frac{1}{2}$ per cent. molybdenum suitable for tubes exposed to high temperature steam. (153).

Esshete 1250. A steel containing 16 per cent. chromium, 10 per cent. nickel and 6 per cent. manganese. An austenitic creep resisting steel. (153).

Essence, Coffee. See COFFEE ESSENCE.

Essence of Ananas. See ANANAS OIL.

Essence of Apples. A solution of isoamyl-isovalerate in rectified spirit. Used for flavouring confectionery, and for other purposes. The term is also used for a mixture of ethyl nitrite, ethyl acetate, amyl valerate, glycerol, aldehyde, chloroform, and alcohol.

Essence of Apricots. A mixture of isoamyl butyrate and isoamyl alcohol. Another essence consists of a mixture of benzaldehyde, amyl butyrate, chloroform, and alcohol.

Essence of Bergamot. Oil of Bergamot.

Essence of Bigarade. Oil of Bitter Orange Peel.

Essence of Bitter Almonds. Benzoic aldehyde, C_6H_5CHO.

Essence of Ginger. See JAMAICA GINGER.

Essence of Greengage. Usually contains ethyl œnanthylate, $C_2H_5.C_7H_{13}O_2$.

Essence of Lemon. Oil of Lemon.

Essence of Melon. Often contains ethyl sebacate, $(C_2H_5)_2.C_{10}H_{16}O_4$.

Essence of Mirbane (Oil of Mirbane). Nitro-benzene, $C_6H_5NO_2$. Used for scenting soap.

Essence of Mulberry. Often contains ethyl suberate, $(C_2H_5)_2.C_8H_{12}O_4$.

Essence of Orange. Oil of Orange.

Essence of Petroleum. Petroleum jelly.

Essence of Pineapple. A mixture of ethyl butyrate, amyl valerate, chloroform, aldehyde, and alcohol.

Essence of Pineapple Oil. Ananas oil (q.v.).

Essence of Portugal. Essence of Sweet Orange Peel.

Essence of Quince. Often contains ethyl pelargonate, $C_2H_5.C_9H_{17}O_2$.

Essence of Ratafia. Approximately Essence of Almonds.

Essence of Resin. See ROSIN SPIRIT.

Essence of Smoke (Cambrian Essence, Westphalian Essence, Smokene). Materials used for imparting a smoky taste to bacon and fish. The liquids are painted on and allowed to dry. One formula contains spirit of tar, wood naphtha, and crude pyroligneous acid.

Essence of Tar. Creosote.

Essence of Turpentine. See OIL OF TURPENTINE.

Essence of Violets, Artificial. Ionone.

Essences. Essential oils in alcohol.

Essential Oil of Bitter Almonds. Benzaldehyde, C_6H_5CHO.

Essential Salt of Lemons. Salt of sorrel (acid potassium oxalate).

Essential Salt of Urine. See MICROCOSMIC SALT.

Essex Powder. An explosive, containing 22–24 per cent. nitro-glycerin, 0·5–1·5 per cent. collodion cotton, 33–35 per cent. potassium nitrate, 33–35 per cent. wheat flour, and 5–7 per cent. ammonium chloride.

Essogen. A proprietary vitamin A concentrate.

Essonite. A mineral. It is a calcium-aluminium-silicate.

Estaflex ATC. A Dutch trade name for acetyl butyl citrate. A plasticiser/stabiliser. (50).

Estabex 2386. Epoxidised monoesters.

Estabex 2307, 2349. Dutch trade names for epoxidised Soyabean oils. (50).

Estabex 2307, 2349. Epoxidised soyabean oils.

Estaflex ATC. Acetyl tributyl citrate. (50).

Estane. A proprietary range of thermoplastic polyurethane moulding and extrusion compounds. (102).

Estar. A proprietary polyester laminating resin. (718).

Ester Copal. An ester gum obtained by the interaction between glycerin and copal.

Esterellite. A mineral. It is a variety of quartz.

Ester Gums (Rosin Ester, Rosin Gum). These gums are mainly prepared by heating rosin with glycerin in an autoclave at 350° C., an ester being formed. The product is treated with a small quantity of manganese dioxide, to give it a light colour. They are used in the manufacture of varnishes. Proprietary brands of ester gum are known by the names Alcho, Glycero, and Zinco. Sometimes copal, kauri, and anime resins, and other hydroxy-compounds such as resorcinol, phenol, or carbohydrates, are used in the production of these bodies.

Esterin. Colouring of polyester resins. (533).

Ester Margarine. A margarine prepared in Germany by churning " ester oil " (ethyl and glycol esters of fatty acids) with refined oil and milk.

Ester Oil. See ESTER MARGARINE.

Esterol. A proprietary brand of benzyl succinate. Used in medicine. Also a proprietary trade name for alkyd synthetic varnish and lacquer resins.

Esterpol. A proprietary synthetic resin.

Estersil. Ethyl and propyl salicylglycollic esters. (515).

Estersol. Solubilised vat dyestuffs. (582).

Estigyn. Tablets of ethinyloestradiol-oestrogen. (182).

Estol. A range of fatty esters. (597).

Estone. See LENICET.

Estopen. A proprietary trade name for Penethamate Hydriodide.

Estoral. Menthyl borate, $B(O.C_{10}H_{19})_3$. Has been recommended in acute and chronic inflammation of the throat. Used as a snuff.

Estragol. p-propenylanisol, $CH_2 : CH$. $CH_2.C_6H_4OCH_3$.

Estralite. See BAKELITE.

Estramadurite. The Spanish variety of apatite.

Estrovis. A proprietary preparation of QUINESTROL used for the suppression of lactation. (262).

Estynox 308. A proprietary trade name for an epoxy type vinyl plasticiser. (104).

Etafedrine. 2 - (Ethylmethylamine) - 1-phenylpropan-1-ol.

Etamiphylline. A smooth muscle relaxant. 7 - (2 - Diethylaminoethyl)theophylline. MILLOPHYLINE is the camsylate.

Etard's Reagent. Anhydrous chromium oxychloride, an oxidising agent.

Eteleen. Trigallic acetal.

Eteline. A proprietary solvent reputed to be perchlorethylene, $CCl_2 : CCl_2$. Its specific gravity is 1·624, boiling range 119–121° C., and is toxic. A solvent for cellulose acetate.

Etenzamide. An analgesic. 2-Ethoxy-benzamide. LUCAMID.

Eternite. A slate-like mass made from 6 parts Portland cement and 1 part asbestos fibre. Used for roofing.

Eterol. A name given to etherised alcohol which has been subjected to a special treatment. A motor fuel.

Ethacol. Pyrocatechin monoethyl ether.

Ethacrynic Acid. 2,3 - Dichloro - 4 - (2 - ethylacryloyl)phenoxyacetic acid. Edecrin.

Ethal. Cetyl alcohol, $C_{16}H_{33}OH$.

Ethalfluralin. A pesticide. N-ethyl-N-(2 - methylallyl) - 2 - 6 - dinitro - 4 - trifluoromethyl = aniline.

Ethambutol. (+) - NN' - Di - (1 - hydroxymethylpropyl)ethylenediamine. Myambutol is the dihydrochloride.

Ethamivan. NN-Diethylvanillamide. Vandid.

Ethamsylate. Diethylammonium 2,5-dihydroxybenzenesulphonate. Dicynene.

Ethanesal. Ether containing 4 per cent. normal butyl alcohol and a small quantity of aldehyde. An anæsthetic.

Ethanite. A proprietary plastic made from ethylene dichloride and calcium polysulphide. It is stated to be resistant to practically all solvents, oils, fats, and greases. It may be used as an ingredient in rubber compounds.

Ethanol. See SPIRITS OF WINE.

Etharsanol. A compound, $C_6H_4(NH. CH_2.CH_2OH).AsO.(OH)_2$, used in the treatment of trypanosomal infections of animals.

Ethasan. A proprietary trade name for a rubber vulcanising ultra accelerator. It is the zinc salt of diethyl dithiocarbamic acid. (57).

Ethavan. A proprietary trade name for ethylvanillin.

Ethazate. Zinc diethyl dithiocarbamate. A proprietary accelerator. (435).

Ethchlorvynol. 1 - Chloro - 3 - ethyl - pent-1-en-4-yn-3-ol. Arvynol; Placidyl; Serensil.

Ethebenecid. 4 - Diethylsulphamoyl - benzoic acid. Urelim.

Ethene. See OLEFIANT GAS.

Ether, Acetic. Ethyl acetate, $CH_3. COOC_2H_5$.

Ether, Arbutin-benzoic. Cellotropin.

Ether, Benzoic. Ethyl benzoate, $C_6H_5 COOC_2H_5$.

Ether, Bromhydric. See HYDROBROMIC ETHER.

Ethereal Oil of Bitter Almonds. Benzaldehyde, $C_6H_5.CHO$.

Ether, Fluoric. Ethyl fluoride, C_2H_5F.

Ether, Formic. Ethyl formate, $HCOOC_2H_5$.

Etherin. See OLEFIANT GAS.

Ether, Methylated. Ether prepared from methylated spirit.

Ether, Propane Disulphinic. See SULPHONAL.

Ether, Pyroacetic. Acetone, $CH_3 COCH_3$.

Ether, Resorcin. See RESORCINYL OXIDE.

Ether, Salicylic. Ethyl salicylate, $HO. C_6H_4.COOC_2H_5$.

Ether, Thymolcarbonic. Thymol carbonate.

Ethiazide. 6 - Chloro - 3 - ethyl - 3,4 -

dihydrobenzo - 1,2,4 - thiadiazine - 7 - sulphonamide 1,1-dioxide.

Ethibute. A proprietary preparation of PHENYLBUTAZONE. An anti-inflammatory drug. (719).

Ethidivin. A proprietary preparation of HOMIDIUM BROMIDE. A veterinary trypanocide.

Ethidol. (1) A proprietary preparation. It is the ethyl ester of iodo-ricinoleic acid. It is stated to be suitable for intraglandular injection.

Ethidol. (2) A proprietary preparation of ethinylœstradiol and phenobarbitone, used for menopausal discomfort. (265).

Ethinamate. 1 - Ethynylcyclohexyl carbamate. Valmid ; Valmidate.

Ethiodan. Ethyl 4-iodophenylundec-10-enoate. Iophendylate—X-ray contrast medium for myclography. (182).

Ethionamide. 2 - Ethylisonicotinthio - amide. Trescatyl.

Ethiopsite. A mineral. It is a black mercurous sulphide.

Ethiop's Mineral. A mixture of black mercuric sulphide and sulphur.

Ethipen. A proprietary preparation of penicillin V. An antibiotic. (719).

Ethipram. A proprietary preparation of IMIPRAMINE. An anti-depressant. (719).

Ethitet. A proprietary preparation of TETRACYCLINE. An antibiotic. (719).

Ethnine. A proprietary preparation containing pholcodine. A cough linctus. (284).

Ethobral. A proprietary preparation of phenobarbitone., butobarbitone and quinalbarbitone. A hypnotic. (245).

Ethocaine. See NOVOCAINE.

Ethocel. A registered trade mark for ethyl cellulose thermoplastics. (64).

Ethodryl. A proprietary trade name for the dihydrogen citrate of diethylcarbamazine.

Ethoduomeen. An ethoxylated long chain aliphatic diamine. (555).

Ethofat. An ethoxylated ester of long chain fatty acids. (555).

Ethofoil. A proprietary trade name for ethyl cellulose film.

Ethoglucid. Triethyleneglycol diglycidyl ether. Epodyl.

Ethoheptazine. Ethyl 1-methyl-4-phenylazacycloheptane-4-carboxylate. Ethyl hexahydro - 1 - methyl - 4 - phenylazepine-4-carboxylate. Zactane.

Ethol. Basic dyestuffs for printing inks. (582).

Ethomeen. Ethoxylated aliphatic amines. (555).

Ethomid. Ethoxylated long chain aliphatic amides. (555).

Ethomulsion. A proprietary trade name for ethyl cellulose lacquer emulsion.

Ethomoxane. 2 - n - Butylamino - methyl - 8 - ethoxybenzo - 1,4 - dioxan.

Ethone. Ethyl-o-formate, $CH(OC_2H_5)_3$. A mild hypnotic and sedative.

Ethopabate. An anti-protozoal present in AMPROLMIX-UK, AMPROL-PLUS and PANCOXIN. It is methyl 4-acetamido-2-ethoxybenzoate.

Ethopropazine. 10 - (2 - Diethylamino - n-propyl)phenothiazine. Profenamine. Lysivane is the hydrochloride.

Ethoquad. Ethoxylated quaternary ammonium salts. (555).

Ethosalamide. Salicylamide 2-ethoxyethyl ether 2-(2-Ethoxyethoxy)benzamide.

Ethosperse. Coloured ethyl cellulose chips for ethyl cellulose printing inks and general ethyl cellulose finishes. (260).

Ethosuximide. α - Ethyl - α - methyl - succinimide. Emeside ; Simatin ; Zarontin.

Ethotoin. 3 - Ethyl - 5 - phenylhydantoin. Peganone.

Ethox. A proprietary trade name for diethoxy-ethyl phthalate, a plasticiser. A vinyl plasticiser. (73).

Ethoxy Acetone. Acetol-ethyl ether, $CH_3.CO.CH_2OC_2H_5$. A solvent having a boiling-point of 128° C.

Ethoxol. A proprietary glycol ether. (2).

Ethoxyethane. Ethyl ether $(C_2H_5)_2O$.

Ethoxytet. A proprietary preparation of OXYTETRACYCLINE. An antibiotic. (719).

Ethrine. Linctus of pholcodine. (510).

Ethulon. Ethyl cellulose film for tracing and industrial purposes. (507).

Ethulose. A proprietary preparation of alcohol and laevulose. A parenteral source of calories. (386).

Ethyl. The trade mark for a fluid for admixture with petrol to prevent or reduce knocking in internal combustion engines. The active ingredient of this fluid is tetraethyl lead, and the amount employed is a few c.c. per gallon of petrol. See ANTI-KNOCK FLUIDS.

Ethyl Acid Violet S4B. A dyestuff. It is an equivalent of Victoria violet 4BS.

Ethylan. A non-ionic detergent and emulsifier. (558).

Ethylbenztropine. 3-Diphenylmethoxy-8-ethylnortropane.

Ethyl Biscoumacetate. Ethyl di-(4-hydroxycoumarine-3-yl)acetate. Pelentan ; Tromexan.

Ethyl Blue. A dyestuff obtained by heating ethyl-diphenylamine with oxalic acid, or by the action of ethyl chloride upon Diphenylamine blue. Dyes silk blue.

Ethyl-butyl-carbinol. Sec-Heptyl alcohol.

Ethyl Carbinol. Propyl alcohol, CH_3. $CH_2.CH_2OH$.

Ethylcyanine. A dyestuff. It is lepidinquinoline-ethyl-cyanine-bromide, a similar dye to Cyanine, using ethyl bromide instead of amyl iodide. Used as a substitute for Cyanine in dyeing.

Ethyl Dibunate. Ethyl 2,6-di-t-butylnaphthalenesulphonate.

Ethylene Blue. A dyestuff of the same class as Methylene blue.

Ethylene Petrol. An oil. It consists of a mixture of hydrocarbons.

Ethyl Eosin. See SPIRIT EOSIN.

Ethyl Eosin Rose JB. A dyestuff, $C_{22}H_{11}O_5Br_4K$, obtained by the alkylation of eosin.

Ethyl Gasoline. Motor gasoline to which Ethyl (q.v.) has been added. An anti-knock gasoline.

Ethyl Green (Methyl Green). A dyestuff. It is the zinc double chloride of ethyl - hexamethyl - pararosaniline-bromide, $C_{27}H_{35}N_3Cl_3BrZn$. Dyes wool mordanted with sodium thiosulphate and sulphuric acid or zinc acetate, and silk and cotton mordanted with tannin, bluish-green. Also see BRILLIANT GREEN.

Ethylmethylthiambutene. 3-Ethylmethylamine - 1,1 - di - (2 - thienyl)but - 1-ene.

Ethyl Mustard Oil. Ethyl-thiocarbimide, $C_2H_5.N.CS$.

Ethyl Nitrite. In commerce, this is usually a solution of ethyl nitrite in alcohol.

Ethyloestrenol. 17α - Ethyloestr - 4 - en-17β-ol. Orbolin.

Ethylphenacemide. Pheneturide.

Ethyl Purple 6B. See ETHYL VIOLET.

Ethyl Pyrophosphate. A preparation used in the treatment of myasthenia gravis. It is tetraethyl pyrophosphate.

Ethyl Red. A dyestuff. It is quinaldine - quinoline - ethyl - cyanine - iodide. Used as a sensitiser for silver bromide gelatin plates.

Ethyl Sulphonal. See TETRONAL.

Ethyl Thiurad. Tetraethyl thiuram disulphide. An ultra accelerator for rubber. (559).

Ethyl Tuads. A proprietary accelerator. Tetraethyl thiuram disulphide. (450).

Ethyl Tuex Powder. A proprietary accelerator. Tetraethyl thiuram disulphide. (435).

Ethyl Violet (Ethyl Purple 6B). A dyestuff. It is the hydrochloride of hexa-ethyl-pararosaniline, $C_{31}H_{42}N_3Cl$. Dyes silk and wool bluish-violet, and cotton mordanted with tannin and tartar emetic.

Ethyl Zimate. A proprietary accelerator. Zinc diethyl dithiocarbamate. (450).

Ethynodiol. A progestational steroid present as the 3,17-diacetate in DEMULEN, FEMULEN, METRULEN. METRULEN M and OVULEN. It is 19-No-17α-pregn-4-en-20-yne-3β, 17-diol.

Etidocaine. A local anæsthetic. (±)-2 - (N - Ethylpropylamino)butyro - 2', 6'-xylidide.

Etifoxine. A tranquilliser currently under clinical trial as " 36–801 ". It is 6 - Chloro - 2 - ethylamino - 4 - methyl - 4 - phenyl - 4H - 3, 1 - benzoxazine.

Etiolin. A yellow modification of chlorophyll. It is formed when plants grow out of light.

Etisazole. A fungicide. 3-Ethylamino-1, 2 - benzisothiazole. ECTIMAR.

Etisul. A proprietary trade name for ditophal. A proprietary preparation of diethyl dithiolisophthalate. An antileprotic. (2).

Etna Powder. An explosive similar in composition to Giant powder.

Etonitazene. 1 -'(2 - Diethylamino - ethyl) - 2 - (4 - ethoxybenzyl) - 5 - nitro - benzimidazole.

Etophylate. A proprietary preparation of acepifylline. A bronchial antispasmodic. (349).

Etophylate P.P. A proprietary preparation of acepifylline, phenobarbitone and papaverine. A bronchospasm relaxant. (349).

Etorphine. 1,2,3,3a,8,9 - Hexahydro - 5 - hydroxy - 2a - [1(R) - hydroxy - 1 - methylbutyl] - 3 - methoxy - 12 - methyl - 3,9a - etheno - 9,9b - imino - ethanophenanthro[4,5 - bcd]furan - 7, 8 - Dihydro - 7a[1(R) - hydroxy - 1 - methylbutyl] - O^6 - methyl - 6,14 - endoethenomorphine. 7α - [1(R) - Hydroxy - 1 - methylbutyl) - 6,14 - endoethenotetrahydro - oripavine - 19 - Propylorvinol. M.99 is the hydrochloride.

Etoxeridine. Ethyl 1-[2-(2-hydroxyethoxy)ethyl] - 4 - phenylpiperidine - 4-carboxylate.

Etrimfos. A pesticide. It is o-6-ethoxy-2 - ethylpyrimidin - 4 - yl oo - dimethyl phosphorothioate.

Etromeyerite. A mineral, $(Ag.Cu)S$.

Etronite. The trade name for a Norwegian synthetic resin-paper product. It is used for electrical insulation.

Etrynit. A proprietary trade name for Propatylonitrate.

Etryptamine. An anti-depressant. 3-(2-Aminobutyl)indole. α - Ethyltryptamine. MONASE is the acetate.

Ettringite. A mineral. It is a calcium-aluminium-silicate, a garnet.

Etymide. α-Ethoxy-N-methyl-N-[2-(Nmethylphenethylamino) - ethyl]di - phenylacetamide. SQ 10629 is the hydrochloride.

Euban. A mineral. It is a variety of quartz.

Eubeco. Vitamin B complex for injection. (510).

Eubornyl. See BROVALOL.

Eucaine. α-Eucaine is *n*-methyl-benzoyl- triacetone - alkamine - carboxylic acid. Betacaine, benzamine, or β-eucaine, is the hydrochloride of benzoyl-vinyl-diacetone-alkamine. Both are used as substitutes for cocaine.

Eucairite (Eukairite, Eukarite). A mineral. It is a selenide of copper and silver, $CuSe.Ag_2Se$.

Eucalomel. A very light calomel.

Eucalyptol. Cineol, $C_{10}H_{18}O$, a terpene alcohol.

Eucalyptosan. An emulsion of eucalyptus oil and caseosan. It is a German preparation, and is used in the treatment of chronic bronchitis.

Eucalyptus Gum (Red Gum, Eucalyptus kino). See KINO, AUSTRALIAN.

Eucalyptus Kino, See KINO, AUSTRALIAN.

Eucaphene. A German product containing phenol, camphor, menthol, eucalyptol, and turpentine. Used in the treatment of affectations of the air passage by pouring it into hot water and inhaling.

Eucasein. Casein-ammonia, obtained by passing ammonia over finely powdered casein. It is an easily digested strengthening substance. Used in stomach and lung trouble.

Eucasol. Eucalyptus-amytol. (Amytols are solutions of phenols in ichthyol.)

Eucatropine. 4 - Mandeloyloxy - 1,2,2, 6-tetramethylpiperidine. Euphthalmine.

Eucerin. A proprietary preparation of wool alcohols. A skin cream. 5 parts of an oxycholesterin, obtained from wool fat with 95 parts paraffin. Intended as an ointment base. (824).

Euchinine. See EUQUININE.

Euchlorine. Prepared by the action of hydrochloric acid upon potassium chlorate. It appears to be a mixture of chlorine peroxide and chlorine, and is explosive.

Euchlorite. A mineral. It is a variety of magnesium mica of a deep-green colour.

Euchlorose. The mineral, chalcophyllite.

Euchroite. A mineral, $CuO.As_2O_5$.

Euchrome. See CAPPAGH BROWN.

Euclase (Prismatic Emerald). A mineral, $2BeO.Al_2O_3.2SiO_2.H_2O$.

Euclidan. A proprietary trade name for the dihydrogen citrate of Nicanutate.

Eucodal. The hydrochloride of dihydro-hydroxy-codeine. A narcotic.

Eucodal. A proprietary trade name for the hydrochloride of Oxycodone.

Eucodeine. See EUCODINE.

Eucodine (Eucodeine). Codeine-methyl-bromide, $C_{18}H_{21}NO_3.CH_3Br$. Used for pacifying irritating coughs.

Eucol (Eucols). Guaiacol acetate, $CH_3O.C_6H_4O.C_2H_3O$. An antiseptic used in medicine.

Eucolite. See EUDIALYTE.

Eucols. See EUCOL.

Eucopine. Disinfectants and antiseptics. (552).

Eucrasite. A mineral. It is altered thorite.

Eucryptite. A mineral. It is an orthosilicate of aluminium and lithium, $LiAlSiO_4$.

Eucupin. Isoamyl-hydro-cuprein. An antiseptic and anæsthetic.

Eucupin Bihydrochloride. Iso-amyl-hydro - cupreine - dihydrochloride $C_{27}H_{39}O_2N_2Cl_2$. Used in medicine.

Eudalene. 6 - methyl -4-isopropyl-naphthalene. It is produced by hydrogenation of endesmol (*q.v.*).

Eudeiolite. A mineral, $(Ca.Ce.Fe.Hg.Na)O.(Nb.Ti.Th)O_2.H_2O$.

Eudemine. A proprietary preparation of DIAZOXIDE used in the treatment of hypertension and hypoglycaemia. (284).

Eudermol. Nicotine salicylate, $C_{10}H_{14}N_2.C_7H_6O_2$. Applied as a 1 per cent. ointment for itching. Used in veterinary practice.

Eudesmol. A sesquiterpene obtained from eucalyptus. Its constitution is 1 - methylene - 5 - methyl -8-isopropyl-8-hydroxy-decahydro-naphthalene.

Eudialite. See EUDIALYTE.

Eudialyte (Eudialite, Eucolite). A mineral. It is a silicate and zirconate of sodium, calcium, and ferrous iron, $NaCl.6Na_2O.6(Fe.Ca)O.2o(Si.Zr)O_2$.

Eudidymite. A mineral of Norway. It is a beryllium-sodium-hydro-silicate, $NaBeHSi_3O_6$.

Eudnophite. A mineral. It is a variety of analcite.

Eudoxin. The bismuth salt of Nosophene (*q.v.*). Used in medicine internally as a disinfectant, and externally as a substitute for iodoform.

Eudralin. A German preparation. It is a combination of adrenalin and Folicaine (*p*-amino-benzoyl-diethyl-amino-ethenol). A local anæsthetic.

Eudrenine (Adreucaine). A solution containing 0·01 gram eucaine and 0·03 milligram adrenalin per c.c. A local anæsthetic used in dentistry.

Euferrol. An iron and gelatin preparation. It is sold in the form of capsules containing 0·012 gram ferrous oxide, and 0·00009 gram arsenic oxide, As_2O_3. Prescribed in cases of blood diseases.

Euflavine (Neutroflavin, Neutral Acriflavine). A proprietary brand of neutral acriflavine (3 : 6 - diamino - acridine - methyl-chloride $H_2NC_6H_3(CH)N(CH_3)$ $Cl.NH_2$. A powerful antiseptic for intravenous use. Also see ACRIFLAVINE.

Euflavine Cadmium (Euflavine Gold, Euflavine Copper). Metal euflavine compounds. Trypanocides.

Euformol. Dextrin-formaldehyde.

Eugallol. See GALLOLS.

Eugenglanz. Synonym for Polybasite.

Eugenoform. Eugenol-carbinol-sodium, formed by the action of formaldehyde upon phenol. An intestinal disinfectant used in cases of typhoid fever.

Eugenol. Allyl-4 : 3-guaiacol, $C_{10}H_{12}O_2$.

Euglissin. A German product. It is an aperient consisting of paraffin, linseed, seeds of *Psyllium*, and extract of sagrada.

Euglucon. A proprietary preparation of GLIBENCLAMIDE used in the treatment of late-onset diabetes mellitus. (443).

Euglycin. A proprietary trade name for Metahexamide.

Euguform. Partly acetylated methylene-diguaiacol. Prescribed in irritating skin diseases in the form of an ointment or a solution in acetone.

Eugynon 30 and 50. Proprietary preparations of ethinyloestrodiol and dinogestrel. Oral contraceptives. (438).

Euka-drya. A Dutch cellulose rayon.

Eukarite. See EUCAIRITE.

Eukinase. A powder obtained by the desiccation of the pancreatic juice of swine, and contains mainly trypsin, an active proteolytic ferment. A digestive substance.

Eukodal. See EUCODAL.

Eulactol. A casein preparation. It contains 33·25 per cent. albumin, 46·3 per cent. fat, 14·3 per cent. carbohydrates, and 4·3 per cent. saline matter.

Eulan. A German preparation. It is absorbed by wool from dilute solution, rendering the fabric moth-proof.

Eulatin. A mixture of *p*-bromo-benzoic acid, *o*-amino-benzoic acid, and antipyrine. Recommended in the treatment of whooping cough.

Eulissin. Decamethonium iodide for injection. (510).

Eulite. A mineral. It is $16[(Fe,Mg)SiO_3]$.

Eulykol. A proprietary preparation. It consists of the phenyl-ethyl esters of of a selected fraction of hydrocarpus oil. Used for intradermal injection.

Eulytine. A mineral, $2Bi_2O_3.SiO_2$.

Eumecon. A 2 per cent. morphine solution of German origin.

Eumenol. A trade mark for a liquid preparation of the Chinese root, Tang-kui. Recommended in the relief of dysmenorrhœa.

Eumictine. A proprietary preparation of hexamine with salol and santalol.

Eumydrin. A proprietary preparation of atropine methonitrate used in whooping cough and pyloric stenosis. See also methyl atropine. (112).

Eunatrol. Pure sodium oleate, used medicinally.

Euosmite. A resin found in the lignite in Bavaria.

Eupad. An antiseptic consisting of bleaching powder and boric acid in equal parts.

Euphorbia Gum. See POTATO GUM.

Euphorine. Phenyl-urethane, $CO(NH. C_6H_5)(O.C_2H_5)$. Used in medicine as an antipyretic and antineuralgic, also externally in the treatment of ulcers.

Euphthalamine. The hydrochloride of phenyl - glycolyl - *n* - methyl - β-vinyl-di-acetone-alkamine. Used for widening the pupil of the eye.

Euphthalmine. A proprietary trade name for Eucatropine.

Euphyllin. A registered trade mark for a mixture of equal parts of primary and secondary theophylline - ethylene - diamine, $C_2H_4(NH_2)_2.C_7H_8N_4O_2+C_2H_4 (NH_2)_2.2C_7H_8N_4O_2$. Used in medicine. (226).

Euphyllite. A mineral, $CaO.K_2O. 3Na_2O.14Al_2O_3.23SiO_2.9H_2O$.

Eupicin. A condensation product of formaldehyde and coal tar. Employed in the treatment of eczema.

Eupinal. A proprietary preparation of caffeine iodide.

Eupion. A distillation product of rubber. It has a boiling-point of from 33–44° C.

Eupnine Vernade. A proprietary preparation containing caffeine iodide, extract of liquorice, cherry laurel water, glycerine and liquid extract of coffee. A cough linctus. (343).

Euphoramin. A proprietary preparation of meprobamate and methylamphetamine hydrochloride. (273).

Euporphine. Apomorphine - methyl - bromide, $C_{17}H_{17}NO_2.CH_3Br.H_2O$. Said to be used in cases of bronchitis.

Eupurgo. A proprietary preparation of phenolphthalein.

Eupyrchroite. See APATITE.

Eupyrine. *p*-Phenetidine-vanillin-ethyl-carbonate, $C_6H_4.(OC_2H_5)N : CH.C_6H_3 (OCH_3)O.CO_2.C_2H_5$. Prescribed as an antipyretic and astyptic.

Euquinine (Euchinine). Quinine-ethyl-carbonate, $C_2H_5O.CO.O.C_{20}H_{23}.N_2O$.

Prescribed for malaria and whooping cough.

Euralite. A chloritic mineral, of Finland.

Eurax. A proprietary trade name for Crotamiton.

Eurax-Hydrocortisone. A proprietary preparation of hydrocortisone and crotamiton for dermatological use. (17).

Eureka Alloy. A trade mark for a copper-nickel alloy used for electrical resistance wires.

Eureka Compound. See BURNT HYPO.

Euresol (Monoresate). Resorcinol-mono-acetate, $C_6H_4.(OH)(O.C_2H_3O)$. Used in medicine in the treatment of seborrhea, sycosis, and other skin diseases.

Eurobin. Chrysarobin-triacetate. Used as a substitute for chrysarobin.

Europhene (Butocresiol). Isobutyl-*o*-cresol-iodide, $(C_6H_3.C_4H_9.CH_3O)_2I$. Used as a substitute for iodoform in the treatment of wounds, either as a powder (mixed with boric acid), or in the form of an ointment.

Europrene. A registered trade mark (19) for general purpose butadiene styrene copolymers coded as follows:
CIS. 1-4 cis-polybutadiene
SS. high styrene copolymers
N. butadiene acrylonitrile copolymers.

Eusapyl. An aqueous solution of chlorcresol in potassium ricinoleate. A disinfectant.

Euscopol. Optically inactive scopolamine hydrobromide, $C_{17}H_{21}NO_4.HBr$.

Eusol. A solution containing 0·54 per cent. hypochlorous acid, 1·28 per cent. calcium biborate, and 0·17 per cent. calcium chloride. It is made by shaking Eupad (*q.v.*) in water, and filtering.

Eusolvan. A proprietary solvent stated to consist of ethyl lactate.

Euspasmin. A German proprietary preparation stated to contain benzyl succinate, papaverine hydrochloride, and atropine-methylo-bromate. Recommended in cases of dysmenorrhœa, etc.

Eustenin. The addition product of theobromine-sodium and sodium iodide, $C_7H_7N_4O_2.NaI$. Used in medicine.

Eusynchite. A mineral, $(Pb.Cu)O.V_2O_5$.

Eutannin. A mixture of gallic acid and milk sugar. An intestinal astringent.

Eutectal. An aluminium alloy containing small amounts of copper, manganese, magnesium silicide, silicon, and titanium.

Euthallite. A mineral. It is a variety of analcite.

Euthatal. Pentobarbitone sodium solution. (507).

Eutirsol. A thiophene-sulphur prepared from the distillation products of Seefeld shale. It is a bright yellow liquid of faint odour, density 0·96, containing 12–13 per cent. sulphur, nearly all in the form of thiophene compounds. In combination with suitable salve bases it is expected to be more useful than ichthyol.

Eutonyl. A proprietary preparation of pargyline hydrochloride. An antihypertensive. (311).

Euvalerol B. A proprietary preparation of valerian root with phenobarbitone. A sedative. (284).

Euvernil. A proprietary trade name for Sulphaurea.

Euvitol. A proprietary preparation of fencamfamin hydrochloride. (284).

Euxamite. A radio-earth found in Brazil.

Euxenite (Loranskite). A rare mineral consisting essentially of columbates and titanates of yttrium and erbium.

EVA. Ethylene vinyl acetate. A flexible polythene-like polymer for moulding and extrusion. Adheres well to metals.

Evadyne. A proprietary preparation of BUTRIPTYLINE hydrochloride. An antidepressant. (467).

Evans Blue. Azovan blue.

Evans' Cement. A metallic cement made by adding cadmium amalgam (74 per cent. mercury) to mercury.

Evansite. A mineral, $Al_3(OH)_6PO_4.6H_2O$.

Evansol. Lysol (*q.v.*). (591).

Evatmine. A combination of pituitary extract, adrenalin, and physiological salt solution, used in asthma.

Eventin. A proprietary trade name for Propylhexedrine.

Everbrite. A proprietary trade name for an alloy of 60 per cent. copper, 30 per cent. nickel, 3 per cent. iron, 3 per cent. silicon, and 3 per cent. chromium.

Everdur. The trade mark for a metal which may be described as a proprietary alloy of copper and silicon with controlled amounts of other elements, most commonly a copper-silicon-manganese alloy. It is stated to be resistant to sulphuric acid, aluminium sulphate, chlorine water, carbolic, malic, and citric acid, salts of metals, and caustic soda. The most widely used alloy melts at 1019° C., has a specific gravity of 8·54 for the rolled metal. It is available in several other compositions suitable for special applications. (134)

Eveready Prestone. The trade marks applied to ethylene glycol antifreeze.

Everite. A trade name for an asbestos-cement board.

Everitt's Salt. Potassium ferrous ferrocyanide, $K_2Fe.Fe.(CN)_6$, formed in the preparation of hydrocyanic acid from prussiate of potash.

Everlastic. A proprietary trade name for a textile incorporating Duprene.

Everlasting Pills. Pills of metallic antimony.

Everseal. A bituminastic liquid applied as a corrosion-resisting material.

Eversoft Plastex. A low-freezing explosive containing a nitrated mixture of glycerin and ethylene glycol, ammonium nitrate, sodium chloride, wood meal, trinitro-toluene, and nitro-cotton.

Eversoft Sea Mex. A low-freezing explosive containing a nitrated mixture of glycerin and ethylene glycol, ammonium nitrate, sodium chloride, and wheat flour.

Eversoft Tees Powder. A low-freezing explosive consisting of a nitrated mixture of glycerin and ethylene glycol, ammonium nitrate, wood meal, and sodium chloride.

Evidorm. A proprietary preparation of hexobarbitone and cyclobarbitone calcium. A hypnotic. (112).

Evigtokite. A mineral. It is Gearksutite.

Evipan. A proprietary preparation. It is stated to be N-methyl-C.C.-cyclohexenyl-methyl-barbituric acid. An intravenous anæsthetic.

Evo-stik 873 Super. A proprietary synthetic rubber latex adhesive having a high content of solids. (720).

Evramycin. A proprietary preparation containing triacetyloeandomycin. An antibiotic. (245).

Ewer and Pick's Acid. Naphthalenedisulphonic acid, $C_{10}H_6(SO_3H)_2$, (1 : 6).

Ewol. See NERADOL.

Exalgin. Methyl-acetanilide, $C_6H_5N(CH_3).CO.CH_3$. Prescribed as an antineuralgic.

Exaltone. A trade name for muskone (cyclo-pentadecanone), the perfuming principle of natural musk.

Exanthalite. Synonym for Exanthalose.

Exanthalose. A mineral. It is $Na_2SO_4.2H_2O$.

Exbepa. A proprietary preparation. It is a desiccated liver extract.

Excelite. See METALITE. Also an American trade name for a thermosetting fibrous plastic.

Excellerex. A proprietary rubber vulcanisation accelerator. It is an aniline derivative.

Excello. An electrical resistance alloy containing 85 per cent. nickel, 14 per cent. chromium, 0·5 per cent. iron, and 0·5 per cent. manganese. It is also the name for a carbon black used in rubber mixings.

Excelo. A proprietary trade name for a hot die steel containing 2·5 per cent.

tungsten, 1·5 per cent. chromium, 0·35 per cent. vanadium, and 0·55 per cent. carbon.

Excelon. A proprietary trade name for acrylic denture material.

Excelsior. A proprietary trade name for carbon black.

Exkins. Anti-skinning agents for paints. (607).

Exlax. A proprietary phenol-phthalein preparation.

Exl-die Steel. A proprietary trade name for a non-shrinking steel containing 1·15 per cent. manganese, 0·5 per cent. chromium, 0·5 per cent. tungsten, and 0·9 per cent. carbon.

Exodin. A mixture of 23 per cent. diacetyl - rufigallic acid - tetramethyl-ether, 30 per cent. rufigallic-hexamethyl-ether, and 47 per cent. acetyl-rufigallic-pentamethyl-ether. A mild aperient.

Exolon. A trade mark for abrasive articles consisting essentially of silicon carbide.

Exonic OT. A proprietary trade name for dioctyl sodium sulphosuccinate.

Exovax. Botulism vaccine for mink. (514).

Exolan. A proprietary preparation of triacetoxyanthracene used as a peeling agent in dermatology. (377).

Expanded Graphite. A substance prepared by covering flake graphite with an oxidising agent, to produce a film of graphitic acid, then heating strongly to cause the particles to become distended.

Expanding Solder. An alloy of 37·5 per cent. lead, 6·75 per cent. bismuth, and 56·25 per cent. antimony. It expands on cooling, and is used for fixing metal into holes.

Expansatol. See BUTABARBITAL. (721).

Expansyl Spansule. A proprietary preparation of trifluoperazine hydrochloride, diphenylpyraline hydrochloride and ephedrine sulphate. (281).

Experatol. An aqueous solution of 4 per cent. formalin, 8 per cent. potash alum, and 10 per cent. pine-needles extract.

Expiral. Pentobarbitone sodium for veterinary euthanasia. (525).

Explosif O3 (Prométhée). An explosive. It consists of chlorate mixed with a little manganese dioxide, and dipped into a mixture of nitro-benzene, turpentine, and naphtha.

Explosifs Favier. See EXPLOSIFS N.

Explosifs N (Explosifs Favier, Grisou-nites). French explosives consisting of mixtures of ammonium nitrate with nitro-naphthalene, paraffin, and resin.

Explosive Antimony. Obtained by electrolysing a solution of antimony trichloride, with one antimony electrode and one platinum electrode. The metallic deposit, consisting of 93·5 per cent. antimony, 6 per cent. antimony trichloride, and 0·5 per cent. hydrochloric acid, is explosive.

Explosive D. Ammonium picrate, the ammonium derivative of trinitrophenol, $NH_4O.C_6H_2(NO_2)_3$.

Explosive Gum. Nitro-glycerin gelatinised with collodion cotton. It contains 96 per cent. of the former, and 4 per cent. of the latter compound.

Expulin. A proprietary preparation of PHOLCODINE, EPHEDRINE hydrochloride, CHLORPHENIRAMINE maleate, glycerin and menthol. A cough medicine. (722).

Exsiccated Alum. This is burnt alum (q.v.).

Exsilite. A proprietary trade name for a pure electrically fused silica in which have been incorporated special colouring agents permanent above 2000° C. (156).

Extil. A proprietary preparation containing carbinoxamine maleate and pseudoephedrine hydrochloride. A cough suppressant. (182).

Extil Compound Linctus. A proprietary preparation containing noscapine, pseudoephedrine, and carbinoxamine maleate. A cough linctus. (182).

Extol. A proprietary trade name for a sulphated compound with solvents used as a detergent.

Exton Bristles. A proprietary trade name for bristles for toothbrushes, etc., made from polymerised cellulose compounds.

Extract, Indigo. See INDIGO CARMINE.

Extract of Gamboge. A compound of gamboge and aluminium oxide. A yellow pigment.

Extract of Ginger. See JAMAICA GINGER.

Extract of Lead. (*Liquor Plumbi Subacetatis Fortis B.P.*) A strong solution of lead acetate.

Extract of Malt (Sugar of Malt). An evaporated infusion of malt, prepared by heating potato starch with water, adding from 1–3 per cent. of malt, cooling, and adding from 4–7 per cent. of green malt. The starch is converted into sugar. It assists enfeebled digestion, and is used mixed with iron, quinine, iodine, or cod-liver oil, as a strengthening agent.

Extract of Scammony. Resin of scammony.

Extract of Vermilion. See SCARLET VERMILION.

Extra Gilder's White. A variety of whiting.

Extra White Metal. An alloy of 50 per cent. copper, 30 per cent. nickel, and 20 per cent. zinc. It is a nickel silver (German silver).

Extrox. Low viscosity paints containing zinc oxide. (610).

Exyphen. A proprietary preparation of BROMPHENIRAMINE maleate, guaiphenesin, phenylephrine hydrochloride and phenyl propanolamine hydrochloride. A cough medicine. (462).

Eyelet Brass. See BRASS.

Eye Ointment (Golden Ointment). *Unguentum Hydrargyri Flavi B.P.* and *Ung. Hyd. Ox. Rubri B.P.* The former consists of yellow mercury oxide mixed with paraffin, and the latter is the red oxide mixed with paraffin.

F

F12. See FREON.

F789. A bactericide. It is the hydroxyethyl derivative of hexamethylene-tetramine.

F790. A bactericide. It is the methiodide of hexamethylene-tetramine.

Fabahistin. A proprietary preparation of mebhydrolin napadisylate. (341).

Fabrethane. A proprietary one-component foamable polyurethane, 100% solids. (404).

Fabrifil. A proprietary trade name for a macerated cotton fabric filler.

Fabrikoid. A trade mark for a fabric coated with pyroxylin.

Fabroil. A proprietary trade name for a synthetic resin.

Fabrolite. A synthetic resin of the phenol-formaldehyde type. (409).

F-acid. See DELTA ACID.

Facteka. A proprietary trade name for a rubber substitute.

Factis. A term applied to rubber substitutes prepared from oils.

Factoprene NS. A proprietary hard factice. (417).

Factoprene Z. A proprietary soft factice. (417).

Fæxin Extract. The fatty acids of yeast.

Fagacid. A product derived from beechwood tar. It is an antiseptic agent used in the preparation of soaps and plasters.

Fagara Silk (Ailanthus Silk). A silk produced from the insect *Attacus atlas*.

Faheyite. A mineral. It is $3[(Mn,Mg,Na)Be_2Fe_2'''(PO_4)_4.6H_2O]$.

Fahlerz. Synonym for Tetrahedrite.

Fahl Ore. See TETRAHEDRITE.

Fahlun Diamonds (Tin Brilliants). Lead-tin alloys, containing about 40 per cent. lead, used for theatre jewellery.

Fahlunite. A mineral, $2FeO.SiO_2.Al_2O_3.2SiO_2.H_2O$.

Fahralloy. A proprietary trade name for chromium-nickel-iron alloys.

Fai Nuts. See OWALA BEANS.

Fairchildite. A mineral. It is $K_2Ca(CO_3)_2$.

Fairco. See CRISCO.

Fairey Metal. A proprietary trade name for an alloy of aluminium with copper and smaller amounts of magnesium. It is of the duralumin type.

Fairfieldite. A mineral. It is $[Ca_2(Mn^{··},Fe^{··})(PO_4)_2 . 2H_2O]$.

Fairprene. A proprietary trade name for a chloroprene polymer for fabric coating.

Faktex. A proprietary trade name for a yellow rubber substitute which can be dispersed in water for addition to latex.

Falapen. A proprietary preparation containing Benzylpenicillin. An antibiotic. (182).

Falcodyl. A proprietary preparation of pholcodine and ephedrine hydrochloride. A bronchial antispasmodic. (350).

Falkaloid. A proprietary trade name for a soft oil modified alkyd resin.

Falkenhaynite. A mineral. It is a sulphide of antimony and copper.

Falkmanite. A mineral. It is $4[Pb_3Sb_2S_6]$.

Falkyd. A proprietary trade name for a soft oil modified alkyd resin.

False Columba (Tree Turmeric). The dried stem of *Coscinium fenestratum*. Used in Ceylon and Southern India as a yellow dye and bitter tonic.

False Gold Bronze. Powdered brass foil.

False Hyacinth. Garnet (*q.v.*).

False Lead. See BLACK JACK.

False Topaz. See CITRINE.

False Tragacanth. See BASSORA GUM.

Famatinite. A mineral. It is a copper-antimony sulphide, $3Cu_2S.Sb_2S_5$.

Famel Syrup. A trade mark for a syrup comprising purified beech wood creosote rendered water soluble by means of lactic acid in combination with calcium lactophosphate, aconite and codeine. Used for treatment of infections of the lungs. See Beatin. (228).

Famprofazone. An analgesic and antipyretic present in GEVODIN. It is 4-isopropyl - 2 - methyl - 3 - [N - methyl - N-(α-methylphenethyl) aminomethyl]-1-phenyl-5-pyrazolone.

Fanasil. A proprietary trade name for Sulphormethoxine.

Fan Blade Brass. See BRASS.

Fangerine. Tablets containing phenolphthalein, chocolate, and sucrose. A laxative.

Fanghidi Sclofani. A yellow powder of volcanic origin, consisting chiefly of sulphur, with small quantities of iron, calcium, and manganese. Used as a remedy for acne rosacea.

Fantan. Phenyl-cinchonoyl-urethane.

Fantasit. A proprietary casein plastic.

Fanthridone. 5 - (3 - Dimethylamino-propyl)phenanthridone. AGN 616 is the hydrochloride.

Faradayin. A low boiling fraction obtained by the distillation of rubber by Himly.

Faraday's Cement (Electrical Cement). A cement for use in fastening brass work to glass tubes. It consists of 5 oz. of resin, 1 oz. of beeswax, and 1 oz. of red ochre.

Faraday's Gold. A colloidal gold solution.

Farallonite. A mineral. It is $Mg_2W_2SiO_9 . nH_2O$.

Faratsihite. A mineral. It is a silicate of aluminium and iron.

Farbruss FW. A trade name for carbon black. (German).

Fard's Spanish White. See BISMUTH WHITE.

Fargite. A mineral. It is a variety of natrolite.

Farina. Flour, or potato starch.

Farine. A term applied in the West Indies to a product obtained by grating fresh cassava root, draining away the juice from the wet pulp, and then heating the residue. It is also known as Cassava meal.

Faringets. A proprietary preparation of myristyl benzalkonium iodide chloride. (112).

Farinose. Starch cellulose, or the outer covering of the starch granule.

Farlite. A proprietary trade name for a phenol-formaldehyde synthetic resin laminated product.

Farmer's Reducer. A photographic reducer. It consists of 100 c.c. of hypo solution (1 : 4), with from 5–10 c.c. of a 10 per cent. solution of potassium ferricyanide.

Farnesol. A sesquiterpene alcohol prepared from nerolidol (*q.v.*). A perfume.

Färoelite. A mineral. It is a variety of thomsonite.

Farrant's Medium. A microscopic medium. It consists of a mixture of equal parts of glycerin and arsenious acid, to which is added powdered gum arabic, allowed to stand and filtered.

Farronic. A heat-resisting nickel-copper alloy.

Fasciculite. A mineral. It is a variety of hornblende.

Faspite. Retarded hemihydrate gypsum. (512).

Fassaite. A pyroxene mineral, (Mg.Fe)(Al.Fe)$_2$SiO$_6$.

Fast Acid Blue B (Intensive Blue). A dyestuff obtained from dimethyl-aniline and α-naphthylamine-disulphonic acid. Intensive blue is chemically identical.

Fast Acid Blue R (Violamine 3B). A dyestuff. It is the sodium salt of di-p-ethoxy - phenyl - m - amino - phenol - dichloro-phthalein-sulphonic acid, C$_{36}$H$_{26}$N$_2$O$_8$SCl$_2$Na$_2$. Dyes wool and silk blue.

Fast Acid Cerise. A dyestuff. It is a British equivalent of Fast acid fuchsine B.

Fast Acid Eosin G. A phthalein dyestuff. It dyes wool in the presence of 10 per cent. sodium sulphate, and 4 per cent. sulphuric acid. Fast phloxin A and Irisamine are similar dyestuffs.

Fast Acid Fuchsine B (Fast acid Magenta B, Fast Acid Cerise, Fast Acid Magenta BL). A dyestuff. It is the sodium salt of benzene-azo-1 : 8-amino-naphthol-3 : 6-disulphonic acid, C$_{16}$H$_{11}$N$_3$S$_2$O$_7$Na$_2$.

Fast Acid Magenta B. See FAST ACID FUCHSINE B.

Fast Acid Magenta BL. A dyestuff. It is equivalent to Fast acid fuchsine B, and is of British manufacture.

Fast Acid Orange S. A dyestuff obtained by treating J acid with phosgene gas.

Fast Acid Phloxine A. See FAST ACID EOSIN G.

Fast Acid Ponceau. See FAST ACID SCARLET.

Fast Acid Red. An acid dyestuff.

Fast Acid Scarlet (Acid Ponceau, Ponceau S for Silk). A dyestuff. It is the sodium salt of sulpho-naphthalene-azo-β-naphthol, C$_{20}$H$_{13}$N$_2$O$_4$SNa. It is isomeric with Fast red A, Fast brown 3B, and Double brilliant scarlet G. Dyes wool red from an acid bath.

Fast Acid Violet A2R (Violamine R, Acid Violet 4R). A dyestuff. It is the sodium salt of di-o-tolyl-m-amino-phenol-phthalein-sulphonic acid, C$_{34}$H$_{24}$N$_2$O$_6$SNa$_2$. Dyes silk and wool reddish-violet.

Fast Acid Violet B (Violamine B). A dyestuff. It is the sodium salt of diphenyl - m - amino - phenol - phthalein-sulphonic acid, C$_{34}$H$_{24}$N$_2$O$_6$SNa$_2$. Dyes wool and silk reddish-violet.

Fast Acid Violet 10B. A dyestuff. It is the sodium salt of benzyl-ethyl-tetramethyl - pararosaniline - disulphonic acid, C$_{32}$H$_{35}$N$_3$O$_7$S$_2$Na$_2$. Dyes wool violet-blue from an acid bath.

Fast Black (Fast Blue-Black). A dyestuff obtained by the action of nitroso-dimethyl-aniline-hydrochloride upon m-oxy - diphenylamine. Dyes tannined cotton blue-black.

Fast Black B. A dyestuff obtained by the action of sodium sulphide in aqueous solution upon 1 : 8-dinitronaphthalene. Dyes cotton black from an alkaline bath.

Fast Black BS. A dyestuff obtained by the action of alkalis upon Fast black B. Dyes cotton deep black.

Fast Blue. See ALKALI BLUE and MELDOLA'S BLUE.

Fast Blue-black. See FAST BLACK.

Fast Blue, Greenish. A soluble induline. See INDULINE.

Fast Blue B, Spirit Soluble. See INDULINE, SPIRIT SOLUBLE.

Fast Blue 2B or R. See NEW BLUE.

Fast Blue 2B for Cotton. See NEW BLUE B or G.

Fast Blue 6B for Wool. A soluble induline. See INDULINE.

Fast Blue R, 2R, 3R, for Cotton. See MELDOLA'S BLUE and INDULINE, SOLUBLE.

Fast Blue R, Spirit Soluble. See INDULINE, SPIRIT SOLUBLE.

Fast Bordeaux O. Cloth red B.

Fast Brown. A dyestuff. It is the sodium salt of bi-sulpho-naphthalene-disazo - resorcinol, C$_{26}$H$_{16}$N$_4$O$_8$S$_2$Na$_2$. Dyes wool brown from an acid bath.

Fast Brown 3B (Acid Brown R (Brotherton & Co., Ltd.)). A dyestuff. It is the sodium salt of sulpho-naphthalene-azo-α-naphthol, C$_{20}$H$_{13}$N$_2$O$_4$SNa. Dyes wool brown from an acid bath.

Fast Brown G (Acid Brown). A dyestuff. It is the sodium salt of bi-sulpho-benzene-disazo-α-naphthol, C$_{22}$H$_{14}$N$_4$O$_7$S$_2$Na$_2$. Dyes wool brown from an acid bath.

Fast Brown N (Acid Brown, Naphthylamine Brown, Azo Brown O, Chrome Brown RO, Lighthouse Chrome Brown R, Acid Brown G (Brotherton & Co., Ltd.), Acid Brown R (British Dyestuffs Corporation)). A dyestuff. It is the sodium salt of p-sulpho-naphthalene-azo-α-naphthol, C$_{20}$H$_{13}$N$_2$O$_4$SNa. Dyes wool brown from an acid bath.

Fast Brown O. See CRUMPSALL DIRECT.

Fast Brown ONT, Yellow Shade. A dyestuff. It is the sodium salt of bi-sulpho-xylene-disazo-α-naphthol, C$_{26}$H$_{22}$N$_4$S$_2$O$_7$Na$_2$. Dyes wool and silk brownish-red. Also used for lakes.

Fast Chrome Cyanine. A dyestuff. It is equivalent to Chrome fast cyanine G.

Fast Chrome Cyanine 2B. A dyestuff. It is equivalent to Palatine chrome blue.

Fast Chrome Yellow C. See ANTHRACENE YELLOW C.

Fast Chrome Yellow GG. An azo dyestuff obtained from *o*-anisidine and salicylic acid.

'ast Cotton Blue B. See NEW BLUE B or G.

'ast Cotton Blue R. See INDOINE BLUE R.

'ast Cotton Brown R. See POLY-CHROMINE B.

'ast Diamine Red. A dyestuff, $C_{29}H_{20}N_5O_7SNa$, obtained from benzidine, salicylic acid, and 2 : 8-amino-naphthol-6-sulphonic acid. Used for dyeing wool and cotton.

'astex. A proprietary trade name for a specially stabilised and purified rubber latex supplied in concentrations of 40 and 60 per cent.

'ast Green (Fast Green Extra, Fast Green Extra Bluish). A dyestuff. It is the sodium salt of tetramethyl-dibenzyl-pseudo-rosaniline-disulphonic acid, $C_{37}H_{37}N_3O_7S_2Na_2$. Dyes wool bluish-green from an acid bath.

'ast Green. See MALACHITE GREEN.

'ast Green Extra. See FAST GREEN.

'ast Green Extra, Bluish. See FAST GREEN.

'ast Green FCF. An American dyestuff prepared by the condensation of 2 molecules of ethyl-benzyl-aniline sulphonic acid with 1 molecule-*p*-hydroxybenzaldehyde-*o*-sulphonic acid followed by oxidation with lead peroxide. Proposed for admission as a food dye.

'ast Green G. See GALLANILIC GREEN.

'ast Green J. See BRILLIANT GREEN.

'ast Green M. A dyestuff. It is probably dimethyl-phenyl-diaminooxy-naphtho-phenazonium-chloride. Employed for printing on cotton in conjunction with tannin.

'ast Green O. See DINITROSO-RESORCIN.

'ast Green S. See BRILLIANT GREEN.

Fast Grey D and S. Sulphide dyestuffs. They dye cotton in a solution containing sodium sulphide, caustic soda, and sodium chloride.

Fast Light Orange G. A dyestuff. It is equivalent to Orange G, and is of British manufacture.

Fast Marine Blue. See MELDOLA'S BLUE.

Fast Marine Blue G, BM, BG. See NEW BLUE B AND G.

Fast Milling Red B. See CLOTH RED B.

Fast Mordant Yellow. See ANTHRACENE YELLOW C.

Fast Myrtle Green. See DINITROSO-RESORCIN.

Fast Navy Blue G, BM, GM. See NEW BLUE B OR G.

Fast Navy Blue R. See MELDOLA'S BLUE.

Fast Navy Blue RM, MM. See MELDOLA'S BLUE.

Fast Neutral Violet B. A dyestuff. It is dimethyl-diethyl-diamino-phenazonium-chloride, $C_{18}H_{23}N_4Cl$. Dyes tannined cotton violet.

Fast New Blue for Cotton. See PARAPHENYLENE BLUE R.

Fastocaine. A proprietary preparation of lignocaine hydrochloride and noradrenaline. A local anæsthetic. (308).

Fast Oil Brown S. A dyestuff. It is equivalent to Sudan brown.

Fast Oil Orange I. A British brand of Sudan I.

Fast Oil Orange II. A dyestuff. It is a British equivalent of Sudan II.

Fast Oil Scarlet III. A British equivalent of Sudan III.

Fast Pink for Silk. See MAGDALA RED.

Fast Ponceau B. See BIEBRICH SCARLET.

Fast Ponceau 2B (Ponceau S extra, Scarlet S). A dyestuff. It is the sodium salt of sulpho-benzene-azo-sulpho-benzene-azo-β-naphthol-disulphonic acid, $C_{22}H_{12}N_4O_{13}S_4Na_4$. Dyes wool scarlet from an acid bath.

Fast Red (Fast Red A, Fast Red HF, Fast Red O, Brilliant Red, Roccellin, Rauracienne, Rubidine, Orseilline No. 3, Orseilline No. 4, Ceracine, Cardinal Red J, Alliance Fast Red A Extra, Fast Red A New, AL, G Extra, KG, Acid Red). A dyestuff. It is the sodium salt of *p*-sulpho-naphthalene-azo-β-naphthol, $C_{20}H_{13}N_2O_4SNa$. Dyes wool red from an acid bath.

Fast Red. See FAST RED D.

Fast Red A. See FAST RED.

Fast Red A New, AL, G Extra, KG. Dyestuffs. They are British equivalents of Fast red.

Fast Red B (Bordeaux B, Bordeaux BL, Bordeaux D, Bordeaux R Extra, Double Brilliant Scarlet G, Scarlet for Silk, Silk Scarlet S). A dyestuff. It is the sodium salt of α-naphthalene-azo-β-naphthol-disulphonic acid, $C_{20}H_{12}N_2O_7S_2Na_2$. Dyes wool red from an acid bath.

Fast Red 7B. See CLOTH RED G.

Fast Red BT (Fast Red SX). A dyestuff. It is the sodium salt of α-naphthalene-azo-β-naphthol-mono-sulphonic acid, $C_{20}H_{13}N_2O_4SNa$. Dyes wool red from an acid bath.

Fast Red C. See AZORUBINE S.

Fast Red D (Bordeaux DH, Bordeaux S, Fast Red, Fast Red NS, Fast Red EB, Naphthol Red O, Naphthol Red S, Amaranth, Azo Acid Rubine 2B, Œnanthin, Wool Red Extra, Victoria Rubine, Brilliant Acid Bordeaux, Brilliant Bordeaux B, 2B, Brilliant Crimson O). A dyestuff. It is the sodium salt of *p*-sulpho-naphthalene-azo-β-naphthol-disulphonic acid. $C_{20}H_{11}N_2O_{10}S_3Na_3$. Dyes wool red from an acid bath.

Fast Red E (Fast Red, Fast Red S, Acid Carmoisine). A dyestuff. It is the sodium salt of p-sulpho-naphthalene-azo-β-naphthol-mono-sulphonic acid, $C_{20}H_{12}N_2O_7S_2Na_2$. Dyes wool red from an acid bath.

Fast Red EB. See FAST RED D.

Fast Red HF. See FAST RED.

Fast Red NS. See FAST RED D.

Fast Red O. See FAST RED.

Fast Red PR Extra. An acid wool dye.

Fast Red RBE Base. A dyestuff. It is 6-benzamino-m-4-xylidine hydrochloride.

Fast Red S. See FAST RED E.

Fast Red SX. A British brand of Fast red BT.

Fast Red TR Base. The hydrochloride of 5-chloro-2-amino-o-toluene.

Fast Scarlet. See BENZOPURPURIN 4B.

Fast Scarlet (Double Scarlet). The Badische Co. A dyestuff obtained by the action of diazotised amino-azo-benzene-sulphonic acid upon β-naphthol, $C_{22}H_{16}N_4O_4S$.

Fast Scarlet B. A dyestuff. It is the sodium salt of sulpho-benzene-azo-benzene-azo-β-naphthol-mono-sulphonic acid, $C_{22}H_{14}N_4O_7S_2Na_2$. Dyes wool scarlet from an acid bath.

Fast Scarlet TR Base. The hydrochloride of 6-chloro-2-amino-toluene.

Fast Sulphone Violet 5BS. A mono-azo dyestuff, giving wool and silk bluish-violet shades from an acid bath.

Fast Sulphone Violet 4R. A dyestuff of the same class as Fast sulphone violet 5BS.

Fast Vat Blue R. A Bohme dyestuff. It contains logwood extract, methylene blue, and a chrome mordant.

Fast Violet. See GALLOCYANINE DH and BS.

Fast Violet B. A dyestuff. It is the sodium salt of sulpho-p-toluene-azo-α-naphthalene - azo - β - naphthol - β - sul - phonic acid, $C_{27}H_{18}N_4O_7S_2Na_2$. Dyes wool violet from an acid bath.

Fast Violet R. A dyestuff. It is the sodium salt of sulpho-benzene-azo-α-naphthalene - azo - α - naphthol - β - sul - phonic acid, $C_{26}H_{16}N_4O_7S_2Na_2$. Dyes wool from an acid bath, or chrome mordanted cotton, reddish-violet.

Fast Wool Blue. Alpine blue ($q.v.$).

Fast Yellow. See ORANGE IV and FAST YELLOW R.

Fast Yellow. See ACID YELLOW, ORANGE IV.

Fast Yellow B. An azo dyestuff. It is amino-azo-toluene-disulphonic acid.

Fast Yellow Extra. See ACID YELLOW.

Fast Yellow G. See ACID YELLOW.

Fast Yellow GL. A dyestuff. It is a British brand of Acid yellow.

Fast Yellow, Greenish. See ACID YELLOW.

Fast Yellow N (Curcumeine, Jaune Solide N, Orange N, Yellow OO). A dyestuff. It is the sodium salt of sulpho-p-toluene-azo-diphenylamine $C_{19}H_{16}N_3O_3SNa$. Dyes wool orange from an acid bath.

Fast Yellow R (Fast Yellow, Acid Yellow R, Yellow W). A dyestuff. It is the sodium salt of amino-azo-toluene disulphonic acid, $C_{14}H_{13}N_3S_2O_6Na_2$ Dyes wool reddish-yellow from an acid bath.

Fast Yellow S. See ACID YELLOW.

Fat, Bromine. See BROMIPIN.

Fat, Chocolate. See VEGETABLE BUTTER.

Fat Clay. Ball clay ($q.v.$).

Fat Colours. Colours used for dyeing wood, leather, candles, soaps, pomades, and butter. They are insoluble in water, but soluble in mineral, plant, and animal oils, and in benzene and naphtha.

Fat Liquor. An emulsion for use with oils to soften leather.

Fat, Mineral. See PETROLEUM JELLY.

Fat, Otoba. See OTOBA BUTTER.

Fat Ponceau. See SUDAN IV.

Fat, Tacamahac. See LAUREL NUT OIL.

Fat, Tangkawang. See BORNEO TALLOW.

Faturan. See BAKELITE.

Faujasite. A mineral, $(Na_2.Ca)O.Al_2O_3.5SiO_2.10H_2O$.

Faunolen. Canine contagious hepatitis vaccine. (514).

Fauserite. A mineral. It is $4[(Mg,Mn)SO_4.7H_2O]$.

Faustite. A mineral. It is $ZnAl_6(PO_4)_4(OH)_8.5H_2O$.

Favas. See PAREDRITE.

Faversham Powder. An explosive. It consists of 85 per cent. ammonium nitrate, 10 per cent. dinitro-benzene, and 5 per cent. of Trench's flame-extinguishing compound.

Faversham Powder No. 2. An explosive containing ammonium nitrate, potassium nitrate, trinitro-toluene, and ammonium chloride.

Favierite No. 1. An explosive consisting of 88 per cent. ammonium nitrate and 12 per cent. dinitro-naphthalene.

Favierite No. 2. An explosive containing 90 per cent. of No. 1 and 10 per cent. ammonium chloride.

Favioxanthic Acids. Stains for biology. (544).

Fayalite. A mineral. It is iron-*o*-silicate, Fe_2SiO_4.

Fax. Edible fats. (563).

Faxola. Cooking oil. (563).

Faxtender. Factices or rubber substitutes. (610).

Fazadinium Bromide. A neuromuscular blocking agent currently undergoing clinical trial as "AH 8165 D". It is 1, 1' - azobis(3 - methyl - 2 - phenyl - imidazo[1,2 - *a*]pyridinium bromide.

Feather Alum (Pickeringite). Aluminium sulphate, $Al_2(SO_4)_3.18H_2O$. Occurs naturally.

Feathered Lead. Granulated lead, obtained by pouring molten metal into water. Used in the production of white lead.

Feathered Tin. Granulated tin.

Feather Ore. See ZINKENITE.

Feather Shot Copper. See BEAN SHOT COPPER.

Febkol Elastomer 110 and 122. Proprietary brands of polysulphide sealant. (723).

Febkol Plastomer 555. A proprietary polysulphide/epoxy sealant and adhesive. (723).

Febplate. A proprietary range of epoxy mortars. (723).

Febramine. The hydrochloride of Cetoxime.

Febrifuge Salt. Potassium chloride, KCl.

Febrilix. A proprietary preparation of paracetamol. An analgesic. (253).

Febset. A proprietary range of epoxy mortars. (723).

Fe-Cap. A proprietary preparation of ferrous glycine sulphate used in the treatment of anaemia caused by deficiency of iron. (249).

Fe-Cap C. A proprietary preparation of ferrous glycine sulphate and ascorbic acid used in the treatment of anaemia. (249).

Fe-Cap Folic. A proprietary preparation of ferrous glycine sulphate and folic acid used in the treatment of anaemia during pregnancy. (249).

Feck (Feech). Methylated spirit.

Fecraloy. A proprietary trade name for an alloy of iron, chromium, aluminium and yttrium under development for use in sintered form as a catalyst to assist in the control of atmospheric pollution by reducing the emission of carbon monoxide and other fumes from automobile exhausts. (944).

Feculose. The name given to various commercial starch esters.

Fedorovite. A mineral. It is $4[CaMgSi_2O_6]$.

Fedralite. A proprietary trade name for a Vinsol resin-treated laminated paper.

Fedrin Lilly. A proprietary preparation of ephedrine sulphate.

Feech. See FECK.

Fefol. A proprietary preparation of ferrous sulphate and folic acid. A hæmatinic. (281).

Fefol-Vit Spansule. A proprietary preparation of ferrous sulphate, folic acid and vitamins C and B complex, used in the prophylaxis of anaemia during pregnancy. (658).

Fehling's Solution. An alkaline solution of potassio-tartrate of copper. It is prepared in two solutions. (*a*) Consisting of copper sulphate, and (*b*) a solution of Rochelle salt (potassium-sodium tartrate) and caustic soda. Used for the identification and determination of sugars.

Fehling's Solution, Neutral. A solution made by adding 25 c.c. of a solution containing 2 grams copper (7·86 grams $CuSO_4.7H_2O$) per litre to 25 c.c. of a solution containing 3·292 grams sodium carbonate and 20 grams Rochelle salt per litre. It is stated to be a much more sensitive reagent for the detection of sugars.

Felamine. A proprietary preparation of cholic acid and hexamine. (267).

Felamine Sandoz. Hexamethylene-tetramine-glyco-cholate.

Feldspath Apyre. Synonym for Andalusite.

Felite. A constituent of blast furnace slag, and of slag cement.

Felixite. An explosive. It is a 42-grain powder, and contains metallic nitrates, nitro-hydrocarbons, and 3 per cent. pet. jelly.

Felopan. A proprietary preparation of bile, pancreatin, used for constipation. (361).

Felsinosima. A trade name for cultures of *Bacillus felsineus*. Used for flax retting.

Felsite. Orthoclase (*q.v.*).

Felsobanyite. A mineral. It is a hydrated basic aluminium sulphate.

Felsol. A proprietary preparation of phenazone, phenacetin, caffeine, iodine, liquid extract of grindel and extract of viscum. A bronchial antispasmodic. (837).

Felspar (Potassium Felspar, Microcline, Orthoclase). A potassium-aluminium silicate, $(K_2O.3SiO_2.) + (Al_2O_3.3SiO_2)$. Used in the manufacture of porcelain, as a building material, and as a fertiliser.

Felspar, Blue. Chessylite.

Felspar, Glassy. See SANIDINE.

Felspar, Lime. See ANORTHITE.

Felspar, Potash. See FELSPAR.

Felspar, Soda. See ALBITE and OLIGOCLASE.

Felt. Wool or hair made into sheet by matting it together.

Feltex. A proprietary trade name for a material consisting of an asphalt impregnated felt.

Felypressin. A vasoconstrictor.

Femapor. A proprietary trade name for base-metal compounds and base-metal mixtures used in the treatment of cast-iron. (899).

Femergin. A proprietary preparation of ergotamine tartrate, used for migraine. (267).

Femerital. A proprietary preparation containing nifuratel. Contains ambucetamide. (249).

Feminal. See GYNOVAL.

Feminor 21. A proprietary preparation of mestranol (16 pink tablets) and mestranol and norethynodrel (5 white tablets). Oral contraceptive. (804).

Feminor Sequential. A proprietary preparation of mestranol (15 pink tablets) and mestranol and norethynodrel (5 white tablets). Oral contraceptive. (804).

Femipausin. A proprietary preparation containing methyltestosterone and ethinylœstradiol. (312).

Femulen. A proprietary preparation of ETHYNODIOL diacetate, used as an oral contraceptive. (644).

Fenafix. A proprietary trade name for a series of modified vinyl-pyrrolidone resins. Used to modify the properties of other vinyl films and also to improve the adhesion of difficult surfaces. (193).

Fenamisal. Phenyl Aminosalicylate.

Fenasprate. 4-Acetamidophenyl O-acetylsalicylate.

Fenbendazole. An anthelmintic currently undergoing clinical trial as " Hoe 881 ". It is methyl 5-(phenylthio) - 1*H* - benzimidazol - 2 - ylcarba - mate.

Fenbutatin Oxide. A pesticide. It is di [tri - (2 - methyl - 2 - phenyl = pro - pyl) tin] oxide.

Fencamfamin. N - Ethyl - 3 - phenyl - bicyclo[2,2,1]hept-2-ylamine. Euvitol is the hydrochloride.

Fenchlorphos. An insecticide. *oo* - Di - methyl *o* - (2, 4, 5 - trichlorophenyl) phosphorothioate. NANCOR, RONNEL, TROLENE.

Fenchyval. The isovaleric ester of fennel oil. Employed for the same purposes as other valerates.

Fenclofenac. An anti-inflammatory currently undergoing clinical trial as " RX 67408 ". It is 2-(2,4-dichlorophenoxy)-phenylacetic acid.

Fenclozic Acid. An anti-inflammatory currently undergoing clinical trial as " ICI 54 450 ". It is 2-(4-chlorophenyl)-thiazol-4-ylacetic acid.

Fenethylline. A cerebral stimulant. 7-[2 - (α - Methylphenethylamino) ethyl] - theophylline.

Fenfluramine. 2 - Ethylamino - 1 - (3 - trifluoromethylphenyl)propane. Ponderax is the hydrochloride.

Fenimide. 3 - Ethyl - 2 - methyl - 2 - phenylsuccinimide. CI 419.

Fenisorex. An anorexigenic. (±)-*cis*-7 - Fluoro - 1 - phenylisochroman - 3 - ylmethylamine.

Fenmetramide. An anti-depressant. 5 - Methyl - 6 - phenyl - 3 - morpholinone.

Fennel. The ripe fruit of *Fœniculum capillaceum*.

Fennite. A fungicide. (509).

Fenobelladine. A proprietary preparation of phenobarbitone and belladonna dry extract. (250).

Fenolac. Cyclised rubber used in adhesives and bonding agents. THERMO-PRENE.

Fenolite. An Italian synthetic resin material for use in electrical insulation.

Fenoprofen. An anti-inflammatory and analgesic currently undergoing clinical trial as " 53 858 ". It is 2-(3-phenoxy-phenyl)propionic acid.

Fenopron. A proprietary preparation of FENOPROFEN calcium. An anti-inflammatory drug. (309).

Fenostil. A proprietary preparation of dimethindine hydrogen maleate. (373).

Fenostil Retard. A proprietary preparation of DIMETHINDENE maleate. An anti-pruritic drug. (724).

Fenoval. Phenoval.

Fenox. A proprietary preparation of phenylephrine hydrochloride. A nasal spray. (253).

Fenpipramide. 2,2-Diphenyl-4-piperidinobutyramide.

Fenpiprane. (1 - (3,3 - Diphenylpropyl) - piperidine.

Fentachol. Tranquillizing and anticholinergic preparations. (510).

Fentanyl. 1 - Phenethyl - 4 - (N - pro - pionylanilino)piperidine. Sublimaze.

Fentazin. A proprietary preparation of perphenazine. A sedative. (284).

Fenthion. An insecticide. *OO*-Dimethyl *O* - 4 - methylthio - *m* - tolyl phosphoro-thioate. TIGUVON.

Fenticlor. Di - (5 - chloro - 2 - hydroxy - phenyl)sulphide. S7.

Fenton's Metal. A bearing metal. It contains about 80 per cent. zinc, 15 per cent. tin, and 5 per cent. copper.

Fenton's Reagent. Hydrogen peroxide and a ferrous salt. Used for the oxidation of polyhydric alcohols.

Fenton's Rubber. A patented rubber substitute made from oils, tar, and pitch, which has been treated with dilute nitric acid.

Fenzol. A proprietary preparation of dichloralphenazone. A hypnotic. (279).

Feospan. A proprietary preparation of dried ferrous sulphate. A hæmatinic. (281).

F.E.P. A proprietary name for Teflon 100. (10).

Feprazone. An analgesic and anti-inflammatory currently undergoing clinical trial as "DA 2370". It is 4-(3-methylbut - 2 - enyl) - 1, 2 - diphenyl - pyrazolidine-3,5-dione.

Feraloy. A nickel-steel-chromium alloy, having a specific gravity of 8·15, and melting at 1480° C.

Feran. A proprietary trade name for a dual metal of aluminium and iron in which the aluminium is made to adhere to the iron.

Fer Ascoli. A compound of iron with nuclein. Used in medicine.

Feraspartyl. A proprietary preparation of ferrous aspartate and ascorbic acid. A hæmatinic. (320).

Feravol. A proprietary preparation of ferrous gluconate and folic acid. A hæmatinic. (246).

Ferbelan. Preparations of B Vitamins with iron. (527).

Ferberite. A black opaque mineral. It consists essentially of iron tungstate, $FeWO_4$.

Fer Cremol. See FERROHEMOL.

Feretene. A trade mark for a polyethylene. (61).

Ferfolic. A proprietary preparation containing folic acid, thiamine hydrochloride, riboflavine, ferrous gluconate, nicotinamide, and ascorbic acid. Iron-vitamin supplement. (259).

Ferfolic " M ". A proprietary preparation of folic acid, ferrous gluconate and ascorbic acid. A hæmatinic. (259).

Ferfolic " SV ". A proprietary preparation of folic acid, ferrous gluconate and ascorbic acid. A hæmatinic. (259).

Ferganite. A mineral. It is a hydrated uranium-vanadate, $U_3(VO_4)_26H_2O$.

Ferghanite. Ferganite, a mineral.

Fergluvite. A proprietary preparation of ferrous gluconate, thiamine hydrochloride, fiboflavine, nicotinamide and ascorbic acid. A hæmatinic. (259).

Fergon. A trade mark. See FERGAN. A proprietary preparation of ferrous gluconate. A hæmatinic. (112).

Fergusonite. A mineral. It is a niobate of yttrium and tantalum, $Y(Nb. Ta)O_4$.

Ferlucon. A proprietary preparation of ferrous gluconate and aneurine hydrochloride. A hæmatinic. (182).

Fermangol. The trade name for a

remedy for chlorosis and nervous complaints. It is said to be an aqueous-alcoholic solution, containing 5 per cent. iron-manganese saccharate, 1·5 per cent. glycerol phosphate, and 14 per cent. cane sugar.

Fermenlactyl. A preparation of lactic acid bacilli.

Fermentation Alcohol. Ethyl alcohol, C_2H_5OH.

Fermentation Amyl Alcohol. See FUSEL OIL.

Ferment Diagnosticum. A solution of glycyl-tryptophan. Used for determining the presence of a peptolytic ferment.

Fermented Oil. The oil obtained from fermented olives.

Fermenticide. An antiseptic compound. (504).

Fermentine. A preparation of yeast containing benzoin. Used in medicine for local applications.

Fermentol. A brown liquid, which contains from 98–99 per cent. glycerin. Used for the manufacture of nitro-glycerin.

Fermet Alloy. An alloy of 74·5 per cent. iron, 18 per cent. nickel, 2·2 per cent. manganese, 0·7 per cent. tungsten, 0·3 per cent. copper, and 0·35 per cent. carbon.

Fermin. A proprietary preparation of cyancobalamin in an oral form. A hæmatinic. (395).

Fermine. A proprietary trade name for methyl phthalate.

Fermocyl. A mixture of dried yeast, pancreas, and sodium phosphate. Used in medicine.

Fermorite. A mineral, $(Cu.Sr)_5(F.OH) [(P.As)O_4]_3$.

Ferna-Col. A fungicidal spray. (511).

Fernambuco Wood. See REDWOODS.

Fernandinite. A mineral. It is a hydrous calcium vanadyl vanadate, $CaO.V_2O_4.5V_2O_5.14H_2O$.

Fernasan. A non-mercurial seed dressing. (511).

Fernasul. A lime and sulphur fungicide. (511).

Fernbach's Culture. A micro-organism of the long rod type which breaks up the complex molecule of starch into acetone and n-butyl alcohol. About 15–24 per cent. acetone and about twice this amount of n-butyl alcohol are produced.

Fern Balsam. See BALSAM OF FERN.

Fernesta. A selective weed killer. (511).

Fernico Alloy. See KOVAR ALLOY.

Fernide. A foliage fungicide. (511).

Fernimine. A selective weed killer. (511).

Fernoxone. A selective weed killer. (511).

Feronia. Wood-apple gum, derived from an Indian tree. It is often used as an adulterant of gum arabic.

Feronuclin. Dry yeast extract.

Feroxal. A granular preparation consisting of ferrous oxalate in combination with sodium phosphate. Recommended as a nerve tonic in anæmia.

Ferox-Celotex. A proprietary trade name for Celotex (*q.v.*) which has to be treated to resist attack by fungi and termites.

Ferozon. A disinfectant.

Ferralbol. Asserted to be a chemical combination of iron with eggs, and said to contain 3 per cent. iron and 1 per cent. lecithin. Used in the treatment of anæmia.

Ferralbumose. Made by treating finely-cut meat, freed from fat, with an artificial gastric juice.

Ferrantigorite. Synonym for Ferro-antigorite.

Ferraplex-B. A proprietary preparation containing ferrous sulphate copper carbonate, ascorbic acid and natural Vitamin B complex. A hæmatinic. (272).

Ferratine. A compound of iron and albumin, containing from 6–8 per cent. iron. It is a form soluble in alkaline liquids. A tonic.

Ferratogen. A compound of nuclein and iron, containing 1 per cent. iron. Used in medicine.

Ferratose. A solution of Ferratin (*q.v.*), containing 0·3 per cent. iron.

Ferrazite. A mineral. It is a hydrated phosphate of lead and barium, $3(Pb.Ba)O.P_2O_5.8H_2O$, found in Brazil.

Ferrhæmine. A compound of fresh ox-blood with iron, wine being added as preservative.

Ferribeidellite. A clay of composition $(Al,Fe''')_2Si_3O_9.4H_2O$.

Ferrical. A proprietary preparation of ferrous iron, calcium, aneurine hydrochloride, and phenolphthalein. A tonic. (350).

Ferrichrompicotite. A mineral. It is $8[(Mg,Fe)(Cr,Al,Fe)_2O_4]$.

Ferrichthol (Ferrichthyol). An iron preparation containing 3·5 per cent. iron, and 96·5 per cent. ichthyol-sulphonic acid. Used medicinally as a hæmatinic.

Ferrichthyol. See Ferrichthol.

Ferricodil. Iron cacodylate.

Ferricopiapite. A mineral. It is $[Fe_5'''(SO_4)_6O(OH).nH_2O]$.

Ferriepidote. A mineral. It is $Ca_2Fe_3'''Si_3O_{12}OH$.

Ferrierite. A mineral. It is a silicate of aluminium, magnesium, and sodium.

Ferrigarnierite. A mineral. It is $(Ni,Mg)_3Si_2O_5(OH)_4$.

Ferrigedrite. A mineral. It is $4[(Mg,Fe''')_7(Si,Fe''')_8(O,OH)_{24}]$.

Ferrikalite. An artificial potassium ferric sulphate.

Ferrimolybdite. A mineral, $Fe_2O_3.7MoO_3.19H_2O$.

Ferrinol. Iron nucleinate, containing 21 per cent. ferric oxide, and 2·5 per cent. phosphorus.

Ferriobenate. See Ferro-Sajodin.

Ferri-Orthoclase. A mineral. It is $KFe'''Si_3O_8$.

Ferri-Purpurite. Synonym for Heterosite.

Ferripyrine. See Ferropyrine.

Ferrisol. A substance said to be derived from cinnamic acid and guaiacol. It has been used intramuscularly in tubercular diseases.

Ferrisul. A trade name for ferric sulphate.

Ferri-symplesite. A mineral of Ontario. It approximates to the formula $3Fe_2O_3.2As_2O_5.16H_2O$.

Ferrite. Nearly pure iron. Phosphorus and sulphur may be present in minute quantities, but the carbon content is not more than 0·05 per cent. It is free α-iron.

Ferrititanbiotite. Synonym for Ferriwotanite.

Ferritungstite. A mineral, $Fe_2O_3.WO_3.6H_2O$. It occurs associated with quartz in the State of Washington.

Ferrivine. Ferric sulphanilate, $Fe(SO_3.C_6H_4.NH_2)_3$. Used in the treatment of syphilis.

Ferriwotanite. A mineral. It is $2[K_2(Mg, Fe'', Fe''', Ti)_5(Al, Ti, Si)_8O_{20}(OH)_4]$.

Ferro-Alumen. Iron alum (*q.v.*).

Ferro-aluminium. See Aluminium-Iron. It contains up to 20 per cent. aluminium with iron, and is used in the preparation of iron and steel. The alloy containing 80 per cent. iron and 20 per cent. aluminium has a specific gravity of 6·3, and melts at 1480° C.

Ferro-Anthophyllite. A mineral. It is $4[Fe_7Si_8O_{22}(OH)_2]$.

Ferro-Antigorite. A mineral. It is $Fe_3Si_2O_5(OH)_4$.

Ferro-argentan. An alloy resembling silver and containing 70 per cent. copper, 20 per cent. nickel, 5·5 per cent. zinc, and 4·5 per cent. cadmium.

Ferroaugite. A mineral. It is $8[(Ca,Mg,Fe'')SiO_3]$.

Ferroaxinite. A mineral. It is $2[Ca_2FeAl_2BSi_4O_{15}OH]$.

Ferro-boron. An alloy of iron and boron, containing from 20–25 per cent. boron. It is added to steel.

Ferrobrucite. A mineral. It is $(Mg,Fe)(OH)_2$.

Ferrocalcite. A mineral containing calcite and ferrous carbonate.

Ferrocap. A proprietary preparation of ferrous fumarate and Vitamin B_1. A hæmatinic. (256).

Ferrocap F 350. A proprietary preparation of ferrous fumarate and folic acid used in the treatment of anæmia during pregnancy. (256).

Ferro-carbon-titanium. An alloy of iron and titanium, containing carbon. Used for making steel.

Ferro-cerium. See AUER METAL.

Ferrochlor. A mixture of ferric chloride and calcium hypochlorite. Used to clarify water.

Ferro-chrome. See CHROME IRON.

Ferro-chromium. A British Chemical Standard Alloy. Low carbon No. 203 contains 69 per cent. and over chromium, 0·08 per cent. carbon, and 0·01 per cent. sulphur, while the high carbon alloy contains 71·4 per cent. and over chromium, 5·09 per cent. carbon, and 0·02 per cent. sulphur.

Ferrochromite. See CHROME IRON ORE.

Ferro-cobalt. An alloy of 70 per cent. cobalt with iron. Used for adding cobalt to steel.

Ferrocobaltite. A mineral. It is a cobaltite containing iron.

Ferrocolumbite. A mineral. It is $4[Fe(Cb,Ta)_2O_6]$.

Ferro-concrete. See REINFORCED CONCRETE.

Ferrocrete. A brand of rapid hardening Portland cement of high strength.

Ferro-cupralium. An alloy of 75–80·5 per cent. copper, 11–12 per cent. aluminium, and 2–13 per cent. iron.

Ferrodic. A proprietary preparation containing ferrous carbonate and ascorbic acid. A hæmatinic. (284).

Ferrodur. A substance containing Nitrolim (q.v.). It is used for case-hardening and tempering iron and steel.

Ferro-Elarsan. Strontium chlor-arseno-behenolate and reduced iron.

Ferroepsomite. A mineral. It is $4[(Mg,Fe)SO_4 . 7H_2O]$.

Ferroferrite. See MAGNETIC IRON ORE.

Ferrogen. An iron degasser, scavenger and deoxidant. (531).

Ferro-glidin. A combination of iron with Glidine (q.v.).

Ferrogoslarite. The mineral Goslarite, $ZnSO_4$, in which part of the zinc is replaced by iron.

Ferrograd C. A proprietary preparation containing ferrous sulphate and ascorbic acid. A hæmatinic. (311).

Ferrograd Folic. A proprietary preparation of ferrous sulphate and folic acid used in the treatment of anæmia during pregnancy. (311).

Ferro-Gradumet. A proprietary preparation of ferrous sulphate in a slow release form. A hæmatinic. (311).

Ferrohemol. A brown powder containing 3 per cent. iron, which is made by the precipitation of blood by a dilute solution of iron, and neutralisation with soda. Fer cremol is a similar preparation.

Ferroids. A proprietary preparation of iron and aminoates. A hæmatinic. (275).

Ferrolon. Sequestering agents. (582).

Ferroilmenite. A mineral. It is columbite.

Ferroludwigite. Synonym for Paigeite.

Ferro-magnesite. Obtained by burning magnesite with iron ore. It is used as a lining for furnaces.

Ferro-Mandets. A proprietary preparation of ferrous fumarate and ascorbic acid. A hæmatinic. (306).

Ferromangan. A solution containing 0·6 per cent. iron, 0·1 per cent. manganese, and 1·5 per cent. peptone. Used in anæmia, chlorosis, and as a general tonic.

Ferromangandolomite. A mineral. It is $Ca(Fe,Mn)(CO_3)_2$.

Ferro-manganese. An alloy of manganese with iron and carbon, usually made in the blast furnace. High-grade alloys contain about 78 per cent. manganese and 8 per cent. carbon. These alloys vary from 50–80 per cent. manganese, 10–42 per cent. iron, 2 per cent. silicon, and 5–8 per cent. carbon.

Ferrominit. A German non-poisonous pigment containing 94 per cent. ferric oxide, Fe_2O_3. It is free from acids and very resistant to carbon dioxide, sulphur dioxide, sulphuretted hydrogen, and ammonia.

Ferro-molybdenum. An alloy of iron with 80 per cent. molybdenum. Used in the place of molybdenum in the manufacture of hard steel.

Ferromyn. A proprietary preparation of ferrous succinate. A hæmatinic. (247).

Ferromyn S. A proprietary preparation of ferrous succinate and succinic acid. A hæmatinic. (247).

Ferron. An alloy of 50 per cent. iron, 35 per cent. nickel, and 15 per cent. chromium. Also a building material prepared from the pickling liquor from steel mills. It consists of precipitated iron oxide.

Ferronatrite. A mineral. It approximates to $3Na_2SO_4.Fe_2(SO_4)_3.6H_2O$.

Ferro-nickel (Nickel Iron). An alloy

of iron and nickel, usually containing 25 per cent. nickel. The alloy of 74·2 per cent. iron, 25 per cent. nickel, and 0·8 per cent. carbon has a specific gravity of 8·1, and melts at 1500° C., and that containing 67·8 per cent. iron, 32 per cent. nickel, and 0·2 per cent. carbon has a gravity of 8, and melts at 1480° C.

Ferronite. A solid solution of about 0·27 per cent. carbon in β-iron.

Ferro-phosphorus. An alloy of iron and phosphorus, used in steel making for thin castings.

Ferrophytin. Neutral colloidal inositol hexaphosphate of iron containing about 7·5 per cent. iron and 6 per cent. phosphorus. A tonic and hæmatopoietic.

Ferropyrine (Ferripyrine). A compound of antipyrine with ferric chloride, $(C_{11}H_{12}N_2O)_3(FeCl_3)_2$. It is taken internally as a remedy for chlorosis and neuralgia.

Ferropyroaurite. Synonym for Eisenbrucite.

Ferro Redoxon. A proprietary preparation of ferrous sulphate and ascorbic acid. A hæmatinic. (314).

Ferrorichterite. A mineral. It is Na_2CaFe_5··$Si_8O_{22}(OH)_2$.

Ferro-sajodin. Basic ferric iodobehenate, containing 5 per cent. iron, and 24 per cent. iodine. Used in medicine.

Ferrosal. A solution of iron saccharate and sodium chloride, containing 0·77 per cent. iron. Employed in anæmia and chlorosis.

Ferro-silicon. Alloys of iron and silicon made in the arc type electric furnace. They are graded upon the silicon content. The ordinary grades containing 25, 45–50, 75, and 95 per cent. silicon are used in steel works. The quality containing 95 per cent. silicon generates hydrogen by treatment with boiling water. Also see SILICOL.

Ferro-silicon-aluminium. An alloy of iron, silicon, and aluminium, containing up to 15 per cent. silicon.

Ferrostabil. A stabilised ferrous chloride preparation used in medicine in the form of tablets and suppositories.

Ferrostibian. A mineral. It is a ferrous manganese antimonate.

Ferrostyptine. The double salt of hexamethylene-tetramine-chloride, and ferric chloride, $(CH_2)_6N_4.HCl.FeCl_3$. Used as an astringent and astyptic in dental practice.

Ferrotantalite. A mineral. It is columbite.

Ferrotellurite. A mineral, $FeTeO_4$.

Ferro-titanium. An alloy of iron and titanium, usually containing from 15–18 per cent. titanium. It, however,

sometimes consists of 23–25 per cent. titanium, 70–72 per cent. iron, and 5 per cent. aluminium. Used as a purifying agent for steels.

Ferrotone. A proprietary preparation containing iron and ammonium citrate, potassium glycerophosphate, copper sulphate, manganese glycerophosphate, glycerin, citric acid, and Vitamin D. A tonic. (380).

Ferrotubes. An iron degassing and scavenging agent. (531).

Ferro-tungsten. An alloy of iron and tungsten. It usually contains from 65–85 per cent. tungsten, and from 1–2 per cent. carbon. Used in the steel industry.

Ferro-uranium. Alloys of iron and uranium, containing from 30–50 per cent. uranium. Used in steel making.

Ferro-vanadium. An alloy of iron and vanadium, containing from 20–40 per cent. vanadium. It is added to steel and iron.

Ferro-vanadium (No. 205). A British Chemical Standard alloy. Vanadium 52·2 per cent. (standard). It also contains carbon 0·16 per cent., sulphur 0·03 per cent. (not standard), silicon 1·18 per cent., phosphorus 0·05 per cent.

Ferrowollastonite. A mineral. It is $6[(Ca,Fe··)SiO_3]$.

Ferrox (Ferrite). Trade names for yellow iron oxides used as paint pigments. They consist of 98–99 per cent. $Fe(OH)_3$, with calcium sulphate.

Ferroxyl Reagent. A gelatin or agar-agar jelly containing phenolphthalein and potassium ferricyanide. When a piece of iron is placed in the jelly, colours are formed at the ends of the metal after a time. Iron ions give a colour of Turnbull's blue with the potassium ferricyanide, and hydroxyl ions a pink colour with the phenolphthalein.

Ferrozell. A proprietary trade name for a synthetic resin.

Ferro-zirconium. A 20 per cent. zirconium alloy with iron. Used to remove nitrogen and oxides from steel.

Ferrozoid (Vestalin). Alloys. They are usually 28 per cent. nickel steels and are used as electrical resistances. The alloy containing 70 per cent. iron and 30 per cent. nickel has a specific resistance of 84 micro-ohms per cm.^{3}, melts at 1490° C., and has a specific gravity of 8·13.

Ferrozone. A saccharated iron and vanadium compound.

Ferrucite. A mineral. It is $4[NaBF_4]$.

Ferrugo. Ferric hydroxide, $Fe(OH)_3$.

Ferrul. An alloy of 54·6 per cent.

copper, 40 per cent. zinc, 5 per cent. lead, and 0·4 per cent. aluminium.

Ferrum. The Latin name for iron.

Ferry Alloy. A trade mark for an electrical resistance alloy of 54 per cent. copper and the balance nickel, a material with a very low temperature coefficient of resistance. (143).

Ferrybar. A proprietary preparation of iron and ammonium citrate, riboflavine, nicotinamide and aneurine hydrochloride, used to prevent iron deficiency anæmia. (273).

Ferry Metal. Alloys used for electrical resistance and containing 40–45 per cent. nickel and 55–60 per cent. copper. The alloy containing 40 per cent. nickel and 60 per cent. copper has a specific resistance of 48 micro-ohms per cm.[3], melts at 1250° C., and has a specific gravity of 8·9. The name appears to be also applied to a bearing alloy and solder containing lead with 2 per cent. barium, 1 per cent. copper, and 0·25 per cent. mercury.

Fersaday. A proprietary preparation of ferrous fumarate. A hæmatinic. (335).

Fersamal. A proprietary preparation of ferrous fumarate. A hæmatinic. (335).

Fersmite. A mineral. It is $(Ca,Ce)(Cb,Ti)_2(O,F)_6$.

Fersolate. A proprietary preparation of ferrous, manganese and copper sulphates. A hæmatinic. (335).

Fersan. A para-nucleine compound containing iron. It contains albumin in a soluble form, also phosphorus. It may be regarded as an iron compound, also a food, and is prepared from ox-blood.

Fertene. A registered trade mark for low density polyethylene. (61).

Fertiliser, Completely Soluble. A Belgian fertiliser which is stated to be a mixture containing nitrogen, phosphorus, potassium, and calcium mixed in a special manner. There are three varieties : (a) Containing 4 per cent. nitrogen, 10 per cent. phosphorus pentoxide, and 4 per cent. potassium ; (b) containing 6 per cent. nitrogen, 10 per cent. phosphorus pentoxide, and 8 per cent. potassium ; and (c) containing 4 per cent. nitrogen, 10 per cent. phosphorus pentoxide, and 8 per cent. potassium.

Fertiliser, Nitrogenous. See SKIN AND LEATHER MEALS.

Fervanite. A mineral. It is a hydrous ferric vanadate, $Fe_2O_3.2V_2O_5.5H_2O$.

Fervine. A meat extract combined with iron compounds.

Fesovit. A proprietary preparation of ferrous sulphate, ascorbic acid and

B vitamins used in the treatment of anæmia. (658).

Festoform. A solid preparation of formaldehyde, obtained by mixing an aqueous solution of formaldehyde with a soda soap solution. An antiseptic disinfectant and deodoriser.

Fetid Aloe. See ALOE.

Fetid Spirit of Ammonia. (Spiritus ammoniæ fetidus B.P.) An alcoholic solution of the volatile oil of asafœtida mixed with a solution of ammonia.

Fetoxylate. An anti-diarrhæal. 2-Phenoxyethyl 1 - (3 - cyano - 3, 3 - diphenyl - propyl) - 4 - phenylpiperidine - 4 - carboxylate. " R 13,558 " is currently under clinical trial as the hydrochloride.

Fetron. A mixture of yellow pet. jelly and pure anilide of stearic acid. An ointment.

Feuerkohle. A brown coal of Germany. It is dark in colour, and is used for making briquettes.

Feuille Morte. A jeweller's alloy, containing 70 per cent. gold and 30 per cent. silver.

Fever Drops. Compound tincture of cinchona.

Feximac. A proprietary preparation of Buteximac in a cream base, used in the treatment of eczema. (271).

Fiberglass. A trade mark for glass in the form of fine fibres. These are twisted into yarn or woven into tapes, etc.

Fiberlac. A proprietary trade name for cellulose nitrate lacquer.

Fiberloid. See VISCOLOID. A proprietary trade name for a cellulose nitrate plastic resistant to oils.

Fiberlon. A proprietary trade name for a phenol-formaldehyde synthetic resin resistant to oils.

Fiberoid. An American grade of vulcanised fibre used for electrical insulation. It is also the name for a celluloid.

Fiberock. A proprietary trade name for a roofing material consisting of asbestos impregnated with asphalt.

Fibestos. A proprietary trade name for a cellulose acetate plastic.

Fibralda. A proprietary trade name for untwisted cellulose acetate fibres.

Fibre B. See KEVLAR.

Fibre, Egyptian. See VULCANISED FIBRE.

Fibre, Grey. See VULCANISED FIBRE.

Fibre, Horn. See VULCANISED FIBRE.

Fibre, Red. See VULCANISED FIBRE.

Fibrestos. A silica-asbestos product used as an insulator.

Fibresul. A sulphur-impregnated fibre product. The fibre is made from wood pulp, and is then immersed in molten sulphur. It is stated to be a suitable material for acid-proof pipes.

Fibre, Vegetable. See Vulcanised Fibre.

Fibre, Whalebone. See Vulcanised Fibre.

Fibrin-carmine. Carmine, 1 gram, is dissolved in 1 c.c. ammonia. Add 400 c.c. water and leave in flask stoppered with cotton wool until the odour of ammonia has nearly gone. Drop washed fibrin into the solution and leave for 24 hours. Strain fibrin from the fluid until the washings are colourless and store in ether.

Fibrino-plastic Substance. See Globulin.

Fibrinoplastin. See Globulin.

Fibro. A proprietary artificial silk product. It is a staple fibre (q.v.) made by the viscose process.

Fibroc. A laminated product consisting of fibre impregnated with a phenol-formaldehyde resin.

Fibro-cement. A term used in Australia for a product manufactured from asbestos and cement in the form of sheet. Usually it contains 1 part asbestos and 5 parts cement. It is fireproof, and is used as a building material.

Fibrofelt. A proprietary trade name for a sound insulating board for walls made from flax, and rye-straw fibres.

Fibro-ferrite. A mineral, $2Fe_2(SO_4)_2$ $(OH)_2.Fe_2SO_4(OH)_4.24H_2O$.

Fibrolite (Bucholzite). A mineral, $Al_2O_3.SiO_2$.

Fibrolysin (Thiosinyl). A trade mark for a 15 per cent. solution of the double salt of thiosinamine and sodium salicylate. Used in medicine for softening scars and wounds.

Fibron. A trade mark for a surfacing material for resurfacing and treatment of floors resulting in a plastic finish.

Fibrotex. A proprietary trade name for a roofing cement consisting of asbestos mixed with oil and gum.

Fibrox. A variety of siloxicon. It is obtained by heating in a gas furnace fragments of silicon, whereby the silicon vapour, reacting with carbon monoxide or dioxide, gives oxycarbide, a soft, elastic material, which is called fibrox. It is a thermal insulator.

Ficel. Blowing agents. (509).

Ficel AC. A proprietary activated azodicarbonamide blowing agent. (725).

Ficel CR. A proprietary blowing agent similar to Ficelac, but developed specifically for expanding neoprene, natural rubber, polyolefines and P.V.C. Its decomposition temperature lies between 320–375° F. (725).

Fi-Chlor. A proprietary trade name for a chlorcyanurate used for rendering wool shrink-resistant. (188).

Fichtelite. Perhydro-retene, $C_{18}H_{32}$, found in fossil coniferous resins.

Ficinite. A mineral. It is a variety of hypersthene.

Ficoid. A range of proprietary skin creams containing fluocortolone used in the treatment of eczema and related dermatoses. (188).

Fi-Cryl. Acrylic copolymer solutions and dispersions. (509).

Fiddle Gum. Gum tragacanth.

Fiedlerite. A mineral, $2PbCl_2.Pb(OH)_2$.

Field's Orange Vermilion. See Scarlet Vermilion.

Fierroso. See Acerado.

Fife or Fife's Snuff. White snuff (q.v.).

Fi-Gard. Rubber and plastics additives. (509).

Filabond. A proprietary polyester laminating resin. (181).

Filair. A proprietary preparation of Terbutaline. A bronchodilator. (275)

Filao Bark. The bark of *Casuarina equisetifolia*. It contains 15 per cent. catechol-tannin.

Filastic. A proprietary rubber textile yarn in which the rubber latex impregnation takes place during spinning. The yarns contain 50 per cent. of rubber.

Filcryl. A proprietary acrylic polymer for dental fillings.

File Bronze. An alloy of 64·4 per cent. copper, 18 per cent. tin, 10 per cent. zinc, and 7·6 per cent. lead.

Filfloc. A proprietary trade name for a cotton flock filler.

Filhos's Caustic. A caustic. It is a mixture of potassium hydroxide, 5 parts, and quick lime, 1 part, fused together.

Filicic Acid. An acid obtained from male fern. It is used in the treatment of worms.

Filicon. A tænicide, consisting of an extract of the rhizome of *Aspidium spinulosum*, combined with castor oil.

Fi-Line. Dyeline chemicals for photocopying. (509).

Filite. A smokeless explosive. It is Ballistite (q.v.), drawn out into cords with the aid of a solvent.

Filivex. A proprietary fish liver extract.

Filled Gold. Rolled gold plate (q.v.).

Fillite. A proprietary inert silicate in the form of spheres. Hard as glass, it is used as a filling material for plastics. (726).

Fillowite. A mineral, $(Mn.Fe.Na_2.Ca)_3$ $(PO_4)_2$.

Filmafalie. Nitrocellulose film scrap.

Filmarone Oil. A 10 per cent. solution of filmarone (present in fern roots) in castor oil. Used as a specific against worms.

Filmite. White oil preparations. (501).

Filmogen (Collosin). A vehicle for the application of medicinal substances to the skin. It is said to be a solution of pyroxylin in acetone, containing a trace of fixed oil.

Filon. A proprietary preparation of phenmetrazine theocalate, and phenbutrazate hydrochloride. An appetite suppressant. (281).

Filt-char. A proprietary brand of bone charcoal, used as a filtering medium.

Filter-cel. A proprietary preparation of infusorial earth, used in filtering.

Filtered Cylinder Oils. See CYLINDER OILS.

Filtrol. A trade mark for a decolorising substance consisting of fine silica with a little aluminium silicate.

Filtros. An acid-proof mineral used for filtering.

Finajet. A proprietary preparation of TRENBOLONE. An anabolic steroid used in veterinary medicine.

Finalgon. A proprietary preparation of nonylic acid vanillymide and butoxyethyl nicotinate. (278).

Fine Black. Aniline Black (q.v.).

Fine Blue. See ANILINE BLUE, SPIRIT SOLUBLE.

Fine Gold. A jeweller's alloy, containing 75 per cent. gold and 25 per cent. silver.

Fine Gold Solder. See GOLD SOLDERS.

Fine Silver. 99·9 per cent. pure silver.

Fine Solder. See PLUMBER'S SOLDER and TINSMITH'S SOLDER.

Finestol. A phloroglucinol dye coupler. (509).

Fining Glue. A glue manufactured from bones.

Finings. The term applied to isinglass dissolved in an acid such as tartaric acid. Used to clarify beer.

Finnemanite. A mineral. It is a lead arsenite found at Langban, and approximates to the formula, $Pb_5Cl(AsO_3)_3$.

Finnish Turpentine. See TURPENTINE.

Fiolax. A trade name for a resistance glass.

Fiorite. A mineral. It is silica, SiO_2.

Fique. A native name for hemp fibre grown in tropical America.

Fir Balsam. See CANADA TURPENTINE.

Fire-air. See EMPYREAL.

Fire-armour. Heat-resisting alloys containing 60–61 per cent. nickel, 18–20 per cent. chromium, 10–20 per cent. iron, 0–1·8 per cent. manganese, and 0·5 per cent. carbon.

Fire Blende (Pyrostilpnite). A rare silver mineral similar to Pyrargyrite.

Firebrake. A fire retardant for forest fires. (548).

Fireclay. Clay containing a considerable amount of free silica.

Firecol. An alginate fire-fighting suspension. (536).

Firecrete. A proprietary trade name for a calcined high alumina clay used as a refractory in furnaces.

Fire-damp. A gas, mainly consisting of methane, which is formed by the conversion of wood into coal, and is therefore often found in coal mines.

Fire Felt. A proprietary trade name for a felt made from asbestos.

Firefoam. A carbon dioxide froth made by mixing solutions of alum and sodium bicarbonate. A protective colloid, such as glue, is used to stabilise the finely dispersed foam. Used for blanketing fires.

Fireite. A trade mark for a heat-resisting, unglazed porcelain.

Fire Marble. A marble containing fossil shells.

Fireproof Cyanite. A fireproof paint. It is a mixture of aluminium and sodium silicates.

Fireproof Glue. A glue obtained by soaking 1 part glue in 8 parts raw linseed oil, and heating. Quicklime, 2 parts, is then stirred in.

Fireproof Paints. Paints containing asbestos with small amounts of sodium tungstate, alum, or ammonium phosphate, in a medium of casein and sodium silicate. Others contain an oil or glue medium with the fireproof base.

Firit. A foundry refractory coating. (531).

Firnagral. A mineral drying oil. Used to replace up to 30 per cent. of the linseed oil in putty. It is stated to give the putty a harder and smoother surface.

Firnis. Linseed oil and driers.

First Runnings. See LIGHT OIL.

Firthite. A proprietary trade name for a material consisting of a mixture of tungsten and other carbides.

Fir Wool. See PINE WOOL.

Fir Wool Oil. The oil of Scotch fir leaves.

Fischerite. A mineral, $2Al_2O_3.P_2O_5.8H_2O$.

Fischer-Langbein Solution. A solution containing 450 grams ferrous chloride, 500 grams calcium chloride, and 750 c.c. water. Used for the electro-deposition of iron, with a current density of up to 120 amps. per sq. ft. A temperature of 60–70° C. is employed.

Fischer's Reagent. A test solution for sugars. It consists of 2 parts phenylhydrazine-hydrochloride and 3 parts sodium acetate in 20 parts of water.

Fischer's Salt. Potassium-cobaltic-nitrite, $K_3Co(NO_2)_6$. Used for the detection and determination of potassium.

Fischer's Yellow. Fisher's salt (q.v.).

Fisetin. Tetroxy-methyl-anthraquinone, $C_{15}H_{10}O_6$. A yellow colouring from the wood of *Quebracho Colorado*, etc.

Fish Albumen. An albumen prepared from fish spawn. It is an impure product.

Fish Berry. The fruit of *Cocculus indicus* and *Anamirata paniculata*. It acts as a powerful fish poison. The active principle is picrotoxin, $C_{30}H_{34}O_{13}$.

Fish Gelatin. See ISINGLASS.

Fish Glue. A material made at fish manure factories, from the water in which the fish has been cooked. The aqueous liquid is separated from the oil and gurry (finely divided fish), and clarified by the addition of alum. The solution is filtered and concentrated to a strength representing 32 per cent. of dry glue, and bleached. The glue is used by shoe manufacturers, bookbinders, makers of boxes, musical instruments, and artificial flowers.

Fishing Salt. A coarse-grained variety of salt (sodium chloride), used in the fish-curing trade.

Fish Manure (Guanos). A manure made from the offal of fish, and from whole fish, by the removal of fats, and disintegration.

Fish Paper. See LEATHER PAPER.

Fisons MCPB. A selective weedkiller. (509).

Fisons P.C.P. A weedkiller. (509).

Fisons 18-15. A selective weedkiller. (509).

Fi-Vi. A lightweight expanded PVC. (509).

Fixanal. Analytical chemicals accurately weighed and sealed, ready for rapid volumetric solution.

Fixanol. A fixing agent for direct dyestuffs. (512).

Fixature. See BANDOLIN.

Fixed Air. The name given by Black to carbon dioxide, CO_2.

Fixed Nitre. Potassium carbonate, K_2CO_3.

Fixed White. Commercial barium sulphate, $BaSO_4$.

Fixin. Aluminium lactate. Recommended as a gastro-intestinal disinfectant.

Fixinvar. An alloy having the same properties as Elinvar, but having greater stability.

Fixol. Adhesives. (587).

Fixopone. See OIL WHITE.

Fix-Sol. A concentrated fixing and hardening solution. (515).

Fizelyite. A mineral. It is an argentiferous lead ore, $5PbS.Ag_2S.4Sb_2S_3$.

Flagstaffite. A material found with resin in the cracks of buried pine trees in Arizona. It is identical with terpene hydrate, $C_{10}H_{20}O_2.H_2O$.

Flagyl. A proprietary preparation containing metronidazole. (336).

Flagyl Compak. A proprietary preparation comprising METRONIDAZOLE, taken orally, and vaginal pessaries containing nystatin. Used in the treatment of vaginitis. (494).

Flajolotite. A mineral, $FeSbO_4.\frac{1}{2}H_2O$.

Flake Lead. See WHITE LEAD.

Flake Litharge. Litharge made by the oxidation of lead.

Flake White (Cremnitz, Kremnitz, Crems White, Blanc D'argent, Silver White, London White, Nottingham White, Cremnitz White). Lead whites. They are all carbonates of lead, and contain varying quantities of hydrated oxide of lead. Flake white is a variety of chamber white lead, obtained in flaky pieces by heating lead plates. A basic bismuth is also known as Flake white. The term Flake or Pearl white is sometimes used for oxychloride of bismuth.

Flake White, Toilet. Bismuth oxynitrate.

Flakice. A form of ice having a large surface area. It is made by freezing a layer of ice to a metal surface, then causing the shape of the metal to change, thereby resulting in the peeling of the ice.

Flaky Arsenic. Native arsenic, often containing silver, as well as a little iron, and sometimes cobalt and nickel.

Flamazine. A proprietary preparation of silver sulphadiazene, used to prevent infection in burns. (824).

Flamco. Mould and core dressings. (531).

Flamenol. A proprietary trade name for a polyvinyl chloride synthetic resin.

Flaming. A decolorising agent for sugar juices.

Flamma. Flame retardants. (572).

Flammex 4BA. A proprietary tetrabromo bisphenol-A. A reactive additive used as a flame-retardant in epoxy and polyester resins. (137).

Flammex 5BP. A proprietary brominated phenol-type reactive flame-retardant additive, for use with phenolic resins. (137).

Flammex AP. A proprietary aqueous solution of inorganic salts used as a flame retardant in cellulose-based materials. (137).

Flammex B10. A proprietary brominated aromatic type of non-reactive flame retardant, used in conjunction with antimony oxide where high temperatures are encountered. (137).

Flammex CA. A proprietary chlorendic anhydride type of flame retardant. It is a di-functional acid anhydride and is used in alkyd, epoxy and polyester resins. (137).

Flammex S. A proprietary brominated phosphate-ester flame retardant used in polystyrene. (137).

Flammex TFI. A proprietary aqueous solution of inorganic and organic salts used as a flame retardant. (137).

Flammocite. A safety explosive. It contains 44 per cent. ammonium nitrate, 16 per cent. sodium chloride, 14 per cent. sodium nitrate, 10 per cent. trinitro-toluene, 6 per cent. nitroglycerin, 5 per cent. ammonium sulphate, and 5 per cent. cellulose.

Flander's Stone. A name applied to graphite when used in pencils.

Flandrac. A decolorising agent for sugar juices.

Flar. A proprietary preparation containing a lactic ferment resistant to antibiotics, thiamine, riboflavine, pyridoxine, cyanocobalamin, nicotinamide, sodium pantothenate, inositol, folic acid and liver extract. It is used in the treatment of antibiotic side-effects. (256).

Flavacid Nitrate. 2 : 7-Dimethyl-6-amino-3-dimethyl-amino-methyl-acridinium nitrate. Used against diphtheria bacilli.

Flavaniline. A quinoline dyestuff. It is 2'-p-amino-phenyl-4'-methyl-quinoline, $C_{16}H_{14}N_2$.

Flavaniline S. A sulphonated flavaniline.

Flavanthrene (Indanthrene Yellow). A dyestuff, $C_{28}H_{14}N_2O_2$, obtained by the oxidation of β-amino-anthraquinone. The leuco-compound dyes cotton blue, becoming yellow upon air oxidation.

Flavaurine. An old yellow dyestuff allied to Alizarin yellow.

Flavazine S (Kiton Yellow, Pyrazine Yellow S). A dyestuff similar to Tartrazine. Employed on wool, in a bath containing sodium hydrogen sulphate.

Flavazol. A dyestuff, $C_{13}H_{12}N_2O_3$, prepared from p-toluidine and salicylic acid. It dyes yellow on chrome mordanted wool.

Flavelix. A proprietary preparation containing mepyramine maleate, ephedrine hydrochloride, ammonium chloride and sodium citrate. A cough linctus. (330).

Flaveosine. A dyestuff obtained by condensing m-acetamino-dimethyl-aniline with phthalic anhydride. It dyes tannined cotton and wool reddish-yellow, and silk golden yellow.

Flavinduline (Induline Yellow). A dyestuff. It is phenyl-phenanthra-phenazonium chloride, $C_{26}H_{17}N_2Cl$. Dyes tannined cotton yellow.

Flavine. Three materials are known by this name : (a) β-Diamino-benzophenone ; (b) a grade of quercitron bark extract ; and (c) diamino-methyl-acridinium chloride.

Flavizid. A derivative of acriflavine for use in medicine.

Flavizide. 6-Amino-3-dimethyl-amino-2 : 7 : 10-trimethyl-acridinium chloride. An antiseptic.

Flavocents. Concentrated flavour compositions. (522).

Flavodine. A proprietary acriflavine solution.

Flavoline. 2-Phenyl-4-methyl-quinoline.

Flavone. β-Phenyl-benzo-pyrone.

Flavophenin. See CHRYSAMIN.

Flavopurpurin (Alizarin No. 10CA, Alizarin FA, Alizarin GB, Alizarin GI, Alizarin RG, Alizarin SDG, Alizarin X, Alizarin YCA, Alizarin CAF, DCA, JCA, YAR). Trioxy-anthraquinone, $C_{14}H_8O_5$. Dyes cotton mordanted with alumina, red.

Flavotint. Food colour and flavour compounds. (522).

Flavoxate. An anti-spasmodic. 2-Piperidinoethyl 3-methylflavone-8-carboxylate. URISPAS is the hydrochloride.

Flaxedil. A proprietary preparation of gallamine triethiodide. A muscle relaxant. (336).

Flax Wax. A wax associated with flax fibre and with the cortical tissues. The air-dried cortex contains as much as 10 per cent. by weight of wax. It is removed by extraction with a volatile solvent. Different varieties of flax yield wax with the following characteristics : saponification value 77–84, iodine value 21–27, acid value 17–23, specific gravity 0·963–0·983, and melting-point 68–69° C. It is suggested as a substitute for beeswax.

Flax, Mountain. See AMIANTHUS.

Flax Seed. Linseed.

Flazalone. An anti-inflammatory. 3-(4-Fluorobenzoyl)-4-(4-fluorophenyl)-1-methylpiperidin-4-ol.

Flectol A. A proprietary trade name for an antioxidant used in rubber. It is a ketone-amine.

Flectol B. A proprietary trade name for a ketone-amine compound used as an antioxidant for rubber.

Flectol H. A proprietary trade name for a rubber antioxidant. It is a polymer of 1,2-dihydro-2,2,4-trimethyquinolate.

Flectol White. A proprietary trade name for a white antioxidant for rubber. It is an aryl-oxy-ketone.

Flemming's Solution. A fixing agent used in microscopic work. It consists of 80 c.c. of a 1 per cent. solution of osmic acid, 15 c.c. of a 10 per cent. solution of chromic acid, 100 c.c. of glacial acetic acid, and 95 c.c. of water.

Flerhenol. See AVIVAN.

Flesh Guano. See FLESH MANURE.

Flesh Manure (Flesh Guano, Meat Meal). A manure obtained from the carcases of horses and other animals. It is the putrid animal refuse of slaughter-houses. The refuse is treated with superheated steam, and the solid matter left is pulverised and dried. It contains from 6–8 per cent. nitrogen, and from 6–15 per cent. phosphorus oxide.

Flesh Sugar. Inosite, $C_6H_{12}O_6.2H_2O$.

Fletcher's Alloy. An alloy of 95·5 per cent. aluminium, 3 per cent. copper, 1 per cent. tin, 0·5 per cent. antimony, and 0·5 per cent. phosphor tin.

Fletcher's Bearing Alloys. Aluminium base alloys. One contains 92 per cent. aluminium. 7·5 per cent. copper, and 0·25 per cent. tin, and another 90 per cent. aluminium, 7 per cent. copper, and 1 per cent. zinc.

Flexachlor 40, 50, 60, 50C. A trade name for chlorinated hydrocarbon secondary plasticisers for vinyl polymers containing 40 per cent., 50 per cent., etc., chlorine. (55).

Flexalyn. A proprietary trade name for a plasticiser. It is diethylene glycol diabietate.

Flexamine. A proprietary trade name for a mixture of 65 per cent. of a diarylamineketone reaction product and 35 per cent. of N-N'-diphenyl-p-phenylene diamine. (163).

Flexane. A proprietary polyurethane elastomer. (727).

Flexazone. A proprietary preparation of phenylbutazone. (137).

Flex Carbon. A proprietary carbon black made by the thermatomic process. It has a specific gravity of 1·75, and is used in rubber mixings.

Flexible Collodion. (*Collodium flexible B.P.*) Consists of 940 c.c. collodion, 40 c.c. Canada turpentine and 20 c.c. castor oil.

Flexible Sandstone. An Indian sandstone, containing angular quartz grains.

Flexibond. Adhesives. (508).

Flexin. A proprietary trade name for Zoxazolamine.

Flexite. A reclaimed rubber which gives an acetone extract of 9·5 per cent., mineral matter 0·52 per cent., sulphur 3·25 per cent., rubber 86·25 per cent., and has a specific gravity of 1·68.

Flexocel. Prepolymer systems for flexible polyurethane foams. (557).

Flexol. A proprietary trade name for vinyl plasticisers.

Flexol Plasticiser 3GH. A proprietary trade name for a plasticiser. It is triethylene glycol-di-2-ethyl butyrate.

Flexol Plasticiser 3GO. A proprietary trade name for a plasticiser. It is triethylene glycol-di-2-ethyl hexoate.

Flexoresin. Proprietary brands of glycol and glyceryl phthalates. Also polymerised terpenes.

Flexricin 59. A proprietary trade name for castor oil used as a vinyl plasticiser additive. (104).

Flexricin 66. A proprietary trade name for isobutyl acetyl ricinoleate. A vinyl plasticiser. (104).

Flexzone 3C. A proprietary anti-oxidant. It is N-isopropyl-N'-phenyl-p-phenylene diamine. (435).

Flexzone 6-H. A proprietary anti-oxidant. It is N-phenyl-N'-cyclohexyl-p-phenylene diamine. (435).

Flinkite. A mineral, $Mn[Mn(OH)_2]_2 AsO_4$.

Flint. A form of silica, SiO_2.

Flint Alloy. A heat and corrosion-resisting alloy containing 83 per cent. iron, 12·5 per cent. chromium, 3 per cent. carbon, and 0·5 per cent. silicon.

Flint Bricks. Bricks made from powdered flint and used as firebricks.

Flintcast. A white iron made in the electric furnace. It resists abrasion.

Flint Glass (Lead Glass, Cristal). A glass composed of lead and potassium silicates. Used for hollow-ware, superior bottles, and optical work. See POTASH-LEAD GLASS.

Flint Metal. A proprietary trade name for an alloy of iron with 4–4·5 per cent. nickel, 1·25–1·75 per cent. chromium, and 3–3·5 per cent. carbon.

Flintshire Stones. See DINAS BRICKS.

Flinty Zinc Ore. Flinty calamine.

Flit. A proprietary insecticide consisting of a mixture of coal-tar oil (light and medium boiling-points), with the addition of refined petroleum. Probably the whole has been chlorinated as the action depends upon the presence of chlorinated benzene and its homologues.

Flo. A fluorescent dye for textiles. (537).

Floating Soap. Made by churning up a pure white soap in a special machine. The soap is beaten up into a froth, and air incorporated with it, to make it lighter than water.

Floatstone. Porous opal.

Float Tin. Cassiterite, occurring in the soil, and formed by the disintegration of tin rocks, is called float tin.

Flocsil. A silicate for water clarification. (546).

Floctafenine. An analgesic currently undergoing clinical trial as "RU 15750". It is 2,3-dihydroxypropyl N-(8 - trifluoromethyl - 4 - quinolyl) - anthranilate.

Floex. A proprietary trade name for a wetting agent for paint, etc. It is a condensation product of higher fatty alcohols.

Flomax. A stabilizer for PVC cadmium/barium complexes. (572).

Flooring Plasters. Plasters made from over-burnt gypsum, and slow hydration.

Floranid. A German fertiliser consisting of urea which is made by compressing a mixture of ammonia and carbon dioxide in a lead-lined autoclave at 135° C. for 2 hours. The solution containing urea with unchanged ammonia and carbon dioxide is evaporated after separation.

Floraquin. A proprietary preparation of di-iodohydroxyquinoline. Vaginal antiseptic tablets. (285).

Florantyrone. γ - Fluoranthen - 8 - yl - γ-oxobutyric acid. Zanchol.

Florence Lake (Vienna Lake, Paris Lake). Lakes produced from cochineal, by precipitating alkaline solutions of cochineal with alum, or with a mixture of alum and tin salts.

Florence Oil. Olive oil imported from Leghorn.

Florence Zinc. Zinc oxide, made by the French method in America.

Florencite. A mineral. It is a basic phosphate of aluminium and cerium, $AlPO_4.CePO_4.Al(OH)_6$.

Florentine Brown (Vandyck Red, Hatchette Brown). A pigment. It is copper ferrocyanide.

Florette Silk. The longer and better quality silk, after spinning and carding, obtained from waste silk.

Floricin (Florizine, Derizine, Dericin). A substance produced by heating castor oil to 300° C., and distilling 10 per cent. of it. There remains the product termed "floricin," which solidifies at −20° C. It is also made by heating castor oil with formaldehyde. It is miscible with ceresin and pet. jelly and is used as a vehicle for menthol and oil of eucalyptus. Unlike castor oil, it is insoluble in alcohol.

Florida Earth. See BLEACHING EARTH.

Florida Phosphates. Mineral phosphates. There are two types, hard rock phosphates, containing 80 per cent. calcium phosphate, and soft clay phosphates, containing 40–60 per cent. calcium phosphate. Fertilisers.

Floridin. A form of fuller's earth found in Florida, U.S.A. It removes the colouring matter from oils and waxes, and is used for this purpose.

Florinef. A proprietary preparation of fludrocortisone 21-acetate. (326).

Florite. A proprietary trade name for a carefully prepared and screened bauxite.

Florizine. See FLORICIN.

Flos Ferri. A mineral. It is calcium carbonate, containing iron.

Flosol. A proprietary trade name for colloidal barium silico-fluoride for horticultural purposes.

Floss, Akund. See AKUND.

Floss Silk. The waste parts of the cocoon after the silk has been reeled.

Floto. See PINE OILS.

Flour. The name for the starch from cereals, legumens, acorns, and chestnuts.

Flour Alum. Minute crystals of alum.

Floured Mercury. A state of metallic mercury when in a fine state of division and a film of foreign matter prevents the globules uniting together.

Flour, Fossil. See INFUSORIAL EARTH.

Flour, Mountain. Infusorial earth.

Flour of Sulphur. Powdered sulphur. It has been powdered by grinding, but is not so finely powdered as flowers of sulphur.

Flour, Self-raising. A common mixture of this type contains 10,000 parts flour, 190 parts cream of tartar, and 85 parts sodium bicarbonate.

Flovic. A proprietary trade name for a vinyl chloride/acetate copolymer rigid foil for vacuum forming. (2).

Flow Blue. See FLOW-POWDER.

Flowers of Antimony. Formed when antimony burns in air. It consists of antimony oxide, Sb_4O_6. Used in medicinal preparations, and as a white pigment.

Flowers of Arsenic. See ARSENIOUS ACID.

Flowers of Benjamin (Flowers of Benzoin). Benzoic acid, C_6H_5COOH.

Flowers of Benzoin. See FLOWERS OF BENJAMIN.

Flowers of Bismuth. Bismuth oxide, Bi_2O_3, obtained by burning bismuth metal at a red heat.

Flowers of Brimstone. See FLOWERS OF SULPHUR.

Flowers of Camphor. Camphor in crystalline form.

Flowers of Copper. The oxide of copper, CuO, produced when copper burns. It is also the name applied to a hair-

like form of cuprous oxide, Cu_2O, a mineral.

Flowers of Sulphur (Flowers of brimstone). Consists of minute crystals of sulphur, obtained by chilling sulphur vapour.

Flowers of Tin. Stannic oxide, SnO_2. A polishing powder.

Flowers of Zinc. See PHILOSOPHER'S WOOL.

Flow-powder. A mixture of white lead and salt which gives off chlorine on heating. It is used for the production of flow blues (cobalt blues) on ceramics. Cobalt chloride is formed, which, being volatile, gives blues of varying intensity.

Flox. A fibrous material made from cellulose sulphate pulp. It is mixed with cotton, wool, etc., and woven. It is an art wool.

Floxapen. A proprietary preparation of FLUCLOXACILLIN. An antibiotic. (364).

Fluanisone. 4' - Fluoro - 4 - [4 - (2 - methoxyphenyl) - 1 - piperanzinyl] - butyrophenone.

Fluanxol. A proprietary preparation of FLUPENTHIXOL. An anti-depressant and tranquilliser. (645).

Fluates. This term is used for fluosilicates. It is also the name for waterproofing compounds consisting of solutions of sodium silicate, or silicofluoride, and other silicofluorides, such as those of zinc, magnesium, and aluminium.

Fluclorolone Acetonide. A corticosteroid. $9\alpha, 11\beta$ - Dichloro - 6α - fluoro - 21 - hydroxy - 16α, 17α - isopropylidenedi - oxypregna - 1, 4 - diene - 3, 20 - dione. TOPILAR.

Flucloxacillin. 3 - (2 - Chloro - 6 - fluorophenyl) - 5 - methylisoxazol - 4 - ylpenicillin. See FLOXAPEN. Present also in MAGNAPEN. An antibiotic.

Flucytosine. An anti-fungal preparation. 4 - Amino - 5 - fluoro - 1, 2 - dihydro - pyrimidin - 2 - one. 5 - Fluorocytosine. ALCOBON. (314).

Fludor Solder. An aluminium solder containing 56·5 per cent. tin, 40 per cent. zinc, 3 per cent. lead, 0·2 per cent. antimony, and 0·1 per cent. copper.

Fludrocortisone. 9α - Fluoro - $11\beta,17\alpha$, 21 - trihydroxypregn - 4 - ene - 3,20 - dione 9α-Fluorohydrocortisone. Alflorone, Florinef and Fludrocortone are the 21-acetate.

Fludrocortone. A proprietary trade name for the 21-acetate of Fludrocortisone.

Fludroxycortide. Flurandrenolone.

Fluellite. A mineral. It is $8[3AlF_3.4H_2O]$.

Flufenamic Acid. N - $(\alpha,\alpha,\alpha$ - Trifluoro - m-tolyl)anthranilic acid.

Flugène 113. A proprietary non-inflammable solvent of low toxicity for cleaning precision equipment. It is trifluorotrichloroethane. (729).

Flugestone. A progestational steroid. 9α - Fluoro - 11β, 17 - dihydroxypregn - 4 - ene - 3, 20 - dione. FLUROGESTONE.

Fluid Gelatin. See OIL PULP.

Fluid Magnesia. (*Liquor Magnesii carbonatis B.P.*) It consists of magnesium carbonate dissolved in water containing carbon dioxide (10 grains per ounce).

Fluinlan. Liquid lanolin. (526).

Flukanide. See RAFOXANIDE.

Flumedroxone. A preparation used in the treatment of migraine. 17α - Hydroxy - 6α - trifluoromethylpregn - 4 - ene - 3, 20 - dione.

Flumerin. The disodium salt of hydroxy-mercuri-fluoresceine, $C_{20}H_{10}O_6$ Na_2Hg. Used in the treatment of syphilis.

Flumethasone. $6\alpha,9\alpha$ - Difluoro - 11β, $17\alpha,21$ - trihydroxy - 16α - methyl - pregna - 1,4 - diene - 3,20 - dione. 6α, 9α - Difluoro - 16α - methylprednisolone. Locorten is the pivalate.

Flumethiazide. 6 - Trifluoromethyl - benzo - 1,2,4 - thiadiazine - 7 - sulphonamide 1,1-dioxide.

Flunitrazepam. A hypnotic currently undergoing clinical trial as " Ro 05–4200 ". It is 5-o-fluorophenyl-1,3-dihydro - 1 - methyl - 7 - nitro $2H$ - 1, 4 - benzodiazepin-2-one.

Fluoborite. A mineral, $Mg(F.OH)_2$.

Fluocerite. A mineral, $(Ce.La.Di)_2$ $O.F_4$.

Fluocinolone. 6α-9α - Difluoro - 11β, $16\alpha,17\alpha,21$ - tetrahydroxypregna - 1,4 - diene - 3,20 - dione. $6\alpha,9\alpha$ - Difluoro - 16α - hydroxyprednisolone. Synalar and Synandone are the acetonide.

Fluocinonide. A corticosteroid. 21-Acetoxy-6α, 9α-difluoro-11β-hydroxy-16α, 17α - isopropylidenedioxypregna - 1, 4 - diene - 3, 20 - dione. FLUOCINOLONE $16\alpha,17\alpha$-acetonide - 21-acetate. METOSYN.

Fluocortin Butyl. An anti-inflammatory currently undergoing clinical trial as " SH K 203 ". It is butyl 6α - fluoro - 11β - hydroxy - 16α - methyl - 3, 20 - dioxopregna-1,4-dien-21-oate.

Fluocortolone. 6α - Fluoro - $11\beta,21$ - dihydroxy - 16α - methylpregna - 1,4 - diene-3,20-dione. Ultralanum is the 21-hexanoate.

Fluoderm. A proprietary preparation of fluorometholone, clioquinol and chlorphenesin for dermatological use. (182).

Fluolite. Fluorescent whitening agents. (512).

Fluomimetite. Obtained by melting lead fluoride and *o*-arsenate. It is lead fluo-tri-*o*-arsenate, $3Pb_3(AsO_4)_2.PbF_2$.

Fluon. A trade mark for polytetra-fluoroethylene (P.T.F.E.). A hard plastics material with a very water repellant surface which has a working temperature range from -200 to $+280°$ C. It has a low high frequency power factor and permittivity and outstanding non-stick properties. It is used in heat resistant glands, packings and bearings. See also TEFLON. (2).

Fluon CD 123. A proprietary coagulated P.T.F.E. dispersion powder used in extrusion operations. (512).

Fluon G 201. A proprietary pre-sintered P.T.F.E. granular extrusion powder. (512).

Fluon G 401. A proprietary unsintered P.T.F.E. granular powder used in ram extrusion operations. (512).

Fluon L 170. A proprietary finely-divided P.T.F.E. powder designed as an additive to improve the lubricity and wear of such materials as printing inks. (512).

Fluon VX2FT. A range of proprietary bronze and graphite-filled P.T.F.E. moulding powders. (512).

Fluon VG 15. A proprietary P.T.F.E. powder loaded as to 15% with glass.

Fluon VG15F. A proprietary free-flowing variant of FLUON VG 15. (512).

Fluon VG 25 and 25F. Proprietary P.T.F.E. moulding powders containing 25% glass, having poor-flow and free-flow characteristics respectively. (512).

Fluon VX2. A bronze-filled PTFE used as a bearing material.

Fluon VX3. A lead oxide filled PTFE used as a bearing material.

Fluopromazine. 10 - (3 - Dimethyl - aminopropyl) - 2 - trifluoromethyl - pheno-thiazine. Triflupromazine. Vespral and Vesprin are the hydrochloride.

Fluor. See FLUORSPAR.

Fluor-Adelite. Synonym for Tilasite.

Fluoram. Ammonium bifluoride.

Fluoran. 9-hydroxy-9-xanthene-*o*-benzoic acid lactone.

Fluor Apatite. A mineral. It is a variety of apatite ($3Ca_3 2PO_4.CaF_2$).

Fluor-Biotite. A mineral. It is $K(Mg,Fe)_3AlSi_3O_{10}F_2$.

Fluorchrome. Chromium fluoride, $CrF_3.4H_2O$. A mordant.

Fluorel. A trade name for a heat resistant fluoroelastomer rated at 400° F. continuously and 600° F. for short periods. It is a vinylidene fluoride-hexafluoropropylene copolymer, 70 : 30. (35).

Fluoremetic. Antimony - sodium -

fluoride, $SbF_3.NaF$. Used in the place of tartar emetic in dyeing processes.

Fluorene. *o*-Diphenyl-methane.

Fluorenone. *o*-Diphenylene-ketone.

Fluoresceine. Resorcinol - phthalein, $C_6H_4.CO.O.C.C_6H_3OH.C_6H_3(OH)O$. It is obtained by heating phthalic anhydride with resorcinol. Its alkali salts dye silk and wool yellow, as does fluoresceine itself, with green fluorescence. See URANINE.

Fluorescent Blue (Resorcin Blue, Iris Blue). A dyestuff. It is the ammonium salt of tetrabromo-resorufin, $C_{12}H_6Br_4N_2O_3$. Dyes silk and wool blue, with brownish fluorescence.

Fluorescent Red 5B. A proprietary organic fluorescent-red dye used for colouring polystyrene polymethylmethacrylate and unplasticised P.V.C. (431).

Fluorfolpet. A proprietary fungicide applied as a paint. It is *N*-(fluordichloromethylthio) phthalimid. (112).

Fluor-Herderite. A mineral. It is $4[CaBePO_4F]$.

Fluoric Acid. Hydrofluoric acid, HF.

Fluoric Ether. Ethyl fluoride, C_2H_5F.

Fluorite. See FLUORSPAR.

Fluormanganapatite. A mineral. It is a variety of apatite containing manganese and no chlorine (MnO = 4·9 per cent.).

Fluormount. A mountant for fluorescent microscopy. (544).

Fluorocarbon-12. Dichlorodifluoromethane. A refrigerant. See Arcton. (2).

Fluorocarbon-22. Chlorodifluoromethane. A refrigerant for low temperature work. See Arcton. (2).

Fluoroform. Methenyl fluoride, CHF_3.

Fluoroformol. A 2·4 per cent. solution (aqueous) of fluoroform, CHF_3. Used in the treatment of tuberculosis.

Fluoro-heavy Spar. A mineral. It contains barium sulphate and calcium fluoride.

Fluoroil. Immersion oil for fluorescent microscopy. (544).

Fluorol. Sodium fluoride, NaF.

Fluorolene. A registered trade mark for polytetrafluoroethylene. (730).

Fluorometholone. 9α - Fluoro - 11β, 17α - dihydroxy - 6α - methylpregna - 1, 4-diene-3,20-dione.

Fluorosint. A registered trade mark for polytetrafluorethylene with a ceramic-like texture suitable for machining.

Fluorouracil. 5-Fluorouracil.

Fluorrheumine. A mixture of fluorine-phenetol and difluor-diphenyl, in the form of an ointment. Said to be used for rheumatism and lumbago.

Fluorspar (Fluor, Fluorite, Derbyshire

Spar). A mineral. It is calcium fluoride, $4[CaF_2]$.

Fluosite. Thermosetting moulding powders based on phenolformaldehyde resins. (61).

Fluothane. A proprietary preparation of halothane. A general anæsthetic. (2).

Fluotrimazole. A pesticide. It is 1-(3-trifluoromethyltrityl)-1,2,4-triazole.

Fluoxymesterone. 9α - Fluoro - 11β, 17β - dihydroxy - 17α - methylandrost - 4-en-3-one. 9α-Fluoro-11β-hydroxy-17α-methyltestosterone. Ultandren.

Flupenthixol. A tranquilliser. 9-{3-[4-(2 - Hydroxyethyl)piperazin - 1 - yl] - propylidene} - 2 - trifluoromethylthio - xanthene. DEPIXOL is the decanoate.

Fluperolone. 9α - Fluoro - $11\beta,17\alpha,21$ - trihydroxy - 21 - methylpregna - 1,4 diene-3,20-dione. 9α-Fluoro-21-methyl-prednisolone. Methral is the acetate. -

Fluphenazine. 10 - {3 - [4 - (2 - Hydroxy-ethyl)piperazin - 1 - yl]propyl} - 2 - trifluoromethylphenothiazine. Moditen and Prolixin are the dihydrochloride.

Fluprednidene. A gluco-corticosteroid. 9α - Fluoro - 11β, 17α, 21 - trihydroxy - 16 - methylenepregna - 1, 4 - diene - 3, 20-dione. " StC 1106 " is currently under clinical trial as the 21-acetate.

Fluprednisolone. 6α - Fluoro - $11\beta,17\alpha$, 21 - trihydroxypregna - 1,4 - diene - 3, 20-dione. Alphadrol.

Fluprofen. An anti-inflammatory and analgesic currently undergoing clinical trial as " R.D. 17345 ". It is 2-(2'-fluorobiphenyl - 4 - yl)propionic acid.

Fluprostenol. A preparation used in cases of infertility. It is $(\pm)$-7-{(1R, 2R, 3R, 5S) - 3, 5 - Dihydroxy - 2 - [(3R) - 3 - hydroxy - 4 - (3 - trifluoro-methylphenoxy)but - 1 - (E) - enyl]-cyclopent - 1 - yl}hept - 5 - (Z) - enoic acid. " ICI 81,008 " or EQUIMATE are the sodium salt, with the former still under clinical trial.

Flurandrenolone. 6α - Fluoro - $11\beta,21$ - dihydroxy - $16\alpha,17\alpha$ - isopropyl - idenedioxypregn - 4 - end - 3,20 - dione. 6α - Fluor - $16\alpha,17\alpha$ - isopropylidene - dioxyhydrocortisone. Fludroxycortide. Cordran ; Drenison ; Haelan.

Flurazepam. A hypnotic. 7-Chloro-1-(2 - diethylaminoethyl) - 5 - (2 - fluoro - phenyl) - 1, 3 - dihydro - 2H - 1, 4 - benzodiazepin - 2 - one. DALMANE is the hydrochloride.

Flurbiprofen. An anti-inflammatory and analgesic. 2 - (2 - Fluorobiphenyl - 4 - yl) propionic acid.

Flurene. A patented fluoro-carbon solution coating for high chemical resistance and low friction properties. (254).

Flurene CE. A proprietary solution coating system containing polychloro-trifluoroethylene. (254).

Flurogestone Acetate. A proprietary progestin. It is 9-fluoro-11β, 17-di-hydroxy - pregn - 4 - ene - 3, 20 - dione, 17 - acetate. (644).

Flurothyl. Di - (2,2,2 - trifluoroethyl) - ether. Indoklon.

Fluscorbin. A proprietary preparation of vitamin C, quinine dihydrochloride and sulphate, phenacetin and caffeine. A cold remedy. (248).

Flush Plate Brass. See BRASS.

Fluspirilene. A tranquilliser. 8 - [4, 4 - Di - (4 - fluorophenyl)butyl] - 1 - phenyl - 1, 3, 8 - triazaspiro[4, 5]decan - 4-one.

Flutec. A series of extremely inert, temperature stable, non-toxic, non-inflammable liquids exhibiting excellent electrical insulating properties and good heat transfer characteristics. (300).

Flutec PP1. A fluorinated hydrocarbon C_6F_{14}. Perfluoro-n-hexane.

Flutec PP2. A fluorinated hydrocarbon C_7F_{14}. Perfluoromethylcyclohexane.

Flutec PP3. A fluorinated hydrocarbon C_8F_{16}. Perfluoro-1,3-dimethylcyclo-hexane.

Flutec PP9. A fluorinated hydrocarbon $C_{11}F_{20}$. Perfluoro-1-methyl decalin.

Flutherite. A mineral. It is urano-thalite.

Flutra. A diuretic. Trichlormethazide.

Fluvia. See JELUTONG.

Flux Oil. An oil obtained from asphalt and used to flux asphalts.

Fluxol. A hardwood pitch prepared from the distillation of hardwood. It has a specific gravity of 1·16–1·19, and melts at 185–203° F. It is used as a rubber softener.

Flux Skimmings. The result of the action of ammonium chloride and atmospheric oxygen upon molten zinc during the galvanising process. A skin, known as flux skimmings, is produced. It consists of chloride and oxide of zinc, with some ammonium chloride, and dirt. Used by smelters.

Fly Blister. Cantharides plaster.

Flyosan. A proprietary preparation used as an insecticide. It is stated to contain petroleum, sassafras oil, and traces of amyl acetate.

Fly Stone. Mercuric chloride in lumps. It is also the name for a native arsenate of cobalt, used for poisoning flies, by grinding it, and putting a small quantity in a saucer with sweetened water.

F.C.N.3 Steel. A proprietary trade name for a 3 per cent. nickel case-hardening steel.

F.C.N.C. Steel. $4\frac{1}{2}$ per cent. nickel chrome for case hardening.

Foam Spar. See APHRITE.

Foam Tannin. Tannin extracted from sumach, or galls, by means of ether.

Fob Metal. See BRASS.

Foerdite. An explosive consisting of 25 per cent. nitro-glycerin, 1·5 per cent. collodion wool, 5 per cent. nitro-toluene, 4 per cent. dextrin, 3 per cent. glycerin, 37 per cent. ammonium nitrate, and 24 per cent. potassium chloride.

Fœtid Quartz. See STINK QUARTZ.

Foil Lead. An alloy of 86·5 per cent. lead, 12·5 per cent. tin, and 1 per cent. copper.

Folæmin. A proprietary preparation of iron aminoates and folic acid. A hæmatinic. (275).

Folcovin. A proprietary preparation containing pholcodine. A cough linctus. (273).

Folex. A proprietary preparation of ferrous gluconate and folic acid. A hæmatinic. (273).

Folex 350. A proprietary preparation containing ferrous fumarate and folic acid used in the treatment of anæmia during pregnancy. (273).

Folgerite. A mineral. It is pentlandite.

Foliac Super Red. A proprietary trade name for a graphite jointing compound. (215).

Foliated Tellurium. See MAGYAGITE.

Folic Acid. Pteroylglutamic acid. Part of the Vitamin B complex. Deficiency produces anæmia in man.

Folicaine. See EUDRALIN.

Folicin. A proprietary preparation of ferrous, copper and manganese sulphates and folic acid. A hæmatinic. (329).

Folin-Dennis Solution. A solution prepared by adding slowly 400 c.c. of a 0·7268 per cent. solution of silver nitrate to a solution containing 10 grams mercuric cyanide and 180 grams caustic soda, in 1,200 c.c. water. Used for the determination of acetone in milk, blood, and urine.

Folin-Dennis Phenol Reagent. To 750 c.c. distilled water 100 grams sodium tungstate are added. This is dissolved and 20 grams phosphomolybdic acid added, and finally 50 c.c. of syrupy 85 per cent. phosphoric acid. The solution is boiled for 2 hours under the reflux and diluted to 1,000 c.c.

Folin-Dennis Uric Acid Reagent. To 750 c.c. distilled water 100 grams sodium tungstate are added and 80 c.c. of 85 per cent. syrupy phosphoric acid. The solution is boiled for 2 hours under the reflux, cooled, and transferred to a 1,000 c.c. flask and made up to 1,000 c.c.

Folinic Acid. Citrovorum factor, an active form of folic acid. N^5-formyl tetrahydrofolic acid.

Folin-McEllroy Lactose Reagents. (a) A saturated picric acid solution (2 grams picric acid in 100 c.c. water). (b) A 20 per cent. sodium carbonate solution.

Folin-McEllroy Sugar Reagents. Qualitative. 100 grams sodium pyrophosphate, $Na_4P_2O_7.10H_2O$, 30 grams crystalline disodium-monohydrogen phosphate, Na_2HPO_4, and 50 grams dry sodium carbonate in 900 c.c. water. Dissolve and add 13 grams copper sulphate dissolved in 200 c.c. water. Quantitative. (a) Acidified copper sulphate solution containing 60 grams copper sulphate in 900 c.c. water. Dissolve, and add 5 c.c. concentrated sulphuric acid. Make up to 1,000 c.c. (b) Phosphate-carbonate-thiocyanate dry mixture prepared by mixing 100 grams dry sodium carbonate, Na_2CO_3, H_2O, and 30 grams potassium thiocyanate mixed in a large mortar.

Folin's Uranium Acetate Mixture. Uric acid reagent. It consists of 500 grams ammonium sulphate, 5 grams uranium acetate, and 6 c.c. glacial acetic acid dissolved in 650 c.c. water and made up to 1 litre.

Folin's Uric Acid Reagent. Add to 160 c.c. water, 50 c.c. syrupy phosphoric acid. Heat to 85° C., and add 100 grams sodium tungstate. Boil for 1 hour under reflux. Place 25 grams lithium carbonate in a beaker, add 50 c.c. syrupy phosphoric acid and 200 c.c. water. Boil 10 minutes, cool, and add first solution. Mix and dilute to 1,000 c.c.

Folin-wu Silver Lactate. Into a 100 c.c. flask place 5 grams silver lactate, $H_3C.CHOH.COO.AgH_2O$, and 5 grams lactic acid (specific gravity 1.21, 85 per cent. $C_3H_6O_3$). Dilute to 100 c.c.

Follicle - Stimulating Hormone (F.S.H.). An extract of human postmenopausal urine containing primarily the follicle-stimulating hormone. PERGONAL.

Follotropin. A follicle-stimulating hormone.

Folosan. Horticultural fungicides. (511).

Folpet. A proprietary fungicide applied as a paint. N-(Trichlormethylthio) phthalimid. (112).

Folvron. A proprietary preparation of folic acid and ferrous sulphate. A hæmatinic. (306).

Fomac. A fungicide for rubber. (511).

Fomescol U and X series. Proprietary non-ionic surfactants based on polyoxyalkylene glycol. (547).

Fomitine. A liquid extract from the

fungi *Fomes cinnamomeus.* Used in medicine for diseases of the bladder.

Fomocaine. A local anæsthetic. 4-(3-Morpholinopropyl(benzyl phenyl ether.

Fomodox. Veterinary compound pessaries. (514).

Fond Rouge (Orcellin, Orcellin Deep Red). A dyestuff. It is picramic acid-azo-resorcin, $C_6H_2.OH(NO_2)_2.N_2.C_6H_3(OH)_2$.

Fontaine's Powder. An explosive consisting of potassium picrate and potassium chlorate.

Food of the Gods. Asafœtida.

Fool's Gold. Iron pyrites (*q.v.*).

Footeite. A mineral, $CuCl_2.8Cu(OH)_2.4H_2O$.

Foots. Matter deposited by oils on standing.

Foral 85. A proprietary trade name for a rosin glycerol ester used in adhesives and hot melts. (93).

Foral 105. A proprietary trade name for a rosin pentaerythritol ester used in adhesives and hot melts. (93).

Forbesite. A mineral, $(Ni.Co)H.AsO_4.3\frac{1}{2}H_2O$.

Forbes Metal. An alloy of 53·5 per cent. zinc, and 46·5 per cent. copper.

Force. A malt preparation.

Forceval. A proprietary preparation of vitamins and minerals. (731).

Forcherite. A mineral. It is a variety of opal of orange-yellow colour.

Forcite. A trade mark for various types of explosives.

Fordath Resins. A proprietary range of phenolic and urea resins used in foundry work. (732).

Fordit. A German explosive consisting of 24 per cent. nitro-glycerin, 1 per cent. nitro-cotton, 34 per cent. nitrotoluene, 2 per cent. flour, 2 per cent. dextrin, 5 per cent. glycerin, and 32 per cent. ammonium nitrate. To this mixture is added 30 per cent. potassium chloride.

Ford's Silicate of Limestone. A stone made by pressing a mixture of burnt lime and sand into a perforated mould. The lime is slaked with boiling water, and the stone hardened by superheated steam.

Fordura 30. A proprietary solvent-free urethane flooring system suitable for use in food plants. (733).

Forestite. A proprietary trade name for a wood flour.

Forgenin. Tetramethyl-ammonium-formate, $H.COON(CH_3)_4$. A stimulant and appetiser.

Forging Brass. See BRASS.

Forlay. Selective weed killers. (501).

Formadermine. Guaiacol methylene ether.

Formagen. A dental cement consisting of two parts : (*a*) A liquid containing creosote, phenol, olive oil, and alcoholic formalin, and (*b*) a powder consisting of aluminium silicate, magnesium and zinc carbonates, and lime. It sets very hard.

Formagene. Paraformaldehyde tablets (591).

Formal. Formaldehyde. See METHYLAL.

Formalbumin. A condensation product of formaldehyde and casein. Used as an external antiseptic.

Formaldehyde. See ADORIN, ALDOFORM, FORMAL, FORMALIN, FORMALITH, FORMATOL, FORMITROL, FORMOLYPTOL, FESTOFORM, METHYLALDEHYDE, and PRESERVALINE.

Formaldehyde-alcohol Fluid. Consists of 100 parts of 70 per cent. alcohol and 5 parts of 40 per cent. formaldehyde. A fixative for plant material in microscopy.

Formaldehyde-ammonia 6 : 4. See HEXAMINE.

Formaldehyde Dust. Formaldehyde mixed with an inert material for use in plant disinfecting.

Formaldehyde Soaps. See FORMALIN SOAPS.

Formalin (Formol, Formol-chloral). A 40 per cent. aqueous solution of formaldehyde. The commercial product often contains from 12–15 per cent. methyl alcohol, to prevent the separation of polymerised compounds.

Formalin Cream. A cream made by adding 10 per cent. liquid formaldehyde to lanoline.

Formalin Sapene. A liquid soap containing formaldehyde. It is employed as a remedy for perspiring feet, and as a preventative of night-sweats.

Formalin Soaps. (*a*) Liquid formalin soap, which is a mixture of olive oil, alcohol, and essential oils with from 10–25 per cent. liquid formaldehyde. (*b*) Solid formaldehyde soaps, consisting of neutral soap, 5 per cent formaldehyde and perfumes.

Formalite. A trade name for phenol-formaldehyde resin moulded material for use in electrical insulation.

Formalith. A formaldehyde solution.

Formaloin. A condensation product of formaldehyde and aloin. It has the therapeutic properties of aloin, and the antiseptic character of formaldehyde.

Formamine. See HEXAMINE.

Formamint. A compound of formaldehyde and milk sugar. Used as a mouth disinfectant, and as a remedy in cases of diseases of the mouth and throat. Each tablet contains 0·01 gram formaldehyde.

Formamol. See HELMITOL.

Formamylum. See AMYLOFORM.

Forman. Chloro-methyl-menthyl-ether, $C_{10}H_{19}O.CH_2Cl$. An antiseptic. It is prescribed as a remedy for colds.

Formanek's Indicator. Alizarin green used as an indicator. It gives a violet colour with $pH0·3$, pink with $pH1·0$, yellow with $pH12$, and brown with $pH14$.

Formanilide. Phenyl - formamide, $HCONHC_6H_5$.

Formaniline. A rubber vulcanisation accelerator. It is anhydro-formaldehyde aniline.

Formanite. A mineral. It is $8[(Yt,Sr)(Cb,Ta)O_4]$.

Formapex. A proprietary phenol-formaldehyde resin varnish.

Formasal. A condensation product of formaldehyde and salicylic acid, claimed to be methylene-disalicylic acid. Several salts are on the market. The caffeine salt (Cafformasal) has been suggested as a means for the hypodermic administration of caffeine. The bismuth salt (Bisformasal) is used as an internal antiseptic. The sodium salt (Sodiformasal) is used as an antirheumatic and antineuralgic.

Formatol. A preparation of formaldehyde, an antiseptic.

Formax. Sodium formate. (507).

Formebolone. An anabolic steroid. 2-Formyl - 11α, 17β - dihydroxy - 17α - methylandrosta - 1, 4 - dien - 3 - one.

Formeston. A preparation of Eston. (See LENICET.) It has germicidal and astringent properties.

Formex. A proprietary trade name for an enamelled wire. The enamel is flexible and heat resisting up to about 185° C. It contains a mixture of polyvinyl acetal and a thermosetting phenol-formaldehyde synthetic resin. See THERMEX.

Formica. A proprietary trade name for phenolic and urea resins.

Formic Ether. Ethyl formate, $HCOO.C_2H_5$.

Formichthol. See ICHTHOFORM.

Formicin. Formaldehyde - acetamide, $CH_3.CO.NH.CH_2OH$ or $CH_3.C(:NH).O.CH_2OH$. Used as an antiseptic by injection, and for wounds.

Formidine. Methylene - disalicylic - iodide, $C_{15}H_{16}O_6I_2$.

Formin. See HEXAMINE.

Forminitrazole. 2 - Formamido - 5 - nitrothiazole. Aroxine.

Formit. See BAKELITE.

Formite. See BAKELITE.

Formitrol. A formaldehyde preparation.

Formobor. An aqueous solution containing 0·4 per cent. formaldehyde, and 1·5 per cent. borax. Recommended as a disinfectant for hair - dressing articles.

Formocortal. A topical corticosteroid. It is 21 - acetoxy - 3 - (2 - chloroethoxy)-9α - fluoro - 6 - formyl - 11β - hydroxy - 16α, 17α - isopropylidenedioxypregna - 3,5-dien-2o-one.

Formo-gelatin. Glutol (*q.v.*).

Formol. See FORMALIN.

Formolactin. A concentrated solution of casein and formaldehyde. Used as a wash, or for painting on paper or other material to make same washable, waterproof, and dustless, as well as disinfecting.

Formol-chloral. See FORMALIN.

Formolide. An antiseptic consisting of an aqueous solution of 15 per cent. alcohol, 2 per cent. boric acid, ½ per cent. sodium benzoate, and ¼ per cent. formaldehyde.

Formolites. Phenol - formaldehyde resins. See BAKELITE, ALBERTOL, and ISSOLIN.

Formolyptol. A formaldehyde preparation.

Formopon. See DISCOLITE.

Formopyrin. Methylene-diantipyrine, $(C_{11}H_{11}N_2O)_2 : CH_2$. It has the medicinal action of formaldehyde and antipyrine.

Formosa Camphor. Camphor from China.

Formose. A mixture of saccharine compounds obtained from formaldehyde.

Formosul. Sodium formaldehyde sulphoxylate. See also hydrosulphite NP. (503).

Formosul F.A. A special compound of acetaldehyde sulphoxylate. (503).

Formotan. See TANNOFORM.

Formrez. A registered trade name for a polyester-based polyurethane elastomer cross-linked with a diamine. (40).

Formural. See FORMUROL.

Formurol. An additive product of hexamethylene-tetramine and sodium citrate, $C_6H_{12}N_4.C_6H_7O_7Na$. Administered in cases of gout and urinary troubles.

Formvar. A trade mark for an acetal type of synthetic resin made from Gelva (*q.v.*) and formaldehyde.

Formyl Violet 6B and 10B. Mixtures of Formyl violet S_4B with Thiocarmine R.

Formyl Violet 5BN. See FORMYL VIOLET S_4B.

Formyl Violet S4B (Acid Violet 6B, Acid Violet 4B Extra, Formyl Violet 5BN, Coomassie Violet, Alizarin Celestol, Benzyl Violet 5BN, Coomassie Violet R). A dyestuff. It is the sodium salt of tetraethyl-dibenzyl-triamino-triphenyl-carbinol-disulphonic acid, $C_{41}H_{45}N_3O_7S_2Na_2$. Dyes wool violet from an acid bath.

Formysole. A disinfectant. It is a glycerin-potash soap, with an addition of from 10–25 per cent. formaldehyde.

Fornacite. A mineral. It is $4[(Pb,Ca)_3\{(Cr,As)O_4\}_2]$.

Fornitrol. A compound containing 2 molecules of formic acid, combined with 1 molecule of endo-anilo-diphenyl-dihydro - triazol, $(H.COOH)_2 \cdot (C_6H_3)_3 \cdot (N_4C_2H)$. A reagent used for the estimation of nitric acid, nitro-compounds, and nitrates.

Foroid. A proprietary preparation of sodium tetraiodo-phenol-phthalein.

Forsterite. A mineral. It is a magnesium-ortho-silicate, Mg_2SiO_4.

Forster Powder. See VON FORSTER POWDER.

Fortafil. A siliceous filler. (512).

Fortagesic. A proprietary preparation of PENTAZOCINE and PARACETAMOL. An analgesic. (685).

Fortex. (1) An explosive containing ammonium nitrate and Tetryl (q.v.).

Fortex. (2) Flavours for confectionery, foodstuffs and beverages. (504).

Fortrex (2). An activated clay for reinforcing natural and synthetic rubber. (189).

Fortification Agate. A term applied to those agates in which strata of jasper and chalcedony have been deposited upon quartz crystals.

Fortior. A proprietary preparation of THIAMINE hydrochloride, RIBOFLAVIN, NICOTINAMIDE, ascorbic acid, ferrous sulphate, copper sulphate and manganese citrate. A hæmatinic. (334).

Fortis. An explosive. It is a mixture of nitre - tar, and sulphur, with small quantities of iron sulphate and glycerin.

Fortisan. A proprietary trade name for a synthetic fibre made from regenerated cellulose. It is three times as strong as natural silk and can be produced in diameters one-tenth that of silk. Its heat resistance is superior to that of cotton. It can be impregnated and is resistant to mould growth.

Fortoin. A methyl-dicotoin, $CH_2(C_{14}H_{11}O_4)_2$. Used in medicine in cases of diarrhœa, also as a local application in badly ulcerated sore throats, and gonorrhœa.

Fortral. A proprietary preparation of pentazocine hydrochloride. An analgesic. (439).

Fosalsil. A proprietary trade name for a natural diatomaceous material made into bricks or used as a cement. It has high heat and sound insulating properties, and is fireproof.

Fosazepam. A hypnotic currently undergoing clinical trial as " HR 930 " and " 48 390 ". It is 7-chloro-1-di methylphosphinylmethyl - 1, 3 - di hydro - 5 - phenyl - $2H$ - 1,4 - benzodia zepin-2-one.

Foseco. Foundry preparations and fluxes. (531).

Fosfaragonite. A lime fertiliser giving on analysis, 39 per cent. CO_2, 5 per cent. SiO_2, 2 per cent. H_2O, and 1 per cent. P_2O_5.

Fosferno. Parathion insecticide. (511).

Fosfestrol. A preparation used in the treatment of carcinoma of the prostate It is *trans*-$\alpha\alpha'$ - diethylstilbene - 4, 4' diol bis(dihydrogen phosphate). HON VAN is the tetrasodium salt.

Fosfor. A proprietary preparation of phosphorylcolamine. A tonic. (256).

Fosfoxyl. A proprietary preparation It is stated to be a colloidal phosphorus product.

Foshaigite. A mineral, $H_2Ca_5(SiO_4)_3$ $2H_2O$, of California.

Fospirate. A pesticide. It is dimethyl 3, 5, 6 - trichloro - 2 - pyridyl phosphate

Fossil Carbon. Peat, lignite, and coal

Fossil Flour. See INFUSORIAL EARTH.

Fossiline. See OZOKERINE.

Fossil Salt. Rock Salt.

Fossil Wax. See OZOKERITE.

Fostacene. A proprietary polystyrene (734).

Fosterite. A mineral. It is magnesium silicate, Mg_2SiO_4.

Fotosensin. A condensation product of phthalic acid and resorcinol con taining small proportions of copper and iron. Small quantities increase the root and stem growth of plants.

Fouadin. A sodium salt of antimony-3 pyrocatechin-disulphonate containing 13.3 per cent. antimony. It is used in the treatment of schistosomaisis.

Fouane. Benzthiazide.

Founders' Type. See TYPE METAL.

Foundry Clay. A clay containing from 80–90 per cent. silica and 15–18 per cent. aluminia.

Foundry Pattern Metal. An alloy of 75 per cent. zinc and 25 per cent. tin.

Fouqueite. A mineral. It is a variety of epidote.

Fourdrinier Wire. A brass containing from 80–85 per cent. copper, and 15–20 per cent. zinc.

Fourmarierite. A uranium mineral from the Belgian Congo. It approximates to the formula $PbO.4UO_3.5H_2O$ and contains small quantities of tellurium and iron oxide.

Fourneau 189. Oxy - amino - phenyl arsinic acid. An antisyphilitic.

Fourneau 190 (Acetarsol, Spirocid Stovarsol). The acetyl derivative of hydroxy - amino - phenyl - arsinic acid (Fourneau 189), $HO.C_6H_3(NHCH_3)As O_3H_2(1:2:4)$.

Fourneau 309. A trypanocide. It is sodium-sym-bis -(1″:4″:6″:8″-naph-thyl-amino-trisulphonate-4′-methyl′-3 - amino-benzoyl-3-amino-benzoyl) urea, $(SO_3Na)_3.C_{10}H_4.NH.CO.C_6H_3(CH_3)NH. CO.C_6H_4.NH.CO.NH.C_6H_4.CO.NH.(C H_3)C_6H_3.CONH.C_{10}H_4(SO_3Na)_3$. It is thought to be identical with Bayer 205. Used in sleeping sickness.

Fourneau 575. 8 - Dimethyl - amino - *n*-propyl-amino-6-methoxy-quinoline. An antimalarial.

Fourneau 710. 8 - Diethyl - amino - *n* - propyl-amine-6-methoxy-quinoline. An antimalarial.

Fourneau 728. 8 - Diethyl - amino - *n* - propyl - amino - quinoline. An anti - malarial.

Fourneau 735. 8 - Diethyl - amino - *n* - amyl-amino-6-methoxy-quinoline. An antimalarial.

Fourneau 769. Ethyl - β - ethyl - butyl - barbituric acid. A hypnotic.

Fousel Oil. Fusel Oil (*q.v.*).

Fowlerite. A mineral, $(Mn.Fe.Ca.Zn. Mg).O.SiO_2$.

Fowler's Solution. (*Liquor arsenicalis B.P.*) A solution of white arsenic, As_4O_6, in potassium carbonate, coloured with tincture of lavender. It contains 1 part of white arsenic in 100 parts of solution.

Foxglove Oil. See OIL OF FOXGLOVE.

FR-28. Sodium borate. (548).

FR-1360. A proprietary flame-retardant material used in the production of flexible polyurethane foam. It is tribromoneopentyl alcohol. (64).

FR-2406. Tris (2,3-dibromopropyl) phosphate. A proprietary fire-retardant additive used in acrylics, epoxies, latices, phenolics, polyesters, polystyrenes, polyvinyl chloride, rayon celluloses and polyurethanes. (735).

Fracton. A refractory dressing. (531).

Fractorite. An explosive containing 90 per cent. ammonium nitrate, 4 per cent. resin, 4 per cent. dextrin, and 2 per cent. potassium bichromate.

Fractorite B. A Belgian explosive consisting of 75 per cent. ammonium nitrate, 2·8 per cent. dinitro-naphthalene, 2·2 per cent. ammonium oxalate, and 20 per cent. ammonium chloride.

Fragarol (Fragasol). The butyl ether of β-naphthol. A synthetic perfume.

Fragaroma. Perfumery products. (507).

Fragasol. See FRAGAROL.

Fragula Bark. The bark of *Rhamnus fragula*.

Fraipontite. A mineral. It is a hy-drated basic silicate of aluminium and zinc.

Fraissite. A tear gas. It is benzyl iodide, $C_6H_5.CH_2.I$.

Fraizone. Aromatic flavouring chemicals. (522).

Framycetin. An antibiotic derived from Streptomyces decaris. Framygen and Soframycin are the sulphate.

Framycort. A proprietary preparation of framycetin sulphate and hydrocortisone used in dermatology as an antibacterial agent. (188).

Framygen. A proprietary preparation of framycetin sulphate. A systemic antibiotic. (188).

Framyspray. An antibiotic aerosol. (509).

Franckeite. A mineral. It is a sulphide ore of tin, $Pb_5FeSn_3Sb_2S_{14}$.

François Reagent. A solution of sulphurous acid containing 1 part rosaniline acetate in 1,000 c.c. To 30 c.c. of this solution are added 3 c.c. of sulphuric acid. The reagent is used to test chloroform by shaking 5 c.c. of it with 5 c.c. of reagent. No colour should result.

Francolite. An impure variety of the mineral Apatite.

Frankfort Black (Drop Black). A pigment prepared by heating vine twigs, bones, ivory, or other materials in closed retorts, washing, and grinding. The ground product is mixed with water, and sent into the market in drops or tears called drop black. The best kinds are obtained by carbonising wine yeast.

Frankincense. Olibanum, a gum resin.

Frankincense, American. Gum thus.

Frankincense, Common. See GUM THUS.

Frankincense, Indian. See OLIBANUM.

Franklandite. A mineral, $Na_2CaB_6O_{11}. 7H_2O$.

Franklinite (Zincoferrite). A mineral. It is a mangano-ferrate of manganese, iron, and zinc, $(Fe.Zn.Mn)O.Fe_2O_3$. It is also stated to be $8[(Zn,Mn,Fe'')(Fe''',Mn''')_2O_4]$.

Franklin Substitute. A rubber substitute. It consists of an oxidised mixture of coal tar and boric acid.

Frankoline. A solution of cuprous chloride and ferric chloride in strong hydrochloric acid, absorbed in kieselguhr. Used for the removal of phosphorus compounds from acetylene.

Frankonite (Silitonite, Tonsil). German bleaching earths obtained from deposits in Germany.

Franocide. Diethylcarbamazine citrate (Veterinary). (514).

Franol. A proprietary preparation of

theophylline, phenobarbitone and ephedrine hydrochloride. A bronchial antispasmodic. (112).

Franol Expectorant. A proprietary preparation containing phenobarbitone, ephedrine, theophylline monohydrate and guaiphenesin. A cough linctus. (112).

Franol Plus. A proprietary preparation of THEOPHYLLINE, PHENOBARBITONE, EPHEDRINE sulphate, and THENYLDIAMINE hydrochloride used as a bronchospasm relaxant. (439).

Franquerite. A mineral. It is $2[(Mg, Fe^{..})(Fe^{...}, Al)_3(SO_4)_4(OH)_3.18H_2O]$.

Frantin. An anthelmintic for the unweaned lamb. (514).

Frary Metal. A calcium-barium-lead alloy containing up to 2 per cent. barium and 1 per cent. calcium. Used as a bearing metal. It melts at 445° C.

Fraude's Reagent. Perchloric acid, $HClO_4$.

Fray-Bentos Guano. A fertiliser. It is produced in a similar way to flesh guano, during the manufacture of Liebig's extract of meat.

Fredo. Calcium hydrosulphite.

Fredricite. A mineral. It is a variety of tennantite.

Freeman's Non-poisonous White Lead. A pigment. It is a mixture of white lead, zinc white, baryta white, and magnesium carbonate.

Freestone. Calcareous sandstone containing particles of sandstone united with calcium carbonate.

Freezine. See PRESERVALINE.

Freezing Salt. Crude sodium chloride (common salt).

Freibergite. A mineral. It is a variety of tetrahedrite, rich in silver.

Freieslebenite. A mineral. $3Ag_2S.4PbS.3Sb_2S_3$, containing 23 per cent. silver.

Freirinite. A rock mineral, $6(Cu.Ca)O.3Na_2O.2As_2O_5.6H_2O$, found in Chile.

Fremontite. Synonym for Natromontebrasite.

Fremy's Salt. Potassium hydrogen fluoride, $KF.HF$.

French Asphalt. See AUVERGNE BITUMEN.

French Automobile Bearing Metal. See BEARING METALS.

French Blue. See ULTRAMARINE.

French Brandy. See COGNAC.

French Cement. A mucilage of gum arabic mixed with powdered starch. Used by naturalists, artificial flower makers, and confectioners.

French Chalk. A variety of Steatite or Soapstone (q.v.). It is a hydrated silicate of magnesium, and is used for marking cloth, removing grease from silk, as a filler, and for other purposes.

French Chrome Yellow. See CHROME YELLOW.

French Fat Oil. Oil of turpentine.

French Gutta-percha. A product obtained from the bark of the birch by boiling in water.

French Maltine. Extract of Malt (q.v.).

French Metal. A term used for ingots of antimony.

French Ochre. See YELLOW OCHRE.

French Pine Resin. See GUM THUS.

French Polish. A polish for wood. It consists of shellac dissolved in alcohol and coloured with dragon's blood.

French Purple (Archil Violet, Red Indigo, Perseo). A dyestuff obtained from lichens. The lichen acids are first extracted by milk of lime, and precipitated by acids. They are then dissolved in ammonia, and exposed to air at ordinary temperatures, until a cherry-red colour is produced. After being boiled, the liquor is placed in large flat dishes, and kept at 70–75° C. until it becomes purplish-violet. The solid dyestuff is precipitated by sulphuric acid or calcium chloride.

French Putty. A putty obtained by boiling 7 parts linseed oil with 4 parts brown umber, $5\frac{1}{2}$ parts chalk, and 11 parts white lead are added. Used for wood.

French Red. See LUTECIENNE.

French Silver. These are nickel-copper-silver alloys containing 25–35 per cent. of nickel, and are used as silver substitutes. Special alloys of this class are Roulz, Odessa, Mousset, and Abe alloys.

French Turpentine (Bordeaux Turpentine). The oleo-resin from *Pinus maritima*. It contains from 15–20 per cent. essential oil and 70–80 per cent. rosin. It contains terebenthene.

French Ultramarine. See ULTRAMARINE.

French Verdigris. The basic acetate of copper having the formula $Cu_2(C_2H_3O_2)_2.(OH)_2$.

French Veronese Green. See CHROMIUM GREEN.

French White. Powdered talc.

French Yellow. See CAMPOBELLO YELLOW.

French Zinc (French Zinc Oxide). Zinc oxide produced by the French method when metallic zinc is volatilised, and the vapours brought into contact with air.

French Zinc Oxide. See FRENCH ZINC.

Frenokone. A selective weedkiller (511).

Frenquel. The hydrochloride of diphenyl-4-piperidylmethanol.

Frenzelite. A mineral. It is guanajuatite, Bi_2Se_3.

Freon (F-12, Kinetic No. 12). A registered trade mark. Dichloro-difluoromethane, CCl_2F_2. It is a colourless, almost odourless gas, boiling-point $-29.8°$ C., non-toxic, non-corrosive, non-irritating, non-inflammable, and is used as a refrigerant.

Freund's Acid. 1-Naphthylamine-3 : 6-disulphonic acid.

Freyalite. A mineral, $ThSiO_2$.

Friable Amber. See GEDANITE.

Friar's Balsam. See TURLINGTON BALSAM.

Frick's Alloys. Nickel silvers containing from 50–69 per cent. copper, 18–39 per cent. zinc, and 5–31 per cent. nickel.

Friedelite. A mineral, Mn_2SiO_4.

Friedländer's Stain. A microscopic stain. It contains 50 grams of a saturated alcoholic solution of gentian violet, 10 grams of glacial acetic acid, and 100 c.c. water.

Frieseite. A mineral. It is a variety of sternbergite.

Frigidite. A mineral. It is a variety of tetrahedrite.

Frishmuth's Aluminium Solder. Alloys. One contains 67 per cent. tin, 27 per cent. lead, and 3 per cent. aluminium. Another consists of 94 per cent. zinc, 4 per cent. aluminium, and 2 per cent. copper.

Frishout Solder. An alloy of 46 per cent. tin, 23 per cent. zinc, 15 per cent. aluminium, 8 per cent. copper, and 9 per cent. silver.

Frisonnets. Waste silk obtained in reeling the cocoons.

Frits. Specially prepared glasses, made by fusing together various constituents of a glaze (usually alkali or carbonate), and some form of silica or clay.

Fritscheite. A mineral. It is a vanadiferous autunite.

Fritzsche's Reagent. Dinitro-anthraquinone, $C_{14}H_6N_2O_6$.

Frohde's Reagent. An alkaloid reagent. It consists of 0·5 gram of sodium molybdate dissolved in 100 c.c. of concentrated sulphuric acid.

Fromager Wood. A wood obtained from *Eriodendron anfractuosum*, of Gaboon. The dry wood contains 68·3 per cent. cellulose, 25·18 per cent. lignin, and 0·62 per cent. fats and waxes.

Frondox. A weed killer preparation. (514).

Frosted Silver. See DEAD SILVER.

Frostine Balsam. A solution of 1 part Bromocoll in 10 parts collodion, with the addition of 1 part alcohol, and ½ part tincture of benzoin. Used for chilblains.

Frostine Salve. A 10 per cent. Bromocoll resorbin ointment. Used for chilblains.

Frost Preservative Solution. This contains 80 grams chloral hydrate, 160 grams potassium acetate, 3,500 grams cane sugar, 80 grams sodium fluoride, and 8,000 grams of a saturated solution of thymol.

Frost Rubber. Sponge rubber.

Froth, Copper. See TYROLITE.

Fructol. A formic acid preservative, used for fruit preparations. It contains from 10–14 per cent. formic acid.

Frugardite. A mineral. It is vesuvianite.

Fruit Sugar. See LEVULOSE.

Fruktin. A honey substitute. It consists of cane sugar, and a small quantity of tartaric acid.

Frusemide. 4 - Chloro - N - furfuryl - 5 - sulphamoylanthranilic acid. Furosemide. Lasix.

Frusid. A proprietary preparation of FRUSEMIDE. A diuretic. (498).

Fuchsia. See METHYLENE VIOLET 2RA. The diethyl-safranines, also the chlorides of α- and β-diamyl-safranines, are known under this name.

Fuchsianite. See MAGENTA.

Fuchsine. See MAGENTA.

Fuchsine Extra Yellow (Fuchsine Scarlet). A dyestuff. It consists of Fuchsine mixed with Auramine, to give a yellower shade than fuchsine alone.

Fuchsine S. See ACID MAGENTA.

Fuchsine Scarlet. See FUCHSINE EXTRA YELLOW.

Fuchsite. A mineral. It is a bright green variety of Muscovite (*q.v.*), and contains chromium.

Fuchsone. Diphenyl-quino-methane.

Fucidin. A proprietary preparation containing sodium fusidate. An antibiotic. (308).

Fucidin-H. A proprietary preparation of sodium fusidate and HYDROCORTISONE acetate used in the treatment of dermatosis. (308).

Fucidin V.P. A proprietary preparation of sodium fusidate and phenoxymethyl-penicillin. An antibiotic. (308).

Fucole. The trade name for a preparation of iodine-containing sea-weeds which are broken up, roasted, powdered, and then mixed with sesamé oil. Suggested as a substitute for cod-liver oil.

Fucosol (Fucusol). A mixture of furfurol and methyl-furfurol.

Fucusin. Tablets containing fucusin (alleged to be the principle of *Fucus vesiculosis*), are recommended as having anti-fat properties.

Fucusol. See FUCOSOL.

Fudow. A proprietary range of phenolic moulding materials. (442).

Fudowlite U. A proprietary range of urea formaldehyde moulding materials. (442).

Fuel, Colloidal. See COLLOIDAL FUEL.

Fuel, Liquid. See COLLOIDAL FUEL.

Fuel Oil. This name includes heavy petroleum distillates and residues, shale, cannel, and coal oils, tar and tar oils. Used as a liquid fuel for Diesel engines. Also see HEAVY OILS.

Fulbent. Bentonite clay for binding, suspending and emulsifying. (513).

Fulbond. A bonding agent for foundry sands. (513).

Fulcat. Clay catalysts. (513).

Fulcin. A proprietary trade name for Griseofaloin.

Fulcin Forte. A proprietary preparation containing griseofulvin. An antibiotic. (2).

Fulgurite. An explosive containing 60 per cent. nitro-glycerin and 40 per cent. wheaten flour and magnesium carbonate.

Fuligo. Soot.

Fulla Panza Nuts. See OWALA BEANS.

Fullerite. A proprietary trade name for a slate powder used as a rubber filler.

Fuller's Earth. A term applied to a sandy loam or argillaceous earth found in Surrey and Kent. It consists of aluminium and magnesium hydrosilicates. A deodoriser. Used for clarifying olive oil, and for decolorising other oils.

Fullonite. See ONEGITE.

Fulmargin. A solution of colloidal silver.

Fulmenit. An explosive consisting of 86·5 per cent. ammonium nitrate, 5·5 per cent. trinitro-toluene, 2·5 per cent. paraffin oil, 1·5 per cent. charcoal, and 4 per cent. guncotton.

Fulminating Gold. A compound having the formula $2AuN_2H_3.H_2O$, prepared by the action of concentrated ammonia on gold hydroxide. It is explosive.

Fulminating Platinum. Explosive compounds formed by acting upon ammonium platinochloride with potash.

Fulminating Silver. A compound of nitrogen and silver, NAg_3. It has been prepared by the action of ammonia on precipitated silver oxide. It is explosive.

Fulöppite. A mineral. It is $4[Pb_3Sb_8S_{15}]$.

Fulton White. A proprietary brand of lithopone (q.v.).

Fulvite. An artificial titanium monoxide.

Fumagillin. An antibiotic produced by certain strains of Aspergillus fumigatus.

Fumigating Vinegar. Usually tincture of benzoin with alcohol, acetic acid and perfume oils. It is evaporated by gentle heat.

Fuming Liquor of Libavius. Stannic chloride, $SnCl_4$.

Fuming Nitric Acid. Nitric acid containing some of the lower oxides of nitrogen.

Fuming Oil of Vitriol. See FUMING SULPHURIC ACID.

Fuming Sulphuric Acid (Oleum, Nordhausen Sulphuric Acid, Pyrosulphuric Acid). It consists of sulphur trioxide dissolved in sulphuric acid. The commonest fuming acid contains 55 per cent. sulphuric acid, and 45 per cent. sulphur trioxide. Nordhausen sulphuric acid has a specific gravity of from 1·86-1·90.

Fumonex. A proprietary gas black for rubber mixing.

Fungex. A copper containing fungicide (501).

Fungilin. A proprietary preparation containing Amphotericin B. An antibiotic. (326).

Fungitex 656. A proprietary trade name for a solution of an organo-mercuric complex. Used as a durable mildewproofing agent for textiles. (18).

Fungizone. A proprietary preparation containing Amphotericin B. An antibiotic. (326).

Furac No. 3. A proprietary rubber vulcanisation accelerator. It is the lead salt of dithio-furoic acid.

Furacin. (1) A trade name for an acid and solvent resistant furane resin cement used for bedding and jointing pipes. (157).

Furacin. (2). A proprietary preparation of nitrofurazone. An antibiotic. (281).

Furacrinic Acid. A diuretic currently undergoing clinical trial as " GP 48 674 ". It is 6-methyl-5-(2-ethylacryloyl)benzofuran - 2 - carboxylic acid.

Furadantin. A proprietary preparation containing nitrofurantoin. A urinary antiseptic. (281).

Furamide. A proprietary preparation of diloxanide 2-furoate. An antiamœbic. (253).

Furan. A proprietary preparation of nitrofurantoin. A genito-urinary antibiotic. (363).

Furanculine. Dried yeast.

Furane (Tetra-phenol, Tetrol, Tetrane). Furfurane, C_4H_4O.

Furazolidone. An antiseptic. 3-(5-Nitrofurfurylideneamino)oxazolidin - 2 - one. FUROXONE.

Furbac. A proprietary accelerator. It is n-cyclohexyl - 2 - benzthiazyl sulphenamide. (290).

Furethidine. Ethyl 4 - phenyl - 1 - [2 - (tetrahydrofurfuryloxy)ethyl] - piperi - dine-4-carboxylate. A narcotic analgesic.

Furfural Resins. Artificial resins obtained by the condensation of furfuraldehyde with phenols, cresols, or other similar bodies.

Furnace-calamine. Masses consisting mainly of zinc oxide, formed during the smelting of zinciferous iron ores.

Furnace Creek Talc. A very pure proprietary magnesium silicate talc supplied in either 325-mesh or micronised form. (293).

Furnacite. Synonym for Fornacite.

Furniture Polishes. These consist chiefly of beeswax or paraffin, soap, turpentine, alcohol, and water, made into an emulsion. Others consist of a solution of gum sandarac and shellac in alcohol.

Furol Green. A dyestuff obtained by condensing furfural with dimethylaniline in the presence of zinc chloride, then converting the leuco base produced into the dyestuff by means of lead peroxide.

Furosemide. Frusemide.

Furoxone. A proprietary preparation of furazolidine. An antibiotic. (818).

Fursatil CS 12. A proprietary cold-setting resin based on urea/furane. (732).

Fursatil CS15. A faster-setting, lower-strength variant of FURSATIL CS12. (732).

Fursatil CS25. A fast-setting, medium-strength variant of FURSATIL CS12 having a low fume level. (732).

Fursatil CS30. A proprietary phenol/furane cold-setting resin. (732).

Fursatil CS40. A proprietary plasticised urea cold-setting resin. (732).

Fursatil CS60. A proprietary phenol/urea cold-setting resin. (732).

Fursatil CS65. See FURSATIL CS60. (732).

Fursatil CS71. A proprietary modified phenolic cold-setting resin. (732).

Fursatil CS81. A proprietary modified urea formaldehyde cold-setting resin. (732).

Furunculin. A yeast preparation

Fusafungine.⁋ An antibiotic produced by Fusarium lateritium 437. Locabiotal.

Fusain. See CLARAIN.

Fusariol. A mercury-formaldehyde preparation. A seed preservative.

Fuscamine Brown. A dyestuff. It is m-amino-phenol.

Fuscanthrene. An anthracene dyestuff, prepared by heating formaldehyde and certain diamino-anthraquinones

with potassium hydroxide. It is a yellow-brown vat dye.

Fuscochlorin. A dark-green pigment from algæ.

Fuscorhodin. A dark-red pigment from algæ.

Fused Cement (Electro-fused Cement). Terms applied to an aluminous cement with a high alumina content.

Fusel Oil (Fermentation Amyl Alcohol, Potato Oil, Grain Oil, Marc Brandy Oil). A by-product in alcoholic fermentation, especially in the preparation of potato spirit, and in the rectification of alcohol. It consists of from 12–24 per cent. water, 15–45 per cent. ethyl alcohol, 6–14 per cent. normal propyl alcohol, 10–25 per cent. isobutyl alcohol, and 10–40 per cent. fermentation amyl alcohol. Used for the preparation of amyl alcohol, and for gelatinising explosives.

Fusible Alloy (Fusible Metal). Alloys of tin, bismuth, lead, and sometimes cadmium, of variable composition. See WOOD'S, KRAFT'S, HOMBERG'S, and ROSE'S METAL.

Fusible Metal. See FUSIBLE ALLOY.

Fusible Salt. See MICROCOSMIC SALT.

Fusible Salt of Urine. This is microcosmic salt.

Fusible White Precipitate. See MERCURAMMONIUM CHLORIDE.

Fusidic Acid. An antibiotic produced by a strain of Fusidium. Fucidin is the sodium salt.

Fussolon. A proprietary range of fluorocarbon and similar resins. (736).

Fustet Wood. The wood of the sumac tree. Used for dyeing mordanted wool reddish-yellow.

Fustic (Old Fustic, Brazil Wood, Yellow Wood, Cuba Wood). The chips or extract from *Morus tinctoria*. It is a natural dyestuff, the dyeing principles being morin (tetraoxy-flavenol), $C_{15}H_{10}O_7$, and maclurin (pentaoxy-benzophenone), $C_{13}H_{10}O_6$. Chiefly used for dyeing wool yellow.

Fustin. The diazo-benzene compound of maclurin (obtained from fustic).

Futurit. A proprietary plastic of the phenol-formaldehyde type.

Fybogel. A proprietary preparation of ispaghula husk with sodium bicarbonate and citric acid. A laxative. (243).

Fybranta. A proprietary preparation of bran and calcium phosphate. A laxative. (286).

Fyrol 6. A dihydroxy-terminated phosphonate ester. An additive for rigid polyurethane foam imparting self extinguishing properties.

G

G. Glycerol-phthalic anhydride resins.

G.62. A proprietary trade name for a vinyl plasticiser based on an epoxy resin. (13).

G124. 6-Monochlor-thymol.

G.500. A proprietary preparation containing hexamine mandelate and methionine. A urinary antiseptic. (319).

Gabbett's Stain. A microscopic stain. It contains 2 grams methylene blue, 25 c.c. sulphuric acid, with water up to 100 c.c.

Gabbro. A coarse, crystalline rock, composed mainly of lime-soda felspar.

Gabian Oil. An inflammable mineral naphtha.

Gabianol. A substance prepared from a natural shale. Asserted to be of value in various diseases of the lungs and throat.

Gaboon Chocolate. See DIKA BREAD.

Gabraster. Polyester resins.

Gabrite. Thermosetting] moulding powders based on ureaformaldehyde resins. (61).

G-acid. Naphthol-sulphonic acid, (OH $=2$, $SO_3H=6$ and 8).

Gadolinite. A mineral, $FeO.2SiO_2$. $2BeO.(Y.Ce.La.Di)_2O_3$.

Gadose. A grease prepared from cod-liver oil and lanoline. Used as a basis for ointments.

Gad's Cement. A mason's cement, consisting of 3 parts clay and 1 part ferric oxide.

Gaduol. An extract containing the alcohol-soluble constituents of cod-liver oil.

Gadusan. Colloidal copper morrhuate, stated to be used with success on tubercular abscesses.

Gagat. A variety of soft coal.

Gageite. A mineral. It is $(Mg,Mn,Zn)_8Si_3O_{14}.2H_2O$.

Gahnite (Zinc Spinel). A mineral, $Al_2O_3.ZnO$.

Gahnospinel. A mineral. It is $8[(Mg,Zn)Al_2O_4]$.

Gahn's Ultramarine. See COBALT BLUE.

Gajacyl. See GUAIACYL.

Gajite. A mineral. It is a hydrated carbonate of calcium and magnesium.

Gala. A proprietary casein plastic.

Galactan (Gelose). A gum, $(C_6H_{10}O_5)n$, from agar-agar.

Galactochloral. A combination of chloral and galactose. Used as a somnifacient.

Galactogen. A soluble casein preparation, made from skim milk. It contains 3·5 per cent. fat, 70 per cent. albumin, and 2 per cent. phosphoric acid.

Galactomin. Proprietary trade name for an artificial milk food with low lactose content. (665).

Galadin. A condensation product of antipyrine and phenyl-methane in 30 per cent. alcoholic solution. It is a German preparation, and is used as an antirheumatic.

Galafatite. A mineral, an aluminium-potassium sulphate.

Galagum. A mixture of modified polysaccharides. It gives a colloidal solution when boiled in water. It is a protective colloid, and is used in making baker's and flavouring emulsions, and in cosmetic and hair lotions.

Galahad A. A proprietary trade name for a steel containing 0·10 per cent. carbon, 0·50 per cent. silicon, 0·04 per cent. sulphur, 0·04 per cent. phosphorus, 0·60 per cent. manganese, 13·00 per cent. chromium and sometimes 1·50 per cent. nickel and 0·25 per cent. molybdenum. It is used in the manufacture of forged steel pressure vessels with good corrosion resistance. (155).

Galalith (Erinoid). The trade name for a horn-like material, obtained by the action of formaldehyde upon casein. It is a substitute for ivory, ebony, and horn, and can be kneaded and moulded. It is not easily inflammable, and can be used as an insulator.

Galalkerite (Proteolite, Zoolite). Italian casein plastics, used for the manufacture of combs, buttons, and umbrella handles.

Galam Butter. Shea butter.

Galanack. A mixture of tar, sulphur, and oil. Used as a cement or lute.

Galangal. The dried rhizome of *Alpinia officinarum*.

Galangin. Dihydro-flavonol, obtained from galanga root, *Alpinia officinarum*.

Galapectite. A variety of the mineral Halloysite.

Galaton. A proprietary preparation containing codeine phosphate and phenylpropanolamine hydrochloride. A cough linctus. (244).

Galaxite. A mineral. It is $8[MnAl_2O_4]$.

Galazin. A liquid produced by placing skimmed cow's milk, with 2 per cent. sugar, and 0·3 per cent. beer yeast, in tightly stoppered bottles, and allowing fermentation to take place for twenty-four hours.

Galbanum. A gum-resin obtained from *Ferula galbanifluum*, of Persia. A gum resin allied to ammoniacum. It is obtained from species of *Peucedanum*, and usually contains from 8–10 per cent. of oil and 54–63 per cent. of resin. The specific gravity varies from 1·11–1·13, the saponification value from

75–225, the acid value from 5–65, and the ash from 1–25 per cent.

Galena (Galenite, Blue Lead Ore, Tesselated Ore, Dice Ore). A mineral. It is lead sulphide, PbS. It often contains silver as sulphide, up to 1 per cent.

Galena, Pseudo. See ZINC BLENDE.

Galenite. A pigment. It is a basic sulphate of lead. The name is sometimes applied to Galena.

Galenobismuthite. A mineral, PbS. Bi_2S_3.

Galenoceratite. A mineral. It is lead chloro-carbonate.

Galenomycin. A proprietary preparation of OXYTETRACYCLINE. An antibiotic. (476).

Galen's Cerate. Cold cream.

Galettame Silk (Ricotti Silk, Neri Silk, Basinetto Silk). The residue of the silk cocoon after reeling.

Galicar. An anti-friction bearing metal containing 83 per cent. tin.

Galigen. A German preparation. It is an alkaloid found in the clover, *Galiga officinalis*. It reduces the blood-sugar, and is used in cases of diabetes.

Galipot. The resin from *Pinus maritima*.

Galismuth. Ethylene-diamine-bismuth gallate. An antisyphilitic.

Gallacetophenone. Trioxy - aceto - phenone, $C_8H_8O_4$. Used medicinally as an antiseptic in skin diseases.

Gallal. Aluminium subgallate, $Al_4(C_7H_2O_5)_3$. An antiseptic and astringent.

Gallamine. 1,2,3, - Tri - (2 - diethyl - aminoethoxy)benzene. Flaxedil is the triethiodide.

Gallamine Blue. A dyestuff. It is the amide of gallocyanine, $C_{15}H_{14}N_3O_4Cl$. Dyes chromed wool blue, also used in calico printing.

Gallanilic Blue (Gallanilic Indigo P and PS, Tannic Indigo). A dyestuff obtained by the action of aniline upon Gallanilic violets R and B. The mark PS is the sulphonic acid of the product, and it dyes silk and wool from an acid bath, or upon a chrome mordant. The brand P gives indigo blue upon a chrome mordant.

Gallanilic Green (Fast Green G, Solid Green G). A dyestuff obtained by the nitration of Gallanilic indigo PS. Dyes chromed wool green.

Gallanilic Indigo Blue P and PS. See GALLANILIC BLUE.

Gallanilic Violets R and B (Gallanilic Violet BS). Dyestuffs. They are the anilides of dimethyl- and diethyl-gallocyanines. BS is the bisulphite compound. They dye metallic mordanted wool or silk, reddish-violet, and give bluer shades from an acid bath.

Gallanol (Gallinol). Gallic acid anilide,

$C_6H_2.CO.NH.C_6H_5(OH)_3$. Used in medicine for chronic eczema.

Gallatite (Lactite, Lactoform, Cornalith, Ingalite, Lactorite, Sicalite, Proteolite). Casein preparations of a similar type to Galalith (*q.v.*).

Gallazine A. A dyestuff obtained by the condensation of Gallocyanine with β-naphthol-sulphonic acid S. Dyes chromed wool indigo blue. It is also employed in printing.

Galleine (Galleine A, Galleine W, Alizarin Violet, Anthracene Violet). A dyestuff. It is pyrogallol-phthalein, $C_{20}H_{12}O_7$. Dyes chrome mordanted wool, silk, or cotton, violet.

Galleine A. See GALLEINE.

Galleine W. See GALLEINE.

Gallicin. Gallic acid methyl ester, $C_6H_2(OH)_3CO_2CH_3$. Used by oculists as an antiseptic in conjunctivitis.

Gallin. 3 : 4 : 5 : 6 - tetrahydroxy - 9 - xanthene-*o*-benzoic acid.

Gallinol. See GALLANOL.

Gallipoli Oil. An olive oil used in the textile industries.

Gallisin. That portion of commercial starch syrup which resists fermentation.

Gallitzenite. See GOSLARITE.

Gallium-Albite. A mineral. It is $NaGaSi_3O_8$.

Gall Nuts. See GALLS.

Gallobromol. Dibromo - gallic acid, $C_6Br_2(OH_3)COOH$. Used in medicine for neurasthenia and epilepsy.

Gallocarmine Blue. A dyestuff allied to Gallocyanine. It is obtained from gallamic acid and nitroso-dimethyl-aniline.

Gallocyanine BD. See LITHOL RED B.

Gallocyanine DH and BS (Fast Violet, Solid Violet, Alizarin, Purple, Gallocyanine RS, BS, and D). A dyestuff. It is dimethyl-amino-dioxy-phenoxazonium-carboxylate, $C_{15}H_{12}N_2O_5$. BS is the bisulphite compound. Dyes chromed wool bluish-violet. Employed in printing.

Gallocyanine RS, BS, and D. See GALLOCYANINE DH and BS.

Gallocyanine S. A dyestuff. It is Gallocyanine sulphonic acid. Dyes chromed wool blue.

Galloflavine. A dyestuff, $C_{13}H_6O_9$, obtained by the moderate oxidation of gallic acid in aqueous or alcoholic alkaline solution, by means of air. Dyes chromed wool yellow. Used in printing.

Galloformin. A combination of gallic acid and formin, $C_6H_2(OH)_3COOH$ $(CH_2)_6H_4$. Used internally in cystitis.

Gallogen. Ellagic acid, $C_{14}H_6O_6$. Recommended as an astringent in cases of tuberculosis and inflammation of the bowels.

Gallols. Chemical combinations of pyrogallol, chrysarobin, and resorcinol, with various acids. Lenigallol is pyrogallol-triacetate, $(CH_3COO)_3C_6H_3$. Eugallol is a 66 per cent. solution of pyrogallol-mono-acetate. Both are used in skin diseases.

Gallophysin. A German product. It was formerly Ictophysin (q.v.).

Gallotannic Acid (Tannic Acid). Digallic acid, $C_6H_2.(OH)_3CO.O.C_6H_2(OH)_2COOH$.

Gallo Violet. A dyestuff. It is a leuco-pyro-gallocyanine.

Galls. Vegetable excrescences formed upon the branches, shoots, and leaves of trees of the oak type, grown in the Levant, by the puncture of certain insects for the purpose of depositing eggs. Varieties are Oak-apple galls, Aleppo galls (Turkey or Levant galls), Chinese galls, Japanese galls, and Acorn galls. These contain 40–70 per cent. tannin, estimated as gallotannic acid. Used as tanning materials. Galls are sold in their natural state or crushed, and also after being roasted. The roasted material yields a darker ink.

Gallstone. A yellow pigment obtained from the gall gladder of oxen.

Galmer. A mineral. It is Hemimorphite (q.v.).

Galorn. A proprietary trade name for a casein plastic.

Galt Glass. Polyester resin bonded glass fibre mouldings with a special surface giving good resistance to weather and chemicals.

Galuchat Leather. A leather made from the Japanese ray fish

Galvanised Iron (Zinced Iron). Iron coated with metallic zinc.

Galvanit. A plating powder consisting of a mixture of the salt of the metal to be deposited (silver for silver plating), and a more electro-positive metal.

Galvano Lac. A mixture of celluloid varnish with powdered metal.

Galvene. An inhibitor used during metal pickling. (512).

Galyl. A derivative of salvarsan, used in the treatment of syphilis. It is 4 : 4'-dihydroxy-arseno-benzene-3 : 3'-phosphamic acid, $C_{24}H_{22}O_8N_4P_2As_4$, containing 35·3 per cent. arsenic, and 7·2 per cent. phosphorus.

Gamagarite. A mineral. It is $Ba_4(Fe,Mn)_2V_4O_{15}(OH)_2$.

Gamaquil. A proprietary trade name for Phenprobamate.

Gambene Extract. A gambier substitute.

Gambier. See CUTCH.

Gambine B. See DIOXINE.

Gambine G. See GAMBINE Y.

Gambine R. See DIOXINE.

Gambine Y (Alsace Green J, Gambine G, Steam Green S, Mulhouse Green). A dyestuff. It is α-nitroso-β-naphthol, $C_{10}H_7NO_2$. Dyes iron mordanted fabrics green.

Gambine Yellow. A dyestuff. It dyes chromed wool yellow.

Gamboge (Gummi Gutta). A gum resin, the product of *Garcinia morella* of Siam. It is a yellow pigment used for water colours, also for colouring spirit and other varnishes. It is poisonous. It has a saponification value of 148, an acid value of 81, and an ash of 0·5 per cent.

Gamboge Butter. A fat obtained from *Garcinia morella*.

Gambo Hemp (Bimli Hemp). A fibre stated to be suitable as a substitute for jute. It is the product of *Hibiscus cannabinus*.

Gambria. See JELUTONG.

Gamma Acid. 2-Amino-8-naphthol-6-sulphonic acid.

Gamma Benzene Hexachloride. An anti-parasitic. γ-1, 2, 3, 4, 5, 6-Hexachlorocyclohexane. LOREXANE, QUELLADA.

Gamma-Col. Insecticidal spray. (511).

Gammaform. See PYROFORM.

Gamma-H.C.H. A pesticide. It is the gamma isomer of H.C.H. Gamma Benzene Hexachloride.

Gammalex. A seed dressing. (511).

Gammalin. Insecticides. (511).

Gamman. A Tunisian dyestuff.

Gammexane. A registered trade name for insecticidal preparations. (512).

Gamolenic Acid. A preparation used in the treatment of hypercholesterolæmia. It is *cis, cis, cis*-octadeca-6,9, 12-trienoic acid.

Gamsigradite. A mineral. It is a variety of hornblende.

Gangue. The earthy portion of an ore which leaves the metal when reduced.

Ganister. A rock mineral with a composition corresponding to a pure silica with about 1/10 of its weight of clay. Used in the manufacture of siliceous fire-bricks, and for lining furnaces.

Ganja. See BHANG and GUAZA.

Ganomalite. A mineral, $Ca_4Pb_4(PbOH)_2(Si_2O_7)_3$.

Ganomatite. A mineral, $Fe_2O_3.As_2O_3.Sb_2O_5.XH_2O$.

Ganophyllite. A mineral. It is an aluminium-manganese silicate, $Mn_7Al_2Si_8O_{26}.6H_2O$.

Gansil. See CHLORAMINE T.

Gant. A barrier cream. (526).

Gantanol. A proprietary preparation of sulphamethoxazole. An antibiotic. (314).

Gantrisin. A proprietary preparation containing sulphafurazole. (314).

Gapite. A mineral. It is morenosite.

Garamycin. A proprietary trade name for Gentamicin.

Garanceux (Spent Garancine). A low quality of Garancine (*q.v.*).

Garancine. A preparation made from madder, by treating it with sulphuric acid. Used as a dye.

Garantose. A benzoyl-sulphonicimide. A sugar substitute used for diabetes.

Garbyite. A mineral. It is enargite.

Garcinia Oil. See KOKUM BUTTER.

Gardan. A German preparation. It is a mixture of novalgin and pyramidone in molecular proportions. An antipyretic, analgesic, and antirheumatic.

Gardenal. See LUMINAL. Phenobarbitone. (507).

Gardenal Sodium. A proprietary preparation of phenobarbitone sodium. A hypnotic. (336).

Gardeniol. Phenyl - methyl - carbinyl acetate, $C_6H_5.CH(OOC.CH_3).CH_3$. A perfume.

Gardinol. A proprietary trade name for wetting agents. They are sodium salts of sulphated higher fatty alcohols.

Gardlite. A proprietary trade name for a synthetic resin, produced from toluene sulphonamide and formaldehyde.

Garganine. A madder extract.

Garget. Poke root.

Garj. A bituminous sandstone. It contains from 6–17 per cent. bitumen.

Garlic. The bulbs of *Allium sativum*.

Garnet. A magnesium, calcium, or iron-aluminium silicate, a gem stone. Also see MAGENTA and CARMINAPH GARNET.

Garnet Brown. A dyestuff. It is the potassium or ammonium salt of isopurpuric acid, $C_8H_4N_5O_6K$. Dyes wool and silk from an acid bath.

Garnet Jade. Synonym for Grossular.

Garnet Lac. See LEMON LAC.

Garnet Red. A pigment. It is usually a red lead coloured with Crocein. Another quality consists of Orange lead coloured with Ponceau 2R and 3R.

Garnierite (Noumeite, Genthite). A mineral. It is a hydrated silicate of magnesium, containing variable amounts of nickel (2–30 per cent.).

Garnsdorffite. See PISSOPHANITE.

Garoin. A proprietary preparation of phenytoin sodium and phenobarbitone sodium. An anticonvulsant. (336).

Garolux 5 and 6. A proprietary trade name for two coarse limestones of exceptional whiteness. Used in spray plasters. (189).

Garomycin. A proprietary preparation of gentamycin sulphate. An antibiotic. (265).

Garona A and F. Proprietary trade names for two fine particle size clays. (189).

Garouille. The root bark of Kermes oak. Used in tanning.

Garrathion. A proprietary preparation of CARBOPHENOTHION. A veterinary insecticide.

Garri. Dried and grated Cassava root. A West African product, which consists mainly of starch.

Gas, Aerogene. See AEROGENE GAS.

Gas, Air. See AIR GAS.

Gas, Alcohol-hydrocarbon. See ALCOHOL-HYDROCARBON GAS.

Gasanol. A proprietary trade name for a motor fuel. It contains 75 per cent. alcohol, 20 per cent. petrol, and 5 per cent. kerosene.

Gas Black (Satin Gloss Black, Hydrocarbon Black, Hydrocarbon Gas Black, Silicate of Carbon, Jet Black, Ebony Black). A carbon black made by the incomplete combustion of natural gas, the carbon being deposited upon a metal surface. The specific gravity approximates to 1·75. Used in rubber mixings.

Gas, Blue. See BLAU GAS.

Gas Blue. See PRUSSIAN BLUE AND SODA BLUE.

Gas Carbon. See RETORT CARBON.

Gas-concrete. A porous building material consisting of lime, cement, and ashes from bituminous alum slate. The ashes are finely ground and mixed with lime, then with cement and a material containing a substance which gives gas with water, such as aluminium or zinc dust, which generates hydrogen. The material is moulded and then heated.

Gas, Generator. See DOWSON GAS and PRODUCER GAS.

Gas, Green Cross. Superpalite (trichloro-methyl-chloro-formate), a military poison gas.

Gasil. Micromised silica gel. (586).

Gasil EBC and EBN. Proprietary compounds of silica used in the paint, resin and plastics industries, *e.g.* in the matting of electron-beam-cured coatings. (205).

Gaskoid. A proprietary trade name for a rubber jointing resistant to oil and petrol.

Gas Lime. Consists chiefly of calcium sulphide or sulphydrate.

Gas Liquor. The aqueous condensation product obtained in the distillation of coal tar, and in the purification of coal gas.

Gas, Mixed. See DOWSON GAS.

Gas Oil. The name applied to all mineral oils intended for the preparation of gas, such as the light oils of brown coal tar, and shale oil. The term is also used for a fraction of Russian petroleum distillation of specific gravity 0·865–0·885, and flashing at 90° C. A burning oil.

Gasol. A gas obtained by the low-temperature distillation of coal. It is a mixture of hydrocarbons, and consists of homologues of methane and ethylene. It is recommended for welding and cutting operations.

Gasoline (Canadol, Kandol). That fraction of the first distillate of crude American or Canadian petroleum, of 80–90° B., and having a boiling-point of from 70–90° C. Used for extracting oil from seeds, and for carburetting coal gas. The name gasolene is also applied to petroleum ether, ligroin, and petrol.

Gas, Pintsch. See OIL GAS.

Gas Salts. Naphthalene, $C_{10}H_{18}$.

Gas, Siemens'. See PRODUCER GAS.

Gas, Suction. See DOWSON GAS.

Gastalar. A proprietary preparation containing aluminium hydroxide, magnesium carbonate and sorbitol. An antacid. (327).

Gastaldite. A mineral. It is a variety of glaucophane.

Gas Tar Asphalt. See GERMAN ASPHALT.

Gasterin. A preparation made from the stomach juices of dogs.

Gastex. A proprietary gas black used in rubber mixings.

Gastrils. A proprietary preparation containing aluminium hydroxide and magnesium carbonate as a co-dried gel. An antacid. (268).

Gastro Caloreen. A proprietary preparation of a polyglucose polymer used as a high-calorie food supplement. (447).

Gastrocote. A proprietary preparation of alginic acid, dried aluminium hydroxide gel, magnesium trisilicate and sodium bicarbonate. An antacid. (249).

Gastronida. A German product. It contains magnesium peroxide, magnesia usta, bismuth subsalicylate, and compound powder of liquorice. Used for stomach and intestinal troubles.

Gastrosan. Bismuth bisalicylate. Used in diarrhœa and gastric troubles.

Gastrosil. A trade mark for a specific against hyperacidity and gastric ulcers, the active component of which is calcium silicate in gel. form. (43).

Gastrotest. A proprietary trade name for Phenazopyridine.

Gastrovite. A proprietary preparation of ferrous glycine sulphate, ascorbic acid and calcium gluconate. A hæmatinic. (249).

Gas, Yellow Cross. Mustard gas (dichloro-diethyl-sulphide).

Gatinar. A proprietary preparation of lactulose and other sugars used in the treatment of constipation. (244).

Gauduin's Fluid. A mixture of finely powdered cryolite, and a solution of phosphoric acid in alcohol. A soldering fluid.

Gauging Metal. An alloy similar to Delta metal, but containing iron.

Gaultheria Oil. Oil of wintergreen.

Gaultheriasalol. Methyl-salicylo-salicylate, $HO.C_6H_4.COO.C_6H_4.COOCH_3$.

Gaultheric Acid. Methyl salicylate, $CH_3.C_7H_5O_3$.

Gauslinite. See BURKEITE.

Gaviscon. A proprietary preparation of alginic acid, sodium alginate, magnesium trisilicate, aluminium hydroxide and sodium bicarbonate. An antacid. (243).

Gavite. A mineral. It is a variety of talc, $H_4(Mg.Fe)_4Si_5O_{16}$, found in Italy.

Gaylussite (Natrocalcite). A mineral. It is sodium-calcium carbonate, $Na_2CO_3.CaCO_3.5H_2O$.

Gazelle. Acid phosphates for the baking trade. (503).

GDL. Glucono delta lactone. (577).

GE. 2557. A proprietary trade name for a polyester vinyl plasticiser. (59).

Gearksutite. A mineral, $Al(F.OH)_3.CaF_2.H_2O$.

Geax. A trade name for a synthetic resin-paper product used as an electrical insulation.

Geblitol. Sodium hydrosulphite. Used as a disinfectant.

Gechophen. A trade name for chlorphenesin B.P. (182).

Gedanite (Friable Amber, Soft Amber). A resin found on the shores of the Baltic. It is a variety of amber, and melts at 150°–180° C. It is also called Soft amber. A variety of amber low in succinic acid.

Gedeflex. A registered trade mark for dibutyl, butyl-benzyl and di-octyl phthalates.

Gedelite. A registered trade mark for phenolic moulding powders and resins.

Gedge's Metal. See AICH METAL.

Gedrite. A yellow resin found with Prussian amber. It is also the name for a mineral, $2(Mg.Fe)SiO_3.MgAl_2SiO_3$ of the amphibole class.

Gedroitzite. A mineral. It is $Na_2Al_2Si_3O_{10}.2H_2O$.

Gefarnate. A preparation used in the treatment of peptic ulcers. It is a mixture of stereoisomers of 3,7-dimethyloc- ta - 2, 6 - dienyl 5, 9, 13 - trimethyl - tetradeca - 4, 8, 12 - trienoate. Geranyl farnesylacetate. GEFARNIL.

Gefarnil. A proprietary preparation of gefarnate. A gastrointestinal sedative. (280).

Gehlenite. A mineral, $3CaO.Al_2O_3. 2SiO_2$.

Geikielite. A mineral. It is an iron magnesium ore containing titanium.

Gelanth. See GELANTHUM.

Gelanthum (Gelanth). A mixture of gelatin, tragacanth, glycerin, and water, with thymol.

Gelatase. Bacterial proteolytic enzymes. (550).

Gelatin. A colourless and odourless glue, obtained either from calves' heads, or the cartilage and skins of young animals. Used as a food, for making jellies, in photography, for preparing negatives, and for various other purposes.

Gelatin-arsphenamine. A combination of salvarsan and gelatin. An antisyphilitic.

Gelatinastralite. An explosive consisting of ammonium nitrate, with some sodium nitrate, up to 20 per cent. dinitro - chlorhydrin, and maximum amounts of 5 per cent. nitroglycerin, and 1 per cent. collodion cotton.

Gelatin, Bengal. See AGAR-AGAR.

Gelatin Carbonite. An explosive consisting of 25 per cent. nitro-glycerin, 0·7 per cent. collodion wool, 7 per cent. gelatin, 25 per cent. sodium chloride, and 42 per cent. ammonium nitrate.

Gelatin, Ceylon. See AGAR-AGAR.

Gelatin, Chinese. See AGAR-AGAR.

Gelatin Dynamites. Explosives. English ones consist of 80 per cent. gelatinised nitro-glycerin (with 3 per cent. collodion cotton), 15 per cent. potassium nitrate, and 5 per cent. wood meal. Also see GELIGNITES.

German. These are mixtures prepared from explosive gum (96 per cent. nitro-glycerin gelatinised with 4 per cent. collodion cotton), and a nitre base as absorbent.

Italian. No. 0 consists of 74 per cent. nitro-glycerin, 5 per cent. collodion cotton, 15½ per cent. sodium nitrate, 5 per cent. wood meal, and ½ per cent. carbonate. No 1 contains 70–72 per cent. nitro-glycerin, and No. 2, 48 per cent. nitro-glycerin.

Gelatin, Fish. See ISINGLASS.

Gelatin, Fluid. See OIL PULP.

Gelatin, Japanese. See AGAR-AGAR.

Gelatin Silk. See VANDURA SILK.

Gelatin, Vegetable. Agar-agar.

Gelatinwetterastralite. An explosive, containing 40 per cent. ammonium nitrate, 7·5 per cent. sodium nitrate, 16 per cent. dinitro-chlorhydrin, 4 per cent. nitro-glycerin, 0·5 per cent. collodion cotton, 0·5 per cent. wood meal, 8 per cent. potato starch, 2 per cent. rape-seed oil, 1 per cent. nitro-toluene, 2 per cent. dinitro-toluene, 14 per cent. salt, and 2·5 per cent. ammonium oxalate.

Gelato-glycerin. Glycerin jelly.

Gelbin. See STEINBUHL YELLOW.

Geldolomite. See GURHOFITE.

Gelignites. Gelatin dynamites containing usually 65 per cent. gelatinised nitro-glycerin, with 35 per cent. of absorbents (75 per cent. nitre, 24 per cent. wood meal, and 1 per cent. soda).

Gellert Green. A variety of cobalt green, obtained by roasting and igniting metallic cobalt with 4–5 parts of salt-petre and 8–10 parts of zinc oxide.

Gelline. A colloidal iodine gel.

Gelose. See GALACTAN.

Gelmagnesite. A mineral. It is an amorphous colloidal form of magnesite.

Gelobel. A proprietary trade name for a gelatine dynamite. An explosive.

Gelofusine. A proprietary preparation of gelatin, calcium chloride and sodium chloride used in the treatment of shock. (256).

Gelose. See GALACTAN.

Gelosine. A mucilaginous material extracted from a Japanese algæ. It is soluble in alcohol and water.

Geloxite. An explosive consisting of 54–64 parts nitro-glycerin, 4–5 parts nitro-cotton, 13–22 parts potassium nitrate, 12–15 parts ammonium oxalate, 0–1 part red ochre, and 4–7 parts wood meal.

Gel Rubber. A term used for the residue of rubber left undissolved when raw rubber is treated with a solvent.

Gelufen. A proprietary preparation containing heavy magnesium carbonate, aluminium glycinate and oil of peppermint. An antacid. (259).

Gelusil. A proprietary preparation containing aluminium hydroxide and magnesium trisilicate. An antacid. (262).

Gelva. A trade mark for a series of polymerised vinyl acetates.

Gelva RA 737, 784, 788, 858. Acrylic multipolymer solutions for use in pressure sensitive adhesives.

Gelvatol. A proprietary brand of polyvinyl acetate. (57).

Gemfibrozil. A preparation used in the treatment of hypercholesterolæmia, currently undergoing clinical trial as " CI-

719 ". It is 2,2-dimethyl-5-(2,5-xyly-loxy)valeric acid.

Gemglo. A proprietary trade name for styrene and methacrylate.

Gemlite. A proprietary trade name for a urea formaldehyde synthetic resin.

Gemloid. Proprietary brands of pyroxylin (*q.v.*) and cellulose acetate.

Gemme. Crude turpentine.

Gemstone. A proprietary trade name for a phenol-formaldehyde cast resin.

Gemstone M.1.2. A proprietary trade name for a phenolic laminating resin.

Genacort. A proprietary preparation of hydrocortisone for dermatological use. (188).

Genal P4300-CM. A proprietary heat-resistant but asbestos-free phenolic moulding compound. (59).

Genappe. A smooth worsted yarn of animal origin. The thread is made more lustrous by burning off the surface fibre.

Genasco. A proprietary trade name for a bituminous softener.

Genasprin. A trade name for acetyl-salicylic acid. See ASPIRIN.

Genatosan Skin Bar. A soap free detergent bar. (509).

Genclor. Chlorinated PVC. (512).

Genexol. A proprietary preparation of triisopropylphenoxypolyethoxy - ethanol. A spermicide. (737).

Genisol. A medicated shampoo. (509).

Genitron. Blowing agents. (509).

Genklene. A proprietary trade name for 1-1-1-trichlorethane. A cleaning fluid with lower toxicity than some other chlorinated solvents.

Genelit. A spongy bronze-like bearing metal prepared from a very finely ground mixture of copper, tin, and graphite. When the mixture is heated the graphite burns away, the copper and tin melt together, leaving behind a porous mass which is able to absorb large quantities of lubricating oil.

Generator Gas. See DOWSON GAS and PRODUCER GAS.

Geneva. Gin.

Genoa Oil. Fine olive oil.

Genochrome. A stabilised photographic colour developer. (507).

Genoform. Methylene - glycol - salicylic ester. It is used in medicine, and is said to split up in the intestines into formaldehyde, salicylic and acetic acids.

Genoxide. A registered trade name for a special quality of hydrogen peroxide for medical purposes. (513).

Genster. A trade mark for a manufacture of carbon.

Gentamicin. An antibiotic produced by Micromonospora purpurea. Garamycin. Genticin is the sulphate.

Genteles Green (Tin Green, Tin-Copper Green). Copper stannate, a pigment.

Gentersal. A proprietary preparation of gentian violet and alkyldimethylbenzalkonium chloride, used to treat vulvo-vaginal candidiasis. (369).

Genthane SR. (GS 338). A proprietary trade name for a polyurethane based moulding compound with a temperature range from −60° C. to +160° C.

Genthelvite. A mineral. It is $(Zn,Fe,Mn)_8Be_6Si_6O_{24}S_2$.

Genthite (Nickel gymnite). A mineral. See GARNIERITE.

Gentian. The dried rhizome and roots of *Gentiana lutea*, or of other Gentiana species.

Gentian Blue 6B. See ANILINE BLUE, SPIRIT SOLUBLE.

Gentiannie. The zinc double chloride of dimethyl-diamino-phenazthionium chloride, $C_{14}H_{14}N_3SCl$. Dyes mordanted cotton bluish-violet.

Gentian Violet. A dyestuff. It is a mixture of penta and hexa-methyl-*p*-rosaniline hydrochlorides. Used as a microscopic stain, and as a bactericidal agent.

Genticin Injectable. A proprietary preparation containing Gentamicin. An antibiotic. (265).

Gentisin. The yellow pigment of *Gentiana lutea*. It is 1 : 3 : 7-trihydroxy-xanthone-3-methyl ether.

Gentisone HC. A proprietary preparation of GENTAMICIN sulphate and HYDROCORTISONEL acetate used in the treatment of infections and inflammations of the ear. (271).

Geoform. See GUAIAFORM.

Geogorite. A mineral. It is a bismuth carbonate.

Geokronite. A mineral, $5PbS.Sb_2S_3$.

Geoline. See OZOKERINE.

Geon 100. A proprietary trade name for polyvinyl chloride.

Geon 200. A proprietary trade name for polyvinylidene chloride (Saran).

Geon 140X31. A proprietary vinyl powder used in the manufacture of fluidised-bed and electrostatic coatings. (102).

Geon 450X23. A proprietary vinyl chloride/acrylic copolymer used in latex paints. (294).

Geon 460X6. A proprietary vinyl chloride copolymer incorporating a synthetic anionic emulsifier. (102).

Geon 590X3. A proprietary dioctyl-phthalate plasticised vinyl-chloride copolymer. (102).

Geon 590X4. A proprietary phosphate ester plasticised vinyl-chloride copolymer. (102).

Geon 590X6. A proprietary phosphate

ester plasticised vinyl chloride copolymer. (102).

Georgiadesite. A mineral, $Pb_5Cl_4(AsO_4)_2$.

Georgia Kaolin. A general filler for paint, rubber and plastics. It has a high brightness, fine particle size and low impurity content. (137).

Geosot. Guaiacol valerianate, C_6H_4O. $CH_3O.CO.C_4H_9$. Used in phthisis, and as a gastro-intestinal antiseptic.

Geranine (Brilliant Geranine, Titan Rose, Patent Atlas Ted, Columbia Red 8B, Direct Scarlet, Thiazine Red, Thiazine Brown). Dyestuffs of a similar nature to Erika B, Diamine rose, and Sultan red.

Geranium. See MAGENTA.

Geranium Crystals. A name which is sometimes used for diphenyl oxide, $C_6H_5.O.C_6H_5$. Used in perfumery.

Geranium Oil. See ROSÉ OIL.

Gerhardite. A mineral. It is copper nitrate, $Cu(NO_3)_2$.

Gerhardt's Caustic. This consists of litharge boiled with potassium hydroxide until it is dissolved, and water is added.

Geriden. A proprietary preparation of leptazol and nicotinic acid. (382).

Gerisom. A proprietary preparation of paracetamol, amylobarbitone, and chlormezanone.

Germalgene. Trichlorethylene, C_2HCl_3.

German Asphalt. Coal tar boiled down to the required consistency.

German Brass. See BRASS.

German Chamomile. The dried flower-heads of *Matricaria chamomilla*.

German Green. See CHROME GREEN.

German Gum. A reclaimed rubber of German manufacture. It is stated to contain 90 per cent. of rubber and to have the physical properties of the best raw rubber.

Germanin. A name for Bayer 205. See FOURNEAU 309.

Germanit. A copper sulphide mineral found at Tsumeb, in South-West Africa. It contains 45 per cent. copper, 7 per cent. iron, 6 per cent. germanium, 3 per cent. zinc, 0·7 per cent. lead, 31 per cent. sulphur, 5 per cent. arsenic, and 0·7 per cent. silica.

German Razor Hones. Fine-grained argillaceous rocks.

German Saltpetre (Artificial Saltpetre, Conversion Saltpetre). A saltpetre made by dissolving 10 parts sodium nitrate, and 9 parts potassium chloride (obtained from the Stassfurt " carnalite," in the mother liquors from previous operations), concentrating, crystallising, and refining. Used in the manufacture of gunpowder.

German Sesamé Oil. See CAMELINE OIL.

German Silver. See NICKEL SILVERS.

German Silver Solder. Usually consists of 5 parts German silver, and 4 parts zinc.

German Tinder. See AMADOU.

German Turpentine. See TURPENTINE.

German Yeast. Dried yeast.

Germisan. A compound having the formula $C_6H_4(ONa)HgCN$. A disinfectant.

Germol. See BACTEROL.

Gersdorffite (Plessite). A mineral. It is a sulpharsenide of nickel, NiAsS.

Gesarol. A proprietary trade name for an insecticide containing 5 per cent. D.D.T. (*q.v.*) with a wetting agent for spray and 3 per cent. D.D.T. (dust) for human lice.

Gesilit. German safety explosives. No. 1 contains 30·75 per cent. nitro-glycerin jelly, 5·25 per cent. dinitro-toluene, 7 per cent. sodium chloride, 18 per cent. sodium nitrate, and 39 per cent. dextrin. No. 2 consists of 30·75 per cent. nitro-glycerin jelly, 5·25 per cent. dinitro-toluene, 22 per cent. ammonium nitrate, 21 per cent. sodium chloride, and 21 per cent. dextrin.

Gestanin. A proprietary preparation of allylestrenol, used to control uterine bleeding and abortion. (316).

Gesteins-tremonit V. A German explosive containing ammonium nitrate, nitro-glycerin, vegetable meal, potassium perchlorate, and nitro-compounds.

Gesteins-Westfalit B and C. Explosives. They are Ammonals containing dinitro-benzene, and dinitro-toluene respectively.

Gestronol. A progestational steroid. 17 - Hydroxy - 19 - norpregn - 4 - ene - 3, 20 - dione. DEPOSTAT is the hexanoate. GESTRONORONE.

Gestronorone. See GESTRONOL.

Gestyl. A proprietary preparation of serum gonadotrophin. (316).

Getah Wax. See JAVA WAX.

Gevodin. A proprietary preparation of famprofazone, paracetamol, isopropylphenazone and caffeine. An analgesic. (386).

Gevral. A proprietary preparation of multivitamins and minerals. (306).

Geyerite. A mineral. It is a variety of lolingite. It is $Fe(As,S)_2$.

Geyserite (Siliceous Sinter). A mineral. It is a variety of opal, found as an incrustation on vegetable matter in hot springs.

Ghati Gum (Ghatti Gum). Bassora gum (*q.v.*).

Ghé Butter. See INDIAN BUTTER.

Ghee. A pure clarified milk fat, made from buffalo, Indian cow, or goat milk.

Ghee Butter. See INDIAN BUTTER.

Giannettite. A mineral. It is $Na_3Ca_3Mn(Zr,Fe)TiSi_6O_{21}Cl$.

Giant Cement. See ACETIC-GELATIN CEMENT.

Giant Powders. Explosives. No. 1 contains 40 per cent. nitro-glycerin, 40 per cent. sodium nitrate, 6 per cent. rosin, 6 per cent. sulphur, and 8 per cent. kieselguhr. No. 2 consists of 36 per cent. nitro-glycerin, 48 per cent. potassium or sodium nitrate, 8 per cent. sulphur, and 8 per cent. rosin.

Gibbsite. See HYDRARGILLITE.

Giemsa's Stain. A microscopic stain for white blood corpuscles. It contains eosin, glycerin, and methanol.

Gies Biuret Reagent. See BIURET REAGENT.

Gieseckite. A mineral. It is a variety of pinite.

Gilbertite. A mineral. It is a variety of mica.

Gilder's Size. Cologne glue (*q.v.*), in thin leaves, bleached by means of chlorine.

Gilder's Whiting. A superior quality of whiting. See CHALK.

Gilding Metal. A jeweller's alloy of 90 per cent. copper, and 10 per cent. zinc. Another alloy contains 70 per cent. copper, 17·5 per cent. brass, and 12·5 per cent. tin.

Gilding Solutions. These generally consist of solutions of gold chloride and potassium carbonate in water. Used for the electro-deposition of gold.

Gilead Balm. See BALM OF GILEAD.

Gillebackite. A mineral. It is a variety of wollastonite.

Gillespite. A mineral. It is a silicate of barium and ferrous iron, $Fe.BaSi_4O_{10}$, found in Alaska.

Gilpinite. A mineral of Colorado. It is represented by the formula RO. $UO_3.SO_3.4H_2O$. $(R=Cu.Fe.Na_2.)$

Gilsonite (Uintahite, Uintaite). A product resembling and closely allied to asphalt, found in Utah. It has a specific gravity of 1·05–1·10, melts at about 130° C., is soluble in carbon disulphide, and yields an ash of 0·75 per cent.

Gin. An alcoholic liquid containing about 52 per cent. alcohol, prepared from grain spirit, either by redistilling it from juniper berries, or by treating it with oil of juniper. Geneva or Hollands is a spirit which is prepared from malted barley, rye and barley, in equal proportions, then distilled from juniper berries.

Gina. A proprietary preparation of propatylnitrate. Used to treat angina pectoris. (112).

Ginal. A purifier for sugars. It contains the sodium compound of alginic acid (from seaweed).

Ginetris Vaginal Tablets. A proprietary preparation containing chloramphenicol, myralact and cloponone. (365).

Gingelly Oil (Gingili Oil, Teal Oil, Teel Oil, Til Oil, Beni Oil, Benne Oil, Beniseed Oil). Sesamé oil obtained from the seeds of *Sesamum indicum* and of *S. orientale*. Used in the manufacture of margarine and soap, and as a burning oil.

Ginger. The dried rhizome of *Zingiber officinale*.

Ginger Grass Oil. See ROSÉ OIL.

Gingerine. The base of ginger tincture or essence, obtained by the extraction of the rhizome of *Zingiber officinale*. It is a mixture of camphene, phellandrene, cineol, citral, borneol, and gingerol.

Gingicain. A proprietary preparation of amethocaine gentisate and chlorbutol in aerosol. A local anæsthetic. (312).

Gingili Oil. See GINGELLY OIL.

Ginilsite. A mineral. It is a silicate of aluminium, calcium, iron, and magnesium.

Ginorite. A mineral. It is $Ca_2B_{14}O_{23}.8H_2O$.

Ginseng. The root of *Panax quinquefolium*.

Gin-shi-bui-chi. A Japanese alloy of 30–50 per cent. silver with copper.

Giobertite. A mineral. It is magnesium carbonate, $MgCO_3$.

Gioddu. A Sicilian fermented milk, similar to Yogurt (*q.v.*).

Giorgiosite. A mineral, $3MgCO_3.Mg(OH)_2.2H_2O$.

Gippon. Antiseptic paint. (541).

Girard's Reagent. Betaine hydrazide hydrochloride.

Girasol Opal. A blue-white variety of opal with red reflection in strong light.

Giroflé. See METHYLENE VIOLET 2RA.

Gismondine. See GISMONDITE.

Gismondite (Gismondine, Zeagonite, Abrazite). A zeolite mineral, $CaO.Al_2O_3.4SiO_2.4H_2O$.

Gitalin. A purified digitalin.

Githagin. See STRUTHIIN.

Giufite. A mineral. It is a variety of milarite.

Giv Tan " F ". It is 2-ethoxyethyl para methoxy cinnamate. Odourless, insoluble in water and glycerol, soluble in alcohol and mineral oil. Used in sun tan lotions.

Glacetex. A proprietary safety glass.

Glacialin. See BOROGLYCERIDE.

Glacialin Rose Extract (Glacialin Salt Mixture). A mixture of boric acid and borax. Intended to preserve meat without altering the colour.

Glacialin Salt Mixture. See GLACIALIN ROSE EXTRACT.

Glacial Phosphoric Acid. See PHOSPHORIC ACID, GLACIAL.

Glacier Blue. The zinc double chloride of dimethyl-diamino-di-o-tolyl-dichlorophenyl-carbinol-hydrochloride, $C_{23}H_{23}N_2Cl_3$. Dyes silk, wool, and tannined cotton, greenish-blue.

Gladiolin O, I, II. Direct cotton dyestuffs, similar to Brilliant congo G.

Gladite. A mineral, $2PbS.Cu_2S.5Bi_2S_3$.

Glagerite. A Bavarian white clay.

Glance Coal. Gas carbon. See RETORT CARBON.

Glance, Cobalt. See COBALTINE.

Glance, Copper. See REDRUTHITE.

Glance Green. See MOUNTAIN GREEN.

Glance, Iron. See HÆMATITE.

Glance, Molybdenum. See MOLYBDENITE.

Glance Ore. See ARGENTITE.

Glance Pitch. See MANJAK.

Glance, Silver. See ARGENTITE.

Glancodot. A mineral. It is a sulpharsenide of cobalt and iron, (Co.Fe) AsS.

Glandiposan. A proprietary preparation of thyroid hormone and amphetamine. An antiobesity agent. (808).

Glanzdiarlin. Cinnamic acid in linseed oil.

Glanzkohle. A form of carbon described as between graphite and the diamond. It is very resistant to chemicals, and is stated to contain 99 per cent. carbon.

Glanzstoff (Sirius Silk, Meteor Silk, Artificial Hair, Meteor Yarn, Sirius Yarn). Pure cellulose, manufactured by dissolving purified cellulose in cuprammonium solution, and forcing the solution through narrow orifices into sulphuric acid. Used for making wigs.

Glasbachite. A mineral. It is a variety of zorgite.

Glascol HA2. A proprietary aqueous solution of acrylic copolymer. The ammonium salt provides hard, water-resistant films. (565).

Glascol HA4. See GLASCOL HA2. (565).

Glascol HN2. A proprietary aqueous acrylic copolymer, the sodium salt of which gives hard and brittle films soluble in water. (565).

Glascol HN4. A proprietary aqueous acrylic copolymer the sodium salt of which gives soft and flexible films soluble in water. (565).

Glascol PA6. A proprietary acrylic copolymer supplied in the form of low-viscosity aqueous solutions. The ammonium salt gives tacky, pressure-sensitive films resistant to water. (565).

Glascol PA8. A proprietary acrylic copolymer supplied in the form of low-viscosity aqueous solutions. The ammonium salt gives soft, tacky, pressure-sensitive films resistant to water. (565).

Glascol PN 8. A proprietary acrylic copolymer supplied in the form of low-viscosity aqueous solutions. The sodium salt gives soft, tacky, pressure-sensitive films soluble in water. (565).

Glaserite. Obtained from the Stassfurt deposits. See APHTHITALITE.

Glaser's Salt. Potassium sulphate and sulphite.

Glass. Amorphous silica, containing embedded in it invisible crystals of silica, or one or more silicates.

Glass, Fluid. See SOLUBLE GLASS.

Glass-gall. See SANDIVER.

Glassite. Magnetic oxide of iron from precipitation of ferrous sulphate with caustic soda. Also called black rouge. It is used for buffing.

Glass, Lead. See POTASH-LEAD GLASS.

Glass Liquor. See SOLUBLE GLASS.

Glass-maker's Soap. Manganese dioxide, MnO_2.

Glass, Muller's. See HYALITE.

Glass, Muscovy. See MUSCOVITE.

Glass of Antimony. See ANTIMONY GLASS.

Glass of Borax. See BORAX GLASS.

Glassona. Cellulose bonded fibre glass bandage. (529).

Glass Ore. See ARGENTITE.

Glass, Potash. See POTASH - LIME GLASS.

Glass, Red Arsenic. See REALGAR.

Glass, Ruby. See RED GLASS.

Glass Sand. A sand containing 98–100 per cent. silica, and no iron oxide.

Glass, Silica. See VITREOSIL.

Glass Silk. Glass wool.

Glass, Smoke. See GREY GLASS.

Glass, Soda. See SODA-LIME GLASS.

Glass Sponge. A patented sponge-like product obtained by mixing glass wool with salt, heating, then dissolving out the salt with water or acid.

Glass, Water. See SOLUBLE GLASS.

Glassy Felspar. See SANIDINE.

Glaubapatite. A mineral. It contains calcium phosphate and sodium sulphate.

Glauberite. A mineral which occurs in the Stassfurt deposits, $Na_2SO_4.CaSO_4$. It is decomposed by water.

Glauber's Salt (Mirabilite). Sodium sulphate, $Na_2SO_4.10H_2O$.

Glaucocerinite. A mineral. It is $Ca_7Zn_{13}Al_8(SO_4)_2(OH)_{60} \cdot 4H_2O$.

Glaucochroite. A mineral. It is $4[(Mn,Ca)_2SiO_4]$.

Glaucodote. A mineral. It is a sulpharsenide of cobalt, $(Co.Fe)AsS$.

Glaucolite. A mineral. It is a variety of scapolite.

Glauconic Acids. Bluish-violet dyestuffs, obtained by the successive action of pyroracemic acid and formaldehyde on aromatic primary amines.

Glauconite (Greensand). A mineral. It is a silicate of aluminium, iron, potassium, magnesium, and calcium.

Glauconite, Soda. See SODA-GLAUCONITE.

Glaucophane. A mineral, $Na_2O.Al_2O_3.4SiO_2.(Fe.Mg.Ca)O.SiO_2$.

Glaucopyrite. A mineral approximating to $Fe_{13}S_2As_{24}$.

Glaucosan. A proprietary preparation of deltanephrin and methyl-amino-aceto-pyrocatechol.

Glaucosiderite. Synonym for Vivianite.

Glaucosil. The siliceous residue obtained by extracting green-sand with mineral acids. It is a by-product in the manufacture of potash from greensand, and consists of practically pure silica. It is used as an absorbent for vapours.

Glauramine. A proprietary solution of specially purified auramine (British Dyestuffs Corporation). A powerful antiseptic for use in the treatment of gonorrhœa, and for other purposes.

Glaurin. A proprietary trade name for a plasticiser. It is diethylene glycol mono-laurate.

Glazier's Salt. Potassium sulphate, K_2SO_4.

Glazing and Parchment Glue. Pure skin glue, resembling gelatin.

Glazy Pig. A whitish crystallo-granular iron resembling grey iron, but brighter and whiter. It contains up to 12 per cent. silicon.

Glekosa. See CHLORAMINE T.

Glendonite. A mineral. It is a variety of calcite.

Gleptoferron. An iron preparation. It is a macromolecular complex of ferric hydroxide and dextranglucoheptonic acid.

Glessite. A variety of amber melting at 250–300° C.

Glibenclamide. An oral hypoglæmic agent. $1 - \{4 - [2 - (5 - Chloro - 2 - methoxybenzamido)ethyl]benzenesulphonyl\} - 3 - cyclohexylurea$. DAONIL, EUGLUCON.

Glibenese. A proprietary preparation of GLIPIZIDE used in the treatment of diabetes. (85).

Glibornuride. An oral hypoglycæmic agent. It is $1 - [(1R) - 2 - endo - hydroxy - 3 - endo - bornyl] - 3 - (toluene - p - sulphonyl)urea$. GLUTRIL.

Glidine. A protein preparation obtained from wheat flour, and asserted to contain 98 per cent. albumin and 1 per cent. lecithin. Used as a nutrient, but also as a substance which combines with stomach irritants.

Glievor Bearing Metals. One alloy contains 76 per cent. lead, 14 per cent. antimony, 8 per cent. tin, and 2 per cent. iron, and another consists of 73 per cent. zinc, 9 per cent. antimony, 7 per cent. tin, 5 per cent. lead, and 4 per cent. copper.

Glimmer. Mica.

Glinkite. See OLIVINE.

Glipizide. An oral hypoglycæmic agent. $1 - Cyclohexyl - 3 - \{4 - [2 - (5 - methyl - pyrazine - 2 - carboxamido)ethyl] - benzenesulphonyl\}urea$. GLIBENESE, MINODIAB.

Gliquidone. An oral hypoglycæmic currently undergoing clinical trial as "ARDF 26". It is 1-cyclohexyl-3-p-$[2 - (3. 4 - dihydro - 7 - methoxy - 4, 4 - dimethyl - 1, 3 - dioxo - 2(1H) - iso-quinolyl)ethyl]phenylsulphonylurea$.

Glisoxepide. An oral hypoglycæmic agent currently undergoing clinical trial as "FB b 4231". It is 3-$[4$-(perhydroazepin - 1 - ylureidosulphonyl)phenethylcarbamoyl]$-$5$-methylisoxazole.

Glist. Mica.

Globe Granite. A stone similar to Ward's stone (*q.v.*).

Globol. A German patented material for killing insects. It consists of chlorine derivatives of benzene.

Globosite. A mineral. It is $Fe_5(PO_4)_4(OH)_3 \cdot 10H_2O$.

Globulin (Fibrino-plastic Substance, Paraglobin, Paraglobulin, Serum Casein). Serum globulin, an albuminoid obtained from blood serum. Crystallin, an albuminoid which occurs in the crystalline lens, is also called globulin.

Glockerite. A mineral. It is a hydrated sulphate of iron.

Glokem. Synthetic waxes. (547).

Glonoine Oil. See NITRO-GLYCERIN.

Gloria. A new white rubber substitute of German manufacture. It is stated to be stable in the presence of accelerators, and to be a completely dechlorinated product. It is free from acidity and completely saponifiable.

Glos. Cellulose artificial silk.

Glossecollite. A variety of the mineral Halloysite.

Glossite. An abrasive, the active material of which is stated to be black oxide of iron.

Gloss Oil. A varnish made from rosin and benzine.

Glover Acid. A 78 per cent. sulphuric acid. It is the acid pumped up to the Glover tower in the lead chamber process.

Glover's Wool. See TANNER'S WOOL.

Glowray. An electrical resistance alloy containing 65 per cent. nickel, 12 per cent. chromium, and 23 per cent. iron, with some manganese. (793).

Gloxazone. An anti-protozoan. 3-Ethoxybutane - 1, 2 - dione bis(thio - semicarbazone). CONTRAPAR.

Gloy. A trade mark. An adhesive said to be a mixture of dextrin and starch, with magnesium chloride.

Glucagon. A polypeptide hormone produced in the alpha cells of the islets of Langerhans in the pancreas. Proprietary preparation. (333).

Glucal. A reducing compound, $C_6H_{10}O_4$, obtained by the reduction of β-aceto-bromo-glucose with zinc dust and acetic acid.

Glucalox. Glycalox.

Glucanal. Proprietary preparations of silver and anthraquinone glucosides.

Glucase. Maltase, an enzyme which converts maltose into glucose.

Glucin (Glucine). A sweetening substance. It is the sodium salt of the mono- and disulphonic acids of a substance having the composition $C_{19}H_{16}N_4$.

Glucina. Beryllium oxide, BeO.

Glucinite. A mineral. It is herderite.

Glucinium. Beryllium, Be.

Gluckauf. A German safety explosive containing ammonium nitrate, wood meal, dinitro-benzene, and copper oxalate.

Glucodin. A proprietary preparation of glucose and ascorbic acid. A food supplement. (738).

Glucophage. A proprietary preparation of metformin hydrochloride. An oral hypoglycæmic agent. (323).

Glucose (d-Glucose, Starch Sugar, Corn Sugar, Grape Sugar, Honey Sugar, Diabetic Sugar). Dextrose, $C_6H_{12}O_6$.

Glucose, Brewing. See GLUCOSE CHIPS.

Glucose Chips (Brewing Glucose). This is prepared by the hydrolysis of starch with acid. It consists mainly of dextrose and is used in brewing.

Glucose D. A preparation of glucose with vitamin D and calcium glycerophosphate.

Glucose or Sugar Vinegar. A vinegar prepared by the conversion of starch substance into sugar, by the action of dilute acids, followed by fermentation and acetification.

Glucose Syrup (Dextrin-Maltose). A partially hydrolysed starch employed in brewing and in confectionery.

Glucotannin. A tannin found in Chinese rhubarb. It is 1-galloyl-β-glucose.

Glucotin. A cement for broken articles. It consists of 4 oz. isinglass, 3 oz. gelatin, 24 oz. water, 4 oz. acetic acid, and 2 oz. spirit of wine.

Glue, Albumen. Partly decomposed gluten prepared from flour washed with water, made to ferment, and dried. It does not contain glue or albumen.

Glue and Gelatin Pearls. Forms of glue or gelatin in bright, hard drops. They are obtained by forcing the glue or gelatin extract through an orifice into a liquid medium. The pearls are purer products, and swell more speedily in water.

Gluferate. A proprietary preparation of ferrous gluconate and ascorbic acid. A hæmatinic. (245).

Glue, Hatter's. See PARIS GLUE.

Glumorin. A proprietary preparation of kallikrein. (341).

Glue, Parchment. See GLAZING GLUE.

Glue, Vegetable. See VEGETABLE GLUE and AGAR-AGAR.

Glurub. A proprietary rubber-glue compound for rubber stiffening.

Gluside. The B.P. name for Saccharin (q.v.).

Gluside, Easily Soluble. See SACCHARIN, EASILY SOLUBLE.

Gluside, Soluble. See SACCHARIN, EASILY SOLUBLE.

Glutannin. A combination of tannic acid and gluten. Used in the treatment of diarrhœa.

Glutazine. Dioxy-amino-pyridine, $C_5H_6N_2O_2$.

Gluteketone. A preparation in the form of a semi-solid jelly, containing glycerin, α-eigon, zinc oxide, salicylic acid, and ichthyol. Employed in eczema and other skin diseases.

Glutenex. A proprietary biscuit free of gluten and milk. (739).

Glutethimide. α - Ethyl - α - phenyl - glutarimide. Doriden.

Glutin. A substance made from curd and a solution of sodium tungstate. It is used as a glaze for dressing curtain fabrics, and also as a medium for colours in dyeing and calico printing.

Glutoform. See GLUTOL.

Glutoid. See GLUTOL.

Glutoid Capsules. Gelatin capsules which have been hardened by exposure to formaldehyde. Used for conveying internal antiseptics.

Glutol (Glutoform, Glutoid, Formo-Gelatin). Formaldehyde gelatin obtained by evaporating a solution of gelatin with one of formaldehyde. Employed as an antiseptic powder for wounds. Glutol is also a name applied to a proprietary synthetic resin.

Glutolin. Methyl cellulose.

Glutone. A nutritive preparation, soluble in water, obtained from gelatin. Recommended as a strengthening agent.

Glutril. A proprietary preparation of GLIBORNURIDE used in the treatment of diabetes mellitus. (314).

Glyakol. Diglyceryl ether tetracetate.

Glycalox. A polymerised complex of glycerol and aluminium hydroxide. Glucalox. Manalox AG. Antacids. (190).

Glycamyl. See GLYCERIN OF STARCH.

Glycarbin. Glyceryl carbonate.

Glycene. A proprietary trade name for an alkyd synthetic resin used for dentures.

Glyceria Wax. A wax formed in the stem of cane grass, *Glyceria ranirgera*, of Australia. It melts at 82° C.

Glycerin. Commercial glycerol, $C_3H_8O_3$.

Glycerin-formal. A condensation product of glycerol and formaldehyde, $CH_2OH.CH(O.CH_2.O)CH_2$.

Glycerin Glue. Obtained by dissolving glue in warm glycerin. Used for hectographic work.

Glycerin Jelly (Gelato-Glycerin, Glyco-Gelatin). Gelatin dissolved in glycerin. Glycerin jelly is usually a mixture of 12 oz. glycerin, 8 oz. white soap, 6 oz. bleached almond oil, 2 drachms oil of thyme, 4 drachms oil of bergamot, and 1 drachm oil of roses.

Glycerin-litharge Cement. A mixture of litharge and glycerin, sometimes with fuller's earth to delay setting. It usually contains 70 per cent. glycerin and 30 per cent. litharge, and sets in about 3 hours.

Glycerin of Borax. This contains 2 oz. borax and 12 oz. glycerin. The borax is triturated with the glycerin in a warmed mortar until it dissolves.

Glycerin of Boric Acid. This contains boric acid and glycerin. It is mixed, and the boric acid dissolved by heating.

Glycerin of Starch (Glycamyl, Plasma). This preparation contains 2 oz. starch, 13 oz. distilled water, and 13 oz. glycerin. These are heated together until a jelly is formed. It is used for chilblains and chapped hands.

Glycerinova. A mixture of calcium chloride, potassium lactate, and a vegetable gum. A glycerin substitute.

Glycerin Soaps. See TRANSPARENT SOAPS.

Glycerin Substitutes. See KIPP'S GLYCERIN SUBSTITUTE, GLYCERINOVA, GLYCERYL, GLYCINAL, MOLLPHORUS, PERGLYCERIN, and PERKAGLYCERIN.

Glycero. See ESTER GUMS.

Glycero-ester. A trade name for ester gum.

Glycerogen. A mixture of polyhydric alcohols obtained by inversion of sugar to hexose, then reduction with hydrogen and vacuum distilled. The final product is 40 per cent. glycerine, 40 per cent. propylene glycol and 20 per cent. hexyl alcohols.

Glycero-piperaz. Basic piperazine-glycerophosphate.

Glyceryl. A mixture of molasses and phosphoric acid. A glycerin substitute.

Glycinal. (1) A mixture of potassium chloride and β-dipyridine. A glycerin substitute.

Glycinal. (2) A proprietary preparation containing aluminium glycinate, magnesium trisilicate and oil of peppermint. An antacid. (250).

Glycine. p-Oxy-phenyl-glycocoll, C_6H_4(OH)(NH.CH$_2$COOH). A photographic developer.

Glycine Blue. A dyestuff. It is the sodium salt of diphenyl-sulphone-disazo-bi-α-naphthyl-glycine, $C_{36}H_{24}N_6O_6SNa_2$. Dyes cotton blue from a soap bath.

Glycine Corinth. A dyestuff. It is the sodium salt of diphenyl-disazo-bi-α-naphthyl-glycine, $C_{36}H_{26}N_6O_4Na_2$. Dyes cotton red from a soap bath.

Glycine Red. A dyestuff. It is the sodium salt of diphenyl-disazo-α-naphthyl-glycine-naphthionic acid, $C_{34}H_{24}N_6SO_5Na_2$. Dyes cotton red from a soap bath.

Glyco. See PARAFFIN, LIQUID.

Glycobiarsol. Bismuth glycollylarsanilate.

Glycoborval. See BORNYVAL, NEW.

Glycobrom. The glyceryl ester of dibromo-hydro-cinnamic acid, C_3H_5(O.CO.(CHBr)$_2C_6H_5$)$_3$.

Glycocaine. See Nirvanine.

Glycochloral. A chloral and glucose compound. A sedative.

Glycodine. A proprietary trade name for Pholcodine.

Glyco-gelatin. See GLYCERIN JELLY.

Glycola. A proprietary preparation of caffeine, calcium glycerophosphate, kola, chloroform, ferrous perchlorate and vermiculi. A tonic. (259).

Glycoline. See PARAFFIN, LIQUID.

Glycoluril. Acetylene-urea, $C_4H_6N_4O_2$.

Glyco Metal. An alloy of 70–74 per cent. lead, 14–16 per cent. antimony, and 8–12 per cent. tin.

Glyconin. An emulsion of glycerin and egg-yolk. Used in medicine, and for toilet purposes.

Glyconyl. A new photographic developer, the active constituent of which is p-hydroxy-phenyl-glycine.

Glycophenol. See SACCHARIN.

Glycopon. A material containing opium alkaloids as glycero-phosphates.

Glycopyrin. See ASTROLIN.

Glycopyrrolate. A proprietary trade name for Glycopyrronium bromide.

Glycopyrronium Bromide. 3 - (α - Cyclopentyl - α - phenylglycolloyloxy) - 1,1 - dimethyl - pyrrolidinium bromide. Glycopyrrolate. Robinul.

Glycosal. The mono-salicylic ester of glycerol, $C_3H_5(OH)_2.O.CO.C_6H_4(OH)$. Used in medicine in the treatment of rheumatism.

Glycosin. See SACCHARIN.

Glycosterine. A proprietary preparation. It is a glycol glyceryl stearate. Used to replace beeswax in certain polishes.

Glycotauro. A proprietary preparation of bile salts freed from bile pigment, and standardised to contain 50 per cent. of the natural mixture of sodium glycocholate and sodium taurocholate. Used in medicine.

Glycothymoline. An antiseptic, usually containing thymol, eucalyptol, menthol, borates, bicarbonates, benzoates, and glycerin. It is often coloured with cudbear.

Glycozone. A proprietary preparation claimed to contain 5 per cent. glyceric acid, and 90 per cent. glycerin. An antiseptic. Also see PEROXOL AND HYDROZONE.

Glycrex C-100. A polyethoxylated glycerol derivative. Used as a humectant. (195).

Glyecin. Ethyl-thio-diglycol. A solvent used in treatment of wool for dyeing.

Glykola. A proprietary preparation of caffeine, calcium glycerophosphate, kola, chloroform and ferrous perchlorate. A tonic. (740).

Glymaxil. A proprietary trade name for sodium glucoldrate.

Glymidine. Sodium salt of 2-benzene-sulphonamido - 5 - (2 - methoxy - ethoxy) pyrimidine. Gondafon; Lycanol.

Glymol. See PARAFFIN, LIQUID.

Glyoxiline. An unsuccessful dynamite made by soaking guncotton and potassium nitrate in nitro-glycerin.

Glyptal 2557, 2559. Proprietary trade names for polyester plasticisers. (59).

Glyptal Resins. Resinous products obtained by the interaction of glycerol and organic acids. A resin of this type is made by reacting upon glycerol with oleic acid and phthalic anhydride.

Glyptanite. A registered trade name for a glyptal bonded mica product used for insulation.

Glysal. The mono-glycol ester of salicylic acid.

Glysantin. A German ethylene glycol preparation for automobiles. It is 100 per cent. strength and is about 30 per cent. higher in anti-freeze protection than high percentage glycerin.

Glysobuzole. Isobusole.

Gmelin's Blue. See TURNBULL'S BLUE.

Gmelinite. A mineral, $(Na_2.Ca)O.Al_2 O_{33}SiO_2$.

GMS 263. A solution of an acrylic resin for use in compounding pressure sensitive adhesives.

Gneiss. A rock containing silica, alumina, iron, lime, soda, potash, and titanium.

Goa Butter. Kokum butter ($q.v.$).

Goa Powder (Chrysarobin, Bahia Powder, Brazil Powder, Ringworm Powder). Araroba powder, a substance which collects in the cavities of a leguminous tree *Andira araroba*. Employed as an ointment for diseases of the skin.

Gobon. A German product. It consists of phenacetin, pyrazolone-salicyl, caffeine-sodium-benzoate, and quinine hydrochloride. Used for stilling pain.

Godfrey's Cordial. A mixture of treacle, water, sassafras oil, alcohol, and 1 per cent. of laudanum.

Goethite. A mineral, FeO.OH.

Gohi Iron. A proprietary trade name for iron containing manganese, sulphur, phosphorus, copper (total less than 0·125 per cent.).

Golaz Wax. A mixture of 3 parts resin, 1 part beeswax, 1 part tallow, and about 4 parts mineral powder. A vacuum wax used for jointing glass and metal.

Gold, Artificial. Stannous sulphide, SnS_2.

Gold-beater's Skin. The thin, light, toughened, outer membrane of the blind gut of the ox. It is separated, cleaned, and dried, and is finished by treating it with camphor or alum, and coating it with egg-albumin. Thin plates of gold are placed between layers of this skin, and beaten to form gold leaf.

Gold, Black. See MALDONITE.

Gold Bronze. See SAFFRON BRONZE.

Gold Brown. See BISMARCK BROWN.

Gold Button Brass. See BRASS.

Gold Coinage. See STANDARD GOLD.

Gold-copper. See CHRYSOCHALK.

Gold Earth. An ochre. See YELLOW OCHRE.

Golden Acorn. Nutmeg.

Golden Antimony Sulphide (Golden Sulphuret of Antimony, Golden Sulphide of Antimony). Antimony pentasulphide, Sb_2S_5. Used in vulcanising rubber, and in the manufacture of matches.

Golden Beryl. A bright yellow variety of beryl.

Golden Hair Water. See AURICOME.

Golden Ochre. See YELLOW OCHRE.

Golden Ointment. See EYE OINTMENT.

Golden Santonin. See CHROMO SANTONIN.

Golden Seal. *Hydrastis Canadensis.* Used as a source of yellow dye.

Golden Sulphide of Antimony. See GOLDEN ANTIMONY SULPHIDE.

Golden Sulphuret of Antimony. See GOLDEN ANTIMONY SULPHIDE.

Golden Syrup (Drip Syrup). This is the product obtained when raw or brown sugar is dissolved and the solution clarified with animal charcoal and the white sugar crystallised from it. The mother liquor is drip or golden syrup.

Golden Topaz (Spanish Topaz). Yellow varieties of quartz are called by these names.

Golden Yellow. See MARTIUS YELLOW and TROPÆOLINE O.

Gold, Euflavine. See EUFLAVINE CADMIUM.

Goldfieldite. A mineral. It is a sulphantimonide of copper, in which some of the antimony is replaced by arsenic, and the sulphur by tellurium.

Gold, Filled. Rolled gold plate (*q.v.*).

Gold, Fool's. Iron pyrites (*q.v.*).

Gold Leaf. Nearly pure gold. It often contains from 1·5–4 per cent. of silver or copper.

Gold Leaf, Imitation. An alloy of from 77–78 per cent. copper, and 22–23 per cent. zinc.

Gold, Mock. See MOSAIC GOLD.

Gold Ochre, Transparent. See YELLOW OCHRE.

Gold Orange. See ORANGE II, III, IV.

Gold Orange for Cotton. See CHRYSOIDINE R.

Gold Purple (Purple of Cassius). A flocculent purple precipitate, produced by the addition of stannous chloride to a solution of gold chloride. Used in the manufacture of artificial gems, and for colouring porcelain.

Goldschmidtite. A mineral, $Au_2Ag.Te_6$.

Gold Size. Consists of a mixture of 1 part yellow ochre, 2 parts copal varnish, 3 parts linseed oil, 4 parts turpentine, and 5 parts boiled oil.

Gold Solders. Various alloys of gold, silver, and copper, sometimes with zinc. An ordinary gold solder contains 43 per cent. gold, 30 per cent. silver, 20 per cent. copper, and 7 per cent. zinc. A fine gold solder consists of 66 per cent. gold, 22 per cent. copper, and 12 per cent. silver, and a hard gold solder contains 37·5 per cent. 18-carat gold, 21 per cent. silver, and 21 per cent. copper.

Gold Solder, Fine. See GOLD SOLDERS.

Gold Solder, Hard. See GOLD SOLDERS.

Gold, Sterling. See STANDARD GOLD.

Gold, Talmi. See ABYSSINIAN GOLD.

Gold Yellow. See TROPÆOLINE O, VICTORIA YELLOW, and MARTIUS YELLOW.

Goluthan. A German salve containing mercuric chloride.

Gomabrea. An exudation from a Chilean tree, offered as a substitute for gum arabic, in France.

Gomaine. A solution of iodoform and camphor, in sesamé oil.

Gomenol. A preparation of some New Caledonian plants.

Gomma. Italian gelatin dynamites. A contains 92 per cent. nitro-glycerin, and 8 per cent. collodion cotton, and B contains 83 per cent. nitro-glycerin, and 5 per cent. collodion cotton.

Gommeline. See BRITISH GUM.

Gommier Resin. A resin collected in Dominica. It is of the elemi type. It is the product of *Dacryodes hexandra.*

Gon. Tablets for children. (518).

Gonacrine. Preparation of 2,8-diamino-10-methylacridinium chloride and di-amino acridine. (507).

Gonadorelin. A luteinising follicle-stimulating hormone which stimulates and releases other hormones. Currently under clinical trial as "LH-RH Hoechst" and "Relefact LH-RH". It is L-pyroglutamyl - L - histidyl - L - tryptophyl - L - seryl - L - tyrosyl-glycyl-leucyl - L - arginyl - L - prolyl - L - gly - cinamide.

Gonadotrophon. A group of proprietary preparations of human gonadotrophins. (330).

Gonal. Santalol, $C_{15}H_{24}O$.

Gonazole. Sulphathiazole with proflavine monohydrochloride. (507).

Gondaphon. A proprietary preparation of glymidine. An oral hypoglycæmic agent. (265).

Gondang Wax. See JAVA WAX.

Gond Babel. See BABAL GUM.

Gond Babul. See BABAL GUM.

Gong Metal. A brass containing 78 per cent. copper and 22 per cent. zinc.

Gonnardite. A mineral. It is a silicate of aluminium, calcium, and sodium.

Gonoflavine. Diammo - methyl - acridinium chloride. Used in cases of gonorrhœa.

Gonorol. Approximately pure sanatol. Used in the treatment of gonorrhœa.

Gonsogolite. A mineral. It is pectolite.

Gooch and Eddy Reagent. A reagent used for precipitating magnesium as magnesium carbonate. It contains 180 parts concentrated ammonia, 800 parts water, and 900 parts absolute alcohol, the solution being saturated with ammonium carbonate.

Goodpasture's Stain. A microscopic stain containing 0·05 gram sodium nitroprusside, 0·05 gram benzidine, 0·1 gram basic fuchsin, and 100 c.c. 95 per cent. alcohol.

Goodrite. A proprietary trade name for vinyl plasticisers coded as follows:
GP-223. Dioctyladipate.
GP-235. Octyldecyl adipate.
GP-236. Didecyl adipate.
GP-261. A phthalate.
GP-265. Octyl decyl phthalate.
GP-266. Didecyl phthalate. (102).

Goongarite. A mineral, $4PbS \cdot Bi_2S_3$, of Western Australia.

Gooroo Nuts. Kola seeds.

Gooseberry-Garnet. Synonym for Grossular.

Go Pain. A proprietary analgesic preparation of salicylamide, potassium salicylate, calcium succinate, p-aminobenzoic acid, vitamins B_1 and C, and aluminium hydroxide. (741).

Gopmann Solder. An alloy of 49·1 per cent. tin, 20·3 per cent. zinc, and 26 per cent. lead.

Gorceixite. A mineral. It is an alumino-phosphate of alkaline and ceria earths, of Brazil.

Gordaite. A mineral. It is a variety of ferronatrite.

Gordal. A German product. It consists of tablets containing bryonia, eucalyptus, quinine, and colloidal iodine. Used for the prevention and cure of influenza.

Gordon Aerolite. A proprietary product. It is a special linen reinforced with phenol-formaldehyde resin. It possesses very high tensile strength.

Gordonite. A mineral. It is $MgAl_2(PO_4)_2(OH)_2 \cdot 8H_2O$.

Gordon Superflex D. A trade name for a graft of stereospecific rubber and polystyrene for high impact. Izod 1·35–1·65. Elongation 27 per cent. Tensile strength 3800 p.s.i. Modulus 291,000 p.s.i. (flex). (65).

Gorgeyite. A mineral. It is $K_2Ca_5(SO_4)_2 \cdot H_2O$.

Gorilla Yarn. A mixture of hair fibres, such as alpaca, sheep's wool, and mohair, with cotton and waste silk.

Gorite. Consists mainly of calcium peroxide, with some hydrate and carbonate. It has some germicidal properties, and is used as an addition to dentifrices, also as an anti-acid body.

Gornerol. The trade name for an essential oil prepared from the leaves of *Melaleuca leucadendron*. It is also known as Niaoul oil.

Gorun. A proprietary preparation stated to contain cinchophen with glycocoll and hexamine.

Goschenite. A mineral. It is a variety of beryl.

Goslarite (Gallitzenite). Native zinc sulphate, $ZnSO_4$.

Gossypetin. A flavanol. $5:7:8:3':4'$-pentahydroxy-flavanol.

Gossypium. Cotton wool.

Gossypol. The colouring matter, $C_{30}H_{28}O_9$, of the cotton seed.

Gothar. A tanning material obtained from *Zizyphus xylopra*, of South India.

Gothite. See NEEDLE IRON ORE.

Gottardite. A mineral. It is probably a variety of dufrenoysite.

Goudron (Asphalt Tar, Artificial Bitumen). Obtained by the distillation of the resinous masses produced from tar by treatment with acid, with varying amounts of creosote oil. Used in the manufacture of impermeable pasteboard for roofing.

Goulac. A proprietary trade name for a concentrated sulphite pulp process waste. A binder.

Goulard Powder. Lead acetate.

Goulard's Cerate. Lead subacetate ointment.

Goulard's Extract. (*Liquor plumbi subacetatis fortis B.P.*) A 24 per cent. solution of lead subacetate.

Goulard's Lotion. See GOULARD WATER.

Goulard Water (Goulard's Lotion, Lead Lotion). *Liquor plumbi subacetatis dilutus B.P.* A solution of lead subacetate, containing 1 part of a 24 per cent. solution of lead subacetate, and 1 part alcohol, in 78 parts water.

Goutine. See CITARIN.

Goyazite. A mineral, $Al_{10}Ca_3P_{23} \cdot 9H_2O$.

G.P.V. A proprietary preparation of penicillin V. (476).

GR Acid. α-Naphthol-disulphonic acid.

Grafene. A proprietary trade name for a lubricant containing graphite and oils.

Grafita. A proprietary trade name for a lubricant containing graphite and grease.

Graftonite. A mineral. It is a phosphate of calcium, iron, and manganese.

Grahamite. An asphaltic substance found in Mexico and Cuba. It is usually associated with mineral matter. It has a specific gravity of 1·17, melts at 175–230° C., up to 45 per cent. mineral matter, and is very soluble in carbon disulphide.

Grain Alcohol. Ethyl alcohol, C_2H_5 OH.

Grain Cartiers. The kernels of the fruit of a Haitian palm, *Pseudophœnix vinifera*. They contain 14 per cent. fat, and 2 per cent. saponin.

Grain Oil. See FUSEL OIL.

Grains D'Ambrette. Musk seeds.

Grains, Guinea. See GRAINS OF PARADISE.

Grains of Kermes. See KERMES.

Grains of Paradise (Guinea Grains). Seeds of *Amomum Melegueta*.

Grain Tin (Dropped Tin). Tin heated until brittle, hammered, and dropped from a height, when it splits up into prismatic fragments. Usually it is pure tin.

Gralmandite. A mineral. It is $8[(Ca,Fe'')_3Al_2Si_3O_{12}]$.

Grammatite. See TREMOLITE.

Grammite. See WOLLASTONITE.

Gramoxone. Paraquat weedkiller preparations. (511).

Gramp's Solder. An alloy of 60·4 per cent. tin, 36·1 per cent. zinc, 3 per cent. copper, 0·25 per cent. lead, and 0·18 per cent. antimony. An aluminium solder.

Gram's Iodine Stain. This consists of 1 gram iodine, 2 grams potassium iodide, and 300 c.c. distilled water.

Gram's Stain. A microscopic stain. It contains 10 c.c. of a saturated alcoholic solution of gentian violet and 90 c.c. aniline water.

Graneodin. A proprietary preparation of neomycin sulphate and gramicidin used in dermatology as an antibacterial agent. (326).

Graney Bronze. An alloy of 76·5 per cent. copper, 9·2 per cent. tin, and 15·2 per cent. lead.

Granilla. An inferior cochineal containing small insects and vegetable matter.

Granite. A mineral. It is composed of three minerals : quartz (silica), felspar (a silicate of aluminium and potassium), and mica (a silicate of aluminium and other bases). A building material.

Granodine. Phosphate pre-treatment for rust prevention. (512).

Granodised Steel. Steel which has been treated with zinc phosphate to give the surface resistance to corrosion.

Granol. Carbonised granulated peat. Used for gas production for power.

Granulite. A mineral found on Dartmoor. It contains silica, potash, and soda, and is used in the manufacture of glass.

Granulose. The inner part of the starch granule is known by this name.

Grape Honey. A concentrated grape juice. That from muscat grapes contains 24 per cent. water ; total sugar (as invert sugar) 72 per cent. ; total protein 0–7·87 per cent.

Grape-nuts. A food made from cereals. It contains 6 per cent. water, 2 per cent. mineral matter, 1·6 per cent. fat, 15 per cent. proteids, 49·4 per cent. soluble carbohydrates and 26 per cent. insoluble carbohydrates.

Grape Rubber. A rubber-like material obtained from the seeds and skins of grapes.

Grape Sugar. See GLUCOSE.

Graphalloy, Silver. A trade mark for a moulded graphite impregnated with silver. It is stated to be a highly efficient conductor, self-lubricating, and durable. It is used in electrical brushes and similar appliances.

Graphic Tellurium. See SYLVANITE.

Graphite (Black Lead, Plumbago). A mineral. It is a form of carbon (crystalline). Used for making electrodes and crucibles, and as a lubricant.

Graphite Metal. An antifriction and fitting metal. It contains 15 per cent. tin, 68 per cent. lead, and 17 per cent. antimony.

Graphite, Retort. See RETORT CARBON.

Graphitic Carbon. The black shiny flakes of carbon present in pig iron.

Graphitic Temper Carbon. The black amorphous carbon present in certain varieties of iron.

Graphitites. Graphites which swell on moistening them with strong sulphuric acid, and then heating them to redness. They are not true graphites.

Grappes. Bunches of green-coloured non-transparent crystals of copper acetate.

Grappier Cements. Hydraulic limes are slaked and passed through sieves. The hard lumps left on the sieve consist of unchanged limestone and calcium silicates. These are finely ground, and are then known as grappier cements. Le Farge cement belongs to this class.

Grass Cloth. A fabric made from ramie fibre.

Grasselerator 101. A trade mark for a rubber vulcanisation accelerator. It is aldehyde ammonia.

Grasselerator 102. A trade mark for a

rubber vulcanisation accelerator. It is hexamethylene-tetramine.

Grasselerator 552. A trade mark for a rubber vulcanisation accelerator. It is piperidine-piperidine-1-carbo-thionolate.

Grasselerator 808. A proprietary rubber vulcanisation accelerator. It is butylidene-aniline.

Grasselerator 833. A trade mark for a liquid aldehyde amine condensation product. It is a low temperature rubber vulcanisation accelerator, and has antioxidant properties.

Grass Tree Gum. See ACAROID BALSAM.

Gratonite. A mineral. It is $Pb_9As_4S_{15}$.

Graupen. See BISMUTH BARLEY.

Gravergol. A proprietary preparation of ergotamine tartrate, caffeine and dimenhydrinate, used for migraine. (255).

Gravitol. A German product. It consists of the hydrochloride of diethyl-amino-ethyl ether of 2-methoxy-6-allyl-phenol, a white crystalline substance soluble in water. Its action is similar to that of ergotamine in its effects upon blood pressure.

Gravol. A proprietary preparation of dimenhydrinate. An anti-emetic. (255).

Gravy Colour. See BURNT SUGAR.

Gravy Salt. See BURNT SUGAR.

Grease, Leather. See DÉGRAS.

Grease Paint. Powdered French chalk tinted with carmine, burnt sienna, burnt umber, or other pigment, mixed with lard, glycerin, or pomade to form a paste.

Grease-proof Paper. A non-absorbent sulphate chemical wood-pulp on cotton paper.

Grease, Tanning. See DÉGRAS.

Grease, Wakefield. See YORKSHIRE GREASE.

Green Acid (Green Sulphonate, Green Sulphonic Acid). Crude mixtures of sulphonic acids from refining petroleum sludge.

Greenalite. A mineral. It is a hydrated iron silicate, $Fe_2(Fe.Mg)_3(SiO_4)_3.3H_2O$.

Green Ashes. See BREMEN.

Greenbank. Caustic soda and bleaching powder. (512).

Green-blue Oxide. An artist's pigment made by calcining a mixture of chromic oxide, aluminium oxide, and a cobalt salt.

Green Butter. See BORNEO TALLOW.

Green Calamine. See AURICHALCITE.

Green, Cinnabar. See CHROME GREEN.

Green Copperas. See IRON VITRIOL.

Green Chrome Rouge. Chromium oxide, CrO, used for buffing steels.

Green Cross Gas. Superpalite (tri-chloro-methyl-chloroformate), a military poison gas.

Green Earth (Veronese Green, Veronese Earth). A natural pigment. It contains ferrous iron, silica, magnesia, alumina, and lime. Used as an absorbent for basic dyestuffs.

Green Ebony. A yellow dyewood formerly employed. It is now out of use.

Greener's Powder. An explosive. It consists of nitrocellulose and nitrobenzene, coloured with lampblack.

Green Glass. Chromium or iron produce green, but the colour is usually made by combining several oxides, such as copper and iron, or chromium and copper.

Green Gold. Alloys of 75 per cent. gold, with from 11–25 per cent. silver, and 4–12 per cent. cadmium.

Greenhart Bark. Nectandra bark, also Bebeeru bark.

Green Iodide of Mercury. Mercurous iodide, HgI. Employed in medicine as an irritant poison for syphilis.

Greenish Blue. A dyestuff. It is the hydrochloride of tri-p-tolyl-rosaniline.

Greenite. A mineral. It is chlorite.

Green John. A green fluorspar.

Greenlandite. A mineral. It is columbite.

Greenland Spar. Cryolite ($q.v.$).

Green Lead Ore. A mineral. It is a variety of Pyromorphite ($q.v.$).

Green Leather. Made in Yorkshire, by first giving the skins a light gambier tannage, and then handling them in a hot and strong solution of salt and alum, and finally, heavily stuffing with sod oil.

Green Malt. Germinated barley.

Green Mordant. Sodium thiosulphate. Used as a mordant for fixing aniline greens on fibre.

Green Naphtha. Obtained from Scotch shale by distillation, and treating the distillate with sulphuric acid and sodium hydroxide, then again distilling.

Green Ochre. Usually a mixture of silica, clay, and ferrous hydroxide. Used as a base for cheap lakes.

Greenockite (Cadmium Blende). Cadmium sulphide, CdS. It occurs naturally, usually as a yellow incrustation on zinc blende.

Green Oils. A fraction of oil obtained from Scotch shale by treating the distillate with sulphuric acid and sodium hydroxide and then again distilling. It is also the name for a fraction of

Yorkshire grease distillation. The term is also applied to a preparation of elder leaves boiled in olive oil and filtered. Also see CHROMOL PAINTS.

Greenovite. Titanite containing manganese. A mineral.

Green Oxide of Chromium. Chromic oxide, Cr_2O_3.

Green Powder. See METHYL GREEN.

Green Ramie. See RAMIE.

Green Sand. See GLAUCONITE.

Green Sand. A mixture of sand with 8–15 per cent. coal dust. Used for moulds for casting.

Green Spar. See MALACHITE.

Green Starch. Impure moist starch, obtained in the preparation of starch.

Greenstone, New Zealand. A mineral. It is a variety of Nephrite (q.v.).

Green Sulphide Oil. A recovered product from the residue of the olive oil press.

Green Sulphonate. See GREEN ACID.

Green Sulphur. See BLUE SULPHUR.

Green Sulphuric Acid. See GREEN ACID.

Green Syrup. The impure mother liquor in the massecuite (concentrated sugar juice), from which the sugar is separated. It is also called Centrifugal Syrup.

Green Tar. Barbadoes tar, a bitumen or mineral tar.

Green Tea. Prepared by drying the fresh green leaves after they have been rolled to express the juice.

Green T-stoff. A lachrymator gas of German origin. It consists of 88 per cent. xylyl bromide and 12 per cent. bromo-acetone.

Green Ultramarine. A pigment produced by heating kaolin, silica, sodium carbonate, sulphur, coal, and rosin, washing and grinding. It is used in water-colours, and is known as Lime green. When heated again with sulphur, it gives blue ultramarine. Also see ULTRAMARINE.

Green Verdigris. See VERDIGRIS.

Green Verditer. See MALACHITE and VERDITER BLUE.

Green Vermilion (English Green, Mineral Green). Pigments. They are usually prepared by treating Prussian blue, rendered soluble by oxalic acid, with a solution of potassium bichromate, and then adding a solution of lead acetate. Varieties of Schweinfurth green pass under this name. Also see CHROME GREEN.

Green Vitriol. See IRON VITRIOL.

Green Wood Spirits. See ACETONE ALCOHOL.

Grefco. A proprietary trade name for a chrome ore cement used as a refractory.

Greggio. Crude Sicilian sulphur. The impurities vary from 2–11 per cent. Also see RAFFINATE.

Gregoderm. A proprietary preparation of NEOMYCIN sulphate, mystatin, POLYMYXIN and HYDROCORTISONE used in the treatment of skin disorders. (731).

Gregory's Powder. Compound powder of rhubarb. (*Pulvis rhei compositus B.P.*) It contains 22 grams rhubarb root, 66 grams light magnesia, and 12 grams ginger.

Grenacher's Alum Carmine. An aqueous solution containing 1–5 per cent. common or ammonia alum, boiled with 0·5–1 per cent. carmine, and filtered. A microscopic stain.

Grenacher's Borax Carmine. A microscopic stain. It contains 3 grams carmine, 4 grams borax, and 100 c.c. distilled water. After dissolving by heat, 100 c.c. of 70 per cent. alcohol added, and the solution filtered.

Grenadine. A dyestuff. It is a British equivalent of Magenta.

Grenadine S. See ACID MAGENTA.

Grenaldine. See MAGENTA.

Grenat S. See MAROON S.

Grenat Soluble (Soluble Garnet). A dyestuff. It is the ammonium salt of isopurpuric acid.

G Resin. A proprietary trade name for cumarone-indene resins.

Grey Acetate. Crude acetate of lime, prepared with distilled pyroligneous acid. It contains from 80–82 per cent. calcium acetate, and 20 per cent. water.

Grey Antimony. Trigonal antimony obtained by allowing molten antimony to cool in a crucible.

Grey Antimony Ore. See ANTIMONITE.

Grey Cast Iron. A cast iron containing much of its carbon in the uncombined state. A typical one contains 94 per cent. iron, 3·5 per cent. carbon, and 2·5 per cent. silicon. It has a specific gravity of 7·0 and melts at 1230° C.

Grey Cobalt. See COBALTINE.

Grey Copper. See TETRAHEDRITE.

Grey Copper Ore. See TETRAHEDRITE.

Grey Fibre. See VULCANISED FIBRE.

Grey Forge Pig. A pig iron usually containing less silicon than other grey irons.

Grey Glass (London Smoke Glass). A glass made by adding to it substances which produce complementary colours.

Grey Gold. An alloy of gold and iron, sometimes with silver. One alloy contains 86 per cent. gold, 8·5 per cent. silver, and 5·5 per cent. iron, and another 83 per cent. gold and 17 per cent. iron.

Grey Lotion. (*Lotio hydrargyri nigra B.P.*) It contains mercurous chloride, glycerin, mucilage of tragacanth, and lime water.

Grey Mixture. A mixture of 7 parts meal powder, with 100 parts saltpetre and sulphur. Used in fireworks.

Grey Ochre. This material usually consists of silica, clay, and carbonaceous matter. Used as a filler for cheap paints.

Grey Oil (Mercurial Cream, Mercurial Oil, Lambkin's Cream). An oil containing mercury, lanoline, paraffin, and phenol. Used as an injection.

Grey Ointment. See MERCURY OINTMENT.

Grey Powder (Alkalised Mercury). (*Hydrargyri cum creta B.P.*) It contains 20 grams mercury and 40 grams chalk.

Grey R and B. See INDULINE, SOLUBLE.

Grey Silver. See TETRAHEDRITE.

Grey Tin. A form of tin obtained by exposing the metal to low temperatures. It reverts back to the ordinary form when heated.

Grey Ultramarine. A pigment prepared by replacing the sodium constituents of ultramarine with manganese.

Griess Reagent. Pure white α-naphthylamine is dissolved in water (0·1 gram in 100 c.c.). Glacial acetic acid (5 c.c.), and sulphanilic acid (1 gram) in 100 c.c. water are added. Used for detecting nitrous acid.

Grifa. The trade name for lithium-acetyl-salicylate.

Griffith's Mixture. (*Mistura ferri composita B.P.*) It is a mixture of ferrous carbonate and sugar.

Griffith's White. See LITHOPONE.

Grignard's Reagent. Magnesium reacts with alkyl and aryl halides, in the presence of ether, forming compounds of the type R.MgX (R=alkyl or aryl radicle, and X=a halogen radicle).

Grilon. A registered trade name for polyamide plastics. See NYLON. (742).

Grimm Aluminium Solder. An alloy of 69 per cent. tin, 29 per cent. lead, 1·5 per cent. zinc, and 0·75 per cent. silver.

Grinding Oils. Drying oils such as linseed, etc., used for grinding pigments for paints.

Grindstone. A sandstone consisting almost entirely of quartz.

Griphite. A mineral. It is $8[(Na,Al,Ca,Fe)_3Mn_2(PO_4)_{2·5}(OH)_2]$.

Gripta. Special road dressing material. (603).

Griqualandite. A mineral. It is a variety of crocidolite.

Griscom's Substitute. A rubber substitute made from a petroleum residue, animal fat, candle tar, and sulphur.

Griseofulvin. 7 - Chloro - 4,6 - di - methoxycoumarin - 3 - one - 2 - spiro - 1'- (2' - methoxy - 6' - methylcyclohex - -en-4'-one). Fulcin ; Grisovin.

Griserin (Tryen). A mixture of Loretine (*m*-iodo-*o*-oxy-quinoline-sulphonic acid), with 6·75 per cent. of sodium hydrogen carbonate. Recommended as an internal antiseptic.

Grisounites. Coal mine explosives. They contain nitro-glycerin, collodion cotton, ammonium nitrate, and sometimes magnesium sulphate in the place of ammonium nitrate. Others contain ammonium nitrate and nitro-naphthalene.

Grisoutine Favier. A French explosive containing 29 per cent. nitro-glycerin, 0·9 per cent. collodion cotton, 69·5 per cent. ammonium nitrate, and 0·5 per cent. sodium carbonate.

Grisoutite. An explosive consisting of 53 per cent. nitro-glycerin, 14·5 per cent. kieselguhr, and 32·5 per cent. magnesium sulphate.

Grisovin. A proprietary preparation containing Griseofulvin. An antibiotic. (335).

Grisuten Silk. Fibrous polyethylene terephthalate used as an insulation for conductors in place of silk.

Grochauite. A mineral. It is $2[(Mg,Fe",Al)_6(Si,Al)_4O_{10}(OH)_8]$.

Grodex. Seed germination indicator. (507).

Grodnolite. A mineral. It approximates to the formula $(2Ca_3(PO_4)_2·CaCO_3·Ca(OH)_2)_4·H_4Al_2Si_2O_9$.

Grofas. A proprietary preparation of QUINDOXIN. A veterinary growth promoter.

Grog. A burned clay, which is added to raw clay to reduce the shrinkage of the latter.

Gromidin. A proprietary preparation of gramicidin. An antibiotic. (802).

Grommett Brass. See BRASS.

Groroilite. A mineral. It is a variety of bog manganese.

Grossin. A thickening material for cream. It consists of 5·5 per cent. lime and 10·5 per cent. sugar.

Grossman's Alloy. An alloy containing 87 per cent. aluminium, 8 per cent. copper, and 5 per cent. tin.

Grossmann Reagent. An ammoniacal solution of dicyandiamidine sulphate, $(C_2H_6ON_4)_2·H_2SO_4$. A reagent for nickel.

Gross Solder. An alloy of 62·5 per cent. tin, 12·5 per cent. zinc, 6·3 per cent. aluminium, 12·5 per cent. lead, and 6·3 per cent. phosphor-tin.

Grossular. A mineral. It is $8[Ca_3Al_2Si_3O_{12}]$.

Grossularite. A mineral. It is a calcium-aluminium-silicate, $Ca_2Al_2Si_3O_{12}$.

Grotan. m-Chloro-cresol compound with an alkali. A disinfectant.

Grothine. A mineral. It is a silicate of aluminium, calcium, and iron.

Grothite. A mineral. It is a variety of sphene.

Ground-nut Oil. A non-drying oil obtained from the ground-nut. It consists chiefly of the glycerides of oleic and linolic acids. It is edible, and when hydrogenated is used in the manufacture of margarine. Lower qualities are employed in soap-making.

Ground Nuts (Pea Nuts, Earth Nuts, Monkey Nuts). Arachis nuts, from *Arachis hypogœa*. The nuts contain about 35 per cent. oil (the kernels about 50 per cent.).

Grout. A mixture of cement and water. sometimes including sand.

Groutite. A mineral. It is 4[MnO.OH].

Growth Hormone. An extract of human pituitaries containing predominantly growth hormone. CRESCORMON.

GR Salt. The sodium salt of GR acid (*q.v.*).

Gruber Solder. An alloy of 60 per cent. tin, 25 per cent. zinc, 2 per cent. aluminium, 10 per cent. copper, and 3 per cent. cadmium.

Grudekok (Lignite Char). Lignite from which about 30–40 per cent. water has been expelled. Used in the manufacture of briquettes.

Gru-gru Fat. The fat from the seeds of *Acrocomia sclerocarpa*, of the palm family.

Grunauite. A mineral. It consists of bismuth and nickel sulphides.

Grünerite. A mineral. It is a variety of pyroxene.

Grüne's Chromoxyd. See CHROMIUM GREEN.

Grünlingite. A mineral, Bi_4S_3Te. Synonym for Joseite.

GT50A. A proprietary preparation of calciferol, aneurin, neostigmine, and carbachol, for use in osteoarthritis. (386).

Guacamphol. Guaiacol camphorate, $C_8H_{14}(COOC_6H_4O.CH_3)_2$. Recommended in the night sweats of tuberculosis.

Guadalcazarite. A mineral. It consists mainly of cinnabar with a little zinc sulphide.

Guadarramite. A mineral. It is a variety of ilmenite.

Guaiacetin. Sodium - pyrocatechin - mono - acetate, $C_6H_4(HO).O.CH_2$. COONa. Recommended in the treat-

ment of tuberculosis and chronic bronchitis.

Guaiachinol (Guaiakinol, Guaiaquinol). Quinine-dibromo-guaiacolate, $C_6H_2Br_2$ $(OCH_3)OH.C_{20}H_{24}N_2O_2$. Used as an antipyretic and a sedative in tuberculosis.

Guaiacodeine. Codeine-o-guaiacol-sulphonate. Used in medicine for its codeine content.

Guaiacodyl. See CACODYLLIAGOL.

Guaiacol Ethyl (Guethol, Thanatol, Ajacol). Pyrocatechin - mono - ethyl ether, $C_6H_4.O.C_2H_5.OH$. Used in cases of neuralgia and neuritis.

Guaiacol Ethylene. $(CH_3O.C_6H_4O)C_2$ $H_4(OC_6H_4.OCH_3)$. Used in the treatment of phthisis.

Guaiacol Iodoform. A solution of iodoform in guaiacol. It has been used in the treatment of tuberculosis.

Guaiacol Salol. Guaiacol salicylate, $C_6H_4OH.COO(C_6H_4OCH_3)$. A compound related to Salol (phenol salicylate). Used in the treatment of phthisis, and as an intestinal astringent.

Guaiaco-phosphal. Guaiacol phosphite, used in medicine.

Guaiac Solution. This is made by dissolving 0·5 gram guaiac resin in 30 c.c. of 95 per cent. alcohol. Used to test for blood in milk and other substances.

Guaiacum Resin. A resin obtained from the wood of *Guaiacum officinale* and *Guaiacum sanctum*. Used in medicine in the treatment of gout and rheumatism. It has a specific gravity of 1·2, an acid value of 46–53, saponification value 167–192, ash 1–5 per cent., and 87–96 per cent. is soluble in alcohol.

Guaiacyl (Gajacyl). Calcium-o-guaiacolsulphonate, $(C_2H_3OH.(OCH_3).SO_3)_2Ca$. Used hypodermically, also as a local anæsthetic.

Guaiaform (Geoform, Pulmoform). Methylene - diguaiacol, $CH_2.(C_6H_3)(OC$ $H_3)(OH)_2$. It is said to contain 95·4 per cent. guaiacol, and is used medicinally as a substitute for creosote, in the treatment of bronchitis and tuberculosis.

Guaiaglycol. See MONOTAL.

Guaiakinol. See GUAIACHINOL.

Guaialin. Methylene-diguaiacol-benzoic ester. Recommended as an antipyretic, and as a tonic in tuberculosis.

Guaiamaltin. A mixture of malt extract and potassium-sulpho-guaiacolate.

Guaiamar. See GUAJAMAR.

Guaiaperol. An addition product of guaiacol and piperidine, $C_5H_{11}N(C_7H_8$ $O_2)_2$. Used in medicine for the treatment of tuberculosis.

Guaiaquin. Quinine - guaiacol - bisulphonate, $(C_6H_4O_2.CH_3.HSO_3)_2.C_{20}H_{24}$

N_2O_2. Used as an antiperiodic in malarial diseases.

Guaiaquinol. See GUAIACHINOL.

Guaiasanol (Gujasanol, Guiasanol). Diethyl - glycocoll - guaiacol - hydro - chloride, $C_6H_4(OCH_3).O.CO.CH_2.N(C_2H_5)_2.HCl$. Used as a subcutaneous injection, and for internal use in tuberculosis of the lungs and larynx.

Guaiathol. Prepared by heating catechol with alcohol, $C_6H_4(OH)O.C_2H_5$. An antiseptic used in medicine.

Guaic. Guaiacol resin.

Guaiol. Tiglic aldehyde, C_5H_8O.

Guaiotyle. Guaiasanol (q.v.).

Guaiphenesin. 3-(2-Methoxyphenoxy)-propane-1,2-diol. Guaiacol glycerol ether. Respenyl.

Guajamar (Guaiamar). Guaiacol-glycerol ester, $C_6H_4(OCH_3)O.C_3H_7O_2$. An intestinal antiseptic for typhoid fever and diarrhœa. Also used as an ointment consisting of 2 parts guajamar, and 1 part lanoline, for rheumatism.

Guamecycline. An antibiotic. N-(4-Guanidinoformimidoylpiperazin - 1 - ylmethyl)tetracycline.

Guanabacoite. A mineral. It is a variety of quartz.

Guanacline. 1 - (2 - Guanidinoethyl) - 1,2,3,6 - tetrahydro-4 - picoline. FBA 1464.

Guanajuatite (Selenobismutite, Frenzelite, Castillite). A mineral, Bi_2Se_3.

Guanapite. See OXAMMITE.

Guanethidine. 1 - (2 - Guanidinoethyl)-azacyclo-octane. Ismelin is the sulphate.

Guanicaine. See ACOIN.

Guanimycin. A proprietary preparation of dihydrostreptomycin sulphate, sulfaguanidine and kaolin. An antidiarrhœal. (284).

Guanite. See STRUVITE.

Guano (Bird Manure). A fertiliser. It consisted of deposits of excrements and skeletons of birds and animals, during past geological epochs, now exhausted. The deposits contained phosphates together with much ammonium urate and oxalate. Also see FISH MANURE.

Guano, Flesh. See FLESH MANURE.

Guano, Whale. See WHALE MEAL.

Guanoclor. [2-(2,6-Dichlorophenoxy)-ethyl]aminoguanidine. Vatensol is the sulphate.

Guanol. A mixture of molasses and peat. A fertiliser.

Guanor Expectorant. A proprietary preparation of ammonium chloride, DIPHENHYDRAMINE hydrochloride, sodium citrate, chloroform and menthol. (368).

Guantal. Diphenylguanidine phthalate rubber accelerator. (559).

Guanoxan. 2 - Guanidinomethylbenzo - 1,4-dioxan. Envacar is the sulphate.

Guapinole Resin. A resin obtained from a Mexican tree of the same name. It is stated to be insoluble in alcohol and petroleum ether, and to be suitable for the manufacture of lacquers and varnishes.

Guara. The ground fruits of a species of *Cæsalpinia*, from Central and South America. A tanning material.

Guarana. A dried paste prepared from the seeds of *Paullinia sorbilis*. An article of food, as cocoa.

Guaranine. See THEINE.

Guardex. A proprietary safety glass.

Guarinite. A mineral, $CaO.TiSiO_4$.

Guatannin. Guaiacol tanno-cinnamate. Employed in the treatment of bronchial catarrh.

Guayacanite. A mineral. See ENARGITE.

Guayale (Durango). A rubber obtained from *Parthenium argentatum*. The dry shrub weighs 2 lb., more mature specimens about 3 lb., of which 10 per cent. is represented by leaves and stem, 45 per cent. by woody fibre and pith containing practically no rubber. The rubber may amount to as much as 20 per cent. of the dry weight, but is 15 per cent. for older shrubs. The rubber is obtained by crushing the shrub in the presence of water, when the particles of rubber float. It is usually mixed with other rubbers for making rubber goods.

Guayarsin. A combination of guaiacol and arsenious oxide. Used medicinally.

Guayule (Durango). A rubber obtained from *Parthenium argentatum*. It is usually mixed with other rubbers for making rubber goods.

Guaza (Ganja, Gunjah). Indian hemp.

Gudmundite. A mineral. It is 8[FeSbS].

Guejarite. A mineral. It is a cuprous antimony sulphide.

Guernsey Blue. See SOLUBLE BLUE.

Guethol. See GUAIACOL ETHYL.

Guhr Dynamite. Ordinary dynamite. Nobel's contains 75 per cent. nitroglycerin and 25 per cent. kieselguhr.

Guiasanol. See GUAIASANOL.

Guettier Metal. An alloy of 62 per cent. copper, 32 per cent. zinc, and 6 per cent. tin.

Guido's Balsam. Liniment of opium.

Guignet's Green. See CHROMIUM GREEN.

Guildite. A mineral. It is an iron and copper sulphate.

Guillaume Alloy. A proprietary trade name for an alloy of 66 per cent. iron and 34 per cent. nickel. It has a low coefficient of expansion.

Guillaume Metal. An alloy of 64 per cent. copper and 36 per cent. bismuth.

Guinea Carmine B. A dyestuff. It dyes wool from an acid bath, red shades with a bluish tinge.

Guinea Grains. See GRAINS OF PARADISE.

Guinea Green B (Guinea Green BV, Acid Green, Acid Green G). A dyestuff. It is the sodium salt of diethyl dibenzyl - diamino - triphenyl - carbinol - disulphonic acid, $C_{37}H_{36}N_2O_7S_2Na_2$. Dyes silk and wool green from an acid bath.

Guinea Green BV. See GUINEA GREEN B.

Guinea Pepper. Capsicum fruit.

Guinea Red. An acid dye for wool and silk.

Guitermanite. A mineral. It is $Pb_{10}As_6S_{19}$.

Guitermannite. A mineral, $PbS.Pb As_2O_4$.

Gujasanol. See GUAIASANOL.

Gum Acacia. See GUM ARABIC.

Gum Accroides. See ACAROID BALSAM.

Gum, Amritsar. An acacia gum from *Acacia modesta*. It is in large brown tears and is used in calico printing.

Gum Angico. See BRAZILIAN GUM.

Gum Animi. See ANIMI RESIN.

Gum, Arabian. See KHORDOFAN GUM.

Gum Arabic (Arabin, Calcium Gummate). Gum acacia from *Acacia senegal* and other species. See GUMS AMRITSAR, AUSTRALIAN, BABAL, CAPE, EAST INDIAN, KHORDOFAN, MOROCCO, PICKED TURKEY, SENEGAL, and SUAKIN. Also HABBAK.

Gum, Artificial. See BRITISH GUM.

Gum, Babal. See BABAL GUM.

Gumbellite. A mineral. It is an aluminium silicate.

Gum, Bengal. See BABAL GUM.

Gum Benguela. A semi-fossil copal used in varnishes.

Gum Benjamin. Benzoin, a balsam.

Gum, Black Boy. See ACAROID BALSAM.

Gumbo. See BENTONITE.

Gum, Bombay. See EAST INDIA GUM.

Gum, Brown Barberry. See MOROCCO GUM.

Gum, Butea. See KINO, BENGAL.

Gum, Cadie. See CAMBOGIA.

Gum, Cambogia (Gutti). See CAMBOGIA.

Gum-carbo. A rubber substitute made from cotton-seed oil.

Gum, Carmania. See BASSORA GUM.

Gum, Catechu. Catechu. See Cutch.

Gum, Copal. See COPAL RESIN.

Gum, Cowdie. See AUSTRALIAN COPAL.

Gum, Cowrie. Gum dammar.

Gum, Cumar. See PARACOUMARONE RESIN.

Gum, Crystallised. See BRITISH GUM, CRYSTALLISED.

Gum D. A gelatin dynamite (French). It contains 69·5 per cent. nitro-glycerin.

Gum, Dhak. See KINO, BENGAL.

Gum Dragon. Gum tragacanth.

Gum E. A gelatin dynamite (French). It contains 49 per cent. nitro-glycerin.

Gum Elemi. Manila elemi. See MANILA ELEMI.

Gum, Fiddle. Gum tragacanth.

Gum, Ghati. Bassora gum (*q.v.*).

Gum, Grass-tree. See ACAROID BALSAM.

Gum, Hemlock. See CANADA PITCH.

Gum, Hog. See BASSORA GUM.

Gum, India. See BASSORA GUM.

Gum Juniper. Gun sandarac.

Gum, Karite. See GUTTA-SHEA.

Gum, Kauri. See AUSTRALIAN COPAL.

Gum Kino. See KINO.

Gum, Kutera. See BASSORA GUM.

Gum Lac. See LAC.

Gum Lini. A gum made from linseed by treatment with water, then treating the mass with 90 per cent. alcohol. The gum is soluble in water and is used as a substitute for gum arabic.

Gum, Locust Bean. See INDUSTRIAL GUM.

Gum, Locust-kernel. See CAROB-SEED GUM.

Gum, Magyey. See CHAGUAL GUM.

Gum MB. A gelatin dynamite (French). It contains 74 per cent. nitro-glycerin.

Gummeline. Dextrin.

Gum, Mesquite. Mesquite gum (*q.v.*).

Gummi Gutta. See GAMBOGE.

Gummite. A mineral. It is an amorphous hydrated uranium oxide, containing 61–74 per cent. UO_3, with some lead, calcium, and barium, and traces of radium. The name has also been applied to a variety of the mineral Halloysite.

Gum, Mogador. See MOROCCO GUM.

Gummoid. An Austrian moulded asbestos product for use in electrical insulation.

Gummolin. A linseed oil preparation used as a coating for protection against rust and weather.

Gummon. A German compressed and impregnated asbestos composition.

Gum, Muccocota. See OCOTA COCOTA GUM.

Gum, Oomra. See AMRAD GUM.

Gum, Para. See BRAZILIAN GUM and JUTAHYCICA RESIN.

Gum, Persian. See INDIA GUM.

Gumption. A mixture of mastic varnish, linseed oil, and lead acetate. Used in oil painting.

Gum, Red. See EUCALYPTUS GUM.

Gum, Red Yacca. See ACAROID BALSAM.

Gums. The French name for gelatin dynamites.

Gum, Seiba. See TUNO GUM.

Gum, Sennaar. See SUAKIN GUM.

Gum, Silk. See CUITE AND SERICINE.

Gum, Starch. See BRITISH GUM.

Gum, Sweet. See AMERICAN STORAX.

Gum, Talca. See SUAKIN GUM.

Gum, Talha. See SUAKIN GUM.

Gum Thus (French Pine Resin). Common frankincense, the crude turpentine from French pine-trees.

Gum, Toonu. See TUNO GUM.

Gum, Touchpong. See POUCKPONG GUM.

Gum Tragacanth. A gum obtained from shrubs of the *Astragalus* family. The qualities include Fiori, white flakes obtained from the first incisions, and Biondo, yellow gum from the second incision.

Gum, Tunu. See TUNO GUM.

Gum, Turkey. See KHORDOFAN GUM.

Gum, Vegetable. See BRITISH GUM.

Gum, Wattle. See AUSTRALIAN GUM.

Gum, White Sennaar. See PICKED TURKEY GUM.

Gum, Wood. See COTTON and TREE GUM.

Gum, Yacca. See ACAROID BALSAM.

Gum, Zapoto. See CHICLE.

Guncotton (Nitro-cotton, Pyroxylin, Collodion Cotton). An explosive. It is nitro-cellulose, made by acting upon cotton with nitric and sulphuric acids.

Gunjah. See GUAZA.

Gun Metal. An alloy containing from 89–91 per cent. copper, 8–11 per cent. tin, and 1–2 per cent. zinc. The alloy, containing 90 per cent. copper, 8 per cent. tin, and 2 per cent. zinc, has a specific gravity of 8·8, and melts at 1010° C.

Gunnarite. A mineral. It is an iron-nickel sulphide.

Gunning's Reagent. A 10 per cent. iodine solution in alcohol. Used for the detection of acetone in urine.

Gunny. A coarse grade of burlap.

Gunpowder (Black Powder). A mixture of saltpetre, carbon, and sulphur, in varying proportions. An average one contains 75 per cent. potassium nitrate, 10 per cent. sulphur, and 15 per cent. carbon. An explosive.

Gunsberg Reagent. A solution containing 2 parts phloroglucinol, 1 part vanillin, and 30 parts alcohol. Used for testing for hydrochloric acid in the gastric juices. A drop of the reagent gives red crystals when evaporated with the gastric juice if hydrochloric acid is present.

Gur. A name applied to jaggery (*q.v.*).

Gurdynamite. See KIESELGUHR DYNAMITE.

Gurhofite (Geldolomite). A mineral. It is an amorphous colloidal form of dolomite.

Gurjun Balsam or Oil (Wood Oil). The oleo-resin from the stems of *Dipterocarpus species*. It has a specific gravity of 0·96–0·985 and an acid value of 10–25. Soluble in alcohol, benzene, amyl acetate, chloroform, ether. Slightly soluble in petroleum ether.

Gurley's Metal. An alloy of 86·5 per cent. copper, 5·4 per cent. zinc, 5·4 per cent. tin, and 2·7 per cent. lead.

Gurney's Bronze. An alloy of 76 per cent. copper, 15 per cent. lead, and 9 per cent. tin.

Gurry. Finely divided fish, obtained in fish manure manufacture.

Guthrie's Eutectic Alloy. An alloy of 47 per cent. bismuth, 20 per cent. tin, 20 per cent. lead, and 13 per cent. cadmium.

Gutta-alco. A patented substitute for gutta-percha. It is made from rubber latex and the salts of fatty acids (palmitic, stearic).

Gutta-gentzsch. An artificial gutta-percha, consisting of rubber and a special kind of palm wax. Other varieties contain resin, as well as wax. An insulator.

Gutta-grip. See GUTTA-SUSU.

Gutta-hangkang. A product of *Palaquium leiocarpum*, of Borneo. It has been used to adulterate the better grade gutta-perchas.

Gutta-Jangkar. An inferior gutta-percha from Sarawak.

Guttaline. A mixture of bitumen, Manila gum, and resin oil.

Gutta-percha. The coagulated latex of species of *Palaquium* and *Payena*, of Malay, Sumatra, and Borneo. The material is plastic when hot and can be moulded, and in this way differs from rubber. As imported into England it is associated with varying amounts of dirt and moisture. The hydrocarbon $(C_{10}H_{16})n$, is also associated with resin in varying proportions, and the amount of gutta hydrocarbon obtainable from the clean material varies from 30–80 per cent. A standard product of much higher value is now obtained from cultivated trees on certain estates, and the gutta is also obtained from the leaves of the trees by mechanical means. Gutta-percha is mainly used as the insulating material for submarine cables, but in addition, it is made into carboys and bottles for use as containers for hydrofluoric acid, for packings for cold-water systems, as a dental stopping, in blocks for use as cutting beds for glove-making, and in the form of thin tissue as an adhesive.

Gutta-percha, Artificial. See GUTTA-GENTZSCH, NIGRITE, VELVRIL, and SOREL's SUBSTITUTE.

Gutta-percha, Indian. See PALA GUM.

Gutta-percha Pitch. A term applied to the resin obtained by the extraction of gutta-percha by solvents and evaporation.

Gutta-percha, Reboiled. Raw gutta-percha, generally of inferior quality to average grades, which is the result of mixing odd lots of raw material by softening in hot water and making it more or less homogeneous.

Gutta-shea (Karite Gum). A product resembling gutta-percha obtained from an African tree, *Bassia Parkii*. The gutta is separated from the fat (Shea butter).

Gutta-Siak. A low grade gutta-percha gum from *Payena Leerii* and other *Payena* species, of Siak, Sumatra. It is a mixed product with a high resin content.

Gutta-soh. A mixed gutta-percha on the Singapore market. It is often coloured red with bark stain.

Gutta-sundik. A gutta-percha from *Payena* species. It is of fairly good quality, but has a high resin content.

Gutta-susu (Assam White, Gutta Grip, Gutta Gerip, Gutta Singarip, Borneo Rubber). A wild rubber mainly obtained from *Willughbeia firma*, of Borneo.

Gutti. See CAMBOGIA.

Guttmann's Nutrient Milk Flour. Consists of casein mixed with carbohydrates (mainly gluten-free oatmeal). It contains 65 per cent. carbohydrates, 20 per cent. albumin, and 4 per cent. fat.

Guvacine. An alkaloid derived from pyridine. It is $1:2:5:6$-tetrahydropyridine-3-carboxylic acid. Its methyl ester is also called Guvacine.

G Varnish. A proprietary trade name for varnish and lacquer resins made from glycerol and phthalic anhydride.

Gyle. The resulting liquid after the wort is treated with yeast, and aerated in the preparation of vinegar. It contains 6–7 per cent. alcohol. The liquid is treated with acetic acid bacteria.

Gymnite. See DEWEYLITE.

Gymnite, Nickel. See GENTHITE.

Gynaflex. A proprietary preparation of noxytiolin, and lignocaine hydrochloride, used in vaginitis. (386).

Gynergen. Ergotamine tartrate. Used to alleviate atonia of the uterus.

Gynocardia Oil. An oil similar to chaulmoogra oil. It is obtained from the seeds of *Taraktogenon Kurzii*.

Gynocardic Acid. A term used for the acids contained in the oil expressed from the seeds of *Gynocardia odorata*.

Gyno-Daktarin. A proprietary preparation of MICONAZOLE nitrate used in the veterinary treatment of candidiasis. (356).

Gynoval (Feminal). Isobornyl-isovalerate, $C_4H_9.COO.C_{10}H_{17}$. Used in medicine for the same purposes as Bornyval (*q.v.*).

Gynovlar. A proprietary preparation of norethisterone acetate and ethinyl-œstradiol. Oral contraceptive. (265).

Gypsite. A deposit consisting of small grains of gypsum, disseminated through an earthy mass. It is used for the production of wall plastics.

Gypsona. Plaster of Paris bandage. (529).

Gypsteel. A proprietary trade name for a steel reinforced gypsum in the form of slabs used for floors.

Gypsum (Alabaster, Selenite, Annaline, Terra Alba, Satinite, Mineral White, Light Spar, Satin Spar, Lenzit). A mineral. It is calcium sulphate, $CaSO_4.2H_2O$. Its specific gravity is 2·3.

Gyrite. A mineral. It is chalybite.

Gyrolite. A calcium silicate mineral.

H

H.11. A proprietary preparation of a polypeptide from male urine, used in the treatment of cancer. (698).

H.400. Descaling solvent for marine heat exchangers. (503).

HA 819. Polyoxyethylene nonyl phenol. A water soluble material of medium chain length. Used as a detergent. (301).

HB 40. Partially hydrogenated terphenyl vinyl plasticiser. (559).

Haarlem Oil (Dutch Drops). Linseed oil, sulphurated and mixed with turpentine. Antiseptic.

Habbak. A generic term applied to acacia gums and gum resins, by the Somalis. Some of them are used as adhesives, others are eaten by the Somalis, whilst the remainder have no commercial value.

H-acid. $1:8$-Amino-naphthol-3:6-disulphonic acid.

Hachimycin. An antibiotic produced by *Streptomyces hachijoensis*, used in the treatment of trichomoniasis. TRICHOMYSIN.

Hackmanite. A mineral. It is $Na_8Al_6Si_6O_{24}(Cl_2,S)$.

Hædensa. A German preparation. It

is an ointment containing menthol, anthrasol, mercury, and silver nitrate. Used for the treatment of piles.

Hæfelyite. A trade name for a shellac varnish-paper product. Used as an electrical insulation.

Haelan. A proprietary preparation of flurandrenalone used in the treatment of skin diseases. (309).

Haelan, Haelan-X. A proprietary preparation of flurandrenalone for dermatological use. (309).

Hæmachates. Agates marked with red jasper.

Hæmafibrite. A mineral, $Mn_6As_2H_{10}O_{16}$.

Hæmalum. A microscopic stain. One gram hæmatoxylin or its ammonium salt is dissolved in 50 c.c. of 90 per cent. alcohol, added to a solution of 50 grams of alum in 1,000 c.c. water, and filtered.

Hæman. A peptone preparation which contains iron sulphocyanide. Said to be used in cases of chlorosis.

Hæmatin. Hæmatoxylin, $C_{16}H_{14}O_6$.

Hæmatite (Specular Iron Ore, Kidney Ore, Micaceous Iron Ore, Iron Glance, Martite). Varieties of anhydrous ferric oxide, Fe_2O_3, which occur naturally.

Hæmatite Pig. See BESSEMER PIG.

Hæmatite, Red. See HÆMATITE.

Hæmato-crystallin. SEE HÆMATO-GLOBULIN.

Hæmatogen. See HÆMOL.

Hæmato-globulin (Hæmato-crystallin). Hæmoglobin, the pigmentary matter of the red corpuscles of the blood.

Hæmatolite. Synonym for Hematolite.

Hæmatophan. A hæmoglobin preparation, obtained from defibrinated blood.

Hæmatostibiite. A mineral, 8MnO.Sb_2O_5.

Hæmatoxylin Solution. Logwood (15 grams), boiled in 100 grams absolute alcohol for 48 hours, then filtered. Used for the detection of iron and copper in drinking water.

Hæmogallol. Prepared by the action of pyrogallol on defibrinated blood solution. Prescribed in cases of chlorosis.

Hæmoglobulin. Has a variety of uses, one of which is as a protective colloid serving as an antioxidant for rubber latex. The latex is coagulated by acid, not by agitation. It can also be used as a filler for rubber particularly in cases of mixes for adhering to metals.

Hæmol (Hæmatogen). Obtained by the reducing action of zinc-dust upon defibrinated blood. Prescribed in cases of chlorosis. It consists of albuminate of iron.

Hæmoplastin (Hæmostatic Serum). A fluid preparation from blood serum. It consists chiefly of prothrombin and thrombokinase.

Hæmosistan. A German hæmostatic. It is stated to be a compound of ethylene-diamine-acetate and calcium chloride.

Hæmostasin. Adrenalin (*q.v.*).

Hæmostatic Serum. See HÆMO-PLASTIN.

Hæmostop. A proprietary preparation of naftazone. Used to stop capillary bleeding by injection. (256).

Hæmovin. A proprietary preparation of titanium dioxide, salicylamide, hexachlorophane and ephedrine hydrochloride. An antipruritic anal cream. (339).

Haffkine's Prophylactic Fluid. Plague vaccine.

Hafnefjordite. A mineral. It is labradorite.

Hagafilm. Volatile filming amine. (503).

Hagatalite. A mineral. It is a variety of zircon.

Hagendorfite. A mineral. It is $(Na,Ca)(Fe^{..},Mn^{..})(PO_4)_2$.

Hager's Reagent. Picric acid (1 gram) dissolved in 100 c.c. water.

Hagevap. Treatment for sea water evaporators. (503).

Hahnmann's Mercury. Black oxide of mercury.

Haidingerite. A mineral. It is dicalcium arsenate, $Ca_2H_2(AsO_4)_2.H_2O$.

Hailcris. A British glass, used for making miners' lamps, and other purposes.

Hailuxo Glass. A British glass used for chemical apparatus.

Haine's Solution. Copper sulphate, 8.314 grams, is dissolved in 400 c.c. water, 40 c.c. glycerol, and 500 c.c. of 5 per cent. potassium hydroxide added. A test for sugar.

Hainite. A mineral. It is a tantalosilicate and titanate of zirconium, calcium, and sodium, of Bohemia.

Hair, Artificial. See GLANZSTOFF.

Hair Copper. See CHALCOTRICHITE.

Hair Pyrites. See MILLERITE.

Hair Salt. See ALUNOGEN AND EPSOM SALTS.

Hair Water, Golden. See AURICOME.

Hakuenka. A Japanese native whiting (calcium carbonate) specially processed for use as a reinforcing agent for rubber. It has very fine particle size, and gives higher tensile, greater resisting power, and greater wetting power than ordinary calcium carbonate.

Hakuenka CCR. A Japanese trade name for precipitated calcium carbonate. (25). See also CALOFIL.

Hakuenka OT. A proprietary trade name for a 30 millimicron average particle size transparent precipitated calcium carbonate used in high gloss offset printing inks. (189).

Halabar. A proprietary preparation of butobarbitone, and mephenisin. A sedative. (830).

Halarsol. A proprietary preparation. It is 3-amino-4-hydroxy-phenyl-dichlorarsine hydrochloride solution.

Halazone. p - Sulphon - dichlor - amino - benzoic acid, $C_6H_4.SO_2.NCl_2.COOH.$ Recommended for sterilising drinking water.

Halberland Metal. A brass containing 87 per cent. copper and 13 per cent. zinc.

Halciderm. A proprietary preparation of HALCINONIDE used in the treatment of skin diseases. (682).

Halcinonide. A corticosteroid currently undergoing clinical trial as " SQ 18 566 ". It is 21-chloro-9α-fluoro-11β-hydroxy - 16α, 17α - isopropylidenedi - oxypregn-4-ene-3,20-dione.

Haldrate. A proprietary trade name for the 21-acetate of Paramethasone. (333).

Halethazole. 5 - Chloro - 2 - [4 - (2 - diethylaminoethoxy)phenyl] - benzo - thiazole. Episol.

Half-ganister (Semi-silica, Quartzite). A material containing 80–85 per cent. silica, made by mixing ganister with a high silica fireclay. A refractory.

Half-stuff. Refined wood cellulose obtained as sulphite pulp and in the form of thick sheets. It is used either alone or mixed with esparto pulp in the manufacture of paper.

Half-wool Black. See UNION BLACK B, R.

Halibol. A proprietary preparation of halibut-liver oil with irradiated ergosterol.

Haliborange. Halibut-oil and orange juice. (510).

Halite. See ROCK SALT.

Haliverol. A proprietary preparation of halibut-liver oil with irradiated ergosterol.

Hallite. A mineral. It is a variety of mica.

Halloysite. A clay mineral. It is an aluminium silicate, $Al_2O_3.2SiO_2.2H_2O.$

Halofenate. A hypolipœmic agent currently undergoing clinical trial as " MK-185 " and " Livipas ". It is 2-acetamidoethyl (4-chlorophenyl)(3-trifluoromethylphenoxy)-acetate.

Haloflex. Chlorinated polythene. (512).

Halofuginone. An anti-protozoan. DL-*trans* - 7 - Bromo - 6 - chloro - 3 - [3 - (3 - hydroxy - 2 - piperidyl) - 2 - ace - tonyl] quinazolin-4-one.

Halon. A proprietary polychlorotri fluoroethylene. See HOSTAFLON C2 KEL-F. (30).

Halopenium Chloride. 4-Bromobenzyl 3 - (4 - chloro - 2 - isopropyl - 5 - methyl phenoxy)propyldimethylammoniun chloride.

Haloperidol. 4 - (4 - Chlorophenyl) - 1 [3 - (4 - fluorobenzoyl)propyl] - piperi din-4-ol. Serenace.

Halopyramine. N - (4 - Chlorobenzyl) N'N' - dimethyl - N - 2 - pyridyl ethylenediamine. Chloropyramine Synopen is the hydrochloride.

Halo Rubber. A synthetic rubber con taining combined chlorine.

Halothane. 2 - Bromo - 2 - chloro - 1,1 1-trifluoroethane. Fluothane.

Halotrichite. See ALUNOGEN. A mineral. It is $Fe^{..}Al_2(SO_4)_4.22H_2O.$

Halowax. Proprietary preparations They consist essentially of chlorin substitution products of naphthalene or other materials.

Halowax 1014. A proprietary trade name for hexachlornaphthalene. (95).

Halowax 4000 B-2. A proprietary trade name for a chlorinated hydrocarbon vinyl plasticiser. (34).

Haloxon. An anthelmintic. Bis (2 - chloroethyl) 3 - chloro - 4 - methyl - coumarin-7-yl phosphate. LOXON.

Halphen Reagent. A solution of sulphur (1 per cent.) in carbon disulphide. Used to test for cotton-seed oil. T 1 c.c. of oil add 1 c.c. of reagent and 1 c.c. of amyl alcohol, and heat. A red colour is given with cotton-seed oil.

Halquinol. A mixture of the chlorinated products of 8-hydroxyquinoline containing about 65 per cent. of 5,7 dichloro-8-hydroxyquinoline. Quixalin

Halumin. A new aluminium alloy with 1·48 per cent. copper, 2 per cent. nickel 2·3 per cent. manganese, 0·47 per cent iron, and 0·09 per cent. silicon. It is specially resistant to corroding agents.

Halva. A food product prepared from " thaban " (roasted sesamé meal).

Halwa. The name used for certain sweetmeats in Turkey. They consist of mixtures of sugar, nuts, saponin, and meal.

Hamamelin. A preparation from witchhazel.

Hamameli Tannin. A tannin originally isolated from *Hamamelis virginica*. It has a formula, $C_{20}H_{20}O_{14}$, and is probably a glucoside of *m*-digallic acid anhydride.

Hamartite. See BASTNASITE.

Hambergite. A mineral, $BeOH.BeBO_3.$

Hamburg Blue. See MOUNTAIN BLUE and PRUSSIAN BLUE.

Hamburg White. A pigment consisting

of 1 part white lead and 2 parts heavy spar.

Hamilton Metal. A brass containing 67 per cent. copper and 33 per cent. zinc. An alloy of 90 per cent. zinc, with small quantities of copper, lead, antimony, and phosphor tin, is also known as Hamilton metal.

Hamlinite (Bowmanite). A mineral. It is an aluminium-beryllium-hydro-fluo-phosphate.

Hammarite. A mineral, $5PbS.3Bi_2S_3$. It is also stated to be $Cu_2Pb_2Bi_4S_9$.

Hammerloid. A proprietary trade name for metal paints.

Hammer Slag. A basic silicate of iron produced and used in the puddling process for iron.

Hammonia Metal. An alloy of 64·5 per cent. tin, 32·2 per cent. zinc, and 3·2 per cent. copper.

Hamonite. A British-made activated carbon, made from peat.

Hampdenite. A mineral. It is a variety of serpentine.

Hampden Steel. A proprietary trade name for a chromium tool steel containing 12·5 per cent. chromium, 0·25 per cent. nickel, 0·25 per cent. manganese, and 2·1 per cent. carbon.

Hancockite. A mineral, $(Ca.Pb)_4(Al.Fe)_6H_2Si_6O_{26}$.

Hang-ge. A drug. It is the root nodule of *Pinellia tuberifera*. Used as an antemetic in Japan and China.

Hanksite. A mineral, $Na_2SO_4.Na_2CO_3$.

Hanleite. A mineral. It is $8[Mg_3Cr_2Si_3O_{12}]$.

Hannayite. A mineral, $Mg_3(NH_4)_2H_4(PO_4)_4.8H_2O$.

Hansa Oil. A proprietary trade name for polymerised marine animal oil for soap manufacture.

Hansa Yellow G (Monolite Yellow, Pigment Fast Yellow Conc. New). A dyestuff obtained by coupling diazotised *m*-nitro-*p*-toluidine with aceto-acetic anilide.

Hanus' Iodine Bromide Solution. Iodine bromide (10 grams), dissolved in 500 c.c. of glacial acetic acid. Used for determining the iodine value of fats and oils.

Hanusite. A mineral. It is $Mg_2Si_3O_7(OH)_2.H_2O$.

Haplotypite. A mineral. It is a variety of ilmenite.

Harbortite. A mineral. It is $Al_3(PO_4)_2(OH)_3.3H_2O$.

Hard Aluminium. An alloy of 77 per cent. aluminium, 11 per cent. zinc, and 11 per cent. magnesium. Another alloy contains copper in the place of zinc.

Hardened Rosins. These are metallic resinates prepared by heating rosin with metallic oxides, usually lime, and to a smaller extent magnesium oxide and zinc oxide. The amount of lime normally used is from 5–11 per cent. as calcium hydrate. The hard rosin formed, which contains 3 per cent. calcium oxide, has an acid value of 106, and that containing 5·3 per cent. calcium oxide gives 84 as the acid value. A zinc resinate is formed by adding about 3 per cent. lime and 5 per cent. zinc oxide and heating to 150° C. for from 4–5 hours. Magnesium oxide and zinc oxide may be added to the rosin at 240° C. Lead and cobalt are sometimes incorporated with rosin as acetates.

Hardened Rubber. See EBONITE.

Hardening Carbon. The form in which carbon exists in hard steels.

Hardenite. A collective name for austenite and martensite of eutectoid composition.

Hardeste. Dressing for core and moulds. (531).

Hard-finish Plaster. Plaster made from oven-burnt gypsum dipped in alum solution, and again calcined.

Hard Gold Solder. See GOLD SOLDERS.

Hard-head. The name by which the impurities obtained from the refining of tin are known.

Hardite. Heat-resisting alloys containing 55–65 per cent. nickel, 15–18 per cent. chromium, 1–4 per cent. silicon, 1–2 per cent. manganese, the balance being iron.

Hardite X. A heat-resisting alloy. It contains 82–86 per cent. nickel, 10–13 per cent. chromium, and 2 per cent. manganese.

Hard Jatoba. A Brazilian copal resin.

Hard Lac. A hard shellac made by the removal of half of the soft constituents by alkali.

Hard Lac Resin. A hard resin obtained by extraction of shellac with alkaline reagent to remove the soft resin.

Hard Lead (Antimony Lead). An alloy of lead with 10–30 per cent. antimony. It is also known as antimony lead, and is used as a type metal.

Hard Metal. The name usually applied to a tin-copper alloy, containing 1 part tin, and 2 parts copper.

Hard Platinum. Platinum containing from about 5–30 per cent. iridium.

Hard Silver Solder. See SILVER SOLDERS.

Hard Solder. An alloy of 86 per cent. copper, 9 per cent. zinc, and 4 per cent. tin.

Hardware Brass. See BRASS.

Hardystonite. A mineral. It is a calcium-zinc silicate.

Hard Zinc. An alloy of 92 per cent. zinc, 5 per cent. iron, and 3 per cent. lead.

Harflex. A proprietary trade name for vinyl plasticisers. (101).

Hargus Steel. A proprietary trade name for a die steel containing 1·0 per cent. manganese, 0·35 per cent. nickel, and 1·0 per cent. carbon.

Harle's Solution. A solution of sodium arsenite.

Harlington Bronze. See HARRINGTON BRONZE.

Harmaline. See MAGENTA.

Harmogen. A proprietary preparation of piperazine œstrone sulphate used in the treatment of menopausal symptoms. (311).

Harmomang A and B. Iron alloys containing carbon. manganese and molybdenum. Proprietary alloys.

Harmonia Bronze. An alloy of 57 per cent. copper, 40 per cent. zinc, 1·8 per cent iron, and 0·4 per cent. lead.

Harmonyl. A proprietary preparation of deserpidine. A sedative and antihypertensive. (311).

Harmotone (Morvenite, Andreolite). A mineral. It is a zeolite, containing barium, $(K_2Ba)O.Al_2O_3.5SiO_2.5H_2O$.

Harrington Bronze (Harlington Bronze). An alloy of 55·75 per cent. copper, 42·5 per cent. zinc, 1 per cent. tin, and 0·75 per cent. iron. A white bearing metal.

Harringtonite. An Irish mineral. It contains 41·4 per cent. SiO_2, 30·2 per cent. Al_2O_3, 11·2 per cent. CaO, 5·2 per cent. Na_2O, and 12·5 per cent. H_2O.

Harris' Hæmatoxylin Stain. A microscopic stain. It contains 1 gram hæmatoxylin, 10 c.c. alcohol, 20 grams alum, 0·5 gram mercuric oxide, and 200 c.c. water.

Harrisite. A mineral. It is a variety of chalcocite.

Harrison's Indicator. A small amount of starch boiled with a few c.cs. of water, adding to it 100 c.c. of a freshly prepared 10 per cent. potassium iodide solution.

Harstigite. A mineral. It is an aluminium-barium-potassium silicate.

Hartin. A white resin found in the lignite in Austria.

Hartite. A similar resin to Hartin, and found with it.

Hartman's Crimson Salt. A disinfectant. It consists of a mixture of manganates or permanganates, with other substances, such as potash alum, borax, and salt.

Hartmetall. See CAMITE.

Hartolan. Wool wax alcohols. (526).

Hartsalz. Occurs in the Stassfurt deposits. It is a mixture of Sylvine, Steinsalz, and Kieserite.

Hartshorn and Oil. Liniment of ammonia.

Hartshorn Oil. See BONE OIL.

Hartshorn Powder. Approximately prepared chalk.

Hartshorn Salt. See SALT OF HARTS-HORN.

Hartshorn Spirit. See SPIRIT OF HARTSHORN and VOLATILE ALKALI.

Hartzburgite. A mineral. It is a mixture of Olivine and Enstatite.

Harvesan. Mercurial seed dressing. (502).

Harvite. A proprietary trade name for a shellac compound.

Harz. Resin (rosin, collophany).

FLS Harz. A trade name for sulphonamide—formaldehyde resin. (3).

KM Harz. A trade name for a series of rosin-maleic-glycerine esters. (3).

Hascrome. A proprietary alloy. It is a manganese-chromium-iron welding rod.

Hashish. See BHANG.

Haskelite. A proprietary trade mark for a laminate of plywood and synthetic resin.

H.A. Solvent. A proprietary solvent. It is cyclohexanyl acetate. Its specific gravity is 0·947–0·98, flash-point 65–69° C., boiling range 170–180° C. It is a solvent for cellulose acetate and nitrate, rosin, rubber, oils, and metallic resinates.

Hastelloy Alloys. A registered trade mark for nickel base alloys. (a) A nickel-molybdenum-iron alloy usually containing 58 per cent. nickel, 20 per cent. molybdenum, 20 per cent. iron, and 2 per cent. manganese, is capable of being forged, cast, rolled into sheet, and welded. It is resistant to most acids except nitric acid, hot or cold. (b) A modification of alloy (a). It contains 58 per cent. nickel, 17 per cent. molybdenum, 6 per cent. iron, 14 per cent. chromium, and 5 per cent. tungsten. An acid-resisting alloy. (c) A nickel base alloy containing appreciable amounts of silicon and small amounts of copper and aluminium. It is resistant to hydrochloric and sulphuric acids.

Hastelloy B. Suitable for boiling HCl and HCl gas and has good resistance to sulphuric acid and phosphoric acid. (Ni/Mo).

Hastelloy C. An alloy capable of resisting corrosion from a wide variety of attacking agents. It can withstand strong oxidising agents and is also resistant to sulphuric, phosphoric, acetic, formic and sulphurous acids. (Ni-Mo-Cr).

Hastelloy C-276. A wrought version of HASTELLOY C, possessing improved working properties. (147).

Hastelloy D. Resists wear, abrasion and galling whether accompanied by corrosion or not. Excellent resistance to all concentrations of sulphuric acid even up to the boiling point. (147).

Hastelloy F. Withstands corrosion from oxidising and reducing acids and both both acid and alkaline solutions. (147).

Hastelloy G. A proprietary columbium-stabilized nickel-base alloy possessing good resistance to hot sulphuric and phosphoric acids. (147).

Hastelloy X. Possesses good creep and rupture properties, excellent forming characteristics and high resistance to oxidation up to 2200° F. (Ni-Cr-Mo-Fe). (714).

Hastingsite. A mineral. It is a variety of amphibole.

Hasting's Naphtha. Wood spirit (methyl alcohol).

Hata. See SALVARSAN.

Hatafa Bark. See BADAMIER BARK.

Hatchette Brown. See FLORENTINE BROWN.

Hatchettine. See BITUMEN.

Hatchettolite. A mineral. It consists chiefly of a tantalo-columbate of uranium, $U(Nb.Ta)O_3.H_2O$.

Hatchite. A mineral. It is a sulpharsenate of lead.

Hatter's Glue. See PARIS GLUE.

Hauchecornite. A mineral, $(Ni.Co)$. $(S.Bi.Sb)_3$.

Hauerite (Hauterite). A mineral. It is manganic sulphide, MnS_2.

Haughtonite. A mineral. It is a variety of meroxene.

Hausmannite (Red Oxide of Manganese). A mineral. It is manganese tetroxide, Mn_3O_4.

Haut de Cote. A term for arachis nuts in shells.

Hautefeullite. See BOBIERRITE.

Hauterite. See HAUERITE.

Hauyine. A mineral, $5(Na_2Ca)O.3Al_2 O_3.6SiO_2.SO_3$. It is also given as $(Na,Ca)_{4-8}Al_6Si_6O_{24}(SO_4)_{1-2}$.

Havapen. A proprietary preparation containing Penamecillin. An antibiotic. (245).

Haveg. A proprietary trade name for a moulded constructional material made from a phenolic type resin and special acid washed asbestos. It has a specific gravity of 1·6, is stated to be resistant to acids, bases, certain solvents and other chemicals and can be used up to 265° F.

Hawaiite. A mineral. It is a variety of olivine, a gem stone.

Haw, Black. The bark of *Viburnum prunifolium*.

Hawke's Anti-ferment (Meat Preserve Crystals). Preparations of sulphites, used as preservatives.

Haydenite. A mineral. It is a variety of chabazite. Also a proprietary trade name for a resin for waterproofing fabrics.

Hayem's Solution. A solution of 5 grams sodium sulphate, 1 gram sodium chloride, and 0·5 gram mercuric chloride in 200 c.c. water. Used in the examination of blood corpuscles.

Hayesin. See BORONATROCALCITE.

Hayphryn. A proprietary preparation of phenylephrine hydrochloride, and thenyldiamine hydrochloride. An anti-allergic nasal spray. (112).

Haylite No. 1. An English explosive. It contains 25–27 per cent. nitroglycerin, 0·5–1·5 per cent. collodion cotton, 19–21 per cent. potassium nitrate, 19–21 per cent. barium nitrate, 12–14 per cent. wood meal, 6–8 per cent. mineral jelly, and 10–12 per cent. ammonium oxalate.

Haylite No. 3. An explosive. It consists of 9·5 per cent. nitro-glycerin, 60 per cent. ammonium nitrate, 5 per cent. wood meal, 19·5 per cent. sodium chloride, and 5 per cent. ammonium oxalate.

Haynes Alloy No. 25. A proprietary trade name for an alloy of cobalt, nickel, tungsten and chromium. Possesses exceptional mechanical properties up to 1800° F

Haynes Metals. Alloys of from 10–75 per cent. iron, 20–30 per cent. chromium, and 5–25 per cent. cobalt. A harder alloy contains 45 per cent. cobalt, 40 per cent. tungsten, and 15 per cent. chromium, and a softer one, 62 per cent. cobalt, 28 per cent. tungsten, and 10 per cent. chromium. They are stated to be non-corrosive.

Haynon. A proprietary preparation of CHLORPHENIRAMINE. An anti-allergic drug. (368).

Hayphryn. A proprietary preparation of PHENYLEPHRINE hydrochloride and THENYLDIAMINE hydrochloride. An anti-allergic nasal spray. (685).

Haystellite. See CAMITE.

Hazeline. The trade name for an essence obtained from the fresh bark of *Hamamelis virginiana*, or witch-hazel. It is an oil used in the treatment of eczema, ulcers, and burns.

Hazol. A proprietary preparation of oxymetazoline hydrochloride. A nasal decongestant. (284).

HB-20. A proprietary trade name for a partially hydrogenated alkyl aryl hydrocarbon. (57).

HB-40. A proprietary trade name for a

hydrogenated terphenyl vinyl plasticiser. (57).

H.C. Copper. Copper containing 99.9 per cent. of the metal. Used for electrical work.

H.C.H. A pesticide containing mixed isomers of 1,2,3,4,5,6-hexachlorocyclo = hexane benzene hexachloride.

Heading. Saponine is used for making " heading " for beverages.

Head-light Oil. A kerosine of 150° C. fire test, and water-white in colour.

Heart Sugar. Inosite, $C_6H_6(OH)_6$.

Heavithane. A proprietary polyester-based polyurethane elastomer cross-linked by diols. (699).

Heavy Benzine. See LIGROIN.

Heavy Benzol. That fraction of the light oils of coal tar which distils over from 140–170° C.

Heavy Carburetted Hydrogen. See OLEFIANT GAS.

Heavy Lead Ore. A mineral. It is lead dioxide, PbO_2.

Heavy Oils (Dead Oils, Creosote Oils, Medium Oils). Fractions of coal tar, distilling over between 230 and 270° C. About 17 gallons are obtained from 1 ton of tar. Heavy oil is sold under the names of Creosote oil, Oil for fuel, Absorption oils, and Carbolineum.

Heavy Spar. See BARYTES.

Heavy Spar, Artificial. A pigment. It is barium white.

Heavy White. See BARYTES.

Heazlewoodite. A mineral. It is a nickel-iron sulphide.

H.E.B. " A ". A proprietary preparation of olive oil, glycerin, wax and water. Used in dermatology. (381).

Hebarol-sodium. A proprietary preparation of hexyl-ethyl barbiturate.

H.E.B. Burn Cream. A proprietary preparation of calamine, hamamelis water, glycerine, olive oil, wool fat, and liquid paraffin. (381).

Heberden's Ink. Mist. Ferri. Aromat.

H.E.B. Lac. A proprietary preparation of emulsified H.E.B. simplex in water. Used in dermatology. (381).

Hebler Powder. A Swiss explosive. It is a mixture of nitre, ammonium nitrate, sulphur, and charcoal.

H.E.B. " M ". A proprietary preparation of sulphur, resorcinol, salicylic acid and titanium dioxide used in dermatology as an antibacterial agent. (381).

Hebronite. A mineral. It is a variety of amblygonite.

H.E.B. Simplex. A proprietary preparation of liquid paraffin, paraffin wax and higher fatty alcohols in an emulsifying base. Used in dermatology. (381).

H.E.B. " S.S.". A proprietary preparation of salicylic acid, sulphur, and diphenyliodonium acetate used in dermatology as an antibacterial agent. (381).

Heckel's Antiseptic Solution. A solution containing sodium sulphite, benzoic acid, and water.

Hecla. A proprietary trade name for alloy steels with the following codings :

Hecla 35 contains 0.15 per cent. carbon, 0.25 per cent. silicon, 0.04 per cent. sulphur, 0.04 per cent. phosphorus and 0.70 per cent. manganese. Used to manufacture forged steel pressure vessels. (155).

Hecla 37 contains 0.04 per cent. carbon, 0.25 per cent. silicon, 0.04 per cent. sulphur, 0.04 per cent. phosphorus and 0.70 per cent. manganese and is used for the manufacture of forged steel pressure vessels. (155).

Hecla 115 contains 0.35 per cent. carbon, 0.25 per cent. silicon, 0.04 per cent. sulphur, 0.04 per cent. phosphorus, 0.7 per cent. manganese and 1.00 per cent. nickel and is used for manufacturing forged steel pressure vessels. (155).

Hecla 135 contains 0.60 per cent. carbon, 0.30 per cent. silicon, 0.30 per cent. manganese, 2.00 per cent. chromium, 2.00 per cent. nickel and 0.45 per cent. molybdenum. Used for the manufacture of forged high tensile steel pressure vessels. (155).

Hecla 138 contains 0.30 per cent. carbon, 0.30 per cent. silicon, 0.30 per cent. manganese, 2.00 per cent. chromium, 2.00 per cent. nickel and 0.45 per cent. molybdenum. It is used for the manufacture of forged high tensile steel pressure vessels. (155).

Hecla 138H contains 0.40 per cent. carbon, 0.30 per cent. silicon, 0.04 per cent. sulphur, 0.04 per cent. phosphorus, 0.60 per cent. manganese, 0.70 per cent. chromium, 2.70 per cent. nickel, 0.50 per cent. molybdenum and 0.25 per cent. vanadium (optional). It is used in the manufacture of forged high tensile steel pressure vessels. (155).

Hecla 155 contains 0.12 per cent. carbon, 0.20 per cent. silicon, 0.04 per cent. sulphur, 0.04 per cent. phosphorus, 0.60 per cent. manganese, 2.30 per cent. chromium, 1.00 per cent. molybdenum. It is used in the manufacture of forged steel pressure vessels for use at higher temperatures. (155).

Hecla 174 contains 0.30 per cent. carbon, 0.30 per cent. silicon, 0.04 per cent. sulphur, 0.04 per cent. phosphorus, 0.50 per cent. manganese, 5.00 per cent. chromium, 1.30 per cent. molybdenum and 0.90 per cent. vanadium. It is used for the manufacture of forged steel pressure vessels for hydrogen service and high tensile purposes. (155).

Hecla 180 contains 0·40 per cent. carbon, 0·30 per cent. silicon, 0·04 per cent. sulphur, 0·04 per cent. phosphorus, 0·60 per cent. manganese, 1·00 per cent. chromium, 3·2 per cent. nickel, 0·50 per cent molybdenum and 0·25 per cent. vanadium. It is used in the manufacture of forged high tensile steel pressure vessels. (155).

Hecla 306 contains 0·30 per cent. carbon, 0·20 per cent. silicon, 0·04 per cent. sulphur, 0·04 per cent. phosphorus, 0·50 per cent. manganese, 3·00 per cent. chromium, 0·50 per cent. molybdenum and 0·20 per cent. vanadium. It is used in the manufacture of forged steel pressure vessels for hydrogen service. (155).

Hecla 307 contains 0·18 per cent. carbon, 0·20 per cent. silicon, 0·04 per cent. sulphur, 0·04 per cent. phosphorus, 0·50 per cent. manganese, 3·00 per cent. chromium, 0·50 per cent. molybdenum and 0·20 per cent. vanadium. It is used in the manufacture of forged steel pressure vessels for hydrogen service. (155).

Hecla Powder. An explosive. It is similar in composition to Giant powder (q.v.).

Hectargyre. A mixture containing 0·1 gram of Hectine (q.v.), and 0·01 gram of mercury oxycyanide in each mil. Used medicinally.

Hectine. Sodium - benzo - sulpho - p - amino-phenyl-arsinate, $C_6H_5.SO_2.NH.C_6H_4.AsO(OH)(ONa)$. Used in the treatment of syphilis.

Hector Bases. Basic substances obtained by the oxidation of thioureas with hydrogen peroxide. They are stated to be good vulcanisation accelerators.

Hectorite Laponite. A sodium magnesium lithium fluoro silicate. A white, iron free suspending and gelling agent for aqueous systems.

Hedaquinium Chloride. Hexadecamethylenedi - (2 - isoquinolinium chloride). Teoquil.

Hedenbergite. A mineral, $CaO.FeO.2SiO_2$.

Hedgehog Crystals. Crystals of ammonium urate found in urinary deposits.

Hediorite (Hediosit). The lactone of α-gluco-heptonic acid, $C_7H_{12}O_7$. Recommended in the treatment of diabetes, in the place of sugar.

Hediosit. See HEDIORITE.

Hedonal. Methyl-propyl-carbinol-urethane, $NH_2.COO.CH(CH_3)(C_3H_7)$. A hypnotic used in cases of hysteria and neurasthenia.

Hedyphane. See MIMETISITE.

Heel Ball. Made by melting together 4 oz. mutton suet, 1 oz. beeswax, and ½ oz. olive oil, then ½ fluid oz. turpentine is added. When cooling add 1 oz. powdered acacia and ½ oz. lampblack.

Hegolit 3. A patented product which is a preparation of higher aliphatic alcohols, with a melting-point of 50° C. A plasticiser.

Hegonon (Argalbose). Ammonio-silver-nitrate-albuminose. It contains 70 per cent. silver, and is used locally in the treatment of gonorrhœa.

Heidenhain's Chrome Hæmatoxylin. A microscopic stain. It is produced by staining the object in 0·33 per cent. solution of hæmatoxylin in water, then soaking in 0·5 per cent. solution of potassium chromate.

Heidenhain's Iron Hæmatoxylin. (a) A solution of 4 grams iron alum in 100 c.c. distilled water. (b) A solution of 0·5 gram hæmatoxylin crystals in 100 c.c. distilled water. It is used as a microscopic stain by placing sections in (a) for 1 hour, washing, and then into (b) for ½ hour.

Heikkolite. Synonym for Heikolite.

Heikolite. A mineral. It is $2[Na_2(Mg,Fe)_3(Al,Fe)_2Si_8O_{22}(OH)_2]$.

Heiloy. A proprietary trade name for stainless steels used for dairy utensils.

Heintzite. A mineral, $Mg_4K_2B_{22}O_{38}.14H_2O$.

Helabon. Formerly Helamon, a mixture of 70 per cent. antipyrine with 16·7 per cent. pyramidone and 13·3 per cent. veronal.

Helcosal. See HELCOSOL.

Helcosol (Helcosal). Basic bismuth pyrogallate, $HO.C_6H_3O_2.Bi(OH)$. An internal antiseptic.

Heleco. A proprietary pyroxylin preparation.

Helenin (Alantin, Menyanthin, Dahlin, Synanthin, Sinistrin). An alanta lactone, $C_{72}H_{124}O_{62}$.

Helenite. See ELATERITE.

Helgoloid. An emulsion paint (German) made from alginic acid. (44).

Helgotan. See TANNOFORM.

Helianthi (Helianti, Salsefis). A fodder plant. It is *Helianthus macrophyllus*.

Helianthin. See ORANGE III. The name is also used for Citronine B (q.v.).

Helianti. See HELIANTHI.

Helicon. See ASPIRIN.

Heligoland Blue 3B. A direct cotton blue.

Heligoland Yellow. A dyestuff. It is the sodium salt of diphenyl-thio-urea-disazo-bi-phenol, $C_{25}H_{18}N_6SO_2Na_2$. Dyes cotton yellow.

Helindone Blue B. A dyestuff prepared by the oxidation of 9-chloro-3(2H)thiophanthrenone.

Helindone Blue 2B. A dyestuff. It is 5 : 5'-dibromo-indigo.

Helindone Fast Scarlet R. A dyestuff. It is 5 : 5'-dichloro-6 : 6'-di-ethoxythio-indigo.

Helindone Grey 2B. A dyestuff. It is 7 : 7'-diamino-thio-indigo.

Helindone Grey BR. A dyestuff. It is dichloro-7 : 7'-diamino-thio-indigo.

Helindone Orange D. A dyestuff. It is dibromo-6 : 6'-diamino-thio-indigo.

Helindone Orange R. A dyestuff. It is 6 : 6'-diethoxy-thio-indigo.

Helindone Pink BN. A dyestuff. It is 6 : 6'-dibromo-dimethyl-thio-indigo.

Helindone Red B. A dyestuff. It is 5 : 5'-dichloro-thio-indigo.

Helindone Red 3B (Durindone Red 3B). A dyestuff. It is 5 : 5'-dichloro-6 : 6'-dimethyl-thio-indigo.

Helindone Red R. A dyestuff prepared by the oxidation of thioindoxyl.

Helindone Scarlet S. A dyestuff. It is 6 : 6'-dithio-ethyl-thio-indigo.

Helindone Violet 2B. A dyestuff. It is dichloro-dimethyl-dimethoxy-thio-indigo.

Helindone Violet D. A dyestuff. It is methyl-indoxy condensed with isatin, and bromination.

Helindone Yellow 3GN. A dyestuff. It is a urea derivative of β-amino-anthraquinone.

Heliochrysin (Sun Gold). The sodium salt of tetranitro-α-naphthol.

Heliodor. A mineral. It is a clear golden-yellow variety of beryl. It occurs in pegmatites, in South-West Africa.

Helio Fast Red RL (Pigment Fast Red HL, Monolite Fast Scarlet R, RN, Sitara Fast Red, Lithol Fast Scarlet R, Permanent Red 4R, Pigment Fast Scarlet 3L Extra). A dyestuff prepared from diazotised m-nitro-p-toluidine, and β-naphthol.

Helio Fast Yellow RL. A dyestuff. It is the dibenzyl derivative of a mixture of 1 : 5 and 1 : 8-diamino-anthraquinones.

Heliolac. A French nitro-cellulose lacquer.

Heliophyllite. A mineral, $Pb_4As_2O_7.2PbCl_2$.

Heliopurpurin 4BL. A dyestuff prepared from β-naphthylamine-3 : 6-disulphonic acid, and α-naphthol-3 : 6-disulphonic acid. Used for the preparation of lakes.

Heliopurpurin 7BL. A dyestuff prepared from β-naphthylamine-1 : 6-di-sulphonic acid, and-β-naphthol-3 : 6-disulphonic acid. Used in the preparation of lakes.

Heliopurpurin GL. A dyestuff prepared from β-naphthylamine-3 : 6-disulphonic acid, and β-naphthol-3 : 6 : 8-trisulphonic acid. Used in the preparation of lakes.

Heliotrope. See BLOODSTONE.

Heliotrope B. A dyestuff. It is the sodium salt of dimethoxy-diphenyl-disazo-bi-ethyl-β-naphthylamine-β-sulphonic acid, $C_{38}H_{32}N_6S_8Na_2$. Dyes cotton reddish-violet from an alkaline bath.

Heliotrope 2B. A dyestuff. It is the sodium salt of diphenyl-disazo-α-naphthol-4 : 8-disulphonic-β-naphthol-8-sulphonic acid, $C_{32}H_{19}N_4S_3O_{11}Na_3$. Dyes cotton violet.

Heliotrope B, 2B. See TANNIN HELIOTROPE.

Heliotropes Modernes. Dyestuffs. They are gallocyanines, prepared from nitroso-mono-alkyl-anilines.

Heliotropine (Piperonal). The methylene ether of protocatechuic aldehyde, $CH_2O.O.CH_3.CHO$. It occurs native in the oil of the flowers of *Spiræa ulmaria*. It is made by converting safrol into isosafrol and oxidising the latter by means of sodium bichromate. It consists of snow-white crystals melting at 37° C., which are soluble in 70 per cent. alcohol to the extent of 5·2 per cent. at 10° C.

Helioxanthin. See ORANGE IV.

Helisen. Hay fever is caused by the inhalation of the pollen of grasses. The extract of different European grasses and flowers, which is injected subcutaneously in gradually increasing quantities, is stated to protect against hay fever. The extract is known as helisen.

Helkomen. The trade name for a basic bismuth-dibromo-hydroxy-naphthoate, containing 73 per cent. bromine. It is used as a substitute for iodoform.

Hellandite. A mineral. It is a silicate of yttrium, erbium, calcium, aluminium, and manganese.

Hellhoffite. An explosive. It is a solution of dinitro-benzene in nitric acid.

Helmacet. Aluminium oxynaphtholate.

Helmatac. A proprietary preparation of PARBENDAZOLE. A veterinary anthelmintic.

Helmet Brass. See BRASS.

Helmezine. Preparations of salts of piperazine. (510).

Helmitol (New Urotropine, Neurotropine, Citramine, Hexamol, Citramine Oxyphen, Formamol, Uropurgol). The anhydro-methylene-citrate of hexamethylene-tetramine, $C_7H_6O_7.(CH_2)_6$ N_4. Prescribed as an internal antiseptic in bladder diseases.

Ielthin. Potassium - amino - naphthol, $KO.C_{10}H_6NH_2$. A reagent for nitrites in water.

Ielvetan. A mineral. It is a variety of mica, found in Switzerland.

Ielvetia Blue. See METHYL BLUE.

Ielvetia Green (Acid Green). A dyestuff, $C_{23}H_{25}NSO_3$. It is prepared by the sulphonation of Malachite green, and is an obsolete acid dye

Ielvine. A mineral. It is $(Mn,Fe,Zn)_8.$ $Be_6Si_6O_{24}S_2$.

Ielvite. A mineral, $3SiO_2.6(Mn.Be.Fe)$ $O.(Mn.Fe)S$.

Helwig's Blood Solvent. A solution of 1 part potassium iodide in 4 parts water. It dissolves dried blood spots.

Hema-Combistix. A proprietary test strip comprised of four separate tests. (1) Methyl red and bromothymol blue for pH. (2) A buffered mixture of glucose oxidase, peroxidase, o-toluidine, and a red dye for glucose. (3) Buffered tetrabromophenol blue for glucose. (4) O-toluidine for blood. Used to test urine. (807).

Hemalbumin. A powder containing hæmatin, hæmoglobulin, serum albumin, paraglobulin, and the natural mineral salts. Employed in tubercular coughs treatment.

Hemastix. A proprietary test-strip containing O-tolidine and an organic peroxide, used to detect the presence of blood in urine. (807).

Hematest. A proprietary test tablet of o-toluidine and an organic peroxide, used to detect blood in fæces and urine. (807).

Hematine Paste and Powder. See LOGWOOD.

Hematite. Hæmatite (q.v.).

Hematolite. A mineral. It is $3[(Mn,Mg)_4(Mn,Al)AsO_4(OH)_8]$.

Hematophanite. A mineral. It is $4[Pb_5Fe_4(Cl,OH)_2O_{10}]$.

Hemcoware. A proprietary trade name for urea mouldings.

Hemellithenol. See HEMIMELLITHENOL.

Hemi-celluloses. These are contained in plant cell walls. They are reserve celluloses and are readily converted into hexoses and pentoses by acids and by the enzyme cytase.

Hemimellithenol (Hemellithenol). Trimethyl-phenol, $C_6H_2(CH_3)_3OH$.

Hemimorphite. A mineral. It is a hydrosilicate of zinc, $3ZnO.SiO_2.H_2O$. It is also given as $2[Zn_4Si_2O_7(OH)_2.$ $H_2O]$.

Heminevrin. A proprietary preparation of chlormethiazole edisylate. A hypnotic. (338).

Hemi-oxide of Copper (Protoxide of Copper, Red Oxide of Copper.) Cuprous oxide, Cu_2O.

Hemisine. See ADRENALIN.

Hemlock Gum or Pitch. See CANADA PITCH.

Hemlock Oil. See OIL OF HEMLOCK.

Hemogallol. A blood derivative made by the action of pyrogallol on the colouring matter of blood.

Hemol. A blood derivative made by the action of zinc-dust upon blood colouring matter. A tonic.

Arsenhemol, Bromhemol, Iodo hemol, Iodomercury hemol, and Copper hemol, are hemol preparations.

Hemol, Copper. See HEMOL.

Hemoplex. A proprietary preparation of Vitamin B complex. (329).

Hemostatin. Adrenalin.

Hemostyl. See COAGULOSE.

Hemp Bast, Artificial. Prepared by mixing zinc white, barium sulphate, or chalk, into a collodion, viscose, or other cellulose solution.

Hemp, Bimli. See GAMBO HEMP.

Hempel's Solution. A solution made by mixing a solution of 120 grams potassium hydroxide in 80 c.c. water with a solution of 5 grams pyrogallol in 15 c.c. water. Used in the absorption pipette for the determination of oxygen.

Hemp, Mauritius. See ALOE FIBRE.

Hemp Resin. Cannabinone, $C_8H_{12}O$.

Hemrids. A proprietary preparation of phenylephrine hydrochloride, amethocaine hydrochloride, bismuth carbonate and tyloxapol. Anæsthetic suppositories. (112).

Hemyphone. A hypnotic containing diacetyl - morphine - diallyl - barbiturate and trichloro - tertiary - butyl - alcohol (chlorbutol).

Hengleinite. A mineral. It is $4[(Ni,Fe,Co)S_2]$.

Henlex. A new cable insulation to replace rubber insulation. It is resistant to acids, alkalis, heat and cold, and has high tensile strength and dielectric resistance.

Henna. Derived from the leaves and roots of Lawsonia inermis or L. alba. Used as a dye, and for staining the hair.

Henrite. An explosive. It is a 33-grain powder.

Henryite. A mineral. It is a mixture of lead telluride and pyrites.

Hensler's Cement. A cement containing 3 parts litharge, 2 parts quick-lime, and 1 part white bole, ground up with boiled oil. Used for china and glass.

Henwoodite. A mineral, Al_4CuH_{10} $(PO_4)_8.6H_2O$.

Hepacon. A proprietary name for a group of preparations containing liver extract and/or vitamins of the B complex. (256).

Hepacort Plus. A proprietary preparation of heparin, hydrocortisone acetate, and methyl parahydroxybenzoate. Local anti-inflammatory cream. (323).

Hepadis. Distemper contagious hepatitis vaccine. (514).

Hepanorm. A proprietary preparation of liver extract and Vitamin B_{12}. (255).

Hepar Sulphur. Potassium sulphide.

Hepastab. A proprietary product. It is a specially prepared extract of liver for intra-muscular injection.

Hepatex. A proprietary liver extract.

Hepatic Aloes. Liver-coloured aloes.

Hepatic Cinnabar. See IDRIALITE.

Hepatite. A mineral. It is a variety of barite.

Heptabarbitone. 5 - (Cyclophept - 1 - enyl)-5-ethylbarbituric acid. Medomin.

Heptachlor. Heptachlorbenzypyrene. A powerful insecticide.

Heptalgin. A proprietary trade name for the hydrochloride of Phenadoxone.

Heptaline. 3 - Methyl - cyclo - hexanol, $CH_3.C_6H_{10}OH$. It has a boiling-point of 170–180° C., and is suggested as a solvent for resins.

Heptaminol. 6-Amino-2-methylheptan-2-ol.

Heptanal. Heptoic aldehyde, CH_3 $(CH_2)_5CHO$.

Heptene. A proprietary rubber vulcanisation accelerator. It is heptaldehyde-aniline.

Heptonal. A proprietary preparation of piperazine theophylline ethanoate, heptaminol hydrochloride and phenobarbitone. A bronchial antispasmodic. (349).

Herabol Myrrh. A true myrrh (a gum-resin). It is obtained from *Balsamodedron* species. Used in medicine as an aromatic.

Herapathite. Quinine sulphate periodide.

Herapath's Salt. Quinine iodo-sulphate.

Heratol. A solution of chromic acid in acetic acid or hydrochloric acid, and absorbed in kieselguhr. Used for the removal of phosphorus compounds from acetylene.

Herbohn Bronze. An alloy of 71 per cent. copper, 26 per cent. tin, and 3 per cent. zinc.

Herco. A proprietary trade name for a pine oil (*q.v.*).

Hercoflex. A proprietary trade name for vinyl plasticisers coded as follows:
Hercoflex 150. Octyl decyl phthalate.
Hercoflex 200. Diisooctyl phthalate.
Hercoflex 290. Octyl decyl sebacate. (93).

Hercofloc. Anionic and cationic types of varying high molecular weight synthetic polyelectrolytes. Used in solid/liquid separation processes. (93).

Hercolyn. A registered trade mark for hydrogenated methyl abietate, $C_{21}H_{34}$ O_2. It is a non-volatile, non-oxidising resin plasticiser.

Hercolyn D. A trade mark for deodorised methyl dihydroabietate. (93).

Hercose AP. A trade mark for cellulose acetate propionate. Used in lacquer manufacture.

Hercose C. A trade mark for cellulose acetobutyrate for use in lacquer manufacture.

Hercosol. A trade mark for a solvent made from pine oil.

Hercules. A trade mark for certain dynamite and gelatinous explosives, sporting and blasting powder, fuses and caps, turpentine and pine oil, rosin, nitro-cellulose, purified cotton, linter pulp, and a high boiling solvent for gums and cellulose esters.

Hercules Cement. See ACETIC-GELATIN CEMENT.

Hercules Metal. A bronze containing 85·5 per cent. copper, 10 per cent. tin, 2·5 per cent. aluminium, and 2 per cent. zinc. Another alloy contains 54 per cent. copper, 36 per cent. zinc, 7·5 per cent. iron, and 2·5 per cent. aluminium. The term is also used for an alloy of copper, nickel, and aluminium.

Hercules Stone. A mineral. It is magnetic iron ore.

Herculite. A proprietary trade name for a plate glass made to withstand high temperatures.

Herculoid. A proprietary manufacture of cellulose nitrate.

Herculoy. A patented alloy. It is a silicon bronze containing tin. It has high tensile strength and resists corrosion. Herculoy 418 contains copper with about 3 per cent. silicon and 0·5 per cent. tin.

Hercynite. A mineral, $FeO.Al_2O_3$.

Herderite (Allogonite, Glucinite). A mineral, $Be(OH.F)CaPO_4$.

Heremetal. A proprietary trade name for chimney liners coated with Heresite (*q.v.*). It is stated to resist acid corrosion encountered in gas firing.

Herepathite. A periodide of quinine sulphate, $(C_{20}H_{24}O_2N_2)_4 . (H_2SO_4)_3 . (HI)_2I_4.6H_2O$.

Heresite. A proprietary trade name for

a phenol-formaldehyde synthetic resin moulding compound.

Herkules. An alloy of 50 per cent. silver and 50 per cent. copper. Used for fuse wire.

Hermannotite. A mineral. It is columbite.

Hermann's Fluid (Platino-Aceto-Osmic Acid). A fixing agent used in microscopy. It contains 15 parts of a 1 per cent. platinum chloride solution, 4 parts of a 2 per cent. osmic acid solution, and 1 part of glacial acetic acid.

Hermesite. A mineral. It is a tetrahedrite containing mercury.

Hermite Fluid. A disinfectant, prepared by an electrolytic method. It contains magnesium oxide and free hypochlorous acid, with from 4–5 per cent. available chlorine.

Hermophenyl. Mercury-sodium-phenol-disulphonate, $(SO_3Na)_2C_6H_2O(Hg)$. It is an antiseptic used in the treatment of syphilis, and contains 40 per cent. mercury.

Heroin (Acetomorphine, Diamorphine). Diacetyl-morphine, $C_{17}H_{17}NO(C_2H_3O_2)_2$. Used as a substitute for morphine and codeine, for pulmonary catarrh.

Herolith. A synthetic resin of the phenol-formaldehyde type. See BAKELITE.

Herpid. A proprietary preparation of IDOXURIDINE and dimethyl sulphoxide used in the treatment of skin infections of the herpes type. (650).

Herrengrundite (Urvölgyite). A mineral, $CaSO_4.3Cu(OH)_2.CuSO_4.3H_2O$.

Herrerite. A mineral. It is a cupriferous variety of zinc carbonate.

Herschelite. A mineral. It is a variety of chabazite.

Herschel's Crystals. Hydrated calcium-tetrahydroxy-trisulphide, $Ca_3(OH)_4.S_3.8H_2O$.

Hertzberg's Stain. For paper. Solution A contains zinc chloride saturated at 70° C. Solution B contains 0·25 gram iodine and 5·25 grams potassium iodide in 12·5 c.c. water. A mixture of 25 c.c. of A and 25 c.c. of B is allowed to stand until clear before using it.

Herzenbergite. A mineral. It is 4[SnS].

Herzfeld's Honey. See ARTIFICIAL HONEY.

Hesperetic Acid. Isoferulic acid, $C_6H_3(OCH_3)(OH).CH.CH.COOH$.

Hesperidene (Citrene, Carvene) + or — Limonene, $C_{10}H_{16}$.

Hesperidine. See AURANTIIN.

Hessenbergite. A mineral. It is similar to Bertrandite.

Hessian Blue. See ANILINE BLUE, SPIRIT SOLUBLE.

Hessian Bordeaux. A dyestuff. It is the sodium salt of disulpho-stilbene-disazo-bi-α-naphthylamine, $C_{34}H_{24}N_6S_2O_6Na_2$. Dyes cotton bordeaux red.

Hessian Brown BB. A dyestuff. It is the sodium salt of diphenyl-disazo-bi-resorcinol-azo-benzene-sulphonic acid $C_{36}H_{24}N_8O_{10}S_2Na_2$. Dyes cotton brown.

Hessian Brown MM. A dyestuff. It is the sodium salt of ditolyl-disazo-bi-resorcinol-azo-benzene-sulphonic acid, $C_{38}H_{28}N_8O_{10}S_2Na_2$. Dyes cotton brown.

Hessian Purple B. A dyestuff. It is the sodium salt of disulpho-stilbene-disazo-bi-β-naphthylamine-sulphonic acid, $C_{34}H_{22}N_6O_{12}S_4Na_4$. Dyes cotton bluish-red from a soap bath.

Hessian Purple D. A dyestuff. It is the sodium salt of disulpho-stilbene-disazo-α-naphthylamine-sulphonic acid, $C_{34}H_{22}N_6O_{12}S_4Na_4$. It is isomeric with Hessian purple B. Dyes cotton bluish-red from a soap bath.

Hessian Purple N. A dyestuff. It is the sodium salt of disulpho-stilbene-disazo-bi-β-naphthylamine, $C_{34}H_{24}N_6O_6S_2Na_2$. Dyes cotton bluish-red from a soap bath.

Hessian Purple P. An azo dyestuff obtained from diazotised diamino-stilbene-disulphonic acid, and naphthionic acid, $C_{34}H_{26}N_6O_{12}S_4$.

Hessian Violet. A dyestuff. It is the sodium salt of disulpho-stilbene-disazo-α-naphthylamine-β-naphthol, $C_{34}H_{23}N_5O_7S_2Na_2$. Dyes cotton violet from a soap bath.

Hessian Yellow. A dyestuff. It is the sodium salt of disulpho-stilbene-disazo-bi-salicylic acid, $C_{28}H_{16}N_4O_{12}S_2Na_4$. Dyes cotton yellow from a neutral or acid bath.

Hessite (Tellurium Silver). A mineral. It is silver telluride, Ag_2Te.

Hessonite (Cinnamon-stone). A mineral. It is a calcium-aluminium garnet (a silicate), and contains in addition, small amounts of ferrous and ferric iron, manganese, and magnesium. A precious stone.

Hesthasulphid. A German proprietary brand of sodium sulphide. Used in leather manufacture.

Hetacillin. 6-(2,2-Dimethyl-5-oxo-4-phenylimidazolidin-1-yl)penicillanic acid.

Hetærolite. A mineral. It is 4[ZnMn_2O_4].

Hetairite. Synonym for Hetærolite.

Heterobrochantite. A mineral. It is a variety of antlerite.

Heterocline. See MARCELINE.

Heterogenite. A mineral. It is a hydrous oxide of cobalt.

Heteromerite. A mineral. It is a variety of vesuvianite.

Heteromorphite. A mineral, PbS. $4Sb_2S_3$.

Heteronium Bromide. 1-Methylpyrrolidin - 3 - ylα - phenyl - α - (2 - thienyl) - glycollate methobromide.

Heteronucleal. α-Naphthol-disulphonic acid, $C_{10}H_5(OH)(SO_3H)_2$.

Heterosite. A mineral. It is $4[(Fe^{\cdots},Mn^{\cdots})PO_4]$.

Hetocresol. The m-cresol ester of cinnamic acid $C_6H_5.CH : CH.COOC_6H_4 CH_3$. A dusting powder, used in the treatment of external tuberculosis.

Hetoform. Basic bismuth cinnamate, $(C_6H_5.CH : CH.CO_2)_3.Bi.Bi_2O_3$.

Hetol. The trade name for synthetic sodium cinnamate, usually in the form of a dilute solution. It is prescribed for pulmonary tuberculosis.

Hetol Caffeine. Caffeine-sodium-cinnamate. A general stimulant.

Hetrazan. A proprietary preparation of diethylcarbamazine dihydrogen citrate. A filaricide. (306).

Hetralin. Resorcinol - hexamethylene - tetramine. Used in the treatment of gonorrhœa.

Hetroform. A hexamine benzoate. Used in medicine.

Hetron. A proprietary polyester laminating resin. (37).

Heulandite. A mineral, $Al_2O_3.3SiO_2$. $CaO.5SiO_2$.

Heusler Alloy. An alloy of 66–68 per cent. copper, 18–22 per cent. manganese, 10–11 per cent. aluminium, and 0–4 per cent. lead.

Heveacrumb. A proprietary compressed rubber crumbled in oil. (700).

Heveaplus MG. A proprietary graft copolymer of methacrylate with natural rubber. (701).

Heveaplus MG-30. A proprietary graft polymer. (702).

Heveaplus MG-49. A proprietary graft polymer. (702).

Heveatex. The trade name to denote a series of preserved, concentrated, or processed Hevea rubber latices.

Hevitan. A German product. It is a vitamin preparation from yeast, and contains vitamin B. Used in the treatment of anæmia.

Hewettite. A mineral. It is a hydrated vanadate of calcium, $CaO.3V_2O_5.9H_2O$.

Hex. See HEXAMINE.

Hexa. A rubber vulcanisation accelerator. It is hexamethylene-tetramine.

Hexacarb. Black dye for leather. (551).

Hexacarbacholine Bromide. Carbolonium bromide.

Hexacert. U.S.A. certified foodstuff colourant. (551).

Hexachlorophane. Di - (3,5,6 - tri - chloro - 2 - hydroxyphenyl)methane. Hexachlorophene.

Hexachlorophene. A proprietary trade name for Hexachlorophane.

Hexacide. Insect and vermin killer. (551).

Hexacol. Foodstuff colourants of guaranteed specification. (551).

Hexaderm. Fast dyes for leather. (551).

Hexadimethrine Bromide. Poly - (NNN′N′ - tetramethyl - N - tri - methylenehexamethylenediammonium dibromide. Polybrene.

Hexagonite. A variety of the mineral tremolite.

Hexal (Hexalet, Sulphexine, Sulphexet). Hexamethylene - tetramine - salicyl - sulphonic acid, $C_6H_{12}N_4 . (OH)C_6H_3 (COOH)SO_3H$.

Hexalan. Fast to light colours for wool. (551).

Hexalet. See HEXAL.

Hexalin. A registered trade name for cyclohexanol, $C_6H_{11}OH$. It has a boiling range of 155–170° C., a flashpoint of 68° C., and a specific gravity at 15° C. of 0·944. A solvent.

Hexalin Acetate. See ADRONAL ACETATE.

Hexallac. Dyes for cellulose lacquers. (551).

Hexamecoll. A preparation of guaiacol and hexamethylene-tetramine. A disinfectant.

Hexamethonium Bromide. Hexamethylenedi(trimethylammonium bromide). Vegolysen.

Hexamethonium Iodide. Hexamethylenedi - trimethylammonium iodide). Hexathide.

Hexamethonium Tartrate. A preparation used in the treatment of low blood pressure. A hypotensive. It is hexamethylene di(trimethylammonium hydrogen tartrate).

Hexamethylamine. Hexamethylene - tetramine, $C_6H_{12}N_4$.

Hexamethyl - para - rosaniline. See CRYSTAL VIOLET.

Hexamic Acid. Cyclamic acid sweetening agents. (Now banned in some countries). (525).

Hexamine (Cystamin, Cystogen, Metramine, Urotropine, Formamine, Formin, Naphthamine, Xametrin, Vesaloin, Urisol, Uritone, Hex, H.M.T., Formaldehyde-ammonia 6 : 4, Carin, Ammonioformaldehyde, Vesalvine). Hexamethylene-tetramine, $C_6H_{12}N_4$. A uric acid solvent.

Hexamine Azurine 5G. A dyestuff. It is a British brand of Brilliant azurine 5G.

examine Hippurate. A 1 : 1 complex of hexamine and hippuric acid.

examol. See HELMITOL.

exanatrin. A combination of hexamethylene-tetramine and acid sodium sulphate.

exanhexol. Mannite, $C_6H_{14}O_6$.

exanitrin. Mannitol-hexanitrate, $CH_2O(NO_2)(CHO.NO_2)_4.CH_2ONO_2$.

exanon. Cyclohexanone.

exapar. Detergent preparations. (551).

exaplas BUT. A proprietary trade name for poly-1 : 3-butylene adipate. A polymeric plasticiser. (2).

exaplas PLA. A medium viscosity modified polypropylene adipate migration resistant plasticiser. (2).

exaplas PPA. A proprietary trade name for polypropylene adipate. A vinyl plasticiser. (2).

exaplas PPS. A proprietary polymeric plasticiser comprising polypropylene sebacate. (2).

exaplus PPL. A proprietary polymeric plasticiser comprising polypropylene laurate. (2).

exapropymate. 1 - (Prop - 2 - ynyl) - cyclohexyl carbamate. Merinax.

exasilane. Silicohexane (hexasilicon-tetradecahydride), Si_6H_{14}.

exasol. Alcohol soluble dyes. (551).

exathide. A proprietary trade name for Hexamethonium iodide.

exatype. Dyes for doubletone printing inks. (551).

exazole. 3 - Ethyl - 4 - cyclohexyl - 1,2,4 - triazole. Azoman.

exela. Cellulose acetate and nylon dyes. (551).

Hexetidine. 5 - Amino - 1,3 - di - (2 - ethylhexyl)hexahydro - 5 - methyl - pyrimidine. Sterisil.

Hexeton. A synthetic camphor compound. It is 3-methyl-5-isopropyl-Δ-2 : 3-cyclo-hexane. It is a liquid suggested for use in medicine in a similar way to camphor.

Hexil. Hexanitro-diphenylamine.

Hexine. α-Diallyl, C_6H_{10}.

Hexnitrol. Leather stains. (551).

Hexo. Alkaline cleaner. (546).

Hexobendine. A coronary vasodilator. 1, 2 - Di - [N - methyl - 3 - (3, 4, 5 - tri - methoxybenzoyloxy) - propylamino] - ethane.

Hexocyclium Methylsulphate. 4 - (β - Cyclohexyl - β - hydroxyphenethyl) - 1,1 - dimethylpiperazinium methylsulphate. Tral.

Hexoil. Oil and varnish dyes. (551).

Hexopal. A proprietary preparation of inositol nicotinate. A peripheral vasodilator. (112).

Hexophan. Hydroxy-phenyl-quinoline-

dicarboxylic acid, $C_{17}H_{11}NO_5.H_2O$. Used in the treatment of gout.

Hexoprenaline. A bronchodilator. NN'-Di - [2 - (3, 4 - dihydroxyphenyl) - 2 - hydroxyethyl] - hexamethylenediamine. "ST 1512" is currently under clinical trial as the dihydrochloride.

Hexaprofen. An anti-inflammatory, anti-pyretic and analgesic currently undergoing clinical trial as "BTS 13 622". It is 2-(4-cyclohexylphenyl) - propionic acid.

Hexoran (Trioran). Emulsions of carbon tetrachloride and trichloro-ethylene. Used for degreasing wool.

Hexsotate. Hexamethylene-tetramine-sodio-acetate.

Hexylresorcinol. A urinary antiseptic.

Heyn's Reagent. The double chloride of copper and ammonia. Used to reveal ferrite in the micro-analysis of carbon steels.

HF(2). A proprietary histidine-free food used in the dietary treatment of histidinæmia. (665).

Hg. 33. 2-Myristoxymercuri-3-hydroxy-benzaldehyde, $(C_6H_3(CHO)(OH)(HgO.CO.C_{13}H_{27}))$. Used for the treatment of syphilis and leprosy.

Hi. Nylon detergent brightener. (552).

Hiawa Gum (Resin of Conima). A resin obtained from *Protium heptaphyllum*. Used for incense.

Hibbenite. A mineral. It is a basic zinc phosphate.

Hibbo. A proprietary aluminium bronze containing iron. It has a high tensile strength, hardness, high fatigue, ductility, and toughness. A corrosion resistant. The metal is produced in a number of grades.

Hibiscrub. A proprietary preparation of CHLORHEXIDINE gluconate in a detergent base. An antiseptic cleansing solution. (2).

Hibitane. A proprietary preparation of chlorhexidene digluconate, an antiseptic cream. (2).

Hibschite. A mineral. It is a silicate of copper and aluminium.

Hibosol. High boiling point solvents. (578).

Hibudine. A proprietary trade name for a synthetic rubber, probably derived from butadiene.

Hicore 90. A nickel-chromium-molybdenum case-hardening steel for heavy motor vehicles and other gears.

Hicoseen. A proprietary preparation of diethylaminoethyl phenylbutyrate, codeine phosphate and guaicol albuminate. A cough linctus. (396).

Hiddenite. A mineral. It is a green variety of spodumene (a silicate of lithium and aluminium). A gem stone.

Hide Glue. A glue made from the fresh

waste from tanneries and slaughter-houses, by first placing scraps of hide, tails, and feet of oxen and sheep, with milk of lime for three weeks, washing, and heating with water to extract glue.

Hide Powder Substitute. Formalin-gelatin, obtained by soaking filter paper in gelatin, drying, then exposing it to formaldehyde.

Hidrofugal. A German preparation. It consists of aluminium sulphate, aluminium chloride, aluminium acetate, resorcinol, and chlorthymol dissolved in water. A yellow, odourless, non-drying liquid used for excessive sweating by rubbing into the skin.

Hiduminium. A registered trade name for a range of aluminium alloys. (232).

Hidurax. A proprietary trade name for a range of corrosion resistant aluminium bronzes. (158).

Hiduron 191. A cupronickel which has been strengthened by the addition of aluminium. (158).

Hieratite. A mineral. It is a potassium-silicon fluoride.

Higginsite. A mineral, $2CuO.2CaO.As_2O_5$, found in Arizona.

High Brass. See BRASS.

Highgate Resin. Copalin, a resin from the blue clay of Highgate Hill.

High-speed Tool Steels. These alloys usually contain in addition to iron, 13–18 per cent. tungsten, 3–5 per cent. chromium, 0·5–2 per cent. vanadium, 0·6–0·77 per cent. carbon, 0·14–0·5 per cent. manganese, 0·1–0·4 per cent. silicon, 0·02 per cent. sulphur, and 0·02 per cent. phosphorus. Some contain 2–5 per cent. cobalt, and a few 0·18 per cent. nickel. Some German steels of this type contain 0·7 per cent. molybdenum. Also see STELLITE.

High Tensile Brass. An alloy of 76 per cent. copper, 22 per cent. zinc, and 2 per cent. aluminium.

Hightensite. A trade name for a proprietary hard rubber composition for use in electrical insulation.

Hi-heet. An American synthetic resin-moulded product for electrical insulation.

Hiirogane. A blood-red coloured metal prepared either by the treatment of copper with a solution of copper sulphate and verdigris, or by heating a copper alloy with a paste containing a salt of copper, borax, and water.

Hilgardite. A mineral. It is $Ca_8B_{18}O_{33}Cl_4.4H_2O$.

Hilgenstockite. A calcium phosphate, $4CaO.P_2O_5$, found in the basic slag of the Thomas-Gilchrist process for the dephosphorisation of iron.

Hillebrandite. A mineral of Mexico. I is stated to be a hydrated calcium silicate

Hills-McCanna Alloy No. 45. A alloy containing 88 per cent. copper 10·5 per cent. aluminium, and 1·5 pe cent. iron.

Hilomid. A proprietary preparation o TRIBROMSALAN. A veterinary anthel mintic.

Himaizol. A proprietary food produc containing vegetable oils used in th treatment of hypercholesterolæmia. (665).

Hingra. An asafœtida sold in th Bombay market.

Hinsdalite. A mineral. It is simila to Alunite, and has the composition $2PbO_3.Al_2O_3.P_2O_5.6H_2O$.

Hiortdahlite. A mineral, $3CaSiO_3.Ca (F.OH)Na.ZrO_3$, of Norway.

Hipernick. A registered trade mark fo an alloy of nickel and iron in equa parts. It has an initial permeability ten times higher than that of iron and a maximum permeability up to 50,000 Used for making cores of audio-fre quency transformers.

Hipersil. A registered trade mark fo high permeability silicon steel. It i used in high-frequency communications equipment.

Hippol. Methylene-hippuric acid, C_6H_5 $CO.N(CH_2).CH_2.CO.O$. Used as an antiseptic in inflammation of the bladder.

Hippo Wine. Ipecacuanha wine.

Hippuryl Amide. N-benzoyl-glycina-mide. $C_6H_5.CO.NH.CH_2.CO.NH_2$. A white powder. A substrate in studies of papain action. (182).

Hiprex. A proprietary preparation of HEXAMINE hippurate. A urinary anti-septic. (275).

Hirathiol. A compound used as a substitute for Ichthyol (q.v.).

Hircalite. A trade name for an asbestos-slate product.

Hirudine. The substance of the leech, which prevents blood from clotting. Used medicinally.

Hirudoid. A proprietary preparation of organo-heparoid. Surgical application. (391).

Hishilen. A preparation containing mercury thio-salicylate. Used in medicine.

Hisingerite. A mineral, $2Fe_2O_3.7H_2O. 3SiO_2$.

Hislopite. A mineral. It is a calcium carbonate coloured with glauconite.

Histadyl E.C. A proprietary preparation containing codeine phosphate, ephedrine hydrochloride, thenylpyramine fumarate, ammonium chloride and chloroform. A cough linctus. (333).

Iistalix. A proprietary preparation of DIPHENHYDRAMINE hydrochloride, ammonium chloride, sodium citrate, chloroform and menthol. An expectorant. (255).

Iistalog. The hydrochloride of 3 - (2 - aminoethyl)pyrazole. (Ametazole hydrochloride). (333).

Iistamine. See ERGAMINE.

Iistantin. A proprietary preparation of chlorcyclizine hydrochloride. (Antihistamine). (277).

Iistazarin. 2 : 3 - Dihydroxy - anthraquinone.

Iistex. A proprietary preparation of diphenhydramine hydrochloride. (279).

Iistidine. α-Amino-β-imino-azole-propionic acid.

Histo-Acryl. A proprietary preparation of ENBUCRILATE. A surgical tissue adhesive.

Histofax. A proprietary preparation of chlorcyclizine and calamine, used for allergic skin disorders. (277).

Histogenol. A registered trade mark for a combination of nucleic acid and sodium-methyl-arsenate. Used in the treatment of tuberculosis. (232).

Histosan. Triguaiacol-albuminate, prepared by the action of an alkaline solution of guaiacol upon egg albumin. Used in the early stages of tuberculosis.

Histostab. The mesylate of Antazoline.

Histrixite. A mineral, $7Bi_2S_3.2Sb_2S_3.5CuFeS_2$.

Histron Balm. A proprietary preparation of histamine dihydrochloride, heparin, methyl nicotinate glycol salicylate, and oleoresin capsicum. An embrocation. (380).

Histryl. A proprietary preparation of diphenpyraline hydrochloride. (281).

Hitchcockite. A mineral. It is a variety of plumbogummite.

Hitenso. A proprietary trade name for a cadmium bronze.

Hitox. A proprietary trade mark for titanium dioxide. (703).

Hittorf's Phosphorus. Black phosphorus (q.v.).

Hi-Zex. A trade mark for Japanese high density polyethylene. See also Rigidex. (60).

Hive Powder. Mercury with chalk.

Hjelmite. A mineral. It consists of the tantalates of calcium, iron, and manganese.

H.M.T. See HEXAMINE.

H.M.T.D. Hexamethylene - triperoxy - diamine. A new detonating explosive.

Hochst New Blue. A dyestuff. It is the calcium salt of the di- and trisulphonic acids of trimethyl-triphenyl-pararosaniline. Trisulphonic acid, C_{40}

$H_{34}N_3O_{10}S_3Na_3$. Dyes wool blue from an acid bath.

Hock. A light wine containing 6–8 per cent. alcohol.

Hodgkinsonite. A mineral. It is a zinc-manganese silicate. It is $MnZn_2SiO_5.H_2O$.

Hoeferite. A mineral. It is a ferric silicate.

Hoenle's Cement. A cement, consisting of 2 parts shellac, and 1 part Venice turpentine.

Hoernesite. A mineral. It is a magnesium arsenate, $Mg_3(AsO_4)_2.8H_2O$.

Hoevelite. A mineral. It is sylvine.

Hoffmannite. A mineral. It is lolingite.

Hoffmann's Anodyne (Anodyne drops). (*Spiritus ætheris compositus B.P.*) Compound spirit of ether, containing a solution of heavy oil of wine (a mixture of ethylene sulphate, ethyl sulphate, and ether), in ether and alcohol.

Hoffner's Blue. See COBALT BLUE.

Hofmann's Blue. A pigment, $KFe(Fe(CN)_6)+H_2O$.

Hofmann's Violet (Iodine Violet, Dahlia, Primula, Violet 4RN, Red Violet 5R Extra, Violet 5R, Violet R, Violet RR). A dyestuff. It is a mixture of the hydrochlorides and acetates of the mono-, di-, or trimethyl (or ethyl)-rosaniline, and pararosanilines. Hydrochloride of triethyl-rosaniline, $C_{26}H_{32}N_3Cl$. Dyes wool, silk, and mordanted cotton violet.

Hog Gum. See BASSORA GUM.

Högomite. A mineral. It is $(MgFe^{..})_7(Al,Fe^{...})_{20}TiO_{39}$.

Hohmannite. A mineral. It is amarantite.

Hokutolite. A mineral. It is a mixture of anglesite and barytes.

Holadin. A dried extract of the entire pancreas, containing all the constituents of the glands, and having the power to digest starch and proteins.

Holdenite. A mineral from Franklin, N.J. It is an arsenate of manganese and zinc, $8MnO_2.4ZnO.As_2O_5.5H_2O$.

Holfos Bronze. An alloy of copper with 11–12 per cent. tin, 0·25 per cent. lead, and 0·1–0·2 per cent. phosphorus.

Holite. A proprietary trade name for a synthetic resin for moulding and laminating.

Holken Silk. An artificial fibre prepared from Maco cotton.

Hollandite. A manganese ore of essentially the same composition as Psilomelane (q.v.).

Holmesite. A mineral. It is clintonite.

Holmite. A mineral. It is a calcium carbonate and silicate.

Holocaine (Phenocaine). p-Diethoxyethenyl - diphenyl - amidine - hydro - chloride, $CH_3 . C(N . C_6 H_4 O . C_2 H_5)$ $(NH . C_6H_4O . C_2H_5)$. Used by oculists as a local anæsthetic, in the place of cocaine.

Holoklastite. See PETROCLASTITE.

Holopon. A trade name for a preparation stated to contain all the alkaloids of opium.

Homac. A proprietary synthetic resin.

Homadrenine. See ARTERENOL.

Homalourea. Dipropyl-barbituric acid.

Homatropine. The tropine ester of mandelic acid, $C_6H_5 . CH(OH) . CO_2$. $C_2H_3 . C_5H_8NCH_3$. Used for dilating the pupil of the eye.

Homberg's Metal. A fusible alloy, containing 3 parts bismuth, 3 parts tin, and 3 parts lead. It has a melting-point of $122°$ C.

Homberg's Phosphorus. Anhydrous calcium chloride, $CaCl_2$, which, when fused, and exposed to the sun, becomes phosphorescent in the dark.

Homberg's Salt. Boric acid, H_3BO_3.

Homesurma. A product consisting mainly of allyl mustard oil, used for the prevention of mould growth on the surface of fodder.

Homidium Bromide. A trypanocide. 3, 8 - Diamino - 5 - ethyl - 6 - phenyl - phenanthridinium bromide. ETHIDI-VIN.

Homilite. See HOMILITH.

Homilith. A mineral, $FeO . 2CaO . B_2O_3$. SiO_2.

Homoantipyrin. $1 : 2 : 3$-Phenyl-methyl-ethyl-pyrazolone, $C_{12}H_{14}ON_2$.

Homobarbital. See PROPONAL.

Homocamphin. Hexeton (q.v.).

Homochlorcyclizine. $1 - (4 - Chlorodi - phenylmethyl) - 4 - methylhomopiper - azine.

Homoflavine. $2 : 7$-Dimethyl-$3 : 6$-di-amino - acridinium - methylo - chloride hydrochloride, $(CH_3C_6H_2NH_2)_2CH.N.$ $CH_3Cl.HCl.$ An antiseptic.

Homoguaiacol. See CREOSOTE-GUAI-ACOL.

Homokol. A sensitiser for silver bromide plates. It is a mixture of Quinoline red with an isocyanine dye.

Homolle's Digitalin. See DIGITALIN, AMORPHOUS.

Homophan. See PARATOPHAN.

Homophosphine G. An acridine dye-stuff, belonging to the same class as Acridine orange R extra.

Homorenon (Paradrin). Ethyl-amino-aceto - pyrocatechol - hydrochloride, $(HO)_2 C_6H_3 . CO . CH_2 . NH(C_2H_5) . HCl.$ Used in medicine as a substitute for adrenaline.

Homosaligenin. Methyl-hydroxy-ben-zene.

Homprenorphine. An analgesic currently undergoing clinical trial as " R & S 5205—M ". It is N-cyclopropyl-methyl - 7, 8 - dihydro - 7α - [$1(R)$-hy-droxy-1-methylpropyl] - O^3O^6-dimethyl-*endo*ethenonormorphine.

Honduras Silk. See ARGHAN FIBRE.

Hondurite. A proprietary moulded composition, with cotton and a vulcanised binder, for electrical insulation.

Honestone. See WHETSTONE.

Honey. A mixture of dextrose and levulose, together with a little sucrose, dextrines, flavouring matter, pollen, and a small quantity of free formic acid.

Honey Balsam. Oxymel of squill.

Honey-dew. A viscid, sugary material, occasionally met with in the leaves of the maple, lime, and black alder.

Honey, Herzfeld's. See ARTIFICIAL HONEY.

Honey Stone. See MELLITE.

Honey Sugar. See GLUCOSE.

Honthin. A Hungarian tannalbin (albu-min tannate). An intestinal astrin-gent.

Honvan. A proprietary preparation of stilbœstrol diphosphate. Treatment of prostatic carcinoma. (319).

Hoochinoo. An alcoholic drink prepared by North American Indians. It contains from 20–25 per cent. alcohol.

Hoof Meal. See HORN MEAL.

Hoof Oil. Neatsfoot oil (q.v.).

Hooker Brass. See BRASS.

Hooker's Green. See CHROME GREENS.

Hoolamite. A mixture of 53–56 per cent. fuming sulphuric acid, 10–12 per cent. iodine pentoxide, and 33–35 per cent. pumice stone. Used for the detection of carbon monoxide, which it oxidises to carbon dioxide, with the liberation of iodine. The iodine reacts with the sulphur trioxide to give a green colour.

Hopcalite I. A mixture of 50 per cent. manganese dioxide, 30 per cent. copper oxide, 15 per cent. cobalt oxide, and 5 per cent. silver oxide. Used as a catalyst to oxidise the carbon monoxide in air. It has been employed in gas masks.

Hopeite. A mineral, $Zn_3PO_4 . xH_2O$.

Hop-flour. A sticky, yellow dust, attached to the lower inside base of the bracts of the hop cone. It is the flavouring material of the hop.

Hopkin's-Cole Reagent. To a litre of a saturated solution of oxalic acid 60 grams of sodium amalgam are added, and the mixture allowed to stand. It is then filtered and diluted with from 2–3 volumes of water. Used for the detection of proteins.

Hopkin's-Cole Reagent (Benedict's

modification). Powdered magnesium (10 grams) are placed in a conical flask and shaken up with enough water to cover the magnesium; 250 c.c. of a cold saturated solution of oxalic acid are added slowly, and the whole shaken and poured on to a filter. The filtrate is acidified with acetic acid and made up to a litre. Used to detect proteins.

Hopkin's-Cole Tyrosin Reagent. This contains mercuric sulphate dissolved in a solution of sulphuric acid.

Hopkin's Lactic Acid Reagent. Thiophene.

Hopogan. The main constituent of this substance is magnesium peroxide (magnesium perhydrol), MgO_2.

Hopol Powder. A proprietary preparation for making liquid metal polish, by adding a turpentine substitute to it.

Hopper Salt. Sodium chloride, which has been caused to crystallise in large hollow cubes which float when alum is added to the bath.

Hop Substitute. Quassiin, a bitter principle, said to be used as a hop substitute. Picric acid is also said to be used for a similar purpose.

Horak Glass. A glass manufactured in Czechoslovakia. It is made from sand, boric acid, potassium carbonate, sodium carbonate, zirconia, and titanium oxide.

Horbachite. A native sulphide of gold, iron, and nickel.

Horco X. A proprietary trade name for a thermosetting polyvinyl butyral synthetic resin.

Hordeine. See HORDENINE.

Hordenine (Hordeine, Anhaline). *p*-Oxy-phenyl-dimethyl-ethylamine, $HO.C_6H_4.CH_2CH_2.N(CH_3)_2$, obtained from barley malt. It retards coagulation of the blood, and raises the blood pressure. Used in dysentery, and as a tonic.

Hormofemin. A proprietary preparation of dienœstrol, in a cream base used to treat senile vaginal conditions. (250).

Hormonal. A preparation made from the spleen of an animal, used in medicine for constipation.

Hormonin. A proprietary preparation of œstriol, œstrone, and œstradiol. Œstrogen replacement. (830).

Hornbergite. A mineral. It is a uranium arsenate.

Hornblende. A mineral. It is a constituent of a variety of rocks, $3(Mg.Fe)O.CaO.SiO_2$.

Hornesite. A mineral, $Mg_3(AsO_4)_2.8H_2O$.

Horn Fibre. See VULCANISED FIBRE.

Horn Lead. See KERASITE.

Horn Meal (Hoof Meal). A nitrogenous manure, obtained from hoofs, horns, and claws, by treating them with superheated steam, drying, and grinding. It contains on an average about 14 per cent. nitrogen.

Horn-quicksilver. A mineral. It is mercurous chloride, HgCl (calomel).

Horn Silver (Chlorargyrite). A mineral. It consists of silver chloride, AgCl.

Hornstone. A mineral. It is a variety of jasper.

Horse Aloe. See ALOE.

Horse Brimstone. Black sulphur (*q.v.*).

Horse-flesh Ore. See PURPLE COPPER ORE.

Horse Oil. A yellow-brown oil from horse fat. Used in the manufacture of soaps.

Horse Tincture. Friar's balsam.

Horsfordite. A mineral, Cu_6Sb.

Hortonolite. A mineral, $(Fe.Mg.Mn)_2SiO_4$. It is also given as $4[(Fe,Mg)_2SiO_4]$.

Hosal. A German preparation. It contains the calcium double salts of polyamic acids, and has a taste similar to that of salt. It is used for addition to food when salt-free diet is necessary, e.g. in epilepsy, kidney, and heart troubles, and high blood pressure. It is not added to butter.

Hostacain. A proprietary preparation of butanilicaine phosphate, procaine phosphate, and adrenaline. A local dental anæsthetic. (312).

Hostadur. A trade name for partially crystalline thermoplastic polyester based on ethylene terephthalate used for construction of rigid components, e.g. gears. (9).

Hostaflon. A trade mark for polytrifluormonochlorethylene. (9).

Hostaflon C2. A proprietary polychlorotrifluoroethylene. (9).

Hostaflon ET. A proprietary ethylenetetrafluoroethylene copolymer. (9).

Hostaflon TF. A trade mark for polytetrafluorethylene. (9).

Hostaform. Highly crystalline acetal copolymer. (9).

Hostaform C. A trade name for acetal plastics. (9). See also DELRIN. (10). CELCON. (11). POLYFYDE. (12).

Hostalen. A trade mark for a high density polythene. (9). See also RIGIDEX. (12).

Hostalen GM. A proprietary polyethylene resin used for covering pipes and wires. (9).

Hostalen GP. A proprietary polyethylene resin used for making film, monofil and blow-moulded containers. (9).

Hostalen OO. A trade mark for isotactic polypropylene. A thermoplastic moderately rigid moulding material. (9).

Hostalen PP. Crystalline polypropylene. (9).

Hostalit. A trade mark for PVC in sheet form.

Hostalit Z. A trade mark for a blend of PVC and chlorinated polyolefin. A thermoplastic used for manufacturing pipes. (9).

Hostamid. A proprietary trade name for a group of nylons. (9).

Hostaphane. A proprietary trade name for polyethylene terephthalate film. (9).

Hotfoil GH. A proprietary waterproof heating tape sheathed in chlorosulphonated polyethylene and carrying 4 watts per foot. It is used for heating pipes carrying resins and other materials which are normally solids at room temperature. (704).

Houghite. See HYDROTALCITE.

Houillite. Anthracite.

Howard's Silver. Mercury fulminate, used for percussion caps.

Howes Algæcide. Algæcide for swimming pools. (535).

Howflex. A proprietary trade name for a group of plasticisers. (705).

Howflex BB. A proprietary plasticiser. Tri (dimethyl cyclohexyl) oxalate. (705).

Howflex CA. Cyclohexyl adipate. (705).

Howflex CBS. A proprietary plasticiser. Cyclohexyl benzene sulphonamide. (705).

Howflex CC. A proprietary plasticiser. Tricyclohexyl citrate. (705).

Howflex CP. A proprietary plasticiser. Cyclohexyl phthalate. (705).

Howflex CS. A proprietary plasticiser. Cyclohexyl stearate. (705).

Howflex DAP. A proprietary plasticiser. Dialkyl phthalate (C7-C9). (705).

Howflex SA. A proprietary plasticiser. Sextyl adipate. (705).

Howflex SB. A proprietary plasticiser. Sextyl sebacate. (705).

Howflex SP. A proprietary plasticiser. Sextyl phthalate. (705).

Howflex SS. A proprietary plasticiser. Sextyl stearate. (705).

Howlite. A mineral, $CaH_3(BO_2)_5.CaH_2SiO_4$.

Howsorb. Sorbitol syrup aqueous solution. (513).

Howtol. Cyclic alcohols. (513).

Hoyle's Metals. Bearing metals, usually containing about 46 per cent. tin, 12 per cent. antimony, and 42 per cent. lead.

Hoyt Metal. A proprietary trade name for an antimonial lead containing 6–10 per cent. antimony.

H-scale. A proprietary trade name for a synthetic pearl essence.

Hsihutsunite. A mineral. It is $10[(Mg,Mn)_7Si_8O_{22}(OH)_2]$.

H.T.S. A salt mixture used as a heat transfer medium. It contains approximately 40 per cent. sodium nitrite, 7 per cent. sodium nitrate, and 50 per cent. potassium nitrate by weight. The temperature limits are 290–1000° F.

Huantajayite. A mineral (Na.Ag)Cl.

Huascolite. A mineral. It is a lead-zinc sulphide.

Huber's Alloy. A pyrophoric alloy containing 85 per cent. cerium and 15 per cent. magnesium.

Huber's Reagent. A solution of ammonium molybdate and potassium ferrocyanide. Used for the detection of free mineral acids.

Hubl's Reagent. (a) Iodine (50 grams), dissolved in 1 litre of 95 per cent. alcohol. (b) Mercuric chloride (60 grams), dissolved in 1 litre of alcohol. Used for obtaining the iodine value of fats and oils.

Hubnerite. A mineral. It consists mainly of a tungstate of manganese, $MnWO_4$.

Hudsonite. A mineral. It is a variety of pyroxene.

Hugel A. A proprietary solution of the sodium salt of an acrylic copolymer, used for thickening natural and synthetic rubber latices. (417).

Hugel AH. A proprietary viscous solution of an acrylic copolymer. (417).

Hugel B. A proprietary trade name for emulsions of acrylic copolymer containing free carboxyl groups. (417).

Hugel BC 10. A proprietary viscous solution, colourless and odourless, of the sodium salt of an acrylic polymer. (417).

Hugel CH14. The sodium salt of a proprietary acrylic polymer. It is very viscous, and takes the form of a short non-stringy yellow gel. (417).

Hügelite. A hydrated zinc-lead vanadate mineral.

Hullite. A black chloritic mineral.

Hulot's Solder. An alloy of 37·5 per cent. tin, 37·5 per cent. lead, and 25 per cent. zinc amalgam. Used for soldering aluminium bronze.

Hulsite. A mineral, $12(Fe.Mg)O.2Fe_2O_3.SnO_2.2B_2O_3.2H_2O$.

Humagel. A proprietary preparation of paromycin sulphate, kaolin and pectin. An antidiarrhœal. (264).

Human Coagulation Fraction (II, IX and XI). A preparation of human blood containing coagulating factors II, IX and XI, used in the treatment of hæmophilia B deficiency.

Humatin. A proprietary preparation containing paromomycin (as the sulphate). An antibiotic. (264).

Humboldtilite. See MELILITE.

Humite. See CHONDRODITE.

Humogen. Bacterised peat.

Humorsol. A proprietary trade name for demecarium bromide.

Hünefeld Solution. This contains 25 c.c. alcohol, 5 c.c. chloroform, 1·5 c.c. glacial acetic acid, and 15 c.c. turpentine. It is used for the detection of blood.

Hungary Blue. Cobalt blue (q.v.).

Hungary Green. Malachite green (q.v.).

Hunterite. A mineral. It is a variety of cimolite.

Huntilite. A mineral, Ag_2As.

Huntite. A mineral. It is $2[Mg_3Ca(CO_3)_4]$.

Huppert's Reagent. A 10 per cent. aqueous solution of calcium chloride used for the detection of biliary pigments in urine.

Hureaulite. A mineral, $5(Mn.Fe)O.2P_2O_5.5H_2O$.

Hurka. See DUKA.

Hurlbutite. A mineral. It is $4[Ca,Be_2(PO_4)_2]$.

Huron. An aluminium alloy containing from 3·5–6·6 per cent. copper and small amounts of manganese, magnesium, and chromium. The name is also used for a chromium steel containing 12·5 per cent. chromium, 1·0 per cent. vanadium, and 0·2 per cent. carbon. It is used for dies.

Hurr Nut. Myrobalans.

Hurtig's Wood Composition. A preparation obtained by mixing sawdust with a strong solution of curd soap dissolved in water, drying, treating with milk of lime, and again drying. It is then mixed with casein and slaked lime, dried, and pressed.

Husman Metal. An alloy of 74 per cent. tin, 11 per cent. antimony, 10·6 per cent. lead, 4 per cent. copper, 0·22 per cent. iron, and 0·18 per cent. zinc.

Hussakite. See XENOTINE.

Hutchinsonite. A mineral. It is a sulph-arsenite of thallium, lead, silver, and copper. $PbS(Tl.Cu.Ag)_2S.2As_2S_3$. A mineral. It is also given as $8[(Tl,Pb)_2AgAs_5S_{10}]$.

Hüttenbergite. A mineral. It is a variety of lolingite.

Huttonite. A mineral. It is $4[ThSiO_4]$.

Huxham's Tincture. Tinct. Cinchonæ Co.

Huyssenite. A mineral. It is $(Mg,Fe)_6B_{14}O_{26}Cl_2$.

HVA-2. NN'-m-phenylenedimaleimide. A free radical regulator used as an auxiliary in the curing of HYPALON. (10).

Hversalt. Synonym for Halotrichite.

Hyacinth (Jacinth, Jargon, Matura Diamond). Minerals. They are clear varieties of zircon (a silicate of zirconium). Gem-stones.

Hyacinth, Ceylon. A mineral. It is garnet.

Hyacinth-geranium Oil. An oil obtained by distilling geraniol over hyacinth flowers.

Hyacinth, False. A mineral. It is garnet.

Hyalase. A proprietary trade name for an enzyme which depolymerises hyaluronic acid. Hyaluronidase.

Hyaline. A mineral. It consists of alumina, Al_2O_3. It is also the name for a mixture of gun-cotton and resins, a vulcanite substitute.

Hyalite (Water Opal, Muller's Glass). A mineral. It is a colourless and transparent variety of opal (q.v.).

Hyalithe. A black glass manufactured in Bohemia.

Hyalophane. A mineral, $BaO.K_2O.Al_2O_3.8SiO_2$.

Hyalosiderite. A mineral. It is a deep olive green olivine used as a gem.

Hyalotekite. A mineral. It is a barium-calcium-lead silicate.

Hyaluronidase. Enzyme which depolymerises hyalyronic acid. Hyalase. Rondase. Wydase.

Hyasorb. A proprietary preparation containing Benzylpenicillin (as the potassium salt). An antibiotic. (366).

Hyb-lum. An alloy of aluminium in which the alloying elements, consisting of about 2 per cent., are mainly nickel and metals of the chromium group. Used for reflectors of therapeutic lamps.

Hycal. A proprietary preparation of dextrose and related compounds, used as a high-calorie diet supplement. (364).

Hycar 1203X17. A proprietary 70 : 30 blend of HYCAR medium acrylonitrile rubber and GEON P.V.C. resin. (102).

Hycar 1204X5. A proprietary 100 : 70 : 120 pre-fluxed blend of HYCAR medium-high acrylonitrile rubber, GEON P.V.C. resin and a phthalate plasticiser. (102).

Hycar 1204X9. A proprietary 100 : 70 : 100 pre-fluxed blend of HYCAR medium-high acrylonitrile rubber, GEON P.V.C. resin and a phthalate plasticiser, protected by non-staining stabilisers. (102).

Hycar 1205X3. A proprietary 50 : 50 : 60 pre-fluxed blend of HYCAR medium-high acrylonitrile rubber, GEON P.V.C. resin and a phthalate plasticiser. (102).

Hycar 1273. A proprietary 70 : 30 blend of carboxy-modified HYCAR acrylonitrile rubber and GEON P.V.C. resin, possessing good resistance to abrasion. (102).

Hycar 1402 H82. A proprietary acrylonitrile-butadiene copolymer in powder form, having a medium-high content of acrylonitrile. (706).

Hycar 1402 H83. A proprietary acrylonitrile-butadiene copolymer in powder form, having a high content of acrylonitrile. (706).

Hycar 1403 H84. A proprietary acrylonitrile-butadiene copolymer having a medium acrylonitrile content. It is supplied in powdered form and used when good behaviour at low temperatures is required. (706).

Hycar 2100. A proprietary range of polyacrylic-solution polymers used as pressure-sensitive adhesives. (706).

Hycar 2550H33. A proprietary reinforced styrene-butadiene copolymer rubber used in the manufacture of foam rubber. (706).

Hycar 2550H5. A proprietary aqueous, anionic dispersion of a cold-polymerised styrene-butadiene copolymer. (706).

Hycar 2550H55. A proprietary aqueous, anionic dispersion of a styrene-butadiene copolymer used in the foam-backing of carpets. (706).

Hycar 2570H28 and 2570H29. A proprietary group of aqueous anionic dispersions of self-reactive styrene-butadiene copolymers, used in the making of carpet backings. (706).

Hycar 2570X5. A proprietary aqueous, anionic dispersion of a carboxy-modified styrene-butadiene copolymer reactive to heat. It is used for leather finishes and adhesives. (706).

Hycar 2671H49. A proprietary aqueous anionic dispersion of a heat-reactive carboxy-modified acrylic polymer used in the making of surgical rubber materials. (706).

Hycar 4021. A proprietary copolymer of ethyl acrylate having a small percentage of 2-chloro-ethyl vinyl ether. (102).

Hycar 4032. A proprietary polyacrylic rubber used to make rubber seals and gaskets. (706).

Hycar 4043. A proprietary acrylic rubber which remains flexible at minus 40° C, but which also gives good resistance to oil at high temperatures. (102).

Hycar 4201. A proprietary copolymer of ethyl acrylate having a small percentage of 2-chlorovinyl ether. (102).

Hycar ATBN. A proprietary amine-terminated butadiene-acrylonitrile liquid polymer used in the curing and modifying of epoxy resins. (102).

Hycar VTBN. A proprietary vinyl-terminated butadiene-acrylonitrile copolymer. It is a liquid polymer used for modifying polyesters, polystyrene, etc. (102).

Hycar EP. A proprietary butadiene-styrene copolymer containing 50 per cent. styrene (cf. 25 per cent. in GR–S). It is fawn in colour. Also known as Hycar S–10. It is not oil resistant.

Hycar OR. A proprietary trade name for oil-resisting synthetic rubber.

Hycathane. A proprietary polyurethane elastomer. (707).

Hychlorite. A disinfectant. It is anti-formin.

Hycol. A disinfectant. It contains phenolic bodies, and gives emulsions with water. (552).

Hycolin. Disinfectants and antiseptics. (552).

Hycoloid. A proprietary trade name for a cellulose nitrate composition used for container manufacture.

Hycon. Photographic developer. (507).

Hycran. A registered trade mark for a group of nitrile resins. (512).

Hydan. Chlorinated organic bleach. (509).

Hydantil. A proprietary preparation of methoin and phenobarbitone. An anticonvulsant. (267).

Hydeltracin. A proprietary preparation of prednisolone and neomycin sulphate, used to treat inflammatory skin lesions. (310).

Hydergine. A proprietary preparation of mesylates of dihydroergocornine, dihydroergocristine and dihydroergokryptine. Peripheral vasodilator. (267).

Hydnestryle. Ethyl hydnocarpate.

Hydnocreol. Ethyl hydnocarpate with 4 per cent. creosote.

Hydracetin. See PYRODINE.

Hydrach. Arsenious oxide, As_4O_6. The inhabitants of Styria eat it under the above name, to increase their endurance.

Hydralin. Methyl-cyclo-hexane, a dry-cleaning agent. The name is also applied to a methyl-hexalin soap. See SAVONADE.

Hydrallazine. 1-Hydrazinophthalazine. Hydralazine.

Hydramin. A combination of p-phenylene-diamine and quinol.

Hydramyl. Pentane, C_5H_{12}.

Hydranthrene Blue RS Paste. A dyestuff. It is a British equivalent of Indanthrene X.

Hydranthrene Dark Blue. A dyestuff. It is a British brand of Indanthrene dark blue BO.

Hydranthrene Olive R. A British equivalent of Indanthrene olive G.

Hydranthrene Yellow AGR. A British dyestuff. It is equivalent to Algol yellow WG.

Hydranthrene Yellow ARN, AG New. Dyestuffs. They are British brands of Algol yellow R.

Hydraphthal. A preparation containing 90 per cent. tetralin, 5 per cent. ammonium oleate, and 3 per cent. water. A wetting-out and scouring agent.

Hydrargaphen. Phenylmercury 2,2'-dinaphthylmethane - 3,3' - disulphonate. Penotrane.

Hydrargillite (Gibbsite, Claussenite). A mineral. It is a hydrated alumina, $Al_2O_3.3H_2O$.

Hydrargol. Mercury succinimide (C_2H_4 $(CO)_2.N)_2Hg$. Used in medicine, in the treatment of pulmonary tuberculosis.

Hydrargotin. Mercury tannate, Hg_2. $3(C_{14}H_9O_9)OH$, containing 50 per cent. mercury. Used in the treatment of syphilis.

Hydrargyrol. The mercury-potassium salt of thymol-p-sulphonic acid, (C_6H_4. $OK.SO_3)_2Hg$. Used in medicine.

Hydrargyroseptol. Quinoline-mercury-sodium chloride. An antiseptic.

Hydrarsan. A solution of a mixture of arsenious, mercuric, and potassium iodides, and antipyrine.

Hydrastis. The dried rhizome and roots of *Hydrastis canadensis*.

Hydrated Cellulose. See ALKALI CELLULOSE.

Hydrated Manganese Ore. See VARVICITE.

Hydrated Zinc Oxide. Zinc hydroxide, $Zn(OH)_2$.

Hydratene. A proprietary preparation of paracetamol and chloral hydrate. A hypnotic. (361).

Hydraulic Bronze. An alloy of 83 per cent. copper, 5 per cent. lead, 5 per cent. zinc, 5 per cent. tin, and 2 per cent. nickel. Another alloy contains 83 per cent. copper, 10·8 per cent. tin, 6 per cent. zinc, and 0·1 per cent. lead. Also see KERN'S HYDRAULIC BRONZE.

Hydraulic Cements or Mortars. These are prepared by calcining mixtures of calcium carbonate with from 10–30 per cent. clay. Tricalcium silicate, $3CaO.SiO_2$, and tricalcium aluminate, $3CaO.Al_2O_3$, are formed.

Hydraulic Limes. Limes containing from 15–30 per cent. clayey matter (aluminium silicate). They are made by burning impure limestones at a low temperature. They slake in water, but show hydraulic properties.

Hydraulic Mortar. See HYDRAULIC CEMENTS.

Hydrazine Yellow. See TETRAZINE.

Hydrea. A proprietary preparation of hydroxyurea, a carcino-chemotherapeutic agent. (326).

Hydrenox. A proprietary preparation of hydroflumethiazide. A diuretic. (253).

Hydrenox-M. A proprietary preparation of hydroflumethiazide. A diuretic. (253).

Hydrex (Betex). Hydrogenated fish oils. They are largely replacing stearic acid for activating and plasticising rubber in the U.S.A.

Hydridin. A substance stated to be identical with hydroxy-dihydro-iso-indigotin.

Hydril. A proprietary trade name for Hydrochorothiazide.

Hydriodic Ether. Ethyl iodide, C_2H_5I.

Hydriodol. See CYPRIDOL.

Hydriol. A proprietary preparation of iodised poppy-seed oil.

Hydro-Adreson. A proprietary preparation of hydrocortisone acetate. (316).

Hydroapatite. A mineral, $3Ca_3(PO_4)_2$. $CaF_2.3H_2O$.

Hydrobasaluminite. A mineral. It is $Al_4SO_4(OH)_{10}.nH_2O$ with $n \backsimeq 36$.

Hydrobol. Water repellent finishes. (543).

Hydroboracite. A mineral which occurs in the Stassfurt deposits, CaB_2O_4. $MgB_2O_4.6H_2O$.

Hydrobromic Ether (Bromhydric Ether). Ethyl bromide, C_2H_5Br.

Hydrobuna. A name applied to hydrogenated rubber.

Hydrocalcite. A mineral. It is a hydrated calcium carbonate.

Hydrocalumite. A mineral. It is $8[Ca_2Al(OH)_7.3H_2O]$.

Hydrocarbon. An artificial bitumen produced by the distillation of asphalt. It is used for the prevention of blooming in rubber compounds.

Hydrocarbon-aldehyde Resins. Resins obtained by the interaction of hydrocarbons such as naphthalene with formaldehyde in the presence of sulphuric acid. Other aldehydes may be used. They are resins of the oil-soluble type.

Hydrocarbon Black. See GAS BLACK.

Hydrocarbon Cement. A cement made by mixing heated pitch or tar with from one to four times its volume of calcium or magnesium sulphate. Used for paving or building purposes.

Hydrocarbon Gas Black. See GAS BLACK.

Hydrocarbon Oil. See PARAFFIN, LIQUID.

Hydrocarbon Rubber. A patented rubber substitute, containing oxidised oil, rosin, and sulphur. The name must not be confused with " rubber hydrocarbon," which is used for pure rubber freed from resin and dirt.

Hydrocerussite. A mineral, $2PbCO_3$. $PbO.H_2O$.

Hydrochinin. Hydroquinone (*q.v.*).

Hydrochloric Ether. Ethyl chloride, C_2H_5Cl.

Hydrochlorthiazide. 6 - Chloro - 3,4 - dihydrobenzo - 1,2,4 - thiadiazine - 7 - sulphonamide-1,1-dioxide. Dichlotride; Direma; Esidrex; Hydril; Hydrodiuril; Hydrosaluric; Salupres.

Hydroclinohumite. A mineral. It is $2[Mg_9Si_4O_{16}(OH)_2]$.

Hydrocodone. A cough suppressant. 7, 8 - Dihydro - O^3 - methylnormor - phine.

Hydrocortisone. $11\beta,17\alpha,21$-Trihydroxy-pregn-4-ene-3,20-dione. 17-Hydroxy-corticosterone. Cortef; Cortifoam; Cortril; Ef-Cortelan; Hydro-Adreson; Hydrocortistab; HydroCortisyl; Hydrocortone.

Hydrocortistab. A proprietary preparation of hydrocortisone. (253).

Hydrocortisyl. A proprietary preparation of hydrocortisone acetate. (307).

Hydrocortone. A proprietary preparation of hydrocortisone. (310).

Hydrocuprite (Vanadium Ochre). A mineral, $Cu_2O.H_2O$.

Hydrocyanic Acid, Powdered. See POWDERED HYDROCYANIC ACID.

Hydrocyanite. A mineral. It is copper sulphate, $CuSO_4$.

Hydroderm. A proprietary preparation of hydrocortisone, neomycin sulphate and bacitracin used in dermatology as an antibacterial agent. (310).

Hydrodiuril. A proprietary trade name for Hydrochlorothiazide.

Hydrodolomite (Hydromagnocalcite). A mineral. It is a hydrated calcium-magnesium carbonate.

Hydroergotinine. Ergotoxine.

Hydrofit. A proprietary polyester-based polyurethane elastomer. (708).

Hydroflumethiazide. 3,4 - Dihydro - 6 - trifluoromethylbenzo - 1,2,4 - thia - diazine-7-sulphonamide 1,1-dioxide. Di-Ademil; Hydrenox; Naclex; Rontyl.

Hydroforsterite. A mineral. It is $MgSiO_4.2H_2O$.

Hydrogen, Dicarburetted. See OLE-FIANT GAS.

Hydrogenite. A mixture of 5 parts ferro-silicon, 90–95 parts silicon, 12 parts sodium hydroxide, and 4 parts slaked lime. When ignited it yields hydrogen.

Hydrogen Peroxide, Solid. See HY-PEROL.

Hydrogen, Phosphoretted. See PHOS-PHINE.

Hydrogen Rubeanide. The amide of dithio-oxalic acid, $CS(NH_2)CS(NH_2)$.

Hydrogen-Uranospinite. A mineral. It is $HUO_2AsO_4.4H_2O$.

Hydro-glockerite. A mineral. It is a basic ferric sulphate, $2Fe_2O_3.SO_3.8H_2O$.

Hydrohæmatite. See TURITE.

Hydrohalite. A mineral. $NaCl.2H_2O$.

Hydroherderite. A mineral. It is $4[CaBePO_4OH]$.

Hydroilmenite. A mineral. It is a partially altered ilmenite of Sweden.

Hydrokinone. Hydroquinone (q.v.).

Hydrolaine. A solution of rubber in turpentine and alcohol. Oil of worm-wood is added. Used for water-proofing.

Hydrolene. See BLOWN PITCH.

Hydrolete. Calcium hydride, CaH_2.

Hydrolites. See WATER-STONES.

Hydrolith. A 90 per cent. calcium hydride which yields hydrogen on con-tact with water. It is used for filling military balloons.

Hydromagnesite. A mineral. It is a basic carbonate of magnesium, $3Mg CO_3.Mg(OH)_2.3H_2O$.

Hydromagnocalcite. See HYDRO-DOLOMITE.

Hydromanganocalcite. A mineral. It is a hydrated calcium carbonate con-taining manganese.

Hydromelanothallite. A mineral, $CuCl_2.CuO.2H_2O$.

Hydromet. A proprietary preparation of methyldopa and hydrochlorothiazide. Used to control hypertension. (310).

Hydromorphinol. 7,8 - Dihydro - 14 - hydroxymorphine. 14-Hydroxydi-hydromorphine. Numorphan Oral.

Hydromorphone. 7,8-Dihydromorphi-none.

Hydromycin-D. A proprietary prepara-tion of prednisolone and neomycin used in dermatology as an antibacterial agent. (253).

Hydronal. (1) A polymerised product of pyridine and chloral. A hypnotic.

Hydronal. (2) Specially pure (+)-hydroxy citronellaldehyde. (504).

Hydronalium. An aluminium alloy containing from 7–9 per cent. magne-sium and small amounts of silicon and manganese. It is resistant to sea-water, soap, and soda.

Hydronaphthol. β-naphthol, $C_{10}H_7OH$.

Hydron Blue. A dark-blue vat dye, obtained by reducing nitroso-phenol with carbazole to form an indophenol, which is heated with sodium sulphide, and subsequently with sulphur. Also used as a trade mark for a range of other dyestuffs. (983).

Hydrone. An alloy of 35 per cent. sodium and 65 per cent. lead, which generates hydrogen by action of water.

Hydronepheline. A mineral, $2Na_2O.3Al_2O_3.6SiO_2.7H_2O$.

Hydronyx. A proprietary trade name for sodium sulphoxylate.

Hydropalat A. A registered trade name

for diethyl-hydro-phthalate, a solvent for nitro-cellulose. It has a specific gravity of 1·0955 and flashes at 131° C.

Hydropalat B. A registered name for dibutyl-hydro-phthalate, a solvent for nitro-cellulose. It has a specific gravity of 1·005 and flashes at 152° C.

Hydrophane. A mineral. It is an opal.

Hydrophilite. A mineral. It is calcium chloride, $CaCl_2$.

Hydrophite. A Swedish dark green Serpentine.

Hydropirin. Sodium-acetyl-salicylate.

Hydroplumbite. A mineral, $3PbO.H_2O$.

Hydropyrine. See APYRON.

Hydroquinone. p-Dihydroxy-benzene, $C_6H_4(OH)_2$. A photographic developer.

Hydro-resin A. A proprietary trade name for a water soluble resin.

Hydro Rubber. A product, $C_{10}H_{20}$, obtained by adding platinum to a benzene solution of rubber, and precipitating it with alcohol. It is a colourless, sticky mass.

Hydros. A registered trade mark for hyposulphite of soda, specially prepared for use as a reducing agent in vat colour dyeing. Sodium hydrosulphite. (503).

Hydros 1. Sodium hydrosulphite and sodium pyrophate. (503).

Hydrosaluric. A proprietary preparation of hydrochlorothiazide. A diuretic. (310).

Hydrosaluric-K. A proprietary preparation of HYDROCHLORTHIAZIDE with potassium chloride. A diuretic with a potassium supplement. (472).

Hydrosept. A preparation of p-toluene-sulphon-chloro-amide (chloramine).

Hydrosol. An aqueous colloidal silver solution.

Hydrospray. A proprietary preparation of hydrocortisone, propadrin and neomycin. A nasal spray. (310).

Hydrosulphite (Rongalite, Hyraldite, Decroline, Redo, Blanchite, Blankit). Hydrosulphites used in dyeing, and for decolorising sugar syrups The active principle is sodium hydrosulphite, $Na_2S_2O_4$, or sodium hydrosulphite-formaldehyde (sulphoxylate), $NaHSO_2$. $CH_2O.2H_2O$.

Hydrosulphite A. A 10 per cent. solution of hydrosulphite NF, or hyraldite. Used for testing dyed fabrics.

Hydrosulphite A.W. See DISCOLITE.

Hydrosulphite B. Prepared by acidifying 200 c.c. of hydrosulphite A with 1 c.c. acetic acid. Used for testing dyed fibres.

Hydrosulphite BASF. A 90 per cent. sodium hydrosulphite, $Na_2S_2O_4$.

Hydrosulphite NF. (Rongalite Conc., Brittalite, Formosul). A condensation product of formaldehyde and sodium hydrosulphite. It consists of a mixture of formaldehyde - sodium bisulphite, $NaHSO_3.CH_2O$, and formaldehyde - sodium - sulphoxylate, $NaHSO_2.CH_2O$. Used as a discharger in calico printing.

Hydrosulphite NF Conc. See HYRALDITE C EXT.

Hydrotalc. See PENNINE.

Hydrotalcite (Houghite, Völkernite). A mineral, $MgCO_3.5Mg(OH)_2.2Al(OH)_3$ $4H_2O$.

Hydroterpin. A mixture of hydrogenated terpene hydrocarbon and tetralin. Used as a wash material for solvents, more particularly for washing benzol.

Hydrothorite. A mineral. It is thorium silicate.

Hydrotitanite. A mineral. It is an altered perowskite.

Hydrous Silica. See OPAL.

Hydrous Wool Fat. See LANOLIN.

Hydrovit. A proprietary trade name for Hydroxocobalamin.

Hydroxal. A proprietary oral antacid consisting of a suspension of aluminium hydroxide. (709).

Hydroxamethocaine. 2-Dimethyl.aminoethyl 4-n-butylaminosalicylate.

Hydroxine Yellow G, L, L Conc- Dyestuffs. They are British equivalents of Tartrazine.

Hydroxocobalamin. α-(5,6-Dimethylbenzimidazolyl)hydroxocobamide. Hydrovit; Neo-Cytamen.

Hydroxyl - Herderite. Synonym for Hydroherderite.

Hydroxyacetone. Acetol. Ketol, $CH_3.CO.CH_2.OH$.

Hydroxyamphetamine. A sympathomimetic and mydriatic. 4-(2-Aminopropyl)phenol. PAREDRINE.

Hydroxyapatite. Calcium - hydroxy - hexaphosphate, $3Ca_3(PO_4)_2.Ca(OH)_2$. It is also given as $2[Ca_5(PO_4)_3OH]$.

Hydroxycarbamide. See HYDROXYUREA.

Hydroxychloroquine. 7 - Chloro - 4 - [4 - (N - ethyl - N - 2 - hydroxyethylamino) - 1 - methylbutylamino] - quinoline. Plaquenil is the sulphate.

Hydroxydione Sodium Succinate. Sodium 21-hydroxypregnane-3,20-dione succinate. Presuren; Viadril.

Hydroxy-Fluor-Apatite. A mineral. It is $2[Ca_5(PO_4)_3(F,OH)]$.

Hydroxymimetite. A basic lead arsenate made artificially.

Hydroxyprogesterone. 17α-Hydroxy-pregn-4-ene-3,20-dione. Primolut Depot is the caproate.

Hydroxypethidine. Ethyl 4-(3-hydroxyphenyl) - 1 - methylpiperidine - 4 - carboxylate.

Hydroxyprocaine. 2 - Diethylaminoethyl 4-aminosalicylate.

Hydroxystilbamidine. 4,4'-Diamidino-2-hydroxystilbene.

Hydroxytetryl. Trinitro - hydroxyphenyl-methyl-nitro-amine.

Hydroxytoluic Acid. An analgesic. 2-Hydroxy - m - toluic acid. 3 - Methyl - salicyclic acid.

Hydroxyurea. An anti-neoplastic agent. HYDROXYCARBAMIDE.

Hydroxyzine. 1 - (4 - Chloro - α - phenyl-benzyl) - 4 - [2 - (2 - hydroxyethoxy) - ethyl]piperazine. Atarax is the hydrochloride.

Hydrozincite (Zinc Bloom, Zinconise, Zinconite, Cegamite). A mineral. It is a carbonate of zinc, $3ZnCO_3.2H_2O$.

Hydrozone (Glycozone, Pyrozone). Trade names for hydrogen peroxide, H_2O_2, used as an antiseptic in dental practice.

Hydryl. The product of the interaction between mercuric oxide and Orsudan (q.v.).

Hydurol. See ASUROL.

Hyflux. Soldering fluxes. (618).

Hy-glo Steel. A proprietary trade name for a stainless steel containing 17·0 per cent. chromium and 0·6 per cent. carbon.

Hygrol. Colloidal mercury.

Hygroton. A proprietary preparation of chlorthalidone. A diuretic. (17).

Hygroton K. A proprietary preparation of CHLORTHALIDONE and potassium chloride. A diuretic with a potassium supplement. (17).

Hylastic. A high-manganese steel containing 1·6–1·8 per cent. manganese and 0·35 per cent. carbon.

Hylene. A proprietary trade name for an organic diisocyanate used in the manufacture of polyurethane foam having a range of rigidities. (10).

Hylene B. A proprietary trade name for di(3,5,5,trimethylhexyl)esters of adipic acid. Vinyl plasticisers. (56).

Hylene C. A proprietary trade name for di(3,5,5,trimethylhexyl) esters of dighyeollic acid. (56).

Hylene D. A proprietary trade name for di-(n-octyl)adipate. A vinyl plasticiser. (56).

Hymatol. Thymol carbonate, $CO(O.C_6H_3(CH_3)C_3H_7)_2$. An antiseptic.

Hyoscin. Scopalamine, $C_{17}H_{21}NO_4$, an alkaloid.

Hyoscine Methobromide. Scopolamine methobromide.

Hypalon. A proprietary trade name for a synthetic rubber (chlorosulphonated polyethylene). It has good resistance to heat, oils, oxidising chemicals, sunlight and weathering. It is colourstable and ozone-proof. (10).

Hypaque. A proprietary trade name for Sodium Diatrizoate.

Hypargyrite. A mineral. It is pyrargyrite.

Hypargyron-Blende. Synonym for Miargyrite.

Hypercal. A proprietary preparation of rauwolfia alkaloids. An antihypertensive. (246).

Hypercal B. A proprietary preparation of rauwolfia alkaloids and amylobarbitone, used in the control of hypertension. (246).

Hyperdol. Reserpine and hydroflumethiazide. An antihypertensive. (502).

Hyperit. This is the same as hyperol.

Hypernick. See HIPERNICK.

Hyperol (Solid Hydrogen Peroxide, Ortizon, Perhydrate, Hyperit). Proprietary names for a compound of hydrogen peroxide and urea, $CO(NH_2)_2.H_2O_2$. It contains 35 per cent. hydrogen peroxide, which is obtained by dissolving in water or ether. One gram in 10 c.c. =a 10-volume strength solution of hydrogen peroxide.

Hypersthene. A pyroxene mineral, $(Mg.Fe)SiO_3$.

Hypertane. A proprietary preparation of rauwolfia alkaloids. An anti-hypertensive. (250).

Hypertane Forte. A proprietary preparation of rauwolfia alkaloids, ethiazide and potassium chloride, used in the control of hypertension. (250).

Hypertensan. A proprietary preparation of rauwolfia root. An antihypertensive. (250).

Hypertensin. A proprietary preparation of angiotensin amide. A vasoconstrictor. (18).

Hyperthermine. See AMBRINE.

Hyperysin. A proprietary preparation of papaverine nitrite, hexamethylenetetramine, dichloralhydrate and carbromal. (396).

Hypnal (Chlorpyrin). Mono-chloral-antipyrine, $C_{11}H_{12}N_2O.CCl_3(OH)_2$. Used medicinally in the place of chloral, as a soporific.

Hypnoacetin. Acetophenone-acetyl-p-amino-phenol ester, $C_6H_4(OCH_2COC_6H_5)(NHCOCH_3)$. A hypnotic and antiseptic.

Hypnogen. A proprietary preparation of barbitone. A hypnotic. See VERONAL. (366).

Hypnone. See PHENYL METHYL KETONE.

Hypnosed. A proprietary preparation of barbitone, calcium aspirin and codeine phosphate. A sedative and hypnotic. (838).

Hypnotal. Diethyl-oxy-acetyl-urea.

Hypnotics. See SOMNOLIN, SONERYL, NEONAL, HYPNAL, HYPNACETIN, HYP-

NOGEN, HYPNONE, VERONAL, PERNOCTON.

Hypo. Sodium thiosulphate, $Na_2S_2O_3$. It dissolves halogen silver salts, and is employed in photography.

Hypo, Black. See BURNT HYPO.

Hypoloid. Sheep immunization products. (514).

Hypon. A proprietary preparation of aspirin, phenacetin, caffeine, codeine phosphate, and phenolphthalein. An analgesic. (247).

Hyponitrous Ether. Ethyl nitrite, $C_2H_5NO_2$.

Hypophysin. The crystalline sulphate obtained from the phosphotungstate acid precipitate of the pituitary gland extract.

Hyporit. The trade name for calcium hypochlorite, containing 80 per cent. available chlorine. An antiseptic.

Hyposclerite. A mineral. It is a variety of Albite.

Hyposiderite. A mineral. It is a variety of limonite.

Hypotan. This is stated to be bromo-choline and α-methyl-acetyl-choline bromide with chloral hydrate.

Hypotonin. A German preparation containing the amino-compound of isovaleric acid. Used in medicine.

Hypovase. A proprietary preparation of PRAZOSIN hydrochloride. An antihypertensive. (85).

Hypromellose. A surface-active agent. It is a partial mixed methyl and hydroxypropyl ether of cellulose. ISOPTO.

Hyraldite. See HYDROSULPHITE.

Hyraldite A. See HYDROSULPHITE NF.

Hyraldite C Ext. (Hydrosulphite NF Conc., Britulite). Sodium formaldehyde-sulphoxylate, $NaHSO_2.CH_2O$. $2H_2O$. Used as a reducing agent in calico printing.

Hyrganol. A proprietary preparation of ethyl chaulmoograte.

Hyrgol. Colloidal mercury used medicinally.

Hyrgolum. The trade name for a form of colloidal mercury.

Hysol. A proprietary polyurethane elastomer. (710).

Hysol MBI—02. A proprietary epoxy-resin moulding powder modified for use as a load-bearing material. (711).

Hysol XC7—W529. A proprietary flexible, one-component epoxy casting and potting compound possessing good thermal shock properties. (712).

Hystatite. A mineral. It is a variety of ilmenite.

Hystazarin. Dioxy - anthraquinone, $C_{14}H_8O_4$.

Hysterol. Another name for Bornyval (*q.v.*).

Hystl. A trade mark registered in the U.S.A. for polybutadiene resins, elastomers and cured thermosetting polybutadiene resins. (713).

Hytemco. A proprietary trade name for an iron-nickel resistance alloy.

Hyten M Steel. A proprietary trade name for a nickel-chromium-molybdenum steel.

Hytens. A trade mark for goods of the abrasive class. The essential constituent is crystalline alumina.

Hy-ten-sl. A proprietary trade name for an alloy of 66 per cent. copper, 19 per cent. zinc, 10 per cent. aluminium, and 5 per cent. manganese.

Hytox. Germicidal detergent. (506).

Hytrel. A trade mark for a group of polyester elastomers used as thermoplastic rubbers. They are graded for hardness as follows:—**Hytrel 4055,** 92A; **Hytrel 5550,** 55D; **Hytrel 6350,** 63D. (10).

Hyvar X. Bromacil. A weed killer. Used as a broad spectrum weed control agent in raspberries. (10).

Hyzone. A term applied to nascent hydrogen.

I

Iachiol. Silver fluoride. AgF.

IA-IA Alloy. An alloy of 60 per cent. copper and 40 per cent. nickel. Used for electrical resistances.

Ialine. A proprietary coal-tar disinfectant which is miscible with water.

Ianthinite. A uranium mineral from Katanga. It is found in pitchblende and contains uranium and iron.

Iatrevin. A condensation product of menthol and isobutyl-phenol. Has been used as an internal antiseptic in the treatment of pulmonary tuberculosis.

Iatrol. Oxy-iodo-methyl-anilide, C_6H_5. $NH.O_2.C_2H_5OI_2$. An antiseptic.

Iberol. A proprietary preparation of ferrous sulphate, B complex vitamins and minerals. (311).

Ibex. Acid calcium phosphate for aerating purposes. (503).

Ibite. A compound of tannin and bismuth oxy-iodide, $C_6H_2(OH)_2(COOH)O$. $CO.C_6H_3(OH)O_2.BiI$. Said to be used as an antiseptic and deodorant powder. It is related to " Airol."

Ibufenac. 4-Isobutylphenylacetic acid. Dytransin.

Ibuprofen. 2 - (4 - Isobutylphenyl) - propionic acid. R.D. 13621.

Ice Colours. Colours formed on the

fibre by treating it with a phenol, and then with a diazotised amine, in the presence of ice. They are also known as Ingrain colours.

Iceland Spar. Purest crystalline calcite, $CaCO_3$, from Iceland.

Iceline. See PRESERVALINE.

Ice-spar. See SANIDINE. Also cryolite, which name means ice-spar.

Ice-stone. See CRYOLITE.

Ichden. See ICHTHYOL.

Ichtammon. See ICHTHYOL.

Ichthaband. A proprietary zinc paste and ICHTHYOL bandage used in the treatment of eczema. (484).

Ichthadone. A proprietary preparation of ammonium ichtho-sulphonate.

Ichthalbine. See ICHTHYOL-ALBUMIN.

Ichthammon. See ICHTHYOL.

Ichthammonium. See ICHTHYOL.

Ichthamol. See ICHTHYOL.

Ichthargan (Silver Ichthyol, Argentichthol). Silver ichthyol-sulphonate, containing 30 per cent. silver. It is used in medicine as a germicide.

Ichthermol. The mercury compound of ichthyol. It contains $\frac{1}{4}$ of its volume of mercury and is used as an antiseptic in the treatment of wounds and burns.

Ichthium. See ICHTHYOL.

Ichthocalcium. Calcium ichthosulphonate.

Ichthoferrum. Iron ichthosulphonate.

Ichthoform (Ichthyoform, Formichthol). A compound of ichthyol-sulphonic acid, and formaldehyde. Given internally as an intestinal antiseptic. Also used externally as a substitute for iodoform for wounds.

Ichtholan. German preparations consisting of lanoline-petroleum jelly salves with 10, 20, and 50 per cent. ammonium-sulpho-ichthyolicum.

Ichthopaste. Zinc paste and ichthamaol bandage. (529).

Ichthosan. See ICHTHYOL.

Ichthosodium (Ichthylolate). Sodium ichthosulphonate.

Ichthosulphol. See ICHTHYOL.

Ichthosulphonic Acid. Produced by the action of sulphuric acid upon crude ichthyol.

Ichthozincum. Zinc ichthosulphonate.

Ichthydrin. A by-product in the manufacture of ichthyol.

Ichthylolate. See ICHTHOSODIUM.

Ichthynal. See ICHTHYOL.

Ichthynat. See ICHTHYOL.

Ichthyocolla. Isinglass (*q.v.*).

Ichthyodine. See ICHTHYOL.

Ichthyoform. See ICHTHOFORM.

Ichthyol. A volatile oil containing sulphur, obtained by the dry distillation of a bituminous schist (stink stone) found in the Tyrol.

The name ichthyol is also applied to the ammonium salt, ammonium-ichthyol-sulphonate, $C_{28}H_{36}S_3O_6(NH_4)_2$. The following are some of the many products on the market which are said to be ichthyol preparations : Ichtammon, Ichthium, Ichthammon, Ichthammonium, Ichthamol, Ichthosan, Ichthyodine, Ichthosulphol, Lithyol, Ichtolithium, Ichden, Ichthynat, Ichthyopon, Isarol, Ichthynal, Piscarol, Pisciol, Subitol, Thiolin, and Isurol.

The salts of ichthyol are used in skin diseases, and internally in tuberculosis, as an antiseptic. Also see THIOL.

Ichthyol-albumen (Ichthalbine, Albichthol). A substance obtained by precipitating an albumin solution with a solution of ichthyol-sulphonic acid. Used in the treatment of intestinal catarrh.

Ichthyolate. A proprietary preparation of sodium ichtho-sulphonate.

Ichthyolidine. The piperazine salt of ichthyol-sulphonic acid. Used in the treatment of gout and uric acid troubles.

Ichthyol-isapogen. See ISAPOGEN.

Ichthyol-salicyl. A mixture of ichthyol and sodium salicylate.

Ichthyol Soap. An antiseptic soap made by the addition of about 5 per cent. sodium ichthosulphonate to soap.

Ichthyophthalmite. A variety of apophyllite, sometimes used as a gem stone.

Ichthyopon. See ICHTHYOL.

Ichthyo-resorcin. A solution of resorcin (10 per cent.) in ichthyol. An antiseptic.

Ichthysmut. Bismuth sub-sulpho-ichthyolate.

Ichtoform. See ICHTHOFORM.

Ichtolithium. See ICHTHYOL.

Icipen V. A proprietary trade name for Phenoxymethylpenicillin Potassium. An antibiotic. (2).

Icipen 300. A proprietary preparation of penicillin V. An antibiotic. (2).

Ickosan. A proprietary preparation of icthammol, zinc oxide, boric acid, bismuth subgallate, sulphur and hexachlorophane. Skin dusting powder. (344).

Icoral. A proprietary preparation stated to be *m* - hydroxy - N - ethyl - diethyl - amino-ethyl-amino-benzene and *m*-hydroxy-phenyl-propanolamine hydrochloride solution. Analeptic.

Icterosan. A German preparation. It is a 10 per cent. solution of the sodium salt of phenyl-quinoline-carboxylic acid, and 0·16 per cent. β-eucaine-hydrochloride.

Ictophysin. A German preparation. It is a combination of hypophysin and icterosan, and is used for the removal of gall-stones.

Ictotest. A proprietary test tablet of *p*-nitrobenzene diazonium, *p*-toluene sulphonate, salicylsulphonic acid and sodium bicarbonate, used for the detection of bilirubin in urine. (807).

Ideal Alloy. An alloy of 53·5 per cent. copper, 45 per cent. nickel, 0·66 per cent. iron, and 0·45 per cent. manganese.

Idemin. A proprietary preparation of meprobamate and benactyzine hydrochloride. Antidepressant. (269).

Idilon. See CARBORA.

Iditol. A proprietary shellac substitute.

Idocrase (Hoboite). A mineral. It is essentially a silicate of calcium and aluminium, but many other elements are also present.

Idokyl. Disinfectant, sanitizer and cleaner. (574).

Idoxuridine. 5-Iodo-2'-deoxyuridine. Kerecid.

Idozan. A proprietary preparation of colloidal iron.

Idrapid-spalter. A synthetic lipolytic agent. It is a sulphonic acid.

Idrialine. See IDRIALITE.

Idrialite (Hepatic Cinnabar). A peculiar variety of cinnabar from India. It is a mixture of cinnabar, HgS, with Idrialine (a hydro-carbon which is present in Idrialite to the extent of as much as 75 per cent.), iron pyrites, clay, and gypsum.

Idryl. Fluoranthrene, $C_{16}H_{10}$.

Igasurine. Impure brucine.

Igazol. A mixture of paraform, iodoform, and terpene hydrate. It has been employed in the treatment of tuberculosis.

Igelit. A German trade name for vinyl polymers. See also VINOFLEX. (3).

Igelit PCU. A proprietary manufacture of polyvinylchloride.

Igelströmite. A mineral. It is a variety of pyroaurite.

Igepal. A wetting agent.

Igepon A. A proprietary product. It is an oleyl derivative of isethionic acid (hydroxy-ethane sulphonic acid), $C_{17}H_{33}.COO.CH_2.CH_2.SO_3.Na$. A detergent.

Igepon T. A proprietary preparation. It is an oleyl derivative of taurine (amino-ethane sulphonic acid), $C_{17}H_{33}.CO.NH.CH_2.CH_2.SO_3.Na$.

Igevin. A trade name for a series of vinyl ether polymers used as plasticisers, e.g., polyvinylisobutyl ether. (3).

Igewsky's Reagent. A solution of 5 per cent. picric acid in absolute alcohol. Used for etching in the microanalysis of carbon steels.

Iglésiasite. A mineral. It is a mixture of zinc and lead carbonates.

Iglite. A mineral. It is a variety of arragonite.

Igmerald. A synthetic emerald.

Ignition Pills. Small pills of platinised asbestos or pumice.

I.G. Wax N & S. Montan wax, freed from resin, bleached and hardened by oxidation.

Ihlenite. A mineral. It is a ferric sulphate, $Fe_2(SO_4)_3.12H_2O$.

Ildeforsite. A mineral. It is a variety of columbite.

Ilesite. A mineral, $MnSO_4.(Zn.Fe)SO_4.H_2O$.

Iletin. See INSULIN.

Iliadin-Mini. A proprietary preparation of OXYMETAZOLINE hydrochloride. Used in the form of nose drops. (310).

Ilinol. Conjugated linseed oil. (506).

Illipé Butter. See BASSIO TALLOW.

Illium. An acid-resisting alloy containing 60·65 per cent. nickel, 21·07 per cent. chromium, 6·42 per cent. copper, 4·67 per cent. molybdenum, 2·13 per cent. tungsten, 1·04 per cent. silicon, 1·09 per cent. aluminium, 0·98 per cent. manganese, 0·76 per cent. iron, and 1·19 per cent. carbon and boron.

Illurin Balsam. The oleo-resin from *Paradaniella oliveri*. Used for adulterating balsam of copaiba. Specific gravity 0·985–0·995, acid value 55–60.

Illyrin. A preparation similar to Ichthyol (*q.v.*).

Ilmenite (Titaniferous Iron Ore, Titanic Iron, Titaniferous Iron). A mineral. It consists of about 52 per cent. titanic oxide, TiO_2, and 48 per cent. ferrous oxide, FeO.

Ilmenorutile. A black variety of the mineral rutile containing more than 10 per cent. ferric oxide. A mineral. It is $Fe(Cb,Ta)_2O_6$.

Iloglandol. A German insulin preparation.

Ilonium. A proprietary preparation of COLOPHONY, pinene, turpentine, camphene, chlorophyll, thymol, phenol dipentine, cineole and borneol, used in the treatment of septic skin disorders. (384).

Ilosone. A proprietary preparation containing erythromycin (as the estolate). An antibiotic. (333).

Ilotycin. A proprietary preparation containing erythromycin, ethyl carbonate. An antibiotic. (333).

Ilsemannite. A blue oxide of molybdenum, Mo_3O_8, found native.

Iltovax. Infectious laryngotracheitis vaccine (Veterinary). (514).

Ilvaite. A mineral, $(Ca.Fe)O.Fe_2O_3.2SiO_2$.

Image Stone. A variety of agalmatolite.

Imbretil. Carbolonium bromide.

Imbrilon. A proprietary preparation of INDOMETHACIN used in the treatment of arthritis. (137).

Imburel. See CHAY ROOT.

Imesatin. β-Imino-isatin, $C_8H_6ON_2$.

Imferon. A proprietary preparation of iron dextran. A hæmatinic. (188).

Imidocarb. An anti-protozoan. 3,3'-Di - (2 - imidazolin - 2 - yl) carbanilide. " 4A65 " is currently under clinical trial as the hydrochloride.

Imiodid. A substance obtained by heating p-ethoxy-phenyl-succinimide with a solution containing potassium iodide and iodine. A powerful antiseptic.

Imipramine. 1 - (3 - Dimethylamino - propyl) - 4,5 - dihydro - 2,3 : 6,7 - di - benzazepine. Berkomine and Tofranil are the hydrochloride.

Imitation Rum. See RUM.

Immadium. An alloy of manganese bronze containing aluminium.

Immedial. A trade mark for a range of dyestuffs. (983).

Immedial Black. A dyestuff obtained by fusing 1-chloro-2 : 4-dinitro-benzene and p-amino-phenol with sodium polysulphide.

Immedial Black N. A dyestuff produced by the sulphurisation of dinitrophenol. See DINITRO-PHENOL BLACK.

Immedial Black V. A dyestuff produced by the fusion of dinitro-oxy-diphenylamine with sodium polysulphide. It dyes cotton black from a sodium sulphide bath. By the oxidation of the colour on the fibre by hydrogen peroxide, it is converted into Immedial blue (indigo blue).

Immedial Blue. A dyestuff produced from hydroxy - dinitro - diphenylamine, and sulphurising at low temperature. It is not Immedial pure blue.

Immedial Bordeaux G. A dyestuff prepared by the action of alkaline sulphides and sulphur on amino-hydroxy-phenazines.

Immedial Bronze. A dyestuff prepared by fusing dinitro-cresol with polysulphides.

Immedial Brown B. A brown sulphide dyestuff, giving yellowish-brown shades on cotton. It is manufactured by boiling 4-hydroxy-4-amino-diphenyl-amine with a solution of sodium hydroxide, and heating the product with sodium polysulphides.

Immedial Cutch (Immedial Direct Blue). A dyestuff belonging to the same class as Immedial indone R.

Immedial Green. A dyestuff manufactured by dissolving 4-dimethyl-amino-4-hydroxy-diphenylamine in a solution of sulphur in crystalline sodium sulphide, copper sulphate being added to the solution. The dye is precipitated with salt. It gives bluish-green shades on cotton.

Immedial Indone R. A dyestuff obtained from the indophenol produced by oxidising o-toluidine with p-amino-phenol by heating it with sodium polysulphide in aqueous or alcoholic solution. It dyes cotton.

Immedial Maroon B (Thionine Redbrown B). A dyestuff prepared by the action of alkaline sulphides and sulphur upon amino-hydroxy-phenazine. See IMMEDIAL BORDEAUX G.

Immedial Orange C. A dyestuff produced by fusing tolylene-2 : 4-diamine with sulphur. The product is fused with sodium sulphide. Dyes cotton orange-brown.

Immedial Orange N. A sulphide dyestuff made by fusing together the same materials as in Immedial yellow D.

Immedial Pure Blue (Sky Blue). A dyestuff manufactured by oxidising a mixture of reduced nitroso-dimethyl-aniline, and an aqueous solution of phenol, to an indophenol. This is converted by reduction into 4-dimethyl-amino-4-hydroxy-diphenylamine, which is added to a hot solution of sulphur and sodium sulphide, then heated.

Immedial Sky Blue. A dyestuff produced by the fusion of dimethyl-p-amino - p - oxy - diphenylamine with sodium polysulphides at 110–115° C. Dyes cotton blue from a sulphide bath.

Immedial Yellow D. A dyestuff obtained from m-tolylene-diamine and sulphur at 190° C.

Immedial Yellow GG. A greenish-yellow dyestuff for cotton, made by fusing together sulphur, benzidine, and dehydro-thio-toluidine.

Immetal. Di-iodo-erucic acid isobutyl ester.

Immunised Cotton. Cotton obtained by partial esterification with toluene-p-sulphon-chloride. It is characterised by its resistance to the usual type of substantive cotton dyestuffs.

Imodium. A proprietary preparation of LOPERAMIDE. An anti-diarrhœal. (356).

Imogen. Sodium-diamino-naphthol-sulphonate, $(NH_2)_2C_{10}H_4(OH)SO_3Na$.

Imogen Sulphite. A combination of several developers with the required amount of sodium sulphite. A photographic developer.

Imolamine. 4 - (2 - Diethylamino-ethyl)5 - imino - 3 - phenyl - 1,2,4 - oxadiazoline. Angolon and LA 1211 are the hydrochloride.

Imperacin. A proprietary preparation

of OXYTETRACYCLINE dihydrate. An antibiotic. (2).

Imperial Black. See DIRECT BLACK.

Imperial Ester. A proprietary trade name for rosin-glycerol varnish and lacquer resins.

Imperial Green. See EMERALD GREEN.

Imperial Jade. A green aventurine quartz used as a gem in China.

Imperial Metal. An alloy of 80 per cent. copper and 20 per cent. nickel.

Imperial Red. See BENZOPURPURIN 4B, EOSIN BN, and INDIAN RED.

Imperial Scarlet. See BIEBRICH SCARLET.

Imperial Schultze Powder. A smokeless powder containing 80 per cent. nitro-lignin, 10 per cent. barium nitrate, and 8 per cent. pet. jelly.

Imperial Stone. A stone similar to Victoria stone (q.v.).

Imperial Violet. See REGINA PURPLE.

Imperial Yellow. See AURANTIA and CHROME YELLOW.

Impervite. See CERESIT.

Impetigo Cream. A proprietary preparation of red mercuric sulphide, and ammoniated mercury in H.E.B. Co. (381).

Impletol. A German preparation. It is a complex compound of caffeine and novocaine, and is used for subcutaneous injection.

Impsonite. An asphaltic substance found in Arkansas and other places. It is difficultly fusible, and is only slightly soluble in carbon disulphide.

Imuran. A proprietary preparation of azathioprine. An immunosuppressive agent used in renal transplantation. (277).

Inactive Tartaric Acid. Mesotartaric acid.

Inalium. An aluminium alloy containing 2 per cent. cadmium, 0·8 per cent. magnesium, and 0·4 per cent. silicon.

Inapassade. A proprietary preparation of sodium para-aminosalicylic acid and isoniazid. Antituberculous drug. (386).

Inca Stone. A pyrite used as a gem stone.

Incarb. Proprietary name for a concentrated liquid form of carbon black. (619).

Incense. Olibanum, a gum resin.

Incense Powders. These usually contain gums benzoin, thus, or olibanum mixed with lavender flowers or musk.

Inco Chrome Nickel (Inconel). Proprietary nickel alloys containing 12–14 per cent. chromium and 6–7 per cent. iron. They resist corrosion by the organic acids met with in food-

stuffs. The alloy containing 80 per cent. nickel, 14 per cent. chromium, and 6 per cent. iron melts at 1390° C.

Inconel Alloy 600. A trade mark for an alloy of 16 per cent. chromium, 7 per cent. iron and 77 per cent. nickel. It has good oxidation resistance at high temperatures. (143).

Inconel Alloy 700. A trade mark for an alloy of 15 per cent. chromium, 29 per cent. cobalt, 3 per cent. molybdenum, 2·25 per cent. titanium, 3·3 per cent. aluminium and the balance nickel. (143).

Inconel Alloy 718. A trade mark for an alloy of 19 per cent. chromium, 3 per cent. molybdenum, 0·8 per cent. titanium, 5 per cent. niobium, 53 per cent. nickel and the balance iron. (143).

Incoloy Alloy 800. A trade mark for an alloy of 20 per cent. chromium, 32 per cent. nickel and 48 per cent. iron. It is resistant to hydrogen/hydrogen sulphide corrosion. (143).

Incoloy Alloy 825. A trade mark for an alloy of 40 per cent. nickel, 21 per cent. chromium, 3 per cent. molybdenum, 2 per cent. copper, 1 per cent. titanium and the balance iron. It is resistant to corrosion in hot oxidising acid conditions. (143).

Incoloy Alloy 901. A trade mark for an alloy of 12·5 per cent. chromium, 5·7 per cent. molybdenum, 2·9 per cent. titanium, 42 per cent. nickel and the balance iron. (143).

Incoloy Alloy DS. A trade mark for an alloy of 18 per cent. chromium, 2·3 per cent. silicon, 37 per cent. nickel and the balance iron. (143).

Inconel Alloy X-750. A trade mark for an alloy of 15 per cent. chromium, 2·5 per cent. titanium, 0·9 per cent. aluminium, 0·9 per cent. niobium, 7 per cent. iron and the balance nickel. (143).

Inconel (Inco Chrome Nickel). A proprietary trade name for an alloy resistant to corrosion and heat, containing 80 per cent. nickel, 14 per cent. chromium, and 6 per cent. iron.

Incrinit. A proprietary insulation.

Inda. A casein preparation similar to gallatite.

Indalitan. A proprietary name for Clorindione.

Indalizarin. A dyestuff formed by the action of sulphites upon gallo-cyanine-sulphonic acid. It represents a series of dyes known as chromo-cyanines.

Indalizarin Green. Nitro-gallo-cyanine-sulphonic acid. Dyes chromed wool green.

Indalizarin R and J. A dyestuff produced by the action of sulphites upon gallocyanine-sulphonic acid. Dyes blue upon a chrome mordant. Used in printing.

Indamine Blue. See INDAMINE BLUE R and B.

Indamine Blue R and B. A dyestuff. It is amino-dianilido-phenyl-phenazonium-chloride, $C_{30}H_{24}N_5Cl$. Dyes tannined cotton bluish-violet.

Indamine 3R (Indamine 6R, Rubramine). A dyestuff prepared by the action of nitroso-dimethyl-aniline-hydrochloride upon o-toluidine, or upon a mixture of o-toluidine and p-toluidine. Dyes tannined cotton reddish to bluish violet.

Indamine 6R. See INDAMINE 3R.

Indanthrene. A dyestuff produced by fusing β-amino-anthraquinone with potassium hydroxide. It is N-dihydro-$1:2:2':1'$-anthraquinone-azine.

Indanthrene Black. A dyestuff. It consists of Indanthrene green chlorinated on the fibre.

Indanthrene Blue GC (Caledon Blue GC, Alizanthrene Blue GC). A dyestuff. It is dibromo-indanthrene.

Indanthrene Blue GCD (Alizanthrene Blue GCP, Caledon Blue GCD, Duranthrene Blue GCD). A dyestuff. It is dichloro-indanthrene.

Indanthrene Blue RC. A dyestuff. It is mono-bromo-indanthrene.

Indanthrene Blue RS. See INDANTHRENE X.

Indanthrene Bordeaux B. A dyestuff allied to Indanthrene red G.

Indanthrene C. A dyestuff. It is a mixture of di- and tri-bromo-indanthrene.

Indanthrene CD. A dyestuff. It is dichloro-indanthrene. Used in calico printing.

Indanthrene Dark Blue BO (Cyanthrene, Alizanthrene Dark Blue O, Caledon Dark Blue O, Duranthrene Dark Blue BO, Hydranthrene Dark Blue). A dyestuff. It is prepared by fusing benzanthrone with alkali.

Indanthrene Gold Orange G. See PYRANTHRONE.

Indanthrene Gold Orange R. A dyestuff. It is the halogen product of Indanthrene gold orange G.

Indanthrene Green. A dyestuff. It is the nitro-derivative of Indanthrene dark blue BO.

Indanthrene Grey B. A dyestuff prepared by the alkaline fusion of $1:5$-diamino-anthraquinone.

Indanthrene Maroon. A dyestuff obtained by the alkaline fusion of the formaldehyde compound of $1:5$-diamino-anthraquinone.

Indanthrene Olive G (Alizanthrene Olive G, Duranthrene Olive GL, Hydranthrene Olive R). A dyestuff prepared by the fusion of anthracene with sulphur at 250° C.

Indanthrene Red G. A dyestuff obtained by condensing $2:6$-dichloro-anthraquinone with α-amino-anthraquinone.

Indanthrene Red Violet 2RN. A dyestuff formed by the oxidation of 5-chloro-$4:7$-dimethyl-thioindoxyl.

Indanthrene S. The leuco compound of indanthrene. Used in printing.

Indanthrene Scarlet G. A dyestuff obtained by the halogenation of Indanthrene gold orange G.

Indanthrene Violet RT. A dyestuff produced by the halogenation of Indanthrene dark blue BO.

Indanthrene X (Indanthrene Blue RS, Duranthrene Blue RD Extra, Hydranthrene Blue RS Paste). A dyestuff. It is anthraquinone-azine, $C_{28}H_{14}N_2O_4$. Dyes cotton from a reduced vat, blue. Employed in printing.

Indanthrene Yellow. See FLAVANTHRENE.

Indanthrene Yellow GK. A dyestuff. It is equivalent to Algol yellow R.

Indazin. A proprietary trade name for a range of products used in the dyeing of textiles. (983).

Indazole. Indazine, $C_7H_6O_2$.

Indazurine B. A dyestuff. It is the sodium salt of dimethoxy-diphenyl-disazo-β-naphthol-disulphonic-dioxynaphthalene-sulphonic acid, $C_{34}H_{23}N_4S_3O_{14}Na_3$. Dyes cotton reddish-blue.

Indazurine BB. A dyestuff. It is the sodium salt of dimethoxy-diphenyl-disazo-dioxy-naphthoic-sulphonic-β-naphthol-disulphonic acid, $C_{35}H_{22}N_4S_3O_{16}Na_4$. Dyes cotton blue.

Indazurine GM. A dyestuff. It is the sodium salt of dimethoxy-diphenyl-disazo-dioxy-naphthoic-sulphonic-α-naphthol-p-sulphonic acid, $C_{35}H_{23}N_4S_2O_{13}Na_3$. Dyes cotton blue.

Indazurine 5GM. A dyestuff. It is the sodium salt of dimethoxy-diphenyl-disazo-dioxy-naphthoic-sulphonic-amino-naphthol-disulphonic acid, $C_{35}H_{23}N_5S_3O_{16}Na_4$. Dyes cotton greenish blue.

Indazurine RM. A dyestuff. It is the sodium salt of ditolyl-disazo-dioxy-naphthoic-sulphonic-α-naphthol-p-sulphonic acid, $C_{35}H_{23}N_4O_{11}S_2Na_3$. Dyes cotton reddish-blue.

Indazurine TS. A dyestuff. It is the sodium salt of ditolyl-disazo-dioxy-naphthoic-sulphonic-amino-naphthol-sulphonic acid, $C_{35}H_{24}N_5S_2O_{11}Na_3$. Dyes cotton reddish-blue.

Inderal A proprietary preparation of propranolol hydrochloride. (2).

Inderborite. A mineral. It is $CaMgB_6O_{11}.6H_2O$.

Inderite. A mineral. It is $Mg_2B_6O_{11}.15H_2O$.

Indestructo. A proprietary safety glass.

India Gum (Persian Gum). A gum resembling gum arabic. It is probably the product of *Prunus bokharensis*. It is not produced in India, but is sent to Bombay. It is probably a Soudan gum. See also BASSORA GUM.

Indian Agate. A moss agate used as a gem stone.

Indianaite. A mineral. It is an aluminium silicate.

Indian Balsam. Peru balsam.

Indian Brandy or Tincture. Consists of 1 part spirit of nitrous ether, 1 part compound tincture of rhubarb, and 1 part syrup.

Indian Butter (Ghé Butter, Ghee Butter). Phulwara butter, the fat expressed from the seeds of *Bassia butyracea*, of India. A foodstuff.

Indian Cornflour (Oswego). The starch from maize.

Indian Fire. A mixture of 7 parts sulphur, 2 parts realgar, and 24 parts nitre. Used for signalling, and in the manufacture of fireworks.

Indian Frankincense. See OLIBANUM.

Indian Gutta-percha. See PALA GUM.

Indian Ink (Chinese Ink). Consists of lampblack held together by means of fish glue. It sometimes contains a little camphor, and also sepia.

Indianite. A cement made up of 100 parts rubber, 15 parts rosin, and 10 parts shellac, dissolved in carbon disulphide. It is also the name for a white clay resembling Halloysite.

Indian Lake. See CARMINE LAKE.

Indian Madder. See CHAY ROOT.

Indian Millet. Sorghum (*q.v.*).

Indian Ochre. A native ferric oxide of North America.

Indian Oil. A drying spirit, distilled from crude mineral oil. Recommended as a solvent and turpentine substitute.

Indian Pipestone. See CATLINITE.

Indian Purple. A pigment. It is prepared by precipitating cochineal extract with copper sulphate.

Indian Red (Venetian Red, Venetian Bole, Rouge, Colcothar, Red Bole, Bole, Armenian Bole, Caput Mortuum, English Red, Angel Red, Chemical Red, Pompeian Red, Berlin Red, Iron Minium, Iron Red, Persian Red, Raddle, Reddle, Red Rudd, Red Ochre, Red Chalk, Red Earth, Terra di Sienna, Mineral Purple, Stone Red, Prussian Red, Italian Red, American Sienna, Mineral Rouge, Crocus, Blood Red, Brown Red, Pale Oxide of Iron, Dark Oxide of Iron, Violet Oxide of Iron, Iron Saffron, Imperial Red, Nuremberg Red, Scarlet Ochre, Prague Red, Red Oxide, Scarlet Red, Rubrica, Sinopis, Lemnos Earth, Vandyck Red, Spanish Oxide, Turkey Red Oxide, Turkey Red, Bright Red Oxide).

Red pigments consisting mainly of ferric oxide, Fe_2O_3, with varying amounts of natural argillaceous compounds. Some are natural products (ochres or earths), burnt or unburnt, but the names are also applied to products obtained by heating ferrous sulphate, or as a red residue from the manufacture of fuming sulphuric acid. This consists chiefly of ferric oxide, together with some basic ferric sulphate. The burnt product comes on to the market as Stone red. The following are some of the raw materials used for the production of these pigments : red iron ore, hæmatite, iron ochre, and limonite. These pigments contain from 12–95 per cent. ferric oxide.

Indian Saffron. See TURMERIC.

Indian Sago. The sago prepared from potato starch.

Indian Sarsaparilla. Hemidesmes root.

Indian Yellow (Piuri, Purree, Pioury). A pigment used in India, obtained from the urine of cows fed on mango leaves. The colouring principle is the magnesium or calcium salt of euxanthic acid, $C_{19}H_{16}O_{11}Mg.5H_2O$. Used as a permanent water and oil colour. The name is also used for Citronine B. See CITRONINE B and AUREOLIN.

Indian Yellow G. See CITRONINE B.

Indiarubber, Artificial. See CAOUTCHOUC, ARTIFICIAL.

Indican. Potassium-indoxyl-sulphate.

Indicator, Universal. A mixture of indicators, usually methyl red, α-naphthol-phthalein, phenol-phthalein, bromo-thymol blue, and cresol red. The colour indicates the pH value when added to the solution.

Indicolite. A mineral from Brazil. It is a blue tourmaline.

Indigen D and F. See INDULINE, SPIRIT SOLUBLE.

Indigo (Indigo Pure BASF, Indigo LL Paste, Powder, LL Vat I, Vat II). Indigotine, $C_{16}H_{10}N_2O_2$. Natural indigo is obtained by steeping the leaves of indigo-bearing plants in water, then oxidising the extract. Synthetic indigo is prepared by several methods. Used for cotton, wool, and silk, by steeping the material in a vat containing the leuco compound, then exposing to air.

Indigo Blue (Indigotine). Indigo, $C_{16}H_{10}N_2O_2$. A mixture of methyl violet and malachite green is also known as indigo.

Indigo Blue N and SGN. Dyestuffs. They are mixtures of cyanol, with green and red dyes.

Indigo Brown. Brown substances found in commercial indigo. Two of these bodies have been called Indiretin and Indirubin. They are secondary decompositions of indican.

Indigo Carmine (Indigo Extract, Indigotine, Indigo Carmine Extra, Indigo Extract, Paste, L Paste). A dyestuff. It is the sodium salt of indigotine-disulphonic acid, or the free acid, $C_{16}H_8N_2O_8S_2Na_2$. Also called Soluble indigo. Dyes wool blue from an acid bath.

Indigo Carmine Extra. A dyestuff. It is a British brand of Indigo carmine.

Indigo Copper. See COVELLITE.

Indigo Extract. See INDIGO CARMINE.

Indigo Extract, Paste, L Paste. British equivalents of Indigo carmine.

Indigo LL Paste, Powder, LL Vat I, Vat II. Dyestuffs. British equivalents of Indigo.

Indigo LL 2R. A British equivalent of Indigo MLB/R.

Indigo MLB. Indigo in powder, or in a 20 per cent. paste.

Indigo MLB/4B. See BROMINDIGO FB.

Indigo MLB/6B (Durindone Blue 6B). Pentabrom-indigo.

Indigo MLB/G. Indigo in powder.

Indigo MLB/R (Indigo LL, 2R). 5-Brom-indigo.

Indigo MLB/RR. A mixture of mono- and di-brom-indigo.

Indigo MLB/T. 7 : 7′-Dimethyl-indigo.

Indigo MLB/Vat L. A 20 per cent. indigo.

Indigo Pure BASF. See INDIGO.

Indigo Purple. A dyestuff. It is the sodium salt of indigotin-sulphonic acid. It blues bleached cotton, thread, and silk in a soap bath.

Indigo RBN and RB. Dibrom-indigo.

Indigo Red. See INDIPURPURIN.

Indigo, Red. See FRENCH PURPLE.

Indigo Salt T. o-Nitro-phenol-lactic-methyl-ketone, $C_{10}H_{11}NO_4$. Employed in calico printing, the compound being applied dissolved in bisulphite and converted into indigo by treatment with alkalis.

Indigosol DH. The sodium salt of the acid disulphuric ester of indigo. Used in dyeing and printing.

Indigosol-O (Blue Salt). A colourless product obtained by adding dry dihydro-indigo at 0° C., in an inert atmosphere, to a reaction mixture obtained by adding chlor-sulphonic acid to dimethyl - aniline in chlor - benzene, warming to 60° C., making alkaline with caustic soda, removing chlorbenzene and dimethyl-aniline by distillation, and salting out the sodium salt of the acid ester. The product is stable in air, but with mild oxidising agents gives indigo. It is suitable for dyeing wool and cotton, no reducing vat being necessary.

Indigo, Soluble. See INDIGO CARMINE.

Indigo Substitutes. Violet-blue liquids used in dyeing. They are prepared from logwood extracts by means of potassium bichromate and sodium bisulphite. They probably consist of the bisulphite compounds of an oxidation product of hæmatoxylin, and a chromium salt.

Mixtures of Patent blue with Victoria violet and Lanacyl violet are also known as indigo substitute. Also Induline, soluble, and Diaminogen blue BB (q.v.).

Indigotine. See INDIGO CARMINE and INDIGO BLUE.

Indigotine P. A dyestuff. It is the sodium salt of indigotine-tetra-sulphonic acid, $C_{16}H_6N_2O_{14}S_4Na_4$. Dyes wool bluish-violet from an acid bath.

Indigo TRG. Indigo.

Indigo Vat. See INDIGO WHITE.

Indigo White (Indigo Vat). Leucindigotine, $C_{16}H_{12}N_2O_2$. Used for the preparation of indigo vats. Fibres such as cotton, wool, or silk, are immersed in an alkaline bath of this compound, and then exposed to air, to precipitate indigo within the fibres.

Indigo Yellow. A material frequently present in commercial indigo, particularly Java indigo. It is stated to be identical with kæmferol, $C_{15}H_{10}O_6$.

Indigo Yellow 3G. Produced by the interaction of indigo and benzoyl chloride in the presence of copper powder, $C_{23}H_{14}N_2O_2$.

Indilitans. An alloy of 36 per cent. nickel, 0·06 per cent. carbon, 0·68 per cent. manganese, 0·09 per cent. silicon, and remainder iron. The alloy has a low coefficient of thermal expansion at medium temperatures, and is suitable for use in clocks and other precision apparatus in which a minimum expansion is important.

Indimulsin. A name which has been applied to the hydrolytic enzyme which brings about the decomposition of indican.

Indin. A substance stated to be identical with iso-indigotin.

Indipurpurin (Indigo Red). Indirubin, $C_{16}H_{10}N_2O_2$, the basis of a number of dyestuffs.

dirubin. Indigo red, $C_{16}H_{10}N_2O_2$, a ed colouring matter associated with ndigo in amount usually from 1–5 per cent. Java indigo often contains up to to per cent. It is produced from the decomposition of indican.

disin. A registered trade mark currently awaiting re-allocation by its proprietors. (983).

docarbon. A trade mark for a range of dyestuffs. (983).

dochromine T. See BRILLIANT ALIZARIN BLUE G and R.

dochromogen S. A dyestuff. It is the sodium salt of sulpho-oxy-indophenol-thiosulphonic acid, $C_{20}H_{17}N_2O_8$ S_2Na_2. It is used in calico printing, giving greenish-blue shades.

docid. A proprietary preparation of indomethacin. (310).

docybin. A proprietary trade name for Psilocybin.

doform. Salicylic methylene-acetate, $C_6H_4(CO_2H)O.CH_2.CO_2.CH_3$. Prescribed for gout and neuralgia.

doil CPD 142 and CPD 143. A proprietary trade name for petroleum type vinyl plasticisers. (100).

doine Blue R (Janus Blue, Naphthindone, Naphthindone Blue BB, Bengaline, Vac Blue, Fast Cotton Blue R, Indole Blue R, Diazine Blue, Diazine Blue BR). A dyestuff. It is safranine-azo-β-naphthol. It dyes unmordanted and tannined cotton indigo blue. It is discharged by stannous chloride to a red pattern on a blue ground.

doklon. A proprietary trade name for Flurethyl.

dole Blue R. See INDOINE BLUE R.

doline. A basic navy blue dye.

domethacin. 1 - (4 - Chlorobenzoyl) - 5 - methoxy - 2 - methylindol - 3 - ylacetic acid. Indocid.

donal. A German soporific. It is diethyl-barbituric acid and *Cannabis indica*.

donex VG. A proprietary trade name for aromatic hydrocarbon vinyl plasticiser. (100).

dophenine. See PARAPHENYLENE BLUE R.

dophenine Extra. See INDULINE, SPIRIT SOLUBLE.

dophenol (α-Naphthol Blue). A dyestuff. It is the oxidation product of dimethyl-*p*-amino-phenyl-*p*-oxy-α-naphthylamine, $C_{18}H_{16}N_2O$. Used in vat dyeing.

dophenol White (Leucindophenol). The tin compound of dimethyl-*p*-aminophenyl-*p*-oxy-α-naphthylamine, $C_{18}H_{18}N_2O$. Used in printing and vat dyeing, in conjunction with indigo.

Indophor. A dyestuff produced by heating phenyl - glucine - *o* - carboxylic acid with an alkali hydroxide. It is indoxylic acid.

Indoramin. An anti-hypertensive agent currently under clinical trial as " Wy-21901 ". It is 3-[2-(4-benzamidopiperidino)ethyl]indole.

Indorm. A proprietary preparation of propiomazine. A hypnotic. (245).

Induline. A mixture of aryl-aminoazines, made by heating together amino-azo-benzene, aniline, and aniline hydro-chloride. Used in making inks.

Induline A, 2B, 5B, BL, 5B Crystals, L 332. Dyestuffs. They are British brands of Induline.

Induline 3B. See INDULINE, SOLUBLE.

Induline 3B, Spirit Soluble. See INDULINE, SPIRIT SOLUBLE.

Induline 3B Opal and 6B Opal. See INDULINE, SPIRIT SOLUBLE.

Induline 6B. See INDULINE SOLUBLE.

Induline 6B, Spirit Soluble. See INDULINE, SPIRIT SOLUBLE.

Induline Black. A dyestuff. It consists of the sulphonic acids of induline, and dyes silk, wool, and cotton.

Induline Opal. See INDULINE, SPIRIT SOLUBLE.

Induline R and B. See INDULINE SOLUBLE.

Induline Scarlet. A safranine dyestuff. It is amino-ethyl-tolu-naphthazonium-chloride, $C_{19}H_{18}N_3Cl$. Dyes tannined cotton scarlet. Chiefly used in printing.

Induline, Soluble (Induline R and B, Induline 3B, Induline 6B, Fast Blue R and 3R, Fast Blue 2R, B and 6B, Sloe-line RS and BS, Fast Blue greenish, Nigrosine, soluble, Solid Blue 2R and B, Blue CB, Grey R and B, Bengal Blue, Induline A, 2B, 5B, BL, 5B Crystals, L332). A dyestuff. It is a mixture of the sodium salts of the sulphonic acids of the various spirit soluble indulines. Dyes wool or silk blue, reddish-blue. or bluish-violet, from an acid bath, Used in silk dyeing, and in the manufacture of inks.

Induline, Spirit Soluble (Induline Opal, Fast Blue R, spirit soluble, Induline 3B, spirit soluble, Induline 6B, spirit soluble, Induline 3B Opal and 6B Opal, Fast Blue B, spirit soluble, Azine Blue, spirit soluble, Indigen D and F, Printing Blue, Acetin Blue, Nigrosine, spirit soluble, Coupier's Blue, Spirit Black, Oil Black, Indophenine Extra, Blue CB, spirit soluble, Pelican Blue, Sloeline, Azo Diphenyl Blue, Violaniline). Mixtures of dianilido, amino, trianilido, and tetra - anilido - phenyl-phenazonium-chlorides. Employed for the preparation of the corresponding water soluble colours. Also used (mixed with chrysoidine), for the preparation of black spirit varnishes and polishes. Employed in calico printing.

Induline Yellow. See FLAVINDULINE.

Indur. A proprietary trade name for a phenol-formaldehyde synthetic resin used for moulding.

Indurated Stone. A stone similar to Victoria stone.

Indurite. An explosive containing 40 per cent. guncotton, freed from lower nitrates, and 60 per cent. mono-nitrobenzene. The name is also applied to a moulding powder of the phenol-formaldehyde condensation product type.

Indusoil. A proprietary product. It is a refined Talleol.

Industal. A proprietary material containing asbestos and portland cement. A roofing material.

Industrial Alcohol. See ALCOHOL, INDUSTRIAL.

Industrial Gum (Tragasol, Locust Bean Gum). Carob bean gum, an ingredient of mucilages. It is also used as a protective colloid.

Industrial Methylated Spirit. See METHYLATED SPIRIT.

Inertex. Protective powder for molten magnesium (531).

Inesite. A mineral. It is a manganese-calcium silicate.

Influvac. A proprietary influenza vaccine. (654).

Infusible White Precipitate. Dimercuri-diamonium chloride ($N_2H_4Hg_2.Cl_2$).

Infusorial Earth (Celite, Fossil Flour, Kieselguhr, Diatomite, Tripolite, Mountain Flour). The siliceous remains of diatoms. Used as a non-conducting material for boilers, an absorbent for liquids and liquid manures, in the preparation of dynamite, and as a filling material in soaps, dyes, and rubber goods. Also see DIATOMITE and FILTER-CEL.

Ingalikite. A mineral. It is $NaKAl_4Si_4O_{15}.2H_2O$.

Ingalite. See GALLATITE.

Ingluvin. An extract of fowl's gizzard. Was used as a digestive ferment, but has proved of little value.

Ingot Iron. A malleable iron prepared in the liquid form.

Ingotol. Non-ferrous chill and ingot dressing. (531).

Ingrain (Primuline Red). An azo dyestuff produced by dyeing the fabric with primuline, diazotising its amino group, and developing in a bath of β-naphthol.

Ingrain Brown (Alkali Brown D, Benzo Brown 5R, Cotton Brown R, Terracotta G, Cotton Orange 6305, Direct

Fast Orange D2G, D2R). A dyestuff It is the sodium salt of primuline-azo phenylene-diamine, being similar t Ingrain, a developing bath of pheny lene-diamine being used.

Ingrain Orange. A dyestuff produce in a similar manner to Ingrain, usin resorcinol instead of β-naphthol.

Injacom. Vitamin injection for animals (540).

Ink. Ordinary writing inks consist prin cipally of iron salts of gallotannic and gallic acids. Indigo carmine or log wood extract is sometimes added.

Ink Blue (Ink Blue 8671, 7567). Dye stuffs. They are British equivalent of Soluble blue.

Ink, Chinese. See INDIAN INK.

Ink Powder. A mixture of 10 oz. pow dered nut galls, 2 oz. zinc sulphide 4 oz. iron sulphate, and 1 oz. gun arabic. Ink is made by adding 1 oz of the powder to ½ pint of water.

Inkretan. This is stated to be a thyroi and pituitary compound.

I.N.N. International recommended Non registered Name.

Inolaxine. A proprietary preparation o STERCULIA. A laxative. (320).

Inositol Nicotinate. meso-Inosito hexanicotinate. Hexopal ; Mesonex.

Inproquone. 2,5 - Di - (aziridinyl) - 3 6-dipropoxy-1,4-benzoquinone.

Insecticides. Pitteliene Sulfosept oi Balbiani's ointment, Disparin Knadolin Nessler's insecticides, Picro-fœtidine Pyrethrum powder. See also INSEC POWDER.

Insect Powder. The powdered un expanded flowerheads of *Pyrethrum cinerariæfolium*. The mixure, styled Dalmatian Insect Powder, usually con tains about 70 per cent. ground pyre thrums, 25 per cent. ground sumach and 5 per cent. yellow ochre.

Insect Wax. See CHINESE WAX.

Insidon. A proprietary preparation o opipramol dihydrochloride. A seda tive. (17).

Insipin (Tasteless Quinine). The digly collic ester of quinine sulphate, $(CH_2COO.C_{20}H_{23}N_2O)_2O.H_2SO_4.3H_2O$. substitute for quinine employed in th treatment of malaria.

Insoluble Algin. Alginic acid, ob tained from seaweed.

Insomnol. A proprietary preparatio of methylpentynol. A hypnotic (250).

Instoms. Indigestion tablets. (509).

Instrument Bronze. An alloy of 8 per cent. copper, 13 per cent. tin, an 5 per cent. zinc.

Insulate. An electrical insulating pro

duct made from shellac, mica, asbestos, barytes, and wood flour.

Insulin. Iletin, $C_{45}H_{69}O_{14}N_{11}S.3HO$, a hormone isolated from the pancreas. Used in the treatment of diabetes.

Insuline. A proprietary trade name for calcined fire-clay.

Insulin Lente. A proprietary trade name for Insulin Zinc Suspension.

Insulin Novo Actrapid. A proprietary trade name for Neutral Insulin Injection.

Insulin Semilente. A proprietary trade name for Insulin Zinc Suspension. (Amorphous).

Insulin Ultralente. A proprietary trade name for Insulin Zinc Suspension. (Crystalline).

Insulin Zinc Suspension. Insuline Lente. Insulin Semilente.

Insulite Mastic Flooring. A flooring made from Elaterite or similar mineral rubber, mixed with asbestos. It is resistant to acids, and has good wearing qualities.

Insullac. A copal spirit varnish with the resin acids neutralised. An insulating material.

Insurok. A proprietary trade name for a phenol-formaldehyde synthetic resin used for moulding compounds and laminated products.

Intal Compound. A proprietary preparation of disodium cromoglycate and isoprenaline sulphate. A bronchial inhalant. (188).

Intarvin. Glyceryl margarate, $(C_{16}H_{33}COO)_3.C_3H_5$. Used in the treatment of diabetes.

Integrin. A proprietary preparation of oxypertine. A tranquilliser. (112).

Intene. A butadiene solution of polymerised rubber. (619).

Intene OE 60. An aromatic oil extended poly-butadiene rubber. (220).

Intense Blue. A pigment. It is indigo refined by precipitation.

Intensive Blue. See FAST ACID BLUE B.

Interacton. A proprietary preparation of cocarboxylase, glutaminase, histaminase, Co-enzyme A, ascorbic and allyl sulphide. (274).

Interenin. A hormone from the suprarenal gland. It contains 30 per cent. Cl.

Interferon. A protein formed by the interaction of animal cells with viruses. It is capable of conferring on animal cells resistance to virus infection.

Internol. See PARAFFIN, LIQUID.

Interol. See PARAFFIN, LIQUID.

Intestinol. A German product. It is a combination of pancreas extract and animal charcoal.

Intex. A proprietary trade name for a

group of styrene-butadiene rubber latices. (619).

Intex 082. A proprietary cold-polymerised butadiene-styrene latex with a low solids content, used for blending with INTEX 181. (619).

Intex 084. A proprietary styrene-butadiene rubber latex with a low solids content. (619).

Intex 105. A proprietary styrene-butadiene rubber latex with a high solids content used in the carpet industry. (619).

Intex 131. A proprietary styrene-butadiene rubber latex with a high solids content. (619).

Intex 151. A proprietary polybutadiene rubber latex. (619).

Intex 164. A proprietary carboxylated styrene-butadiene rubber used in the carpet industry. (619).

Intex 166. A proprietary carboxylated styrene-butadiene rubber latex used in the manufacture of carpets. It has good resistance to fraying and its tuft-locking properties are tough. (619).

Intex 167. A stiffer variety of INTEX 166. (619).

Intex 168. A proprietary styrene-butadiene rubber latex with high styrene content. (619).

Intex 178. A proprietary carboxylated styrene-butadiene rubber latex. (619).

Intex 181. A proprietary styrene-butadiene-vinyl pyridine polymer latex used in the formulation of dip compositions to promote rubber-to-textile adhesion. (619).

Intex 191. A proprietary polystyrene rubber latex used in the carpet industry. (619).

Intol. An emulsion type of styrene-butadiene synthetic rubber. (619).

Intolan. A proprietary range of ethylene-propylene rubbers. (619).

Intolene. A proprietary medium-vinyl polybutadiene. (619).

Intradex. A proprietary preparation of dextran in saline or dextrose solution. An intravenous infusion used to restore blood volume. (335).

Intraflodex. A proprietary preparation of dextran in saline or dextrose, used to improve capillary circulation. (335).

Intralgin. A proprietary preparation of benzocaine and salicylamide. (275).

Intra-Lipid-Vitrum. A proprietary preparation of soya bean oil. (329).

Intralloy. A proprietary trade name for base metal alloys and base metal mixtures used in the treatment of metal melts, especially of cast-iron. (899).

Intramine. Di-o-amino-phenyl-disulphide, $NH_2.C_6H_4.S.S.C_6H_4.NH_2$. Used in the treatment of syphilis. Now discontinued. (182).

Intraval. A proprietary preparation of thiopentone sodium. Used as a short duration anæsthetic or for induction of anæsthesia. (336).

Introcid. A German preparation. It is an iodine compound of cerium in aqueous solution, and is used for injection into operable tumours and in tuberculosis.

Introsolvan HS. A mixture of isohexyl and isoheptyl alcohol mixtures.

Inulase. An enzyme which decomposes inulin.

Invaderm C9B. A sodium aryl disalphonate in powder form used as an anionic level dyeing assistant for leather. (18).

Invadine. A proprietary trade name for a wetting agent containing a sodium alkyl-phenylene sulphonate.

Invar. An alloy of 36 per cent. nickel and 64 per cent. steel (0·2 per cent. carbon in steel). It has very little expansion on heating, and is used for delicate instruments for measuring. It has a specific gravity of 8·0, and melts at 1500° C. Its specific heat is 0·120 cal./gm. at 15–100° C.

Invariant. An alloy of 47 per cent. nickel and 53 per cent. iron. Has similar properties to Permalloy.

Invaro Steel. A proprietary trade name for a tool steel containing 1·15 per cent. manganese, 0·5 per cent. tungsten, and 0·9 per cent. carbon.

Invenol. Carbutamide.

Inverarite. A mineral found in mines formerly worked at Inverary, in Argyllshire. It is similar in composition to Pentlandite.

Inversal. A proprietary preparation of ambazone. (341).

Inversine. A proprietary preparation of mecamylamine hydrochloride. An antihypertensive. (310).

Invertase. See SUCRASE.

Invertin. See SUCRASE.

Invert Sugar. A mixture of molecular proportions of dextrose and levulose, obtained in the hydrolysis of cane sugar by acids. It is used to improve wines, also in the manufacture of liqueurs, fruit preserves, and honey substitutes.

Invisible Inks. Various. (a) A solution of lead acetate, made visible by exposure to sulphuretted hydrogen ; (b) a solution of potassium ferrocyanide, made visible by means of a solution of an iron salt ; (c) a solution of rice starch, made visible by vapours of iodine.

Inyoite. A mineral. It is calcium borate, $2CaO.3B_2O_3.13H_2O$.

Iobenzamic Acid. N - (3 - Amino - 2,4, 6 - tri - iodobenzoyl) - N - phenyl - β - alanine. Osbil.

Iocarmic Acid. A radio-opaque substance. 5,5′-(Adipoyldiamino(bis-(2,4, 6 - tri - iodo - N - methylisophthalamic acid). Dimer X is a sterile solution of the meglumine salt.

Iocetamic Acid. A contrast medium. 3 - (N - 3 - Amino - 2, 4, 6 - tri - iodo - phenyl) acetamido - 2 - methylpropionic acid.

Iodagol. An iodine preparation used as an antiseptic.

Iodal. A substance prepared by the action of iodine upon a mixture of alcohol and nitric acid.

Iodalbacid. Albumin iodate, containing 10 per cent. iodine, used in the treatment of syphilis.

Iodalbin. A compound of iodine and protein, containing 21·5 per cent. iodine. Used as a substitute for alkali iodides.

Iodalgin. A soluble preparation, containing 50 per cent. iodine.

Iodalia. A saccharated iodine compound, containing tannic acid. It is used to administer iodine to children.

Iodalose. A proprietary preparation stated to be iodine with peptone.

Iodamide. A contrast medium. α, 5 - Di(acetamido) - 2, 4, 6 - tri - iodo - m - toluic acid.

Iodamylene. Valerylene hydriodide, C_5H_8HI. Used in medicine.

Iodanisol. o-Iodo-anisol, $C_6H_4.OCH_3.I$. An antiseptic.

Iodantifebrin. Iodo-acetanilide, C_6H_4I. $NH.(C_2H_3O)$.

Iodanytol. A 10 per cent. solution of iodine in anytin (a 33 per cent. solution of ichthyol).

Iodargol (Iodéol). Colloidal iodine.

Iodargyrite (Iodyrite). A mineral. It is silver iodide, AgI.

Iodaseptine. A trade mark. Iodo-benzomethyl-formine.

Iodatol. A proprietary preparation stated to be an iodised vegetable oil.

Iodeigon. An iodine derivative of peptone (if soluble), or of albumin (if insoluble). See EIGONES.

Iodeikon. Sodium tetra-iodo-phenolphthalein.

Iodembolite. A mineral. It is $4[Ag(Cl,Br,I)]$.

Iodéol. See IODARGOL.

Iodeosin B. See ERYTHROSIN.

Iodeosin G. See ERYTHROSIN G.

Iodesin. Tetra-iodo-fluoresceine, $C_{20}H_8I_4O_5$.

Iodex. A proprietary ointment containing 4% organically-combined iodine in a petroleum jelly base. (281).

Iodferratin. An iodine compound of Ferratin (q.v.), containing 6 per cent. each of iodine and iron.

Iodgallicin. Bismuth-oxy-iodo-methyl-gallol, $C_6H_2(COOCH_3)(OH)_2OBi(OH)I$. It contains 23 per cent. iodine, and is used as a disinfectant and wound antiseptic.

Iodgorgon. A German preparation. It is an amino acid containing iodine.

Iodicin. A proprietary preparation of calcium-iodo-ricinoleate. An internal antiseptic.

Iodicyl. Di-iodo-salicylic-methyl ester.

Iodin. Iodised arachis oil.

Iodinated Glycerol. An expectorant. A mixture of iodinated dimers of glycerol. ORGANIDIN.

Iodine Blister. (*Ungentum hydrargyri iodidi rubri B.P.*) It consists of mercuric iodide mixed with benzoated lard.

Iodine Eigones. See EIGONES.

Iodine Green (Night Green, Pomona Green, Light Green, Metternich Green). A dyestuff. It is the zinc double chloride of heptamethyl - rosaniline chloride, $C_{27}H_{35}N_3Cl_4Zn$. Dyes silk green.

Iodine Ointment. A mixture of 4 per cent. iodine and 4 per cent. potassium iodide with glycerin and benzoated lard.

Iodine-potassium Iodide Solution. See LUGOL'S SOLUTION.

Iodine Red (Iodine Scarlet, Brilliant Scarlet, Scarlet Red, Royal Scarlet). A pigment. It is mercuric iodide, HgI_2.

Iodine Scarlet. See IODINE RED.

Iodine Soap. Usually made by the addition of a solution of potassium iodide to Castile soap.

Iodine Violet. See HOFMANN VIOLET.

Iodine-zinc-starch Solution. Starch (5 grams), and zinc chlorate (20 grams), are boiled with 100 c.c. of water until clear, when 2 grams of zinc iodide are added, and the whole made up to 1 litre.

Iodinol. See IODIPIN.

Iodipamide. NN'-Di-(3-carboxy-2,4,6-tri-iodophenyl)adipamide. Biligrafin and Endografin are the bis-meglumine salt.

Iodipin (Iodinol). A compound of iodine and sesamé oil, sold as 10 per cent. and 25 per cent. Iodipin (containing 10 per cent. and 25 per cent. iodine). It is used in the treatment of syphilis.

Iodisalin. Methylene - iodo - disalicylic acid. Used in medicine.

Iodisan. A German preparation. It is hexamethyl - diamino - isopropanol - di - iodide. Used for injection.

Iodised Almond Oil. See IODISED OIL and IODURETTED OIL.

Iodised Arachis Oil. See IODIN.

Iodised Casein. See IODOMENIM.

Iodised Chloroform Water. Tincture of iodine added to chloroform water to increase antiseptic action.

Iodised Ether, Magendi's. See MAGENDI'S IODISED ETHER.

Iodised Gelatin. See IODYLOFORM.

Iodised Oil. Almond oil containing iodine. Used as a liniment.

Iodised Oil of Wintergreen. See IODOZEN.

Iodised Quinine Hydriodide. Quinine-iodo-hydriodide.

Iodised Salt. Common salt, NaCl, containing a very small amount of sodium iodide. Goitre in certain districts is said to be associated with lack of iodide in the water supply.

Iodised Starch. Iodide of starch, used in medicine.

Iodistol (Iodosol, Iodohydromol, Iothymol, Iosol). Thymol iodide, $C_6H_2(CH_3)(OH)(C_3H_7)I$. Used in medicine.

Iodite. An ore of silver, AgI, containing 46 per cent. silver.

Iodival (Iovalurea). Mono-iodo-isovaleryl-urea, $(CH_3)_2.CH.CHI.CO.NH.CO.NH_2$. Used in medicine as a substitute for iodine and iodides.

Iodivical. A proprietary preparation of alphidine and aspirin. (366).

Iodoacetone. A substance obtained by dissolving iodine in acetone, $CH_3.CO.CH_2I$. Used in the treatment of carbuncles.

Iodoantipyrine. A compound, $C_{11}H_{11}ION_2$, formed from iodine and antipyrine in the presence of sodium acetate. It melts at 160° C., and is used in medicine as a source of iodine and as an antipyretic.

Iodo-atoxyl. The sodium salt of iodophenyl-arsenic acid. Used as a substitute for Atoxyl in the treatment of syphilis.

Iodo-bismitol. Sodium iodo-bismuthite.

Iodo-bismuth Erce. An oil suspension of quinine and bismuth iodide.

Iodobromite. A mineral, Ag(Cl.Br.I).

Iodo-caffeine. Sodium-caffeine-iodide.

Iodochlorhydroxyquin. A proprietary name for Clioquinol.

Iodochlorhydroxyquinoline. Clioquinol.

Iodocol. A substance prepared from guaiacol and iodine in sodium iodide. An antiseptic used for the treatment of eczema and other diseases.

Iodocrase. A mineral. It is vesuvianite.

Iodocresol (Traumatol). Cresol iodide $C_6H_3I.(CH_3)OH$. It contains 55 per cent. iodine, and is used locally as an antiseptic in syphilitic ulcers and eczema. Also used as a substitute for iodoform.

Iodocrol. Carvacrol iodide, $C_{10}H_{13}OI$. An antiseptic used in surgical dressings.

Iodo-elarsan. Strontium - chlorarseno-behenolate and iodine.

Iodo-eosin. Erythrosin or Pyrosin. Used as an indicator.

Iodo-eugenol. Eugenol iodide.

Iodofan. Mono-iodo-dihydroxy-benzene-formaldehyde, $C_6H_3I(OH)_2H.COH$. A condensation product of formaldehyde and iodo-resorcin. Used as a substitute for iodoform as a dusting powder for wounds.

Iodoform. Tri-iodo-methane, CHI_3.

Iodoformal. A compound of iodoform and hexamethylene - tetramine-ethyl iodide, $C_6H_{12}N.C_2H_5I.CHI_3$. Used as a substitute for iodoform.

Iodoform, Deodorised. A preparation of iodoform mixed with coumarin or a similar substance to mask the odour of the iodoform.

Iodoformin (Iodoformolin). Hexamethylene - tetramine - iodoform, CH_3I $(CH_2)_6N_4$. Used in surgery as an odourless substitute for iodoform.

Iodoformogen. Iodoform albuminate. An antiseptic.

Iodoform Ointment. A mixture of 10 per cent. iodoform with benzoated lard.

Iodoformolin. See IODOFORMIN.

Iodoformosol. A German preparation. It is a red-brown, odourless, water-soluble iodoform preparation used for the treatment of wounds.

Iodoform Substitutes. Aristol, Iodocresol, Iodofan, Iodoformal, Iodoformin, Iodosyl, Iodoterpene, Iodylin, Iodyloform.

Iodo-gallicin. Bismuth oxyiodo-methyl-gallol.

Iodoglidine. A preparation of wheat gluten, containing 10 per cent. iodine. Used in medicine.

Iodoglobin. Di-iodo-tyrosine, $HO.C_6$ $H_2I_2.CH_2.CH(COOH)NH_2$.

Iodogluten. A preparation of vegetable albumin containing 8 per cent. iodine.

Iodohemol. See HEMOL.

Iodohydromol. See IODISTOL.

Iodol. Tetra-iodo-pyrrol, C_4I_4NH. Used externally as a substitute for iodoform, and internally as a substitute for alkali iodides.

Iodolein. A proprietary preparation of iodised poppy-seed oil.

Iodolen. A combination of Iodol (tetra-iodo-pyrrol), and albumin. It contains 30 per cent. Iodol.

Iodolin. Quinoline-chloro-methyl-iodo-chloride, $(C_9H_7N)(CH_3Cl)ClI$. Used in medicine.

Iodolysin (Sinetide). A preparation similar to Tiodine. It contains 43 per cent. thiosinnamine, and 47 per cent. iodine, and is used in medicine.

Iodomenim. Iodised casein, containing bismuth. It contains 10 per cent. iodine, and is intended as a substitute for iodides in the treatment of syphilis.

Iodomenthol. Iodopyrrole with 1 per cent. menthol.

Iodomercury Hemol. See HEMOL.

Iodomethane. Methyl iodide, CH_3I.

Iodonascin. An iodine antiseptic in solid or liquid condition. It splits off iodine in contact with organic or inorganic acids.

Iodophen. See NOSOPHENE.

Iodophenin. Tri-iodo-phenacetin. An antiseptic.

Iodoprotein. An albumin preparation containing 10 per cent. iodine.

Iodopyrine. Iodine antipyrine, $C_{11}H_{11}$ IN_2O. Prescribed as an antipyretic and antineuralgic.

Iodo-ray. Sodium tetra-iodo-phenolphthalein.

Iodose. A red powder stated to be a compound of iodine and a nucleo-protein. It contains 10 per cent. iodine, and is recommended in the treatment of goitre and glandular enlargement.

Iodosol. See IODISTOL.

Iodosyl. A red powder, $C_6H_3I.OI.$ COOH, used as a substitute for iodoform in the treatment of wounds, ulcers, and ringworm.

Iodoterpene. An ointment-like substance composed of iodine and terpine hydrate. Said to be a substitute for iodoform.

Iodotetragnost. Sodium tetra - iodo - phenol-phthalein.

Iodothion. An antiseptic. It is 1-iodo-2 : 3-dihydroxypropane, $CH_2I.CH(OH).$ CH_2OH.

Iodothiouracil. 4 - Hydroxy - 5 - iodo - 2 - mercaptopyrimidine. Itrumil is the sodium derivative.

Iodothymoloform. Iodised thymoloform.

Iodothyrine. An iodine compound of the thyroid gland. Prescribed in goitre and eczema.

Iodoval. Mono - iodo - isovaleryl - urea, $(CH_3)_2CH.CHI.CO.NH.CO.NH_2$.

Iodoxamic Acid. A contrast medium. NN' - (1, 16 - Dioxo - 4, 7, 10, 13 - tetra-oxahexadecane - 1, 16 - diyl) di - (3 - amino-2,4,6-tri-iodobenzoic acid).

Iodozen. An iodised oil of wintergreen. It is an American substitute for Sanoform.

Iodozol. Di - iodo - p - phenol - sulphonic acid, $C_6H_2I_2(OH)SO_3H.3H_2O$.

Iodtriferrin. Iodised paranucleate of iron. It contains 15 per cent. iron and 8·5 per cent. iodine. An antiseptic.

Iodurase. This is stated to be yeast albumin with iodides.

oduretted Oil. A 0·5 per cent. solution of iodine in almond oil.

odylin. Bismuth-iodo-salicylate. Said to be used as an odourless substitute for iodoform.

odyloform. Iodised gelatin, containing 10 per cent. iodine. Used as a substitute for iodoform.

odyrite. See IODARGYRITE.

ogen. A substance produced by the action of iodine upon the anhydride of phthalic acid. It is antiseptic and germicidal, and is used in oil, for the treatment of nose and throat affections.

oglycamic Acid. A radio-opaque substance. $\alpha\alpha'$ - Oxydi - 3 - acetamido - 2, 4, 6 - tri - iodobenzoic acid). BILIGRAM is the meglumine salt.

ohydrin (Iothion). Di-iodo-isopropyl-alcohol, $CH_2I.CH(OH).CH_2I$. Used as a substitute for alkali iodides. When mixed with olive oil, pet. jelly or lanoline, it forms a preparation containing iodine in an easily absorbable form.

olanthite. A trade name for a mineral resembling jasper.

olase. Stated to be a yeast albumin with iodine.

olite (Dichroite, Cordierite). A mineral, $4(Mg.Fe)O.4Al_2O_3.10SiO_2.H_2O$.

onac ECP-88. A proprietary acrylic polymer used in the formulation of coatings applied by electrostatic spray. (743).

onagar. Pure standardised agar. (620).

onamin. A proprietary preparation of PHENTERMINE in a resin complex, used in the treatment of obesity. (744).

onex. Luboil additives. (553).

onol. A proprietary trade name for a phenolic antioxidant possessing a symmetrical structure thus giving a low power loss at high frequencies when used in dielectrics. (99).

onomer Resins. A name given to thermoplastic resins containing both covalent and ionic bonds. Carboxyl groups are located along the polymer chain by copolymerization to provide the anionic portion of the ionic cross links. Metal ions constitute the cationic part of the links. Sodium, potassium, magnesium and zinc are examples of ions used. See also SURLYN a proprietary trade name. (10).

onox 99. General antioxidants. (99).

onox 220. 4,4'-methylene-bis-2,6-di-*tert*. butyl phenol. An antioxidant of 98 per cent. minimum purity. (303).

onox 330. Non-toxic antioxidants for polymeric systems, thermoplastic rubbers and fatty acid distillates. (99).

onox 901. An ultra violet light stabiliser for polymeric systems. (99).

Iopanoic Acid. α - (3 - Amino - 2,4,6 - tri-iodobenzyl)butyric acid. Telepaque.

Iopax. The sodium salt of 2-oxy-5-iodo-pyridine-N-acetic acid.

Iophendylate. Ethyl 10-(4-iodophenyl) - undecanoate. Ethiodan ; Myodil.

Iopropane. Iodopropanol.

Iopydol. N - (2,3 - Dihydroxypropyl) - 3,5-di-iodo-4-pyridone.

Iopydone. 3,5-Di-iodo-4-pyridone.

Iosal. A proprietary preparation of alphidine, thyroid and aspirin. (366).

Iosol. See IODISTOL.

Iothalamic Acid. 5-Acetamido-2,4,6-tri - iodo - N - methylisophthalamic acid. Angio-Conray is the sodium salt ; Conray is the meglumine salt.

Iothalen. See NOSOPHENE.

Iothion. See IOHYDRIN.

Iothymol. See IODISTOL.

Iotifix. An iodine compound containing 8 per cent. iodine.

Iovalurea. See IODIVAL.

Ioxantin. A proprietary preparation of alphidine, guaicum resin and precipitated sulphur. (366).

Ioxynil. 1,4-diiodo-3-cyanophenol. A specific herbicide. Used in cereal crops.

Ipecac (Ipecacuanha). The dried root of *Uragoga ipecacuanha*.

Ipecacuanha. See IPECAC.

Ipecacuanha Wine. A mixture of 1 part by volume of liquid extract of ipecacuanha with 19 parts of sherry.

Ipecine. Emetine, $C_{30}H_{40}O_5N_2$, an alkaloid.

Ipesandrine. A proprietary preparation of alkaloids of opium and ipecacuanha, and ephedrine hydrochloride. A cough linctus. (267).

Ipesumman Syrup. A German product. It contains ipe-summan (0·006 gram of the total alkaloids from 0·3 gram of ipecacuanha root), codeine phosphate, benzoic acid (about 0·05 gram), and syrup. An expectorant.

Ipexon. A proprietary preparation of guaiphenesin, ephedrine hydrochloride, sucrose and glycerine. A cough linctus. (182).

Iphaneine. Specially pure isobutyl phenyl acetate. (504).

Ipomic Acid. Sebacic acid, $CO_2H(CH_2)_8$ CO_2H.

Iporka. A trade name for foamable urea resins. (49).

Ipral. A trade mark. Calcium ethyl-isopropyl - barbiturate, $Ca(C_9H_{13}O_3$ $N_2)_2.3H_2O$. Hypnotic.

Ipratropium Bromide. A bronchodilator currently undergoing clinical trial as " Sch. 1000 " and " Atrovent ". It is $(8r)$ - 8 - isopropyl - 3 - ($\pm$) - tro - poyloxy - $1\alpha H$, $5\alpha H$ - tropanium bro - mide N-Isopropylatropinium bromide.

Iprindole. An anti-depressant. 5-(3-

Dimethylaminopropyl) - 6, 7, 8, 9, 10, 11-hexahydrocyclooct[*b*]indole. PRON-DOL is the hydrochloride.

Iproclozide. N - 4 - Chlorophenoxy-acetyl-N'-isopropylhydrazine.

Iproniazid. N - Isonicotinoyl - N' - iso-propylhydrazine. Marsilid.

Ipronidazole. An antiprotozoal currently undergoing clinical trial as " M & B 16905 " and " Ro 7-1554 ". It is 2 - isopropyl - 1 - methyl - 5 - nitro - imidazole.

I.P.S. *Iso*-propylalcohol. (553).

Ipsel. A proprietary ointment containing GLYCERYL trihydroxyoleate, zinc undecanoate, undecanoic acid and zinc oxide. It is used in the treatment of nappy rash. (745).

Ipsiform. A German product. It is a liquid formaldehyde soap and is used as a disinfectant.

Ipsilene. A disinfecting gas made by heating ethyl chloride and iodoform under pressure.

Iragcet. Solvent soluble dyes. (543).

Iranolin. A proprietary trade name for aromatic extracts from petroleum suitable for use as extenders for vinyl plasticisers. PDL 18 and 2241 T are examples. (108).

Ircogel. A proprietary trade name for thixotropic agents for PVC plastosols and organosols for cold dipping and general use to prepare non-drip compounds. They are metallo-organic complexes high in calcium content, having a particle size of 0·01 micron.

Ireson's Packing Compound. A mixture of rubber, ultramarine, and silicate of magnesium.

Iretol. 1 : 2 : 3-Trihydroxy-5-methoxy-benzene, $C_7H_8O_4$, used in perfumery.

Irgaclarol. Wetting and scouring agents. (543).

Irgafin. Predispersed pigments for plastics polymers. (543).

Irgafiner. Predispersed pigments for polyolefines. (543).

Irgalan. Premetallised dyes. (543).

Irgalite. Pigment dispersions for emulsion paints and other organic pigments. (543).

Irgalite Blue GST. Beta form copper phthalocyanine blue pigment (Cl Pigment Blue 15) with excellent texture, dispersibility, gloss and high strength for letterpress and lithographic inks. (17).

Irgalite Dispersed. Pigment plasticiser dispersions. (543).

Irgalite MPS. Multipurpose pigment stainers for paints. (543).

Irgalite PDS. Predispersed pigment powders for paints. (543).

Irgalite PR. Predispersed pigment powders for rotogravure. (543).

Irgalite Yellow BGW. Metaxylidide *bis*-arylamide yellow pigment (Cl Pigment Yellow 13) with excellent dispersibility and improved gloss and transparency for letterpress and lithographic inks. (17).

Irgalite Yellow F4G. A proprietary monoazo pigment of the arylamide type. It has a greenish hue which makes it suitable for use in letterpress, offset litho, flexographic and gravure printing inks. (746).

Irgalon. Chelating agents. (543).

Irganox. A trade mark for a range of speciality antioxidants of the hindered phenol type developed initially for the stabilisation of polyolefines for use at high frequencies. (17).

Irgaphor. Rubber masterbatch pigments. (543).

Irgapyrol. Flame proofing agents. (543).

Irgasan DP300/ic. A proprietary bacteriostat for use in plastics and rubbers. It is 2,4,4'-trichloro-2-hydroxy-diphenyl ether. It is active against both Gram-negative and Gram-positive bacteria. (519).

Irgatron. Premetallised dyes for polyamides. (543).

Irgazin. A trade mark for a range of organic pigments derived from isoindolinone and dioxazine. Typical colours are Yellow 2GLT, Yellow 3RLT, Orange RLT, Red 2BLT and Violet BLT. (17).

Iridin (Irisin). The powdered extract of iris.

Iridium. A term used for alloys containing 77–83 per cent. zinc, 1·1–1·25 per cent. copper, and 15–22 per cent. tin.

Iridium Black. A pigment for china. It is an oxide of iridium.

Iridium Steel. A German steel containing 4 per cent. cobalt, 16 per cent. tungsten, 3–5 per cent. chromium, 0·67 per cent. vanadium, 0·8 per cent. molybdenum, and 0·6 per cent. carbon.

Iridosmine (Osmiridium). An alloy of iridium and osmium containing 40–77 per cent. iridium, and 20–50 per cent. osmium. If there is more iridium, the alloy is called Nevyanskite, and Siserskite if the content of osmium is high.

Irigenin. 5 : 7 : 3'-trihydroxy-6 : 4' : 5'-trimethoxy-iso-flavone.

Iriphan (Triphan). The strontium salt of atophan.

Irisamine G (Rhodine 3G). A dyestuff. It is the ethyl ether of unsymmetrical dimethyl - homo - rhodamine, $C_{25}H_{25}$ N_2O_3Cl. Dyes tannined cotton, silk and wool, red.

Iris Blue. See FLUORESCENT BLUE.

Iris Green. See SAP GREEN.

Irish Diamond. An Irish quartz crystal sometimes used as a gem stone.

Irish Pearl Moss. Caragheen moss, a gelatinous seaweed, *Chondrus crispus*. It contains carrageenin allied to pectin. It is employed as a substitute for isinglass, as a size for thickening colours in calico printing, and for stiffening silk.

Irish Peat Wax (Montana Wax, Montanin Wax). Waxes extracted from Irish peat are sold under these names. They resemble Montan wax.

Irisin. See IRIDIN.

Irisol. Irone, $C_{13}H_{20}O$, a ketone found in orris oil. Used in perfumery.

Iris Violet. See AMETHYST VIOLET.

Irocaine. See NOVOCAINE.

Irofol C. A proprietary preparation of ferrous sulphate, folic acid, sodium ascorbate and vitamin C. A hæmatinic. (311).

Iroline. A French motor fuel of unknown composition.

Iron A. A British chemical standard. It is a cast iron containing 0·734 per cent. combined carbon, 1·989 per cent. silicon, 0·047 per cent. sulphur, 0·049 per cent. phosphorus, 0·688 per cent. manganese, 0·042 per cent. arsenic, 0·052 per cent. titanium, and 2·387 per cent. graphitic carbon.

Ironac. An acid-resisting alloy of iron and silicon. It contains 13 per cent. silicon, 84 per cent. iron, 0·77 per cent. manganese, 1·08 per cent. carbon, and 0·78 per cent. phosphorus.

Iron, Alginoid. See ALGIRON.

Iron, Alibated. See ALIBATED IRON.

Iron, Alum (Ferro-alumen). Ferric ammonium sulphate, $Fe_2(SO_4)_3.(NH_4)_2 SO_4.24H_2O$.

Iron, Aluminium. See ALUMINIUM IRON.

Iron Amalgams. One amalgam is formed by rubbing powdered iron with mercuric chloride and water, and others are formed by electrolysis.

Iron-Andradite. Synonym for Skiagite.

Iron, Armco. See ARMCO IRON.

Iron, Arsenical. See ARSENICAL IRON.

Iron Arsenite, Soluble. Ammoniated citro-arsenite of iron.

Iron B. A British chemical standard. It is a cast iron containing 0·39 per cent. combined carbon, 0·031 per cent. sulphur, 0·026 per cent. phosphorus, 0·031 per cent. arsenic, 0·108 per cent. titanium, and 2·67 per cent. graphitic carbon.

Iron Berlinite. A mineral. It is $3[FePO_4]$.

Iron Black. Antimony precipitated as a fine powder, by action of zinc upon an acid solution of an antimony salt. It is used for giving the appearance of polished steel to papier maché and plaster of Paris. See BLACK ANTIMONY.

Iron, Bonderised. See BONDERISED IRON.

Iron-Boracite. Synonym for Huyssenite.

Iron Brass. See BRASS, IRON.

Iron Buff (Nankin Yellow). A mineral colour. It is ferric hydroxide, and is used for colouring cotton by impregnating the fibre with green vitriol (ferrous sulphate), and passing it through caustic soda.

Iron, Burnt. See BURNT IRON.

Iron, Busheled. See BUSHELED IRON.

Iron Carbonate, Saccharated. See SACCHARATED IRON CARBONATE.

Iron, Cast. The molten iron from the blast furnace.

Iron Cement (Rusting Cement). A cement used for drain pipes and other purposes. It usually contains 85 per cent. iron filings, 10 per cent. flowers of sulphur, and 5 per cent. sal-ammoniac. The mixture is stirred into a paste with water. The iron filings rapidly oxidise, and the cement becomes stone hard. See IRONITE and EISEN-PORTLAND CEMENT.

Iron Chamois. See IRON BUFF.

Iron, Charcoal. See CHARCOAL IRON.

Iron, Chateaugy. See CHATEAUGY IRON.

Iron-chlorite. See DELESSITE.

Iron, Chrome. See CHROME IRON.

Iron, Cold Short. See COLD SHORT IRON.

Iron-Cordierite. A mineral. It is $4[Fe_2''Al_4Si_5O_{18}]$.

Iron, Cromodined. See CROMODINED IRON.

Iron D (Phosphoric D). A British chemical standard. It contains 11·8 per cent. phosphorus.

Iron D2. A grey phosphoric cast iron in the form of fine turnings. It contains 1·31 per cent. silicon, 1·07 per cent. phosphorus, and 1·64 per cent. manganese. The approximate quantities of other elements are 2·5 per cent. graphitic carbon, 0·8 per cent. combined carbon, and 0·03 per cent. sulphur. It is a British chemical standard, and takes the place of Iron D in regard to phosphorus.

Iron, Delhi Rustless. See DELHI RUSTLESS IRON.

Iron, Dialysed. See DIALYSED IRON.

Iron Earth, Blue. See BLUE IRON EARTH.

Iron, Electro-granodised. See ELECTROGRANODISED IRON.

Iron Flint. An opaque variety of quartz containing iron.

Iron Froth. A spongy variety of hæmatite.

Iron G. A new standard cast iron containing 1·82 per cent. graphite carbon, 0·86 per cent. combined carbon, 1·3 per cent. silicon, 0·41 per cent. manganese, 0·125 per cent. sulphur, and 0·45 per cent. phosphorus.

Iron, Galvanised. See GALVANISED IRON.

Iron Glance. See HÆMATITE.

Iron Glycerophosphate. A ferric glycerophosphate containing approximately 15 per cent. iron.

Iron Gohi. See GOHI IRON.

Ironite. A trade mark applied to a product used with cement for hardening floors and waterproofing walls. Another variety is used for waterproofing only without cement. The processes depend upon the rapid oxidation of the material in the presence of moisture. Also a proprietary trade name for chromium-nickel-vanadium cast iron. (842).

Iron L (Nickel-chromium-copper-Austenitic Iron L). A British Chemical Standard. It contains 3·06 per cent. total carbon, 2·26 per cent. silicon, 1·01 per cent. manganese, 0·119 per cent. phosphorus, 13·45 per cent. nickel, 3·96 per cent. chromium, 4·73 per cent. copper, and 0·031 per cent. sulphur, the remainder being iron.

Iron Liquor. See PYROLIGNITE OF IRON.

Iron-Leucite. A mineral. It is $KFe^{\cdots}Si_3O_8$.

Iron, Manganese. See FERRO-MANGANESE.

Iron Mica. See LEPIDOMELANE.

Iron Minium. See INDIAN RED.

Iron-Monticellite. A mineral. It is $4[CaFe^{\cdots}SiO_4]$.

Iron Mordant. Ferrous sulphate and ferric nitrate are both used as mordants.

Iron, Nickel. See FERRO-NICKEL.

Iron Ore A, Hæmatite Type. A standard iron ore containing 58·19 per cent. iron, 0·056 per cent. phosphorus, 8·14 per cent. silica, and 0·066 per cent. sulphur. Used as a standard for checking analyses of iron ore.

Iron Ore, Arsenical. See MISPICKEL.

Iron Ore, Blue. See BLUE IRON EARTH.

Iron-ore Cement. Cements in which a large proportion of the alumina is replaced by ferric oxide. Also see SIDERO CEMENT.

Iron Ore, Red. See BOG IRON ORE.

Iron Ore, Spathic. See CHALYBITE.

Iron Ore, Specular. See HÆMATITE.

Ironorm. A proprietary preparation of iron DEXTRAN complex. A hæmatinic. (747).

Iron-Orthoclase. Synonym for Ferriorthoclase.

Iron Paranucleinate. An iron casein compound containing 22 per cent. iron.

Iron Pill (Blaud's Pill). *Pilula ferri, B.P.* Prepared from ferrous sulphate and sodium carbonate.

Iron-plating Solutions. Usually consist of solutions of ferrous sulphate and ammonium chloride in water. Used for the electro-deposition of iron.

Iron Portland Cement. See EISEN PORTLAND CEMENT.

Iron Putty. A mixture of ferric oxide and boiled linseed oil. Used for joints in iron pipes.

Iron Pyrites (Mundic). A mineral. It is iron sulphide, FeS_2.

Iron Pyrites, White. See MARCASITE.

Iron Pyrochroite. A mineral. It is $(MnFe)(OH)_2$.

Iron Pyrolignite. See PYROLIGNITE OF IRON.

Iron Red. See INDIAN RED.

Iron-Rhodonite. Synonym for Pyroxmargite.

Iron Rubber. A mixture of iron filings and petroleum pitch.

Iron Saffron. See INDIAN RED.

Iron Sand. A heavy black sand of metallic lustre. It consists mainly of magnetic ore, but often contains a considerable quantity of titanium. It is also called Iserine or Menaccanite.

Iron-Schefferite. A mineral. It is $4[(Ca,Mn,Fe)(Mg,Fe,Mn)Si_2O_6]$.

Iron Scurf. A mixture of particles of stone and iron, produced by the wear of siliceous grindstones in grinding gun-barrels. Blue bricks are glazed by sprinkling with this material.

Iron Sinter. A mineral, $Fe_2O_3.As_2O_5$.

Iron Somatose. See SOMATOSE.

Iron Spinel. See PLEONASTE.

Iron, Titanic. See ILMENITE.

Iron, Titaniferous. See ILMENITE.

Iron Vitriol (Green Vitriol, Copperas, Green Copperas). Ferrous sulphate, $FeSO_4.7H_2O$. Used in dye-works, in the manufacture of ink, and as a disinfectant.

Iron Wood. The wood of the hornbeam. It is said to be tonic and antiperiodic.

Iron, Wrought. See MALLEABLE IRON.

Iron, Zinced. See GALVANISED IRON.

Iron-zinc Spar. See CAPNITE.

Irosorb 59. Diagnostic test for anæmia. (525).

Irox. A proprietary trade name for a synthetic yellow iron oxide.

Irrigal. A substance obtained by the dry distillation of wood. It contains various phenolic bodies, and is germicidal.

Irvingia Butter. See COCHIN-CHINA WAX.

Irvingite. A lithium mineral.

Isabellite. A mineral. It is richterite.

Isalax. See ACETPHENOLISATIN. (696).

Isamoxole. A drug used in the treatment of asthma. N-Butyl-2-methyl-N-(4-methyloxazol-2-yl) propionamide.

Isapogen. A German preparation. It is a viscous soap containing 6 per cent. iodine and 6 per cent. camphor, and is rubbed in for rheumatism. There is also salicyl-isapogen containing 15 per cent. salicylic acid, and ichthyol-isapogen containing 10 per cent. ichthyol in addition.

Isarit. A bleaching earth.

Isarol. See ICHTHYOL.

Isatin Yellow. A dyestuff. It is sodium - isatin-phenyl-hydrazone-p-sulphonate, $C_{14}H_{10}O_4N_3SNa$. Dyes wool and silk greenish-yellow.

Isatophan. See METHOPHAN.

Isaverine. A proprietary preparation of METHINDIZATE hydrochloride. A veterinary spasmolytic.

Isceon. A proprietary range of halogen derivations of aliphatic hydrocarbons used as refrigerants and propellants. (748).

Ischelite. A mineral. It is polyhalite.

Iserine. See IRON SAND.

Ishikawaite. A mineral. It is (U,FeYt,Ce)(Cb,Ta)O₄.

Isinglass (Fish Gelatin). Consists of the dried inner skin of the swimming bladder of the sturgeon, and similar fish. Used for the clearing of liquids, such as beer and wine. Also see THAO.

Isinglass, Bengal. See AGAR-AGAR.

Isinglass, Ceylon. See AGAR-AGAR.

Isinglass, Chinese. See AGAR-AGAR and THAO.

Isinglassine. An isinglass substitute made from calves' feet and other materials.

Isinglass, Japan. See AGAR-AGAR and THAO.

Island Cacao. Theobroma seeds.

Ismelin. A proprietary preparation of guanethidine. An antihypertensive. (18).

Ismelin-Navidrex K. A proprietary preparation of guanethidine sulphate, cyclopenthiazide and potassium chloride in a slow release core. (18).

Isn. A liquid preparation of iron with 0·2 per cent. of ferrous saccharate. It is really a mixture of ferrous citrate (containing oxide), and sugar.

Isoaminile. 4 - Dimethylamino - 2 - isopropyl-2-phenylvaleronitrile. Dimyril is the citrate.

Iso-anthraflavic acid. 2 : 7-Dioxyanthraquinone, $C_{14}H_6O_2.(OH)_2+H_2O$.

Iso-Autohaler. A proprietary preparation of ISOPRENALINE sulphate as a metered aerosol, used in the treatment of asthma. (275).

Isobronchisan. A proprietary preparation of isoprenaline sulphate, ephedrine chloride and theophylline. A cough linctus. (839).

Isobutad. A mineral rubber or bitumen.

Isobutyl Carbinol. Amyl alcohol, $C_5H_{11}OH$.

Isobuzole. 5 - Isobutyl - 2 - (4 - methoxy-benzenesulphonamido)-1,3,4-thiadiazole. Glysobuzole. Stabinol.

Iso-Brovon. A proprietary preparation of isoprenaline sulphate, theophylline, and methylephedrine hydrochloride. A bronchial antispasmodic. (339).

Iso-Brovon Inhalant. A proprietary preparation of isoprenaline and atropine. A bronchial inhalant. (339).

Isocaine. A local anæsthetic. It is p-amino - benzoyl - di - isopropyl - amino-ethanol-hydrochloride.

Isocarboxazid. 3 - N - Benzylhydrazino-carbonyl-5-methylisoxazole. Marplan.

Isoclase. A mineral, $Ca(CaOH)PO_4$. $2H_2O$.

Isoclasite. A mineral. It is a calcium phosphate.

Iso-Cornox. Selective weedkillers. (502).

Isodiphenyl Black. A dyestuff. It is the sodium salt of benzene-disazo-naphthol-sulphonic acid-azo-m-phenylene-diamine-resorcin, $C_{28}H_{21}N_8O_6SNa$. Dyes cotton black.

Isodulcite (Rhamnodulcite). Rhamnose, $C_6H_{12}O_5+H_2O$.

Isodurindine. Amino-tetramethyl-benzene, $C_6H(CH_3)_4.NH_2$.

Isoetharine. 1 - (3,4 - Dihydroxy-phenyl)-2-isopropylaminobutan-1-one.

Isoflupredone. An adrenocortical steroid. 9α - Fluoro - 11β, 17α, 21 - trihydroxypregna - 1, 4 - diene - 3, 20 - dione. 9α-Fluoroprednisolone.

Isoform (Oxiosol). p-Iodo-anisol, C_6H_4(OCH₃)IO₂. Used as an antiseptic. Because of its explosive properties, it is mixed with calcium phosphate for use as an internal and external antiseptic.

Isogalithe. A casein product. It resembles gallalith.

Isogel. A proprietary preparation derived from the husks of mucilaginous seeds. A laxative. (284).

Isokite. A mineral. It is 4[CaMgPO₄F].

Isokol. A similar preparation to Perikol (q.v.).

Isol. (1) An oil forming a permanent emulsion with hot or cold water. Used for oiling textiles.

Isol. (2) An antiskinning agent. (564).

Isolantite. A proprietary trade name for a ceramic material made from steatite with binders.

Isoleucine. α - Amino - β - methyl-ethyl-propionic acid, $(C_2H_5)(CH_3)CH.CH(NH_2)COOH$.

Isolevin. A proprietary trade name for the hydrogen tartrate of Isoprenaline.

Isolierstahl. See BAKELITE.

Isolit (Isolose). Cements for fastening porcelain insulators to metal supports.

Isolose. See ISOLIT.

Isometamidium. An anti-protozoan. 8 - m - Amidinophenyldiazoamino - 3 - amino - 5 - ethyl - 6 - phenylphenan - thridinium chloride. SAMORIN is the hydrochloride.

Isomethadone. 6 - Dimethylamino - 5 - methyl - 4,4 - diphenylhexan - 3 - one. Isoamidone.

Isomidone. A proprietary trade name for Isomethadone.

Isomin. A proprietary melamine formaldehyde moulding compound. (749).

Isomist. A proprietary preparation of isoprenaline sulphate. A bronchial antispasmodic. (271).

Isonaphthol. β-Naphthol, $C_{10}H_7OH$.

Isonian. A proprietary preparation of guaiphenesin, theophylline, ephedrine hydrochloride and phenobarbitone. A bronchial decongestant and antispasmodic. (840).

Isoniazid. Isonicotinoylhydrazine.

Isopal. A proprietary urea formaldehyde moulding material. (749).

Isopar. A trade mark for a mixture of isoparaffin s containing C_{10}, C_{11} and C_{12} hydrocarbons. (621).

Isophane. Synonym for Franklinite.

Isophane Insulin. A hypoglycæmic agent. Isophane Insulin (NPH).

Isophylline. A proprietary trade name for Diprophylline.

Isoplac. A proprietary insulation.

Isopral. An internationally-registered trade mark. (983).

Isoprednidene. An A.C.T.H. inhibitor. It is 11β, 17α, 21 - trihydroxy - 16 - methylenepregna - 4, 6 - diene - 3, 20 - dione.

Isoprenaline ($\pm$) - 1 - (3,4 - Dihydroxy-phenyl) - 2 - isopropylaminoethanol Isopropylnoradrenaline. Aleudrin, Neo-drenal, Neo-Ephinine and Norisodrine are the sulphate ; Isoleven is the hydrogen tartrate ; Aerotrol and Isupren are the hydrochloride.

Isopropamide Iodide. (3-Carbamoyl-3, 3-diphenylpropyl)di-isopropyl-methyl-ammonium iodide. Darbid ; Tyrimide. Present in Stelabid.

Isopropyl Carbinol. Isobutyl alcohol, $(CH_3)_2.CH.CH_2.OH$.

Isopropylnoradrenaline. Isoprenaline.

Isoptin. A proprietary trade name for Veraparnil.

Isopto Atropine. A proprietary preparation of atropine sulphate solution for ocular use. (374).

Isopto Carbachol. A proprietary pre-

paration of carbachol solution for ocular use. (374).

Isopto Carpine. A proprietary preparation of pilocarpine hydrochloride solution for ocular use. (374).

Isoptocetamide. A proprietary preparation of sodium sulphacetamide solution. An ocular antiseptic. (374).

Isopto Eserine. A proprietary preparation of physostigmine salicylate solution for ocular use. (374).

Isopto Frin. A proprietary preparation of phenylephrine hydrochloride solution for ocular use. (374).

Isopto Homatropine. A proprietary preparation of homatropine hydrobromide solution for ocular use. (374).

Isopto Hydrocortisone. A proprietary preparation of hydrocortisone for ocular use. (374).

Isopto Hyoscine. A proprietary preparation of hyoscine hydrobromide solution for ocular use. (374).

Isopto P-ES. A proprietary preparation of pilocarpine hydrochloride and physostigmine salicylate solution for ocular use. (374).

Isopto Alkaline. A proprietary preparation of HYPROMELLOSE for ocular use. (374).

Isopto Plain. A proprietary preparation of HYPROMELLOSE for ocular use. (374).

Isopurpurin. See ANTHRAPURPURIN.

Isoquinoline Red. See QUINOLINE RED.

α-Isorcin. See CRESORCIN.

Isordil. A proprietary preparation of isosorbide dinitrate used in the treatment of angina pectoris. (467).

Isorubine. See NEW MAGENTA.

Iso Soap. A solid sulphonic derivative of castor oil, soluble in hot water. Used as a bleaching, washing, and dressing agent in the textile industries.

Isostannite. A mineral. It is $[Cu_2FeSnS_4]$.

Isotachiol. Silver silico-fluoride, $Ag_2 SiF_6$. Used for sterilising water.

Isotense. A proprietary preparation of syrosingopine. An antihypertensive. (271).

Isothane. Urethane elastomeric products. (557).

Isothionine. See BERNTHSEN'S VIOLET.

Isothipendyl. An anti-histamine. 10-(2 - Dimethylaminopropyl) - 10H - pyrido[3, 2 - b] - [1, 4]benzothiazine. NILERGEX and THERUHISTIN are the hydrochloride.

Isotonic Salt Solution. See PYHSIO-LOGICAL SALT SOLUTION.

Isovanat. A proprietary preparation of vanadium in isotonic and isobaric solution.

Isovon. A proprietary preparation of

isoprenaline hydrochloride in an aerosol. A bronchial antispasmodic. (339).

Isoxarsone. 8-Acetyl-amino-3-hydroxy-1 : 4-benz-isoxazine-6-arsinic acid.

Isoxsuprine. A peripheral vasodilator. 1 - (4 - Hydroxyphenyl) - 2 - (1 - methyl - 2 - phenoxyethylamino) pro-pan-1-ol.

Isoxyl. A proprietary trade name for Thiocarbide.

Issolin. A phenol-formaldehyde resin, which is soluble in alcohol.

Issolith. See BAKELITE.

Issue Peas. Orange berries.

Istin (Istizin). Proprietary preparations of 1 : 8-dihydroxy-anthraquinone.

Istizin. See ISTIN.

Isupren. A proprietary trade name for the hydrochloride of Isoprenaline.

Isurol. See ICHTHYOL.

Isutan. See BISMUTAN.

Itabirites. Minerals. They are hæmatite schists.

Itacolumite. A flexible sandstone of Brazil.

Italian Earth. A pigment. It is a sienna.

Italian Green. A sulphide dyestuff obtained by heating *p*-nitro-phenol with copper sulphate in water mixed with caustic soda solution and sulphur. Dyes cotton dull green.

Italian Pink. See YELLOW CARMINE.

Italian Red. See INDIAN RED.

Itramin Tosylate. A vasodilator. 2-Nitratoethylamine toluene-*p*-sulphonate.

Itrol. Silver citrate, $C_6H_5O_7Ag_3$. Used as an antiseptic dust for wounds.

Itrumil. A proprietary trade name for the sodium derivation of Iodothiouracil.

Ittiolo. An Italian preparation of ammonium ichthyolate. A substitute for Ichthyol.

Ivaleur. A proprietary trade name for pyroxylin (cellulose nitrate).

Ivax. A proprietary preparation of neomycin sulphate, sulphaguanidine and light kaolin. An antidiarrhœal. (253).

Iversal. Ambazone.

Iviglite. A mineral. It is a variety of mica.

Ivoride. A casein product used as an electrical insulation.

Ivory, Artificial. Stearic acid made into a paste with gypsum.

Ivory Black (Cassel Black). A black pigment made by charring ivory cuttings in closed retorts. Animal charcoal, consisting of practically pure carbon, is also known as ivory black. Also Bone black and Lampblack.

Ixolite. A resin found in Austria.

Izal. A distillate from coke residues. It is a proprietary disinfectant.

Izarine. A proprietary trade name for artificial suède made from a base of rubberised fabric with cotton fibres specially treated and vulcanised.

J

Jaborandi. The native name for several drugs of a sudorific and salivating character, obtained from the leaves and twigs of various species of *Pilocarpus*.

Jabroc. See PERMALI.

Jacana Metal. An alloy of 70 per cent. lead, 20 per cent. antimony, and 10 per cent. tin.

J-acid. 2 : 5 - Amino - naphthol - 7 - sulphonic acid.

Jacinth. A mineral. It is hyacinth.

Jacksonite. A mineral. It is prehnite.

Jackson's Button Brass. See BRASS.

Jackwood. A tree cultivated in India. In Europe, it is used as a furniture wood, but in India, as the source of a yellow dye for colouring the robes of priests.

Jacobsite (Magnoferrite). A mineral, $MnO.Fe_2O_3$. Another formula is $8[(Mn,Fe,Mg)(Fe,Mn)_2O_4]$.

Jacoby Metal. An alloy of 85 per cent. tin, 10 per cent. antimony, and 5 per cent. copper.

Jacquemart's Reagent. An aqueous solution of mercuric nitrate with nitric acid. Used as a test for ethyl alcohol.

Jade. A mineral. Both Nephrite and Jadeite are known by this name.

Jadeite (Soda-spodumene, Chloromelanite). A mineral. It is a silicate of sodium and aluminium, $NaAl(SiO_3)_2$.

Jadit. A proprietary preparation of bulclosamide and salicylic acid. A skin fungicide. (312).

Jadit H. A proprietary preparation containing BUCLOSAMIDE, salicylic acid and HYDROCORTISONE used in the treatment of inflammatory and fungal skin diseases. (312).

Jaft. See DCHIT.

Jaggery. A sugary material obtained by evaporating toddy before fermentation has commenced, 8 gallons of toddy yielding 2 gallons of jaggery. In Java, jaggery is fermented, and the product distilled for alcohol.

Jahnite. An explosive. It is a mixture of nitre, lignite coal, and sulphur, with very small quantities of picric acid, potassium chlorate, and calcined soda.

Jaipurite. A mineral. It is a cobalt sulphide.

Jalap. The roots and tubers of certain convolvulaceous plants which yield purgative resins. The resin has an acid value of 13–17.

Jalapin. The active principle of jalap resin, obtained by decolorising tincture of jalap with animal charcoal and evaporating.

Jalapurgin. Jalapin, the chief constituent of jalap resin.

Jalcase. A steel with a high resistance to wear with a soft core. It has forging properties.

Jalovis. Hyaluronidase. An enzyme.

Jalpaite. A mineral. It is approximately a sulphide of silver and copper, $(Ag.Cu)_2S$.

Jalupap. A proprietary analgesic. (750). ACETAMINOPHEN.

Jamaica Ginger (Essence of Ginger, Extract of Ginger). Tincture of ginger, U.S.P., containing the alcohol-soluble matter of 200 grams of ginger per litre in 93 per cent. alcohol.

Jamaica Pepper. See ALLSPICE.

Jamaica Wood. See LOGWOOD.

Jambu Assu. The root of *Piper jaborandi*. Used by the natives of Brazil as a sudorific, diuretic, and febrifuge.

Jambul. The seeds of *Eugenia jambolana*.

James Fever Powder. A similar product to *Pulvis antimonalis B.P.*

Jamesonite. See ZINKENITE.

Jamesonite, Silver. See OWYHEEITE.

James Powder. (*Pulvis antimonalis B.P.*) It consists of antimonious oxide mixed with twice its weight of calcium phosphate (purified bone earth).

Jamrosin. A preparation asserted to be the fluid extract of jambul. It is said to possess remedial qualities in the treatment of diabetes.

Janosite. A mineral. It is a hydrated ferric sulphate.

Janthone. A synthetic perfume obtained by condensing citral or lippial with mesityl oxide. It has a violet odour.

Janus Blue. See INDOINE BLUE R.

Janus Brown B (Janus Yellow R). An azo dyestuff produced by diazotising *m*-amino-phenyl-trimethyl-ammonium chloride, coupling the product with α-naphthylamine, diazotising, and finally coupling with chrysoidine, $N(CH_3)_3Cl . C_6H_4 . N_2 . C_{10}H_6N_2 . C_6H_2 (NH_2)_2.N_2.C_6H_5$.

Janus Dyes. Basic azo dyestuffs. Yellow, red, and brown dyes are produced from the diazo derivatives of amino-phenyl-ammonium, and amino-benzyl-amine compounds, whilst blue and green dyes are derivatives of various safranines. They dye half-woollens.

Janus Green B and G. See DIAZINE GREEN.

Janus Grey. A dyestuff prepared by the action of phenol upon diazo-safranine. It is identical with Diazine black (Kalle).

Janus Red (Janus Brown B). A dyestuff. It is the chloride of trimethyl-amino-benzene-azo-*m*-toluene-azo-β-naphthol, $C_{26}H_{26}N_5OCl$. Dyes cotton from an acid bath.

Janus Yellow. See JANUS BROWN.

Japan Agar. See AGAR-AGAR.

Japan Black. See BLACK JAPAN.

Japan Camphor (Laurel Camphor). Ordinary camphor, $C_{10}H_{16}O$, which separates from the essential oil of *Laurus camphora*. Used for making celluloid and in explosives.

Japan Drier. See TEREBINE.

Japan Earth. See CUTCH.

Japanese Acid Clay (Kambara Earth). A clay having the formula $Al_2O_3.6SiO_2.XH_2O$ (X is larger than 6). It has powerful adsorptive and decolorising properties. The dried clay has strong dehydrating action. This action is used in a commercial product named Adsol, and is used for drying air in theatres and storehouses.

Japanese Antimony. See BLACK ANTIMONY.

Japanese Bell Metal. An alloy of 60·5 per cent. copper, 18·5 per cent. tin, 12 per cent. lead, 6 per cent. zinc, and 3 per cent. iron.

Japanese Brass. See BRASS and SINCHU.

Japanese Bronze. An alloy of from 81–83 per cent. copper, 10 per cent. lead, 4·6 per cent. tin, and 0–1·8 per cent. zinc.

Japanese Cement. A paste made of fine rice flour, well boiled and ground.

Japanese Drops. Japanese peppermint oil.

Japanese Gelatin. See AGAR-AGAR.

Japanese Lac. See LAC, JAPANESE.

Japanese Silver. An alloy of 50 per cent. aluminium and 50 per cent. silver.

Japanese Steel. See K.S. MAGNET STEEL.

Japanese Wax. See CHINESE WAX.

Japanese Wood Oil. An oil obtained from *Elæococca vernica* or *Aleurites cordata*, of Japan. It is stated to be of less value than Chinese wood oil, gelatinising less readily.

Japan Isinglass. See AGAR-AGAR.

Japan Lac. See LAC, JAPANESE.

Japan Sago. The starch from *Cycas revoluta*.

Japan Tallow (Sumach Wax, Vegetable Wax). Japan wax, derived from *Rhus succedanea*, *Rhus vernicifera*, and *Rhus sylvestric*. It melts at from 48–55° C., has an acid value of 11–33, a saponi-fication value 207–237, and an iodine value 8–15.

Japan Varnishes. These are obtained by blending asphalt varnishes with dark coloured copal or amber varnishes.

Japan Wax. See JAPAN TALLOW.

Jara Jara. β-Naphthol-methyl ether.

Jargoon (Jargon). A mineral. It is a green or yellow variety of Zircon (q.v.). A precious stone.

Jargon. See JARGOON.

Jargonelle Pear Essence. A solution of isoamyl acetate in ethyl alcohol. Used for flavouring confectionery.

Jarische's Ointment. An ointment containing 1 part pyrogallic acid, and 7 parts lard.

Jarlite. A mineral. It is $8[NaSr_2Al_2(F,OH)_{11}]$.

Jarosite. A hydrated basic sulphate of potassium and ferric iron, $K_2O.3Fe_2O_3.4SO_3.6H_2O$.

Jasmacyclene. A perfumery speciality. (549).

Jasmal. That fraction of the essential oil of jasmine flowers distilling at 100° C.

Jasmin. See CITRONINE B.

Jasmolide. A perfumery chemical. (549).

Jasper. A mineral. It consists of grains of quartz intermixed with clayey material, and with red and yellow oxides and hydroxides of iron.

Jasper, Opal. See OPAL JASPER.

Jatex. A proprietary brand of pure concentrated rubber latex, 60 per cent.

Jatobá Duro. A hard copal obtained from Ceará and Northern Bahia, Brazil. Specific gravity 1·033 at 15° C., melting-point 110–130° C., ash 0·08 per cent. Used in varnishes.

Jatobá Lagrima (Trapocá Resin). A rather soft copal from the Jatobá tree in Brazil. Specific gravity 1·05 at 15° C., ash 0·02–0·06 per cent., acid value 84–96. Used in spirit varnish.

Jatobá Resin. Brazilian copals from *Hymenæa Courbaril* and *Hymenæa Parvifolia*. There are hard and soft qualities. The soft is called jatobá tean and Trapocá.

Jatrevine. A condensation product of menthol camphol and isobutyl phenol. Recommended as an inhalation in cases of catarrh.

Jaulingite. A mineral resin obtained from a species of lignite.

Jaune Acide. Acid yellow (q.v.).

Jaune Anglais. Victoria yellow (q.v.).

Jaune Brilliant. See CADMIUM YELLOW.

Jaune de fer. Mars yellow (q.v.).

Jaune de mars. Mars yellow (q.v.).

Jaune d'or. Martius Yellow (q.v.).

Jaune N. See FAST YELLOW N.

Jaune Soleil. See SUN YELLOW.

Jaune Solide N. See FAST YELLOW N.

Jaune Yellow. See CITRONINE B.

Java Olives. See OLIVES OF JAVA.

Java Wax (Sumatra Wax, Gondang Wax, Kondang Wax, Getah Wax). A wax obtained from the milky juice which runs from cuts made in the bark of the gondang (wild fig) tree, *Ficus variegata*. It melts about 60° C., and is soluble in benzene, chloroform, carbon disulphide, etc.

Jaydalene. A proprietary soldering paste consisting of o-phosphoric acid with a base which vaporises without decomposition, e.g., aniline, etc.

Jecovol. An emulsion of cod-liver oil with the glycerophosphates of sodium, calcium, and iron.

Jectofer. A proprietary preparation of iron-sorbitol-citric acid complex. A hæmatinic. (338).

Jectoral. A proprietary preparation of ferrous glycine sulphate and SORBITOL. A hæmatinic. (338).

Jectoral F. A proprietary preparation of ferrous glycine sulphate, sorbitol and folic acid. A hæmatinic. (338).

Jectothane. A proprietary polyester-based polyurethane thermoplastic injection-moulding compound. (290).

Jeffamine D-230 and D-400. Trade marks for polyoxypropyleneamine. (751).

Jefferisite. A mineral. It is a hydrated mica.

Jeffersonite. A pyroxene mineral, $(Mg.Fe.Zn)(Ca.Mn)(SiO_3)_2$.

Jellettite. A mineral. It is a calcium-iron garnet.

Jellitac. A prepared starch, gluten, sold as a powder. An adhesive.

Jelly, Glycerin. See GLYCERIN JELLY.

Jelly, Lieberkuhn's. See LIEBERKUHN'S JELLY.

Jelly, Mineral. See PETROLEUM JELLY.

Jelly, Petroleum. See PETROLEUM JELLY.

Jelly Rock. See WILKINITE.

Jelly, Toothache. See TOOTHACHE JELLY.

Jelly, Vegetable. See VEGETABLE JELLY.

Jelonet. Paraffin gauze dressing. (529).

Jelutong (Pontianac, Bresk, Dead Borneo, Fluvia, Gambia). A resinous latex yielded by *Dyera costulata*. It contains from 19–24 per cent. of rubber and 75–80 per cent. of resin, and is used for mixing with rubber and for other purposes.

Jenkinsite. A mineral. It is $(Mg,Fe)_4Si_3O_{10}.3H_2O$.

Jenner's Stain. A microscopic stain for white blood corpuscles. It consists of (*a*) a solution of water-soluble, yellowish eosin, 0·5 gram, in 100 c.c. methyl alcohol, and (*b*) a solution of methyl blue, 0·5 gram, in 100 c.c. methyl alcohol. For use 25 c.c. of (*a*) are mixed with 20 c.c. of (*b*).

Jenzschite. A mineral. It is a variety of silica similar to opal.

Jeppel's Oil. Bone oil (*q.v.*).

Jequirty (Prayer Beads, Jumble Beads). Abrus, the seeds of *Abrus precatorius*, an Indian and Brazilian shrub.

Jeremejewite. A mineral, $AlO.BO_2$.

Jerntalk. A mineral. It is $Mg_3Si_4O_{10}(OH)_2$.

Jeromite. A mineral, $As(S.Se)_2$.

Jersey Lily White. A pigment. It is a lithopone.

Jersey Stone. A felspathic rock, similar to Cornish stone.

Jesuit's Balsam. The oleo-resin copaiba.

Jesuit's Bark. Cinchona bark.

Jesuit's Tea. See MATÉ.

Jet. A mineral. It is a fossilised wood, and falls between lignite and coal. Used for ornaments.

Jet Black. See GAS BLACK.

Jet Black G. A dyestuff. It is the sodium salt of disulpho-toluene-azo-α-naphthalene-azo-phenyl-α-naphthyl-amine. It dyes greener shades than the R mark.

Jet Black R. A dyestuff. It is the sodium salt of disulpho-benzene-azo-α-naphthalene-azo-phenyl-α-naphthyl-amine, $C_{32}H_{21}N_5S_2O_6Na_2$. Dyes wool bluish-black from an acetic acid or a salt bath.

Jeunite. Copper fungicide. (501).

Jevreinovite. A mineral. It is a variety of vesuvianite.

Jewelith. See ABALAK.

Jeweller's Borax. Octahedral borax, $Na_2B_4O_7.5H_2O$.

Jeweller's Brass. See BRASS.

Jeweller's Rouge. The finest calcined ferric oxide or hæmatite.

Jeweller's Salt. Finely calcined iron oxide.

Jew's Pitch. See BITUMEN.

Jeyes Disinfectant. A disinfectant containing creosote, rosin, caustic soda, and water. It forms emulsions with water.

Jezekite. A mineral. It is $Na_4CaAl_2(PO_4)_2OF_2(OH)_2$.

Jiconger Nuts. The fruit of *Telfairia pedata*.

Jicwood. See PERMALI.

Joaquinite. A mineral. It is $4[NaBa(Ti,Fe)_3Si_4O_{15}]$.

Jobramag. A German medicinal preparation containing iodine and bromine.

Jocketan. A mineral. It is an iron carbonate.

Jodisan. Hexamethyl-diamino-isopropanol-di-iodide.

Jodival. A proprietary preparation of iovalurea (iodival), iodo-isovaleryl-urea.

Joha. A solution of 40 per cent. salvarsan in iodinol.

Johachidolite. A mineral. It is $Na_2Ca_3Al_4B_6O_{14}F_5(OH)_5$.

Johannite. A hydrated sulphate of uranium and copper, containing 68 per cent. UO_3. A mineral. It is $4[Cu(UO_2)_2(SO_4)_2(OH)_2.6H_2O]$.

Johannsenite. A mineral. It is $4[CaMnSi_2O_6]$.

Johnite. A mineral. It is a variety of turquoise.

Johnsonite. A mineral. It is a variety of fibrous alum.

Johnstonite. A mineral. It is a mixture of lead sulphide, lead sulphate, and sulphur.

Johnstonotite. A mineral. It is a variety of garnet.

Johnstrupite. A mineral. It is a silicotitanate of cerium, yttrium, aluminium, magnesium, calcium, and sodium, with fluorine.

Joiner's Glue. Either skin or bone glue.

Jollyte. A mineral. It is an aluminium-iron-magnesium silicate.

Jonite. A proprietary wood fibre product.

Jordanite. A mineral, $4PbS.Bi_2S_3$. The term has also been applied to a mineral, $4PbS, As_2S_3$.

Jordisite. A mineral. It is a colloidal form of molybdenum sulphide, MoS_2.

Jordoset. A proprietary trade name for a thermosetting acrylic resin used as a baking varnish on motor cars. (189).

Joseite. A mineral, $Bi_3Te(S.Se)$. It is also given as Bi_4TeS_2.

Josephinite. A mineral. It is a nickel-iron alloy, Fe_2Ni_5.

Jossaite. A mineral. It contains lead and zinc chromates.

Jothion. A proprietary preparation of iodo-propanol.

Journal Box Brass. See BRASS.

JSR-10. A proprietary A.B.S. material possessing high impact strength. (752).

JSR-12. A proprietary A.B.S. material possessing high impact strength. (752).

JSR-21. A proprietary A.B.S. resin capable of giving good surface finish in moulding operations. (752).

J-Thane. A trade mark for two-component polyurethane systems (polyols and isocyanates) used for mixing in-place polyurethanes. (751).

Juchten Red. Fuchsine mixed with chrysoidine, to give a yellower shade.

Judson Powder. An American blasting powder, consisting of sodium nitrate, sulphur, cannel coal, and nitro-glycerin.

Juglon. 5-α-Oxy-naphthoquinone, HO. $C_{10}H_5O_2$.

Julianite. A mineral. It is a variety of tennantite.

Julienite. A mineral. It is a chloro-nitrate of cobalt.

Julin's Chloride. Hexachlor-benzene, C_6Cl_6.

Julol Violet. A dyestuff, obtained by acting upon keto-methyl-julololine with phosphorus pentachloride, containing a little oxychloride.

Jumble Beads. See JEQUIRTY.

Junijot. A German preparation. It is a spirit extract of *Retmispera plumosa*. Used as a substitute for tincture of iodine for antiseptic purposes.

Juniper Berries. The dried, ripe fruit of *Juniperus communis*.

Juniper, Gum. Sandarac.

Juniper Tar. An anti-eczematic.

Juniper Tar Oil (Oil of Cade). *Oleum cadium B.P.* It is the product of the distillation of *Juniperus oxycedrus*.

Jupiter Oil. See OIL OF JUPITER.

Jurinite. A variety of the mineral brookite.

Jurupaite. A mineral. It is a hydrated calcium-magnesium silicate, $H_2(Ca.Mg)_2.SiO_7$, of California.

Justite. A substance approximating to the formula, $(Ca.Mg.Fe.Zn.Mn)_3Si_2O_7$, found in furnace slag. The name was formerly applied to the mineral Koenite.

Jutahy. See JUTAHYCICA.

Jutahycica (Jutahy). A copal from Brazil. It is obtained from the roots of *Hymenæa Courbaril* and *Hymenæa Parvifolia*. It has a specific gravity of 1·04 at 15° C., a melting-point of 125–135° C. and an acid value of 107.

Jutahycica Resins (Paragum, Resina Animé). Brazilian copal resins from *Hymenæa Courbaril* and *Hymenæa Parvifolia*.

Jute Scarlet R. A dyestuff. It is a British equivalent of Azococcin 2R.

Juvel. A proprietary preparation of Vitamins A, D, C, thiamine, riboflavine, pyridoxine and nicotinamide. Vitamin supplements. (272).

Juvelith. See BAKELITE.

Juvenin. A German preparation. It consists of tablets containing small quantities of yohimbine-methyl-arsi-nate and strychnine-methyl-arsinate.

Juvidur. A proprietary polyvinyl chloride. (753).

Juvocaine. Novocaine. An anæsthetic.

K

K34. See HEXACHLOROPHANE.

K285. Absorbable dusting powder. (502).

Kalar. A proprietary name for a range of copolymers of polyisobutylene and isoprene of low molecular weight, compounded with fillers, extenders and plasticisers. They are used as elasto-meric sealers and caulking compounds. (482).

K.A. Alloy. An aluminium alloy resembling duralumin.

Kabaite. A mineral wax of the ozo-kerite type.

Kabikinase. A proprietary preparation of streptokinase. A fibrinolytic agent. (318).

Kachin. A photographic developer. Its active constituent is pyrocatechol, $C_6H_4(OH)_2.(I : 2)$.

K-acid. 1 : 8-Amino-naphthol-4 : 6-di-sulphonic acid.

Kacynoids. Specified weight potassium cyanide. (512).

Kade Oil. Cade oil (*q.v.*).

Kadox. A very fine zinc oxide used as a rubber filler. It is in a specially active colloidal condition, with average particle size of 0·15 microns diameter. There are three grades : Black label containing lead as litharge not exceeding 0·1 per cent., blue label, 0·1–0·25 per cent., and red label, 0·25–1 per cent.

Kaempferol. The colouring matter of the blue flowers of *Delphinium consolida*. It is a trihydroxy-flavonol.

Kærsutite. A mineral. It is $2[(Na, K, Ca)_{2-3}(Mg, Fe''Ti)_5(Al, Si)_8O_{22}(OH)_2]$.

Kagoo Oil. See KORUNG OIL.

Kahlerite. A mineral. It is $Fe''(UO_2)_2(AsO_4)_2.12H_2O$.

Kainite. A salt found in the Stassfurt deposits, consisting mainly of potassium magnesium sulphate and magnesium chloride, $K_2Mg(SO_4)_2.MgCl_2.6H_2O$. The crude material consists of a mixture of kainite and rock salt, and contains 23 per cent. of potassium sulphate. Used as potash fertiliser, and as a source of potassium sulphate.

Kainosite (Cenosite). A mineral, $CaY_2(SiO_3)_4.CaCO_3.2H_2O$ (Y=yttria metals) of Norway.

Kairine A. The corresponding ethyl compound to Kairine M (*q.v.*).

Kairine M. 8-Hydroxyl-1-methyl-tetra-hydro-quinoline hydrochloride, $C_9H_9(OH)N.CH_3.HCl$. A febrifuge.

Kairoline A. *n*-Ethyl-tetrahydro-quino-line.

Kairoline B or M. *n*-Methyl-quinoline-tetrahydride, $C_{10}H_{13}N$. It is similar to Kairine. The sulphate is the commercial salt.

Kairoline, Nitro. See NITRO-KAIRO-LINE.

Kaiser Green. See EMERALD GREEN.

Kaiserling Solution. A solution used

for preserving tissue. It contains 3 grams potassium acetate, 1 gram potassium nitrate, 75 c.c. water, and 30 c.c. formaldehyde.

Kaiser Red. See EOSIN BN.

Kaiserroth. See EOSIN BN.

Kaiser Yellow. An old name for Aurantia (q.v.).

Kakachlore. A mineral. It is asbolane.

Kaki. Ebony wood of Japan.

Kakishibu. A Japanese product. It is an aqueous extract of unripe Kaka fruit. Used for preserving fishing nets, and waterproofing paper.

Kaki-sibu. An antiseptic product obtained from the juice of a plant. Used in Japan.

Kakodyljacol. See CACODYLIAGOL.

Kakoxene. A mineral, $2Fe_2O_3.P_2O_5.12H_2O$.

Kaladana. The seeds of *Ipomœa lederacea*.

Kalaite. Turquoise (q.v.).

Kalammon. A new German fertiliser containing 17 per cent. nitrogen and 30 per cent. calcium carbonate.

Kalaqua. A Belgian tonic containing caffeine.

Kalene. A proprietary name for a range of copolymers of polyisobutylene and isoprene of low molecular weight, used as linings for tanks containing acids. (482).

Kaleoilris. A proprietary filling compound. It has a specific gravity of 1·024.

Kalgoorlite. See COOLGARDITE.

Kali (Skokian). An intoxicating drink made from the roots of the S. African plant, *Raphionacme divaricata*.

Kaliammon Saltpetre. A potassium-ammonium nitrate prepared by mixing equivalent molecular proportions of solid potassium chloride and ammonium nitrate in the presence of a little water. A fertiliser.

Kaliborite. A hydrated borate of magnesium and calcium, $K_2Mg_4B_{22}O_{38}14H_2O$.

Kalicine (Kalicinite). A mineral. It is potassium hydrogen carbonate, $KHCO_3$.

Kalicinite. See KALICINE.

Kalif. A proprietary copper-lead bearing alloy. It melts at 952° C., and has a tensile strength of 10,000 lb. per sq. in. at 21° C.

Kali, Lemon. Sherbet.

Kalinite. A mineral. It is potash alum, $K_2SO_4.Al_2O_3.SO_3.24H_2O$. It is also given as $KAl(SO_4)_2.11H_2O$.

Kali Nuts (Cali Nuts). The seed of *Mucuna urens*.

Kaliohitchcockite. A mineral. It is $KAl_3(PO_4)_2(OH)_4.2H_2O$.

Kaliophilite. A mineral. It is a potassium-aluminium silicate.

Kalio-tonal (Tonol). Potassium-glyceryl-phosphate. A proprietary trade name for a preparation used medicinally for nervous complaints. See TONAL.

Kalipol 18. A proprietary potassium polyphosphate solution used in the manufacture of liquid detergents. (503).

Kalite. A proprietary form of chalk prepared by a special process whereby it has a very small particle size, and the particles are coated with a calcium soap. A rubber filler.

Kalithomsonite. A mineral. It is a variety of thomsonite containing potassium.

Kalium. Potassium, K.

Kalium Durules. A proprietary preparation of potassium chloride in a sustained-release tablet, used to prevent potassium depletion. (338).

Kaliuzoto. A new Italian fertiliser containing nitrogen, potassium, and organic matter. It is manufactured from residual molasses by allowing kieselguhr to absorb molasses of Bé 32–36° until the consistency of the mass is nearly solid. It is sold as a granular powder.

Kalkammon. A fertiliser. It is a mixture of ammonium chloride with 30 per cent. chalk.

Kalkeisenolivin. Synonym for Iron monticellite.

Kalkowskite. A mineral, $(Fe.Ce)_2O_3.4(Ti.Si)O_2$.

Kalkowskyn. A mineral. It is $Fe_2Ti_3O_9$.

Kalkstickstoff. See NITROLIM.

Kalle's Acid. α-Naphthylamine-disulphonic acid, $C_{10}H_5(SO_3H)_2NH_2.(1:2:7)$.

Kalle's Indigo Salt T. See INDIGO SALT T.

Kalle's Salt. Acetone bisulphite, $CH_3.CO.CH_3.NaHSO_3$.

Kallidinogenase. A vasodilator. An enzyme that splits the kinin known as kallidin from kininogen. GLUMORIN, DEPOT-GLUMORIN.

Kallikrein. A proprietary preparation. It is a solution of a hormone preparation in biologically standardised form. It gives diminution in blood pressure temporarily.

Kallilite. A mineral, NiBiS.

Kallodent. Registered trade name for a methyl methacrylate thermoplastic material used for moulding dentures.

Kallodoc. Acrylic powder for artificial teeth and eyes. (512).

Kalmadol. A proprietary preparation of phenpropamine. A gastrointestinal antispasmodic. (357).

Kalmex. Thermit compound for feeding ingots and castings. (531).

Kalmm. A proprietary preparation of MEPROBAMATE. (754).

Kalmopyrin (Soluble Aspirin, Tylcalsin, Solupyrine, Analutos, Kalsetal, Calcium Aspirin). Calcium acetyl - salicylate, $(C_2H_3O.OC_6H_4.COO)_2Ca$. Used as an antirheumatic and analgesic.

Kaloempang Beans. See OLIVES OF JAVA.

Kaloo Nuts. The nuts of *Aleurites Fordii*. They furnish tung oil.

Kalpak. A scented mixture of petroleum jelly and wood tar.

Kalprotholite. A mineral, $Cu_6Bi_4S_9$.

Kalsetal. See KALMOPYRIN.

Kalsomines. See DISTEMPERS.

Kalspray. Polyurethane prepolymer spray formulation. (557).

Kaluszite. See SYNGENITE.

Kalvan. A proprietary trade name for calcium carbonate for use in rubber to give wear resistance. It has an ultra-fine particle size.

Kalzana. A proprietary calcium-sodium lactate.

Kalzose. A casein preparation containing calcium.

Kamala. See KAMELA.

Kamaresite (Kamarezite). A native basic copper sulphate. A mineral. It is $Cu_3SO_4(OH)_4.6H_2O$.

Kambara Earth. See JAPANESE ACID CLAY.

Kambe Wood. See REDWOODS.

Kamela (Kamala). A dyestuff obtained from the seeds or fruits of *Mallotus phillpenis* or *Rottlera tinctoria*. Used in India as medicine, and for dyeing silk orange.

Kamforite (Carboniet). A proprietary fertiliser and insecticide. It is stated to supply nitrogen, phosphorus, and potash to the soil, and to destroy insects.

Kamillosan. A proprietary preparation of chamomile oil, hexylresorcinol and extract of chamomile, used in the treatment of sore gums. (370).

Kampa. Synthetic waxes and surface active agents. (622).

Kamptulicon. A floor covering made from rubber, jute, and fillers.

Kampylite. See MIMETISITE.

Kanamycin. An antibiotic produced by *Streptomyces kanamyceticus* in which KANNASYN and KANTREX are the sulphate. It is also present as the sulphate in KANFOTREX and KANTREXIL.

Kanda Haree. A variety of asafœtida in the Bombay market.

Kandol. See GASOLINE.

Kaneite. A mineral, $MnAs_2$.

Kanelstein. Synonym for Grossular.

Kane's Salt. A salt prepared by Kane by dissolving mercuric nitrate in a boiling solution of ammonium nitrate. The crystals have the formula $Hg(NO_3)_2.2NH_3.2H_2O.NH_2.HgNO_3$.

Kanfotrex. A proprietary preparation of kanamycin, amphymycin and hydrocortisone used in dermatology as an antibacterial agent. (324).

Kanga Butter. See LAMY BUTTER.

Kanigen. Chemical nickel plate. (503).

Kanja Butter. See LAMY BUTTER.

Kankar. See KUNKUR.

Kannasyn. A proprietary preparation containing kanamycin (as the sulphate). An antibiotic. (112).

Kanten. A variety of agar-agar from red Tegusa seaweed of Japan.

Kanthal Alloy. A trade mark for an iron alloy containing aluminium, cobalt, and chromium, and having a high degree of resistance to heat, a low electrical conductivity, and good hot and cold working properties.

Kanthosine J. See TOLUYLENE ORANGE G.

Kanthosine R. See TOLUYLENE ORANGE R.

Kantrex. A proprietary preparation containing kanamycin sulphate. An antibiotic. (324).

Kantrexil. A proprietary preparation containing kanamycin sulphate, pectin, bismuth carbonate and activated attapulgite. An antidiarrhoeal. (324).

Kaodene. A proprietary preparation of codeine phosphate and kaolin used as an anti-diarrhœal. (280).

Kaogel. A proprietary preparation containing kaolin and pectin. An antidiarrhoeal. (264).

Kaolin. See CHINA CLAY.

Kaolinase. A purified kaolin.

Kaolin Cataplasm (Antiphlogistine, Denver Mud). A substance obtained by mixing kaolin, glycerin, and boric acid with small quantities of thymol, methyl salicylate, and peppermint oil. Used in medicine.

Kaolinite (Myelin). A crystalline mineral similar to kaolin, $H_4Al_2Si_2O_9$. A mineral. It is also given as $2[Al_2Si_2O_5(OH)_4]$.

Kaomycin. A proprietary preparation containing neomycin sulphate and kaolin. An antidiarrhœal. (325).

Kaopectate. A proprietary preparation containing kaolin, bentonite and pectin. An antidiarrhœal. (325).

Kaovax. A proprietary preparation of succinylsulphathiazole, and kaolin. An antidiarrhœal. (350).

Kapak. A material made from the mineral rubber Elaterite. Used in rubber mixings.

Kapazang Oil. A fat obtained from the seeds of *Hodgsonia heteroclita*.

Kapeloid. Kaolin poultice. (591).

Kapilon Soluble. A proprietary trade name for Menadoxime.

Kapithamia Piscum. See WOOD-APPLE GUM.

Kapnite. See CAPNITE.

Kapok. A cotton-like down produced in the seed-pods of the kapok tree. Used in making life-saving jackets.

Kaporie Tea. The leaves of *Epilobium angustifolium*. Largely used in Russia as a beverage.

Kapsol. A proprietary trade name for a methoxy-ethyl-oleate. Used as a plasticiser.

Kapsovit. Multi-vitamin capsules. (510).

Kapton. A proprietary trade name for polyimide resin in the form of film. (10).

Kapur Kachri. See SANNA.

Karachaite. A mineral. It is $MgSiO_3.4H_2O$.

Kara-haue. A Japanese alloy containing copper, zinc, tin, and lead.

Karakane. An alloy of 62·5 per cent. copper, 25 per cent. tin, 9·4 per cent. zinc, and 3·1 per cent. iron.

Karaya Paste. A proprietary preparation of STERCULIA in isopropyl alcohol, used as a dressing around colostomies. (311).

Karbolite (Carbolite). A Russian artificial resin made from phenols, formaldehyde, and naphthol-sulphonic acid.

Karbolon. A proprietary thermoplastic of Soviet origin made by blending polycarbonate with polysulphone to give better heat resistance. (424).

Karbos. A char made from charred sawdust, purified by acid treatment, and mixed with animal charcoal. A decoloriser and filtering medium.

Karbozit. A fuel prepared from lignite, wood, or peat, by heating the raw material to 450° C.

Karelinite. A mineral. It is an oxysulphide of bismuth, $Bi_2S.Bi_2O_3$.

Karetnja. A bituminous insulation used for leadless cables. It is stated to be impermeable to water. It consists mainly of asphalt, with an aluminium stearate binder.

Karite Gum. See GUTTA-SHEA.

Karite Oil. An oil obtained from the seeds of *Bassio paski*.

Karkade Tea. A drug consisting of the dried flowers of *Hibiscus Sabdariffa*. It contains 5·85 per cent. nitrogenous matter and 94 per cent. dry matter.

Karma Metal. An alloy of 80 per cent. nickel and 20 per cent. chromium. It is a heat-resisting alloy and melts at 1415° C.

Karmarsch Metal. An alloy containing 88·8 per cent. tin, 7·4 per cent. antimony, and 3·7 per cent. copper.

Karnak. A preparation containing 16 per cent. ethyl alcohol, 0·132 gram anthraquinones, 0·008 gram alkaloids, and 0·55 gram glycyrrhizic acid per 100 c.c.

Karolith. A casein preparation similar to Galalith, used for the manufacture of buttons and other objects.

Karpholite. A mineral, $(Mn.Fe)O.(Fe.Al)_2O_3.SiO_2.2H_2O$.

Karstenite. A mineral. It is a variety of anhydrite which occurs in the Stassfurt deposits.

Karu. See DUKA.

Karumga Gum. A gum of Nigeria, from *Acacia seyal*.

Karvol. (1) A 5 per cent. solution of chlorcarvacrol.

Karvol. (2) A proprietary preparation of menthol, chlorbutol, chlorothymol and essential oils. A decongestant inhalant. (280).

Karyinite. See CARYINITE.

Karyocerite. See CARYOCERITE.

Kaschkawal. A Bulgarian cheese prepared mostly from sheep's milk, and rarely from goat's milk. It contains about 29 per cent. fat, and 29 per cent. nitrogenous matter.

Käseleim. A cementing material made from cheese and slaked lime.

Käseleim-pulver. A cementing material made from lime and skim milk.

Kasolite. A mineral. It is a hydrated silicate of uranium and lead, $3PbO.3UO_3.3SiO_2.4H_2O$, found in the Belgian Congo.

Kaspine Leather. Leather goods prepared from oil tannage, and treated with dilute formaldehyde made alkaline with sodium carbonate.

Kastel. A registered trade mark for ion exchange resins. (61).

Kastel A 100. A medium basic exchanger with amine and quaternary ammonium active groups for industrial water treatment. (61).

Kastel A 101. A highly porous, weakly basic anion exchanger, monofunctional (tertiary amine active groups) for industrial water treatment. (61).

Kastel A 105. A weakly basic exchanger with tertiary amine active groups for sugar juices treatment. (61).

Kastel A 300. A strongly basic exchanger (type 11) with dialkyl-alkanol amine active groups for industrial water treatment. (61).

Kastel A 300P. A porous strongly basic exchanger (type 11) with dialkylalkanol amine active groups for industrial water treatment. (61).

Kastel A 500. A strongly basic exchanger (type 1) with trialkylamine active groups for industrial water treatment. (61).

Kastel A 500P. A porous strongly basic exchanger (type 1) with trialkylamine active groups for industrial water treatment. (61).

Kastel A510. A highly absorbent resin with strongly porous matrix and strongly basic active groups: specially recommended as scavenger for industrial water treatment. (61).

Kastel A 501 D. A decolorising resin with a highly porous matrix and strongly basic active groups for sugar juices treatment. (61).

Kastel C100. A weakly acidic exchanger with carboxylic active groups for industrial water treatment. (61).

Kastel C101. A weakly acidic cation exchanger with carboxylic active groups and higher exchange capacity or industrial water treatment. (61).

Kastel C300. A strongly acidic exchanger with sulphonic active groups for industrial water treatment. (61).

Kastel C300P. A porous strongly acidic exchanger with sulphonic active groups for industrial water treatment. (61).

Kastel C300 AGR. A porous highly crosslinked exchanger with sulphonic active groups for industrial water treatment. (61).

Kastel C300 AGR/P. A porous highly crosslinked exchanger with sulphonic active groups for industrial water treatment. (61).

Kastle-Meyer Reagent. A 2 per cent. solution of phenol-phthalein in 20 per cent. aqueous caustic potash decolorised by boiling with zinc powder. It gives a pink colour with copper salts when 4 drops of the reagent are added to 10 c.c. of the solution to be tested, and 1 drop of hydrogen peroxide (5 vols.).

Kat. See ARABIAN TEA.

Katalabu Gum. A gum of Nigeria, from *Acacia Sieberiana*. An adhesive.

Katamen. Pyrazolone-phenyl-dimethyl-sulphamine-benzoate.

Katangite. A mineral. It is a hydrated copper silicate, $CuH_2SiO_4.H_2O$, found in the Belgian Congo.

Katanol. A trade mark for a range of mordants used in the dyeing of textiles. (983).

Katapleite. A mineral, $(Na_2Ca)O.ZrO_2.$ $3SiO_2.2H_2O$.

Katarsit. A calcium sulphite pellet for use as a dechlorinating agent for water.

Katavel Oil. The oil from *Hydnocarpus wightiana*.

Katbél-ki-gond. See WOOD-APPLE GUM.

Katchung Oil. Peanut oil.

Katechu. See CUTCH.

Kath. See ARABIAN TEA.

Katharin. A proprietary trade name for carbon tetrachloride used as a grease remover.

Katharite. A mineral. It is alunogen.

Kathro. Semi-refined cholesterol. (526).

Katigen. A registered trade mark currently awaiting re-allocation by its proprietors to cover a range of dyestuffs. (983).

Katonium. A proprietary preparation of ammonium and potassium polystyrene sulphonate. A diuretic. (112).

Katorin. A proprietary preparation of potassium gluconate. A potassium supplement. (253).

Katzenstein Bearing Metal. See BEARING METALS.

Kauaiite. A mineral. It is an aluminium-potassium-sodium sulphate.

Kau Drega. See TALOTALO GUM.

Kauk Catalyst. A proprietary trade name for a spherical catalyst of 5 mm. diameter consisting of potassium salts and vanadium on a porous silica carrier. V_2O_5 content 6·5 per cent. It is used for converting SO_2 into SO_3.

Kauresin. A trade name for phenol formaldehyde resins. (49). (See also BAKELITE.)

Kauri Gum. See AUSTRALIAN COPAL.

Kaurit. A trade name for urea formaldehyde resins. (49). (See also BEETLEWARE.)

Kava-kava Resin. A mixture of resins and resin acids from the dried roots of *Piper methysticum*.

Kawasaki Hakkinko. A proprietary Japanese steel containing 0·19 per cent. carbon, 1·8 per cent. silicon, 1·0 per cent. manganese, 17·0 per cent. nickel, 25·0 per cent. chromium, and 0·2 per cent. molybdenum. Offers resistance to hydrogen embrittlement.

Kawrie Gum. See AUSTRALIAN COPAL.

Kaydox. 1,4-dichlorobenzene paste. (501).

Kaylene. A proprietary preparation of colloidal aluminium silicate.

Kaylene-ol. A proprietary preparation of colloidal aluminium silicate with liquid paraffin.

Kaynitro. Concentrated nitrogen/potash fertiliser. (512).

Kayserite. A mineral. It is aluminium hydroxide.

K-Contin. A proprietary preparation of potassium chloride in a delayed-release capsule. It is used as a potassium supplement in diuretic therapy. (334).

Keatingine. A mineral. It is fowlerite.

Keeleyite. A mineral from Bolivia, approximating to the formula $2PbS$, $3Sb_2S_3$. Synonym for Zinckenite.

Keene's Alloy. See NICKEL SILVERS.

Keene's Cement. The name for a number of different plasters prepared by various manufacturers. It is usually obtained by calcining the plaster stone to plaster of Paris, dipping this into a solution of alum or aluminium sulphate, and recalcining.

Kefaldol. A chemical combination resulting from the interaction between phenetidin, quinine, citric acid, and salicylic acid. It has been used as an antipyretic and antirheumatic.

Keffekilite. A fuller's earth.

Keffekill. Synonym for Sepiolite.

Kefir. See KEPHIR.

Keflex. A proprietary preparation containing cephalexin monohydrate. An antibiotic. (333).

Kefglycin. See CEPHALOGLYCIN.

Keflin. A proprietary preparation of CEPHALOTHIN sodium. An antibiotic. (463).

Kefzol. A proprietary preparation of CEPHAZOLIN sodium. An antibiotic. (463).

Keilhauite. A mineral, $CaYt(Ti.Al.Fe)SiO_5$.

Kekuna Oil. Bakoly oil (candlenut oil)

Kel-F. A proprietary extrusion and moulding material resistant to high temperatures. Polychlorotrifluoroethylene. (35).

Kel-F-Elastomer 3700. Polychlorotrifluoroethylene-vinylidene fluoride, 30 : 70. A proprietary synthetic rubber resistant to high temperatures. (35).

Kel-F Elastomer 5500. Chlorotrifluoroethylene-vinylidene fluoride, 50 : 50. A proprietary synthetic rubber resistant to high temperatures. (471).

Keleastol. A proprietary trade name for a ricinoleate type of vinyl plasticiser. (109).

Kelene. Ethyl chloride, C_2H_5Cl.

Kelenmethyl. A mixture of ethyl and methyl chlorides. An anæsthetic.

Kelferron. A proprietary preparation of ferrous glycine sulphate. A hæmatinic. (249).

Kelfizine W. A proprietary preparation of sulphametopyrazine. An antibiotic. (353).

Kelfolate. A proprietary preparation of ferrous glycine sulphate, and folic acid. A hæmatinic. (249).

Kelgin. A proprietary trade name for a sodium alginate.

Kelgum. A linseed oil rubber substitute.

Kellite. A proprietary trade name for a synthetic resin.

Kelly's Paint. A benzoated collodion containing tincture of benzoin, glycerin, and collodion. Used for painting on abrasions of the skin.

Kelocyanor. A proprietary preparation of cobalt tetracemate in a glucose solution. An antidote for cyanide poisoning. (323).

Kelo-form. A proprietary trade name for ethyl amino-benzoate.

Kelp (Varec). The ash from certain algæ, called kelp in Scotland, varec in Normandy. The seaweed and the ash from the seaweed are known as kelp. A source of iodine.

Kelp Ash. The ash from seaweed. Used as a source of iodine.

Kelpchar. A decolorising carbon obtained from seaweed by carbonising in two stages and extracting with water and dilute hydrochloric acid.

Kelp Salt. A mixture of potassium chloride, with some alkaline sulphates and carbonates, formed in the preparation of potassium chloride from kelp.

Keltan. A proprietary name for a range of EPDM terpolymers having differing Mooney viscosities. (755).

Kemadrin. A proprietary preparation of procyclidine hydrochloride. An antiparkinsonism agent. (277).

Kematal. A trade mark for acetal copolymers. A moulding material possessing outstanding dimensional stability, high strength and rigidity, low coefficients of friction and good electrical insulating properties. (756).

Kemgo. A proprietary trade name for inks for use with heat.

Kemick. Heat-resisting paint. (512).

Kemicetine. A proprietary preparation of CHLORAMPHENICOL. An antibiotic. (365).

Kemisuede. A proprietary leather shoe lining made by coating sheeting with rubber latex.

Kemite. A ceramic material of uniform texture made by patented processes. Its pores are filled with carbon in the form of coke. It is hard, dense and suitable for table tops. It is apparently unaffected by acids except hydrofluoric. It has a specific gravity of 1·87.

Labstone is the name given to the material when used for laboratory bench tops.

Kemithal. A proprietary preparation of thialbarbitone sodium. Intravenous anæsthetic. (2).

Kemlet Metal. An alloy consisting mainly of zinc, with aluminium and copper.

Kemp. The shorter fibres of Mohair.

Kempite. A mineral found in the so-called alum rock of California. It approximates to the formula, $Mn_4Cl_2O_6$.$3H_2O$.

Kempol. A proprietary trade name for vulcanisable vegetable oil polymers.

Kempy Wool. Wool prepared from sheep badly fed, or subjected to exposure. It dyes badly, and in streaks, and is used for making horse rugs.

Kemsol. Saltcake of special quality. (512).

Kemthal. A proprietary trade name for Thialbarbitone.

Kenalog. A proprietary preparation of TRIAMCINOLONE acetonide. (682).

Kenbond 300 Series. A proprietary name for a range of general-purpose epoxy resin adhesives. (418).

Kencast 700 Series. A proprietary name for a range of general-purpose epoxy resin potting compounds. (418).

Kendex 0869X. A proprietary trade name for a petroleum type of vinyl plasticiser/extender. (110).

Kenflex. A proprietary trade name for a vinyl plasticiser. The condensation products of alkyl naphthalenes. (111).

Kenflex A. A proprietary trade name for a vinyl plasticiser. A dimethyl naphthalene derivative. (111).

Kenmag. A proprietary dispersion of magnesium oxide. (757).

Kenngottite. Synonym for Miargyrite.

Kennigottite. A mineral. It is a variety of miargyrite.

Kenseal 600 Series. A proprietary name for a range of general-purpose potting, casting and encapsulating resins. (418).

Kenstrip 902. A proprietary non-flammable, thermosetting resin-remover. (418).

Kenthane 102. A proprietary general-purpose epoxy casting, potting and encapsulating resin compatible with polyurethane resins. (418).

Kentish Rag. A siliceous limestone, used as an adulterant of Portland cement.

Kentite. An explosive. It contains from 32–35 per cent. ammonium nitrate, 32–35 per cent. potassium nitrate, 16–18 per cent. ammonium chloride, and 14–16 per cent. trinitro-toluene.

Kentrolite. A mineral, $Mn_4Pb_3Si_3O_{15}$.

Keottigite. A mineral, $(Zn.Co.Ni)_3$ $(AsO_4)_2$.

Kephaldol. A preparation made from phenetidine, citric and salicylic acids.

Kephir (Kefir). An imitation of fermented mare's milk (Koumiss). It is cow's milk fermented by the addition of a special ferment, *Saccharomyces kephiri*, made in Western Europe. It is a drink, and contains 2 per cent. alcohol. Also see KOUMISS.

Kephos. Non-aqueous phosphating solution. (512).

Ker. A Polish trade name for a butadiene copolymer.

Kerament. Slabs of concrete covered with a special glaze.

Keramite. Synonym for Mullite.

Keramo. Tiles of coloured glass, used for floors and walls.

Keramohalite. See ALUNOGEN.

Keramonite. A skeleton of metal embedded in a ceramic mass. It is very resistant to temperature changes. Also see THERMONIT.

Keramyl. A solution of hydrofluosilicic acid, H_2SiF_6.

Keraphen. Sodium tetraiodo-phenol-phthalein.

Kerargyrite. See CERARGYRITE.

Kerasin. See MENDIPITE.

Kerasite (Cerasine, Horn Lead, Cromfordite, Phosgenite). A mineral. It is a compound of lead chloride and carbonate, $PbCl_2.PbCO_3$.

Kerasol. Tetraiodo-phenol-phthalein, $C_{20}H_{10}O_4I_2$.

Keratite. A name applied to a vulcanite.

Keratol. A cellulose waterproofing compound.

Kerchenite. A mineral. It is $Fe''Fe_2'''$ $(PO_4)_2(OH)_2.6H_2O$.

Kerecid. A proprietary preparation of idoxuridine. An ocular antiseptic. (281).

Kerimid 500. A proprietary polyamide-imide polymer in solution form. (758).

Kerimid 501. A proprietary polyamide-imide polymer in film form. (758).

Kerimid 502. A proprietary polyamide-imide polymer in the form of a green paste, comprising a thermosetting polymer and an aluminium powder filler. (758).

Kerimid 503. A proprietary polyamide-imide polymer in film form, coloured green. (758).

Kerimid 601. A proprietary thermosetting polyimide used in the manufacture of glass-fibre laminates. (758).

Kerite. A rubber substitute containing vegetable oil, waxes, bitumen, coal tar, sulphur, and a little tannin.

Kerites. See PETROLITE.

Kerman. A German disinfectant containing 22.5 per cent. fluosilicic acid.

Kermes (Alkermes, Kermes Berries, Scarlet Corns, Grains of Kermes). The dried females of the shield louse, *Coccus ilicis*. It contains a red dye related to carmine, and is used for dyeing in Turkey. The name kermes is also applied to an impure antimony sulphide, Sb_2S_3, containing antimony oxide, Sb_2O_3. Specific gravity is 4.5. See KERMES MINERAL.

Kermes Berries. See KERMES.

Kermes Grains. See KERMES.

Kermesin Orange. See ORANGE T.

Kermesite. A mineral. It is $8[Sb_2S_2O]$. See RED ANTIMONY.

Kermes Mineral (Brown-Red Antimony Sulphide). A mixture of trisulphide and trioxide of antimony, containing alkali. Used in medicine.

Kermesome. Synonym for Kermesite.

Kernite. A new sodium borate mineral found in California.

Kerute. A new sodium borate mineral found in California.

Kern's Hydraulic Bronze. An alloy of 78 per cent. copper, 12 per cent. tin, and 10 per cent. zinc.

Kerocain. A local anæsthetic, said to be identical with German novocaine.

Kerogen. The organic matter contained in shale. It amounts to from 20–27 per cent. and splits up upon distillation into water, ammonia, gas, and oil.

Kerol. A proprietary disinfectant. It contains phenol, and forms emulsions with water.

Keromask. A proprietary cosmetic preparation containing titanium oxide and ochre pigments. (759).

Keronyx. A proprietary trade name for a casein plastic material used for the manufacture of combs, etc.

Kerosene (Paraffin Oil, Astral Oil, Coal Oil). A refined distillate of petroleum, 150–300° C. An illuminating oil.

Kerosene-Lime. A garden insecticide prepared by mixing lime with kerosene and water.

Kerotenes. A name suggested for asphaltic materials which are insoluble in carbon disulphide, and which yield hydrocarbon products when distilled.

Kerstenite. A mineral. It is a lead selenite.

Kesso Oil. Japanese valerian oil. It is distilled from the roots of *Valeriana officinalis*, and is used in pharmacy.

Kest. A proprietary preparation of magnesium sulphate and PHENOLPHTHALEIN used in the treatment of constipation. (137).

Kester. Fatty acid esters. (526).

Ketalar. A proprietary preparation of KETAMINE hydrochloride. An anæsthetic agent.

Ketamine. An anæsthetic. 2-(2-Chlorophenyl) - 2 - methylaminocyclohexa - none. KETALAR is the hydrochloride.

Kethamed. A proprietary preparation for pemoline. A central nervous stimulant. (250).

Ketobemidone. A narcotic analgesic. 4 - (3 - Hydroxyphenyl) - 1 - methyl - 4 - propionylpiperidine. CLIRADON.

Keto-Diastix. A proprietary test-strip used to detect ketones and glucose in urine. (807).

Ketol. The trade name for a series of products with boiling-points ranging from 60–200° C., obtained by the rectification of the crude distillate from the dry distillation of calcium butyrate. It is a mixture of ketones. The calcium butyrate is obtained by the saccharification of sawdust and selective butyric fermentation of the sugars. The ketols are used as industrial solvents.

Ketone Base. Tetramethyl - diaminobenzophenone. An intermediate for dyes.

Ketone Blue 4BN. A dyestuff. It is the sulphonic acid of ethoxy-trimethylphenyl - triamino - triphenyl - carbinol, $C_{30}H_{31}N_2OCl$. Dyes wool and silk blue.

Ketone Blue G and R. Dyestuffs of the same type as Patent blue V, N.

Ketone Chloroform. Chloroform prepared from acetone, and not from alcohol.

Ketone Musk. An artificial musk perfume. It is dinitro-tertiary-butylxylyl-methyl ketone.

Ketonone. A proprietary trade name for benzoic acid derivatives used as plasticisers for cellulose acetate and cellulose nitrate.

Ketonone B. A proprietary trade name for butyl-benzoyl benzoate. A plasticiser.

Ketonone E. A proprietary trade name for ethyl *o*-benzoyl benzoate. A plasticiser.

Ketonone M. A proprietary trade name for methyl *o*-benzoyl benzoate. A plasticiser.

Ketonone M.O. A proprietary trade name for methyl-ethyl benzoyl benzoate. A plasticiser.

Keto Resins. Artificial resins obtained by the polymerisation of aldehydeketone condensation products.

Ketostix. A proprietary test strip of buffered sodium nitroprusside and glycine, used for the detection of ketones in urine, serum or milk. (807).

Ketovite. A proprietary multi-vitamin preparation. (330).

Ketrax. See LEVAMISOLE.

Kevlar 49. A proprietary aromatic polyamide fibre of great strength. See FIBRE B and ARAMID FIBRE. (10).

Keweenawite. A mineral, $(Cu.Ni.Co)_2As$.

Key Alloy. A nickel-silver containing 60–65 per cent. copper, 20–26 per cent. zinc, 12 per cent. nickel, 1–2 per cent. lead, and 0–0.4 per cent. iron.

Keystone. Adhesives. (617).

K.G.S. Powder. An explosive. It is a 33-grain powder.

K-Gutta. A patented thermoplastic, flexible insulating material designed especially to meet the needs of the modern high-speed submarine telegraph and telephone cable. It is a mixture of the hydrocarbon from gutta-percha or balata, or both, with purified petroleum jelly, stabilised with an anti-oxidant. The material has extremely good electrical properties at telephone frequencies and sea-bottom conditions.

Khadi. An intoxicating beverage made by the natives of the Transvaal, from the roots of *Mesembryanthemum Mahoni*.

Khaki. A colouring matter produced on the fibre. The material is dipped in chrome alum, ferrous sulphate, and pyrolignite of iron, and then passed through a solution of sodium silicate.

Khaki Brown C. See KHAKI YELLOW C.

Khaki, Mineral. See MINERAL KHAKI.

Khaki Yellow C (Khaki Brown C, Cross Dye Black FNG). Sulphur dye-stuffs.

Khari Salt. A native salt of India consisting chiefly of sodium sulphate. Used for curing skins.

Kharophen. A proprietary preparation. It is 3-acetyl-amino-oxyphenyl-arsinic acid.

Kharsin. An arsenic compound formerly used in the treatment of syphilis. It has now become obsolete.

Kharsivan. A proprietary name for Salvarsan (*q.v.*).

Kharsulphan. A proprietary preparation. It is sulpharsenobenzene.

Khodnevite. A mineral. It is Na_2AlF_5.

Khoharite. A mineral. It is $8[Mg_3Fe_2'''Si_3O_{12}]$.

Khordofan Gum (Arabian Gum, Turkey Gum). Gum arabic. It is the picked gum derived from *Acacia senegal*, and is considered the finest gum arabic obtainable.

Khotinski Cement. An American preparation made by macerating gum lac in 10 per cent. of its weight of North Carolina oil of tar on a water bath. Used for jointing glass and metal.

Kibdelophane. A mineral. It is a variety of ilmenite.

Kickh. A Lebanon food product. It is a mixture of bourghoul and lében (*q.v.*). It contains 12 per cent. H_2O, 0·99 per cent. P_2O_5, 7·05 per cent. fat, 1·72 per cent. lactose, and 50·56 per cent. starch.

Kick Plate Brass. See BRASS.

Kidnamin. A proprietary preparation of essential aminoacids used as a dietary supplement. (760).

Kidney Cotton. Peruvian cotton.

Kidney Iron Ore. A mineral. It is a massive, dull red variety of hæmatite.

Kidney Stone. A mineral. It is Nephrite (*q.v.*). It was worn by the ancients as a charm against kidney disease.

Kiel Compound. An insulating material containing rubber, sulphur, and mineral oil. It sometimes contains pumice and beeswax.

Kienmeyer's Amalgam. An amalgam consisting of 2 parts mercury, 1 part tin, and 1 part zinc. Used as a coating for frictional electrical machines.

Kien Oil. Turpentine oil obtained by the dry distillation of resinous wood. The refined oil has a specific gravity of 0·86–0·88 at 15° C., a refractive index of 1·466–1·481 at 15° C., and boils at 150–180° C.

Kienol. Russian oil of turpentine.

Kieselgalmei. Synonym for Hemimorphite.

Kieselguhr. See INFUSORIAL EARTH.

Kieselguhr Dynamite (Gurdynamite, Dynamite). Ordinary dynamite, consisting of nitro-glycerin absorbed by kieselguhr.

Kieserite. Occurs in the Stassfurt deposits. It is magnesium sulphate, $MgSO_4.H_2O$, and is a source of Epsom salts.

Kietyöite. A mineral. It is a variety of apatite.

Kievite. A mineral. It is a variety of hornblende.

Kil. Insecticides. (509).

Kilbrickenite. A mineral, $Pb_6Sb_2S_9$.

Kilbrush. Concentrated organic acids—brushwood killers. (589).

Kilianite. A proprietary synthetic resin product.

Killas. The local name in Cornwall for clay-slates.

Killed Spirits. A solution of zinc chloride. Prepared by dissolving zinc in commercial hydrochloric acid until action ceases.

Killgerm. Disinfectants and antiseptics. (574).

Kilmacooite. A mineral. It is (Zn,Pb)S.

Kilmet. Selective weedkillers. (507).

Kiltree. Concentrated organic acid—tree killers. (589).

Kimitotantalite. A mineral. It is a variety of tantalite.

Kinar. A proprietary modified polyvinylidene fluoride. (424).

Kinectine. Quinine-benzo-sulphon-*p*-amino-phenyl-arsinate.

Kinel 5502. A proprietary polyimide casting, potting and encapsulating resin. (450) (758).

Kinel 5514. A proprietary polyimide moulding composition reinforced with glass fibre. (450) (758).

Kinel 5517. A proprietary free-sintering, self-lubricated, heat-resistant polyimide moulding powder. (450) (758).

Kinetic No. 12. See FREON.

Kinetine. A combination of quinine and bectine.

Kinetite. An explosive. It is a mixture of the jelly formed by dissolving guncotton in nitrobenzene, with potassium chlorate or potassium nitrate and sulphur.

Kineurine. Quinine glycero-phosphate $C_3H_7O_3 . H_2PO_3 . (C_{20}H_{24}N_2O_2)_2.4H_2O$. An antiperiodic.

Kingley's Alloy. An alloy used in dental work. It contains 94 per cent. tin and 6 per cent. bismuth.

King's Blue. See SMALT and COBALT BLUE.

King's Green. Oil Green (q.v.).

Kingston Bronze. An alloy of 85 per cent. copper, 12 per cent. zinc, 2·5 per cent. tin, and 0·05 per cent. iron.

King's Yellow. A pigment. It is arsenic sulphide, and occurs naturally as Orpiment (q.v.). Also see CHROME YELLOW.

Kinic Acid (Chinic acid). Quinic acid, $C_7H_{12}O_6$.

Kinidin Durules. A proprietary preparation of quinidine bisulphate. (338).

Kinite. A proprietary trade name for a steel containing 12·5–14·5 per cent. chromium, 1·5 per cent. carbon, 1·1 per cent. molybdenum, 0·7 per cent. cobalt, 0·55 per cent. silicon, 0·5 per cent. manganese, and 0·4 per cent. nickel.

Kino (Kino Gum). The dried juice obtained from incisions in the trunk of *Pterocarpus marsupium*. It resembles catechu, and is used in dyeing and in medicine.

Kino, Australian (Kino, Eucalyptus, Kino, Botany Bay). The dried exudation of *Eucalyptus* species.

Kino, Bengal (Kino, Madras, Dhak Gum). Butea gum, from *Butea frondosa*.

Kino, Botany Bay. See KINO, AUSTRALIAN.

Kino, Cochin (Kino, Malabar). *Kino B.P.*

Kino, Eucalyptus. See KINO, AUSTRALIAN.

Kinonglas. A safety glass consisting of two plates of glass with cellulose nitrate between them.

Kino Gum. See KINO.

Kino, Madras. See KINO, BENGAL.

Kino, Malabar. See KINO, COCHIN.

Kinradite. A trade name for a quartz mineral resembling jasper.

Kipp's Glycerin Substitute. A mixture of quince seed, sugar, and boric acid.

Kips. The skins of small cattle, imported mostly from India

Kipushite. A mineral of Katanga having the composition $(Cu.Zn)_3(PO_4)_2 + 3(Cu.Zn)(OH)_2. + 3H_2O$.

Kirchberger Green. A pigment. It has the same composition as Scheele's green.

Kirkite, New. See VARVICITE.

Kirschwasser. A spirit obtained by the distillation of the fermented juice of cherries. Imitation Kirsch is prepared from grain spirit flavoured with peach stones, or cherry laurel leaves.

Kish. Crystalline graphite formed in blast furnace slag during iron smelting.

Kischtimite. A mineral. It contains fluo-carbonates of the cerium metals and is found in the Urals.

Kiton. (1) A mixture of tar and clay. A dust-laying composition for roads.

Kiton. (2) Acid wool dyestuffs. (530).

Kiton Fast Orange G. A dyestuff. It is a British equivalent of Orange G.

Kiton Yellows. British equivalents of Flavazine S.

Kittool Fibre. A fibre obtained from the leaves of a Ceylon palm, *Caryota urens*. Used in the manufacture of brushes.

Kiwit. A German explosive.

Kjerulfine. A mineral. It is a variety of wagnerite.

Klaprothite. See CHESSYLITE.

Klaprotholite. A mineral. It is a copper-bismuth sulphide.

Klee's Salt. Acid potassium oxalate, $KHC_2O_4.H_2O$.

Kleinenberg's Fat Mixture. A solution of cacao butter and spermaceti in castor oil. Used as an embedding material in microscopy.

Kleinenberg's Fixative. Used in microscopy. It consists of 100 c.c. of a saturated aqueous solution of picric acid, 3 c.c. of sulphuric acid, and 300 c.c. of water.

Kleinenberg's Stain. A microscopic stain. It consists of a saturated solution of alum and calcium chloride in alcohol (70 per cent.) diluted with six times its volume of alcohol (70 per cent.), to which is added an alcoholic solution of hæmatoxylin until the colour is violet blue.

Kleinite. A mineral. It is a mercury-ammonium-chloride.

Klein's Reagent. A saturated solution of cadmium borotungstate, $2(Cd_2H_2W_8O).7(WO_3)B_2O_3.H_2O$. Used for the separation of minerals.

Klementite. A mineral. It is an aluminium-iron - magnesium - manganese silicate.

Kletzinsky's Wood Paste. An artificial wood prepared from gelatin, alumina, and sawdust.

Klingerite. An asbestos preparation containing 80 per cent. asbestos, 2 per cent. flax, and 17 per cent. rubber and balata (the agglutinant). A material of German manufacture, used as a packing for steam-pipe joints.

Klinophaite. A mineral, $(Fe.Al)_2(Fe.Mg.Ni.Ca)$ $(K.Na)_2(OH)_6(SO_4)_5.5H_2O$.

Klipsteinite. A mineral. It is a manganese-iron silicate.

Kljakite. A mineral. It is a colloidal aluminium hydroxide.

Klochmannite. A new copper selenide mineral of Argentine and Sweden.

Klorax. A solution of Chloramine T. An antiseptic.

Kloref. A proprietary preparation containing betaine hydrochloride and potassium bicarbonate. (248.)

Kluchol. Anethol benzoate.

Klucine. A proprietary waterproofing compound

" K "-Monel. A new proprietary alloy. It contains nickel and copper in approximately the same ratio as in monel metal with the addition of 4 per cent. aluminium.

Knadolin. A Swiss insecticide. It contains 58 per cent. amyl alcohol, 39 per cent. soft soap, 2 per cent. nitrobenzene, and 1 per cent. sodium xanthogenate.

Knapp's Solution. Mercurous chloride (10·8 grains) are treated with potassium cyanide solution until the addition of caustic soda causes no precipitate. Caustic soda solution (100 c.c. of specific gravity 1·145), added, and the whole diluted to 1 litre. Used for the estimation of glucose.

Knebelite. A mineral. It is an iron-manganese-magnesium silicate.

Kneiss Alloy. An alloy of 42 per cent. lead, 40 per cent. zinc, 15 per cent. tin, and 3 per cent. copper. Used for machine bearings. Another alloy contains 50 per cent. zinc, 25 per cent. tin, and 25 per cent. lead.

Knight's Patent Zinc White. See LITHOPONE.

Knoch's Patent Insulating Material. A heat insulator. It is prepared from wood, jute, wool, argillaceous shale, and glue.

Knopite. A mineral. It is a variety of perovskite (calcium titanate, $CaTiO_3$), containing cerium and yttrium metals, and iron.

Knoppern (Acorn Galls). Galls (q.v.) produced in the immature acorns of various species of oak. A tanning material.

Knoxvillite. A mineral. It is a chromium-iron-aluminium sulphate.

Kobaltchrom. Cobaltcrome (q.v.).

Kobaltfahlerz. A mineral. It is $(Cu,Co)_3SbS_3$.

Kobaltnickelpyrit. Synonym for Hengleinite.

Kobeite. A mineral. It is $(Yt,Fe,U)(Ti,Cb,Ta)_2(O,OH)_6$.

Kobellite. A mineral, $3PbS(Bi.Sb)_2S_3$.

Kochelite. A mineral of Silesia. It is a niobate of the rare metals with ThO_2, SiO_2, and Ca.

Kochenite. A fossil resin resembling amber.

Kocher-Fonio. See COAGULEN CIBA.

Kochite. A Japanese mineral, $2Al_2O_3.3SiO_2.5H_2O$.

Kochlin's Bearing Bronze. An alloy of 90 per cent. copper and 10 per cent. tin.

Koch's Acid. 1-Naphthylamine-3 : 6 : 8-trisulphonic acid.

Kodaflex DMP, DEP, etc. A trade mark for the following plasticisers.

DMP. Dimethyl phthalate.
DEP. Diethyl phthalate.
DBP. Dibutyl phthalate.
DIBP. Diisobutyl phthalate.
DMEP. Di - (2 - methoxyethyl) phthalate.
DOP. Dioctyl phthalate.
OIDP. Octyl isodecyl phthalate.
DIDP. Diisodecyl phthalate.
DOA. Di-octyl adipate.
DIDA. Diisodecyl adipate.
DOZ. Dioctyl azelate.
DBS. Dibutyl sebacate.

Kodaloid. A proprietary trade name for a cellulose nitrate. It is made in the form of sheets.

Kodapak. A proprietary trade name for transparent cellulose acetate film. Used for making packets.

Koenerite. A mineral, $Al_2O_3.3MgO.3MgCl_2.6H_2O$. It occurs at Stassfurt.

Koechlinite. A native bismuth-molybdenum oxide, $Bi_2O_3.MoO_3$. A mineral. It is also given as $4[(BiO_2)_2MoO_4]$.

Koenenite. A mineral. It is $Mg_5Al_2Cl_4(OH)_{12}.2H_2O$.

Koenig Solder. An alloy of 60 per cent. tin, 30 per cent. aluminium, and 10 per cent. antimony.

Koerzit. An alloy for permanent magnets made by Krupp, containing 1·1 per cent. carbon, 3·5 per cent. manganese, 36 per cent. cobalt, 4·8 per cent. chromium, the remainder being iron.

Koerzit, I., II., III. Proprietary cobalt steels containing 10, 20, and 30 per cent. cobalt respectively.

Kohlensilesia A4. A German safety explosive, consisting of 80 per cent. potassium chlorate, 16 per cent. resin, and 4 per cent. nitrated resin.

Kohlenspath. Synonym for Whewellite.

Kohlerite. A mineral. It is onofrite.

Koka Seki. A variety of pumice stone found in the Nūjima Islands, near Tokio. It is used as a building material, and in the production of reinforced concrete.

Koken. A proprietary synthetic resin.

Koko. The leaves of *Celastrus buxifolia*. Used in Natal as a sumac substitute for tanning.

Kokowai. A variety of rouge used by the Maori.

Koktaite. A mineral. It is $2[(NH_4)_2 Ca(SO_4)_2 . H_2O]$.

Kokum Butter (Garcinia Oil, Concrete Oil of Mangosteen). A fat obtained from the seeds of *Garcinia indica* or *G. purpurea*. It is composed of stearine, myristicine, and oleine.

Kolanticon. A proprietary preparation of aluminium hydroxide, magnesium oxide, DICYCLOMINE hydrochloride and DIMETHICONE used as a gastro-intestinal sedative. (690).

Kolantyl. A proprietary preparation containing dicyclomine hydrochloride, aluminium hydroxide, magnesium hydroxide and methyl cellulose. An antacid. (263).

Kolantyl-NV. A proprietary preparation of DICYCLOMINE hydrochloride, aluminium hydroxide, magnesium hydroxide and magnesium trisilicate. An antacid. (690).

Kola Nut (Kola Seeds). The seeds of *Cola acuminata* and *C. vera*.

Kola Seeds. See KOLA NUT.

Kolax. An explosive of the same type as Carbonite.

Kolbeckine. Synonym for Herzenbergite.

Kolbeckite. A new mineral of Saxony of specific gravity 2·39. It appears to be a silico phosphate of beryllium.

Kolen-Carbonite. A Belgian explosive containing 25 per cent. nitro-glycerin, 34 per cent. potassium nitrate, 1 per cent. barium nitrate, 38·5 per cent. flour, 1 per cent. tan meal, and 0·5 per cent. sodium carbonate.

Kolinit. An artificial resin product made from coal. The coal is treated with a phenol, e.g. cresol, the product washed with, benzene, milled, dried, and moulded. Used for the manufacture of buttons, dishes, and electrical fittings.

Kol-kol Gum. A gum of Nigeria, from *Acacia senegal*.

Kollag. Oildag. See OILDAG AND AQUADAG.

Kollargol. See COLLARGOL.

Kollofast. Pathological museum specimens colour preservative. (544).

Kolm. A variety of bituminous coal found in Sweden. The ash contains from 1–3 per cent. of uranium oxide, U_3O_8.

Koloran. See AVIVAN.

Kolosorukite. A mineral. It is a variety of jarosite.

Kolovratite. A mineral. It is a nickel vanadate.

Kolpon. A proprietary preparation of œstrone in pessary form. A treatment for vaginitis. (316).

Kolton. A proprietary trade name for Piprinhydrinate.

Kombé Arrow Poison. Strophanthus, the seed of *Strophanthus hispidus*.

Kombé Seeds. See KOMBÉ ARROW POISON.

Komed. A proprietary preparation of sodium thiosulphate, salicylic acid, resorcinol, menthol, camphor, isopropyl alcohol and alumina. Acne lotion. (265).

Kommoid. A sulphurised corn oil rubber substitute.

Konakion. A proprietary trade name for Phytomenadione. A proprietary preparation of Vitamin K. (314).

Kondang Wax. See JAVA WAX.

Konel. Proprietary nickel-cobalt-iron alloys containing about 2·5 per cent. titanium. They are high temperature resisting alloys and possess high tensile strength at elevated temperatures.

Kongsbergite. A natural silver amalgam found in Norway.

Konichalcite. A mineral, (Cu.Ca)(CuOH) (As.P.V)O_4.

Konilite. A silica in powder form.

Koninokite. A mineral. It is a ferric phosphate.

Konnan Bark. Obtained from *Cassia fistula* of Southern India. A tanning material.

Kon Oil (Kusum Oil). Macassar oil obtained from the seeds of *Schleichera trijuga*. It has a saponification value of 215–230, an iodine value of 48–69, and an acid value of 6–35.

Konstrastin. Zirconium basic acetate, sold for weighting silk.

Konstructal. An aluminium alloy containing 1 per cent. copper or 8 per cent. zinc.

Kontakt. A purified form of the Twitchell reagent, used for the hydrolysis of fatty glycerides.

Kontrastin. Pure zirconium oxide, ZrO_2. Used for defining the intestines in X-ray photographs.

Konzentrole. A term used for essential oils free from terpenes and sesquiterpenes. Used for flavouring.

Koodilite. A mineral. It is a variety of thomsonite.

Kopan. See BAKELITE and ABALAK.

Kopols. Commercial products consisting of esterified copal resins. Used in varnishes.

Koppeschaar Solution. A bromine solution of N/10 strength.

Koppite. A mineral, (Ca.Ce.Fe.Hg. Na$_2$)O.NbO$_2$.H$_2$O.

Korad A. A registered trade name for acrylic film. (13).

Koraton. A proprietary trade name for a synthetic resin.

Koreon. A basic chromium sulphate, Cr(OH)SO$_4$. Used in the tanning industry.

Korginite. A mineral. It is caryinite.

Kornelite. A mineral. It is Fe$_2$(SO$_4$)$_3$. 9H$_2$O.

Kornerupine. A mineral. It is an aluminium-magnesium silicate.

Kornite. An artificial horn made from horn scraps.

Kornka. A proprietary name for granulated lime for agricultural use. (899).

Koro. A proprietary trade name for an alloy of 98 per cent. copper and 2 per cent. nickel.

Korogel. A proprietary trade name for a soft Koroseal (q.v.).

Korolac. A proprietary trade name for solutions of Koroseal (q.v.). It is used in acid-resisting tank linings.

Koron. A proprietary trade name for polyvinyl chloride.

Koronit. A German explosive.

Koronium Bromide. The trade name for strontium bromide, SrBr$_2$.6H$_2$O.

Koroplate. A proprietary synthetic paint in which Koroseal (q.v.) is the base. It is extremely resistant to acid fumes.

Koroseal. A proprietary trade name for a rubber-like thermoplastic varying in hardness, from soft jellies to hard rubber. It is detained by treating highly polymerised vinyl chloride with plasticisers at high temperatures and cooling. It can be worked like rubber when hot but requires rather higher temperatures. It is resistant to light, water, oils, and most other chemicals. It is used for impregnating and coating paper, fabrics, and metals for the manufacture of tubing for corrosive materials, and cable sheathing. Other proprietary names for this rubber substitute are Welvic, Telcovin. See also under POLYVINYL CHLORIDE.

Korrelkool. An activated charcoal in grains made by the Allgemeine Norit Maatschappij.

Korteite. Synonym for Koenenite.

Korum. See ACETAMINOPHEN. (761).

Korung Oil (Kagoo Oil). Pongam oil, obtained from the fruits of *Pongamia glabra*.

Koryfin. See CORYFIN.

Korynite. A mineral, (Ni.Fe)(As.Sb)S.

Kosam Seeds. The fruit of *Brucea sumatrana* of China. Used in dysentery and diarrhœa.

Kosmos Black, 3XB, BB, and F4. A proprietary gas black used in rubber mixings. Also used as a black pigment.

Kosmos Fibre (Artificial Wool). A product made from the waste obtained from the spinning of jute mixed with similar waste from linen and hemp.

Kostil. Styrene-acrylonitrile copolymer.

Kotonin. Cottonised flax (q.v.).

Köttigite. A mineral. It is a zinc arsenate.

Koumiss (Kumys, Kumiss). A Tartar drink, prepared by the action of a peculiar ferment known as kephir (kepir), (called by the Mohammedans " the Prophet's millet ") upon mare's milk. It contains 1·7 per cent. alcohol. In England, it is a preparation of cow's milk, and is used as a therapeutic food.

Kourbatoff's Reagents. (a) A 4 per cent. solution of nitric acid in isoamyl alcohol. (b) A 20 per cent. solution of hydrochloric acid in isoamyl alcohol, with the addition of one-third of its volume of a saturated solution of nitraniline or nitrophenol in alcohol. (c) Consists of 1 pint of a 4 per cent. solution of nitric acid in acetic anhydride, 1 part methyl alcohol, 1 part ethyl alcohol, and 1 part isoamyl alcohol. (d) Contains 3 parts of a saturated solution of nitrophenol, and 1 part of a 4 per cent. solution of nitric acid in ethyl alcohol. Used as etching agents in the micro-analysis of carbon steels.

Kovar Alloy (Fernico Alloy). A registered trade mark for an alloy of iron with 23–30 per cent. nickel, 17–30 per cent. cobalt and 0·6–0·8 per cent. manganese. Various compositions will give alloys which have coefficients of expansion practically identical with all commercial hard glasses. Kovar is stated to form a seal with glass by means of solution of its surface oxides in the glass.

Kowree Gum. See AUSTRALIAN COPAL.

KP-23. A proprietary trade name for a plasticiser consisting of butoxy-ethyl stearate.

KP 90. A proprietary trade name for a vinyl plasticiser of the epoxy type.

KP-140. A proprietary trade name for a plasticiser consisting of tributoxyethylphosphate. (73).

KP 201. A proprietary trade name for a vinyl plasticiser. Dicyclohexyl phthalate.

KP 220. A proprietary trade name for a vinyl plasticiser. Capryl glycollate. (73).

KP 555. A proprietary trade name for a

vinyl plasticiser. Bis(dimethylbenzyl) ether.

Kra. A term used in the Malay for certain pegmatites containing tourmaline, muscovite, and cassiterite.

Kraft Paper. A paper produced by the sulphate pulp process.

Kraft's Metal. A fusible alloy containing 5 parts bismuth, 3 parts lead, and 1 part tin. It melts at 104° C.

Kramerite. Synonym for Proberite.

Krantzite. A variety of Retinite.

Kratites. Explosives. They are mixtures of perchlorates with nitro-glycerin and nitro-cellulose.

Kraton. A proprietary name for a range of thermoplastic rubbers used for footwear and as adhesives. (99).

Kraurite. A mineral, $Fe_2(OH)_3PO_4$.

Krausite. A mineral, $K_2SO_4.Fe_2(SO_4)_3.2H_2O$, of California.

Kraut's Reagent. See DRAGENDORF'S REAGENT.

Krebisote. Creosote-bismuth.

Krelos. A mixture of tar distillates and a rosin soap solution.

Kremersite. A mineral, $2KCl.2NH_4Cl.2FeCl_3.3H_2O$.

Kremnitz. See FLAKE WHITE.

Kremser White. The purest form of white lead. See WHITE LEAD.

Krems White. See FLAKE WHITE.

Krennerite. See WHITE TELLURIUM.

Kreolin. The trade name for a preparation of crude carbolic acid. Antiseptic.

Kreosal. See TANOSAL.

Kresamin (Cresamol). A mixture of 25 per cent. tricresol with ethylene-diamine, $H_2N.CH_2.CH_2.NH_2$. A powerful antiseptic.

Kresatin. m-Cresol-acetate.

Kresegol. See EGOLS.

Kresival. A German preparation. It contains the water-soluble calcium salts of the sulphonic acids of the cresols.

Kreso. See BACTEROL.

Kresolig. See BACTEROL.

Kresolin. See KRESOPOLIN.

Kresol Red. See CRESOL RED.

Kresopolin (Kresolin). Preparations of crude carbolic acid. Disinfectants.

Kresosolvin. See CREOLIN.

Kresosteril. See CRESOSTERIL.

Kreuzbergite. A mineral. It is a hydrated aluminium phosphate, found in Bavaria.

Kriegr-o-dip. A proprietary trade name for liquid dyes for plastics. S—standard chemical dye. A—for cellulose acetate. W—powder dye for use in hot water. V—for polystyrene.

Krist-o-kleer. A proprietary trade name for a plasticiser containing 50 per cent. dextrose and 50 per cent. levulose.

Krisuvigite. A mineral. It is a basic copper sulphate.

Kröhnkite. A mineral, $Na_2SO_4.CuSO_4.2H_2O$.

Krokoloy. A proprietary trade name for a steel containing 14 per cent. chromium with some cobalt.

Krokydolite. A mineral, $Na_2OFe_2O_3.4SiO_2FeO.SiO_2$.

Kromax. An electrical resistance alloy of 80 per cent. nickel and 20 per cent. chromium.

Kromore. An alloy of nickel with 15 per cent. chromium. Used for the heating elements in wire-wound electric furnaces. It has a specific resistance of 98 micro-ohms. cm. at 0° C.

Kronisol. A proprietary trade name for dibutoxy-ethyl-phthalate, a plasticiser. (73).

Kronit. See TENACIT.

Kronitex AA or Kronitex 1 or K3. Proprietary trade names for vinyl plasticisers. Tricresyl phosphate. (73).

Kronitex MX. A proprietary trade name for cresyl phenyl phosphate. A vinyl plasticiser. (73).

Krugite. A mineral, $K_2SO_4.MgSO_4.4CaSO_4.2H_2O$. It occurs in the Stassfurt deposits, and is decomposed by water.

Krung Sap. The oleo-resin of *Dipterocarpus crinetus*.

Krupp Bearing Metal. See BEARING METAL, KRUPP'S.

Kruppin. An electrical resistance alloy containing 28 per cent. nickel and the rest iron.

Kryalith. A proprietary trade name for a synthetic cryolite.

Krylene 606. A registered trade mark for a cold polymerised, alum coagulated non-staining butadiene-styrene rubber.

Krylene 608. A registered trade mark for a cold polymerized styrene butadiene rubber. (230).

Krynac 27.50. A proprietary acrylonitrile rubber. (230).

Krynac 34.35. A proprietary cold-polymerised gel-free oil-resistant butadiene/acrylonitrile rubber. (230).

Krynac 34.50. A proprietary cold-polymerised gel-free oil-resistant butadiene/acrylonitrile rubber. (230).

Krynac 34.60 SP. A proprietary nitrile rubber capable of withstanding temperatures up to 135° C. (230).

Krynac 34.80. A proprietary cold-polymerised gel-free oil-resistant butadiene/acrylonitrile rubber. (230).

Krynac 34.140. A proprietary acrylonitrile rubber possessing high viscosity. (230).

Krynac 823X2. A registered trade mark for a copolymer of acrylonitrile and butadiene containing a medium level of bound acrylonitrile. (230).

Krynac 833. A proprietary isoprene acrylonitrile rubber containing 31·0% bound acrylonitrile. Its Mooney viscosity is 70. (230).

Krynac 843. A proprietary medium nitrile rubber containing 50% dioctyl phthalate as an extender. (230).

Krynac 850. A registered trade mark for a vinyl-modified nitrile rubber. (230).

Krynac 881 and 882. Proprietary names for synthetic rubbers of the ethylacrylate type. (230).

Krynac 882X1. A registered trade mark for a low temperature resistant acrylic rubber for oil seals. (230).

Kryofine (Cryofine, Methoxetin). Methylglycollic-phenetidin, $C_6H_4(OC_2H_5)NH.CO.CH_2OCH_3$. Recommended as an analgesic for pains of nervous origin, and as an antipyretic in various fevers.

Kryogen Blacks, G, BG, B, N. Dyestuffs produced from the condensation of dinitro-*m*-dichlor-benzene with *p*-amino-phenol and its sulphonic and carboxylic acids. On thionation they give black, greenish-black, and bluish-black shades respectively.

Kryogene Blue G and R. (Kryogene Brown). Dyestuffs obtained by fusing with sodium polysulphides, the product obtained by the reduction of 1 : 8-dinitro-naphthalene with sodium sulphide in the presence of sodium sulphite. Dyes cotton blue or brown.

Kryogene Brown. See KRYOGENE BLUE G and R.

Kryogene Yellow G. A dyestuff obtained from the thiourea derivative of *m*-toluylene-diamine, mixed with benzidine.

Kryogene Yellow R. A dyestuff obtained by fusing *m*-toluylene-dithiourea with sulphur.

Kryogenin. See CRYOGENIN.

Kryptocyanines. A series of dyestuffs obtained by dissolving lepidine-ethyl iodide in boiling alcohol and adding a solution of sodium ethoxide and formaldehyde, with exclusion of air. They are purple-black in colour, and are used as photo-sensitising dyes.

Kryptol. See CRYPTOL.

Krysolgan. The sodium salt of a complex 4-amino-2-aurophenol-1-carboxylic acid. Stated to be used in the treatment of tuberculosis.

Krystallazurin. A fungicide consisting of ammoniacal copper sulphate.

Krystallin. See KYANOL.

Krystallos. Quartz.

Krytox. A proprietary name for a range of fluorinated greases used as lubricants in aircraft and missiles. (10).

K-Slag. Potassium basic slag. (618).

K.S. Magnet Steel. A cobalt steel containing 35 per cent. cobalt. It is suitable for short magnets. Also see PERMANITE, COBALT STEEL, and JAPANESE STEEL.

K.S. Powder. An explosive. It is a 42-grain powder.

K.S. Seewasser Alloy (K.S. Seawater Alloy). A German aluminium alloy containing 3 per cent. manganese, 2·5 per cent. magnesium, and 0·5 per cent. antimony. It is stated to resist seawater very well, and to resist attack by fatty acids.

Ktenasite. A mineral. It is $7(Cu,Zn)_3 SO_4(OH)_4.17H_2O$.

Ktypeite. A mineral similar to Aragonite.

Kubeite. Synonym for Rubrite.

Kuhne Phosphor Bronze. An alloy of 78 per cent. coppper, 10·6 per cent. tin, 10·45 per cent. lead, 0·57 per cent. phosphorus, and 0·26 per cent. nickel.

Kühnite. A mineral. It is berzelite.

Kukkersite. An oil shale of Esthonia, of specific gravity 1·2–1·4. It contains about 55 per cent. volatile matter, and when distilled at 500° C. yields from 70–80 gallons per ton of oil of specific gravity 0·92–0·93.

Kuk-Seng. A Chinese drug. It is the dried root of *Sophora fiavescens*, and contains the alkaloid matrine.

Kummel. A Russian liqueur obtained by distilling brandy with Dutch cumin seeds, and dissolving sugar in the distillate.

Kumys. See KOUMISS.

Kundaite. A variety of grahamite.

Kunerol. See CUNEROL.

Kunheim Metal. A pyrophoric alloy containing cerium earth metals. It consists of hydrides of these metals, together with a certain proportion of aluminium and magnesium, usually 85 per cent. cerium earth hydrides, 10 per cent. magnesium, and 1 per cent. aluminium.

Kunkur (Kankar). An argillaceous limestone used in India for making mortar.

Kunstharz AP. A trade name for polyphenylvinyl ketone. (3).

Kunstharz AP 100. A trade name for isobutyl phenol formaldehyde condensate. (3).

Kunstharz AP 200. A trade name for isohexyl phenol formaldehyde condensate. (3).

Kunstharz AP 300. A trade name for isooctyl phenol formaldehyde condensate. (3).

Kunstharz AW2. A trade name for cyclohexanone. (3).

Kunstharz FX. A trade name for an

acetaldehyde condensate used as a shellac substitute. (3).

Kunstharz HW. An ammonia condensed phenol formaldehyde resin melting at 55° C. (3), (4).

Kunstharz W60. See LUPOLEN B.60. A trade name for a synthetic resin composed of butyl phenolformaldehyde condensate, trimethylolpropane adipic acid ester, ethylcellulose, butanol, methyl glycol acetate. (3).

Kunststein. An artificial stone made from magnesite.

Kunzite. A transparent violet variety of Spodumene (q.v.).

Kuoxam. A cellulose solvent prepared by dissolving 50 grammes of copper sulphate in 300 c.c. water and adding ammonia until all the copper hydroxide is precipitated. The precipitate after filtration is dissolved in 25 per cent. ammonia solution.

Kupferdermasan. A salicyl-copper soap preparation containing 2 per cent. copper. A bactericide.

Kupfernickel. See NICCOLITE.

Kupferrite. A mineral. It is a variety of anthophyllite. It is $4[(Mg,Fe)_7Si_8O_{22}(OH)_2]$.

Kupferuranit. Synonym for Metazeunerite.

Kuracap. A proprietary accelerator. 2 - mercaptobenzthiazole + dibenzthia - zole disulphide. (197).

Kurade. Accelerator for rubber. (583).

Kuranol. A proprietary preparation of zinc oxide, mercurous chloride, bismuth oxychloride, camphor, phenol, and witch hazel. Rectal ointment. (391).

Kurgantaite. A mineral. It is (Sr,Ba) $B_2O_4 \cdot \frac{1}{2}H_2O$.

Kurnakovite. A mineral. It is $Mg_2B_6O_{11} \cdot 13H_2O$.

Kurchi. The root bark of *Holarrhena antidysenteriea*. A febrifuge.

Kurofan. A registered trade mark for polyvinylidene chloride. (49).

Kurofan D. A registered trade mark for a polyvinylidene chloride dispersion. (49).

Kuroko (Black Ore). A complex copper ore of Japan. It contains zinc blende and Galena.

Kurom 1. A jewellery alloy of copper with tin and cobalt.

Kuromoji Oil. An oil from *Lindera* species. It is used in Japan for perfuming soaps and oils and contains 1-α-phellandrene, nerolidol, linaloöl, and geraniol.

Kurrodur. A proprietary trade name for an alloy of copper with 0·75 per cent. nickel and 0·5 per cent. silicon.

Kurrol Salts. Alkaline metaphosphates insoluble in water, but soluble in pyrophosphate solutions. They are produced by heating sodium trimetaphosphate or ethyl sodium phosphate.

Kurskite. A mineral. It contains calcium phosphate, fluoride, and carbonate.

Kusambi Nuts (Pacca Nuts). Nuts obtained from *Schleichera trijuga*.

Kusum Oil. See KON OIL.

Kutch. See CUTCH.

Kutera Gum. See BASSORA GUM.

Kutnohorite. A mineral. It is a manganiferous dolomite.

Küttner Silk. An artificial silk prepared by the viscose process.

Ku-Zyme. A proprietary preparation of amylolytic, lipolytic, proteolytic and cellulolytic enzymes, used to replace natural enzymes after gastrectomy or pancreatitis. (320).

Kyanite (Cyanite, Disthene, Rhoetizite). A mineral. It is an aluminium silicate, $Al_2O_3 \cdot SiO_2$.

Kyanol (Krystallin, Anilin, Benzidam). Aniline, $C_6H_5NH_2$. Kyanol is the term used by Runge in 1834. Also see ANILINE OIL.

Kyapootic Oil. Cajuput oil, obtained from the leaves of *Melaleuca leucadendron*.

Kylindrite. A mineral, $Pb_6Sn_6Sb_2S_{21}$.

Kymrie Green. The Solway Dyes Co. Alizarin Cyanine Green (q.v.).

Kynar. A proprietary trade name for polyvinylidene fluoride. (762).

Kynar 500. A proprietary polyvinylidene fluoride resin. (762).

Kynazac. A slate dust used as a filler.

Kynex. A proprietary preparation of sulphamethoxypyridazine. An antibiotic. (306).

Kynite. An explosive containing 24–26 per cent. nitro-glycerin, 2–3 per cent. wood pulp, 32–32½ per cent. starch, 31–34 per cent. barium nitrate, and 0–0·5 per cent. calcium carbonate.

Kynoch's Smokeless Powder. An explosive powder containing 52 per cent. nitro-cotton, 19·5 per cent. dinitrotoluol, 1·4 per cent. potassium nitrate, 22 per cent. barium nitrate, and 2·7 per cent. wood meal.

Kynol. A highly cross-linked amorphous phenolic polymer. It resists temperatures up to 2500°C.

Kynurin. *p*-Hydroxy-quinoline, $C_9H_6(OH)N$.

Kypfarin. Warfarin. (589).

Kypnix 40. Nicotine sulphate. (589).

Kyrock. A rock asphalt consisting of sand with about 7 per cent. bitumen. Originates from Kentucky. A paving material.

Kyrosite. A mineral. It is a variety of marcasite containing arsenic.

Kysite. A proprietary trade name for a phenol-formaldehyde synthetic resin with a fibre filler.

L

LA-12. A trade name for HYDROXO-COBALAMIN. (763).

Labarraque's Solution. A solution of chlorinated soda, containing $2\frac{1}{2}$ per cent. of available chlorine.

Labdanum (Ladanum). A resinous substance obtained from various species of *Cistus*. A stimulant expectorant.

Laben Raieb. A fermented milk product of Egypt, similar in composition to Miciurata (*q.v.*).

Labetalol. A beta adrenergic blocking agent currently undergoing clinical trial as " AH 5158A " and " Sch 15719W ". It is 5 - [1 - hydroxy - 2 - (1 - methyl - 3 - phenylpropylamino)ethyl] salicylamide.

Labiton. A proprietary preparation of aneurine hydrochloride, extract of Kola nuts, syrup and glycerophosphoric acid. (352).

Labophylline. A proprietary preparation of theophylline and lysine. A respiratory and cardiac stimulant. (352).

Laboprin. A proprietary preparation of aspirin and LYSINE. An analgesic. (352).

Labordin. See ANALGEN.

Labosept. A proprietary preparation of DEQUALINIUM CHLORIDE. An antibacterial lozenge taken orally. (352).

Labradorite (Hafnefjordite, Labrador-spar). A mineral. It is a member of the felspar group, and consists of a mixture of 1 part of albite, and 3-6 parts of anorthite (*q.v.*), $Al_2O_3.2SiO_2$. $Na_2O.SiO_2$.

Labrador-Spar. See LABRADORITE.

Labrol RO-O. A refined oleamide lubricant for polyolefines. (2).

Labstix. A proprietary test-strip used for the detection of pH, protein, glucose, ketones and blood in urine. (807).

Labstone. See KEMITE.

Lac (Gum Lac, Lacca, Button Lac, Sheet Lac, Shellac). The resinous excretion of the lac insect, *Laccifer lacca*, cultivated in India, Burma, and Siam. The insects living on the twigs become surrounded with the lac, and in this form it is known as stick-lac. The stick-lac is made into seed-lac, by removing the twigs, insect bodies, etc., by crushing, winnowing, and washing, and is finally purified by heating it in a cloth bag and forcing the lac through the cloth by twisting the bag. This is shellac. Rosin and orpiment are often added to give a light appearance to the material. The hot resin is stretched into sheets or run into buttons. Modern mechanical methods designed to give standard products have to an extent superseded the native process, and at least one firm can supply shellac to specification in regard to purity and constant quality. Shellac is sold as " lemon," " fine orange," " orange," " garnet," and " dewaxed." The best quality shellac contains no rosin or orpiment, and has from 3·5–4·5 per cent. wax. Garnet lacs contain from 1–20 per cent. rosin. Shellac softens at about 68° C. and melts at about 78° C. It has a specific gravity of 1·14–1·18, a saponification value of 240–250, and an acid value of 75–80. It is soluble in many organic solvents, particularly the lower alcohols, in ammonia, borax, and in solutions of alkaline carbonates and hydrates. Shellac has many applications, being used as a constituent of polishes, varnishes, and lacquers, in electrical insulation, and in the manufacture of gramophone records and sealing waxes. A variety of Japanese wax obtained from the wood of *Rhus Vernicifera* of Japan is called Lac. It is used in lacquer.

Lacajolin. A German product. It is a new name for Lakajol, a lactic acid guaiacol preparation in the form of syrup and pills, the latter containing very small amounts of guaiacol lactate and cod-liver oil. Used in diseases of the respiratory organs.

Lacanite. A proprietary trade name for a shellac compound.

Lacarnol. A proprietary preparation. It is an extract of the skeletal muscles of animals, free from albumin, histamine, and adrenalin. It contains a new hormone and is recommended in angina pectoris, etc.

Lacca. See LAC.

Laccain. A phenol-formaldehyde resin made with the aid of hydroxy acids, such as tartaric acid. It is brittle, melts at 90–100° C., and is soluble in alcohol, acetone, and ethyl acetate.

Laccase. An oxidising enzyme.

Lac-Dye (Lack-lack, Dyer's Lac). The colouring matter derived from Lac (*q.v.*). Stick-lac contains 10 per cent. of colouring matter. Used for dyeing wool mordanted with aluminium or tin salts.

Lacessan. An aqueous solution of calcium benzoate. It is a German preparation, and is injected for arthritis.

Lac, Garnet. See LEMON LAC.

L-Acid. See LAURENT'S ACID.

Lacimoid. A proprietary trade name for

a synthetic resin. It is used in laminated form for walls, etc.

Lacitin Red B. The calcium lake of Lithol Red R.

Lacitin Red R. The barium lake of Lithol Red R.

Lac, Japanese. The lac obtained from *Rhus vernicifera*. It is a natural varnish or lacquer, and contains 85 per cent, of urushic acid.

Lack-lack. See LAC-DYE.

Lackmoid. The blue colouring matter obtained by heating resorcinol with sodium nitrite. Used as an indicator in alkalimetry.

Lac Lake. See CARMINE LAKE.

Lacmoid. See LACKMOID.

Lacmus. Litmus.

Lac, Oil. See OIL LACS.

Lac, Orange. See LEMON LAC.

Lacorene. A proprietary trade name for a polystyrene moulding resin.

Lacqran. A proprietary trade name for an ABS (*q.v.*) moulding resin.

Lacqran E. Antistatic acrylonitrile butadiene styrene.

Lacqrene 550. A proprietary polystyrene used in extrusion and injection moulding. (291).

Lacqrene 635, 811, 835 and 836. Proprietary polystyrenes used to produce extrusions of differing tensile strengths. (291).

Lacqrene 740. An impact and heat resistant polystyrene suitable for use at 90° C. (291).

Lacqrene E. Antistatic polystyrene. (291).

Lacqsan. A proprietary trade name for a styrene-acrylonitrile copolymer moulding compound. Exhibits mechanical strength, transparency, high gloss finish, high impact resistance and good thermal resistance. Resists hydrocarbons, oils, greases, alcohol, bases and most acids. (185).

Lacqsan E. Antistatic styrene acrylonitrile.

Lacqsan 125 and 125L. Proprietary copolymers of styrene and acrylonitrile used in extrusion and injection moulding. (291).

Lacqtene 1070 MN20. A proprietary low-density polyethylene used in injection moulding. (291).

Lacqtene 1200 MN26. A proprietary low-density polyethylene used in injection moulding. (291).

Lacquer. Shellac dissolved in alcohol, and coloured with saffron, annatto, or dragon's blood.

Lacrinite. A proprietary casein-phenol-formaldehyde product.

Lacroixite. A mineral. It is Na_4 $(Ca,Mn)_4Al_3(PO_4)_3(OH)_{12}$.

Lac Sulphur. Sulphur precipitated from some of its combinations. It is very light in colour.

Lac Sulphuris. Milk of sulphur.

Lactagol. The powdered extract of cotton seed. Used for increasing the quality and quantity of milk in nursing mothers. It causes an increase in the amount of casein and fat.

Lactalox. A proprietary trade name for an aluminium antacid. (190).

Lactannin. Bismuth-dilacto-mono-tannate. Used in tubercular and other chronic diarrhœa in children.

Lactarin. Also called technical casein. It is the dried casein obtained from skim milk, and is a foodstuff containing 0·4 per cent. fat, 78 per cent. casein, and 10 per cent. water.

Lactase. An enzyme which converts lactose into *d*-glucose and *d*-galactose.

Lacteol (Lactigen, Lactilloids, Lactobacilline, Lactone). Preparations of lactic acid bacilli.

Lactigen. See LACTEOL.

Lactilloids. See LACTEOL.

Lactin. Lactose, a sugar.

Lactine. See VEGETABLE BUTTER.

Lactinium. A German preparation of neutral aluminium lactate. An astringent and disinfectant.

Lactite. See GALLATITE.

Lactitis. A casein preparation containing borax and lead acetate. It is an artificial ivory.

Lactobacilline. See LACTEOL.

Lactoform. See GALLATITE.

Lactoid. A casein preparation.

Lactoiod. A compound of iodine and casein, containing 8 per cent. iodine. Used for the administration of iodine.

Lactol (Lactonaphthol). The lactic acid ester of β-naphthol, $CH_3CHOH.COO.$ $C_{10}H_7$. An intestinal astringent.

Lactoleum. A kind of liquid linoleum made from cork powder and cement. Used for making floors.

Lactolin (Antimonin). The double salts of antimony lactate with alkalis, alkaline earths, and zinc salts. A convenient means for the transport of lactic acid. Also used as a substitute for tartar emetic in dyeing.

Lactolith. A casein preparation.

Lactoloid. A proprietary casein product.

Lactomaltose. A preparation of sour milk, to which maltose has been added, prior to fermentation.

Lactonaphthol. See LACTOL.

Lactone. See LACTEOL.

Lactophenine. *p*-Lactyl-phenetidine, $C_6H_4(OC_2H_5)NH.CO.CH(OH).CH_3$. Given as an antipyretic and analgesic.

Lactoprene. A patented vulcanisable synthetic rubber made from emulsified methyl or ethyl acrylate copolymerised with small quantities of a polyfunctional monomer such as butadiene, isoprene or allyl maleate. The copolymer is compounded with sulphur and accelerator and cured.

Lactorite. See GALLATITE.

Lactose Molasses. Molasses obtained from the preparation of milk sugar.

Lactoserve. A butter-milk in the form of powder, made by subjecting sterilised milk to the action of lactic acid bacteria until a certain degree of acidity is obtained. The product is dried, and ground with sugar, wheat flour, and Roborate (*q.v.*).

Lactucarium (Lettuce Opium). The dried juice of *Lactuca virosa*.

Lactulose. A preparation used in the treatment of hepatic coma and chronic constipation. It is 4-*O*-β-D-galactopyranosyl - D - fructose. DUPHALAC.

Lac, White. See WHITE INSECT WAX.

Ladanum. See LABDANUM.

Ladelloy. Ferro-alloy ladle additions. (531).

Laettbentyl. A motor fuel containing 25 per cent. alcohol obtained from waste sulphite liquor and 75 per cent. petrol.

Lævadosin. A proprietary preparation of adenosine phosphate, adenosine, inosine, guanosine, guanosine monophosphate and uridine. (247).

Lævo-Glucose. See LEVULOSE.

Lævo-Menthone. See APINOL.

Lævo-Pinene. Terebenthene, a terpene.

Lævoral. A proprietary preparation of fructose, used in liver disease and shock. (247).

Lævosan. A proprietary preparation of lævulose solution for intravenous therapy. (247).

Lævotonin. A proprietary preparation of fructose, sodium hydrogen phosphate, manganese, strychnine and caffeine. A tonic. (247).

Laevuflex. A proprietary preparation of lævulose used as a parenteral calorie supplement. (386).

Lævuline Blue. A solution of induline in acetin.

Lafond's Bearing Metal. See BEARING METALS.

Lafond's Bronzes. Alloys of 80–98 per cent. copper, 2–18 per cent. tin, 0–2 per cent. zinc, and 0–0·5 per cent. lead.

Lafou's Reagent. A sulphuric acid solution of ammonium selenite. An alkaloidal reagent.

Lager Beer. A beverage containing 3·5–4 per cent. alcohol.

Lagonite. A borate of iron, found in Tuscany.

Lagoriolite (Soda-garnet). Sodium dialuminium triorthosilicate, $Na_6Al_2(SiO_4)_3$.

Lagos Silk Rubber. A rubber obtained from *Funtumia Elastica* of Africa.

Lake Bordeaux B. A dyestuff. It is prepared by diazotising β-naphthylamine-α-sulphonic acid, and condensing the product with β-hydroxy-naphthoic acid.

Lake Copper. An American class of copper containing 99·8 per cent. of the metal.

Lake, Florentine. See FLORENCE LAKE.

Lake, Indian. See CARMINE LAKE.

Lake, Lac. See CARMINE LAKE.

Lake Ore. See BOG IRON ORE.

Lake, Paris. See FLORENCE LAKE.

Lake Red C. A dyestuff prepared from 5 - amino - 2 - chlor - toluene - 4 - sulphonic acid.

Lake-red Ciba B. An insoluble pigment, made by the interaction between indigo and phenyl-acetyl-chloride. The soluble dyestuff is obtained by sulphonating the product.

Lake Red D. A dyestuff. It consists of diazotised anthranilic acid condensed with β-naphthol.

Lake Red F. A dyestuff. It is an equivalent of Lake red P.

Lake Red P (Lake Red F, Monolite Red P). A dyestuff obtained by diazotising *p*-nitraniline-*o*-sulphonic acid, and condensing it with β-naphthol.

Lake, Rubine. See MADDER LAKE.

Lakes. Compounds of inorganic bodies with organic colouring matters, usually aluminium oxide or other metallic oxide.

Lake's Indiarubber Compound. A compound consisting of saponified resin and vulcanised oil, incorporated with indiarubber or gutta-percha.

Lake, Vienna. See FLORENCE LAKE.

Lake, Venetian. Crimson Lake (*q.v.*).

Lake, Yellow. See YELLOW CARMINE.

Lalona Bark. The bark of *Weinmannia bojeriana*. It contains 13·75 per cent. tannin, and is used for tanning.

Lambertite. A uranium mineral of Wyoming.

Lambkin's Cream. See GREY OIL.

Lamellon. A proprietary trade name for a range of unsaturated polyester resins used for reinforced plastics and coatings. (195).

Lamepon. A proprietary trade name for the condensation product of a fatty acid chloride and a protein decomposition product. It is a wetting agent.

Lamex 173/FR. A self extinguishing type of the above resins complying with BS 476 Part I (Class II). (66).

Lamex 185. A trade name for a flexible

amine preaccelerated polyester resin used for motor car body repairs. (66).

Lamex 186. A trade name for a rigid amine preaccelerated polyester resin used for motor car body repairs. (66).

Lamicoid. A proprietary trade name for a phenol-formaldehyde synthetic resin with a mica filler used for laminated products.

Laminac EPX-176. A trade mark. A self extinguishing (flame resistant) polyester resin. Suitable for manufacture of reinforced plastics in the transportation industry.

Laminac EPX-196. A trade mark. A polyester resin derived from isophthalic acid. It meets requirements in repeated contact with food. (56).

Laminated Talc. See MICA.

Laming Mixture. Made by mixing 160 parts lime, 180 parts sawdust, and 30 parts ferrous sulphate, moistening, and turning over, until the ferrous sulphate is converted into ferrous hydroxide, and then into ferric hydroxide. Calcium sulphate is also formed. Used for fixing ammonium salts, hydrogen, and other sulphides in the purification of coal gas.

Laminic. A nickel-iron alloy.

Laminol E. A phenolic mixture for resin manufacture. Used for laminating paper and cloth. (2).

Lamitex. A proprietary trade name for a hard vulcanised fibre.

Lamotte's Drops. Tr. Ferri Perchlor Ether.

Lamotte Standard Indicators. These are prepared by mixing 0·5 c.c. of the prepared commercial solutions of the following indicators with 10 c.c. of the special buffer solution M5: Thymol Blue (acid range), Bromophenol Blue, Bromocresol Red, Bromocresol Purple, Bromothymol Blue, Phenol Red, Cresol Red, Thymol Blue (alkaline range).

Lampadite. An earthy variety of manganese dioxide, containing copper oxide.

Lampblack. This is carbon in a fine condition prepared by the incomplete combustion of tar, colophony, vegetable oils, and the pitch or heavy oils from tar. Other organic materials are also used for its production. The British specification No. 287 (1927) states that it should consist of soft dry powder free from added adulterant and colouring matter. The loss on heating at 95–98° C. should not exceed 1 per cent., the ash should not be more than 4 per cent., not more than 2 per cent. should be soluble in methylated ether, and not more than 1 per cent. soluble in water.

Lamprene. A proprietary preparation of CLOFAZIMINE. An anti-leprotic. (17).

Lamprophyllite. A mineral. It is $4[Na_2SrTiSO_2O_8]$.

Lamy Butter (Kanja Butter, Kanga Butter, Sierra Leone Butter). The fat obtained from the seeds of *Pentadesma butyracea* of West Africa. It melts at 32–42° C., has an acid value of 16–26, a saponification value of 186–199, and an iodine value of 42–68.

Lana Batu. Citronella oil.

Lanacron. Wool dyestuffs. (530).

Lanacyl Blue BB (Indigo Substitute). An azo dyestuff. It is prepared by coupling 1 : 8 : 3 : 6-amino-naphthol-disulphonic acid with 1 : 5-amino-naphthol. An acid wool dye.

Lanacyl Blue R. A dyestuff belonging to the same group as Lanacyl Blue BB.

Lanacyl Navy Blue B. A dyestuff belonging to the same group as Lanacyl Blue BB.

Lanacyl Violet B. A dyestuff. It is the sodium salt of disulpho-oxy-naphthalene-azo-ethyl-α-naphthylamine, $C_{22}H_{17}N_3S_2O_7Na_2$. Dyes wool from an acid bath.

Lanacyl Violet BB. A dyestuff. It is the sodium salt of disulpho-oxy-naphthalene - azo - amino - naphthol, $C_{20}H_{13}N_3S_2O_8Na_2$. Dyes wool blue from an acid bath.

Lanadin. An alcoholic soap solution containing 87 per cent. trichlor-ethylene. A wetting-out and scouring agent.

Lanafuchsine S.B. See SORBIN RED.

Lanain. See LANOLIN.

Lanalin. See LANOLIN.

Lanamar. See LANELLA.

Lana Philosophica. Philosopher's wool, zinc oxide, ZnO.

Lanarkite (Dioxylite). A mineral. It is a double sulphate and carbonate of lead, $PbSO_4.PbCO_3$.

Lanasol. Reactive dyes for wool and silk. (530).

Lanatoside C. A myocardial stimulant. 3 - (3′ - Acetyl - 4′ - β - D - glucosyltri - digitoxosido) digoxogenin. CEDILANID.

Lanbitum. Black bituminous paint. (580).

Lanbritol. Self emulsifying waxes. (562).

Lancashire Brass. See BRASS.

Lancasterite. A mineral. It is stated to be a mixture of brucite and hydromagnesite.

Lancaster Yellow. A colouring matter obtained from picramic acid and phenol, $C_{12}H_8N_4O_6$. It dyes wool and silk from an acid bath, but is no longer used.

Lancastrine Oil. A compound oil used for lubrication. It has a specific gravity of 0·867, a flash-point of 353° F., and is a spindle oil.

Lancegaye. A proprietary safety glass.

Landevanite. A mineral. It is a variety of montmorillonite.

Land Plaster. Ground gypsum, $CaSO_4.2H_2O$, a soil dressing.

Landshoff and Meyer's Acid. 2-Naphthylamine-disulphonic acid (1 : 2 : 5).

Laneite. A mineral. It is a variety of amphibole.

Lanella (Lanamar). Woolly types of fibre obtained from a kind of seaweed, *Posidonia oceania*, found in the Pacific.

Lanesin. See LANOLIN.

Lanette Self emulsifying waxes. (562).

Lanette Wax. A proprietary trade name for a mixture of cetyl and stearyl alcohols.

Lanette Wax Ester. A proprietary trade name for palmitic acid ester of cetyl and stearyl alcohols used in emulsions.

Lanette Wax SX. A proprietary trade name for an emulsified mixture of cetyl and stearyl alcohols.

Langbanite. A mineral containing manganese silicate and iron antimonate.

Langbeinite. A double salt of potassium and magnesium sulphates, $K_2SO_4.2MgSO_4$. It is a mineral, but has been prepared artificially. It is also written $4[K_2Mg_2(SO_4)_3]$.

Lange Solution. A colloidal gold solution.

Langford. See CATALPO.

Langite. A mineral. It is a basic copper sulphate, $CuSO_4.3Cu(OH)_2.2H_2O$.

Lanichol. See LANOLIN.

Laniol. See LANOLIN.

Lanital. A proprietary trade name for a casein textile fibre made by dissolving casein in a dilute alkaline solution and extruding the viscous compound in the form of thin filaments. These are treated with acid and rendered insoluble by means of formaldehyde. The modern process employs the use of a dispersion of casein and the addition of carbon disulphide or similar reagent.

Lanitop. See MEDIGOXIN.

Lankro Mark. Stabilisers for PVC. (558).

Lankol. Coal tar disinfectant. (580).

Lankroflex ED3. A proprietary epoxy plasticiser. Octyl epoxy stearate. (558).

Lankroflex ED6. A proprietary epoxy plasticiser. (558).

Lankroflex GE. A proprietary epoxidised vegetable oil used as a plasticiser or as an extender. See ABRAC A. (558).

Lankrol. Sulphated oils. (558).

Lankroline. Sulphated oils and pigment finishes. (558).

Lankroplast. Synthetic resin dispersions. (558).

Lankropol. Anionic wetting agents. (558).

Lanoid. See MULSOID.

Lanolac. Lanolin paint. (526).

Lanolin (Hydrous Wool Fat, Lanalin, Lanain, Laniol, Lanesin, Lanichol). Purified wool fat which has absorbed a large quantity of water. Used as a basis for ointments and cosmetics.

Lanolin, Anhydrous. Purified wool fat.

Lanolin Cream. Usually a mixture of wool fat, soft paraffin, balsam of Peru, and rose oil.

Lanolin Substitute. An American sample of this material contains 65 per cent. pet. jelly, 20 per cent. paraffin wax, 10 per cent. spermaceti, and 5 per cent. lanolin, the whole emulsified in 100 parts water.

Lanoplast. A proprietary cellulose acetate.

Lanoresin. A resin obtained from the washing of wool.

Lanosol. Lanolised mineral oil. (526).

Lanoxin. A proprietary preparation of digoxin. (277).

Lanoxine-PG. A proprietary preparation of digoxin. (277).

Lansfordite. A mineral, $Mg_2(OH)_2.CO_3.21H_2O$.

Lanthana. Lanthanum oxide.

Lanthanite. A mineral. It is hydrated carbonate of lanthanum, didymium, and cerium.

Lantol. A name applied to colloidal rhodium, and also to lanthopine, $C_{23}H_{25}O_4N$, an alkaloid of opium.

Lanvis. A proprietary preparation of THIOGUANINE used in the treatment of leukæmia. (277).

Lapaquin. A proprietary preparation containing chloroquine phosphate and chlorproguanil hydrochloride. An antimalarial drug. (2).

Lapintine. See PICROFOETIDINE.

Lapis Albus. A mineral. It is a calcium-silico-fluoride.

Lapis Amiridis. Emery (*q.v.*).

Lapis Baptista. Talc.

Lapis Cæruleus. Copper sulphate.

Lapis Calaminaris. Calamine (*q.v.*).

Lapis Causticus. Fused sodium or potassium hydroxide.

Lapis Imperialis. Silver nitrate.

Lapis Infernalis. Fused silver nitrate.

Lapis-Lazuli. A mixture of minerals consisting of a crystalline limestone impregnated with lazurite, sodalite, and iron pyrites. Used for ornaments, and for the preparation of ultramarine. See ULTRAMARINE, GENUINE.

Lapis Lunaris. Silver nitrate.

Lapis-Ollario. See STEATITE.

Lapis Smiridis. Emery. (A trade mark.)

Lapix. A proprietary trade name for a flux used in steel moulding. It contains carbon and clay.

Lapparentile. Synonym for Tamarugite.

Lapparentite. A mineral. It is $AlSO_4$ $OH.4\frac{1}{2}H_2O$.

Lapudrine. A proprietary preparation containing chlorproguanil hydrochloride. An antimalarial drug. (2).

Larch Extract. An extract of the bark of *Pinus larix*. Used for tanning.

Larch Sugar. Melezitose, a trisaccharide, $C_{18}H_{32}O_{16}$.

Larch Turpentine. See VENICE TURPENTINE.

Lard. Purified hog's fat.

Lard Compound. A substance prepared by thickening edible cotton-seed oil with oleo-stearine.

Larderellite. A mineral. It is ammonium borate, and is found in Tuscany.

Lardine Oil. An oil prepared from cotton-seed oil. It has a specific gravity of 0·967–0·980, a saponification value of 194–215, and a flash-point of 400° F.

Lardite. A green variety of the mineral Agalmatolite.

Lard Oil. The oleine of lard, a lubricant. An oil obtained from rape seed is also known as lard oil.

Lard Substitutes. Mixtures of lard or lard stearine, with cotton-seed stearine, maize oil, coco-nut oil, and similar oils.

Largactil. A proprietary preparation of chlorpromazine hydrochloride. A sedative. (336).

Largin (Argonine, Agyrol, Protargol, Protein Silver Salt). Compounds of silver with albumin, casein, and wheat gluten. Used as bactericides.

Largon. A proprietary trade name for Propiomazine.

Larixin. Larixinic acid, $C_{10}H_{10}O_5$, obtained from larch bark.

Larnite. A new mineral. It is a calcium silicate.

Larnol. Rust preventative. (526).

Larodopa. A proprietary preparation of LEVODOPA used in the treatment of Parkinson's disease. (314).

Laroxyl. A proprietary preparation of amitriptyline hydrochloride. An antidepressive agent. (314).

Larsenite. A mineral. It is a silicate of lead and zirconium.

Larvacide. A proprietary trade name for chloropicrin used as an insecticide.

Larvex. A proprietary clothes-moth remedy. It is a solution of sodium fluosilicate.

Lasilso. Sodium metasilicate. (513).

Lasionite. A mineral. It is wavellite.

Lasix. A proprietary preparation of frusemide. A diuretic. (312).

Lasix + K. A proprietary preparation of FRUSEMIDE and potassium chloride. A diuretic containing a potassium supplement. (312).

Lasonil. A proprietary preparation of heparinoid, and hyaluronidase. Surgical application. (341).

Lassallite. A mineral. It is $Mg_3Al_4Si_{12}$ $O_{38}8H_2O$.

Lassolatite. A mineral. It is a variety of opal.

Lastane. A proprietary-based polyurethane elastomer cross-linked with diamine. (764).

Lastex. A proprietary product. It is rubber latex threads spun with fibre.

Lastil. A fungicide for bonded cork. (182).

Lastilac. A proprietary moulding compound.

Lasurite. See CHESSYLITE.

Latene. A proprietary rubber vulcanisation accelerator. It is a solution of trimene base in rubber latex.

Latensol AP8. A non-ionic surfactant. (49).

Laterite. A material consisting of hydroxides of iron and aluminium, with sand and clay. It is used in India as a building stone.

Latex. The milky emulsion, containing minute suspensions of rubber, which flows from incisions in the bark of rubber trees, such as *Hevea brasiliensis*.

Latex Foam. A proprietary trade name for a type of cellular sponge rubber made from latex by a special method.

Latex, Reversible. See REVERTEX.

Latialite. Synonym for Haüyne.

Latiumite. A mineral. It is K_2Ca_6 $(Si,Al)_4O_{25}(SO_4,CO_3)$.

Latri's Bois Durci. An artificial wood prepared from sawdust and blood albumen.

Lattbentyl. A Swedish motor spirit. See LAETTBENTYL.

Laubanite. A mineral. It is a calcium-aluminium silicate.

Laubmannite. A mineral. It is $Fe_3\cdots$ $Fe_6\cdots(PO_4)_4(OH)_{12}$.

Laudanon. A mixture of opium alkaloids.

Laudanum. Tincture of opium (*Tinctura Opii B.P.*).

Laudexium Methylsulphate. A neuromuscular blocking agent. Decamethylenedi-{2-[1-(3,4-dimethoxybenzyl)-1,2,3,4-tetrahydro-6,7-dimethoxy-2-methylisoquinolinium methylsulphate]}.

Laudolissin. See LAUDEXIUM METHYLSULPHATE.

Laueïte. A mineral. It is $MnFe_2(PO_4)_2$ $(OH)_2.8H_2O$.

Laughing Gas. Nitrous oxide, N_2O.

Laumonite. A mineral. It is a zeolite, $CaO.Al_2O_3.4SiO_2.2H_2O$.

Laundros. See Burmol.

Laurel Camphor. See Japan Camphor.

Laurel Nut Oil (Domba Oil, Alexandrian Laurel Oil, Poonseed Oil, Tacamahac Oil, Njamplung Oil, Calaba Oil, Dilo Oil, N dilo oil, Pinnay Oil, U dilo oil). Calophyllum oil, from the nuts of *Calophyllum* species. The natives use it for medicinal and illuminating purposes. It is not an edible oil, being poisonous.

Laurel Oil. Bayberry oil, obtained from the berries of the laurel tree, *Laurus nobilis*.

Laurel Wax (Myrtle Wax, Bayberry Tallow). Myrtle berry wax, obtained from *Myrica cerifera*.

Laurent's Acid (L-acid). 1-Naphthylamine-5-sulphonic acid.

Laurent's Aluminium Solder. The hard solder contains 63–74 per cent. zinc and 19–30 per cent. tin. The soft solder contains 60–70 per cent. zinc, 16–27 per cent. tin, and 12 per cent. lead.

Laurent's Naphthalidinic Acid. See Laurent's Acid.

Laureol. See Vegetable Butter.

Laurex. The zinc salt of lauric acid. A light-grey soft powder used as a softener and stabiliser for rubber in the place of stearic acid. The zinc content is about 17 per cent. and its gravity is 1·1.

Laurex. A proprietary trade name for a series of primary fatty alcohols. (210).

Laurionite. A mineral, $PbCl_2.Pb(OH)_2$.

Laurite. A mineral found in Borneo. It contains sulphides of ruthenium and osmium.

Laurodin. Bactericidal preparations of laurolinium acetate. (510).

Laurol. Laurene, $C_{11}H_{16}$.

Laurolinium Acetate. A surface-active agent. 4 - Amino - 1 - dodecylquinaldinium acetate. Laurodin.

Laurydol. Trade name for Lauroyl Peroxide, used for the polymerisation of vinyl chloride and acrylics, also for curing polyester resins.

Laurydol B-50. Paste of Laurydol in a phthalate plasticiser.

Lausenite. A mineral. It is $Fe_2(SO_4)_3$. $6H_2O$.

Lausofan. A hexamethylene ketone used for destroying insects of the vermin type.

Lautal. A proprietary trade name for an aluminium alloy containing 4 per cent. copper and 2 per cent. silicon. It has a specific gravity of 2·74, an electrical conductivity of 40 per cent. that of copper, and 70 per cent. that of aluminium. It is more resistant to corsion than aluminium.

Lautarite. A mineral. It is calcium iodate, $Ca(IO_3)_2$.

Lauth's Violet (Thionine). Amino-imino-imino - diphenyl - sulphide, $C_{12}H_9N_3S$. Used as a microscopic stain.

Lautite. A mineral. It is $4[CuAsS]$.

Lauxite. A proprietary product. It is a water soluble synthetic resin consisting of a zinc chloride-urea-formaldehyde condensation product, and is used as an adhesive. Now applied generally to synthetic resin glues. (559).

Lava. A material from volcanoes. It consists essentially of ferrous, calcium, and aluminium silicates.

Lavalloy. A proprietary trade name for a ceramic product made from mullite and alumina.

Lavarock. An American heat-treated steatite.

Lavasan. See Chlorthymol.

Lavasteril. *p*-Chlor-*m*-cresol and *p*-chlor-*m*-thymol.

Lavasul. A mixture of 40 per cent. coke dust with sulphur. Used for tank linings, as it resists dilute acids.

Lavender Drops. Compound tincture of lavender.

Lavender, Red. See Red Drops.

Lavendol. Impure linalol.

Lavendulanite. A mineral, $Cu_3(AsO_4)_2$. $2H_2O$.

Lavenite. A mineral, $(Mn.Ca.Fe)O$. $(ZrO.F)Na_2O.SiO_2$.

Laveran's Stain. (*a*) Eosin 1 gram, distilled water 1,000 c.c. (*b*) A few crystals of silver nitrate are dissolved in 60 c.c. water and sodium hydroxide added in excess. Silver oxide is precipitated. This is collected and added to a saturated solution of methylene blue. For use 4 c.c. of solution (*a*) are mixed with 6 c.c. water and 1 c.c. solution (*b*) added.

Lavite. A heated steatite product of American origin.

Lavodermin. A soap preparation containing about 3 per cent. of the mercury compound of casein.

Lavrovite. A mineral. It is $4[MgCaSi_2O_6]$.

Lawrencite. A mineral, $FeCl_2$.

Lawrowit. Synonym for Lavrovite.

Lawsone. Hydroxy-naphthoquinone, $C_{10}H_6O_3$, the colouring matter of henna leaves.

Lawsonite. A mineral, $CaO.Al_2O_3.2SiO_2$. $2H_2O$.

Laxans. See Purgen.

Laxatin. See Purgen.

Laxatol. See Purgen.

Laxatoline. See PURGEN.

Laxen. See PURGEN.

Laxiconfect. See PURGEN.

Laxin. See PURGEN.

Laxmannite. A mineral. It is a phospho-chromate of lead and copper, $2(Pb.Cu)CrO_4.(PbCu)_3.(PO_4)_2$. It is probably identical with vauquelinite.

Laxoberal. A proprietary preparation of SODIUM PICOSULPHATE. A laxative (650).

Laxoin. See PURGEN.

Laxophen. See PURGEN.

Laxothalen. A phenol phthalein preparation.

Laybourn's Stain. A microscopic stain. It consists of two solutions: (a) Toluidine blue 0·15 gram, malachite green 0·2 gram, glacial acetic acid 1 c.c., alcohol (95 per cent.) 2 c.c., and water 100 c.c. (b) Iodine 2 grams, potassium iodide 3 grams, and water 300 c.c.

Layor Carang. See AGAR-AGAR.

Laytex. A proprietary insulating material derived from rubber latex. It is prepared by removing proteins, sugar, and water-soluble substances. It is used for coating conductors by the dip or pass method by moving the conductor through a bath. It is stated to have high tensile strength and insulation resistance.

Lazialite. Synonym for Haüyne.

Lazulite. A mineral, $2AlPO_4.(Fe.Mg)(OH)_2$. The name is sometimes applied to the mineral Chessylite.

Lazurapatite. A mineral. It is an impure apatite.

Lazurite. A mineral. It is a silicate and sulphide of sodium and aluminium. It is present in lapis-lazuli. The name is also applied to the mineral Chessylite.

L.C. Pulver. A German explosive.

Lead, Antimony. See HARD LEAD.

Lead Arsenate Paste 50 per cent. A preparation of basic lead arsenate with water and containing 50 per cent. dry salt.

Lead Ashes. The skimmings formed during the melting of lead. It consists mainly of oxide.

Lead, Black. See GRAPHITE.

Lead Bottoms (Bottom Salts). An impure lead sulphate obtained by adding sulphuric acid to impure liquid containing lead sulphate obtained as a by-product by the textile printer. The precipitate is sold as lead bottoms. It is a poor type of pigment for use in paints.

Lead Bronze. An alloy of from 70–90 per cent. copper, 6–16 per cent. lead, and 4–13 per cent. tin.

Lead Chamber Crystals. See NITROSYL-SULPHURIC ACID.

Lead Dust. Metallic lead in a finely divided condition.

Lead Earth. See CERUSSITE.

Leaded Bronze. An alloy of 88·5 per cent. copper, 10 per cent. zinc, and 1·5 per cent. lead. Another source gives the following proportions: 80 per cent. copper, 10 per cent. tin, and 10 per cent. lead. It has a melting-point of 945° C.

Leaded Gun Metal. An alloy of 85·5 per cent. copper, 2 per cent. zinc, 9·5 per cent. tin, and 3 per cent. lead. It melts at 980° C.

Leaded Zinc Oxides. Pigments containing 20–35 per cent. lead sulphate and 60–80 per cent. zinc oxide. They are of American manufacture.

Lead Extract. See EXTRACT OF LEAD.

Lead, False. See BLACK JACK.

Lead Flake. White Lead (q.v.).

Lead Fume. In the smelting of galena a considerable quantity of the lead is carried off in the form of vapour (lead fume). This is condensed by passing the gases through long flues.

Lead Glass. See POTASH-LEAD GLASS and FLINT GLASS.

Leadhillite (Maxite, Susanite). A mineral, $PbSO_4.2PbCO_3.Pb(OH)_2$.

Lead Horn. See KERASITE.

Lead Lotion. See GOULARD WATER.

Lead Malachite (Plumbomalachite). A mineral, $2CuCO_3.PbCO_3.Cu(OH)_2$.

Lead Ochre. Lead monoxide, PbO, found naturally.

Lead Ore, Black. See ALQUIFON.

Lead Ore, White. See CERUSSITE.

Lead Oxide, Puce. See BROWN OXIDE OF LEAD.

Lead Oxide, Red. See RED LEAD.

Lead Plaster. Lead oleate.

Lead Rope. See LEAD WOOL.

Lead, Shot. See SHOT METAL.

Lead Soap. Lead resinate.

Lead Solder. An alloy of 50 per cent. lead and 50 per cent. tin used for soldering lead.

Lead Spar, Yellow. A mineral. It is wulfenite.

Lead Styphnate. Lead trinitro-resorcinate. Used in explosives.

Lead, Tempered. An alloy. It is Noheet metal.

Leadtex. A proprietary trade name for a lead-coated copper in sheet form.

Lead Tungate. A preparation obtained from lead acetate and tung oil. Used as a drier in the preparation of paints.

Lead Vinegar. A basic lead acetate, $Pb(C_2H_3O_2)_2.PbO.2H_2O$.

Lead Vitriol. See Anglesite.

Lead Water. A 1 per cent. solution of basic lead acetate.

Lead Wool. Lead metal made into fine threads by melting the metal and allowing it to run through tiny holes. The threads are twisted into a lead rope. A harder product is shredded lead. Used for caulking joints.

Lead Yellow. Lead chromate, $PbCrO_4$.

Leaf Gold. Gold which occurs in leaflets.

Leaf Green. See CHROME GREEN, CHLOROPHYLL, and OIL GREEN.

Lean Coal. Coal of the poorest quality, used for lime and brick kilns.

Leantin. A proprietary trade name for a bearing alloy of lead and tin.

Leather Black CT. A colour used for dyeing leather by the chrome process.

Leather Blonde. See LEATHER YELLOW AND BLONDE.

Leather Brown. K. Oehler. A dyestuff. It is the hydrochloride of bi-*p*-amino - benzene - disazo - *m* - phenylene-diamine, $C_{18}H_{19}N_8Cl$. Dyes leather and jute brown. Also see BISMARCK BROWN and PHOSPHINE.

Leather Cloth. See AMERICAN CLOTH.

Leather Grease. See DÉGRAS.

Leatherine. A thin leather substitute made from rubber, antimony sulphide, iron oxide, sulphur, zinc sulphide, magnesium carbonate, and barium sulphate.

Leatherlubric. A proprietary trade name for a sulphonated sperm oil.

Leather Meal. A nitrogenous manure obtained from scrapped leather, by steaming, drying, and grinding. It contains from 6–11 per cent. nitrogen.

Leatheroid (Leather Paper). Terms used for thin grey vulcanised fibre. Used for electrical insulation. See VULCANISED FIBRE.

Leather Paper (Fish Paper). A cotton-rag paper chemically treated with zinc chloride. The term Leather Paper is also used for thin vulcanised fibre. See LEATHEROID and VULCANISED FIBRE.

Leather Yellow. See PHOSPHINE.

Leather Yellow and Blonde. A Bohme dyestuff. It contains fustic extract and a chrome mordant.

Lebal. A German product obtained by combining cod-liver oil with albumin. It contains 33·3 per cent. cod-liver oil, and is used in cases of rickets.

Lebalca. A German preparation containing cod-liver oil, albumin, and calcium lactate.

Lebalpho. A German product containing cod-liver oil, albumin, and calcium glycero-phosphate.

Lebanon No. 34. A proprietary trade name for an alloy of 20 per cent. chromium, 30 per cent. nickel, 3·25 per cent. molybdenum, 5 per cent. copper, and 3·25 per cent. silicon. It is stated to be resistant to hydrochloric and sulphuric acid.

Lebanon No. 48. A proprietary trade name for an alloy of 30 per cent. chromium, 30 per cent. nickel, 0·4 per cent. carbon, and the remainder iron.

Lebbin Salt. A proprietary mixture containing nitrite. Used for curing meat.

Lében. A food product of Lebanon. It is a specially coagulated milk similar to youghhourt.

Lechatelierite. A mineral. It is silica in a fused state.

Lecibrin. A preparation of lecithin from the brain. It contains 33·3 per cent. lecithin with nucleo-protein.

Lecin. An iron albuminate, stated to be an easily assimilable form of iron. It contains 20 per cent. albumin and 0·6 per cent. iron.

Lecipon. A 10 per cent. lecithin in a soluble form.

Lecithan. A trade name for lecithin.

Lecithcerebrin. Lecithin prepared from brain substance.

Lecithin (Yolk Powder). A substance made from egg yolks. It is a compound of choline, glycerol, phosphoric, and various fatty acids. Used as a component of invalid foods.

Lecithmedullan. A lecithin preparation from bone marrow.

Lecithol. An emulsion of brain lecithin containing 1·5 per cent. lecithin.

Leclo. Lead chloride. (589).

Lecontite. A mineral. It is a sulphate of ammonium, potassium, and sodium.

Lectricon. A proprietary trade name for alkyl ammonium compounds dispersed in mixed aromatic/aliphatic oils with film strength agents. For electrostatically spraying onto form work for the casting of concrete. (221).

Lectro 80. A trade name for a basic lead chlorosilicate-sulphate complex in fine powder form, insoluble in water but slightly soluble in organic solvents. It is used as a stabiliser for wire insulation at 60°–100° C. (47).

Lectro Cast. A proprietary trade name for an alloy of iron with 2·75 per cent. nickel, and 0·7 per cent. chromium.

Lecutyl. A combination of lecithin and copper cinnamate, containing 1·5 per cent. copper.

Ledac. Lead acetate. (589).

Ledca. Lead carbonate. (589).

Ledclair. A proprietary preparation of SODIUM CALCIUMEDETATE used in the treatment of heavy-metal poisoning. (637).

Leddel Alloy. An alloy of 86 per cent. zinc, 9·5 per cent. antimony, and 4·5 per cent. copper. Another alloy used for bearings consists of 87·5 per cent. zinc, 6·25 per cent. copper, and 6·25 per cent. aluminium.

Ledeburite. Austenite-cementite eutectic.

Ledebur's Metal. Bearing metals. One contains 85 per cent. zinc, 10 per cent. antimony, and 5 per cent. copper. Another consists of 77 per cent. zinc, 17·5 per cent. tin, and 5·5 per cent. copper.

Ledercort. A proprietary preparation of triamcinolone. (306).

Ledercort-D. A proprietary preparation of triamcinolone acetonide for dermatological use. (306).

Lederite. A mineral. It is a variety of sphene.

Lederkyn. A proprietary preparation containing sulphamethoxypyridazine. (306).

Ledermycin. A proprietary preparation containing Demethylchlortetracycline. An antibiotic. (306).

Lederplex. A proprietary preparation of vitamin B complex. (306).

Lederspan. A proprietary preparation of TRIAMCINOLONE hexacetonide. (306).

Lederstatin. A proprietary preparation of dimethylchlortetracycline, and nystatin. An antibiotic. (306).

Ledfo. Lead formate. (589).

Ledmin LPC. A proprietary trade name for lauryl pyridinium chloride. An emulsifier for waxes giving bacteriostatic polishes. (9).

Ledni. Lead nitrate. (589).

Ledrite Brass. A proprietary trade name for a leaded brass containing 60–63 per cent. copper and 2·5–3·7 per cent. lead.

Leedsite. A mineral. It is a mixture of calcium and barium sulphates.

Leefex. A hop defoliant. (511).

Leelite. A mineral. It is a variety of orthoclase of a red colour.

Lees. Yeast and various suspended matters of the must produced during the fermentation of grape juice.

Leesbergite. A mineral, $Mg_2Ca(CO_3)_3$.

Le Farge Cement. A Grappier Cement (*q.v.*).

Leffmann and Beam's Glycerol Reagent. For Reichert-Meissl number. It is a mixture of 180 c.c. pure glycerol and 20 c.c. of 50 per cent. sodium hydroxide solution.

Legrandite. A mineral. It is $Zn_{14}(AsO_4)_9OH.12H_2O$.

Leguval. A proprietary polyester laminating resin. (112).

Lehner's Cement. See ELASTIC CEMENT.

Lehner Silk. A similar silk to collodion silk.

Lehrbachite. A mineral, PbSe.HgSe.

Leidyite. A mineral. It is an aluminium-iron-calcium-magnesium silicate.

Leifite. A mineral. It is a fluo-silicate of aluminium and sodium

Leinsaat Oils. Linseed oil obtained by extracting the residues from the presses.

Leipsic Yellow. See CHROME YELLOW.

Leitch's Blue. See CYANINE.

Leithner's Blue. See COBALT BLUE.

Lekutherm. A proprietary epoxide resin. (112).

Lemarquand's Alloy. An alloy made from 75 per cent. copper, 14 per cent. nickel, 2·0 per cent. cobalt oxide, 1·8 per cent. tin, and 7·2 per cent. zinc. It is stated to be very resistant to oxidation.

Lembergite. An artificial hydrous aluminium-sodium silicate.

Lemberg's Solution. A solution of aluminium chloride and extract of logwood. It colours calcite mineral surfaces violet.

Lemco. A proprietary meat extract.

Lemery Salt. See SALT OF LEMERY.

Lemery's White Precipitate. See WHITE PRECIPITATE.

Lemnian Earth (Lemnos Earth). A red, yellow, or grey, earthy substance consisting of a hydrated silicate of aluminium, found at Lemnos. It is an ochre, and is used as a pigment.

Lemnos Earth. See LEMNIAN EARTH.

Lemol. Lemon oil substitute for flavouring. (504).

Lemolac. A proprietary preparation. It is a very light mercurous chloride.

Lemonades. Drinks containing sugar and citric or tartaric acids.

Lemonal. A name for linalyl acetate.

Lemon Balm Oil. See OIL OF BALM.

Lemon Chrome. See BARIUM YELLOW.

Lemon Essence. Oil of lemon.

Lemongrass Oil. See OIL OF LEMONGRASS.

Lemon Juice, Artificial. A coloured aqueous solution of citric acid.

Lemon Lac (Orange Lac, Garnet Lac). Terms referring to the colour of lac, determined to some extent upon the tree from which it is obtained. See LAC.

Lemon, Salt of. See SALT OF SORREL.

Lemon Yellow. See CHROME YELLOW and ZINC YELLOW.

Lengenbachite. A mineral. It is lengenbuchite.

Lengenbuchite. A mineral, 6PbS.$(Ag.Cu)_2S.2As_2S_3$.

Lenicet (Estone). The trade name for a basic aluminium acetate, $Al_2(OH)_2(C_2H_3O_2)_4.H_2O$. Used in the treatment of wounds, and for perspiring feet. It is sold as powder, ointment, or cream.

Lenigallol. See GALLOLS.

Lenirobin. Chrysarobin tetracetate, used as a substitute for chrysarobin.

Lenitive Electuary. Confection of senna.

Lenium. A proprietary preparation of

selenium sulphide and bithionol. A skin cleanser. (112).

Lenka. Strongly alkaline detergent powder. (553).

Lensex. Midly alkaline detergent pastes. (553).

Lensine. Alkali detergent powders. (553).

Lentana. A red earthy ironstone of the type commonly called Laterite. Beads andornaments are made in Sokota of this material. It is also used for reddening pottery.

Lenticillin. A proprietary preparation containing Procaine Penicillin. An antibiotic. (336).

Lentin (Doryl). The urethane of choline, $(CH_3)_3NCl.CH_2.CH_2.OCONH$. A medicinal.

Lentizol. A proprietary preparation of AMITRYPTYLINE hydrochloride. An anti-depressant. (262).

Lentulite. A mineral. It is a variety of liroconite.

Lenzit. See GYPSUM.

Lenzite. A variety of the mineral Halloysite.

Lenzol. A pale cedar-wood oil of known viscosity and refractive index. Used for oil immersion in microscopy.

Leo K. A proprietary preparation of potassium chloride used in potassium-replacement therapy. (308).

Leonhardite. A mineral. It is $2[Ca_2Al_4Si_8O_{24}.7H_2O]$.

Leonil. See NEKAL.

Leonite (Kaliastrakanite). A mineral. It is a double salt of potassium and magnesium sulphates, $K_2SO_4.MgSO_4.4H_2O$.

Leopardite. A mineral. It is a variety of quartz rock containing stains of manganese.

Lepargylic Acid. See ANCHOIC ACID.

Lepidine. 4-Methyl-quinoline. It boils at 261–263° C., and resembles quinoline in its antipyretic and antiseptic properties.

Lepidocrocite. A mineral. It is a hydrated ferric oxide.

Lepidokrokite. A mineral. See GOTHITE.

Lepidolamprite. A mineral. It is franckeite.

Lepidolite (Lithiferous Mica, Lithium Mica, Lithia Mica). A mineral of variable composition, usually $K_4Li_4(Al.Fe)_3.Al_4(Si_3O_8)(SiO_4)_3$. Used for the preparation of lithium salts, and lithia water.

Lepidomelane (Iron Mica). A mineral, $(KH)_2O.(Fe.Al)_2O_3.2(Fe.Mg)O . 3SiO_2$. It is black in colour.

Lepidomorphite. A mineral. It is a mica.

Lepidone Violet. A dyestuff obtained from methyl-lepidone and phosphorus pentachloride.

Lepidophaeite. A mineral. It is a variety of wad containing copper.

Lepro. Lead peroxide. (589).

Leptinol. An alcoholic extract of *Leptotænia dissecta*. It has a therapeutic value.

Leptonematite. A mineral. It is a variety of braunite.

Leptynol. Colloidal palladous hydroxide suspended in sesamé oil. Employed as an injection in the treatment of obesity.

Lerastan. A German preparation. It is stated to be a tin preparation to be rubbed in pyodermia, furunculosis, and staphylococcus diseases.

Lerbachite. See LEHRBACHITE.

Lergine. A proprietary preparation containing tricyclamol chloride. An antispasmodic. (277).

Lergoban. A proprietary preparation of DIPHENYLPYRALINE. An anti-allergic drug. (275).

Leridine. The phosphate of Anileridine.

Leroy's Insulating Mixture. A heat insulator. It is composed of clay, hemp, coco-nut fibre, wool refuse, paper pulp, charcoal, sawdust, flour, and tar.

Le Sage Cement (Plaster Cement). A natural cement obtained from nodules found at Boulogne-sur-Mer.

Lesleyite. A mineral. It is $K_2Al_9Si_4O_{20}(OH)_4$.

Lessbergite. A mineral of the dolomite class. It is probably a mixture.

Lethalbine. A lecithin albuminate, containing 20 per cent. lecithin.

Lethane. Trade mark for insecticide concentrates based on beta butoxy beta—thio cyanodiethyl ether. Supplied in petroleum distillate. Used in industrial insecticide sprays and mosquito larvicides.

Lethi. Lead thiosulphate. (589).

Lethidrone. A proprietary preparation of nalorphine hydrobromide, used to counteract overdosage of morphine and similar drugs. (284).

Letovicite. A mineral. It is $(NH_4)_3H(SO_4)_2$.

Lettuce Opium. Lactucarium (*q.v.*).

Letusin. A proprietary trade name for the napsylate of Levopropoxyphene.

Leucaniline. Triamino-diphenyl-tolyl-methane, $(H_2N.C_6H_4)_2.CH.C_7H_6.NH_2$.

Leucarsone. 4-carbaminophenyl arsonic acid. (507).

Leucaugite. A mineral. It is a variety of augite.

Leucaurin. Trioxy-triphenyl-methane, $CH(C_6H_4.OH)_3$.

Leuchtenbergite. A mineral. It is a variety of Clinochlore.

Leuchtol. A proprietary synthetic resin.

Leucindophenol. See INDOPHENOL WHITE.

Leucine. α-Amino-isobutyl-acetic acid, $(CH_3)_2CH.CH_2CH(NH_2)COOH$.

Leucite (Amphigene). A mineral. It is a double silicate of potassium and aluminium, $K_2O.Al_2O_3.4SiO_2$, obtained from lava deposits in Italy. A source of potassium. It is also used as a fertiliser and water softener.

Leucite-Ferrique. Synonym for Iron-leucite.

Leucoargilla. See LITHOMARGE.

Leucobenzaurin. Di-p-oxy-triphenyl-methane, $CH.C_6H_5(C_6H_4OH)_2$.

Leucochalcite. A mineral, $Cu_2(OH)AsO_4.H_2O$.

Leucocylite. A mineral. It is a variety of apophyllite.

Leucofermantin. A substance stated to be obtained from the blood serum of animals immunised against trypsin. It has been used in the treatment of abscesses and boils.

Leucogallothionines. Dyestuffs obtained by condensing alkyl-diamino-thio-sulphonic acids with gallic acid or derivatives.

Leucogen. Sodium hydrogen sulphite, $NaHSO_3$, used in bleaching and paper making.

Leucol. An impure quinoline.

Leucoline (Chinoline). Quinoline, $C_6H_4.CH : CH.N : CH$.

Leuco-Malachite Green. Di-methyl-amino-phenyl-phenyl-methane, $C_6H_5CH(C_6H_4N(CH_3)_2)_2$.

Leucomanganite. A mineral. It is fairfieldite.

Leuco-Methylene-Blue. Tetra-methyl-diamino - imino - diphenyl - sulphide, $C_{16}H_{19}N_3S$.

Leuconate. Triphenylmethane triiso-cyanate.

Leucone. o-Silico-formic acid, $SiH(OH)_3$.

Leuconine (Leukonin). An antimony preparation containing 98 per cent. of sodium metantimoniate. Recommended as a substitute for tin oxide for enamelling.

Leucopetrite. A deposit between a wax and a resin in properties. It is an oxygenated hydrocarbon.

Leucophane. See LEUCOPHANITE.

Leucophanite (Leucophane). A mineral, $NaF.CaO.BeO.SiO_2$.

Leucophœnicite. A mineral. It is $4[Mn_7Si_3O_{12}(OH)_2]$.

Leucophosphite. A mineral. It is $(K,NH_4)Fe_2^{\cdots}(PO_4)_2OH.2H_2O$.

Leucophyll. A material which can be transformed into chlorophyll.

Leucophyllite. A mineral. It is a variety of mica.

Leucopyrite. See ARSENICAL IRON.

Leucosphenite. A mineral, $BaO.2Na_2O.2(Ti.Zr)O_2 10SiO_2$, of Greenland.

Leucotrope O. Dimethyl - phenyl - benzyl-ammonium chloride. An indigo discharger giving yellow shades on indigo dyed cloth.

Leucotrope W. The calcium salt of disulphonated Leucotrope O. An indigo discharger, giving yellow colours soluble in alkalis.

Leucovorin. A proprietary preparation of calcium folinate used as an antidote to METHOTREXATE. (306).

Leukarion. See OIL WHITE.

Leukeran. A proprietary preparation of chlorambucil. A carcino-chemo-therapeutic agent. (277).

Leukol. Runge's name for quinoline.

Leukon. A proprietary synthetic resin moulding powder.

Leukonin. See LEUCONINE.

Leuna Gas. A proprietary trade name for a compressed propane for use as a fuel. This type of fuel is also sold under the name Propagas.

Leunaphos B.A.S.F. A new fertiliser containing 20 per cent. nitrogen, 15 per cent. phosphoric acid, P_2O_5, and lime. It is a mixture of phosphate, nitrate, and sulphate of ammonia.

Leunaphoska B.A.S.F. A fertiliser specially prepared for Eastern markets. It contains 13 per cent. nitrogen, 10 per cent. phosphoric acid, P_2O_5, and 13 per cent. potash, K_2O.

Leuna Saltpetre B.A.S.F. A double salt of ammonium sulphate and nitrate. A fertiliser similar to Chilean nitrate in its action.

Leupoldite. A mineral, KCl.

Levallorphan. A narcotic antagonist present as the hydrogen tartrate in PETHILORFAN. It is $(-)$-N-allyl-3-hydroxymorphinan. LORFAN is the hydrogen tartrate.

Levamisole. An anthelmintic. $(-)$-2, 3, 5, 6 - Tetrahydro - 6 - phenylimidazo - [2, 1 - b] thiazole. KETRAX.

Levamphetamine. $(-)$ - α - Methyl-phenthylamine. $(-)$-2-Aminopropyl-benzene. $(-)$-Amphetamine. Cydril is the succinate.

Levant Berries. *Cocculus indicus*, the dried fruits of *Anamirta paniculata*.

Levapren. An ethylene-vinyl acetate copolymer. (112).

Leverrierite. A mineral, $2Al_2O_3.5SiO_2.5H_2O$. See also BEIDELLITE.

Levigated Litharge. The hard masses of litharge formed by the oxidation of lead after grinding.

Leviglianite. A mineral. It is a variety of cinnabar containing zinc.

Levius. A proprietary preparation of sustained-release aspirin. An analgesic. (353).

Levodopa. A preparation used in the treatment of the Parkinsonian syndrome. It is (—)-3-(3,4-dihydroxy-phenyl)-L-alanine. BERKDOPA, BROCADOPA, LARODOPA, LEVOPA, VELDOPA.

Levomethorphan. (—) - 3 - Methoxy - N-methylmorphinan.

Levopa. A proprietary preparation of LEVODOPA used in the treatment of Parkinson's disease. (765).

Levophed. A proprietary preparation of noradrenaline acid tartrate, used to raise blood pressure. (112).

Levophenacylmorphan. (—) - 3 - Hydroxy-N-phenacylmorphinan.

Levopropoxyphene. α - (—) - 4 - Di - methylamino-3-methyl-1,2-diphenyl-2-propionyloxybutane. Letusin is the napsylate.

Levormoramide. (—) - 1 - (β - Methyl - γ - morpholino - αα - diphenylbutyryl) - pyrrolidine.

Levorphan. A proprietary trade name for Levorphanol.

Levorphanol. (—) - 3 - Hydroxy - N - methylmorphinan. Levorphan. Dromoran is the hydrogen tartrate.

Levuline. A proprietary preparation used in the textile industry for finishing.

Levulose (Fruit Sugar, Diabetin, Lævo glucose, Sucro-levulose). d-Fructose, $C_6H_{12}O_6$.

Lévurargyre. A mercury - nucleo-proteid. An antiseptic.

Levuretine (Levurin, Levurinose). A proprietary form of dried beer yeast, used medicinally.

Levurin. See LEVURETINE.

Levurinose. See LEVURETINE.

Levyn. A mineral, $CaO.Al_2O_3.3SiO_2.5H_2O$.

Levynite. A mineral. It is a calcium-aluminium silicate.

Lewisite (Dew of Death). A military poison gas. It is β-chloro-vinyl-di-chloro-arsine, $CHCl : CH.AsCl_2$. The name Lewisite is also used for a mineral, $5CaO.2TiO_2.3SbO_3$.

Lewis Metal. An alloy of 1 part tin and 1 part bismuth having the property of expanding when cooling. It has a melting-point of 138° C. and is used for sealing and holding die parts.

Lewisol. Synthetic resins for varnishes and lacquers. (593).

Lexan. A registered trade name for polycarbonate resins. Used for electrical moulding purposes where good transparency is required. 92 per cent. transparency and a refractive index of

1·58 are obtained. Very high creep resistance is exhibited under load. (59).

Lexan 130. Proprietary name for a range of high-viscosity polycarbonates. (766).

Lexan 140. A proprietary range of low-viscosity polycarbonates used in the moulding of intricate parts. (766).

Lexan 141R. A proprietary polycarbonate resin used in injection moulding. It possesses intrinsic properties which facilitate mould-release. (766).

Lexan 145. A proprietary free-flowing granular polycarbonate used in the casting of films from solution and for dip-coating. (766).

Lexan 700. A proprietary polycarbonate resin used in rotational moulding. (766).

Lexan 700 SE1 and SE2. Self-extinguishing grades of LEXAN 700. (766).

Lexan 3312. A proprietary polycarbonate resin containing 20% glass reinforcement. (766).

Lexan 3314. A proprietary polycarbonate resin containing 40% glass reinforcement. (766).

Lexan LS. A proprietary polycarbonate resin used for moulding lenses. (766).

Lexan RP 700. A proprietary polycarbonate resin used in the rotational moulding of containers and road-lighting fittings. (766).

Lexel. A registered trade mark for a cellulose acetate-butyrate insulating tape. It is flame retardant, has low moisture absorption and has a high dielectric strength.

Lextron. A proprietary preparation of liver extract, and ferric ammonium citrate. (333).

Ley - Cornox. Selective weedkillers. (502).

Leytosan. Organomercurial seed dressings. (572).

Leyden Blue. See COBALT BLUE.

LG Wax. A montan wax derivative used in the production of non-ionic and ionic dry-bright emulsions for floor polishes and similar applications. (6).

Lherzolite. A mineral. It is a mixture of olivine, diallage, and hypersthene.

Liantral. An extract of coal tar, obtained by treatment with benzene, filtering, and evaporating. It has been recommended as a substitute for coal and wood tar in the treatment of eczema, and is employed as a 10 per cent. solution with olive oil.

Libanol. An ethereal oil obtained from the wood of *Cedrus atlantica*. It is said to be a valuable substitute for sandal-wood oil, in the treatment of gonorrhœa.

Libavius' Fuming Spirit. A solution of stannic chloride, $SnCl_4$, obtained by the distillation of a mixture of tin or tin amalgam and mercuric chloride.

Libethenite. A mineral, $4CaO.P_2O_5.H_2O$.

Libollite. A variety of asphaltum.

Libraxin. A proprietary preparation of chlordiazepoxide, and clidinium bromide. Gastro-intestinal sedative. (314).

Librium. A proprietary preparation of chlordiazepoxide hydrochloride. A sedative. (314).

Licareol. A name applied by Morin to a *l*-linalol.

Licella Yarn. A product made from narrow strips of wood-pulp paper. A jute substitute.

Lichen Blue. See LITMUS.

Lichen Starch. Lichenin, $C_6H_{10}O_5$. A carbohydrate derived from Iceland moss.

Lichen Sugar. Erythrite, $C_4H_6(OH)_4$.

Lichner's Blue. A variety of smalt. It is a silicate of cobalt and potassium.

Lichtenberg's Metal. An alloy of 50 per cent. bismuth, 30 per cent. lead, and 20 per cent. tin, melting at $91.5°C$.

Liconite. A rubber substitute made from bitumen and oils.

Lidanil. See MESORIDAZINE.

Lidocaine. A proprietary trade name for Lignocaine.

Lidoflazine. A cardiac stimulant. 4-[3-(4, 4′ - Difluorobenzhydryl) propyl] piperazin - 1 - ylacet - 2′, 6′ - yxlidide. CLINIUM.

Lidothesin. A proprietary preparation of lignocaine hydrochloride, a local anæsthetic. (348).

Lidothesin with Noradrenaline. A proprietary preparation containing Lignocaine hydrochloride and noradrenaline 1 in 80,000. A local anæsthetic. (348).

Liebenerite. A mineral. It is a variety of pinite.

Lieben Solution. A solution of iodine in potassium iodide.

Lieberkuhn's Jelly. The jelly formed by mixing egg serum with about one-third of its volume of twice normal sodium hydroxide.

Liebigite. A mineral found near Adrianople, Turkey, and at Joachimsthal corresponding to the formula $CaCO_3.(UO_2)CO_3.20H_2O$.

Liebmann and Studer's Acid. α Naphthol-sulphonic acid (1 : 7).

Lievrite. A mineral, $(Ca.Fe)O.Fe_2O_3.SiO_2$.

Life Balsam. See BALSAM OF LIFE.

Ligdynite. A coal mine explosive consisting of 25–27 per cent. nitro-glycerin, 27–29 per cent. sodium nitrate, 10–12 per cent. sodium chloride, and 30–33 per cent. wood meal.

Ligdyns. South African explosives similar to American straight dynamite (a mixture of nitro-glycerin, wood pulp, sodium nitrate, and calcium or magnesium carbonate).

Light Aniline. See ANILINE OIL.

Light Benzine. A fraction of Russian petroleum, distilling above $130°C$. of specific gravity $0.700–0.717$.

Light Benzol. That fraction of the light oils from coal tar (after purification), which distils over from $70–140°C$.

Light Blue. See METHYL BLUE and ANILINE BLUE, SPIRIT SOLUBLE.

Light Carburetted Hydrogen. See MARSH GAS.

Light Fast Yellow 3G. A dyestuff belonging to the same class as Tartrazine.

Light Green. See METHYL GREEN and IODINE GREEN.

Light Green SF, Bluish (Acid Green, Acid Green, bluish, Acid Green M). A dyestuff. It is the sodium salt of dimethyl - dibenzyl - diamino - triphenyl-carbinol-trisulphonic acid, $C_{35}H_{31}N_2O_{10}S_3Na_3$. Dyes wool and silk green from an acid bath.

Light Green SF, Yellowish (Acid Green, Acid Green extra conc., Acid Green GG, Acid Green D, Acid Green JJ, Acid Green SOF, Acid Green Yellowish). A dyestuff. It is the sodium salt of diethyl-dibenzyl-diamino - triphenyl - carbinol - trisulphonic acid, $C_{37}H_{35}N_2O_{10}S_3Na_3$. Dyes wool and silk green.

Lighthouse Chrome Blue B. A dyestuff. It is a British equivalent of Chromotrope 2R.

Lighthouse Chrome Brown R. A dyestuff. It is a British brand of Fast brown N.

Lighthouse Chrome Cyanine R. A British dyestuff. It is equivalent to Palatine chrome blue.

Lighthouse Chrome Yellow O. A British equivalent of Milling yellow.

Lightning Powder. A double base smokeless rifle powder adapted for use in a large class of sporting cartridges such as the 25–35, 30–30, 32 special, 30–40 high power and like cartridges.

Light Nitrobenzene. See NITROBENZENE, LIGHT.

Light Oil (First Runnings, Crude Oil). That fraction of coal tar distilling up to $170°C$., of specific gravity $0.91–0.95$. Approximately 12 gallons are obtained from 1 ton of tar.

Light Oil of Wine. Wine oil, an oil obtained in the preparation of ether by the distillation of alcohol with sulphuric acid.

Light Red. A pigment prepared by calcining Oxford or yellow ochre.

Light Red Silver Ore. See PROUSTITE.

Light Spar. See GYPSUM.

Light White. See OIL WHITE.

Lignax. A proprietary preparation of lignocaine hydrochloride, extract of hamamelis, bismuth subgallate, zinc oxide, balsam of Peru and menthol. Suppositories for anal conditions. (348).

Lignin Dynamites. Explosives used in rock blasting. They consist of nitroglycerin absorbed by a mixture of woodpulp, and a nitrate, usually sodium nitrate.

Lignite. A fuel found in Devonshire. It is of a more recent date than true coal.

Lignite Char. See GRUDEKOK.

Lignite Wax (Mineral Wax). Montan wax (q.v.).

Lignizik. A charcoal made by heating wood waste by means of gases containing no oxygen. It is a hard product.

Lignocaine. N - (Diethylaminoacetyl) - 2,6-xylidine. Lidocaine. Duncaine; Lignostab; Xylocaine; Xylotox.

Lignol. An oily or tarry substance, obtained by the dry distillation of bituminous fossilised wood. It is asserted to have considerable bactericidal powers and is recommended in the treatment of eczema and acne.

Lignolite. See SOREL CEMENT.

Lignolite Sheets. A building material. It consists of sheets of wood, preferably a special quality of birch plywood, which have been treated with an impregnating material and subjected to long heat treatment at a high temperature. The sheets of wood have an applied surface of veneer and are pressed between metal plates at high temperatures. The material can be worked, but is very resistant to mechanical damage.

Lignorosin. Calcium lignosulphonate, a by-product in the manufacture of paper pulp. It is a dark-brown syrup, and is used as an assistant in mordanting wool with chrome.

Lignostab. A proprietary trade name for Lignocaine.

Lignostone. See PERMALI.

Lignosulfin. See LIGNOSULFIT.

Lignosulfit (Lignosulfin). A by-product in the manufacture of cellulose. The active ingredient is sulphurous acid. It is employed in pulmonary complaints, by the inhalation of the vapour.

Lignum Vitæ. Guaiacum wood.

Ligro. A proprietary trade name for a crude pine fatty acid mixture.

Ligroin (Heavy Benzine). A term rather loosely applied. It usually denotes a refined distillate of petroleum oil having a boiling-point of 120–135° C., of specific gravity 0·73. Used as a polishing oil, and as a turpentine substitute in varnishes. The term is sometimes applied to such fractions as benzoline.

Ligulin. The dyestuff of privet berries.

Ligurite. A mineral. It is a variety of sphene.

Lilacin (Lilicin, Terpilenol). Terpineol, $C_{10}H_{17}OH$, a terpene alcohol.

Lilaminox 10. A proprietary trade name for a dimethyl alkyamine oxide of general formula $R(Me_2)N{\rightarrow}O$ where R is lauryl, myristyl or cetyl. A completely biodegradable detergent. (203).

Lilaminox 100. A proprietary trade name for a bis-(2-hydroxyethyl)alkylamine oxide of the general formula $R(C_2H_4OH)_2N{\rightarrow}O$ where R is lauryl, myristyl or cetyl. Used similarly to Lilaminox 10. (203).

Lilianite. A mineral, $3(Pb.Ag_2)S.Bi_2S_3$.

Lilicin. See LILACIN.

Lillhammerite. A mineral. It is pentlandite.

Lillite. An earthy mineral resembling Glauconite, of Bohemia.

Lily of Valley, Artificial. Terpineol, $C_{10}H_{17}OH$.

Lilyolene. See PARAFFIN, LIQUID.

Limanol. A preparation of salt-marsh mud used for rheumatism.

Lima Oil. See EARTH OIL.

Lima Wood. See REDWOODS.

Limarsol. See ACETARSONE.

Limbachite. A Serpentine mineral.

Limbritrol. A proprietary preparation of AMITRYPTYLINE and CHLORDIAZEPOXIDE. An anti-depressant. (314).

Limbux. A registered trade name for a form of mechanically slaked lime. It is used in agriculture, building and construction, metallurgical and chemical industries. (2).

Limclair. A proprietary preparation of DISODIUM EDETATE used in the treatment of hypercalcæmia. (637).

Lime (Burnt Lime, Quicklime, Caustic Lime). Calcium monoxide, CaO, produced by calcining calcium carbonate. Sometimes a mixture of calcium and magnesium oxides is sold under this name.

Lime-Arsenic Greens. See SODA GREENS.

Lime, Ashby. See ASHBY LIME.

Lime Blue. A pigment. It consists of a mixture of copper hydroxide and lime, and is prepared by precipitating a solution of copper sulphate with milk of lime, with the addition of ammonium chloride. Sold in rectangular lumps is known as Neuwied blue. The name Lime blue is also applied to the limefast coal-tar colour lakes, and to ultramarine cheapened by the addition of gypsum.

Lime, Burnt. See LIME.

Lime, Caustic. See LIME.

Lime Chrome. A pigment. It is calcium chromate, $CaCrO_4$. It is little used.

Lime, Durham. See DURHAM LIME.

Lime Felspar. See ANORTHITE.

Lime Flux. Limestone. See CALCITE.

Lime Green (Farth Green). Lime Green was the name originally given to a pigment obtained by precipitating copper sulphate solution with chalk or milk of lime. Also see GREEN ULTRAMARINE.

Limeite. A cement containing rubber, tallow, and lime. The addition of vermilion causes the mixture to harden.

Lime-Liniment. See CARRON OIL.

Lime Malachite (Calciomalachite). A mineral containing calcium and copper carbonates.

Lime Mortar. Mixtures of slaked lime and sand.

Lime Mud. See CHANCE MUD.

Lime Nitrate (Lime Saltpetre). Calcium nitrate, $Ca(NO_3)_2.2H_2O$. A fertiliser.

Lime Nitrogen. See NITROLIM.

Limeolivine. Calcium orthosilicate $2CaO.SiO_2$.

Lime Precipitate. A fertiliser obtained by soaking bones in 8 per cent. hydrochloric acid, in which the phosphate is dissolved. This is precipitated by lime as dicalcium phosphate.

Lime, Pyrolignite. See PYROLIGNITE OF LIME.

Lime Saltpetre. See LIME NITRATE.

Lime Soap. Calcium resinate.

Limestone. See CALCITE.

Limestone, Magnesium. See DOLOMITE.

Lime-Sulphur Dips. Sheep dips for the treatment of scab. They contain flowers of sulphur, lime, and water.

Lime Water. A solution of calcium hydroxide, $Ca(OH)_2$, in water. It contains 13 grains calcium hydroxide in 1 pint of water at 60° F.

Limnite. A mineral. It is a hydrated ferric oxide, $Fe_2O_3.3H_2O$.

Limo (Sablon, Tartrate of Lime). The raw materials obtained by the precipitation of tartaric acid in tartar works or wine distilleries.

Limonite (Marsh Ore). A mineral. It is a brown hydrated oxide of iron, $2Fe_2O_3.3H_2O$.

Linalux. Marine gloss finishes. (512).

Linarite. A mineral. It is a basic sulphate of lead and copper, $PbSO_4.Cu(OH)_2$.

Linocin. A proprietary preparation containing lincomycin hydrochloride. An antibiotic. (325).

Lincomycin. An antibiotic produced by Streptomyces lincolnensis var. lincolnensis. Lincocin and Mycivin are the hydrochloride.

Linctavit. A proprietary preparation of pholcodine, ephedrine hydrochloride, ipecacuanha extract, paracetamol and vitamins B and C. (397).

Linctified Expectorant. A proprietary preparation of TRIPROLIDINE, PSEUDO-EPHRINE, CODEINE phosphate and guaiphenesin. An expectorant. (277).

Lindackerite. A mineral, $Ni_3Cu_6(OH)_4(AsO_4)_4(SO_4).5H_2O$.

Lindene. Indene coumarone resins. (623).

Lindenol. Pure alpha terpineol. (504).

Lindesite. Synonym for Urbanite.

Lindol. A proprietary trade name for tricresyl phosphate. A vinyl plasticiser. (83).

Lindstromite. A mineral, $2PbS.Cu_2S.3Bi_2S_3$.

Linen. A fabric manufactured from flax fibre.

Lingraine. A proprietary preparation of ergotamine tartrate, used for migraine. (112).

Lining Metal. Alloys of 70–90 per cent. lead, 2–20 per cent. tin, and 5–20 per cent. antimony. They are used for car bearings.

Linnæite. See LINNALITE.

Linnalite (Linnæite). Native cobalt sulphide, Co_3S_4.

Linnets. See BROWN LEAD ORE.

Linofelt. A proprietary sound and heat insulating material made from flax fibres and paper.

Linolana. The trade name of a cottonised flax tow. A cotton substitute.

Linoleum. A material consisting chiefly of cork and wood dust, oxidised linseed oil, mineral earth colours, rosin, various siccatives, and jute. It is prepared by oxidising linseed oil by blowing hot air through it, after the addition of dryers, for 18–20 hours. After adding 30 per cent. of colophony, the whole is converted into a paste with cork dust, and the mass is pressed hot on to a strong textile previously varnished. It is coloured with mineral colouring matters. Used as floor coverings. Also see LINOLEUM CEMENT.

Linoleum Cement. A step in the manufacture of linoleum. It is obtained by melting solidified linseed oil with various gum-resins (usually kauri gum and rosin), and mixing it with cork flour, various pigments, and fillers.

Linoleum, Inlaid. Prepared from coloured sheets by cutting these to form various patterns, and stamping them on to canvas; or different coloured compositions are moulded to various shapes, and placed on canvas. Wood flour, in addition to cork dust, is used for inlaid linoleum.

Linotype Metal. An alloy of 13·5 per cent. antimony, 2 per cent. tin, and 84·5 per cent. lead. See also TYPE METAL.

Linoxyn. See SOLIDIFIED LINSEED OIL.

Linseed Meal. Ground oil-cake.

Linseed Oil, Artist's. Raw linseed oil which has been allowed to stand for weeks, then treated with litharge, and finally bleached by exposure.

Linseed Oil, Refined. Raw linseed oil which has been treated with a 1 per cent. solution of sulphuric acid.

Linseed Oil Soap. A potashsoap.

Linters. The short fibres removed mechanically from the " fuzz " remaining in the cotton seed after the lint has been taken.

Linter's Starch. A soluble starch prepared by mixing raw starch with 7·5 per cent. hydrochloric acid in water, allowing it to stand for several days, with stirring. The solution is decanted, and the starch washed with water and dried.

Lintex A10. A proprietary aqueous emulsion of a styrene copolymer internally plasticised with a copolymerised ester. It contains crosslinkable groups and is used as a binder for emulsion paints. (29).

Lintonite. A mineral. It is a variety of thomsonite.

Lionite. A mineral. It is an impure native tellurium. The name is also used as a trade name for an abrasive consisting essentially of fused alumina.

Lioresal. A proprietary preparation of BACLOFEN used to reduce muscle spasticity. (519).

Liothyronine. L - α - Amino - β - [4 - (4 - hydroxy - 3 - iodophenoxy) - 3,5 - di - iodo-phenyl] propionic acid.   (−)-Tri-iodothyronine. Cynomel and Tertroxin are the sodium derivative.

Lipal. A proprietary name for a range of polyoxyethylene esters and ethers. (767).

Liparite. A mineral. It is a variety of talc.

Lipases. Enzymes which split up fats.

Lipiodol. A proprietary preparation. It is iodised poppy-seed oil.

Lipiphysan. A proprietary preparation of cotton seed oil. (313).

Lipobromol. Brominated poppy-seed oil, containing 33 per cent. bromine.

Lipoflavonoid. A proprietary preparation of choline bitartrate, inositol, di-methionine, ascorbic acid, lemon bioflavinoid, thiamine, riboflavine, nicotinamide, pyridoxine, panthenol and cyanocobalamin. Used for neurosensory deafness. (252).

Lipofor. A stabilised fat, hydrolysed without alkali, which exerts such a strong solvent action upon fats, tars, and resins that a clear solution is produced in its presence.

Lipoiodine " Ciba." Ethyl ester of di-iodobrassidic acid. An organic odourless and tasteless preparation containing 41 per cent. of iodine. Indicated in all afflictions calling for iodine therapy.

Lipotriad. A proprietary preparation of tricholine citrate, inositol, DL-methionine, cyanocobalamin, thiamine hydrochloride, riboflavine, nicotin-amide, pyridoxine hydrochloride and pantothenyl alcohol. (252).

Lipowitz's Alloy. An alloy of 50 per cent. bismuth, 27 per cent. lead, 13 per cent. tin, and 10 per cent. cadmium. It has a specific heat of 0·0345 calories per gram at from 5–50° C., and melts at 65° C. It is used for automatic sprinklers and other purposes.

Liqueur de Ferraile. See PYROLIGNITE OF IRON.

Liqueur de Goudron. A solution of Norwegian tar.

Liqueur de van Swieten. Consists of 1 part mercuric chloride, 100 parts alcohol, and 900 parts water.

Liqueurs. Alcoholised wines. They are usually mixtures of alcohol, water, and sugar, with essential oils, vegetable extracts, and essences, and contain 40–70 per cent. alcohol.

Liquibor. Corrosion control additives. (548).

Liquibor 524. A borax-glycol condensation product which is used at a concentration of 1–2 per cent. as a corrosion inhibitor in synthetic hydraulic fluids.

Liquidambar (Copalm Balsam). A balsam obtained from a large Mexican tree. It contains cinnamyl cinnamate and styrene. It is also erroneously called liquid storax.

Liquid, Bleaching.   See BLEACH LIQUOR.

Liquid Bronzes. Varnishes in which bronze colours are suspended.

Liquid Camphor (Camphor Oil). The essential oil from the camphor laurel, from which camphor is obtained. Bornene, an oil accompanying Borneo camphor in the tree, is also known by this name.

Liquid Chloride of Lime. See BLEACH LIQUOR.

Liquid Drier. See TEREBINE. Also a name given to a concentrated solution of calcium chloride.

Liquid Fuel. See COLLOIDAL FUEL.

Liquid Glue. Dextrin is known by this name. A liquid glue is also obtained by the protracted heating of glue with its own weight of water, and one fourth of its weight of hydrochloric, acetic, or nitric acid. A solution of shellac in alcohol is sold under this name. Also see DUMOULIN'S GLUE and SPAULDING'S GLUE.

Liquid Gold. Contains about 10 per cent. of gold (as chloride), resin, lavender oil, and bismuth. Used for painting china.

Liquid Gutta-percha. This is usually a solution of gutta-percha in chloroform.

Liquid Phosphoretted Hydrogen. Phosphorus dihydride, P_2H_4.

Liquid, Pink Cutting. See TIN SPIRITS.

Liquid Pitch Oil. See OIL OF TAR.

Liquid Resins (Polyterpene, Sulphate Resin, Talleol). Semi-resinous compounds obtained as by-products in the manufacture of wood pulp by the sulphite and sulphate processes of papermaking. The terms are more particularly applied to products obtained when pinewood is used. They have little technical importance, and are used to some extent in the manufacture of detergents. Also see TALLOEL.

Liquid Storax. Storax, a balsam. Also see LIQUIDAMBAR.

Liquid Sulphur. There are two forms of liquid sulphur, Sλ and Sμ, in dynamic equilibrium when sulphur is melted. Sλ is mobile and Sμ is viscous. Sμ is formed on heating Sλ.

Liquitalis. A substance obtained by the extraction of digitalis leaves with cold water. It contains digitalin, but no digitoxin or digitonin. A cardiac remedy.

Liquix. A proprietary analgesic and anti-pyretic. ACETAMINOPHEN in a 120 mg/5 ml. solution. (768).

Liquix C. A preparation of LIQUIX with the addition of 8 mg. codeine per cubic centimetre. (768).

Liquoid. A proprietary preparation of sodium polyanethol sulphonate. (314).

Liquor, Gas. See GAS LIQUOR.

Liquor, Glass. See SOLUBLE GLASS.

Liquorice. The dried root of *Glycyrrhiza glandulifera*.

Liquorice Juice. Extract of liquorice in sticks.

Liquor, Iron. See PYROLIGNITE OF IRON.

Liroconite. A mineral, $2Al_2O_3.2As_2O_5$. $7Cu(OH)_2.2CuO.20H_2O$.

Liskeardite. A mineral. It is a basic aluminium-ferric arsenate, $(Fe.Al)AsO_4$. $8H_2O$.

Lissamine. Acid wool levelling dyestuffs. (512).

Lissamine Green B. See WOOL GREEN S.

Lissapol. A synthetic detergent. (512).

Lissapol L S. A synthetic detergent. An anionic wetting agent, it is oleyl *p* -amsidine sulphonate. $R.CoNH.C_6H_3$ $(OCH_3)SO_3$, where R is the oleyl radical (2).

Lissapol N. A synthetic detergent and wetting agent. It is either a polyglycol-alkylphenol ether or an alkylphenol ethylene oxide adduct. (2).

Lissatan. A bleaching and mordant agent for leather. (512).

Lissephen. A proprietary trade name for Mephensin carbamate.

Lissokor. A trade mark for artificial leather (made from latex), leather and leather substitutes.

Lissolamine. Azoic and vat stripping agent. (512).

Listerine. A proprietary antiseptic containing boric acid, benzoic acid, thymol, and essential oils of eucalyptus, gaultheria, and others.

Listica. A proprietary preparation for the relief of tension. (327). Hydroxyphenamate.

Litalbin. Lecithin albuminate.

Litaner Balsam. See DAGGET.

Litchi Nut. The Chinese hazel nut. Used in the East for medicinal and edible purposes.

Litharge (Massicot). Pigments consisting of lead monoxide, PbO. Litharge is obtained in silver refining, and has a more reddish colour than Massicot, which is made by roasting lead.

Lithargrite. A mixture of oxide of lead and calcined magnesia. Used as a rubber filler.

Lithex. A proprietary rubber vulcanisation accelerator. It is lead dithiobenzoate precipitated on an inert base such as clay.

Lithia. A mixture of lithium carbonate and citric acid. It dissolves in water, and is used as a remedy for gout.

Lithia Mica. See LEPIDOLITE.

Lithic Acid. Uric acid, $C_5H_4O_3N_4$.

Lithiferous Mica. See LEPIDOLITE.

Lithionpsilomelane. Synonym for Lithiophorite.

Lithiophorite. A mineral. It is $2[(Al,Li)MnO_2(OH)_2]$.

Lithiophyllite. A mineral, $Li(Fe.Mn)PO_4$.

Lithiopiperazine. A preparation of piperazine and lithium salts. A solvent for uric acid.

Lithite. See PETALITE.

Lithium Aspirin. Lithium - acetylsalicylate. See APYRON.

Lithium-Diuretin. An additive product of theobromine-lithium and lithium salicylate, $C_7H_7N_4O_2Li . OH . C_6H_4 . COOLi$.

Lithium Iron Mica. See ZINNWALDITE.

Lithium Mica. See LEPIDOLITE.

Litho-Carbon. A material resembling asphalt, found in Texas.

Lithoclastite. An explosive. It is a dynamite.

Lithocolla. See LITHOMARGE.

Lithofor. Dyestuffs for printing inks. (512).

Lithoform. A pretreatment for zinc. (512).

Lithofracteur. An explosive of the dynamite class. It contains 54 per cent. nitro-glycerin, 15 per cent. barium nitrate, 17 per cent. kieselguhr, 2 per cent. wood meal, 1 per cent. bran, 4 per cent. sulphur, 2 per cent. manganese dioxide, and 2 per cent. sodium carbonate.

Lithographer's Varnish. See STAND OIL.

Lithographic Stones. Limestones prepared and polished.

Lithol. Ammonium ichtho-sulphonate.

Lithol Fast Scarlet R. See HELIO FAST RED RL.

Lithol Red B (Gallocyanine BD, Monolite Red R, Signal Red). Sulphonaphthalene-azo-β-naphthol. It forms red lakes.

Lithol Red R. A dyestuff from diazotised β-naphthylamine-1-sulphonic acid and β-naphthol.

Lithol Rubine B (Permanent Red 4B). A dyestuff obtained from p-toluidine-o-sulphonic acid and β-hydroxy-naphthoic acid.

Lithomarge (Lithocolla, Leucoargilla). A compact clay found in rock fissures.

Litho-oil (Stand oil, Polymerized oil). Raw linseed oil heated in such a manner that practically no oxidation occurs, thus thickening the oil through polymerization alone.

Lithophone. See LITHOPONE.

Lithopone (Beckton White, Charlton White, Duresco, Enamel White, Fulton White, Griffith's White, Jersey Lily White, Knight's Patent Zinc White, Lithophone, Marbon White, Nevin, Orr's White, Oleum White, Pinolith, Porcelain White, Ross's White, Zinc-Baryta White, Zincolith, Zinc Sulphide White, Sulphide White). A pigment. It is a mixture of barium sulphate and zinc sulphide, with some zinc oxide, containing from 11–42 per cent. zinc sulphide. The Dutch red seal quality contains 70·1 per cent. barium sulphate, 26·6 per cent. zinc sulphide, and 3·2 per cent. zinc oxide; German red seal contains 69·9 per cent. barium sulphate, 29·5 per cent. zinc sulphide, and 0·5 per cent. zinc oxide. A yellow seal variety contains about 15 per cent. zinc sulphide, a green seal about 30 per cent. zinc sulphide, a white seal about 26 per cent., and a blue seal about 22 per cent. zinc sulphide. The specific gravity varies from about 3·6 to 4·25, and the oil absorption approximates to 8 per cent.

Lithopone, Titanium. See TI-TONE.

Lithyol. See ICHTHYOL.

Litmoryrine. See APYRON.

Litmus (Lichen Blue, Lacmus, Tournesol, Turnsole). The colouring matter produced by the action of air and alkalis upon certain colourless principles, as orcin, $C_6H_3(OH)_2CH_3$, derived from different species of lichens. The extract is mixed with gypsum or chalk, and made into tablets. Used as an indicator. Also see ARCHIL.

Litmus-Milk Solution. Used to test for pancreatic lipase. It consists of 1 part of powdered litmus to 50 parts of dried milk. One part to 9 parts of water is used to make the solution.

Litnum Bronze. An alloy of from 80–85 per cent. copper, 10–15 per cent. aluminium, and 4 per cent. iron.

Lito-Silo. See SOREL CEMENT.

Littoral. A decolorising agent for sugar juices.

Liveingite. A mineral, $Pb_5As_8S_{17}$.

Liver of Antimony. The name applied to the impure double sulphides of antimony, obtained by heating antimony sulphide, Sb_2S_3, with various metallic sulphides, more especially with the alkali and alkaline earth sulphides.

Liver of Sulphur. See SULPHURATED POTASH.

Liveroid. A proprietary liver extract.

Liver-Opal. See MENILITE.

Liver Ore. See CINNABAR.

Liver Resin. A name suggested for the unsaponified matter obtained from the oil of the Jewfish. When dried at 100° C., it resembles colophony.

Liver Sugar. Glycogen ($C_6H_{10}O_5$).

Livingstonite. A mineral. It is Antimonite containing 14 per cent. mercury.

Livonal. A 20 per cent. alcoholic solution of benzyl benzoate. It is a German preparation used in the treatment of asthma.

Livox. A proprietary preparation of liver extract, thiamine hydrochloride, riboflavine, nicotinamide, ferrous gluconate, copper sulphate and manganese sulphate. (334).

Llama. See ALPACA.

Llanca. Chrysocolla, a copper ore. It is the name used in Chile.

Llicteria. Synonym for Franckeite.

Lloyd's Reagent. A hydrous aluminium silicate prepared from fuller's earth. It absorbs alkaloids.

Loadstone. See MAGNETIC IRON ORE.

Loaisite. A mineral. It is a variety of scorodite.

Loalin. A proprietary trade name for a polystyrene moulding compound.

Loams. Natural mixtures of clay and sand. Used for the manufacture of bricks and tiles.

Lobak. A proprietary preparation of chlormezanone, and paracetamol. (112).

Lobelin. A German preparation. It is a mixture of alkaloids obtained from the North American plant *Lobelia Inflata*. The preparation is stated to be suitable for combating carbon monoxide poisoning.

Lobidan. A proprietary preparation of lobeline sulphate, magnesium carbonate, and tribasic calcium phosphate. (137).

Loboite. A mineral. It is idocrase.

Lobosol. Blended low boiling solvents. (578).

Locabiotal. A proprietary trade name for Fusafungine.

Locamphen. An antiseptic made from iodine, phenol, and camphor.

Locan. A proprietary preparation of amethocaine, amylocaine, bismuth subnitrate and zinc oxide. Anæsthetic anal suppositories. (182).

Locasol. A proprietary milk preparation with a low content of calcium. (665).

Locke's Solution. A saline solution containing dextrose and used for injections.

Locoid. A proprietary preparation of HYDROCORTISONE 17-butyrate used as a steroid skin cream. (317).

Locorten. A proprietary preparation of flumethasone pivalate for dermatological use. (18).

Locorten-N. A proprietary preparation of flumethasone pivalate and neomycin sulphate used in dermatology as an antibacterial agent. (18).

Locorten-Vioform. A proprietary preparation of flumethasone pivalate and CLIOQUINOL. A steroid and anti-infective skin cream. (519).

Locoum. A gum-like mass prepared from starch paste and sugar.

Locron. A proprietary trade name for a urea synthetic resin.

Loctite. A trade mark. Single component structural adhesives and thread locking materials possessing the unique property of setting when air is excluded. Some are based on oxygenated methacrylic molecules of patented formulation.

Loctite 904. A proprietary trade name for a single component cyanoacrylate adhesive.

Locust Bean (St. John's Bread). Carob bean, a food.

Locust Bean Gum. See INDUSTRIAL GUM.

Locust-kernel Gum. See CAROB-SEED GUM.

Loda. An artificial fibre stated to be suitable for use in the place of wool.

Lodal. A proprietary preparation of 6 : 7 - dimethoxy - 2 - methyl - 3 : 4 - dihydro-isoquinoline chloride. It causes a rise in blood pressure.

Lodestone. See MAGNETIC IRON ORE.

Loestrin 20. A proprietary preparation of ethinylœstradiol and norethisterone acetate. An oral contraceptive. (264).

Lo-ex. An aluminium alloy containing 14 per cent. silicon, 2 per cent. nickel, 1 per cent. copper, and 1 per cent. magnesium.

L.O.F. A proprietary safety glass.

Lofenalac. Proprietary name for an infant milk feed low in PHENYLALANINE, used in the dietary treatment of phenylketonuria. (324).

Lofendazam. A tranquilliser. 8-Chloro-4, 5 - dihydro - 1 - phenyl - $3H$ - 1, 5 - benzodiazepin-2-one.

Lofepramine. An anti-depressant. 5 - {3 - [N - (4 - Chlorophenacyl)methyl - amino]propyl} - 10, 11 - dihydrodi - benz[b, f]azepine. LOPRAMINE.

Löffler's Methylene Blue. A stain for bacteria. It consists of 100 parts of a solution of sodium hydroxide (1 in 10,000) and 30 parts of a saturated alcoholic solution of methylene blue.

Löffler Stain for Flagella. To 10 c.c. of a 20 per cent. solution of tannin are added 5 c.c. of a cold saturated solution of ferrous sulphate and 1 c.c. of a solution of fuchsine or methyl violet.

Lofoam. Textile detergents. (537).

Lofotol. Cod-liver oil charged with carbon dioxide.

Loftine. See MULSOID.

Lofton Merritt's Stain. (a) A solution of 2 grams of malachite green in 100 c.c. water. (b) A solution of 1 gram of basic fuchsine in 100 c.c. water. The test solution contains 1 part of (a) and 1 part of (b), and is used to distinguish between unbleached sulphate and unbleached sulphite fibres, the former giving a blue or blue-green colour, the latter a purple or lavender.

Loganite. A mineral. It is a variety of amphibole.

Logwood (Campeachy Wood, Jamaica Wood, Hematine Paste and Powder, Steam Black). A natural dyestuff from the wood of *Hæmatoxylon campechianum*. It is sold as chips or extract. The wood contains hæmatoxylon, $C_{16}H_{14}O_6$. $3H_2O$, the dyeing principle, which is converted into hæmatin, $C_{16}H_{12}O_6$, the dyestuff, by oxidation. Used for dyeing black with a chrome mordant, for wool; with an iron mordant for silk, and with a chrome, or an iron and aluminium mordant, for cotton.

Lohys Steel. A mild steel having a high magnetic permeability.

Lokain. Lokaonic acid.

Lokandi. See VENTILAGO MADRASPANTA.

Lokas. See SAP GREEN.

Lolingite (Lollingite). A mineral. It is ferric arsenide, $FeAs_2$, or Fe_2As_3.

Loman Steel. An abbreviation of " low-manganese steel." It contains from 7–10 per cent. manganese.

Lomodex. A proprietary preparation of dextrans in saline or dextrose solution, used to improve capillary circulation. (188).

Lomonosovite. A mineral. It is $Na_5Ti_2Si_2PO_{13}$.

Lomotil. A proprietary preparation of DIPHENOXYLATE hydrochloride and atropine sulphate. An anti-diarrhœal. (644).

Lomotil with Neomycin. A proprietary preparation containing diphenoxylate hydrochloride, atropine sulphate and neomycin sulphate. An anti-diarrhœal. (285).

Lomudase. A proprietary preparation of chymotrypsin, and isoprenaline sulphate. Treatment of bronchitis. (188).

Lomupren. A proprietary preparation of isoprenaline sulphate. A bronchial antispasmodic. (188).

Lomusol. A proprietary preparation of sodium cromoglycate as a nasal spray in the treatment of allergic rhinitis. (509).

Lonarit. An acetyl cellulose product. It can be moulded and coloured.

Lonchidite. A mineral. It is a variety of marcasite.

Londal. A proprietary trade name for certain aluminium alloys.

London Blue Extra. See SOLUBLE BLUE.

London Paste. A paste made by adding a third of the weight of water to a mixture of equal parts of sodium hydroxide and powdered lime. A caustic.

London Purple. A waste product used as a poisonous insecticide. It usually contains about 40 per cent. arsenious oxide, 25 per cent. lime, the rest being iron, alumina, and dyestuff.

London Smoke Glass. See GREY GLASS.

London White. See FLAKE WHITE.

Longifene. The hydrochloride of buclizine.

Long Life Balsam. Friar's balsam.

Long Oil Varnishes. A classification of varnishes. They contain about 1 part of the solid constituents to $1\frac{1}{2}$ parts of drying oil. Short oil varnishes contain 1 part of solid to $\frac{1}{2}$ part oil.

Lontar. Creosote. (573).

Lonzoid. A proprietary cellulose acetate.

Loparite. A rare earth mineral, given as $11Ce(TiO_3)_2 . 6(Di . La . Y)_2(TiO_3)_3 . 6CaTiO_3 . 9(Na . K(TiO_3)$. It is also given as $8[(Ce,La,Na)(Ti,Cb,Ta)O_3]$.

Loperamide. An anti-diarrhœal. 4-(4-p - Chlorophenyl - 4 - hydroxypiperi - dino) - NN - dimethyl - 2, 2 - diphenyl - butyramide. " R 18,553 " is currently under clinical trial as the hydrochloride.

Lopezite. A mineral. It is $4[K_2Cr_2O_7]$.

Lophine. Triphenyl-glyoxaline.

Lopion. A German preparation. It is the sodium salt of auro-allyl-thiourea-benzoic acid, containing 40 per cent. gold. Used for intramuscular application in cases of tuberculosis.

Lopox. A registered trade mark for epoxy resins.

Lopramine. See LOFEPRAMINE.

Lopresor. A proprietary preparation of METOPROLOL tartrate used in the treatment of angina pectoris. (17).

Lorandite. A mineral. It is a sulph-arsenite of thallium, $Tl_2S.As_2S_3$.

Loranskite. See EUXENITE.

Lorazepam. A tranquilliser. 7-Chloro-5 - (2 - chlorophenyl) - 1, 3 - dihydro - 3 - hydroxy - 2H - 1, 4 - benzodiazepin - 2 - one. ATIVAN.

Lorenit. p-Iodo-ana-oxyquinoline-o-sulphonic acid. An isomer of Loretine ($q.v.$).

Lorenzenite. A mineral. It is a titano-silicate of sodium and zirconium, found in Greenland.

Loretine (Sulphiolinic Acid). Iodo-hydroxy-quinoline-sulphonic acid, $C_9H_4NI(OH)SO_3H$. A germicide used as a substitute for iodoform for treating wounds. The sodium and bismuth salts are used medicinally, the bismuth one being employed in the treatment of diarrhœa.

Loretine Bismuth. Bismuth-iodo-oxyquinoline-sulphonate. Has been used in diarrhœa, and as a powder for ulcers.

Lorettoite. A mineral. It is lorettorite.

Lorettorite. A mineral, $13PbO.2PbCl_2$.

Lorexane. A proprietary preparation of gamma benzene hexachloride. BHC. γ-benzene hexachloride. An insecticide. Antiparasitic hair lotion. (2).

Lorfan. A proprietary preparation of levallorphan tartrate, used in narcotics overdosage and in obstetrical anæsthesia. (314).

Lorival. A proprietary electrical insulation made from a synthetic resin of the phenol-formaldehyde type.

Lorival R5, R25 and R200. Proprietary names for a range of depolymerised rubbers. (770).

Lorol. A trade mark for a mixture of

alcohols produced by the reduction of coco-nut oil. Sulphated fatty alcohols. (562).

Lorsban. A proprietary preparation of CHLORPYRIFOS. An insecticide.

Losan. A proprietary preparation. It consists of quinophan and sodium bicarbonate.

Losetic. A proprietary procaine hydrochloride and adrenalin preparation.

Loseyite. A mineral. It is $4[(ZnMn)_7 (CO_3)_2(OH)_{10}]$.

Losite. Synonym for Vishnevite.

Losophane. Tri-iodo-m-cresol, $C_6H_3(I_3)OHCH_3$. Used externally in skin diseases, as an antiseptic and astringent.

Lossenite. A mineral, $9Pb(FeOH)(AsO_4)_6SO_4.12H_2O$.

Lost (Mustard Gas). An exceedingly toxic and vesicant poison gas. It is $\beta\beta'$-dichloro-diethyl-sulphide, $(CH_2Cl.CH_2)_2S$. It was also called Yperite during the 1914–1918 war.

Losungsmittel 9C. A trade name for glycolmono-acetate. (3).

Losungsmittel E13. A trade name for a solvent mixture composed of ethyl acetate, methanol and a small amount of butanol. (3).

Losungsmittel E33. A trade name for 70 parts of methylacetate and 30 parts methanol. (3).

Losungsmittel T. A trade name for tetrahydrofurane. (3).

Lotens. See CARBORA.

Lotion, Copal. See ADHÆSOL.

Lotion, Goulard's. See GOULARD WATER.

Lotion, Grey. See GREY LOTION.

Lotion, Lead. See GOULARD WATER.

Lotion, Yellow Mercurial. See YELLOW WASH.

Lotol. A new rubber product prepared from rubber latex by a special process. It is used as an adhesive to replace rubber cements.

Louderbackite. A mineral. It is $Fe''(Fe''',Al)_2(SO_4)_4.12H_2O$.

Loughlinite. A mineral. It is $MgSi_2O_5.nH_2O$.

Louisville Cement. A cement made from cement rock obtained at Louisville, in Kentucky.

Lovenite. A mineral, $(Zr.Ca.Na_2)O.SiO_2$.

Lovol. A mixture of aliphatic alcohols formed by the high pressure hydrogenation of coco-nut oil.

Lovozerite. A mineral. It is $(Na,K)_2(Mn,Ca)ZrSi_6O_{16}.3H_2O$.

Low Brass. See BRASS.

Löweite. A mineral. It is $Na_2Mg(SO_4)_2.2\frac{1}{2}H_2O$.

Lowerite. Strippable coating compositions. (526).

Löwigite. A mineral. It is an aluminium-potassium sulphate.

Lowroff Phosphor Bronze. Alloys of 70–90 per cent. copper, 4–13 per cent. tin, 5–16 per cent. lead, and 0·5–1 per cent. phosphorus.

Low Setting-point Oils. A class of oils manufactured from non-paraffinoid crude oils of petroleum. They are valuable for lubrication at low temperatures.

Lox. An explosive consisting of liquid oxygen. Used in mining.

Loxa Bark. Pale cinchona bark.

Loxapine. A tranquilliser currently undergoing clinical trial as " SUM 3170 ". It is 2 - chloro-11-(4-methylpiperazin - 1 - yl)dibenz[b,f][1, 4]oxazepine.

Loxiol G-70, G-71, G-72 and G-73. Proprietary names for a range of multifunctional polyesters of high molecular weight, used as additives to P.V.C. compounds to reduce surface tackiness. (771).

Loxoclase. An orthoclase mineral.

Loxon. A proprietary preparation of HALOXON. A veterinary anthelmintic.

Loza. A material consisting mainly of anhydrous calcium sulphate. It is found above the caliche beds in Chile.

L.P. Aerosol. A proprietary preparation of fusafungine. (313).

L.P.D. An ultra-accelerator for rubber vulcanisation. It is the lead salt of piperidine - pentamethylene - dithio - carbamate.

L.Q.3. Blended alkaline liquid detergents for bottle washing. (546).

L.Q.4. Blended alkaline liquid detergents for bottle washing. (546).

L.Q.5 and L.Q.6. Blended alkaline liquid detergents for bottle washing. (546).

L.Q.12. General purpose liquid detergents. (546).

L.Q.14. Medium alkaline liquid detergents. (546).

L.Q.21. Acid liquid detergents. (546).

LSD. Lysergic acid diethylamide.

Luargol (Preparation 102). A compound of salvarsan with silver bromide and antimony, $(C_{12}H_{12}O_2N_2As_2)_2AgBr.SbO(H_2SO_4)_2$. Used in the treatment of syphilis. It is diamino - dihydroxy - arsenobenzene - silver bromide - antimonyl sulphate.

Luatol A. An aqueous glucose solution of sodium - potassium - tartro - bis - muthate. An antisyphilitic. Also see TARBISOL.

Luatol B (Tarbisol 1, Trépol). An oily suspension of sodium-potassium-tartrobismuthate. An antisyphilitic.

Luban Matti. A name applied to an

elemi resin obtained from *Boswellia Freriana*, of Africa.

Lubeckite. A mineral, $8CuO.Co_2O_3.2Mn_2O_3.8H_2O$.

Lublinite. A mineral. It is a variety of calcite.

Lubricating Oils, Dark. Residues of crude petroleum oils or concentrations obtained by the redistillation of lubricating oil base, or a mixture of residue oils with oils of lower viscosity. Used in cheap lubricants.

Lubrico. A proprietary trade name for an alloy of 75 per cent. copper, 20 per cent. lead, and 5 per cent. tin.

Lubrol MB. A fatty alcohol ethoxylate with a neutral reaction. Used for the production of self-emulsifiable blends of oils. (2).

Lucamid. See ETENZAMIDE.

Lucanthone. 1-(2-Diethylaminoethyl)-amino-4-methylthiazanthone. Miracil D and Nilodin are the hydrochloride.

Lucca Oil. Olive oil.

Lucerno. An alloy containing 67·9 per cent. nickel, 27·5 per cent. copper, 2·4 per cent. iron, and 2·2 per cent. manganese.

Lucianite (Auxite). A mineral. It is a colloidal magnesium silicate.

Lucidol (Benzoyl superoxide, Sallidol). A registered trade mark for benzoyl peroxide. Recommended as an antiseptic for use in burns and skin diseases. Also for use as a bleaching agent, improver for flour, and as a polymerisation catalyst. (564). Cf. LUPERCO.

Lucidril. A proprietary trade name for the hydrochloride of Meclofenoxate. (358).

Lucinite. A native aluminium phosphate, $Al_2O_3.P_2O_5.4H_2O$.

Lucipal. Benzolyl peroxide compositions. (564).

Lucite (Perspex). Proprietary trade names for a methyl methracrylate and acrylate synthetic resins. The material is more transparent than glass and has a very high refractive index. (10).

Lucitone. See VERNONITE.

Luckite. A mineral. It is a ferrous sulphate containing manganese.

Lucofen S A. A proprietary preparation of chlorphentermine hydrochloride. Antiobesity agent. (262).

Lucovyl. A proprietary polyvinyl chloride. (772).

Lucullite. A mineral. It is a black marble coloured by carbonaceous matter.

Ludenscheidt's Button Brass. See BRASS.

Ludigol. Sodium-*m*-nitro-benzene-sulphonate. Used as an assistant in the treatment of fabrics.

Ludiomil. A proprietary preparation of MAPROTILINE hydrochloride. An antidepressant. (519).

Ludlamite. A mineral, $Fe_3(PO_4)_2$.

Ludlum Alloy. A heat-resisting alloy of iron with from 13–17 per cent. chromium, 1 per cent. silicon, 1 per cent. molybdenum, and 0·4 per cent. carbon.

Ludlum No. 602 Steel. A proprietary trade name for a steel containing 1·7 per cent. silicon, 0·7 per cent. manganese, 0·4 per cent. molybdenum, 0·12 per cent. vanadium, and 0·48 per cent. carbon.

Ludwigite. A mineral, $2MgO.B_2O_3.Fe_3O_4$.

Ludyl. Benzene-*m*-3′ : 3′-disulphamino-*bis*-3-amino-4 : 4′-dihydroxy-arseno-benzene. Used in the treatment of syphilis.

Luesan. A glidine preparation containing mercury. Used in the treatment of syphilis.

Lueside. A preparation containing glucose and metallic mercury. An antisyphilitic.

Luftseide (Celta, Soie Nouvelle, Tubulated Silk). Artificial silks of the rayon type made with hollow central spaces. They are formed by adding gas-evolving materials to the viscous solution.

Luglas. A proprietary safety glass.

Lugo. A patented rubber substitute made from oxidised oil, rubber, and potassium permanganate.

Lugol's Solution. Iodine-potassium iodide solution. Iodine (5 grams) are triturated with 5 grams of potassium iodide, and 100 c.c. of water, and diluted to 1 litre.

Lugo's Powder. Powdered cinchona bark.

Luizym. A proprietary preparation of cellulase, hemicellulase, amylase, and proteases. (391).

Lulea Tar. A variety of Stockholm tar.

Lumapone. A proprietary trade name for a glazing material consisting of a plastic reinforced with heavy duty wire.

Lumarith EC. A registered trade mark for an ethyl cellulose thermoplastic resistant to oils.

Lumarith ER. A registered trade mark for ethyl rubber.

Lumbang Oil. A drying oil obtained by pressing the seeds of *Aleurites moluccana*. Used in the manufacture of paints and soap.

Lumen Alloy 11-C. A proprietary trade name for an alloy containing copper with 10 per cent. aluminium and 1 per cent. iron.

Lumen Bronze. An alloy of 86 per cent. zinc, 10 per cent. copper, 4 per cent. aluminium, and 0·1 per cent. magnesium. A softer variety contains 88 per cent. zinc, 8 per cent. aluminium, and 4 per cent. copper.

Lumilux. A proprietary trade name for a methacrylate plastics. (71).

Lumina. A proprietary safety glass.

Luminal. A proprietary preparation of phenobarbitone. A hypnotic. (112). See also Gardenal, Phenylbarbital Phenyl ethyl malonyl urea.

Luminal Sodium. The sodium salt of Luminal (*q.v.*).

Luminous Paint. This is usually a mixture of zinc sulphide or barium sulphide in lacquer.

Lumite. An aluminium alloy containing 5·6 per cent. nickel and 1 per cent. iron.

Lummer's Solution. Hydrogen platinochloride, H_2PtCl_6 (3 grams in 100 c.c. water), with 0·02 gram lead acetate. Used for coating platinum electrodes with finely divided platinum.

Lumnite Cement. An alumina cement consisting of about 40 per cent. alumina, 40 per cent. lime, 15 per cent. iron oxide, 5 per cent. silica, and magnesia. The material is made from bauxite, and is stated to be stronger than Portland cement.

Lumolit. A proprietary wood fibre product.

Lump-Lac. See LAC.

Lunar Caustic. Silver nitrate, $AgNO_3$, fused and cast into sticks or rods. An energetic caustic for wounds and sores.

Lund's Oil (Catheter Oil). Lubricant Oil (*Oleum lubricans B.P.C.*).

Lunebergite (Luneburgite). A mineral. It is a magnesium borate and phosphate, $Mg(BO_2)_2.2MgHPO_4.7H_2O$.

Lunnite. A mineral. It is a basic copper sulphate.

Lunosol. A proprietary preparation. It is a colloidal silver chloride.

Luo-calcite. A name given to the calcium bicarbonate occurring in solution in natural waters.

Luo-Chalybite. A name given to the ferrous bicarbonate occurring in solution in natural waters.

Luo-Diallogite. A name given to the manganese bicarbonate occurring in solution in natural waters.

Luo-Magnesite. A name given to the magnesium bicarbonate occurring in solution in natural waters.

Lupeose (Mannotetrose). Stachyose, $C_{24}H_{12}O_{21}$, a tetrasaccharose. It occurs in the tubes of *Stachy tubifera*.

Luperco. A registered trade mark for organic peroxide compounds for polymerisation catalysis, drying accelerator, oxidizing agent and bleaching applica-

tions. A consists of benzoyl peroxide with an inorganic filler and AC consists of benzoyl peroxide with an organic filler.

Luperco AFR. A trade mark for 50 per cent. benzoyl peroxide in plasticiser. (52).

Lupersol DDM. A trade mark for 60 per cent. methyl ethyl ketone peroxide in dimethyl phthalate. (52).

Lupersol 227. A registered trade mark for a 50 per cent. solution of diisobutyryl peroxide in mineral spirits. An organic peroxide for polymerising. (52).

Lupersol DEL. A trade mark for 60 per cent. methyl ethyl ketone peroxide in dimethyl phthalate. (52).

Lupersol DSW. A proprietary trade name for methyl ethyl ketone peroxide. A liquid fire resistant peroxide containing 11·5 per cent. active oxygen. (201).

Lupetazin. Dimethyl-piperazine, $(C_2H_3(CH_3)NH)_2$. A uric acid solvent.

Luphen. A trade name for phenolic resins. (German). (3).

Luphen AM. A trade name for 7 parts phenol-formaldehyde and 3 parts adipic acid trimethylolpropane. (3).

Luphen AT. A trade name for 5 parts phenol-formaldehyde and 5 parts of adipic acid-trimethylolpropane. (3).

Luphen AW. A trade name for 4 parts phenol-formaldehyde and 6 parts adipic acid-trimethylolpropane. (3).

Luphen H. A trade name for 45 per cent. phenol formaldehyde, 5 per cent. acetylene dimethylolurea, 0·1 per cent. ethyl cellulose plus 50 per cent. ethanolethylglycol mixture. (3).

Luphen L. A trade name for pure phenol formaldehyde. (3).

Luphen SH. A trade name for a xylenol formaldehyde condensate plasticised with adipic acid-hexanetriol ester. (3).

Luphen SM. A trade name for a 70 per cent. xylenol formaldehyde and 30 per cent. trimethylol-propane-adipic acid ester dissolved to 60 per cent. solids in ethanol-ethylene glycol mixture. (3).

Luphen SW. A trade name for 40 per cent. xylenol formaldehyde and 60 per cent. trimethylolpropane adipic acid ester dissolved to 70 per cent. solids in ethanol-ethylene glycol mixture. (3).

Luphenharter L2. A trade name for naphthalene sulphonic acid. (3).

Lupinit. A proprietary casein product.

Lupolen. A registered trade name for polyethylenes. (3).

Lupolen 200. A proprietary low-density polyethylene used in injection moulding. (49).

Lupolen 804H and 1814H. Proprietary low-density polyethylenes used in the manufacture of liners and barriers. (49).

Lupolen 1800 H/M/S. A proprietary low-density polyethylene used in injection moulding. (49).

Lupolen 1810E. A proprietary low-density polyethylene used in the manufacture of milk containers. (49).

Lupolen 1810H. A proprietary low-density polyethylene used in blow moulding. (49).

Lupolen 1812D and 1812EH. Proprietary polyethylenes used in the manufacture of wires and cables. (49).

Lupolen 1814E. A proprietary low-density polyethylene used in the manufacture of milk containers. (49).

Lupolen 1852E/H. A proprietary low-density polyethylene used in the manufacture of pipes. (49).

Lupolen 2040EX and 2410DX. Proprietary low-density polyethylenes used in the manufacture of bags. (49).

Lupolen 2410S. A proprietary low-density polyethylene used in injection moulding. (49).

Lupolen 2424H and 2425K. Proprietary low-density polyethylenes used in the manufacture of transparent packaging materials. (49).

Lupolen 2430H. A proprietary low-density polyethylene used in blow moulding. (49).

Lupolen 2452 E. A proprietary low-density polyethylene used in the extrusion of pipes. (49).

Lupolen 3010 S. A proprietary low-density polyethylene used in injection moulding. (49).

Lupolen 3020 D. A proprietary low-density polyethylene used in blow moulding. (49).

Lupolen 3020 KX and 3025 KX. Proprietary low-density polyethylenes used as over-wrappings. (49).

Lupolen 4261 AX. A proprietary high-density polyethylene of high molecular weight. (49).

Lupolen 5011 K. A proprietary high-density polyethylene used in injection moulding. (49).

Lupolen 5052 C. A proprietary high-density polyethylene used in the extrusion of pipes. (49).

Lupolen 6011 K. A proprietary high-density polyethylene used in injection moulding. (49).

Lupolen V-2524EX and V-3510K. Proprietary low-density polyethylene copolymers. (49).

Luran. A trade name for polystyrene and styrene-acrylonitrile copolymers. (49).

Luran 368R. A proprietary styrene-acrylonitrile copolymer used in injection moulding. (49). LURAN 534.

Luran 378P. A proprietary styrene-acrylonitrile copolymer used in injec-tion moulding and extrusion. LURAN KR2551. (49).

Luran 757R. A proprietary acrylonitrile-styrene-acrylonitrile copolymer used in injection moulding. (49).

Luran 776S. A proprietary acrylonitrile-styrene-acrylonitrile used in extrusion and injection moulding. (49).

Luran KR 2517. A proprietary acrylonitrile-styrene copolymer containing 35% glass fibre. (49).

Luran S. Acrylonitrile-styrene-acrylic terpolymer. (49).

Lurgi Metal. An alloy of lead with 2 per cent. barium.

Luron. Nylon monofilament for fishing line. (512).

Lussatite. A mineral. A form of silica.

Lustilac. A proprietary moulding compound.

Lustra-Cellulose. Artificial silk.

Lustran. A trade mark for an ABS (*q.v.*) moulding and extrusion material for piping for chemical resistance. (57).

Lustran A. A trade mark. A styrene acrylonitrile copolymer suitable for the moulding of battery boxes. (57).

Lustran ABS. A registered trade name for ABS (*q.v.*) moulding and extrusion materials.

Lustran A.B.S. 244. A proprietary acrylonitrile-butadiene-styrene copolymer used in the injection moulding of automobile components. (559).

Lustran I, 200 Series. A proprietary range of acrylonitrile-butadiene-styrenes of high tensile strength and medium impact, used in extrusion and injection moulding. (559).

Lustran I, 400 Series. A proprietary range of acrylonitrile-butadiene-styrenes of medium-to-high impact strength, used in the moulding of automobile parts and household goods. (559).

Lustran I, 600 Series. A proprietary range of acrylonitrile-butadiene-styrenes possessing high impact strength, used in injection moulding. (559).

Lustran I, 700 Series. A proprietary range of acrylonitrile-butadiene-styrenes of high impact strength used for the extrusion of sheet to be formed into luggage, pipe fittings, automobile parts, etc. (559).

Lustran HR 850. A proprietary heat-resistant acrylonitrile - butadiene - styrene copolymer used in moulding the front grilles of automobiles. (559).

Lustran PG 299. A proprietary acrylonitrile-butadiene-styrene copolymer used for plating and in the production of automobile trim. (559).

Lustran SAN 21. A proprietary acrylonitrile copolymer used in the injection moulding of automobile components. (559).

Lustran SAN UV. A proprietary styrene-acrylonitrile copolymer used in the injection moulding of automobile components. (559).

Lustrasol. A proprietary trade name for acrylic copolymers for surface coatings. (181).

Lustrex. Polystyrene moulding powder. (559).

Lustrex Latex. Polystyrene dispersion. (559).

Lustron. See CELANESE. Also a proprietary trade name for polystyrene moulding compounds.

Lustrose. A proprietary compound used in the textile industry for sizing.

Lutate. A proprietary preparation of HYDROXYPROGESTERONE caproate. (773).

Lutécienne (French Red). A mixture of Fast red A with Orange II (β-naphthol orange, the sodium salt of p-sulpho-benzene-azo-β-naphthol). Also see EOSIN BN.

Lutecin. See NICKEL SILVERS.

Lutecite. A mineral. It is a variety of chalcedony (crypto-crystalline quartz), SiO_2. Also see QUARTZINE.

Lutéienne. An old name for Eosin BN (*q.v.*).

Lutein. A preparation of the corpus luteum of the hog.

Lutensit An 10. A surfactant. (49).

Luteol. Oxychlor-diphenyl-quinoxalin. An indicator.

Luteolin. See WELD. Also the sodium salt of *m* - xylidine - sulphonic - azo - diphenylamine. An orange-yellow dyestuff.

Lutidine. Dimethyl-pyridine, C_7H_9N.

Lutofan. A trade mark for polyvinyl chloride. (49).

Lutofan D. A registered trade mark for a polyvinyl chloride dispersion. (49).

Lutol. See CUTOL.

Lutonal. See also IGEVIN. A proprietary trade name for a series of vinyl-ether resins. (3).

Lutonal. A registered trade name for polyvinyl ether. (49).

Lutonal LC. A registered trade name for polyvinyl isobutyl ether. (49).

Lutonal A. A proprietary trade name for polyvinyethyl ether. (3).

Lutonal D. A registered trade mark for a polyvinyl ether dispersion. (49).

Lutonal J. A proprietary trade name for polyvinylisobutyl ether. (3).

Lutorcin. See CRESORCIN.

Luvican M170. A trade name for polyvinyl carbazole. (49).

Luxene. A proprietary synthetic resin of the phenol-formaldehyde class.

Luxene 44. A proprietary denture compound made from a vinyl copolymer.

Lux Mixture. Consists of the residue obtained in the manufacture of aluminium hydroxide from bauxite. I contains 51 per cent. Fe_2O_3, with som sawdust and lime. Used in the purifi cation of coal gas.

Luxol Colours. A new series of spiri colours of American manufactur They are probably aryl-guanidine salts Some are soluble in the higher alcohol and are useful for producing transparen coloured pyroxylin lacquers fast t light.

Luxor 950 NL. A proprietary trad name for an oil-free aluminium powde used for manufacturing expanded con crete in alkaline environments. (189).

Luxullianite. A Cornish rock containin, tourmaline associated with quartz anc felspar.

Luzerne. A proprietary trade name for a hard rubber.

Luzidol. Benzoyl peroxide, $C_{14}H_{10}O_4$.

Luzonite. See ENARGITE.

Lyargol. Silver proteinate.

Lycanol. A proprietary trade name fo Glymidine.

Lycetol (Tetradine). Dimethyl-piper azine-tartrate, $C_6H_{14}N_2.C_7H_6O_6$. A solvent for uric acid.

Lycin. See ACIDOL.

Lycine. The base of *Lycium barbarum* It is identical with betaine.

Lycopodium. A pale yellow powde consisting of the spores of *Lycopodium clavatum.*

Lycopon (Vatrolite). A proprietary trade name for sodium hyposulphite.

Lyddite (Melinite, Pertite, Ecrasite, Shimose). Picric acid, an explosive. Lyddite (English), contains 87 per cent. picric acid, 10 per cent. nitro-benzene, and 3 per cent. pet. jelly. Other mixtures contain ammonium picrate, and ammonium salts of trinitro-cresol, sometimes with the addition of potassium nitrate.

Lydian Stone (Lydite, Touchstone). A siliceous slate containing about 84 per cent. silica, 5 per cent. alumina, and 1 per cent. ferric oxide. Used for testing gold by rubbing it upon the stone. and testing the streak of metal produced with acid.

Lydin. A name given to Mauve or Aniline purple. See MAUVEINE.

Lydite. See LYDIAN STONE.

Lye Glycerin. Glycerin obtained from soap liquor. Crude soap-lye glycerin contains 81 per cent. glycerin, has a specific gravity of 1·278, a total residue at 160° C. of 12 per cent., inorganic salts 9·5 per cent., and arsenic 1 part in 1,000,000.

Lygosine. Di-o-dioxy-dibenzal-acetone, $(HO.C_6H_4.CH : CH)_2CO$. It is prepared in the form of the sodium salt. Both this salt and the quinine salt are used in the treatment of wounds.

Lymecycline. N^2-{[(+)-5-Amino-5-carboxypentylamino]methyl}tetracycline. Armyl; Mucomycin; Tetralysal.

Lyndiol. A proprietary preparation of lynestrenol and mestranol. Oral contraceptive. (316).

Lyndiol 2.5. A proprietary preparation of lynestranol and mestranol. Oral contraceptive. (316).

Lyndochite. A new mineral of the euxenite-polycrase group from Ontario (U.Th.Ce.La.Di.Yt.Er.Zr.Ta and Cb= 73 per cent.).

Lynite. A trade mark for an alloy of 88 per cent. aluminium, 10 per cent. copper, 1·5 per cent. iron, and 0·25 per cent. magnesium. It has a specific gravity of 2·95.

Lynoestrenol. 17α-Ethynyloestr-4-en-17β-ol. An œstrogen present in LYNDIOL, MINILYN, ORGATULON, OVANON, and SISTOMETRIL.

Lynoral. A proprietary preparation of ethinylœstradiol, used in menopausal disorders. (316).

Lyofix 363. A proprietary trade name for a modified urea formaldehyde resin precondensate used in the crease-resistance finishing of cellulosic textiles. (18).

Lyofix F. A proprietary trade name for a melamine-formaldehyde resin precondensate used in finishing paper makers' felts. (18).

Lyonore. A proprietary trade name for a steel containing 0·2 per cent. copper with some chromium and nickel.

Lyons Blue. See ANILINE BLUE, SPIRIT SOLUBLE, and SOLUBLE BLUE.

Lyons Sugar. Sucramine, the ammonium salt of saccharine.

Lyopan. A proprietary preparation of phenazone theophyllin and urethane.

Lyophrin. A proprietary preparation of adrenaline hydrogen tartrate with sodium sulphate as a preservative. Treatment of glaucoma. (374).

Lyovac Saluric. A proprietary preparation of chlorthiazide sodium. Emergency intravenous treatment of heart failure. (310).

Lypressin. 8-Lysinevasopressin. Syntopressin.

Lyptol. Oil of peppermint and oil of eucalyptus dissolved in alcohol with chinasol. An antiseptic and disinfectant.

Lysan. A saponaceous alcoholic solution of a disinfectant which is the result of an action between formaldehyde and terpene.

Lysargine. A colloidal silver, containing 60 per cent. of silver. An antiseptic.

Lysenyl. See LYSURIDE.

Lysergide. NN-Diethyl-lysergamide. Lysergic acid diethylamide. LSD. Delysid.

Lysidine (Piperazenyl). Ethylene-ethenyl-diamine, $CH_2.CH_2.NH.N.C CH_2$. A solvent for uric acid.

Lysine. A preparation containing inorganic salts and essences in an acid solution of formalin. An antiseptic and bactericide.

Lysinex. A proprietary preparation of L-lysine hydrochloride and stanolone. An anabolic agent. (347).

Lysitol. See LYSOL.

Lysivane. A proprietary preparation of ethopropazine hydrochloride, used for Parkinsonism. (336).

Lysochlor. Chloro-m-cresol, $C_6H_3Cl (CH_3)OH$. A disinfectant. It is a similar preparation to Eusapyl (q.v.).

Lysoform. A scented solution of formaldehyde in alcoholic potash soap solution. A disinfectant much like Lysol.

Lysol (Lysitol). Consists of crude cresols mixed with a soft-soap solution. A disinfectant used in surgery, and for cleaning floors and walls. Also see NEOLYSOL, NIZOLYSOL, and SAPOCARBOL.

Lysophan. Cresol-tri-iodide, $C_6HI_3(CH_3)OH$. An antiseptic.

Lystonol. A zinc chloride preparation. It is used for preventing fermentation in sugar juices.

Lysulfol. Lysol with 10 per cent. of sulphur. Recommended for use in scabies and acne.

Lysuride. A preparation used in the prophylaxis of migraine. It is 9-(3, 3-diethylureido) - 4, 6, 6a, 7, 8, 9 - hexahydro - 7 - methyl - indolo[4, 3 - f, g] - quinoline.

Lytensium. A proprietary trade name for Pentamethonium Bromide.

Lythol Oil. A commercial phenolated oil used as a disinfectant.

Lytrol. An alcoholic solution of potassium-β-naphthol. A disinfectant.

Lytron 820. Thickening agents. (559).

Lytron 886 and Lytron 887. Flocculating agents. (559).

Lytta. Cantharides.

M

M33, MN3. Polyimide resins. (88).

MA 20. Soap powder for launderettes. (506).

Maali Resin. A yellowish-white resin resembling elemi of Samoa.

M & B 639. A proprietary preparation of sulphapyridine. An antibiotic. (336).

M & B Antiseptic Cream. A proprietary preparation of propamidien isethionate in a cream base. (336).

Maakite. Synonym for Hydrohalite.

Maalox. A proprietary preparation containing magnesium aluminium hydroxide gel. An antacid. (266).

Maboula Nuts. See OWALA BEANS.

Macadamite. An alloy of 72 per cent. aluminium, 24 per cent. zinc, and 4 per cent. copper.

Macaja Butter. Mocaya oil.

Macarite. A Belgian high explosive. It is a mixture of trinitro-toluene and lead nitrate.

Macaroni. A flour preparation made from Italian wheat.

Macassar Copal. See COPAL RESIN.

Macassar Nutmeg Butter (Papua Nutmeg Butter). The fat from *Myristica argentæ*.

MacDougall's Disinfecting Powder. Made by adding a certain quantity of crude carbolic acid to crude sulphide of calcium.

Mace. Arillus of the nutmeg.

Mace Butter. Nutmeg butter, obtained from the seeds of *Myristica officinalis*, of the East.

Mace Butter, American. See OTOBA BUTTER.

Mace Oil. See OIL OF MACE.

MacFarland's Alloy. Heat-resisting alloys. (*a*) Contains 59 per cent. nickel, 30 per cent. chromium, and 11 per cent. copper. (*b*) Contains 46 per cent. nickel, 43 per cent. chromium, and 11 per cent. copper.

Macfarlanite. A mineral. It is a complex silver ore.

Macgill Metal. An alloy containing 88 per cent. copper, 7 per cent. nickel, 4·5 per cent. iron, and traces of tin and lead. It resists corrosion.

Macgovernite. A mineral. It is $Mn_7(AsO_3,AsO_4)SiO_4(OH)_7$.

Machacon Juice. An alkaline decoction of the juice of the root of a plant having the same name. It is used to coagulate rubber latex.

Machine Bronze. Variable alloys containing 50–90 per cent. copper, 25 per cent. nickel, 0–30 per cent. tin, and 0·8 per cent. lead. Some alloys contain no nickel, and also contain zinc. Another alloy of this type consists of 83 per cent. copper, 16 per cent. zinc, and 1 per cent. tin.

Machine Oil. A lubricating oil. It is that fraction of specific gravity 0·895–0·910 obtained from the residue from Russian petroleum distillation. It flashes at 185–215° C.

Machinery Brass. See BRASS.

Mach's Metal. An alloy of aluminium with 2–10 per cent. magnesium.

Macht's Metal. An alloy of 57 per cent.

copper and 43 per cent. zinc, used for castings.

M-Acid. 1-Amino-5-naphthol-7-sulphonic acid.

Mackayite. A mineral. It is $Fe_2(TeO_3)_3 \cdot nH_2O$.

Mackechnie's Bronze. Usually an alloy of 57 per cent. copper, 41 per cent. tin, 1 per cent. zinc, 1 per cent. iron, and 0·5 per cent. lead.

Mackenzie's Amalgam. Bismuth (2 parts) and lead (4 parts) are melted separately in crucibles, and each poured into 1 part mercury. These amalgams are then rubbed together.

Mackenzie's Metal. An alloy of 70 per cent. lead, 17 per cent. antimony, and 13 per cent. tin; also 68 per cent. lead, 16 per cent. antimony, and 16 per cent. bismuth. Electrotype metals.

Mackintoshite. A mineral. It is a hydrous uranium-thorium-silicate found in Texas.

Mack's Cement. Prepared by adding calcined sodium sulphate or potassium sulphate to dehydrated gypsum.

Maclureite. A mineral. It is a variety of amphibole.

Maclurin. A tanning material contained in concentrated extract of yellow wood (old fustic).

Macquer's Salt. Potassium arsenate, KH_2AsO_4. Used in calico printing.

Macrisalb. An injection of human albumen treated with iodine (^{131}I) and used in the examination of pulmonary perfusion.

Macrodantin. A proprietary preparation of NITROFURANTOIN. An antibiotic. (818).

Macrodex. A proprietary preparation of dextran in saline or dextrose solution, used to replace plasma volume in shock or hæmorrhage. (337).

Macrogol 400. Polyethylene glycol 400. Polyoxyethylene glycol 400.

Macrogol 4000. Polyethylene glycol 4000. Polyoxyethylene glycol 4000.

Macrogol Stearate. Polyoxyl 40 stearate. Polyoxyl 8 stearate.

Maculanin. Potassium amylate.

Madagascar Copal. See COPAL RESIN.

Madanite. A product used as a binding material made from 2 parts petroleum jelly and 1 part rubber.

Madar (Mudar). Calotropis bark. Has been used as a remedy for eczema.

Madar Fibre. A bast fibre, known in India by this name, obtained from *Calotropis procera* and *C. gigantea*.

Madder. The powdered root of the plant, *Rubia tinctorum*. The chief constituent is ruberythric acid. This acid is a glucoside, and is split up into alizarin and a sugar by the action of acids.

Madder Carmine. That preparation of madder which contains the greatest amount of colouring matter and the smallest amount of base.

Madder, Indian. See CHAY ROOT.

Madder Indian Red (Tuscan Red). Red oxides of iron improved in colour by the addition of an alizarin lake. Pigments.

Madder Lakes (Rose Madder, Pink Madder). Lakes prepared by precipitating the colouring matter of the root of the madder plant with alumina.

Madder, Pink. See MADDER LAKES.

Madder, Rose. See MADDER LAKES.

Maddrell Salts. Alkaline metaphosphates which are insoluble in water, and saline solvents. They are made by heating mono-sodium phosphate at 245–250° C.

Madecassol. A proprietary preparation extract of *Centella asiatica* in an ointment, used for skin protection. (323).

Madol Oil. An oil obtained from the seeds of *Garcinia echinocarpa*. It is a burning oil, but is also used medicinally as a vermifuge.

Madopar. A proprietary preparation of LEVODOPA and BENSERAZIDE used in the treatment of Parkinson's disease. (314).

Madras Blue. The name applied to a mixture of gallocyanine and logwood extract.

Madras Kino. See KINO, BENGAL.

Madribon. A proprietary preparation containing sulphadimethoxine. (314).

Madzonn (Matzoon). A fermented milk prepared by the Armenians.

Mafenide. An anti-bacterial agent. α-Aminotoluene - *p* - sulphonamide. MARFANIL is the hydrochloride, SULFOMYL the propionate and SULFAMYLON the acetate.

Mafura Fat. A fat obtained from the seeds of *Mafureira oleifera*. It contains from 92–95 per cent. fatty acids, has an acid value of 31–32, and a saponification value of 202–207. Used in the manufacture of soaps.

Magadi Soda. An East African soda. It contains sodium carbonate and sodium bicarbonate.

Magdala Red (Naphthalene Red, Naphthalene Scarlet, Naphthalene Rose, Naphthylamine Red, Naphthaline Red, Naphthylamine Pink, Sudan Red, Fast Pink for Silk). A dyestuff. It is a mixture of amino-naphthyl-naphthazonium chloride and diamino-naphthyl-naphthazonium chloride, $C_{30}H_{20}N_3Cl$ and $C_{30}N_{21}N_4Cl$. Dyes silk from a soap bath pink.

Magecol. A proprietary lampblack suitable for rubber goods.

Magendi's Iodised Ether. A solution of 1 gram of iodine in 15 grams of ether.

Magenta (Roseine, Fuchsine, Aniline Red, Rubine, Magenta Red, Magenta Roseine, Rubesine, Rubianite, Fuchsianite, Magenta Crystals, Amethyst, Ponceau, Maroon, Grenaldine, Geranium, Cerise, Garnet, Russian Red, Azaleine, Chestnut, Solferino, Erythrobenzine, Rubianin, Harmaline, Grenadine, Russian Red G, 967, Cerise B, BB, G, 2YS). A dyestuff. It consists of a mixture of the hydrochlorides or acetates of pararosaniline (triamino-triphenyl-carbinol), and rosaniline (triamino-diphenyltolyl-carbinol). Hydrochlorides $= C_{19}H_{26}N_3ClO_4$, and $C_{20}H_{28}N_3ClO_4$. Acetates $= C_{21}H_{21}N_3O_2$, and $C_{22}H_{23}N_3O_2$. Dyes silk, wool, leather, and cotton mordanted with tartar emetic and tannin, red.

Magenta Base. See ROSANILINE.

Magenta BB Powder, Crystals FF, CV Conc., L, Superfine Powder. Dyestuffs. They are British equivalents of New magenta.

Magenta Bronze (Violet Bronze). Tungsten potassium bronze, $K_2W_4O_{12}$, prepared by the addition of tungsten trioxide to fused potassium tungstate, then heating in a current of hydrogen, and digesting in water and acids. The residue is magenta bronze.

Magenta Crystals. See MAGENTA.

Magenta Red. See MAGENTA.

Magenta Roseine. See MAGENTA.

Magenta S. See ACID MAGENTA.

Magisal. Magnesium-acetyl salicylate, $(CH_3CO.OC_6H_4COO)_2Mg$. Used for similar purposes as aspirin.

Magister of Bismuth. Basic bismuth nitrate, $BiO.NO_3.H_2O$. Used in medicine, and as a constituent of enamels and cosmetics.

Magister of Sulphur. Sulphur precipitated in an amorphous condition from solutions of hyposulphites or polysulphides, by acids.

Magistery of Lead. White lead (*q.v.*).

Magistral. An impure copper sulphate containing ferric oxide, sodium sulphate, and sodium chloride.

Maglite Y. A trade mark for magnesium oxide. (896).

Magmilor Oral Tablets. A proprietary preparation containing nifuratel. (247).

Magmilor Vaginal Tablets. A proprietary preparation containing nifuratel. (247).

Magnafloc LT. A polyacrylamide flocculant used for the clarification of potable water and sugar juice. (207).

Magnafloc R351, R352, R455. Polyacrylamide flocculants available in both mining and industrial grades for mineral processing and effluent clarification (207).

Magnalite. (1) An alloy of 94·2 per cent. aluminium, 2·5 per cent. copper, 1·5 per cent. nickel, 1·3 per cent. magnesium, and 0·5 per cent. zinc. It has a specific gravity of 2·8.

Magnalite. (2) A mineral. It is $Mg_4Al_6Si_{12}O_{37}.13H_2O$.

Magnalium. Alloys of magnesium and aluminium. One contains from 1–2 per cent. of magnesium. It is lighter than aluminium, and is as hard as brass. Another alloy contains 10 per cent. magnesium, and is used for parts of machinery, for cooking utensils, and for optical mirrors. One alloy containing 90 per cent. aluminium and 10 per cent. magnesium has a specific gravity of 2·8.

Magnapen. A proprietary preparation of AMPICILLIN and FLUCLOXACILLIN. An antibiotic. (364).

Magnesia. Magnesium oxide, MgO.

Magnesia Alba. A basic magnesium carbonate of variable composition.

Magnesia Alum. Synonym for Pickeringite.

Magnesia Bleaching Liquid. Magnesium oxychloride, $Mg(OCl)_2$. Used for bleaching.

Magnesia Cement. See SOREL CEMENT.

Magnesia-Citrate Mixture. Citric acid (20 grams), is dissolved in 20 per cent. ammonium hydrate, and mixed with 1 litre of magnesia mixture (q.v.). Used in the determination of phosphorus in manures.

Magnesia-Cordierite. A mineral. It is $4[Mg_2,Al_4,Si_5O_{18}]$.

Magnesia-Goslarite. A mineral. It is $4[(Zn,Mg)SO_4.7H_2O]$.

Magnesia Covering. A heat insulator. It usually contains about 85 per cent. hydrated magnesium carbonate and 15 per cent. asbestos fibre.

Magnesia, Heavy. Magnesium oxide, MgO, obtained by heating heavy magnesium carbonate. See MAGNESIUM CARBONATE HEAVY.

Magnesia, Light. Magnesium oxide, MgO, obtained by heating light magnesium carbonate.

Magnesia Mixture. Magnesium sulphate (1 part), and ammonium chloride (1 part) are dissolved in 8 parts of water with the addition of 4 parts of ammonia (specific gravity 0·96), and filtered. Used for the determination of phosphorus in manures.

Magnesia White. Both magnesium oxide and magnesium carbonate are known by this name.

Magnesiochromite. A mineral. It is $8[(Mg,Fe)(Al,Cr)_2O_4]$.

Magnesio-Copiapite. A mineral. It is $MgFe_4^{\cdots\cdots}(SO_4)_6(OH)_2.nH_2O$.

Magnesio-Cummingtonite. A mineral. It is $2[(Mg,Fe)_7Si_8O_{22}(OH)_2]$.

Magnesioferrite. A mineral, M $(FeO_2)_2$.

Magnesio-Magnetite. A mineral. I is $[(Fe,Mg)Fe_2O_4]$.

Magnesiosussexite. A mineral. It $(Mg,Mn)_2B_2O_5.H_2O$.

Magnesio-Wüstite. A mineral. It $4[(Mg,Fe)O]$.

Magnesite. A mineral. It is magnesiu carbonate, $MgCO_3$.

Magnesite Spar. A mineral. It is rhombohedral form of magnesite.

Magnesium Base Alloys. The mag nesium in these alloys usually varie from 90–95 per cent.

Magnesium Carbonate, Heavy. Mag nesium carbonate obtained by mixin concentrated solutions of magnesiur sulphate and sodium carbonate, evap orating to dryness, then removing th sodium sulphate.

Magnesium Carbonate, Light. Mag nesium carbonate prepared by addin, a solution of sodium carbonate to solution of magnesium sulphate, an boiling. The precipitate is the car bonate.

Magnesium Chalcanthite. Synonyn for Pentahydrite.

Magnesium-Chlorophœnicite. A min eral. It is $(Mg,Mn)_5(AsO_4)(OH)_7$.

Magnesium Dust. Finely powdered magnesium. Used in fireworks.

Magnesium-Halotrichite. Synonym for Pickeringite.

Magnesium Iron Mica. See BIOTITE.

Magnesium Limestone. See DOLO MITE.

Magnesium Mica. A mineral. It is Phlogopite (q.v.).

Magnesium-Monel. A proprietary trade name for an alloy of 50 per cent. magnesium and 50 per cent. monel metal.

Magnesium-Pectolite. Synonym for Vivianite.

Magnesium-Perhydrol. See BIOGEN.

Magnesium Phytinate. The mag nesium salt of anhydro-methylene-phos phoric acid.

Magnesium Sulphate, Effervescent. See EFFERVESCENT EPSOM SALTS.

Magnesium Superoxyl. Magnesium peroxide, MgO_2.

Magnesium-Zinc-Spinel. Synonym for Gahnospinel.

Magnetic Black. A proprietary trade name for a finely ground magnetic iron oxide for use as a pigment.

Magnetic Iron Ore (Loadstone, Lode stone, Magnetite, Black Oxide of Iron, Ferroferrite). A black ore of iron. It is a ferroso-ferric oxide, $FeO.Fe_2O_3$, and contains over 72 per cent. iron. It is a source of Swedish iron.

Magnetic Oxide of Iron. See MAGNETIC IRON ORE.

Magnetic Pyrites (Pyrrhotine, Pyrrhotite, Magnetic Sulphide of Iron). A mineral). In composition it approximates to ferrous sulphide, FeS, or ferroso-ferric sulphide, Fe_3S_4, but often contains nickel.

Magnetic Sulphide of Iron. Magnetic iron ore.

Magnetite. See MAGNETIC IRON ORE.

Magnetoplumbite. A mineral from Långben. It corresponds to PbO $2Fe_2O_3$. It is also given as $4[(Pb,Mn^{\cdot}.Mn)^{\cdots}(Fe^{\cdots},Mn^{\cdots},Ti)_6O_{10}]$.

Magnets. The following alloy has been used for permanent magnets: 91 per cent. iron, 5·3 per cent. tungsten, 2·5 per cent. aluminium, 0·25 per cent. manganese, 0·65 per cent. carbon, and 0·2 per cent. chromium.

Magnet Steel. Alloys of iron with from 5–50 per cent. cobalt, 5–18 per cent. nickel, 0–7 per cent. manganese, 1–12 per cent. tungsten, and 0–12 per cent. chromium.

Magniotriplite. Synonym for Talktriplite.

Magno. An electrical resistance alloy containing 95 per cent. nickel and 5 per cent. manganese.

Magnocarbon. A proprietary preparation of charcoal, magnesium peroxide and extract of belladonna. Gastrointestinal sedative. (359).

Magnochromite. A mineral. It is a chromite containing magnesium.

Magnocid. Magnesium oxychloride,

$Mg(OH)(OCl)$. It has bleaching properties.

Magnodat. See BIOGEN.

Magnoferrite. See JACOBSITE.

Magnofranklinite. A mineral. It is a magnetic variety of franklinite.

Magnogene. A proprietary preparation of magnesium chloride, bromide, fluoride and iodide used in the treatment of prostatic irritability. (802).

Magnojacobsite. A mineral. It is $(Mn,Mg)Fe_2O_4$.

Magnolan. Calcium-anhydro-oxy-diamine-phosphate. Used in treatment of *Diabetes mellitus.*

Magnolax. A proprietary preparation of magnesium hydrate with liquid paraffin.

Magnolia Metal. A bearing metal consisting mainly of antimony and lead, and sometimes tin, with small quantities of iron and bismuth. (*a*) Contains 83½ per cent. lead and 16½ per cent. antimony; (*b*) consists of 79 per cent. lead and 21 per cent. antimony. Other alloys contain 78–80 per cent. lead, 15–16 per cent. antimony, 5–6 per cent. tin, and 0–0·25 per cent. bismuth.

Magnolite. A mineral, Hg_2TeO_4.

Magnolium. An alloy of 90 per cent. lead and 10 per cent. antimony.

Magno Masse. A calcined variety of the mineral dolomite sold in certain particle sizes.

Magnophorite. A mineral. It is NaK $CaMg_5Si_8O_{23}OH$.

Magnuminium. A proprietary trade name for magnesium-aluminium alloys having the following exemplary analyses:—

Alloy	Aluminium	Zinc	Manganese	Silicon	Copper	Magnesium
	per cent. up to 0·2	per cent. up to 0·2	per cent. up to 2·5	per cent. up to 0·4	per cent. up to 0·2	
133	up to 0·2	up to 0·2	up to 2·5	up to 0·4	up to 0·2	rest
166	,, 11·0	,, 1·5	,, 1·0	—	—	rest
166A	,, 11·0	,, 1·5	,, 1·0	—	—	rest
177	,, 8·5	,, 3·5	,, 0·5	up to 0·4	up to 0·4	rest
199	,, 8·5	,, 3·5	,, 0·5	,, 0·4	,, 0·4	rest
220	9–11	,, 3·5	,, 0·5	Fluorine 1·0	Lead 0·4	rest

Magnus Green Salt. An ammoniacal platinum compound, $Cl_2Pt.2NH_3$.

Magtran. A proprietary trade name for optical grades of magnesium fluoride for lens blooming, hot-pressing and crystal growing. (182).

Magyagite (Black Tellurium, Foliated Tellurium). A mineral $(Au.Pb)_2(TeS.Sb)_3$.

Magyey Gum. See CHAGUAL GUM.

Mahogany Acid. Crude mixtures of sulphonic acid for the refining of petroleum sludge.

Mahogany Brown. A sienna which has been ignited, ground wet, and made up in the form of pieces, and dried.

Mahogany Nuts. See ACAJOU BALSAM. Njave butter.

Ma Huang. A Chinese drug. It contains an alkaloid ephedrine. The alkaloid is stated to have valuable properties in the treatment of Addison's disease.

Mahura. Bael fruit.

Mahwa Butter. See BASSIO TALLOW.

Maillechort. A nickel silver containing copper, zinc, nickel, and iron. The Paris variety contains 66·3 per cent. copper, 13·4 per cent. zinc, 16·4 per cent. nickel, and 3·2 per cent. iron. The Vienna type consists of 66·6 per cent. copper, 13·6 per cent. zinc, 19·3 per cent. nickel, and 0·5 per cent. iron. German maillechort contains 65·4 per cent. copper, 13·4 per cent. zinc, 16·8 per cent. nickel, and 3·4 per cent. iron. Also see NICKEL SILVERS.

Maize. See SUN YELLOW.

Maizena. A proprietary trade mark for corn starch.

Maize Oil. See CORN OIL.

Maize Pro-Gen. A proprietary preparation of ARSANILAC ACID. A growth promoter for veterinary use.

Maizite. A proprietary trade name for a preparation of alcohol soluble protein.

Maizolith. A material somewhat resembling hard rubber in appearance and properties. It is obtained by the hydration of the corn-stalk or corn-cob by chemical and mechanical treatment usually by shredding, heating with sodium hydroxide, and beating the pulp when a jelly-like mass is produced. This is dried and machined into the desired shape. It is suggested as an insulating material and for noiseless gears and wheels.

Majamin. Sodium-β-tetralin sulphonate. Majamin-kalium, the potassium salt, and Majammonium, the ammonium salt, are similar products. They are added to soap and soap powders to increase lathering power.

Majammonium. Ammonium-β-tetralin-sulphonate.

Majeptil. A proprietary trade name for the mesylate of Thioproperazine.

Majolica. A pottery enamelled with a tin oxide enamel.

Majunga Noir. A rubber yielded by *Landolphia Perrieri*.

Makalot. A proprietary trade name for a phenol-formaldehyde synthetic resin, used for moulding.

Makhonine. A motor fuel obtained by a new process of cracking petroleum, starting from crude oil. Its specific gravity is 0·95.

Makite. A mineral. It is thenardite.

Makrolon. A trade mark for a poly-carbonate resin.

Makrolon GV. A trade mark. Glass-filled polycarbonate resin.

Malabar Kino. See KINO, COCHIN.

Malabar Tallow. A fat obtained from the seeds of *Vateria indica*. It melts at 37·5° C., has a saponification value of 188·7–189·3, and an iodine value of 37·8–39·6.

Malacca Primers. Primers for bonding materials such as rubber, wood, fabric, and metals. They usually have a rubber base.

Malachite (Mountain Green, Green Verditer, Copper Rust, Green Spar). A mineral used as a pigment. It is a hydrated basic copper carbonate, $CuCO_3.Cu(OH)_2$, and when ground is sold as Mountain green and Mineral green.

Malachite, Blue. See CHESSYLITE.

Malachite Green (Malachite Green B, New Victoria Green, New Green, Fast Green, Vert Diamant, Bitter Almond Oil Green, Benzaldehyde Green, Benzal Green, Diamond Green B, Victoria Green, Solid Green, Solid Green Crystals, Solid Green O, Benzoyl Green, Dragon Green, China Green, Hungary Green). A dyestuff. It is the zinc double chloride, oxalate, or ferric double chloride of tetra-methyl-p-amino-triphenyl-carbinol. Zinc double chloride = $(C_{23}H_{25}N_2Cl)_3 + 2ZnCl_2 + 2H_2O$. Oxalate = $(C_{23}H_{24}N_2)_2(C_2H_2O_4)_3$. Dyes wool, silk, jute, and leather bluish green, also cotton mordanted with tannin and tartar emetic. The name Malachite Green is also used at times for Malachite (*q.v.*). A specially purified form of this dyestuff has been used as an antiseptic dressing for wounds.

Malachite Green B. See MALACHITE GREEN.

Malachite Green G. See BRILLIANT GREEN.

Malachite Green, Spirit Soluble. The picrate of tetramethyl-di-p-amino-triphenyl-carbinol. Used for colouring spirit varnishes.

Malacolite. A mineral. It is White Augite (*q.v.*).

Malacone. A mineral. It is an altered form of zircon, $ZrSiO_4$.

Malakin (Phenosalin). Salicyl-phenetidine, $C_6H_4(OC_2H_5)N : NCH.C_6H_4(OH)$. An antipyretic.

Malapaho. Oil of Panao, collected from *Dipterocarpus vernicifluus*.

Malarine. Acetophenone-phenetidine, $C_6H_5.C(CH_3) : N.C_6H_4.OC_2H_5$. Used as an antiseptic and antineuralgic.

Malatex. A proprietary preparation of propylene glycol, malic acid, benzoic acid and salicylic acid. A de-sloughing agent. (462).

Malathion. An insecticide which is rapidly detoxified by the mammalian liver. Diethyl 2 - (dimethoxyphos - phinothioylthio)succinate. CYTHION.

Malay Camphor. See CAMPHOL.

Malazide. Plant growth inhibitor. (618).

Maldene 285. A proprietary copolymer of butadiene and maleic anhydride supplied as a 25% solution in acetone for use as an intermediate. (434).

Maldene 286. A copolymer similar to MALDENE 285 save that it is a 25% solution of the partial ammonium salt in water. (434).

Maldene 288. A copolymer similar to MALDENE 285 save that the solids content in the solution is 35%. (434).

Maldene 289. A proprietary copolymer of butadiene and maleic anhydride partially ethyl-esterified and dissolved to 25% in ethyl alcohol. (434).

Maldene 292. A proprietary copolymer of butadiene and maleic anhydride supplied as a partially-butyl-esterified 25% solution in butanol. (434).

Maldene 293. A copolymer similar to MALDENE 285 save that it is a 25% solution of the partial amide-ammonium salt in water. (434).

Maldene 300. A copolymer similar to MALDENE 285 save that it is a 25% solution of the partial octyl ester in toluene. (434).

Maldene 631. A copolymer similar to MALDENE 285 save that it is an 18% solution of a zinc-ammonium-complexed form of MALDENE 286 in water. (434).

Maldonite (Black Gold). A natural alloy of bismuth and gold, found at Maldon, Australia. It has the formula Au_2Bi.

Male Fern. The rhizome of *Aspidium felix-mas*.

Malenite. A material containing an antimony double salt, $SbF_3.Na_2SO_4$, in addition to sodium fluoride, and the sodium compound of dinitro-phenol or dinitro-*o*-cresol. Used for impregnating wood.

Malethamer. Maleic anhydride-ethylene polymer.

Malidone. Aloxidone.

Malinowskite. A mineral. It is a variety of tennantite containing lead.

Malladrite. Sodium fluosilicate, Na_2SiF_6.

Mallardite. A mineral. It is a manganous sulphate, $MnSO_4.7H_2O$.

Malleable Iron (Wrought Iron). Practically pure iron, through which are scattered particles of slag or oxide.

Malleable Nickel. Nickel commercially refined, and treated with a deoxidising agent such as magnesium, and cast into ingots. It is suitable for hot or cold working.

Mallebrein. Aluminium chlorate. An antiseptic and astringent.

Mallet Alloy. A brass containing 74·6 per cent. zinc and 25·4 per cent. copper.

Mallet Bark. The bark of *Eucalyptus occidentalis*. It contains from 35–52 per cent. tannin.

Mallophene. Phenyl-azo-α-α-diamino-pyridine hydrochloride.

Malloydium. An alloy of 61 per cent. copper, 23 per cent. nickel, 14 per cent. zinc, and 1 per cent. iron. It is stated to be acid-resisting.

Malmal. Myrrh of Somaliland. There are two kinds, Ogo malmal, and Gubar malmal.

Malonal. A preparation intended to replace veronal.

Malonoben. A pesticide. It is 2-(3,5-di-*tert*-butyl-4-hydroxybenzylidene) = malononitrile.

Malonurea. See VERONAL.

Maloprim. A proprietary preparation of DAPSONE and PYRIMETHAMINE used in the prophylaxis of malaria. (277).

Malotte's Alloy. An alloy of 46 per cent. bismuth, 20 per cent. lead, 34 per cent. tin. Melting-point is 203° F.

Malouren. See VERONAL.

Malros. Fortified rosin size. (575).

Malt. Barley which has been allowed to germinate slightly. Also see BYNE, BYNIN, SITOGEN, and STAPHYLASE.

Malta Grey. See NEW GREY.

Maltase. See GLUCASE.

Maltesite. A mineral. It is a variety of chiastolite.

Malt Extract. See EXTRACT OF MALT.

Maltha. A variety of mineral tallow or wax found in Finland. It is also the name applied to certain types of soft bitumen.

Malthactite. A clay of the fuller's earth type.

Malthenes. See PETROLENES.

Malthite. A name for viscous bitumens.

Maltine. Extract of Malt (*q.v.*).

Maltine, French. Diastase. See AMYLASE.

Maltobiose (Amylon). Maltose, $C_{12}H_{22}O_{11}.H_2O$, a sugar.

Maltol. 3-Hydroxy-2-methyl-1:4-pyrone.

Maltose. Malt sugar, an isomer of cellobiose. α'-4-glucosido-glucose.

Malt, Sugar of. See EXTRACT OF MALT.

Malt Vinegar. See BROWN VINEGAR.

Maltyl. Dry malt extract. It contains about 90 per cent. soluble carbohydrates.

Maltzyme. Diastase, an enzyme.

Maluminium. An alloy of 87 per cent. aluminium, 6·4 per cent. copper, 4·8 per cent. zinc, 1·4 per cent. iron, and traces of manganese, silicon, and lead.

Mamanite. A mineral resembling Polyhalite.

Mamarron Nuts. Nuts from a species

of *Attalea*, of Columbia. The kernels contains 70 per cent. fat.

Manaca. The dried root of *Brunfelsia hopeana*, of Brazil. An extract is used as a diuretic and diaphoretic.

Manaccanite. A mineral. It is a variety of titaniferous magnetic oxide of iron.

Manal. Manganese acetate. (589).

Manalox. Additives for oils, resins and paints. (534).

Manalox 30A. A proprietary trade name for a white spirit solution of aluminium tri-isopropoxide modified with a higher molecular weight alcohol. (190).

Manalox 205. A proprietary trade name for a polymeric organic aluminium compound used as a structuring agent in paints and inks. (190).

Manalox 402. A proprietary trade name for a poly-oxo aluminium tallate. (190).

Manalox 403. A proprietary trade name for a poly-oxo aluminium stearate used as a waterproofing agent. (190).

Manalox AG. A proprietary trade name for Glycalox.

Manalox AS. A proprietary trade name for Sucralox.

Manandonite. A mineral. It is a lithium boro-silicate.

Manasseïte. A mineral. It is $Mg_6Al_2 CO_3(OH)_{16}.4H_2O$.

Manchester Brown. See BISMARCK BROWN.

Manchester Brown EE (Bismarck Brown R, Manchester Brown PS, Vesuvine B). A dyestuff. It is the hydrochloride of toluene-disazo-*m*-toly-lene-diamine, $C_{21}H_{26}N_8Cl_2$. Dyes wool, leather, and tanned cotton reddish-brown.

Manchester Brown PS. See MANCHESTER BROWN EE.

Manchester Yellow. See MARTIUS YELLOW.

Mancopper. A pesticide. It is an ethylene bisdithiocarbamate-mixed metal complex containing about 13·7% manganese and about 4% copper.

Mandarin G. See ORANGE II.

Mandarin G Extra. See ORANGE II.

Mandarin GR. See ORANGE T.

Mandarin Orange. See ORANGE III.

Mandelamine. A proprietary preparation containing hexamine mandelate. A urinary antiseptic. (244).

Mandelin's Reagent. An alkaloidal reagent, consisting of 1 gram of ammonium vanadate dissolved in 200 c.c. of concentrated sulphuric acid.

Manderite. A preparation of asbestos.

Mandrax. A proprietary preparation of methaqualone and diphenhydramine hydrochloride. A hypnotic. (307).

Mandurin. A proprietary preparation

of hexamine mandelate. A urinary antiseptic. (331).

Mangabeira Rubber. A rubber from the small tree, *Hancornia speciosa* cultivated in Paraguay and Venezuela.

Mangadiaspore. A mineral. It is $4[(Al,Mn)O.OH]$.

Mangal. An aluminium alloy containing 1·5 per cent. manganese. It resists corrosion and has a tensile strength of 18–25 kg. per sq. mm.

Mangaloy. An alloy of nickel containing iron and manganese.

Manganaluminium Bronze. An alloy containing from 9–10 per cent. manganese, 85·5–86 per cent. copper, and $4\frac{1}{2}$–5 per cent. aluminium.

Mangan-Ankerite. A mineral. It is $[Ca(Fe,Mg,Mn)(CO_3)_2]$.

Manganapatite. A mineral. It is $2[(Ca,Mn)_5(PO_4)_3F]$.

Manganar. Manganese arsenate.

Manganated Linseed Oil. Linseed oil which has been boiled with manganese dioxide to increase its drying properties.

Manganberzellite. A mineral. It is $8[(Ca,Na)_3Mn_2(AsO_4)_3]$.

Manganbrucite. A mineral. It is $(Mg,Mn)(OH)_2$.

Manganepidote. Synonym for Piemontite.

Manganese-Aluminium Brass. An alloy of 56 per cent. copper, 40 per cent. zinc, 3 per cent. manganese, and 1 per cent. aluminium.

Manganese Alums. Compounds of manganese having the formula R_2SO_4. $Mn_2(SO_4)_3.24H_2O$. $(R=(NH_4)_2 : K.)$

Manganese Amalgam. A manganese-mercury alloy prepared by electrolysing a solution of manganous chloride, using a negative pole of mercury.

Manganese, Battery. See MANGANESE BLACK.

Manganese Bistre. A mineral colour used for textile purposes, which consists of manganese peroxide. The cotton fabric to be coloured is first impregnated with a solution of manganese sulphate, and dried. It is then run through soda lye, which precipitates manganese hydroxide upon the fibre. The colour is finally developed by oxidation with bleaching powder or potassium bichromate.

Manganese Black (Battery Manganese). A pigment. It is native manganese dioxide, MnO_2.

Manganese Blende. Native manganese sulphide, MnS.

Manganese Blue. A pigment obtained by calcining a mixture of china clay (2 parts), oxides of manganese (3 parts), and barium nitrate, (8 parts).

Manganese, Bog. See WAD.

Manganese Boron. An alloy of manganese and boron. It is used for making other alloys.

Manganese Brass. Variable alloys, usually containing 51–69 per cent. copper, 1–4 per cent. manganese, 0–3 per cent. iron, 29–40 per cent. zinc, 0–2 per cent. tin, 0–2 per cent. nickel, and sometimes aluminium. Alloys containing higher amounts of manganese approximate to 50–84 per cent. copper, 12–25 per cent. manganese, and 4–15 per cent. zinc.

Manganese Bronze. An alloy made by adding ferro-manganese or manganese to bronze. It usually contains from 82–83.5 per cent. copper, 8 per cent. tin, 5 per cent. zinc, 3 per cent. lead, and 0.5–2 per cent. manganese, but an alloy containing 59 per cent. copper, 39 per cent. zinc, 1.5 per cent. manganese, and 0.5 per cent. iron, melting at 900° C. and having a specific gravity of 8.6, is also known as manganese bronze. The term is also used for manganese copper. The name is sometimes applied to a colouring matter produced on the fibre by the oxidation of manganese hydroxide.

Manganese Brown. A mineral colour used for dyeing cotton. The material is impregnated with a solution of manganous chloride, then passed through a hot solution of sodium hydroxide, and exposed to oxidation by air.

Manganese-Chalcanthite. A mineral. It is $2[MnSO_4.5H_2O]$.

Manganese Copper. An alloy used in the place of manganese in the manufacture of non-ferrous alloys. The manganese may be as high as 40 per cent., but is usually 4 per cent., 10 per cent., 15 per cent., or 30 per cent. An alloy containing 95 per cent. copper and 5 per cent. manganese, has a specific gravity of 8.8 and melts at 1060° C. These alloys are sometimes called manganese bronzes.

Manganese Cupro. See MANGANESE COPPER.

Manganese Cupro Nickel. This is usually an alloy containing from 65–83 per cent. copper, 15–30 per cent. manganese, and 2–8 per cent. nickel.

Manganese German Silver. An alloy of 80 per cent. copper, 15 per cent. manganese, and 5 per cent. zinc.

Manganese Green. See ROSENTHIEL'S GREEN.

Manganese Iron. See FERRO-MANGANESE.

Manganese Nickel. An alloy of from 51–82 per cent. copper, 14–31 per cent. manganese, and 3–16 per cent. nickel.

An alloy containing 95 per cent. nickel and 5 per cent. manganese, and melting at 1420° C. is also known as manganese nickel.

Manganese Nickel Brass. Alloys containing 51–65 per cent. copper, 5–40 per cent. zinc, 0–2.78 per cent. iron, 1.5–3.24 per cent. manganese, and 2–18 per cent. nickel.

Manganese Nickel-silver. Alloys containing 50–70 per cent. copper, 9–40 per cent. zinc, 1–20 per cent. manganese, and 2–20 per cent. nickel.

Manganese Ore A. A standard manganese ore. It contains 51.3 per cent. manganese, 14.3 per cent. available oxygen, 6.5 per cent. silica, 1.3 per cent. iron, and 0.22 per cent. phosphorus.

Manganese Ore, Brown. See MANGANITE.

Manganese Ore, Hydrated. See VARVICITE.

Manganese Ore, Spathic. See RHODOCROSITE.

Manganese Oxide, Red. See HAUSMANNITE.

Manganese, Recovered. See WELDON MUD.

Manganese, Red. See RHODOCROSITE.

Manganese Saccharate 10 per cent. A sweetened form of manganese oxide containing 10 per cent. cane sugar.

Manganese, Silico. See SILICO-MANGANESE.

Manganese Silver. See MANGANESE GERMAN SILVER.

Manganese Spar. See RHODOCROSITE.

Manganese Steel. An alloy of manganese and steel containing up to 20 per cent. manganese. Commercial manganese steel contains 11–14 per cent. manganese, 1–1.3 per cent. carbon, 0.3 per cent. silicon, 0.05–0.08 per cent. sulphur, and 0.05–0.08 per cent. phosphorus. An alloy containing 86 per cent. iron, 13 per cent. manganese, and 1 per cent. carbon. It has a specific gravity of 7.81 and melts at 1510° C. Used for making points and crossings, and rails.

Manganese, Thermit. See THERMIT MANGANESE.

Manganese Tin. An alloy of manganese and tin, containing 50 per cent. manganese. It is used in the place of manganese in the manufacture of non-ferrous alloys.

Manganese Titanium. An alloy of manganese and titanium, containing from 30–35 per cent. titanium. Used for removing oxygen and nitrogen from copper alloys.

Manganese Velvet Brown. See UMBER.

Manganese Violet (Nuremberg Violet, Mineral Violet, Permanent Violet). A pigment prepared by fusing a mixture of phosphoric acid, and either powdered pyrolusite or residues from the preparation of chlorine. The mass is allowed to cool, and is then heated with ammonia or ammonium carbonate to boiling, when manganese hydroxide, $Mn(OH)_2$, is precipitated out. The liquid is filtered and evaporated to dryness. It is allowed to cool, and boiled with water, when a violet powder is produced which constitutes manganese violet. It is a manganic phosphate, and is used as a mineral colour.

Manganese White. A pigment. It is manganous carbonate, $MnCO_3$.

Manganese Zinc. An alloy containing 20 per cent. manganese, and zinc. Used in the place of manganese in the manufacture of non-ferrous alloys.

Manganic. Nickel containing a small percentage of manganese.

Manganin. Alloys usually containing 70–86 per cent. copper, 4–25 per cent. manganese, and 2–12 per cent. nickel. One of the best varieties contains 83.6 per cent. copper, 13.6 per cent. manganese, 2.5 per cent. nickel, and 0.3 per cent. iron. Later alloys consist of 80–81 per cent. copper, 17–18 per cent. manganese, and 1.5–2 per cent. nickel. The specific resistance of an alloy containing 84 per cent. copper, 12 per cent. manganese, and 4 per cent. nickel, is 42 micro-ohms . cm. at 0° C., and an alloy of 82 per cent. copper, 15 per cent. manganese, and 3 per cent. nickel has a specific gravity of 8.5. These alloys are used for electrical resistances.

Manganite (Brown Manganese Ore). A mineral. It is a hydrated oxide of manganese, $Mn_2O_3.H_2O$. It is also a name for a war gas which consisted of a mixture of hydrocyanic acid and arsenic trichloride. Synonym, ACERDESE.

Mangan-Monticellite. Synonym for Glaucochroite.

Mangan-Neusilber. A nickel silver containing manganese. It contains from 59–72 per cent. copper, 5–20 per cent. zinc, 10–18 per cent. nickel, and 2–20 per cent. manganese.

Manganocalcite. A mineral, $(Mn.Ca.Mg)CO_3$.

Manganochlorite. A mineral. It is a variety of clinochlore containing 2.5 per cent. MnO.

Manganocolumbite. A mineral. It is columbite.

Manganoferrite. Synonym for Jacobsite.

Manganomossite. A mineral. It is $2[Mn(Cb,Ta)_2O_6]$.

Manganosite. A mineral, MnO.

Manganospinel. A mineral. $(Mg.Mn)(Al.Fe)(O_2)_2$. Also given as $8[(Mn,Mg)(Al,Mn)_2O_4]$.

Mangano Steel. A proprietary trade name for a non-shrinking steel containing 1.6 per cen,t. manganese, 0.2 per cent. chromium and 0.95 per cent. carbon.

Manganostibite. A mineral. It is $Mn_{10}(Sb,As)O_{15}$.

Manganostibnite. A mineral, $Mn_{10}Sb_2O_{15}$.

Manganostilbite. A mineral. It consists of manganese sulphide, MnS, with some oxides of arsenic and antimony.

Manganotantalite. A mineral containing manganese oxide, MnO, tantalum oxide, TaO_5, with some tin and tungsten.

Mangano-Titanium. This is usually an alloy of manganese with 30 per cent. titanium, and is used as a deoxidiser in making bronze and brass castings.

Mangol. A powder consisting of basic magnesium hypochlorite. Used for testing alkaloids.

Mangonic. A manganese-nickel alloy containing about 3 per cent. manganese.

Manguinite. A war poison gas. Cyanogenchloride CN.Cl.

Manhardt's Aluminium Bronze. An alloy of 83.3 per cent. aluminium, 6.25 per cent. copper, 10.13 per cent. tin, 0.16 per cent. antimony, 0.05 per cent. magnesium, and 0.08 per cent. phosphorus.

Manhatton Spirits. See ACETONE ALCOHOL.

Manila Copal. See COPAL RESINS.

Manila Elemi. Elemi from *Canarum commune.*

Manila Paper. A sulphate wood-pulp paper which sometimes contains manila hemp and jute fibre. Pure manila paper contains manila only.

Manjak (Glance Pitch). A bitumen found in Mexico, South America, and West Indies. It has a specific gravity of 1.1–1.15, fuses at 250–375° F., and is very soluble in carbon disulphide. Used as a paint, as a roofing material, and in connection with drilling for oil.

Manlianite. An explosive consisting of 72 per cent. ammonium perchlorate, 14.75 per cent. charcoal, and 13.25 per cent. sulphur.

Manna. A sugary exudation which occurs in the rising sap of *Fraxinus ornus*, and *F. rotundifolia.* The crude material contains from 12–13 per cent. water, 10–15 per cent. sugar, 32–42 per cent. mannitol, 40–41 per cent. mucilaginous substances, organic acids, and nitrogenous matter. Australian manna is obtained from *Myoporum platycorpum,* and contains as much as 90 per cent. of mannitol. Manna is used as a purgative for children.

Mannheim Gold. A brass containing

80 per cent. copper and 20 per cent. zinc. A jeweller's alloy. The term is also applied to a bronze consisting of 83·7 per cent. copper, 9·3 per cent. zinc, 7 per cent. tin, with a little phosphorus.

Mannin. See ORTHOFORM.

Mannite, Inactive. α-Acrite.

Mannol. Ethyl-acetanilide. A febrifuge. It is also used as a softening agent for cellulose esters.

Mannolit. See CHLORAMINE T.

Mannomustine. 1,6-Di-(2-chloroethyl-amino) - 1,6 - dideoxy - D - mannitol. Degranol is the dihydrochloride.

Mannotetrose. See LUPEOSE.

Manofast. A proprietary trade name for thiourea dioxide. Textile reducing compounds. (190).

Man Oil. Bone oil (*q.v.*)

Manolene. A trade mark for a range of high-density polyethylenes. (672).

Manolene 5203. A hexene copolymer.

Manomet 50. A proprietary trade name for a solution of a complex of calcium in a high boiling petroleum distillate used as a PVC stabiliser. (190).

Manomet 105. A proprietary trade name for a PVC stabiliser based on a blend of calcium, zinc and aluminium compounds. (190).

Manomet 201, 220. Proprietary trade names for a liquid mixture of cadmium, calcium and zinc acylate complexes used as a stabiliser for p.v.c. plastisols for leathercloth. (190).

Manomet 280. A proprietary trade name for a PVC kicker/stabiliser consisting of metal organic complexes based on the metals—cadmium, zinc, zirconium and aluminium. (190).

Manomet 360. A proprietary trade name for a liquid lead acylate in a diluent used as a heat stabiliser for p.v.c. plastisols for carpet backing and conveyor belting. (190).

Manomet 361. A proprietary trade name for a liquid lead stabiliser for PVC. (190).

Manomet 450. A proprietary trade name for a barium/cadmium/zinc liquid stabiliser for PVC. (190).

Manomet 451. A proprietary trade name for a barium/cadmium/zinc liquid stabiliser for PVC. (190).

Manomet 452. A proprietary trade name for a barium-cadmium/zinc liquid stabiliser for PVC. (190).

Manomet 453. A proprietary trade name for a barium/cadmium/zinc liquid stabiliser for PVC. (190).

Manomet 455. A proprietary trade name for a barium/liquid co-stabiliser for PVC. (190).

Manomet 456. A barium/calcium/zinc liquid stabiliser for PVC. (190).

Manomet Zinc 22. A proprietary trade name for a zinc organic compound as PVC kicker/stabiliser. (190).

Manosec. High metal paint driers. (534).

Manosec Calcium 6 per cent. A proprietary trade name for a synthetic acid based calcium drier used in paints and printing inks. (190).

Manosec Cobalt 18 per cent. A proprietary trade name for a high metal synthetic acid based cobalt drier used in paints and printing inks. (190).

Manosec Lead 36 per cent. A proprietary trade name for a high metal synthetic acid based lead drier used in paints and printing inks. (190).

Manosec Manganese 10 per cent. A proprietary trade name for a high metal synthetic acid based manganese drier used in paints and printing inks. (190).

Manosec Zinc 17 per cent. A proprietary trade name for a high metal synthetic acid based zinc drier used in paints and printing inks. (190).

Manosec Zirconium 18 per cent. A proprietary trade name for a high metal synthetic acid based zirconium drier used in paints and printing inks. (190).

Manosil. Auxiliaries for paint plastic and rubber industries. (534).

Manosil AS7 & AS9. A proprietary trade name for aluminium silicates used in the rubber industry. (190).

Manosil VN2/VN3. A proprietary trade name for pure precipitated hydrated silicas used in the rubber industry. (190).

Manosperse. A proprietary trade name for processing aids in rubber compounds. (190).

Manox. Ferrocyanide blues and chemicals. (534).

Manox B.S.3. Wire drawing lubricant. (534).

Manoxol " IB ". A proprietary trade name for a sodium di-isobutyl sulphosuccinate wetting agent. (190).

Manoxol " MA ". A proprietary trade name for a sodium di-methylamyl sulphosuccinate. (190).

Manoxol " N ". A proprietary trade name for a sodium dinonyl sulphosuccinate. (190).

Manoxol " OT ". A proprietary trade name for a sodium dioctyl sulphosuccinate wetting agent. (190).

Manoxol " OT/P ". A pharmaceutical grade of the above. (190).

Manoxol " TR ". A proprietary trade name for a sodium bis-tridecyl sulphosuccinate. (190).

Manoxolot. A proprietary trade name for dioctyl sodium sulphosuccinate.

Manqueta (Manquta). African names for a fossil gum resin, resembling copal.

Mansfieldite. A mineral. It is $8[AlAsO_4.2H_2O]$.

Mansjöite. A variety of the mineral diopside, containing fluorine, found in Sweden.

Manson Blue. See BLUE OF MANSON.

Mansu. Manganese sulphate. (589).

Manucol. A registered trade name for sodium alginate. (Sodium salt of a polysaccharide.) An emulsifier. Algin is extracted from seaweed from the Atlantic coast of Scotland and marketed as the sodium derivative under the trade name above. It is used in the manufacture of cosmetics, plastics (typewriter rollers), transparencies, and textiles. It is also used in dried milk and cocoa to prevent sedimentation and to provide greater solubility respectively.

Manucol Ester. Propylene glycol alginate. (536).

Manucol Ester EX/LL. A proprietary trade name for propylene glycol alginate. A food grade emulsifying agent. (204).

Manus. Dishwashing detergent and sanitizer. (537).

Mapé. A coarse starch obtained from the fruit of *Inocarous edulis*.

Maphenide. 4-Sulphanomoylbenzylammonium chloride. Marfanil.

Mapico Browns. Proprietary trade names for iron oxide browns.

Maple Sugar Sand. A by-product in the manufacture of maple sugar. The sap from the maple is evaporated in pans, and a precipitate forms when the water content is about 35 per cent. This precipitate is maple sugar sand. The chief constituent is calcium malate (60–80 per cent.), and malic acid is easily prepared from it. The sand has an ash of 50–60 per cent., containing 37–46 per cent. CaO, 25–36 per cent. CO_2, 11–31 per cent. SiO_2, and 2·3–3·3 per cent. MnO.

Maprenal. A trade mark for a range of synthetic lacquers. They are resins on malamine or benzoguanamine bases. (983).

Maprofix. A proprietary trade name for a sulphonated fatty alcohol used as a wetting agent and detergent.

Mapromin. A proprietary trade name for a sulphated fatty alcohol used as a wetting agent.

Maprotiline. An anti-depressant. 3-(9, 10 - Dihydro - 9, 10 - ethanoanthracen - 9 - yl)propylmethylamine. " 34276-Ba " is currently under clinical trial as the hydrochloride.

Marabout Silk. A white silk which still contains its gum. It is dyed and used for the manufacture of imitation feathers.

Maranite. A variety of the mineral Andalusite.

Maranta. Arrowroot starch.

Maranyl. A proprietary trade name for nylon. (2).

Maranyl A 100. A proprietary brand of Nylon 66. (2).

Maranyl A 101. A proprietary heat-stabilised form of Nylon 66. (2).

Maranyl A 102. A fast processing grade of type 66 nylon. (2).

Maranyl A 108. Proprietary name for a grade of Nylon 66 containing graphite and molybdenum disulphide. (2).

Maranyl A 150. Proprietary name for a high-viscosity grade of Nylon 66. (2).

Maranyl A 190. Glass filled nylon 66, (2).

Maranyl B 100. A proprietary grade of Nylon 610. (2).

Maranyl B 102. A proprietary grade of Nylon 610 stabilised for both heat and light. (2).

Maranyl C 109. A proprietary range of N-substituted soluble Nylons. (2).

Maranyl DA. A proprietary grade of Nylon 66/610/6. (2).

Maranyl F. A proprietary grade of Nylon 6. (2).

Maranyl LA 29. A transparent grade of Nylon 66/610. (2).

Maranyl LA 145. A grade of Nylon 66/610 possessing high viscosity. (2).

Maraphos. Bone marrow, malt and phosphate. (591).

Maraschino. An alcoholic liquor produced from small Zara cherries.

Marbadal. A proprietary trade name for Sulphatolamide.

Marble. Crystallised limestone, $CaCO_3$. See CALCITE.

Marble, Fire. See FIRE MARBLE.

Marble, Serpentinous. See OPHIOLITE.

Marblette. A proprietary trade name for a phenol-formaldehyde cast resin.

Marbo. A proprietary trade name for a chlorinated rubber.

Marbolith. See BAKELITE.

Marbon B. A proprietary trade name for a cyclo-rubber.

Marbon Latex. Marmix reactive S.B.R. ABS latices. (608).

Marbon White. A proprietary brand of Lithopone (*q.v.*).

Marboran. A proprietary preparation containing methisazone. (277).

Marcain. A proprietary preparation of bupivacaine and adrenaline. Long acting local anæsthetic. (182).

Marcasite (White Iron Pyrites, Coxcomb Pyrites, Radiated Pyrites). A mineral. It is disulphide of iron, FeS_2. The term is also occasionally applied to bismuth.

Marcasol. See MARKASOL.

Marc Brandy Oil. See FUSEL OIL.

Marceline (Heterocline). A mineral, $Mn_2O_3.MnSiO_3$.

Marchies. See MARGINES.

Marcoumar. A proprietary preparation of phenprocoumon. An anticoagulant. (314).

Marcs. The name given to the residue from wine factories, consisting of the stems and skins of grapes. It is used for making verdigris.

Marcylite. A mineral. It is a variety of atacamite.

Mareepa. Kernels of the fruits of the cokerite palm of British Guiana.

Marennin. A green colouring matter from certain oysters on the French coast. It is a chlorophyll derivative.

Maretin. m-Tolyl-semi-carbazide, $CH_3.C_6H_4.NH.NH.CO.NH_2$. Said to be a good febrifuge.

Marevan. A proprietary preparation of warfarin sodium. An anticoagulant. (182).

Marezzo Marble. An artificial marble produced from oxychloride cement. It is used for building.

Marfanil. A preparation of benzylamine sulphonamide. It is stated to be superior to many sulphonamides in its bactericidal action.

Marfanil. A proprietary trade name for Maphenide.

Margalite. A phenol-formaldehyde resin product. It is used in the manufacture of varnishes and insulators.

Margarin. A term used for beef suet free from stearine.

Margarine (Butter, artificial, Butterine, Dutch Butter, Butter Surrogate). A butter substitute. Oleo oil, neutral lard, coco-nut oil, earth-nut oil, sesamé oil, and cotton-seed oil are all used in various grades of margarine. Soured milk is churned up with the fats. The higher grades contain 10 per cent. oil, 70 per cent. oleo oil, and 20 per cent. coco-nut oil, whilst in inferior qualities, cotton-seed oil is used. Margarine is coloured with butyro-flavine (dimethyl-amino-azo-benzene), and a little cholesterol is added to the milk to give the same odour, whilst frying, as butter. In some countries no milk is used, the margarine being flavoured with strong cheese. Also see CUNEROL and BUTTIROL.

Margarine Cheese. In this material a cheaper fat is substituted for the valuable butter fat extracted from the milk from which the cheese is made.

Margarite (Diphanite, Emeryllite, Pearl mica). A mineral, $CaO.2Al_2O_3.2SiO_2.H_2O$.

Margarodite. A mica having an appearance similar to talc.

Margaron. A white, odourless, tasteless ointment base, obtained by distilling beef suet with lime.

Margarosanite. A mineral, $PbO.2CaO.3SiO_2$.

Margines (Marchies). The residues obtained from the manufacture of olive oil. Used as a manure, but is also recommended as an insecticide.

Margol. A mixture of volatile fatty acids, used as a flavouring material for margarine, to give it the taste of butter.

Margosa Bark. Indian azadirach.

Margraff Alloy. An alloy consisting of 58 per cent. copper, 28 per cent. tin, and 14 per cent. zinc.

Maria-glass. A mineral. It is a variety of gypsum.

Marialite. A mineral, $2NaCl.3Na_2O.3Al_2O_3.18SiO_2$.

Maricol. Magnesium ricinoleate, $(HO.C_{17}H_{32}.CO_2)_2Mg$.

Marignacite. A mineral. It is a variety of pyrochlore.

Marignac's Salt. Potassium-stanno-sulphate.

Marine Acid. Hydrochloric acid, HCl.

Marine Blue (New Methylene Blue NX). Mixtures of Methylene blue 2B with Methyl violet. Also see SOLUBLE BLUE.

Marine Blue B. See BAVARIAN BLUE DSF.

Marine Fibre. A fibre obtained by dredging the shallow water of a gulf in South Australia. It is a hydrated ligno-cellulose.

Marine Glue. Various mixtures are known by this name, but it usually consists of 1 part rubber dissolved in 12 parts turpentine or paraffin, to which is added 2 parts asphalt or shellac. A waterproof adhesive.

Marine Oil. A mixture of blown rape oil and a mineral oil. Used for marine engines.

Marine Salt. Sodium chloride, NaCl.

Marine Soap. A soap made from coco-nut oil, which is soluble in fresh and sea water.

Marine Sperm Oil. A burning oil made from Scotch shale. It has a specific gravity of 0·830 and a flash-point of 230° F.

Marionite. A mineral. It is a hydrated zinc carbonate.

Mark 1330. A proprietary sulphur-containing organotin stabilised for use in P.V.C. for injection moulding and pipe extrusion. (97).

Mark 1414. A proprietary organotin mercaptide stabiliser for P.V.C. used in the extrusion of rigid pipe. (97).

Markasol (Marcasol). Bismuth borophenate. Used as a substitute for iodoform.

Marking Inks. The juices obtained from *Coriaria thymifolia, Anacardium orientale,* and *Anacardium occidentale* are natural products. Chemical marking inks usually contain silver nitrate, which is reduced on the fibre to a black deposit.

Markus Alloy. See NICKEL SILVERS.

Marlex. A proprietary trade name for high density polyethylenes coded as follows:

TR.885 for injection moulding thin walled containers. It has a density of 0·965 gm/c.c. and a melt index of 30.

TR.610 is for use as a wire and cable insulation. (891).

Marlex 1708. A proprietary trade name for a low density polyethylene suitable for the extrusion of heavy duty film. Type 1 Class A Grade 4 resin has a density of 0·917 and a melt index of 0·8. (891).

Marlex. Proprietary name (891) for a range of polyethylenes. Grades include the following:—

55250. Used in the injection moulding of household goods, toys and containers with thin walls.

6003. High-density. Used in the blow moulding of articles requiring a high degree of stiffness such as dustbins and similar large containers.

BHB 5003. A high-performance, high-density polyethylene copolymer used in the blow-moulding of detergent bottles and similar containers.

BHB 5012. Used in the injection moulding of large articles such as petrol tanks, industrial containers, school seating and outdoor furniture.

BHP TR-201. Used for the coating of wires and cables.

BHP TR-203. Used in the insulation of power cables.

BHP TR-551. A resin treated to resist attack by rodents.

BHM 5002. A high-density copolymer used in the blow-moulding of detergent bottles.

BHM 5402. Used in the blow-moulding of light-duty bottles for bleaches and detergents.

BHM 5603. A resin used in the blow-moulding of bottles and tanks.

BHV TR-204. Possessing good resistance to stress cracking, it is used in the coating by extrusion of wire and cables.

BHV TR-553. A compound used in the extrusion coating of wire and cable. Specially treated to give protection against attack by rodents.

BMB 5040, 5065 and *5095.* Used in the injection moulding of household goods and toys.

BMN 5565. Combining superior toughness, high impact strength, good resistance to stress cracking and easy processing characteristics, it is used in the injection moulding of industrial containers, boxes and crates.

BMN TR-880. A narrow distribution of molecular weight gives high impact strength and good resistance to warping. Used in the injection moulding of large flat surfaces.

BMN TR-980. Used in the rotational moulding of air ducts, etc.

BMN TR-995. Good impact strength. Used in the rotational moulding of bins, etc.

BXM 43065. Used to make sheets for forming into trays and similar large, thin, rigid or semi-rigid products.

CL-100-35. Used in the rotational moulding of agricultural and chemical sewage tanks and small engine and automobile fuel tanks, it cross-links during moulding.

EHB 6002. Used in the extrusion of film having a high degree of mechanical performance.

EHB 6007. A high-density resin used in the blow-moulding of bottles for milk, fruit juices and soft drinks.

EHB 6009. Of high density. Used in the blow-moulding of pharmaceutical bottles and industrial containers.

EHB 6009-MT. A variant of EHB 6009 giving off little odour.

EHM 6001. Used in the manufacture of barrier film.

EHM 6006. Of high density. Used in the blow-moulding of milk and water bottles.

EMB 6035. Having good impact strength and rigidity and a fast processing cycle, it is used in the injection moulding of high-quality components and containers.

EMB 6050. Giving a high-gloss finish and possessing easy processing characteristics, it is used in the injection moulding of large household articles and toys.

EMN 6065. Used in the injection moulding of large articles requiring a high degree of stiffness.

EMN TR-885. Used in the injection moulding of thin-walled containers when good product rigidity and a high production rate are required.

EMN TR-960. Used in the rotational moulding of air ducts and large bottles.

EXM 55035. Used to make sheet for the thermo-forming of large components requiring maximum stiffness.

GF-830. Glass-reinforced and of high density, it is used in the injection moulding of articles requiring rigidity, ability to resist heat, and good bearing qualities.

HGR-120-01. A flame-retardant grade of polypropylene used in the manufacture of radio components and electrical mouldings.

HHM 5003. Used in the thermo-forming of intricate components requiring to be accurately reproduced from a moulding.

HHM 5202 and *5502.* High density polyethylenes used in the blow-moulding of bleach and detergent bottles.

Proprietary rights in the name Marlex are held by another manufacturer (159) for the following grades:—

HHM TR-210. An ethylene hexene-1 copolymer producing a high-density polyethylene which acts as a tough coating resin.

HHM RE-415. A polyethylene resin used in the extrusion of water pipes. It has good resistance to ultra-violet light.

HHM TR-416. Easily coiled. Used in the making of unitised electrical conduits and water pipes.

HHM TR-418. A polyethylene used in the manufacture of tube for gas-distribution and drinking-water pipes.

HHM TR-550. A polyethylene used in the extrusion of wires and cables (especially telephone and high-voltage cables) which need to be resistant to the attack of rodents.

HRV-120-01. A flame-retardant polypropylene used in the manufacture of carpeting and industrial fabrics.

J-403. A polyethylene resin of medium density used to coat wires and cables, especially for the sheathing of multi-core cables.

M 0100 A. A polyethylene prepared as a foaming concentrate used in the injection moulding of articles made of closed-cell expanded polyethylene of either high or low density.

Marlie's Alloy. An alloy containing 10 per cent. iron, 35 per cent. nickel, 25 per cent. brass, 20 per cent. tin, and 10 per cent. zinc, which has been quenched in a mixture of acids.

Marls. Natural mixtures of clay and chalk (aluminium silicate and calcium carbonate). Used in the manufacture of cements. The term is also applied to friable earths which are devoid of chalk, such as those of Staffordshire.

Marmalite. Also called Marmatite. It is a mineral consisting of zinc blende with ferrous sulphide.

Marmatite. A mineral. It is a variety of zinc blende containing ferrous sulphide. $4[(Zn,Fe)S]$.

Marme's Reagent. Consists of 10 parts cadmium iodide, CdI_2, and 20 parts potassium iodide, KI, dissolved in 80 parts water. Used for testing for alkaloids.

Marmite. A yeast extract. It is a food preparation resembling meat extract.

Marmo Bardiglio de Bergamo. Vulpinite, a variety of anhydrite, mixed with silica. Used for ornamental purposes.

Marmolite. Serpentine (*q.v.*) with a lamellæ structure.

Maroon. See MAGENTA.

Maroon S (Grenat S, Acid Cerise, Cardinal Red S, Acid Maroon). Impure qualities of Acid Magenta (*q.v.*).

Marphos. Phosphoric acid. (608).

Marplan. A proprietary preparation of isocarboxazid. An antidepressant. (314).

Marquardt Material. A fireproof substance resembling porcelain, which may be used in the manufacture of pieces of apparatus in chemical engineering and in the laboratory. It is of German origin.

Marquis's Reagent. A solution of formalin in sulphuric acid. An alkaloidal reagent.

Marrubin. A proprietary preparation of the glycerin extract of red bone marrow.

Marsala. A liqueur wine. It is manufactured from Trapani wine with 13 per cent. alcohol, to which is added the must of mature white grapes, concentrating, and then adding the must from ripe white grapes.

Mars Brown. See BURNT SIENNA.

Marseilles Soap. Olive oil soap.

Marseilles Vinegar. Vinegar or dilute acetic acid macerated with the leaves of peppermint, rosemary, rue, sage, or lavender flowers.

Marsh Gas (Light Carburetted Hydrogen). Methane, CH_4.

Marshite. A mineral. It is $4[(CuI)]$.

Marsh Ore. A mineral. It is limonite.

Marsilid. A proprietary preparation of iproniazid. An antidepressant. (314).

Mars Orange. A pigment obtained by the calcination of an artificial ochre, such as Mars Yellow (*q.v.*).

Mars Purple. A pigment produced by the calcination of an artificial ochre.

Mars Red (Rouge de Mars). A pigment. It is an artificial ochre, prepared by calcining an artificial ochre such as Mars yellow. The term is also applied to an acid red dye. See AZORUBINE S.

Mars Violet (Purple Ochre). A pigment. It is prepared by calcining an artificial ochre such as Mars Yellow.

Mars Yellow. A pigment. It is an artificial preparation of yellow ochre and is obtained by dissolving ferrous sulphate in water, warming, and introducing strips of zinc. The bright yellow precipitate is Mars yellow. The average composition is from 51–53 per cent. ferric hydroxide, 23–24 per cent. calcium oxide, 18–18·5 per cent. carbon dioxide, and 3·5 per cent. water. The yellow variety contains calcium sulphate.

Martensite. A solid solution of carbon in iron, and is a characteristic constituent of steel which has been tempered at a temperature a little above the transformation point.

Martin Steel. A proprietary trade name for a steel containing 12–14 per cent. chromium, 0·85–1·25 per cent. molybdenum, 0·35 per cent. vanadium, 0·8 per cent. cobalt, and 1·4–1·6 per cent. carbon.

Martinite. A mineral, $Ca_5H_2(PO_4)_4$.

Martino's Alloys. Alloys containing 17·25 per cent. pig iron, 3–4·5 per cent. ferro-manganese, 1·5–2 per cent. chromium, 5·25–7·5 per cent. tungsten, 1·25–2 per cent. aluminium, 0·5–0·75 per cent. nickel, 0·75–1 per cent. copper, and 65–70 per cent. wrought iron. Used for drilling and cutting tools.

Martin's Cement. A similar cement to Keene's, except that potassium carbonate solution is used instead of alumina.

Martin Steel (Open Hearth Steel). Steel obtained in the Martin process by melting from 75 per cent. of cast iron in a reverberatory furnace with the necessary quantity of wrought iron to obtain the required amount of carbon.

Martite. See HÆMATITE.

Martius Yellow (Naphthol Yellow, Naphthylamine Yellow, Gold Yellow, Primrose, Jaune D'Or, Manchester Yellow, Naphthalene Yellow, Naphthaline Yellow). A dyestuff It is the ammonium, sodium, or calcium salt of dinitro-α-naphthol, $C_{10}H_9N_3O_5$, or $C_{10}H_5N_2O_5Na.H_2O$, or $C_{20}H_{10}N_4O_{10}Ca$. Dyes wool golden yellow from an acid bath.

Martocirite. A mineral, $3FeS.4Sb_2S_3$, related to berthierite.

Martonite. A lachrymator gas used in the Great War. It contains 80 per cent. bromo-acetone and 20 per cent. chloroacetone.

Martourite. A mineral. It is an iron-antimony sulphide.

Marzine. A proprietary trade name for the hydrochloride of Cyclizine.

Marzipan Masse. Consists of cane sugar with crushed almonds.

Mascagnine. A mineral $(NH_4)_2SO_4$.

Masonite. A mineral. It is chloritoid.

The name is also applied to a proprietary waste-wood product obtained by the treatment of waste wood from sawmills with steam and pressure, then releasing the pressure to produce fibres. It is made into sheets.

Massaranduba (Balata Rans, Brittle Balata). A pseudo gutta-percha derived from the sap of the Brazilian cow tree, *Mimusops elata*.

Masse. A proprietary preparation of 9-aminoacridine base, allantoin, 8-hydroxyquinolone sulphate in an emollient base used in dermatology as an antibacterial agent. (369).

Massecuite. The boiled mass of beet-sugar syrup. It is a semi-solid mass formed during the evaporation of the sugar juice, and consists of sugar crystals and a thick syrup. It contains from 3·5–7 per cent. water.

Massicot. See LITHARGE.

Masteril. A proprietary preparation of drostanolone propionate. Treatment of mammary carcinoma. (805).

Mastic. The name for an important resin obtained from *Pistachia lentiscus*, from various parts of the Mediterranean coast. It has a specific gravity of 1·04–1·07, softens at 100° C., and melts below 110° C. It has an acid value of 50–71, a saponification value of 73–89, an iodine value of 64, and an ash of 0·1–0·2 per cent. It is soluble in amyl alcohol, amyl acetate, benzene, and ether, and slightly soluble in ethyl alcohol, chloroform, and carbon tetrachloride. It is used in the manufacture of spirit varnishes. The term Mastic is also applied to a mixture of asphalt rock and Trinidad pitch, used for paving.

Masuron. A proprietary trade name for a cellulose acetate plastic.

Masut. See MAZUT.

Mat. 4-*p*-Toluene-azo-acetanilide-3-mercuric-hydroxide. An antiseptic.

Matali. Round strips of rubber obtained from the roots of various species of *Apocynaceae*.

Mataziette. An explosive, containing nitro-glycerin, sand, and chalk.

Maté (Jesuit's Tea, St. Bartholomew's Tea). Paraquay tea, the dried leaves and shoots of an evergreen, *Ilex paraguayensis*. It contains caffeine, theobromine, *i*-inositol, a tannin, and a dihydroxy-sterol (matesterin, $C_{28}H_{46}O_3$).

Matezite. A saccharin substance obtained from Madagascar caoutchouc. It is identical with mono-methyl-*d*-inositol, an isomeride of bornesite. Also see PINITE.

Mathesius Metal. A proprietary trade

name for an alloy of lead with calcium and strontium in small amounts.

Matic. A proprietary preparation of guaicol carbonate, theophylline, ephedrine hydrochloride, and butabarbital. A bronchial antispasmodic. (830).

Matico-camphor. A camphor from the Peruvian matico.

Matildite (Peruvite, Morocochite). A mineral, $Ag_2S.Bi_2S_3$.

Matlockite. A mineral. It is an oxychloride of lead, $PbCl_2.PbO$.

Matrix Alloy. An alloy of 48 per cent. bismuth, 28·5 per cent. lead, 14·5 per cent. tin, and 9 per cent. antimony. It expands on cooling and is used to hold tools in position.

Matrix Brass. See BRASS.

Matromycin. A proprietary trade name for Oleandomycin.

Matt Blues. Cobalt aluminate colours used in the ceramic industry. They are prepared by calcining cobalt oxide with ammonia alum, with varying amounts of zinc oxide.

Matte. See COARSE METAL.

Mattenccite. A mineral. It is Na $HSO_4.H_2O$.

Matteucinol. 6 : 8-dimethyl-5 : 7-dihydroxy-4'-methoxy-flavanone.

Matthieu-Plessy Green. See CHROMIUM GREEN.

Matt Salt. Acid ammonium fluoride, $NH_4F.HF$.

Matura Diamond. See HYACINTH.

Matzoon. See MADZOON.

Maucherite. A mineral, Ni_3As_2.

Mauritius Elemi. Elemi from *Canarum mauritianium*.

Mauritius Hemp. See ALOE FIBRE.

Mauve. See MAUVEINE.

Mauve Dye. See MAUVEINE.

Mauveine (Mauve, Mauve Dye. Chrome Violet, Aniline Violet. Aniline Mauve. Aniline Purple, Perkin's Violet, Perkin's Purple, Violine, Violeine, Rosolane, Indisin, Anileine, Phenamin, Phenamein Purpurin, Rosein, Tyraline, Lydin). A dyestuff prepared by the oxidation of a mixture of aniline and toluidine by potassium chromate. It is phenyl-phenosafranine, $NH_2.C_6H_3NN$ $(C_6H_5)(Cl)C_6H_3.NH.C_6H_5$. Dyes silk reddish-violet. It is also employed for colouring postage stamps. Many of the above names are now obsolete.

Mauzeliite. A mineral, $4(Ca.Pb)O.TiO_2$. $2Sb_2O_3$.

Mawele. A millet of East Africa.

Maw Seeds. Black poppy seeds.

Maxhete. A steel containing nickel, chromium, tungsten, copper, and silicon.

Maxidex. A proprietary preparation of dexamethasone, benzalkonium chloride and phenylmercuric nitrate. Anti-inflammatory eye drops. (374).

Maxilon. Modified basic dyes. (543).

Maxilvry Steel. A proprietary high nickel-chromium steel containing copper. It is stated to be corrosion resisting, and particularly resistant to attack by cider.

Maximite. An explosive. It is similar to cordite.

Maxim-Schupphaus Powder. An explosive containing 80 per cent. of insoluble nitro-cellulose, 19·5 per cent. of soluble nitro-cellulose, and 0·5 per cent. of urea.

Maxim's Powder. An explosive containing soluble and insoluble nitrocellulose, nitro-glycerin, and castor oil.

Maxipen. A proprietary preparation of potassium phenethicillin. An antibiotic. (817).

Maxite. See LEADHILLITE.

Maxitrol. A proprietary preparation of DEXAMETHASONE, NEOMYCIN and POLYMYXIN B used as eye-drops. (374).

Maxium Metal. Castings of magnesium metal.

Maxolon. A proprietary preparation of metoclopramide hydrochloride. An anti-emetic. (364).

Maxyntan. See NERADOL. (508).

Mayari Iron. An iron made from Cuban ores. Small amounts of vanadium and titanium are present which give strength to the metal obtained from these ores.

Mayari Steel. A Cuban low nickel-chromium steel.

Mayer's Albumen. A fixing agent used in microscopy. It consists of 50 c.c. white of egg, 50 c.c. glycerin, and 1 gram sodium salicylate.

Mayer's Carmalum. Carmalum (*q.v.*).

Mayer's Hæmalum. Hæmalum (*q.v.*).

Mayer's Paracarmine Stain. Paracarmine (*q.v.*).

Mayer's Picrocarmine Stain. Picrocarmine (*q.v.*).

Mayer's Solution. Mercuric iodide dissolved in aqueous potassium iodide. An alkaloidal reagent.

May Green (Mignonette Green). Pigments. They are mixtures of Chrome yellow and Paris blue, with yellow predominating. Also see CHROME GREEN.

Maypon UD. A proprietary trade name for undecyleryl polypeptide. An anionic protein based detergent. Used in hair shampoos and conditioners. (222).

Maytee. Fœnugreek seeds.

Mazam. A dextrin of high molecular weight.

Mazapilite. A mineral, $Fe_4(OH)_6.Ca_3$ $(AsO_4)_4.3H_2O$.

Mazarine Blue. See WILLOW BLUE.

Mazindol. An appetite suppressant currently undergoing clinical trial as " AN 448 ". It is 5-(4-chlorophenyl)-2, 5 - dihydro - 3H - imidazo[2, 1 - a] - isoindol-5-ol.

Mazum. An alcoholic liquid prepared from fermented milk. It is made in Armenia from the milk of sheep, goats, or buffaloes.

Mazut (Astatki, Ostatki). Crude Russian petroleum deprived of its volatile constituents by exposure to air or distillation up to 280° C. It has a specific gravity above 0·880, and is used as a fuel, and in the manufacture of oils. The Tartar name is Masut or Mazut, and the Russian terms Astatki or Ostatki.

MBMC. Mono - teriary - butyl - meta - cresol. (167).

Mbocaya Fibre. A fibre obtained from the South American palm.

M.B. General chemicals and pharmaceuticals. (507).

M.B. Powder No. 1. Modified black powder No. 1. A black gunpowder in which part of the potassium nitrate is replaced by potassium chlorate.

M.B. Powder No. 2. An explosive powder consisting of a mixture of ammonium perchlorate, sodium, potassium, or barium nitrate, charcoal, and sulphur.

MBS. A terpolymer of methylmethacrylate, butadiene and styrene. Its properties include rigidity, hardness, high impact strength, heat resistance and good clarity.

MCPB. 3-phenoxybutyric acid. A systemic fungicide active against chocolate spot disease in broad beans.

McCrorie's Stain. A microscopic stain. Solution A contains 1 gram tannic acid, 1 gram potash alum dissolved in 40 c.c. distilled water. Solution B consists of 0·5 gram night blue dissolved in 20 c.c. absolute alcohol. The solutions are mixed and filtered.

McGill Metal. A proprietary trade name for a group of copper-aluminium-iron alloys, one of which contains 89 per cent. copper, 9 per cent. aluminium, and 2 per cent. iron.

McGovernite. A mineral from Sterling Hill, N.J. Its analysis gives the following: SiO_2, 9 per cent.; MnO_2, 43 per cent.; FeO, 1·5 per cent.; MgO, 11 per cent.; ZnO, 10 per cent.; As_2O_3, 4·5 per cent. ; As_2O_5, 12·5 per cent.; and H_2O, 9 per cent.

Mead. An alcoholic liquor prepared from honey by fermentation. Water is added to honey, mixed and sterilised by boiling. As the liquid cools, flavouring is added and then the mixture is fermented with brewers' yeast.

Meadow Green. See EMERALD GREEN.

Meadow Ore. Bog Iron Ore (*q.v.*).

Meadowsweet. Salicylic aldehyde, $C_6H_4(OH).CHO$. This aldehyde is the chief constituent of the essential oil of meadowsweet, *Spiræa ulmaria*.

Meal, Hoof. See HORN MEAL.

Meal, Skin. See SKIN and LEATHER MEALS.

Mearlmaid. A proprietary trade name for a pearl essence.

Mearthane. A proprietary polyurethane elastomer. (673).

Measac. Ethanolamine sesquisulphite. (503).

Meat Flour. A powder made from the waste products in the manufacture of meat extracts. Used as a feeding-stuff for animals.

Meat Meal. See FLESH MANURE.

Meat Preserve Crystals. See HAWKE'S ANTI-FERMENT.

Meat Sugar. Inosite, $C_6H_6(OH)_6$.

Mebadin. A proprietary trade name for dehydroemetine.

Mebanazine. α-Methylbenzylhydrazine. Actomol.

Mebeverine. 7-(3,4-Dimethoxybenzoyloxy) - 3 - ethyl - 1 - (4 - methoxyphenyl) - 2-methyl-3-azaheptane. 4-[N-Ethyl-2-(4 - methoxyphenyl) - 1 - methylethyl - amino]-butyl 3,4-dimethoxybenzoate. Duphaspasmin is the hydrochloride.

Mebezonium Iodide. 4,4′-Methylenedi(cyclohexyltrimethylammoniumiodide). Present in T.61.

Mebhydrolin. 9-Benzyl-3-methyl-1,2,3,4-tetrahydro-γ-carboline. Fabahistin is the napadisylate.

Mebinol. A proprietary trade name for Clefamide.

Mebryl. A proprietary preparation of embramine hydrochloride. An antiallergic. (281).

Mebutamate. 2,2-Di(carbamoyloxymethyl)-3-methylpentane. Capla.

Mecadox. A proprietary preparation of CARBADOX. A veterinary anti-bacterial.

Mecamylamine. 3-Methylaminoisocamphane. Inversine and Mevasine are the hydrochlorides.

Mecca Balsam. An oleo-resin obtained from *Balsamodendron gileadense*, of Arabia. It is known in India as " Balsan-katel," and is imported there under the name of " Duhnul-balasan." Balm of Gilead is another name for Mecca balsam.

Mecholyl. Acetyl - β - methyl - choline chloride. A parasympathetic stimulant.

Mecholyl Chloride. A proprietary trade name for Methacholine Chloride.

Mecillinam. An antibiotic currently undergoing clinical trial as " FL 1060 ". It is (2S, 5R, 6R) - 6 - (perhydroazepin - 1 - ylmethyleneamino) penicillanic acid.

Meclofenamic Acid. An anti-inflammatory. N - (2, 6 - Dichloro - m - tolyl) - anthranilic acid.

Meclofenoxate. Dimethylaminoethyl-4-chlorophenoxyacetate. ANP 235 and Lucidril are the hydrochlorides.

Meclozine. 1-(4-Chloro-α-phenylbenzyl)-4-(3-methylbenzyl)-piperazine. Ancolan is the dihydrochloride.

Meco. A cupro-nickel alloy.

Mecobalamin. A preparation used in the treatment of vitamin B_{12} deficiency. It is α - (5, 6 - dimethylbenzimidazol - 1 - yl) cobamide methyl.

Meconium. Opium.

Mecpa. Selective weedkillers. (501).

Mecufix. An adhesive for polyester film. (507).

Meculon. Metallised polyester film. (507).

Mecysteine. Methyl Cysteine.

Medal Metal. An alloy of 84 per cent. copper and 16 per cent. zinc.

Medal Bronze. An alloy of from 92–97 per cent. copper, 1–8 per cent. tin, and 0–2 per cent. zinc.

Medang Losoh Oil. An oil from the wood of *Cinnamonum parthenoxylon*. It consists mainly of safrole.

Medarsed. See BUTABARBITAL. (674).

Medazepam. A tranquilliser. 7-Chloro-2, 3 - dihydro - 1 - methyl - 5 - phenyl - 1H-1,4-benzodiazepine. NOBRIUM.

Medigoxin. A preparation used in the treatment of cardiac insufficiency. It is 3β - [O - (2, 6 - dideoxy - 4 - O - methyl - D - *ribo* - hexopyranosyl) - (1→4) - O - (2, 6 - dideoxy - D - *ribo* - hexopyrano - syl) - (1→4) - (2, 6 - dideoxy - D - *ribo* - hexopyranosyloxy] - 12β, 14 - dihydro - xy - 5β, 14β - card - 20(22) - enolide. LANITOP.

Medihaler-Duo. A proprietary preparation of ISOPRENALINE and PHENYLEPHRINE as a metered aerosol in the treatment of asthma. (275).

Medihaler-EPI. A proprietary preparation of adrenaline acid tartrate in an aerosol form. A bronchial antispasmodic. (275).

Medihaler-Ergotamine. A proprietary preparation of ergotamine tartrate in an aerosol form, used for migraine. (275).

Medihaler-Iso. A proprietary preparation of isoprenaline sulphate in an aerosol form. A bronchial antispasmodic. (275).

Medina Cement. A variety of Roman cement, made from a stone found in the Isle of Wight.

Medinal. A proprietary preparation of barbitone sodium. A hypnotic. (265). See VERONAL SODIUM.

Medium Oils. See HEAVY OILS.

Medjidite. A mineral. It is a calcium-uranium sulphate.

Medocodene. A proprietary prepara-

tion of paracetamol, codeine phosphate, and phenolphthalein. (250).

Medodorm. A proprietary preparation of chlorhexadol. A hypnotic. (250).

Medol. A combination of cresols and iodine. Recommended as an anti-parasitic in skin troubles.

Medolit. A phenol-formaldehyde condensation product. It is a resinous material, and is recommended as a shellac varnish substitute.

Medomet. A proprietary preparation of METHYLDOPA. An anti-hypertensive. (498).

Medomin. A proprietary preparation of heptabarbitone. A hypnotic. (17).

Medro-Cordex. A proprietary preparation of methylprednisolone and aspirin. An antirheumatic agent. (325).

Medrol. A proprietary trade name for Methylprednisolone.

Medrone. A proprietary preparation of methylprednisolone. (325).

Medrone Medules. A proprietary preparation of methylprednisolone. (325).

Medrone Veriderm. A proprietary preparation of methylprednisolone acetate for dermatological use. (325).

Medroxyprogesterone. 17α-Hydroxy-6α - methylpregn - 4 - ene - 3,20 - dione. Provera is the acetate.

Medusa. A waterproofing compound mixed with cement.

Meehanite. A proprietary trade name for a close-grained, pearlitic, sorbitic iron with properties superior to cast iron. It has good casting and machining properties.

Meena Harma. A name given to an opaque variety of bdellium gum-resin.

Meerschaluminite. See SIMLAITE.

Meerschaum (Sepiolite). A mineral. It is a hydrated silicate of magnesium, $Mg_2Si_3O_8.2H_2O$.

Mefenamic Acid. N-2,3-Xylylanthranilic acid. Ponstan.

Mefruside. 4 - Chloro - N' - methyl - N' - (2 - methyltetrahydrofurfuryl) - benzene-1,3-disulphonamide. FBA 1500.

Megabasite. A mineral. It is a variety of hubnerite.

Megabromite. A mineral. It is a mixture of silver bromide and chloride.

Megaclor. A proprietary preparation containing Clomocycline. An antibiotic. (330).

Megadyne. A Belgian explosive consisting of a mixture of ammonium perchlorate, aluminium powder, and paraffin wax.

Meganite. An explosive containing nitro-glycerin, and dinitro-cellulose, to which has been added a nitro mixture to ensure complete combustion.

Megaperm 4510. A magnetic alloy

containing 45 per cent. nickel, 45 per cent. iron, and 10 per cent. manganese.

Megaperm 6510. A magnetic alloy containing 65 per cent. nickel, 25 per cent. iron, and 10 per cent. manganese.

Megapren C 150. A proprietary chloroprene rubber used in the manufacture of mouldings and extrusions resistant to sunlight, weathering and ozone. (421).

Megapren Si 10, 20, 30 and 60. A proprietary range of materials based on silicone rubber. (421).

Megapren U225. A proprietary polyurethane rubber. (421).

Megasil. Silicone dielectric materials. (614).

Megass. See BEGASSE.

Megestrol. 17α-Hydroxy-6-methylpregna-4,6-diene-3,20-dione. Present in Serial 28 and Volidan as the acetate.

Megilp. A mixture of linseed oil and mastic varnish. Used in artist's oil paints.

Megimide. A proprietary preparation of bemegride. (271).

Meglumine. N-Methylglucamine.

Megomit. A mica product used as an electrical insulator.

Megum 661. A proprietary solution of a number of halogenated polymers dispersed in chlorinated and aromatic hydrocarbons. (675).

Megum 3220, 3225 and 3230. Proprietary name for a range of adhesives comprising a solution of chlorinated polymers, used to bond chlorinated rubbers to metal. (675).

Megum 3287. A proprietary siliconebased bonding agent used to bond silicone rubber and VITON to metals. (675).

Mejonite. A mineral, $4CaO.3Al_2O_3.6SiO_2$.

Mekad. Antiskinning agent for paints. (518).

Mekong Yellow G. A dyestuff. It is the sodium salt of bi-phenyl-tetra-kisazo-dioxy-diphenyl-methane-bi-salicylic acid, $C_{51}H_{34}N_8O_8Na_2$. Dyes cotton greenish-yellow from a soap bath.

Mekong Yellow R. A dyestuff. It is the sodium salt of bi-tolyl-tetra-kisazo-dioxy-diphenyl-methane-bi-salicylic acid, $C_{54}H_{42}N_8O_8Na_2$. Dyes cotton yellow.

Mekoxim " MX ". A proprietary trade name for methyl ethyl ketoxime used as antiskinning agent in paints; printing inks. (190).

Melaconite (Black Copper). A mineral. It is copper oxide, CuO.

Melacos. Bottle washing detergents. (512).

Melafix DM. A proprietary trade name for a melamine formaldehyde resin.

Used as a shrink-resistant in wool. (18).

Melalith. A steatite-porcelain product.

Melamine, Normal. Normal cyanuramide, $C_3H_6N_6$.

Melamit 200. A proprietary melamine formaldehyde cellulose moulding powder. (6).

Melampyrite. Dulcite, $CH_2(OH)(CH.OH)_4.CH_2OH$.

Melan Asphalt. Albertite (q.v.).

Melanchroite (Melanochroite). Native lead chromate, $PbCrO_4$.

Melanex. A proprietary trade name for Metahexamide.

Melaniline. Diphenyl-guanidine, $NH:C(NH.C_6H_5)_2$.

Melanite. A lime alumina garnet, $6CaO.3SiO_2.2Al_2O_3.3SiO_2$.

Melanocerite. A mineral. It consists mainly of cerium and yttrium fluosilicates.

Melanochalcite. A mineral. It is a basic copper silico-carbonate.

Melanochroite. See MELANCHROITE.

Melanogen Black. See SULPHUR BLACK T.

Melanogen Blue B, BG, D. Dyestuff obtained by dissolving 1 : 5-dinitronaphthalene in concentrated sulphuric acid, treating with sulphuretted hydrogen or metallic sulphide, and precipitating as the zinc chloride compound. This substance is heated with sulphur and sodium sulphide to form the dyestuff. It dyes cotton greenish-blue, but subsequent treatment with copper sulphate gives blue-black, with chromium, nickel, or cobalt salts, blue-black, and with zinc, cadmium, or aluminium salts, bright blue shades.

Melanoid. A colloidal bituminous paint material. It is used as a preservative paint for metal which is in contact with corrosive gases.

Melanolite. A mineral. It is an iron silicate.

Melanosiderite. A mineral, $Fe_2O_3.H_2O.SiO_2$.

Melanostibian. A mineral. It is a manganese-iron antimonite.

Melanotekite. A mineral, $Pb_3Fe_4Si_3O_{15}$.

Melanothallite. A mineral, $CuCl_2.CuO.H_2O$.

Melanovanadate. A mineral. It is calcium-vanadium vanadate, $2CaO.2V_2O_4.3V_2O_5$, found in Peru.

Melanterite. Native ferrous sulphate, $FeSO_4.7H_2O$.

Melanthrene B. An anthracene dyestuff obtained by heating diaminoanthraquinone with caustic potash. Dyes cotton grey.

Melanthrene BH. A dyestuff. It is a British equivalent of Diamine black BH.

Melarsonyl Potassium. Dipotassium

2 - [4 - (4,6 - diamino - 1,3,5 - triazin - 2 - ylamino)phenyl] - 1,3,2 - dithiarsolan-4,5-dicarboxylate.   Trimelarsan.

Melarsoprol.   2-[4-(4,6-Diamino-1,3,5-triazin - 2 - ylamino)phenyl] - 4 - hy - droxymethyl-1,3,2-dithiarsolan. Mel B.

Mel B.   A proprietary trade name for Melarsoprol.

Melatix.   A proprietary melamine formaldehyde moulding compound. (676).

Melco.   A synthetic milk made from the peanut.

Melcret.   A proprietary melamine-formaldehyde condensation product containing sulphonic acid groups, used for improving concrete, cement, lime and gypsum. (899).

Melcril 4079.   A trade mark for tetrahydrofurfuryl acrylate. (677).

Melcril 4083.   A trade mark for an ethylene glycol acrylate phthalate. (677).

Melcril 4085.   A trade mark for a benzyl acrylate. (677).

Melcril 4087.   A trade mark for a phenoxy ethyl acrylate. (677).

Melcril 5919.   A trade mark for a melamine acrylate. (677).

Meldola's Blue (New Blue; Naphthylene Blue R, in Crystals; Fast Blue R, 2R, 3R, for Cotton, in Crystals; Cotton Blue R; Fast Navy Blue R, RM, MM; Naphthol Blue R, D; Cotton Fast Blue 2B; β-Naphthol Violet; New Fast Blue, for Cotton; New Fast Blue R Crystals, for Cotton; New Blue R; Fast Marine Blue; Phenylene Blue). A dyestuff.   It is the zinc double chloride of dimethyl-amino-naphtho-phenoxazonium-chloride, $C_{18}H_{15}N_2$OCl. Dyes cotton mordanted with tannin and tartar emetic indigo blue. Used as a substitute for indigo.

Meldrite.   A proprietary melamine formaldehyde moulding powder. (61).

Melengestrol.   17α-Hydroxy-6-methyl-16 - methylenepregna - 4,6 - diene - 3,20 - dione.

Melflock.   A proprietary aminoplast condensate used as a hardening and modifying agent in resins (especially laminating resins) and as a purifying agent for metal surfaces. (899).

Melflux.   A proprietary aminoplast condensate used as a surface-active agent having dispersive properties. (899).

Melgan.   Plasticisers for cement mortars (512).

Melibiase.   An enzyme which splits melibiose into d-glucose and d-galactose.

Meligrin.   A condensation product of dimethyl-oxyquinoline and methyl-phenyl-acetamide.

Melilite (Humboldtilite).   A mineral, $6(Ca.Mg.Na_2)O.(Al.Fe)_2O_3.S.SiO_2$.

Melilot.   p-Methyl-acetophenone, C_6H_4(CH$_3$).CO.CH$_3$.   It imparts the honey-

like fragrance of sweet clover, and is used for perfuming soap.

Melilot, Methyl.   See METHYL MELILOT.

Melinex.   A proprietary trade name for polyethylene terephthalate film. An extremely tough material used for cable lapping, motor insulation and capacitors, valve diaphragms and conveyor belting. (2) See also MYLAR (10).

Melinite.   The French name for Lyddite (q.v.).   It consists of 70 per cent. picric acid and 30 per cent. collodion cotton.

Melinite.   A yellow clay found in Bavaria.

Melinophane.   See MELIPHANITE.

Melinose.   A mineral.   It is wulfenite.

Melioform.   A ruby-red liquid containing 25 per cent. formaldehyde and 15 per cent. aluminium acetate. Used as a disinfectant for the hands, and for wounds.

Meliphanite (Melinophane).   A mineral, $2(Mg.Fe)O.SiO_2$, with some beryllium.

Melit.   Melamine resins.

Melitase.   A proprietary preparation of CHLORPROPAMIDE used in the treatment of diabetes. (137).

Melite.   A mineral.   It is an aluminium-iron silicate.

Melkbosch.   A fibre obtained from Asclepias fructicosa of South Africa.

Mellane.   Plasticisers for synthetic resins. (503).

Melleril.   A proprietary preparation of thioridazine. A sedative. (267).

Melling Powder.   An explosive containing from 51–55 per cent. ammonium nitrate, 11–13 per cent. sodium nitrate, 5–7 per cent. trinitro-toluene, 3–5 per cent. wood meal, 4–6 per cent. nitroglycerin, and 18–20 per cent. ammonium oxalate.

Mellite.   (1) (Artificial Amber, Honey Stone).   A mineral which consists chiefly of the aluminium salt of mellitic acid.   Mellitic acid is benzene-hexa-carboxylic acid, $C_6(COOH)_6$.

Mellite.   (2) A registered trade mark for organotin stabilisers, metallic soaps and complexes for p.v.c. (24).

Mellonite.   A mineral.   It is a pseudo-cotunnite mixed with chlorides and sulphates of sodium, potassium, copper, and lead.

Mellotte's Alloy.   A fusible alloy used in dental work.   It contains 50 per cent. bismuth, 31·25 per cent. tin, and 18·75 per cent. lead.

Melmac.   A proprietary trade name for a melamine-formaldehyde synthetic resin and adhesive.

Melment.   Proprietary name for a range of melamine-formaldehyde condensation products containing sulfonic acid groups, used for improving the quality of concrete, cement, lime, gypsum and other building materials. (899).

Melmex. Melamine moulding powder for tableware. (570).

Melnetz. A proprietary aminoplast condensate used as a hardening and modifying agent in resins (especially laminating resins) and in additives for paper, textiles, leather, glass mat, woods and synthetics. (899).

Melocol. A proprietary trade name for a polyamide-formaldehyde product.

Melogene Blue BH (Diamine Beta Black). A dyestuff. It is the sodium salt of diphenyl-disazo-p-xylene-azo-bi-amino-naphthol-disulphonic acid, $C_{40}H_{28}N_8O_{14}S_4Na_4$. Dyes cotton black-blue.

Melon Essence. See ESSENCE OF MELON.

Melonite. A mineral, NiTe.

Melon Oil. The name given to the mass of fat from the head of the dolphin. It has the shape of a half water-melon. It weighs about 25 lb. and yields about 6 quarts of oil.

Melpers. A proprietary name for a range of modifying agents used in laminate aminoplast resins, especially elasticising agents, accelerators and release compounds. (899).

Melphalan. 4 - Di - (2 - chloroethyl) - amino - L - phenylalanine. Alkeran.

Melsed. A proprietary preparation of methaqualone. A hypnotic. (253).

Melsedin. A proprietary preparation of methaqualone hydrochloride. A hypnotic. (253).

Melsir. Melamine moulding powder. (678).

Meltrol. A proprietary preparation of PHENFORMIN hydrochloride used in the treatment of diabetes. (137).

Melubrin (Sulphantipyrin). Sodium-1-phenyl - 2 : 3 - dimethyl - pyrazolone - 4 amino-methane-sulphonate, $C_{11}H_{11}NO_2NH.CH_2.SO_3Na$. Used in medicine as an antipyretic and analgesic.

Melurac. A proprietary trade name for a melamine-urea-formaldehyde laminating synthetic resin.

Membranit. A proprietary trade name for an emulsion paint binder composed of alkyd resin and casein. (3).

Memilite. See RANDANITE.

Menaccanite. A variety of titaniferous magnetic oxide of iron, found in Cornwall. Synonyms: Menacconite, Menakanite, Manaccanite, Manakeisenstein.

Menadione. A synthetic analogue of Vitamin K. 2-methyl-1,4-naphthoquinone.

Menadoxime. Ammonium salt of 2-methylnaphthaquinone-4-oxime O-carboxymethyl ether. Kapilon Soluble.

Mendeleeffite. A radio-active mineral.

It is a urano-titano-columbate of calcium, containing 23·5 per cent. U_3O_8.

Mendipite (Kerasin). A mineral. It is an oxychloride of lead, $PbCl_2.2PbO$, found in Somerset.

Mendozite. Sodium alum, $Na_2SO_4.Al_2O_3(SO_4)_3.24H_2O$, found in South America.

Meneghinite. A mineral. It is $Pb_{13}Sb_7S_{23}$. Also stated to be $4PbS.Sb_2S_3$.

Menhaden Oil. An oil prepared from the heads and intestines of fish, especially of the menhaden. Used in the leather, paint, and rope trades.

Menilite (Liver-opal). A mineral. It is a brown opaque variety of opal.

Menispermin. An extract of Canadian moon-seed.

Mennige. A mineral, Pb_3O_4.

Menoformon. A proprietary preparation of œstrone in a cream base. Treatment of senile vaginal changes. (316).

Menolet Sublets. A proprietary preparation of ethynylœstradiol and methyltestosterone. (357).

Menopax. A proprietary preparation of ethinylœstradiol, carbomal and bromvaletone used in the treatment of menopausal symptoms. (271).

Menopax Forte. A proprietary preparation of ethinyloestradiol, methyltestosterone, CARBROMAL, BROMVALETIN and MEPHENESIN used in the treatment of menopausal disorders. (271).

Menosal. Menthyl - salicylic - methyl ester.

Mensalin. Diphenyl-dioxy-carbonate of dimethyl - pyrazolon - hexahydrocymolvalerate. A nervine.

Menstrogen. A proprietary preparation of ethinylœstradiol and ethisterone. Treatment of secondary amenorrhœa. (316).

Menthiodol. A substance prepared by heating 4 parts of menthol, and adding 1 part of iodol. Used as a local application for neuralgia.

Menthival. See VALIDOL.

Menthofax. A proprietary preparation of compound methyl salicylate ointment containing methyl salicylate, menthol, eucalyptol, oil of cajuput, beeswax, and wool fat.

Mentholeate. Menthol (200 grains) in 4 fluid drachms oleic acid. Used as an outward application for neuralgia.

Menthol-Y. A synthetic menthol. It is isomeric with menthol, but differs from it in that it melts at 22° C., and is optically inactive.

Menthophenol. A mixture of 1 part phenol and 3 parts menthol.

Menthorol. A mixture of p-chlorophenol and menthol. Used in 5–15 per cent. solution in the treatment of ulceration of the throat.

Menthospirin. Menthol-acetyl-salicylate. Used in the treatment of colds and catarrh.

Mentopin. A German preparation. It is a solution of menthol, thymol, and terpichin. Used as an intramuscular injection against influenza and bronchitis.

Mentrinol. A proprietary preparation of norethinodrel and mestranol. Treatment of menstrual disorders. (806).

Menyanthin. See HELENIN.

Me Oil. Illipé butter, from the seeds of *Illipé malabrorum*, is known by this name in Ceylon.

Mep 4. A proprietary preparation of meprobamate. A sedative. (362).

Mepacrine. A British quinine substitute. See also ATEBRIN (German), ATABRINE (U.S.A.). It is stated to be 8-(8 - diethylamino - α - methyl - butylamino)-6-methoxy-acridine.

Mepavlon. A proprietary preparation of meprobamate. A sedative. (2).

Mepedyl. A proprietary trade name for Piprinhydrinate.

Mepenzolate Bromide. 1-Methyl-3-piperidyl benzilate methobromide. Cantil.

Mephaneine. Specially pure methyl phenylacetate. (504).

Mephenesin. 3-(2-Methylphenoxy)propane-1,2-diol. 3-o-Tolyloxypropane-1, 2-diol. Lissephen; Myanesin; Tolseram is the carbamate.

Mephenesin Quinineco. A proprietary preparation of mephenesin, quinine sulphate, ascorbic acid and acetomenaphthone. (398).

Mephentermine. Nαα - Trimethylphenethylamine. Mephine is the sulphate.

Mephenytoin. Methoin.

Mephine. A proprietary preparation of mephentermine sulphate. Used to raise blood pressure. (245).

Mephitic Air. Nitrogen, so-called by Rutherdorf and Priestley.

Mephosol. A proprietary preparation of mephenesin and salicylamide. An antirheumatic. (280).

Mephyton. A proprietary trade name for Phytomenadione.

Mepilin. A proprietary preparation of ethinylœstradiol and methyltestosterone. Treatment of senile and menopausal conditions. (182).

Mepiprazole. A psychotrophic agent. 1 - (3 - Chlorophenyl) - 4 - [2 - (5 - methylpyrazol - 3 - yl)ethyl]piperazine. " H 4007 " is currently under clinical trial as the dihydrochloride.

Mepivacaine. A local anæsthetic. 1 - Methyl - 2 - (2, 6 - xylylcarbamoyl) - piperidine. CARBOCAINE.

Mepol. A proprietary melamine formaldehyde moulding compound. (679).

Meprate. A proprietary preparation of meprobamate. A sedative. (363).

Meprobamate. 2,2-Di(carbamoyloxymethyl)pentane. EQUANIL; KALMM; MEPAVLON; MILTOWN.

Meprochol. N-(2-Methoxyprop-2-enyl)-trimethylammonium bromide. Esmodil.

Meprothixol. An anti-inflammatory and analgesic. 9-(3-Dimethylaminopropyl)-9 - hydroxy - 2 - methoxythiaxanthen.

Meptazinol. An analgesic. 3-Ethyl-3-(3-hydroxy phenyl) - 1 methylhexahydroazepine. " Wy 22811 " is currently under clinical trial as the hydrochloride.

Mepyramine. N-4-Methoxybenzyl-N'N' - dimethyl - N - 2 - pyridyl - ethyl - enediamine. Anthisan and Neo-antergan are the hydrogen maleate.

Meral. An aluminium alloy containing 3 per cent. copper, 1 per cent. nickel, 0·8 per cent. magnesium, and 0·3 per cent. manganese.

Meralluride. Mixture of N-(3-hydroxymercuri-2-methoxypropyl-carbamoyl) succinamic acid and theophylline. Mercardan; Mercuhydrin.

Meraneine. Specially pure geranyl acetate. (504).

Meratran. A proprietary trade name for the hydrochloride of Pipadol.

Merbaphen. Novasurol (q.v.).

Merbentyl. A proprietary preparation containing dicyclomine hydrochloride. An antispasmodic. (263).

Mercallite. A mineral. It is $KHSO_4$.

Mercamin. See SUBLAMINE.

Mercaptan. Ethyl sulphydrate, C_2H_5HS.

Mercaptomerin Sodium. Disodium salt of N-(3-carboxymethylthiomercuri-2-methoxypropyl)camphoramic acid. Thiomerin Sodium.

Mercaptopurine. 6-Mercaptopurine. Puri-nethol.

Mercarbolide. A preparation containing 0·1 per cent. o-hydroxyphenyl-mercuric chloride. A germicide.

Mercardan. A proprietary trade name for Meralluride.

Mercarsan. A combination of arsenic and mercury, $C_{17}H_{17}O_3As_2Hg$. Used for the treatment of syphilis.

Mercazole. A proprietary trade name for Methimazole.

Mercedan. Sodium-mercury p-nucleinate.

Mercerised Cotton. Cotton which has been immersed in a solution of sodium hydrate (30–35° Bé.). It has a lustrous appearance.

Mercer's Liquor. A solution containing potassium ferricyanide. Used for etching.

Merchloran. Chlormerodrin.

Merclor D. A proprietary trade name for a solution of sodium hypochlorite, NaOCl. A bleaching agent.

Mercoid. A suspension of mercury-salicyl-sulphonate and calomel. Used medicinally.

Mercolloid. A colloidal mercury sulphide.

Mercoloy. A proprietary trade name for a nickel bronze containing 60 per cent. copper, 25 per cent. nickel, 10 per cent. zinc, 1 per cent. tin, 2 per cent. lead, and 2 per cent. iron.

Mercotan. See MERGAL.

Mercresin. A proprietary trade name for a solution of 0·1 per cent. *o*-hydroxy-phenyl-mercuric chloride and 0·1 per cent. sec-amyl-tricresol in 50 per cent. alcohol, 10 per cent. acetone, and water. A germicide.

Mercuhydrin. A proprietary trade name for Meralluride.

Mercuramine. Mercury-ethylene-dia-mine-citrate. It has bactericidal properties.

Mercurammonium Chloride (Fusible White Precipitate, Ammonio-mercuric-chloride). Mercuri-diammonium chloride, $N_2H_6Hg.Cl_2$.

Mercuranine. See MERCUROCHROME-220 SOLUBLE.

Mercurgan. Sodium-hydroxy-mercuric-methoxy - propyl - carbamyl - phenoxy-acetate.

Mercurial Cream. See GREY OIL.

Mercurialine. The name given by Reichardt to methylamine found in *Mercurialis annua* and *M. perensis*.

Mercurial Lotion, Yellow. See YELLOW WASH.

Mercurial Oil. A substance containing 90 per cent. mercury in the form of an amalgam, the vehicle being lanoline and almond oil.

Mercurial Soap. A germicidal soap prepared by adding an alcoholic solution of mercuric chloride to soap.

Mercurichrome. Dibromo - oxy - mer - curic-fluoresceine. A bactericide.

Mercuricide. Lithio-mercuric-iodide, $3LiI + HgI_2$. A germicide.

Mercuric Iodide, Red. See RED IODIDE OF MERCURY.

Mercuric Potassium Iodide. See MAYER'S SOLUTION.

Mercuriocoleols. A double stearate of cholesterol and mercury.

Mercuriol. A mercury amalgam with aluminium.

Mercurit. A German explosive containing potassium chlorate and mineral oil.

Mercuritol. See AFRIDOL.

Mercurochrome-220 Soluble (Mercurochrome, Mercuranine, Mercurome, Planochrome). The disodium salt of di-bromo - hydroxy - mercury - fluores-

ceine. It is used as an antiseptic for the genito-urinary tract. A skin disinfectant is prepared by dissolving 2 grams of mercurochrome in 35 c.c. water and adding 55 c.c. of 95 per cent. alcohol and 10 c.c. acetone.

Mercurol. A compound of nuclein and mercury. Used internally and locally as a substitute for corrosive sublimate.

Mercurome. See MERCUROCHROME-220 SOLUBLE.

Mercurophen. Sodium-oxy-mercuric-*o*-nitro-phenoxide, $C_6H_3.HgOH.ONa.NO_2$. A bactericide.

Mercurophylline Sodium. Mixture of the sodium salt of N-(3-hydroxymer-curi - 2 - methoxypropyl)camphoramic acid and theophylline. Mercuzanthin.

Mercurosal. The disodium salt of hydroxy - mercuric - *o* - carboxy - phen-oxy-acetic acid, $C_6H_3(OCH_2COONa)(COONa)HgOH$. Used in the treatment of syphilis.

Mercury, Alkalised. See GREY POWDER.

Mercury, Ammoniated. See WHITE PRECIPITATE.

Mercury Blende. See CINNABAR.

Mercury, Cosmetic. See WHITE PRECIPITATE.

Mercury Iodide, Green. See GREEN IODIDE OF MERCURY.

Mercury Ointment (Blue Ointment, Grey Ointment). *Unguentum hydrargyri B.P.* An emulsion of 50 per cent. finely divided mercury with 10 per cent. lard and 35 per cent. suet.

Mercury Oxide, Red. See RED OXIDE OF MERCURY.

Mercury Pill (Blue Pill). *Pilula hydrargyri B.P.*

Mercury Resorbin. Consists of 1 part of mercury and 2 parts of resorbin.

Mercutin. A proprietary preparation of mercury with talc.

Mercuzanthin. A proprietary trade name for Mercurophylline Sodium.

Mergal (Mercotan). A combination of mercury cholate, $(C_{24}H_{39}O_5)_2Hg$, and albumin tannate. Has been used in the treatment of syphilis.

Mergol. Mercury cholate, $(C_{24}H_{39}O_5)_2$ Hg. Prescribed in capsules with tannin and egg albumin for syphilis.

Mérimée's Yellow. See ANTIMONY YELLOW.

Merinax. A proprietary trade name for Hexapropymate.

Meriodine. See MERJODIN.

Meritol. A photographic developer. (515).

Merjodin (Meriodine). Mercury-di-iodo-phenol-*p*-sulphonate (mercury so-zoidol), $(HO.C_6H_2I_2.SO_3)_2Hg$. Prescribed for syphilis.

Merlon. A registered trade name for polycarbonates.

Merlon M-39. A proprietary general-purpose polycarbonate moulding resin, transparent and with good flow properties. (680).

Merlon M-40 and M-50. Proprietary polycarbonate resins used for moulding and extrusion. (680).

Merlon M-60. A proprietary polycarbonate resin used in profile extrusion and blow-moulding. (680).

Merlon SE-2100. A proprietary thermoplastic polycarbonate resin prepared in the form of pellets for use in injection moulding and extrusion. (680).

Merlon SF-600. Proprietary trade name for an expandable polycarbonate. (680).

Merlon T-70. A proprietary general-purpose polycarbonate resin used in injection moulding and extrusion. (680).

Merlusan. A mercury compound of tyrosine, $C_9H_9O_3.N.Hg$.

Merocets. A proprietary preparation of cetylpyridinium chloride. Throat lozenges. (263).

Merochinol. Mercury-hydroxy-quinoline-sulphonate, $(HO.C_9H_5N.SO_3)_2Hg$.

Meroxene. A mineral, $(KH)_2O.(Fe.Al)_2O_3.2(Fe.Mg)O.3SiO_2$.

Merpentine. A proprietary trade name for a sodium alkyl naphthalene sulphonate product used as a wetting agent.

Merpol. A proprietary trade name for a wetting agent.

Merpol DSR. A non-ionic softener for use in conjunction with acrylic type soil release agents in durable press finishes on textiles. (10).

Merpol SH. A nonionic long chain alcohol-ethylene oxide condensate. A highly efficient biodegradable surfactant used in the preparation of textile fabrics for dyeing and finishing. (10).

Merquinox. Fungicide for industrial applications. (518).

Merrillite. A mineral. It is a calcium phosphate.

Mersalyl. Sodium hydroxy-mercuric-methoxy - propyl - carbamyl - phenoxy acetate.

Mersalyl BDH. Injection and tablets. A proprietary diuretic (mercurial). (182).

Mersil. Turf fungicide. (507).

Mersol. A proprietary trade name for a solvent.

Mertec. A proprietary trade name for a chlorinated rubber-base paint.

Merthiolate. A proprietary preparation of thiomersal. A skin antiseptic. Sodium ethyl-mercurithio-salicylate. (333).

Merwinite. A mineral. It is a calcium-magnesium silicate, $Ca_3Mg(SiO_4)_2$, found in California.

Mesabite. A mineral. It is a variety of gothite.

Mesamoll-Verdingrin. A proprietary trade name for a paraffin sulphonated ester. A vinyl plasticiser. (112).

Mescal (Tequila). Distilled liquors produced from a species of *agave* in Mexico.

Mescal Buttons. The seeds of *Anhalonium lewinii*.

Mesgamma. Combined insecticide fungicide and seed dressing. (511).

Mesicerin. Tri-ω-hydroxy-mesitylene, $C_6H_3(CH_2OH)_3$.

Mesidine. 2, 4, 6-trimethylaniline. A dyestuff intermediate.

Mesitine Spar. A mineral, $Mg.Fe(CO_3)$.

Mesole. A mineral. It is a variety of thomsonite.

Mesoline. A mineral. It is $(Na_2Ca)Al_2Si_4O_{12}.5H_2O$.

Mesolite. A mineral, $Na_2O.Al_2O_3.3SiO_2 . 2H_2O . Ca(OH)_2 . Al_2O_3 . 3SiO_2 . 2H_2O$.

Mesolitine. A mineral. It is thomsonite.

Mesonex. A proprietary preparation of inositol nicotinate. A peripheral vasodilator. (279).

Mesontoin. A proprietary preparation of methoin. An anticonvulsant. (267).

Mesoridazine. A tranquilliser. 10-[2-(1 - Methyl - 2 - piperidyl)ethyl] - 2 - (methylsulfinyl)phenothiazine. LIDANIL, SERENTIL.

Mesotan (Ericin, Ulmarine, Methosal). The methyl-oxy-methyl ester of salicylic acid, $C_6H_4(OH).COOCH_2.O.CH_3$. Mixed with olive oil, it is used as an ointment in rheumatic diseases.

Mesotanol. A proprietary preparation. It is a mixture of mesotan and olive oil.

Mesotartaric Acid. Inactive tartaric acid, $CO_2H.CH(OH).CH(OH).CO_2H$.

Mesothorium. The name applied to the first product of the disintegration of thorium. It has similar therapeutic properties to radium.

Mesotype. See NATROLITE.

Mesquite Gum. A gum found on the mesquite tree, *Prosopis Juliflora*, of Mexico. It exudes from the stem and branches of the tree. When hydrolysed the gum yields 1-arabinose.

Mesquitelite. A mineral. It is $(Mg,Ca)Al_4Si_9O_{25}.5H_2O$.

Messelite. A mineral, $(Ca.Fe.Mg)_3(PO_4)_2.2\frac{1}{2}H_2O$.

Messingite. A mineral. It is aurichalcite.

Mestanolone. 17β - Hydroxy - 17α -

methyl-5α-androstan-3-one. Androstalone.

Mestarine. A proprietary preparation. It is a derivative of theobromine, and is a water-soluble diuretic.

Mesterolone. An androgen. 17β-Hydroxy - 1α - methyl - 5α - androstan - 3 - one. PRO-VIRON.

Mestinon. A proprietary preparation containing Pyridostigmine Bromide. (314).

Mestranol. 17α-Ethynyl-3-methoxyoestra-1,3,5(10)-trien-17β-ol.

Mesulphen. 2,6-Dimethylthianthren. Mitigal ; Sudermo.

Mesurol. A German preparation. It is a bismuth salt of dihydroxy-benzoic methyl ether, and is used in the treatment of syphilis by intramuscular injection as a 20 per cent. emulsion with olive oil.

Mesuximide. Methsuximide.

Meta (Meta Fuel). Pure metaldehyde produced by the polymerisation of acetaldehyde. A solid fuel of Swiss manufacture used in the place of methylated spirit.

Metabisulphite. Sodium metasulphite, $Na_2S_2O_5$.

Metabrushite. A mineral, $HCaPO_4$. H_2O.

Metacetamol. An analgesic currently undergoing clinical trial as " BS 749 ". It is 3-acetamidophenol.

Metacetone. The name given by Frémy to an oil which occurs among the products of the distillation of sugar, starch, or gum, with quicklime. It consists of propionic aldehyde, dimethyl-furfurane, and hydrocarbons.

Metachalcophyllite. A mineral. It is $Cu_{18}Al_2(AsO_4)_3(SO_4)_3(OH)_{27}$.

Metachrome. Chrome dyestuffs. (541).

Metachrome Brown B (Alizarin Brown M, Metachrome Brown G). A dyestuff prepared from picramic acid and m-tolylene-diamine.

Metachrome Brown G. A dyestuff. It is a British equivalent of Metachrome brown B.

Meta Chrome Mordant. The name given to a mixture of potassium, bichromate and ammonium acetate. Used as a mordant.

Metacinnabarite. A mineral having the same composition as cinnabar, but black in colour.

Metacon. Rust remover. (526).

Metacortandracin. A proprietary trade name for Prednisone.

Metacortandralone. A proprietary trade name for Prednisolone.

Metacrylene. A proprietary trade name for a styrene-methyl methacrylate copolymer.

Metacrylene BS. A proprietary trade name for an MBS terpolymer.

Metaethyl. A mixture of methyl and ethyl chlorides. An anæsthetic.

Metaferrin. An iron albumin compound containing 10 per cent. iron. Used in the treatment of anæmia.

Metaform. A plastic packing material for packing stuffing-boxes, and consisting of powdered white metal, graphite, cylinder oil, and asbestos fibre.

Meta Fuel. See META.

Metahewettite. A mineral. It is a diamorphous form of hewettite.

Metahexamide. N-(3-Amino-4-methyl-benzenesulphonyl) - N' - cyclohexylurea. Euglycin ; Melanex.

Metahohmannite. A mineral. It is $Fe_4\cdots(SO_4)_5(OH)_2.17H_2O$.

Metahydroboracite. Synonym for Inderborite.

Metaiodin. A German product. It is an alcoholic solution of bromine salts with thiocyanates and iron preparations. It is an antiseptic used as a substitute for tincture of iodine.

Metakaline. A solid disinfectant. It is a cresol soap preparation containing 80 per cent. of a crystallised double compound of potassium m-cresolate, and 3 molecules of m-cresol, with 20 per cent. of soap.

Metal Argentum. An alloy consisting of 85½ per cent. tin and 14½ per cent. antimony.

Metal-furnace Slag. A slag formed and used in the preparation of copper metal. It is essentially a silicate of iron, and contains about 4 per cent. copper.

Metalife 2ZS. A proprietary zinc-powder-loaded ethyl zinc silicate used to give protection against corrosion in the offshore industries. (681).

Metalite (Durundum, Idilite, Exelite, Orelite). Proprietary trade names for aluminium oxide abrasives.

Metallic Carbon. See RETORT CARBON.

Metallichrome. A finish for motor bodies. (512).

Metalline. An alloy consisting of 35 per cent. cobalt, 30 per cent. copper, 10 per cent. iron, and 25 per cent. aluminium. Used in jeweller's work.

Metallised Rubber. A rubber having a film of metal on its surface. It is made by combining the sulphur of the rubber with a metal such as zinc, nickel, tin, silver, or gold, and then reducing the sulphide produced.

Metal Polishes. These preparations generally contain abrasives and a detergent. Most of them contain soap, oil, whiting, chalk, or fine silica, and a colouring agent.

Metal Soaps. Salts of the heavy metals with fatty acids.

Metalyn. A proprietary trade name for the distilled methyl ester of tall oil. A vinyl plasticiser. (93).

Metamine Blue B and G. See NEW BLUE B or G.

Metamitron. A pesticide. 4-amino-4, 5 - dihydro - 3 - methyl - 6 - phenyl - 1, 2, 4 - triazin - 5 - one.

Metamfepramone. International recommended Non-registered Name (I.N.N.) for DIMEPROPION.

Metamsustac. A proprietary preparation of methylamphetamine hydrochloride, in a slow release tablet. An antiobesity agent. (330).

Metamucil. A proprietary preparation containing psyllium mucilloid, dextrose, benzyl benzoate, sodium bicarbonate, potassium biphosphate and citric acid. A laxative. (285).

Metaniline Grey. See NEW GREY.

Metanil Orange I, II. Dyestuffs prepared by the action of diazotised m-amino-benzene-sulphonic acid upon α- and β-naphthol respectively. They are acid dyes.

Metanil Yellow (Orange MN, Tropæoline G, Yellow GA, Victoria Yellow, Extra Conc., Orange MNO, Metanil Yellow Y). A dyestuff. It is the sodium salt of m-sulpho-benzene-azodiphenylamine, $C_{18}H_{14}N_3O_3SNa$. Dyes wool orange-yellow from an acid bath. Used in paper-staining, carpet printing, and for colouring varnishes. It is poisonous.

Metanil Yellow S (Acid Yellow 2G). The sulphonated product of Metanil Yellow ($q.v.$).

Metanil Yellow Y. A dyestuff. It is a British equivalent of Metanil yellow.

Metanium. A proprietary preparation of titanium dioxide, peroxide, salicylate and tannate. Skin cream. (802).

Metaphen. (1) A trade mark for the dimercury salt of 4-nitro-2-cresol. C_6H_2. $CH_3.ONa(HgOOCCH_3)_2$, a bactericide.

Metaphen. (2) A proprietary preparation of nitromersal solution, a sterilising fluid. (311).

Metaphenylene Blue B. A dyestuff. It is dimethyl-tolyl-diamino-tolyl-phenazonium chloride, $C_{28}H_{27}N_4Cl$. Dyes cotton mordanted with tannin indigo blue.

Metaphlorone. See PHLORONE.

Metaquest. Sequestering agents. (618).

Metaraminol. (−) - 2 - Amino - 1 - (3 - hydroxyphenyl)propan-1-ol. Aramine is the hydrogen (+)-tartrate.

Metargon. A name given by W. Ramsey and M. W. Travers to a solid obtained when argon is liquefied.

Metarossite. A Colorado mineral, $CaO.V_2O_5.2H_2O$.

Metarseno-Argenticum. A proprietary preparation. It is a sodium salt of silver-4 : 4′-dihydroxy-3 : 3′-diamino-arseno - benzene - N - diformaldehyde bisulphite. It is prepared for use by intramuscular injection in the treatment of disseminated sclerosis and syphilis.

Metarsenobillon (Sulphostab). Proprietary names for the disodium salt of dioxy - diamino - arseno - benzene - diformaldehyde bisulphite.

Metasideronatrite. A mineral. It is $Na_2Fe'''(SO_4)_2OH. 1\frac{1}{2}H_2O$.

Metasol. (1) m-Cresol-anytol. (Anytols are solutions of phenol in ichthyol.)

Metasol. (2) Spirit soluble colours. (512).

Metaspirine. A proprietary preparation. It consists of acetyl-salicylic acid and caffeine.

Metastab. A proprietary trade name for Methylprednisolone.

Metastibnite. A mineral, Sb_2S_3.

Metastrengite. Synonym for Phosphosiderite.

Metastyrene. See POLYSTYRENE.

Metastyrol. See POLYSTYRENE.

Metatone. A proprietary preparation containing thiamine hydrochloride, and calcium, potassium, sodium, manganese, and strychnine glycerophosphates. A tonic. (264).

Meta-Torberite. A mineral from Utah, containing 36 per cent. UO_3, 11 per cent. P_2O_5, 36 per cent. SiO_2, 6 per cent. CuO, water, CaO, and arsenic.

Metatyuyamunite. A mineral. It is $Ca(UO_2)_2(VO_4)_25-7H_2O$.

Metavariscite. A mineral. It is $4[AlPO_4.2H_2O]$.

Metavauxite. A mineral. It is $2[Fe''Al_2(PO_4)_2(OH)_2.8H_2O]$.

Metaxalone. 5-(3,5-Xylyloxymethyl)-oxazolidin-2-one. Skeladin.

Metaxite. A Serpentine mineral with a fibrous structure.

Metazeunerite. A mineral. It is $[Cu(UO_2)_2(AsO_4)_2.8H_2O]$.

Metazocine. 1,2,3,4,5,6-Hexahydro-8-hydroxy - 3,6,11 - trimethyl - 2,6 - methano - 3 - benzazocine.

Metco 450. A nickel/aluminium composite material for building up metal surfaces by spraying. Similar materials are Metco 451 (nickel-chromium), Metco 44 (nickel base/chromium) and Metco 51 (aluminium bronze).

Meteorite. An alloy of aluminium with from 1–2 per cent. zinc and 1–4 per cent. phosphorus.

Meteor Oil. A fraction of Russian petroleum distillation, having a specific gravity of 0·806–0·810. A burning oil.

Meteor Silk. See GLANZSTOFF.

Meteor Yarn. See GLANZSTOFF.

Meterdos-Iso. A proprietary preparation of isoprenaline sulphate in an aerosol. Treatment of asthma. (839).

Metformin. N′N′-Dimethyldiguanide. Glucophage is the hydrochloride.

Methacetin. Methoxy - acetamino - phenol, $C_6H_4(OCH_3)NH.COCH_3$. An antipyretic.

Methacholine Chloride. Acetyl-β-methylcholine chloride. Amechol; Mecholyl Chloride.

Methacrifos. A pesticide. o-2-methoxy-carbonylprop - 1 - enyl oo - dimethyl phosphorothioate. Methyl 3 - dimeth - oxy = phosphinothioyloxy - 2 - methyl-acrylate.

Methacycline. An antibiotic. 4-Di-methylamino - 1, 4, 4a, 5, 5a, 6, 11, 12a - octahydro - 3, 5, 10, 12, 12a-penta - hydroxy - 6 - methylene - 1, 11 - di - oxonaphthacene - 2 - carboxamide. RONDOMYCIN is the hydrochloride.

Methadol. A proprietary trade name for Dimepheptanol.

Methadone. 6-Dimethylamino-4,4-diphenylheptan - 3 - one. Amidone. Adanon; Physeptone is the hydrochloride.

Methadyl Acetate. O-Acetate of 6-dimethylamino - 4,4 - diphenylheptan - 3-ol. Acetylmethadol.

Methaform. Chloretone (*q.v.*).

Methallenoestril. 3-(6-Methoxy-2-naphthyl)-2,2-dimethylpentanoic acid. Vallestril.

Methallibure. N-Methylthiocarba-moyl - N′ - (1 - methylallylthio - car - bamoyl)hydrazine.

Methalone. A proprietary trade name for Drostanolone.

Methamphazone. 4-Amino-6-methyl-2-phenylpyridaz-3-one.

Methampyrone. A proprietary trade name for Dipyrone.

Methanal. See METHYLALDEHYDE.

Methandienone. 17β-Hydroxy-17α-methylandrosta-1,4-dien-3-one. Diana-bol.

Methanthelinium Bromide. 2-Diethyl-aminoethyl xanthen - 9 - carboxylate metho-bromide. Banthine Bromide.

Methanol (Carbinol, Wood Spirit, Methyl Hydrate, Wood Alcohol, Wood Naphtha, Pyroxylic Spirit, Pyroligneous Spirit). Methyl alcohol, CH_3OH. The term has been officially adopted by the U.S. Public Health Service as representing wood alcohol.

Methaphenilene N′N′-Dimethyl-N-phenyl - N - 2 - thenylethylenediamine. Diatrin is the hydrochloride.

Methapyrilene. NN-Dimethyl-N′-(2-pyridyl) - N′ - (2 - thenyl)ethylene - diamine. Present in Vortel as the hydrochloride.

Methaqualone. 2 - Methyl - 3 - o - tolyl-quinazolin-4-one. 2-Methyl-3-o-tolyl-4-quinazolone. Melsedin is the hydrochloride.

Metharbital. A proprietary preparation of methabitone. An anticonvulsant. (311).

Metharbitone. 5,5-Diethyl-1-methyl-barbituric acid.

Methazolamide. 5-Acetylimino-4-methyl - 1,3,4 - thiadiazoline - 2 - sulphonamide. Neptazane.

Metharsan. An isotonic solution of disodium-methyl-arsinate.

Methasan. Proprietary ultra rubber vulcanising accelerator. It is the zinc salt of dimethyl dithiocarbamic acid. (57).

Methazate. A proprietary preparation of zinc dimethyl dithiocarbamate. (435).

Methdilazine. 10-(1-Methylpyrrolidin-3-ylmethyl)phenothiazine. Dilosyn is the hydrochloride.

Methedrine. A proprietary preparation of methylamphetamine hydrochloride. A stimulant. (277).

Methenolone. 17β-Hydroxy-1-methyl-5α-androst-1-en 3-one. Primobolan is the acetate; Primobolan-Depot is the enanthate.

Metheph. A proprietary preparation of methylephedrine hydrochloride. A bronchial antispasmodic. (339).

Methergin. A proprietary preparation of methylergometrine maleate, used in obstetrics to stimulate uterine contraction. (267).

Methethyl. A mixture of ethyl and methyl chlorides. A local anaesthetic.

Methetoin. 5-Ethyl-1-methyl-5-phenyl-hydantoin.

Methicillin. 6 - (2,6 - Dimethoxybenz-amido)penicillanic acid. Celbenin and Staphcillin are the sodium salts.

Methimazole. 2-Mercapto-1-methylimi-dazole. Thiamazole. Mercazole; Tapazole.

Methindizate. A spasmolytic. 2-(1-Methyloctahydroindol - 3 - yl)ethyl benzilate. It is present in ISAVERINE as the hydrochloride.

Methiodal Sodium. A radio-opaque substance. Sodium iodomethanesulphonate.

Methisazone. 1-Methylindoline-2,3-dione 3-thiosemicarbazone. Marboran.

Methisul. A proprietary preparation of SULPHAMETHISOLE. An antibiotic. (368).

Methixene. 9-(1-Methylpiperid-3-ylmethyl)thiaxanthen. Termonil is the hydrochloride.

Methocarbamol. 2-Hydroxy-3-(2-methoxyphenoxy)propyl carbamate. Robaxin.

Methocel. A proprietary trade name for methyl cellulose.

Methocidin. A proprietary preparation of hydroxymethylgramicidin ephedrine and cetylpyridinium chloride. Throat lozenges. (323).

Methofas P. A trade mark. Methyl Hydroxypropyl Cellulose. (2).

Methohexitone. α-5-Allyl-1-methyl-5-(1-methylpent-2-ynyl)barbituric acid. Brevital; Brietal.

Methoidal Sodium. Sodium iodomethanesulphonate. Skiodan Sodium.

Methoxone. Selective weed killers. (512).

Methoin. 5-Ethyl-3-methyl-5-phenyl-hydantoin. Mephenytoin. Mesontoin.

Methonal. Dimethyl - sulphone - di-methyl-methane, $(CH_3)_2C(SO_2CH_3)_2$. A hypnotic.

Methone. 5 : 5-Dimethyl-resorcinol.

Methophan (Isatophan, Methoxy-ato-phan). 8-Methyl-2-phenyl-quinoline-4-carboxylic acid. Used in medicine for gout.

Methosal. See MESOTAN.

Methoserpidine. 10-Methoxydeserpi-dine. Decaserpyl.

Methotrexate. N-{4-[N-(2,3-Diaminop-teridin - 6 - ylmethyl) - N - methyl - amino]benzoyl}-glutamic acid. Ame-thopterin.

Methotrimeprazine. 10-(3-Dimethyl-amino - 2 - methylpropyl) - 2 - methoxy-phenothiazine. Veractil.

Methox. A proprietary trade name for dimethoxy ethyl phthalate. A vinyl plasticiser. (73).

Methoxamine. 2-Amino-1-(2,5-dime-thoxyphenyl)propan-1-ol. Vasoxine and Vasylox are the hydrochloride.

Methoxetin. See KRYOFINE.

Methoxsalen. An aid to dermal pigmen-tation. 9-Methoxy-7H-furo[3,2-g]chro-men-7-one.

Methoxy-Atophan. See METHOPHAN.

Methoxyflurane. 2,2-Dichloro-1,1-fluoroethyl methyl ether. Penthrane.

Methoxyphenamine. 1-(2-Methoxy-phenyl)-2-methylaminopropane. Orth-oxine.

Methozin. See ANTIPYRINE.

Methral. A proprietary preparation of fluperolone acetate. Topical anti-in-flammatory agent. (315).

Methscopolamine Bromide. A pro-prietary trade name for Hyoxine Metholeromide.

Methsuximide. Nα-Dimethyl-α-phenyl-succinimide. Mesuximide. Celontin.

Methutin. See TRIGEMINE.

Methyclothiazide. 6-Chloro-3-chloro-methyl - 3,4 - dihydro - 2 - methyl - benzo - 1,2,4 - thiadiazine - 7 - sulpon-amide 1,1-dioxide. Enduron.

Methyl Acetone. A crude fraction of wood distillation. Its principal con-stituents are acetone, methyl alcohol, and methyl acetate. Some samples contain from 70–80 per cent. acetone. Used as a rubber solvent. Methyl ethyl ketone, $CH_3CO.CH_2CH_3$, is also known by this name.

Methylal (Formal). Methylene-di-methyl ester $CH_2(O.CH_3)_2$. Used medicinally, externally as an ointment, internally as a hypnotic, and inhaled as an anæsthetic.

Methylaldehyde (Methanal). Formalde-hyde, H.CHO.

Methyl Alizarin. Dioxy-methyl-anthraquinone, $C_6H_4 : C_2O_2 : C_6HCH_3$ $(OH)_2$, 1 : 6 : 4 : 3 : 2.

Methyl Alkali Blue (Alkalin, Alkali Blue D, Alkali Blue 6B). A dyestuff. It is the sodium salt of triphenyl-p-rosaniline-monosulphonic acid, $C_{37}H_{30}$ N_3SO_4Na. Dyes wool from an acid bath.

Methyl Aniline Green. See METHYL GREEN.

Methyl Aniline Violet. See METHYL VIOLET B.

Methylanone. Methyl-cyclohexanone, a solvent for resins.

Methyl Aspirin. See METHYL RHODIN.

Methyl-Aspirodine. A proprietary preparation. It is methyl-acetyl-iodo-salicylate.

Methylated Chloroform. Chloroform made from methylated spirit.

Methylated Spirit. New regulations (1926) state that Industrial methylated spirit shall consist of 95 volumes of alcohol, with 5 volumes wood naphtha. Industrial methylated spirit (pyridi-nised), to contain 95 volumes alcohol, 5 volumes wood naphtha, with the ad-dition of 0·5 volume of pyridine to every 100 parts of mixture. Mineralised methylated spirit to consist of 90 volumes alcohol, 9·5 volumes wood naphtha, 0·5 volume pyridine, and ⅜ of a gallon mineral naphtha or petroleum oil, and $\frac{1}{40}$ oz. methyl violet to every 100 gallons of the mixture.

Methyl-Atophan. See PARATOPHAN.

Methyl Atropine. Two salts are on the market under this name. Methyl-atropine-bromide, $C_{16}H_{20}O_3N(CH_3)_2Br$, and methyl-atropine-nitrate or Eumy-drine, $C_{16}H_{20}O_3N(CH_3)_2NO_3$. They are used in the place of atropine.

Methylbenzethonium Chloride. Ben-zyldimethyl - 2 - {2 - [4 - (1,1,3,3 - tetra -

methylbutyl)cresoxy]ethoxy}ethyl-ammonium chloride.

Methyl Benzene (Retinaphtha, Phenyl-Methane). Toluene, $C_6H_5CH_3$.

Methyl Benzoquate. An anti-protozoan. Methyl 7 - benzyloxy - 6 - butyl - 1, 4 - dihydro - 4 - oxoquinoline - 3 - carboxyl-ate. NEQUINATE, STATYL.

Methyl Blue (Brilliant Cotton Blue, greenish, XL Soluble Blue, Diphenyla-mine Blue, Bavarian Blue DBF, Soluble Blue 8B, 10B, Helvetia Blue, Methyl Blue for Cotton, Light Blue, Pararosaniline Blue, Paris Blue, Night Blue). A dyestuff. It is the sodium salt of triphenyl-pararosaniline-trisul-phonic acid, $C_{37}H_{26}N_3O_9S_3Na_3$. Dyes silk and mordanted cotton blue.

Methyl Blue for Cotton. See METHYL BLUE.

Methyl Blue for Silk MLB. See BAVARIAN BLUE DSF.

Methyl Blue M. A dyestuff, which consists chiefly of the sodium salt of triphenyl - pararosaniline - trisulphonic acid (Methyl blue).

Methyl Blue, Water Soluble. See BAVARIAN BLUE DSF.

Methyl Carbinol. Ethyl alcohol, C_2H_5OH.

Methyl Catechol. Guaiacol, $C_6H_4(OH)$ (OCH_3), 1 : 2, the chief constituent of beech-wood creosote.

Methyl Cellosolve. A proprietary name for the monomethyl ether of ethylene glycol. It is a colourless and nearly odourless liquid boiling at $124.5°$ C., and has the lowest boiling-point and greatest rate of evaporation of all avail-able glycol ethers. It is a solvent for cellulose acetate, nitro-cellulose, and hydrocarbons.

Methyl Cellulose. See TRYLOSE S and METHOCEL.

Methylchromone. 3-Methylchromone. Crodimyl.

Methyl Cysteine. Methyl-α-amino-β-mercaptopropionate. Mecysteine. Ac-drile is the hydrochloride.

Methyl-Dambose. Bornesite, $C_7H_{14}O_6$, a volatile substance which occurs in the caoutchouc of Borneo.

Methyldesorphine. 6-Methyl-Δ^6-deoxy-morphine.

Methyldopa. (−)-β-(3,4-Dihydroxy-phenyl)-α-methylalanine. Aldomet.

Methyldopate. Ethyl (−)-2-amino-3-(3,4 - dihydroxyphenyl) - 2 - methyl - propionate. Aldomet Injection is the hydrochloride.

Methylene-Azure. Methylo-iodide of trimethyl - amino - imino - imino - di - phenyl-sulphone, $C_{16}H_{18}N_3SO_3I$.

Methylene Blue. A dyestuff. It is the methylo-chloride of trimethyl-amino-imino - imino - diphenyl - sulphide,

$C_{16}H_{18}N_3SCl$. Used in dyeing, calico printing, and as a staining material in bacteriological work. It is used in medicine for malaria and neuralgia.

Methylene Blue A Extra. See METHY-LENE BLUE B and BG.

Methylene Blue B and BG (Methylene Blue BB in Powder Extra D, Methy-lene Blue A Extra, Methylene Blue 2B, 2B Conc, 2BZF, FZP, G, GS, R Conc, ZF, GSF). Dyestuffs. The chloride (Methylene blue BG, BB in powder extra D, printing blue), or zinc double chloride (Methylene blue B, BB in powder extra, dyeing blue), of tetra-methyl - diamino - phenazthionium. Chloride, $C_{16}H_{18}N_3SCl$; zinc double chloride, $(C_{16}H_{18}N_3SCl)_2 + ZnCl_2 + H_2O$. They dye cotton mordanted with tannin blue.

Methylene Blue 2B, 2B Conc., 2BZF, FZP, G, GS, R Conc., ZF, GSF. Dyestuffs. They are British equiva-lents of Methylene blue B and BG.

Methylene Blue BB in Powder Extra. See METHYLENE BLUE B and BG.

Methylene Blue BB in Powder Extra D. See METHYLENE BLUE B and BG.

Methylene Blue, Löffler's. See LÖFFLER'S METHYLENE BLUE.

Methylene Blue T50. A dyestuff It is a British equivalent of Toluidine blue O.

Methylene Green. A mono-nitro derivative of methylene blue. It dyes unmordanted cotton bluish-green shades.

Methylene Green G Conc., Extra Yellow Shade. Nitro-methylene blue, $C_{16}H_{17}N_4O_2SCl$. Dyes cotton bluish-green.

Methylene Grey. See NEW GREY.

Methylene Red. The methylo-chloride of dimethyl-amino-imino-phenyl-disul-phide, $(C_8H_9N_2S_2Cl)_2$. A by-product in the manufacture of methylene blue, found in the mother liquor.

Methylene Violet. A dyestuff. It is oxy - imino - dimethyl - amino - di - phenylene-sulphide, $C_{14}H_{12}N_2SO$. Dyes silk and cotton.

Methylene Violet 2RA, 3RA (Fuchsia, Safranine MN, Clemantine, Giroflé, Brilliant Heliotrope 2R Conc.). A dyestuff. It is dimethyl-diamino-phenyl-phenazonium chloride, $C_{20}H_{19}N_4Cl$. It is employed in printing reddish-violet.

Methyl Eosin. See ERYTHRIN.

Methylergometrine, (+)-1-(Hydroxy-methyl)propylamide of (+)-lysergic acid. Methergin is the hydrogen maleate.

Methyl-Erythrine. The methyl ester of tetrabromo - fluoresceine, $\overset{.}{C}_{20}H_7CH_3$ Br_4O_5.

Methyl-Glycocoll (Sarcosine). Methyl-amino-acetic acid, $CH_3 . NH . CH_2 . CO_2H$. It is a derivative of creatine, and is used in rheumatism.

Methyl Green (Paris Green, Light Green, Double Green SF, Green Powder, Methyl Aniline Green). A dyestuff. It is the zinc double chloride of hepta-methyl-pararosaniline chloride, $C_{26}H_{33}N_3Cl_4Zn$. Dyes silk green from a soap bath. Also see ETHYL GREEN.

Methylhexalin. A registered trade name for a mixture of three isomeric methyl-cyclo-hexanols. It has a specific gravity of 0·930, boiling range of 170–180° C., and a flash-point of about 68° C. A solvent for fats, resins, oils, and waxes.

Methyl Hydrate. See METHANOL.

Methylia. A name formerly used for methylamine.

Methyl Indigo B. A dyestuff. It is o-methyl-indigotin, $C_{18}H_{14}N_2O_2$. Dyes cotton greenish-blue from a reduced vat.

Methyl Indigo R. A dyestuff. It is p-methyl-indigotin, $C_{18}H_{14}N_2O_2$. Dyes cotton reddish-blue from a reduced vat.

Methylindone. A dyestuff. It is an indoine blue prepared with amino-naphthol.

Methyl-Ketole Yellow. Potassium-2-methylindyl - 2 - methyl - indolidene - phenyl - methane - o - carboxylate. It imparts a yellow colour to silk and wool, and the colour is stable towards acids.

Methyl Melilot. Dimethyl-aceto-phenone, $C_6H_3(CH_3)_2CO.CH_3$. It has a similar odour to Melilot (q.v.).

Methyl Morphine. Codeine, $C_{18}H_{21}NO_3$, an alkaloid.

Methyl Mustard Oil. Methyl iso-thiocyanate, $CH_3.NCS$.

Methyl-oil. A distillation product from cellulose boilers.

Methyl Orange. See ORANGE III.

Methylpentynol. 3-Methylpent-1-yn-3-ol. Atempol; Oblivon; Somnesin; Oblivon-C is the carbamate.

Methyl Phenidate. Methyl-α-phenyl-α-2-piperidylacetate. Ritalin is the hydrochloride.

Methyl-phenol. Cresol, $C_6H_4(CH_3)OH$

Methyl Phloxin (Cyanosin). A dyestuff obtained by the alkylation of phloxin.

Methylprednisolone. 11β,17α,21-Tri-hydroxy - 6α - methylpregna - 1,4 - diene-3,20-dione. 6α-Methylpredniso-lone. Medrol ; Medronel Metastab.

Methyl Pulvate. Vulpic acid, $C_{19}H_{14}O_5$.

Methyl Rhodin. (Methyl Aspirin). Methyl - acetyl - salicylate, $C_2H_3O.O.C_6H_4COOCH_3$. Used in medicine in the treatment of rheumatism and neu-ralgia.

Methyl Salol. The phenyl ester of p-cresotinic acid, $CH_3.C_6H_3(OH)COOC_6H_5$. Used for the relief of muscular pain.

Methyl Sulphonal. See TRIONAL.

Methylsulphonatum. Diethyl-sulpho-methyl-ethyl-methane.

Methyl Tuads. A proprietary prepara-tion of tetramethyl thiuram disulphide. (450).

Methyl Violet B (Paris Violet, Methyl Aniline Violet, Direct Violet, Dahlia, Pyoctanin, Violet 3B Extra). A dye-stuff. It consists chiefly of the hydro-chlorides of penta- and hexa-methyl-pararosaniline, $C_{24}H_{28}N_3Cl$. Dyes silk and wool violet, and cotton mordanted with tannin and tartar emetic.

Methyl Violet 5B. See BENZYL VIOLET.

Methyl Violet 6B. See BENZYL VIOLET.

Methyl Violet 6B Extra. See BENZYL VIOLET.

Methyl Violet 10B. A British dyestuff. It is equivalent to Crystal violet.

Methyl Water Blue. See BAVARIAN BLUE DSF.

Methyl Zimate. A proprietary prepara-tion of zinc dimethyl dithiocarbamate. (450).

Methyprylone. 3,3-Diethyl-5-methyl-2,4-dioxopiperidine. Noludar.

Methyridine. An anthelmintic. 2-(2-Methoxyethyl)pyridine. PROMINTIC.

Methysergide. 1-(Hydroxymethyl)-propylamide of 1-methyl-(+)-lysergic acid. Deseril.

Metiamide. A histamine H_2-receptor blocking agent. 1 - Methyl 3 - [2 - (5 - methylimidazol - 4 - ylmethylthio) - ethyl] thiourea.

Metiguanide. A proprietary prepara-tion of METFORMIN hydrochloride used in the treatment of diabetes. (365).

Metilar. A proprietary trade name for the 21-acetate of Paramethazone. (805).

Metillure. An acid-resisting alloy con-sisting of 17 per cent. silicon, 81 per cent. iron, 0·9 per cent. manganese, 0·25 per cent. aluminium, 0·6 per cent. carbon, and 0·17 per cent. phosphorus.

Metiloil. A registered trade mark for methyl esters of ricinoleic acid and other fatty acids in castor oil. Used for tan-ning and in textiles and for the manu-facture of sulphonated emulsifying agents. (291).

Metoclopramide. 4-Amino-5-chloro-N - (2 - diethylaminoethyl) - 2 methoxy - benzamide. Maxolon is the hydro-chloride.

Metofoline. 1-(4-Chlorophenethyl)-1,2,3,4 - tetrahydro - 6,7 - dimethoxy - 2 - methylisoquinoline.

Metol. Mono-methyl-p-amino-m-cresol-sulphate, $(C_6H_3(OH)CH_3(NH)CH_3)_2H_2$

SO₄. A developing agent used in photography.

Metolazone. A diuretic. 7-Chloro-1,2,3,4 - tetrahydro - 2 - methyl - 4 - oxo - 3 - *o* - tolyl - 6 - quinazolinesulphona - mide. ZAROXOLYN.

Metolhydroquinone (Metol-quinone). A photographic developer. It contains Metol (*q.v.*), hydroquinone, sodium phosphate, sodium sulphite, and potassium carbonate.

Metol-Quinone. See METOLHYDRO-QUINONE.

Metopimazine. 10-[3-(4-Carbamoyl-piperidino)propyl] - 2 - methanesul - phonylphenothiazine. 9965 R.P.

Metopirone. A proprietary preparation of metyrapone, used to test pituitary gland function. (18).

Metopon. 7,8-Dihydro-5-methylmorphi-none.

Metoprolol. A beta adrenergic blocking agent. (±) - 1 - Isopropylamino - 3 - *p* - (2 - methoxyethyl)phenoxypropan - 2 - ol. " CGP 2175 " is the L-(+) - tar - trate.

Metoserpate. A tranquilliser. Methyl *O* - methyl - 18 - epireserpate. AVICALM is the hydrochloride.

Metosyn. A proprietary preparation of FLUOCINONIDE. A steroid skin cream. (512).

Metozin. See ANTIPYRINE.

Metramine. See HEXAMINE.

Metrazol. Pentamethylene-tetrazole.

Metriphonate. An insecticide and anthelmintic. Dimethyl 2,2,2-trichloro-1-hydroxyethylphosphonate. TRICHLOR-PHON, DIPTEREX, DYVON, NEGUVON, TUGON.

Metronidazole. 1-(2-Hydroxyethyl)-2-methyl-5-nitroimidazole. Flagyl.

Metro - nite. A refined natural mineral composed of calcium carbonate, and carbonates and silicates of magnesium. Used in the paint industry as a pigment.

Metrotect. A proprietary bitumen paint for protecture treatment of iron-work, etc.

Metrotonin. A mixture stated to contain sparteine and acetyl-choline. Used in medicine.

Metrulen. A proprietary preparation containing ethynodiol and mestranol. (285).

Metrulen M. A proprietary preparation containing ethynodiol and mestranol. (285).

Metsapol. A degreasing detergent. (546).

Metso. Sodium metasilicate. (546).

Metso 22, 66, 99. Trade names for proprietary cleaning agents containing alkaline silicates.

Metso 99. A patented crystalline sodium silicate. It contains 3 parts Na₂O, 2 parts SiO₂, and 11 parts H₂O on a molecular basis. A cleaning material.

Metternich Green. See IODINE GREEN.

Metycaine. A proprietary trade name for the hydrochloride of Piperocaine.

Metyrapone. 2-Methyl-1,2-di-(3-pyridyl)propan-1-one. Metopirone.

Metyrin. A German preparation. It is papaverine-yohimbine-tartrate with the addition of amino-phenazone. Used in cases of dysmenorrhœa.

Metyzoline. A vasoconstrictor. 3-(2-Imidazolin - 2 - ylmethyl) - 2 - methyl - benzo[*b*]thiophene. Eunasin is the hydrochloride.

Mevasine. A proprietary trade name for the hydrochloride of Mecamylamine.

Mevilin-L. A proprietary live measles vaccine. (444).

Mexenone. 2-Hydroxy-4-methoxy-4'-methylbenzophenone. Uvistat 2211.

Mexican Blue. The colouring matter of *Sericographis mohite.* Dyes wool and cotton purplish-blue direct.

Mexican Elemi. Elemi from *Amyris elemifera.*

Mexican Onyx. A variety of calcite.

Mexican Turpentine. See TURPEN-TINE.

Mexican Whisk. See RAIZ DE ZACATON.

Mexico Seeds. Castor oil seeds.

Mexiletine. An anti-arrhythmic. 1-Methyl - 2 - (2, 6 - xylyloxy)ethylamine. " Kö 1173 " is currently undergoing clinical trial as the hydrochloride.

Mexphalte (Spramex, Shelspra). Trade marks for varieties of bitumen, used for road dressing and other purposes.

Meyerhofferite. An artificially prepared mineral. It is calcium borate.

Meyersite. A mineral. It is a hydrated aluminium phosphate.

Meyer's Solution. Mercury-potassium iodide solution, obtained by dissolving 13·35 grams mercuric chloride and 49·8 grams potassium iodide separately in water, mixing the solutions and diluting to 1 litre.

Meymacite. A native hydrated tungsten oxide, WO₃.H₂O.

Mezcaline. 3 : 4 : 5-Trimethoxyphenyl-methylamine. Used in medicine.

Mezlocillin. An antibiotic. 6-[D-2-(3-Methylsulphonyl - 2 - oxoimidazolidine-1 - carboxamido 2 - phenylacetamido] - penicillanic acid. " BAY f 1353 " is currently undergoing clinical trial as the sodium salt.

M.F.C. Materials for chromatography. (527).

M.F. Stain. Biological stain for keratin. (544).

MG. Blue. Polychrome dye for protein electrophoresis. (544).

Mgoa Rubber. Commercial name for the rubber from *Muscarenhasia elastica* of East Africa.

Mianin. See CHLORAMINE T.

Mianserin. An anti-depressant. 1,2,3, 4, 10, 14b - Hexahydro - 2 - methyldi - benzo[*c*,*f*,] pyrazino [1, 2 - α] azepine. BOLVIDON.

Miargyrite. See PYRARGYRITE HYPARGYRITE. A mineral. It is 8[AgSbS₂].

Miascite. A mineral. It is a mixture of calcite and strontianite.

Miazine. Metadiazine.

Mica (Laminated Talc, Glimmer Glist). This material consists mainly of a double silicate of aluminium, and sodium or potassium. It also contains magnesium and iron silicates. Used for making fireproof window-panes and lamp-chimneys, also as an electrical insulating material. Mica has an induction capacity of 6·6 compared with ebonite 1·92, and a specific gravity of 2·7–3·15. Also see BIOTITE, MUSCOVITE, PARAGONITE, LEPIDOLITE, and PHOLGOPITE.

Mica, Biaxial. See MUSCOVITE.

Mica, Black. See BIOTITE.

Micabond. A proprietary trade name for a material consisting of mica, shellac, and resin.

Mica Cambric. See MICANITE CLOTH.

Micaceous Iron Ore. A mineral. It consists of thin plates or scales of hæmatite. See HÆMATITE.

Mica, Common. See MUSCOVITE.

Mica, Copper. See TAMARITE.

Micafil B. A proprietary synthetic resin-varnish-paper product used in electrical insulation.

Micafil G. A proprietary electrical insulator made in the form of tubes from shellac, coated paper, and mica.

Micafil S. A proprietary trade name for a shellac varnish-paper product used as an electrical insulation.

Micafolium. A general name for electrical insulators made from mica splittings and paper.

Mica Iron. See LEPIDOLITE.

Mica-Kote. A proprietary trade name for a roofing felt made from asphalt-impregnated felt.

Micalex. See MYCALEX.

Mica, Lithia. See LEPIDOLITE.

Mica, Lithiferous. See LEPIDOLITE.

Mica, Lithium. See LEPIDOLITE.

Mica, Lithium Iron. See ZINNWALDITE.

Mica, Magnesium Iron. See BIOTITE.

Micanite. A mica material built up of small plates of mica with an insulating material such as shellac, or on a foundation of paper or cloth. Used as an electrical insulating material.

Micanite Cloth (Mica Cambric, Toile Micanite). Products used for electrical insulation, made from mica splittings on a cotton-cambric backing.

Micaphilit. Synonym for Andalusite.

Mica, Potash. See MUSCOVITE.

Mica Powder. An explosive consisting of ground mica and nitro-glycerin.

Micarta. A trade name for a range of varnished paper and fabric products using natural and synthetic resin varnishes. They are used for electrical insulation.

Micarta Folium. A similar product to micafolium.

Mica Schist. A foliated rock consisting of quartz and mica.

Mica Silk. An electrical insulating tape made from mica splittings on a silk cloth.

Mica, Soda. See PARAGONITE.

Mica, Uranium. See AUTUNITE.

Mica, Vanadium. See ROSCOELITE.

Mica, White. See MUSCOVITE.

Michaelite. A mineral. It is a variety of fiorite.

Michaelsonite. See ERDMANNITE.

Michel-Lévyte. A mineral. It is barite.

Michler's Hydrol. *p*-*p*'-Bisdimethylamine - benzhydrol, $HO . CH[C_6H_4N(CH_3)_2]_2$.

Michler's Ketone. Tetramethyl-diamino-benzophenone, $[(CH_3)_2N.C_6H_4]_2CO$. Used for making dyestuffs.

Michrome. Microscopic stains and biological reagents. (544).

Miciuratu. A fermented milk product of Sardinia, prepared by boiling milk (cow's, ewe's, or goat's, or a mixture of the three) for about 6 minutes, then cooling it to 37° C., and introducing a ferment. After 7 hours, the milk forms a doughy mass, and is ready for consumption.

Mi-Col. A mildew fungicide. (511).

Micoren. A proprietary preparation of cropropamide and crotethamide. A respiratory stimulant. (17).

Micralax. A proprietary preparation containing sodium citrate, sodium alkyl sulphoacetate and sorbic acid. An enema. (281).

Micro-asbestos. An Austrian asbestos of short fibre unsuitable for the ordinary uses of asbestos. It is a homogeneous bluish-green to white, fine matte powder, and is similar in composition to Canadian asbestos containing approximately 50 per cent. SiO_2, 5 per cent. Al_2O_3, 15 per cent. CaO, and 23 per cent. MgO. The disintegrated, washed, and dried material is resistant to alkalis, acids, heat, and atmosphere, is a bad conductor of heat and electricity, and is strongly absorbent. It is suitable as an insulation for pipe lines, and as a refractory for furnaces. It is used in the production of " stone-wood," a wood substitute, as a filler in the rubber industry, as an ingredient in blotting-paper, and for electrical insulation boards. With gypsum it makes a mould-lining composition for casting metals, and is also used as an ingredient of fire-proof paints.

Microbin. A German product containing 45 per cent. sodium-p-chlor-benzoate and 55 per cent. sodium-o-chlor-benzoate. A preservative.

Microbromite. A mineral, a mixture of silver, bromide, and chloride.

Microcal. A registered trade mark for calcium silicate. (546).

Microcal ET. A trade mark for a series of synthetic calcium silicates of controlled particle size used as an extender in emulsion paint. (205).

Microcidin (Beta-Naphthol Sodium). Varying compounds of sodium naphtholate with other naphtholates and phenates. Surgical antiseptics.

Microcline. See FELSPAR.

Microcosmic Salt (Phosphorus Salt, Fusible Salt of Urine, Fusible Salt, Essential Salt of Urine). Sodium-ammonium-hydrogen phosphate, $Na(NH_4)H.PO_4.4H_2O$.

Microdol (Extra). A trade name for ground Dolomite.

Microgynon 30. A proprietary preparation of ethinyloestradiol and α-norgestrel. An oral contraceptive. (438).

Microlan. Powdered lanolin. (526).

Microlite. A mineral, $2(Ca.Mn.Fe.Mg)O.(Ta.Nb)_2O_5$.

Micromeritol. An alcoholic solution of yerba buena.

Micromet. A slowly soluble sodium metaphosphate. (503).

Micromya. A French name for a wet-ground whiting from Champagne. (26). See also OMYA BSH.

Micronazole. An anti-fungal agent. $1-[2, 4 - Dichloro - \beta (2, 4 - dichlorobenzyloxy)phenethyl]imidazole.$

Micronex. A proprietary gas black for use in rubber mixings.

Micronor. A proprietary preparation of NORETHISTERONE. An oral contraceptive. (369).

Micropore. A proprietary surgical tape of rayon with a hypoallergenic adhesive. (471).

Microporite. An insulating concrete made from ground silica and lime, hardened by treatment with steam.

Microsan. A copper fungicide. (589).

Microsol. A soluble disinfectant for stables and drains, consisting of 75 per cent. copper sulphate, and sulphocarbolate. It contains some free sulphur dioxide.

Microthene F. A registered trade name for ultra fine polyolefine powder dispersions for coating purposes.

Microx. Zinc oxide for special enamels. (610).

Micryston. A proprietary name for a group of hormone preparations used for replacement therapy. (352).

Mictine. 1-Allyl-6-amino-3-ethylpyrimidine-2,4-dione.

Midamor. A proprietary preparation of AMILORIDE hydrochloride used to conserve potassium in the course of diuretic therapy. (472).

Midas Gold. A proprietary trade name for hydrated transparent iron oxide dispersions. Used for metallic motor-car finishes. (189).

Middle Oils (Carbolic Oils). That fraction of coal tar distilling at 170–230° C. Approximately 20 gallons are obtained from 1 ton of tar. It is usually split up into (a) carbolic oil, with a specific gravity up to 1·000; (b) naphthalene oil, specific gravity 1·000–1·025; and (c) heavy oil, the residue.

Middletonite. A resin found near Leeds.

Middlings. See BRAN.

Midicel. A proprietary preparation containing sulphamethoxypyridazine. (264).

Midrid. A proprietary preparation of isometheptene mucate, DICHLORALPHENAZONE and PARACETAMOL used in the treatment of migraine. (830).

Midrol. The iodo-methylate of phenylpyrazolone. A mydriatic, i.e. a drug causing dilatation of the pupil of the eye.

Midvale Alloys. Heat-resisting alloys A.T.V. alloy contains from 33–39 per cent. nickel, 10–12 per cent. chromium, 1·1–1·8 per cent. manganese, and the balance iron. B.T.G. alloy contains 60–62 per cent. nickel, 10–11 per cent. chromium, 1·2–1·5 per cent. manganese, and the balance iron.

Midvaloy H.R. A proprietary nickel-tungsten-chromium alloy. It resists corrosion.

Midzuame. Glucose prepared in Japan by the diastase conversion of the starch of millet and rice.

Miedziankite. A mineral. It is a variety of tennantite containing zinc, and corresponds to the formula $2Cu_3As_2S_3.ZnS$.

Miemite. A mineral, $CaCO_3.MgCO_3$.

Miersite. Cubic crystals of Iodite (silver iodide).

Miesite. A mineral. It is a variety of pyromorphite.

Migafar AL. A proprietary trade name for an aqueous emulsion of fatty products suitable for the removal of chafe marks from dyed fabrics. (18).

Migen. A proprietary extract of house dust mite used in the desensitisation treatment of asthma and allergic rhinitis. (272).

Mignonette-geranium Oil. An oil obtained by distilling geraniol over mignonette flowers.

Mignonette Green. See MAY GREEN.

Migraineine. A mixture of caffeine citrate and antipyrine. An antipyretic and antineuralgic.

Migra Iron. A special pig iron for high quality castings obtained by a special heat treatment before casting, which results in a remarkably fine grain.

Migraleve. A proprietary preparation of BUCLIZINE dihydrochloride, PARACETAMOL, CODEINE phosphate and dioctylsodium sulphosuccinate tablets with tablets of codeine, paracetamol and dioctyl sodium sulphosuccinate, used in the treatment of migraine. (852).

Migralgine. A mixture of 88 per cent. antipyrine, 9 per cent. caffeine, and 3 per cent. salicylic acid, fused together.

Migraneol. A clear brown liquid used externally for headache. It is a 10 per cent. solution of menthol in acetic ester, to which camphor and essential oils have been added.

Migril. A proprietary preparation of ergotamine tartrate, cyclizine hydrochloride and caffeine, used for migraine. (277).

Migrol. A compound of pyrocatechin and pyramidone.

Migrophene. A mixture of lecithin and quinine. Prescribed for headaches.

Mikado Brown B, 3GO, M. A dyestuff obtained by the action of alkalis upon p-nitro-toluene-sulphonic acid, in the presence of an oxidising substance. Dyes cotton brown.

Mikado Gold Yellow 2G, 4G, 6G, 8G. See MIKADO YELLOW.

Mikado Orange G, 4R (Direct Orange 2R, Direct Orange G, Chloramine Orange, Stilbene Orange 4R, Paramine Fast Orange D, G, Direct Fast Orange 4R, RL, 2RL, Chlorazol Fast Orange D). A dyestuff obtained by the action of alkaline reducing agents upon Direct yellow. Dyes cotton yellow-orange to reddish-orange.

Mikado Yellow (Mikado Gold Yellow 2G, 4G, 6G, 8G; Direct Yellow 2G, 4G, Diazamine Fast Yellow H, Direct Fast Yellow 2GLO, Paramine Fast Yellow 3G, Stilbene Yellow 2G, 3G, 8G). A dyestuff obtained by the treatment of the condensation product of p-nitro-toluene-sulphonic acid and sodium hydroxide with oxidising agents. Dyes cotton yellow from a salt bath.

Mikamycin. An antibiotic produced by Streptomyces mitakaensis. Mikamycin B is Ostreogrycin B.

Mykamycin B. Ostreogrycin B.

Mikolite. A proprietary material. It is a vermiculite which has been expanded by calcination giving a very fine product. It is used in paints.

Mikrobin. Sodium-p-chlor-benzoate. A preservative for wines.

Mikronal. Menthol-formalin combined with tetra-pyridine-carbonate. A 1 per cent. solution is used for inhalation in the treatment of catarrh and nasal affections.

Milanite. A mineral. A variety of halloysite.

Milanol. Basic bismuth trichloro-butyl-malonate.

Milarite. A mineral, $HKCa_2Al_2(SiO_3)_{12}$.

Milbis. Bismuth glycollylarsanilate.

Mild Alkali. Sodium carbonate, Na_2CO_3.

Mild Lime. Calcium carbonate (chalk), $CaCO_3$, is known in agriculture as mild lime.

Milfoil or Yarrow. The flowering plant of *Achillea millefolium*.

Milid. See PROGLUMIDE.

Milk Albumen, Riegel's. See RIEGEL'S MILK ALBUMEN.

Milk Glass. A soda or flint glass rendered opaque by the addition of a mineral phosphate.

Milk of Almonds. *Mistura amygdalæ* B.P.

Milk of Asafœtida. A 4 per cent. emulsion of asafœtida in water. Used as a sedative and carminative.

Milk of Barium. A suspension of barium hydrate, $Ba(OH)_2$.

Milk of Lime. Slaked lime and water in a thin cream.

Milk of Magnesia. A suspension of magnesium hydroxide, $Mg(OH)_2$.

Milk of Sulphur (Lac Sulphuris). A soluble amorphous sulphur obtained by the action of acids upon solutions of polysulphides. Employed medicinally.

Milk, Pasteurized. Milk which has been heated for 30 minutes to 60° C.

Milk Powder. The name given to dried milk, or artificial mixtures such as casein, butter, milk sugar, and salts. Often skim milk is dried *in vacuo*, and powdered. This contains no fat, therefore does not become rancid, and if a little calcium saccharate is added, it dissolves in water to give skim milk.

Milk Somatose. See SOMATOSE.

Milkstone. A mixture of milk salts and protein obtained from milk.

Milk Sugar. Lactose, $C_{12}H_{22}O_{11}+H_2O$.

Milk Sugar Rennet. See PEGNIN.

Milk Tree Wax. Cow tree wax.

Milky Quartz. A quartz consisting of innumerable cavities containing liquid.

Millaloy. A proprietary trade name for a nickel-chromium steel containing 4 per cent. nickel, 1·5 per cent. chromium, and 0·4 per cent. carbon.

Millektrol. A German sodium carbohydrate preparation. A styptic.

Millerite (Hair Pyrites, Capillary Pyrites, Capillose, Nickel Blende, Nickel Pyrites, Beyrichite). A mineral. It is nickel sulphide, NiS.

Millicarb. A proprietary trade name for crystalline calcite with a mean particle size of 2·4 microns. (189).

Millicorten. A proprietary trade name for dexamethasone.

Milling Blue. A dyestuff. It is the sodium salt of a sulphonic acid of di-phenyl - diamino - phenyl - naphthazonium chloride. Dyes chromed wool blue.

Milling Green. An acid dyestuff, giving bluish-green shades on wool.

Milling Green S. A quinone-oxime dyestuff. The iron salt of Gambine B is probably present in the dye.

Milling Orange. A dyestuff. It is the sodium salt of sulpho-benzene-azo-benzene-azo-salicylic acid, $C_{19}H_{12}N_4SO_6Na_2$. Dyes chromed wool orange-red from an acid bath.

Milling Red B, FFG, G, FR, R. Acid mordant dyestuffs. They dye direct from an acid bath, or on chrome mordanted wool.

Milling Silver. A nickel silver. It is an alloy of 56 per cent. copper, 27·5–31 per cent. zinc, 12–16 per cent. nickel, and 0·5–1 per cent. lead.

Milling Yellow (Chrome Yellow D, Anthracene Yellow BN, Mordant Yellow O, Chrome Fast Yellow, Monochrome Yellow Paste, Powder, Phorochrome Yellow Y, Chrome Fast Yellow O, Alizarin Yellow L, WS, Lighthouse Chrome Yellow O, Solochrome Yellow Y). A dyestuff. It is the sodium salt of sulpho-naphthalene-azo-salicylic acid, $C_{17}H_{10}N_2O_5SNa_2$. Dyes wool yellow from an acid bath.

Millon's Base. Hydroxy-dimercuro-ammonium hydroxide, $OH.Hg_2NH_2O$. Used medicinally, for colouring porcelain, and for coating or varnishing the keels of vessels, to prevent shellfish and marine plants from adhering to them.

Millon's Reagent. Mercury dissolved in an equal weight of nitric acid (specific gravity 1·41), and the solution diluted to twice its volume. After standing, the liquid is decanted from the precipitate. Used as a test for albumin.

Millophyline. A proprietary preparation of diethylaminoethyltheophylline and camphorsulphonate, used as a cardiac and respiratory stimulant. Also as a bronchial antispasmodic. (320).

Mills Plastic. A proprietary trade name for a vinylidene chloride synthetic resin.

Milonorm. A proprietary preparation of MEPROBAMATE. A tranquilliser. (255).

Milontin. A proprietary preparation of phensuximide. An anticonvulsant. (264).

Milorganite. An American fertiliser obtained from sewage and trade wastes. It contains 5·4 per cent. of nitrogen and 3·08 per cent. phosphoric acid. The nitrogen is readily available, and the material compares favourably with dried blood, fish scrap, and cotton-seed meal.

Milori Blue. See PRUSSIAN BLUE.

Milori Green. See CHROME GREENS.

Milory Green. See CHROME GREENS.

Miloschite. A mineral. It is an aluminium silicate containing chromium.

Milososin. A mineral containing hydrated oxides of chromium, aluminium, and silicon.

Milowite. A proprietary amorphous silica for paint, polishing, and chemical trades. It is very white in colour, and 90 per cent, is below 0·01 mm. particle size.

Milpath. A proprietary preparation of meprobamate and tridihexethyl chloride. Gastro-intestinal sedative. (255).

Milton. The trade mark for a disinfectant, the active principle of which is sodium hypochlorite.

Miltown. A proprietary preparation of meprobamate. A sedative. (255).

Milvan Steel. A proprietary trade name for a high speed tool steel containing 19 per cent. tungsten, 4 per cent. chromium, and 2 per cent. vanadium.

Milwaloy. A proprietary trade name for a chromium-vanadium steel.

Mimea. See MOMEA.

Mimetisite (Kampylite, Hedyphane, Mimetite). A mineral. It is lead arseniate, $3Pb_3(AsO_4)_2PbCl_2$.

Mimetite. See MIMETISITE.

Mimico. See CARBORA.

Mimosa. See CLAYTON YELLOW and WATTLE BARK.

Minac. A proprietary preparation of aspirin, magnesium carbonate, and dihydroxyaluminium aminoacetate. An analgesic. (348).

Minadex. A proprietary preparation containing Vitamins A and D, ferric ammonium citrate, calcium and potassium glycerophosphates, and manganese and copper sulphates. (335).

Minafen. A proprietary artificial milk feed low in PHENYLALANINE used in the dietary treatment of phenylketonuria. (665).

Minamino. A proprietary preparation of vitamins, minerals and aminoacids used as a dietary supplement. (256).

Minargent. An alloy of copper, nickel, and aluminium.

Minargentatum. An alloy of 56·82 per cent. copper, 39·77 per cent. nickel, 2·84 per cent. tungsten, and 0·57 per cent. aluminium.

Minasragrite. A mineral. It is a hydrated acid vanadyl sulphate, V_2O_4. $3SO_3.16H_2O$.

Mindererus's Spirit. A solution of ammonium acetate (*Liquor ammonii acetatis B.P.*).

Minelite. An explosive containing chlorates and paraffin wax.

Minepentate. A drug used in treatment of the Parkinsonian syndrome, currently undergoing clinical trial as " UCB 1549 ". It is 2-(2-dimethylaminoethoxy)ethyl 1 - phenylcyclopentanecarboxylate.

Mineral Acid. An inorganic acid.

Mineral Alkali. Sodium hydroxide, NaOH.

Mineral Black. A pigment. It is a shale found naturally, and contains 70 per cent. silica and 30 per cent. carbonaceous matter. Other types of mineral black are coal black, from coal, and slate black. Also a name for ground coal in the form of soft dry powder. British Standard Specification, page 288/1927, states that on heating to constant weight at 95–98° C., the loss shall not exceed 2·5 per cent. On ashing not more than 50 per cent. ash shall remain. Not more than 2 per cent. shall be soluble in ether, and not more than 1 per cent. in water.

Mineral Blue. See MOUNTAIN BLUE, PRUSSIAN BLUE, and VERDITER BLUE.

Mineral Brown. See CAPPAGH BROWN and UMBER.

Mineral Burning Oils (Belmontine Oil, Cazeline Oil, Colzarine Oil, Mineral Colza Oil, Mineral Seal Oil, Mineral Sperm Oil, Pyronaphtha). Burning oils obtained from petroleum.

Mineral Butters. A term formerly used for several of the metallic chlorides, such as those of antimony, arsenic, bismuth, and zinc.

Mineral Caoutchouc. See ELATERITE.

Mineral Carbon. Anthracite.

Mineral Chameleon. Potassium manganate, K_2MnO_4, which is decomposed by water, the green solution becoming violet through the formation of potassium permanganate. Sodium manganate behaves similarly, and is also known by this name.

Mineral Colza Oil. See MINERAL BURNING OILS.

Mineral Cotton. See SLAG WOOL.

Mineral Fat. See PETROLEUM JELLY.

Mineral Flour. A Florida clay used in rubber mixings.

Mineral Glycerin. See PARAFFIN, LIQUID.

Mineral Green. See MALACHITE, SCHEELE'S GREEN, VERDITER BLUE, and GREEN VERMILION.

Mineral Grey. The ash from lapislazuli after the extraction of ultramarine. Also see SLATE GREY.

Mineral Gum. Soluble glass (*q.v.*).

Mineral Indigo. The blue oxide of molybdenum is known by this name.

Mineralised Methylated Spirit. See METHYLATED SPIRIT.

Mineral Jelly. See PETROLEUM JELLY.

Mineral Kermes. See KERMES MINERAL.

Mineral Khaki. A mineral colour produced on the fibre by impregnating cotton with a mixture of ferrous and chromic acetates, drying, and then steaming. Mixtures of basis ferric and chromic acetates are formed on the material, which are fixed by passing the fibre through solutions of sodium carbonate and sodium hydroxide.

Mineral Lake. A basic chromate of tin, prepared by adding potassium chromate solution to stannous chloride solution. Used for colouring paper, and in oil painting.

Mineral Lard. Soft paraffin.

Minerallic 20 and 78. Filling compounds.

Mineral Oil. See EARTH OIL.

Mineral Oils. Natural oils of the petroleum series.

Mineral, Orange. See ORANGE LEAD.

Mineral Pitch. See BITUMEN.

Mineral Pulp. See AGALITE.

Mineral Purple. See INDIAN RED. Purple of Cassius is also sometimes called by this name.

Mineral Rouge. See INDIAN RED and JEWELLER'S ROUGE.

Mineral Rubber. Bitumens of the gilsonite type.

Mineral Seal Oil. See MINERAL BURNING OILS.

Mineral Sperm Oil. See MINERAL BURNING OILS.

Mineral Superphosphate. See SUPERPHOSPHATE.

Mineral Syrup. See PARAFFIN, LIQUID.

Mineral Tallow. See BITUMEN.

Mineral, Turbith. See TURPETH MINERAL.

Mineral Umber. See UMBER.

Mineral Violet. See MANGANESE VIOLET.

Mineral Wax. See OZOKERITE and LIGNITE WAX.

Mineral White. See GYPSUM and BARIUM WHITE.

Mineral Wool. See SLAG WOOL.

Mineral Yeast. Torula, a yeast-like organism, used for fodder production.

Mineral Yellow. See TURNER'S YELLOW and YELLOW OCHRE.

Miner's Friend. An explosive similar in composition to Giant powder.

Minervite. A mineral. It is a hydrated phosphate of alkalis and aluminium, $2AlPO_4.(K.NH_4.H)_3PO_4.7H_2O$.

Mine Tin. See BLOCK TIN.

Minilyn. A proprietary preparation of ethinyloestradiol and lynestrenol. An oral contraceptive. (316).

Minims. A proprietary range of eye-drops for use in a number of ophthalmic applications. (824).

Minite. A similar explosive to Kohlencarbonite (q.v.), without barium nitrate.

Minium. See RED LEAD.

Minium, Iron. See INDIAN RED.

Minium Tego. A high dispersed red lead marketed in Germany.

Minlon. A proprietary brand of Nylon 66 reinforced with 40% mineral filler. (10).

Minocin. A proprietary preparation of MINOCYCLINE hydrochloride. An antibiotic. (306).

Minocycline. An antibiotic. 4,7-Bis-(dimethylamino) - 1, 4, 4a, 5, 5a, 6, 11, 12a - octahydro - 3, 10, 12, 12a - tetra - hydroxy - 1, 11 - dioxonaphthacene - 2 - carboxamide. MINOCIN.

Minodiab. A proprietary preparation of GLIPIZIDE used in the treatment of diabetes. (365).

Minofor. Alloys used by jewellers. They contain from 9–64 per cent. antimony, 20–84 per cent. tin, 2–10 per cent. copper, and 1–10 per cent. zinc.

Minol. See PARAFFIN, LIQUID.

Minolite Antigrisouteuse. A Belgian explosive containing 72 per cent. ammonium nitrate, 23 per cent. sodium nitrate, 3 per cent. trinitro-toluene, and 2 per cent. trinitro-naphthalene.

Minolith. A wood preservative against dry rot. (562).

Minovlar. A proprietary preparation of norethisterone acetate and ethinyl-œstradiol. Oral contraceptive. (265).

Mint Camphor. Menthol, $C_{10}H_{20}O$, occurring to the extent of 80 per cent. in peppermint oil, which is obtained from the leaves and stems of *Mentha piperita*.

Mintezol. A proprietary preparation of thiabendazole. An antihelmintic. (310).

Mintite. A patent finish for rubber surfaces. It consists of powdered mica.

Minuric. See BENZBROMARONE.

Minyak Kerung. An oleo-resin obtained from *Dipterocarpus* species of Malay. It is obtained as a viscous liquid.

Minyulite. A mineral. It is $2[KAl_2(PO_4)_2(OH,F).4H_2O]$.

Miocarpine. A proprietary preparation of pilocarpine hydrochloride solution for ocular use. (329).

Miocarta. A synthetic varnish-paper product used for electrical insulation.

Miochol. A proprietary preparation of acetylcholine chloride and mannitol fused as eye-drops in the treatment of miosis. (473).

Mio-Pressin. A proprietary preparation of rauwolfia serpentina, protoveratrine and phenoxybenzamine hydrochloride. An antihypertensive. (281).

Miotine. The methyl-urethane of α-m-hydroxy - phenyl - ethyl - dimethyl - amine. A miotic.

Mipolam. A proprietary trade name for polyvinyl chloride which, when plasticised, has properties resembling soft rubber. It is used in moulding and for sheathing cables. It is resistant to ozone and other chemicals, and has a specific gravity of 1·4.

Mipor-Scheider. A German product. It is a micro-porous rubber used as diaphragms in the accumulator industry.

Miraculoy. A proprietary trade name for a steel containing 1·25 per cent. nickel, 0·65 per cent. chromium, 0·4 per cent. molybdenum, and 1·55 per cent. manganese.

Mirabilite. See GLAUBER'S SALT.

Miracil D. A proprietary trade name for the hydrochloride of Lucanthone.

Miradon. Anisindione.

Miralite. A light aluminium alloy which can be cast or rolled, and drawn into wire. It contains 12 per cent. copper and 2 per cent. tin.

Miramant. A cutting alloy of heat-resisting metals with a definite fraction of stable and hard carbides, especially molybdenum and tungsten carbides in eutectic proportions. It enables work to be done on materials that were regarded as unworkable, such as austenitic 12–14 per cent. manganese steel, highly alloyed austenitic nickel and chromium-nickel steels, constructional steels hardened to 150 kilos per sq. cm., and hard cast iron.

Mira Metal. Acid-resisting alloys. One contains 74·7 per cent. copper, 16·3 per cent. lead, 6·8 per cent. antimony, 0·91 per cent. tin, 0·62 per cent. zinc, 0·43 per cent. iron, and 0·24 per cent. nickel. Another consists of 75 per cent. copper, 16 per cent. lead, 8 per cent. tin, and 1 per cent. nickel.

Mirasol. A proprietary trade name for alkyd varnish and lacquer resins.

Mirbane Essence. See ESSENCE OF MIRBANE.

Mirbane Oil. See ESSENCE OF MIRBANE.

Mirion. A proprietary preparation. It is iodo-hexamine.

Miriquidite. A mineral, $PbO.Fe_2O_3$. $As_2O_5.P_2O_5H_2O$.

Mirror Bronze. A copper and tin alloy, containing 28–35 per cent. tin. It sometimes contains a little nickel.

Misch Metal. A pyrophoric alloy containing 35 per cent. iron and 65 per cent. cerium earth metals. The cerium earth metals are obtained from the cerium earths, which contain about 45 per cent. cerium sesquioxide, 25 per cent. lanthanum oxide, 15 per cent. neodymium oxide, and some yttrium and samarium oxides. It is a by-product in the manufacture of thorium nitrate for incandescent mantles. Used in the production of pyrophoric alloys such as Auer metal. Synonym: MIX METAL.

Mischzinn. A tin alloy. Theoretically it is an eutectic containing 63 per cent. tin and 37 per cent. lead, but in practice the tin is at least 55 per cent., antimony and copper must be more than 3·5 and 0·5 per cent. respectively, and zinc is present in traces. It is prepared from metal scrap, chiefly bearing alloys.

Miscible Carbon Disulphide. A mixture of carbon disulphide with castor oil, caustic potash, denatured alcohol, and water. An insecticide for destroying the Japanese beetle in the soil without serious damage to the plant.

Miscible Wood Naphtha. Wood naphtha recovered from the brown acetate of lime, prepared from the destructive distillation of wood.

Misco. An alloy of 57·5 per cent. iron, 15 per cent. chromium, 25 per cent. nickel, 1·5 per cent. silicon, 0·5 per cent. manganese, and 0·5 per cent. carbon.

Miscrome. A proprietary trade name for a corrosion-resisting alloy of iron with 28 per cent. chromium.

Misenite. A mineral. It is potassium hydrogen sulphate, $KHSO_4$.

Miserables. A trade name for a mildly stimulating cocoa tea made from cacao shell.

Miskeyite. A serpentine mineral. It is a precious stone.

Mispickel (Arsenical Pyrites, Arsenical Iron Ore, White Mundic, Arsenopyrite, Plinian). A mineral. It is a sulph-arsenide of iron, $FeAsS$.

Mistletoe Rubber. A rubber obtained from the fruit of certain *Loranthaceae* as parasites on the coffee tree.

Mist-o-Matic. A liquid seed dressing. (501).

Misylite. A mineral. It is copiapite.

Mitchalloy A. A proprietary trade name for an alloy of iron with 2·5 per cent. nickel and 0·9 per cent. chromium.

Mitchellite. A mineral. It is a variety of chromite containing magnesium.

Mithramycin. An antibiotic produced by *Streptomyces argillaceus*, *Streptomyces plicatus* and *Streptomyces tanashiensis*.

Mitigal. A proprietary preparation of dimethyl-diphenylene-sulphide.

Mitigal. A proprietary trade name for Mesulphen.

Mitigated Caustic (*Argenti nitras mitigatus B.P.*). It is a fused mixture of 1 part silver nitrate with 2 parts potassium nitrate.

Mitin. Moth proofing agents. (543).

Mitine. A base for ointments prepared from an emulsion which is superfatted with a non-emulsifying fat. Wool fat is used as the fat, and milk as the serum-like liquid to the extent of 50 per cent.

Mitis Green. See EMERALD GREEN.

Mitobronitol. An anti-neoplastic agent. 1, 6 - Dibromo - 1, 6 - dideoxy - D - mannitol. MYELOBROMOL.

Mitoclomine. An anti-neoplastic agent. NN - Di - (2 - chloroethyl) - 4 - methoxy-3 - methyl - 1 - naphthylamine.

Mitopodozide. An anti-neoplastic agent. 2′ - Ethylpodophyllohydrazide.

Mitotenamine. 5-Bromo-3-[N-(2-chloroethyl)ethylaminoethyl]benzo[b]thiophen.

Mitscherlichite. A mineral. It is a potassium-copper chloride.

Mitschlich's Ammoniacal Salt. Probably a double compound of mercur-oxy-ammonium nitrate, and mercuri-ammonium nitrate $(NH_2.Hg_2O)NO_3.(NH_2.Hg)NO_3.H_2O$.

Mittel AEP. A proprietary softening agent for cellulose esters. It is ethyl-*p*-toluene-sulphonate.

Mittel HNA. A proprietary solvent stated to be acetophenone, $C_6H_5.COCH_3$. It is a solvent for cellulose nitrate and acetate, shellac, etc.

Mittel KP. Cresyl-*p*-toluene-sulphonate. A softening agent for cellulose esters.

Mittel L. A solvent resembling turpentine.

Mittler's Green. See CHROMIUM GREEN.

Mixed Acid (Nitrating Acid). Any mixture of nitric and sulphuric acids used for nitrating. Usually it consists of 37 per cent. nitric acid and 63 per cent. sulphuric acid. Used for nitrating dyes and explosives.

Mixed Ether. An ether containing two

different alkyl radicles, as in ethyl-methyl ether, $C_2H_5.O.CH_3$, as opposed to one containing two similar ones, as in ethyl ether, $C_2H_5OC_2H_5$.

Mixed Gas. See DOWSON GAS.

Mixed Metal. A term used for alloys of cerium, lanthanum, and praseodymium.

Mixed Vitriol (Salzburg Vitriol). Cupric-ferrous sulphate, $CuSO_4.3FeSO_4.28H_2O$.

Mixite. A mineral, $Cu_2O.As_2O_5.H_2O$, with some Bi_2O_3.

Mix Metal. Misch metal.

Mixogen. A proprietary preparation of ethinylœstradiol and methyltestoster-one. Treatment of menstrual disorders. (316).

Mixol. A timber insecticide. (511).

Mixtamycin. A proprietary preparation of streptoduocin. An antibiotic. (309).

MN Powder (Maxim-Nordenfelt Powder). An American guncotton powder gelatinised with ethyl acetate.

M.N.T. Mono-nitro-toluene, $CH_3.C_6.H_4.NO_2$, an intermediate in the preparation of trinitro-toluene.

Moac. Very finely divided form of mica which, when used as a dusting agent on rubber, gives a silky lustre.

Mobilsol 44. Mobilsol 66. Registered trade marks for modifying diluents for epoxy resins. They give flexibility. (160).

Mocasco Iron. A proprietary trade name for a nickel-chromium-molybdenum cast iron containing 1–1·35 per cent. nickel, 0·25–0·3 per cent. chromium, and 0·75 per cent. molybdenum.

Mocaya Oil (Macaja Butter). Paraguay palm oil obtained from the kernels of *Acrocomia scelerocarpa*. It has a saponification value of 189·8, an iodine value of 77·2, and an acid value of 110.

Mocha-stone. Agates of white or brown chalcedony from India, with markings due to oxides of iron and manganese.

Mock Epsoms. Needle crystals of sodium sulphate, Na_2SO_4.

Mock Gold. See MOSAIC GOLD.

Mock Lead. Both tungsten ore found in Cornwall and zinc blende are known by this name.

Mock Silver. An alloy of 84 per cent. aluminium, 10 per cent tin, 5·5 per cent. copper, and 0·1 per cent. phosphorus.

Mock Turkey Red. Barwood red on cotton.

Mock Vermilion. Lead chromate.

Modderite. A mineral. It is an arsenide of cobalt.

Moddite. An explosive. It is a variety of Cordite MD, but is made with a nitro-cellulose partially soluble in ether alcohol.

Modecate. A proprietary preparation of FLUPHENAZINE decanoate used in the treatment of psychotic disorders. (682).

Modern Blue. See MODERN VIOLET.

Modern Blue CVI (Modern Cyanines). Dyestuffs. They are derivatives of gallocyanine.

Modern Heliotrope PH. A dyestuff obtained by condensing nitroso-mono-ethyl-*o*-toluidine, with gallamide, and reducing.

Modern Violet (1900 Blue, Modern Violet R, Chromazol Violet). Dyestuffs. They are leuco derivatives of gallo-cyanines.

Modern Violet N. A dyestuff. It is the leuco compound of a pyrogallo-cyanine.

Modern Violet R. A dyestuff. It is a British equivalent of Modern violet.

Modified Soda. A mixture of sodium carbonate and bicarbonate used as a cleaning agent in laundries.

Modified Butacite. A proprietary trade name for thermosetting polyvinyl butyral synthetic resin.

Modified Vinylite X. A proprietary trade name for thermosetting polyvinyl butyral synthetic resin.

Modinal: T. A proprietary trade name for a wetting agent consisting of a long chain alcohol sulphate.

Moditen. A proprietary preparation of fluphenazine hydrochloride. A sedative. (326).

Moditen Enanthate. A proprietary preparation of FLUPHENAZINE enanthate used in the treatment of psychotic disorders. (682).

Modulac. A thixotropic alkyd resin. (512).

Modulac 135W. The proprietary trade name for a typical drying oil alkyd. (2).

Modulex. A trade mark for a carbon black for use as a pigment.

Modumite. A mineral. It is skutte-rudite.

Moduretic. A proprietary preparation of AMILORIDE hydrochloride and HYDROCHLORTHIAZIDE. A diuretic. (472).

Moellon. See DÉGRAS, ARTIFICIAL.

Moerner's Reagent. This consists of 1 c.c. formalin with 45 c.c. distilled water and 55 c.c. of 50 per cent. sulphuric acid.

Mogadon. A proprietary preparation of nitrazepam. A hypnotic. (314).

Mogador Gum. See MOROCCO GUM.

Mohair. A material made from the hair of the Angora goat.

Mohawkite. See ALGODONITE.

Mohawk Steel. A proprietary trade name for a hot die steel containing about 14 per cent. tungsten, 3·5 per cent. chromium, 0·7 per cent. vanadium, and 0·45 per cent. carbon.

Mohrah Butter. See BASSIO TALLOW.

Mohr's Salt. Ferrous ammonium sulphate, $FeSO_4.(NH_4)_2.SO_4.6H_2O$. Used in volumetric analysis for the standardisation of permanganate solution.

Mohsite. A mineral. It is a variety of ilmenite.

Moissanite. A mineral. It is carbon silicide, CSi, and is identical with carborundum.

Molaschar. A decolorising carbon used for sugar juices.

Molascuit. A cattle food. It is the fine fibre of the sugar cane or begasse, with cane molasses absorbed by it.

Molasocarb. A decolorising black made from molasses.

Molasses. The non-crystallisable residue from sugar. Cane molasses contain 55 per cent. sugar, 20 per cent. water, and 9 per cent. ash, whilst beet molasses consists of 50 per cent. sugar, 10 per cent. salts, 20 per cent. water, 10 per cent. nitrogenous matter, and 10 per cent. non-nitrogenous matter. Used as a cattle food.

Molassine Meal. A mixture of molasses and peat moss. A cattle food.

Molcer. A proprietary preparation of DIOCTYL SODIUM SULPHOSUCCINATE used as ear-drops for the removal of wax. (255).

Molcose. A food preparation containing 22·7 per cent. proteids, obtained from milk. Recommended for debility in children, and for convalescents.

Moldarta. A proprietary trade name for wood-flour plastics.

Moldesite. Proprietary name for a range of phenolic moulding materials. (683).

Moldovite. A mineral. It is a variety of obsidian.

Molengraaffite. Synonym for Lamprophyllite.

Moler. A Danish diatomaceous earth containing 82·6 per cent. silica, 5·33 per cent. aluminium oxide, a small proportion of ferric oxide, and organic matter. It is very light in weight, and is used in the manufacture of heat-insulating materials.

Molera. A heat insulator obtained by mixing fine clay with cork dust, and firing.

Molindone. An anti-psychotic. 3-Ethyl - 6, 7 - dihydro - 2 - methyl - 5 - (morpholinomethyl) indol - 4(5H) - one.

Molivate. A proprietary preparation of CLOBETASONE butyrate used in the treatment of eczema. (335).

Mollin. A base for ointments. It is a soft soap containing 17 per cent. of uncombined fat.

Mollit. A proprietary trade name for a polystyrene synthetic resin.

Mollit B. A proprietary trade name for glyceryl tribenzoate.

Mollit I. Diethyl-diphenyl-urea, a softening agent for cellulose esters.

Mollite. See CHESSYLITE.

Moloie. A proprietary trade name for a manganese-molybdenum steel.

Mollphorus. A glycerin substitute consisting of raw and invert sugar.

Moloid Tablets. A German preparation. It is a mixture of nitro-bodies in 1 mg. tablets. It is chewed.

Molybdate Red. A lead chromate pigment consisting of mixed crystals of lead chromate, lead sulphate, and a small proportion of lead molybdate. Its colour varies from reddish-orange to scarlet.

Molybdenite (Molybdenum Glance). A mineral. It is molybdenum disulphide, MoS_2.

Molybdenum Blue. A blue pigment, $4MoO_3+MoO_2$, a product of the reduction of molybdic acid.

Molybdenum, Chromium. See CHROMIUM MOLYBDENUM.

Molybdenum Glance. See MOLYBDENITE.

Molybdenum Indigo. A molybdenum oxide, Mo_5O_7, used for colouring rubber.

Molybdenum Nickel. An alloy of 75 per cent. molybdenum and 25 per cent. nickel. Used in the manufacture of saws.

Molybdenum Ochre. See MOLYBDIC OCHRE.

Molybdenum Permalloy. A proprietary trade name for an alloy of 81 per cent. nickel, 17 per cent. iron, and 2 per cent. molybdenum. It has a higher permeability than Standard Permalloy (q.v.).

Molybdenum Steel. A variable alloy. It usually contains from 0·06–1·73 per cent. carbon and 0·23–15 per cent. molybdenum.

Molybdic Ochre (Molybdine, Molybdite). A mineral, MoO_3.

Molybdine. See MOLYBDIC OCHRE.

Molybdite. See MOLYBDIC OCHRE.

Molybdoferrite. A mineral, $FeMoO_4$.

Molybdomenite. A mineral. It is a lead selenite.

Molybdophyllite. A mineral. It is $(Pb,Mg)_2SiO_4.H_2O$.

Molybdoscheelite. Synonym for Seyrigite.

Molybdosodalite. A variety of sodalite containing nearly 3 per cent. MoO_3. It is green in colour.

Molybdurane. A mineral, $UO_2.UO_3.2MoO_4$.

Molysite. A mineral, $FeCl_3$.

Molyte. A trade name for a patented

mixture of calcium and molybdenum oxides with a flux.

Momea (Mimea). A hemp preparation made in Tibet.

Momordicine. Elaterin, $C_{20}H_{28}O_5$.

Monacetin. Glyceryl monoacetate, $CH_2OH.CHOH.CH_2OOCCH_3$.

Monachit. An explosive containing 12 per cent. trinitro-xylene, 1 per cent. charcoal, and 1 per cent. collodion cotton.

Monamath. A saline earth used as a manure in India.

Monarsone. Sodium-dimethyl-arsonate. A substitute for salvarsan.

Monase. A proprietary trade name for the acetate of Etryptamine.

Monastral Colours. A proprietary trade name for copper phthalo-cyanine or its derivatives.

Monazite. A mineral composed essentially of phosphates of the cerium and lanthanum earths, together with a small and variable proportion of thoria. Also see THORITE and URDITE.

Mond 70 Alloy. An alloy of 70 per cent. nickel, 26 per cent. copper, and 4 per cent. manganese.

Mond Gas. A combustible gas produced by passing air and steam over heated coal or peat. It consists of a mixture of carbon monoxide, hydrogen, and nitrogen.

Monel 400. A trade mark. Copper 30, Manganese 1, iron 2·5 max. nickel balance. A general engineering alloy with good resistance to corrosion by sea water, sulphuric, hydrochloric and phosphoric acids. (208).

Monel Alloy 400. A trade mark for an alloy of 30 per cent. copper, 1 per cent. manganese, 2·5 per cent. (max) iron and the balance nickel. (143).

Monel 414. Monel 400 with a high carbon content. Improved machining properties. (208).

Monel Alloy 414. As for alloy 400 but with high carbon content. (143).

Monel Alloy K-500. A trade mark for an alloy of 30 per cent. copper, 3 per cent. aluminium, 0·5 per cent. titanium and the balance nickel. (143).

Monel, Ebonised. See EBONISED MONEL.

Monel Metal. An alloy. The cast metal usually contains from 68–70 per cent. nickel, 28 per cent. copper, 2 per cent. iron, 1 per cent. silicon, and 0·25 per cent. manganese; and the forged alloy consists of 68 per cent. nickel, 28 per cent. copper, 2 per cent. iron, 1·5 per cent. manganese, and 0·2 per cent. silicon. There are sometimes traces of carbon, phosphorus, and zinc. An alloy of this class containing 60 per cent. nickel, 33 per cent. copper, and

7 per cent. iron, has a specific gravity of 8·9 and melts at 1360° C. The average Monel Metal contains 67 per cent. nickel, 28 per cent. copper, and 5 per cent. other metals (2 per cent. iron and 1·5 per cent. manganese). It melts at 1360° C. and has a specific gravity of 8·82 and a coefficient of expansion of 0·0000137 per °C. Resistivity 48 micro-ohms . cm. Heat conductivity 1/15 that of copper. Hardness, cast 120–140 kg. per sq. mm. Hot rolled 150–190 kg. per sq. mm. Cold drawn 200–217 kg. per sq. mm. Monel metal is used in the manufacture of chemical plant.

Monensin. An anti-protozoan. 4 - {2 - [2 - Ethyl - 5′ - (tetrahydro - 6 - hydroxy - 6 - hydroxymethyl - 3, 5 - dimethylpyran - 2 - yl) - 3′ - methylbitetrahydro - 2, 2′ - furyl - 5 - yl] - 9 - hydroxy - 2, 8 - dimethyl - 1, 6 - dioxaspiro [4, 5] dec - 7 - yl} - 3 - methoxy - 2 - methylvaleric acid. COBAN is the sodium salt.

Monesia. A South American vegetable extract, said to be obtained from the bark of *Lucuma glycyphlœa*. An astringent used in diarrhœa.

Monesin. A saponin-like substance extracted from the bark of the South African plant, *Chrysophyllum viridifolium*.

Monetite. A calcium phosphate, $CaHPO_4$, found in guano.

Monex. A proprietary trade name for a rubber vulcanisation accelerator. It is tetramethyl-thiurum-monosulphide, a yellow powder with a melting-point of 105° C. (435).

Monheimite. A variety of the mineral zinc spar (zinc carbonate) containing over 20 per cent. iron carbonate.

Monimolite. A mineral, $(Pb.Fe.Mn)_3$ $(SbO_4)_2$.

Monistat. A proprietary preparation of MICONAZOLE nitrate used in the treatment of vaginal candidiasis. (369).

Monite. A proprietary plastic.

Monkey Nuts. See GROUND NUTS.

Monnex. A compound obtained by reacting urea with potassium carbonate. Used as a fire extinguisher. It is non-toxic. (2).

Monobel (A2 Monobel). A trade mark for a smokeless powder. It is a mixture of 9–11 parts nitro-glycerin, 56–61 parts ammonium nitrate, 8–10 parts wood meal, 0·5–1·5 parts magnesium carbonate, and 18·5–21·5 parts potassium chloride.

Monocast. A trade mark for a cast nylon obtainable in large diameter cylinders.

Monochin. A German preparation. It

is a mixture of quinine, codeine, aspirin, and pyramidone.

Monochlorinated Dutch Liquid. See DUTCH LIQUID MONOCHLORINATE.

Monochrome Yellow Paste, Powder. Dyestuffs. They are British brands of Milling yellow.

Monodral. A proprietary preparation containing penthienate methobromide. An antispasmodic. (263).

Monofil. A proprietary brand of artificial horse hair.

Monoformin. The formyl derivative of glycerin, $C_3H_5(OH)_2(OCHO)$.

Monogermane. Germanium tetrahydride, GeH_4.

Monoglyme. Ethylene glycol dimethyl ether.

Monol. Calcium permanganate, Ca $(MnO_4)_2.4H_2O$. Used in the textile industry.

Monolite. Insoluble lake colours. (512).

Monolite Dyestuffs. A registered trade name for certain dyestuffs.

Monolite Fast Scarlet R. See HELIO FAST RED RL.

Monolite Fast Scarlet RN. A dyestuff. It is a British equivalent of Helio fast red RL.

Monolite Red P. A British dyestuff. It is equivalent to Lake red P.

Monolite Red R. See LITHOL RED B.

Monolite Yellow. See HANSA YELLOW G.

Monomethyl Methylene Blue. Phenyltrimethyl-thionine.

Monopar. A proprietary trade name for Stilbazium Iodide.

Monoplas. Monomeric plasticisers. (542).

Monoplas 7L and 9L. Proprietary aliphatic phthalate plasticisers used in P.V.C. (94).

Monoplas 204. A proprietary plasticiser. Dibutylphthalate. (542).

Monoplas 208. A proprietary plasticiser. Di-2-ethyl hexyl phthalate. (542).

Monoplas 209. A proprietary plasticiser. Dinonyl phthalate. (542).

Monoplas 210. A proprietary plasticiser used in PVC. Di-iso-decyl phthalate. It has a viscosity of 81 centipoise. (94).

Monoplas 214. A proprietary plasticiser. Di-isobutyl phthalate. (542).

Monoplas 218. A proprietary plasticiser. Diisooctyl phthalate. (542).

Monoplas 230. A proprietary adipic ester plasticiser. (542).

Monoplas 279. A proprietary dialkyl (C_7-C_9) phthalate plasticiser.

Monoplas 410. A proprietary low-temperature monomeric plasticiser used in PVC. (94).

Monoplas 468. A proprietary plasticiser. Isobutyl nonyl phthalate. (542).

Monoplas 513. A proprietary plasticiser. Ditridecyl phthalate. (542).

Monoplas 524. A proprietary plasticiser. Triisooctyl trimellitate. (542).

Monoplas 530. A proprietary monomeric plasticiser for PVC of the trimellitate type. It has a viscosity of 102 centipoise. (94).

Monoplas 1000. A proprietary C_6-C_{10} aliphatic alcohol phthalate plasticiser. (542).

Monoplex 5. A proprietary trade name for dibenzyl sebacate. A vinyl plasticiser.

Monoplex II. A proprietary trade name for the dialkyl ester of a synthetic long chain dibasic acid. (13).

Monoplex 16. A proprietary trade name for a monomeric nitrile. A vinyl plasticiser. (13).

Monoplex S71. A proprietary trade name for hexyl epoxystearate. A vinyl plasticiser. (13).

Monopoline. A liquid motor fuel containing 90 per cent. petrol and 10 per cent. absolute alcohol.

Monopol Oil. See TURKEY RED OILS.

Monopol Soap (Avirol). Sulphonated oils similar to Turkey red oil. Used as wetting-out agents.

Monopyrate. See GALLOLS.

Monoresate. See EURESOL.

Monosol. Soluble lake colours. (512).

Monosulfiram. A parasiticide. Tetraethylthiuram monosulphide.

Monotah. A name for guaiacol-methylglycollate.

Monotal. Guaiacol - methyl - glycol - late, $CH_2(OCH_3) . COOC_6H_4 . OCH_3$. Used in cases of neuralgia, and as a local anæsthetic.

Monotheamin and Amytal Pulvules. A proprietary preparation of theophylline monoethanolamine and amylobarbitone. (333).

Monotheamin Pulvules. A proprietary preparation of theophylline monoethanolamine. (333).

Mono-Thiurad. An ultra accelerator for rubber. It is tetramethyl thiuram monosulphide. (559).

Monotype Metal. See TYPE METAL.

Monox. A product containing mainly silicon monoxide, with some silicon, silicon dioxide, and small quantities of silicon carbide. It is obtained by heating sand with silicon, carborundum, or coke, in the electric furnace. A good thermal and electrical insulator. It decomposes water with the evolution of hydrogen.

Monphytol. A proprietary preparation of boric acid, chlorbutol, methyl salicylate, salicylic acid and undecylenic acid. Antifungal skin powder. (352).

Monrepite. A mineral. It is a mica containing iron.

Monrolite. A mineral. It is a variety of sillimanite.

Monsanto Salt. A proprietary trade name for o-chloro-p-toluene sodium sulphonate, $Cl.C_7H_6.SO_3.Na$.

Monsell's Salt. Basic ferric sulphate, $Fe_4O(SO_4)_5$. Used in medicine.

Montacel. A special preparation of montan wax and pitch for paper and cardboard.

Montaclere. Styrenated phenol rubber antioxidant. (559).

Montago. A German light alloy with a specific gravity lower than aluminium.

Montana Gold. An alloy of 89 per cent. copper, 10·5 per cent. zinc, and 0·5 per cent. aluminium.

Montana Wax. See IRISH PEAT WAX.

Montanine. A liquid containing 31 per cent. hydrofluosilicic acid. Recommended as a disinfectant for the walls of breweries and distilleries. It is obtained from by-products in the pottery industry.

Montanin Wax. See IRISH PEAT WAX.

Montanite. A mineral, $(Bi(OH)_2)_2$. TeO_4.

Montan Pitch. The residue from the production of montan wax. The crude material gives an ash of 1·7 per cent., has an acid value of 3, and a saponification value of 6.

Montan Wax. A wax obtained by distilling in steam the bitumen prepared by extracting sulphurous brown coal with benzine. The crude material melts at 81° C., has a specific gravity of 1, a saponification value of 58, an acid value of 25, and contains 15 per cent. soluble in ether. The ash is about 0·5 per cent. The white purified material consists of montanic acid, $C_{28}H_{58}O_2$, and an alcohol. It is used as a substitute for ceresin in the candle industry, also as a substitute for carnauba wax in polishes, and as an insulating material.

Montar. Synthetic pitchy resins. (559).

Montax. A proprietary trade name for a filler for rubber, etc. It is a mixture of hydrated magnesium carbonate and silica.

Montbrayite. A mineral. It is $12[Au_2Te_3]$.

Montebrasite. See AMBLYGONITE.

Monthier's Blue. A coloured compound obtained by the oxidation of the precipitate formed by the action of ammoniacal ferrous chloride upon potassium ferrocyanide, $(Fe_2)_2(Fe(CN)_6)_3$. $6NH_3.9H_2O$.

Montesite. A mineral. It is $PbSn_4S_5$.

Montgomeryite. A mineral. It is $2[Cu_4Al_5(PO_4)_6(OH)_5.11H_2O]$.

Monticellite (Scacchite). A mineral, $2(Ca.Mg)O.SiO_2$.

Montmartite. A mineral. It is a variety of gypsum.

Montmorillonite. A mineral, Al_2O_3. SiO_2. It is a colloidal clay similar to Bentonite, and is often combined with alkalies or alkaline earths. Used for decolorising oils.

Monto. A lubricating oil additive. (559).

Montopore. Foamable polystryene beads. (559).

Montorate TU. A proprietary accelerator. Trimethyl thiourea. (57).

Montothene. Ethylene vinyl acetate plastics. (539).

Montothene G50. A proprietary trade name for an ethylene vinyl acetate copolymer. It is translucent, nontoxic and has good mechanical properties. It is used for film injection and blow moulded articles and extrusions. (75).

Montpelier Yellow. See TURNER'S YELLOW.

Montreal Potash. Commercial potassium carbonate.

Montronite. A mineral. It is a hydrated ferric silicate.

Montroseite. A mineral. It is $4[(V,Fe)O.OH]$.

Montroydite. A mineral, HgO.

Monzaldon. A proprietary preparation of REVATRINE hydrochloride. A relaxant of the uterus used in veterinary work.

Moogrol. A proprietary preparation. It is a mixture of the acids of the chaulmoogric series. Used as a therapeutic agent in leprosy.

Moon-stone. A mineral. It is an opalescent variety of orthoclase (q.v.).

Moore's Ointment. Resin ointment.

Moore's Teejel. A proprietary preparation of CHOLINE salicylate and CETALKONIUM CHLORIDE used in the treatment of mouth ulcers. (487).

Moplen. A registered trade mark for a brand of polypropylene. A flexible hard, tough hydrocarbon thermoplastic used for moulding domestic ware and for electrical purposes. (61).

Moplen-RO. A registered trade mark for polyethylene, high density. (61).

Morantel. An anthelmintic. (E)-1,4,5, 6 - Tetrahydro - 1 - methyl - 2 - [2 - (3 - methyl - 2 - thienyl) vinyl] - pyrimidine. "UK 2964–18" and "CP 12,009–18" are undergoing clinical trial as the tartrate.

Moranyl. A proprietary preparation. It is sym-disodium-m-amino-benzoyl-m - amino - p - methyl - benzoyl - l - naphthyl - amino - 4 : 6 : 8- trisulphonate-urea.

Morat White (Moudan White). A white pigment. It is a clay found in Switzerland.

Morazone. 2,3-Dimethyl-4-(3-methyl-2-phenylmorpholinomethyl) - 1 - phenyl - pyrazol-5-one.

Mordant, Green. See GREEN MORDANT.

Mordant, Red. See RED LIQUOR.

Mordant Rouge. See RED LIQUOR.

Mordant Yellow O. See MILLING YELLOW.

Moreau Marble. A marble prepared by immersing soft amorphous limestone in a bath of zinc sulphate, and drying.

Moreau's Solution. A solution of thymic acid and guaiacol in olive oil. Used for intramuscular injection.

Morell's Solution. A disinfecting solution containing arsenious acid, caustic soda, and a small quantity of phenol, dissolved in water.

Morencite. A mineral. It is essentially a ferric silicate.

Morenosite (Nickel Vitriol, Gapite, Epimillerite, Pyromeline). A mineral. It is nickel sulphate, $NiSO_4.7H_2O$.

Moresnetite. A mineral. It is a mixture of clay and calamine.

Morflex 100. A proprietary trade name for di-iso-octyl phthalate. A vinyl plasticiser.

Morflex 125. A proprietary plasticiser. n-Octyl n-decyl phthalate. (85).

Morflex 130. A proprietary trade name for didecyl phthalate. A vinyl plasticiser. (115).

Morflex 175. A proprietary trade name for octyl decyl phthalate. A vinyl plasticiser.

Morflex 200. A proprietary trade name for di-iso-octyl sebacate. A vinyl plasticiser. (115).

Morflex 210. A proprietary plasticiser. Diethyl hexyl phthalate. (85).

Morflex 240. A proprietary plasticiser. Dibutyl phthalate. (85).

Morflex 310. A proprietary plasticiser. Di-2 ethyl hexyl adipate. (85).

Morflex 325. A proprietary plasticiser. n-octyl n-decyl adipate. (85).

Morflex 330. A proprietary trade name for di-decyl adipate. A vinyl plasticiser. (85).

Morflex 375. A proprietary trade name for n-octyl decyl adipate. A vinyl plasticiser. (115).

Morflex 410. A proprietary plasticiser. Di-2-ethyl hexyl azelate. (85).

Morflex 510. A proprietary plasticiser. Tri-2-ethyl hexyl trimellitate. (85).

Morflex 525. A proprietary plasticiser. Tri (n-octyl n-decyl) trimellitate. (85)

Morflex 530. A proprietary plasticiser. Triisodecyl trimellitate. (85).

Morflex P50. A proprietary n-alkyl phthalate plasticiser. (85).

Morganite. A trade name for a beryl of rose colour.

Morhal Resin. A resin obtained from *Vatica lanceæfolia*, of India.

Morhulin. A proprietary preparation of cod-liver oil, zinc oxide and Dakin's solution. A wound cleanser. (487).

Morin. Tetrahydroxy-flavonol.

Morin. A product obtained by extracting rasped fustic with boiling water containing 2 per cent. sodium carbonate, and concentrating the solution to 1·041 specific gravity.

Morintannic Acid. Maclurin (*q.v.*).

Morion. A mineral. It is a dark variety of Cairngorm stone.

Morland's Salt. $(Cr(NH_3)_2(SCN)_4)$ HNH. The guanidinium salt of the same complex as Reinecke's salt, formed as a by-product in the preparation of the latter salt.

Mornidine. A proprietary trade name for Pipamazine.

Moroccan Olive Oil. Argan oil from *Arganum sideroxylon*. It has a specific gravity at 15° C. of 0·919, a saponification value of 192, an iodine value of 95·9, and an acid value of 0·18.

Morocco Gum (Mogador Gum, Brown Barberry Gum). A variety of gum acacia in the form of tears.

Morocochite. See MATILDITE.

Moronal. Aluminium-formaldehyde-sulphite. An antiseptic and astringent.

Moronite. A mineral. It is a variety of jarosite.

Moroxite. An impure variety of the mineral Apatite.

Moroxydine. N-(Guanidinoformimidoyl)morpholine.

Morpan NBB. A proprietary trade name for a 50 per cent. active composition of octyldimethyl benzyl ammonium bromide. A low forming flocculating agent used as a filtering aid.

Morphacetin. Acetomorphine. See HEROIN.

Morpheridine. Ethyl 1-(2-morpholinoethyl) - 4 - phenylpiperidine - 4 - car - boxylate. Morpholinoethylnorpethidine.

Morphia. Morphine, $C_{17}H_{19}NO_3$.

Morpholine. Tetrahydro-1 : 4-oxazine. It has a boiling-point of 128° C.

Morpholinoethylnorpethidine. Morpheridine.

Morphosan. Morphine-methyl-bromide, $C_{17}HN_{19}O_3.CH_3Br.H_2O$. Suggested as a substitute for morphine.

Morsep. A proprietary preparation of cetrimide, cod-liver oil and Dakin's solution used to treat nappy rash. (487).

Mortar. Ordinary mortars consist of a mixture of lime and sand, sometimes with the additions of small quantities of other materials. Cement mortars are made with Portland, natural, or slag cement.

Mortar, Cement. See CEMENT MORTAR and MORTAR.

Mortar, Hydraulic. See HYDRAULIC CEMENT.

Mortar, Lime. See LIME MORTAR and MORTAR.

Mortha. A proprietary preparation of morphine hydrochloride and tetrahydroaminacrine hydrochloride. An analgesic. (319).

Morto. Weedkiller and potato haulm destroyer. (501).

Morton's Fluid. A solution containing iodine, potassium iodide, and glycerin.

Morubiline. An extract of the fresh liver of the cod.

Moruette. A proprietary preparation. It is a vitamin cod-liver oil preparation.

Morvenite. A mineral. It is harmotone.

Mosaic Gold (Mock Gold, Cat's Gold, Bronzing Powder, Tin Bronze). A flaky yellow form of disulphide of tin, SnS_2. It was formerly used for gilding and imitating bronze. A substance also called Mosaic gold is made from an amalgam of tin and mercury, with ammonium chloride and sulphur. Used as a bronzing material for plastics.

Mosaic Gold. An alloy of 50 per cent. copper, and 50 per cent. zinc. One containing 65 per cent. copper and 35 per cent. zinc is also known as Mosaic gold.

Mosaic Silver. An alloy of tin and bismuth.

Mosandrite. A mineral, $3(CaO.SiO_2. TiO_2)(Ce.La.Di)_4O_3.SiO_2.TiO_2$.

Moscavon. A German proprietary preparation stated to contain the active principle of nutmeg in sodium stearate. A remedy for heartburn.

Mosesite. A mineral. It is a mixture of ammonium mercuro-chloride with mercurous sulphate.

Moskonfyt. A South African name for a grape syrup prepared by boiling the clarified must until it contains 68–69 per cent. sugar.

Moss Agate. An agate consisting of chalcedony enclosing twisted filaments, usually green, but sometimes red or brown.

Moss Alcohol. Ethyl alcohol prepared from reindeer moss and Iceland moss.

Mossbunk Oil. See POGY OIL and MENHADEN OIL.

Moss Copper. During the refining of " coarse metal " for the production of copper the oxide of copper used for refining is in excess of that required; copper metal separates out in the form of velvety filaments known as moss copper.

Moss Green. See OIL GREEN.

Mossite. A mineral. It is an iron-columbo-tantalate containing 82·92 per cent. Cb_2O_5 and Ta_2O_5.

Moss Starch. Lichenin, $(C_6H_{10}O_5)x$.

Mos-Tox. Moss eradicant. (507).

Mota. Tablets of metaldehyde.

Motalin. A motor fuel containing 0·4 per cent. motyl (a mixture of equal volumes of iron carbonyl and benzine). See MOTYL.

Mother of Coal. Charcoal.

Mother of Pearl. This consists of alternate layers of calcium carbonate and chitin.

Mother of Pearl Sulphur (Nacreous Sulphur). A form of monoclinic sulphur obtained by heating sulphur with benzene at 140° C. It is unstable.

Mother of Vinegar. See VINEGAR PLANT.

Motival. A proprietary preparation of FLUPHENAZINE hydrochloride and NOR-TRIPTYLINE hydrochloride. A tranquilliser. (682).

Motor Alcohol. Motor spirits obtained by fermentation of carbohydrates, or distillation, and mainly consisting of alcohols.

Motor Benzol. Motor spirit derived from coal gas, or its condensation products.

Motor Fuel, National. See NATIONAL MOTOR FUEL.

Motor Spirit. The saturated aliphatic hydrocarbons of American oil, the poly-methylenes from Baku oil, or the unsaturated hydrocarbons derived from shale oil, as well as benzene, C_6H_6, and alcohol, C_2H_5OH, are employed in internal combustion engines. They usually have a boiling-point below 120° C.

Mots. See SAKÉ.

Mottled Soaps. Soaps to which have been added insoluble colouring matters, such as ferrous sulphate, Venetian red, ferric oxide, or ultramarine.

Mottramite. A mineral. It is a vanadate of lead and copper, $(Pb.Cu)_3(VO_4)_2 +2PbCu(OH)_2$.

Motung Steel. A proprietary trade name for a high speed steel containing 7·5–8·5 per cent. molybdenum, 1·25–2 per cent. tungsten, 3·5–4·5 per cent. chromium, 0·9–1·5 per cent. vanadium, 0·8 carbon, 0·2–0·4 per cent. manganese, and 0·25–0·5 per cent. silicon.

Motyl. A German anti-knocking agent consisting of equal volumes of benzine and iron carbonyl, $Fe(CO)_5$. It is not sold as it is,· but in motalin, an anti-knock petrol. Motalin contains 0·4 per cent. motyl.

Motyline. Petrol containing 1 per cent.

ron-carbonyl. An anti-knock compound for internal combustion engines.

Moudan White. See MORAT WHITE.

Mou-iéou (Pi-yu). Chinese vegetable tallow.

Mouldensite. A proprietary synthetic resin product used for electrical insulation.

Mouldrite. Thermosetting moulding powders. (570).

Mountain Blue (Azure Blue, Mineral Blue, Copper Blue, Hamburg Blue, English Blue). These names have been transferred from the mineral azurite to the artificial product, which is prepared from cupric chloride. This substance is treated with lime paste, which produces green copper oxychloride, and the latter is then converted into the blue colour by adding potassium carbonate and milk of lime. It is a basic copper carbonate. A pigment used by painters.

Mountain Butter. A hydrated aluminium sulphate found in fibrous masses.

Mountain Cork (Elastic Asbestos). An asbestos which floats on water.

Mountain Flax. See AMIANTHUS.

Mountain Flour. See INFUSORIAL EARTH.

Mountain Green. A pigment prepared by precipitating a boiling solution of alum and copper sulphate with a hot solution of sodium or potassium sulphate. Varieties of mountain green blended with white clay, or heavy spar, are known under the names of Alexander green, Glance green, Napoleon's green, and Neuwieder green. The mineral malachite was formerly mined under the name of mountain green, but modern mountain greens are the artificial copper pigments.

Mountain Leather (Mountain Paper). Thin, tough types of asbestos.

Mountain Milk. An earth similar to infusorial earth, used as a rubber filler.

Mountain Paper. See MOUNTAIN LEATHER.

Mountain Soap. See STEATITE.

Mountain Tallow. See BITUMEN.

Mountain Wood. A variety of asbestos.

Mountford's Paint. A waterproof paint. It consists of asbestos, ground in water, potassium or sodium aluminate, and potassium or sodium silicate sometimes with oil, and zinc white.

Mourey's Aluminium Solder. An alloy of 82 per cent. zinc and 18 per cent. aluminium.

Mousset Alloy. See FRENCH SILVER.

Moussett's Alloy. An alloy of 60 per cent. copper, 27·5 per cent. silver, 9·5 per cent. zinc, and 3 per cent. nickel.

Mouth Glue. Bone glue, scented with lemon essence, and sweetened with sugar.

Movelat. A proprietary preparation of adrenocortical extract, salicylic acid and mucopolysaccharide polysulphuric acid ester used in the treatment of arthritis. (453).

Mowilith. A trade mark for a series of vinyl acetate, polymers and copolymers sold in solution and in emulsion form. (3).

Mowlith 15, 20, 50, 70. Trade marks for polyvinyl acetate powders or solutions of increasing K value. (3).

Mowlith D. A trade mark for a 50 per cent. dispersion of polyvinylacetate in water without plasticiser.

Mowlith 200. A trade mark for a copolymer of vinyl acetate and vinyl benzoate dispersed in water. (3).

Mowilith ABC. A trade mark for a copolymer of vinyl acetate, vinylbenzoate and crotonic acid in solution— 70 per cent solids. (3).

Mowilith D32. A trade mark for 38·7 per cent. polyvinyl acetate, 1·9 per cent. polyvinyl alcohol, 11·6 per cent. tricresyl phosphate, 7·7 per cent. dibutyl phthalate, 3·3 per cent. alcohol and 36·8 per cent. water. (3).

Mowilith D300. A trade mark for 70 per cent. vinylacetate and 30 per cent. vinyl chloride copolymerised and dispersed in water, dioctyl phthalate and dibutylphthalate added as plasticiser. (3).

Mowiol. A proprietary brand of polyvinyl alcohol. (3).

Mowital. A proprietary brand of polyvinylacetal. (3).

Moxaverine. An anti-spasmodic. 1-Benzyl - 3 - ethyl - 6, 7 - dimethoxyiso - quinoline. Eupaverin is the hydrochloride.

Moxipraquine. A protozoacid currently undergoing clinical trial as " 349C 59 ". It is 8 - {6 - [4 - (3 - hydroxybutyl) - piperazin - 1 - yl] hexylamino} - 6 - methoxyquinoline.

Moxisylyte. Thymoxamine.

MPE 750 and 770. Proprietary highdensity polyethylene injection moulding compounds. (559).

MPE 771. An antistatic form of MPE-770. (559).

MPS 500. A chlorinated fatty acid ester. A vinyl plasticiser. (37).

MR-1, MR-1A, MR-17A, MR-17B. Proprietary nomenclature for allyl resins.

M S 4. A silicone electrical insulating compound. (614).

M.S.S. Sodium metasilicate. (537).

Mucaine. A proprietary preparation of oxythazaine, aluminium hydroxide gel, and magnesium hydroxide. Treatment of œsophagitis. (245).

Muccocota Gum. See OCOTA COCOTA GUM.

Mucicarmin. A solution of 2 parts carmine, 1 part aluminium chloride, and 4 parts water. A staining solution.

Mucidan. A German preparation containing the mucilage and pus-dissolving salts of thiocyanates. Used in the treatment of catarrh.

Mucilage. Mucilage of acacia.

Mucilage Oil. See OIL OF MUCILAGES.

Mucodyne. A proprietary preparation of CARBOXYMETHYLCYSTEINE. An expectorant. (137).

Mucogen (Mukogen). Dimethyl-phenyl-p - ammonium - β - oxy - naphthox - azine - chloride, $C_{18}H_{15}N_2O_2Cl$. Prescribed for the removal of mucus in the intestines.

Mucomycin. A proprietary trade name for Lymecycline. (182).

Mudar. See MADAR.

Mudar Gum. A material obtained from *Calotropis giganteas*. It resembles gutta-percha, and contains about 20 per cent. of a rubbery material.

Mud, Denver. See KAOLIN CATAPLASM.

Mudge's Speculum Metal. An alloy of 69 per cent. copper and 31 per cent. tin.

Mud, Lime. See CHANCE MUD.

M.U.F. A rubber antioxidant for white and lightly coloured rubber.

Muflin. A proprietary preparation containing DEXTROMETHORPHAN hydrobromide, sodium citrate, PHENIRAMINE maleate and citric acid. An expectorant. (633).

Muga Silk. The product of the caterpillar, *Antheraca assama*, of Assam.

Muhlhaus White. A pigment. It is lead sulphate, $PbSO_4$.

Mukogen. See MUCOGEN.

Mulberry Essence. See ESSENCE OF MULBERRY.

Muldan. An orthoclase mineral.

Mule Gum. A name sometimes applied to Ceara rubber.

Mulene. See AMBRINE.

Mulhouse Green. A dyestuff. It is Gambine Y.

Mullanite. A mineral, $5PbS.2Sb_2S_3$.

Mullerine. A mineral of the Sylvanite type, containing antimony and lead.

Muller's Fluid. A solution of phosphoric acid in alcohol. A soldering fluid for brass and copper. The term is also used for a hardening agent used in microscopy. It consists of 30 grains potassium bichromate and 15 grains sodium sulphate dissolved in 100 c.c. distilled water.

Muller's Glass. See HYALITE.

Mullex. A proprietary refractory material made from mixtures containing various proportions of clay and mullite.

Mullfrax 301. A mullite-alumina product. (292).

Mullicite. A mineral. It is blue iron earth.

Mullite. (1) A refractory material formed by heating sillimanite to a temperature of 1550° C. It is also made from the minerals andalusite, dumortierite, and Indian cyanite, and by the electric fusion of alumina and silica. Mullite has a melting-point of about 1800° C. and a low coefficient of expansion.

Mullite. (2) A mineral. It is $Al_9Si_3O_{19.5}$.

Mulsivin. A proprietary preparation containing bromoform, codeine phosphate, tincture of aconite, belladonna alkaloids, ipecacuanha alkaloids, benzoin, balsam of tolu, storax and liquid paraffin. A cough linctus. (273).

Mulsoid (Lanoid, Loftine). Trade names for Bentonite detergents, usually consisting of soap with from 25–50 per cent. Bentonite.

Multaglut. A mixture of persulphate and calcium phosphate. Used to improve flour.

Multamine. Dyestuff for nylon. (512).

Multibionta. A proprietary multi-vitamin preparation. (310).

Multibrol. An organic combination of bromine consisting primarily of the sodium derivative of brom-oleate with a bromine content of 16 per cent.

Multilan. Dyes for wool, viscose and nylon carpets. (512).

Multistix. A proprietary test strip for the detection of pH, protein, glucose, ketones, bilirubin, blood and urobilinogen in urine. (807).

Multivite. A proprietary preparation of Vitamins A, D, C and thiamine hydrochloride. Vitamin supplement. (182).

Multrathane. A proprietary range of polyester and polyether-based polyurethane elastomers cross-linked with diols, castor oil, triols or diamines. (680).

Mulukilivary. A gum-resin obtained from *Balsamodedron Berryi*. A myrrh substitute.

Mumetal. A patented nickel-iron-copper alloy having the highest permeability of all known commercial materials. Its exceptional magnetic properties and low losses make it invaluable for cable loading, instrument transformers, relays, magnetic shields, etc.

Mummy. A bituminous product mixed with animal remains. It is used in oil as a brown pigment.

Mumpsvax. A proprietary live vaccine for mumps. (472).

Mundic. See IRON PYRITES.

Mundic, White. See MISPICKEL.

Mungo Fibres. Short fibres of shoddy (q.v.).

Munjeet. The root of *Rubia munjista*. It contains purpurin, and is an important Indian dyestuff.

Munjistin (Purpuroxanthic Acid). Dioxy - anthraquinone - carboxylic acid, $C_{15}H_8O_6$.

Munkforssite. A mineral. It is an aluminium-calcium phosphate and sulphate.

Muntz Metal. A brass containing 60 per cent. copper and 40 per cent. zinc, having a specific gravity of 8·4. It was originally used for sheathing ships, but is now used under the name of Yellow metal for the cheaper varieties of brass tube, wire, and sheet.

Murac. An insulating material stated to be made by treating the latex of *Sapotaceæ* species.

Murald. Aldrin insecticides. (501).

Murcurite. A mercury fungicide. (501).

Murdiel. Dieldrin insecticides. (501).

Murex. A proprietary trade name for a manganese steel containing 3 per cent. manganese, 1 per cent. carbon, and 0·85 per cent. nickel.

Murexan. See DIALURAMIDE.

Murexide (Naples Red). An obsolete red basic dyestuff, obtained by the action of nitric acid upon guano, and subsequently treating the product with ammonia.

Murfixtan. A mercury fungicide. (501).

Murfly. An insecticide. (501).

Murfoleos. Organo-phosphorus insecticides. (501).

Murfotox. Organo-phosphorus insecticides. (501).

Murfume. Pesticidal smoke generators. (501).

Muriacite. See ANHYDRITE.

Muriate of Ammonia. See SAL-AMMONIAC.

Muriate of Potash. Potassium chloride, KCl.

Muriate of Soda. Sodium chloride, NaCl.

Muriatic Acid. Hydrochloric acid, HCl.

Muriatic Acid Gas, Dephlogisticated. See DEPHLOGISTICATED MURIATIC ACID GAS.

Muriatic Ether. Hydrochloric Ether (q.v.).

Muripsin. A proprietary preparation of glutamic acid hydrochloride and pespin. (286).

Murman's Alloy. One contains 92 per cent. aluminium, 4·4 per cent. zinc, and 3·6 per cent. magnesium. Another consists of 72 per cent. aluminium,

14·5 per cent. zinc, and 13·5 per cent. magnesium.

Muromonite. A mineral. It contains about 30 per cent. berylla.

Murpherin. Rodenticides. (501).

Murphex. Disinfestation products. (501).

Murphicol. Pesticide suspensions. (501).

Murphos. Parathion insecticides. (501).

Musa Rubber. See BANANA RUBBER.

Muscarine (Campanuline). A dyestuff. It is the dihydroxy derivative of Meldola's blue, and is dimethyl-amino-oxynaphtho-phenoxazonium-chloride, $C_{18}H_{15}N_2O_2Cl$. Dyes cotton mordanted with tannin and tartar emetic blue. Employed for calico printing.

Muscle Sugar. Inositol, $C_6H_6(OH)_6$.

Muscovite (Potash Mica, Common Mica, White Mica, Biaxial Mica, Muscovy Glass). A mineral. It is a potassium-aluminium silicate, $K_2O.3Al_2O_3.4SiO_2$. A source of potassium.

Muscovy Glass. Clear sheets of the mineral muscovite. See MUSCOVITE.

Musculin. Syntonin, an albuminoid.

Müsenite. Synonym for Selenolinnacite.

Mushet Steel (Self-hardening Steel). Steels containing from 0·7–1·2 per cent. carbon and 2–3 per cent. tungsten. They require no quenching or tempering.

Mushroom Sugar. Mannite.

Musiv Gold. An alloy of from 66–70 per cent. copper and 30–34 per cent. zinc.

Musk. The dried animal secretion of the musk deer. It has been practically superseded by synthetic compounds.

Musk Ambrette. See AMBRENE.

Musk, Artificial. See MUSK BAUR, KETONE MUSK, XYLENE MUSK and AMBRENE.

Musk B. See MUSK BAUR.

Musk Baur (Musk B, Tonquinol). An artificial musk. It is 2 : 4 : 6-trinitro-*l*-methyl - 3 - tertiary - butyl - toluene, $C_6H(CH_3)(NO_2)_3.C(CH_3)_3$. Used for soap and toilet purposes, also as an antispasmodic and nervine.

Musk C. Ketone musk (q.v.).

Musk, Ketone. See KETONE MUSK.

Musk, Xylene. See XYLENE MUSK.

Mussanin. An extract from *Albizzin anthelmintica*. A vermifuge.

Must. Grape juice.

Mustard (*Sinapis B.P.*) A powdered mixture of reddish-brown mustard seeds from *Brassica nigra*, and white mustard seeds from *B. alba*.

Mustard Bran. Mustard seed husks ground with a small proportion of the seeds.

Mustard Gas (Yperite, Yellow Cross Gas). Dichloro-diethyl-sulphide. A military poison gas.

Mustard Oil. Ethyl-isothiocyanate, C_2H_5NCS. Also see OIL OF MUSTARD.

Mustard Oil, Methyl. See METHYL MUSTARD OIL.

Mustine. NN-Di-(2-chloroethyl)methylamine. Chlormethine. A proprietary preparation of mustine hydrochloride. A carcino-chemotherapeutic agent. (253).

Mustone. A Japanese chloroprene polymer. See NEOPRENE.

Muthanol. An oil suspension of radioactive bismuth hydroxide. An antisyphilitic.

Muthmannite. A mineral, (Ag.Au)Te.

Muthmann's Liquid. Acetylene-tetrabromide, $CHBr_2.CHBr_2$. A solvent.

Muthol. See PARAFFIN, LIQUID. Also see MUTHMANN'S LIQUID.

Muthydral. A combination of bismuth and mercury. An antisyphilitic.

M'Varavara. The roots of *Securidaca longipedunculata*. Employed medicinally.

MXM-7500. A proprietary epoxy putty used for filling in difficult radii and depressions in epoxy-fibre glass components. (684).

Myambutol. A proprietary preparation containing ethambutol. An antituberculous agent. (306).

Myanesin. A proprietary trade name for Mephenesin carbamate—muscle relaxant and tranquilliser. (182).

Mybasan. A proprietary trade name for Isoniazid.

My-B-Den. Adenosine Phosphate.

Mycalex. A registered trade mark for an insulating material used for the production of radio-frequency insulators. It consists essentially of mica dust bonded with boro-silicate glass. It can be extruded, and can be produced in the form of rods, sheets, or tubes. It has a high tensile strength and is unaffected by moisture. (39).

Mycardol. A proprietary trade name for Pentaerythritol Tetranitrate.

Mycifradin. A proprietary preparation of neomycin sulphate. An antibiotic. (325).

Myciguent. A proprietary preparation of neomycin sulphate used in dermatology as an antibacterial agent. (325).

Mycil. A proprietary trade name for Chlorphenesin. Preparations of chlorphenesin (=" Gecophen "). Antifungal. (182).

Mycivin. A proprietary preparation containing lincomycin hydrochloride. An antibiotic. (253).

Mycocide. A slime control agent. (624).

Mycoderma Aceti (Mycoderma Vini). Names given to the acetic acid bacteria. It is now called Bacteria aceti.

Mycoderma Vini. See MYCODERMA ACETI.

Mycodermine. A yeast preparation used in medicine.

Mycolactine. A proprietary preparation of dried yeast, ox bile, lactic principles, frangula, aloes and belladonna. A laxative. (781).

Mycose. Trehalose, $C_{12}H_{22}O_{11}.H_2O$, a sugar.

Mycostatin. A proprietary trade name for Niptatin.

Mycota. A proprietary preparation of undecenoic acid and zinc undecenoate. Treatment of fungal infections of the skin. (253).

Mycozol. A proprietary preparation of chlorbutol, salicylic acid, benzoic acid and malachite green. Treatment of fungal infections of the skin. (268).

Mydocalm. See TOLPERISONE.

Mydriacyl. A proprietary trade name for Tropicamide.

Mydriasine. Atropine-methyl-bromide, $C_{17}H_{23}NO_3.CH_3Br$. A mydriatic.

Mydrilate. A proprietary trade name for Cyclopentolate.

Mydrol. Iodo-methyl-phenyl-pyrazolone, a mydriatic.

Myelin. A white, fatty substance obtained from various animal and vegetable tissues.

Myelobromol. A proprietary preparation of MITOBRONITOL. An anti-leukæmic agent. (137).

Myer's Naphthol Green (Standard). For cholestrol determination. It consists of 50 mg. naphthol green B dissolved in 1,000 c.c. water. The solution is standardised against 5 c.c. of a chloroform solution of cholesterol containing 0·4 mg. cholesterol.

Mylanta. A proprietary preparation containing magnesium hydroxide, aluminium hydroxide and methylpolysiloxane. An antacid. (264).

Mylar. A proprietary trade name for a polyester film used for electrical insulation, cable lapping, magnetic tape. It has a very high tensile strength. (10). See also MELINEX. (2).

Myleran. A proprietary preparation of busulphan (1,4-dimethane sulphonoxy butane). Treatment of chronic myeloid leukæmia. (277).

Mylocon. A proprietary preparation containing methyl polysiloxane in flavoured vehicle. (264).

Mylodex A. A proprietary preparation of amylobarbitone and dexamphetamine. A sedative. (354).

Myloin. See ACIDOL.

Mylol. An insect repellant. (502).

Mylomide. A proprietary preparation of amylobarbitone and megemide. A sedative. (271).

Mynah. A proprietary preparation of ETHAMBUTOL and ISONIAZID used in the therapy of tuberculosis. (306).

Myoarsphenamine. The sodium salt of dihydroxy - diamino - arseno - benzene - diformaldehyde-sulphonic acid. It has the chemotherapeutic activity of neoarsphenamine.

Myocardol. A proprietary preparation of pentaerythritol tetranitrate. A vasodilator used in angina pectoris. (112).

Myocrisin. A proprietary preparation of sodium aurothiomalate. (336).

Myodil. Ethyl iodophenylundecylate. (520).

Myogen. A food preparation, consisting of pure albumin obtained from blood serum of cattle. It contains 13 per cent. nitrogenous matter, corresponding to 83 per cent. albumin.

Myolgin. A proprietary preparation of aspirin, paracetamol, codeine phosphate, caffeine citrate and acetomenaphthone. An analgesic. (248).

Myo-salvarsan. A German product closely related to neosalvarsan.

Myotonine. A proprietary preparation of bethanechol chloride, a smooth muscle stimulant. (810).

Myrabola Oil. A German soap-making material. It is an oil stated to be a mixture of different fatty acids and their glycerides. It is obtained from fat waste, and has an acid value of from 13–30, a saponification value of 166–217, and contains 0·8–2·2 per cent. unsaponifiable matter.

Myralact. (N-(2-Hydroxyethyl)tetradecylammonium lactate.

Myrcia Spirit. See BAY RUM.

Myrickite. A trade name for a chalcedony.

Myrilos. Calcium borogluconate solution (Veterinary). (514).

Myristica. Nutmeg.

MYRJ 45. A proprietary trade name for Polyoxyl 8 stearate.

MYRJ 52. A proprietary trade name for Polyoxyl 40 stearate.

Myrmekite. A mineral. It is quartz.

Myrobalans. The fruit of *Terminalia chebula*, of India. This is the chief variety of this product, but there are at least five varieties of the commercial article which are named after the district where they are marketed. They contain from 24–39 per cent. tannin and from 2·2–3·1 per cent. ash. A solid extract is made containing from 50–60 per cent. tannin.

Myrophine. Tetradecanoyl 6-ester of 3-benzylmorphine.

Myrosinase. The ferment of black mustard seed.

Myrrh. The true myrrh is that known as Herabol myrrh, a gum resin obtained from various species of *Balsamodendron* and *Cammiphora*. The acid value varies from 60–70; 24–40 per cent. is soluble in alcohol, and 5–8 per cent. in petroleum ether. Bisabol myrrh is derived principally from *Balsamea erythrea* and has an acid value of about 42. Myrrh is used mainly in toilet preparations, perfumery, and incense.

Myrrh, Substitute. See MULUKILIVARY.

Myrtle Green. See DINITROSORESORCIN.

Myrtol. A refined myrtle oil. It is an essential oil containing myrtenol, pinene, and cineol. It is used in medicine for bronchial and pulmonary affections, and as an antiseptic.

Mysoline. A proprietary preparation of primidone. An anticonvulsant. (2).

Mysorine. A mineral. It is a basic copper carbonate.

Myspamol. A proprietary trade name for Proquamezine.

Mysteclin Syrup. A proprietary preparation containing Tetracycline and Amphotericin. An antibiotic. (326).

Mysteclin Tablets. A proprietary preparation containing Tetracycline and Nystatin. An antibiotic. (326).

Mystery Gold. An alloy of 1 part platinum and 2 parts copper, with a little silver.

Mystic Metal. An alloy of 88·7 per cent. lead, 10·8 per cent. antimony, 0·4 per cent. iron, and 0·1 per cent. bismuth.

Mystin. A mixture of formaldehyde and sodium nitrite. A preservative.

Mytelase. Ambenonium chloride.

Myuizone. *p*-acetylaminobenzaldehyde thiosemicarbazone.

N

N33. An alloy of cast iron with additions of nickel, copper, and chromium.

N.A.B. Neosalvarsan (*q.v.*).

Nacarat. See AZO RUBINE S.

Nacconates. A trade mark for diisocyanates for use in urethane foams. (30).

Nacconol. A proprietary trade name for a wetting agent containing a sodium alkyl aryl sulphonate or of similar constitution.

Naclex. A proprietary preparation of hydroflumethiazide. A diuretic. (335).

Nacreous Sulphur. See MOTHER OF PEARL SULPHUR.

Nacrite. A name applied to two minerals: (*a*) a variety of mica, and (*b*) a variety of kaolinite.

Nactisol. A proprietary preparation of poldine methylsulphate and *sec*butobarbitone. (364).

Nacton. A proprietary preparation containing poldine methylsulphate. A gastrointestinal sedative. (272).

Nadinsulan AF. A heat resistant insulation withstanding a temperature of 600° C under continuous use. A specially prepared calcium silicate with glass fibre. (407).

Nadide. An antagonist to alcohol and to narcotic analgesics. Nicotinamide adenine dinucleotide. 1-(3-Carbamoylpyridinio - β - D - ribofuranoside 5 - (adenosine - 5' - pyrophosphate. ENZOPRIDE.

Nadisan. A 10 per cent. suspension of bismuth-methyl tartrate in olive oil. It is a German preparation, and is used in the treatment of syphilis.

Nadisan. Carbutamide.

Nadolol. A beta adrenergic blocking agent currently undergoing clinical trial as " SQ 11 725 ". It is $(2R, 3S)$ - 5 - (3 - *tert* - butylamino - 2 - hydroxypropoxy) - 1, 2, 3, 4 - tetrahydronaphthalene - 2, 3- diol.

Nadorite. A mineral, PbO.SbOCl.

Naëgite. A mineral. It is a silicate of zirconium, $ZrSiO_4$, with yttrium, thorium, uranium, and niobium, of Japan.

Nafcillin. 6-(2-Ethoxy-1-naphthamido)-3,3 - dimethyl - 7 - oxo - 4 - thia - 1 - azabicyclo[3.2.0]heptane - 2 - carboxylic acid.

Nafenopin. A hypolipidæmic. 2-Methyl - 2 - [4 - (1, 2, 3, 4 - tetrahydro -1 - naphthyl) phenoxyl] propionic acid.

Nafisal-ovula. The hydrochloride of sec. octyl-hydrocupreicin for use in urinary cancer.

Naftalan. Naphthalane (*q.v.*).

Naftazone. β-Naphthaquinone 2-semicarbazone.

Naftidrofuryl. A vasodilator. 2-Diethylaminoethyl 2 - (1 - naphthyl - methyl) - 3 - (tetrahydro - 2 - furyl) - propionate. PRAXILENE is the oxalate.

Naftolen. A proprietary trade name for an unsaturated hydrocarbon prepared from the higher fractions of petroleum. It is used as an extender, modifying agent, and plasticiser.

Naftolen R 100, 510, 530, 550, 570, X413, X414, X10, 134. Vinyl plasticisers. (116).

Naganol. A veterinary " Bayer 205."

Naga Red. An azo dyestuff obtained from benzidine. Used as a trypanocide in the treatment of nagana.

Nagatelite. A mineral. It is $Ca_2(Ce,La)_2Al_4Fe_2(Si,P)_6O_{15}OH$.

Nageli's Solution. A solution containing a mixture of zinc chloride and iodide. A disinfectant.

Nagyagite. A mineral, $10PbS.Sb_2S_3.2AuTe_3$ or $2AuS$.

Nahcolite. A native sodium bicarbonate.

Naheran. Cinchophen with glycocoll and hexamine.

Nail-head Spar. A mineral. It is a variety of calcite, $CaCO_3$.

Nairit P. MERCAPTAN-modified polychloroprene.

Nakrite. A mineral, $Al_2O_3.2SiO_2.H_2O$.

Nalcite. A proprietary trade name for a water softener containing an organic zeolit type exchanger material.

Nalicine. A mixture containing 1 per cent. nitro-glycerin solution, thymol, water, sodium chloride, alcohol, formaldehyde, phenol, and 1 gram of cocaine to every 100 grams of the mixture. A local anæsthetic used in dentistry.

Nalidixic Acid. An anti-bacterial agent. 1 - Ethyl - 7 - methyl - 4 - oxo - 1, 8 - naphthyridine - 3 - carboxylic acid. NEGRAM.

Nalorphine. N-Allylnormorphine.

Naloxone. An antagonist to narcotics. $(-)$ - 17 - Allyl - 4, 5α - epoxy - 3, 14 - dihydroxymorphinan - 6 - one. NARCAN is the hydrochloride.

Namaqualite. A mineral. It is $Cu_2Al(OH)_7.2H_2O$.

Nametal. Foil wrapped metallic sodium. (531).

Nancic Acid. Lactic acid, $C_3H_6O_3$.

Nandrolone. 17β-Hydroxyoestr-4-en-3-one. 17β-Hydroxy-19-norandrost-4-en-3-one. Nortestosterone. Deca-Durabolin is the decanoate; Durabolin is the phenylpropionate.

Nangawhite. A non-staining antioxidant for rubber. (556).

Nankin. See PHOSPHINE.

Nankin Yellow. See IRON BUFF and PHOSPHINE.

Nankor. A proprietary preparation of FENCHLORPHOS. An insecticide.

Nansa. Alkyl aryl sulphonates. (568).

Nansa UC. A range of detergent powders containing a sodium alkyl benzene sulphonate as the active ingredient.

Nansen. A high-speed tungsten steel containing 18 per cent. tungsten.

Nantokite (Nantoquite). A mineral. It is cuprous chloride, $CuCl$.

Nantoquite. See NANTOKITE.

Nantusi. A compound used in rubber mixings. It is stated to be a mixture of paraffin and sulphur, and to preserve rubber against atmospheric action.

Napalite. A dark-red wax which melts at 42° C. It is a hydrocarbon, C_6H_4, and occurs naturally.

Napalm. An aluminium soap consisting of a mixture of oleic naphthenic and coconut fatty acids. Makes petrol thicken and gel. Used in flame throwers and fire bombs.

Napeline. Benzaconine, $C_{32}H_{43}NO_{10}$.

Naphalane. A crude naphtha product containing soap. It is similar to Naphthalane (*q.v.*).

Naphazoline. A vasoconstrictor. 2-(1-Naphthylmethyl) - 2 - imidazoline. PRIVINE is the nitrate. Present in COLUVAL, DELTA-FENOX, and NOMAZE as the nitrate, and in VASOCON-A and VASOZINC as the hydrochloride.

Naphoxytol. See EPICARIN.

Naphtalin. See TAR CAMPHOR.

Naphtha (Petroleum Naphtha). The less volatile portion obtained in redistilling Benzine (*q.v.*), boiling from about 95–100° C. The term is loosely applied, and is synonymous with mineral naphtha. Solvent naphtha is not a petroleum product. Danforth's oil, B.P. 80–110° C., is a similar material. Used as a solvent, and for burning. Also see B-PETROLEUM NAPHTHA, EARTH OIL, and BENZOLINE.

Naphthadol. Synthetic tanning extracts. (508).

Naphthalamine. See NAPHTHALIDAM.

Naphthalane (Naftalan). A greasy mass prepared from the high boiling constituents and residues of the distillation of a naphtha from certain Caucasian naphtha wells. The naphtha is treated with 3–4 per cent. anhydrous soap. Used for skin diseases and burns. Also see NAPHALANE.

Naphthalene Acid Black. A dyestuff obtained by coupling diazotised metanilic acid with α-naphthylamine-sulphonic acids (1 : 6 and 1 : 7), diazotising, and coupling with α-naphthylamine.

Naphthalene Black 12B. A dyestuff. It is a British equivalent of Naphthol blue-black.

Naphthalene Blue B. An acid dye for wool.

Naphthalene Blue-black. A British brand of Naphthol blue-black.

Naphthalene Green Conc. See NAPHTHALENE GREEN V.

Naphthalene Green V (Naphthalene Green Conc.). A diphenyl-naphthylmethane dyestuff, prepared by condensing naphthalene-disulphonic acid with tetramethyl-diamino-benzhydrol, and oxidising the leuco compound. Dyes wool from an acid bath.

Naphthalene Red. See MAGDALA RED.

Naphthalene Rose. See MAGDALA RED.

Naphthalene Scarlet. See MAGDALA RED.

Naphthalene Yellow. See MARTIUS YELLOW.

Naphthalidam (Naphthalidine). α-Naphthylamine, $C_{10}H_7.NH_2$.

Naphthalidine. See NAPHTHALIDAM.

Naphthalin. Naphthalene, $C_{10}H_8$. See TAR CAMPHOR.

Naphthaline Red. See MAGDALA RED.

Naphthaline Yellow. See MARTIUS YELLOW.

Naphthalit. A German explosive.

Naphthalol (Betol). Naphthol salicylate.

Naphthalophos. An anthelmintic. N-(Diethoxyphosphinyloxy) naphthalimide. RAMETIN.

Naphthamine. See HEXAMINE.

Naphtharene Orange G. A dyestuff. It is a British equivalent of Orange G.

Naphtharene Orange R. A British brand of Scarlet GR.

Naphtharene Scarlet 2R. A British dyestuff. It is equivalent to Ponceau.

Naphthase. α-β-Naphthazine, $C_{20}H_{12}N_2$.

Naphtha, Shale. See SHALE SPIRIT.

Naphtha, Vinegar. See ACETIC ESTER.

Naphtha, Wood. See METHANOL.

Naphthazarin. 5, 6-dihydroxy-1, 4 naphthoquinone, $C_{10}H_4(OH)_2O_2$.

Naphthazarin S. See ALIZARIN BLACK S.

Naphthazine Blue. A dyestuff. It is the sodium salt of the disulphonic acid of dimethyl - β - naphthyl - diamino -β - naphthyl-phenazonium, $C_{34}H_{25}N_4O_6S_2$ Na. Dyes wool blue from an acid bath.

Naphthazine Blue-black 6B. A British equivalent of Naphthol blue-black.

Naphthazoline. 2-(1-Naphthylmethyl)-2-imidazoline. Privine is the nitrate.

Naphthazurin B, 2B, R. Basic dyes, giving navy blue shades on tanned cotton.

Naphthindone. See INDOINE BLUE R.

Naphthindone, Blue BB. See INDOINE BLUE R.

Naphthindone Red. See ARCHIL SUBSTITUTE V.

Naphthine Brown-α. See SULPHAMINE BROWN A.

Naphthine Brown-β. See SULPHAMINE BROWN B.

Naphthiomate-T. A proprietary trade name for Tolnaftate.

Naphthionic Acid. 1-Naphthylamine-4-sulphonic acid.

Naphthionic Red (Archil Substitute). A dyestuff. It is nitro-benzene-azo-α-naphthylamine-α-sulphonic acid, $NO_2.C_6H_4.N_2.C_{10}H_5.NH_2.HSO_3$.

Naphthite. Trinitro-naphthalene. Employed in explosives.

Naphthochrome. Wool dyestuffs. (530).

Naphthocyanine. Fast acid wool dyestuffs. (541).

Naphthocyanole. A homologue of Pinacyanol (a red sensitiser for silver bromide plates), prepared by the condensation of β-naphtho-quinaldine ethiodide with formaldehyde in the presence of alcoholic potash. Used as a red sensitiser for photographic plates.

Naphthoformol. A product of α-naphthol and formaldehyde. Used as a dusting powder.

Naphthol Aristol. Iodo-naphthol, $C_{10}H_6I_2O_2$. Used as an antiseptic.

Naphthol AS. β-Oxy-naphthoic-acid-anilide, $C_{17}H_{13}NO_2$. Used in the manufacture of dyestuffs.

Naphthol AS-BG. A dyestuff. It is the 2 : 5-dimethoxy-anilide of β-hydroxy-naphthoic acid.

Naphthol AS-BO. A dyestuff. It is β - hydroxy - naphthoic - α - naphthyl - amide.

Naphthol AS-BS. A dyestuff. It is β-hydroxy-naphthoic-m-nitranilide.

Naphthol AS-D. The o-toluidide of β-hydroxy-naphthoic acid.

Naphthol AS-G. A dyestuff. It is diaceto-aceto-tolidide.

Naphthol AS-OL. A dyestuff. It is the o-anisidide of β-hydroxy-naphthoic acid.

Naphthol AS-RL. A dyestuff. It is β-hydroxy-naphthoic-p-anisidide.

Naphthol AS-TR. 5-Chloro-o-toluidide of β-hydroxy-naphthoic acid.

Naphthol A-SW. A dyestuff. It is β - hydroxy - naphthoic - β - naphthyl - amide.

Naphthol-bismuth. Bismuth beta-naphtholate.

Naphthol Black. See BLUE-BLACK B.

Naphthol Black B (Brilliant Black B). A dyestuff. It is the sodium salt of disulpho - β - naphthalene - azo - α - naphthalene - azo - β - naphthol - disulphonic acid, $C_{30}H_{16}N_4S_4O_{13}Na_4$. Dyes wool blue-black from an acid bath.

Naphthol Black 6B (Acid Black 6B). A dyestuff. It is the sodium salt of disulpho - naphthalene - azo - α - naphthalene - azo - β - naphthol - disulphonic acid R, $C_{30}H_{16}N_4S_4O_{13}Na_4$. Dyes wool blue-black from an acid bath.

Naphthol Black 12B. See NAPHTHOL BLUE-BLACK.

Naphthol Blue B. See NEW BLUE B or G.

Naphthol Blue G, R. Acid dyes giving navy blue shades on wool.

Naphthol Blue-black (Naphthol Black 12B, Acid Black HA, Acid Blue Black, 428, Naphthalene Black 12B, Naphthalene Blue Black, Naphthalene Blue-Black 6B, Naphthol Blue-Black B, 10,

L, S Conc., Coomassie Blue-Black). A dyestuff. It is the sodium salt of p-nitro - benzene - azo - disulpho - amino - naphthol - azo - benzene, $C_{22}H_{14}N_6O_9S_2Na_2$. Dyes wool black from an alkaline bath.

Naphthol Blue-Black, B, 10, L, S Conc. Dyestuffs. They are British equivalents of Naphthol blue-black.

Naphthol Blue R and D. See MELDOLA'S BLUE.

Naphthol-camphor. A mixture of β-naphthol (1 part) and camphor (2 parts). It has been used as an injection into tubercular glands.

Naphthol Green B. A dyestuff. It is the ferrous sodium salt of nitroso-β-naphthol-β-mono-sulphonic acid, $C_{20}H_{10}N_2O_{10}S_2FeNa_2$. Dyes wool from an acid bath containing an iron salt.

Naphtholite. A proprietary trade name for a light petroleum distillate used as a solvent, etc.

Naphtholith. A bituminous shale.

Naphthol Orange. See ORANGE I and II.

Naphthol Red O. See FAST RED D.

Naphthol Red S. See FAST RED D.

Naphtholsalol. See BETOL.

Naphthol Scarlet 3R. An acid dyestuff.

Naphthol Violet. See MELDOLA'S BLUE.

Naphthol Yellow. See MARTIUS YELLOW and NAPHTHOL YELLOW S.

Naphthol Yellow FY. A dyestuff. It is a British brand of Naphthol yellow S.

Naphthol Yellow RS. See BRILLIANT YELLOW.

Naphthol Yellow S (Naphthol Yellow, Sulphur Yellow S, Citronine A, Acid Yellow S, Naphthol Yellow FY). A dyestuff. It is the calcium, ammonium, sodium, or potassium salt of dinitro - α - naphthol - β - mono - sulphonic acid, $C_{10}H_4N_2O_8SNa_2$. Dyes wool and silk yellow from an acid bath.

Naphthoresorcin. 1 : 3-Dioxy-naphthalene.

Naphthoride. A proprietary preparation of naphthalene tetrachloride.

Naphthorubine. See PALATINE RED. It is isomeric with Buffalo rubine.

Naphthosalol. See BETOL.

Naphthosultone. The anhydride of 1 : 1'-naphthol-sulphonic acid, $C_{10}H_6(SO_2)O$.

Naphthothiam Blue. A dyestuff obtained by warming a solution of 1 : 8-nitro-naphthalene-sulphinic acid with zinc dust and potassium sulphite. A vat dye.

Naphthylamine Black 4B. A dyestuff similar to Naphthylamine black D.

Naphthylamine Black 6B. A dyestuff similar to Naphthylamine black D.

Naphthylamine Black D (Acid Black N). A dyestuff. It is the sodium salt of disulpho-naphthalene-azo-α-naph-thalene-azo-α-naphthylamine, $C_{30}H_{19}N_5O_6S_2Na_2$. Dyes wool and silk black from an acid bath, or from a neutral bath containing salt.

Naphthylamine Bordeaux. See CAR-MINAPH GARNET.

Naphthylamine Brown. See FAST BROWN N.

Naphthylamine Pink. See MAGDALA RED.

Naphthylamine Red. See MAGDALA RED.

Naphthylamine Violet. A dyestuff obtained from α-naphthylamine-hydro-chloride by oxidation with potassium chlorate.

Naphthylamine Yellow. See MARTIUS YELLOW.

Naphthyl Blue. A sulphonated milling blue. It dyes silk violet-blue with red fluorescence.

Naphthyl Blue 2B. A dyestuff. It is the sodium salt of dicarboxy-diphenyl-disazo - bi - benzoyl - amino - naphthol - sulphonic acid, $C_{48}H_{28}N_6S_2O_{14}Na_4$. Dyes cotton blue from a salt bath.

Naphthyl Blue Black N. An acid dye-stuff, giving navy blue shades on wool.

Naphthylene Blue. See MELDOLA'S BLUE.

Naphthylene Blue R in Crystals. See MELDOLA'S BLUE.

Naphthylene Red. A dyestuff. It is the sodium salt of naphthalene-disazo-bi-naphthionic acid, $C_{30}H_{20}N_6O_2S_6Na_2$. Dyes cotton red from a boiling alkaline bath.

Naphthylene Violet (Diamine Cutch). A dyestuff. It is the sodium salt of disulpho - naphthalene - disazo - bi - α - naphthylamine, $C_{30}H_{20}N_6S_2O_6Na_2$. Dyes wool and silk direct. It is used to produce diamine cutch (cutch brown), by treatment on the fibre with nitrous acid, and heating.

Naples Red. See MUREXIDE.

Naples Yellow (Paris Yellow). A pigment. It is an antimonate of lead, $PbO.Sb_2O_5$. A mixture of this body with carbonate and chromate of lead is also sold under this name. Cadmium sulphide, CdS, and a pale yellow ochre have been called by this term.

Naplithin. Lithium-β-hydroxy-naph-thalene-α-monosulphonate.

Napliwi. A sand cemented together by the rubber latex from the roots of the Chondrilla plant in Russia. The roots are attacked by the larvæ of certain insects, when the latex exudes and runs into the sandy soil where it coagulates. The sand usually contains about 2–2·5 per cent. rubber and 10 per cent. resin.

Napoleonite. See CORSITE.

Napoleon's Green. See MOUNTAIN GREEN.

Napolite. Synonym for Haüyne.

Naprosyn. A proprietary preparation of NAPROXEN used in the treatment of arthritis. (809).

Naproxen. An anti-inflammatory, anti-pyretic and analgesic currently under-going clinical trial as " RS-3540 ". It is (+) - 2 - (6 - Methoxy - 2 - naphthyl) - propionic acid.

Napsalgesic. A proprietary prepara-tion of dextropropoxyphene and acetyl salicylic acid. (309).

Narcan. A proprietary preparation of NALOXONE hydrochloride used as an antidote to narcotics. (685).

Narceine. The sodium bisulphite com-pound of p-sulpho-benzene-azo-β-naph-thol, $C_{16}H_{12}N_2O_7S_2Na_2$. Used in calico printing.

Narceol. A synthetic perfume of honey-suckle. It is p-tolyl-acetate, $CH_3.C_6H_4.O.COCH_3$.

Narcodeon. Double salts of narcotine and codeine with di- and poly-basic acids.

Narcoform. See SOMNOFORM.

Narcophin. A double salt of morphine and narcotine, $C_{22}H_{23}O_7N.C_{17}H_{19}O_3N.C_7H_4O_7 + H_2O$.

Narcotil. A mixture of ethyl and methyl chlorides. Used as an anæsthetic. Methylene chloride, CH_2Cl_2, used as a local anæsthetic, is also known by this name.

Narcotine. A proprietary trade name for Noscapine.

Narcyl. Ethyl-narceine hydrochloride, $C_{23}H_{26}(C_2H_5)NO_8.HCl$. Used in medi-cine to relieve coughing.

Narcylen. An anæsthetic, the active principle of which is purified acetylene.

Nardil. A proprietary preparation of phenelzine dihydrogen sulphate. An antidepressant. (262).

Narex. A proprietary preparation of PHENYLEPHRINE and CHLORBUTOL used as a nasal spray. (462).

Nargentol. A protein compound of silver, containing 24 per cent. silver. An antiseptic.

Nargol. The silver salt of nucleic acid. It contains 10 per cent. silver, and is used in medicine as a substitute for silver nitrate for local application.

Naridan. A proprietary trade name for the hydrochloride of Oxyphencyclimine.

Nari Oil. See NJAVE BUTTER.

Narki. An acid-resisting silicon-iron alloy.

Narphen. A proprietary preparation of phenazocine hydrobromide. An anal-gesic. (268).

Narsarsukite. A mineral. It is a titano-silicate of iron and sodium.

Nartine. Nartic acid, $C_{20}H_{16}N_2O_6$.

Nasagon. A German preparation containing anæsthesin, paracaine, menthol borate, hypernephrine. Used against colds.

Naseptin. A proprietary preparation of chlorhexidene hydrochloride and neomycin sulphate. (2).

Nasidan. An oil suspension of potassium-bismuthyl-tartrate.

Nasoflu. A proprietary live influenza vaccine. (658).

Nasonite. A mineral, $Ca_4Pb_6Cl(Si_2O_7)_3$.

Nasrol. Sodium-caffeine-sulphonate, $C_8H_9N_4O_2SO_3Na$. A diuretic.

Natal Aloe. See ALOE.

Natalite. A substitute for petrol manufactured from fermented molasses. It is a mixture of alcohol and ether. The name is also used as a trade mark for abrasive and refractory articles consisting essentially of alumina.

Nataloin. The aloin from Natal aloes.

Natalon. See CARBORA.

Natamycin. An antibiotic produced by Streptomyces natalensis. Pimafucin. Pimaricin.

Natcom. A proprietary mechanically-comminuted rubber. (955).

Nateina. A German preparation consisting of a mixture of vitamins A, B, C, and D of vegetable origin, to which calcium phosphate and milk sugar are added.

Natene. A trade mark for Ziegler-type polyethylenes. (686).

National Motor Fuel. A mixture of petrol and alcohol containing a small amount of a solvent such as cyclohexanol, to ensure miscibility. It usually contains 10–20 per cent. of 90 per cent. alcohol.

Natirose. A proprietary preparation of GLYCERYL trinitrate, ethylmorphine hydrochloride and hyoscyamine hydrobromide used in the treatment of angina pectoris. (781).

Natisedine. A proprietary preparation of quinidine phenylethylbarbiturate used as a cardiac sedative. (781).

Native Guano. A manure prepared by pressing the sludge from sewage, and forming cakes of the product.

Nativelle's Digitalin. This is stated to consist almost entirely of digitoxin.

Native Paraffin. Ozokerite (q.v.).

Native Prussian Blue. Synonym for Vivianite.

Native Smalt. See CHESSYLITE.

Native Ultramarine. A blue pigment. It is lapis-lazuli. Also see ULTRA-MARINE, GENUINE.

Natramblygonite. Synonym for Natromontebrasite.

Natricite. A mineral, $Mg_2SiO_4.H_2O$.

Natrium. Sodium, Na.

Natroalunite. A mineral. A sodium variety of Alunite (q.v.). It is $NaAl_3(SO_4)_2(OH)_6$.

Natroamblygonite. A mineral. It approximates to $NaAl(OH)PO_4$ with a little lithium.

Natroborocalcite. Synonym for Ulexite.

Natrocalcite. A mineral. It is a basic sodium-copper sulphate, $Na_2SO_4.Cu(OH)_2.3CuSO_4.3H_2O$. The term is also applied to Gaylussite ($Na_2CO_3.CaCO_3.5H_2O$).

Natrochalcite. A mineral. It is $2[NaCu_2(SO_4)_2OH.H_2O]$.

Natrohitchcockite. A mineral. It is $NaAl_3(PO_4)_2(OH)_4.2H_2O$.

Natrojarasite. A mineral. It is Na $Fe_3^{...}(SO_4)_2(OH)_6$.

Natrolite (Mesotype). A zeolite mineral, $Na_2O.Al_2O_33SiO_2.2H_2O$.

Natrolith. A water-softening material said to consist of granulated clay which removes lime and magnesium salts from hard water when used as a filter. It is stated to be unaffected by carbonic acid.

Natromontebrasite. A mineral. It is $2[(Na,Li)AlPO_4(OH,F)]$.

Natron. See TRONA.

Natrophyllite. A mineral, $Na(Mn.Fe)PO_4$.

Natrum. See TRONA.

Natronsanidine. A mineral. It is $4[(K,Na)AlSi_3O_8]$.

Natrundum. See CARBORA.

Natulan. A proprietary preparation of procarbazine. An antimitotic. (314).

Natural Asphalt. See BITUMEN.

Natural Gas. A mixture of gaseous hydrocarbons found in nature, usually associated with petroleum deposits. Used as a fuel, and for the manufacture of carbon black.

Natural Varnishes. Two plants, belonging to the *Anacardiaceæ*, yield a sap which is used in the East as a natural varnish or lacquer. *Rhus* for Japanese or Chinese lacquer, and *Melanorrhœa*, which yields black Burmese lacquer.

Naubuc. A nickel-silver. It contains 58 per cent. copper, 25 per cent. nickel, 16·25 per cent. zinc, and 0·75 per cent. iron.

Nauganlite. A proprietary anti-oxidant. Alkylated phenol. (556).

Nauganlite Powder. A proprietary name for alkylated bis-phenol. (556).

Naugard 445. A proprietary amine antioxidant which does not cause discoloration. It is used in polymers and lubricants. (287).

Naugatex 219. A proprietary trade name for a dispersion in water of B.L.E. (q.v.). (163).

Naugatex 259. A proprietary trade

name for an acetone-hydroquinone reaction product. An antioxidant. (163).

Naugatex 269. A proprietary trade name for a paste of Betanox (q.v.). (163).

Naugatex 279. A proprietary trade name for a paste of BLE powder (q.v.). (163).

Naugatex 289. A proprietary trade name for a water dispersion of Aminox (q.v.). (163).

Naugatex 505A. A proprietary trade name for a 50 per cent. paste of Aminox (q.v.). (163).

Naugatex 506. A proprietary trade name for a 50 per cent. paste of Betanox (q.v.). (163).

Naugatex 510. A proprietary trade name for a 50 per cent. paste of Aranox (q.v.). (163).

Naugatex 519. A proprietary trade name for a 60 per cent. paste of BLE (q.v.). (163).

Naujakasite. A mineral. It is $2[Na_9Al_9Si_{16}O_{50}.4\frac{1}{2}H_2O]$.

Nauli Gum. An oleo-resin from a tree found in the Solomon Islands. It contains 10 per cent. volatile oil, 8 per cent. resin, and 3 per cent. water-soluble matter containing anisic acid.

Naumannite. A mineral. It is a selenide of silver and lead, $(Ag_2Pb)Se$.

Nauruite. Synonym for Francolite.

Nautisan. A German proprietary product. It is a combination of trichlorisobutyl-alcohol (chloretone) with trimethyl-xanthine (caffeine). It is said to reduce sickness after narcotics.

Navac. Vacuum processed metallic sodium. (531).

Naval Brass. See BRASS.

Naval Bronze. An alloy of 88·1 per cent. copper, 9·74 per cent. tin, and 2·04 per cent. zinc.

Navane. A proprietary preparation of thiothixene. A sedative. (85).

Navidrex. A proprietary preparation of cyclopenthiazide and potassium chloride in slow release core. A diuretic. (18).

Navidrex-K. A proprietary preparation of CYCLOPENTHIAZIDE and slow-release potassium chloride. A diuretic. (18).

Navy Blue. See SOLUBLE BLUE.

Navy Blue B. See BAVARIAN BLUE DSF.

Navy Green Paint. A mixture of barium sulphate, lead chromate, and an organic blue.

Naxogin and Naxogin 500. Proprietary preparations of NIMORAZOLE used in the treatment of trichomonas infections. (365).

NBC. A proprietary name for nickel dibutyl dithiocarbamate. (10).

N.C.T. Nitro-cellulose tutular, a pyrocollodion powder, made from a gelatinised nitro-cellulose, pressed in the form of rods.

N.C.T.3 Alloy. See ALLOY N.C.T.3.

Ndilo Oil. See LAUREL NUT OIL.

Nealbarbitone. 5-Allyl-5-neopentylbarbituric acid. Censedal; Nevental.

Nealpon. See PANTOPON.

Neapolitan Ointment. Mercury ointment.

Neatsfoot Oil. A fixed oil obtained by boiling ox or cow's feet in water. It is also the name for a mixture of 1 part lard and 3 parts colza oil.

Neblah. See BABLAH.

Nebcin. A proprietary preparation of TOBRAMYCIN. An antibiotic. (463).

Necaron. A German product. It is a silver preparation of equal parts of silver-sodium cyanate and sodium cholate.

Necket. A proprietary medical preparation consisting of mandelic acid granules.

Necol. Cellulose lacquers, organic enamels and adhesives. (512).

Necolin S4070E. A proprietary trade name for a modified soya oil for use in chlorinated rubber paints. Used for traffic paints. (195).

Necoloidine. A high-grade purified nitrocellulose prepared for use for diffusion membranes. (512).

Necronite. A mineral. It is a variety of orthoclase.

Nectandra Bark. Bebeera bark.

Nectrianin. A French product from the Nectria ditissima of apple and pear trees.

Needle Bronze. An alloy of 84·5 per cent. copper, 8 per cent. tin, 5·5 per cent. zinc, and 2 per cent. lead.

Needle Iron Ore (Gothite, Lepidokrokite, Pyrrhosiderite). An ore of iron. It is an oxyhydride, $FeO(OH)$.

Needle Ore. See ACICULITE.

Needle Spar. A variety of arragonite (q.v.).

Needle Stone. Natrolite (q.v.).

Needle Tin Ore. Acute pyramidal crystals of cassiterite.

Neem Bark. Indian azadirach.

Neem Oil. See VEEPA OIL.

Nefomolit. A proprietary trade name for a plastic made from mineral oil, formaldehyde, etc.

Nefopam. A muscle relaxant. 3,4,5,6-Tetrahydro - 5 - methyl - 1 - phenyl - 1H2, 5 - benzoxazocine.

Nefranutrin. A proprietary preparation of essential amino acids used in the dietary treatment of renal failure. (386).

Nefrolan. A proprietary preparation of clorexolone. A diuretic. (336).

Negomel. Antistatic agent for PVC and rubber. (512).

Negram. A proprietary preparation containing nalidixic acid. A urinary antiseptic. (112).

Negro-heads. A mineral. It is a tourmaline from the island of Elba. The name is also applied to a variety of rubber. See SERNAMBY.

Negrolin. Creoline (*q.v.*).

Negro Powder (Nigger Powder). An explosive consisting of 86–91 per cent. ammonium nitrate, 9–11 per cent. trinitro-toluene, and 1–3 per cent. graphite.

Neguvon. A proprietary preparation of METRIPHONATE. An insecticide.

Neillite. A proprietary phenol-formaldehyde synthetic resin moulding compound.

Neisser's Stain. A microscopic stain. (*a*) Solution contains 0·1 gram methylene blue, 2 c.c. alcohol, 5 c.c. glacial acetic acid, and 95 c.c. water. (*b*) Solution contains 0·2 gram bismarck brown in 100 c.c. boiling water.

Nekal (Leonil, Oranit, Neomerpin N). Wetting-out agents probably derived from alkylated naphthalene sulphonic acids. Used to assist the penetration of textiles by liquids. Also see TETRACARNIT and TETRAPOL.

Nelco. A proprietary trade name for whiting.

Nelio Resin. A proprietary trade name for a purified wood resin.

Nelson's Patent Gelatin. A proprietary gelatin containing a small quantity of chalk to render it opaque. It is stated to be prepared by washing skin parings and other materials and digesting them in caustic soda for some days. They are bleached and washed, and the gelatin obtained by heating in water.

Nemafax. A proprietary preparation of THIOPHANATE. A veterinary anthelmintic.

Nemalite. See BRUCITE.

Nemaphyllite. A mineral. It is a magnesium silicate containing iron and sodium.

Nematolite. See BRUCITE.

Nembudonnal. A proprietary preparation of pentobarbitone sodium, and belladonna dry extract. (311).

Nembutal. A proprietary preparation of pentobarbitone sodium. A hypnotic. (311).

Neoacryl. Sodium Succinanilomethylamide-*p*-arsonate. Used as a trypanocide, $Na_2O_3As . C_6H_4 . NHCO . CH_2 . CH_2 . CONH . CH_3$.

Neo-Antergan. A proprietary trade name for the hydrogen maleate of Mepyramine.

Neoarsaminol. Salvarsan (*q.v.*).

Neoarsphenamine. Salvarsan (*q.v.*).

Neoarsycodyl. Sodium methyl-arsenate.

Neobacrin. A proprietary preparation of neomycin sulphate and zinc bacitracin. An antibiotic skin ointment. (335).

Neobistovol. A proprietary preparation. It is the bismuth salt of stovarsol. Used in the treatment of syphilis.

Neobor. Trade name for disodium tetraborate pentahydrate. It is borax partially dehydrated to effect economy of transport and handling. (183).

Neobornyval. Bornyval, new (*q.v.*).

Neocaine. Novocaine (*q.v.*).

Neocaine-surrénine. A proprietary preparation. It consists of ethocaine hydrochloride with adrenalin.

Neo-Cantil. A proprietary preparation of MEPENZOLATE BROMIDE and NEOMYCIN sulphate. An anti-diarrhœal. (249).

Neo-cardyl. A proprietary preparation. It is an oil solution of bismuth-butyl-thio-laurate.

Neo Cortef. A proprietary preparation of hydrocortisone acetate and neomycin sulphate used in dermatology as an antibacterial agent. (325).

Neochrome Blue-black B. A dyestuff. It is equivalent to Chrome fast cyanine G.

Neochrome Blue-black R. A dyestuff. It is equivalent to Chrome blue.

Neocianite. A mineral, $CuO.SiO_2$.

Neocid. A proprietary preparation containing 5 per cent. D.D.T. (*q.v.*). An insecticide.

Neocinchophen. Novatophan (*q.v.*).

Neocinchophen. Ethyl 6-methyl-2-phenylcinchonate.

Neocryl. A proprietary preparation of sodium succinanilo-methylamide-*p*-arsonate. Used in the treatment of syphilis and sleeping-sickness.

Neocryl. A proprietary trade name for acrylic and acrylic copolymer emulsions used for floor polishes. (91).

Neocryl BT-8. A proprietary acrylic copolymer emulsion used in flexographic inks. (687).

Neocryl NP-4-C. A 35% solution of an acrylic polymer in a 9/1 toluene/butanol mixture, used as a proprietary petrol-resistant lacquer on automobiles. (687).

Neocutrene. A German preparation. It is a suspension of the bismuth salts of iodo-*o*-oxy-quinoline sulphonic acid, and salicylic acid.

Neocyanite. A material found at Vesuvius. It is a copper silicate. See NEOCIANITE.

Neo Cytamen. A proprietary preparation of hydroxycobalamin. (335).

Neodermin. A 5 per cent. ointment of fluor-pseudo-cumol. Used as a local application in cases of sciatica.

Neodiarsenol. Salvarsan (q.v.).

Neodorm. α-Bromo-α-isopropyl-butyramide. A soporific.

Neodrenal. A proprietary trade name for the sulphate of Isoprenaline.

Neo-Epinine. A proprietary trade name for the sulphate of Isoprenaline.

Neoferrum. A proprietary preparation of ferric hydroxide. A hæmatinic. (280).

Neoform. The bismuth compound of tri-iodo-phenol, $C_6H_2I_3OBi(OH)_2.Bi_2O_3$. Suggested as a dusting-powder for ulcers and wounds.

Neogen. An alloy of 58 per cent. copper, 12 per cent. nickel, 27 per cent. zinc, 2 per cent. tin, and 0·5 per cent. aluminium. It is a nickel silver (German silver). Also see NICKEL SILVERS.

Neogest. A proprietary preparation of dl-norgestrel. An oral contraceptive. (438).

Neo-Hetramine. A proprietary trade name for Thonzylamine.

Neohexal (Neosulphexine). Secondary hexamethylene-salicyl-sulphonate. Used in medicine as a urinary antiseptic.

Neohydrin. Chlormerodrin.

Neo-hydriol. A proprietary preparation of an iodised poppy-seed oil.

Neo-iopax. Sodium-3 : 5-di-iodo-1-methyl-1 : 4-dihydro-4-pyridone-2 : 6-dicarboxylate.

Neokharsivan. A proprietary name for sodium-3 : 3'-diamino-4 : 4'-dihydroxy-arseno - benzene - formaldehyde - sul - phoxylate, $NH_2C_6H_3(OH).As.As.C_6H_3(OH)NH(CH_2O)SONa$, (Neosalvarsan). One of the most efficient remedies for the treatment of syphilis.

Neolan. Wool dyestuffs. (530).

Neoleptol. Triformyl-trimethylene-triamine.

Neoleukorit. A proprietary synthetic resin of the phenol-formaldehyde class.

Neolin. Benzathine Penicillin.

Neolite. A mineral, $(Mg.Fe)O.Al_2O_3.SiO_2.H_2O$.

Neolyn. Synthetic resins. (593).

Neolyn 24. A proprietary trade name for a rosin alkyd in toluene 75 per cent. (93).

Neolysol. Lysol made with chlor-cresol. An antiseptic.

Neo-Medrone Acne Lotion. A proprietary preparation of sulphur, aluminium chlorhydroxide, methylprednisolone acetate and neomycin sulphate used in dermatology for acne. (325).

Neomedrone Veriderm. A proprietary preparation of methylprednisolone and neomycin used in dermatology as an antibacterial agent. (325).

Neomercazole. A proprietary preparation of carbimazole. Used in treatment of thyrotoxicosis. (265).

Neomerpin N. See NEKAL.

Neomin. A proprietary preparation containing neomycin sulphate. An antibiotic. (335).

Neomycin. An antibiotic produced by a strain of Streptomyces fradiae. Mycifradin, Myciguent, Neomin, and Nivemycin are the sulphate.

Neo-Naclex. A proprietary preparation of bendrofluazide. A diuretic. (335).

Neo-Naclex-K. A proprietary preparation of BENDROFLUAZIDE and potassium chloride. A diuretic. (335).

Neonal. An explosive. It contains potassium perchlorate, nitro-glycerin wood meal, ammonium oxalate, and a little nitro-toluene. The name is also applied to 5-butyl-5-ethyl-barbituric acid, $C_{10}H_{14}O_3N_2$, a hypnotic. Also see SONERYL.

Neonalium. An alloy of aluminium with 6–14 per cent. copper, 1 per cent. nickel, and small amounts of other metals.

Neonite. A 33-grain sporting rifle powder. It contains 10 per cent. barium or potassium nitrate, 6 per cent. pet. jelly, and insoluble nitro-cellulose.

Neopelline. An alkaloid, $C_{32}H_{45}NO_8.8H_2O$, obtained from Aconitum napellus.

Neopen SS. A proprietary trade name for a wetting agent containing sodium abietene sulphonate.

Neopentyl. A proprietary name for Clemizole penicillin.

Neophax. A proprietary trade name for brown factice-vulcanised vegetable oils.

Neophax FA. A proprietary brown factice added to polychloroprene, nitrile rubber, HYPALON and polyurethane when maximum resistance to oil is required. (417).

Neophenoquin. A proprietary preparation of lithium phenyl-cinchoninate.

Neophryn. A proprietary trade name for the hydrochloride of Phenylephrine.

Neopine. Hydroxy-codeine, $C_{18}H_{20}NO_3(OH)$.

Neoplen. A proprietary preparation of powdered polyethylene and polyethylene liquids and emulsions. Expanded polyethylene. (49).

Neoprene (GR-M). A generic term for synthetic rubbers made by polymerising chloroprene, the latter being obtained from acetylene and hydrogen chloride. It has a greater resistance to oils and ozone than rubber. It is used particularly for belts in machinery where oil resistance is required.

Neo-protosil. A proprietary colloidal

silver preparation. It contains about 20 per cent. silver iodide, AgI, combined with a protein base, and is a germicide.

Neopyrin (Valinopyrin). Valeryl-amino-antipyrine. An antipyretic.

Neopurpurite. Synonym for Hetero-site.

Neoquinine. Quinine glycerophosphate.

Neoquinophan. 6-Methyl-2-phenyl-quinoline-4-carboxylic ethyl ester. Used in medicine.

Neoresit. A phenol-formaldehyde resin.

Neosaccharin. See SACCHARIN.

Neosalvarsan (Ehrlich 914, 914, Neo-kharsivan, Novarsenobenzol, Novar-senobillon, Neotreparsenan, Novarsan, Novarsenol, Rhodarsan). A yellow powder soluble in water. It is sodium-3 : 3′-diamino-4 : 4′-dihydroxy-arseno-benzene - formaldehyde - sulphoxylate, $NH_2 . C_6H_3(OH)As . As . C_6H_3(OH)NH(CH_2O)SONa$. One of the most efficient remedies for the treatment of syphilis.

Neo-silver Salvarsan. Sodium silver arsenobenzol and novarsenobenzol combined.

Neosiode. Iodo-catechin, $(C_{15}H_{14}O_6 . 3H_2O)_3I$. Used for the administration of iodine.

Neosote. The phenoloids of blast-furnace tar. It contains a small quantity of phenol, and a large amount of cresols.

Neospectra. A proprietary trade name for carbon black.

Neosporin. A proprietary preparation of polymixin B sulphate, neomycin sulphate and gramicidin. Ocular antibiotic solution. (277).

Neostam. A proprietary preparation. It is stibamine glucoside (the nitrogen glucoside of sodium-p-amino-phenyl-stibonate).

Neo-stibosan. A proprietary preparation of dimethyl-amine-p-amino-phenyl-stibinate.

Neosulfazon. A proprietary preparation of NEOMYCIN sulphate, KAOLIN, phtha-lylsulphathiazole and PECTIN. An anti-diarrhœal. (255).

Neosulphexine. Neohexal (hexamethy-lene - tetramine - salicyl - sulphonate).

Neosyl. A registered trade mark for an amorphous silica (SiO_2) of very low bulk density and high absorption power, prepared by a patented process. It has a bulk density of 0·8–0·2 gm. per c.c. (205).

Neosyl C. A registered trade mark for a pure precipitated silica specially adapted for hard rubber.

Neosyl MH. A registered trade mark for a modified form of precipitated silica containing about 5% of magnesium as oxide. A new white reinforcing agent for rubber.

Neosynephrine. A proprietary trade name for the hydrochloride of Phenyle-phrine.

Neotantalite. A mineral. It is a variety of tantalite.

Neotestite. A mineral. It is a man-ganese-magnesium silicate.

Neothane. A proprietary polyester-based polyurethane elastomer cross-linked with diamine. (688).

Neotreparsenan. NEOSALVARSAN.

Neó-trépol. A 10 per cent. suspension of metallic bismuth in a very finely divided state. An antisyphilitic.

Neo-triplex. A proprietary safety glass involving the use of cellulose acetate.

Neo-trivalin. A German preparation. It differs from the old trivalin (which contained the valerates of morphine, caffeine, and cocaine) in that an atro-pine derivative is substituted for the cocaine. A narcotic.

Neotulle. A proprietary preparation of paraffin tulle with neomycin sulphate, zinc bacitracin, and polymixin B sul-phate. Wound dressing. (188).

Neotype. A mineral. See ALSTONITE.

Neovax. A proprietary preparation of neomycin sulphate, succinylsulphathia-zole and kaolin. An antidiarrhœal. (350).

Neovit. A proprietary preparation of thiamine hydrochloride, riboflavine, pyridoxine hydrochloride, nicotinamide, manganese, sodium and potassium glycerophosphates. A tonic. (273).

Neozone A. A proprietary antioxidant for rubber. It is phenyl-$α$-naphthyl-amine.

Neozone B. A proprietary trade name for meta-toluylene diamine. An anti-oxidant. (2), (10).

Neozone C. A proprietary antioxidant for rubber. It resembles neozone standard, but contains less m-toluylene-diamine.

Neozone D. A proprietary antioxidant for rubber. It is pure phenyl-$β$-naph-thylamine.

Neozone E. A proprietary trade name for an antioxidant for rubber containing 75 parts of phenyl-beta-naphthylamine and 25 parts of meta toluylene diamine. (2), (10).

Neozone L. A proprietary anti-oxidant for both natural and synthetic rubber. It is the product of reaction between acetone and diphenylamine. (10).

Neozone Standard. A proprietary trade name for an antioxidant for rub-ber containing 50 parts phenyl-alpha-naphthylamine, 25 parts meta toluylene diamine and 25 parts stearic acid. (2), (10).

Nepaulite. A mineral. It is tetra-hedrite.

Nepheline (Elæolite). A mineral, $3(Na_2 K_2)O.4Al_2O_3.9SiO_2$.

Nephridine. Adrenaline (*q.v.*).

Nephril. A proprietary preparation of polythiazide. A diuretic. (85).

Nephrite. A mineral. It is a variety of amphibole.

Nephthelite. A mineral, $NaAlSiO_4$.

Nepouite. A mineral. It is a nickel-magnesium silicate.

N. E. Powder. A 36-grain powder, containing metallic nitrates, and nitro-hydrocarbons.

Nepresol. A proprietary trade name for the mesylate or sulphate of Dihydrallazine.

Neptal. A proprietary preparation of mercuramide and theophylline. A diuretic. (336).

Neptazane. A proprietary trade name for Methazolamide.

Neptune Green (Disulphine Green B). A dyestuff formed by the condensation of *o*-chloro-benzaldehyde and benzyl-ethyl-aniline, oxidation and conversion into the sodium salt.

Neptune Green S. An acid dyestuff, giving bluish-green shades on wool.

Neptunite. A mineral. It is a titano-silicate of iron and alkalis.

Nequinate. Methyl benzoquate.

Neradol (Maxyntan, Paradol, Cresyntan, Ordoval G, 2G, Ewol). Synthetic tannins generally prepared by the condensation of phenol-sulphonic acids with formaldehyde, under conditions that only water-soluble products are formed. Also see Esco Extract and Corinal.

Neradol D. A condensed sulphonated cresol. A synthetic tannin.

Neradol N. A synthetic tannin made from sulphonated naphthalene.

Neral. β-Citral, $C_9H_{15}CHO$.

Neraltein. See Neuraltein.

Neran. A leather glazing finishing agent. (512).

Nerco. Solvents and detergents. (550).

Nercol. Soluble cutting oils. (550).

Nerfinol. Textile finishing agent. (550).

Nergandin. An alloy containing 70 per cent. copper, 28 per cent. zinc, and 2 per cent. lead. Used for condenser tubes.

Neri Silk. See Galettame Silk.

Nerisol. Germicides. (522).

Nerloate. Paint dryers. (550).

Nerogene D. A developer for Zambesi black.

Nerol Blacks B, BB. Dyestuffs prepared from *p*-amino-diphenyl-amino-sulphonic acid, α-naphthylamine, and α-naphthol-sulphonic acid, and β-naphthol-3 : 6-disulphonic acid respectively. Both brands contain an admixture of a reddish-brown dye. Used for dyeing wool from a bath containing sodium sulphate and acetic acid.

Nerolidol. Methyl-vinyl-homo-geranyl carbinol. A perfume.

Nerolin (Yara-yara). β-Naphthol-methyl ester, $C_{10}H_7.O.CH_3$. β-Naphthol-ethyl ester is also known under this name.

Nerolin II (Bromelia). β-Naphthol ethyl ester, $C_{10}H_7.O.C_2H_5$. A synthetic perfume.

Neroli Oil. See Oil of Orange Flowers.

Nervagenin. A German preparation. It is a combination of sodium-diethyl-barbiturate with a valerian preparation. Used in sea-sickness.

Nervan. Wetting agents and finishing oils. (550).

Nervanaid. Metal sequestering agents. (550).

Nerve Oil. Neatsfoot oil (*q.v.*).

Nervin. An extract of meat.

Nervocidine. A yellow powder which is said to be an alkaloid obtained from an East Indian plant, *Gasu Basu*. A local anæsthetic.

Nesfield's Triple Tablets. A water steriliser. It consists of (*a*) a tablet containing an iodide and an iodate; (*b*) a tablet containing citric or tartaric acid; and (*c*) a tablet containing sodium sulphite. The addition of (*a*) to (*b*) liberates iodine from the iodate, and (*c*) removes free iodine.

Neslite. A mineral. It is a variety of opal.

Nesquehonite. A mineral, $MgCO_3.3H_2O$.

Nessler's Insecticides. These consist of soft soap and amyl alcohol with either tobacco extract or potassium sulphide.

Nessler's Reagent. An alkaline solution of mercuric iodide in potassium iodide. Employed as a delicate test for ammonia.

Nestargel. A proprietary preparation containing carob seed flour and calcium lactate. A gastrointestinal sedative. (282).

Nestorite, Nestor. A proprietary range of urea and phenolic moulding materials. (689).

Nestosyl. A proprietary preparation of benzocaine, butylaminobenzoate and resorcin. Used to relieve itching. (802).

Nethaprin Dospan. A proprietary preparation of Etafedrine hydrochloride, Bufylline, Doxylamine succinate and Phenylephrine hydrochloride used in the treatment of bronchospasm. (690).

Nethaprin Expectorant. A proprietary preparation of Etafedrine hydrochloride, Bufylline, Doxylamine succinate and GLYCERYL guaiacol ate. An expectorant. (690).

Netté Meal. A product from the fruit of *Parkia biglobosa*. Valued as a food by the natives of Africa.

Neu-antiluetin. A double salt of potassium - ammonium - antimonyl - tartrate and the neutral sodium sulphonate of mercury-salicylic acid $[SbO(C_4H_4O_6)]_3$ $K.NH_4.C_7H_3O_6SNa.Hg.2H_2O$. A salvarsan substitute used in the treatment of syphilis.

Neu-camphrosal. A white crystalline powder melting at $136°$ C., easily soluble in organic solvents and fairly soluble in hot water. It is used to make cellulose esters elastic.

Neu-cesol. The methyl ester of N-methyl-nipecotic acid metho-chloride.

Neudorfite. A resinous hydrocarbon found in Bavarian coal pits.

Neueuguform. A 50 per cent. solution of Euguform (*q.v.*), in acetone.

Neukirchite. A mineral. It is a variety of manganite.

Neulactil. A proprietary preparation of pericyazine. A sedative. (336).

Neumandin. A proprietary trade name for Isoniazid.

Neuraltein (Neraltein). Sodium-*p*-ethoxy - phenyl - amino - methane - sulphonate, $C_2H_5O.C_6H_4.NH.CH_2.SO_3$ Na. It has analgesic properties, and is used in cases of rheumatism.

Neuramag. A German preparation consisting of tablets containing quinine-acetyl salicylate, codeine phosphate, phenacetin, and acetanilide.

Neurene. Borneol isovalerianate.

Neurine. The methylo-hydroxide of dimethyl - vinyl - amine, $N(CH_3)_3$ $(C_2H_3)OH$.

Neurithrit. A German preparation consisting of acetyl-salicylic acid, dimethyl-amino - phenyl - dimethyl - pyrazolone, and other materials. Antineuralgic.

Neurodine (Acetoxane). (1) Acetyl-*p*-phenyl - urethane, $C_6H_4(CO_2.CH_3)NH$. $CO_2.C_2H_5$. Used as an antineuralgic and antipyretic.

Neurodyne. (2) A proprietary preparation of paracetamol and codeine phosphate. An analgesic. (340).

Neurolecithin. A trade name for lecithin prepared from the brain and spinal cord of animals.

Neurolite. A fibrous variety of the mineral agalmatolite.

Neuronal (Acetobromal). Diethyl-bromo-acetamide, $(C_2H_5)_2CBr.CO.NH_2$. Used in cases of insomnia, headache, and epilepsy.

Neuro-Phosphates. A proprietary preparation of calcium glycerophosphate, sodium glycerophosphate and strychnine. A tonic. (658).

Neuroplex. A proprietary preparation of calcium, sodium, potassium and ferric glycerophosphates and thiamine and strychnine hydrochlorides. A tonic. (399).

Neuro-Transentin. A proprietary preparation of adiphenine hydrochloride and phenobarbitone. (18).

Neurotropine. See HELMITOL.

Neusidonal. A preparation asserted to be the inner anhydride of quinic acid, but consists of quinine and quinic acid. Used in the treatment of gout.

Neu-silver. See NICKEL SILVERS.

Neustab. A proprietary trade name for Thiacetazone. See also *p*-ACETYLAM-INOBENZALDEHYDE THIOSEMICARBAZONE.

Neutex. A proprietary safety glass.

Neutradonna. A proprietary preparation containing aluminium sodium silicate with belladonna alkaloids. An antacid. (265).

Neutradonna Sed. A proprietary preparation containing aluminium sodium silicate, belladonna amylobarbitone and ascorbic acid. An antacid. (265).

Neutral Acriflavine. See EUFLAVINE.

Neutralaleisen. A Swedish silicon-iron alloy, which is stated to resist the action of acids.

Neutral Alum. A neutral basic alum obtained by the addition of sodium hydroxide to a solution of alum, until the precipitate produced is just re-dissolved.

Neutral Blue. A dyestuff. It is di-methyl - amino - phenyl - pheno - naph-thazonium chloride, $C_{24}H_{20}N_3Cl$. Dyes tanned cotton blue.

Neutral Insulin Injection. A solution of insulin buffered at pH 7. INSULIN NOVO ACTRAPID, NUSO.

Neutralite. An asphaltic material made in Germany.

Neutral Oils. The name given to the lightest lubricating oils from American petroleum. The term is also applied to refined coal-tar oils.

Neutralol. See PARAFFIN LIQUID.

Neutralon. An aluminium silicate recommended for use in the treatment of gastric ulcers.

Neutral Orange (Pentley's Neutral Orange). A pigment. It is a compound of Cadmium yellow and Venetian red.

Neutral Phosphate. A fertiliser prepared by digesting mineral phosphate, bone meal, or a mixture of both, with small amounts of sulphuric acid. This renders the P_2O_5 more available. The product contains 20–25 per cent. P_2O_5, and is neutral.

Neutral Quinine Sulphate. See QUININE SULPHATE, SOLUBLE.

Neutral Red (Toluylene Red). A dyestuff. It is the hydrochloride of dimethyl - diamino - toluphenazine, $C_{15}H_{17}N_4Cl$. Dyes cotton mordanted with tannin and tartar emetic, bluishred.

Neutral Tartar. Potassium tartrate.

Neutral Violet. A dyestuff. It is the hydrochloride of dimethyl-diaminophenazine, $C_{14}H_{15}N_4Cl$. Dyes cotton mordanted with tannin and tartar emetic, reddish-violet.

Neutrapen. A proprietary preparation of penicillinase. (275).

Neutraphylline. A proprietary preparation of diprophylline. A bronchial antispasmodic. (351).

Neutroflavine. See EUFLAVINE.

Neutrolactis. A proprietary preparation containing aluminium hydroxide gel, magnesium trisilicate, calcium carbonate and milk solids. An antacid. (267).

Neuwestfalit. An explosive consisting of 70 per cent. ammonium nitrate, 11 per cent. dinitro-toluene, 2 per cent. flour, and 17 per cent. sodium chloride.

Neuwied Blue. See LIME BLUE.

Neuwieder Green. See MOUNTAIN GREEN.

Nevada Silver. See NICKEL SILVERS.

Nevastain. An alloy of 86 per cent. iron, 9·5 per cent. chromium, 4 per cent. silicon, and 0·43 per cent. carbon. A non-corrosive alloy.

Nevastain R.A. A stainless steel alloy containing iron with approximately 16 per cent. chromium, 1 per cent. copper, 1 per cent. silicon, 0·4 per cent. manganese, 0·03 per cent. phosphorus and sulphur, and 0·1 per cent. carbon (maximum).

Nevental. A proprietary trade name for Nealbarbitone.

Nevidene. A proprietary trade name for a coumarone resin.

Nevile and Winther's Acid. α-Naphthol-mono-sulphonic acid (1 : 4).

Nevillac. A proprietary trade name for a phenol-indene-coumarone resin. Used in varnishes and paints.

Neville. A proprietary trade name for coumarone-indene resins.

Nevillite. A proprietary trade name for a hydrocarbon.

Nevin. A proprietary brand of lithopone (q.v.).

Nevindene. A proprietary trade name for coumarone-indene resins.

Nevinol. A proprietary coumarone plasticising oil. A viscous liquid polymer practically non-drying at room temperature.

Nevraltein. A name for sodium-p-phenetidine-methane-sulphonate.

Nevrosthénine. An alkaline solution of the glycero-phosphates of sodium, potassium, and magnesium.

Nevyanskite. See IRIDOSMINE.

Newagit. A trade mark for abrasive and refractory materials consisting essentially of alumina.

Newaloy. A proprietary trade name for a steel containing copper.

Newberryite. A mineral, $MgHPO_4.3H_2O$.

Newboldtite. A mineral. It is a zinc blende.

New Blue. See ULTRAMARINE, PRUSSIAN BLUE, and MELDOLA'S BLUE.

New Blue B or G (Fast Blue 2B for Cotton; Fast Cotton Blue B; Fast Navy Blue G, BM, GM; Naphthol Blue B; Fast Marine Blue G, BM, BG). A dyestuff. It is dimethyl-amino-dimethylamino - anilido - naphtho - phenoxazo - nium chloride, $C_{26}H_{25}N_4OCl$. Dyes cotton mordanted with tannin and tartar emetic, blue.

New Blue R. Meldola's blue (q.v.).

New Brunswick Greens. See CHROME GREENS.

New Cacodyl. See ARRHENAL.

New Coccine (Brilliant Ponceau, Brilliant Ponceau 4R, Scarlet F, Brilliant Scarlet, Cochineal Red A, Croceine Scarlet 4BX, Brilliant Ponceau 5R, Brilliant Scarlet 4R, Acid Scarlet R, Rainbow Scarlet 4R, Confectionery Red). A dyestuff. It is the sodium salt of p - sulpho - naphthalene - azo - β-naphthol-disulphonic acid, $C_{20}H_{11}N_2O_{10}S_3Na_3$. Dyes wool red from an acid bath.

New Coccine R. See CRYSTAL SCARLET 6R.

New Fast Blue F and H. A navy blue dyestuff for tannined cotton.

New Fast Blue for Cotton. See MELDOLA'S BLUE.

New Fast Blue R Crystals for Cotton. See MELDOLA'S BLUE.

New Fast Green 3B. See VICTORIA GREEN 3B.

New Fast Grey. A basic dyestuff giving reddish-grey shades on tannined cotton.

New Fuchsine. See NEW MAGENTA.

New Gold. A Bohme dyestuff. It contains quercitron extract, and a chrome mordant.

New Green. A dyestuff. It is dimethyl - diamino - naphthyl - diphenyl - carbinol-hydrochloride. Used in calico printing. Also see MALACHITE GREEN and EMERALD GREEN.

New Grey (Nigrisine, Methylene Grey, New Methylene Grey, Malta Grey, Alsace Grey, Direct Grey, Nigramine, Metaniline Grey, Special Grey R). A dyestuff obtained by boiling nitroso-dimethyl-aniline hydrochloride with water and alcohol, or by the oxidation of dimethyl-*p*-phenylene-diamine. Dyes cotton mordanted with tannin silver-grey or blackish-grey.

Newkirkite. See VARVICITE.

New Legumex. Selective weedkillers. (509).

Newloy. An alloy of 64 per cent. copper, 35 per cent. nickel, and 1 per cent. tin.

New Magenta (New Fuchsine, Isorubine, Magenta BB Powder, Crystals FF, CV Conc, L, Superfine Powder, New Magenta Crystals, New Roseine O). A dyestuff. It is the hydrochloride of triamino-tritolyl-carbinol, $C_{22}H_{24}N_3Cl$. Dyes wool, silk, leather, and tannined cotton, red.

New Magenta Crystals. A dyestuff. It is a British equivalent of New magenta.

Newmastic. An artificial rubber containing glue and glycerin.

New Methylene Blue. A greenish-blue dyestuff used for dyeing mordanted cotton and silk. It is obtained by condensing Meldola's blue with diphenyl-amine.

New Methylene Blue GG. A dyestuff. It is tetramethyl - diamino - naphtho-phenoxazonium chloride, $C_{20}H_{20}N_3OCl$. Dyes tannined cotton greenish-blue. It dyes silk from a killed soap bath.

New Methylene Blue N. A dyestuff. It is diethyl-diamino-toluphenaz-thionium chloride, $C_{18}H_{22}N_3SCl$. Dyes tannined cotton blue.

New Methylene Blue NX. See MARINE BLUE.

New Methylene Grey. See NEW GREY.

New Patent Blue B, 4B. See PATENT BLUE V, N.

New Phosphine G. A dyestuff. It is *exo* - dimethyl - amino - toluene - azo - resorcin, $C_{15}H_{17}N_3O_2$. Dyes leather and tannined cotton, yellow.

New Pink. See PHLOXINE P.

New Printing Black SS, NR, NGR. Logwood extracts, which have an odour of acetic acid, mixed with sodium chlorate. Employed directly for printing cotton.

New Red. See BIEBRICH SCARLET.

New Red L. See BIEBRICH SCARLET.

New Roseine O. A British brand of New magenta.

New Solid Green (New Victoria Green Extra). A dyestuff. It is analogous in composition to Malachite green, but is prepared with dichlor-benzaldehyde, instead of benzaldehyde. See MALACHITE GREEN and BRILLIANT GREEN.

New Solid Green BB and 3B. Victoria green 3B (*q.v.*).

Newtex. A proprietary safety glass.

Newtonite. A mineral, $Al_2O_3.2SiO_2.5H_2O$, of Arkansas.

Newton's Alloy. An alloy of 50 per cent. bismuth, 31·2 per cent. lead, and 18·7 per cent. tin. It is a fusible alloy, and melts at 94·5° C.

New Urotropine. See HELMITOL.

New Verdone. Selective weedkiller. (511).

New Victoria Black B. See VICTORIA BLACK B.

New Victoria Black Blue. See VICTORIA BLACK B.

New Victoria Blue R (Victoria Blue R). A dyestuff. It is the hydrochloride ethyl - tetramethyl - triamino - α - naph-thyl - disphenyl - carbinol, $C_{29}H_{32}N_3Cl$. Dyes silk, wool, and tannined cotton, blue.

New Victoria Green Extra. See NEW SOLID GREEN.

New White. See BARIUM WHITE.

New White Lead. A sulphate of lead.

New-wrap. A proprietary viscose packing material.

New Yellow. See CHROME YELLOW and ORANGE IV.

New Yellow L. See ACID YELLOW.

New Zealand Dammar. A name given to Kauri copal resin, obtained from *Dammara Australis*.

New Zealand Greenstone. Nephrite (*q.v.*).

Nez. A proprietary preparation of phenylephrine hydrochloride, paracetamol, caffeine and ascorbic acid. Cold remedy. (273).

Ngai Camphor. A camphor obtained from *Blumea balsamifera*. It is closely related to borneol.

N'hangellite. An elastic bitumen.

Naicin. See NICOTINAMIDE.

Niagara Blue. This is the same as trypan blue.

Nialamide. N-(2-Benzylcarbamoyl-ethyl) - N′ - isonicotinoylhydrazine. Niamid.

Niamid. A proprietary preparation of nialamide. An antidepressant. (315).

Niaoul Oil. See GORNEROL.

Nibiol. A proprietary preparation of nitroxoline. (320).

Nibren Wax. Chlorinated naphthalene.

Nibrite. Barrel bright nickel plating process. (528).

Nicametate. 2-Diethylaminoethyl nicotinate.

Nicar. Nickel carbonate. (589).

Nicaragua Wood. See REDWOODS.

Nicat. Nickel catalysts. (586).

Nicat 101. A proprietary brand name for a nickel-aluminium alloy 200 mesh for hydrogenation of readily reducible groups (ethylenic and acetylenic) and for desulphurisation and dehalogenation. (205).

Nicat 102. Similar to 101 except that it is used where a more specific catalyst is required such as for the hydrogenation of aldehydes, ketones and aromatic nuclei. (205).

Nicat K17. A proprietary brand name for a finely divided nickel supported on kieselguhr supplied under hardened oil with a nominal nickel content of 17 per cent. (205).

Nicat NP/AC60PT. As NP/AC60P but in the form of $\frac{1}{8}''$ pellets.

Nicat NP/K50P. A proprietary brand name for finely divided nickel mounted on kieselguhr in an active but stabilized powder form containing 50 per cent. nickel. (205).

Nicat NP/K50PT. As NP/K50P but in the form of $\frac{1}{8}''$ pellets. (205).

Nicat NP/SC60P. A proprietary brand name for finely divided nickel mounted on silica in an active but stabilised powder form containing 55 per cent. nickel. (205).

Nicat RFF12. A proprietary brand name for a finely divided nickel catalyst prepared from nickel formate supplied under hardened oil. The nominal nickel content is 12 per cent. (205).

Nicat RFF25. A proprietary brand name for a finely divided nickel catalyst prepared from nickel formate supplied in hardened oil. The nominal nickel content is 25 per cent. (205).

Nicat S17. A proprietary brand name for a finely divided nickel supported on silica supplied under hardened oil with a nominal nickel content of 17 per cent. (205).

Niccolite (Nickeline, Nickelite, Arsenical Nickel, Speiss, Copper Nickel, Kupfernickel). A mineral. It is an arsenide of nickel, NiAs. The arsenic is largely replaced by antimony in many varieties.

Nice Oil. See AIX OIL.

Nicergoline. A vasodilator. 8β-(5-Bromonicotinoyloxymethyl) - 10 - methoxy - 1, 6 - dimethylergoline.

Niceritrol. A preparation used in the treatment of hypercholesterolæmia. Pentærythritol tetranicotinate.

Nicetal. A proprietary preparation of isoniazid. An antituberculous drug. (244).

Ni-chillite. A proprietary trade name for a nickel-chromium-molybdenum cast iron.

Nicfo. Nickel formate. (589).

Nicholsonite. A mineral. It is a variety of Arragonite, containing less than 10 per cent. zinc.

Nicholson's Blue. See ALKALI BLUE. Also a mixture of Alkali blue and Water blue.

Nichroloy. Electrical resistance alloys containing 23–75 per cent. nickel, 7–20 per cent. chromium, 7–50 per cent. iron, and 1–3 per cent. manganese. They are stated to resist corrosion by steam and dilute acids.

Nichrome. A registered trade mark. Alloys of 54–80 per cent. nickel, 10–20 per cent. chromium, 7–27 per cent. iron, 0–11 per cent. copper, 0–5 per cent. manganese, 0·3–4·6 per cent. silicon, and sometimes 1 per cent. molybdenum, and 0·25 per cent. titanium. Nichrome I is stated to contain 60 per cent. nickel, 25 per cent. iron, 11 per cent. chromium, and 2 per cent. manganese, and Nichrome II 75 per cent. nickel, 22 per cent. iron, 11 per cent. chromium, and 2 per cent. manganese. They are used as electrical resistance metals, and are stated to resist acids. Nichrome I is stated to have a resistivity of 110 microohms . cm. and Nichrome II a resistivity of 113 micro-ohms . cm. at 0° C. Nichrome III—a heat-resisting alloy containing 85 per cent. nickel and 15 per cent. chromium.

Nichrosi. Alloys. (*a*) Contains 25–30 per cent. chromium, 16–18 per cent. silicon, balance nickel. (*b*) Contains 15–25 per cent. chromium, 16–18 per cent. silicon, balance nickel.

Nickeladium. A proprietary trade name for a nickel-vanadium cast steel.

Nickel 200. A grade of nickel containing 99·0 per cent. nickel. (143).

Nickel 201. A grade of nickel containing 99·0 per cent. nickel (min) and 0·02 per cent. carbon (max). (143).

Nickel 204. A nickel containing 4 per cent. cobalt. (143).

Nickel 205. Nickel containing 99·0 per cent. min. nickel and low carbon. (143).

Nickel 211. Nickel containing 5 per cent. manganese. (143).

Nickel 212. Nickel containing 2 per cent. manganese. (143).

Nickel 213. A nickel with improved machining properties. Nickel 96 per cent. min., Manganese 2 per cent. High carbon and silicon content. (208).

Nickel 222. Nickel containing 99·5 per cent. min. nickel, 0·06–0·09 per cent. magnesium and very low impurity levels. (143).

Nickel 223. Nickel containing 99·5 per cent. min. nickel, 0·035–0·065 per cent. magnesium and very low impurity levels. (143).

Nickel 229. Nickel containing 97·5 per cent. min. nickel, 1·8–2·2 per cent. tungsten, 0·35–0·65 per cent. magnesium and 0·02–0·04 per cent. aluminium. (143).

Nickel 270. Nickel 99·9 per cent. pure. (143).

Nickel Aluminium Bronze. An alloy of 10–40 per cent. nickel, 10–88 per cent. copper, and 2–30 per cent. aluminium. One alloy contains 20 per cent. tin.

Nickel, Arsenical. See NICCOLITE.

Nickel Babbitt. A proprietary trade name for a tin-copper-nickel alloy used as a bearing metal for high speeds.

Nickel Blende. See MILLERITE.

Nickel Bloom. See ANNABERGITE.

Nickel Brass. A nickel silver. One alloy contains 55 per cent. copper, 43 per cent. zinc, and 2 per cent. nickel, and another 50 per cent. copper, 34 per cent. zinc, 15 per cent. nickel, and 0·1 per cent. aluminium.

Nickel Brass, Krupp. A nickel silver. It contains 48·5 per cent. copper, 24·3 per cent. zinc, 24·3 per cent. nickel, and 2·9 per cent. iron.

Nickel Bronze. A nickel silver. It usually contains from 20–30 per cent. nickel, 50–86 per cent. copper, and 8–25 per cent. tin, but other alloys contain 11–18 per cent. zinc, and 0–18 per cent. lead.

Nickel - chromium - aluminium. An alloy consisting mainly of nickel, with aluminium and chromium.

Nickel-chromium-copper. An alloy of from 80–85 per cent. nickel, 20–25 per cent. chromium, and 15–20 per cent. copper.

Nickel-chromium-molybdenum Steel " B." A British Chemical Standard. The composition is 3·05 per cent. nickel, 0·68 per cent. chromium, 0·34 per cent. molybdenum, 0·35 per cent. carbon, 0·25 per cent. silicon, 0·034 per cent. sulphur, 0·024 per cent. phosphorus, and 0·61 per cent. manganese.

Nickel - chromium - molybdenum Steels. Steels containing from 0·1–0·7 per cent. carbon, traces–5 per cent. molybdenum, 0·5–5 per cent. nickel, and 0·25–2·5 per cent. chromium.

Nickel-chromium Steels. These alloys are usually made in types containing 1·5, 2·5, and 3·5 per cent. nickel. One alloy contains 0·6–0·7 per cent. chromium, and 1·25–1·5 per cent. nickel, and another 1·1–1·25 per cent. chromium, with 2·5–3 per cent. nickel. They are used for making armour plate, projectiles, and in automobile construction. These steels resist the action of sea water, acetic, boric, citric, malic, nitric, oleic, oxalic, phosphorous, picric, and pyrogallic acids, acetic anhydride, camphor, paraffin, petrol, and oils. They are unaffected by ammonium salts, calcium oxychloride, hydrogen peroxide, formaldehyde,

iodine, and sodium salts but are attacked by hydrocyanic, hydrofluoric and chloroacetic acids and bromine, chlorine, carbon tetrachloride, and ferric chloride.

Nickel Coinage. An alloy of 25 per cent. nickel and 75 per cent. copper. Used in U.S.A., Belgium, Brazil, Britain, Germany, and other countries.

Nickel Copper. See NICCOLITE.

Nickel-copper Alloys. Various. Alloys containing 2–10 per cent. nickel and 90–98 per cent. copper are known as nitro-copper. An alloy of 15 per cent. nickel and 85 per cent. copper is termed Benedict metal. Cupro-nickel contains 80 per cent. copper and 20 per cent. nickel, with a specific gravity of 8·9, and alloys containing 40–45 per cent. nickel and 55–60 per cent. copper are known as Ferry metal and Constantan. An alloy of 70 per cent. nickel and 30 per cent. copper is known as Corronil.

Nickel, Cupro. See NICKEL-COPPER ALLOYS.

Nickelene. A name suggested for nickel silver (German silver) alloys.

Nickel Glance. A mineral, $Ni(AsS)_2$.

Nickel Gymnite. See GENTHITE.

Nickelin. Electrical resistance alloys of nickel and copper, usually with zinc. One alloy contains 55·3 per cent. copper, 31 per cent. nickel, 13 per cent. zinc, 0·4 per cent. iron, and 0·2 per cent. lead; another 68 per cent. copper and 32 per cent. nickel; and a third 74·5 per cent. copper, 25 per cent. nickel, and 0·5 per cent. iron.

Nickeline. Synonym for Niccolite.

Nickeling Solutions. These usually consist of solutions of nickel-ammonium sulphate and ammonia or ammonium sulphate in water. Used for the electro-deposition of nickel.

Nickel Iron. See FERRO-NICKEL.

Nickelite. See NICCOLITE.

Nickel-linnæite. See POLYDYMITE.

Nickel Manganese Bronze. An alloy containing 2·5 per cent. nickel, 53·4 per cent. copper, 39 per cent. zinc, 1·7 per cent. manganese, and 2·6 per cent. tin, with small quantities of aluminium and lead.

Nickel-manganese-copper. An alloy of 73 per cent. copper, 24 per cent. manganese, and 3 per cent. nickel. Used for electrical resistances.

Nickel-molybdenum Steels. Alloys containing 0·13–0·54 per cent. carbon, 0·12–4·4 per cent. molybdenum, and 1·8 per cent. nickel.

Nickeloid. A proprietary trade name for a dual metal. It is zinc faced with nickel.

Nickel Oreide. A nickel silver. It is

an alloy of from 63–87 per cent. copper, 6–33 per cent. zinc, and 2–7 per cent. nickel.

Nickeloy. An alloy of 1·5 per cent. nickel, 4 per cent. copper, and 94 per cent. aluminium.

Nickel Pyrites. See MILLERITE.

Nickel Salipyrin. A salt of nickel, salicylic acid, and antipyrin. An antiseptic.

Nickel - silicon - molybdenum Steels. Alloys containing from 0·37–0·66 per cent. carbon, 0·27–0·78 per cent. molybdenum, 0·7–3·5 per cent. nickel, and 0·7–2·5 per cent. silicon.

Nickel Silvers. Ternary alloys of copper, nickel, and zinc, the standard of which is determined by the nickel content. The alloy containing 30 per cent. nickel is known as BB, 25 per cent. nickel as A1, and 20, 15, 12, 10, and per cent. nickel as first, seconds, thirds, fourths, and fifths respectively. English nickel silvers usually contain 61·3 per cent. copper, 19·1 per cent. nickel, 19·1 per cent. zinc. Another English alloy consists of 57·4 per cent. copper, 25 per cent. zinc, 13 per cent. nickel, and 3 per cent. iron. Birmingham nickel silvers vary from 50–62 per cent. copper, 20–36 per cent. zinc, and 7–30 per cent. nickel, the best containing 50 per cent. copper, 29 per cent. zinc, 21 per cent. nickel. Nickel silver for plate contains 57 per cent. copper, 36 per cent. zinc, and 7 per cent. nickel. Extra White contains 50 per cent. copper, 20 per cent. zinc, and 30 per cent. nickel.

A Sheffield nickel silver contains 57·4 per cent. copper, 26·5 per cent. zinc, 13 per cent. nickel, 3 per cent. lead, and another 57 per cent. copper, 24 per cent. nickel, and 19 per cent. zinc. Chinese nickel silver contains 40·4–41 per cent. copper, 25·4–26·5 per cent. zinc, 30·8–31·6 nickel, and 2·6–2·7 per cent. iron. Another name for these alloys is German silver, but this term is used for alloys containing varying proportions of copper, zinc, and nickel. A good quality German silver contains 46 per cent. copper, 34 per cent. zinc, and 20 per cent. nickel, and a common one 55 per cent. copper, 25 per cent. zinc, and 20 per cent. nickel. Others vary between 46–63 per cent. copper, 18–36 per cent. zinc, and 6–31 per cent. nickel. The resistivity of nickel silver containing 56 per cent. copper, 28 per cent. nickel, and 16 per cent. zinc is 21 micro-ohms . cm. at 0° C., and one containing 60 per cent. copper, 15 per cent. nickel, and 25 per cent. zinc, is 30 micro-ohms . cm. This name is also applied to those containing in

addition a little lead, and iron, and in rare cases manganese and tin. The following are some of the trade names of nickel silver alloys. Albatra, Alfenide, Alpacca, Amberoid, Ambrac, American silver, Aphit, Argentan, Argentin, Argiroide, Argentan solder, Argyroide, Argyrolith, Aterite, Benedict plate, Bismuth bronze, Bismuth brass, Bolster silver, Brazing solder, China silver, Carbondale silver, Charcoal nickel silvers, Christofle, Chromax bronze, Colorado silver, Craig gold, Electroplate, Electrum, Elner's German silver, Frick's alloys, Keene's alloy, Key alloy, Lutecin, Maillechort, Markus alloy, Milling silver, Naubuc, Neogen, Neu-silver, Nevada silver, Nickel brass, Nickel bronze, Nickelin, Nickel oreide, Packfong, Packtong, Platinoid, Potosi silver, Rheotan, Ruolz alloys, Silverite, Silveroid, Spoon metal, Sterlin, Sterline, Suhler white copper, Toncas metal, Tuc-tur metal, Tungsten brass, Tutenag, Victoria silver, Victor metal, Virginia silver, Wessel's silver, White button alloy, White copper, White metal, White solder. Alfenide, Argyroide, and Christofle are plated German silvers. These alloys are used in the manufacture of table ware, and for other purposes.

Nickel Silver, English. An alloy usually containing 61·3 per cent. copper 19·1 per cent. nickel, and 19·1 per cent. zinc.

Nickel Silver, Sheffield. An alloy of 57 per cent. copper, 24 per cent. nickel, and 19 per cent. zinc.

Nickel Silver Solder. An alloy of 35 per cent. copper, 57 per cent. tin, and 8 per cent. nickel.

Nickel Steel. An alloy of nickel with steel, usually containing from 3–5 per cent. of nickel, but sometimes a larger amount. One alloy contains 30 per cent. of nickel, 1 per cent. manganese, and 1 per cent. chromium. Nickel steels are used for armour plates, ships' screws, boiler plates, cable wires, and gun barrels. Also see STEELS O and T.

Nickel Stibine. See ULLMANITE.

Nickel-Skutterudite. A mineral. It is 8NiAs₃.

Nickel Tantalum. Alloys of nickel and tantalum. An alloy containing 30 per cent. tantalum is stated to resist the action of boiling aqua regia.

Nickel Tungsten. Alloys containing 25–50 per cent. nickel and 50–75 per cent. tungsten.

Nickel - vanadium - molybdenum Steels. Alloys containing traces–0·8 per cent. carbon, 0·25–3 per cent.

molybdenum 1–5 per cent. nickel, and traces–0·6 per cent. vanadium.

Nickel Vitriol. See MORENOSITE.

Nickel, White. See CHLOANTITE.

Nickel Yellow. A pigment obtained by treating nickel sulphate with sodium phosphate, and calcining the product.

Nickel Zirconium. An alloy of 86·4 per cent. nickel, 6 per cent. aluminium, 6 per cent. silicon, and 1·5 per cent. zirconium. Used for cutting tools.

Niclad. A proprietary trade name for a duplex metal in which nickel or nickel alloy is deposited on steel or iron.

Niclofolan. An anthelmintic. 2,2′-Bis-(4-chloro-6-nitrophenol). BILEVON.

Niclosamide. 5-Chloro-N-(2-chloro-4-nitrophenyl)salicylamide. Yomesan.

Nicocodine. 6-O-Nicotinoylcodeine.

Nicodicodine. An anti-tussive. 7,8-Di-hydro - O^3 - methyl - O^6 - nicotinoyl - morphine.

Nicolane. A proprietary trade name for Noxapine.

Nicolle's Carbol-thionin Blue. A microscopic stain. It consists of a mixture of 10 c.c. of a saturated solution of thionin blue in 50 per cent. alcohol, and 100 c.c. of a 2 per cent. carbolic acid solution.

Nicolube. Trade mark for a fatty-acid soap-based lubricant containing additives such as graphite, molybdenum disulphide, PTFE and a precipitating agent. (691).

Nicomorphine. 3,6-Di-O-nicotinoyl-morphine.

Nicon. An alloy of 70 per cent. iron and 30 per cent. nickel.

Nicoschwab. See UBA.

Nicotinamide. Nicotinic acid, niacin. One of the substances comprising the Vitamin B complex. Deficiency in the diet is partly the cause of pellagra, a skin disorder.

Nicotine Humate. See NICOTINE PEAT. Nicotine humate is the *water soluble* product formed on mixing nicotine and peat, and is isolated by evaporation. It contains 28–34 per cent. nicotine and is used as an insecticide.

Nicotine Peat. A *water-insoluble* insecticide formed by mixing nicotine and peat in water.

Nicotinyl Alcohol. 3-Pyridylmethanol.

Nicoumalone. 3-[2-Acetyl-1-(4-nitrophenyl)ethyl] - 4 - hydroxycoumarin. Acenocoumarol. Sinthrome.

Nicral Alloys. Aluminium alloys containing varying percentages of nickel, chromium, and copper.

Nicro-copper. An alloy of 98 per cent. copper and 2 per cent. nickel.

Nicroman. A proprietary trade name for a tool steel containing 1 per cent. chromium, 1·65 per cent. nickel, 0·35

per cent. copper, and 0·7 per cent. carbon. An oil-hardening hob steel.

Nicrosil. A proprietary trade name for an alloy of iron with 18 per cent. nickel and 4–6 per cent. silicon.

Nicrosilal. An alloy of 71·2 per cent. iron, 18 per cent. nickel, 6 per cent. silicon, 2 per cent. chromium, 1·8 per cent. carbon, and 1 per cent. manganese. It is a non-magnetic grey cast iron which is resistant to staining and has great ductility.

Nicu Steel. A nickel steel containing 2·13 per cent. nickel, 0·2 per cent. copper, 0·51 per cent. manganese, 0·03 per cent. sulphur, 0·03 per cent. silicon, and 0·006 per cent. phosphorus.

Nidrin. A proprietary preparation containing aluminium hydroxide and magnesium carbonate in a co-dried gel. An antacid. (268).

Niello Silver (Russian Tula, Blue Silver). An alloy of silver, copper, lead, and bismuth with a bluish colour.

Nifedipine. A coronary vasodilator currently undergoing clinical trial as " BAY a 1040 ". It is dimethyl 1,4-dihydro - 2, 6 - dimethyl - 4 - (2 - nitrophenyl) pyridine - 3, 5 - dicarboxylate.

Nifenazone. 2,3-Dimethyl-4-nicotinamido-1-phenyl-5-pyrazolone. Thylin.

Niferex. A proprietary preparation of a polysaccharide-iron complex used in the treatment of anæmia. (781).

Nifuratel. A drug used in the treatment of trichomoniasis. It is 5-methylthiomethyl - 3 - (5 - nitrofurfurylidene - amino) oxazolidin - 2 - one.

Nifursol. An anti-protozoan. 3, 5 - dinitro - N - (5 - nitrofurfurylidene) salicylohydrazide. SALFURIDE.

Nifurtimox. A drug used in the treatment of trypanosomiasis. It is tetrahydro - 3 - methyl - 4 - (5 - nitrofurfurylideneamino) - 1, 4 - thiazine 1, 1 - dioxide.

Nigagin. Methyl-*p*-hydroxy-benzoate. A preservative.

Nigella (Black Carraway, Black Cumin). The dried seeds of *Nigella sativa*. A condiment. It is also used in snuff.

Nigerite. A mineral. It is $(Zn,Fe^{\cdot\cdot},Mg)(Sn,Zn)_2(Al,Fe^{\cdot\cdot\cdot})_{12}O_{22}(OH)_2$.

Nigger Powder. See NEGRO POWDER.

Night Blue. A dyestuff. It is the hydrochloride of *p*-tolyl-tetraethyl-triamino - diphenyl - α - naphthyl - carbinol, $C_{38}H_{42}N_3Cl$. Dyes silk and wool greenish-blue. Also see METHYL BLUE.

Night Blue B. A dyestuff. It is the sodium salt of *o*-chloro-*m*-nitro-diethyldibenzyl - diamino - triphenyl - carbinol-disulphonic acid, $C_{37}H_{34}N_3ClS_2O_9Na_2$. Dyes wool and silk bluish-green from an acid bath.

Night Green. See IODINE GREEN.

Night Green 2B. A dyestuff. It is the sodium salt of chloro-diethyl-dibenzyl-diamino - triphenyl - carbinol - disulphonic acid, $C_{37}H_{35}N_2S_2O_7ClNa_2$. Dyes wool and silk bluish-green from an acid bath.

Nigramine. See NEW GREY.

Nigraniline. Aniline black, $C_{30}H_{25}N_5$.

Nigre. The impure soap remaining after the good soap has been removed by running out. It contains iron soaps, caustic soda, and sodium chloride.

Nigrine. A mineral. It is a variety of rutile.

Nigrisine. See NEW GREY.

Nigrite. A gutta-percha substitute. It consists of rubber mixed with the residue from the distillation of ozokerite.

Nigrol. The residue obtained after the removal of kerosene, gasoline, and light oils from petroleum naphtha at Baku.

Nigrosalin. See DIRECT BLACK.

Nigrosine ABKS, G, L, SG, SS, 1471, 7600. Dyestuffs. They are British equivalents of Water-soluble nigrosine.

Nigrosine Crystals, B, BP, G, JB, R, W, WM, 79694. Dyestuffs. They are British brands of Water-soluble nigrosine.

Nigrosine, Soluble. See INDULINE, SOLUBLE.

Nigrosine, Spirit Soluble. See INDULINE, SPIRIT SOLUBLE.

Nigrosulphine. A sulphur black dye made from 2-hydroxy-*m*-phenylenediamine and sulphur.

Nigroth Metal. A proprietary trade name for a heat-resisting nickel-chromium cast iron.

Nigrotic Acid. Dihydroxy-sulpho-naphthoic acid (2 : 8 : 3 : 6).

Ni-hard. Alloys of iron with from 4–5 per cent. nickel, 1·5 per cent. chromium, and varying amounts of silicon and carbon.

Nihil. See NIL.

Nikalgin. A proprietary preparation of quinine and urea hydrochloride.

Niketol. Ethyl-*p*-amino-benzo-phthalamate hydrochloride. A dental anæsthetic.

Nikoteen. An American product containing 26 per cent. nicotine. A fumigant.

Nikrome. Proprietary trade name for nickel-chromium steels. Nikrome M contains 2·25 per cent. nickel, 1 per cent. chromium, 0·45 per cent. molybdenum, and 0·4 per cent. carbon.

Nikro-trimmer Steel. A proprietary trade name for a nickel-chromium steel containing 0·3 per cent. nickel, 0·55 per cent. chromium, and 0·85 per cent. carbon.

Nil (Nihil). Zinc oxide, ZnO.

Nile Blue A. A dyestuff. It is diethyl-diamino - naphtho - phenoxazonium sulphate, $(C_{20}H_{20}N_3O)_2SO_4$. Dyes tannined cotton blue.

Nile Blue 2B. A dyestuff. It is diethyl-benzyl - diamino - phenoxazonium chloride, $C_{27}H_{26}N_3OCl$. Dyes tannined cotton greenish-blue.

Nile Blue R. A dyestuff allied to Nile blue A and 2B, of a more reddish colour.

Nilergex. A proprietary preparation of isothipendyl hydrochloride. (2).

Nilevar. A proprietary preparation of norethandrolone. An anabolic steroid. (285).

Nilex. A proprietary 36 per cent. nickel steel used for pendulums, etc., on account of its low coefficient of expansion.

Nilo. A trade mark for controlled expansion alloys. Their compositions are coded as follows:

    Alloy 36. 36 per cent. nickel, balance iron.

    Alloy 42. 42 per cent. nickel, balance iron.

    Alloy K45. 32 per cent. nickel, 13 per cent. cobalt, 54 per cent. iron.

    Alloy P50. 50 per cent. nickel, 50 per cent. iron. (143).

    Alloy 475. 47 per cent. nickel, 5 per cent. chromium, balance iron.

    Alloy 48. 48 per cent. nickel, balance iron.

    Alloy 51. 51 per cent. nickel, balance iron.

    Alloy K. 29 per cent. nickel, 17 per cent cobalt, balance iron. (143).

Nilodin. A proprietary trade name for the hydrochloride of Lucanthrone.

Nilomag. A trade mark for magnetic alloys made by a powder metallurgy process. Their compositions are as follows:

    Alloy 471. 47 per cent. nickel, 50 per cent. iron, 3 per cent. molybdenum.

    Alloy 475. 47 per cent. nickel, 5 per cent. chromium, balance iron.

    Alloy 48. 48 per cent. nickel, balance iron.

    Alloy K. 29 per cent. nickel, 17 per cent. cobalt, balance iron.

    Alloy 51. 51 per cent. nickel, balance iron. (143).

Nilstim. A proprietary preparation of microcrystalline cellulose and methylcellulose used in the treatment of obesity. (692).

Nilverm. A proprietary preparation of TETRAMISOLE. A veterinary anthelmintic.

Nimocast Alloy 242. A patented alloy of 21 per cent. chromium, 10 per cent. cobalt, 10·5 per cent. molybdenum and the balance nickel. (143). A trade mark.

Alloy 771. 77 per cent. nickel, 14 per cent. iron, 5 per cent. copper, 4 per cent. molybdenum. (143).

Alloy 713. 13·4 per cent. chromium, 4·5 per cent. molybdenum, 1 per cent. titanium, 6·2 per cent. aluminium, 2·3 per cent. niobium and the balance nickel. (143).

Alloy PE10. 20 per cent. chromium, 6 per cent. molybdenum, 6·5 per cent. niobium, 2·5 per cent. tungsten, and the balance nickel. (143).

Alloy PK24. 10 per cent. chromium, 15·2 per cent. cobalt, 3 per cent. molybdenum, 5·2 per cent. titanium, 5·5 per cent. aluminium, and the balance nickel. (143).

Nimol. An alloy of cast iron with 20 per cent. monel metal and 2–4 per cent. chromium. It is non-magnetic and has high resistance to corrosion by acid and sea water.

Nimonic Alloy PE 11. A trade mark for an alloy of 18 per cent. chromium, 5·2 per cent. molybdenum, 2·3 per cent. titanium, 0·8 per cent. aluminium, 38 per cent. nickel, and the balance iron. (143).

PE 13. 22 per cent. chromium, 1·5 per cent. cobalt, 9 per cent. molybdenum, 18·5 per cent. iron, 0·6 per cent. tungsten, and the balance nickel. (143).

PK 31. 20 per cent. chromium, 14 per cent. cobalt, 4·5 per cent. molybdenum, 2·3 per cent. titanium, 0·4 per cent. aluminium, 5 per cent. niobium, and the balance nickel.

PK33. 19 per cent. chromium, 14 per cent. cobalt, 7 per cent. molybdenum, 2 per cent. titanium, 2·0 per cent. aluminium, and the balance nickel.

Alloy 75. 20 per cent. chromium, 0·4 per cent. titanium, and the balance nickel. (143).

Alloy 80A. 20 per cent. chromium, 2·3 per cent. titanium, 1·3 per cent. aluminium, and the balance nickel. (143).

Alloy 90. 20 per cent. chromium, 17 per cent. cobalt, 2·5 per cent. titanium, 1·5 per cent. aluminium, and the balance nickel. (143).

Alloy 93 is the same as alloy 90 except that closer control is maintained.

Alloy 105. 15 per cent. chromium, 20 per cent. cobalt, 5 per cent. molybdenum, 1·2 per cent. titanium, 4·7 per cent. aluminium, and the balance nickel. (143).

Alloy 108 has a closer compositional control. (143).

Alloy 118. A fully vacuum-melted and cast version of alloy 115. (143).

Alloy 115. 15 per cent. chromium,

15 per cent. cobalt, 3·5 per cent. molybdenum, 4 per cent. titanium, 5 per cent. aluminium, and the balance nickel. (143).

Nimorazole. A drug used in the treatment of trichomoniasis. It is 4-[2-(5-nitroimidazol-1-yl) ethyl] morpholine. Nitrimidazine. NAXOGIN, NULOGYL.

Nimox. Nickel molybdenum oxides on alumina. (513).

Ninhydrin (Triketol). Tri-keto-hydrindene-hydrate. Employed to demonstrate the presence of albumin.

Niobe Oil. See OIL OF NIOBE.

Niobite. A mineral, $(Fe.Mn)O.(Nb.Ta)_2O_3$.

Nioform. See VIOFORM.

Nipabenzyl. Benzyl 4-hydroxybenzoate. (625).

Nipabutyl. Butyl 4-hydroxybenzoate. (625).

Nipacombin. Compounded sodio-4-hydroxybenzoate esters. (625).

Nipagina. A world wide registered trade mark for the ethyl ester of p-hydroxybenzoic acid. A preservative. (625).

Nipagin M. A world-wide registered trade mark for the methyl ester of p-hydroxybenzoic acid. A preservative. (625).

Nipasol M. A world-wide registered trade mark for the propyl ester of p-hydroxybenzoic acid. A preservative. (625).

Nipantiox. Butylated hydroxyanisole. (625).

Nipa Salt. A material obtained by ignition of the plant *Nipa fructicans*. Used to coagulate rubber latex.

Nipasept. Compounded esters of 4-hydroxybenzoic acid. (625).

Nipasol. Propyl-4-hydroxybenzoates. (625).

Nipholite. Synonym for Khodnevite.

Nipol. A proprietary synthetic rubber. (693).

Nipro. Caprolactam. A trade mark. (694).

Ni-resist. A corrosion and heat-resisting cast iron containing 12–15 per cent. nickel, 5–7 per cent. copper, and 1·5–4 per cent. chromium.

Nirex. A proprietary trade name for an alloy of 80 per cent. nickel, 14 per cent. chromium, and 6 per cent. iron.

Niridazole. 1-(5-nitro-2-thiazolyl)-2-imidazolidinone. Used in the treatment of bilharaziasis. Available as Ambilhar.

Nirolex Expectorant Linctus. A proprietary preparation containing guaiphenesin, ephedrine sulphate and mepyramine maleate. A cough linctus. (253).

Nirostaguss. A non-rusting and heat-resisting 34 per cent. chromium cast iron.

Nirvanine (Glycocaine). The hydrochloride of the methyl ester of diethyl-glycocoll-*p*-amino-*o*-oxybenzoic acid. A local anæsthetic.

Nirvanol. Phenyl-ethyl-hydantoin, $C_6H_5.C_2H_5.C.CO.NH.NH.CO.$ A hypnotic.

Nisapas. A proprietary preparation of sodium para-aminosalicylic acid and isoniazid. An antituberculous agent. (841).

Nisentil. The hydrochloride of Alpha-prodine.

Nispan Alloy C-902. A trade mark for an alloy of 42 per cent. nickel, 5·2 per cent. chromium, 2·4 per cent. titanium, 0·6 per cent. aluminium, and the balance iron. (143).

Nisser Powder. An explosive containing 10·5 per cent. potassium perchlorate, 44·5 per cent. potassium nitrate, 2 per cent. potassium bichromate, 1·5 per cent. potassium ferrocyanide, 15·5 per cent. sulphur, 19·5 per cent. charcoal, and 6·5 per cent. vegetable substance.

Nissex. A parasiticide. It is stated to contain the active principles of insect powder with hydroxy-quinoline dissolved in a petroleum-fat-acetone mixture.

Nitolac. A proprietary polyurethane flooring material. (695).

Nitoman. A proprietary trade name for Tetrabenazine. (314).

Nitragin. A pure culture of bacteria which, when spread over the soil renders the ordinary fixation of nitrogen more active.

Nitral. A trade name for moist nitrous oxide used as a bactericide.

Nitralloy. A nitrided aluminium-chromium-molybdenum steel.

Nitram. Ammonium nitrate. (2).

Nitrammite. A native ammonium nitrate.

Nitrammomkalk. A mixture of ammonium and calcium nitrates in a granular form of Norwegian manufacture. A fertiliser.

Nitraniline N. Nitraniline mixed with sufficient sodium nitrite necessary for its diazotisation.

Nitraniline Red. It has the formula, $C_{16}H_{11}N_3O_5$, but is sold as nitraniline and β-naphthol to make it.

Nitraphen. See DINITRA.

Nitrapo. A product obtained from crude caliche by crystallisation. It contains about 66 per cent. sodium nitrate, 29 per cent. potassium nitrate, and a little sodium chloride. It is used as a fertiliser.

Nitrated Oils. Thick syrupy liquids obtained by treating castor oil or linseed oil with a mixture of concentrated sulphuric and nitric acids. They form homogenous mixtures with nitro-cellulose. Dissolved in acetone, these oils form varnishes, which are used for enamelling leather or similar material, and mixing paints.

Nitrate of Tin. A mixture of stannous and stannic chlorides. Used by dyers.

Nitratine. A mineral. It is $2[NaNO_3]$. See CHILE SALTPETRE.

Nitrating Acid. See MIXED ACID.

Nitrazepam. 1,3-Dihydro-7-nitro-5-phenyl-1,4-benzodiazepin-2-one. Mogadon.

Nitrazine Yellow. A dyestuff prepared from dioxy-tartaric acid and nitroxyl hydrazine-sulphonic acid. It is now out of use.

Nitrazol. See PARANITRANILINE RED.

Nitre. See SALTPETRE.

Nitre Balls (Throat Balls). Balls of fused potassium nitrate (sal prunella).

Nitre Cake. The residue from the manufacture of nitric acid. It consists of a mixture of normal and acid sodium sulphate.

Nitre, Chile. See CHILE SALTPETRE.

Nitre, Cubic. See CHILE SALTPETRE.

Nitre, Soda. See CHILE SALTPETRE.

Nitre Spirit. See SPIRIT OF NITRE.

Nitre, Sweet. See SPIRIT OF SWEET NITRE.

Nitrex. A proprietary name for nitrofurantoin. (696).

Nitric Ether. Ethyl nitrate, $C_2H_5NO_3$.

Nitrided Steel. Steel which has been treated with ammonia gas at a temperature of 950° C., whereby nitrogen is absorbed on the surface giving a hard, non-brittle surface. Steels containing from 0·5–2 per cent. aluminium and 0·5–4 per cent. of other elements are used. See NITRALLOY.

Nitrified Borax. See BORAX GLASS.

Nitrimidazine. See NIMORAZOLE.

Nitrite Rubber. A product obtained by treating latex with a nitrite and coagulating with an acid.

Nitro-26. A nitrogeneous fertiliser. (618).

Nitrobacterine. A culture of bacteria for soil inoculation.

Nitrobarite. A mineral. It is $4[Ba(NO_3)_2]$.

Nitrobenzene, Heavy. A mixture of nitro-benzene with the higher homologues.

Nitrobenzene, Light. Almost pure nitro-benzene.

Nitrobenzene, Very Heavy. Commercial nitro-toluene is known under this name.

Nitrobenzide. The name given to nitrobenzene by Mitscherlich.

Nitrobenzol for Red. A mixture of nitro-benzene with the two nitro-toluenes.

Nitrocalcite (Wall saltpetre). Calcium nitrate, $Ca(NO_3)_2$.

Nitro-cellulose Varnishes. Varnishes used in the manufacture of artificial leather. They consist of nitro-celluloses dissolved in amyl acetate, and coloured. Used also for painting iron-work.

Nitro-chalk. A proprietary fertiliser consisting of an intimate mixture of chalk and ammonium nitrate in the form of a fine powder. There are two grades containing 15·5 and 10 per cent. nitrogen respectively. The former grade is for export. The nitrogen exists in two forms, one half in the form of nitrate and the other in the form of ammonia. The chalk content of the 15·5 per cent. grade is 52 per cent. and that of the 10 per cent. grade 66 per cent. (512).

Nitro-chloroform. See CHLORO-PICRIN.

Nitrocine. A proprietary preparation of glyceryl trinitrate. A vasodilator used in angina pectoris. (250).

Nitro-cobalt. The name given to a compound, $Co_2(NO_2)$, formed by passing nitrogen peroxide over reduced cobalt. It is decomposed by water, and is explosive.

Nitro-copper. A compound of copper and nitrogen peroxide, Cu_2NO_2. Also see NICKEL-COPPER ALLOYS.

Nitro-cotton. See GUNCOTTON.

Nitro-dextrin. A similar product to Nitro-starch. Used in explosives.

Nitrodracrylic Acid. p-Nitro-benzoic acid, so-called since it is formed from dragon's blood by oxidation with nitric acid.

Nitro-erythrite. Erythrol-tetranitrate.

Nitrofer. Mixed cultures of *Azotobacter* used for soil inoculation to increase the nitrogen fixation.

Nitroferrite. An explosive. It contains 93 per cent. ammonium nitrate, 2 per cent. trinitro-naphthalene, 2 per cent. potassium ferricyanide, and 3 per cent. sugar.

Nitroform. Tetranitro-methane.

Nitrofurantoin. 1-(5-Nitrofurylidine-amino)hydantoin. Berkfurin. Furadantin.

Nitrofurazone. An anti-infective drug. 5 - Nitro - 2 - furaldehyde semicarbazone. FURACIN.

Nitroganic. A fertiliser material from sewage. It is similar to milorganite and contains 65 per cent. organic matter and 35 per cent. mineral matter. The organic nitrogen amounts to 5·4 per cent., the total phosphoric acid to 2·4 per cent. (available phosphoric acid 2 per cent.), and the potash to 0·3 per cent.

Nitrogenina (Soluble Nitrogenous Organic Fertiliser). A fertiliser stated to be composed of animal skins, bones, and blood. It contains 6 per cent. organic nitrogen and 0·16 per cent. ammoniacal nitrogen.

Nitrogen, Lime. See NITROLIM.

Nitrogenous Fertiliser. See SKIN and LEATHER MEALS.

Nitroglauberite. A mineral, $2Na_2SO_4 \cdot 6NaNO_3 \cdot 3H_2O$.

Nitroglycerin (Glonoine Oil, Pyroglycerin). Glycerol nitrates, usually trinitro-glycerol, $C_3H_5(O.NO_2)_3$.

Nitro-iron. A compound, Fe_2NO_2, formed by allowing finely divided iron to absorb nitrogen peroxide.

Nitro-kairoline. Nitro-methyl-quinoline-tetrahydride, $C_6H_3(NO_2)CH_2.CH_2.N.CH_3.CH_2.$

Nitrolac. A German nitro-cellulose lacquer.

Nitrolan. A proprietary preparation of glyceryl trinitrate used in the treatment of angina pectoris. (487).

Nitro-leuco-malachite Green. Nitro-tetramethyl- di-p-amino- triphenyl-methane, $C_6H_4(NO_2).CH(C_6H_4N(CH_3)_2)_2$.

Nitroleum. See BLASTING OIL.

Nitrolignin. Wood which has been nitrated.

Nitrolim (Lime Nitrogen). A fertiliser. Commercial nitrolim contains 57–63 per cent. calcium cyanamide, 20 per cent. lime, 14 per cent. graphite, and 7–8 per cent. silica, iron oxide, and alumina.

Nitrolite. An explosive. It contains nitro-glycerin.

Nitrol-nickel. A compound formed when nitrogen peroxide is passed over reduced nickel. It is decomposed by water, and is explosive.

Nitromagnesite. A mineral, $Mg(NO_3)_2$.

Nitromagnite (Dynamagnite). An explosive made in a similar way to dynamite, using magnesia alba as the absorbent.

Nitro-malachite Green. Nitro-tetramethyl - di - p - amino - triphenyl - carbinol, $C_6H_4(NO_2) . C(OH)[C_6H_4N(CH_3)_2]_2$.

Nitromannite. Mannitol hexanitrate.

Nitro-methylene Blue. Methylene green.

Nitro-muriate of Tin. See TIN SPIRITS.

Nitro-muriatic Acid. See AQUA REGIA.

Nitron. 1: 4-Diphenyl-3: 5-endanilo-dihydro-triazole, $CH . N(C_6H_5) . C: N . (N . C_6H_5)_2$. A base which forms a nitrate almost insoluble in water. It is used for the determination of nitric acid. It is also used as a rubber vulcanisation accelerator. Also a proprietary trade name for a cellulose nitrate plastic.

Nitrophenine (Paramine Yellow GG, Thiazol Yellow R). The sodium salt of diazo - dehydro - thio - toluidine - sulphonic acid-p-nitraniline, $C_{20}H_{14}N_5O_5$ SNa. Dyes cotton greenish-yellow.

Nitrophoska. Fertilisers of German origin. They are mixtures of potassium nitrate, ammonium chloride, and ammonium phosphate produced by adding diammonium phosphate and a high-grade (over 50 per cent. K_2O) potassium chloride or sulphate to ammonium nitrate. The ammonium nitrate is heated to fusion, the other salts added, the pasty mass stirred, cooled, and ground. There are several brands containing from 9·7–13·4 per cent. nitrogen in the form of ammonia, 1·6–6·4 per cent. nitrogen in the form of nitrate, 11–30 per cent. phosphoric acid, and 15–26·5 per cent. potash.

Nitrophosphate. A fertiliser sometimes wrongly called ammonium superphosphate. It is prepared by mixing calcium superphosphate with ammonium sulphate. Some mixtures contain ammonium phosphate and calcium sulphate.

Nitropropiol. Sodium-o-nitro-phenyl-propiolate.

Nitrosamine Red. A dyestuff. It is diazotised nitraniline.

Nitroscleran. A German preparation. It consists of sodium nitrite, sodium chloride, sodium phosphate, and potassium phosphate, in water. Used for injection to lower blood pressure.

Nitro Silk. See TUBIZE YARN.

Nitrosin Saltpetre. A proprietary preparation containing nitrite, used for curing meat.

Nitro Base. p-Nitroso-dimethyl-aniline. An intermediate for dyes.

Nitroso Blue. See RESORCIN BLUE.

Nitro-starch (Xyloidin). A nitric ester of starch, probably the octonitrate $C_{12}H_{12}(NO_2)_8O_{10}$. Used in America for blasting explosives, either alone, or by mixing 10 per cent. of it with a mixture of sodium nitrate and carbonaceous material.

Nitrosulphate. Ferric sulphate $Fe_2(SO_4)_3$. Sold as a mordant for dyeing.

Nitrosyl. See NITROTYL.

Nitrosyl Silver. Silver hyponitrite. $Ag_2N_2O_2$.

Nitrosyl Sulphuric Acid (Lead Chamber Crystals). Nitro-sulphonic acid, $NO_2(SO_2.OH)$.

Nitrotyl (Nitrosyl, Nitroyl). The group NOH. Nitrosyl is used for the radicle

NO, therefore nitrotyl or nitroyl has been recommended.

Nitrous Ether. Ethyl nitrite, $C_2H_5NO_2$.

Nitrous Gas or Air. Nitrogen dioxide, NO_2.

Nitrous Sulphuric Acid. A solution of nitrosyl-sulphuric acid (Weber's acid), in sulphuric acid.

Nitrovin. A growth promoter. 1,5-bis-(5 - nitro - 2 - furyl) penta - 1, 4 - dien - 3 - one amidinohydrazone. PAYZONE and PANAZON are the hydrochloride.

Nitroxan. A catalyst used for the conversion of ammonia into nitric acid. It is stated to be a compound of barium metaplumbate and barium manganate. The ammonia is oxidised to nitric acid, which is retained as barium nitrate.

Nitroxoline. 8-Hydroxy-5-nitroquinoline. Nibiol.

Nitroxynil. An anthelmintic. 4-hydroxy - 3 - iodo - 5 - nitrobenzonitrile. TRODAX is the eglumine salt.

Nitroyl. See NITROTYL.

Nivan. See OIL WHITE.

Nivaquine. A proprietary preparation of chloroquin sulphate. An antimalarial. (336).

Nivar. A proprietary trade name for Invar ($q.v.$).

Nivebaxin. A proprietary preparation of neomycin sulphate, bacitracin zinc, polymyxin B sulphate, calcium phosphate, glycine and starch. (253).

Niveite. A mineral. It is a variety of copiapite.

Nivembin. A proprietary preparation of chloroquin sulphate and di-iodohydroxyquinoline. An antiamœbic. (336).

Nivemycin. A proprietary preparation containing noemycin sulphate. An antibiotic. (253).

Nivenite. A mineral. It is a variety of uraninite.

Nixenoid. See VISCOLOID.

Nixon C/A. A proprietary brand of cellulose acetate.

Nixon C/N. A proprietary brand of cellulose nitrate.

Nixon E/C. A proprietary brand of ethyl cellulose.

Nixonite. A proprietary trade name for a cellulose acetate plastic.

Nixonoid. A proprietary trade name for a cellulose nitrate plastic.

Nizin. A proprietary preparation of zinc sulphanilate, $(NH_2.C_6H_4.SO_3)_2Zn.4H_2O$. An astringent and antiseptic.

Nizolysol. A lysol preparation, having a more agreeable odour than lysol.

Njamplung Oil. See LAUREL NUT OIL.

Njatuo Tallow. A fat obtained from *Palaquium oblongifolium*.

Njave Butter (Djave Butter, Nari Oil, Noumgou Oil, Adjab Fat). A fat obtained from the seeds of *Mimusops njave* or *djave*, also from *Bassia toxisperma* and *Bassia djave*. The nuts are known as Abeku, Bako, or Mahogany Nuts. The fat has a saponification value of 182–188 and an iodine value of 56–65.

N-Labstix. A proprietary test-strip used to detect pH, protein, glucose, ketones, blood and nitrite in urine. (807).

NLA-10. Dibutylphthalate. A vinyl plasticiser. (117).

NLA-20. Di-2-ethylhexylphthalate. A vinyl plasticiser. (117).

NLA-30. Di-iso-decylphthalate. A vinyl plasticiser. (117).

NLA-40. Di-decylphthalate. (117).

NLF-32. A mixed adipate vinyl plasticiser. (117).

NLF-33. A modified adipate vinyl plasticiser. (117).

Nobecutane. A proprietary preparation of acrylic resin dissolved in acetic esters used in aerosol form as a plastic wound dressing. (477).

Nobel Ardeer Powder. A dynamite containing 33 per cent. nitro-glycerin, 49 per cent. magnesium sulphate, 13 per cent. kieselguhr, and 5 per cent. potassium nitrate.

Nobel Polarite. An explosive. It is a mixture of potassium perchlorate and ammonium nitrate, with trinitrotoluene, a little starch, and wood meal.

Nobel's Explosive Oil (Trinitrine), Nitro-glycerin (trinitro-glycerol), C_3H_5 $(O.NO_2)_3$.

Nobilite (Phyllinglanz). A variety of the mineral Nagyagite.

Noble Metals. The rare metals, such as platinum, compared with the commoner metals.

Noblen. A trade mark for Japanese polypropylene. (60). See also PROPATHENE (2) and MOPLEN (61).

Nobrium. A proprietary preparation of MEDAZEPAM. A tranquilliser. (314).

Nocco. A bismuth-lead alloy used by dentists.

Nocerite. A mineral, $(Mg.Ca)_3OF_4$.

Noctal. A German soporific. It is β-brom - propenyl - isopropyl - malonyl - urea, $(CH_3.CBr.CH)(C_3H_7).C.CO.CO.$ $NH.CO.NH.$

Noctec. A proprietary preparation of chloral hydrate. A hypnotic. (682).

No-Del. A proprietary preparation of allyl isothiocyanate, ethyl nicotinate, methyl salicylate, eugenol, oil of turpentine, and cholesterol in a hydrophillic base. (273).

Nohaesa. A proprietary preparation of camphor, menthol, calcium chloride and chloral hydrate used as an ointment in the treatment of hæmorrhoids. (370).

Noheet Metal (Tempered lead). An alloy of 98·4 per cent. lead, 1·4 per cent. sodium, 0·11 per cent. antimony, and 0·08 per cent. tin.

Nohlite. A mineral. It is a variety of Samarskite, of Sweden.

Noil. An alloy of 80 per cent. copper and 20 per cent. tin.

Noiret's Aluminium Solder. An alloy of 80 per cent. zinc and 20 per cent. tin.

Nolascite. A mineral. It is a variety of galenite.

Nolu. A proprietary trade name for a bearing material. It is an oil impregnated wood.

Noludar. A proprietary preparation of methyprylone. A hypnotic. (314).

Nolvadex. A proprietary preparation of TAMOXIFEN used in the treatment of breast cancer. (2).

No-mag. A non-magnetic alloy of 77 per cent. iron, 12 per cent. nickel, 6 per cent. manganese, 3 per cent. carbon, and 2 per cent. silicon. It has a high specific resistance, is close-grained, and of good mechanical properties.

No-max. A proprietary trade name for a high speed molybdenum-tungsten steel for cutting tools.

Nomaze. A proprietary preparation of ephedrine and naphazoline. (188).

Nomex. A proprietary trade name for nylon fibre specially fabricated to withstand exposure to 500° F. It will not melt or drip. (10).

Nomifensine. A thymoleptic stimulating the central nervous system. It is 8 - amino - 1, 2, 3, 4 - tetrahydro - 2 - methyl - 4 - phenylisoquinoline. "36-984" is currently under clinical trial as the hydrogen maleate.

Nonad Tulle. Paraffin gauze dressing. (510).

Noncorrodite. A proprietary trade name for a chromium steel.

Nongo. A gum resembling tragacanth, obtained from *Albizzia brownei*, of Uganda.

Non-gran Metal. A bronze containing 87 per cent. copper, 11 per cent. tin, and 2 per cent. zinc.

Nonidet. Ethylene oxide condensate. (553).

Non-inflammable Wood. Wood rendered fireproof by treatment with steam and vacuum to remove air and moisture. It is then impregnated with fire-resisting chemicals, and again dried, leaving crystals of the chemicals in the wood.

Non-mordant Cotton Blue. See ALKALI BLUE XG.

Nonox. A registered trade mark for a group of anti-oxidants. (512).

Nonox B. Proprietary name for a condensation product of acetone and diphenylamine. (512).

Nonox BL. Proprietary name for a condensation product of acetone and diphenylamine. (512).

Nonox BLW. A proprietary acetone/diphenylamine liquid-condensate type of antioxidant prepared in the form of a dustless powder and used in rubber. (2).

Nonox CC. A proprietary anti-oxidant. Phenol sulphide. (512).

Nonox CGP. A proprietary anti-oxidant. Mercaptobenzimidazole + NONOX CI. (512).

Nonox CI. A registered trade mark for N - N' - Di - β - naphthyl - p - phenylene diamine, a high temperature antioxidant for rubber and plastics. (2).

Nonox CNS. A proprietary anti-oxidant. Mercaptobenzimidazole + Nonox WSP. (512).

Nonox D. A proprietary anti-oxidant. Phenyl-β-naphthylamine. (512).

Nonox DED. A proprietary anti-oxidant. N,N'-diphenyl ethylene diamine. (512).

Nonox DN. A proprietary anti-oxidant. Phenyl-β-naphthylamine. (2).

Nonox DPPD. A registered trade mark for N-N'-Diphenyl-p-phenylene diamine, an antioxidant for rubber and plastics. (2).

Nonox EX. A proprietary anti-oxidant. A condensation product of phenol. (512).

Nonox EXN. A proprietary anti-oxidant. A condensation product of phenol. (512).

Nonox EXP. A proprietary anti-oxidant. A condensation product of phenol. (512).

Nonox HFN. A proprietary blend of aryl amines. (512).

Nonox GP. A proprietary anti-oxidant. Styrenated phenol. (512).

Nonox NS. A trade mark for an antioxidant for rubber. It is a phenolaldehyde-amine reaction product. (2).

Nonox NSN. Similar to Nonox NS.

Nonox OD. A proprietary anti-oxidant. Octylated diphenylamine. (512).

Nonox S. A trade mark for a mixture of the condensation products of aldol with α and β-naphthylamines. (2).

Nonox TBC. A trade mark for 4-methyl-2,6,-tert. butyl phenol, an antioxidant. (2).

Nonox WMP. A phenolic anti-oxidant used in thermoplastics for its non-staining properties. (2).

Nonox WSL. A trade mark for a liquid substituted phenol antioxidant. (2).

Nonox WSO. A trade mark for a non staining white antioxidant of the phenolic type. It has a melting point of 168° C. and is suitable for olefins and other polymers. (2).

Nonox WSP. A trade mark for 2-2'-methylene-bis(6-(1-methyl cyclo hexyl)-4-methyl phenol. An antioxidant for polythene with relatively non staining properties. (2).

Nonox ZA. A proprietary anti-oxidant. 4-Isopropylamine diphenylamine. (512).

Nonoxal AW (Nonox CI). A proprietary trade name for an antioxidant. It is di-β-naphthyl-p-phenylene-diamine. See also AGE-RITE WHITE. (2).

Nonoxan. A proprietary anti-oxidant phenyl-alpha-naphthylamine. (512).

Nonoxol CM. A proprietary preparation of CATECHOL and anhydrous methanol. (2).

Nonoxol DCP. A proprietary trade name for an antioxidant for polythene. It is di-cresylolpropane. (2).

Non-pareil Metal. An alloy of 78 per cent. lead, 17 per cent. antimony, and 5 per cent. tin.

Non-poisonous White Lead (Patent White Lead). A pigment. It is lead-sulphate, $PbSO_4$.

Non-slip Stone. A concrete made of crushed York stone chippings and Portland cement.

Nontronite. A mineral, $Fe_2O_3.3SiO_2.5H_2O$.

Noors Honey. A type of honey derived from several species of *Euphorbia*. It is bitter to the taste.

Nootropyl. See PIRACETAM.

Nopalin. See EOSIN BN.

Nopirine. This material is stated to consist of aspirin with caffeine and phenacetin.

Noracymethadol. α-4-Acetoxy-1,N-dimethyl-3,3-diphenylhexylamine.

Noradran. A proprietary preparation of ephedrine hydrochloride, theophylline and papaverine hydrochloride. A bronchial antispasmodic. (270).

Noradrenaline. (—)-2-Amino-1-(3,4-dihydroxyphenyl)ethanol. Levophed.

Noralite. A mineral. It is a variety of amphibole.

Noratex. A proprietary preparation of talc, kaolin, zinc oxide, cod liver oil and wool fat used as a protective skin cream. (462).

Norbergite. A mineral, $Mg_2SiO_4.Mg(F.OH)_2$.

Norbide. A proprietary trade name for an amorphous boron carbide. An abrasive.

Norbo. A proprietary trade name for phenolic resin.

Norbutrine. 2-Cyclobutylamino-1-(3,4-dihydroxyphenyl)ethanol. N-Cyclobutylnoradrenaline.

Norcodeine. A narcotic analgesic. N-Dimethyl-O^3-methylmorphine.

Nordel. A proprietary ethylene-propylene synthetic rubber. (10).

Nordel 2744. A trade mark for a fast-curing EPDM hydrocarbon rubber possessing high green strength. (10).

Nordhausen Acid. See FUMING SULPHURIC ACID.

Nordenskiöldine. A mineral. It is $CaSnB_2O_6$.

Noreplast. A proprietary trade name for laminated thermosetting, plastic thermosetting and thermoplastic synthetic resins.

Norepol. A proprietary trade name for vulcanisable vegetable polymers.

Noreseal. A proprietary trade name for a cork substitute made from low cost domestic raw materials. It is stated to be equal in strength to cork.

Norethandrolone. 17α-Ethyl-17β-hydroxyoestr - 4 - en - 3 - one. 17α - Ethyl - 17β - hydroxy - 19 - norandrost - 4-en-3-one. Nilevar.

Norethisterone. 17β - hydroxy- 19 - norpregn - 4 - en - 20 - yn - 3 - one. 17α - Ethynyl - 17β - hydroxyoestr - 4 - en - 3 - one. 17α- Ethynyl - 19 - nor - testosterone. Present as the acetate in ANOVLAR, CONTROVLAR, GYNOVLAR 21, MINOVLAR, NORIDAY, NORINYL-1, NOR-INYL-2, NORLESTRIN, ORLEST 28, ORTHO-NOVIN, and PRIMODOS. It is a hormone used in many contraceptive preparations.

Norethynodrel. 17α-Ethynyl-17β-hydroxyoestr-5(10)-en-3-one.

Norfer. A proprietary preparation of ferrous fumarate used in the treatment of anæmia. (462).

Norflex. A proprietary preparation of orphenadrine citrate, a muscle relaxant. (275).

Norge, Saltpetre. Air saltpetre (*q.v.*).

Norgesic. A proprietary preparation of orphenadrine citrate and paracetamol. A muscle relaxant. (275).

Norgestomet. A progestational steroid. 11β - Methyl - 3, 20 - dioxo - 19 - nor - pregn-4-en-17α-yl acetate.

Norgestrel. 13β-Ethyl-17-hydroxy-18, 19 - dinor - 17α - pregn - 4 - en - 20 - yn - 3-one. Present in OVRAN.

Norgine. The sodium-ammonium salt of laminaric acid from seaweed, *Laminaria digitata* and *Saccharinus digitatus*. It is used in the treatment of textiles.

Norgotin. A proprietary preparation of ephedrine, amethocaine, chlorohexidine acetate and propylene glycol used as eardrops. (286).

Noriday. A proprietary preparation of

Norethisterone. An oral contraceptive. (809).

Norinyl-1, Norinyl-2. A proprietary preparation of norethisterone and mestranol. Oral contraceptive. (805).

Norisodrine. A proprietary preparation of isoprenaline sulphate and calcium iodide. A bronchial antispasmodic. (311).

Norit. A registered trade mark for a purified charcoal made from birch and used as a decolorising agent for sugar juices and other purposes.

Norite. A rock consisting of the minerals labradorite and hypersthene.

Norlestrin. A proprietary preparation of norethisterone acetate and ethinyl-œstradiol. Treatment of menstrual disorders. (264).

Norlestrin 21. A proprietary preparation of norethisterone acetate and ethinyl œstradiol. Oral contraceptive. (264).

Norleusactide. See PENTACOSACTRIDE.

Norlevorphanol. $(-)$-3-Hydroxy-morphinan.

Norlutin " A ". A proprietary preparation of NORETHISTERONE acetate used in the control of menstrual irregularity. (264).

Normacol. A proprietary preparation containing sterculia. A laxative. (286).

Normal Powder. A gelatinised gun-cotton powder.

Normal Salt Solution. See PHYSIOLOGICAL SALT SOLUTION.

Normannite. A mineral. It is a basic bismuth carbonate.

Normax. A proprietary preparation containing dioctyl sodium sulphosuccinate and 1-8-dihydroxyanthraquinone. A laxative. (269).

Normethadone. 6-Dimethylamino-4,4-diphenylhexan-3-one.

Normogastrine. A proprietary product. It is a specially prepared aluminium silicate having the property of reducing the acidity of the stomach without rendering it alkaline.

Normorphine. N-Demethylmorphine.

Norolen. A proprietary preparation of norethinodrel and mestranol. Oral contraceptive. (806).

No-Roma. A proprietary preparation of paraformaldehyde, sodium carboxymethyl cellulose, sodium hydroxide and methylene blue. A medical deodorant. (697).

Norpipanone. 4,4-Diphenyl-6-piperidinohexan-3-one.

Norpramine. A proprietary preparation of IMIPRAMINE hydrochloride. An anti-depressant. (462).

Norsed. A proprietary preparation of

cyclobarbitone and amylobarbitone. A hypnotic. (350).

Norsip. Special industrial pitch. (623).

Norskalloy. A Norwegian brand of electric furnace pig iron produced from ore containing a quantity of vanadium and titanium, and this iron has the following typical analyses :—

	Standard. per cent.	Refined. per cent.
Total carbon	4–4·5	2·5–3·5
Silicon . .	0·5–1·5	1·7–2·5
Manganese .	0·2	0·2
Sulphur . .	Trace	Trace
Phosphorus .	0·2–0·25	0·15–0·2
Vanadium .	0·3–0·4	0·2–0·3
Titanium .	0·4–0·8	0·3–0·5

Norsodyne. A registered trade mark for unsaturated polyester resins. (487).

Norsolene. A registered trade mark for coumarone and polyindene resins. (892).

Norsomix. A registered trade mark for polyester compounds.

Nortestosterone. Nandrolone.

Northovan. Sodium-o-vanadate.

Northupite. A mineral, $NaCl.Na_2CO_3 MgCO_3$.

Nortran. A proprietary preparation of TRIFLUOMEPRAZINE. A veterinary tranquilliser.

Nortriptyline. 3-(3-Methylaminopropylidene)-1,2 : 4,5-dibenzo cyclohepta-1,4-diene. Allegron and Aventyl are the hydrochloride.

Norval. A proprietary preparation o-dioctyl-sodium sulphosuccinate in gelatine. A fæcal softener. (269).

Norvinyl. A proprietary polyvinyl chloride. (893).

Norwegian Saltpetre. See AIR SALTPETRE.

Noryl. A proprietary trade name for polyphenylene oxide polymer. It has a tensile strength of 10,000 lbs./sq. in., an elongation at break of 500 per cent., a Youngs modulus of 350,000 lbs./sq. in., an Izod impact of 1·3, a thermal expansion coefficient of 6·7° C. 10⁵ and a Rockwell hardness of R 119. (59).

Noryl SE-100. A registered trade mark for self extinguishing ABS resins. (59).

Noryl SE-1000. Self-extinguishing acrylonitrile butadiene styrene plastic. (59).

Noscapine. Narcotine. Coscopin. Nicolane.

Nosean (Noselite, Nosite, Nosian, Nosin). A mineral. It is a sodium-aluminium silicate containing sodium sulphate, $2Na_2O.Al_2O_33SiO_2.Na_2SO_4$.

Noselite. See NOSEAN.

Nosian. See NOSEAN.

Nosiheptide. A veterinary antibiotic. It is a peptide obtained from cultures of *Streptomyces actuosus 40037* or the same

substance obtained by any other means.

Nosin. See NOSEAN.

Nosite. See NOSEAN.

Nosophene (Iodophen, Iothalen). Tetra-iodo-phenol-phthalein, $(C_6H_2I_2OH)_2C (C_6H_4)CO.O$.

Three salts are on the market. (a) Antinosine, the sodium salt, $(C_6H_2I_2 ONa)_2CO.C_6H_4O$; (b) Eudoxine, the bismuth salt, $(C_6H_2I_2OBi)_2C.O.C_6H_4O$; and (c) Apallagin, the mercury salt. It is mercury tetra-iodo-phenol-phthalein.

They are used medicinally, internally as intestinal astringents, and externally as antiseptics.

Nostyn. A proprietary trade name for Ectylurea.

Notack. A tack remover for rubber. (610).

Nottingham White. See FLAKE WHITE.

Noumeite. See GARNIERITE.

Noumgou Oil. See NJAVE BUTTER.

Noury Aun. 2,2'-azo-bis-(2,4-dimethyl) valeronitrile. A very reactive azo compound used as a polymerisation initiator. (90).

Noury AIBN/c. Azo-*bis-iso*-butyronitrile. A polymerisation initiator. (90).

Nouveau Textile. A proprietary wool substitute obtained from the treatment of jute.

Nova. A bright nickel plating process. (605).

Novacetoform. A solid preparation of aluminium acetate. It is of German origin.

Novaculite. A quartz rock used as an abrasive.

Nováčekite. A mineral. It is $2[Mg(UO_2)_2(AsO_4)_2.12H_2O]$.

Novacryl. Cationic dyes for acrylic fibres. (508).

Novacyl. A preparation containing mainly magnesium acetyl salicylate with some free acetyl salicylic acid.

Novadelox. A mixture of benzoyl peroxide and acid calcium phosphate. It has been used as a bleaching agent for flour.

Novain (Carnitine). γ-Trimethyl-β-oxybutyro-betaine, $C_7H_{15}O_3N$.

Novalak Resins. Synthetic resins of the formaldehyde-phenol type, which possess the characteristics of the natural resins in that they are fusible, and soluble in certain solvents.

Novalgin. Sodium - phenyl - dimethyl - pyrazolone-methyl-amino-methane sulphonate.

Novaloy 6521. A proprietary one-component epoxy resin supplied in powder form for the coating of electronic components. (774).

Novamidon. See PYRAMIDONE.

Novargan (Omorol). A protein preparation containing 10 per cent. silver. It is used as a germicide in the treatment of gonorrhœa.

Novarsan. Neosalvarsan (*q.v.*).

Novarsenobenzol. A French product. See NEOSALVARSAN.

Novarsenobillon. See NEOSALVARSAN.

Novarsenol. Neosalvarsan (*q.v.*).

Novasorb. Magnesium trisilicate. (591).

Novaspirin. Methylene-anhydro-citryl-salicylic acid, $C.CO.O.CH_2(CH_2.CO.OC_6H_4.CO_2H)_2$. Used in medicine for rheumatism and neuralgia.

Novasurol (Salyrgan). A combination of mercury-*o*-chloro-phenoxy-acetic acid, C_6H_3 (OCH_2COONa) Cl . HgOH, and barbital. An antisyphilitic drug.

Novatophan. Ethyl-6-methyl-2-phenyl-quinoline-4-carboxylate, $C_6H_5.C_9H_4N$ $(CH_3)COOC_2H_5$. It is the ethyl ester of Paratophan (methyl-atophan).

Novemol. A trade name for a wetting agent containing sulphonated terpene alcohols.

Novatropine. *n*-Methyl-atropine nitrate amygdalic acid ester.

Noveloid. Proprietary products of cellulose esters and ethers.

Noveril. A proprietary trade name for the hydrochloride of Dibenzepin.

Novesine. A proprietary trade name for the hydrochloride of Oxybuprocaine.

Novex. Benzal-bis-dimethyl-dithio-carbamate. It has a specific gravity of 1·365, is a white crystalline powder, and melts at 175° C. It is used as a rubber vulcanisation accelerator.

Noviform (Betaform). The bismuth compound of tetrabromo-pyrocatechol. Used as a surgical dressing.

Novirudin. A new compound belonging to the melanin acid group. It hinders the coagulation of blood.

Novitane. A proprietary polyurethane elastomer. (102).

Novite. A proprietary trade name for an alloy of iron with about 1·5 per cent. nickel and 0·5 per cent. chromium.

Novobiocin. An antibiotic produced by Streptomyces niveus and Streptomyces spheroides. Streptonivicin. Albamycin is the calcium salt and Biotexin and Cathomycin are the sodium salts.

Novocaine (Ethocaine, Irocaine, Neo-caine, Syncaine, Juvocaine). *p*-Amino-benzoyl-diethyl-amino-ethanol hydro-chloride, $NH_2.C_6H_4.CO.O.CH_2.CH_2.$ $N(C_2H_5)_2.HCl$. A local anæsthetic. It is used as a substitute for cocaine.

Novocodine. Hexamethylene-tetra-mine-di-iodide.

Novocol. Sodium-mono-guaiacol phosphate.

Novocrete. A proprietary name or af

new concrete containing mineralised sawdust.

Novoiodine. Hexamethylene-tetramine di-iodide. A mixture of this substance with powdered talc, in equal proportions, is used as a dusting powder for wounds.

Novol. High purity oleyl alcohol. (597).

Novolac. A name applied by Baekeland to fusible and soluble phenol-formaldehyde resins.

Novolen. A proprietary trade name for polypropylene. (49).

Novolith. A photographic developer. (507).

Novon 700. A trade name for PVC foils for cladding for decorative purposes. (2).

Novonal. A hypnotic. It is a commercial preparation of diethyl-allyl-acetamide, $(C_2H_5)_2(C_3H_5)C.CO.NH$.

Novonasco. A proprietary trade name for a wetting agent containing modified sodium alkyl naphthalene sulphonate.

Novoplas. A proprietary trade name for an organic polysulphide synthetic elastic material.

Novoprotin. A proprietary preparation of crystalline vegetable albumen.

Novorenal. A solution of novocaine and adrenalin. Sold as an anæsthetic.

Novostab. A proprietary preparation. It is salvarsan.

Novotextil. A French proprietary artificial silk stated to be wool-like.

Novo-zirol. Procaine hydrochloride and adrenalin preparations.

Novozone. A German registered name for a magnesium peroxide, MgO_2, prepared for medical purposes. An antiseptic used internally, and externally as an ointment, for wounds and gatherings. Also see BIOGEN.

Novutox. A proprietary preparation of procaine. A local anæsthetic. (803).

Noxiptyline. An anti-depressant. 3-(2-Dimethylaminoethyloxyimino) diben-zo[*a*, *d*]cyclohepta-1,4-diene. BAY 1521 is under clinical trial as the hydrochloride.

Noxyflex. A proprietary preparation of NOXYTHIOLIN used to irrigate the bladder. (386).

Noxythiolin. N-Hydroxymethyl-N'-methylthiourea.

N.P.L. Alloy. An alloy of 94·5 per cent. aluminium, 2·5 per cent. nickel, and 1·5 per cent. magnesium.

Nsa-sana Seeds. The seeds of *Ricinodedron heudolitii*.

N.S.B. A proprietary preparation of noxytiolin, and vinyl pyrrolidone/vinyl acetate copolymer in an aerosol. Wound dressing. (386).

N.S. Fluid. A mixture of sodium chloride, aluminium chloride, and iron chloride.

NSM. A proprietary tobacco substitute. (982).

N.T. See NOUVEAU TEXTILE.

Nuade. Painting grinding aid. (607).

Nuba. A proprietary trade name for a thermoplastic coal-tar pitch and a cumarone-indene resin for paints.

Nubrite. A bright nickel plating process. (605).

Nubun. A proprietary trade name for a synthetic rubber latex insulation for power and communication cables. It is made from a special modification of buna S synthetic rubber.

Nucerite. A proprietary trade name for metal ceramic composite materials. They comprise ceramic fused to a metal base and have outstanding corrosion, abrasion and impact resistance.

Nuchar. A decolorising agent of American origin. It is recommended for purifying alkaloids and medicinals.

Nucin. Oxy-α-naphtho-quinone.

Nucleant. An aluminium alloy grain refiner. (531).

Nucleic Acid. An organic acid obtained from nuclein by the action of alkalis or by tryptic digestion.

Nuclein. A nucleo-protein obtained by peptic digestion or treatment with dilute acid.

Nucleogen. A compound of iron with nuclein and arsenic. Used in medicine.

Nucleol. A pure nuclein obtained from yeast.

Nucleosil. A silver compound of nucleic acid. Used in medicine.

Nucoline. See VEGETABLE BUTTER.

Nuelin. A proprietary preparation of THEOPHYLLINE. A bronchodilator. (275).

Nufome. A sequestering agent; EDTA type. (537).

Nujol. A registered trade mark for a liquid paraffin or medicinal oil and other articles.

Nulacin. A proprietary preparation containing milk, dextin, maltose base with magnesium trisilicate, calcium carbonate, magnesium oxide and oil of peppermint. An antacid. (269).

Nulla Panza Nuts. See OWALA BEANS.

Nulogyl. A proprietary preparation of NIMORAZOLE used in the treatment of trichomonas infections. (324).

Nuloid. See BAKELITE.

Nulomoline. A solution of partly inverted sugar. Used for some purposes as a substitute for glycerin.

Numoquin. Ethyl-hydro-cupreine.

Numorphan. A proprietary trade name for the hydrochloride of Oxymorphone. Injection of oxymorphone—Analgesic.

Numorphan Oral. A proprietary trade name for Hydromorphinol. Tablets of hydromorphinol—Analgesic. (182).

Numotac. A proprietary preparation of ISOETHARINE hydrochloride used in the treatment of bronchospasm. (275).

Nuocure. Stabilised stannous octanoate. (607).

Nuocure 935. A proprietary range of tin-organic complexes used as catalysts for polyurethanes. The tin content is 23·6%. (516).

Nuocure 938. See NUOCURE 935, but with the tin content reduced to 20%. (516).

Nuoc-mam (Salted Fish Water). A condiment prepared by the fermentation of the solution obtained from a mixture of salt and small fish. The concentrated solution is known as Nuoc-nhut, and is diluted to form Nuoc-mam It contains sodium chloride, tryptophan, and tyrosine.

Nuoc-nhut. See NUOC-MAM.

Nuodex 72. A fungicide for paints, ropes and cordage. (607).

Nuodex 84. Non-toxic fungicide. (607).

Nuodex 87. A wall washing compound. (607).

Nuodex 321 Extra. A mercury paint fungicide. (607).

Nuodex A.F.7. An antifoaming agent. (607).

Nuodex N.A. A pigment dispersant for vinyl resins. (607).

Nuodex P.M.A.18. A soluble mercury fungicide. (607).

Nuogel A.O. An aluminium soap. (607).

Nuolates. Tallate based paint driers. (607).

Nuomix. Oil soluble wetting agents. (607).

Nuostabe 1317. A proprietary trade name for a Ba/Cd/Zn stabiliser for PVC. (223).

Nuostabe 1374. A proprietary trade name for a non sulphide staining stabiliser for PVC used in blown foams. (223).

Nuostabe 1515. A barium/cadmium/zinc complex in liquid form designed for use as a stabiliser in plastisol and organisol applications. (304).

Nuostabe V-1515. A barium-cadmium-zinc complex. An efficient stabiliser for plastisols for high processing temperatures. (304).

Nuostabe 1602. A calcium-zinc stabiliser for use in PVC plastisols. (304).

Nuostabe 1605. A stabiliser for PVC based on a barium, cadmium zinc complex. (223).

Nuostabe 1619. A barium-cadmium-zinc complex. A stabiliser for plasticised PVC injection moulding compounds and extrusion compounds. Good heat stability. (304).

Nuosyn. Synthetic acid paint driers. (607).

Nupa-Sal. A proprietary trade name for Salinazid.

Nupercainal. A proprietary preparation of cinchocaine hydrochloride. An anæsthetic ointment. (18).

Nupercaine. A proprietary preparation of cinchocaine hydrochloride. Surface anæsthetic. (18).

Nuprin. A proprietary trade name for Sulphamoxole.

Nupyrin. See ASPIRIN.

Nurac. A proprietary rubber vulcanisation accelerator. It is stated to be diphenyl-guanidine. It has also been said to be thiocarbanilide.

Nureco GK. A proprietary epoxyresin. (775).

Nuremberg Gold. An alloy of 90 per cent. copper, 7·5 per cent. aluminium, and 2·5 per cent. gold. A jeweller's alloy.

Nuremberg Green (Victoria Green, Permanent Green). Pigments. They are mixtures of chromium green with zinc yellow.

Nuremberg Red. See INDIAN RED.

Nuremberg Violet. See MANGANESE VIOLET.

Nurexform. A preparation of lead arsenate, having a high covering power, and remaining in suspension for a long period of time.

Nusat. A satin finish nickel plating process. (605).

Nu-Seals Aspirin. A proprietary preparation of aspirin in an enteric coated form. An analgesic. (333).

Nu-Seals Potassium Chloride. A proprietary preparation of potassium chloride in enteric coated tablets. (333).

Nu-Seals Sodium Salicylate. A proprietary preparation of sodium salicylate used as an analgesic and antipyretic. (463).

Nuso. A proprietary trade name for Neutral Insulin Injection.

Nuso Neutral Insulin. A proprietary preparation of neutral insulin. (182, 277, 253).

Nussierite. A mineral. It is an impure variety of pyromorphite.

Nut, Abeku. See NJAVE BUTTER.

Nut, Chop. See ORDEAL BEAN.

Nut, Gall. See GALLS.

Nut, Gooroo. See KOLA NUT.

Nut, Hurr. Myrobalans.

Nutmeg Butter. Expressed oil of nutmeg.

Nut Oil. A term used in China for Tung oil (Chinese wood oil). The same name is also used for Walnut and Arachis oils.

Nut Oil, Sweet. See BUTTER OIL.

Nut, Physic. See PHYSIC NUT.

Nut, Pontianak Illipé. See BORNEO TALLOW.

Nut, Purging. See BARBADOES NUTS.

Nutramigen. A proprietary artificial lactose-free infant milk food. (324).

Nutregen. A proprietary gluten-free wheat starch used in the dietary treatment of coeliac disease. (776).

Nutrient Gelatin. A gelatin prepared by adding specially made beef broth or bouillon (about 10 per cent.) to gelatin. Used for cultivating bacteria.

Nutrincreosote. See CREOSOTE ALBUMINATE.

Nutrium. A foodstuff prepared from casein, salt, and milk sugar.

Nutrizym. A proprietary preparation of pancreatin, BROMELAINS and ox-bile used as a pancreatic supplement in the treatment of cystic fibrosis. (310).

Nutrose (Plasmon, Caseon). The sodium compound of casein. The casein of fresh cow's milk is precipitated, and converted into nutrose by treatment with sodium carbonate. It contains 62 per cent. albuminous matter, 20 per cent. nitrogen-free matter, 4 per cent. mineral matter, and 10 per cent. water. It is asserted to be a highly nutritious food preparation.

Nuts, Arachis. See GROUND NUTS.

Nuts, Attawa. See OWALA BEANS.

Nuts, Bako. See NJAVE BUTTER.

Nuts, Caco Babassa. See BASSOBA NUTS.

Nuts, Cognito. See BASSOBA NUTS.

Nuts, Coquilho. See BASSOBA NUTS.

Nuts, Coquilla. See BASSOBA NUTS.

Nuts, Curcia. See BASSOBA NUTS.

Nuts, Earth. See GROUND NUTS.

Nuts, Fai. See OWALA BEANS.

Nuts, Fulla Panza. See OWALA BEANS.

Nuts, Maboula. See OWALA BEANS.

Nuts, Mahogany. See NJAVE BUTTER and ACAJOU BALSAM.

Nuts, Monkey. See GROUND NUTS.

Nuts, Nulla Panza. See OWALA BEANS.

Nuts, Odu. See OWALA BEANS.

Nuts, Pacca. See KUSAMBI NUTS.

Nuts, Panza. See OWALA BEANS.

Nuts, Pea. See GROUND NUTS.

Nut, Split. See ORDEAL BEAN.

Nuts, Vaua-assu. See BASSOBA NUTS.

Nuttallite. A mineral. It is a variety of scapolite.

Nuvacon. A proprietary preparation of megestrol acetate and ethinyl œstradiol. Oral contraceptive. (182).

N.W. Acid. See NEVILLE'S AND WINTHER'S ACID.

Nyanza Black B. A dyestuff. It is the sodium salt of amino-benzene-azonaphthalene - azo - γ - amino - naphthol-sulphonic acid, $C_{26}H_{19}N_6SO_4Na$. Dyes wool and cotton direct from an acid bath.

Nyctal. Adalin (*q.v.*)

Nydrane. A proprietary preparation of beclamide. An anticonvulsant. (323).

Nylac. A patented nylon solution coating for metalwork. (254).

Nylander's Reagent. An alkaline solution of bismuth subnitrate and Rochelle salt obtained by dissolving 40 grams Rochelle salt and 20 grams bismuth subnitrate in 1,000 c.c. of 8 per cent. caustic soda. It is used for the detection of glucose in urine by boiling 5 parts of the glucose solution with 1 part of the reagent when reduction occurs and a black precipitate is produced.

Nylatron. Nylon moulding compounds. (230).

Nylatron GS. A trade mark for a bearing material manufactured from nylon 66 filled with molybdenum disulphide. (240).

Nylestriol. An œstrogen currently undergoing clinical trial as " 49 825 ". It is 3 - cyclopentyloxy - 19 - nor - 17α - pregna - 1, 3, 5 (10) - trien - 20 - yne - 16α,17β-diol.

Ny-lite. A proprietary trade name for a luminous pigment.

Nylofanol. Atophan (*q.v.*).

Nylomine. Dyestuffs for nylon and polyamide fibres. (512).

Nylon. A generic term for polyamide products prepared from adipic and related acids and hexamethylene and related diamines by condensation. It has a protein-like chemical structure resembling the proteins of the animal kingdom. It is fabricated into bristles, fibres, and sheets.

Nypene. A proprietary trade name for a polyterpene hydrocarbon resin used in adhesives, paints, and varnishes.

Nystadermal. A proprietary preparation of NYSTATIN and TRIAMCINOLONE acetonide used in the treatment of fungal and eczematous skin disorders. (682).

Nystaform. A proprietary preparation of nystatin and clioquinol used in dermatology as an antibacterial agent. (372).

Nystaform-HC. A proprietary preparation of nystatin, clioquinol and hydrocortisone used in dermatology as an antibacterial agent. (372).

Nystan. A proprietary preparation containing nystatin. (326).

Nystatin. An antibiotic produced by *Streptomyces noursei*. MYCOSTATIN NITACIN, NYSTAM. Also present in LEDERSTATIN, MYSTECLIN, NYSTAN TA and SILTETRIN.

Nystavescent. A proprietary preparation of NYSTATIN in an effervescent

pessary used in the treatment of vaginal candidiasis. (682).

Nytrazid. A proprietary trade name for Isoniazid.

Nyxolan. A proprietary preparation of aluminium hydroxyquinoline. An antihelmintic. (396).

O

Oak Bark. A tanning material containing 10–14 per cent. tannin. Oak wood contains 6 per cent. tannin and is used in the manufacture of bark extracts. The extracts contain from 25–37 per cent. tannin.

Oak Moss Resin. A resin obtained from lichens. Used as a fixative in perfumery.

Oak Red. A colouring matter, phlobaphene, $C_{28}H_{22}O_{11}$, obtained by the hydrolysis of quercitannic acid.

Oakum. Hemp fibre obtained by untwisting rope. It is usually impregnated with tar or pitch.

Oakwood Extract. A tanning material manufactured from the wood of the oak. The extract contains from 26–28 per cent. tannin.

Oba Oil. See DIKA BUTTER.

Obermayer's Reagent. A solution of ferric chloride in concentrated hydrochloric acid (4 gm. ferric chloride dissolved in 1 litre of concentrated hydrochloric acid). It is used for the detection of indoxyl in urine, indigo being formed if this substance is present.

Oblivon. A proprietary preparation of methylpentynol. (265).

Oblivon-C. A proprietary trade name for the carbamate of Methylpentynol.

Obracin. A proprietary preparation of TOBRAMYCIN sulphate. An antibiotic. (309).

Obsidene. A plastic residue from the distillation of petroleum.

Obsidian. See PUMICE STONE.

Obsidianite. A proprietary brand of fireproof and acid-proof material, used in the construction of Glover and Gay-Lussac towers.

Obturin. Soluble fluoresceine.

Occidental Topaz. See CITRINE.

Occidine. A fungicide. It contains copper sulphate, iron sulphate, sulphur, naphthalene, and calcium carbonate.

Occultest. A proprietary test tablet of *o*-toluidine, strontium peroxide, calcium acetate, tartaric acid and sodium bicarbonate, used for the detection of blood in urine. (807).

Ocenol. The mixture of fatty alcohols derived from sperm oil. Also a proprietary trade name for technical oleic acid.

Ochran. A yellow bole (or clay-earth).

Ochre (Yellow Ochre, Oxide Yellow, Chinese Yellow). A natural pigment consisting of hydrated oxides of iron and manganese, mixed with clay and sand. Specific gravity, 3·5. The term is frequently restricted to a pale, yellowish-brown variety.

Ochre, Artificial. See MARS YELLOW and SIDERINE YELLOW.

Ochre, Bismuth. See BISMUTH OCHER.

Ochre, Brown. See YELLOW OCHRE.

Ochre, Burnt. See INDIAN RED.

Ochre, Cobalt. A mineral. It is cobalt bloom.

Ochre, Golden. See YELLOW OCHRE.

Ochrematite. A mineral, $3(3Pb(AsO_4)_2 \cdot PbCl_2)4Pb_2MoO_5$.

Ochre, Orange. Burnt Yellow ochre (q.v.).

Ochre, Purple. See MARS VIOLET.

Ochre, Red. See INDIAN RED.

Ochre, Roman. See YELLOW OCHRE.

Ochre, Spruce. Yellow ochre (q.v.).

Ochre, Tellurium. See TELLURITE.

Ochre, Transparent Gold. See YELLOW OCHRE.

Ochre, Vanadium. See HYDROCUPRITE.

Ochroite (Ochrolite). A mineral, $Pb_4Sb_2O_7.2PbCl_2$.

Ochrolite. See OCHROITE.

Ocota Cocota Gum (Muccocota Gum). Names applied in West Africa to varieties of copal.

Ocre de Ru. Brown ochre.

Octacosactrin. A corticotrophic peptide.

Octaflex. A proprietary preparation of octaphen resin-acrylate and methacrylate polymers. Plastic wound dressing. (319).

Octaflex. A proprietary anti-oxidant mixture of diphenylamine and *p*-phenylene-diamine. (556).

Octahedrite. See ANATASE.

Octamine. An antioxidant. A reaction product of diphenylamine and diisobutylene. (556).

Octaphen. A proprietary trade name for Octaphonium chloride.

Octaphonium Chloride. Benzyldiethyl-2 - [4 - (1,1,3,3 - tetra methylbutyl) - phenoxy]-ethyl ammonium chloride. Octaphen. Phenoctide.

Octatropine Methylbromide. 8-Methyl - O - (2 - propylvaleryl)tropinium bromide. Anisotropine Methylbromide. Valpin.

Octaverine. An antispasmodic. 6,7-dimethoxy - 1 - (3, 4, 5 - triethoxyphenyl) - isoquinoline.

Octin. The hydrochloride or acid tartrate of methyl-amino-6-methyl-2-heptene ($C_8H_{15}NHCH_3$). It is prescribed

for convulsive conditions and for ulcers of the gastro-intestinal canal.

Octoil. A proprietary trade name for a plasticiser. Dioctylphthalate.

Octoil S. A proprietary trade name for a plasticiser. Dioctylsebacate.

Ocuba Wax. Ocuba fat, from *Myristica ocuba*.

Ocusol. A proprietary preparation of sodium sulphacetamide, zinc sulphate and cetrimide used as eye-drops. (502).

Odessa Alloy. See FRENCH SILVER.

Odite. A mineral. It is a variety of muscovite.

Odontolite. See BONE TURQUOISE.

Odorine. A volatile base found in bone oil. It is probably impure picoline.

Odu Nuts. See OWALA BEANS.

Œllacherite. A mineral. It is $2[(K_2,B_2)Al_6Si_6O_{20}(OH)_4]$.

Œnanthin. See FAST RED D.

Œnolin. The red colouring matter of wine, precipitated by lime or basic lead acetate.

Oerelin. A German product containing oil of lavender, oil of nutmeg, thymol, camphor, and alcohol. It is used as a pain-stilling material for rubbing-in in cases of rheumatism.

Œstrad. A proprietary preparation of ethinylœstradiol, glyceryl trinitrate and sodium bromide. Treatment of menstrual disorders. (354).

Œstradin. A proprietary preparation of ethinylœstradiol, phenobarbitone, sodium bromide and glyceryl trinitrate. (350).

Oestriol Succinate. Oestra-1,3,5(10)-triene - 3,16α,17β-triol 16,17 - di(hydrogen succinate).

Oestroform. Injection, oestradiol benzoate. Tablets, oestradiol-oestrogenic hormone. (182).

Offretite. A mineral. It is a potassium-calcium-aluminium silicate.

OFHC Copper. A trade mark for oxygen-free, high conductivity copper. The composition per cent. is: Iron 0·0005, sulphur 0·0025, silver 0·0010, nickel 0·0006, antimony 0·0005, arsenic 0·0003, selenium 0·0002, tellurium 0·0001, lead 0·0006, tin 0·0002, manganese 0·0005, bismuth 0·0001, oxygen 0·0001–0·0007, copper 99·9997. (33). Conductivity 101 per cent. IACS.

Ogwin. A mixture of lime and starch used to increase the rate of sedimentation of solids in water.

O-hi-o. A proprietary trade name for a die steel containing 12 per cent. chromium, 1·55 per cent. carbon, 0·85 per cent. vanadium, 0·4 per cent. cobalt, and 0·8 per cent. manganese.

Ohio Oil. See EARTH OIL.

Ohmal. A resistance alloy similar in composition to manganin. It usually contains 87·5 per cent. copper, 9 per cent. manganese, and 3·5 per cent. nickel.

▶hmlac kapak. A refined elaterite.

▶hmoid. A proprietary trade name for a phenol-formaldehyde synthetic resin laminated product used for electrical insulation.

▶hm Oil. A mineral oil which has been treated or contains in solution an anti-oxidant, thereby stabilising the oil and increasing its electrical resistivity.

▶hopex Q-10. A proprietary trade name for octyl fatty phthalic acid esters. (73).

▶il, Alexandrian Laurel. See LAUREL-NUT OIL.

▶il, Andiroba. See CRAB WOOD OIL.

▶il and Hartshorn (Oil Ananas) (*Linimentum ammoniæ B.P.*). It consists of olive oil mixed with ammonia and almond oil.

▶il, Animal. See BONE OIL.

▶il Asphalt. A thick fluid remaining after distilling crude petroleum. Used for roofing materials, and for paving when mixed with natural asphalt.

▶ilatum Application. A proprietary preparation of arachis oil and polyvinyl pyrrolidine. Used in dermatology. (379).

▶ilatum Emollient. A proprietary preparation of liquid fraction of acetylated lanolin alcohols in a semi-dispersing emollient base. Used in dermatology. (379).

▶il Babulum. Neatsfoot oil (*q.v.*).

▶il, Bagasses. See PYRENE OIL.

▶il, Bakoby. See BAKOLY OIL.

▶il, Bari. See AIX OIL.

▶il, Batana. See PATAVA OIL.

▶il, Beniseed. See GINGELLY OIL.

▶il, Black. See INDULINE, SPIRIT SOLUBLE.

▶il Blue. A pigment prepared by introducing copper filings into boiling sulphur, and after cooling, boiling the mass with sodium hydroxide to remove excess of sulphur. It is applicable as an oil colour only. Also see PRUSSIAN BLUE.

▶il Brown D. A dyestuff. It is equivalent to Sudan brown.

▶il, Cade. See JUNIPER TAR.

▶il Cakes. The residue obtained after pressing out the oil from oil-containing seeds. Used as a feeding stuff for animals.

▶il, Calaba. See LAUREL-NUT OIL.

▶il, Camphor. See LIQUID CAMPHOR.

▶il, Candlenut. See BAKOLY OIL.

▶il, Carbolic. See MIDDLE OILS.

▶il, Catheter. See LUND'S OIL.

▶il, Cazeline. See MINERAL BURNING OILS.

▶il, Chinese. Peru balsam.

Oilcloth. Fabrics coated with linseed oil, whiting, and pigments, and painted in oil colours. Also see LINOLEUM.

Oil, Coal. See KEROSENE.

Oilcoals. A name suggested for a mixture of Aspthalenes and Kerotenes (*q.v.*).

Oil, Coco-nut. See COCO-NUT BUTTER.

Oil, Cold Drawn. See VIRGIN OIL.

Oil, Colzarine. See MINERAL BURNING OILS.

Oildag. A registered trade mark for colloidal graphite in mineral oil used as a special lubricant and as an additive to lubricating oils and greases. (207).

Oil, Danforth's. See NAPHTHA.

Oil, Dark Cylinder. See CYLINDER OILS.

Oil Die. A proprietary trade name for tool steel containing 1·6 per cent. chromium, 0·45 per cent. tungsten, and 0·9 per cent. carbon.

Oil, Dika. See DIKA BUTTER.

Oil, Dilo. See LAUREL-NUT OIL.

Oil, Dippel's. See BONE OIL.

Oil, Disinfection. See SAPROL.

Oil, Dodder. See CAMELINE OIL.

Oil, Domba. See LAUREL-NUT OIL.

Oil, Dutch. See HAARLEM OIL.

Oil, Duty. See OIL OF RHODIUM.

Oiled Silk. Thin silk fabric which has been impregnated with oil, usually linseed.

Oil, Ester. See ESTER MARGARINE.

Oil, Filtered Cylinder. See CYLINDER OILS.

Oil for Fuel. See HEAVY OILS.

Oil, Fousel. Fusel oil (*q.v.*).

Oil, French Almond. See ALMOND OIL, FRENCH.

Oil, Garcinia. See KOKUM BUTTER.

Oil Gas (Pintsch Gas). A gas made by the destructive distillation of mineral oils, usually mineral seal oil, Scotch shale oil, and Russian solar oil. It is a mixture of methane, ethylene, acetylene, and benzene.

Oil, German Sesamé. See CAMELINE OIL.

Oil, Ginger Grass. See ROSÉ OIL.

Oil, Gingili. See GINGELLY OIL.

Oil, Glonoine. See NITROGLYCERIN.

Oil, Grain. See FUSEL OIL.

Oil Green (Maple Green, Leaf Green, Moss Green, Emerald Green). Pigments. They are mixtures of Chrome yellow and Paris blue, containing more Paris blue, so that they have a blue shade. Also see CHROME GREEN.

Oil Gutta-percha. A product obtained by heating together 100 parts gutta-percha, 10 parts olive oil, and 2 parts stearin. Used for moulding purposes.

Oil, Hoof. See NEATSFOOT OIL.

Oil, Hydrocarbon. See PARAFFIN LIQUID.

Oil, Jeppel's. Bone oil (q.v.).

Oil, Kagoo. See KORUNG OIL.

Oil, Kekuna. See BAKOLY OIL.

Oil, Kusum. See KON OIL.

Oil Lacs. These are obtained by adding to almost boiling oil varnish, fused copal or other resin, and diluting with oil of turpentine when using.

Oil, Lima. See EARTH OIL.

Oil, Linseed. See LINSEED OIL ARTIST'S and LINSEED OIL REFINED.

Oil, Maize. See CORN OIL.

Oil, Medium. See HEAVY OILS.

Oil, Mineral. See EARTH OIL.

Oil, Mineral Colza. See MINERAL BURNING OILS.

Oil, Mineral Seal. See MINERAL BURNING OILS.

Oil, Mineral Sperm. See MINERAL BURNING OILS.

Oil, Monopol. See TURKEY RED OILS.

Oil, Nari. See NJAVE BUTTER.

Oil Ndilo. See DILO OIL and LAUREL-NUT OIL.

Oil, Neem. See VEEPA OIL.

Oil, Neroli. See OIL OF ORANGE FLOWERS.

Oil, Nerve. Neatsfoot oil (q.v.).

Oil, Niaoul. See GORNEROL.

Oil, Nice. See AIX OIL.

Oil, Njamplung. See LAUREL-NUT OIL.

Oil, Noumgou. See NJAVE BUTTER.

Oil, Oba. See DIKA BUTTER.

Oil of Absinthe. Wormwood oil.

Oil of Acetone. Crude impure acetone. Also see ACETONE OILS.

Oil of Adders. See OIL OF VIPERS.

Oil of Ajava. Ajowan oil, the essential oil from the seeds of *Carum copticum*.

Oil of Akee. A yellow non-drying butter-like fat from *Blighia sapida*.

Oil of Allspice. Oil of Pimento.

Oil of Aloes. The oil obtained from Socotrine aloes.

Oil of Amber. An oil distilled from amber. The oils obtained from copal or dammar are also called by this name.

Oil of Anthos. Rosemary oil.

Oil of Ants. Olive oil, in which ants have been digested.

Oil of Ants, Artificial. Furfural, $C_4H_3(CHO)O$.

Oil of Apple. Amyl valerate, C_4H_9COO C_5H_{11}, used in the manufacture of fruit essences and medicinally as a sedative.

Oil of Asarabacca. An oil obtained from the roots of *Asarum europæum*.

Oil of Asphaltum. The oil obtained from asphaltum.

Oil of Aspic. Oil of spike (q.v.).

Oil of Balm (Oil of Lemon Balm). The volatile oil from *Melissa officinalis*. Used as a diaphoretic.

Oil of Bay. A volatile oil obtained from the leaves of *Myrcia acris*.

Oil of Bay Berries. The oil expressed from the berries of *Laurus nobilis*.

Oil of Been (Oil of Behn, Oil of Ben). The oil expressed from the seeds of *Moringa aptera*.

Oil of Behn. See OIL OF BEEN.

Oil of Benjamin. The oil obtained from benzoin, after the sublimation of benzoic acid.

Oil of Benné. See GINGELLY OIL.

Oil of Bergamot, Artificial. See BERGAMINOL.

Oil of Birch. The volatile oil from *Betula lenta*, the sweet birch. It consists mainly of methyl salicylate. It is also the name for the oil from *Betula alba*, the white birch.

Oil of Bitter Almonds. The essential oil of bitter almonds. The name is also applied to benzaldehyde and nitro-benzene.

Oil of Bones. See BONE OIL.

Oil of Box. The oil obtained from box-wood.

Oil of Bricks. The oil obtained by heating bricks to redness, and quenching in olive oil. The term is also applied to a mixture of 1 part turpentine and 4 parts linseed oil, coloured with alkanet or tar.

Oil of Cabbage (Oil of Elder). Olive oil in which elder leaves have been boiled.

Oil of Cacao. See OIL OF COCOA.

Oil of Cade. See JUNIPER TAR OIL.

Oil of Caoutchouc. Dipentene (q.v.).

Oil of Cassia (Oil of Chinese Cinnamon). The oil distilled from the bark, leaves, and twigs of *Cinnamomum cassia*, of China and Java. The yield is from 0·5–2 per cent. The specific gravity of the oil varies from 1·05–1·065, the refractive index from 1·585–1·605. It contains from 75–95 per cent. cinnamic aldehyde, $C_6H_5CH : CH.CHO$. It is a stimulant and germicide.

Oil of Cedrat. The oil obtained from citron peel.

Oil of Chinese Cinnamon. See OIL OF CASSIA.

Oil of Cinnamon. The oil obtained from the bark of *Cinnamomum zeylanicum*, of Ceylon. The yield varies from 0·5–1 per cent. The specific gravity of the oil varies from 1–1·04, and the refractive index from 1·565–1·582. It contains from 55–75 per cent. cinnamic aldehyde, $C_6H_5.CH : CH.CHO$.

Oil of Cinnamon Leaf. An oil distilled from the leaves of *Cinnamomum zeylanicum*. It contains from 75–90 per cent. eugenol and safrole, and only traces of cinnamic aldehyde. Its specific gravity is 1·045–1·065, and refractive index 1·533–1·536.

Oil of Citronella. The oil obtained from *Andropogon nardus*. Its chief constituent is citronellal, $C_{10}H_{18}O$.

Oil of Cocoa (Oil of Cacao). Oil of theobroma (cacao butter, *q.v.*).

Oil of Cognac. A mixture of different esters. It consists partly of rectified wine fusel oil, and partly of an artificial grape juice. Artificial oil of cognac is made from coco-nut oil and alcohol, and from palargonium oil, and castor oil.

Oil of Colza. See COLZA OIL.

Oil of Cus-cus. See CUS-CUS OIL.

Oil of Daget. Oil of birch tar.

Oil of Dog-fish. Shark oil.

Oil of Dolphin. Porpoise oil.

Oil of Duty. See OIL OF RHODIUM.

Oil of Earth Worms. A mixture of olive oil and white wine, in which earth-worms have been boiled.

Oil of Elder. See OIL OF CABBAGE.

Oil of Ergot. The residue left when an ethereal solution of tincture of ergot is evaporated.

Oil of Exeter. Oil of elder, mixed with euphorbium and mustard.

Oil of Flaxseed. Linseed oil.

Oil of Foxglove. Olive oil, in which fresh leaves of the foxglove have been digested.

Oil of Garlic. Allyl sulphide, $(C_3H_5)_2S$.

Oil of Geranium. See ROSÉ OIL.

Oil of Geranium, East Indian or Turkish. See ROSÉ OIL.

Oil of Gingelli. See GINGELLY OIL.

Oil of Ginger Grass. See ROSÉ OIL.

Oil of Gourd. Cucumber oil.

Oil of Grain. See FUSEL OIL.

Oil of Hartshorn. See BONE OIL.

Oil of Hemlock. The volatile oil from *Pinus canadensis*, the hemlock spruce. The term is also applied to olive oil in which fresh leaves of *Conium maculatum* have been digested.

Oil of Infernal Regions. Very impure olive oil.

Oil of Japanese Mint. This is not a true mint oil. It is derived from *Mentha arvensis* and is a partly de-mentholised oil.

Oil of Jupiter. Oil of juniper. It contains pinene, $C_{10}H_{16}$, cadinene, $C_{15}H_{24}$, and juniper camphor.

Oil of Lemon Balm. See OIL OF BALM.

Oil of Lemon Grass. The oil obtained from *Andropogon citratus*.

Oil of Liquid Pitch. See OIL OF TAR.

Oil of Mace. The name erroneously given to expressed oil of nutmeg.

Oil of Mirbane. See ESSENCE OF MIRBANE.

Oil of Mosoi Flowers. Ylang-ylang oil (*q.v.*).

Oil of Mucilages. Olive oil boiled with a decoction of marshmallow root, linseed, and fœnugreek seeds.

Oil of Mustard (Allyl Mustard Oil). Isoallyl-thio-carbimide, $C_3H_5N:CS$.

Oil of Neroli. See OIL OF ORANGE FLOWERS.

Oil of Nerves. Neatsfoot oil (*q.v.*).

Oil of Niobe. The methyl ester of benzoic acid. Used in the soap industry.

Oil of Orange Flower (Oil of Neroli). The oil distilled from the flowers of the bitter orange tree.

Oil of Origanum. Oil of thyme. Also oil from *Origanum* species.

Oil of Palma Christi (Ricinus oil). Castor oil.

Oil of Palmarosa. See ROSÉ OIL.

Oil of Paper. The oil obtained by burning paper on a tin plate.

Oil of Partridge Berry. Oil of wintergreen.

Oil of Pear. See PEAR OIL.

Oil of Pelargonium. See ROSÉ OIL.

Oil of Pennyroyal (Oil of Poley). Oil of pulegium. It consists chiefly of pulegone, $C_{10}H_{18}O$.

Oil of Peter (Oil of Petre). Rock oil, or a mixture of 1 part of oil rosemary, 4 parts turpentine, and 4 parts of Barbados tar.

Oil of Petitgrain. The oil obtained from the leaves of the bitter orange tree.

Oil of Petre. See OIL OF PETER.

Oil of Poley. See OIL OF PENNY-ROYAL.

Oil of Pompilion. An ointment of poplar seeds, also green elder ointment.

Oil of Portugal. Oil of sweet orange peel.

Oil of Ptychotis. Oil of ajowan.

Oil of Rhodium (Oil of Duty). The oil obtained from the root of *Genista canariensis*. Also a mixture of sandal-wood oil, and otto of rose, or oil of rose geranium.

Oil of Rose Geranium. See ROSÉ OIL.

Oil of St. John Wort. The oil obtained by digesting the flowering tops of *Hypericum perforatum* in warm olive oil.

Oil of Scorpions. Oil in which scorpions have been digested.

Oil of Smoke. Creosote (*q.v.*).

Oil of Spike. The volatile oil from *Lavandula spica*. It is also the name for a mixture of lavender oil, and oil of turpentine, coloured with alkanet.

Oil of Sweet Birch. Oil of betula, from *Betula lenta*.

Oil of Tar (Oil of Liquid Pitch). Creosote, the greenish liquid distilled from tar.

Oil of Tartar. Deliquescent potassium carbonate, K_2CO_3.

Oil of Tea. The oil obtained from the seeds of *Camellia species*.

Oil of Theobroma. Cacao butter (*q.v.*).

Oil of Three Ingredients. A mixture of the oils of turpentine, lavender, and brick, in equal parts.

Oil of Turpentine (Spirits of Turpentine, Essence of Turpentine, Turps). Derived from the pine, *Pinus palustris* and *P. Taeda,* and from the Scotch fir, *Pinus sylvestris.* The exudation from the bark is vacuum distilled in a current of steam, when the volatile turpentine oils distil over. The main constituent is pinene, $C_{10}H_{16}$. A solvent for resins and oils. It has a specific gravity of 0·86–0·875 and a boiling-point of 160° C.

Oil of Turpentine, Oxidised. See DECAMPHORISED OIL OF TURPENTINE.

Oil of Verbena. The oil obtained from *Verbena triphylla.* Also the name for oil of lemongrass.

Oil of Vetiver. See CUS-CUS OIL.

Oil of Vipers (Oil of Adders). The fat of oil from *Peluis berus,* the viper or adder. An approximation is 3 parts lard oil and 1 part bone oil.

Oil of Vitriol (O.V., Vitriolic Acid). Ordinary concentrated sulphuric acid, H_2SO_4.

Oil of Vitriol, Brown. See BROWN ACID.

Oil of Wax. See WAX BUTTER.

Oil of Wheat. The oil obtained from bruised wheat.

Oil of Wintergreen. Oil of gaultheria, obtained by distilling the leaves of *Gaultheria procumbens,* or from the bark of *Betula lenta.* The chief constituent is methyl salicylate, $C_6H_4(OH)COO.CH_3$.

Oil, Ohio. See EARTH OIL.

Oil Orange, E. Dyestuffs. They are British equivalents of Sudan I.

Oil Orange O. A dyestuff. It is a British brand of Sudan G.

Oil, Oxidised. See BLOWN OILS.

Oil, Oxidised Linseed. See SOLIDIFIED LINSEED OIL.

Oil Paalsgaard. See SCHOU OIL.

Oil, Palmichristi. See OIL OF PALMA CHRISTI.

Oil, Paraffin. See KEROSENE.

Oil, Peanut. See EARTHNUT OIL.

Oil, Physic Nut. See PURGING NUT OIL.

Oil, Pinnay. See LAUREL-NUT OIL.

Oil, Pogy. See MOSSBUNK OIL.

Oil, Poonseed. See LAUREL-NUT OIL.

Oil Pulp (Thickener, Fluid Gelatin, Viscom). A gelatinous material made by heating aluminium oleate with mineral oil. It is sold to give increased viscosity to mineral oils.

Oil, Rangoon. See RANGOON OIL and EARTH OIL.

Oil Red Base. A British equivalent of Rosaniline.

Oil-red-O-pyridine. A fat stain made by adding 3–5 grams oil-red-O to 100 c.c. of 70 per cent. solution of pyridine in water.

Oil Red S. An oil-soluble colour. It is toluene-azo-toluene-azo-β-naphthol.

Oil, Ricinus. See OIL OF PALMA CHRISTI.

Oil, Riviera. See AIX OIL.

Oil, Rock. See EARTH OIL.

Oil, Roshé. See ROSÉ OIL.

Oil, Safety. See C-PETROLEUM NAPHTHA.

Oil, Scarlet (Red B Oil, Soluble Extra Conc., Ponceau 3B). A dyestuff. It is toluene-azo-toluene-azo-β-naphthol, $C_{24}H_{20}N_4O$. Used for colouring oils and varnishes, also for cotton fibre.

Oil, Scarlet AS. A dyestuff. It is a British brand of Sudan III.

Oil Scarlet L, Y. Dyestuffs. They are British equivalents of Sudan II.

Oil, Seconds. See VIRGIN OIL.

Oil Shale. A sedimentary rock which yields from 12–60 gallons of shale oil per ton on distillation.

Oil, Sherwood. See PETROLEUM ETHER.

Oil Skin. A waterproof material made by impregnating cotton or other fabric with hardening oils.

Oil, Sod. See DÉGRAS.

Oil, Soluble. See TURKEY RED OILS.

Oil, Soluble Castor. See BLOWN OILS.

Oil-soluble Resins. Synthetic resins of the phenol-formaldehyde type obtained by a fixing process, using rosin in the mixing. They dissolve in hydrocarbon solution and are no longer soluble in alcohol. Also see HYDROCARBON-ALDEHYDE RESINS.

Oilstone. Whetstone (*q.v.*).

Oil-stone. See STINK-STONE.

Oil, Sublime Salad. See SUBLIME OLIVE OIL.

Oil Sugars. They are obtained by triturating volatile oils with sugar until the product is homogeneous. They are used for flavouring medicines.

Oil, Sulphated. See TURKEY RED OILS.

Oil, Sulphocarbon. See SANSE.

Oil, Sulphonated. See TURKEY RED OILS.

Oil, Sulphurated. Balsam of sulphur (*q.v.*).

Oil, Sulphur Olive. See SANSE.

Oil, Sweet. See COLZA OIL.

Oil, Sweet Nut. See BUTTER OIL.

Oil, Teal. See GINGELLY OIL.

Oil, Teel. See GINGELLY OIL.

Oil, Theobroma. See COCOA BUTTER.

Oil, Thickened. See BLOWN OILS.

Oil, Til. See GINGELLY OIL.

Oil, Touloucouna. See CRAB WOOD OIL.

Oil, Turkey Geranium. See ROSÉ OIL.

Oil Udilo. See LAUREL-NUT OIL.

Oil, Var. See AIX OIL.

Oil Varnishes. Solutions of resins in linseed oil.

Oil, Veppam. See VEEPA OIL.

Oil Vermilion. A dyestuff. It is a British brand of Sudan R.

Oil, Vetiver. See CUS-CUS OIL.

Oil, Vulcanised. See THINOLINE.

Oil, Wax. See WAX BUTTER.

Oil White (Light White, Leukarion, Albanol, Diamond White, Edelweiss, Snow White, Anti-White Lead, Blenda, Condor, Fixopone, Nivan). White lead substitutes. They consist chiefly of lithopone mixed with white lead or zinc white, also with whiting, gypsum, magnesia, or silica.

Oil, Wild Mango. See DIKA BUTTER.

Oil Yellow. See BUTTER YELLOW.

Oil Yellow. A dyestuff. It is a British equivalent of Sudan G.

Ointment, Blue. Mercury ointment.

Ointment, Golden. See EYE OINTMENT.

Ointment, Grey. See MERCURY OINTMENT.

Oisanite. A mineral. It is anatase.

Oiticica Oil. An oil obtained from the Brazilian plant *Conepia grandifolia*. It resembles Tung oil in its properties.

Okenite. A mineral. It consists of calcium-tetrahydro-disilicate, $CaH_4Si_2O_7$.

Okol. A disinfectant consisting of an emulsion containing phenols. It is miscible with water.

Okonite. A mineral. It is a calcium silicate, $CaH_2Si_2O_6$. The name has also been applied to an insulator consisting of rubber, lampblack, zinc oxide, litharge, and sulphur. Another variety is stated to contain vulcanised rubber and ozokerite.

Okstan X3. A trade name for a butyl tin mercaptide, a stabiliser for PVC with an exceptionally high heat stability. (67).

Okstan XO. A di-*n*-octyl tin mercaptide for stabilisation of non toxic PVC compounds. (66).

Olafite. A mineral. It is an albite of Norway.

Olaquindox. A growth promoter and bactericide currently undergoing clinical test as " Bay Va 9391 ". It is 2-(2-hydroxyethylcarbamoyl) - 3 - methyl - quinoxaline 1,4-dioxide.

Olcotrop Leather. A leather made from a species of shark skin.

Old Fustic. See FUSTIC.

Oldhamite. A mineral of meteoritic origin. It is calcium sulphide 4[CaS].

Old Scarlet. See BIEBRICH SCARLET.

Oleanodyne. A preparation containing oleic acid, aconitine, atropine, and veratrine.

Oleandomycin. An antibiotic produced by certain strains of Streptomyces antibioticus. Matromycin. Romicil.

Olefiant Gas (Heavy Carburetted Hydrogen, Dicarburetted Hydrogen, Elayl, Ethene, Etherin). Ethylene, C_2H_4.

Olein. The tri-oleyl derivative of glycerol, $C_3H_5(OC_{18}H_{33}O)_3$. The term is applied commercially to any liquid oil obtained from solid fats by pressure, to crude oleic acid, and to the potassium, sodium, or ammonium salt of the sulphonate of oleic acid. Elaine is another name given to commercial oleic acid. Also see TURKEY RED OILS.

Olein of Saponification. Commercial oleic acid prepared by the saponification of pure fats, and the separation from stearine, by pressing.

Oleite. Sodium-sulpho-ricinoleate.

Oleo (Premier Jus). The oil expressed from beef-fat. Used in margarine manufacture.

Oleobismuth. An oil suspension of bismuth oleate.

Oleochrysine. Calcium auro-thioglycerol sulphonate. A new French proprietary gold preparation for therapeutic use. The calcium content increases the tolerance to gold.

Oleocreosote. Creosote oleate. Used as an antiseptic in catarrhal condition of the respiratory organs, in the treatment of scrofula and phthisis.

Oleoformine. Cholic acid with hexamine and sodium oleate.

Oleogen. See PAROGEN.

Oleoguaiacol. Guaiacol oleate, $CH_3O.C_6H_4.O.CO.C_{17}H_{23}$. Used as an antiseptic.

Oleomargarine. The liquid fat obtained from tallow (the fat of oxen and sheep), after clarification and squeezing out the stearine. It consists of about 55 per cent. triolein, 35 per cent. tripalmitin, and 10–15 per cent. tristearin. Used in the manufacture of margarine (*q.v.*).

Oleo-resin Copaiba (Balsam of Copaiba, Balsam of Copaiva, Balsam of Capivi). Copaiba, an oleo-resin.

Oleo-resins. Resins mixed with a volatile oil.

Oleo-sanocrysin. A proprietary preparation of gold-sodium thiosulphate in oil.

Oleosol. Oildag. See OILDAG and AQUADAG.

Olesal. A bismuth compound of iminoglyceryl-benzoic acid, $(OH)_2BiO.C_6H_3(COOH)NHC_3H_5(OH)_2$. Used in medicine.

Oleum. See FUMING SULPHURIC ACID.

Oleum Spirits. A petroleum distillate.

Oleum White. A pigment. It is a sulphide of zinc mixed with a small proportion of barium sulphate (a lithophone). Used as a rubber filler.

Olex. Fat soluble powdered flavours. (504).

Olibanum (Indian Frankincense, Salai-gugl). A gum-resin obtained from *Boswellia* species. It contains from 8–18 per cent. of essential oil, 55–57 per cent. of resin, and 20–23 per cent. gum (carbohydrates). It has a specific gravity of 1·2, an acid value of 45–88, and a saponification value of 65–120.

Oligoclase (Soda Felspar, Sun Stone, Avanturine Felspar). A mineral. It is a felspar consisting of 6 parts albite, and 1 part anorthrite, $3Al_2O_3.3SiO_2.$ $2(Na_2O.CaO)3SiO_2.$

Oligonite (Pelosiderite). A mineral. It is a variety of siderite containing manganese.

Oliolase. An iodised vegetable oil.

Olive Green. See BRONZE GREEN and DE WINT'S GREEN.

Olive Green Oxide of Uranium. Uranoso-uranic oxide, $U_3O_8.$

Olive Lake. A pigment. Originally it was prepared from green ebony, but is now exclusively a mixture.

Olivenite. A mineral, $4CuO.As_2O_5.$

Olive Oil. See BOTTLE-NOSE OIL, FLORENCE OIL, GALLIPOLI OIL, GENOA OIL, LUCCA OIL, OIL OF INFERNAL REGIONS, PROVENCE OIL, RED OIL, SALAD OIL, SICILY OIL, and SPANISH OIL.

Olive Oils, Sulphur. See SANSE.

Olives of Java (Kaloempang Beans, Beligno Seeds, Sterculia Kernels). Seeds of *Sterculi fœtida*, the source of Sterculia oil.

Olivieraite. A mineral. It is a hydrated titanate of zirconium, $3ZrO_2.2TiO_2.$ $2H_2O$, of Brazil.

Olivine (Chrysolite, Peridot, Glinkite). A mineral. It is a silicate of iron and magnesium. A clear transparent variety is a precious stone. Also see ANDRADITE. The term is used for a family of ortho-silicates, as well as for a particular mineral.

Olivita. A proprietary preparation of olive oil containing vitamin D.

Olivite. A substance having a rubber base. Used as an acid-proofing material in pumps.

Ollite. A mineral. It is an impure steatite.

Olminal. A trade name for a commercial aluminium oleate. It contains 1·5 per cent. aluminium.

Olobintin. A German preparation. It is a 10 per cent. solution of a mixture of various rectified turpentine oils.

Olutkombul. The gelatinous sap of the plant *Abroma angustum*. Used in India in dysmenorrhœa.

Olympic Bronze. A proprietary trade name for an alloy of copper with 3 per cent. silicon and 1 per cent. zinc.

Olympic Bronze G. A proprietary trade name for an alloy of copper with 22 per cent. zinc and 1 per cent. silicon.

Olyntholite. Synonym for Grossular.

Omal. Trichloro-phenol.

Omarsan. A detergent and sterilizer. (522).

Omeire. A drink resembling Koumiss.

Omnadin. A German preparation. It is a mixture of reactive albumen bodies and lipoids, from gall and animal fats.

Omnex. A textile and hard surface detergent. (537).

Omnipen. A proprietary preparation of ampicillin. An antibiotic. (245).

Omnopon. A proprietary preparation of papaveretum. An analgesic. (314). See PANTOPON.

Omorol. See NOVARGAN.

Omphacite. A mineral. It is $8[(Ca,Mg,Fe^{..},Al)(Si,Al)O_3].$

Omya BLR3. A trade name for dry-ground calcite surface treated with a calcium salt of aliphatic acids. (French). (26).

Omya BSH. A French trade name for a wet-ground whiting from Champagne surface treated with the calcium salt of aliphatic acids. (26).

Omyastab. A secondary stabiliser for chlorinated polyester resins. It is a colourless cyclic organic compound containing ether linkages which are compatible with the polyester resin.

Onadox-118. A proprietary preparation of aspirin and dihydrocodeine bitartrate. An analgesic. (182).

Oncor. Silica-cored anticorrosive pigments. (572).

Oncosine. A mineral. It is a variety of muscovite.

Oncovin. A proprietary trade name for the sulphate of Vincristine.

Ondoita. A proprietary synthetic resin moulding powder.

Onegite (Fullonite). Minerals. They are varieties of gothite.

Onion's Alloy. A fusible alloy containing 50 per cent. bismuth, 30 per cent. lead and 20 per cent. tin.

Onofrite (Köhlerite). A mineral approaching the composition, $HgSe.4HgS$ (mercury selenosulphide).

Ontariolite. A mineral. It is a variety of scapolite.

Ontario Steel. A proprietary trade name for a non-shrinking steel containing 11 per cent. chromium, 0·75 per cent. molybdenum, 0·25 per cent. vanadium, 0·35 per cent. silicon, 0·30 per cent. manganese, and 1·45 per cent. carbon.

O.N.V. A proprietary trade name for a rubber vulcanization accelerator. It is diphenyl-carbamyl-dimethyl-dithiocarbamate. It has a specific gravity of 1·9, a melting-point of 183–185° C.

Onyx. Consists chiefly of silica.

Onyx Marble. A marble containing fossil shells.

Onyx of Tecali. A variety of alabaster. The colour varies from milk-white to pale yellow and pale green.

Oolitic Limestone. A massive variety of calcium carbonate, used for building purposes.

Oomra Gum. See AMRAD GUM.

Opachala Beans. See OWALA BEANS.

Opacin. A proprietary preparation of the sodium salt of tetra-iodo-phenol-phthalein. It is used in cholecystography.

Opacite. Stannic chloride, $SnCl_4$. It fumes in air and produces a corrosive smoke.

Opacol. A proprietary preparation of sodium tetra-iodo-phenol-phthalein.

Opal (Hydrophane, Hydrous Silica). A mineral. It consists of hydrated colloidal silica.

Opal Blue. See ANILINE BLUE, SPIRIT SOLUBLE, and FINE BLUE.

Opal Blue 6B. See ANILINE BLUE, SPIRIT SOLUBLE.

Opal Blue XL. See DIPHENYLAMINE BLUE SPIRIT SOLUBLE.

Opalescent Glass. Produced by the addition of arsenious oxide and calcium phosphate.

Opaline. Spirit lacquers. (512).

Opalite. A proprietary trade name for an amorphous silica.

Opal Jasper. It is silica, resembling jasper.

Opal, Liver. See MENILITE.

Opalon 740. A trade name for a graft copolymer used as a semi rigid wire insulation.

Opalon. A proprietary trade name for phenol-formaldehyde cast resins.

Opal Violet. See SPIRIT PURPLE.

Opal, Water. See HYALITE.

Opalwax. Hydrogenated castor oil.

Opaque Oxide of Chromium. A green pigment. It is sesquioxide of chromium.

Open Hearth Steel. See MARTIN STEEL.

Operidine. A proprietary preparation of phenoperidine hydrochloride. Analgesic supplement in anæsthesia. (356).

Ophicalcite. See OPHIOLITE.

Ophiolite (Ophicalcite, Serpentinous Marble). A mineral. It is serpentine associated with limestone or dolomite. It is usually variegated, and is used for ornamental work.

Ophorite. An ignition powder for projectiles. It consists of magnesium powder and potassium chlorate.

Ophthaine. A proprietary trade name for the hydrochloride of Proxymetacaine.

Opilon. A proprietary preparation of thymoxamine hydrochloride. A vasodilator. (262).

Opipramol. 1-{3-[4-(2-Hydroxyethyl)-piperazin - 1 - yl]propyl}-2,3 : 6,7 - di - benzazepine.

Opium. The dried juice from the unripe capsules of *Papaver somniferum*. It contains morphine, codeine, narcotine, narceine, thebane, papaverine, and meconin.

Opodeldoc. A solution of soap, camphor, and volatile oils in alcohol. Used externally for rheumatism.

Opoidine. A British proprietary preparation of opium containing the total alkaloids of opium in a soluble form. It contains about 50 per cent. of morphine, the remainder being the other alkaloids of opium, narcotine, narceine, codeine, papaverine, and thebaine, and is standardised for its alkaloidal content. It is a hypnotic, sedative, and analgesic, and has been used in the production of " twilight sleep," in surgical practice, and for the relief of pain caused by various diseases. It is administered orally, hypodermically, or rectally.

Opol. Chemical polishers for plastics. (513).

Opon. The trade name for a mixture of the hydrochlorides of opium alkaloids except morphine.

Opopanax. The dried juice from the roots of *Pastinaca opopanax*. It is used in perfumery, and medicinally as an antispasmodic.

Oppanol B. A registered trade name for polyisobutylene. (49). Its properties depend on the molecular weight, i.e. the stage of polymerisation. Molecular weight varies from about 3,000 (a viscous liquid) to 200,000, a fairly hard rubbery material. It has very good H.F. electrical properties, and is very resistant to water absorption and the action of acids. It is used as a synthetic rubber in cable manufacture and general insulation.
Synonyms : P.I.B., Isolene, Vistanex. (3).

Oppanol D. A registered mark for a polyisobutylene dispersion. (49).

Opsan. Antibacterial eye drops. (502).

Optalidon. A proprietary preparation of isoallyl-ethyl-barbituric acid with aminopyrin and caffeine.

Optannin. Basic calcium tannate.

Optarson. A proprietary preparation of ammonium heptin-chlor-arsinate.

Opthaine. A proprietary preparation of proxymetacaine hydrochloride solution. An ocular anæsthetic. (326).

Ophthalmidine. A proprietary preparation of IDOXURIDINE eye-drops used in the treatment of eye infections of the herpes type. (777).

Optical Bronze An alloy of 89 per cent. copper, 6·5 per cent. zinc, and 4·5 per cent. tin.

Optimax. A proprietary preparation of *l*-tryptophan, PYRIDOXINE and ascorbic acid. An anti-depressant. (451).

Optinol. Dye carrier for polyester textiles. (508).

Optipol. Cerium oxide glass polishing powder. (626).

Optisal. A German product. It is stated to be calcium-sodium citrate, a tasteless lime preparation used in rickets and asthma.

Optochal. Calcium-α-isobutyrate, Ca $(C_4H_7O_3)_2$. Used medicinally for the injection of calcium.

Optochin (Optoquin). Ethyl hydrocupreine. Used in medicine. It acts upon the pneumonia microbe.

Optoquin. See OPTOCHIN.

Optran. High grade optical chemicals. (527).

Optrex Eye Ointment. A proprietary preparation of gramicidin and aminacrine hydrochloride in wool fat-paraffin base. (228).

Opulets. A proprietary preparation of atropine sulphate in gelatine, used as a long acting mydriatic as an aid in diagnosis, and in treatment of uveitis. (330).

Orabase. A proprietary preparation containing sodium carboxy-methyl cellulose, pectin and gelatin in a liquid paraffin polyethylene base. (326).

Orabilix. The sodium salt of 3-(3-Butyramido - 2,4,6 - tri - iodophenyl) - 2-ethylacrylic acid.

Orabolin. A proprietary preparation of ethylœstrenol. An anabolic agent. (316).

Oracet. Organic solvent soluble dyestuffs. (530).

Oradexon. A proprietary preparation of dexamethasone. (316).

Oragulant. A proprietary trade name for diphenadione.

Orahesive. A proprietary preparation containing sodium carboxy-methyl cellulose, pectin and gelatin. (326).

Oralcer. A proprietary preparation of CLIOQUINOL and ascorbic acid used in the treatment of mouth ulcers. (779).

Oraldene. A proprietary preparation of HEXETIDINE. An antiseptic mouth wash. (262).

Oralith. Organic pigments.

Oranabol. A proprietary preparation of oxymesterone. An anabolic steroid. (336).

Oranabol. A proprietary trade name for oxymesterone.

Orange I (Tropæoline ooo No. 1, α-Naphthol Orange, Naphthol Orange, Orange B). A dyestuff. It is the sodium salt of *p*-sulpho-benzene-azo-α-naphthol, $C_{16}H_{11}N_2O_4SNa$. Dyes wool orange from an acid bath. It is now little used.

Orange II (Tropæoline ooo No. 2, Mandarin G, β-Naphthol Orange, Mandarin G Extra, Chrysaureine, Gold Orange, Orange Extra, Atlas Orange, Orange A, Acid Orange, Orange P, Acid Orange G, Orange II Conc, Special). A dyestuff. It is the sodium salt of *p*-sulpho - benzene - azo - β - naphthol, $C_{16}H_{11}N_2O_4SNa$. Dyes wool and silk orange from an acid bath.

Orange II Conc., Special. Dyestuffs. They are British equivalents of Orange II.

Orange III (Orange No. 3, Methyl Orange, Porrier's Orange III, Dimethylaniline Orange, Helianthin, Tropæoline D, Gold Orange, Mandarin Orange). A dyestuff. It is the sodium salt of *p*-sulpho-benzene-azo-dimethylaniline, $C_{14}H_{14}N_3SO_3Na$. Dyes wool orange from an acid bath. Used as an indicator in alkalimetry.

The sodium salt of *m*-nitro-benzene-azo - β - naphthol - disulphonic acid-$C_{16}H_9N_3O_9S_2Na_2$, is also called Orange III. It dyes wool orange from an acid bath.

Orange No. 3. See ORANGE III.

Orange IV (Diphenylamine Orange, Tropæoline oo, Orange M, Diphenylamine Yellow, Fast Yellow, Orange W, Orange GS, New Yellow, Diphenyl Orange, Helioxanthin, Gold Orange, Orange N, Acid Yellow D). A dyestuff. It is the sodium salt of *p*-sulpho-benzene-azo-diphenylamine, $C_{18}H_{14}N_3O_3S$ Na. Dyes wool orange-yellow from an acid bath.

Orange A. See ORANGE II.

Orange B. See ORANGE I.

Orange Chrome. See CHROME RED.

Orange ENL. See PONCEAU 4GB.

Orange Essence. Oil of orange.

Orange Extra. See ORANGE II.

Orange Flower Oil. See OIL OF ORANGE FLOWERS.

Orange G (Orange GG, Orange Yellow, Patent Orange, Kiton Fast Orange G, Fast Light Orange G, Naphtharene Orange G). A dyestuff. It is the sodium salt of benzene-azo-β-naphthol-disulphonic acid G, $C_{16}H_{10}N_2S_2O_7Na$. Dyes wool orange-yellow from an acid bath.

Orange GG. See ORANGE G.

Orange GRX. See PONCEAU 4GB.

Orange GS. See ORANGE IV.

Orange GT (Orange RN, Orange O, Orange N, Brilliant Orange O, Brilliant Orange OL, Crocein Orange (Brotherton & Co.)). A dyestuff. It is the sodium salt of toluene-azo-β-naphthol-sulphonic acid, $C_{17}H_{13}N_2O_4SNa$. Dyes wool from an acid bath.

Orange L. See SCARLET GR.

Orange Lac. See LEMON LAC.

Orange Lead (Orange Mineral, Orange Red, Sandix, Saturn Red). A red lead obtained by calcining powdered white lead. It is a better red lead than crystal minium.

Orange M. See ORANGE IV.

Orange Mineral. See ORANGE LEAD.

Orange MNO. A British brand of Metanil yellow.

Orange N. See ORANGE IV. See ORANGE GT. Also see FAST YELLOW N and SCARLET GR.

Orange O. See ORANGE GT.

Orange Ochre. Burnt yellow ochre (q.v.).

Orange Oxide of Uranium (Yellow Sesquioxide of Uranium, Uranium Yellow). Sodium uranate, $Na_2U_2O_7$, has been called by these names.

Orange P. See ORANGE II.

Orange R. See PONCEAU 2G, ORANGE T, ORANGE RR, and ALIZARIN YELLOW R.

Orange Red. See ORANGE LEAD.

Orange Red I. See BRILLIANT DOUBLE SCARLET G.

Orange RL and RRL. Identical with Resorcin yellow.

Orange RN. See ORANGE GT.

Orange RR (Orange R). A dyestuff. It is the sodium salt of sulpho-xylene-azo-β-naphthol, $C_{18}H_{15}N_2O_4SNa$. Dyes wool orange from an acid bath.

Orange Russet. Rubens madder.

Orange Spirit. See TIN SPIRITS.

Orange T (Mandarin GR, Orange R, Kermesin Orange). A dyestuff. It is the sodium salt of sulpho-o-toluene-azo-β-naphthol, $C_{17}H_{13}N_2O_4SNa$. Dyes wool orange from an acid bath.

Orange TA. A direct cotton dyestuff.

Orange Tungsten. Saffron bronze, tungsten-sodium tungstate, $Na_2WO_4 \cdot W_2O_5$.

Orange Vermilion. See SCARLET VERMILION.

Orange Vermilion, Field's. See SCARLET VERMILION.

Orange W. See ORANGE IV.

Orange Wine. A wine made by the fermentation of a saccharine solution, to which fresh bitter orange peel has been added.

Orange Yellow. See ORANGE G.

Orangite. See THORITE.

Oranit. See NEKAL.

Oranium Bronze (Dirigold). An aluminium bronze. It contains from 87–97 per cent. copper and 3–11 per cent. aluminium.

Orap. A proprietary preparation of PIMOZIDE. A tranquilliser. (356).

Orapen V-K. A proprietary preparation containing Phenoxymethylpenicillin Potassium. An antibiotic. (362).

Orargol. Colloidal gold and silver.

Orarsan. A proprietary preparation of 3-acetylamino-4-oxy-phenyl arsinic acid. See FOURNEAU 190.

Orasan. See ACETARSONE.

Orasol. Organic solvent soluble dyestuffs. (530).

Oratrol. A proprietary preparation of dichlorphenamide. An ocular antiseptic. (374).

Orasecron. A proprietary preparation of ethisterone and ethinyloestradiol. Treatment of menstrual disorders. (265).

Orastrep. A proprietary preparation of streptomycin sulphate, sulphadimidine and kaolin. An antidiarrhœal. (309).

Orbenin. A proprietary preparation containing Cloxacillin. An antibiotic. (364).

Orca. A French synthetic resin prepared from acrolein. It is used for electrical insulation.

Orcellin. See FOND ROUGE.

Orcellin Deep Red. See FOND ROUGE.

Orchidee (Sanfoin). The isoamyl ester of salicylic acid or o-oxy-benzoic acid, $C_6H_4(OH)COOC_5H_{11}$. Used in perfumery.

Orchil. See ARCHIL.

Orchil Extract G. See ARCHIL SUBSTITUTE G.

Orchil Extract N Extra. See APOLLO RED.

Orchil Extract V. See ARCHIL SUBSTITUTE V.

Orchil Extract 3VN. See ARCHIL SUBSTITUTE 3VN.

Orchil Red A (Union Fast Claret). A dyestuff. It is the sodium salt of xylene-azo-xylene-azo-β-naphthol-disulphonic-acid, $C_{26}H_{22}N_4O_7S_2Na_2$. Dyes wool red from an acid bath.

Orchil Substitute G. See ARCHIL SUBSTITUTE G.

Orchil Substitute N Extra. See APOLLO RED.

Orchil Substitute V. See ARCHIL SUBSTITUTE V and ARCHIL SUBSTITUTE 3VN.

Orchil Substitute 3VN. See ARCHIL SUBSTITUTE 3VN.

Orchindone. Isobutyl-salicylate. Used in perfumery.

Orcin. Orcinol, $C_6H_3 \cdot CH_3(OH)_2$.

β-Orcin. Betorcinol, $C_6H_2(CH_3)_2(OH)_2$.

Orciprenaline. 1-(3,5-Dihydroxy-phenyl)-2-isopropylaminoethanol. Alupent is the sulphate.

Ordeal Bark. The bark of *Erythrophlœum guincense*.

Ordeal Bean (Chop Nut, Split Nut). Calabar bean, a source of eserine.

Ordonezite. A mineral. It is $2[ZnSb_2O_6]$.

Ordoval. A synthetic tannin made from sulphonated anthracene.

Ordoval G.2G. See NERADOL.

Orea. See RESINS, ACROLEIN.

Ore, Burnt. See BLUE BILLY.

Ore, Fahl. See TETRAHEDRITE.

Ore, Feather. See ZINKENITE.

Ore, Flinty Zinc. Flinty Calamine.

Ore-furnace Slag. A slag obtained when roasted copper sulphide ore is mixed with oxidised ores and slag and fused for the production of " coarse metal." It often contains unfused quartz and less than 1 per cent. copper, and is mainly a silicate of iron.

Ore, Glance. See ARGENTITE.

Ore, Glass. See ARGENTITE.

Oregon Balsam. The true oleo-resin is obtained from *Pseudotsuga mucronata*, but another product sold under the same name consists of a mixture of rosin and turpentine.

Ore, Horseflesh. See PURPLE COPPER ORE.

Oreide. A yellow alloy resembling gold. It usually contains from 80–90 per cent. copper, 10–14·5 per cent. zinc, and 0–4·5 per cent. tin.

Ore, Lake. See BOG IRON ORE.

Orelite. See METALITE.

Ore, Liver. See CINNABAR.

Orellin. A yellow colouring matter found in annatto. It is probably an oxidation product of bixin, another colouring matter of annatto.

Ore, Meadow. Bog iron ore (*q.v.*).

Ore, Needle. See ACICULITE.

Oreoselone. Oreoselin, $C_{14}H_{12}O_4$.

Ore, Potter's. See ALQUIFON.

Ore, Red Copper. See CUPRITE.

Ore, Red Silver. See PYRARGYRITE.

Ore, Ruby Silver. See PYRARGYRITE.

Ore, Soft. See ARGENTITE.

Ore, Spathic Iron. See CHALYBITE.

Ore, Spathic Manganese. See RHODOCROSITE.

Ore, Spathic Zinc. See CALAMINE.

Ore, Specular Iron. See HÆMATITE.

Ore, Titaniferous. See ILMENITE.

Ore, Wolfram. See WOLFRAMINE.

Orexine (Phenzoline, Cedrarine, Tannexin). The name formerly given to orexin-hydro chloride (phenyl-dihydro-quinazoline hydrochloride), $C_{14}H_{12}N_2$ ($C_{14}H_{10}O_9$). The tannic acid is now employed. An aperitive.

Ore, Yellow Copper. Copper pyrites.

Orgaluton. A proprietary preparation of mestranol and lynestrenol used in the treatment of menstrual disorders. (316).

Orgametril. A proprietary trade name for Lynoestrenol.

Orgamide R. A proprietary trade name for nylon 6.

Organidin. A proprietary preparation of 2,3-(2- and 3-indopropylidene-doxy)-propanol in solution. A cough linctus. (382).

Organy. Pennyroyal, also Origanum.

Organzine Silk (Warp Silk). The reeled-off fibres from a number of silk cocoons. It is the best quality raw silk obtained from the most perfect cocoons. Used for warps in weaving.

Orgol. A special tar for tar/epoxy combination. (627).

Orgotein. An anti-inflammatory comprising a group of soluble metalloproteins isolated from liver, red blood cells and other mammalian tissues.

Orgrafin Sodium. Sodium Ipodate.

Orgraine. A proprietary preparation of ergotamine tartrate, caffeine, hyoscyamine sulphate, atropine sulphate and phenacetin, used in the treatment of migraine. (316).

Oriental Agate. An agate used as a gem stone. It is translucent.

Oriental Blue. Ultramarine (*q.v.*).

Oriental Emerald. A green variety of corundum (*q.v.*). It is a green sapphire.

Oriental Hyacinth. A rose-coloured corundum.

Oriental Powder. An explosive used in fireworks. It is a mixture of gamboge and potassium nitrate.

Oriental Ruby. A mineral, Al_2O_3.

Oriental Sapphire. A blue variety of corundum.

Oriental Sweet Gum. Storax (*q.v.*).

Oriental Topaz. Aluminium oxide with metallic oxides. Also the name given to a yellow sapphire.

Orientite. A mineral. It is a silicate of manganese and calcium, $4CaO.2Mn_2O_3.5SiO_2.4H_2O$, found in Cuba.

Orient Yellow. See CADMIUM YELLOW.

Origanum Oil. See OIL OF ORIGANUM.

Original Green. See EMERALD GREEN.

Orinase. A proprietary trade name for Tolbutamide.

Orinol. A colloidal copper silicate containing 0·1 per cent. copper. It is a German preparation, and is employed in solution and salves for the treatment of catarrh and colds.

Oriol Yellow (Cotton Yellow R, Alkali Yellow). A dyestuff. It is the sodium salt of primuline-azo-salicylic acid. Dyes cotton yellow from an alkaline bath.

Orion Blue 2B. A dyestuff. It is a British equivalent of Diamine blue 2B.

Orion Blue R33. A dyestuff. It is a British brand of Trisulphone blue R.

Orion Brown B. A dyestuff. It is a British equivalent of Benzo brown B.

Orion Brown G. A British dyestuff. It is equivalent to Benzo brown G.

Orion Green B. A dyestuff. It is a British brand of Diamine green B.

Orion Green G. A British equivalent of Diamine green G.

Orion Sky-blue. A dyestuff. It is a British equivalent of Diamine sky-blue.

Orion Violet. A British dyestuff. It is equivalent to Trisulphone violet B.

Orisol. A solution of berberine acid sulphate.

Orisulf. A proprietary preparation of sulphaphenazole. An antibiotic. (18).

Orkast. A proprietary polyester laminating resin. (627).

Orlean. See ANNATTO.

Orleans Vinegar. See WINE VINEGAR.

Orleans White. Chalk (q.v.).

Orlest 28. A proprietary preparation of ethinylœstradiol and NORETHISTERONE acetate. An oral contraceptive. (264).

Ormolu. One alloy contains 58 per cent. copper, 25·3 per cent. zinc, and 16 per cent. tin, and another consists of 90·5 per cent. copper, 3 per cent. zinc, and 6·5 per cent. tin.

Ornalith. See BAKELITE.

Ornithine. α-δ-Diamino-valeric acid, $(H_2N)CH_2.CH_2.CH_2.CH(NH_2)COOH$.

Ornithite. A tricalcic phosphate, $Ca_3P_2O_8.2H_2O$.

Oroglas. A proprietary grade of polymethylmethacrylate. (13).

Oronite. A registered trade mark for a variety of chemical products. (231).

Oronite (R). The trade mark for a line of industrial chemicals, including detergent intermediates, drying-oil extenders, sodium sulphonates, naphthenic acids, flotation agents, etc. (231).

Oropon. A bate for leather. It is composed of the enzymes of pancreas absorbed in sawdust or kieselguhr, and intimately mixed with ammonium chloride or boric acid.

Orovite. A proprietary preparation of thiamine, riboflavine, pyridoxine, nicotinamide and ascorbic acid. Vitamin supplement. (272).

Orphenadrine. 2-Dimethylaminoethyl 2-methyldiphenylmethyl ether Disipal is the hydrochloride; Norflex is the citrate. Present in Norgestic as the citrate.

Orphole. A bismuth-β-naphthol compound, $Bi_2O_2(OH).(C_{10}H_7O)$. Used internally as an intestinal antiseptic, and externally as a substitute for iodoform.

Orpiment (Yellow Sulphide of Arsenic). A mineral, As_2S_3. Also see KING's YELLOW.

Orpiment, Red. Realgar (q.v.).

Orris Camphor. Essential oil of orris.

Orris Root. The rhizome of Iris florentina.

Orr's White. See LITHOPONE.

Orseille. See ARCHIL.

Orseilline BB. A dyestuff. It is the sodium salt of sulpho-toluene-azo-toluene-azo-α-naphthol-p-sulphonic acid, $C_{24}H_{18}N_4O_7S_2Na_2$. Dyes wool red from an acid bath.

Orseilline No. 3. See FAST RED.

Orseilline No. 4. See FAST RED.

Orsudan. Sodium-methyl-acetyl-p-amino-phenyl-arsenate, $C_2H_3O.NH.C_6H_3(CH_3).AsO(OH)(ONa)$. It contains 25·4 per cent. of arsenic, and has been used in medicine in the treatment of syphilis, now obsolete.

Ortamine Brown. A dyestuff. It is o-dianisidine.

Orthesin. See ANÆSTHESIN, BENZOCAINE.

Orthin. o-Hydrazine-p-oxy-benzoic acid, $C_6H_3(OH)(COOH)(NH.NH_2)$. It has a feeble antipyretic action.

Orthite. A mineral containing thorite.

Orthocaine. Orthoform, new (q.v.).

Orthochrome T. p-Toluquinaldine p-toluquinoline-ethyl-cyanine bromide. A red sensitiser for silver bromide plates.

Ortho-Chrysotile. A mineral. It is $2[Mg_3Si_2O_5(OH)_4]$.

Orthoclase. See FELSPAR.

Orthocoll. See THIOCOLL.

Ortho-Creme. A proprietary preparation of ricinoleic acid, boric acid, and sodium lauryl sulphate. A spermicidal cream for use with a vaginal diaphragm. (359).

Orthodol. Synthetic tanning extract. (508).

Orthoferrosilite. A mineral. It is $16[FeSiO_3]$.

Orthoform (Mannin). The methyl ester of p-amino-m-hydroxy-benzoic acid, $C_6H_3(NH_2)(OH)COOCH_3$. A local anæsthetic.

Orthoform, New. Methyl-m-amino-p-hydroxy-benzoic acid, $C_6H_3(NH_2)(OH)COOCH_3$. A local anæsthetic.

Orthoforms. A proprietary preparation of nonylphenoxypolyethoxyethanol and benzethonium chloride. Contraceptive pessaries. (369).

Ortho-Gynol. A proprietary preparation of ricinoleic acid and p-diisobutyl-phenoxypolyethoxyethanol. Vaginal spermicidal cream for use with a diaphragm. (369).

Orthonovin 1/50, 1/80. Proprietary preparations of norethisterone and mestranol. Oral contraceptives. (369).

Orthonovin 2 mg. A proprietary preparation of norethisterone and mestranol. Oral contraceptive. (369).

Orthonovin SQ. A proprietary preparation of norethisterone and mestranol. Oral contraceptive. (369).

Orthosichol. A German product. It is a proprietary preparation which is stated to contain the isolated active principle of *Orthosiphon stamineus*. For bile and kidney complaints.

Orthosil. A proprietary trade name for an anhydrous sodium orthosilicate. A detergent.

Orthoxicol. A proprietary preparation containing methoxyphenamine hydrochloride, codeine phosphate and sodium citrate. A cough linctus. (325).

Orthoxine. A proprietary preparation of methoxyphenamine. A bronchial antispasmodic. (325).

Orth's Acid Alcohol. This mixture contains 3 c.c. hydrochloric acid, 70 c.c. alcohol, and distilled water up to 100 c.c.

Orth's Stain. A microscopic stain. It contains 1 gram lithium carbonate and 2·5 grams carmine in 100 c.c. water.

Ortison. Tablets containing a solid preparation of hydrogen peroxide for use as a mouth disinfectant.

Ortizon. See HYPEROL.

Ortol. A photographic developer. Methyl-o-amino-phenol, $C_6H_4(OH)$ $(NHCH_3)$ (2 mols.), combined with hydroquinone (1 mol.), forms the basis of this developer.

Ortolan. A proprietary insulation.

Ortosol. A mixture for dry cleaning. It consists of 10 per cent. chlor-benzene, 88 per cent. o- and m-dichlor-benzenes, and 2 per cent. p-dichlor-benzene.

Orudis. A proprietary preparation of ketoprofen used in the treatment of arthritis. (507).

Oruetite. A mineral. It is a bismuth sulpho-telluride, $Bi_2Te_3Bi_2S_3$.

Orvillite. A mineral. It is a hydrated silicate of zirconium, $8ZrO_2.6SiO_2.5H_2O$.

Orvus WA. A proprietary trade name for a wetting agent. It is sodium lauryl sulphate.

Osage Orange. A material obtained from the bark of the Osage orange tree, containing 25 per cent. of tannin. Used as a dyestuff.

Osanite. A mineral from Alter Pedroso, Portugal. It contains 47·5 per cent. SiO_2, 22·1 per cent. FeO, 17·6 per cent. Fe_2O_3, 7·6 per cent. Na_2O, and small amounts of TiO_2, MnO, MgO, CaO, and K_2O.

Osarsal. See ACETARSONE.

Osarsan. A derivative of phenyl-glycine-arsinic acid. Used in the treatment of sleeping sickness.

Osbil. A proprietary preparation of isobenzamic acid, used for oral cholangiography. (336).

Osbornite. A mineral of meteoritic origin. It is 4[TiN].

Oscodal. A proprietary preparation of vitamin cod-liver oil product.

Oskalsan. A double salt of calcium chloride and calcium lactate, containing 33 per cent. calcium chloride. It is used medicinally for bone development in children.

Osmal Black. Activated carbon which has absorbed ammonia. It can replace part of the red lead in rust-preventive pigments. It is used particularly for exposure to sewer gases.

Osmelite. A mineral of Bavaria. It is similar to Pectolite ($q.v.$).

Osmic Acid. The name formerly used for osmium tetroxide, OsO_4.

Osmiridium. See IRIDOSMINE.

Osmo-calamine. Colloidal calamine.

Osmo-kaolin. A preparation of kaolin obtained by a patented electro-osmosis process. It has a high covering power and clinging properties, and is used in toilet powders.

Osmondite. The stage in the transformation of austentite, at which the solution in dilute sulphuric acid reaches its maximum rapidity.

Osmo-sil. A dye absorbent. It is a very pure form of silica, SiO_2.

Osnol. A proprietary preparation of liquid paraffin with vitamin D.

Ospolot. A proprietary preparation of sulthiame. An anticonvulsant. (341).

Osram. An alloy of osmium and tungsten.

O.S.S. Sodium orthosilicate. (537).

Ossalin. A fat obtained from the marrow of ox bone. Used in ointments.

Ossein. A variety of gelatin prepared from bones.

Os Sepiæ (White Fish-Bone). The calcareous shell lying within the back of the cuttle fish. It consists mainly of calcium carbonate, and is used as a tooth powder.

Ossivite. A proprietary preparation of bonemeal and Vitamins A and D. (245).

Ossopan. A proprietary preparation of calcium phosphate and bone fluorides used in the treatment of osteoporosis. (780).

Ossophyt. Sodium-glycocoll-phosphate. It is a German preparation, and is stated to aid bone formation.

Ostan. A trade name for pure sodium and potassium hydroxides in disc form

OST 527 OWA

(5 mm. diameter), suitable for analytical work and convenience in weighing.

Ostatki. See MAZUT.

Ostelin. A proprietary preparation. It is a vitamin cod-liver oil product.

Osteolite. A mineral. It is calcium orthophosphate, $Ca_3(PO_4)_2.2H_2O$.

Ostranite. See ZIRCON.

Ostreocin. A proprietary trade name for a mixture of Ostreogrycins B and G.

Ostreogrycin. Antimicrobal substances produced by Streptomyces ostreogriseus (specific substances are designated by a terminal letter; thus, Ostreogrycin B). Ostreocin is a mixture of Ostreogrycins B and G.

Osumilite. A mineral. It is $[(K,Na,Ca)(Mg, Fe\cdots)_2(Al, Fe\cdots, Fe\cdots)_3(Si, Al)_{12}O_{30}.H_2O]$.

Osvarsan. See ACETARSONE.

Oswego. A proprietary trade mark for cornflour.

Otalgan. A proprietary preparation of 5 per cent. phenazone in glycerin. Used for all painful affections of the ear.

Otamidyl. A proprietary preparation of dibromopropamidine isethionate and diamidino-diphenylamine dihydrochloride used as ear-drops. (507).

Otavite. A mineral. It is a cadmium carbonate.

Otaylite. A Californian bentonite found near Otay.

Ote Seeds (Acoomo Seeds). Seeds of *Myristica angolensis*, of Nigeria. The seeds yield Kombe fat.

Otita. A solution containing 1·8 grains quinine dihydrochloride per fluid ounce in glycerin and water.

Otoba Butter (Otoba Wax, American Mace Butter). Otoba fat, obtained from the fruit of *Myristica otoba*.

Otoba Fat. See OTOBA BUTTER.

Otoba Wax. See OTOBA BUTTER.

Otopred. A proprietary preparation of CHLORAMPHENICOL, PREDNISOLONE and THIOMERSAL used as anti-infective ear-drops. (355).

Otoseptil. A proprietary preparation of NEOMYCIN undecylenate, TYROTHRICIN, HYDROCORTISONE and MACROGOL used as anti-bacterial ear-drops. (334).

Otosporin. A proprietary preparation of POLYMYXIN B sulphate, NEOMYCIN sulphate and HYDROCORTISONE used as anti-infective ear-drops. (277).

Ototrips. A proprietary preparation of POLYMYXIN B, BACITRACIN and TRYPSIN used as anti-infective ear-drops. (256).

Otreon. A German preparation containing papaverine, magnesium carbonate, and bismuth carbonate. It is used against excessive acidity of the stomach.

Otrivine. A proprietary trade name for a preparation used as nasal decongestant drops. It is the hydrochloride of Xylometazoline.

Otto. Essential oil.

Ouabaine Arnaud. A proprietary preparation of ouabaine used in cases of threatened heart failure. (781).

Oulopholite. A mineral. It is a variety of gypsum.

Ounce Metal. A bronze consisting of 85 per cent. copper, 5 per cent. tin, 5 per cent. zinc, and 5 per cent. lead.

Ouralpatti. See VENTILAGO MADRASPANTA.

Ourari. Curare.

Outremer. Ultramarine (q.v.).

Ouvarovite. A mineral. It is a lime-chrome-garnet.

O.V. See OIL OF VITRIOL.

Ovac. A proprietary trade name for a rubber vulcanising accelerator. It is a blend of the thiazole derivatives of two specially selected aldehyde-amines.

Ovanon. A proprietary preparation of MESTRANOL or MESTRANOL-plus-lynestrol in tablet form. An oral contraceptive. (316).

Ovastrol. A proprietary preparation of mestranol. Treatment of menopausal disorders and senile vaginal changes. (806).

Overite. A mineral. It is $2[Ca_3Al_8(PO_4)_8(OH)_6.15H_2O]$.

Overnite. A slug killer. (501).

Ovestrin. A proprietary preparation of œstriol. Treatment of senile vaginal changes and menstrual irregularities. (316).

Ovicide. An insecticide. (511).

Ovoferrin. Brand Colloidal Iron Tonic. Readily assimilable colloidal iron. For use for all those conditions for which iron is indicated. Does not stain the teeth, upset digestion or constipate.

Ovogall. A combination of ox-gall and egg-albumin. It dissolves in the intestines, and increases secretion of bile.

Ovol. A proprietary preparation of dicyclomine hydrochloride and dimethicone used in the treatment of infant colic. (782).

Ovolecithin. Lecithin (q.v.).

Ovran. A proprietary preparation of ethinylœstradiol and d-norgestrel. An oral contraceptive. (245).

Ovranette. A proprietary preparation of ethinyl œstradiol and d-norgestrel. An oral contraceptive. (245).

Ovulen 1 mg. A proprietary preparation of ethynodiol diacetate and mestranol. Oral contraceptive. (285).

Ovulen 50. A proprietary preparation of MESTRANOL and ETHYNODIOL diacetate. An oral contraceptive. (644).

Owala Beans (Opachala Beans, Atta Beans, Maboula Nuts, Panza Nuts,

Attawa Nuts, Odu Nuts, Fai Nuts, Fulla Panza Nuts, Nulla Panza Nuts). The seeds of *Pentaclethra macrophylla*, of West Africa. They are the source of Owala oil.

Owenite. A mineral. It is a variety of thuringite.

Owyheeite (Silver Jamesonite). A mineral. It is a sulphantimonite of lead and silver, $5PbS.Ag_2S.3Sb_2S_3$, found in Idaho.

Oxacalcite. Native calcium oxalate. Synonym for Whewellite.

Oxacillin. 3,3-Dimethyl-6-(5-methyl-3-phenyl - 4 - isoxazolecarboxamido) - 7 - oxo - 4 - thia - 1 - azabicyclo[3.2.0]hep - tane-2-carboxylic acid.

Oxaf. A proprietary accelerator. Zinc mercaptobenzthiazole. (435).

Oxalan. Oxaluramide, $C_3H_5N_3O_3$.

Oxalantin. Leucoturic acid, $C_6H_6N_4O_6$.

Oxalate Blasting Powder. A safety explosive powder containing 71 per cent. nitre, 14 per cent. charcoal, and 15 per cent. ammonium oxalate.

Oxalic Ether. Ethyl oxalate $(C_2H_5.COO)_2$.

Oxalite. A mineral. It is ferrous oxalate.

Oxalumina. An abrasive consisting of small crystals of alumina.

Oxamine Black BR. A dyestuff. It is the sodium salt of dimethoxy-diphenyl disazo - phenylene - diamine - oxamic - α-naphthol-sulphonic acid, $C_{32}H_{24}N_6SO_9Na_2$. Dyes cotton black.

Oxamine Black MB. A dyestuff. It is the sodium salt of diphenyl-disazo-*m*-phenylene - oxamic - acid - azo - bi - amino-naphthol-sulphonic acid, $C_{40}H_{26}N_9O_{11}S_2Na_3$. Dyes cotton black.

Oxamine Black MD. A dyestuff. It is the sodium salt of dimethoxy-diphenyl - disazo - phenylene - oxamic - acid - azo - bi - amino - naphthol - sul - phonic acid, $C_{42}H_{30}N_9O_{13}S_2Na_3$. Dyes cotton black.

Oxamine Black MT. A dyestuff. It is the sodium salt of ditolyl-disazo-phenylene - oxamic - acid - azo - bi - amino-naphthol-sulphonic acid, $C_{42}H_{29}N_9O_{11}S_2Na_3$. Dyes cotton black.

Oxamine Blue B. A dyestuff. It is the sodium salt of dimethoxy-diphenyl-disazo - α - naphthol-sulphonoic-amino-naphthol-sulphonic acid, $C_{34}H_{25}N_5S_2O_{10}Na_2$. Dyes cotton indigo blue.

Oxamine Blue BB. A dyestuff. It is the sodium salt of dimethoxy-diphenyl-disazo - amino - benzene - azo - β - naph-thol-α-naphthol-p-sulphonic acid, $C_{40}H_{30}N_7O_7SNa$. Dyes cotton blue from an alkaline salt bath.

Oxamine Blue BT. A dyestuff. It is the sodium salt of dimethoxy-diphenyl-disazo - phenylene - oxamic acid - azo - *m* - phenylene - diamine - β - naphthol

disulphonic acid, $C_{38}H_{28}N_9O_{12}S_2Na_3$. Dyes cotton dark reddish-blue from a salt bath.

Oxamine Blue MD. A dyestuff. It is the sodium salt of dimethoxy-diphenyl-disazo - phenylene - oxamic - acid - azo - bi - β - naphthol - disulphonic acid, $C_{42}H_{26}N_7O_{19}S_4Na_5$. Dyes cotton blue from an alkaline bath.

Oxamine Blue 3R. A dyestuff. It is the sodium salt of ditolyl-disazo-α-naphthol - sulphonic - amino - naphthol-sulphonic acid, $C_{34}H_{25}O_5S_2O_8Na_2$.

Oxamine Maroon. A dyestuff. It is the sodium salt of diphenyl-disazo-salicylic acid-1 : 5-amino-naphthol-sul-phonic acid, $C_{29}H_{19}N_5SO_7Na_2$.

Oxamine Orange G. A dyestuff. It is the sodium salt of diphenyl-disazo-phenol-*m*-tolylene-diamine-oxamic acid, $C_{27}H_{21}N_6O_4Na$. Dyes cotton orange from a salt bath.

Oxamine Red. A dyestuff. It is the sodium salt of diphenyl-disazo-salicylic acid-2 : 5-amino-naphthol-sulphonic acid, $C_{29}H_{19}N_5SO_7Na_2$. Dyes cotton dark red.

Oxamine Red B. A dyestuff. It is the sodium salt of diphenyl-disazo-α-naph-thol-sulphonic acid-*m*-phenylene-dia-mine-oxamic acid, $C_{30}H_{20}N_6SO_7Na_2$. Dyes cotton bluish-red.

Oxamine Red MT. A dyestuff. It is the sodium salt of ditolyl-disazo-pheny-lene-oxamic acid-azo-bi-resorcin, $C_{34}H_{26}N_7O_7Na$. Dyes cotton brownish-red from an alkaline salt bath.

Oxamine Scarlet B. A dyestuff. It is the sodium salt of diphenyl-disazo-naphthionic - *m* - phenylene - diamine - oxamic acid, $C_{30}H_{21}N_7SO_6Na_2$. Dyes cotton scarlet from a salt bath.

Oxamine Violet. A dyestuff. It is the sodium salt of diphenyl-disazo-bi-amino-naphthol-sulphonic acid, $C_{32}H_{22}N_6S_2O_8Na_2$. Dyes cotton reddish-violet.

Oxamine Violet BBR. A dyestuff. It is the sodium salt of ditolyl-disazo-phenylene - oxamic - azo - β - naphthol - α-naphthol-sulphonic acid, $C_{42}H_{29}N_7O_8SNa_2$. Dyes cotton violet from a salt bath.

Oxamine Violet GR. A dyestuff. It is the sodium salt of ditolyl-disazo-*m*-phenylene - diamine - oxamic - α - naph-thol-sulphonic acid, $C_{32}H_{24}N_6SO_7Na_2$. Dyes cotton dark reddish-violet.

Oxamine Violet GRF. A dyestuff. It is the sodium salt of diphenyl-disazo-*m*-phenylene-oxamic acid-azo-phenylene-diamine-β-naphthol-disulphonic acid $C_{36}H_{24}N_9O_{10}S_2Na_3$. Dyes cotton red-dish-violet from a salt bath.

Oxamine Violet MT. A dyestuff. It is the sodium salt of ditolyl-disazo-phenylene-oxamic acid-azo-bi-β-naphthol-disulphonic acid, $C_{42}H_{26}N_7O_{17}S_4Na_5$. Dyes cotton violet from a salt bath.

Oxamine Violet 4R. A dyestuff. It is the sodium salt of diphenyl-disazophenylene-oxamic acid-azo-bi-α-naphthol-p-sulphonic acid, $C_{40}H_{24}N_7O_{10}S_2Na_3$. Dyes cotton violet from a salt bath.

Oxammite (Guanapite). A mineral. It is ammonium oxalate.

Oxamniquine. A preparation used in the treatment of schistosomiasis, currently undergoing clinical trial as " UK-4271 ". It is 6-hydroxymethyl-2-isopropylaminomethyl - 7 - nitro - 1, 2, 3, 4-tetrahydroquinoline.

Oxandrolone. 17β - Hydroxy - 17α - methyl - 2 - oxa - 5α - androstan - 3 - one. Anavar.

Oxantel. An anthelmintic. 1,4,5,6-tetrahydro - 1 methyl - 2 - (*trans* - 3 - hydroxystyryl) pyrimidine. " CP-14, 445-16 " is currently under clinical trial as the pamoate.

Oxantin. A synthetic sugar (dioxyacetone ($CH_2OH . CO . CH_2OH)_2$), for diabetic patients. It is of German origin.

Oxaphor. Oxycamphor, $C_9H_{15}.CO_2H$ (*q.v.*), in 50 per cent. solution. Used internally for all kinds of respiratory diseases.

Oxaprozin. An anti-inflammatory currently undergoing clinical trial as " Wy 21/743 ". It is 3-(4,5-diphenyloxazol-2-yl)propionic acid.

Oxatets. A proprietary preparation containing Oxytetracycline. An antibiotic. (248).

Oxazepam. 7-Chloro-1,3-dihydro-3-hydroxy - 5 - phenyl - 1,4 - benzodiaze - pin-2-one.

Oxeladin. 2-(2-Diethylaminoethyoxy)-ethyl-$\alpha\alpha$-diethylphenylacetate 2-(2-Diethylaminoethoxy)ethyl α - ethyl - α - phenylbutyrate. Pectamol is the citrate.

Oxethazaine. 2-Di-[($\alpha\alpha$N-trimethylphenethylcarbamoyl)methyl] - aminoethanol.

Oxilube. Polyoxyalkylene diols and derivatives. (553).

Oxitex. Textile fibre lubricant. (553).

Oxitol. 2-Ethoxyethanol. (553).

Oxford Ochre. See YELLOW OCHRE.

Oxford Yellow. See YELLOW OCHRE.

Oxidation Black. Aniline black (*q.v.*).

Oxide of Copper, Red. See HEMI-OXIDE OF COPPER.

Oxide of Iron, Dark. See INDIAN RED.

Oxide of Iron, Magnetic. See MAGNETIC IRON ORE.

Oxide of Iron, Pale. See INDIAN RED.

Oxide of Iron, Violet. See INDIAN RED.

Oxide of Lead, Red. See RED LEAD.

Oxide of Manganese, Red. See HAUSMANNITE.

Oxide, Red. See INDIAN RED.

Oxide, Resorcinyl. See RESORCINYL OXIDE.

Oxide Yellow. See OCHRE.

Oxidised Linseed Oil. See SOLIDIFIED LINSEED OIL.

Oxidised Oil of Turpentine. See DE-CAMPHORISED OIL OF TURPENTINE.

Oxidised Oils. See BLOWN OILS.

Oximony. A proprietary red oxide of iron used as a rubber pigment.

Oxine. 8-Hydroxyquinoline. An analytical reagent for metal analysis.

Oxiosol. See ISOFORM.

Oxi-tan. A proprietary tanning compound.

Oxolin (Perchoid). A patented material made from oxidised oil, jute fibre, and sulphur. It is a rubber substitute.

Oxolinic Acid. An anti-proteus agent. 5 - ethyl - 5, 8 - dihydro - 8 - oxo - 1, 3 - dioxolo [4, 5 - g] quinoline - 7 - carboxylic acid.

Oxone (Oxolin). Compressed sodium peroxide, Na_2O_2. Used in washing powders.

Oxonite. An explosive. It is made from 54 per cent. of nitric acid (specific gravity 1·5), and 46 per cent. of picric acid.

Oxpentifylline. A vasodilator. 3,7-Dimethyl - 1 - (5 - oxohexyl) xanthine. PENTOXIFYLLINE, TRENTAL.

Oxprenolol. A beta-adrenergic receptor blocking agent. It is 1-(*o*-allyloxyphenoxy) - 3 - isopropylaminopropan -2-ol. TRASICOR is the hydrochloride.

Oxyacetone (Ketol, Hydroxyacetone). Acetol, $CH_3CO.CH_2OH$.

Oxyalizarin. Purpurin, $C_{14}H_5O$.

Oxyammonia. Hydroxylamine, NH_2OH.

Oxyanthracene. Anthrol, $C_{14}H_{10}O$.

Oxybuprocaine. 2-Diethylaminoethyl 4 - amino - 3 - butoxybenzoate. Novesine is the hydrochloride.

Oxycamphor. Campholenic acid, $C_9H_{15}.CO_2H$.

Oxycarbonate of Bismuth. See BISMUTH SUBCARBONATE.

Oxycel. A proprietary preparation of oxidised cellulose. Used to stop hæmorrhage. (264).

Oxychinaseptol (Diaphtherine). A compound of hydroxy-quinoline with phenol-sulphonic acid. An antiseptic.

Oxychloride. A disinfectant. It is a solution of sodium hypochlorite, containing 10–12 per cent. available chlorine.

Oxychloride of Tin. See TIN SALTS.

Oxycholine. Muscarine, $C_5H_{15}O_3N$.

Oxycinchophen. 3-Hydroxy-2-phenyl-cinchonic acid.

Oxyclozanide. An anthelmintic. $3,3',5,5',6$ - Pentachloro - $2'$ - hydroxysali - cylanilide. ZANIL.

Oxycodone. 7,8-Dihydro-14-hydroxy-codeinone. Dihydrohydroxycodeinone. Eucodal is the hydrochloride. Proladone is the pectinate.

Oxycymol. Carvacrol, $C_{10}H_{14}O$.

Oxydase. An oxidising enzyme.

Oxydasine. Consists mainly of a 0·05 per cent. solution of vanadic acid. An antiseptic recommended for wounds.

Oxydiamine Black N. See DIAMINE BLACK-BLUE B.

Oxydiamine Black NR. See DIAMINE BLACK-BLUE B.

Oxydiamine Black SOOO. See DIAMINE BLACK-BLUE B.

Oxydiamine Deep Black N. See DIAMINE BLACK-BLUE B.

Oxydiamine Orange G and R. Polyazo dyestuffs. They are substantive cotton dyes, but can also be employed on wool and silk.

Oxydimorphine. See DEHYDROMORPHINE.

Oxydislin. A compound of silicon having the formula Si_2H_2O, prepared by treating calcium silicide with cold dilute alcoholic hydrochloric acid in the dark. It is a white solid spontaneously inflammable in air.

Oxydon. A proprietary preparation containing Oxytetracycline. An antibiotic. (368).

Oxydpech. Oxide pitch obtained as a residue from the distillation of the fatty acids obtained from the oxidation of paraffin.

Oxydrot F-140. See also EISENOXYDROT. A synthetic red oxide of iron. A pigment.

Oxyfedrine. A coronary vasodilator. L - 3 - [(β - Hydroxy - α - methyl - phenethyl) aminol] - $3'$ - methoxypro - piophenone.

Oxygen, Active. Ozone, O_3.

Oxygenated Oil. Olive oil through which chlorine has been passed for several days.

Oxygenated Paraffin. Parogen (q.v.).

Oxygenated Water. An old name for hydrogen peroxide, H_2O_2.

Oxygen Cubes. Made by mixing sodium peroxide and bleaching powder together and compressing into tablets. They contain 100 parts of bleaching powder (33–35 per cent. available chlorine), and 39 parts of sodium peroxide. On contact with water, oxygen is evolved.

Oxygenite. A mixture of perchlorates or nitrates with a substance. When ignited the mixture produces oxygen, and the material is used for this purpose.

Oxygen Powder. Sodium peroxide, Na_2O_2.

Oxygen, Solid. Oxygen cubes (q.v.).

Oxyguard. A proprietary anti-oxidant. 2, 6 - di - tertiary - butyl - p - cresol. (435).

Oxyhæmoglobin. Hæmoglobin.

Oxyliquit. A blasting explosive. It is formed by rapidly mixing liquid air, rich in oxygen, with powdered charcoal, petroleum residues, or cotton wool.

Oxylith. A compressed powder. It is a mixture of sodium peroxide and bleaching powder, which evolves oxygen on treatment with water. Also see OXYGEN CUBES.

Oxymel. Clarified honey (80 per cent.), mixed with acetic acid (10 per cent.), and water (10 per cent.).

Oxymesterone. 4,17β-Dihydroxy-17α-methylandrost - 4 - en - 3 - one. 4 - Hydroxy-17α-methyltestosterone. Oranabol.

Oxymetazoline. 2 - (4 - t - Butyl - 3 - hydroxy - 2,6 - dimethylbenzyl) - 2 - imidazoline. Afrin and Hazol are the hydrochloride.

Oxymetholone. 17β - Hydroxy - 2 - hydroxymethylene - 17α - methyl - 5α - androstan-3-one. 4,5α-Dihydro-2-hydroxymethylene - 17α - methyltestosterone. Adroyd. Anapolon.

Oxymicin. A proprietary preparation containing Oxytetracycline Dihydrate. An antibiotic. (363).

Oxymorphine. See DEHYDROMORPHINE.

Oxymorphone. 7,8 - Dihydro - 14 - hydroxymorphinone. Dihydroxymorphinone. Numorphan is the hydrochloride.

Oxymuriate. Chlorate.

Oxymuriate of Lime. See BLEACHING POWDER.

Oxymuriate of Tin. See BUTTER OF TIN and TIN SPIRITS.

Oxymuth Saca. A colloidal suspension of bismuth oxyhydrate.

Oxyneurine. Betaine, $C_5H_{11}NO_2$.

Oxynitro Zanibeletti. Colloidal hydrated manganese peroxide. It is used in the treatment of cholera.

Oxynone. A proprietary trade name for a rubber antioxidant. It is 2, 4-diamino-diphenylamine.

Oxyntin. A combination of albumin and 5 per cent. of hydrochloric acid.

Oxypertine. 1-[2-(5,6-Dimethoxy-2-methylindol - 3 - yl)ethyl] - 4 - phenyl - piperazine.

Oxyphenbutazone. 4-n-Butyl-2-(4-hydroxyphenyl) - 1 - phenylpyrazoli - dine-3,5-dione.

xyphencyclimine. 1,4,5,6-Tetrahydro-1-methylpyrimidin-2-ylmethyl α-cyclohexyl-α-hydroxy-α-phenyl-acetate. Daricon and Naridan are the hydrochloride.

xyphenine. Direct cotton dyestuffs. (530). See CHLORAMINE YELLOW.

xyphenine Gold. Similar to Chlormine yellow (q.v.)

xyphenisatin. 3,3-Di-(4-hydroxyphenyl)oxindole. Bydolax and Contax are the diacetate.

xyphenonium Bromide. 2-Diethylaminoethyl α-cyclohexyl-α-hydroxy-α-phenyl-acetate methobromide.

xypurinol. An inhibitor of xanthine oxidase currently undergoing clinical trial as " B.W.55–5 ". It is 1H-pyrazolo [3, 4 - d] pyrimidine - 4, 6 - diol.

xyquin. See QUINOSOL.

xytetracycline. 4-Dimethylamino-1, 4,4a,5,5a,6,11,12a - octahydro - 3,5,6,10, 12,12a - hexahydroxy - 6 - methyl - 1, 11 - dioxonaphthacene - 2 - carboxamide. An antibiotic. OXYDON, TERRAMYCIN. It is present as the hydrochloride in BISOLVOMYCIN.

xytoluol. Cresol, $CH_3C_6H_4OH$.

xytri. Polymerised β-hydroxy-trimethylene sulphide.

xzone. Hydrogen peroxide. (513).

yamalite. A mineral. It is a variety of zircon.

zalid. A copying paper of German manufacture. It gives positive prints (dark lines on white ground), which is an advantage over blue prints.

zamin. Benzopurpurin.

zarkite. A mineral. It is a variety of thomsonite.

Ozark White. A pigment used in mixed paints. It usually consists of approximately 60 per cent. zinc oxide and 40 per cent. lead sulphate.

Ozia Gum. Copal gum, derived from *Daniella ogea*.

Ozoform. A formaldehyde soap obtained from sulpho-ricinoleic acid. A disinfectant.

Ozogen. Hydrogen peroxide, H_2O_2.

Ozokerine (Adepsine, Chrisma, Chrismaline, Chrysine, Cosmoline, Fossiline, Geoline, Petrolina, Saxoline, Paraffin Jelly). Trade names applied to varieties of soft paraffin. Also see PETROLEUM JELLY.

Ozokerite (Mineral Wax, Earth Wax, Cerasin, Cerosin, Cerin, Fossil Wax). The solid residue left when petroleum evaporates, which occurs mixed with earth in most oil-bearing districts. It consists of hydrocarbons of the olefine series, and is chiefly used for the preparation of ceresine. The crude wax melts at 79·5° C., has no saponification

value, no acid value, and an iodine value of 7·8.

Ozole. A term applied to volatile aromatic odours contained in certain dextrines and plant extracts. " Dextrinozole " is the name applied to the body giving the scent of commercial dextrin.

Ozonic Ether (Ethereal Solution of Hydrogen Peroxide). A solution of hydrogen peroxide, H_2O_2, in ether. It contains 1·2 per cent. H_2O_2, and has been used in scarlet fever and whooping cough.

Ozonised Wool. Wool which has been treated with slightly ozonised air after it has been treated with dilute ammonia or other suitable solution. The affinity of the wool for dyestuffs at low temperatures is increased and the strength and lustre improved.

Otonite. A trade mark for a preparation containing soap, soda, sodium ailicate, and perborate. A washing agent.

P

P3. A proprietary degreasing material. It is a mixture of water-glass and tri-sodium phosphate in solid form. It is used for cleaning metals, glass, and textiles. It is stated to have no corrosive action on aluminium, aluminium alloys, tin, zinc, and brass.

P 13 N. A trade mark for a range of polyimide varnishes and coatings. (783).

P-51. A proprietary name for dibasic lead stearate used as a heat stabiliser for PVC. (706).

P-289. A proprietary name for a PVC plasticiser of the lead stearate type. (894).

P-2003-K and P-2020-T. Low-density polyethylenes, of Soviet origin.

Paalsgaard Oil. See SCHOU OIL.

Pabracort. A proprietary preparation of HYDROCORTISONE acetate. A nasal decongestant spray. (330).

P.A.C. A proprietary manufacture of formaldehyde.

Pacatal. A proprietary trade name for the hydrochloride or the acetate of Pecazine. A sedative. (262).

Pacca Nuts. See KUSAMBI NUTS.

Pacherite. A mineral, $BiVO_4$.

Pachnolite. A mineral. It is a fluoride of aluminium, calcium, and sodium, $NaF.CaF_2.AlF_3.H_2O$.

Pacific Blue. A dyestuff, $C_{58}H_{49}N_6$. Dyes wool and cotton greenish-blue.

Pacite. A mineral. It approximates to $Fe_5S_2As_8$.

Packfong. A nickel silver. It contains from 26–44 per cent. copper,

16–37 per cent. zinc, and 32–41 per cent. nickel. One alloy contains 40·4 per cent. copper, 25·4 per cent. zinc, 31·6 per cent. nickel, and 2·6 per cent. iron. See NICKEL SILVERS.

Packtong. See NICKEL SILVERS.

Paco (Pacos). A Peruvian term for a ferruginous earth containing small quantities of metallic silver.

Pacolin and Pacoline. Disinfectants and antiseptics. (552).

Pacolol. A preparation of the lysol class.

Pacos. See PACO.

Pacyl. A German product. It is a choline derivative for reducing the blood pressure in arterosclerosis.

Pædo Sed. A proprietary preparation of dichloralphenazone and paracetamol. A hypnotic. (330).

Pæonine (Coralline Red, Aurine Red, Rosophenoline, Aurine R, Coralline). A dyestuff. It is the sodium salt of the reaction product of alcoholic ammonia upon aurine. Dyes silk and wool.

Pagodite. See AGALMATOLITE.

Paigeite. A mineral. It is $Mg_2Fe^{...}BO_5$.

Painterite. A mineral. It is a variety of vermiculite.

Painter's Naphtha. A petroleum distillate. It has a boiling-point of 105–200° C.

Paint, Waterproof. See MOUNTFORD'S PAINT.

Paint, White. See BISMUTH WHITE.

Paisamel Bituminous Solution. Coating for ships bottoms and bunkers. (581).

Paisbergite. A mineral. It is a manganese silicate.

Pakfong. See NICKEL SILVERS.

Pakolin. A proprietary trade name for a synthetic varnish paper product used for electrical insulation.

Palacheite. A mineral. It is datolite.

Paladac. A proprietary preparation containing Vitamin A palmitate, calciferol, thiamine, riboflavine, pyridoxine, nicotinamide and ascorbic acid. (264).

Pala Gum (Indian Gutta-percha). A product from a Ceylon tree. The coagulated juice resembles gutta-percha.

Palaite. A mineral. It is a hydrated manganese phosphate.

Palamoll 632. A polyester of adipic acid and propanediol. A plasticiser for PVC. (49).

Palamoll 644 and 646. A polyester of adipic acid and butanediol. A plasticiser for PVC. (49).

Palamoll 645 and 647. Proprietary polyadipates having viscosities of 6000 mPa.S and 10,000 mPa.S at 20° C respectively. They are used as polymeric plasticisers. (49).

Palamoll 855. A proprietary polymeric plasticiser having a viscosity of 5000 mPa.S. (49).

Palao Amarillo. A rubber obtained from the Mexican *Euphorbia fulva*. It has a high resin content.

Palaprin. A proprietary preparation of aloxiprin. An analgesic. (265).

Palatal. A trade name for an unsaturated polyester resin. (49).

Palatal KR 1397. A proprietary unsaturated polyester based on chlorendic acid dissolved in monostyrene. It is used in the manufacture of articles made of glass-reinforced plastics. (149).

Palatal P5. A proprietary unsaturated polyester resin dissolved in monostyrene. It has low viscosity and is used in the manufacture of articles made of glass-reinforced plastics. (49).

Palatal P8, P50T and P52TL. Proprietary polyester resins used in the manufacture of articles made of glass-reinforced plastics. (49).

Palatal S333. A proprietary polyester resin used in the manufacture of articles made of glass-reinforced polyesters. (49).

Palatine Black (Wool Black 4B and 6B, Toluylene Black G). A dyestuff. It is the sodium salt of p-sulpho-benzene-azo-disulpho-amino-naphthol-azo-naphthalene, $C_{26}H_{16}N_5S_3O_{10}Na_2$. Dyes wool and silk black.

Palatine Chrome Black. A dyestuff. It is equivalent to Palatine chrome blue.

Palatine Chrome Blue (Eriochrome, Blue-black R, Chrome Fast Cyanine B, BN, Alliance Chrome Blue-black R, Fast Chrome Cyanine 2B, Palatine Chrome Black, Stellachrome Black L757, Diadem Chrome Blue-black P6B, Lighthouse Chrome Cyanine R, Era Chrome Dark Blue B). A dyestuff. It is obtained by combining diazotised 1-amino-2-naphthol-4-sulphonic acid with β-naphthol.

Palatine Chrome Brown. A dyestuff prepared from diazotised o-amino-phenol-p-sulphonic acid, and m-phenylene-diamine.

Palatine Orange. A dyestuff. It is the ammonium salt of tetranitro-γ-diphenol. Dyes wool and silk from an acid bath.

Palatine Red (Naphthorubine). A dyestuff. It is the sodium salt of α-naphthalene-azo-α-naphthol-disulphonic acid, $C_{20}H_{12}N_2S_2O_7Na_2$. Dyes wool bluish-red.

Palatine Scarlet (Cochineal Scarlet PS, Brilliant Cochineal). A dyestuff. It is the sodium salt of m-xylene-azo-naphthol-disulphonic acid, $C_{18}H_{14}N_2S_2O_7Na_2$. Dyes wool scarlet from an acid bath.

alatinit. A mixture of sodium hydrosulphite (Blankit) and zinc dust. A bleaching agent.

alatinol. A trade mark for PVC plasticisers. (49).

alatinol A. A registered trade mark for plasticiser for cellulose lacquers. It is stated to be diethyl-phthalate, $C_6H_4(COO.C_2H_5)_2$. It dissolves cellulose nitrate, ester gum, coumarone, etc., has a specific gravity of $1 \cdot 12 - 1 \cdot 13$, and a boiling range of $290 - 300°$ C.

alatinol AH. Di-2-ethyl benzylphthalate and dioctylphthalate. PVC plasticisers for general application. (49).

alatinol BB. A proprietary trade name for benzyl butyl phthalate. (49). (3).

alatinol C. Di-butyl phthalate. A plasticiser.

alatinol 1C. Di-isobutyl-o-phthalate. A non-volatile softening agent for cellulose esters. A trade mark.

alatinol D10. Di-iso-octylphthalate. A plasticiser for PVC. (49).

alatinol DN. Di-iso-nonyl phthalate. A plasticiser for PVC. (49).

alatinol K. Di-butylglycol phthalate. A plasticiser for PVC. (49).

alatinol M. A non-volatile softening agent for cellulose esters. It is di-methyl-o-phthalate. A trade mark.

alatinol O. Dimethyl glycol phthalate. (3).

alatinol Z. Di-iso-decyl phthalate. A plasticiser for PVC. (49).

Palatone. A proprietary trade name for an acrylic denture material.

Palau. A platinum substitute. It is an alloy of gold and palladium, usually 80 per cent. gold and 20 per cent. palladium. Another alloy termed Palau contains 60 per cent. nickel, 20 per cent. platinum, 10 per cent. palladium, and 10 per cent. vanadium.

Pale Acid. Nitric acid containing less than $0 \cdot 1$ per cent. nitrogen oxides.

Pale Cadmium. See CADMIUM YELLOW.

Pale Catechu. See CUTCH.

Pale Cinchona. See CINCHONA, PALE.

Pale Cobalt Violet. A pigment. It is cobalt arsenite.

Pale Lemon-yellow. A pigment. It is a chromate of barium.

Pale Oils. A name applied to a distillate from the residue of petroleum which has been treated with acid and soda, washed or filtered to a certain degree of refining or colour. They have a light and medium viscosity and are employed as lubricants for rapidly moving machinery.

Pale Oxide of Iron. See INDIAN RED.

Pale Smalt. See SMALT.

Pale Yellow Gold. Alloys. One contains $91 \cdot 6$ per cent. gold and $8 \cdot 3$ per cent. silver, and another $91 \cdot 6$ per cent. gold and $8 \cdot 3$ per cent. iron.

Paleva. A proprietary preparation of aspirin and PARACETAMOL. An analgesic. (633).

Palfium. A proprietary preparation of dextromoramide. An analgesic. (249).

Palite. Chloro-methyl-chloro-formate, $COCl.OCH_2Cl$. A military poison gas.

Palladium Asbestos. An asbestos coated with palladium used in gas analysis for the absorption of hydrogen.

Palladium Black. A finely divided palladium used as a catalyst in the hydrogenation of oils.

Palladium Gold (White Gold). An alloy of 90 per cent. gold and 10 per cent. palladium. An alloy of 40 per cent. copper, 31 per cent. gold, 19 per cent. silver, and 10 per cent. palladium, is also known by this name.

Palladium Red. Ammonio-chloride of palladium. A red pigment.

Pallamine. A colloidal solution of palladium. Recommended in the treatment of gonorrhœa.

Pallicid. Sodium tribismuthyl-tartrate.

Pallite. A mineral. It is $Ca_3Al_{13}(PO_4)_8(OH)_{18}.6H_2O$.

Palma Christi Oil. See OIL OF PALMA CHRISTI.

Palmarosa Oil. See ROSÉ OIL.

Palm Butter. Palm oil.

Palmerite. A mineral. It is an aluminium-potassium phosphate.

Palmer's Wood Composition. A preparation consisting of sawdust, mixed dried blood, bone dust, and glue solution.

Palmetto. A palm-like shrub used to produce a tannin extract called Palmetto extract which contains from 5–12 per cent. tannin.

Palmiacol. See CETYLGUAIACYL.

Palmierite. A mineral, $K_4Na_2Pb_4(SO_4)_7$. It is also given as $[(K, Na)_2 Pb(SO_{42}]$.

Palmine. See VEGETABLE BUTTER.

Palmitin. Commercial palmitic acid is incorrectly called by this name.

Palm Oil, Para. See PARA PALM OIL.

Palm Pitch. A pitch obtained by the treatment of palm oil with sulphuric acid.

Palm Spirit. Arrack (q.v.).

Palm Wax. A yellow wax from *Ceroxylon andicola*. Used as a beeswax substitute.

Palm Wine. The fermented juice of the sugar palm.

Palorium. A platinum substitute. It is a white alloy of gold and platinum only distinguished from platinum with difficulty. It is a ductile, homogeneous alloy with a melting-point of $1310°$ C., and it remains stronger than platinum on heating. It is not quite so resistant as platinum to dilute acids and alkalis, but is more resistant to concentrated sulphuric acid and caustic melts, and it is 45 per cent. lower in price.

Paludrine. A proprietary preparation of proguanil hydrochloride. An anti-malarial. (2).

α-Palygorskite. Synonym for Lassallite.

Pamergan. A proprietary preparation of pethidine hydrochloride and promethazine hydrochloride. An analgesic. (336).

Paminal. A proprietary preparation of hyoscine methobromide and phenobarbitone. (325).

Pamine. A proprietary preparation of methscopolamine bromide. A gastro-intestinal sedative. (325).

Pamn. The methonitrate of Pramine.

Pamol. A proprietary preparation of paracetamol. An analgesic. (357).

Panacur. A proprietary preparation of FENBENDAZOLE. A veterinary anthelmintic.

Panadeine CO. A proprietary preparation of paracetamol and codeine phosphate. An analgesic. (112).

Panadol. A proprietary preparation of paracetamol. An analgesic. (112).

Panadonin. A preparation of *Adonis davurica*, a Japanese herb. Advocated as a substitute for digitalis.

Panaflex BN-1. A proprietary trade name for a hydrocarbon type vinyl plasticiser. (118).

Panaflex BN-2. A proprietary trade name for a petroleum derivative used as a vinyl plasticiser. (118).

Panama Bark. Quillaia bark.

Panama Crimson. A colouring matter obtained from the leaves of a vine called " china." Employed for dyeing straw hats.

Panar. A proprietary preparation containing pancreatin. (327).

Panase. A combination of digestive enzymes of the pancreas, derived from the pancreatic glands of the pig.

Panasorb. A proprietary preparation of paracetamol in a sorbitol base. An analgesic. (112).

Panazon. A proprietary preparation of NITROVIN hydrochloride. A veterinary growth promoter.

Panclastite. Explosives. They are mixtures of liquid nitrogen tetroxide with carbon disulphide, ether, benzene, or nitro-benzene, petroleum, and other hydrocarbons. The name is also used for certain potassium chlorate mixtures.

Pancortex. An extract of the suprarenal cortex intended for the treatment

of disturbances of the suprarenal func-tion.

Pancoxin (1). A proprietary preparatio containing sulphaquinoxaline. A vetei inary coccidiostat.

Pancoxin (2). A proprietary preparatio of AMPROLIUM. An anti-protozoan fc veterinary use.

Pancoxin (3). A proprietary preparatio of ETHOPABATE. A veterinary anti protozoal.

Pancreas Diastase. Amylopsin, a enzyme.

Pancreatin Granules. A proprietar preparation containing pancreatin trip] strength. (328).

Pancreatokinase. A mixture o Eukinase (*q.v.*), and pancreatin.

Pancreol. A trypsin preparation, use(for bating skins.

Pancreon. A pancreatic preparation, ii which the pancreatin is not destroye(by the pepsin of the stomach, becaus it is converted into a form insoluble ii water and acids by treatment with tan nic acid. It contains 8 per cent. tanni(acid, and is said to have trypsilytic amylolytic, and emulsifying properties

Pancreozymin. A hormone obtained from duodenal mucosa.

Pancrex. A proprietary preparation o pancreatin used in the treatment o cystic fibrosis. (329).

Pancrex V. A proprietary preparatior of concentrated pancreatin used in the treatment of cystic fibrosis. (329).

Pancuronium Bromide. A neuro-muscular blocking agent. 3α, 17β-Di-acetoxy - 2β, 16β - dipiperidino - 5α -androstane dimethobromide. PAVULON

Pandermite. A mineral which is a source of borax. It has the formula $Ca_2O.2B_2O_3.4H_2O$, and is used in ceramic frits.

Pandex. A selenium preparation for rubber vulcanisation.

Panelyte. A proprietary trade name for phenol-formaldehyde laminated products and paper, fabric, wood veneer, fibre glass, and asbestos base ther-mosetting plastics for structural work.

Panflavin. A remedy for influenza having as an active ingredient trypaflavine.

Panidazole. An amœbicide. 2-Methyl-5 - nitro - 1 - [2 - (4 - pyridyl) ethyl] -imidazole.

Panilax. A registered trade name for materials made from aniline-formalde-hyde synthetic resin. They are ther-moplastic but have a softening point about 100° C.

Panitrin. Papaverine nitrate in acetyl-diethyl-amide.

Pankreon. See PANCREON.

Panmycin. A proprietary preparation of tetracycline. An antibiotic. (325).

ennantite. A mineral. It is Mn_9Al_6 $Si_5O_{20}(OH)_{16}$.

annetier Green. See CHROMIUM GREEN.

anno-di-morti Marble. A black marble containing fossil shells.

anok. A proprietary trade name for Paracetamol.

an Scale. The calcium sulphate, containing some sodium chloride, which settles out during the crystallisation of salt from brine. It is sold as " salt lick " for cattle, also for manuring purposes.

anoxyl 5 and 10. Proprietary preparations of benzoyl peroxide used in the treatment of acne. (379).

ansecretin. See DUODENIN.

antakaust. A fuel similar to meta.

antal. An aluminium alloy containing 0·8–2 per cent. magnesium, 0·4–1·4 per cent. manganese, 0·5–1 per cent. silicon, and 0·3 per cent. titanium. It resists corrosion, and has a tensile strength of from 18–33 kg. per sq. mm.

antal. An aluminium containing 0·2 per cent. titanium.

antarol. A proprietary product for the protection of metal. It is applied by brushing, spraying, and dipping. It resists the action of light, sea air, steam, acid fumes, but is destroyed by concentrated acids and alkalis.

anteric. A proprietary preparation containing pancreatin (triple B.P. strength). (264).

anthenol. A preparation used in the treatment of paralytic ileus and postoperative distension. It is ($\pm$)-2,4-dihydroxy - N (3 - hydroxypropyl) - 3, 3 - dimethylbutyramide.

antothenic Acid. Co-enzyme A. Part of the Vitamin B complex. No deficiency disease known.

anthesin. A salt of p-amino-benzoyl-N-diethyl-leucinol.

antocaine. A local anæsthetic. It is 4 - butyl - amino - benzoic acid - β - di - methyl - amino - ethyl ester hydrochloride, $C_4H_9 . NH . C_6H_4 . CO . O . CH_2.CH_2.N(CH_3)_2$.

antolit. A proprietary synthetic resin of the phenol-formaldehyde type.

antopon (Omnopon, Nealpon). Mixtures of the soluble hydrochlorides of opium alkaloids.

antosept. A German antiseptic in which the active agent is hypochlorous acid.

antothenyl Alcohol. Panthenol.

anturon. A proprietary preparation containing atropine sulphate, papaverine hydrochloride phenobarbitone, aluminium hydroxide gel, kaolin, magnesium carbonate, magnesium trisilicate and oil of peppermint. An antacid. (270).

Panza Nuts. See OWALA BEANS.

Paoferro. The inner bark of the Brazilian ironwood tree. Used as an antidiabetic.

Papaw. The seeds of *Asimina tribola*. An emetic.

Papain. See PAPAYOTIN.

Papaveroline. A vasodilator. 1-(3,4-Dihydroxybenzyl) - 6, 7 - dihydroxyiso - quinoline. It is present in UTEN CAPSULES, and in UTEN INJECTION as the 6'-sulphonic acid.

Papayotin (Papain, Papoid). Papain, a vegetable digestive ferment obtained from the unripe fruit of the papaw tree. Has been used for dyspepsia.

Papelac. A trade name for an air drying vinyl coating for packaging and other papers. (254.)

Paper-clay. See BENTONITE.

Paper-coal. A synonym for Dysodil (*q.v.*).

Paper, Fish. See LEATHER PAPER.

Paperhanger's Alum. Aluminium sulphate, used for sizing paste.

Paperine. A starch product used in paper manufacture.

Paper, Leather. See LEATHER PAPER and LEATHEROID.

Paper, Mountain. See MOUNTAIN LEATHER.

Paper Oil. See OIL OF PAPER.

Paper Scarlet Blue Shade. See BRILLIANT CROCEINE M.

Paper-spar. A mineral. It is a variety of calcite.

Paper, Tetra-base. See TETRA-PAPER.

Paper, Vulcanised. See VULCANISED FIBRE.

Papier-maché. A material made from paper pulp and binding substances.

Papite. A tear gas. It is acrolein with stannic chloride.

Papoid. See PAPAYOTIN.

Paposite. A mineral. It is a ferric sulphate.

Pappenheim's Stain (Pyronin Stain). A microscopic stain. It consists of 1 part of a concentrated solution of pyronin and 3 parts of a concentrated solution of methyl green.

Paprika. Cayenne pepper.

Papua Nutmeg Butter. See MACASSAR NUTMEG BUTTER.

Papyrine. See PARCHMENT PAPER.

Papyrus. A proprietary casein product.

Par. See CATALPO.

Para Arrowroot. Tapioca.

Parabal. A proprietary preparation of phenobarbitone and dihydroxyaluminiumamino acetate. (137).

Parabayldonite. A mineral. It is a basic lead-copper arsenate, of Southwest Africa.

Para Bismut. Bismuth paranucleinate.

Para Blue. A basic induline dyestuff,

obtained by heating Spirit blue with p-phenylene-diamine. Dyes tannined cotton greyish-blue.

Parabutlerite. A mineral. It is Fe''' SO$_4$OH.2H$_2$O.

Para Butter. See PARA PALM OIL.

Paracarmine. A microscopic stain. It contains 1 gram carminic acid, 0·5 gram aluminium chloride, 4 grams calcium chloride, and 100 c.c. 70 per cent. alcohol.

Paracelsian. A mineral. It is 4[BaAl$_2$Si$_2$O$_8$].

Paracetamol. 4-Acetamidophenol. Calpol. Eneril. Febrilix. Panadol. Panok. Tabalgin.

Parachlor. Parachlormetacresol. p-chlor-m-cresol. A germicide.

Para, Coarse. See SERNAMBY.

Paracodeine. Dihydro-codeine hydro-chloride.

Paracodol. A proprietary preparation of paracetamol and codeine phosphate. An analgesic. (188).

Paracolline. A rubber solution.

Paracolumbite. A mineral. It is a variety of ilmenite.

Paracon. A generic name for polyester elastomers. See PARAPLEX.

Paracoquimbite. A mineral. It is a hydrated ferric sulphate, from Troja, near Prague (Fe$_2$O$_3$=21·7 per cent.).

Paracoto Bark. Coto bark.

Paracoumarone Resin (Cumar Resin, Cumar Gum, Coumarone Resin, Benzo-Furane Resin, Cumar). A synthetic resin produced from coal-tar distillates. Solvent naphtha distilling between 150 and 200° C. is used, and is polymerised with sulphuric acid, or aluminium chloride, tin chloride, ferric chloride, and phosphoric acid. It contains p-coumarone, p-indene, and polymers of other hydrocarbons. It is unaffected by most acids and alkalis, and if an excess of sulphuric acid is used in its preparation it gives an insoluble and infusible resin. A medium hard cumar resin melts at 104·5° C., has an acid value of 0, a saponification value of 0, and an iodine value of 53·4. The resin is used in the production of varnishes, polishes, artificial leather, and linoleum.

Paracril J4940. A proprietary buta-diene-acrylonitrile rubber used in the manufacture of oil hydraulic hoses for use in automobiles to contain a special stabiliser system to reinforce the material against the effects of heat and oil. (287).

Paradene. A proprietary trade name for coumarone-indene resins.

Paradione. A proprietary preparation of paramethadione. An anticonvulsant. (311).

Paradise Grains. See GRAINS OF PARADISE.

Paradol. See NERADOL.

Paradoxite. A mineral. It is a variety of orthoclase.

Paradrin. See HOMORENON.

Paradura. A proprietary trade name for a phenolic synthetic resin for varnish and lacquers.

Paradyn. Antipyrine (2:3-dimethyl-1-phenyl-5-pyrazolon). Used in the treatment of fever, neuralgia, and rheumatism.

Parafecol. A paste containing phenol. A disinfectant.

Paraffagar. A proprietary preparation of liquid paraffin and agar-agar.

Paraffin, Brilliant. See BRILLIANT PARAFFIN.

Paraffin Jelly. See OZOKERINE.

Paraffin, Liquid. A mixture of liquid hydrocarbons obtained by the distillation of the liquid remaining after the lighter hydrocarbons have been removed from petroleum. It is decolorised and purified, and has a specific gravity of 0·86–0·89, and boils about 360–390° C. A refined paraffin melts at 51·7° C. Specific gravity, 0·774. Saponification value, 0. Acid value, 0. The following names (proprietary and otherwise) are some of those used for liquid paraffin and similar preparations. Adepsine Oil, Alboline, Alboline, Amilee, Atoleine, Atolin, Bakurol, Blandine, Crysmalin, Deeline, Glycoline, Glyco, Glymol, Hydrocarbon oil, Interol, Internol, Lilyolene, Mineral glycerin, Mineral syrup, Minol, Muthol, Neutralol, Nujol, Paroleine, Paroline, Petralol, Petro, Petrolax, Petrolia, Petronol, Petrosio, Russol, Saxol, Seneprolin, Stanolax, Stanolind, Terraline, Terroline, Usoline.

Paraffin, Native. Ozokerite (q.v.).

Paraffin Oil. See KEROSENE.

Paraffin, Oxygenated. Parogen (q.v.).

Paraffin Scale. Crude paraffin wax.

Paraffinum Molle. See PETROLEUM JELLY.

Paraffin Wax. The wax obtained from petroleum, and.from bituminous shales.

Parafilm. A proprietary trade name for a rubber composition.

Paraflex. A proprietary preparation of chlorzoxazone. (369).

Paraflow. A synthetic lubricating oil prepared by the condensation in the presence of anhydrous aluminium chloride of chlorinated wax with aromatic hydrocarbons.

Parafon Forte. A proprietary preparation of chlorzoxazone, and paracetamol. (369).

Paraform. See TRIFORMAL. It is also the name for a new local anæsthetic consisting of the butyl ester of p-amino-benzoic acid, $H_2NC_6H_4CO_2C_4H_9$.

araforme. See SCUROFORM.

araformol. See TRIFORMAL.

arafuchsine (Para-magenta). A dye-stuff. It is the hydrochloride of tri-amino - triphenyl - carbinol, $C_{19}H_{26}N_3$ ClO_4. Dyes wool, silk, and leather red, and cotton mordanted with tannin and tartar emetic, red.

aragearksutite. A mineral. It is $CaAl(F,OH)_5 \cdot \frac{3}{4}H_2O$.

aragesic. A proprietary preparation of PARACETAMOL, caffeine and PSEUDO-EPHRINE hydrochloride. A decongest-ant and analgesic. (478).

aragite. A mineral. It is an apatite containing iron.

araglobin. See GLOBULIN.

araglobulin. See GLOBULIN.

aragol. A rubber substitute made from oxidised oil.

aragonite. A mineral. It is sodium-mica, and consists of snow-white scales.

aragon Steel. A proprietary trade name for a non-shrinking steel contain-ing 1·55 per cent. manganese, about 0·6 per cent. chromium, and 0·25 per cent. vanadium.

araguanajuatite. A mineral. It is $3[Bi_2(Se,S)_3]$.

araguay Tea (Yerba maté). The leaves of *Ilex paraguariensis*.

ara Gum. See BRAZILIAN GUM and JUTAHYCICA RESINS.

aragutta. A patented insulating com-pound for use in the manufacture of submarine telegraph and telephone cables. It is made from deproteinised rubber obtained by the heat treatment of rubber latex to hydrolyse the protein and washing. This rubber has reduced water-absorbing properties and is mixed with gutta-percha or balata and suitable waxes to produce paragutta.

arahilgardite. A mineral. It is $2[Ca_8B_{18}O_{33}Cl_4 \cdot 4H_2O]$.

arahopeite. A mineral. It is a hydrated zinc phosphate.

Para-Hypon. A proprietary prepara-tion of paracetamol, caffeine, codeine phosphate and phenolphthalein. An analgesic. (247).

arailmenite. A mineral. It is a variety of ilmenite.

Parajamesonite. A mineral. It is Pb_4 $FeSb_6S_{14}$.

Parake. A proprietary preparation of PARACETAMOL and codeine phosphate. An analgesic. (476).

Paralac. Synthetic resin for varnish, enamel and water paints (phenol formaldehyde type). (512).

Paralactic Acid. Sarco-lactic acid.

Paralaudin. Diacetyl-dihydro-morphine.

Paralaurionite. A mineral. It is $4[PbClOH]$. See RAFAELITE.

Paralene and Paralex. Synthetic tan-nin. (508).

Paralgin. A proprietary preparation of PARACETAMOL, caffeine and codeine phosphate. An analgesic. (462).

Paraloid K12ON and KM-228. A trade mark for acrylic modifiers for PVC. (13).

Paralysol (Solid Cresol). A combination of cresol and potassium, $C_6H_4(CH_3)OK$. $3C_6H_4(CH_3)OH$. A bactericide.

Param. Cyanoguanidine, $C_2H_4N_4$.

Paramagenta. See PARAFUCHSINE.

Paramar. A mineral rubber used in rubber mixing.

Paramelaconite. A mineral. It is $16[CuO]$.

Paramet Ester Gum. A proprietary trade name for rosin-glycerol synthetic resin for lacquer and varnish manufac-ture.

Paramethadione. 5 - Ethyl - 3,5 - di - methyloxazolidine - 2,4 - dione. Para-dione.

Paramethasone. 6α - Fluoro - $11\beta,17\alpha$, 21 - trihydroxy - 16α - methylpregna - 1,4-diene-3,20-dione.

Parametol. Emetine oleate solution in liquid paraffin.

Paramine Black BH. A dyestuff. It is a British equivalent of Diamine black BH.

Paramine Black HW. A dyestuff. It is a British brand of Diamine black HW.

Paramine Blue 2B New. A dyestuff. It is equivalent to Diamine blue 2B, and is of British manufacture.

Paramine Brilliant Black FB, FR. Dyestuffs. They are British equiva-lents of Columbia black FF extra.

Paramine Brown. A dyestuff. It con-sists of p-phenylene-diamine oxidised on the fibre.

Paramine Fast Brown B. A British equivalent of Diamine brown B.

Paramine Fast Brown M. A dyestuff. It is a British brand of Diamine brown M.

Paramine Fast Orange D, G. Dye-stuffs. They are British equivalents of Mikado orange.

Paramine Fast Orange S. A British dyestuff. It is equivalent to Benzo fast orange S.

Paramine Fast Red F. A British brand of Diamine fast red.

Paramine Fast Scarlet 4BS. An equivalent of Benzo fast scarlet 4BS.

Paramine Fast Violet N. A dyestuff. It is a British equivalent of Diamine violet N.

Paramine Fast Yellow 3G. A British brand of Mikado yellow.

Paramine Green B, G. British equivalents of Diamine green G.

Paramine Orange G, R. British brands of Benzo orange R.

Paramine Yellow GG. A British dyestuff. It is equivalent to Nitrophenine.

Paramine Yellow R, 2R, Y. Dyestuffs, They are equivalent to Direct yellow R. and are of British manufacture.

Paramisan. A proprietary preparation of sodium aminosalicylate used in the treatment of tuberculosis. (824).

Paramol. o-Amino-m-hydroxy-benzyl alcohol, $C_6H_3(OH)(CH_2OH)NH_2$. A photographic developer.

Paramol-118. A proprietary preparation of paracetamol and dihydrocodeine bitartrate. An analgesic. (182).

Paramontmorillonite. A mineral. It is a hydrated aluminium silicate.

Paramorphan. Dihydro-morphine hydrochloride.

Paramorphine. Thebaine, $C_{19}H_{21}O_3N$.

Paranephrine. Adrenaline (q.v.).

Paranitraniline Red (Azophor Red, Para Red, Nitrazol, Discharge Lake R and RR). A dyestuff. It is p-nitro-benzene-azo-β-naphthol, $C_{16}H_{11}N_3O_3$. Used for dyeing cotton, and in the preparation of lakes for paper-staining.

Paranol. A proprietary trade name for a phenol-formaldehyde synthetic resin.

Paranolin. A soya-bean casein.

Paranoval. A German preparation. It is an almost tasteless veronal, consisting of equimolecular proportions of diethyl-barbituric acid (veronal), and trisodium phosphate.

Paranox. Detergent-inhibitor lubricating oil additives. (621).

Paranthine. See SCAPOLITE.

Para Palm Oil (Para Butter). Pinot oil, a semi-drying oil from the seeds of *Euterpe oleracea*.

Paraphenylene Blue G, R, and B. Dyestuffs formed by heating p-phenylene-diamine with certain amino-azo compounds.

Paraphenylene Blue R (Fast New Blue for Cotton, Indophenine). A dyestuff obtained by the action of p-phenylene-diamine upon the hydrochloride of amino-azo-benzene. Dyes cotton mordanted with tannin and tartar emetic, blue.

Paraphenylene Violet. A dyestuff obtained by heating amino-azo-naphthalene or benzene-azo-α-naphthylamine with p-phenylene-diamine. Dyes tannined cotton violet.

Paraphthalein. A preparation of phenol-phthalein.

Paraplex. A trade mark for a synthetic resin and a glycol sebacate and other polyester plasticisers. (13).

Paraplex 962. A trade mark. An epoxy type vinyl plasticiser. (13).

Paraplex G57. A proprietary polyester plasticiser for plastics used as refrigerator gaskets. (13).

Paraplex G59. A proprietary polyester plasticiser for plastics used as insulation tapes. (13).

Paraplex X 100. A trade mark for a curable polyester elastomer. It is stated to have good oil and petrol resistance when cured. (13).

Parapoid. Gear oil additives. (621).

Paraquat. It is a herbicide which functions by stifling the chlorophyll uptake by plants. It is 1·1'-dimethyl-4-4'-di-pyridium dichloride. It is fatal to humans.

Pararammelsbergite. A mineral. It is $8[NiAs_2]$.

Para Red. See PARANITRANILINE RED.

Pararosaniline. Triamino-triphenyl-carbinol. Pararosaniline chloride dyes wool and silk, purple-red, and cotton with mordants.

Pararosaniline Base. An equivalent of Rosaniline.

Pararosaniline Blue. See METHYL BLUE.

Pararosilic Acid. Aurine (q.v.).

Paraschœpite. A mineral. It is $10UO_3.19H_2O$.

Para-Seltzer. A proprietary preparation of PARACETAMOL and caffeine. An analgesic. (244).

Parasepiolite. A mineral. It is $Mg_2Si_3O_6(OH)_4$.

Parashade and Parasheen. Dyes. (621).

Paraset 26–31. Proprietary trade names for close-cut high boiling petroleum fractions with an aromatic content giving a Kauri-Butanol value in the 25–30 range. They have freedom from "tail" which ensures greater control of the drying rate. (196).

Parasiticine. A fungicide containing 57 per cent. copper sulphate and sodium carbonate and bicarbonate.

Parasulphurine S. Sulphanil yellow (q.v.).

Paratacamite. A mineral. It is a basic cupric chloride.

Paratartaric Acid. Racemic tartaric acid, $C_4H_6O_6$.

Paratex. A proprietary preparation of thiamine hydrochloride, calcium and strychnine glycerophosphates, and sodium hypophosphate. A tonic. (395).

Parathion. O,O-diethyl O-p-nitrophenyl phosphorothioate. Diethoxy, nitrophenoxy phosphorothioate. A powerful insecticide.

Para-Thor-Mone. A proprietary preparation of parathyroid hormone. (463).

Para Toner. Paranitraniline red. Used as a toner for lakes.

Paratophan (Homophan, Methyl-atophan). 6-Methyl-2-phenyl-quinoline-4-carboxylic acid. Used in medicine.

Paratoxin. Asserted to be an extract of the gall substance, with subsequent distillation with benzine.

Paraurichalcite. A mineral. It is a basic carbonate of copper and zinc, $3(Cu.Zn)CO_3.4(Cu.Zn)(OH)_2$, of Southwest Africa.

Paravauxite. A mineral of Bolivia, $5FeO.4Al_2O_3.5P_2O_5.26H_2O$.

Paravivianite. A mineral. It is $(Fe^{..},Mn,Mg)_3PO_4.8H_2O$.

Parawallastonite. A mineral. It is $12[CaSiO_3]$.

Paraxin. Chloramphenicol.

Paraxin. Dimethyl-amino-1 : 7-dimethylxanthine, $(CH_3)_2N.C(NC.NH.CO)N(CH_3)C.CO.N(CH_3)$. A diuretic.

Parazol. Crude dinitro-dichloro-benzene. It contains m-dinitro-p-dichloro-benzene, o-dinitro-p-dichloro-benzene, and p-dinitro-p-dichloro-benzene. It is used as a high explosive.

Parazolidin. A proprietary preparation of PHENYLBUTAZONE and PARACETAMOL used in the treatment of arthritis. (17).

Parazone. A proprietary trade name for paraphenyl-phenol. An antioxidant. (2).

Parbendazole. An anthelmintic. Methyl 5 - butylbenzimidazol - 2 - ylcarba - mate. HELMATAC.

Parchment Glue. See GLAZING GLUE.

Parchment Paper (Papyrine, Vegetable Parchment). Made by dipping white unsized paper for a few seconds into concentrated sulphuric acid, and washing free from acid in dilute ammonia. Used for jam and pickle jar covers.

Pardale. A proprietary preparation of paracetamol, codeine phosphate and caffeine. An analgesic. (320).

Paredrine. See HYDROXYAMPHETAMINE.

Paredrite (Favas). A mineral. It is a variety of titanium dioxide.

Paregoric (Paregoric Elixir). Compound tincture of camphor (*Tinctura camphoræ composita B.P.*). It contains 50 c.c. tincture of opium, 5 grams benzoic acid, 3 grams camphor, 3 c.c. oil of anise, and sufficient alcohol (60 per cent.) to produce 1,000 c.c.

Paregoric Elixir. See PAREGORIC.

Pareira. The dried root of *Chondrodendron tomentosum*. Used medicinally.

Parel 58. A proprietary self-vulcanisable copolymer of propylene oxide and allyl glycidyl ether. (393).

Parenol. Consists of 65 per cent. soft paraffin, 15 per cent. wool fat, and 20 per cent. distilled water.

Parenol, Liquid. Consists of 70 per cent. liquid paraffin, 5 per cent. white beeswax, and 25 per cent. distilled water.

Parenterovite. A proprietary preparation containing thiamine, riboflavine, pyridoxine, nicotinamide, dextrose, sodium ascorbate, benzyl alcohol and chlorocresol. (272).

Parfenac. A proprietary preparation of BUFEXAMAC used as a skin cream. (306).

Pargasite (Syntagmatite). An amphibole mineral, $(Mg.Fe)_2(Al.Fe)_4(SiO_6)_2$.

Pargonyl. A proprietary trade name for the sulphate of Paromomycin.

Pargyline. N-Benzyl-N-methylprop-2-ynylamine. Eutonyl is the hydrochloride.

Parian Cement. A cement which is similar to Keene's cement, except that a solution of borax is used instead of alum.

Parianite. An asphaltum from the pitch lake at Trinidad.

Parian Marble. A marble obtained from the Isle of Paros.

Parian Ware. Unglazed porcelain, used for the production of statuary.

Parietic Acid. Chrysophanic acid (*q.v.*).

Parilene. A proprietary trade name for polyparaxylyene. A plastics material used for film manufacture for electrical purposes.

Paris Black. Bone black (*q.v.*).

Paris Blue. Finest Prussian blue (*q.v.*). Also see METHYL BLUE. The term is generally applied to a mixture of Prussian blue, Turnbull's blue, and Willow blue.

Paris Cement. Consists of 5 parts gum arabic, 2 parts sugar candy, and white lead. Used for mending sea-shells.

Parisepsin. A proprietary preparation of diphenylchlorethanes and diphenyliodonium chloride. (381).

Paris Glue (Hatter's Glue). Dark brown, soft varieties of glue, used in hat making.

Paris Green. See EMERALD GREEN and METHYL GREEN.

Parisilon. A proprietary preparation of PREDNISOLONE sodium phosphate. (275).

Parisite. A mineral $(CeF)(CaF_2)CeCO_3$, with lanthanum and didymium.

Paris Lake. Carmine lake (*q.v.*).

Paris Red. See RED LEAD. A variety of rouge employed in polishing is also sold under this name.

Paris Salts. A disinfectant containing 50 parts zinc sulphate, 50 parts ammonia alum, 1 part potassium permanganate, and 1 part lime, perfumed with a little thymol.

Paris Violet. See METHYL VIOLET B.

Paris Violet 6B. See BENZYL VIOLET.

Paris White. See CHALK.

Paris Yellow. See CHROME YELLOW, TURNER'S YELLOW, and NAPLES YELLOW.

Parkerised Steel. A patented process for the treatment of steel with iron and manganese phosphates to give the surface resistance to corrosion.

Parkerite. A mineral. It is $Ni_3B_2S_2$.

Parker's Cement. See ROMAN CEMENT.

Parkesine. Celluloid (q.v.).

Parlodion. A trade mark for a shredded form of pure collodion.

Parlon. A proprietary trade name for a chlorinated rubber compound, rubber derivatives and rubber-like resins for use as a base for concrete paint and alkyd enamels.

Parma. See ANILINE BLUE, SPIRIT SOLUBLE.

Parmentine. A mixture of glycerin, gelatin, dextrine, sodium sulphite, and zinc sulphate. Used for sizing and finishing cotton, wool, and silk.

Parme R. See PRUNE, PURE.

Parmetol (Parol). A material containing 40 per cent. chloro-*m*-cresol in 8 per cent. caustic soda. A 5 per cent. solution is used as a disinfectant in tuberculosis expectoration.

Parmr. The trade name for a blown bitumen residue. It is a mineral rubber for use in the rubber industry. Grade I melts as from 190–310° F., and Grade II at above 300° F.

Parnate. A proprietary preparation of tranylcypromine. An antidepressant. (281).

Paroa-caxy Oil. The seed-oil of *Pentaclethra filamentosa*.

Parodyne. Antipyrine (q.v.).

Parogen (Oleogen, Oxygenated Paraffin, Vasogen). Consists of 2 parts liquid paraffin, 2 parts oleic acid, and 1 part ammoniated alcohol (5 per cent.).

Parogen, Thick. Consists of 6 parts hard paraffin, 24 parts liquid paraffin, 15 parts oleic acid, and 5 parts ammoniated alcohol (5 per cent.).

Parol. See PARMETOL.

Par-o-lac. An impregnating compound.

Paroleine. See PARAFFIN, LIQUID.

Paroline. See PARAFFIN, LIQUID.

Paromomyicn. An antibiotic produced by Streptomyces rimosus forma paromomycinus. D-Glucosaminedeoxystreptamine D-ribosediaminohexose. Humatin and Pargonyl are the sulphate.

Parosan. A proprietary preparation. It is a new arsinic acid derivative of low toxicity. It is 8-acetyl-amino-3-hydroxy-1 : 4-benzisoxazine-Δ-6-arsinic acid. It has been used with good results in the treatment of disseminated sclerosis.

Paroven. A proprietary preparation of hydroxyethylrutosides used in the treatment of varicose veins. (724).

Paroxyl. See ACETARSONE.

Parpanit. The hydrochloride of Caramiphen.

Parpevoline. Dimethyl-ethyl-pyridine-hexahydride, $C_9H_{19}N$.

Parr M-4592. A proprietary mineral-filled melamine-formaldehyde moulding compound. (778).

Parraynite. A trade name for a rubber compound which is used by X-ray operators to protect them from injury by exposure to the rays.

Parresine. See AMBRINE.

Parrot Coal. Cannel coal.

Parrot Green. See EMERALD GREEN and ZINC GREENS.

Parr's Alloys. Anti-corrosion alloys. One alloy contains 80 per cent. nickel. 15 per cent. chromium, and 5 per cent, copper. Another alloy contains 66·6 per cent. nickel, 18 per cent. chromium, 8·5 per cent. copper, 3·3 per cent. tungsten, 2 per cent. aluminium, and 1 per cent. manganese.

Parsettensite. A mineral. It is a manganese silicate.

Parsley Camphor (Camphre de Persil). Crystallised apiol.

Parsley Fruit. The fruit of *Carum petroselinum*.

Parsonite. A radioactive mineral of the Belgian Congo. It approximates to the formula, $2PbO.UO_3.P_2O_5.H_2O$.

Parson's Alloy. A proprietary trade name for an alloy of 56 per cent. copper, 41·5 per cent. zinc, 1·2 per cent. iron, 0·7 per cent. tin, 0·1 per cent. manganese, and 0·46 per cent. aluminium. It has a specific gravity of 8·4.

Parstelin. A proprietary preparation of tranylcypromine, and trifluoperazine. (281).

Partagon. A German preparation. It consists of rods containing silver chloride and sodium-silver chloride.

Partigene. A German preparation obtained from lactic acid and tubercule bacilli.

Partinium (Victoria Aluminium). An aluminium alloy. It varies in composition, and often contains tungsten, copper, tin, zinc, and magnesium. One alloy contains 96 per cent. aluminium, 2·4 per cent. antimony, 0·8 per cent. tungsten, 0·64 per cent. copper, and 0·16 per cent. tin. Another alloy consists of 88·5 per cent. aluminium, 7·4 per cent. copper, 1·7 per cent. zinc, 1·3 per cent. iron, and 1·1 per cent. silicon.

Partridge Berry Oil. See OIL OF PARTRIDGE BERRY.

Partridgeite. A mineral. It is $16[Mn_2O_3]$.

Partzite. A mineral. It is a hydrated antimonious oxide mixed with copper, lead, and silver oxides.

Parvol. Contraceptive jelly, containing polyoxyethyleneothylcresol. (182).

Parvoline. Dimethyl - ethyl - pyridine, $C_9H_{13}N$.

Parylene N. A plastic material used to make thin film membranes, 2–1000 Angstroms thick. (34).

Pasade. A proprietary preparation of sodium aminosalicylate. An antituberculous agent. (824).

Pascoite. A mineral. It is a hydrated vanadate of calcium, $2CaO.3V_2O_5.XH_2O$.

Pashets. Cachets of sodium amino salicylates. (529).

Pasilex. A proprietary trade name for an aluminium silicate powder 35μ particle size. Used in the paper trade. (190).

Pasinah. A proprietary preparation containing sodium aminosalicylate, isoniazid. An antituberculous agent. (244).

Paskalium. A proprietary preparation of potassium paraaminosalicylate used in the treatment of tuberculosis. (810).

Passauite. A mineral. It is a variety of scapolite.

Passini's Solution. An aqueous solution of mercury and sodium chlorides and glycerin. Used to preserve animal tissue.

Passow's Slag Cement. Prepared by blowing into liquid slag as it issues from the blast furnace, when it becomes granulated. It is then finely ground.

Passyite. A mineral. It is a variety of quartz.

Pastaccio. A residue from the manufacture of calcium citrate. It consists of vegetable cellulose with some hydrocarbon, $C_{10}H_{16}$.

Paste. See STRASS.

Paste blue. See PRUSSIAN BLUE.

Pasteurised Milk. A milk that has been heated to 65° C. for not less than 30 minutes, becoming partially sterilised.

Patagosite. A mineral. It is a form of calcite.

Patava Oil (Batana Oil). Coumou oil, a semi-drying oil obtained from the kernels of the Brazilian palm tree, *Œnocarpus batava*. It has a specific gravity of 0·925, a saponification value of 190–192, an acid value of 1–1·4, and an iodine value of 78–80.

Patchouli. An Indian herb, *Pogostemon patchouly*. Used in perfumery.

Patent Alum. See ALUM CAKE.

Patent Atlas Red. See GERANINE.

Patent Bark. Commercial quercetin.

Patent Black. An acid dyestuff. It is a substitute for logwood.

Patent Blue A (Disulphine Blue A). A dyestuff. It is the calcium salt of the disulphonic acid of *m*-oxy-diethyl-dibenzyl - diamino - triphenyl - carbinol $(C_{37}H_{35}N_2S_2O_7)_2Ca$. Dyes wool greenish-blue.

Patent Blue JOO. A dyestuff. It is a mixture of Patent blue and violet.

Patent Blue V, N, Superfine, and Extra (New Patent Blue B and 4B). Dyestuffs. The calcium, magnesium, or sodium salt of the disulphonic acid of *m*-oxy-tetra-allyl-diamino-triphenyl-carbinol, $C_{27}H_{31}N_2S_2O_7Na$. Dyes wool greenish-blue.

Patent Chrome Green. An azo dyestuff obtained by coupling diazo-salicylic acid with α-naphthylamine, diazotising the product, then coupling with K-acid and aniline.

Patent Fustin (Wool Yellow). Dyestuffs. They are condensation products of fustic extracts and diazo compounds. They dye wool and cotton. The brand K is only partially oxidised, and is recommended for use with oxidising mordants. The brand E is fully oxidised.

Patent Glue. A dark brown bone glue.

Patent Green. See EMERALD GREEN.

Patent Greens O and V. Dyestuffs. They are mixtures of Patent blue and Acid green.

Patent Mixture. A mixture of ether and ammonia. (*Mistura ætheris cum ammonia B.P.*)

Patent non-poisonous White Lead. A pigment. It consists of lead sulphate with 25 per cent. zinc oxide.

Patent Orange. See ORANGE G.

Patent Phosphine. See PHOSPHINE.

Patent Red. See CINNABAR.

Patent Rock Scarlet. See ST. DENNIS RED.

Patents. The small portion of very white flour obtained from wheat. It is poor in proteins, and is used for fancy breads.

Patent White Lead. See NON-POISONOUS WHITE LEAD.

Patent Yellow. See TURNER'S YELLOW.

Patent Zinc White. A pigment made by adding a soluble sulphide to a zinc chloride or zinc sulphate solution, filtering off the precipitate, drying it, and then calcining it. It has the composition, $5ZnS.ZnO$.

Pateriate. A mineral. It is a molybdate of cobalt.

Paternoite. A mineral. It is a hydrated magnesium tetraborate, $MgO.4B_2O_3.4H_2O$, of Sicily.

Patersonite. See THURINGITE.

Pathibamate-200. A proprietary preparation of tridihexethyl chloride and meprobamate. (306).

Patina. The green film which forms on

copper and bronze mouldings. It consists of basic copper carbonate.

Patrinite. See ACICULITE.

Patronite (Rizopatronite). A mineral. It is a sulphide of vanadium, and approximates to V_2S_9.

Pattern Metal. An alloy of 83 per cent. copper, 10 per cent. zinc, 4 per cent. tin, and 3 per cent. lead.

Pattern Metal, Light. An alloy of 72 per cent. aluminium, 16 per cent. zinc, and 2 per cent. copper.

Pattes de Lièvre. See ÉDRÉDON VÉGÉTALE.

Pattinson's White Lead. A pigment. It is basic lead chloride, $PbCl_2.2Pb(OH)_2$.

Paulite. A mineral. It is similar to Hypersthene.

Paul Veronese Green. See BREMEN.

Pauly Silk. Glanzstoff (*q.v.*).

Pavacol. A proprietary preparation containing pholcodine, papaverine hydrochloride, balsam of tolu, oil of clove, tincture of ginger, oil of aniseed, tincture of capsicum, oil of peppermint, glycerin, alcohol, chloroform and treacle. A cough linctus. (319).

Pavacol Diabetic Cough Syrup. A proprietary preparation containing agents as in Pavacol without carbohydrates. A cough linctus. (319).

Paviin. Fraxin, $C_{16}H_{18}O_{10}$. A substance which occurs in the bark of the common ash.

Pavon. The total alkaloids of opium in a soluble form.

Pavulon. A proprietary preparation of PANCURONIUM BROMIDE. A muscle relaxant. (316).

Pavy's Solution. A modified Fehling's solution used for the determination of sugar. The solution consists of (*a*) 4·158 grams copper sulphate in 500 c.c. water, and (*b*) 20·4 grams sodium-potassium tartrate, 20·4 grams potassium hydroxide, and 300 c.c. strong ammonia, made up to 500 c.c. Equal parts of (*a*) and (*b*) are used.

Paxbestos. Asbestos products bonded with hydraulic cement and impregnated with bitumen. Used as insulating materials.

Paxidorm. A proprietary preparation of methaqualone hydrochloride. A hypnotic. (255).

Paxolin. A synthetic resin bonded paper product used for insulating purposes.

Payne's Grey. An oil and water colour prepared from black alizarin, madder, and indigo.

Payne's Solution. A solution of sodium hypobromite.

Paynocil. A proprietary preparation of aspirin and glycine. An analgesic. (272).

Paynocil Junior Tablets. A proprietary preparation of aspirin and glycine. An analgesic. (272).

Payzone. A proprietary preparation of NITROVIN hydrochloride. A veterinary growth promoter.

Pazite. A mineral. It is $Fe(As,S)_2$.

Pazo. An insecticide. (501).

P.B.N. Phenyl-beta-naphthylamine. A rubber antioxidant. See NEOZONE D.

P.C.M. A proprietary preparation of paracetamol. An analgesic. (334).

PDA-10. A proprietary anti-oxidant blend of polyarylamines. (784).

PE-209. A proprietary ethylene vinyl acetate copolymer used in the manufacture of extruded film. (785).

PE-210. A proprietary ethylene vinyl acetate resin used for liquid packaging and in the manufacture of extruded film. (785).

Peach Black. A variety of carbon black similar to lampblack.

Peach Bordeaux Mixture. Consists of 3 lb. copper sulphate, 9 lb. lime, and 50 gallons water.

Peach Essence. A flavouring material. It contains 100 parts amyl isovalerate, 100 parts amyl butyrate, 20 parts amyl acetate, 10 parts benzaldehyde, and 770 parts alcohol.

Peach Wood. See REDWOODS.

Peacock Blue. A dyestuff. It is a mixture of Methyl violet and Malachite green.

Peacock Blue Lake. The barium lake of Patent blue. Used in printing inks.

Peacock Copper Ore. An iridescent copper pyrite, produced by the partial decomposition of the yellow mineral, $Cu_2S.Fe_2S_3$.

Pea Iron Ore. See BOG IRON ORE.

Pealite. A mineral. It is a variety of geyserite.

Peanut Oil. See EARTHNUT OIL.

Peanut Ore. A mineral. It is a variety of wolframite.

Pea Nuts. See GROUND NUTS.

Pearceite. A mineral $(Ag.Cu)_{16}(Sb.As)_2S_{11}$.

Pear Essence. A flavouring material. It consists of 200 parts amyl acetate, 100 parts ethyl nitrite, 50 parts ethyl acetate, and 645 parts alcohol.

Pearl. A calcareous secretion obtained chiefly from the oyster. The term Pearl is also a registered trade mark for a particular product.

Pearl Alum. A specially prepared aluminium sulphate used in the paper industry.

Pearl Ash. A variety of potassium carbonate, K_2CO_3.

Pearl Dust. A registered trade name for

a form of potassium carbonate, K_2CO_3, used as a filler.

Pearl Essence. A product made from fish scales. Used on materials to give a lustrous finish, especially for imitation pearls.

Pearl-hardening. Calcium sulphate, $CaSO_4$, used as a loading for paper.

Pearlite. Iron carbide eutectoid, consisting of alternate masses of ferrite and cementite.

Pearl-mica. A mineral. It is margarite.

Pearl Powder. Bismuth oxychloride. Used as a cosmetic.

Pearl Sinter. A variety of opal, SiO_2, found in volcanoes.

Pearl Spar. A double carbonate of magnesium and calcium, $Mg.Ca(CO_3)_2$.

Pearl Stone. A felspathic mineral with a pearly lustre.

Pearl White (Flake White). Bismuth oxychloride, $BiOCl$, used as a cosmetic, and in enamels (blanc de perle). A basic bismuth nitrate, $Bi(OH)_2NO_3$, is also known as pearl white.

The term is sometimes used in connection with a white lead which has been tinted with Paris blue or indigo.

A preparation of mother of pearl is called by this name.

Also see BISMUTH WHITE and FLAKE WHITE.

Pear Oil. Isoamyl acetate, $CH_3.COO.C_5H_{11}$. Used in the manufacture of fruit essences for flavouring confectionery.

Pearsol. Chemical closet fluids. (552).

Pearson's Cerate. Consists of 4 parts lead plaster, 1 part beeswax, and 3 parts almond oil.

Pearson's Solution. A solution of dried sodium arsenate 1 per 100 to 1 per 1,000.

Pea-stone. See PISOLITE.

Peat. The partially decayed remains of plants. Used as fuel.

Peat Coal. An intermediate between peat and lignite.

Pecan Oil. Oil obtained from the seed of the North American walnut, *Juglans niger*.

Pecazine. 10-(1-Methyl-3-piperidyl-methyl)phenothiazine. Pacatal is the hydrochloride or the acetate.

Pechman Dyes. Coloured dehydration products of β-benzoyl-acrylic acid or its homologues.

Pecilocin. An antibiotic produced by Paecilomyces varioti banier var. antibioticus. Variotin.

Peckhamite. A mineral. It is a magnesium-iron silicate.

Pectamol. A proprietary preparation containing oxeladin citrate, in flavoured vehicle containing menthol, chloroform and glycerin. A cough linctus. (182).

Pectase. A clotting enzyme, which produces vegetable jellies.

Pectin. A substance soluble in water, which is a constituent of many fruits, such as apples, pears, and gooseberries, also of carrots and beetroot. Its aqueous solution gelatinises on cooling. Four forms may occur: protopectin or pectinogen, pectin, pectinic acid, and pectic acid. Protopectin is the mother substance and is similar in composition to the carbohydrates. It is insoluble in water, but as the fruit ripens it becomes hydrolysed and pectin and cellulose are formed. The pectin is soluble in water giving a jelly, and is the jellifying principle of jams. Little is known of the chemistry of the substance, and its formula is stated to be $2C_6H_{10}O_5.H_2O$, but Ehrlich describes it as the calcium-magnesium salt of a complex anhydro-arabino-galactose-methoxyl-tetragalacturonic acid. It is sold in a concentrated liquid form containing 4–4·5 per cent. pectin, a pure pectin, and a mixture of pectin and sugar of variable strength.

Pectine. Sodium-benzene-sulphonyl-*p*-arsanilate. Used in sleeping-sickness.

Pectolite (Gonsogolite, Ratholite). A mineral, $4CaO.Na_2O.6SiO_2.H_2O$.

Pectomed. A proprietary preparation containing ipecacuanha liquid extract, squill liquid extract, tolu solution, strong ammonium acetate, solution and cherry syrup. A cough linctus. (250).

Pediamycin. A proprietary preparation of erythromycin ethyl succinate. An antibiotic. (816).

Peerless Alloy. A heat-resisting alloy containing 78·5 per cent. nickel, 16·5 per cent. chromium, 3 per cent. iron, and 2 per cent. manganese.

Pegamoid. A leather substitute prepared from nitro-cellulose and camphor, with some solvent. It is also the name for a brand of aluminium paint.

Peganite. See VARISCITE.

Peganone. A proprietary preparation of ethotoin. An anticonvulsant. (311).

Pegmatite. A felspathic rock, similar to Cornish stone.

Pegmatolite. A mineral. It is felspar (*q.v.*).

Pegnin. A preparation of lactose and rennet, which yields a finely divided curd from cows' milk. Used in infant food.

Pegu. See CUTCH.

Pegu Brown. A direct cotton dyestuff.

Pegu Catechu. See CUTCH.

Peka Glas (PK Glas). A proprietary safety glass.

Pelagite. A mineral, $MnO_2.Fe_2O_3$.

Pelagosite. A mineral. It is an impure calcium carbonate.

Pelargone. A trade name for Nylon 9.

Pelargonium Oil. See ROSÉ OIL.

Pelaspan GP. A trade mark. General purpose expandible polystyrene.

Pelaspan 333FR. A trade mark. A flame retardant expandible polystyrene. (64).

Pelaspan PAC. A trade mark. Polystyrene foam strands.

Pelentan. A proprietary trade name for ethyl biscoumacetate.

Pelhamite. A mineral. It is a hydrated mica.

Pelican Blue. An induline dyestuff. See INDULINE, SPIRIT SOLUBLE.

Peligot Blue. A pigment. It is a hydrated copper oxide.

Peligotite. Synonym for Johannite.

Peliom. A mineral. It is a variety of iolite.

Pelionite. A coal of the cannel type.

Pellidol (Alborubol). Diacetyl-amino-azo-toluene. Used in surgery.

Pellitory Root. Pyrethrum root.

Pellote. Mescal buttons (*q.v.*).

Pellurin. Hexamethylene-tetramine hydrochloride.

Pelokonite. A mineral. It is a variety of lampadite.

Pelosiderite. See OLIGONITE.

Pelosine. Berberine.

Pelouze's Green (Prussian Green). Ferroso-ferric-ferricyanide, $Fe_3Fe_4(FeCy_6)_6$.

Pembrite. Cologne rottweil (*q.v.*).

Pemoline. 2-Imino-5-phenyloxazolidin-4-one. Kethamed. Pioxol Ronyl.

Pempidine. 1,2,2,6,6-Pentamethyl-piperidine. Perolysen and Tenormal are the hydrogen tartrate.

Penacol. Resorcinol (technical). (170).

Penacolite G–1124 and G–1131. Proprietary trade names for resorcinol-formaldehyde glues.

Penak Dammar. See DAMMAR RESIN.

Penamecillin. Acetoxymethyl 6-phenyl-acetamidopenicillanate.

Penaryl A. A proprietary trade name for a plasticiser. It is mono-amyl-diphenyl.

Penaryl B. A proprietary trade name for a plasticiser. It is a diamyl-diphenyl.

Penavlon V. A proprietary preparation containing Phenoxymethylpenicillin Potassium. An antibiotic. (2).

Penbenemid. A proprietary preparation containing Benzylpenicillin (as the potassium salt) and Probenecid. An antibiotic. (310).

Penbritin. A proprietary preparation containing Ampicillin. An antibiotic. (364).

Penbritin K-S. A proprietary preparation of ampicillin, sulphadimidine and kaolin. An antidiarrhœal. (364).

Penbutolol. A beta adrenergic blocking agent currently undergoing clinical trial as " 39 893d " and as " Hoe 893d ". It is (—) - 1 - *tert* - butylamino - 3 - (2 - cyclopentylphenoxy)propan - 2 - ol.

Pencatite. A mineral. It is a mixture of calcite and brucite.

Penchlor. A proprietary acid-proof cement made from cement powder and sodium silicate solution. Used for lining tanks.

Pencil Ore. A mineral. It is a variety of hæmatite.

Pencils for Glass. Usually obtained by stirring Prussian blue with a mixture of 8 parts white wax and 2 parts tallow, and when nearly cold, rolling into a pencil.

Pencil Stone. A mineral. It is Agalmatolite.

Pendare. A name for Venezuelan chicle.

Pendecamaine. A surface-active agent present in TEGO-BETAINES. It is NN-dimethyl - (3 - palmitamidopropyl) - glycine betaine.

Pendiomide. Azamethonium Bromide.

Penethamate Hydriodide. Benzylpenicillin 2-diethylaminoethyl ester hydriodide. Estopen.

Penetrodine. A new iodine compound having a stimulating antiseptic action.

Penetrol. As insecticide against aphides. It is a sulphonated oxidation product of petroleum.

Penetrol. A compound used as a textile detergent.

Penfieldite. A mineral, $2PbCl_2.PbO$.

Penfluridol. A neuroleptic. 4 - (4 - Chloro - 3 - trifluoromethylphenyl) - 1 - [4, 4 - di - (4 - fluorophenyl)butyl] piperidin-4-ol.

Penicals. A proprietary preparation containing Phenoxymethylpenicillin (as the calcium salt). An antibiotic (308).

Penicals 333. A proprietary preparation containing Phenoxymethylpenicillin Calcium. An antibiotic. (308).

Penicillamine. A chelating agent. (—)-3, 3 - Dimethylcysteine. D - Penicillamine. CUPRIMINE, DISTAMINE.

Penicillin. The name given to the antibiotic principle of the mould *penicillium notatum* by Fleming. The material is now prepared by special fermentation processes in large quantities for the destruction of staphylococci and germs of pneumonia, diphtheria, gas gangrene and meningitis.

Penicillin - G - Sodium Salt. A proprietary preparation containing Benzylpenicillin Sodium. An antibiotic. (2).

Penicillin N. Adicillin.

Penicillin V Pulvules. A proprietary preparation containing Phenoxymethylpenicillin. An antibiotic. (333).

Penicillin V Sulpha. A proprietary

preparation of phenoxymethylpenicillin sulphadiazine, sulphamerazine and sulphadimidine. An antibiotic. (333).

Penicillinase. An enzyme obtained from cultures of Bacillus cereus which hydrolyses benzylpenicillin to penicilloic acid. Neutrapen.

Penidural All Purpose. A proprietary preparation containing Benzathine Penicillin, Benzylpenicillin and Procaine Penicillin. An antibiotic. (245).

Penitriad. A proprietary preparation of phenoxymethylpenicillin potassium, sulphadiazine, sulphadimidine and sulphathiazole. An antibiotic. (336).

Pen Metal Brass. See BRASS.

Pennalene White Oil. A colourless, odourless oil, used for medical purposes.

Pennettier's Green. See CHROMIUM GREEN.

Pennine (Penninite). A mineral, $4H_2O$. $(Al.Fe_2)O_3.5(Mg.Fe)O.3SiO_2$.

Pennite. A mineral. It is a hydrated calcium-magnesium carbonate.

Pennsalt TD-5032. Hexamethylditin. A contact and systemic insecticide.

Pennyroyal Oil. See OIL OF PENNYROYAL.

Penotrane. A proprietary preparation containing hydrargaphen. (319).

Penroseite. A mineral. It is a copper-lead-cobalt selenide, $3CuSe.2PbSe_2.5(Ni.Co)Se_2$.

Pensa's Rubber. A rubber substitute made from coal tar, petroleum tar, oil of turpentine, and boric or phosphoric acids.

Penspek. A proprietary trade name for the potassium salt of Phenbenicillin.

Pensulate. Proprietary insulating materials.

Pentabor. Sodium pentaborate. (548)

Pent-acetate. An American amyl acetate. It has a specific gravity of 0·86–0·87. A lacquer solvent.

Pentacizers. Proprietary trade names for plasticisers.

Pentacosactride. A corticotrophic peptide.

Pentacosane. A hydrocarbon, $C_{25}H_{52}$, obtained from beeswax.

Pentacresol. Sec-amyl-tricresol. A germicide.

Pentacynium Methyl Sulphate. N^1-(5 - Cyano - 5,5 - diphenylpentyl) - $N^1N^1N^2$ - trimethyl - ethylene - 1 - ammonium-2-morpholinium bismethyl sulphate. Presidal.

Pentaerythritol Tetranitrate. 2,2-Bishydroxymethylpropane-1,3-diol tetranitrate. Pentaerythrityl Tetranitrate. Mycardol. Peritrate.

Penta-erythritol Tetrastearate (PET). A proprietary release agent used in injection-moulding processes. (10).

Penta G.P. 79. Rust preventatives. (526).

Pentagastrin. A polypeptide resembling gastrin. It is N-t-butyloxycarbonyl-β-alanyl-L-tryptophyl-L-methionyl-L-aspartyl-L-phenylalanine amide. PEPTAVLON.

Pentahydrite. A mineral. It is $MgSO_4$. $5H_2O$.

Pental. Trimethyl-ethylene, C_5H_{10}. A hypnotic.

Pentalamide. 2-Pentyloxylbenzamide. O-Pentylsalicylamide.

Pentaline. Pentachlorethane, $CHCl_2$. CCl_3. It boils at 159° C., is of specific gravity 1·7, and has a limited use in the varnish industry.

Pentalyn. Synthetic resins for varnish, inks and adhesives. (593).

Pentamethonium Bromide. Pentamethylenedi(trimethylammonium bromide). Lytensium.

Pentamethonium Iodide. Pentamethylenedi(trimethylammonium iodide). Antilusin.

Pentamidine. 1,5 - Di - (4-amidinophenoxy)pentane.

Pentanol. Amyl alcohol, $CH_3(CH_2)_4OH$.

Pentaphane. A proprietary trade name for a film made from a chlorinated polyether (polymerised 3,3-bis(chloromethyl)oxetane. (173).

Pentapiperide. 1-Methyl-4-(3-methyl-2-phenylvaleryloxy)piperidine. Quilene is the methyl sulphate.

Pentaquine. 8-(5-Isopropylaminopentylamino)-6-methoxyquinoline.

Pentaryl A. A proprietary trade name for a monoamyl diphenyl plasticiser. (119).

Pentaryl B. A proprietary trade name for a diamyl diphenyl plasticiser. (119).

Pentasilane. Silicopentane (penta-silicon-dodecahydride), Si_5H_{12}.

Pentasol. A mixture of pure amyl alcohols. It is stated to contain 75 per cent. primary alcohol and 25 per cent. secondary alcohol, and is obtained from pentane fraction of gasoline. It has a specific gravity of 0·812–0·820 and a boiling range of 116–136° C. It has no acid value and is used as a varnish and lacquer solvent.

Pentazocine. 1,2,3,4,5,6-Hexahydro-8-hydroxy - 6,11 - dimethyl - 3 - (3 - methylbut - 2 - enyl) - 2,6 - methano - 3-benazocine.

Pentek. A proprietary trade name for a technical grade of pentaerithritol used in synthetic resins and in the paint and varnish industry.

Pentenel. Cyclo-pentethylene-pentenylethyl-barbituric acid.

Penthienate. 2-Diethylaminoethyl α-cyclopentyl - α - hydroxy - α - (2 -

thienyl)acetate. Monodral is the methobromide.

Penthrane. A proprietary preparation of methoxyfluorane. Inhalation anæsthetic, used in obstetrics. (311).

Penthrichloral. A sedative and hypnotic. 5, 5 - Di(hydroxymethyl) - 2 - trichloro - methyl-1,3-dioxan.

Penthrinit. An explosive. It is a plastic mixture of 80 per cent. penta-erythritol-tetranitrate and 20 per cent. nitroglycerin.

Penthrit. Pentaerythrite tetranitrate.

Pentifylline. A vasodilator. 1-Hexyl-3,7-dimethylxanthine. COSALDON.

Pentlandite (Zillhammerite, Folgerite). A mineral. It is a sulphide of nickel and iron, usually corresponding to the formula (Fe.Ni)S.

Pentley's Neutral Orange. See NEUTRAL ORANGE.

Pentol. Timber fungicides. (511).

Pentolinium Tartrate. NN'-Pentamethylenedi - (1 - methylpyrrolidinium hydrogen tartrate). Ansolysen.

Penton. A trade mark for a chlorinated polyether (the polymer of 3,3-bis (chloromethyl) oxetane). Useful as a corrosion protective coating for mild steel. (93).

Pentosalen. An aid to dermal pigmentation. 9 - (3 - Methylbut - 2 - enyloxy) - furo[3,2-g]chromen-7-one.

Pentostam. A proprietary trade name for Sodium Stibogluconate.

Pentothal. A proprietary preparation of thiopentone sodium. Intravenous anæsthetic. (311).

Pentovis. A proprietary preparation containing Quinestradol. (262).

Pentoxifylline. See OXPENTIFYLLINE.

Pentoxylon. A proprietary preparation of rauwolfia alkaloids, and pentaerythrityl tetranitrate. An antihypertensive. (275).

Pentral 80. A proprietary preparation of pentaerythritol tetranitrate in a sustained-release form, used in the treatment of angina pectoris. (633).

Pentrexyl. A proprietary preparation of AMPICILLIN. An antibiotic. (324).

Pentrium. A proprietary preparation of chlordiazepoxide and pentaerythritol tetranitrate, used in angina pectoris. (314).

Pentrone ON. A proprietary trade name for a 33 per cent. active composition of sodium-2-ethyl hexyl sulphate. A high purity, low fat content material used as a surface tension reducing agent for caustic soda solutions. (194).

Pentyl. Amyl, C_5H_{11}.

Pen Vee Dural. A proprietary preparation containing Phenoxymethylpenicillin Potassium and Benzathine Penicillin. An antibiotic. (245).

Penwithite. A mineral. It is a hydrated manganese silicate.

Penyl. Polamide.

Penzold's Reagent. A solution of diazo-benzo-sulphonic acid and potassium hydroxide. A reagent for sugar in urine.

Peonine. See PÆONINE.

Peonol. The aqueous distillate from the root of the Japanese *pæonia moutan*. It is a ketone, and is said to have the formula, $C_9H_{10}O_3$.

PEP-1. A polyethylene of Soviet origin.

Pepo. Melon pumpkin seeds.

Pepper Bark (Winter's Bark). The bark of *Drimys winteri*.

Peppermint Camphor. Menthol, $C_{10}H_{19}OH$. Used in medicine.

Peppermint Water. (*Aqua menthæ B.P.*) It consists of 1 c.c. oil of peppermint and 1,500 c.c. water, distilled to give 1,000 c.c.

Pepsalin. A preparation of pepsin and sodium chloride.

Pepsin. A proteolytic enzyme of the mucous membrane of the stomach. It decomposes albuminous bodies into peptone.

Peptacol-10. A proprietary preparation of homatropine methylbromide and phenobarbitone. (330).

Peptard. A proprietary preparation of *l*-hyoscyamine sulphate used to reduce gastric acidity. (275).

Peptavlon. A proprietary preparation of pentagastrin used for the clinical testing of gastric secretion. (2).

Peptic Salt. A mixture of sodium chloride and pepsin.

Peptofer. A proprietary preparation of iron chlor-peptonate.

Pepton 65. A proprietary trade name for zinc 2-benzamide thiophenate. M.p. 200–300 C. A peptizer for natural rubber. SBR and synthetic polyisoprene above 65° C. (162).

Peptonal. A purified peptone.

Peptoneigone. See EIGONES.

Peptone Paste. Beef peptone.

Peptonised Iron. A compound of iron and peptone made soluble by sodium citrate.

Peptonising Tablets. Pancreatic tablets.

Peptorub. Pre-plasticised comminuted rubber.

Per-abrodil. A proprietary preparation. It is stated to be a 35 per cent. solution of the diethanolamine salt of 3 : 5-di-iodo-4-pyridone-N-acetic acid.

Peradinol. Textile dyeing auxiliaries. (521).

Peralga. A material stated to be a mixture of amidopyrin and diethyl-barbituric acid. An analgesic.

Peralit. A German insulating material

to replace fibre, hard rubber, and other insulating materials. It is stated to have a high electrical resistance and not to be affected by oil, air, or moisture, and only by certain acids.

Peralvex. A proprietary preparation of anthraquinone glycosides and salicylic acid used in the treatment of mouth ulcers. (286).

Perandren. The first synthetic chemically pure testicular hormone indicated in disturbances or insufficiency of the male sex hormone.

Peratizole. An anti-hypertensive agent. 1 - [4 - (2, 4 - Dimethylthiazol - 5 - yl) - butyl] - 4 - (4 - methylthiazol - 2 - yl) - piperazine.

Perborax. Sodium perborate, $NaBO_3$. $4H_2O$, a washing and bleaching agent.

Perborin. Sodium perborate, $NaBO_3$, a constituent of washing powders.

Perborin M. A mixture of soap, soda, and sodium perborate. A dry soap.

Perborol. Perborax (q.v.).

Perbunan N. A proprietary trade name for a synthetic rubber consisting of butadiene-acrylonitrile copolymer. It is oil-resistant. (112).

Percainal. An antipruritic and analgesic ointment containing 1 per cent. percaine.

Percaine. Hydrochloride of α-butyloxycinchoninic acid diethylethylenediamide. A powerful local anæsthetic for infiltration, spinal and surface anæsthesia.

Perchlorethylene. Tetrachlorethylene, C_2Cl_4.

Perchloron. A technical calcium hypochlorite containing 68·1 per cent. available chlorine.

Perchoid. See OXOLIN.

Per-clene. A proprietary trade name for perchlorethylene (q.v.).

Percorten " M " Crystules. A proprietary preparation of deoxycortone pivalate. (18).

Percresan. A mixture of cresols, soap, and water. Used as a disinfectant in 1–2 per cent. solution.

Percussion Cap Brass. See BRASS.

Percylite. A mineral, $Pb.Cu.Cl(OH)_2$.

Perdeca. A proprietary preparation of dexamethasone and cyproheptadine hydrochloride. (310).

Perdilatal. The hydrochloride of buphenine.

Perdolat. A German preparation. It is a compound of phenyl-cinchonine-carboxylic acid (40 per cent.), dimethyl-amino-phenazolone (50 per cent.), and caffeine (10 per cent.). The tablets are used for rheumatism, influenza, and headache.

Perduren. An organic polysulphide synthetic rubber.

Perdynamine. A compound of albumen and hæmoglobin.

Perecot. A copper fungicide. (511).

Pereiro Bark. The bark of *Geissospermum vellosii*. A Brazilian febrifuge.

Pereman. A copper fungicide. (511).

Perenox. A copper fungicide. (511).

Perenyi's Fluid (Chromo-nitric Acid). A fixing agent used in microscopy. It contains 3 parts of 92 per cent. alcohol, 4 parts of 10 per cent. nitric acid, and 3 parts of 0·5 per cent. chromic acid. The objects are treated with alcohol after fixing.

Perezol (Perezone). A 0·5 per cent. alcoholic solution of pipitzahoic acid, $C_{15}H_{20}O_3$. It is a compound obtained from the Mexican plant, *Perezia adnata*, and is used as an indicator.

Perezone. See PEREZOL.

Perflex. A proprietary trade name for unstretched vinylidene chloride (Saran q.v.).

Perfumed Formosyls. Saponaceous solutions of origanum and other oils. They are antiseptic, and are used in medicine.

Perfumery Oil. Refined Russian petroleum, of specific gravity 0·880–0·885. Used in perfumery.

Pergalen. A proprietary preparation of sodium apolate, and benzyl nicotinate. Resolution of bruises and local trauma. (312).

Pergament. Consists of calf-skins and the flesh splits of sheep-skins, well rubbed with pumice stone and chalk paste, and dried in a stretched condition.

Pergamin (Pergamyn). A grease-proof paper made from cellulose pulp.

Pergamyn. See PERGAMIN.

Pergenol. A mixture of sodium perborate and bitartrate. It gives hydrogen peroxide on the addition of water.

Perglow. A bright nickel plating process. (605).

Perglycerol. An aqueous solution of sodium lactate. Used as a substitute for glycerol for medical and cosmetic purposes.

Pergonal. A proprietary trade name for a follicle-stimulating hormone.

Pergut. See CHLORKAUTSACHUK and ALLOPRENE. A proprietary trade name for chlorinated rubber.

Perhexiline. A drug used in the treatment of angina pectoris. 2-(2,2-Dicyclohexylethyl)piperidine. PEXID is the hydrogen maleate.

Perhydrate. See HYPEROL.

Perhydrit. See HYPEROL.

Perhydrol. Hydrogen peroxide, H_2O_2, one volume of 30 per cent. hydrogen peroxide giving 100 volumes of oxygen. Used for bleaching, also as a disinfectant.

Perhydrol of Magnesia. See BIOGEN.

Peri Acid. 1-Naphthylamine-8-sulphonic acid.

Periactin. A proprietary preparation of cyproheptadine hydrochloride. (310).

Perichthol. A proprietary preparation of ammonium ichthosulphonate.

Periclase. A mineral, MgO.

Pericline. See ALBITE.

Pericyazine. 2-Cyano-10-[3-(4-hydroxy-piperidino)propyl]phenothiazine. Neulactil.

Perideca. A proprietary preparation of DEXAMETHASONE and CYPROHEPTADINE hydrochloride. An anti-allergic drug. (472).

Peridot. A mineral. It is a variety of olivine (q.v.), used as a gem stone.

Perifenil. A proprietary preparation of pheneizine, and pentaerythritol tetranitrate. (262).

Perihemin. A proprietary preparation of vitamin B_{12}, ferrous fumarate, folic acid and ascorbic acid. A hæmatinic. (306).

Perikol. A sensitiser for silver bromide plates, prepared by treating the addition product of toluquinaldine and the ethyl ester of toluene-sulphonic acid with alcoholic potassium hydroxide.

Perilla Oil. An oil obtained from the seed of the Asiatic mint, *Perilla ocymoides*. It is made in Japan and used as a drying oil in substitution for linseed oil. It has the following characteristics: specific gravity at 15° C., 0·928–0·933; refractive index at 15° C., 1·483–1·485; saponification value 187–197, and iodine value 180–200.

Perin. A proprietary trade name for Piperazine Calcium Edetate.

Peristaltin. A cascara preparation containing the water-soluble glucosides extracted from the bark of *Rhamnus purshiana*. It stimulates peristalsis without drastic purgative action.

Peristerite. Albite (q.v.), with quartz.

Peritrate. A proprietary preparation of pentaerythritol tetranitrate. A vasodilator used for angina pectoris. (262).

Perkadox SE 9. A proprietary trade name for pelargonyl peroxide. A polymerisation initiator for polyolefines. (90).

Perkadox SE.10. A proprietary trade name for decanoyl peroxide, a low critical temperature solid replacement for lauroyl and benzoyl peroxides. Used as a catalyst for the polymeriza-

tion of ethylene, styrene, vinyl chloride and for the medium and high temperature polymerization of unsaturated polyesters. (192).

Perkaglycerol. An aqueous solution of potassium lactate. Used as a substitute for glycerol for medical and cosmetic purposes.

Perkin's Base. *p*-Tolylamino-ditolyl-*p*-toluquinone-di-imine.

Perkin's Purple. See MAUVEINE.

Perkin's Violet. See MAUVEINE.

Perklone. Perchloroethylene. A solvent for dry cleaning. (512).

Perlankrol. Sulphated fatty alcohols. (558).

Perlapine. A hypnotic. 6-(4-Methyl-piperazin-1-yl)-11*H*-dibenz[*b*, *e*] azepine.

Perlate Salt. See SALT PERLATE.

Perlite. A eutectic product resulting from an alloy of ferrite and cementite in steel.

Perlka. A proprietary name for granulated calcium cyanamide, a nitrogen fertilizer having herbicidal and fungicidal properties. (899).

Perlygel. A registered trade mark for benzoyl peroxide.

Permabond. A proprietary cyanoacrylate adhesive. ARON ALPHA, EASTMAN 910, LOCTITE 904. (786).

Permalba. A composite pigment consisting mainly of barium sulphate. An artist's colour.

Permali (Jicwood, Jabroc, Durisol, Lignostone). A proprietary trade name for laminated products containing wood or paper impregnated with synthetic resin. Some are made from thin wood coated with synthetic resin solution and compressed under heat, others are impregnated under pressure, solvent removed and then compressed.

Permalloy. A registered trade mark for alloys of nickel and iron containing more than 30 per cent. nickel. They are prepared by certain heat treatment and show unusual magnetic properties, giving a high initial permeability. One of the best alloys contains 78·5 per cent. nickel and 21·5 per cent. iron. Another alloy of this class contains 78·5 per cent. nickel, 18 per cent. iron, 3 per cent. molybdenum, and 0·5 per cent. manganese. A typical analysis gives 78·23 per cent. nickel, 21·35 per cent. iron, 0·04 per cent. carbon, 0·03 per cent. silicon, 0·035 per cent. sulphur, 0·22 per cent. manganese, 0·37 per cent. cobalt, 0·1 per cent. copper and traces of phosphorus.

Permalon. A proprietary trade name for stretched vinylidene chloride (Saran q.v.).

Permalux. A proprietary trade name of the di-orthotolylguanidine salt of diatechol borate. (2).

Permanent Blue. See ULTRAMARINE.

Permanent Green. See CHROMIUM GREEN, NUREMBERG GREEN, and CHROME GREEN.

Permanent Orange R. A dyestuff prepared from 5-chloro-aniline-2-sulphonic acid.

Permanent Red 4B. See LITHOL RUBINE B.

Permanent Red 4B. See HELIO FAST RED RL.

Permanent Vermilion. A pigment. It is usually Orange mineral (q.v.), tinted with p-nitraniline.

Permanent Violet. See MANGANESE VIOLET.

Permanent White. See BARIUM WHITE and ZINC WHITE.

Permanent Yeast. See ZYMIN.

Permanent Yellow. See BARIUM YELLOW.

Permanite. A cobalt steel which has very high magnetic properties. Also see K.S. MAGNET STEEL and COBALT STEEL.

Permapen. Benzathine Penicillin.

Permaplex. An ion exchange membrane. (609).

Permapruf T. A proprietary wood preservative containing tri-n-butyltin oxide made soluble with quaternary ammonium compound. (425).

Permatol A. A proprietary trade name for a preservative for wood. It contains pentachlorphenol in oil.

Permax. A nickel steel containing 76 per cent. nickel, of French manufacture. It has magnetic properties.

Permeatin. A German preparation. It is a camphor-lanolin ointment with guaiacol valerate and formic acid. It is used for the percutaneous treatment of phthisis.

Permidan. Dimethyl-amino-pyrazolone, $C_5H_9N_3O$.

Perminvars. Proprietary alloys having exceptional magnetic properties. They are particularly suited for use in electrical communication circuits. One alloy contains 45 per cent. nickel, 25 per cent. cobalt, and 30 per cent. iron, and has a high initial permeability. These properties are developed by heat treatment. For magnetising forces below 1·7 gauss the permeability of the alloys is practically constant. These alloys also have very small hysteresis loss in the range of magnetising forces and flux densities in which the permeability is constant.

Permite. An explosive containing ammonium perchlorate, zinc dust, pet. jelly, potassium chlorate, pitch, and sulphur. Also a proprietary trade name for aluminium alloys of variable composition.

Permonite. An explosive used in mines. It is a mixture of potassium perchlorate and ammonium nitrate, with trinitrotoluene, a little starch, and wood meal.

Permutite. An artificially made zeolite, prepared by igniting together china clay (aluminium silicates), and (sometimes) quartz or sand, with alkali carbonates. Used for removing calcium and magnesium salts, sodium and potassium salts, and manganese and iron, from water.

Pernambuco Wood. See REDWOODS.

Pernax. The modern form of Gutta-Gentzsch (an artificial gutta-percha made from rubber, wax, and rosin).

Pernazene. A proprietary preparation of tymazoline hydrochloride. Anti-allergic nasal spray. (320).

Pernivit. A proprietary preparation of nicotinic acid and acetomenaphthone. (182).

Pernocton. The sodium salt of secbutyl-α-bromoallyl-barbituric acid. A hypnotic.

Pernomol. A proprietary preparation of chlorobutol, phenol, camphor, tannic acid and spirit Chilblain paint. (352).

Perocide. A fungicide. It consists essentially of the sulphates of the cerite earths (cerium, lanthanum, and neodymium), and is a by-product in the manufacture of the incandescent gas mantles. It contains from 43–47 per cent. of cerite earth oxides.

Perofskite (Perovskite, Perowskite). A mineral $(Ca.Fe)TiO_3$.

Peroidin. A proprietary preparation of potassium perchlorate used in the treatment of hyperthyroidism. (786).

Peroline. A preservative for iron and steel. (512).

Perolysen. A proprietary preparation of pempidine hydrogen tartrate. An antihypertensive. (336).

Perone. A proprietary trade name for pure hydrogen peroxide.

Peronine. Benzyl - morphine - hydro - chloride, $C_{17}H_{18}NO_2.O.C_6H_5CH_2.HCl$. Used as a substitute for morphine and codeine, to alleviate coughs.

Peronoid. A trade name for a mixture of copper sulphate and lime. A fungicide.

Peroxal. A trade name for hydrogen peroxide, H_2O_2.

Peroxide RH–2. A proprietary trade name for a high melting, stable, aromatic organic peroxide used as a polymerisation catalyst. (Compare LUPERCO, LUCIDOL.)

Peroxol (Pyrozone, Glycozone). Disinfectants. They are solutions containing hydrogen peroxide, sometimes mixed with other disinfectants.

Peroxtik. A proprietary preparation of solid hydrogen peroxide. See HYPEROL.

Peroxydol. Sodium perborate, $NaBO_3$. $4H_2O$. An antiseptic, deodorant, and bleaching agent.

Peroxyl. Hydrogen peroxide. (507).

Perparin. 6 : 7-Diethoxy-1-(3 : 4-diethoxy-phenyl)-isoquinoline hydrochloride. A compound analogous to papaverine with similar pharmacological effects.

Perpentol. A tetralin preparation used for cleaning wool.

Perphenazine. 2-Chloro-10-{3-[4-(2-hydroxyethyl)piperazin - 1 - yl]propyl} - phenothiazine. Fentazin ; Trilafon.

Perrierite. A mineral. It is $4[(Ce,La,Ca,Th)_2(Ti,Fe'')_2Si_2O_{11}]$.

Perry. See CIDER.

Persantin. A proprietary preparation of dipyridamole. (278).

Perseo. See FRENCH PURPLE.

Persian Balsam. Compound tincture of benzoin.

Persian Berries (Yellow Berries, Rhamnine). A natural dyestuff obtained from the dried unripe fruits of various species of *Rhamnus.* The dyeing principles are rhamnetin, or quercetin-mono-ethyl ether, $C_{16}H_{12}O_7$, rhamnazin or quercetin-dimethyl-ether, $C_{17}H_{14}O_7$, and quercetin, all as glucosides. Used for cotton printing with tin, chrome, or aluminium mordants, giving yellow to orange shades.

Persian Berry Carmine (Dutch Yellow). A pigment consisting of the aluminium and calcium lakes of the Persian berry colouring matters.

Persian Gum. See INDIA GUM.

Persian Powder. Insect powder (*q.v.*).

Persian Red. See INDIAN RED and CHROME RED.

Persian Yellow. A dyestuff. It is nitrotoluene - azo - nitro - salicylic acid, $C_{14}H_{10}N_4O_7$. Dyes chromed wool yellow. Also used in cotton printing giving yellow shades with chromium acetate. Also see CHINESE YELLOW.

Persil. A mixture of soap, soda, sodium silicate, and sodium perborate. A washing powder.

Persio. See ARCHIL and FRENCH PURPLE.

Persionin. An acetone extract of cudbear.

Persodine. A mixture of ammonium and potassium persulphates. Used in medicine as an aperitive and eupeptic.

Persoz's Reagent. Zinc Oxide (2 grams), is added to a solution of zinc chloride (10 grams), in 10 c.c. water. It dissolves silk, and detects silk in the presence of wool.

Perspex. A trade mark for acrylic (methyl methacrylate) resins in sheet form. The material can be used as a constructional material for electrical equipment where good insulation resistance is required. The material is also used for aircraft turrets, radomes, roof lights, machine guards and goggles. See also DIAKON, PLEXIGLAS, LUCITE, RESIN M. (2).

Perstoff (Diphosgene). A poison gas. Trichloromethyl-chloroformate, $ClCO.OCCl_3$.

Persulphocyanogen Yellow. See CANARIN.

Perthane. Trade mark for an agricultural insecticide based on diethyl diphenyl dichloro ethane, supplied as a wettable powder or emulsifiable concentrate. Control insects on plants and livestock, also used as a moth protection for textiles.

Perthite. A mineral. It is a potassium-sodium felspar.

Pertinax. A German synthetic varnish-paper product used for electrical insulation.

Pertinit. A proprietary synthetic resin of the urea-formaldehyde type.

Pertite. An Italian explosive. The main constituent is picric acid. See LYDDITE.

Pertofran. A proprietary trade name for the hydrochloride of desipramine.

Pertonal. *p*-Acetyl-amino-ethoxy-benzene. An antipyretic.

Petrothene. A trade mark for polyethylene. (788).

Pertusa. A proprietary preparation containing ephedrine hydrochloride, tincture of belladonna, liquid extract of ipecacuanha, syrup of tolu, honey, citric acid and sodium benzoate. A cough linctus. (253).

Peru Balsam, Synthetic. See PERUGEN.

Perugen (Synthetic Peru Balsam). A synthetic Peru balsam made by mixing benzyl benzoate with storax, benzoic, and tolu balsams.

Peruol. A registered trade mark currently awaiting re-allocation by its proprietors to cover a range of pharmaceuticals. (983).

Peru Saltpetre. See CHILE SALTPETRE.

Peruscabin. Benzyl benzoate, $C_6H_5.CO_2.CH_2.C_6H_5$. It is the active constituent of Peru balsam, and is used in the same manner, and for the same purposes as Peruol.

Peru Silver. See CHINESE SILVER.

Peruvian Balsam. The oleo-resin of *Myroxylon Pereiræ*, of Central America.

Peruvian Bark. Cinchona bark.

Peruvin. Cinnamic alcohol, C_6H_5CH : $CHCH_2OH$.

Peruvite. See MATILDITE.

Pervon. A proprietary trade name for a fully reacted polyurethane/pitch coating. (174).

Pescola Oil. An oil used in the tanning industry.

Pesillite. A mineral. It is a manganese oxide with silica.

Pesta. An insecticide. (562).

Pestex. An insecticide. (618).

Petal 17/50. A proprietary trade name for a close cut aliphatic hydrocarbon containing 50 per cent. aromatics. (196).

Petalite (Lithite, Berzelite). A mineral. It is a silicate of sodium, lithium, and aluminium. A source of lithium.

Peteraite. A mineral, $FeCoMo_2O_8$.

Peter Oil. See OIL OF PETER.

Pethilorphan. A proprietary preparation of pethidine hydrochloride and levallorphan tartrate. An analgesic. (314).

Petitgrain Oil. See OIL OF PETITGRAIN.

Petracin A and B. Dyestuffs obtained from petroleum by treatment with sulphuric acid, then with halogens in the presence of oxidising agents. Upon neutralising, the solution gives a precipitate of Petracin B, the solution containing Petracin A. A dyes silk and wool; B when heated with nitric acid, then treated with alkalis, dyes silk, cotton, and wool.

Petralol. See PARAFFIN, LIQUID.

Petralon. A German name for a preparation of wood tar. An antiseptic.

Petrasul. (**Transite.** A trade mark.) A synthetic stone material made from asbestos and Portland cement, which is sulphur-impregnated. The sulphur content varies from 15–35 per cent., and the material can be coloured, and is suitable for counter-tops, or similar purposes.

Petre. Potassium nitrate, KNO_3, used in pyrotechny.

Petre Oil. See OIL OF PETER.

Petrex. A proprietary trade name for a polybasic acid used in synthetic resin manufacture, the essential constituent of which is 3-isopropyl-6-methyl-3 : 6-endo-ethylene - $\triangle_4$ - tetrahydrophthalic anhydride.

Petrified Asafœtida. An inferior type of asafœtida.

Petro. See PARAFFIN, LIQUID.

Petroacid. A proprietary trade name for a mixture of fatty acids obtained from petroleum distillates.

Petrobenzol. A petroleum distillate. A solvent. It has a boiling-point of 61–96° C.

Petro-Bond. A bonding agent for foundry sands. (572).

Petroclastite (Petroklastite, Holoklastite). An explosive. It contains potassium nitrate, sulphur, coal tar pitch, and potassium bichromate.

Petrofracteur. An explosive resembling Kinetite in composition, but differing in that it contains no gunpowder. It consists of 10 per cent. nitro-benzene, 67 per cent. potassium nitrate, 20 per cent. potassium chlorate, and 3 per cent. antimony pentasulphide.

Petrogens (Petroxolins). Similar to Vasogen (*q.v.*).

Petrohol. A propyl alcohol obtained from the waste gases formed in the Burton petroleum-cracking still by removing the sulphuretted hydrogen from the mixture of propylene and sulphuretted hydrogen, passing the propylene through sulphuric acid, then recovering the propyl alcohol by distillation.

Petroklastite. See PETROCLASTITE.

Petrol. A product of the distillation of petroleum. The term is synonymous with gasoline and petroleum spirit. Other names for the same product are naphtha, petroleum naphtha or mineral naphtha, benzoline, benzine, and carburine. The material usually boils between 40–190° C., and has a specific gravity of 0·65–0·72.

Petrolagar. A proprietary emulsion of liquid paraffin and agar-agar.

Petrolatum. See PETROLEUM JELLY.

Petrolatum Wax. A wax contained in the residual stock from the refining of petroleum waxes. It is one of them, the others being slop wax from the heavier wax distillate, and paraffin from the lighter wax distillate.

Petrolax. See PARAFFIN, LIQUID.

Petrolenes (Malthenes). Constituents of bitumens which are soluble in hexane.

Petroleum Benzine. See C-PETROLEUM NAPHTHA.

Petroleum Ether (Gasoline, Solene). A distillate from petroleum oil, of boiling-point 40–70° C., and specific gravity 0·64–0·65. It consists essentially of pentane and hexane, and is a solvent for resins. Petroleum oil of boiling-point 70–90° C. (Sherwood oil), is called petroleum ether, also that fraction of boiling-point 120–140° C.

Petroleum Jelly (Mineral Jelly, Paraffinum Molle, Petrolatum). Varieties of soft paraffin. They consist of the yellow, semi-solid, purified residue left when petroleum is distilled, and contains several hydrocarbons. The mixture has a specific gravity of 0·87–0·90. An artificial product has been prepared by dissolving 1 part paraffin or cerasin in 4 parts liquid paraffin. Also see OZOKERINE.

Petroleum Naphtha. A term very loosely applied. It often denotes the first fraction of boiling-point up to 150° C., obtained from the distillation of crude petroleum oil, but is sometimes applied to any low boiling petroleum product. See BENZENE.

A-Petroleum Naphtha (Cleaning Oil). That fraction of petroleum distilling at 120–150° C.

B-Petroleum Naphtha (Ligroine, Naphtha, Cleaning Oil). That fraction of petroleum distilling at 100–120° C., of specific gravity 0·707–0·772.

C-Petroleum Naphtha (Petroleum Benzine, Safety Oil). That fraction of petroleum distilling at 80–100° C., of specific gravity 0·667–0·707.

Petroleum Pitch. Asphalt.

Petroleum Spirit (Light Petroleum). Both benzoline and naphtha are sold under this name. They are used as motor spirits, and for dry-cleaning cloths. See BENZINE, BENZOLINE, and NAPHTHA, with all of which the terms are synonymous.

Petroleum, Stockholm. Stockholm tar.

Petrolia. See PARAFFIN, LIQUID.

Petrolina. See OZOKERINE.

Petroline. A fraction of petroleum distillation of boiling-point 120–150° C., of specific gravity 0·722–0·737. Used for defatting, or cleaning. The term is also used for a volatile oil yielded by asphalt, when it is distilled with water.

Petrolit. A German explosive containing potassium chlorate and mineral oil.

Petrolite. A name suggested for solid bitumens. " Kerites " is proposed as a term for those bitumens insoluble in carbon disulphide, and " Asphaltites " for those soluble in carbon disulphide. It is also the name for a Hungarian explosive. It is a mixture of nitre, wood pulp, coke-dust, and sulphur.

Petromor. Petroleum sulphonates. (628).

Petronol. See PARAFFIN, LIQUID.

Petropul. A proprietary trade name for a synthetic resin.

Petrosapol. A brown ointment-like compound of soap and the by-products of petroleum. Used as a vehicle for ointments.

Petrosio. See PARAFFIN, LIQUID.

Petrosulfol. See THIOL.

Petrothene XL. Nos. 6301, 6310, 6311, 6312. A trade mark for cross-linkable grades of high density polyethylene. (41).

Petroxolins. See PETROGENS.

Petroxylin. An inhibitor for petrol and oils. (552).

Petterdite. A mineral. It is an oxychloride of lead.

Pettkoite. A mineral. It is voltaite.

Petzite. A mineral (Ag_2Te+Au_2Te), containing 3·2–25·6 per cent. Au, and 40·8–46·8 per cent. Ag.

Peucedanin. Imperatorin, $CH_3.O.C_6H_4$ $O.C_6H_4.O.CH_2.O.CH_3$. It occurs in the root of masterwort.

Pevafix. An adhesive for polyvinyl alcohol film. (507).

Pevalon. Polyvinyl alcohol film. (507).

Pevidine. A proprietary preparation of POVIDONE iodine used as a skin disinfectant. (137).

Pewter. A variable alloy of from 73–89 per cent. tin, 1·6–6·7 antimony, 1–6·8 per cent. copper, and 0–20·5 per cent. lead, and sometimes zinc. The alloy containing 80 per cent. tin and 20 per cent. lead has a specific gravity of 10 and melts at 200° C. Another alloy of this class consists of 88 per cent. tin, 7 per cent. antimony, 3 per cent. copper, and 1 per cent. zinc. A harder variety contains 75 per cent. tin and 25 per cent. lead.

Pewter, Berthier's. An alloy of 72 per cent. copper, 25 per cent. zinc, 2 per cent. lead, and 1 per cent. tin.

Pewter Solder. See TINSMITH'S SOLDER.

Pexid. A proprietary preparation of PERHEXILINE maleate used in the treatment of angina pectoris. (690).

Pexol. Fortified rosin size. (593).

Peyron's Chloride. Platino-semi-diamine-chloride, $Pt(NH_3.NH_3Cl)Cl$.

Peyton Powder. An explosive. It is a nitro-cellulose and nitro-glycerin powder, containing 20 per cent. ammonium picrate.

Pfaudlon 301. A water suspension of Penton (q.v.).

Pfeilringspalter. A catalyst used in the decomposition of fats. It is prepared by treating with sulphuric acid a mixture of hydrogenated ricinoleic acid and naphthalene.

Pfeufer's Green. A blue-green dye obtained from Chlorospenium æruginosum, a fungus.

Phacelite. A mineral. It is a potassium-aluminium silicate.

Phacolite. See CHABAZITE.

Phanodorm. A German preparation. It is cyclo-hexo-phenyl-ethyl-barbituric acid, and is for use as a general hypnotic.

Phanodorm. A proprietary preparation of cyclobarbitone. A hypnotic. (112).

Phanquinone. Phanquone.

Phanquone. 4,7-Phenanthroline-5,6-quinone. Phanquinone. Entobex.

Phanteine. Specially pure linalyl acetate. (504).

Phantol. Specially pure linalol. (504).

Pharaoh's Serpent Eggs. Made by preparing pills of mercuric thiocyanate, $(NC.S)_2Hg$, and gum. They swell up when heated.

Pharmacolite. A mineral. It is dicalcium arsenate, $Ca_2H_2(AsO_4)_2.5H_2O$. It is also given as $4[CaHAsO_4.2H_2O]$.

Pharmacosiderite. A mineral. It is $[R.Fe_4(AsO_4)_3(OH)_4.6H_2O]$ where R is an exchangeable base. See TYROLITE.

Pharmagel. A proprietary trade name for pure gelatin.

Phasal. A proprietary preparation of lithium carbonate in slow-release form used in the treatment of mania. (330).

Phaseomannite. Inosite, $C_6H_{12}O_6$.

Phazyme. A proprietary preparation of simethicone, PEPSIN, DIASTASE and pancreatin used in the treatment of dyspepsia. (374).

P.H.D. A proprietary trade name for a plasticising oil.

Phecine. Sulpho-*m*-dihydro-oxy-benzene. Recommended as a dermal antiseptic.

Phemaloid. A proprietary trade name for a resin-bonded plywood.

Phemox. Mercury fungicide. (501).

Phenac. A proprietary trade name for a phenolic synthetic resin for varnish and lacquer.

Phenacaine. Holocaine (*q.v.*).

Phenacemide. (Phenylacetyl)urea.

Phenacetin. Acetyl-*p*-phenetidine (Ethoxy-acetamino-benzene), $C_6H_4(O.C_2H_5)$ $NH.CO.CH_3$.

Phenacite. A mineral. It is beryllium silicate, $BeSiO_4$.

Phenactropinium Chloride. N-Phenacylhomatropinium chloride. Trophenium.

Phenadoxone. 6-Morpholino-4,4-diphenylheptan-3-one. Heptalgin is the hydrochloride.

Phenaglycodol. 2-(4-Chlorophenyl)-3-methylbutane-2,3-diol. Ultran.

Phenaldine. A proprietary trade name for a rubber vulcanisation accelerator. It is diphenylguanidine.

Phenald Resins. A general term for phenol-formaldehyde resins.

Phenalgin. Phenol-phthalein, $C_{20}H_{14}O_4$.

Phenalgin. See AMMONAL.

Phenalin. A proprietary trade name for a phenol-formaldehyde synthetic resin.

Phenamein. See MAUVEINE.

Phenamin. See MAUVEINE.

Phenamine. Amino-acet-*p*-phenetidine hydrochloride, $C_2H_5O.C_6H_4.NH.CO.$ $CH_2.NH_2.HCl$.

Phenampromide. N-(1-Methyl-2-piperidinoethyl)-N-propionylaniline.

Phenanthraquinone Red. A dyestuff. It is sodium-phenanthraquinone-di-α-naphthyl-osazone-disulphonate $C_{34}H_{22}$

$O_6N_4S_2Na_2$. Dyes wool red from an acid bath.

Phenaseptin. See ASPIROPHEN.

Phenazocine. 2′-Hydroxy-5,9-dimethyl-2-phenethyl-6,7-benzomorphan. 1,2,3,4,5,6-Hexahydro-8-hydroxy-6,11-dimethyl-3-phenethyl-2,6-methano-3-benzazocine. Narphen is the hydrobromide.

Phenazone. See ANTIPYRINE.

Phenazopirin. See ACOPYRIN.

Phenazopyridine. 2,6-Diamino-3-phenylazopyridine. Gastrotest. Pyridium is the hydrochloride. Present in Mezuran and Uromide as the hydrochloride.

Phenbenicillin. α-Phenoxybenzylpenicillin. Penspek is the potassium salt.

Phenbutrazate. 2-(3-Methyl-2-phenylmorpholino)ethyl α-phenyl-butyrate.

Phenchizine. Phenyl-dihydro-quinazoline.

Phencyclidine. 1-(1-Phenylcyclohexyl)-piperidine. Sernylan is the hydrochloride.

Phendimetrazine. (+)-3,4-Dimethyl-2-phenylmorpholine. Plegine is the tartrate.

Phenedin. Phenacetin (*q.v.*).

Phenegol. The mercury-potassium salt of nitro-*p*-phenol-sulphonic acid. A bactericide.

Phenelzine. Phenethylhydrazine.

Phenergan. A proprietary preparation of promethazine hydrochloride. (336).

Phenester. A proprietary trade name for a synthetic resin of the coumarone-indene type.

Phenethicillin. 6-(α-Phenoxypropionamido)penicillanic acid. (1-Phenoxyethyl)penicillin. Broxil is the potassium salt.

Phenethyl Alcohol. 2-Phenylethanol.

Phenethyl Mustard Oil. *p*-Ethylphenylthiocarbimide, $SCN.C_6H_4.(C_2H_5)$.

Phenetidine. Amino-phenyl ethyl ether, $C_6H_4(OC_2H_5)NH_2$. Antipyretic.

Phenetidine Red. An azo dyestuff obtained from nitro-phenetidine and β-naphthol.

Phenetol. Phenyl-ethyl ether, $C_6H_5.O.C_2H_5$.

Phenetole Red (Coccinin). An azo dyestuff, $C_2H_5.O.C_6H_4.N_2.C_{10}H_{14}(HSO_3)_2.$ OH. It is homologous with Anisole red.

Phenetsal. Acetyl-amino-salol. (Salophene (*q.v.*).)

Pheneturide. (α-Phenylbutyryl)urea. Ethylphenacemide Benuride.

Phenex. A proprietary trade name for a rubber vulcanisation accelerator. An aldehyde-amine.

Phenformin. N^1-Phenethyldiguanide. D.B.I. and Dibotin are the hydrochlorides.

Phengite. Muscovite (q.v.).

Phenglutarimide. α-2-Diethylamino-ethyl-α-phenylglutarimide. Aturbane is the hydrochloride.

Phenic Acid (Phenic Alcohol). Phenol, C_6H_5OH.

Phenic Alcohol. See PHENIC ACID.

Phenicienne. See PHENYL BROWN.

Phenicine. See PHENYL BROWN.

Phenidex. A proprietary preparation of tyrothricin, cetyl pyridinium chloride and benzocaine. Throat lozenges. (364).

Phenin. A proprietary preparation of phenacetin.

Phenindamine. 1,2,3,4-Tetrahydro-2-methyl - 9 - phenyl - 2 - azafluorene. Thephorin is the hydrogen tartrate.

Phenindione. 2-Phenylindane-1,3-dione. Phenylindanedione. Dindevan.

Pheniprazine. α-Methylphenethyl-hydrazine. Cavodil is the hydrochloride.

Pheniramine. 3-Dimethylamino-1-phenyl - 1 - (2 - pyridyl)propane. Daneryl is the 4-aminosalicylate. Trimeton is the maleate.

Phenistix. A proprietary preparation of ferric ammonium sulphate, magnesium sulphate and cyclohexylsulphamic acid impregnated on a test strip, used for the detection of phenylketonuria and ingestion of salicylates. (807).

Phenmetrazine. Tetrahydro-3-methyl-2-phenyl-1,4-oxazine. 3-Methyl-2-phenylmorpholine. Preludin is the hydrochloride.

Phenobarbitone. Ethyl phenyl malonyl urea.

Phenobutiodil. α - (2,4,6 - Tri - iodo - phenoxy)butyric acid. Biliodyl.

Phenocaine. See HOLOCAINE.

Phenocitrain. A proprietary preparation of phenolic analogues, citral and LIGNOCAINE used in the treatment of varicose ulcers. (789).

Phenocitrin. See CITROPHEN.

Phenoco. Said to be a mixture of coal-tar creosote and higher phenol homologues, suspended in a soap solution. An antiseptic.

Phenocoll. Amino-acet-phenetidine (α-amino-phenacetin), $C_6H_4.O.C_2H_5.NH.$ $CO.CH_2.NH_2$, prepared by the introduction of NH_2 group into phenacetin. Antipyretic and antineuralgic.

Phenoctide. A proprietary trade name for Octaphonium chloride.

Phenocyanines. A series of dyestuffs produced by reacting upon gallocyanines with resorcin.

Phenocyanine TC. A dyestuff. It is diethyl - amino - oxy - phenoxazone - oxyphenyl ether, $C_{22}H_{20}O_5N_2$. Dyes chromed wool blue.

Phenocyanine TV. A dyestuff. It is the sulphonic acid of diethyl-amino-oxy-phenoxazone-oxy-phenyl ether. Dyes chromed wool and silk blue. Employed in printing.

Phenocyanine VS. A dyestuff. It is diethyl-amino-dioxy-phenoxazine-oxy-phenyl-ether, $C_{22}H_{22}N_2O_5$. Used for printing blue upon chromed cotton.

Phenoflavine. A dyestuff. It is the sodium salt of m-sulpho-benzene-azo-amino-phenol-sulphonic acid, $C_{12}H_9N_3$ $S_2O_7Na_2$. Dyes wool yellow from an acid bath.

Phenoform. See BAKELITE.

Phenokol. Amino-p-phenetidine, $C_6H_4.$ $O.C_2H_5.NH.CO.CH_2.NH_2$.

Phenolax. See PURGELLA.

Phenolene and Phenolene Supra. Tin plating bath brighteners. (543).

Phenol-Bismuth. A combination of 80 per cent. bismuth with 19 per cent. phenol. Has been used in gastro-intestinal catarrh.

Phenol Blue. Dimethyl-amino-phenyl-imide, $C_{14}H_{14}N_2O$.

Phenol Camphor (Carbolic Camphor). Phenol with camphor (Phenol cum camphoræ B.P.).

Phenol Coralline. See CORALLINE PHTHALEIN.

Phenol-Formaldehyde and Allied Synthetic Resins. See under A.K. BAKELITE, BECKACITE, BE X, BONNYWARE, CHRISTOLIT, COGELBI, CRISTALOID, DURCOTON, ERICON, FORMICA, FUTURITE, HAVEG, KILIANIT, LACRINITE, LORIVAL, LUXENE, NEOLEUKORITE, PANTOLIT, RAMOS, RESINOX, REVOLITE, ROCKITE.

Phenolic Sisal Plastics. Phenol-formaldehyde synthetic resins containing sisal fibres as fillers.

Phenoline. A disinfectant identical with Lysol. It is a cresol made soluble in water by saponification.

Phenolite. A proprietary trade name for a phenol-formaldehyde synthetic resin laminated product.

Phenol-mercury. Mercury carbolate, $Hg(C_6H_5O)_2$. An antiseptic and antisyphilitic.

Phenolphthalein. The lactone of dioxy-triphenyl - carbinol - carboxylic acid, $C_{20}H_{14}O_4$. Used as an indicator in alkalimetry.

Phenolphthalein Synthetic Resin. Alkalit.

Phenol Red. Phenol-sulphonphthalein. An indicator.

Phenol Red. Tetrabrom-phenol-sulphonphthalein. An indicator.

Phenol Soda. Compound solution of sodium carbolate (Liquor sodii carbolatis compositus B.P.C.).

Phenol-sodium Sulphoricinate 25 per cent. This material contains 25 per cent. phenol and 75 per cent. of the sodium salt of sulphonated castor oil.

Phenomauveine. A dyestuff prepared from nitraniline and diphenyl-m-phenylene-diamine.

Phenomet. A proprietary preparation of phenobarbitone and ipecacuanha. A hypnotic. (360).

Phenomine. See TYRAMINE.

Phenomorphan. 3-Hydroxy-N-phenethylmorphinan.

Phenoperidine. Ethyl 1-(3-hydroxy-3-phenylpropyl) - 4 - phenylpiperidine - 4-carboxilate.

Phenopreg. A proprietary trade name for phenolic impregnated fabrics and papers.

Phenopyrine. A mixture of antipyrine and phenol. Used externally as an analgesic in rheumatism and neuralgia, and sometimes internally as an antipyretic.

Phenoquin. An equivalent of atophan. Used in the treatment of rheumatism.

Phenosafranine (Safranine B Extra). A dyestuff. It is diamino-phenylphenazonium chloride, $C_{18}H_{15}N_4Cl$. Dyes cotton mordanted with tannin and tartar emetic, red.

Phenosal. Mono-phenetidide of salicylacetic acid, COOH.C_6H_4.O.CH_2.CO.NH. C_6H_4.OC_2H_5. Used in the treatment of rheumatism.

Phenosalin. See MALAKIN.

Phenosalyl. A mixture of phenol, salicylic acid, menthol, and lactic acid. An antiseptic.

Phenostal. Diphenyl - oxalic ester, asserted to be a combination of oxalic acid and phenol. A germicide.

Phenoval (Bromovaletin, Fenoval). Bromovaleryl - phenetidine, $(CH_3)_2$.CH. CHBr.CO.NH.C_6H_4.O.C_2H_5. An antipyretic.

Phenox. A proprietary trade name for phenyl-mercury-hydroxide.

Phenoxetol. Monoaryl ethers of aliphatic glycols. (625).

Phenoxin. Carbon tetrachloride, CCl_4.

Phenoxybenzamine. 2-(N-Benzyl-2-chloroethylamino) - 1 - phenoxypropane. Dibenylene and Dibenzyline are the hydrochloride.

Phenoxylene Plus. A selective weed killer. (618).

Phenoxymethylpenicillin. A biosynthetic penicillin formed by fermentation with suitable precursors, of Penicillium Notatum. Penicillin V. Apsin VK. Calcipen-V. Compocillin-VK. Crystapen V. Distaquaine V. Eskacill in V. Icipen V. Penavlon V. Penicals. Stabillin V-K. and V-Cil-K are the free acid or its salts.

Phenoxypropazine. (1 - Methyl - 2 - phenoxyethyl)hydrazine. Drazine is the hydrogen maleate.

Phenprobamate. 3-Phenylpropyl carbamate. Gamaquil.

Phenprocoumon. 4-Hydroxy-3-(1-phenylpropyl)coumarin. Marcoumar.

Phensedyl Linctus. A proprietary preparation containing promethazine hydrochloride, codeine phosphate and ephedrine hydrochloride. A cough linctus. (336).

Phensuximide. N - Methyl - α - phenyl-succimide. Milontin.

Phentermine. $\alpha\alpha$-Dimethylphenethylamine. Duromine (an ion-exchange resin complex).

Phentolamine. 2-N-(3-Hydroxyphenyl)-p-toluidinomethyl-2-imidazoline. Rogitine is the hydrochloride or the mesylate.

Phenurone. A proprietary trade name for Phenacimide.

Phenyform. An antiseptic powder prepared from phenol and formaldehyde. Used as an indicator, also medicinally, and for denaturing purposes.

Phenyl-acetamide. See ANTIFEBRIN.

Phenyl Acetone. Benzyl-methyl-ketone, C_6H_5.CH_2.CO.CH_3.

Phenylalanine. α-Amino-β-phenyl-propionic acid, C_6H_5.CH_2CH(NH$_2$)COOH.

Phenylamine. Aniline, $C_6H_5NH_2$.

Phenyl Aminosalicylate. Phenyl 4-aminosalicylate. Fenamisal. Pheny-PAS-Tebamin.

Phenylaspriodine. Acetyl-iodo-salol.

Phenylbarbital. See LUMINAL.

Phenyl Brown (Phenicine, Phenicienne, Rotheine). A yellow-brown powder consisting of dinitro-phenol, and an amorphous brown substance. It dyes wool and silk, but explodes on heating, and is now obsolete.

Phenylbutazone. 4 - n - Butyl - 1,2 - diphenylpyrazolidine-3,5-dione. Butazolidin.

Phenylene Black. See ANTHRACITE BLACK B.

Phenylene Blue. See MELDOLA'S BLUE.

Phenylene Brown. See BISMARCK BROWN.

Phenylenefil. A registered trade name for flame retardant modified polyphenylene-oxide. (402).

Phenylephrine. $(-)$-1-(3-Hydroxy-phenyl)-2-methylaminoethanol. Neophryn and Neosynephrine are the hydrochloride.

Phenyl-Gamma Acid. 2-Phenyl-amino-8-naphthol-6-sulphonic acid.

Phenylindanedione. Phenindone.

Phenyl Methane. See METHYL BENZENE.

Phenyl-methyl-ketone (Acetyl-Benzene, Hypnone). Acetophenone, C_6H_5. CO.CH_3. A hypnotic.

Phenylon. See ANTIPYRINE.

Phenyl-peri Acid. 1-Phenyl-naphthylamine-8-sulphonic acid.

Phenylpropanolamine. A sympathomimetic. 2-Amino-1-phenylpropan-1-ol. PROPADRINE. Present in ESKACEF, GALATON and RINUREL as the hydrochloride.

Phenyl-sedaspirin. Acetyl-bromo-salol.

Phenyltoloxamine. An anti-histamine. 2-Benzylphenyl 2-dimethylaminoethyl ether. Present in RINUREL as the citrate.

Phenyl Violet. See REGINA PURPLE.

Pheny-Pas-Tebamin. A proprietary preparation of phenyl aminosalicylate. An antituberculous agent. (821).

Phenyramidol. α-(2-Pyridylaminomethyl)benzyl alcohol. Analexin is the hydrochloride.

Phenzoline. See OREXINE.

Pheophytin. The brownish derivative obtained by treating chlorophyll with acid.

Philadelphia Yellow G. See PHOSPHINE.

Philadelphite. A mineral. It is a variety of vermiculite.

Philanised Cotton. Cotton material which has been treated with concentrated nitric acid to convert it into a wool-like fabric.

Philgas. Liquefied petroleum gas distributed in the United States of America. It consists chiefly of propane and butane.

Philipstadite. A mineral. It is a variety of hornblende.

Phillipite. A mineral, $CuSO_4.FeSO_4$.

Phillipsite (Spangite). A zeolite containing potassium.

Philonin. A German preparation. It is an ointment for wounds containing copper iodo-o-oxy-quinolate, silver sulphate, boric acid, balsam of Peru, trypaflavine, and zinc paste.

Philosopher's Wool (Flowers of Zinc). The zinc oxide produced in a flocculent condition by burning zinc.

Phisomed. A proprietary preparation of HEXOCHLOROPHENE in a detergent base used as a skin cleanser. (439).

Phlobaphenes. Red or brown colouring matter from barks, usually oak bark.

Phloba-tannins. Tannins which give the phlobaphene reaction.

Phlogin. A German product. It consists of phenyl-cinchonine-carboxylic (75 per cent.), tannic acid (18 per cent.), and alumina (6·3 per cent.). Tablets are used for rheumatism and neuralgia.

Phlogopite. A mineral. It is a magnesia mica, $AlMg_3KH_2Si_3O_{12}$.

Phloroglucinol. 1 : 3 : 5-trihydroxybenzene, $C_6H_3(OH)_3$.

Phlorol. o-Ethyl-phenol, $C_6H_4.C_2H$ (OH).

Phlorone (Metaphlorone). p-Xyl(quinone, $C_6H_2(CH_3)_2O_2$.

Phloxine (Phloxine TA, Eosine 10F Cyanosine). The sodium salt of tetra bromo-tetrachloro-fluoresceine, $C_{20}H$ $Cl_4Br_4O_5Na_2$. See PHLOXINE P.

Phloxine P (Phloxine, New Pink, Erythrosin BB). Dyestuffs. The alkali salt of tetrabromo - dichloro - fluorescein $C_{20}H_4Cl_2Br_4O_5K_2$. Dyes wool bluish red, and cotton mordanted with tin alumina, or lead.

Phloxine TA. See PHLOXINE.

Phobotex. Water repellent products (530).

Phobotex FTN. A proprietary trade name for a fat modified melamine resin with outstanding water-repellancy. It i fast to repeated washing in soap and soda and provides a durable water repellant finish for textiles. (18).

Phocenic Acid. Chevreul's name for isovaleric acid.

Phocil. A proprietary preparation con taining pholcodine, liquid extract o cocillana, euphorbia, squill, senega and cascara and potassium acid tartrate menthol and glycerin. A cough linctus (344).

Phœnicite. Basic lead chromate.

Phœnicochroite. A mineral, $PbCrO_4$.

Phœnix Alloy. An electrical resistance alloy containing 25 per cent. nickel and 75 per cent. iron.

Phœnixite. A proprietary pyroxylin product.

Phœnix Powder. An explosive containing 28–31 per cent. nitro-glycerin, 0–1 per cent. nitro-cotton, 30–34 per cent. potassium nitrate, and 33–37 per cent. wood meal.

Pholcodine. 3-(2-Morpholinoethyl)-morphine. Ethnine. Glycodine.

Pholcomed Forte Diabetic Linctus. A proprietary preparation containing pholcodine and papaverine hydrochloride in a sugar free base. A cough linctus. (250).

Pholcomed Forte Linctus. A proprietary preparation containing pholcodine and papaverine hydrochloride. A cough linctus. (250).

Pholedrine. 4-(2-Methylaminopropyl)-phenol. Veritain. Veritol.

Pholerite. A mineral. It is aluminium silicate.

Pholin's Alloy. An alloy of 77 per cent. tin, 19 per cent. bismuth, and 4 per cent. copper.

Pholtex. A proprietary preparation containing pholcodine and phenyltoloxamine. A cough linctus. (275).

Phono Bronze. A proprietary trade name for certain copper alloys containing about 1·25 per cent. tin and small amounts of silicon and cadmium.

honolite. A mineral. It contains silica, alumina, iron, calcium, potassium and sodium.

horan. See CAMITE.

horochrome Yellow Y. A dyestuff. It is a British equivalent of Milling yellow.

hosbrite. Chemical polishing solutions for metals. (503).

hos-Chek P/30. A proprietary trade name for a fire retardant material for paint. It is ammonium pyrophosphate. (57).

hosclere. Stabilisers and antioxidants. (583).

hos-copper. A proprietary welding alloy composed essentially of copper with from 5–10 per cent. phosphorus. It melts at 700° C., becoming extremely fluid at 750° C.

hosene. Synanthene, $C_{14}H_{10}$.

hosflex 300 and 400. Proprietary mixed triaryl phosphate ester plasticisers containing halogen and possessing better flame-retardant properties in PVC than tricresyl phosphate. (401).

hosforme. Monoethyl-phosphoric acid.

hosgard. Flame retardants. (559).

hosgard 2XC-20. A proprietary organic phosphate ester plasticiser containing halogen, used as a non-reactive flame retardant in polyurethane foams. (57).

hosgene Gas. Carbonyl chloride, $COCl_2$.

hosgene Spar. A mineral. It is kerasite.

hosgene Toluol Solution 20 per cent. A 20 per cent. solution of phosgene, $COCl_2$, in toluene.

hosgenite (Phosgene Spar). See KERASITE.

hoso-guaiacol. Guaiacol phosphite, $P(OC_6H_4OCH_3)_3$. Used for the administration of guaiacol.

hosote. See PHOSPHOTE.

hospham. Phosphorus imidonitride, PN_2H or $N : P : NH$.

hosphammite. Native ammonium phosphate.

hosphate, Basic. See BASIC SLAG.

hosphate, Rock. See PHOSPHORITE.

hosphate-Sandoz. A proprietary preparation of sodium acid phosphate and sodium and potassium bicarbonates used in the treatment of hypercalcæmia. (478).

hosphate, Somberero. See SOMBERITE.

hosphate, Thomas. See BASIC SLAG.

hosphate, Vesta. See RHENANIA PHOSPHATE.

Phosphatic Guano. A manure found on islands in the Atlantic Ocean.

Phosphatol. See CREOSOTE PHOSPHITE.

Phosphazote. A fertiliser. It is an intimate mixture of superphosphate and urea containing 4–11 per cent. nitrogen and 10–14 per cent. P_2O_5.

Phosphin. 3-Amino-9-*p*-amino-phenyl-acridine. Used in the treatment of malaria.

Phosphine (Leather Yellow, Xanthine, Vitoline Yellow 5G, Patent Phosphine, Nankin, Philadelphia Yellow G, Chrysaniline, Acridine Yellow, Acridine Orange, Leather Brown, Phosphine II, N, P). The nitrate of chrysaniline (unsym - diamino - phenyl - acridine, $C_{16}H_{26}N_4O_3$), and homologues. Dyes leather reddish-yellow.

Phosphine (Phosphoretted Hydrogen). Phosphorus trihydride, PH_3.

Phosphine II. See PHOSPHINE.

Phosphine N. See PHOSPHINE.

Phosphine P. See PHOSPHINE.

Phosphine, Patent. See PHOSPHINE.

Phosphocalcite. See PSEUDOMALACHITE.

Phosphocerite. A mineral. It is $(La,Ce)PO_4$. See RHABDOPHANE.

Phosphochalcite. See PSEUDOMALACHITE.

Phosphochromite. A mineral. It is a lead phospho-chromate.

Phosphoferrite. A mineral. It is an acid phosphate of ferrous iron, magnesium, and calcium, found in Bavaria.

Phospho-gélose. A German clarifier for sugar juice. It consists of 70 per cent. phosphate of lime, and 30 per cent. kieselguhr.

Phospholine Iodide. Ecothiopate iodide.

Phospholutein. Lecithin.

Phosphomort Organo-phosphorous insecticides. (501).

Phosphophyllite. A mineral. It is a hydrated phosphate and sulphate of ferrous iron, magnesium, calcium, potassium and aluminium, found in Bavaria.

Phosphorated Oil. (*Oleum phosphoratum B.P.*) It contains 98 per cent. almond oil, 1 per cent. phosphorus, and 1 per cent. oil of lemon.

Phosphor Bronze. A bearing metal. It is an alloy of from 70–97 per cent. copper, 3–13 per cent. tin, 0–16 per cent. lead, and 0·1–1·0 per cent. phosphorus and sometimes a little zinc. The alloys for casting usually contain 85–92 per cent. copper, 7–13 per cent. tin, 0·3–1·0 per cent. phosphorus, and traces of lead and zinc. For malleable phosphor bronze the alloys consist of 94–97 per cent. copper, 3–5 per cent. tin, and 0·1–0·35 per cent. phosphorus. One alloy contains 79·7 per cent. copper, 10 per cent. tin, 9·5 per cent. antimony, and 0·8 per cent. phosphorus.

It has a specific gravity of 8·8. The average specific resistance of an ordinary phosphor bronze at 0° C. is 8·5 microhms. cm.

Another contains 94 per cent. copper, 6 per cent. tin, and traces of phosphorus. It melts at 1050° C. and has a specific gravity of 8·94.

The alloy of 92 per cent. copper, 8 per cent. tin, and up to 0·3 per cent. phosphorus has a melting-point of 1000° C. and a specific gravity of 8·68.

Also see AJAX BRONZE, KUHNE PHOSPHOR BRONZE, and LOWROFF PHOSPHOR BRONZE.

Phosphor Copper. An alloy of copper with from 5–15 per cent. of phosphorus. Used as an addition to other metals, and in the manufacture of phosphor bronze.

Phosphorerdenepidot. Synonym for Nagatelite.

Phosphoretted Hydrogen. See PHOSPHINE.

Phosphoric Acid, Glacial. Metaphosphoric acid, HPO_3. Usually it contains some sodium phosphate.

Phosphoric D. Iron D (q.v.).

Phosphoric Ether. Triethyl-phosphate, $OP(OC_2H_5)_3$. Also see SULPHURIC ETHER.

Phosphorite (Rock Phosphate). A mineral. It is calcium phosphate, $Ca_3(PO_4)_2$.

Phosphorochalcite. A mineral, $Cu(OH)_3 PO_4$.

Phosphoro - Orthite. Synonym for Nagatelite.

Phosphorous Ether. Triethyl-phosphite, $(OC_2H_5)_3P$.

Phosphor Resin. A resin prepared from triphenyl phosphamide by heating and passing carbon dioxide through the compound.

Phosphor-Rösslerite. A mineral. It is $8[MgHPO_4.7H_2O]$.

Phosphor Steel. An alloy of steel with phosphorus.

Phosphor Tin. An alloy of tin and phosphorus, containing up to 10 per cent. phosphorus.

Phosphorus, Amorphous. See RED PHOSPHORUS.

Phosphorus, Hittorf's. Black phosphorus (q.v.).

Phosphorus Paste. A rat poison made from phosphorus and flour.

Phosphorus Pill. (*Pilula phosphori B.P.*) It contains 1 gram of phosphorus, 40 grams of oil of theobromine, 11 grams of wool fat, 16 grams of kaolin, 32 grams of sodium sulphate, and 20 grams of carbon disulphide.

Phosphorus Salt. See MICROCOSMIC SALT.

Phosphorus, Schenk's. See SCARLE PHOSPHORUS.

Phosphoscorodite. A mineral. It $8[Fe(As,P)O_4.8H_2O]$.

Phosphosiderite. A mineral. It $4[FePO_4.2H_2O]$.

Phosphosol (Phosphotex). A proprietary trade name for tetrapotassium pyrophosphate.

Phosphotal. Creosote phosphite. It contains 90 per cent. creosote, and is used in the treatment of phthisis.

Phosphote (Phosote). Creosote phosphate, $PO_4(C_6H_7)_3$. It contains 80 per cent. creosote, and is antiseptic.

Phosphotex. See PHOSPHOSOL.

Phosphuranylite. A mineral. It is hydrated phosphate of uranium, $(UO_2 (PO_4)_2.6H_2O$, and occurs as an incrustation on quartz, felspar, and mica in North Carolina.

Phosphuret of Baryta. A brown-red mixture of barium phosphide an phosphate.

Phosvichin. A quinine-lecithin solution containing 10 per cent. quinine.

Photal. A photographic developer.

Photodyn. A proprietary trade name for a hæmato-porphyrin preparation which is active only in the light. It is most sensitive in yellow light. It is used therapeutically.

Photogen. A variety of benzine, similar to that from petroleum, but obtained by the distillation of wood, lignite, and coal. It is used in the purification of paraffin, in carburetting lighting gas and for cleaning.

Photophor. A registered trade name for a calcium phosphide, Ca_2P_2. Used for signal fires.

Photoxylin. See CELLOIDIN.

Phoxim. An insecticide and anthelmintic currently undergoing clinical trial as " Bayer 9053 ". It is α-diethoxyphosphinothioyloxyimino - α - phenylaceto - nitrile.

Phoxim-Methyl. A pesticide. O-α-cyanobenzylideneamino oo-dimethyl phosphorothioate.

Phrenotropin. A proprietary trade name for the hydrochloride of Prothipendyl.

Phtalopal. A proprietary trade name for a series of phthalic acid-alcohol condensates or esters. (3).

Phtalopal PP. A proprietary trade name for a phthalic anhydride-pentaerythritol condensate used as a substitute for shellac. (3).

Phtalopal SEB. Similar to 1156 except that it is made to a higher degree of condensation. (3).

Phtalopal BU. A proprietary trade name for 1,3-butylene glycol phthalate. (3).

Phtalopal 1156. A proprietary trade name for a phthalic anhydride-trimethyl propane condensate used as a shellac substitute. (3).

Phthalate 79. A mixed heptyl-nonyl phthalate vinyl plasticiser. (190).

Phthalofyne. See WHIPCIDE.

Phulwa. The fat obtained from the seeds of *Bassia butyracea*. Also see BASSIO TALLOW.

Phycite. See ERYTHROMANNITE.

Phyldrox. A proprietary preparation of aminophylline suppositories. A bronchial antispasmodic. (246).

Phyldrox-G. A proprietary preparation of theophylline sodium glycinate, ephedrine hydrochloride and phenobarbitone. A bronchial antispasmodic. (246).

Phyllinglanz. See NOBILITE.

Phyllite. See CHLORITOID.

Phyllocontin. A proprietary preparation of aminophylline used in the treatment of bronchospasm. (334).

Phyllol. Specially pure eugenol. (504).

Physalite. A mineral, $Al_2O_3.SiO_2$.

Physeptone. A proprietary preparation of methadone hydrochloride. An analgesic. (277).

Physeptone Linctus. A proprietary containing methadone hydrochloride. A cough linctus. (277).

Physic Nut. The seed of *Jatropha curcas*. It contains curcas oil, which has purgative and emetic properties.

Physic Nut Oil. See PURGING NUT OIL.

Physiological Salt Solution (Normal Salt Solution, Isotonic Salt Solution, Surgical Solution). This consists of 8·5 grams of sodium chloride in 1,000 c.c. of distilled water. It is sterilised and used for intravenous injection.

Physiol Soaps. Soaps of German manufacture in which all or part of the fatty acids in them have been replaced by polysaccharides.

Physostigma. See CALABAR BEAN.

Physostigmine. Eserine, $C_{15}H_{21}N_3O_2$, an alkaloid from the calabar bean.

Physostol. A sterile solution of 1 per cent. of an eserine salt in olive oil. Used in the treatment of certain eye diseases.

Phytex. ˆA proprietary preparation of boratannic complex in alcohol in ethylacetate solvent. Used in dermatology. (330).

Phytic Acid (Phytinic Acid). Inositolhexaphosphoric acid, $C_6H_6(OPO_3H_2)_6$.

Phytin. Calcium magnesium salt of inositol hexaphosphoric acid; contains about 22 per cent. phosphorus, 12 per cent. calcium, and 1·5 per cent. magnesium. A powerful nerve and general tonic.

Phytocil. A proprietary preparation of 3-phenoxypropanol, 2-P-chlorophenoxyethanol, salicylic acid, menthol and glycerin. (340).

Phytodermine. A proprietary preparation of methyl hydroxybenzoate and salicylic acid. A skin fungicide. (507).

Phytoforol. Vitamin E capsules. (182).

Phytol. A primary alcohol, $C_{20}H_{39}OH$, obtained by the decomposition of chlorophyll, the colouring matter of plants.

Phytomenadione. 2-Methyl-3-phytyl-1, 4-naphthaquinone. Vitamine K. Konakion. Mephyton. A synthetic analogue of Vitamin K. Available as Aqua-mephyton.

P.I.B. (1) A proprietary preparation of isoprenaline hydrochloride and atropine methonitrate in an aerosol. A bronchial antispasmodic. (339).

P.I.B. (2) An abbreviation for polyisobutylene.

Picamar. Propyl - pyrogallol - dimethyl ether, $C_{11}H_{16}O_3$. Used in perfumery.

Piccolastic A-5, A-25. Proprietary trade names for vinyl plasticisers based upon polymerised styrene and homologues. (121).

Piccolyte. A proprietary trade name for thermoplastic terpene resins.

Piccoumaron. A proprietary trade name of terpene varnish and lacquer resins.

Piccoumarone Resins. Proprietary trade name for coumarone-indenes.

Pichurim Beans. Sassafras nuts, the seeds of *Nectandra puchury*. An aromatic.

Pichurim Camphor. A substance resembling laurel camphor obtained from pichurim beans.

Picite. A mineral. It is a hydrated phosphate of iron.

Picked Turkey Gum (White Sennaar Gum). The best variety of gum acacia.

Pickeringite. A mineral. It is $MgAl_2(SO_4)_4.22H_2O$. See FEATHER ALUM.

Pickle Alum. Aluminium sulphate, $Al_2(SO_4)_3$. Used for packing and preserving.

Pickle Green. A commercial variety of Scheele's green.

Pickling Acid. Acetic acid, CH_3COOH, is called by this name. Sulphuric acid, H_2SO_4, is also known by this term, and is used for treating iron and steel.

Picloxydine. 1,4-Di-(4-chlorophenylguanidinoforminidoyl)-piperazine.

Picoleum. A rubber-tar product used for road or pavement surfacing.

Picoline. Methyl-pyridine, C_6H_7N.

Pi-cone. A proprietary lithopone containing 15 per cent. titanium oxide, 25 per cent. zinc sulphide, and 60 per cent. precipitated barium sulphate. It is stated to have a much higher covering power than ordinary lithopone.

Picotite (Chromospinel). A mineral, $MgO.Al_2O_3$, with iron and chromium.

Picraalluminite. Synonym for Piero-allumogene.

Picrasmin. Quassin, $C_{10}H_{12}O_3$.

Picrastol. Dimethylol - formyl - me - thenyl - tetramethylene - pentamine $C_9H_{17}N_5O_4$.

Picrate Powder. The name given to explosive powders, in which the main constituent is the potassium or ammonium salt of picric acid.

Picratol. Silver picrate, $C_6H_2O(NO_2)_3$ $Ag+H_2O$. An antiseptic used in gonorrhœa.

Picric Acid. Trinitro-phenol, C_6H_2. $OH(NO_2)_3$. Used for making explosives and dyes. A solution is employed in the treatment of burns, erysipelas, eczema, and gonorrhœa. Also see BRANDOL, CARBAZOTIC ACID, IGEW-SKY'S REAGENT, and PICRONTIC ACID.

Picric Powder. An explosive consisting of ammonium picrate and potassium nitrate.

Picroallumogene. A mineral. It is $Mg_2Al_2(SO_4)_5.28H_2O$.

Picro-aniline Blue. A stain used in microscopy. It is prepared by adding aniline blue to a saturated solution of picric acid in 92 per cent. alcohol until the liquid becomes deep blue-green in colour.

Picrocarmine. A microscopic stain obtained by mixing 1 gram carmine in 10 c.c. water and 3 c.c. strong ammonia solution, and adding the mixture to 200 c.c. of a saturated solution of picric acid. Allow to stand and filter.

Picrocrichtonite. Synonym for Picroilmenite.

Picrofoetidine (Lapintine, Pomoline). Trade names for insecticides usually consisting of putrid fish or animal oils. Used for the protection of fruit trees.

Picro-formol (Bouin). A fixing agent used in microscopy. It contains 75 parts of a saturated aqueous solution of picric acid, 25 parts of 40 per cent. formaldehyde, and 5 parts of glacial acetic acid.

Picroilmenite. A mineral. It is $2[(Mg,Fe)TiO_3]$.

Picroknebelite. Synonym for Talk-nebelite.

Picrol. Di - iodo - resorcin - potassium - mono-sulphonate, $C_6HI_2(OH)_2SO_3K$. An antiseptic.

Picrolite. Serpentine (q.v.), of a fibrous structure.

Picromerite. A mineral, $(K_2Mg)SO_4$.

Picronigrosine. An alcoholic solution of picric acid and nigrosin. A microscopic stain.

Picrontric Acid. Picric acid (q.v.).

Picropharmacolite. A mineral, $(Ca.Mg)_3(AsO_4)_2.6H_2O$.

Picro-sulphuric Acid. A liquid made by adding to 100 volumes water, 2 volumes sulphuric acid, and about 0.25 per cent. picric acid. Used in microscopy as a fixing agent.

Picrotanate. Synonym for Picroilmenite.

Picrotitanite. A mineral. It is a magnesium-iron titanate. Synonym for Picroilmenite.

Picryl Brown. A colouring matter obtained by sulphonating the nitro derivatives of secondary and tertiary aromatic amines. Dyes silk and wool yellow from an acid bath.

Pictet Crystals. White crystals, SO_2. XH_2O, formed when liquid sulphur dioxide evaporates.

Pictet's Fluid. Liquid carbon dioxide, used for freezing machines.

Pictet's Liquid. A mixture of liquid carbon dioxide and sulphur dioxide. Used for producing low temperature.

Pictite. A variety of the mineral sphene or titanite, $CaTiSiO_5$.

Pictolin. A mixture of liquid carbon dioxide and sulphur dioxide.

Piedmontite. A mineral, $4CaO.3(Al,Mn)_2O_3.6SiO_2.H_2O$.

Piemontite (Piedmontite). A mineral. It is $2[Ca_2(Al,Fe,Mn''')_3Si_3O_{12}OH]$.

Pierrot Metal. An alloy consisting mainly of zinc, with smaller amounts of copper, tin, antimony, and lead.

Pifenate. Ethyl 2,2-diphenyl-3-(2-piperidyl)propionate.

Pigeon Berry. Phytolacca.

Pigeonite. A mineral. It is $8[(Mg,Fe,Ca)SiO_3]$.

Pig Iron. Cast iron in pieces of D section are called pigs. It is an impure form of iron, usually containing from 92-93 per cent. iron, and at least 2.5 per cent carbon as graphitic carbon. It is classified into six classes, Nos. 1, 2, 3, 4 mottled, and white, according to the appearance of the fracture. No. 1 contains the largest flakes of graphite, and white the smallest.

Pig Iron, Glazy. See GLAZY PIG.

Pig Iron, Grey Forge. See GREY FORGE PIG.

Pig Lead. Lead is obtained from galena by heating in a reverberatory furnace with a silica flux, then heated with coke and sometimes lime in a cupola furnace. The lead drawn off is called pig lead.

Pigment 40-40-20. An American pigment. It contains 40 per cent. zinc oxide, 40 per cent. lithopone, 10 per cent. silica, and 10 per cent. asbestine.

Pigmentar. A proprietary trade name for standardised pine tar prepared for use as a rubber softener in rubber compounding. It is made in three consistencies having the following characteristics:—

	S.G. at 25° C.	Sp. viscosity at 25 °C.	Distillation
Thin	1·065	75	85 per cent. between 178 and 335° C.
Medium	1·075	275	75 per cent. between 178 and 338° C.
Heavy	1·085	3500	60 per cent. between 211 and 346° C.

Practically nothing distils over below 170° C. from any one of these.

Pigment Brown. See SUDAN BROWN.

Pigment Chrome Yellow L. A dyestuff made from diazotised *o*-toluidine, and 1-phenyl-3-methyl-5-pyrazolone.

Pigment Fast Orange L Extra. A dyestuff. It is a British equivalent of Pigment orange R.

Pigment Fast Red HL. See HELIO FAST RED RL.

Pigment Fast Scarlet 3L Extra. A dyestuff. It is a British brand of Helio fast red RL.

Pigment Fast Yellow Conc., New. See HANSA YELLOW G.

Pigment Fast Yellow GRL Extra. A dyestuff obtained by coupling diazotised *m*-nitro-*p*-toluidine with acetoacetic toluidide.

Pigment Orange R (Pigment Fast Orange L Extra). A dyestuff obtained from diazotised *p*-nitro-*o*-toluidine coupled with *β*-naphthol.

Pigment Purple. A dyestuff formed by diazotising *o*-anisidine and coupling with *β*-naphthol.

Pigment Scarlet 3B. A dyestuff obtained by diazotising anthranilic acid and coupling it with R salt.

Pig, Silicon. See SILICON-EISEN.

Pilarite. A mineral approximating to the formula, $CaCu_5Al_6Si_{12}O_{39}.24H_2O$.

Pilasonite. A mineral, Bi_3Te_2.

Pilbarite. A mineral, $PbO.UO_3.ThO_2.2SiO_2.2H_2O$.

Pilinite. A mineral. It is an aluminium-calcium silicate.

Pili Nuts. The seeds of *Canarium luzonicum*. Used in America for dessert.

Piliophen. Sodium-tetra-iodo-phenol-phthalein.

Pilite. A mineral. It is a variety of amphibole.

Pilocarp-phenol. Pilocarpine carbolate.

Pilocarpus. Jaborandi leaves.

Pilolite. A mineral. It is a hydrated aluminium and magnesium silicate, $4MgO.Al_2O_3.10SiO_2.15H_2O$.

Pilsenite. A mineral. It is tetradymite.

Pimafucin. A proprietary preparation of NATAMYCIN used in the treatment of fungal infections. (317).

Pimaricin. A proprietary trade name for Natamycin.

Pimeleine. Petrolatum.

Pimelite. A mineral. It is meerschaum containing nickel.

Piminodine. Ethyl 4-phenyl-1-(3-phenylaminopropyl)piperidine-4-carboxilate.

Pimozide. A tranquilliser. 1-{1-[4,4-Bis-(4-fluorophenyl)butyl]-4-piperidyl}benzimidazolin-2-one. ORAP.

Pimple Metal. A term used for a type of copper metal produced from the "coarse metal" obtained from sulphide ores which have been fused with an excess of copper oxide in their purification.

Pinachrom. *p*-Ethoxy-quinaldine-*p*-methoxy-quinoline-ethyl-cyanine-bromide. A red sensitiser for silver bromide plates.

Piña Cloth. A Philippine cloth made from pineapple fibre.

Pinacoline. Methyl-*ter*-butyl-ketone, $CH_3.CO.C(CH_3)_3$.

Pinacyanol. A red sensitiser for silver bromide plates obtained by treating quinaldinium salts with formaldehyde followed by alkali.

Pinaflavol. A basic dye used as a green sensitiser in photography.

Pinakiolite. A mineral, $(Mg.Mn)_2O(BO_2)MnO_2$.

Pinakol. A pyrogallol photographic developer in which part of the alkali usually employed is replaced by sodium-amino-acetate.

Pinakol P. A photographic developer. Pyrogallol is the developing substance, but it also contains pinakol salt N.

Pinakol Salt N. A 20 per cent. solution of sodium-amino-acetate, $CH_2.NH_2.COONa$. It replaces the alkali in organic developers.

Pinakryptol. A green dye that is used as a photographic sensitiser.

Pinaverdol. A green sensitiser for silver bromide plates. It is *p*-tolu-quinaldine-quinolinium-methyl-cyanine bromide.

Pinchbeck. An alloy of from 83–93 per cent. copper and 6–17 per cent. zinc. A brass.

Pincoffin. Commercial alizarin.

Pindolol. A beta adrenergic receptor blocking agent currently undergoing clinical trial as " LB46 " and " Visken ". It is 1-(indol-4-yloxy)-3-isopropylaminopropan-2-ol.

Pineapple Essence. See ESSENCE OF PINEAPPLE.

Pineapple Oil, Artificial. Ethyl butyrate dissolved in alcohol. Used in confectionery and perfumery. Also see ESSENCE OF PINEAPPLE.

Pineapple Oil Essence. See ANANAS OIL.

Pine Camphor. Sombrerone, a terpene, $C_{10}H_{16}O$.

Pine Gum (Pine Resin, White Pine Resin, Cypress Pine Resin). Names applied to Australian sandarac resin, obtained from *Callitris quadrivalis* and *C. calcarata*. See SANDARAC RESIN.

Pine Needle Ether. A mixture of 120 parts pine oil, 250 parts absolute alcohol 40 parts ether, 10 parts lavender oil, and 10 parts juniper oil. Another mixture contains 1 part pine-needle oil and 9 parts petroleum ether.

Pine-needle Oil, Artificial. Bornyl acetate.

Pine Oil. The name was originally applied to turpentine oils obtained from pine trees. The term is used in America to designate turpentine obtained by distilling pine wood. It is also used for the lighter oils of pine tar, a refined rosin oil, and a by-product in the manufacture of wood pulp by the sulphite process. Proprietary brands of pine oil obtained from waste wood and the stumps of yellow pine by steam distillation are "Herco," recommended as an alchohol denaturant, "Risor," a crude brand used as a reagent in the flotation of some ores, and is also recommended for use in the manufacture of rubber goods, "Floto," and "Yarmor," the latter for use as a frothing agent in the flotation of ores and in the manufacture of disinfectants, etc. The proprietary brands are obtained from waste wood and stumps of yellow pine by steam distillation, e.g. "Herco": specific gravity 0·925–0·935, refractive index 1·479. Residue after polymerisation less than 3·5 per cent. 90 per cent. distils below 220° C. Moisture less than 0·5 per cent. Flashpoint above 60° C. Used as an alcohol denaturant.

Pine Resin. See PINE GUM.

Pine Resin, Cypress. See PINE GUM.

Pine Resin, French. See GUM THUS.

Pine Resin Shellac. Pine resin is steam distilled to remove turpentine, and the residue extracted with turpentine. The remainder is extracted with alcohol and the extract evaporated to give pine resin shellac.

Pine Tar. See STOCKHOLM TAR.

Pine Wool (Fir Wool). Leaves of *Pinus sylvestris* broken down into a woolly condition. Used for making vermin-repelling blankets.

Piney Tallow. Malabar tallow, an edible fat obtained from the seeds of *Vateria indica*, of East Indies.

Pingos d'Agoa. The Brazilian name for topaz.

Pinguin. Alantol, $C_{10}H_{16}O$, obtained from the roots of *Inula elecampane*.

Pinite. A mineral, $2Al_2O_3.K_2O.5SiO_2.3H_2O$. It is also the name given to monomethyl-cyclohexanol, $C_6H_6OCH_3(OH)_5$, discovered in *Pinus lambertiana*. It has been isolated as sennite from senna leaves and as matezite from the juice of *Mateza roritina*.

Pink. An old name for safranine.

Pink Cutting Liquid. See TIN SPIRITS.

Pink Madder. See MADDER LAKES.

Pinko and **Pink-o-perf.** Proprietary trade names for absorbent paper for plastic lamination.

Pink Root. The root of *Spigelia marilandica*.

Pink Salt. Ammonium-stannic-chloride $SnCl_4.2NH_4Cl$. Formerly used as a mordant for dyes.

Pinna Silk. Byssus silk (*q.v.*).

Pinnay Oil. See LAUREL-NUT OIL.

Pinnoite. A mineral, $Mg(BO_2)_2.3H_2O$, which occurs at Stassfurt. It is decomposed by water. It is also given as $4[MgB_2O_4.3H_2O]$.

Pinol. Sombreron, $C_{10}H_{16}O$. A tar distillate obtained from the black spruce and used as a disinfectant is known by this name. A trade mark.

Pinolin. A name for rosin oil (the first distillate from rosin, boiling at from 78–250° C.).

Pinolith. See LITHOPONE.

Pinosal. A distillate of tar about the consistency of honey. Used as a local application in various skin diseases.

Pintadoite. A mineral. It is a hydrated calcium vanadate, $2CaO.V_2O_5.9H_2O$.

Pintsch Gas. See OIL GAS.

Pintsch Oil. A product obtained by compressing Pintsch gas. It contains 50 per cent. benzene, and 25 per cent. toluene.

Pinwire Brass. An alloy of from 66–73 per cent. copper and 27–34 per cent. zinc.

Pioneer Alloy. An alloy of 20 per cent. copper, 38 per cent. nickel, 4 per cent. silicon, 3 per cent. molybdenum, 2 per cent. tungsten, and the remainder iron.

Pioury. See INDIAN YELLOW.

Pioxol. A proprietary trade name for Pemoline.

Pipadone. A proprietary preparation of dipipanone hydrochloride. An analgesic. (277).

Pipamazine. 10-[3-(4-Carbamoylpiperidino)propyl] - 2 - chlorophenothiazine. Mornidine.

Pipamperone. 1-[3-(4-Fluorobenzoyl)-

propyl] - 4 - piperidinopiperidine - 4 - carboxamide.

ipanol. A proprietary preparation of benzhexol hydrochloride, used in parkinsonism. (112).

ipazethate. 2-(2-Piperidinoethoxy)-ethyl 10-thia-1,9-diaza-anthracene-9-carboxilate. Selvigon is the hydrochloride.

ipeclay. A mineral. It is an aluminium silicate. An abrasive. See CHINA CLAY.

ipenzolate Bromide. 1-Ethyl-3-piperidyl benzilate methobromide. Piptal.

iperazenyl. See LYSIDINE.

iperazidine. Diethylene-diamine (piperazine) $C_4H_{10}N_2$.

iperazine Calcium Edetate. A chelate produced by reacting ethylenediamine - NNN'N' - tetra - acetic acid with calcium carbonate and piperazine. Perin.

iperidolate. 1-Ethyl-3-piperidyl diphenylacetate. Dactil is the hydrochloride.

iperocaine. 3-(2-Methylpiperidino) - propyl benzoate. Metycaine is the hydrochloride.

Piperonal. Heliotropin (q.v.).

Piperonyl Butoxide. An acaricide. 5-[2 - (2 - Butoxyethoxy)ethoxymethyl] - 6-propyl-1,3-benzodioxole.

Piperoxan. 2-P peridinomethylbenzo-1, 4-dioxan. Diphenyl-2-piperidylmethanol.

Pipestone, Indian. See CATLINITE.

Pipothiazine. A neuroleptic. 2-Dimethylsulphamoyl - 10 - {3 - [4 - (2 - hydroxyethyl)piperidino]propyl}phenothiazine.

Pip Pip. An abbreviated name for piperidinium - pentamethylene - dithiocarbamate. An accelerator for rubber vulcanisation.

Pipoxolan. An anti-spasmodic. 5,5-Diphenyl - 2 - (2 - piperidinoethyl) - 1, 3 - dioxolan - 4 - one. Rowapraxin is the hydrochloride.

Pipradol. Diphenyl-2-piperidylmethanol. Meratran is the hydrochloride.

Piprelix. Worm expellent. (502).

Piprinhydrinate. Diphenylpyraline salt of 8-chlorotheophylline. 4-Diphenylmethoxy-1-methylpiperidine salt of 8-chlorotheophylline. Kolton. Mepedyl.

Piptal. A proprietary preparation containing pipenzolate methobromide. A gastrointestinal sedative. (249).

Piptalin. A proprietary preparation of pipenzolate bromide and simethicone used as a gastro-intestinal sedative. (249).

Piracetam. A cerebral stimulant. 2-Oxopyrrolidin - 1 - ylacetamide. NOOTROPYL.

Piral. Pyrogallol, $C_6H_3(OH)_3$.

Pirazoline. Phenazone (antipyrin).

Pirbuterol. A bronchodilator. 2-tert-Butylamino - 1 - (5 - hydroxy - 6 - hydroxymethyl - 2 - pyridyl)ethanol. " CP-24,314-1 " is under clinical trial as the dihydrochloride.

Piria's Naphthionic Acid. α-Naphthylamine-sulphonic acid (1 : 4).

Piridoxilate. A preparation for the treatment of angina pectoris currently undergoing clinical trial as " GLYO-6 ". It is the reciprocal salt of (5-hydroxy-4-hydroxymethyl - 6 - methyl - 3 - pyridyl)methoxyglycolic acid with [4, 5 - bis(hydroxymethyl) - 2 - methyl - 3 - pyridyl] oxyglycolic acid(1 : 1).

Piriex. A proprietary preparation containing chlorpheniramine maleate, ammonium chloride and sodium citrate. A cough linctus. (284).

Piriton. A proprietary preparation of chlorpheniramine maleate. (284).

Piritramide. 4-(4-Carbamoyl-4-piperidinopiperidine) - 2,2 - diphenyl - butyronitrile.

Pirsch-Baudoin's Alloy. An alloy of 71 per cent. copper, 16·5 per cent. nickel, 1·75 per cent. cobalt oxide, 2·5 per cent. tin, and 7 per cent. zinc.

Pirssonite. A mineral, $Na_2Ca(CO_3)_2$. $2H_2O$.

Pirtoid. A proprietary trade name for a synthetic resin varnish-paper product used as an electrical insulator.

Pisanite. A mineral. It is an iron sulphate containing copper $(FeCu)SO_4$. $7H_2O$.

Piscarol. See ICHTHYOL.

Pisciol. See ICHTHYOL.

Pisekite. A radioactive mineral found in Bohemia. It contains Cb, Ta, Ti, U, Ce, Yt, Yb, Th, also Si, Al, K, Ca, and Mg.

Pisolite (Pea-stone). A mineral. It is a form of Arragonite (q.v.).

Pissophanite (Garnsdorffite). A mineral. It is a basic ferrous aluminium sulphate.

Pistacite. A mineral. A variety of Epidote (q.v.).

Pitankite. A mineral. It is $(Ag,Cu,Pb)_4$ Bi_2S_5.

Pitayin. Quinidine, $C_{20}H_{24}N_2O_2$. An alkaloid used in medicine.

Pitch Barm. A cement made from casein, water-glass, and caustic lime.

Pitch Blende. A black mineral consisting essentially of uranium oxide, together with varying amounts of rare earths, lead, lime, and bismuth. It is a source of radium, and is also known as Uraninite.

Pitch, Candle. See STEARIN PITCH.

Pitch, Glance. See MANJAK.

Pitch, Hemlock. See CANADA PITCH.

Pitch, Jew's. See BITUMEN.

Pitch Mineral. See BITUMEN.

Pitch Oil, Liquid. See OIL OF TAR.

Pitch-Opal. An opal with a pitchy lustre.

Pitchphalte. Special pitch preparation for grouting. (573).

Pitibilin. A standardised solution of the pituitary body extract, specially prepared for hypodermic use. Also see HYPOPHYSIN and PITUITRIN.

Pit-ite No. 2. An explosive consisting of 23–25 per cent. nitro-glycerin, 28–31 per cent. potassium nitrate, 33–36 per cent. wood meal, and 7–9 per cent. ammonium oxalate.

Pitocin. A proprietary preparation of oxytocin used to induce labour. (264).

Piton Snuff. A proprietary preparation of posterior pituitary extract. Antidiuretic hormone. (316).

Pitralon. A German antiseptic. It is a wood tar derivative.

Pitressin. A proprietary preparation of vasopressin. Antidiuretic hormone. (264).

Pitressin Tannate. A proprietary preparation of vasopressin tannate. Antidiuretic hormone. (264).

Pittaccal. Eupittonic acid, $C_{25}H_{26}O_9$.

Pitteliene. A mixture of coal tar and oil. An insecticide.

Pitti. See VENTILAGO MADRASPANTA.

Pitticite. See SCORODITE.

Pittinite. A mineral. It is a variety of gummite.

Pittylen. A mixture of pine tar with formaldehyde. Used in skin diseases.

Pituitrin. COLUITRIN. Preparations of the extract of the pituitary body, obtained by the extraction of the infundibular portion of fresh glands (usually of the ox). Also see PITIBILIN and HYPOPHYSIN. A proprietary preparation of posterior pituitary extract. Antidiuretic hormone. (264).

Piuri. See INDIAN YELLOW.

Pivampicillin. An antibiotic. Pivaloyloxymethyl 6 - [D(—) - α - amino - phenylacetamido]penicillinate.

Pivazide. See PIVHYDRAZINE.

Pivhydrazine. N-Benzyl-N′-pivaloyl-hydrazine.

Pivmecillinam. An antibiotic currently undergoing clinical trial as "FL 1039". It is pivaloyloxymethyl $(2S,5R,6R)$ - 6 - (perhydroazepin - 1 - ylmethyl - eneamino)penicillanate.

Pivofax. A dried yeast.

Pix. Pitch.

Pixol. A form of wood tar soluble in water, made from tar and soap. A disinfectant.

Pix Solubilis (Soluble Pitch). A soluble modification of the tar obtained by sulphonating the tar obtained from peat.

Pixtonet. A trade name for a slate substitute. It is used as an electrical insulator.

Pi-yu. See MOU-IÉON.

Pizotifen. 4-(9, 10-Dihydrobenzo[4,5] cyclohepta[1, 2 - b]thien - 4 - ylidene) 1-methylpiperidine.

Placadol. A proprietary preparation of papaverine hydrochloride, homatropin methylbromide and codeine phosphate (319).

Placet Alloy. An alloy of 60 per cent nickel, 20 per cent. iron, 15 per cent chromium, and 5 per cent. manganese. An electrical resistance.

Placidyl. A proprietary trade name for ethchlorvynol.

Placodin. Contains Aloxiprin.

Plagionite. A mineral, $5PbS.4Sb_2S_3$.

Planadalin. A proprietary preparation. It is identical with Adalin (diethyl bromo-acetyl-urea).

Planavin. A trade mark for a nitrogenous weed-killer. (790).

Plancheite. A variety of the mineral Chrysocolla, $15CuO.12SiO_2.5H_2O$.

Planerite. A mineral. It is a contain aluminium phosphate, iron, and copper

Planidets. A proprietary preparation of dibromopropamidine embonate, CHLOR PHENOCTIUM AMSONATE and butyl aminobenzoate used in the treatment of mouth ulcers. (507).

Planocaine. (1) A proprietary preparation of ethocaine hydrochloride.

Planocaine. (2) A proprietary preparation of procaine hydrochloride, adrenaline, sodium chloride, potassium sulphate, in water. Local anæsthetic Type " P " contains no adrenaline (336).

Planochrome. Mercurochrome. (507)

Planoferrite. A mineral. It is Fe_2 $SO_4(OH)_4.13H_2O$.

Planovin. Trade mark for a German contraceptive containing megestrolacetate and ethinyloestradiol. (791).

Plantex. A proprietary weed-killer containing sodium chlorate. (829).

Plant Indican. Indican, $C_{26}H_{31}NO_{17}$.

Plantowoll. A Swiss cottonised jute obtained by splitting the fibre. A wool substitute.

Plantvax. A proprietary systemic fungicide based on oxycarboxin. (438).

Plaquenil. A proprietary preparation of hydroxychloroquin sulphate. An antimalarial. (112).

Plasgon. A proprietary trade name for plastic gasket and joint cement.

Plaskon. A trade mark for a range of urea-formaldehyde and melamine resins and for Nylon 6. (30).

Plaskon 2201. A proprietary high-density polyethylene used in the manufacture of blown film. (30).

Plaskon 8200. A proprietary nylon claiming superior mechanical properties. (30).

Plaskon 8200-P. A proprietary nylon with high impact resistance. (30).

Plaskon 8201-HS. A proprietary nylon used in extrusion and moulding operations. (30).

Plaskon 8201-P. A proprietary nylon similar to PLASKON 8201-HS but possessing higher impact resistance. (30).

Plaskon 8202. A proprietary nylon used for injection moulding. (30).

Plaskon 8205. A proprietary nylon possessing high melt viscosity, used in extrusion and blow-moulding operations. (30).

Plaskon 8206. A proprietary nylon of good flexibility used in extrusion and moulding operations. (30).

Plaskon 8226. A proprietary nylon material with high impact resistance used in the moulding of such articles as hammer heads, etc. (30).

Plaskon 8229. A proprietary flexible grade of nylon. (30).

Plaskon 8230. A proprietary nylon with low melt viscosity used in the moulding of intricate sections. (30).

Plaskon 8231, 8233 and 8234. Proprietary names for a range of glass-filled nylons. (30).

Plaskon 8250, 8251 and 8253. A proprietary range of copolymer grades of nylon possessing high impact strength. (30).

Plaskon AA60-003. A proprietary high-density polyethylene resin used in blow-moulding and extrusion operations. (30).

Plaskon AA60-007. A proprietary high-density polyethylene used in the blow-moulding of food and bleach bottles. (30).

Plaskon AB50-003. A proprietary ethylene vinylacetate copolymer used in the moulding of detergent bottles. (30).

Plaskon AC 1220 and 1221. A proprietary range of polyethylene resins with high molecular weights. (30).

Plaskon AC Copolymer 400. A proprietary ethylene vinyl acetate copolymer of low molecular weight used in the coating of paper. (30).

Plaskon A.C. (Polyethylene 1220). A proprietary polyethylene resin of very high molecular weight. (30).

Plaskon Alpha 8200 C, 8202 C and 8203 C. A proprietary range of alpha-type nylon moulding compounds used in the production of stiff mouldings. (30).

Plaskon C 1012. A proprietary high-density polyethylene possessing officially approved self-extinguishing properties. (30).

Plaskon PP 60-002. A proprietary polyethylene used in such operations as the blow-moulding of large containers. (30).

Plaskon SS 50-035. A proprietary general-purpose polyethylene of high density. (30).

Plaskon SS 50-050 and SS 50-090. Proprietary high-density polyethylenes used in injection moulding. (30).

Plaskon SS 50-250. A proprietary high-density polyethylene with good flow characteristics used in the making of intricate mouldings. (30).

Plaskon SS 55-100. A proprietary high-density polyethylene used for injection moulding. (30).

Plaskon SS 55-180. A proprietary high-density polyethylene possessing good melt-flow properties. (30).

Plaskon SS 60-035. A proprietary high-density polyethylene with moderate flow properties used in injection moulding. (30).

Plaskon SS 60-050. A proprietary high-density polyethylene with moderately good flow properties used in the moulding of toys and articles for household use. (30).

Plaskon SS 60-090. A proprietary high-density polyethylene resin with a narrow distribution of molecular weight giving it good resistance to impact and warping. (30).

Plaskon SS 60-300. A proprietary polyethylene combining high flow properties and high density with a narrow distribution of molecular weight. (30).

Plasma. A dark leek-green chalcedony, SiO_2. Also see GLYCERIN OF STARCH.

Plasmin. The proteolytic enzyme derived from the activation of plasminogen.

Plasminogen. The specific substance derived from plasma which, when activated, has the property of lysing fibrinogen, fibrin and other proteins.

Plasmocain. A proprietary preparation of German manufacture. It is a remedy for malaria, the active principle of which is N-diethyl-amino-isopentyl-8-amino-6-methoxy-quinoline.

Plasmon. See NUTROSE.

Plasmoquin (Beprochin). A synthetic quinine substitute. It is stated to be 8 - diethyl - amino - isoamyl - amino - 6 - methoxy-quinoline. Used in the treatment of malaria.

Plasmosan. A proprietary trade name for Povidone.

Plastacele. A proprietary trade name for a plasticised cellulose acetate compound.

Plastalac. A trade name for an air drying waterproof sealing coating for orthopædic plaster casts. (254).

Plastammone. An explosive containing ammonium nitrate, glycerin, mono-nitro-toluene, and nitro-semicellulose.

Plastamol. A registered trade mark for plasticers for PVC. (49).

Plastazote. This is the registered trade mark for an expanded formvar thermoplastic, having an exceptionally high impact strength and modulus of elasticity under compression. Compression strength is 170 lb. per sq. in. at the standard density of 6 lb. per sq. ft. At this density the thermal conductivity is 0·24 B.Th.U. and modulus of elasticity is greater than 10,000 lb. per sq. in.

Plaster Board. Sheets employed for walls and ceilings, consisting of gypsum mixed with about 15 per cent. fibre.

Plaster Cement. Cements made from gypsum. See LE SAGE CEMENT.

Plaster of Paris. A partially dehydrated gypsum, $2(CaSO_4).H_2O$. It is made from gypsum by heating the latter from 212–400° F., when 3 parts of the water of crystallisation is given off.

Plasteryl. An explosive. It is a mixture of 99·5 per cent. trinitrotoluene and 0·5 per cent. resin.

Plastic A. A proprietary trade name for glyceryl tribenzoate.

Plasticalk. A proprietary trade name for a plastic resin used as an adhesive and filler.

Plastic Bronze. An alloy of 64 per cent. copper, 30 per cent. lead, 5 per cent. tin, and 1 per cent. nickel.

Plasticede. A proprietary trade name for a plasticiser for clay. It contains tannins and lignin.

Plastic Iron. A jointing metal made by heating iron compounds, such as magnetite, below the melting-point of iron to give a porous mass. This is impregnated with bitumen, bonded with iron wire and fashioned into strips for making rings. When compressed the material loses plasticity. It is used as a substitute for lead joints.

Plasticiser 13. The neutral esterification of p-oxybenzoic acid and 2-ethyl hexanol. A plasticiser for polyamides. (49).

Plasticiser 28P. Dioctyl phthalate. A vinyl plasticiser.

Plasticiser E. A proprietary trade name for a chlorinated paraffin plasticiser.

Plastic Metal. An alloy of 80·5 per cent. tin, 9·5 per cent. copper, 8·6 per cent. antimony, and 1·4 per cent. iron.

Plastic Plant Product. See PLASTIC WOOD.

Plastic Steel. Trade mark for a paste containing 80% powdered steel and 20% epoxy resin capable of being hardened. (792).

Plastic Sulphur. Prepared by heating sulphur to 225° C. Used to a limited extent as a material for preparing moulds for electrotyping.

Plastic Wood (Plastic Plant Product). A material prepared by cooking vegetable matter with neutral salt solution followed by mechanical treatment to break down intercellular binding material. The name is also used to describe wood cellulose in solution with certain additional solvents such as ether or acetone. Used for filling up holes in wood or other materials.
" Plastic wood " is a registered trade mark for a brand of cellulose fibre filler.

Plastic X. A proprietary plasticiser. It is tricresyl phosphate, and has a specific gravity of from 1·177–1·18.

Plastifix. Trade mark for a brand of polychloroprene. (112).

Plastikon. A proprietary trade name for a rubber putty for filling and adhesive purposes.

Plastite. A name applied to a vulcanite.

Plastitube. A proprietary trade name for cellulose acetate tubing.

Plastokyd 310. A proprietary trade name for a short oil linseed alkyd resin for rapid drying trichlorethylene paints. (66).

Plastolein. A registered trade name for vinyl plasticisers identified as follows. (123).
3049R. A polyester.
8058. Di-2-ethylhexyl azelate.
9050. Di-2-ethyl butyl azelate.
9055. Di-ethyl glycol dipelargonate.
9056. Di-2-ethyl hexyl azelate.
9057. Di-iso-octyl azelate.
9058. Dioctyl azelate.
9250. Tetrahydrofurfuryl oleate.
9404. Triethylene glycol dipelargonate.
9722. A polymeric plasticiser.

Plastolin I. A proprietary solvent. It is reported to be benzyl acetate, $C_6H_5.CH_2.OOC.CH_3$.

Plastomenite. A German explosive powder made by incorporating 1 part nitro-lignin with 5 parts fused dinitrotoluene, and granulating. It may contain barium nitrate.

Plastomoll BMB. N-Butylbenzene sulphonamide. A plasticiser for polyamides. (49).

Plastomoll DMA. A proprietary trade name for the adipic ester of hydrogenated higher cyclic alcohols. A vinyl plasticiser. (49).

Plastomoll DOA. Di-2-ethylhexyl adipate. A plasticiser for PVC. (49).

Plastomoll NA. Di-iso-nonyl adipate. A plasticiser for PVC. (49).

Plastomoll TAH. A proprietary trade name for the thiodibutyric acid ester of a synthetic octyl alcohol. A vinyl plasticiser. (49).

Plastomoll WH. A proprietary trade name for the ester of an aliphatic dicarboxylic acid. A vinyl plasticiser. (49).

Plastomoll 34. A proprietary trade name for a vinyl plasticiser. An aliphatic acid ester mixture. (49).

Plastone. A proprietary rubber vulcanisation accelerator. It is methylenediphenyl-diamine mixed with a small amount of stearic acid or with naphthalene or naphthalene oil.

Plastone A. A proprietary trade name for a phenolic moulding compound containing cotton seed hull.

Plastone B. A proprietary trade name for an inorganic and phenolic moulding compound.

Plastopal. A proprietary trade name for a series of urea formaldehyde resins. (3).

Plastopal 11. A modified urea-formaldehyde condensation product in the form of a 65 per cent. butanol—white spirit solution. (49).

Plastopal H. A proprietary trade name for a urea formaldehyde condensate of high quality. (3).

Plastopal ATX. A proprietary trade name for a condensate of urea formaldehyde and adipic acid hexanetriol. (3).

Plastopal CB. A proprietary trade name for a condensate between 100 parts of urea formaldehyde and 3 parts of adipic acid-trimethylolpropane added before condensation. (3).

Plastopal AT. A proprietary trade name for a condensate between 27 parts of urea formaldehyde and 23 parts of adipic acid. (3).

Plastopal AW. A proprietary trade name for a condensate of 1 part urea formaldehyde and 4 parts adipic acid-trimethylolpropane. (3).

Plastopal W. A proprietary trade name for a urea formaldehyde condensate in a solvent. (3).

Plastoprene No2/LV. A proprietary trade name for an isomerised rubber resin used for printing inks and chemical resistant coatings. (66).

Plastorub. A proprietary brand of peptised rubber. (895).

Plastose. Proprietary synthetic resin moulding powders.

Plastosol. An antiseptic liquid plaster containing copper guaiacol-sulphonate, and penetrodine dissolved in a volatile organic solvent.

Plastosperse. A range of pigments dispersed in plasticisers for use in plastics. (260).

Plastosperse 40. 40 per cent. carbon black dispersed in a plasticiser. (260).

Plastplate. A proprietary trade name for moulded plastics plated with chromium, copper, gold, or nickel.

Plastrotyl. An explosive. It is a plastic product prepared from trinitrotoluene, resin, collodion cotton, and crude dinitro-toluene. Sometimes larch turpentine is used.

Plastules with Folic Acid. A proprietary preparation of ferrous sulphate, folic acid, yeast and liver extract. A hæmatinic. (245).

Plastules with Liver. A proprietary preparation of ferrous sulphate, yeast and liver extract. A hæmatinic. (245).

Plastylene. A French trade name for polyethylene. (69). See POLYTHENE, ALKATHENE (Trade Mark), etc. A registered trade mark for LD polyethylene and ethylene—vinylacetate copolymers.

Plastyrol E6X. A proprietary trade name for a styrenated epoxide ester resin. A rapid air drying resin. (66).

Plastyrol S88X. A proprietary trade name for a styrenated alkyd resin for rapid drying finishes. (66).

Plastyrol S-77X. A proprietary trade name for a styrenated alkyd resin. A very fast drying resin. (66).

Platalargan. An alloy of aluminium and silver with some platinum. It is similar to Alargan, except that it contains platinum. Used as a platinum substitute.

Platamid. Polyamides for extrusion and injection moulding.

Plate Black. A dyestuff of the same class as Diphenyl fast black.

Plate Glass. A glass which is mainly a silicate of sodium and calcium, but it contains, in addition, a considerable quantity of potassium silicate.

Plate Pewter. An alloy of 90 per cent. tin, 6 per cent. antimony, and 2 per cent. each bismuth and copper.

Plate Powder. Bone ash of which calcium phosphate, $Ca_3(PO_4)_2$, forms 80 per cent., is sold as a non-mercurial plate powder under the name of White rouge.

Plate Sulphate. The double sulphate, $K_2SO_4.Na_2SO_4$, is called plate sulphate. It crystallises from hot water, a flash of light accompanying the separation of each crystal.

Platina. An alloy of 53·5 per cent. tin, and 46·5 per cent. copper.

Platinating Solution. These generally consist of solutions of platinum chloride with a sodium, potassium or ammonium salt. Used for the electrodeposition of platinum.

Platine. A brass containing 43 per cent. copper and 57 per cent. zinc.

Platine-autitre. A proprietary trade name for a platinum substitute containing 65–83 per cent. silver and platinum.

Platinised Asbestos. Loosely fibred asbestos moistened with a concentrated solution of platinum chloride, dried, dipped into ammonium chloride solution, again dried, and brought to a red heat. It usually contains 8–8·5 per cent. platinum, and is used in the manufacture of sulphuric anhydride in the contact process for sulphuric acid.

Platinite. A proprietary trade name for a nickel steel containing 46 per cent. nickel, and 0·15 per cent. carbon. It has a low coefficient of expansion, and can be sealed in glass. It has a specific gravity of 8·2 and a melting-point of 1470° C.

Platino. An alloy of 11 per cent. platinum and 80 per cent. gold. It is resistant to fused potassium nitrate, and to alkalis.

Platino-aceto-osmic Acid. See HERMANN'S FLUID.

Platinoid. An alloy of 60 per cent. copper, 24 per cent. zinc, 2 per cent. tungsten, and 14 per cent. nickel. It is used as the material which connects filaments with outside wires of electric lamps. It has the same coefficient of expansion as glass. Also employed as an electrical resistance. One alloy (a nickel-silver containing 1 per cent. tungsten) has a resistivity of 41·7 microhms . cm. at 0° C. Another alloy contains 54 per cent. copper, 25 per cent. nickel, 20·5 per cent. zinc, 0·47 per cent. iron, 0·15 per cent. lead, and 0·15 per cent. manganese.

Platinor. An alloy of 2 parts platinum, 5 parts copper, 1 part silver, and 1 part nickel.

Platinum Black. If platinum be precipitated in the metallic state from solutions, it is obtained in the form of a powder called platinum black, which possesses the power of promoting combination with oxygen in the highest proportions. This form of platinum may be obtained by boiling a solution of platinic chloride with Rochelle salt, when the platinum black is precipitated.

Platinum Bronze. An alloy consisting of 90 per cent. nickel, 9 per cent. tin, and 1 per cent. platinum.

Platinum Gold. Variable alloy containing from 12–81 per cent. copper, 9–58 per cent. platinum, 0–70 per cent. gold, 0–37 per cent. silver, and 0–4 per cent. zinc.

Platinum Grey. See ZINC GREY.

Platinum Iridium. An alloy usually containing 90 per cent. platinum and 10 per cent. iridium.

Platinum Silver. An alloy containing 66·6 per cent. silver and 33·3 per cent. platinum. It has a specific resistance of 24·33 microhms per cm.³ at 0° C.

Platinum, Soft. See SOFT PLATINUM.

Platinum Solder. An alloy usually consisting of 73 per cent. silver and 27 per cent. platinum.

Platinum, Spongy. Ammonium platinochloride, $(NH_4)_2PtCl_6$, is heated to redness, when all constituents except platinum are expelled, the latter being left in a porous condition called spongy platinum.

Platinum Substitute. An alloy of 72 per cent. nickel, 23·6 per cent. aluminium, 3·7 per cent. bismuth, and 0·7 per cent. gold.

Platinum Yellow. A barium chloroplatinate or other alkaline chloroplatinate. Used as a coating for fluorescent screens in X-ray work.

Platnam. An alloy consisting of 56 per cent. nickel, 31 per cent. copper, 12 per cent. lead, 0·48 per cent. iron, and 0·32 per cent. aluminium.

Platnik. An alloy of nickel and platinum. A platinum substitute.

Plattnerite. A mineral. It is a brown oxide of lead, PbO_2.

Platynite. A mineral, $PbS.Bi_2S_3$.

Plazolite. A mineral. It is a hydrated silicate and carbonate of calcium and aluminium, $3CaO.Al_2O_3.2(SiO_2.CO_2)2H_2O$, found in California.

Plecavol. Consists chiefly of p-aminobenzoyl-eugenol, $NH_2.C_6H_4.CO.O.C_6H_3.(C_3H_5)O.CH_3$, tricresol (a mixture of the three cresols), and formaldehyde. Employed in dentistry.

Plegine. A proprietary trade name for the tartrate of Phendimetrazine.

Pleiapyrin. A condensation product of benzamide and phenyl-dimethyl-purazolone. Antipyretic and analgesic.

Pleistopon. Similar to Pantopon, except that it contains no narcotin.

Plenargyrite. A mineral. It is a silver sulpho-bismuthite.

Pleogen. A proprietary polyester laminating resin. (794).

Pleonaste (Iron Spinel). A mineral. $MgO.Fe_2O_3$.

Plesmet. A proprietary preparation of ferrous glycine sulphate. A hæmatinic (361).

Plesmet FA. A proprietary preparation of ferrous glycine sulphate and folic acid used in the treatment of anæmia of pregnancy. (334).

Plessite. A German explosive containing potassium chlorate and mineral oil. It is also the name for the mineral Gersdorffite (NiAsS).

Plessy's Green. Chromic phosphate, $Cr(PO_3)_3$, a pigment.

Plex-Hormone. A proprietary preparation of methyltestosterone, deoxycortone acetate, ethinyloestradiol and vitamin E used to treat male androgen deficiency. (256).

Plexiglas. A trade mark for a thermoplastic transparent sheet of acrylic synthetic resin. (13). See under PERSPEX.

Plexiglo. A proprietary trade name for a polish and cleaner for transparent plastics.

Plexigum. A proprietary trade name for a series of methyl methacrylate polymers used for glazing including laminated glass. (13).

Plexigum M315. A proprietary trade name for a copolymer comprising 60 parts methyl methacrylate, 20 parts ethyl methacrylate and 20 parts butyl methacrylate. (13).

Plexon. A proprietary trade name for a synthetic resin coated yarn.

Pliabrac 810. A proprietary C_8-C_{10} aliphatic phthalate plasticiser. (503).

Pliabrac 987. A proprietary adipic ester plasticiser. (503).

Pliabrac 989. A proprietary adipic ester plasticiser. (503).

Pliabrac 990. A proprietary adipic ester plasticiser. (503).

Pliabrac DIOZ. A proprietary plasticiser. Diisooctyl azelate. (120).

Plialite. A proprietary product stated to be rubber resin.

Plimmer and Paine's Stain. A microscopic stain. It contains 10 grams tannic acid, 18 grams aluminium chloride, 18 grams zinc chloride, 1·5 grams rosaniline hydrochloride, and 40 c.c. 60 per cent. alcohol.

Plimmer's Salt. Sodium antimony tartrate.

Plinian. A mineral. It is mispickel.

Plinthite. A red clay from Ireland.

Pliofilm. A trade mark for a preparation of rubber hydrochloride in the form of transparent sheet, used for wrapping material. Rubber derivatives and rubber-like resins.

Plioflex. A proprietary trade name for polyvinyl chloride.

Plioform. A proprietary product. A type of rubber plastic obtained by the action of halogenated acids on rubber.

Pliolite. A proprietary trade name for modified, isomerised rubber, rubber derivatives and rubber-like resins.

Plitex. A proprietary trade name for a wood and phenolic resin.

Plomb. See BLEI.

Plumbago. See GRAPHITE.

Plumbago Grease. A mixture of plumbago and tallow. Used for lubricating.

Plumbago Gummite. A mineral, $PbO.Al_2O_3$.

Plumber's Solder. Usually a mixture of lead and tin, sometimes with a little antimony. Coarse solder contains 75 per cent. lead and 25 per cent. tin, and melts at 250° C. Ordinary solder (slicker solder) usually consists of 67 per cent. lead and 33 per cent. tin, and melts at 227° C. It has a specific gravity of 9·4. Plumber's fine solder, or soft solder, contains 50 per cent. lead and 50 per cent. tin, and melts at 188° C.

Plumber's White Alloy. An alloy of from 54–58 per cent. copper, 25–27 per cent. zinc, 13–17 per cent. nickel, 1–7 per cent. lead, and sometimes 1 per cent. tin and 1 per cent. iron.

Plombierite. A mineral. It is a calcium silicate.

Plumbinnite. A mineral. It is dufrenoisite.

Plumbiodite. A mineral. It is schwartzembergite.

Plumbionite. A mineral containing 46 per cent. Cb_2O_5.

Plombit. A German acid-resisting material made from hard rubber, oleic acid, sulphuric acid, and sulphur. It is stated to be an artificial asphalt.

Plumbo-Argentojarasite. A mineral. It is $(Pb,Ag_2)Fe_6(SO_4)_4(OH)_{12}$.

Plumbobinnite. Synonym for Dufrenoysite.

Plumbobismuth Glance. A mineral, $Pb_2Bi_2S_5$.

Plumbocalcite. See TARNOWITZITE.

Plumbocuprite. See CUPROPLUMBITE.

Plumboferrite. A mineral $(PbO.FeO.CuO)Fe_2O_3$.

Plumbogummite. A mineral. It is a lead phosphate and aluminate.

Plumbojarosite. A mineral. It is a hydrous sulphate of lead and iron, $Pb(Fe_6)(OH)_{12}(SO_4)_4$.

Plumbomalachite. See LEAD MALACHITE.

Plumbomanganite. A mineral containing manganese, lead, and sulphur.

Plumbonacrite. A mineral. It is a basic lead carbonate.

Plumboniobite. A mineral. It consists of Samarskite containing lead oxide.

Plumbostannite. A mineral, $Pb_2(Fe.Zn)_2.Sn_2.Sb_2.S_{11}$.

Plumbostibnite. A mineral. $Pb_3Sb_2S_6$.

Plumboxan. A compound or solid solution of sodium manganate and sodium metaplumbate, of the composition, $Na_2MnO_4.Na_2PbO_3$. It gives up oxygen when treated with steam.

Plumbum. See BLEI.

Plumite. A mineral. It is plumosite.

Plummer's Pill. (*Pilula hydrargyri subchloridi composita B.P.*) It contains

1 gram gum arabic, 1 gram gum tragacanth, 20 grams mercurous chloride, 20
grams sulphurated antimony, 10 grams
glucose syrup, and 40 grams guaiacum
resin.

Plumose Mica. A variety of muscovite.

Plumosite. A mineral. It is an antimonial lead sulphide, $2PbS.Sb_2S_3$, which
occurs in the Isle of Man.

Plum Spirits. See Tin Spirits.

Plusbrite. A chemical for bright nickel
plating. (503).

Plush. A name for woven fabrics having
a pile. The material can be cotton, silk,
or wool. At one time the pile consisted
of mohair.

Plush Copper Ore. A Cornish name for
the mineral chalcotrichite.

Plusol Y30. A proprietary trade name
for a short oil alkyd resin for stoving
finishes dissolved in xylol/ethylene
glycol. It has outstanding adhesion to
light metals. (189).

Pluto Black. A dyestuff of the same class
as Diphenyl fast black See Diamine
Black-blue B.

Pluviusin. A proprietary trade name for
a synthetic resin of the urea type.

Plybond. Proprietary products. They
consist of thin absorbent paper impregnated with a solution of phenol-
formaldehyde resin. They are used for
bonding plywood, etc.

Plymax. A laminated wood product consisting of three-ply wood with a metal
covering. Used for disc wheels and
vehicle bodies.

Plymetl. A proprietary trade name for a
plywood faced with galvanised steel.

Plymul 98-759. A range of polyvinyl
acetate thermosetting emulsions used
for bonding cellulose to cellulose. (48).

Plyophen. A proprietary trade name for
a phenolic laminating resin and varnish.

Plyothene. A proprietary phenolic
moulding material. (48).

Plywood. A laminated wood built up by
layers with the grain in each layer at
right angles to adjacent layers. The
layers are usually bonded with glue
and pressure. Recently, proprietary
plywoods have been made in which the
bonding agent is one or other of the
thermosetting plastics. Dielectric heating has increased the speed of gluing.

P.M.G. Metal. A proprietary trade name
for an alloy of copper with 3–4 per cent.
silicon, 2 per cent. iron, and 2 per cent.
zinc.

P.M.T. Alloy. A proprietary alloy made
as a substitute for Admiralty gun metal.
It contains 88 per cent. copper, 2 per
cent. zinc, and 10 per cent. silicon,
manganese, and iron.

Pneulec Core Gum. A proprietary product. It is a linseed oil and wood ex-

tract material, and is used as a binder
for the sand for cores in metal casting.

Pneumarol. A German preparation containing dimethyl- and trimethyl-
xanthin, suprarenal extract, papaverine, digitalis, strophanthus, and
nitroglycerin. Used in asthma.

Pneumatogen. A mixture of the peroxides of potassium and sodium.

Pneumin. A condensation product of
creosote and formaldehyde. It is recommended in the treatment of tuberculosis and pulmonary catarrh.

Pneumitren. A German preparation. It
contains creosotal, Peru balsam, ichthyol, oil of peppermint, etc. Used
for the treatment of nasal catarrh and
bronchitis, influenza and colds.

Pneumosan. Amyl-thio-trimethylamine.

P.N.P. *p*-Nitro-phenol. Used as fungicide
in the rubber industry.

Poco Oil. The oil from *Mentha aquatica*.
Used in Java as a remedy for headache.

Podolite. See Dahllite.

Podophyllum. The dried rhizome of
Podophyllum peltatum. Used in medicine.

Pod Pepper. Capsicum.

Pogy Oil. See Mossbunk Oil.

Poilite. A trade name for an asbestos
cement product used for building work.

Poirrier's Orange III. See Orange III.

Poison Flour. Arsenious oxide, As_4O_6.

Poison Wheat. Wheat grains poisoned
with arsenic and dyed green with
malachite green. Used to destroy field
mice. Sometimes the poison is strychnine.

Poivrette. The ground stones of olive
fruit. Used as an adulterant and toning
agent in spices.

Polar Dynobel (Polar Monobel No. 2,
Polar Rex, Polar Saxonite, Polar Sto-
monal, Polar Super Cliffite, Polar
Thames Powder, Polar Viking). Proprietary low freezing explosives containing a mixture of nitrated glycerin
and polyglycerin or glycerin and ethylene glycol, ammonium nitrate, sodium
chloride, wood meal, etc.

Polaroid H-Glass. A patented product
consisting of laminations in select drawn
glass of a linear high-polymeric plastic
containing oriented molecules which
have been treated to render them light-
polarising.

Polarwhite. A headless white paint.
(541).

Polastor. A proprietary polyester laminating resin. (795).

Poldine Methylsulphate. 2-Benziloyl-
oxymethyl - 1 - methylpyrrolidine
methylsulphate. Nacton.

Polectron. A proprietary trade name for
a vinyl carbosol resin. (796).

Poley Oil. See Oil of Pennyroyal.

Polianite. A mineral, MnO_2.

Polidene. Vinylidene chloride copolymer emulsions. (542).

Polidene 528F. A proprietary polyvinylidene chloride emulsion. (797).

Polidexide. An agent used to restrict the increase of cholesterol in the blood beyond normal limits. It is DEXTRAN cross-linked with epichlorhydrin and o-substituted with 2-diethylaminoethyl groups, some of them quaternised with diethylaminoethyl chloride.

Polidine. A proprietary trade name for emulsions of polyvinylidine chloride. (94).

Polidine 901/55. A proprietary trade name for a polyvinylidene chloride emulsion in water. Used as a water and oil barrier coating for paper. (94).

Poligen MMV. A proprietary aqueous dispersion of styrene/acrylic copolymers. (49).

Poligen PE. An aqueous solvent-free dispersion of a polyethylene of average molecular weight. (49).

Polishing Oil. A fraction of petroleum oil, having a boiling-point of 130–160° C., and a specific gravity of 0·74–0·77. Used as a turpentine substitute.

Polish Turpentine. See .TURPENTINE.

Politint. Proprietary dyes for plastics. No. 1 for methylmethacrylate. No. 2 for cellulose acetate, cellulose acetate butyrate, and ethyl cellulose.

Pollack's Cement. A cement made from glycerin, litharge, and red lead.

Pollards. See BRAN.

Pollinex. A proprietary vaccine for treating by desensitisation pollen-induced asthma and hay-fever. (272).

Pollopas. A synthetic resin. It is a condensation product of urea and formaldehyde, prepared by a special process. It is stated to be suitable as a glass substitute for lenses. (169).

Pollucite (Pollux). A rare mineral. It is a silicate of aluminium and cæsium, $H_2Cs_4Al_4(SiO_3)_9$.

Pollux. See POLLUCITE.

Polmerite. Synonym for Taranakite.

Polnec. A proprietary polyester laminating resin. (683).

Pologol Tarsone. A proprietary preparation of hydrocortisone and coal tar for dermatological use. (375).

Poloxalkol. A polymer of ethylene oxide, propylene oxide and propylene glycol.

Poloxyl Lanolin. A polyoxyethylene condensation-product of anhydrous lanolin. Aqualose.

Poly 1F4 and 2F4. Proprietary brands of synthetic rubber capable of withstanding high temperatures. Fluoracrylates. (471).

Polyalk. A proprietary preparation of DIMETHICONE and aluminium hydroxide gel. An antacid. (476).

Polyall. A proprietary melamine formaldehyde moulding resin. (647).

Polyargite. A mineral. It is a variety of pinite.

Polyargyrite. A mineral, $12Ag_2S.Sb_2S_3$.

Polyarsenite. See SARKINITE.

Polybactrin. A proprietary preparation of POLYMYXIN B sulphate and zinc BACITRACIN. A topical antibiotic. (247).

Polybasite. A silver ore, $As_2S_3.9(Ag_2 \cdot Cu_2)S$ or $(Cu_2S.Ag_2S)_9(Sb_2S_3.As_2S_3)$. A mineral. It is also given as $16[Ag_{16}Sb_2S_{11}]$.

Polyblack. A proprietary 50 : 50 mixture of nitrile rubber and carbon black used as a master batch. (288).

Polybor. Trade name for a prepared product intermediate between borax and boric acid in its chemical properties. It is, however, more soluble in water than either of these and is therefore useful for obtaining sodium borate rich solutions which are almost neutral. (183).

Polybrene. A proprietary trade name for Hexadimethrine Bromide.

Polycarbafil. A registered trade name for flame retardant polycarbonate materials.

Polycat 20, 21 and 22. A proprietary range of amine catalysts. (311).

Polychloral. See VIFERRAL.

Polychrest Salt. An old name for normal potassium sulphate, K_2SO_4. The term is also applied to Rochelle salt.

Polychrome Blue of Unna. A dyestuff prepared by the action of potassium carbonate on Methylene blue. Used in microscopy.

Polychrome, Methylene Blue. See TERRY'S STAIN.

Polychromine. See PRIMULINE.

Polychromine B (Fast Cotton Brown R, Direct Brown R). A dyestuff obtained by boiling equal molecules of p-nitrotoluene-sulphonic acid and p-phenylenediamine with aqueous sodium hydroxide. Dyes cotton orange-brown.

Polycillin. A proprietary preparation of ampicillin. An antibiotic. (324).

Polycin. A proprietary polyester-based polyurethane elastomer cross-linked with diamine or polyols. (76).

Polycizer. A proprietary trade name for vinyl plasticisers as follows :
Coded 162-dioctyl phthalate.
Coded 332-dioctyl adipate. (121).

Polyco 2140. A proprietary trade name for polyvinyl acetate. (168).

Polycrase. A mineral similar to Euxenite.

Polycroit. The colouring matter of saffron.

Polycrol. A proprietary preparation containing methylpolysiloxane, aluminium hydroxide gel and magnesium hydroxide. An antacid. (271).

Polydymite (Nickel - linnæite). A mineral, $(Ni.Co)_4S_5$.

Polyeite. A proprietary polyester laminating resin. (48).

Polyestal. A proprietary polyester laminating resin. (798).

Polyestradiol Phosphate. An œstrogen currently undergoing clinical trial as " Leo 114 " and " Estradurin ". It is a polyester of œstra-1,3,5(10)-triene-3, 17β-diol and phosphoric acid.

Polyethylene. See POLYTHENE, ALKATHENE.

Polyethysol. A patented low density polyethylene solution coating for metalwork. (254).

Polyfax. A proprietary preparation of POLYMYXIN B sulphate and zinc BACITRACIN. A topical antibiotic. (277).

Polyflex (1). A registered trade mark for a flexible polystyrene sheet and fibre.

Polyflex (2). A proprietary anti-oxidant. It is 6-ethoxy-2,2,4-trimethyl-1,2-dihydroquinoline. (435).

Polyflon. A proprietary brand of polytetrafluoroethylene (PTFE). (799).

Polyfyde. A trade name for acetal plastics. (12). See also CELCON (11), DELRIN (10), and HOSTAFORM C (9).

Polygalin. See STRUTHIIN.

Polygallic Acid. See STRUTHIIN.

Polygard. A stabiliser for SBR and plastics. A mixture of alkylated aryl phosphites. (556).

Polygeline. A polymer of urea and polypeptides derived from denatured gelatin. Present in Haemacel.

Polyglactin. A synthetic suture capable of being absorbed by the patient's body. It is a mixture of lactic acid polyester with glycolic acid.

Polyglandin. A solution of the autacoid principles of the thyroid, parathyroid, ovary, testic, and pituitary gland substances.

Polyglycolic Acid. A synthetic suture capable of being absorbed by the patient's body. Poly(oxycarbonylmethylene). DEXON.

Polyhalite (Isobelite). A mineral which occurs in the Strassfurt deposits. It is a crystalline mixture of the sulphates of calcium, magnesium, and potassium, and is found with rock salt. It has the formula, $K_2SO_4.MgSO_4.2CaSO_4.2H_2O$.

Polyhexanide. An anti-bacterial. Poly-(1 - hexamethylenebiguanide hydrochloride). Present in TEATCOTE PLUS.

Polyhydrite. A mineral. It is a silicate of iron.

Polyisobutylene (P.I.B., Isolene, Oppanol, Vistanex). Polymers of isobutylene. See under OPPANOL B. It is sold under the above trade names.

Polylite. (1) A mineral. It is a variety of pyroxene.

Polylite. (2) A registered trade name for polyester resins for moulding purposes.

Polylite (3). A proprietary anti-oxidant. Alkylated diphenylamine. (48).

Polylithionite. A mineral. It is a variety of zinnwaldite containing lithium.

Polymaster. A proprietary polyester laminating resin. (800).

Polymerised Oils. See BLOWN OILS.

Polymignite. A mineral, $4(Ca.Ce.Fe)O.(Ti.Zr)O_2.CaO.Nb_2O_5$.

Polymon Green 69S. Phthalocynine green for plastics. (2).

Polymon Green GN 500. Phthalocyamine green for plastics. (2).

Polymyxin. Antimicrobial substances produced by Bacillus polymyxa. Specific substances are designated by a terminal letter, e.g., Polymyxin B sulphate. AEROSPORIN is Polymyxin B sulphate. Present in DIPLOMYCIN, FRAMYSPRAY, POLYFAX and TRIBACTRIC as the sulphate.

Polynoxylin. Poly[methylenedi(hydroxymethyl)urea]. Anaflex Ponoxylan.

Polyox. A range of water soluble, high molecular weight polymers of ethylene oxide. They are used in adhesives, binders, pharmaceuticals and lubricants.

Polyoxyl 8 Stearate. Polyoxyethylene 8 stearate. Macrogol stearate 400. Myrj 45.

Polyoxyl 40 Stearate. Polyoxyethylene 40 stearate. Macrogol stearate 2000. Myrj 52 and 52S.

Poly-pale. A proprietary trade name for a polymerised resin having a 15-20° C. higher melting-point than gum or wood resin.

Poly-Pale Esters. Polymerised rosin esters. (593).

Polypentek. A proprietary trade name for polypentaerythritol.

Polyphenyl Black. A polyazo dyestuff of the same class as Chloramine green B. It dyes cotton black.

Polyphenyl Yellow R. A direct cotton dyestuff.

Polyprene. Weber's name for rubber.

Polysil. Silicone rubber gums and compounds. (614).

Polysolvan E. A proprietary trade name for a solvent mixture comprising the acetates of propyl, isobutyl and amyl alcohols.

Polysolvan O. A proprietary trade name for a solvent composed of the ester of isobutyl alcohol with glycollic and butyl glycollic acids.

Polysolvan SHS. A proprietary trade name for acetic acid esters of alcohols up to C_{11}.

Polysorbate 20. Polyoxyethylene 20 sorbitan monolaurate. Sorbimacrogol laurate 300. Tween 20.

Polysorbate 40. Polyoxyethylene 20 sorbitan monopalmitate. Sorbimacrogol palmitate 300. Tween 40.

Polysorbate. 60. Polyoxyethylene 20 sorbitan monostearate. Sorbimacrogol stearate 300. Tween 60.

Polysorbate 65. Polyoxyethylene 20 sorbitan tristearate. Sorbimacrogol tristearate 300. Tween 65.

Polysorbate 80. Polyoxyethylene 20 sorbitan mono-oleate. Sorbimacrogol oleate 300. Tween 80.

Polysorbate 85. Polyoxyethylene 20 sorbitan trioleate. Sorbimacrogol trioleate 300. Tween 85.

Polysphærite. A mineral, $(Pb.Ca)_3 (PO_4)_2 (Pb.Ca)_2 Cl(PO_4)$.

Polystab. Stabilizers for PVC. (558).

Polystat. Antistatic agents. (597).

Polystyrene. A synthetic thermoplastic formed by the polymerisation of monomeric styrene $C_6H_5CH : CH_2$. It is used in moulding parts for high frequency insulation and in the preparation of lacquers. Its high-frequency electrical properties are among the best. Trade names for the material are: Polystyrol, Metastyrol, Metastyrene, Resoglass, Vistron, Trolitul, Superstyrex, B.P. Polystyrene. (288). Many of these materials vary in properties such as flexibility and hardness according to the particular use for which they are required.

Polystyrol 143E. A proprietary polystyrene having easy melt-flow and good mechanical properties. (49).

Polystyrol 165H. A proprietary polystyrene similar to POLYSTYROL 143E but possessing greater mechanical strength. (49).

Polystyrol 168N. A proprietary polystyrene stabilised against ultra-violet light. (49).

Polystyrol 427M. An impact-resistant polystyrene with good resistance to deformation at high temperatures. (49).

Polystyrol 432F. A proprietary impact-resistant styrene-butadiene copolymer. (49).

Polystyrol 466 I. A proprietary styrene/butadiene copolymer with high impact resistance. (49).

Polystyrol 472 D. A proprietary styrene/butadiene copolymer offering high impact resistance. (49).

Polystyrol 473 E. A proprietary styrene-butadiene copolymer offering high-impact resistance and easy flow properties. (49).

Polystyrol 475 K. A proprietary styrene-butadiene copolymer offering high resistance to impact. (49).

Polystyrol KR 2536. A proprietary styrene/butadiene copolymer offering good resistance to impact and to deformation at high temperatures. (49).

Polystyrol KR 253 and KR 2538. Proprietary styrene-butadiene copolymers offering very high resistance to impact at low temperatures. (49).

Polytar. A proprietary preparation of liquid paraffin, tar CADE OIL, coal tar and ARACHIS OIL extract of coal tar used in the treatment of skin diseases. (846).

Polytelite. A mineral, $(Pb.Ag_2)_4Sb_2S_7$, with $(Zn.Fe)_4Sb_2S_7$.

Polyterpene. See LIQUID RESINS.

Polythene. The general term for a range of solid polymers of ethylene. See ALATHON, ALKATHENE, BAKELITE POLYETHYLENE, MONSANTO POLYETHYLENE, PETROTHENE, TENITE POLYETHYLENE, HI-FAX, HOSTALEN, ROTENE, VESTOLEN, FORTIFLEX, FORTILENE, MARLEX, RIGIDEX and CARLONA.

Polythiazide. 6-Chloro-3,4-dihydro-2-methyl - 3 - (2,2,2 - trifluoroethylthio - methyl)benzo - 1,2,4 - thiadiazine - 7 - sulphonamide 1,1-dioxide. Nephril. Renese.

Polytuf 200. A proprietary flame-resistant terpolymer of vinyl chloride, ethylene and vinyl acetate, used as a moulding and extrusion compound. (659).

Polyvidone. See POVIDONE. Polyvinylpyrrolidone.

Polyvinox. Poly(butyl vinyl ether). Shostakovsky Balsam.

Polyvinyl Chloride. PVC., Breon, Carina, Chlorovene, Corvic, Flamenol, Geon 100, Koron, Koroseal, Mipolam, Plioflex, Vinylite Q, and Welvic are trade names for polyvinyl chloride which may or may not be plasticised.

Polyvite. A proprietary preparation of vitamin A, CALCIFEROL, THIAMINE, RIBOFLAVIN, PYRIDOXINE, ascorbic acid calcium pantothenate and nicotinamide. A multi-vitamin supplement. (250).

Polyvon. A proprietary polyester-based polyurethane elastomer cross-linked with diols. (847).

Poly-zole AZDN. A trade mark. Azodiisobutyronitrile. A vinyl polymerization catalyst. Gives freedom from side reactions and is not readily poisoned. (53).

Polyzote. A proprietary trade name for nitrogen-expanded synthetic resin plastics.

Pomace. The residue from the extraction of apple juice in cider manufacture. A cattle food.

Pomade en Crème. Cold cream.

Pomaret's 132. See EPARSENO.

Pombe. A beer made from Sorghum millet.

Pomoline. See PICROFOETIDINE.

Pomoloy. A proprietary trade name for a cast iron made by a special process.

Pomona Green. See IODINE GREEN.

Pompeian Red. See INDIAN RED.

Pompey Red. Ferric oxide, Fe_2O_3.

Pompholix. Zinc oxide, ZnO.

Pompilion Oil. See OIL OF POMPILION.

Ponceau. Synonym for scarlet. See MAGENTA.

Ponceau B. See BIEBRICH SCARLET.

Ponceau B Extra. See BIEBRICH SCARLET.

Ponceau 3B. See OIL SCARLET.

Ponceau BO Extra. See BRILLIANT CROCEINE M.

Ponceau G. See PONCEAU R.

Ponceau 2G (Orange R, Brilliant Ponceau GG, Ponceau JJ, Acid Scarlet G, Ponceau GL). A dyestuff. It is the sodium salt of benzene-azo-β-naphthol-disulphonic acid R, $C_{16}H_{10}N_2O_7S_2Na_2$. Dyes wool reddish-orange from an acid bath.

Ponceau 3G (Scarlet 3G). A dyestuff. It is the sodium salt of sulpho-anisoil-azo-β-naphthol, $HSO_3.OCH_3.C_6H_3N : N.C_{10}H_6ONa.$

Ponceau 4GB (Croceine Orange, Orange ENL, Brilliant Orange, Brilliant Orange G, Orange GRX, Brilliant Orange GL, Rainbow Orange). A dyestuff. It is the sodium salt of benzene-azo-β-naphthol-β-sulphonic acid, $C_{16}H_{11}N_2O_4SNa$. Dyes wool orange-yellow from an acid bath.

Ponceau GL. A dyestuff. It is a British equivalent of Ponceau 2G.

Ponceau GR. See PONCEAU R.

Ponceau GT. A dyestuff. It is toluene-azo-β-naphthol-disulphonic acid.

Ponceau J. See PONCEAU R.

Ponceau JJ. See PONCEAU 2G.

Ponceau R, 2R, G, GR (Ponceau J, Xylidine Red, Xylidine Scarlet, Scarlet G, Brilliant Ponceau G, Ponceau 2RX, Naphtharene Scarlet 2R, Rainbow Scarlet G, Scarlet 2RS, Scarlet 2R, 2RL, R, 3R). A dyestuff. It is the sodium salt of xylene-azo-β-naphthol-disulphonic acid, $C_8H_9.N_2.C_{10}H_4(HSO_3)_2ONa$. Dyes wool scarlet from an acid bath.

Ponceau 2R. See PONCEAU R.

Ponceau 2RX. A British brand of Ponceau.

Ponceau 3R (Cumidine red). A dyestuff. It is the sodium salt of ethyl-dimethyl-benzene - azo - β - naphthol - disulphonic acid, $C_2H_5 . (CH_3)_2 . C_6H_2 . N_2 . C_{10}H_4 (HSO_3)_2ONa$. Dyes wool bluish-scarlet from an acid bath. See BIEBRICH SCARLET.

Ponceau 3R (Ponceau 4R, Cumidine Red, Cumidine Ponceau). A dyestuff. It is the sodium salt of ψ-cumene-azo-β-naphthol-disulphonic acid, $C_{19}H_{16}N_2O_7S_2Na_2$. Dyes wool bluish-scarlet from an acid bath.

Ponceau 4R. See PONCEAU 3R.

Ponceau 5R (Erythrin X, Scarlet 5R). A dyestuff. It is the sodium salt of benzene - azo - benzene - azo - β - naphthol-trisulphonic acid, $C_{22}H_{13}N_4O_{10}S_3 Na_3$. Dyes wool bluish-red from an acid bath.

Ponceau 6R. See SCARLET 6R.

Ponceau 3RB. See BIEBRICH SCARLET.

Ponceau 4RB. See CROCEINE SCARLET 3B.

Ponceau 6RB. See CROCEINE SCARLET 7B.

Ponceau 10RB. An acid dyestuff. It dyes wool or silk bluish-crimson from an acid bath.

Ponceau RT. A dyestuff. It is the sodium salt or toluene-azo-β-naphthol-disulphonic acid, $C_{17}H_{12}N_2O_7SNa$. Dyes wool orange-red from an acid bath. It is isomeric with Ponceau GT from R salt.

Ponceau S Extra. See FAST PONCEAU 2B.

Ponceau S for Silk. See FAST ACID SCARLET.

Ponderax. A proprietary preparation of fenfluramine hydrochloride. An anti-obesity agent. (313).

Ponderax PA. A proprietary preparation of FENFLURAMINE hydrochloride in a sustained-release capsule. An appetite suppressant. (848).

Pondermite. A mineral, $Ca_2B_6O_{11}. 4H_2O$. A source of boric acid.

Ponder's Stain. A microscopic stain. It consists of 0·02 gram toluidine blue, 1 c.c. glacial acetic acid, and 2 c.c. absolute alcohol in 100 c.c. distilled water.

Pondicherry Oil (Nut Oil). Arachis oil.

Ponite. A mineral. It is a variety of rhodocrosite containing iron.

Ponolith (Sunolith, Superlith). Lithopone pigments.

Ponoxylan. A proprietary trade name for Polynoxylin.

Ponsital. The registered trade name for a neuroleptic agent. It is 3-chloro-10-{γ - [N' - β' - (1″ - methyl - 2″ - oxo - imi-dazolidyl - 3″) - ethyl - N - piperazinyl] -

propyl}-phenothiazine dihydrochloride. (844).

Ponstan. A proprietary preparation of mefenamic acid. An analgesic. (264).

Pontianac. Jelutong (*q.v.*).

Pontianac Copal. See COPAL RESIN.

Pontianak Illipé Nuts. See BORNEO TALLOW.

Ponticin. See RHAPONTICUM.

Poonac. Coconut cake, a cattle food. The term is also used for the residue from castor oil seeds after cold and hot pressing and solvent extraction. Used for caulking timber.

Poonahlite. A mineral. It is a hydrated aluminium-calcium silicate.

Poonseed Oil. See LAUREL-NUT OIL.

Pope's Solution. A solution of 1 part in 10,000 of a mixture of 10 parts 2 : 7-dimethyl-3 : 6-diamino-acridinium-me-thylo-chloride hydrochloride and 1 part crystal violet. An antiseptic for wounds.

Poppy Capsules or Heads. The dried, immature fruit of *Papaver somniferum*.

Populin. Benzoyl-salicin, $C_{20}H_{22}O_8$.

Porcelain. A mixture of clay, quartz, and felspar. A normal mix consists of 50 per cent. clay, 25 per cent. quartz, and 25 per cent. felspar.

Porcelain Clay. Synonym for Kaolinite. See CHINA CLAY.

Porcelain Earth. See CHINA CLAY.

Porcelain, Semi. See SEMI-PORCELAIN.

Porcelain White. See LITHOPONE.

Porcelanite. A fused clay and shale found in burned coal seams.

Porcelave. A proprietary trade name for a ceramic material.

Porfiromycin. An antibiotic. 6-Amino-8 - carbamoyloxymethyl - 1, 1a, 2, 8, 8a, 8b-hexahydro- 8a - methoxy - 1, 5 - dimethylazirono [2', 3' : 3, 4] pyrrolo[1, 2-*a*]indole-4,7-dione.

Porocel. A proprietary trade name for a carefully prepared and screened bauxite.

Poron. A trade mark for expanded polystyrene plastics. (239).

Porous Alum. Sodium aluminium sulphate (soda alum), $Al_2(SO_4)_3.Na_2SO_4.$ $24H_2O$.

Porpezite. A native alloy of gold and palladium.

Porphyry. A building stone having the same composition as felspar.

Porporino. An alloy of mercury, tin, and sulphur. Used for decorating purposes.

Porrier's Orange III. See ORANGE III.

Portagen. A proprietary artificial infant food containing medium chain triglycerides and non-lactose carbohydrates, for use in cases of intolerance of fat and lactose. (324).

Porter. A beverage. London porter contains 5·4 per cent. alcohol.

Portland Arrowroot. The starch from *Arum maculatum*.

Portland Cement. Made by heating an intimate mixture of argillaceous and calcareous substances, such as lime and clay, and pounding the product. The material does not slake with water, and has energetic hydraulic properties.

Portland Stone. A limestone.

Portsmouth Accelerator No. 3. A proprietary rubber vulcanisation accelerator. It is phenyl - *o* - tolyl - guanidine.

Portugallo Oil. Essential oil of orange peel.

Portugal Oil. See OIL OF PORTUGAL.

Portuguese Turpentine. See TUR-PENTINE.

Portyn. Benzilonium bromide.

Portyn Kapseals. A proprietary preparation containing benzinolinium bromide. An antispasmodic. (264).

Porzite. Synonym for Mullite.

Posalfilin. A proprietary preparation of podophyllin and salicylic acid used in dermatology as a peeling agent. (370).

Poskine. A depressant of the central nervous system. *O*-Propionylhyoscine. PROSCOPINE is the hydrobromide.

Potaba. A proprietary preparation of potassium *p*-aminobenzoate and envule. An anti-inflammatory. (810).

Potaba + 6. A proprietary preparation of potassium *p*-aminobenzoate and PYRI-DOXINE. An anti-inflammatory. (810).

Potarite. A mineral (Pd.Hg).

Potash. Potassium hydroxide, KOH. Potassium carbonate is also called potash. Another name for the carbonate is Salts of tartar.

Potash Alum. A double sulphate of potassium and aluminium, $Al_2(SO_4)_3.$ $K_2SO_4.24H_2O$.

Potash Bordeaux Mixture. Contains 6 lb. copper sulphate, 2 lb. potassium hydroxide, and 50 gallons water.

Pot-ashes. Impure potassium carbonate.

Potash Felspar. See FELSPAR.

Potash Glass. A glass containing silicate of potassium.

Potash-lead Glass. A glass usually containing from 40–50 per cent. SiO_2, 28–53 per cent. PbO, 8–11 per cent. K_2O, and 1 per cent. Al_2O_3 and Fe_2O_3.

Potash-lime Glass (Bohemian Glass). A glass usually containing from 72–76 per cent. SiO_2, 12–15 per cent. K_2O, 8–10 per cent. CaO, 1 per cent. Al_2O_3 and Fe_2O_3, and 0–3 per cent. Na_2O.

Potash Lozenges. Potassium chlorate lozenges.

Potash Mica. See MUSCOVITE.

Potash Pellets. Compressed tablets of potassium chlorate. See POTASH LOZENGES.

Potash Salts. See ABRAUM SALTS.

Potash Water-glass. A mixture of potassium silicates.

Potassalumite. A mineral. It is a potash alum.

Potassic Superphosphate. A manure made by combining calcium superphosphate with potash salts.

Potassium Amalgam. An alloy of potassium and mercury, formed by the combination of the elements.

Potassium Cadmium Iodide. See MARME'S REAGENT.

Potassium Felspar. See FELSPAR.

Potassium Menaphthosulphate. Dipotassium 2-methyl-1,4-disulphato-naphthalene dihydrate. Vikastab.

Potassium Muriate. See MURIATE OF POTASH.

Potassium Prussiate, Red. See RED PRUSSIATE OF POTASH.

Potassium Prussiate, Yellow. See YELLOW PRUSSIATE OF POTASH.

Potato Flour. Potato starch.

Potato Gum (Almadina, Euphorbia Gum). Almeidina gum, stated to be derived from *Euphorbia rhipsaloides*, of West Africa. The latex contains about 10 per cent. rubber, 32 per cent. water, 51 per cent. resin, 1 per cent. protein, 6 per cent. insoluble matter, and gives an ash of 2·5 per cent. The dry material contains 14·3 per cent. rubber and 75·8 per cent. resin.

Potato Oil or Spirit. The alcohol obtained from potato starch. Also see FUSEL OIL.

Potato Rubber. See POTATO GUM.

Potato Stones. Agate from the marls of Somersetshire.

Potazote. A French fertiliser containing 14 per cent. nitrogen, as ammonium chloride, and 20 per cent. potassium oxide as potassium chloride.

Potensan. A proprietary preparation of yohimbine hydrochloride, dexamphetamine sulphate, strychnine hydrochloride and amylobarbitone. (250).

Potensan Forte. A proprietary preparation of yohimbine hydrochloride, methyltestosterone, pemoline and strychnine hydrochloride. (250).

Potentite. See TONITE.

P.O.T.G. Phenyl-*o*-tolyl-guanidine, a rubber vulcanisation accelerator.

Potin. An alloy of 72 per cent. copper, 25 per cent. zinc, 2 per cent. lead, and 1 per cent. tin.

Potinjaune. See POTIN.

Pot Metal. An alloy of lead and copper.

Potosi Silver. See NICKEL SILVERS.

Potstone. An impure steatite (*q.v.*).

Potter's Clay. Pipeclay (*q.v.*).

Potter's Ore. See ALQUIFON.

Pouckpong Gum (Touchpong Gum). A rubber gum of British Guiana.

Poudre B (Vieille Powder). A French explosive. It is a smokeless powder made from a mixture of soluble and insoluble nitro-cellulose, thoroughly gelatinised with a mixture of ether and alcohol, rolled into sheets, and cut into strips.

Poudre EF. A French explosive made from nitro-cellulose and binding material.

Poudre J. A French explosive containing 83 per cent. guncotton and 17 per cent. potassium bichromate.

Poudre Pyroxulée. A French sporting powder. It consists of insoluble nitrocellulose, with 35 per cent. barium and potassium nitrates.

Poudre Savory. Seidlitz powder.

Poulenc 309. A proprietary preparation. It is sym-di-sodium-*m*-amino-benzoyl-*m*-amino-*p*-methyl-benzoyl-1-naphthyl-amino-4 : 6 : 8-trisulphonate-urea.

Pounce. Powdered sandarac.

Poutet's Reagent. Consists of 1 c.c. mercury dissolved in 12 c.c. nitric acid, specific gravity 1·42. Used for testing oils.

P.O.V. Purified oil of vitriol (sulphuric acid containing 93–96 per cent. H_2SO_4).

Povidone. Poly(vinylpyrrolidone). Plasmosan.

Povidone-Iodine. A complex produced by reacting iodine with poly(vinylpyrrolidone). Betadine.

Powder 19/04/15H Black 904. A proprietary polyethylene used in rotational moulding and carpet-backing applications. It can be used in contact with foodstuffs.

Powder 22/04/00A 400. A proprietary polyethylene powder of micron size having a low melting point. It is used for making interliners for fabrics and as carpet backing.

Powder 26/04/00. A proprietary polyethylene powder possessing good rigidity, used in rotational moulding.

Powder 215 Natural. A proprietary 400-micron powder used in flame-retardant rotational mouldings.

Powdered Hydrocyanic Acid. The name applied to a calcium cyanide prepared from calcium carbide and hydrocyanic acid. It evolves hydrocyanic acid with moisture, hence the name. A fumigator.

Powder of Algaroth (Basic Chloride, English Powder, Powder of Algarotti). A mixture of antimony oxychloride, SbOCl, and antimony oxide, Sb_2O_3. Used in the preparation of tartar emetic.

Powder of Algarotti. See POWDER OF ALGAROTH.

Powellite. A mineral. It is calcium molybdate, $CaMoO_4$.

Pozzolana Cement. See EISEN PORT-LAND CEMENT.

P.P.D. Piperidine - pentamethylene - dithio-carbamate. A rubber vulcanisation accelerator.

PPO. Polyphenylene oxide.

P.P.S. A proprietary polyphenylene sulphide. A cross-linkable aromatic thermoplastic with a high modulus used as a coating material capable of withstanding temperatures in the range 200°–260° C. (730).

P.R. Spray. A proprietary preparation of trichlorofluoromethane and dichlorodifluoromethane used as an analgesic spray. (502).

Practolol. A beta adrenergic receptor blocking agent. It is 4-(2-hydroxy-3-isopropylaminopropoxy)acetanilide.

Praenitrona. A proprietary trade name for Trolnitrate Phosphate.

Prage Alizarin Yellow G. A dyestuff. It is m-nitro-benzene-azo-resorcylic acid, $C_{13}H_9N_3O_6$. Dyes chrome mordanted cotton pure yellow, and chromed wool brownish-yellow.

Prage Alizarin Yellow R. A dyestuff. It is p-nitro-benzene-azo-resorcylic acid, $C_{13}H_9N_3O_6$. Dyes chromed wool and cotton orange-yellow.

Pragmatar. A proprietary preparation of cetyl alcohol and coal tar distillate with sulphur and salicylic acid used in the treatment of dandruff. (658).

Pragmoline. A proprietary preparation of acetyl-choline bromide.

Prague Red. See INDIAN RED.

Pralidoxime. N-Methylpicolinaldoxime. Protopam is the iodide.

Prajmalium Bitartrate. A preparation used in treatment of arrhythmia of the heart. N-Propylajmalinium hydrogen tartrate.

Pramidex. A proprietary preparation of TOLBUTAMIDE used in the treatment of diabetes. (137).

Praminil. A proprietary preparation of IMIPRAMINE hydrochloride. An anti-depressant. (363).

Pramiverine. 4, 4 - Diphenyl - N - iso - propylcyclohexylamine.

Pramoxine. 4-(3-(4-Butoxyphenoxy) - propyl]morpholine.

Prampine. O-Propionylatropine. PAMN is the methonitrate.

Prase. A mineral, SiO_2.

Praseolite. Similar to Cardierite.

Praxilene. A proprietary preparation of NAFTIDROFURYL used in the treatment of cerebrovascular disease. (849).

Prayer Beads. The seeds of *Abrus precatorius*. See JEQUIRTY.

Prazepam. A muscle relaxant. 7-Chloro - 1 - (cyclopropylmethyl) - 1, 3 - dihydro - 5 - phenyl - $2H$ - 1, 4 - benzo - diazepin-2-one.

Praziquantel. An anthelmintic currently undergoing clinical trial as " EMBAY 8440 ". It is 2-cyclohexylcarbonyl-1, 3, 4, 6, 7, 11b - hexahydro - $2H$ - pyra-zino[2,1-a]isoquinolin-4-one.

Prazitone. An anti-depressant. 5-Phenyl - 5 - (2 - piperidyl)methylbar - bituric acid.

Prazosin. An anti-hypertensive. 1-(4-Amino - 6, 7 - dimethoxyquinazolin -2 - yl) - 4 - (2 - furoyl)piperazine.

Preceptin. A proprietary preparation of p-diisobutylphenoxypolyethoxyetha-nol and ricinoleic acid. A spermicidal gel for use with a vaginal diaphragm. (369).

Precipitated Calomel. See CALOMEL.

Precipitated Phosphate. Insoluble calcium phosphate.

Precipitate, Red. See RED OXIDE OF MERCURY.

Precipitate, White Fusible. See MER-CURAMMONIUM CHLORIDE.

Precortisyl. A proprietary preparation of prednisolone acetate. (307).

Predazzite. A mineral. It is a mixture of calcite and brucite.

Predef 2X. A proprietary preparation containing ISOFLUPREDONE 21-acetate. A veterinary steroid.

Prednelan. A proprietary preparation of prednisolone. (335).

Prednesol. A proprietary preparation of PREDNISOLONE disodium phosphate. (335).

Prednisolamate. Prenisolone 21-di-ethylaminoacetate.

Prednisolone. 11β,17α-21-Trihydroxy-pregna-1,4-diene-3,20-dione. 1,2-Dehy-drocortisone. Metacortandralone. Co-delcortone. Delta-Cortef. Delta-cortril. Delta-Stab. Di-Adreson-F. Precortisyl. Ultracorten-H. Prednelan is the acetate. Predsol is the disodium phosphate.

Prednisone. 17α-21 - Dihydroxypregna-1,4 - diene - 3,11,20 - trione - 1,2 - De-hydrocortisone. Metacortandracin. De-cortisyl. Deltacortone. Di-Adreson. Ultracorten. Delta-Cortelan is the ace-tate.

Prednylidene. 11β,17α,21-Trihydroxy-16 - methylenepregna - 1,4 - diene - 3,20-dione. Dacortilene. Decortilen.

Predsol. A proprietary preparation of prednisolone disodium phosphate. (335).

Predsol-N. A proprietary preparation of prednisolone sodium phosphate and neomycin sulphate. (335).

Pregaday. A proprietary preparation of ferrous fumarate and folic acid. Hæmatinic for use in pregnancy. (335).

Pregamal. A proprietary preparation

of folic acid and ferrous fumarate. A hæmatinic. (335).

Pregfol. A proprietary preparation of folic acid and ferrous sulphate. A hæmatinic. (245).

Pregl's Solution. A solution of potassium iodide and sodium iodate with a little sodium chloride and bicarbonate.

Pregnavite. A proprietary preparation of Vitamins A, D, C, thiamine, nicotinamide, ferrous sulphate and calcium phosphate. (272).

Pregnavite Forte. A proprietary preparation of Vitamins A, D, C, thiamine, riboflavine, pyridoxine, nicotinamide, ferrous sulphate and calcium phosphate. (272).

Pregnavite Forte (F). As for Pregnavite Forte but with folic acid. (272).

Pregnenolone. 3β-Hydroxypregn-5-en-20-one.

Pregnyl. A proprietary preparation of chlorionic gonadotrophin used in the treatment of delayed puberty and anovulatory sterility. (316).

Pregolan. A chlorine compound giving 65–72 per cent. available chlorine.

Pregrattite. A mineral. It is a variety of paragonite.

Pregwood. A proprietary synthetic resin impregnated wood, made by impregnating and then subjecting the wood to heat and pressure.

Prehensol. A proprietary preparation of zinc salicylate and lecithin used in the treatment of detergent dermatitis. (377).

Prehnite (Jacksonite). A mineral. It is a silicate of aluminium and calcium, $2CaO.Al_2O_3.3SiO_2$.

Preludin. A proprietary trade name for the hydrochloride of Phenmetrazine.

Premarin. A proprietary preparation of conjugated œstrogens. (467).

Premier Alloy. A heat-resisting alloy containing 61 per cent. nickel, 11 per cent. chromium, 25 per cent. iron, and 3 per cent. manganese.

Premier Jus. See OLEO.

Prenatal Dri-Kaps. A proprietary preparation of Vitamins A, C, D, K, B_{12}, thiamine, riboflavin, folic acid, niacinamide, calcium hydrogen phosphate, ferrous and manganese sulphates. (306).

Prenite. A proprietary trade name for an asbestos sheet bonded with neoprene. A packing material.

Prenomiser. A proprietary preparation of isoprenaline sulphate in aerosol form. A bronchial antispasmodic. (188).

Prenylamine. N - (3,3 - Diphenylpropyl)-α-methyl phenethylamine. Segontin and Synadrin are the lactate.

Prepacol. A proprietary preparation containing magnesium hydroxide, hydroxyethylcellulose and dioctyl sodium sulphosuccinate. A laxative. (321).

Prepagen WK. A proprietary trade name for 75 per cent. distearyl dimethyl ammonium chloride in isopropanol. Used as a softener in the laundry trade. (9).

Preparation 102. See LUARGOL.

Prepared Bark. See QUERCITRON.

Prepared Calamine. Obtained by calcining and powdering negative zinc carbonate or calamine, and freeing the product from gritty particles. It consists of zinc carbonate with some oxide of iron.

Prepared Chalk. (*Creta præparata B.P.*) Washed chalk or whiting.

Prepared Cobalt Oxide. Cobalt oxide, CoO, obtained by heating the black oxide, Co_2O_3. Used in the ceramic industry.

Preparing Salt. Sodium stannate, $Na_2SnO_3.3H_2O$. Used as a mordant in dyeing and calico printing.

Prescollan. A proprietary polyester-based polyurethane elastomer cross-linked with diols. It is VULKOLLEN made under licence in the U.K. (850).

Presdwood. A material made in a similar way to masonite (*q.v.*). The boards are thinner.

Preservaline (Freezine, Iceline). Names for formaldehyde used as a preservative for milk.

Preservol. Creosote, partially emulsified with pyroligneous acid. Used for preserving wood.

Presidal. A proprietary trade name for Pentacynium Methylsulphate.

Preslite. See TSUMEBITE.

Presol W. Mercury fungicide solution. (624).

Press-cake. The mill-cake formed by mixing the ingredients of gunpowder in the incorporating mill, is subjected to a high pressure to make press-cake.

Pressed Amber. See AMBROID.

Pressimmune. A proprietary preparation of anti-human lymphocyte globulin used in tissue transplants to produce immuno-suppression. (312).

Pressolith (Sillimanith). Earthenware-porcelain products.

Pressphan. A German name for press-boards made from wood pulp. Used as insulating materials.

Presszell. A German synthetic resin varnish-paper product used as an electrical insulator.

Prest-o-lite. A proprietary brand of acetylene gas compressed in cylinders.

Prestone. The trade mark of National Carbon Company Inc. applied to ethylene glycol anti-freeze.

Preston Salts (Smelling Salts). Consist of acid ammonium carbonate, NH_4HCO_3.

Presto Steel. A proprietary trade name for a steel containing 1·4 per cent. chromium.

Presuren. A proprietary trade name for Hydroxydione sodium succinate.

Pretamazium Iodide. A preparation used in the treatment of enterobiasis. It is 4 - (biphenyl - 4 - yl) - 3 - ethyl - 2 - [4 - (pyrrolidin - 1 - yl) styryl] thiazo - lium iodide.

Preventol. A proprietary preparation. It is stated to be a chlorinated phenol dissolved in organic bases. It is used as a preservative for adhesives, etc., liable to attack by moulds or bacteria.

Previson. A proprietary preparation of norethinodrel and mestranol. Oral contraceptive. (307).

Priadel. A proprietary preparation of lithium carbonate used in the treatment of mania. (349).

Pribramite. A mineral. It is a variety of sphalerite.

Pricite (Bechilite). A mineral. It is a calcium borate $3CaO.4BO_3.6H_2O$.

Priderite. A mineral. It is $(K,Ba)_{1.3}$ $(Ti,Fe''')_8O_{16}$.

Prilocaine. N-(α-Propylaminopropio-nyl)-o-toluidine. Citanest.

Primal. A solution of toluylene-diamine with neutral sulphite. Recommended as a hair dye. Also a proprietary trade name for a synthetic resin of the acrylate type used for leather finishes.

Primaquine. 8-(4-Amino-1-methyl-butylamino)-6-methoxyquinoline.

Primer Gilding Brass. See BRASS.

Primex. See CRISCO.

Primidone. 5 - Ethylhexahydro - 5 - phenylpyrimidine - 4,6 - dione. Mysoline.

Primobolan. A proprietary preparation of methenolone acetate. An anabolic agent. (898).

Primobolan Depot. A proprietary preparation of methenolone enanthate. An anabolic agent. (898).

Primodos. A proprietary preparation of ethinylœstradiol and NORETHISTERONE acetate used in the treatment of secondary amenorrhœa. (438).

Primogyn Depot. A proprietary preparation of œstradiol valerate used in the treatment of amenorrhœa and prostatic carcinoma. (438).

Primolut Depot. A proprietary trade name for the caproate of Hydroprogesterone. (898).

Primolut N. A proprietary trade name for Norethisterone. (898).

Primor. Lubricating and industrial oils. (628).

Primoteston Depot. A proprietary preparation of testosterone used in the treatment of male osteoporosis or sterility. (438).

Primperan. A proprietary preparation of METOCLOPRAMIDE hydrochloride. An anti-emetic. (137).

Primrose. See MARTIUS YELLOW and SPIRIT EOSIN.

Primrose Smokeless. A smokeless 42-grain powder.

Primrose Soluble. See ERYTHROSIN.

Primrose Soluble in Alcohol. See ERYTHRIN.

Primula. See HOFMANN'S VIOLET. Also a mixture of Methyl violet and Fuchsine.

Primuline (Polychromine, Carnotine, Thiochromogen, Aureoline, Sulphine, Primuline Extra). A dyestuff. It is the sodium salt of the mono-sulphonic acids of the dehydro-thionated condensation products of dehydro-thio-toluidine (mixed with some sodium-dehydro-thio-toluidine-sulphonate). It is chiefly $C_{28}H_{17}N_4O_3S_4Na$. Dyes cotton primrose yellow from an alkaline or neutral bath, and is employed for the production of ingrain colours. It also has application in photography.

Primuline Base. p-Toluidine heated with sulphur.

Primuline Bordeaux. Fabric dyed with primuline ($q.v.$), then passed through a solution of ethyl-β-naphthylamine.

Primuline Brown. Fabric dyed with primuline ($q.v.$), then passed through a solution of m-phenylene-diamine.

Primuline Extra. A British equivalent of Primuline.

Primuline Orange. Fabric dyed with primuline ($q.v.$), then passed through a solution of resorcinol.

Primuline Red. See INGRAIN.

Prinalgin. A proprietary preparation of alclofeniac used in the treatment of arthritis. (137).

Prince Rupert's Metal. See PRINCE'S METAL.

Prince's Blue. A mineral. It is a blue variety of Sodalite ($q.v.$). The slabs are polished for ornamental purposes.

Prince's Metal (Prince Rupert's Metal). An alloy. It is a variety of brass containing from 61–83 per cent. copper and 17–39 per cent. zinc. Another alloy, also called Prince's metal, consists of 84·75 per cent. tin and 15·25 per cent. antimony.

Prince's Metallic. See PRINCE'S MINERAL.

Prince's Mineral (Prince's Metallic). A clay containing about 40 per cent. oxides of iron.

Principen. A proprietary preparation of ampicillin. An antibiotic. (326).

Printer's Acetate. Aluminium acetate, $Al(C_2H_3O_2)_3$.

Printer's Iron Liquor. A deep black solution of ferrous acetate, containing some ferric acetate. It contains about 10 per cent. iron.

Printing Black for Wool. A dyestuff produced by the reduction of a mixture of 1 : 5- and 1 : 8-dinitro-naphthalene by means of glucose in alkaline solution in the presence of sodium sulphite. Dyes wool violet black from an acid bath. Employed in printing.

Printing Blue. See INDULINE, SPIRIT SOLUBLE, and METHYLENE BLUE B and BG.

Printing Blue for Wool. A dyestuff obtained by the reduction of 1 : 8-dinitro-naphthalene with sodium sulphide in the presence of sodium sulphite and sodium hydroxide. Dyes cotton violet-blue. Used in wool printing.

Printing Green. See CHROME GREENS.

Printing Inks. Inks consisting of pigments incorporated with varnish made by heating linseed oil.

Prioderm. A proprietary preparation of MALATHION used in the treatment of infestation by lice. (334).

Priolenes. A range of oleines. (597).

Pripsen. A proprietary preparation of piperazine phosphate, and senna. An antihelmintic. (322).

Priscol. A proprietary preparation of tolazoline hydrochloride. A vasodilator. (18).

Prismatic Emerald. See EUCLASE.

Prismatic Nitre. Potassium nitrate, KNO_3, so-called from the form of the crystals.

Pristane. Iso-octadecane, $C_{18}H_{38}$.

Pristerene 59, 60, 61, 62, 63, 65 and 67. Proprietary brands of stearine used for lubrication. (597).

Pristinamycin. An antibiotic produced by Streptomyces pristina spiralis.

Privine. A proprietary trade name for the nitrate of Naphazoline.

Pro-Actidil. A proprietary preparation of TRIPROLIDINE hydrochloride. An anti-allergic drug. (277).

Proban. Antiflame finish for textiles. (503).

Pro-Banthine Bromide. A proprietary trade name for Propantheline Bromide.

Probenecid. 4 - (Di - n - propylsulpha - moyl)benzoic acid. Benemid.

Proberite. A mineral. It is $2[Na,Ca, B_8O_9.5H_2O]$.

Probilin. A preparation of phenolphthalein.

Probnal. Propyl-butyl-acet-urethane.

Probucol. An agent used to control the increase of cholesterol in the blood beyond normal limits. It is 4,4'-(iso-propylidenedithio) bis - (2, 6 - di - t - butylphenol.

Procaine. Novocaine (q.v.).

Procainamide. N-(4-Aminobenzoyl)-2-diethylaminoethamine. 4-Amylino-N-(2-diethylaminoethyl)benzamide. Pronestyl.

Procarbazine. N - 4 - Isopropylcarbamoylbenzyl - N' - methylhydrazine. Natulan is the hydrochloride.

Process Butter. See BUTTER, RENOVATED.

Procion. Dyestuffs, chemically reactive with cellulosic and protein fibre. (512).

Proclonol. $\alpha\alpha$-Di-(4-chlorophenyl) - cyclopropylmethanol.

Prochlorite. A mineral similar to chlorite.

Prochlorperazine. 2-Chloro-10-[3-(4-methylpiperazin - 1 - yl) - propyl]phenothiazine. Compazine is the dimaleate or the edisylate.

Proctofoam H.C. A proprietary preparation of HYDROCORTISONE acetate and PRAMOXINE hydrochloride as an aerosol, used in the treatment of anorectal inflammation. (374).

Proctosedyl. A proprietary preparation of HYDROCORTISONE, cinchocaine hydrochloride and FRAMYCETIN used in the treatment of ano-rectal inflammation. (443).

Procyclidine. 1-Cyclohexyl - 1 - phenyl - 3 - pyrrolidinopropan - 1 - ol. Kemadrin is the hydrochloride.

Procythol. A liver extract.

Prodag. A trade mark for a colloidal form of graphite in water.

Prodexin. A proprietary preparation containing aluminium glycinate and magnesium carbonate. An antacid. (272).

Prodoraqua. A concrete hardener. (561).

Prodorcote. Synthetic resin coating materials. (594).

Prodorite. An acid-resisting material. It is a concrete with a hardened pitch binder. The mineral part is carefully graded and mixed with the pitch. It is stated to be suitable for plants containing corrosive gases.

Prodorkitt. An acid resisting bituminous compound. (594).

Prodorlac. Bituminous paint. (594).

Prodorohalte. An acid resisting mastic. (594).

Prodorplast. A plastisol coating material. (594).

Prodox. High contrast photographic developer. (507).

Prodoxol. A proprietary preparation of OXOLINIC ACID. A urinary antiseptic. (262).

Producer Gas (Air Gas, Siemen's Gas, Generator Gas). Any gas fuel obtained from solid fuel in a gas producer in which the solid fuel is as completely as possible consumed by partial oxidation. It is usually prepared by passing air together with a spray of water over red hot carbon, and contains 28–33 per cent. carbon monoxide, 0·8–4 per cent. carbon dioxide, and 62–64 per cent. nitrogen. Used for heating and power. Air gas, water gas, and semi-water gas are varieties of producer gas.

Profadol. 1 - Methyl - 3 - propyl - 3 - (3-hydroxyphenyl)pyrrolidine.

Profenamine. Ethopropazine.

Profil. A registered trade name for flame retardant polypropylene. (851).

Profil PP-14BX. A proprietary antistatic polypropylene reinforced with glass fibre. (851).

Profil PP-14WX. A proprietary polypropylene copolymer reinforced with glass fibre for use where resistance to heat is required. (851).

Profil PP-15H5. A proprietary glass-reinforced polypropylene copolymer possessing a high degree of resistance to heat and oxidation. (851).

Profil PP-15S5. A proprietary glass-reinforced polypropylene copolymer used in the moulding of automobile accessories. It has high impact resistance. (851).

Profil PP51HX. A proprietary glass-reinforced polypropylene resistant to ultraviolet light and oxidation, used in the extrusion of fibres. (851).

Profil PP-23S. A proprietary polypropylene copolymer used in blow-moulding. (851).

Proflavine. 3 : 6-Diamino-acridine-sulphate, $C_{13}H_{11}N_3.H_2SO_4.H_2O$. An antiseptic.

Progallin. Esters of gallic acid. (625).

Proganol. See PROTARGENTUM.

Pro-Gen. Arsanilic acid and its salts. (525).

Progene. Liquid detergents. (506).

Progilite. A proprietary trade name for phenol-formaldehyde.

Proglumide. An anti-gastrinic. 4-Benzamido *NN* - dipropylglutaramic acid. MILID.

Progressite. An explosive containing 89 per cent. ammonium nitrate, 4·7 per cent. aniline hydrochloride, 6 per cent. ammonium sulphate, and 0·2 per cent. colouring matter.

Progynova. A proprietary preparation of œstradiol valerate used in the treatment of menopausal symptoms. (438).

Proheptazine. Hexahydro-1,3-dimethyl-4 - phenyl - 4 - propionyloxy - azepine.

Proidonite. A mineral, SiF_4.

Proil. Rust preventatives. (526).

Proiodin. A combination of iodine with protein, containing 4·4 per cent. iodine.

Prokayvit Oral. Tablets of acetomenaphthone (vitamin K). (182).

Proketazine. The dimaleate of Carphenazine.

Prokliman Ciba. A German product. It consists of tablets containing ovarial hormone, peristaltin, nitro-glycerin, pyramidone, and a sodium salt of caffeine.

Proladone. A proprietary preparation of oxycodone pectinate. An analgesic. (280).

Prolaurin. A proprietary trade name for propylene glycol mono-laurate.

Prolein. A proprietary trade name for propylene glycol mono-oleate.

Proline. α-Pyrolidine-carboxylic acid.

Prolintane. 1-(α-Propylphenethyl)pyrrolidine. 1-Phenyl-2-pyrrolidinopentane. Villescon.

Prolixin. A proprietary trade name for the dihydrochloride of Fluphenazine.

Prolugen. Propylene glycol. (518).

Promazine. 10-(3-Dimethylamino-propyl)phenothiazine. SPARINE is the embonate or the hydrochloride.

Prometal. A variety of cast iron, used in the construction of furnace parts.

Promethazine. 10-(2-Dimethylamino-*n*-propyl)phenothiazine. Phenergen is the hydrochloride.

Promethazine Theoclate. Promethazine salt of 8-chlorotheophylline. Promethazine chlorotheophyllinate.

Prométhée. See EXPLOSIF O3.

Promethoestrol. 3,4-Di-(4-hydroxy-3-methylphenyl)hexane. Methoestrol.

Promethus. A blasting powder. It contains potassium chlorate, manganese dioxide, iron oxide, mono-nitro-benzene, turpentine oil, and naphtha.

Promicrol. Ultrafine grain photographic developer. (507).

Prominal. A proprietary preparation of N-methyl-ethyl-phenyl-malonyl-urea. (methylphenobarbitone). A hypnotic. (112).

Promintic. A proprietary preparation of METHYRIDINE. A veterinary anthelmintic.

Promoloid. A fertiliser containing colloidal magnesium silicate. It is a Japanese product.

Promoxolan. 4-Hydroxymethyl-2,2-di-isopropyl-1,3-dioxolan.

Pronalys. Analytical grade reagents. (507).

Prondol. A proprietary preparation of iprindole. An antidepressant. (245).

Pronestyl. A proprietary preparation of procainamide hydrochloride. (326).

Pronethalol. 2-Isopropylamino-1-(2-naphthyl)ethanol. Alderlin is the hydrochloride.

Proofite. A proprietary trade name for a product similar to aquatec.

Proof Spirit. A term originally intended to denote alcohol, that was just strong enough to ignite gunpowder, when burnt upon it. It is alcohol containing 49·24 parts of alcohol to 50·76 parts of water by weight, or 100 volumes of alcohol to 81·82 volumes of water. It has a specific gravity of o·920 at 15° C.

Proof Vinegar. A vinegar containing 66 per cent. acetic acid.

Propaderm. A proprietary preparation of beclomethasone dipropionate for dermatological use. (284).

Propaderm A. A proprietary preparation of beclomethasone dipropionate and chlortetracycline hydrochloride for dermatological use. (284).

Propaderm C. A proprietary preparation of beclomethasone dipropionate and clioquinol for dermatological use. (284).

Propaderm N. A proprietary preparation of beclomethasone dipropionate and neomycin sulphate for dermatological use. (284).

Propadrine. A proprietary trade name for Phenylpropanolamine.

Propæsin (Propocaine). The propyl ester of p-amino-benzoic acid, $NH_2.C_6H_4.COO.C_3H_7$. A local anæsthetic.

Propafilm. Biaxially oriented polypropylene film. (512).

Propafilm C 90/300 and C 110/255. Polyvinylidene chloride coated Propathene film. (2).

Propal. See PROPONAL.

Propalanin. Amino-butyric acid, $C_4H_9O_2N$.

Propamidine. 1,3-Di(4-amidinophenoxy)-propane.

Propamine. Catalysts for polyurethane foams. (558).

Propamine D. A liquid aliphatic tertiary amine catalyst miscible with water and organic liquids. It is tetramethyl ethylene diamine. (77).

Propanal. Propionic aldehyde, CH_3CH_2CHO.

Propane Disulphinic Ether. See SULPHONAL.

Propanidid. Propyl 4-diethylcarbamoylmethoxy - 3 - methoxyphenyl - acetate. Epontol. FBA 1420.

Propanol. Normal propyl alcohol, $CH_3.CH_2.CH_2OH$.

Propanthelline Bromide. 2-Di-isopropylaminoethyl xanthen-9-carboxylate methobromide. Pro-Banthene Bromide.

Proparacaine. A proprietary trade name for Proxymetacaine.

Propathene. A trade mark for polypropylene. The lightest of the thermoplastics. It has good rigidity and tensile strength which are retained at elevated temperatures. It has excellent resistance to chemicals and has no tendency to environmental stress cracking. Used for wire and cable covering, moulding and tank lining. (2). See also NOBLEN.

Propathene GY 621M. A proprietary propathene copolymer used in injection moulding when difficult sections are involved. (512).

Propathene HW 70 GR. A proprietary glass-reinforced polypropylene offering high strength, high rigidity, low creep and resistance to distortion at high temperatures. (512).

Propathene LW 604 M. A proprietary polypropylene copolymer with low stabiliser content to give low levels of taste and odour. It is used for the injection moulding of such articles as cups. (512).

Propathene PXC 4717. A proprietary polypropylene giving good adhesion to plating. (512).

Propathene PXC 5563. A proprietary polypropylene containing glass fibre, treated to give good properties of adhesion. (512).

Propatylnitrate. 1,1,1-Trisnitratomethylpropane Etrynit. Gina.

Propellum. Marine paint for stern plates. (512).

Propenol. Allyl alcohol, $CH_2 : CH.CH_2OH$.

Properidine. A narcotic analgesic. Isopropyl 1 - methyl - 4 - phenylpiperidine - 4-carboxylate.

Proper-Myl. A proprietary preparation of lyophilised yeasts, for injection. (256).

Propetamphos. A pesticide. Z-o-2-isopropoxycarbonyl - 1 - methylvinyl o - methyl ethyl phosphoramido = thioate.

Propezite. A natural alloy of palladium and gold, containing 7 per cent. gold.

Prophet's Millet. See KOUMISS.

Propicillin. 6 -(α-Phenoxybutyramido) - penicillanic acid. (1-Phenoxypropyl)-penicillin. Brocillin and Ultrapen are the Potassium salt.

Propinol. Propargyl alcohol, $CH : C.CH_2OH$.

Propiofan. A trade name for polyvinyl propionate. (49).

Propiofan D. A registered trade mark for a polyvinyl propionate dispersion. (49).

Propiolactone. β-Propiolactone. Betaprone.

Propiolic Acid. o-Nitro-phenyl-propiolic acid, $C_6H_4(NO_2)C : C.CO_2H$., is known commercially by this name. It is in the form of a thin paste.

Propiomazine. 10-(2-Dimethylaminopropyl) - 2 - propionylphenothiazine. Dorevane Indorm Largon.

Propione. Diethyl-ketone $(C_2H_5)_2.CO$. A hypnotic and anæsthetic.

Propionylhyoscine. Poskine.

Propiram Fumarate. N-(1-Methyl-2-piperidinoethyl) - N - (2 - pyridyl) - pro - pionamide fumarate. FBA 4503.

Proplatinum. An alloy of 72 per cent. nickel, 23·6 per cent. silver, 3·7 per cent. bismuth, and 0·7 per cent. gold.

Proponal (Propal, Homobarbital). Di-propyl-malonyl urea or dipropyl-barbituric acid, $(C_3H_7)_2.C(CO_2).(NH)_2CO$. A soporific.

Proponesin. A proprietary trade name for the hydrochloride of Tolpronine.

Propoquad. Propoxylated quaternary ammonium salts. (555).

Proposote. Creosote-phenyl-propionate. A stimulating expectorant.

Propoxyphene. See DEXTROPROPOXY-PHENE.

Propranolol. 1-Isopropylamino-3-(1-naphthyloxy)propan-2-ol. Inderal is the hydrochloride.

Propylan. Polyethers for polyurethane foams. (558).

Propylan A350. A proprietary amine-initiated polyether used in the manu-facture of polyurethane foam. (77).

Propylan G600. A proprietary poly-oxypropylene triol of low molecular weight used in the production of rigid urethane foams and other urethane compositions, including elastomers. (77)

Propylan RF55. A proprietary modified sorbitol-based polyether for making rigid, flame-proof urethane foams. (77).

Propyl Docetrizoate. Propyl 3-dia-cetylamino - 2,4,6 - tri - iodobenzoate. Pulmidol.

Propyl Ether. Dipropyl-oxide, $(CH_2.CH_2.CH_3)_2O$.

Propylhexedrine. 1-Cyclohexyl-2-methylaminopropane. Benzedrex. Eventin.

Propyliodone. N-Propyl 3,5-di-iodo-4-pyridone-N-acetate. 3,5-Diodo-1-pro-poxycarbonylmethylpyrid - 4 - one. Dionosil.

Propyphenazone. An analgesic. 4-Isopropyl - 2, 3 - dimethyl - 1 - phenyl - 5 - pyrazolone. Present in GEVODIN.

Propyrin. Sodium thymol-benzoate.

Propytal. Dipropyl-barbituric acid. Proponal (q.v.).

Proquamezine. 10-(2,3-Bisdimethyl-aminopropyl)phenothiazine. Myspamol.

Proresid. A proprietary preparation of mitopopozide. An antimitotic. (267).

Proscillaridin. 3β,14β-Dihydroxybufa-4,20,22-trienolide 3-rhamnoside. Talu-sin.

Proscopine. A proprietary trade name for the hydrochloride of Poskine.

Prosobee. A proprietary artificial baby milk derived from soya, used in cases of intolerance of cows' milk. (324).

Prosol. A proprietary high protein food. (665).

Prosopite. A mineral, $Ca(F.OH)_2.Al_2(F.OH)_3$.

Prosparol. A proprietary preparation of ARACHIS OIL and water emulsion. A high calorie food. (444).

Prostaphlin. A proprietary prepara-tion of sodium oxacillin. An antibiotic. (324).

Prostigmin. A proprietary preparation of neostigmine bromide, used in myes-thenia gravis and to counteract the effect of curare-like drugs. (314).

Protagon. Lecithin.

Protamyl. A proprietary preparation of promethazine hydrochloride and amylobarbitone. A hypnotic. (336).

Protan. See ALBUTANNIN.

Protargentum (Protargin, Proganol). A compound of gelatin and silver. It contains 8 per cent. silver, and is used in aqueous solution in medicine.

Protargin. See PROTARGENTUM.

Protargol. See LARGIN.

Protargolgranulat. A German product. It consists of 1 part protargol and 2 parts urea. It has the advantage of easy solubility.

Protars. A dry arsenical fungicide pre-pared from talc, lime, and arsenic oxide.

Protectoid. A proprietary trade name for a cellulose acetate plastic in the form of a non-inflammable film.

Protectol. A brown, syrupy liquid used for the protection of animal products such as hair, wool, silk, skin, and leather from the action of alkaline liquids. It contains sodium lignin sulphonate.

Protectolex. Water displacing protec-tive oil. (545).

Protectyl. A solution containing 0·2 per cent. mercury, 1 per cent. salicylic acid, 3 per cent. glycerin, and 95·8 per cent. water. A disinfectant.

Protegin X. A base containing petro-latum and oxycholesterin.

Proteids. The same as albuminoids.

Protein Silver Salt. See LARGIN.

Proteol. A combination of casein with formaldehyde. An antiseptic dusting-powder.

Proteolite. See GALALKERITE.

Proteryl. Emetine-bismuth iodide cap-sules.

Protex. A proprietary safety glass.

Protheite. A mineral. It is a variety of pyroxene.

Prothiaden. A proprietary preparation of DOTHIEPIN. An anti-depressant. (280).

Prothidium. A proprietary preparation

of PYRITHIDIUM BROMIDE. A veterinary anti-protozoan.

Prothionamide. 2-Propylisonicotinthioamide. Trevintix.

Prothipendyl. 9-(3-Dimethylaminopropyl) - 10 - thia - 1,9 - diaza - anthracene. Phrenotropin and Tolnate are the hydrochloride.

Protirelin. A hormone for the release of thyrotrophin, currently undergoing clinical trial as "TRH Roche" and "Abbott 38 579". It is 1-[N-(5-oxo-L-prolyl)-L-histidyl]-L-prolinamide.

Protobastite. A mineral. It is a variety of enstatite.

Protoflavine Oleate. The oleic acid salt of 3 : 6-diamino-acridine. A dressing for wounds.

Protokylol. 3,4 - Dihydroxy - α - [(α - methyl - 3,4 - methylenedioxyphen - ethylamino)methyl benzyl]alcohol. Caytine is the hydrochloride.

Protopam. A proprietary trade name for the iodide of Pralidoximine.

Protosal. Salicylic acid glycerol-formal ester, $HO.C_6H_4.CO.O.CH_2.CH(O.CH_2O)CH_2$. Used as a means for the endermic administration of salicylic acid in the treatment of rheumatism.

Protosil. A silver protein compound containing 20 per cent. silver.

Protoxide of Copper. See HEMI-OXIDE OF COPPER.

Protriptyline. 7-(3-Methylaminopropyl) - 1,2 : 5,6 - dibenzocycloheptatriene. Concordin is the hydrochloride.

Prouera. A proprietary trade name for the acetate of Medroxyprogesterone.

Proustite (Light Red Silver Ore, Arsenical Silver Blende). A mineral. It is a silver sulpharsenite, Ag_3AsS_3 or $3Ag_2S.AsS_3$, containing 65·5 per cent. silver.

Provasine. A proprietary preparation of menthol, camphor, chlorbutol, phenylephrine hydrochloride and naphazoline nitrate. (380).

Provence Oil. The finest (Aix) olive oil.

Provera. A proprietary preparation containing Medroxyprogesterone acetate. (325).

Providoform. Tribrom-naphthol. An antiseptic.

Provinite. A stabilizer for PVC consisting of cadmium/barium complexes. (572).

Pro-Viron. A proprietary preparation of MESTEROLONE used in the treatment of androgen deficiency. (438).

Provol. A compound used for electrical insulation. It is a mixture of pitch, bitumen or similar materials, and mineral matter.

Proxy. A hydrogen peroxide solution.

Proxyl. A proprietary trade name for a pyroxylin denture material.

Proxymetacaine. 2-Diethylaminoethyl 3 - amino - 4 - propoxybenzoate. Proparacaine. Ophthaine is the hydrochloride.

Proxyphylline. 7-(2-Hydroxypropyl) - theophylline. Brontyl.

Prozine. A proprietary preparation of meprobamate, and promazine hydrochloride. (245).

Prune, Pure (Parme R). A dyestuff. It is the methyl ether of gallocyanine (q.v.), $C_{16}H_{15}N_2O_5Cl$. Dyes tanned cotton, and wool bluish-violet. Used in calico printing.

Prussian Black. A pigment prepared by calcining Prussian blue. It consists of carbon and oxide of iron.

Prussian Blue (American Blue, Antwerp Blue, Berlin Blue, Bronze Blue, Chinese Blue, Erlangen Blue, Gas Blue, Hamburg Blue, Milori Blue, Mineral Blue, New Blue, Oil Blue, Paris Blue, Paste Blue, Steel Blue). Prussian blue is a collective name for several blues, the basis of which is Prussian blue, ferric ferrocyanide, $Fe_4(FeC_6N_6)_3$. In commerce these blues consist of Prussian blue mixed with clay, gypsum, heavy spar, zinc white, and magnesia. The British Standard Specification No. 283 (1927) for Prussian blue for paints states that (1), the sum of the basic iron (expressed as iron), and iron-cyanogen complex (expressed as $Fe(CN)_6$) is not less than 54 per cent. The loss on heating for 2 hours in a water oven at 95–98° C. not to exceed 4 per cent. Not more than 2 per cent. to be soluble in water, and the acidity of the water extract not to exceed the equivalent of 0·1 per cent. sulphuric acid calculated on the material.

Prussian Brown. A pigment composed of sesquioxide of iron and alumina.

Prussian Green. A pigment. It consists of Prussian blue and gamboge, with Prussian blue largely in excess. Also see PELOUZES GREEN.

Prussian Red. See INDIAN RED.

Prussiate Black. A carbonaceous pigment obtained as a by-product in the manufacture of potassium ferrocyanide.

Prussic Acid. Hydrocyanic acid HCN.

Pruvagol. A proprietary preparation of acid fuchsin and borax, used as a vaginal pessary in the treatment of vaginitis. (370).

Prystal. A French proprietary rubber vulcanisation accelerator. It is a formaldehyde-urea condensation product. Also a proprietary trade name for a cast, clear phenolic moulding.

Prystaline. A proprietary moulding compound of the urea-formaldehyde type of synthetic resin.

Przibramite. A mineral. It is a cadmium-zinc blend.

Psatyrin. Hartin, $C_{10}H_{16}O$.

P.S.E. No. 15 Powder. An explosive. It is a mixture of ammonium perchlorate and rosin.

Pseubrookite. A mineral, $2Fe_2O_3.TiO_2$.

Pseudo-alums. Double sulphates of aluminium and another metal containing a bivalent metal sulphate instead of a monovalent one. A type is $MnSO_4$. $Al_2(SO_4)_3.24H_2O$. They are not isomorphous with the alums.

Pseudoapatite. An impure variety of the mineral Apatite.

Pseudobolite. A mineral, $5PbCl_2.4CuO$.

Pseudobrookite. A diamorphous modification of the mineral ilmenite.

Pseudocamphylite. A mineral. It is a variety of pyromorphite.

Pseudochalcedonite. A mineral. It is a variety of silica.

Pseudocotunnite. A mineral, $2KCl$. $PbCl_2$.

Pseudodeweylite. A mineral. It is a hydrated magnesium silicate.

Pseudoephedrine. (+)-2-Methylamino-1-phenylpropan-1-ol (a stereoisomer of ephedrine).

Pseudoeucryptite. An artificial lithium-aluminium silicate.

Pseudo-galena. See ZINC BLENDE.

Pseudoglaucophane. A mineral. It is a variety of amphibole.

Pseudojadeite. A mineral. It is a variety of albite.

Pseudolaumontite. A mineral. It is a hydrated silicate of aluminium, iron, magnesium, and potassium.

Pseudolibethenite. A mineral, Cu $(CuOH)PO_4.\frac{1}{2}H_2O$.

Pseudo-malachite (Phosphochalcite, Dihydrite, Ehlite, Phosphocalcite). A mineral, $Cu_3(PO_4)_2$.

Pseudo-mendipite. A mineral, $6PbO$. $2PbCl_2$.

Pseudomorphine. Dehydro-morphine.

Pseudonepheline. A mineral. It is a variety of nepheline.

Pseudoparisite. A mineral. It is cordylite.

Pseudopyrochroite. A mineral. It is bäckströmite.

Pseudopyrophyllite. A mineral. It is a hydrated silicate of aluminium and magnesium.

Pseudorcin. See ERYTHROMANNITE.

Pseudo-steatite. An impure variety of the mineral Halloysite.

Pseudotridymite. A mineral. It is a variety of tridymite.

Pseudotriplite. Synonym for Heterosite.

Pseudo-wavellite. A mineral from Bavaria. It is a hydrous aluminium phosphate containing also CaO, SrO, BaO, and rare earths to the extent of 2–3 per cent.

Pseudowollastonite. Obtained by heating wollastonite above $1800°$ C.

Psicain. The acid tartrate of d-ψ-cocaine, a drug having the same use as cocaine hydrochloride.

Psicain New. The hydrochloride of benzoyl-d-ψ-ergonin propylester. A local anæsthetic.

Psicobenyl. A German preparation. It is an emulsion of psicaine, anæsthesin, and paraffin. Used in the treatment of inflammation of the neck.

Psiconal. A German preparation containing 0.0075 part psicaine, 0.00005 part adrenalin, 0.007 part sodium chlorate, made up with water to 1 part. A local anæsthetic.

Psilocybin. 3 - (2 - Dimethylaminoethyl)indol-4-yl dihydrogen phosphate. Indocybin.

Psilomelane (Carbronigrite). A mineral. It is a hydrated manganese dioxide, $RO_4MnO_2(R=Mn, Ba, K_2)$.

Psittacinite. A mineral. It is a basic vanadate of copper and lead, $(Pb.Cu)_3$. $V_2O_8.(Pb.Cu)(OH)_2$.

Psorialan. A rose-coloured ointment obtained by the action of margaric acid upon yellow mercuric oxide. Used as a salve for diseases of the skin.

Psoriderm. A proprietary preparation of LECITHIN and coal tar used in the treatment of psoriasis. (377).

Psoriderm S. A proprietary preparation of LECITHIN, salicylic acid and coal tar used in the treatment of psoriasis. (377).

Psorigallol. A German product. It is stated to be a combination of pyrogallol and tar (lithantrol). Used in cases of psoriasis.

Psorox. A proprietary preparation of coal tar extract and allantoin for dermatological use. (188).

P.T.D. A rubber vulcanisation accelerator. It is dipentamethylene-thiuram-disulphide.

Pterolite. A mineral. It is a variety of mica, found in Norway.

P.T.F.E. Polytetrafluoroethylene.

P.T.M. A rubber vulcanisation accelerator. It is dipentamethylene-thiuram-monosulphide.

Ptychotis Oil. Ajowan oil.

Puce Oxide of Lead. See BROWN LEAD OXIDE.

Puce Spirit. See TIN SPIRIT.

Pucherite. A mineral. It is a bismuth vanadate, $Bi_2O_3.V_2O_5$.

Pudding-stone Marble. A type of marble containing pebbles cemented together.

Pufahlite. A mineral. It contains 41.9

per cent. tin, 37·4 per cent. lead, 6·3 per cent. zinc, 13·5 per cent. sulphur, and silver in small quantities.

Puflerite. A mineral. It is a variety of stilbite.

Puknos. A mixture of sodium hyposulphite and silver chloride tinted with eosin. It has been tried for plant diseases.

Pularin. A proprietary preparation of heparin. An anti-coagulant. (591).

Pulleite. A mineral. It is a variety of apatite.

Pulmadil. A proprietary preparation of RIMITEROL hydrobromide used in the treatment of bronchospasm. (275).

Pulmidol. A proprietary trade name for Propyl Docetrizoate.

Pulmodrine. A proprietary preparation containing guaiacol glyceryl ester and N-methyl ephedrine hydrochloride. A cough linctus. (250).

Pulmofluid. A German preparation. It is an extract from herbs, *Equisetum, Polygonatum, Galeopsis, Pulmonaria, Plantago*, etc. A cure for colds.

Pulmoform. See GUAIAFORM.

Pulmonal. A preparation containing *Succus liquiritiæ* and other extracts, sucrose, alcohol, sodium salicylate, anise oil, and menthol, in water.

Pulp, Asbestine. See AGALITE.

Pulp, Mineral. See AGALITE.

Pulque. An intoxicating beverage used in Mexico, obtained from fermentation caused by young flower-heads of *Agave Americana*.

Pulvatex. A trade mark for rubber (raw or partly prepared) for manufacture.

Pulverent Ammonium Dynamite. See AMMONIA DYNAMITE, PULVERENT.

Pumice Stone (Obsidian). A volcanic mineral (lava froth), consisting mainly of aluminium silicate. Specific gravity 2·2–2·5. An abrasive.

Punicin. A colouring matter obtained from *Purpura capillus* and other shellfish.

Pumiline. A proprietary preparation of pine oil.

Purac. Decolourising and absorptive activated carbon.

Purapen G. A proprietary preparation containing Benzylpenicillin (sodium salt) from which a highly allergic fraction has been removed. An antibiotic. (364).

Puratylene. A mixture of lime and bleaching powder. Used to purify acetylene.

Purbeck Stone. A limestone used in building.

Purdox. A proprietary trade name for high purity recrystallised alumina. (175).

Purdurum. See CAMITE.

Pure Air. See VITAL AIR.

Pure Blue. See SOLUBLE BLUE.

Pure Chrysoidine RD, YD. British equivalents of Chrysoidine.

Pure Scarlet. Mercuric iodide, HgI_2.

Pure Soluble Blue. A British brand of Soluble Blue.

Purex. A pigment. It is basic lead sulphate, and varies between $2PbSO_4$. PbO and $3PbSO_4$.PbO.

Purganol. A phenol-phthalein preparation.

Purgatin (Purgatol). Anthrapururin-diacetate, $C_{14}H_5O_2(OH)(O.C_2H_3O)_2$. An aperient.

Purgatol. Consists mainly of the anhydrides and lactones of fatty acids. Used for dressing hides. Also see PURGATIN.

Purgen (Laxans, Laxatin, Laxatol, Laxatoline, Laxen, Laxiconfect, Laxin, Laxoin, Laxophen, Phenolax, Purgella, Purgen, Purgo, Purgylum). Proprietary preparations of phenolphthalein, $C_6H_4(OC)C.(C_6H_4.OH)_2O$, sometimes with malic acid.

Purging Nut Oil (Physic Nut Oil). Curcas oil, from the seeds of *Jatropha Curcas*. Used in soap-making, and for lubricating.

Purging Nuts. See BARBADOS NUTS.

Purging Salt, Tasteless. Sodium dihydrogen phosphate, $NaH_2PO_4.H_2O$.

Purgo. See PURGELLA.

Purgolade. See PURGELLA.

Purgylum. See PURGELLA.

Puri-Nethol. A proprietary preparation of MERCAPTOPURINE used in the treatment of leukæmia. (277).

Purlboard. Polyurethane insulating board. (512).

Purochem. Organotin bactericides. (583).

Puromycin. An antibiotic. 3'-(2-Amino-*p* - methoxyhydrocinnamido) - 3' - de-oxy-*NN*-dimethyladenosine.

Purone. 2 : 8-Dioxy-1 : 4 : 5 : 6-tetra-hydro-purine.

Purosol. Hydro refined tolulene, xylene and naphthas. (580).

Puroverine. A proprietary preparation of protoveratrine and PANTHESIN used in the treatment of eclampsia. (478).

Purozone. A proprietary preparation. It consists of an alcoholic solution of the sodium salts and acids of wood tar or similar materials. A disinfectant.

Purple Carmine. Murexide (*q.v.*).

Purple Copper Ore (Chaleomiclite, Erubescite, Horse-flesh Ore, Bornite). A mineral, it is a double sulphide of copper and iron, $Fe_2S_3.3Cu_2S$, found in Cornwall, Ireland, and Chile.

Purple Glass. A glass in which the colour is produced by manganese dioxide.

Purple Madder. A pigment. It is a lake formed by the precipitation of the colouring matter of madder in combination with a metallic oxide.

Purple Ochre. See MARS VIOLET.

Purple of Cassius. See GOLD PURPLE.

Purple Spirit. See TIN SPIRIT.

Purpurblende. A mineral. It is kermesite.

Purpurin (Alizarin No. 6). Trioxy-anthraquinone, $C_{14}H_8O_5$. Dyes cotton mordanted with alumina, red, and with chromium, reddish-brown. Also see MAUVEINE.

Purpurite. A mineral. It is a ferric-manganic phosphate.

Purpuroxanthic Acid. See MUNJISTIN.

Purpuroxanthin (Xanthopurpurin). *m*-Dioxy-anthraquinone, $C_{14}H_8O_4$.

Purree. See INDIAN YELLOW.

Purrenone (Purrone). Euxanthon, $C_{13}H_8O_4$, obtained from purree, a yellow dyestuff.

Purrone. See PURRENONE.

Pursennid. A proprietary preparation containing sennosides A and B (as calcium salts). A laxative. (267).

Purub. Hydrofluoric acid, formerly used as a coagulating agent for rubber latex. The term " Purub " is also used for a patented flooring composition, consisting of vulcanised rubber.

Purus. See CHLORAMINE T.

Puschkinite. A mineral. It is a variety of epidote.

Putrescine. Tetramethylene-diamine, $NH_2(CH_2)_4.NH_2$. A base found in ergot.

Putridel. A German product. It is a yellowish-green liquid containing phenols of the type of pyrocatechol, and naphthalene derivatives. A disinfectant.

Putty. Consists of whiting with 18 per cent. raw linseed oil, sometimes with the addition of white lead.

Putty Powder. An impure stannic oxide, SnO_2, used for polishing glass. It is also used sometimes in rubber mixing and has a specific gravity of 6·6.

Puzzolana Cements. See EISEN-PORTLAND CEMENT.

Puzzuolana. A volcanic material found in various parts of Italy, especially Puzzuoli. It is employed for the conversion of pure lime into a hydraulic lime.

P.V.A. Polyvinyl alcohol.

P.V.C. Polyvinyl chloride.

PVF2. A polyvinyl fluoride. See KYNAR.

PX 104. Dibutyl phthalate.

PX 108. Di-iso-octyl phthalate.

PX 114. Decyl butyl phthalate.

PX 120. Di-iso-decyl phthalate.

PX 138. Dioctyl phthalate.

PX 208. Di-iso-octyl adipate.

PX 209. Dinonyl adipate.

PX 404. Dibutyl sebacate.

PX 408. Di-iso-octyl sebacate.

PX 658. Tetrahydrofurfuryl oleate.

PX 916. Triphenyl phosphate.

PX 917. Tricresyl phosphate.

The above codes are proprietary trade names. (124).

Pycamisan. A proprietary preparation containing sodium aminosalicylate, isoniazid. An antituberculous agent. (268).

Pycasix. A proprietary preparation containing sodium aminosalicylate, isoniazid. An antituberculous agent. (268).

Pycazide. A proprietary preparation containing isoniazid. An antituberculous agent. (268).

Pycnite. A mineral. It is a variety of topaz.

Pycnophyllite. A mineral. It is a variety of muscovite.

Pyelocystin. A German product. It consists of hexamethylene-tetramine-hydrargyrum-chloratum.

Pyelognost. Sodium iodide urea.

Pylkrome. A composite alloy steel. The base is mild steel, and it has a corrosion-resisting surface of high chrome or high chromium-nickel-iron alloys.

Pylura. A proprietary preparation of adrenaline, benzyamine lactate and phenol used in the treatment of hæmorrhoids. (330).

Pyoctanin (Pyoktanin). The name given to different coal-tar colours. (1) Yellow pyoctanin (auramine, *q.v.*), and (2) Blue pyoctanin (methyl violet, *q.v.*). Used in surgery as bactericides.

Pyoctanin Blue. See PYOCTANIN.

Pyoctanin Mercury. A combination of mercury with pyoctanin blue. Used as an antiseptic.

Pyoctanin Yellow. See PYOCTANIN.

Pyoktanin. See PYOCTANIN.

Pyopen. A proprietary preparation containing Carbenicillin (as the disodium salt). An antibiotic. (364).

Pyorox. A proprietary preparation of ACETARSOL, aminacrine and sodium ricinoleate used as a tooth-paste in the treatment of gingivitis. (802).

Pyotropin. A German preparation. It is stated to contain calcium carbonate, potassium phenolate, and phenol, in liquid and salve form.

Pyracetosalyl. An addition product of antipyrine and acetyl-salicylic acid, $C_{11}H_{12}N_2O + C_2H_3O.O_6H_4COOH$.

Pyracine. Phenazone.

Pyradiolin. A proprietary synthetic plastic used as a dielectric material in wireless telegraphy, and for other purposes. It is a modified pyroxylin plastic.

Pyradone. A proprietary preparation of amidopyrin.

Pyralin. (1) A registered trade mark for a celluloid product. Available in transparent, translucent, opaque coloured and colourless forms. Resistant to hydrocarbons and oils.

Pyralin. (2) A registered trade name for polyimide high temperature resistant materials. (10).

Pyraloxin. Oxidised pyrogallic acid, a dark brown powder obtained by oxidation with air and ammonia. Used in the treatment of skin diseases.

Pyramid. Sodium and potassium silicates. (546).

Pyramidol Brown BG. A dyestuff. It is the sodium salt of diphenyl-disazo-bi-resorcin, $C_{24}H_{18}N_4O_4Na$. Dyes cotton red.

Pyramidol Brown T. A dyestuff. It is the sodium salt of ditolyl-disazo-bi-resorcin, $C_{26}H_{22}N_4O_4Na$. Dyes cotton brownish-red.

Pyramidone (Amidopyrine, Novamidon). Di-methyl-amino-antipyrine, $C_{13}H_{27}N_{30}$. Used to allay high temperature of typhoid fever, and of syphilis, also in neuralgia and rheumatism.

Pyramine Orange 3G. A dyestuff. It is the sodium salt of diphenyl-disazo-nitro-m-phenylene-diamine-m-phenylene-diamine-disulphonic acid, $C_{24}H_{19}N_9S_2O_8Na_2$. Dyes cotton yellowish orange.

Pyramine Orange R. A dyestuff. It is the sodium salt of disulpho-diphenyl-disazo-bi-nitro-m-phenylene-diamine, $C_{24}H_{18}N_{10}O_{10}S_2Na_2$. Dyes cotton orange-red.

Pyramine Orange 2R. A dyestuff. It is the sodium salt of diphenyl-disazo-nitro-m-phenylene-diamine-β-naphthylamine-disulphonic acid, $C_{28}H_{20}N_8S_2O_8Na_2$. Dyes cotton reddish-orange.

Pyramol. Silica sols for industry. (546).

Pyranil Black. A sulphide dyestuff.

Pyranol. Sodium acetyl salicylate, $C_2H_3O.OC_6H_4COONa$.

Pyrantel. An anthelmintic. 1,4,5,6-Tetrahydro-1-methyl-2-[trans-2-(2-thienyl)vinyl]pyrimidine.

Pyranthrone (Indanthrene gold orange G, Duranthrene Gold Yellow Y). A dyestuff prepared by treating 2 : 2′-dimethyl-1 : 1′-dianthraquinonyl with alkali or zinc chloride. Dyes orange tints.

Pyrantimonite. See RED ANTIMONY.

Pyrantine. p-Ethoxy-phenyl-succinimide $(CH_2.CO)_2N.C_6H_4.OC_2H_5$. It has properties similar to acetphenetidine.

Pyrantine, Soluble. The sodium salt of pyrantine (q.v.), used in acute rheumatism.

Pyranum (Pyrenol). Thymol - sodium -

benzoyl-oxybenzoate. An antipyretic and antineuralgic.

Pyrargyrite (Dark Red Silver Ore, Ruby Silver Ore, Miargyrite, Stephanite, Brittle Silver Ore, Black Silver). Minerals. They are sulpho-antimonites of silver of varying composition Ag_2S. Sb_2S_2, $3Ag_2S.Sb_2S_3$, and $5Ag_2S.Sb_2S_3$.

Pyrasteel. A proprietary trade name for a heat-resisting alloy of iron with 25 per cent. nickel, 14 per cent. chromium, and 2·5–3·0 per cent. silicon.

Pyraton. A proprietary trade name for diacetone alcohol.

Pyrax Talcs A and B. Qualities of pure white talc mineral from deposits in America. Used in rubber, textile, and ceramic industries.

Pyrazinamide. Pyrazinoic acid amide. Zinamide.

Pyrazine. See ANTIPYRINE.

Pyrazine Yellow S. A British equivalent of Flavazine S.

Pyrazole Blue. A blue compound, $C_{20}H_{16}O_2N_4$, obtained when bis-phenyl-methyl-pyrazolone is treated in alkaline solution with excess of sodium nitrite, and the mixture poured into sulphuric acid.

Pyrazoline. See ANTIPYRINE.

Pyrazolon. A German preparation containing derivatives of barbituric acid, belladonna, ammonium benzoate, phenacetin, and uzara.

Pyreneite. A mineral. It is an iron-lime garnet.

Pyrene Oil (Begasses Oil). An inferior olive oil obtained from the twice pressed marc, which is stored and allowed to ferment, ground, stirred up with boiling water, and subjected to heavy pressure.

Pyrenol. See PYRANUM.

Pyrethane. A trade mark for Urethane-methyl-pyrazo-carbonate.

Pyrethrum Powder. An insecticide made from the powdered flowers of some species of pyrethrum plants. The active principle is rotenone (q.v.).

Pyrex. A trade mark for a glass having a very low coefficient of expansion. It contains 80 per cent. silica, and 12 per cent. boric oxide. (237).

Pyrgom. A mineral. It is a variety of augite.

Pyrgos. See CHLORAMINE T.

Pyribenzamine. A proprietary trade name for the citrate or the hydrochloride of Tripelennamine.

Pyricit. Sodium boro-fluoride, $NaBF_4$. A food preservative and antiseptic.

Pyricol. Acopyrin (q.v.).

Pyridium. A proprietary preparation of phenazopyridine hydrochloride. (262).

Pyridinised Industrial Methylated Spirit. See METHYLATED SPIRIT.

Pyridium. A proprietary preparation of phenyl-azo-α-α-diamino-pyridine.

Pyridium. A proprietary trade name for the hydrochloride of Phenazopyridine.

Pyrido Rubber. A polymerised acrolein-methyl-amine. A rubber-like material.

Pyridostigmine. 3-Dimethylcarbamoyloxy - 1 - methylpyridinium. Mestinon is the bromide.

Pyridoxine. Vitamin B$_6$. C$_8$H$_{11}$O$_3$N.

Pyrilin. A proprietary preparation of pyridine-ethyl-phosphinate.

Pyrimethamine. 2,4-Diamino-5-(4-chlorophenyl) - 6 - ethylpyrimidine. Daraprim.

Pyrimithate. An insecticide. o-2-Dimethylamino - 6 - methylpyrimidin - 4 - yl oo - diethyl phosphorothioate. DIOTHYL.

Pyrites, Arsenical. See MISPICKEL.

Pyrites, Capillary. See MILLERITE.

Pyrites, Cobalt-nickel. See COBALT-NICKEL BLENDE.

Pyrites, Cockscomb. See COCKSCOMB PYRITES and MARCASITE.

Pyrites, Coxcomb. See COCKSCOMB PYRITES and MARCASITE.

Pyrites, Efflorescent. See WHITE PYRITES.

Pyrites, Hair. See MILLERITE.

Pyrites, Nickel. See MILLERITE.

Pyrites, Radiated. See MARCASITE

Pyrites, Spear. See COCKSCOMB PYRITES.

Pyrites, White Iron. See MARCASITE.

Pyrithidium Bromide. An anti-protozoan. 3 - Amino - 8 (2 - amino - 1, 6 - di - methylpyrimidinium - 4 - ylamino) - 6 - (4 - aminophenyl) - 5 - methylphenan-thridinium dibromide. PROTHIDIUM.

Pyrithione Zinc. Zinc bis(pyridine-2-thiol 1-oxide). Zinc Omadine.

Pyritinol. Di-(5-hydroxy-4-hydroxy-methyl - 6 - methyl - 3 - pyridylmethyl)-disulphide. Encephabol is the dihydrochloride.

Pyro. Pyrogallol, C$_6$H$_3$(OH)$_3$.

Pyro-acetic Spirit. Acetone, CH$_3$.CO.CH$_3$.

Pyro Alcohol. Methyl alcohol, CH$_3$OH.

Pyro-Antimonite. Synonym for Kermesite.

Pyroaurite. A mineral, 3Mg(OH)$_2$.Fe(OH)$_3$.3H$_2$O.

Pyrobel. Heat and flame resisting finishes and dopes. (512).

Pyrobelonite. A mineral. It is a vanadate of lead and manganese, 4PbO.7MnO.2V$_2$O$_5$.3H$_2$O, found in Sweden.

Pyro-bitumen. Bitumen which is insoluble in carbon tetrachloride. It is synonymous with carbenes.

Pyrobromone. Bromo-dimethyl-amino-antipyrine.

Pyrocain. Guaiacol-benzyl-ester, OCH$_3$.C$_6$H$_4$.OCH$_2$.C$_6$H$_5$. It possesses the therapeutic properties of guaiacol, and is less caustic.

Pyrocast. A proprietary trade name for a nickel-chromium cast iron.

Pyrocatechin (Catechol). Pyrocatechol, C$_6$H$_4$(OH)$_2$.

Pyrocatechol Arsenic Acid. o-Hydroxy - phenyl - arsenate, O : As(O.C$_6$H$_4$OH)$_3$. A reagent for alkaloids.

Pyrochinin. The quinine amidopyrin double salt of camphoric acid.

Pyrochlor. Fire-resistant and wood preservative. (562).

Pyrochlore. A mineral, (Ca.Fe.Ce)O.(Nb.Ti.Th)O$_2$.

Pyrochroite. A mineral. It is a manganese hydrate.

Pyroclasite. Synonym for Hydroxy-fluor-apatite.

Pyroclastite. A mineral, 6CaHPO$_4$.Ca$_3$(PO$_4$)$_2$.H$_2$O.

Pyrocollodion. A soluble nitrocellulose containing the highest practicable percentage of nitrogen, about 12·5 per cent.

Pyroconite. A mineral. It is a hydrated fluoride of aluminium, calcium, and sodium.

Pyro Cotton. A nitrated cellulose, not so fully nitrated as guncotton.

Pyrodialite. An explosive. It contains 80–88 per cent. potassium chlorate, 5–6 per cent. charcoal, 10–18 per cent. gas tar, and 3–4 per cent. sodium and ammonium bicarbonates.

Pyrodine (Hydracetin). Phenyl-hydrazine-acetyl, (C$_6$H$_5$)HN.NH(CO.CH$_3$).

Pyro-emerald. See CHLOROPHANE.

Pyrofax. The trade mark of Carbide and Carbon Chemicals Corporation applied to compressed gas for heating, lighting, and power purposes.

Pyroform (Gammaform). Bismuth-oxy-iodo-dipyrogallate, used in medicine.

Pyrofulmin. A yellow substance obtained by heating mercuric culminate. It is probably a mixture of mercuric oxycyanide and oxide.

Pyrogallic Acid. Pyrogallof C$_6$H$_3$(OH)$_3$. It absorbs oxygen, and is used in gas analysis.

Pyrogene Blacks and Blues. See SULPHUR BLACK T.

Pyrogene Blues and Greys. Dyestuffs produced by heating hydroxy-dinitro-diphenylamine and indophenols under pressure with polysulphides in alcoholic solution.

Pyrogene Brown D. A dyestuff obtained by sulphurising sawdust or bran.

Pyrogene Dark Green. A dyestuff produced by sulphurising p-amino-phenol and its substitution derivatives with sodium sulphide and sulphur, in the presence of copper.

Pyrogene Green. A sulphide dyestuff prepared by the fusion of various indophenols with polysulphides, in the presence of copper compounds.

Pyrogene Indigo. A dyestuff obtained by heating indophenol, $C_2H_5.NH.C_6H_4$ $N.C_6H_4O$, with polysulphides.

Pyrogene Olive N (Pyrogene Yellow M). Dyestuffs prepared by heating various methyl-amino and nitro-amino compounds with sodium sulphide, sulphur, and alkalis.

Pyrogene Yellow M. See PYROGENE OLIVE N.

Pyroglycerin. See NITRO-GLYCERIN.

Pyroguanite. Synonym for Hydroxy-fluor-apatite.

Pyroligneous Acid. Crude acetic acid.

Pyroligneous Liquor (Crude Wood Vinegar). The aqueous layer from the distillation of wood in retorts. It contains about 10 per cent. acetic acid, 1–2 per cent. methyl alcohol, and 0·1–0·5 per cent. acetone.

Pyroligneous Spirit. See METHANOL.

Pyrolignite of Iron (Iron Liquor, Black Mordant, Black Liquor, Bouillon Noir, Liqueur de Ferraile). Ferrous acetate, $Fe(C_2H_3O_2)_2$, prepared by the action of pyroligneous acid upon iron turnings. The solution also contains ferric acetate. Used in calico printing, and in dyeing, for the preparation of blue, violet, black, and brown colours.

Pyrolignite of Lime. Diacetate of lime, $Ca(C_2H_3O_2)_2$.

Pyrolusite. A mineral. It is manganese dioxide, MnO_2.

Pyromeline. A mineral. It is morenosite.

Pyromic. A particularly pure form of induction melted nickel-chromium used for electric furnaces, heaters, ovens, etc.

Pyromorphic Phosphorus. Prepared by heating red phosphorus with a trace of iodine at 280° C., or *in vacuo*.

Pyromorphite (Green Lead Ore). A mineral. It is lead phosphate with lead chloride, $3(Pb_3(PO_4)_2)+PbCl_2$.

Pyronaphtha. A fraction of Russian petroleum distillation. It has a specific gravity of 0·855–0·865, and flashes at 98° C., or above. A burning oil. Also see MINERAL BURNING OILS.

Pyronil. A proprietary trade name for the phosphate of Pyrrobutamine. (333).

Pyronine B. A dyestuff. It is tetraethyl-diamino-xanthenyl-chloride, $C_{19}H_{23}N_2$ OCl. Dyes cotton, wool, and silk, red.

Pyronine G (Casan Pink). A dyestuff. It is tetramethyl-diamino-xanthenyl-chloride, $C_{17}H_{19}N_2OCl$. Dyes cotton, wool, and silk, red.

Pyronin Stain. See PAPPENHEIM'S STAIN.

Pyronite. Tetryl (*q.v.*).

Pyronium. A proprietary substitute for

tin oxide in enamels. It is used in conjunction with tin oxide.

Pyrope (Vogesite). A mineral. It is a magnesium aluminium silicate, Mg_3Al_2 Si_3O_{12}, a precious stone.

Pyrophan. A combination of pyrogallol and dimethylamine.

Pyrophane. A mineral. It is a variety of opal.

Pyrophanite. A mineral. It is a manganese titanate, $MnTiO_3$.

Pyrophori. Certain reduced metals such as iron, nickel, cobalt, and copper, when poured into oxygen, spontaneously ignite, and when in this finely divided state are termed pyrophoric.

Pyrophoric Alloy. See MISCH METAL, AUER METAL, and KUNHEIM METAL.

Pyrophoric Lead. The mixture of lead and carbon left when lead tartrate is heated.

Pyrophosphorite. A mineral, $Mg_2P_2O_7$. $4(Ca_2P_2O_7.Ca_3(PO_4)_2)$.

Pyrophyllite (Pencil Stone). A mineral. It is a hydrated aluminium silicate.

Pyrophysalite. A mineral. It is a variety of topaz.

Pyropissite. A lignite now almost exhausted, obtained from deposits of oily wood, extracted from mines in Saxony.

Pyroretin. A brown resin found in the lignite in Bohemia.

Pyros. A paramagnetic alloy consisting of nickel with 7 per cent. chromium, 5 per cent. tungsten, 3 per cent. manganese, and 3 per cent. iron. It is suitable for expansion pyrometers.

Pyrosal. See ACOPYRIN.

Pyrosine B. See ERYTHROSIN.

Pyrosine G. See ERYTHROSIN G.

Pyrosin J. Di-iodo-fluoresceine was sold and used for a time under this name.

Pyrosin R. A dyestuff. It is a mixture of Pyrosine B and G.

Pyrosmaltite. A mineral, $H_5(Fe.Mn)_4Si_4$ $O_{16}Cl$.

Pyrostibite. Synonym for Kermesite.

Pyrostibnite. See RED ANTIMONY.

Pyrostilpnite. See FIRE BLENDE.

Pyrosulphuric Acid. See FUMING SULPHURIC ACID.

Pyrotin. An azo dyestuff. It is isomeric with Double scarlet extra S, and is prepared from β-naphthylamine-mono-sulphonic acid, and α-naphthol-α-mono-sulphonic acid.

Pyrotin RRO. A dyestuff. It is the sodium salt of sulpho-naphthalene-azo-α-naphthol-mono-sulphonic acid, $C_{20}H_{12}$ $N_2O_7S_2Na_2$. Dyes wool red from an acid bath.

Pyroxene. A mineral, $CaO.MgO.2SiO_2$, with magnesium, iron, and aluminium silicates. The term is applied to a group of metasilicate minerals.

Pyroxmangite. A mineral. It is $16[(Mn,Fe)SiO_3]$.

Pyroxylic Spirit. See METHANOL.

Pyroxylin. See GUNCOTTON.

Pyroxylin Plastics. See AETHROL, ALCOLITE, CAMPHORLOID, DROTT, HELECO, PHOENIXITE, TRELIT, TROLIT F, VENITE.

Pyrozone. See PEROXOL.

Pyrrhite. A mineral. It is a variety of microlite.

Pyrrhoarsenite. A mineral. It is $8[Ca_3(Mg,Mn)_2(As,Sb)_3O_{12}]$.

Pyrrhodid. Aminopyrin sulphocyanide.

Pyrrhol (Pyrroline). Pyrrole, C_4H_5N.

Pyrrhosiderite. See NEEDLE IRON ORE.

Pyrrhotine. See MAGNETIC PYRITES.

Pyrrhotite. See MAGNETIC PYRITES.

Pyrrobutamine. 1-[4-(4-Chlorophenyl)-3 - phenylbut - 2 - enyl pyrrolidine. Pyronil is the phosphate.

Pyrrocaine. N - (pyrrolidin - 1 - ylace - tyl)-2,6-xylidine. Endocaine.

Pyrrodiazole. Triazole, $C_2H_3N_3$.

Pyrrol Black. See SULPHUR BLACK T.

Pyrrole Blue A. An indigo-blue compound, $C_{24}H_{16}O_3N_4$, obtained by adding pyrrole to a solution of isatin in dilute sulphuric acid.

Pyrrole Blue B. A compound, $C_{24}H_{16}O_2N_4$, obtained by adding pyrrole in glacial acetic acid to isatin in acetic and sulphuric acids cooled to 0° C.

Pyrrole Red. A polymer of pyrrole, produced by hot hydrochloric acid.

Pyrroline. See PYRRHOL.

Pyrvinium Pamoate. Viprynium Embonate.

Pyxol. An emulsion of coal-tar acids with soap. A disinfectant.

Q

Quantrovanil. A name applied to ethyl-protocatechuic aldehyde.

Quartz. A mineral, silica, SiO_2. In the main there are three types. (1) Crystalline, such as Tridymite and Cristobalite. (2) Crypto-crystalline, such as Chalcedony, and (3) hydrated silica or Opal.

Quarzal. An alloy of aluminium with 15 per cent. copper, 6 per cent. manganese, and 0·5 per cent. silicon. Used for the cylinders of internal combustion engines.

Quartz, Fœtid. See STINK QUARTZ.

Quartz Glass. Fused silica glass.

Quartzilite. A new metallic carbide formed by action of silica and carbon at 2000–3000° C. It is suitable for electrical resistances.

Quartzine. A mineral. It is a variety of chalcedony (crypto-crystalline quartz), SiO_2. Also see LUTECITE.

Quartzite. A rock composed of quartz and siliceous cement. Also see HALF-GANISTER.

Quartz, Yellow. See CITRINE.

Quaternary Steels. Steels containing two special elements in addition to the iron and carbon.

QE-12904. A polyurethane-based adhesive giving vinyl-to-leather adhesion. (900).

QE 13910, 13920 and 13930. A proprietary range of polyurethane-based adhesives. (900).

Quebrachite (Quebrachitol). The monomethyl ether of lævo-inositol, $C_6H_6(OH)_5(OCH_3)$. It is found in the latex of *Hevea brasiliensis* (the rubber tree) to the extent of 1–2 per cent.

Quebrachitol. 1-methyl-inositol.

Quebracho. *Loxopteryngium lorenzii*. The wood of this tree contains about 20 per cent. of tannin, and is used in the form of an extract for tanning.

Queensland Arrowroot. See TOUS-LES-MOIS STARCH.

Queensland Fever Bark. The bark of *Alstonia constricta*.

Queen's Metal. A jeweller's alloy. It is very variable in composition. It contains from 50–85 per cent. tin, 7–16 per cent. antimony, 0–16 per cent. lead, 0–3·5 per cent. copper, and 1–12 per cent. zinc.

Queen's Yellow. See TURPETH MINERAL.

Quellada. A proprietary preparation of gamma benzene hexachloride. A skin fungicide. (376).

Quenselite. A mineral, $2PbO.Mn_2O_3.H_2O$.

Quenstedtite. A mineral. It is a hydrated ferric sulphate.

Quercetin. 1 : 3 : 3′ : 4′ - Tetroxy - flavonol.

Quercitron. A colouring matter, sold as chips, or as a coarse powder, obtained by grinding the bark of *Quercus tinctoria* and *Q. nigra*. The dyeing principle is quercitin or flavin, $C_{15}H_{10}O_7$, which forms yellow lakes with aluminium and tin salts. Used for calico printing and wool dyeing, for yellows and browns. Flavin, Patent bark, and Prepared bark are commercial preparations.

Quercyite. A mineral. It is a variety of phosphorite.

Questran. A proprietary preparation of cholestyramine chloride used in the treatment of hypercholesterolæmia. (324).

Quetenite. A mineral. It is a hydrated magnesium-iron silicate.

Quevenne's Iron. Reduced iron.

Quickening Liquid. A solution of mercuric nitrate or cyanide. Used in electro-plating.

Quicklime. See LIME.

Quick-pach. A plastic fire-clay for making monolithic linings and quick repairs.

Quicksilver. Mercury, Hg. A pigment, which consists of sulphide of mercury, is also known under this name.

Quietol. Dimethyl-amino-dimethyl-iso-valeryl-propyl-ester-hydrobromide.

Quilene. A proprietary trade name for the methyl sulphate of Pentapiperide.

Quillaic Acid. An acid obtained from the inner bark of *Quillaja saponaria* (soap bark). Used as an expectorant.

Quimbo. A German product. It is an ointment for chapped hands and contains potassium sozoidolate.

Quinaband. A proprietary bandage coated with zinc paste, CALAMINE and IODOCHLOROHYDROXYQUINOLONE used in the treatment of leg ulcers. (484).

Quinacillin. 3-Carboxyquinoxalin-2-ylpenicillin.

Quinacol. A chemical combination of quinine and guaiacol, containing 50 per cent. of each compound. Recommended in influenzal pneumonia.

Quinalbarbitone Sodium. The mono-sodium derivative of 5-allyl-5-(1-methyl-butyl)barbituric acid. Seconal Sodium.

Quinaldine. 2-Methyl-quinoline.

Quinalgen. Analgen (*q.v.*).

Quinalizarin. See ALIZARIN BORDEAUX B, BD.

Quinalspan. A proprietary preparation of quinalbarbitone sodium. A hypnotic. (330).

Quinanalgen. See ANALGEN.

Quinanil. An antiseptic prepared from 2-*p*-dimethyl-amino-anil-6-methyl-quinoline methochloride.

Quinaphenin (Chinaphenin). Phenetidine-quinine-carbonic ester, $NH.C_6H_4$. $OC_2H_5.CO.O.C_{20}H_{23}N_2O$. Used in medicine.

Quinaphthol. See CHINAPHTHOL.

Quinaseptol. See DIAPTHOL.

Quinbi. A trade name for quinine-iodo-bismuthate. Used as an anti-syphilitic.

Quinby. Bismuth and quinine iodide in oil suspension.

Quincite. A mineral. It is a silicate of magnesium and iron.

Quinconal. Quinine-diethyl-barbiturate.

Quindoxin. A growth promoter. Quin-oxaline 1,4-dioxide. GROFAS.

Quineine. A standard solution of *Cinchona succiruba*, containing 7 per cent. of alkaloids, 5 per cent. of which is quinine.

Quineonal. See CHINEONAL.

Quinestradol. 3-Cyclopentyloxyoestra-1,3,5,(10)-triene-16α,17β-diol. Pentovis.

Quinestrol. An œstrogen. 3-Cyclopentyloxy - 19 - nor - 17α - pregna - 1, 3, 5(10)-trien-20-yn-17-ol. ESTROVIS.

Quinethazone. 7-Chloro-2-ethyl-1,2-dihydro - 6 - sulphamoylquinazolin - 4 - one. Aquamox.

Quinetum. A mixture of 3–22 per cent. quinine, 24–60 per cent. cinchonidine, 18–54 per cent. cinchonine, and 3–5 per cent. quinidine.

Quingestanol. A progestational steroid. 3 - Cyclopentyloxy - 19 - nor - 17α - pregna-3,5-diene-20-yn-17-ol.

Quinicardine. Quinidine sulphate, $(C_{20}H_{24}O_2N_2)_2.H_2SO_4$.

Quinimuthol Saca. Colloidal quinine-iodo-bismuthate.

Quinine. Quinine sulphate, $(C_{20}H_{24}N_2O_2)_2.H_2SO_4.8H_2O$.

Quinine Acetosalate (Quinine Aspirin, Quinine Salacetate, Xaxaquin). Quinine-acetyl-salicylate, $CH_3.CO.O.C_6H_4$. $COOH.C_{20}H_{24}O_2N_2$. Has been used in pleurisy and influenza.

Quinine Aspirin. See QUININE ACETOSALATE.

Quinine Lygosinate. Dihydrodibenzal-acetone - quinine, $CO(CH : CH.C_6H_4$. $C_{20}H_{24}N_2O_2)_2$.

Quinine Salacetate. See QUININE ACETOSALATE.

Quinine Sulphate, Neutral. See QUININE SULPHATE, SOLUBLE.

Quinine Sulphate, Soluble (Quinine Sulphate, Neutral). Acid quinine sulphate, $C_{20}H_{24}N_2O_2(H_2SO_4)_2.7H_2O$.

Quinine, Tasteless. See INSIPIN.

Quinine Troposan. A proprietary preparation. It is a quinine salt of 2-hydroxy-5-acetyl-amino-phenyl-arsinic acid, and has been found to be of considerable value in the treatment of benign tertian malaria.

Quinio - stovarsol. Quinine - acetyl-amino-oxyphenyl-arsinate.

Quinisal. A German preparation. It is quinine disalicylate. An antipyretic.

Quinisan. A proprietary trade name for the quinine salt of bisalicylo-salicylic acid. It is stated to be effective in influenza and tonsillitis.

Quinisocaine. Dimethisoquin.

Quinitol. *p*-dihydroxy-hexamethylene, $C_6H_{10}(OH)_2$.

Quinizarine Blue. A dyestuff. It is the sodium salt of anilido-oxy-anthraquinone-sulphonic acid. Dyes wool reddish-blue from an acid bath, and chromed wool, a greenish-blue.

Quinizarine Greens. See ALIZARIN CYANINE GREENS.

Quinn's Rubber. A patented rubber substitute made from rapeseed oil, petroleum, and chloride of sulphur.

Quinoderm. A proprietary preparation of potassium hydroxyquinoline sulphate used in the treatment of acne. (901).

Quinoform (Chinoform). Quinine formate, $C_{20}H_{24}N_2O_2.H.COOH$. A condensation product of cinchotannic acid and formaldehyde is also known by this name. It is used as an internal astringent and antiseptic.

Quinoformine. A compound of quinic acid and hexamethylene - tetramine. Used as a solvent for uric acid.

Quinoidine (Chinoidine). Quinoline-chlor-iodide, a substitute for iodoform.

Quinol. Hydroquinone (p-dihydroxy-benzene), $C_6H_4(OH)_2$. Used as a photographic developer.

Quinoline Blue. See CYANINE.

Quinoline Green. A dyestuff prepared by the action of phosphorus oxychloride upon a mixture of quinoline and tetra-methyl-diamino-benzophenone.

Quinoline-Hydroquinone. Dioxy-quinoline, $C_9H_7NO_2$.

Quinoline Red (Isoquinoline Red). A dyestuff, $C_{26}H_{19}N_2Cl$, obtained by the action of benzo-trichloride upon a mixture of quinaldine (methyl-quinoline), and isoquinoline. Dyes wool and silk rose-red. Also used for isochromatising photographic plates.

Quinoline Yellow (Quinoline Yellow, Water Soluble). A dyestuff. It is the sodium salt of the sulphonic acids (chiefly the disulphonic acid), of quino-phthalone, $C_{18}H_9NO_8S_2Na_2$. Dyes silk and wool greenish-yellow from an acid bath.

Quinoline Yellow, Spirit Soluble. Quino-phthalone, $C_{18}H_{11}NO_2$. Used for colouring spirit varnishes and waxes. See SPIRIT SOLUBLE QUINOLINE YELLOW.

Quinoline Yellow, Water Soluble. See QUINOLINE YELLOW.

Quinoped. A proprietary preparation of benzoyl peroxide and potassium hydro-xyquinolone sulphate used in the treatment of fungal foot infections. (901).

Quinophan. See ATOPHAN.

Quinophthol. Quinine-β-naphthol-sul-phonate. Used in medicine.

Quinopyrin. A combination of 2 parts antipyrine and 3 parts quinine hydrochloride.

Quinosol (Chinosol, Oxyquin, Sunoxol). The potassium salt of oxy-quinoline-sulphonic acid, $C_9H_6.NO.SO_3K$. A valuable non-corrosive and practically non-poisonous antiseptic, used for disinfecting the hands, and for wounds.

Quinostab. A proprietary preparation of bismuth and quinine iodide in oil suspension.

Quinotropine. A compound of urotropine. It is hexamethylene-tetramine-quinate, and is used as a solvent for uric acid, in cases of gout.

Quinovabitter (Quinovic acid). Quino-vin, $C_{30}H_{48}O_8$.

Quinovasugar. Quinovite, $C_6H_9O(OH)_3$.

Quinovic Acid. See QUINOVABITTER.

Quinoxyl. A proprietary preparation of m-iodo-oxyquinoline-sulphonic acid and sodium bicarbonate.

Quinquenal. A German soporific containing luminal, adalin, codeine phosphate, trional, and phenacetin.

Quintiofos. An insecticide. o-Ethyl o-8-quinolyl phenylphosphonothioate. BACDIP.

Quintolan. Water repellent finishing agent for textiles, leather and paper. (512).

Quirogite. A mineral, $Pb_{23}Sb_6S_{32}$.

Quisqueite. An asphaltum-like compound containing much sulphur. It is found in Peru.

Quixalin. A proprietary preparation containing halquinol. A gastrointestinal sedative. (326).

Quotane. A proprietary trade name for the hydrochloride of DIMETHISOQUIN. An anti-pruritic cream.

Quram 3365. A proprietary organic ammonium silicate used with zinc-rich primers. (902).

R

R-2 Crystals. A proprietary trade name for an ultra accelerator for latex, etc. It is the reaction product of carbon disulphide with methylene dipiperidine.

Rabbittite. A mineral. It is $Ca_3Mg_3(UO_2)_2(CO_3)_6(OH)_4.18H_2O$.

Rabdionite. A mineral. It is a hydrate of copper, manganese, cobalt, and iron, and is similar to asbolite.

Rabdite. A mineral. It is an iron-nickel phosphide.

Rabdophanite. A mineral. It is a hydrated phosphate of rare earth metals.

Rabro. A proprietary preparation containing liquorice, bismuth subnitrate, magnesium carbonate, sodium bicarbonate, alder buckthorn bark and calamus rhizome. An antacid. (273).

Racahout. A starch food prepared from corn. It is sweetened and flavoured.

Racemethorphan. ($\pm$) - 3 - Methoxy - N-methylmorphinan.

Racemoramide. ($\pm$) - 1 - (β - Methyl - γ - morpholino - $\alpha\alpha$ - diphenylbutyryl) - pyrrolidine.

Racemorphan. ($\pm$) - 3 - Hydroxy - N - methylmorphinan.

Racewinite. A mineral. It is a hydrated silicate of aluminium and iron.

R-Acid. 2-Naphthol-3 : 6-disulphonic acid.

2R-Acid. 2 : 8-Amino-naphthol-3 : 6-disulphonic acid.

Rackarock. A blasting explosive, consisting of 79 per cent. potassium chlorate and 21 per cent. nitro-benzene, mixed sometimes with picric acid or sulphur.

Rackarock Special. An explosive similar to Rackarock, but containing 12–16 per cent. picric acid.

Radarsan. See RAWSTOL.

Radauite. A mineral. It is a variety of labradorite.

Radax. A proprietary accelerator formed from a blend of thiazole and thiuram. (57).

Raddle. See INDIAN RED.

Radial Yellow 3G. A dyestuff prepared by heating 2 molecules of the hydrazine of 2-amino-3-chloro-toluene-5-sulphonic acid with dioxy-tartaric acid.

Radiant Yellow. See CADMIUM YELLOW.

Radicle Vinegar. Acetic acid, glacial.

Radiolite. A mineral. It is a natrolite of Norway.

Radio-malt. A proprietary preparation consisting of malt extract with irradiated ergosterol (radiostol). It contains vitamins A, B, and D.

Radiometal. A nickel-iron-copper alloy having high incremental permeability and low losses. It is largely used for radio transformers, relays, etc.

Radiopaque. A special barium sulphate.

Radiophan. A German product. It is stated to contain radium chloride and atophan. Used for intramuscular injection in cases of neuralgia, rheumatism, and gout.

Radiophyllite. A mineral. It is a hydrated calcium silicate.

Radiose. A French nitro-cellulose lacquer.

Radiostol. A proprietary preparation of calciferol, vitamin D_2. (182).

Radiostoleum. A proprietary preparation of vitamins A and D in oil.

Radiotine. A mineral. It is $Mg_3Si_2O_5(OH)_4$.

Radmolite. An electrical insulating material for supporting heating coils. It largely consists of diatomaceous earth, and having a slower rate of heat absorption, replaces fireclay.

Radoform. A brewery disinfectant containing sodium hypochlorite and caustic soda. Also see ANTIFORMIN.

Radumine. A synthetic oxalic acid.

R.A.E. 57 Alloy. An alloy of aluminium with 4 per cent. copper, 2 per cent. iron, and 0·5 per cent. magnesium. Specific gravity 2·8.

Rafaelite (Paralaurionite). A mineral, PbCl(OH). Also a proprietary bitumen.

Raffia. A fibre peeled from the leaves of a Madagascar palm.

Raffinade Wafers. Cane sugar, with ethereal oils and colouring matter.

Raffinase. An enzyme.

Raffinate. A refined Sicilian sulphur with impurities amounting to about 0·5 per cent. Also see GREGGIO and CALCARONI.

Raffinose. Mellitose, $C_{12}H_{22}O_{11}$, a sugar.

Rafoxanide. An anthelmintic. 3'-Chloro-4'-p-chlorophenoxy-3,5-di-iodo-salicylanilide. FLUKANIDE.

Raidmondite. A mineral. It is a hydrated ferric sulphate, $2Fe_2O_3.3SO_3.7H_2O$.

Raimondite. A mineral. It is a hydrated ferric sulphate.

Rainbow Orange. A dyestuff. It is a British equivalent of Ponceau 4GB.

Rainbow Red RB, NB. Dyestuffs. They are British brands of Azorubine S.

Rainbow Scarlet G, 2RS. British equivalents of Ponceau.

Rainbow Scarlet 4R. A dyestuff which is equivalent to New coccine. It is of British manufacture.

Rainbow Ware. A proprietary synthetic resin of the urea-formaldehyde type.

Raiz del Indio. See CANAIGRE.

Raiz de Zacaton (Chiendent, Mexican Whisk). The roots of the grass, *Epicampes macroura*, used in the manufacture of brushes.

Rakel's Alloy. An aluminium bronze. It contains 87·5 per cent. copper, 10·5 per cent. aluminium, 1 per cent. manganese, and 1 per cent. lead or zinc.

Raki. See ARRACK.

Ralstonite. A mineral, $3Al(OH.F)_3(Na_2Mg)F_2.2H_2O$.

Ramarite. A mineral. It is descloizite.

Rambufaside. A cardiac glucoside. 14-Hydroxy-3-(4-O-methyl-α-L-rhamnopyranosyloxy)-14β-bufa-4,20,22-trienolide. 4'-o-Methylproscillaridin.

Ramenti Ferri. Iron filings.

Ramet. A proprietary cutting material for steel alloys, cast iron, etc. It consists of tantalum carbide with nickel, and melts at 4100° C.

Rametin. A proprietary preparation of NAPHTHALOPHOS. A veterinary anthelmintic.

Ramie (Rhea, Green Ramie). A fibre obtained from *Bœhmeria tenacissima*.

Ramie, Green. See RAMIE.

Ramie, White. See CHINA GRASS.

Raminal. A proprietary preparation. It consists of theobromine-calcium salicylate with chlorophyll and iron phosphate.

Ramix. A proprietary magnesite refractory.

Rammelsbergite. A mineral. It is nickel diarsenide, $NiAs_2$.

Ramos. A proprietary product. It is a phenol-formaldehyde resin.

Ramsdellite. A mineral. It is $4[MnO_2]$.

Ramseyite. A mineral, $Na_2O.2SiO_2.2TiO_2$.

Ranciéite. A mineral. It is (Ca,Mn'') Mn_4····O_9·$3H_2O$.

Ranciérite. Synonym for Ranciéte.

Randanite (Ceyssatite, Memilite). Varieties of hydrated silica or opal.

Randite. A mineral, $Ca_5U_2C_6O_{20}·3H_2O$.

Raney Nickel. A form of nickel hydride, etc., used in hydrogenating certain organic compounds.

Rangoon Oil. The name for a mineral lubricating oil for rifles. Also see EARTH OIL.

Rankinite. A mineral. It is $4[Ca_3Si_2O_7]$.

Ransome's Stone. An artificial stone made by mixing sand with sodium silicate and a little chalk, or other similar material. The product is moulded to shape, and immersed in a solution of calcium chloride.

Ransomite. A mineral. It is a hydrated copper-iron sulphate.

Rape Cake (Rape Dust). Fertilisers made from rapeseed which has not been completely oil-extracted. The best quality contains 6·5 per cent. of nitrogen.

Rape Dust. See RAPE CAKE.

Rapeseed Meal. A fertiliser made from oil-extracted rapeseed.

Raphilite. A mineral. It is a variety of Tremolite.

Raphite. A mineral. It is boronatrocalcite.

Rapic Acid. A name which has been applied to the fatty acids of rape oil. It appears to be identical with oleic acid.

Rapid Fast Red B. A dyestuff. It consists of the anilide of β-hydroxynaphthoic acid mixed with 5-nitro-o-anisidine. Used in calico printing.

Rapid Fast Red BB. A dyestuff. It consists of a mixture of p-nitro-anilide of β-hydroxy-naphthoic acid and the nitrosamine salt of 5-nitro-o-anisidine.

Rapid Fast Red GG. A dyestuff. It consists of the anilide of β-hydroxynaphthoic acid mixed with p-nitroaniline. Used in calico printing.

Rapid Fast 3G. A dyestuff. It consists of the anilide of β-hydroxy-naphthoic acid mixed with p-chloro-o-nitro-aniline. Used in calico printing.

Rapid Fast Red GL. A dyestuff. It consists of the anilide of β-hydroxynaphthoic acid mixed with o-nitro-p-toluidine. Used in calico printing.

Rapid Fast Red 3GL. A dyestuff. It is a mixture of the aniline of β-hydroxynaphthoic acid and p-chloro-o-nitroaniline. Used in calico printing.

Rapidal. A proprietary preparation of CYCLOBARBITONE calcium. A sedative. (250).

Rapidine. A product of the distillation of mineral oil, from 100–125° C. It is

purified, and benzene added, so that it may be used for internal combustion engines.

Rapitard. A proprietary preparation of beef insulin crystals in a rapid acting solution of neutral pork insulin. (182).

Rar (Bu-shol). Beans from a species of *Canavalia*, of Burma.

Rarical. A proprietary preparation of ferrous calcium citrate and tricalcium citrate. A hæmatinic. (369).

Rarical " F ". A proprietary preparation of ferrous calcium citrate, tricalcium citrate and folic acid. A hæmatinic. (369).

Raschit. A preservative for latex. It is p-chloro-m-cresol.

Rashleighite. A mineral. It is $Cu(Al,Fe)_6(PO_4)_4(OH)_8·5H_2O$.

Rasorite. A trade name for kernite (hydrated sodium borate), and in general boron ores and products. (548).

Raspberry Spar. A mineral. It is a variety of rhodochrosite.

Raspite. A mineral. It is a form of lead tungstate, $PbWO_4$.

Rastik. A perfumed mixture of powdered galls, henna, alum, sugar, and a trace of copper sulphate or iron sulphate. A Turkish hair dye.

Rastinon. A proprietary preparation of tolbutamide. An oral hypoglycæmic agent. (312).

Rastone. Aromatic flavouring chemicals. (522).

Ratafia. Essence of almonds.

Rathite. A mineral, $3PbS.2As_2S_3$.

Ratholite. See PECTOLITE.

Ratofkite. A mineral. It is a variety of fluorite.

Rattlesnake Root. The root of *Polygala senega*.

Raudixin. A proprietary preparation of rauwolfia. An antihypertensive. (326).

Rauracienne. See FAST RED.

Rautrax. A proprietary preparation of rauwolfia, HYDROFLUMETHIAZIDE and potassium chloride used as an antihypertensive agent. (682).

Rautrax Sine K. A proprietary preparation of rauwolfia and HYDROFLUMETHIAZIDE used as an anti-hypertensive agent. (682).

Rauvite. A mineral of Utah, containing 38·3 per cent. V_2O_4 and V_2O_5, 24 per cent. SiO_2, water, CaO, and Fe_2O_3.

Rauwiloid. A proprietary preparation of rauwolfia alkaloids. An antihypertensive. (275).

Rauwiloid Veriloid. A proprietary preparation of alseroxylon and alkavervir. (275).

Rauxite. Proprietary trade name for urea-formaldehyde varnish and lacquer resins.

Rauxone. A proprietary trade name for alkyd varnish and lacquer resins.

Rauzene. A proprietary trade name for a phenolic varnish and lacquer resin.

Rauzene Ester. A proprietary trade name for an ester gum.

Ravolen. Plasticiser and rubber extender. (628).

Ravolen 11(T). A proprietary trade name for a decolourized petroleum aromatic extract. (125).

Raw Palmira Root Flour. See TALIPOT.

Raw Sienna. A yellow pigment. It is a native ferruginous earth.

Rawstol (Radarsan). Vermicides. They consist of solutions of fluosilicic acid.

Raw Turkey Umber. See UMBER.

Raw Umber. See UMBER.

Raybar. A proprietary preparation of barium sulphate. (979).

Rayo. An electrical resistance alloy consisting of 85 per cent. nickel and 15 per cent. chromium.

Rayon. A name proposed for artificial silk.

Rayox. A proprietary trade name for titanium dioxide, TiO_2.

Razoxane. An anti-neoplastic currently undergoing clinical trial as " ICI 59,118 " and " ICRF159 ". It is 1,2-bis (3, 5-dioxopiperazin-1-yl)propane.

R.B.C. A proprietary preparation of phenylmercuric nitrate, iso-butyl para-amino benzoate, N-butyl para-amino-benzoate, benzocaine, cholesterol and calamine. Antipruritic. (273).

Reactivan. A proprietary preparation of FENCAMFAMIN hydrochloride, thiamine, PYRIDOXINE, CYANOCOBALAMIN and ascorbic acid. A tonic. (896).

Realgar (Ruby Sulphur, Red Orpiment, Red Arsenic, Red Arsenic Glass, Ruby Arsenic, Arsenic Orange). Arsenic disulphide, As_2S_2.

Realgarite. Synonym for Realgar.

Real Ultramarine. See ULTRAMARINE GENUINE.

Réamur's Alloy. An alloy of 70 per cent. antimony and 30 per cent. iron.

Reargon. A colloidal silver preparation. Used in the treatment of gonorrhœa.

Reasec. A proprietary preparation of DIPHENOXYLATE hydrochloride and atropine sulphate used in the control of diarrhœa. (356).

Reaumerite. A compound, $(Ca.Na_2)O.3SiO_2$, obtained by heating glass at its softening temperature.

Réboulet's Solution. An aqueous solution of calcium chloride, potassium nitrate, and alum. Used to preserve anatomical specimens.

Reclaim Acid. See RECLAIMED RUBBER.

Reclaim, Alkali. See RECLAIMED RUBBER.

Reclaimed Rubber. Rubber which has been recovered from used rubber goods. It is obtained by means of acid or alkali treatment and heat to render it a pliable product.

Reconstituted Rubies. Splinters of rubies cemented together with a lead-containing flux.

Reconstructed Milk. A product obtained by emulsifying unsalted butter in a solution of skimmed milk powder.

Reconstructed Stone. A stone manufactured from the debris of limestone quarries. This is crushed and mixed with dolomitic lime, and heated in retorts.

Recoura's Sulphate. A chromium hexa-hydrated sulphate, $Cr_2(SO_4)_3.6H_2O$.

Recovered Grease. The oil used to lubricate wool during spinning is recovered from the wash-water. It is used to manufacture a low-grade stearin.

Recovered Manganese. See WELDON MUD.

Recresal. A proprietary preparation of acid sodium phosphate in tablets.

Rectified Spirit S.V.R. A specially rectified ethyl alcohol 68–69 over proof, containing 96–97 per cent. ethyl alcohol by volume. It is used in perfumery and in pharmaceutical extracts and tinctures.

Rectorite. A mineral, $Al_2O_3.2SiO_2.H_2O$, of Arkansas.

Recto-serol. A German preparation containing extract of humamelid, alum, acet. tartar, formaldehyde, balsam of Peru, and novocaine, finely divided in serol (serol is a water-soluble protein body). Used for hæmorrhoids.

Red Acid. Nitric acid of 40° Bé., or stronger. It contains dissolved nitrogen oxides.

Red Algar. Arsenic disulphide, As_2S_2.

Redalloy. A proprietary trade name for brass containing 85 per cent. copper, 14 per cent. zinc, and 1 per cent. tin.

Red Aniline. See ANILINE OIL FOR RED.

Red Antimony (Antimony Blende, Pyr-antimonite, Pyrostibnite, Kermesite, Antimony Cinnabar). A mineral. It is an oxysulphide of antimony, $Sb_2O_3.2Sb_2S$. Antimony cinnabar, or red antimony, is also obtained by treating antimony chloride with sodium thio-sulphate in aqueous solution. Used as a pigment to replace ordinary cinnabar.

Red Argol. See ARGOL.

Red Arsenic. See REALGAR.

Red Arsenic Glass. See REALGAR.

Red B. See SUDAN II.

Red Balsam or Xanthorrhœa. See ACAROID BALSAM.

Red B Oil-soluble Extra Conc. See OIL SCARLET.

Red Bole. See INDIAN RED.

Red Brass. A brass containing 90 per cent. copper and 10 per cent. zinc. Also Tombac (*q.v.*), which has been pickled in acid. The following particulars are also given: a brass containing 85 per cent. copper, 5 per cent. tin, 5 per cent. lead has a melting-point of 970° C., and one containing 82 per cent. copper, 3 per cent. tin, and 5 per cent. lead has a melting-point of 980° C. (See also BRASS.)

Red Bricks. Bricks coloured with oxide of iron, made by the addition of ferric oxide to the clay.

Red C. See SUDAN III.

Red Calamine. A mineral which consists chiefly of Smithsonite, containing cadmium, and is usually ferruginous. It contains from 28–35 per cent. zinc.

Red Chalk. See INDIAN RED.

Red Charcoal. A wood charcoal made at low temperature. It contains hydrogen and oxygen.

Red Chromate of Potash. Potassium bichromate, $K_2Cr_2O_7$.

Red Chrome. See CHROME RED.

Red Cobalt. A mineral. It is erythrite.

Red Copper. See VIOLET COPPER.

Red Copper Ore. See CUPRITE.

Red Coralline. See PÆONINE.

Red Cotton Spirits. See TIN SPIRITS.

Red Crocus. Ferric oxide, Fe_2O_3.

Red Developer. β-Naphthol, used for developing red on fibre which has been treated with Primuline yellow.

Reddingite. A mineral, $3(Mn.Fe)O.P_2O_5$.

Reddle. See INDIAN RED.

Red Drops (Red Lavender). Compound tincture of lavender.

Red Earth. See INDIAN RED.

Red Fibre. See VULCANISED FIBRE.

Red Fire. Usually consists of strontium nitrate, sulphur, charcoal, potassium chlorate, and antimony sulphide.

Red Glass (Ruby Glass). A glass, the colour of which is produced by gold, selenium, or copper.

Red Gold. A jeweller's alloy containing 75 per cent. gold and 25 per cent. copper.

Red Gum. See EUCALYPTUS GUM.

Red Hæmatite. Native ferric oxide, Fe_2O_3. See HÆMATITE.

Redicote. Specialised cationic bitumen emulsifiers. (555).

Red Indigo. See FRENCH PURPLE.

Redingtonite. A mineral. It is a hydrated chromic sulphate.

Red Iodide of Mercury. Mercuric iodide, HgI_2. Employed in medicine as an irritant poison for syphilis, also as a germicide for washing wounds.

Red Iron Ore. A mineral. It is a massive variety of hæmatite.

Red Lavender. See RED DROPS.

Red Lavender Spirit. See SPIRIT OF RED LAVENDER.

Red Lead (Red Lead Oxide, Minium, Paris Red, Saturn Red). A pigment. It is oxide of lead, Pb_3O_4, made by heating litharge, PbO. There are several kinds on the market distinguished by their colour and amount of lead dioxide they contain. This varies between 18 and 34 per cent. A good quality for a paint contains 25 per cent. lead dioxide.

Red Lead Oxide. See RED LEAD.

Red Lead, True. See CRYSTAL MINIUM.

Red Liquor (Mordant Rouge). A solution corresponding to the formula, $Al_2(C_2H_3O_2)_6$, which appears to consist of a diacetate of aluminium, and acetic acid. Red liquor is largely used in dyeing and calico printing, especially for the production of red colours, for the manufacture of dense lakes, and for waterproofing woollen fabrics.

Red M. An azo dyestuff prepared from benzidine, *m*-amino-phenol, and naphthionic acid.

Red Manganese. See RHODOCROSITE.

Redmanol. See BAKELITE.

Red Metal. A term usually applied to an alloy of 90 per cent. copper, and 10 per cent. zinc.

Red Mordant. See RED LIQUOR.

Red Nickel Ore. A mineral. It is niccolite or nickeline.

Redo. See HYDROSULPHITE.

Red Ochre. See INDIAN RED.

Red Oil. Acid-treated distillates or residual oils of petroleum which are finished by a soda wash or by a clay treatment after the acid tar produced by sulphuric acid treatment has been allowed to settle. These oils are used for purposes of general lubrication. The term is also used for olive oil containing the red colouring matter of alkanet and for commercial oleic acid. Also see TURKEY RED OILS.

Redonda Phosphate. A mineral. It consists mainly of calcium phosphate.

Redondite. A mineral. It is a hydrated phosphate of aluminium and iron.

Red Orpiment. See REALGAR.

Red Oxide. See INDIAN RED.

Red Oxide of Chromium. Chromium trioxide, CrO_3.

Red Oxide of Copper. See HEMI-OXIDE OF COPPER.

Red Oxide of Lead. See RED LEAD.

Red Oxide of Manganese. See HAUSMANNITE.

Red Oxide of Mercury (Red Precipitate). Mercuric oxide, HgO. Used in porcelain painting and in medicine.

Redoxon. A proprietary preparation of ascorbic acid. (314).

Red Pepper. Capsicum.

Red Phosphorus (Amorphous phosphorus). A modification of phosphorus.

Red Pitch. The sticky sublimate obtained when Saxon lignite tar is distilled at high temperature. Crackene, $C_{24}H_{18}$, has been obtained from this material.

Red Poppy Petals. Obtained from *Papaver rhœas.*

Red Poppyseed Oil. The oil from poppyseeds, pressed hot.

Red Precipitate. See RED OXIDE OF MERCURY.

Red Prussiate of Potash. Potassium ferricyanide, $K_3Fe(CN)_6$.

Redray. An electrical resistance alloy containing 85 per cent. nickel and 15 per cent. chromium. The resistivity 93 microhms. cm. The melting-point is 1380° C., and the specific gravity is 8·28 (Henry Wiggin & Co.).

Red Rice (Brown Rice). Undermilled rice, retaining the proteid layer and some of the pericarp, which contain the salts.

Red Rudd. See INDIAN RED.

Redruthite (Copper Glance, Chalcosine, Vitreous Copper). A variety of cuprous sulphide, Cu_2S, found in Cornwall.

Red Salts. Both crude sodium acetate and crude sodium carbonate, coloured red by ferric oxide, are known as red salts.

Red Saunderswood. See REDWOODS.

Red Silver Ore. See PYRARGYRITE.

Red Soda. A solution of red ink containing a little gum arabic and sodium carbonate. It is used as a marking ink for " blue prints."

Red Spirit. See TIN SPIRITS.

Red Star Powder. An explosive. It is a 33-grain smokeless powder containing metallic nitrates, nitro-hydrocarbons, and pet. jelly.

Red Storax (Solid Storax). An artificial product obtained by mixing poor storax with sawdust, and pressing the mixture. Used for fumigating candles and powders.

Reduced Oils. Crude American petroleum, which, after the sunning process (see SUNNED OILS), has been partially evaporated or concentrated in stills. Used as lubricating oils.

Reduced Turpentine. A mixture of turpentine oil with petroleum.

Reducin. Triamino-resorcin, $C_6H(NH_2)_3$ $(OH)_2$. Used in photography as a developer.

Red Ultramarine. Obtained from blue ultramarine by the action of dry hydrochloric acid gas and oxygen, at 150–180° C. See ULTRAMARINE.

Red Violet 5R Extra. See HOFMANN'S VIOLET.

Red Violet 4RS (Acid Violet 4RS). A dyestuff. It is the sodium salt of dimethyl-rosaniline-trisulphonic acid, $C_{22}H_{22}N_3O_{10}S_3Na_3$. Dyes wool from an acid bath.

Red Violet 5RS (Acid Violet 5RS). A dyestuff. It is the sodium salt of ethyl-rosaniline-sulphonic acid, $C_{22}H_{22}N_3O_2$ S_3Na_3. Dyes wool bluish-red from an acid bath.

Red Vitriol (Botryogen). A native ferroso-ferric sulphate from Sweden.

Red Wash. A zinc sulphate solution containing red colouring matter.

Red Water Bark. Sassy bark.

Redwoods (Red Dye Woods). These dye woods are divided into two classes: (1) Soluble, which comprise Brazil, Pernambuco or Fernambuco wood, Peach wood, Lima wood, Sapan wood, Bimas redwood, and Nicaragua wood. All of them contain the colouring principle brazilin, $C_{16}H_{14}O_5$, which, by oxidation, is converted into brazilein, $C_{16}H_{12}O_5$, which gives purple shades with chrome mordants, and crimson with alum.

2) Insoluble redwoods, consisting of Camwood or Cambe wood, Barwood, Saunderswood, Santalwood, Sandlewood or Sandelwood, Bresille wood, and Caliatur wood. The dyeing principle is santaline. These woods have a limited application for dyeing wool with alumina, chrome, tin, or iron mordants.

Red Yacca Gum. See ACAROID BALSAM.

Red Zinc Ore. A mineral. It consists of zinc oxide, containing up to 12 per cent. of manganese, as oxide.

Redeptin. A proprietary preparation of FLUSPIRILINE used in the treatment of schizophrenia. (281).

Reed Brass. See BRASS.

Reedmergnerite. A mineral. It is Na BSi_3O_8.

Reese's Alloy. An alloy used in dental work. It contains 87 per cent. tin, 8·6 per cent. silver, and 4·4 per cent. gold.

Rees's Thionin Stain. A microscopic stain. It consists of 1·5 grams thionin and 10 c.c. alcohol in 100 c.c. of 5 per cent. solution of carbolic acid. It is used at the rate of 5 c.c. in 20 c.c. water.

Refagan. A proprietary preparation of salicylamide, phenacetin, caffeine and mebhydrolin napadisylate. (341).

Refikite. A resin found in lignite.

Refined Silver. A silver usually containing from 99·7–99·9 per cent. metal.

Refined Tartar. See CREAM OF TARTAR.

Reflite. A proprietary synthetic resin moulding powder.

Reflorit. Picric acid. It has been tried for the disinfection of seed-corn.

Reform Phosphate. Rock phosphate which has been treated with small quantities of dilute acid to render it more porous, converting calcium carbonate into calcium hydrogen carbonate.

Refrax. Bricks made from recrystallised silicon carbide. A refractory material.

Regalox. A sintered material comprising 88 per cent. alumina.

Regenerated Turpentine. A product of synthetic camphor manufacture. It boils at 170° C.

Regepyrin. A trade name for aspirin (q.v.).

Regianin. Sec-oxy-α-naphthoquinone.

Regina Purple (Regina Violet, Violet Imperial Rouge, Violet Phenylique, Phenyl Violet, Imperial Violet). A dyestuff. It is the acetate of o-tolyl-p-rosaniline, $C_{28}H_{27}N_3O_4$. Dyes wool reddish-violet.

Regina Spirit Purple. See SPIRIT PURPLE.

Regina Violet. See REGINA PURPLE.

Regina Violet, Water Soluble. A dyestuff. It is the trisulphonic acid of Regina purple. Dyes wool reddish-violet from an acid bath.

Reglykol. A German preparation. It contains glycerin, phosphoric acid, chrysophan, arbutin, emetin, and capsacin. Used in diabetes.

Regnaud's Anæsthetic. Contains 4 parts chloroform and 1 part methyl alcohol.

Regnolite. A mineral. A sulpharsenite of copper, iron, and zinc.

Regulin. Broken-up agar-agar, mixed with 25 per cent. of cascara extract. Used in medicine.

Regulus Metal. This is usually a 5–12 per cent. antimony with lead. According to the British Engineering Standards Association No. 335 (1928) there are three types: (1) Containing lead with from 6–8 per cent. antimony, recommended for the construction of tanks, acid eggs, and pans. (2) From 8–10 per cent. antimony, for valves, taps, acid pumps, and plugs. (3) With from 10–12 per cent. antimony, for small parts such as screws, jets, and nozzles. The specification gives the limits of impurities (As, Zn, Cu, Sn, and S), and the physical tests include the tensile strength, Brinell hardness, and bend test.

Regulus of Antimony. Produced by heating antimony ore, Sb_2S_3. It contains about 10 per cent. of iron.

Regulus of Venus. An alloy of copper and antimony, $SbCu_2$.

Reichardite. A mineral. It is magnesium sulphate, $MgSO_4.7H_2O$, and occurs in the Stassfurt deposits.

Reichite. A mineral. It is calcite.

Reich's Bronze. An aluminium bronze containing 85.2 per cent. copper, 7.52 per cent. iron, 6.6 per cent. aluminium, 0.5 per cent. manganese, and 0.15 per cent. lead.

Reicolit. A proprietary insulation.

Reinecke's Salt. A metal chromammine, $Cr(NH_3)_2.(CNS)_4.NH_4$, produced when ammonium cyanate is melted and ammonium bichromate added.

Reinforced Concrete (Armoured Concrete, Ferro-Concrete). Concrete in which iron rods, wire-netting, or perforated iron plates are buried.

Reinforced Glass. Plate glass, in which wire-netting is embedded.

Reinforced Lead. A patented material consisting of lead sheet strengthened by embedding iron or steel gauze or perforated sheet within the lead.

Reinforced Wine. Wine to which alcohol has been added.

Reinite. A mineral. It is an iron tungstate, $FeWO_4$, and occurs in Japan.

Reiset's First Base. Plato-diamine-hydroxide, $Pt(NH_3.NH_3.OH)_2$.

Reiset's First Chloride. Plato-diamine-chloride, $Pt(NH_3.NH_3.Cl)_2.H_2O$.

Reissacherite. A mineral. It is a variety of wad.

Reissite. A mineral. It is a variety of epistilbite.

Reith Alloy. An alloy of 75 per cent. copper, 10 per cent. tin, 10 per cent. lead, and 5 per cent. antimony.

Relcryl 100 and 200. A trade mark for a range of acrylic enamel finishes for metal. Relcryl 100 is a one-coat finish; 200 a two-coat finish. (980).

Releasil. Silicone release agents. (614).

Relugan GT. An aldehyde tanning agent. It is an aqueous solution of glutaraldehyde.

Remarcol. Sodium fluoride, NaF, a preservative for sweet wines and beer.

Remeflin. A proprietary trade name for the hydrochloride of Dimefline.

Remiderm. A proprietary preparation triamcinolone acetonide and halquinol used in dermatology as an antibacterial agent. (326).

Remingtonite. A mineral. It is a hydrated cobalt carbonate.

Remolinite. See ATACAMITE.

Remnos. A proprietary preparation of NITRAZEPAM. A hypnotic. (498).

Remotic. A proprietary preparation of HALQUINOL and TRIAMCINOLONE acetonide used as a local anti-infective agent. (682).

Renaglandin. An extract of the suprarenal gland.

Renaleptine. Synthetic adrenalin.

Renalina. Adrenalin (q.v.).

Renarcol. Tribromethyl alcohol.

Renardite. A uranium mineral, $PbO.4UO_3.P_2O_5.9H_2O$, from Chinkolobwe.

Renault Alloy. An aluminium alloy containing 88 per cent. aluminium, 10 per cent. zinc, and 2 per cent. copper.

Rendells. A proprietary preparation of nonoxinol used in the form of contraceptive pessaries. (737).

Rendrock. An explosive. It is a modification of Lithofracteur, consisting of 40 per cent. potassium nitrate, 40 per cent. nitro-glycerin, 13 per cent. wood pulp, and 7·0 per cent. paraffin or pitch.

Renese. A proprietary trade name for Polythiazide.

Reniformite. A Japanese mineral, $5PbS.As_2S_3$.

Rennet. A clotting enzyme, which coagulates milk by precipitating the casein. It is the aqueous or alcoholic infusion of the dried stomach of the calf.

Rennet, Milk Sugar. See PEGNIN.

Rennin. A solid form of rennet.

Rennselaerite. A mineral. It is Talc (q.v.).

Renoform. Adrenalin (q.v.).

Renostypticin. Adrenalin (q.v.).

Renostyptin. Adrenalin (q.v.).

Renotrat. A German preparation. It is a proprietary concentrated extract of sarsaparilla recommended for kidney trouble.

Renoxin. A rodenticide. (535).

Renyx. A proprietary trade name for an alloy of aluminium with nickel, copper, and silicon.

Reochlor (LF and 54). A trade mark for an extender-plasticiser for PVC. They are chlorinated paraffins. (17).

Reoflam 20, 40 and 60. A range of proprietary plasticisers used with PVC. (18).

Reofos. A trade name for a range of synthetic organic phosphates. (903).

Reolube. A trade name for a range of synthetic organic phosphates. (903).

Reolube FAD. A projectory trade name for a long chain fatty acid mixture, the principal components being C_{14}, C_{16}, and C_{18} acids. (17).

Reomol. A trade mark (17) for vinyl plasticisers as follows: Reomol DBS dibutyl sebacate. DCP dicapryl phthalate. DOS di-2-ethyl-hexyl sebacate. D79S a mixture of heptyl and nonyl sebacates.

Reomol 4PG. A proprietary plasticiser. Butyl phthalyl butyl glycollate. (17).

Reomol BCF. A proprietary plasticiser. Butyl carbinol formal. (17).

Reomol P. A proprietary trade name for dimethoxy ethyl phthalate. A chemical bonding agent for cellulose acetate staple fibre. (17).

Reomol PBPS. A sebacic acid polyester and a small proportion of non polymeric ester. (17).

Reomol TC9. A proprietary trade name for a chemical bonding agent for Terylene and cellulose triacetate fibres.

Reoplex 200, 220, 300. A trade mark. Vinyl plasticisers of the polyester type. (17).

Reoplex 901. A trade mark for a plasticiser for PVC sheeting intended for manufacture of surgical and electrical tapes. (17).

Reoplex 902. A plasticiser with good resistance to extraction by petroleum. (17).

Reostene. A nickel-iron alloy.

Repelit. A proprietary synthetic resin varnish-paper product used for electrical insulation.

Resacetophenone. 4 : 2 : 1-Dioxy-acetophenone, $C_6H_3(OH)_2.CO.CH_3$.

Resaldol. Resorcyl-benzoic-acid-ethyl ester, $(HO)_2.C_6H_3.CO.C_6H_4.CO_2.C_2H_5$. An antiseptic.

Resalgin. Antipyrine - resorcylate, $(C_{11}H_{12}N_2O_4)_2.C_7H_6O_4$. An antipyretic.

Resamine. A proprietary trade name for formaldehyde resins. (4).

Resan. See BAKELITE.

Resanite. A mineral. It is $Cu_5Fe_2Si_{10}O_{28}.30H_2O$.

Resart. A proprietary range of phenolic moulding materials. (904).

Resazoin. See DIAZORESORCIN.

Rescinnamine. Methyl O-(3,4,5-trimethoxycinnamoyl)reserpate. Anaprel.

Reserpine. An alkaloid obtained from Rauwolfia serpentina Benth. Serpasil.

Resilia. A proprietary trade name for a silico-manganese spring steel.

Resilla. A proprietary trade name for a special silicon-manganese spring steel.

Resin. See COLOPHONY.

Resin, Acaroid. See ACAROID BALSAM.

Resins, Acrolein. Resins obtained by the polymerisation of acrolein by means of inorganic and organic bases or salts of iron and lead. Orea is a trade name for a resin of this type. Acrolein also condenses with phenols to form resins.

Resins, Alkyd. Resins resulting from the interaction of a polyhydric alcohol and a polybasic acid. See BECKOSOL, ALKYDAL, and REZYL.

Resin, Benzo-furane. See PARACOUMARONE RESIN.

Resin Blende. Zinc blende, ZnS, of a yellow colour, is sometimes called by this name.

Resin Cerate. See BASILICON.

Resin, Coumarone. See PARACOUMARONE RESIN.

Resin, Cumar. See PARACOUMARONE RESIN.

Resin, Cypress Pine. See PINE GUM.

Resineon. A volatile oil obtained from back tar by distillation. It is said to be free from phenol, and is used as an antiseptic in skin affections.

Resin Essence. See ROSIN SPIRIT.

Resin Ether L. A proprietary synthetic resin for use as a cellulose-lacquer plasticiser. It is stated to be non-drying, not susceptible to atmospheric oxidation, and to have a low acid value. It is soluble in all the usual solvents except industrial methylated spirit, but soluble in anhydrous alcohol. It has a boiling-point of 500° C. and a flash-point of 600° F.

Resinette. A synthetic resin obtained from phenol and formaldehyde.

Resinite. See BAKELITE.

Resin Lutea (Acaroid Balsam). A name applied to yellow acaroid balsam, a yellow resin obtained from *Xanthorrhœa Hastile*.

Resin M. See PERSPEX.

Resin M.S.2. Cyclohexanone condensation products. (513).

Resin of Botany Bay. See ACAROID BALSAM.

Resin of Conima. See HIAWA GUM.

Resin Oil. That fraction of the distillation of resin (colophony), which distils over from 300–400° C. It consists principally of terpineol, $C_{10}H_7OH$. Used as a lubricant.

Resinol. A varnish substitute obtained by the dehydrogenation of petroleum, distillation, and polymerisation.

Resinous Silica. A variety of hydrated silica or opal.

Resinox. A proprietary synthetic resin moulding powder of the phenol-formaldehyde type.

Resin, Saliretin. See SALIRETINS.

Resin, Spiller's. See SPILLER'S RESIN.

Resin Spirit. See ROSIN SPIRIT.

Resin, Sulphate. See LIQUID RESINS.

Resin, Urea. See UREA and THIOUREA RESINS.

Resin WP. A proprietary trade name for a melamine based thermosetting resin used for crease-resisting finishes. (18).

Resin, West Indian Anime. See TACAMAHAC RESIN.

Resin, White Pine. See PINE GUM.

Resiosol. See ANUSOL.

Resipol DL. A proprietary trade name for a plasticiser. It is glyceryl dilactate.

Resipol ML. A proprietary trade name for a plasticiser. It is glyceryl monolactate.

Resisco. A proprietary trade name for an alloy of 91 per cent. copper, 7 per cent. aluminium, and 2 per cent. nickel.

Resista. A glass similar in composition to Pyrex glass. It contains 70 per cent. silica and 13·5 per cent. boric oxide.

Resistac. A proprietary trade name for an alloy of copper with 9 per cent. aluminium and 1 per cent. iron.

Resistal. A heat-resisting alloy containing 63·5 per cent. iron, 16·6 per cent. nickel, 15 per cent. chromium, 4·5 per cent. silicon, and 0·3 per cent. carbon.

Resistance Bronze. A term for an alloy of from 84–86·5 per cent. copper, 11·5–13·5 per cent. manganese, and 2 per cent. iron.

Resista Steel. An alloy of iron, nickel, and manganese, which is ductile at low temperatures.

Resistin. An electrical resistance alloy containing 84–86 per cent. copper, 2 per cent. iron, and 11–13 per cent. manganese.

Resistoflex. A proprietary trade name for polyvinyl alcohol synthetic resins.

Resistopen. A proprietary preparation of sodium oxacillin. An antibiotic. (326).

Resistox. A proprietary antioxidant for rubber. It is an aldehyde-amine condensation product.

Resochin. A proprietary preparation of chloroquin phosphate. An antimalarial. (341).

Resoflavin. A dyestuff obtained by the oxidation of *m*-dioxy-benzoic acid in sulphuric acid solution, by means of ammonium persulphate. Dyes wool mordanted with chromium or alumina, yellow.

Resoglass. See POLYSTYRENE.

Resoglaz. A proprietary trade name for a polymerised styrene.

Resol. A disinfectant said to be made by saponifying wood tar with caustic potash, and adding wood spirit.

Resonium-A. A proprietary preparation of sodium polystyrene sulphonate, used in hyperkalæmia. (112).

Resopol. A proprietary polyester laminating resin. (905).

Resopyrin. A mixture of solutions of antipyrine and resorcinol.

Resorbin. A registered trade mark currently awaiting re-allocation by its proprietors to cover a range of pharmaceuticals. (983).

Resorcin. Resorcinol, $C_6H_4(OH)_2$.

Resorcinal. A mixture of equal parts of resorcinol and iodoform. Used as an antiseptic dusting-powder.

Resorcin Blue. See FLUORESCENT BLUE. The name resorcin blue is also used for Lackmoid (*q.v.*).

Resorcin Blue. Nitroso Blue. A dyestuff. It is the tannin compound of dimethylamino-phenoxazone, $C_{14}H_{12}N_2O_2$, and is produced on the fibre, giving indigo blue shades.

Resorcin Brown, A Conc., G, R, RBW. British equivalents of Resorcin brown.

Resorcin Brown (Resorcin Brown A Conc., G, R, RBW, Resorcinol Brown). A dyestuff. It is the sodium salt of xylene-azo-resorcin-azo-benzene-p-sulphonic acid, $C_{20}H_{17}N_4O_5SNa$. Dyes wool brown from an acid bath.

Resorcin Ether. See RESORCINYL OXIDE.

Resorcin Green. See DINITROSO-RESORCIN.

Resorcinoform. An antiseptic used in medicine. It is obtained by dissolving resorcinol in formaldehyde, adding hydrochloric acid to cause precipitation, and filtering.

Resorcinol Brown. A dyestuff. It is a British equivalent of Resorcin brown.

Resorcinol Green. See DINITROSO-RESORCIN

Resorcinopyrin. Resorcin-phenazone.

Resorcin-percutol. A German preparation of 33 per cent. resorcin and 66·5 per cent. salicylic ester.

Resorcin Yellow. See TROPÆOLINE O.

Resorcin Yellow O Extra. A British brand of Tropæoline O.

Resorcinyl Oxide (Resorcin Ether). The anhydride of resorcinol, $O(C_6H_4.OH)_2$.

Resorcylalgin. Antipyrine resorcylate.

Resorufin. Oxyphenazone, $C_{12}H_7O_3N$.

Resotren. A proprietary trade name for cloquinate.

Resovin. A proprietary synthetic resin of the vinyl type. A denture.

Respenyl. A proprietary preparation containing guaiaphenesin (guaiacol glyceryl ether). A cough suppressant. (280).

Resticel. Expansible polystyrene.

Restil. SAN copolymers.

Restiran. ABS copolymers.

Resyl. A completely absorbable guaiacol preparation, expectorant and antiseptic in acute and chronic affection of the respiratory organs.

Restirolo. Polystyrenes.

Retariox. A proprietary trade name for an aldehyde-amine condensation product.

Retcin. A proprietary preparation of ERYTHROMYCIN. An antibiotic. (498).

Retene. Methyl-isopropyl-phenanthrene, $C_{18}H_{18}$.

Retenema. A proprietary enema containing BETAMETHASONE valerate used in the treatment of proctitis. (335).

Retin-A. A proprietary preparation of TRETINOIN used in the treatment of acne vulgaris. (906).

Retinalite (Vorhauserite). A mineral. It is a variety of serpentine.

Retinaphtha. See METHYL BENZENE.

Retin Asphalt. See RETINITE.

Retinite (Retin Asphalt, Walchowite). A fossil resin found in brown coal. It occurs in Derbyshire and in Walchow. The material found near Walchow is a polymeric resin made up chiefly of sesquiterpenes. It has an acid value of o, saponification value 42, and an ash of 3 per cent. An analysis gives 79·4 per cent. carbon and 9·99 per cent. hydrogen.

Retinol. Vitamin A alcohol.

Retinol (Codoil, Rosinol, Rosin Oil). A product obtained by the distillation of rosin. An antiseptic and a vehicle for ointments.

Retnolite. A proprietary trade name for a phenol-formaldehyde synthetic resin.

Retort Carbon (Gas Carbon, Metallic Carbon, Retort Graphite). Formed by the decomposition of the hydrocarbons which are evolved during the distillation of coal in gas works. It is found as a deposit on the walls of the retorts, and is used for the manufacture of graphite crucibles, and carbon electrodes.

Retort Graphite. See RETORT CARBON.

Retz Alloy. An alloy of 75 per cent. copper, 10 per cent. lead, 10 per cent. tin, and 5 per cent. antimony.

Retzbanyite. A mineral, Pb.BiS.

Retzian. A mineral. It is a hydrated arsenate of manganese and calcium, with cerium and yttrium metals, of Sweden.

Reuniol. See ROSEOL.

Reussinite. A reddish-brown resin found in certain coal deposits.

Revalon. A proprietary trade name for an alloy of 76 per cent. copper, 22 per cent. zinc, and 2 per cent. aluminium.

Revatol S. A proprietary trade name for sodium m-nitro-benzene-sulphonate.

Revatrine. A relaxant for the uterus. It is 1-(3,4-dimethoxyphenyl)-4-phenyl-butyldimethylamine. NN-Dimethyl-α-3-phenylpropylveratrylamine. MONZALDON is the hydrochloride.

Reversible Latex. See REVERTEX.

Revertex (Reversible Latex). A trade mark for a highly concentrated rubber latex produced by a patented process, in which the latex is concentrated in the presence of an alkaline protective colloid.

Revolite. A proprietary phenol-formaldehyde synthetic resin impregnated cloth.

Revona. A proprietary trade name for a water-soluble aminoplast. A very effective pitch dispersant in papermaking. (18).

Revonal. A proprietary preparation of METHAQUALONE. A sedative. (310).

Revultex. A trade mark for vulcanised

rubber latex of any concentration produced from revertex.

Rewdanskite. A mineral. It is a nickel-magnesium silicate, containing up to 18 per cent. nickel.

Rex. A trade mark for abrasive goods consisting essentially of alumina.

Rex 95. A proprietary trade name for a cobalt steel containing 5 per cent. cobalt, 14 per cent. tungsten, and 4 per cent. chromium, 2 per cent. vanadium, and 0·5 per cent. molybdenum.

Rex-blak. A special preparation of carbon black, containing carbon, rubber, and glue, in varying proportions. Used in rubber mixings.

Rexenite. A proprietary trade name for a cellulose acetate butyrate plastic.

Rexhide. A rubber-glue stock material for use in rubber mixings.

Rexine. An imitation leather made by mixing nitro-cellulose with a drying oil and colouring matter, and rolling. The name is a trade mark.

Rexite. A blasting explosive, containing 6·5–8·5 per cent. nitro-glycerin, 64–68 per cent. ammonium nitrate, 13–16 per cent. sodium nitrate, 6·5–8·5 per cent. trinitro-toluene, and 3–5 per cent. wood meal.

Rexol. An explosive. It is a mixture of ammonium perchlorate, potassium chlorate, rosin, zinc or aluminium, and mineral oil or wax.

Rexolite. A cross-linked polystyrene. (408).

Rexoll Black. A sulphide dyestuff.

Rexotan. Methylene-tannin-urea. Used as an internal antiseptic and astringent.

Rextox. A proprietary trade name for a material consisting of copper with a layer of cuprous oxide formed on the surface of the metal at high temperatures.

Rextrude. A proprietary trade name for a cellulose-acetate-butyrate plastic.

Rezbanyite. A mineral. It is a sulpho-bismuthite of lead.

Rezistal. Corrosion and heat-resisting steels consisting of iron with up to 0·4 per cent. carbon, 1–5·5 per cent. silicon, 8–26 per cent. chromium, and 7–35 per cent. nickel.

Rezyl. A proprietary trade name for an alkyd synthetic resin.

R G-Acid. α-Naphthol-disulphonic acid.

R G-Salt. The sodium salt of R G-acid.

Rhabdophane (Phosphocerite). A mineral, $(Ce.La.Di)_3(PO_4)_2$.

Rhadoonit. A German slate and marble substitute.

Rhagite. A mineral, $Bi(BiO)_9.(AsO_4)_4.8H_2O$.

Rhamnetin. Quercetin-3-methyl-ether.

Rhamnine. See PERSIAN BERRIES.

Rhamnodulcite. See ISODULCITE.

Rhaponticin (Rheic Acid, Rheumin, Rhubarbaric Acid, Rhubarbarin). Chrysophanic acid, $C_{15}H_{10}O_4$, found in rhubarb root.

Rhaponticum (Rhapontin, Ponticin). The crystalline substance from the common English rhubarb.

Rhapontin. See RHAPONTICUM.

Rhatany. The dried root of *Krameria triandra* and *K. argentea*.

Rhea. See RAMIE.

Rheic Acid. See RHAPONTICIN.

Rhemattan. See ATROPHAN.

Rhenania Phosphate (Vesta Phosphate). Prepared by sintering together in a furnace at 1200–1300° C., a mixture of raw phosphate, limestone, and alkali silicate. The resulting product approximates to the formula $Ca_2KNa(PO_4)_2$. The German phosphate contains 23–31 per cent. soluble phosphoric acid and 40 per cent. lime, and is made by treating raw phosphate with soda at high temperatures.

Rhenanit V. A German explosive containing nitro-glycerin, ammonium nitrate, vegetable meal, nitro-compounds, and potassium perchlorate.

Rhenish Dynamite. A solution of 75 per cent. nitro-glycerin in naphthalene, 2 per cent. chalk or barium sulphate, and 23 per cent. kieselguhr.

Rheomacrodex. A proprietary preparation of dextran in saline or dextrose solution, used as a blood volume expander in shock and hæmorrhage. (337).

Rheonine. A dyestuff. It is the hydrochloride of tetramethyl - triamino - phenyl-acridine, $C_{23}H_{24}N_4$. Dyes tannined cotton and leather brownish-yellow.

Rheostene. A nickel-iron alloy. It has a specific resistance of 77 microhms per cm.[3] at 0° C.

Rheotan I. An electrical resistance alloy containing 84 per cent. copper, 12 per cent. manganese, and 4 per cent. zinc.

Rheotan II. An alloy for electrical resistances. It contains 25 per cent. nickel, 52 per cent. copper, 5 per cent. iron, and 18 per cent. zinc. This alloy or a similar one has been called Rheostan.

Rhesal. Glyceryl monosalicylate.

Rheukomen. A German preparation. It consists of the ester of iodo-salicylic acid with a small addition of formic acid. It is used in the form of ointment for neuralgia and rheumatism.

Rheumajecta. A proprietary preparation of sulphurylsulphokinase, cholin-acetylase and catalase, used in the treatment of rheumatic diseases. (907).

Rheumasan. A superfatted cream soap to which 10 per cent. of salicylic acid has been added. Recommended for rubbing in cases of rheumatism.

Rheuma-sensit. An analgesic containing 40 per cent. of soap, 15 per cent. petroleum jelly, 5 per cent. lanoline, 10 per cent. free fatty acids, 10 per cent. salicylic acid as potassium salicylate, 5 per cent. essential oils of camphor and menthol, and 15 per cent. water.

Rheumasol. A dark brown liquid consisting of 10 parts salicylic acid, 10 parts ichthyol, or similar compound, and 80 parts salicylic ointment. Used as a paint in cases of skin disease and rheumatism.

Rheumatine. Salicyl-quinine-salicylate, $C_6H_4(OH)CO_2 . C_{20}H_{23}N_2O . C_6H_4(OH)COOH$. Recommended as a substitute for salicylic acid in rheumatism and neuralgia.

Rheumin. See RHAPONTICIN.

Rhexite. Rexite (*q.v.*).

Rhiamer PC 7066. A clear polycarbonate sheet material for vandal-resistant lighting fittings. (15).

Rhigolene. See CYMOGEN.

Rhinamid. A proprietary preparation of EPHEDRINE hydrochloride, sulphanilamide and butacaine sulphate, in the form of decongestant nasal drops. (802).

Rhine Metal. An alloy of 97 per cent. tin and 3 per cent. copper. It has a specific gravity of 7·35 and melts at 300° C.

Rhinestone. A highly refractive glass cut as a gem stone.

Rhodaform. An addition product of hexamethylene-tetramine and methyl-thio-cyanate, $C_6H_{12}N_4.CH_3.CNS$. Used in medicine.

Rhodallin. Thiosinamine (*q.v.*).

Rhodamine B (Rhodamine O, Safraniline). A dyestuff. It is the hydrochloride of diethyl-*m*-amino-phenol-phthalein, $C_{28}H_{31}N_2O_3Cl$. Dyes wool and silk bluish-red with fluorescence, also tanned cotton, violet-red.

Rhodamine 3B (Anisoline). A dyestuff. It is the ethyl ester of tetraethyl-rhodamine, $C_{30}H_{35}N_2O_3Cl$. Dyes wool, silk, and mordanted cotton, bluish-red.

Rhodamine G and G Extra. Dyestuffs, consisting chiefly of triethyl-rhodamine, $C_{26}H_{27}N_2O_3Cl$. Dyes wool, silk, and tanned cotton, red.

Rhodamine 6G (Trianisoline). A dyestuff. It is the ethyl ester of sym-diethyl-rhodamine, $C_{26}H_{27}N_2O_3Cl$. Dyes silk and mordanted cotton red or pink shades.

Rhodamine 12GM. A dyestuff. It is the ethyl ether of dimethyl-amino-ethoxy-rhodamine, $C_{25}H_{24}NO_4Cl$. Dyes silk and tanned cotton yellowish-red.

Rhodamine O. See RHODAMINE B.

Rhodamine S. A dyestuff. It is the

hydrochloride of dimethyl-*m*-amino-phenol-succineine, $C_{20}H_{23}N_2O_3Cl$. Dyes cotton red, and is used for dyeing half-silk goods, and for colouring paper pulp and wool.

Rhodanate. Potassium thiocyanate. Potassium sulphocyanate.

Rhodapurin. A German preparation consisting of rhodan and caffeine.

Rhodarsan. Neosalvarsan (*q.v.*).

Rhodazil. A proprietary preparation. It is stated to be benzyl benzoate, and to be used as an antispasmodic.

Rhodazines. Dyestuffs prepared by the action of phenyl-hydrazine upon rosilic acid. They have no technical importance.

Rhodeoretin. Jalapin, the chief constituent of Jalap resin.

Rhoderite. A mineral. It is a silicate of manganese, $MnSiO_3$.

Rhodester. A proprietary trade name for polyester resins. (88).

Rhodialite. A proprietary cellulose acetate.

Rhodiaseta. A proprietary artificial silk.

Rhodine. A proprietary preparation of acetyl-salicylic acid.

Rhodine 2G. A dyestuff. It is the ethyl ester of dimethyl-ethyl-rhodamine, $C_{26}H_{27}N_2O_3Cl$. Dyes silk, wool, and tanned cotton, red.

Rhodine 3G. See IRISAMINE G.

Rhodine 12GF. A dyestuff, $C_{49}H_{38}N_4O_8Cl_2$, obtained by the action of formaldehyde upon the etherified condensation product of dimethyl-amino-benzoyl-benzoic acid and resorcin. Dyes tanned cotton and silk yellowish-red, and is used for printing on cotton and silk.

Rhodinol. A terpene alcohol prepared from the oils of rose, geranium, and citronella. It is practically pure geraniol.

Rhodione. See VIOLETTON.

Rhodite. See RHODIUM GOLD.

Rhodium Gold (Rhodite). A native alloy of from 57–66 per cent. gold and 34–43 per cent. rhodium.

Rhodizite. A borate of lime, $3CaO.4B_2O_3$, imported from the West Coast of Africa.

Rhodochrome. See CHROMOCHLORITE.

Rhodocrosite (Spathic Manganese Ore, Manganese Spar, Diallogite, Red Manganese). A mineral. It is manganese carbonate, $MnCO_3$.

Rhodoform. Rhodaform (*q.v.*).

Rhodoid. A proprietary preparation. It is a plastic material made from cellulose acetate.

Rhodole. Intermediate products between fluoresceine-phthalein and the rhodamines.

Rhodolite. A mineral. It is a pink gemstone, and is intermediate in composition between pyrope and almandine.

Rhodonite. A mineral. It is manganese silicate, $MnSiO_3$. See RHODERITE.

Rhodopas. A proprietary trade name for polyvinyl acetates. (88).

Rhodopas AX. A proprietary trade name for polyvinyl acetate/polyvinyl chloride copolymers. (88).

Rhodopas X. A proprietary trade name for polyvinyl chlorides. (88).

Rhodophosphite. A mineral. It contains calcium, manganese, and iron phosphate, chloride, and sulphate.

Rhodoviol. A proprietary polyvinyl alcohol. (88).

Rhoduline Reds G and B (Rhoduline Violets, Brilliant Rhoduline Red). Alkylated safranines, which dye in a similar way to safranine.

Rhoduline Violets. See RHODULINE REDS G and B.

Rhodusite. A mineral. It is a variety of glaucophane.

Rhoetizite. A mineral. See KYANITE.

Rhoféine. A proprietary preparation of acetyl-salicylic acid and caffeine.

Rhomboclase. A mineral. It is $HFe^{\cdot\cdot}$ $(SO_4)_2.4H_2O$.

Rhomb-spar. A mineral. It is a variety of calcite.

Rhometal. A complex nickel-iron alloy having high electrical resistivity and retaining its permeability up to very high frequencies. It is used for television transformers, special radio transformers, H.F. alternators, etc.

Rhomnogyre. A similar preparation to Mercurol (q.v.).

Rhomnol. A trade name for nucleinic acid.

Rhönite. A mineral. It is a variety of amphibole. Also a proprietary trade name for a urea resin.

Rhoplex. A proprietary trade name for an acrylic resin for textile finishes.

Rhotanium. A series of alloys, consisting mainly of gold (60–90 per cent.), and palladium, and in some cases with a small proportion of rhodium. They are said to be more resistant to hot concentrated sulphuric acid and fused caustic soda than lead.

Rhovinal B. A proprietary trade name for polyvinyl butyrals. (88).

Rhovinal F. A proprietary trade name for polyvinyl formals. (88).

Rhodoviol. A proprietary trade name for polyvinyl alcohols. (88).

Rhubarb. The dried rhizome of *Rheum palmatum*.

Rhubarbaric Acid. See RHAPONTICIN.

Rhubarbarin. See RHAPONTICIN.

Rhyolite. A volcanic rock. It usually contains lime and iron.

Riblene. A trade name for polyethylene. (20).

Riboflavin. Vitamin B_2. Lactoflavin.

Ribostamycin. An antibiotic. O-2,6-Diamino - 2, 6 - dideoxy - α - D - gluco - pyranosyl - (1→4,) - O - [β - D - ribo - furanosyl - (1→5)] - 2 - deoxy - D - streptamine.

Rice Paper. A paper made from plant pith, particularly in China and Japan.

Rice Rubber. An elastic cellulose product made from Japanese rice.

Rice's Bromide Solution. A solution containing 125 parts bromine, 125 parts sodium bromide, and 1,000 parts water. Used in the determination of urea.

Richardite. See EPSOMITE.

Richard's Aluminium Solder. An alloy of 71·5 per cent. tin, 25 per cent. zinc, and 3·5 per cent. aluminium.

Richardson's Speculum Metal. An alloy of 65·3 per cent. copper, 30 per cent. tin, 2 per cent. silicon, 2 per cent. arsenic, and 0·7 per cent. zinc.

Riché Gas. A gas obtained during the dry distillation of wood. It contains, on an average, about 60 per cent. carbon dioxide, 25 per cent. carbon monoxide, 15 per cent. methane, and a very small quantity of hydrogen.

Richellite. A mineral, $(Fe.Ca)_2(F_6(PO_4)_2$.

Richmondite. A mineral. It has a lead sulphantimonate. The name is also applied to a mineral related to gibbsite.

Richterite (Isabellite). A mineral, $(K_2.Na_2.Mg.Ca.Mn)SiO_3$.

Ricin. The name given to the toxic constituents of castor beans.

Ricinose. Castor oil.

Ricinus Oil. See OIL OF PALMA CHRISTI.

Rickardite (Sanfordite). A mineral. It is copper telluride, Cu_4Te_3.

Ricolite. A mineral. It is a green serpentine.

Ricolite. See BAKELITE.

Ricons. A proprietary range of polybutadiene homopolymers and copolymers of butadiene/styrene. (908).

Ricotti Silk. See GALETTAME SILK.

Ridzol. A proprietary preparation of RONIDAZOLE. A veterinary anti-protozoal.

Riebeckerite. A mineral, $Na_2O.Fe_2O_3$. $4SiO_2.4FeO.4SiO_2$ Also see CROCIDOLITE.

Riegel's Milk Albumen. A casein preparation, containing 86 per cent. albumin, 8·2 per cent. water, 0·31 per cent. fat, and 5·3 per cent. mineral matter.

Riemannite. See ALLOPHANE.

Riems. A South African leather, made by cutting a strip of raw hide and suspending it with a heavy weight on one end, after coating it with oil. It is afterwards repeatedly oiled and twisted.

Rifadin. A proprietary preparation of RIFAMPICIN used as an antibiotic in the treatment of tuberculosis. (909).

Rifamide. An antibiotic. RIFAMYCIN B diethylamide. RIFOCIN-M.

Rifampicin. An antibiotic. 3-(4-Methyl-piperazin - 1 - yliminomethyl)rifamycin SV. RIFADIN, RIMACTANE.

Rifamycin. Antibiotics isolated from a Streptomyces mediterranei. (Specific substances are designated by a terminal letter, e.g., Rifamycin B).

Rifinah. A proprietary preparation of RIFAMPICIN and ISONIAZID used in the treatment of tuberculosis. (909).

Rifleite. Explosive. It is a nitrocellulose gelatinised by acetone.

Rifocin-M. See RIFAMIDE.

Rifomycin. An alternative spelling of RIFAMYCIN.

Riga Balsam. Consists of 1 part each of oils of lavender, cloves, cinnamon, thyme, mace, and lemon; 4 parts balsam of Peru, ½ part oil of sage, 2½ parts tincture of saffron, and 250 parts 90 per cent. alcohol. It is a medicine of Riga, and is recommended as a pick-me-up and cold cure.

Rigidex 3. A registered trade mark for high density polyethylene copolymer having a density of 0·946, melt index 0·3. (288).

Rigidex X4RR. A registered trade mark for a high density polyethylene having a density of 0·946. (288).

Rigidex 9. A registered trade mark for a high density polyethylene with a density of 0·960 and a melt index of 0·9. (288).

Rikospray Antibiotic. A proprietary preparation of neomycin sulphate, bacitracin zinc and polymixin B sulphate. An antibiotic dermatological aerosol. (275).

Rikospray Balsam. A proprietary preparation of benzoin and prepared storax. A dermatological aerosol. (275).

Rikospray Silicone. A proprietary preparation of aluminium dihydroxyallantoinate, CETYLPYRIDINIUM chloride and silicone used in the treatment of skin soreness. (275).

Rilsan 11 and 12. Proprietary names for Nylon 11 and 12. (910).

Rilan Wax. Waxy acids and vegetable oils hardened by hydrogenation. It is obtained from the higher alcohols of the fat series.

Rilata. Vulcanised bitumen linseed oil mixture containing 16 per cent. sulphur.

Rimactane. A proprietary preparation of rifampicin. An antibiotic and anti-tuberculous agent. (18).

Rimactazid. A proprietary preparation of RIFAMPICIN and ISONIAZID used in the treatment of tuberculosis. (18).

Rimifon. A proprietary trade name for Isoniazid.

Rimiterol. A bronchodilator. It is erythro - 3, 4 - dihydroxy - α - (2 - piper - idyl)benzyl alcohol.

Rimpylite. A mineral. It is a variety of amphibole.

Rinex. A registered trade mark applied to a remedy for respiratory diseases.

Ringer Solution. An isotonic solution containing 0·7 per cent. sodium chloride, 0·03 per cent. potassium chloride, and 0·025 per cent. calcium chloride in water. Used in physiological experiments.

Ringworm Powder. See GOA POWDER.

Rinkite. A mineral, $3NaF.4CaO.6(Ti.Si)O_2.Ce_2O_3$.

Rinmann's Green. See COBALT GREEN.

Rinneite. A mineral. It is an anhydrous chloride of potassium, sodium, and ferrous iron, $FeCl_2.3KCl.NaCl$.

Rinurel. A proprietary preparation of paracetamol, phenacetin, phenylpropanolamine hydrochloride and phenyltoloxamine dihydrogen citrate. (262).

Rio Arrowroot. Tapioca.

Riodene. A 66 per cent. solution in oil of the iodised glycerin ester of ricinoleic acid.

Riolite. A mineral. It is a variety of onofrite.

Rionite. A bismuth fahlore.

Riopan. Registered trade mark for a brand of monalium hydrate antacid tablets taken orally to give relief in cases of gastric hyperacidity. (226). (467).

Rio Resin. A proprietary amorphous resin. (911).

Ripodolite. A mineral, $(Al.Cr)_2O_3.5(Mg.Fe)O.3SiO_2$.

Ripping Ammonal. An explosive. It contains 84–87 per cent. ammonium nitrate, 7–9 per cent. aluminium, 2–3 per cent. charcoal, and 3–4 per cent. potassium bichromate.

Rippite. An English explosive. It contains 56–63 per cent. nitro-glycerin gelatinised with a small quantity of collodion cotton, potassium nitrate, wood meal, castor oil, ammonium oxalate, with the addition of calcium or magnesium carbonate, and a pet. jelly.

Risigallum. Synonym for Realgar.

Riso. A material consisting of ammonium carbonate ground in a mixture of mineral and vegetable oils (Cycline oil). Used in the manufacture of sponge rubber.

Risor. See PINE OILS.

Risörite. A mineral. It is a yttria niobate, of Norway.

Risseite. A mineral. It is aurichalcite.

Rissicol. A castor oil powder, containing 49 per cent. castor oil and 36 per cent. of inorganic matter, mainly magnesia.

Ristin. A 25 per cent. solution of ethylene-glycol monobenzol ester.

Ristocetin. Anti-microbial substances produced by Nocardia lurida. Spontin contains Ristocetin A and Ristocetin B.

Risunal. A proprietary preparation of β-diethylaminobutyric acid, aniline hydrochloride, isopropylphenazone, and ethyl and benzyl nicotinate. A rubefacient skin ointment. (386).

Ritalin. A proprietary trade name for the hydrochloride of Methyl Phenidate.

Ritex. A proprietary trade name for magnesite brick for use as a refractory.

Ritha. An alkaline deposit found on the land in India. Used as a soap substitute. Also see SAJJI.

Ritodrine. A relaxant for the uterus. It is *erythro*-1-(4-hydroxyphenyl)-2-(4-hydroxyphenethylamino)propan-1-ol.

Rittingerite. A mineral, $Ag_3\cdot As(Se.S)_4$.

Rivaite. A mineral. It is a calcium-sodium silicate.

Rivalit P. A German explosive containing nitro-glycerin, ammonium nitrate, vegetable meal, nitro-compounds, and potassium perchlorate.

Rivanol (Vucine). 2-Ethoxy-6 : 9-diamino-acridine-hydrochloride. A powerful antiseptic.

Riversideite. A mineral. It is a hydrated calcium metasilicate. It is $Ca_5Si_6O_{16}$ $(OH)_2\cdot 2H_2O$.

Riviera Oil. See AIX OIL.

Rivet Metal. An alloy of copper and tin, to which zinc is sometimes added.

Rivoren. A proprietary preparation of ammonium heptinchlorarsenate and hexamethyl - diamino - isopropanol - di-iodide.

Rivotite. A mineral, $CuCO_3.Sb_2O_4$.

Rivotril. A proprietary preparation of clonazepam. An anti-convulsant. (314).

Rivtex. A proprietary trade name for a cotton fibre, wood, and phenolic-resin plastic.

Rizopatronite. A mineral. It is patronite.

Roachban. A cockroach insecticide. (501).

Road Oil. A heavy residue from petroleum oil used on roads.

Roanoid. A proprietary urea-formaldehyde moulding compound.

Roaster Slag. A slag produced in the purification of copper metal. It contains from 17–40 per cent. copper as silicate and metal.

Ro-a-vit. A proprietary preparation of vitamin A. (314).

Robac. Trade mark for a range of proprietary accelerators (554), of which the following are constituent members:—

Robac 22. Ethylenethiourea.

Robac 44. *p-p*-diaminodiphenyl methane. (554).

Robac Alpha. A rubber accelerator.

Robac Cu.D.D. Copper dimethyl dithiocarbamate.

Robac C.P.D. Cadmium pentamethylane dithiocarbamate.

Robac D.B.U.D. Dibutylammonium dibutyl dithiocarbamate.

Robac D.B.V.D. An anti-oxidant. Dibutylammonia dibutyl dithiocarbamate.

Robac D.E.T.U. N-N'-diethyl thiourea.

Robac D.F.T.U. A substituted urea.

Robac L.M.D. Lead dimethyl dithiocarbamate.

Robac L.P.D. Lead pentamethylene dithiocarbamate.

Robac M.Z.1. Zinc mercaptobenzthiazole.

Robac M.Z.2. A blend of 91% zinc mercaptobenzthiazole and 9% di-*ortho*-tolyl guanidine.

Robac P.P.D. Piperidinium pentamethylene dithiocarbamate.

Robac P.T.D. Dipentamethylene thiuram disulphide.

Robac P.T.M. Dipentamethylene thiuram monosulphide. Used in the rapid vulcanisation of rubber.

Robac S.P.D. Sodium pentamethylene dithiocarbamate.

Robac T.B.U.T. Tetrabutyl thiuram disulphide.

Robac T.E.T. Tetraethyl thiuram disulphide.

Robac Thiuram P25. Dipentamethylene thiuram tetrasulphide.

Robac T.M.S. Tetramethyl thiuram monosulphide.

Robac T.M.T. Tetramethyl thiuram disulphide.

Robac Z.B.E.D. Zinc dibenzyl dithiocarbamate.

Robac Z.B.U.D. Zinc dibutyl dithiocarbamate. (554).

Robac Z.B.U.D. Extra. A zinc dibutyl dithiocarbamate-dibutylamine complex.

Robac Z.D.C. Zinc diethyl dithiocarbamate.

Robac Z.I. Zinc dimethyl pentamethylene dithiocarbamate.

Robac Z.I.X. Zinc isopropyl xanthate.

Robac Z.M.D. Zinc dimethyl dithiocarbamate.

Robac Z.P.D. Zinc pentamethylene dithiocarbamate.

Robac Z.P.D. Extra. A zinc pentamethylene dithiocarbamate-piperidine complex.

Robac T.B.Z. A trade name for zinc thiobenzoate. A low temperature peptising agent for natural rubber.

Robalate. A proprietary preparation containing dihydroxy aluminium aminoacetate. An antacid. (258).

Robaxin. A proprietary preparation of methocarbamol, a muscle relaxant. (258).

Robellazite. A mineral. It contains aluminium, iron, manganese, tungsten, vanadium, niobium, and tantalum.

Robenidine. An anti-protozoan. 1,3-Bis(4 - chlorobenzylideneamino)guanidine. CYCOSTAT is the hydrochloride.

Robertson Alloy. Consists of 1 part gold, 3 parts silver, and 2 parts tin. It is mixed with mercury for use as a dental filler.

Robert's Reagent. For proteids. It consists of 1 volume of pure nitric acid with 5 volumes of a 40 per cent. (saturated) solution of magnesium sulphate.

Robinsonite. A mineral. It is $Pb_7Sb_{12}S_{25}$.

Robinul. A proprietary trade name for Glycopyrronium bromide.

Robitussin. A proprietary preparation containing guaiphenesin. A cough linctus. (258).

Robitussin A.C. A proprietary preparation containing codeine phosphate, guaiphenesin and pheniramine maleate. A cough linctus. (258).

Robol. An amylolytic and proteolytic ferment.

Roborate. A foodstuff prepared from corn. It consists of pure vegetable proteid.

Roborin. A preparation made from ox or calf blood. It is an amylolytic and proteolytic ferment.

Robur. An Italian motor fuel containing 30 per cent. ethyl alcohol and 22 per cent. methyl alcohol in petrol.

Roburine. A French impregnated asbestos board containing mica.

Roburite. An explosive used in mines. It consists of 86 per cent. ammonium nitrate and 14 per cent. chloro-dinitrobenzene.

Roburite I. An explosive containing 87·5 per cent. ammonium nitrate, 7 per cent. dinitro-benzene, 0·5 per cent. potassium permanganate, and 5 per cent. ammonium sulphate.

Roburite III. An explosive containing 87 per cent. ammonium nitrate, 11 per cent. dinitro-benzene, and 2 per cent. chloro-naphthalene.

Roccal. A proprietary preparation of benzalkonium chloride. A pre-operative skin cleanser. (112).

Roccelline. See FAST RED.

Rochdale Salt. See ROCHELLE SALT.

Roche Alum (Rock Alum). An impure native variety of alum containing iron. A mixture of common alum and ferric oxide is sold under this name.

Rochelle Salt (Rochdale Salt, Tartrated Soda, Seignette's Salt). Sodium-potassium-tartrate, $C_4H_4O_6Na.K.4H_2O$. Used to reduce the silver salts in the silvering of mirrors, and in medicine as a mild aperient. It is the active constituent of seidlitz powders.

Rock Alum. See ROCHE ALUM.

Rock Ammonia. Ammonium carbonate $(NH_4)_2CO_3$.

Rock Asphalt. Limestone or other material and found naturally, impregnated with bitumen.

Rockbridgeite. A mineral. It is $(Fe ,Mn)Fe_4'''(PO_4)_3(OH)_5$.

Rock Cork. A variety of asbestos. Also see ROCK WOOL.

Rock Crystal. Transparent and colourless quartz.

Rock Dammar. A variety of dammar resin derived from *Hopea odorata*, of Burma.

Rockite. A trade mark for a series of phenol formaldehyde synthetic resin moulding compounds. (12).

Rock Kieserite. A mixture of carnallite and kieserite.

Rock Meal. A mineral. It is calcium carbonate, $CaCO_3$.

Rock Milk. Finely ground rock meal (*q.v.*).

Rock Oil. See EARTH OIL.

Rock Phosphate. See PHOSPHORITE.

Rock Salt (Halite). Sodium chloride, NaCl, from sea water.

Rock Scarlet BS. See ST. DENIS RED.

Rock Scarlet YS. A dyestuff. It is the sodium salt of azoxy-toluene-disazo-β-naphthol-α - naphthol - mono - sulphonic acid, $C_{34}H_{25}N_6O_6SNa$. Dyes wool scarlet from an acid bath.

Rocksil. A proprietary trade name for rockwool insulating materials. It withstands 760° C. (138).

Rock Tallow. See BITUMEN.

Rocktex. A proprietary trade name for rock wool.

Rock Wool (Rock Cork). A furnace product made from self-fluxing siliceous and argillaceous dolomite in which the basic and acidic constituents are present in proportions that their fluxing action is nearly balanced. The molten rock at 2800–3000° F. is atomised by a blast of steam under pressure. The wool treated with a binder is sold as rock cork. A heat insulator and corrosion-resisting packing.

Rocol P.R. A proprietary silicone-based spray used for mould release. (912).

Rocol R.S.7. A proprietary non-silicone-based wet film spray used for mould release. (912).

Rocou. See ANNATTO.

Rocsol. Industrial waxes. (521).

Rodagen. A dried antithyroid preparation.

Rodinal. A photographic developer. The active constituent is *p*-amino-phenol hydrochloride.

Rod Wax. A wax-like mass deposited on the drill rods in many petroleum oil wells.

Rœbaryt. A barium sulphate prepared for use in X-ray work.

Roeblingite. A mineral, $5CaH_2SiO_4$. $2(Pb.Ca)SO_4$.

Roemerite. A mineral, $Fe_3(SO_4)_4.12H_2O$.

Roepperite. A mineral. It is $4[(Fe,Mn,Zn)_2SiO_4]$.

Roesch's Aluminium Solder. An alloy of 50·2 per cent. zinc, 50 per cent. tin, 0·7 per cent. antimony, and 0·2 per cent. copper.

Roesslerite. A mineral, $MgHAsO_4.\frac{1}{2}H_2O$.

Rogersite. A mineral. It is an yttrium niobate, of Carolina.

Roghan (Afridi Wax). Obtained by boiling safflower oil for 2 hours, then putting it into vessels partly filled with water.

Rogitine. A proprietary preparation of phentolamine mesylate. (18).

Rohagit SH 150 and SHV 5 per cent. A proprietary trade name for 70 parts methyl acrylic acid and 30 parts methyl acrylate copolymerized. (13).

Rohagum 5N, 6N, 7H, 7N, 8H. A range of proprietary acrylic polymers. (13).

Rohn Alloys. Heat-resisting alloys containing nickel, chromium, and iron, sometimes with the addition of manganese and molybdenum. Also see CHROMAN.

Rohrbach's Solution. A solution of barium and mercuric iodides (100 grams barium iodide and 130 grams mercuric iodide heated with 20 c.c. water to 150–200° C.). The solution is allowed to cool, when a double salt is deposited. The liquid is decanted. The salt is used in the separation of the heavy metals by gravity as it has a specific gravity of 3·58.

Rolicton. Amisometradine.

Rolicypram. An anti-depressant. (+)-5 - Oxo - *N* - (*trans* - 2 - phenylcyclo - propyl) - L - pyrrolidine - 2 - carbox - amide.

Rolitetracycline. N-(1-Pyrrolidinyl-methyl)tetracycline. Tetrex PMT.

Rolled Gold Plate. This usually consists of base metal partially or completely covered with a layer of gold alloy considerably thicker than is usually applied by plating. Articles of jewellery are often made from stock rolled or drawn from billets consisting of base metal alloy and gold alloy which have been soldered or brazed together.

Roll Sulphur. Sulphur which has been melted and poured into moulds.

Rol-man Steel. A proprietary trade name for a high-manganese steel containing 11–14 per cent. manganese and 1–1·4 per cent. carbon.

Roman Alum. Potash alum, $Al_2(SO_4)_3$. $K_2SO_4.24H_2O$, prepared at Tolfa, near Rome, from Alumite, $(SO_4)_2Al.K.Al (OH)_3$.

Roman Bronze. An alloy of 90 per cent. copper and 10 per cent. tin.

Roman Cement (Parker's Cement). A natural cement made by calcining the nodules of argillaceous limestone mixed with calcareous spar, which occurs in London and other clays.

Roman Chamomile. The flower-heads of *Anthemis nobilis*.

Romanechite. A mineral. It is a variety of psilomelane.

Romanite. See ROUMANITE.

Romanium. An alloy of 97·43 per cent. aluminium, 1·75 per cent. nickel, 0·25 per cent. copper, 0·25 per cent. antimony, 0·17 per cent. tungsten, and 0·15 per cent. tin.

Roman Ochre. See YELLOW OCHRE.

Roman Ointment. A mixture of extract of opium, extract of belladonna, glycerin, and resin ointment.

Romanowsky's Stain. A microscopic stain. It consists of (A) methylene blue 2 grams, distilled water, 200 c.c. N/10 caustic potash 10 c.c. This is boiled, cooled, and 10 c.c. of N/10 sulphuric acid added. (B) Eosin 1 gram, distilled water 1,000 c.c. For use mix 1 c.c. of (A) with 6 c.c. of (B).

Roman Sepia. A brown pigment. It consists of sepia mixed with yellow-browns.

Roman Vitriol. See BLUE COPPERAS.

Roman Yellow. See YELLOW OCHRE.

Romanzovite. A mineral. It is a variety of grossularite.

Romcite. A mineral, $CaO.Sb_2O_5$.

Romeite. A mineral, $CaSb_2O_4$.

Romicil. A proprietary trade name for Oleandomycin.

Romilar. A proprietary trade name for the hydrochloride of Dextromethorphan.

Romite. An explosive. It is ammonium nitrate mixed with a solid, melted hydrocarbon (paraffin or naphthalene), gelatinised with a liquid hydrocarbon (paraffin oil), and contains gelatinised potassium chlorate.

Romotal. A proprietary trade name for the hydrochloride of Tacrine.

Rompel's Alloy. An antifriction metal, containing 62 per cent. copper, 10 per cent. zinc, 10 per cent. tin, and 18 per cent. lead.

Romperit G. A German explosive containing nitro-glycerin, ammonium nitrate, vegetable meal, nitro-compounds, and potassium perchlorate.

Rompun. A proprietary preparation of XYLAZINE used as a veterinary analgesic.

Rondase. A proprietary preparation of hyaluronidase, used to facilitate absorption of injected fluids. (182).

Rondomycin. A proprietary preparation containing Methacycline. An antibiotic. (85).

Ronfusil Steel. A proprietary trade name for a manganese-steel containing 12 per cent. manganese.

Rongalite. See HYDROSULPHITE.

Rongalite C. A combination of the sodium salt of the unstable sulphoxylic acid and formaldehyde. A reducing agent used in the dye industry, and as a photographic developer. See DISCOLITE.

Rongalite, Concentrated. See HYDROSULPHITE NF.

Ronia Metal. A brass containing small quantities of cobalt, manganese, and phosphorus.

Ronicol. A proprietary trade name for the tartrate of Nicotinyl alcohol.

Ronicol Timespan. A proprietary preparation of NICOTINYL tartrate used in the treatment of circulatory disorders. (314).

Ronidazole. An anti-protozoal. (1-Methyl - 5 - nitroimidazol - 2 - yl) methyl carbamate. RIDZOL.

Ronilla. A proprietary trade name for polystyrene. (3).

Ronnel. A proprietary preparation of FENCHLORPHOS. An insecticide.

Röntgenite (Roentgenite). A mineral. It is $9[Ca(Ce,La)_3(CO_3)_5F_3]$.

Rontyl. A proprietary trade name for Hydroflumethiazide.

Ronyl. A proprietary trade name for Pemoline.

Ronozol Salts. See SOZOIODOL.

Rooseveltite. A mineral. It is $4[BiAsO_4]$.

Rosamine. A dyestuff obtained from benzo - trichloride and dimethyl - *m* - amino-phenol, $C_{23}H_{23}N_2OCl$.

Rosaniline (Magenta Base, Rosaniline Base, O, SF, Pararosaniline Base, Oil Red Base, Brilliant Oil Crimson). Triamino-tolyl-diphenyl-carbinol.

Rosaniline Base, O, SF. Equivalents of Rosaniline.

Rosaniline Blue. See ANILINE BLUE, SPIRIT SOLUBLE.

Rosanthrenes O, R, A, B, CB. Tetrazo dyestuffs, which produce shades similar to Turkey Red. They are prepared by coupling diazo compounds with *m*-

amino-benzoyl-amino-naphthol-sulphonic acids, and dye from a bath of sodium sulphate, sodium carbonate, and soap.

Rosarin. See AZOCARMINE G.

Rosasite. A mineral. It is a basic zinc-copper carbonate, $(Zn.Cu)CO_3.Cu(OH)_2$. It is also given as $4[(Cu,Zn)_2 CO_3(OH)_2]$.

Rosaurine. Rosilic acid, $C_{21}H_{16}O_3$.

Rosazine. See AZOCARMINE G.

Rosazurine B. A dyestuff. It is the sodium salt of ditolyl-disazo-bi-ethyl-β-naphthylamine-sulphonic acid, $C_{33}H_{34}N_6O_6S_2Na_2$. Dyes cotton bluish-red from an alkaline bath.

Rosazurine G. A dyestuff. It is the sodium salt of ditolyl-disazo-ethyl-β-naphthylamine - sulphonic - β - naphthylamine-sulphonic acid, $C_{36}H_{30}N_6O_6S_2Na_2$. Dyes cotton bluish-red from an alkaline bath.

Roscherite. A mineral. It is an aluminium - calcium - manganese - iron hydrated basic phosphate.

Roscoelite (Vanadium Mica). A mineral. It is a muscovite mica, in which part of the aluminium is replaced by vanadium, and is practically an aluminium vanadate, with potassium silicate. It contains up to 24 per cent. of vanadium pentoxide.

Rose B. See ERYTHROSIN.

Rose Bengal (Rose Bengal N, Rose Bengal AT, Rose Bengal G, Bengal Red). A dyestuff. It is the alkaline salt of tetra - iodo - dichloro - fluoresceine, $C_{20}H_4Cl_2I_4O_5K_2$. Dyes wool bluish-red. Also see ROSE BENGAL 3B.

Rose Bengal AT. See ROSE BENGAL.

Rose Bengal B. See ROSE BENGAL 3B.

Rose Bengal 3B (Rose Bengal B, Rose Bengal). A dyestuff. It is the potassium salt of tetra-iodo-tetrachloro-fluoresceine, $C_{20}H_2Cl_4I_4O_5K_2$. Dyes wool bluish-red.

Rose Bengal G. See ROSE BENGAL.

Rose Bengal N. See ROSE BENGAL.

Rose, Cobalt. See COBALT RED.

Rose de Benzoyl. See BENZOYL PINK.

Rose Ester. A trade name for trichloromethyl phenyl carbinyl acetate. M.Pt. 85–87° C. A rose scent used in the perfumery industry. (197).

Rose-Geranium Oil. An oil obtained by distilling geraniol over rose flowers. Also see ROSÉ OIL.

Rosein. An alloy of 44·4 per cent. nickel, 33·3 per cent. aluminium, 11·1 per cent. silver, and 11·1 per cent. tin. Another alloy contains 40 per cent. nickel, 30 per cent. aluminium, 20 per cent. tin, and 10 per cent. silver. Used by jewellers.

Roseine. See MAGENTA and MAUVEINE.

Roseine Crystals SF. A dyestuff. It is a British equivalent of New magenta.

Rose JB. See SPIRIT EOSIN.

Rose JB, Alcohol Soluble. See SPIRIT EOSIN.

Roselite. A mineral, $3(Ca.Co.Mg)O.As_2O_5.2H_2O$.

Rosellane (Rosite). A mineral. It is decomposed anorthite.

Roselle Fibre. A Malay fibre similar to jute.

Rose Madder. See MADDER LAKES.

Rosenbuschite. A mineral. It is a titano-silicate of calcium, zirconium, sodium, cerium, yttrium, iron, and manganese.

Rosendale Cement. A cement made from argillaceous magnesium limestone, found along the Appalachian range.

Rosenite. A mineral. It is another name for plagionite.

Rosenöl. A synonym for rose otto.

Rosenthiel's Green (Baryta Green, Cassel Green, Manganese Green). A pigment. It is a manganate of barium, obtained by heating a mixture of oxides of manganese, barium nitrate, and heavy spar or kaolin.

Rosé Oil (Roshé Oil, Oil of Geranium, Oil of Rose-geranium, Oil of Pelargonium, Ginger Grass Oil, Turkish Geranium Oil, Oil of Palmarosa). Andropogon oils, obtained from a grass, *Andropogon nardus*. They contain geraniol, $C_{10}H_{18}O$.

Rose Oil, Artificial. An oil made from citronellol, $C_{10}H_{20}O$, and from geraniol, $C_{10}H_{18}O$.

Roseol (Reuniol). Names applied to citronellol, $C_9H_{17}.CH_2OH$, or to mixtures of this body with geraniol. Perfumes.

Rose Petals. Petals of the pale or cabbage rose, which are obtained from *Rosa centifolia*.

Rose Pink. A pigment prepared by dyeing whiting with Brazil wood. Used in paper-staining.

Rose Quartz. A mineral. It is a variety of quartz, SiO_2, which is stated to owe its colour to manganese.

Rose's Metal. Fusible bismuth alloys. (1) Consists of 42 per cent. bismuth, 42 per cent. lead, and 16 per cent. tin. It melts at 79° C. (2) Consists of 33·3 per cent. bismuth, 33·3 per cent. lead, and 33·3 per cent. tin. It melts at 93° C. (3) Consists of 48·9 per cent. bismuth, 27·5 per cent. lead, and 23·6 per cent. tin. It has a specific heat of 0·0552 cal. per gm. between 20–89° C.

Rosette Copper. This is obtained in thin films by throwing water on to the surface of molten copper and removing the crusts formed.

Rose Vitriol. Cobalt sulphate.

Rosewater. (*Aqua rosæ B.P.*) Made by distilling the flowers of *Rosa damascena*. The rosewater of commerce is a saturated solution of the volatile oil of fresh rose flowers.

Roshé Oil. See ROSÉ OIL.

Rosilic Acid. See AURINE.

Rosin. See COLOPHONY.

Rosin Blende. See RESIN BLENDE.

Rosinduline 2B. See AZOCARMINE B.

Rosinduline G. A dyestuff. It is the sodium salt of rosindone-mono-sulphonic acid, $C_{22}H_{13}N_2SO_4Na$. Dyes wool and silk scarlet, and is chiefly used for printing.

Rosindulone. Rosindone, $C_6H_4.CO.CH : C.C : N.NC_6H_5.C_6H_4$.

Rosindulone 2G. A dyestuff. It is the sodium salt of rosindone-β-mono-sulphonic acid, $C_{22}H_{13}N_2SO_4Na$. Dyes silk and wool orange from an acid bath.

Rosin Ester. See ESTER GUMS.

Rosin Grease. A combination of rosin oil (*q.v.*) with lime, $13(C_{10}H_{16}).Ca(OH)_2$.

Rosin Gum. See ESTER GUMS.

Rosin, Hardened. See HARDENED ROSINS.

Rosinjack. See ZINC BLENDE.

Rosin Oil. An oil obtained by the distillation of rosin from 300–400° C. Used as an adulterant for olive oil, also as a lubricant for iron bearings. See RETINOL.

Rosinol. See RETINOL.

Rosin Pitch. Obtained by the distillation of rosin. It is the residue, and amounts to about 16 per cent.

Rosin Soap. Sodium resinate.

Rosin Spirit (Essence of Resin, Resin Spirit). The first distillate from rosin, 78–250° C. It is a complex mixture of hydrocarbons somewhat resembling turpentine. Used as a substitute for turpentine.

Rosin Tin. A yellow variety of the mineral Cassiterite or Tinstone.

Rosite. See ROSELLANE.

Rosolane. See MAUVEINE.

Rosolane B, R, O, T. A dyestuff. It is phenyl-diamino-phenyl-toluphen-azonium chloride, $C_{24}H_{21}N_4Cl$. Dyes silk violet-pink.

Rosolene. A rosin oil obtained by the distillation of rosin.

Rosoli. A liqueur.

Rosolic Acid. See AURINE.

Rosophenine 4B. See ST. DENIS RED.

Rosophenine 10B (Rosophenine Pink, Thiazine Red R, Chlorazol Pink Y). A dyestuff. It is the sodium salt of sulpho-benzenyl-amino-thiocresol-azo-α-naphthol-sulphonic acid, $C_{24}H_{16}N_3O_7S_3Na$. Dyes cotton pink to red shades.

Rosophenine Geranine. A direct cotton colour.

Rosophenine Pink. See ROSOPHENINE 10B.

Rosophenoline. See PÆONINE.

Ross Alloy. A bronze containing 68 per cent. copper and 32 per cent. tin.

Rossite. A Colorado mineral, CaO.$V_2O_5.4H_2O$.

Rosslerite. A mineral, $MgHAsO_4.H_2O$.

Ross's White. See LITHOPONE.

Rosstrevorite. A mineral. It is a variety of epidote.

Rosterite. A mineral. It is a variety of beryl.

Rotenone. A crystalline material found in the roots of derris, a plant grown in the rubber plantations of the Malay Peninsula. It is also found in the South American " cube " plant. It occurs up to 5·5 per cent. in derris and up to 7 per cent. in " cube." It is stated to be a contact and stomach poison for use as an insecticide.

Rotene. A trade mark for polyethylene. (61).

Roter. A proprietary preparation containing bismuth subnitrate, magnesium carbonate, sodium bicarbonate and frangula. An antacid used in the treatment of peptic ulcers. (274).

Rotercholon. A proprietary preparation of turmeric, ox-bile extract, peppermint oil, fennel oil, caraway oil and methyl salicylate used in the treatment of biliary disorders. (913).

Rotersept. A proprietary aerosol spray containing CHLORHEXIDINE digluconate used in the prophylaxis of puerperal mastitis. (913).

Rotheine. See PHENYL BROWN.

Rothoffite. A mineral, 6(Mn.Ca)O.$3SiO_2.2Fe_2O_3.3SiO_2$.

Rothspiesglaserz. Synonym for Kermesite.

Rotnickelkies. Synonym for Niccolite.

Rotoxit. A resistant high-silicon-copper alloy. It is stated to be resistant to uric, fluosilicic, and fatty acids, to dilute hydrochloric, sulphuric, and acetic acids, to 30 per cent. phosphoric acid, hydrogen peroxide, ammonia, lyes, and sulphates, but not to chromic, nitric, and lactic acids.

Rotra Bark. The bark of *Rotra fotsy* and *R. meno.* The bark contains 12·6 per cent. tannin.

Rotten Stone. A soft, friable aluminium silicate, containing a little organic matter. It is a disintegrated rock, found in Derbyshire, and is used as a polishing material. The term is also sometimes applied to Tripoli (*q.v.*). Specific gravity is 1·98.

Rottisite. A mineral, $2NiO.3SiO_2.2H_2O$.

Rouen White. A pigment. It is a clay found near Rouen.

Rouge. Good qualities of rouge consist of very fine iron oxide, Fe_2O_3, and are used as abrasives. The finest rouge is prepared from safflower. See INDIAN RED.

Rouge de Mars. See MARS RED.

Rouge Flambé (Aventurine, Sang de Bœuf). Red colours produced by the addition of copper oxide, CuO, to glazes, for pottery.

Rouge, Mineral. See JEWELLER'S ROUGE.

Rouge, Mordant. See RED LIQUOR.

Rouge, Toilet. A mixture of carmine and chalk.

Rouge, Vegetable. See VEGETABLE ROUGE and SAFFLOWER.

Rouge, White. See PLATE POWDER.

Rouille. See CHAMOIS.

Roulz Silver. See FRENCH SILVER.

Roumanite (Romanite). The amber of Roumania, which much resembles the Prussian variety.

Roussel's Solution. A solution of sodium phosphate.

Roussin's Black Salt. A compound, $Fe_3S_5H_2N_4O_4$, obtained by adding a solution of ferrous or ferric chloride slowly to the mixed solution of potassium nitrite and ammonium sulphide, and then boiling. The filtered liquid yields black crystals called Roussin's black salt.

Roussin's Red Salt. A salt, $Fe_2S_4N_2O_2Na_2.H_2O$, obtained by treating the sodium salt of Roussin's black salt with excess of acid after boiling, and then evaporating.

Roussin's Salts. Salts of the type $KFe_4(NO)_7S_3$, obtained when nitric oxide is passed through a suspension of ferrous sulphide in a sulphide solution.

Roux's Stain. A microscopic stain. It contains 0·5 gram gentian violet, 1·5 grams methyl green, and 200 c.c. distilled water.

R.O.V. Rectified oil of vitriol (sulphuric acid containing 93–96 per cent. H_2SO_4).

Rovamycin. A proprietary trade name for Spiramycin.

Rowachol. A proprietary preparation of menthol, methone, α-4-β-pinenes, camphene, eucalyptol, borneol and olive oil, for biliary and hepatic disorders. (811).

Rowatinex. A proprietary preparation of pinenes, α-camphene, ANETHOL fenchone and EUCALYPTOL in olive oil. A urinary antiseptic. (811).

Roweite. A mineral. It is 4[Ca(Mn,Mg,Zn)B₂O₅.H₂O].

Rowlandite. A mineral. It is a yttrium silicate, containing cerium, lanthanum, and thallium.

Roxamine. A dyestuff. It is the sodium

salt of dioxy-azo-naphthalene-sulphonic acid, $C_{20}H_{13}N_2O_5SNa$.

Roxarsone. An anti-protozoal and growth promoter. It is 4-hydroxy-3-nitrophenylarsonic acid.

Roxite. A synthetic resin product used for electrical insulation.

Roxon. A proprietary synthetic resin of the phenol-formaldehyde type.

Roxotit. A copper-silicon acid-resisting alloy.

Royal Blue (Saxon Blue, Azure Blue). Varieties of Smalt (*q.v.*).

Royal Powder. Pulv. Scammon. Co.

Royal Scarlet. See IODINE RED.

Royal Yellow. See CHROME YELLOW and CHINESE YELLOW.

Royalene. A proprietary ethylene propylene synthetic rubber. (287).

Roydalox. A proprietary trade name for alumina porcelain with good resistance to corrosion and thermal shock. It is used in ball mills and grinding balls. (176).

Roydazide. A proprietary trade name for silicon nitride. It is used for the manufacture of turbine blades and generally where a temperature of up to 1650° C. is present. (176).

Roylar. A proprietary polyester-based polyurethane thermoplastic used in injection moulding. (287).

Roylar A-863. A proprietary polyurethane elastomer used in cable work. It is a polytetramethylene - ether glycol - methylene bis (4-phenyl diisocyanate) base. (287).

Roylar A-863FR. A proprietary fire-resistant polyurethane elastomer used in cable manufacture. (287).

R.R. 53 Alloy. An aluminium alloy containing 91·85 per cent. aluminium, 2·25 per cent. copper, 1·3 per cent. nickel, 1·5 per cent. magnesium, 1·5 per cent. iron, 1·5 per cent. silicon, 0·1 per cent. titanium. Used for die casting pistons.

R.R.V. A proprietary antioxidant for rubber. It is resorcylidene.

R-Salt. The sodium salt of R-acid (*q.v.*).

Rubalt. A proprietary compound of rubber, bitumen, and benzene. It is a waterproof, rust-proof, and acid-resisting paint, and is stated to be highly resistant to mineral acids, alkalis, chlorine, ammonia, and salt solution.

Rubber. An elastic material contained in the latex of certain plants. The most important plant is *Hevea Brasiliensis*, of South America, which yields the Para rubber of commerce. Other plants yielding similar latices are *Castilloa elastica*, of Central America; *Castilloa Ulei*, of Peru; and *Manihot Glaziovii*, of Brazil. African rubbers are obtained from *Funtumia elastica* and *Landolphia owariensis*, Asiatic rubber from *Ficus*

elastica, and Guayule rubber from *Parthenium argentatum*, of Texas and Mexico. Plantation rubber is obtained from cultivated *Hevea Brasiliensis*. The latex is treated with a coagulating agent, usually acetic acid, and is then marketed in the form of smoked sheet, or pale crêpe. The latex itself is now shipped, coagulation being prevented by the addition of ammonia or sodium carbonate. It has been used in the preparation of paper, and for other purposes. Coagulated latex, as crêpe, is used for soling shoes and other purposes, and the vulcanised material (with or without accelerators and fillers) is used for tyres, mats, and many other purposes. Rubber as received contains resin in varying amounts, from 2 per cent. for the finest pale crêpe to 15 per cent. for Guayule and 70–80 per cent. for Jelutong.

Rubber, Abba. A low-grade African rubber, probably from *Ficus vogelii*. It is a red rubber containing 55 per cent. rubber and 45 per cent. resin.

Rubber, Artificial. See CAOUTCHOUC, ARTIFICIAL.

Rubber Cements. These are made by dissolving rubber in suitable solvents, such as coal-tar naphtha or carbon disulphide. Sometimes rosin or turpentine is added. Rubber cement is often a mixture of rubber and sulphur dissolved in oil.

Rubbered Asphalt. A combination of rubber and asphalt or bitumen obtained by dissolving the rubber in a suitable solvent that will also dissolve the asphalt. The powdered asphalt is put into a mixer, and the solution of rubber added gradually, when the solvent evaporates. The product is used for paving.

Rubbered Concrete. A combination of rubber, coal tar, and cement. (Rubber =2–5 per cent.) A paving material.

Rubber Formolite. A product obtained by the action of formaldehyde on a petroleum ether solution of a pale crêpe rubber to which has been added concentrated sulphuric acid.

Rubber, Frost. A name for sponge rubber.

Rubberite. An artificial rubber made from asphalt, oxidised oil, pet. jelly, and sulphur.

Rubber Lead No. 4. A vulcanisation accelerator for rubber. It contains organic and mineral constituents, and is manufactured by precipitating a substituted guanidine on a rubber lead base.

Rubberlene. A white refined petroleum product. It is used as a solvent, and can be substituted for carbon disulphide for dissolving rubber. It boils at 145–300° F.

Rubber, Mineral. See ELATERITE.

Rubber, Musa. See BANANA RUBBER.

Rubber Powder. Rubber produced in powder form by means of spraying rubber latex on to a heated endless belt. The dry rubber is removed in the form of powder.

Rubber, Reclaimed. This is obtained from scrap vulcanised rubber which is treated so as to remove free sulphur and fibre, etc., thus rendering the rubber plastic. There are two main processes, in which the material is first heated with either caustic soda or sulphuric acid, and then plasticised mechanically with admixture of softening and plasticising agents. The products are referred to as acid reclaim or alkali reclaim according to the method used.

Rubber, Silk. See CUITE, SERICINE, and SILK RUBBER.

Rubber Substitute. See CAOUTCHOUC, ARTIFICIAL.

Rubber-sulphur. Amorphous plastic sulphur, obtained from the Kobui sulphur mine, Japan.

Rubbone. A patented composition stated to be rubber resin prepared by oxidising rubber catalytically. It has an acid value of 5, an iodine value of 305, a saponification value of 60, and a specific gravity of 0·966 at 27° C. It is used in paints, varnishes, etc., in electrical insulation, and in the impregnation of coils.

Rubelix. A proprietary preparation containing pholcodine and ephedrine hydrochloride. A cough linctus. (330).

Rubellan. A mineral. It is a variety of mica.

Rubellite (Apyrite, Siberian Ruby). A mineral. It is a red tourmaline.

Rubel Metal. An alloy of 55 per cent. copper, 40 per cent. zinc, and 5 per cent. aluminium-iron-manganese-nickel-alloy. Another alloy contains 51 per cent. copper, 40 per cent. zinc, and 5 per cent. aluminium-iron-manganese-nickel alloy, and 4 per cent. ferro-manganese.

Ruben's Brown. See VANDYCK BROWN.

Ruben's Madder. A pigment. It is a preparation of madder.

Rubeosine. A nitro-chloro-fluoresceine, obtained by the action of nitric acid upon aureosin.

Ruberite. A name for red copper ore, Cu_2O.

Rub-erok. A proprietary trade name for hard rubber for electrical insulation.

Rub-er-red. A proprietary pigment for rubber. It is a red iron oxide of fine particle size, which is acid and alkali free, and contains no soluble salts.

The manganese and copper content are claimed to be below any rubber specification. It has a specific gravity of 5·15, and contains over 99 per cent. Fe_2O_3.

Rubesine. See MAGENTA.

Rubianic Acid. Ruberythric acid, $C_{26}H_{28}O_4$.

Rubianin. See MAGENTA.

Rubianite. See MAGENTA.

Rubicelle. Spinel (q.v.).

Rubidine. See FAST RED.

Ribidium-Microcline. A mineral. It is $4[(K,Rb)AlSi_3O_8]$.

Rubiesite. A mineral, $8Bi_2S_3.Sb_2S_3.Bi_2(Te.Se)_3$.

Rubinblende. Synonym for Miargyrite.

Rubine. See MAGENTA.

Rubine S. See ACID MAGENTA.

Rubini's Essence. A saturated solution of camphor in alcohol.

Rubio Ore. A brown ore of iron, from Bilbao, in Spain.

Rubmag. A proprietary trade name for a light magnesium carbonate used for rubber reinforcing.

Rubramine. See INDAMINE 3R.

Rubrax. A mineral rubber containing 98–99 per cent. of material soluble in chloroform, with an ash less than 0·5 per cent. It has a specific gravity of 1·01.

Rubrene. Phenyl - ethinyl - diphenyl - methane. $C_{21}H_{20}$.

Rubrescin. An indicator prepared from resorcinol and chloral hydrate.

Rubrica. A red pigment. It is a natural burnt ochre, containing varying quantities of iron. See INDIAN RED.

Rubric Lake. See MADDER LAKE.

Rubriment. A proprietary preparation of nicotinic acid benzyl ester and capsicin. An embrocation. (269).

Rubrite. A mineral. It is $Mg_3Fe_4(SO_4)_6(OH)_6.27H_2O$.

Rubrophen. Dihydroxy-trimethoxy-triphenyl-methane.

Rub-tex. A proprietary trade name for a hard rubber.

Ruby. A mineral. It is alumina, Al_2O_3, coloured with metallic oxides, and used as a precious stone.

Ruby, Almandine. Spinel (q.v.).

Ruby Arsenic. See REALGAR.

Ruby, Chrome. See CHROME RED.

Ruby Copper. See CUPRITE.

Ruby Glass. See RED GLASS.

Rubyl. Quinine-bismuth iodide, a proprietary medicinal preparation for the treatment of syphilis.

Ruby Ore. Cuprous oxide, Cu_2O.

Ruby Powder. A sporting 42-grain powder containing 50 per cent. nitrocellulose, metallic nitrate, 8 per cent. nitro-hydrocarbon, and 6 per cent. starch.

Ruby, Siberian. See RUBELLITE.

Ruby Silver. See PYRARGYRITE.

Ruby Silver Ore. See PYRARGYRITE.

Ruby, Spinel. See BALAS RUBY.

Ruby Sulphur. See REALGAR.

Ruby Tin. A red variety of the mineral Cassiterite, or tinstone.

Ruddle. See INDIAN RED.

Ruelene. A proprietary preparation of CRUFOMATE. A veterinary insecticide.

Rufigallol. Hexa-oxy-anthraquinone, $C_{14}H_8O_8$. Dyes chromed wool brown.

Rufiopin. Tetra - oxy - anthraquinone, $C_{14}H_8O_6$.

Rufocromomycin. An antibiotic produced by Streptomyces rufuchromogenus.

Rufol. β-Dioxy-anthraquinone, C_6H_3 (OH) : C_2H_2 : C_6H_3(OH).

Rufus Pill. Pil. Aloes et Myrrhæ.

Rugar. A proprietary preparation of barium sulphate. (914).

Ruge's Solution. A solution containing 1 c.c. glacial acetic acid, 2 c.c. formalin, and 100 c.c. distilled water.

Rule Brass. See BRASS.

Rum. An alcoholic liquor. Genuine rum exists in two qualities, one being Jamaica rum, and the other Demerara rum. The first is obtained by the slow fermentation (10 to 12 days) of cane sugar molasses, and the second by rapid fermentation (36 to 48 hours). They contain from 65–73 per cent. of alcohol. Imitation rum, made in Germany, is obtained from grain or beet spirit, and flavoured with artificial essences.

Rumasal. A proprietary preparation of basic aluminium acetylsalicylate. An analgesic. (257).

Rum Essence. Essential oils used to flavour rum.

Rumicin. A preparation from yellow dock. It consists of chrysophanic acid, $C_{15}H_{10}O_4$.

Rumpff's Acid. See BAYER'S ACID.

Rumpfite. A mineral. It is a basic silicate of aluminium and magnesium.

Rumongite. Synonym for Ilmenorutile.

Runa. Rutile type titanium dioxide. (513).

Runge's Madder Orange. Rubiacin, a yellow, crystalline substance obtained from the madder root.

Ruolz Alloys. See NICKEL SILVERS.

Rupel Alum. Roche alum.

Rupert's Drops. Hardened glass drops, obtained by dropping molten glass into hot oil or hot water. A scratch will cause a drop to break into many pieces.

Ruselite. A proprietary trade name for an alloy of 94 per cent. aluminium, 4 per cent. copper, 2 per cent. chromium, and 2 per cent. molybdenum. It is stated to be resistant to corrosion.

Rusma. A mixture of arsenic sulphide (orpiment), As_2S_3, with lime. It is made into a paste with water, and used for unhairing skins prior to tanning them.

Ruspini's Solution. A styptic containing tannic acid, rose water, alcohol, and water.

Russet Rubiate. A pigment. It is a preparation of madder.

Russian Cast Brass. See BRASS.

Russian Glue. A variety of Cologne glue (q.v.), rendered opaque by means of white lead, zinc white, or chalk.

Russian Green. See DINITROSORESOR-CIN.

Russian Red. Fuchsine mixed with safranine to give it a yellow shade. See MAGENTA.

Russian Red G, 967. Dyestuffs. They are British equivalents of Magenta.

Russian Tallow. A mixture of beef and mutton fat.

Russian Tula. See NIELLO SILVER.

Russian Turpentine. The oleo-resin from *Pinus sylvestris* and *P. Ledebourii*.

Russian White Lead. A white lead having the composition, $5PbCO_3.2Pb(OH)_2.PbO$.

Russol. See PARAFFIN, LIQUID.

Rusting Cement. See IRON CEMENT.

Rustless Iron. A rustless steel containing about 0·1 per cent. carbon. It is made in the electric furnace by means of practically carbon-free ferro-chrome.

Rustless Steel. See CHROME STEELS.

Ruthenium Red. Ammoniated ruthenium oxychloride, $Ru_2(OH)_2Cl_4.7(NH_3).3H_2O$. Used as a microscopic stain, and as a reagent for pectin, plant mucin, and gum.

Rutherfordine. A mineral. It is a yellow uranium carbonate, UO_2CO_3.

Ruthmol. A proprietary preparation of potassium chloride, lactose and gluten-free starch. A sodium-free table salt. (787).

Rutile (Cajuelite). A mineral containing 98–99 per cent. titanium oxide, TiO_2, and 1–2 per cent. ferric oxide, Fe_2O_3.

Rutiox. A rutile titanium dioxide pigment. (513).

Rutonal. Phenyl-methyl-malonyl-urea. A proprietary preparation of phenyl-methylbarbituric acid. An anticonvulsant. (336).

R.X.X.L. High boiling tar acids. (602).

Rybaferrin. A proprietary preparation of ferrous sulphate, manganese, copper, strychnine hydrochloride, thiamine, hydrochloride and nicotinic acid. A tonic. (273).

Rybarex. A proprietary preparation of chloroxylenol, papaverine hydrochloride, atropine methonitrate, methyl salicylate, menthol, benzocaine, pituitary extract, adrenaline, tri-iodophenol and a saline base. (273).

Rybarvin. A proprietary preparation of atropine methonitrate, ADRENALIN, papaverine hydrochloride, benzocaine and posterior pituitary extract, used in the treatment of asthma. (273).

Rybronsol. A proprietary preparation of phenazone iodoantipyrine, caffeine, and butethamate citrate. (273).

Rymel. A proprietary preparation containing ipecacuanha liquid extract, acetic acid, squill liquid extract, sodium citrate and glycerin. A cough linctus. (273).

Rynabond. A proprietary preparation containing phenylephrine tannate, pheniramine tannate and mepyramine tannate. (188).

Rynacrom. A proprietary preparation of sodium cromoglycate used in the treatment of allergic rhinitis. (188).

Ryotol. A proprietary preparation of phenoxyethanol, phenylmercuric nitrate and phenazone. (273).

Ryspray. A proprietary preparation of atropine methonitrate and isoprenaline hydrochloride in an aerosol. A bronchial antispasmodic. (273).

Rythmodan. A proprietary preparation of DISOPYRAMIDE used in the treatment of cardiac arrhythmias. (443).

Ryton. Polyphenylene sulphide (PPS). A thermoplastic moulding resin used for structural components and the encapsulation of semiconductors.

S

Saamite. A mineral. It is $2[(Ca,Sr,R*)_5 (PO_4)_3(F,O)]$ where R* signifies rare earth metals.

Sabadilline. See CEVADILLINE.

Sabeco Metal. A proprietary trade name for copper with 21 per cent. lead and 9 per cent. tin.

Sablon. See LIMO.

Sabromine (Bromobehenate). Calcium-dibromo-behenate, $(C_{22}H_{41}O_2Br_2)_2Ca$, containing 29 per cent. bromine. It has therapeutic value similar to bromides.

Sabugalite. A mineral. It is [HAl $(UO_2)_2PO_4 . 24H_2O]$.

Sabulite. An explosive containing ammonium nitrate, charcoal, and calcium silicide.

Saccharase. See SUCRASE.

Saccharated Iron Carbonate. (*Ferri carbonas saccharatus B.P.*) A mixture of ferrous carbonate and sugar.

Saccharated Iron Phosphate. (*Ferri phosphas saccharatus B.P.*) Ferrous phosphate, more or less oxidised, mixed with glucose.

Saccharated Solution of Lime. (*Liquor calcis saccharatus B.P.*) A solution of 2 oz. of sugar and 188 grains of slaked lime, in 1 pint of water.

Sacchareines. Dyestuffs similar to rhodamines, obtained by condensing saccharin with dialkylated *m*-aminophenol.

Saccharin (Gluside, Glycophenol, Glycosin, Neosaccharin, Sycorin, Sykose). *o*-Anhydro - sulphamide - benzoic acid (benzoic sulphimide), $C_6H_4(CO)(SO_2)$. NH. A sweetening substance which is 500 times sweeter than cane sugar. Used for diabetic patients.

Saccharin, Easily Soluble (Soluble Gluside, Crystallose). The sodium salt of saccharin (*q.v.*).

Saccharinol. See SACCHARIN.

Saccharinose. See SACCHARIN.

Saccharol. See SACCHARIN.

Saccharosolval. A preparation of oxybenzoic acid, trypsin, and the spinal marrow of oxen. It is administered in the treatment of diabetes.

Sacholith (Sachtolith). A German pigment stated to be a specially prepared zinc sulphide.

Sachsse's Solution. A solution containing 18 grams mercuric iodide, 25 grams potassium iodide, and 80 grams potassium hydroxide in a litre. Used for the determination of reducing sugars.

Sachtolith. A proprietary trade name for zinc sulphide used as a pigment.

S-acid. 1 : 8 - Amino - naphthol - 4 - sulphonic acid.

2 S-acid. 1 : 8-Dihydroxy-naphthalene-2 : 4-disulphonic acid.

Sacred Bark. The bark of *Rhamnus purshianus.*

S.A.E. No. 50 Alloy (Dow H. Alloy). A magnesium alloy containing 6 per cent. aluminium, 3 per cent. zinc, and 0·2 per cent. manganese. An aircraft alloy.

Safapryn. A proprietary preparation of aspirin and PARACETAMOL. An analgesic. (85).

Safapryn-Co. A proprietary preparation of aspirin, PARACETAMOL and codeine phosphate. An analgesic. (85).

Safetex. A proprietary safety glass.

Safety Dynamite. An explosive consisting of 24 per cent. nitro-glycerin, 1 per cent. guncotton, and 75 per cent. ammonium nitrate.

Safety Glass. A glass made from sheets of glass laminated with cement, usually one of the transparent plastics.

Proprietary trade names for some safety glasses are: Acetex, Aerolite, Aeroplex, Citalo, Duplate, Glacetex, Guardex, Indestructo, Kinonglas, Lancegaye, Lumina, L.O.F., Neo-triplex,

Neutex, Newtex, Peka Glas, Protex, Safetex, Securex, Sigla, Simplex, Splintex, Supertex, Thorax A and N, Trilob, Triplex, Tyrex, Veracetex, Vis, Xetal.

Safety Matches. Usually consist of a mixture of potassium chlorate and sulphur, with additions, to reduce violence of ignition. They are ignited upon a surface coated with a mixture of red phosphorus, antimony sulphide, and powdered glass.

Safety Nitro-powder. An explosive similar in composition to Giant powder.

Safety Oil. See C-PETROLEUM NAPHTHA.

Safex. A proprietary super-accelerator for rubber vulcanisation. It is stated to be dinitro-phenyl-dimethyl-dithio-carbamate. It melts at $140°$ C., and has a specific gravity of $1·54$. It is used with zinc oxide.

Safflor. See SAFFLOWER.

Safflorite. A mineral. It is a cobalt arsenide, $CoAs_2$.

Safflower (Bastard Saffron, Safflor, Safflower Extract). A natural dyestuff. Carthamine, $C_{14}H_{14}O_7$, is the colouring principle, and it dyes silk and cotton red. It is sold under the names of Rouge végétable and Safflower Carmine, and is used as a cosmetic and pigment.

Safflower Carmine. See SAFFLOWER.

Safflower Extract. See SAFFLOWER.

Saffron. A colouring matter obtained from the dried and powdered flowers of the saffron plant, *Crocus sativus*. Used for colouring confectionery. See SULPHURATED ANTIMONY.

Saffron, Bastard. See SAFFLOWER.

Saffron Bronze (Gold Bronze). Tungsten-sodium bronze, $Na_2W_3O_9$. Used as a pigment. The corresponding potassium salt is known as Violet bronze or Magenta bronze.

Saffron, Dyer's. See SAFFLOWER.

Saffron, Indian. See TUMERIC.

Saffron, Iron. See INDIAN RED.

Saffron of Antimony. See SULPHURATED ANTIMONY.

Saffron Oil. Safflower oil from the seeds of *Carthamus tinctorius*. It has an acid value of $9·8$ and a saponification value of $197·3$.

Saffron Substitute. See VICTORIA YELLOW.

Saffron Sugar. Crocase, $C_6H_{12}O_6$.

Saffron Surrogate. See VICTORIA YELLOW.

Saflex. A proprietary trade name for a synthetic resin of the polyvinyl acetate or polyvinyl butyrate type. It is used in the interleaving of safety glass.

Safraniline. See RHODAMINE B.

Safranine (Safranine T, Safranine Extra G, Safranine S, Safranine FF Extra, Safranine Conc., Safranine GGS, Safranine AG, AGT, OOF, Safranine GOO, Aniline Rose, Aniline Pink). A dyestuff. It is a mixture of diamino-phenyl and tolyl-tolazonium-chlorides, $C_{21}H_{21}N_4Cl$ and $C_{20}H_{19}N_4Cl$. Dyes cotton mordanted with tannin and tartar emetic, red. Used in calico printing.

Safranine AG, AGT, OOF. See SAFRANINE.

Safranine B. Phenosafranine (*q.v.*).

Safranine B Extra. Phenosafranine (*q.v.*).

Safranine Conc. See SAFRANINE.

Safranine Extra G. See SAFRANINE.

Safranine FF Extra. See SAFRANINE.

Safranine GGS. See SAFRANINE.

Safranine GOO. See SAFRANINE.

Safranine MN. See METHYLENE VIOLET 2RA.

Safranine RAE (Tolusafranine). Dyes tannined cotton.

Safranine S. See SAFRANINE.

Safranine Scarlet. A dyestuff. It is a mixture of auramine and safranine.

Safranine T. See SAFRANINE.

Safranisol. Methoxy-safranine.

Safrol. The methylene ether of allyl-pyrocatechol, $C_6H_3.C_3H_5.(O.OCH_2)$. It is found in oil of sassafras, and is obtained from red oil of camphor. Used in the place of oil of sassafras.

Safrosine. See EOSIN BN.

Sagenite. A mineral. It consists of silica containing crystals of rutile.

Sago. A prepared starch obtained from several varieties of palm. An artificial sago is prepared by mixing dark potato starch with a little dextrin.

Sagrotan. Described as a molecular mixture of a chlor-xylenol and a chlor-cresol. An antiseptic.

Sahamalite. A mineral. It is $2[(Mg,Fe··)(Ce,La)_2(CO_3)_4]$.

Sahlinite. A mineral. It is $Pb_{14}(AsO_4)_2O_9Cl_4$.

Sahli's Reagent. A mixture of equal parts of a 48 per cent. solution of potassium iodide and an 8 per cent. solution of potassium iodate. Used to test for free hydrochloric acid in stomach contents.

Sahli's Stain. A solution of borax and methylene blue in water. Used to stain nervous tissues and cell nuclei.

Sahlite. A pyroxene mineral, $(Mg.Fe)Ca(SiO_3)_2$.

Sailor's Pepper. Cubebs.

St. Bartholomew's Tea. See MATÉ.

St. Denis Black. A dyestuff obtained by the fusion of *p*-phenylene-diamine with sodium polysulphides. Dyes cotton greyish-blue to black.

St. Denis Red (Dianthine, Rosophenine 4B, Trona Red, Rock Scarlet BS, Patent Rock Scarlet). A dyestuff. It is the sodium salt of azoxy-toluene-disazo-bi-α-naphthol-sulphonic acid, $C_{34}H_{24}N_6O_9S_2Na_2$. Dyes cotton red from an alkaline bath.

St. Helen's Powder. An explosive consisting of 92–95 per cent. ammonium nitrate, 2–3 per cent. aluminium powder, and 3–5 per cent. trinitro-toluene.

St. Ignatius Bean. The seed of *Strychnos ignatii*.

St. John's Bread. See LOCUST BEAN.

St. John Wort Oil. See OIL OF ST. JOHN WORT.

Saiodine (Sajodin, Calioben). The calcium salt of mono-iodo-behenic acid, $(C_{22}H_{42}O_2I)_2Ca$. It contains 26 per cent. of iodine, and is used medicinally as an iodine preparation.

Sajji. An alkaline deposit found on the land in India. It is used as a soap substitute. Also see RITHA.

Sajodin. See SAIODINE.

Sakaloid. A new synthetic resin. It is a polymerised sugar product obtained from sugar, dextrose, levulose, etc. It can be used for varnishes and lacquers and, when extruded, as an artificial silk.

Saké. An alcoholic drink made from rice by the Japanese and Chinese. The rice is saccharified by means of *Aspergillus oryzæ*, and the liquid fermented by a yeast mould called Mots. It contains 10–15 per cent. of alcohol, 0·4–0·95 per cent. sugar, 0·24–0·5 per cent. dextrin, and 0·95–1·08 per cent. glycerol.

Saké Oil. This consists chiefly of rice oil, and is found floating on the surface of the liquid after the fermentation of saké, a Japanese drink prepared from rice. The oil has a saponification value of 179, an acid value of 22·6, and an iodine value of 101·6.

Sakoa Oil. An oil obtained from the seeds of *Sclerocarpa caffra*. It is a non-drying oil, and has a saponification value of 193·5.

Salabrose. A diabetic food preparation. It is stated to be obtained by heating glucose with petroleum jelly, a small quantity of an iron salt, and a catalyst.

Sal Absinthii. Salt of wormwood, potassium carbonate, K_2CO_3.

Salacetin. Aspirin (*q.v.*).

Salacetol (Salicyl-Acetol, Salantol, Acetomesal). Acetol-salicylic ester, $C_6H_4(OH).CO_2.CH_2.CO.CH_3$. Prescribed for articular rheumatism.

Sal Acetosella. Acid potassium oxalate.

Salactol. A proprietary preparation of salicylic acid and lactic acid used in the treatment of warts. (377).

Salad Oil. This usually consists of olive oil, but any edible oil is called by this name.

Salad Oil, Sublime. See SUBLIME OLIVE OIL.

Sal Aeratus. Potassium bicarbonate, $KHCO_3$.

Salaeratus. An American baking powder, consisting of sodium carbonate, sodium chloride, and cream of tartar or tartaric acid.

Salaigugl. A local name for Olibanum (*q.v.*).

Sal Alembroth. See SALT OF ALEMBROTH.

Sal Alkali Minerale. Sodium carbonate, Na_2CO_3.

Sal Alkali Vegetable (Sal Tartari). Potassium carbonate, K_2CO_3.

Salamac. Compressed blocks of ammonium chloride. (512).

Sal Amarum (Sal Catharticum, Sal Anglicum, Sal Seidlitense). Magnesium sulphate, $MgSO_4$.

Sal Ammoniac (Salmiak, Muriate of Ammonia). Ammonium chloride, NH_4Cl.

Sal Anglicum. See SAL AMARUM.

Salantin. See ASPIRIN.

Salantol. See SALACETOL.

Salaphene. A proprietary preparation of resorcinol acetate, salicylic acid, bithionol and methyl hydroxybenzoate used in dermatology as a peeling agent. (378).

Salargyl. A protein-silver preparation.

Salarmoniac. An ammonium chloride obtained from volcanoes of Central Asia.

Salaspin. See ASPIRIN.

Sal Auri Philosophicum. Potassium bisulphate, $KHSO_4$.

Salazolon. See SALIPYRIN.

Salazopyrin. A proprietary preparation of sulphasalazine. (337).

Salazosulphadimidine. 4'-(4,6-Dimethylpyrimidin - 2 - ylsulphamoyl) - 4 - hydroxyazobenzene - 3 - carboxylic acid. Azudimidine.

Salbromalide. Salicyl-brom-anilide.

Salbutamol. A bronchodilator. 1-(4-Hydroxy - 3 - hydroxymethylphenyl) - 2 - (t - butylamino)ethanol. VENTOLIN.

Sal Carolinum. Carlsbad salt (*q.v.*).

Sal Catharticum. See SAL AMARUM.

Salcatonin. A component of natural salmon CALCITONIN currently undergoing clinical trial as " SCT 1 ", with " SMC 20–051 " as the hydrated polyacetate, for use in the treatment of hypercalcæmia and Paget's disease.

Sal Chalybdis. Iron sulphate.

Sal Commune. Sodium chloride, NaCl.

Sal Cornu Cervi. Ammonium carbonate, $(NH_4)_2CO_3$.

Sal Culinaris. Sodium chloride, NaCl.

Sal de Duobus (Sal Polycrest). Potassium sulphate, K_2SO_4.

Sal Digestnum Sylvii. Potassium chloride, KCl.

Sal Diureticum. Potassium acetate, $(C_2H_3O_2)K$.

Sal Duobus. See SAL DE DUOBUS.

Salecolen. See SALENE.

Salem Copal. See BOMBAY COPAL.

Salene (Salecolen). A mixture of ethyl and methyl glycollic acid esters of salicylic acid. It is an oily liquid, said to be used in cases of rheumatism.

Salenixon. Crude potassium sulphate, obtained in the manufacture of nitric acid.

Saleratus. Sodium - hydrogen - carbonate, $NaHCO_3$.

Salesite. A mineral. It is $4[CuIO_3OH]$.

Salesthin. Methylene chloride, CH_2Cl_2.

Sal-ethyl. Ethyl salicylate, used in medicine.

Saletin. Aspirin (q.v.).

Salforcose. A mixture of 90 per cent. carbon disulphide with 10 per cent. of 50 per cent. alcohol. Used to destroy bed insects.

Salfuride. A proprietary preparation of NIFURSOL. A veterinary anti-protozoan.

Salge Metal. An alloy of 4 per cent. copper, 9·9 per cent. tin, 1·1 per cent. lead, and 85 per cent. zinc.

Salhar Gum. A gum-resin obtained from *Boswellia serrata*.

Salhypnone. Benzoyl-methyl-salicylic ester, $C_6H_4O(CO.C_6H_5).CO.OCH_3$. A mild antiseptic.

Salibromine. Methyl-dibromo-salicylate, $C_6H_2Br_2(OH)COOCH_3$. Prescribed as an antirheumatic and antipyretic.

Salicaine. Saligenin (q.v.).

Saliceral. Glyceryl monosalicylate.

Salicine Red B. A dyestuff obtained from nitro-benzidine, salicylic acid, and β-naphthol, and sulphonating the product. Gives yellowish-red to bluish-red shades on wool.

Salicine Red G, 2G. Dyestuffs similar to Salicine red B.

Salicine Yellow. A dyestuff from nitrobenzidine, and two molecules of salicylic acid. An acid mordant dye for wool.

Salicitrin. Citrosalic acid.

Salicolen. Salene, a mixture of ethyl and methyl glycollic acid esters of salicylic acid.

Salicor. See BLANQUETTE.

Salicreol. See SALOCRESOL.

Salicresol. Creosote salicylate. See SALOCRESOL.

Salicyl-acetol. See SALACETOL.

Salicylamide. An analgesic and antipyretic. It is 2-hydroxybenzamide. SALIMED.

Salicylanilide. See SALIFEBRIN.

Salicylazosulphapyridine. Sulphasalazine.

Salicylic Orange. A dyestuff obtained from sulpho-salicylic acid by bromination. Dyes wool or silk dark golden-yellow or orange.

Salicylic Yellow. A dyestuff obtained from sulpho-salicylic acid by nitration. Dyes wool and silk yellow.

Salicyl-isapogen. See ISAPOGEN.

Salicylsuccinate. See DIASPIRIN.

Salidol. Lucidol (q.v.).

Salifebrin (Salicylanilide). A product obtained by the fusion of salicylic acid and acetanilide.

Saliformin. A salicyl derivative of hexamethylene-tetramine, $C_8H_{12}N_4.C_6H_4OH.COOH$. Used as a uric acid solvent, and as a genito-urinary antiseptic.

Saligallol. Pyrogallol-disalicylate. Used externally in skin diseases.

Saligenin (Saligenol). Salicyl alcohol (o - oxy - benzyl - alcohol), $C_6H_4OH.CH_2OH$. Used in the treatment of rheumatism.

Saligenol. See SALIGENIN.

Saliglycol. See SPIROSAL.

Salimed. A proprietary preparation of salicylamide. An analgesic. (250).

Salimed Compound. A proprietary preparation of salicylamide, mephenesin and amylobarbitone. (250).

Salimenthol (Samol). Menthol salicylate, $HO.C_6H_4.COOC_{10}H_9$. Used externally and internally as an antiseptic and as an anodyne in cases of toothache and rheumatism.

Salinaphthol. See BETOL.

Salinazid. N-Isonicotinyl-N′-salicylidenehydrazine. Nupa-sal.

Saliphen (Saliphenin). Salicyl-p-phenetidine, $C_6H_4(O.C_2H_5).NH.C_6H_4(OH)CO$. Has antifebrile properties.

Saliphenin. See SALIPHEN.

Salipyrazolon. See SALIPYRINE.

Salipyrine (Salipyrazolon, Salazolon). Antipyrine salicylate, $C_{18}H_{18}N_2O_4$. It reduces fever, and is also used for rheumatism.

Saliretin Resins. See SALIRETINS.

Saliretins (Saliretin Resins). Resins obtained from saligenin by either heating or treating it with formaldehyde. They are similar to phenol-formaldehyde resins.

Salit. Bornyl salicylate, $C_{10}H_{17}O.CO.C_6H_4OH$. Prescribed in cases of rheumatism. Equal parts of Salit and olive oil are used for rubbing.

Salitannol. A condensation product of salicylic and gallic acids, $OH.C_6H_4.CO.O.CO.C_6H_2(OH)_3$. A surgical antiseptic and substitute for salol.

Salite. A mineral. It is a variety of hedenbergite.

Salithymol. The thymol ester of salicylic acid, $C_6H_3(CH_3)(C_3H_7)O.CO.C_6H_4OH$. An antiseptic.

Salitre. Sodium nitrate, $NaNO_3$.

Saliva Diastase. Ptyalin, the enzyme in saliva.

Salizone. See AGATHINE.

Salkowski's Solution. A solution of phospho-tungstic acid. For albumen in urine.

Sallit's Speculum Metal. An alloy of 64·6 per cent. copper, 31·3 per cent. tin, and 4·1 per cent. nickel.

Sally Nixon. Fused nitre cake (acid sodium sulphate).

Sal Martis. Ferrous sulphate, $FeSO_4$.

Salmefamol. 1 - (4 - Hydroxy - 3 - hy - droxymethylphenyl) - 2 - (4 - methoxy - α - methylphenethylamino)ethanol.

Salmester. Methoxy - methyl - sali - cylate.

Salmiak. See SAL AMMONIAC.

Sal Mineral. Ferric oxide, Fe_2O_3.

Sal Mirabil. Sodium sulphate, Na_2SO_4.

Salmite. A mineral. It is a variety of chloritoid containing magnesium.

Salmocid. A proprietary preparation of polynoxylin. (386).

Salmon Pink. See EOSIN ORANGE.

Salmon Red. The Badische Co. A dye-stuff. It is the sodium salt of diphenyl-urea-disazo-bi-naphthionic acid, $C_{33}H_{24}$ $N_8O_7S_2Na_2$. Dyes cotton flesh colour to brownish-orange from a boiling alkaline bath.

Salmon Red. Farbwerk, Griesheim, Notzel, Istel & Co. A dyestuff. It is the sodium salt of diphenyl-thiourea-bi-naphthionic acid, $C_{33}H_{24}N_8O_6S_3Na_2$. Dyes cotton orange-red.

Salmon Red. Berlin Aniline Co. (En-bico Direct Fast Pink Y Conc.). A dyestuff. It is the sodium salt of methyl-benzenyl - amino - thioxylenol - azo - β - naphthylamine-disulphonic acid, $C_{26}H_{20}$ $N_4O_6S_3Na_2$. Dyes cotton direct salmon red.

Salmonsite. A mineral. It is a mag-nesium-iron hydrated phosphate.

Sal Nitre. Potassium nitrate, KNO_3.

Salochinine. See SALOQUININE.

Salocoll. Phenocoll salicylate, C_6H_4 $(OC_2H_5).NH.CO.CH_2.NH_2.C_6H_4(OH)$. COOH. An antirheumatic.

Salocreol. See SALOCRESOL.

Salocresol (Salicresol, Salicreol, Salo-creol). Creosote salicylate. Used ex-ternally in erysipelas and rheumatism.

Salodine. A proprietary preparation. It is an iodised salt.

Salol. Phenyl salicylate, $HO.C_6H_4.CO_2$. C_6H_5, melting-point 43° C. Used in medicine as a substitute for sodium salicylate, and is also recommended as an antipyretic and antiseptic.

Salol Camphor. A mixture of 3 parts salol and 2 parts camphor. An anti-septic for boils.

Salol Red. A reddish-brown powder,

$C_{19}H_{16}O_4$, obtained by heating salicyl-metaphosphoric acid with phenol. It dyes wool.

Salophene (Cetosalol). Acetyl-p-amino-phenyl-salicylate, $C_6H_4(OH).COOC_6H_4$. $NH.CO.CH_3$. An antirheumatic, anti-neuralgic, and antiseptic.

Saloquinine (Salochinine). Salicyl-quinine, $HO.C_6H_4.COOC_{20}H_{23}N_2O$. An antipyretic and antiperiodic.

Sal Polycrest. See SAL DE DUOBUS.

Sal Prunella. Potassium nitrate, KNO_3, in balls.

Sal Rupellensis. Sodium-potassium tartrate.

Sal Saturni. Lead acetate, $Pb(C_2H_3O_2)_2$.

Sal Sedativus. Boric acid, H_3BO_3.

Salsefis. See HELIANTHI.

Sal Siedlitense. See SAL AMARUM.

Sal Soda. Sodium carbonate, Na_2CO_3.

Sal Succini. Succinic acid, $HOOC.CH_2$. $CH_2.COOH$.

Salt, Abraum. See ABRAUM SALTS.

Salt, Amido-G. See AMIDO-G-ACID.

Salt, Amido-R. See AMIDO-R-ACID.

Sal Tartari. See SAL ALAKALI VEGE-TABLE.

Salt, Bitter. See EPSOM SALTS.

Salt Cake. Crude sodium sulphate, Na_2SO_4, produced in the Leblanc soda process.

Salt, Common. Sodium chloride, NaCl.

Salted Fish Water. See NUOC-MAM.

Salt, Fossil. See ROCK SALT.

Salt, Fusible. See MICROCOSMIC SALT.

Salt, Gravy. See BURNT SUGAR.

Salt, Hair. See ALUNOGEN and EPSOM SALTS.

Salt Lick. See PAN SCALE.

Salt, Lister's. See LISTER'S ANTISEPTIC.

Salt, Mining. A mixture of sodium bromate and bromide. It was formerly used in the extraction of gold from its ores.

Salt of Alembroth (Salt of Wisdom, Sal-Alembroth). A compound of mercuric chloride and ammonium chloride, $2NH_4$ $Cl.HgCl_2.H_2O$. Used as a surgical dressing.

Salt of Amber. Succinic acid, C_2H_4 $(COOH)_2$.

Salt of England. See EPSOM SALTS.

Salt of Hartshorn. Ammonium car-bonate, $(NH_4)_2CO_3$.

Salt of Lemery. Potassium sulphate, K_2SO_4.

Salt of Lemon. See SALT OF SORREL.

Salt of Norton. Platinum tetrachloride, $PtCl_4.5H_2O$.

Salt of Saturn (Sugar of Saturn). Nor-mal lead acetate, $Pb(C_2H_3O_2)_2$, was formerly known under these names.

Salt of Soda. Sodium carbonate, Na_2CO_3.

Salt of Sorrel (Salts of Sorrel, Salts of Lemon). The two acid salts of potas-sium oxalate, C_2O_4HK, and C_2O_4KH.

$C_2H_4O_2.2H_2O$, are both sold under these names, for removing ink stains.

Salt of Steel. Ferrous sulphate, $FeSO_4$.

Salt of Tartar. See POTASH.

Salt of Tin. Stannous chloride, $SnCl_4$.

Salt of Urine, Fusible. See MICROCOSMIC SALT.

Salt of Wisdom. See SALT OF ALEMBROTH.

Salt of Wormwood (Sal Absinthii). Impure potassium carbonate, K_2CO_3, made from plant ash.

Salt Perlate. Sodium phosphate, HNa_2PO_4.

Saltpetre (Nitre). Potassium nitrate, KNO_3.

Saltpetre, Artificial. See GERMAN SALTPETRE.

Saltpetre, Conversion. See GERMAN SALTPETRE.

Saltpetre, Cubic. See CHILE SALTPETRE.

Saltpetre Flour. Minute crystals of refined saltpetre, KNO_3, used in the manufacture of gunpowder.

Saltpetre, Lime. See LIME NITRATE.

Saltpetre, Norwegian. See AIR SALTPETRE.

Saltpetre, Peru. See CHILE SALTPETRE.

Saltpetre Rot. Calcium nitrate, $Ca(NO_3)_2.4H_2O$. It causes the rapid disintegration of mortar.

Saltpetre, Soda. See CHILE SALTPETRE.

Saltpetre Superphosphate. A fertiliser made by mixing nitre with calcium superphosphate.

Saltpetre, Wall. See NITROCALCITE.

Salt, Phosphorus. See MICROCOSMIC SALT.

Salt, Pinakol. See PINAKOL SALT N.

Salt, Potash. See ABRAUM SALTS.

Salt, Protein Silver. See LARGIN.

Salt, Ronozol. See SOZOIODOL.

Salt, Roussin's. See ROUSSIN'S BLACK AND RED SALTS.

Salts. See EPSOM SALTS.

Salt, Seignette's. See ROCHELLE SALT.

Salt, Smelling. See ENGLISH SALT.

Salt Spar, Bitter. See DOLOMITE.

Salt, Stassfurt. See ABRAUM SALTS.

Salt, Stripping. See ABRAUM SALTS.

Salt, Wonderful. Sodium sulphide, Na_2S.

Salubrol. Tetrabromo-methylene-di-antipyrine, $C_{23}H_{24}O_2N_4Br_4$. An antiseptic.

Salufer. The sodium salt of hydrofluosilicic acid, Na_2SiF_6. An antiseptic.

Salumin (Salumen). Aluminium salicylate, $Al(C_6H_4OH.COO)_3$. An antiseptic and astringent employed in nasal and throat affections.

Salunol. An aqueous solution of sodium hypochlorite. A disinfectant.

Salupres. A proprietary trade name for Hydrochlorothiazide.

Salurene. Hexamethylene-tetramine salicylate.

Salurheuma. A German preparation containing menthol, anæsthesin, and salicylic acid. Used as a rubbing agent for rheumatism.

Saluric. A proprietary preparation of chlorothiazide. A diuretic. (310).

Salvacid. A German product. It consists of pastilles containing the terpene thujone and the gall acids of ox gall. Used in the treatment of stomach troubles.

Salvadorite. A mineral. It is a copper-iron hydrated sulphate.

Salva-glas. A proprietary safety glass.

Salvamin. A German product. It is the lactone of ethanol-amino-gallic acid hydrochloride, and is similar in its action to adrenalin, but is less poisonous. Tablets containing it are used against asthma and hay fever.

Salvandra. A German product. It is a compound which forms nascent oxygen and chlorine by dissociation of chloric acid, $HClO_3$.

Salvarom. The diethyl ester of phthalic acid. A solvent.

Salvarsan (Hata, Kharsivan, Arsenobenzol, Arsenobillon, (A.B.) Arsphen, Ehrlich 606, Arsphenamine, Ehrlich-Hata 606, 606). Dihydroxy-diaminoarseno-benzene-di-hydrochloride, $HO(NH_2.HCl).C_6H_3.As.As.C_6H_3(NH_2.HCl)OH.2H_2O$. It contains 31·6 per cent. arsenic, and is one of the most efficient remedies for the treatment of syphilis. Also see ARSAMINOL, ARSEN-PHENOLAMINE, DIARSENAL, NEOARSAMINOL, NEOARSPHENAMINE, and NEODIARSENOL.

Salvarsan, New. See NEOSALVARSAN.

Sal Vegetable. Potassium tartrate, $K_2C_4H_4O_6.\frac{1}{2}H_2O$.

Salvochin. A German product. It is a 25 per cent. aqueous solution of quinine for intramuscular injection in cases of pneumonia.

Sal Volatile. A spiritous solution of $1\frac{1}{4}$ per cent. ammonia, $3\frac{1}{2}$ per cent. ammonium carbonate, with oils of nutmeg and lemon (Spiritus ammoniæ aromaticus, B.P.). Commercial ammonium carbonate, $(NH_4)_2CO_3$, is also called Sal volatile.

Salvosal-lithia. Lithium-salol-o-phosphite, the lithium salt of salo-o-phosphorous acid, $C_6H_4(O.PO.(OH)(OLi)).COO.C_6H_5$. Used in influenza and gout.

Salyran. The sodium salt of allylamido-phenoxy-acetic acid, $C_3H_5.NH.CO.C_6H_4.OCH_2COONa$. Used in medicine.

Salyrgan. A German preparation. It is a mercury compound of the sodium salt of salicyl-allyl-amide-o-acetic acid. An antiluetic and diuretic.

Salysal. Salicyl-salicylate.

Salzburg Vitriol. See MIXED VITRIOL.

Salzone. A proprietary preparation of PARACETAMOL. An analgesic. (255).

Samarskite (Uranotantalite). A tantalum mineral. It consists mainly of the niobates and tantalates of iron, calcium, yttrium, and cerium earths.

Sambol. A light yellow refined mineral oil from the Dutch Indies.

Samiresite. A mineral. It is a hydrated uranyl columbate containing 45·8 per cent. Cb_2O_5 and 3·7 per cent. Ta_2O_5.

Samite. A trade name for a carborundum product. An abrasive.

Samli. A clarified butter from East Africa.

Samna (Samn). Arabic terms for ghee (*q.v.*).

Samoite. A mineral of Samoa. It is Allophane.

Samol. See SALIMENTHOL.

Samorin. A proprietary preparation of ISOMETAMIDIUM hydrochloride. A veterinary anti-protozoan.

Samsonite. An explosive containing nitro-glycerin, collodion cotton, potassium nitrate, wood meal, and ammonium oxalate. Used in coal mines.

Samson Steel. A proprietary trade name for nickel-chromium steel containing 1·25 per cent. nickel and 0·6 per cent. chromium.

Samsu. A fermented drink made from rice. It is made and used in Eastern Asia.

Sanalgin. Phenazone-caffeine citrate and phenacetin.

Sanatogen. A registered trade mark for a food preparation, consisting of 95 per cent. casein and 5 per cent. sodium-glycero-phosphate.

Sanatol. See CREOLINE.

Sancos. A proprietary preparation of pholcodine, menthol and glycerin. A cough linctus. (267).

Sancos Compound. A proprietary preparation of pholcodine, pseudoephedrine hydrochloride and chlorpheniramine maleate. A cough linctus. (267).

Sand Acid. Hydrofluosilicic acid, H_2SiF_6.

Sandalwood. See REDWOODS.

Sandalwood, Red. See REDWOODS.

Sandalwood, White. The wood of *Santalum album*.

Sandaracha. Synonym for Realgar.

Sandarac Resin. A resin obtained from the N.W. African tree *Callitris quadrivalis*. It melts at from 135–160° C., has an acid value of from 90–154 (usually 140–154), a saponification value of 145–157, an iodine value of 112–134, and an ash of 0·04–0·1 per cent. Its specific gravity varies from 1·04–1·09. It is soluble in amyl alcohol, acetone,

and ether, and partly soluble in chloroform. It is used in the manufacture of spirit varnishes. Pine gum or Australian sandarac is obtained from *Callitris* species in Australia, and resembles the African variety.

Sandbergite. A mineral. It is a variety of tennantite.

Sand Cement. A cement consisting of equal parts of Portland cement and sand.

Sand, Chemical. See CHEMICAL SAND.

Sandel Red. The red resinous colouring principle of Sanderswood.

Sandelwood. See REDWOODS.

Sanderite. A mineral. It is $MgSO_4$. $2H_2O$.

Sander's Blue. Anhydrous basic copper carbonate.

Sanderswood. See REDWOODS.

Sand, Green. See GLAUCONITE.

Sandiver (Glass Gall). The scum formed on the surface of molten glass. It consists of calcium and sodium sulphates, with about one-tenth of its weight of glass.

Sandix. See ORANGE LEAD.

Sandocal. A proprietary preparation of calcium lactate gluconate and sodium and potassium bicarbonate, used in the treatment of osteoporosis. (267).

Sandoce. Methyl-saccharin, $C_6H_3(CH_3)$. $CO.SO_2.NH$. A sweetening substance.

Sando-K. A proprietary preparation of potassium chloride. (267).

Sandoptal. A German product. It is a new hypnotic isobutyl-allyl-barbituric acid. By replacing the diethyl group in hypnotics of the barbituric acid group by higher radicles, like the allyl complex, a higher hypnotic effect is produced in sandoptal.

Sandosten. A proprietary trade name for the tartrate of Thenalidine.

Sandoz Effervescent Compound. A proprietary preparation of caffeine, pseudoephedrine hydrochloride and paracetamol. An analgesic. (267).

Sandscale. Impurities formed in the pan during the concentration of brine. It consists of calcium carbonate, $CaCO_3$.

Sandstone. A stone consisting of grains of sand cemented together by a cementing material, the most common being silica, carbonate of iron, and oxide of iron. It is called calcareous sandstone when the grains are united together with calcium carbonate, argillaceous sandstone when a clayey cement unites the particles, micaceous when scales of mica are present, felspathic when grains of felspar are there, and ferruginous when ferric oxide or hydroxide is present.

Sanescol. A proprietary preparation of belladonna dry extract, kaolin, thiamine hydrochloride, riboflavine, nicotinamide and ascorbic acid. (334).

Sanfoin. See ORCHIDEE.

Sanfordite. A mineral. It is rickardite.

Sangajol. A trade mark for a fraction of Borneo petroleum distillate boiling at 160–170° C. It contains cyclic hydrocarbons, and is used as a turpentine substitute and resin solvent.

Sang de Bœuf. See ROUGE FLAMBÉ.

San-gri-na. A proprietary preparation containing 33 per cent. sulphur, 21 per cent. potassium hydrogen tartrate, $KHC_4H_4O_6$, 15 per cent. vegetable extract, 11 per cent. calcium carbonate, 11 per cent. arrowroot starch, 4 per cent. talc, 2 per cent. phenol-phthalein, and water. Sold as a remedy for obesity.

Sanguial. A preparation of blood and iron, containing 10 parts hæmoglobin, 44 parts muscle albumin, and 46 parts blood salts. Prescribed for chlorosis.

Sanguinarine Nitrate. Nitrates of the mixed alkaloids of *Sanguinaria canadensis*.

Sanguinella Oil. See DOGWOOD OIL.

Sanguinite. A mineral. It is a silver sulpharsenite.

Sanidine (Ice-Spar, Glassy Felspar). A mineral. It is a variety of Felspar (*q.v.*).

Sanitas. An aqueous liquid prepared by blowing air through warm oil of turpentine, in contact with water. It contains hydrogen peroxide and thymol, and is used as a disinfectant.

Sanmartinite. A mineral. It is $2[ZnWO_4]$.

Sanna (San Nai, Kapur Kachri, Sitruti, Sheduri). A Chinese drug. It is the dried roots and stems of *Hedychium spicatum*.

San Nai. See SANNA.

Sanochinol (Sanschinol). A quinine compound, said to be the product of the action of ozone upon a solution containing 4 per cent. quinine hydrochloride. It is recommended in the treatment of malaria.

Sanocrysin. A colloidal gold solution, containing sodium aurous thiosulphate $Na_3Au(S_2O_3)_2$. It has been known as Fordos and Geles salt. Used in the treatment of tuberculosis.

Sanoform. A disinfectant consisting of a mixture of the disinfecting constituents of various tar oils, with calcium chloride and magnesium chloride in a saponified form. The term is also used for the methyl ester of di-iodo-salicylic acid, an iodoform substitute.

Sanoleum. A mixture of crude cresols with hydrocarbons, used in disinfecting urinals.

Sanomigran. A proprietary preparation of PIZOTIFEN hydrogen maleate, used as a preventative of migraine. (244).

Sanoscent (Camphortar). Preparations containing camphor as the main ingredient. Disinfectants.

Sanose. A powder containing 80 per cent. pure casein and 20 per cent. albumose from white of egg.

Sanovite. A proprietary foodstuff having a high content of roughage, claimed to be similar to bread. (739).

Sanschinol. See SANOCHINOL.

Sanse. The residual cakes obtained from pressed Italian olives. When dried, and the oil extracted with carbon disulphide, it gives the so-called sulphocarbon oil, or sulphur olive oils. Used in the manufacture of green soap for use in the textile industries.

Sanse Oil. See SANSE.

Sansol. A proprietary trade name for a water thinnable emulsion paint based upon epoxy resin. (154).

Santal. The red colouring matter of Sanderswood.

Santal Oil. Oil of sandal wood.

Santalwood. See REDWOODS. It contains 16 per cent. of santalin, $C_{15}H_{14}O_5$. The extract is used to colour confectionery and liqueurs.

Santheose. Theobromine, $C_5H_2(CH_3)_2N_4O_2$.

Santiago New Yellows E, K. Dyestuffs. They are fustic preparations.

Santicizer. Proprietary products. (1) *p*-toluene-sulphon-anilide. It is used as a plasticiser in cellulose lacquers. (2) Ethyl-*p*-toluene-sulphon-amide. (3) Ethyl-*m*-toluene-sulphon-amide.

Santicizer 8. A proprietary trade name for a plasticiser. A mixture of *o*- and *p*-toluene ethyl sulphonamides. It is much used in the plasticising of nylon. (57).

Santicizer 9. A proprietary trade name for a plasticiser. A mixture of *o*- and *p*-toluene-sulphonamides.

Santicizer 10. A proprietary trade name for a plasticiser. *o*-Cresyl-*p*-toluene-sulphonate.

Santiciser. A proprietary trade name for vinyl plasticisers coded as follows :

B-16. Butyl phthalyl butyl glycollate. (57).

LOY. Dioctyl phthalate. (57).

E-15. Ethyl phthalyl ethyl glycollate. (57).

603. Butyl decyl phthalate. (57).

606. A mixed phthalate. (57).

SC. The triglycol ester of a vegetable oil fatty acid. (121).

M-17. Methyl phthalyl ethyl glycollate.

1-H. N-cyclohexyl paratoluenesulphonamide.

107. Di-2-ethyl hexyl phthalate.
140. Cresyl diphenyl phosphate.
141. Octyl diphenyl phosphate.
160. Butyl benzyl phthalate.
601. A 50/50 blend of di-*n*-octyl and *n*-decyl phthalate and di-iso-octyl phthalate. (57).
602. A 50/50 blend of di-iso-decyl phthalate and di-iso-octyl phthalate. (57).

Santobrite. A proprietary trade name for sodium pentachlorphenate. A preservative used in paints, adhesives, etc.

Santocel. A proprietary trade name for silica gel (*q.v.*), a porous form of silica. It is used as a heat insulator, drying agent, etc.

Santochlor. A proprietary trade name for *p*-dichloro-benzene. Used as a deodoriser, moth preventative, etc.

Santocure. A proprietary trade name for benzothiazyl-2-monocyclohexyl sulphonamide and amino derivatives of 2-thiobenzo-thiazole. N-cyclohexyl-2-benzthiazyl sulphenamide. They are used as accelerators in the vulcanisation of rubber. (57).

Santocure Mor. It is 4-(Benzo-thiazole-2-sulphenyl) morpholine. A rubber accelerator. (559).

Santocure Mor 90. A proprietary compound of 90% 2-(4-morpholinyl mercapto)-benzthiazole and 10% dibenzthiazyl disulphide. (57).

Santocure NS. A proprietary accelerator. N-*tert*.-butyl-2-benzthiazyl sul-phenamide. (559).

Santoflex 75. A proprietary blend of 25% SANTOFLEX DD and 75% N,N'-diphenyl paraphenylene diamine. (57).

Santoflex 9010. A proprietary anti-oxidant. Phenyl - β - naphthylamine plus a small proportion of diphenyl-*p*-phenylene diamine. (559).

Santoflex A. A proprietary trade name for a mixed ketone-amine and diphenyl-*p*-phenylene-diamine. A rubber anti-oxidant.

Santoflex AW. A proprietary anti-oxidant. 6 - ethoxy - 2, 2, 4 - trimethyl - 1, 2-dihydroquinoline. (559).

Santoflex B. A proprietary trade name for the condensation product of acetone and *p*-amino-diphenyl. A rubber anti-oxidant.

Santoflex BX. A proprietary trade name for a constant composition blend of Santoflex B and diphenylparaphenylenediamine. A rubber antioxidant.

Santoflex CP. A proprietary anti-oxidant. N-cyclohexyl-N-phenyl-*p*-phenylene diamine. (559).

Santoflex DD. A proprietary anti-oxidant. 6 - dodecyl - 2, 2, 4 - trimethyl - 1,2-dihydroquinoline. (559).

Santoflex DPA. A proprietary anti-oxidant. An acetone/diphenylamine reaction product. (559).

Santoflex I.P. A proprietary anti-oxidant. N-phenyl-N-isopropyl-*p*-phenylene diamine. (559).

Santolite. A proprietary trade name for synthetic resins of the sulphonamide-aldehyde type, e.g. toluene sulphonamide-formaldehyde, for use in lacquers, etc.

Santomerse. A proprietary trade name for an alkylated aryl sulphonate. Used as a wetting agent.

Santonin, Chromo. See CHROMO SANTONIN.

Santonin, Golden. See CHROMO SANTONIN.

Santonin Wafers (Worm Wafers). Consist of cane sugar with santonin and egg albumin.

Santoperonin. A proprietary preparation stated to be used in the treatment of worms. It is said to contain copper, salicylic acid, and other materials.

Santophen 20. A proprietary trade name for pentachlorphenol. A preservative for paints, wood, adhesives, etc.

Santoresin. A proprietary trade name for a synthetic resin.

Santorin Earth. A volcanic ash found in the island of Santorin. Used to convert lime into hydraulic lime.

Santosite. A proprietary trade name for anhydrous sodium sulphite. A reducing agent.

Santotan KR. A proprietary trade name for a basic chromium sulphate, $Cr_2(SO_4)_3.(OH)_2$. A tanning agent.

Santovar A. A proprietary trade name for an alkylated polyhydroxyphenol. (57).

Santovar-O. Insoluble sulphur 60.

Santovar O. See SANTOVAR A.

Santowhite. A proprietary trade name for a dialkyl phenol sulphide. (57).

Santowhite CI. A proprietary anti-oxidant. N, N - dinaphthyl-*p*-phenylene diamine. (559).

Santyl. Santalol salicylate, $C_6H_4(OH).COO.C_{15}H_{23}$. An oil prescribed for gonorrhœa.

Sanusin. A proprietary compound of resorcin, boric acid, balsam of Peru, zinc and bismuth carbonates. It has antiseptic and healing properties.

Sanyan. A silk from a wild silkworm of Nigeria.

Sapamine. A proprietary trade name for diethyl - amino - ethyl - oleyl - amino ace-tate and similar compounds. Used in conjunction with dyes.

Sapan Wood. See REDWOODS.

Saparoform. A solution of 3–5 per cent. of paraformaldehyde in liquid potash soap. Said to be used medicinally.

Sapene. A liquid soap, a vehicle for medicaments.

Sap Green (Buckthorn Green, Vegetable Green, Bladder Green, Chinese Green, Lokas, Iris Green). The colouring matter obtained by evaporating to dryness a mixture of lime, or sometimes a little alum, and indigo carmine with the juice of the berries of buckthorn. It is employed in China for giving green shades on silk.

Sapin. A mixture of Japan wax with heavy mineral oil (soft or liquid paraffin). A superfatting agent for soaps.

Sapocarbol. Lysol (q.v.).

Sapodermin. A soap preparation containing about 3 per cent. of the mercury compound of casein.

Sapoform. A product containing oleic acid, alcohol, potassium hydroxide, formalin, and distilled water.

Sapogenin. A decomposition product $(C_{14}H_{22}O_2)$ of saponin.

Saponia. A German soap substitute. It is a mixture of soap, with sodium carbonate, and clayey matter.

Saponification Glycerin. That glycerin obtained in stearine manufacture. Crude saponification glycerin contains 90·55 per cent. glycerol, has a specific gravity of 1·255, gives a residue at 160° C. of 0·6 per cent., and contains 0·53 per cent. inorganic salts and 2 parts per 1,000,000 of arsenic.

Saponification Olein. See OLEIN OF SAPONIFICATION.

Saponified Cresol. See BACTEROL.

Saponine. A name usually applied to the active constituent of Panama bark, which is used instead of soap, for washing and producing a lather. The term is also used for a boring and cutting oil.

Saponite. See STEATITE.

Sapphire. Alumina, Al_2O_3, coloured with metallic oxides.

Sapphire-quartz. A variety of quartz having a sapphire colour.

Sapphire, White. See WHITE SAPPHIRE.

Sapphirine. A mineral, $5MgO.6Al_2O_3.2SiO_2$.

Saprol (Disinfection Oil). A mixture of crude cresols, hydrocarbons, and pyridine bases. Used for disinfecting lavatories.

Sap Yellow. A vegetable dyestuff obtained from the half-ripe berries of various species of Rhamnus. Used in the form of a lake in painting.

Saralasin. A drug used in the diagnosis of renin-dependent hypertension. It is [1 - (N - methylglycine), 5 - L - valine, 8 - L - alanine]angiotensin. P-113 is under clinical trial as the hydrated acetate.

Saran. A trade mark for a range of polyvinylidene chloride plastics. (613).

Saratoga Steel. A proprietary trade name for a non-shrinking steel containing small quantities of manganese, chromium, tungsten, and carbon.

Sarawakite. A mineral. It is an antimony chloride.

Sarcine (Sarkine). Hypoxanthine, $C_5H_4N_4O$.

Sarcinite. A mineral. It is a basic manganese arsenate.

Sarco. A material made from elaterite (a mineral rubber). Used in rubber mixings.

Sarcocoll. A gum resin from Penœa sarcocolla, of Africa.

Sarcolite. A mineral, $3(Ca.Na_2)O.Al_2O_3.3SiO_2$.

Sarcopside. A mineral. It is $(Fe,Mn,Ca)_7(PO_4)_4F_2$.

Sarcosine. See METHYL-GLYCOCOLL.

Sard. A mineral. It is a brown variety of Chalcedony (q.v.).

Sardinian (Sardinianite). Minerals. They are varieties of anglesite, $PbSO_4$.

Sardonyx. An agate containing alternating bands of red and white chalcedony.

Saridone. A proprietary preparation of phenacetin with caffeine and phenyl-dimethyl-isopropyl-pyrazolone. Also given as a proprietary preparation of isopropylantipyrine, phenacetin and caffeine. An analgesic. (314).

Sarkalyt. A solution of the sodium salt of adenosin-phosphoric acid.

Sarkine. See SARCINE.

Sarkinite (Polyarsenite). A mineral, $Mn(MnOH)AsO_4$.

Saroten. A proprietary preparation of amitriptyline hydrochloride. An antidepressant. (262).

Sarpol. A preparation of crude phenol (carbolic acid). A disinfectant.

Sarsaparilla. The dried root of Smilax officinalis and other species. A tonic.

Sarton. A preparation made from the seeds of Glycine soja. Used in the treatment of diabetes.

Sartorite. A mineral, $PbS.As_2S_3$.

Sase. A viscose artificial silk fibre.

Sassafras. The dried bark of the root of Sassafras variifolium. The extract is used as a carminative.

Sassolin. Tuscan boric acid.

Sassolite. A mineral. It is $4[B(OH)_3]$.

Sassy Bark (Saucy Bark). The bark of Erythrophleum guineanse.

Satco Metal. A proprietary trade name for a lead-base bearing alloy modified by the addition of tin, calcium, magnesium, mercury, aluminium, potassium lithium, all in very small quantities except tin which may rise to 1 per cent.

Sateen. An imitation satin made from mercerised cotton.

Satin. A heavy silk fabric woven in a special manner.

Satin-gloss Black. See GAS BLACK.

Satin Green. See CHROME GREENS.

Satinite. See GYPSUM.

Satin Rouge. A variety of lamp-black used for polishing.

Satin Spar. A mineral. It is a variety of Gypsum (*q.v.*). Used for making beads.

Satin White. A pigment. It consists of gypsum mixed with alumina. A mixture of calcium sulphate with aluminium sulphate is also known under this name.

Sativic Acid. Trioxy-stearic acid, $C_{17}H_{31}(OH)_3.COOH$.

Satrapol. A photographic developer containing mono-methyl-*p*-amino-phenol-sulphate.

Saturn Red. See RED LEAD and ORANGE LEAD.

Sauconite. A clay containing zinc.

Saucy Bark. See SASSY BARK.

Sauerin. A preparation of lactic acid bacilli.

Saunderswood, Red. See REDWOODS.

Saurol. A distillation product of Meride shale. It resembles ichthyol in its therapeutic properties.

Saussurite. A mineral. It is impure labradorite.

Saventrine. A proprietary preparation of isoprenaline hydrochloride, used as a cardiac stimulant. (330).

Savlon. A proprietary preparation of CHLORHEXIDINE and cetrimide, used as a skin cleanser. (2).

Savol. A medicated soap. It contains salol (phenyl salicylate) with perfumes.

Savonade (Texapon, Texalin, Hydralin). Liquid hexalin and methyl-hexalin soaps.

Savonette Oil. A mixture of vegetable fatty acids and resin acids, a by-product of paper manufacture. It has a specific gravity of 0·965, a saponification value of 155–170, and an acid value of 150–165. It is recommended as a substitute for oleic acid in soap manufacture.

Savore. A preparation of milk, cereal proteins, albuminoses, and carbohydrates.

Saxifragin. An explosive mixture containing 76 per cent. barium nitrate, 2 per cent. potassium nitrate, and 22 per cent. charcoal.

Saxin. A proprietary preparation of saccharin.

Saxol. See PARAFFIN, LIQUID.

Saxoline. See OZOKERINE.

Saxon Blue. The term is usually applied to a solution of indigo in sulphuric acid, but it is also a synonym for smalt.

Saxon Green. A pigment. It is a green earth found in Saxony.

Saxonite. An explosive similar to Samsonite in composition. The term is also used for a mineral which is a mixture of olivine and enstatite.

Saxon Verdigris. A pigment which approaches Brunswick green in composition, and is prepared by precipitating a mixture of copper sulphate and sodium chloride with milk of lime. It also contains gypsum or Vienna white.

Saxony Blue. See SMALT.

Sayawer. See CHAY ROOT.

Saynite. A mineral. It is a mixture of polydymite, Ni_4S_5, and bismuthinite.

Scabiol. A liniment containing balsam of Peru, storax, alcohol, and castor oil.

Scacchite. See MONTICELLITE.

Scadomex 320. A proprietary trade name for a reactive melamine resin used in stoving enamels. (195).

Scadonal 84. A proprietary trade name for a non yellowing long oil alkyd resin used in brushing enamels. (195).

Scadonal 150-X-70. A proprietary trade name for alkyd resin based on synthetic fatty acids. Used in appliance finishes. (195).

Scadonoval 14-X-60. A proprietary trade name for a styrenated alkyd resin. It has quick drying properties and high hardness. It is used in industrial paints. (195).

Scadonoval 15-ML-55. A proprietary trade name for a vinyl toluene modified alkyd resin. It has good gloss and brushing properties. (195).

Scadoplast RA3L, RA350. A proprietary trade name for an adipic acid polyester vinyl plasticiser. (126).

Scadoplast RS 20, RS 150. A proprietary trade name for a sebacic acid polyester vinyl plasticiser. (126).

Scadoplast S 1209. A proprietary trade name for a pentaerythritol ester plasticiser with low volatility and good heat resistance. It is used for vinyl covered heating wire for electric blankets. (195).

Scadoset 620-XB-60. A proprietary trade name for a thermosetting acrylic resin. (195).

Scæchite. A mineral, $MnCl_2$.

Scagliola. A stone manufactured from Keene's cement mixed with colouring matter, to which is added water containing dissolved glue or isinglass.

Scale. Crude Scotch paraffin wax is known in commerce by this name.

Scalol. A photographic developer containing methyl-*p*-amino-phenol, $C_6H_4(OH)(NH.CH_3)$, as the active constituent.

Scammonium Resin. A resin obtained from *Convolvulus scammonia*, of Aleppo and Smyrna. Used in pharmacy. Acid value 14–21. Iodine value 10–15.

Scammony Extract. See EXTRACT OF SCAMMONY.

Scandia. Scandium oxide, Sc_2O_3.

Scandonoval. 16-ML-50. A proprietary trade name for a vinyl toluene modified alkyd resin. (195).

Scapolite (Paranthine, Wernerite). A mineral, $4CaO.3Al_2O_3.6SiO_2$, or $2Na_2O.Al_2O_3.9SiO_2$, with chlorine and calcium tungstate.

Scarab. A proprietary urea formaldehyde moulding material. (915).

Scarat. A proprietary trade name for a synthetic resin of the urea type.

Scarlet B. See BIEBRICH SCARLET.

Scarlet 3B, 3R, 4R. Varieties of Biebrich Scarlet (q.v.).

Scarlet Corns. See KERMES.

Scarlet EC. See BIEBRICH SCARLET.

Scarlet F. See NEW COCCINE.

Scarlet for Cotton. A dyestuff. It is a mixture of Chrysoidine with Safranine.

Scarlet for Silk. See FAST RED B.

Scarlet G. See PONCEAU R.

Scarlet GR (Scarlet R, Orange N, Orange L, Brilliant Orange R, Xylidine Orange, Naphtharene Orange R, Scarlet RL). A dyestuff. It is the sodium salt of xylene-azo-β-naphthol-mono-sulphonic acid, $C_{18}H_{15}N_2O_4SNa$. Dyes wool yellowish-red from an acid bath.

Scarlet GT. A dyestuff obtained from p-toluidine and β-naphthol-sulphonic acid.

Scarlet J, JJ. See EOSIN BN.

Scarlet Lake. A pigment. It is a carmine lake with a scarlet hue, which is imparted to it by mixture with vermilion.

Scarlet Ochre. See INDIAN RED.

Scarlet OOO. See CROCEINE 3BX.

Scarlet Phosphorus (Schenk's Phosphorus). Obtained by heating 70 parts phosphorus with 30 parts phosphorus tribromide, PBr_3.

Scarlet R. See SCARLET GR.

Scarlet 2R. See CARMINAPH GARNET.

Scarlet 5R. See PONCEAU 5R.

Scarlet 6R (Ponceau 6R, Amaranth). A dyestuff. It is the sodium salt of p-sulpho-naphthalene-azo-β-naphthol-trisulphonic acid, $C_{20}H_{10}N_2O_{13}S_4Na_4$. Dyes wool red from an acid bath.

Scarlet Red. See IODINE RED and BIEBRICH SCARLET.

Scarlet RL. A dyestuff. It is a British equivalent of Scarlet GR.

Scarlet 2R, 2RL, R, 3R. British equivalents of Ponceau.

Scarlet S. See FAST PONCEAU 2B.

Scarlet Spirit (Bowl Spirit). A Tin Spirit (q.v.), used by wool dyers for producing cochineal scarlet.

Scarlet Vermilion (Extract of Vermilion, Chinese Vermilion, Orange Vermilion, Field's Orange Vermilion). Varieties of vermilion, see CINNABAR.

Scatole (Skatole). Methyl-indole, $C_6H_4(C.CH_3)(NH)CH$.

Scawtite. A mineral. It is $Ca_7Si_6O_{16}CO_3(OH)_4$.

Schaeffer's Acid (Baum's Acid, Armstrong Acid). $1:2$-α-Naphthol-sulphonic acid, $C_{10}H_6(OH)(SO_3H)$, also $2:6$-β-naphthol-sulphonic acid.

Schaeffer's Salt. The sodium salt of $2:6$-β-naphthol-sulphonic acid.

Schafarzikite. A mineral. It is a ferrous phosphite, found in Hungary.

Schairerite. A mineral of California. It approximates to the formula, $Na_2SO_4.Na(F.Cl)$.

Schallerite. An arseno-silicate found in the Franklin furnace. It approximates to the formula, $9MnSiO_3.Mn_3As_2O_8.7H_2O$.

Schambit. A proprietary trade name for a heat resistant coating employing phosphoric acid and reactive pigments. (3).

Schapbachite. A mineral, $(Pb.Ag_2)_2Bi_2S_5$.

Scharizerite. A black asphalt-like mineral. It gives 35 per cent. carbon, 4·5 per cent. hydrogen, and 8 per cent. nitrogen, and an ash of 17 per cent.

Scheele's Acid. A 4 per cent. solution of hydrocyanic acid, HCN.

Scheele's Green (Mineral Green, Swedish Green). A pigment consisting of copper arsenite, $CuHAsO_3$.

Scheeletine (Scheelinite). A tungstate of lead, $PbWO_4$.

Scheelinite. See SCHEELETINE.

Scheelite (Calciosheelite). See TUNGSTENITE.

Scheerite. A mineral wax resembling ozokerite.

Schefferite. A pyroxene mineral, $(Mg.Fe)(Ca.Mn)(SiO_3)_2$.

Scheiber Oil. The glyceride of dehydrated ricinoleic acid. It is suitable for varnishes.

Scheibler's Reagent. Sodium-phospho-tungstate, obtained by dissolving 100 grams sodium tungstate and 70 grams sodium phosphate in 500 c.c. water, and acidifying with nitric acid. Used as a testing reagent for alkaloids.

Scheiderite. A mixture of trinitro-naphthalene and ammonium nitrate. An explosive.

Schellan Solution. A colloidal solution of a synthetic resin made from urea and formaldehyde, and kept from gelatinising by means of sodium acetate. Used as a dressing material for textiles.

Schenk's Phosphorus. See SCARLET PHOSPHORUS.

Schericur. A proprietary preparation of hydrocortisone and clemizole-hexachlorophane for dermatological use. (898).

Schering I and II. See MINERAL TABLETS.

Scheriproct. A proprietary preparation of PREDNISOLONE, cinchocaine hexachlorophane and CLEMIZOLE undecylenate used in the treatment of hæmorrhoids. (438).

Schernikite. A mineral. It is a variety of muscovite.

Schertelite. A mineral, $Mg(NH_4)_2.H_2(PO_4)_2.4H_2O$. It occurs in bat guano.

Scheteligite. A mineral. It is $(Ca,Mn,Sb,Bi)_2(Ti,Ta,Cb,W)_2(O,OH)_7$.

Schiff's Reagents. (1) Consists of a solution of rosaniline hydrochloride, decolorised by sulphur dioxide, and is used to test for aldehydes. (2) Furfuraldehyde and hydrochloric acid, employed for testing for urea. (3) Concentrated sulphuric acid, followed by ammonia, a test for cholesterin.

Schillerspar (Bastite). A mineral. It is an alteration product of Enstatite or Bronzite.

Schimose. See LYDDITE.

Schirmerite. A mineral, $3(Ag_2Pb)S.2Bi_2S_3$.

Schlempe. Beet sugar waste. It is the thick brown liquor remaining after the extraction of all possible sugar. It is also called Vinasse.

Schleretinite. A resin found in the coal mines of Wigan.

Schlerospathite. A mineral. It is $(Fe''',Cr)SO_4OH.7H_2O$.

Schlippe's Salt. Sodium-thio-antimonate, $Na_3SbS_4.9H_2O$. Used for the preparation of golden sulphide of antimony.

Schnapps. German potato spirit, an alcoholic liquor.

Schneebergite. A mineral. It approximates to the formula, $CaSbO_3$.

Schneiderite. An explosive. It contains 88 per cent. ammonium nitrate, 11 per cent. dinitro-naphthalene, and 1 per cent. resin. Used for high explosive shells.

Schnitzer's Green. A pigment similar to Arnaudon's green, except that crystallised sodium phosphate is used instead of ammonium phosphate in its preparation. Chromium phosphate is also known by this name.

Schoellkopf's Acids. α-Naphthol-disulphonic acid (1 : 4 : 8), and α-naphthylamine-sulphonic acid (1 : 8).

Schœnanthe. Oil of lemon grass.

Schoenite (Schonite). Potassium-magnesium sulphate, $K_2SO_4.MgSO_4.6H_2O$. It occurs in the Stassfurt deposits.

Schoepite. A new uranium mineral found in the Belgian Congo.

Scholine. A proprietary preparation of suxamethonium chloride. Short-acting muscle relaxant. (284).

Schollpopf Acid. See SCHOELLKOPF'S ACIDS.

Scholzite. A mineral. It is $2[Ca_3Zn(PO_4)_2(OH)_2.H_2O]$.

Schonberg's Alloy. A die-casting alloy containing 87 per cent. zinc, 10 per cent. tin, and 3 per cent. copper.

Schonite. See SCHOENITE.

Schorl (Shorle). A black tourmaline.

Schorlomite. A mineral, $Ca(Ti.Fe).SiO_5$.

Schorl Rock. An aggregate of black tourmaline and quartz.

Schou Oil (Paalsguard Oil). An emulsifier made from soya-bean oil.

Schraufite. A fossil resin found in Carpathian sandstone.

Schreibersite (Dyslytite). An iron-nickel phosphide found in meteorites. A chromium sulphide has also been called Schreibersite.

Schröckingerite. A mineral. It is $4[NaCa_3UO_2SO_4(CO_3)_3F.10H_2O]$.

Schrötterite. A mineral. It is a hydrated aluminium silicate.

Schultenite. A mineral of S.W. Africa with a composition corresponding to $2PbO.As_2O_5.H_2O$.

Schultzenite. A mineral, $CuO.2CoO.Co_2O_3.4H_2O$.

Schultze's Reagents. (1) Phosphoantimonic acid, made from sodium phosphate and antimony pentachloride, an alkaloidal reagent. (2) Consists of 25 parts dry zinc chloride, 8 parts potassium iodide, $8\frac{1}{2}$ parts water, and iodine. It gives a blue colour with cellulose.

Schultze's Stain. A microscopic stain. It consists of equal parts of a 2 per cent. solution of β-naphthol sodium and a 2 per cent. solution of dimethyl-p-phenylene-diamine hydrochloride. The solutions are mixed and filtered.

Schultze's Smokeless Powder. Consists of 62 per cent. nitro-lignin, 2 per cent. potassium nitrate, 26 per cent. barium nitrate, 5 per cent. pet. jelly, and $3\frac{1}{2}$ per cent. starch.

Schungite. A mineral. It is carbon in an amorphous form.

Schuppenglanz. Synonyn for Franckeite.

Schutzenberger's Salt. Sodium hydrosulphite, $NaHSO_2$.

Schwartzenbergite (Plumbiodite). A mineral, $Pb(I.Cl)_2.2PbO$.

Schwatzite. A mineral. It is a mercury tetrahedrite.

Schweinfurth Green. See EMERALD GREEN.

Schweitzer's Reagent. A solution of copper hydrate, $Cu(OH)_2$, in strong ammonia. A solvent for cellulose.

Schweizerite. See ZERMATTITE.

Schwelkohle. A brown coal of Germany. It is light brown in colour.

Scian Turpentine. See CHIAN TURPENTINE.

Sciatago. A proprietary preparation of cinchophen with glycocoll and hexamine.

Scillin. See SCILLIPICRIN.

Scillipicrin (Scillotoxin, Scillin). Commercial names for pharmaceutical preparations of squill, the fleshy bulb of *Urginea scilla.*

Scillotoxin. See SCILLIPICRIN.

Sclair. A range of proprietary polyethylenes. (10). Constituent members of the range include the following:—
Sclair 17A and 19A. Used in the manufacture of film.
Sclair 30. Used for the extrusion of pipe.
Sclair 44A. A black compound used in the extrusion of wire coverings.
Sclair 44B. Black in colour, it is used in the extrusion of high-grade telephone cable coverings.
Sclair 44F. A natural polyethylene resin used for insulating coaxial cables.
Sclair 44G. A black polyethylene compound used to insulate military field wire.
Sclair 46A. A weather-resistant polyethylene compound, black in colour, used for the covering of line wires.
Sclair 47C. A natural polyethylene.
Sclair 56A. Used in the blow-moulding of containers.
Sclair 57. Used in the making of bottles for detergents.
Sclair 58 and 59. Used in blow-moulding applications.
Sclair 61A. Of low density. Used in extrusion coating processes.
Sclair 63B. Of medium density. Used in extrusion coating processes.
Sclair 79B. Used in the forming of sheet and extrusions.
Sclair 79C. Used in pressure-forming applications.
Sclair 79D. Its easy processing properties make it suitable for use in low-power extruders.
Sclair 96A. Used in rotational moulding applications.
Sclair 97B. An easily-processed grade of *Sclair 96A.*
Sclair 99B. Used in the preparation of fibrillated string.
Sclair 2000. A range of moulding resins having densities varying from 0·920 to 0·960.

Scleroclase. A mineral, $PbAs_2S_4$.

Sclerolac. A suggested name for hard lac resin (*q.v.*).

Scleron Alloys. See SELERON.

Sclevoveine. A solution of pure sodium salicylate.

Scolaban. A proprietary preparation of BUNAMIDINE. A veterinary anthelmintic.

Scolecite. A mineral, $CaO.Al_2O_3.3SiO_2.2H_2O$.

Scoline. A proprietary trade name for Suxamethonium chloride.

Scon C 7180 and C 7200. A proprietary vinyl chloride/vinyl acetate copolymer with 12–15% vinyl acetate content. (916).

Scon C 7260. A proprietary vinyl chloride/vinyl acetate copolymer with 9–11% vinyl acetate content. (916).

Scon C 7400. A proprietary vinyl chloride/vinyl acetate copolymer with a vinyl acetate content of 3–5%. (916).

Scon P 9470. A proprietary vinyl chloride homopolymer of high molecular weight used in the production of PVC plastisols. (916).

Scon S 5260. A proprietary vinyl chloride homopolymer of low molecular weight used in moulding applications. (916).

Scon S 5300. A proprietary vinyl chloride homopolymer of low molecular weight used for calendering. (916).

Scon S 5350. A proprietary general-purpose vinyl chloride homopolymer of intermediate molecular weight used in calendering, extrusion and moulding applications. (916).

Scon S 5380. A proprietary vinyl chloride homopolymer of intermediate molecular weight used for general-purpose fabrication. (916).

Scon S 5410. A proprietary general-purpose vinyl chloride homopolymer of high molecular weight. (916).

Scon S 5520. A proprietary general-purpose vinyl chloride homopolymer of very high molecular weight, used in the production of heavy-duty footwear. (916).

Scon S 8340. A proprietary vinyl chloride homopolymer of low molecular weight and with dense and fine particle size, used as a filler polymer in PVC plastisols. (916).

Scopacron. A proprietary trade name for thermosetting acrylic resins modifiable by means of epoxy resins. (216).

Scopacron 50, 75 and 80. A proprietary trade name for a thermo-setting acrylic resin capable of cross-linking with amino and epoxy compounds. Primarily intended for use with melamine-formaldehyde resin for motor car top coats. (198).

Scopacryl. A proprietary trade name

for thermoplastic acrylic resin solutions used for wall paints and road marking applications. (198).

Scopasol 550. A proprietary trade name for a water-dilutable thermosetting acrylic resin. Used for high performance white gloss coatings to be applied by electrophoresis techniques. (198).

Scopol 58M, 58SP. A proprietary trade name for a vinyl toluene modified alkyd resin. (198).

Scopol 85X. A proprietary trade name for a styrene modified alkyd resin. Used for quick drying coatings with exceptional adhesion properties. (198).

Scopolamine. Hyoscine, $C_{17}H_{21}NO_4$, an alkaloid.

Scopolamine Methobromide. Hyoscine methobromide.

Scopoline. Oscine, $C_8H_{13}O_2N$. It results from the hydrolysis of hyoscine.

Scopolux 221SP. A proprietary trade name for a medium oil alkyd based on linseed oil. (216).

Scopomannite. A solution of scopolamine hydrobromide in water containing 10 per cent. mannite.

Scopomorphine. A sterilised solution containing, per c.c., 0·0006 gram Euscopal (*q.v.*) and 0·015 gram morphine hydrochloride.

Scorbital. Tablets of phenobarbitone with ascorbic acid. (182).

Scorodite (Pitticite). A mineral, $FeAsO_4$. $2H_2O$.

Scorpion Oil. See OIL OF SCORPIONS.

Scorzalite. A mineral. It is $2[Fe^{..},Mg]$ $Al_2(PO_4)_2(OH)_2]$, a variety of Lazulite.

Scotch Cement. A cement prepared from feebly hydraulic limes, by the addition of 5 per cent. plaster of Paris, and grinding.

Scotch Foundry Pig. A pig iron made for foundry purposes from Scotch clayband or black-band ores. It usually contains from 0·7–1 per cent. phosphorus, and 2·5 per cent. silicon.

Scotch Gin. Spirit of sweet nitre (*q.v.*).

Scotch Pebbles. Agates, found chiefly in Forfarshire and Perthshire.

Scotch Soda. Impure sodium carbonate, Na_2CO_3.

Scotch Topaz. Golden topaz, a yellow variety of quartz.

Scotiolite. A mineral. It is a variety of hisingerite.

Scott's Dressing. Compound ointment of mercury.

Scott's Liniment. Lin. Hydrarg.

Scouring Slag. A slag produced in making spiegel. It is black in colour and contains up to 8 per cent. of oxide of iron.

Scovillite. A mineral. It is a variety of rhabdophanite.

Scrap Rubber. Formed by the drying of the latex on the bark at the tapping cut. It is variable in quality and colour.

Screen Plate Brass. See BRASS.

Screw Brass. See BRASS.

Screw Bronze. An alloy of 93·5 per cent. copper, 5 per cent. zinc, 1 per cent. tin, and 0·5 per cent. lead.

Scurane V. A proprietary trade name for polyurethane varnishes. (88).

Scurenaline. A proprietary preparation. It is stated to be a solution of synthetic adrenalin. A hæmostatic.

Scurocaine. Stated to be the chlorhydrate of Ethocaine (*p*-amino-benzoyl - diethyl - amino - ethanol hydro - chloride). A local anæsthetic.

Scuroforme (Butesin, Paraforme). Proprietary preparations of butyl-*p*-aminobenzoate, $NH_2C_6H_4.COOC_4H_9$. An antiseptic dusting powder or ointment for wounds.

S.D.V. A proprietary vaccine used in the de-sensitisation treatment of allergic asthma. (272).

SE-458. A proprietary silicone rubber compound used for bonding to unprimed surfaces during the curing process. (59).

Sea-Legs. Meclozine hydrochloride tablets—travel sickness remedy. (182).

Sealing Wax. Made from shellac, turpentine, and vermilion. Other varieties contain beeswax and rosin, the rosin being sometimes replaced by shellac, and various colouring matters are used.

Sealite. A liquid containing glucose, corn starch, glycerol, calcium chloride, and glue. Used to prevent evaporation from oil storage tanks.

Seamanite. A mineral. It is $4[Mn_3 PO_4BO_3.3H_2O]$.

Sea Onion. Squill.

Searlesite. A mineral, $Na_2O.B_2O_3$. $4SiO_2.2H_2O$, of California.

Sea Silk. See BYSSUS SILK.

Sea-water Bronze (Sheathing Bronze). An alloy of 32·5 per cent. nickel, 45 per cent. copper, 5·5 per cent. zinc, 16 per cent. tin, and 1 per cent. bismuth. It resists sea water.

Sebaclen. A proprietary preparation of xenysalate hydrochloride. A skin cleanser. (382).

Sebastine. A dynamite explosive.

Sebesite. A mineral. It is a variety of tremolite.

Sebkanite. A crude potassium chloride obtained by the evaporation and crystallisation of the water of the salt lake in Tunis.

Seboderm. A proprietary preparation of cetrimide. A skin cleanser. (383).

Secaderm Salve. A proprietary preparation of phenol, TEREBENE, melaleuca and turpentine oil used as a local anti-infective agent. (340).

Secaline. Trimethyl-amine, $CH_3.CH_3.CH_3.N$.

Secbutobarbitone. 5-Ethyl-5-s-butyl barbituric acid.

Secholex. A proprietary preparation of POLIDEXIDE used in the treatment of hypercholesterolæmia. (337).

Seclomycin. A proprietary preparation containing streptomycin, benzylpenicillin and procaine penicillin. An antibiotic. (335).

Seconal Sodium. A proprietary preparation of quinalbarbitone sodium. A hypnotic. (333).

Secondary Vermilion. A pigment. It is vermilion mixed with heavy spar.

Seconds Oil. See VIRGIN OIL.

Seconesin. A proprietary preparation of quinalbarbitone and mephenesin. A hypnotic. (280).

Secretan. An alloy of from 91–95 per cent. copper, 5–9 per cent. aluminium, 1·5 per cent. magnesium, and 0·5 per cent. phosphorus.

Secretin. A hormone obtained from duodenal mucosa.

Secretol. A fat-splitting material similar and equal to Twitchell's reagent in its action.

Secrodyl. Tablets dimethisterone with ethinyloestradiol — in gynaecological disorders. (182).

Secrosteron. A proprietary trade name for Dimethisterone. (182).

Sectral. A proprietary preparation of ACEBUTOLOL hydrochloride used in the treatment of cardiac arrhythmias. (507).

Securex. A proprietary safety glass.

Securite. A safety explosive for mines. It is a mixture of 26 per cent. m-dinitro-benzene and 74 per cent. ammonium nitrate. It sometimes contains dinitro-naphthalene and potassium nitrate.

Securitol. This is essentially a sodium silicate used to hasten the setting of cements.

Sedacao. A German sedative. It contains calcium bromide and cocoa.

Sedacol. A proprietary preparation of clioquinol and phanquone. Antidiarrhœal. (373).

Sedan Blue. See SOLUBLE BLUE.

Sedasprin. A compound formed from acetyl-salicylic acid (aspirin) and bromine. A medicinal preparation.

Sedatin. See ANTIPYRINE and VALERYDINE.

Sedative Salt. Boric acid, H_3BO_3.

Sedatussin. A proprietary preparation of cephæline hydrochloride, sodium benzoate, syrup of squill and syrup of tolu. A cough linctus. (333).

Sedeff. An effervescent preparation containing opium, bismuth, and digestive ferments.

Sedestran. A proprietary preparation of stilbœstrol and PHENOBARBITONE used in the treatment of menopausal disorders. (17).

Sednine. A proprietary preparation of pholcodine and pseudoephedrine hydrochloride. A cough linctus. (284).

Sedonan. A proprietary preparation of PHENAZONE and CHLORBUTOL used in the treatment of infections of the ear. (917).

Seebachite. A mineral, $4PbSe.4CuSe.Cu_2Se.HgSe$.

Seed, Kola. See KOLA NUT.

Seed, Kombé. See KOMBÉ ARROW POISON.

Seed-lac. Stick-lac, after washing free from the colouring matter soluble in water. See LAC and LAC-DYE.

Seekay Pitch. A registered trade name for chlorinated naphthalene products available in various grades. They have a melting-point from 65–70 to 120–125° C. One brand melts at 200°F., and has a specific gravity of 1·5–1·55.

Seelandite. A mineral. It is a variety of feather alum.

Segetan. A silver cyanide with a copper complex. A seed preservative.

Segoldus. A pigment consisting of 90–92 per cent. zinc oxide and 6 per cent. lead.

Segontin. A proprietary trade name for the lactate of Prenylamine.

Sehta. Indian jeweller's name for cobaltite.

Seiba Gum. See TUNO GUM.

Seidlitz Powder (Effervescent Tartrated Soda Powder). Consists of 3 parts rochelle salt with 1 part sodium bicarbonate in the blue paper, and 1 part tartaric acid in the white paper.

Seidlitz Powder, Double. Contains a double dose of rochelle salt.

Seidlitz Salt. A name applied to magnesium sulphate, $MgSO_4.7H_2O$ (Epsom salts), found in the mineral waters of Seidlitz.

Seidschütz Salt. A name applied to native magnesium sulphate, $MgSO_4.7H_2O$.

Seifert Solder. An alloy of 73 per cent. tin, 21 per cent. zinc, 5 per cent. lead, 0·5 per cent. phosphorus and 0·5 per cent. tin.

Seignette's Salt. See ROCHELLE SALT.

Seladon Green. See BOHEMIAN EARTH.

Selazate. A proprietary accelerator. Selenium diethyl dithiocarbamate. (435).

Sel d'Angleterre. Magnesium sulphate, $MgSO_4$.

Sel de Barnit. Zinc tannate.

Sel de Sagesse (Sel de Science). Salt of alembroth (q.v.).

Selek. A proprietary trade name for powders for jointing metal to metal.

Selektan. A proprietary preparation of 2-hydroxyl-5-iodo-pyridine.

Selenac. A proprietary trade name for selenium diethyl-dithio-carbamate.

Selenhydric Acid. See SELENIETTED HYDROGEN.

Selenietted Hydrogen (Seleniuretted Hydrogen, Hydroselenic Acid, Selenhydric Acid, Selenion Hydride). Hydrogen selenide, H_2Se.

Seleniol. Colloidal selenium.

Selenite. Crystals of gypsum, $CaSO_4$. $2H_2O$. Used in optical instruments.

Seleniuretted Hydrogen. See SELENIETTED HYDROGEN.

Selenobismutite. See GUANAJUATITE.

Selenocosalite. A mineral. It is $2[Cu Pb_7Bi_8(S,Se)_{22}]$.

Selenojarosite. A mineral. It is $[KFe_3'''\{(S,Se)O_4\}_2(OH)_6]$.

Selenolinnæite. A mineral. It is $8[(CO_3S,Se)_4]$.

Selenolite. A mineral. It is lead selenide covered with selenium dioxide.

Selenopyrine. Synonym for seleno-antipyrine. Melting-point 168° C.

Selenoxene. Dimethyl - selenophene, C_6H_8Se.

Selenpalladium. A mineral. It is allopalladium.

Seleron (Aeron). A group of aluminium alloys containing 85 per cent. aluminium, with copper, nickel, zinc, manganese, silicon, and lithium, as the other ingredients. The specific gravity is 2·8–3, and the melting-point about 600° C. They are claimed to be useful for electrical apparatus.

Selex. See CRISCO.

Self-hardening Steel. See MUSHET STEEL.

Seligmannite. A mineral, $PbCuAsS_3$.

Seliwanoff's Reagent. A solution of 0·05 gram resorcinol in 100 c.c. dilute (1 : 2) hydrochloric acid. It gives a red colour with fructose.

Sellaite. Native magnesium fluoride, A mineral. It is $2[MgF_2]$.

Selora. A proprietary preparation of potassium chloride used as a substitute for table salt. (439).

Selsun. A proprietary preparation of selenium sulphide and a detergent used as a treatment for dandruff. (311).

Seltzers. Usually consist of 25 parts sodium carbonate, 5 parts sodium chloride, 6 parts sodium sulphate, and 1,000 parts water.

Selvadin. A proprietary preparation of a readily soluble calcium salt of catechol-disulphonic acid, which injected into the blood stream causes an increase in the calcium content by rendering the sparingly soluble calcium compounds more soluble.

Selvigon. A proprietary preparation of pipazethate hydrochloride, a cough linctus. (281).

Selwynite. Yellow Ochre (q.v.).

Semi-benzide. See CRYOGENIN.

Semicoke. A fuel made from coal by low temperature carbonisation. It is a smokeless fuel with a low ash.

Seminose. Mannose, $C_6H_{12}O_6$, a sugar.

Semi-opal. An impure opaque opal. See OPAL.

Semi-porcelain. A type of stoneware made from a white plastic clay to which ground quartz or flint is added. It is afterwards glazed.

Semi-refractory Material. A heat insulating material. It is a mixture of diatomaceous earth and lime, heated, treated with water and steam, then pressure applied.

Semi-silica. See HALF-GANISTER.

Semi-steel. A metal having properties between cast iron and cast steel. Used for filter-press plates. The term is applied to grey cast irons of low carbon content.

Semi-water Gas. A gaseous fuel produced by blowing air and steam through red hot coke or other material. It contains from 16–20 per cent. carbon monoxide, 8–19 per cent. hydrogen, 1–3 per cent. methane, 5–11 per cent. carbon dioxide, and 50–60 per cent. nitrogen. See DOWSON GAS and PRODUCER GAS.

Semseyite. A mineral, $9PbS.4Sb_2S_3$.

Senaite. A mineral, $(Fe.Mn.Pb)TiO_3$.

Senarmontite. A mineral. It is $16[Sb_2O_3]$. See ANTIMONY BLOOM.

Sendust. A proprietary trade name for an iron-silicon-aluminium alloy.

Seneca Oil. A name given to American petroleum, used in medicine.

Senega. The dried rhizome and roots of Polygala senega.

Senegal Gum (West African Gum). A gum arabic ranking second to Khordofan gum. It is derived from Acacia Senegal and other species of Acacia. It gives a good adhesive mucilage.

Senegin. See STRUTHIIN.

Seneprolin. See PARAFFIN, LIQUID.

Sengierite. A mineral. It is $2[Cu (UO_2)_2(VO_4)_2.8H_2O]$.

Sengite. An American explosive. It has a guncotton base and is similar to Tonite (q.v.), except that sodium nitrate replaces barium nitrate. The name " sengite " is derived from the initial letters of " substituted explosive, no glycerin," and the ending of the word dynamite.

Senna. The dried leaflets of *Cassia acutifolia.*

Sennaar Gum. See SUAKIN GUM.

Sennatin. A preparation containing the water-soluble principle of senna leaves.

Sennax. A preparation containing the water-soluble glucosides of senna leaves.

Sennite. See PINITE.

Senokot. A proprietary preparation containing sennosides A and B. A laxative. (322).

Sensitol, Red and Green. The German Pinacyanol and Pinaverdol (*q.v.*).

Seominal. A proprietary preparation of reserpine, phenobarbitone and theobromine. (112).

Sepia. A brownish-black pigment derived from the ink-bag of the cuttlefish. Used as water colour.

Sepiolite. A mineral. It is $Mg_3Si_4O_{11}$. $5H_2O$. See MEERSCHAUM.

α-Sepiolite. Synonym for Parasepiolite.

Septacrol. A double salt of silver and dimethyl-amino-methyl-acridine. A powerful antiseptic.

Septamid. Magnesium - *p* - toluol - sulphonic chloramide.

Septex No. 1. A proprietary skin cream containing boric acid, zinc oleate and zinc oxide. (462).

Septex No. 2. A proprietary skin cream containing boric acid, zinc oxide, zinc oleate and sulphathiazole. (462).

Septicemine. Iodo-benzo-methyl-diformine.

Septicidin. The active principle of a serum used for swine fever.

Septoform. A condensation product of formaldehyde with members of the terpene, naphthalene, and phenol groups. The chief one is dioxy-naphthyl-methane, $(C_{10}H_7O)_2.CH_2$. It is germicidal, and is used in veterinary practice.

Septovince. A proprietary preparation of trichloro-acetyl-di-phenol-di-iodide. An antiseptic.

Septrin. A proprietary preparation of trimethoprim and sulphamethoxazole. An antibiotic. (277).

Sequens. A proprietary preparation of mestranol (15 white tablets) and mestranol and chlormadinone acetate (5 peach tablets). Oral contraceptive. (333).

Seracelle. A proprietary cellulose acetate packing material.

Seractide. A corticotrophic peptide. Ala^{26}-Gly^{27}-Ser^{31}-$α^{1-39}$-corticotrophin.

Serandite. A mineral. It is Na_6 $(Ca,Mn)_{15}Si_{20}O_{58}.2H_2O$.

Serc. A proprietary preparation of betahistine hydrochloride used as an anti-emetic. (654).

Serenace. A proprietary preparation of haloperidol. A sedative. (285).

Serenesil. A proprietary preparation of ETHCHLORVYNOL. A hypnotic. (311).

Serenid-D. A proprietary preparation of oxazepam. A sedative. (245).

Serenid Forte. A proprietary preparation of OXAZEPAM. A tranquilliser. (245).

Serendibite. A mineral containing aluminium, iron, calcium, and magnesium silicates and borates.

Serenid-D. A proprietary trade name for Oxazepam.

Serentil. See MESORIDAZINE.

Seretin. Carbon tetrachloride, CCl_4.

Serge Blue. A lower quality of Soluble Blue (*q.v.*).

Serial 28. A proprietary preparation of ethinylœstradiol (16 red tablets) and megestrol acetate and ethinylœstradiol (5 white tablets). Sequential oral contraceptive. (182).

Sericine (Silk Size, Silk Rubber). The gum surrounding the silk from the silk spinner. See CUITE.

Sericite. A mineral. It consists of flaky muscovite.

Sericose. Cellulose acetate, used for making artificial silk and dope.

Serilene. Disperse dyes for polyester fibres. (508).

Serilic. Disperse dyes for acrylic fibres. (508).

Serimet. Dispersed premetallised dyes. (508).

Serine. α-Amino-β-hydroxy-propionic acid, $CH_3(OH)CH(NH_2)COOH$.

Serinyl. Disperse dyes for polyamide fibres. (508).

Seriplas. Dyestuffs for plastics. (508).

Serizyme. A proprietary trade name for an enzyme used in desizing acetate fabrics and similar materials which contain protein.

Sernamby (Coarse Para, Negroheads). Scraps of self-coagulated rubber from the collecting vessels, made into balls.

Sernylan. A proprietary trade name for the hydrochloride of Phencyclidine.

Seroden. A proprietary preparation of iodine with serum protein.

Seroden. A proprietary trade name for Thiacetazone.

Serol. See RECTO-SEROL.

Seromycin. A proprietary preparation of cycloserine. An antibiotic.

Serosine. Bromanilid.

Serpasil. A proprietary preparation of phenobarbitone and theobromine (Reserpine). An antihypertensive. (336).

Serpasil-Esidrex. A proprietary preparation containing RESERPINE and HYDROCHLORTHIAZIDE. An anti-hypertensive. (519).

Serpasil-Esidrex-K. A proprietary preparation of RESERPINE, HYDROCHLORTHIAZIDE and potassium chloride. An anti-hypertensive. (519).

Serpatonil. A proprietary preparation of reserpine, and methyl phenidate hydrochloride. (18).

Serpentine. A mineral. It is a rock consisting mainly of a hydrous magnesium silicate, $Mg_3Si_2O_8H_2+H_2O$, with more or less iron.

Serpentinous Marble. See OPHIOLITE.

Serpierite. A mineral. It is a basic copper-zinc sulphate.

Serum-casein. See GLOBULIN.

Sesamé Oil, German. See CAMELINE OIL.

Sesseralite. A mineral. It is a variety of corundum.

Setalana. A proprietary name for a natural nest silk produced by worms of the genus *Anaphe*, introduced into Germany from Africa. The fibre resembles tussah silk, but is stated to be not so strong.

Setarol. A proprietary polyester laminating resin. (918).

Sethadil. A proprietary trade name for Sulphaethidole.

Setilose. A French cellulose acetate artificial silk.

Setocyanine (Brilliant Glacier Blue, Acronol Brilliant Blue). A dyestuff. It is the hydrochloride of diethyldiamino - o - chloro - phenyl - ditolyl - carbinol, $C_{25}H_{28}N_2Cl$. Dyes silk and tannined cotton greenish-blue.

Setoglaucine. See VICTORIA GREEN 3B.

Setopaline. A triphenyl-methane dyestuff closely related to Erioglaucine A.

Sevin. 1-naphthyl N-methylcarbamate. A proprietary preparation of carbaryl used as a veterinary insecticide.

SE Wax. A Montan Wax ester containing an emulsifier used in the preparation of non-ionic self polishing emulsions for floors. (6).

Sextate. Cyclohexanol (hexahydrophenol) acetate. It boils at 170–195° C., has a specific gravity of 0·94–0·96, and a flash point of 155° F. Also stated to be methyl cyclohexyl acetate. (513).

Sextol. Commercial cyclohexanol. A mixture of cyclohexanol (*q.v.*) and the three methyl-cyclohexanols. Boiling-point 160–180° C. Specific gravity 0·93–0·94, flash-point 155° F. Solubility in water 3 per cent. It is incorporated in soap to increase the detergent action.

Sextol Z. Dimethyl cycohexanol. (513).

Sextone. See CYCLOHEXANONE.

Sextone B. Methyl-cyclohexanone. A

solvent boiling between 160–170° C., having a density of 0·93 and a flash-point of 130° F.

Seybertite. A mineral, 6(Fe.Ca)O. 9(Fe.Al)$_2$O$_3$.5SiO$_2$.

Seymourite. A proprietary trade name for an alloy of 64 per cent. copper, 18 per cent. nickel, and 18 per cent. zinc.

Seyrigite. A mineral. It is 8[Ca (W,Mo)O$_4$].

SF147 (Sandoz). A German product. It is a local anæsthetic, N-diethyl-leucinol ester of *p*-amino-benzoic acid. It is stated to be less toxic than tutocaine, acts longer, and in lower concentration.

SH 420. A proprietary preparation of NORETHISTERONE acetate used in the treatment of breast cancer. (438).

Shadocol. Sodium - tetraiodo - phenol - phthalein.

Shaku-do. A Japanese alloy. It usually contains 94–96 per cent. copper, 3·76–4·16 per cent. gold, and 0·08–1·55 per cent. silver.

Shale. A dark-grey or black mineral containing 73–80 per cent. mineral matter and 20–27 per cent. organic matter. It is a source of oil for lubricating purposes.

Shale Motor Spirit. See SHALE SPIRIT.

Shale Naphtha. See SHALE SPIRIT.

Shale Oil. The tarry oil obtained by the distillation of certain bituminous shales. It contains unsaturated hydrocarbons.

Shale Spirit (Shale Naphtha, Shale Motor Spirit). A spirit obtained from Scotch shale, by treating the distillate from the shale with sulphuric acid and caustic soda, and redistilling. Used as a substitute for turpentine, and as a motor spirit. It consists chiefly of paraffins and olefines.

Shandite. A mineral. It is [Ni$_3$Pb$_2$S$_2$].

Shanghai Oil. A variety of Colza Oil (*q.v.*).

Shanyavskite. A mineral. It is an aluminium hydroxide.

Sharpite. A mineral. It is (UO$_2$)$_6$ (CO$_3$)$_5$(OH)$_2$.7H$_2$O.

Sharps. See BRAN.

Shattuckite. A rare mineral, CuH$_2$S$_2$O$_7$.

Shawinigan's Black. See ACETYLENE BLACK.

Shea Butter (Bambuk Butter). The fat obtained from the seeds of *Butyrospernum parkii* or *Bassio parkii*. It has a saponification value of 175–190 and an iodine value of 54–66.

Shea Gutta. See GUTTA-SHEA.

Shear Steel. Blister Steel (*q.v.*) which has been sheared, reheated, and rolled into bars, to render it more homogeneous.

Sheathing Bronze. See SEA-WATER BRONZE.

Shé-Chuang-Tzu. A Chinese drug. It is the fruit of *Selinum monnieri*, and contains an essential oil which contains *l*-pinene, camphene, and bornyl-isovalerate.

Sheduri. See SANNA.

Sheet Brass. See BRASS.

Sheet-lac. See LAC.

Shellac. See LAC.

Shellac, Arizona. See SONORA GUM.

Shellac Cement. A cement suitable for glass and metals made by heating together 50 parts shellac, 5 parts wood creosote, 2 parts terpineol, and 1 part strong ammonia by weight.

Shellackose. An alcohol-soluble phenolformaldehyde resin. Used in the preparation of lacquers.

Shellac Substitute. See IDITOL.

Shellac Varnish. This often contains up to 25 per cent. by weight of shellac. Various other resins are usually added, such as Venice turpentine, sandarac, mastic, and Manila copal.

Shellac Water Varnish. This consists of shellac dissolved in a borax or ammonia solution. It is used as a varnish for stiffening hats and for leather.

Shell-head Brass. See BRASS.

Shellite. An explosive. It is a mixture of ammonium perchlorate and paraffin wax.

Shell-lac (Shellac). See LAC.

Shell Limestone. A variety of calcium carbonate in massive form.

Shellsol. High boiling hydrocarbon solvents. (553).

Shelspra. See MEXPHALTE.

Sherwood Oil. See PETROLEUM ETHER.

Shibu-ichi. A Japanese alloy containing 51–67 per cent. copper, 32–49 per cent. silver, and traces of gold and iron.

Shibuol. A phenolic compound contained in Kakishibu (*q.v.*).

Shikimole. Safrole, $C_{10}H_{10}O_2$, the chief constituent of oil of sassafras.

Shikon. The dried roots of *Lithospermum erythrorhizon*.

Shilajatu. An Indian mineral gum.

Shimose. See LYDDITE.

Shimosite. A Japanese explosive, the chief constituent of which is picric acid.

Shinkle Plastic. See AXF PLASTIC.

Shinnamu. A vegetable dye obtained from a species of maple found in Korea.

Shio Liao. A Chinese cement for marble. porcelain, etc., made from 54 per cent, slaked lime, 6 per cent. alum, and 40 per cent. blood.

Shipley's Solutions. Solutions of pyrogallol and caustic soda in water, usually 10 c.c. of 1 : 1 caustic soda solution, 1 and 4 c.c. water, and 2 and 10 grams

pyrogallol. Used for the absorption of oxygen.

Shirlan. A trade mark for salicylanilide. It is used to protect fabric from moth and mildew. (Shirlanised material) (Shirley Institute). (2).

Shoddy. The recovered and broken up wool of old cloth.

Shoemaker's Black. Ferrous sulphate, $FeSO_4$.

Shoemaker's Paste. A paste made by allowing the gluten from flour to putrefy, rolling it out thin, and making it into a paste. Used for securing leather to leather, paper, or other material.

Shoe Nail Brass. See BRASS.

Sho-ju. See SOY.

Shorle. See SCHORL.

Shortite. A mineral. It is $2[Na_2Ca_2(CO_3)_3]$.

Short Oil Varnishes. See LONG OIL VARNISHES.

Shostakovsky Balsam. Polyvinox. A A proprietary preparation of synthetic vinyl butyl ether. (400).

Shot Lead. See SHOT METAL.

Shot Metal (Shot Lead, Bullet Metal). An alloy of lead with not more than 3 per cent. arsenic. One alloy contains 99·8 per cent. lead and 0·2 per cent. arsenic.

Shoya. See SOY.

Shredded Lead. See LEAD WOOL.

Sialonite. A mineral, Be_8Se_3.

Siberian Ruby. See RUBELLITE.

Siberite. A mineral. It is a variety of tourmaline.

Sibley Alloy. An alloy of 67 per cent. aluminium and 33 per cent. zinc.

Sibor. A proprietary safety glass.

Sibo Wood. A French West African wood. It is obtained from *Sarcocephalus osculentus*, and the dried wood contains 0·96 per cent. fats and waxes, 67·4 per cent. cellulose, and 31·36 per cent. lignin.

Sical. An alloy of from 22–29 per cent. aluminium, 50–51 per cent. silicon, 2–4 per cent. titanium, 1 per cent. calcium, 0·2–0·3 per cent. carbon, and the remainder iron.

Sicalite. See GALLATITE.

Siccative. Manganese borate, MnB_4O_7. Used as a siccative mixed with linseed oil and resin, for impregnating leather.

Siccolam. Compound titanium dioxide paste. Desiccant for exudatory dermatoses. (182).

Sicily Oil. Inferior olive oil.

Sicklerite. A lithium mineral.

Sicoflex MBS. A thermoplastic material based on methyl methacrylate, butadiene and styrene. (8).

Sicoflex 80. A proprietary A.B.S. ter-polymer possessing very high flow pro-perties. (8).

Sicoflex 85. A proprietary A.B.S. ter-polymer possessing high flow properties. (8).

Sicoflex 90. A proprietary A.B.S. ter-polymer having high impact strength. (8).

Sicoflex 93. A proprietary general-purpose A.B.S. terpolymer. (8).

Sicoflex 95. A proprietary A.B.S. ter-polymer possessing high tensile strength. (8).

Sicoflex 99. A proprietary A.B.S. ter-polymer offering high resistance to heat. (8).

Sicoid. A cellulose acetate product similar to cellon (*q.v.*).

Sicromo Steel. A proprietary trade name for a chromium-silicon-molyb-denum steel containing from 2·25–2·75 per cent. chromium, 0·5–1·0 per cent. silicon, 0·4–0·6 per cent. molyb-denum and up to 0·15 per cent. carbon.

Sicron. Suspension polyvinyl chloride homopolymers. (919).

Siderac. A German preparation. It is stated to be an " active " oxide of iron to be taken in cases of anæmia.

Sideranatrite. A term used for both Autunite and Urasite (*q.v.*).

Sideraphthite. An alloy resembling silver. It contains 64·5 per cent. iron, 22·5 per cent. nickel, 4·5 per cent. each aluminium and copper, and 4 per cent. tungsten. It is stated to be non-oxidisable.

Siderazote (Siderazotite). A mineral Fe_5N_2, of volcanic origin.

Siderazotite. See SIDERAZOTE.

Siderine Yellow. A basic chromate of iron, used to a small extent as a water colour, and mixed with waterglass as a paint.

Siderite. See CHALYBITE.

Sidero Cement. A cement in which iron ores are wholly or partly substituted for the clay.

Siderochalcite. See CLINOCLASTITE.

Siderochrome. A mineral. It is chromite.

Sideroconite. A mineral. It is a calcite containing iron oxide.

Sideronatrite. A mineral. It is an iron-sodium hydrated sulphate.

Sideroplen. A German product. It is a water-soluble iron preparation for intra-muscular injection. The preparation differs only slightly from ferrum oxydat saccharate.

Sideroplesite. A mineral. It is $2[(Fe,Mg)CO_3]$.

Siderotile. A mineral, $FeSO_4.5H_2O$.

Sideroxene. A similar mineral to Bertrandite.

Sidonal. See UROL.

Sidonal, New. A mixture of quinic acid, $C_6H_7(OH)_4COOH$, and its anhydride.

Sidot's Blende. A phosphorescent zinc sulphide.

Sidros. A proprietary preparation con-taining ferrous gluconate and ascorbic acid. A hæmatinic. (269).

Siemen's Gas. See PRODUCER GAS.

Siemensite. A refractory material pro-duced by fusing a mixture of chromite, bauxite, magnesite, and a reducing agent in the arc furnace to obtain a slag containing from 20–40 per cent. Cr_2O_3, 25–45 per cent. Al_2O_3, 18–30 per cent. MgO, and 8–14 per cent. other constituents.

Sienna. A pigment. It consists of hy-drated oxide of iron, mixed with a little manganese, and clay. It contains from 50–70 per cent. Fe_2O_3, 8–12 per cent. SiO_2, 2–8 per cent. Al_2O_3, 2–5 per cent. $CaSO_4$ or CaO, and water.

Sienna, American. See BURNT SIENNA and INDIAN RED.

Sierra Leone Butter. See LAMY BUTTER.

Sierra Leone Copal. See COPAL RESIN.

Sifbronze. A proprietary trade name for a brass containing some ferroman-ganese and tin.

Siflural. A trade name for a solution of aluminium fluosilicate. A disinfectant.

Sigal. A proprietary alloy of 10 per cent. Si and 90 per cent. Al. A pig-ment.

Sigla. A proprietary safety glass.

Sigmaform. See XEROFORM.

Sigmalium. A proprietary trade name for an alloy of aluminium containing 1 per cent. silicon, 4 per cent. copper, and 0·7 per cent. magnesium.

Sigmamycin. A proprietary prepar-tion containing oleandomycin and tetra-cycline. An antibiotic. (85).

Signal Red. A dyestuff. It is a British equivalent of Lithol red B.

Sigtesite. A mineral. It is a variety of albite.

Silage. A cattle food made by cutting up and storing green fodder.

Silajit. A preparation containing ben-zoic and hippuric acids, gums, albumin-oids, resin, and fatty acids.

Silal. A proprietary trade name for a grey iron with 5 per cent. silicon and 2·5 per cent. total carbon. It is stated to resist oxidation, growth, and scaling up to 750° C.

Sil-al. Aluminium hydro-silicate.

Silanca. A stainless silver with a high silver content.

Silane (Silicane). Silicon tetrahydride, SiH_4.

Silani. *Vigna marina*, a forage crop of the Philippines.

Silanox. Trade mark for a fumed oxide rendered hydrophobic used as a water-resistant thickening agent in greases and inks. (63).

Silanox 101. A proprietary fine-particle fumed silica rendered hydrophobic and added to powdered polymers to improve their flow properties. (63).

Silantox. A proprietary colloidal silicon dioxide.

Silaonite. A mineral. It is a mixture of bismuth and bismuth trisulphide.

Silar-1. A proprietary copolymer, black in colour, consisting of heteroarylene and siloxane blocks. (424).

Silargel. A German product. It is a silver chloride-silica gel preparation, a white odourless powder containing 5 per cent. silver. It is an adsorbent and disinfectant for the external treatment of burns.

Silastomer. Silicone rubbers. (614).

Silastoseal. Room temperature curing silicone rubber sealants. (614).

Silbamine. Silver fluoride, AgF.

Silbe. A proprietary preparation of EPHEDRINE hydrochloride, THEOPHYLLINE and calcium benzylphthalate used in the treatment of asthma. (137).

Silbe Inhalant. A proprietary preparation of PAPAVERINE hydrochloride, ADRENALIN tartrate, atropine methonitrate, posterior pituitary extract and HYOSCIN hydrobromide, used in the treatment of asthma. (137).

Silbephylline. A proprietary preparation of diprophylline. A bronchial antispasmodic. (137).

Silberit. A jewellery alloy. It contains aluminium, nickel, and silver.

Silberol. Silver - phenol - sulphonate, $C_6H_4(OH)SO_3Ag$. an antiseptic used in the treatment of gonorrhœa and ophthalmic inflammation.

Silcar. A proprietary trade name for a pigment comprising a mixture of silicon dioxide and silicon carbide.

Silchrome. A heat-resisting alloy containing 86 per cent. iron, 9·5 per cent. chromium, 4 per cent. silicon, and 0·5 per cent. carbon.

Silchrome 46M. A proprietary trade name for a chromium steel containing 4–6 per cent. chromium, 0·5 per cent. molybdenum, and 0·2 per cent. carbon.

Silchrome R.A. A proprietary trade name for a steel containing 16 per cent. chromium, 1 per cent. silicon, 1 per cent. copper, and 0·12 per cent. carbon.

Silchrome Wire. An alloy of iron with 18 per cent. chromium, 3 per cent. silicon, 3 per cent. tungsten, and 0·3 per cent. carbon.

Silcocell. Cell control agents for polyurethane foams. (512).

Silcoloid. Transparent silicone potting compound. (512).

Silcoset. Rubber curing agent. (512).

Silcoset 101. A proprietary self-curing silicone rubber. (512).

Silcote. Grease eliminator for the bakery trade equipment. (503).

Silcron. Registered trade mark for a fine-particle silica. (920).

Silderm. A proprietary preparation of TRIAMCINOLONE acetonide, NEOMYCIN sulphate and undecanoic acid. A steroid skin cream. (306).

Sildura. Registered trade mark for a range of silicone-rubber compositions. curable by the application of heat. (59).

Silene. A proprietary trade name for a precipitated calcium silicate. Used in rubber mixes to give wear-resistance.

Silent Spirit. See SPIRIT OF WINE.

Silesia Powder. An explosive. It is a mixture of 75 per cent. potassium chlorate, with pure or nitrated resin, and a little castor oil.

Silesite. A tin silicate with 55 per cent. tin, found in the Bolivian tin deposits.

Silester OS. A proprietary trade name for a mixture of tetraethyl orthosilicate and ethyl polysilicate. (57).

Silex. A name applied to silica (SiO_2). It is used also for tripoli employed as a filler in paints. A ground flint is also known as silex.

Silexon. See CARBORA.

Silfbergite. A mineral. It is a variety of dannemorite.

Sil-fos. A proprietary trade name for a phosphor-silver brazing solder containing 80 per cent. copper, 15 per cent. silver, and 5 per cent. phosphorus.

Silfrax (Silicised Carbon). A product obtained by the action of silicon on carbon and consisting of carbon with a coating of silicon carbide and carbon. It is stated to be tougher and stronger than carborundum, and is used as a refractory material in the manufacture of pyrometer tubes for electrical heating elements.

Silica Gel. The name applied to a colloidal form of silica, prepared by treating sodium silicate with acetic or hydrochloric acid, washing the gelatinous silica, and drying. It is highly absorbent, and is used to absorb vapours such as benzene or ether from air. It has also been applied to the recovery and refining of petroleum oil.

Silica Glass. See VITREOSIL.

Silica, Hydrous. See OPAL.

Silica SM 111. A proprietary silica with

a particle size of 1 micron, used as an anti-blocking agent in plastic film and to promote matting in vinyl solutions and acrylic coil coatings. (205).

Silical. A term used to denote the radicle Si_2OH.

Silicam. Silicon imido-nitride, Si_2N_3H, formed when silicon di-imide is heated to 900° C. in an atmosphere of dry nitrogen.

Silicane. See SALINE.

Silicargol. A colloidal silver preparation for wound treatment.

Silicate Cotton. See SLAG WOOL.

Silicated Soap. A soap to which water-glass (sodium silicate) has been added. A detergent.

Silicate of Carbon. See GAS BLACK.

Silicate Ore. A zinc ore. It is a mixture of Hemimorphite and Smithsonite.

Siliceous Calamine. A mineral. It is a hydrated ortho-silicate of zinc.

Siliceous Sinter. See GEYSERITE.

Silicic Ether. Ethyl - o - silicate, $(C_2H_5)_4$ SiO_4.

Silicised Carbon. See SILFRAX.

Silicite. A mineral. It is a variety of labradorite.

Silicium. A proprietary trade name for silicon used as a pigment in the Atephen system (q.v.). A chemically resistant coating.

Silico-carnotite. Calcium - silico - phosphate, $3CaO.P_2O_5 + 2CaO.SiO_2$, found in the basic slag of the Thomas Gilchrist process for the dephosphorisation of iron. It is also given as a mineral: $4[Ca_5SiP_2O_{12}]$.

Silicol. Ferro-silicon, usually containing 84 per cent. silicon, for use in the preparation of hydrogen by the action of caustic soda.

Silicolloid. A natural siliceous material free from iron. It is a New Zealand product, and is in the form of a very fine powder. Suitable for use in paper manufacture, cleansers, and tooth pastes.

Silico-manganese. An alloy of silicon and manganese made in the electric arc type of furnace. It contains 60–75 per cent. manganese, 20–25 per cent. silicon, and the rest iron.

Silicon Brass. An alloy of 81 per cent. copper, 14 per cent. zinc, and 3 per cent. silicon.

Silicon Bronze (Silicum Bronze). An alloy of 97·37 per cent. copper, 1·32 per cent. tin, 1·24 per cent. zinc, and 0·7 per cent. silicon.

Silicon Copper. Alloys of copper with small amounts of silicon. Used for the manufacture of telephone and telegraph wires. An alloy with 10 per cent. silicon is also called silicon-copper.

Silicone. See CHRYSEONE.

Silicon-Eisen (Silicon Pig). A pig iron containing from 5–15 per cent. silicon.

Silicones. A generic name for compounds prepared with consistencies varying from greases to tough solids, in which silicon atoms are linked together in long chains by alternate oxygen atoms and in which alkyl or aryl groups fill the third and fourth valencies of the silicon atoms. The compounds are unique in that their physical properties are almost independent of temperature.

Silicone compounds of which details have been disclosed include the following:

E300 and *E301*: Polydimethyl siloxane gums; *E302* and *E303*: Polymethyl-vinyl siloxane gums; *E350* and *E351*: Polymethyl phenyl vinyl siloxane gums; *E367*: A partially-filled silicone gum.

Silicon, Ferro. See FERRO-SILICON.

Silicon Nickel Brass. An alloy of 81 per cent. copper, 14 per cent. zinc, 3 per cent. silicon, and 2 per cent. nickel.

Silicon Pig. See SILICON-EISEN.

Silicon Steel. A steel made by melting steel and ferro-silicon in crucibles. It is used for making sheets, springs, and acid-resisting plants.

Silicoset. Silicone rubbers for curing at room temperature. (2).

Silico-spiegel. An alloy of 20 per cent. manganese, 12 per cent. silicon, and the rest iron.

Silico-superphosphate. A preparation made by mixing superphosphate with kieselguhr or precipitated silicic acid. It is stated to give better results on medium and light soils.

Silico-titanium. A titanium-silicon alloy used in the steel industry.

Silicum Bronze. Silicon bronze.

Silicures. Metallic soaps, silicone resin catalysts. (607).

Siline. A compound of hexamethylene-tetramine, citric acid, and silicic acid. Used in uric acid troubles.

Silistren. Silicic acid tetra-glycollic ester.

Silit. A material made by exposing mixed silicon, silicon carbide, and carbon, to the action of carbon monoxide at 1500° C. It is made in three qualities. (1) A material for resistance subjected to permanent losses, (2) for electric heating work up to 1400° C., and (3) a fireproof material capable of withstanding violent changes of temperature.

Silitonite. See FRANKONITE.

Silk, Ailanthus. See FAGARA SILK.

Silk, Anaphe. The silk obtained from a caterpillar in German East Africa. It has a specific gravity of 1·282.

Silk, Basinetto. See GALETTAME SILK.

Silk, Bemberg. See BEMBERG YARN.

Silk Blue. See SOLUBLE BLUE.

Silk Blue O. A dyestuff. It is a British equivalent of Soluble blue.

Silk, Cuprate. See CUPRAMMONIUM SILK.

Silk, Degummed. See BOILED - OFF SILK.

Silk, Gelatin. See VANDURA SILK.

Silk Grass. A term applied to pineapple fibre, obtained from the pineapple plant. Used for making cloth in the Philippine Islands.

Silk Green. See CHROME GREENS.

Silk Grey. An azine dyestuff obtained by oxidising the product of the interaction of dimethyl- or diethyl-phenosafranine and formaldehyde. Dyes silk from an acid bath.

Silk Gum. See CUITE and SERICINE.

Silkin. A proprietary cellulose nitrate silk.

Silk, Meteor. See GLANZSTOFF.

Silk, Neri. See GALETTAME SILK.

Silkons. See AZONINES.

Silk, Pauly's. See GLANZSTOFF.

Silk, Ricotti. See GALETTAME SILK.

Silk Rubber. A rubber from an African tree, *Kickxia elastica*, also called *Funtumia elastica*. See CUITE and SERICINE.

Silk Scarlet S. A dyestuff. It is a British equivalent of Fast red B.

Silk, Sea. See BYSSUS SILK.

Silk, Sirius. See GLANZSTOFF.

Silk Size. See CUITE and SERICINE.

Silk, Tasar. See TUSSAR SILK.

Silk, Tubulated. See LUFTSEIDE.

Silk Wadding. The waste from the spinning of silk.

Silk, Warp. See ORGANZINE.

Sillénite. A mineral. It is $12[Bi_2O_3]$.

Sillimanite. A mineral. It is a silicate of aluminium $Al_2O_3.SiO_2$.

Sillimanith. See PRESSOLITH.

Sillitin N. A general purpose filler for rubber. It is a natural, finely divided mixed product of silicic acid and kaolin with a particle size of less than 20μ.

Sillman Bronze. An alloy of 86 per cent. copper, 10 per cent. aluminium, and 4 per cent. iron.

Silman Steel. A proprietary trade name for a silicon steel containing 2·1 per cent. silicon, 0·85 per cent. manganese, 0·3 per cent. vanadium, 0·25 per cent. chromium, and 0·55 per cent. carbon.

Silocalm. A proprietary preparation of aluminium hydroxide, simethicone and PROPANTHELLINE BROMIDE. An antacid. (921).

Sil-o-cel. A brand of kieselguhr, also a heat insulator made from kieselguhr.

Siloid. A registered trade mark for micron-sized silica gels.

Siloxicon. A fireproof material and a resistant to the action of acids and alkalis. It is made by heating powdered silica with a small quantity of carbon in the electric furnace, the composition approximating to Si_2C_2O. It is produced with silicon carbide in the carborundum furnace. Employed alone, or with binding materials, for making crucibles or muffles.

Siloxide. A mixture of silica with a little titanium, or zirconium oxide.

Siloxyl. A proprietary preparation of aluminium hydroxide gel and DIMETHICONE. An antacid. (921).

Silumin (Alpax). Proprietary alloys of aluminium and silicon containing 12 per cent. silicon. They have a specific gravity of 2·63–2·65. Also see ALUDUR, ALUMINAC, WILMIL and SIGAL.

Siluminite. An electric insulator consisting of 75 per cent. of mineral matter (asbestos, calcium silicate, and aluminium silicate), with pitch as the binding material.

Silumin-Y. A proprietary aluminium-silicon alloy with small additions of manganese and magnesium. It has high corrosion resistance.

Silundum. A product similar to carborundum. Articles, such as crucibles and tubes, are made by shaping pieces of graphite, embedding them in carborundum, and subjecting them to the action of silicon vapour at high temperatures in the electric furnace. It has a high electrical resistance, and is used for making electrodes.

Silva. A name used mainly in Germany for a type of artificial silk.

Silvaz. A proprietary trade name for an alloy used for the manufacture of steel. It contains iron with 40–45 per cent. silicon, 6·0–6·5 per cent. vanadium, 6·0–6·5 per cent. aluminium, and 6·0–6·5 per cent. zirconium.

Silvel. A proprietary trade name for an alloy containing 67·9 per cent. copper, 16 per cent. zinc, 6·5 per cent. nickel, 0·5 per cent. lead, 2·2 per cent. iron, and 6·8 per cent. manganese.

Silver Alum. An aluminium-silver sulphate, $Al_2(SO_4)_3.Ag_2SO_4.24H_2O$.

Silver Amalgam. An alloy of mercury and silver. It occurs as a mineral, but is also prepared artificially.

Silver, Antimonial. See DYSCRASITE.

Silver Arsphenamine. The sodium salt of silver - diamine - dihydroxy - arseno - benzene (silver-salvarsan). Used in the treatment of syphilis and in malaria.

Silver Atoxylate. Silver - p - amino - phenyl-arsonate.

Silver Bell Metal. An alloy of 40–42 per cent. copper and 58–60 per cent. tin.

Silver, Black. See PYRARGYRITE.

Silver, Blue. See NIELLO SILVER.

Silver Bronze. An alloy of 64 per cent. copper, 17 per cent. manganese, 13 per cent. zinc, 5 per cent. silicon, and 1 per cent. aluminium. An electrical resistance alloy.

Silver, China. See ARGYROLITH.

Silver Coinage. An alloy of 50 per cent. silver, 40 per cent. copper, and 10 per cent. nickel. Before 1920 the coinage contained 92·5 per cent. silver and 7·5 per cent. copper. See NICKEL COINAGE.

Silver Copper Glance. See STROH-MEYERITE.

Silver, Crede's. See COLLARGOL.

Silver Foil. An alloy of from 90–97 per cent. tin, 0–2·5 per cent. copper, and 0–10 per cent. zinc, is known by this name.

Silver, Frosted. See DEAD SILVER.

Silver, German. See NICKEL SILVERS.

Silver Glance. See ARGENTITE.

Silver Grain. The cochineal insect killed in an oven at three months old is called silver grain.

Silver Grey. A Bohme dyestuff. It contains extracts of logwood and redwood, together with a chrome and iron mordant. Also see SLATE GREY and ZINC GREY.

Silver, Grey. See TETRAHEDRITE.

Silver-hansa Plaster. A German product. It is a proprietary wound dressing containing metallic silver.

Silver Ichthyol. See ICHTHARGAN.

Silverine. An alloy of 77 per cent. copper, 17 per cent. nickel, 2 per cent. iron, 2 per cent. zinc, and 2 per cent. cobalt.

Silvering Solutions. Usually consist of solutions of silver cyanide and ammonium cyanide in water. Used for the electro-deposition of silver.

Silver Ink. A mixture of gum arabic and ground white mica. Employed for inlaying buttons.

Silverite. See NICKEL SILVERS.

Silver Jamesonite. See OWYHEEITE.

Silver Leaf. An alloy of 91 per cent. tin, 8 per cent. zinc, 0·35 per cent. lead, and 0·2 per cent. iron. Another alloy contains 91 per cent. tin, 8·25 per cent. zinc, and 0·4 per cent. antimony.

Silver Metal. An alloy of 66·5 per cent. zinc and 33·5 per cent. silver. Also see ALUMINIUM SILVER.

Silver Methylene Blue. The silver salt of methylene blue. A germicide.

Silver, Nevada. See NICKEL SILVERS.

Silver Nitrate Points and Sticks. These articles contain 95 per cent. silver nitrate and 5 per cent. potassium nitrate.

Silver Nucleinate. A product obtained by dissolving silver oxide in nucleic acid. It contains about 10 per cent. silver.

Silveroid. An alloy of 45 per cent. nickel, 54 per cent. copper, and 1 per cent. manganese. It has the merest traces of impurities and has maximum whiteness of all nickel-copper alloys. It polishes like silver, is not susceptible to sulphur, and does not oxidise below 260° C.

Silver Ore, Brittle. See PYRARGYRITE.

Silver Ore, Dark Red. See PYRARGYRITE.

Silver Ore, Red. See PYRARGYRITE.

Silver Ore, Ruby. See PYRARGYRITE.

Silver Ore, Vitreous. See ARGENTITE.

Silver, Oxidised. Silver covered with a thin film of sulphide, by immersion in a solution obtained by boiling sulphur with potash.

Silver Percylite. A mineral. It is a percylite containing silver chloride.

Silver, Peru. See CHINESE SILVER.

Silver, Potosi. See NICKEL SILVERS.

Silver Quinaseptolate. Silver-oxychino-line-sulphonate.

Silver, Refined. See STANDARD SILVER.

Silver, Ruby. See PYRARGYRITE.

Silver-salt. Pure sodium - anthra - quinone-mono-sulphonate, obtained in alizarin manufacture.

Silver Saltpetre. A name which has been applied to silver nitrate, $AgNO_3$.

Silver-salvarsan. See SILVER ARS-PHENAMINE.

Silver Sand. Quartz sand.

Silver Solder. Variable alloys. Usually they contain silver, copper, and zinc. A soft silver solder contains 67 per cent. silver and 33 per cent. brass, and is suitable for sheet. A hard silver solder consists of 80 per cent. silver and 20 per cent. copper. Some alloys contain smaller amounts of silver, and an ordinary one of this type is composed of 47 per cent. copper, 47 per cent. zinc and 6 per cent. silver. Sometimes a little tin is present.

Silver, Sterling. See STANDARD SILVER.

Silver, Tellurium. See HESSITE.

Silver Ultramarine. See YELLOW ULTRAMARINE.

Silver, Victoria. See NICKEL SILVERS.

Silver, Virginia. See NICKEL SILVERS.

Silver White. See FLAKE WHITE.

Silvestrite. A mineral. It is siderazote.

Silvialite. A mineral. It is $Ca_4Al_6Si_6O_{24}SO_4$.

Silvikrin. A German product. It is a shampoo powder said to contain sulphur-albuminoses as effective constituents. Used for preventing baldness.

Silzin Bronze. An alloy of copper with 10–20 per cent. zinc and 4·5–5·5 per cent. silicon.

Simatin. A proprietary trade name for Ethosuximide.

Simeco. A proprietary preparation of aluminium hydroxide, sucrose and simethicone used as a gastro-intestinal sedative. (245).

Simetite. Sicilian amber of wine-red to garnet-red colour.

Similex. A proprietary product. It is a flexible material, transparent as glass, and is a phenol-formaldehyde compound. It is stated to be acid, fire, and sea-water proof. It is a non-conductor of heat and an electrical insulator.

Similor. A rich-coloured brass. It usually contains from 80–89 per cent. copper, 9–20 per cent. zinc, and 0–7 per cent. tin.

Simlaite (Meerschaluminite). An aluminium silicate mineral found in India.

Simonellite. A hydrocarbon, $C_{15}H_{20}$, found in the lignite, in Tuscany.

Simonyite. See ASTRAKANITE.

Simplene. A proprietary preparation of ADRENALIN used in the treatment of glaucoma. (824).

Simplex. A proprietary safety glass.

Simplex Steel. A proprietary trade name for a nickel-chromium steel containing 1·25 per cent. nickel and 0·6 per cent. chromium.

Sinamine. Allyl-cyanamide, $C_4H_6O_2$.

Sinapoline. Diallyl-urea, $(C_3H_5NH)_2CO$.

Sinaxar. A proprietary preparation of STYRAMATE used in the treatment of muscle spasm. (327).

Sincaline. See CHOLINE.

Sin-Chu (Japanese Brass). An alloy of 66·5 per cent. copper, 33·4 per cent. zinc, and 0·1 per cent. iron.

Sincosite. A mineral. It is a hydrated phosphate of calcium and vanadium, $CaO.V_2O_4.P_2O_5.5H_2O$, found in Peru.

Sindanyo. A trade name for proprietary asbestos products.

Sinecain. Quinine hydrochloride and antipyrine for hypodermic use.

Sinemet. A proprietary preparation of CARBIDOPA and LEVODOPA used in the treatment of Parkinson's disease. (472).

Sinequan. A proprietary preparation of doxepin. A tranquillizer and antidepressant. (85).

Sinetens. A proprietary preparation of PRAZOSIN hydrochloride used as an anti-hypertensive. (365).

Sinetide. See IODOLYSIN.

Sinflavin. A German product. It is stated to be 3 : 6-dimethoxy-10-methyl-acridinium chloride. It is a wound disinfecting agent which does not dye or irritate. It is used as a 10–20 per cent. ointment, a 1 in 300 solution, or as a powder for the treatment of wounds.

Singapore Copal. See COPAL RESIN.

Singer's Electric Cement. A cement for fixing glass to brass. It contains 20 parts resin, 4 parts beeswax, 4 parts red ochre, and 1 part plaster of Paris.

Single Muriate of Tin. An acid solution of stannous chloride, used as a mordant.

Single Nickel Salt. Nickel sulphate, $NiSO_4.7H_2O$, used in the plating trade.

Singoserp. A proprietary trade name for Syrosingopine.

Sinhalite. A mineral. It is $4[MgAlBO_4]$.

Sinigrin. Potassium myronate, $KC_{10}H_{16}NS_2O_9$. A constituent of black mustard seed.

Sinistrin. See HELENIN.

Sinnodin. A German product. It is a solution of trimethyl-xanthin-sodium oxybenzoate, hexamethylene tetramine, and phenazone. It is used for intra-muscular and intravenous injection in cases of neuralgia, rheumatism, and gout.

Sinodor. A basic magnesium acetate, containing an excess of magnesium hydrate. Used for disinfecting purposes.

Sinopis. See INDIAN RED.

Sinox. A proprietary trade name for sodium dinitro-o-cresylate. A highly toxic and selective herbicide. It is activated by ammonium sulphate and sodium hydrogen sulphate. 1 gal./100 gals. water/acre is used for Charlock in cereals.

Sin Red. Potassium permanganate, $KMnO_4$.

Sinter. Incrustations, usually on rocks, from mineral waters.

Sinter-corundum. A proprietary preparation. It is a ceramic material produced from pure alumina at a temperature of about 1800° C. The thermal conductivity at 16° C. is about twenty times as high as that of porcelain, and it is stated to be not attacked by hydrofluoric acid or hot alkali.

Sinterit. A proprietary trade name for a form of sponge iron (*q.v.*) used for coupling packings.

Sinterloy. A proprietary trade name for a steel powder.

Sinter, Siliceous. See GEYSERITE.

Sinthrome. A proprietary preparation of nicoumalone. An anticoagulant. (17).

Sintisone. A proprietary preparation of PREDNISOLONE stearoylglycolate. A steroid skin cream. (365).

Sintisone-C. A proprietary preparation of CHLORQUINALDOL and PREDNISOLONE stearoylglycolate. An anti-fungal steroid skin cream. (365).

Sintol. A proprietary trade name for a plasticiser, dimethylthioanthrene. (3). See T-OL or WEICHMACHER T.

Sintox. A proprietary trade name for an impervious alumina ceramic. It has a high degree of chemical inertness. (177).

Sioglur. A proprietary trade name for a glass with a narrow melting range. It contains boron and may be used with sodium carbonate to replace borax in enamels.

Siomine. Hexamethylene - tetramine - tetraiodide, $(CH_2)_6.N_4I_2$, an antiseptic.

Sionon. A proprietary sugar substitute. It is d-sorbite.

Siopel. A proprietary preparation of DIMETHICONE and cetrimide. A barrier skin cream. (2).

Sipalin. A proprietary trade name for the cyclohexyl esters of palmitic and stearic acids. (45).

Sipalin AOC. A proprietary solvent for cellulose nitrate. It is dicyclohexanyl adipate. It boils at $212°$ C., has a specific gravity of 1·03, and a flash-point of $185°$ C.

Sipalin AOM. A proprietary solvent for cellulose nitrate and plasticiser for rubber. It is dimethyl-cyclohexanyl adipate. It boils at from $225-232°$ C., has a specific gravity of 1·011, and a flash-point of $189°$ C.

Sipalin MOM. A proprietary solvent and plasticiser. It is dimethyl-cyclohexanyl-β-methyl-adipate. It boils from $216-224°$ C., has a specific gravity of 1·009, and flashes at $195°$ C.

Sipeira. Bebeeru bark.

Sipilite. See BAKELITE.

Sipon. A registered trade mark currently awaiting re-allocation by its proprietors. (983).

Sipylite. A mineral, $(Y.Er)NbO_4$, with cerium.

Sirene. Dodecylbenzene.

Sirester. A proprietary polyester laminating resin. (715).

Sirfen. Phenolic moulding powders and resins. (119).

Sirflex. Alkylates for textiles.

Sirit. Urea resins.

Siritle. A proprietary urea formaldehyde moulding material. (715).

Sirius Silk. See GLANZSTOFF.

Sirius Yarn. See GLANZSTOFF.

Sirius Yellow G. A dyestuff. It is 1 : 2-benzanthraquinone.

Sirmasse. Polyester premixes. (715).

Sirpol. Acetovinylic resins.

Sirtene. L-D polyethylene.

Sisal Kraft. A proprietary trade name for a waterproof building paper made from sisal.

Siserskite. See IRIDOSMINE.

Sismondite. A mineral. It is a variety of chloritoid.

Sistomensin. The physiologically standardised ovarian hormones. Regularises the phenomenon of menstruation and stimulates the development of the female genital organs.

Sitaparite. A mineral. It is $16[(Mn, Fe)_2O_3]$.

Sitara Fast Red. See HELIO FAST RED RL.

Sitilan. The methyl-cyclohexyl ester of adipic acid. A solvent for cellulose and rubber. It is also used in the manufacture of leather oils and varnishes because of the softening effect which remains even at low temperatures. It can be used as a celluloid adhesive.

Sitogen. A malt preparation.

Sitol. A proprietary trade name for the sodium salt of m-nitro benzene sulphonic acid. Used in dyeing.

Sit-ruti. See SANNA.

Size. Usually consists of a starch solution containing small amounts of tallow or oil, and China clay or French chalk.

Size, Silk. See CUITE and SERICINE.

Sjögrenite. A mineral. It is $Mg_6Fe_2^{\cdots}(CO_3)(OH)_{16}.4H_2O$.

Sjogrufvite. A mineral, $Fe.Mn_3(AsO_4)_3.3H_2O$. It is also given as $(Ca,Mn)_3 Fe^{\cdots}(AsO_4)_3.3H_2O$.

SK-65. A proprietary preparation of DEXTROPROPOXYPHENE. An analgesic. (658).

S.K.A. A butadiene polymer of Soviet origin, derived from petroleum.

Skatole. See SCATOLE.

S.K.B. A butadiene polymer of Soviet origin, derived from alcohol.

Skefron. A proprietary preparation of dichlorodifluoromethane and trichlorofluoromethane. An analgesic spray. (658).

Skeladin. A proprietary trade name for Metaxalone.

Skelleftea. A variety of Stockholm tar.

Skellysolve. A proprietary trade name for a series of petroleum solvents.

Skemmatite. A mineral. It is an iron-manganese oxide.

Skiagite. A mineral. It is $8[Fe_3^{\cdots}Fe_2^{\cdots}Si_3O_{12}]$.

Skiargan. A colloidal silver solution.

Skin and Leather Meals (Nitrogenous Fertiliser). A fertiliser little used, consisting of trimmings and waste ground to a meal. It is sold under the name of nitrogenous fertiliser, and contains from 4–6 per cent. of nitrogen.

Skin Wool. The wool obtained from slaughtered sheep.

Skiodan Sodium. A proprietary trade name for Methiodal Sodium.

Skleron. A proprietary trade name for an aluminium alloy containing 12 per cent. zinc, 3 per cent. copper, 0·6 per cent. manganese, 0·25 per cent. silicon, and a small amount of nickel.

Sklodowskite. A radio-active mineral

found in the Belgian Congo. The main constituents are uranium oxide and silica. (Compare CHINKOLOB-WITE.)

Skogbölite. A mineral. It is a variety of tapiolite containing 84·4 per cent. Cb_2O_5 and Ta_2O_5.

Skokian. See KALI.

Skopyl. A proprietary preparation of methylscopolamine nitrate used in the treatment of infantile pyloric stenosis. (337).

Skunk. A name applied to an American petroleum which contains sulphur compounds, thereby having a considerable odour.

Skutterudite (Modumite). A mineral. It is cobalt triarsenide, $CoAs_3$.

Sky Blue. See IMMEDIAL PURE BLUE, WILLOW BLUE, and CŒRULEUM.

Skybond. A trade mark for a polyimide insulation resin suitable for use up to 700° F. (57).

Skydrol. A hydraulic fluid used in aircraft.

Slack Wax. A soft paraffin wax from the pressing of paraffin distillate.

Slag A. A British chemical standard. It is a basic slag containing 44·5 per cent. CaO, 16·15 per cent. SiO_2, 12·93 per cent. P_2O_5., 8·97 per cent. Fe, and 6·9 per cent. MgO.

Slag, Belgian. See BASIC SLAG.

Slagbestos. A similar product to slag wool (blast furnace slag).

Slag Cement. See EISEN-PORTLAND CEMENT.

Slag Sand. Blast furnace slag is run out of the furnace to fall into a running stream of water, when it is broken up into a fine sand.

Slag, Thomas. See BASIC SLAG.

Slag Wool (Mineral Cotton, Silicate Cotton, Mineral Wool). Blast furnace slag (essentially a glass composed of silicates of aluminium and calcium), which has had air blown through it. It resembles spun glass, and is used for packing steam pipes.

Slaked Lime. Calcium hydrate, $Ca(OH)_2$.

Slate. A mineral. It is a silicate of aluminium and magnesium.

Slate Black. See MINERAL BLACK.

Slate Dust (Slate Filler). A ground slate used as a filler in rubber mixings. It usually has a specific gravity of 2·7–2·8.

Slate Grey (Stone Grey, Silver Grey, Mineral Grey). Grey pigments obtained by grinding and levigating special kinds of grey slate, which occur in Germany. Used as priming paint, and for the preparation of putty. They are imitated by mixtures of white clay, blacks, ochres, and ultramarine.

Slate Lime. A mixture of 60 per cent.

lime with 40 per cent. of calcined slate powder used in the manufacture of porous concrete.

Slate Spar. A type of the mineral calcite, $CaCO_3$.

Slavikite. A mineral of Bohemia, $(Na. K)_2SO_4.2Fe_5(OH)_3(SO_4)_6.63H_2O$.

Slicker Solder. See PLUMBER'S SOLDER.

Slix. Heat resisting refractory cement. (531).

Sloeline. See INDULINE, SPIRIT SOLUBLE.

Sloeline RS, BS. See INDULINE, SOLUBLE.

Slop Wax. The wax present in the heavier wax distillates obtained in the refining of petroleum waxes. It is commonly considered unpressable and therefore different from the paraffin wax pressed from lighter wax distillates.

Slow-Fe. A proprietary preparation of ferrous sulphate in a slow release base. Iron supplement. (18).

Slow-Fe-Folic. A proprietary preparation of ferrous sulphate and folic acid. A hæmatinic. (18).

Slow-K. A proprietary preparation of potassium chloride in a slow release core. (18).

Slow-Sodium. A proprietary preparation of sodium chloride in sustained release form. (18).

Sludge Acid. Sulphuric acid which has been used in the refining of petroleum.

Sludge Gas. This gas, approximating to 69 per cent. methane, 29 per cent. carbon dioxide, and 2 per cent. nitrogen, is produced by the digestive process (anaerobic fermentation) of sewage sludge. It is used for gas engines aggregating 950 H.P.

S.M.A. A proprietary milk feed for babies. (245).

SMA 2625 A. A proprietary styrene maleic anhydride copolymer of low molecular weight, used as a levelling resin in floor polishes. (922).

SMA 3840. A proprietary styrene-maleic anhydride copolymer of low molecular weight used for coating cans and drums. (922).

SMA 5500. A proprietary copolymer of styrene and maleic anhydride, partially esterified and of low molecular weight, used as a vehicle for thermosetting electro-deposited coatings. (922).

SMA 17352 A. A proprietary copolymer of styrene and maleic anhydride, of low molecular weight, used as a levelling agent in polishes. (922).

Smalt (Saxony Blue, Saxon Blue, King's Blue, Royal Blue, Zaffer, Zaffre, Bleu D'Azure, Bleu de Saxe, Azure Blue). A potash glass containing oxide of cobalt. It is prepared by mixing zaffre (cobalt oxide) with powdered quartz and potassium carbonate and heating. The resulting product is a double silicate of cobalt and potassium, containing about 6 per cent. cobalt oxide. Used as a pigment. Ash blue and pale smalt are finer qualities of smalt. Also see COBALT BLUE, ESCHEL, and STREWING SMALT.

Smalt Blue. See CHESSYLITE.

Smalt F, M, O. Varieties of smalt (F = fine, M = medium, and O = ordinary).

Smaltine (Smaltite, Arsenical Cobalt). A mineral. It is cobalt arsenide, $CoAs_4$. Used for the preparation of smalt.

Smaltite. See SMALTINE.

Smalt, Native. See SMALTINE.

Smalt, Pale. See SMALT.

Smaragdgreen. See BRILLIANT GREEN.

Smaragdine. A trade name for a solidified alcohol consisting of alcohol and gun-cotton, coloured with malachite green.

Smaragdite. A mineral. It is a green variety of amphibole.

Smectite. A mineral. It is a variety of the mineral Halloysite.

Smegmatite. A mineral. It is steatite.

Smelite. Synonym for Kaolinite.

Smelling Salts. See ENGLISH SALT.

Smelling Salts, Violet. See ENGLISH SALT.

Smelling Salts, White. See ENGLISH SALT.

Smithite. A mineral. It is a silver sulpharsenite.

Smithsonite. A mineral. See CALAMINE.

Smitter-Lénian. An alloy containing 72 per cent. copper, 12·75 per cent. nickel, 9·75 per cent. zinc, 2·3 per cent. iron, 2·25 per cent. tin, and 1 per cent. bismuth.

Smoke Black. A carbon black used as a pigment. It contains 99·75 per cent. carbon.

Smoke Blue. A Bohme dyestuff. It contains logwood extract and a chrome mordant.

Smokeless Diamond Powder. A 33-grain powder consisting of insoluble nitro-cellulose, with 15 per cent. metallic nitrates, 6 per cent. charcoal, and 3 per cent. pet. jelly.

Smokene. See ESSENCE OF SMOKE.

Smoking Deterrent. A proprietary preparation of magnesium carbonate, LOBELIN sulphate and tribasic calcium phosphate. (923).

Smoking Salts. Impure hydrochloric acid.

Smoky Quartz. A quartz containing organic matter or hydrocarbons. It is usually brown in colour.

S Monel. A Monel metal with 3·75 per cent. silicon used in valves, etc., which are subject to corrosion.

SN-20PM. A proprietary copolymer of styrene and acrylonitrile, of Soviet origin. (424).

Snake Root. Senega root.

Snake Stone (Water-of-Ayr Stone). A Scotch stone used for rubbing down the surfaces of other stones, and copper plates.

Sneezing Gas. A poison gas. It is blue cross gas.

Sniafil. An Italian synthetic wool substitute. It is a product obtained in a similar way to Viscose silk, but differs from it in the treatment of the viscose solution.

Sniamid. A trade mark for superpolyamides 6 and 66 for injection moulding, extrusion and blow moulding. (87).

Snowcal 5SW. Dryground whiting from Swanscombe Kent. (27).

Snowcal 7ML. Dryground whiting from Hessle (Yorks). See also OMYA BLR3. (27).

Snow White. See BARIUM WHITE, ZINC WHITE, and OIL WHITE.

Snuff, White. See WHITE SNUFF.

Soa. A proprietary trade name for sucrose octa-acetate. It is used as a plasticiser.

Soamin. A proprietary preparation. It is the mono-sodium salt of p-aminophenyl-arsonic acid, $NH_2.C_6H_4.AsO(OH)(ONa)$. It is used in the treatment of sleeping sickness, syphilis, and tuberculosis. Other names for this compound are Atoxyl and Arsamin.

Soap. Metallic or inorganic salts of fatty acids.

Soap, Animal. Curd soap.

Soap Bark. The bark of *Quillaja saponaria*, of Chile. The commercial product consists of the layer of bast with the dead bark removed. The active principle is saponin. It is used to clean clothes.

Soap Balsam. Soap liniment.

Soap, Clay. See BENTONITE.

Soap, Dry. See SOAP POWDERS.

Soap, Formaldehyde. See FORMALIN SOAPS.

Soap, Glycerin. See TRANSPARENT SOAPS.

Soap, Mountain. See STEATITE.

Soap Powders (Dry Soaps, Washing Powders). Usually consist of soap and soda ash, sometimes with the addition of fillers. Often 1 per cent. of turpentine is present, and occasionally sal-ammoniac is added.

Soap Root. Ordinary soap root is composed of the stems and root of the soapwort, *Saponaria officinalis*. The white soap root is the root of species of *Gypsophila*.

Soap, Soft. This is usually a potash soap

of linseed oil or oleine, but cotton-seed, colza, sesamé, palm, or fish oil is often used.

Soap, Stearin. Curd soap.

Soapstone. See STEATITE.

Soap, Tallow. Curd soap.

Sobee. A proprietary baby feed based on soya, used in cases when an infant is intolerant of milk. (324).

Sobenite. A sodium benzoate/nitrite corrosion inhibitor. (503).

Sobiacrin. A proprietary preparation of mepacrin hydrochloride, chloroquin phosphate, para-aminobenzoic acid, pyridoxine hydrochloride and calcium pantothenate. (320).

Sobita. A proprietary preparation of neutral sodium bismuthyl-tartrate.

Soborol. Methyl-p-hydroxyl-benzoate.

Sobralite. Synonym for Pyroxmargite.

Socaloin (Zanaloin). Aloin from Socotrine or Zanzibar aloes.

Socotrine Aloe. See ALOE.

Soda. Sodium carbonate and bicarbonate are both known by this term.

Soda Ash. Practically anhydrous sodium carbonate, Na_2CO_3. A commercial variety of soda ash used for softening boiler feed water is known as 58 per cent. soda ash and contains 58 per cent. Na_2O.

Soda Ash, Ammonia. Sodium carbonate prepared by the Solvay ammonia process.

Soda Blue (Gas Blue). Impure Prussian blues prepared by using sodium ferrocyanide instead of the potassium salt.

Soda Bordeaux Mixture. Made with 6 lb. copper sulphate, 2 lb. caustic soda, and 50 gallons water. See BORDEAUX MIXTURE.

Soda, Chlorinated. See EAU DE JAVELLE.

Sodacopperas. Synonym for Natrojarasite.

Soda Crystals (Washing Soda, Washing Crystals). Sodium carbonate, Na_2CO_3. $10H_2O$.

Soda Felspar. See ALBITE and OLIGOCLASE.

Soda-garnet. See LAGORIOLITE.

Soda Glass. See SODA-LIME GLASS.

Soda Glass, Soluble. See SOLUBLE GLASS.

Soda-glauconite. A mineral. It is a variety of glauconite in which part of the potassium has been replaced by soda.

Soda Greens. Pigments. They are arsenic greens, obtained by neutralising the mother liquor containing white arsenic, acetic acid, and dissolved emerald green, which is produced in the preparation of the latter with sodium carbonate. If the neutralisation is

carried out with milk of lime, " lime-arsenic greens " are produced.

Sodaheterosite. Synonym for Heterosite.

Soda-jadeite. See JADEITE.

Soda-lime Glass. A glass usually containing from 71–78 per cent. SiO_2, 12–17 per cent. Na_2O, 5–15 per cent. CaO, 1–4 per cent. Al_2O_3 and Fe_2O_3, and 0–2 per cent. K_2O.

Soda-lime, Sofnol. See SOFNOL SODALIME G.

Sodalite. A mineral. It is a silicate and chloride of sodium and aluminium, $3(Al_2O_3.SiO_2.Na_2O.SiO_2)_2.NaCl$.

Sodalumite. A soda alum in cubic form.

Soda Lye. Obtained by boiling a solution of sodium carbonate with slaked lime.

Soda-Melilite. A mineral. It is $Na_2Si_3O_7$.

Soda Nitre. See CHILE SALTPETRE.

Soda-olein. A sulphonated castor oil.

Soda Powders. (*Pulveres effervescens B.P.*) They contain 33 grains sodium bicarbonate and 25 grains tartaric acid.

Soda Pulp. Wood pulp obtained by means of caustic soda.

Soda-Richterite. A mineral. It is $2[Na_2(Mg,Mn,Ca)_6Si_8O_{22}(OH)_2]$.

Soda Saltpetre. See CHILE SALTPETRE.

Soda-spodumene. See JADEITE.

Soda Tar. The name applied to an alkaline solution which has been used to purify petroleum oils after they have been treated with sulphuric acid. The alkali used has a specific gravity of 1·3, and is agitated with the oil and then allowed to settle. Also see ACID TAR.

Soda, Tartrated. See ROCHELLE SALT.

Sodatol. An agricultural explosive. It is a mixture of sodium nitrate and trinitro-toluene.

Soda Ultramarine. See ULTRAMARINE.

Soda, Vitriolated. Sodium sulphate, Na_2SO_4.

Soda, Washing. See SODA CRYSTALS.

Soda Waste. See ALKALI WASTE.

Soda Water Glass. A mixture of sodium silicates.

Soddite. A radio-active mineral. It contains 7·83 per cent. SiO_2, 85·33 UO_3, and 6·23 per cent. H_2O, and corresponds to the formula $12UO_3.5SiO_2.14H_2O$.

Soderseine. Colloidal bismuth.

Sodiformasal. See FORMASAL.

Sodinoc. A proprietary trade name for the sodium derivatives of dinitro-o-cresol. A herbicide. See SINOX.

Sodium Acetrizoate. Sodium 3-acetamido-2,4,6-tri-iodobenzoate. Diaginol.

Sodium Alum. Aluminium-sodium sulphate, $Al_2(SO_4)_3.Na_2SO_4.24H_2O$.

Sodium Amalgam. An alloy of mercury and sodium containing from 1 part sodium in 100 parts mercury to 1 part sodium in 80 parts mercury. Harder alloys can be made with less mercury. Used as a reducing agent.

Sodium Amidotrizoate. Sodium diatrizoate.

Sodium Ammonia. A solution of sodium in ammonia. It is a blue liquid.

Sodium Amytal. Sodium isoamyl-ethylbarbiturate. A proprietary preparation of amylobarbitone sodium. A hypnotic. (333).

Sodium Anoxynaphthonate. Sodium 4′ - anilino - 8 - hydroxy - 1,1′ - azo - naphthalene - 3,6,5′ - trisulphonate. Coomassie Blue.

Sodium Antimonylgluconate. A preparation used in the treatment of schistosomiasis. It is the sodium salt of a trivalent antimony derivative of gluconic acid. TRIOSTAM.

Sodium Apolate. Poly(sodium ethylenesulphonate). Sodium Lyapolate. Pergalen.

Sodium Borobenzoate. A product obtained by dissolving boric acid in sodium benzoate solution.

Sodium Calciumedetate. The calcium chelate of the disodium salt of ethylenediamine - NNN′N′tetra - acetic acid. Calcium Disodium Versenate.

Sodium Citrotartrate. A mixture of sodium citrate and tartrate.

Sodium Citrotartrate, Effervescent. A mixture of sodium bicarbonate, citric, and tartaric acids, and sugar.

Sodium Cyclamate. Sodium cyclohexylsulphamate. Sucaryl.

Sodium Diatrizoate. Sodium 3,5-diacetamido - 2,4,6 - tri - iodobenzoate. Sodium Amidotrizoate. Hypaque.

Sodium Dibunate. Sodium 2,6-di-*t*-butylnaphthalenesulphonate.

Sodium Diprotrizoate. Sodium 3,5-dipropionamido - 2,4,6 - tri - iodo - benzoate.

Sodium Edetate. Disodium dihydrogen ethylenediamine - NNN′N′ - tetra - acetate.

Sodium Glucaldrate. Sodium gluconatodihydroxyaluminate. Glymaxil.

Sodium Glucaspaldrate. Octasodium tetrakis(gluconato)bis(salicylato) μ-diacetatodialuminate. III dihydrate.

Sodium Hydrosulphite Formaldehyde. See HYDROSULPHITE.

Sodium Ipodate. Sodium β-(3-dimethylaminomethylenamino- 2,4,6 - tri - iodophenyl)propionate. Biloptin. Orgrafin Sodium.

Sodium Ironedetate. The iron chelate of the monosodium salt of ethylenediamine - NNN′N′ - tetra - acetic acid. Sytron.

Sodium Lactophosphate. A mixture of sodium lactate and sodium acid phosphate.

Sodium Metrizoate. Sodium 3-acetamido - 2,4,6 - tri - iodo - 5 - N - methyl - acetamidobenzoate. Triosil.

Sodium Mica. See PARAGONITE.

Sodium Morrhuate. The sodium salt of the fatty acids of cod-liver oil.

Sodium Muriate. See MURIATE OF SODA.

Sodium Permutite. A sodium zeolite, made artificially.

Sodium Phosphate, Effervescent. A mixture of sodium phosphate and bicarbonate and citric and tartaric acids.

Sodium Picosulphate. A laxative currently undergoing clinical trial as " La 391 " and " DA 1773 ". It is *di*sodium 4, 4′ - (2 - pyridyl) methylenedi (phenyl sulphate). LAXOBERAL.

Sodium Salvarsan. Prepared by precipitating a solution of salvarsan in aqueous caustic soda.

Sodium Sesquicarbonate. See CREX.

Sodium Stannite. Sodium chloride, caustic soda, sodium nitrate, and tin, are heated together to obtain a dry powder, which is called sodium stannite. Used in dyeing and calico printing.

Sodium Tyropanoate. A contrast medium currently undergoing clinical trial as " WIN 8851–2 " and " Bilopaque ". It is sodium 2-(3-butyramido-2, 4, 6 - tri - iodobenzyl) butyrate.

Sodium Versenate. A proprietary preparation of trisodium edetate used in the treatment of hypercalcæmia. (275).

Sod Oil. See DÉGRAS.

Sodos. A mixture of sodium dihydrogen phosphate and sodium bicarbonate. Used in medicine.

Sofanate. A fungicide for fruit storage. (511).

Sofnolite. See SOFNOL SODA-LIME G.

Sofnol Soda-lime G (Sofnolite). A proprietary form of soda-lime containing a little manganic acid. It is stated to absorb much more carbon dioxide than ordinary soda-lime, and to change colour as the degree of saturation is approached.

Sofracort Skin Spray. A proprietary preparation of framycetin sulphate, gramicidin and hydrocortisone. An antibacterial aerosol used in dermatology. (307).

Sofradex. A proprietary preparation of DEXAMETHASONE, FRAMYCETIN and gramicidin used in the form of anti-infective ear-drops. (443).

Soframycin. A proprietary preparation of FRAMYCETIN sulphate and gramicidin. A topical antibiotic. (443).

Sofratulle. A proprietary gauze dressing impregnated with FRAMYCETIN. (443).

Soft Amber. See GEDANITE.

Soft Copal. A name applied to varieties of Australian sandarac resin.

Softex. Proprietary trade name for pure red oxide (*q.v.*).

Soft Ore. See ARGENTITE.

Soft Platinum. Commercially pure platinum, containing about 1 per cent. iridium.

Softrite. A proprietary rubber softener. It has a zinc laurate base.

Soft Solder. See PLUMBER'S SOLDER.

Soie Nouvelle. See LUFTSEIDE.

Soilime. A lime residue from cyanamide manufacture. It contains 50 per cent. of lime.

S-Oils. Sulphur-containing oils obtained by the distillation of crude petroleum oil in the presence of sulphur. They have a strong antiseptic action against wood-destroying fungi.

Soimonite. A mineral. It is a variety of corundum.

Soja Bean Oil. See SOYA BEAN OIL.

Solac. A synthetic milk made from the soya bean, which is ground and stirred with an alkaline solution. It is then filtered, and the oil separated. Arachis and sesamé oils are added, and the whole emulsified. It resembles cow's milk.

Solacen. A proprietary preparation of tybamate. A sedative. (255).

Solactol. A proprietary trade name for ethyl lactate.

Solæsthin. Methylene chloride, CH_2Cl_2, an anæsthetic.

Solapsone. Tetrasodium salt of bis-[4-(3 - phenyl - 1,3 - disulphopropylamino)-phenyl]sulphone. Solasulphone. Sulphetrone.

Solargentum. A compound of silver and gelatin, containing from 10–23 per cent. of silver in a colloidal form. Used in medicine.

Solargyl. A compound of silver and protein containing 30 per cent. silver. An antiseptic.

Solar Oil. The name given to various hydrocarbons obtained as by-products in the treatment of brown coal tar in paraffin works. It has a specific gravity above 0·85, but not exceeding 0·88, and a flash-point not below 176° F.

Solar Oil, Light. A fraction of Russian petroleum distillation, of specific gravity 0·885–0·895, and flashing at 136° C. A burning oil.

Solar Salt. Salt (sodium chloride) obtained by the evaporation of sea-water by sunshine.

Solarson. The ammonium salt of chloro-heptenyl-arsonic acid, $CH_3.(CH_2)_4.CCl : CH.AsO(OH)_2$.

Solar Stearin. Lard stearin.

Solar Steel. A proprietary trade name for a silicon steel containing 1 per cent.

silicon, 0·5 per cent. molybdenum, 0·4 per cent. manganese, and 0·5 per cent. carbon.

Solasulphone. Solapsone.

Solatol. A preparation of crude phenol (carbolic acid). A disinfectant.

Solazzi Juice. A variety of liquorice in sticks.

Sol-bi. Bismuth - campho - carbonate solution in oil.

Solbrol. Methyl - *p* hydroxy - benzoate. Used in the preservation of food.

Solcod. A proprietary trade name for sulphonated cod oil.

Solcornol. A proprietary trade name for a sulphonated corn oil.

Solder. The various alloys or mixtures which constitute solder are usually classified as hard or soft, according to their melting-point. Hard solder includes brazing solder, silver solder, and gold solder, whilst the soft solders usually consist of tin and lead, and melt below 300° C. The addition of cadmium and bismuth tends to lower the melting point, and antimony to raise it.

Solder, Coarse. See PLUMBER'S SOLDER and TINSMITH'S SOLDER.

Solder, Common. Contains equal parts of lead and tin.

Solder, Fine. See PLUMBER'S SOLDER and TINSMITH'S SOLDER.

Soldering Acid. Hydrochloric acid, HCl.

Soldering Pastes. Various. Many consist of tallow and rosin, sometimes with the addition of sal ammoniac.

Soldering Powders. Various. Many consist of rosin only, whilst others contain rosin, sal ammoniac, and zinc sulphate, or rosin, sal ammoniac, and borax.

Soldering Salt. Ammonium and zinc chlorides.

Soldering Solution. A solution of zinc chloride. Also see GAUDUIN'S FLUID and MULLER'S FLUID.

Solder, Pewter. See TINSMITH'S SOLDER.

Solder, Slicker. See PLUMBER'S SOLDER.

Solder, Soft. See PLUMBER'S SOLDER.

Solder, Tinned Iron. See TINSMITH'S SOLDER.

Soldier's Ointment. Ung. Hydrarg. Mite.

Soldis. A disinfectant containing phenolic and cresylic bodies. It is miscible with water.

Soldo. A flux used for tinning metals. It is mixed with powdered tin.

Soldona. A trade name for a preparation containing formaldehyde and hydrogen peroxide. Used as a milk preservative.

Solene. See PETROLEUM ETHER.

BRITISH STANDARD SPECIFICATIONS FOR SOLDERS
In addition to lead :

Code letter	Tin, per cent.	Antimony, per cent.	Main impurities		Uses
			Fe, per cent.	As, per cent.	
A	64–66	1 max.	0·02	0·05	Low melting-point steel tube joint.
B	49–51	2·5–3·0	0·02	0·05	Tinsmith's and coppersmith's work. Hand soldering.
C	39–41	2·0–2·4	0·02	0·05	General hand soldering.
D	29–31	1·0–1·7	0·02	0·05	Plumbers' wiped joints.
E	94·5–95·5	0–0·5	0·02	0·05	Special electrical purposes.
F	49–51	0–0·5	0·02	0·05	Machine soldering. General electrical purposes. For zinc and galvanised iron.
G	41–43	0–0·4	0·02	0·05	Dipping baths. Zinc and galvanised iron. Tinning electrical joints.
H	34–36	0–0·3	0·02	0·05	Lead cable wiped joints.
J	29–31	0–0·3	0·02	0·05	Dipping baths.
K	59–61	0–0·5	0·02	0·05	Special machine soldering.

Solenhofen Stone. A porous limestone containing clay.

Solenite. An explosive. It is an Italian smokeless powder, and contains 30 per cent. nitro-glycerin, 40 per cent. " insoluble," and 30 per cent. " soluble " nitro-cellulose.

Solferino. See MAGENTA.

Solganal. A proprietary preparation of sodium - 4 - sulpho - methyl - amino - 2 - auro-mercapto-benzol-1-sulphonate.

Solicum. A material made from waste rubber and oil.

Solidago. The dried herb of *Solidago odora*. It is used medicinally as a stimulant, carminative, and diuretic.

Solid Alcohol. A soapy mass containing about 20 per cent. water, 20 per cent. sodium stearate, and 60 per cent. alcohol.

Solid Ammonia. Prepared by adding to a mixture of from 3–5 parts sodium stearate (dissolved in 10 parts aqueous ammonia, or 80 per cent. spirits of wine), about 85–90 parts ammonia solution, containing 25–33 per cent. ammonia, NH_3.

Solid Blue 2R, B. See INDULINE, SOLUBLE.

Solid Cresol. See PARALYSOL.

Solid Green. See DINITROSORESORCIN and MALACHITE GREEN.

Solid Green Crystals. See MALACHITE GREEN.

Solid Green G. See GALLANILIC GREEN.

Solid Green J, TTO. See BRILLIANT GREEN.

Solid Green O. See MALACHITE GREEN.

Solid Hydrogen Peroxide. See HYPEROL.

Solidified Alcohol. A name applied to a solution of nitrocellulose in ethyl alcohol for use in heaters, etc., as a fuel.

Solidified Linseed Oil (Oxidised Linseed Oil, Linoxyn). A flexible solid mass obtained when linseed oil is exposed to oxidation.

Solidified Sulphuric Acid (Consolidated Sulphuric Acid). Sulphuric acid mixed with kieselguhr for the purpose of transporting it.

Solidite. A proprietary trade name for a range of moulded products made from shellac, bitumen, and fillers. Electrical insulation.

Solidonia Fibre. A proprietary name for a vegetable fibre used mainly in the form of combed yarn.

Solid Storax. See RED STORAX.

Solid Violet. See GALLOCYANINE DH and BS.

Solid Yellow S. See ACID YELLOW.

Soligen. A proprietary trade name for certain metallic naphthenates used as paint driers.

Solithane. A proprietary polyester-based polyurethane elastomer cross-linked with castor oil or diamine. (129).

Soliwax. A proprietary preparation of dioctyl sodium sulphosuccinate used to remove ear-wax. (921).

Sollacaro's Aluminium Solder. An alloy of 64 per cent. zinc, 30 per cent. tin, and 6 per cent. lead.

Solochrome. A registered trade name for certain dyestuffs.

Solochrome Black 6B. A dyestuff. It is equivalent to Chrome fast cyanine G.

Solochrome Black F. A dyestuff. It is a British equivalent of Diamond black F.

Solochrome Yellow Y. A dyestuff. It is a British brand of Milling yellow.

Soloform. A proprietary product. It is a tri-iodo-phenol preparation.

Solox. A proprietary trade name for an alcohol-type solvent, a fuel for alcohol lamps, blow torches, portable stoves, etc. It contains ethyl alcohol mainly, with small quantities of ethyl acetate and petrol.

Solozone. A proprietary trade name for a sodium peroxide containing 20·5 per cent. available oxygen.

Solpadeine. A proprietary preparation of PARACETAMOL, CODEINE phosphate and caffeine. An analgesic. (439).

Solprene. A thermoplastic elastomer.

Solprin. A proprietary preparation of aspirin, calcium carbonate, citric acid and saccharin. An analgesic. (243).

Sol Rubber. The portion of rubber which enters solution when unmilled raw rubber is treated with a solvent.

Soltercin. A proprietary preparation of soluble aspirin, phenacetin and buto-barbitone. (248).

Soluble Algin. Alginate of soda, obtained from seaweed.

Soluble Anhydrite. Obtained from plaster of Paris by heating it to 200° C.

Soluble Aspirin. See KALMOPYRIN.

Soluble Blue (Water Blue, Water Blue 6B Extra, Navy Blue, Serge Blue, Lyons Blue, Marine Blue, Pure Blue, Blackley Blue, Sedan Blue, Silk Blue, Guernsey Blue, China Blue, London Blue Extra, Cotton Blue, Blue, Acid Blue, Soluble Blue B, Conc. NS, L, 2R, Silk Blue O, Pure Soluble Blue, Water Blue B, R, Ink Blue, Ink Blue 8671, 7567). Dyestuffs. They consist of the ammonium, sodium, or calcium salts of the trisulphonic acid (with some disulphonic acid), of triphenyl-rosaniline, and triphenyl-pararosaniline. Free acids, $C_{38}H_{31}N_3O_9S$, and $C_{37}H_{29}N_3O_9S$.

They dye silk and mordanted cotton blue. Also see ALKALI BLUE and METHYL BLUE.

Soluble Blue B, Conc. NS, L, 2R. British equivalents of Soluble blue.

Soluble Blue 8B, 10B. See METHYL BLUE.

Soluble Blue XG. See ALKALI BLUE XG.

Soluble Blue XL. See METHYL BLUE.

Soluble Calomel. See CALOMELOL.

Soluble Castor Oils. See BLOWN OILS.

Soluble Cream of Tartar. Cream of Tartar (potassium bitartrate, $C_4H_5O_6K$) dissolved in a solution of boric acid or borax. It is potassium boro-tartrate, and is used medicinally.

Soluble Glass (Soluble Soda-Glass, Water Glass, Glass Liquor). A syrupy solution containing 50 per cent. of sodium silicate, Na_4SiO_4, and Na_2SiO_3. It is used to impregnate articles to render them fire-resistant, as an adhesive for glass and porcelain, and as adulterant in soap, in dyeing, and for egg-preserving.

Soluble Gluside. See SACCHARIN, EASILY SOLUBLE.

Soluble Indigo. See INDIGO CARMINE.

Soluble Nitrogenous Organic Fertiliser. See NITROGENINA.

Soluble Oil. See TURKEY RED OILS.

Soluble Oils. See CUTTING OILS.

Soluble Phenyle. A fluid containing coal-tar creosote, rosin oil, potassium oleate, and caustic soda. It gives an emulsion with water, and is used as a sheep dip.

Soluble Pitch. See PIX SOLUBILIS.

Soluble Potash Glass. Potassium silicate, K_2SiO_3.

Soluble Primrose. See ERYTHROSIN.

Soluble Prussian Blue. A blue pigment. It is potassium ferrous ferrocyanide, $FeK.Fe(CN)_6$.

Soluble Regina Purple. A dyestuff obtained by the sulphonation of Spirit violet.

Soluble Saccharin. See SACCHARIN, EASILY SOLUBLE.

Soluble Salumin. Aluminium ammonium salicylate, $Al_2(C_6H_4(ONH_4)CO_2)_6 2H_2O$. An astringent wash for nose and throat.

Soluble Soda Glass. See SOLUBLE GLASS.

Soluble Starch (Amylodextrin). Obtained by heating starch with glycerin and adding alcohol. An emulsifying agent. A soluble starch is also made by boiling starch in water and adding a little caustic soda to clear it.

Soluble Tannalum. Obtained by the treatment of Tannalum (q.v.) with tartaric acid, $Al(C_4H_4O_6)(C_{14}H_9O_9)+3H_2O$. Used medicinally.

Soluble Tartar. Potassium tartrate, $(CHOH)_2(COOK)_2$.

Soluble Tartaric Acid. Tartrelic acid, $C_4H_4O_5$.

Solubor. A highly soluble form of sodium borate $Na_2B_8O_{13}.4H_2O$. It is used to correct boron deficiency in plants by applying either as a foliar spray, in nutrient feeds or with herbicides. (183).

Solucortef. A proprietary preparation of hydrocortisone sodium succinate. (325).

Solumedrone. A proprietary preparation of METHYLPREDNISOLONE used in the treatment of shock. (325).

So-luminum. An aluminium solder It is an alloy of 55 per cent. tin, 33 per cent. zinc, 11 per cent. aluminium, and 1 per cent. copper.

Solupen. A proprietary preparation containing Benzylpenicillin Sodium. An antibiotic. (309).

Solupen Buffered. A proprietary preparation containing Benzylpenicillin Sodium. An antibiotic. (309).

Solupyrin. See KALMOPYRIN.

Solurol. Thyminic acid, $C_{30}H_{46}N_4O_{15}(P_2O_5)_2$. It dissolves uric acid, and is said to be used for gout.

Solusalvarsan. A proprietary trade name for sodium 3 : 4'-diacetamino-4-hydroxyarsenobenzene-2'-glycollate.

Soluseptacine. A proprietary trade name for disodium-p-(γ-phenyl-propyl-amino)-benzene-sulphonamide-$\alpha\gamma$-disulphonate.

Solution of Chlorinated Lime. Obtained by shaking 1 lb. bleaching powder with 1 gallon distilled water and filtering.

Solution SBR. See UNIDENE.

Solutol. An alkaline solution of sodium cresol in an excess of cresol, obtained by treating cresol with caustic soda. A disinfectant.

Soluvone. A proprietary preparation of penicillin G and streptomycin sulphate. An antibiotic. (309).

Solux. A proprietary trade name for a rubber antioxidant. It is p-hydroxy-phenyl-morpholine.

Solvaloid A. A proprietary trade name for a naphthenic petroleum type vinyl plasticiser. (127).

Solvaloid C. An aromatic petroleum type of vinyl plasticiser. (127).

Solvaloid L. A naphthenic petroleum type of vinyl plasticiser. (127).

Solvaloid N. An aromatic petroleum type of vinyl plasticiser. (127).

Solvatone. A mixture of approximately 80 per cent. acetone, 10 per cent. isopropyl alcohol, and 10 per cent. toluene. A solvent for lacquers.

Solvay Soda. A registered trade name for a sodium carbonate for water softening.

Solvene. A proprietary solvent. It is a heavy grade coal-tar naphtha. It has a boiling range of from 160–190° C., a specific gravity of 0·88–0·91, and flashes at 100° F. It is a solvent for ester gum, pitches, etc.

Solvenol. A proprietary product obtained from the steam distillation of waste wood and stumps of the Southern yellow pine. It is stated to be a terpene with properties similar to turpentine but superior in its solvent power. It is a volatile thinner and has the following characteristics : Specific gravity at 15·5° C., 0·8565; refractive index at 20° C., 1·476; unpolymerised residue, 2·5 per cent.; initial boiling-point, 167° C., and moisture less than 1 per cent.

Solvent Naphtha. A fraction of coal-tar distillation of specific gravity 0·875. Also the wood naphtha recovered from grey acetate of lime, prepared from the distillation of wood.

Solventol. See TERPURILE.

Solventol. Bamipine. (*q.v.*).

Solveol. Cresols made soluble in water by the addition of sodium cresotinate. A disinfectant and substitute for guaiacol and creosote.

Solvesso. A proprietary trade name for petroleum solvents.

Solvochin. A proprietary preparation of basic quinine, 25 per cent. solution.

Solvol. A proprietary solvent. It is tetrahydro-naphthol acetate.

Solvosal. Salol-phosphinic acid, C_6H_4. $COOC_6H_5.O.PO(OH)_2$, used medicinally. The term Solvosal appears to be also applied to a German product stated to contain theobromine, calcium salicylate, adrenalin, quinine hydrochloride, camphor, oil of eucalyptus, and oil of aniseed. A cough mixture.

Solway Blue. Alizarin Sapphirol (*q.v.*).

Solway Purple. Alizarin Irisol (*q.v.*).

Somali Gum. An acacia gum from *Acacia glaucophylla* and *Acacia abyssinica*.

Somatose. A food preparation made from meat. The chief ingredients are albuminose (78 per cent.) and peptone (3 per cent.). Prescribed in cases of chlorosis and indigestion. Iron somatose containing 2 per cent. of iron, and milk somatose, are similar preparations.

Somatose, Iron. See SOMATOSE.

Somatose, Milk. See SOMATOSE.

Somberero Phosphate. See SOMBRERITE.

Sombrerite (Somberero Phosphate). A

mixed calcium and aluminium phosphate, which occurs on Sombrero Island.

Somilan. Chloral Betaine.

Sommairite. A mineral. It is 8[(Fe,Zn)SO₄.7H₂O].

Somnal. A solution of a mixture of chloral hydrate and urethane in alcohol. Used as a hypnotic.

Somnesin. A proprietary trade name for Methylpentynol.

Somnifene. A diethyl-dipropenyl-barbiturate of diethyl-amine. A hypnotic. No longer commercially available.

Somnoform (Narcoform). A mixture of 60 per cent. ethyl chloride, 35 per cent. methyl chloride, and 5 per cent. ethyl bromide. Said to be used as a local anæsthetic.

Somnolin. A mixture of chloral hydrate and acetyl-*p*-amino-phenol. A hypnotic.

Somnos. A glycerin solution of trichlorethidene - propenyl - ether, CCl₃CHO (OH)₃.C₃H₅.

Somnosal. A proprietary preparation of brom-isovalerianyl-urea combined with amidopyrin.

Sonalgin. A proprietary preparation of butobarbitone, codeine phosphate and phenacetin. (336).

Sonatin. A solution of benzoyl-benzoic acid, C₆H₅.CO.C₆H₄COOH, in castor oil. A substitute for balsam of Peru in the treatment of skin diseases.

Sonergan. A proprietary preparation of promethazine hydrochloride and butobarbitone. A hypnotic. (336).

Soneryl (Neonal). Butyl-ethyl-malonyl-urea. A proprietary preparation of butobarbitone. A hypnotic. (336).

Songanal B. A gold-thioglucose compound used as an anti-tubercle reagent.

Sonilyl. A proprietary preparation of sulphchlorpyridazine. An antibiotic. (819).

Sonnenschein's Reagent. An alkaloidal reagent prepared by adding phosphoric acid to a warm solution of ammonium molybdate in nitric acid, boiling the precipitate produced in *agua regia*, evaporating to dryness, and dissolving in 10 per cent. nitric acid.

Sonomaite. A mineral, Mg₃Al₂(SO₄)₆.33H₂O.

Sonora Gum (Arizona Shellac). A variety of shellac obtained from *Larrea Mexicana*.

Soot. A deposit formed when the products of combustion, carrying smoke, are brought into contact with a cool surface. It usually contains carbon, 16–23 per cent. ash, and 5–15 per cent. hydrocarbons.

Soot Brown. Bistre (*q.v.*).

Sophol (Argophol). Silver - formo -

nucleinate, containing 20 per cent. silver. A germicide.

Sorban. A strong solution of sorbitol, HO.CH₂.(CH.OH)₄.CH₂.OH.

Sorbidel. A proprietary preparation containing sorbitol. A laxative. (323).

Sorbide Nitrate. 1,4 : 3,6-Dianhydrosorbitol dinitrate. Isosorbide. Dinitrate. Vascardin.

Sorbimacrogol Laurate 300. Polysorbate 20. (See under other Polysorbates for equivalent Sorbimacrogol esters.)

Sorbin. See SORBINOSE.

Sorbinose (Sorbin). Sorbose, C₆H₁₂O₆, a sugar.

Sorbin Red (Azogrenadine S, Lanafuchsine SB). A dyestuff prepared from acetyl-*p*-phenylene-diamine and α-naphthol-3 : 6-disulphonic acid. Lanafuchsine SB has similar properties, but gives a yellow shade of red.

Sorbismal. A German preparation of finely divided bismuth in oil.

Sorbitan Monolaurate. A surface-active agent currently undergoing clinical trial as " Sorbester P12 " and " Span 20 ".

Sorbitan Mono-oleate. A surface-active agent currently undergoing clinical trial as " Sorbester P17 " and " Span 80 ".

Sorbitan Monopalmitate. A surface-active agent currently undergoing clinical trial as " Sorbester P16 " and " Span 40 ".

Sorbitan Monostearate. A surface-active agent currently undergoing clinical trial as " Sorbester P18 " and " Span 60 ".

Sorbitan Sesquioleate. A surface-active agent currently undergoing clinical trial as " Arlacel C " and " Arlacel 83 ".

Sorbitan Trioleate. A surface-active agent currently undergoing clinical trial as " Sorbester P37 " and " Span 85 ".

Sorbitan Tristearate. A surface-active agent, currently undergoing clinical trial as " Sorbester P38 " and " Span 65 ".

Sorbite. A constituent of iron, which is formed in the transformation of austenite, the stage following trootsite and osmondite, and preceding pearlite.

Sorbitol (EGIC). A proprietary preparation of sorbitol used in intravenous nutrition. (925).

Sorbolene. The registered trade mark for a fat liquor for the leather trade. It is used in the tanning and dyeing process to give greater elasticity to the leather, and enables fuller shades to be obtained in dyeing.

Soredur. A proprietary polyester laminating resin. (924).

Soreflon. Polytetrafluoroethylene (PTFE).

Sorel Cement (Lignolite, Xylolite Xylolith, Magnesia Cement, Lito-silo). A magnesium oxychloride cement, made from magnesite (magnesium carbonate, $MgCO_3$) and magnesium chloride, $MgCl_2$.

Mixtures of magnesite or magnesium oxide, magnesium chloride, and wood-dust are made in varying proportions. The composition varies between $MgCl_2$.$5MgO.13H_2O$, at the commencement, to $MgCl_2.2MgO.9H_2O$, after some days. Used for floorings. Also a trade name for a dental cement. 1 part zinc oxide, $\frac{1}{2}$ part fine sand. Add zinc chloride of specific gravity 1·26. It hardens at once.

Sorel's Gutta-percha Substitutes. Substitutes containing rosin, pitch, rosin oil, slaked lime, and gutta-percha. Some are filled with china clay, and in others coal tar is used.

Sorensen's Salt. Sodium phosphate, $Na_2HPO_4.2H_2O$.

Sorghum Beer. A beer made from Sorghum or Kaffir corn. It contains about 2·9 per cent. alcohol.

Soricin. Sodium ricinoleate.

Sormodren. See BORNAPRINE.

Sornyl. Benzoyl-benzyl-succinic ester.

Sorrel Salt. See SALT OF SORREL.

Sorrel's Alloy. An alloy of 98 per cent. zinc, 1 per cent. iron, and 1 per cent. copper. Another alloy contains 80 per cent. zinc, 10 per cent. iron, and 10 per cent. copper.

Sotacor. A proprietary preparation of SOTALOL hydrochloride used in the treatment of angina pectoris. (324).

Sotalol. A beta adrenergic blocking agent. It is $(\pm)$-4'-(1-hydroxy-2-iso-propylaminoethyl) methanesulphonanilide. BETA-CARDONE and SOTACOR are the hydrochloride.

Soubieran's Ammonical Salt. Mercury-ammonium-nitrate, $(NH_2.Hg_2O)NO_3$.

Soude Douce. See BLANQUETTE.

Souesite. A nickel-iron alloy which occurs naturally.

Soulan's Cement. Consists of 7 parts resin, 10 parts ether, 15 parts collodion, and aniline red. A semi-transparent varnish, used for sealing corks into bottles.

Soumansite. A mineral. It is an aluminium-sodium fluophosphate.

Souple Silk. A silk which has been deprived of a considerable proportion of its sericin or silk gum, usually to the extent of from 8–12 per cent. of its weight.

South African Bush Tea. The dried leaves of several species of *Cyclopia*. It is used as a substitute for tea, but contains no theine.

South African Jade. Synonym for Grossular.

Southalite. A phenol-formaldehyde condensation product, with a filler of paper, used for insulating purposes.

Souzalite. A mineral. It is $(Fe^{..},Mg)_3$ $Al_4(PO_4)_4(OH)_6.2H_2O$.

Sovprene. A Russian chloroprene synthetic rubber obtained by the polymerisation of acetylene to form divinyl-acetylene and then the formation of chloroprene by treating with hydrogen chloride, followed by polymerisation.

Soxhlet's Solution. A modified Fehling's solution. It consists of (1) a solution of 34·639 grams copper sulphate in 500 c.c. water and (2) 50 grams caustic soda and 173 grams potassium sodium tartrate in 500 c.c. water. Used for the determination of sugars.

Soy (Shoya, Sho-ju). Names for a sauce made from the soya bean. It is made by grinding the beans with wheat, boiling with water, and allowing the mixture to ferment.

Soyabean Glue. The extracted soya bean meal from which the flour has been removed, giving a vegetable protein which is used as a vegetable glue.

Soya Bean Miso. A Japanese food product made from fermented rice and mixing with steamed soya beans, salt, and water.

Soya Bean Oil. An oil obtained from the seeds of *Soja hispida* by expression or extraction with a solvent. It is used as an edible oil, and is also employed in soap-making, paints and varnishes, and in the linoleum industry. It is also known as Chinese bean oil. Specific gravity at 15° C. 0·924–0·929. Refractive index at 15° C. 1·472–1·480. Saponification value 188–194.

Soy Oil. A mixture of soya bean oil and wheat oil, obtained as a by-product in the brewing of soy.

Soyolk. A proprietary preparation made from the soya bean. A food product containing high fat and protein content.

Sozal. Aluminium-phenol-sulphonate, $Al_2(C_6H_4HSO_4)_6$. Recommended as a substitute for iodoform.

Sozionic Acid. See SOZOIODOL.

Sozoiodalic Acid. See SOZOIODOL.

Sozoiodol (Sozoiodalic Acid, Sozoinic Acid, Ronozol Salts). Commercial names given to certain di-iodo-*p*-phenol-sulphonates. Easily soluble sozoiodol is the sodium salt, whilst the potassium salt is difficultly soluble. Di-iodo-*p*-phenol-sulphonic acid has the formula $C_6H_2.I_2.OH.SO_2OH$. The salts are used as substitutes for iodoform.

Sozolic Acid. o-Phenol-sulphonic acid, $C_6H_4.OH.SO_3H$. An antiseptic. Also see ASEPTOL.

Sozonon. A German dusting powder for wounds. It contains 5 per cent. of the sodium salt of sozioidol and 95 per cent. talc.

Spadaite. A mineral, $H_2Mg_5Si_6O_{18}$, of Spain.

Spalerite. Zinc blende, ZnS.

Span 20. A proprietary trade name for Sorbitan Monolaurate.

Span 60. Sorbitan Monostearate.

Span 80. Sorbitan Mono-oleate.

Span 40. Sorbitan Monopalmitate.

Span 85. Sorbitan Trioleate.

Span 65. Sorbitan Tristearate.

Spandofoam. Rigid polyurethane foam-in-place plastic. (539).

Spandoglue. An adhesive for polyurethane and polystyrene. (539).

Spaneph Spansules. A proprietary preparation of EPHEDRINE sulphate in a sustained release form, used in the treatment of bronchospasm. (658).

Spangite. This is phillipsite, a zeolite containing potassium.

Spangolite. A mineral. It is an aluminium-copper basic sulphate.

Spaniolite. Tetrahedrite containing mercury.

Spanish Chalk. See ACTINOLITE.

Spanish Flies. Cantharides, the dried insects.

Spanish Juice. Liquorice in sticks.

Spanish Moss. The fibre of the plant *Tillandsia Usneoides* of tropicalAmerica. Used as a packing material.

Spanish Ochre. Burnt Roman ochre (Yellow Ochre $(q.v.)$).

Spanish Oil. Inferior olive oil.

Spanish Oxide. A natural red pigment. It is a red oxide of iron, and contains over 80 per cent. Fe_2O_3. Also see INDIAN RED.

Spanish Pepper. Capsicum $(q.v.)$.

Spanish Soap. An olive oil soap.

Spanish Topaz. See GOLDEN TOPAZ.

Spanish Turpentine. See TURPENTINE.

Spanish White. See CHALK and BISMUTH WHITE.

Spanish White, Fard's. See BISMUTH WHITE.

Spanish Yellow. See CHINESE YELLOW.

Spanoilite. A mineral. It is a cinnabar ore.

Sparable Tin. Acute crystals of cassiterite found in Cornwall.

Spar, Bismuth. See BISMUTOSPHÆRITE.

Spar, Bitter Salt. See DOLOMITE.

Spar, Blue. See CHESSYLITE.

Spar, Bolognian. A mineral. It is barite.

Spar, Brown. See BROWNSPAR.

Spar, Carmine. See CARMINITE.

Spar, Chlorite. See CHLORITOID.

Spar, Derbyshire. See FLUORSPAR.

Spar, Dolomite. Dolomite, $MgCa(CO_3)_2$.

Spar, Foam. See APHRITE.

Spar, Green. See MALACHITE.

Spar, Greenland. Cryolite $(q.v.)$.

Spar, Heavy. See BARYTES.

Spar, Heavy-fluoro. See FLUORO-HEAVY SPAR.

Spar, Ice. See SANIDINE.

Sparine. A proprietary sedative. It is either the embonate or the hydrochloride of PROMAZINE. (245).

Spar, Iron-zinc. See CAPNITE.

Sparkaloy. A proprietary trade name for silicon-manganese-nickel alloy used for spark-plug wire.

Spar, Labrador. See LABRADORITE.

Spar, Light. See GYPSUM.

Spar, Manganese. See RHODOCROSITE.

Spar, Tabular. See WOLLASTONITE.

Spartaite. A mineral consisting of calcspar and manganese spar. It is $2[(Ca,Mn)CO_3]$.

Spartalite. A mineral. It is zinc oxide ZnO.

Spar, Yellow Lead. A mineral. It is wulfenite.

Spar, Zinc. See CALAMINE.

Spar, Zinc-iron. See CAPNITE.

Spasmine. A proprietary preparation. It is stated to be benzyl succinate, $C_2H_4(COOC_6H_5CH_2)_2$.

Spasmodin. A proprietary preparation. It is stated to be benzyl benzoate, $C_6H_5.CH_2.O.COC_6H_5$.

Spasmonal. A proprietary preparation of alverine citrate used in the treatment of colonic disorders. (286).

Spasmotropin. A German preparation. It is a brom-calcium-glucose prepared for the treatment of cardiospasms.

Spasmyl. Camphor-benzyl-valerianate.

Spasticine. A proprietary preparation of atropine methyl bromide, benzyl succinate, and papaverine hydrochloride combined.

Spastipax. A proprietary preparation of hyoscyamine sulphate, atropine sulphate, hyoscine hydrobromide and amylobarbitone. (271).

Spathic Iron Ore. See CHALYBITE.

Spathic Manganese Ore. See RHODOCROSITE.

Spathic Zinc Ore. See CALAMINE.

Spathiopyrite. A mineral. It is a variety of safflorite.

Spathose Iron Ore. Ferrous carbonate, $FeCO_3$, containing also carbonates of calcium, manganese, and magnesium.

Spaulding's Glue. A liquid glue prepared from glue by means of strong vinegar, instead of water.

Spauldite. A proprietary trade name for a phenol-formaldehyde synthetic resin laminated product.

S.P.D. A rubber vulcanisation accelerator. It is sodium-pentamethylene-dithiocarbonate.

Spear Pyrites. See COCKSCOMB PYRITES and MARCASITE.

Special Grey R. See NEW GREY.

Specpure. Spectrographically standardised metals and chemicals. (615).

Spectacles. See DENTELLES.

Spectinomycin. An antibiotic produced by Streptomyces spectabilis. Trobicin.

Spectraban. A proprietary preparation of isoamyl-p-N, N-dimethylaminobenzoate in ethanol. A lotion used to protect skin from ultra-violet light. (379).

Spectroflux. Buffer mixtures for spectrographic analysis. (615).

Spectrosol. Solvents for spectrophotometry. (527).

Specular Iron Ore. See HÆMATITE.

Specularite (Specular Hæmatite). Iron oxide, Fe_2O_3, with a bright metallic lustre.

Speculite. A mineral. It is a gold-silver telluride containing 36·1–36·6 per cent. gold and 3·5–4·5 per cent. silver.

Speculum Metal. An alloy of 66 per cent. copper and 34 per cent. tin, with a little arsenic. An alloy of 64 per cent. copper, 32 per cent. tin, and 4 per cent. nickel. It has a specific gravity of 8·6 and a melting-point of 750° C. Used for making mirrors of reflecting telescopes.

Speed X Accelerator. A proprietary rubber vulcanisation accelerator containing 60 per cent. diphenylguanidine and 40 per cent. zinc oxide.

Speiss. See NICCOLITE.

Spelter. Zinc used in galvanising. The term is also used for hard solder.

Spence Metal. A material obtained by melting ferrous sulphide with sulphur. Used as a jointing material.

Spencerite. A mineral, $Zn_3(PO_4)_2$. $Zn(OH)_2.3H_2O$.

Spent Acid. Mixed acid (nitric and sulphuric acids), used in nitrating organic substances, in which it is partly deprived of nitric acid.

Spent Garancine. See GARANCEUX.

Spent Oxide. Obtained from gasworks. It consists mainly of a mixture of sulphur, hydrated oxides of iron, some undecomposed sulphides, and sawdust.

Spent Wash. The wash from corn and potato distilleries, resulting after the alcohol is distilled off. Used as a food for cattle.

Speriden. A proprietary preparation of hesperiden ascorbate. (332).

Spermaceti. A wax obtained from the head of the sperm whale. The crude product is obtained by chilling the head and blubber oils. It is a crystalline, dark brown, translucent substance, but when purified is white and lustrous, and melts at 44° C., has a saponification value of 121·3, an acid value of 2·8, and an iodine value of 5·8. It consists principally of cetyl palmitate, $C_{16}H_{33}$ $OCOC_{15}H_{31}$.

Spermaceti Ointment. Consists of 5 parts spermaceti, 1¾ parts white wax, and 14 parts olive oil.

Spermaceti Sugar. A material prepared by moistening a mixture of spermaceti and sugar with alcohol and evaporating.

Spermaceti, Vegetable. See CHINESE WAX.

Spermolin. A linseed oil product. It is a proprietary binder for sands used as cores in metal casting.

Spermoline Oil. A compound spindle oil used for lubrication. It has a specific gravity of 0·879 at 60° F. and a flash-point of 376° F.

Sperrylite. A rare mineral. It is an arsenide of platinum, $PtAs_2$. It also contains rhodium and antimony.

Spessartite. A mineral. It is a manganese-aluminium-silicate, $Mn_3Al_2Si_3O_{12}$.

Sphaerite. A mineral, $Al_5(OH)_9(PO_4)_2$. $12H_2O$.

Sphaero-cobaltite. A mineral. It is a cobalt carbonate.

Sphagni. An insulating material prepared from the white moss found on the Swedish peat moors.

Sphagnol. See CORBA OIL.

Sphalerite. A mineral. It is a Zinc Blende (q.v.).

Sphene (Titanite). A mineral. It is a calcium-titanium silicate, $Ca.Ti.SiO_5$.

Spiauterite. A mineral. It is a variety of wurtzite.

Spiegeleisen. A ferro-manganese alloy containing from 10–35 per cent. manganese, 60–85 per cent. iron, 1 per cent. silicon, and 4–5 per cent. carbon.

Spiegler Jolle's Reagent. A solution of 2 grams mercuric chloride, 4 grams succinic acid, and 4 grams sodium chloride in 100 c.c. water. A reagent for albumin in urine.

Spiegler's Reagent. This consists of 40 grams mercuric chloride and 20 grams tartaric acid dissolved in 500 c.c. water. To this solution is added 100 grams glycerol and 50 grams, sodium chloride, and the whole made up to 1,000 c.c. It is used for proteids.

Spiller's Resin. The oxidation products of rubber are sometimes called by this name.

Spinel (Talc-spinel). A mineral. It is a magnesium aluminate, $MgO.Al_2O_3$.

Spinel, Chromo. See PICOTITE.

Spinel, Iron. See PLEONASTE.

Spinel Ruby. See BALAS RUBY.

Spinning Brass. See BRASS.

Spinocain. β-Dimethyl-amino-ethyl N-butyl-amino-benzoate, $C_4H_9NH.C_6H_4.CO.O.CH_2.CH_2.N(CH_3)_2$.

Spinthere. A variety of the mineral sphene.

Spiperone. 8-[3-(4-Fluorobenzoyl)-propyl] - 1 - phenyl - 1,3,8 - triazaspiro [4,5]decan-4-one.

Spiracin. Methyl-carboxyl-salicylic acid. Used in medicine for the same purposes as salicylates.

Spiramycin. An antibiotic produced by Streptomyces ambofaciens. Rovamycin.

Spirarsyl (Ehrlich 418). The sodium salt of arseno-phenyl-glycine, $As_2(C_6H_4.NH.CH_2.CO_2Na)_2$. Has trypanocidal and spirillicidal properties.

Spirilene. 8 - [4 - (4 - Fluorophenyl) - pent - 3 - enyl] - 1 - phenyl - 1,3,8 - tria - zaspiro[4,5]decan-4-one.

Spirit Black. SEE INDULINE, SPIRIT SOLUBLE.

Spirit Blue O. See ANILINE BLUE, SPIRIT SOLUBLE.

Spirit, Cotton. See TIN SPIRITS.

Spirit, Crimson. See TIN SPIRITS.

Spirit Eosin (Ethyl Eosin, Eosin S, Eosin BB, Rose JB, Spirit Primrose, Primrose). A dyestuff. It is the potassium salt of tetrabromo-fluoresceine ethyl ether, $C_{22}H_{11}Br_4O_5K$. It dyes wool yellowish-red with slight fluorescence. Also see ERYTHRIN.

Spirit, Green Wood. See ACETONE ALCOHOL.

Spirit, Manhatton. See ACETONE ALCOHOL.

Spirit of Alum. Sulphuric acid, H_2SO_4.

Spirit of Copper. Acetic acid obtained from copper acetate.

Spirit of French Wine. See BRANDY.

Spirit of Hartshorn. A solution of ammonia. See VOLATILE ALKALI.

Spirit of Myrcia. See BAY RUM.

Spirit of Nitre. Spirit of nitrous ether. (See SPIRIT OF SWEET NITRE.) The term is also applied to nitric acid.

Spirit of Red Lavender. Compound tincture of lavender.

Spirit of Salt. Strong impure hydrochloric acid.

Spirit of Sal Volatile. Aromatic spirit of ammonia. It consists of 100 parts ammonium carbonate, 200 c.c. strong ammonia, 15 c.c. oil of nutmeg, 20 c.c. oil of lemon, 3,000 c.c. alcohol, and 1,500 c.c. water.

Spirit of Sulphur. Sulphurous acid, H_2SO_3.

Spirit of Sweet Nitre. Spirit of nitrous ether, consisting of a solution of 1·52–2·66 per cent. of ethyl nitrite in alcohol.

Spirit of Sweet Wine. Ethyl chloride, C_2H_5Cl, a local anæsthetic.

Spirit of Tar. See OIL OF TAR.

Spirit of Tin. Stannic chloride, $SnCl_4$.

Spirit of Turpentine. See OIL OF TURPENTINE.

Spirit of Verdigris. Acetic acid, CH_3COOH.

Spirit of Vinegar. Dilute acetic acid.

Spirit of Vitriol. Sulphuric acid, H_2SO_4.

Spirit of Vitriol, Sweet. Spirit of ether.

Spirit of Wood. Methyl alcohol, CH_3OH.

Spirit Oil. The first fraction from the distillation of Yorkshire Grease (q.v.). Used for making black varnish.

Spirit, Orange. See TIN SPIRITS.

Spirit, Palm. Arrack (q.v.).

Spirit, Plum. See TIN SPIRITS.

Spirit, Potato. See POTATO OIL.

Spirit Primrose. See SPIRIT EOSIN.

Spirit, Puce. See TIN SPIRITS.

Spirit Purple (Spirit Violet, Opal Violet, Regina Spirit Purple). Dyestuffs which consist of diphenylated rosanilines.

Spirit, Purple. See TIN SPIRITS.

Spirit, Pyroligneous. See METHANOL.

Spirit, Pyroxylic. See METHANOL.

Spirit, Red. See TIN SPIRITS.

Spirit, Red Cotton. See TIN SPIRITS.

Spirit Red III. See SUDAN IV.

Spirit, Resin. See ROSIN SPIRIT.

Spirit, Silent. See SPIRITS OF WINE.

Spirit, Sky Blue. See DIPHENYLAMINE BLUE, SPIRIT SOLUBLE.

Spirits of Wine (Ethanol, Silent Spirit). Ethyl alcohol, C_2H_5OH. Commercial spirits of wine contain 84 per cent. by weight of alcohol.

Spirit Soluble Blue. See ANILINE BLUE, SPIRIT SOLUBLE.

Spirit Soluble Eosin. See ERYTHRIN.

Spirit Soluble Quinoline Yellow. A quinoline dyestuff prepared from quinaldine, phthalic anhydride, and zinc chloride, $CO.O.C_6H_4.C : CH.C_9H_6N$.

Spirit, Standard Wood. See ACETONE ALCOHOL.

Spirittine. Soft wood tar creosote. A wood preservative.

Spirit Varnishes. Prepared by mixing resins with such solvents as methylated spirit or turpentine.

Spirit Vinegar. Made from potato or grain spirit. It contains up to 12 per cent. of acetic acid.

Spirit Violet. See SPIRIT PURPLE.

Spirit, Wood. See METHANOL.

Spirit Yellow. See ANILINE YELLOW.

Spirit Yellow I. A dyestuff. It is a British brand of Sudan I.

Spirit Yellow R (Yellow Fat Colour). A dyestuff prepared from o-toluidine, $C_{14}H_{15}N_3$. Used for colouring butter.

Spirobismol. A proprietary preparation of sodium-potassium-bismuthyl-tartrate and quinine-bismuth iodide in oil suspension.

Spirocid. See FOURNEAU 190.

Spirocid. An antileutic stated to be identical with stovarsol.

Spiroform. See VESIPYRINE.

Spironal. Sodium citro-bismuthate.

Spirone. A spray solution employed for asthma, and said to contain potassium iodide and acetone.

Spironolactone. β-(7α-Acetylthio-17β-hydroxy - 3 - oxoandrost - 4 - en - 17α-yl)propionic acid lactone.

Spirosal (Saliglycol). The mono-glycol ester of salicylic acid, $HO.C_6H_4.COO.CH_2CH_2OH$. Used in medicine for rheumatism.

Splintex. A proprietary safety glass.

Split Nut. See ORDEAL BEAN.

Spodiosite. A mineral, $Ca_3(PO_4)_2.CaF_2$.

Spodium, Black. Animal Charcoal (q.v.).

Spodium, White. Bone Ash (q.v.).

Spodumene (Hiddenite). A pyroxene mineral, $(Li.Na)Al(SiO_3)_2$.

β-Spodumene. Spodumene. A lithum-aluminium silicate ore, heated and freed from associated minerals.

Spodumene, Soda. See JADEITE.

Spondite. A glass substitute. It consists of cellulose acetate dope on wire netting.

Sponge Iron. A finely porous form of iron obtained by reducing iron oxide at a temperature where no sintering or fusion takes place. A reagent for the precipitation of copper, lead, and other metals from solution.

Spoon Metal. See NICKEL SILVERS.

Sporocide. A wood preservative mainly consisting of potassium-o-dinitro-cresylate.

Sporogelite. A mineral. It is aluminium oxide, $Al_2O_3.H_2O$, in a colloidal form.

Sporting Ballistite. A smokeless powder consisting of 37·6 per cent. nitroglycerin and 62·3 per cent. nitrocotton.

Spramex. See MEXPHALTE.

Sprengel's Explosives. Cakes of potassium chlorate which have absorbed combustible liquids.

Sprengsalpeter. An explosive consisting of 75 per cent. sodium nitrate, 15 per cent. brown coal, and 10 per cent. sulphur.

Sprilon. A proprietary preparation of DIMETHICONE, zinc oxide, soft paraffin, wool fat and alcohols, cetyl alcohol and DEXTRAN, used as an aerosol spray for skin protection. (337).

Spring Brass. See BRASS.

Sprödglaserz. Synonym for Polybasite.

Spruce Ochre. Yellow Ochre (q.v.).

Spumagen. A saponin product. It is

aphrogen (q.v.), made by an improved process, and is suitable for use in beverages.

Spurrite. A mineral. It is a calcium silico-carbonate.

Sputamin. A German preparation. It is a powder containing 80 per cent. chloramine. An antiseptic.

S.Q.D. Synthetic microscopical mountant. (544).

Squaw Root. Caulophyllum.

Squill. Bulb of Urginea maritima.

S.R.A. Dyestuffs. A series of colours used for dyeing of acetyl silk. They are of British manufacture. The letters " S.R.A." are the initial letters of the words " sulpho-ricinoleic acid," and the dyestuffs are treated with this acid to render them in suitable condition (colloidal dispersion) for use in dyeing acetyl silk.

S.S.F. Sodium silico-fluoride, Na_2SiF_6.

S.S.S. Sodium sesqui silicate. (537).

S.T. 137. A solution of hexyl-resorcinol in glycerin and water. A germicide.

Stabilarsan. A preparation of salvarsan. It is a combination of salvarsan with glucose, and is used in the treatment of syphilis.

Stabilator A.R. A proprietary antioxidant for rubber. It is phenyl-β-naphthylamine. It melts at 108° C., and is recommended for white mixings.

Stabiliser No. 1. A proprietary trade name for 1 : 3 : 5-isopropyl-cresol.

Stabilite. A proprietary rubber vulcanisation accelerator. It is diphenylethylene-diamine. The term is also applied to a German product used for electrical insulation. It is made from cotton, rubber, and fillers. (926).

Stabilite Alba. A proprietary rubber vulcanisation accelerator. It is di-o-tolyl-ethylene-diamine.

Stabilite L. A proprietary anti-oxidant. Diphenyl propylene diamine. (926).

Stabillin V-K. A proprietary preparation containing Phenoxymethylpenicillin Potassium. An antibiotic. (253).

Stabilite White. A proprietary antioxidant. Diphenyl ethylene diamine. (926).

Stabinol. A proprietary trade name for Isobuzole.

Stabismol. A proprietary preparation. It is a solution of α-carbonyl-cyclohexanyl acetate in olive oil.

Stablex. A stabilised bitumen used for protective coatings to resist acids.

Stabochlor. A proprietary chloride of lime specially prepared.

Stacol. A complex sodium borophosphate, an inorganic water-soluble resin stable to acids and alkalis.

Staff. A mixture of plaster-of-Paris and tow. Used for mouldings.

Staffelite. A mineral. It contains calcium phosphate with calcium chloride or fluoride.

Staflene Ho. Polyethylene.

Staflex 1XA. A proprietary trade name for a highly acetylated methyl cellosolve ricinoleate. A vinyl plasticiser (128).

Staflex CP. A mixed alkyl phthalate. A vinyl plasticiser.

Staflex KA. A polymeric type vinyl plasticiser. (128).

Staflex LA. A sebacate. A vinyl plasticiser. (128).

Staffordshire All Mine Pig. A pig iron made in Staffordshire from ore. It contains about from 0·5–0·75 per cent. of phosphorus.

Stag Brand. Paints, enamels and jointing pastes. (512).

Stagnine. An astyptic obtained from the spleen.

Stahl's Sulphur Salt. Potassium sulphite, $K_2SO_3.2H_2O$.

Stainless Iron. This is really stainless steel, and usually contains from 0·1–0·2 per cent. carbon, 12–27 per cent. chromium, and up to 0·5 per cent. silicon.

Stainless Invar. A Japanese alloy of 54 per cent. cobalt, 36·5 per cent. iron, and 9·5 per cent. chromium. It has a low coefficient of expansion and withstands corrosion well.

Stainless Silver. This is usually an alloy of 92·5 per cent. silver with copper and antimony, and is used for table ware.

Stainless Steel (Rustless Steel). A chromium-steel alloy containing 12–15 per cent. chromium, and not more than 0·45 per cent. carbon. Used for cutlery, acid pumps, turbine blades, and exhaust valves for engines. Some alloys contain 12·18 per cent. chromium, 8–12 per cent. nickel, 74–76 per cent. iron, sometimes with a little tungsten and carbon. Also see CHROME STEELS.

Stalactites. Deposits of calcium carbonate in the form of icicles, formed when water containing calcium carbonate drips from the roofs of caves.

Stalagmites. Similar deposits to stalactites, except that they are formed on the floors of caves.

Stalloy. A proprietary trade name for an alloy containing 3·5–4·0 per cent. silicon and 0·1–0·2 per cent. aluminium. It has a specific resistance of about 55 michroms. cm. Its magnetic hysteresis is much lower than that of pure iron. It is used in the construction of cores for field and armature magnets.

Stamford Powder. An explosive containing from 68–72 per cent. ammonium nitrate, 21–23 per cent. sodium nitrate, 3–4 per cent. trinitrotoluene, and 3½–4½ per cent. ammonium chloride.

Stamglan. A registered trade mark for L.D. polyethylene. (294).

Stamping Brass. See BRASS.

Stamylan. A registered trade mark for high density polythene manufactured in Holland. (22).

Stanaprin. A proprietary preparation of choline salicylate. An analgesic. (358).

Stanclere. Organotin stabilizers for PVC. (583).

Standacol. United Kingdom food stuffs colours. (533).

Standard Benzine. Light petroleum spirit of specific gravity 0·695–0·705 at 15° C., of which 95 per cent. boils between 65° and 95° C. Used for the determination of asphalt in oils.

Standard Copper. See COPPER, STANDARD.

Standard Gold (Sterling Gold). It is 22-carat gold containing 91·6 per cent. gold with 8·4 per cent. other metals, usually copper, to render it harder. American standard gold contains 90 per cent. gold and 10 per cent. copper. This latter alloy has a specific gravity of 17·7 and melts at 940° C.

Standard Silver (Sterling Silver). Silver, 92·5 per cent., with another metal, usually copper, to harden it. American standard silver contains 90 per cent. silver and 10 per cent. copper.

Standard Wood Spirit. See ACETONE ALCOHOL.

Stand Oil (Standöl Varnish, Dicköl Varnish, Lithographer's varnish). Linseed oil boiled strongly, and allowed to burn until it has the desired thickness.

Standöl Varnish. See STAND OIL.

Staniform. A proprietary preparation of methyl stannic iodide. A remedy for boils, carbuncles, small wounds, and injuries.

Stanley Red. See CLAYTON CLOTH RED.

Stannekite. A resinous hydrocarbon, $C_{20}H_{22}O_3$, found in coal deposits in Bohemia.

Stannicide. Fungicides, bactericides and algicides. (583).

Stannine. See TIN PYRITES.

Stanniol. An alloy of 96·2 per cent. tin, 2·4 per cent. lead, 1 per cent. copper, 0·3 per cent. nickel, and 0·1 per cent. iron.

Stannite. A mineral. It is tin pyrites.

Stannoxyl. A proprietary preparation of tin powder and stannous oxide. (801).

Stannoxyl Liquid. A solution of tin chloride, $SnCl_2$, in glycerin. Used as a lotion in the treatment of boils.

Stannum. Tin, Sn. A proprietary trade name for a bearing metal. A tin-lead alloy.

Stanolax. See PARAFFIN, LIQUID.

Stanolind. See PARAFFIN, LIQUID.

Stanolone. 17β-Hydroxy-5α-androstan-3-one. Anabolex.

Stanozolol. 17β-Hydroxy-17α-methyl-5α-androstano[3,2-c]-pyrazole. Stromba.

Stantienite. A brown resin found with Prussian amber.

Stanzaite. A mineral. It is a variety of Andalusite.

Staphcillin. A proprietary preparation of sodium methicillin. An antibiotic. (324).

Staphylase. A malt preparation.

Staple Artificial Silk. See STAPLE FIBRE.

Staple Fibre (Staple Artificial Silk, Artificial Wool, Artificial Chappe). This fibre consists of artificial threads of cellulose or cellulose compounds possessing a definite medium length. It is worked up by ordinary spinning machinery and is suitable for mixing with cotton or wool.

Staralox. A trade mark for abrasive goods made essentially of alumina.

Star Antimony. Pure antimony, Sb.

Star Bowls. Antimony metal obtained by refining with iron. The metal containing about 91 per cent. antimony with about 7 per cent. iron is mixed with crude antimony and salt and heated. The product is known as star bowls. It contains about 99·5 per cent. antimony.

Starch Cellulose. See FARINOSE.

Starch Glazes. Made by adding borax, powdered stearic acid, or paraffin to potato starch.

Starch Glue. Prepared by adding 3 pints water and ½ lb. nitric acid to 2½ lb. starch, warming, then heating.

Starch Gum. See BRITISH GUM.

Starch Paste. Mucilage of starch.

Starch Sugar. Glucose.

Starch Syrups. Glucose mixed with dextrine. They are used in the place of sugar for various purposes.

Starkey's Soap. Turpentine soap.

Starlite. A proprietary synthetic resin.

Starpass. A proprietary urea - for - maldehyde synthetic resin.

Stasite. A mineral. It is a hydrated phosphate of uranium and lead, $8UO_3$. $4PbO.3P_2O_5.12H_2O$, found in Katanga.

Stassanised Milk. Milk which has been heated in cylindrical tubes 1 mm. diameter, so that a great surface is exposed, thereby killing all germs.

Stassfurtite (Crystalline boracite). $2Mg_3$ $B_8O_{15}.MgCl_2$. It occurs in the Stassfurt deposits, and is used in the preparation of boric acid. See BORACITE.

Stassfurt Salts. See ABRAUM SALTS.

Staszicite. A mineral from Meidzianka, containing 39 per cent. As_2O_5, 26·5 per cent. CuO, 20·8 per cent. CaO, and 7·3 per cent. ZnO.

Statox. A proprietary solvent-based organotin wood preservative. (425).

Statuary Bronze. A variable alloy. It usually contains from 75–95 per cent. copper, 1–10 per cent. tin, 0–5 per cent. zinc, 0–6 per cent. lead, 0·12–0·34 per cent. phosphorus, and 0·19–0·7 per cent. nickel.

Statuary Marble. Marble, $CaCO_3$, with a crystalline or saccharoid structure.

Statyl. A proprietary preparation of METHYL BENZOQUATE. A veterinary anti-protozoan.

Staurolite. A mineral. It is a basic aluminium ferrous iron silicate, $HFeAl_5$ Si_2O_{13}.

Staurotide. See STAUROLITE.

Staybelite. A proprietary trade name for a hydrogenated rosin and plasticiser consisting of diethylene glycol ester of hydrogenated rosin. Hydrogenated rosin esters for adhesives. (593).

Staybrite. A proprietary trade name for stainless steels containing chromium and nickel. They usually contain 18 per cent. chromium, 8 per cent. nickel, 74 per cent. iron, sometimes with molybdenum and occasionally with titanium and tungsten. They possess extreme malleability and are very resistant to corrosion. See also ANKA STEEL.

Staycept. A proprietary preparation of nonoxinol. A contraceptive pessary. (809).

Stcherbokov's Solder. An aluminium solder. It contains 49 per cent. zinc, 46 per cent. tin, and 1·5 per cent. aluminium.

S.T.D. (1) A proprietary preparation of sodium tetradecyl sulphate, used for injection of varicose veins. (812).

S.T.D. (2). A trade name for a proprietary anti-oxidant. (784). It is phenyl-β-naphthylamine.

S.T.D.X. Phenyl-β-naphthylamine plus dinaphthyl-p-phenylene diamine. (784).

Steadite. Iron-phosphorus eutectic, consisting of about 61 per cent. iron-phosphide, Fe_3P, with iron, a constituent of cast iron. The same name has been applied to a basic calcium-silicophosphate, $3(CaO.P_2O_5).2CaO(2CaO. SiO_2)$, found in the basic slag of the Thomas-Gilchrist process for the dephosphorisation of iron.

Stead's Reagent. A reagent consisting of 100 c.c. methyl alcohol, 18 c.c. water, 2 c.c. concentrated hydrochloric acid,

1 gram copper chloride ($CuCl_2.2H_2O$), 4 grams magnesium chloride ($MgCl_2.6H_2O$). An etching reagent used in the examination of steels.

Steam Black. See LOGWOOD.

Steamed Bone Meal. A fertiliser. It consists of crushed bones, which have been treated with superheated steam and benzene, to remove fat and glue. It contains about 1 per cent. nitrogen.

Steam Glue (Russian Steam Glue). A preparation of glue made by treating glue with nitric acid.

Steam Orange, Green, and Olive. Dyestuffs prepared from Persian berries, and used in calico printing. See GAMBINE Y.

Steapsin. Lipase, a lipolytic ferment.

Stearex. A trade mark for a standardised stearic acid. It is a commercially pure, free fatty acid prepared for rubber manufacture. There are two grades: (a) Double pressed stearic acid, and (b) single pressed stearic acid. The (b) quality contains more oleic acid than (a).

Steargillite. A variety of the mineral Montmorillonite.

Stearin (Tristearin). The tristearyl-derivative of glycerol, $C_3H_5(O.C_{18}H_{35}O)_3$. The term is, however, principally used to indicate the solid part of any fat from which the liquid portion has been expressed, and the solid white cakes of stearic and palmitic acids, obtained by the hydrolysis of fats, melting at 56–56·5° C., are known as stearin. When pressed again to remove the final portion of oleic acid, and melting at 57·5° C., it is known as Double stearin.

Stearin, Double. See STEARIN.

Stearin Pitch (Candle Pitch, Candle Tar). A pitch obtained in the sulphuric acid treatment of fats. After distillation in steam of the washed acids (stearic, palmitic, and oleic), stearin pitch remains to the extent of 2 per cent.

Stearin Soap. Curd soap.

Stearite. A proprietary trade name for synthetic stearic acid produced by hydrogenation of certain oils.

Stearodine. Calcium iodostearate.

Stearopodis. Magnesium stearate, used in the preparation of soap and face creams.

Stearosan. A compound of santalol and stearic acid.

Steatite (Soapstone, Pot-stone, Lapis Ollaris, French Chalk, Mountain Soap, Saponite, Bowlingite, Smegmatite). A mineral. It is a massive variety of Talc (q.v.), and is used for gas burners, as an insulator, and as a source of magnesium salts.

Steclin. A proprietary preparation containing tetracycline. An antibiotic. (326).

Stecsolin. A proprietary preparation of OXYTETRACYCLINE dihydrate. An antibiotic. (927).

Steel. Iron containing combined carbon up to 1·5 per cent. High carbon steels contain from 0·5–1·5 per cent. carbon, and mild steels from a trace to 0·5 per cent. carbon. Steel containing 1 per cent. carbon has a specific gravity of 7·8, and melts at 1430° C. The specific heat of a steel containing 0·004 per cent. carbon is 0·107 cal./gm./° C. at 20° C. and 0·117 cal./gm./° C. at 100° C.

Steel A2. A British chemical standard. It is a carbon steel containing 0·037 per cent. carbon, 0·034 per cent. silicon, 0·020 per cent. sulphur, 0·008 per cent. phosphorus, 0·043 per cent. manganese, 0·031 per cent. arsenic, 0·059 per cent. nickel, 0·013 per cent. chromium, 0·067 per cent. copper, 0·04 per cent. oxygen, and 99·72 per cent. iron.

Steel B4. A British chemical standard. It is a carbon steel containing 0·400 per cent. carbon, 0·026 per cent. silicon, 0·046 per cent. sulphur, 0·103 per cent. phosphorus, 0·735 per cent. manganese, and 0·140 per cent. arsenic.

Steel, Blue. See PRUSSIAN BLUE.

Steel, Bonderised. See BONDERISED IRON and STEEL.

Steel Bronze. See UCHATIUS BRONZE.

Steel C. A British chemical standard. It is a carbon steel containing 0·093 per cent. carbon.

Steelcrete. Rapid hardening Portland cement. (512).

Steel, Crucible Cast. See CAST STEEL.

Steel Drops. Tinct. Ferri Perchlor.

Steel E. A standard steel containing 0·115 per cent. carbon and 0·491 per cent. manganese. It is used as the colorimetric standard for the determination of carbon in steels containing more than 0·100 per cent. carbon.

Steel, Electro-granodised. See ELECTROGRANODISED IRON and STEEL.

Steel F. A German steel containing 0·67–1·1 per cent. silicon and 0·1–0·14 per cent. carbon.

Steel H. A British chemical standard. It is a carbon steel, and contains 0·428 per cent. carbon, 0·047 per cent. sulphur, and 0·035 per cent. phosphorus.

Steel I. A carbon steel containing 0·521 per cent. carbon and 0·726 per cent. manganese. It is a British chemical standard.

Steelite. An explosive consisting of potassium chlorate, mixed with oxidised resin, and a little castor oil.

Steel, Japanese. Magnet steel (*q.v.*).

Steel M. A British chemical standard. It is a carbon steel containing 0·228 per cent. carbon and 0·057 per cent. silicon.

Steel Mixture. Mist. Ferri. Co.

Steel N. A carbon steel containing 0·17 per cent. carbon, 0·117 per cent. silicon, 0·034 per cent. sulphur, 0·037 per cent. phosphorus, 0·432 per cent. manganese, and 0·029 per cent. arsenic. It is a British chemical standard.

Steel N1. A carbon steel containing 0·153 per cent. carbon, 0·176 per cent. silicon, 0·050 per cent. sulphur, 0·036 per cent. phosphorus, 0·527 per cent. manganese, 0·030 per cent. arsenic, 0·260 per cent. nickel, and 0·04 per cent. copper. It is a British chemical standard.

Steel O. A British chemical standard. It is a nickel steel containing 0·325 per cent. carbon, 0·590 per cent. manganese, and 3·985 per cent. nickel.

Steel O1. A carbon steel containing 0·333 per cent. carbon, 0·162 per cent. silicon, 0·032 per cent. sulphur, 0·031 per cent. phosphorus, 0·617 per cent. manganese, 0·024 per cent. arsenic, 0·162 per cent. nickel, 0·017 per cent. chromium, and 0·037 per cent. copper. It is a British chemical standard.

Steel, Open Hearth. See MARTIN STEEL.

Steel Ore. A variety of cinnabar containing 75 per cent. mercury.

Steel P. A high silicon and phosphorus steel. It is a British chemical standard.

Steel R. A British chemical standard. It is a carbon steel containing 0·786 per cent. carbon, 0·053 per cent. sulphur, and 0·914 per cent. manganese.

Steel, Rustless. See CHROME STEELS.

Steel S1. A British chemical standard. It is a carbon steel containing 0·921 per cent. carbon and 0·051 per cent. phosphorus.

Steels, Aircraft Construction (Ternary and Quaternary). These are usually iron with from 0·25–1 per cent. carbon, 0·15–3·25 per cent. nickel, and 0·45–1·25 per cent. chromium.

Steel, Self-hardening. See MUSHET STEEL.

Steel, Stainless. See CHROME STEELS.

Steel T. A nickel steel containing 3·367 per cent. nickel. It is a British chemical standard.

Steel U. A carbon steel containing 1·203 per cent. carbon, 0·472 per cent. manganese, and 0·608 per cent. nickel. It is a British chemical standard.

Steel V. A British chemical standard. It is an alloy steel containing 0·548 per cent. carbon, 0·161 per cent. silicon, 0·063 per cent. sulphur, 0·024 per cent. phosphorus, 0·542 per cent. manganese,

0·861 per cent. chromium, and 0·273 per cent. vanadium.

Steel V2A. A rustless steel containing iron with 20 per cent. chromium, 7 per cent. nickel, and 0·2 per cent. carbon.

Steel, V2A. See ANKA STEEL; also STAYBRITE.

Steel W. A British chemical standard. It is an alloy steel containing 0·695 per cent. carbon, 0·187 per cent. silicon, 0·075 per cent. sulphur, 0·028 per cent. phosphorus, 0·101 per cent. manganese, 0·44 per cent. nickel, 3·01 per cent. chromium, 0·791 per cent. vanadium. 4·76 per cent. cobalt, and 16·21 per cent. tungsten.

Steel W2. A British chemical standard high-speed alloy steel. It contains 0·17 per cent. carbon, 0·14 per cent. silicon, 0·051 per cent. sulphur, 0·220 per cent. manganese, 3·29 per cent. chromium, 0·82 per cent. vanadium, 16·12 per cent. tungsten, 4·35 per cent. cobalt, 0·43 per cent. nickel, and 0·55 per cent. molybdenum.

Steel Wine. See WINE OF IRON.

Steel Wool. Long and fine steel fibres used for abrasive purposes.

Steenstrupine. A mineral. It is a silicate of cerium, yttrium, iron, sodium, thorium, manganese, aluminium, and titanium, of Greenland.

Steigerite. A vanadium mineral, Al_2O_3. $V_2O_5.6.5H_2O$, similar to fervanite, found in deposits of the Gypsum Valley Colorado. Bright yellow in colour, decomposed by mineral acids.

Steinbuhl Yellow (Gelbin Yellow Ultramarine). A pigment. A chromate of calcium is sold under these names, but Barium yellow also frequently passes under the same terms.

Steinmannite. A mineral. It is a variety of galena containing antimony and arsenic.

Steinmark. Synonym for Kaolinite.

Stelabid. A proprietary preparation of trifluoperazine dihydrochloride and isopropamide iodide. Gastrointestinal sedative. (281).

Steladex. A proprietary preparation of trifluoperazine dihydrochloride and dexamphetamine sulphate. (281).

Stelazine. A proprietary preparation of trifluoperazine or its hydrochloride. A sedative. (281).

Stellachrome Black L757. A dyestuff. It is equivalent to Palatine chrome black.

Stellerite. A mineral. It is a calcium-aluminium silicate.

Stellin. A distillate of Indian petroleum. It contains unsaturated hydrocarbons, and no benzene.

Stellite. An alloy of 59·5 per cent. cobalt, 22·5 per cent. molybdenum,

10·77 per cent. chromium, 3·11 per cent. iron, 2·04 per cent. manganese, 0·87 per cent. carbon, 0·77 per cent. silicon, 0·08 per cent. sulphur, and 0·04 per cent. phosphorus. Another alloy containing 75 per cent. cobalt, 20 per cent. chromium, and 5 per cent. tungsten, is also called Stellite. Used for cutting tools, being very hard and non-rusting. The term is also used for a mineral, probably Pectolite, of Stirlingshire.

Stellited Metal. Metals treated with an alloy consisting chiefly of chromium, tungsten, and cobalt (stellite). The metals treated are usually steel, cast iron, malleable iron, and semi-steel. It is an economical method for these treated metals are rendered suitable for wear-resisting parts of machinery.

Stelznerite. A mineral. It is antlerite.

Stemetil. A proprietary preparation of prochlorperazine dimaleate or mesylate. An anti-emetic and sedative. (336).

Stenediol. A proprietary preparation of methandriol used in the treatment of osteoporosis. (316).

Stenol. A proprietary trade name for technical stearyl alcohol.

Stenosin. Di - sodium - mono - methyl - arsonate.

Stentor Steel. A proprietary trade name for a non-shrinking steel containing 1·6 per cent. manganese, 0·25 per cent. silicon, and 0·9 per cent. carbon.

Stephanite. See PYRARGYRITE.

Ster 5. A proprietary brand of PREDNISOLONE, available in 5 mg. tablets. (754).

Sterbon. See CARBORA.

Stercorite. Sodium ammonium hydrogen phosphate. It occurs in guano.

Sterculia Gum. Indian tragacanth.

Sterculia Kernals. See OLIVES OF JAVA.

Stereosine Grey. A dyestuff which dyes wool and mixed goods bluishbrown from a neutral bath.

Stereotype Plate. An alloy of 85 per cent. lead and 14 per cent. antimony, sometimes with the addition of a little tin. See TYPE METAL.

Steresol. An antiseptic varnish made by dissolving 270 parts purified shellac, 10 parts benzoin, 10 parts balsam of tolu, 100 parts phenol, 6 parts oil of cinnamon, and 6 parts saccharine in alcohol, to make 1,000 parts.

Steriform. See STERISOL.

Sterilised Butter. See BUTTER, RENOVATED.

Sterinovo. A proprietary preparation of procaine hydrochloride and adrenalin preparations.

Steriosol. See STERISOL.

Sterisil. A proprietary trade name for Hexetidine.

Sterisol (Steriosol). A liquid containing formaldehyde, milk sugar, sodium chloride, potassium phosphate, and water. Used as an antiseptic. The terms Sterisol and Steriform are used for a combination of milk sugar and formaldehyde, used as an antiseptic dusting powder. A trade mark.

Sterline. An alloy of 68 per cent. copper, 17–18 per cent. nickel, 13–14 per cent. zinc, 0·75–0·8 per cent. iron, and 0–0·8 per cent. lead. It is a nickel silver (German silver).

Sterling Brass. See BRASS.

Sterling Gold. See STANDARD GOLD.

Sterling Silver. See STANDARD SILVER.

Sterling Solder. An alloy of 61·6 per cent. tin, 15·2 per cent. zinc, 11·2 per cent. aluminium, 8·3 per cent. lead, 2·5 per cent. copper, and 1·2 per cent. antimony.

Sterlite. A proprietary trade name for a nickel brass containing 25 per cent. nickel, 20 per cent. zinc, and small amounts of iron, manganese, silicon, and carbon.

Sterlith. A trade mark for materials of the refractory and abrasive type. They consist essentially of crystalline alumina.

Sternbergite. A mineral, $(Ag_2Fe)S$.

Sternite. A proprietary trade name for phenolic and polystyrene moulding materials. SPF 5092 is mineral filled. (928).

Sterogyl-15. A proprietary preparation of calciferol. (307).

Steroxin. A proprietary trade name for Chlorquinaldol.

Steroxin-Hydrocortisone. A proprietary preparation of chlorquinaldol and hydrocortisone used in dermatology as an antibacterial agent. (17).

Sterpon. A proprietary polyester laminating resin. (929).

Sterrettite. A mineral. It is $2[Al_6(PO_4)_4(OH)_6 . 5H_2O]$.

Sterro Metal. See AICH METAL.

Ster-Zac. A proprietary preparation of HEXACHLOROPHANE. A topical antiseptic. (346).

Stetefeldite. A mineral containing silver, copper, iron, antimony, and sulphur.

Stevensite. A mineral. It is a variety of talc.

Stewartite. A mineral. It is a manganese phosphate. The term is also used for a variety of bort.

Sthenosised Cotton. Cotton which has been treated with formaldehyde. It becomes resistant to alkalis and cannot be mercerised. It has a greatly decreased affinity for direct dyestuffs.

Stibacetin. Sodium - p - acetyl - amino - phenyl - stibinate, $C_6H_5NH.C_6H_4.SbO(ONa)OH$. Used in medicine.

Stibamine. The sodium salt of 4-amino-phenyl-stibinic acid. Used in medicine. Other derivatives, such as urea-stiba-mine and metachloro-stibacetin, have been used in the treatment of kala-azar.

Stibenyl. The sodium salt of acetyl-p-amino-phenyl-stibinic acid. Used in medicine.

Stiberite. A variety of boronatrocalcite found as a soft coating on gypsum. Synonym for Ulexite.

Stibianite. A mineral, $Sb_2O_5.H_2O$.

Stibiated Tartar. Tartar emetic ($q.v.$).

Stibiatil. A mineral. It is a ferrous manganese antimonate.

Stibiconite. A mineral. It is a variety of antimony tetroxide, Sb_2O_4.

Stibine. See ANTIMONIURETTED HYDROGEN.

Stibiobismuthinite. A mineral, $(Sb.Bi)_4 S_7$.

Stibioferrite. A mineral. It contains Sb_2O_5 with Fe_2O_3, SiO_2, and H_2O.

Stibiogalenite. A mineral. It is bind-heimite.

Stibiotantalite. A mineral, $(SbO)_2 Ta(Cb)_2O_6$.

Stibium. Antimony, Sb.

Stiblite. A mineral, $Sb_2O_4.H_2O$.

Stibnite. See ANTIMONITE.

Stibocaptate. Antimony (III) sodium meso-2,3-dimercaptosuccinate.

Stibosan. A proprietary preparation of sodium - m - chloro - p - acetyl - amino-phenyl-stibinate.

Stichtite (Chromobrugnatellite). A mineral, $MgCO_3.5Mg(OH)_2.2Cr(OH)_3. 4H_2O$.

Stick-lac. See LAC and LAC-DYE.

Stickstoffoxydbaryt. Barium nitrite, $Ba(NO_2).H_2O$.

Stie-Lasan. A range of proprietary skin pastes containing dithranol and sali-cylic acid in a zinc oxide, starch and soft paraffin base, used in the treatment of psoriasis. (379).

Stilbamidine. A preparation used in the treatment of tripanosomiasis. It is 4,4'-diamidinostilbene.

Stilbazium Iodide. 1-Ethyl-2,6-di-[4(pyrrolidin - 1 - yl)styryl]pyridinium iodide. Monopar.

Stilbene Orange 4R. A dyestuff. It is a British equivalent of Mikado orange.

Stilbene Red. A dyestuff prepared from diamino-stilbene, $C_{34}H_{23}N_6O_9S_3Na_3$.

Stilbene Yellow G, 4G, 6G, 8G. Dye-stuffs. They are alkaline condensation products of dinitro-dibenzyl-disulphon-ic acid and dinitroso-stilbene-disul-phonic acid. Dyes cotton greenish-yellow from a salt or sodium sulphate bath.

Stilbene Yellow 2G, 3G, 8G. Dye-

stuffs. They are British brands of Mikado yellow.

Stilbite. A zeolite mineral. It is a cal-cium aluminium silicate, $CaO.Al_2O_3. 6SiO_2.6H_2O$.

Stil de Grain. See BROWN PINK.

Stillingia Oil. An oil obtained by crushing the kernel of Stillingia sebifera. Also see CHINESE TALLOW.

Stilpnomelane. A mineral, $2FeO. SiO_2(Ca.Al)$.

Stimmi. A mineral. It is an antimony sulphide, Sb_2S_3.

Stimplete. A proprietary preparation of phenobarbitone dexamphetamine sulphate, thiamine hydrochloride, ribo-flavine, pyridoxine hydrochloride, nico-tinamide and alcohol. (245).

Stink Quartz (Fœtid Quartz). A quartz which has a bad odour, due to organic matter.

Stink-stone (Oil-stone). A bituminous schist found in the Tyrol. A source of ichthyol.

Stipolac. A proprietary preparation. It is a sodium - tetraiodo - phenol - phthalein.

Stirimazole. A drug used in the treat-ment of amœbiasis, trichomoniasis and trypanosomiasis. It is 2-(4-carboxy-styryl)-5-nitro-1-vinylimidazole.

Stirlingite. Synonym for Roepperite.

Stirling Metal. See STERLING BRASS.

Stirling's Gentian Violet. A micro-scopic stain. It contains 5 grams gentian violet, 10 c.c. 95 per cent. alcohol, 2 c.c. aniline, and 88 c.c. water.

Stiven's Solution. A rectal anæsthetic consisting of 1 part olive oil and 3 parts ether.

Stockalite. A proprietary product. It is a very highly refined china clay used as a filler for tyres, cables, and high grade mixes.

Stockholm Petroleum. See PETROLEUM, STOCKHOLM.

Stockholm Pitch. Pine-wood tar pitch. It is soluble in alkalies, and is used in the preparation of varnishes, in the rubber and gutta-percha trades, and in the preparation of impervious cements.

Stockholm Tar (Pine Tar). A tar obtained principally from pine-wood distillation. It is obtained from Pinus sylvestris and other species of Pinus. There are several qualities. A light Stockholm tar has a specific gravity of 1·065, a medium one of 1·075, and a heavy one of 1·085. Umea, lulea, shelleftea, and wasa tars are varieties of Stockholm tar. It is used as a pre-servative paint for ships and roofing and as a rubber softener. It has anti-oxidant properties.

Stoco. A proprietary bituminous plastic.

Stoddard Solvent. A proprietary trade name for a refined petroleum product for dry cleaning.

Stoffertite. A calcium phosphate, $CaHPO_4.5H_2O$. It occurs in guano.

Stoic Metal. An alloy similar in composition to Invar.

Stokesite. A mineral. It is a hydrated silicate of tin and calcium, $H_4CaSn Si_3O_{11}$.

Stoke's Reagent. A reducing agent prepared by dissolving 30 grams ferrous sulphate and 20 grams tartaric acid in 1 litre of water. When required for use, strong ammonia is added until the precipitate first formed is redissolved.

Stolpenite. A mineral. It is a variety of montmorillonite.

Stolzite (Raspite). A mineral. It is lead tungstate, $PbWO_4$.

Stomahesive. A proprietary preparation containing gelatin Pectin, carboxymethyl cellulose and polyisobutylene on a protective film, used for the protection of skin around surgical stomata. (927).

Stomosan. Ethylamine phosphate, used in medicine.

Stone Black. Animal charcoal.

Stone, China. See Cornish Stone.

Stone, Cinnamon. See Hessonite.

Stone, Coal. Anthracite.

Stone, Cross. See Chiastolite.

Stone, Flintshire. See Dinas Bricks.

Stone, Green. See Bohemian Earth.

Stone, Grey. See Slate Grey.

Stone, Honey. See Mellite.

Stone, Ice. See Cryolite.

Stone, Mercury. Mercuric chloride, $HgCl_2$, in lumps.

Stone, Needle. Natrolite (q.v.).

Stone, Oil. See Stink-stone.

Stone, Pea. See Pisolite.

Stone, Pencil. See Pyrophyllite.

Stone, Pot. See Steatite.

Stone Red. See Indian Red.

Stone Root. Collinsonia.

Stone's Bronze. An alloy of 87 per cent. copper, 11 per cent. tin, and 2 per cent. phosphor-copper.

Stone, Soap. See Steatite.

Stone, Sun. See Oligoclase.

Stone, Tin. See Stream Tin.

Stone, Touch. Lydian Stone (q.v.).

Stone, Water-of-Ayr. See Snake Stone.

Stone Wax. A name applied to carnauba wax.

Stone Yellow. See Yellow Ochre.

Stonite. An explosive consisting of 68 per cent. nitro-glycerin, 20 per cent. kieselguhr, 8 per cent. potassium nitrate, and 4 per cent. wood meal.

Stoodite. A proprietary trade name for a high manganese steel.

Stora. A proprietary trade name for a Swedish charcoal iron used for making malleable iron.

Storalon. See Carbora.

Storax (Styrax). An oleo-resin, the product of the tree *Liquidambar orientalis*. The crude material contains 20–30 per cent. water and fragments of bark, etc. Prepared storax is used as a drug, and is soluble in alcohol. It has a specific gravity of $1 \cdot 109$–$1 \cdot 125$, a saponification value of 130–230, an acid value of 35–175, and gives an ash of 0–2 per cent.

Storax Calamita. The powdered bark of *Liquidambar styraciflua*, most of the resin being first extracted. The product has no connection with storax, an oleo-resin.

Storax, Liquid. See Liquid Storax and Liquidambar.

Storax, Solid. See Red Storax.

Stout. An alcoholic beverage. Dublin stout contains 7 per cent. of alcohol.

Stovaine (Amylocaine). A registered trade mark for amylocaine hydrochloride (dimethyl - amino - benzoyl - dimethyl-ethyl-carbinol-hydrochloride), $C_6H_5 . COO . C(CH_3)(C_2H_5)CH_2N(CH_3)_2 HCl$. A local anæsthetic.

Stovarsol. A proprietary preparation containing acetarsol. (336). See Fourneau 190.

Stovarsolan. p - Hydroxy - m - acetyl - amino-arsenious acid.

Stove Polishes. Blacklead, formerly considerably used, consists of graphite. Others contain graphite mixed with water-glass and glycerin. Aniline black is also used.

Stowite. An explosive containing 58–61 per cent. nitro-glycerin, $4 \cdot 5$–5 per cent. nitro-cotton, 18–20 per cent. potassium nitrate, 6–7 per cent. wood meal, and 11–15 per cent. ammonium oxalate.

Stowmarket Powder. A 33-grain powder.

Straight Dynamite. An explosive consisting of nitro-glycerin, wood pulp, sodium nitrate, and calcium or magnesium carbonate.

Strass (Paste). A kind of glass used to imitate precious stones. It is made from 100 parts sand, 40 parts minium, 24 parts potassium carbonate, 20 parts borax, and 12 parts potassium nitrate. This gives a colourless product, and various oxides are added to colour it.

Strassburg Turpentine. The oleo-resin from the silver fir, *Pinus picea*.

Strasser Solder. An alloy of 62 per cent. tin, 12 per cent. zinc, 4 per cent. aluminium, 8 per cent. lead, 5 per cent. copper, 5 per cent. bismuth, and 4 per cent. cadmium.

Stratyl. A proprietary polyester laminating compound. (686).

Straus Metal. See CAMITE.

Strawberry Essence. Usually contains 9 parts amyl formate, 9 parts butyric acid, 18 parts isovaleric acid, 27 parts acetic acid, 13 parts ethyl acetate, 9 parts essence of violets, and 915 parts alcohol.

Strawstone. A mineral. It is a variety of carpholite.

Stream Tin (Wood Tin, Toad's Eye, Stone Tin). Varieties of Tinstone (*q.v.*).

Strenes Metal. A proprietary trade name for a nickel-chromium-molybdenum cast iron.

Strengite. A mineral, $FePO_4.2H_2O$.

Strepolin. A proprietary preparation containing streptomycin sulphate. An antibiotic. (335).

Strepsils. A proprietary preparation of dichlorobenzyl alcohol and AMYLMETACRESOL taken as a mouth lozenge. (502).

Streptaquaine. A proprietary preparation containing streptomycin. An antituberculous agent. (309).

Streptets. A proprietary preparation of zinc bacitracin, polymixin B sulphate, and neomycin sulphate. Antibiotic lozenges. (245).

Streptodornase. An enzyme obtained from cultures of various strains of Streptococcus haemolyticus and capable of catalysing the depolymerisation of polymerised deoxyribonucleoproteins.

Streptoduocin. A mixture of equal parts of Dihydrostreptomycin. Sulphate and Streptomycin Sulphate. Dimycin. Mixtamycin.

Streptohydrazid. See STREPTONICOZID.

Streptokinase. An enzyme obtained from cultures of various strains of Streptococcus haemolyticus and capable of changing plasminogen into plasmin. Kabikinase.

Streptonicozid. A compound, used in the treatment of tuberculosis formed by interaction of a suitable streptomycin salt and isoniazid. STREPTOHYDRAZID.

Streptonivin. A proprietary trade name for Novobiocin.

Streptotriad. A proprietary preparation of streptomycin, sulphadiazine, sulphathiazole and sulphadimidine, used in the treatment of dysentery. (507).

Stresnil. A proprietary preparation of AZAPERONE. A veterinary tranquilliser.

Strewing Smalt. The coarsest powdered Smalt (*q.v.*).

Striatran. A proprietary trade name for emylcamate.

Stripping Salt. See ABRAUM SALTS.

Strobane. A trade mark for an insecticide and acaricide. It is based on polychlorinated terpine and contains 66% chlorine

Strohmeyerite (Stromeyerite, Argentiferous Copper Glance, Silver Copper Glance). A mineral, $Cu_2S.Ag_2S$, containing 53 per cent. silver.

Stromba. A proprietary preparation of stanozolol. An anabolic agent. (112).

Stromeyerite. See STROHMEYERITE.

Strontia. Strontium oxide, SrO.

Strontianite. A mineral. It is strontium carbonate, $SrCO_3$.

Strontian White. A pigment. It consists of strontium sulphate, $SrSO_4$.

Strontian Yellow (Yellow Ultramarine). A pigment. Originally it was a chromate of strontium, $SrCrO_4$, but a more durable pigment is now sold under this name.

Strontiobarite. A mineral. It is $4[(Ba,Sr)SO_4]$.

Strontiocalcite. A calcium-strontium carbonate, $CaSr(CO_3)_2$.

Strontisal. Strontium salicylate. An antirheumatic.

Strontium Apatite. Synonym for Saanute.

Strontiuran. A preparation of strontium chloride and urea. A solvent and reagent for arsphenamine and neoarsphenamine.

Strophantofix. A German preparation consisting of tablets containing tincture of Strophantii.

Strumolysin. A German product. It is an adrenalin-hypophysin-iodine compound, and is used in the treatment of primary struma.

Struthiin (Githagin, Polygalin, Polygallic Acid, Senegin). Saponin, $C_{19}H_{30}O_{10}$, a glucoside found mainly in the common soapwort.

Struverite. A titanium mineral.

Struvite (Guanite). A mineral, $(NH_4)MgPO_4$.

Strychnine Arsenite. A mixture of 77 per cent. strychnine and 23 per cent. arsenious oxide.

Stryphnon. Methyl-amino-aceto-pyrocatechol. It has a physiological action similar to adrenalin.

Strytopyrin. A German preparation. It is a compound of cotarnine and amidoantipyrine.

ST-Size. Modified rosin emulsion size. (593).

Stuart's Granolithic Stone. A stone similar to Ward's stone.

Stucco. A specially hard plaster which can be polished. There are two kinds: (1) made from plaster-of-Paris; and (2) made from lime. They are usually mixed with size.

Studerite. A mineral. It is a variety of tetrahedrite.

Stugeron. A proprietary preparation of CINNARIZINE. An anti-nauseant. (356).

Stupp. A mercurial soot condensed in

the chambers during the treatment of mercury ores. It contains about 20 per cent. mercury as metal, and sulphate.

Sturcal. Calcium carbonate. (524).

Stutzite. A mineral. It is a silver telluride, Ag_4Te.

Stybial. Antimony-sodium tartrate solution in ampoules.

Stylotypite. A mineral, Cu_3SbS_3. It is also stated to be $(Cu,Ag,Fe)_{12}Sb_4S_{13}$.

Styphen I. A proprietary anti-oxidant. It is a mixture of styrenated phenols. (930).

Styphnic Acid. Trinitro-resorcinol (2 : 4-dihydroxy-1 : 3 : 5-trinitro-benzene), $C_6H_3O_8N_3$. Used in explosives.

Stypol. A proprietary polyester laminating resin. (931).

Stypticine. Cotarnine hydrochloride, used medicinally as a styptic.

Styptirenal. Adrenalin (q.v.).

Styptogan. A paste of vaseline with permanganate. An astyptic used externally.

Styptol. A compound of phthalic acid with cotarnine, $(C_{12}H_{13}NO_3)_2.C_6H_4(COOH)_2$. Used to stop bleeding.

Styracin. Cinnamyl cinnamate, $C_6H_5.CH : CH.CH_2.O.CO.CH : CH.C_6H_5$.

Styracol. The cinnamic acid ester of guaiacol, $C_6H_5.CH : CH.CO_2.C_6H_4.CH_3$. Used as a substitute for guaiacol in the treatment of phthisis and chronic gastro-intestinal catarrh.

Styrafil. A registered trade name for flame retardant polystyrene.

Styraloy 22, 22A. A proprietary trade name for an elastomeric styrene derivative.

Styramate. β-Hydroxyphenylethyl carbamate.

Styramic H.T. and M.T. Proprietary trade names for polystyrene thermoplastics possessing a higher softening point than usual (about 236° F.). They are stated to be polydichlorstyrenes.

Styrocell. Expanded and expansible polystyrene. (553).

Styrocolor. A proprietary trade name for two colour high impact polystyrene laminate. (71).

Styrofan D. A registered trade mark for a polystyrene dispersion. (49).

Styroflex. A proprietary trade name for a synthetic resin said to be a flexible polymer of styrene.

Styrofoam. Trade mark for an extruded polystyrene foam. (64).

Styroform. A proprietary trade name for high impact polystyrene. (71).

Styrogallol. Dioxy - anthracoumarin, $C_{16}H_8O_5$, a yellow dyestuff.

Styrol (Styrolene, Cinnamene, Cinnamol). Styrene, $C_6H_5.CH : CH_2$. Styrol is

also the name for a colloidal silver preparation.

Styrolene. See STYROL.

Styrolyl Alcohol (Styryl Alcohol). Phenyl-glycol, $C_6H_5.CHOH.CH_2OH$. Used in perfumery.

Styron. A proprietary trade name for polystyrene (q.v.).

Styrone. Cinnamyl alcohol, $C_6H_5.CH : CH.CH_2OH$.

Styropor. A trade name for expandable polystyrene. (49).

Styrowood. A proprietary trade name for imitation wood high impact polystyrene. (71).

Styvarene. A proprietary trade name for impact resistant polystyrene. (240).

Styxol. See UBA.

Suakin Gum (Talca Gum, Talka Gum, Sennaar Gum). A brittle variety of gum acacia from *Acacia fistula*. It gives a ropy mucilage.

Suanite. A mineral. It is $4[Mg_2B_2O_5]$.

Suari Nuts. See BUTTER NUTS.

Subacetate of Lead. Monobasic lead acetate, $(C_2H_3O_2)_2Pb + PbO + H_2O$.

Subcutin (Soluble Anæsthesin). Ethyl-p-amino-benzoate salt of p-phenolsulphonic acid, $NH_2.C_6H_4COOC_2H_5.HO.C_6H_4.SO_3H$. Used for hypodermic injections.

Suberite. A cork substitute made from cork chips and an adhesive.

Subeston. A preparation of Estone (see LENICET). It is a double basic acetate of aluminium.

Subitine. A proprietary preparation of equal parts of ammonium ichthosulphonate and glycerin.

Subitol. See ICHTHYOL.

Sublamine (Mercamin). Mercuric-ethylene-diamine-sulphite, $HgSO_4.2C_2H_4(NH_2)_2 + H_2O$. A germicide.

Sublimaze. A proprietary preparation of fentanyl. An analgesic. (356).

Sublimed Blue Lead. A pigment produced by heating mixed ores of zinc and lead in a furnace with an air blast. It usually contains 50 per cent. lead sulphate, 20 per cent. lead oxide, PbO, 11 per cent. lead sulphide, PbS, 8 per cent. lead sulphite, and 3 per cent. zinc oxide.

Sublimed Calomel. See CALOMEL.

Sublimed White Lead. A white lead manufactured from mixed ores of galena and zinc blende. They are roasted in the presence of an air blast, and the lead sulphate, lead oxide, and zinc oxide formed is collected in large chambers. The average composition is 75 per cent. lead sulphate, 20 per cent. lead oxide, and 5 per cent. zinc oxide. A pigment.

Sublime Olive Oil (Sublime Salad Oil). The best olive oil.

Sublimoform. A mercury-formaldehyde preparation. A seed preservative.

Sublimo-phenol. Chloro-phenolate of mercury. Used in antiseptic surgery.

Subox. A protective coating paint consisting of a suspension of colloidal lead in linseed oil.

Substitute Gutta-percha. See GUTTA-GENTZSCH, NIGRITE, VELVRIL, and SOREL'S SUBSTITUTES.

Substitute of Tartar. See SUPERARGOL.

Substitute Rubber. See CAOUTCHOUC, ARTIFICIAL.

Substitute Saffron. See VICTORIA YELLOW.

Succinellite. Succinic acid obtained from amber.

Succinite. Baltic amber which contains succinic acid. Also the name for a mineral, a lime-aluminium garnet.

Succinol. An oil obtained by the distillation of amber.

Succinoxate. See ALPHOGEN.

Suchar. A decolorising carbon used for sugar juices. It is prepared from waste sulphite-cellulose liquors.

Suclofenide. An anti-convulsant. N-(2 - Chloro - 4 - sulphamoylphenyl) - 2 - phenylsuccinimide. SULFALEPZINE.

Sucralox. A polymerised complex of sucrose and aluminium hydroxide. Manalox AS. A proprietary trade name for an aluminium antacid. (190).

Sucramine (Lyons Sugar). The ammonium salt of saccharin, used in France as a sweetening substance.

Sucrase (Saccharase, Invertin). Invertase, an enzyme which decomposes saccharose into glucose and levulose.

Sucrate of Hydrocarbonate of Lime. See SUCRO-CARBONATE OF LIME.

Sucrene. See DULCINE.

Sucrets. A proprietary preparation of HEXYLRESORCINOL in sugar, taken as a mouth lozenge. (932).

Sucro-carbonate of Lime (Sucrate of Hydrocarbonate of Lime). A complex compound of lime, calcium sucrate, and calcium carbonate formed in the production of sugar from the beet, when carbon dioxide gas is passed into a solution of sucrate of lime.

Sucrol. See DULCINE.

Sucro-levulose. See LEVULOSE.

Sucrose (Cane Sugar, Beet Sugar). Saccharose, $C_{12}H_{22}O_{11}$.

Suction Gas. See DOWSON GAS.

Suction Gum (Suction Powder). Powdered gum tragacanth.

Sucuaryl. A proprietary trade name for sodium cyclamate.

Sudan I (Carminaph, Fast Oil Orange I, Oil Orange, E, Spirit Yellow I). A dyestuff. It is benzene-azo-β-naphthol, $C_{16}H_{12}N_2O$. Used for colouring oils and varnishes.

Sudan II (Red B, Scarlet G, Fast Oil Orange II, Oil Scarlet L, Y). A dyestuff. It is xylene-azo-β-naphthol, $C_{18}H_{16}N_2O$. Used for colouring oils and varnishes.

Sudan III (Red C, Cerasine Red, Fast Oil Scarlet III, Oil Scarlet AS, Spirit Red III). A dyestuff. It is benzene-azo-benzene-azo-β-naphthol, $C_{22}H_{16}N_4O$. Used for colouring oils and varnishes.

Sudan IV (Fat Ponceau, Biebrich Scarlet R Medicinal, Spirit Red III). A dyestuff. It is o-toluene-azo-o-toluene-azo-β-naphthol. The medicinal variety has been recommended for wound treatment.

Sudan Brown (Pigment Brown, Fast Oil Brown S, Oil Brown D). A dyestuff. It is α-naphthalene-azo-α-naphthol, $C_{20}H_{14}N_2O$. Used for colouring oils and soap.

Sudan Coffee. The roasted seeds of *Parkia Africana.*

Sudan G (Carminaph J, Cerasine Orange G, Oil Orange O, Oil Yellow). A dyestuff. It is dioxy-azo-benzene, $C_{12}H_{10}N_2O_2$. Used for colouring oils and varnishes.

Sudan Glycerin. The dyestuff Sudan III (0·01 gram) is dissolved in 5 c.c. 90 per cent. alcohol, and 5 c.c. glycerin added. Used as a microscopic stain.

Sudan R (Oil Vermilion). A dyestuff similar to Sudan III.

Sudan Red. See MAGDALA RED.

Sudbury Ore. A mineral. It is a magnetic iron pyrites containing from 3–8 per cent. nickel.

Sudermo. A proprietary trade name for Mesulphen.

Sudoformal. A 10 per cent. formalin soap recommended for perspiring feet.

Sudoxicam. An anti-inflammatory drug currently undergoing clinical trial as "CP-15,973". It is 4-hydroxy-2-methyl - N - thiazol - 2 - yl - $2H$ - 1, 2 - benzothiazine-3-carboxamide 1,1-dioxide.

Suède. A chrome-tanned soft-finished leather made from kid and calf skin.

Suet Substitutes. Mixtures of suet and cotton-seed oil or cotton-seed stearin.

Suffettil. The white seed from which the best Indian sesamé oil is obtained. See TILLIE and BIGARRÉ.

Sugamo. A Japanese seaweed suggested for use in paper-making.

Sugar, Beechwood. Wood sugar (*q.v.*).

Sugar, Beet. See SUCROSE.

Sugar Candy. Crystallised sugar, obtained by allowing a sugar solution to slowly evaporate.

Sugar Cane Wax. A wax obtained from the dried filter press cake from sugar mills by benzine extraction. The African cake contains 14–17 per cent.

of wax, and the Java cake 4 per cent. The wax obtained is not a pure product, but is a mixture of wax and fatty material with 7 per cent. glycerin. The pure wax is obtained by crystallisation and distillation. The crude wax contains about 7 per cent. glycerin, 61 per cent. free and combined acids, and 28 per cent. unsaponifiable matter. The wax has been used in the manufacture of polishes and gramophone records. It melts at 76–77° C., and has an acid value of 1 and a saponification value of 34.

Sugar Charcoal (Lampblack). Amorphous carbon.

Sugar Colouring. See BURNT SUGAR.

Sugar, Corn. See GLUCOSE.

Sugar, Date-tree. See DATE SUGAR.

Sugar, Diabetic. See GLUCOSE.

Sugar, Fruit. See LEVULOSE.

Sugar, Grape. See GLUCOSE.

Sugar, Honey. See GLUCOSE.

Sugar House Black. A bone black pigment. It is a by-product of the sugar mills.

Sugar Lime. An aqueous solution containing 4 per cent. lime and from 17–28 per cent. sucrose. Used for correcting the acidity and improving the thickness of cream and milk.

Sugar of Gelatin. Glycocoll (amino-acetic-acid), $CH_2.NH_2.COOH$.

Sugar of Lead. Normal lead acetate, $(CH_3COO)_2Pb + 3H_2O$. Used as a mordant in dyeing and printing, and for the preparation of lead salts and paints.

Sugar of Malt. See EXTRACT OF MALT.

Sugar of Milk. Lactose, $C_{12}H_{22}O_{11} + H_2O$.

Sugar of Saturn. See SALT OF SATURN.

Sugar Sand. See MAPLE SUGAR SAND.

Sugar, Starch. See GLUCOSE.

Sugar Vinegar. See GLUCOSE VINEGAR.

Suhler White Copper. An alloy of 40 per cent. copper, 32 per cent. nickel, 25 per cent. zinc, 2·6 per cent. tin, and 0·6 per cent. cobalt. It is a nickel silver.

Suicalm. A proprietary preparation of AZAPERONE. A veterinary tranquilliser.

Suine. A butter substitute made from pig's fat.

Suint. A fatty substance in sheeps' wool. It contains a quantity of potash, and is used for soap-making, and in the preparation of lanoline.

Suleo. A proprietary shampoo containing CARBARYL used to treat infestation by lice.

Sulfacytine. An anti-bacterial. 4-(4-Aminobenzenesulphonamido) - 1 -ethyl - 1, 2 - dihydropyrimidin - 2 - one. 1 - Ethyl-*N*-sulfanilylcytosine.

Sulfadoxine. A sulphonamide. 4-(4-Aminobenzenesulphonamido) - 5. 6 - dimethoxypyrimidine. FANASIL.

Sulfalepzine. See SUCLOFENIDE.

Sulfamerazine. An anti-diarrhœal, present in Kaobiotic V. It is N^1-(4-Methylpyrimidin-2-yl)sulphanilamide.

Sulfametin. N′- (5 - methoxy - 2 - pyrimidinyl) sulphanilamide. SULPHA - METHOXYDIAZINE.

Sulfametopyrazine. A sulphonamide. 2 - (4 - Aminobenzenesulphonamido) - 3 - methoxypyrazine. KELFIZINE.

Sulfamylon. A proprietary preparation of MAFENIDE acetate in a cream, used in the treatment of infected burns. (685).

Sulfan. A proprietary trade name of a stabilised liquid sulphur trioxide. (190).

Sulfapyrazole. A sulphonamide. N^1-(3 - Methyl - 1 - phenylpyrazol - 5 - yl) sulphanilamide. VESULONG.

Sulfaquinoxaline. A coccidiostat, present in PANCOXIN. It is N^1-quinoxalin-2-ylsulphanilamide. EMBAZIN is the sodium salt.

Sulfarine. A mixture of magnesium sulphate with 15 per cent. sulphuric acid. It is used against potato scab.

Sulfarsenol. A combination of salvarsan base with formaldehyde bisulphite. It is used in the treatment of syphilis.

Sulfidal. See SULFOID.

Sulfil. A registered trade name for flame retardant polysulphone.

Sulfoderm. Silicic acid with 1 per cent. colloidal sulphur.

Sulfofix. A powder which gives off sulphur dioxide when in contact with wounds. It is a German preparation, and is suggested for use in veterinary practice.

Sulfogenol. A crude mineral oil obtained from bituminous shale, is saturated with sulphur, and sulphonated. Sulfogenol is the ammonium salt of the sulphonated product. It has similar properties to ichthyol.

Sulfoid (Sulphoid, Sulfidal). A preparation of colloidal sulphur (80 per cent., sulphur), with an albuminous substance as protective colloid.

Sulfoiodol. Colloidal sulphur.

Sulfoliquid. A German preparation. It is a solution of sulphur dioxide which gives off active sulphur dioxide.

Sulfomyl. A proprietary preparation of MAFENIDE propionate, used in the form of anti-infective eyedrops. (439).

Sulfonitrate. A fertiliser containing ammonium-calcium-sulpho-nitrate.

Sulfophosphite. Consists of zinc, phosphorus, and sulphur. Used in Germany for making matches.

Sulfopone. See SULPHOPONE.

Sulfosept Oil. The next higher fraction to thiosept oil (*q.v.*). Used as an insecticide.

Sulfosin Leo. Sulphur, 1 per cent. in olive oil.

Sulglycotide. A preparation used in the treatment of peptic ulcers. It is the sulphuric polyester of a glycopeptide isolated from pig duodenum.

Sulindac. An anti-inflammatory drug. (Z) - 5 - Fluoro - 2 - methyl - 1 - (4 - methylsulphinylbenzylidene) indene - 3 - acetic acid.

Sulphachlorpyridazine. A sulphon - amide drug. N^1 - (6 - Chloropyridazin - 3 - yl) sulphanilamide. COSULID; VETISULID.

Sulphadimethoxine. 6-(4-Aminobenzenesulphonamido) - 2,4 - dimethoxypyrimidine. Madribon.

Sulphaethidiole. 5-(4-Aminobenzenesulphonamido) - 2 - ethyl - 1,3,4 - thiadiazole. Sethadil.

Sulphafurazole. 5-(4-Aminobenzenesulphonamido) - 3,4 - dimethylisoxazole. Gatrisin.

Sulphalite (Sulphohalite). A mineral, $3Na_2SO_4.2NaCl$.

Sulphaloxic Acid. 2-[4-(Hydroxymethylureidosulphonyl)phenylcarbamoyl]benzoic acid. Enteromide is the calcium salt.

Sulphamagna. A proprietary preparation of streptomycin sulphate, phthalylsulphathiazole, sulphadiazine and activated attapulgite. (245).

Sulphamethizole. 2-(4-Aminobenzenesulphonamido) - 5 - methyl - 1,3,4 - thiadiazole. Urolucosil.

Sulphamethoxazole. 3-(4-Aminobenzenesulphonamido) - 5 - methylisoxazole. Gantanol.

Sulphamethoxydiazine. 2-(4-Aminobenzenesulphonamido) - 6 - methoxypyrimidine.

Sulphamethoxypyridazine. 3-(4-Aminobenzenesulphonamido) - 5 - methoxy-pyridazine.

Sulphamezathine. A proprietary preparation containing sulphadimidine sodium. (2).

Sulphamine Brown A (Naphthine Brown-α). A dyestuff obtained from diazotised α-naphthylamine and the sodium bisulphite compound of nitroso-β-naphthol. Dyes chromed wool brown.

Sulphamine Brown B (Naphthine Brown-β). A dyestuff obtained by the action of β-naphthylamine upon the sodium bisulphite compound of nitroso-β-naphthol. Dyes chromed wool chocolate-brown.

Sulphaminol. Thio-oxy-diphenylamine, $C_6H_4(S_2NH)C_6H_3OH$. An antiseptic dusting-powder.

Sulphamipyrin. Melubrin (*q.v.*).

Sulphammonium. A solution of sulphur in liquid ammonia to form a purple solution.

Sulphamoprine. 2 - (4 - Aminobenzenesulphonamido) - 4,6 - dimethoxypyrimidine.

Sulphamoxole. 2 - (4 - Aminobenzenesulphonamido) - 4,5 - dimethyloxazole. Nurin.

Sulphan Blue. Sodium 4-(4-diethylaminobenzylidene) cyclohexa - 2, 5 - dienylidenediethylammonium - α - benzene-2,4-disulphonate.

Sulphanil Black. See SULPHUR-BLACK T.

Sulphanil Brown. A dyestuff obtained by converting 2 : 4-dinitro-4-aminodiphenylamine into its sulphonic acid, and heating this product with sodium sulphide and sulphur.

Sulphanilic Acid. Aniline-*p*-sulphonic acid, $C_6H_4NH_2SO_3H$.

Sulphanil Yellow (Parasulphurine S). An azo dyestuff obtained from diazotised benzidine and sulphanilic acid, $C_{24}H_{20}N_6S_2O_6$. Dyes cotton greenish-yellow.

Sulphantipyrin. See MELUBRIN.

Sulphaphenazole. 5 - (4 - Aminobenzenesulphonamido) - 1 - phenyl-pyrazole. Orisulf.

Sulphapred. A proprietary preparation of sulphacetamide and PREDNISOLONE used in the form of anti-inflammatory eye drops. (255).

Sulphaproxyline. N'-(4-Isoproxybenzoyl) - 4 - aminobenzenesulphonamide.

Sulpharsenol. Sulfarsenol (*q.v.*).

Sulphasalazine. 4-Hydroxy-4'-(pyrid-2 - ylsulphamoyl)azobenzene - 3 - carboxylic acid.

Sulphasan R. A rubber accelerator 4,4'-dithiomorpholine. (559).

Sulphasomidine. 6-(4-Aminobenzenesulphonamido) - 2,4 - dimethyl - pyrimidine. Elkosin.

Sulphasomizole. 5-(4-Aminobenzenesulphonamido) - 3 - methylisothiazole. Bidizole.

Sulphasuxidine. A proprietary preparation of succinyl sulphathiazole. An antibiotic. (932).

Sulphate Meionite. Synonym for Silvialite.

Sulphate-Scapolite. Synonym for Silvialite.

Sulphated Oils. See TURKEY RED OILS.

Sulphate Pulp. Wood pulp obtained by the treatment of wood with alkali liquors containing sodium sulphate.

Sulphate Resin. See LIQUID RESINS.

Sulphate Ultramarine. Artificial ultramarine in which sodium sulphate is used as a constituent. See ULTRAMARINE.

Sulphathiourea. 4 - Aminobenzenesulphonylthiourea. Badional.

Sulphatic Cancrinite. Synonym for Vishnevite.

Sulphatine. A fungicide. It is a mixture of 73 per cent. sulphur, 20 per cent. lime, and 7 per cent. copper sulphate. It is used against black rot.

Sulphato. A name suggested for the acidic group SO_4H.

Sulphatolamide. 4-Aminobenzene-sulphonylthiourea salt of 4-sulpha-moylbenzylamine. Marbadal.

Sulphatriad. A proprietary preparation of sulphathiazole, sulphadiazine and sulphamerazine. An antibiotic. (507).

Sulphaurea. 4-aminobenzenesulphonylurea. Sulfacarbamide. Euvernil.

Sulphesatyd. A substance stated to be 3-thio-oxindole.

Sulphetrone. A proprietary preparation of solapsone. An antileprotic. (277).

Sulphex. A proprietary preparation of sulphathiazole and hydroxyamphetamine hydrobromide. (281).

Sulphexet. See HEXAL.

Sulphexine. See HEXAL.

Sulphide Dyestuffs. A class of dyestuffs prepared by the fusion of organic amines and other substances with sulphur and sodium sulphide. They are used for cotton dyeing, and are usually fixed by oxidising agents.

Sulphide of Arsenic, Yellow. See ORPIMENT.

Sulphide White. See LITHOPONE.

Sulphiformin. Formaldehyde - sulphurous acid, $HO.CH_2.SO_3H$. An antiseptic. A 1 per cent. solution has been used for spraying vines.

Sulphine. See PRIMULINE.

Sulphine Brown (Cattu Italiano). A dyestuff obtained by the action of sodium polysulphides upon oils, fats, or fatty acids. Dyes cotton dark brown, changing to reddish-brown by oxidation.

Sulphinpyrazone. 1,2-Diphenyl-4-(2-phenylsulphinylethyl)pyrazolidine - 3, 5-dione. Anturan.

Sulphiolinic Acid. See LORETIN.

Sulphite Carbon. A decolorising carbon used for sugar juices. It is prepared from sulphite-cellulose liquors.

Sulphite Pulp. Wood pulp obtained by means of calcium bisulphide. It is made by digesting the disintegrated wood under pressure with the calcium bisulphite, which gives a mass of cellulose fibres free from lignocellulose amounting to about 45 per cent. of the wood.

Sulphite Turpentine. A by-product obtained from the pulping of spruce by the sulphite process, 0·36 to 1 gallon is obtained from 1 ton of pulp. The main constituent is p-cymene.

Sulphite Turpentine Oil (Cellulose Turpentine Oil). A by-product obtained in the manufacture of cellulose. When decolorised it resembles turpentine oil.

Sulpho Blacks. See CROSS DYE BLACKS.

Sulphoborite. A mineral, $Mg_6H_4O_{12}$ $(SO_4)_2.7H_2O$. It is also stated to be $Mg_3SO_4B_2O_5.4\frac{1}{2}H_2O$.

Sulphocarbon Oil. See SANSE.

Sulphocol. See THIOCOLL.

Sulphoform. Triphenyl - stibine - sulphide, $(C_6H_5)_3SbS$. Used in eczema and skin troubles.

Sulphoguaiacin. Quinine methyl-sulphoguaiacate.

Sulphohalite. See SULPHALITE.

Sulphoid. See SULFOID.

Sulphomyxin. Penta - (N - sulpho - methyl)polymyxin B. Thiosporin is the sodium salt.

Sulphonal (Sulphonmethane). Diethyl-sulphone - dimethyl - methane, $(CH_3)_2.$ $C(SO_2)_2.(C_2H_5)_2$. A hypnotic.

Sulphonal, Methyl. See TRIONAL.

Sulphonamide-P. p-Amino-benzene-sulphonamide. Used in the chemotherapeutic treatment of haemolytic streptococcal and other infections.

Sulphonated Oils. See TURKEY RED OILS.

Sulphoncyanines. Dyestuffs obtained by diazotising metanilic acid, coupling with α-naphthylamine, diazotising the product, and coupling it with phenyl- and tolyl-1-naphthylamine-8-sulphonic acids. Dye cotton direct; also used in wool dyeing.

Sulphone Acid Blue B. A tetrazo dyestuff. It dyes wool from a bath containing sodium sulphate and acetic acid.

Sulphoneazurin. A dyestuff. It is the sodium salt of disulpho-diphenyl sulphone - disazo - bi - phenyl - β-naphthylamine, $C_{44}H_{28}N_6O_8S_3Na_2$. Dyes wool blue from a sodium sulphate bath, and cotton blue from a neutral or soap bath.

Sulphonmethane. Sulphonal (q.v.).

Sulphophenol. Aseptol (q.v.).

Sulphophone. (Sulfopone). A trade mark for a mixture of zinc sulphide and calcium sulphate. It is an analogous product to lithopone.

Sulphophosphite. See SULFOPHOSPHITE.

Sulphopyrine. See BETA-SULPHOPYRINE.

Sulphormethoxine. 6 - (4 - Aminobenzenesulphonamido) - 4,5 - dimethoxy - pyrimidine. FANASIL. SULFADOXINE.

Sulphosalicylic Acid. Salicyl-sulphonic acid, $C_6H_3OH(SO_3H)COOH$. Used in the treatment of articular rheumatism.

Sulphosot. Potassium creosote sulphonate. Used in pulmonary disorders.

Sulphostab. See METARSENOBILLON.

Sulphourea. Thiourea, CH_4N_2S.

Sulphoxyl-salvarsan. Sodium-p-arseno-phenyl - dimethyl - amino - pyra - zolone-methylene-sulphoxylate.

Sulphurated Antimony (Antimony Crocus, Saffron of Antimony). A mixture of antimony pentasulphide, Sb_2S_5, with a little oxide, Sb_4O_6, and some free sulphur. (*Antimonium sulphuratum B.P.*) Formerly used in making tartar emetic.

Sulphurated Lime (Calcic Liver of Sulphur). A mixture of calcium sulphide, CaS, and sulphate, $CaSO_4$, containing 50 per cent. of the sulphide. (*Calx sulphurate B.P.*)

Sulphurated Oil. Balsam of sulphur (*q.v.*).

Sulphurated Potash (Liver of Sulphur). A mixture of sulphides, mainly $K_2S_2O_3$, and K_4S_3, obtained by heating potassium carbonate with one-half its weight of sulphur. When fresh and carefully prepared it is the colour of liver, and was called liver of sulphur. (*Potassa sulphurata B.P.*) It is sometimes used in the form of ointment. Calcium sulphide, CaS, is also called liver of sulphur, and is used in the leather industry as a depilating agent.

Sulphur Auratum. Antimony sulphide.

Sulphur Black T (Thional Black, Katigene Black, Pyrrol Black, Thiogene Blue, Sulphanil Black, Pyrogene Blacks and Blues, Melanogen Black). Dyestuffs of the same class as Immedial black V. They are prepared by the action of sodium polysulphides upon various amino-oxy-derivatives of diphenylamine. They dye from a sulphide bath.

Sulphur Black T Extra. See DINITROPHENOL BLACK.

Sulphur Blue B (Sulphur Blue L Extra; Sulphur Brown G, 2G; Sulphur Cutch R, G; Sulphur Corinth B; Sulphur Indigo B). Dyestuffs of a similar type to sulphur black T (*q.v.*).

Sulphur Blue L Extra. See SULPHUR BLUE B.

Sulphur Brown G, 2G. See SULPHUR BLUE B.

Sulphur Corinth B. See SULPHUR BLUE B.

Sulphur Cutch R, G. See SULPHUR BLUE B.

Sulphuretted Hydrogen. Hydrogen monosulphide, H_2S.

Sulphur, Flour. See FLOUR OF SULPHUR.

Sulphur Gold. Antimony pentasulphide, Sb_2S_5. Used for vulcanising and imparting a red colour to rubber.

Sulphur, Green. See BLUE SULPHUR.

Sulphur Hypochlorite. A mixture of sulphur and sulphur chloride. Used in rubber vulcanising.

Sulphuric Acid, Consolidated. See SOLIDIFIED SULPHURIC ACID.

Sulphuric Ether (Phosphoric Ether). Names formerly given to ethyl oxide (ether), $C_2H_5.O.C_2H_5$. An anæsthetic. Also used in the manufacture of colloidal and artificial silk.

Sulphur Indigo B. See SULPHUR BLUE B.

Sulphurion. Colloidal sulphur.

Sulphurite. A name applied to a sulphur from Java, which contained 29 per cent. arsenic.

Sulphur, Liver of. See SULPHURATED POTASH.

Sulphur, Nacreous. See MOTHER OF PEARL SULPHUR.

Sulphur Ointment. A mixture of 20 per cent. sublimed sulphur and 10 per cent. potassium carbonate with benzoated lard. Antiparasitic.

Sulphur Olive Green. See THIOCHEM SULPHUR GREEN.

Sulphur Olive Oils. A name for the oil dissolved out from residual olive oil cake by means of carbon disulphide. It is also called sulphocarbon oil. It is rich in stearin, and has a saponification value of 183, and an acid value of 25–70 per cent. oleic acid. Also see SANSE.

Sulphur, Ruby. See REALGAR.

Sulphur Soap. Usually a yellow medicated soap to which has been added about 10 per cent. powdered sulphur.

Sulphur Waste. The residue from the distillation of iron pyrites.

Sulphur Yellow S. See NAPHTHOL YELLOW S.

Sulpiride. An anti-depressant and anxiolytic. It is N-(1-ethylpyrrolidin-2-ylmethyl) - 2 - methoxy - 5 - sulpha - moylbenzamide. DOGMATYL.

Sulporex. A proprietary trade name for iron sponge containing magnesium, used in the de-sulphurisation of iron melts. (899).

Sulsol. A proprietary trade name for a colloidal sulphur preparation for horticultural purposes.

Sultan Red 4B. See BENZOPURPURIN 4B.

Sulthiame. 4-(Tetrahydro - 2H - 1,2 - thiazin - 2 - yl)benzene - sulphonamide. SS-dioxide.

Sultrin. A proprietary preparation of sulphathiazole, sulphacetamide, and N-benzoyl sulphanilamide, used in the treatment of vaginitis. (369).

Sulvanite. A mineral, $Cu_3S.V_2S_5$.

Sulzin. An inorganic rubber vulcanisation accelerator. It approximates to the composition $ZnSO_4.5NH_3$.

SUM 36. A symmetrical urea of m-benzoyl - p - amino - benzoyl - amino -

naphthol-3 : 6-sodium sulphonate. A proprietary medicinal preparation allied to Bayer 205.

SUM 468. Sym - di - p - benzoyl - p - amino - benzoyl - 1 - naphthylamine - 4 : 6 : 8-sodium-sulphonate urea.

Sumac. See SUMACH.

Sumacel. A diatomaceous earth containing 80 per cent. SiO_2, 5·3 per cent. Fe_2O_3 and Al_2O_3, 2·02 per cent. CaO, and 8·16 per cent. H_2O. It is stated to be suitable as a filtering medium for sugars.

Sumach (Sumac). The dried and finely powdered leaves and shoots of species of *Rhus*. Used for tanning leather, also for dyeing and printing. Sicilian sumach consists of the leaves of *Rhus coriaria*. The material is often imported in the form of powder containing from 25–28 per cent. tannin. It is often adulterated with leaves of *Pistacia lentiscus*. Venetian or Turkish sumach consists of the leaves of *Rhus cotinus* (the wood of this tree gives young fustic), and contains 17 per cent. tannin. American sumach is obtained from varieties of *Rhus*, chiefly *Rhus glabra*, and usually contains about 25 per cent. tannin. Virginian sumach consists of the leaves of *Rhus typhina*. French sumach is obtained from *Coriari myrtifolia*, and contains 15·6 per cent. tannin. Cape sumach is from the leaves of *Colpoon compressum*, and contains 23 per cent. tannin, and Russian sumach comes from *Arctostaphylos uva-ursi*, and contains 14 per cent. tannin.

Sumach Wax. See JAPAN TALLOW.

Sumalban. Alban obtained from Sumatra gutta-percha.

Sumaphos. A mixture of diatomaceous earth and acid phosphate, containing 36·22 per cent. P_2O_5. Recommended for use in the manufacture of white sugar.

Sumatra Camphor. See CAMPHOR.

Sumatra Wax. See JAVA WAX.

Sumet Processed Lead. An alloy of 70–80 per cent. copper and 15–30 per cent. lead.

Sumikon. A proprietary range of Phenolic moulding materials. (933).

Sumner's Reagents. For glucose determination. 3 : 5 - Dinitro - salicylic acid (10 grams) are dissolved in 500 c.c. warm water in a 1,000-c.c. flask and made up to 1,000 c.c. Sodium hydroxide (13·5 grams) is dissolved in 300 c.c. cold water in a 2-litre beaker, then 880 c.c. of the salicylic acid solution are added and mixed; 225 grams of potassium sodium tartrate (Rochelle salt) are added and dissolved by stirring, and the whole transferred to a bottle. Standard iron-alum solution

is made by weighing 345 mg. violet ferric ammonium sulphate and transferring to a 1,000-c.c. flask, adding 1 gram 3 : 5-dinitro-salicylic acid and then 500 c.c. water. The colour of the solution is equivalent to a glucose solution containing 1 gram treated by the Sumner method.

Sumycin. A proprietary preparation of tetracycline base buffered with potassium metaphosphate. An antibiotic. (326).

Sun Bronze. An alloy of from 40–60 per cent. copper, 30–40 per cent. tin, and 10 per cent. aluminium. Used in jeweller's work. The name is also used for an alloy of from 50–60 per cent. cobalt, 30–40 per cent. copper, and 10 per cent. aluminium.

Sundora. A proprietary trade name for cellulose acetate.

Sundtite. A mineral. It is a variety of andorite.

Sun Gold. See HELIOCHRYSIN.

Sunned Oils. Crude American petroleum oil which has been spread out over warm water in tanks, and exposed to the sun to improve the gravity and consistency.

Sunoco Spirits. A proprietary trade name for petroleum solvents.

Sunolith. A proprietary trade name for a pigment containing 71 per cent. barium sulphate and 29 per cent. zinc sulphide. See also PONOLITH.

Sunoxol. Quinosol (*q.v.*).

Sun Stone. Oligoclase (*q.v.*).

Suntei Tallow. A white sweetish fat expressed from the seeds of *Palaquium oleosum*.

Sun Yellow (Jaune Soleil, Curcumine S, Maize). A dyestuff. It is the sodium salt of the so-called azoxy-stilbene-disulphinic acid, $C_{14}H_8N_2O_7S_2Na_2$. Dyes wool and silk reddish-yellow from an acid bath.

SUP 36. A proprietary product. It is the symmetrical urea of p-benzoyl-p-amino - benzoyl - amino - naphthol - 3 : 6-sodium sulphonate. Used formerly in the treatment of colds and influenza by injection. Product now discontinued.

SUP 468. A proprietary product. It is sym - di - p - benzoyl - p - amino - benzoyl-1-naphthylamine-4 : 6 : 8-sodium-sulphonate urea. Product now discontinued.

Suparac Standard. A proprietary trade name for rubber vulcanisation accelerator. It contains 25 per cent. piperidine piperidine-1-carbothionolate and 75 per cent. colloidal clay. Suparac 2 contains 25 per cent. piperidine piperidine-1-carbothionolate and 75 per cent. zinc oxide.

Superam. A fertiliser obtained by neutralising the acids of ordinary super-phosphate with ammonia gas.

Superargol (Tartar Cake, Substitute of Tartar). Preparations containing simply acid sodium sulphate. Others contain oxalates, and a few tartaric acid and sulphuric acid.

Super-ascoloy. A ferrous alloy containing 8 per cent. nickel and 18 per cent. chromium.

Super-ba. An Italian synthetic resin used for making varnish and varnished paper for electrical insulation.

Superba. A proprietary trade name for a carbon black.

Superbasique Metal. A modification of cast iron. It is resistant to alkalis.

Superbeckacite. A proprietary trade name for pure phenolic varnish and lacquer resins.

Super Bronze. An alloy of from 57–69 per cent. copper, 1·2–5·1 per cent. aluminium, 1·3–2 per cent. iron, 21–37 per cent. zinc, and 3–3·2 per cent. manganese.

Super Cement. An ordinary Portland cement to which has been added a waterproofing material.

Super-cliffite. Explosives. No. 1 contains 10 per cent. nitro-glycerin, 1 per cent. collodion cotton, 60 per cent. ammonium nitrate, 16 per cent. sodium chloride, 11 per cent. ammonium oxalate, and 6 per cent. wood meal. No. 2 has the sodium chloride increased to 20 per cent., and the ammonium oxalate reduced to 6 per cent.

Super Die. A proprietary trade name for a tool steel containing 10·5 per cent. chromium, 1 per cent. tungsten, and 1 per cent. silicon.

Super-excellite. An explosive containing 73·5–77 per cent. ammonium nitrate, 6·5–8 per cent. potassium nitrate, 2–4 per cent. wood meal, 3·5–5 per cent. nitro-glycerin, and 9–11 per cent. ammonium oxalate.

Superez. A proprietary cellulose acetate denture.

Superfiltchar. A proprietary product. It is an active decolorising carbon made from sawdust.

Superforcite. A Belgian gelatin dynamite containing 64 per cent. nitro-glycerin.

Super Hartolan. Wool wax alcohols. (526).

Superior Alloy. A heat-resisting alloy containing 78 per cent. nickel, 19·5 per cent. chromium, 2 per cent. manganese, and 0·5 per cent. iron.

Superite. An explosive consisting of 80–84 per cent. ammonium nitrate, 9–11 per cent. potassium nitrate, 2–5

per cent. starch, and 3·5–4·5 per cent. nitro-glycerin.

Super-karma. An alloy wire containing 80 per cent. nickel and 20 per cent. chromium.

Super-kolax No. 2. An explosive containing nitro-glycerin, collodion cotton, potassium nitrate, barium nitrate, wood meal, starch, and ammonium oxalate.

Superlan. Level dyeing, fast to light wool dyestuffs. (582).

Super-ligdynite. A coal mine explosive containing from 15–17 per cent. nitro-glycerin, 15–17 per cent. ammonium nitrate, 23–25 per cent. sodium nitrate, 10–12 per cent. flour, 19–21 per cent. wood pulp, and 9–11 per cent. sodium chloride.

Superlith. See PENOLITH.

Superneutral Metal. A silicon-iron alloy. Used for nitric acid plants.

Super Nickel. A proprietary trade name for alloys of 20–30 per cent. nickel with 70–80 per cent. copper. They are corrosion resisting.

Superol. o-Oxy-quinoline sulphate. An antiseptic.

Superlit. A proprietary synthetic resin.

Superpalite (Diphosgene, Green Cross Gas). Trichloro-methyl-chloro-formate, $Cl.COO.CCl_3$. It has a specific gravity of 1·6525, and boils at 127·5–128° C. A military poison gas.

Superphosphate (Mineral Superphosphate, Superphosphate of Lime). A fertiliser. It consists of mono-calcium phosphate, $CaH_4(PO_4)_2$, mixed with calcium sulphate, and contains 25–28 per cent. soluble phosphate.

Superphosphate, Ammonium. See NITRO-PHOSPHATE.

Superphosphate, Mineral. See SUPER-PHOSPHATE.

Superphosphate of Lime. See SUPER-PHOSPHATE.

Super-rippite. A smokeless powder containing from 51–53 parts nitroglycerin, 2–4 parts nitro-cotton, 13·5–15·5 parts potassium nitrate, 15·5–17·5 parts dried borax, and 7–9 parts potassium chloride.

Super-rippite No. 2. An explosive for coal mines containing 51 per cent. nitro-glycerin, 3 per cent. nitro-cotton, 11 per cent. potassium perchlorate, 24 per cent. borax, and 10 per cent. potassium chloride.

Supersoy. A high grade soya bean flour.

Superston. A proprietary alloy of aluminium manganese and bronze used in the manufacture of wood-grained mould for injection moulding. (934).

Superstyrex. See POLYSTYRENE.

Super-sulphur. Thiuramdisulphide, a vulcanisation accelerator.

Super Sulphur No. 1. A proprietary

rubber vulcanisation accelerator. It is the oxidised zinc salt of dimethyl-dithio-carbamic acid.

Super Sulphur No. 2. A proprietary rubber vulcanisation accelerator. It is lead dimethyl-dithio-carbamate.

Supertex. A proprietary safety glass.

Superthane. A proprietary polyester-based polyurethane elastomer cross-linked with diols. (935).

Supertone. A proprietary range of urea formaldehyde moulding materials. (936).

Superturpentine. Spirits of turpentine specially rectified *in vacuo*. It boils at 155° C., and distils completely below 160° C.

Suprac. Decolorizing and absorptive activated carbon. (629).

Supracapsulin. Adrenalin (*q.v.*).

Supracet. Dispersive dyestuffs for synthetic fibres. (582).

Supradin. Adrenalin (*q.v.*). A trade mark. No longer commercially available.

Suprafrax. A clay with a high percentage of alumina. It is used as a furnace lining.

Suprasec. Isocyanates for general application. (512).

Suprasec 4275. A proprietary solution of toluene diisocyanate-polyol adduct in ethyl acetate. It is used as the isocyanate component of two-pack polyurethane coatings. (512).

Supramica. A trade mark for ceramoplastics. (39).

Supranephrane. Adrenalin (*q.v.*).

Suprarenaline. Adrenalin (*q.v.*).

Suprarenine. Adrenalin (*q.v.*).

Supraresen. The residue obtained when dammar is prepared for use in lacquers and is soluble in hydrocarbons. Used in the varnish industry.

Suprex. See CATALPO.

Suprexcel. Fast to light direct cotton dyestuffs. (582).

Suprex White. A highly purified precipitated calcium carbonate for use as a rubber filler in the place of blanc fixe. It has a specific gravity of 2·7.

Supronic B10, B25, B50, B75 and B100. A proprietary range of low-foam surface-active agents. They are polyoxy-alkylated polyalkylene glycols. (547).

Supronic E800. A solid, non-ionic surface-active agent. It is a polyoxy-ethylene polyoxypropylene condensate.

Suprox. Zinc oxide for paints and enamels. (516).

Surahwa Nuts. See BUTTER NUTS.

Surbex T. A proprietary multi-vitamin preparation. (311).

Surfathesin. A proprietary trade name for cyclomethycaine.

Surfil. Precipitated calcium carbonate. (542).

Surgical Solution. See PHYSIOLOGICAL SALT SOLUTION.

Surlyn. A registered trade mark (10) for a range of ionomer resins. Constituents of the range include the following:—

Surlyn 1555. Possesses good flow properties. Used in injection moulding.

Surlyn 1558. A 25-mesh resin used in rotational moulding.

Surlyn 1559. Used in injection moulding.

Surlyn 1560. Used in injection moulding when maximum clarity is required.

Surlyn 1603. Used in film extrusion when good slip is required.

Surlyn 1605. Used as an extruded coating on paper.

Surlyn 1652. Used as an extruded coating on foil.

Surlyn 1707. Used in the extrusion of high-clarity sheet and for blow moulding.

Surlyn 1800. Used as a tough coating for wires and cables.

Surmontil. A proprietary preparation of trimipramine. An antidepressive drug. (336).

Surophosphate (Dasag). A fertiliser made from sewage, other waste material, and peat. It is of German origin.

Surparil. A mixture of Novotropin and Perparin. Suparil-strong tablets also contain dimethylaminoantipyrin and phenylethylmalonylurea. The preparation is an antiseptic for the stomach and intestine. It is a trade mark.

Surrenine. Adrenalin (*q.v.*).

Surrogat. A by-product obtained from the refining of Russian kerosene. Used in soap-making.

Surrogate, Butter. See MARGARINE.

Surrogate, Saffron. See VICTORIA YELLOW.

Sursassite. A mineral. It is an aluminium-magnesium silicate.

Survon. Nylon monofilament for sports racquet strings. (512).

Susanite. See LEADHILLITE.

Suscardia. A proprietary preparation of isoprenaline hydrochloride, used as a cardiac stimulant. (330).

Susini. An alloy of aluminium containing from 1·5–4·5 per cent. copper, 0·5–1·5 per cent. zinc, and 1–8 per cent. manganese.

Suspensol. A proprietary colloidal bismuth cream.

Sussexite. A hydrated borate of manganese and magnesium, $(Mg.Mn)B_4O_7$, found in the United States of America.

Sustac. A proprietary preparation of

glyceryl trinitrate. A vasodilator used in angina pectoris. (330).

Sustamycin. A proprietary preparation of TETRACYCLINE hydrochloride. An antibiotic. (249).

Sustanon. A proprietary preparation of TESTOSTERONE propionate, phenylpropionate and isocaproate, used as an androgen supplement. (316).

Sutermeister's Stain. For paper (a) Contains 1·3 grams iodine and 1·8 grams potassium iodide in 100 c.c. water, and (b) consists of a clear saturated solution of calcium chloride.

Sutron. Nylon monofilament for surgical sutures.

Suxamethonium Bromide. Bis-2-dimethylaminoethyl succinate bismethobromide. Bevidil M.

Suxamethonium Chloride. Bis-2-dimethylaminoethyl succinate bismethochloride. Anectine. Brevidil M. Scoline.

Suxethonium Bromide. Bis-2-dimethylaminoethyl succinate bisethobromide. Brevidil E.

Svabite. See SVAVITE.

Svavite (Svabite). A mineral, $Ca_5 FAs_3O_{12}$.

S.V.C. A proprietary preparation of ACETARSOL used in the treatment of vaginitis. (507).

Swale Powder. An explosive containing potassium perchlorate, nitro-glycerin, collodion cotton, ammonium oxalate, wood meal, and a little nitrotoluene.

Swalite. An explosive for coal mines, similar to Swale powder (q.v.).

Swanbergite. A mineral, $Na_3(CaOH) (AlO)_6(PO_4)_2(SO_4)_2.3H_2O$.

Swarf. The scrapings of soft iron castings. It is used in the preparation of aniline.

Swartzite. A mineral. It is $2[CaMg UO_2(CO_3)_3.12H_2O]$.

Swedelec. A Swedish charcoal iron. It has a high magnetic permeability.

Swedenborgite. A new mineral from Långban. It approximates to the formula, $Na_2O.2Al_2O_3.Sb_2O_5$.

Swedish Factory Tar. A tar obtained from waste wood in charcoal kilns, as a by-product in charcoal burning.

Swedish Green. See SCHEELE'S GREEN.

Swedish Liquid Resin. See TALLOEL.

Swedish Turpentine. See TURPENTINE.

Sweet Bark (Sweet Wood Bark, Eleuthera Bark). Cascarilla, used for extracting cascarilla oil and as an ingredient in insecticides, etc.

Sweet Gum. See AMERICAN STORAX.

Sweet Nitre. See SPIRIT OF SWEET NITRE.

Sweet Nut Oil. See BUTTER OIL.

Sweet Oil. See COLZA OIL.

Sweet-water. Consists of glycerin and water, obtained in the distillation of crude glycerol.

Sweet Wine Spirit. See SPIRIT OF SWEET WINE.

Swinbourne's Gelatin. A patented gelatin obtained from skins and hides by reducing to shavings or slices, soaking in cold water, changing at intervals, then heating with water and straining.

Syanthrose. Levulin, $C_6H_{10}O_5$.

Sychnodymite. A mineral. It is a sulphide of cobalt and copper, $(Co.Cu)_4S_5$.

Sycorin. See SACCHARIN.

Sycose. See SACCHARIN.

Syenite. An igneous rock consisting of potash-felspar and hornblende. This is called hornblende-syenite. Augite sometimes replaces hornblende and is then called augite-syenite. If biotite replaces hornblende, it is termed micasyenite. They resemble granite and are used for building and ornamental work.

Syepoorite. A mineral. It is a cobalt sulphide, CoS.

Sykose. Saccharin (q.v.).

Syl. A proprietary preparation of dimethicone 350, benzalkonium solution and nitrocellulose used in dermatology as an antibacterial agent. (347).

Syloid 72. A proprietary trade name for a silica gel for addition (2 per cent.) to plasticised vinyls to prevent plate out.

Syloid 83. A proprietary trade name for a silica aerogel. A flatting agent which may be stirred into varnishes giving improved suspension and flatting efficiency. (189).

Sylopal. A proprietary preparation of DIMETHICONE, magnesium oxide and aluminium hydroxide. A gastro-intestinal sedative. (462).

Sylphrap. A proprietary trade name for a regenerated cellulose transparent sheet.

Sylplast. A proprietary trade name for urea-formaldehyde resins, lacquers, and coatings. (937).

Sylvan. α-Methyl-furane, C_5H_6O. A constituent of wood tar.

Sylvania Cellophane. A proprietary trade name for regenerated cellulose.

Sylvanite (Graphic Tellurium, Calaverite). A mineral, $(Ag.Au)Te_3$.

Sylvic Acid. Impure abietic acid.

Sylvid. A registered trade mark for a range of silica fillers for plastics processing. (404).

Sylvine (Sylvite, Hoevelite). A mineral. It is potassium chloride, KCl, found in the Stassfurt deposits.

Sylvinite. A mineral. It is a mixture of sylvine (q.v.) with rock salt.

Sylvite. See SYLVINE.

Symikite. A mineral. It consists of

mono-hydrated manganous sulphate, $MnSO_4.H_2O$.

Symmetrel. A proprietary preparation of AMANTADINE hydrochloride, used in the treatment of parkinsonism. (17).

Sympathol. A German product. It is stated to be a hydrochloric acid combination of p-methyl-amino-ethanol-phenol. It is similar to adrenalin, but differs from it in that it has lost an OH group. It has rather weaker action than adrenalin, and can be sterilised.

Sympatol. A proprietary preparation of oxedrine tartrate used in the treatment of cardiac disorders. (252).

Symphoral. See SYMPHOROL.

Symphorol (Symphoral). Lithium sodium, or strontium sulphonates. Diuretics.

Symplesite. A mineral, $Fe_3(AsO_4)_2$. $8H_2O$.

Synacthen. A proprietary preparation of tetracosactrin. (18).

Synadelphite. A mineral, (Mn.Al) $[Mn(OH)_2]_5.(AsO_4)_2$.

Synadryn. A proprietary preparation of prenylamine lactate. Treatment of angina pectoris. (312).

Synalar, Synalar Forte, Synandone. A proprietary preparation of fluocinolone acetonide for dermatological use. (2).

Synalar-N. A proprietary preparation of fluocinolone acetonide and neomycin sulphate used in dermatology as an antibacterial agent. (2).

Synandone. A proprietary trade name for the acetonide of Fluocinolone.

Synandone-N. A proprietary preparation of fluocinolone acetonide and neomycin sulphate used in dermatology as an antibacterial agent. (2).

Synanthin. See HELENIN.

Synasol. A proprietary trade name for a denatured ethyl alcohol.

Syncaine. Novocaine (*q.v.*).

Syncillin. A proprietary preparation of potassium phenethicillin. An antibiotic. (324).

Syncurine. A proprietary trade name for decamethonium iodide.

Syndite. An explosive consisting of 10–22 per cent. nitro-glycerin, 0·1–0·3 per cent. collodion cotton, 45–49 per cent. ammonium nitrate, 7–9 per cent. sodium nitrate, 2–5 per cent. glycerin, 2–5 per cent. starch, and 26–28 per cent. sodium chloride.

Synephrin Tartrate. Hydroxy-phenyl-methyl-amino-ethanol tartrate.

Synergel. A proprietary preparation of aluminium phosphate gel, PECTIN and AGAR, used in the treatment of dyspepsia. (938).

Syngenite (Kaluszite). A mineral, $CaK(SO_4)_2.H_2O$.

Synkavit. A proprietary preparation of menadiol sodium diphosphate. A Vitamin K supplement. (314).

Synmould. A proprietary phenolic moulding powder. (939).

Synobel. Insulating varnishes. (512).

Synocryl 820S and 821S. Proprietary trade names for hydroxyacrylics curing at 120° C. with melamine. Used for thermosetting flow enamels for the car industry. (91).

Synocure 867S. A proprietary acrylic resin used for coating metals, particularly aluminium. It is hydroxyl-functional. (91).

Synocure 868S. A proprietary flexible acrylic resin used for coating rigid surfaces. It is hydroxyl-functional. (91).

Synocure 869S. A proprietary fast-drying acrylic resin used as a commercial wood finishing. (91).

Synogist. A proprietary preparation of sodium sulphosuccinate and undecylenic monoalkyl amide, used as a shampoo for dandruff. (940).

Synolate MFF. A proprietary trade name for dibutoxy ethyl phthalate. A high molecular weight fast gelling plasticiser with low volatility and low water extraction used for polyvinyl chloride. (189).

Synolite. A proprietary polyester laminating resin. (941).

Synopen. A proprietary trade name for the hydrochloride of Halopyramine.

Synourin. A proprietary trade name for a type of castor oil used in paints. Dehydrated castor oil. (564).

Synpron 1032 and 1033. A proprietary range of liquid-antimony mercaptides used as heat stabilisers. (942).

Synresin RD 461. A proprietary blocked, one-component polyurethane resin. It is thermosetting and is used as a rubber flock adhesive. (943).

Syntagmatite. A mineral. It is pargasite.

Syntamol V. An auxiliary for tanning and dyeing chrome leather. It is a mixture of the neutral salts of aromatic sulphonic acids.

Syntan. See NERADOL.

Syntetrin. A proprietary preparation of rolitetracycline. An antibiotic. (324).

Syntex. A proprietary trade name for an oil modified alkyd resin (*q.v.*).

Syntex Menophase. A proprietary preparation of MESTRANOL used in the treatment of menopausal symptoms. (809).

Synthalin. Piperonyl-quinoline-carboxylic-acid-methyl ester, $CH_2O_2.C_6H_3.C_9H_5N.COOCH_3$. It has a similar action to insulin.

Synthalin B. A proprietary preparation.

It is dodecamethylene-guanidine. It is used in diabetes.

Sythamica. A trade mark for synthetic mica. (39).

Synthane. A proprietary trade name for phenol-formaldehyde synthetic resin laminated products and other plastics.

Synthaprufe. A proprietary waterproofing and jointing material consisting of a mixture of rubber and tar.

Synthargol. A colloidal silver product consisting of 90 per cent. bile salts and 10 per cent. colloidal silver. A germicide.

Synthawax. Hydrogenated castor oil. (506).

Synthecite. A rubber softener. It is a distillate from vulcanised rubber containing vegetable oils and waxes.

Synthe-plastic. A reaction product of a terpene base. It is a rubber plastic, and is stated to contain no pitches or waxes.

Synthetic Peru Balsam. See PERUGEN.

Synthetic Tallow. See CANDELITE.

Synthetic Tannin. Neradol (*q.v.*).

Synthin. A product obtained by heating Synthol (*q.v.*), at 400° C. in an autoclave. A liquid results which contains saturated hydrocarbons and sulphuric acid. A liquid fuel.

Synthite. A proprietary condensation product of phenol.

Synthite. A proprietary trade name for formaldehydes. (89).

Synthocarbone. A specially prepared charcoal for use as a fuel.

Synthol. A liquid fuel containing hydrocarbons, acids, alcohols, aldehydes, and esters. It is obtained by reducing carbon monoxide in water gas at high temperatures and under pressure, using iron borings coated with potassium carbonate as contact material. It contains 10 per cent. fatty acids, 20 per cent. water-soluble alcohols, aldehydes, and ketones (1·5 per cent. methyl alcohol, 14·5 per cent. ethyl alcohol, and 5 per cent. acetone), 11 per cent. oils partly miscible with water, 48 per cent. oils volatile in steam, and 2 per cent. non-volatile oils. Rectified synthol on distillation gives 87 per cent. of a spirit suitable as a light fuel. It has a calorific value of 8200 calories per kilogram. It is a German product, and is mixed with benzol for use as a motor fuel.

Syntholvar. A proprietary trade name for extruded polyvinyl chloride.

Syntocinon. A proprietary preparation containing oxytocin. Used for promoting uterine contraction during and following labour. (267).

Syntometrin. A proprietary preparation of ERGOMETRINE maleate and oxytocin, used in the treatment of postpartum hæmorrhage. (267).

Syntopressin. A proprietary preparation of lypressin. (267).

Synvaren. A proprietary trade name for a phenol formaldehyde resin adhesive.

Synvarol. A proprietary trade name for an urea formaldehyde resin adhesive.

Syrgol. Colloidal silver. A bactericide.

Syrian Asphalt. A natural asphalt containing about 100 per cent. bituminous matter. It has a specific gravity of about 1·06, a melting-point of about 100° C., and practically no mineral matter.

Syringa Vulcanine. An organic rubber vulcanisation accelerator.

Syrosingopine. Methyl O-(4-ethoxycarbonyloxy - 3,5 - dimethoxy - benzoyl)reserpate. Isotense. Singoserp.

rtussar. A proprietary preparation of dextromethorphan hydrobromide, phenylpropanolamine hydrochloride, sodium citrate, citric acid and chloroform. A cough linctus. (327).

Syrup, Mineral. See PARAFFIN, LIQUID.

Systogen. *p* - Hydroxy - phenyl - ethylamine.

Sytam. Systemic organo phosphorous insecticide. (501).

Sytron. A proprietary preparation of sodium iron edetate. A hæmatinic. (264).

Szabdite. A mineral. It is a variety of hypersthene.

Szaibeyite. See BOROMAGNESITE.

Szájbelyite. A mineral. It is $MgBO_2OH$.

Szmikite. A mineral, $MnSO_4.H_2O$.

Szomolnokite. A mineral, $FeSO_4.H_2O$.

T

Taamya. An Egyptian food product. It is prepared by grinding beans, mixing with flour, and frying in cottonseed or sesamé oil. It contains about 15 per cent. protein, 25 per cent. fat, 27 per cent. moisture, and 23 per cent. cellulose and starch.

Tabalgin. A proprietary preparation of paracetamol. An analgesic. (137).

Tabbyite. See WURTZILLITE.

Tabergite. A mineral. It is $(Na,K)_{0·8}$ $(Mg,Fe)_{3·3}(Si,Al)_4O_9(OH)_3$.

Tablasthma. A German product. It consists of tablets containing extract of belladona, potassium iodide, tincture of lobelia and adrenalin. It is used against asthma.

Table Salt. Sodium chloride, NaCl, usually containing some added calcium phosphate to prevent lumpiness and dampness.

Tabora Black. A direct cotton dyestuff. It is used direct, or developed upon the

TAB 677 **TAL**

fibre with toluylene-diamine. See DIA-MINE BLACK-BLUE B.

Tab Rybar Co. A proprietary preparation of isoprenaline sulphate, methyl-ephedrine hydrochloride, butethamate citrate, and theophylline. A bronchial antispasmodic. (273).

Tabular Oxylepider. The lactone of γ-oxy-tetraphenyl-crotonic acid, C.C$_6$H$_5$.C.C$_6$H$_5$.C(C$_6$H$_5$)$_2$.CO.O.

Tabular Spar. See WOLLASTONITE.

Tacamahac Oil. See LAUREL NUT OIL.

Tacamahac Resin (West Indian Anime Resin). A resin obtained from various plants, usually from *Calophyllum* species.

Tace. A proprietary preparation containing chlorotrianisene. (263).

Tachalgan. A German product. It is a compound of caffeine, antipyrine, salicylic acid, and hexamethylene-tetra-mine. An antineuralgic for intravenous injection.

Tachhydrite (Tachydrite). A mineral. It is a calcium-magnesium chloride, CaCl$_2$.2MgCl$_2$.12H$_2$O. It occurs in the Stassfurt deposits.

Tachostyptan. A proprietary preparation of THROMBOPLASTIN used to control bleeding. (256).

Tachiol (Tachyol). Silver fluoride, AgF. An antiseptic used for diseases of the urinary organs.

Tachyaphaltite. A mineral. It is a variety of zircon, ZrSiO$_4$.

Tachydrite. See TACHHYDRITE.

Tachylite. A dark volcanic glass.

Tachyol. Tachiol (*q.v.*).

Tacitin. A proprietary preparation of BENZOCTAMINE hydrochloride. A tranquilliser. (18).

Tackol. A mixture of oils and resins used as a rubber plasticiser.

Taconite. A magnetic ore consisting of magnetite, quartz, and amphibole.

Tacrine. 9-Amino-1,2,3,4-tetrahydro-acridine. Romotal is the hydrochloride.

Tacuasonte Balsam. See BALSAM OF CASCARA.

Taeniol. A mixture of sebirol (obtained from various species of *Embelia*), dithymol salicylate, and turpentine oil. Used as a specific for the ankylo-stomiasis of miners, and for intestinal worms.

Taeniolite. A mineral. It is a mica.

Taffy. A residue from the neutralisation of the mixed organic acids produced by the fermentation of kelp-seaweed in the production of acetone. It consists chiefly of calcium propionate.

Taflite 900. A proprietary range of weather-resistant high-impact poly-styrenes made from an EPDM graft polymerised with styrenes and dispersed

as spherical microgels in polystyrene phases. (948).

Tagilite. A mineral, Cu(CuOH)PO$_4$.H$_2$O.

Tahasskite. A mineral. It is 4[Fe$_2$$^{\cdot\cdot}SiO_4$].

Ta-Hong. A lead glass containing ferric oxide. Used by the Chinese as a red enamel on porcelain.

Tailor's Chalk. This material consists of French chalk (magnesium silicate) mixed with a little China clay.

Tainiolite. A Greenland mineral H$_2$(Li.Na.K)$_2$(Mg.Fe)$_2$(SiO$_3$)$_3$.

Tak. Mould sealing compound. (531).

Taka-diastase. A registered trade mark applied to an enzyme from the fungus *Eurotium orzæ*, grown on rice. A proprietary preparation containing aspergillus oryzæ enzymes. An antacid. (264).

Takazyma. A proprietary preparation containing takadiastase (*q.v.*) magnesium carbonate, bismuth carbonate, ginger and calcium carbonate. An antacid. (264).

Takatol. *p*-Amino-phenol, C$_6$H$_4$OH.NH$_2$.

Takizolit. A red micro-crystalline kaolin found in Japan, and having the composition, 2Al$_2$O$_3$.7SiO$_2$.7H$_2$O. It also contains appreciable amounts of yttrium, lanthanum, didymium, scandium, cerium, ytterbium, and ruthenium, and possibly also masurium.

Ta-Kong. A lead glass containing ferric oxide and used by the Chinese as a red enamel.

Taktene 1252. A registered trade mark for a high cis-1,4-polybutadiene rubber containing 37·5 parts per hundred of a highly aromatic oil. (230).

Taktic. See AMITRAZ.

Talampicillin. An antibiotic currently undergoing clinical trial as " BRL 8988 ". It is phthalidyl 6-[D(-)-α-aminophenylacetamido] penicillanate.

Talatrol. A proprietary trade name for Trometamol.

Talbor's Powder. Cinchona bark in powder form.

Talc. A hydrated magnesium silicate, H$_2$Mg$_3$Si$_4$O$_{12}$, specific gravity 2·7. A lubricating agent.

Talc-Chlorite. A mineral. It is (Mg,Fe$^{\cdot\cdot}$,Al)$_6$(Si,Al)$_4$O$_{10}$(OH)$_8$.

Talca Gum. See SUAKIN GUM.

Talcapatite. An impure variety of apatite.

Talc, Laminated. See MICA.

Talcoid. A variety of talc. A mineral. It is Mg$_3$Si$_5$O$_{12}$(OH)$_2$.

Talcose Slate. A rock containing some talc.

Talcosite. A mineral. It is an aluminium silicate.

Talc Schist. A slaty talc.

Talc-spar. Magnesite (*q.v.*).

Talc-spinel. See SPINEL.

Talgol. See CANDELITE.

Talide. A proprietary trade name for a tungsten carbide material.

Talipot (Raw Palmira Root Flour). A starch obtained from a palm, *Corypha umbraculifera*.

Talite. A siliceous earth containing 84 per cent. silica with small quantities of oxides of iron and aluminium. It is used as a rubber filler. See also TALITOL.

Talitol (Talite). An alcohol, CH_2OH $(CHOH)_4CH_2OH$.

Talka Gum. See SUAKIN GUM.

Talkeisenerz. Synonym for Magnesio-magnetite.

Talk-Knebelite. A mineral. It is $4[(Mn,Fe,Mg)_2SiO_4]$.

Talktriplite. A mineral. It is $8[(Mn,Fe,Mg,Ca)_2PO_4F]$.

Tallingite. A mineral. It is a copper oxychloride, $CuCl_2.4Cu(OH)_2.4H_2O$, found in Cornwall.

Talloel. A Swedish liquid resin obtained as a by-product in the production of cellulose from Swedish fir by the soda process. It is stated to consist mainly of resin acids, and is closely related to rosin. It contains 87·5 per cent. resin acids, 8 per cent. unsaponifiable matter, and 3 per cent. oxy-acids. It has an acid value of 171. Also see LIQUID RESINS.

Tall Oil. A by-product of sulphate pulp manufacture. It contains 2·2 per cent. of material soluble in petroleum ether, 12·4 per cent. unsaponifiable matter, 30·4 per cent. resin acid, and 54·9 per cent. fatty acids. The resin acid consists of abietic acid, and the fatty acids contain oleic, linoleic, and linolenic acids.

Tallow. The solid fat of oxen (beef tallow), and sheep (mutton tallow). It consists of tristearin, tripalmitin, and triolein.

Tallow Clays. Clays containing varying proportions of zinc silicate.

Tallow, Mineral. See BITUMEN.

Tallow, Mountain. See BITUMEN.

Tallow, Rock. See BITUMEN.

Tallow Seed Oil. Stillingia oil, obtained from the seeds of *Stillingia sebifera*.

Tallow Soap. Curd soap.

Tallow, Synthetic. See CANDELITE.

Talotalo Gum (Kau Drega). A gum somewhat resembling gutta-percha, from Fiji.

Ta-Lou. The Chinese term for a glass flux used for enamelling on porcelain. It is mainly a silicate of lead with a little copper.

Talmi Gold. See ABYSSINIAN GOLD.

Taloximine. A respiratory stimulant. 4 - (2 - Dimethylaminoethoxy) - 1, 2 - dihydro-1-hydroxyiminophthalazine.

Talusin. A proprietary trade name for Proscillaridin.

Talwaan. A tanning material. It is the root of *Elephantorrhiza burchelli*.

Tamarac. The dried bark of *Larix laricina*, an American larch. The extract is used as an astringent and stimulant.

Tamarinds. The fruit of *Tamarindus indica* preserved with sugar. The pulp contains about 13 per cent. tartaric acid, and is a mild purgative.

Tamarite (Chalcophyllite, Copper Mica). A mineral. It is a hydrated copper arsenate, $Cu_3As_2O_8.5Cu(OH)_2+H_2O$, found in Cornwall.

Tamarugite. A mineral, $Na_2SO_4.Al_2O_3.$ $12H_2O$.

Tambac. See TOMBAC.

Tambookie Grass. The product of *Hyperrhenice glauca*. It is stated to be suitable for paper-making.

Tamclad 7200. A proprietary PVC organosol. (949).

Tamguard 840, 840H and 840S. A proprietary range of PVC plastisols used for coating electroplating racks. Their Shore hardnesses are A90, D35 and A70 respectively. (949).

Tammite. A mineral containing iron, manganese, and tungsten.

Tamol. A combination of formaldehyde and naphthalene-sulphonic acid. Used as a precipitant for the production of lakes, from basic dyestuffs.

Tamoxifen. A drug used in the treatment of mammary carcinoma. It is 1 - *p* - β - dimethylaminoethoxyphenyl - *trans* - 1, 2 - diphenylbut - 1 - ene. NOLVADEX is the citrate.

Tampicin. A resin, $C_{34}H_{54}O_{14}$, obtained from *Ipomœa simulans*.

Tampovagan. A proprietary vaginal pessary containing stilbœstrol and lactic acid used in the treatment of atrophic vaginitis. (370).

Tampovagan N. A proprietary vaginal pessary containing NEOMYCIN, used in the treatment of candidiasis. (370).

Tampovagan P.S.S. A proprietary vaginal pessary containing penicillin, sulphanilamide and sulphathiazole, used in the treatment of vaginal infections. (370).

Tamtam. An alloy of 78 per cent. copper and 22 per cent. tin.

Tanacetone. Thujone, $C_{10}H_{16}O$.

Tanacetyl. Tannin acetic ester.

Tanalith. A wood preservative against insects and fungus. (562).

Tanargan (Tannargentan). A combination of tannin and silver albuminate. Used as an intestinal astringent.

Tanargentan. See TANARGAN.

Tanatarite. A name proposed for a mineral occurring in crevices in the chromite near Kairakty, Russian Central Asia. It contains 74·25 per cent. Al_2O_3, 1·44 per cent. Fe_2O_3, 1·72 per cent. MgO, 3·09 per cent. CaO, 3·72 per cent. SiO_2, and 15 per cent. H_2O. It has previously been called Kayserite.

Tanatol. See UBA.

Tancolin. A proprietary preparation of dextromethorphan hydrobromide, theophylline, sodium citrate, citric acid, ascorbic acid and glycerin. A cough linctus. (345).

Tandacote. A proprietary preparation of OXYPHENBUTAZONE. An anti-inflammatory drug. (17).

Tandalgesic. A proprietary preparation of OXYPHENBUTAZONE and PARACETAMOL. An inti-inflammatory and analgesic drug. (17).

Tanderil. A proprietary preparation of oxyphenbutazone. (17).

Tanekaha. The bark of *Phyllocladus trichomanoides*. Used in tanning leather.

Tanformal. Claimed to be a chemical combination of tannin, phenol, and formaldehyde. An intestinal astringent and antiseptic.

Tangeite. A crystalline mineral, $2CuO.2CaO.V_2O_5.H_2O$, from Tyuya-Muyun, Fergana. The name Turkestan Volborthite is retained for colloidal varieties.

Tangkawang Fat. See BORNEO TALLOW.

Taninol. A mordant for basic dyestuffs. (512).

Tanked Oil. Linseed oil from which the moisture and other matter has settled out. It has a higher value than the ordinary oil.

Tank Waste. See ALKALI WASTE.

Tannacetin. See ACETANNIN.

Tannafax. Tannic acid jelly. (631).

Tannal (Tannalum). A basic aluminium tannate, $Al(OH)_2 . (C_{14}H_9O_9) + 5H_2O$. An astringent used as a dusting-powder.

Tannalbine. An albumen compound of tannic acid. Used as an intestinal astringent in catarrh.

Tannaline Films. Gelatin films hardened by formaldehyde, used for photographic purposes.

Tannalum. See TANNAL.

Tannaphthol. A condensation product of tannin albuminate and benzonaphthol.

Tannargentan. See TANARGAN.

Tannate, Creosote. See TANOSAL.

Tannenite. A mineral. It is a bismuth-copper sulphide, $Cu_2S.Bi_2S_3$, found in the Erzgebirge.

Tanner's Wool (Glover's Wool). A wool pulled from the carcases of slaughtered sheep with the assistance of lime. It does not dye well.

Tannexin. See OREXINE.

Tannia. Tannic acid substitute. (631).

Tannic Acid. See GALLOTANNIC ACID.

Tannic Indigo. See GALLANILIC BLUE.

Tannigen. Triacetyl-tannin, $C_{14}H_7O_9$ $(CO.CH_3)_3$. Used as an intestinal astringent in chronic diarrhœa.

Tanning Grease. See DÉGRAS.

Tannin Heliotrope (Heliotrope B, 2B, Basic Heliotrope B). A dyestuff. It is dimethyl - di - amino - xylyl - xylo, phenazonium - chloride, $C_{24}H_{27}N_4Cl$. Dyes tannined cotton reddish-violet and is employed in calico printing.

Tannin Indigo. See GALLANILIC BLUE.

Tannin Orange R. A dyestuff. It is exo-dimethyl - amino - toluene - azo - β - naphthol, $C_{19}H_{19}N_3O$. Dyes leather and tannined cotton orange.

Tanninphenolmethane. A combination of tannin, formaldehyde, and phenol.

Tannin, Synthetic. See NERADOL.

Tanninthymolmethane. See TANNOTHYMAL.

Tannismuth. Bismuth bitannate.

Tannisol. Methylene-ditannin, $(C_{14}H_9O_9)_2CH_2$. An astringent and antiseptic.

Tannobromine. A compound obtained from formaldehyde and dibrom-tannic acid. Prescribed internally for diseases of the stomach.

Tannocasum. A combination of casein and tannin. An intestinal astringent.

Tannochrome. A compound of chromium oxide, tannin, and resorcin. Said to be used for the skin.

Tannocol (Tannogelatin). A compound of glue and tannin, used as an intestinal astringent.

Tannocreosoform. A condensation product of tannin, formaldehyde, and creosote. An antiseptic.

Tannocreosote. A condensation product of creosote and tannin.

Tannoform (Helgotan, Formotan). A condensation product of tannin and formaldehyde, $CH_2(C_{14}H_9O_9)_2$. An astringent and antiseptic.

Tannogelatin. Tannocol (*q.v.*).

Tannoguaiaform. A condensation product of tannin, formaldehyde, and guaiacol. An intestinal astringent and antiseptic.

Tannone (Tannopine). Hexamethylene-tetramine-tannin, $(C_{14}H_{10}O_9)_3.(CH_2)_6N_4$. An intestinal astringent.

Tannopine. See TANNONE.

Tannothymal (Tanninthymolmethane). A condensation product of thymol, formaldehyde, and tannin. An intestinal astringent.

Tannoxyl. A compound of tannin and oxychlor-casein.

Tannurgyl. Vanadium and manganese albuminate.

Tannyl. Oxy-chloro-casein-tannate. Used in diarrhœa.

Tanocol. See TANNOCOL.

Tanolin. A proprietary trade name for a basic chromium chloride for use in chrome tanning baths.

Tanosal (Creosote Tannate, Creosal, Kreosal). The tannic acid ester of creosote, obtained by the action of phosgene upon a mixture of tannin and creosote. Used as a substitute for creosote, especially in the treatment of consumption.

Tanret's Reagent. To a solution of 1·35 grams mercuric chloride in 25 c.c. water is added a solution of 3·32 grams potassium iodide in 25 c.c. water. This is made up to 60 c.c. with water and 20 c.c. glacial acetic acid.

Tansel. A specially prepared salt for curing hides.

Tantalite. A mineral, $FeO.Ta_2O_5$, but usually some of the tantalum is replaced by niobium, and some of the iron by manganese. It is a source of tantalum.

Tantalohatchettolite. A mineral. It is $8[(Ca,U)_2(Ta,Cb)_2(O,OH)_7]$.

Tantalorutile. Synonym for Ilmenorutile.

Tantalum Cassiterite. See AINALITE.

Tantcopper. A copper alloy analogous to tantiron.

Tantiron. An alloy of 84 per cent. iron, 15 per cent. silicon, and 1 per cent. carbon. It has a specific gravity of 6·8 and is acid-resisting.

Tantnickel. A nickel alloy analogous to tantiron.

Tanzite. A mineral. It is bismuth arsenantimonate.

Tao. A proprietary preparation of tri-aceto-oleandromycin. An antibiotic. (817).

Taoffeite. A mineral. It is $4[BeMg Al_4O_8]$.

Tapalpite. A mineral. It is a silver-bismuth sulpho-telluride, $Ag_3Bi(S.Te)_3$.

Tapazole. A proprietary trade name for Methimazole.

Tap Cinder. The basic silicate of iron constituting the slag, and flowing through the tap-hole of the puddling furnace.

Taphosote. Tanno-phosphate of creosote. An antiseptic.

Tapiolite. A mineral. It consists mainly of iron tantalite. $2[Fe(Ta,Cb)_2O_6]$.

Tara. The tannin from the pods of *Cæsalpinia tinctoria*.

Taractan. A proprietary trade name for Chlorprothixene.

Taramellite. A mineral. It is a basic silicate of barium and iron.

Taramite. A mineral. It is $(Ca,Na,K)_3 (Fe'',Fe''')_5(SiAl)_8O_{22}(OH)_2$.

Taranakite. A mineral. It is $KAl_3 (PO_4)_3OH.8\frac{1}{2}H_2O$.

Tarapacaite. A mineral. It is potassium chromate, K_2CrO_4.

Tar, Asphalt. See GOUDRON.

Taraspite. A mineral. It is a variety of dolomite.

Tarband. A proprietary bandage impregnated with zinc oxide and coal tar paste, used in the treatment of eczema. (484).

Tarbisol I. See LUATOL B.

Tarbisol II. An aqueous solution of diethylamine-tartro-bismuthate. An antisyphilitic.

Tar, Bone. See BONE OIL.

Tarbuttite. A mineral, $Zn_2(OH)PO_4$.

Tar Camphor (Naphthalin, Naphtalin). Naphthalene, $C_{10}H_{18}$.

Tar, Candle. See STEARIN PITCH.

Tarcortin. A proprietary preparation of coal tar and hydrocortisone for dermatological use. (376).

Tardrox. A proprietary preparation of chlorhydroxyquinolone and tar for dermatological use. (246).

Targesin. A German preparation containing 6 per cent. silver. Used in medicine.

Targionite. A mineral. It is a variety of galena containing antimony.

Targite. A mineral. It is a hydrated ferric oxide, $2Fe_2O_3.H_2O$.

Tari. See WHITE TAN.

Tarmac. A proprietary preparation of blast furnace slag, refined tar, and other ingredients. Used for road dressing.

Tarmex. A proprietary name for a combination of prepared tar and Mexphalte. Used for road dressing.

Tarnovicite. See TARNOWITZITE.

Tarnowitzite (Plumbocalcite). A mineral, $(Ca.Pb)CO_3$.

Tar Oil. See OIL OF TAR.

Tarola. A coal-tar product used as a sheep dip.

Tarragon Vinegar. Tarragon leaves are macerated with vinegar, allowed to stand, and filtered. It is used as a flavouring agent in salads and sauces.

Tar, Regenerated. A mixture of pitch and anthracene oil which does not crystallise.

Tar, Skelleftea. See SKELLEFTEA.

Tarslag. A proprietary preparation of cold blast slag which has been treated with a bituminous compound. Used as a road dressing.

Tar Soap. Soaps of this type are made with the addition of tar, usually juniper tar.

Tar Spirit. Benzene. See OIL OF TAR.

Tartar. See ARGOL.

Tartar Cake. See SUPERARGOL.

Tartar, Chalybeated. Tartrated iron (q.v.).

Tartar, Crude. See ARGOL.

Tartar Emetic (Tartrated Antimony).

Potassium-antimonyl-tartrate, $C_4H_4O_6$ (SbO)$K+\frac{1}{2}H_2O$. Used in medicine as an emetic, and in dyeing as a mordant.

Tartar Emetic (Tartar Emetic Substitute, Antimony Mordant). Mixtures of tartar emetic and zinc sulphate.

Tartar Emetic Substitute. See TARTAR EMETIC POWDER.

Tartarised Borax. Potassium borotartrate.

Tartarline. Potassium bisulphate, used as a substitute for tartaric acid for industrial purposes.

Tartar, Refined. See CREAM OF TARTAR.

Tartars. Raw materials which contain more than 40 per cent. tartaric acid are termed tartars.

Tartar, Salt of. See POTASH.

Tartar, Stibiated. Tartar emetic (q.v.).

Tartar Substitute. See SUPERARGOL.

Tartar, Vitriolated. Potassium sulphate, K_2SO_4.

Tartar Yellow FS, FS Conc. Dyestuffs. They are British equivalents of Tartrazine.

Tar Tea. Tar water.

Tartrachromin G.G. See ALIZARIN YELLOW 5G.

Tartrated Antimony. See TARTAR EMETIC.

Tartrated Iron. Iron and potassium tartrate. Used medicinally.

Tartrated Soda. See ROCHELLE SALT.

Tartrated Soda Powder, Effervescent. See SEIDLITZ POWDER.

Tartrate of Lime. See LIMO.

Tartratol Yellow L. A British brand of Tartrazine.

Tartrazine (Acid Yellow 79210, Hydroxine Yellow G. L. L Conc, Tartar Yellow FS, FS Conc, Tartratol Yellow L, Tartrine Yellow O). A dyestuff. It is the sodium salt of dihydroxytartrate-diphenyl-osazone-disulphonic acid, $C_{16}H_{10}O_{10}N_4S_2Na_4$. Dyes silk and wool.

Tartrine Yellow O. A British equivalent of Tartrazine.

Tarvia. A proprietary trade name for a specially refined coal-tar.

Tarwar. The bark of *Cassia auriculate*. A tanning material.

Tasar Silk. See TUSSAR SILK.

Tasajo. A salted and pressed meat containing 23 per cent. albuminoids and 45 per cent. fat.

Tasch. An abbreviation for tuberculin-antibody-Scheitlin, a proprietary substance which, according to Scheitlin, contains a small amount of Koch tuberculin, antituberculous serum, and $1:2:3$-$C_6H_3SO_3H.OH.OCH_3$.

Tasprin. A proprietary preparation of soluble aspirin. (434).

Tasteless Quinine. See INSIPIN.

Tasteless Salts. Sodium phosphate, HNa_2PO_4.

Taumasthman. A proprietary preparation of THEOPHYLLINE, PHENAZONE, caffeine, EPHEDRINE and atropine used in the treatment of asthma. (255).

Taurin. Amino-ethane sulphonic acid.

Tauriscite. A mineral. It is $4[FeSO_4.7H_2O]$.

Tauroflex. See TAUROLIN.

Taurolin. An antibacterial agent. $4\text{-}4'\text{-}$ Methylenedi(tetrahydro-1, 2, 4-thiadiazine-1-dioxide). TAUROFLEX.

Taurultam. An antibacterial and antifungal agent. It is tetrahydro-1,2,4-thiadiazine 1,1-dioxide.

Tautocline. A mineral. It is a variety of Dolomite.

Tavegil. A proprietary preparation of CLEMASTINE hydrogen fumarate, used in the treatment of allergic rhinitis. (244).

Tavistockite. A mineral, $Al_2Ca_3(OH)_6$ $(PO_4)_2$.

Tavorite. A mineral. It is $2[Li,Fe\text{'''} PO_4OH]$.

Tawmawite. A mineral. It is a variety of epidote.

Taxol. A proprietary preparation of pancreatin, bile salts, aloes and AGAR used in the treatment of constipation. (248).

Taylor. A proprietary trade name for a phenol-formaldehyde synthetic resin laminated product.

Taylorite. See BENTONITE.

Taylorite. A mineral. It is $4(K,NH_4)_2 SO_4]$. See BENTONITE.

Taylor Oil. A patented binding material obtained by boiling raw linseed oil with driers (litharge), then forcing air through the oil when heated to $300°$ F., and finally heating it for some time at 500–$600°$ F. It is a suitable cement for cork carpets.

Taylor Solder. An alloy of 60 per cent. tin, 12 per cent. lead, 12 per cent. silver, 8 per cent. zinc, 4 per cent aluminium, and 4 per cent. copper.

Tazoline. A proprietary preparation of antazoline hydrochloride, octaphonium chloride, titanium dioxide and calamine. (273).

T.B.T.O. A registered trade mark for tri-n-butyltin oxide used as a timber preservative. (24).

T.C.A. A rubber vulcanisation accelerator. It is thiocarbanilide.

Tcha-Lau. A blue powder containing copper. Used by the Chinese for obtaining a blue colour on porcelain.

TC-Harz. A proprietary trade name for a polymer of coumarone-indene and cyclopentadiene. Used as a substitute for shellac. (3).

T.C.P. An aqueous solution of trichloro-phenyl-iodo-methyl-salicyl. An antiseptic and germicide.

T.C.P. (Plastic X, Plastol X). Trade names for tricresyl phosphate, a plasticiser for cellulose lacquers, and polyvinylchloride. It has a specific gravity of $1 \cdot 185$–$1 \cdot 189$, a boiling range of 430–440° C., and a flash-point of 215° C. The term T.C.P. appears to be also applied to an aqueous solution of trichloro-phenyl-iodo-methyl-salicyl, an antiseptic and germicide.

Tea, Abyssinian. See ARABIAN TEA.

Teaberry Oil. Methyl salicylate, C_6H_4 OH.COO.CH$_3$.

Tea, Carolina. See APPALACHIAN TEA.

Tea, Jesuit's. See MATÉ.

Tea-Lead. An alloy of from 97–99 per cent. lead and 1–3 per cent. zinc. Also an alloy of lead with 2 per cent. tin used for wrapping tea.

Teallite. A mineral, PbS.SnS. It is also given as 2[PbSnS$_2$].

Teal Oil. See GINGELLY OIL.

Tea Oil. See OIL OF TEA.

Tea, St. Bartholomew. See MATÉ.

Teatcote Plus. A proprietary preparation of POLYHEXANIDE. A veterinary antibacterial.

Tebelon. A German commercial preparation, the chief constituent of which is an iodine derivative of thymol. It is stated to be a valuable therapeutic agent in pulmonary tuberculosis.

Tebethion. p-acetylaminobenzaldehyde thiosemicarbazone.

Tec. A proprietary trade name for cellulose acetate varnish resins.

Teca. A proprietary trade name for an acetate fibre of special form.

Tecaldrine. A preparatory preparation of methapyrilene hydrochloride, dextromethorphan hydrobromide, ammonium chloride, ipecacuanha, menthol and syrup. A cough linctus. (311).

Tecali Onyx. See ONYX OF TECALI.

Técarine. Sodium - theobromine - N - acetate.

Tec-Char. A proprietary trade name for a granular charcoal.

Technical Casein. See LACTARIN.

Teclothiazide. 6-Chloro-3,4-dihydro-3-trichloromethylbenzo - 1,2,4 - thiadiazine-7-sulphonamide 1,1-dioxide. Deplet is the potassium derivative.

Tecquinol. A proprietary anti-oxidant. Hydroquinone. (242).

Tedlar. A proprietary trade name for a clear or pigmented polyvinyl fluoride film. It has high resistance to weathering and is generally chemically inert. (10).

Tedral. A proprietary preparation of theophylline, ephedrine hydrochloride, and phenobarbitone. A bronchial antispasmodic. (262).

Teejel. A proprietary preparation of choline salicylate and cetalkonium chloride. (334).

Teel Oil. See GINGELLY OIL and TECHNICAL CASEIN.

Teepleite. A mineral. It is 2[Na$_2$BO$_2$ Cl.2H$_2$O].

Teerlack. Coal tar pitch.

Teevax. A proprietary preparation of CROTAMITON and HALOPYRAMINE used in the treatment of pruritis. (17).

Teflon. A proprietary polytetrafluoro-ethylene (P.T.F.E.) plastic material having good resistance to high temperatures. (10).

Teflon F.E.P. A proprietary range of hexafluoropropylene copolymers. (10).

Tefzel. A copolymer of ethylene and tetrafluoroethylene for wire and cable insulation.

Tefzel 200. A proprietary ETFE fluoropolymer resin extruded for use as wire insulation. (10).

Tegin. A patented preparation. It is a neutral ester closely related to the natural fats. It is wax-like and melts at 57° C. Recommended as a salve base.

Teglac. A proprietary trade name for an alkyd synthetic varnish and lacquer resin.

Tegmin. An emulsion made from 1 part yellow wax, 2 parts acacia, and 3 parts water. It also contains 5 per cent. zinc oxide, and a little wool fat. Used as a surgical dressing.

Tego Films. Proprietary products. They are tough films produced by impregnating very thin transparent paper with adhesive material along with cresol and formaldehyde. The papers are placed between layers of wood and subjected to heat and pressure. Condensation reactions take place, and very tough films are produced.

Tegoglätte. A litharge having smaller particles than the ordinary type. Used in rubber mixings.

Tegofan. See ALLOPRENE.

Tegopen. A proprietary preparation of cloxacillin. An antibiotic. (324).

Tegretol. A proprietary preparation of carbamezepine. An analgesic. (17).

Tegul. A proprietary sulphur jointing compound for bell and spigot pipes. It contains sulphur and sand.

Teineite. A mineral. It is Cu(Te,S)O$_4$. 2H$_2$O.

Teka Oil. A proprietary trade name for an extract from stand oil ($q.v.$) from which bases and acids have been removed.

Teknon. A proprietary trade name for polyurethane moulding materials. (241).

Telconax. A patented insulating compound made from selected bitumen, waxes, and rubber. It has extremely good dielectric properties, which are unaffected by moisture, mild acids, alkalis, or other chemical liquors. It resists well attack by ozone and ultra-violet light, and is useful for insulation or as a protective sheathing.

Telconite. A proprietary insulating material made in various colours.

Telconstan. A non-magnetic nickel-copper alloy prepared in induction furnaces and having exceptional purity and very low temperature coefficient of resistance. It is used for resistances where standard of resistance with temperature is important.

Telcothene. A registered trade name for polythene powder, tube and sheet. (178).

Telcovin. A registered trade name for polyvinyl chloride tube and sheet. (178).

Teleblock. A proprietary range of thermoplastic rubbers. (159).

Telegraph Bronze (Telegraph Metal, Electric Metal). An alloy of 80 per cent. copper, 7·5 per cent. lead, 7·5 per cent. zinc, and 5 per cent. tin.

Telemarkite. A mineral. It is a variety of grossularite.

Telepaque. A proprietary trade name for Iopanoic acid.

Telloy. A proprietary product. It is stated to be a form of elementary tellurium specially pulverised and purified for use as a rubber vulcanising agent. It has a specific gravity of 6·27.

Tellurbismuth. A mineral. It is $[Bi_2Te_3]$.

Telluretted Hydrogen. Hydrogen telluride, H_2Te.

Telluric Bismuth. See TETRADYMITE.

Tellurite (Tellurium Ochre). A mineral. It is tellurium oxide, $8[TeO_2]$.

Tellurium, Black. See MAGYAGITE.

Tellurium, Graphic. See SYLVANITE.

Tellurium Lead. An alloy containing 0·05 per cent. tellurium with lead. It resists sulphuric acid.

Tellurium Ochre. See TELLURITE.

Tellurium Silver. See HESSITE.

Tellurobismuth. A mineral, Bi_2Te_3.

Telmid. A proprietary trade name for the iodide of Dithiazanine.

Telsit. A gelatin explosive containing from 10–15 per cent. dinitro-toluene or liquid trinitro-toluene.

Teluran. A registered trade name for acrylonitrile - butadiene - styrene polymers and related products. (49).

Temazepam. A tranquilliser. 7-Chloro-1, 3 - dihydro - 3 - hydroxy - 1 methyl - 5 - phenyl - $2H$ - 1, 4 - benzodiazepin - 2-one.

Temex. A barium/zinc complex plasticiser for PVC. (572).

Temiskamite. A mineral, Ni_4As_3, found in mines in Ontario.

Temlock. A proprietary trade name for a board made from wood fibres impregnated with resin and subjected to pressure.

Tempaloy. A patented alloy of approximate composition, 95 per cent. copper, 4 per cent. nickel, and 1 per cent. silicon.

Temper. Alloys of arsenic and lead or copper, and tin. Used as hardening materials for shot or pewter.

Tempered Lead. Noheet metal (q.v.).

Temperite. A proprietary trade name for calcium chloride anti-freeze for concrete.

Temperite Alloys. A proprietary trade name for alloys of lead, tin, and cadmium.

Tempo. Proprietary cellulose esters.

Tenacit (Kronit). Ebonite substitutes made from asbestos with a gum binder.

Tenacite. See BAKELITE.

Tenacity. Fluxes for silver alloy brazing. (615).

Tenalan. An insulating material for power currents. It is stated to have properties equal to bakelite. It is a German synthetic resin made by a protected dry process from resorcinol and paraform.

Tenamine 1. A proprietary anti-oxidant. N-butylated-p-aminophenol. (242).

Tenamine 2. A proprietary anti-oxidant. N,N¹-di-sec-butyl-p-phenylene di-amine. (242).

Tenamine 3. A proprietary anti-oxidant. 2,6-di-$tert$-butyl-p-cresol. (242).

Tenasco. A proprietary trade name for synthetic fibre resembling Nylon.

Tenasco Fibre. A proprietary trade name for a fibre obtained by stretching viscose fibre when in a plastic condition.

Tenavoid. A proprietary preparation of bendrofluazide and meprobamate. Prophylaxis for premenstrual syndrome. (308).

Tenaxatex VA 632. A proprietary trade name for a high molecular weight vinyl acetate homopolar water emulsion containing 55 per cent. solids. Used as an adhesive base. (199).

Tenaxatex VA 956. A proprietary trade name for a vinyl acetate/acrylate co-polymer emulsion containing 55 per cent. solids. A medium viscosity adhesive base. VA 957 and VA 959 are as above but are low viscosity materials. (199).

Tenax Metal. A zinc alloy containing from 0·35–2·56 per cent. copper, 0·2–4·42 per cent. aluminium, 0–0·35 per cent. iron, and up to 1·2 per cent. lead. Used for the manufacture of guide rings.

Tenazit. A proprietary trade name for laminated bakelite or similar synthetic resin.

Tenebryl. Di-iodo-methane-sodium sulphonate.

Tenefoil S. A proprietary trade name for ionomer resin (*q.v.*). (71).

Tenephrol. A proprietary preparation of lithium iodide (31 per cent. solution).

Tengerite. A mineral which is said to be yttrium carbonate.

Tenite 1. A proprietary cellulose acetate moulding compound.

Tenite 2. A proprietary trade name for a cellulose acetate-butyrate moulding compound.

Tenite 7 D.R.D. A proprietary polyethylene terephthalate resin used for injection moulding. (214).

Tennal. A proprietary trade name for certain aluminium alloys for casting purposes.

Tennanite. A mineral, $4Cu_2S.As_2S_3$, with some iron.

Tennant's Salt. Chlorinated lime.

Tennessee Phosphates. Mineral phosphates containing from 60–70 per cent. calcium phosphate. Fertilisers.

Tenorite. A mineral. It consists of copper oxide, CuO.

Tenormal. A proprietary trade name for the hydrogen tartrate of Pempidine.

Tenosin. A solution containing 0·0005 gram β-iminazolyl-ethylamine and 0·02 gram p-hydroxy-phenyl-ethylamine per c.c.

Tenox T.B.H.Q. A proprietary antioxidant used to stabilise oils, fats and foods against oxidative deterioration. (214/242).

Tensanyl. A proprietary preparation of bendrofluazide, reserpine, and potassium chloride. Antihypertensive. (308).

Tensilac 39. A proprietary rubber vulcanisation accelerator. It is a soft form of ethylidene-aniline.

Tensilac 40. A proprietary rubber vulcanisation accelerator. It is a resinous condensation product.

Tensilac 41. A proprietary rubber vulcanisation accelerator. It is a hard form of ethylidene-aniline.

Tensilite. An aluminium bronze. It contains from 64–67 per cent. copper, 3·1–4·4 per cent. aluminium, 0–1·2 per cent. iron, 2·5–3·8 per cent. manganese, and 24–29 per cent. zinc.

Tensilon. A proprietary preparation of edrophonium chloride used in the diagnosis of myasthenia gravis. (314).

Tensloy. A proprietary trade name for an alloy of iron with approximately 1·5 per cent. nickel and 0·5 per cent. chromium.

Tensol. (1) A proprietary trade name for a dispersing and emulsifying agent containing a sulphonated ether.

Tensol. (2) Cements for vinyl and acrylic sheets. (512).

Tensorub. A proprietary fast-vulcanising rubber. (895).

Tentor. A proprietary preparation of phenylbutazone. (368).

Tenuate. A proprietary preparation of diethylpropion hydrochloride. An antiobesity agent. (263).

Tenuate Dopan. A proprietary preparation of diethylpropion hydrochloride in a slow release base. (263).

Teoquil. A proprietary trade name for Hedaquinium chloride.

Tephroite. A mineral. It is manganese silicate, $2MnO.SiO_2$.

Tepperite. A proprietary polystyrol denture.

Tequila. See MESCAL.

Terbufos. A pesticide. S-*tert*-butylthiomethyl OD-diethyl phosphoro = dithioate.

Terbutaline. A bronchodilator. 1-(3, 5 - Dihydroxyphenyl) - 2 - (t - butyl - amino)ethanol. BRICANYL is the sulphate.

Tercin. A proprietary preparation of aspirin, phenacetin and butobarbitone. An analgesic. (248).

Tercoda G.W. A proprietary preparation of codeine phosphate, terpin hydrate, eucalyptol, menthol and oil of peppermint. A cough linctus. (259).

Tercod Bricks. A special type of silicon carbide brick with a borosilicate glaze to prevent oxidation. A refractory material.

Terebene. French or American turpentine is mixed with 5 per cent. sulphuric acid. It is left some time until the action on polarised light is destroyed, and then steam distilled. The distillate is washed and redistilled. It has a specific gravity of 0·862–0·866, and boils at 156–180° C. It consists of a mixture of dipentene and other hydrocarbons, and is used as a substitute for turpentine.

Terebenthene. Pinene, $C_{10}H_{16}$.

Terebine (Liquid Drier, Japan Drier). Made by heating oxides of lead and manganese with linseed oil or rosin, or mixtures of the oil and rosin, and thinning with turpentine or turpentine substitute. A drier for paints. It is not to be confused with Terebene.

Terephane. A proprietary name for polyethylene terephthalate film. (950).

Terephthal Brilliant Green. A dyestuff, $C_{48}H_{60}N_4Cl_2.3ZnCl_2$, prepared from phthalyl chloride, diethyl-aniline, and zinc chloride. Dyes wool and silk yellowish-green.

Terephthal Green. A dyestuff, $C_{40}H_{44}N_4Cl_2.3ZnCl_2$, prepared from phthalyl chloride, dimethyl-aniline, and zinc chloride. Dyes wool and silk yellowish-green.

Tergitol S. A proprietary trade name for a series of biodegradable non-ionic intermediates comprising ethoxylates and ethoxysulphates of linear secondary alcohols. Used in the production of biodegradable detergents. (34).

Tergitols. A proprietary trade name for wetting agents consisting of the sodium salts of the sulphates of higher alcohols.

Terlinguaite. A mineral. It is mercury oxychloride, Hg_2ClO.

Terlon. A proprietary aromatic polyamide used in the forming of film. (424).

Terluran 846 L. A proprietary A.B.S. of medium rigidity and toughness used for injection moulding, extrusion and thermoforming. (49).

Terluran 886. A tough grade of A.B.S. (49).

Terluran 8760 Galvano. A special grade of A.B.S. used for electroplating. (49).

Termierite. A mineral, $Al_2O_3.6SiO_2.18H_2O$.

Ternary Steels. Alloy steels containing one special element in addition to the iron and carbon.

Terne Metal. An alloy of 80 per cent. lead, 18 per cent. tin, and 2 per cent. antimony.

Terne Plate. An alloy of lead and tin, coated on iron plate, and intended for use in roofing.

Ternovskite. A mineral. It is $2[Na_2(Mg,Fe^{..})_3Fe_2^{...}Si_8O_{22}(OH)_2]$.

Teronac. A proprietary preparation of MAZINDOL used in the treatment of obesity. (244).

Terpalin. A proprietary preparation of EUCALYPTOL, menthol, TERPINE hydrate and codeine phosphate. A cough syrup. (462).

Terpaln. A proprietary trade name for modified synthetic polyterpenes used as resin tackifiers. (93).

Terpestrol. A powder containing lactose with 5 per cent. oil of turpentine.

Terpex D,K-3,S. A proprietary trade name for terpene vinyl plasticisers. (122).

Terpigol. A proprietary trade name for terpinyl monoethylene glycol ether. (197).

Terpilenol. See LILACIN.

Terpine. A turpentine substitute. It is a product of the distillation of Borneo petroleum. It has similar physical properties to turpentine, a specific gravity of 0·81 at 15° C., and a boiling point of 142–188° C.

Terpinol. A mixture of terpenes containing terpinene, dipentene, also terpineol and cineol.

Terpoin. A proprietary preparation of eucalyptol, terpin hydrate, codeine phosphate, menthol and guaiphenesin. A cough linctus. (346).

Terposol No. 3. A proprietary trade name for a solvent consisting of terpene methyl ethers.

Terposol No. 8. A proprietary trade name for a solvent consisting of terpene glycol ethers.

Terpurile (Cycloran, Solventol). Wetting-out agents consisting of soaps with organic solvents.

Terra Alba. See GYPSUM.

Terra-Bron. A proprietary preparation of oxytetracycline, ipecacuanha and ephedrine hydrochloride. (85).

Terra Cariosa. Rotten Stone (q.v.).

Terra Catechu. See CUTCH.

Terra-Cortril. A proprietary preparation containing oxytetracycline hydrochloride, hydrocortisone acetate and polymixin B sulphate. (85).

Terra-Cotta. A building material made from clay.

Terra-Cotta F. A dyestuff. It is the sodium salt of primuline-azo-phenylene-diamine - azo - naphthalene - sulphonic acid. Dyes cotton brown from a neutral or alkaline bath.

Terra-Cotta G. See INGRAIN BROWN.

Terra-Cotta R. See ALIZARIN YELLOW R.

Terra di Sienna. See INDIAN RED and YELLOW OCHRE.

Terra Fullonica. Fuller's earth.

Terra Japonica. See CUTCH.

Terraline. See PARAFFIN, LIQUID.

Terra Merita. See TURMERIC.

Terramycin. A proprietary preparation containing Oxytetracycline, Oxytetracycline Hydrochloride. An antibiotic. (85).

Terra Nobilis. Bergmann's name for the diamond.

Terra Ponderosa. Barium sulphate, $BaSO_4$.

Terrar. A preparation from earthy zirconia, in Brazil. Used as an opacifying agent in enamels and glazes.

Terra-Systam. Systemic organo-phosphorus insecticides. (501).

Terra Verte. A pigment. It is a green earthy material found in the Mendip Hills. It consists of a species of ochre, and is essentially silica with oxide of iron and small quantities of other oxides. See also BOHEMIAN EARTH.

Terroline. See PARAFFIN, LIQUID.

Terry's Stain. A microscopic stain. It contains 20 c.c. of a 1 per cent. aqueous solution of methylene blue, 20 c.c. of a 1 per cent. solution of potassium carbonate, and 60 c.c. of water. This is boiled, cooled, and 10 c.c. of a 10 per cent. solution of acetic acid added, and the whole made up with water to 100 c.c.

Tertroxin. A proprietary preparation of liothyronine sodium. Thyroid hormone preparation. (335).

Tertschite. A mineral. It is $Ca_4B_{10}O_9$ $20H_2O$.

Tervelite. A mineral. It is a variety of dolomite.

Terylene. A synthetic polyester textile fibre, resistant to most dry cleaning solvents, possesses good wear resistance. It is polyethylene terephthalate produced from dimethyl terephthalate and ethylene glycol. A trade mark. (512).

Teschemacherite. A mineral, NH_4 HCO_3.

Tes P.P. A proprietary preparation of testosterone phenylpropionate. (316).

Tessalon. Benzonatate.

Tessan. A proprietary trade name for animal or vegetable fibre treated with gelatin.

Tesselated Ore. See GALENA.

Tesselite. A mineral. It is a variety of apophyllite.

Tessilite. A proprietary phenolic moulding material. (61).

Testalin. An aluminium soap, made by treating ordinary soap with aluminium sulphate. Used for the cementing together of sandstone to form a solid block.

Testifas Oil. A fraction of Russian petroleum distillation. It has a specific gravity of 0·820–0·823, and flashes at 38° C., or above. Used as a burning oil.

Testijodyl. A combination of coagulated blood, albumin, and iodine.

Testoral. A proprietary preparation of testosterone. Male sex hormone supplement. (316).

Tetalite. A mineral. It is a variety of calcite containing manganese. Synonym for Spartaite.

Tetanol. A proprietary preparation of calcium lævulinate.

Tetiothalein. Sodium-tetraiodo-phenolphthalein.

Tetjamer. An aluminium bronze. It contains from 86–93 per cent. copper, 5–10 per cent. aluminium, 1–3 per cent. silicon, and 0·72–0·98 per cent. iron.

Tetmosol. A proprietary preparation of MONOSULFIRAM in alcohol, used in the treatment of scabies. (2).

Tetra. See TETRAPHOSPHATE.

Tetra-Base-Paper. See TETRA-PAPER.

Tetrabenazine. 1,3,4,6,7,11b-Hexahydro - 3 - isobutyl - 9,10 - dimethoxy - benzo[a]quinolizin-2-one. Nitoman.

Tetrabid. A proprietary preparation of TETRACYCLINE hydrochloride. An antibiotic. (316).

Tetracarnit. A mixture of pyridine and its homologues with Turkey red oil or similar substances. It is used as a wetting-out agent to assist the penetration of textiles by liquids. Also see NEKAL and AVIVAN.

Tetrachel. A proprietary preparation of tetracycline hydrochloride. An antibiotic. (137).

Tetracol. A proprietary product. It is a specially purified carbon tetrachloride manufactured for internal administration.

Tetracosactide. See TETRACOSACTRIN.

Tetracosactrin. A corticotrophic peptide. β^{1-24}-Corticotrophin. CORTROSYN; COSYNTROPIN; SYNACTHEN; TETRACOSACTIDE.

Tetracycline. 4-Dimethylamino-1,4,4a, 5,5a,6,11,12a, octahydro - 3,6,10,12,12a-pentahydroxy - 6 - methyl - 1,11 - dioxonaphthacene-2-carboxamide. CLINETRIN, ECONOMYCIN, Telotrex, TETRACHEL, Tetragen and TOTOMYCIN are the hydrochloride. TETRACYCLINE is an antibiotic, present also in SIGMAMYCIN and Siltetrin, and in MYSTECLIN and Servicin as the hydrochloride.

Tetracycline Phosphate Complex. An antibiotic. A sparingly soluble complex of sodium metaphosphate and TETRACYCLINE. BRISTREX; TETREX. Present also in UROPOL.

Tetracyn. A proprietary preparation containing Tetracycline Hydrochloride. An antibiotic. (85).

Tetracyn P. A proprietary preparation containing Tetracycline Hydrochloride, Sodium Hexametaphosphate. An antibiotic. (85).

Tetracyn S.F. A proprietary preparation of TETRACYCLINE with vitamin supplements. An antibiotic. (85).

Tetradine. See LYCETOL.

Tetradymite (Telluric Bismuth). A mineral, Bi_2Te_3.

Tetraethyl Lead. A mobile liquid $Pb(C_2H_5)_4$. It is mixed with petrol to lower the rate of explosion in internal combustion engines, thus reducing the knock tendency of the fuel. See ETHYL (Trade Mark).

Tetraflon. A proprietary polytetrafluoroethylene (P.T.F.E.). (951).

Tetraform. A specially pure carbon tetrachloride.

Tetrahedral Garnet. Synonym for Helvine.

Tetrahedrite (Grey Copper Ore, Fahlore, Grey Silver, Nepaulite). A mineral. It is a sulphantimonite of copper, Cu_3SbS_3. The copper may be replaced by silver, iron, zinc, or mercury, and the antimony by arsenic. E.g. $8[Cu_3AsS_3]$.

Tetrahydrozolone. 2-(1,2,3,4-Tetrahydro-1-naphthyl) - 2 - imidazolone. Tetryzoline.

Tetra-Isol. A preparation of carbon tetrachloride, CCl_4, completely soluble in water. It is used as a cleaning agent in the textile industry.

Tetralin. Tetrahydro - naphthalene, $C_{10}H_{12}$. It is a solvent for gums, oils, waxes, and resins, and is used as a substitute for turpentine. It has a specific gravity of 0·975 at 20° C. and a boiling point of 206–208° C. Its melting-point is −27° to −30° C. and the flash-point is 78° C. The refractive index is 1·535–1·55.

Tetraline. An old name for tetrachlorethane.

Tetralin Extra. A mixture of Tetralin (*q.v.*) and Dekalin (*q.v.*).

Tetralitbenzol. A mixed fuel for internal combustion engines. It contains 50 per cent. benzol, 25 per cent. tetralin, and 25 per cent. of 95 per cent. alcohol.

Tetralite. See TETRYL.

Tetralol. Tetrahydro-β-naphthol. An antiseptic.

α-Tetralone. A synonym for α-keto-tetrahydro-naphthalene.

Tetralysal. A proprietary preparation containing Lymecycline. An antibiotic. (365).

Tetramethyl Base. Tetramethyl-diamino-diphenyl-methane, $(CH_3)_4.N_2.(C_6H_4)_2.CH_2$.

Tetramisole. An anthelmintic. (±)-2, 3, 5, 6 - Tetrahydro - 6 - phenyl - imidazo[2,1-*b*]thiazole. NILVERM is the hydrochloride.

Tetrane. See FURANE.

Tetranitrin. See TETRA-NITROL.

Tetra-nitrol (Tetranitrin, Butane Tetrol). Erythrol-tetranitrate, $C_4H_6(NO_3)_4$. Has been used in angina.

Tetranyl. An explosive. It is 2 : 3 : 4 : 6-tetranitro-aniline.

Tetra-paper (Tetra-base-paper). Paper which has been treated with dimethyl- or tetramethyl-*p*-phenylene-diamine. It is used in testing for ozone.

Tetraphenol. See FURANE.

Tetraphosphate (Tetra). Produced by mixing natural phosphate rock powder with 6 per cent. of a powder containing equal parts of the carbonates of calcium, sodium, and magnesium, with a little sulphate of soda. The mixture is roasted at from 600–800° C. After treating the product with cold phosphoric acid, it is then reduced to the required strength by mixing with dry earth and sand. It is a fertiliser free from acidity, causticity, and deterioration.

Tetrapol. See AVIVAN.

Tetrasilane. Silicobutane (tetrasilicon-decahydride), Si_4H_{10}. See SILICONES.

Tetrathal. Tetrachlorophthalic anhydride. A flame retardant resin intermediate used for manufacturing polyester, urethane and alkyd resins. (57).

Tetrazets. Antibiotic throat lozenges. A proprietary preparation of BACITRACIN, NEOMYCIN, TYROTHRICIN and benzocaine. (932).

Tetrazine (Hydrazine Yellow). A dyestuff. It is the sodium salt of benzene-azo - pyrazolone - carboxy - disulphonic acid, $C_{16}H_{10}N_4S_2O_9Na_2$. Dyes wool and silk yellow from an acid bath.

Tetrex. A proprietary preparation containing tetracycline phosphate complex. An antibiotic. (324).

Tetrex Bidcaps. A proprietary preparation containing Tetracycline Phosphate Complex. An antibiotic. (324).

Tetrex PMT. A proprietary trade name for Rolitetracycline.

Tetrol. See FURANE.

Tetron. Sodium pyrophosphate for detergents. (503).

Tetronal (Ethyl sulphonal). Diethyl-sulpho-diethyl-methane, $(C_2H_5)_2.C.(SO_2.C_2H_5)_2$. A hypnotic allied to Sulphonal.

Tetrone A. A proprietary trade name for a rubber vulcanisation accelerator. It is dipentamethylene-thiuram tetrasulphide.

Tetrophan. Dihydro-naphtho-acridine carbonic acid.

Tetryl (Tetralite). A trade name for trinitro - phenyl - methyl - nitramine, $C_6H_2(NO_2)_3.NCH_3.NO_2$, used in explosives.

A detonator known as " tetryl " contains 0·4 gram tetranitro-phenyl-methyl-nitramine, and 0·3 gram of a mixture of 87·5 per cent. mercury fulminate and 12·5 per cent. potassium chlorate.

Texaco BQ. An insecticide having a petroleum base. It is used for killing the boll weevil.

Texalin. See SAVONADE.

Texanol. A trade mark. 2,2,4-trimethyl - 1,3 - pentanediolmonoisobutyrate. An intermediate for the manufacture of plasticisers, surfactants, urethanes and pesticides. (242).

Texapon. See SAVONADE.

Texicote. (1) A proprietary range of vinyl emulsions used to provide surface coatings. (542).

(2) A self-cross-linking vinyl/acetate acrylic copolymer available in several grades as a binder in floor coverings, etc.

Texicote 105X and 134X. A proprietary trade name for acrylic copolymer emulsions. (94).

Texicryl. A proprietary range of self-cross-linking styrene/acrylate copolymers used in carpet manufacture. (94).

Texigel ET8. A proprietary trade name for a sodium polyacrylate dispersion. (94).

Texilac. Copolymer solutions for surface coatings. (542).

Texilac 230. A proprietary trade name for an acrylic copolymer solution in ethyl acetate. Used as a waterproofing agent for nylon, an adhesive for polyurethane foam and polyester film. (94).

Texileather. A proprietary trade name for pyroxylin-coated leather cloth.

Texilose Yarn. See XYLOLIN YARN.

Texin. A proprietary polyester-based polyurethane. (680).

Texiprint. Textile printing pastes. (542).

Texoderm. A cellulose product. It is an imitation leather.

Texofor A and B. A proprietary range of higher fatty alcohol-based polyoxyalkylene condensates used as non-ionic surfactants. (547).

Texofor C. A proprietary range of unsaturated fatty acid-based polyoxyalkylene condensates used as non-ionic surfactants. (547).

Texofor D. A proprietary range of glyceride oil-based polyoxyalkylene condensates used as non-ionic surfactants. (547).

Texofor E and ED. A proprietary range of saturated fatty acid-based polyoxyalkylone condensates used as non-ionic surfactants. (547).

Texofor FN and FP. A proprietary range of alkyl phenol-based polyoxyalkylene condensates used as non-ionic surfactants. (547).

Texofor G. A proprietary unsaturated fatty acid-based polyoxyalkylene condensate used as a non-ionic surfactant. (547).

Texofor J4. A proprietary bio-degradable non-ionic emulsifier. It is a linear fatty alcohol ethoxylate. (547).

Texofor M. A proprietary range of unsaturated fatty acid-based polyoxyalkylene condensates used as non-ionic surfactants. (547).

Texofor N. A proprietary range of fatty alcohol-based polyoxyalkylene condensates used as non-ionic surfactants. (547).

Texofor P. A proprietary range of complex amide-based polyoxyalkylene condensates used as non-ionic surfactants. (547).

Texofor T. A proprietary range of higher fatty alcohol-based polyoxyalkylene condensates used as non-ionic surfactants. (547).

Texowax. Polyoxyethylenes. (547).

Textase. A diastase preparation.

Textulite. A proprietary trade name for phenol formaldehyde laminated synthetic resin and moulded compounds.

Tfol. An argillaceous earth containing free gelatinous silica. Used in North America as a soap.

TG-8. Tri-ethylene glycol di-caprylate. (106).

T-gas. The commercial mixture of ethylene oxide and carbon dioxide. Used as an insecticide.

T.H.A. A proprietary preparation of TACRINE hydrochloride used as a narcotic antagonist. (650).

Thaban. See HALVA.

Thalackerite. A Greenland mineral. It is an anthophyllite.

Thalamonal. A proprietary preparation of fentanyl, and droperidol, used as a premedication in anæsthesia. (356).

Thalassan. A German product. It consists of dial and belladona, and is recommended against sea-sickness.

Thalazole. A proprietary preparation containing phthalylsulphathiazole. (336).

Thalenite. A mineral, $2Y_2O_3.4SiO_2.H_2O$.

Thalidomide. α-Phthalimidoglutarimide. Distaval.

Thalline. 6-Methoxy-tetrahydro-quinoline sulphate, $C_9H_9(OCH_3)NH.H_2SO_4$. A febrifuge.

Thallite. A mineral. It is a variety of epidote.

Thallium Alum. A double sulphate of thallium and aluminium, $Tl_2SO_4.Al_2(SO_4)_3.24H_2O$.

Thallochlore. Knop's and Schnedermann's name for the green colouring matter of lichens.

Thalo Blue No. 1. An alpha, solvent sensitive, red shade phthalocyanine blue pigment. Used for solventless printing inks. (243).

Thalo Blue No. 2. A beta, solvent stable, green shade phthalocyanine blue pigment. (243).

Thalo Green No. 1. A halogenated copper phthalocyanine green. Extremely light and heat-fast. (243).

Thanatol. See GUAIACOL ETHYL.

Thanite. A mineral. It is a mixture of halite and kainite.

Thapsia Resin. The resin of *Thapsia garganica* root. It contains caprylic acid and thapsic acid.

Thao. A gelatinous preparation made in Cochin China from seaweed. It has frequently appeared in England under the names of Japanese or Chinese isinglass, and is used for the same purposes as isinglass.

Tharandite. A mineral. It is a variety of Dolomite.

Thaumasite. A mineral. It is calcium carbonate, silicate, and sulphate, $CaSiO_3 . CaCo_3 . CaSO_4 . 15H_2O$. It is also given as $[Ca_6(CO_3)_2(SO_4)_2Si_2O_4 (OH)_4 . 27H_2O]$.

Thawpit. A preparation of carbon tetrachloride used for cleaning materials.

Thean 500. A proprietary preparation of proxyphylline. A bronchial antispasmodic. (338).

Thebacon. 7,8-Dihydrocodeinone enol acetate. Acetyldihydrocodeinone.

Theic. A proprietary preparation of tris-(2-hydroxyethyl)-isocyanurate used in the manufacture of heat-resistant wire lacquers. (49).

Theine (Guaranine). Caffeine, $C_8H_{10}N_4 O_2$, the principal alkaloid of tea and coffee. It is used in medicine.

Themalon. A proprietary trade name for Diethylthiambutene.

Thenalidine. 1-Methyl-4-N-(2-thenyl) - anilinopiperidine. Sandosten is the tartrate.

Thenardite (Makite). A mineral. It is anhydrous sodium sulphate, Na_2SO_4.

Thenard's Blue. See COBALT BLUE.

Thenfadil. A proprietary trade name for the hydrochloride of Thenyldiamine.

Thenium Closylate. Dimethyl-(2-phenoxyethyl) - 2 - thenylammonium 4-chlorobenzenesulphonate.

Thenyldiamine. N′N′-Dimethyl-N-2-pyridyl - N - 3-thenyl - ethylenediamine. Thenfadil is the hydrochloride.

Theobroma Oil. See COCOA BUTTER.

Theobromine. Dimethyl-xanthine, $C_7 H_8O_2N_4$. The active principle of the cocoa bean.

Theobromose. The lithium compound of theobromine, $C_7H_7N_4O_2Li$.

Theocal. A double salt of calcium theobromine and calcium lactate. A diuretic.

Theocalcin. A proprietary preparation of theobromine and calcium salicylate.

Theocine (Theophylline). 1 : 3-Dimethyl-xanthine, $C_7H_8N_4O_2$, a diuretic.

Theocine Sodium Acetate. Theophylline-sodium-acetate, a diuretic.

Theocyl. Theobromine - acetyl - sali - cylate.

Theocylene. A condensation product of aspirin and theobromine.

Theodrenaline. An analeptic drug. 7 - [2 - (3, 4, β - Trihydroxyphenethyl - amino)ethyl]theophylline.

Theodrox. A proprietary preparation of aminophylline and aluminium hydroxide used in the treatment of bronchospasm. (275).

Theoform. A proprietary name for a condensation product of theobromine and formaldehyde, with the addition of citric acid as preservative.

Theogardenal. A proprietary preparation of a combination of phenyl-ethyl-barbituric acid and theobromine.

Theograd. A proprietary preparation of THEOPHYLLINE. A bronchodilator. (311).

Theolactin (Theosate). An addition product of theobromine-sodium and sodium lactate.

Theominal. A proprietary preparation of phenobarbitone and theobromine. An antihypertensive. (112).

Theonacet. See AQUIRIN.

Theonar. A proprietary preparation of THEOPHYLLINE and NOSCAPINE. A bronchodilator. (249).

Theonasal. A preparation of theobromine and sodium salicylate.

Theonyl. A proprietary preparation of proxyphylline, ephedrine hydrochloride and caffeine. A bronchial antispasmodic. (338).

Theophen. A proprietary preparation of butethamate citrate, amylobarbitone, ephedrine hydrochloride and theophylline. A bronchial antispasmodic. (273).

Theophorin. An addition product of theobromine-sodium and sodium-formate, $C_7H_7N_4O_2Na.HCO_2Na.H_2O$. A diuretic.

Theophylline. See THEOCINE.

Theophysem. Iodo-ethyl-allyl-thioxyurea.

Theosal. A preparation of theobromine and sodium salicylate.

Theosalin. An addition product of theobromine-sodium and sodium-sulpho-salicylate.

Theosate. See THEOLACTIN.

Theo-sod-acet. Theobromine-sodium-odio-acetate.

Theo-sod-form. Theobromine-sodium-sodio-formate.

Theo-sod-sal. Theobromine and sodium salicylate.

Theosol. A proprietary preparation of theobromine-calcium salicylate.

Thephorin. A proprietary preparation of phenindamine hydrogen tartrate. (314).

Theranol. See THERMIOL.

Therlo. An electrical resistance alloy containing 85 per cent. copper, 13 per cent. manganese, and 2 per cent. aluminium. It has a low temperature coefficient and a resistivity of 46·7 microhms . cm. at 0° C.

Thermalene. An intimate mixture of acetylene and vaporised oils. It is used for the production of high temperatures, in the cutting and welding metals.

Thermalloy. A patented form of thermit containing 50 per cent. iron oxide, 27 per cent. aluminium, and 23 per cent. sulphur. The name appears to be applied also to an alloy containing 66·5 per cent. nickel, 30 per cent. copper, and 2 per cent. iron. It has a magnetic permeability which decreases at higher temperature. An alloy containing 75–85 per cent. iron, 10–20 per cent. chromium, 2–6 per cent. silicon, 0·5–1 per cent. manganese, 0·5–1 per cent, tungsten, and 0·2–2 per cent. carbon is also known as Thermalloy.

Thermalloy A. A proprietary trade name for an alloy containing 67·5 per cent. nickel, 0·15 per cent. carbon, 0·15 per cent. silicon and 30 per cent. copper.

Thermalloy B. A proprietary trade name for an alloy containing 57·8 per cent. nickel, 0·15 per cent. carbon, 0·15 per cent. silicon, and 40 per cent. copper.

Thermatomic Carbon. A fine carbon produced by " cracking " natural gas into carbon and hydrogen by passing the gas over heated brickwork. Used as a rubber filler.

Thermax. A thermatomic carbon used in rubber mixings. It is a proprietary product. Also a proprietary trade name for an insulating board made from wood fibre with a fire-resisting magnesia cement. Also a trade name for heat resisting chromium nickel alloys.

Thermazote. A trade mark for an expanded thermosetting plastic, manufactured in densities between 7 and 30 lb. per cubic foot. It is noninflammable and odourless and withstands temperatures as high as 300° C. It has a low thermal conductivity and is used in building construction, etc.

Thermex. A proprietary trade name for a polyvinyl acetal thermosetting phenolic synthetic wire enamel possessing flexibility and resistance to temperatures of the order of 185° C. See FORMEX.

Thermifugin. Sodium-methyl-trihydroxy-quinoline-carbonate, $C_9H_8(CH_3)$. N.COONa. An analgesic and antipyretic.

Thermine. Tetrahydro-β-naphthylamine-hydrochloride, $C_{10}H_7.H_4.NH_2$. HCl. It increases body temperature.

Thermiol (Theranol). A 25 per cent. solution of sodium-phenyl-propiolate, $C_6H_5.C:C.CO_2Na$. Recommended for inhalation in cases of laryngitis.

Thermisilio. A proprietary trade name for a chemical-resisting iron-silicon alloy in which the brittleness has been diminished.

Thermisilio Extra. Similar to the above but has a higher silicon content, is more resistant to acids and has greater hardness and brittleness.

Thermisilizid. A Swedish iron-silicon alloy of the acid-resisting type.

Thermit. A world-wide registered trade mark for (a) Alumino-thermic mixtures consisting essentially of nearly equal parts of powdered aluminium and metal oxides, usually iron or manganese oxides. These mixtures burn with a high temperature, and are used for welding metals. They are also used as an ingredient in incendiary bombs. (b) A bearing metal containing lead, antimony (20 per cent.), and small quantities of tin, nickel and copper.

Thermit Manganese. Manganese metal made by the Thermit reduction method. It contains approximately 98 per cent. manganese.

Thermit Metal. A German bearing metal containing : 14–16 per cent. antimony, 5–7 per cent. tin, 0·8–1·2 per cent. copper, 0·7–1·5 per cent. nickel, 0·3–0·8 per cent. arsenic, 0·7–1·5 per cent. cadmium and 72–78·5 per cent. lead.

Thermlo F. A proprietary rubber vulcanisation accelerator. It is an organic polysulphide.

Thermocast. A proprietary trade name for an ethyl cellulose composition.

Thermo-couples. These consist of bars or wires of two different metals or alloys soldered together. If the junction of the metals is heated, an electric force is set up, which is a measure of the temperature. Platinum with an alloy of platinum with 10 per cent. iridium or rhodium is usually used for temperatures from 300–1500° C. Iron with a nickel-iron alloy is often used for temperatures up to 1000° C. For lower temperatures copper and Constantan is used, and for still lower temperatures bismuth and antimony (not above 276° C.), or an alloy of 66 per cent. antimony with 33 per cent. zinc, and tin, is employed (not above 260° C.).

Thermodin (Uracetin). The acetyl derivative of p-ethoxy-phenyl-urethane,

$C_6H_4(OC_2H_5).NH.COOC_2H_5$. An antipyretic.

Thermoflex. A proprietary trade name for di-para-methoxy-diphenylamine. An antioxidant. (10), (2).

Thermoflex A. A proprietary trade name for an antioxidant. It contains 50 parts of phenyl-beta-naphthylamine, 25 parts of methoxy-diphenylamine and 25 parts of diphenyl-para-phenylenediamine. (10), (2).

Thermoflex C. Similar to Thermoflex A.

Thermokalite. A mixture of trona, thenardite, thermonatrite, and sodium bicarbonate.

Thermol. Acetyl - salicyl - phenetidin, $C_{17}H_{17}NO_4$. An antipyretic and analgesic.

Thermolastic. A registered trade name for extrudable thermoplastic rubber-like materials based upon styrene butadiene copolymers not requiring vulcanisation. (99).

Thermonatrite. Mono-hydrated sodium carbonate. A mineral. It is $4[Na_2CO_3 . H_2O]$.

Thermonit (Keramonit). A refractory cement made in the electric furnace. It is stated to be used as a paint or mortar, and to resist high temperatures. Keramonit is the cement reinforced with metal mesh.

Thermophyllite. A mineral. It is a variety of serpentine, found in Finland.

Thermoprene. Products obtained by heating rubber with either an organic sulphonyl chloride or an organic sulphonic acid at 125–135° C. for several hours. p-Toluene-sulphonyl chloride and p-toluene-sulphonic acid are suitable reagents. One product is a protective paint which is resistant to acids, alkalis, and corrosive gases, and has a low permeability to water. Products resembling gutta-percha, balata, and shellac are also obtained. See also CYCLORUBBERS.

Thermotex. A proprietary trade name for a material used for coating steam pipes. It is an emulsified asphalt mixed with asbestos.

Thermozine. See AMBRINE.

Theruhistin. See ISOTHIPENDYL.

Thiabendazole. 2-(Thiazol-4-yl)benzimidazole. Mintezol. Thibenzole.

Thiacetazone. 4-Acetamidobenzaldehyde thiosemicarbazone. Berculon A. Neustab. Seroden. Thioparamizone.

Thiacril 44. A proprietary polymer rubber with an ethyl-acrylate base. (129).

Thialbarbitone. 5-Allyl-5-(cyclohex-2-enyl) - 2 - thiobarbituric acid. Kemithal.

Thial. Hexamethylene-tetramine-oxymethyl-sulphonate. An antiseptic.

Thialion. A preparation containing 26 per cent. sodium sulphate, 56 per cent. sodium citrate, 3 per cent. potassium citrate, 3 per cent. lithium citrate, 3 per cent. sodium chloride, and 9 per cent. water.

Thialon. A registered trade mark for a proprietary product.

Thiamazole. Methimazole.

Thiambutozine. N-(4-Butoxyphenyl)-N'-(4-dimethylaminophenyl) - thiourea. Ciba 1906.

Thiamine. Vitamin B_1. The quaternary salt of methyl-hydroxy-ethyl-thiazole. $C_{12}H_{17}ON_4SCl,HCl$.

Thiamin Yellow. A direct cotton dyestuff produced by the action of formaldehyde upon primuline.

Thiarsol. A colloidal suspension of arsenic sulphide, As_2S_3, proposed for the treatment of cancer.

Thiate E. A proprietary accelerator. Trimethyl thiourea. (450).

Thiaver. A proprietary preparation of alkavervir, and epithiazide. (275).

Thiazamide. A proprietary preparation of sulphathiazole. An antibiotic. (507).

Thiazesim. 5-(2-Dimethylaminoethyl)-2,3 - dihydro - 2 - phenylbenzo - 1,5 - thiazepin-4-one.

Thiazina. A proprietary preparation containing thiacetazone, isoniazid. An antituberculous agent. (268).

Thiazine Brown G. A dyestuff similar to the R mark.

Thiazine Brown R. A direct cotton dyestuff, dyeing shades of red-brown.

Thiazine Red G. An azo dyestuff prepared from dehydro-thio-toluidine-sulphonic acid. A direct cotton colour.

Thiazine Red R. See ROSOPHENINE 10 B.

Thiazoline (Thiazylamine). Aminothiazole, $C_3H_4N_2S$.

Thiazol Yellow. See CLAYTON YELLOW.

Thiazol Yellow G. A dyestuff. It is equivalent to Clayton yellow.

Thiazol Yellow R. A dyestuff. It is equivalent to Nitrophenin.

Thiazylamine. See THIAZOLINE.

Thibenzole. A proprietary trade name for Thiabendazole.

Thickened Mineral Oils. Mineral oils which have been thickened by dissolving soap, usually aluminium soap, in them.

Thickened Oils. See BLOWN OILS.

Thickener. See OIL-PULP.

Thiel-Stoll Solution. A saturated solution of lead chlorate, $Pb(ClO_4)_2$. It has a density of 2·6 and is used for the determination of the specific gravity of minerals.

Thiethylperazine. 2-Ethylthio-10-[3-(4-methyl piperazin-1-yl)propyl]-phenothiazine. Torecan is the base or its salts.

Thierschite. A mineral. It is calcium oxalate.

Thiersch's Antiseptic Solution. A solution containing salicylic acid and boric acid.

Thiery's Solution. A solution of picric acid for the treatment of burns.

Thiet-sie. A resinous substance used as a varnish by the Burmese.

Thigan. A preparation of silver with thiogenol (q.v.).

Thilamin. A preparation said to be made by the action of sulphur upon lanoline. Used in skin eruptions.

Thilaven. Ammonium ichthosulphonate.

Thilocologne. Ethyl chloride and eau-de-Cologne.

Thimerosal. A proprietary trade name for Thiomersal.

Thinner No. 22. An industrial solvent containing approximately 55 per cent. terpene hydrocarbon and 45 per cent. gasoline.

Thinoline (Vulcanised Oils). Vulcanised linseed oil, used in rubber mixings.

Thinolite. A mineral. It is calcium carbonate.

Thioantipyrin. See Thiopyrine.

Thiobismol. Sodium bismuth thioglycollate.

Thiocamf. A liquid formed by exposing camphor to the action of sulphur dioxide. Used as a disinfectant as it evolves sulphur dioxide on exposure to air.

Thiocarlide. NN'-Di-(4-isopentoxy-phenyl)thiourea. Isoxyl.

Thiocarmine R. A dyestuff. It is the sodium salt of diethyl-dibenzyl-diamino-phenazthionium-disulphonic acid, $C_{30}H_{28}N_3O_6S_3Na$. Dyes wool and silk indigo-carmine shades from an acid bath.

Thiocatechine (Thiocatechine S). A dyestuff obtained by the fusion of p-diamines or acetyl-nitramines with sodium polysulphide. The S mark is the sulphite compound. Dyes cotton brown.

Thiocatechine S. See Thiocatechine.

Thiochem Sulphur Green (Sulphur Olive Green). A dyestuff produced by the fusion of benzene-azo-phenol with a copper salt at 180–200° C.

Thiochromogen. See Primuline.

Thiochrysine. A gold-sodium thiosulphate.

Thio Cotton Black. A dyestuff obtained by the fusion of a mixture of dinitro-phenol and p-amino-phenol-sulphonic acid with sodium polysulphide. Dyes cotton black.

Thiocyanosin. An eosin dyestuff. It is

tetrabromo-thio-dichloro-fluoresceine-methyl ether.

Thiodin. Thiosinamine-ethyl-iodide.

Thiodotoxyl. A solution of 10 per cent. each of atoxyl and thiosinamine-ethyl-iodide.

Thiofide. A proprietary rubber vulcanisation accelerator. It is 2-2-dithio-bisbenzothiazole.

Thiofide E.P. A proprietary accelerator. Dibenzthiazyl disulphide. (57).

Thioflavine S (Chromine G). A dyestuff. It is the sodium salt of methylated primuline. Dyes cotton, silk, and half-silk goods greenish-yellow from an alkaline bath.

Thioflavine T (Acronol Yellow T.). A dyestuff. It is dimethyl-dehydro-thio-toluidine-methylo-chloride, $C_{17}H_{19}N_2SCl$. Dyes tannined cotton greenish-yellow, and silk yellow with green fluorescence.

Thioform (Deltaform). Bismuth-dithiosalicylate, $S.C_6H_3(OH).COO.Bi_2O_3+Bi_2O_3+H_2O$. Used externally in the place of iodoform, and taken internally for intestinal catarrh.

Thiofurfuran. Thiophene, C_4H_4S.

Thiofurfurane. Thiophene, C_4H_4S.

Thiogene Black (Thiogene Purple; Thiogene Dark Red G, R; Thiogene Rubine O). Sulphide dyestuffs obtained by the action of sulphur upon amino-hydroxy-phenazines.

Thiogene Blue. See Sulphur Black T.

Thiogene Dark Red, G, R. See Thiogene Black.

Thiogene Purple. See Thiogene Black.

Thiogene Rubine O. See Thiogene Black.

Thiogene Violet V. A sulphide dyestuff manufactured by heating together sulphur and phenosafranine, then heating the product with sodium sulphide. It is further heated with sodium polysulphide.

Thioguanine. An anti-neoplastic agent. 2-Aminopurine-6-thiol. Lanvis.

Thioindigo B. See Thioindigo Red B.

Thioindigo Grey B. A dyestuff. It is 7 : 7'-diamino-thioindigo.

Thioindigo Orange R. A dyestuff. It is 6 : 6'-diethoxy-thioindigo.

Thioindigo Pink BN. A dyestuff. It is 6 : 6'-dibromo-dimethyl-thioindigo.

Thioindigo Red B (Vat Dye B, Thioindigo B, Durindone Red B). An indigo dyestuff, $C_{16}H_8O_2S_2$, prepared from phenyl-thio-glycol-o-carboxylic acid, by boiling it with alkalis, heating the product with acids, and oxidising the thioindoxyl produced. It is a yellow-red dyestuff.

Thioindigo Red 3B. A dyestuff. It is 5 : 5′-dichloro-6 : 6′dimethyl-thioindigo.

Thiodigo Red BG. A dyestuff. It is 5 : 5′-dichloro-thioindigo.

Thioindigo Scarlet. An indigo dyestuff, $C_{16}H_9O_2NS$, prepared by condensing thioindoxyl with isatin. A yellow vat dye.

Thioindigo Scarlet S. A dyestuff. It is 6 : 6′-dithioxyl-thioindigo.

Thioindigo Violet 2B. A dyestuff. It is dichloro-dimethyl-dimethoxy-thioindigo.

Thiokol. A registered trade mark for a range of polysulphide polymers made by reacting bis-2-chloroethyl formal with solids which are highly resistant to oil, grease, petroleum and other solvents. Products covered by the THIOKOL trademark include clutch and brake facings and linings; tracked vehicles for the transport of personnel and goods; rocket motors and parts and accessories therefor; organic sulphur-containing compounds such as mercaptans, sulphides, polysulphides, thiosulphates and thioketones; polyesters and polyurethanes; plasticisers of the ester and acetyl types; rocket propellants; and products deriving from polytetrafluoroethylene. (129).

Thiokol A. A proprietary ethylene dichloride sodium sulphide condensate. (129).

Thiokol F.A. A proprietary ethylene dichloride and dichloroethyl formal condensed with sodium sulphide. (129). THIOKOL ZR300.

Thiokol LP2 and LP3. Proprietary grades of di-2-chloroethyl formal sodium sulphide condensate. (129).

Thiokol LP5. A proprietary mercaptan-terminated alkyl polysulphide. (129).

Thiokol LP8, LP31 and LP33. Proprietary grades of di-2-chloroethyl formal condensed with sodium sulphide. (129).

Thiokol LP32. A proprietary polymer of bis (ethylene oxy) methane containing disulphide linkages. (129).

Thiokol RD. A proprietary trade name for an oil-resistant synthetic rubber. It is an interpolymer of butadiene and acrylonitrile with a third unspecified component. It contains 4 per cent. nitrogen.

Thiokol ST. A proprietary di-2-chloroethyl formal sodium sulphide condensate. (129).

Thiokol TP-90. A proprietary trade name for a high molecular weight polyether. (129).

Thiokol TP-90-B. and TP-95. See TP-90.

Thiokol ZR 300. See THIOKOL F.A.

Thiol (Tumenol, Petrosulfol). Names given to artificial substitutes for ammonium ichtho-sulphonates. Thiol is a product similar to ichthyol (*q.v.*), and is obtained by sulphonating gas oil (brown coal tar oil), with sulphur, treating the product with sulphuric acid, and pouring the whole into water. Used medicinally in the same way as ichthyol.

Thiolin. See ICHTHYOL.

Thioline. See ABIETENE.

Thiolite. An insulator prepared from formaldehyde, cresol, and sulphur chloride. The sulphur improves the insulating properties. The material is prepared by condensing formaldehyde and cresol to form a product soluble in alcohol and acetone, which is then treated with sulphur chloride. The hydrogen chloride is eliminated and the product contains up to 12 per cent. sulphur. It is purified by precipitation. It softens at 80° C. and if pressure is applied, hardening takes place. Insulation resistance is 300×10^6 megohms per cm. Dielectric constant 4·5.

Thiomerin Sodium. A proprietary trade name for Mercaptomerin sodium.

Thiomersal. Sodium 2-(ethylmercurithio)benzoate. Sodium ethyl mercurithiosalicylate. Thimerosal : Thiomersalate. Merthiolate.

Thiomersalate. A proprietary trade name for Thiomersal.

Thiomesterone. $1\alpha,7\alpha$-Bis(acetylthio)-17β - hydroxy - 17α - methylandrost-4-en-3-one. Embadol.

Thiomicid. *p*-acetylaminobenzaldehyde thiosemicarbazone.

Thional Black. See SULPHUR BLACK T.

Thional Bronze. A sulphide dyestuff obtained by the fusion of β-hydroxy-naphthoquinone-anilide with sodium polysulphides.

Thional Brown R. A sulphide dyestuff which dyes unmordanted cotton.

Thional Green. A sulphide dyestuff.

Thionalide. A commercial name for thioglycollic acid, β-amino-naphthalide, $HS.CH_2CO.NH.C_{10}H_7$. An analytical reagent.

Thion Black. A dyestuff prepared by heating at 140–180° C. for from 2–3 hours, sodium tetrasulphide and sodium dinitro-phenoxide.

Thion Blue B. A sulphide dyestuff derived from *p*-nitro-*o*-amino-*p*-hydroxdiphenylamine.

Thionex. Tetramethyl-thiuram-monosulphide. An ultra-accelerator for rubber vulcanisation.

Thion Green. A sulphide dyestuff obtained by the action of sodium hydroxide upon *p*-hydroxy-phenyl-thio-carbamide.

Thionhydrol. Colloidal sulphur.

Thionine. See LAUTH'S VIOLET.

Thionine Blue G, O Extra. A dye-stuff. It is the zinc double chloride of trimethyl - ethyl - diamino - phenaz-thionium-chloride, $(C_{17}H_{23}N_3SCl)_2Zn$ Cl_2. Dyes tannined cotton blue.

Thionine Red-brown B. A British brand of Immedial maroon B.

Thionite A. A synthetic rubber of Japan. It is a polymerized compound of ethylene-di-glycoside and sodium tetrasulphide.

Thionol. A registered trade name for certain dyestuffs.

Thionol Black. See SULPHUR BLACK T.

Thionoline. Oxy-amino-imino-diphenyl-sulphide, $C_{12}H_8N_2OS$.

Thionone Brilliant Green GG Conc. A sulphide green dyestuff obtained from phenyl (or tolyl)-α-naphthyl-amine - 8 - sulphonic acid, *p* - amino-phenol, and sodium hypochlorite, re-ducing with sodium sulphide, and boiling the product with sodium poly-sulphide and a copper salt.

Thionotol. A material containing men-thol, eucalyptol, chloral camphor, chlorothymol, azulin, ephedrine, and vasogen. Recommended against colds.

Thion Yellow. A dyestuff obtained by heating thio-*m*-tolylene-diamine with sodium sulphide solution.

Thioparamizone. A proprietary trade name for THIACETAZONE. It is *p*-acetylaminobenzaldehyde thiosemi-carbazone.

Thiophanate. An anthelmintic. 4,4′-*o* - Phenylenebis(ethyl 3 - thioallo - phanate). 4, 4′ - *o* - Phenylenebis (3 - ethoxycarbonyl - 2 - thiourea). NEMA-FAX.

Thiophen Green. An analogue of malachite green. It has a formula, $C(OH) . C_4H_3S .^!C_6H_4 . N(CH_3)_2 . C_6H_4 . N(CH_3)_2$.

Thiophenol. Phenyl-mercaptan, C_6H_5 SH.

Thiophenol Black T Extra. A dye-stuff prepared by the fusion of dinitro-phenol with sodium polysulphides. (Dinitrophenol Black.)

Thiophloxin. An eosin dyestuff. It is tetrabromo-thio-dichloro-fluoresceine.

Thiophor. A registered trade mark currently awaiting re-allocation by its proprietors to cover a range of dyestuffs. (983).

Thiophor Bronze 5G. A dyestuff ob-tained by the fusion of *p*-phenylene-diamine and *p*-amino-acetanilide with sulphur.

Thiophor Indigo. A dyestuff obtained by heating the indophenol derivative

from α-naphthol and *p*-amino-dimethyl-aniline with sodium sulphide and sulphur.

Thiophor Indigo CJ. A leuco com-pound of 4-hydroxy-3-thiol-*p*-dimethyl-amino-phenyl-α-naphthylamine.

Thiophorine. Theobromine - sodium acetate.

Thiophosgene. Thiocarbonyl-chloride, $CSCl_2$.

Thiophosphine J. See CHLORAMINE YELLOW.

Thiophysem. See TIODINE.

Thiopinol. A pure petroleum product of German origin.

Thioprene-48. An elastomeric mercap-tan-terminated polymer used for sealing glass. A registered trade mark. (483).

Thiopropazate. 10-{3-[4-(2-Acetoxy-ethyl)piperazin - 1 - yl]propyl} - 2 - chlorophenothiazine. Dartalan is the hydrochloride.

Thioproperazine. 2-Dimethylsulpha-moyl - 10 - [3 - (4 - methylpiperazin - 1 - yl) - propyl]phenothiazine. Majeptil is the mesylate.

Thiopyrin (Thioantipyrin). 1-Phenyl-2 : 3-dimethyl -2: 5-thiopyrazole. An antipyretic.

Thioquin. Dihydroxy - trimethoxy - quinine-phenyl-sulphonate.

Thioridazine. 10 - [2 - (1 - Methyl - 2 - piperidyl)ethyl] - 2 - methylthio - phenothiazine. Melleril is the hydro-chloride.

Thiorubin. An azo dyestuff. It is thio-*p*-toluidine diazotised and com-bined with β-naphthol-disulphonic acid, $S(C_7H_6NH_2)C_7H_6 . N_2 . C_{10}H_{14}(HSO_2)_2$ OH.

Thiosaccharine. A compound, C_6H_5 $(CS)SO_2.NH$, obtained by heating a mixture of saccharin and phosphorus pentasulphide at 220° C., and extracting with hot benzene.

Thiosal. Disodium-dithio-salicylate.

Thiosan. A proprietary accelerator. A thiazole-dithiocarbamate blend. (57).

Thiosept. A product containing sulphur obtained from the Tyrol oil shale. It is used in salves.

Thiosept Oil. A distillation product of Tyrol shale oil. It contains sulphur and has a boiling range of 100–350° C.

Thiosinamine (Rhodallin). Allyl-thio-urea, $C_6H_5.NH.CS.NH_2$. An anti-septic.

Thiosinyl. See FIBROLYSIN.

Thiosporin. A proprietary preparation containing sulphomyxin sodium. (277).

Thiostab. A proprietary pure sodium thiosulphate in ampoules.

Thiotax. A proprietary semi-ultra rub-ber vulcanisation accelerator. It is 2-mercapto-benzo-thiazole. (57).

Thiotepa. Tri - 1 - aziridinylphosphine sulphide. Triethylene thiophosphoramide.

Thiothixine. NN-Dimethyl-9-[3-(4-methylpiperazin - 1 - yl)propylidene] - thiozanthen-2-sulphonamide.

Thiovanic. Vacuum distilled mercaptoacetic acid. (591).

Thioxine Blacks. Dyestuffs for cotton.

Thioxine Orange (Thioxine Yellow G). Yellow sulphide dyestuffs.

Thioxine Yellow G. See THIOXINE ORANGE.

Thioxolone. 6-Hydroxybenz-1,3-oxathiol-2-one. Camyna.

Thioxydant Lumière. Ammonium persulphate, $(NH_4)_2S_2O_8$.

Thiozin. Ammonium ichthosulphonate.

Thissirol. An aqueous solution of about 57 per cent. castor oil soap and 29 per cent. chloroxylenol mixture. A bactericide.

Thitsi (Burma Black Varnish). A natural lacquer. It is the sap of the black varnish tree, *Melanorrhœa visitata*.

Thiurad. A proprietary trade name for a rubber ultra accelerator. It is tetramethyl thiuram disulphide. (57).

Tholaform. A mixture of equal parts of menthol and trioxy-methylene.

Thomaite. A mineral. It is an iron carbonate.

Thomas Ammonia Phosphate Lime. Thomas slag mixed with ammonium sulphate. A fertiliser.

Thomasite. A compound, $6CaO.P_2O_5.Fe_2SiO_4$. It is a constituent of the basic slag of the Thomas-Gilchrist process for the dephosphorisation of iron.

Thomas Meal. Ground slag obtained from the Thomas process for iron. Used as a fertiliser. See BASIC SLAG.

Thomas Phosphate. See BASIC SLAG.

Thomas Slag. See BASIC SLAG.

Thomsenolite. A mineral, $NaF.CaF_2.AlF_3.H_2O$.

Thomsonite (Mesolitine). A mineral, $2(Ca.Na_2)O.2Al_2O_3.4SiO_2.5H_2O$.

Thonzylamine. N-4-Methoxybenzyl-N'N' - dimethyl - N - pyrimidin - 2 -yl - ethylenediamine. Neohetramine.

Thoracin. A proprietary preparation of phenylethyl nicotinate, guaiacol furoate, tetrahydrofurfuryl salicylate, camphor and eucalyptol. (347).

Thoragol. A proprietary preparation of bibenzonium bromide. A cough linctus. (347).

Thoran. An alloy of 96 per cent. tungsten with 4 per cent. carbon.

Thorax A. A proprietary safety glass involving the use of cellulose acetate.

Thorax N. A proprietary safety glass involving the use of pyroxylin.

Thoren. A technical diamond substitute. It is an alloy made from tungsten and tungsten carbide, and has a hardness of 9·8 compared with 10 for the diamond, and melts at about 3000° C.

Thoria. Thorium dioxide, ThO_2.

Thorianite. A mineral. It consists of about 70 per cent. of thorium dioxide, and from 10–30 per cent. of uranium oxide, U_3O_8.

Thorite (Orangite, Monazite). A mineral. It consists essentially of thorium silicate, $ThO_2.SiO_2$.

Thoroclear. A silicone-based water-repellent coating for limestone. A trade mark. (952).

Thorogummite (Chlorothorite). A mineral. It is a hydrous silicate of uranium and thorium, $UO_3.ThO_2.3SiO_2.6H_2O$.

Thoron. A term used for the emanation of thorium.

Thorosheen. An acrylic paint for masonry. A registered trade mark. (952).

Thorotrast. A proprietary colloidal thorium dioxide preparation.

Thorotungstite. A mineral of the Federated Malay States approximating to the formula, $(ThO_2.Ce_2O_3.ZrO_2).H_2O.+2WO_3.H_2O$.

Thortveitite. A scandium mineral containing about 36 per cent. scandia.

Thoryl. The Edisylate of Caramiphen.

Thoulet's Solution. A concentrated solution of potassium and mercury iodides in water. Used to determine the density of minerals.

Thovaline. A proprietary preparation of talc, kaolin, zinc oxide and cod-liver oil,'used in dermatology. (384).

Thowless Solder. An alloy of tin and zinc with small amounts of aluminium and silver.

Thresh's Reagent. Potassium - bismuthic-iodide. Used for testing alkaloids.

Throat Balls. See NITRE BALLS.

Thrombase. A clotting enzyme. It coagulates blood.

Thrombolite. A mineral, $10CuO.3Sb_2O_3.19H_2O$.

Thromboplastin. An extract of the brain of cattle dissolved in salt solution. A hæmostatic.

Throphleol. A 50 per cent. solution of erythrophleine hydrochloride (from casca bark) in eugenol.

Throsil. A proprietary preparation of CETALKONIUM CHLORIDE and amethocaine hydrochloride. An antiseptic mouth lozenge. (248).

Thsing-Hoa-Liao. The Chinese name for a cobaltiferous aluminic silicate. Used in the manufacture of porcelain.

Thucholite. A mineral found in the vicinity of Parry Sound, Ontario. It occurs, associated with uraninite and calciosamarskite, embedded in felspar, quartz, or mica. It contains about 50 per cent. carbon, about 13 per cent. water, and about 28 per cent. of ash, of which one-half is thorium dioxide.

Thuenite. A mineral. It is a variety of ilmenite.

Thujol (Absinthol). Tanacetone, $C_{10}H_{16}O$.

Thulite. A mineral. It is a calcium-aluminium silicate.

Thunderite. An explosive containing 92 per cent. ammonium nitrate, 4 per cent. trinitro-toluene, and 4 per cent. flour.

Thurfyl Nicotinate. Tetra hydrofurfuryl nicotinate. Trafuril.

Thuringite (Patersonite). A mineral, $4(AlFe)_2O_3.7FeO.6SiO_2.9H_2O$.

Thurston's Alloy. An alloy of 80 per cent. zinc, 14 per cent. tin, and 6 per cent. copper.

Thurston's Brass. See BRASS.

Thus, Gum. See GUM THUS.

Thwaites' Solution. A mixture of alcohol, creosote, and chalk in water. Used to preserve animal tissues.

Thyangol. A German product. It contains anæsthesin, phenacetin, thymol, menthol, and eucalyptol.

Thylin. A proprietary preparation of nifenazone. (137).

Thymacetin. Thymol - phenacetin, $C_6H_2CH_3.C_3H_7(O.C_2H_5).NH.C_6H_3O$. A hypnotic and analgesic.

Thymacetol. A condensation product of acetone and o-thymic acid. An intestinal astringent.

Thymatol. Thymol carbonate, $CO(O.C_{10}H_{13})_2$, an anthelmintic used in medicine.

Thyme Camphor (Thymic acid). Thymol, $C_6H_3(OH).CH_3.C_3H_7$.

Thymegol. The mercury-potassium salt of thymol-sulphonic acid. Used in medicine. See EGOLS.

Thymene. The residual oils obtained from the preparation of thymol. Used as a cheap perfume for soaps.

Thymic Acid. See THYME CAMPHOR.

Thymidol. Methyl-propyl-phenol-menthol. An antiseptic wash.

Thymine. 5-Methyl-uracil, $C_5H_6O_2N_2$, obtained by the hydrolysis of nucleic acid.

Thymiodin. See THYMOIDE.

Thymiodol. See THYMOIDE.

Thymodrosin. A preparation containing the active principles of thyme, aconite, belladonna, hyoscyamus, ipecacuanha, and other plant extracts. Another type contains about 3 per cent. potassium-guaiacol-sulphonate.

Thymoform. See THYMOLOFORM.

Thymoide (Thymotol, Thymiodin, Thymiodol). Thymol iodide.

Thymol Blue. Thymol-sulphon-phthalein. An indicator.

Thymolcarbonic Ether. See ETHER, THYMOLCARBONIC.

Thymolin. A mixture of 18 parts naphthalene, 1 part camphor, and 1 part thymol.

Thymoloform (Thymoform). A condensation product of formaldehyde and thymol, $CH_2 : [C_6H_3(CH_3)(C_3H_7)O]_2$. An antiseptic dusting-powder.

Thymolol. Thymol iodide, $(C_{10}H_{13}OI)_2$.

Thymotal (Tyratol). Thymol carbonate. See THYMATOL.

Thymotol. See THYMOIDE.

Thymoxamine. 4-(2-Dimethylaminoethoxy) - 5 - isopropyl - 2 - methyl - phenyl acetate. Moxisylyte. Opilon is the hydrochloride.

Thyol. A substitute for Ichthyol ($q.v.$), obtained by treating tar oils with sulphur.

Thyreol. See THYRESOL.

Thyresol (Thyreol). Santalol-methylester, $C_{15}H_{23}O.CH_3$. Used as a substitute for santalol in the treatment of urethritis.

Thyroidectin. Dried antithyroid serum.

Thyrocalcitonin. See CALCITONIN.

Thyrodex. A proprietary preparation of dexamphetamine sulphate and thyroid. (329).

Thyropit. A proprietary preparation of thyroid and anterior pituitary glands used in the treatment of obesity. (250).

Thyropurin. A German preparation. It contains the effective constituents of the thyroid gland in tablet form. Each tablet (0·25 gram) contains 0·0005 grain thyroxin.

Thyrotrophic Hormone. Thyrotrophin.

Thyrotrophine. Thyrotrophic hormone. Thytropar. Thytrophin.

Thyroxin. A complex protein present in the thyroid gland. It is di-iodohydroxy-phenyl ether of di-iodo-tyrosine.

Thytropar. A proprietary preparation of thyrotropin. (327).

Thytrophin. A proprietary trade name for Thyrotrophin.

Tiamulin. It is 11-hydroxy-6,7,10,12-tetramethyl - 1 - oxo - 10 - vinylperhydro-3a, 7 - pentanoinden - 8 - yl (2-diethyl aminoethylthio)acetate. A veterinary antibiotic.

Tiargirio. A compound, $C_{14}H_9O_6Hg_2ClS_2$, containing 11·18 per cent. sulphur, 34·93 per cent. mercury, and 48·25 per cent. salicylic acid.

Tibergite. A mineral. It is $(Ca,Na)_3(Mg,Fe^{···})_5(Si,Al)_8O_{22}(OH)_2$.

Tibric Acid. A preparation used in the treatment of hyperlipæmia currently undergoing clinical trial as "CP-18,524". It is 2-chloro-5-(*cis*-3,5-dimethylpiperidinosulphonyl) benzoic acid.

Ticapur. A proprietary trade name for carbide metals in the form of pure metal-carbides, mixed metal-carbides and metal powders; and for raw materials in the form of carbides, silicides, borides and nitrides. (899).

Ticarbodine. An anthelmintic currently undergoing clinical trial as " EL-974 ". It is 2, 6 - dimethylpiperidino - 3' - (tri - fluoromethyl)thioformanilide.

Ticarcillin. An antibiotic. 6-[2-Carboxy - 2 - (3 - thienyl)acetamido] - penicillanic acid.

Ticelgesic. A proprietary preparation of PARACETAMOL. An analgesic. (434).

Ticevite. A proprietary preparation of Vitamins A, D, E and B complex. (434).

Ticillin V.K. A proprietary preparation of Penicillin V. An antibiotic. (434).

Ticipect. A proprietary preparation of DIPHENHYDRAMINE hydrochloride, ammonium chloride, sodium citrate, menthol and chloroform. A cough syrup. (434).

Tico. An electrical resistance alloy containing 67·5 per cent. iron, 30·5 per cent. nickel, and small quantities of manganese and copper.

Ticomplex. Organo-titanium complexes. (534).

Tidolith. See CRYPTONE.

Tiemannite. Mineral, HgSe.

Tiemonium Iodide. An anti-spasmodic and anti-cholinergic drug. It is 4-[3-hydroxy - 3 - phenyl - 3 - (2 - thienyl) - propyl] - 4 - methylmorpholinium iodide.

Tienilic Acid. An anti-hypertensive drug and diuretic. [2,3-Dichloro-4-(2-thenoyl)phenoxy]acetic acid.

Tiers Argent. An alloy containing 66·6 per cent. aluminium and 33·3 per cent. silver.

Tiff. Barytes, $BaSO_4$. A local name for the mineral barite (*q.v.*).

Tifolic. A proprietary preparation of ferrous fumarate and folic acid, used in the treatment of anæmia in pregnancy. (434).

Tigerez. Argentiferous galena found in the limestone in Austria.

Tiger's Eye. A mineral. It is quartz having a golden-yellow sheen.

Tigloidine. Tiglylpseudotropeine. Tiglyssin is the hydrobromide.

Tiglylpseudotropeine. A proprietary trade name for Tigloidine.

Tiglyltropeine. A proprietary trade name for Tropigline.

Tiglyssin. A proprietary trade name for the hydrobromide of TIGLOIDINE, used in the treatment of muscular spasm. (444).

Tiguvon. A proprietary preparation of FENTHION. A veterinary insecticide.

Tikhvinite. A mineral. It is an aluminium-strontium phosphate and sulphate.

Tikitiki. A preparation of rice-polishing extract used as a cure for infantile beri-beri.

Til. Organic compounds of titanium. (632).

Til Alkyl. Alkyl titanates. (632).

Til Aryl. Aryl titanates. (632).

Tilasite. A mineral. It is stated to be $4[CaMgAsO_4F]$.

Tilbutyl. Butyl titanates. (632).

Tilcom. Organic compounds of titanium. (632).

Tile Ore. An earthy variety of native cuprous oxide.

Tiletamine. An anti-convulsant and anæsthetic. 2-Ethylamino-2-(2-thienyl)-cyclohexanone.

Tilidate. An analgesic. Ethyl 2-dimethylamino - 1 - phenylcyclohex - 3 - ene-1-carboxylate.

Tilkerodite. A mineral. It is a lead selenide.

Tillantin B. A copper-arsenic compound used in a 0·2 per cent. solution against smut of cereals.

Tilleyite. A mineral. It is $Ca_5Si_2O_7$ $(CO_3)_2$.

Tillie. The black seed containing the largest amount of sesamé oil. See SUFFETTIL and BIGARRE.

Tillit. A Swedish rubber paint for use as a protective coating for iron and concrete. It is stated to consist of rubber and bitumen dissolved in benzol, the bitumen being first artificially oxidised.

Tilly Drops. Dutch drops.

Til Oil. See GINGELLY OIL.

Tima. A medicinal preparation imported from Tampico. It is prepared from the fruit of *Crescentia edulis*, by boiling with sugar. It is said to be a remedy for phthisis.

Timang Steel. A proprietary trade name for a high manganese steel.

Timbo. The root rind of a variety of *Conchocarpus*. A narcotic.

Timbor. A specially soluble form of di-sodium octaborate tetrahydrate (Na_2 $B_8O_{13}.4H_2O$). It is highly toxic to wood destroying insects and decay fungi and is used for the preservation of building timber by diffusion impregnation at the sawmill. (183).

Timodine. A proprietary preparation of NYSTATIN, HYDROCORTISONE, BENZALKONIUM CHLORIDE and DIMETHICONE used in the treatment of infected eczema. (347).

Timolol. A beta adrenergic blocking agent currently undergoing clinical trial as "MK-950". It is (—)-1-*tert*-butyl-amino - 3 - (4 - morpholino - 1, 2, 5 - thiadiazol - 3 - yloxy) - propan - 2 - ol. BLOCADREN is the maleate.

Timonox. A specially prepared antimony oxide. Used as a pigment.

Timonox Blue Star. A proprietary preparation of pure antimony oxide. The arsenic amounts to 0·0018 per cent.

Tinaderm. A proprietary preparation of tolnaftate. A skin fungicide. (335).

Tin Amalgam. A tin-mercury alloy containing 44–51 per cent. tin. It is prepared by electrolysis.

Tin Ash. Stannic oxide, SnO_2. A polishing powder.

Tin Brilliants. See FAHLUN DIAMONDS.

Tin Bronze. An alloy of 89 per cent. copper and 11 per cent. tin.

Tincal (Tinkal). An impure borax imported into Europe in the form of yellowish-white crystals. It consists of borax mixed with lime, magnesia, alumina, chlorides, and sulphates of sodium and calcium, as well as a greasy substance.

Tincalconite. A mineral. It is a Californian borax.

Tin-copper Green. See GENTELES GREEN.

Tin Crystals. See TIN SALTS.

Tinder, German. See AMADOU.

Tinder Ore. A mineral, $Pb_7Sb_8S_{19}$.

Tin, Dropped. See GRAIN TIN.

Tineafax. A proprietary preparation of zinc undecanoate and naphthenate used in the treatment of fungal skin infections. (535).

Tinegal AC. A cationic dyeing agent for acrylics.

Tin Foil. Consists of tin, but the common variety usually contains lead in considerable quantities. One alloy contains 87·5 per cent. tin, 8 per cent. lead, 4 per cent. copper, and 0·5 per cent. antimony.

Tin Green. See GENTELES GREEN.

Ting-yu. See TSÉ-IÉOU.

Tinidazole. An anti-trichomonal agent. It is ethyl 2-(2-methyl-5-nitroimidazol-1-yl)ethyl sulphone.

Tinkal. See TINCAL.

Tinkalzit. Synonym for Ulexite.

Tin, Mine. See BLOCK TIN.

Tinned Iron Solder. See TINSMITH'S SOLDER.

Tinnevelly Senna. Indian senna.

Tinning Solutions. Usually consist of solutions of tin bichloride and potassium cyanide in water. Used for the electro-deposition of tin.

Tin, Nitro-muriate. See TIN SPIRITS.

Tinol. A proprietary preparation of PARACETAMOL and DIPHENHYDRAMINE hydrochloride. An analgesic. (434).

Tinopal PCRP. A registered trade mark for an optical whitener. (17).

Tin Ore. A mineral. It is tinstone, SnO_2.

Tin Oxychloride. See TIN SPIRITS.

Tin, Phosphor. See PHOSPHOR TIN.

Tin Plate. Iron plated with tin.

Tin Prepare Liquor. Sodium stannate, $Na_2SnO_3.3H_2O$. Used as a mordant for dyes in calico, printing.

Tin Pyrites (Stannine, Bell Metal Ore, Stannite). A mineral found in Cornwall. It is a sulpho-stannate of copper and iron, sometimes containing zinc, $SnS_2.Cu_2S.FeS.ZnS$.

Tin Salts (Tin Crystals). Stannous chloride, $SnCl_2$. Used as a wool mordant for dyeing cochineal scarlet, for dyeing blacks on silk, for weighting silk, and for calico printing.

Tinsel. An alloy of 60 per cent. tin and 40 per cent. lead.

Tinsmith's Solder. Usually a mixture of tin and lead in varying proportions. Coarse solder contains 60 per cent. tin and 40 per cent. lead, and melts at 168° C., and fine solder consists of 67 per cent. tin and 33 per cent. lead, and melts at 171° C. A solder suitable for cans contains 59 per cent. tin, 37 per cent. lead, and 4 per cent. bismuth, and one for tinned iron, melting at 275° C., consists of 87·5 per cent. lead and 12·5 per cent. tin. A solder for pewter contains 50 per cent. bismuth, 25 per cent. tin, and 25 per cent. lead, and melts at 96° C.

Tin Spirits. Solutions of stannous salts, employed in wool dyeing. Yellow, orange, scarlet, amaranth, bowl, purple, plum, and puce spirits are prepared by adding sulphuric acid to stannous chloride or by treating tin with a mixture of sulphuric and hydrochloric acids.

Red, purple, and aniline spirits are produced by dissolving tin in a mixture of hydrochloric and nitric acids.

Cotton spirits consist of stannic salts with admixtures.

Such mordants are used under the names oxychloride of tin, nitro-muriate of tin, oxymuriate of tin, crimson, barwood, plum, red cotton, purple, and cotton spirits, and pink cutting liquid.

Tinstone (Cassiterite). A mineral. It consists mainly of tin dioxide, SnO_2.

Tin, Stone. See STREAM TIN.

Tin Tacks. Tinned iron tacks.

Tinticite. A mineral. It is $Fe_3(PO_4)_2$ $(OH)_3.3\frac{1}{2}H_2O$.

Tinuvin. A trade mark for ultraviolet light absorbers for incorporation in plastics materials. (746).

Tinuvin P. A substituted benzotriazole derivative having a peak absorption at

340 mμ. It is recommended for PVC, polystyrene and acrylics. (17).

Tinuvin 770. A proprietary ultra-violet-light stabiliser used in the making of polyolefin plastics. It is a modified hindered amine. (746).

Tin White. A pigment. It is stannic hydroxide, $Sn(OH)_4$. Used in enamel and glass-making.

Tin White Cobalt. A mineral consisting of an arsenide of cobalt, nickel, and iron.

Tin, Wood. See STREAM TIN.

Tinzenite. A mineral. It is an aluminium-calcium-manganese silicate.

Tiodine (Thiophysem). Thiosinamine-ethyl-iodide, $C_3H_5.NH.CS.NH_2.C_2H_5I$.

Tiona. A trade mark for a proprietary titanium oxide containing 98–100 per cent. titanium dioxide. A white pigment.

T.I.P. A proprietary preparation. It is tetra-iodo-phenol-phthalein for use in cholecystography.

Tiprenolol. A beta adrenergic receptor blocking agent. It is 3-isopropylamino-1 - [2 - (methylthio)phenoxy]propan - 2-ol.

Tiquinamide. A drug used in the treatment of peptic ulcer. It is 5, 6, 7, 8 - tetrahydro - 3 - methylquinoline - 8 - thiocarboxamide. " Wy 24081 " is under clinical trial as the hydrochloride.

Tirodite. A mineral. It is $2[(Mg,Mn)_8 Si_8O_{22}(O,OH)_2]$.

Tirolite. A mineral, $CaCO_3.2Cu(OH)_2. 2Cu_3(PO_4)_2.2H_2O$.

Tirucalli Gum. A product of an Indian plant of the *Euphorbia* species. It somewhat resembles gutta-percha.

Tisco Steel. A proprietary trade name for a high manganese steel containing up to 15 per cent. manganese.

Tised. A proprietary preparation of MEPROBAMATE. A sedative. (434).

Tissier's Metal. An alloy of 97 per cent. copper, 2 per cent. zinc, and 1 per cent. arsenic.

Titanaugite. A mineral. It is a titaniferous augite.

Titanbiotite. A mineral. It is a titaniferous biotite.

Titan Blue 3B, R. A direct cotton dyestuff.

Titan Brown O, Y, R. Direct cotton dyestuffs. The O brand gives a yellowish, and Y and R reddish shades.

Titan Cements. Cements obtained by fusing a mixture of titaniferous iron ore, limestone, and coke in an electric or blast furnace. Pig iron and a slag is produced. The ground slag constitutes Titan cement. It consists essentially of calcium titanate ($CaTiO_3$) with small amounts of ferrites, aluminates and calcium silicate, together

with from 2–10 per cent. ferric oxide. These cements have a specific gravity varying from 3·35–3·55. They are quick hardening and chemically resistant. These cements are stated to be very resistant.

Titan Como G, R, S. Direct cotton dyestuffs, giving blue shades.

Titaneisen. A mineral. It is menaccanite.

Titaneisstein (Iserin). A titaniferous iron sand from Iserwiesl and Riesengbirge.

Titanellow. Titanium oxalate, $Ti_2 (C_2O_4)_3.10H_2O$. A mordant.

Titan Grey. A direct cotton dyestuff.

Titaniferous Iron. See ILMENITE.

Titaniferous Iron Ore. See ILMENITE.

Titanital. The trade name for a proprietary titanium white. The golden seal brand contains from 95–98 per cent. titanium oxide. TiO_2, and the silver seal grade is a mixture of 80 per cent. titanium dioxide with 20 per cent. zinc oxide.

Titanite. See SPHENE. Also a proprietary aluminium-manganese alloy containing titanium.

Titanite No. 1. An explosive consisting of 85–88 per cent. ammonium nitrate, 6–8 per cent. trinitro-toluene, and 4·5–6·5 per cent. charcoal.

Titanium Alloy. A ferro-titanium is called by this name.

Titanium-calcium Pigment. A pigment similar to titanox, but the titanium oxide is precipitated on calcium sulphate instead of barium sulphate. It is stated to have a greater pigmenting power than titanox and is lighter, the gravity being 3·19 as against 4·3. It is stated to be not quite so good as titanox in elasticity, toughness, and resistance to abrasion.

Titanium Green. Titanium ferro-cyanide.

Titanium Lithopone. See TI-TONE.

Titanium White. A pigment with varying amounts of titanium dioxide and barium sulphate, but usually consisting of 25 per cent. titanium oxide with 75 per cent. barium sulphate. One quality known in trade as Extra X contains 70 per cent. titanium oxide, 10 per cent. barium sulphate, and 20 per cent. calcium phosphate.

Titanmelanite. A mineral. It is $8[Ca_3 (Fe,Ti)_2Si_3O_{12}]$.

Titan Navy. A dyestuff giving bluish and reddish-navy blue on cotton.

Titanoferrite. A variety of the mineral ilmenite.

Titanolith. See CRYPTONE.

Titanolivine. A mineral, $(Mg.Fe)_2 (Si.Ti)O_4$.

Titanomagnetite. A mineral, Ti_3O_4.

Titanomorphite. A mineral. It is an alteration product of sphene or titanite, $CaTiSiO_5$.

Titanox. A trade mark designation for a variety of titanium pigment products. Individual products bear designations such as A, B, C, etc., connoting different materials: for instance, A represents a substantially pure titanium dioxide : B represents a composite pigment containing about 75 per cent. barium sulphate : C represents a composite pigment containing about 70 per cent. calcium sulphate. (235).

Titanox-B-30. A pigment containing 30 per cent. titanium dioxide instead of 25 per cent. in " B."

Titanox RA-39. A proprietary trade name for a stearate coated titanium dioxide pigment easily dispersible in polystyrene and polyolefines. (289).

Titanpigeonite. A mineral. It is $(Ca,Mg,Fe'',Fe''',Ti)(Si,Ti)O_3$.

Titan Pink 3B. A bluish-pink dyestuff for cotton.

Titan Red. See GERANINE.

Titan Red for Wool. A dyestuff for cotton or wool.

Titan Rose. See GERANINE.

Titan Scarlet C, S. Direct cotton dyestuffs. The brand C gives yellowish, and S bluish-scarlet shades.

Titanspinel. Synonym for Ulvospinel.

Titan Yellow. A dyestuff. It is equivalent to Clayton yellow.

Titan Yellow GG, G, R, Y. Direct cotton dyestuffs. The brand GG gives a greenish-yellow, and R a dull reddish-yellow shade.

Titanweiss (C, Extra T, Standard T, Standard A). Trade names for titanium dioxide pigments extended with calcium or barium sulphates.

Titite. A proprietary rubber cement, partly made from rubber. It is waterproof, and is used for mending cloth, paper, rubber, leather, and wood.

Ti-tone. A titanium lithopone containing 15 per cent. titanium dioxide, 25 per cent. zinc oxide, and 60 per cent. barium sulphate. Its specific gravity is 4·25, and it is stated to have a covering power 60 per cent. greater than ordinary lithopone.

Titralac. A proprietary preparation containing calcium carbonate and glycine. An antacid. (275).

Tixo K100. A cyanoacrylate adhesive.

Tixylix. A proprietary preparation of promethazine hydrochloride pholcodine citrate and phenylpropanolamine hydrochloride. A cough linctus. (336).

Tiza. A name for boronatrocalcite (*q.v.*), found in South America.

Tizit. An alloy of 40–80 per cent.

tungsten, 4–15 per cent. titanium, 4 per cent. chromium, 2–4 per cent. carbon, 1–5 per cent. cerium, and 3–40 per cent. iron.

T. Metal. An alloy of 95 per cent. aluminium, 4 per cent. magnesium, 0·5 per cent. silicon, 0·5 per cent. iron, and 0·1 per cent. copper.

T.N.A. Tetranitro-aniline.

T.N.B. Trinitro-benzene.

T.N.T. See TROTYL.

T.N.X. Tetranitro-xylene.

Toad's Eye. See STREAM TIN.

Tobacco Extract. A dark-brown viscid liquid preparation from tobacco leaves or stems. A 1–2 per cent. solution is used for parasites on domestic animals, also for destroying insects on plants.

Tobacco Water. An aqueous infusion of tobacco.

Tobermorite. A calcium silicate mineral, of Tobermory.

Tobias Acid. 2-Naphthylamine-1-sulphonic acid.

Tobin Bronze. Alloys of 59–83 per cent. copper, 3–48 per cent. zinc, 0·9–12·4 per cent. tin, 0·31–2·14 per cent. lead, and 0·1–0·8 per cent. iron. One alloy contains 58·79 per cent. copper, 40·43 per cent. zinc, and 0·88 per cent. tin.

Tobramycin. An antibiotic obtained from cultures of *Streptomyces tenebrarius*, or the same substance obtained by any other means. It is O^6 - (3 - amino - 3 - deoxy - α - D - glucopyranosyl) - O^4 - (2, 6 - diamino - 2, 3, 6 - trideoxy - α - D - ribopyranosyl) - 2 - deoxystreptamine.

Toce. A proprietary trade name for Diethadione.

Tochlorine. A proprietary preparation. It is chloramine-T.

Tocopherol. See VITAMIN E.

Toddite. A mineral. It is a variety of columbite.

Toddy. The fermented sap of the coconut palm.

Todralazine. A drug used as an antihypertensive agent. It is ethyl 3-phthalazin-1-ylcarbazate. BINAZINE.

Tofenacin. A drug used in the treatment of the Parkinsonian syndrome. It is N - methyl - 2 - (2 - methylbenzhydryl-oxy)ethylamine. ELAMOL is the hydrochloride.

Tofranil. A proprietary preparation of imipramine hydrochloride. An antidepressant. (17).

Toile Micanite. See MICANITE CLOTH.

Toilet Vinegar. Dilute acetic acid with odorants.

Toisin's Solution. A microscopic stain used for staining white blood corpuscles. It consists of a mixture of $\frac{1}{2}$ grain methyl violet, 1 oz. glycerin, 15 grains sodium chloride, 2 drachms sodium sulphate, and 8 oz. distilled water.

T-ol. Dimethylthioanthrene. See SIN-TOL. A plasticiser.

Tolamine. See CHLORAMINE T.

Tolamolol. A beta adrenergic receptor blocking agent. It is 4-[2-(2-hydroxy-3 - o - tolyloxypropylamino)ethoxy] - benzamide.

Tolanase. A proprietary preparation of tolazamide. An oral hypoglycæmic agent. (325).

Tolane. Diphenyl-acetylene, $C(C_6H_5)$: $C(C_6H_5)$.

Tolane Red. A dyestuff. It is the sodium salt of benzene-azo-1 : 8-amino-naphthol-4 : 6-disulphonic acid, $C_{16}H_{11}N_3S_2O_7Na_2$. Dyes wool a brilliant red from an acid bath.

Tolantipyrine. p - Tolyl - dimethyl - pyrazole.

Tolazamide. N'N'-Hexamethylene-N^4 - toluene-p-sulphonyl-semicarbazide. Tolanase.

Tolazoline. 2-Benzyl-2-imidazoline. Priscol is the hydrochloride.

Tolbutamide. N - Butyl - N' - toluene - p-sulphonylurea. Artosin. Orinase. Rastinon.

Toldimfos. A source of phosphorus. 4-Dimethylamino - o - tolylphosphinic acid. TONOPHOSPHAN is the sodium salt.

Toledo Blue V. A polyazo dyestuff. It dyes cotton direct from a sodium sulphate and sodium carbonate bath, or a soap bath.

Tolex. A proprietary trade name for a vinyl resin coated leather cloth.

Tolite. See TROTYL.

Tollac Solvent. A proprietary trade name for a toluene substitute.

Tollen's Reagent. A solution of ammoniacal silver nitrate containing free caustic soda. It is prepared when required by mixing (1) 10 per cent. caustic soda with (2) ammoniacal silver nitrate, obtained by dissolving sufficient silver nitrate to yield a 10 per cent. solution, in a mixture of equal volumes of concentrated ammonia and distilled water. It is used to test for aldehydes and other reducing substances.

Tolmesoxide. An anti-hypertensive drug. 4, 5 - Dimethoxy - o - tolyl methyl sulphoxide.

Tolnaftate. O-2-Naphthyl N-methyl-m-tolylthiocarbamate. Naphthiomate-T. Tinaderm.

Tolnate. A proprietary preparation of prothipendyl hydrochloride. A sedative. (281).

Toloy 45. An alloy of 45 per cent. nickel and 20 per cent. chromium with other materials. Used where stress corrosion

resistance is required. The material conforms to BS.1648 Grade H.

Tolpentamide. N-Cyclopentyl-N'-toluene-p-sulphonylurea.

Tolperisone. A muscle relaxant. 2-Methyl - 3 - piperidino - 1 - p - tolyl - propan-1-one. MYDOCALM.

Tolpiprazole. A tranquilliser. 5-Methyl - 3 - [2 - (4 - m - tolylpiperazin - 1-yl)ethyl]pyrazole.

Tolpronine. 1-(1,2,3,6-Tetrahydropyridino)-3-o-tolyl oxypropan-2-ol. Proponesin is the hydrochloride.

Tolpropamide. NN-Dimethyl-3-phenyl-3-p-tolylpropylamine. Tylagel is the hydrochloride.

Tolseram. A proprietary preparation of mephenesin carbamate. Relief of spasticity in paralysed limbs. (326).

Tolu Balsam. The oleo-resin of *Myroxylon toluifera*, of South America.

Toluidine Blue O (Methylene Blue T50). A dyestuff. It is the zinc double chloride of dimethyl-diamino-toluphenazthionium-chloride, $(C_{15}H_{16}N_3SCl)_2.ZnCl_2$. Dyes tanned cotton blue.

Toluol. Commercial toluene, $C_6H_5CH_3$. It contains traces of benzene, xylene, paraffin, and thio-toluene. It is employed in the explosive industry, in the preparation of saccharin, and in the synthesis of indigo. Specific gravity at 15° C. is 0·87. Refractive index 1·489. Boiling-point is 108–112° C. Ignition-point is about 5° C.

Tolupyrine (Tolypyrine). Tolyl-antipyrine (1-o-tolyl-2 : 3-dimethyl-pyrazolone), $CH_3.C_6H_4N[N(CH_3)C.CH_3][CO.CH]$.

Tolusafranine. See SAFRANINE RAE.

Toluylene. Stilbene, $C_6H_5.CH : CH.C_6H_5$.

Toluylene Black G. See PALATINE BLACK.

Toluylene Blue (Witt's Toluylene Blue). $C_{15}H_{18}N_4.HCl$. An indamine dyestuff obtained by the oxidation of dimethyl-p-phenylene-diamine and m-toluylene-diamine, or by the combination of nitroso-dimethyl-aniline-hydro-chloride and m-toluylene-diamine. Dyes cotton.

Toluylene Brown G. A dyestuff. It is the sodium salt of sulpho-toluene-disazo-m-phenylene-diamine, $C_{13}H_{11}N_6SO_3Na$. Dyes cotton yellowish-brown.

Toluylene Brown R. A dyestuff. It is the sodium salt of sulpho-toluene-disazo-bi-m-phenylene - diamine - azo-naphthalene-sulphonic acid, $C_{39}H_{29}N_{12}O_9S_3Na_3$. Dyes cotton brown from a soap bath.

Toluylene Orange G. (Kanthosine J). A dyestuff. It is the sodium salt of ditolyl - disazo - o - cresol - carboxylic -m,

toluylene - diamine - sulphonic acid-
$C_{29}H_{26}N_6O_6SNa_2$. Dyes cotton orange.

Toluylene Orange R (Kanthosine R).
A dyestuff. It is the sodium salt of
ditolyl-disazo-bi-m-toluylene - diamine -
sulphonic acid, $C_{28}H_{28}N_8O_6S_2Na_2$.
Dyes cotton reddish-orange.

Toluylene Orange RR. A dyestuff.
It is the sodium salt of sulpho-toluene-
disazo-bi-β-naphthylamine. $C_{27}H_{21}N_6$
O_3SNa. Dyes cotton reddish-orange
from a soap bath.

Toluylene Red Neutral red $(q.v.)$.
Dianol brilliant red $(q.v.)$.

Toluylene Yellow. A dyestuff. It is
sulpho - toluene-disazo - bi - nitro - m -
phenylene-diamine, $C_{19}H_{17}N_{10}O_7SNa$.
Dyes cotton yellow.

Tolycaine. Methyl 2-diethylaminoaceta-
mido-m-toluate. Baycain is the hydrochloride.

Tolyl Peri Acid. 1-Tolyl-naphthyl-
amine-8-sulphonic acid.

Tolypyrine. See TOLUPYRINE.

Tolysal. Tolypyrine salicylate (p-tolyl-
dimethyl-pyrazole), $C_{12}H_{14}N_2 . O . C_7$
H_6O_3. Used in rheumatism.

Tolysin. The ethyl ester of p-methyl-
phenyl-cinchoninic acid. An antipyretic.

Tombac. An alloy usually containing
89 per cent. copper, 5·5 per cent. zinc,
and 5·5 per cent. tin.

Tombac, Red (Red Brass, Tambac). A
brass containing less than 18 per cent.
zinc. It usually consists of 90 per cent.
copper and 10 per cent. zinc, and is
used for rolling into leaf.

Tombac, White (White Copper). An
alloy containing 75 per cent. copper
and 25 per cent. tin.

Tombasil. A proprietary trade name
for an alloy consisting mainly of tombac
metal with silicon.

Tomophan. A proprietary viscose packing material.

Tonal. Salts of glyceryl - phosphoric
acid, $C_3H_5(OH)_2.O.PO(OH)_2$. Used in
medicine as a rapidly assimilated form
of phosphorus.

Tonca (Tonka). Tonquin bean.

Toncan. A corrosion-resisting alloy
containing pure iron, copper, and
molybdenum.

Toncas Metal. An alloy of 29 per
cent. nickel, 36 per cent. copper, 7·1 per
cent. iron, 7·1 per cent. zinc, 7·1 per cent.
lead, 7·1 per cent. tin, and 7·1 per cent.
antimony. Used for ornamental work.

Tonic Cups. Cups made from quassia
wood.

Toniron. A proprietary preparation of
ferrous sulphate used in the treatment of anæmia. (250).

Tonite (Potentite). An explosive. It
consists of mixtures of granulated guncotton and barium nitrate. No. 1
contains 50–52 parts guncotton and
40–47 parts barium nitrate. No. 2
contains charcoal also. No. 3 consists
of 18–20 parts guncotton, 67–70 parts
barium nitrate, and 11–31 parts dinitro-
benzene. Also a name for chloracetone,
$CH_3COCH_2.Cl$.

Tonivitan Capsules. A proprietary preparation of Vitamins A, C, D, thiamine
hydrochloride, nicotinic acid and dried
yeast. A tonic. (250).

Tonivitan A & D. A proprietary preparation of Vitamins A, D, iron and
ammonium citrate, calcium and manganese glycerophosphate and copper
sulphate. A tonic. (250).

Tonivitan B. A proprietary preparation
of thiamine hydrochloride, riboflavine,
pyridoxine hydrochloride, nicotinamide,
and calcium, manganese and strychnine
glycerophosphates. A tonic. (250).

Tonka. See TONCA.

Tonka-bean Camphor. Coumarin,
$C_6H_4OCOCH : CH$, found in species of
Dipteryx.

Tonnenite. A mineral. It is double
sulphide of copper and bismuth,
$CuBiS_2$.

Tonol. See KALIO-TONAL and TONAL.

Tonophosphan. A proprietary preparation of TOLDIMFOS sodium. A
source of phosphorus used for veterinary purposes.

Tonox. A proprietary trade name for
para-para-diamino-diphenylmethane. A
material for increasing the abrasion
resistance of rubber. It is soluble in
organic solvents and has a melting
point of 60° C., a specific gravity of 1·1,
is non volatile and is used in amounts of
0·25 to 1·0 per cent. It also acts as a
rubber anti-oxidant. (163).

Tonox D. A pure form of Tonox. (163).

Tonquin Bean. Tonka bean, used to
flavour snuff.

Tonquinol. Trinitro-butyl-toluene.

Tonsil. See Frankonite.

Tonsillin. A proprietary preparation of
BENZALKONIUM CHLORIDE and penicillin V used in the treatment of tonsillitis.
(439).

Tool Steel. See HIGH SPEED TOOL
STEELS.

Toonu Gum. See TUNO GUM.

Toothache Jelly. Consists of equal
parts of collodion and phenol.

Toothache Seed. Henbane seed.

Topanol CA. A proprietary trade name
for an antioxidant for polypropylene
and other olefines and vinyls against
thermal attack during processing or in
service. It is approved by the U.S.
Food and Drugs Administration for use

in food wrappings. It is a phenolic material. (2).

Topanol M. N, N¹ - di - *sec* - butyl - *p* - phenylene diamine. (512).

Topanol O, OC. A proprietary trade name for 4-methyl-2,6-tert. butyl phenol, an antioxidant. (2).

Topaz. A mineral. It is an aluminium-hydroxy-fluo-silicate, $(AlF)_2SiO_4$. A gem stone. Specific gravity 3·55 and refractive index 1·61–1·63.

Topaz, False. See CITRINE.

Topaz, Occidental. See CITRINE.

Topazolite. A mineral. It is a lime-aluminium-garnet.

Topaz, Scotch. Golden topaz (*q.v.*).

Topaz, Spanish. See GOLDEN TOPAZ.

Töpfer's Reagent. Dimethyl-amino-azo-benzene (0·5 gram) in 100 c.c. of 95 per cent. alcohol. It is used to test acidity in stomach contents.

Tophet. An electrical resistance alloy containing 61 per cent. nickel, 10 per cent. chromium, 26 per cent. iron, and 3 per cent. manganese.

Tophet A. A proprietary trade name for 80 per cent. nickel, 20 per cent. chrome resistance wire.

Tophet C. A proprietary trade name for nickel chrome iron resistance wire.

Tophosote. Creosote tanno-phosphate.

Topilar. A proprietary preparation of FLUCLOROLONE ACETONIDE in a cream base used in the treatment of eczema and psoriasis. (809).

Toracsol. A proprietary preparation of cetyltrimethyl ammonium bromide, BENZALKONIUM and CETYLPYRIDINIUM bromides used in the treatment of acne. (953).

Toramin. The ammonium salt of tri-chlor-butyl-malonate. A substitute for opium in medicine.

Toray C.M.1001. A proprietary brand of Nylon 6. (954).

Toray C.M.2001. A proprietary brand of Nylon 610. (954).

Toray X3001. A proprietary brand of Nylon 66. (954).

Torbanite. A variety of cannel coal.

Torberite. See CUPROURANITE.

Torbernite. A mineral. It is a phosphate of uranium and copper.

Torecan. A proprietary trade name for Thiethylperazine or its salts. An antiemetic. (267).

Torendrikite. A mineral. It is $2[(Ca,Na)_2(Mg,Fe\dotdot,Fe\dotdotdot)_5Si_8O_{22}(OH)_2]$.

Torfoleum. A sound- and heat-insulating material made from peat moss.

Toril. A variety of beef extract.

Torlon 2000. A range of poly(amide-imide) resins variously filled for moulding purposes. (956).

Tormentil. The dried rhizome of *Poten-*

tilla tormentilla. Used for tanning and as an astringent.

Tormol. A nickel-chromium-molybdenum steel highly resistant to shock and fatigue.

Törnebohmite. A mineral. It is $Ce_3Si_2O_8OH$.

Tornesit. A trade name for a proprietary protective coating base prepared by the chlorination of rubber. It is a white powder of specific gravity 1·5, is chemically inert and non-inflammable. It is soluble in benzene, toluene, xylene, in esters, in chlorinated hydrocarbons, and in vegetable oils. It is insoluble in water, alcohol, glycerin, mineral oils, paraffin, and fatty acids. It is stated to be very resistant to acids, and is used in the paint and varnish industry.

Toron. A sulphur-terpene compound prepared by heating turpentine with sulphur. It is a black viscid liquid or semi-solid. Used for waterproofing cloth, preparing rubberised cloth, and for attaching or coating metal surfaces with rubber.

Torrensite. A mineral. It is a manganese-silico-carbonate.

Torreyite. A mineral. It is $(Mg,Zn,Mn)_7SO_4(OH)_{12}.4H_2O$.

Torrington's Drops. Tinct. Benz. Co.

Torula. A variety of yeast found in milk and beer.

Tosmilen. A proprietary preparation of DEMECARIUM BROMIDE as eye-drops, used in the treatment of glaucoma. (477).

Totaigite. A mineral. It is $(Mg,Ca)_2SiO_4.H_2O$.

Totaquina. Cinchona alkaloids 70 per cent. crystallisable.

Totin Glue. A glue made from gelatin.

Totolin. A proprietary preparation of PHENYLPROPANOLAMINE hydrochloride and GUAIPHENESIN. A cough linctus. (476).

Totomycin. A proprietary preparation containing Tetracycline Hydrochloride. An antibiotic. (253).

Toucas Metal. An alloy of 35·75 per cent. copper, 28·56 per cent. nickel, 7·1 per cent. zinc, 7·2 per cent. tin, 7·1 per cent. lead, 7·2 per cent. antimony, and 7·1 per cent. iron.

Touch-paper. Paper which has been dipped in a solution of potassium nitrate, and dried. Used in fireworks.

Touchpong Gum. See POUCKPONG GUM.

Touchstone. See LYDIAN STONE.

Tough Copper. Commercial copper, containing impurities such as arsenic.

Toughened Caustic. Consists of 95 per cent. silver nitrate and 5 per cent.

potassium nitrate, fused together. (*Argenti nitras induratus, B.P.*)

Touloucouna Oil. See CRAB WOOD OIL.

Toulouron Oil. An oil obtained from *Pagarus latro*. Used in Senegal for the treatment of rheumatism.

Tourmaline. A mineral. It is an alkali-calcium-aluminium-silicate.

Tournant Oil. A commercial brand of olive oil obtained from fermented marc of expressed olives. It contains free fatty acids, and is used as a Turkey-red oil.

Tournay's Metal. An alloy of 82·5 per cent. copper and 17·5 per cent. zinc.

Tournesol. See LITMUS.

Tournesol en Drapeaux. A blue colouring matter allied to litmus, manufactured from *Croton tinctorium*. It is used for colouring cheese wrappers to detect ripeness in them. When lactic acid is formed the wrapping changes to a red colour.

Tous-les-mois Starch (Queensland Arrowroot). The starch from the rhizomes of *Canna edulis*.

Tova. A proprietary preparation of cyclical hormones consisting of 16 white tablets of ethinyl œstradiol and 5 pink tablets of ethinyl œstradiol and di-methisterone. Used for control of menstruation. (182).

Towanite. See COPPER PYRITES.

Tower Acid. See CONTACT ACID.

Toxynone. Sodium - *m* - acetyl - amino - mercuri-benzoate, $HgOH.C_6H_3(NH.C_2H_3O)CO_2Na$. Used in the treatment of syphilis.

T.P.G. Triphenyl-guanidine.

TPX. A proprietary trade name for 4-methyl pentene-1 polymers. They are characterised by high clarity, a melting point of over 200° C., excellent electrical properties, resistance to many chemicals and a specific gravity lower than any other known thermoplastics. (2).

Trabuk. An alloy containing 87·5 per cent. tin, 5·5 per cent. nickel, 5 per cent. antimony, and 2 per cent. bismuth. It resists vegetable acids.

Trachyte. A volcanic rock composed of felspar with some hornblende and mica.

Tracumin. Copper trichloro - butyl - malonate. Used for diseases of the eye as a 7–10 per cent. salve.

Trafuril. A proprietary trade name for Thurfyl Nicotinate. (Tetrahydrofurfuryl nicotinate). (519).

Tragacanth, Arrehbor. See ARREHBOR TRAGACANTH.

Tragacanth, False. See BASSORA GUM.

Tragasol. A gum obtained by steep-ing locust-bean kernels in water. Used as a binding material. See also INDUSTRIAL GUM.

Trainite. A mineral. It is an impure variety of varicite.

Train Oil. Whale Oil.

Tral. A proprietary trade name for Hexocyclium methylsulphate.

Tramazoline. 2-(5,6,7,8-Tetrahydro-1-naphthylamino)-2-imidazoline.

Tram Silk. A silk obtained from less perfect cocoons than organzine silk.

Trancopal. A proprietary preparation of chlormezanone. A sedative. (112).

Tran-cor. A proprietary trade name for high silicon steel used in transformers.

Tranexamic Acid. trans-4-Amino-methylcyclohexanecarboxylic acid. Amikapron.

Tranquo - Adamon. The registered trade name for an antispasmodic and tranquilliser. It is a mixture of Adamon (*q.v.*) and Meprobamate (*q.v.*). (844).

Transargan. A German product. It is a silver-sodium-thiosulphate.

Transformer Oils. Usually the higher petroleum fractions, and more or less identical with lubricating oils. They are free from moisture and acids. Formerly rosin oil was used.

Transite Board. (Transite is a trade mark.) See PETRUSAL.

Transithal. Buthalitone soduim.

Translink 37. The proprietary name for a specially-treated and processed complex aluminium silicate pigment used in the making of cross-linkable polyethylene and ethylene-propylene rubbers. (957).

Transpar. Lactic acid and buffered lactic acid mixtures.

Transparent Gold Ochre. See YELLOW OCHRE.

Transparent Oxide of Chromium. Sesquioxide of chromium, a green pigment.

Transparent Soaps (Glycerin Soaps). Made from mixtures of decolorised tallow, with castor, linseed, and coconut oils, with the addition of glycerin, and also of from 20–30 per cent. saccharose or glucose. At one time transparent soaps were made by evaporating an alcoholic solution of the soap.

Transpex 1. A proprietary trade name for an unplasticised polymethylmethacrylate. It is a plastic said to be equivalent to optical crown glass.

Transpex 2. A proprietary trade name for unplasticised polystyrene. It is a plastic said to be equivalent to optical flint glass.

Transpulmin. (1) A solution of 3 parts quinine and 2 parts camphor in 100 parts ethereal oils. Used as an injection for bronchitis and pneumonia.

Transpulmin. (2) A proprietary preparation of chinindium, camphor, menthol, eucalyptus and olive oil, a smoking deterrent. (370).

Transvaal Jade. Synonym for Grossular.

Transvasin. A proprietary preparation of tetrahydrofurfuryl salicylate, ethyl nicotinate, N-hexyl nicotinate and ethyl *p*-aminobenzoate. (347).

Tranxene. A proprietary preparation of potassium clorazepate. A sedative. (278).

Tranylcypromine. ($\pm$)-trans-2-Phenylcyclopropylamine. Parnate is the sulphate.

Trap. An igneous rock containing SiO_2, Al_2O_3, Fe, MnO, CaO, MgO, K_2O, and Na_2O.

Trapocá Resin. Jatoba Lagrima.

Trasentine. An antispasmodic combining the actions of atropine and papaverine without the unpleasant side-effects of the former, indicated in all spastic conditions of the smooth musculature.

Trasicor. A proprietary preparation of OXPRENOLOL hydrochloride used in the treatment of angina pectoris and hypertension. (18).

Trass. A volcanic material found on the bank of the Rhine. It is used in Holland as an addition to lime, to convert it into hydraulic lime.

Trasulphane. A name for ammonium-sulpho-ichthiolicum.

Trasylol. A proprietary preparation of APROTININ used to treat shock. (112).

Traumatic Balsam. Compound tincture of benzoin.

Traumaticin. Consists of 1 part gutta-percha and 10 parts chloroform, by weight.

Traumatol. See IODOCRESOL.

Trautwinite. Synonym for Uravorite.

Traversellite. A mineral. It consists of augite with some alumina.

Traversoite. A mineral of Sardinia, $2(Cu.Ca)O.Al_2O_3.2SiO_2.12H_2O$.

Travertine (Calc Sinter, Calcareous Tufa). A limestone deposited by calcareous springs.

Travis Silk. A proprietary viscose silk.

Trazodone. A stimulant for the central nervous system. It is 2-[3-(4-*m*-chlorophenyl - 1 - piperazinyl)propyl] - 1, 2, 4 - triazolo[4, 3 - *a*]pyridin - 3 - (2*H*)-one.

Treacle. When refined sugar liquor has been repeatedly boiled and crystallised an uncrystallisable residue is obtained. This is treacle and corresponds to the molasses obtained at the raw sugar works.

Treacles. See GOLDEN SYRUPS.

Trebizond Opium. Persian opium.

Treble Superphosphate. This consists of mono-calcium phosphate, containing 48–49 per cent. P_2O_5 (41–42 per cent. water soluble P_2O_5). A fertiliser which does not tend to take up water from the atmosphere.

Trecator. A proprietary preparation of ethionamide. An antituberculous agent. (820).

Tree Copal. A name applied to white Zanzibar copal.

Tree Gum (Wood Gum). Xylane, $C_6H_{10}O_5$.

Tree Turmeric. See FALSE COLUMBA.

Tree Wax. See CHINESE WAX.

Trefol. A name applied to amyl salicylate.

Trehalase. An enzyme. It splits trehalose.

Trelit. A proprietary pyroxylin plastic.

Tremerad. A proprietary preparation of CLIOXANIDE. A veterinary anthelmintic.

Tremolite (White Amphibole, Grammatite). A mineral, $CaMg_3.(SiO_3)_4$. It is also given as $2[Ca_2Mg_5Si_8O_{22}(OH)_2]$.

Tremonil. A proprietary preparation of methixene hydrochloride, used in Parkinsonism. (244).

Trenbolone. An anabolic steroid. 17β-Hydroxyœstra - 4, 9, 11 - trien - 3 - one. FINAJET is the acetate.

Trench's Flameless Explosive. An explosive containing ammonium nitrate.

Trenimon. A proprietary preparation of triaziquone. An antimitotic. (341).

Trentadil. A proprietary preparation of bamifylline hydrochloride. (327).

Trental. A proprietary preparation of oxypentifylline used in the treatment of peripheral vascular disease. (312).

Trent Sand. A sand found in the Trent, Severn, and other rivers. Used for polishing.

Trényline. A proprietary trade name for rubber vulcanisation accelerator. It is triphenyl-guanidine.

Treosulfan. L - Threitol 1, 4 - dimethanesulphonate.

Treparsol. A formyl derivative of *m*-amino - hydroxy - *p* - phenyl-arsonic-acid. It is used in the treatment of syphilis.

Trépol. See LUATOL B.

Tréposan. A preparation of bismuth succinate. An antisyphilitic.

Trescatyl. A proprietary preparation of ethionamide. An antituberculous agent. (336).

Trescazide. A proprietary preparation of ethionamide and isoniazid. An antituberculous agent. (336).

Tretamine. 2,4,6-Tri-(1-aziridinyl)-1,3, 5-triazine. Triethylene Melamine. TEM. A proprietary preparation of triethanomelamine. A carcino-chemo-therapeutic agent used for leukæmias. (2).

Trethinium Tosylate. 2-Ethyl-1,2,3, 4 - tetrahydro - 2 - methylisoquinolinium toluene-*p*-sulphonate.

Trethylene. Trichlorethylene, C_2HCl_3.

Tretinoin. 3, 7 - Dimethyl - 9 - (2, 6, 6 - trimethylcyclohex - 1 - enyl)nona - 2, 4, 6, 8 - all - *trans* - tetraenoic acid. RETIN-A.

Tret-o-Lite. A patented preparation for the destruction of petroleum emulsions. It consists of 83 per cent. sodium oleate, 5·5 per cent. sodium resinate, 5·0 per cent. sodium silicate, 4·0 per cent. phenol, and 1·5 per cent. paraffin wax.

Trevintix. A proprietary preparation containing prothionamide. An antituberculous agent. (336).

Trevorite. A mineral, (Ni.Fe)O.FeO₃.

Tri. An abbreviation for trichlorethylene, C_2HCl_3, a solvent with a specific gravity of 1·4726 and a boiling-point of 86·7° C.

Triacetin. Glyceryl-triacetate, $C_9H_{14}O_6$.

Triacetyloleandomycin. The triacetyl ester of oleandomycin. Evramycin.

Tri-Adcortyl. A proprietary preparation of triamcinolone acetonide, neomycin sulphate, gramicidin and nystatin used in dermatology as an antibacterial agent. (326).

Triadimefon. A pesticide. 1-(4-Chlorophenoxy) - 3, 3 - dimethyl - 1 - (1, 2, 4 - triazol - 1 - yl) butan - 2 - one.

Triamcinolone. 9α-Fluoro-11β,16α,17α, 21 - tetrahydroxypregna - 1,4 - diene - 3,20-dione.   9α-Fluoro-16α-hydroxy-prednisolone.   Adcortyl.   Aristocort. Ledercort. Adcortyl-A is the acetonide.

Triamterine. 2,4,7-Triamino-6-phenyl-pteridine. Dytac.

Trianisoline. See RHODAMINE 6G.

Triatox. See AMITRAZ.

Triatrix. See AMITRAZ.

Triaziquone. Tri-(1-aziridinyl)-1,4-benzoquinone. Trenimon.

Tribase. Lead based stabilizer for PVC. (572).

Tribase XL. A trade mark for modified tribasic lead sulphate. A stabiliser for vinyl polymers. (47).

Tribromsalan. An anthelmintic. 3, 4', 5-Tribromosalicylanilide. HILOMID.

Tricaderm. A proprietary preparation of TRIAMCINOLONE acetonide, salicylic acid and BENZALKONIUM CHLORIDE used in the treatment of eczema. (682).

Tricalcine. A trade mark for a preparation containing the active properties of the tribasic phosphates and carbonates of calcium and magnesium. Recommended in tuberculosis.

Tricalcol. A colloidal combination of tricalcium phosphate and albumin.

Trichalcite. A mineral, $Cu_3(AsO_4)_2$. $5H_2O$.

Trichlormethine. Trimustine.

Trichlorphon. See METRIPHONATE.

Trichorad. Acinitrazole.

Tri-clene. A proprietary trade name for trichlorethylene used in dry-cleaning. See also TRIKLONE.

Triclofenol Piperazine. Piperazine di-(2,4,5-trichlorphenoxide).

Triclofos. 2,2,2-Trichloroethyldihydrogen phosphate. Tricloryl is the mono-sodium salt.

Tricloryl. A proprietary preparation of triclofos sodium. (The monosodium salt of trichlorethyl phosphate.). A sedative. (335).

Triclosan. A bactericide currently undergoing clinical trial as " GP 41,353 ". It is 5-chloro-2-(2,4-dichlorophenoxy)-phenol.

Tricoid. Cine film cement. (507).

Tricresol. A purified mixture of the three cresols. It contains about 35 per cent. ortho, 20 per cent. meta, and 25 per cent. para-cresol.

Tricyclamol Chloride. 1-(3-Cyclohexyl-3 - hydroxy - 3 - phenylpropyl) - 1 - methylpyrrolidinium chloride. Elorine Chloride. Lergine.

Tridesilon. A proprietary preparation of DESONIDE used in the treatment of eczema and psoriasis. (651).

Tridia. A proprietary preparation of neomycin sulphate, clioquinol and kaolin. An antidiarrhœal. (280).

Tridihexethyl Chloride. 1-Cyclohexyl-3 - diethylamino - 1 - phenylpropan - 1 - ol ethochloride. (3 - Cyclohexyl - 3 - hydroxy - 3 - phenyl)triethylammonium chloride.

Tridione. A proprietary preparation of troxidone. An anticonvulsant. (311).

Tridymite. A mineral. It is quartz, SiO_2.

Trieline. A term for trichlor-ethylene, C_2HCl_3.

Triethanolamine. Tri - β - hydroxyl - ethylamine. The industrial material contains from 20–25 per cent. of secondary, 75–80 per cent. of tertiary, and 0·5 per cent. primary amines.

Triethanomelamine. Tretamine. Triethylene Melamine. TEM.

Triferrin. See FERRINOL.

Trifluomeprazine. A tranquilliser. *NN*-Dimethyl - 2 - (2 - trifluoromethyl - phenothiazin - 10 - ylmethyl propyl-amine). NORTRAN.

Trifluoperazine. 10 - [3 - (4 - Methyl - piperazin - 1 - yl)propyl] - 2 - trifluoro-methylphenothiazine. Stelazine is the base or the dihydro chloride.

Trifluperidol. A neuroleptic. $1-[3-(4-$Fluorobenzoyl)propyl] - 4 - (3 - tri - fluoromethylphenyl)piperidin - 4 - ol. TRIPERIDOL.

Triformal (Paraform, Triformol, Paraformol). Paraformaldehyde, $(CH_2O)_3$, a polymer of formaldehyde. A disinfectant.

Triformin. Glyceryl formate.

Triformol. See TRIFORMAL.

Trigemine (Methutin, Asciatine). Dimethyl-amino-antipyrine-butyl-chloral-hydrate, $C_{17}H_{24}O_3N_3Cl_3$. Prescribed for headache and neuralgia, also as a mild antipyretic. Its solutions are antiseptic.

Trigonite. A mineral. It is an acid arsenite of lead and manganese, Pb_3 $MnH(AsO_3)_3$, of Sweden.

Trigonox A-75. A proprietary catalyst used in the polymerisation of vinyl monomers, unsaturated melamine resins and unsaturated polyesters; also in the preparation of graft polymers. It is composed as to 75 per cent. of *tert.* butyl hydroperoxide and as to 25 per cent. of *di-tert.* butyl peroxide. (90).

Trigonox B. A proprietary vulcanisation and cross-linking agent. It is *di-tert.* butyl peroxide. (90).

Trigonox B.P.I.C. A proprietary liquid monoperoxycarbonate used in the polymerisation of acrylics, olefins, styrene and vinyl acetate. It is *tert.* butyl isopropyl carbonate. (90).

Trigonox C. A proprietary vulcanising and cross-linking agent. It is *tert.* butyl perbenzoate. (90).

Trigonox FM-50. A proprietary catalyst for the polymerisation of unsaturated polyester resins. It is 50 per cent. *tert.* butyl peracetate in dimethyl phthalate. (90).

Trigonox T. A proprietary vulcanising and cross-linking agent. It is *tert.* butyl cumyl peroxide. (90).

Trigonox 21. A proprietary catalyst for polyester resins, curing at temperatures of 80° to 100° C. It is *tert.* butyl peroxy-2-ethyl hexanoate. (90).

Trigonox 25/75. A 75 per cent. solution of *tert.*-butyl perpivalate in white hydrocarbon oil. It is used as an initiator in the production of polyethylene by the high-temperature process. (90).

Trigonox 29/40. A proprietary polyfunctional peroxide in calcium carbonate used in the cross-linking of olefin copolymers, silicones and both natural and synthetic rubbers. (90).

Trigonox 36. A proprietary preparation used as an initiator in the polymerisation of ethylene, vinyl chloride, styrene and acrylates; also for the curing of unsaturated polyester resins. It is

nonanoyl peroxide. 3, 5, 5 trimethyl-hexanoyl peroxide. It is supplied as either a 75 per cent. or a 50 per cent. solution in white hydrocarbon oil. (90).

Trigonox 40. A proprietary liquid ketone peroxide used in the curing of unsaturated polyester resins in combination with cobalt accelerators. (90).

Trigonox 44. A proprietary accelerator for the curing of unsaturated polyester resins. It is a paste based on a liquid ketone peroxide, and is used in combination with cobalt accelerators. (90).

Trihexyphenidyl. Benzhexol.

Trihydrocalcite. A mineral, $CaCO_3$. $3H_2O$.

Triketol. See NINHYDRIN.

Triklone. A proprietary trade name for trichlorethylene.

Trikresol. See BACTEROL and TRICRESOL.

Trilactine. A preparation of lactic acid bacilli.

Trilafon. A proprietary trade name for Perphenazine.

Trilaurin. Glyceryl - ceryl - laurate, $C_3H_5(OOC.C_{11}H_{23})_3$.

Trilene. A trade mark for trichlorethylene, a general anæsthetic. (2).

Trilene M. A proprietary solid polyisocyanate used in non-discolouring surface coatings and adhesives. (539).

Trilite. See TROTYL.

Trillat's Reagent. An acetic acid solution of tetramethyl-diamino-di-phenylmethane, $[(CH_3)_2N.C_6H_4]_2.CH_2$. The reagent is prepared by dissolving 5 grams of the base in 100 c.c. of 10 per cent. acid. Lead peroxide gives a blue colour with the reagent, but the same colour is given by manganese dioxide and other oxidising agents.

Trillekamin. A proprietary trade name for the hydrochloride of Trimustine.

Trilob. A proprietary safety glass.

Trimelarsan. A proprietary trade name for Melarsonyl Potassium.

Trimellitic Acid. 1 : 2 : 4 - Benzene - tricarboxylic acid, $C_6H_3(COOH)_3$.

Trimene. A proprietary rubber vulcanisation accelerator. It is the stearic acid salt of the condensation product of ethylamine and formaldehyde (trimene base). It is sold in the concentrated form as trimene base, dissolved in latex, as latene, and in a paste form as trimene. (435).

Trimene Base. A proprietary rubber vulcanisation accelerator. It is triethyl-trimethylene-triamine. (435).

Trimeperidine. 1,2,5-Trimethyl-4-phenyl-4-propionyloxypiperidine.

Trimeprazine. 10-(3-Dimethylamino-2-methylpropyl)phenothiazine. Alimemazine. Vallergan is the tartrate.

Trimerite. A mineral, $Be(Mn.Ca.Fe)SiO_4$.

Trimesitinic Acid. Trimesic acid, (1 : 3 : 5-benzene-tricarboxylic acid), $C_6H_3(COOH)_3$.

Trimetaphan. 4,6-Dibenzyl-5-oxo-1-thia - 4,6 - diaza - tricyclo[6,3,0,0³,⁷] - undecanium.

Trimetaphan Camsylate. A hypotensive. 1, 3 - Dibenzyldecahydro - 2 - oxoimidazo[4, 5 - c]thieno[1, 2 - a] - thiolium (+) - camphor - 10 - sulphonate. ARFONAD.

Trimetazidine. 1-(2,3,4-Trimethoxybenzyl)piperazine. Vastarel is the dihydrochloride.

Trimethadione. Troxidone.

Trimethol. Trimethylo - methoxy - phenol. An antiseptic.

Trimethidinium Methosulphate. (+)-3 - (3 - Dimethylaminopropyl) - 1,8,8 - trimethyl - 3 - azabicyclo[3,2,1]octane di(methyl methosulphate). Baratol.

Trimethoprim. 2,4-Diamino-5-(3,4,5-trimethoxybenzyl)pyrimidine.

Trimethylborine. See BOROMETHYL.

Trimeton. A proprietary trade name for the maleate of Peniramine.

Trimipramine. 1-(3-Dimethylamino-2-methyl propyl)-4,5-dihydro-2,3 : 6,7-dibenzazepine. Surmontil.

Trimustine. Tri-(2-chloroethyl)amine. Trichlormethine. Trillekamin is the hydrochloride.

Trinidad Asphalt. A natural asphalt obtained from the Trinidad pitch lake. The crude pitch contains from 40–46 per cent. bitumen, 24–30 per cent. mineral matter (clay), and 21–29 per cent. water. It also contains mud, sticks, and roots. The refined material contains about 50 per cent. bitumen or asphalt, and 43 per cent. mineral matter, with small amounts of water. The asphalt is soluble in carbon disulphide. Used for paving. The melting-point of the refined product is about 85° C. and the specific gravity is 1·4.

Trinitrine. See NOBEL'S EXPLOSIVE OIL.

Trinitrol. Erythrol tetranitrate, $C_4H_6(NO_3)_4$.

Trinol. A proprietary preparation of benzhexol, used for Parkinsonism. (279). See TROTYL.

Trinuride. A proprietary preparation of phenylethylacetylurea, phenytoin and phenobarbitone. An anticonvulsant. (802).

Triogesic. A proprietary preparation of phenylpropanolamine hydrochloride and paracetamol. An analgesic. (244).

Triolein. See OLEIN.

Triolin. A material similar to linoleum.

Triominic. A proprietary preparation containing phenylpropanolamine hydrochloride and mepyramine and pheniramine maleate. (244).

Trional (Dithan, Methyl Sulphonal). Diethyl-sulpho-methyl-ethyl-methane, $C(CH_3)(C_2H_5)(SO_2)_2(C_2H_5)_2$. A soporific.

Trioran. See HEXORAN.

Triosil. A proprietary trade name for Sodium Metriozoate.

Triostam. A proprietary trade name for sodium antimonyl gluconate.

Triotussic. A proprietary preparation of phenylpropanolamine hydrochloride, mepyramine maleate, pheniramine maleate, noscapine, terpin hydrate and paracetamol. (244).

Trioxymethylene. The solid, polymerised form of formaldehyde, $(HCHO)_3$. See TRIFORMAL.

Trip. Ferric oxide, Fe_2O_3.

Triparanol. 2-(4-Chlorophenyl)-1-(4-diethylaminoethoxyphenyl) - 1 - p - tolylethanol. MER-29.

Tripelennamine. N-Benzyl-N′N′-dimethyl - N - 2 - pyridylethylenediamine. Pyribenzamine is the citrate or the hydrochloride.

Triperidol. A proprietary preparation of trifluperidol. A sedative. (356).

Tripestone. Anhydrite (q.v.).

Triphal. A German preparation. It is the sodium salt of auro-thio-benziminazole-carboxylic acid. Used in the treatment of tuberculosis.

Triphan. See IRIPHAN.

Triphane. The mineral Spodumene, $LiAl(SiO_3)_2$, is known by this name.

Triphenine. Propionyl-p-phenetidine, $C_6H_4(OC_2H_5)NH(CO.CH_2.CH_3)$. An antipyretic and antineuralgic.

Triphyllite. A mineral, $Li.Fe.Mn.PO_4$.

Triplastic. An explosive. It is prepared by mixing trinitro-toluene, together with some lead nitrate and chlorate, with a gelatin made from dinitro-toluene and nitro-cellulose.

Triplastite. An explosive consisting of a mixture of di- and tri-nitrotoluene 70 parts, guncotton 1·2 parts, and lead nitrate 28·8 parts.

Triple Salts (Trisalytes). Used in the electro-deposition of metals. They consist of the cyanide of the metal to be deposited, potassium cyanide, and potassium sulphide.

Triplex. A proprietary safety glass involving the use of pyroxylin.

Triplite. A mineral, $Fe.Mn_2.F.PO_4$.

Triploidite. A mineral, $(Fe.Mn)(FeHO.Mn)PO_4$.

Triplopen. A proprietary preparation containing Benethamine Penicillin, Procaine Penicillin and Benzylpenicillin Sodium. An antibiotic. (335).

Tripoli (Tripoli Powder, Rotten Stone). A mineral. It consists mainly of silica associated with small quantities of alumina and iron oxide, but the composition varies. The variety of tripoli powder found in Derbyshire, is called Rotten stone. Used as an abrasive.

Tripoli Powder. See TRIPOLI.

Tripolite. See INFUSORIAL EARTH.

Trippkeite. A mineral, $CuAs_2O_4$.

Triprolidine. trans.-1-(2-Pyridyl)-3-pyrrolidino - 1 - p - tolylprop - 1 - ene. Actidil is the hydrochloride.

Tripsa. Tribasic phosphate of soda. Used for the prevention of incrustation on boilers.

Triptafen. A proprietary preparation of amitriptyline hydrochloride and perphenazine. An antidepressant. (284).

Triptecol. A proprietary preparation of antazoline hydrochloride, pholcodine and ephedrine hydrochloride. A cough linctus. (348).

Tripuhyite. A mineral, $Fe_2Sb_2O_7$.

Tripyrate. See LENIGALLOL (Gallols).

Trisalytes. See TRIPLE SALTS.

Triscal. A proprietary preparation of calcium and magnesium carbonates. An antacid. (271).

Trisilane. Silicopropane (Trisilicon-octohydride), Si_3H_8.

Tristearin. See STEARIN.

Trisulphone Blue B, R. See TRISULPHONE VIOLET B.

Trisulphone Brown B (Chlorazol Brown LF). A polyazo dyestuff obtained from benzidine, salicylic acid, 2-amino-8-naphthol-3 : 6-disulphonic acid, and m-phenylene-diamine. Dyes cotton in a bath containing 2 per cent. sodium sulphate and 20 per cent. sodium chloride. Also dyes half-wool and half-silk goods in a neutral bath.

Trisulphone Brown G, 2G (Chlorazol Brown 2G). Polyazo dyestuffs of similar composition to the B brand, except that tolidine and dianisidine respectively are used instead of benzidine. They give rather yellower shades than the B quality.

Trisulphone Violet B (Trisulphone Blue B, R, Orion Violet, Orion Blue R33). Dyestuffs obtained from tetrazo compounds of benzidine, tolidine, or dianisidine, combined with α-naphthol-disulphonic acid S, and β-naphthol-sulphonic acid B.

Tritane. Triphenyl-methane, $(C_6H_5)_3$ CH.

Tritheon. Acinitrazole.

Trithion. A proprietary preparation of CARBOPHENOTHION. A veterinary insecticide.

Tritochlorite. A mineral. It is a variety of descloizite.

Tritole. See TROTYL.

Tritolo. See TROTYL.

Tritomite. A mineral. It is a fluosilicate of cerium, yttrium, thorium, calcium, with zirconium, and sodium, of Norway.

Triton. See TROTYL.

Tritonite. A mineral. It contains 3·63 per cent. Ta_2O_5, ZrO, etc.

Trivalin. See NEO-TRIVALIN.

Trivax. A proprietary vaccine for protection against diphtheria, tetanus and pertussis. (277).

Trixidin. An emulsion of antimony trioxide, containing 30 per cent. Sb_2O_3.

Trizma. A proprietary trade name for Trometamol.

Trobicin. A proprietary trade name for SPECTINOMYCIN. An antibiotic. (325).

Trodax. A proprietary preparation of NITROXYNIL. A veterinary anthelmintic.

Trogamid T. A registered trade mark for a transparent polyamide. It is an amorphous polycondensate of terephthalic acid or dimethyl terephthalate and an alkyl substituted hexamethylene diamine. Polyamide injection moulding compounds. (169).

Trogerite. A mineral, $(UO_2)_3.As_2O_5.$ $8H_2O$.

Troilite. Haidinger's name for the ferrous sulphide which occurs in meteorites.

Troisdorf Powder. An explosive. It is a gelatinised nitro-cellulose flake powder.

Trolen. A trade mark for polyethylene. (883).

Trolene. A proprietary preparation of FENCHLORPHOS. An insecticide.

Trolitan. A proprietary range of phenolic moulding materials. (169).

Trolitan 6520. A registered trade mark for glass reinforced moulding materials. (169).

Trolitan H. A urea-formaldehyde condensation product. A synthetic resin. (169).

Trolitan Super P. A proprietary trade name for a phenolic moulding compound. (169).

Trolitax. A proprietary phenol-formaldehyde resin laminated board.

Trolite. A synthetic resin of the phenol-formaldehyde type. It is a term also applied to trinitro-toluene. See TROTYL.

Trolitul. A proprietary trade name for polystyrene. (169).

Trolleite. A mineral, $Al_4(OH)_3(PO_4)_3$.

Trolit F. A proprietary pyroxylin product.

Trolit S and Special. Proprietary phenol-formaldehyde resin moulding compounds.

Trolit W. A proprietary cellulose acetate product.

Trolnitrate Phosphate. Triethanolamine OO'O''-trinitrate diorthophosphate. Praenitrona.

Trolon. A proprietary phenol-formaldehyde synthetic resin. See BAKELITE.

Troluoil. A proprietary trade name for a petroleum solvent.

Trombovar. A proprietary preparation of sodium tetradecyl sulphate, used for injection of varicose veins. (813).

Trometamol. 2-Amino-2-hydroxymethyl propane-1,3-diol. Tromethamine. Talatrol. Trizma.

Tromethamine. A proprietary trade name for Trometamol.

Tromexan. A proprietary preparation of ethyl bicoumacetate. An anticoagulant. (17).

Trona (Urao, Natron, Natrum). Trona is the name given to a salt deposit in Egypt, and Urao is the term given in Columbia to the same substance. It has the formula, $Na_2CO_3.CO_3HNa.3H_2O$. A mineral. It is $4[Na_3H(CO_3)_2.2H_2O]$.

Trona Red. See ST. DENIS RED.

Troostite. A mineral, $2(Zn.Mn)O.SiO_2$. It is also the name for a constituent of steel tempered at a high temperature. It occurs in the transformation of austenite, the stage following martensite, and preceding sorbite.

Troosto-Sorbite. A constituent of steel. It is similar to Troostite.

Tropacocaine. Benzoyl-pseudo-tropine, $C_5H_9.N(CH_2).CH.O.COC_6H_5(CH_2)$. A local anæsthetic.

Tropæoline D. See ORANGE III.

Tropæoline G. See METANIL YELLOW.

Tropæoline O (Tropæoline R, Tropæoline Y, Acid Yellow RS, Acme Yellow, Yellow T, Chrysoine, Resorcin Yellow, Chrysoline, Chryseoline Yellow, Chryseoline, Gold Yellow, Golden Yellow, Chrysoine Extra, Resorcin Yellow O Extra). A dyestuff. It is the sodium salt of dioxy-benzene-azo-benzene-p-sulphonic acid, $C_6H_3(OH)_2.N_2.C_6H_4.SO_3Na$. Dyes wool reddish-yellow from an acid bath.

Tropæoline OO. See ORANGE IV.

Tropæoline OOO No. 1. See ORANGE I.

Tropæoline OOO No. 2. See ORANGE II.

Tropæoline OOOO. A dyestuff. It is benzene-azo-α-naphthol-sulphonic acid, $C_6H_5.N_2.C_{10}H_5.HSO_3.OH$.

Tropæoline R. See TROPÆOLINE O.

Tropæoline Y. See TROPÆOLINE O.

Trophenium. A proprietary trade name for Phenactropinium Chloride.

Trophysan. A proprietary preparation of aminoacids, minerals and vitamins in SORBITOL used for intravenous feeding. (925).

Tropicamide. N-Ethyl-N-(4-pyridylmethyl)tropamide. Mydriacyl.

Tropigline. Tiglytropeine.

Tropium. A proprietary preparation of CHLORDIAZEPOXIDE. A tranquilliser. (498).

Tropon. A food preparation consisting of pure coagulated albumin from animal and vegetable residues, especially from meat flour. Usually it is a mixture of one-third animal and two-thirds vegetable proteid. Also see NUTROSE.

Troposan. A proprietary preparation. It is 2 - oxy - 5 - acetyl-amino-phenyl-arsinic acid.

Troposan-Quinine. The quinine salt of 2 - oxy - 5 - acetyl - amino - phenyl - arsinic acid.

Trotter Oil. Neatsfoot oil.

Trotyl (Trolite, Trilite, Tritolo, Trinol, Tolite, Triton, Tritole, T.N.T.). Trinitro-toluene, $CH_3.C_6H_2(NO_2)_3$. An explosive constituent.

Trovidur HT. A proprietary trade name for a chlorinated PVC. (169).

Troxerutin. 7,3',4' - Tri[o(2 - hydroxyethyl)]rutin.

Troxidone. 3,5,5-Trimethyloxazolidine-2,4-dione. Trimethadione. Tridione.

Troxonium Tosylate. Triethyl-2-(3,4, 5 - trimethoxybenzoyloxy) - ethylammonium p-toluenesulphonate.

Troxypyrrolium Tosylate. N-Ethyl-N - 2 - (3,4,5 - trimethoxybenzoyloxy) - ethylpyrrolidinium p-toluenesulphonate.

Troyes White. A white pigment consisting of calcium carbonate.

T.R.S. Rubber. A proprietary air-dried fast-curing rubber. (945).

Trudellite. A mineral, $Al_2(SO_4)_3.4AlCl_3.4Al(OH)_3.30H_2O$.

True Red Lead. See CRYSTAL MINIUM.

Truscottite. A mineral. It is a calcium silicate.

Tryen. See GRISERIN.

Trygase. Dried yeast cells, used in medicine.

Trylose S. Methylcellulose.

Trypaflavine. See ACRIFLAVINE.

Trypan Blue. An azo dyestuff derived from tolidine and naphthalene. It destroys the trypanosome of the cattle disease " piroplasmosis."

Trypanotoxyl. A combination of the reduction products of Atoxyl with fresh liver substance. Used in medicine as a trypanocide.

Trypan Red. A dyestuff prepared from tetrazotised benzidine-sulphonic acid, and 2 molecules of sodium-naphthyl-amino-disulphonate, $(SO_3H)C_6H_3.N_2.C_{10}H_4(SO_3Na)_2NH_2.C_6H_4.N_2.C_{10}H_4(SO_3Na)_2.NH_2$. It destroys the trypanosome of the South American horse-disease " mal-de-caderas."

Tryparosan. Chlorinated *p*-fuchsine.

Tryparsamide. The sodium salt of N-phenyl-glycine-amino-*p*-arsonic acid (OH)(ONa)AsO.C$_6$H$_4$.NH.CH$_2$CONH$_2$. Used in the treatment of sleeping-sickness and syphilis.

Tryposafrol. A derivative of pheno-safranine. A trypanocide.

Trypoxyl. A proprietary preparation of sodium-amino-arsonate.

Trypsin. The proteolytic ferment of the pancreas.

Tryptizol. A proprietary preparation of amitriptyline hydrochloride. An antidepressant. (310).

Tryptophane. Indole-α-amino-propionic acid.

Trypure Novo. A proprietary preparation of trypsin. Used to promote wound healing and as an inhalation to reduce sputum viscosity. (182).

Tscheffkinite. (Chevkinite) A mineral. It is a titano-silicate of cerium, yttrium, thorium, iron, and calcium.

Tschermakite. A mineral. It is a variety of albite.

Tschermigite. Ammonium alum, (NH$_4$)$_2$SO$_4$.Al$_2$O$_3$(SO$_3$)$_3$.24H$_2$O, found in Bohemia.

Tschewkinite. A mineral. It is a silico-titanite of iron and cerium, lanthanum and didymium, and is found in the Ilmen Mountains.

Tsé-Hong. A mixture of white lead, aluminia, ferric oxide, and silica, used by the Chinese for painting on porcelain.

Tsé-Iéou (Ting-yu). An oil expressed from Chinese tallow seeds (seeds of *Sapium sebiferum*).

T-Siloxide. A trade name for a product of silica fused with 0·1–2 per cent. titania. A silica glass.

Tsing-Lieu. A red pigment used in porcelain painting. It consists of a mixture of stannic and plumbic silicates, with copper oxide or cobalt, and gold.

T-Size. Rosin emulsion size. (593).

T-Stoff. A mixture of benzyl and xylyl bromide. A lachrymator gas used in warfare.

Tsumebite (Preslite). A mineral. It is a basic phosphate of copper and lead.

Tuads. Tetramethyl - thiuram - disulphide. A vulcanisation accelerator.

Tuamine. A proprietary trade name for Tuaminoheptane.

Tuaminoheptane. 1-Methylhexyl-amine. Tuamine.

Tubania. Jeweller's alloys of varying composition, usually containing copper or brass, antimony, tin, and bismuth. The English alloy contains 12 parts brass, 12 parts tin, 12 parts antimony and 12 parts bismuth.

German Tubania consists of 4 parts copper, 3¼ parts tin, and 42 parts antimony.

Tubarine. A proprietary preparation of tubocurarine chloride. A muscle relaxant. (247).

Tube Brass. See BRASS.

Tubercumet. A water-soluble extract of the tuberculosis organism. It is used for the diagnosis and prognosis of tuberculosis.

Tub Glue. A liquid glue which is preserved by a disinfectant.

Tubize Yarn (Nitro Silk). A proprietary synthetic fibre made by the cellulose nitrate process.

Tubomel. A proprietary trade name for Isoniazid.

Tubulated Silk. See LUFTSEIDE.

Tuc-tur Metal. A nickel silver. It contains from 59–61 per cent. copper, 21–28 per cent. zinc, 12–18 per cent. nickel, and 0·3 per cent. iron.

Tucum Oil. Aouara oil, obtained from a palm, *Astrocaryum vulgare*. Used for culinary purposes.

Tudenza Silk. A proprietary brand of viscose silk with a matt appearance.

Tuesite. A mineral. It is a variety of lithomarge.

Tuex. A proprietary trade name for a rubber vulcanisation accelerator. It is tetramethyl - thiuram - disulphide. (435).

Tufa. A mineral. It is calcium carbonate, CaCO$_3$.

Tufnol. A proprietary trade name for laminated plastics materials bonded with synthetic resins incorporating fillers such as cotton fabric, paper, asbestos fabric and paper. The materials are used for construction in the electrical and chemical industry. (179).

Tufseal. A trade mark for a range of polymerisable mixtures of asphalt, polyols and isocyanates used as adhesives. (946).

Tugon. A proprietary preparation of METRIPHONATE. An insecticide.

Tuinal. A proprietary preparation of quinalbarbitone sodium and amylobarbitone sodium. A hypnotic. (333).

Tula Metal. An alloy of silver, copper, and lead, made at Tula, in Russia.

Tully's Powder. Powdered liquorice, morphine sulphate, camphor, and calcium carbonate.

Tumelina. A product manufactured from beet molasses.

Tumenol. See THIOL.

Tumenol Powder. Tumenol-sulphonic acid. See THIOL.

Tumeson. A proprietary preparation of prednisolone, sulphonated distillate of shale oils and titanium dioxide for dermatological use. (312).

Tunga Resin. A neutral glycerol-rosin ester, made with the aid of tung oil as esterifying catalyst.

Tung Oil (Chinese Wood Oil). The oil obtained by pressure from the seeds of *Aleurites cordata* and *Aleurites fordii*, of China and Japan. The seeds contain from 40–53 per cent. of oil. The oil has a specific gravity of 0·936–0·943, a saponification value of 190–195, an iodine value of 150–176, and an acid value of 0·5–1·2.

Tung Oil, Black. See BLACK TUNG OIL.

Tungsten Blue. A colloidal solution of the blue oxide of tungsten (ditungsten pentoxide). It may be used for dyeing silk.

Tungsten Brass (Wolfram Brass). An alloy of 60 per cent. copper, 22 per cent. zinc, 14 per cent. nickel, and 4 per cent. tungsten. An alloy containing 60 per cent. copper, 34 per cent. zinc, 2·8 per cent. aluminium, 2 per cent. tungsten, 0·7 per cent. manganese and 0·15 per cent. tin is also known by this name.

Tungsten Bronze (Wolfram Bronze). An alloy made by fusing potassium tungstate with pure tin. Used for decorative purposes.

It is also the name for an alloy of 95 per cent. copper, 3 per cent. tin, and 2 per cent. tungsten. An alloy containing 90 per cent. copper and 10 per cent. tungsten is also known by this term.

The term is also applied to sodium ditungstate and tungsten dioxide, $Na_2W_3O_7(Na_2W_2O_5+WO_2)$, in which tungsten amounts to from 70–85 per cent.

Tungsten Iron. An alloy of iron and tungsten.

Tungstenite (Scheelite). A mineral found in Utah. It is essentially a sulphide of tungsten WS_2.

Tungsten Powder. Metallic tungsten.

Tungsten Powellite. A mineral. It is $4[Ca(Mo,W)O_4]$.

Tungsten Steel. A very hard alloy of steel and tungsten. It usually contains from 5–8 per cent. tungsten, often 4 per cent. chromium, and 1·25 per cent. carbon. Used for armour plates, projectiles, firearms, and high speed tools. Tool steels contain 1–4 per cent. tungsten, and a rifle-barrel steel contains 3–6 per cent. tungsten.

Tungsten Steel, High. These steels usually contain from 80–85 per cent. iron with more than 14 per cent. tungsten. Some alloys contain from 77–81 per cent. iron, 15–18 per cent. tungsten, 3–4 per cent. chromium, and 0·15–0·35 per cent. silicon.

Tungsten Steel, Low. These alloys usually contain about 96 per cent. iron

with 1·5–2 per cent. tungsten, 0·5–1 per cent. chromium, and 0·15–0·35 per cent. silicon.

Tungsten Yellow. A pigment. It consists of tungstic acid, and is prepared by decomposing wolframite with sodium carbonate, reacting upon the alkali tungstate with calcium chloride, then adding the calcium tungstate to warm hydrochloric or nitric acid, and washing the precipitated tungstic acid.

Tungstic Ochre. A tungsten mineral, WO_3.

Tungstite. A mineral. It is H_2WO_4. See WOLFRAMINE.

Tuno Gum (Seiba Gum, Toonu Gum, Tunu Gum). A gum obtained from a tree in Nicaragua. The coagulated gum, or latex, is a sticky product, and is mixed with balata for belting. See also CHICLE.

Tunu Gum. See TUNO GUM.

Turacine. A red colouring matter contained in the feathers of the turaco birds of Africa. The colouring matter contains 8 per cent. copper.

Turanite. A mineral. It is a hydrated copper vanadate, $5CuO.V_2O_5.2H_2O$.

Turbadium Bronze. An alloy of 46 per cent. copper, 44 per cent. zinc, 5 per cent. lead, 2 per cent. nickel, 1·5 per cent. manganese, and small quantities of tin and aluminium. Used for propeller castings.

Turbax. A German synthetic varnish-canvas product used for electrical insulation.

Turbine Brass. See BRASS.

Turbiston Bronze. An alloy containing 55 per cent. copper, 41 per cent. zinc, 2 per cent. nickel, 1 per cent. aluminium, 0·84 per cent. iron and 0.16 per cent. manganese. It resists sea water.

Turbith. Mineral. See TURPETH MINERAL.

Turbomika. A German micanite used for electrical insulation.

Turbonit. A German synthetic varnish-paper product used for electrical insulation.

Turbuli. See CHAY ROOT.

Turex. A proprietary trade name for a vinyl resin coated leather.

Turgite. A brown ore of iron, $2Fe_2O_3.2H_2O$. See also TURITE.

Turgoids. A name applied to substances such as textile fibres, hide, tissue, leather, and wood fibres, which swell in water but do not dissolve.

Turicin. A compound of tannin (21 per cent.) and glutenin. An intestinal astringent.

Turin Yellow. See TURNER'S YELLOW.

Turite (Turgite, Hydrohæmatite). A mineral. It is a hydrated ferric oxide, $2Fe_2O_3.H_2O$.

Turkestan Volborthite. See TANGEITE.

Turkey Blue. A dyestuff of the same group as Chrome violet, obtained by the condensation of the tetramethyl-amino-benzhydrol with *p*-nitro-toluene.

Turkey Gum. See KHORDOFAN GUM.

Turkey Red. See ALIZARIN and INDIAN RED.

Turkey Red Oils (Sulphonated Oils, Monopol Oil, Soluble Oil, Sulphated Oil, Red Oil, Oleine). Viscid, transparent liquids, used in the preparation of cotton fibre for printing Turkey red (Alizarin red).

Concentrated sulphuric acid is run into castor oil, stirred, the product washed with water, and finally, ammonia or soda added to give an emulsion. Turkey red oil F is completely soluble in water, and Turkey red oil S is partially soluble in water.

Turkey Red Oxide. See INDIAN RED.

Turkey Rhubarb. Rhubarb root.

Turkey Umber. See UMBER.

Turkish Geranium Oil. See ROSÉ OIL.

Turlington's Balsam (Friar's Balsam). Contains benzoin, prepared storax, balsam of tolu, aloes, and alcohol.

Turmali. A Singalese name for zircon.

Turmeric (Indian Saffron, Terra Merita, Curcuma). A natural dyestuff obtained from the underground stems of rhizome of *Curcuma longa* and *C. rotunda*. The dyeing principle is curcumine, $C_{21}H_{20}O_6$. It dyes cotton greenish-yellow, and is also a colouring matter for wool, silk, oil, butter, cheese, curry powder, wood, and wax.

Turmeric Oil. A thick orange-coloured aromatic oil present to the extent of 3 per cent. in turmeric.

Tumeric, Tree. See FALSE COLUMBA.

Turmerine. See CLAYTON YELLOW.

Turnbull's Blue (Gmelin's Blue). Ferrous ferricyanide, $Fe_3[Fe(CN)_6]_2$.

Turnerite. A mineral. It is a variety of monazite.

Turner's Black. Animal charcoal.

Turner's Cement. Consists of 2 parts beeswax, a part resin, and 1 part pitch. These are all melted, and fine brick-dust is stirred in.

Turner's Cerate. Ung. Calaminæ.

Turner's Yellow (Patent Yellow, Cassel Yellow, Verona Yellow, Montpelier Yellow, Mineral Yellow, Veronese Yellow, Turin Yellow, Paris Yellow, English Yellow). Pigments which consist of oxychlorides of lead, usually $3PbO.PbCl_2$.

Turnsole. See LITMUS.

Turpenteen. See TURPENTYNE.

Turpentine. The exudation from incisions made in certain varieties of pine, fir, and larch. The terms American, French, German, Mexican, Portuguese, and Spanish Turpentine are usually used for the balsam turpentine ; German, Finnish, Polish, Russian, and Swedish Turpentine often refer to wood turpentine (*q.v.*), but German, Finnish, or Swedish Oil can refer to refined sulphite turpentine oil.

Turpentine, Artificial. Camphene, $C_{10}H_{16}$.

Turpentine, Bordeaux. See FRENCH TURPENTINE.

Turpentine, Chio. See CHIAN TURPENTINE.

Turpentine, Chios. See CHIAN TURPENTINE.

Turpentine, Cyprian. See CHIAN TURPENTINE.

Turpentine, English. See AMERICAN TURPENTINE.

Turpentine, Essence. See OIL OF TURPENTINE.

Turpentine, Finnish. See TURPENTINE.

Turpentine, German. See TURPENTINE.

Turpentine, Larch. See VENICE TURPENTINE.

Turpentine, Mexican. See TURPENTINE.

Turpentine Oil, Cellulose. See SULPHITE TURPENTINE OIL.

Turpentine, Polish. See TURPENTINE.

Turpentine, Portuguese. See TURPENTINE.

Turpentine, Scian. See CHIAN TURPENTINE.

Turpentine, Spanish. See TURPENTINE.

Turpentine Spirit. See OIL OF TURPENTINE.

Turpentine, Swedish. See TURPENTINE.

Turpentyne (Turpenteen). A turpentine substitute composed of rosin spirit, shale spirit, petroleum spirit, and coal tar naphtha.

Turpeth Mineral (Turbith Mineral, Queen's Yellow). A yellow basic sulphate of mercury, $HgSO_4.2HgO$.

Turps. See OIL OF TURPENTINE.

Turquoise. A mineral. It is a hydrated aluminium phosphate, coloured with copper and iron compounds. A gem stone.

Turquoise Blue. A dyestuff prepared by the condensation of tetramethyl-diamino - benzhydrol and *p* - nitro - toluene. Used in calico printing, and in the manufacture of lakes.

Turquoise Green. A colour used chiefly in porcelain painting. It is usually prepared by heating a mixture of aluminium hydroxide, chromium hydroxide, and cobalt carbonate.

Tusadin. A proprietary product. It is an agent for protection against frost in motor engines. It does not attack metals and does not change in composition when hot.

Tusana. A proprietary preparation of liquid extracts of cocillana, ipecacuanha, squill, senega and senna and glycerin and dextromethorphan hydrobromide. A cough linctus. (253).

Tuscan Red. See MADDER INDIAN RED.

Tusputol. A German product. Like ufinol it consists of oleic acid and phenols, and is used for the disinfection of tubercular sputum.

Tussar Silk (Tasar Silk). The product of the caterpillar of *Antheraca paphia*, of India.

Tussiex. A proprietary preparation of ammonium chloride, sodium citrate, ephedrine hydrochloride, pholcodine and menthol. A cough linctus. (280).

Tussifan S. A proprietary preparation of belladonna extract, potassium citrate, ipecacuanha, squill, anise oil and chloroform spirit. A cough linctus. (462).

Tussipect. A German product. It contains the ammonium salt of primulic acid, the saponin from the root of the primrose. An expectorant.

Tussol. A registered trade mark currently awaiting re-allocation by its proprietors to cover a range of pharmaceuticals. (983).

Tutania. Alloys. An English one contains 91 per cent. tin, 8 per cent. lead, 0·7 per cent. copper, and 0·3 per cent. zinc ; and another 80 per cent. tin, 16 per cent. antimony, 2·7 per cent. copper, and 1·3 per cent. zinc. A German alloy contains 92 per cent. antimony, 7 per cent. tin, and 1 per cent. copper ; and another consists of 62 per cent. antimony, 31 per cent. copper and 7 per cent. tin.

Tutenag (Tutenague, Tutenay). A nickel silver. It consists of from 44–46 per cent. copper, 16–40 per cent. zinc, and 15–40 per cent. nickel.

Tutenague. See TUTENAG.

Tutia. See TUTTY POWDER.

Tutocaine (Tutokain). An anæsthetic of German origin. It is the hydrochloride of γ - dimethyl - amino - α - β-dimethyl - propyl - p - amino - benzoate, $NH_2.C_6H_4.CO_2.CHCH_3.CHCH_3.CH_2N(CH_3)_2.HCl$. A local anæsthetic to replace cocaine.

Tutol. A registered trade name for certain explosives.

Tutol No. 2. An explosive similar to Rexite (*q.v.*). It contains sodium nitrate instead of potassium nitrate, and 12 per cent. of the explosive base is replaced by sodium chloride.

Tutorol. See DR. HALLER'S TUTOROL.

Tutty Powder (Tutia). An impure oxide of zinc, formed during the smelting of lead ores containing zinc.
  Sometimes a mixture of blue clay and copper filings is sold under this name.

Tuxtilite. A mineral. It is a variety of pyroxene.

Tween 20, 40, 60, 65, 80, 85. A proprietary trade name for Polysorbate 20, 40, 60, 65, 80, and 85.

Twitchells Reagent. Benzene-stearosulphonic acid. Used in the decomposition of fats.

Tybamate. 2 - Methyl - 2 - propyltri - methylene butylcarbamate. Benvil. Solacen.

Tychite. A mineral, $Na_6Mg_2(CO_3)_4.SO_4$.

Tyformin. An oral hypoglycæmic agent. 4-Guanidinobutyramide.

Tygafluor. A proprietary trade name for an aqueous dispersion of PTFE (*q.v.*) with a curing temperature of 90–140° C.

Tygon F. A proprietary trade name for a furan resin.

Tykarin. A proprietary preparation of nikethamide. A respiratory stimulant. (255).

Tylac. A proprietary range of butadiene copolymer rubbers. (481).

Tylagel. A proprietary trade name for the hydrochloride of Tolpropamine.

Tylan. A proprietary preparation of TYLOSIN. A veterinary antibiotic.

Tylarsin. A proprietary preparation. It is sodium-p-acetyl-amino-phenyl-arsonate.

Tylcalsin. Calcium - acetyl - salicylate, $(C_2H_3O.OC_6H_4.COO)_2Ca$. See KALMOPYRIN.

Tyloxapol. Oxyethylated *t*-octylphenol formaldehyde polymer. Present in ANS Suppositories and Hemrids.

Tyllithin. See APYRON.

Tylmarin. Acetyl - o - coumaric acid, $C_2H_3O.OC_6H_4.CH : CH.CO_2H$. An internal antiseptic.

Tylnatrin Sodium - acetyl - salicylate, $C_2H_3O.OC_6H_4.COONa$.

Tylose. A proprietary trade name for water soluble cellulose ethers for use as substitutes for gum tragacanth, glue, etc., in the finishing of textiles. It is stated to be resistant to mildew.

Tylosin. An antibiotic derived from an actinomycete resembling *Streptomyces fradiæ*. TYLAN.

Tymahist. A proprietary preparation of CHLORPROPHENPYRIDAMINE. (947).

Tymazoline. 2-Thymoloxymethyl-2-imidazoline. Pernazene is the hydrochloride.

Type Metal. A variable alloy of lead and antimony, frequently with the addition of tin, and sometimes copper or bismuth. The lead is present to the extent of from 50–93 per cent., the antimony 4–30 per cent., the tin 2–40 per cent., copper 0–5 per cent., and bismuth 0–29 per cent. A German type metal contains 60 per cent. lead, 12 per cent. tin, 18 per cent. antimony, 4·7 per cent copper, 4·7 per cent. nickel, and 1 per cent. bismuth. Founder's type usually contains from 20–25 per cent. tin, 25 per cent. antimony, and the rest lead, for hand type. Alloys poorer in tin are used for linotype and often contain from 2–5 per cent. tin, 10–12 per cent. antimony, and the rest lead. Monotype metal contains from 6–10 per cent. tin, 15 per cent. antimony, and the rest lead.

Typewriter Metal. An alloy of 57 per cent. copper, 20 per cent. nickel, 20 per cent. zinc, and 3 per cent. aluminium.

Typing Inks. Usually contain a soluble colour, such as methyl violet, dissolved in glycerol and water.

Tyraline. See MAUVEINE.

Tyramine (Phenomine). p-Hydroxy-phenyl-ethylamine-hydrochloride, HO. $C_6H_4.C_2H_4.NH_2.HCl$. The base is a constituent of ergot. It is used to raise blood pressure, to contract pregnant uterus, and is also administered in cases of shock or collapse.

Tyratol. See THYMATOL.

Tyrenka. A proprietary trade name for a synthetic fibre resembling Nylon.

Tyrex. A proprietary safety glass.

Tyrian Purple. The colouring matter obtained from the shell-fish, *Murex trecuculus*, and from *M. brandaris*. That from *M. trecuculus* contains indigo.

Tyril 767. A registered trade mark for styrene acrylonitrile copolymers. (64).

Tyrimide. A proprietary preparation containing isopropamide iodine. An antispasmodic. (281).

Tyrite. A mineral containing complex columbates of yttrium, cerium, uranium, calcium, and iron.

Tyrol-2, 32B, 6, CEF. Flame retardant materials for plastics. (401).

Tyrolean Earth. Bohemian earth (*q.v.*).

Tyrolite (Pharmacosiderite, Cupriferous Calamine, Copper Froth). A basic copper arsenate of green colour, found in the Tyrol. It has the formula, $Cu_3As_2O_8.2Cu(OH)_2.7H_2O$.

Tyrosal. Salipyrin (*q.v.*).

Tyrosinase. An oxidising enzyme.

Tyrosine. α-Amino-β-hydroxy-phenyl-propionic acid, $HO.C_6H_4.CH_2.CH(NH_2)$ COOH.

Tyrosolvon. A proprietary preparation of tyrothricin, benzocaine and cetyl pyridinium chloride. (244).

Tyrothricin. An antibiotic produced by a strain of Bacillus brevis.

Tyrozets. A proprietary preparation of TYROTHRICIN, benzocaine and CETYL-PYRADINIUM CHLORIDE. A throat lozenge. (262).

Tysonite. (1) A mineral. It is a fluoride of cerium metallic elements with thorium.

(2) A blend of Gibsonite and vulcanised vegetable oils. (911).

Tyuyamunite. A mineral of Utah. It contains 17·62 per cent. V_2O_5, 52·22 per cent. UO_3, 5·36 per cent. CaO, 8·51 per cent. SiO_2, water, Fe_2O_3, K_2O, and BaO.

Tyzanol. A proprietary trade name for the hydrochloride of Tetrahydrozolone.

U

Uacolite. A trade name for a proprietary asbestos-board used as an electrical insulator.

Uba (Styxol, Nicoschwab, Tanatol). Preparations containing sodium fluosilicate, Na_2SiF_6, as the main ingredient.

Ubisindine. An anti-arrhythmic drug. 2 - (2 - Diethylaminoethyl) - 3 - phenyl - isoindolin-1-one.

Ubretid. A proprietary trade name for DISTIGMINE BROMIDE, used in the treatment of urinary retention. (137).

UBS. Dry cleaning soap and paint remover. (523).

Ucaflex. A proprietary polyester laminating compound. (958).

Uchatius Bronze (Steel Bronze). An alloy containing 92 per cent. copper and 8 per cent. tin.

Ucinite. A proprietary trade name for a phenol-formaldehyde resin laminated product.

Ucon. Synthetic lubricants and fluids. (592).

Ucrete. A proprietary cement-modified polyurethane resin used for flooring. (512).

Ucuhuba Fat. A fat obtained from the seeds of *Myristica bicuhyba*. It contains 92 per cent. fatty acids, has an acid value of 30, and a saponification value of 224.

Uddevallite. A mineral. It is a variety of menaccanite.

Udel. A proprietary polysulphone. A high-performance, high-temperature thermoplastic resin used for injection moulding and extrusion. (34).

Udenam. A proprietary preparation containing bismuth subnitrate, magnesium carbonate, sodium bicarbonate, magnesium carbonate, calcium carbonate, frangula, calamus, aneurine hydrochloride and ascorbic acid. An antacid. (276).

Udilo Oil. See LAUREL NUT OIL.

Uffelmann's Reagent. A lactic acid reagent, prepared by adding a ferric chloride solution to a 2 per cent. phenol solution until it is of a violet colour. The colour of the reagent is changed to deep yellow by the addition of lactic acid.

Ufinol. A German product. It is phenol made into an emulsion with oleic acid, with addition of a deodoriser. It yields about 54 per cent. crude phenols on distillation.

Uformite. A proprietary trade name for an urea-formaldehyde synthetic resin.

Uganda Aloe. See ALOE.

Ugikral RA, RB and SN. Proprietary trade names for ABS terpolymers.

Ugitex S. A proprietary styrene-butadiene rubber. (240).

Uhligite. A mineral. It is a titanate of zirconium, calcium, and aluminium, found in East Africa.

Uintahite. See GILSONITE.

Uintaite. See GILSONITE.

Ulbreval. Buthelitone sodium.

Ulcatite. A proprietary trade name for a rubber vulcanisation accelerator containing hexamethylene-tetramine, benzthiazyl disulphide and diphenyl-guanidine.

Ulcedal. A proprietary preparation of deglycyrrhizinised liquorice extract used in the treatment of peptic ulcers. (959).

Ulcerex. A German product. It is a wound powder consisting of a bismuth preparation and sterilised clay. It is used in the treatment of chronic eczema.

Ulco. A metal used as a substitute for Babbitt metal. It contains 98–99 per cent. lead, the rest being barium and calcium.

Ulcony. An alloy of 65 per cent. copper and 35 per cent. lead.

Ulcusin. Capsules containing colloidal gold salt, extract of belladonna, and sodium phosphate.

Uleron. A proprietary trade name for p-amino-benzene- sulphonyl-p' - amino - benzene-sulphone-dimethyl-amide, $NH_2.C_6H_4.SO_2.NH.C_6H_4.SO_2N(CH_3)_2$.

Ulesi. A millet of East Africa.

Ulexine. Cytisine, $C_{10}H_{14}N_2O$, an alkaloid found in laburnum and furze.

Ulexite. A mineral. It is $2[NaCaB_5O_9 . 8H_2O]$. See BORONATROCALCITE.

U-lite. A proprietary urea formaldehyde moulding material. (718).

Ullmanite (Nickel Stibine). A mineral, NiSbS.

Ulmal. A proprietary trade name for an alloy containing aluminium with

10 per cent. magnesium, 1 per cent. silicon and 0·5 per cent. manganese.

Ulmaren. Amyl salicylate. An antipyretic.

Ulmarine. See MESOTAN.

Ulmin Brown. See VANDYCK BROWN.

Ulmite. A name proposed for the dark-coloured fibre covering the grains of sandstone found on the coast of New South Wales. In its properties it resembles those of humus, obtained from brown peat.

Ulokint. A proprietary phenolic moulding material. (960).

Ulon. A proprietary polyurethane elastomer. (140).

Ulrichite. A mineral. It is uranium oxide.

Ultandren. A proprietary preparation of fluoxymesterone. (18).

Ulto Accelerator. A proprietary rubber vulcanisation accelerator. It is a zinc salt of a complex dithiocarbamate. Specific gravity about 1·5.

Ultrabasite. A mineral, $Sb_4Pb_{28}Ag_{22}Ge_3S_{55}$, found in Saxony.

Ultracene. A proprietary rubber vulcanisation accelerator. It is a guanidine derivative.

Ultracorten. A proprietary trade name for Prednisone.

Ultracorten-H. A proprietary trade name for Prednisolone.

Ultracortenol. A proprietary preparation of prednisolone pivalate. (18).

Ultradil. A proprietary preparation of fluocortolone pivalate and hexanoate used in the treatment of eczema. (438).

Ultra-DMC. A proprietary trade name for a vulcanisation accelerator. It is dimethylamine - dimethyl - dithiocarbamate.

Ultraferran. A colloidal iron.

Ultralanum. A proprietary preparation of flucortolone, flucortolone caproate and clemizole hexachlorophane for dermatological use. (265).

Ultra-light Alloys. Alloys having a specific gravity below 2 are known by this name. Magnesium - aluminium - zinc and magnesium-copper are alloys of this type.

Ultralin. A rosin-fatty acid reaction product.

Ultralumin. A jeweller's alloy. It contains more than 90 per cent. aluminium, with nickel, copper, and some rare earth metals of the thorium group. It is specially resistant to sea water.

Ultramarine (Lapis - Lazuli Blue, Oriental Blue, Brilliant Ultramarine, French Blue, New Blue, Permanent Blue, French Ultramarine, Soda Ultramarine). A blue colouring matter formerly prepared from the rare mineral lapis-lazuli, by powdering and washing.

It is now prepared artificially by fusing together kaolin, sulphur, with soda, or with a mixture of sodium sulphate and charcoal. Also see LAPIS-LAZULI.

Ultramarine, Acid-proof. An ultramarine which resists the action of alum, due to an excess of silica in its composition. It does not resist true acids, but because alum in solution gives an acid reaction it is called acid-proof, and can be used for colouring paper, fabric, and soap where alum is also used.

Ultramarine, Artificial. See ULTRA-MARINE.

Ultramarine Ash. In obtaining ultramarine from lapis-lazuli, a blue product is first yielded, then a pale blue, and finally a pale bluish-grey material, which is called Ultramarine ash.

Ultramarine Blue. See WILLOW BLUE, ULTRAMARINE, and GREEN ULTRA-MARINE.

Ultramarine, Brilliant. See ULTRA-MARINE.

Ultramarine, Cobalt. See COBALT BLUE.

Ultramarine, French. See ULTRA-MARINE.

Ultramarine, Gahn's. See COBALT BLUE.

Ultramarine, Genuine. A pigment. It is native ultramarine, obtained by isolating the colouring matter of lapis-lazuli, a stone found in China and Thibet. The stone consists mainly of alumina, silica, sulphur, and soda. Also see ULTRAMARINE.

Ultramarine, Green. See CHROME GREENS and GREEN ULTRAMARINE.

Ultramarine, Real. See ULTRAMARINE, GENUINE.

Ultramarine, Silver. See YELLOW ULTRAMARINE.

Ultramarine, Soda. See ULTRAMARINE.

Ultramarine, Sulphate. See SULPHATE ULTRAMARINE and ULTRAMARINE.

Ultramarine Yellow. See BARIUM YELLOW, STEINBUHL YELLOW, STRONTIAN YELLOW and ZINC YELLOW.

Ultramid. A trade mark for a wide range of Nylons. (49). Grades of ULTRAMID include the following:

Ultramid A. Nylon 66.

A3K and *A3W.* Stabilised 6·6 nylon copolymers. *A3WG5.* A stabilised 6·6 nylon copolymer, 25 per cent. glass-loaded.

A3WG6. As above, but 30 per cent. glass-loaded.

A3WG7. As above, but 35 per cent. glass-fibre-loaded.

A3WG10. As above, but 50 per cent. glass-fibre-loaded.

Ultramid A4. A stabilised 6·6 nylon copolymer.

A4H. As above, but pigmented brown.

AEHG5. A4H 25 per cent. loaded with glass fibre.

Ultramid B. Nylon 6.

B3. A general-purpose nylon 6 copolymer with low viscosity.

B3K. A stabilised version of B3 above.

B3W G5. A nylon 6 copolymer 25 per cent. loaded with glass fibre.

B3W G6. As above, but stabilised and 30 per cent. loaded with glass fibre,

B3W G7. As above, but stabilised and 35 per cent. loaded.

B3W G10. As above, but stabilised and 50 per cent. loaded.

B4. A nylon 6 copolymer with medium viscosity.

B4K. A stabilised grade of *B4.*

B5. A nylon 6 copolymer with high melt viscosity.

B6. As above, but with very high melt viscosity.

B35. As above, but with low-to-medium viscosity.

B35W. A stabilised grade of *B35.*

Ultramid S. Nylon 6.10.

S3. A general-purpose nylon 6·10 copolymer with medium viscosity.

S3K. A stabilised grade of *S3.*

S4. A modified stabilised 6·10 nylon copolymer.

See also GRILON; MARANYL.

Ultran. A proprietary trade name for Phenaglycodol.

Ultrapen. A proprietary preparation containing Propicillin (as potassium salt). An antibiotic. (85).

Ultraplas. A proprietary melamine formaldehyde moulding material. (169).

Ultraproct. A proprietary suppository containing fluocortolone pivalate and hexanoate, cinchocaine hydrochloride, CLEMIZOLE undecanoate and HEXA-CHLOROPHANE used in the treatment of hæmorrhoids. (438).

Ultrasol. Solvents doubly distilled in glass. (527).

Ultrathene. A trade mark for ethylene-vinyl acetate copolymer. Used for manufacturing bottle seals. (41).

Ultra Violet Dyestuffs. Quinhydrones, obtained by the condensation of a leuco-gallocyanine with a gallocyanine.

Ultravon AN. A proprietary trade name for a fatty acid amide derivative. It emulsifies oils and fats more effectively and maintains the handle and colour of wool better than conventional detergents. A special detergent primarily for wool scouring. (18).

Ultra-zeozon. See ZEOZON.

Ultra Zinc DMC. A proprietary trade

name for a rubber vulcanisation accelerator. It is zinc-dimethyl-dithiocarbamate.

Ultroil. A proprietary trade name for a wetting agent for textiles. It is a sulphonated vegetable oil.

Ultryl 6010. A proprietary non-stabilised PVC resin of the suspension type.

Ultryl 6500. A proprietary PVC polymer of the suspension type, used in the production of glass-clear film, tube, etc. (961).

Ultryl 6800. A proprietary plasticiser-free PVC resin used in the manufacture of pipe and profiles. (961).

Ultryl 7100. A proprietary PVC resin of the suspension type with easy processing properties. (961).

Ultryl 7150. A proprietary PVC resin of the suspension type containing additives to give high clarity. (961).

Ulvio Cocoa. A German proprietary food material prepared by exposing cocoa to ultra-violet radiation.

Ulvospinel. A mineral. It is Fe_2TiO_4.

Umangite. A mineral. It is a copper selenide Cu_3Se_2.

Umber (Umber Brown, Mineral Brown, Velvet Brown, Chestnut Brown, Manganese Velvet Brown, Burnt Umber). Mineral varieties of umber. They are ochres coloured brown by oxides of manganese, and containing varying amounts of clay. The best is obtained from Cyprus, and is known as Raw Turkey Umber. Umber contains MnO_2 or Mn_3O_4, to the extent of from 6–12 per cent., Fe_2O_3 25–40 per cent., and SiO_2, 16–32 per cent. The term umber has also been applied to brown earthy products which contain lignite as the chief constituent. The so-called " Cologne earth," Coal brown, and Cassel brown belong to this class. Also see VANDYCK BROWN. Burnt umber is the calcined product of umber. Also see BURNT SIENNA.

Umber Brown. See UMBER.

Umber, Mineral. See UMBER.

Umbrathor. A solution of thorium dioxide.

Umbrenal. A 25 per cent. solution of lithium iodide in ampoules.

Umbrite A. An explosive containing 49 per cent. nitro-guanidine, 38 per cent. ammonium nitrate, and 13 per cent. silicon.

Umbrite B. An explosive containing 37·5 per cent. nitro-guanidine, 49·5 per cent. ammonium nitrate, and 13 per cent. silicon.

Umburana Seed. The product of *Amburana Claudii*. Used in Brazil for perfuming tobacco.

Umea Tar. A pale Swedish pine-wood tar. It is a good variety of Stockholm tar produced in the Umea district.

Unads. A proprietary preparation of tetramethyl thiuram monosulphide. (450).

Unal. A photographic developer. It is Rodinal in a solid form, containing, besides *p*-amino-phenol, the ingredients necessary for solidification.

Unburn. A proprietary anti-burn cream containing benzocaine, HEXACHLOROPHANE, orthophenyl phenol, menthol and lanolin. (962).

Uneprene. A proprietary thermoplastic rubber based on polyethylene. (619).

Ungemachite. A mineral. It is [$Na_9 K_3Fe^{\cdots}(SO_4)_6(OH)_3 \cdot 9H_2O$].

Unguentum. A proprietary preparation of silicilic acid, liquid paraffin, soft paraffin, cetostearyl alcohol, polysorbate-40, glycerol, oil, sorbic acid and propylene glycol. A protective skin cream. (896).

Unidene. A proprietary synthetic rubber. SOLUTION S.B.R. (619).

Unidiarea. A proprietary preparation of NEOMYCIN sulphate, CLIOQUINOL and attapulgite. An anti-diarrhœal. (731).

Uniflu. A proprietary preparation of DIPHENHYDRAMINE hydrochloride, PARACETAMOL, caffeine, PHENYLEPHRINE hydrochloride and codeine phosphate. A remedy for colds. (731).

Unigest. A proprietary preparation of DIMETHICONE, aluminium hydroxide and magnesium hydroxide and carbonate. An antacid. (731).

Unihepa. A proprietary preparation of di-methionine, choline, inositol, thiamine, pyridoxine, biotin, vitamin E, cyanocobalamin, panthenol and folic acid, used to counteract senile muscular degeneration. (731).

Uniloy Chrome Steels. A proprietary trade name for alloys containing 4–6 per cent. chromium, 0·1–0·25 per cent. carbon, up to 0·6 per cent. manganese and 0·4–0·6 per cent. molybdenum and 1·0–1·25 per cent. tungsten.

Unimycin. A proprietary preparation of OXYTETRACYCLINE hydrochloride. An antibiotic. (731).

Union Black B, 2B, S. Direct cotton or union dyestuffs.

Union Black B, R. Union dyestuffs which dye from a bath containing sodium sulphate. The brand B gives a deep black, and R a reddish-black.

Union Fast Claret. See ORCHIL RED A.

Unionite. A mineral. It is a variety of ziosite.

Unioptal. A proprietary preparation of polymixin B sulphate, neomycin sulphate and tyrothricin. An ocular antiseptic. (137).

Unipen. A proprietary preparation of sodium nafcillin. An antibiotic. (245).

Uniplast. A proprietary trade name for phenol-formaldehyde moulding compound.

Uniroid. A proprietary preparation of NEOMYCIN sulphate, POLYMIXIN B sulphate, HYDROCORTISONE and cinchocaine hydrochloride, used in the treatment of hæmorrhoids. (731).

Unithane 640 W and 641 W. A proprietary range of urethane oils used in the manufacture of tough chemical-resistant coatings such as floor finishes, etc. (91).

Univan. A nickel-vanadium steel.

Universal Balsam. Consists of 1 part camphor, 6 parts lead acetate, 16 parts beeswax, and 48 parts rape oil.

Univol U308. A proprietary trade name for 90 per cent. caprylic acid. (70).

Univol U310. A proprietary trade name for 90 per cent. capric acid. (70).

Univol U312. A proprietary trade name for a mixture of caprylic and capric acids. (70).

Univol U314/a/b. A proprietary trade name for 90/98 per cent. lauric acids. (70).

Univol U320. A proprietary trade name for 90 per cent. myristic acid. (70).

Univol U332. A proprietary trade name for 90 per cent. palmitic acid. (70).

Univol U334. A proprietary trade name for 90 per cent. stearic acid. (70).

Univol U304. A proprietary trade name for a mixture of distilled C20/22 acids. (70).

Univol U344. A proprietary trade name for 85/90 per cent. behenic acid. (70).

Univol U342. A proprietary trade name for 85/90 per cent. erucic acid. (70).

Unna's Caustic Paste. A caustic consisting of caustic potash, quicklime, soft soap, and distilled water in equal parts. Glycerin is also added to prevent drying, and sometimes morphine to relieve pain.

Unna's Stain. A microscopic stain. It contains 0·15 gram methyl green, 0·5 gram pyronin, 5 c.c. 95 per cent. alcohol, 20 c.c. glycerin, and the whole made up to 100 c.c. with 2 per cent. carbolic acid solution.

Unna's Zinc Paste. A paste made from gelatin, zinc oxide, glycerin, and water.

U.O.P. 88. A proprietary anti-oxidant. N-N¹ dioctyl-p-phenylene diamine. (51).

U.O.P. 288. A proprietary anti-oxidant. N,N¹-bis-(1-methyl/peptyl)-p-phenylene diamine. (51).

Upas. An arrow poison obtained from the *Upas antjar* and *U. radja*. Used in the East Indies.

U.P.E. 770, 772, 774, 775, 778 and 779. A proprietary range of low-density polyethylene homopolymers used in rotomoulding applications. (785).

U.P.E. 783. A proprietary ethylene vinyl acetate (E.V.A.) copolymer of low density used in rotomoulding applications. (785).

Uplees Powder. An explosive containing 62–65 per cent. ammonium nitrate, $12\frac{1}{2}$–$14\frac{1}{2}$ per cent. sodium nitrate, 4–6 per cent. trinitro-toluene, $13\frac{1}{2}$ per cent. ammonium chloride, and 2–4 per cent. starch.

Uraband. A proprietary bandage impregnated with zinc paste, urethane and ICHTHAMOL used in the treatment of eczema. (484).

Urac. A proprietary trade name for urea-formaldehyde adhesives.

Uracetin. See THERMODIN.

Uracil. 2 : 6 - Dioxy - pyrimidine-tetrahydride, $C_4H_3N_2O(OH)$. A hydrolytic product of nucleic acid.

Uracil Mustard. A proprietary preparation of URAMUSTINE used in the treatment of leukæmia. (325).

Uraconite. See URANIUM OCHRE.

Uracryl. A trade mark for a range of acrylic synthetic resins in emulsion form. (506).

Uradal. See ADALIN.

Uradil. A proprietary range of resins dispersible in water. (963). *Uradil 580/585* are used for air-drying and storing. *Uradil 587/588* are acrylic resins cross-linked by water-thinnable amino resins. *Uradil 503 and 415* are non-oxidising oil-free polyesters.

Uralane. A proprietary polyurethane elastomer. (670).

Uralian Emerald. See ANDRADITE.

Uraline. Chloral-urethane, $CCl_3.CH(OH).NH.CO_2.C_2H_5$. A hypnotic.

Uralite (1). An insulating incombustible product which is prepared from a mixture of asbestos and Spanish white, with gelatinous siliceous earth. It is rolled into plates, which plates are soaked in a solution of water-glass, dried, and dipped into a solution of sodium bicarbonate, dried, and finally treated with calcium chloride. A registered trade mark. (981).

Uralite (2). A proprietary trade name for urea-formaldehyde.

Uralium. Chloral - urethane, $CCl_3.CH(OH).NH.CO_2.C_2H_5$.

Uralysol. A proprietary preparation of hexamine-lysidine-thyminic acid.

Uramil. Amido-malonyl-urea, $C_4H_5N_3O_3$.

Uramino-Antipyrine. Antipyryl-urea, $NH_2.CO.NH.C_{11}H_{11}N_2O$.

Uramon. A proprietary trade name for a fertiliser containing 43 per cent. nitrogen in the form of urea or similar compounds.

Uramustine. 5-Di(2-chloroethyl)amino-uracil.

Urania Blue. A dyestuff obtained by the conjoint oxidation of di-β-naphthyl-m-phenylene-diamine-sulphonic acid and dimethyl-p-phenylene-diamine-thiosulphonic acid. Dyes wool and silk blue from an acid bath.

Uranine (Fluoresceine). The sodium or potassium salt of fluoresceine, $C_{20}H_{10}O_5Na_2$. Dyes silk and wool yellow.

Uraninite. See PITCHBLENDE.

Uranite. See AUTUNITE.

Uranium Blue. A mineral colour. The fibre is impregnated with uranium nitrate, and the colour is developed by means of heat.

Uranium Mica. See AUTUNITE.

Uranium Ochre (Uraconite). A mineral. It consists mainly of uranium oxide, U_2O_3, and contains radium.

Uranium Red. Prepared by passing sulphuretted hydrogen into a solution of uranium nitrate, containing caustic potash. The orange precipitate formed is treated with potassium carbonate, when it is converted into uranium red.

Uranium Yellow. Hydrated sodium uranate, $Na_2U_2O.6H_2O$. Used for painting and staining glass and porcelain, and for making fluorescent uranium glass. Hydrated ammonium uranate is also known by this name.

Uranium, Yellow Sesquioxide. See ORANGE OXIDE OF URANIUM.

Uranocalcite. A mineral. It is $(Ca,Cu)_2 UO_2(SO_4)(OH)_2.11H_2O$.

Uranocircite. (Bariumuranite). A mineral. It is a hydrous phosphate of uranium and barium, $BaO.2UO_3.P_2O_5$.

Uranophane (Uranotil). A mineral. It is a hydrous silicate of uranium and calcium, $CaO.2UO_3.2SiO_2.6H_2O$.

Uranopilite. A similar or identical mineral to Gilpinite. It is probably $CaU_8S_2O_3$, and is found in Cornwall.

Uranospathite. A Cornish mineral similar to autunite.

Uranosphærite. A mineral, $U_2O_7 (BiO_2)_3$, containing radium.

Uranospinite. A mineral $Ca(_{22})UO (AsO_4)_28H_2O$. Compare AUTUNITE.

Uranotantalite. See SAMARSKITE.

Uranothallite (Flutherite). A mineral. It contains the carbonate of uranium, $Ca_2U(CO_3)_410H_2O$.

Uranothorianite. A mineral. It is $4[(Th,U)O_2]$.

Uranothorite. A mineral. It is a variety of thorite.

Uranotil. A mineral. It is urano-phane.

Uranotile. A mineral. It is $2[Ca(UO_2)_2 Si_2O_7.6H_2O]$.

Uranox. A high nickel stainless steel.

Urantoin. A proprietary preparation of NITROFURANTOIN. An antibiotic. (498).

Uranvitriol. A mineral. It consists of uranium sulphate, containing copper. Synonym for Johannite.

Urao. See TRONA.

Urapurgol. See HELMITOL.

Urari. See CURARE.

Urasite (Sideranatrite). A mineral, $2Na_2O.Fe_2O_3.4SO_3.7H_2O$.

Urasol (Afsal). Acetyl-methylene-di-salicylic acid, $CH_2.(C_6H_3.OC_2H_3O. COOH)_2$. Used in the treatment of muscular rheumatism.

Urazine. A proprietary preparation. It is stated to be piperazine-citro-salicylate, and to be an antirheumatic.

Urbanite. A mineral. It is $4[(NaFe''', CaMg)Si_2O_6]$.

Urdite. A mineral. It is monazite.

Urea and Thiourea Resins. Resins obtained by the reaction between urea or thiourea and formaldehyde. Trade names for resins of this class are :— Bonnyware, Ciba Formica, Pertinit Pollopas, Prystaline, Rainbow Ware, Roanoid, Starpass.

Urea-Bromine. A combination of urea and calcium bromide, $4CO(NH_2)_2. CaBr_2$, and containing 36 per cent. bromine.

Urea Glue. A glue formed from the condensate of urea and formaldehyde. It is used in conjunction with a hardener.

Ureaphil. A proprietary preparation of urea used as an osmotic diuretic and as an abortifacient. (311).

Ureaphos. A fertiliser containing phosphate of ammonia and urea.

Urea-potash-phosphor, B.A.S.F. A German fertiliser containing 28 per cent. nitrogen, 14 per cent. phosphoric acid, and 14 per cent. potash.

Urease. An enzyme which splits up urea into carbon dioxide and ammonia.

Urease Solution. This is prepared by grinding 5 grams soya bean or jack bean (whole bean or meal) with 50 c.c. of a 0·6 per cent. solution of potassium dihydrogen phosphate. Allow to stand and filter. The solution is active for 2 days.

Urea-Stibamine. p-Amino-phenyl-stibinic carbamide, $CO(NH_2).H_2NC_6H_4 SbO(OH)_2$. Used in the treatment of kala-azar.

Urecoll. Urea-formaldehyde condensation products. (49).

Ureit. A German name for urea-formaldehyde resins.

Ureka. A mixture of diphenyl-guanidine and mercapto-benzo-thiazole. A rubber vulcanisation accelerator. Also a constant composition blend of diphenyl guanidine and 2, 4-dinitro-phenyl-thiobenzo-thiazole.

Ureka B. A blend similar to Ureka with a portion of D.P.G. replaced by Guantal. (59).

Ureka C. Benzothiazyl-thiobenzoate. Rubber vulcanisation accelerator. (59).

Ureka DD and HR, and Ureka Base. A range of proprietary accelerators based on 2 - (2, 4 - dinitrophenyl thio)-benzthiazole. (57).

Ureka White, White F and White FM. A range of proprietary accelerators based on blends of thiazole guanidine. (57).

Urelim. A proprietary trade name for ethebenecid.

Urepan. A proprietary range of poly-ester-based urethane rubbers. (112).

Uresin B. A proprietary trade name for a soft plasticising resin made from butyl urethane-formaldehyde. (3).

Uresine. Lithium - urotropine - citrate. A diuretic.

Uresol 60. A proprietary trade name for a one pack 60 per cent. solids air drying polyurethane resin. Used for floor and marine varnishes. (189).

Urethane. Ethyl carbamate, $NH_2.COOC_2H_5$.

Urethol. Urethane oils for paint manufacture. (508).

Urethon. Plastic film for sunblinds. (507).

Uretix. A proprietary urea formaldehyde moulding material. (964).

Uretrol. Sodium-dimethyl-amino-azo-benzene-*m*-sulphonate.

Urgon. A German preparation consisting of formaldehyde, calcium chlorate, zinc oxide, methyl salicylate, and glycerin.

Uricedin. Lithium, succinate, $Li_2C_4H_4O_4 3H_2O$.

Urifluine. Lithium succinate, $Li_2C_4H_4O_4.3H_2O$.

Urisol. See HEXAMINE.

Urispas. A proprietary preparation of FLAVOXATE hydrochloride used as an urinary anti-spasmodic. (809).

Uristix. A proprietary test strip impregnated with a citrate buffer, tetra-bromophenol blue, glucose oxidase, peroxidase and *o*-toluidine, used for the detection of protein and glucose in urine. (807).

Uritone. See HEXAMINE.

Urobilistix. A proprietary test strip impregnated with *p*-dimethylamino - benzaldehyde in an acid buffer, used to detect urobilinogen in urine. (807).

Urobinyl. Tablets containing benzoyl-hexamethylene-tetramine.

Urocitral. Sodium-theobromine-citrate, $C_7H_7N_4O_2Na.C_3H_4(OH)(COONa)_3$. A diuretic.

Urodonal. A granular preparation of hexamethylene-tetramine, Sidonal (piperazine quinate), and Lysidine (methyl-glyoxalidine).

Uroformine. Hexamine (hexamethylene-tetramine), $C_6H_{12}N_4$.

Urogenin. An addition product of theobromine and lithium hippurate, $C_7H_7N_4O_2.C_6H_5CO.NH.CH_2.COOLi$.

Urogosan. A compound of gonosan and hexamethylene-tetramine. Recommended as a disinfectant for the bladder.

Uro-Hexene. Hexamine and lithium benzoate.

Uro-Hexoids. A proprietary preparation of hexamine and lithium benzoate tablets. Product now discontinued. (182).

Urokinase. A plasminogen activator isolated from human urine.

Urol (Sidonal). The urea salt of quinic acid, $C_7H_{12}O_6.2[CO(NH_2)_2]_2$. Recommended for gout.

Urolucosil. A proprietary preparation containing sulphamethizole. A urinary antiseptic. (262).

Urolysin. Iodo-*p*-phenetidine-aceto-colchinine.

Uromat PE. A proprietary trade name for an aqueous dispersion based on titanium dioxide used for delustring synthetic fibre fabrics. (18).

Urometin. See HEXAMINE (*q.v.*).

Uromide. A proprietary preparation of sulphacarbamide and PHENAZOPYRIDINE. A urinary anti-spasmodic. (256).

Uronal. A proprietary preparation. It is veronal.

Uronidal. A German diuretic containing hexamethylene-tetramine, sodium phosphate, sodium salicylate, ext. valerianæ, ext. uvæ ursi, and ext. herniaria.

Uronovan. A German product. It contains hexamethylene-tetramine and methylene phosphate.

Uropherin B. Lithium-theobromine-salicylate, $LiC_7H_7O_2N_4+LiC_7H_5O_3$. A diuretic.

Uropherin S. Lithium - theobromine - benzoate, $LiC_7H_7O_2N_4+LiC_7H_5O_2$. A diuretic.

Uroplas. A proprietary range of urea formaldehyde moulding materials. (683).

Uropol. A proprietary preparation of TETRACYCLINE phosphate complex, SULPHAMETHIZOLE and PHENAZOPYRIDINE hydrochloride. A urinary antibiotic. (324).

Uropurgol. See HELMITOL.

Uroselectan. A German proprietary preparation. It is 2-hydroxy-5-iodo-pyridine-N-acetic acid (containing 42 per cent. iodine), or its sodium salt. It enables Röntgen photographs to be taken after intravenous injection.

Uroselectan B. A proprietary trade name for sodium-3 : 5-di-iodo-4-hydroxy-pyridine-2 : 6-dicarboxylate.

Urosine. A 50 per cent. aqueous solution of the lithium salt of quinic acid, $C_6H_7(OH)_4COOLi$. Prescribed for gout.

Urotropine (Urometin). See HEXAMINE.

Urotropine, New. See HELMITOL.

Ursal. Urea salicylate. Used as a substitute for sodium salicylate in medicine.

Ursol D (Ursol P, Ursol DD). Dyestuffs. They consist of the hydrochlorides of p-phenylene-diamine, p-amino-phenol, and diamino-diphenylamine respectively. Used for dyeing fur, feathers, and hair, brown to black.

Ursol DD. See URSOL D.

Ursol P. See URSOL D.

Urtal. An acrylonitrile-butadiene-styrene resin.

Urtenol. A combination of oils for the textile trades. It possesses great penetration and detergent properties and is recommended for all textile materials.

Urvölgyite. See HERRENGRUNDITE.

Urystamine. Lithium-hexamethylene-tetramine-benzoate. Recommended for gout.

Usbekite. A mineral, $Cu_3(VO_4)_2.3H_2O$.

Usco. A special variety of cast iron for use as fire-bars.

Usébe Green. See ALDEHYDE GREEN.

Usga. A combustion engine fuel. It contains alcohol, ether, and a small quantity of castor oil. It is made in Pernambuco.

Usoline. See PARAFFIN, LIQUID.

Uspulun. A material containing sodium sulphate, sodium hydroxide, aniline, and mercury-chloro-phenol. A fungicide. Also a trade name for a German product recommended as a dissipating agent in connecting with rubber tapping work. Its active constituent is monochlor-mercuro-phenolate.

Ussingite. A mineral. It is a sodium aluminium hydro-trimeta-silicate, $Na_2HAl(SiO_3)_3$, of Greenland.

Usunify. A root found in the Sudan. It is used locally as a food.

Utahite. A mineral, $3Fe_2O_3.3SO_3.4H_2O$.

Utahlite. A mineral. It is a variety of variscite.

Uteplex. A proprietary preparation of uridine-5-triphosphoric acid used to relieve muscle spasm. (323).

Uteramin p - Hydroxy - phenyl-ethyl-amine.

Utica Steel. A proprietary trade name for a die steel. It contains 1·4 per cent. tungsten, 1·25 per cent. carbon, 0·4 per cent. chromium and 0·2 per cent. vanadium.

Uticillin. A proprietary preparation of sodium CARFECILLIN. An antibiotic. (364).

Uvaleral. See BROMURAL.

Uvalysat. A preparation of the dialyzate from *Folia uvæ ursi*. Used as a urinary antiseptic.

Uvanite. A mineral. It is hydrated uranium vanadate, $2UO_3.3V_2O_5.15H_2O$.

Uvarovite (Chrome Garnet). A mineral. It is a calcium-iron-silicate, $Ca_3Fe_2Si_3O_{12}$. It is also given as $8[Ca_3Cr_2Si_3O_{12}]$.

Uvinol D50. A registered trade mark for 2,2' - 4,4' - tetrahydroxybenzophenone. An ultra violet absorber for plastics. (229).

Uvinul 400. A registered trade mark for 2,4-dihydroxybenzophenone. An ultra violet absorber for plastics. (229).

Uvistat (1). A proprietary preparation of mexenone, used in dermatology. (319).

Uvistat 12, 24, 247, 2211. A proprietary trade name for a series of additives for protecting plastics against ultra violet light. They have the general formula $R_1.C_6H_5-CO-C_6H_4OH-OR$. Uvistat 247 is particularly effective for the stabilisation of polyolefines. (72).

Uvistat 247. A proprietary trade name for 2 - hydroxy - 4 - n - heptoxybenzophenone. A virtually non-toxic ultra violet light absorber. (72).

Uvistat 2211. A proprietary trade name for Mexenone.

Uvitex. A proprietary trade name for a series of fluorescent brighteners for incorporation in soap-based and synthetic detergents as follows : Uvitex SFC is a stilbenic derivative giving high intensity whites on cellulosic fibres. Uvitex SK is a benzoazole derivative effective on a wide variety of fibres, stable to hypochlorite and chlorisocyanate. Uvitex SOF is also a benzoxazole derivative. Uvitex ERN CONC P is a benzoxazole derivative—a fluorescent brightener for polyester fibres. (18).

Uvitex MA. A proprietary trade name for an imidazole derivative fluorescent brightener for acrylic fibres. It is applied in the dope before spinning. (18).

Uvitex MP. A proprietary trade name for a heterocyclic-stilbene type fluorescent brightener for polyamid fibres. (18).

Uwarowite. A mineral, $6CaO.3SiO_2.2Cr_2O_3.3SiO_2$.

Uzara. The powdered root of an African plant. Used in diarrhœa.

Uzbekite. A mineral. It is a copper vanadate.

V

Vacancin Blue. A basic dyestuff giving a dark blue shade on tannined cotton.

Vacancin Scarlet. An azo red produced on the fibre.

Vac Blue. See INDOINE BLUE R.

Vacsol. A trade name for a semi-rigid PVC quick air drying solution coating for steel work. (254).

Vactran. High grade chemicals for vacuum evaporation. (527).

Vacuum Salt. A pure salt, NaCl, obtained by boiling brine under a vacuum.

Vacuum Silicon Iron. Alloys containing about 0·15 or 3·4 per cent. silicon, made by melting *in vacuo*. They are annealed at 1100° C., and contain about 0·01 per cent. carbon. They have remarkable magnetic properties.

Væsite. A mineral. It is $4[NiS_2]$.

Vagadil-Alk. A proprietary preparation of aspirin, phenacetin, codeine phosphate and aluminium hydroxide. An analgesic. (359).

Vagintus. A colloidal aluminium preparation that forms a foam and carbon dioxide in solution. It is a German product.

Valamin. The isovaleryl ester of tertiary amyl alcohol, $(CH_3)_2.C_2H_5.COOC_5H_9$. A sedative.

Valbornyl. A proprietary preparation of borneol isovalerianate.

Valearin. Valeryl-trimenthyl-ammonium chloride.

Valencianite. A mineral. It is a variety of orthoclase.

Valentinite. See ANTIMONY BLOOM.

Valeramyl. Amyl valerianate, $C_4H_9CO.OC_5H_{11}$.

Valerene. Bourneene, a liquid hydrocarbon, isomeric with oil of terpentine.

Valerian. The dried rhizome of *Valeriana officinalis*.

Valerianic Ether. Ethyl valerianate.

Valeridin. See VALERYDINE.

Valerobromine. A mixture of sodium bromide and sodium valerate in molecular proportions. Used in the treatment of nervous maladies.

Valerone. Di - isobutyl - ketone, $C_4H_9.CO.C_4H_9$.

Valerophen. The menthyl ester of phenol-phthalein. A cathartic.

Valerydine (Valeridin, Sedatin). Isovaleryl - *p* - phenetidine, $C_6H_4(NH.CO.C_4H_9)OC_2H_5$. Prescribed in cases of neuralgia and headaches.

Valide. See VALYL.

Validol (Menthival). Menthyl-valerate, $CH_3.CH_2.CH_2.COO(C_{10}H_{19})$. Used in medicine as an antispasmodic.

Valimyl. Diethyl - amide - isovalerianic acid.

Valine. α-Amino-isovaleric acid, $(CH_3)_2CH.CH(NH_2)COOH$.

Valinopyrin. See NEOPYRIN.

Valisan. See BROVALOL.

Valium. A proprietary preparation of diazepam. A sedative. (314).

Valledrine. A proprietary preparation of trimezaprine tartrate, pholcodine citrate and ephedrine hydrochloride. A cough linctus. (336).

Valleite. A mineral. It is a variety of amphibole.

Vallergan. A proprietary preparation of trimeprazine tartrate. (336).

Vallestril. A proprietary trade name for Methallenoestril.

Vallet's Pills. Iron pills (*q.v.*).

Vallex. A proprietary preparation of trimeprazine tartrate, menthol, phenylpropanolamine hydrochloride, guaiphenesin, citric acid, sodium citrate and ipecacuanha. A cough linctus. (336).

Valmid. A proprietary trade name for Ethinamate.

Valmidate. A proprietary preparation of ethinamate. A sedative. (333).

Valofin. A combination of peppermint extract with valeryl-ethylate and ammonium valerate. A nerve sedative.

Valoid. A proprietary trade name for the hydrochloride or lactate of cyclizine.

Valonia. The acorn cups of *Quercus ægilops*. They contain about 35 per cent. of tannin, and are used in the leather industry.

Valothalein. See APERITOL.

Valpin. A proprietary trade name for Octatropine Methylbromide.

Valproic Acid. An anti-convulsant. 2-Propylpentanoic acid.

Valray Alloy 1. A trade mark for an alloy of 20 per cent. chromium and the balance nickel with controlled manganese, carbon and silicon. (143).

Valsol. Vasogen (*q.v.*).

Valuevite. A mineral. It is a variety of xanthophyllite.

Valve Brass. See Brass.

Valve Bronze. An alloy of from 83–89 per cent. copper, 4–5 per cent. tin, 3–7 per cent. zinc, and 3–6 per cent. lead.

Valvoline. A registered trade name for lubricating oils. They are American petroleum products.

Valyl. A registered trade mark currently awaiting re-allocation by its proprietors. (983).

Valzin. Mono-*p*-phenetol-carbamide.

Vam. A proprietary preparation of vinyl and ethyl ethers. General anæsthetic. (336).

Vamin. A proprietary preparation of

aminoacids and minerals with glucose or fructose, used for intravenous feeding. (330).

Vanadenbronzite. A mineral. It is $4[MgCaSi_2O_6]$.

Vanadinaugite. Synonym for Lavrovite.

Vanadinite. A mineral. It is a chlorovanadate of lead, $9PbO.3V_2O_5.PbCl_2$.

Vanadioardennite. A mineral. It is $2[Mn_5Al_5VO_4Si_5O_{20}(OH)_2.2H_2O]$.

Vanadiol. Vanadium hypochlorite. Used in the treatment of anæmia.

Vanadite. A mineral, $3Pb_3V_2O_8+PbCl_2$.

Vanadium Alum. An ammonium-vanadium sulphate, $(NH_4)_2SO_4.V_2(SO_4)_3.24H_2O$.

Vanadium Brass. An alloy of 70 per cent. copper, 29·5 per cent. zinc, and 0·5 per cent. vanadium.

Vanadium Bronze. Metavanadic acid, HVO_3. Used as a pigment in the place of gold bronze. It is also the name for an alloy of 61 per cent. copper, 38·5 per cent. zinc, and 0·5 per cent. vanadium.

Vanadium-Manganese Brass. An alloy containing 58·56 per cent. copper, 38·54 per cent. zinc, 1·48 per cent. aluminium, 1 per cent. iron, 0·48 per cent. manganese, and 0·03 per cent. vanadium.

Vanadium Mica. See ROSCOELITE.

Vanadium-Molybdenum Steels. Alloys containing from 0·1-1·0 per cent. carbon, 0·52-6·0 per cent. molybdenum and 0·1-1·0 per cent. vanadium.

Vavadium Ochre. See HYDROCUPRITE.

Vanadium Steel. An alloy of steel with vanadium.

Vanadium-tin Yellow. A pigment. It is a mixture of vanadium pentoxide and tin oxide, and is used in the manufacture of yellow glass.

Vanadous Acmite. A mineral. It is $4[Na(Fe,V)Si_2O_6]$.

Vanair. A proprietary preparation of benzoyl peroxide and sulphur used in the treatment of acne. (782).

Vanbeenol. Ethyl vanillin. (504).

Vancocin. A proprietary trade name for the hydrochloride of Vancomycin.

Vancomycin. An antibiotic produced by Streptomyces orientalis. Vancocin is the hydrochloride.

Vancure D.A.A. A proprietary accelerator. N-cyclohexyl-2-benzthiazyl sulphonamide. (450).

Vandenbrandeite. A mineral. It is $2[CuVO_4.2H_2O]$.

Vandex. A proprietary selenium compound used in the rubber industry for imparting ageing properties and high abrasion. Specific gravity 4·3.

Vandid. (1) A proprietary trade name

for Ethamivan. (2) A proprietary preparation of vanillic acid diethylamide. Respiratory stimulant. (275).

Vandiestite. A mineral. It is a silver-gold-lead-bismuth telluride.

Vandike P.360. A proprietary trade name for a vinyl acetate-dioctyl maleate copolymer emulsion used in water-based adhesives. (200).

Vandike 7085. A proprietary trade name for a vinyl acetate-butyl acrylate copolymer emulsion used in emulsion paints. (200).

Vandike 7086. A proprietary trade name for a vinyl acetate-butyl acrylate copolymer emulsion. Used in the manufacture of "non-drip" emulsion paints. (200).

Vandura Silk (Gelatin Silk). An artificial silk prepared from gelatin and formaldehyde. It is also the name for a silk made from casein and formaldehyde.

Vandyck Red. See FLORENTINE BROWN and INDIAN RED.

Vandyke Brown (Rubens Brown, Cassel Brown, Ulmin Brown, Cologne Brown). Brown pigments of vegetable origin prepared from peat, cotton, or soot. The modern vandyke brown is a purified bituminous ochre. Genuine vandyke brown is a natural earth containing a large proportion of organic matter and with an ash usually amounting to 10 per cent. The artificial product generally consists of a carbon black with red oxide and possibly yellow ochre.

Van Ermengem's Stain. A microscopic stain. Solution (A) contains 1 gram of osmic acid, 20 grams tannin, 150 c.c. distilled water, and 8 drops glacial acetic acid. Solution (B) is a 0·25-0·5 per cent. silver nitrate solution. Solution C contains 6 grams tannin, 1 gram gallic acid, 20 grams sodium acetate, and 700 c.c. water.

Vanexane. A proprietary preparation of gamma benzene hexachloride. A lotion for removal of head lice. (399).

Vanilla. The cured unripe fruit of *Vanilla planifolia*. Used as an aromatic and flavouring material.

Vanillal. See BOURBONAL.

Vanillin. Proto-catechuic-aldehyde-methyl-ether, $C_6H_3.(COH)(OH).O.CH_3$. A perfume.

Vanilloes. Vanilla pods.

Vanitox. Selective weed killer. (507).

Vankalite. A proprietary trade name for a beryllium-copper alloy used for setting diamonds in drills.

Vanoxite. A mineral. It is $(VO)_4V_2O_9.8H_2O$.

Vanquelinite. A mineral, $2PbO.CuO.2CrO_3$.

Vanquin. A proprietary preparation

containing viprynium embonate. An anthelmintic. (264).

Van Swieten's Solution. See LIQUEUR DE VAN SWIETEN.

Vanthoffite. A mineral, $MgSO_4.3Na_2SO_4$. It occurs in the Stassfurt deposits.

Vanuxemite. A mineral. It is a clay containing zinc.

Vapona. A proprietary preparation of DICHLORVOS. An insecticide.

Vaporine. Naphthalene-eucalypto-camphor. It is mixed with water and inhaled. Recommended for whooping-cough.

Varac. See KELP.

Varech. See VRAIC.

Variamine Blue B Base. A dyestuff. It is 4-amino-4'-ethoxy-diphenyl-amine sulphate.

Varidase. A proprietary preparation STREPTOKINASE and STREPTODORNASE. A fibrinolytic drug. (306).

Variderm. A proprietary preparation of placental extract, used to treat ulcers. (329).

Variegated Lead Ore. Pyromorphite (q.v.).

Varifos. Trademark for a range of phosphate ester surfactants. (966).

Variotin. A proprietary trade name for PECILOCIN, used in the treatment of fungal skin infections. (308).

Variscite (Peganite). A mineral, $Al_2O_3.P_2O_5$.

Varitox. Sodium trichloroacetate. (507).

Varley's Cement. Consists of 16 parts of dried whiting, 16 parts of black resin, and 1 part of beeswax. Used as a cement for fixing brass to glass.

Varnish. A mixture of boiled oil with various gum resins, and oil of turpentine.

Varnish, Burma Black. See THITSI.

Varnish, Lithographer's. See STAND OIL.

Varnish, Short Oil. See LONG OIL VARNISHES.

Varnish, Standöl. See STAND OIL.

Varnish, Victoria. See ZAPON VARNISH.

Varnodag. A trade mark for a varnish made from phenol-formaldehyde synthetic resin with colloidal graphite.

Varnoline. A petroleum distillate used as a lubricant.

Var Oil. See AIX OIL.

Varon K. A proprietary trade name for a hydrophobic coated, fine particle size clay which does not absorb moisture. Used in sealants and mastic. (189).

Varsol. A proprietary trade name for a petroleum solvent.

Varulite. A mineral. It is $(Na,Ca)(Mn_{...},Fe'')(PO_4)_2$.

Varvicite (Newkirkite, Hydrated Manganese Ore). Minerals. They are hydrated varieties of pyrolusite, MnO_2.

Vasal. A German product. It was formerly known as Vesipyrine. It consists of phenyl-acetyl-salicylate. An antirheumatic and antiseptic.

Vasano. A German product. It is a combination of 1-scopolamine-camphor-combination of 1-scopolamine - camphorate and 1 - hyoscyamine - camphorate. Used to prevent sea and air sickness.

Vascardin. A proprietary preparation of sorbide nitrate. A vasodilator used for angina pectoris. (271).

Vascoloy-Ramet D. A proprietary trade name for a corrosion-resisting alloy of 80 per cent. tantalum carbide, 20 per cent. tungsten and nickel.

Vasconite BT. A proprietary anti-corrosion agent added to poor-quality boiler fuels. It is a suspension of magnesium compounds and combustion catalysts. (965).

Vasculit. The sulphate of 2-n-Butyl-amino - 1 - (4 - hydroxyphenyl)ethanol (bamethan sulphate). (278).

Vascutonix. A proprietary preparation of diethylamine salicylate and glycol salicylate used in the treatment of arthritis. (247).

Vaseline. The trade mark for a line of products among which is petroleum jelly. (234).

Vashegyite. A mineral. It is an aluminium phosphate. $Al_4(PO_4)_3(OH)_3.13\frac{1}{2}H_2O$.

Vasocidin. A proprietary preparation of chloramphenicol, polymixin B sulphate in a buffered solution. An ocular antiseptic. (329).

Vasocon-A. A proprietary preparation of antazoline phosphate and naphazoline hydrochloride solution for ocular use. (329).

Vasoconstrictin. See ADRENALIN.

Vasocort. A proprietary preparation of HYDROCORTISONE, HYDROXYAMPHETAMINE hydrobromide and PHENYLEPHRINE hydrochloride, used as a nasal spray in cases of allergic rhinitis. (281).

Vasodex. A proprietary preparation of dexamethasone sodium phosphate and phenylephrine hydrochloride in a buffered vehicle. An anti-inflammatory oculant. (329).

Vasogen. (1) A proprietary preparation of DIMETHICONE, zinc oxide and calamine used as a protective cream for the skin. (330).

(2) A trade mark for an ointment vehicle—an oxygenated petroleum. PAROGEN.

Vasolastine. A proprietary preparation of lipoxydase citrogenase, amino acid oxydase and tyrosinase complex. (274).

Vasoliment. Parogen (q.v.).

Vasopred. A proprietary preparation of prednisolone acetate and phenylephrine in a buffered vehicle. An anti-inflammatory oculant. (329).

Vasosulf. A proprietary preparation of sodium sulphacetamide solution. An ocular antiseptic. (329).

Vasotonin. A proprietary preparation of a yohimbine and urethane combination. It has the property of producing a fall in blood pressure.

Vasotran. A proprietary preparation of ISOXSUPRINE hydrochloride used in the treatment of peripheral vascular disease. (324).

Vasoxine. A proprietary trade name for the hydrochloride of Methoxamine.

Vasozinc. A proprietary preparation of zinc sulphate and naphazoline hydrochloride in a buffered vehicle. An ocular antiseptic. (329).

Vassy Cement. A cement similar to Roman cement.

Vastarel. A proprietary preparation of trimetazidine dihydrochloride. A vasodilator. (313).

Vasylox. A proprietary trade name for the hydrochloride of Methoxamine.

Vatensol. A proprietary preparation of guanoclor sulphate. An antihypertensive drug. (85).

V2A Steel. See ANKA STEEL.

Vat Dye B. See THIOINDIGO RED B.

Vat Dyes. Dyestuffs insoluble in water which must be treated with a reducing agent, and dissolved in an alkaline solution for dyeing. The colour is produced by oxidation. Indigo is a type of these colours.

Vaterite. A mineral. It is a modification of calcium carbonate, but is distinct from arragonite and calcite.

Vat Indigo. See INDIGO WHITE.

Vat Red Paste. An acid dyestuff used to replace barwood as the ground colour for indigo vat blue.

Vaua-Assu Nuts. See BASSOBA NUTS.

Vaucher's Alloy. An alloy of 75 per cent. zinc, 18 per cent. tin, 4·5 per cent. lead, and 2·5 per cent. antimony.

Vauquelinite. A mineral probably identical with laxmannite. It is a phospho-chromate of lead and copper, $2(Pb.Cu)CrO_4.(Pb.Cu)_3.(PO_4)_2$, found in the Urals.

Vauquelin's Salt. A compound obtained by treating palladium chloride with ammonia, $[Pd(NH_3)_4]Cl_2.PdCl_2$.

Vauxite. A mineral, of Bolivia, $4FeO.2Al_2O_3.3P_2O_5.24H_2O$.

Vayrynenite. A mineral. It is BeMn $PO_4(OH,F)$.

Vazadrine. A proprietary trade name for Isoniazid.

V-Cil-K. A proprietary preparation containing Phenoxymethylpenicillin Potassium. An antibiotic. (333).

V-Cil-K Sulpha. A proprietary preparation of phenoxymethylpenicillin potassium, and sulphadimidine. An antibiotic. (333).

Veatchite. A mineral. It is $16[Sr_3B_{16} O_{27}.5H_2O]$.

Vec. A proprietary trade name for a vinylidene chloride synthetic resin.

Vecopyrin. Phenazone-iron chloride.

Vedar. A proprietary trade name for a calcite (q.v.) of exceptional whiteness giving good hiding power in high p.v.c. emulsion paints and wallpaper. (189).

Vedar. A French trade name for precipitated calcium carbonate. (26). See also CALOFIL.

Vedril. A proprietary transparent plastic used in moulding applications. Polymethylmethacrylate. (61).

Veepa Oil (Veppam Oil, Neem Oil). Margosa oil, obtained from the seeds of *Melia Azadirachta*.

Vega Brown R. See ACID BROWN R.

Veganin. A proprietary preparation. It is a combination of acetyl-salicylic acid, codeine phosphate, and phenacetin.

Vega Red S. A dyestuff. It is a British brand of Brilliant sulphone red B.

Vegasite. A mineral. It is a lead-iron sulphate.

Vegetable Alkali. Potassium hydroxide, KOH.

Vegetable Black. A very light lampblack containing 99 per cent. carbon. The British Standard Specification No. 286 (1927) states that the loss on heating at 95–98° C. should not be more than 1 per cent., the ash not more than 0·5 per cent. Not more than 1·5 per cent. should be soluble in methyl ether, and not more than 1 per cent. soluble in water.

Vegetable Butter (Lactine, Vegetaline, Cocoaline, Laureol, Nucoline, Albene, Palmine, Cocose, Chocolate Fat). Names for an edible fat prepared from coco-nut oil and palm-nut oil. Used in chocolate manufacture as a substitute for cocoa-butter.

Vegetable Calomel. The resin of *Podophyllum*.

Vegetable Casein. Legumin, found in leguminous seeds.

Vegetable Ethiops. A form of charcoal obtained by the incineration of *Fuci*.

Vegetable Fibre. See VULCANISED FIBRE.

Vegetable Gelatin. Agar-agar.

Vegetable Glue (Aparatine). A glue obtained by treating starch with alkali. An adhesive. Also see AGAR-AGAR.

Vegetable Green. See SAP GREEN.

Vegetable Gum. See BRITISH GUM.

Vegetable Ivory (Corajo). Tagua nut, the fruit of *Phytelephas macrocarpa*, of South America.

Vegetable Jelly. Pectin (*q.v.*), found in vegetable juices.

Vegetable Mercury. Manacine, the alkaloid from manaca. It is used as a diuretic and diaphoretic.

Vegetable Milk. An emulsion of fats derived from the soya bean.

Vegetable Parchment. See PARCHMENT PAPER.

Vegetable Rouge. Carthamin, the colouring matter of *Carthamus tinctorius* mixed with French chalk. Used as a cosmetic. See SAFFLOWER.

Vegetable Salt. Potassium tartrate.

Vegetable Soda. The general name for the ash of soda plants (land plants).

Vegetable Spermaceti. See CHINESE WAX.

Vegetable Sulphur. Lycopodium.

Vegetable Tallow. The name applied to vegetable fats similar to tallow, such as Chinese tallow and Malabar tallow. See CHINESE TALLOW.

Vegetable Wax. See JAPAN TALLOW.

Vegetable Wool. A product obtained from green pine and fir cones by processes of fermentation, washing, and disintegration. It is mixed with cotton for the production of yarns.

Vegetaline. The name given to a preparation of lactic acid, used in tanning processes for the removal of lime. It is obtained from the drainage water of preserve manufacture by evaporation, and contains from 8·6–9·6 per cent. lactic acid.

Vegolysen. A proprietary preparation of hexamethonium bromide. (336).

Vegolysin T. A proprietary preparation of hexamethonium tartrate. (336).

Velactin. A proprietary artificial baby feed for use in cases of allergy to cow's milk and of galactosæmia. (244).

Velampishin. See WOOD-APPLE GUM.

Velan. A proprietary product. It is a complex organic compound soluble in water which renders fabric fibres water-repellant.

Velardenite. A mineral, $2CaO.Al_2O_3.SiO_2$.

Velbe. A proprietary trade name for the sulphate of VINBLASTINE, used in the treatment of malignant diseases. (463).

Veldopa. A proprietary preparation of LEVODOPA used in the treatment of Parkinsonism. (824).

Velkor. A trade mark for an artificial leather made from plant fibres and rubber latex.

Velocite. A rubber vulcanisation accelerator. It is thiocarbanilide.

Velon. A proprietary trade name for a vinylidene chloride synthetic resin.

Velosan. A proprietary rubber vulcanisation accelerator. It is aldehyde-ammonia.

Velosef. A proprietary preparation of CEPHRADRINE. An antibiotic. (682).

Velpeau's Caustic Powder. A caustic consisting of burnt alum and powdered savin tops.

Velvet. A silk fabric with the pile on one side, made by carrying some of the thread over wires and cutting the loops.

Velvet Black. A variety of gas carbon black.

Velvet Brown. Umber (*q.v.*).

Velveteen. An imitation velvet made from cotton.

Velvetex. A proprietary carbon black (thermatomic carbon) in a soft form used in rubber mixings.

Velvet Red. A pigment. It is a reddish brown powder consisting of ferric oxide coloured by a mixture of spirit soluble rosaniline blue and fuchsine.

Velvril. A mixture of collodion cotton and nitrated castor or linseed oil. It is a proposed gutta-percha substitute.

Vendril. Polymethylmethacrylate.

Venetian Bole. A pigment. It is Indian red (*q.v.*).

Venetian Lake. Crimson lake (*q.v.*).

Venetian Red. See INDIAN RED.

Venetian White. A white lead pigment. It is white lead with barium sulphate, and is sold as No. I, II, or III, containing respectively 20, 40, and 60 per cent. barium sulphate.

Venice Soap. An olive oil soap.

Venice Turpentine. The oleo-resin of the larch, *Pinus larix*. It contains 20–22 per cent. essential oil and 74–80 per cent. rosin. It has a specific gravity of 1·1–1·2, a saponification value of 75–125, and an acid value of 75–100. It is soluble in amyl alcohol, amyl acetate, benzene, chloroform, and ether. A substance is often sold under this name consisting of a mixture of rosin oil, rosin, and turpentine. Used in the varnish industry.

Venice White. Venetian white (*q.v.*).

Venite. A proprietary pyroxylin product.

Ventilago Madraspanta (Ouralpatti, Pitti, Lokandi). An Indian dyestuff obtained from a climbing shrub.

Ventolin. A proprietary preparation of salbutamol. A broncho-dilator. (284).

Ventromil. A proprietary preparation

of ORPHENADRINE hydrochloride, aluminium hydroxide and magnesium carbonate, used in the treatment of peptic ulceration. (275).

Veova 911. A proprietary trade name for the vinyl ester of versatic acid. $CH_3COOCH=CH_2$. It is used as a modifying monomer in emulsion lattices. (99).

Veppam Oil. See VEEPA OIL.

Veracetex. A proprietary safety glass.

Veracolate. A proprietary preparation of cascara, bile salts and phenolphthalein. A laxative. (262).

Veractil. A proprietary preparation of methotrimeprazine maleate. A tranquillizer. (336).

Veracur. A proprietary preparation of formaldehyde used to treat warts. (355).

Veramon. A combination of 2 molecules dimethyl-amino-phenyl-dimethyl-pyrazolone and 1 molecule diethyl-barbituric acid. An antineuralgic.

Verapamil. 5-[N-(3,4-Dimethoxyphenethyl)methylamino] - 2 - (3,4 - di - methoxyphenyl) - 2 - isopropylvaleronitrile. Isoptin.

Veratrine. A mixture of various alkaloids obtained from the seeds of *Veratrum sabadilla*. It causes sneezing and irritation when inhaled.

Veratrole. Pyrocatechin-dimethyl-ester, $C_6H_4(OCH_3)_2$. It is applied locally, mixed with tincture of iodine, in the treatment of intercostal neuralgia, and also internally for tuberculosis.

Verazide. N-Isonicotinoyl-N'-veratrylidenehydrazine.

Verbena Oil. See OIL OF VERBENA.

Vercazol. Furfuramide, an accelerator used in rubber vulcanisation.

Verchon. Cerium - quinoline - hydro - chloride.

Verdigris. Basic copper acetate. It is usually a mixture of mono-, di-, and tri-acetates of copper. Green verdigris consists chiefly of the basic acetate, $2(C_2H_3O_2)_2Cu_2O$. Blue verdigris consists mainly of the basic acetate, $(C_2H_3O_2)_2Cu_2O$. The various forms of verdigris are used in dyeing and calico-printing, and for the preparation of oil and water colours.

Verdigris, Crystallised. See CRYSTALS OF VENUS.

Verdigris Green. See EMERALD GREEN.

Verdigris Green. See VERDIGRIS.

Verdite. A gem stone consisting essentially of hydrated magnesium silicate and coloured by chromium. Used for ornamental purposes.

Verditer Blue (Verditer Green, Bremen Green, Mineral Blue, Bremen Blue). An anhydrous basic copper carbonate, produced by the addition of sodium

carbonate to a hot solution of copper sulphate or nitrate. Verditer green is an intermediate product. Used for paper-staining. Copper hydrate and copper carbonate are both sold under these names.

Verditer Green. See VERDITER BLUE.

Verditers. Highly basic copper carbonates.

Verdiviton. A proprietary preparation of sodium, calcium, potassium and manganese glycerophosphates, and vitamin B complex. (682).

Veridian. See CHROMIUM GREEN.

Verilite. An alloy of 96 per cent. aluminium, 2·5 per cent. copper, 0·7 per cent. nickel, 0·4 per cent. silicon, and 0·3 per cent. manganese.

Veriloid. A proprietary preparation of alkaverir. (275).

Veriloid V.P. A proprietary preparation of alkaverir and phenobarbitone. (275).

Veritain. A proprietary trade name for Pholedrine.

Veritol. A proprietary trade name for Pholedrine.

Verjuice. The old name for the very sour juice of unripe green grapes, and of crabapples. It contains tartaric, racemic, and malic acids.

Vermiculite. A mineral, $3MgO.(Fe.Al)_2O_{33}SiO_2$.

Vermilion. See CINNABAR.

Vermilion, American. See CHROME RED.

Vermilion, Chinese. See SCARLET VERMILION.

Vermilionettes. Red pigments. They are combinations of white lead or zinc white, and eosin.

Vermilion, Field's Orange. See SCARLET VERMILION.

Vermilion, Orange. See SCARLET VERMILION.

Vermouth. A wine made from muscat wine by the addition of a vinous infusion of aromatic drugs in which wormwood predominates, and which also contains sweet flag, juniper, and gentian. Alcohol is added to make 15–18 per cent. Sparkling vermouth is made by saturating with carbon dioxide.

Vernadite. A mineral. It is MnO_2. nH_2O.

Vernadskite. A mineral. It is a basic copper sulphate.

Vernisol Z60. A filling compound with a specific gravity of 1·028. Brand Z61 has a gravity of 1·04.

Vernonite (Lucitone). Proprietary trade names for acrylic synthetic resins for denture bases, etc.

Veroform. A soap solution containing formaldehyde. Also see VERTOFORM.

Verona Brown. A pigment obtained by calcining a ferruginous earth.

Veronacetin. Phenacetin-veronal.

Veronal (Malouren, Malonurea, Hypnogen, Barbital, Barbitone, Dormonal, Deba, Uronal). Diethyl-barbituric acid, $C(C_2H_5)_2.(CO.NH)_2.CO$. A soporific used in seasickness and vomiting.

Veronal Sodium (Sodium Veronal, Medinal). Sodium-diethyl-barbiturate, $Na(C_8H_{11}O_3N_2)$. Used as a substitute for veronal.

Verona Yellow. See TURNER'S YELLOW.

Veronese Earth. See GREEN EARTH.

Veronese Green. See GREEN EARTH.

Veronese Yellow. See TURNER'S YELLOW.

Veronida. A liquid preparation containing a hypnotic with aromatics.

Verophene. A German product. It is a combination of veronal and phenacetin. A soporific.

Veropyrin. A mixture of 0·01 part Dionine (ethyl-morphine-hydrochloride) 0·2 part Veronal sodium, and 0·5 part Kalmopyrin.

Versalon 1140. A registered trade name for a polyamide resin used as an adhesive between plasticised vinyl resins and metal.

Versamid. A trade mark for polyamide curing agents for epoxy resins. (114). See also EPICURE.

Verschon. Cerium - quinoline - hydro - chloride. A disinfectant used in a similar way to tincture of iodine.

Versene. Registered trade mark for a range of chelating agents based on ethylene diamine tetra-acetic acid. (64).

Versiflex. A proprietary trade name for a transparent vinyl chloride-acetate.

Verstarktes Chromammonit. An explosive containing 70 parts ammonium nitrate, 10 parts potassium nitrate, 12·5 parts trinitro-toluene, 7 parts chromium-ammonium alum, and 6 parts pet. jelly.

Vert D'eau. A jeweller's alloy containing 60 per cent. gold and 40 per cent. silver.

Vert Diamant. See MALACHITE GREEN.

Vert d'Usebe. See ALDEHYDE GREEN.

Vertigon. A proprietary preparation of PROCHLORPERAZINE maleate. An antiemetic. (281).

Vertoform. A German preparation. It was formerly called Veroform, and consists of basic lead formate. An odourless antiseptic.

Vert Paul Veronese Green. See EMERALD GREEN.

Verton. A proprietary preparation of chlormadinone acetate. (326).

Vert Sulpho B. See ACID GREEN.

Vert Sulpho J. See ACID GREEN.

Verv. Calcium stearyl-2-lactylate. A dough conditioner for yeast-leavened bakery products.

Vesaloin. Hexamine (q.v.).

Vesalvine. Hexamine (q.v.).

Vesalvine B. A hexamine benzoate. Used in medicine.

Vesalvine S. Hexamine salicylate. Used in medicine.

Vesbine. A mineral. It is $PbCuVO_4OH.3H_2O$.

Vesipyrine (Spiroform, Acetyl-salol). Phenyl-acetyl-salicylate, $CH_3.CO.O.C_6H_4.COOC_6H_5$. An antirheumatic and bladder disinfectant.

Vespel. A proprietary trade name for polyimide in the form of prefabricated parts. (10).

Vespral. A proprietary trade name for the hydrochloride of Fluopromazine.

Vesprin. A proprietary trade name for the hydrochloride of Fluopromazine.

Vestalin. See FERROZOID.

Vestamid. A proprietary trade name for nylon 12. A polyamide used for extrusion, film making and injection moulding. (224).

Vesta Phosphate. See RHENANIA PHOSPHATE.

Vestinol AH. A proprietary trade name for a vinyl plasticiser. Dioctyl phthalate. (29).

Vestinol C. Dibutylphthalate. (29).

Vestolen A. High density polyethylene. (29).

Vestolen AS. A trade mark for a flame resistant high density polyethylene building material. (29).

Vestolen BT. A proprietary trade name for an isotactic polybutene-1. Used for the extrusion of piping, cable sheathing and corrosion resistant coatings. (224).

Vestolit. PVC. (29).

Vestopal. Unsaturated polyester resins. (29).

Vestoran. Styrene acrylonitrile. (29).

Vestorian Blue. See EGYPTIAN BLUE.

Vestyron. Polystyrene. (29).

Vestyron 550. A trade mark. Polystyrene containing low residual monomer. A packaging material. (29).

Vestyron 551. A trade mark. Polystyrene packaging materials with low residual monomer and exceptional stress cracking resistance in contact with oils. (29).

Vestyron X984 and X1260AK. Trade marks. Polystyrenes with high impact resistance. (29).

Vesulong. A proprietary preparation of sulphapyrazole. A veterinary antimicrobial agent.

Vesuvianite (Frugardite). Idocrase (q.v.).

Vesuvian Salt. A mineral. It is a variety of aphthitalite.

Vesuvine. See BISMARCK BROWN.

Vesuvine B. See MANCHESTER BROWN EE.

Vesypin. Vesipyrine ($q.v.$).

Veszelyite. A mineral. It is a copper-zinc phospho-arsenate.

Vetisulid. A proprietary preparation of SULPHACHLORPYRADAZINE. A veterinary anti-microbial agent.

Vetiver Oil. See CUS-CUS OIL.

Vetol. A proprietary pure vegetable oil palm product for use as a lard substitute.

V.G.B. A brown resinous powder consisting of acetaldehyde-α-naphthylamine. An antioxidant for rubber. (2) A proprietary acetaldehyde/aniline reaction product. (435).

Viacutan. A proprietary preparation of methargen in a cream base. Antiseptic skin cream. (319).

Viadril. A proprietary trade name for Hydroxydione sodium succinate.

Viandite. A mineral. It is a variety of geyserite.

Via Rasa. An insecticide. It is the calcium salt of p-toluene-chloro-sulphonamide.

Vibazine. A proprietary preparation of buclizine hydrochloride. An antiemetic. (315).

Vibrac. A nickel chrome steel.

Vibrac Steel. A nickel - chromium - molybdenum steel.

Vibramycin. A proprietary preparation of calcium doxycyline (syrup), or doxycyline hydrochloride (capsules). An antibiotic. (315).

Vibrathane. A proprietary polyurethane elastomer. (435).

Vibriomune. A proprietary cholera vaccine. (444).

Vibrocil. A proprietary preparation of dimethindine hydrogen maleate, phenylephrine hydrochloride and neomycin sulphate. (373).

Vicalloy. A proprietary trade name for a high permeability alloy containing iron with 36–62 per cent. cobalt and 6–16 per cent. vanadium.

Vichy Salt. Sodium hydrogen carbonate, $NaHCO_3$.

Viclan. Vinylidene copolymer resins and latices. (512).

Vicmos Powder. A smokeless 33-grain powder.

Vicron. Highly refined calcite. (85).

Victor Bronze. An alloy of 58·5 per cent. copper, 38·5 per cent. zinc, 1·5 per cent. aluminium, 1 per cent. iron, and 0·03 per cent. vanadium.

Victoria Aluminium. See PARTINIUM.

Victoria Black B (New Victoria Black B, New Victoria Black-blue).
A dyestuff. It is the sodium salt of sulpho - benzene - azo - naphthalene-azo - dioxy-naphthalene-sulphonic acid, $C_{26}H_{16}N_4S_2O_8Na_2$. Dyes wool bluish-black from an acid bath.

Victoria Black G, 5G. Dyestuffs of the same class as Victoria Black B.

Victoria Blue B. A dyestuff. It is the hydrochloride of phenyl-tetra-methyl-triamino-diphenyl α-naphthyl-carbinol, $C_{33}H_{32}N_3Cl$. Dyes silk and wool blue from an acid bath, and cotton mordanted with tannin.

Victoria Blue R. See NEW VICTORIA BLUE R.

Victoria Blue 4R. A dyestuff. It is the hydrochloride of phenyl-penta-methyl-triamino-diphenyl-α-naphthyl - carbinol, $C_{34}H_{34}N_3Cl$. Dyes silk and wool reddish-blue.

Victoria Green. See MALACHITE GREEN.

Victoria Green 3B (Setoglaucine, New Fast Green 3B, New Solid Green 3B). A dyestuff. It is the hydrochloride or zinc double chloride of tetramethyl-diamino - dichloro - triphenyl - carbinol. Hydrochloride $= C_{23}H_{23}N_2Cl_3$. Dyes silk and wool, and cotton mordanted with tannin and tartar emetic, bluish-green.

Victoria Orange. See VICTORIA YELLOW.

Victoria Red. See BENZOPURPURIN 4B and CHROME RED.

Victoria Rubine. See FAST RED D.

Victoria Silver. See NICKEL SILVERS.

Victoria Stone. An artificial stone. It consists of 66·6 per cent. powdered granite, the particles being cemented together with 33·3 per cent. Portland cement, and the block hardened by immersion in a solution of sodium silicate for some weeks.

Victoria Varnish. See ZAPON VARNISH.

Victoria Violet 4BS (Ethyl Acid Violet S4B, Coomassie Violet AV). A dyestuff. It is the sodium salt of p-amino-benzene-azo-1 : 8 - dioxy - naphthalene-disulphonic acid, $C_{16}H_{11}N_3S_2O_8Na_2$. Dyes wool bluish-violet from an acid bath.

Victoria Violet 8BS. A dyestuff of the same class as Victoria Violet 4BS.

Victoria Yellow (Gold Yellow, Saffron Surrogate, English Yellow, Victoria Orange, Saffron Substitute, Aniline Orange). A dyestuff. It is a mixture of the potassium or ammonium salts of dinitro-o-cresol, and dinitro-p-cresol, $C_7H_5N_2O_5K$. Dyes wool and silk orange, and is employed for colouring butter. A 0·1 per cent. solution is used for spraying trees infected with the caterpillar of *Liparis monacha*. It is explosive when dry.

Victoria Yellow Extra Conc. See METANIL YELLOW.

Victorite. A mineral. It is Enstatite. Also a proprietary trade name for an impregnated paper specially made for gaskets.

Victorium. A proprietary trade name for lignin thermoplastic materials.

Victor Metal. An alloy of 50 per cent. copper, 34·3 per cent. zinc, 15·4 per cent. nickel, 0·28 per cent. iron, and 0·11 per cent. aluminium. Used for sand castings and marine work.

Victor Powder. A trade mark for a smokeless powder containing nitro-ammonium nitrate, wood meal, and potassium chloride.

Victron. A proprietary trade name for polystyrene and vinylite resins.

Viczsal. An ammoniacal solution of copper and zinc phenolates. A wood preservative.

Vidal Black (Vidal Black S). A dye-stuff obtained by the fusion of *p*-amino phenol (or of *p*-amino-phenol and other compounds) with sodium polysulphide. Vidal Black S is the bisulphite compound. Cotton is dyed greenish- to bluish-black.

Vidal Black S. See VIDAL BLACK.

Vidal's Caustic Powder. A caustic consisting of burnt alum and powdered savin tops.

Vidar. A proprietary grade of poly-vinylidene fluoride used in the production of fluor-polymer resins, powders, pastes, liquids, emulsions, dispersions, pellets and chips. (899).

Vidarabine. An anti-viral drug. 9-β-D-Arabinofuranosyladenine.

Vi-Daylin. A proprietary preparation containing Vitamins A, D, C, thiamine, riboflavine, nicotinamide and pyridoxine. (311).

Videnal. A proprietary preparation containing magnesium carbonate, bismuth subnitrate, sodium carbonate, frangula, calamus, aneurine hydrochloride and ascorbic acid. An antacid. (276).

Vidopen. A proprietary preparation of AMPICILLIN. An antibiotic. (137).

Vidry. See BIDRY.

Vieirine. A substance obtained from the bark of *Remijia vellozii*. Used in Brazil as a tonic and antiperiodic in the place of quinine.

Viellaurite. A mineral. It is a silicate and carbonate of manganese.

Vielle Powder. See POUDRE B.

Vienna Blue. See COBALT BLUE.

Vienna Caustic. Potassium hydroxide with lime. See VIENNA PASTE.

Vienna Cement. A metallic cement made from 86 per cent. copper and 14 per cent. mercury. An imitation gold.

Vienna Green. See EMERALD GREEN.

Vienna Lake. See FLORENCE LAKE.

Vienna Mixture. Consists of 3 parts ether and 1 part chloroform by weight.

Vienna Paste. A mixture of lime and potash. Used as a caustic to extirpate malignant growths.

Vienna Red. See CHROME RED.

Viennese Tombac. An alloy of 97 per cent. copper and 2·8 per cent. zinc.

Vierzonite. A mineral. It is a variety of melinite.

Viferral (Polychloral). A polymerised product of chloral and pyridine. A hypnotic.

Vigantol. A German proprietary product. It consists of ergosterol that has been radiated with ultra-violet light, and 0·001 gram is stated to be equal in effect of vitamin content to 20 grams cod-liver oil. It is supplied in three forms : Vigantol oil (1 per cent. solution), 1 c.c.=25 drops, contains 10 mg. vigantol, vigantol pastilles with sugar coating, one pastille containing 2 mg. vigantol, and vigantol pills with chocolate coating, one pill containing 4 mg.

Vigantol Cod-liver Oil. Cod-liver oil containing biologically controlled vitamin A and D content, with the addition of vigantol.

Vigel. A proprietary preparation containing aluminium glycinate, calamus powder, frangula, bismuth subnitrate, magnesium carbonate and sodium bicarbonate. An antacid. (305).

Vigogne. See ALPACA.

Vigorite. A safety explosive for mines. It consists of 30 per cent. nitroglycerin, 49 per cent. potassium chlorate, 7 per cent. potassium nitrate, 9 per cent. wood pulp and 5 per cent. magnesium carbonate.

Vigorite. See BAKELITE.

Vijochin. Quinine-iodo-bismuthate. An antisyphilitic.

Vikastab. A proprietary trade name for Potassium Menaphthosulphate.

Vikro. Proprietary alloys containing from 63–65 per cent. nickel, 13–23 per cent. chromium, 0·5–1 per cent. silicon, up to 1 per cent. manganese and carbon, and the balance iron.

Vilatéite. A mineral. It is $(Fe''',Mn)'''$ $PO_4.2H_2O$.

Villamaninite. A mineral. It is a disulphide of nickel, copper, cobalt and iron $(Cu.Ni.Co.Fe)S_2$, with small quantities of selenium. Found in Spain.

Villarsite. A mineral. It is $(Mg,Fe)_{1·8}$ $SiO_{3·6}(OH)_{0·14}$.

Villescon. A proprietary preparation of prolintane with vitamins. (278).

Villiaumite. A mineral. It is 4[NaF].

Viloxazine. A drug used in the treatment of mental disease. It is 2-(2-ethoxy-phenoxymethyl)tetrahydro - 1, 4 - oxazine. VIVALAN is the hydrochloride.

Viluite. Synonym for Grossular.

Vimlite. A proprietary trade name for a cellulose acetate plastic.

Vimopyrine. *p*-Phenetidine-tartrate.

Vinaccia Tartar. Tartar obtained from the manufacture of wines.

Vinaconic Acid. Trimethylene-1 : 1-dicarboxylic acid (ethylene - malonic acid) $CH_2 : CH.CH.(COOH)_2$.

Vinacryl R3929, R3940. Proprietary trade names for a 55 per cent. concentrated vinyl-acrylic copolymer emulsions used for non-fray carpet backings. (92).

Vinacryl 4001/B. A proprietary trade name for a 50 per cent. concentrated acrylic copolymer emulsion used as a cement additive. (92).

Vinacryl 4005. A proprietary acrylic copolymer emulsion soluble in alkali. (92).

Vinacryl 4152, 4500/X, 4501/X. Proprietary trade names for vinyl acrylic copolymer emulsions used as adhesives. (92).

Vinacryl 4160. A proprietary trade name for an acrylic copolymer emulsion used as a 46·5 per cent. concentrate for a binder in paper board manufacture. (92).

Vinacryl 4260. A proprietary emulsion. Poly - 2 - ethoxyethyl methacrylate. (92).

Vinacryl 4290. A proprietary polybutyl methacrylate emulsion. (92).

Vinacryl 4320. A proprietary self-reactive vinyl acrylic copolymer emulsion used in the finishing of textiles. (92).

Vinacryl 4322. A proprietary self-reactive vinyl acrylic copolymer emulsion. (92).

Vinacryl 4450. A proprietary trade name for a vinyl-acrylic copolymer. Used in crack filling compounds. (92).

Vinacryl 4512. A proprietary acrylic polymer emulsion. (92).

Vinacryl 7170, 7172 and 7175. A proprietary range of styrene acrylic copolymer emulsions. (92).

Vinal. A proprietary trade name for a synthetic vinyl resin.

Vinalak 5150. A proprietary self-reactive acrylic polymer solution in iso-propyl acetate. (92).

Vinamul 3240. A proprietary vinyl acetate-ethylene copolymer emulsion. (92).

Vinamul 3250. A proprietary vinyl acetate-ethylene copolymer emulsion containing a non-ionic emulsifying system. (92).

Vinamul 6000. A proprietary vinyl acetate emulsion of the unsaturated acid copolymer type, soluble in alkali. (92).

Vinamul 6050. A proprietary vinyl acetate-vinyl caprate-unsaturated acid terpolymer emulsion, internally plasticised. (92).

Vinamul 6208. A proprietary trade name for a 50 per cent concentrated vinyl-acrylic copolymer emulsion used for wallpaper grounding, printing and overcoating. (92).

Vinamul 6275. A proprietary trade name for a 55 per cent. concentrated vinyl-acrylic copolymer emulsion used in the production of emulsion paints. (92).

Vinamul 6705. A proprietary ethylene grafted vinyl acetate copolymer emulsion. (92).

Vinamul 6888. A proprietary modified vinyl acetate-acrylate copolymer emulsion. (92).

Vinamul 6930. A proprietary trade name for 52 per cent. concentrated vinyl acetate—VeoVa 911 copolymer emulsion used for emulsion paints. (92).

Vinamul 7700. A proprietary polystyrene emulsion. (92).

Vinamul 7715. A proprietary polystyrene emulsion containing 15 per cent. dibutyl phthalate in a non-volatile plasticiser. (92).

Vinamul 8400. A proprietary trade name for a 50 per cent. emulsion of polyvinyl acetate. Used for adhesives. (92).

Vinamul 8430. A proprietary polyvinyl acetate emulsion. (92).

Vinamul 8460 and 9000. Proprietary polyvinyl acetate emulsions. (92).

Vinapol 1000. A proprietary polyvinyl acetate powder dispersible in water. (92).

Vinapol 1030. A proprietary polyvinyl acetate powder plasticised with 10 per cent. dibutyl phthalate and dispersible in water. (92).

Vinapol 1070. A proprietary finely-divided polyvinyl acetate powder. (92).

Vinapol 1088. A proprietary acrylic processing aid used in the processing of rigid PVC compounds. (92).

Vinapol R3626. A proprietary trade name for a water-dispersible alkali soluble vinyl acetate powder. (92).

Vinapol R.3800, R.3863, R10, 030. A proprietary trade name for a water-dispersible polyvinyl acetate powder. (92).

Vinasse. See DISTILLERS WASH.

Vinbarbitone. 5 - Ethyl - 5 - (1 - methyl-but-1-enyl)barbituric acid. Delvinal.

Vinblastine. An alkaloid extracted from Vinca rosea. Velbe is the sulphate.

Vincennite. A poison gas. It was hydrocyanic acid mixed with stannic chloride.

Vinchel 11. A proprietary metal-free liquid organic complex with a stabiliser system of barium and cadmium soaps, used as a chelating agent in PVC compounds. (92).

Vinchel 20. A proprietary metal-free liquid organic complex with a stabiliser system of barium and cadmium soaps and tribasic barium sulphate, used as a chelating agent in PVC compounds. (92).

Vinchel 22. A proprietary zinc-based chelating agent for PVC compounds. A basic lead carbonate stabilising system is employed. (92).

Vinchel 35. A proprietary zinc-based chelating agent for PVC compounds. A stabiliser system of barium and cadmium soaps, and barium and cadmium liquids is employed. (92).

Vinco 99A. A proprietary stabiliser for PVC and PVA of the liquid-barium/cadmium-zinc complex type. (967).

Vinco 99G. A stabiliser similar to VINCO 99A used with paste-grade resins in rotational moulding. (967).

Vinco 248. A proprietary stabiliser of the liquid-barium and cadmium-complex type, used with PVC polymers sensitive to zinc. (967).

Vinco 249C. A proprietary stabiliser for PVC with a liquid barium, cadmium and zinc base. (967).

Vinco 265. A proprietary stabiliser for PVC pastes with a liquid barium, cadmium and zinc base. (967).

Vinco A33. A proprietary liquid potassium/zinc complex used as a stabiliser and initiator in the production of expanded PVC. (967).

Vinco A183. A proprietary liquid complex used as a stabiliser and initiator in the production of expanded PVC. (967).

Vincristine. An alkaloid obtained from Vinca rosea. Oncovin is the sulphate.

Vinegar. Essentially a dilute solution of acetic acid, containing from 4–10 per cent. acetic acid. It is formed by the acetic fermentation of saccharine liquids which have undergone alcoholic fermentation. such as wine, beer, or cider, and is mainly manufactured from malt or a mixture of malt grain and sugar.

Vinegar 24. A vinegar containing 5·5 per cent. of acetic acid. The number 24 refers to the number of grains of sodium carbonate required to neutralise the acid in 1 fluid ounce of vinegar.

Vinegar, Aromatic. See AROMATIC VINEGAR.

Vinegar, Artificial. Vinegars prepared from purely alcoholic liquids or from acetic acid and colouring material such as caramel. They contain no cream of tartar and the acetic acid sometimes amounts to 12 per cent.

Vinegar, Crystal. See DISTILLED VINEGAR.

Vinegar Essence. A fluid containing usually from 25–30 per cent. acetic acid, but occasionally 60 per cent. acetic acid. It is derived mainly from pyroligneous acid.

Vinegar, Martial. Ferric acetate.

Vinegar, Mother of. See VINEGAR PLANT.

Vinegar Naphtha. See ACETIC ESTERS.

Vinegar, Orleans. See WINE VINEGAR.

Vinegar Plant (Mother of Vinegar). A thick skin produced by the activity of the acetic acid bacteria upon dilute alcoholic liquids. It consists principally of cellulose.

Vinegar, Radicle. Glacial acetic acid, CH_3COOH.

Vinegar Salts. Calcium acetate. $Ca(C_2H_3O_2)_2.H_2O$.

Vinegar, Sugar. See GLUCOSE OR SUGAR VINEGAR.

Vinegar, White. See DISTILLED VINEGAR.

Vinegar, White Wine. See WINE VINEGAR.

Vinescol 23. A fluorinated synthetic rubber.

Vinesthene. A proprietary preparation of vinyl ether. A general anæsthetic. (336).

Vinethen. An anæsthetic. A substitute for ethyl chloride. It is vinyl ether, to which has been added 3·5 parts absolute alcohol and 0·01 part of an antioxidant.

Vinic Ether. Ethyl ether, $C_2H_5O.C_2H_5$.

Vinnapas. A proprietary vinyl resin.

Vinnapas AD. A proprietary trade name for a polyvinyl acetate dispersion (50 per cent. solids). (46).

Vinnathen. A proprietary trade name for a vinyl acetate/ethylene copolymer which can be cross-linked with peroxides. It can be used for cable jackets.

Vinnol. A proprietary grade of polyvinyl chloride (PVC). (46).

Vinoflex. A registered trade name for polyvinyl chloride polymers. (49).

Vinoflex 377. A proprietary PVC-emulsion homopolymer used in the production of PVC film. (49).

Vinoflex 516. A proprietary PVC suspension-type homopolymer. It is an easily-flowing powder used in the extrusion of rigid PVC. (49).

Vinoflex 526. A proprietary PVC homopolymer similar to VINOFLEX 516. (49).

Vinoflex 534. A proprietary PVC suspension-type homopolymer used in the extrusion of high-quality cables. (49).

Vinoflex 535. A proprietary PVC suspension-type homopolymer in the form of an easy-flowing powder with porous particles, used in the making of soft, calendered products. (49).

Vinoflex 719. A proprietary PVC suspension-type polymer used in the making of rigid, tough, weather-resistant products. (49).

Vinoflex PCU. Unchlorinated polyvinylchloride. (3).

Vinoflex N. Chlorinated polyvinylchloride. (3).

Vinoflex PC. Highly chlorinated polyvinylchloride. (3).

Vinoflex PC 820. High viscosity chlorinated polyvinyl chloride. (3).

Vinoflex S3. Normal polyvinal chloride of low viscosity. (3).

Vinoflex MP 400. A copolymer of 75 per cent. vinyl chloride and 25 per cent. vinylisobutyl ether. (3).

Vinoline. A mixture of different coaltar red dyes, usually the salts of rosaniline. Used for colouring Italian wines.

Vinopyrine. The tartaric acid compound of p-phenetidine. Used for headaches.

Vinous Alcohol. Ethyl alcohol, C_2H_5OH.

Vinsol. A proprietary trade name for the black residue from the extraction of rosin with solvents. Used as an insulating varnish.

Vinsol Resin. A proprietary trade name for a resin produced by the distillation of wood.

Vinychlon. A registered trade mark for a series of Japanese vinylchloride polymers. (60).

Vinylbitone. 5 - (1 - Methylbutyl) - 5 - vinylbarbituric acid.

Vinyl Acetate Resins. See ALVA, GELVA.

Vinyl Resins. See VYDON, VINNAPAS, RESOVIN, KORON, POLYVINYLCHLORIDE.

Vinylite. A proprietary trade name for polyvinyl acetate, polyvinyl chloride-acetate and polyvinyl chloride synthetic resins.

Vinylite X. A proprietary trade name for polyvinyl chloride-acetate.

Vinylite V. A proprietary name for an inter-polymer of P.V.C. and P.V.A. (*q.v.*).

Vinyloid. A proprietary trade name for a polyvinyl acetate resin.

2-Vinyl Pyridine. A similar compound to styrene except that the benzene ring is replaced by pyridine.

Vinylseal. A proprietary trade name for vinyl acetate resin adhesives.

Vinyon. A proprietary trade name for vinyl resins for textile fibres.

Vinyon Fiber. A proprietary trade name for a material manufactured from polyvinyl chloride-acetate.

Vinyzene. A fungicide and bactericide. BROMCHLORENONE.

Viocin. A proprietary preparation containing viomycin sulphate. An antituberculous agent. (85).

Vioflor. A proprietary preparation of volatile hydrocarbons. Used to deodorise turpentine substitutes.

Vioform (Nioform). Iodochloroxyquinolin. An odourless, non-irritant, innocuous, sterilisable substitute for iodoform.

Vioform. A proprietary name for Cliquinol.

Vioform-Hydrocortisone. A proprietary preparation of clioquinol and hydrocortisone used in dermatology as an antibacterial agent. (18).

Violaite. A mineral. It is $4[Ca(Mg,Fe)Si_2O_6]$.

Violamine B. See FAST ACID VIOLET B.

Violamine 3B. See FAST ACID BLUE R.

Violamine G. See ACID ROSAMINE A.

Violamine R. See FAST ACID VIOLET A2R.

Violan. A mineral. It is $4[CaMgSi_2O_6]$.

Violaniline. A blue colouring matter belonging to the induline class. See INDULINE, SPIRIT SOLUBLE.

Violanthrene. A dyestuff prepared from anthranol by heating it with glycerin.

Violarite. A mineral. NiS_2. It is also given as $8[FeNi_2S_4]$.

Violeine. See MAUVEINE.

Violet 3B Extra. See METHYL VIOLET B.

Violet 5B. See BENZYL VIOLET.

Violet 6B. See BENZYL VIOLET.

Violet 7B Extra. See CRYSTAL VIOLET.

Violet C. See CRYSTAL VIOLET.

Violet R, RR, 5R. Hofmann's violet (*q.v.*).

Violet 4RN. See HOFMANN'S VIOLET.

Violet Black. A dyestuff. It is the sodium salt of benzene-disazo-α-naphthylamine-α-naphthol-sulphonic acid, $C_{26}H_{18}N_5O_4SNa$. Dyes cotton and wool violet-black.

Violet Bronze. See MAGENTA BRONZE and SAFFRON BRONZE.

Violet Carmine. A pigment from the root of *Anchusa tinctoria*.

Violet Copper (Red Copper). Reduced copper, prepared by reducing copper oxide (prepared by the action of sodium hydroxide on pure copper oxide, and washing), in a slow current of hydrogen. When reduced in a rapid current of hydrogen at 200° C., a red copper is produced. The violet variety has a greater oxidising activity as a catalyst.

Violet Essence, Artificial. See Essence of Violets, Artificial.

Violet Imperial Rouge. See Regina Purple.

Violet Moderne. See Blue 1900.

Violet Phenylique. See Regina Purple.

Violet Phosphorus. The coarse-grained red variety is metallic or violet phosphorus.

Violet Powder. Perfumed starch powder.

Violet Root. Orris root.

Violet Smelling Salts. See English Salt.

Violettol. A mixture of 10 per cent. ionone and 90 per cent. salicyl aldehyde. Used to strengthen natural violet perfume.

Violetton (Rhodione). Trade names for the commercial ionones. Used as perfumes.

Violet Tungsten. Potassium tritungstate, $K_2W_3O_9.W_2O_5$. A pigment.

Violet Ultramarine. A pigment obtained from blue ultramarine by the action of dry hydrochloric acid gas and oxygen at 150–180° C. See Ultramarine.

Violin. An emetic substance contained in the common violet, supposed to be identical with emetine.

Violine. See Mauveine.

Violite. A mineral. It is a variety of copiapite.

Violuric Acid. Isonitroso-barbituric acid, $C_4H_3N_3O_4$.

Viomycin. An antibiotic produced by certain strains of Streptomyces griseus var. purpureus. Viocin is the sulphate ; Vionactane is the pantothenate-sulphate.

Viomycin Sulphate. A proprietary preparation containing viomycin sulphate. An antituberculous agent. (264).

Vionactane. A proprietary preparation of viomycin pantothenate and sulphate. An antituberculous agent. (18).

Viosterol. The officially approved name for preparations of irradiated ergosterol. Approved by the Council on Pharmacy and Chemistry of the American Medical Association.

Vipla. Emulsion polyvinyl chloride homopolymers.

Viprynium Embonate. 6-Dimethylamino - 2 - [2 - (2,5 - dimethyl - 1 - phenyl - 3 - pyrrolyl)vinyl] - 1 - methyl - quinolinium 4,4' - methylenedi - (3 - hydroxy - 2 - naphthoate). Pyrivinium Pamoate. Vanquin.

Virginiamycin. An antibiotic produced by Streptomyces virginiæ.

Virginia Silver. See Nickel Silvers.

Virgin Oil. The term usually applied to the oil pressed cold from seeds such as castor or olive. The first pressing gives " virgin " or " cold drawn " oil, and after breaking up the cakes and again pressing, as " seconds " oil.

Virgo Fibre Half-stuff. A fibre produced from the short fibres or so-called seed-lint by treatment with caustic soda.

Viridian (French Veronese Green). A green pigment. It is hydrated sesquioxide of chromium, $Cr_2O_3.2H_2O$. See Chromium Green.

Viridine. See Alkali Green.

Virogen. A preparation of casein with glycerophosphates, including sodium glycerophosphate. Used in the treatment of general weakness.

Virol. A proprietary preparation of vitamins A, D and B complex, with iron and iodine. A vitamin and mineral supplement. (968).

Virugon. A proprietary preparation of N',N'-anhydrobis (betahydroxyethyl) bignamide hydrochloride, atropine methonitrate and hyoscine methonitrate. (112).

Virvina. A proprietary preparation of Thiamine, Riboflavin, Pyridoxine, and calcium, potassium and manganese glycerophosphates. A tonic. (310).

Vis. A proprietary safety glass.

Visca. An artificial horse-hair made from viscose.

Viscanite. A vegetable glue.

Viscalex EP 30. A proprietary acrylic copolymer emulsion used as a thickener in water-based paints. (565).

Visclair. A proprietary preparation of Carboxymethylcysteine. A nasal spray. (740).

Viscogen. A solution of cane sugar in lime water. It is used as a cream thickener.

Viscoid. A mixture of viscose and clay with powdered horn, or zinc oxide.

Viscolane. An ointment base prepared from the bark of the mistletoe.

Viscolith. The hard mass obtained when a viscose solution coagulates.

Viscoloid (Nixenoid, Fiberloid). A proprietary trade name for pyroxylin plastics.

Viscom. See Oil Pulp.

Viscose. The sodium salt of cellulose xanthate, obtained by the action of carbon bisulphide and alkali upon cellulose. In thin sheets, it is used as a substitute for glass and celluloid. It is also employed as a thickening and dressing substance, as a partial substitute for resin glue in paper manufacture, and in the production of artificial silk. Also see Dextran. Other synonyms are : Cellushi, Clar-Apel, Crystex, New-wrap, Sidac, Tomophan, Zellwonet.

Viscose Silk. An artificial silk produced when Viscose (*q.v.*) is forced through narrow orifices into ammonium chloride solution.

Viscosin. A proprietary refined oil tar.

Viscosine. See VALVOLINE.

Viscovasin. A German proprietary preparation stated to contain the isolated active principles of *Viscus album, Cratægus oxyacantha*, and *Sedum acre*. It reduces blood pressure.

Visem. A food product containing lecithin and salts of glycero-phosphoric acid.

Viseite. A mineral. It is $2[NaCa_5Al_{10}Si_3P_5O_{30}(OH,F)_{18}.16H_2O]$.

Vishnevite. A mineral. It is $(NaKCa)_{6-8}Al_6Si_6O_{24}(SO_4,CO_3).1-5H_2O$.

Vi-Siblin. A proprietary preparation of ispaghula husks and THIAMINE. A laxative. (264).

Visken. A proprietary preparation of PINDOLOL used in the treatment of angina pectoris. (267).

Visnadine. 10-Acetoxy-9,10-dihydro-8, 8 - dimethyl - 9 - methylbutyryloxy - 2H, 8H - benzo[1,2 - ; 3,4 -] - dipyran - 2 - one. Cardine.

Visor. A proprietary preparation of carbromal and bromvaletone. A sedative. (344).

Vistanex. A proprietary trade name for polyisobutylene. A synthetic rubber. See also OPPANOL, ISOLENE, P.I.B. Hydrocarbon resin.

Vistex. Trademark for a blend of cellulosic polymers and calcium carbonates used to increase viscosity and loss of fluidity in liquids. (969).

Vistra. An artificial silk which is made in a similar way to viscose silk.

Vistron. See POLYSTYRENE and other trade names under the latter.

Vita-E. A proprietary preparation of D-α-tocopherol succinate used in the treatment of vascular disorders. (398).

Vita-E-Gels. A proprietary preparation of D-α-tocopherol acetate used in the treatment of vascular disorders. (398).

Vitafer. A preparation of milk and the glycero-phosphates of calcium and magnesium. The name has also been applied to a mixture of magnesium carbonate and sodium sulphate.

Vita Glass. A synthetic resin which allows passing of ultra-violet rays.

Vital Air (Pure Air, Dephlogisticated Air). Old names for Oxygen.

Vital Red. See BRILLIANT CONGO R.

Vitalum. A trade mark for materials of the abrasive class and consisting essentially of alumina.

Vitamalt. A proprietary preparation containing vitamins A, B, C, and D.

Vitamin A. A highly unsaturated alcohol, lack of which in the diet produces night blindness and cheilosis at the corners of the mouth in man.

Vitamina. A German product composed of phosphatides, fat, phytosterin, and water. Half the phosphatide is as lecithin. The material is added to foodstuffs to put in vitamins which are lacking.

Vitamin B Complex. Comprises Vitamins B_1, B_2, B_6, B_{12}, biotin, folic acid, pantothenic acid and nicotinamide. Most of these compounds are thought to be coenzymes concerned with cellular oxidation reactions or with protein synthesis.

Vitamin B_1. Thiamine, aneurin. Deficiency in the diet in man produces Beri-beri and peripheral neuritis.

Vitamin B_2. Riboflavin. Deficiency in man produces minor symptoms only. Also known as lactoflavin.

Vitamin B_6. Pyridoxine. No known deficiency disease in man.

Vitamin B_{12}. Cobalamin, cyanocobalamin. Deficiency in man causes pernicious anæmia and neural degeneration.

Vitamin C. Ascorbic acid. Deficiency in man causes scurvy. It probably plays a part in the production of collagen in the tissues.

Vitamin D. A fat soluble vitamin, deficiency of which causes rickets in children and osteomalacia in adults.

Vitamin D_2. Calciferol. A synthetic vitamin produced by the irradiation of ergosterol with ultra-violet light. Available as Sterogyl-15, and in many multivitamin preparations.

Vitamin D_3. Naturally occurring Vitamin D. Derived from fish oils.

Vitamin D_4. A synthetic vitamin derived form 22-dihydroergosterol by irradiation with ultra-violet light. It has less activity than D_2 and D_3.

Vitamin E. Tocopherol. A fat soluble vitamin not known to have any function in human metabolism. Deficiency in rats produces sterility. Available as Ephynal for use as a vasodilator.

Vitamin G. See RIBOFLAVIN.

Vitamin K_1. Phytomenadione. A fat soluble vitamin. 2-methyl-3-phytyl-1, 4-naphthaquinone. Deficiency gives rise to hæmorrhage.

Vitamin K_2. A fat soluble vitamin. 2-methyl-3-difarnesyl-1,4-napthaquinone. Synthesised in the gut by bacteria.

Vitamin M. Folic acid.

Vitamin P. See QUERCITIN.

Vitamin U. A vitamin extracted from cabbage.

Vitavel. A proprietary preparation containing Vitamins A, D, C, thiamine, hydrochloride and glucose. (272).

Vitoline Yellow 5G. See PHOSPHINE.

Viton. A proprietary trade name for a fluoroelastomer resistant to attack by corrosive chemicals up to 400° F. (10).

Viton A. A proprietary vinylidene fluoride-hexafluoropropylene copolymer possessing high powers of resistance to heat, oil and solvents. Its Mooney viscosity at 100° C. is 67. (10).

Viton A35. A copolymer similar to VITON A but possessing easier processing qualities. Its Mooney viscosity at 100° C. is 35. (10).

Viton AHV. A copolymer similar to VITON A but possessing greater strength at high temperatures. Its Mooney viscosity at 100° C. is 180. (10).

Viton E-430. A proprietary fluoroelastomer with good processing and storage properties. (10).

Viton LN. A waxy semi-solid fluoroelastomer of the VITON type used as a plasticiser to improve moulding and extrusion characteristics. (10).

Vitrain. See CLARAIN.

Vitralene. Polypropylene.

Vitrathene. Polythene.

Vitre-colloid. A proprietary cellulose acetate.

Vitreo-colloid. A proprietary trade name for cellulose acetate plastic.

Vitreon. A porcelain used in the condensing plant for nitric acid.

Vitreosil (Silica Glass). Fused Silica, SiO_2.

Vitreous Copper. See REDRUTHITE.

Vitreous Silver Ore. See ARGENTITE.

Vitreous White Arsenic. See ARSENIC GLASS.

Vitriolated Magnesia. Magnesium sulphate, $MgSO_4$.

Vitriolated Soda. Sodium sulphate, Na_2SO_4.

Vitriolated Tartar. Potassium sulphate, K_2SO_4.

Vitriol, Blue. See BLUE COPPERAS.

Vitriol, Copper. See CHALCANTHITE.

Vitriol, Elixir of. See ACID ELIXIR OF VITRIOL.

Vitriol, English. Ferrous sulphate, $FeSO_4$.

Vitriol, Green. See IRON VITRIOL.

Vitriolic Acid. See OIL OF VITRIOL.

Vitriolised Bones. Bones which have been treated for a long time with sulphuric acid to obtain the phosphates in a more soluble form. A fertiliser.

Vitriol, Nickel. See MORENOSITE.

Vitriol Ochre. A mineral, $FeSO_4$. $2Fe(OH)_3.H_2O$.

Vitriol Oil. See OIL OF VITRIOL

Vitriol, Roman. See BLUE COPPERAS.

Vitriol, Salt of. Zinc sulphate, $ZnSO_4$.

Vitriol, Salzburg. See MIXED VITRIOL.

Vitriol Stone. Impure ferric sulphate obtained by the oxidation of pyrites.

Used in the manufacture of fuming sulphuric acid.

Vitrite. A poison gas containing a mixture of cyanogen chloride and arsenic trichloride.

Vitrite. A Swiss compressed asbestos product used for electrical insulation.

Vitrone. PVC.

Vivalan. A proprietary preparation of VILOXAZINE hydrochloride. An antidepressant. (512).

Vivianite. A mineral. It is $2[Fe_3(PO_4)_2 .8H_2O]$. See BLUE IRON EARTH.

Vivonex. A proprietary preparation of aminoacids, glucose, essential fats, vitamins and minerals used as a dietary aid in cases of mal-absorption of the gut. (818).

Vladimirite. A mineral. It is $Ca_3 (AsO_4)_2.4H_2O$.

Vleminck's Solution. Lime sulphurated solution, B.P.C.

V.L. Yarn. A German artificial wool fibre.

V.M. and P. Naphtha. Varnish-maker's and painter's naphtha, a deodorised petroleum product which is practically a gasoline (benzine) of 100–160° C. boiling range and specific gravity 0·730.

Vogan. A vitamin A concentrate.

Vogel's Alloy. An alloy containing 8 parts copper, 1 part zinc, 2 parts tin, and 1 part lead. Used for polishing steel.

Vogesite. A mineral. It is pyrope.

Voglianite. A mineral. It is a basic sulphate of uranium.

Voglite. A Joachimsthal mineral containing carbonates of calcium, uranium, and copper.

Voidox 100 per cent. A registered trade name for a food grade antioxidant. It is a modified fatty acid derivative of a substituted phenol. (54).

Vol. Ammonium carbonate, $(NH_4)_2CO_3$.

Volatile Alkali (Alkaline Air, Spirit of Hartshorn). Ammonia, NH_3.

Volatile Liniment. Liniment of ammonia.

Volatile Oil of Bitter Almonds. Benzaldehyde, $C_6H_5.CHO$.

Volatile Salt. Ammonium carbonate, $(NH_4)_2CO_3$.

Volborthite. A mineral, $4(Cu.Ca).9V_2O_5$.

Volborthite, Turkestan. See TANGEITE.

Volcanite. A mixture of selenium and sulphur found in the Lipari Islands.

Völckerite. A mineral, $3Ca_3(PO_4)_2.CaO$.

Volckmann's Solution. A solution containing thymol, alcohol, glycerin, and water.

Volenite. A rubber substitute made from rosin, oil, and some fibrous material.

Volgerite. A mineral consisting mainly of antimony tetroxide, Sb_2O_4.

Volidan and Volidan 21. Megestrol acetate with ethinyloestradiol tablets—oral contraceptives. (182).

Volital. A proprietary preparation of pemoline. Central nervous stimulant. (352).

Völkernite. See HYDROTALCITE.

Volkite. A German moulded rubber product used for electrical insulation.

Vollsalz. A table salt sold in Switzerland and Austria. It contains about 5 mg. of potassium iodide per 1,000 grams of salt. Used as a preventative agent against goitre.

Volomite. An abrasive consisting of tungsten carbide. It is stated to be as hard as the diamond.

Volpar. Spermicidal contraceptive containing phenylercuric acetate. Gels, paste and foaming tablets. (182).

Voltaite (Pettkoite). A mineral, $2(Fe.Al)_2O_3 . 5(Mg.Fe.Na_2.K_2)(OH)_2SO_4$.

Voltalef 300. Polytrichlorethylene. (406).

Voltoids. A registered trade name for compressed tablets of ammonium chloride used in the preparation of voltaic cells.

Voltol Oils. Prepared by treating oil with high tension electric current at 60–80° C. Vegetable and marine oils are changed by this treatment to highly viscous oils having great stability.

Voltzine. See VOLTZITE.

Voltzite (Voltzine). A mineral, $4(ZnS).ZnO$.

Voluntal. Trichloro-ethyl-urethane. A hypnotic.

Vomiting Salt. Zinc sulphate, $ZnSO_4$.

Vondiestite. A mineral, $(Ag.Au)_5BiTe_4$.

Von Forster Powder. A gelatinised nitro-cellulose flake powder, with a little calcium carbonate.

Vonges Dynamite. An explosive containing 75 per cent. nitro-glycerin, 20·8 per cent. randanite (decomposed felspar), 3·8 per cent. quartz, and 0·4 per cent. magnesium carbonate.

Vonsenite. A mineral, $(3Fe.Mg)O.B_2O_3.+FeO.Fe_2O_3$, of California.

Von Vetter's Solution. An aqueous solution of glycerin, sugar, and potassium nitrate. Used to preserve anatomical specimens.

Vorane. A proprietary polyether-based polyurethane elastomer cross-linked with diamine. (64).

Vorhauserite. A mineral. It is retinalite.

Vorobyevite. A mineral. It is a variety of beryl.

Vortel. A proprietary preparation of chlorprenaline, ethomoxane and methapyrilene. (333).

Vosol. A proprietary preparation of 1,2-propanediol diacetate, acetic acid and benzethonium chloride in propylene glycol. Ear drops for otitis externa. (382).

Vraic (Varech). French names for Kelp (q.v.).

Vrbaite. A thallium mineral, $TlAs_2SbS_5$.

Vredenburgite. A mineral, $3Mn_3O_4.2Fe_2O_3$.

Vucine. See RIVANOL.

Vuelite. A proprietary trade name for a cellulose acetate plastic. Vuelite-reinforced is a transparent cellulose acetate sheet reinforced with wire mesh.

Vuepak. A proprietary trade name for an acetate wrapping material.

Vulcabond. Adhesive for bonding rubber to metal. (Mixed isocyanates). (512).

Vulcaflex. A registered trade mark for an antioxidant which offers protection against flex-cracking in rubber materials. It is a complex substituted secondary amine of the dimethoxy-diphenylamine type.

Vulcaflex A. A proprietary trade name for a mixture of Vulcaflex and other antioxidants. (2).

Vulcafor. A registered trade name for a range of proprietary products, including rubber vulcanisation accelerators and colours for rubbers and plastics. (2). Products in the range include the following:

Vulcafor I. A proprietary rubber vulcanisation accelerator. See ACCELERENE.

Vulcafor II. Diphenyl-guanidine, $NH C(NHC_6H_5)_2$, a rubber vulcanisation accelerator. Also see D.P.G. SALT.

Vulcafor III. Triphenyl-guanidine, $NHC_6H_5 : C(NHC_6H_5)_2$, a rubber vulcanisation accelerator.

Vulcafor IV. Thio-carbanilide. A rubber vulcanisation accelerator.

Vulcafor V. A rubber vulcanisation accelerator. It consists of aldehyde ammonia, $C_2H_4O.NH_3$.

Vulcafor VI. Zinc-diethyl-dithio-carbamate. A rubber vulcanisation accelerator.

Vulcafor VI R. A proprietary rubber vulcanisation accelerator. It is di-ethylamine-diethyl-dithio-carbamate.

Vulcafor VII. Tetraethyl-thiuram-disulphide. A rubber vulcanisation accelerator.

Vulcafor VIII. Diethylamine-diethyl-dithio-carbamate. A rubber vulcanisation accelerator.

Vulcafor IX. A rubber vulcanisation accelerator. It is zinc isopropyl xanthate.

Vulcafor XII. Di-o-tolyl-guanidine. A

proprietary rubber vulcanisation accelerator.

Vulcafor XIII. Di-*o*-xylyl-guanidine. It is one of the more recent proprietary vulcanisation accelerators of the Vulcafor series.

Vulcafor XIV. A rubber vulcanisation accelerator. It is methylene-*p*-toluidine.

Vulcafor BA. A medium rubber vulcanisation accelerator. It is butyraldehyde-aniline, and is a liquid.

Vulcafor BSO. An accelerator. Benzthiazol sulphen *tert*-octylamide.

Vulcafor BT. An accelerator. It is a thiazole-thiuram mixture.

Vulcafor DAU. An accelerator. It is a dibenzthiazyl disulphide + thiuram disulphide.

Vulcafor DAW. An accelerator. It is a blend of thiazole and guanidine.

Vulcafor DDC. A rubber vulcanisation accelerator. It is diethyl-ammonium-diethyl-dithio-carbamate.

Vulcafor DDCN. An accelerator. Diethylammonium diethyl dithiocarbamate.

Vulcafor DHC. An accelerator. It is a blend of thiazole and dithiocarbamate.

Vulcafor DOTG. A medium rubber vulcanisation accelerator. It is di-*o*-tolyl-guanidine.

Vulcafor DPG. A rubber vulcanisation accelerator. It is diphenyl-guanidine.

Vulcafor EFA. An accelerator, produced from ethyl chloride, formaldehyde and ammonia.

Vulcafor F. An accelerator. Dibenzthiazyl disulphide + diphenyl guanidine.

Vulcafor FN. An accelerator. Dibenzthiazyl disulphide + diphenyl guanidine + hexamethylene tetramine.

Vulcafor HBS. An accelerator. It is N - cyclohexyl - 2 - benzthiazole sulphenamide.

Vulcafor MA. A rubber vulcanisation accelerator. It is a slow accelerator, and is a condensation product of formaldehyde and aniline.

Vulcafor MBT. A semi-ultra rubber vulcanisation accelerator. It is mercapto-benzo-thiazole.

Vulcafor MBTS. A semi-ultra rubber vulcanisation accelerator. It is di-benzo-thiazyl-disulphide.

Vulcafor MPT. A rubber softener. It is methylene-*p*-toluidine.

Vulcafor MS. A super rubber vulcanisation accelerator. It is tetra-methyl-thiurammono-sulphide.

Vulcafor MT. A rubber vulcanisation accelerator. It is a slow accelerator, and is condensation product of formaldehyde and *p*-toluidine.

Vulcafor PT. A semi-ultra rubber vulcanisation accelerator. It is a complex aldehyde-amine.

Vulcafor Resin. A registered trade mark for a rubber vulcanisation accelerator. It is probably an acetaldehyde-aniline condensation product.

Vulcafor RN. A rubber vulcanisation accelerator obtained by polymerising ethylidene-aniline.

Vulcafor SDC. A water-soluble rubber vulcanisation accelerator. It is sodium diethyl-dithio-carbamate.

Vulcafor SPX. A rubber vulcanisation accelerator. It is sodium isopropyl xanthate.

Vulcafor TC. A slow rubber vulcanisation accelerator. It is thiocarbanilide.

Vulcafor TET. A super rubber vulcanisation accelerator. It is tetraethyl-thiuram-disulphide.

Vulcafor TMT. A super rubber vulcanisation accelerator. It is tetramethyl-thiuram-disulphide.

Vulcafor TPG. A slow vulcanisation accelerator. It is triphenyl-guanidine.

Vulcafor ZDC. A super rubber vulcanisation accelerator. It is zinc di-ethyl-dithio-carbamate.

Vulcafor ZEP. An accelerator. Zinc ethyl/phenyl dithiocarbamate.

Vulcafor ZIX. A rubber vulcanisation accelerator. It is zinc-isopropyl-xanthate.

Vulcafor ZMBT. An accelerator. Zinc mercaptobenzthiazole.

Vulcaid. A proprietary litharge of small particle size.

Vulcaid 27. A proprietary trade name for a rubber vulcanisation accelerator. It is zinc butyl xanthate.

Vulcaid 28. A proprietary trade name for a rubber vulcanisation accelerator. It is dibenzylamine.

Vulcaid 33. A proprietary trade name for an antioxidant. A liquid amine condensation product.

Vulcaid 44. A proprietary trade name for an antioxidant. It is a naphthol-amine reaction product.

Vulcaid 55. A proprietary trade name for an antioxidant. It is an acetaldehyde-aniline condensation product.

Vulcaid 111. A proprietary trade name for a rubber vulcanisation accelerator. It is butyraldehyde-aniline.

Vulcaid 222. A proprietary trade name for a rubber vulcanisation accelerator. It is tetramethyl-thiuram monosulphide.

Vulcaid 444B. A proprietary trade name for a rubber vulcanisation accelerator. It is heptaldehyde-aniline.

Vulcaid DPG. A proprietary trade name for a rubber vulcanisation accelerator. It is diphenyl-guanidine.

Vulcaid LP. A proprietary vulcanisation accelerator. It is lead-penta-methylene-dithio-carbamate.

Vulcaid P. A rubber vulcanisation accelerator. It is piperidine-penta-methylene-dithio-carbamate. It melts at 172° C., and is soluble in water, alcohol, and benzene.

Vulcaid ZP. A proprietary rubber vulcanisation accelerator. It is zinc penta-methylene-dithio-carbamate.

Vulcamel. A proprietary rubber vulcanisation accelerator. It is butyr-aldehyde-ammonia.

Vulcan-asbestos. An ebonite substitute made by adding a mixture of asbestos and sulphur to a solution of rubber in benzene, evaporating the solvent, and vulcanising the mass in moulds.

Vulcan Bronze. A proprietary trade name for a bearing bronze containing 1·0 per cent. silicon with iron and nickel.

Vulcanex. A proprietary rubber vulcanisation accelerator. It is an alde-hyde-amine.

Vulcaniline. A proprietary trade name for a rubber vulcanisation accelerator. It is paranitroso-dimethylaniline.

Vulcaniline A. See ACCELERENE.

Vulcanine. A patented material made from rubber, asbestos, litharge, lime, sulphur, and zinc oxide.

Vulcanised Fibre. A material made by treating sheets of paper with zinc chloride solution and subjecting the gelatinised sheets to pressure. The paper is sometimes mixed with glycerin and vulcanised oils. Used for making valve discs, brake blocks, and tubes. Other names for this and similar products are Hard Fibre, Red Fibre, Grey Fibre, Vegetable Fibre, Whalebone Fibre, Egyptian Fibre, Fiberoid, Leatheroid, and Horn Fibre. The dielectric strength of certain types of vulcanised fibre varies from 9,000–16,000 volts per mm., the tensile strength from 9,000–20,000 lbs. per sq. in., the density 1·3–1·5 grams per c.c., and water absorbed in 24 hours from 26–45 per cent. Acetone, alcohol, benzol, carbon disulphide, and ether have no permanent effect upon it, but strong alkalis attack it, also strong acids.

Vulcanised Oils. See THINOLINE.

Vulcanised Paper. See VULCANISED FIBRE.

Vulcanite. See EBONITE. It is also the name for a nitro-glycerin explosive.

Vulcanol. A proprietary rubber vulcanisation accelerator. It is thio-keto-phenyl-dipropyl-propylidene-dihydro-thiazine.

Vulcan Powder. An explosive containing 30 per cent. nitro-glycerin, 52·5 per cent. sodium nitrate, 7 per cent. sulphur, and 10·5 per cent. charcoal.

Vulcan Red. A pigment for rubber. It is stated to be a dye deposited upon a base.

Vulcan Red MO. A pigment for rubber. It is said to be a highly dispersed iron oxide, of German origin.

Vulcaplas. A proprietary trade name for an organic polysulphide synthetic elastic material.

Vulcapont. A proprietary rubber vulcanisation accelerator. It consists of equal parts of thionex and vulcanol.

Vulcaprene. Synthetic rubbers. (512).

Vulcasbeston. A mixture of rubber and asbestos. A heat and electrical insulator.

Vulcase. A proprietary preparation of colloidal sulphur.

Vulcastab. Stabilizer for rubber latex mixing. (512).

Vulcatard. Retarding agent for rubber vulcanization. (512).

Vulcatex Colours. A proprietary trade name for rubber colours containing the vulcafor colours with rubber and stearic acid.

Vulcatex Fast Green GS. A phthalo-cyanine green. (2).

Vulcatex Fast Orange GS. A pigment of the azoic class. A bright mid orange. (2).

Vulcazol. Furfuramide, a vulcanisation accelerator.

Vulco-asbestos. A similar material to amianite, and used for electrical insulation.

Vulcoferran. A trade mark for linings principally of rubber or ebonite for chemical apparatus.

Vulcogene. A proprietary trade name for a rubber vulcanisation accelerator. It is thiocarbanilide.

Vulcogene ND. A proprietary trade name for a rubber vulcanisation accelerator. It is diphenyl-guanidine.

Vulcoid. A proprietary trade name for a phenolic resin impregnated vulcanised fibre.

Vulcone (Du Pont Accelerator No. 19). A trade mark for a hard synthetic resin suitable for use in rubber mixings. It is produced by the condensation of aliphatic aldehydes with aniline, and melts at 80° C. and over.

Vulconex. A proprietary trade name for a rubber vulcanisation accelerator. It is ethylidine-aniline.

Vulkacit. A proprietary trade name for o-tolyl guanidine. A rubber vulcanisation accelerator.

Vulkacit A. Aldehyde-ammonia, a vulcanisation accelerator.

Vulkacit BP. A rubber vulcanisation accelerator. It is a paste, and consists of a mixture of bases.

Vulkacit CA. A rubber vulcanisation accelerator. It is thio-carbanilide. Also see VULCAFOR IV.

Vulkacit CT. A proprietary trade name for a rubber vulcanisation accelerator. It is crotonaldehyde-aniline.

Vulkacit D. Diphenyl-guanidine, a vulcanisation accelerator.

Vulkacit DM. A proprietary trade name for a rubber vulcanisation accelerator. It is benzthiazyl disulphide.

Vulkacit FP. A proprietary trade name for a rubber vulcanisation accelerator. It is methylene-p-toluidine.

Vulkacit H. Hexamethylene-tetramine, a vulcanisation accelerator.

Vulkacit M. A proprietary rubber vulcanisation accelerator. It is mercapto-benzo-thiazole.

Vulkacit Mercapto. A proprietary trade name for a semi-ultra rubber vulcanisation accelerator. It is mercapto-benzo-thiazole.

Vulkacit NP. 5-Methylhexahydro-1,3,5-triazine-2-thione.

Vulkacit P. Piperidine-piperidyl-dithioformate, a vulcanisation accelerator.

Vulkacit P Extra. The zinc salt of ethyl-phenyl-dithio-carbaminic acid, a rubber vulcanisation accelerator.

Vulkacit Thiuram. A rubber vulcanisation accelerator. It is tetramethyl-thiuram-disulphide.

Vulkacit TR. A rubber vulcanisation accelerator. It is a mixture of the free bases of polyamines of ethylene.

Vulkacit 774. A proprietary trade name for an ultra-accelerator for rubber. It is the dithiocarbamate of cyclohexylethylamine. Used for the vulcanisation of thin-walled rubber articles (1 per cent. aq. solution at 80° C. for 30 minutes).

Vulkacit 470. A rubber vulcanisation accelerator. It is a condensation product of homologous acroleins with aromatic bases.

Vulkacit 576. A condensation product of homologous acroleins with aromatic bases. A rubber vulcanisation accelerator most suitable for regenerated rubber.

Vulkacit 1000. A rubber vulcanisation accelerator. It is o-tolyl-biguanide.

Vulkide A. A proprietary trade name for a hard and rigid material based on acrylonitrile/butadiene styrene resin. Used where hard treatment results, e.g., in luggage. (2).

Vulkollan. A proprietary polyester-based polyurethane casting elastomer cross-linked with diols. (112).

Vulnodermol. An iodine preparation stated to be suitable as an iodoform substitute.

Vulnojod. A German product. It is an antiseptic plaster which gives off iodine after application to the skin.

Vulpinite. A variety of anhydrite mixed with silica.

Vultac 1, 2 and 3. A proprietary trade name for an antioxidant comprising the sulphides of phenols and substituted phenols. (119).

Vultex. A trade mark for vulcanised rubber latex preserved with ammonia.

Vuzin. Octyl-hydro-cupreine. A bactericide used in medicine.

Vuzin Bihydrochloride. Iso-octyl-hydro-cupreine-dihydrochloride, $C_{27}H_{45}O_2N_2Cl_2$. Used in medicine.

Vydon. A proprietary vinyl resin denture.

Vyflex NT80S. A proprietary PVC powder coating material. Containing no toxic metals and supporting no microbiological growth, it can be safely brought into contact with drinking water. (146).

Vygen. A trade mark for PVC resins.

Vynamon. Pigments for polyvinyl chloride. (512).

Vynamon Green G-FW. A proprietary bluey-green phthalocyanine pigment for use in plastics. It is claimed to comply with French regulations for food packaging. (512).

Vyncolite. A proprietary range of phenolic moulding materials. (970).

W

Wachenrodite. A mineral. It is a variety of wad containing lead.

Wackenroder's Solution. A solution obtained by passing sulphuretted hydrogen through an aqueous solution of sulphurous acid. The solution contains $H_2S_4O_6$, $H_2S_3O_6$, $H_2S_5O_6$, and colloidal sulphur.

Wackerschellak. An artificial shellac obtained by the condensation and polymerisation of acetaldehyde.

Wad (Bog Manganese). An earthy variety of hydrated manganese oxide, MnO_2.

Wadeite. A mineral. It is $2[K_2CaZr Si_4O_{12}]$.

Wade's Drops. Tinct. Benzoin Co.

Wade's Secaderm Salve. A proprietary preparation of phenol, terebene melaleuca oil, turpentine oil, colophony and chlorophyll, used in dermatology. (340).

Wafer, Worm. See SANTONIN WAFERS.

Wagite. A mineral. It is a variety of calamine.

Wagnerite. A mineral, Mg_2FPO_4.

Wagner's Reagent. A solution of 2 grams potassium iodide in 100 c.c. water.

Wahnerit. A German synthetic varnish-paper product used for electrical insulation.

Wahoo Bark. The root bark of *Euonymus atropurpureus*.

Waite's Fluid. A proprietary preparation of procaine hydrochloride and adrenalin.

Wakefield Grease. See YORKSHIRE GREASE.

Walchowite. See RETINITE.

Waldheimite. A mineral. It is $Na_{3.4}Ca_{1.7}Mg_2FeSi_8O_{22}(OH)_2$.

Walkerite. A clay of the Fuller's earth type. A mineral. It is $NaCa_2Si_3O_8$ OH.

Walker's Earth. Fuller's earth.

Wallerian. A mineral. It is a black variety of hornblende.

Wall Saltpetre. See NITROCALCITE.

Walmstedtite. A mineral. It is a variety of magnesite containing manganese.

Walpurgite. A mineral. It is a basic arsenate of uranium and bismuth.

Walsrode. Powder. A proprietary smokeless powder containing 98·6 per cent. nitro-cotton and 1·4 per cent. volatile matter.

Waltherite. A mineral. It is a bismuth carbonate.

Wando Steel. A proprietary non-shrinking steel containing 1·05 per cent. manganese, 0·5 per cent. chromium, 0·5 per cent. tungsten, and 0·95 per cent. carbon.

W.A. Powder. An American smokeless powder. It is a guncotton-nitroglycerin powder, with barium and potassium nitrates.

Wapplerite. A mineral, $(Ca.Mg)HPO_4$. $3H_2O$.

Waras (Wars, Warrus). A resinous powder which covers the seed pods of *Flemingia congesta*, of India. It is used in Arabia as a dye.

Warburg's Tincture. A quinine preparation containing aloes, rhubarb, gentian, camphor, and oils. Used in India for the treatment of malaria.

Wardite. A mineral. It is a hydrated aluminium phosphate.

Ward's Essence. Lin. Camph. Ammon.

Ward's Paste. Confection of pepper.

Ward's Stone. A concrete composed of limestone and Portland cement. Used for ornamental stairways.

Warfarin. 4 - Hydroxy - 3 - (3 - oxo - 1 - phenylbutyl)coumarin 3 - (2 - Acetyl - 1 - phenylethyl) - 4 - hydroxycoumarin. Coumadin : Marevan is the sodium derivative.

Wargonin. A proprietary trade name for a Swedish concrete additive. It is a lignosulphonate powder. (137).

Warm Sepia. Sepia, warmed by mixing it with a redder brown. A pigment.

Warne's Metal. An alloy of 26 per cent. nickel, 37 per cent. tin, 26 per cent. bismuth, and 11 per cent. cobalt.

Warp Silk. See ORGANZINE SILK.

Warrenite (Domingite). A mineral, $Pb_2Sb_4S_9$.

Warringtonite. See BROCHANTITE.

Warrus. See WARAS.

Wars. See WARAS.

Wartext. A proprietary phenolic moulding material. (960).

Warwickite. A mineral. It is $(Mg,Fe)_3TiB_2O_8$.

Wasa Tar. A wood tar of a similar type to Stockholm tar, and sometimes sold under this name.

Wash. The fermented wort of the distilleries.

Wash, Dry. Dry spent wash. See SPENT WASH.

Washer Brass. See BRASS.

Washing Blue. See BLUES, LAUNDRY.

Washing Crystals. See SODA CRYSTALS.

Washing Powder. See SOAP POWDERS.

Washing Soda. See SODA CRYSTALS.

Washington Bleach. See CHLORAMINE T.

Waste, Tank. See ALKALI WASTE.

Watchmaker's Alloy. An alloy of 59 per cent. copper, 40 per cent. zinc, and 1·2 per cent. lead.

Water Blue. See NICHOLSON'S BLUE and SOLUBLE BLUE.

Water Blue 6B Extra. See SOLUBLE BLUE.

Water Blue B, R. Dyestuffs. They are British equivalents of Soluble blue.

Water-dag. See AQUADAG.

Water-eosines. Dyestuffs. They are the alkali salts of eosin and are water soluble. Used in the paper-staining industry.

Water Gas. The general name for a mixture of gases obtained by the decomposition of steam by incandescent carbon. It usually contains from 43–44 per cent. carbon monoxide, 48–49 per cent. of hydrogen, 3–4 per cent. of carbon dioxide, and 3–4 per cent. of nitrogen. Used for heating and lighting. Also see PRODUCER GAS.

Water-glass. See SOLUBLE GLASS.

Water-glass, Sodium. See SOLUBLE GLASS.

Water Mica. Clear transparent Muscovite (potash mica).

Water Nigrosine Crystals W. A

British equivalent of Water-soluble nigrosine.

Water of Ammonia. A solution of ammonia, NH_4OH.

Water-of-Ayr Stone. See SNAKE STONE.

Water of Saturn. A dilute solution of lead subacetate.

Water Opal. See HYALITE.

Water Paints. Paints which contain saponified oil and colouring matter.

Waterproof Paints. See MOUNTFORD'S PAINT.

Water Soluble Eosin. See EOSIN.

Water Stones (Hydrolites, Enhydros). Agates found in England, consisting of a rind of chalcedony, lined with quartz crystals and containing liquid (water) movable within.

Water Varnishes. Varnishes made by dissolving gums or glue in water.

Waterwax. Embedding waxes for histology. (544).

Watsonite. A proprietary material used as a mica substitute. It is made from scrap mica with a binding agent.

Wattevillite. A mineral, $(Na.K)_2(Ca.Mg)(SO_4)_2.2H_2O$.

Wattle Bark (Mimosa). A tanning material obtained from species of *Acacia*. The amount of tannin varies from 12–49 per cent.

Wattle Gum. See AUSTRALIAN GUM.

Watt's and Li's Solution. A solution used for the electro-deposition of iron. It contains 150 grams ferrous sulphate, 75 grams ferrous chloride, 120 grams ammonium sulphate, and 1,000 c.c. water, and is used at a current density of 10 amps. per sq. ft.

Wavellite (Lasionite). A mineral. It is a hydrated basic phosphate of aluminium, $Al_2(PO_4)_2 + Al_2(HO)_6.9H_2O$.

Wax, Brazil. See BRAZIL WAX.

Wax Butter (Wax Oil). A thick oil obtained by the distillation of beeswax. It consists mainly of cerotene C_2H_{54}; melissine $C_{30}H_{60}$, and palmitic acid, formerly used medicinally externally and internally.

Wax C. A high melting point amide wax. Recommended as an internal lubricant in processing ABC, PVC and polystyrene. It is N,N'-distearyl ethylenediamine. (9).

Wax, Earth. See OZOKERITE.

Waxemul. Wax emulsions. (575).

Waxene. A proprietary rubber vulcanisation accelerator. It is a dispersion of accelerator in wax.

Wax, Fossil. See OZOKERITE.

Wax, Getah. See JAVA WAX.

Wax, Insect. See CHINESE WAX.

Wax, Japanese. See CHINESE WAX.

Wax, Mineral. See OZOKERITE, MINERAL WAX, and LIGNITE WAX.

Wax, Montana. See IRISH PEAT WAX.

Wax, Montanin. See IRISH PEAT WAX.

Wax, Oil. See WAX BUTTER.

Waxolene. A registered trade mark for oil and wax soluble dyes. (2).

Waxsol. A proprietary preparation of dioctyl sodium sulphosuccinate. Ear drops for wax removal. (286).

Wax Spirit. A colourless watery distillate from beeswax. It contains acetic and propionic acids.

Wax Tailings. The remaining petroleum distillate after the paraffin wax has been removed.

Wax, Tree. See CHINESE WAX.

WB 200SL. A single layer microcracked chromium process. (605).

" WB " Warfarin. A proprietary preparation of warfarin sodium. An anticoagulant. (319).

Weatherite. A proprietary kraft building paper which has been waterproofed.

Webert Alloy. A proprietary copper-silicon alloy containing small amounts of manganese.

Web-Neb. See BABLAH.

Webnerite. A mineral. It is a variety of andorite.

Websterite. See ALUMINITE.

Webskyite. A serpentine mineral.

Wedel's Oil. Consists of 1 part bergamot, 4 parts camphor, and 32 parts oil of almonds.

Wedl's Stain. A microscopic stain containing 1 gram orseille, 20 c.c. absolute alcohol, 5 c.c. 60 per cent. acetic acid, and 40 c.c. water.

Wehrlite. A mineral containing bismuth, tellurium, and sulphur.

Weibullite. A mineral, $PbS.Bi_2S_3.PbS, Bi_2Se_3$.

Weichhaltungsmittel PA. o-Phthalic ester of ethyl-glycol. A softening agent for cellulose esters.

Weichhaltungsmittel PM. The o-phthalic ester of methyl-glycol. A softening agent for cellulose esters.

Weichharz 398A. A proprietary trade name for a non-drying alkyd made from adipic acid and trimethylene glycol. (4).

Weichmacher ABG. A polyester formed from 1,3 butylene glycol and adipic acid. (3).

Weichmacher T. Dimethylthioanthrene. See SINTOL.

Weichmacher 90. A polymer from acetylene, glycerine and ethylene oxide.

Weichmacher 238S. A proprietary trade name for an ester formed from adipic acid and C_4–C_9 synthetic fatty acids with pentaerythritol. (4).

Weichmacher 333A. As for 238 above except that it has a lower fatty acid content. (4).

Weichmacher AG. A polyester formed from adipic acid and ethylene glycol. (3).

Weigert's Stain. A microscopic stain. It is made by dissolving 2 grams fuchsine and 4 grams resorcin in 200 c.c. water. This heated and 25 c.c. ferric chloride (10 per cent. iron) added, and the solution boiled. The precipitate is collected and added to 200 c.c. boiling alcohol, and the solution cooled. It is filtered, and 4 c.c. 25 per cent. hydrochloric acid is added to the filtrate, and it is then made up to 200 c.c. with 90 per cent. alcohol.

Weighted Silk. Silk impregnated with various inorganic and organic substances in order to increase the weight. Tannin and metallic salts are often used.

Weinschenkite. A mineral of Bavaria. It contains 52·5 per cent. rare earth oxides, 30 per cent. P_2O_5, 16·5 per cent. water, and small amounts of Fe_2O_3.

Weisalloy. A proprietary sheet aluminium alloy.

Weissigite. A mineral. It is a variety of orthoclase.

Weissite. A mineral, Cu_5Te_3. It is also stated to be $8[Cu_2Te]$.

Weiss-Kupfer. See NICKEL SILVERS.

Wekaform. Camphyl-amino-azo-toluene.

Weld (Luteolin). A yellow dyestuff obtained from the dried stalks and leaves of the herbaceous plant known as *Reseda luteola*. Luteolin is the colouring principle. It dyes wool and silk. Weld green is produced on silk by adding indigo carmine and sulphuric acid to the weld dye-bath.

Weldanka A. A proprietary trade name for a casting stainless steel 18 chromium and 8 titanium. (135).

Weldanka B. A proprietary trade name for a stainless steel containing 18 parts chromium and 10 parts titanium. (135).

Weldanka CB. A proprietary trade name for a stainless steel containing 18 parts chromium and 8–10 parts niobium. It is corrosion and acid resisting. (135).

Weldanka KK. A free cutting 18 chromium 8 titanium steel. (135).

Weld Extract. An extract of Weld (*q.v.*). The dyeing principle is luteolin (tetra-oxy-flavone), $C_{15}H_{10}O_6$. It is employed to a small extent for dyeing silk and wool mordanted with tin or alumina.

Weld Green. See WELD.

Weldite. A mineral. It is an aluminium-sodium silicate.

Weldon Mud (Recovered Manganese). A material obtained in the Weldon process for the production of chlorine.

In this process the chlorine is produced by the action of manganese dioxide upon hydrochloric acid, when chlorine and manganese chloride are the products. The manganese chloride solution is treated with milk of lime, when Weldon mud is produced. The mud contains calcium manganite, $CaO.2MnO_2$, and manganese manganite, $MnO.MnO_2$. It is separated from the calcium chloride solution and used again in the process for obtaining chlorine.

Welgum. Alginates for ceramics, electrodes and water treatment. (536).

Welldorm. A proprietary preparation of dichloralphenazone. A hypnotic. (268).

Wellsite. A mineral. It is an aluminium-barium-calcium-potassium silicate.

Welmet. A proprietary chromium-nickel-molybdenum steel.

Welvic. A registered trade-mark for a range of plasticised and unplasticised polyvinyl chlorides used in the manufacture of cables, flooring, pipes, etc. (2).

Welvic MRO/652. A proprietary grade of rigid PVC used in the extrusion of profiles on conventional equipment. (2).

Werderol. A formic acid preservative used for fruit preparations.

Wernerite. See SCAPOLITE.

Werthemanite. A mineral. It is a hydrated sulphate of aluminium.

Wescodyne. A proprietary preparation of iodophor compound with combined iodine. Germicide. (802).

Weslienite. A mineral from the Langban mines. It approximates to the formula, $Na_2O.FeO.3CaO.2Sb_2O_5$.

Wessell's Silver. An alloy of 51–65 per cent. copper, 19–32 per cent. nickel, 12–17 per cent. zinc, and 2 per cent. silver. It is a nickel silver.

West African Copaiba. Illurin balsam, an oleo-resin, is known by this name. It is used as a substitute for balsam of copaiba.

West African Gum. A gum arabic resembling Senegal gum, obtained from *Acacia nilotica*.

Westfalite No. 3. An explosive consisting of 58–61 per cent. ammonium nitrate, 13–15 per cent. potassium nitrate, 4–6 per cent. trinitro-toluene, and 20–22 per cent. ammonium chloride.

West Indian Anime Resin. See TACAMAHAC RESIN.

Weston 618. A proprietary anti-oxidant and process stabiliser added to polyolefins, PVC and high-impact grades of polystyrene to protect them against colour degradation and ultra-violet light. It is di(stearyl) pentaerithritol diphosphate. (433).

Westoran. A registered trade name for a cleaning agent for cotton. It contains emulsified hydrocarbons, and is also used as an insecticide.

Westphalian Essence. See ESSENCE OF SMOKE.

Westphalite I. A safety explosive for mines, consisting of 95 per cent. ammonium nitrate and 5 per cent. resin.

Westphalite II. (Westphalite, Improved). A safety explosive for mines, containing 92 per cent. ammonium nitrate, 3 per cent. potassium nitrate, and 5 per cent. resin.

Westphalite, Improved. See WESTPHALITE II.

Westrol. A registered trade name for a cleaning liquid for cotton. It contains oils with a solvent. Soaps containing trichlorethylene. A degreasing agent.

Westropol. A registered trade name for a cleaning and degreasing agent.

Westrosol. A registered trade name for a preparation of trichlor-ethylene, $CHCl : CCl_2$. It boils at $87 \cdot 4°$ C., and has a specific gravity of $1 \cdot 49$.

Westrumite. A soluble mineral oil. The oil is rendered soluble by heating it, and allowing fine jets of compressed air to be forced in. It is then mixed with a little caustic soda and saponified resinous matter, and air again forced in. Used for sprinkling the streets to lay the dust.

Weta Material. A porcelain substitute consisting of fine, uniformly distributed carborundum particles with silicates and metals of the iron series, cobalt and nickel, and sinters after firing at $1400°$ C.

Wetanol. A proprietary trade name for a wetting agent for textiles, etc. It is a modified sulphated fatty acid ester.

Wetexi 623. A proprietary resin in a hydrocarbon solvent applied by brush or spray to seal concrete and stone from atmospheric pollution by acids. (971).

Wetherillite. Synonym for Hetærolite.

Wetter-Dynamite. A safety explosive for mines, consisting of 53 per cent. nitro-glycerin, 14 per cent. kieselguhr, and 33 per cent. magnesium sulphate.

Wetter-Dynammon. An Austrian explosive containing 94 per cent. ammonium nitrate, 2 per cent. potassium nitrate, and 4 per cent. charcoal.

Wetteren Powder. A guncotton powder, containing a little calcium carbonate, gelatinised with amyl acetate.

Wetter-Fulminite. An explosive containing ammonium nitrate.

Wetting Agents. See ALKANOL, ALPHASOL, AMALGOL, ARESKAP, ARESKET, ARYLENE, AVITEX, BENSAPOL, BETASOL OT-A, BOZETOL, COREKAL A, DAINTEX, DUPONOL LS, ME, WA, ELEMITE, EMULSAMIN, EXTOL, FLOEX, GARDINOL, IGEPAL, IGEPON, INVADINE, LAMEPON, MAPROFIX, MAPROMIN, MERPENTINE, MERPOL, MODINAL T, NACCONOL, NOVONASCO, NOVEMOL, ORVUS WA, SANTOMERSE, TERGITOLS, ULTROIL, WETANOL.

Weyl and Zeitler's Solution. A solution used to absorb oxygen in a pipette. It consists of $2 \cdot 5$ grams pyrogallol in 100 c.c. sodium hydroxide solution of $1 \cdot 05$ specific gravity.

Whalebone Fibre. See VULCANISED FIBRE.

Whale Guano. See WHALE MEAL.

Whaleite. See WOODITE.

Whale Meal (Whale Guano). A fertiliser made from whale refuse by extraction of the oil, and disintegration.

Whartonite. A mineral. It is $(Fe,Ni)S_2$

Wheat Germ Oil. A vitamin E concentrate.

Wheat Oil. See OIL OF WHEAT.

Wheat-stone. See CERUSSITE.

Wheel Brass. See BRASS.

Wheeler's Solution. A mixture of 2 volumes of a solution of 1 oz. pyrogallol in 100 c.c. boiling water, with 5 volumes of a solution containing 1 Ib. potassium hydroxide in 500 c.c. boiling water. For the absorption of oxygen in the Hempel pipette, 175 c.c. are used.

Wheel Ore. A mineral. It is bournonite.

Wheelswarf. The refuse from under the grindstones, consisting of particles of rusted iron and sand.

Wherryite. A mineral. It is $Pb_4Cu(SO_4)_2CO_3(Cl,OH)_2O$.

Whetstone (Oilstone, Honestone). Hard rocks, usually siliceous in character, used for sharpening tools. Suitable rocks include hornstone, sandstone, slate, lydian stone, schist, etc.

Whetstone, Chocolate. See CHOCOLATE WHETSTONE.

Whewellite. A mineral. It is $CaC_2O_4.H_2O$. Calcium oxalate, found in plants

Whipcide. PHTHALOFYNE. It is mono-(1 - ethyl - 1 - methyl - 2 - propynyl) phthalate. (972).

Whiskey. An alcoholic liquid distilled from the fermented mash of malt alone, or of malt and grain of various kinds.

White Acid. A mixture of hydrofluoric acid and ammonium fluoride. Used for etching glass.

White Agate. Chalcedony (q.v.).

White Alkali. Refined sodium carbonate from the Le Blanc soda process.

White Alloy. An alloy of 10 per cent. cast iron, 10 per cent. copper, and 80 per cent. zinc. This name is also applied to alloys containing 49–53 per cent. copper, 23–24 per cent. zinc, 22–24 per cent. nickel, and 2 per cent. iron. They are nickel silvers.

White Amphibole. See TREMOLITE.

White Antimony. See ANTIMONY BLOOM.

White Argol. See ARGOL.

White Arsenic. See ARSENIOUS ACID.

White Arsenic, Vitreous. See ARSENIC GLASS.

White Bengal Fire. Consists of 2 parts realgar, 7 parts sulphur, and 24 parts potassium nitrate. Used for signal purposes.

White Bole. See CHINA CLAY.

White Brass. Variable alloys containing from 2–45 per cent. copper, 33–80 per cent. zinc, 0–81 per cent. tin, 0–13 per cent. lead, and 0–11 per cent. aluminium.

White Bricks. Bricks made of clay containing very little iron. Sand is usually mixed with it, and sometimes chalk.

White Button Alloy. A nickel-silver containing from 49–53 per cent. copper, 23–24.5 per cent. zinc, 22–24 per cent. nickel, and 2–2.5 per cent. iron.

White Calamine. A mineral. It consists chiefly of silicates, and may carry up to 45 per cent. zinc.

White Cast Iron. A good variety of cast iron. It usually contains 97 per cent. iron and 3 per cent. carbon, mainly in the uncombined state. Such an iron has a specific gravity of 7.6 and melts at 1150° C.

White Caustic. Colourless sodium hydroxide.

White Cerate. Spermaceti ointment.

White Clay. See CHINA CLAY.

White Copper. A nickel silver usually containing 70 per cent. copper, 18 per cent. zinc, and 12 per cent. nickel. See NICKEL SILVERS.

White Copperas (Cinquinolite. A mineral. It is ferric sulphate, $Fe_2(SO_4)_3.9H_2O$. The name White copperas is also used for zinc sulphate.

White Cosmetic. A trade name for basic nitrate or mixture of basic nitrates obtained by adding water to bismuth nitrate.

White Dammar. Manila copal resin, obtained from *Vateria indica*, is known by this name.

White Drying Oil. Linseed oil which has been bleached.

White Fish-bone. See Os SEPIÆ.

White Gold. Various alloys are known by this term. A jeweller's alloy of gold whitened by means of silver, is called white gold. An alloy of 90 per cent. gold, and 10 per cent. palladium, and a platinum substitute, containing 59 per cent. nickel, and 41 per cent. gold, are both known under this name. Other alloys consisting of from 70–85 per cent. gold, 8–10 per cent. nickel, and 2–9 per cent. zinc, are also sold under this term.

White Gunpowder. A mixture of 2 parts potassium chlorate and 1 part each potassium ferrocyanide and sugar. An ingredient of explosives.

White Indigo. See INDIGO WHITE.

White Insect Wax (White Lac.). Arjun wax of India produced by the insect *Ceroplastes ceriferus.*

White Iron Pyrites. See MARCASITE.

White Lac. Shellac which has been bleached. Also see WHITE INSECT WAX.

White Lead (Ceruse). A pigment. It is a basic carbonate of lead, the composition of which is variable. The most generally approved formula is $3PbCO_3 . Pb(OH)_2$. Specific gravity about 6.7. Also see KREMSER WHITE and FLAKE WHITE.

White Lead Colours. Mixtures of lead chloride and sulphate. Pigments.

White Lead, Freeman's. See FREEMAN'S NON-POISONOUS WHITE LEAD.

White Lead Ore. See CERUSSITE.

White Lead, Patent. See NON-POISONOUS WHITE LEAD.

White Metal. An alloy of 54 per cent. copper, 24 per cent. nickel, and 22 per cent. zinc. It is a nickel silver. (See NICKEL SILVERS.) It is also the name applied to bearing metals. (See ANTI-FRICTION METALS and BABBITT'S METALS.) The term is also used for Matte Copper, consisting mainly of copper sulphide.

White Metal A. A British standard alloy. It contains 82.58 per cent. lead 12.05 per cent. antimony, 4.64 per cent. tin, 0.34 per cent. copper, 0.08 per cent. zinc, 0.07 per cent. iron, 0.06 per cent. arsenic, and 0.03 per cent. bismuth.

White Mica. See MUSCOVITE.

White Mundic. See MISPICKEL.

White Nickel. See CHLOANTHITE.

White Ochre. Ordinary clay is known by this name.

White Oils (Egg Oils). A liniment usually containing turpentine, acetic acid, and eggs. Sometimes ammonia and camphor are added.

White Paste. Copper sulphocyanide, $Cu_2(CNS)_2$.

White Phosphorus. The yellow poisonous variety is called by this name.

White Pine Resin. See PINE GUM.

White Poppyseed Oil. The oil obtained from poppy seeds pressed cold.

White Portland Cement. A Portland cement in which iron compounds are absent.

White Precipitate (Ammoniated Mercury, Lemery's White Precipitate). Mercury - ammonium - chloride, NH_2 HgCl. Used for the preparation of cinnabar, and in medicine.

White Precipitate, Fusible. See MERCURAMMONIUM CHLORIDE.

White Precipitate, Lemery's. See WHITE PRECIPITATE.

White Pyrites (Efflorescent Pyrites). A variety of iron pyrites, FeS.

White Ramie. See GLASS GRASS.

White Rouge. See PLATE POWDER.

White Sapphires. Consist of pure alumina, AI_2O_3.

White Sennaar Gum. See PICKED TURKEY GUM.

White Smelling Salts. See ENGLISH SALT.

White Snuff. Menthol and cocaine snuff.

White Solder. An alloy of 10 per cent nickel, 45 per cent. copper, and 45 per cent. zinc. A soldering alloy. Also see BUTTON SOLDER.

White Spirit. A turpentine substitute. It is usually a petroleum product, having flash-point and degree of evaporation similar to turpentine. It is prepared by distilling petroleum distillate of 0·798–0·810 specific gravity, in steam. The distillate boiling at 150–250° C., is used as white spirit. Another source gives the following constants : Specific gravity at 15° C. is 0·79–0·82. Refractive index 1·38–1·45 Boiling-point 120–180° C. Ignition point about 21° C.

Whitestuff. See CARBORUNDUM FIRESAND.

White Swan. Anhydrous lanolin. (526).

White Tan (Tari, Teri). *Cæsalpinia digyna*, containing 30–50 per cent. tannin.

White Tar. Naphthalene.

White Tellurium (Krennerite, Bunsenine). A mineral. It contains from 25–29 per cent. gold, 2·7–14·6 per cent. silver, and 2·5–19·5 per cent. lead, as tellurides.

White Tung Oil. Tung oil obtained by the cold pressing method.

White Ultramarine. Obtained by heating aluminium silicate, sodium carbonate, sulphur, and carbon, in the absence of air.

White Vinegar. See DISTILLED VINEGAR.

White Vitriol. Zinc sulphate, $ZnSO_4$. $7H_2O$.

White Wash. A dilute solution of lead subacetate. It is also called Goulard's lotion and Goulard's water.

Whitewash. A mixture of lime and water.

White-water. The technical name given to the waste from pulp-paper mills. It contains fine particles of cellulose.

White Wax. White beeswax.

White Wine Vinegar. See WINE VINEGAR.

Whitewood Bark. See CANELLA.

Whitex. A filler for plastics. Calcined clay.

White Zinc. Zinc carbonate, $ZnCO_3$.

Whiting (Chalk). Calcium carbonate It has a specific gravity of 2·2–2·5.

Whitlockite. A mineral. It is $7[Ca_3 (PO_4)_2]$.

Whitneyite. See DARWINITE.

Whitworth's Steel. Steel which has been subjected to high pressures to eliminate blow-holes.

Whole Latex Rubber. Rubber which contains all the solid constituents of latex except any which may be volatile with water vapour. It is obtained by evaporating the water from latex.

Wiborg Phosphate. A German fertiliser made by heating mineral phosphate with soda. It consists mainly of a tetraphosphate.

Wichmann's Substitute. A material made from casein and albumen.

Wiegold Alloy. A dental alloy. It is a brass containing aluminium, and resembles gold in appearance. It is said to consist of 67·73 per cent. copper, 32 per cent. zinc, and 0·27 per cent. aluminium, but some analysts state that it contains 0·25–0·5 per cent. lead.

Wiikite. A mineral. It is a titano-tantalo-silicate of zirconium, thorium, cerium, iron and uranium, of Finland.

Wij's Solution. Iodine trichloride (9·4 grams) and iodine (7·2 grams) are dissolved separately in glacial acetic acid, and the solutions added together. Used for the determination of the iodine value of fats and oils.

Wilcoloy. A proprietary tungsten carbide material.

Wild Ginger Oil. The oil of *Asarum canadense*.

Wild Mango Oil. See DIKA BUTTER.

Wilhelmit. A German explosive containing potassium chlorate and mineral oil.

Wilkeite. A mineral, $3Ca_3(PO_4)_2.3Ca_2 SiO_4.3CaSO_4.CaCO_3.CaO$.

Wilkinite (Jelly Rock). A colloidal clay. It is suggested as a substitute for china clay as a paper filter.

Wilkinson's Ointment. Ung. Sulph. Co.

Willemite (Williamsite). A mineral. It is zinc ortho-silicate, Zn_2SiO_4.

Willesden Fabrics. Vegetable textiles are passed through a solution of cuprous hydrate in concentrated ammonia (Cuprammonium or Willesden solution). They become coated with a film of gelatinised cellulose containing copper oxide, as the solution dissolves cellulose. It renders paper or other vegetable textiles waterproof and antiseptic.

Williamsite. A mineral. It is variety of Serpentine. See also WILLEMITE.

Williamsonite. An apple-green translucent serpentine (*q.v.*).

Williamson's Blue or Violet. A pigment, $KFe[Fe(CN)_6] + H_2O$, produced from Everitt's salt by treatment with dilute nitric acid, and warming.

Willow Blue (Mazarine Blue, Ultramarine Blue, Celeste, Sky Blue) Preparations of cobalt blue used for colouring pottery.

Willyamite. A mineral. It is a sulphantimonide of cobalt and nickel, $CoS_2.NiS_2.CoSb_2.NiSb_2$.

Wilmil. Alloys similar in composition to silumin.

Wilmot's Aluminium Solder. An alloy of 86 per cent. tin and 14 per cent. bismuth.

Wilnite. See WOLLASTONITE.

Wilouite. Synonym for Grossular.

Wilson's Ointment. Ung. Zinci.

Wiluite. A mineral. It is a variety of grossularite.

Winchite. A mineral. It is $2[NaCa(Mg,Fe\cdot\cdot)_5Si_8O_{21}(OH)_3]$.

Windolite. A name applied to an acetyl-cellulose glass substitute, which is reinforced by wire netting. Suggested for use for open-air shelters.

Wine. An alcoholic liquid obtained by the spontaneous fermentation of the must of fresh grapes. With more than 25 per cent. of sugar sweet wines are obtained, with less than 25 per cent. dry wines.

Winebergite. A mineral. It is a basic sulphate of aluminium.

Wine, Fortified. Wine to which has been added raw spirit, to increase the alcohol content.

Wine of Iron (Steel Wine). *Vinum ferri B.P.* Made by digesting iron wire in sherry wine. It contains a little iron, potassium tartrate, and other iron salts.

Wine Vinegar (White Wine Vinegar, Orleans Vinegar). Produced by the acetous fermentation of grape wine. Its odour is partly due to ethyl acetate

and aldehyde, and it contains from 5–9 per cent. acetic acid.

Wingstay 100. A proprietary anti-oxidant. It is an alkyl aryl amine. (688).

Wingstay S. A proprietary anti-oxidant. It is styrenated phenol. (688).

Wingstay T. A proprietary anti-oxidant. It is a blend of substituted phenols. (688).

Winnofil S. A trade name for precipitated calcium carbonate surface treated with calcium stearate. (2).

Winnowed Sulphur. Prepared by grinding sulphur between stones, having at the same time a current of air passing. The finest sulphur is carried away to the farthest point of a closed chamber. Used for viticultural purposes.

Winogradsky's Solution. It contains 1000 c.c. water, 1·0 gram potassium phosphate, 0·5 gram magnesium sulphate, 0·01–0·02 gram each sodium chloride, ferrous, sulphate and manganese sulphate. From 1–4 per cent. of sugar is added.

Wintergreen Oil. See OIL OF WINTERGREEN.

Winter Oils. Lubricating oils which remain liquid at low temperatures.

Winter's Bark. Pepper bark.

Wipla Metal. V2A steel (see ANKA STEEL). Used for dental purposes.

Wire Brass. See BRASS.

Wischnewite. Synonym for Vishnevite.

Wisdom, Salt of. See SALT OF ALEMBROTH.

Wiserite. A mineral. It is $(Mn\cdot\cdot,Mg)_2 Mn_8 CO_3O_{12}(OH_2.8H_2O$.

Wismolen. A German bismuth preparation used in the treatment of syphilis.

Wismulen. A proprietary preparation. It is the bismuth-ammonium compound of oxytricarb-allylic acid.

Witcizer. Proprietary trade names :
100. Butyl oleate.
312. Dioctyl phthalate.
313. Di-iso-octyl phthalate.
412. Dioctyl adipate.

Withamite. A mineral. It is a variety of epidote.

Witherite. A mineral. It is barium carbonate, $BaCO_3$.

Withnell Powder. An explosive containing 88–92 per cent. ammonium nitrate 4–5 per cent. trinitro-toluene, and 4–6 per cent. flour.

Wittenburg Weather Dynamite. An explosive, consisting of 25 per cent. nitro-glycerin, 34 per cent. potassium nitrate, 38·5 per cent. rye meal, 1 per cent. wood meal, 1 per cent. barium nitrate, and 0·5 per cent. sodium bicarbonate.

Wittichenite. A mineral, Cu_3BiS_3.

Wittite. A mineral, $5PbS.3Bi_2(S.Se)_3$.

Wittol Wax. A wax possessing similar properties to beeswax. It is a proprietary material, and is suitable for acid-proof linings

Witt's Phenylene Blue. An indamine dyestuff, $C_{14}H_{16}N_4HCl$. It is the dimethyl derivative of Witt's phenylene violet.

Witt's Phenylene Violet. An indamine dyestuff, $C_{12}H_{12}N_4HCl$, obtained by the oxidation of p-phenylene-diamine with m-phenylene-diamine.

Witt's Toluylene Blue. See TOLUYLENE BLUE.

Woad. A dark, clay-like preparation made from the leaves of the woad plant, *Isatis tinctoria*. It is used for the purpose of exciting fermentation in the indigo vat.

Wocheinite. A mineral. It is a variety of bauxite.

Wodanite. A mineral. It is a biotite rich in titanium, $(TiO_2 = 12 \cdot 5$ per cent.). Found in Baden.

Wohlerite. A mineral. $5Na_2O.20CaO.2Nb_2O_5.20SiO_2.3ZrO_2$.

Wolchite. A mineral. It is a variety of bournonite.

Wolchonskoite. A mineral containing hydrated oxides of chromium, aluminium, and silicon.

Wolfachite. A mineral. It is Ni(As,Sb)S.

Wolfamid IG. A proprietary polyamide resin soluble in alcohol, used as a base for varnishes. (967).

Wolfeite. A mineral. It is $16[(Fe,Mn)_2PO_4OH]$.

Wolfert. A rubber substitute consisting of felt which has been impregnated with a vulcanised oil.

Wolfram. Tungsten. Also see WOLFRAMITE.

Wolfram Brass. See TUNGSTEN BRASS.

Wolfram Bronze. See TUNGSTEN BRONZE.

Wolframine (Wolfram Ochre, Wolfram Ore, Tungstite). A mineral, WO_3.

Wolframite (Wolfram). A mineral. It is a tungstate of iron and manganese, $(Mn.Fe)WO_4$.

Wolframium. An alloy of 98 per cent. aluminium, $1 \cdot 4$ per cent. antimony, $0 \cdot 4$ per cent. copper, $0 \cdot 1$ per cent. tin, and $0 \cdot 04$ per cent. tungsten.

Wolfram Ochre. See WOLFRAMINE.

Wolfram Ore. See WOLFRAMINE.

Wolfram White. A pigment. It is barium tugstate.

Wolfsbergite. See ANTIMONIAL COPPER GLANCE.

Wolftonite. A mineral. It is a manganese-zinc oxide.

Wollastonite. . (Wilnite, Tabular Spar, Grammite). A mineral. It is calcium

metasilicate, $CaSiO_3$. A filler for plastics.

Wollaston's Cement. Consists of 1 part beeswax, 4 parts resin, and 5 parts plaster of Paris. Used for fossils

Wolle. An abbreviation for Collodiumwolle nitrocellulose in various viscosities.

Wolmanite and Wolmanol. Wood preservative against fungus and insects. (562).

Wongshy (Wongsky). Chinese names for the pods of *Gardenia grandiflora*, which contain large quantities of crocin for saffron. Dyes silk and wool yellow.

Wongsky. See WONGSHY.

Wood Alcohol. See METHANOL.

Wood, Alligator. See ALLIGATOR WOOD.

Wood-apple Gum (Katbél-ki-gond, Velampishin, Kapithamia Piscum). The gum of *Feronia elephantum*, sold in Madras.

Wood Arsenate. See WOOD COPPER.

Wood, Bitter. Quassia.

Wood-cloth. Strips of wood treated with sulphurous acid or alkaline bisulphite, making the fibre stronger.

Wood Copper (Wood Arsenate). A variety of the mineral olivenite.

Wood, Cuba. See FUSTIC.

Wood Ether. Methyl ether, CH_3OCH_3.

Woodex. A proprietary machine bearing material. It is an impregnated wood, usually Maple impregnated with oils.

Wood, Fernambuco. See REDWOODS.

Wood Flour (Wood Meal). Finely powdered wood, usually white pine. Used as a rubber, linoleum, or soap filler.

Wood Gas. A gas obtained from pinewood by distillation. It is used for lighting.

Wood-Gum. See COTTON GUM. The name is also applied to tree gum, which is xylane, $C_6H_{10}O_5$.

Wood Hæmatite. A hæmatite resembling wood in being made up of alternate light and dark layers.

Woodhouseite. A mineral. It is $CaAl_3PO_4SO_4(OH)_6$.

Woodine. A name suggested for wood alcohol (Methyl alcohol, CH_3OH).

Woodite (Whaleite). A patented rubber compound made from rubber, asbestos, earth wax, charcoal, ground whalebone, and sulphur.

Wood, Jamaica. See LOGWOOD.

Wood, Kambe. See REDWOODS.

Wood, Lima. See REDWOODS.

Wood Naphtha. See METHANOL.

Wood, Nicaragua. See REDWOODS.

Wood Oil. The final fractions obtained in the distillation of wood spirit, containing high boiling ketones. Also see

CHINESE WOOD OIL and GURJUN BALSAM or OIL.

Wood-Opal. Wood fossilised with opaline silica.

Wood, Peach. See REDWOODS.

Wood, Pernambuco. See REDWOODS.

Wood Potash. Potash salts obtained from the ash of certain woods.

Wood, Red. See REDWOODS.

Wood Rosin. Rosin obtained from the stumps and top wood of felled trees which are useless as timber. It is usually of yellow pine. It is cut up into chips, shredded, and distilled in steam to remove turpentine. The residue is extracted with petroleum ether which extracts rosin and pine oil. The pine oil is removed by distillation. A proprietary brand of wood rosin is ruby-red in colour, has an acid value of 150–155, a saponification value of 168–175, unsaponifiable matter 6–10 per cent., and gives an ash of less than 0·2 per cent.

Wood's Alloys. Alloys of bismuth, tin and lead, usually containing cadmium. One alloy contains 50 per cent. bismuth, 27 per cent. lead, 13 per cent. tin and 10 per cent. cadmium while another contains 50 per cent. bismuth, 25 per cent. lead, and 25 per cent. tin. A third contains 50 per cent. bismuth, 25 per cent. lead, 12·5 per cent. tin and 12·5 per cent. cadmium. This has a specific gravity of 9·7 and a melting-point of 65·5° C.

Woodruffite. A mineral. It is $(Zn,Mn^{..})_2Mn_5^{....}O_{12}.4H_2O$.

Wood, Sandal. See REDWOODS.

Wood, Sapan. See REDWOODS.

Wood Spirit. See METHANOL.

Wood Spirit, Crude. Contains 80 per cent. of methyl alcohol. See METHANOL.

Wood Spirits, Green. See ACETONE ALCOHOL.

Wood Spirit, Standard. See ACETONE ALCOHOL.

Woodsteel. Steel actually coated with wood in the form of a veneer. It is made with the use of a special binder, treatment of the wood and metal, and a process of joining. For furniture.

Wood Stone. Xylolite (q.v.).

Wood Sugar. Xylose, $C_6H_{10}O_5$.

Wood Tin. See STREAM TIN.

Wood Turpentine. Turpentine obtained from the stumps and waste wood of the pine by reducing the wood to chips, and either steam-distilling or solvent-extracting the chips. A proprietary brand of steam-distilled wood turpentine has a specific gravity of 0·860–0·890, a refractive index of 1·465–1·469, a boiling range of 150–185° C. (90 per cent. distilling below 170° C.).

Wood Turpentine Oils. See WOOD TURPENTINE.

Wood Vinegar. A vinegar which owes its acetic acid to refined pyroligneous acid.

Wood Vinegar, Crude. See PYROLIGNEOUS LIQUOR.

Woodwardite. A mineral, $CuSO_4.Al_2(SO_4)_3$. It is also given as $Cu_4Al_2SO_4(OH)_{12}.2–4H_2O$.

Wood Wool. A product obtained from the long bleached fibres of wood.

Wood, Yellow. See FUSTIC.

Wool, Artificial. See STAPLE FIBRE and KOSMOS FIBRE.

Wool Black. A dyestuff. It is the sodium salt of sulpho-benzene-azo-sulpho-benzene-azo-p-tolyl-β-naphthylamine, $C_{29}H_{21}N_5O_6S_2Na_2$. Dyes wool bluish-black from an acid bath.

Wool Black 4B. See PALATINE BLACK.

Wool Black 6B. See PALATINE BLACK.

Wool Blue S. A dyestuff. It is a mixture of Acid violet 7B with Blue green S.

Wool Fat, Hydrous. See LANOLIN.

Wool Fat Ointment. Oily lanoline ointment (*Unguentum lanolini oleosum B.P.C.*)

Wool, Glover's. See TANNER'S WOOL.

Wool Grease. The exudation of sheep which collects on the wool.

Wool Green S (Lissamine Green B). A dyestuff. It is the sodium salt of tetra-methyl-diamino-diphenyl-β-oxy-naphthyl-carbinol-disulphonic acid, $C_{27}H_{26}N_2O_8S_2Na_2$. Dyes silk and wool sea-green shades.

Wool Grey B, G, and R. A dyestuff obtained by the action of aniline (or p-toluidine) upon the condensation product from nitroso-dimethyl-aniline and β-naphthol-sulphonic acid S. Dyes wool grey.

Wool Milk. An emulsion, obtained from the treatment of wool fat with caustic soda, and dilution. Lanolin is obtained from this wool milk.

Wool, Mineral. See SLAG WOOL.

Wool Oil. An impure oleic acid used for oiling wool, and for making lubricants and soaps.

Wool Oils (Cloth Oils). Oils used by woollen manufacturers for lubricating wool before spinning. The best varieties consist of pure fatty oils, such as lard, olive, etc. Cheap wool oils include distilled grease olein and manufactured oils.

Wool Pitch. A pitchy material obtained as a residue after the distillation of wool grease (Yorkshire grease).

Wool Red Extra. See FAST RED D.

Wool Scarlet R. A dyestuff. It is the sodium salt of xylene-azo-α-naphthol-disulphonic acid, $C_{18}H_{14}N_2O_7S_2Na_2$. Dyes wool red from an acid bath.

Wool, Vegetable. See LANELLA.

Wool Violet S. A dyestuff. It is the sodium salt of dinitro-benzene-azo-diethyl-metasulphanilic acid, $C_{16}H_{16}N_5SO_7Na$. Dyes wool reddish-violet from an acid bath.

Woolyarna. A new Italian artificial wool.

Wool Yellow. See PATENT FUSTIN.

Woorali. See CURARE.

Woorara. See CURARE.

Woorari. See CURARE.

Wootz. An Indian name for steel.

Work-lead. Crude lead containing a considerable amount of impurity.

Wormseed. The flower-heads of *Artemisia maritima*, used in medicine.

Wormseed, American. The fruit of *Chenopodium ambrosoides*.

Wormseed Oil, American. See AMERICAN WORMSEED OIL.

Worm Wafer. See SANTONIN WAFERS.

Wormwood. Absinthium, the dried leaves and flowering-tops of *Artemisia absinthium*.

Wormwood, Salt of. See SALT OF WORMWOOD.

Wort. Malt is crushed and heated with water until the starch is converted into sugar by the diastase in the malt. The resulting liquid is known as wort.

Worthite. A mineral. It is a variety of sillimanite.

Wovco SP. Trade mark for a range of polyethylene plastics reinforced with carbon fibre. (973).

W.P.C./89. The German mark for Ballistite.

Wresinol. A proprietary polyester laminating resin. (974).

Wright's Stain. A microscopic stain for white blood corpuscles. It consists of 1 gram methylene blue eosin mixture in 600 c.c. methyl alcohol.

Wright's Vaporiser. A proprietary fluid containing 90 per cent. cresols used as an inhalant for the relief of bronchitis.

Wrought Iron. See MALLEABLE IRON.

Wulfenite (Melinose, yellow lead spar). A mineral. It is a molybdate of lead, $PbMoO_4$.

Wurster's Blue. An oxidation product of tetramethyl-*p*-phenylene-diamine. An indicator.

Wurster's Red. An oxidation product of *p*-amino-dimethyl-aniline.

Wurtzillite (Tabbyite, Aegenite, Aeonite). An asphaltic mineral, soluble hot in water.

Wurtzite. A mineral. It is sulphide of zinc.

Wydase. A proprietary trade name for Hyaluronidase.

Wyovin. A proprietary preparation containing dicyclomine hydrochloride. An antispasmodic. (245).

Wytox ADP. A registered trade name for alkylated diphenylamine. An antioxidant for rubber for protection against heat ageing and flex cracking. (53).

Wytox BHT. A registered trade name for alkylated *p*-cresol. A non-staining antioxidant for plastics. (53).

Wytox LT. A registered trade name for dilauryl thiodipropionate. An antioxidant suitable for use in plastics of the polyolefine and ABS type in contact with food. (53).

Wytox 312. A registered trade name for tris-nonylphenyl phosphite, a non-staining, non-discolouring, low volatility antioxidant for polyolefins, vinyl chloride polymers, high impact polystyrenes, etc. (53).

Wytox 335. A registered trade name for a modified polymeric phosphite stabiliser for emulsion type styrene-butadiene polymers. It has outstanding suppression of gel build-up. It is exceptionally resistant to hydrolysis. (53).

W2 Beta. A proprietary name for a special Angelo shellac. (171).

X

X2B. A proprietary hard rubber for radio insulation.

Xala. A name for borax.

Xalostocite. A mineral. It is a pink variety of grossularite.

Xametrin. Hexamine (*q.v.*).

Xanol. Caffeine-sodio-salicylate.

Xanthein. Fremy and Cloez name for the soluble yellow colouring matter of flowers.

Xanthine. See PHOSPHINE.

Xanthinol Nicotinate. 7-{2-Hydroxy-3-[N - (2 - hydroxyethyl)methylamino] - propyl}theophylline nicotinate. Complamin.

Xanthoarsenite. A mineral. It is a manganese arsenate.

Xanthochroite. A mineral. It is an amorphous cadmium sulphide.

Xanthocillin. Antibiotics obtained from the mycelium of Penicillium notatum (Xanthocillin X is 2,3-Di-isocyano-1,4-di-(4-hydroxyphenyl)buta-1,3-diene).

Xanthoconite. A mineral. It is a silver-arsenic-sulphide, $3Ag_2S.As_2S_5$.

Xanthoisite. A mineral. It is $Ni_3(AsO_4)_2$.

Xanthophyll. The yellow pigment of leaves. It has the formula, $C_{40}H_{56}O_2$.

Xanthophyllite. A mineral, $14(Mg.Ca)O.8(Al.Fe)_2O_3.5SiO_2$.

Xanthopicrin. A yellow colouring matter from the bark of *Xanthoxylum caribœum*.

Xanthopicrite. A yellow resin from *Xanthoxylum* species.

Xanthopone. Zinc-ethyl-xanthate. A rubber vulcanisation accelerator.

Xanthopurpurin. See PURPUROXANTHIN.

Xanthorrhœa Balsam. See ACAROID BALSAM.

Xanthorthite. A mineral. It is a variety of orthite.

Xanthosiderite. A mineral. It is a hydrated ferric oxide, $Fe_2O_3.2H_2O$.

Xanthoxenite. A mineral. It is a basic phosphate of ferric iron, with some manganese and calcium, found in Bavaria.

Xaxa. A name formerly applied to Aspirin. It has now become obsolete.

Xaxaquin. See QUININE ACETOSALATE.

Xeneisol 133. A proprietary trade name for the 133 isotope of XENON. (819).

Xenolite. A mineral. It is a variety of silimanite, an aluminium silicate.

Xenon. A heavy inert gaseous element present in minute quantities in the atmosphere.

Xenon XE 133. See XENEISOL 133.

Xenotime (Hussakite). A mineral. It is mainly a yttrium phosphate, YPO_4, but is sometimes associated with cerium and with La, Dy, U, Ti, Zr, Fl, Fe, and Si.

Xenysalate. 2 - Diethylaminoethyl 3 - phenylsalicylate. Biphenamine. Sebaclen is the hydrochloride.

Xerase. A mixture of yeast with kaolin, glucose, and salts. Used for local application as a bactericide.

Xeroform (Sigmaform). Bismuth-tribromo-phenol, $Bi_2O_3.OH(O.C_6H_2Br_3)$. Taken internally as an antiseptic, and used externally as a substitute for iodoform in the treatment of wounds.

Xerol. A proprietary trade name for glyceryl mono-stearate.

Xerumenex. A proprietary preparation containing triethanolamine oleyl polypeptide condensate and chlorbutol in propylene glycol. (334).

Xetal. A proprietary safety glass.

X.F.L.X. A substituted ethylene mixed with secondary aromatic amines. (172).

Xiphonite. A mineral. It is a variety of amphibole.

X-Ite. A proprietary alloy containing 37–39 per cent. nickel and 17–19 per cent. chromium with iron.

X.L. High boiling tar acids. (602).

XL Carmoisine 6R. A dyestuff. It is a British equivalent of Chromotrope 2R.

X.L.O. A rubber vulcanisation accelerator consisting of magnesia and diphenylguanidine.

XL Opal Blue. See DIPHENYLAMINE BLUE, SPIRIT SOLUBLE.

XL Soluble Blue. See METHYL BLUE.

Xonotlite (Eakleite). A mineral of Mexico. It is a hydrated calcium silicate.

X-Prep. A proprietary preparation containing sennosides A and B. An enema. (324).

XX 601. A proprietary brand of zinc oxide used in rubber compounding.

Xylamidine Tosylate. N-2-(3-Methoxyphenoxy)propyl - *m* - tolyacetamidine *p*-toluenesulphonate. BW 545C64.

Xylan 330. A proprietary aerosol form of PTFE used as a mould-releasing agent in plastics processing. (975).

Xylan 1052. A proprietary lubricant based on PTFE, for use under extreme pressures and in extremes of temperature. (975).

Xylazine. An analgesic. N-(5,6-Dihydro - 4H - 1, 3 - thiazin - 2 - yl) - 2, 6 - xylidine. ROMPUN.

Xylene Musk. An artificial musk perfume. It is trinitro-tertiary-butyl-*m*-xylene.

Xylenol Blue. 1 : 4 - Dimethyl - 5 - hydroxy-benzene-sulphon-phthalein. An indicator used in biochemistry.

Xylidine Orange. See SCARLET GR.

Xylidine Red. See PONCEAU R.

Xylidine Scarlet. See PONCEAU R and AZOCOCCIN 2R.

Xylite. (1) A proprietary rubber vulcanisation accelerator. It is a tarry diphenylguanidine.

Xylite. (2) A mineral. It is $(Ca,Mg)_{22} Fe_4\cdots Si_{6.5}O_{19}(OH)_5$.

Xylite Oil. Xylitone, $C_{12}H_{18}O$.

Xylocaine. A proprietary preparation of lignocaine hydrochloride. A local anæsthetic. (338).

Xylocard. A proprietary preparation of LIGNOCAINE hydrochloride used in the treatment of cardiac arrhythmias. (477).

Xylochloral. A compound of chloral and xylose, (wood gum, $C_6H_{10}O_5$). A hypnotic analogous to Chloralose.

Xylock 225. A proprietary condensation product of phenols with an aryl alkyl ether, used as a high-performance, heat-stable moulding resin. (503).

Xylodase. A proprietary preparation of lignocaine and hyaluronidase gel. Mucosal anæsthesia. (338).

Xyloidin. Braconnot's name for an explosive white powder, $C_6H_8(NO_2)O_5$, obtained by the action of nitric acid upon starch or wood.

Xyloidine. See NITRO-STARCH.

Xylol. Commercial xylene. It consists of a mixture of about 60 per cent. *m*-xylene, 10–25 per cent. *o*- and *p*-xylene, ethyl-benzene, and small quantities of trimethyl-benzene, paraffin, and thioxene. Xylene has a flash-point of about 78° F. Specific gravity at 15° C. is 0·868. Refractive index at 15° C. is 1·496. Boiling-point 130–135° C. Ignition point about 20° C.

Xylolin Yarn (Texilose Yarn). Products made from an unsized paper by cutting into strips, rolling, and twisting.

Xylolite (Xylolith). A cement composed of sawdust mixed with Sorel cement (q.v.).

Xylolith. See XYLOLITE and SOREL CEMENT.

Xylon. Wood cellulose.

Xylon FR. A proprietary nylon containing a flame-retarding additive. (942).

Xylonite. See CELLULOID.

Xylopal. Wood opal.

Xyloproct. A proprietary preparation of LIGNOCAINE, aluminium acetate, zinc oxide and HYDROCORTISONE acetate used in the treatment of hæmorrhoids. (477).

Xyloquinone. Dimethyl-quinone, $C_6H_2(CH_3)_2O_2$.

Xylorcinol. Dimethyl - orcinol, $C_6H_2(CH_3)_2.(OH)_2$.

Xylose. A proprietary benzyl cellulose denture.

Xylotile. A mineral. It is $(Mg,Fe)_3 Fe_2'''Si_7O_{20}10H_2O$.

Xylol. A disinfectant containing formaldehyde. It has strong bactericidal properties.

Xylometazoline. 2-(4-*t*-Butyl - 2,6 - dimethylbenzyl-2-imidazoline. Otrivine is the hydrochloride.

Xylotox. A proprietary preparation of lignocaine. A local anæsthetic. (348).

Y

Yaba Bark. The bark of *Andira excelsa*.

Yacca Gum, Red. See ACAROID BALSAM.

Y Alloy. An aluminium alloy containing 4 per cent. copper, 2 per cent. nickel, and 1·5 per cent. magnesium with small amounts of iron and silicon. It has a specific gravity of 2·8. It is used for die-cast pistons, etc.

Yama-Mai Silk. A silk produced by the Japanese oak caterpillar of Japan, China, and India.

Yaowite. See YOGHOURT.

Yara-Yara. See NEROLIN.

Yarmor. See PINE OIL.

Yarn, Chemical. Artificial silk.

Yarn, Filastic. See FILASTIC.

Yarn, Meteor. Glanzstoff (q.v.).

Yarn, Sirius. See GLANZSTOFF.

Yarn, Texilose. See XYLOLIN YARN.

Yarrow. The dried leaves of *Achillea millefolium*. A tonic.

Yatren. Iodo-oxy-quinoline-sulphonic acid, $C_9H_6O_4NSI$. Used in the treatment of dysentery.

Yeast, Beer *Fœx medicinalis*.

Yeast, Cider. Cidrase.

Yeast Fat. A brown liquid obtained from yeast by extraction with a solvent. It contains a sterol which appears to be identical with ergosterol from ergot.

Yeatmanite. A mineral. It is $[(Mn,Zn)_{16}Sb_2Si_4O_{19}]$.

Yeast, Permanent. See ZYMIN.

Yeast Powder. See BAKING POWDER.

Yellow AB. A coal-tar colour. It is benzene-azo-β-naphthylamine, and is allowed in food products by the U.S. Department of Agriculture.

Yellow Acid. 1 : 3-Dihydroxy-naphthalene-5 : 7-disulphonic acid.

Yellow Arsenic. A form of arsenic obtained by distilling black arsenic *in vacuo* at 450° C. in a dark chamber, with a red light. Heat or light changes it into the black variety.

Yellow Bark. The bark of *Cinchona calisaya*.

Yellow Basilicon. Resin ointment.

Yellow Berries. See PERSIAN BERRIES.

Yellow Brass. See BRASS.

Yellow Carmine (Italian Pink, Yellow Lake). Pigments prepared by precipitating the glucoside quercitrin with alumina.

Yellow Catechu. See CUTCH.

Yellow, Chinese. See CHINESE YELLOW and OCHRE.

Yellow, Cobalt. See AUREOLIN.

Yellow, Cologne. See CHROME YELLOW.

Yellow Copperas. Copiapite (q.v.).

Yellow Copper Ore. See COPPER PYRITES.

Yellow Cross Gas. Mustard gas (q.v.).

Yellow Earth. See OCHRE.

Yellow Fast to Soap. A dyestuff. It is the sodium salt of *m*-carboxy-benzene-azo - diphenylamine, $C_{19}H_{14}N_3O_2Na$. Dyes cotton orange with a chrome mordant.

Yellow Fat Colour. See SPIRIT YELLOW R.

Yellow GA. See METANIL YELLOW.

Yellow Gold. An alloy of 53 per cent. gold, 25 per cent. silver, and 22 per cent. copper.

Yellow Lake. See YELLOW CARMINE.

Yellow Lead Ore. A mineral. It is wulfenite.

Yellow Lead Spar. A mineral. It is wulfenite.

Yellow Liquors. The drainage from alkali waste heaps.

Yellow Mercurial Lotion. See YELLOW WASH.

Yellow Metal. See MUNTZ METAL.

Yellow OB. A coal-tar colour. It is o-toluene-azo-β-naphthylamine, and is allowed by the U.S. Department of Agriculture in food products.

Yellow Ochre (Roman Ochre, Transparent Gold Ochre, Brown Ochre, Terra di Sienna, Stone Yellow, Roman Yellow, Mineral Yellow, Oxford Yellow, Golden Ochre). Yellow pigments. Specific gravity 5·0. They are native earths consisting chiefly of silica and alumina coloured by hydrated ferric oxide. French ochre contains from 52–65 per cent. SiO_2, 12–22 per cent. Fe_2O_3, 8–14 per cent. Al_2O_3, and up to 3 per cent. CaO. Spanish ochre contains approximately 56 per cent. Fe_2O_3, 19 per cent. $CaCO_3$, 11 per cent. SiO_2, and 6 per cent. Al_2O_3. Some ochres are adulterated with calcium carbonate and barium sulphate, and occasionally chrome yellow or a yellow dyestuff is used to improve colour. Also see LIMONITE.

Yellow Oil. A material containing higher alcohols, such as hexyl alcohol, obtained during the production of butanol from corn.

Yellow OO. See FAST YELLOW N.

Yellow Paste. See CONGO YELLOW.

Yellow Precipitate. Yellow mercury oxide, HgO. Ammonium phosphomolybdate is also given this name.

Yellow Prussiate of Potash (Ferro-prussiate of potassium). Potassium ferrocyanide, $K_4Fe(CN)_6$.

Yellow Quartz. See CITRINE.

Yellow Sesquioxide of Uranium. See ORANGE OXIDE OF URANIUM.

Yellow Soda Ash. A soda ash (sodium carbonate) containing traces of iron oxide.

Yellow Spirit. A tin spirit (see TIN SPIRITS), used in conjunction with yellow dyewoods, for dyeing yellow.

Yellow Sulphide of Arsenic. See ORPIMENT.

Yellow T. See TROPÆOLINE O.

Yellow Ultramarine (Silver Ultramarine, Barium Yellow). A pigment prepared by replacing the sodium constituent of ultramarine with silver. A yellow ultramarine is also prepared by treatment of red ultramarine with hydrochloric acid above 360° C.

Yellow W. See FAST YELLOW R.

Yellow Wash (Yellow Mercurial Lotion). It consists of 4·6 grains of mercuric chloride dissolved in 100 c.c. of a solution of lime. (*Lotio hydrargyri flava B.P.*)

Yellow Wax. A viscous, semi-solid, difficultly volatile substance obtained by the distillation of the still residues of petroleum. It contains anthracene and other hydrocarbons.

Yellow Wood. See FUSTIC.

Yellow WR. See BRILLIANT YELLOW S.

Yenite. A mineral. It is Lievrite (*q.v.*).

Yenshee. The dregs and carbonised opium which remains after smoking. It contains from 1–10 per cent. morphine.

Ylang-Ylang Oil. Orchid oil.

Yerba Mate. Paraguay tea.

Yeremeyevite. Synonym for Eremeyevite.

Yocca. A crystalline alkaloid obtained from the bark and stems of the *Yocca* plant of South Columbia. It agrees in composition and properties with caffeine.

Yoghourt (Yaowite). A fermented milk prepared by the Bulgarians. The milk is soured by *Bacillus bulgaricus*, and the characteristic constituent is lactic acid.

Yohydrol. Yohimbine hydrochloride.

Yolk Powder. See LECITHIN.

Yoloy. A proprietary alloy. It is a steel containing 1 per cent. copper, 2 per cent. nickel, and up to 0·2 per cent. carbon.

Yomesan. A proprietary preparation of niclosamide. An anthelmintic. (112).

Yonckite. A Belgian explosive consisting of ammonium perchlorate, ammonium nitrate, sodium nitrate, and trinitro-toluene or nitro-naphthalene.

Yorkshire Grease (Wakefield Grease). The recovered fatty acids from wool grease.

Young Fustic (Cotinin). A natural dyestuff the dyeing principle of which is fisetin or trioxy-flavenol, $C_{15}H_{10}O_6$. It has a limited use for dyeing wool orange or scarlet (chrome or tin mordant), and for dyeing leather.

Young-Ki-Shih. A Chinese mineral consisting essentially of a magnesium-iron silicate containing small quantities of calcium, aluminium, chromium, and manganese.

Yperite. See MUSTARD GAS.

Ytter-Garnet. A mineral. It is $8[Ca_3Fe_2Si_3O_{12}]$.

Yttrialite. A mineral, $Y_2O_3.2SiO_2$.

Yttrium-Apatite. A mineral. It is $2[(Ca,Yt)_5(PO_4)_3(O,F)]$.

Yttroalumite. A mineral. It is $Yt_3Al_5O_{12}$.

Yttrocerite (Yttrofluorite). A mineral, $(Y.Er.Ce)F_3.5CaF_3$.

Yttrocrasite. A mineral. It is $Yt_2Ti_4O_{11}$.

Yttrofluorite. A mineral. It is yttrocerite.

Yttrogarnet. A mineral. It is $8[Yt_3Al_5O_{12}]$.

Yttrotantalite. A mineral. It contains the niobates and tantalates of yttrium and cerium earths, iron, and uranium.

Yttrotitanite. A mineral, $CaO.TiO_2.SiO_2.(Y.Al.Fe)_2O_3.SiO_2.$

Yukalon. Trademark for a proprietary grade of polyethylene. (663).

Yukonite. A mineral. It is a hydrated arsenate of calcium and iron.

Yuksporite. A mineral. It is a calcium-potassium-sodium-iron silicate.

Yxin. A German product. It was formerly Alfa-Antar. It is a product of silver oxide and starch containing 5 per cent. silver. It is a brown powder, and is used as a wound antiseptic.

Z

Zaccatila. See COCHINEAL.

Zackingummi. A rubber substitute, the basis of which is glycerin and paraffin.

Zactane. A proprietary trade name for ethoheptazine.

Zactipar. A proprietary preparation of ethoheptazine citrate and paracetamol. An analgesic. (245).

Zactirin. A proprietary preparation of ethoheptazine citrate, aspirin, and calcium carbonate. An analgesic. (345).

Zaffer. See SMALT.

Zaffre. See SMALT.

Zakin Rubber. A rubber-like substance prepared from glue or similar material.

Zala. Borax.

Zamak Alloys. Proprietary alloys of zinc with aluminium and sometimes small amounts of copper and magnesium. Copper-aluminium-zinc alloys, suitable for die castings contain : 3·9–4·3 per cent. aluminium, 0·9–2·9 per cent. copper, 0·003–0·06 per cent. magnesium, remainder zinc. They have a specific gravity of 6·64–6·7.

Zambesi Black B, BR, D, F, R. Direct cotton dyestuffs, which are diazotised, and developed with β-naphthol. See DIAMINE BLACK-BLUE B.

Zambesi Blue BX, B, RX, R. Direct cotton colours which are developed with β-naphthol.

Zambesi Brown G, GG. Direct cotton colours which are diazotised and developed. The GG mark gives yellowish, and the G mark dark-brown shades.

Zambesi Grey B. A direct cotton colour. It dyes cotton or wool bluish-grey, and is diazotised and developed on the fibre.

Zambesi Indigo Blue. A tetrazo dyestuff, which dyes cotton direct, and when diazotised on the fibre and developed with β-naphthol, gives a reddish-blue colour.

Zam Metal. A proprietary alloy of zinc with aluminium and magnesium.

Zanaloin. See SOCALOIN.

Zanchol. A proprietary trade name for Florantyrone.

Zanil. A proprietary preparation of OXYCLOZANIDE. A veterinary anthelmintic.

Zant. A disinfectant based on chlorxylenol. (591).

Zanzibar Aloe. See ALOE.

Zanzibar Animi. See COPAL RESINS.

Zanzibar Copal. See COPAL RESINS.

Zapon Varnish (Brassoline, Cristalline, Victoria Varnish). Celluloid varnishes.

Zapoto Gum. See CHICLE.

Zaratite. See EMERALD NICKEL.

Zarontin. A proprietary preparation of ethosuximide. An anti-convulsant. (264).

Zaroxolyn. A proprietary preparation of METOLAZONE. An anti-hypertensive. (744).

Zaspis. See DAPICHO.

Zauberin. A material containing 10 per cent. of the sodium compound of p-toluene-sulpho-chloramide (Chloramine T). A detergent and bleaching agent of German origin. Also see CHLORAMINE T. Mannolit, Gansil, Glekosa, Purus, and Washington Bleach are names for other washing and bleaching agents, the active principle of which is Chloramine T.

ZBX. See ACCELERATOR ZBX.

Zeagonite. See GISMONDITE.

Zeanin. A maize flour used in the production of beer. It is practically free from oil and proteins.

Zeasorb. A proprietary preparation of microporous cellulose hexachlorophane, chloroxylenol, aluminium dihydroxy-allantoinate and purified talc, used in dermatology. (379).

Zebedassite. A mineral. It is an aluminium-magnesium silicate.

Zebinide. Horticultural fungicide. (511).

Zebromal. Dibromo - cinnamic - ethyl - ester, $C_6H_5 . CHBr . CHBr.COOC_2H_5.$ Used as a substitute for alkaline bromides in epilepsy.

Zedox. Zirconium oxides. (590).

Zeiodelite. A mixture obtained by stirring 24 parts powdered glass into 20 parts melted sulphur. Used as a cement, and for taking casts.

Zeise's Salt. A salt, $[Pt(C_2H_4)Cl_3]K$, formed when potassium chloride is added to a solution of platinous chloride saturated with ethylene.

Zelco Metal. An alloy of 83 per cent.

zinc, 15 per cent. aluminium, and 2 per cent. copper.

Zeller's Ointment. Ammoniated mercury ointment.

Zellner's Paper. Fluoresceine paper.

Zellwonet. A proprietary viscose packing material.

Zelulone. A proprietary artificial yarn made from wood pulp.

Zenker's Fluid. A solution containing 2·5 grams potassium chromate, 1 gram sodium sulphate, 5 grams mercuric chloride, 5 c.c. glacial acetic acid, and 100 c.c. water.

Zentralin. Dimethyl - diphenyl - urea, $(CH_3)_2(C_6H_5)_2CON_2$. Used for explosives.

Zentralit I. Diethyl-diphenyl-carbamide.

Zentralit II. Dimethyl-diphenyl-carbamide.

Zeo-Karb. Cation exchange materials. (609).

Zeolite Mimetica. Synonym for Dachiardite.

Zeolites. Hydrated aluminium silicates containing alkali or alkaline earth metals. They occur naturally.

Zeolox. A pure grade molecular sieve. (513).

Zeopan. A proprietary complex of aluminium, magnesium and silicone oxide in the form of a porous amorphous powder, mainly used as a binding agent in tablets. (976).

Zeo-Sorb. A technical grade molecular sieve. (513).

Zeotokol. A coarse dolerite (an igneous rock composed essentially of labradorite and anorthite, with augite and sometimes olivine) ground up and used as a fertiliser.

Zeozon (Ultra-zeozon). A paste, said to contain an oxy derivative of esculin. It has medicinal properties.

Zepharovichite. A mineral, $AlPO_4$. $3H_2O$.

Zephiran. Benzalkonium chloride.

Zephrol. A proprietary preparation of ephedrine hydrochloride. A bronchial antispasmodic. (336).

Zeppelin Alloys. Alloys used in the construction of Zeppelin airships. They contain 88–89 per cent. aluminium, 0·13–9 per cent. zinc, 0·06–0·7 per cent. copper, 0·36–0·49 per cent. silicon, 0·38–0·45 per cent. iron, 0·45 per cent. manganese, and 0–0·15 per cent. tin.

Zeranol. A non-steroidal œstrogen drug. It is $(3S,7R)$-3,4,5,6,7,8,9,10,11,12 - decahydro - 7, 14, 16 - trihydroxy- - 3 - methyl - $1H$ - 2 - benzoxa - cyclotetradecin-1-one.

Zerefil G-700/30. Trade-mark for a flame-retardant vinyl polymer reinforced with long glass. (415).

Zerefil J-700/30. Trade-mark for a flame-retardant vinyl polymer reinforced with medium-length glass. (415).

Zerex. A proprietary trade name for a polyvinyl alcohol antifreeze compound.

Zermattite (Schweizerite). A serpentine mineral.

Zerofil. A proprietary rock wool (q.v.) which has been treated with asphalt.

Zeroform. Tribromphenol-bismuth.

Zerone. A proprietary trade name for a methanol and polyvinyl alcohol antifreeze product.

Zerox. Hydrazine for boiler water de·oxygenation. (618).

Zettnow's Stain. A microscopic stain. Solution (A) contains 10 grams tannic acid to which has been added 30 c.c. of a 5 per cent. solution of tartar emetic. Solution (B) contains 1 gram silver sulphate in 250 c.c. water. Take 50 c.c. and add ethylamine until precipitate redissolves.

Zeugite. A mineral. It is a variety of metabrushite.

Zeunerite. A mineral. It is an arsenate of copper and uranium, $Cu(UO_2)_2$ $(AsO_4)_2$. Synonym for Metazeunerite.

Zeus. An alloy of 20 per cent. silver and 80 per cent. copper. Used for fuse wire.

Zewaphosphate. A product of German origin. It is made by treating a suspension of chalky Belgian mineral phosphate with sulphur dioxide, and then with hot gases from the sulphur furnaces to allow the sulphur trioxide to react with more of the calcium carbonate. The phosphoric acid content rises from 12–20 per cent. A fertiliser.

Zewa Powder. Sodium lignin sulphonate obtained by evaporation of waste sulphite lyes. It has detergent and water-softening properties.

Zeyringite. A mineral. It is a calcareous sinter containing nickel.

Zide. See HYDROCHLOROTHIAZIDE. (750).

Ziegelite. A mineral. It is a variety of cuprite.

Ziehl's Stain. See CARBOLFUCHSINE.

Zigueline. A red oxide of copper.

Zilloy. A proprietary zinc alloy containing zinc with 1 per cent. copper, 0·01 per cent. magnesium, and lead and cadmium in addition. The rolled sheets are stiffer than zinc, and are suitable for corrugated sheets for building purposes.

Zimalium. Alloys containing 74–93·5 per cent. aluminium, 2·8–14·8 per cent. zinc, and 3·7–11·2 per cent. magnesium. They are harder and cheaper than aluminium, but are not so resistant to chemical action, and not such good electrical conductors.

Zimate. A trade name for the oxidised

zinc salt of dimethyl-dithio-carbamic acid. A rubber vulcanisation accelerator. Also see SUPER-SULPHUR No. 1.

Zimphen. Sodium - m - oxy -cinnamate. An antiseptic used in medicine.

Zinalin. An obsolete nitro compound.

Zinamide. A proprietary preparation containing pyrazinamide. An antituberculous agent. (310).

Zincalium. See ZIMALIUM.

Zincaluminite. A mineral, $2ZnSO_4$. $4Zn(OH)_2.6Al(OH)_3.5H_2O$.

Zinc, Aluminium. See ALZEN and ZISCON.

Zinc Ancap. A proprietary accelerator. (290). It is zinc mercaptobenzthiazole.

Zinc Anhydride. A variety of lithopone, also known as Zinc Barytes. It consists of a mixture of calcium and barium sulphates and zinc oxide.

Zincazol. A proprietary rubber vulcanisation accelerator. It is zinc-α-phenyl-biguanide.

Zinc Azurite. A mineral. It is azurite combined with zinc sulphate.

Zinc-Baryta White. See LITHOPONE.

Zinc Barytes. See ZINC ANHYDRIDE.

Zinc Blende (Black Jack, Blende, Sphalerite, Rosinjack, Pseudo-galena). A mineral. It is zinc sulphite, ZnS. A pigment.

Zinc Bloom. See HYDROZINCITE.

Zinc Bronze. See ADMIRALTY GUN METAL.

Zinc Chrome (Citron Yellow). A pigment. It is zinc chromate, $ZnCrO_4$. Also see ZINC YELLOW.

Zinc Chrome Yellow. See ZINC YELLOW.

Zinc-Copper-Melanterite. A mineral. It is $8[(Zn,Cu)SO_4.7H_2O]$.

Zincdibraunite. A mineral, $ZnO.2MnO_2$. $2H_2O$.

Zinc Dust. See BLUE POWDER.

Zinced Iron. See GALVANISED IRON.

Zinc-Fausterite. A mineral. It is $(Mn,Mg,Zn)SO_4.7H_2O$.

Zinc Flowers. See PHILOSOPHER'S WOOL.

Zinc Formosul. A basic zinc-formaldehyde-sulphoxylate. It is employed in fat-splitting.

Zincfrin. A proprietary preparation of zinc sulphate and PHENYLEPHRINE hydrochloride, used as eye-drops. (374).

Zinc Fume. See BLUE POWDER.

Zinc Greens. The name is now applied exclusively to mixtures of Zinc yellow and Prussian blue, that is Zinc-yellow greens. The palest varieties are known as Parrot greens. The term was formerly applied to Cobalt green (q.v.).

Zinc Grey. The name originally used for zinc dust employed for painting on

iron. The term is now used for finely ground zinc blende. A mixture of zinc oxide with finely divided charcoal is sold under this name. It is produced in the manufacture of zinc. Other names for this product are Diamond grey, Silver grey, and Platinum grey.

Zinc-Hogbomite. A mineral. It is $(Zn,Mg,Fe)_7(Al,Fe)_{20}TiO_{39}$.

Zincing Solutions. Usually consist of solutions of zinc sulphate and ammonium chloride in water. Used for the electro-deposition of zinc.

Zinc-Iron Spar. See CAPNITE.

Zincite (Zinkite). A mineral. It is zinc oxide, ZnO.

Zinckenite. A mineral. It is $12[Pb_6 Sb_{14}S_{27}]$. See ZINKENITE.

Zinc-Magnesia-Chalcanthite. A mineral. It is $2[(Ca,Zn,Mg)SO_4.5H_2O]$.

Zinc-Manganese-Cummingtonite. A mineral. It is $2[(Mg,Fe,Zn,Mn)_7Si_8O_{22} (OH)_2]$.

Zinco. See ESTER GUMS.

Zincocalcite. A calcite containing zinc carbonate.

Zincoferrite. See FRANKLINITE.

Zincogen. Zinc peroxide preparations. (513).

Zincolith. See LITHOPONE.

Zinc Omadine. A proprietary trade name for Pyrithione Zinc.

Zinc Omadine. The zinc salt of 1-hydroxypyridine - 2 - thione. A broad spectra anti-bacterial and anti-fungal compound. A preservative for shampoos, cosmetics and paints. (201).

Zinconal. See EKTOGAN.

Zinconise. See HYDROZINCITE.

Zinconite. See HYDROZINCITE.

Zincopyrin. Phenyl-dimethyl pyrazolone-zinc chloride, $(C_{11}H_{12}N_2O)_2ZnCl_2$. Used as a caustic for certain cancers.

Zincore, Flinty. See FLINTY ZINC ORE.

Zinc Ore, Spathic. See CALAMINE.

Zincorhodochrosite. A mineral. It is $2[(Mn,Zn)CO_3]$.

Zincosite. A mineral $ZnSO_4$.

Zinc Oxide, French. See FRENCH ZINC.

Zinc Oxide Ointment. A mixture of 20 per cent. zinc oxide with petroleum jelly or benzoated lard. Antiseptic and astringent.

Zinc Perhydrol. See EKTOGAN.

Zinc Powder. Zinc oxide, ZnO.

Zinc-Römerite. A mineral. It is $(Zn,Fe\text{''})Fe_2\text{'''}(SO_4)_4.12H_2O$.

Zinc-Schefferite. A mineral. It is $4[Ca(Mg,Mn,Zn)Si_2O_6]$.

Zinc Spar. See CALAMINE.

Zinc Spinel. See GAHNITE.

Zinstabe. A proprietary range of powdered-zinc compounds used as activators/stabilisers in PVC foam materials. (977).

Zinc Sulphide Grey (Calamine White).

A pigment. It is a dense zinc oxide used for painting iron, and is artificially made by tinting lithopone with ochres and charcoal.

Zinc Sulphide White. See LITHOPONE.

Zinc Vitriol. Zinc sulphate, $ZnSO_4$.

Zinc White (Chinese White, Permanent White, Snow White). A pigment. It is zinc oxide, ZnO. Chinese white is a very dense oxide, and snow white a very pure one. The following brands of zinc oxide are sold : White seal, Green seal, Red seal, Yellow seal, and Grey seal. White seal is the purest mark, and contains 99 per cent. zinc oxide. The Grey seal contains metallic zinc.

Zinc Yellow (Zinc Chrome Yellow, Buttercup Yellow, Lemon Yellow). A pigment consisting of zinc chromate, $ZnCrO_4$.

Zinc Yellow Greens. See ZINC GREENS.

Zineb. Zinc dithiocarbamate. A fungicide.

Zinkan. A proprietary combination of aluminium coated with zinc. It is obtained by rolling at elevated temperatures.

Zinkenite (Jamesonite, Boulangerite, Feather Ore, Zinckenite). A mineral. It is a lead sulphantimonite, $PbS.Sb_2S_3$, and is found in France.

Zinkglas. Synonym for Hemimorphite.

Zinkgrau. Cheap, off colour zinc oxide pigment.

Zinkhausmannit. Synonym for Hetærolite.

Zinkite. See ZINCITE.

Zinkkieselerz. Synonym for Hemimorphite.

Zinkweiss Weissiegel. Zinc oxide, white seal, best quality zinc white.

Zinnal. A proprietary dual metal consisting of aluminium sheet coated on both sides with tin.

Zinntitanite. A mineral. It is $Ca(Ti,Sn)SiO_5$.

Zinnwaldite. A mineral. It is a mica, and has the composition, $4(Na.Li.K)F.$ $10SiO_2.Fe_2O_3.3Al_2O_3$.

Zinol. Aluminium - β - naphthol - disulphonate-zinc acetate, $[C_{10}H_5OH(CO_3)_2 Al_2]+Zn(C_2H_3O_2)_2$. Used in medicine as an antiseptic and astringent.

Zinox. A paint pigment. It is a hydrated zinc oxide made in France and sold in the form of ready-mixed enamel or as a semi-paste.

Zinsser's Insulating Wax. (1) Consists of beeswax; (2) consists of shellac, rosin, and oxide of iron.

Zintox. A proprietary trade name for an agricultural spray containing basic zinc arsenate.

Zippeite. A mineral of Utah, containing 72·5 per cent. UO_3, 3·73 per cent.

P_2O_5, 11·11 per cent. SO_3, 1·96 per cent. SiO_2, water, CaO, and CuO.

Zippo. A trade name for an aluminium solder for joining aluminium to itself, to copper, zinc, tin, or brass. It is stated to be unaffected by the atmosphere, a good heat and electrical conductor, to contain no bismuth, and to melt at 280° C.

Zircomplex. An organo zirconium complex. (534).

Zircomplex FW. A proprietary trade name for a complex of zirconium with organic acids used as a textile water repellent. (190).

Zircomplex PA & PN. A proprietary trade name for a complex of zirconium with organic acids used as a textile water repellent. Used for thixotropy in P.V.A. emulsion paints. (190).

Zircomplex R. A proprietary trade name for a complex of zirconium with organic acids used as a textile water repellent. A release agent for adhesive tapes. (190).

Zircon (Engelhardtite, Ostranite). A mineral. It is zirconium-o-silicate.

Zirconia. Zirconium dioxide, ZrO_2.

Zircosil. Zirconium silicates. (590).

Zirkelite. A mineral. It is essentially a zirconate, titanate, and thorate of calcium and iron, $(Ca.Fe)O.2(Zr.Ti.Th)O_2$.

Zirkite. Brazilian deposits of Baddelyite, ZrO_2.

Zirklerite. A mineral. It is $(Fe,Mg,Ca)_9 Al_4Cl_{18}(OH)_{12}.14H_2O$.

Zirlite. A mineral. It is aluminium hydroxide, $Al(OH)_3$.

Ziscon (Ziskon). An alloy of 60 per cent. aluminium and 40 per cent. zinc.

Zisium. An alloy of from 82–83 per cent. aluminium, 1–3 per cent. copper, 15 per cent. zinc, and 0–1 per cent. tin.

Ziskon. See ZISCON.

ZL-694. A proprietary thermoplastic polyurethane lacquer used in the preparation of abrasion-resistant coatings on fabrics, leather plastics and metals. (129).

ZL-736. A proprietary thermosetting polyurethane lacquer otherwise similar to ZL 694. (129).

Z–M–L. A proprietary rubber vulcanising accelerator. It is the zinc salt of mercapto - benzene - thiazole with laurex.

Zoalene. A proprietary preparation of DINITOLMIDE. A veterinary anti-protozoan.

Zoamix. A proprietary preparation containing DINITOLMIDE. A veterinary anti-protozoan.

Zoesite. A mineral. It is a variety of fibrous silica.

Zoisite. A mineral, $Ca_2Al_3(OH)(SiO_4)_3$.

Zômol. A soluble meat paste.

Zonarez. Registered trade-mark for a range of polyterpene resins. (978).

Zonite. A proprietary trade name for sodium hypochlorite solution for disinfecting and antiseptic uses.

Zonolite. A proprietary registered trade mark for a mineral product made by the heat treatment of vermiculite, a mineral which is similar to a crude mica. The mineral weighs 144 lb. per cub. ft., but after the heat treatment, only 10 lb. per cub. ft. It is used in the manufacture of building materials and for use as high temperature insulation.

Zoolite. See GALALKERITE.

Zootic Acid. Hydrocyanic acid, HCN.

Zootinsalz. Synonym for Nitratine.

Zopaque. A proprietary form of titanium oxide for rubber mixing.

Zophirol. An aqueous solution of a long chain benzyl alkyl dimethyl ammonium chloride for use as a disinfectant in obstetrics. It does not attack instruments.

Zorgite. A mineral, $(Cu_2Pb)Se$.

Zorite. A proprietary alloy containing 35 per cent. nickel, 15 per cent. chromium, 1·75 per cent. manganese, and 0·5 per cent. carbon with iron.

Zoxazolamine. 2 - Amino - 5 - chloro-benzoxazole. Flexin.

Z.P.D. The zinc salt of pentamethylene-dithio-carbamic acid. An ultra-rubber vulcanisation accelerator.

Z-Siloxide. A trade name for a compound obtained by fusing silica with 0·1–2 per cent. zirconia. It is a silica glass.

Z-Tron G. A trade name for strip polyphenylene oxide sometimes copper clad for use in high frequency printed circuits. (59).

Zulite. A bituminous paint similar to Melanoid but for use as a preservative for wood.

Zunyite. A mineral, $Al_2[Al(OH.F.Cl)_2]_6$ of Colorado.

Zwickau Yellow. See CHROME YELLOW.

Zwieselite. A mineral. It is a variety of triplite.

Zycloform. *p*-Amino-benzoic isobutyl ester.

Zygadite. A mineral. It is a variety of albite.

Zyklon. A proprietary hydrogen cyanide fumigant for use against insect and vermin pests.

Zyloric. A proprietary preparation of allopurinol. (277).

Zymase. An alcohol-producing enzyme secreted by yeast cells. It decomposes grape sugar.

Zymin (Permanent Yeast). A product obtained by partially drying ordinary yeast, immersing it in acetone for fifteen minutes, which kills yeast, drying on filter paper, and washing with ether It produces alcohol from grape sugar.

Zymocasein. A phospho-protein obtained from yeast. It is similar to Caseinogen.

Zymogen. A commercial product. It consists of a nitrogenous substance, to provide food for yeast in fermentation.

Zymoidan. A powder said to be composed of zinc oxide, bismuth oxide, aluminium oxide, iodine, boric acid, phenol, gallic acid, salicylic acid, quinine, and other ingredients. An antiseptic dusting-powder.

Zypanar. A proprietary preparation of pancreatin used in the treatment of cystic fibrosis. (327).

Zytel. A registered trade-mark (10) for a range of nylon resins coded as follows:—

42. A grade used for the extrusion of tubes, with high melting-point, good resistance to abrasion and impact and a high degree of stiffness.

58 HS-L. A flexible grade used for general-purpose jacketing.

63. A grade soluble in alcohol.

70G. Nylon 66 reinforced with short glass fibres to give good dimensional stability.

70 GHR. Similar to *70G*, but stabilised against hydrolysis.

71 G. A series of modified Nylon 66 reinforced with short glass fibres.

77 G. A series of Nylon 6.12 reinforced with glass fibres.

91 HS. A heat-stabilised and plasticised grade used to make flexible tubing.

101. A nylon moulding compound.

105 BK-10. A black nylon composition with good resistance to weathering.

109 L. A chemically-modified nylon used in the moulding of heavy sections.

122. Zytel 101 modified to give enhanced resistance to hydrolysis.

131 L. Zytel 101 modified to permit fast moulding cycles.

141. A nylon with extra toughness and strength.

151 L. A grade of Nylon 6.12 with low absorption of moisture and good dimensional stability.

153 HS-L. A grade of Nylon 6.12 modified to provide heat stability.

158 L. A grade of Nylon 6.12 possessing higher melt viscosity and extra toughness.

408. A modified natural Nylon 66, tough but easily processed.

408 HS. A heat-stabilised grade of Nylon 66.

410 BK-10. A black Nylon 66 with good resistance to weathering.

3606. A grade of nylon stabilised against heat and weather.

ST 801. A modified 66 polyamide claiming outstanding toughness.

INDEX OF MANUFACTURERS

(The numbers refer to figures bracketed in the text).

1. Kinnis and Brown.
2. Imperial Chemical Industries.
3. I. G. Farben (Ludwigshaven).
4. Chemische Werke, Albert.
5. Algemeene Kunstzijde Unie N.V.
6. Bush Beach and Segner Bagley, London, for Dégussa, Frankfurt.
7. Shawinigan.
8. S.I.C. Plastics, Epsom. Mazzacchelli Celluloide S.P.A. Castilione Olona, Italy.
9. Hoechst Chemicals.
10. Dupont.
11. Celanese Corporation.
12. British Resin Products.
13. Rohm and Haas.
14. Deutsche Advance Production G.M.B.H.
15. M and B Plastics.
16. Allied Products Corporation, Connecticut.
17. Geigy (U.K.).
18. Ciba.
19. Anic Gela, Milan.
20. ABCD Petrochimica.
21. Rhodiaceta, Lyon.
22. Staatsmijnen, Holland.
23. John and E. Sturge, Birmingham.
24. Albright and Wilson.
25. Shiraishi Calcium, Kaisha, Japan.
26. Omya S.A. Paris.
27. The Cement Marketing Co.
28. AS Norwegian Talc, Bergen.
29. Chemische Werke Hüls.
30. Allied Chemical Corporation.
31. Louis Blümer, Zwickau.
32. Beckacite Kunstharzfabrik, Hamburg.
33. American Metal Climax.
34. Union Carbide.
35. Minnesota Mining.
36. Hughron Chemical.
37. Hooker.
38. Goodrich-Gulf Chemicals, Ohio.
39. Mycalex Corporation of America.
40. Witco Chemical, New York.
41. U.S. Industrial Chemicals Division of National Distillers and Chemical Corporation.
42. International Nickel Co.
43. VEB Artzneimittelwerk Dresden, Radebeul, Germany.
44. Dr. Kurt Herberts.
45. Deutsche Hydrierwerke.
46. Dr. Alexander Wacker.
47. The National Lead Company.
48. Reichhold Chemicals Inc.
49. Badische Anilin und Soda-Fabrik.
50. Noury und van der Lande.
51. UOP Chemical, N.J.
52. Wallace and Tiernan, N.Y.
53. National Polychemicals, Wilmington, Mass.
54. The Guardian Chemical Corporation.
55. Pearson Chemical Corporation, N.J.
56. American Cyanamid Co.
57. Monsanto.
58. Diamond Alkali Co.
59. General Electric.
60. Mitsui Chemical Industry, Japan.
61. Montecatini.
62. Catalin Corporation.
63. Cabot Corporation.
64. Dow Chemical Company.
65. Gordon Chemical Co.
66. Plastanol.
67. Chemische Werke München Otto Bärklocker.
68. Centre technique d'applications, France.
69. Ethylene Plastique.
70. Universal Oil Co.
71. Sordelli, Italy.
72. Ward, Blenkinsop & Co.
73. Ohio Apex Inc.
74. Archer Daniels Midland Co.
75. Armour & Co.
76. Baker Castor Oil Co.
77. Lankro Chemicals.
78. Columbia Chemical Division.
79. Tennessee Products and Chemical Corporation.
80. British Industrial Solvents.
81. Koninklijke Nederlandsche Gist-En Spiritusfabriek.
82. Godfrey L. Cabot Inc.
84. Diamond Alkali Co.
85. Pfizer International Service Co. Inc.
86. Armour Hess Chemicals.
87. Snia Viscosa, Milan.
88. Rhone-Poulenc.
89. Synthite Ltd.
90. Novadel Ltd.
91. Cray Valley Products Ltd.
92. Vinyl Products, Carshalton.
93. Hercules Powder Co.
94. Scott Bader.
95. Bakelite Corporation.
96. Briggs and Townsend.
97. Argus Chemical Corporation.
98. U.S. Stoneware Co.
99. Shell Chemicals.
100. Standard Oil Co.
101. Hardesty Chemical Co. Inc.
102. B.F. Goodrich Chemical Co.

103. Consortium de Products Chemiques et de Synthese.
105. Carbide and Carbon Chemicals Corporation.
106. Rubber Corporation of America.
107. Swift and Co.
108. Irano Products.
109. Spencer Kellogg and Sons.
110. Kendall Refining Co.
111. Kenrich Corporation.
112. Bayer & Co.
113. Atlas Chemical Industries Inc.
114. General Mills.
115. Morton Withers Chemical Co.
116. Wilmington Chemical Co.
117. National Lead Co.
118. Pan American Refining Co.
119. Sharples Chemical Inc.
120. A Boake Roberts and Co. Ltd.
121. Harwick Standard Chemical Co.
122. Glidden Co.
123. Energy Industries Inc.
124. The Pittsburgh Coke & Chemical Co.
125. Petromar Ltd.
126. Scado Kunsthars Industrie.
127. Socony Vacuum Oil Co. Inc.
128. Deecy Products Co.
129. Thiokol Corporation.
130. F.M.C. Corporation.
131. Imperial Aluminium.
132. Alcan.
133. Associated Lead.
134. Kynoch.
136. Brown Bayley.
137. Berk.
138. Cape Insulation.   Cape Asbestos Co.
139. Leicester Lovell.
140. Unitex.
141. Sheepbridge Stokes.
142. English Steel.
143. Wiggin.
144. Balfour and Darwins.
145. Powell Duffryn.
146. Plastic Coatings.
147. Samuel Osborn.
148. Detel Products.
149. Dexine.
150. Haworth.
151. Dunlop Rubber.
152. Turner Brothers Asbestos.
153. Samuel Fox.
154. Corrosion.
155. Hadfields.
156. Thermal Syndicate.
157. Prodorite.
158. Langley Alloys.
159. Phillips.
160. Mobil.
161. Agawan Chemicals.
162. Anchor Chemical Co.
163. U.S. Rubber Co.
164. Binney and Smith and Ashby.
165. Japan Dyestuffs.
166. Soc. Anon. des matieres colorantes et produits chimiques de Saint-

Denis.   Paris.   (Allied Colloids Bradford).
167. Hoppers.
168. Borden Chemical.
169. Dynamit Nobel AG.
170. Pennsylvania Coal Products Co.
171. Zinsser, N.Y.
172. C.P. Hall (Akron).
173. British Cellophane.
174. E. & F. Richardson.
175. Morgan Refractories.
176. Doulton.
177. S. Smith.
178. Telcon Plastics.
179. Tufnol.
180. William Pearson.
181. Beck, Koller.
182. British Drug Houses.
183. Borax Consolidated.
184. Sturge.
185. Aquitaine-Fisons.
186. Standard Telecommunications Laboratories.
187. Polychemi AKU-GE Holland.
188. Fisons.
189. Croxton and Garry.
190. Hardman and Holden.
191. Croda.
192. Novadel.
193. Fine Dyestuff.
194. Glovers.
195. Rex Campbell.
196. Carless, Capel and Leonard.
197. Cocker.
198. Styrene Copolymers.
199. H. A. Smith.
200. British Oxygen.
201. Kingsley and Keith.
202. S and D Chemical Manufacturing.
203. Guest Industrials.
204. Alginate Industries Ltd.
205. Joseph Crosfield and Sons Ltd.
206. British Ceca.
207. Acheson Colloids.
208. Wiggin.
209. Honeywell and Stein.
210. Marchon Products.
211. Distillers.
212. J. H. Little.
213. British Celanese.
214. Eastman Chemical International AG.
215. Graphite Products.
216. Styrene Co.-Polymers Ltd.
217. The Pyrene Co.
218. Thomas Swan.
219. Fisons.
220. International Synthetic Rubber.
221. British Solvent Oils.
222. Revai.
223. Durham Raw Materials.
224. Greenham.
225. Barronia.
226. Byk-Gulden Lomberg.   Chemische Fabrik GmbH.
227. Knoll A.G.

228. Optrex.
229. Antara Chemicals.
230. Polymer Corporation.
231. The Oronite Chemical Co.
232. High Duty Alloys.
233. Mouneyrat & Cie.
234. Chesebrough-Pond's Inc.
235. Lafarge Aluminous Cement Co.
236. National Lead Co.
237. Fermerich et Cie.
238. James A. Jobling.
239. Poron Insulation.
240. Plastugil (France).
241. Dowty Seals.
242. Eastman Chemical Products.
243. Reckitts.
244. Wander.
245. Wyeth.
246. Carlton Laboratories.
247. Calmic.
248. Cox-Continental.
249. M.C.P. Pure Drugs.
250. Medo-Chemicals.
251. Tillots Laboratories.
252. Lewis Laboratories.
253. Boots.
254. Chemical Coatings.
255. Wallace.
256. Consolidated Chemicals.
257. Phillips, Scott, Turner.
258. Robins.
259. Agprolin.
260. International Colloids.
261. Oerlikon-Buhrle & Co.
262. Warner.
263. Richardson Merrell.
264. Parke-Davis.
265. British Sheering.
266. Rorer.
267. Sandoz.
268. Smith and Nephew.
269. Horlicks.
270. Norma.
271. Nicholas.
272. Bencard.
273. Rybar.
274. F.A.I.R. Laboratories.
275. Riker.
276. Modkem.
277. Burroughs Wellcome & Co.
278. Boehringer.
279. Therapharm.
280. Crookes Laboratories.
281. Smith Kline and French.
282. Nestlés.
283. Trufood.
284. Allan & Hanbury.
285. G. O. Searle.
286. Norgine.
287. Uniroyal Chemical.
288. B. P. Chemicals.
289. Laporte.
290. Anchor Chemical Co.
291. Aquitaine-Organico.
292. Carborundum Co.
293. Compounding Ingredients.

294. AKU Holland.
294a. Staatsmijnen/DSM Holland.
295. Novadel.
296. Allied Colloids.
297. Rex Campbell.
298. Bavon.
299. BIP Chemicals.
300. Imperial Smelting.
301. Honeywell Atlas.
302. Ashland Chemical Company.
303. Shell International.
304. Nuodex.
305. Vigel Products.
306. Lederle.
307. Roussell Laboratories.
308. Leo Laboratories.
309. Dista Products.
310. Merck Sharp & Dohme.
311. Abbott Laboratories.
312. Hoechst Pharmaceuticals.
313. Selpharm Laboratories.
314. Roche Products.
315. Harvey Pharmaceuticals.
316. Organon Laboratories.
317. Brocades.
318. Kabi Pharmaceuticals.
319. Ward-Blenkinsop.
320. Dales Pharmaceuticals.
321. Damancy Co.
322. Westminster Laboratories.
323. Rona Laboratories.
324. Bristol Laboratories.
325. Upjohn.
326. E. R. Squibb & Sons.
327. Armour Pharmaceuticals.
328. Clay & Abraham.
329. Paines & Byrne.
330. Pharmax.
331. Harker Stagg.
332. Dunster Laboratories.
333. Eli Lilly & Co.
334. H. R. Napp.
335. Glaxo.
336. Pharmaceutical Specialities (M & B).
337. Pharmacia (G.B.).
338. Astra Hewlett.
339. Moore Medicinal Products.
340. Wade Pharmaceuticals.
341. F.B.A. Pharmaceuticals.
342. Charnell.
343. Wilcox Fozeau & Co.
344. Ayrton Saunders.
345. Maws Pharmacy Suppliers.
346. Hough, Hogeason.
347. Lloyd, Hamol.
348. Willows Francis.
349. Delandale Laboratories.
350. H. N. Norton Co.
351. A. Cox & Co.
352. Laboratory For Applied Biology.
353. Farmitalia.
354. Ucal.
355. Typharm.
356. Janssen Pharmaceuticals.
357. Marshalls Pharmaceuticals.
358. Lloyd's Pharmaceuticals.

359. Gedeon-Richter.
360. British Ethical Proprietaries.
361. Coates & Coops.
362. Independent Research Labs.
363. Chelsea Drug & Chemical Co.
364. Beecham Research Laboratories.
365. Carlo Erba (U.K.).
366. Oppenheimer & Sons.
367. Co-Caps (Coded Capsules).
368. R.P. Drugs.
369. Ortho Pharmaceuticals.
370. Camden Chemical Co.
371. M. A. Steinhard.
372. Miles Laboratories.
373. Zyma (U.K.).
374. Alcon Universal.
375. Christie, George & Co.
376. Stafford-Miller.
377. Dermal Laboratories.
378. Linfield Laboratories.
379. Stiefel Laboratory.
380. Reynolds & Branson.
381. H.E.B. Pharmaceuticals.
382. Denver Laboratories.
383. Priory Laboratories.
384. Ilon Laboratories.
385. Multipax Chemicals.
386. Geistlich Sons.
387. Phillip Harris Medical.
388. Richards & Appleby.
389. Aero Smoke.
390. Radiol Chemicals.
391. Inter-Alia Pharmaceuticals Services.
392. Gerhardt-Penick.
393. Knoll.
394. Alysin & Barrett.
395. Albion Laboratories.
396. Hommel Pharmaceuticals.
397. Pharmaceuticals Developments.
398. Bioglan Laboratories.
399. R. Sumner & Co.
400. Leopold Charles & Co.
401. Stauffer Chemical.
402. Roxall Chemical Co., Indiana.
403. Standard Brands Chemical Industries.
404. W. R. Grace, Massachusetts.
405. Cenetron Nourg Corp.
406. K.W. Chemicals.
407. Joseph Nadin.
408. Plasttrading
409. Associated Electrical Industries (G.E.C.)
410. Emmerson and Cummings Inc.
411. Lennig Chemicals.
412. Heyden Newport Chemical Corp.
413. Sindar Corp.
414. B. F. Goodrich Canada, Kitchener, Ontario.
415. Dart Industries Inc., Evansville, Ind.
416. Cyanamid of Canada, W. Montreal, Quebec.
417. Hubron Rubber Chemicals, Failsworth, Manchester.
418. Kenics Corp., Danvers, Mass.

419. Cincinnati Milacron Chemicals Inc., Reading, Ohio.
420. Biwax Corp., Des Plaines, Illinois.
421. Rhein-Chemie Rheinau, U.K. and Mannheim.
422. International Colloids, Widnes, Lancs.
423. Allaco Products Inc., Braintree, Mass.
424. Soviet Origin.
425. Wykamol.
426. Canada Colors and Chemicals.
427. Canadian Hoechst.
428. Streetley Refactories, Hartlepool, Co. Durham.
429. International Dioxide Inc., N.Y.
430. Poly-Version Inc., Tulsa, Oklahoma.
431. Farbwerke Hoechst, Frankfurt-am-Main.
432. Alma Paint and Varnish Co., Ontario.
433. Borg Warner Corp., Washington.
434. Unichem.
435. Naugatuck (U.S. Rubber).
436. Lederle Laboratories, P.O. Box 500, Pearl River, N.Y.
437. Jenkins Laboratories Inc., Union, N.J.
438. Schering Corp., Kenilworth, N.J.
439. Winthrop.
440. Glyco Chemicals Inc.
441. Les Usines de Melle, France.
442. Fuchow, Japan.
443. Roussel.
444. Duncan, Flockhart.
445. Péchiney-St.-Gobain, France.
446. Cariosta.
447. Scientific Hospital Supplies.
448. Lovelock.
449. Alma S.A., Switzerland.
450. Greeff Chemicals.
451. Cambrian.
452. Ashe Chemicals.
453. Luitpold-Werke.
454. Pacific Anchor Chemical Corp.
455. Ansul Co.
456. Societé Indochine de Plantations d'Hévéas (S.I.P.H.), Ivory Coast.
457. Arden.
458. Enka Glanzstoff.
459. Toagosie Chemical Co.
460. Artrite Great Britain.
461. Arveta S.A., Basel, Switzerland.
462. Norton.
463. Lilly.
464. Australian Synthetic Rubber Co.
465. Avon Rubber.
466. Carlosta.
467. Ayerst.
468. British Xylonite Co.
469. Ethnor Medical Products Inc., N.J.
470. Distillers Co.
471. Minnesota Mining and Manufacturing Co.
472. M.S.D.
473. Knox.

474. Borregaard, Norway.
475. Bambrini Parodi Delfine, Italy.
476. Galen.
477. Astra.
478. Sandoy.
479. Columbian Carbon Co.
480. Bunawerke Hülls GmbH.
481. Revertex, Harlow, Essex.
482. Hardman Inc., Bellville, N.J.
483. Polymeric Systems Inc., Valley Forge, Pa.
484. Seton.
485. Elastomer Products.
486. Pharmaceutical Manufacturing Co.
487. Of French origin.
488. Shell Chemicals, Ireland.
489. Chimimport, Bucharest.
490. Borden Inc., N.Y.
491. Longfield Chemicals, Warrington.
492. H. L. Blachford, Montreal.
493. Chemcell Resources, Quebec.
494. May and Baker.
495. Potter and Clarke.
496. Industrial Chem-Tech. Co., Cambridge, Mass.
497. Butese, Italy.
498. D.D.S.A.
499. American Rubber and Chemicals Co.
500. Citrox, S.A.
501. Murphy Chemical Co.
502. Boots Pure Drug Co.
503. Albright and Wilson.
504. Bush Boake Allen
505. Abril Industrial Waxes
506. Unilever.
507. May and Baker.
508. Yorkshire Dyeware and Chemical Co.
509. Fisons Pharmaceuticals.
510. Allen and Hanbury.
511. Plant Protection (Subsidiary of I.C.I.).
512. Imperial Chemical Industries.
513. Laporte Industries.
514. Burroughs Wellcome and Co.
515. Johnsons of Hendon.
516. Durham Chemicals.
517. Thomas Ness.
518. Ward Blenkinsop and Co.
519. Ciba Laboratories.
520. Glaxo Laboratories.
521. Fine Dyestuffs and Chemicals.
522. Food Industries.
523. S and D Chemical Manufacturing Co.
524. John and E. Sturge.
525. Abbott Laboratories.
526. Croda
527. British Drug Houses.
528. Hanshaw Chemicals.
529. T. J. Smith and Nephew.
530. Clayton Aniline Co.
531. Foseco (F.S).
532. Burts and Harveys.
533. Williams (Hounslow).
534. Hardman and Holden.
535. Gerhardt-Penick.
536. Alginate Industries.

537. Alcock (Peroxide).
538. Dista Products.
539. Baxenden Chemical Co.
540. Roche Products.
541. J. C. Bottomley and Emerson.
542. Scott Bader and Co.
543. Geigy (U.K.).
544. Edward Gurr.
545. B.H. Chemicals.
546. Joseph Crosfield and Sons.
547. Glover (Chemicals).
548. Borax Consolidated.
549. Proprietary Perfumes.
550. A.B.M. Industrial Products.
551. L. J. Pointing and Sons.
552. William Pearson.
553. Shell Chemicals U.K.
554. Robinson Brothers.
555. Armour Hess Chemicals.
556. Rubber Regenerating Co.
557. Baxenden Chemical Co.
558. Lankro Chemicals.
559. Monsanto Chemicals.
560. Staveley Chemicals.
561. Prodorite.
562. Hickson and Welch.
563. Marfleet Refining Co.
564. Novadel.
565. Allied Colloids.
566. Carless, Capel and Leonard.
567. Rex Campbell and Co.
568. Hickson and Welch.
569. Y.B. Baron
570. BIP Chemicals (U.K.).
571. Barium Chemicals.
572. Berk.
573. North Thames Gas Board.
574. Mirvale Chemical Co.
575. Bulter Malros.
576. British Hydrological Corporation.
577. Pfizer.
578. BP Chemicals (U.K.).
579. Macfarlane Smith
580. Lancashire Tar Distillers.
581. Scottish Tar Distillers.
582. L. B. Holliday and Co.
583. Akzo Chemie GmbH., Düren.
584. British Geon.
585. Bridge Chemicals.
586. Joseph Crosfield & Sons.
587. Associated Adhesives.
588. Alcoch (Peroxides).
589. Mechema.
590. Associated Lead Manufacturers.
591. Evans Medical.
592. Union Carbide.
593. Hercules Powder Co.
594. Prodorite.
595. Thorium.
596. A. H. Marks and Co.
597. Prices Chemicals.
598. Lancashire Chemical Works.
599. James Robinson and Co.
600. Cocker Chemical Co.
601. Dorman Long (Chemicals).
602. Coalite and Chemical Products.

603. Tar Distillers.
604. Chas. Lowe and Co. (Manchester).
605. Hanshaw Chemicals.
606. Cyclo Chemicals.
607. Nuodex.
608. Marbon Chemical Division of Borg-Warner.
609. Permutit Co.
610. Durham Chemicals.
611. Imperial Smelting Corporation (Alloys).
612. Walker Chemical Co.
613. Dow Chemical Co. (U.K.)
614. Midland Silicones.
615. Johnson Matthey Chemicals.
616. Midland-Yorkshire Tar Distillers.
617. Associated Adhesives.
618. Fisons Industrial Chemicals.
519. International Synthetic Rubber Company.
620. British Dyewood Company.
621. Esso Chemicals.
622. Skip Co.
623. Lane Bros. (Tar Distillers).
624. Lunevale Products.
625. Nipa Laboratories.
626. British Rare Earths.
627. United Coke and Chemical.
628. Burmah Oil Trading.
629. Lancashire Chemical Works.
630. British Titan Products.
631. British Dyewood Company.
632. Titanium Intermediates.
633. Concept.
634. Gemini.
635. Coperbo, Brazil.
636. Montédison, U.K.
637. Sinclair.
638. Dayton Tyre and Rubber.
639. Koppers.
640. de Beers Laboratories Inc., Broadview, Ill.
641. Societá San Giuliano, Italy.
642. Enjay.
643. Degussa Wolfgang Kunststoff Betriebe, Hanan-am-main.
644. Searle.
645. Lunbeck.
646. Dermalex.
647. Mesa Plastics, U.S.A.
648. Asalin Chemical Industry Co., Japan.
649. Disogrin Industries.
650. W.B.P.
651. Dome.
652. Simpla.
653. Synthetic Resins, Speke, Liverpool.
654. Duphar.
655. L.R. Industries.
656. General Tyre and Rubber Co.
657. Alpha Chemical and Plastics Corp., Newark, N.J.
658. Smith Kline and French.
659. Diamond Shamrock Chemical Co., N.J.
660. Dynasil Corp. of America, N.J.
661. Guthrie Estate Agencies.

662. Norsk Spraengstofindsestre.
663. Mitsubishi Petrochemical Co.
664. Hental and Cie., Düsseldorf.
665. Cow and Gate.
666. Acushnet Process.
667. Industrias Químicas Electro-Cloro, Brazil.
668. Birkby's.
669. Monmouth Plastics Inc.
670. Furane Plastics Inc.
671. Lawrence Adhesives, U.S.A.
672. Normandie.
673. Mears Inc.
674. Medar Laboratory Inc., Chicago.
675. Metallgesellschaft A.G., Frankfurt-am-Main.
676. Nisshin Boseki.
677. Danbert Chemical Co., Oak Brook, Ill.
678. Societá Italiana, Italy.
679. Stockholms Superfosfor, Sweden.
680. Mobay Chemical Co., Pittsburgh.
681. Metalife International, Harrogate.
682. Squibb.
683. S.P.R.E.A. S.p.a., Italy.
684. Fiberite West Coast Corp., Orange, California.
685. Winthrop.
686. Péchiney, France.
687. Polyvinyl Chemie. Holland NV, Waaldijk.
688. Goodyear Tyre and Rubber.
689. James Ferguson.
690. Merrell.
691. Nicotex Inc.
692. Trentham.
693. The Japanese Geon Co.
694. Nipro Inc., Augusta, Georgia.
695. Tercol, Shifnal, Shropshire.
696. Vale Chemical Co., Allentown, Pa.
697. Salt.
698. Standard Laboratories.
699. HMC Wheels.
700. RRI Malaya.
701. Natural Rubber Producers Research Association, Welwyn Garden City.
702. Plantation Agencies, Penang.
703. Bernlite Corp. of America.
704. Hotfoil, Northampton.
705. Howards.
706. NV Chemische Industrie AKU-Goodrich, Arnhem.
707. FPT Industries.
708. Carl Frendenberg.
709. Blue Line Chemical Co., St. Louis, Missouri.
710. Hysol (Canada).
711. Dexter Corp., Oleon, N.Y.
712. Dexter Corp., Oleon, N.Y. & Hysol Sterling, London.
713. Hystl Development Co., Redondo Beach, California.
714. Rezolin Inc., U.S.A.
715. Societá Italiana Rosine, Italy.
716. Mobil Chemicals.
717. Plastimer 92, Clichy, France.

718. Tokyo Koatsu, Japan.
719. Ethigel.
720. Evode, Stafford.
721. Merit Pharmaceutical Co. Inc., Houston, Texas.
722. Galen.
723. FEB (Great Britain), Manchester.
724. Zyma.
725. Sobin Chemicals Inc., Boston, Mass.
726. Fillite (Runcorn).
727. Devon Corp.
728. Robinsons.
729. Rhône-Poulenc-Chemie Fine.
730. Liquid Nitrogen Processing Corp., Melven, Pa.
731. Unigreg.
732. Fordath Engineering Co., West Bromwich.
733. Armalux Flooring, Stratford-upon-Avon.
734. Foster Grant, U.S.A.
735. Dow Chemicals, Midland, Michigan.
736. Daikin Kogyo Co., Osaka.
737. Rendell.
738. Farley.
739. Liga.
740. Sinclair.
741. De Pree Co., Holland, Michigan.
742. Emser Werke (Switzerland).
743. Ionac Chemical Co., Birmingham, N.J.
744. Pennwalt.
745. Sale.
746. Ciba-Geigy (U.K.).
747. Wallace and Norma.
748. ISC Chemicals, Avonmouth, Bristol.
749. Stanska Attikfabriken, Sweden.
750. Tutag, Detroit, Michigan.
751. Jefferson Chemical (Texaco).
752. Japan Synthetic Rubber Co., Tokyo.
753. Jugovinil, Jugoslavia.
754. Scrip, Peoria, Ill.
755. D.S.M., P.O. Box 65, Haarlem, Holland.
756. Amcel.
757. Kenrich Petrochemicals Inc., Bayonne, N.J.
758. Societé des Usines Chimiques, Rhône-Poulenc.
759. Innoxa.
760. Keibi-Vitrum.
761. Geneva Drugs, Nanvet, N.Y.
762. Pennsalt Chemical Corp.
763. Hyrex-Key.
764. Zeta Inc.
765. I.C.N.
766. Canadian General Electric Co., Ontario.
767. Pacific Vegetable Oil Corp.
768. Elder, Bryan, Ohio.
769. Intercontinental Chemical Corp.
770. United Ebonite and Loriŏal.
771. Hental Inc., N.J.
772. Péchiney-St. Gobain, France.
773. Savage Laboratories, Missouri City, Texas.

774. Rogers Corp., Rogers, Mass.
775. Nureco Inc., U.S.A.
776. Energen.
777. S.A.S.
778. Parr Moulding Compounds Corp.
779. Vitabiotics.
780. Welbeck.
781. Wilcox.
782. Carter-Wallace.
783. T.R.W. Inc., Redondo Beach, California.
784. Benson.
785. Rexane Polymers Co., N.J.
786. Pearl Chemical Co.
787. Woodward.
788. U.S. Industrial Chemicals.
789. Whaley.
790. Shell Chemie GmbH, Germany.
791. Novo, Germany.
792. Devson Corp., Danvers, Mass.
793. Henry Wiggin and Co.
794. Mol-Rez, U.S.A.
795. Astor.
796. General Aniline.
797. A. E. Stanley Manufacturing Co., Decatur, Ill.
798. A/S Denofar, Norway.
799. Daikin Kogyo, Japan.
800. C. & A. Mitchell, G.B.
801. Continental Laboratories.
802. Bengue.
803. Pharmacological Manufacturing Co.
804. London Rubber.
805. Syntex.
806. W. T. Rendell.
807. Ames.
808. Gidion Richter.
809. Sepe.
810. Glenwood.
811. Rowa.
812. Pharmaceutical Research.
813. Laboratories T. Laurin.
814. Commercial Drug Co.
815. Anglo-French.
816. Ross Laboratories.
817. T. B. Roerig & Co.
818. Eaton Laboratories.
819. Mallinckrodt Chemical Works.
820. Ives Laboratories.
821. Purdue Frederick & Co.
822. E.G.H. Laboratories.
823. Phillips Yeast Products.
824. S.N.P.
825. Unipharma.
826. Martindale Samoore.
827. Ingasetter.
828. B.M. Laboratories.
829. Schacht.
830. Carnrick.
831. Quinoderm.
832. Baxter Laboratories.
833. Evans Medical.
834. Hatrick.
835. Nativelle.
836. Mayfair Chemicals.
837. British Felsol.

838. Legat.
839. West Stilton.
840. S. G. Duncan (Pharm).
841. Antigen Laboratories.
842. Ironite Co.
843. Clevite Corporation of Cleveland, Ohio.
844. Asta-Werke.
845. Celotex Corporation.
846. Stiefel.
847. Rubber Plastics.
848. Servier.
849. Lipha.
850. John Bull Rubber.
851. Canadian Fiberfil.
852. International.
853. McNeil Laboratories, Fort Washington, Pa.
854. Alcon Laboratories Inc., Fort Worth, Texas.
855. A.V.P. Pharmaceuticals, N.Y.
856. City Chemical Corp., N.Y.
857. Philips Roxane, Columbus, Ohio.
858. Warner-Chilcott, Morris Plains, N.J.
859. O'Neal, Jones and Feldman, St. Louis, Mo.
860. Meyer.
861. Ascher, Kansas City, Mo.
862. Baker Laboratories, Miami, Florida.
863. Macsil Inc., Philadelphia, Pa.
864. Westwood Pharmaceuticals, Buffalo, N.Y.
865. Century Pharmaceuticals Inc., Indiana.
866. Misemar Pharmaceuticals Inc., Springfield, Mo.
867. King Research.
868. Buffington, Cambridge, Mass.
869. Cole Pharmacal Co., St. Louis, Mo.
870. Sheryl Pharmaceutical Inc., St. Louis, Mo.
871. Commerce Drug Co. Inc., Farmingdale, N.Y.
872. USV Pharmaceuticals Corp., Tuckahoe, N.Y.
873. Barnes-Hind Pharmaceuticals Inc., Sunnydale, California.
874. Societé Polyterpene Résines, Terpeniques.
875. Wampole.
876. Cook-Waite.
877. Jenkins Laboratories Inc., Auburn, N.Y.
878. Maumée.
879. Hill Dermaceuticals, Tampa, Florida.
880. Dainippon Celluloid, Japan.
881. Kengate.
882. Dynamit AG, Troisdorf, Germany.
883. Dynamit Nobel, Germany.
884. Badger Pharmacal Inc., Cedarburg, Wisc.
885. Zemmer Co. Inc., Oakmont, Pa.
886. Mead Johnson Laboratories, Evansville, Ind.
887. Enzymes.
888. Morton Chemical Co.

889. Ulmer Pharmacal Co., Minneapolis.
890. Standex Laboratories, Columbus, Ohio.
891. Pacific Petroleums (Quebec).
892. C.d.F. Chimie, France.
893. Norsk Hydro-Elektrisk, Norway.
894. NV Chemische Fabriek with Dr. A. Haagen, Roermond, Holland.
895. Socfin Co.
896. E. Merck, Darmstadt.
897. The Ansul Co., Marinette, Wisc.
898. Schering Chemicals, Schering AG, Germany.
899. Süddeutsche Kalkstickstoff-Werke AG, Trostberg.
900. Cornelius Produce Co., London.
901. Quinoderm.
902. Philadelphia Quartz Co.
903. Ciba-Geigy.
904. Resart, Germany.
905. S. A. Shéby, France.
906. Ortho.
907. Enzypharm.
908. Colorado Chemical Specialities.
909. Lepetit.
910. Atochemie-Organo, France.
911. Vanderbilt.
912. Rocol.
913. Fair.
914. McKesson Laboratories, Bridgeport, Conn.
915. B.I.P.
916. Vinatex, Havant, Hampshire.
917. Napp.
918. Kunsthars Fabrik Synthese, Holland.
919. Sicedison, Italy.
920. S.C.M. Corp., Cleveland, Ohio.
921. Concept.
922. Arco Chemical Co.
923. Campana Corp., Batavia, Ill.
924. Svenska Oljeslageri, Sweden.
925. Servier.
926. Hall.
927. Squibb.
928. Sterling Moulding Machinery.
929. Ets. G. Couvent, S.A.R.C., France.
930. Corning.
931. Roberts-Thain.
932. M.S.D.
933. Sumitomo Bakelite, Japan.
934. Abex Corp., N.Y.
935. Flexello Castors and Wheels.
936. Lacrinoid.
937. American Viscose.
938. Biotherax.
939. Moulding Powders.
940. Maltown.
941. NV Chem. Ind. Synres, Holland.
942. Dart Industries Inc., Cleveland, Ohio and Evansville, Ind.
943. Synres International NV, Holland.
944. Atomic Energy Establishment, Harwell.
945. Plantations des Terres Rouges, Vietnam.

946. Robertson Co., Pittsburgh, Pa.
947. Mason Pharm., Inc., Sacramento, California.
948. Mitsui Toatsu, Japan.
949. Tamite Industries Inc., Miami, Fla.
950. French Origin.
951. Nitto Chemical, Japan.
952. Standard Dry Wall Products Inc., Miami, Fla.
953. Torbet.
954. Tojo Rayon, Japan.
955. Banstead Buttery Estate Agency.
956. Amoco Chemicals Corp., Chicago, Ill.
957. Freeport Kaolin Co., N.Y.
958. Un. Chem. Belge.
959. Boehrringer.
960. Giech Export, Poland.
961. Philips Petroleum International.
962. Leeming/Pacquin, N.Y.
963. Synthetic Resins.
964. Nisshin Roseki, Japan.
965. Gamlon Chemical Co.

966. Northern Petroleum Co.
967. V. Woolf, Manchester, M11 4RR.
968. Slimcea.
969. Texas Brine Corp., Houston, Texas.
970. Usine Belge Vynckier, Belgium.
971. Chemical Building Products, Hemel Hempstead, Herts.
972. Pitman-Moore, Englewood, N.J.
973. Worcester Valve Co., Haywards Heath.
974. Resinous Chemical (GB).
975. Whitfield Plastics, Runcorn, Cheshire.
976. Fuji Chemical Industry, Kobe, Japan.
977. New Jersey Zinc Co.
978. Arizona Chemical Co.
979. Fleet C.B. Co. Inc., Lynchburg, Va.
980. Reliance Universal Inc.
981. British Uralite.
982. New Smoking Materials Ltd.
983. Cassella Farbwerke Mainkur A.G.